Table of Atomic Masses and Numbers

Based on the 1989 Report of the Commission on Atomic Weights and Isotopic Abundances of the International Union of Pure and Applied Chemistry and for the elements as they exist naturally on earth. Scaled to the relative atomic mass of carbon-12. The estimated uncertainties in values, between ±1 and ±9 units in the last digit of an atomic mass, are in parentheses after the atomic mass. (From *Journal of Physical and Chemical Reference Data*, Vol. 20 (1991), pp. 1313–1325. Copyright © 1991 IUPAC.)

Element	Symbol	Atomic number	Atomic mass	
Actinium	Ac	89	227.0278	(L)
Aluminum	Al	13	26.981539(5)	
Americium	Am	95	243.0614	(L)
Antimony	Sb	51	121.757(3)	
Argon	Ar	18	39.948(1)	(g, r)
Arsenic	As	33	74.92159(2)	
Astatine	At	85	209.9871	(L)
Barium	Ba	56	137.327(7)	
Berkelium	Bk	97	247.0703	(L)
Beryllium	Be	4	9.012182(3)	
Bismuth	Bi	83	208.98037(3)	
Boron	B	5	10.811(5)	(g, m, r)
Bromine	Br	35	79.904(1)	
Cadmium	Cd	48	112.411(8)	(g)
Calcium	Ca	20	40.078(4)	(g)
Californium	Cf	98	251.0796	(L)
Carbon	C	6	12.011(1)	(r)
Cerium	Ce	58	140.115(4)	(g)
Cesium	Cs	55	132.90543(5)	
Chlorine	Cl	17	35.4527(9)	(m)
Chromium	Cr	24	51.9961(6)	
Cobalt	Co	27	58.93320(1)	
Copper	Cu	29	63.546(3)	(r)
Curium	Cm	96	247.0703	(L)
Dysprosium	Dy	66	162.50(3)	(g)
Einsteinium	Es	99	252.083	(L)
Erbium	Er	68	167.26(3)	(g)
Europium	Eu	63	151.965(9)	(g)
Fermium	Fm	100	257.0951	(L)
Fluorine	F	9	18.9984032(9)	
Francium	Fr	87	223.0197	(L)
Gadolinium	Gd	64	157.25(3)	(g)
Gallium	Ga	31	69.723(4)	
Germanium	Ge	32	72.61(2)	
Gold	Au	79	196.96654(3)	
Hafnium	Hf	72	178.49(2)	
Helium	He	2	4.002602(2)	(g, r)
Holmium	Ho	67	164.93032(3)	
Hydrogen	H	1	1.00794(7)	(g, m, r)
Indium	In	49	114.82(1)	
Iodine	I	53	126.90447(3)	
Iridium	Ir	77	192.22(3)	
Iron	Fe	26	55.847(3)	
Krypton	Kr	36	83.80(1)	(g, m)
Lanthanum	La	57	138.9055(2)	(g)
Lawrencium	Lr	103	262.11	(L)
Lead	Pb	82	207.2(1)	(g, r)
Lithium	Li	3	6.941(2)	(g, m, r)
Lutetium	Lu	71	174.967(1)	(g)
Magnesium	Mg	12	24.3050(6)	
Manganese	Mn	25	54.93805(1)	
Mendelevium	Md	101	258.10	(L)
Mercury	Hg	80	200.59(2)	
Molybdenum	Mo	42	95.94(1)	(g)
Neodymium	Nd	60	144.24(3)	(g)
Neon	Ne	10	20.1797(6)	(g, m)
Neptunium	Np	93	237.0482	(L)
Nickel	Ni	28	58.6934(2)	
Niobium	Nb	41	92.90638(2)	
Nitrogen	N	7	14.00674(7)	(g, r)
Nobelium	No	102	259.1009	(L)
Osmium	Os	76	190.2(1)	(g)
Oxygen	O	8	15.9994(3)	(g, r)
Palladium	Pd	46	106.42(1)	(g)
Phosphorus	P	15	30.973762(4)	
Platinum	Pt	78	195.08(3)	
Plutonium	Pu	94	244.0642	(L)
Polonium	Po	84	208.9824	(L)
Potassium	K	19	39.0983(1)	
Praseodymium	Pr	59	140.90765(3)	
Promethium	Pm	61	144.9127	(L)
Protactinium	Pa	91	231.03588(2)	
Radium	Ra	88	226.0254	(L)
Radon	Rn	86	222.0176	(L)
Rhenium	Re	75	186.207(1)	
Rhodium	Rh	45	102.90550(3)	
Rubidium	Rb	37	85.4678(3)	(g)
Ruthenium	Ru	44	101.07(2)	(g)
Samarium	Sm	62	150.36(3)	(g)
Scandium	Sc	21	44.955910(9)	
Selenium	Se	34	78.96(3)	
Silicon	Si	14	28.0855(3)	(r)
Silver	Ag	47	107.8682(2)	(g)
Sodium	Na	11	22.989768(6)	
Strontium	Sr	38	87.62(1)	(g, r)
Sulfur	S	16	32.066(6)	(g, r)
Tantalum	Ta	73	180.9479(1)	
Technetium	Tc	43	98.9072	(L)
Tellurium	Te	52	127.60(3)	(g)
Terbium	Tb	65	158.92534(3)	
Thallium	Tl	81	204.3833(2)	
Thorium	Th	90	232.0381(1)	(g)
Thulium	Tm	69	168.93421(3)	
Tin	Sn	50	118.710(7)	(g)
Titanium	Ti	22	47.88(3)	
Tungsten	W	74	183.85(3)	
Unnilennium	Une	109	(266)	(L, n, s)
Unnilhexium	Unh	106	263.118	(L, n)
Unniloctium	Uno	108	(265)	(L, n, s)
Unnilpentium	Unp	105	262.114	(L, n)
Unnilquadium	Unq	104	261.11	(L, n)
Unnilseptium	Uns	107	262.12	(L, n)
Uranium	U	92	238.0289(1)	(g, m)
Vanadium	V	23	50.9415(1)	
Xenon	Xe	54	131.29(2)	(g, m)
Ytterbium	Yb	70	173.04(3)	(g)
Yttrium	Y	39	88.90585(2)	
Zinc	Zn	30	65.39(2)	
Zirconium	Zr	40	91.224(2)	(g)

(g) Geologically exceptional specimens of this element are known that have different isotopic compositions. For such samples, the atomic mass given here may not apply as precisely as indicated.

(L) The atomic mass is the relative mass of the isotope of longest half-life. The element has no stable isotopes.

(m) Modified isotopic compositions can occur in commercially available materials that have been processed in undisclosed ways, and the atomic mass given here might be quite different for such samples.

(n) Name and symbol are assigned according to systematic rules developed by the IUPAC.

(r) Ranges in isotopic compositions of normal samples obtained on earth do not permit a more precise atomic mass for the element, but the tabulated value should apply to any normal sample of the element.

(s) Element was not listed in the 1989 report but has been added here.

FUNDAMENTALS OF GENERAL, ORGANIC, AND BIOLOGICAL CHEMISTRY

FUNDAMENTALS OF GENERAL, ORGANIC, AND BIOLOGICAL CHEMISTRY

FIFTH EDITION

JOHN R. HOLUM
Augsburg College

JOHN WILEY & SONS, INC.
New York Chichester Brisbane Toronto Singapore

Acquisitions Editor	Nedah Rose
Marketing Manager	Catherine Faduska
Production Editors	Savouia Amanatidis/Charlotte Hyland
Designer	Kevin Murphy
Manufacturing Manager	Andrea Price
Photo Researcher	Lisa Passmore
Illustration Coordinator	Sigmund Malinowski
Cover Photo	Sidney Moulds/Science Photo Library/Photo Researchers, Inc.

This book was set in ITC Souvenir by Progressive Typographers and printed and bound by Von Hoffman Press, Inc. The cover was printed by The Lehigh Press, Inc.

Recognizing the importance of preserving what has been written, it is a policy of John Wiley & Sons, Inc. to have books of enduring value published in the United States printed on acid-free paper, and we exert our best effort to that end. The paper in this book was manufactured by a mill whose forest management programs include sustained yield harvesting of its timberlands. Sustained yield harvesting principles ensure that the number of trees cut each year does not exceed the amount of new growth.

Holum, John R.
    Fundamentals of general, organic, and biological chemistry / John
R. Holum. — 5th ed.
        p.   cm.
    Includes index.
    ISBN 0-471-57949-1
    1. Chemistry   I. Title.
QD31.2.H62   1994
540 — dc20   93-33263
                                                        CIP

Printed in the United States of America

10 9 8 7 6 5 4 3 2 1

PREFACE

Two types of students study the material offered in this book. A number intend careers in the health sciences other than that of a physician. Many others see their futures outside of science but are interested in how nature and human life work at their molecular levels. As in its previous editions, the Fifth Edition addresses two common requirements of a program of study for these students: that they take just one year of chemistry and that this study include several topics in biochemistry and physiological chemistry. Students have not necessarily taken one year of high-school chemistry. A number of features contribute to the continuing success of this book in conveying the breadth of topics required by this course.

THE MOLECULAR BASIS OF LIFE AS A CENTRAL THEME

The overarching, organizing theme of this book is, in this edition as in previous editions, that life in both health and disease has a molecular basis. No other text remains so consciously faithful to a consistent theme. From the start, students sense that this text is always concerned about what is relevant to their professional needs and that nothing is studied that does not have an application either as background for later topics or for their professional careers. I have allowed no topics that do not fit the theme or serve as background material for topics that do.

Whether the book can be used in its entirety in one year depends somewhat, of course, on both admissions standards and the unique interests that teachers may have in particular areas. An appreciation of these facts has tempered the writing. I have placed much material in **Special Topics,** which can be assigned or omitted as desired without complicating the study of later topics. Some whole chapters, such as Chapter 10 ("Oxidation–Reduction Equilibria") can be omitted, although its more quantitative aspects are desirable, but not essential, background for parts of Chapter 25 ("Biochemical Energetics"). Chapter 29 ("Nutrition"), while fitting the theme, is a candidate for omission, particularly when students take a separate course on this subject. Parts of chapters might also be omitted. The chapters on metabolism, 25–28, for example, generally start with broad overviews, which, in some classes might be all to which time can be allotted.

THE STRUCTURE OF THE BOOK

Laboratory-Related Organization The vocabulary of terms and the background of concepts often required for good laboratory exercises develop too slowly under a traditional text organization. This is why I have *introduced* a number of topics early in the book, while knowing that their *full treatment* can only come later. Many teachers, for example, want to use solutions early in the first term. This is why I have defined a number of words relating to solutions in Chapter 3, although a fuller study comes in Chapter 7. To enable the use of acids and bases in the lab, I briefly introduce them in Chapter 4, but their major study occurs in Chapters 8 and 9. Similarly, the concept and the basic vocabulary of the redox reaction appears early in Chapter 5, while the larger treatment is in Chapter 10.

The mole concept and molarity are fundamental to much of experimental chemistry, so I have brought the study of stoichiometry forward from Chapter 5 to Chapter 3.

Chapters 1 – 11: Traditional General Chemistry Topics   We cannot study the molecular basis of life without learning about molecules as well as several fundamental concepts concerning the structure and properties of matter in general. As long as students retain confidence that a topic, no matter how seemingly remote, relates somehow to their interest in life and health, they will be motivated. Acids, bases, and buffers are studied, for example, because the acid – base status of the body is a matter of life and death. Any teacher realizes, of course, that acids, bases, and buffers cannot be studied without a good background in formulas, structures, equations, solutions, and equilibria.

Chapters 12 – 18: Topics of Organic Chemistry Essential to the Study of Biochemistry   The wedding between the theme of the course and the limitation of time results in a very abbreviated survey of organic chemistry. Some of the major topics developed in even a one-term course of organic chemistry are excluded, topics such as the theory of resonance, nucleophilic substitution reactions, the Grignard synthesis, and many others. I have stressed only those functional groups that occur widely among the molecules of life and their reactions with four kinds of compounds: acids, bases, oxidizing agents, and reducing agents. I have provided some mechanisms because organic reactions otherwise seem too much like magic, and their learning becomes merely rote. Two major reactions that form carbon – carbon bonds, the aldol condensation and the Claisen condensation, are studied in **Special Topics** only, and these occur not among the organic chemistry chapters, but untraditionally nearest their potential applications to metabolism.

Chapters 19 – 29: Illustrations of The Molecular Basis of Life   Carbohydrates, lipids, and proteins begin this closing section of the book. Because of their importance to all that follows, I next take up enzymes, hormones, neurotransmitters, and the extracellular fluids of the body. A study of nucleic acids completes the survey of the chief kinds of substances in the body and their particular settings.

The citric acid cycle and the respiratory chain serve the metabolism of all types of biochemicals, so these pathways are studied next. Then come treatments of the metabolism of carbohydrates, lipids, and proteins. Finally, after developing a knowledge of the structures, properties, and uses of substances in cells, I close with nutrition and the sources of nutrients.

CHANGES TO EARLIER EDITIONS

Most of the **Practice Exercises** that require a calculation or a structure and many of those asking for an equation are new. Beyond that (and more importantly), there have been several changes in organization and content. In the general chemistry chapters (1 – 11), the following can be mentioned.

- Section 1.6 ("Accuracy, Error, Uncertainty, and Precision") is revised thanks to a paper by Professor Charles Guare in the *Journal of Chemical Education* (August 1991, p. 649).

- The chapter on quantitative aspects of chemical reactions (Chapter 3, formerly Chapter 5) has been brought forward, as I discussed earlier.

- The periodic table (studied now at the *start* of Chapter 4) becomes supporting evidence for atomic structure. Condensed representations of electron configurations are introduced.

- The study of the individual laws of gases (in Chapter 6) has been considerably shortened, with no loss of coverage, to help make room for other things.

- The concept of an equilibrium is introduced in Chapter 6 ("States of Matter and the Kinetic Theory") with *physical* equilibria between states of matter. This enables students to be familiar with the terminology associated with equilibria before studying chemical equilibria and lets us use the concept in connection with the states of matter. The study of water and the hydrogen bond has been moved into this chapter from 7 (formerly "Water, Solutions, and Colloids").

- The Brønsted concept of acids and bases now dominates Chapter 8 ("Acids, Bases, and Ionic Compounds") where before it was brought in later in the chapter. This chapter now also has special topics on solubility product equilibria and K_{sp} values and another on the greenhouse effect.

- Kinetic theory as it can be applied to the factors that affect reaction rates and to *chemical equilibria* now starts Chapter 9 ("Reaction Kinetics and Chemical Equilibria: Acid – Base Equilibria"). Before, kinetic theory as applied to chemical reactions was presented earlier, at the close of the chapter on the gas laws. The new arrangement strengthens and significantly enhances the treatment of chemical equilibria. A new Special Topic, "Acid Rain" appears in this chapter.

A number of new special topics enrich the organic chemistry chapters (Chapters 12 – 18):

- New Special Topics such as "Organic Fuels," "The Alkene Double Bond in Nature," and "The Primary Chemical Event That Lets Us See" represent applications.

- Another new Special Topic emphasizes an explanation but then gives a major application: "The Chlorination of Methane — A Free Radical Chain Reaction with Counterparts at the Molecular Level of Life."

Considerable reorganization and updating have occurred to the biochemistry chapters. Some reorganizations occur *within* chapters, such as the earlier placement of the study of fatty acids in Chapter 20 ("Lipids"). Similarly, the section in Chapter 25 on the citric acid cycle now (at the behest of many) comes *before* the section on the respiratory chain. In addition, the Chapters have been put in a different order.

- Chapters 20 ("Lipids") and 21 ("Proteins") now offer a much more extensive treatment of cell membranes than ever before. The lipid components are in Chapter 20 and are used to introduce membrane structure, but the equally important glycoprotein components are in Chapter 21 where cell membranes are revisited. The glycoproteins provide recognition sites for hormones, neurotransmitters, and certain drugs, so Chapter 21 has a new application in a special topic, "Mifepristone (RU 486) — Receptor Binding of a Synthetic Anti-pregnancy Compound."

- Chapter 21 ("Proteins"), besides new material on cell membranes, also includes for the first time information on how cartilage works and how gap junctions between cells enable the synchronized actions by many cells in certain tissues (for example, the heart).

- Chapter 22 ("Enzymes, Hormones, and Neurotransmitters;" old chapter 24) now immediately follows the chapter on proteins. The concepts of a catalyst in general and of an enzyme in particular have been used often. However, once proteins have been introduced, a full-fledged study of enzymes becomes both possible and desirable. New to this chapter is a section on the kinetics of simple enzyme – substrate interactions. A low-key look at the Michaelis – Menten equation in this section is self-contained and can be omitted, but some teachers have requested it.

    Recent developments described in Chapter 22 include retrograde chemical messengers such as nitric oxide and carbon monoxide. A new special topic — "Molecular Complementarity and Immunity, AIDS, Even the ABO Blood Groups" — apply the concepts of this chapter and earlier studies of cell membranes to topics of general interest.

- Chapter 23 ("Extracellular Fluids of the Body;" old Chapter 25) remains about the same. *For future nurses, this chapter is the most important single chapter in the entire book.* Three sections concern the acid – base status of the blood. This topic is virtually the only topic studied in the course that reaches well beyond background applications for other nursing courses to the lifetime careers of nurses. Nearly all medical emergencies involve serious changes in the acid–base status of the blood. The terminology and quantitative values associated with this status are widely encountered both in professional nursing work and in the literature nurses ought to be reading throughout their careers. Where nursing students make up the majority of a class, the acid–base status of the blood must be taught as thoroughly as possible. People going into inhalation therapy or into any aspect of sports medicine also need this material. If presented in the right way, liberal arts students with any interest in sports will be fascinated by the subject (and so motivated to learn it). I have given January interim courses on the subject where all of the students entered the class simply (and reluctantly) to satisfy a general education requirement. But some were athletes and many enjoyed strenuous outdoor activities, such as mountain skiing or wilderness trekking. Grudgingly, many became "believers," and one senior was a bit rueful over missing the opportunity to "go into chemistry." Imagine that! (If you can get *first year* liberal arts majors to take the course for which this book is designed, you may recruit some into chemistry or the life sciences.)

- Chapter 24 ("Nuclei Acids;" old Chapter 22) updates the terminology of the field, the factors that stabilize duplex DNA, the recent developments in gene therapy, the defective gene in cystic fibrosis, and briefly describes the Human Genome Project. The Special Topic on DNA typing ("genetic fingerprinting") has been updated, but the legal controversy is unresolved.

- Chapter 25 ("Biochemical Energetics") contains an updated but somewhat simpler treatment of the respiratory chain. Exciting details on how ATP forms and is then released from its enzyme are included.

- Chapter 26 ("Metabolism of Carbohydrates") has a new Special Topic on the aldol condensation and a shortened but updated special topic on diabetes.

- Chapter 27 ("Metabolism of Lipids") has an updated treatment of lipoprotein complexes and "good" and "bad" cholesterol.

- Chapter 28 ("Metabolism of Nitrogen Compounds") and 29 ("Nutrition;" old Chapter 23) remain about the same.

FEATURES OF THIS EDITION

Many features of this new edition will be familiar to users of the fourth edition. There are frequent **margin comments** to restate a point, offer data, or simply remind.

Key terms are highlighted in boldface at those places where they are defined and then discussed. A complete **glossary** of these terms plus a few others appears at the end of the book. The *Study Guide* that accompanies this book also has individual chapter glossaries.

Each section of a chapter begins with a **headline.** This is *not* a one-sentence summary of the section but rather a lead-in to the beginning of the section that tries to state the section's major point.

Each chapter has a **Summary** that uses key terms in a narrative manner. The summaries are not necessarily organized in the same order in which the material occurs in the various chapter sections. The summaries assume that the sections have been studied and students are familiar with the needed vocabulary.

The chapters in the first two-thirds of the book have several **worked examples.** In those involving calculations, the **factor-label** method is used. New features of these examples, which have always had "Problem" and "Solution" parts, are "Analysis" and "Check" comments. Thus, immediately after the statement of the problem comes the *analysis.* What is the problem really asking? In a multistep solution, what must be done first? Then comes the *solution.* Following is the *"Check"* section of an example. "Does the *size* of the answer make sense? " This takes the student back over the problem and encourages the use of the mind (as opposed to a mechanical use of factor-labels) to see the sense of the analysis and the solution.

Nearly all worked examples are followed by **Practice Exercises,** which encourage immediate reinforcements of skills learned in the examples. Answers to all Practice Exercises are in Appendix D. A copious number of **Review Exercises** closes each chapter, including some that are "additional;" they are not identified by topic. Many additional exercises require the use of material from earlier chapters. Thus you will find stoichiometry problems scattered throughout the book.

As a new feature of this edition, I have introduced three **icons** that will draw the attention of a student to places that emphasize various skills that should be mastered or that point out topics of particular interest from either a health or an environmental view.

This icon, which suggests either a measurement or the application of a skill, draws attention to discussions of chemical calculations, balancing equations, or similar skills.

This is my "map sign" icon, and it appears almost exclusively in the organic chemistry chapters. I draw an analogy between the representations of functional groups in organic structures (like an alkene group) and the symbols used by map makers. We need only a few map symbols to enable us to read almost any map. Similarly, we can see a functional group symbol as representing a relatively short list of properties conferred on the substance. By knowing structural "map signs," we can "read" structural formulas like a map and predict some properties surprisingly well.

This icon, suggesting not only planet earth but also all people on it, draws attention to topics whereby applications of chemical knowledge are made to matters of health- or earth-care.

SUPPLEMENTARY MATERIALS FOR STUDENTS AND TEACHERS

The complete package of supplements that are available to help students study and teachers plan includes the following:

Laboratory Manual for Fundamentals of General, Organic, and Biological Chemistry, **fifth edition** This is a thoroughly revised edition prepared by Dr. Sandra Olmsted, Augsburg College. An *Instructor's Manual* to this laboratory manual is a section of the general Teacher's Manual described below.

Study Guide for Fundamentals of General, Organic, and Biological Chemistry, **fifth edition** This softcover book contains chapter objectives, chapter glossaries, additional worked examples and exercises, sample examinations, and the answers to all of the Review Exercises.

Teachers' Manual for Fundamentals of General, Organic, and Biological Chemistry, **fifth edition.** This softcover supplement is available to teachers. It contains all the usual services for *both the text and the laboratory manual*.

Test Bank

Available in both hard copy and software (Macintosh© and IBM© compatible) versions, this test resource contains roughly 1000 questions.

Transparencies

Instructors who adopt this book may obtain from Wiley, without charge, a set of color transparencies that duplicate key illustrations from the text.

ACKNOWLEDGMENTS

My wife Mary has been my strongest supporter, and I am deeply grateful to this wonderful woman. My daughters, Liz, Ann, and Kathryn, now grown, also have been strong champions, and I thank them for what they have meant to Mary and me.

At Augsburg College, I have always enjoyed unstinting support from the Chair of the Chemistry Department, Dr. Earl Alton; from the Academic Dean, Dr. Ryan LaHurd; and from the President, Dr. Charles Anderson. Dr. Arlin Gyberg, Dr. Joan Kunz, and Dr. Sandra Olmsted of the Chemistry Department have been important sources of suggestions and corrections.

Extraordinarily nice people are all over the place at John Wiley & Sons. I think particularly of my Chemistry Editor, Nedah Rose, her Administrative Assistant, Eric Stano, my Development Editor, Cathleen Petree, and my Supplements' Editor, Joan Kalkut.

The overall design was the responsibility of Kevin Murphy with whom I have worked with immense pleasure on this and other books. Sigmund Malinowski, a worthy successor to John Balbalis, has been skillful, artistic, and faithful in handling the line drawing art work. Lisa Passmore, Associate Photo Editor, produced such a rich supply of outstanding choices for photographs that my choosing became difficult, yet exciting and pleasurable. My copy editor, Peggy Brigg, handled her assignment with grace. Most of the production was supervised by Savoula Amanatidis, and her job is surely one of the most difficult in textbook publishing. Coordinating copy editing, art work, photos, the setting of galleys, proofreading, the preparation of page dummies and final pages, and printing and binding, without letting anything fall between the slats, requires the patience of Job, the accuracy of a computer, and the discipline of a wagon master. Savoula rose to all of the challenges without ever raising her voice! Near the end of the production, she handed the work over to Charlotte Hyland, who stepped forward with the same aplomb I now take for granted among Wiley's Production Editors.

Two outstanding proofreaders saved me from innumerable embarrassments — Dr. Sandra Olmsted and Connie Parks. Dr. Melinda Lee (St Cloud State University) checked the answers to all of the Practice and Review Exercises, and she also prevented many glitches. It's hard to imagine that any errors remain, but, based on experience, no doubt some do. They are now entirely my responsibility. Please send a letter to my Chemistry Editor to let me know about them.

The professional critiques of many teachers are part of the process of preparing a manuscript. I am most pleased to acknowledge and to thank the following people for their work. The following people reviewed the manuscript for this edition.

Margaret Asirvatham
University of Colorado/Boulder

Larry Jackson
Montana State University

Lorraine Brewer
University of Arkansas

Herman Knoche
University of Nebraska — Lincoln

Jack Dalton
Boise State University

Nancy Paisley
Montclair State University

Donald Harriss
University of Minnesota/Duluth

Kent Thomas
Kansas-Newman College

In addition, I'd like to thank the following users/reviewers of *Fundamentals* 4/e.

Robert Ake
Old Dominion University

Fred Schell
University of Tennessee/Knoxville

Muriel Bishop
Clemson University

Justine Walhout
Rockford University

John Meisenheimer
Eastern Kentucky University

Finally, I'd like to thank the following people for reviewing the table of contents for this edition.

Arlin Gysberg
Augsburg College

Atilla Tuncay
Indiana University Northwest

Ram Singhal
Wichita State University

Leslie Wynston
California State University/Long Beach

CONTENTS

Appendices

Glossary

Photo Credits

Index

Index to Special Topics

FUNDAMENTALS OF GENERAL, ORGANIC, AND BIOLOGICAL CHEMISTRY

GOALS, METHODS, AND MEASUREMENTS

We human beings have exquisite methods for adapting to the external environment, mechanisms with a molecular basis as beautiful as this scene from the Towers of Paine National Park, Patagonia. Our aim in this book is to see the beauty of life at the molecular level.

1.1

CHEMISTRY AND THE MOLECULAR BASIS OF LIFE

The theme of this book is the molecular basis of life.
Many centuries ago, people surely saw that different animals drink at the same water holes, breathe the same air, eat the same kinds of food, and enjoy the same salt licks. Ancient farmers knew that the droppings of animals nourish plants. Many animals prosper by eating plants, and some animals can eat weaker animals and grow. At the most basic level of existence, a common pool of parts must occur for all living creatures, plants, and animals.

■ Well over 9 million chemical substances are known.

The shared parts are not organs and tissues but things much smaller, extremely tiny particles called molecules made of even smaller particles called atoms. All of life, whether plant or animal, has a *molecular* basis, and chemistry has been the route to this discovery. *Chemistry* is the study of that part of nature that bears on substances, their compositions and structures, and their abilities to be changed into other substances. There are so many different substances, however, that we must develop a plan of study.

Molecules, Like Maps, Can Be Read When the Keys or Map Signs Are Known Life at the molecular level involves molecules and chemical reactions that often are complicated. The symbols we use for them, however, are actually less complex than many symbol systems you have already mastered. You learned how to read and understand dozens of maps by mastering just a few map symbols, for example. Symbols for molecules are like maps because the same pieces of molecules, like molecular "map signs," occur over and over again. It will be a good idea, therefore, before we study some of the most complicated molecules in nature (Chapters 19 through 29), to learn these "signs" among simpler substances. Our chapters on organic compounds (Chapters 11 through 18) do this.

■ The atoms of all matter are made of varying combinations of three extremely tiny particles: electrons, protons, and neutrons.

Molecules, as we said, are made of atoms, so to understand molecules we must first learn about atoms and how their own (even tinier) parts get reorganized into molecules. The study of atomic and molecular structure occurs mainly in the first third of the book, together with essential background about a variety of substances such as acids, bases, salts, and solutions. All of these studies rest on experimental evidence involving measurements. In this chapter we'll learn about some of the measured quantities that have been useful in chemistry.

Our plan, therefore, is to use Chapters 1 through 10 to study some of the basic principles that underlie all of chemistry. Then, in Chapters 11 to 18, we will learn the molecular "map signs" of the major substances in living systems. In Chapters 19 through 29 we'll apply all this to a broad study of how nature uses substances to sustain living processes. Expect to learn a lot about how nature works, and be prepared to be surprised at how enjoyable this can be. Expect also to discover that chemistry stands in service to a large number of careers and professions besides research in chemistry: agriculture, forestry, home economics, engineering, medicine, nursing, dentistry, veterinary science, dietetics, nutrition, inhalation therapy, physical therapy, public health, science education, pharmacy, clinical lab work, crime lab work, consumer products safety, and many others. Little wonder that chemistry is often referred to as the *central science*. It's at the heart of understanding how nature works, and the knowledge of chemistry is vital to many ways of helping people and also enjoying the work.

1.2
FACTS, HYPOTHESES, AND THEORIES IN SCIENCE

Scientific theories speak to one general question: How does nature work?
One of the most common activities of scientists is the gathering of information and facts by observing nature. Some facts are reproducible and some aren't. Both kinds are important in science. A *reproducible fact* is one that can be observed over and over by independent observers. It might be a chemical reaction that can be carried out again and again, or it might be a property of a substance, like temperature, that can be checked often.

Just because a fact isn't reproducible does not make it unimportant. We cannot rerun yesterday just to check weather measurements, for example. But if we are confident about our instruments and our ability to keep records, yesterday's weather maps can be trusted and can even be used to forecast tomorrow's weather.

■ In science, the most reliable facts are those that can be observed in repeated observations or measurements.

Hypotheses and Theories Are Used to Explain Facts Scientists are forever interested in underlying causes. "How does nature work?" "What must be true about what we cannot *see* in nature to account for what we do *see*?" Facts and observations, therefore, are seldom interesting all by themselves. They are valued, instead, as building blocks for *hypotheses*.

A **hypothesis** is a conjecture that appears to explain a set of facts in terms of a common cause. A hypothesis also serves as the basis for designing additional tests or experiments that could disclose the truth about the hypothesis. Is the hypothesis right or wrong? A mechanic might listen to a funny noise in your car engine, do some tests, and conclude—make a hypothesis—that a spark plug is defective. This suggests a test: Replace the plug and see if the sound disappears. In medicine, a preliminary diagnosis is a hypothesis based on observed or reported facts. A preliminary diagnosis usually suggests what new information should be sought, perhaps by further questions of the patient or by additional lab tests.

■ The aim in testing a hypothesis is not to prove the hypothesis but to discover the truth about it. Is the hypothesis right or wrong?

Sometimes hypotheses are considered on a large scale. In the history of nutrition science, for example, a number of quite different ailments simply did not fit the notion that *all* diseases are caused by germs. In searching for other causes, trace substances in certain foods were discovered that are vital to health, and the vitamin theory developed. After much research, this theory became so solidly based on experiments and tests that everybody now regards it as simply another fact: We need various vitamins to be healthy. What clinched the promotion from vitamin *theory* to fact were the discoveries of the ways that many vitamins chemically react inside cells at the molecular level of life.

A *theory* differs in scope from a hypothesis. A **theory** is an explanation for a large number of facts, observations, and hypotheses in terms of one or a few basic assumptions of what the world is like. One of the broadest, grandest theories about how our world works is that all matter is made of tiny, invisible particles called atoms and that the way a substance behaves depends on the identities and arrangements of its atoms.

The *Scientific Method* Bases Conclusions on Evidence The use of isolated facts and human reason to construct testable hypotheses and theories has occurred often in the history of science, and it is popular now to call this approach to questions the **scientific method.** The scientific method, however, is as much an attitude of mind as a procedure, an attitude of trying as hard as possible to let observations and facts control reason. Reason and logic are wondrous tools for constructing hypotheses and theories. Reason and logic might strongly suggest what is true, but neither discovers facts. Experience, observations, and experiments do that.

■ It was once perfectly logical to believe that the earth does not move and that the sun moves around it.

The scientific method is not some mechanical, cut-and-dried operation applied in a uniformly orderly way to problems. Chance discoveries, hunches, lucky guesses, false leads, and wishful thinking are found repeatedly in the history of science along with many wrong theories, beautifully logical in their day but now discarded. There is seldom anything tidy about a scientific study, but underlying it will be the will to base conclusions — whether hypotheses or grand theories — on observations, measurements, and facts.

"How?" versus "Why?" In any scientific study, asking the right question of nature is critical all in itself. Asking "*How* does nature work?" leads, for example, to more progress in understanding nature than asking "*Why* does nature work?" If we ask, for example, "*Why* do we get sick?" instead of "*How* do we get sick?" we can get bogged down in speculations that people have never resolved to everyone's satisfaction. Not that asking "Why?" isn't important, but the value of the *"How?"* question is that it gets at *mechanism*. Knowledge of the physical or chemical *mechanisms* of various illnesses doesn't answer all serious questions, but it has certainly helped to reduce pain and suffering and to raise living standards. Very few scientists would claim that science addresses *all* important questions, but science has been extremely successful with questions that ask "How?" Scientists, of course, still use the language of "Why?" After all, both "Why?" and "How?" are ways of asking "What causes . . . ?" But almost always "Why?" means "How?" in science.

To do experiments that enable us to know and describe *how* nature works involves taking measurements and using the results. This is why we must learn more about measurements very early in our study.

1.3

PROPERTIES AND PHYSICAL QUANTITIES

A physical property differs from a chemical property by being observable without changing a substance into a different substance.

A **property** is any characteristic of something that we can use to identify and recognize it when we see it again. The observations of some properties, however, change an object or a sample of a substance into something else. We can measure, for example, how much gasoline it takes to drive a car 100 miles, but this measurement uses up the gasoline. As it burns it changes into water and carbon dioxide (the fizz in soda pop). A property that, when observed, causes a substance to change into new substances is called a **chemical property,** and what is being observed is called a **chemical reaction.** A chemical property of iron, for example, is that it rusts in moist air; it changes slowly into a reddish solid, iron oxide, quite unlike metallic iron. **Chemistry** is the study of these kinds of changes in substances, how they occur, and how atoms and molecules become reorganized as they happen.

Properties such as color, height, or mass that can be observed without changing the object into something different are called **physical properties.** We usually rely on such properties to recognize and name things. For example, some physical properties of liquid water are that it is colorless and odorless; that it dissolves sugar and table salt but not butter; that it makes a thermometer read 100 °C (212 °F) when it boils (at sea level); and that if it is mixed with gasoline it will sink, not float. If you were handed a glass containing a liquid having these properties, your initial hypothesis undoubtedly would be that it is water. Think of how often each day you recognize things (and people) by simply observing physical properties.

Notice how much a description of water's properties depends on human senses, our abilities to see, taste, feel, and to sense hotness or coldness. Our senses, however, are limited, so inventors have developed instruments that extend the senses and make possible finer and sharper observations. These devices are equipped with scales or readout panels, and the data we obtain by using measuring instruments are called *physical quantities.*

A **physical quantity** is a property to which we can assign both a numerical value *and a unit.* Your own height is a simple example. If it is, say, 5.5 feet, its numerical value, 5.5, and its unit, feet, together tell us at a glance how much greater your height is than an agreed-upon reference of height, the foot.

The unit in a physical quantity is just as important as the number. If you said that your height is "two," people would ask, "Two what?" If you said "two yards," they would know what you meant (provided they knew what a yard is). (But they might ask, "*Exactly* two yards?") This example shows that we can't describe a physical property by a physical quantity without giving both a number and a unit.

$$\text{Physical quantity} = \text{number} \times \text{unit}$$

■   2 = a number
2 yards = a physical quantity

Physical Quantities Are Obtained by Measurements   A **measurement** is an operation by which we compare an unknown physical quantity with one that is known. Maybe, as you were growing up, someone measured your height by comparing it with how many sticks, perhaps one-foot rulers, it took to equal your height. Usually the number of sticks did not match your height exactly, so fractions of sticks called inches (each with their own fractions) were also used. Somebody has decided what the inch, the foot, and the yard are, and the rest of us have agreed to the definitions. That's all they are, *definitions*. We'll learn those that are the most useful in chemistry in the next section.

1.4

UNITS AND STANDARDS OF MEASUREMENT—THE INTERNATIONAL SYSTEM OF UNITS

The most fundamental quantities of measurement are called *base quantities* and each has an official *standard of reference* for one unit.

Mass, Length, Time, and Temperature Are *Base Quantities*   The most fundamental measurements in chemistry are those of mass, volume, temperature, time, and quantity of chemical substance.

Mass is the measure of the inertia of an object. Anything said to have a high inertia, such as a train engine, a massive boulder, or an oceanliner, is very hard to get into motion or, if it is in motion, it is difficult to slow it down or make it change course. It is this inherent resistance to

Large inertia goes with large mass.

■ *Quantitative* describes something expressible by a number and a unit.

A traditional two-pan balance showing a small container on the left pan and some weights on the right pan.

Volume of a cube = $(l)^3$

■ The other two SI base quantities are *electric current* and *luminous intensity.* Their base units are called the *ampere* and the *candela,* respectively.

■ An alloy is a mixture of two or more metals made by mixing them in their molten states.

any kind of change in motion that we call **inertia,** and *mass* is our way of describing inertia quantitatively. A large inertia means a large *mass.*

A large mass doesn't always mean a large *weight.* Your mass does not depend on where you are in the universe, but your weight does. The **weight** of an object is a measure of the gravitational force of attraction that the earth makes on the object. The gravitational force is less on the moon, which is a smaller object than the earth — about six times less. Thus, an astronaut's *weight* on the moon is 1/6th of its value on earth, but the astronaut's *mass,* the fundamental resistance to any change in motion, is the same in both locations.

When we use a laboratory balance to *weigh* something, we are actually measuring mass because we are comparing two weights *at the same place on the earth* and, therefore, under the same gravitational influence. One weight is the quantity being measured, and the other is a "weight" (or set of weights) built into the weighing balance. Although we commonly call the result of the measurement a "weight," we'd more properly call it the *mass* of the object or the sample. (We'll generally speak of masses, not weights, in this book.)

The **volume** of an object is the space it occupies, and space is described by means of a more basic physical quantity, length. The volume of a cube, for example, is the product of (length) × (length) × (length), or (length)³. **Length** is a physical quantity that describes how far an object extends in a direction, or it is the distance between two points.

A fundamental quantity such as mass or length is called a **base quantity,** and any other quantity such as volume that can be described in terms of a base quantity is called a **derived quantity.**

Another base quantity in science is **time** — our measure of how long events last. We need this quantity to describe how rapidly the heart beats, for example, or how fast some chemical reaction occurs.

Still another important physical quantity is **temperature,** which we use to describe the hotness or coldness of an object.

All of these base quantities are necessary to all sciences, but chemistry has a special base quantity called the *mole* that describes a certain amount of a chemical substance. (We'll study it in Chapter 3.)

Every Base Unit Has a *Reference Standard* of Measurement To measure and report an object's mass, its temperature, or any of its other base or derived physical quantities we obviously need some units and some references. By international treaties among the countries of the world, the reference units and standards are decided by a diplomatic organization called the General Conference of Weights and Measures, headquartered in Sèvres, a suburb of Paris, France. The General Conference has defined a unit called a **base unit** for each of seven base quantities, but at this time we need units only for the five that we have already mentioned: mass, length, time, temperature, and mole. We also need some units for the derived quantities, for example, for volume, density, pressure, and heat. The standards and definitions of base and derived quantities and units together make up what is now known as the **International System of Units** or the **SI** (after the French name, *Système Internationale d'Unites*).

Each base unit is defined in terms of a **reference standard,** a physical description or embodiment of the base unit. Long ago, the reference (such as it was) for the *inch* was "three barleycorns, round and dry, laid end to end." Obviously, which three barleycorns were picked had a bearing on values of length under this "system." And if the barleycorns got wet, they sprouted. You can see that a reference standard ought to have certain properties if it is to serve the needs of all countries. It should be entirely free of such risks as corrosion, fire, war, theft, or plain skulduggery, and it should be accessible at any time to scientists in any country. The improvement that the SI represents over its predecessor, the *metric system,* is not so much in base units as in their reference standards.

The SI base unit of length is called the **meter,** abbreviated **m,** and its reference standard is called the *standard meter.* Until 1960, the standard meter was the distance separating two thin scratches on a bar of platinum – iridium alloy stored in an underground vault in Sèvres. This bar, of course, could have been lost or stolen, so the new reference for the meter is based on a property of light, something available everywhere, in all countries, and that obviously

TABLE 1.1
Some Common Measures of Length[a]

SI	U.S. Customary
1 kilometer (km) = **1000** meters (m)	1 mile (mi) = **5280** feet (ft)
1 meter = **100** centimeters (cm)	= **1760** yards (yd)
1 centimeter = **10** millimeters (mm)	1 yard = **3** feet (ft)
	1 foot = **12** inches (in.)

Other Relationships
1 meter = 39.37 inches   1 inch = **2.54** centimeters

[a] Numbers in boldface are exact.

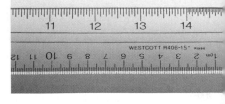

Common ruler marked in inches and centimeters. Notice that there are ten 1-millimeter spaces marked between the centimeter divisions.

can't be lost or damaged. This change in reference did not change the actual length of the meter. It only changed its official reference.[1]

In the United States, older units are now legally defined in terms of the meter. For example, the yard (yd), roughly nine-tenths of a meter, is defined as 0.9144 m (exactly). The foot (ft), roughly three-tenths of a meter, is defined as 0.3048 m (exactly).

In chemistry, the meter is usually too long for convenience, and submultiples are often used, particularly the **centimeter,** or **cm**, and the **millimeter,** or **mm**. Expressed mathematically, these are defined as follows:

$$1 \text{ m} = 100 \text{ cm} \quad \text{or} \quad 1 \text{ cm} = 0.01 \text{ m}$$
$$1 \text{ m} = 1000 \text{ mm} \quad \text{or} \quad 1 \text{ mm} = 0.001 \text{ m}$$
$$1 \text{ cm} = 10 \text{ mm} \quad \text{or} \quad 1 \text{ mm} = 0.1 \text{ cm}$$

Paper clip   0.4 g

Penny   3.1 g

Notice that the subunits are in fractions based on 10. The millimeter, for example, is one-tenth of a centimeter. As we'll often see, this makes many calculations much easier than they are under other systems (for example, where the inch is one-twelfth of a foot, and the foot is one-third of a yard).

The inch (in.) is about two and a half centimeters; more exactly,

$$1 \text{ in.} = 2.54 \text{ cm} \quad \text{(exactly)}$$

Table 1.1 gives several relationships between various units of length.

The SI base unit of mass is named the **kilogram,** abbreviated **kg**, and its reference is named the *standard kilogram mass.* This is a cylindrical block of platinum–iridium alloy housed at Sèvres under the most noncorrosive conditions possible (Fig. 1.1). This is the only SI reference

FIGURE 1.1
The SI standard kilogram mass kept at the International Bureau of Weights and Measures in France. It is made out of a very corrosion-resistant alloy of platinum and iridium.

[1] The SI now defines the standard meter as how far light will travel in 1/299,792,458th of a second. It is thus based on the speed of light as measured by an "atomic clock."

1 kg of butter

TABLE 1.2

Some Common Measures of Mass[a]

SI

1 kilogram (kg) = **1000** grams (g)
1 gram = **1000** milligrams (mg)
1 milligram = **1000** micrograms (μg, γ, or mcg)[b]

U.S. Customary (avoirdupois)[c]

1 short ton = **2000** pounds (lb avdp)
1 pound = **16** ounces (oz avdp)

Other Relationships

1 kilogram = 2.205 lb   1 lb avdp = 453.6 grams

---

[a] Numbers in boldface are exact.

[b] The microgram is sometimes called a *gamma* in medicine and biology.

[c] These are the common units in the United States.

that could still be lost or stolen, but no alternative has yet been devised. Duplicates made as much like the original as possible are stored in other countries. One kilogram has a mass roughly equal to 2.2 pounds, and Table 1.2 gives some useful relationships among various quantities and units of mass.

The most often used units of mass in chemistry are the kilogram (kg), the **gram (g),** the **milligram (mg),** and the **microgram (μg).** These are defined as follows:

$$1 \text{ kg} = 1000 \text{ g} \quad \text{or} \quad 1 \text{ g} = 0.001 \text{ kg}$$
$$1 \text{ g} = 1000 \text{ mg} \quad \text{or} \quad 1 \text{ mg} = 0.001 \text{ g}$$
$$1 \text{ mg} = 1000 \text{ }\mu\text{g} \quad \text{or} \quad 1 \text{ }\mu\text{g} = 0.001 \text{ mg}$$

■ Unless otherwise specified, the U.S. customary (avoirdupois or avdp) system will be meant when we refer to *common* mass units in this text.

Lab experiments in chemistry usually involve grams or milligrams of substances.

The SI unit of volume, one of the important derived units, is the cubic meter, m^3, but this is much too large for convenience in chemistry. An older unit, the **liter,** abbreviated **L,** is accepted as a *unit of convenience.* The liter occupies a volume of 0.001 m^3 (exactly), and one liter is almost the same as one liquid quart; 1 quart (qt) = 0.946 L.

■ One cubic meter holds a little over 250 gallons.

Even the liter is often too large for convenience in chemistry, and two submultiples are used, the **milliliter (mL)** and the **microliter (μL).** These are related as follows.

$$1 \text{ L} = 1000 \text{ mL} \quad \text{or} \quad 1 \text{ mL} = 0.001 \text{ L}$$
$$1 \text{ mL} = 1000 \text{ }\mu\text{L} \quad \text{or} \quad 1 \text{ }\mu\text{L} = 0.001 \text{ mL}$$

One drop of water is about 60 mg.

TABLE 1.3

Some Common Measures of Liquid Volume[a]

SI

1 cubic meter (m^3) = **1000** liters (L)
1 liter = **1000** milliliters (mL)
1 milliliter = **1000** microliters (μL)

U.S. Customary

1 gallon (gal) = **4** liquid quarts (liq qt)
1 liquid quart = **2** liquid pints (liq pt)
1 liquid pint = **16** liquid ounces (liq oz)

Other Relationships

1 cubic meter = 264.2 gallons   1 liter = 1.057 liquid quarts
1 liquid quart = 946.4 milliliters   1 liquid ounce = 29.57 milliliters

---

[a] Numbers in boldface are exact.

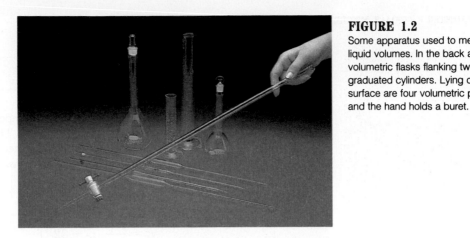

In routine chemistry work, the milliliter is by far the most common unit you will encounter. Table 1.3 gives several other relationships among units of volume. Figure 1.2 shows apparatus used to measure volumes in the lab.

The SI unit of time is called the **second,** abbreviated **s.** The *duration* of the second is 1/18,400 of a mean solar day. (Although the duration is basically the same, the actual SI *definition* of the second is different from this, but it involves complexities of atomic physics unneeded in our study.) Decimal-based multiples and submultiples of the second are used in science, but so are such deeply entrenched old units as minute, hour, day, week, month, and year.

The SI unit for degree of temperature is called the **kelvin, K.** (Be sure to notice that the abbreviation is K, not °K.) This degree used to be called the **degree Celsius (°C),** and even earlier the *degree centigrade* (also °C). Then it was defined as 1/100 the interval between the freezing point of water (named 0 °C) and the boiling point of water (named 100 °C). The most extreme coldness possible is −273.15 °C, and this is named 0 K on the **Kelvin scale.** The kelvin is the name of the degree on the Kelvin scale, and it is identical in size with the Celsius degree. Only the *numbers* assigned to points on the scales differ. See Figure 1.3, where the scales are compared.

Because 0 K corresponds to −273.15 °C, we have the following simple relationships between kelvins and degrees Celsius (where we follow common practice of rounding 273.15 to 273).

$$°C = K - 273$$
$$K = °C + 273$$

■ The *kelvin* is named after William Thomson, Baron Kelvin of Largs (1842–1907), a British scientist.

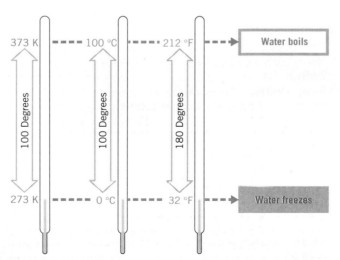

FIGURE 1.3
Relationships among the Kelvin, Celsius, and Fahrenheit scales of temperature.

PRACTICE EXERCISE 1

Normal body temperature is 37 °C. What is this in kelvins?

■

The Kelvin scale is used in chemistry mostly to describe temperatures of gases. The Celsius scale is more popular for most other uses, including medicine. The **degree Fahrenheit (°F)** is five-ninths the size of the degree Celsius. To convert a Celsius temperature, t_C, to a Fahrenheit temperature, t_F, we can use either of the following equations.

$$t_C = \frac{5\ °C}{9\ °F}\ (t_F - 32\ °F)$$

$$t_F = \frac{9\ °F}{5\ °C}\ t_C + 32\ °F$$

Table 1.4 gives some common temperatures in both °C and °F.

1.5

SCIENTIFIC NOTATION

Scientific notation expresses very large or very small numbers in exponential form to make comparisons and calculations easier.

The typical human red blood cell has a diameter of 0.000008 m. Whether we want to write it, say it, or remember it, 0.000008 m is an awkward quantity, and to make life easier scientists have developed a method called **scientific notation** for recording very small or very large numbers. In scientific notation (sometimes called exponential notation), a number is written as the product of two numbers. The first is a decimal number with a value usually between 1 and 10, although sometimes a wider range is used. Following this number is a times (✕) sign and then the number 10 with an exponent or power. For example, we can write the number 4000 as follows.

■ Appendix A has a review of exponential numbers.

$$4000 = 4 \times 1000 = 4 \times 10 \times 10 \times 10$$
$$= 4 \times 10^3$$

■ When the decimal point is omitted, we assume that it is after the last digit in the number.

Notice that the exponent 3 is the number of places to the left that we have to move the decimal point in 4000 to get to 4, which is a number in the desirable range.

$$4\ \underset{3\quad 2\quad 1}{0\ 0\ 0}$$

TABLE 1.4
Some Common Temperature Readings in °C and °F

	°F	°C
Room temperature	68	20
Very cold day	−20	−29
Very hot day	100	38
Normal body temperature	98.6[a]	37
Hottest temperature the hands can stand	120	49

[a] A revision in this value is currently under way. In some healthy people, the normal temperature is as low as 98.2 °F and in others as high as 99 °F.

If our large number is 42,195, the number of meters in a marathon distance, we can rewrite it as follows after figuring out that we have to move the decimal point four places to the left to get a decimal number between 1 and 10.

$$42{,}195 \text{ m} = 4.2195 \times 10^4 \text{ m}$$

In rewriting numbers smaller than 1 in scientific notation, we have to move the decimal point to the *right* to get a number in the acceptable range of 1 to 10. This number of moves is the value of the *negative* exponent of 10. For example, we can rewrite 0.000008 as

$$0.000008 = 8 \times 10^{-6}$$

You should not continue until you are satisfied that you can change large or small numbers into scientific notation. For practice, do the following exercises.

PRACTICE EXERCISE 2

Express each number in scientific notation.[2] Let the decimal part be a number between 1 and 10.

(a) 545,000,000   (b) 5,670,000,000,000   (c) 6454
(d) 25   (e) 0.0000398   (f) 0.00426
(g) 0.168   (h) 0.00000000000987 (See footnote 2.)

Prefixes to the Names of SI Base Units Are Used to Specify Fractions or Multiples of These Units If we rewrite 3000 m as 3×10^3 m and try to pronounce the result, we have to say "three times ten to the third meters." There's nothing wrong with this, but it's clumsy. This is why the SI has names for several exponential expressions — not independent names but prefixes that can be attached to the name of any unit. For example, 10^3 has been assigned a prefix name of *kilo-*, abbreviated *k-*. Thus 1000 or 10^3 meters can be called 1 kilometer. Abbreviated, this becomes 10^3 m = 1 km.

With just a few exceptions, the prefixes defined by the SI go with exponentials that involve powers of 3, 6, 9, 12, 15, and 18 or powers of $-3, -6, -9, -12, -15$, and -18. These are all divisible by 3. Table 1.5 has a list of the SI prefixes and their symbols. Those given in boldface are so often encountered in chemistry that they should be learned now.

Notice that there are four prefixes that do not go with powers divisible by 3. The SI hopes their usage will gradually fade away, but this hasn't happened yet. The two in boldface have to be learned. However, *centi* is used almost entirely in just one physical quantity, the centimeter. *Deci* is limited almost completely to another physical quantity, the deciliter (100 mL or 1/10 L), and you won't see it often in strictly chemical situations. (Clinical chemists often use the deciliter because it saves space on clinical report sheets to abbreviate 100 mL to dL.)

To take advantage of the SI prefixes, we sometimes have to modify a rule used in converting a large or small number into scientific notation. The goal in this conversion will now be to get the exponential part of the number to match one with an SI prefix even if the decimal part of the number isn't between 1 and 10. For example, we know that the number 545,000

■ $4\,\underbrace{2\,1\,9\,5}$.
   4 3 2 1

■ $0.\underbrace{0\,0\,0\,0\,0\,8}$
   1 2 3 4 5 6

■ 1 dL = 1×10^{-1} L = 1/10 liter
   But 1/10 liter = 100 mL.

Therefore, **1 dL = 100 mL**

---

[2] Some of the numbers in this exercise illustrate a small problem that the SI is trying to get all scientists to handle in a uniform way. In part (h), for example, you might become dizzy trying to count closely spaced zeros. The SI recommends — and most European scientists have accepted the suggestion — that the digits in numbers having four or more digits be grouped in threes separated by thin spaces. For large numbers, just omit the commas. Thus 545,000,000 would be written as 545 000 000. The number 0.00000000000987 becomes 0.000 000 000 009 87. You will not soon see this as common usage in the United States, but when you do you'll know what it means. Incidentally, European scientists use a comma instead of a period to locate the decimal point. You might see this yourself soon when you first weigh something in the lab. If the weighing balance was made in Europe, a reading such as 1,045 g means 1.045 g.

TABLE 1.5
SI Prefixes for Multiples and Submultiples of Base Units[a]

Relationship	Prefix	Symbol
$1\ 000\ 000\ 000\ 000\ 000\ 000 = 10^{18}$	exa	E
$1\ 000\ 000\ 000\ 000\ 000 = 10^{15}$	peta	P
$1\ 000\ 000\ 000\ 000 = 10^{12}$	tera	T
$1\ 000\ 000\ 000 = 10^{9}$	giga	G
$1\ 000\ 000 = 10^{6}$	**mega**	**M**
$1\ 000 = 10^{3}$	**kilo**	**k**
$100 = 10^{2}$	hecto	h
$10 = 10^{1}$	deka	da
$0.1 = 10^{-1}$	**deci**	**d**
$0.01 = 10^{-2}$	**centi**	**c**
$0.001 = 10^{-3}$	**milli**	**m**
$0.000\ 001 = 10^{-6}$	**micro**	**μ**
$0.000\ 000\ 001 = 10^{-9}$	nano	n
$0.000\ 000\ 000\ 001 = 10^{-12}$	pico	p
$0.000\ 000\ 000\ 000\ 001 = 10^{-15}$	femto	f
$0.000\ 000\ 000\ 000\ 000\ 001 = 10^{-18}$	atto	a

[a] The most commonly used prefixes and their symbols are in bold-face. Thin spaces instead of commas are used to separate groups of three zeros to illustrate the format being urged by the SI (but not yet widely adopted in the United States).

can be rewritten as 5.45×10^{5}, but 5 isn't divisible by 3, and there isn't an SI prefix to go with 10^{5}. If we counted 6 spaces to the left, however, we could use 10^{6} as the exponential part.

$$5\ 4\ 5\ 0\ 0\ 0 = 0.545 \times 10^{6}$$
$$\quad 6\ \ 5\ \ 4\ \ 3\ \ 2\ \ 1$$

■ We usually put a zero in front of a decimal point in numbers that are less than 1, such as in 0.545. The zero helps us to remember that the decimal point is there.

Now we could rewrite 545,000 m as 0.545×10^{6} m or 0.545 Mm (megameter), because the prefix *mega*, abbreviated M, goes with 10^{6}.

We also could have rewritten 545,000 as 545×10^{3}, and then 545,000 m could have been written as 545 km (kilometers) because *kilo* goes with 10^{3}.

EXAMPLE 1.1
Rewriting Physical Quantities Using SI Prefixes

Bacteria that cause pneumonia have diameters roughly equal to 0.0000009 m. Rewrite this using the SI prefix that goes with 10^{-6}.

ANALYSIS In straight exponential notation, 0.0000009 m is 9×10^{-7} m, but -7 is not divisible by 3 and no SI prefix goes with 10^{-7}. If we move the decimal six places instead of seven to the right, however, we get 0.9×10^{-6} m, and -6 is divisible by 3.

SOLUTION The prefix for 10^{-6} is *micro* with the symbol μ, so

$$0.0000009\ \text{m} = 0.9 \times 10^{-6}\ \text{m} = 0.9\ \mu\text{m}$$

The diameter of one of these bacteria is 0.9 micrometers (0.9 μm).

PRACTICE EXERCISE 3

Complete the following conversions to exponential notation by supplying the exponential parts of the numbers.

(a) $0.0000398 = 39.8 \times$ _____   (b) $0.000000798 = 798 \times$ _____
(c) $0.000000798 = 0.798 \times$ _____   (d) $16500 = 16.5 \times$ _____

PRACTICE EXERCISE 4

Write the abbreviation of each of the following.

(a)   milliliter   (b)   microliter   (c)   deciliter
(d)   millimeter   (e)   centimeter   (f)   kilogram
(g)   microgram   (h)   milligram

PRACTICE EXERCISE 5

Write the full name that goes with each of the following abbreviations.

(a)   kg   (b)   cm   (c)   dL   (d) μg
(e)   mL   (f)   mg   (g)   mm   (h) μL

PRACTICE EXERCISE 6

Rewrite the following physical quantities using the standard SI abbreviated forms to incorporate the exponential parts of the numbers.

(a) 1.5×10^6 g   (b) 3.45×10^{-6} L   (c) 3.6×10^{-3} g
(d) 6.2×10^{-3} L   (e) 1.68×10^3 g   (f) 5.4×10^{-1} m

PRACTICE EXERCISE 7

Express each of the following physical quantities in a way that uses an SI prefix.
(a)   275,000 g   (b)   0.0000625 L   (c)   0.000000082 m

1.6

ACCURACY, ERROR, UNCERTAINTY, AND PRECISION

The way in which the number part of a physical quantity is expressed says something about the uncertainty of the measurement but nothing about its accuracy or precision.

Most people use the terms *accuracy* and *precision* as if they meant the same thing, but they don't. The same is true about the terms *error* and *uncertainty*. **Accuracy** refers to the closeness of a measurement to the true value. In an accurate measurement, the instrument is faithful; its scale, gauge, or needle is very steady; and the experimenter has good eyesight and knows how to use the instrument. The **error** in a measurement is the difference between an experimental value and the correct value. In an accurate measurement, the error is small.

We assume here that you are able to begin the test with your body exactly opposite the 0-mile post and that your body is exactly opposite the 5-mile post when you take your odometer reading. We also assume that the highway department has accurately positioned the posts. You can see that there are many sources of uncertainty in even an ordinary measurement such as this.

Gauges, readout panels, and scales can vary among instruments meant for measuring the same physical quantity, like volume or mass. The **uncertainty** in a measurement is expressed by the *range* in values that must be recorded because of the need to estimate the last digit being read. If you have ever tried to test the *accuracy* of an automobile or bicycle odometer (mileage gauge) against mile posts set up by the local highway department for such tests, you have sensed the problem—the *uncertainty*—of estimating the reading beyond the first decimal place. You might have to record the odometer reading as 5.1 ± 0.1 mi as you pass the 5-mile post, because you judge that you cannot read the odometer more closely. The symbol $\pm$ stands for "plus or minus," and what follows this symbol indicates how much uncertainty is carried in the last digit. By recording 5.1 ± 0.1 mi, the mileage is said to be between 5.0 and 5.2, so the range of uncertainty is 2 in the tenths position. The tenths position has the first uncertain digit. You can see that *uncertainty* does not directly supply information about the *accuracy* of the measurement. *Uncertainty* is only an estimate of how finely the number could be read at the time it was taken. *Accuracy,* we repeat, is the closeness of a measurement to the true value, and only the care of the highway department in planting its mile posts, your own care in reading, and the steadiness of the odometer bear on the true value of the mileage.

When a measurement can be repeated, the experimenter takes several measurements as carefully as possible. This strategy produces data needed to calculate the *precision* with which a physical quantity is known from the measurements. **Precision** is a measure of how *reproducible* the measurements are when several measurements are taken. The results of the measurements are averaged, and the *precision* is calculated by some index, such as the average absolute deviation from the mean or by a standard deviation. These indices are described in the field of statistics; they will no longer concern us because in routine laboratory work involving measurements of mass or volume, almost never is more than one measurement taken of the same quantity. It is important to realize, however, that *precision* and *accuracy* are not the same concepts. One could have an average of several measurements all agreeing very closely with each other (and so of high *precision*), but still have an average value grossly different from the true value (and so of great *error*) because the instrument happens to have been inaccurately manufactured.

Figure 1.4 illustrates the difference between accuracy and precision in the measurement of someone's height. Each dot represents one measurement. In the first set, the dots are tightly clustered close to or exactly at the true value, and obviously a skilled person was at work with a carefully manufactured meter stick. This set illustrates both high precision and great accuracy. In the second set, a skilled person, without realizing it, used a faulty meter stick, one mislabeled by a few centimeters. The precision is as great as that shown by the first set, because the successive measurements agree well with each other. But they're all untrue, so the accuracy is poor. In the third set of measurements, someone with a good meter stick did careless work. Only by accident do the values average to the true value, so the accuracy, in terms of the average, turned out to be high, but the precision is terrible and no one would really trust the average. The last set displays no accuracy and no precision.[3]

No matter what physical quantities we use, we want to be able to judge how accurately they were measured, but this presents a problem. When we read the value of some physical quantity in a report or a table, we have no way of telling *from it alone* if it is the result of an accurate measurement. Someone might write, for example, "4.5678 mg of antibiotic," but in spite of all its digits we can't tell from this report alone if the balance was working or if the person using it knew how to handle it and read it correctly. A skilled and careful experimenter frequently checks the instruments against references of known accuracy. Thus the question of *accuracy* is a human problem. We learn to trust the *accuracy* of data by employing trained people, giving them good instruments, requiring that they prove they are doing consistently accurate work, and rewarding consistently good results.

---

[3] Suggested additional reading: Charles J. Guare, "Error, precision, and uncertainty," *Journal of Chemical Education,* August 1991, p. 649.

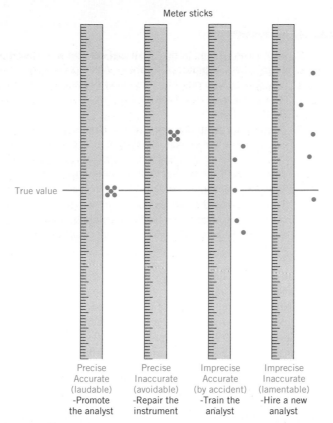

Meter sticks

FIGURE 1.4
Accuracy and precision

True value ————

| Precise
Accurate
(laudable)
-Promote
the analyst | Precise
Inaccurate
(avoidable)
-Repair the
instrument | Imprecise
Accurate
(by accident)
-Train the
analyst | Imprecise
Inaccurate
(lamentable)
-Hire a new
analyst |

The Number of Significant Figures in a Physical Quantity Is the Number of Digits That Are Known to Be Accurate plus One More When we do not wish (or need) to specify the *range* of uncertainty in a measurement, we need an understanding about this range. Then we can round off the numerical part of a physical quantity to leave it with a particular number of *significant figures*. The number of **significant figures** in a physical quantity is the number of digits known with complete certainty to be accurate plus one more. Suppose, for example, that you read a report that refers to "4.56 mg of antibiotic." You would be in a quandary because you'd wonder about the range of uncertainty in this quantity, about what should be in place of the question mark in "4.56 ± ?" We solve the problem in this text by assuming that the question mark is replaceable by *one unit of the last decimal place shown*; in our example, the reported quantity would, by this rule, mean 4.56 ± 0.01 mg. The quantity "4.56 mg," therefore, has three significant figures. The first two, the 4 and the 5, are known to be accurate, but in the last digit the analyst is acknowledging a small uncertainty (± 0.01). This report means that the actual measurement of the mass of this one sample was closer to 4.56 mg than to 4.55 mg or 4.57 mg. If the mass is reported as 4.560 mg, then it carries four significant figures. The 4, 5, and 6 are certainly accurate, but there is some uncertainty in the last digit, 0. The measurement, as reported, means a value closer to 4.560 mg than to 4.559 mg or 4.561 mg. Thus "4.560 mg" implies a greater certainty or fineness of measurement than 4.56 mg.

Figuring out how many significant figures there are in a number is easy, provided we have an agreement on how to treat zeros. Are all the zeros counted as *significant* in such quantities as 4,500,000 people, or 0.0004500 L, or 400,005 m? We will use the following rules to decide.

Rules Governing Significant Figures

1. Zeros sandwiched between nonzero digits are always counted as significant Thus both 400,005 and 400.005 have six significant figures. Both 4056 g and 4506 g have four significant figures.

■ The other zeros are needed to locate the decimal points in these numbers, and they definitely are important in this sense. They just have nothing to do with precision.

2. **Zeros that do no more than set off the decimal point on their *left* are never counted as significant figures**   Although such zeros are necessary to convey the general *size* of a quantity, they don't say anything about the *certainty* of the measurement. Thus such quantities as 0.045 mL, 0.0045 mL, and 0.00045 mL all have only two significant figures.
3. **Trailing zeros to the *right* of the decimal point are always significant**   Trailing zeros are any that come to the right of a decimal point at the very end of the number, as in 4.56000. This number has three trailing zeros, and because they are to the right of the decimal point, all are significant. The number 4.56000 has six significant figures and represents considerable certainty or fineness of measurement.
4. **Trailing zeros that are to the *left* of the decimal point are counted as significant only if the author of the book or article has somewhere said so**   *In this text, if we leave trailing zeros before the decimal point, count them as significant figures.*

Rule 4 is really the only tricky rule. The zeros in 4,500,000, for example, are trailing zeros to the left of the decimal point, but are they significant? Suppose this number stands for the population of a city. A city's population changes constantly as people are born and die, and as they move in and out. No one could claim to know a population is *exactly* 4,500,000 people—not 4,499,999 and not 4,500,001, but 4,500,000 people. Most scientists handle this problem by restating the number in scientific notation so that any desired trailing zeros can be placed *after* the decimal point. By doing this, as many or as few of such zeros can be given to convey the proper degree of certainty. If the census bureau feels that the population is known to be closer to 4,500,000 people than to 4,400,000 or to 4,600,000 people, and that no better certainty than this is possible, then only two significant figures should be in the result, for example, 4.5×10^6 people. Giving the population as 4.50×10^6 people indicates a greater certainty—to three significant figures. There are four significant figures in 4.500×10^6.

Not everyone agrees with this way of handling trailing zeros that stand to the *left* of the decimal point, so you have to be careful. Some say that they aren't significant unless the decimal point is actually given, as in 45,000. L. When the decimal point is the last item in a number, however, it is easily forgotten at the time of making the record. This problem is avoided by switching to scientific notation so that all trailing zeros come after the decimal point. This is the practice we will usually follow in this book, unless noted otherwise, or unless the context makes the intent very clear.

A Few Rules Govern the Rounding Off of Calculated Physical Quantities   When we mathematically combine the values of two or more measurements, we usually have to round the result so that it expresses the same amount of uncertainty allowed by the data. Normally, such rounding is done at the *end* of a calculation (unless specified otherwise) to minimize the errors that can accumulate and grow when we round at intermediate steps in a multistep calculation. We will use four simple rules for rounding calculated quantities.[4]

Rules for Rounding Calculated Results

1. When we multiply or divide quantities, the result is allowed no more significant figures than carried by the least certain quantity (the one with the fewest significant figures).
2. When we add or subtract numbers, the result is allowed no more decimal places than are in the number having the fewest decimal places.
3. When the first of the digits to be removed by rounding is 5 or higher, round the digit to its left *upward* by one unit. Otherwise, drop it and all others after it.
4. Treat *exact numbers* as having an infinite number of significant figures.

[4] When the actual range in uncertainty is known from experiment for each piece of data, the uncertainty in a calculated result can be expressed with greater care. See the reference in footnote 3.

An **exact number** is any that we define to be so, and we usually encounter exact numbers in statements relating units. For example, all of the numbers in the following expressions are exact and, for purposes of rounding calculated results, have an infinite number of significant figures.

$$1 \text{ in.} = 2.54 \text{ cm} \qquad \text{(exactly, as } defined \ by \ law\text{)}$$
$$1 \text{ L} = 1000 \text{ mL} \qquad \text{(exactly, by the } definition \text{ of mL)}$$

■ We use the period in the abbreviation of inch (in.) to avoid any confusion with the preposition *in*, which has the same spelling.

The significance of having an infinite number of significant figures lies in our not letting such numbers affect how we round results. It would be silly to say that the "1" in "1 L" has just one significant figure when we intend, by definition, that it be an exact number.

EXAMPLE 1.2
Rounding the Result of a Multiplication or a Division

A floor is measured as 11.75 m long and 9.25 m wide. What is its area, correctly rounded by our rules?

SOLUTION

$$\text{Area} = (\text{length}) \times (\text{width})$$
$$= 11.75 \text{ m} \times 9.25 \text{ m}$$
$$= 108.6875 \text{ m}^2 \qquad \text{(not rounded)}$$

But the measured width, 9.25 m, has only three significant figures whereas the length, 11.75 m, has four. By our rules, we have to round the calculated area to three significant figures.

$$\text{Area} = 109 \text{ m}^2 \qquad \text{(correctly rounded)}$$

■ Resist the impulse that some owners of new calculators have of keeping all the digits they paid for.

EXAMPLE 1.3
Rounding the Result of an Addition or a Subtraction

Samples of a medication having masses of 1.12 g, 5.1 g, and 0.1657 g are mixed. How should the total mass of the resulting sample be reported?

SOLUTION The sum of the three values, obtained with a calculator, is 6.3857 g, which shows four places following the decimal point. However, one mass is precise only to the first decimal place, so by our rules we have to round to this place. The final mass should be reported as 6.4 g. Notice that the value of the second sample mixed, 5.1 g, says nothing about the third or fourth decimal places. We don't know whether the mass is 5.101 g or 5.199 g, or what; the sample just wasn't measured precisely. This is why we can't know anything beyond the first decimal place in the sum.

PRACTICE EXERCISE 8

The following numbers are the numerical parts of physical quantities. After the indicated mathematical operations are carried out, how must the results be expressed?

(a) 16.4×5.8
(b) $5.346 + 6.01$
(c) 0.00467×5.6324
(d) $2.3000 - 1.00003$
(e) $16.1 + 0.004$
(f) $(1.2 \times 10^2) \times 3.14$
(g) $9.31 - 0.00009$
(h) $\dfrac{1.0010}{0.0011}$

1.7

THE FACTOR–LABEL METHOD IN CALCULATIONS

In calculations involving physical quantities, the units are multiplied or canceled as if they were numbers.

Many people have developed a mental block about any subject that requires the use of mathematics. They know perfectly well how to multiply, divide, add, and subtract, but the problem is in knowing *when,* and no pocket calculator tells this. We said earlier that the inch is defined by the relationship, 1 in. = 2.54 cm. This fact has to be used when a problem asks for the number of centimeters in some given number of inches, but for some people the problem arises in knowing whether to divide or multiply.

Science teachers have worked out a method called the *factor-label* method for correctly setting up such a calculation and *knowing* that it is correct. The **factor-label method** takes a relationship between units stated as an equation (such as 1 in. = 2.54 cm), expresses the relationship in the form of a fraction, called a **conversion factor,** and then multiplies some given quantity by this conversion factor. In this multiplication, identical units (the "labels") are multiplied or canceled as if they were numbers. If the units that remain for the answer are right, then the calculation was correctly set up. We can learn how this works by doing an example, but first let's see how to construct conversion factors.

The relationship, 1 in. = 2.54 cm, can be restated in either of the following two ways, and both are examples of conversion factors.

$$\frac{2.54 \text{ cm}}{1 \text{ in.}} \quad \text{or} \quad \frac{1 \text{ in.}}{2.54 \text{ cm}}$$

If we read the divisor line as "per," then the first conversion factor says "2.54 cm per 1 in." and the second says "1 in. per 2.54 cm." These are merely alternative ways of saying that "1 in. equals 2.54 cm." Any relationship between two units can be restated as two conversion factors. For example,

$$1 \text{ L} = 1000 \text{ mL} \qquad \frac{1000 \text{ mL}}{1 \text{ L}} \quad \text{or} \quad \frac{1 \text{ L}}{1000 \text{ mL}}$$

$$1 \text{ lb} = 453.6 \text{ g} \qquad \frac{453.6 \text{ g}}{1 \text{ lb}} \quad \text{or} \quad \frac{1 \text{ lb}}{453.6 \text{ g}}$$

PRACTICE EXERCISE 9

Restate each of the following relationships in the forms of their two possible conversion factors.

(a) 1 g = 1000 mg   **(b)** 1 kg = 2.205 lb

Suppose we want to convert 5.65 in. into centimeters. The first step is to write down what has been given, 5.65 in. Then we multiply this by the one conversion factor relating inches to centimeters that lets us cancel the unit no longer wanted and leaves the unit we want.

$$5.65 \text{ in.} \times \frac{2.54 \text{ cm}}{1 \text{ in.}} = 14.4 \text{ cm} \qquad \text{(rounded correctly from 14.351 cm)}$$

Notice how the units of "in." cancel. Only "cm" remains, and it is on top in the numerator where it has to be. Suppose we had used the wrong conversion factor.

$$5.65 \text{ in.} \times \frac{1 \text{ in.}}{2.54 \text{ cm}} = 2.22 \frac{\text{in.}^2}{\text{cm}} \qquad \text{(correctly rounded)}$$

■ Some call the factor-label method the cancel-unit or the factor-unit method.

■ When we divide both sides of the equation 2.54 cm = 1 in. by 2.54 cm, we get

$$\frac{2.54 \text{ cm}}{2.54 \text{ cm}} = \frac{1 \text{ in.}}{2.54 \text{ cm}}$$

This only restates the relationship of the centimeter and the inch; it doesn't change it. The use of a conversion factor just changes *units,* not actual quantities.

■ The arithmetic is correct, but the result is still all wrong.

Surprised? We *must* do to the units exactly what the times sign and the divisor line tell us, and (in.) times (in.) equals (in.)^2 just as $2 \times 2 = 2^2$. Of course, the units in the answer, (in.)^2/cm, make no sense, so we know with certainty that we can't set up the solution this way. The reliability of the factor-label method lies in this use of the units (the "labels") as a guide to setting up the solution. Now let's work an example.

EXAMPLE 1.4
Using the Factor-Label Method

How many grams are in 0.230 lb?

ANALYSIS From Table 1.2, we find that 1 lb = 453.6 g, so we have our pick of the following conversion factors.

$$\frac{453.6 \text{ g}}{1 \text{ lb}} \quad \text{or} \quad \frac{1 \text{ lb}}{453.6 \text{ g}}$$

To change 0.230 lb into grams, we want "lb" to cancel and we want "g" in its place in the numerator. Therefore, we pick the first conversion factor; it's the only one that can give this result.

SOLUTION

$$0.230 \text{ lb} \times \frac{453.6 \text{ g}}{1 \text{ lb}} = 104 \text{ g} \quad \text{(correctly rounded)}$$

There are 104 g in 0.230 lb. (We rounded from 104.328 g to 104 g because the given value, 0.230 lb, has only three significant figures. Remember that the "1" in "1 lb" has to be treated as an exact number because it's in a definition.)

PRACTICE EXERCISE 10

The *grain* is an old unit of mass still used by some pharmacists and physicians, and 1 grain = 0.0648 g. How many grams of aspirin are in a tablet containing 5.00 grain of aspirin?

Often there is no single conversion factor that does the job, and two or more have to be used. For example, we might want to find out how many kilometers are in, say, 26.22 miles, but our tables don't have a direct relationship between kilometers and miles. However, if we can find in a table that 1 mile = 1609.3 m and that 1 km = 1000 m, we can still work the problem. We'll see in the next example how we can string two (or more) conversion factors together before doing the calculation that gives the final answer.

EXAMPLE 1.5
Using the Factor-Label Method. Stringing Conversion Factors

How many kilometers are there in 26.22 miles, the distance of a marathon race? Use the fact that 1 mile equals 1609.3 meter and any other relationships that are available.

ANALYSIS The fact that 1 mile equals 1609.3 meter gives us the following conversion factors.

$$\frac{1 \text{ mile}}{1609.3 \text{ m}} \quad \text{or} \quad \frac{1609.3 \text{ m}}{1 \text{ mile}}$$

We're given 26.22 mile, so we know that if we use the second conversion factor, the following calculation would convert miles into meters.

$$26.22 \ \text{mile} \times \frac{1609.3 \ \text{m}}{1 \ \text{mile}}$$

If we paused to carry out this calculation, the answer would not be in kilometers (km). Therefore, *before doing this calculation,* we look for another conversion factor that, if possible, directly relates meters to kilometers and so would let us cancel "m" and replace it by "km." Knowing that "kilo-" stands for 1000, we know that the relationship, 1 km = 1000 m, is what we need.

SOLUTION

$$26.22 \ \text{mile} \times \frac{1609.3 \ \text{m}}{1 \ \text{mile}} \times \frac{1 \ \text{km}}{1000 \ \text{m}} = 42.20 \ \text{km} \qquad \text{(correctly rounded)}$$

The marathon distance is 42.20 km.

PRACTICE EXERCISE 11

Using the relationships between units given in the exercises or in tables in this chapter, carry out the following conversions. Be sure that you express the answers in the correct number of significant figures.

(a)   How many milligrams are in 0.324 g (the aspirin in one normal tablet)?

(b)   A long-distance run of 10.0×10^3 m is how far in feet? (This is the 10-km distance.)

(c)   A prescription calls for 5.00 fluidrams of a liquid. What is this in milliliters? (The "fluidram" is a measure of volume in the old apothecary system; 8 fluidram = 1 liquid ounce.)

(d)   One drug formulation calls for a mass of 10.00 drams. If only an SI balance is available, how many grams have to be weighed out? (16 drams = 1 ounce).

(e)   How many microliters are in 0.00478 L?

TABLE 1.6
Densities of Some Common Substances at 25°C

Substance	Density (g/cm³)
Aluminum	2.70
Bone	1.7–2.0
Butter	0.86–0.87
Cement, set	2.7–3.0
Cork	0.22–0.26
Diamond	3.513
Glass	2.4–2.8
Gold	19.3
Iron	7.86
Marble	2.6–2.8
Mercury	13.534
Milk	1.028–1.035
Wood, balsa	0.11–0.14
ebony	1.11–1.33
maple	0.62–0.75
teak	0.98

On page 10, equations were given relating degrees Celsius and degrees Fahrenheit. The use of these equations illustrates further examples of how to cancel units no longer wanted, as you can demonstrate by using those equations to work the following Practice Exercises.

PRACTICE EXERCISE 12

A child has a temperature of 104 °F. What is this in degrees Celsius?

PRACTICE EXERCISE 13

If the water at a beach is reported as 15 °C, what is this in degrees Fahrenheit? (Would you care to swim in it?)

1.8

DENSITY

One of the important physical properties of a liquid is its density, its amount of mass per unit volume.

Properties Are Called Extensive or Intensive According to Their Dependence on Sample Size Both the mass of a chemical sample and its volume are examples of **extensive properties,** those that are directly proportional to the *size* of the sample. Length is also an extensive property.

An **intensive property** is independent of the sample's size. Temperature and color are such intensive properties, for example. Generally, intensive properties disclose some essential quality of a substance that is true for any sample size, and this is why scientists find intensive properties particularly useful.

An Object's Density Is the Ratio of Its Mass to Its Volume One useful intensive property of a substance, particularly if it is a fluid, is its density. **Density** is the mass per unit volume of a substance.

$$\text{Density} = \frac{\text{mass}}{\text{volume}}$$

The density of mercury, the silvery liquid used in most thermometers, is 13.5 g/mL (at 25 °C), making mercury one of the most dense substances known. In contrast, the density of liquid water at 25 °C is 1.0 g/mL. Table 1.6 gives the densities of several common substances.

The density of a substance varies with temperature, because for most substances the volume of a sample but not the sample's mass changes with temperature. Most substances expand in volume when warmed and contract when cooled, but the effects of such changes in volume on densities isn't great for liquids or solids. Table 1.7 gives the density of water at several temperatures. Notice that, when rounded to two significant figures, the density of water is 1.0 g/mL in the (liquid) range of 0 °C to 30 °C (32 to 86 °F).

Don't make the mistake of confusing *heaviness* with *density*. A pound of mercury is just as heavy as a pound of water or a pound of feathers, because a pound is a pound. But a pound of mercury occupies only 1/13.5th the volume of a pound of water.

One of the uses of density is to calculate what volume of a liquid to take when the problem or experiment specifies a certain mass. Often it is easier (and sometimes safer) to measure a volume than a mass, as we will note in the next example.

TABLE 1.7
Density of Water at Various Temperatures

Temperature (°C)	Density (g/mL)
0	0.9987
3.98	1.0000
10	0.9973
20	0.99823
25	0.99707
30	0.99567
35	0.99406
45	0.99025
60	0.98324
80	0.97183
100	0.95838

■ The density of mercury changes only from 13.60 g/mL to 13.35 g/mL when its temperature changes from 0 °C to 100 °C, a density change of only about 2%.

EXAMPLE 1.6
Using Density to Calculate Volume from Mass

Concentrated sulfuric acid is a thick, oily, and very corrosive liquid that no one would want to spill on the pan of an expensive balance—to say nothing of the skin. It is an example of a liquid that is usually measured by volume instead of by mass, but suppose an experiment called for 25.0 g of sulfuric acid. What volume (in mL) should be taken to obtain this mass? The density of sulfuric acid is 1.84 g/mL.

ANALYSIS The given value of density means that 1.84 g acid = 1.00 mL acid. This gives two possible conversion factors:

$$\frac{1.84 \text{ g acid}}{1 \text{ mL acid}} \quad \text{or} \quad \frac{1 \text{ mL acid}}{1.84 \text{ g acid}}$$

The "given" in our problem, 25.0 g acid, should be multiplied by the second of these conversion factors to get the unit we want, mL.

SOLUTION

$$25.0 \text{ g acid} \times \frac{1 \text{ mL acid}}{1.84 \text{ g acid}} = 13.6 \text{ mL acid}$$

SPECIAL TOPIC 1.1
SPECIFIC GRAVITY

The **specific gravity** of a liquid is the ratio of the mass contained in a given volume to the mass of the identical volume of water at the same temperature. If we arbitrarily say that the "given volume" is 1.0 mL, then the water sample has a mass of 1.0 g (or extremely close to this over a wide temperature range). This means that dividing the mass of some liquid that occupies 1.0 mL by the mass of an equal volume of water is like dividing by 1, but all the units cancel. Specific gravity has no units, and a value of specific gravity is numerically so close to its density that we usually say they are numerically the same. This fact has resulted in a rather limited use of the concept of specific gravity, but one use occurs in medicine.

In clinical work, the idea of a specific gravity surfaces most commonly in connection with urine specimens. Normal urine has a specific gravity in the range of 1.010 to 1.030. It's slightly higher than water because the addition of wastes to water usually increases its mass more rapidly than its volume. Thus the more wastes in 1 mL of urine, the higher is its specific gravity.

Figure 1 shows the traditional method to measure the specific gravity of a urine specimen, by using a urinometer. Its use, however, has largely been supplanted by the use of a refractometer, which needs only one or two drops of urine for the measurement. (The refractometer is an instrument that measures the ratio of the speed of light through air to its speed through the sample being tested. This ratio can be correlated with the concentration of dissolved substances in the urine. *How* the refractometer does this is beyond the scope of our study.)

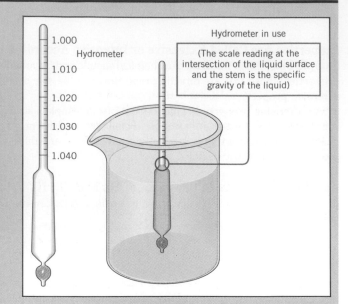

One of the important functions of the kidneys is to remove chemical wastes from the bloodstream and put them into the urine being made. The kidneys' mechanism for doing this does not remove those substances from the blood that ought to remain in the blood. The clinical significance, therefore, of a change in the concentration of substances dissolved in the urine is that it indicates a change in the activity of the kidneys. This might be the result of a kidney disease causing substances that should stay in the blood to leak into the urine being made. Or it might mean that wastes are being generated somewhere else in the body more rapidly than the kidneys can remove them.

Thus if we measure 13.6 mL of acid, we will obtain 25.0 g of acid. (The pocket calculator result is 13.58695652, but our rules require us to round this result to three significant figures.)

PRACTICE EXERCISE 14

An experiment calls for 16.8 g of methyl alcohol, the fuel for fondue burners, but it is easier to measure this by volume than by mass. The density of methyl alcohol is 0.810 g/mL, so how many milliliters have to be taken to obtain 16.8 g of methyl alcohol?

PRACTICE EXERCISE 15

After pouring out 35.0 mL of corn oil for an experiment, a student realized that the mass of the sample also had to be recorded. The density of the corn oil is 0.918 g/mL. How many grams are in the 35.0 mL?

Specific gravity is a property of a fluid that is very similar to density. It is not used very often in chemistry, but in clinical work the specific gravity of liquid specimens (such as urine) helps to reveal the nature of an illness, as described in Special Topic 1.1.

SUMMARY

Chemistry and the molecular basis of life Down at the level of nature's tiniest particles, we find the "parts" — molecules — that nature shuffles from organism to organism in the living world. One of the many ways of looking at life is to examine its molecular basis, the way in which health depends on chemicals and their properties.

Scientific facts and physical quantities Scientific facts are usually capable of being checked by independent observers, and these facts often are physical quantities that pertain to physical properties, whether these are extensive or intensive properties. Physical properties — those that can be studied without changing the substance into something else — include mass, volume, time, temperature, color, and density. For our purposes, the important base quantities are (with the names of the SI base units given in parentheses) mass (kilogram), length (meter), time (second), temperature degree (kelvin), and quantity of chemical substance (mole). Except for temperature, all are extensive properties, those that depend on the size of a sample.

The liter is an important derived unit for volume, and volume is another extensive property. An important intensive physical property is the density of something, the ratio of its mass to its volume. It is often reported in units of grams per milliliter (g/mL).

Special prefixes can be attached to the names of the base units to express multiples or submultiples of these units. To select a prefix we have to be able to put very large or very small numbers into scientific notation.

Accuracy, error, uncertainty, and precision An accurate measurement is one that is true. The error in the measurement is the difference between the true value and what is observed. The range of uncertainty in a direct measurement is determined by the quality of the measuring device, how accurately it is set, how reliable (or steady) is its gauge, scale, or readout panel, and how acute is the observer. In this text, the range of uncertainty is assumed to be plus or minus one unit in the last decimal place allowed by the number of significant figures. The number of significant figures in a physical quantity equals the number of digits known to be true plus one more. When we add or subtract physical quantities, the decimal places in the result can be no more than the least number of decimal places among the original quantities. When we multiply or divide physical quantities, we have to round the result to show the same number of significant figures as are in the least precise original quantity.

Factor–label method The units of the physical quantities involved in a calculation are multiplied or canceled as if they were numbers. To convert a physical quantity into its equivalent in other units, we multiply the quantity by a conversion factor that permits the final units to be correct. The conversion factor is obtained from a defined relationship between the units.

REVIEW EXERCISES

The answers to Review Exercises whose numbers are in color are found in Appendix C. The answers to the other Review Exercises are found in the Study Guide that accompanies this book. The more challenging questions are marked with asterisks.

Molecular Basis of Life

1.1 On which aspects of nature do chemists focus most?

1.2 What kinds of observations of nature led people to suppose that living things must have many common features?

1.3 Concerning living things, at what level of existence are "parts" most freely exchanged between members of different species? What are these parts, in the most general terms?

Facts, Hypotheses, and Theories

1.4 In the most general terms, scientists develop a theory to answer what main questions?

1.5 What is it that makes a fact a *reproducible* fact?

1.6 When you have had some particularly nice success, you no doubt feel very happy. It is a *fact* of your life. Yet, this fact is not described as *reproducible*. Why?

1.7 To someone using a scientific approach, whether a chemist, a physician, or a nurse tending a bedside respirator, an observed fact has value for what purpose?

1.8 When someone constructs a *hypothesis* that involves several facts or observations, what is sought by the hypothesis?

*1.9 When a hypothesis has been created, which is the better question to ask next,
(a) Is the hypothesis true or false?
(b) How can I prove the hypothesis (to be true)?
Explain your choice.

1.10 You are talking by long distance phone to a good friend when suddenly the line "goes dead." What is the most likely explanation that you would make, your first hypothesis to explain this observation of a dead line? Your hypothesis might be logical *based on previous experiences,* but is it therefore *necessarily* correct or incorrect? What would you do to test your hypothesis? Why would it be pretentious to dub your hypothesis a *theory* (a question having to do with how best to use the term)?

1.11 The lack of trace substances, vitamins, in the diet was initially called the vitamin *theory.*

(a) What initially made the use of the term *theory* a better word choice than *hypothesis*?

(b) In general terms only, what kind of experiences led scientists to stop referring to a *theory* concerning vitamins but to a *fact*, instead?

1.12 What is the chief role of reason or logic in the application of the scientific method?

1.13 The question "How do we get sick?" is more often successfully put to nature than the question, "Why do we get sick?" Explain.

Physical Quantities, Properties, and Measurements

1.14 When we speak of the *properties* of some substance, what is meant?

1.15 What is the basis for distinguishing between *chemical* and *physical* properties?

1.16 What marks the difference between a *physical property* and a *physical quantity?*

1.17 In a word, what operation must be done to obtain a value for a physical quantity?

1.18 What is meant by the *inertia* of an object, and what is the name of the physical quantity used to describe this property?

1.19 Your *weight* on the moon would be less than your weight on earth, yet your *mass* is the same at both locations. Explain.

1.20 What makes it possible for us to say that we determine the *mass* of an object when the actual operation we use is *weighing* (and the instrument is a two-pan balance)?

1.21 What is the general name we give to those fundamental quantities in terms of which all other physical quantities are defined? Name five of these physical quantities that are defined or mentioned in this chapter.

1.22 In order for a physical quantity to have any meaning or any usefulness, what must be defined for it?

1.23 What is the *name* of the base unit for the following physical quantities?
(a) length (b) time (c) mass (d) temperature (e) quantity of chemical substance

1.24 In general terms, the *definition* of a base unit involves what kind of a standard? What organization has been responsible for these definitions?

1.25 Why is *volume* not called a *base* quantity?

1.26 What are the *names* of the SI base unit of length and the corresponding *reference standard?*

1.27 What are the *names* of the SI base unit of mass and its corresponding *reference standard?*

1.28 Scientists regard the SI reference standard for length as far more satisfactorily specified than the SI reference standard for mass? Give the reasons.

1.29 Examine each pair of quantities and state which is larger.
(a) meter and yard   (b) inch and centimeter
(c) gram and ounce   (d) millimeter and centimeter
(e) pound and kilogram (f) kilogram and ton
(g) liter and quart   (h) microliter and milliliter
(i) ounce and pound   (j) gram and kilogram

1.30 How many milliliters are in 1 liter?

1.31 How many micrograms are in 1 milligram?

1.32 How many grams are in 1 kilogram?

Degrees and Scales of Temperature

1.33 What is the value given to the point on a mercury-filled thermometer where the mercury level eventually comes to rest after the thermometer is immersed in an ice-water slush on each of the following scales of temperature?
(a) Celsius (b) Fahrenheit (c) Kelvin

1.34 When a mercury-filled thermometer is immersed into a container of boiling water (at sea level), what is the value given to the level the mercury eventually reaches on each of the following scales?
(a) Celsius (b) Fahrenheit (c) Kelvin

1.35 How many degree divisions (arbitrarily) separate the mercury levels for the freezing and the boiling points of water on each of the following scales?
(a) Celsius (b) Fahrenheit (c) Kelvin

1.36 Which is the larger degree, the Celsius degree or the Fahrenheit degree? By how much is it larger?

1.37 Which is the larger degree, the kelvin or the Fahrenheit degree? By how much is it larger?

1.38 When expressed in degrees Celsius, the zero point on the Kelvin scale has what value?

1.39 What is true about a temperature of 0 K?

1.40 A Canadian weather report said that the temperature at one remote reporting station was −40 °C. What is this in °F?

1.41 If you read a Kelvin thermometer in your room as 278 K, would you be comfortable without a coat or sweater? (First convert 278 K into °C and then convert the answer into °F. As part of your answer, give the results of these calculations.)

1.42 In testing an American recipe, a French baker had to decide how to set the French oven for a recipe specification of 320 °F. What oven setting in °C was needed?

1.43 An American visitor to Germany wanted to set a room thermostat for the degree Celsius equivalent of 68 °F. What should the setting be in °C?

*1.44 A clinical thermometer was used to take the temperature of a patient, and it registered 40 °C. Did the patient have a fever? (Do the calculation. Normal body temperature has traditionally been taken to be 98.6 °F, but many healthy individuals have normal temperatures slightly lower or slightly higher than this.)

Scientific Notation and SI Prefixes

1.45 Rewrite the following physical quantities with their units abbreviated.
(a) 2.5 deciliters of solution
(b) 31 milligrams of medication
(c) 46 centimeters in length
(d) 110 kilometers in distance
(e) 35 microliters of solution
(f) 75 micrograms of progestin
(g) 25 millimeters wide

1.46 Rewrite the following physical conditions with their units written out in full.
(a) 110 mL of Ringer's solution
(b) 150 mg of sugar
(c) 16 km to the airport
(d) 50 μg of vitamin K
(e) 1.5 dL of saline solution
(f) 2.5 kg of salt
(g) 75 μL of serum

*1.47 Restate the following physical quantities in scientific notation in which the decimal part of the number is between 1 and 10.
(a) 5230 g (b) 0.0450 L
(c) 1562 m (d) 0.0000093 g

*1.48 How would the following physical quantities be re-expressed in scientific notation in which the decimal part of the number is between 1 and 10?
(a) 0.130 L (b) 3568.5 m
(c) 0.0000042 g (d) 0.0045 g

*1.49 Use a suitable SI prefix to express each of the quantities in Exercise 1.47.

*1.50 Use a suitable SI prefix to restate each of the quantities in Exercise 1.48.

Significant Figures

1.51 The number of tickets sold for a concert was 25342. Restate this number in scientific notation but retain only three significant figures.

1.52 The population of the world constantly changes, but at one moment it was 5154689 people. Re-express this in scientific notation retaining two significant figures.

*1.53 Study the following numbers.
(A) 4.55×10^8 (B) 0.0455 (C) 45,500
(D) 0.00455 (E) 4550 (F) 4.550×10^{-3}
(G) 4.550 (H) 0.45500 (I) 4.5500×10^7
(a) Which of these numbers has three significant figures? (Identify them by their letters.)
(b) Which has four significant figures?
(c) Which has five significant figures?

*1.54 Rewrite the following number according to the number of significant figures specified in each part. Express your answers in scientific notation.

$$16,560,010.01$$

(a) one (b) two (c) three (d) four (e) five (f) six

*1.55 Rewrite the following number according to the number of significant figures specified in each part. Give your answers in scientific notation.

$$199,898.9091$$

(a) three (b) four (c) five (d) six (e) eight (f) nine

1.56 The relationship between the milliliter and the microliter is given by
$$1 \text{ mL} = 1000 \ \mu\text{L}$$

How many significant figures are considered to be in each number?

Precision and Accuracy

*1.57 Consider that the following mathematical operations are calculations that involve physical quantities. (The units have been omitted.) Determine the significant figures that can be retained in the answer in each part according to our rules, and express the results of the calculations in the proper way. Use scientific notation in which the decimal part of the number is between 1 and 10.
(a) $4.665 \times 3.2 \times 10^{-5}$ (b) $6.3 \times 5.6000 \times 10^3$
(c) $4.005 \times 6.23 \times 10^{23}$ (d) $4.5 + 62.003$
(e) $6.004 - 3.2$ (f) $45.0023 + 0.023$
(g) $90.00 \div 3.0$ (h) $0.00050 \div 0.005$
(i) $6.40 \div 3.200$

*1.58 The following mathematical operations are calculations that involve physical quantities. (The units have been omitted.) Determine how many significant figures can be retained, and express the results of the calculations in the proper way. Use scientific notation in which the decimal part of the number is between 1 and 10.
(a) $9600.00 \div 320.0000$ (b) $45.0 \div 1.50$
(c) $45.0 + 1.50$ (d) 45.0×1.50
(e) $45.0 - 1.50$ (f) 0.000009×1.1

*1.59 When a scale was used to take six successive measurements of a person's mass, the following data were recorded.

59.85 kg, 59.70 kg, 59.91 kg, 59.73 kg, 59.94 kg, 59.91 kg

The balance had earlier been tested against a set of official reference standard masses and found to be working exceptionally well. The true value of the mass was verified as 59.86 kg.
(a) Can the measurements be described as *accurate*? Explain.
(b) What, if anything, do the data disclose about their *uncertainty*?

1.60 What is the specific problem when a measurement is known to be in *error*?

Converting between Units

1.61 Conversion factors relate identical physical amounts but that are expressed in different units. Write each of the following relationships between units in the forms of two conversion factors.
(a) 39.37 in. = 1 m (b) 1 L = 1.057 quart
(c) 1 g = 1000 mg (d) 1 kg = 2.205 lb
(e) 1 oz = 23.35 g

*1.62 Given the relationships expressed in Review Exercise 1.61, which of the two quantities that follow the symbol, $\approx$ ("equals approximately"), most closely matches the quantity in the first column. You should develop the skill to make these kinds of judgments without doing an actual calculation using conversion factors.
(a) 0.50 m $\approx$ 20 in. or 80 in.
(b) 0.5 lb $\approx$ 4.4 kg or 0.23 kg
(c) 6 g $\approx$ 140 oz or 0.25 oz
(d) 4500 mg $\approx$ 4.5 g or 4.5×10^6 g

*1.63 Given the relationships that can be found among the tables in this chapter, what quantity most nearly matches what is given in a different unit before the symbol, $\approx$?
   (a) 1 cup $\approx$ 50 mL or 200 mL (1 quart = 4 cup)
   (b) 1 mile $\approx$ 1.6 km or 0.66 km
   (c) 10 mm $\approx$ 2.5 in. or 0.25 in.
   (d) 1000 mL $\approx$ 1 μL or 1 L

1.64 Make the following conversions using relationships found in tables in this chapter. Do all calculations to three significant figures.
   (a) Convert 163 cm into inches (the height of an adult female).
   (b) Convert 154 lb into kilograms (the mass of an adult male).

1.65 Make the following conversions using relationships found in tables in this chapter. Do all calculations to three significant figures.
   (a) Convert 111.5 lb into kilograms (the mass of an adult female).
   (b) Convert 192 cm into inches (the height of an adult male).

1.66 The normal content of cans of popular soft drinks is 12.0 liquid ounces. How much is this in milliliters (to three significant figures)?

1.67 The popular-size bottles of mineral water hold 296 mL. How much is this in liquid ounces (to the proper number of significant figures)?

1.68 The gasoline tank of a small car holds 12.0 U.S. gallons. How much is this in liters?

*1.69 While driving on a country road in a European country you come to a bridge limited to 1.4×10^3 kg. Your vehicle has a mass of 4.5×10^3 lb. Should you cross? (Do the calculation.)

*1.70 A physician prescribed 0.50 g of valinomyocin. The pharmacy dispenses valinomyocin in 250-mg tablets. How many tablets are needed for one prescribed dose?

*1.71 Valium is available in tablets containing 5 mg of this medication. How many tablets must be administered to give a dose of 0.015 g of Valium?

*1.72 An IV (intravenous) solution contains 25 mg of a drug per 5.0 mL of solution. You are to administer 0.75 g of the drug. What volume of the solution should be used?

*1.73 A vial of a medication carries a label instructing the user to add 7.50 mL of water to the contents of the vial to obtain a solution containing 25.0 mg of the active drug per mL of the solution. How many milliliters of this solution must be taken to obtain 0.175 g of the drug?

1.74 The highest mountain in the world, Mount Everest in Nepal, is 8847.7 m. What is this in feet?

1.75 The highest mountain in the United States is Alaska's Mount McKinley, 20322 ft. How high is it in meters?

1.76 One pound of butter can be made into 128 equal-sized pats of butter. What is the mass of each pat in grams?

1.77 A diamond rated as 2.50 carats has a mass of how many grams? (1 carat = 200 mg)

Density

1.78 Mass is called an extensive property, but density is an intensive property. Explain the difference.

*1.79 This book has dimensions of about $8.25 \times 10.0 \times 1.60$ in. Its mass is about 3.75 lb. Calculate its density in units of g/cm^3. What would be the mass of this book in kilograms and in pounds if it were made of solid lead? (Assume that the density of lead is 11.50 g/cm^3.)

1.80 The density of aluminum is 2.70 g/cm^3. A block of aluminum with a volume of 250 mL (about 1 cup) has a mass of how many grams? Pounds?

1.81 Corrosive chemical solutions are usually more safely measured by volume than by mass. To obtain 25.0 g of sulfuric acid (density, 1.84 g/mL), how many milliliters should be measured?

*1.82 Corn oil has a density of 7.60 lb/gal.
   (a) Calculate its density in g/mL.
   (b) How many milliliters of corn oil must be taken to obtain 250 g?

1.83 Liquids and solutions generally expand in volume as they are warmed.
   (a) Assuming no loss by evaporation, does the mass of a sample change as it is warmed?
   (b) When a liquid sample is warmed, does its density increase, decrease, or stay the same?

Specific Gravity (Special Topic 1.1)

1.84 Specific gravity is defined such that the numerical value of something's density is virtually the same as its specific gravity. Explain.

1.85 If a urine specimen has an abnormally low value of specific gravity, what is known about the specimen?

1.86 A urinometer float rides higher in what kind of fluid: one with a high density or one with a low density?

1.87 Scarcely any change in volume occurs when 3.00 g of sugar is dissolved in 100 mL of water. Assuming no volume change, what is the specific gravity of this solution?

Additional Exercise

*1.88 Rehydration therapy is a life-saving procedure for victims of cholera, who typically lose large amounts of fluid. An English physician, Thomas Latta, was the first to use this procedure during the cholera epidemic in London in 1832. The solution he had the victims drink contained 3.0 drachmas of sodium chloride and 2.0 scruples of sodium bicarbonate per 6.0 pints of water. [1 drachma = 60 grains; 1 ounce (avdp) = 480 grains; 1 ounce (avdp) = 28.35 g; 1 scruple = 20 grains]
   (a) How many milligrams of sodium chloride were in each 6-pint unit of the solution? How many would be in 1 L of solution?
   (b) How many milligrams of sodium bicarbonate were in each 6-pint unit of the solution? How many would be in 1 L of solution?

MATTER AND ENERGY

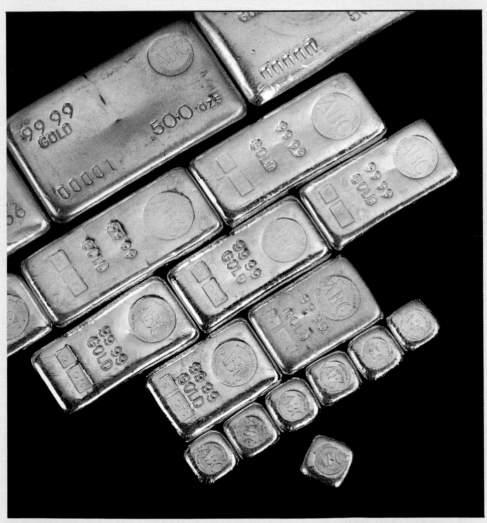

If a large ingot of gold is successively subdivided into smaller and smaller ingots, would there ever be one last "cut" that could not be made and still produce two smaller pieces of gold? Ancient Greek philosophers said "No," and we now call such an uncuttable piece an *atom,* after the Greek *atomas* for "not-cuttable." We introduce the concept of atoms in this chapter.

2.1
STATES AND KINDS OF MATTER
2.2
ATOMS, CHEMICAL SYMBOLS, AND CHEMICAL
EQUATIONS

2.3
KINETIC AND POTENTIAL ENERGY
2.4
HEAT AND THE THERMAL PROPERTIES OF MATTER

2.1

STATES AND KINDS OF MATTER

Elements and compounds have definite compositions, but mixtures do not.
Matter is anything that occupies space and has mass. It includes literally everything, and because there is such a huge variety of matter, making sense out of it might seem impossible. Throughout history, however, we humans have had a powerful impulse to sort and classify whenever we face what is very complex. Biologists, for example, created kingdoms, phyla, species, subspecies, and groups for plants and animals. One reason for sorting, classifying, and naming is simply to be able to find and recognize different things again. Another reason is to focus our minds on possible underlying causes for any patterns in what we see around us.

The *States* of Matter are Solids, Liquids, and Gases One useful way to classify matter is according to the **states of matter,** meaning the possible physical conditions of aggregation. We recognize three physical states, solid, liquid, and gas, which are illustrated very familiarly by ice, liquid water, and steam. Each state is characterized by the ability of a sample to hold its shape and have a definite volume, as Figure 2.1 illustrates. **Solids** have both definite shapes and volumes. **Liquids** have definite volumes but indefinite shapes; they take the shapes of their containers. **Gases** have no definite shapes or volumes. They fill whatever space holds them.

Having pointed out the almost obvious about the three states, we ask: What's behind them? What force makes solids have definite shapes but which seems to be weaker in liquids and still weaker in gases? Such questions try to get to the bottom of things. This course is about getting to the bottom of things in nature, and we'll do this for the states of matter in a later chapter. We have more sorting and classifying to do first.

The Three *Kinds* of Matter Are Elements, Compounds, and Mixtures Any sample of matter consists of one or more *pure substances* of which there are basically two broad kinds, *elements* and *compounds*.[1] When two or more elements or compounds are mixed together (without chemically reacting), we have the third kind of matter, the *mixture.* In this chapter we'll briefly survey what defines elements, compounds, and mixtures and return to them for a deeper study in later chapters.

Elements Cannot Be Changed to Simpler Pure Substances As their name implies, elements are elementary. An **element** is a pure substance that cannot be broken down into simpler pure substances. Among the familiar elements are aluminum, copper, gold, iron, and chromium, as well as the oxygen and nitrogen in air. Water is not an element because we can break water down into two elements, oxygen and hydrogen, which we can cause to recombine again to give water. Hydrogen burns in oxygen, and water forms.

A table of the 109 known elements is inside the front cover of this book. We'll be concerned with only about a dozen, however. Of the bulk mass of the human body, 99% consists of substances made from only four elements: carbon, nitrogen, hydrogen, and oxygen. The remainder is made from at least 21 other elements (possibly 24), but all are vitally important. Many are the "trace elements" of nutrition and must be regularly replenished by the diet.

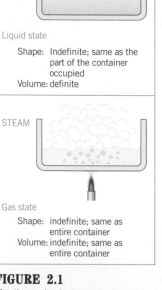

ICE

Solid state
  Shape: definite
  Volume: definite

WATER

Liquid state
  Shape: Indefinite; same as the
      part of the container
      occupied
  Volume: definite

STEAM

Gas state
  Shape: indefinite; same as
      entire container
  Volume: indefinite; same as
      entire container

FIGURE 2.1
The three physical states of matter —solid, liquid, and gas—are illustrated by ice, liquid water, and steam.

---

[1] To chemists (but not always to nonchemists), the word "pure" in *pure substance* is unnecessary. When chemists say "substance," they mean *pure* substance. However, we will often use the two-word term for the sake of emphasis.

Ninety elements occur naturally; the rest have been made by scientists using special equipment. The synthetic elements plus a few that occur naturally exhibit **radioactivity,** the emission of one kind of radiation or another. We will return to their special properties in Chapter 11.

At room temperature, 2 elements are liquids, 11 are gases, and the rest are solids. All but about 20 elements are metals. **Metals** have shiny surfaces (when polished), can be hammered into sheets and drawn into wires, and are good conductors of electricity and heat.

Sometimes two or more metals are melted, mixed together, and allowed to cool to give a solid mixture of metals called an **alloy.** Steel, for example, is actually the name for a family of alloys in which the principal element is iron. The many iron alloys include chromium steel and nickel steel, which have exceptional strength and resistance to corrosion. A few alloys are used to replace bones or to strengthen them, and they must be unusually resistant to corrosion.

Several of the solid elements, like carbon and sulfur, plus all the gaseous elements are **nonmetals.** Nonmetals cannot be worked into sheets or wires, and they do not conduct electricity or heat as well as the metals. Metals and nonmetals have somewhat opposite *chemical* properties, and large numbers of compounds, like table salt, are made by combining metals with nonmetals.

■ Some naturally occurring elements, such as uranium and radium, are radioactive, too.

■ Mercury is a metallic element but a liquid at room temperature. It's still the working fluid of many thermometers, but this use is dwindling.

■ The current U. S. nickel coin is actually an alloy of copper (75%) and nickel (25%).

■ Diamonds consist of one form of pure carbon.

Compounds Are Made from Elements   Chemical **compounds** are pure substances made from two or more elements *always* combined in a proportion by mass that is both definite and unique for the compound. When water, for example, is broken down into hydrogen and oxygen, these two elements are invariably obtained in a mass ratio of 2.0 g of hydrogen to 16.0 g of oxygen. Yet water is not just a *mixture* of hydrogen and oxygen taken in this mass ratio. Such a mixture is in the gaseous state at room temperature and is extremely unstable; given the slightest spark or exposure to ultraviolet light, it explodes (and water forms). The vitamin C in orange juice, the citric acid, and the fruit sugar are all chemical compounds. Their constituent elements are carbon, hydrogen, and oxygen, but the proportions are different for each. Table salt is also a compound. Its elements are sodium and chlorine. Interestingly, both are dangerous elements; sodium (Fig. 2.2), a shiny metal only when freshly cut, combines vigorously with both the oxygen and the moisture in humid air, and chlorine (Fig. 2.3) is a greenish-yellow, poisonous gas. If mixed, sodium and chlorine combine violently, as you can see in Figure 2.4, to give salt, sodium chloride, a compound needed by all animals. How can sodium chloride be so different from its elements? We'll see in the next chapter. We still have sorting and classifying to do.

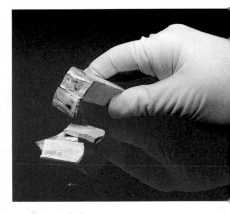

FIGURE 2.2
The shiny, metallic luster of sodium will soon fade, because this soft, easily cut metal reacts quickly with both oxygen and moisture in air.

FIGURE 2.4
A small piece of sodium metal in the metal spoon bursts into flame when it is thrust into chlorine. The reaction produces light and heat as sodium chloride forms.

FIGURE 2.3
Chlorine, a pale yellowish-green gas, is a poison. Warring sides used it in World War I as a weapon.

Mixtures Have Variable Compositions A **mixture** consists of two or more pure substances that are present in a proportion which can vary considerably. Each pure substance contributes something to the mixture's overall properties. Freshly squeezed orange juice, for example, consists of water, vitamin C, a little fruit sugar, citric acid, and other substances. If you remove any of them, you would notice the change (although the removal of some minor components, like vitamin C, would not be noticed right away). You probably can sense that if we want to "get to the bottom" of any mixture, like orange juice, we have to study its individual pure substances.

■ In chemistry, the opposite of *pure* is not *impure,* because *impure* carries an extra biological meaning, such as "dangerous to human health." The opposite of *pure substance* in chemistry is *mixture.*

To recapitulate, there are three *states* of matter: solid, liquid, and gas. There are three *kinds* of matter: elements, compounds, and mixtures. There are two kinds of *pure substances:* elements and compounds.

Chemical Reactions Change Substances into Other Substances Compounds are made by **chemical reactions,** events in which substances called **reactants** change into different substances called **products.** Reactants and products almost always have at least some physical properties that are quite different. Sodium, chlorine, and sodium chloride certainly illustrate this. Chemical reactions always feature at least some changes in physical properties as reactants change over into products.

■ Huge quantities of both sodium and chlorine are made annually by passing electricity through molten salt (sodium chloride).

Laws of Chemical Combination Govern Chemical Composition Sodium chloride, as we said, is a *compound,* not just a mixture of the elements. A stable *mixture* of sodium metal and chlorine gas cannot actually be prepared. Yet it is possible to break down sodium chloride into its elements. (It's done by passing a current of electricity through *molten* salt in such a way that the sodium and chlorine emerge into separate containers.) The breakdown of sodium chloride is also a chemical reaction, because different substances are made. Every time sodium chloride is changed into its elements, or every time sodium and chlorine are allowed to react to form sodium chloride, the ratio of chlorine to sodium is invariably 1.5421 g of chlorine to 1.0000 g of sodium. A sample of sodium chloride obtained from any place in the world would have this ratio of chlorine to sodium. No matter in what proportion you might initially mix sodium and chlorine, the reacting proportion would be exactly 1.5421 g of chlorine to 1.0000 g of sodium. You might have either chlorine or sodium left over, depending on how carefully you mixed these elements, but for every 1.0000 g of sodium that reacts, 1.5421 g of chlorine would react as well, to form 2.5421 g of sodium chloride.

Compounds are like this, so much so that we will now make this property an essential part of the *definition* of a compound. A **compound** is a substance made from two or more elements chemically combined in a definite proportion by mass. This is, in fact, the first of the scientific laws that we study, one of the laws of chemical combination, namely the **law of definite proportions.**

Law of Definite Proportions In a given chemical compound, the elements are always combined in the same proportion by mass.

Elements also obey this law, because an element consists of 100% of itself. Chemists call elements and compounds **pure substances** because their compositions obey the law of definite proportions. The expression, *pure substance,* is thus actually redundant. When we call something (in this text) a *substance,* we mean a *pure* substance (element or compound), one that obeys the law of definite proportions.

Mixtures, in contrast, can be prepared in widely varying proportions of their components. For example, we can prepare mixtures (solutions) of sugar and water, two compounds, in almost any proportion we please. Moreover, we can separate the two simply by letting the water evaporate — a physical change because only a change in physical state occurs. Mixtures, in general, require only physical changes or operations to be separated into their components.

Lemonade is little more than lemon-flavored sugar-water, and its sweetness can be varied from nearly sour to syrupy sweet.

Another law of chemical combination was suggested when we pointed out that 2.5421 g of sodium chloride forms when 1.0000 g of sodium combines with 1.5421 of chlorine. The mass of the product is the sum of the masses of the reactants. This mass relationship between reactants and products has always been observed for chemical reactions, and these observations are behind the **law of conservation of mass.**

> **Law of Conservation of Mass** In any chemical reaction, the sum of the masses of the reactants always equals the sum of the masses of the products.

Any widespread regularity in nature generally suggests some very basic truth about the world in which we live, and the laws of chemical combination are examples of a regularity that draw us to the question: What must be true about substances to explain these laws?

2.2

ATOMS, CHEMICAL SYMBOLS, AND CHEMICAL EQUATIONS

John Dalton

Dalton saw that the laws of chemical combination virtually compel the belief that atoms exist.

John Dalton (1766 – 1844), an English scientist, was the first to find a reasonable explanation for both definite proportions in compounds and the conservation of mass in reactions.

Dalton's Atomic Theory Proposed Indestructible Atoms Dalton reasoned that matter must be made of very tiny, individual particles that undergo a variety of chemical reactions *without breaking apart or losing any mass.* In order to explain the *definite* compositions observed for compounds, he said that these tiny particles simply cannot exist as major fragments of themselves. Each particle is an unbreakable unit.

The idea of a tiny, invisible, unbreakable particle had been around for centuries, because ancient Greek philosophers had proposed it. The Greek word for "not cut" is *atomos,* and from this term came our word *atom.* Dalton revived this ancient belief in "not cuttable" particles with the enormously important difference that he had solid evidence: the laws of chemical combination.

The chief postulates of **Dalton's atomic theory** are the following.

1. Matter consists of definite particles called **atoms.**
2. Atoms are indestructible.
3. All atoms of one particular element are identical in mass.
4. Atoms of different elements have different masses.
5. By becoming bound together in different ways, atoms form compounds in definite ratios *by atoms.*

Not all of Dalton's postulates turned out to be correct, as we'll see in Chapters 3 and 11. However, his theory was never meant to explain everything about atoms and compounds, only something. The only way, said Dalton, that we can observe definite ratios *by mass* in compounds is that they possess definite ratios of atoms, each kind of atom having its own unique mass. Let's see how Dalton's theory works for compounds of iron and sulfur.

The elements iron and sulfur can be made to combine in the ratio of 1.000 g of iron to 0.574 g of sulfur to form one kind of iron sulfide with the formal name iron(II) sulfide. Dalton

■ Sulfur forms more than one compound with iron, so we need a special name to designate one of them. The "(II)" in the formal name, iron(II) sulfide, uniquely defines the sulfide described here.

FIGURE 2.5

Mass ratio versus atom ratio in iron(II) sulfide. If the mass of one sulfur atom is 0.574 times the mass of one iron atom, then combining sulfur and iron atoms in a simple 1 to 1 ratio (by atoms), regardless of how much this is scaled up, must result in a constant ratio by mass.

55.8g Fe
(6.02×10^{23} atoms)

32.1g S
(6.02×10^{23} atoms)

87.9g FeS
(6.02×10^{23} formula units)

would have explained this definite mass ratio as illustrated in Figure 2.5. If we assume that the *atoms* of these two elements combine in a ratio of 1 atom of iron to 1 atom of sulfur, each atom with its own mass, *then the mass ratio cannot help but be a constant.*

The Discovery of the Law of Multiple Proportions Compelled Belief in Atoms

Powerful evidence for Dalton's atomic theory came from the study of different compounds that can be made from the same elements but in different mass ratios. The mineral pyrite is another compound that can be made from iron and sulfur, but the ratio by mass in pyrite is 1.000 g of iron to 1.148 g of sulfur. In iron(II) sulfide, described earlier, the mass ratio is 1.000 g of iron to 0.574 g of sulfur. Notice that 1.148 is exactly twice the size of 0.574. There is a *whole-number* ratio between the grams of sulfur combined with 1.000 g of iron in the two compounds. If Dalton is right, there *must* be a whole-number ratio because atoms combine as whole units, as whole particles. Figure 2.6 shows why this is so. It shows several possible combinations of intact iron and sulfur atoms. Three of the four compounds in Figure 2.6 are known, namely the first, second, and fourth.

Two compounds of tin and oxygen further illustrate the special relationship among compounds made from the same elements. In one compound of tin and oxygen, 1.000 g of oxygen is combined with 3.710 g of tin. In another, 1.000 g of oxygen is combined with 7.420 g of tin. Now compare 7.420 g of tin with 3.710 g of tin.

$$\frac{7.420}{3.710} = 2$$

The ratio of the quantities of tin in the two compounds that combine with the same mass of oxygen is a ratio of small, whole numbers, 2 to 1. These and several other examples led to the third law of chemical combination, the **law of multiple proportions.**

■ Pyrite is often found as golden crystals embedded in rock samples. Many a novice gold miner felt an increased heartbeat on finding what more experienced miners called "fool's gold."

■ Tin is the metal used to coat the inner surfaces of "tin" cans. Tin, unlike iron and less expensive steels, won't corrode in an environment of food juices.

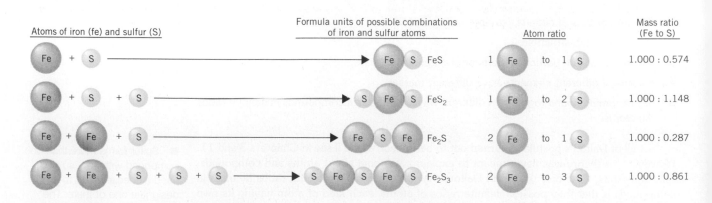

FIGURE 2.6

Possible multiple proportions for combinations of iron atoms and sulfur atoms. If atoms remain essentially intact (and suffer no detectable loss in mass) when they form compounds, then they must assemble in *whole-number ratios by atoms,* regardless of the masses of the individual atoms.

> **Law of Multiple Proportions** Whenever two elements form more than one compound, the different masses of one that combine with the same mass of the other are in the ratio of small whole numbers.

The examples illustrating this law that we described, plus many others, combined with Dalton's astute interpretations, virtually compelled scientists to believe that atoms exist. Since Dalton's time, so much additional evidence for atoms has accumulated that their existence is taken as fact, not theory.

Chemical Formulas Give a Substance's Composition Chemists use special symbols for compounds called **chemical formulas** that tell us at a glance which elements are combined in the compound and in what ratio by atoms. To construct these formulas, each element has been assigned an **atomic symbol** consisting of one or two letters. Those with which we will most often work are given in Table 2.1, and the complete list of atomic symbols appears in a table inside the front cover of this book.

Many elements, like those in the first column of Table 2.1, have single-letter symbols, usually (but not invariably) the capitalized first letter. Because there are more elements than letters, several elements have names beginning with the same letter — for example, carbon, calcium, chlorine, chromium, cobalt, and copper. Many atomic symbols, therefore, consist of the first two letters with the first letter *always* capitalized and the second letter *always* in lower case. Examples are in the second column of Table 2.1. The third column shows how the first letter and a letter occurring beyond the second place in the name are combined to make a symbol. Thus chlorine has the symbol Cl and chromium is Cr. The last column of Table 2.1 lists some elements named long ago when Latin was the almost universal language of educated people, and so the symbols of some elements were derived from Latin names, as shown. Students often find the symbols for sodium (Na) and potassium (K) the trickiest, so be sure to take some extra time to fix these firmly in mind.

Empirical Formulas Give the Ratios of Atoms Combined in Compounds Chemists employ more than one kind of chemical formula, the simplest being the *empirical formula*. In an **empirical formula** the atomic symbols of the elements making up the compound are used in the *smallest* possible whole number ratio that corresponds to the actual ratio by atoms. The formula FeS for iron(II) sulfide is an example of an empirical formula. The ratio that distinguishes this compound from all other iron–sulfur compounds is 1 Fe atom to 1 S atom. Notice that the two atomic symbols, one each of Fe and S, are written without spaces between them in FeS.

In pyrite, another iron–sulfur compound, the combining ratio is 1 atom of iron to 2 atoms of sulfur. Thus the empirical formula of pyrite is written as FeS_2. The 2 is called a *subscript*, and subscripts in formulas always *follow* the symbols to which they refer. In Fe_2S_3, the

■ Among the trickier, but still common, pairs of chemical symbols are

P = phosphorus
K = potassium
S = sulfur
Na = sodium
I = iodine
Fe = iron

■ The formal name of Fe_2S_3 is iron(III) sulfide. Why the "(III)" is used will be explained in a later chapter.

TABLE 2.1
Names and Symbols of Some Common Elements[a]

C	Carbon	Al	Aluminum	Cl	Chlorine	Ag	Silver *(argentum)*
H	Hydrogen	Ba	Barium	Mg	Magnesium	Cu	Copper *(cuprum)*
O	Oxygen	Br	Bromine	Mn	Manganese	Fe	Iron *(ferrum)*
N	Nitrogen	Ca	Calcium	Pt	Platinum	Pb	Lead *(plumbum)*
S	Sulfur	Li	Lithium	Zn	Zinc	Hg	Mercury *(hydrargyrum)*
P	Phosphorus	Si	Silicon	As	Arsenic	K	Potassium *(kalium)*
I	Iodine	Co	Cobalt	Cs	Cesium	Na	Sodium *(natrium)*
F	Fluorine	Ra	Radium	Cr	Chromium	Au	Gold *(aurum)*

[a] The names in parentheses in the last column are the Latin names from which the atomic symbols were derived.

empirical formula of still another compound of iron and sulfur (given in Fig. 2.6), the ratio is 2 atoms of Fe to 3 atoms of S. You can see that empirical formulas give definitive chemical information about a compound's composition.

The subscript 1 is always "understood" in chemical formulas. We write the formula of iron(II) sulfide as FeS, not as Fe_1S_1. We do not write the formula as Fe_2S_2 or as Fe_3S_3 either, although both 2 : 2 and 3 : 3 are equivalent to the 1 : 1 ratio of Fe to S in FeS. By convention, however (not by law of nature), empirical formulas have the *smallest* whole numbers that specify the ratio by atoms. Always remember, however, that empirical formulas disclose *ratios*, not absolute numbers. The actual numbers of combined atoms in FeS even in a speck barely visible under a microscope are extremely large because atoms are exceedingly tiny.

We might have conveyed the identical information about FeS by writing the formula as SFe, but the symbols of metal elements are generally placed before those of nonmetals.

Often compounds are made of two nonmetals, so we will learn a rule that applies when one of the nonmetals is carbon. Its symbol comes first, as in carbon monoxide, CO, and methane, CH_4 (natural gas).

It is too early in our study to go more deeply into the rules for writing formulas and names of compounds, so don't worry about this phase of our study yet. For the present, just be sure you can spot the difference between say, FeS and fes, or that you can tell that CO can't possibly be the symbol for an element (the second letter is not lowercased) but that Co might be. Let's now work an example to illustrate how to use the rules and to become more familiar with chemical formulas and the use of subscripts.

■ There are also *molecular formulas* and *structural formulas,* each differing in the amount of chemical information they carry.

■ CO is the formula of carbon monoxide, a compound made of carbon, C, and oxygen, O, and Co is the atomic symbol of cobalt.

EXAMPLE 2.1
Writing Chemical Formulas from Atomic Compositions

Aluminum, a metal with the symbol Al, and sulfur, a nonmetal with the symbol S, form a compound in which the atom ratio is 2 atoms of Al to 3 atoms of S. Write the formula.

ANALYSIS The symbol for aluminum has to come first because aluminum is a metal.

SOLUTION The formula is Al_2S_3 in which "2" goes with Al and "3" with S. ■

EXAMPLE 2.2
Writing Chemical Formulas from Atomic Compositions

Carbon, a nonmetal with the symbol C, and chlorine, another nonmetal but with the symbol Cl, form a compound in which the ratio of atoms is 1 of C to 4 of Cl. Write the formula.

ANALYSIS By convention, the symbol of carbon comes first.

SOLUTION The formula is CCl_4. (The subscript, 1, of carbon is understood.) ■

PRACTICE EXERCISE 1

Write the formula of the compound between sodium, a metal, and sulfur in which the atom ratio is 2 atoms of sodium to 1 atom of sulfur. ■

PRACTICE EXERCISE 2

Give the names of the elements that are present and the ratio of their atoms in K_2CO_3. ■

A Formula Unit Is a Real or Hypothetical Particle Having the Composition Given by the Formula of the Compound If Fe stands for an atom, what kind of particle does FeS stand for? The most *general* name we can give to the particle with the composition of the formula of a compound is **formula unit.** Special names such as *molecule* or *set of ions* or *ion group* will be used when we are farther along in our study, but *formula unit* embraces all of them. One formula unit of FeS is made of *one* iron atom and *one* sulfur atom. One formula unit of FeS$_2$ is made of *one* iron atom and *two* sulfur atoms.

We can extend the term *formula unit* to include atoms. The chemical formula for sodium, for example, is its atomic symbol, Na; hence, we can say that one formula unit of sodium is one atom of sodium.

Chemical Equations Use Formulas to Describe Reactions Now that we know something about chemical formulas we can learn how to use them in describing chemical reactions by means of chemical equations. A **chemical equation** is a special shorthand description of a reaction; it groups the symbols of the reactants, separated by plus signs, on one side of an arrow and places the symbols of the products, also separated by plus signs, on the arrowhead side of the arrow. A very simple example is the equation for the formation of FeS from iron and sulfur. (The "translation" of the equation appears beneath it.)

$$\text{Fe} + \text{S} \longrightarrow \text{FeS}$$

Iron reacts with sulfur in a ratio of 1 atom of Fe to 1 atom of S   to give   iron(II) sulfide

Read the + sign separating reactants as "reacts with;" the arrow means "to give."

A more complicated example of an equation describes the formation of aluminum sulfide, Al$_2$S$_3$, from aluminum and sulfur.

$$2\text{Al} + 3\text{S} \longrightarrow \text{Al}_2\text{S}_3$$

Aluminum reacts with sulfur in a ratio of 2 Al atoms to 3 S atoms   to give   aluminum sulfide

The numbers in front of the formulas are called **coefficients.** They specify the proportions of the formula units involved in the reaction. As with subscripts, whenever a coefficient is 1, the 1 isn't written; it is understood.

Our objective here is simply to recognize equations and translate them, not to write them. (That will come later.) An essential feature of any chemical equation, however, is that it be a **balanced equation,** one in which all atoms present in the reactants occur somewhere among the products. For example, in the equation for the formation of Al$_2$S$_3$, there are 2 Al atoms on the left and 2 on the right in Al$_2$S$_3$. Similarly, there are 3 S atoms on the left and 3 on the right. The unbreakability of atoms by chemical means and the conservation of mass in chemical reactions ensure this kind of balance. As Dalton said so long ago, when reactions occur the atoms of the reactants rearrange; they do not break up or disappear.

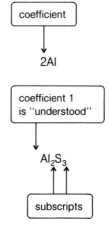

Chemical Reactions Always Convert Substances into Different Substances The absolutely necessary and sufficient condition that determines whether an event is a chemical reaction is that substances change into other substances. To tell if this has happened, we rely on changes in physical appearances or physical properties. As a chemical reaction occurs, some or all of the physical properties associated with the reactants disappear and those of the products emerge. Such changes might be changes in color, odor, or physical state — a gas might bubble out, or a solid might separate. If, once started, the event occurs without any

■ Always be *very* cautious when checking odors of materials. A deep enough whiff of the fumes of ammonia or hydrogen sulfide ("rotten eggs" odor), for example, could hurt you.

further intervention and heat is released, then the change is usually *chemical*. Most chemical reactions that "go by themselves" release heat. If none of these tests decides the matter, then more sophisticated measures must be used, steps that positively identify different compounds.

2.3

KINETIC AND POTENTIAL ENERGY

Chemical energy is a form of potential energy that can be changed to other forms when reactions occur.

We are interested in energy in our study of chemistry, because almost all reactions either require energy in order to occur or they release energy as they happen. **Energy** is the ability to cause change. We say that things "have energy" when they have this ability. Energy can be possessed in two ways: as *kinetic energy* and as *potential energy*.

■ *Kinetic* is from the Greek *kinetikos,* meaning "of motion."

Moving Objects Have Kinetic Energy **Kinetic energy** is the energy associated with motion. A moving car has kinetic energy; so does an avalanche, or a falling star, or a running child. The kinetic energy (KE) of a moving object is calculated by the equation

$$KE = \tfrac{1}{2}mv^2 \tag{2.1}$$

■ Velocity, a derived unit, is distance per period of time, and in the SI its units are meters per second (m/s).

where m is the mass of the object and v is its velocity. Thus the kinetic energy of a moving object is directly proportional to its mass and to the *square* of its velocity. If you double the velocity of a moving car, its kinetic energy increases by a factor of 4 (because $4 = 2^2$). The energy quadruples, not doubles, and this is why a small increase in vehicle velocity can be much more dangerous than the numbers might indicate.

The SI unit of energy, the *joule,* is based on Equation 2.1. An object with a mass of exactly 2 kg moving at a velocity of exactly 1 m/s (meter per second) has one **joule (J)** of kinetic energy.

$$1 \text{ J} = \frac{1}{2} \times (2 \text{ kg}) \times (1 \text{ m/s})^2$$

■ The *joule* is named in honor of J. P. Joule (1818–1889), an English physicist.

The joule is a small amount of energy. If you dropped a 4.4-pound object only 4 inches, it would acquire 1 joule of energy. The **kilojoule (kJ)** is often used for larger amounts of energy.

$$1 \text{ kJ} = 1000 \text{ J}$$

■ Unlike kinetic energy, no simple equation exists for calculating the amount of potential energy in something.

Potential Energy Is Stored Energy There are several ways to put energy into storage. When you wind the spring of a windup toy, you transfer some of your own energy to the coiled spring. There this energy remains until you release the spring and the stored energy appears as the kinetic energy of the moving toy. The stored energy is **potential energy** because it is potentially available to be changed to kinetic energy when you switch on the toy.

When water is pumped into a town's water tank, the stored water has potential energy because of gravitational attraction. Some of this stored energy enables a flow of water that moves wastes through the pipes of the town's sanitary system.

■ Electrons are present in all atoms and so in all matter. They are extremely tiny particles, and when electricity "flows," electrons are moving in one direction.

When battery makers arrange the right chemicals in a battery, the package contains potential energy. At the turn of a switch, chemical reactions occur in the battery that give kinetic energy to electrons to push them through a wire. Thus *chemicals themselves, simply by virtue of their compositions, have potential energy,* usually called **chemical energy.** When you put gasoline into a car, you know that you will be able to use the chemical energy in the gasoline, released when it burns in the engine, to make the car move (and so have kinetic energy).

A piece of paper has stored or chemical energy because of the chemical nature of paper. You can hold a crumpled newspaper in your hands comfortably, until someone puts a match to it. Now the paper burns, giving off heat and light (and maybe a little sound). (Have you ever thought of how remarkable this is? How can the paper have all this energy?)

When you eat, you replenish your reserves of chemicals having chemical energy. Some of the chemical energy might be changed into the mechanical energy of moving arms and legs; some might be transformed into sound energy for speaking; some will appear as heat needed to maintain a steady body temperature.

Total Energy Is Conserved as Changes Occur When you toss a ball into the air you give its mass a certain velocity, so the ball has kinetic energy. As the ball rises, however, its velocity decreases and it loses kinetic energy. Eventually, the velocity becomes zero, and the ball momentarily has no kinetic energy. Where did the energy go? At the moment of no more upward motion, all of the ball's kinetic energy has changed to potential energy. As the ball falls, the reverse process occurs; the ball's potential energy changes into increasing kinetic energy as the velocity increases.

The experience with the ball illustrates one of the most important of all natural laws, the **law of conservation of energy.**

> **Law of Conservation of Energy** The total energy of the universe is constant and can neither be created nor destroyed; it can only be transformed.

The total energy of the universe is the sum of all potential and kinetic energies. This law is also called the *first law of thermodynamics.*

The law of conservation of energy is an example of a *law of nature.* Like all such laws, this law is not *known* to be true *without exception.* It would be impossible to carry out measurements everywhere to see if there are exceptions or violations. The law, instead, expresses the universal experience of those who have made measurements.

The substances in a sandwich or in a piece of paper have chemical energy not because of their locations in the universe, like the tossed ball, but because of their chemical nature. It is in the nature of the way in which the atoms of food, paper, and oxygen are organized that they can undergo energy-releasing chemical reactions and liberate their chemical or stored energy in other forms such as heat.

The chemical energy in paper becomes heat energy.

A great deal of play energy comes from the chemical energy in a sandwich.

2.4

HEAT AND THE THERMAL PROPERTIES OF MATTER

Heat can change the temperature or the physical state of something.
Heat is the energy that transfers from one object to another when the two are at different temperatures and in some kind of contact. We say that heat *flows* from the object with the higher temperature to the one with the lower temperature. If left to itself, the flow continues until both objects reach the same intermediate temperature. *Heat is thus a temperature-changing capacity possessed by an object.* To get heat to flow, all we have to do is put the object next to one with a lower temperature.

Heat is also a physical-state-changing capacity. A block of ice at 0 °C in contact with a warm radiator will not itself undergo a change in temperature. The ice will simply melt — change its physical state from solid to liquid. As long as the freshly melted water is in contact with some ice, its temperature is the same as that of the ice. The temperature at which a solid changes into a liquid is called the **melting point** of the solid.

■ Turning up a burner on a pan of boiling water won't cook the potatoes faster, because it doesn't raise the temperature. It just boils the water away faster.

If you put a pan of water at 100 °C on a hot burner, the flame's higher temperature won't raise the temperature of the water. Instead, the water will boil at a constant temperature and change its state from liquid to gas (vapor). The temperature at which the change from liquid to solid occurs is called the **boiling point.** Thus when an object at a higher temperature is in contact with one at a lower temperature, either a change in state or a change in temperature occurs. In either case, heat flows.

The Calorie Is a Unit of Energy Commonly Used for Heat   The **thermal properties** of matter are physical properties related to the ability of something to handle heat without undergoing a chemical change. One thermal property is **specific heat:** the quantity of energy required to change the temperature of a 1-gram sample of something by 1 °C. It is useful to think of specific heat as the heat-holding ability, the heat-absorbing ability, or the heat-delivering ability of 1 g of a substance.

Water is so common that its heat-absorbing ability was used originally to *define* an older unit of heat, the **calorie (cal):** 1 cal is the amount of heat that, when added to 1 g of water at 14.5 °C, makes the temperature increase to 15.5 °C.[2] If the 1-g sample of water is cooled so that its temperature decreases from 15.5 to 14.5 °C, then the water has given up 1 cal.

One degree is a small change, and one gram of water isn't much — about 16 drops — so the calorie is a small quantity of heat. Scientists, therefore, often use a multiple of the calorie called the **kilocalorie, kcal.**

■ In nearly all popular books on nutrition and diet, the word *calorie* actually means *kilocalorie*. The nutritionist's "Calorie" (capital C) is 1 kcal.

$$1 \text{ kcal} = 1000 \text{ cal}$$

The relationship between the calorie and the joule is known because kinetic energy can be changed quantitatively into heat. The kinetic energy possessed by a paddle rotating below the surface of a sample of water, for example, is changed entirely to heat, which increases the temperature of the water. Roughly 4.2 J are the equivalent of 1 cal. More exactly, the SI now *defines* the calorie in terms of the joule.

$$1 \text{ cal} = 4.184 \text{ J} \qquad \text{(exactly)}$$

Increasingly the joule and the kilojoule (kJ) are replacing the calorie and the kilocalorie as the preferred units for any form of energy, including heat. The change, however, is occurring slowly, particularly in the life sciences, so we will normally use the calorie and the kilocalorie as our units of energy.

PRACTICE EXERCISE 3

The relationship between the calorie and the joule can be restated in the form of two conversion factors.

(a) What are these factors?

(b) Use the proper conversion factor to calculate how many joules and how many kilojoules of energy are equivalent to the 79.7 cal needed to melt 1.00 g of ice.

■ The symbol Δ is the Greek capital *delta*. Pronounce Δ*t* as "delta tee." When Δ is in front of any other symbol, it means a *change* in the value of whatever the other symbol represents. Thus Δ*E* refers to a *change* in energy; Δ*m* is a change in mass.

Now that we have energy units, we can put specific heat on a more general basis. Specific heat is defined by.

$$\text{Specific heat} = \frac{\text{cal}}{\text{g } \Delta t_C} \qquad (2.2)$$

where cal = calories, g = mass in grams, and Δt_C = the *change* in temperature in Celsius degrees. The units in Equation 2.2 give us the common units for specific heat, cal/g °C (calories per gram per degree Celsius).

[2] Although the specific Celsius degree, the one between 14.5 and 15.5 °C, is specified in this formal definition, a one-degree transition anywhere between the 0 °C and 100 °C marks requires virtually an identical quantity of heat.

TABLE 2.2
Specific Heats of Some Substances

Substance	Specific Heat (cal/g °C)[a]
Ethyl alcohol	0.58
Gold	0.031
Granite	0.192
Iron	0.12
Olive oil	0.47
Water (liquid)	1.00

[a] These values are valid in the temperature range of several degrees Celsius on either side of room temperature.

The specific heats of several substances are given in Table 2.2. Notice that metals have very low values; the specific heat of iron, for example, is only about 0.1 cal/g °C or only one-tenth that of water. Only 0.1 cal is needed to make the temperature of a 1-g sample of iron increase by 1 °C. Put another way, the 1 cal that raises the temperature of only 1 g of water by just 1 °C can raise the temperature of the same mass of iron by 10 °C, ten times as much. Iron thus undergoes a much larger change in temperature than the same mass of water by the gain or loss of a relatively small amount of heat.

■ Specific heat could also be in the units of J/(g °C).

■ To two significant figures, water's specific heat is 1.0 cal/g °C over the entire range of 0 to 100 °C.

EXAMPLE 2.3
Using Specific Heat Data

When a 25.4-g piece of iron at 20.0 °C receives 115 cal of heat, what does its temperature change to?

ANALYSIS What the question really calls for is a value of Δt, and we can calculate Δt using Equation 2.2 using the specific heat of iron, 0.12 cal/g °C (Table 2.2). *It is essential that we carry along all of the units as we place data into Equation 2.2, because we have to be sure that the units will cancel properly to leave the answer in the correct unit.* NEVER OMIT THE UNITS OF PHYSICAL QUANTITIES DURING CALCULATIONS UNTIL THEY CANCEL OR MULTIPLY PROPERLY TO GIVE THE CORRECT FINAL UNITS. THE UNITS ARE OUR PRIMARY CHECK ON SETTING UP A SOLUTION CORRECTLY.

■ A large iron nail has a mass of about 25.4 g.

SOLUTION Using the data and Equation 2.2, we have

$$\text{Specific heat of iron} = \frac{0.12 \text{ cal}}{\text{g °C}} = \frac{115 \text{ cal}}{25.4 \text{ g} \times \Delta t}$$

To solve this for Δt we have to cross-multiply. If you think this mathematical procedure is something you can't do, turn to Appendix A where it is described using this problem. Cross multiplication gives us

$$\Delta t = \frac{(115 \text{ cal})(\text{g °C})}{(25.4 \text{ g}) \times (0.12 \text{ cal})}$$

Notice that we cross multiply units just like numbers. Notice also that all of the units cancel, except °C. After doing the arithmetic, we get

$$\Delta t = 38 \text{ °C (rounded by our rules from 37.72965879)}$$

In other words, 115 cal of heat will *increase* the temperature of a 25.4-g piece of iron by 38 °C. Therefore its new temperature is 20.0 °C + 38 °C = 58 °C.

■ The specific heat is known from Table 2.2 to only two significant figures, so we round the answer accordingly.

PRACTICE EXERCISE 4

Suppose that the same amount of heat used in Example 2.3, 115 cal, was absorbed by 25.4 g of water instead of iron, with the initial temperature also 20.0 °C. What will be the final temperature of the water in degrees Celsius? This exercise demonstrates the superior ability of water to absorb heat, as compared to iron, without experiencing a large change in temperature.

The *Heat Capacity* of Something Depends on Both Its Mass and Specific Heat

The **heat capacity** of an object is the quantity of heat that the entire object must absorb (or release) to undergo a 1-degree change in Celsius temperature without changing its physical state.

$$\text{Heat capacity} = \frac{\text{heat}}{\text{degree temperature change}}$$

In terms of the calorie, heat capacity can be calculated by the following equation.

$$\text{Heat capacity} = \frac{\text{cal}}{\Delta t} \qquad (2.3)$$

■ Notice the difference between *specific heat* and *heat capacity*. Specific heat is the heat capacity of a substance *per gram*.

■ If your body's core temperature changes even a few degrees from normal, which is 37.0 °C, you could die.

■ A person using 2000 Calorie (the nutritionist's Calorie being 1000 cal) generates 2×10^6 cal per day.

Heat capacity is an *extensive* thermal property, because *its value is proportional to the total mass of the object*. Specific heat is an intensive property not depending on size. Thus the *specific heat* of the water in Lake Superior, the largest Great Lake, is the same as the specific heat of a small beaker of lake water. The lake, however, has an immense *heat capacity* (a huge heat-holding ability), because of its large size.

Because the adult body is about 60% water and because of the relatively high specific heat of water, *our bodies have a substantial heat capacity*, estimated to be 50×10^3 cal/°C or 50 kcal/°C for a 70-kg adult. This large heat capacity means that considerable heat can be absorbed or released by the body without causing more than a one-degree change in body temperature. The body thus has a thermal "cushion" to help it maintain a steady temperature despite major swings in outside temperature. Nonetheless, the heat generated by the body is considerable; roughly half of the food "calories" taken in each day appears as heat. (The body uses the rest of the energy to run itself and perform the tasks you ask of it.) If you eat, say, 2400 food "calories," meaning 2400 kcal, then about 1200 kcal of heat must be released from the body. If all of this heat stayed in the body, the body temperature would soar by 24 °C!

PRACTICE EXERCISE 5

Show, using Equation 2.3, that the temperature change of the body would be 24 °C if its heat capacity is 50 kcal/°C and the body receives 1200 kcal of extra heat.

The *evaporation* of water from the surfaces of the skin and the lungs is an important mechanism the body uses to remove heat, as explained next.

■ The word *vapor* is usually limited to the gaseous form of something that is a liquid or a solid *at ordinary temperatures.* Thus we speak of "water vapor," but we don't refer to air as a vapor.

Heats of Vaporization and Fusion Are the Energies Needed to Boil or to Melt Substances

The change of a liquid to its gaseous or vapor state is called either **vaporization** or **evaporation.** The opposite change, the conversion of a vapor to its liquid form, is called **condensation.** To vaporize a liquid requires a constant addition of heat, as you no doubt have experienced when you have boiled water on the stove. The heat used to maintain the boiling of a liquid all goes to cause the change of physical state; none goes to change the liquid's temperature as long as both liquid and vapor coexist.

A liquid, of course, does not have to be at its boiling point to evaporate. Wet clothing does dry out, and if it is next to the skin, some of the heat needed for the evaporation is taken from the body. You've also experienced this, no doubt, whenever you've noticed how cold wet

TABLE 2.3
Heats of Vaporization of Some Substances at Their Boiling Points

Substance	Heat of Vaporization[a] (cal/g)
Benzene	94.1
Chloroform	59.0
Ethyl alcohol	204
Ethyl chloride	93
Diethyl ether	84
Gasoline	76–80[b]
Water	539.6

[a] These are values of the heats of vaporization at the boiling points of the substances.

[b] The range in values for the individual compounds in gasoline.

jeans can feel. The conversion of water from liquid to vapor below its boiling point, at body temperature (37 °C), uses up considerable body heat and helps to export heat from the body to the surrounding air.

The heat needed to change 1 g of a substance from its liquid to its gaseous form is called its **heat of vaporization.** Table 2.3 gives the heats of vaporization for several substances at their boiling points. At lower temperatures, the heats of vaporization are higher. The heat of vaporization of water at its boiling point, for example, is 539.6 cal/g; at body temperature (37 °C), the value is about 580 cal/g. Water's heat of vaporization is unusually high among substances, and it accounts for the ability of the body to get rid of considerable heat by letting a little water evaporate from the skin. Special Topic 2.1, which carries this discussion further, shows how the loss of water by evaporation carries away about 40% of the heat that must be released each day from the body.

■ Gaseous water, below 100 °C, can be called *water vapor*, but when the temperature of gaseous water is about 100 °C (water's boiling point), it's called *steam*.

EXAMPLE 2.4
Using Heat of Vaporization Data

How much heat in calories is needed to convert 10.0 g of liquid water to vapor at 37 °C? At this temperature, the heat of vaporization of water is 580 cal/g.

ANALYSIS What we need is a conversion factor that relates grams of water to calories, and water's heat of vaporization at 37 °C, 580 cal/g, gives us the following two choices.

$$\frac{580 \text{ cal}}{1 \text{ g}} \quad \text{or} \quad \frac{1 \text{ g}}{580 \text{ cal}}$$

(Treat the "1" in these conversion factors as an exact number.) Therefore, to get the answer in the right units, we have to multiply 10.0 g by the first factor.

SOLUTION

$$10.0 \text{ g} \times \frac{580 \text{ cal}}{1 \text{ g}} = 5800 \text{ cal} \quad \text{(unrounded)}$$

Because 10.0 has only three significant figures, we have to round the answer and use scientific notation to express the result as 5.80×10^3 cal. This is the same as 5.80 kcal, because

$$5.80 \times 10^3 \text{ cal} \times \frac{1 \text{ kcal}}{10^3 \text{ cal}} = 5.80 \text{ kcal}$$

SPECIAL TOPIC 2.1
METABOLISM AND BODY TEMPERATURE

Basal Metabolism The body's *basal activities* are the minimum activities inside the body that must take place just to maintain muscle tone, control body temperature, circulate the blood, breathe, make compounds or break them down, and otherwise operate tissues and glands during periods of rest. The sum total of the chemical reactions that supply the energy for basal activities is called the body's *basal metabolism.* The rate at which chemical energy is used for basal activities is called the *basal metabolic rate,* which is customarily in units of kcal/min or kcal/kg/h (kilocalories per kilogram of body weight per hour).

Measurements of basal metabolic rates are taken when the person is lying down, has done no vigorous exercise for several hours, has eaten no food for at least 14 hours, and is otherwise awake but at complete rest. A 70-kg (154 lb) adult male has a basal metabolic rate of 1.0 to 1.2 kcal/min. The rate for a 58-kg (128 lb) woman is 0.9 to 1.1 kcal/min. When a person is more active, the metabolic rate is higher than the basal rate. It can go as high as 12 kcal per minute when carrying a heavy load uphill or sprinting in a race.

Mechanisms for Losing Heat from the Body The higher the metabolic rate, the more heat the body must release to preserve its temperature. As explained in Section 2.4, the relatively high heat of vaporization of water at body temperature makes the evaporation of water an important vehicle for this release.

The daily water exchange of a typical adult male is given in the following table.

Water Budget of the Human Body

Water Intake		Water Outgo	
Source	Quantity	Mechanism	Quantity
As drink	1.2 L	Evaporation	
In food	1.0 L	from the skin	0.5 L
Made by metabolism	0.3 L	from the lungs	0.5 L
		Via the urine	1.4 L
		Via the feces	0.1 L
Total intake	2.5 L	Total outgo	2.5 L

The amount of water lost by evaporation is a total of 1 L, or 40% of the 2.5 L of outgo. This evaporated water carries with it 5.8×10^2 kcal of heat per day (as calculated in Practice Exercise 6). The daily intake of food "calories" is normally between 2000 and 4000 kcal per day. Roughly half of this actually appears as *heat.* If you take in 2000 kcal/day of food energy, and half of the energy must be exported as heat, you have to get rid of 1000 kcal of heat per day. The evaporation of water from the skin and the lungs, which removes 5.8×10^2 kcal/day, thus carries away 58% of 1000 kcal/day.

Evaporation occurs two ways. When the sweat glands work and beads of perspiration emerge, the evaporation

TABLE 2.4
Heats of Fusion of Some Substances at Their Melting Points

Substance	Heat of Fusion (cal/g)
Benzene	30
Ethyl alcohol	24.9
Gold	15.0
Iron	65.7
Sulfur	10.5
Water	79.67

PRACTICE EXERCISE 6

The body loses an average of about 1.0 L of water by evaporation each day. How much heat in kilocalories is needed to evaporate 1.0 L of water from the body at 37 °C. The heat of vaporization of water at this temperature is 5.8×10^2 cal/g. (The density of water is 1.0 g/mL; the given volume of water must first be converted to its corresponding mass.)

The change of state from a solid to a liquid also requires a characteristic quantity of heat, the *heat of fusion*. The **heat of fusion** is the heat needed to change 1 g of a solid substance to its liquid state at the melting point. Table 2.4 gives heats of fusion for some common substances, and notice again that water has an unusually high value.

The icepack is a common application of the advantage given by the high heat of fusion of water. Ice is much more effective in an icepack than liquid water, even when the liquid form is at essentially the same temperature. The specific heat of liquid water is roughly 1.0 cal/g °C; its heat of fusion is about 80 cal/g. Thus, if *liquid* water at about 0 °C is to absorb 80 cal and yet increase in temperature only 1 °C, then 80 g of liquid are needed. But if *solid* water at roughly 0 °C is used to absorb 80 cal, then only 1 g of ice is needed.

is called *sensible perspiration.* Evaporation that does not involve the sweat glands is called *insensible perspiration* because we do not notice that it happens. Both forms help to cool the body, but in hot weather and during strenuous exercise, sensible perspiration accelerates. Either form, of course, cannot be sustained without the intake of ample fluids.

Radiation, Conduction, and Convection As Other Mechanisms of Heat Loss The body has other mechanisms besides perspiration for releasing heat. One is by *radiation,* the same heat transfer that occurs at a radiator or a hot clothing iron. Radiation from the body is like light radiation except it isn't visible light but infrared radiation. The uncovered head in cold weather radiates as much as half of the body's heat production. This is why experienced mountaineers say, "If your feet are cold, put on your hat." The hat helps the entire body retain heat.

Conduction is the direct transfer of heat from a warmer body to a colder object. It happens, for example, when we place an icepack on an inflamed area of the skin, or when we sit on a cold surface, or when we put bare hands onto cold machinery or tools.

Finally, *convection* is another mechanism for losing heat from the body. This happens whenever we let the wind or a draft sweep away the warm, thin layer of air next to the skin. Both waffle-weave undergarments and heavy wool slacks or sweaters are filled with tiny pore spaces that trap this layer of warm air. Air is a very poor conductor of heat; so as long as the warm air layer is held close to the skin, little heat is lost by convection.

Body Temperature, Metabolic Rate, Hyperthermia, and Hypothermia One reason why the body tries to maintain a steady temperature is that even small changes in temperature affect the rates of chemical reactions, including those of metabolism. If the body's core temperature increases—a condition called *hyperthermia*—metabolic processes speed up. To sustain this, the body needs more oxygen—about 7% more for every one degree Fahrenheit increase. To deliver this oxygen, the heart must work harder, so a sustained condition of hyperthermia creates problems for the heart.

The opposite of hyperthermia is *hypothermia*—a condition of a lower than normal body temperature. Under this condition, the rates of metabolism eventually slow down, including those reactions that keep vital functions working normally. The initial response of the body to a decrease in its temperature is uncontrolled shivering, however, which *accelerates* the metabolic rate and releases thermal energy from exothermic reactions. This response, however, cannot overcome more than a 2 to 3 °F decrease in the temperature of the body's core. At ever larger decreases, amnesia sets in, the muscles become more rigid, the heart rate becomes erratic, the victim loses consciousness and eventually dies. To prevent this, a victim *must* be made dry, given shelter, and warmed—by drinking warm fluids if conscious, and by blankets or sleeping bags in any case. Give no alcoholic beverages; alcohol dilates (enlarges) capillaries, which enables the heart rapidly to force a flood of chilled blood from the capillaries at the skin's surface into the body's core.

EXAMPLE 2.5
Using Heat of Fusion Data

How much heat in calories is needed to melt an ice cube with a mass of 30.0 g (about 1 ounce) if the ice temperature is 0 °C?

ANALYSIS We need a conversion factor that relates mass to energy for melting ice, and the heat of fusion of ice, 79.67 cal/g, makes available a choice of two.

$$\frac{79.67 \text{ cal}}{1 \text{ g}} \quad \text{or} \quad \frac{1 \text{ g}}{79.67 \text{ cal}}$$

If we multiply the given, 30.0 g, by the first factor, the remaining unit will be "cal."

SOLUTION

$$30.0 \text{ g} \times \frac{79.67 \text{ cal}}{1 \text{ g}} = 2390.1 \text{ cal} \quad \text{(unrounded)}$$

We have to express the answer as 2.39×10^3 cal to show the right number of significant figures: three. Thus one ice cube, by melting, can absorb considerable heat. If this heat were removed from 250 mL of liquid water—about one glassful—with an initial temperature of 25 °C (77 °F), the water temperature would drop by about 10 °C and become 15 °C (59 °F).

PRACTICE EXERCISE 7

An icepack was prepared using 375 g of ice at 0 °C. As this ice melts, how much heat in calories and in kilocalories will be removed from the surroundings by the melting of the ice?   ■

One important additional fact must be noted concerning the thermal properties of matter. If a certain quantity of heat must be absorbed to cause a change of state or a change of temperature in one direction, then exactly the same quantity has to be released to change the temperature or state in the opposite direction. *The law of conservation of energy requires this.* For example, if 80 cal have to be absorbed by ice at 0 °C to *melt* one gram, then to *freeze* one gram of water at the same temperature requires that we *remove* 80 cal from the sample. Similarly, if 540 cal must be absorbed by liquid water at 100 °C to vaporize one gram, then when the same mass of steam at 100 °C condenses it releases the identical quantity of heat. This is why steam is so much more dangerous in contact with the skin than very hot water, although both are life-threatening. When *steam* contacts the much cooler skin, it condenses and all its considerable heat of vaporization is suddenly released, some into the skin itself.

PRACTICE EXERCISE 8

If 16.4 g of steam change to the liquid state at 100 °C, how many kilocalories of heat are released?   ■

Chemical Reactions that Give off Heat Are Said to Be *Exothermic*   Three common sources of heat are solar energy coming from the sun, nuclear energy coming from the unique properties of radioactive elements, and chemical reactions. All of them supply heat and all can be used to generate electricity. Combustion or burning is widely used; it releases the chemical energy in the fossil fuels — gas, oil, and coal — or the stored energy in wood, charcoal, and wastes. **Metabolism,** all of many reactions inside cells, likewise includes many reactions that convert chemical energy into other forms, about half as heat as we said. Both combustion and metabolism consist of *spontaneous* chemical reactions and, in a very broad sense, chemical reactions are either spontaneous or they are not. Spontaneous events are those that, once arranged or started, continue with no further human intervention. Even a small bolt of lightning or the flame from a tiny match can initiate a gigantic forest fire. A reaction such as combustion that continuously releases heat is called an **exothermic** reaction, and most (but not all) spontaneous reactions are exothermic.

Many chemical reactions can be made to take place if we continuously supply them with heat or other forms of energy. Reactions that require a continuous input of heat are called **endothermic** reactions. The process of *photosynthesis* in plants is, overall, energy-consuming, and the necessary energy to sustain it comes from the sun. **Photosynthesis** is the use of solar energy to convert simple compounds with little chemical energy — carbon dioxide and water — into large molecules with considerable chemical energy, like starch. Even the chemical energy in the fossil fuels originated in the sun, because these fuels are the altered remains of plants that flourished in sunlight eons ago.

■ *exo,* out
endo, in
therm, heat

SUMMARY

Matter   Matter, anything with mass that occupies space, can exist in three physical states: solid, liquid, and gas. Broadly, the three kinds of matter are elements, compounds, and mixtures. Elements and compounds are called pure substances, and they obey the law of definite proportions. Mixtures, which do not obey this law, can be separated by operations that cause no chemical changes, but to separate the elements that make up a compound requires chemical reactions.

A chemical reaction is an event in which substances change into different substances with different formulas. Elements can be classified as metals or nonmetals.

Dalton's atomic theory The law of definite proportions and the law of conservation of mass in chemical reactions led John Dalton to the idea — now regarded as well-established fact — that all matter consists of discrete, noncuttable particles called atoms. The atoms of the same element, said Dalton, all have the same mass, and those of different elements have different masses. When atoms of different elements combine to form compounds, they combine as *whole* atoms; they do not break apart. Dalton realized that when different elements combine in different proportions by atoms, the resulting compounds must display a pattern now summarized by the law of multiple proportions.

Symbols, formulas, and equations Every element is given a one- or two-letter symbol, and it can stand either for the element or for one atom of the element. The symbol for a compound is called a formula. The empirical formula (one kind of formula) consists of the symbols of the atoms in one formula unit, with the smallest whole-number subscripts used to show the proportions of the different atoms present.

To describe a chemical reaction, the symbols of the substances involved as reactants, separated by plus signs, are on one side of an arrow that points to the symbols of the products, also separated by plus signs. The equation is balanced when all atoms showing in the formulas of the reactants are present in like numbers in the formulas of the products. Coefficients, numbers standing in front of formulas, are employed as needed to achieve the correct balance. In both formulas and equations, the numbers used for subscripts and coefficients are generally the smallest whole numbers that show the correct proportions.

Forms of energy When something is able to cause a change in motion, position, illumination, sound, or chemical composition, it has energy of one form or another — kinetic energy (energy of motion), light, sound, potential energy, and chemical energy (a special form of potential energy). Energy is neither created nor does it disappear into nothing; it can only be transformed from one type into another or from one place to another. Spontaneous chemical reactions are usually exothermic — they release heat — but many reactions can be made to occur by continuously heating the reactants. These are endothermic reactions.

Heat energy Heat is the form of energy that transfers because of temperature differences. This transfer either causes changes in the temperatures, or causes changes in physical state, such as melting or freezing, boiling or condensing, or it causes chemical change.

The heat that changes the temperature of one gram of a substance by one degree Celsius is called the substance's specific heat. When the substance is water, this quantity of heat is defined as the calorie.

The heat capacity of an object — an extensive property — is the heat that the whole object will absorb or release as its temperature changes by one degree Celsius. The body has a high heat capacity. Its temperature doesn't change much as heat is absorbed or released.

When a change in physical state occurs, heat is released (or absorbed) without any change in temperature. The evaporation of a liquid requires a definite quantity of heat per gram, the heat of vaporization. The value for water is higher than for any other common substance, and the evaporation of water from the body at the skin and the lungs is an important means by which the body gets rid of excess heat. To melt a solid requires a specific quantity of heat per gram, the heat of fusion.

REVIEW EXERCISES

The answers to Review Exercises whose numbers are in color are found in Appendix C. The answers to the other Review Exercises are found in the Study Guide that accompanies this book. The more challenging questions are marked with asterisks.

States and Kinds of Matter

2.1 It is important to distinguish between the *states* of matter and the *kinds* of matter.
(a) What are the names of the states of matter?
(b) What are the names of the kinds of matter?

2.2 Among which, the *states* of matter or the *kinds* of matter, are the distinctions basically in terms of *physical* properties?

2.3 Ice, water, and steam do not look like each other at all, yet we say that they are *chemically* the same. What does this mean?

2.4 It is easy to poke one's finger through the surface of liquid water but not through frozen water. We haven't studied a reason for this yet, but offer a possible explanation (a hypothesis) that explains this at the "molecular level."

2.5 Sulfur is an element that can be obtained as bright yellow, shiny pieces. They are brittle, however, and they do not conduct electricity. Is it a metal or a nonmetal?

2.6 Tungsten, a solid element, can be polished to a shiny finish, and it is a good conductor of electricity. Is it a metal or a nonmetal?

2.7 One element is a liquid at room temperature, and its surface is bright and shiny, like silver. It conducts electricity well. Is it a metal or a nonmetal? What is unusual about the properties described here?

2.8 Changes can be either physical or chemical. What must happen if the change is to be classified as *chemical*? Will physical changes also occur? Explain.

2.9 Which of the following events are chemical changes (as well as being physical changes)?
(a) When heated in a pan, sugar turns brown (caramelizes).
(b) When stirred in water, table salt seems to disappear.
(c) When struck with a hammer, an ice cube shatters.
(d) A bleaching agent causes a colored fabric to lose its color.

2.10 A glass of water and a solution of sugar in water look identical, but the word *substance* properly belongs only to one. Which? Why?

2.11 In broad terms, how many different kinds of *substances* are there, and what are their names?

2.12 Which kind of substance is generally considered to be *simpler*, and why? Which kind of substance occurs in fewer examples?

2.13 Some elements are said to be *radioactive*. What is true about them to earn this title?

2.14 At room temperature, in which *state* are most of the elements found?

2.15 Many metallic materials are *alloys*. What does this mean?

2.16 What is the composition of a diamond?

2.17 How many elements are liquids at room temperature: 0, 1, 2, 20, or 40?

2.18 Roughly how many elements are there: 50, 100, 150, 200, or 250?

2.19 How are compounds different from elements?

2.20 How are compounds different from mixtures?

2.21 How are the terms *substance* and *mixture* used differently in our study?

2.22 Sodium reacts with chlorine to give sodium chloride (table salt).
(a) Physical changes obviously occur (as seen in Figs. 2.2–2.4), so why is this event also called a *chemical* reaction?
(b) Which substances are the *reactants* and which are the *products*?
(c) What is always true about the *ratio of the masses* of sodium and chlorine that combine to form sodium chloride?

2.23 The fact that 2.0 g of hydrogen combines with 16.0 g of oxygen, no more, no less, to give 18.0 g of water illustrates what important law of chemical combination?

2.24 What law of chemical combination most surely distinguishes compounds from mixtures?

Dalton's Atomic Theory

2.25 What are the postulates of Dalton's atomic theory?

2.26 What law of chemical combination contributed most to Dalton's postulate that compounds include their elements in definite ratios by atoms?

2.27 The fact that the ratio of the elements *by mass* in compounds made of just two elements is almost never 1 : 1 suggested what postulate of Dalton's?

*2.28 If the ratio *by mass* of two elements, X and Y, in the hypothetical compound XY were exactly 1 : 1, what would be true in Dalton's theory about the relative masses of the *atoms* of X and Y?

2.29 Dalton's postulate that atoms react as *whole* units and do not break up into pieces of atoms when they react is strongly supported by which law of chemical combination?

*2.30 Copper can combine with oxygen to give two compounds. In one, called cuprite, 1.00000 g of oxygen is combined with 7.94454 g of copper. In the other, called tenorite, the ratio is 1.00000 g of oxygen to 3.97265 g of copper. Do these ratios illustrate the law of multiple proportions? (Do the appropriate calculation.)

*2.31 Hydrogen and oxygen form two compounds, water and hydrogen peroxide. In water, 1.00000 g of hydrogen is combined with 7.93655 g of oxygen. In hydrogen peroxide, 1.00000 g of hydrogen is most likely combined with how much oxygen, 9.45386 g of oxygen, 13.0521 g of oxygen, or 15.8731 g of oxygen? How can you tell?

Chemical Symbols, Formulas, and Equations

2.32 Of the symbols NO and No, which stands for an element and which for a compound?

2.33 Write the symbols of the following elements.
(a) phosphorus (b) calcium (c) bromine
(d) platinum (e) carbon (f) barium

2.34 What are the symbols of the following elements?
(a) lead (b) mercury (c) fluorine
(d) potassium (e) hydrogen (f) iron

2.35 Write the symbols of the following elements.
(a) sulfur (b) iodine (c) manganese
(d) sodium (e) copper (f) magnesium

2.36 What are the symbols of the following elements?
(a) nitrogen (b) oxygen (c) silver
(d) zinc (e) lithium (f) chlorine

2.37 What are the names of the elements with the following symbols?
(a) I (b) Pb (c) Li
(d) N (e) Zn (f) Ba
(g) C (h) Ca (i) Cl
(j) F (k) Cu (l) Fe

2.38 Write the names of the elements with these symbols.
(a) H (b) Mg (c) Al
(d) Hg (e) Mn (f) Na
(g) O (h) Ag (i) Br
(j) P (k) K (l) Pt

2.39 The symbol Fe represents the *element* iron. It also represents what else?

2.40 One formula unit of a constituent of smog consists of one atom of nitrogen and two atoms of oxygen. Which of the following formulas best represents it?
(a) N_2O (b) $2NO$ (c) NO_2 (d) $2NO_2$ (e) N_2O_4

2.41 The formula of sodium chloride is NaCl.
(a) Why would it not be a violation of anything fundamental about sodium chloride to write its formula as Na_2Cl_2?

(b) What is improper about writing Na_2Cl_2 as the formula of sodium chloride?

(c) What kind of formula is represented by NaCl?

(d) What is the name we use for the particle made out of just 1 sodium atom and 1 chlorine atom?

2.42 The balanced equation for the formation of NaCl from sodium and chlorine, as we will see in the next chapter, is properly written as follows.

$$2Na + Cl_2 \rightarrow 2NaCl$$

(a) The *formula unit* of sodium is evidently how many atoms of sodium?

(b) The *formula unit* of chlorine is apparently made out of how many atoms of chlorine?

(c) What is the general name for the numbers that stand in front of formulas in equations?

(d) How many Cl atoms are represented by 2NaCl?

2.43 The following sentences describe chemical reactions. Convert them into balanced equations.

(a) Calcium reacts with sulfur to give calcium sulfide, CaS.

(b) Sodium reacts with sulfur to give sodium sulfide, Na_2S.

Energy

2.44 Compare and contrast the definitions of *matter* and *energy*.

2.45 What is the name of the form of energy associated with the motion itself of a moving object? What is the equation for this form of energy?

2.46 What is the name and symbol of the unit of energy based on the equation that defines kinetic energy?

*2.47 If the *mass* of an object moving at a velocity of 5 m/s is doubled (with no change in velocity), by what factor is the energy of motion of this object changed?

*2.48 If the *velocity* of a moving object changes from 8 m/s to 16 m/s, by what factor is its energy of motion changed?

*2.49 Consider the units used for energy.

(a) Using the equation that defines kinetic energy, what are the *units* of the joule?

(b) How much kinetic energy in joules is possessed by a vehicle with a mass of 2.00×10^3 kg (about the mass of a normal-sized station wagon) when it travels with a velocity of 25.0 m/s (about 56 mi/h)?

(c) When expressed in calories, how much energy does this vehicle have? How much in kilocalories? (1 J = 4.184 cal)

(d) What would the velocity of the vehicle have to become (in m/s) to have twice the kinetic energy it has in part (b)?

*2.50 Use conversion factors made from relationships among units given in Chapter 1 for this question.

(a) What is the equivalent in miles per hour of a velocity of 30.0 m/s?

(b) How does a velocity of 30.0 m/s compare with the normal walking pace of most people (generally taken to be 3.00 mi/h)?

(c) How much energy in kilojoules is possessed by a 70.0-kg adult moving at 30.0 m/s in a vehicle? How much is

possessed by the same person moving at a walking pace of 3 mi/h?

2.51 When an earthquake occurs, a huge amount of kinetic energy is generated. If energy is supposed to be conserved, what happens to this energy?

2.52 Where do you suppose that the light energy released when you turn on a flashlight existed before you closed the switch? (In what form was this energy?)

2.53 When you pick up a book you have to make muscles move. In what form did the kinetic energy of these muscles exist before you caused this to happen?

Thermal Properties of Matter and Heat

2.54 If no heat transfers between two objects in contact, what must be true about them?

2.55 Suppose that two objects, A and B, are in contact and that A has a higher temperature than B.

(a) In which direction, A to B or B to A, will heat transfer?

(b) If by this heat transfer the temperature of B does *not* change, what might be the explanation?

2.56 What name do we give the temperature reading:

(a) At which a solid changes to its liquid state?

(b) At which a liquid changes to its vapor state?

2.57 What is the name we give the amount of heat needed to raise the temperature of 1 g of water by 1 °C? If your finger received this much heat all at once, do you think the skin would be burned?

2.58 When you read that a certain food portion has, say, "100 calories," what does the term "calorie" mean in this context?

*2.59 Referring to Table 2.2, which sample has more heat energy at 25 °C, 100 g of iron or 100 g of water? How can you tell?

*2.60 Enough heat was added to a sample of granite to change its temperature from 20.0 °C to 30.0 °C. What additional information, if any, do we need in order to calculate how much heat was added?

*2.61 Based on data in Table 2.2, which sample would release more heat when cooled from 100 °C to 25 °C, 100 g of iron or 100 g of olive oil?

*2.62 How much heat in kilocalories and in kilojoules is needed to raise the temperature of 250 g of water from 25.0 °C to 75.0 °C?

*2.63 A 250-g bar of iron was cooled from 125 °C to 25.0 °C. How much heat was emitted (in kilocalories and in kilojoules)?

2.64 A large mass of water can absorb a considerable quantity of heat *without any change in its temperature* under two circumstances. What are they?

2.65 A plastic bag filled with 1 kg of water at only 1 °C can be used as an "icepack" on the forehead, but it is far more effective if filled with 1 kg of ice at 0 °C. Explain.

*2.66 Steam at 100.1 °C is regarded as far more dangerous to the skin than an equal mass of water at 100 °C. Explain.

*2.67 Sometimes a gas is called a *vapor*. We do not refer to air, a mixture of gases, as a vapor. Explain.

2.68 What happens to the heat that goes to melt a cube of ice when the meltwater refreezes?

2.69 What terms can be used to describe the change of liquid water to water vapor?

2.70 What term is used for the change of a vapor back to its liquid form?

2.71 Under what circumstance is gaseous water called *water vapor*, and when is it better to call it *steam?*

2.72 Classify each change as *endothermic* or *exothermic*.
(a) Steam condenses
(b) Water freezes
(c) Water evaporates
(d) Ice melts
(e) Water becomes warmer
(f) Paper burns

*2.73 A bar of gold with a mass of 100.0 g at 25.0 °C is given the same quantity of heat as is needed to vaporize 1.00 g of water at 100 °C. What is the new temperature of the gold bar? The specific heat of gold is 0.031 cal/g °C.

*2.74 A bar of iron with a mass of 10.0 g at 20.0 °C is given the same quantity of heat as is needed to melt 1.00 g of ice at 0 °C. What is the new temperature of the iron? The specific heat of iron is 0.119 cal/g °C.

*2.75 One way to prevent pain at a specific spot on the skin is to cool the area rapidly to a temperature low enough to make the chemical reactions associated with pain too slow to matter. A spray of ethyl chloride is often used for this, because it has a low enough boiling point (12 °C) for heat from the skin to boil it. Let us assume that the specific heat of the skin to be sprayed and its underlying flesh is the same as that of water, that the area is initially at body temperature, 37.0 °C, and that the mass affected is 10.0 g. Further suppose that only the heat in this mass (and not from the surrounding air) goes into vaporizing the ethyl chloride. Estimate by a calculation what will be the final temperature of the area if 2.50 g of liquid ethyl chloride is sprayed onto it. The heat of vaporization of ethyl chloride is 93 cal/g.

2.76 When we say that the human body has a "substantial thermal cushion," what is meant? What property of water contributes most to this thermal cushion?

2.77 Which has a higher value for a liter of water, the specific heat of the water or the heat capacity of this sample?

Special Topic 2.1 Metabolism and Body Temperature

2.78 What is the difference between *basal metabolism* and *basal metabolic rate?*

2.79 How do *basal* activities differ from other kinds?

2.80 Describe the basal activities of the body.

2.81 What is the average basal metabolic rate in kcal/min for each (give ranges)?
(a) adult females
(b) adult males

2.82 Calculate how much energy (in kilocalories) an adult with a basal metabolic rate of 1.00 kcal/min expends over a 24.00-hour period.

2.83 Water becomes part of the human body in three ways. What are they? Calculate the average percentage contributed by each route.

2.84 Water leaves the body by three routes. What are they? Calculate the average percentage contributed by each route.

2.85 The body loses water by evaporation chiefly from which areas?

2.86 When the body loses water by evaporation what else does the body lose?

2.87 What are the three nonevaporative means whereby the body gets rid of excess heat?

2.88 An icepack carries heat away from the body by what mechanism?

2.89 What property of air makes waffle-weave or woolen garments effective methods for helping the body retain heat? The use of these fabrics helps particularly to prevent the loss of body heat by what mechanism?

2.90 Covering the head helps to reduce the loss of heat from the body chiefly by preventing what mechanism of heat loss?

2.91 When the body loses heat more rapidly than it can replace the losses, what condition concerning the body's core temperature results?

2.92 In a hyperthermic condition, the body usually breathes more rapidly. Why?

2.93 Why does the heart beat faster (initially) when hyperthermia occurs?

Additional Questions

*2.94 The National Academy of Sciences uses the following conversion factors for the energy content of foods:

proteins	4.0 kcal/g
carbohydrates	4.0 kcal/g
food fat	9.0 kcal/g

One serving of a commercial preparation of baked beans contains 16 g of protein, 49 g of carbohydrate, and 8.0 g of food fat. Calculate the kilocalories of energy in this serving.

*2.95 One serving of elbow macaroni (prepared without cheese sauce) contains 7.0 g of protein, 41 g of carbohydrate, and 1.0 g of food fat. Using the information given in the preceding question, calculate the kilocalories of energy in this serving.

*2.96 A large box of buttered popcorn bought at a theater contains 8.0 g of protein, 32 g of carbohydrate, and 25 g of food fat. Using the information in Review Exercise 2.94:
(a) What is the energy content of this popcorn in kilocalories?
(b) If you walk at a speed of 3.5 mi/h and need 5.0 kcal/min to sustain this activity, how many hours must you walk to "work off" the kilocalories in the popcorn? How many miles long must the walk be?

QUANTITATIVE RELATIONSHIPS IN CHEMICAL REACTIONS

3

Chemists often work with substances dissolved in solutions, many of which have beautiful colors as seen here in an analytical laboratory in Oregon. We'll learn about reacting amounts of substances and their solutions in this chapter.

3.1

THE MOLE CONCEPT

The SI unit for quantity of chemical substance is the mole.

Stoichiometry is the study of the ratios by *atoms* of the elements in compounds and the ratios by *formula units* of the reactants and products in chemical reactions. More information about the kinds of formula units that exist among substances will help our study of stoichiometry.

■ "Stoy-kee-ah-meh-tree" is obtained from the Greek *stoicheion,* meaning "element," and *-metron,* meaning "measure."

The *Formula Units* of Elements and Compounds Vary with the Substance and Can Be Atoms, Molecules, or Ion Groups It is quite adequate to regard the formula units of most metallic elements as being individual **atoms,** which can thus be represented by atomic symbols, like Al (aluminum), Cu (copper), and Ag (silver).

The atoms of most nonmetallic elements do not exist separately. Several have formula units consisting of independent particles made of two atoms, so their formulas have "2" as a subscript. Examples are the elements hydrogen (H_2), oxygen (O_2), nitrogen (N_2), and chlorine (Cl_2). (Atoms of these elements can be generated in special reactions, but they combine as pairs almost instantly.) Small particles capable of independent existence and made of two or more atoms are called **molecules.** When a molecule is made of two atoms it is called a *diatomic molecule.* Many compounds also consist of molecules that can exist as independent particles, only now the molecules are made of atoms from at least two different elements. Examples are water (H_2O), ammonia (NH_3), and methane (CH_4). We say that these compounds are *molecular compounds.*

■ Other elements with diatomic molecules are

fluorine	F_2
bromine	Br_2
iodine	I_2

Many compounds, called *ionic compounds,* consist not of discrete molecules but rather of huge regularly organized arrays of electrically charged particles called *ions.* Sodium chloride (table salt), NaCl, is the most familiar example of an ionic compound. Its crystals consist of enormous numbers of orderly arranged particles, the sodium *ion* and the chloride *ion, in a ratio of 1 to 1, exactly as indicated by the formula NaCl.* Although this formula is not of a discrete *individual* particle—no such independent particle (molecule) of NaCl exists—the formula NaCl does give the elements that make it and in the ratio *of atoms* used. There are times, however, when it is useful to *imagine* a discrete particle corresponding to NaCl, a particle consisting of one Na and one Cl. When necessary we'll call such a formula unit an **ion group,** not a molecule. In summary, the formula units to which we will often refer in our study of stoichiometry can be atoms, molecules, or ion groups.

■ Methane is the chief substance in natural gas, which many communities pipe into homes to be used in "gas" stoves and furnaces and in labs to operate bunsen burners.

Formula Units Enter into Chemical Reactions as Whole Units To simplify the introduction of basic principles of stoichiometry, let's forget for the moment that hydrogen and oxygen are made of diatomic molecules, and let's use only their atoms. Consider their combining to produce H_2O. When we prepare H_2O from H and O with no H or O left over, we are compelled by the formula of water, H_2O, to bring atoms of H and O together in a ratio of exactly 2 to 1. The smallest scale on which we could imagine this reaction is

$$2 \text{ atoms H} + 1 \text{ atom O} \longrightarrow 1 \text{ molecule } H_2O$$

If we mistakenly begin with a 3-to-1 ratio of H to O atoms, one H atom will be left over. Thus,

$$3 \text{ atoms H} + 1 \text{ atom O} \longrightarrow 1 \text{ molecule H}_2\text{O} + 1 \text{ H atom (remaining)}$$

At a larger scale, say 1 dozen molecules of H_2O, then we need the following reaction, but the *ratio* of H to O is the same, 2 to 1.

$$2 \text{ dozen atoms H} + 1 \text{ dozen atoms O} \longrightarrow 1 \text{ dozen molecules H}_2\text{O}$$

If we want 1 gross (144) molecules of H_2O, then we need

$$2 \text{ gross atoms H} + 1 \text{ gross atoms O} \longrightarrow 1 \text{ gross molecules H}_2\text{O}$$

You can see the point. Regardless of the *scale* of the reaction, whether the target is one, a dozen, or a gross of H_2O molecules, the combining ratio of atoms to make H_2O, with nothing left over, is always 2 H to 1 O. We could say that this is the *stoichiometric law* for water. Even when we correct for the diatomic natures of hydrogen and oxygen, the key ratio is still 2 to 1, only this time 2 *molecules* of H_2 and 1 *molecule* of O_2.

$$2 \text{ molecules H}_2 + 1 \text{ molecule O}_2 \longrightarrow 2 \text{ molecules H}_2\text{O}$$

Alternatively,

$$2H_2 + 1O_2 \longrightarrow 2H_2O$$

In this balanced equation we can count 4 atoms of H plus 2 atoms of O on the left, a ratio of $4:2$ or $2:1$.

Formula Units Are Too Small to Be Counted Every chemical substance, regardless of the nature of its formula units, has its own unique stoichiometric law, disclosed by its chemical formula. It seems simple then; for "clean" reactions that make compounds, just count out the necessary atoms (assuming they react as we want). But there's another fact about nature, a catch. Atoms and formula units are too small to be counted like so many corks and test tubes. The number of H_2O molecules in a half a teaspoon of water is well over 10^{23}! No one really comprehends the magnitude of 10^{23}. If you had 10^{23} dollars and distributed dollars equally among the roughly 6×10^9 people of the earth's population, each person could receive over $5000 *per second* for an entire century before the money ran out.

Avogadro's Number Is a Chemical "Counting Unit" The smallness of atoms and formula units necessitates a reference number larger than dozen or gross, and chemists have adopted 6.02×10^{23} for the reference number of formula units. It is called **Avogadro's number** to honor one of the early scientists interested in stoichiometry.

$$\text{Avogadro's number} = 6.02 \times 10^{23}$$

Avogadro's number seems unnecessarily awkward, but there is sense to it. It's scaled large enough so that a sample of Avogadro's number of hydrogen atoms, the element with the lightest atoms, will weigh at least 1 g.

Avogadro's Number Provides One Mole When we have a sample of any pure chemical substance that contains Avogadro's number of its formula units, we have a quantity of substance called a **mole** (abbreviated **mol**). The first of two important meanings of the word *mole* is that it stands for Avogadro's number. The following equations for a much larger scale synthesis of H_2O are identical, therefore.

$$2 \text{ mol H atoms} \quad + \quad 1 \text{ mol O atoms} \quad \longrightarrow \quad 1 \text{ mol H}_2\text{O molecules}$$
$$2 \times [6.02 \times 10^{23} \text{ H atoms}] + 1 \times [6.02 \times 10^{23} \text{ O atoms}] \longrightarrow 1 \times [6.02 \times 10^{23} \text{ H}_2\text{O formula units}]$$

■ Avogadro's number is known to 8 significant figures: 6.0221367×10^{23}, but we'll round to 6.02×10^{23}.

■ Amedeo Avogadro (1776–1856) was an Italian scientist.

As we said, however, it's impossible to count a mole of formula units or even a tiny fraction (yet large enough to be weighed). No doubt you are aware, however, of indirect ways to count large numbers of the same items. If you have a sock full of pennies, for example, you can get a good estimate of their *number* by weighing them and then multiplying the mass by the premeasured pennies per unit mass.

$$\text{mass of pennies} \times \frac{\text{number of pennies}}{\text{total mass}} = \text{number of pennies}$$

The conversion factor of this equation is the connection between numbers and mass of pennies. What we now need is the connection — the conversion factor — between the *number* of formula units of a given chemical substance and the *mass* of the sample. For this, we must learn more about atoms.

3.2

ATOMIC, FORMULA, AND MOLECULAR MASSES

Atoms of one form of carbon, called the carbon-12 isotope, are used to define the mole in mass units.

Since the time of Dalton's theory, atom "smashing" machines have shown that atoms are made of still smaller particles called electrons, protons, and neutrons. The best place to seek a conversion factor that relates *numbers* of formula units to *masses* of substances is with the masses of these tiny particles. Only two, however, protons and neutrons, actually furnish significant mass to an atom. These two have about the same mass, 1.67×10^{-24} g (which is about as unimaginably small as 10^{23} is large). The electron's mass is only about 1/1837th of a proton's mass. The consequence is that electrons contribute almost nothing to the masses of even the heaviest atoms, so we ignore electrons when figuring atomic masses. The masses of atoms stem from their protons and neutrons. The atoms of each element have *unique* numbers of these mass-supplying particles, and we need just a little more information about them.

■ Electrons, protons, and neutrons are the *subatomic* particles of atoms. They are studied in greater detail in the next chapter.

All of the atoms of each individual element have identical numbers of protons. Oxygen atoms all have 8 protons, for example. No other element has atoms with exactly 8 protons. So unique is the number of protons to the atoms of a given element, that this number *defines* an element and is called the element's **atomic number.**

■ The table of the elements inside the front cover of the text gives the atomic numbers of all of the elements.

> **Atomic number** = number of protons per atom

The atomic number of oxygen is 8; that of hydrogen is 1; that of carbon is 6. Element number 11 is sodium; each of its atoms has 11 protons.

■ The isotopes of one element, hydrogen, even have their own names and atomic symbols — protium (H), deuterium (D), and tritium (T). Protium atoms have no neutrons, only 1 proton (and 1 electron) each.

The number of neutrons in the atoms of any given element is not as unique as the number of protons. Thus, although *all* atoms of carbon contain 6 protons, they don't all contain the same numbers of neutrons. Most atoms of naturally occurring carbon have 6 neutrons, but a small percentage of carbon atoms have 7. Nearly all elements are like this; samples of any given element, as it occurs naturally, usually contain atoms with small variations in the numbers of their neutrons. The constituents of an element whose atoms have different numbers of neutrons (but identical numbers of protons) are called the **isotopes** of the element. Like "element," the term "isotope of an element" refers to a *substance,* not to a formula unit. It is possible to separate the isotopes of an element and have samples of them in bottles.

To uniquely define one particular isotope, we use both the atomic number and the *mass number.* The **mass number** of an isotope is the sum of the protons and neutrons in one of its atoms.

> **Mass number** = number of protons + number of neutrons

For every 10,000 naturally occurring carbon atoms, 9889 have the mass number 12. These atoms constitute the isotope of carbon called carbon-12. Each atom of carbon-12 has 6 protons and 6 neutrons. The remaining carbon atoms, 111, have a mass number of 13, from 6 protons + 7 neutrons, and so this isotope is called carbon-13. The same atomic symbol is used for both isotopes because they have the same atomic number and *because they both have identical chemical reactions.*[1]

■ In terms of percentages, naturally occurring carbon is 98.89% carbon-12 and 1.11% carbon-13.

The SI Definition of the *Mole* Is Based on Carbon-12
The SI uses the most abundant isotope of carbon, carbon-12, as the basis for *defining* Avogadro's number or one mole.

> One **mole** of any element or compound has the identical number of formula units (Avogadro's number) as there are atoms in 0.012 kg or 12 g (exactly) of carbon-12.

Thus if we weigh out exactly 12 g of carbon-12, we have Avogadro's number of atoms. This explains why Avogadro's number is awkward in itself. If, purely for our convenience, we want a sample of carbon-12 atoms to have a mass *numerically equal to something familiar about this isotope,* namely its mass number, we are stuck with whatever number of atoms are in such a sample.

The average mass of the atoms of an element as they occur in their natural abundance is the **atomic mass** of the element. Because the masses in grams of individual atoms are so small, a unit of mass new to our study has been defined to express atomic masses. This unit is called the *atomic mass unit,* abbreviated u. The atomic mass of carbon, for example, is 12.011 u. This is the average mass of the carbon atoms as this element occurs in nature. The enormous size of Avogadro's number guarantees that any sample of carbon obtained directly from nature has the proportions of carbon-12 and carbon-13 atoms having an *average* mass of 12.011 u. Most carbon atoms, as we said, are those of carbon-12 with a mass of 12 u (exactly). It's the small percentage of carbon-13 atoms that lifts the average mass, the *atomic mass,* of naturally occurring carbon slightly above 12. Just as 12 g (exactly) of carbon-12 gives us 1 mol of carbon-12 atoms, so 12.011 g of carbon as it occurs naturally also gives us 1 mol of carbon atoms — mostly those of carbon-12, but some of carbon-13. It's like having a sack of pennies in which most of them are old and worn, but a small percentage are brand new. The presence of the new pennies, with their slightly higher mass, lifts the average mass of the pennies slightly above that of just well-worn pennies. Whether worn or new, they have the same name, "pennies," and they have identical "reactions" with vendors when we buy something.

■ $1 u = 1.6656520 \times 10^{-24}$ g

Until recently, the atomic mass unit was abbreviated amu instead of u.

Naturally occurring chlorine (element 17) consists of two isotopes, chlorine-35 (75.53%) and chlorine-37 (24.47%). The average mass of the atoms of these isotopes as they occur in nature is 35.45 u, the atomic mass of chlorine. The average chlorine atom is thus heavier than the carbon-12 atom, 35.45/12 times as heavy, so if we take Avogadro's number or 1 mole of average chlorine atoms the sample has a greater mass, 35.45 g, than that of 1 mole of carbon-12 atoms.

■ An atom of chlorine-35 has 17 protons and 18 neutrons. An atom of chlorine-37 has 17 protons and 20 neutrons. Notice that the average mass, 35.45, is closer to 35 than 37, which reflects the higher percentage of chlorine-35.

$$1 \text{ mol Cl} = 35.45 \text{ g Cl}$$

Thus we have the second, far more important, meaning of *mole* than just "Avogadro's number." *When we have 1 mol of an element we not only have Avogadro's number of its*

---

[1] Dalton did not know about isotopes when he said that *all* atoms of the same element have *identical* masses. Remarkably, this didn't affect the usefulness of his theory because of two facts about elements. First, the proportions of the isotopes of any given element are constant throughout the earth, so wherever you isolate a sample of an element its atoms have identical *average* masses. Second, the isotopes of any given element have the same chemical properties, so chemical reactions giving information about combining masses cannot detect differences among the isotopes of an element. Thus elements in chemical reactions actually do behave as if all their atoms have identical masses. We still treat an element this way, only now we visualize this mass as the *average* mass of the atoms of its isotopes.

atoms, we have a quantity of the element equaling its atomic mass in grams. Figure 3.1 pictures samples of four elements, each sample having Avogadro's number of atoms or 1 mol. The *masses* of the samples are different, however, because the atomic masses are different.

The table of the elements inside the front cover gives the atomic masses of the elements. The footnotes to this table explain why the numbers of significant figures in atomic masses vary so much among the elements. So before we work an example, we need a policy for rounding atomic masses when they are used in calculations.

> **Policy on Rounding Atomic Masses** Round atomic masses to the first decimal place *before* using them in any calculation. (Round the atomic mass of H, however, to 1.01.)

EXAMPLE 3.1
Relating Masses to Avogadro's Number

How many carbon atoms are in 6.00 g of naturally occurring carbon?

ANALYSIS This is our first encounter with a *chemistry* calculation. Like almost all such problems, it basically calls for a transformation of units. The units in this problem, "atoms" and "grams," suggest what we need, a conversion factor that connects numbers of atoms to numbers of grams. You have just learned that the connection is Avogadro's number. *Virtually all problems in chemistry can be analyzed by returning to basic definitions and connections. Whenever you are stuck and don't know what to do, go back to definitions and look for connections among units. These connections are sources of conversion factors.*

We solve this problem by working with the basic meaning of Avogadro's number. Applied to elements, Avogadro's number gives us the number of atoms present in however many grams of the element equal its atomic mass. So we first look up the atomic mass of carbon (12.011), and round it by our rule to the first decimal place, 12.0. Now we can write the connecting relationship between mass and atoms of carbon.

$$12.0 \text{ g of C} = 6.02 \times 10^{23} \text{ atoms of C}$$

As we said, whenever we have a specific connection between two quantities like this, we can devise a choice of two conversion factors, the following in this example.

$$\frac{6.02 \times 10^{23} \text{ atoms C}}{12.0 \text{ g C}} \quad \text{or} \quad \frac{12.0 \text{ g C}}{6.02 \times 10^{23} \text{ atoms C}}$$

SOLUTION If we multiply what is given, 6.00 g C, by the first conversion factor, the units "g C" will cancel, and our answer will be in atoms of carbon.

$$6.00 \text{ g C} \times \frac{6.02 \times 10^{23} \text{ atoms C}}{12.0 \text{ g C}} = 3.01 \times 10^{23} \text{ atoms C}$$

Thus 6.00 g of carbon contains 3.01×10^{23} atoms of carbon.

CHECK Does the answer make sense? Yes; 6.00 g of C is *less* than a whole mole of C atoms (exactly half as much, in fact). So the answer should be *less* than Avogadro's number (exactly half, in fact), and 3.01×10^{23} is half of 6.02×10^{23}. We will sometimes make "sense checks" like this; it's a habit that you will want to develop not just to avoid embarrassing mistakes but also to further your understanding.

FIGURE 3.1
Avogadro's number or 1 mole of atoms are present in each quantity of these elements: 1 mol of iron (55.8 g; the paper clips), 1 mol of liquid mercury (200.6 g), 1 mol of copper (63.5 g; the wire), and 1 mol of sulfur (32.1 g).

PRACTICE EXERCISE 1

How many atoms of gold are in 1.00 oz of gold? Note, 1.00 oz = 28.4 g.

The *Formula Mass* or *Molecular Mass* of a Compound Is the Sum of the Atomic Masses in the Compound's Formula Because atoms lose no weighable quantity of mass when they combine to form compounds, we can expand the idea of an atomic mass to that of a *formula mass* for every compound. The **formula mass** of a compound is simply the sum of the atomic masses of all of the atoms present in one formula unit, no matter what kind of formula unit it represents. For example, the formula mass of ordinary salt, NaCl, is calculated as follows from its formula. We round atomic masses by our rule. We will also follow a common practice of letting the unit of *u* be "understood."

1 atom of Na in NaCl gives	23.0
1 atom of Cl in NaCl gives	35.5
Formula mass of NaCl	58.5

Thus the formula mass of NaCl is 58.5, and one formula unit of NaCl has a mass of 58.5 u. This means that

$$1 \text{ mol NaCl} = 58.5 \text{ g NaCl}$$

The idea of a *formula mass* is general; it applies to anything with a definite formula, including elements as well as compounds. The formula of sodium, for example, is Na, so we can just as well say that its formula mass is 23.0 as to say that this is its atomic mass. The formula of chlorine is Cl_2 so its formula mass is twice the atomic mass of Cl, or $2 \times 35.5 = 71.0$. Thus *subscripts in formulas are multipliers of the atomic masses of the atoms to which the subscripts belong.*

Many compounds have parentheses within their formulas because of the existence of small groups of atoms that often occur together as chemical units. The formula for calcium nitrate, $Ca(NO_3)_2$, is an example. The "2" outside the parenthesis is a multiplier for everything within the parentheses, so in one formula unit of $Ca(NO_3)_2$ there are 2 N, (3×2) or 6 O, and 1 Ca.

There is a synonym for formula mass, namely **molecular mass,** used by many scientists, particularly for molecular compounds. However, *formula mass* is the more general term because it applies to any substance with a chemical formula whether the substance consists of atoms, ion groups, or molecules. You should also be aware that we are in a transition period in chemistry during which a switch is being made from *weight* to *mass*. During this period—it will probably last a generation—you will very often see and hear the terms *atomic weight, formula weight,* and *molecular weight* used for atomic mass, formula mass, and molecular mass.

Chemists who study substances with unusually large formula masses, like those of proteins and genes, frequently attach the word *daltons* to the formula mass. The **dalton,** abbreviated **D,** is a synonym of *atomic mass unit* but is easier to say. A gene might be described as having a mass of 2.5×10^9 daltons or 2.5×10^9 D, for example.

EXAMPLE 3.2
Calculating a Formula Mass

Some baking powders contain ammonium carbonate, $(NH_4)_2CO_3$. Calculate its formula mass.

ANALYSIS A formula mass is the sum of the atomic masses all of the atoms expressed in the formula. So we must look up and write down the atomic masses of all of the elements present, rounding each to the first decimal place (except H is 1.01).

N, 14.0   H, 1.01   C, 12.0   O, 16.0

Each atomic mass must be multiplied by the number of times the atom appears in the formula, and then the resulting numbers are added.

FIGURE 3.2
Avogadro's number or 1 mole of formula units are present in each of these samples of compounds: white sodium chloride (1 mol NaCl = 58.5 g), blue copper sulfate hydrate (1 mol CuSO₄·5H₂O = 185.7 g), yellow sodium chromate (1 mol Na₂CrO₄ = 162.0 g), and liquid water (1 mol H₂O = 18.0 g).

■ *Mol* stands for both the plural and the singular.

■ The moles-to-millimoles conversion factors are

$$\frac{1000 \text{ mmol}}{1 \text{ mol}} \quad \text{and} \quad \frac{1 \text{ mol}}{1000 \text{ mmol}}$$

SOLUTION In each formula unit of $(NH_4)_2CO_3$, N occurs 2 times, H occurs 8 times, C occurs 1 time, and O occurs 3 times. Therefore,

$$2\,N \quad\quad +8\,H \quad\quad +1\,C \quad\quad +3\,O \quad\quad = (NH_4)_2CO_3$$
$$2 \times 14.0 \quad +8 \times 1.01 \quad +1 \times 12.0 \quad +3 \times 16.0 \quad = 96.1 \quad \text{(correctly rounded)}$$

The formula mass of ammonium carbonate is 96.1. This means that one formula unit has a mass of 96.1 u, and it means that Avogadro's number of these units has a total mass of 96.1 g. It also means that

$$1 \text{ mol } (NH_4)_2CO_3 = 96.1 \text{ g } (NH_4)_2CO_3$$

PRACTICE EXERCISE 2

Calculate the formula masses of the following compounds.

(a) $C_9H_8O_4$ (aspirin)   **(b)** $Mg(OH)_2$ (milk of magnesia)

(c) $Fe_4[Fe(CN)_6]_3$ (ferric ferrocyanide or Prussian blue, an ink pigment)

Think of the Mole as the *Lab-Sized Unit* for Amount of Chemical Substance The mole is a quantity of a substance that can be manipulated experimentally and can be taken in fractions or in multiples. For example, the formula mass of H_2O is 18.0, so 1 mol of H_2O has a mass of 18.0 g. If we wished, we could weigh out a smaller sample, say 1.80 g, and then we would have 0.100 mol of water, because 1.80 is one-tenth of 18.0. Or we could take 36.0 g of H_2O, and then have 2.00 mol, because 36.0 is 2 times 18.0. Figure 3.2 pictures 1-mole samples of four compounds. Each sample has Avogadro's number of formula units, but the masses of the samples vary according to their formula masses.

One of the SI prefixes is often used for the mole, the prefix *milli-* signifying one thousandth. The abbreviation of *millimole* is *mmol*.

$$1 \text{ mmol} = 0.001 \text{ mol}$$
$$1000 \text{ mmol} = 1 \text{ mol}$$

Two kinds of calculations are particularly helpful in fixing the mole concept in mind. They are also the two most often used calculations in experimental work — calculating grams from moles and moles from grams.

EXAMPLE 3.3
Converting Moles to Grams

About 21% of the air we breathe is oxygen, O_2. How many grams of O_2 are in 0.250 mol of O_2?

ANALYSIS The problem comes down to the following question.

$$0.250 \text{ mol } O_2 = ? \text{ g } O_2$$

What *always* connects grams to moles is a formula mass. So we first must calculate the formula mass of O_2. It is two times the atomic mass of O, 16.0, or $2 \times 16.0 = 32.0$. This tells us that

$$1 \text{ mol } O_2 = 32.0 \text{ g } O_2$$

This connection between moles and grams of O_2 gives us the following conversion factors.

$$\frac{1 \text{ mol } O_2}{32.0 \text{ g } O_2} \quad \text{or} \quad \frac{32.0 \text{ g } O_2}{1 \text{ mol } O_2}$$

If we now multiply what is given, 0.250 mol of O_2, by whichever conversion factor lets us cancel "mol O_2" we'll have the answer to the question in the desired final unit, "g O_2." The second conversion factor is what we need.

SOLUTION

$$0.250 \cancel{\text{mol } O_2} \times \frac{32.0 \text{ g } O_2}{1 \cancel{\text{mol } O_2}} = 8.00 \text{ g } O_2$$

Thus 0.250 mol of O_2 has a mass of 8.00 g of O_2.

CHECK Does the size of the answer, 8.00 g O_2, make sense? Of course. A whole mole of O_2 has a mass of 32.0 g, so a quarter of a mole of O_2 is 1/4th of 32.0 g, or 8.00 g.

PRACTICE EXERCISE 3

An experiment calls for 24.0 mol of NH_3. How many grams is this?

The next worked example shows how to use a formula mass to convert grams to moles.

EXAMPLE 3.4
Converting Grams to Moles

A student was asked to prepare 12.5 g of NaCl. How many moles is this?

ANALYSIS Here the question is

$$12.5 \text{ g NaCl} = ? \text{ mol NaCl}$$

The connection between grams and moles of NaCl is given by the formula mass of NaCl, which we have already calculated to be 58.5. This tells us that

$$58.5 \text{ g NaCl} = 1 \text{ mol NaCl}$$

Therefore we have these two possible conversion factors.

$$\frac{58.5 \text{ g NaCl}}{1 \text{ mol NaCl}} \quad \text{or} \quad \frac{1 \text{ mol NaCl}}{58.5 \text{ g NaCl}}$$

If we multiply the given, 12.5 g NaCl, by the second ratio, the units will cancel properly and the result will be the moles of NaCl in 12.5 g NaCl.

SOLUTION

$$12.5 \cancel{\text{g NaCl}} \times \frac{1 \text{ mol NaCl}}{58.5 \cancel{\text{g NaCl}}} = 0.214 \text{ mol of NaCl} \qquad \text{(from 0.2136752137)}$$

Thus 12.5 g of NaCl consists of 0.214 mol of NaCl.

CHECK Is the answer reasonable? Yes. The sample is about 2/10th of one mole (58.5 g NaCl), so the answer must certainly be *less*, not more, than 1 mol NaCl.

PRACTICE EXERCISE 4

A student was asked to prepare 6.84 g of aspirin, $C_9H_8O_4$. How many moles is this? How many millimoles?

3.3

BALANCED CHEMICAL EQUATIONS AND STOICHIOMETRY

The coefficients of a balanced equation give the ratios by moles in which the substances react.

One of the most important properties of substances involves their mole relationships when they react. We have just learned that atoms combine only in definite *ratios* by atoms or by moles when they form compounds. Likewise, compounds react only in definite *ratios* by formula units or moles. The coefficients of the reaction's balanced equation give us such ratios. We learned to read a balanced equation in the previous chapter. We'll now carry this another step forward and learn how to write and balance equations when we are given the formulas of the reactants and products.

Balanced Equations Use Formulas and Coefficients to Tell What Reacts, What Forms, and in What Proportions by Moles   Recall that a chemical equation is a **balanced equation** when all the atoms given among the reactants appear in identical numbers among the products. In most equations one or more of the formulas is multiplied by some whole number in order to show the correct balance. These multipliers of formulas in chemical equations are called **coefficients.** For example, the formation of water from hydrogen and oxygen has the following equation (in which we now use the formulas H_2 and O_2.)

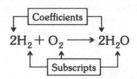

Because of the coefficient of 2 in $2H_2O$, there are $2 \times 2 = 4$ atoms of H on the right side of the equation, and $2 \times 1 = 2$ atoms of O. These figures match the 4 H atoms and 2 O atoms on the left. The equation is thus balanced.

We are never allowed to change subscripts, once we have the right formulas, just to get an equation to balance. For example, changing H_2O to H_2O_2 makes a change from the formula for water to the formula for hydrogen peroxide, an entirely different substance. We can adjust coefficients, however, to balance an equation, as we will see in the next worked example.

■ H_2O_2 cannot be made by a direct combination of H_2 and O_2.

EXAMPLE 3.5
Balancing a Chemical Equation

Sodium, Na, reacts with chlorine, Cl_2, to give sodium chloride, NaCl. Write the balanced equation for this reaction.

ANALYSIS   The first step is to set down all of the correct formulas in the format of an equation. *Never worry about the coefficients until the correct formulas are down.* Then, *never change the formulas.*

SOLUTION After the first step, we have

$$Na + Cl_2 \longrightarrow NaCl \qquad \text{(unbalanced)}$$

So far, we see two chlorine atoms on the left (in Cl_2) but only one on the right. We can't fix this by writing $NaCl_2$, because this isn't the correct formula for sodium chloride. The only way we are allowed to get two Cl atoms on the right is to put a coefficient of 2 in front of NaCl.

$$Na + Cl_2 \longrightarrow 2NaCl \qquad \text{(unbalanced)}$$

Of course, writing 2 in front of NaCl makes it a multiplier for both Na and Cl, so now we have two Na atoms on the right and just one on the left. To fix this, we write a 2 before the Na.

$$2Na + Cl_2 \longrightarrow 2NaCl \qquad \text{(balanced)}$$

Notice particularly how we used a subscript, the 2 in Cl_2, to suggest a coefficient for another formula on the other side of the arrow. This is standard strategy in balancing equations.

■ $NaCl_2$ doesn't even exist.

EXAMPLE 3.6
Balancing a Chemical Equation

Iron, Fe, can be made to react with oxygen, O_2, to form an oxide with the formula Fe_2O_3. Write the balanced equation for this reaction.

ANALYSIS We first write down the correct formulas in the format of an equation. Then we use the subscripts to suggest coefficients.

SOLUTION After the first step, the unbalanced equation is

$$Fe + O_2 \longrightarrow Fe_2O_3 \qquad \text{(unbalanced)}$$

Next we exploit subscripts to suggest coefficients. Oxygen has a subscript of 2 in O_2 and a subscript of 3 in Fe_2O_3. To get a balance, we use the 3 as a coefficient for O_2 and the 2 as a coefficient for Fe_2O_3. This is cross-switching the numbers.

$$Fe + 3O_2 \longrightarrow 2Fe_2O_3 \qquad \text{(unbalanced)}$$

Now there are six oxygen atoms on the left (in $3O_2$) and six on the right (in $2Fe_2O_3$). Of course, the coefficient of 2 in the formula on the right also means that there are 4 Fe atoms on the right. To fix this, we simply use a coefficient of 4 on the left, for Fe.

$$4Fe + 3O_2 \longrightarrow 2Fe_2O_3 \qquad \text{(balanced)}$$

PRACTICE EXERCISE 5

In the presence of an electrical discharge like lightning, oxygen, O_2, can be changed into ozone, O_3. Write the balanced equation for this reaction.

PRACTICE EXERCISE 6

Aluminum, Al, reacts with oxygen to give aluminum oxide, Al_2O_3. Write the balanced equation for this change.

Physical States Are Often Included in Equations   Following the formula of a substance in an equation, a chemist often adds a symbol in parentheses to specify whether the substance is in the gaseous, liquid, or solid state or is in solution in water. The symbols are shown in the margin. Thus, the equation for the reaction of iron with oxygen (Example 3.6) could be written as follows.

■ (g)   gas
 (l)   liquid
 (s)   solid
 (aq)   aqueous solution
       (solution in water)

$$4Fe(s) + 3O_2(g) \longrightarrow 2Fe_2O_3(s) \qquad \text{(balanced)}$$

The practice is not always used, but we get more chemical information when it is.

Sometimes, when we adjust coefficients to balance an equation, we get an equation with coefficients all divisible by the same whole number. Suppose, for example, that we had obtained the following as we tried to balance the equation for the reaction of sodium with chlorine in Example 3.5.

$$4Na + 2Cl_2 \longrightarrow 4NaCl$$

The equation is surely balanced and all of its formulas are correct, so there is nothing basically wrong with it. Chemists, however, generally (but not always) write balanced equations using the set of *smallest* whole numbers as coefficients. We will normally follow this rule. In the equation above, the coefficients are all divisible by 2.

When the formulas in an equation include groups of atoms inside parentheses, and when it is obvious that the groups do not themselves change, treat the groups as whole units in balancing equations.

EXAMPLE 3.7
Balancing Equations Involving Polyatomic Groups

When water solutions of $(NH_4)_2SO_4(aq)$ and $Pb(NO_3)_2(aq)$ are mixed, a white solid separates that has the formula $PbSO_4(s)$. The other product is $NH_4NO_3(aq)$, but it remains dissolved as indicated by the (aq). Represent this reaction by a balanced equation, showing the physical states.

ANALYSIS   As usual, we start by simply writing the correct formulas in the format of an equation.

$$(NH_4)_2SO_4(aq) + Pb(NO_3)_2(aq) \longrightarrow PbSO_4(s) + NH_4NO_3(aq)$$

Because polyatomic groups are involved, we next examine the formulas to see whether any such groups change or if they all appear to react as whole units. We can see here that they do remain as intact units. The subscript of 2 in $(NH_4)_2SO_4$ suggests that we use 2 as the coefficient in the formula on the right where NH_4 occurs.

SOLUTION   Placement of 2 as a coefficient for $NH_4NO_3(aq)$ gives

$$(NH_4)_2SO_4(aq) + Pb(NO_3)_2(aq) \longrightarrow PbSO_4(s) + 2NH_4NO_3(aq)$$

This automatically brought into balance the units of NO_3 on each side of the arrow. The equation is now balanced.

PRACTICE EXERCISE 7

Balance each of the following equations.

(a) $Ca + O_2 \longrightarrow CaO$

(b) $KOH + H_2SO_4 \longrightarrow H_2O + K_2SO_4$

(c) $Cu(NO_3)_2 + Na_2S \longrightarrow CuS + NaNO_3$

(d) $AgNO_3 + CaCl_2 \longrightarrow AgCl + Ca(NO_3)_2$

(e) $Al + H_2SO_4 \longrightarrow Al_2(SO_4)_3 + H_2$

(f) $CH_4 + O_2 \longrightarrow H_2O + CO_2$

An Equation's Coefficients Provide Connections among the Mole Quantities of Reactants and Products With the concept of a mole, we can now think about the coefficients in a balanced equation at two levels at the same time. For example, to return to an equation we used in Chapter 2 to study the laws of chemical combination, the reaction of Fe with S, notice that each formula has a coefficient of 1. Beneath each formula in the equation we can see various ways of interpreting these coefficients.

Fe	+ S	$\longrightarrow$ FeS
1 atom of Fe	+ 1 atom of S	$\longrightarrow$ 1 formula unit of FeS
1 dozen atoms of Fe	+ 1 dozen atoms of S	$\longrightarrow$ 1 dozen formula units of FeS
6.02×10^{23} atoms Fe	+ 6.02×10^{23} atoms S	$\longrightarrow$ 6.02×10^{23} formula units of FeS
1 mol of Fe	+ 1 mol of S	$\longrightarrow$ 1 mol of FeS

Notice that the proportions all remain the same, provided we work with *formula units* of one kind or another. The most important relationship in all of stoichiometry is the following.

> Equal numbers of moles contain equal numbers of formula units.

All that changes as we move through the equations for the formation of FeS is the *scale* of the reaction — the actual numbers of formula units, not their proportions in relationship to each other. *The coefficients in a balanced equation give us the proportions of substances in moles.* Thus, to use an equation that we have employed before:

$$2Na + Cl_2 \longrightarrow 2NaCl$$

we can now interpret this to mean that for every *2 mol* of Na that reacts, *1 mol* of Cl_2 also reacts and *2 mol* of NaCl forms.

One kind of stoichiometric calculation is to use an equation's coefficients to find how many moles of one substance must be present if a certain number of moles of another are involved. *An equation's coefficients are the connection between moles of one substance and moles of another.* Each connection provides two conversion factors, as we'll see. First, we must introduce a substitute for the equals sign when we want to specify a moles-to-moles connection between two *different* substances. We let $\Leftrightarrow$ stand for "is chemically equivalent to." You'll see how this symbol is used in the next example.

EXAMPLE 3.8
Using the Mole Concept with Equations

How many moles of oxygen are needed to combine with 0.500 mol of hydrogen in the reaction that produces water by the following equation?

$$2H_2 + O_2 \longrightarrow 2H_2O$$

ANALYSIS We need the connection between moles of H_2 (the given) and moles of O_2. The coefficients tell us that 2 mol of H_2 combines with 1 mol of O_2 in this particular reaction. They tell us that in this reaction 2 mol of H_2 is *chemically equivalent* to 1 mol of O_2. So we can symbolize this connection as follows.

$$2 \text{ mol } H_2 \Leftrightarrow 1 \text{ mol } O_2$$

It would not be right to say that "2 mol of H_2 *is equal to* 1 mol of O_2 and so write 2 mol $H_2 = 1$ mol O_2. Hydrogen and oxygen are not "equal." Their ability to react in a 2 to 1 ratio, however, tells us that 2 mol of H_2 *requires* exactly 1 mol of O_2, so *for this reaction only* we know that 1 mol of O_2 is the chemical equivalent of 2 mol of H_2. A connection, of course, is a connection, whether it is symbolized by "=" or "↔," and we can use it to construct conversion factors. The connection we just made, 2 mol H_2 ↔ 1 mol O_2, lets us construct and select between the following conversion factors.

$$\frac{2 \text{ mol } H_2}{1 \text{ mol } O_2} \quad \text{or} \quad \frac{1 \text{ mol } O_2}{2 \text{ mol } H_2}$$

We have to choose one of these ratios to multiply by the given quantity, 0.500 mol of H_2, to find out how much O_2 is needed.

SOLUTION The correct conversion factor is the second.

$$0.500 \text{ mol } H_2 \times \frac{1 \text{ mol } O_2}{2 \text{ mol } H_2} = 0.250 \text{ mol } O_2$$

In other words, 0.500 mol of H_2 requires 0.250 mol of O_2 for this reaction.

CHECK Is the size of the answer sensible? Yes. We can see by the balanced equation that half as many moles of O_2 are needed for the given moles of H_2, and half of 0.500 mol is 0.250 mol.

PRACTICE EXERCISE 8

How many moles of H_2O are made from the 0.250 mol of O_2 in Example 3.8? How many millimoles of H_2O are thus made?

PRACTICE EXERCISE 9

Nitrogen and oxygen combine at high temperature in an automobile engine to produce nitrogen monoxide, NO, an air pollutant. The equation is $N_2 + O_2 \rightarrow 2NO$. To make 8.40 mol of NO, how many moles of N_2 are needed? How many moles of O_2 are also needed?

PRACTICE EXERCISE 10

Ammonia, an important nitrogen fertilizer, is made by the following reaction: $3H_2 + N_2 \rightarrow 2NH_3$. In order to make 300 mol of NH_3, how many moles of H_2 and how many moles of N_2 are needed?

With the ability to make the mole calculations involving a balanced equation, we move to study a very common problem that arises in the laboratory—how many grams of one substance are needed to make a given mass of another according to some equation? The next worked example illustrates how this is handled.

EXAMPLE 3.9
Mole Calculations Using Balanced Equations

How many grams of aluminum are needed to make 24.4 g of Al_2O_3 by the following equation?

$$4Al + 3O_2 \longrightarrow 2Al_2O_3$$

■ Aluminum oxide, Al_2O_3, is sometimes used as a white filler for paints.

ANALYSIS *All problems of this nature in stoichiometry must first be worked at the mole level, because the coefficients refer to moles, not masses.* Thus we must first find out how many *moles* are in 24.4 g of Al_2O_3. Then we use the connection, given by the coefficients,

$$2 \text{ mol } Al_2O_3 \Longleftrightarrow 4 \text{ mol } Al$$

to calculate how many *moles* of Al are chemically equivalent to this much Al_2O_3 *for this reaction.* When we know the number of moles of Al, we use the connection,

$$1 \text{ mol } Al = 27.0 \text{ g } Al$$

to calculate the number of grams of Al needed. Figure 3.3 outlines the calculation "flow." In short, our calculation "trip" is

$$24.4 \text{ g of } Al_2O_3 \longrightarrow ? \text{ mol of } Al_2O_3 \longrightarrow ? \text{ mol of } Al \longrightarrow ? \text{ g of } Al$$

Knowing that we'll be needing formula masses, it's usually a good idea at the start of a problem such as this to compute any needed formula masses. The atomic mass of Al is 27.0; the formula mass of Al_2O_3 is 102.0.

SOLUTION To determine the number of moles of Al_2O_3 in 24.4 g Al_2O_3, we devise a conversion factor based on the connection between moles and mass: 1 mol Al_2O_3 = 102.0 g Al_2O_3. Then we carry out the following calculation.

$$24.4 \text{ g } Al_2O_3 \times \frac{1 \text{ mol } Al_2O_3}{102.0 \text{ g } Al_2O_3} = 0.239 \text{ mol of } Al_2O_3$$

Now we can use the connection, given by the equation's coefficients, between the number of moles of Al_2O_3 and the number of moles of Al, 2 mol $Al_2O_3 \Longleftrightarrow 4$ mol Al, to calculate how many moles of Al are needed. The connection gives us the following conversion factors.

$$\frac{4 \text{ mol } Al}{2 \text{ mol } Al_2O_3} \quad \text{or} \quad \frac{2 \text{ mol } Al_2O_3}{4 \text{ mol } Al}$$

So we multiply 0.239 mol of Al_2O_3 by the first factor.

$$0.239 \text{ mol } Al_2O_3 \times \frac{4 \text{ mol } Al}{2 \text{ mol } Al_2O_3} = 0.478 \text{ mol } Al$$

The problem called for the answer in grams of Al, not moles of Al, so we next have to convert 0.478 mol Al into grams of Al. We use the formula mass of Al, 1 mol Al = 27.0 g Al, to devise the correct conversion factor.

$$0.478 \text{ mol } Al \times \frac{27.0 \text{ g } Al}{1 \text{ mol } Al} = 12.9 \text{ g } Al$$

This is the answer; it takes 12.9 g of Al to prepare 24.4 g of Al_2O_3 according to the given equation.

■ The *equals* sign in 1 mol Al = 27.0 g Al is suitable because both sides refer to a quantity of the identical substance but in different units. It's like writing, 1 in. = 2.54 cm; both sides refer to the identical length but in different units.

■ 2Al: $2 \times 27.0 =$ 54.0
3O: $3 \times 16.0 =$ 48.0
Formula mass Al_2O_3 = 102.0

FIGURE 3.3

All calculations involving masses of reactants and products that participate in a chemical reaction must be worked out at the mole level. There is no one-step route from grams of one substance to grams of another, although one can always use a string of conversion factors.

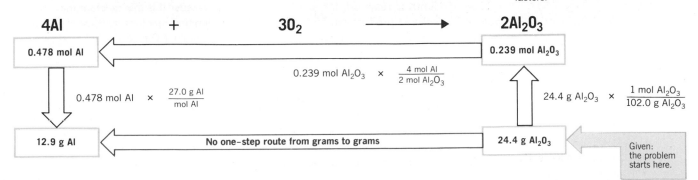

CHECK It's harder to do a "head check" of the sense of this answer, but we can ask if the *size* of the answer makes sense. The answer we found is close to *half* a mole of Al. The equation's coefficients tell us that half again of this amount of Al is chemically equivalent to Al_2O_3, which has a formula mass of 102.0. In other words, half a mole of Al is chemically equivalent to a quarter mole of Al_2O_3. Is this what we began with? Yes, a quarter of a mole of Al_2O_3 is about 25 g, close to the 24.4 g of Al_2O_3 actually given.

PRACTICE EXERCISE 11

How many grams of oxygen are needed for the experiment described in Example 3.9? Use a diagram of the solution in the style of Figure 3.3 as you work out the answer.

PRACTICE EXERCISE 12

If 28.4 g of Cl_2 are used up in the following reaction, how many grams of Na are also used up, and how many grams of NaCl form?

$$2Na + Cl_2 \longrightarrow 2NaCl$$

When the Ideal Stoichiometric Ratio of Reactants Is Not Used, One Reactant *Limits* How Much of the Products Can Be Made Exact stoichiometric ratios by moles of the reactants, the ratios given by the coefficients, are seldom supplied among the reactions occurring in nature and in our cells. A living cell, for example, may have to operate in a short supply of oxygen, or its protein-making machinery shuts down because one "building block" molecule (an amino acid) is unavailable when needed.

To illustrate what "short supply" means with a simple example, consider again the reaction in Example 3.9 and in Practice Exercise 11.

$$4Al + 3O_2 \longrightarrow 2Al_2O_3$$

The ideal stoichiometric ratio of the reactants, Al to O_2, is 4 to 3, 4 mol Al to 3 mol O_2, because 4 and 3 are the respective coefficients. But suppose that in the lab we actually measured out the reactants in a ratio of only 2 mol Al to 3 mol O_2. What then? Now we do not have enough Al for all of the O_2 supplied, so the O_2 taken cannot all be used up. Some O_2 will remain left over from the reaction. (How many moles of O_2 will this leftover amount be?) An excess of O_2 is present, so all of the Al will be used up.

When the ideal mole ratio of reactants is not taken, the reactant that can be entirely used up is called the **limiting reactant.** It's said to be "limiting" because it is this reactant that limits how many moles of product can form. When Al is the limiting reactant, and we start with 2.0 mol of it, by the equation's coefficients we can only obtain 1.0 mol of Al_2O_3, because

$$2.0 \text{ mol Al} \times \frac{2 \text{ mol } Al_2O_3}{4 \text{ mol Al}} = 1.0 \text{ mol } Al_2O_3$$

Even if we took 3.0 mol of O_2 for the 2.0 mol of Al, we could not obtain 2.0 mol of Al_2O_3 — from 3.0 mol $O_2 \times (2 \text{ mol } Al_2O_3/3 \text{ mol } O_2)$ — because we have not begun with sufficient aluminum to make 2.0 mol of Al_2O_3. The concept of a limiting reactant is important in all areas of chemistry, and we'll see it again when we discuss nutritional needs where a short supply of vitamins or amino acids can have injurious or fatal consequences.

3.4

REACTIONS IN SOLUTION

Virtually all of the chemical reactions studied in the lab and that occur in living systems take place in an aqueous solution.
If the particles of one substance are to react with those of another, they must have enough freedom to move about to find each other. Such freedom exists in the gaseous and liquid states but not in the solid state. To get one solid to react with another, chemists usually dissolve them in something. This puts them into a liquid state, and their particles can move about. In order to learn in the next section about mass relationships when reactants are in solution, we must first learn the common terms used to describe solutions.

A Solution Is Made of a Solvent and One or More Solutes A **solution** is a uniform mixture of particles that are exceedingly small, at the atomic size or somewhat larger depending on the formula units involved. A minimum of two substances is needed to have a solution. One is called the *solvent* and all of the others are called the *solutes.*

The **solvent** is the medium into which the other substances are mixed or dissolved. The solvent is usually a liquid, like water. Unless we state otherwise, we will always be dealing with aqueous solutions; *aqueous* designates water as the solvent as we have indicated.

A **solute** is anything that is dissolved by the solvent. In an aqueous solution of sugar, the solute is sugar and the solvent is water. The solute can be a gas. Club soda is a solution of carbon dioxide in water. The solute can be a liquid. Antifreeze, for example, is mostly an aqueous solution of the liquid ethylene glycol.

■ Both solids and gases can function as solvents. Air is a solution chiefly of oxygen and nitrogen. An *alloy* is a hardened solution made by stirring together two or more molten metals.

Solutions Can Be Dilute or Concentrated Several terms are used to describe a solution. A **dilute solution** is one in which the ratio of solute to solvent is very small, such as a few crystals of sugar dissolved in a glass of water. Most of the aqueous solutions in living systems have more than two solutes and are dilute in all of them. In a **concentrated solution,** the ratio of solute to solvent is large. Syrup, for example, is a concentrated solution of sugar in water.

Solutions Can Be Unsaturated, Saturated, or Supersaturated Some solutions are **saturated solutions,** which means that it isn't possible to dissolve more of the solute in them (assuming that the temperature of the solution is kept constant). If more solute is added to a solution already saturated in this solute, the extra solute will just remain separate. If the solute is a solid, it will generally sink to the bottom and lie there.

An **unsaturated solution** is one in which the ratio of solute to solvent is lower than that of the corresponding saturated solution. If more solute is added to an unsaturated solution, at least some of it will dissolve.

It isn't easy, but sometimes a **supersaturated solution** can be made. This is an unstable system in which the ratio of dissolved solute to solvent is actually higher than that of a saturated solution. We can sometimes make a supersaturated solution by carefully cooling a saturated solution. The ability of most solutes to dissolve in water decreases with temperature, so when a saturated solution is cooled some of the now excess solute should separate. But this doesn't always happen. If the excess solute does not separate, then we have a supersaturated solution. If we now scratch the inner wall of the container with a glass rod, or if we add a crystal—a "seed" crystal—of the pure solute to the system, the excess solute will usually separate immediately. This event can be dramatic and pretty to watch (see Figure 3.4). The separation of a solid from a solution is called **precipitation,** and the solid is referred to as the **precipitate.**

FIGURE 3.4
Supersaturation. A seed crystal has been added to a supersaturated solution, (left photo), and whatever solute was present in solution in excess quickly separates. Any solution that still remains in contact with the crystals (right photo) is now a saturated solution.

Solubility Is Sometimes Reported as Grams of Solute per 100 g of Solvent The amount of solute needed to give a saturated solution in a given quantity of solvent at a specific temperature is called the **solubility** of the solute in the given solvent. It is the maximum quantity of the solute that can dissolve and form a stable solution at the given temperature in the given quantity of solvent. If you take more than this maximum by even the smallest quantity, like a single crystal of a solid, the extra will simply remain undissolved in the system. The presence of an extra amount of solute, in fact, is a sign that the solution is saturated (provided enough time has been allowed for the solution to form). If a known extra amount of solute has *all* gone into solution, it's a sign that the solution is supersaturated (and so is unstable).

Solubilities vary widely from substance to substance, as the data in Table 3.1 show. Notice particularly that a saturated solution can still be quite dilute. For example, only a very small mass of barium sulfate can dissolve in 100 g of water.

The solubilities of most solids increase with temperature, as you can see in Table 3.1. All gases become less and less soluble in water as the temperature increases, assuming that the measurements are made under the same pressure.

TABLE 3.1
Solubilities of Some Substances in Water

Solute	Solubilities (g/100 g water)			
	0 °C	20 °C	50 °C	100 °C
Solids				
Sodium chloride, NaCl	35.7	36.0	37.0	39.8
Sodium hydroxide, NaOH	42	109	145	347
Barium sulfate, $BaSO_4$	0.000115	0.00024	0.00034	0.00041
Calcium hydroxide, $Ca(OH)_2$	0.185	0.165	0.128	0.077
Gases				
Oxygen, O_2	0.0069	0.0043	0.0027	0
Carbon dioxide, CO_2	0.335	0.169	0.076	0
Nitrogen, N_2	0.0029	0.0019	0.0012	0
Sulfur dioxide, SO_2	89.9	51.8	4.3	1.8[a]
Ammonia, NH_3			28.4	7.4[b]

[a] At 90 °C.

[b] At 96 °C.

3.5

MOLAR CONCENTRATION

The unit of *moles per liter* is the most useful unit of concentration when working with the stoichiometry of reactions in solution.

The **concentration** of a solution is the ratio of the quantity of solute to some given unit of the solution. The units can be anything we wish, but for the stoichiometry of reactions in solution, the best units are those of moles of solute per liter of solution. This ratio is called the solution's **molar concentration,** or **molarity,** abbreviated M. The molar concentration of a solution, its molarity, is the number of moles of solute per liter of solution.

$$M = \frac{\text{mol solute}}{\text{L solution}} = \frac{\text{mol solute}}{1000 \text{ mL solution}}$$

■ M = moles/liter
mol = moles

A bottle might, for example, have the label "0.10 M NaCl." If so, we know that the solution in this container has a concentration of 0.10 mol of NaCl per liter of solution (or per 1000 mL of solution). There might be a small or large amount of solution in the bottle, but in any case the *ratio* is the same, 0.10 mol NaCl per liter of solution. A value of molarity always gives two conversion factors. In our example, they are

$$\frac{0.10 \text{ mol NaCl}}{1000 \text{ mL NaCl solution}} \quad \text{and} \quad \frac{1000 \text{ mL NaCl solution}}{0.10 \text{ mol NaCl}}$$

A Volumetric Flask Is Used to Make a Solution of Known Molarity Figure 3.5 shows how to make a solution having a known molarity. The mass of solute corresponding to the moles of solute we want is weighed out. The sample is then placed in a *volumetric flask,* a special piece of glassware pictured in Figure 3.5. These flasks are available in several fixed capacities, so the one selected must allow a final volume of solution that gives the solute the molar concentration we want. The solvent is added until the solute all dissolves and the liquid level exactly reaches the etched mark on the flask, as described in the figure legend.

The concept of molarity will become clearer by studying how to do some of the calculations associated with it. In the next worked example we'll see what kinds of calculations have to be done in order to go into the lab and prepare a certain volume of a solution that has a given molar concentration.

FIGURE 3.5

The preparation of a solution of known molarity. The volumetric flask has an etched line on its neck that marks the liquid level at which the flask will hold the specified volume. (*a*) The solute, accurately weighed, has been placed in the flask. (*b*) Some water (distilled or deionized) is added. (*c*) The flask is agitated so that the solute dissolves. (*d*) Enough water is added to bring the level to the etched line. (*e*) After the flask is stoppered, it is shaken so that the solution will be uniform.

(a)

(b)

(c)

(d)

(e)

EXAMPLE 3.10
Preparing a Solution of Known Molar Concentration

■ Sodium bicarbonate is the "bicarb" of home medicine cabinets. It is also an ingredient in baking powders.

How many grams of sodium bicarbonate, $NaHCO_3$, is needed to prepare 500 mL of 0.125 M $NaHCO_3$?

ANALYSIS   The 0.125 M indirectly refers to *moles,* but the question asks for the answer in grams. Before we can calculate the grams needed, we have to find out how many moles of $NaHCO_3$ are required. Here is where the given concentration provides what we need most, a conversion factor to calculate the moles of $NaHCO_3$ in the given volume, 500 mL of 0.125 M $NaHCO_3$. The molarity gives us the following conversion factors.

$$\frac{0.125 \text{ mol NaHCO}_3}{1000 \text{ mL NaHCO}_3 \text{ solution}} \quad \text{or} \quad \frac{1000 \text{ mL NaHCO}_3 \text{ solution}}{0.125 \text{ mol NaHCO}_3}$$

If we multiply the given volume, 500 mL of $NaHCO_3$ solution, by the first conversion factor, the volume units will cancel and we will learn how many moles of $NaHCO_3$ are needed. (Draw the cancel lines yourself.)

SOLUTION

■ From here on, as you judge their usefulness, draw in the cancel lines in calculations involving conversion factors.

$$500 \text{ mL NaHCO}_3 \text{ solution} \times \frac{0.125 \text{ mol NaHCO}_3}{1000 \text{ mL NaHCO}_3 \text{ solution}} = 0.0625 \text{ mol NaHCO}_3$$

In other words, the 500 mL of solution has to contain 0.0625 mol of $NaHCO_3$. So we have to convert 0.0625 mol of $NaHCO_3$ into grams of $NaHCO_3$. To do this we need one of the conversion factors that the formula mass of $NaHCO_3$, 84.0, makes available.

$$\frac{84.0 \text{ g NaHCO}_3}{1 \text{ mol NaHCO}_3} \quad \text{or} \quad \frac{1 \text{ mol NaHCO}_3}{84.0 \text{ g NaHCO}_3}$$

If we multiply 0.0625 mol of $NaHCO_3$ by the first of these factors, then the units of mol $NaHCO_3$ will cancel and our answer will be in what we want, grams.

$$0.0625 \text{ mol NaHCO}_3 \times \frac{84.0 \text{ g NaHCO}_3}{1 \text{ mol NaHCO}_3} = 5.25 \text{ g NaHCO}_3$$

Thus to prepare 500 mL of 0.125 M $NaHCO_3$, we have to weigh out 5.25 g of $NaHCO_3$, dissolve it in some water in a 500-mL volumetric flask, and then carefully add water until its level reaches the mark, making sure that the contents become well mixed.

CHECK   Is the *size* of the answer, 5.25 g $NaHCO_3$, reasonable? Yes. The molarity is 0.125 M $NaHCO_3$, and the formula mass of $NaHCO_3$ is 84 g. If the molarity were 0.100 M instead of 0.125 M, then we'd need 8.4 g of solute for a whole liter, and 4.2 g for the specified half a liter. But the molarity is larger than 0.100 M, so we need a somewhat larger mass of $NaHCO_3$ than 4.2 g, which is what the answer is. (By checking the *size* of the answer, you get an idea if you multiplied when you should have divided something, or vice versa.)

PRACTICE EXERCISE 13

How many grams of each solute are needed to prepare the following solutions?
(a)   250 mL of 0.100 M H_2SO_4 **(b)**   100 mL of 0.500 M glucose ($C_6H_{12}O_6$)

Another calculation is to find the volume of a solution of known molar concentration that will deliver a certain quantity of its solute. The next worked example shows how this is done.

EXAMPLE 3.11
Using Solutions of Known Molar Concentration

In an experiment to see whether mouth bacteria can live on mannitol ($C_6H_{14}O_6$), a student needed 0.100 mol of mannitol. It was available as a 0.750 M solution. How many milliliters of this solution must be used in order to obtain 0.100 mol of mannitol?

ANALYSIS The two conversion factors that are provided by the given concentration are

$$\frac{0.750 \text{ mol mannitol}}{1000 \text{ mL mannitol solution}} \quad \text{and} \quad \frac{1000 \text{ mL mannitol solution}}{0.750 \text{ mol mannitol}}$$

We must multiply the given, 0.100 mol of mannitol by the second conversion factor to calculate the answer.

SOLUTION

$$0.100 \text{ mol mannitol} \times \frac{1000 \text{ mL mannitol solution}}{0.750 \text{ mol mannitol}} = 133 \text{ mL mannitol solution}$$

Thus 133 mL of 0.750 M mannitol solution holds 0.100 mol of mannitol.

CHECK If 1000 mL of solution holds 0.750 mol of solute, we need roughly one-seventh as much (0.100/0.750 is about one-seventh) to hold 0.100 mol, and one-seventh of 1000 is about 130 mL.

■ Mannitol is the sweetening agent in some sugarless chewing gums.

PRACTICE EXERCISE 14

To test sodium carbonate, Na_2CO_3, as an antacid, a scientist needed 0.125 mol of Na_2CO_3. It was available as 0.800 M Na_2CO_3. How many milliliters of this solution are needed for 0.125 mol of Na_2CO_3?

Once solutions of known molar concentration have been prepared, then the most common kind of calculation involves the stoichiometry of some reaction when at least one reactant is in solution. In the next worked example, we'll see how we can do stoichiometric calculations for such a reaction.

EXAMPLE 3.12
Stoichiometric Calculations That Involve Molar Concentrations

Potassium hydroxide, KOH, reacts with hydrochloric acid as follows:

$$\text{HCl}(aq) \quad + \text{KOH}(aq) \longrightarrow \text{KCl}(aq) \quad + \text{H}_2\text{O}$$

Hydrochloric acid · Potassium hydroxide · Potassium chloride

How many milliliters of 0.100 M KOH are needed to react with the acid in 25.0 mL of 0.0800 M HCl?

ANALYSIS The volume and the molarity of the HCl solution will enable us to calculate how many moles of HCl were used. The equation's coefficients will let us relate the number of moles of HCl to the number of moles of KOH. Finally, we will use the number of moles of KOH and the molarity of the KOH solution to calculate the volume needed.

■ We usually do not specify the liquid state for water, as in $H_2O(l)$, because this is the state water normally is in.

■ "Stomach acid" is roughly 0.1 M hydrochloric acid.

■ soln = solution

SOLUTION We can find the number of moles of HCl in 25.0 mL of 0.0800 M HCl by using the first of the following conversion factors obtained from the fundamental meaning of M, moles per liter or moles per 1000 mL.

$$\frac{0.0800 \text{ mol HCl}}{1000 \text{ mL HCl soln}} \quad \text{or} \quad \frac{1000 \text{ mL HCl soln}}{0.0800 \text{ mol HCl}}$$

We multiply the given, 25.0 mL of HCl solution, by the first factor:

$$25.0 \text{ mL HCl soln} \times \frac{0.0800 \text{ mol HCl}}{1000 \text{ mL HCl soln}} = 0.00200 \text{ mol HCl}$$

The next step is to find out how many moles of KOH are needed to react with 0.00200 mol of HCl. The coefficients of the balanced equation tell us that the mole ratio is 1 : 1, which means 1 mol of HCl ⟺ 1 mol of KOH in this reaction. Thus 0.00200 mol of HCl requires 0.00200 mol of KOH.

Finally, we have to calculate the volume (in mL) of the KOH solution that contains 0.0020 mol of KOH. The molarity of the KOH solution gives us the option of the following conversion factors:

$$\frac{0.100 \text{ mol KOH}}{1000 \text{ mL KOH soln}} \quad \text{or} \quad \frac{1000 \text{ mL KOH soln}}{0.100 \text{ mol KOH}}$$

If we multiply 0.00200 mol of KOH by the second conversion factor, we'll have the right units:

$$0.00200 \text{ mol KOH} \times \frac{1000 \text{ mL KOH soln}}{0.100 \text{ mol KOH}} = 20.0 \text{ mL KOH soln}$$

Thus 20.0 mL of 0.100 M KOH solution provides exactly the right amount of KOH to react with the acid in 25.0 mL of 0.0800 M HCl. Check this out yourself. ■

The next worked example shows how to solve a problem in which the relevant mole ratio in the equation is not 1 : 1.

EXAMPLE 3.13
Stoichiometric Calculations That Involve Molar Concentrations

■ Sodium hydroxide, NaOH, is commonly known as "lye," and is an ingredient in some oven cleaners.

Sodium hydroxide, NaOH, reacts with sulfuric acid by the following equation:

$$H_2SO_4(aq) + 2NaOH(aq) \longrightarrow Na_2SO_4(aq) + 2H_2O$$

Sulfuric acid   Sodium hydroxide   Sodium sulfate

How many milliliters of 0.125 M NaOH provide enough NaOH to react completely with the sulfuric acid in 16.8 mL of 0.118 M H$_2$SO$_4$ by the given equation?

ANALYSIS We start by asking how many moles of sulfuric acid are in 16.8 mL of 0.118 M H$_2$SO$_4$. Then, we can relate this number of moles to the number of moles of NaOH that match it according to the coefficients. For this equation, we know from its coefficients that 1 mol H$_2$SO$_4$ ⟺ 2 mol NaOH. Finally, we will find out how many milliliters of the NaOH solution hold this calculated number of moles of NaOH. Figure 3.6 provides a pictorial summary—a calculation flowchart—of the steps for solving this problem.

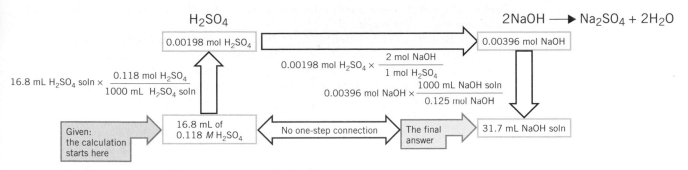

FIGURE 3.6
The calculation flow diagram for Example 3.13.

SOLUTION   First, the number of moles of H_2SO_4 that react:

$$16.8 \text{ mL } H_2SO_4 \text{ soln} \times \frac{0.118 \text{ mol } H_2SO_4}{1000 \text{ mL } H_2SO_4 \text{ soln}} = 0.00198 \text{ mol } H_2SO_4$$

Next, the moles of NaOH that chemically match 0.00198 mol of H_2SO_4 based on the fact that 1 mol $H_2SO_4 \Leftrightarrow$ 2 mol NaOH:

$$0.00198 \text{ mol } H_2SO_4 \times \frac{2 \text{ mol NaOH}}{1 \text{ mol } H_2SO_4} = 0.00396 \text{ mol NaOH}$$

Finally, the volume of 0.125 M NaOH solution that holds 0.00396 mol NaOH:

$$0.00396 \text{ mol NaOH} \times \frac{1000 \text{ mL NaOH soln}}{0.125 \text{ mol NaOH}} = 31.7 \text{ mL NaOH soln}$$

Thus 31.7 mL of 0.125 M NaOH solution is needed to react with all of the sulfuric acid in 16.8 mL of 0.118 M H_2SO_4. As a check, look again at the conversion factors used and if units have properly canceled.

PRACTICE EXERCISE 15

Sodium bicarbonate reacts with sulfuric acid as follows:

$$2NaHCO_3(aq) + H_2SO_4(aq) \longrightarrow Na_2SO_4(aq) + 2CO_2(g) + 2H_2O$$

How many milliliters of 0.112 M H_2SO_4 will react with 21.6 mL of 0.102 M $NaHCO_3$ *according to this equation?*

3.6

PREPARING DILUTE SOLUTIONS FROM CONCENTRATED SOLUTIONS

The dilution of a fixed volume of a concentrated solution changes only the concentration, not the moles of solute.

Often in the lab we have a relatively concentrated solution of known molarity, but we'd rather use a more dilute solution in some experiment. We'll study here how to do the calculations needed for the preparation of dilute solutions of known molarity from concentrated solutions. This or a similar calculation sometimes occurs in a clinical situation when a medication must be diluted. One basic idea guides these calculations. The mass or moles of solute in the final

■ To save shipping costs, acids are often shipped in very concentrated solutions which are then diluted to the desired concentration at the laboratory.

volume of the dilute solution will be the same as were present in the concentrated solution. We add only *solvent,* not solute.

The calculation begins with three facts. We know what final concentration we want, we know what volume we wish to have, and we know the concentration of the concentrated solution to be diluted. The calculation is to tell us what volume of the concentrated solution must be taken. Because the moles of actual solute are the same in both solutions, we can calculate this amount in the usual way from the volume and molarity data *for both solutions.* In the dilute solution,

$$\text{mol solute} = \text{liters}_{\text{dil soln}} \times \frac{\text{mole solute}}{\text{liter}_{\text{dil soln}}} = \text{liters}_{\text{dil soln}} \times M_{\text{dil soln}}$$

In the concentrated solution,

$$\text{mol solute} = \text{liters}_{\text{concd soln}} \times \frac{\text{mole solute}}{\text{liter}_{\text{concd soln}}} = \text{liters}_{\text{concd soln}} \times M_{\text{concd soln}}$$

These two expressions for the moles of solute equal each other, as we have said. Therefore,

$$\text{liters}_{\text{dil soln}} \times M_{\text{dil soln}} = \text{liters}_{\text{concd soln}} \times M_{\text{concd soln}}$$

Actually, the unit of liters isn't required. We can use any volume unit that we please provided that it is the same unit on both sides of the equation. Normally the mL unit is used in the lab, so our equation can be restated as follows.

$$\text{mL}_{\text{dil soln}} \times M_{\text{dil soln}} = \text{mL}_{\text{concd soln}} \times M_{\text{concd soln}} \qquad (3.1)$$

The following example shows how this equation is used.

EXAMPLE 3.14
Doing the Calculations for Making Dilutions

■ Hydrochloric acid can be purchased in hardware stores as "muriatic acid."

Hydrochloric acid can be purchased at a concentration of 1.00 M HCl. How can we prepare 500 mL of 0.100 M HCl?

ANALYSIS What the question really asks is how many milliliters of 1.00 M HCl would have to be diluted to a final volume of 500 mL to make a solution with a concentration of 0.100 M HCl? Equation 3.1 is meant for this problem.

SOLUTION   We first assemble the known data:

$$\text{mL}_{\text{dil soln}} = 500 \text{ mL} \qquad \text{mL}_{\text{concd soln}} = ?$$
$$M_{\text{dil soln}} = 0.100 \ M \qquad M_{\text{concd soln}} = 1.00 \ M$$

Now we use the equation:

$$\text{mL}_{\text{dil soln}} \times M_{\text{dil soln}} = \text{mL}_{\text{concd soln}} \times M_{\text{concd soln}}$$
$$500 \text{ mL} \times 0.100 \ M = \text{mL}_{\text{concd soln}} \times 1.00 \ M$$

Rearranging terms to solve for mL$_{\text{concd soln}}$ gives us

$$\text{mL}_{\text{concd soln}} = \frac{500 \text{ mL} \times 0.100 \ M}{1.00 \ M} = 50.0 \text{ mL}$$

In other words, if we take 50.0 mL of 1.00 M HCl, place this in a 500-mL volumetric flask, and add water to the mark, we will have 500 mL of 0.10 M HCl. Figure 3.7 illustrates how a dilution is carried out.

(a)

(b)

(c)

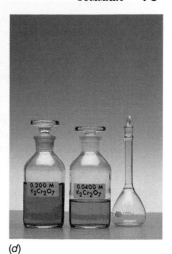
(d)

FIGURE 3.7

Preparing a dilute solution by diluting a concentrated solution. The long glass tube with a bulge in its middle is a volumetric pipet. Like a volumetric flask, it has an etched line on its upper narrow section so that when the liquid level is at this line, the pipet contains the volume printed on it. Notice that suction is being supplied by a suction bulb, not by mouth. The calculated volume of the more concentrated solution is withdrawn (a) and (b) placed in a volumetric flask. This flask would already contain some of the additional water to be added if the solution in the pipet were a concentrated acid or base. (c) Now additional water is added slowly as the new solution is swirled to promote mixing until the final volume is reached. (d) Then the new solution is transferred to a dry bottle and labeled.

PRACTICE EXERCISE 16

Calculate the volume of $0.200\ M\ K_2Cr_2O_7$ needed to prepare 100 mL of $0.0400\ M\ K_2Cr_2O_7$. Figure 3.7 shows the steps for doing this dilution.

PRACTICE EXERCISE 17

The concentrated sulfuric acid that can be purchased from chemical supply houses is $18\ M$ H_2SO_4. How could we use this to prepare 250 mL of $1.0\ M\ H_2SO_4$?

SUMMARY

Mole concept The number of atoms in 12 g (exactly) of the carbon-12 isotope is called Avogadro's number, and to three significant figures it equals 6.02×10^{23} atoms. This many formula units of any pure chemical substance constitutes one mole of the substance. Equal numbers of moles contain identical numbers of formula units.

Most elements consist of small numbers of isotopes. An isotope is completely defined when both its atomic number (number of protons) and its mass number (protons plus neutrons) are specified. The small mass of an atom makes the gram unit awkward, so masses of atoms—the atomic masses—are registered in atomic mass units (u). The atomic masses are not whole numbers for two reasons. First the masses of the mass-bearing particles of an atom, its protons and neutrons, even in atomic mass units are not whole numbers. The second reason is the existence of isotopes. Atomic masses must be regarded as the average masses of the atoms of the element's isotopes as the isotopes occur in their natural relative abundances.

Formula masses The sum of the atomic masses of the atoms shown in a formula is the formula mass (formula weight) of the substance. For a substance of exceptionally large formula mass, the unit *dalton* is sometimes added to the number.

Stoichiometry For calculation purposes, the relationship

1 mol of an element = atomic mass of element in grams

connects quantity of an element by particles with quantity by mass. The coefficients in a balanced equation give the proportions of the chemical involved either in formula units or in moles. A quantity of a substance equal to its formula mass taken in grams is one mole of the substance. When working problems involving balanced equations and quantities of substances, solve them at the mole level where the coefficients can be used. Then, as needed, convert moles to grams.

When reactants are mixed in mole ratios that do not conform to the reaction's coefficients, the reactant that can be entirely used up is called the limiting reactant because its mole quantity limits the moles of products.

Solutions A solution has a solvent and one or more solutes, and its concentration is the ratio of quantity of solute to some unit quantity of solvent or of solution.

A solution can be described as dilute or concentrated according to its ratio of solute to solvent being small or large. Whether a solution is unsaturated, saturated, or supersaturated depends on its ability to dissolve any more solute (at the same temperature).

Each substance has a particular solubility in a given solvent at a specified temperature, and this is often expressed as the grams of solute that can be dissolved in 100 g of the solvent. The solubility is the maximum quantity of solute that can dissolve and form a stable solution at the given temperature in the given solvent.

Molar concentration The ratio of the moles of solute per liter (or 1000 mL) of solution is the molar concentration or the molarity of the solution. When we have to prepare one solution by diluting a more concentrated solution, the equation that we use is

$$\text{mL}_{\text{dil soln}} \times M_{\text{dil soln}} = \text{mL}_{\text{concd soln}} \times M_{\text{concd soln}}$$

REVIEW EXERCISES

The answers to Review Exercises whose numbers are in color are found in Appendix C. The answers to the other Review Exercises are found in the Study Guide that accompanies this book. The more challenging questions are marked with asterisks.

The Mole Concept and Avogadro's Number

3.1 What specifically is meant by the term *stoichiometry?*

3.2 What is it about chemical substances and chemical reactions that makes the study of the *numbers* of formula units in a given mass of a substance important?

3.3 When rounded to one significant figure, what is Avogadro's number?

3.4 When applied to a chemical substance, what is another name for Avogadro's number?

3.5 Describe the stoichiometric relationship given by the formula Al_2O_3.

Atomic Masses

3.6 What three particles make up atoms? Which account for an atom's mass?

3.7 If the atomic number of an element is 16, how many protons does one of its atoms contain? What is the name and symbol of the element? How can you tell?

3.8 Elements and isotopes are *substances,* not tiny particles, although they consist of such particles. What is the distinction between the terms *element* and *isotope of an element?*

3.9 What is the name of an isotope consisting of atoms having 3 protons and 4 neutrons? What is its mass number?

3.10 What fact about the isotopes of any given element makes it possible to use identical atomic symbols for them?

3.11 In what sense is an atomic mass of an element an *average* mass?

3.12 What is the (major) reason why an atomic mass for any given element can be used anywhere in the world?

*3.13 Suppose that an element consists of two isotopes in exactly a 1 : 1 atom ratio. The atomic mass of one isotope is 24.0 u and that of the other is 22.0 u.
  (a) Which isotope, if either, has more protons?
  (b) Which isotope, if either, has more neutrons?
  (c) What value for the atomic mass of this element would be recorded in a table of atomic masses?

3.14 The atomic mass unit has approximately what size, 10^{-230} g, 10^{-23} g, 10^{-3} g, or 10^{23} g?

3.15 Avogadro's number is actually known to eight significant figures. Why did chemists pick such an ungainly number to stand for one unit quantity of chemical substance?

3.16 How many atoms are in 4.00 g of helium? (Express the answer in two ways.)

3.17 How many atoms are in 5.89 g of cobalt?

3.18 How many atoms are in 1.00 oz of silver? (1 oz = 28.4 g)

Formula Masses[2]

3.19 What law of chemical combination permits us simply to add atomic masses to calculate formula masses?

3.20 Calculate the formula masses of the following substances.
(a) HCl   (b) KOH   (c) $MgBr_2$
(d) HNO_3 (e) $NaHCO_3$ (f) $Ba(NO_3)_2$
(g) $(NH_4)_2HPO_4$ (h) $Ca(C_2H_3O_2)_2$ (i) $C_6H_{12}O_6$

3.21 Calculate the formula masses of the following compounds.
(a) Na_2CO_3 (b) H_2SO_4 (c) $(NH_4)_3PO_4$
(d) $Mg_3(PO_4)_2$ (e) $Al(C_2H_3O_2)_3$ (f) $Ca(ClO_4)_2$
(g) $(NH_4)_2SO_3$ (h) $K_2Cr_2O_7$ (i) $Fe_4(OH)_2(SO_4)_5$

Moles of Chemical Substsnces

*3.22 The *mole* is the SI base unit for amount of chemical substance. All other SI base units, like the meter and the kilogram mass, have constant values, but the mass of one mole varies from substance to substance. Explain. Is there any feature about a mole that is *constant* from substance to substance?

*3.23 What is the fundamental reason why we have to calculate the amount of *mass* in a mole in connection with running chemical reactions in the laboratory?

3.24 How is the quantity of mass in one mole of some substance calculated?

3.25 Chemical laboratory balances generally read in *grams*, not in *moles*. Explain why.

3.26 After calculating the formula mass of sodium hydroxide, NaOH, what two conversion factors can we prepare for chemical calculations?

3.27 Calculate the number of grams in 0.125 mol of each of the substances in Review Exercise 3.20.

3.28 How many grams are in 0.750 mol of each of the compounds in Review Exercise 3.21?

3.29 Calculate the number of moles in 50.0 g of each of the compounds in Review Exercise 3.20.

3.30 Calculate the number of moles in 1.50 g of each of the compounds in Review Exercise 3.21.

3.31 How many *molecules* of N_2 are there in 1.00 g of N_2, roughly the amount of nitrogen in one liter of air?

3.32 How many *molecules* of water are in one drop, which we can assume has a volume of 0.0625 mL and a density of 1.00 g/mL?

*3.33 At a level of only 0.5 μg of ozone, O_3, in one cubic meter of air, the air is considered dangerous for active children to breathe. How many molecules of ozone are in 0.5 μg?

3.34 A "5-grain" aspirin tablet holds about 180 mg of aspirin ($C_9H_8O_4$). How many moles and how many molecules are in 180 mg of aspirin?

Balanced Equations

3.35 State the information given in the following equation in words.

$$S + O_2 \longrightarrow SO_2 \quad \text{(sulfur dioxide)}$$

3.36 Write in your own words what the following equation says.

$$2NO + O_2 \longrightarrow 2NO_2$$
$$\text{(nitrogen} \qquad\qquad \text{(nitrogen}$$
$$\text{monoxide)} \qquad\qquad \text{dioxide)}$$

3.37 The following is a balanced equation, but what would be a more acceptable way to write it?

$$4H_2SO_4 + 8NaOH \longrightarrow 4Na_2SO_4 + 8H_2O$$

3.38 Balance the following equations.
(a) $N_2 + O_2 \rightarrow NO$
(b) $MgO + HNO_3 \rightarrow Mg(NO_3)_2 + H_2O$
(c) $CaBr_2 + AgNO_3 \rightarrow Ca(NO_3)_2 + AgBr$
(d) $HI + Mg(OH)_2 \rightarrow MgI_2 + H_2O$
(e) $CaCO_3 + HBr \rightarrow CaBr_2 + CO_2 + H_2O$

3.39 Balance each of the following equations.
(a) $P + O_2 \rightarrow P_4O_{10}$
(b) $Fe_3O_4 + H_2 \rightarrow Fe + H_2O$
(c) $Al_2S_3 + H_2SO_4 \rightarrow Al_2(SO_4)_3 + H_2S$
(d) $HNO_3 \rightarrow N_2O_5 + H_2O$
(e) $KHCO_3 + H_2SO_4 \rightarrow K_2SO_4 + CO_2 + H_2O$

Stoichiometry and Its Use with Balanced Equations

*3.40 The natural gas piped to homes for heating and cooking purposes is generally methane, CH_4. When it burns in a plentiful supply of oxygen, the products are CO_2 and H_2O.
(a) Write the balanced equation for this reaction.
(b) What pairs of conversion factors express the mole relationships between the following?
     CH_4 and O_2
     CH_4 and CO_2
     CH_4 and H_2O

3.41 In one brand of stomach antacid the active ingredient is calcium hydroxide, $Ca(OH)_2$. The stomach acid is hydrochloric acid, HCl, which is neutralized (destroyed) by the following reaction:

$$Ca(OH)_2 + 2HCl \longrightarrow CaCl_2 + 2H_2O$$

What conversion factors describe the *mole* relationship between $Ca(OH)_2$ and HCl?

3.42 Aluminum metal is vigorously attacked by sulfuric acid according to the equation:

$$2Al + 3H_2SO_4 \longrightarrow Al_2(SO_4)_3 + 3H_2$$

What conversion factors describe the mole relationships between the following pairs?
(a) Al and H_2SO_4 (b) Al and H_2
(c) H_2SO_4 and $Al_2(SO_4)_3$

---

[2] Remember that the policy is to round values of atomic masses to the first decimal point before starting any calculations (except that we round the atomic mass of H to 1.01).

*3.43 The high explosive TNT (trinitrotoluene) decomposes (breaks down) according to the following equation when it explodes.

$$2C_7H_5N_3O_6 \longrightarrow 3N_2 + 7CO + 5H_2O + 7C$$

(a) How many moles of N_2 are produced for each mole of TNT that decomposes?
(b) How many moles of CO per mole of TNT are produced?
(c) The water forms as high-temperature water vapor. How many moles of water per mole of TNT are produced?
(d) What is the total number of moles of gases of all three kinds, N_2, CO, and H_2O vapor, produced from one mole of TNT? (The carbon forms as finely divided soot.)
(e) TNT is a solid at room temperature, and 1.00 mol of TNT occupies a volume of about 0.14 L. If the gases produced by the explosion occupy a volume of roughly 31 L/mol, what is the ratio of the volume of gases produced to the volume of TNT taken? (This huge expansion, occurring suddenly, is what makes TNT such an effective explosive. Rapidly produced and expanding gases push things out of the way!)

3.44 When glucose ($C_6H_{12}O_6$) is used in the body as a source of energy, the overall reaction is with oxygen, and the products are CO_2 and H_2O.
(a) Write the balanced equation for the reaction of glucose with oxygen.
(b) What conversion factors express the mole relationship between glucose and oxygen?

3.45 Gasohol is a fuel consisting of various hydrocarbons and ethyl alcohol, C_2H_6O. The ethyl alcohol burns in oxygen to give only carbon dioxide and water.
(a) Write the balanced equation for this reaction.
(b) The burning of 5.00 mol of ethyl alcohol uses up how many moles of oxygen?
(c) How many moles of carbon dioxide are produced by the burning of 5.00 mol of ethyl alcohol?

*3.46 The rusting of iron involves the reaction of oxygen with iron. Although the process is complicated, the following equation can be used to represent the overall results.

$$4Fe + 3O_2 \longrightarrow 2Fe_2O_3$$

(a) If 0.556 mol of iron is changed in this way, how many moles of oxygen are consumed?
(b) How many moles of Fe_2O_3 are produced from 0.556 mol of iron?

3.47 Butane, C_4H_{10}, the fuel in lighters, burns according to the following equation.

$$2C_4H_{10} + 13O_2 \longrightarrow 8CO_2 + 10H_2O$$

(a) If 3.00 mol of O_2 are to be consumed by this reaction, how many moles of butane will be used up?
(b) To produce 1.15 mol of CO_2 by this reaction requires how many moles of butane and how many moles of oxygen?

*3.48 Aluminum metal is made industrially by passing a current of electricity through a solution of aluminum oxide, Al_2O_3, in a special solvent. The other product is molecular oxygen.
(a) Complete and balance the following equation for this reaction.

$$Al_2O_3 \xrightarrow{\text{electric current}}$$

(b) How many grams of aluminum can be made from 100 g of aluminum oxide?
(c) How many grams of oxygen are produced from 100 g of aluminum oxide?
(d) What is the total mass of aluminum plus oxygen produced from 100 g of aluminum oxide? Compare the answer with the amount of aluminum oxide used. What law of chemical combination is illustrated by this?

3.49 One chemical reaction that is used industrially to make iron from iron oxide is the reduction of iron(III) oxide by carbon monoxide according to the following equation.

$$Fe_2O_3 + 3CO \longrightarrow 2Fe + 3CO_2$$

(a) How many grams of iron can be made from 750 g of Fe_2O_3?
(b) How many grams of carbon monoxide are needed to reduce 750 g of Fe_2O_3 by this reaction?
(c) How many grams of carbon dioxide are produced by this reaction from 750 g of iron(III) oxide?

*3.50 The chemical reaction that causes silver to tarnish is between silver metal, oxygen in the air, and traces of hydrogen sulfide, also in the air. The black tarnish consists of silver sulfide.

$$4Ag + 2H_2S + O_2 \longrightarrow 2Ag_2S + 2H_2O$$

(a) If 4.68 mg of Ag tarnish by this reaction, how many milligrams of hydrogen sulfide are needed?
(b) How many milligrams of silver sulfide form from 4.68 mg of silver?

*3.51 The deeply purple colored permanganate ion, MnO_4^-, can be made by the oxidation of the Mn(II) ion using sodium bismuthate, $NaBiO_3$, and nitric acid, HNO_3,

$$2Mn(NO_3)_2 + 5NaBiO_3 + 14HNO_3 \longrightarrow$$
$$2NaMnO_4 + 5Bi(NO_3)_3 + 3NaNO_3 + 7H_2O$$

Consider an experiment in which 12.6 g of sodium permanganate, $NaMnO_4$, are to be made by this experiment.
(a) How many grams of manganese(II) nitrate, $Mn(NO_3)_2$, are needed?
(b) How many grams of sodium bismuthate are required?
(c) This preparation will also require how many grams of nitric acid?
(d) How much bismuth(III) nitrate (in grams) will also be produced?

3.52 When a small amount of an acid is accidentally spilled onto a laboratory bench, it should be promptly destroyed (neutralized) before further cleanup is tried. One common way to do this that poses little danger is to sprinkle the acid spill with sodium carbonate until the fizzing caused by escaping carbon dioxide stops. Sulfuric acid, for example, reacts as follows with sodium carbonate.

$$H_2SO_4 + Na_2CO_3 \longrightarrow Na_2SO_4 + CO_2 + H_2O$$

Suppose that a spill of 30.0 g of sulfuric acid occurs. What is the minimum number of grams of sodium carbonate needed to destroy the acid?

Solutions

3.53 A well-stirred mixture of finely divided sand in water is not properly called a *solution.* Explain.

3.54 When water is the dissolving medium for something like sugar, the water itself is designated in what way? How is the sugar designated? What general term can be used to describe any solution for which water is the dissolving medium?

3.55 Sugar (sucrose) is very soluble in water; 100 g of water will dissolve 200 g of sugar.
(a) A solution made up at this concentration would be described as *supersaturated, saturated,* or *unsaturated?*
(b) A solution made up at this concentration would be described as *dilute* or *concentrated?*

3.56 Using the information in a table in this chapter, name a compound (not a gas) that can form the most dilute solution that could still be called saturated.

3.57 What laboratory operation *not* involving the use of any added solute or solvent could be used to convert a saturated solution of sodium hydroxide into a supersaturated solution? Into an unsaturated solution?

3.58 Suppose that for a series of experiments you needed to have on hand a saturated solution of sodium chloride in water. How could such a solution be prepared in the certain knowledge that it is saturated without actually weighing out the solute?

Molar Concentrations

3.59 What is another term for *molar concentration?*

3.60 Distinguish between the terms *molar concentration, molarity, mole,* and *molecule.*

3.61 The units of molar concentration refer to moles per liter of solvent or moles per liter of solution?

*3.62** Calculate the number of moles and the number of grams of solute needed to prepare the given volumes of the following solutions.
(a) 500 mL of 0.125 M NaCl
(b) 250 mL of 0.100 M $C_6H_{12}O_6$
(c) 100 mL of 0.250 M H_2SO_4
(d) 125 mL of 0.500 M Na_2CO_3

*3.63** How many moles and how many grams of solute are needed to prepare the stated volume of each of the following solutions?
(a) 500 mL of 0.200 M $NaC_2H_3O_2$
(b) 250 mL of 0.125 M HNO_3
(c) 100 mL of 0.100 M NaOH
(d) 50.0 mL of 0.250 M $NaHCO_3$

3.64 How many milliliters of 0.150 M HNO_3 contain 0.0100 mol of HNO_3?

3.65 To obtain 0.125 mol of H_2SO_4, how many milliliters of 0.440 M H_2SO_4 would have to be taken?

3.66 An experiment called for 0.100 mol of Na_2CO_3, which was available in a solution with a concentration of 0.250 M. How many milliliters of this solution are required?

3.67 The stock solution of hydrochloric acid is 0.500 M HCl. If 100 mL of this solution is taken, how many moles of HCl are taken?

3.68 A student obtained 50.0 mL of 6.00 M Na_2CO_3 solution. How many moles of Na_2CO_3 are in this quantity?

3.69 How many moles of glucose, $C_6H_{12}O_6$, are in 100 mL of 0.100 M glucose solution?

*3.70** If a stock solution of nitric acid, HNO_3, has a concentration of 1.00 M, how many milliliters of this solution are needed to obtain 5.00 g of HNO_3?

*3.71** The stock supply of sulfuric acid, H_2SO_4, has a concentration of 0.500 M. To obtain 1.00 g of sulfuric acid in the form of this solution, how many milliliters have to be taken?

3.72 To obtain 10.0 g of HCl, how many milliliters of 12.0 M HCl have to be taken?

Stoichiometry of Reactions in Solution

3.73 The label on a reagent bottle reads "0.250 M HCl." What two conversion factors are available from this information? (Base these factors on the milliliter unit for the volume.)

*3.74** Barium sulfate, the ingredient in a "barium cocktail" given to patients about to undergo an X-ray of the intestinal tract, can be made by the following reaction.

$$Ba(NO_3)_2(aq) + Na_2SO_4(aq) \longrightarrow BaSO_4(s) + 2NaNO_3(aq)$$

The desired product, as you can see, is a water-insoluble compound that can be separated from the other substances by filtration (letting the mixture flow through filter paper). To prepare 1.00 g of $BaSO_4$, how many milliliters of 0.100 M $Ba(NO_3)_2$ and how many milliliters of 0.150 M Na_2SO_4 must be mixed together?

*3.75** Gold is attacked by very few chemicals. A mixture of concentrated nitric acid and hydrochloric acid, called *aqua regia* ("royal water"), however, dissolves gold by the following equation.

$$Au(s) + 3HNO_3(aq) + 4HCl(aq) \longrightarrow$$
$$HAuCl_4(aq) + 3NO_2(g) + 3H_2O$$

To dissolve 28.4 g of Au (1.00 oz) by this reaction, what is the minimum number of milliliters of 12.0 M HCl and of 16.0 M HNO_3 needed?

*3.76** Vinegar is a 5.00% solution of acetic acid ($HC_2H_3O_2$) in water, which corresponds to 0.837 M $HC_2H_3O_2$. If mixed with aqueous sodium carbonate (Na_2CO_3), the following reaction occurs.

$$Na_2CO_3(aq) + 2HC_2H_3O_2(aq) \longrightarrow$$
$$2NaC_2H_3O_2(aq) + CO_2(g) + H_2O$$

(a) If the acetic acid in 40.0 mL of vinegar is to react entirely with 0.500 M Na_2CO_3 by this equation, how many milliliters of the sodium carbonate solution are needed?

(b) If solid sodium carbonate is used instead of the solution, how many grams of this solid are needed to react with the acetic acid in a spill of 50.0 mL of 0.837 M acetic acid?

*3.77 The active ingredient in milk of magnesia, an over-the-counter antacid, is finely divided magnesium hydroxide slurried in water. The acid it destroys in the stomach by the following equation is 0.1 M HCl. (A minimum recommended dose of milk of magnesia is 2 tablespoons or 30 mL.)

$$Mg(OH)_2(s) + 2HCl(aq) \longrightarrow MgCl_2(aq) + 2H_2O$$

How many milliliters of 0.100 M HCl can be destroyed by 30.0 mL of milk of magnesia when this medication contains 1.20 g of solid $Mg(OH)_2$ per 15.0 mL of milk of magnesia slurry? (Normally, nearly 2 L of 0.1 M HCl is secreted per day into the stomach.)

*3.78 The concentration of sodium bicarbonate in pancreatic juice, one of the digestive juices, can reach up to 0.120 M $NaHCO_3$. It reacts with the hydrochloric acid delivered in the stomach contents as they move into the upper intestinal tract by the following reaction.

$$NaHCO_3(aq) + HCl(aq) \longrightarrow NaCl(aq) + CO_2(g) + H_2O$$

How many liters of 0.120 M $NaHCO_3$ solution provide enough solute to react with the solute in 1.25 L of 0.100 M HCl?

Preparing Dilute Solutions from Concentrated Solutions

3.79 The commercially available concentrated phosphoric acid is 15.0 M H_3PO_4. How many milliliters of this acid are needed to prepare 500 mL of 1.00 M H_3PO_4? Describe how one would prepare this dilute solution.

3.80 The concentrated aqueous ammonia that is commercially available is 15.0 M NH_3. If you dissolved 50.0 mL of this solution in water and made up the final volume to equal 250 mL, what would be the molarity of the resulting solution?

Additional Exercises

3.81 Which has more mass, 0.50 mol of NH_3 or 2.0 mol of He?

*3.82 Iodized salt contains a trace amount of calcium iodate, $Ca(IO_3)_2$, to help prevent a thyroid condition called *goiter*. How many moles of iodine, I, are in 0.500 mol of $Ca(IO_3)_2$?

*3.83 Baking soda is sodium bicarbonate, $NaHCO_3$. If a sample of baking soda is large enough to contain 2.50 g of C, then how many grams of Na, H, and O are also present?

*3.84 An acid spill involved 25.60 mL of 12.0 M HNO_3, concentrated nitric acid. In an attempt to neutralize the acid, 20.0 g of Na_2CO_3 was stirred into the spill. The equation is

$$2HNO_3 + Na_2CO_3 \longrightarrow 2NaNO_3 + H_2O + CO_2$$

(a) Was enough Na_2CO_3 added to neutralize all of the acid?

(b) What is the formula of the limiting reactant in this situation?

(c) Which reactant was left over? How many moles of this reactant remained?

(d) How many moles and how many grams of CO_2 formed?

ATOMIC THEORY AND THE PERIODIC SYSTEM OF THE ELEMENTS

4

Many things in nature are periodic. The locations of the crests of these ocean waves are periodic with distance, and they arrive at the beach at intervals that are periodic with time. The properties of the elements also vary periodically, but not with distance or time. We'll learn how in this chapter.

4.1
THE PERIODIC LAW AND THE PERIODIC TABLE
4.2
THE NUCLEAR ATOM
4.3
WHERE ATOMS' ELECTRONS ARE — AN OVERVIEW

4.4
ATOMIC ORBITALS
4.5
ELECTRON CONFIGURATIONS

4.1

THE PERIODIC LAW AND THE PERIODIC TABLE

When the elements are arranged in their order of increasing atomic number, several properties recur periodically.

If each of the 109 elements was completely unlike the others, chemistry would be far more complex. Fortunately, the elements can be grouped into a small number of families whose members have much in common. Dimitri Mendeleev, a Russian scientist, took particular notice of this as he wrote a chemistry textbook, published in 1869.

Many Properties of the Elements Vary Periodically with Their Atomic Numbers

Mendeleev observed that the physical and chemical properties of the elements known at the time seem to go through cycles as we move through the elements beginning with those of lowest atomic mass. (Atomic numbers were not known in his day.) The *boiling points* of the elements, for example, roughly illustrate what Mendeleev noted, as seen in Figure 4.1*a*, a plot of boiling points through element 20. The temperatures do not continuously increase but, instead, fluctuate with higher and higher elements. The fluctuations aren't perfect, but there are definite ups and downs. One general improvement came with the discovery of atomic numbers a few decades after Mendeleev's work. The properties of the elements fit into a periodic pattern better if atomic numbers instead of atomic masses are the basis for arranging the elements in order.

The *ionization energies* of the elements display a similar rising and falling in values (see Fig. 4.1*b*). An element's ionization energy is the energy needed to remove an electron from each atom in one mole of its atoms. Notice that helium, neon, and argon are at the bottoms of cycles in the plot of boiling points versus atomic numbers. They are at the tops in the plot of ionization energies. In other words, helium, neon, and argon seem to form a set of elements that have similar properties.

The *combining abilities* of the atoms of one element for atoms of another also go through a cyclical rise, fall, rise, fall pattern. For example, most of the first 20 elements form binary compounds with hydrogen. A **binary compound** is one made of only two elements.

■ A substance's boiling point is the temperature at which it boils.

■ The product of *ionization,* besides the electron, is an electrically charged particle called an *ion.* Thus the ionization of sodium, Na, produces a sodium ion, symbolized by the atomic symbol of sodium followed by a plus charge as a right superscript, Na^+.

■ CH_4 is methane, the natural gas used for heating and cooking. NH_3 is ammonia.

An atom of atomic number:	3	4	5	6	7	8	9	10
can bind these many H atoms:	1	2	3	4	3	2	1	0
The formulas are:	LiH	BeH_2	BH_3	CH_4	NH_3	H_2O	HF	—

This pattern repeats itself as we go to still higher atomic numbers.

An atom of atomic number:	11	12	13	14	15	16	17	18
can bind these many H atoms:	1	2	3	4	3	2	1	0
The formulas are:	NaH	MgH_2	AlH_3	SiH_4	PH_3	H_2S	HCl	—

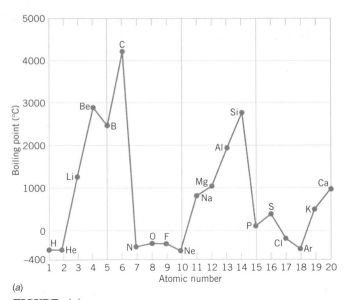

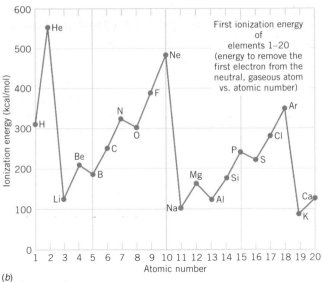

FIGURE 4.1
Two properties of elements 1 through 20 that show periodic fluctuations. *(a)* Boiling points versus atomic numbers. *(b)* Ionization energies versus atomic numbers.

In both series, we see an increase in the number of hydrogens from 1 to 4 and then we see the number fall back again. The number of hydrogens does not keep increasing to ever higher values as we go to elements of higher atomic numbers.

Just as helium (2) neon (10) and argon (18) seem to be similar with respect to boiling points and ionization energies, they also similarly form no compound with hydrogen. Notice that atoms of elements 6 and 14, are alike in their ability to bind four hydrogen atoms, and these two elements similarly share high boiling points (Fig. 4.1*a*). In these examples we thus see that periodically, with increasing atomic number, physical and chemical properties recur — more or less. This is the essence of the **periodic law,** one of the important laws of nature.

■ *Periodic* means the recurrence of something in a regularly repeating way, such as lampposts or sunsets.

> **Periodic Law** The properties of the elements are a periodic function of atomic numbers.

The Periodic Table Organizes the Elements to Show Off Their Periodic Properties

The heart of Mendeleev's discovery was that elements of similar properties line up in vertical columns when horizontal rows, made of their symbols (arranged in order of increasing atomic *mass* — Mendeleev's criterion), were interrupted by breaks at the right places. The result was a table of the elements called the **periodic table.** Its modern form is shown inside the front cover of this book. Today, the table incorporates elements discovered since the time of Mendeleev, and they are arranged in order of their atomic *numbers,* not atomic masses. Each horizontal row in the periodic table is called a **period** and each vertical column is called a **group.**

In constructing his periodic table, Mendeleev had the boldness to leave blanks in the columns whenever this seemed necessary to get elements to line up vertically in families. He even went so far as to declare that these blanks represented elements that had not yet been discovered — and he was right. Such a blank space prompted the discovery of the element germanium (atomic number 32), for example. In order to achieve the best vertical sorting into families, Mendeleev even switched some pairs of elements from their order of increasing

Dimitri Mendeleev (1834–1907)

■ Notice in the periodic table that Co (at. no. 27) has a higher atomic mass than Ni (at. no. 28).

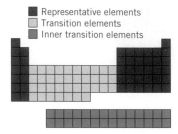

□ Representative elements
□ Transition elements
□ Inner transition elements

■ The IUPAC is an organization made of representatives from the world's several chemical societies, like the American Chemical Society.

■ Hydrogen is a nonmetal, physically and chemically unlike the alkali metals, and it could be left to stand alone in the periodic table. It is put in group IA only because of its electron configuration (as we will soon see).

atomic masses. He listed, for example, tellurium (atomic mass 127.6) *before* iodine (atomic mass 126.9), because iodine seemed to fit far better with fluorine, chlorine, and bromine in group VIIA than with oxygen, sulfur, and selenium in group VIA. When atomic numbers were discovered, it was gratifying to learn that placing tellurium *before* iodine had put these elements in the correct order according to increasing atomic numbers.

The periods in the periodic table are not all the same length; several are broken. This is necessary if the highest priority is to be the chemical similarities of the elements in vertical columns or groups. Thus period 1 is very short, containing only hydrogen and helium, and like the next two periods it is separated into two parts.

The groups have both numbers and letters. Some groups have roman numerals followed by the letter A — IA, IIA, IIIA, and so forth up to VIIA. These groups plus group 0 are called the **representative elements.** The other groups, clustered near the middle of the periodic table, use roman numerals followed by the letter B (except for a cluster in the middle designated as VIII). The B series plus group VIII are called the **transition elements.** There are 10 such elements in each of periods 4, 5, and 6. The two rows of elements placed outside the table are the **inner transition elements.** (The table would not fit well on the page if the inner transition elements were not handled in this way.) Elements 58 through 71 constitute the *lanthanide series,* named after element 57, lanthanum, which just precedes this series. The series of elements 90 to 103 is the *actinide series,* named after actinium, element 89. Each of these two series has 14 elements.

We are using here the column labels still widely employed in the United States. The International Union of Pure and Applied Chemistry or IUPAC has adopted different designations for the vertical columns, but chemists in the United States have resisted the new numbers because they do not correlate electron configurations as easily with the column numbers. In the periodic table, the IUPAC numbers for the groups are given in parentheses beneath the designations we will normally use. Thus group VIIA is the same as the IUPAC group number 17. (The IUPAC group number will sometimes be given in parentheses in this chapter.)

Several of the groups among representative elements also have names. Except for hydrogen, the elements in group IA (IUPAC 1) are called the **alkali metals,** because they all react with water to give an alkaline or caustic (skin-burning) solution.

The elements in group IIA (IUPAC 2) are called the **alkaline earth metals,** because they are commonly found in "earthy" substances. Calcium carbonate or $CaCO_3$, for example, is a compound of calcium, an alkaline earth metal of group IIA, and $CaCO_3$ is the chief substance in limestone.

The elements in group VIIA (IUPAC group 17) are called the **halogens** after a Greek word signifying salt-forming ability. Chlorine of group VIIA, for example, is present in a chemically combined form in table salt, NaCl (sodium chloride).

The elements in group 0 (IUPAC 18), are all gases discovered after Mendeleev's first table. Except for a few compounds that xenon and krypton form with fluorine and oxygen, these elements chemically react with nothing. For this reason they are called the **noble gases** (*noble* signifying, perhaps, aloofness from change).

Other groups of representative elements are named simply after the first member — for example, the **boron family** (group IIIA, IUPAC 3), the **carbon family** (group IVA, IUPAC 14), the **nitrogen family** (group VA, IUPAC 15), and the **oxygen family** (VIA, IUPAC 16).

PRACTICE EXERCISE 1

Referring to the periodic table, pick out the symbols of the elements as specified.

(a) A member of the carbon family: Sr, Sn, Sm, S

(b) A member of the halogen family: C, Ca, Cl, Co

(c) A member of the alkali metals: Rn, Ra, Ru, Rb

(d) A member of the alkaline earth metals: Mg, Mn, Mo, Md

(e) A member of the noble gas family: Ac, Al, Am, Ar

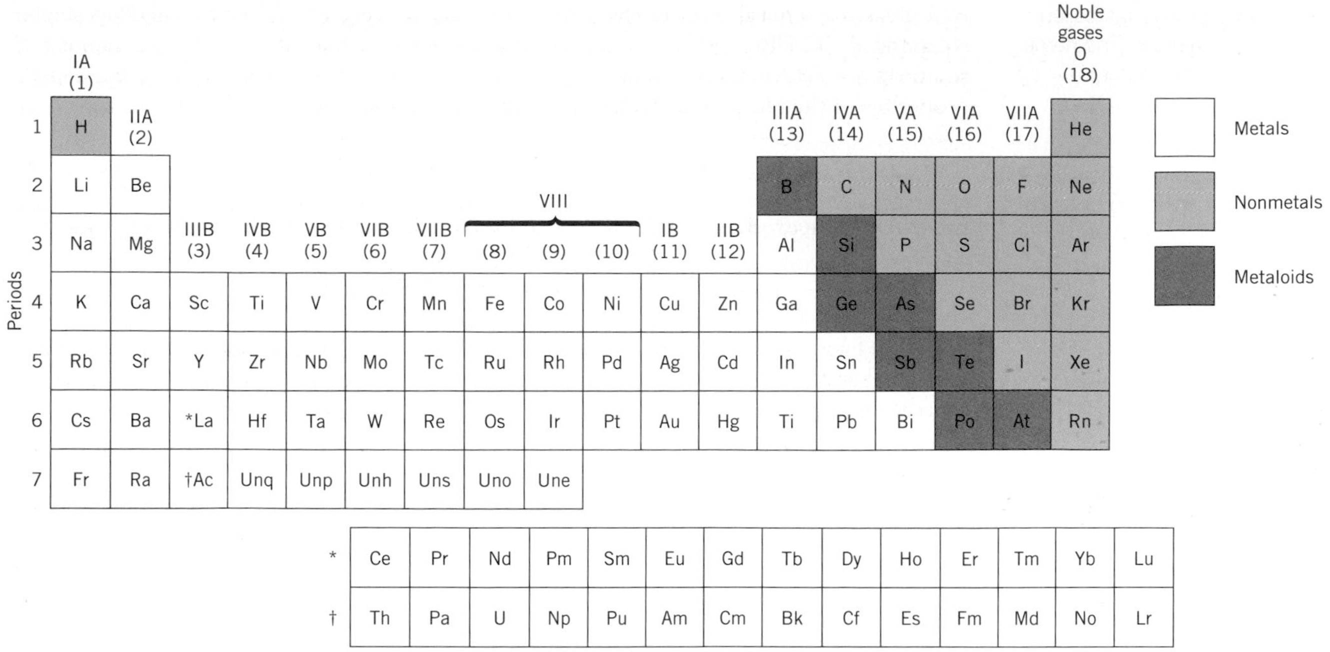

FIGURE 4.2
Locations of metals, nonmetals, and metalloids in the periodic table.

Metals and Nonmetals Are Separated in the Periodic Table

An interesting feature of the periodic table is the location of the metals and nonmetals (see Figure 4.2). As you can see, the great majority of all elements are metals and that, except for hydrogen, the nonmetals are all clustered in the upper right-hand corner. A few elements lie along the borderline between metals and nonmetals. Sometimes called **metalloids,** they have properties that are partly metallic and partly nonmetallic.

The Members of a Group in the Periodic Table Form Compounds with Similar Formulas and Chemical Properties

Perhaps the most noteworthy common property of the elements in the same group, particularly among the representative elements, is that their compounds have like formulas and properties. For example, all of the alkali metals of group IA react with water in the same way. If we let M represent any alkali metal, the general equation for the reaction is

$$2M + 2H_2O \longrightarrow 2MOH + H_2$$

Alkali   Alkali metal
metal   hydroxide

When M is sodium, for example, sodium hydroxide forms.

$$2Na(s) + 2H_2O \longrightarrow 2NaOH(aq) + H_2(g)$$

Sodium   Sodium hydroxide

■ NaOH flakes are an ingredient in one kind of drain cleaner. Use it very carefully, wearing protective gloves, and keep it out of reach of children.

All of the group IA metal hydroxides, MOH, have the common property of being caustic substances capable of causing a chemical burn to the skin. Thus, if we had never handled potassium hydroxide, KOH, we would be very careful with it. It is the hydroxide of an alkali metal, potassium (atomic number 19), and so it must be considered caustic (which it is).

All of the binary compounds between the halogens (group VIIA) and hydrogen have the common formula HX (where X can be F, Cl, Br, or I). If we know that an aqueous solution of

■ Pure HCl is a gas called *hydrogen chloride*. The name of its solution in water is *hydrochloric acid*, HCl(aq).

■ We have much more to learn about acids, so consider the definition here to be a partial definition.

HCl gives a chemical burn to the skin, we would be very careful when handling similar solutions of HF, HBr, and HI. We would expect them to be like HCl. All of the aqueous HX solutions are *acids*. **Acids** are substances capable of destroying or *neutralizing* the caustic properties of the alkali metal hydroxides according to the following general equation:

$$HX(aq) + MOH(aq) \longrightarrow MX(aq) + H_2O$$

For example, hydrochloric acid, HCl(aq) and sodium hydroxide, NaOH(aq), react as follows:

$$HCl(aq) + NaOH(aq) \longrightarrow NaCl(aq) + H_2O$$

Water and sodium chloride (table salt) form in this reaction, and we know that neither can cause a chemical burn to the skin. Because the reaction of an acid with an alkali destroys a characteristic property of any alkali and any acid, it is an important reaction and deserves a special name. We call it **neutralization.** Substances that can neutralize acids are generally called **bases,** so the reaction of hydrochloric acid with sodium hydroxide is an example of an *acid–base neutralization*. Similarly, hydrobromic acid, HBr(aq), reacts with and neutralizes potassium hydroxide, KOH(aq).

$$HBr(aq) + KOH(aq) \longrightarrow KCl(aq) + H_2O$$

These examples show how useful the periodic table is. One overall purpose of our study is to learn the chemical properties of important substances, and the periodic table helps us organize chemical information for easier study. We will often be able, for example, to write *general* equations that apply to several reactions, instead of having to learn each and every reaction, because members of the same family of elements have similar properties.

The periodic law and the periodic table made possible by this law of nature beg one huge question. Why? Why do the elements in the same group have similar properties? Why are the properties periodic? Why are certain periods broken up? Why are there transition elements and why 10 of them per period? Why are there 14 elements in each of the sets of inner transition elements? Answers to these questions lie in atomic structure.

4.2

THE NUCLEAR ATOM

The protons and electrons of an atom occur together in a core or *atomic nucleus*, and the electrons are arranged outside the core.

After Mendeleev, scientists became increasingly interested in what might lie behind the periodic law. The quest led to a deeper knowledge of atomic structure.

■ The proton and the neutron appear to be made of still smaller particles that physicists have named *quarks*.

The Major Subatomic Particles Are Electrons, Protons, and Neutrons Atoms are made of smaller, *subatomic particles,* which we introduced in Chapter 3. Several have been identified, but only three are needed to account for the masses, structures, periodic relationships, and chemical properties of atoms. The three **subatomic particles** are the **electron, proton,** and **neutron.**

As we learned in Chapter 3, only protons and neutrons contribute to an atom's mass to any significant extent. We also learned that the number of protons in the atoms of any given element, the *atomic number* of the element, is identical for all of the element's atoms, including those of its isotopes. The *mass number* of an atom is the sum of its protons and neutrons. An element's *atomic mass* is the average mass of its atoms in atomic mass units taking into account the element's relative isotopic abundances.

■ No known atom has more than 109 electrons, and they contribute only about 0.02% to the mass of such an atom.

Two of the Subatomic Particles, the Proton and the Electron, Carry Electrical Charge You no doubt have experienced an electrical charge. When you receive a shock after walking across a carpet (or touching a bare wire!), the spark causing the shock is a movement of electrical charge.

There are fundamentally two kinds of charge. They give rise to two important phenomena, and you have probably experienced both. Have you ever tried to flick away a small thin piece of plastic, the kind used to wrap albums, only to have it stick stubbornly to your fingers? Most annoying! You shake your hand harder, but it still sticks. This is because you are momentarily carrying one kind of electrical charge, and the plastic has picked up the opposite kind. *Opposite charges attract.* This is the first of two rules of behavior of electrical charges that are among *the most important rules in all of chemistry.* Maybe you have seen your hair stand on end after blow-drying it. The individual hairs act as if they repel each other, because each hair has picked up the *same* kind of charge. *Like charges repel* — this is the second of the two rules about electrical charges.

Attracting and repelling are opposites, so we designate one kind of charge as positive and the other as negative, and give them $+$ and $-$ signs. The proton has one unit of positive charge, $1+$. The electron has one unit of negative charge, $1-$. Electrons repel each other, because they are like-charged; protons also repel each other for the same reason. Electrons and protons attract each other, however, because they are oppositely charged. The neutron has no charge — hence its name. When we need symbols, we'll use p^+ for the proton and e^- for the electron.

All Atoms Have Nuclei That Hold Their Protons and Neutrons In 1911, British scientists working under Ernest Rutherford found that when streams of certain subatomic particles from a radioactive element were allowed to strike a very thin metal foil, most of the particles sailed right through with no change in course. Only a few bounced back, and many went through with various angles of deflection. It was as if the metal foil were mostly empty space, like chainlink fencing, but that at some places there were particles massive enough to bounce the subatomic "bullets" back again (see Fig. 4.3). By studying the angles at which many of the particles careened through the foil, Rutherford deduced that the massive particles in the foil were positively charged and contained virtually all the mass of an atom. Thus, he concluded, an atom must be mostly empty space around one dense, massive inner core, to which he gave the name **nucleus.**

We now know that any given atom has only one nucleus and that all of the atom's protons and neutrons are located in it. The protons cause the positive charge on the nucleus, and the *number* of protons equals the size of this charge.

Atomic number = + charge on nucleus = number of protons

Now we must learn where the electrons are in atoms. This knowledge will help us understand many chemical properties of substances at the molecular level of life.

Like charges repel.

■ The numerical values of charge, 1+ and 1−, are *relative* values, not the values in SI units. The charges are *equal* but *opposite.*

Ernest Rutherford (1871 – 1937)

■ Rutherford's discovery of the nucleus earned him the 1908 Nobel prize in chemistry and, in 1930, the title Baron Rutherford of Nelson.

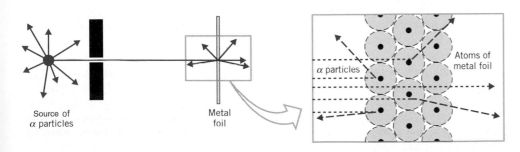
Source of α particles

Metal foil

α particles

Atoms of metal foil

FIGURE 4.3
The discovery of the atomic nucleus. The α (alpha) particles consist of two protons and two neutrons, a "package" that some radioactive elements emit in high-energy streams called alpha rays. Most of the α particles passed through the foil, although some were deflected. A few particles bounded right back as though they had hit something far more massive than an electron.

4.3

WHERE ATOMS' ELECTRONS ARE—AN OVERVIEW

The electrons in an atom are confined to particular energy shells outside the nucleus.

When elements combine to form compounds, some electrons of the elements' atoms become relocated with respect to atomic nuclei. To understand this, we have to learn where an atom's electrons are *initially*. Only then can we think about how they might relocate to give the more stable arrangements of electrons and nuclei in compounds.

The *number* of electrons to arrange must equal the number of protons in an atom, because atoms are electrically neutral.

> Atomic number = number of protons = number of electrons

Thus, the nucleus of an atom with six protons has a total positive charge of $6+$. To balance this charge so that the atom is electrically neutral, there must be a total of $6-$ of negative charge. This requires six electrons, because each electron has a charge of $1-$.

An atom's electrons do not occur simply at random in the space near its nucleus. They are constrained to particular patterns, and the specific arrangement of electrons about a nucleus is called the atom's **electron configuration.** When we know it, we can understand a great deal about the chemical properties of an element.

Niels Bohr (1885–1962)

■ Fireflies use *chemical* energy, not heat, to promote electrons to excited states in molecules.

Light quantum emitted

The Bohr Atom Was an Early Atomic Model In 1913, only two years after Rutherford's discovery of the nucleus, Niels Bohr (1885–1962), a Danish physicist, offered two postulates about electron configurations. Bohr's first postulate was that electrons are confined to what came to be called *allowed energy states.* This says that electrons cannot be just anywhere, like buzzing mosquitoes. They can only be in particular places much as tennis balls on a stairway can only be on the steps, not suspended in air between them. Bohr's allowed energy states, in fact, are commonly called *energy shells* or *energy levels.* And like a stairway, the lower energy states are more stable places to be than are the higher states. In other words, each of the allowed energy states or energy levels of an atom corresponds to some different value of energy.

Bohr's second postulate was that as long as the atom's electrons remain in allowed energy states, the atom neither radiates nor absorbs any energy associated with the electrons' movements. Not that it isn't possible to make an electron move from one allowed state to another. This is what happens, for example, when an iron fireplace poker is heated until it glows in the dark. Before being heated, the iron atoms are nearly all in their *ground state* arrangement of electrons; all their electrons are in the lowest energy states available. The heat causes electrons in iron atoms to shift to higher energy states called *excited states.* This is partly how the iron actually soaks up heat energy. As soon as atoms have become excited in this way, the electrons begin to shift back to lower states, *and the difference in energy between the excited state and the lower state is emitted as light.*

Scientists had known long before Bohr that the emitted light did not possess every conceivable value of light energy, but had only certain values. This is what led Bohr to believe that atoms had only certain allowed energy states.

The precise quantity of energy that is emitted when one electron changes from a higher to a lower energy state is called a **quantum** of energy, and sometimes a **photon** of energy. Sometimes electricity rather than heat is used to generate excited atoms that then emit light. Sodium vapor lamps or mercury vapor lamps along highways work this way. And you know that the sodium vapor lamps have a characteristic color. Excited sodium atoms don't emit all colors of light, just yellow light, and the yellow light from sodium lamps is characterized by photons of just two, very nearly equal values of energy.

To help people understand his postulates, Bohr suggested an analogy. He pictured an atom's electrons as being in very rapid motion around the nucleus, and that they follow paths, called *orbits,* much as planets move in orbits around the sun. This picture of an atom is an example of a scientific **model,** a mental construction—often involving a picture or drawing—used to explain a number of facts. The **Bohr model** of the atom was quickly dubbed the "solar system" model, and it is still commonly used in the communications media to accompany almost any discussion of things atomic.

The solar system picture of the atom is now obsolete; it could not explain a number of observations about higher elements. Yet Bohr's two fundamental postulates behind the model are still firmly fixed in atomic theory.

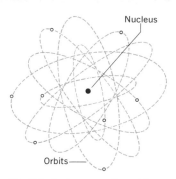

The Bohr "solar system" model

Heisenberg's Uncertainty Principle Gives Us Nature's Limit on What We Can Know about an Atom's Electrons

The Bohr model worked well only for one element, hydrogen. The problem, as the German physicist Werner Heisenberg (1901–1976) soon realized, is that calculations based on the model assume that the location of an electron and its energy can *both* be precisely known at the *same* instant. This is untrue.

The electron is small enough that any act of measuring its location gives it a nudge and changes its energy, and any attempt to find its energy changes its location. This is why the location and energy of an electron cannot *both* be precisely known at the same time. There is always some uncertainty about either the location of an electron or its energy, and this is one way of stating what came to be called the **Heisenberg uncertainty principle.**

Because of Heisenberg's insight, scientists gave up the idea that an electron moves in a fixed orbit. Instead, the locations of electrons are described in terms of *probabilities* of their being at certain places. The question became, In what particular parts of the space surrounding a nucleus is it *likely* that an electron will be? It's like asking: If I go out a certain distance from the nucleus in one particular direction, what are the chances or the probability of an electron being there? The answer depends on what energy state the electron is in.

■ Heisenberg won the 1932 Nobel prize in physics.

■ The idea that the very act of measuring something actually alters what is being measured has profound implications for the fields of psychology, sociology, poll taking, and even television news.

Electrons in Atoms Are Confined to Principal Energy Levels

The specific energy states in which electrons can be are organized in the atom's **principal energy levels.** They roughly correspond to certain successive distances from the atom's center. These levels are given numbers, n, called the *principal quantum number.* The value of n is limited to whole numbers beginning with $n = 1$ for the lowest energy level. This level happens to be the one nearest the nucleus, but it will serve our needs better if you learn to associate a lower value of n with a lower value of energy rather than a location. An electron in level 1 has the least quantity of energy it can have in the atom. Just as nature has a powerful tendency to take up positions having the lowest possible energy, so an atom's electrons nearly always are in the lowest allowed levels—in the ground state.

There Are Limits on the Numbers of Electrons in the Principal Energy Levels

Not all electrons can crowd into level 1. Electrons, remember, are like-charged and so they repel each other. Each principal energy level has a limit to its number of electrons, and the limit for level 1 is only two. At $n = 2$, the limit is eight electrons. Table 4.1 summarizes the maximum number of electrons that can be in the various principal energy levels.

TABLE 4.1
The Principal Energy Levels and the Number of Subshells

Principal Level Number[a]	1	2	3	4	5	6	7
Maximum number of electrons actually observed in nature	2	8	18	32	32	18	8
Number of subshells in theory	1	2	3	4	5	6	7

[a] Principal quantum number, n.

The Principal Energy Levels Are *Electron Shells*   If you've ever studied a whole onion, you know that it exists in sections, each one a hollow sphere with thick walls, each successively larger than the one just inside it. The sections all have a common center, so they are called *concentric* spheres. We can roughly think of an atom's principal energy levels as concentric spheres, too, each with a definite thickness. This is why the principal energy levels are often called **electron shells.** The first principal energy level is thus the first shell.

To summarize, the electrons of an atom reside in principal energy levels or shells that occur concentrically around the atom's nucleus. It's a rather simple picture, and it lacks some details required to correlate an atom's structure with its chemical properties. These details are the concern of the next section.

4.4
ATOMIC ORBITALS

The principal energy levels have *subshells* made up of regions called *atomic orbitals*.

Electron Shells Have *Subshells*   The thickness of an electron shell allows for some fine structure. All electron shells except the first have a small number of **subshells,** the allowed number equaling the value of n for the main level. When $n = 1$, there is actually no *sub*shell. (We could call the entire shell a whole subshell). When $n = 2$, there are two subshells; at $n = 3$, there are three subshells, and so forth. It's as though an atom is an apartment house for electrons (with the nucleus in the basement) and each floor is a principal energy level or shell. Each floor can have one, two, three, four, or five apartments called subshells.

Subshells Have Regions Called *Orbitals*   An atom's electrons are in subshells but these, like apartments, have further structure. Each subshell has a particular number of spaces, called *atomic orbitals,* where individual electrons can reside. **Atomic orbitals** are particularly shaped spaces than can hold up to two electrons apiece, no more. It's as though each apartment (subshell) on a given floor (main shell) has a certain number of rooms (orbitals) for electrons, but no room can hold more than two electrons. To specify the location of an electron, we must name its main energy shell, its subshell, and its orbital. When we do this, we specify all that we can know about an electron's location relative to the atomic nucleus.

Heisenberg said that if we give up wanting to know precisely *where* an electron is within an orbital, we can know what is more important, the *energy* of the electron. Energy is important because of a major fact about our world and the way it works: *nature, given the opportunity, tends to change in whichever direction results in a more stable, lower energy arrangement of things.* When we know that one arrangement of electrons and nuclei is more stable (has lower energy) than another, we then know which arrangement nature prefers. When something gets into the less stable arrangement, it sooner or later will change (perhaps undergo a chemical reaction) to what is more stable. These basic principles of the way of nature are at the heart of understanding chemical reactions, because they exemplify nature's preference for the more stable arrangement of its parts.

An Orbital Can Hold Two Electrons Only if They Are Spinning in Opposite Directions   The final complexity in electron configurations is that electrons are *spinning* particles. Like the earth, electrons spin about an axis. Unlike the earth, electrons have the options of spinning in either of two opposite directions. When an atomic orbital holds two electrons, they can be present only if one electron spins in a direction opposite that of the other electron. Part of the reason is that the actual spinning makes an electron behave as a tiny magnet. By spinning *oppositely,* the two electrons have a magnetic *attraction* for each other. This helps to overcome the electrical repulsion between two like-charged electrons when they are in the same space.

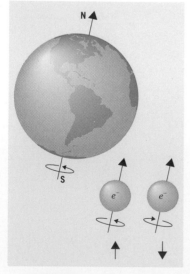
Electron spin

Wolfgang Pauli (1900–1958), an Austrian-born physicist, was the first to realize the limitations on the number and spins of electrons in the same orbital, and we call the rule the **Pauli exclusion principle.**

> **Pauli Exclusion Principle** An orbital can hold as many as two electrons, but only if they have opposite spins.

■ Wolfgang Pauli won the 1945 Nobel prize in physics.

Orbitals Have Unique Shapes Now let's look more closely at the kinds of "rooms," the atomic orbitals, available in the atom's electron apartments (subshells). As we said, when $n = 1$, we're at the first main level where there is only one subshell. The subshell at $n = 1$ has only one orbital. It is named the $1s$ orbital, where "1" is the value of n, and "s" comes from a German word of no interest here. Think of "s" as meaning "spherical."

■ The orbital, subshell, and level are identical when $n = 1$.

Each atomic orbital has a particular shape. The shapes have been deduced by theoreticians who have calculated the probabilities of finding electrons at definite points relative to the atom's nucleus. The shape of an orbital is simply what you get when you wrap an imaginary envelope around enough of a particular space to enclose a region of high probability — say, 90% — of having a particular electron somewhere in it.

Figure 4.4 shows the shape of the $1s$ orbital. It looks like a sphere when viewed from the outside. The nucleus is in its center. The surface of the sphere encloses a space within which the probability of finding an electron belonging to level 1 is greater than 90%. An electron in a $1s$ orbital moves about — very rapidly, in fact. But we can't know exactly how or where because we would rather know about the $1s$ electron's energy (at least its energy relative to other locations). It is far more important to know if an electron is in a $1s$ state than to known where within the state it is.

■ The atomic orbitals that we are describing are based on calculations done on the hydrogen atom. Ample evidence exists that the orbitals of other atoms are like those of hydrogen.

The rapid movement of an electron within an orbital gives us another useful image, that of an **electron cloud.** An electron moves so rapidly that the influence of its negative charge is distributed throughout its orbital much as water molecules are distributed throughout a cloud.

Principal level 2 has two subshells. One consists of only one orbital named the $2s$ orbital. It corresponds to slightly less energy than that of the other subshell at $n = 2$. The shape of the $2s$ orbital looks like that of the $1s$ orbital when "viewed" from the outside, which explains why "s" is used in the name $2s$. The second subshell at $n = 2$ holds three orbitals called the $2p$ orbitals. They have identical energies, and each has two lobes. The long axes of these orbitals,

■   —— —— ——

    $2p_x$ $2p_y$ $2p_z$

  ——

  $2s$

Subshells at main level 2

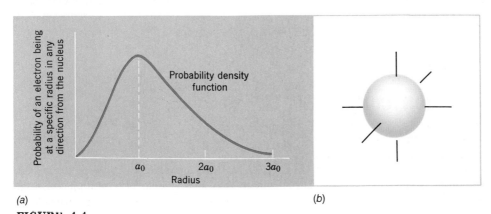

(a)   *(b)*

FIGURE 4.4

The $1s$ orbital. *(a)* Imagine that the space around a nucleus is made up of layer upon layer of extremely thin, concentric shells. At each distance away from the nucleus there is a point on the curve that indicates the probability of finding a $1s$ electron at this distance. Notice that the probability is zero for a zero distance — the electron is not on the nucleus. The probability reaches a maximum at the radius marked a_0, which is 52.9 pm (1 pm = 10^{-12} m). *(b)* One of the thin spheres described in part *(a)* encloses a space within which the total probability of finding an electron is large, say, 90%. This sphere is the "envelope" discussed in the text, and its shape is the shape of the $1s$ orbital.

FIGURE 4.5
The orbitals at principal energy level 2. From the outside, the 2s orbital looks like a 1s orbital, seen in Figure 4.4. However, its radius is larger. Each of the 2p orbitals has the same shape and energy as the others, but they are oriented along the three different axes.

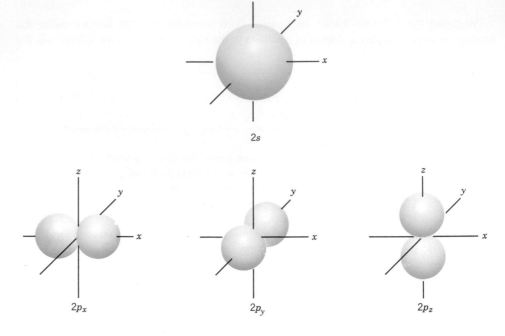

$2s$

$2p_x$ $2p_y$ $2p_z$

each axis symmetrically piercing both of the orbital's two lobes, project at right angles along the x, y, and z coordinate axes (see Fig. 4.5). Each 2p orbital is characterized by a subscript; we write $2p_x$, $2p_y$, and $2p_z$ to indicate the direction of projection. In atoms with two or more electrons, a p orbital electron has more energy than an s orbital electron in the same main level. Thus in atoms that normally have an electron at level two, the electron is in a slightly more stable location in the 2s orbital than in any of the 2p orbitals.

When $n = 3$, we're at main level 3, so it has three subshells. One is an orbital all by itself, the 3s orbital. From the outside, it looks like the other s orbitals, only it has a larger radius. Another subshell at $n = 3$ has three orbitals of identical energy but with right angle orientations, the $3p_x$, $3p_y$, and $3p_z$. They resemble in shape the two-lobed p orbitals at main level 2. An electron in a 3p orbital has more energy than one in the 3s state. The third subshell at $n = 3$ consists of five orbitals, called the 3d orbitals, which correspond to more energy than the 3p orbitals. At $n = 4$, the fourth main level, there are 4 subshells, the 4s (1 orbital), 4p (3 orbitals), 4d (5 orbitals), and 4f (7 orbitals). We will not study the shapes and names of d or f orbitals. Now that we know the main accommodations available to electrons, we are ready to describe the *electron configurations* of the elements.

■ 3d _ _ _ _ _
  3p _ _ _
  3s _
Subshells at main level 3

■ 4f _ _ _ _ _ _ _
  4d _ _ _ _ _
  4p _ _ _
  4s _
Subshells at main level 4

4.5

ELECTRON CONFIGURATIONS

The electrons of an atom fill into the lowest energy atomic orbitals available, spreading out among orbitals of the same energy.

The electron configurations of atoms follow a relatively simple pattern based on the **aufbau principle:** starting with hydrogen (atomic number 1), each additional proton placed in the nucleus is accompanied by placing an electron in whichever of the *available* orbitals corresponds to the lowest energy. What constitutes an "available orbital" is determined by only a few rules. The Pauli exclusion principle is one of them.

■ Aufbau means "building up" in German.

Electrons Spread Out among Orbitals of the Same Subshell Another rule concerning "available orbitals" governs where electrons go when orbitals *of the same energy* are available at some value of n, like the three p orbitals at one of the main levels. **Hund's rule** handles this question.

> **Hund's Rule** Electrons *at the same subshell* spread out among the subshell's orbitals as much as possible.

This rule makes sense because electrons are like-charged and they tend to be as far from each other as possible if it makes no difference in terms of the orbital energies. It is also true that when electrons do spread out like this, they have the same spins. Thus, the distribution on the left is more stable than the one on the right.

$$\underset{2p_x}{\uparrow} \quad \underset{2p_y}{\uparrow} \quad \underset{2p_z}{\_} \qquad \underset{2p_x}{\uparrow\downarrow} \quad \underset{2p_y}{\_} \quad \underset{2p_z}{\_}$$

This configuration with unpaired spins is more stable.

This configuration with paired spins is less stable.

As a symbol for the direction of spin, we use an arrow, which can point either down ($\downarrow$) or up ($\uparrow$). However, because two electrons in the same orbital *must* have opposite spins, we seldom need such arrows. Usually, the symbol used for a pair of electrons in the same orbital is a right superscript as in $1s^2$

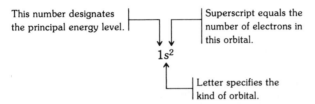

This number designates the principal energy level.

Superscript equals the number of electrons in this orbital.

$$1s^2$$

Letter specifies the kind of orbital.

Beginning with the filling of orbitals at the $3d$ subshell and higher, what constitutes a lower energy subshell does not correlate well with the value of n. For example, the $4s$ subshell is used *before* any electrons are put into any orbitals of the $3d$ subshell, because the energy of the $4s$ subshell is less than that of the $3d$ despite its having a larger value of n. Figure 4.6 displays the

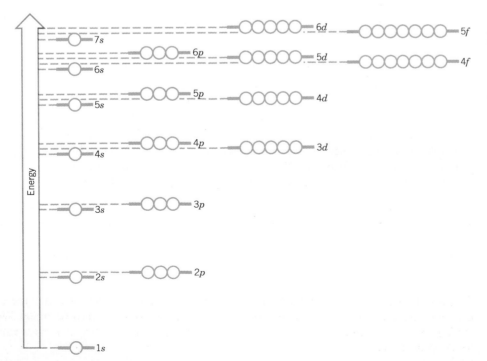

FIGURE 4.6

Approximate relative energy levels for atoms with two or more electrons. (Adapted from J. E. Brady and J. R. Holum, *Chemistry. The Study of Matter and Its Changes,* 1993. John Wiley & Sons, Inc., New York. Used by permission.)

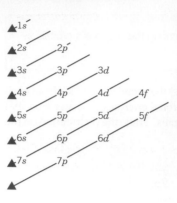

FIGURE 4.7
The order in which sublevels are filled. Each horizontal row names the subshells at a given main energy level. In a many-electron atom, the filling of sublevels begins with the 1s subshell (first slanting arrow at the top); then the 2s subshell receives electrons (second slanting arrow from the top). Hereafter, to find the further order of subshell filling, follow the successive slanting arrows in the directions they point. The order is thus $1s \rightarrow 2s \rightarrow 2p \rightarrow 3s \rightarrow 3p \rightarrow 4s \rightarrow 3d \rightarrow 4p \rightarrow 5s \rightarrow 4d \rightarrow$ etc.

relative energies of all of the subshells, and you can see several overlaps of subshells among different main levels. One easily reconstructed device for knowing the order in which subshells are filled is given in Figure 4.7. Just follow the arrows, starting at the top. Thus, Figure 4.7 tells us that the order of filling orbitals is first 1s, then 2s, 2p, 3s, 3p, 4s, 3d, 4p, 5s, 4d, etc.

The Aufbau Principle Lets Us Construct the Electron Configurations of Elements   A hydrogen atom has one electron. We have learned that the 1s orbital has the lowest associated energy, so this is where the electron resides.

■ This is the electron configuration of all three isotopes of hydrogen.

$$\text{H} \quad 1s^1$$

meaning $1s$ ↑

Helium has atomic number 2 and therefore its atoms have two electrons. Both can (and must) go into the 1s orbital.

■ Helium is a gas used to fill dirigibles such as the Goodyear blimp. It is much less dense than air, and it won't burn.

$$\text{He} \quad 1s^2$$

meaning $1s$ ↑↓

Lithium, atomic number 3, has three electrons in each atom. The first two fill the 1s orbital. According to Figure 4.6, the third electron must go into the next lowest orbital, the 2s.

■ Think of the three lines at the 2p subshell as representing, in order, the $2p_x$, $2p_y$, and $2p_z$ orbitals.

$$\text{Li} \quad 1s^2 2s^1$$

meaning $2p$ __ __ __
$2s$ ↑
$1s$ ↑↓

Beryllium, atomic number 4, has four electrons per atom. The first two fill the 1s orbital, and the last two fill the 2s orbital. None enters a 2p orbital, because these orbitals are at a higher energy level (Fig. 4.6), and we must fill the lower energy orbitals first.

$$\text{Be} \quad 1s^2 2s^2$$

meaning $2p$ __ __ __
$2s$ ↑↓
$1s$ ↑↓

■ Borax, a cleansing agent, contains boron.

Boron, atomic number 5, has five electrons per atom. The first four fill the 1s and the 2s orbitals, exactly as in beryllium (number 4), and boron's fifth electron enters a 2p orbital. We don't know (or need to know) which of the three takes the electron, so we just arbitrarily assign it to the $2p_x$ orbital.

$$\textbf{B} \quad 1s^2 2s^2 2p_x{}^1$$

meaning $2p$ ↑ __ __

               $2s$ ⇅

               $1s$ ⇅

The next element, carbon (atomic number 6), is of central importance at the molecular level of life, because its atoms make up most if not quite all of the "backbones" of molecules, other than water, that are in living cells.

■ All but a handful of the several million compounds of carbon are classified as organic compounds.

$$\textbf{C} \quad 1s^2 2s^2 2p_x{}^1 2p_y{}^1$$

meaning $2p$ ↑ ↑ __

               $2s$ ⇅

               $1s$ ⇅

Notice that carbon illustrates the application of Hund's rule. The last two electrons go into *separate* orbitals at the $2p$ subshell.

Nitrogen, number 7, is another very important element among biological chemicals. It also illustrates Hund's rule.

■ Air is about 79% nitrogen in the form of N_2.

$$\textbf{N} \quad 1s^2 2s^2 2p_x{}^1 2p_y{}^1 2p_z{}^1$$

meaning $2p$ ↑ ↑ ↑

               $2s$ ⇅

               $1s$ ⇅

Oxygen (atomic number 8) illustrates that we don't start with level 3 until level 2 is filled. Oxygen's eighth electron goes into a $2p$ orbital, not the $3s$.

■ Air is about 21% oxygen in the form of O_2.

$$\textbf{O} \quad 1s^2 2s^2 2p_x{}^2 2p_y{}^1 2p_z{}^1$$

meaning $2p$ ⇅ ↑ ↑

               $2s$ ⇅

               $1s$ ⇅

Fluorine (atomic number 9) has nine electrons, and we continue to fill the $2p$ orbitals.

■ Fluorine is so reactive that it burns with water.

$$\textbf{F} \quad 1s^2 2s^2 2p_x{}^2 2p_y{}^2 2p_z{}^1$$

meaning $2p$ ⇅ ⇅ ↑

               $2s$ ⇅

               $1s$ ⇅

With the next element, neon (atomic number 10), we complete the filling of all the atomic orbitals at level 2.

■ Neon is the gas in "neon" lights.

$$\textbf{Ne} \quad 1s^2 2s^2 2p_x{}^2 2p_y{}^2 2p_z{}^2$$

meaning $2p$ ⇅ ⇅ ⇅

               $2s$ ⇅

               $1s$ ⇅

Level 2 now has its maximum of eight electrons, two at the $2s$ subshell and six at the $2p$ subshell.

With element 11, sodium, we start to fill the third main level.

$$\textbf{Na} \quad 1s^2 2s^2 2p_x^2 2p_y^2 2p_z^2 3s^1$$

meaning $3s$ ↑

$2p$ ↑↓ ↑↓ ↑↓

$2s$ ↑↓

$1s$ ↑↓

We could also write the configuration of the Na atom as $1s^2 2s^2 2p^6 3s^1$. When *all* orbitals at a p subshell are filled, we can save space by writing p^6.

The process described for elements $1-11$ continues through the remainder of period 3, just as we have shown. Period 3 ends with argon (atomic number 18), with the configuration

$$\text{Argon} \quad 1s^2 2s^2 2p^6 3s^2 3p^6$$

As we move into period 4, remember (from Figs. 4.6 and 4.7) that the $4s$ orbital accepts electrons *before* the $3d$. Therefore the electron configurations of potassium and calcium, the first two elements in period 4, are

$$\text{Potassium} \quad 1s^2 2s^2 2p^6 3s^2 3p^6 4s^1$$
$$\text{Calcium} \quad 1s^2 2s^2 2p^6 3s^2 3p^6 4s^2$$

EXAMPLE 4.1
Writing an Electron Configuration

Phosphorus, atomic number 15, is another element important at the molecular level of life. Write its electron configuration.

ANALYSIS With 15 electrons, we know that both levels 1 and 2 are filled; they take $2 + 8 = 10$ electrons. The remaining five must be in level 3.

SOLUTION $1s^2 2s^2 2p^6 3s^2 3p_x^1 3p_y^1 3p_z^1$

meaning $3p$ ↑ ↑ ↑

$3s$ ↑↓

$2p$ ↑↓ ↑↓ ↑↓

$2s$ ↑↓

$1s$ ↑↓

PRACTICE EXERCISE 2

Using the one-line representation rather than the system with arrows, write the electron configuration of each of the following elements. Use the table inside the front cover to find out their atomic numbers.

(a) aluminum **(b)** chlorine **(c)** silicon **(d)** calcium

Abbreviated Electron Configurations Focus Attention on the Electrons of the Highest Occupied Level   Still another way to represent the electron configurations of elements like sodium and the others in the third period of the periodic table is in the abbreviated form illustrated for Na and the next element, magnesium, as follows.

Sodium   [Ne]$3s^1$

Magnesium   [Ne]$3s^2$

"[Ne]" stands for $1s^2 2s^2 2p^6$, the electron configuration of neon. In other words, we let "[Ne]" represent all of the inner electrons, the **core electrons,** those in main shells with n less than that of the highest *occupied* shell. Following the symbol for the core electrons, we finish the configuration by specifying in more detail the **outside shell electrons,** those in orbitals associated with the highest value of n at which electrons occur in the atom. The outside shell electrons are generally the electrons involved in chemical reactions; the core electrons often are undisturbed.

Elements in period 4, elements 19 through 35 following argon, have the argon configuration for the inner core electrons, $1s^2 2s^2 2p^6 3s^2 3p^6$. This allows us to write the following abbreviated configurations for potassium and calcium.

Potassium   [Ar]$4s^1$

Calcium   [Ar]$4s^2$

The $3d$ orbitals of potassium and calcium are empty like those of argon.

After calcium, we enter the first series of ten transition elements; we begin to place electrons in the $3d$ orbitals, which correspond to lower energy than the $4p$ state (Fig. 4.6). From element 21 (scandium) through element 30 (zinc), the $3d$ orbitals fill until at zinc the $3d$ orbitals hold 10 electrons. Only after the $3d$ orbitals have filled do electrons go into the $4p$ orbitals, beginning with element 31 (gallium). The filling of the $3d$ orbitals is a general pattern for transition elements in any period. As we move through any series of transition elements from left to right in the periodic table, d orbitals are filling to their maximum of 10 electrons. Because it takes a series of *ten* elements to fill the five d orbitals, each set of transition elements numbers 10.

■ The d orbitals come in sets of five, each capable of holding 2 electrons, so 5 × 2 or 10 is the maximum number of electrons that can be held by a d orbital subshell.

The period 5 elements have a krypton core; krypton is the noble gas element immediately *preceding* period 5. Therefore, the first element in period 5, rubidium (an alkali metal) has the electron configuration [Kr]$1s^1$. Period 6 elements all have xenon cores, and the elements in period 7 have the radon core. We will not carry this study to greater detail; Appendix B gives the electron configurations of all of the elements. You may wish to look at them to note that the atoms of the inner transition elements correspond to the filling of the f orbitals. Because there are seven f orbitals at any value of n, it takes 14 electrons to fill them, so there are 14 elements in each series of inner transition elements. Notice also that whether the d or the f orbitals are filling, the number of electrons in the atom's outside shell is 1, 2, or 3. The elements of most common significance at the molecular level of life are found among the first 20, although a number of transition elements are vitally important as the "trace elements" of nutrition.

When We Write the Electron Configuration of an Element, We Do So for Any of Its Isotopes   Atoms of the different isotopes of the same element differ only in their numbers of neutrons, but these are in the nucleus and have nothing to do with electron configurations. *It is because all isotopes of any given element have the same electron configurations, that they all have the same chemical properties.*

Elements in the Same Family Have the Same Outside Shell Configurations
Among the representative elements of groups IA – VIIA, the roman numeral of the group equals the number of outside shell electrons in the atoms. Thus, all atoms of group IA have one

■

Group Number and Name	Total Outside Shell Electrons
IA Alkali metals	1
IIA Alkaline earth metals	2
IIIA Boron family	3
IVA Carbon family	4
VA Nitrogen family	5
VIA Oxygen family	6
VIIA Halogen family	7
0 Noble gases	8

■ We note in passing that somehow an outside level of 8 electrons confers unusual chemical stability (to be continued in Chapter 5).

electron in the outside shell. The atoms of the group IIA elements have two outside shell electrons, and so on.

Group IA		Group IIA	
Lithium	$[He]1s^1$	Beryllium	$[He]1s^2$
Sodium	$[Ne]2s^1$	Magnesium	$[Ne]2s^2$
Potassium	$[Ar]3s^1$	Calcium	$[Ar]3s^2$
Rubidium	$[Kr]4s^1$	Strontium	$[Kr]4s^2$
Cesium	$[Xe]5s^1$	Barium	$[Xe]5s^2$
Francium	$[Rn]6s^1$	Radium	$[Rn]6s^2$

Similarly, the group IIIA atoms all have 3 outside shell electrons. The atoms in group IVA have 4. Helium, the first element in group 0, has just two electrons in its outside shell, but this is shell 1 and so cannot hold more than two anyway. Otherwise, after helium, all noble gas elements have 8 outside shell electrons. For understanding the chemical properties that we will be studying, the *total* number of electrons in the outside shell is the important number obtained from electron configurations. This number can be told at a glance at the periodic table when the element is a representative element, because it's the same as the roman numeral of the group.

EXAMPLE 4.2
Finding Information in the Periodic Table

How many electrons are in the outside shell of an atom of iodine?

ANALYSIS We need the atomic number of iodine to find it most easily in the periodic table, which will then tell us if iodine is a representative element. A search of the Table of Atomic Masses and Numbers (inside the front cover) tells us that the atomic number of iodine is 53. Now we use the periodic table, and find that iodine is in group VIIA.

SOLUTION Being one of the A-type elements, we know that iodine is a *representative* element, which means that its group number is the same as the number of outside shell electrons, 7.

PRACTICE EXERCISE 3

How many electrons are in the outside shell of an atom of each of the following elements?

(a) potassium **(b)** oxygen **(c)** phosphorus **(d)** chlorine

Most Elements With 4–8 Outside Shell Electrons Are Nonmetals If you compare the locations of the elements in Figure 4.2 with their electron configurations, you will see that all of the nonmetals, except hydrogen and helium, have four to eight electrons in their outside shells. All the metallic elements have atoms with one, two, or three outside shell electrons, occasionally 4. Only among some elements with high atomic numbers do we find metals with more than three outside-shell electrons. Tin and lead of group IVA are common examples.

SUMMARY

Periodic properties Because many properties of the elements are periodic functions of atomic numbers, the elements fall naturally into groups or families organized as vertical columns in a periodic table. The atoms of a particular group of representative elements in groups IA–VIIA have the same number of outside-shell electrons —a number that corresponds to the group number itself. Of the group 0 elements, all except helium have outside shells with 8 electrons; helium atoms have 2.

The horizontal rows of the periodic table are called periods. In the long periods, there are transition and inner transition elements which involve the systematic filling of inner d or f orbitals. The nonmetallic elements are in the upper right-hand corner of the periodic table, and the metals—the great majority of the elements —make up the rest of the table. (At the border between metals and nonmetals occur the metalloids, which have both metallic and nonmetallic properties.)

The metal hydroxides formed from the group IA alkali metals have the general formula MOH and are all bases. They neutralize acids, like hydrochloric acid or the other HX acids of group VIIA, in the same way.

Atomic structure Atoms, which are electrically neutral particles, are the smallest representatives of an element that can display the element's chemical properties. Each atom has one nucleus—a hard inner core—surrounded by enough electrons to balance the positive charge on the nucleus. All the atom's protons and neutrons are in the nucleus. The proton has a charge of $1+$, the electron's charge is $1-$, and the neutron is electrically neutral. In atoms, the atomic number equals both the number of protons and the number of electrons outside the nucleus.

Atomic Orbitals The places where electrons can be in an atom are organized as principal energy shells, which consist of subshells made up of orbitals. An atomic orbital is a volume of space near the nucleus where there is a high probability of finding a particular electron. The shape of an orbital comes from wrapping an imaginary envelope about that much of the space within which the overall probability of finding an electron is at least 90%.

Principal level or shell 1 has only one subshell. (It is its own subshell.) And this subshell has just one orbital. (It is its own orbital.) This orbital is the 1s orbital, the 1 standing for principal level 1, and the s specifying the shape of the orbital—spherical. At level (shell) 2, there are two subshells—the s and the p types. Any s subshell, no matter at which principal level, has just one orbital. A p-type subshell always has three orbitals, designated as p_x, p_y, and p_z, to correspond to the three coordinate axes along which the three point. Each p orbital consists of two lobes of equal size and shape. Level 3 has s-, p-, and d-type subshells. (Subshell d consists of 5 orbitals.) Levels 4 and higher have all these plus an f-type subshell (with 7 orbitals).

Each occupied orbital can be viewed as an electron "cloud." It isn't possible to obtain precise information simultaneously about an electron's location and energy, but the knowledge of where electrons most likely are—information obtained from electron configurations—is enough for understanding chemical properties.

As long as electrons remain in their orbitals, an atom neither absorbs nor radiates energy. However, an atom can absorb the energy that corresponds exactly to the difference in energy between two orbitals, provided the orbital of higher energy has a vacancy. The absorbed energy makes an electron move to the higher orbital. When it drops back down, this energy is radiated as a quantum or photon of light.

Electron configurations To write an electron configuration of an atom, we use the aufbau principle and follow certain rules. We place the atom's electrons one by one into the available orbitals, starting with the one of lowest energy, the 1s orbital. According to the Pauli exclusion principle, each orbital can hold two electrons (if their spins are opposite), but where two or more orbitals are available at the same subshell, electrons spread out (Hund's rule).

REVIEW EXERCISES

The answers to Review Exercises whose numbers are in color are found in Appendix C. The answers to the other Review Exercises are found in the Study Guide that accompanies this book. The more challenging questions are marked with asterisks.

The Periodic Table

4.1 In general terms, how do the boiling points of the first 20 elements change as the atomic number increases?

4.2 What is meant by *first ionization energy* and, in general terms, how does this energy change as you move among the first 20 elements from atomic number 1 to 20?

4.3 The binary hydrides of the first 20 elements display a periodic change in what specific way?

*4.4 What are the likely formulas of the binary hydrides of the following elements?
(a) potassium (b) calcium (c) gallium (d) germanium (e) arsenic (f) selenium (g) bromine

4.5 To what do elements in the same vertical column in the periodic table belong? What name is given to a horizontal row in the periodic table?

4.6 In the form of the periodic table used in this textbook, vertical columns are assigned roman numerals followed by either A or B.
(a) Which represents a family of *representative* elements, IVA or IVB?

(b) Are the elements in group 0 regarded as representative or transition elements?

(c) Are the elements in the B groups likely to be metals or nonmetals?

(d) Are the elements in period 2 representative or transition elements or a mixture of both?

4.7 In what way would period 5 of the periodic table be different if the elements were arranged in their order of increasing atomic mass instead of increasing atomic number? Why did Mendeleev insist on the arrangement we presently see despite the fact that it violated his version of the periodic law?

*4.8 Suppose element X forms a compound XOH.

(a) To what family does X most likely belong?

(b) Is XOH an acid or a base?

(c) What reaction will XOH give with HBr? Write the equation.

(d) What *name* is given to the reaction of part (c)?

4.9 Give the group number and the chemical family name of the set of elements to which each of the following belongs.

(a) potassium (b) fluorine (c) selenium (d) barium

4.10 Give the group number and the chemical family name of the set of elements to which each of the following belongs.

(a) chlorine (b) phosphorus (c) calcium (d) sodium

Atomic Structure

4.11 Dalton said that atoms are indestructible, but we now know that they can be broken up into three (actually more) particles.

(a) What is the *general* name for such particles?

(b) What are the names and electrical conditions of the three particles of greatest interest in understanding the chemical properties of substances?

(c) Two of these particles attract each other. What are their names?

(d) Two simply stated rules govern the behaviors of electrically charged particles toward each other. What are they?

*4.12 In Chapter 5 we will learn that particles having the following compositions exist.

Particle X: 11 protons, 12 neutrons, and 10 electrons
Particle Y: 17 protons, 18 neutrons, and 18 electrons

(a) What is the *net* electrical charge carried by particle X?

(b) What is the *net* electrical charge carried by particle Y?

(c) Would the particles X and Y attract each other, repel each other, or be indifferent to each other? How can you tell?

(d) What are the mass numbers of X and Y?

The Bohr Model of the Atom

4.13 Who discovered the existence of atomic nuclei, and what did he conclude about the nature of an atom?

4.14 What causes the positive charge on an atomic nucleus, and why is it always a *whole* number?

4.15 When we know the specific arrangement of the electrons around an atomic nucleus, what do we know about the atom? (What short term is used?)

4.16 Niels Bohr suggested two postulates about the arrangements of electrons in atoms. What are they? Are they still true?

4.17 When light is emitted from, say, a red-hot bar of iron, what happens at the atomic level? What do we call one unit of emitted light?

4.18 What seemed particularly strange to earlier workers in the field concerning the light energy emitted by a sample of an "excited" element? (It was this strange behavior that suggested one of his postulates to Bohr.)

4.19 What analogy did Bohr suggest to illustrate his view of atomic structure? Do scientists still use this analogy?

Where Electrons Are in Atoms

4.20 What two facts about an electron in an atom cannot be known precisely *at the same time?* Who first realized this? How was the Bohr model affected?

4.21 Instead of trying to describe the orbits taken by electrons in atoms, we instead say that a given electron is in a particular *energy state.* Each such allowed energy state has what two general names (synonyms of each other)?

4.22 When all of the electrons of an atom are in their lowest energy states, what name do we give to this condition or state?

4.23 What does *principal quantum number* refer to? What relationship exists between the number of subshells and the corresponding principal quantum number?

4.24 Complete the following table to describe the *numbers* of subshells and orbitals associated with principal energy levels.

Principal Energy Level Number	Number of Subshells	Number of Orbitals
1	——	——
2	——	——
3	——	——
4	——	——

4.25 What are the symbols used for the individual orbitals at each location?

(a) Principal level 1, subshell 1

(b) Principal level 2, subshell 1

(c) Principal level 2, subshell 2

(d) Principal level 3, subshell 2

4.26 When all lower orbitals are filled and one electron is in principal level 3, what subshell and what orbital is it normally in? Give the combination symbol that summarizes this.

*4.27 When we describe the geometry of the 1s atomic orbital as that of a sphere:

(a) What is said about the region inside the sphere?

(b) What is true about the surface of the sphere?

(c) Is it possible for a 1s electron to be outside its 1s sphere?

4.28 What does the cross section of a 2p orbital look like, roughly?

4.29 Of the three p orbitals at principal level 2, what is different about them?

Electron Configurations

4.30 In this chapter we looked ahead to the use we will make of an electron configuration. In general terms, what is this?

4.31 What is the maximum number of electrons that are allowed in each?

(a) Principal energy levels 1, 2, and 3

(b) The p orbitals of principal levels 2 and 3

(c) The s orbitals of principal levels 1, 2, and 3

(d) The d orbitals of principal levels 4 and higher

(e) The f orbitals where they occur

4.32 What is Hund's rule, and what is a likely reason for it?

4.33 Given the opportunity and freedom to change, nature generally tends to assume a posture of lowest energy. What rule used in connection with the aufbau principle reflects this?

4.34 What does the Pauli exclusion principle tell us about writing electron configurations?

4.35 How many electrons are represented in the following electron configurations, and what is the name of the associated element?

(a) $1s^2 2s^2 2p_x^2 2p_y^1 2p_z^1$

(b) $1s^2 2s^2 2p_x^2 2p_y^2 2p_z^2 3s^2 3p_x^1$

4.36 How many electrons are represented in the following electron configurations? Name the associated element.

(a) $[Ne]\ 3s^2 3p_x^1$

(b) $[Ar]\ 4s^2$

4.37 Write the electron configuration of phosphorus in each of the following ways.

(a) The unabbreviated, one-line mode

(b) The condensed, one-line mode

4.38 Write the electron configuration of potassium in each of the following ways.

(a) The unabbreviated, one-line mode

(b) The condensed, one-line mode

*4.39 The electron configuration of an atom of the metal zinc is $1s^2 2s^2 2p^6 3s^2 3p^6 3d^{10} 4s^2$. Using only this information, answer the following questions.

(a) What is the atomic number of zinc? How can you tell?

(b) Is principal energy level number 3 completely filled?

(c) Does the zinc atom have electrons with unpaired spins? How can you tell?

(d) How would the electron configuration of zinc be written in the condensed mode? (Let your answer reflect the subshell filling order of Figure 4.7; i.e., 4s fills before 3d.)

(e) Using the condensed mode, what would be the electron configuration of an element with four more electrons than zinc? What is its atomic number? What is its atomic symbol?

(f) In answering part (e), how was Hund's rule used?

(g) In answering part (e), how was the Pauli exclusion principle used?

4.40 Using the condensed mode, write the electron configuration of neon, atomic number 10.

*4.41 Examine the following possible electron configurations and answer the questions about them.

$$1s^2 2s^2 2p_x^2 2p_y^2 2p_z^2 3s^2 3p_x^1 \qquad 1s^2 2s^2 2p_x^2 2p_y^2 2p_z^2 3s^2 4s^1$$
$$\mathbf{1} \qquad\qquad\qquad\qquad \mathbf{2}$$

(a) Are these electron configurations for atoms of the same element or of different elements? How can you tell?

(b) Which is the more stable configuration, **1** or **2**? How can you tell?

(c) Which configuration corresponds to the *higher* energy state for the atoms? Why?

(d) What has to happen to change an atom from one configuration to the other?

*4.42 Using only the following atomic numbers as well as the aufbau principle, write the electron configurations of the following atoms in both the noncondensed and abbreviated forms.

(a) 4 (b) 6 (c) 12 (d) 16

*4.43 Using only the following atomic numbers as well as the aufbau principle, write the electron configurations of the following atoms in both the noncondensed and abbreviated forms.

(a) 5 (b) 7 (c) 14 (d) 17

*4.44 Suppose that an atom has the following electron configuration.

$$1s^2 2s^2 2p^6 3s^2 3p^6 3d^{10} 4s^2 4p^3$$

Without consulting the periodic table, answer the following questions.

(a) To what family does it belong (using the roman numeral–letter designation)? How can you tell?

(b) Is it likely a metal or a nonmetal? How can you tell?

Additional Exercises

*4.45 Suppose that an atom has the following electron configuration.

$$1s^2 2s^2 2p^6 3s^2 3p^6 3d^{10} 4s^2 4p^6 4d^{10} 5s^2 5p^2$$

Without consulting the periodic table, answer the following questions. When you write condensed electron configurations, let your answers reflect the subshell filling order of Figure 4.7; i.e., 5s fills before 4d.)

(a) What is the group number of this element? How can you tell without referring to the periodic table?

(b) When written in the condensed form, what is its electron configuration?

(c) Using the condensed form, write the electron configuration of the element standing immediately to its right in the periodic table.

(d) Using the condensed form, write the electron configuration of the element standing immediately to its left in the periodic table.

(e) Using the condensed form, write the electron configuration of the element standing immediately above it in the periodic table. (The group 0 element nearest this one is argon, Ar.)

*4.46 The following table is a section from the periodic table where the numbers are atomic numbers. The numbers of one row have been given hypothetical atomic symbols. You should be able to answer the following questions without referring to the actual periodic table.

5	6	7	8	9
13 a	14 b	15 g	16 d	17 e
31	32	33	34	35

(a) Give the atomic numbers of the elements in the same period as g.

(b) What are the atomic numbers of the elements in the same group as b?

(c) Give the atomic numbers of the elements in the same family as e.

(d) Above each box of the top row of elements, write the group numbers of the elements, including the A or B designation.

(e) How many electrons are in the highest occupied principal energy level of the element that would stand immediately to the left of a?

(f) How many electrons would be in the outside shell of the element standing immediately below d in the periodic table?

(g) Element g forms a binary hydride with hydrogen with the formula gH_3. What are the likely formulas of the binary hydrides of elements 33 (give it the symbol X) and 7 (give it the symbol Z)?

(h) Which element is more likely to be a nonmetal, element 9 or 31?

CHEMICAL COMPOUNDS AND CHEMICAL BONDS

5

These beautiful blue-green crystals of microcline ($KAlSi_3O_8$) from Colorado have such precise shapes because their atomic sized building blocks are organized in repeating patterns. The natural constraint among many solid chemicals to regularity in shape rather than chaos is explained in this chapter.

5.1

ELECTRON TRANSFERS AND IONIC COMPOUNDS

Strong forces of attraction exist between oppositely charged ions in a large number of chemical compounds.

In this chapter we ask a fundamental question: How can things hold together? Why is rock salt hard, water mobile, and why does air move so readily out of our way? Atoms, we have learned, are electrically *neutral*, so how can they become stuck together strongly enough to account for the existence of compounds? The answer, in brief, is that atoms can undergo reorganizations of their electrons and nuclei into new particles in which net electrical forces of attraction, called **chemical bonds,** exist.

- *Molecule* is from a Greek term meaning *little mass.*

- Molecular *elements* include the diatomic elements, H_2, O_2, and N_2.

The Two Principal Kinds of Compounds Are Ionic and Molecular   There are two important ways by which electrons and nuclei can become reorganized relative to each other. One way leads to a new kind of small particle called a *molecule,* which is made of two or more atomic nuclei located among enough electrons to make the whole particle electrically neutral. *Molecular compounds* are those whose molecules have nuclei from *different* elements, like water (H_2O), ammonia (NH_3), sugar ($C_{12}H_{22}O_{11}$), vitamin C ($C_6H_6O_6$), and cholesterol ($C_{27}H_{46}O$).

The other way to reorganize atoms into compounds produces tiny particles of opposite electrical charge called *ions,* and these strongly attract each other. Compounds consisting of oppositely charged ions are called *ionic compounds,* which we will study first.

Electron Transfers between Atoms Can Produce Ions   Sodium chloride (table salt) is a typical ionic compound. We learned in Section 2.1 that the parent elements of this compound, sodium and chlorine, cannot be stored in each other's presence because they react violently. Their atoms undergo the following changes in their electron configurations. These are the *overall* changes—the net results. To simplify this discussion, we're ignoring the fact that the element chlorine consists of molecules, Cl_2, instead of individual atoms. The circles stand for the atomic nuclei, which contain protons (p^+) and neutrons (n).

- For purposes of illustration, we have picked the sodium-23 and chlorine-35 isotopes.

$$\left(\begin{array}{c} 11\ p^+ \\ 12\ n \end{array}\right) 1s^2 2s^2 2p^6 3s^1 + \left(\begin{array}{c} 17\ p^+ \\ 18\ n \end{array}\right) 1s^2 2s^2 2p^6 3s^2 3p_x^2 3p_y^2 3p_z^1 \longrightarrow \left[\left(\begin{array}{c} 11\ p^+ \\ 12\ n \end{array}\right) 1s^2 2s^2 2p^6\right]^+ + \left[\left(\begin{array}{c} 17\ p^+ \\ 18\ n \end{array}\right) 1s^2 2s^2 2p^6 3s^2 3p_x^2 3p_y^2 3p_z^2\right]^-$$

One sodium *atom*	One chlorine *atom*	Outer octet / One sodium *ion*	Outer octet / One chloride *ion*
Na	Cl	Na^+	Cl^-

It must be emphasized that the new particle with the sodium nucleus is no longer a sodium *atom*, because it no longer is electrically neutral. Although it has the sodium nucleus with its

$11+$ charge, the particle now has only 10 electrons. These can cancel only $10+$ of the nuclear charge, so this new particle has a net charge of $1+$.

Similarly, the new particle with a chlorine nucleus isn't an atom either. It carries a net electrical charge of $1-$. The chlorine nucleus has 17 protons for $17+$ but surrounding it are now 18 electrons for $18-$, so the net charge on this new particle is $[(17+) + (18-)]$ or $1-$.

Electrically charged particles at the atomic level of size are called **ions.** Positively charged ions are classified as **cations,** and negatively charged ions are **anions.** Ions of either positive or negative charge having *one* atomic nucleus are **monatomic ions.** In the reaction of sodium with chlorine, electrons relocate relative to atomic nuclei to give sodium ions and chloride ions. These particles, not intact atoms, constitute sodium chloride. Before we see how ions organize themselves into crystals, it will be useful to learn how monatomic ions such as these are named.

■ *Ion* is from the Greek *ienai,* to go or to move. Ions, unlike atoms, can move in response to electrical forces.

Monatomic Ions Are Named after the Parent Elements All ions derived from metals have the same name as the element (plus the word *ion*). Some particularly important metal ions are the sodium ion, Na^+, the potassium ion, K^+, the magnesium ion, Mg^{2+}, and the calcium ion, Ca^{2+}. Notice that the formula of an ion, when set separately, always includes the electrical charge as a right superscript.

All metals form monatomic *cations,* and transition metals generally are able to form more than one cation that differ in the amount of charge. The two ions of iron, for example, are Fe^{2+} and Fe^{3+}. The amount of charge is indicated by a roman numeral written into the ion's name. Fe^{2+} is named the iron(II) ion, and Fe^{3+} is the iron(III) ion. These are the modern names. The iron ions were once (and still often are) called the ferrous ion (Fe^{2+}) and the ferric ion (Fe^{3+}). The two ions that copper can form are Cu^+, the copper(I) ion (cuprous ion), and Cu^{2+}, the

TABLE 5.1
Important Monatomic Ions

Group	Element	Symbol for Neutral Atom	Symbol for Its Common Ion	Name of Ion
IA	Lithium	Li	Li^+	Lithium ion
	Sodium	Na	Na^+	Sodium ion
	Potassium	K	K^+	Potassium ion
IIA	Magnesium	Mg	Mg^{2+}	Magnesium ion
	Calcium	Ca	Ca^{2+}	Calcium ion
	Barium	Ba	Ba^{2+}	Barium ion
IIIA	Aluminum	Al	Al^{3+}	Aluminum ion
VIA	Oxygen	O	O^{2-}	Oxide ion
	Sulfur	S	S^{2-}	Sulfide ion
VIIA	Fluorine	F	F^-	Fluoride ion
	Chlorine	Cl	Cl^-	Chloride ion
	Bromine	Br	Br^-	Bromide ion
	Iodine	I	I^-	Iodide ion
Transition Elements	Silver	Ag	Ag^+	Silver ion
	Zinc	Zn	Zn^{2+}	Zinc ion
	Copper	Cu	Cu^+	Copper(I) ion (cuprous ion)[a]
			Cu^{2+}	Copper(II) ion (cupric ion)
	Iron	Fe	Fe^{2+}	Iron(II) ion (ferrous ion)
			Fe^{3+}	Iron(III) ion (ferric ion)

[a] The names in parentheses are older names but are still in use.

copper(II) ion (cupric ion). Notice that in the older names, the *-ous* ending goes with the ion of the lower charge and the *-ic* ending is for the ion with the higher charge. (In formulas having these ions, the charges are "understood.")

Monatomic ions derived from nonmetals are generally *anions,* and their names end in *-ide,* as in *chloride ion,* whose parent element is chlorine. Other important monatomic anions are the fluoride ion, F^-, bromide ion, Br^-, iodide ion, I^-, oxide ion, O^{2-}, and sulfide ion, S^{2-}.

The names, symbols, and electrical charges of several common ions are given in Table 5.1, *and they must be learned now.* (A list of the principal ions mentioned in this book is in Appendix C.)

■ The *metal* elements in groups IVA and VA can form positively charged ions.

Ionic Compounds Have Ionic Bonds between Symmetrically Arrayed Ions

It isn't physically possible to arrange a chemical meeting between just one atom of sodium and one of chlorine. Any visible sample, even the tiniest speck, has upward of at least 10^{18} atoms. Therefore when actual samples of these elements are mixed, a storm of electron transfers occurs, and countless numbers of oppositely charged ions form.

■ A 0.1-mg sample of NaCl has about 10^{18} pairs of ions.

The new sodium ions repel each other, of course, because like charges repel. The chloride ions also repel each other for the same reason. Sodium ions and chloride ions, however, attract each other, because unlike charges attract. Out of all these attractions and repulsions, the storm of new ions subsides into firm, hard, and regularly shaped crystals of sodium chloride. Spontaneously, the unlike charged ions, Na^+ and Cl^-, nestle together as closest neighbors, and like-charged ions stay just a little farther apart (see Fig. 5.1). If the ions are to make maximum use of their forces of attraction and to minimize their forces of repulsion, *they must assemble in an array as symmetrical as allowed by ionic sizes, shapes, and ratios.* This is why crystals of sodium chloride have regular shapes.

Crystal of sodium chloride

What forms as Na^+ and Cl^- ions come together is an example of an *ionic compound.* **Ionic compounds** are orderly aggregations of oppositely charged ions, and the force of attraction between these ions is called the **ionic bond.** Ionic bonds are very strong, and crystals of ionic compounds are rigid up to high temperatures, after which the crystals melt. Ionic crystals, such as salt crystals, can be pulverized without too much difficulty, however. A sharp blow makes one layer of ions shift over so that like-charged ions suddenly become closest neighbors. Now the net force — at least along this layer in the crystal — is one of repulsion, not attraction, and the crystal splits apart.

■ Sodium chloride melts at 804 °C.

The ionic bond does not extend in any single, unique direction. The force of attraction that Na^+ has for a negative charge radiates equally in all directions, like light from a light bulb. One particular Na^+ ion doesn't belong to any particular Cl^- ion. Except in our imagination, there is no separate, discrete molecule consisting of just one Na^+ ion and one Cl^- ion belonging exclusively together. We mention this because the formula used for sodium chloride, NaCl, might be incorrectly interpreted this way.

■ *Empirical* is from the Latin *empiricus,* something experienced (here signifying "from experimental data").

The formula for sodium chloride or for any ionic compound is meant only to disclose ions and the ratio in which they occur. The ratio in sodium chloride is 1 : 1. It *must* be 1 : 1, because the net electrical charge of a compound is zero — always — and the 1 + charge of one sodium ion is balanced exactly by the 1 − charge of one chloride ion. As we said, ionic compounds are represented by formulas that give only the *ratios* of the particles present, so the chemical formulas of ionic compounds are always *empirical* formulas. **Empirical formulas** give the compound's elements and their ratios by atoms expressed in the smallest whole numbers. (We have to make this distinction now because we'll soon learn about other kinds of formulas.)

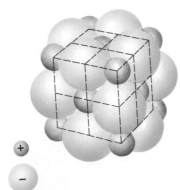

Binary Ionic Compounds Are Named after Their Ions

Generally speaking, expect a compound to be *ionic* if its formula includes the atomic symbol of a metal.[1] Chemists have developed rules for naming ionic compounds, and we'll introduce them for the simplest types, binary ionic compounds. A *binary ionic compound* is one made of the ions of just two elements, like sodium chloride, NaCl, or calcium oxide, CaO. As these examples suggest, binary ionic

FIGURE 5.1
The structure of a sodium chloride crystal. The sodium ions are surrounded by chloride ions as nearest neighbors, and like-charged ions are just a little farther apart.

---

[1] One common exception involves a cation made of N and H with the formula NH_4^+, the ammonium ion. Other exceptions are found among the compounds that we will study in Chapter 12 and following.

compounds are easy to name. We just write the name of the cation, then the name of the anion, omitting the "ion," of course. A binary compound of the iron(III) ion and the bromide ion is thus called iron(III) bromide. (The older name is ferric bromide.)

The ratio of the ions in an empirical formula must be one that permits all opposite charges to cancel, because compounds are electrically neutral, as we said. When calcium ions, Ca^{2+}, and oxide ions, O^{2-}, form an ionic compound, they combine *only* in a 1 : 1 ratio. Only in this ratio can the 2 + on the calcium ion cancel the 2 − on the oxide ion. Thus the formula is CaO. By convention, the cation is written first, and the charges are omitted. They are "understood" in empirical formulas.

Calcium chloride, a compound of calcium ions and chloride ions, must have *two* Cl^- ions for every Ca^{2+} ion because it takes two Cl^- ions to give enough negative charge to cancel the charge of 2 + of a calcium ion. To show this ratio, the formula of calcium chloride is written as $CaCl_2$. Recall that the 2 is called a *subscript,* that subscripts are used to specify ratios of ions (or atoms) present, and that they follow and are placed half a line below the associated atomic symbols. (The subscript 1 is always "understood;" by convention it is not used in chemical formulas.)

■ CaO is also called "quicklime" and is an ingredient in cement.

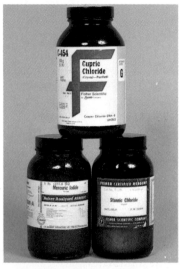

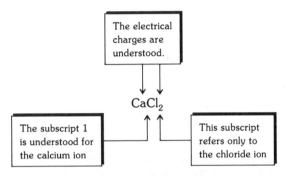

The older system of naming compounds still can be found on the bottles of lab chemicals.

The subscripts used are generally the smallest whole numbers that express the correct ratio. We don't write Ca_2Cl_4 for calcium chloride, even though it gives the correct ratio because 2 : 4 is equivalent to 1 : 2.

EXAMPLE 5.1
Writing Formulas from Names of Ionic Compounds

Write the formula of aluminum oxide.

ANALYSIS We must use the name, aluminum oxide, to infer the ions *and their electrical charges.* Unless this shows that a 1 : 1 ratio works, we use the charges to determine their lowest common multiple and thus to see what is the smallest number of each ion that will have a total charge numerically equal to this multiple. Finally, we assemble the ions in the correct order for a formula using the numbers just determined as coefficients.

SOLUTION From the name "aluminum oxide," we infer that the ions are Al^{3+} and O^{2-}. We know that 3 + isn't canceled by 2 −, so we can't simply write AlO. When a 1 : 1 ratio of ions cannot work, we use the lowest common multiple of the ionic charges to obtain subscripts. With charges of 3 + and 2 −, the least common multiple (ignoring the signs) is 6 ($2 \times 3 = 6$). An electrically neutral balance is thus obtainable if we have 6 + total charge balanced by 6 − total charge. So we can pick the smallest number of aluminum ions that give a total of 6 + and the smallest number of oxide ions that give a total of 6 −. This requires 2 Al^{3+}, because $[2 \times (3+) = 6+]$, and it requires 3 O^{2-}, because $[3 \times (2-) = 6-]$. The ratio in aluminum oxide, therefore, is 2Al^{3+} to 3O^{2-}, and writing the aluminum first in the formula gives us the answer:

$$Al_2O_3$$

■ When a ratio of 1 : 1 does work, the calculation of a least common multiple is obviously unnecessary.

■ Aluminum oxide is a buffing powder for polishing metals.

The strategy in Example 5.1 to find and use the lowest common multiple of (numerically unequal) charges on the ions always works. Try Practice Exercise 1 to develop experience in using this approach.

PRACTICE EXERCISE 1

Write the formulas of the following compounds.

(a) Silver bromide (a light-sensitive chemical used in photographic film)

(b) Sodium oxide (a very caustic substance that changes to lye in water)

(c) Iron(III) oxide (the chief component in iron rust)

(d) Copper(II) chloride (an ingredient in some laundry-marking inks)

EXAMPLE 5.2
Writing Names from Formulas of Ionic Compounds

Write the name of the ionic compound $FeCl_2$, using both the modern and the older forms.

ANALYSIS The symbol Fe stands for one of the two ions of iron, but which? What charge does the iron ion bear in $FeCl_2$? Here we must use our knowledge of the charge on one ion, Cl^-, to figure out the charge on the other. Because there are two Cl^- ions in $FeCl_2$, there must be a total negative charge of $2-$ in each formula unit of $FeCl_2$. The lone Fe part of $FeCl_2$ must therefore provide $2+$ in charge, so we must be dealing with the Fe^{2+} ion—named either the iron(II) ion or the ferrous ion.

SOLUTION The compound is named either iron(II) chloride or ferrous chloride.

PRACTICE EXERCISE 2

Write the names of each of the following compounds. When a name can be given in a modern form or an older form, write both.

(a) CuS **(b)** NaF **(c)** FeI_2 **(d)** $ZnBr_2$ **(e)** Cu_2O

The technique used in Example 5.2 of finding the electrical charge on one ion from the known charge on the other in an ionic compound greatly reduces the number of ions that have to be memorized. The next Exercise gives some practice in this.

PRACTICE EXERCISE 3

What are the charges on the *metal* ions in each of these substances?

(a) Cr_2O_3 (a green pigment in stained glass)

(b) HgS ("Chinese red," a bright, scarlet-red pigment)

(c) $CoCl_2$ (an ingredient in invisible ink)

Many Common Substances Are Ionic Compounds Many familiar substances besides sodium chloride are ionic compounds. Sodium bicarbonate in baking soda, barium sulfate in "barium X-ray cocktails," sodium hydroxide in lye and drain cleaners, and calcium sulfate in plaster and plaster of paris are other examples. When prepared foods are advertised as "low in sodium," it means that they are low in sodium *ion,* certainly not the extremely reactive sodium atom. This, incidentally, hints at how profoundly chemical properties differ between atoms and their ions. Something "rich in calcium" is rich in calcium *ions,* not calcium atoms. Every fluid in every living thing, whether plant or animal, contains dissolved ions. At the molecular level of life, ions are everywhere. People in professional health care fields speak of the "electrolyte status" of this or that body fluid. An **electrolyte** is any substance that can furnish ions in a solution in water, and all body fluids include dissolved electrolytes.

- The principal ions in blood are Na^+ and Cl^-, but many others are also present.

- Ca^{2+} ions are essential not just to build bones but to the chemical reactions by which nerve signals are transmitted.

Ions Have Different Radii than Their Parent Atoms At the molecular level of life, *the shapes and sizes of ions and molecules are as important as anything else about them.* We will need a background in the *geometries* of particles to understand the workings of enzymes, hormones, neurotransmitters, genes, many medications, and many poisons. The simplest principles governing size and geometry involve monatomic ions. The geometry of an ion or a molecule is determined by several factors: which atomic nuclei are involved, what electron clouds they have, the electron configurations, and the kind and amount of any net electrical charge.

Sodium ions and chloride ions have quite different sizes (see Fig. 5.2). In general, when a cation forms from a metal atom, the remaining electron cloud shrinks because the positive nuclear charge now numerically *exceeds* the negative charge of the electron cloud. The excess positive charge in the cation pulls the electron cloud more tightly inward, so *metal ions are always smaller than their parent atoms* (see Fig. 5.2). The sodium *ion* has a radius (95 pm) only about half as large as the sodium *atom* (186 pm). Several similar examples are in Figure 5.2.

- Monatomic ions are spherical, and the *radius* of a sphere is the distance from its center to its surface.

- The unit "pm" is the *picometer;* 1 pm = 10^{-12} meter.

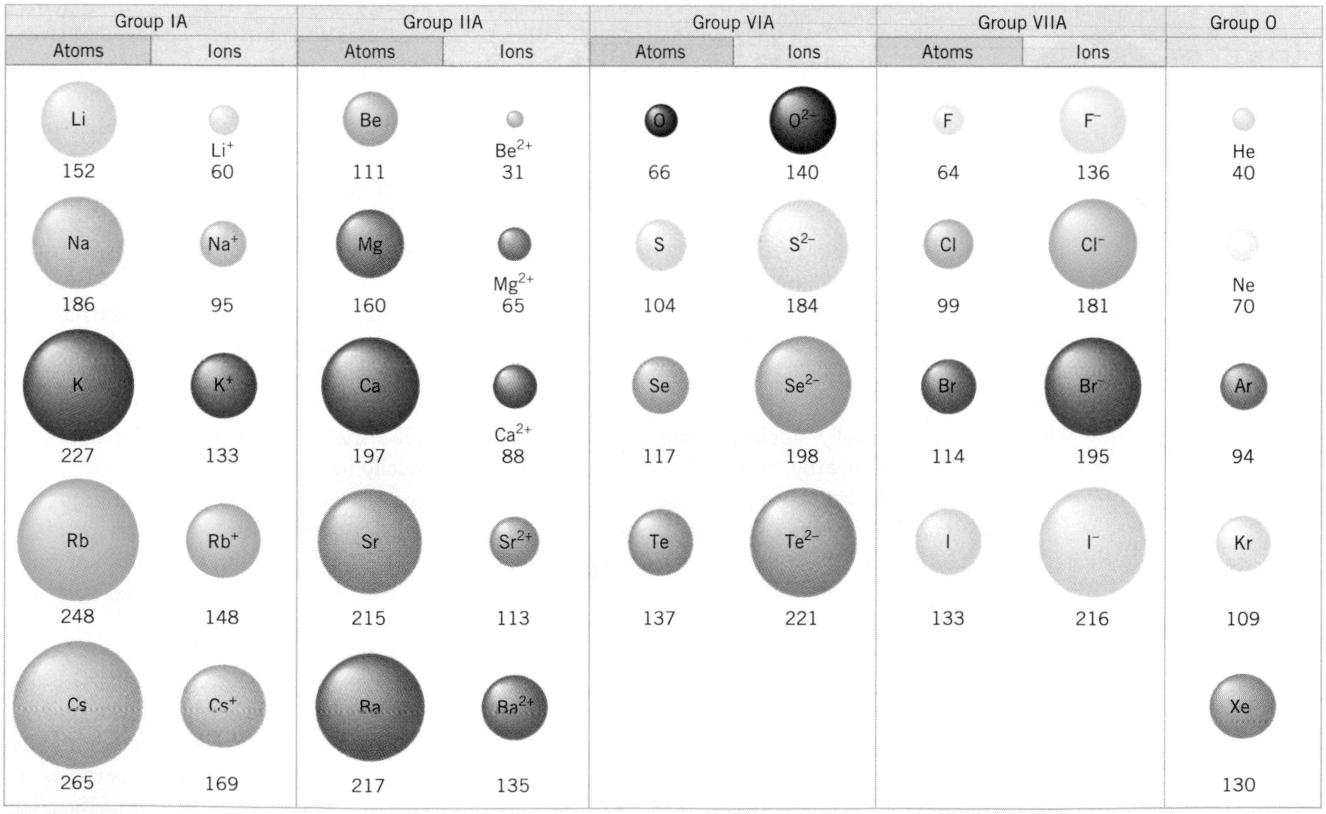

Group IA		Group IIA		Group VIA		Group VIIA		Group 0
Atoms	Ions	Atoms	Ions	Atoms	Ions	Atoms	Ions	
Li 152	Li^+ 60	Be 111	Be^{2+} 31	O 66	O^{2-} 140	F 64	F^- 136	He 40
Na 186	Na^+ 95	Mg 160	Mg^{2+} 65	S 104	S^{2-} 184	Cl 99	Cl^- 181	Ne 70
K 227	K^+ 133	Ca 197	Ca^{2+} 88	Se 117	Se^{2-} 198	Br 114	Br^- 195	Ar 94
Rb 248	Rb^+ 148	Sr 215	Sr^{2+} 113	Te 137	Te^{2-} 221	I 133	I^- 216	Kr 109
Cs 265	Cs^+ 169	Ba 217	Ba^{2+} 135					Xe 130

FIGURE 5.2
Atomic and ionic radii of some representative elements. The radii are given in picometers, pm (10^{-12} meter).

When an anion forms from an atom, one or more electrons are added to the atom's electron cloud. *Monatomic anions are thus always larger than their parent atoms.* The radius of the Cl⁻ ion, 181 pm, is almost twice as large as the radius of the Cl atom, 99 pm. You can see similar increases in Figure 5.2 among the group VIA and VIIA elements.

The sizes of the Na⁺ and Cl⁻ ions are such that the best "fit" using a 1 : 1 ratio causes the final shape of the sodium chloride crystal to be a cube. We say that these ions form a *cubic crystal.* So do K⁺ and Cl⁻ ions. Other combinations of ions form crystals with different but still regular shapes. The shapes *must* be regular because ions can maximize interionic attractions, cations to anions, only this way.

5.2

MONATOMIC IONS AND THE OCTET RULE

Atoms and ions whose outside energy levels hold eight electrons are substantially more stable than those that do not.

The formation of sodium and chloride ions raises several questions. Does sodium always form singly charged ions, Na⁺? Why not Na²⁺ ions? We could ask similar questions about chlorine. Indeed, we could ask, why do they form ions at all? Not all elements do. In this section we'll learn about a pattern in nature that answers these questions in terms of the general principle that nature "moves" toward what is the more stable, lower energy condition in a given circumstance.

Atoms of the Noble Gases Have the Most Stable Electron Configurations In the previous chapter we learned that the group 0 elements, the noble gases, are the least reactive, most stable of all elements. Noble gases form neither positive nor negative ions. There is evidently something quite stable about the two kinds of electron configurations found among noble gas atoms. One configuration is the **outer octet,** eight electrons in whichever principal level happens to be the *outside* level or shell. The other is a *filled* level 1 when it is the *outside* level (as in helium). These two noble gas configurations, conditions of unusual chemical stability, will help us understand why ions have the charges they do and why the elements in the same chemical family, if they form ions at all, form ions of the *same* charge.

Monatomic *Ions* of the Representative Elements Have Noble Gas Configurations A sodium atom, in group IA, has one electron in its outside level — level 3 — but after losing this electron, it has a new outside level: the former inner level number 2, which has eight electrons. As the sodium *ion,* the particle has an outer octet, and it also has the electron configuration of a member of the noble gas family — neon. Whatever causes an outer octet to lend stability to the neon *atom* evidently is at work in the sodium *ion,* because the sodium ion is an unusually stable particle, too. (It's stable provided that there is something oppositely charged nearby, so that the overall system is electrically neutral.)

A chlorine atom, in group VIIA, has seven electrons in its highest occupied energy level, number 3. When this level accepts one electron, the Cl atom acquires an outer octet and becomes Cl⁻. Thus the chloride *ion* has the same electron configuration as an atom of one of the noble gases, argon. The chloride ion, provided that an oppositely charged system is nearby, is also a particle of unusual chemical stability.

When we put sodium and chlorine together, the electron transfers between sodium and chlorine generate particles with noble gas configurations. In other words, these two elements — given the opportunity in terms of each other's actual presence — spontaneously change in a direction that leads to greater stability for both. They change from particles that don't have noble gas configurations to those that do. It does not matter that the new particles are electrically charged. The atoms sacrifice neutrality for increased stability.

We have to emphasize that we don't mean that ions are stable in isolation. It isn't possible to have a bottle of just sodium ions. Matter in bulk must *always* be electrically neutral to be

■ Just why noble gas configurations are so stable is still not fully understood.

■ The *atoms* in sodium metal react violently with water. The sodium *ion* is exceptionally stable, gives no reaction with water (or with much of anything else for that matter), and is present in all body fluids.

■ The *element* chlorine is a poisonous gas. The chloride *ion* is unusually stable, has very few reactions, and is present in all body fluids.

stable. As the crystal of sodium chloride forms, ions of one charge are surrounded by ions of the opposite charge. It is in such an environment that we can say that the ions are exceptionally stable.

We haven't explained *why* a noble gas configuration is stable. We are only pointing out that for some reason it is. The pattern we noted for the formation of sodium chloride is so general for the reactions of the representative metals and nonmetals that it almost amounts to a law of nature traditionally called the **octet rule** but perhaps better named the **noble gas rule.**

> **Octet Rule (Noble Gas Rule)** The atoms of the reactive representative elements tend to undergo those chemical reactions that most directly give them electron configurations of the nearest noble gas.

■ G. N. Lewis (1875–1946) was chiefly responsible for the development of the octet rule.

The Charges on Monatomic Ions of Representative Elements Correlate with the Periodic Table
Notice the phrase "most directly" in the octet rule. What does it mean, and why is it there? To answer these questions, consider another way by which a sodium atom could acquire a noble gas configuration. We could imagine a sodium atom *receiving* seven more electrons to give it an outer octet at level 3, instead of giving up one electron to get an outer octet at level 2. But the sodium nucleus, with a charge of $11+$ and already (in the atom) hanging onto eleven electrons, cannot possibly attract and hold an extra seven electrons. We cannot expect a particle (nucleus) with a charge of $11+$ to attract and hold 18 electrons with a total charge of $18-$. All metal atoms have just a few outside level electrons, so changing to ions "most directly" for all metal atoms means *losing* a few electrons rather than gaining many. Thus all metals form *positively* charged ions by giving up electrons. The metals of the *representative* elements always give up only as many electrons as needed to strip the electron configuration to that of the noble gas nearest the metal (and preceding it) in the periodic table. Notice that the "action" leading to a monatomic cation involves the outside *level* of the original atom. Sometimes "level" might be taken to refer to a specific orbital, however, so a better term for use in this connection is the **valence shell,** meaning an atom's highest occupied principal energy level. For the representative elements, the number of valence shell electrons equals the group number. Therefore, *all members of the same representative family of elements gain or lose electrons in the same way and form ions of identical charges.*

For representative metal elements, *the positive charge on the ion is the same as the group number.* For example, the ions of the group IA metals all have charges of $1+$, Li^+, Na^+, K^+, Rb^+, and Cs^+. Two ions, Na^+ and K^+, are particularly important in the fluids of living systems. The ions of the group IIA metals have charges of $2+$, for example, Mg^{2+}, Ca^{2+}, and Ba^{2+}. We will be interested in only one element in group IIIA, aluminum, and its ion has a charge of $3+$, Al^{3+}. The electrical charge on an ion—both the number and the sign of the charge—is sometimes called the ion's *electrovalence* because this charge ("electro-") is responsible for the ionic bonding ("valence") in ionic crystals.

For the ions of the transition metals there is no simple correlation between their group numbers and the size of their positive charges. Generally, the charges vary from $1+$ to $3+$, and several transition metals can exist as ions of more than one charge, as we have noted.

Nonmetal atoms of the *reactive* elements (not the noble gas elements) have 4, 5, 6, or 7 valence shell electrons. Being already close to valence shell octets, atoms with valence shells of 6 or 7 electrons can achieve noble gas configurations most directly by *gaining* 2 or 1 electrons than by losing 6 or 7. Group VIA or VIIA nonmetal atoms thus *accept* electrons to acquire octets. Atoms of group VIA elements accept two electrons and form monatomic ions with charges of $2-$, like O^{2-} and S^{2-} (see Table 5.2). Atoms of group VIIA elements, with valence shells of seven electrons, need just one more electron for outer octets. So their monatomic ions all have charges of $1-$, for example, F^-, Cl^-, Br^-, and I^-.

The group IVA and VA elements evidently have valence shells too far from octets. Their atoms would have to gain more electrons (four or three) than their nuclear charges can attract *and* hold. The result is that the nonmetal elements in groups IVA and VA exist as anions in so few compounds that we will ignore them. When we encounter an exception, we will note it.

■ "Valence" is from the Latin *valere,* meaning "to be strong." Here it refers to "combining strength" or "combining ability."

■ One of the many roles of Mg^{2+} in the body is to activate enzymes, substances that control how rapidly reactions occur.

■ The nitride ion (N^{3-}), a group VA anion, occurs with ions of group IIA metals in saltlike compounds, e.g., Mg_3N_2, magnesium nitride.

The noble gas (octet) rule, the periodic table, and the aufbau principle enable us easily to figure out the electron configurations and electrical charges of the ions of the representative elements. The following examples show how.

EXAMPLE 5.3
Using the Noble Gas (Octet) Rule

When nutritionists speak of the *calcium requirement* of the body, they always mean the calcium *ion* requirement. Calcium has atomic number 20. What charge does the calcium ion have, and what is the symbol of this ion?

ANALYSIS There are two methods for solving this kind of problem, and you should learn both. The first is to write the electron configuration of the atom, determine the number of electrons in its valence shell, and then decide how many electrons must be gained or lost to acquire an outer octet in the most direct manner.

The second approach to the solution is to use the periodic table to find out in which *group* the element is. Then the group number tells how many electrons must be gained or lost to acquire an outer octet in the most direct manner.

SOLUTION Using the rules, we write the electron configuration of element 20, remembering that $4s$ fills before electrons go into $3d$.

$$1s^2 2s^2 2p^6 3s^2 3p^6 4s^2$$

The atom has two electrons in the valence shell (level 4) and eight electrons in the next lower level (level 3 with two $3s$ electrons plus six $3p$ electrons). Only by losing *both* of the $4s$ electrons can the calcium atom get a new outside level that holds an octet. Losing *one* electron won't do. Neither will losing three or more. Therefore the only stable ion that calcium forms has the following electron configuration.

$$1s^2 2s^2 2p^6 3s^2 3p^6$$

By losing two electrons the net charge becomes $2+$, so the symbol of the calcium ion is Ca^{2+}.

Using the second approach, we locate calcium in group IIA in the periodic table. All group IIA atoms have two valence shell electrons, so all most directly acquire configurations of the nearest noble gases by losing *two* electrons. Thus all of the group IIA ions bear charges of $2+$: Be^{2+}, Mg^{2+}, Ca^{2+}, Sr^{2+}, Ba^{2+}, and Ra^{2+}.

EXAMPLE 5.4
Using the Noble Gas (Octet) Rule

Oxygen exists as the oxide ion in such substances as calcium oxide, an ingredient in stucco and mortar. What is the symbol for the oxide ion, including its electrical charge?

ANALYSIS As in Example 5.3, we will solve this from both the electron configuration of the oxygen atom and the location of oxygen in the periodic table.

SOLUTION First, the electron configuration of an oxygen atom (atomic number 8) is

$$1s^2 2s^2 2p_x^2 2p_y^1 2p_z^1$$

The valence shell—level 2—has a total of six electrons, just two short of an octet and a neon configuration. We could also say that the oxygen atom has six too many electrons to have the helium configuration ($1s^2$). However, gaining two electrons to become like neon ($1s^2 2s^2 2p^6$) is much simpler than losing six, so oxygen achieves a noble gas configuration most directly by accepting two electrons from some metal atom donor. The configuration of the anion of oxygen, then, is

$$1s^2 2s^2 2p_x^2 2p_y^2 2p_z^2 \quad \text{or} \quad 1s^2 2s^2 2p^6$$

The two extra electrons give the anion a charge of 2−, so the symbol of the oxide ion is O^{2-}.

Using the periodic table, we locate oxygen in group VIA, so its valence shell has six electrons. It must pick up two electrons, not just one and not more than two, to have an outer octet. These two extra electrons give the particle a charge of 2−, so we can write O^{2-} directly.

EXAMPLE 5.5
Using the Noble Gas (Octet) Rule

Hardly any element is involved in more compounds than carbon. (Roughly six million carbon compounds are known.) Can carbon atoms change to ions? If so, what is the symbol of the ion?

ANALYSIS When we find carbon's place in the periodic table, we see that it is in group IVA, so carbon atoms have four valence shell electrons. A carbon atom either must lose these four electrons (and become helium-like) or gain four electrons (and become neon-like) to achieve a noble gas configuration. In one or two very rare situations, carbon can do the latter—become the C^{4-} ion. Because this is so rare we ignore it.

SOLUTION By ignoring anions of groups IVA and VA, we need not write their configurations or symbols.

■ The methanide ion, C^{4-}, apparently occurs in Be_2C, beryllium methanide, a brick-red solid.

PRACTICE EXERCISE 4

Write the electron configuration of an atom of each of the following elements, and deduce from it the charge on the corresponding ion. If the atom isn't expected to have a corresponding ion, state so. The numbers in parentheses are atomic numbers. Do not use the periodic table for this practice exercise.

(a) potassium (19) **(b)** sulfur (16) **(c)** silicon (14)

PRACTICE EXERCISE 5

Write the electron configurations of the *ions* of the elements in Practice Exercise 4 that can form ions.

PRACTICE EXERCISE 6

Relying on their locations in the periodic table, write the symbols of the ions of each of the following elements. Always remember that no symbol of an ion is complete without its electrical charge. (The numbers in parentheses are atomic numbers.)

(a) cesium (55) **(b)** fluorine (9) **(c)** phosphorus (15) **(d)** strontium (38)

Electron Transfer Reactions Are Examples of Redox Reactions The reaction between sodium and chlorine that we studied earlier illustrates one very general and important family of reactions, the *redox reaction,* or **oxidation – reduction reaction.** These are at the heart of the sets of reactions whereby we use oxygen from air to obtain chemical energy from the compounds obtained by the partial breakdown of food molecules. You may even carry out oxidation – reduction reactions in the lab soon, so we'll pause here only long enough to introduce some of the major terms concerning them. This is only a bare introduction meant to give us some useful terms; the reactions will be studied in more depth in Chapter 10.

A **redox reaction** is one in which oxidation numbers change. **Oxidation numbers** are numbers assigned to each atom in a formula according to a set of rules. The rules are simple for elements and monatomic ions. The oxidation number of any *element* is zero. The oxidation number for any *monatomic ion* is the same as the charge on the ion. Thus the oxidation number of Na in sodium metal is 0, but the oxidation number of sodium in NaCl is $+1$ because there is a charge of $1+$ on the sodium ion. The oxidation number of chlorine in NaCl is -1 because the chloride ion has a charge of $1-$, but Cl has an oxidation number of 0 in elemental Cl_2. In $FeCl_3$, the oxidation number of iron is $+3$. The oxidation numbers in MgS are $+2$ for magnesium and -2 for sulfur.

When a sodium atom transfers an electron to a chlorine atom, the oxidation number of Na changes from 0 to $+1$. Any change that makes an oxidation number more positive is defined as an **oxidation.** We say that the sodium atom is *oxidized* to the sodium ion. Of course, this occurs here only because an electron actually leaves the Na atom, so in electron-transfer reactions involving elements or monatomic systems, *oxidation means the loss of electrons.*

When a chlorine atom accepts an electron and changes from Cl to Cl⁻, the oxidation number of chlorine changes from 0 to -1. We define any change that makes an oxidation number more negative as a **reduction.** This can't take place without the particle accepting an electron, so for elements and monatomic systems *reduction means the gain of electrons.*

A reduction cannot occur without an oxidation of something else. Electrons don't just leave from or go to outer space; they *transfer*. A reaction that involves a reduction also *must* involve an oxidation, and any such reaction is a *redox reaction.*

The reactant that causes a reduction is called the **reducing agent.** Sodium is the reducing agent in the reaction with chlorine because it causes the reduction of Cl to Cl⁻. The reactant that oxidizes something else is called an **oxidizing agent.** Chlorine is the oxidizing agent in the reaction with sodium because it causes the oxidation of Na to Na⁺. Using the correct formula for chlorine, Cl_2, we have

$$2Na \;+\; Cl_2 \longrightarrow 2Na^+ + 2Cl^-$$

Reducing agent   Oxidizing agent

Notice that an oxidizing agent is always itself reduced in a redox reaction, and that a reducing agent is always itself oxidized.

- The chemical reactions involved in respiration, photosynthesis, and rusting all include oxidation – reduction reactions.

- The earliest examples of oxidation involved oxygen itself as the oxidizing agent; hence the name.

- People used to call the conversion of ores, like iron ore, to the metal a *reduction* of the ore to the metal. This is where we got the general name, reduction.

EXAMPLE 5.6
Analyzing a Redox Reaction

The reaction of calcium with sulfur is a redox reaction in which calcium ions and sulfide ions form to make calcium sulfide, CaS.

$$Ca + S \longrightarrow CaS$$

Determine the oxidation numbers of calcium and sulfur in CaS, and decide what is the oxidizing agent and what is the reducing agent in this redox reaction.

ANALYSIS The reactants are both elements, so they both have oxidation numbers of 0. From Table 5.1 (and soon, it must be said, from memory) we know that the charge on Ca in CaS is 2+ and that the charge on S in CaS is 2−. Calcium, therefore, has an oxidation number of +2 in CaS, and sulfur's oxidation number in this compound is −2.

SOLUTION We can see that calcium's oxidation number becomes more positive (0 to +2), so calcium is oxidized. Sulfur's oxidation number becomes more negative (0 to −2), so sulfur is reduced. Calcium is the reducing agent and sulfur is the oxidizing agent. ■

PRACTICE EXERCISE 7

Identify by their chemical symbols what is oxidized and what is reduced in the following reactions. Also identify what is the oxidizing agent and what is the reducing agent.

(a) $Mg + S \rightarrow MgS$

(b) $CuCl_2 + Zn \rightarrow ZnCl_2 + Cu$

(*Hint:* Both $CuCl_2$ and $ZnCl_2$ are compounds of the chloride ion. From this information, you should be able to deduce the charges on the metal ions in these compounds.)

5.3

ELECTRON SHARING AND MOLECULAR COMPOUNDS

When electron density becomes sufficiently concentrated between two atomic nuclei there is a chemical bond — a covalent bond — between them.
Nature has ways besides electron transfers to develop net forces of attraction, chemical bonds, between atoms, and our purpose in this section is to learn about a second method. It most often occurs when atoms of nonmetals form compounds.

Compounds of Nonmetals Are Not Ionic Several million compounds, probably over 90% of all compounds, involve only nonmetal atoms. These cannot be ionic compounds, however. To have *ionic* compounds, ions of both positive and negative charge are needed; but atoms of nonmetals do not become positive ions because too many electrons have to be lost from their valence shells to achieve noble gas configurations. The nonmetals in groups IVA and VA, moreover, almost never become negative ions, because they would have to gain too many electrons to get outer octets. The result is that compounds made entirely of nonmetals, instead of being ionic, consist of particles called *molecules.* A **molecule,** as we have said, is a small, electrically neutral particle consisting of at least two nuclei and enough electrons to make the whole system neutral. A compound that consists of molecules is called a **molecular compound,** and we have noted that several elements consist of molecules, too. Generally speaking, if the formula of a compound does *not* contain the atomic symbol of a metal, it is a molecular compound. (An exception is NH_4^+; see footnote 1.) Our question now is, "What holds molecules together?"

Electron Clouds Can Become Dense between Suitable Atoms We'll consider the simplest molecule, H_2, first. Imagine two isolated hydrogen *atoms* moving directly toward each other (see Fig. 5.3a). As the H atoms draw closer, the electron cloud of the $1s$ electron of each atom begins to sense the presence of the positively charged nucleus of the other atom

■ A region of *high electron density* is one with a thick or dense electron cloud.

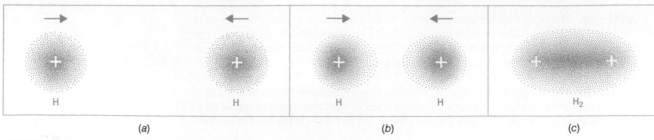

(a)   *(b)*   *(c)*

FIGURE 5.3

Formation of the covalent bond in hydrogen, H_2. (From J. E. Brady and J. R. Holum, *Chemistry. The Study of Matter and Its Changes,* 1993, John Wiley and Sons, New York. Used by permission.)

and experiences an attraction toward it. The attraction distorts both of the two electron clouds, and they bulge toward the sides of the atoms that are nearing each other (see Fig. 5.3*b*). As the atoms continue toward collision, their electrons spend more and more of their time on facing sides of the atoms. Each nucleus, however, begins more and more to sense the like-charged nature of the other, so a collision never fully develops to the touching of the nuclei. Instead, the atoms brake so that their two nuclei are a short distance apart (see Fig. 5.3*c*).

The atoms do not rebound, like two billiard balls, because the electron density between their two nuclei is now too great. The nuclei are attracted toward this concentrated electron cloud, and the cloud itself is attracted to the two nuclei. Because the pair of electrons have become concentrated more or less between the nuclei, the two nuclei now have a bond between them, and a molecule is born, H_2.

The language used to describe such a pair of electrons is *shared pair.* Neither of the shared electrons any longer belongs exclusively to just one of the nuclei, as each did before the atoms came together. The two electrons now belong to both nuclei, and we say that the nuclei *share* the two electrons. The bond created by this sharing of electron pairs between atomic nuclei is called the **covalent bond,** sometimes an *electron pair bond.*

■ *Co-* comes from cooperative; *-valent* from the Latin *valere,* to be strong, signifying strong binding.

The whole new package, the two protons and two electrons of H_2, is electrically neutral, as molecules *must* be. Unlike a single pair of ions, like Na^+ and Cl^- in NaCl, a molecule is a discrete and separate particle that can enjoy independent existence. It can move around as a unit.

The covalent bonds in larger molecules are like those in H_2. They involve the electron density of a pair of electrons that becomes concentrated between two nuclei. The nuclei are attracted into this region and held there at a very small distance apart. We'll see in the next section how the stability of a noble gas configuration influences the *number* of covalent bonds that an atom can form.

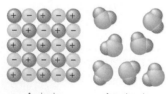

An ionic substance   A molecular substance

5.4

LEWIS STRUCTURES AND THE OCTET RULE

The electron pairs of covalent bonds are counted with the other valence electrons in computing octets.

Like atoms and ions, molecules need symbols, and one useful symbol represents a molecule by an *electron-dot structure.*

Electron-Dot Structures Give Valence Shell Electrons in Atoms   **Electron-dot structures** are meant to draw attention to valence shell electrons by representing them as dots placed around the atomic symbol. The dots are kept separated and at the sides of an imaginary square until there are four dots. The fifth, sixth, seventh, and eighth electrons are placed so as to create pairs of electrons. Because of the great contributions of G. N. Lewis, an American chemist, to the theory of the covalent bond, electron-dot structures are often called **Lewis structures.**

The group numbers of the representative elements tell us how many electrons are in their valence shells and thus how many electron dots must be used. All group IA elements require just one electron dot. Those in group VIIA need seven. (Remember that those of group 0, the noble gases, need eight, with the exception of helium, which needs two.) The electron-dot structure of hydrogen is simply H· and of helium is He: . The electron-dot or Lewis structures of the atoms in the second period of the periodic table are as follows.

$$Li\cdot \quad \cdot Be\cdot \quad \cdot \overset{\cdot}{B}\cdot \quad \cdot \overset{\cdot}{\underset{\cdot}{C}}\cdot \quad \cdot \overset{\cdot}{\underset{\cdot}{N}}\cdot \quad \cdot \overset{\cdot\cdot}{\underset{\cdot\cdot}{O}}\cdot \quad :\overset{\cdot\cdot}{\underset{\cdot\cdot}{F}}\cdot \quad :\overset{\cdot\cdot}{\underset{\cdot\cdot}{Ne}}:$$

■ It doesn't matter which sides of the symbols get the pairs.

Notice how, after the first four dots have gone into place, additional dots are paired with others. More importantly, notice how the number of dots of atoms in groups IA – VIIA always equals the group number of the atom.

EXAMPLE 5.7
Writing the Electron-Dot Symbol of a Representative Atom

What is the electron-dot symbol of sulfur, S?

ANALYSIS   Sulfur is in group VIA of the periodic table, so its atoms must have six electrons in their valence shells. We write the first four electron dots at the sides of an imaginary square, and we add the last two to create two pairs. (It doesn't matter which of the possible pairs are created.)

SOLUTION

$$\cdot \overset{\cdot\cdot}{\underset{\cdot\cdot}{S}}\cdot$$

PRACTICE EXERCISE 8

Write the electron-dot or Lewis symbols for the atoms of the period 3 elements of the periodic table.

PRACTICE EXERCISE 9

Write the electron-dot or Lewis symbol for antimony, Sb, atomic number 51.

The Formation of Ions Can Be Shown by Electron-Dot Symbolism   The electron transfer that occurs when sodium and chlorine react can be represented by electron-dot symbolism as follows.

$$Na\overset{\curvearrowright}{\cdot} + \cdot \overset{\cdot\cdot}{\underset{\cdot\cdot}{Cl}}: \longrightarrow Na^+ + \left[:\overset{\cdot\cdot}{\underset{\cdot\cdot}{Cl}}:\right]^-$$

The valence shell of sodium loses its electron, so no dot remains. (Only valence shells, the *original* outside levels, are given electron dots.) The valence shell of chlorine accepts this electron and an octet forms. To show that this electron is fully the property of chlorine, we put brackets about the symbol of the ion.

The reaction between magnesium and chlorine can similarly be represented.

$$:\ddot{\underset{..}{Cl}} \cdot \; + \; \odot Mg \odot + \; \cdot \ddot{\underset{..}{Cl}}: \longrightarrow Mg^{2+} + 2\left[:\ddot{\underset{..}{Cl}}:\right]^{-}$$

Each chlorine atom has room in its valence shell for only one more electron, so two chlorine atoms are needed to take care of the two electrons of one magnesium atom.

EXAMPLE 5.8
Representing a Reaction in Electron-Dot Symbolism

Sodium sulfide is an ionic compound of the sodium ion, Na^+ and the sulfide ion, S^{2-}. How could its formation be represented in electron-dot symbolism?

ANALYSIS   We first write the Lewis symbols for sodium and sulfur.

$$Na\cdot \qquad \cdot\ddot{\underset{..}{S}}\cdot$$

To achieve an octet, S needs two electrons, so we need two Na atoms for one S atom.

SOLUTION   We can therefore write (remembering the brackets),

$$Na\odot + \cdot\ddot{\underset{..}{S}}\cdot + \odot Na \longrightarrow 2Na^+ + \left[:\ddot{\underset{..}{S}}:\right]^{2-}$$

PRACTICE EXERCISE 10

Represent the formation of calcium oxide by electron-dot symbolism. It consists of Ca^{2+} and O^{2-} ions.

Lewis Structures of Molecules Display Noble Gas Configurations for the Atoms Present   The formation of H_2 can be represented in electron-dot symbolism as follows:

$$H\cdot + \cdot H \longrightarrow H:H$$

The shared pair of electrons is shown between the two atomic symbols. Since they are shared, they both count for each hydrogen when assessing whether a noble gas configuration is achieved. Each hydrogen atom acquires the helium configuration by this sharing of the electron pair. What we see here is a general rule: Noble gas configurations dominate the formation of covalent bonds as they did of ions. Atoms that form covalent bonds tend, by the sharing of electron pairs, to acquire as many electrons as needed to achieve noble gas configurations.

We can represent the formation of the diatomic halogen molecules in electron-dot symbolism as follows. We will also carry the symbolism one step further in these examples. It is customary to represent any shared pair of electrons — any covalent bond — by a short line, a "dash" bond. Any pairs of valence shell electrons that are not used for covalent bonds are called **unshared pairs.** Unshared electron pairs are often omitted from structures when they are not involved in chemical reactions. The structures following the arrows, next, thus illustrate these options.

Fluorine, F_2 $:\ddot{F}\cdot + \cdot\ddot{F}: \longrightarrow :\ddot{F}:\ddot{F}:$ or $:\ddot{F}-\ddot{F}:$ or $F-F$

Chlorine, Cl_2 $:\ddot{Cl}\cdot + \cdot\ddot{Cl}: \longrightarrow :\ddot{Cl}:\ddot{Cl}:$ or $:\ddot{Cl}-\ddot{Cl}:$ or $Cl-Cl$

Bromine, Br_2 $:\ddot{Br}\cdot + \cdot\ddot{Br}: \longrightarrow :\ddot{Br}:\ddot{Br}:$ or $:\ddot{Br}-\ddot{Br}:$ or $Br-Br$

Iodine, I_2 $:\ddot{I}\cdot + \cdot\ddot{I}: \longrightarrow :\ddot{I}:\ddot{I}:$ or $:\ddot{I}-\ddot{I}:$ or $I-I$

■ Remember, only the valence shell electrons are shown. The core electrons are "understood."

By counting the shared electron pairs of the covalent bonds for either atom, we can see that each halogen atom in these diatomic molecules has achieved a noble gas configuration, an outer octet. Each molecule has one covalent bond.

Hydrogen chloride, HCl, is a diatomic molecule with one covalent bond, and its formation can be represented as follows.

$$H\cdot + \cdot\ddot{Cl}: \longrightarrow H:\ddot{Cl}:$$

Again we see that noble gas configurations are achieved for H and Cl by the sharing of an electron pair.

The atoms of the nonmetals of groups IVA, VA, and VIA must form more than one covalent bond to acquire octets. Of particular importance at the molecular level of life are atoms of carbon, nitrogen, oxygen, and sulfur. Their Lewis symbols are

$$\cdot\overset{\cdot}{\underset{\cdot}{C}}\cdot \quad\quad \cdot\overset{\cdot\cdot}{N}\cdot \quad\quad \cdot\overset{\cdot\cdot}{\underset{\cdot\cdot}{O}}\cdot \quad\quad \cdot\overset{\cdot\cdot}{\underset{\cdot\cdot}{S}}\cdot$$

You can see at a glance from these symbols how many electrons each must get by sharing to acquire octets. Carbon has four electrons so it needs four more, and if it combines with hydrogen (which has just one electron per atom to share), it *must* have four hydrogen atoms to get the necessary four additional electrons.

H·   ↶·C·↷   ·H ⟶   H:C:H   or   H—C—H

with H above and H below

Methane

■ Methane is the chief constituent of natural gas.

Similarly, a nitrogen atom, which has five electrons and needs a share of three more to have an octet, combines with three hydrogen atoms. And oxygen and sulfur combine with two.

H:N:H   H:O: H:S:

or   or   or

H—N—H   H—O: H—S:

Ammonia   Water   Hydrogen sulfide

■ Hydrogen sulfide is what gives rotten eggs their odor.

The symbols for methane, ammonia, water, and hydrogen sulfide that show the sequence in which the atoms are joined together are called **structural formulas** or simply **structures**. A formula of a molecular substance, such as H_2O, which gives the composition of one molecule is called a **molecular formula**. You can see that structures are more informative than other kinds of formulas.

The kinds of formulas we used for ionic compounds, empirical formulas, are almost never used for molecular compounds. This is because when molecules react, they often change in only one part of their structure, and we need the fuller information of a structural formula to show this. Thus the *empirical* formula of butane, a familiar lighter fluid, is C_2H_5. This tells us nothing more than that butane is made of carbon and hydrogen in the indicated $2:5$ ratio by atoms; these numbers are the smallest *whole* numbers we could use. But one molecule of butane actually has the composition of C_4H_{10}. This is its **molecular formula,** one that gives the actual composition of a whole molecule of a molecular compound. The fourteen atoms in C_4H_{10} are organized in the butane molecule as shown in the following structural formula or structure.

$$
\begin{array}{ccccccc}
 & H & & H & & H & & H \\
 & | & & | & & | & & | \\
H- & C & - & C & - & C & - & C & -H \\
 & | & & | & & | & & | \\
 & H & & H & & H & & H
\end{array}
$$

Butane, structural formula

Sometimes the empirical and molecular formulas of a molecular compound are the same, as in water, ammonia, and methane.

EXAMPLE 5.9
Using the Octet Rule and Electron-Dot Structures to Figure Out Structural Formulas

■ Phosphine is present in the very unpleasant odor of decaying fish.

Phosphine is a very poisonous compound of phosphorus and hydrogen. What is its most likely structure?

ANALYSIS We need the electron-dot structures for the atoms first in order to see how outer octets are to be satisfied. Phosphorus, like nitrogen, is in group VA.

$$ H \cdot \qquad \cdot \ddot{\underset{\cdot}{P}} \cdot $$

Phosphorus needs three more electrons for an outer octet. To get them using H atoms, P *must* have *three* H atoms.

SOLUTION The structural formula of phosphine, therefore, must be

$$
H \colon \ddot{P} \colon H \quad \text{or} \quad H - \ddot{P} - H \\
\ \ \ddot{H} \qquad\qquad\quad | \\
\qquad\qquad\qquad\qquad H
$$

PRACTICE EXERCISE 11

Silicon and hydrogen form a simple compound called silane that has one Si atom per molecule. What is the structural formula of silane?

■ Ethylene is used to make polyethylene plastics.

In Many Molecules, Two or Three Pairs of Electrons Are Shared in Double or Triple Bonds Ethylene has the molecular formula C_2H_4. Its structure is

$$
\begin{array}{cc}
H & \qquad\quad H \\
\ \ \ \ddot{C} :: \ddot{C} & \quad\text{or} \\
H & \qquad\quad H
\end{array}
\qquad
\begin{array}{c}
H \qquad\qquad H \\
\ \diagdown \qquad \diagup \\
\ \ \ C = C \\
\ \diagup \qquad \diagdown \\
H \qquad\qquad H
\end{array}
$$

Each bond with one shared pair, like the bond from H to C, is called a **single bond.** Each carbon supplies four electrons, as its location in group IVA requires, and each carbon atom has an octet. This would not have been possible without placing two pairs of electrons between the carbon symbols. When two pairs of electrons are shared, the result is called a **double bond.** Such bonds are extremely prevalent in nature.

The nitrogen molecule has a **triple bond,** because three pairs of electrons are shared between the nitrogen nuclei. We can think of N_2 forming as follows.

$$:\!N\!\cdot \quad \cdot\!N\!: \longrightarrow \;:\!N\!:\!:\!:\!N\!: \quad \text{or} \quad \;:\!N\!\equiv\!N\!:$$

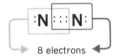

8 electrons

All six valence shell electrons between the two nitrogen nuclei count toward the octet of each N. The triple bond is not as common as single and double bonds, but acetylene, C_2H_2, the fuel for oxyacetylene torches, has one.

$$H\!:\!C\!:\!:\!:\!C\!:\!H \quad \text{or} \quad H\!-\!C\!\equiv\!C\!-\!H$$
Acetylene

Many molecules have two double bonds. Carbon dioxide, CO_2, is a common example. Its Lewis structure is

$$:\!\ddot{O}\!:\!:\!C\!:\!:\!\ddot{O}\!: \quad \text{or} \quad :\!\ddot{O}\!=\!C\!=\!\ddot{O}\!:$$
Carbon dioxide

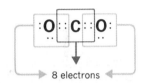

8 electrons

Note how each atom in CO_2 has an octet.

The *Covalence Number* of a Nonmetal Is the Number of Covalent Bonds Its Atoms Have in Molecules

We can summarize part of the foregoing discussion in terms of the bond-forming abilities of the nonmetal elements. Each of these elements can be assigned a number called its **covalence number** that equals the number of bonds its atoms have in molecules. The covalence number of an element thus also equals the number of additional electrons that its atoms must acquire by sharing to have a noble gas configuration. The common covalence numbers of several elements are given in Table 5.2; they apply to the atoms given as they occur in electrically neutral particles. Some elements have more than one covalence number.

We have studied bond-*forming* events, but covalent bonds can be broken if enough energy is supplied. Such reactions occur, for example, in the stratosphere where bond-breaking reactions provide us with protection from the sun's ultraviolet radiation (see Special Topic 5.1).

TABLE 5.2
Common Covalence Numbers of Nonmetals

Periodic Table Group Number				
IA	IVA	VA	VIA	VIIA
H 1	C 4	N 3	O 2	F 1
	Si 4	P 3	S 2	Cl 1
				Br 1
				I 1

OZONE IN THE STRATOSPHERE AND THE OZONE "HOLE"

Photochemical Reactions The covalent bonds in molecules can be broken if the substance is heated strongly enough, which is what starts to happen as sugar caramelizes in candy-making. The molecular fragments are generally too unstable to last, and they recombine often in new ways to form different substances. Covalent bonds are also broken when a molecule absorbs light of the proper energy. The molecular basis of the frightening news stories about the ozone "hole" concern light-induced, bond-breaking reactions that occur in the rarified regions of our stratosphere.

The *stratosphere* is an envelope of space surrounding the earth between altitudes of 10 and 50 km (6 to 31 miles) and immediately above the *troposphere,* the region of the atmosphere in which we live. Ultra-small concentrations of a number of substances are present in the stratosphere, including oxygen, nitrogen, methane, nitrogen dioxide, and ozone. Ozone is a form of oxygen in which the oxygen atoms exist as triatomic molecules, O_3. Ozone is an extremely reactive substance, and a level in air of only 1 μg/L is very dangerous to health. Smog contains some ozone (usually much less than 1 μg/L), and this is but one reason why breathing smoggy air is unhealthy. Ozone in the stratosphere, on the other hand, is one of the silent, unseen protectors of all life.

Light or electromagnetic radiation can be considered as a stream of tiny massless particles called *photons,* each photon being a packet of energy. Photons that make up what we call *ultraviolet* or UV light, which is not visible to our eyes, have more energy than photons of visible light. UV photons have enough energy to break chemical bonds.

Roughly 7% of the solar energy entering our outermost atmosphere, the outer stratosphere, is carried by UV photons. This may seem small, but it is probably safe to say that neither plant nor animal life, including human life, would be possible without the natural mechanism existing in the stratosphere for removing nearly all of the solar UV radiation. Because of the high energies in the UV kind of photons, reactions not otherwise observed occur to the trace chemicals in the stratosphere.

Traditionally, to describe a particular kind of photon, an associated wavelength, symbolized as λ (Greek lambda), is given. We won't develop this concept in detail, except to say that the energy of a photon is inversely proportional to λ. Therefore, the *smaller* the value of λ, the more energy the photon has. UV photons have values of λ ranging from 200 to 400 nm (where nm means nanometer, 10^{-9} meter). Sunlight that reaches the earth's surface at sea level includes some UV energy, principally in the region of 280 nm to 315 nm (peaking at 300 nm). Sunburn occurs upon exposure to such light. With sufficiently prolonged

or severe exposure, damage occurs to genes in the cells of the skin. This can lead to one of two kinds of skin cancer, the chief consequence in humans of exposure to UV radiation with wavelengths of 315 nm or shorter. Little skin damage is caused by wavelengths longer than 320 nm. (The visible region starts at about 380 nm.) Fortunately, nearly all incoming solar UV radiation below 320 nm is absorbed in the stratosphere, but a small amount does survive to reach us at the earth's surface.

Most of the higher energy (shorter λ) photons of the UV are removed by chemical reactions in the stratosphere. Such photons—the optimum λ is 242 nm—have sufficient energy to break oxygen molecules into oxygen atoms.

$$O_2 \xrightarrow[\text{(λ = 242 nm or lower)}]{\text{UV radiation}} 2O \qquad (1)$$

Stratospheric Ozone Cycle The removal of nearly all of the incoming UV radiation is accomplished in the stratosphere by a series of reactions called the *stratospheric ozone cycle.* Overall, the cycle converts UV radiation into heat. By "cycle" we mean a *chemical chain reaction* whereby the reactants needed for an early step are continuously generated by a later step, thus making possible further occurrences of the early step.

The ozone cycle is initiated by the generation of oxygen atoms according to Equation 1. This reaction itself absorbs some of the incoming UV energy. Generally one or both of these oxygen atoms is *electronically excited,* meaning with a valence-shell electron in a higher than normal energy state, thus making the O atom a reactive species. Ozone is made when an electronically excited oxygen atom—we'll symbolize it by O*—combines with an oxygen molecule, during a collision of the two at the surface of a neutral particle, *M. M* can be a molecule of O_2 or N_2, for example. This is the first step of the ozone cycle itself.

$$O^* + O_2 + M \longrightarrow O_3 + M + \text{heat} \qquad (2)$$

(The function of *M* is to absorb some of the energy of the collision between O and O_2 so that the new O_3 molecule has insufficient energy within itself to split apart as soon as it forms.)

The newly formed ozone is broken apart by the absorption of UV radiation, and the second step of the cycle occurs.

$$O_3 + \text{UV energy} \xrightarrow[\text{(λ = 242–320 nm)}]{} O_2 + O^* \qquad (3)$$

Notice that the O* *product* of the second reaction (3) is the necessary *reactant* for the first (2). Reaction 3, therefore, supplies what is needed for 2. When 2 occurs again, reaction 3 can take place once more. Thus 2 and 3 together constitute a *chemical chain reaction.* The cycle repeats itself sometimes hundreds of times before other events terminate a chain. If we add Equations 2 and 3, canceling identical species on opposite sides of the arrows, the net result is simply

$$UV\ energy \longrightarrow heat \qquad (4)$$

This is the net effect of the ozone cycle in the stratosphere, the conversion of dangerous UV radiation into heat. The region of the stratosphere where the ozone cycle occurs is called the *ozone layer.*

One of the many reactions that can terminate a chain of the ozone cycle is the recombination of two oxygen atoms to form molecular oxygen. When any O* made by Equation 3 is removed by such a reaction, then Equation 2 cannot generate the ozone needed for further occurrences of 3. The cycle is broken.

The overall result of the chain initiation, the cycle itself, and any natural chain terminations is a somewhat steady-state concentration of ozone in the stratosphere. This concentration varies rather widely, however, with latitude, altitude, and the month of the year, so no single figure can be cited to describe the stratospheric ozone level. At latitudes reaching 60° north or south from the equator, the ozone concentration in 1974 varied roughly between 260 and 360 Dobson units, reaching as high as 500 Dobson units. The *Dobson unit* equals 2.7×10^{16} molecules of O_3 in a column of air 1 cm^2 in area at its base (the earth's surface) and extending through the stratosphere.

Chlorofluorocarbons and the Ozone Cycle The chlorofluorocarbons (CFCs) are a family of volatile, nonflammable, chemically stable, and essentially odorless and tasteless compounds. CFC-11, for example, is $CFCl_3$, boiling at 24 °C, and CFC-12 is CCl_2F_2, boiling at -30 °C. Once marketed under the trade name Freon, they have been widely used as the fluids in air conditioners, refrigerators, and freezers; as cleaning solvents for computer parts; and as aerosol propellants. At one time, $CFCl_3$ was used in 50–60% of all aerosol cans sold.

Eventually, all of the CFCs migrate throughout the entire atmosphere, become globally distributed, and work their way into the stratosphere. In 1974, chemists M. J. Molina and F. S. Rowland (California) warned that the CFCs could reduce our protection from UV radiation by interfering with the stratospheric ozone cycle. The theory was very logical when it first appeared, but was it true? (There's a difference.) The theory has survived strenuous experimental challenges, and by early 1991 it appeared to be generally accepted. According to Molina and Rowland, the CFCs in the stratosphere absorb UV radiation, which breaks their C—Cl bonds and generates chlorine *atoms.*

$$CCl_3F \xrightarrow{\text{UV radiation}} CCl_2F + Cl$$
$$CCl_2F_2 \xrightarrow{\text{UV radiation}} CClF_2 + Cl$$

Atomic chlorine is able to destroy ozone and so disrupt the ozone cycle by the following chain reaction. (There are other Cl-based cycles besides this one.)

$$Cl + O_3 \longrightarrow ClO + O_2 \qquad (5)$$
$$\underline{ClO + O \longrightarrow Cl + O_2} \qquad (6)$$
$$Net: \quad O_3 + O \longrightarrow 2O_2$$

This is a *chain reaction,* so the breakup of only one CFC molecule can initiate the destruction of thousands of ozone molecules.

Because of their inertness, CFCs endure for several decades. Their influence on the stratospheric ozone level will continue throughout the next century, therefore, even if no more are released. The threat is serious, so in 1987, 36 nations signed a treaty called the Montreal Protocol that called for cutting the worldwide CFC production in half by 1998. Two years later, more deeply alarmed by additional research, 80 nations agreed to ban CFC production by the year 2000. Other processes can also interfere with the ozone layer, but the CFCs constitute a major threat that humans can control.

The Antarctic Ozone "Hole" The British Antarctic Survey headquartered at Halley Bay on the Antarctic continent has been measuring the ozone density over the Antarctic since the 1960s. In 1985 it reported that a puzzling and disturbing trend became apparent in the data from 1970 on. A decline in the ozone concentration occurred during each Antarctic spring. (Spring in the southern hemisphere occurs during the fall of the northern hemisphere.) The ozone level reaches its lowest levels in October each year. In 1984, the loss by mid-October was 30%; in 1989, the reduction was 70%. The result is that a huge, continent-broad column of the atmosphere over the Antarctic becomes significantly less able to destroy UV radiation during this precipitous decline in ozone level. The ozone-poor column of the atmosphere over the south pole is called the *Antarctic ozone hole.* A similar "hole" occurs seasonally over the north pole, but (so far) it has not been as severe.

Slowly, as spring turns to summer between November and March in the southern hemisphere, a significant recovery in the ozone level in the Antarctic ozone hole occurs, but another decline takes place the following spring. Global, atmospheric air-mixing processes bring air richer in ozone into the Antarctic stratosphere so that, overall, the stratospheric ozone level all over the globe has been steadily declining at a rate of 2.3% per decade.

Why over the Antarctic? Why in the Antarctic Spring? A circulating wind pattern called the *Antarctic vortex* develops over the south pole and the Antarctic continent during the Antarctic winter. See Figure 1. It is generated by the rotation of the earth and the sharply uneven heating of the atmosphere. Wind velocities as high as 80 m s^{-1} (180 mph) are seen in the jet stream encircling

(Continued)

the south pole, but very little wind is in the center of the vortex. These winds help to confine the chemical reactions that destroy ozone largely to this vortex. The circulation pattern thus answers the question, "Why over the Antarctic (or over the Arctic)?"

Following summer and fall, the Antarctic winter sets in again, and it is a period of *total* darkness during June, July, and August. The stratospheric temperature drops below 195 K (-78 °C), low enough for vast yet thin clouds to form from traces of water vapor and other substances. These polar stratospheric clouds (PSCs) include microcrystals of water containing sulfur oxides (SO_2 and SO_3) injected into the stratosphere from volcanos (or migrating as *anthrogenic* — human produced — air pollutants). Another type of cloud consists of crystals of nitric acid trihydrate, $HNO_3 \cdot 3H_2O$, formed from water and nitrogen oxides, particularly NO_2 (of both natural and anthrogenic origin). (Hydrates are further discussed on page 186.) NO_2 reacts with water to give HNO_3 (nitric acid) and HNO_2 (nitrous acid).

$$2NO_2 + H_2O \longrightarrow HNO_3 + HNO_2$$

During the Antarctic winter, the polar stratospheric clouds collect and store chlorine, Cl_2, which forms by the reaction of hydrogen chloride with chlorine nitrate (a rare, molecular compound made in the stratosphere).

$$HCl + ClNO_3 \longrightarrow Cl_2 + HNO_3$$

The $ClNO_3$ is made by the reaction of ClO, from Reaction 5, with NO_2.

$$ClO + NO_2 \longrightarrow ClNO_3$$

The hydrogen chloride forms from the reaction of *atomic* chlorine (taken from the ozone-destroying cycle, Reactions 5 and 6) with CH_4 to give CH_3 (methyl) and HCl.

$$CH_4 + Cl \longrightarrow HCl + CH_3$$

(Methane is an air pollutant produced both naturally and by pipeline leaks. Methyl, CH_3, is another very unstable species that can exist only in a low-pressure situation, and then not for long.)

The chlorine and nitric acid molecules stay on the cloud crystals until the Antarctic winter is over and the sun reappears. Now the Cl_2 is released from this reservoir and is broken apart by UV radiation.

$$Cl_2 + UV \longrightarrow 2Cl$$

Just as sunlight returns, therefore, and the Antarctic spring commences, a large supply of chlorine atoms is released. They quickly destroy ozone by Equations 5–6, and the Antarctic ozone hole once again appears. Thus we see why the severity of the ozone depletion — ozone hole — peaks in the Antarctic *spring*. In October of 1993, the ozone level in the Antarctic ozone hole dropped to an all-time low, 90 Dobson units. (For that time of the year, the normal value is 225 Dobson units.)

Between November and March, the polar vortex breaks up and air richer in ozone migrates into the Antarctic stratosphere from mid-latitudes. The result is an overall decrease in the stratospheric ozone level over the southern hemisphere. Similar events are occurring over the north pole. The stratospheric ozone level is believed to have declined by about 3% during the last three decades at the latitude of New York City and by about 5% at the latitude of Buenos Aires, Argentina. The U.S. Environmental Protection Agency issued a prediction in 1991 that the increased human exposure to UV radiation would cause an *additional* 200,000 deaths from skin cancer in the United States over the next 50 years.

FIGURE 1

The Antarctic polar vortex and the ozone hole on October 7, 1989. The wind speeds are indicated by color code in the upper part of the figure (the cylinder missing a wedge). The darkest blue indicates essentially zero wind velocity and the red areas correspond the velocities up to 80 m/s. Beneath the cylinder, the surface plot is color-coded for total ozone where the darkest blue colors are around the south pole and signify values below 200 Dobson units. Latitudes are drawn in at 30° and 60° south. Extending to the lower right is the Andes chain of mountains of South America. Extending to the lower left is the tip of Africa. (Used by permission from M. R Schoeberl and D. L. Hartmann, *Science,* January 4, 1991, page 47.)

5.5

STRUCTURES OF POLYATOMIC IONS

Many important ions are electrically charged clusters of atoms held together by covalent bonds.

A **polyatomic** ion is a cluster of atoms held together by covalent bonds, but which has a net electrical charge. The ammonium ion, NH_4^+, is a particularly important example, because it and substances like it have functions at the molecular level of life. To understand this ion, we must expand our understanding of the covalent bond.

A Shared Electron Pair Can Originate from One Atom   The electron-dot structure of ammonia, NH_3, is

$$\begin{array}{cc} H\!:\!\overset{\displaystyle ..}{\underset{\displaystyle \overset{..}{H}}{N}}\!:\!H & \quad or \quad H\!-\!\overset{\displaystyle ..}{\underset{\displaystyle |}{N}}\!-\!H \\ & \qquad\qquad H \end{array}$$

Now notice that the nitrogen atom in this molecule has one unshared pair in the valence shell. Both electrons of this pair hold a fourth nucleus of hydrogen in the ammonium ion. The nucleus of a hydrogen atom, of course, is just a bare proton, and we can symbolize it here as H^+, a *hydrogen ion,* because it's a hydrogen atom minus its $1s$ electron. Then we can visualize the formation of NH_4^+ from NH_3 and H^+ as follows.

$$H\!-\!\underset{\displaystyle \underset{\text{H}}{|}}{\overset{\displaystyle ..}{N}}\!-\!H + H^+ \longrightarrow \left[H\!-\!\underset{\displaystyle \underset{\text{H}}{|}}{\overset{\displaystyle ..}{N}}\!-\!H \right]^+ \quad or \quad \left[H\!-\!\underset{\displaystyle \underset{\text{H}}{|}}{\overset{\displaystyle \overset{\text{H}}{|}}{N}}\!-\!H \right]^+ \quad or \quad NH_4^+$$

Ammonia   Hydrogen ion   Ammonium ion

The new bond to the fourth hydrogen is a covalent bond like the other bonds, because it's an electron-pair bond. Sometimes it's useful, however, to have a special name for a covalent bond for which *both* shared electrons come from one atom. When we want to indicate this, we call the bond a **coordinate covalent bond.**

The ammonium ion bears a net charge of $1+$ because we have added the $1+$ charge of H^+ to a neutral particle (zero charge), NH_3. The resulting cluster of atoms, all held together by covalent bonds, is therefore an example of a *polyatomic ion.* It can exist in solution and move around as one unit, or it can be present as a unit in a solid, ionic compound, like ammonium chloride, NH_4Cl.

In the structure of the ammonium ion, the nitrogen still has an octet, only now all four electron pairs of the octet are involved in covalent bonds. All four of these bonds are equivalent. The molecule cannot remember which bond formed in which way, so you can see that a coordinate covalent bond and a covalent bond are not different *once they have formed.*

Before considering other polyatomic ions we want to say more about two families of compounds, *acids* and *bases,* that are not just sources of polyatomic ions, but are also involved in supplying or in combining with the hydrogen ion. We do this to facilitate some of your laboratory work, but we have much more to learn about acids and bases later. (In fact, the acid–base status of body fluids is one of the most important topics that we study.)

Compounds That Furnish Hydrogen Ions Are Called Acids   If you're wondering where we can obtain a hydrogen ion, H^+, to make NH_4^+, it is supplied by any of a large family of **acids,** briefly introduced on page 84. For example, hydrochloric acid is actually a $1:1$ mixture of hydrogen ions and chloride ions in water. Sulfuric acid, H_2SO_4, in water furnishes hydrogen ions and both the hydrogen sulfate ion, HSO_4^-, and the sulfate ion, SO_4^{2-}. Nitric acid provides, besides the hydrogen ion, the nitrate ion, NO_3^-.

■ Even dilute solutions of nitric, sulfuric, and hydrochloric acids will quickly eat holes in blue jeans.

Dilute solutions of these acids all have very tart tastes (but don't experiment with them unless your instructor shows you what to do — some acids are poisons and can harm teeth!). The tartness of lemon juice is caused by citric acid, and the tartness of vinegar is caused by acetic acid. Another property common to acids is that they react with most metals such as iron. The reason why acids have properties in common is that all acids are sources of hydrogen ions.

Bases are compounds or ions that combine with hydrogen ions. Ammonia, which can react with and bind hydrogen ions, is an example of a base, but it's just one of several. If we represent hydrochloric acid by its separated ions, H^+ and Cl^-, then its reaction with ammonia, a base, can be written as follows:

■ Ammonia, a gas under ordinary conditions, is available in the lab as solutions in water called *aqueous ammonia*.

$$NH_3 \quad + [H^+ + Cl^-] \longrightarrow [NH_4^+ + Cl^-]$$

Ammonia   Hydrochloric acid   Ammonium chloride

Generally, when an acid and a base react, one product is a *salt*. This is a general term; a **salt** is an ionic compound formed from any cation except H^+ and any anion except OH^- or O^{2-}. Sodium chloride is thus just one of thousands of known salts. Ammonium chloride (above) is another salt.

Parentheses Are Sometimes Needed in Chemical Formulas Involving Polyatomic Ions Table 5.3 lists several of the important polyatomic ions. *Their names and formulas should be learned now.* In many of the reactions of their compounds, the polyatomic ions stay together as intact units. So they are shown as units in formulas, as in the following examples:

NH_4Cl	ammonium chloride, an ingredient in smelling salts
$NaOH$	sodium hydroxide, a raw material for making soap
NH_4NO_3	ammonium nitrate, a fertilizer
$NaNO_2$	sodium nitrite, a preservative in bacon and bologna
Na_3PO_4	sodium phosphate, a powerful cleaning agent
Na_2CO_3	sodium carbonate, washing soda
$NaHCO_3$	sodium bicarbonate, baking soda (not baking *powder*)

All are examples of salts, except NaOH, because its anion is OH^-.

TABLE 5.3
Important Polyatomic Ions

Name	Formula	Name	Formula
Ammonium ion	NH_4^+	Dihydrogen phosphate ion	$H_2PO_4^-$
Hydronium ion[a]	H_3O^+	Nitrate ion	NO_3^-
Hydroxide ion	OH^-	Nitrite ion	NO_2^-
Acetate ion	$C_2H_3O_2^-$	Hydrogen sulfite ion[d]	HSO_3^-
Carbonate ion	CO_3^{2-}	Sulfite ion	SO_3^{2-}
Bicarbonate ion[b]	HCO_3^-	Cyanide ion	CN^-
Sulfate ion	SO_4^{2-}	Permanganate ion	MnO_4^-
Hydrogen sulfate ion[c]	HSO_4^-	Chromate ion	CrO_4^{2-}
Phosphate ion	PO_4^{3-}	Dichromate ion	$Cr_2O_7^{2-}$
Monohydrogen phosphate ion	HPO_4^{2-}		

[a] This ion is known only in a water solution.

[b] Formal name: hydrogen carbonate ion.

[c] Common name: bisulfate ion.

[d] Common name: bisulfite ion.

Whenever a formula has more than one polyatomic ion, we place parentheses about it and put a subscript *outside* the closing parenthesis. One example is ammonium sulfate.

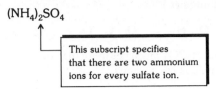

$(NH_4)_2SO_4$

This subscript specifies that there are two ammonium ions for every sulfate ion.

Other compounds whose formulas require parentheses are the following (and all are salts).

$Ca(NO_3)_2$	calcium nitrate
$Mg_3(PO_4)_2$	magnesium phosphate
$Al_2(SO_4)_3$	aluminum sulfate

Appendix C contains a summary of the rules for naming the kinds of compounds we have been using thus far, but notice from the above examples that nothing new has been added to what we have already covered. We make names of ionic compounds from the names of their ions (omitting *ion*) in the usual way.

PRACTICE EXERCISE 12

Spend some time learning the names and formulas of the polyatomic ions that your instructor has assigned, and then drill yourself by writing the formulas of the following compounds.

(a) potassium bicarbonate

(b) sodium monohydrogen phosphate

(c) ammonium phosphate

PRACTICE EXERCISE 13

Write the name of each of the following compounds.

(a) $NaCN$ **(b)** KNO_3 **(c)** $NaHSO_3$ **(d)** $(NH_4)_2CO_3$ **(e)** $NaC_2H_3O_2$

5.6
WRITING LEWIS STRUCTURES OF MOLECULES AND POLYATOMIC IONS

As much as possible, Lewis structures are written so that all atoms have valence shell octets.

Learning how to write Lewis structures is one way to gain a better sense of where electrons are in molecules and polyatomic ions. This is useful to our study, because chemical reactions rearrange valence electrons as bonds break and reform.

Chemists have reduced the writing of a Lewis structure to a few simple steps. The first step is to write a *skeletal structure,* one that roughly groups the atoms of the molecular formula as they are arranged in the structure, but without any valence electrons. This is often the trickiest step, and sometimes you have to make an educated guess. For example, sometimes the molecular formula suggests what might be the *central atom,* the atom around which the others are grouped. Hydrogen is never the central atom because once a covalent bond extends from H, no more bonds can extend from it. (The valence shell of H is principal energy level 1 and this level can hold only two electrons, those of the one covalent bond from H.) Thus in H_2O,

■ NO_2 is the air pollutant that gives the red-brown color to smog.

oxygen must be the central atom. Otherwise, oxygen isn't often the central atom. Here are the correct skeletal structures for H_2O, CO_2, and NO_2. (For these, the central atom is the one that has the highest covalence number.)

H O H   O C O   O N O

Now consider carbonic acid, H_2CO_3 which has more than three atoms. Which of the following is the correct grouping simply is not obvious.

■ Carbonic acid forms by a reaction of CO_2 with water. It isn't a stable acid, and it cannot be isolated and made pure.

<table>
<tr><td>O
H O C O H or</td><td>H O C O O H or</td><td>H
O C O O H</td></tr>
<tr><td>Correct</td><td>Incorrect</td><td>Incorrect</td></tr>
</table>

What we need is a guideline, a rule of thumb that can help us out in most situations like this and lead us to the identity of the central atom. The key word in carbonic acid is *acid*, plus the fact that its formula, H_2CO_3, contains both oxygen and hydrogen besides another nonmetal. Such acids are classified as *oxoacids*, and in all those of our study, H is always joined covalently to O and O is never the central atom in oxoacids. At least, we will assume that the atom other than H and O in an oxoacid is the central atom. The central atom in carbonic acid is thus C; around it can be grouped all of the O atoms of H_2CO_3.

EXAMPLE 5.10
Writing Skeletal Structures

Write the skeletal structure of nitric acid, HNO_3.

ANALYSIS This is an acid with both H and O, so we know that one oxygen atom holds the hydrogen atom and that the central atom is N. We can now write the skeletal structure by grouping the O atoms around the N and placing H near one of the O atoms. (It does not matter which one.)

SOLUTION
                    O
         H O N O

PRACTICE EXERCISE 14

Write the skeletal structures of sulfuric acid, H_2SO_4, probably the most important acid used in industry, and phosphoric acid, H_3PO_4.

The second step in writing a Lewis structure is to add up all of the valence electrons. The total equals the number of dots that must be used. Use the locations of the elements in the periodic table to find the number of their valence electrons.

EXAMPLE 5.11
Adding Up Valence Electrons

How many dots must appear in the Lewis structures of H_2CO_3, HNO_3, and H_2SO_4?

ANALYSIS We use the group number of the element in the periodic table to determine quickly how many electrons one of its atoms has in its valence shell. Then we add up the valence shell electrons over all atoms in the formula.

SOLUTION

H_2CO_3	H has 1 valence electron, so for 2 H	$2 \times 1 = 2$
	C (group IVA) has 4, so for 1 C	$1 \times 4 = 4$
	O (group VIA) has 6, so for 3 O	$3 \times 6 = \underline{18}$
	Total valence electrons	24
HNO_3	H has 1 valence electron, so for 1 H	$1 \times 1 = 1$
	N (group VA) has 5, so for 1 N	$1 \times 5 = 5$
	O (group VIA) has 6 each, so for 3 O	$3 \times 6 = \underline{18}$
	Total valence electrons	24
H_2SO_4	H has 1 electron, so for 2 H	$2 \times 1 = 2$
	S (group VIA) has 6, so for 1 S	$1 \times 6 = 6$
	O (group VIA) has 6, so for 4 O	$4 \times 6 = \underline{24}$
	Total valence electrons	32

PRACTICE EXERCISE 15

How many valence electrons are in H_3PO_4, phosphoric acid?

In the third step, we start to place the electrons *by pairs* into the skeletal structure. Do this in the following order. Put *one* pair in each bond. Then use pairs to complete the octets of all atoms attached to the central atom. Finally, if necessary, complete the octet of the central atom using as many pairs as needed. (Remember to limit the electrons by any H atom to two.)

EXAMPLE 5.12
Placing Electron Pairs into Skeletal Lewis Structures

Fill in the electron pairs for the skeletal structure of sulfuric acid.

ANALYSIS In Practice Exercise 14, the skeletal structure was found to be the following.

$$
\begin{array}{c}
\text{O} \\
\text{H O S O H} \\
\text{O}
\end{array}
$$

(You might have placed the H atoms by different oxygens. This would not matter.) In Example 5.11 we calculated that there are 32 valence electrons — 32 dots — to distribute.

SOLUTION First we put electron pairs into each bond.

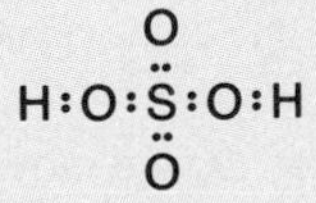

$$
\begin{array}{c}
\text{O} \\
\ddot{} \\
\text{H:O:}\overset{\displaystyle ..}{\text{S}}\text{:O:H} \\
\ddot{} \\
\text{O}
\end{array}
$$

This used up 12 electrons. Now we finish the octets for the oxygen atoms.

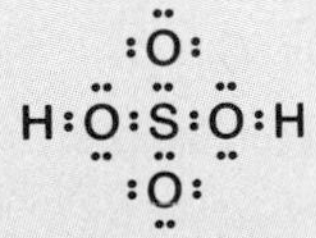

$$
\begin{array}{c}
:\ddot{\text{O}}: \\
\text{H:}\ddot{\text{O}}\text{:}\ddot{\text{S}}\text{:}\ddot{\text{O}}\text{:H} \\
:\ddot{\text{O}}:
\end{array}
$$

■ Lewis structures for the sulfuric acid and phosphoric acid systems (and their ions) will be modified in Chapter 8 (see page 211).

This uses up 20 more electrons. We have used the 32 electrons we were required to place, 12 + 20. We see that the central atom has an octet. So we have answered the problem, but we can still change this structure to one with dash bonds and write the Lewis structure of sulfuric acid as follows.

$$H-\overset{\cdot\cdot}{\underset{\cdot\cdot}{O}}-\overset{:\overset{\cdot\cdot}{O}:}{\underset{:\overset{\cdot\cdot}{O}:}{S}}-\overset{\cdot\cdot}{\underset{\cdot\cdot}{O}}-H$$

PRACTICE EXERCISE 16

One of the oxoacids of chlorine is chloric acid, $HClO_3$. Chlorine is its central atom. Write its skeletal structure, calculate the total number of valence electrons, fill them in (remembering to complete the octet of the central atom), and display the Lewis structure using dash bonds.

In many applications of the steps just given, you run out of electrons before you complete all of the octets. This is a sign that *double or triple bonds have to be created*. (Beryllium and boron are exceptions, as will be discussed on page 131.)

EXAMPLE 5.13
Creating Double Bonds in Writing a Lewis Structure

Write the Lewis structure of carbon dioxide, CO_2.

ANALYSIS   Its correct skeleton was given earlier,

$$O \quad C \quad O$$

Its total number of valence electrons is 16. (Two O atoms contribute $2 \times 6 = 12$ electrons, and the C contributes 4 electrons for a total of 16.) We proceed to place these 16 electrons around the atomic symbols of the skeleton structure, and check to see that each atom has an octet. If it does, we have solved the problem; otherwise, we must consider an additional procedure.

SOLUTION   Filling in the electrons according to Steps 2 and 3 gives us the following.

$$:\overset{\cdot\cdot}{O}:C:\overset{\cdot\cdot}{O}:$$

We have now used up all 16 electrons, but the central atom does not have an octet. There are only 4 electrons around C, and we've run out of valence electrons.

The procedure now is to move a *nonbonding* pair of electrons on an adjacent oxygen atom into an existing carbon–oxygen bond.

$$:\overset{\cdot\cdot}{O}:C:\overset{\cdot\cdot}{O}: \longrightarrow :O::C:\overset{\cdot\cdot}{O}:$$

But carbon still doesn't have an octet. So we move another nonbonding pair, taking it from the other oxygen, and the result is two double bonds in CO_2.

$$:\overset{\cdot\cdot}{O}::C:\overset{\cdot\cdot}{O}: \longrightarrow :O::C::\overset{\cdot\cdot}{O}: \quad \text{or} \quad :\overset{\cdot\cdot}{O}=C=\overset{\cdot\cdot}{O}:$$

Carbon dioxide

Here O is the "adjacent atom." These shifts of electron pairs cannot be done when a halogen atom is the adjacent atom. Halogen atoms (F, Cl, Br, or I) generally do not engage in double or triple bonds. S and N, however, can. ∎

PRACTICE EXERCISE 17

Write the Lewis structure for sulfur trioxide, SO_3, an air pollutant.

Lewis Structures of Polyatomic Ions Also Show Noble Gas Configurations We need only a small modification of our preceding steps to construct Lewis structures of polyatomic ions. These modifications concern the total number of valence shell electrons that are to be inserted into a skeletal structure.

1. For each negative charge on the ion, we have to add an electron to the count. A charge of 1− means adding 1 more electron. A charge of 2− means adding 2 more electrons, and so forth.

2. For each positive charge on an ion, we subtract an electron.

Let's see how this works with the sulfate ion, SO_4^{2-}.

EXAMPLE 5.14
Writing the Lewis Structure of a Polyatomic Negative Ion

What is the Lewis structure of SO_4^{2-}?

ANALYSIS   We first have to total the valence shell electrons and then draw a skeletal structure, using S as the central atom. Next we insert electron pairs into the covalent bonds, place remaining electrons so as to provide for complete octets, and then see if double bonds are needed.

SOLUTION   First, we total the valence shell electrons.

For S (group VIA), 6 electrons	$1 \times 6 = 6$
For each O (group VIA) 6 electrons	$4 \times 6 = 24$
For each negative charge, add 1 electron	$2 \times 1 = \underline{2}$
Total valence shell electrons	32

Next we write a reasonable skeletal structure. We use S as the central atom and arrange the O atoms around it symmetrically.

$$\begin{array}{c} \text{O} \\ \text{O S O} \\ \text{O} \end{array}$$

Next, we insert electron pairs into the covalent bonds. This will use up 8 electrons out of the 32.

$$\begin{array}{c} \text{O} \\ \text{O:S:O} \\ \text{O} \end{array}$$

Now we place electron pairs around the oxygens to give each an octet. This will use up all the remaining 24, so this gives the final structure of the sulfate ion. We don't have to create double bonds. To show that the ion carries a charge of 2−, we'll place brackets about it and place the 2− charge as a right superscript.

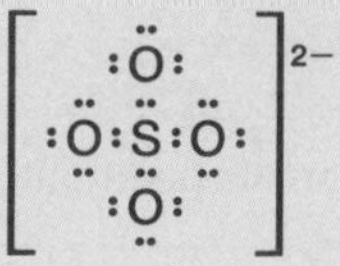

$$\left[\begin{array}{c} :\ddot{O}: \\ :\ddot{O}:S:\ddot{O}: \\ :\ddot{O}: \end{array}\right]^{2-} \qquad \text{The sulfate ion, } SO_4^{2-}$$

PRACTICE EXERCISE 18

Write the Lewis structures of the following ions.

(a) H_3O^+ (b) OH^-

PRACTICE EXERCISE 19

Write the Lewis structures of the following ions. Each will require that you form a double bond. Remember, it won't matter which electron pair is moved into position as the second bond of the double bond so long as proper octets are realized.

(a) CO_3^{2-} (b) NO_3^-

Sometimes the Octet Rule Fails The octet rule is not really a law of nature because it fails sometimes. We will just cite, by structures, some examples of failures without making anything big out of them. We will seldom encounter failures of the octet rule when only the row 1 and the row 2 nonmetal elements of the periodic table are involved. This means that H, C, N, and O will never give us trouble on this score. Their valence shells can never hold more than eight electrons (just two for hydrogen, of course).

Some failures of the octet rule involve more than eight valence shell electrons in the structure. These can occur when a "central" nonmetal atom is from period 3 (or higher) of the periodic table. These atoms have their valence electrons at principal energy level 3 (or higher), *which can hold more than eight electrons.* Phosphorus pentachloride, PCl_5, and sulfur hexafluoride, SF_6, are examples of octet rule failures of this type.

■ SF_6 is a colorless, odorless, tasteless, nonflammable, nontoxic, and unusually stable gas used to insulate high-voltage generators and switches.

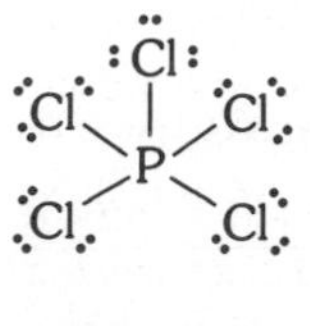

10 electrons around P and 8 electrons around each Cl

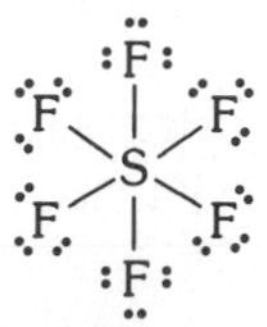

12 electrons around S and 8 electrons around each F

A few failures of the octet rule involve fewer than eight electrons around a central atom. The classic examples are beryllium chloride, $BeCl_2$, and boron trichloride, BCl_3. The Lewis structure of Be is simply ·Be·, because Be is in group IIA. We can represent the formation of $BeCl_2$ as follows.

$$:\overset{..}{\underset{..}{Cl}}\cdot\underset{\rightarrow}{+}\cdot Be\cdot\underset{\rightarrow}{+}\cdot\overset{..}{\underset{..}{Cl}}:\longrightarrow :\overset{..}{\underset{..}{Cl}}:Be:\overset{..}{\underset{..}{Cl}}:$$

BeCl$_2$; four
electrons
around Be

Only four valence electrons are around the Be atom in this structure. Recall that we cannot shift electron pairs from Cl to make multiple bonds and so give Be an octet in BeCl$_2$. Halogen atoms do not participate in multiple bonds.

The formation of BCl$_3$ is represented as follows.

$$\cdot B\cdot + 3\cdot\overset{..}{\underset{..}{Cl}}:\longrightarrow :\overset{..}{\underset{..}{Cl}}:\overset{:\overset{..}{Cl}:}{B}:\overset{..}{\underset{..}{Cl}}:$$

BCl$_3$; B has
six outside-
level electrons.

■ Boron is the first member of group IIIA

Boron has only six electrons in its valence shell in BCl$_3$.

5.7

SHAPES OF MOLECULES

As valence-shell electron pairs act to stay as far apart as possible, they set the overall shapes of molecules.

We continue here what we began on page 107 with monatomic ions, a study of the shapes of particles and their significance. The water molecule, for example, is not a linear molecule, with all its atoms lined up in a straight line. It's a bent molecule, and the angle is known, 104.5°.

■ *Linear* means in a straight line.

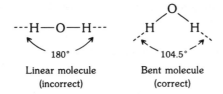

Linear molecule
(incorrect)

Bent molecule
(correct)

The angle formed by two bonds from the same atom is called the **bond angle.**

The VSEPR Theory Is a Simple Explanation for Bond Angles The easiest way to explain the bond angle in water is by a theory with a long but descriptive name — the **valence-shell electron-pair repulsion theory,** or the **VSEPR theory,** for short. The *valence shell* is where covalent bonds originate. *Electron pair* refers to the electrons of the valence shell. *Repulsion* refers to the effect that an electron cloud of one electron pair has on the electron cloud of another in the same valence shell.

VSEPR theory says that the shape of a molecule is largely determined by the efforts of the valence shell electron clouds to stay out of each other's way as much as they can. Imagine that each electron cloud is a balloon with an imaginary line or axis running from where it's tied off to the opposite surface. (See the figure in the margin.) Now imagine how four identical balloons *must* arrange themselves most comfortably (and so most stably) if we tie them all close together at one point (see Figure 5.4). The balloons are least crowded when their axes make angles of 109.5° with each other. This array is called **tetrahedral** because the four axes point to the corners of a regular tetrahedron.

Figure 5.5 shows how VSEPR theory explains the bond angle in the water molecule. The valence shell of O in H$_2$O has four electron pairs. Two pairs are unshared, and the other two carry and hold one hydrogen nucleus apiece. These hydrogen nuclei shrink the associated

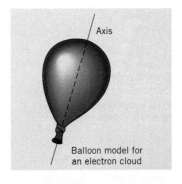

Axis

Balloon model for
an electron cloud

FIGURE 5.4
When four identical balloons are tied off at at common point, their axes will point to the corners of a tetrahedron.

FIGURE 5.5

Shapes of molecules that have four pairs of electrons around a central atom.

Number of Electron Pairs in Bonds	Number of Lone Pairs	Structure	
4	0		Tetrahedral (Example, CH_4) All bond angles are 109.5°.
3	1		Trigonal pyramidal (Pyramid-shaped) (Example, NH_3)
2	2		Nonlinear, bent (Example, H_2O)

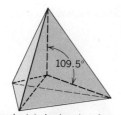

A regular tetrahedron is a four-sided space bounded by identical equilateral triangles. Any two lines from the corners to the midpoint of the space make an angle of 109.5°.

clouds somewhat, so the clouds repel each other less. Consequently, the actual bond angle in H_2O is not quite as large as the true tetrahedral angle of 109.5°, but it is very close.

The central atom is N in NH_3. Once again the VSEPR theory predicts a tetrahedral bond angle of 109.5°, because there are four valence-shell pairs of electrons. The actual bond angle in ammonia is 107.3°, also quite close (see Fig. 5.5).

In methane, CH_4, C is the central atom. VSEPR theory again predicts bond angles of 109.5° for each of the H—C—H bonds (see Fig. 5.5), and the bond angles in methane are all 109.5°.

VSEPR Theory Is Unusually Successful for Systems Like $BeCl_2$, BCl_3, PCl_5, and SF_6, Too   What if there are fewer than four electron pairs in the valence shell, as in $BeCl_2$ and BCl_3? What if there are more than four pairs as around P in PCl_5 and around S in SF_6?

Figure 5.6 illustrates the shapes of molecules with anywhere from two to six valence-shell electron pairs. Each shape is exactly what VSEPR theory predicts. When only two electron pairs are in the valence shell — think of just two balloons tied together — the axes of their clouds must point oppositely to give the most comfortable room, so the bond angle in $BeCl_2$ is 180°. When there are just three electron clouds, their axes will be in a plane and point to the corners of a regular triangle, so the bond angle in BCl_3 is 120°. The bond angles and shapes of the rest of the systems in Figure 5.6 follow from the same arguments.

5.8

POLAR MOLECULES

Even electrically neutral molecules can attract each other if they are polar.
If molecules are neutral, how can they adhere together? Sugar molecules, for example, stack together naturally to make beautiful crystals that are not easy to melt. What holds sugar molecules together in such crystals? We'll introduce the answer here.

Number of Electron Pairs	Shape	Example

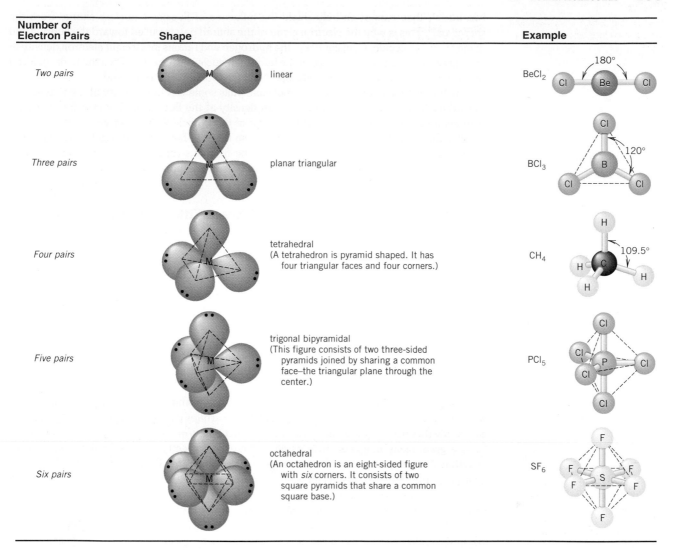

Two pairs	linear	BeCl₂ ... 180° Cl—Be—Cl
Three pairs	planar triangular	BCl₃ ... 120°
Four pairs	tetrahedral (A tetrahedron is pyramid shaped. It has four triangular faces and four corners.)	CH₄ ... 109.5°
Five pairs	trigonal bipyramidal (This figure consists of two three-sided pyramids joined by sharing a common face—the triangular plane through the center.)	PCl₅
Six pairs	octahedral (An octahedron is an eight-sided figure with *six* corners. It consists of two square pyramids that share a common square base.)	SF₆

Shared Pairs between Unlike Atoms Are Usually Not Equally Shared

When atoms have *unequal* nuclear charges, the electron cloud of the shared pair usually is drawn toward the nucleus with the larger charge. This is particularly true when the bond holds atoms of the *same period* in the periodic table. Such atoms have the same number and arrangement of core electrons, so atomic nuclei from the same period are shielded in identical ways by their core electrons. Hence, the stronger the nuclear charge becomes as we move from left to right across a period, the greater will be the ability of the nucleus to attract a valence-shell electron cloud. Thus the nucleus with the larger positive charge pulls some electron density of the shared electron cloud away from the other nucleus, *somewhat exposing its positive charge influence.* The positive charge of the nucleus of *lower* charge is therefore not entirely canceled *where this nucleus is in the molecule;* its associated electron cloud is too thin in negative charge. In other words, where the electron cloud is too thin, a fraction of a nuclear positive charge is still exerting whatever attracting or repelling influences any charge can exert. A fractional charge is called a *partial charge,* and we use the Greek lowercase letter delta, δ, to stand for *partial.* Thus a partial positive charge has the symbol $\delta+$.

The hydrogen fluoride molecule, for example, has a $\delta+$ charge at its hydrogen end. The hydrogen nucleus has only a charge of $1+$, but the fluorine nucleus has a charge of $9+$, nine times as great. Moreover, because fluorine's other electrons are not concentrated between the

FIGURE 5.6

Molecular shapes to be expected for different numbers of pairs of valence shell electrons. (From J. E. Brady and J. R. Holum, *Chemistry. The Study of Matter and Its Changes,* 1993, John Wiley and Sons, New York. Used by permission.)

two nuclei, they cannot entirely shield the influence of fluorine's nuclear charge from the shared pair. This is why the electron cloud of the shared pair is pulled toward the fluorine end of the H—F molecule. Consequently, the hydrogen end retains insufficient electron density to neutralize fully the positive charge of the hydrogen nucleus *where it is.* Thus the hydrogen end of the H—F molecule has a $\delta+$ charge. Now let's look at the other end.

The fluorine nucleus of the H—F molecule pulls some electron density of the shared pair toward itself. This causes the total electron density at the fluorine end of H—F to be more than enough to neutralize the positive charge of the F nucleus. In other words, the electron cloud is thicker than necessary at F in H—F, so this end has a partial negative charge, $\delta-$.

We can't say precisely what the sizes of the fractions represented by $\delta+$ and $\delta-$ are. However, the molecule as a whole is electrically neutral, so the algebraic sum of $\delta+$ and $\delta-$ must be zero.

Polar Bonds Have Opposite Partial Charges at Either End   When a covalent bond has a $\delta+$ at one end and a $\delta-$ at the other, it is called a **polar bond.** We can symbolize the electrical polarity of the bond in hydrogen fluoride in either of two ways, as seen in structures **1** and **2**.

$$\overset{\delta+ \quad \delta-}{\text{H—F}} \qquad \overset{\longleftrightarrow}{\text{H—F}}$$

$$\mathbf{1} \qquad\qquad \mathbf{2}$$

In **2**, the arrow points toward the end of the bond that is richer in electron density. At the other end, there is a hint of the positive character by the merger of the arrow with a plus sign. Because there are two partial, opposite charges in H—F, this molecule is sometimes said to have an **electrical dipole.**

A magnet with two *magnetic* poles — labeled north and south — is a good analogy for describing the consequences of *electrical* polarity. Perhaps you have played with toy magnets and know that two magnets can stick to each other *if they are lined up properly.* In fact, if you have a great many magnets, and line all of them up correctly, you can make them all cling together. You just have to make sure that poles of opposite kind are nearest neighbors and that poles that are alike are as far apart as possible.

Molecules that are electrically polar can stick to each other just like magnets. Given the freedom to move, polar molecules will line up automatically the way magnets can (see Fig. 5.7). This is how neutral molecules are able to adhere to each other. How tightly they stick depends on the sizes of the partial charges and on the shapes of the molecules. In some large molecules with complex shapes, partial charges can be shielded by neighboring sections of the molecules. This interferes with the ability of $\delta+$ and $\delta-$ sites on different molecules to get close to each other and so weakens forces of attraction between the molecules.

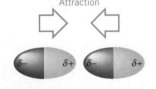

Attraction

Two polar molecules can attract each other.

FIGURE 5.7
Polar molecules attract each other in a crystal of a molecular substance.

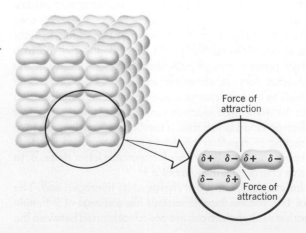

Force of attraction

Force of attraction

When Bonded Atoms Have Different Electronegativities, the Bond Is Polar The relative ability of an element's atoms to draw away electron density of a covalent bond is called the **electronegativity** of the element.

Fluorine has the highest electronegativity of all of the elements, because its atoms have the highest nuclear positive charge while being shielded by only level 1 and level 2 electrons. Oxygen, which stands just to the left of fluorine in the periodic table, has the next highest electronegativity. Its atoms have one less charge than fluorine atoms, and also are shielded by only level 1 and level 2 electrons. The element with the third highest electronegativity, you might now guess, lies just to the left of oxygen. It is nitrogen with atoms that have one less charge on their nuclei than oxygen atoms.

Figure 5.8 shows the relative electronegativities of several elements, metals and nonmetals, and their locations in the periodic table. Notice that carbon isn't the element with the fourth highest electronegativity; chlorine ranks fourth. The chlorine atom has a large positive nuclear charge, $17+$. This doesn't make chlorine even more electronegative than fluorine, however, evidently because chlorine has a sufficient number of additional core electrons to shield the chlorine nucleus. Moreover, the covalent bond in a molecule such as H—Cl is a longer bond than it is in H—F. The shared pair has its electron density farther from the chlorine nucleus to start with, and this also makes it harder for this nucleus to pull electron density toward itself. On balance, chlorine is less electronegative than fluorine, but more so than carbon.

Notice in Figure 5.8 that metals have the lowest electronegativities. In fact, the general trend is that as you move to the right in the same period or as you move upward in the same group, the electronegativities become larger. The most electronegative element, as we said, is fluorine; and the least electronegative is cesium, the last element in group IA (not pictured in Figure 5.8, but below rubidium, Rb). The *trends* should be learned, but not specific values of electronegativities. We'll be working so often with oxygen, nitrogen, carbon, and hydrogen, however, that you should memorize the order of their electronegativities, $O > N > C > H$. We'll see shortly how knowing this can be useful.

FIGURE 5.8
Relative electronegativities.

IA						
H 2.20	IIA	IIIA	IVA	VA	VIA	VIIA
Li 0.97	Be 1.47	B 2.01	C 2.50	N 3.07	O 3.50	F 4.10
Na 1.01	Mg 1.23	Al 1.47	Si 1.74	P 2.06	S 2.44	Cl 2.83
K 0.91	Ca 1.04				Se 2.48	Br 2.74
Rb 0.89	Sr 0.99				Te 2.01	I 2.21

■ Group IA and IIA elements are so weakly electronegative that they give up electrons entirely to group VIA or VIIA elements and form ions, not polar bonds.

Polar Bonds Make Molecules Polar if the Bond Polarities Do Not Cancel It's easy to tell if a *bond* is polar; it *always* is if the atoms that the bond joins have different electronegativities. For diatomic molecules such as H—F and H—Cl that have only one bond, when the bond is polar so is the molecule. Because such diatomics as H—H and F—F involve identical atoms and one bond, we can tell right away that these molecules are nonpolar.

Whether larger molecules are polar in an overall sense depends not just on the presence of polar bonds but also on the *geometry* of the molecule. It's possible for the polarities of individual bonds to cancel each other. Consider, for example, the carbon dioxide molecule, **3**.

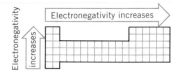

In electronegativity, metals are lowest and nonmetals are highest.

$$O{=}C{=}O \qquad \overset{\displaystyle O}{H \qquad H}$$

3   **4**

Because oxygen is more electronegative than carbon, each carbon–oxygen double bond must be polar. But the two dipoles in CO_2 point in exactly opposite directions, so in an overall sense they cancel each other. This leaves the CO_2 molecule overall nonpolar. The water molecule, **4**, on the other hand, is bent. Its two individual O—H bond polarities thus cannot cancel, and the water molecule is polar, quite polar, in fact.

Another useful way of thinking about the polarity of a molecule uses the idea of a *center of density of charge,* something like a "balance point" for electrical charge. A **polar molecule** is one in which the center of density of positive charge is not at the same place as the center of density of negative charge. We don't have to be able to pinpoint these centers exactly to know if they are in the same place or not. The *symmetry* will usually tell us. For example, the symmetry of the carbon dioxide molecule, **3**, tells us that all the positive charges on the three nuclei balance around the center of the carbon nucleus. Similarly, all of the negative charges contributed by all the electrons must also balance *around the identical point.* Because these

■ Nonbonding valence-shell electrons are often omitted from Lewis structures in general discussions.

two centers are in the same place, the molecule is nonpolar. Wherever these two centers are in the water molecule, **4**, we know that the molecule's angularity prevents their being at the same place. Hence, we know that the H_2O molecule must be polar.

EXAMPLE 5.15
Predicting Molecular Polarity

Place $\delta+$ and $\delta-$ signs at the correct ends of each of the bonds in the following structures (whose correct geometries are shown). Then decide whether each molecule as a whole is polar or nonpolar. (Use information in Figure 5.8 as needed.) A three-dimensional view of the carbon tetrachloride molecule is given. This molecule is entirely symmetrical.

I—Br   O / F ... F 105°   Cl—C—Cl / Cl / Cl 109°

Iodine bromide   Oxygen difluoride   Carbon tetrachloride

ANALYSIS There are six nonmetal atoms involved in these structures, and we must use their relative electronegativities to decide where partial charges are to be placed. Because Br stands *above* I in group VIIA, Br is more electronegative than I. This makes the Br end of the bond in I—Br the site with the $\delta-$ charge. Because F stands to the right of O in period 2, F is more electronegative than O and must have the $\delta-$ charge in the bonds of oxygen difluoride. From Figure 5.8 we learn that Cl is more electronegative than C, so we must place $\delta-$ at the Cl ends of each C—Cl bond in CCl_4. Once we have the partial charges correctly placed, we consider the geometry of each molecule to decide if it has an overall polarity.

$\overset{\delta+}{I}\!\!-\!\!\overset{\delta-}{Br}$ $\overset{\delta+}{O}$ / $\overset{\delta-}{F}$ ... $\overset{\delta-}{F}$ 105° CCl_4 structure with $\delta-$ on Cl and $\delta+$ on C 109°

Iodine bromide   Oxygen difluoride   Carbon tetrachloride

SOLUTION
The linear iodine bromide molecule has a polar bond and so must be a polar molecule. Because the oxygen fluoride molecule is angular, the centers of density of positive and negative charge cannot be at the same location, so the molecule must be (and is) polar. Although each bond in CCl_4 is polar, this molecule is *symmetrical.* Therefore, the balance point—the center of charge density—for all positive charge has to be at the center of the carbon nucleus. Similarly, the center of all negative charge density has to be in the identical place—the symmetrical disposition of the chlorine atoms about this center guarantees this result. Hence, the CCl_4 molecule as a whole is not polar.

PRACTICE EXERCISE 20

The structure of chloroform is just like that of carbon tetrachloride (in Example 5.15, above) *except* that one Cl has been replaced by H. Using the structure of carbon tetrachloride as a model, make the needed changes to draw a structure of chloroform and then place $\delta+$ and $\delta-$ signs by each atom. Finally, decide whether the molecule as a whole is polar.

■ OF_2 is a colorless, poisonous gas and is one product when fluorine burns(!) in water.

5.9

MOLECULAR ORBITALS—ANOTHER VIEW OF THE COVALENT BOND

A shared electron pair resides in a molecular orbital formed by the partial overlapping of the spaces of two atomic orbitals.

We earlier described the formation of a hydrogen molecule, H—H, by thinking of two hydrogen atoms, each with a $1s$ electron, approaching each other (Fig. 5.3, page 114). In a widely held view of the covalent bond, the $1s$ orbitals partially merge to create a new space for the electrons to be shared. The partial merging of atomic orbitals from different atoms is called the **overlapping of orbitals.** We will call the space created by such overlapping of atomic orbitals a **molecular orbital,** and it surrounds two nuclei (sometimes more). Like an atomic orbital, a molecular orbital can hold a maximum of two electrons, provided their spins are opposite. The shared electron pair of a covalent bond thus resides in a molecular orbital whose space encloses both nuclei held by the covalent bond.

The atomic orbitals that overlap are generally those of valence shell electrons. Figure 5.9 shows how orbital overlapping happens between two fluorine atoms. On the left in the figure, we see two separated fluorine atoms, but only their half-filled p_z orbitals are pictured. Their nuclei are centered at the *nodes,* where pairs of lobes touch. Imagine that these two atoms move toward each other. Eventually, the spaces occupied by facing lobes of the p_z orbitals start to overlap. A molecular orbital forms, and the former two p_z electrons take up residence in it, become the shared pair, and thus electron density becomes concentrated between the two nuclei. The covalent bond in F—F has formed.

■ The electron configuration of F is

$$1s^2 2s^2 2p_x^2 2p_y^2 2p_z^1$$

The Geometries of Atomic Orbitals Determine Bond Angles

The bond angle in the hydrogen sulfide molecule is 92°, almost a right angle (90°). Because sulfur is in the same family as oxygen, shouldn't the angle in H_2S be like that in H_2O, nearly a tetrahedral angle? VSEPR theory would predict this, since the S atom in H_2S, like the O atom in H_2O, has four pairs of electrons, two pairs nonbonding and two bonding. Let's see how molecular orbital theory handles this. Sulfur is in group VIA, and its electron configuration is

S $1s^2 2s^2 2p^6 3s^2 3p_x^2 3p_y^1 3p_z^1$

■ The atomic number of sulfur is 16, so we have to make places for 16 electrons.

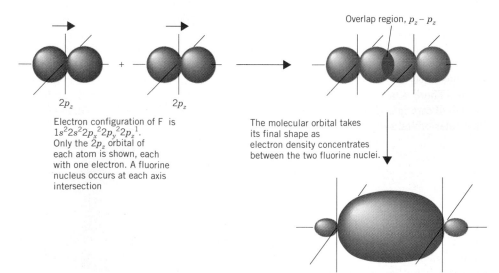

Electron configuration of F is $1s^2 2s^2 2p_x^2 2p_y^2 2p_z^1$. Only the $2p_z$ orbital of each atom is shown, each with one electron. A fluorine nucleus occurs at each axis intersection

Overlap region, $p_z - p_z$

The molecular orbital takes its final shape as electron density concentrates between the two fluorine nuclei.

FIGURE 5.9
The covalent bond in the F—F molecule.

FIGURE 5.10

Bonding in H_2S. The H atoms must position themselves so that their $1s$ orbitals overlap with the half-filled $3p$ orbitals of S.

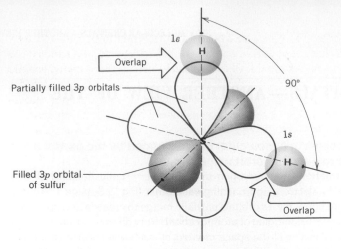

Thus a sulfur atom has two half-filled $3p$ orbitals, *and their axes make an angle of $90°$ with each other* (see Fig. 5.10). Therefore when two hydrogen atoms approach a sulfur atom to form the S—H bonds, the only effective way in which their $1s$ orbitals can overlap with sulfur's $3p$ orbitals is along the two perpendicular axes—also shown in Figure 5.10. Only by this approach can *maximum* overlapping result, and therefore only in this way can the strongest bonds form. Hence, the two resulting S—H bonds should make an angle of $90°$. The actual bond angle, as we said, is $92°$, which gives nice support to the theory that *the chief cause of a particular bond angle is the angle at which the axes of the atomic orbitals of the central atom cross.*

Second Row Nonmetals, O, N, and C, Don't Fit the Simple H_2S Model   Because oxygen is in the same family as sulfur, what works well for H_2S ought to work equally well for H_2O, but it doesn't. As we have learned, the bond angle in H_2O is $104.5°$, not $90°$, not a good fit.

A similar difficulty arises with the bond angles in NH_3. The electron configuration of nitrogen, atomic number 7, is

$$N \qquad 1s^2 2s^2 2p_x^1 2p_y^1 2p_z^1$$

The nitrogen atom has three half-filled $2p$ orbitals, so their axes cross at angles of $90°$. Three hydrogen atoms, each with a $1s$ electron, coming up to overlap with the $2p$ orbitals of N should produce bond angles in NH_3 of $90°$, but they don't. The angles are $107.3°$, very close to what simple VSEPR theory predicts.

So far things are going poorly in extending the molecular orbital model of H_2S to the hydrides of O and N. It becomes a total disaster when applied to CH_4, whose central atom is carbon with the electron configuration

$$C \qquad 1s^2 2s^2 2p_x^1 2p_y^1$$

Here we have only two half-filled $2p$ orbitals, an empty $2p$ orbital (the $2p_z$), and a filled $2s$ orbital. There is no way we can bring the $1s$ electrons of four hydrogen atoms up to these orbitals of carbon and obtain the bond angles of $109.5°$ known to be in CH_4. Yet the idea of a molecular orbital seems "right," because atomic orbitals were right.

FIGURE 5.11

Formation of sp^3 hybrid orbitals. *(a)* The empty atomic orbitals mix (hybridize) to give four new orbitals, sp^3 hybrid orbitals. The four have identical energies, just slightly more than an s orbital and slightly less than the p orbitals. *(b)* Cross-sectional shape of an sp^3 hybrid orbital. *(c)* Arrangement in space of four sp^3 orbitals (showing only their larger lobes). How many electrons will be placed in these four orbitals will depend on the atomic number of the central atom. See Figure 5.12.

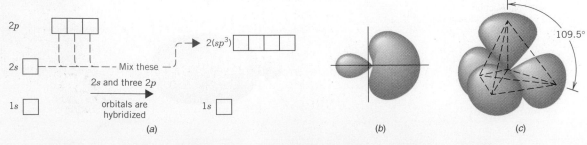

Hybrid Atomic Orbitals Are Used by Some Second Row Elements to Make Molecular Orbitals

The mismatches between fact and theory, just described, led scientists to reconsider whether the simple atomic orbitals of the *isolated* atoms of O, N, and C are the actual atomic orbitals used when these atoms form single bonds. Perhaps it is possible for some ordinary atomic orbitals to change as they engage in the overlapping that creates molecular orbitals. What theoreticians developed was a theory based on the overarching principle that nature takes up the lowest energy arrangements possible in a given situation. The calculations that became part of the theory indicated that lower energy arrangements would indeed occur if the simple atomic orbitals changed as overlapping took place. Out of these calculations came new orbital pictures (and energies). The equations behind simple *s* and *p* orbitals were modified. It is as if the *s* and *p* orbital shapes were "mixed" to obtain pictures of the modified atomic orbitals.

The mixing of atomic orbitals is called **orbital hybridization.** This term uses the word *hybrid* from biology, because the new **hybrid orbitals,** like hybrids anywhere, have some resemblances to their "parents." Hybridization occurs because it leads to better overlapping, to stronger bonds, and therefore to more stable molecules. So nature, not being too interested in our efforts to theorize, does its usual thing — it arranges for the most stable systems it can.

The *sp*³ Hybrid Orbitals Are One Important Set

One way to form hybrid atomic orbitals for C, N, or O is to mix all the orbitals at level 2, the *2s* and the three *2p* orbitals. An important rule about hybridization is that the total number of orbitals is conserved. When we mix four orbitals, we get four new ones.

The mixing of an *s* orbital with three *p* orbitals is called *sp*³ hybridization. It produces four new hybrid orbitals named ***sp*³ hybrid orbitals** (see Fig. 5.11). The four are identical in shape, and their axes make angles of 109.5°. They have a tetrahedral array, in other words, exactly what VSEPR theory would also suggest. Notice that each hybrid orbital has two lobes, like *p* orbitals, but one lobe is much larger, a contribution of the *s* orbital "parent."

According to the theory of hybrid orbitals, atoms of carbon, nitrogen, and oxygen can all form *sp*³ orbitals, but the number of electrons in them depends on which atom is involved. The new electron configurations for the *bonding* states, not the isolated atom states, of C, N, and O are as follows whenever these atoms participate in the formation of just single bonds (see Fig. 5.12).

C	$1s^2 2(sp^3)^1 2(sp^3)^1 2(sp^3)^1 2(sp^3)^1$	Four half-filled orbitals
N	$1s^2 2(sp^3)^2 2(sp^3)^1 2(sp^3)^1 2(sp^3)^1$	Three half-filled orbitals
O	$1s^2 2(sp^3)^2 2(sp^3)^2 2(sp^3)^1 2(sp^3)^1$	Two half-filled orbitals

Figure 5.13 shows how we can imagine the formation of molecules of methane, ammonia, and water. A 1s orbital of a hydrogen atom, which holds one electron, overlaps with an *sp*³ bonding-state orbital of a central atom, also with one electron, to form the molecular orbital for

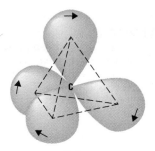

Carbon
Half-filled orbitals, 4
Filled orbitals, 0

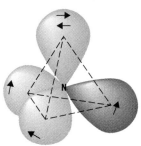

Nitrogen
Half-filled orbitals, 3
Filled orbitals, 1

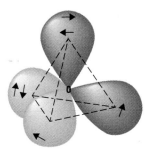

Oxygen
Half-filled orbitals, 2
Filled orbitals, 2

FIGURE 5.12
The bonding orbitals for atoms of carbon, nitrogen, and oxygen when they are involved only in single bonds.

FIGURE 5.13
The sigma bonds in *(a)* methane, *(b)* ammonia, and *(c)* water.

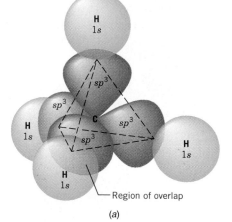

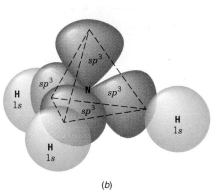

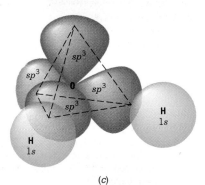

(a)   Region of overlap

(b)

(c)

139

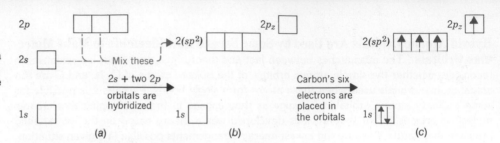

FIGURE 5.14

Formation of sp^2 hybrid orbitals at a carbon atom. (a) The empty atomic orbitals at level 2. (b) Three of these atomic orbitals have been hybridized to create three sp^2 hybrid orbitals, leaving a p orbital unhybridized. (c) Carbon's six electrons are fed into the available orbitals to create the bonding state of carbon at a double bond.

each bond to H. The bond angles in all these compounds should have the same tetrahedral value, 109.5°. They are a trifle smaller in both ammonia and water, presumably because their nonbonding electron clouds exert *squeezing* forces.

The sp^3 hybrid orbitals permit the formation of stronger bonds than the original nonhybrid orbitals because the larger lobes of sp^3 orbitals at level 2 extend a little farther from the nucleus than either the 2s or any of the 2p orbitals. Therefore strong overlapping between an sp^3 orbital of C, N, or O and a 1s of H can occur without bringing the atomic nucleus of H so close to the nucleus of the central atom. Thus repulsions between nuclei are less, and the system is more stable.

The single covalent bonds that we have just described are named **sigma bonds (σ bonds).** Sigma bonds have the symmetry of a long, straight sausage.

■ The single bonds in H—H, F—F, Cl—Cl, Br—Br, and I—I are also *sigma* bonds.

Sometimes Only Two p Orbitals Mix with an s Orbital
Sigma bonds are also *single* bonds. What about double (or triple bonds)? We'll use ethylene (page 118) to answer this.

The bond angles in ethylene are almost exactly 120°, and all the atoms lie in the same plane. Clearly sp^3 hybrid orbitals will not account for these facts. Moreover, the carbon atoms

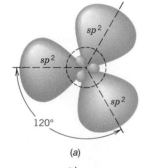

(a)

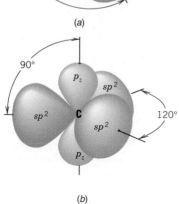

(b)

FIGURE 5.15

The shapes and arrangements of the bonding orbitals at level 2 of carbon when it is in an sp^2 hybridized state. (a) The three sp^2 hybrid orbitals have their axes in a plane and at angles of 120°. (b) The unhybridized p orbital fits with the three sp^2 orbitals.

FIGURE 5.16

The sigma bond network in ethylene. (The 2p orbital at each carbon is omitted so that the others can be seen better.)

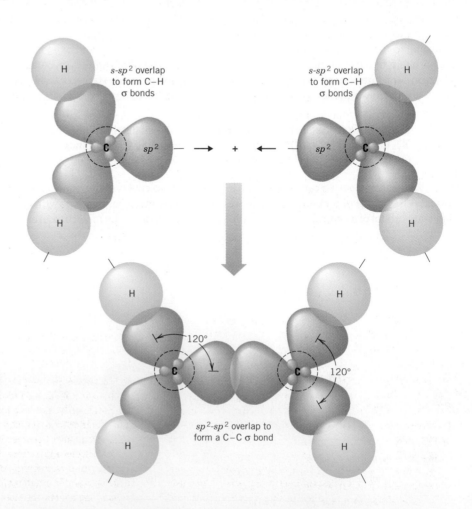

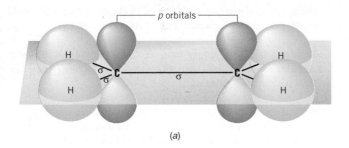

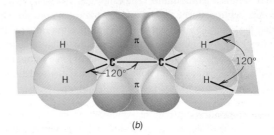

p orbitals

(a)

(b)

FIGURE 5.17

How the pi bond forms in ethylene. *(a)* Just before the *p* orbitals overlap side to side, showing the sigma bonds in place. *(b)* This side-to-side overlap of two *p* orbitals creates two banana-shaped lobes of the pi bond. One pair of electrons resides in this complex space.

at a double bond cannot use sp^3 orbitals because each atom holds three other groups, not four. Different hybrid orbitals are used at a double bond, and these form by the mixing or hybridizing of the 2*s* and just *two* of the 2*p* atomic orbitals (see Fig. 5.14). The third 2*p* orbital is left unchanged, but only for a moment.

Because one *s* and two *p* orbitals are mixed, the new hybrid orbitals are called sp^2 **orbitals.** Each carbon of the double bond develops three of them, and it also has one unhybridized 2*p* orbital. Thus there are four orbitals at each carbon with one electron apiece.

The shapes and the distributions of the sp^2 orbitals at carbon are shown in Figure 5.15. Notice how their axes come out at angles that VSEPR theory would have predicted. The axes of these orbitals lie in the same plane, and between them are angles of $120°$. The axis of the unhybridized 2*p* orbital is perpendicular to and intersects the plane at the point where the three sp^2 orbitals intersect.

Figure 5.16 illustrates how a molecular orbital forms from the overlapping of two sp^2 orbitals, one from each carbon at the double bond. Two electrons are in this molecular orbital and are shared by the two carbons. The result is one of the bonds of the double bond, and it is a sigma bond.

The two groups of CH_2 atoms held so far by this bond would be free to rotate with respect to each other about this bond. Such a rotation would not reduce the degree of overlap and weaken the bond. However, at one stopping point during such a rotation, the two 2*p* orbitals become perfectly aligned, side by side. They are so close to each other that they now overlap *side by side* to create a new kind of molecular orbital, one with two banana-shaped halves (see Fig. 5.17). This molecular orbital now holds another shared pair of electrons so it also is a bond, the second bond of the double bond. A covalent bond made by the side-to-side overlap of two *p* orbitals is called a **pi bond,** or π **bond.**

The triple bond of acetylene (page 119) has two pi bonds and a sigma bond, but because the triple bond is so rare among molecules of biochemical interest we will not study its molecular orbitals further.

■ Carbon has *sp* hybrid orbitals at a triple bond.

SUMMARY

Ionic bonds and ionic compounds A reaction between a metal and a nonmetal usually goes by the transfer of an electron from the metal atom to the nonmetal atom. The metal atom changes into a positively charged ion (cation), and the nonmetal atom becomes a negatively charged ion (anion). The oppositely charged ions aggregate in whatever whole-number ratio ensures that the product is electrically neutral. The electrical force of attraction between the ions is called an ionic bond, and compounds made of ions are called ionic compounds.

Octet rule (noble gas rule) The ions of the representative elements have electron configurations of the nearest noble gases, usually outer octets. Metal atoms lose exactly enough valence shell

electrons to achieve (new) noble gas configurations, and nonmetal atoms gain exactly as many valence shell electrons to acquire noble gas configurations.

Redox reactions Electron transfers mean changes in oxidation numbers and are called redox reactions. Elements have zero oxidation numbers. Monatomic ions have oxidation numbers that equal their electrical charges. An atom that loses electrons and whose oxidation number becomes more positive is oxidized, and one that gains electrons and a more negative oxidation number is reduced. Anything that causes an oxidation is called an oxidizing agent; a reducing agent is anything that causes a reduction. Metals tend to be reducing agents and nonmetals are oxidizing agents.

Formulas of ionic compounds The empirical formulas of ionic compounds begin with the symbol of the cation (without the sign indicating the charge). Subscripts are used to give the ratios of the ions. When a formula must include two or more polyatomic ions, parentheses enclose their symbols. The names of ionic compounds are based on the names of the ions (except that the word *ion* is omitted). The cation is named first, then the anion.

Molecular compounds Atoms of nonmetals form molecules by sharing valence shell electrons in pairs. Each shared pair constitutes one covalent bond, and each pair is counted as the joint property of both atoms when the structure is checked to see if it adheres to the octet rule. The shared pair creates a region of relatively high electron density between the two atoms toward which the nuclei are electrically attracted, and this attraction is called a covalent bond. The covalent-bond-forming abilities of atoms are their covalence numbers.

Lewis structures Molecules have unique structures, which display the sequence in which the atoms are joined. Lewis structures (electron-dot structures) display the valence shell electrons, both the shared and unshared pairs, in molecules. (The shared pairs are usually replaced by dash bonds.) In nearly all molecules the atoms have noble gas configurations, but a few have fewer than four electron pairs and a few have more than four.

To write a Lewis structure, the skeletal structure must be known as well as the total number of valence shell electrons present. First, electron pairs are placed into the bonds. Next, additional electron pairs are used to give noble gas configurations to all atoms, doing this for the central atom last. If there aren't enough for the central atom, double (or triple) bonds have to be made by moving electrons pairs from other atoms, like C, N, or O (but not halogens), toward the central atom.

Polyatomic ions The atoms of polyatomic ions are joined by covalent bonds, but the overall numbers of electrons and protons do not balance. In these ions, coordinate covalent bonds can also occur, bonds whose electron pairs both came from one of the bonded atoms.

The Lewis structures of polyatomic ions can be figured out by the same rules for making other electron-dot structures. In deter-mining the total number of valence shell elections, add 1 for each negative charge and subtract 1 for each positive charge.

Acids and bases Acids are substances that can furnish H^+ ions, and bases are compounds that can combine with H^+ ions. Acids and bases react to neutralize each other.

Molecular shapes and the VSEPR theory Valence shell electron pairs in molecules stay out of one another's way as much as possible. When there are four pairs, the result is a tetrahedral geometry, or close to it. If only three pairs are present, the geometry is that of a triangle with the three atoms at the corners.

Polar molecules Even though molecules are electrically neutral, they can still be polar. If individual bond polarities, caused by electronegativity differences between the joined atoms, do not cancel, the molecule is polar and can adhere to adjacent polar molecules much as magnets can stick together.

Molecular orbitals Molecular orbitals form by the partial overlapping of atomic orbitals. This creates a space between the two nuclei in which the electron pair of the single covalent bond resides. When the molecular orbital has the symmetry of a straight sausage, it is a sigma bond.

According to one theory, when atoms of carbon, nitrogen, and oxygen form covalent bonds, they use hybrid atomic orbitals instead of the simple atomic orbitals of their isolated atoms. When an *s* orbital and three *p* orbitals of the same main level mix (hybridize), four new, equivalent sp^3 hybrid orbitals form whose axes point to the corners of a regular tetrahedron. When a carbon atom has a double bond, it uses sp^2 hybrid orbitals made by mixing an *s* and two *p* atomic orbitals. This leaves the third *p* orbital of carbon unhybridized. One of the two bonds at the double bond is a sigma bond formed by the overlapping of the sp^2 hybrid orbitals of the two carbons. Then the side-to-side overlap of the unhybridized *p* orbitals creates the other bond, called a pi bond, a double-banana shaped space above and below the plane of the sigma bond. One pair of electrons resides in this unusual space.

REVIEW EXERCISES

The answers to Review Exercises whose numbers are in color are found in Appendix C. The answers to the other Review Exercises are found in the Study Guide that accompanies this book. The more challenging questions are marked with asterisks.

Ions, Ionic Compounds, and Their Names and Formulas

5.1   What natural law concerning electrically charged particles is particularly important for understanding the nature of the chemical bond?

5.2   In general terms, what parts of the atoms become reorganized relative to each other when a chemical bond forms between the two atoms? Once the bond forms, can we say that the two *atoms* are still present? Explain.

5.3   What two fundamental kinds of compounds are recognized, based on the way that electrons and nuclei become reorganized?

5.4   What is the name of the fundamental particle that carries all of the chemical properties of a molecular compound?

5.5   Which kinds of elements largely make up molecular compounds?

5.6   Why don't nonmetal atoms readily form positively charged ions?

5.7 Assuming that sodium atoms interact with fluorine atoms in the same way that they react with chlorine atoms, use changes in electron configurations to illustrate how Na and F react. Use electron configurations (including the compositions of nuclei) as we did in Section 5.1 for the reaction of sodium with chlorine to show how a sodium atom and a fluorine atom can change to particles that can attract each other. What are the names of the particles that form?

5.8 In terms of their fundamental structures, what is the difference between a sodium atom and a sodium ion?

5.9 Write the symbols with the correct charges for the following ions.
(a) iodide ion   (b) sodium ion
(c) silver ion   (d) calcium ion
(e) zinc ion   (f) aluminum ion
(g) sulfide ion   (h) lithium ion
(i) bromide ion   (j) chloride ion
(k) potassium ion   (l) barium ion
(m) oxide ion   (n) cupric ion
(o) fluoride ion   (p) copper(I) ion
(q) ferric ion   (r) magnesium ion
(s) iron(II) ion

5.10 What are the names of the following ions? (Give both the older and the modern name for ions with both.)
(a) K^+ (b) S^{2-} (c) Al^{3+} (d) Fe^{2+} (e) Br^-
(f) Fe^{3+} (g) Ba^{2+} (h) Na^+ (i) Cl^- (j) Mg^{2+}
(k) Li^+ (l) Cu^+ (m) Zn^{2+} (n) Ca^{2+} (o) Ag^+
(p) O^{2-} (q) I^- (r) F^- (s) Cu^{2+}

5.11 Two ions of tin are commonly known, Sn^{4+} and Sn^{2+}. Which is the stannous ion and which is the stannic ion? (These common names stem from *stannum*, Latin for tin.)

5.12 What would be the modern names for the following ions for which common names are given in parentheses?
(a) Pb^{2+} (plumbous ion)   (b) Au^{3+} (auric ion)
(c) Hg_2^{2+} (mercurous ion)   (d) Pb^{4+} (plumbic ion)

5.13 What are the formulas of the following compounds?
(a) cuprous sulfide   (b) barium oxide
(c) aluminum oxide   (d) calcium chloride
(e) ferrous bromide   (f) sodium iodide

5.14 Write the formulas of each of the following compounds.
(a) magnesium fluoride   (b) lithium oxide
(c) cupric sulfide   (d) ferric chloride
(e) sodium bromide   (f) calcium oxide

5.15 Write the names of the following compounds. Where two names are possible, one based on modern terminology and the other a common name, write both names.
(a) AgI   (b) $FeCl_2$ (c) Al_2O_3
(d) $BaCl_2$ (e) CaS   (f) KI

5.16 What are the names of the following compounds? If both a modern and a common name are possible, write both.
(a) NaF   (b) Li_2O (c) $CuBr_2$
(d) $MgCl_2$ (e) ZnO   (f) $FeBr_3$

5.17 Which has the larger radius, the atom or the ion of a given metal? Explain.

5.18 Sulfur and oxygen are in the same family in the periodic table. Which atom has the larger radius, S or O? Explain.

5.19 Which has the larger radius, the oxygen atom or the oxide ion? Explain.

5.20 Which kinds of compounds are more likely to be *electrolytes*, ionic or molecular compounds? Explain.

5.21 In certain life-threatening situations, an attending physician or nurse may comment that the *potassium level* of the blood has changed. To what does this term specifically refer?

Ions and the Octet Rule

5.22 Show how calcium atoms and chlorine atoms can interact to form ions that will aggregate in the correct ratio. (Follow the directions given in Review Exercise 5.7.)

5.23 Write diagrams that show how lithium atoms and oxygen atoms can cooperate to form particles that aggregate in a particular ratio to form an ionic compound. (Follow the directions given in Review Exercise 5.7.)

5.24 If M is the symbol of a representative element and M^{2+} is the symbol of its ion, to what group in the periodic table does this element most likely belong?

5.25 An atom of a representative nonmetal X can, in some circumstances, accept three electrons and become an ion. To what group in the periodic table does X most likely belong?

5.26 What electric charge is carried by the ions of the elements in the following groups?
(a) Group IA   (b) Group IIA   (c) Group IIIA
(d) Group VIIA   (e) Group VIA

5.27 The atoms of which group in the periodic table have exceptionally stable electron configurations? Describe these configurations in your own words.

5.28 Can the atom with the following electron configuration be changed to a stable ion? If it can, what would be the electrical charge on the ion?

$$1s^2 2s^2 2p_x^2 2p_y^2 2p_z^1$$

5.29 Study the following electron configuration. Can the atom with this configuration be changed to a reasonably stable ion? If so, what is the charge on the ion?

$$1s^2 2s^2 2p^6 3s^2 3p^6 4s^1$$

5.30 Can an atom with the following electron configuration be changed to a stable ion? If it can, what would be the charge on this ion?
$$1s^2 2s^2 2p_x^2 2p_y^1 2p_z^1$$

5.31 Study the following electron configuration. Can the atom with this configuration be changed to a reasonably stable ion? If so, what is the charge on the ion?

$$1s^2 2s^2 2p_x^1 2p_y^1$$

5.32 Can an atom with the following electron configuration be changed to a stable ion? If it can, what would be the charge on this ion?
$$1s^2 2s^2 2p^6 3s^2 3p^6 3d^{10} 4s^2 4p_x^2 4p_y^2 4p_z^2$$

5.33 Are nonrepresentative elements likelier to have positively or negatively charged ions? Explain.

*5.34 Using the atomic numbers only (and the corresponding electron configurations), without consulting the periodic table, write the symbols and the electron configurations of the *ions* of the following elements. (The atomic symbols are hypothetical.)
(a) *M*, atomic number 19
(b) *Q*, atomic number 3
(c) *Z*, atomic number 16

5.35 In the compound NaH, sodium hydride, sodium exists as its normal ion, Na$^+$. What is the charge on H in this ionic compound? Write the electron configuration of this ion. In what way does this ion agree with the noble gas rule?

Oxidation and Reduction

5.36 What are the oxidation numbers of the following ions?
(a) potassium ion (b) aluminum ion (c) oxide ion
(d) copper(II) ion (e) chloride ion (f) mercury(II) ion

5.37 Determine the oxidation number of the metal component in each of the following compounds.
(a) $AuCl_3$ (b) $GaCl_3$ (c) Cr_2O_3
(d) PbF_4 (e) $ZrOCl_2$ (f) V_2O_5

5.38 What would be the formula of an oxide of manganese, Mn, if the oxidation number of Mn is $7+$? (The oxidation number of oxygen is always $2-$.)

5.39 Tungsten, W, has an oxidation state of $5+$ in one of its oxides. Oxygen always has an oxidation number of $2-$ in oxides, so what is the formula of this tungsten oxide?

5.40 The following reaction between aluminum and oxygen seals an invisible coating of aluminum oxide, an ionic compound, over any aluminum surface exposed to air.

$$4Al + 3O_2 \longrightarrow 2Al_2O_3$$

Write the chemical symbol for the following species.
(a) The substance reduced
(b) The substance oxidized
(c) The reducing agent
(d) The oxidizing agent

*5.41 Calcium reacts with chlorine to form calcium chloride, an ionic compound, as follows.

$$Ca + Cl_2 \longrightarrow CaCl_2$$

Write the chemical symbol for the following substances in this reaction.
(a) The substance oxidized
(b) The oxidizing agent
(c) The substance reduced
(d) The reducing agent

*5.42 Iron(III) oxide reacts with hydrogen as follows.

$$3H_2 + Fe_2O_3 \longrightarrow 2Fe + 3H_2O$$

Write the chemical symbols of the following species in this reaction.
(a) The species whose oxidation number becomes more positive
(b) The species whose oxidation number becomes less positive
(c) The substance reduced

(d) The reducing agent
(e) The substance oxidized
(f) The oxidizing agent

Molecules, Molecular Compounds, and Lewis Structures

5.43 In what ways are a molecule and an atom alike and in what ways are they different?

5.44 In what ways are a molecule and an ion alike and how do they differ?

5.45 How do molecular elements and molecular compounds differ?

5.46 How do molecular compounds and ionic compounds differ?

5.47 What kind of force of attraction holds the two nuclei in one hydrogen molecule, H_2, quite near each other? What special name do we give to this force of attraction when it operates within molecules?

5.48 In your own words, describe how the force of attraction arises as two hydrogen atoms combine to form a hydrogen molecule.

5.49 How does the octet rule work with molecular compounds?

5.50 The diatomic molecule, He_2, does not exist. Offer an explanation for its failure to form.

5.51 Using the periodic table, write the electron-dot structures of the following.
(a) Rubidium (atomic number 37)
(b) Strontium (atomic number 38)
(c) Gallium (atomic number 31)
(d) Tellurium (atomic number 52)

5.52 Use Lewis structures to diagram the formation of $CaCl_2$ from neutral atoms.

5.53 Use Lewis structures to diagram the formation of the following molecules from neutral atoms.
(a) HF (b) H_2S (c) SiH_4

5.54 The terms *bond length* and *bond distance* are used in connection with the (indirectly) measured distance between two atomic nuclei in a covalent bond. Consider now the distance between two carbon nuclei in ethane and ethylene.

Ethane   Ethylene

In which molecule is the carbon–carbon bond distance shorter? Explain.

Polyatomic Ions and Formulas of Compounds Involving Them

5.55 Write the names of the following ions.
(a) OH^- (b) NH_4^+ (c) CN^-
(d) MnO_4^- (e) HSO_4^- (f) HSO_3^-
(g) $H_2PO_4^-$ (h) H_3O^+ (i) $C_2H_3O_2^-$
(j) HCO_3^- (k) SO_4^{2-} (l) NO_3^-
(m) HPO_4^{2-} (n) CrO_4^{2-} (o) CO_3^{2-}
(p) NO_2^- (q) PO_4^{3-} (r) $Cr_2O_7^{2-}$

5.56 Write the formulas of the following ions.
    (a) sulfite ion   (b) acetate ion
    (c) nitrite ion   (d) bicarbonate ion
    (e) hydroxide ion   (f) ammonium ion
    (g) carbonate ion   (h) nitrate ion
    (i) phosphate ion   (j) cyanide ion
    (k) hydronium ion   (l) monohydrogen phosphate ion
    (m) hydrogen sulfate ion   (n) dihydrogen phosphate ion
    (o) dichromate ion   (p) hydrogen sulfite ion
    (q) chromate ion   (r) sulfate ion

5.57 Write the formulas of the following compounds.
    (a) potassium phosphate
    (b) sodium carbonate
    (c) calcium sulfate
    (d) ammonium cyanide
    (e) lithium nitrite
    (f) sodium hydrogen sulfite
    (g) calcium dichromate
    (h) magnesium acetate

5.58 What are the formulas of the following compounds?
    (a) sodium monohydrogen phosphate
    (b) ammonium carbonate
    (c) sodium hydrogen sulfate
    (d) ammonium dihydrogen phosphate
    (e) sodium permanganate
    (f) aluminum hydroxide
    (g) lithium bicarbonate
    (h) calcium nitrate

5.59 Write the names of the following compounds.
    (a) Na_2CO_3 (b) NH_4NO_3
    (c) $Mg(OH)_2$ (d) $BaSO_4$
    (e) $KHCO_3$ (f) $Ca(C_2H_3O_2)_2$
    (g) $NaNO_2$ (h) $(NH_4)_3PO_4$

5.60 What are the names of the following compounds?
    (a) $KHSO_4$ (b) Li_2HPO_4
    (c) $Ca(CN)_2$ (d) $Na_2Cr_2O_7$
    (e) Na_2SO_3 (f) $BaCrO_4$
    (g) $Al_2(SO_4)_3$ (h) $KMnO_4$

5.61 What is the total number of atoms of all kinds in one formula unit of each of the following compounds?
    (a) $(NH_4)_2CO_3$ (b) $Al(C_2H_3O_2)_3$ (c) $Ba(H_2PO_4)_2$

5.62 One formula unit of each of the following compounds has how many atoms of all kinds?
    (a) $Al_2(CO_3)_3$ (b) $Ca(NO_3)_2$ (c) $(NH_4)_3PO_4$

5.63 The *Merck Index,* an encyclopedia of chemicals, drugs, and biologicals, gives the formula of ferrous gluconate, a hematinic agent (promotes the formation of red blood cells), as $Fe[HOCH_2(CHOH)_4CO_2]_2$. How many atoms of all kinds are in one formula unit of this compound?

Acids and Bases

5.64 Solutions in water of any of the common acids all contain what ion? (Give its name and formula.)

5.65 What is the formula of nitric acid, and what are the principal ions present in a solution of this in water? (Give their names and formulas.)

5.66 What is the formula of sulfuric acid? In a solution of this acid in water, what are the names and formulas of the principal ions present?

5.67 Hydrochloric acid is an aqueous solution in which the principal ions are what? (Give their names and formulas.)

5.68 What are two properties of solutions that contain acids? What chemical species is responsible for both?

5.69 What is the general family name for the substances that can react with acids to neutralize them? When these substances neutralize acids, do they react with the positive ion, the negative ion of the acid, or both?

5.70 Using the structural formula of the ammonia molecule in which each bond is represented by a line and the unshared pair of outer-level electrons on nitrogen is shown by a pair of dots, diagram how the ammonia molecule neutralizes a hydrogen ion.

5.71 An unshared pair of electrons on the oxygen atom of a water molecule can form a coordinate covalent bond to a hydrogen ion. The product is the hydronium ion. Using the structural formula of the water molecule in which each bond is shown by a line and each unshared pair of electrons on the oxygen is given by a pair of dots, diagram how the water molecule forms this new bond to H^+ to give H_3O^+.

5.72 Using the periodic table, determine the group numbers of arsenic (As), selenium (Se), and germanium (Ge). On the basis of these locations, write the most likely molecular formulas and the Lewis structures of the compounds of these elements with hydrogen.

5.73 Write the skeletal structures of the following species.
    (a) $PbCl_4$ (b) OF_2 (c) NCl_3 (d) NO_2
    (e) HSO_4^- (f) HSO_3^-

5.74 How many outside level electrons must be represented in the structures of Review Exercise 5.73?

5.75 Draw the Lewis structures of the compounds of Review Exercise 5.73.

*5.76 Draw the Lewis structure of the acetylide ion, $C_2{}^{2-}$.

*5.77 Dinitrogen tetroxide, N_2O_4, is a constituent in smog.
    (a) What is its empirical formula?
    (b) Propose a Lewis structure for it. (It is a symmetrical molecule with a single bond between the two nitrogen atoms.)

VSEPR Theory and the Shapes of Molecules

5.78 What, in general terms, is true at the molecular level of life that makes the study of bond angles useful?

5.79 What does the acronym VSEPR represent? In general terms, what factor is used in VSEPR theory to explain bond angles?

5.80 Describe the geometric arrangements assumed by the *axes* of the electron clouds of the valence shell of a central atom in a molecule under each circumstance given.
    (a) When the atom has four valence-shell electron pairs.
    (b) When the atom has three valence-shell electron pairs.
    (c) When the atom has two valence-shell electron pairs.

5.81 The central atom, Be, in $BeCl_2$ holds two atoms just like the central atom, O, in H_2O. The bond angles at these two central atoms are different, however.
(a) What are the two bond angles?
(b) Why are these angles different?

5.82 What bond angles would be predicted by VSEPR theory for silane, SiH_4?

Polar Molecules

5.83 What is the underlying cause of the polarity of a covalent bond?

5.84 Atoms X and Y are of elements in the same period of the periodic chart. Atom X has the larger atomic number. Which has the higher electronegativity? Briefly explain.

5.85 Suppose that X and Y form a diatomic molecule X—Y, and that X is less electronegative than Y.
(a) Is the X—Y molecule polar?
(b) If so, where are the $\delta+$ and the $\delta-$ charges located?

5.86 Describe the ways in which relative electronegativities change in each circumstance.
(a) Within the same family in the periodic table
(b) Within the same period in the periodic table
(c) As one moves across the table from left to right

Molecular Orbitals

5.87 In what ways is a molecular orbital different from an atomic orbital?

5.88 In what ways is a molecular orbital similar to an atomic orbital?

5.89 How do we envision the formation of a molecular orbital from atomic orbitals?

5.90 How do we envision the formation of the covalent bond in the hydrogen molecule, H_2?

5.91 What atomic orbitals partially overlap to form the covalent bond in the chlorine molecule, Cl_2?

5.92 What atomic orbitals partially overlap to form the covalent bond in the hydrogen fluoride molecule?

5.93 How is the bond angle of nearly 90° explained for hydrogen sulfide, H_2S?

5.94 Why cannot we use two of the simple $2p$ atomic orbitals of an oxygen atom to form molecular orbitals with separate hydrogen atoms (using their $1s$ orbitals) to explain the bonds in the water molecule?

5.95 To explain bond angles in methane, ammonia, and water, what is imagined concerning the atomic orbitals of C, N, and O before bonds to H atoms form?

5.96 What is meant when the C—H single bond in CH_4 is called a sigma bond?

5.97 Where are the unpaired electrons in the valence shells of N and O in molecules of NH_3 and H_2O?

5.98 The hydride of phosphorus is PH_3 and is called phosphine. The H—P—H bond angles in phosphine are all 93.7°. What atomic orbitals of P, simple or sp^3 hybrid, are most likely used to make PH_3?

5.99 What orbitals of carbon are "mixed" to make the hybrid orbitals used for the bonds in ethylene, $H_2C = CH_2$?

5.100 Describe the way in which the four axes of the orbitals lie at a carbon atom involved in sp^2 hybridization.

5.101 What are the names of the two bonds that make up a carbon–carbon double bond?

5.102 How does a pi bond form, in molecular orbital terms?

Ozone in the Stratosphere (Special Topic 5.1)

5.103 Write the equation by means of which an ozone cycle is initiated.

5.104 What two equations represent the ozone cycle? Why is the cycle called a chemical chain reaction?

5.105 What happens to the UV radiation as a result of the ozone cycle?

5.106 By means of equations, explain how CFC-11 can reduce the stratospheric ozone level.

5.107 What is the polar vortex? The Antarctic ozone hole?

5.108 Polar stratospheric clouds provide reservoirs for the subsequent release of chlorine atoms. How is this done? (Include equations.)

5.109 How does the reappearance of the sun at the onset of the Antarctic spring initiate the decline in ozone levels in the ozone hole?

Additional Exercises

5.110 Using the atomic numbers only (and the corresponding electron configurations), without consulting the periodic table, write the symbols and the electron configurations of the ions of the following elements. (The atomic symbols are hypothetical.)
(a) X, atomic number 20
(b) Y, atomic number 17
(c) Z, atomic number 10

5.111 Write the electron-dot structure of indium. What would be a reasonable electrovalence for it?

*5.112 Boron, B, is in group IIIA of the periodic table.
(a) What is the electron configuration of an isolated boron atom?
(b) If its valence-shell atomic orbitals underwent sp^2 hybridization and each new hybrid orbital held one electron, what would be the geometry of the axes of these hybrid orbitals?
(c) In this hybridized state, does there remain any p orbitals at level 2 that hold an electron?
(d) Boron forms a compound with chlorine, BCl_3. Describe the geometry of this molecule.

STATES OF MATTER AND THE KINETIC THEORY

6

A temperature that melts rock also gives explosive pressure to gases, enough to hurl the molten rock far into the night sky, as in this scene of an eruption of Kilauea volcano, Hawaii.

6.1

THE GASEOUS STATE

The four properties that can be used to completely define the physical state of any gas are pressure, temperature, volume, and quantity.

We now move from a study of the *kinds* of matter — elements and compounds — to the *states of matter* — gases, liquids, and solids — a study we introduced in Section 2.1. We begin with gases because they are the simplest. Our goal is to deepen our understanding of the huge gas mixture in which we live — the atmosphere — as well as the gases involved in respiration, oxygen and carbon dioxide.

All Gases Obey the Same Set of Laws   What makes gases simple is that, unlike liquids and solids, gases all follow the same physical laws. Gas volume, temperature, pressure, and quantity all interact in predictable ways. One of the remarkable facts about *all* gases is that *when we know any three of these four measurable properties of a gas sample, the fourth can have only a single value,* a value that we can calculate.

All gases, by a process called **diffusion,** spontaneously fill out whatever space they are given. Therefore, the volume of a confined gas, V, is the total volume of its container and is given in liters (L) or in milliliters (mL).

Gas temperature, T, must be in kelvins (K) to make gas law calculations simple. When the temperature of a gas in a rigid container changes, its pressure also changes. If the container isn't rigid — perhaps it has a movable "wall" like an automobile cylinder — a change in gas temperature results in a change in volume. Thus temperature, pressure, and volume all interact for a given amount of gas.

The pressure of a gas, P, can be given in a variety of units. You became aware of pressure by pumping up bicycle tires or blowing up balloons, so you know that pressure affects gas volume.

The amount of gas in a sample is usually stated in moles, n, although the mass (in grams or milligrams) is sometimes used. One mole of a gas at some particular pressure and temperature occupies a certain volume, *the same volume for any gas.*

Pressure Is the Ratio of the Force Acting to the Area on Which It Acts   To understand pressure, we need to learn about the relationships among weight, force, and pressure. The *weight* of an object is a measure of a natural force, the force of gravity. Thus an object's weight (on earth) is the force it can exert because of the gravitational attraction of the earth. But the effect of a given weight on you, such as that of a book backpack, depends on how the weight is distributed over some area of your body. **Pressure** is force per unit area.

$$\text{Pressure} = \frac{\text{force}}{\text{area}}$$

■ Liquids and solids, of course, do not diffuse like gases.

■ If the gas temperature is known in degrees Celsius, it must be converted to kelvins before being used in gas law calculations; the converting equation is

$$K = {}^\circ C + 273$$

Imagine balancing a ski pole sideways on your palm. Its weight is distributed over a fairly large area of the hand, so the ratio — force (weight) divided by area — is relatively small and you feel little pressure. However, if you could balance the pole by its tip on your finger, the entire force (weight) would now be concentrated on a very small area. The pressure you now would feel would be much greater and much more painful. It could be as much as 100,000 times greater, as you can see by the margin calculation. Thus the distinction between force (or weight) and pressure is important, and *pressure* is what is relevant to gases.

The Pressure Exerted by the Atmosphere Led to the First Units of Pressure

Because air is matter, it has weight, which means that it is pulled toward the earth by the earth's gravitational attraction. Imagine, now, a tall cylinder of still air with invisible walls. The area at its bottom is 1 square inch, it rests on an ocean beach, and the cylinder of air reaches to outer space. This column of air has a weight at sea level of about 14.7 lb when the temperature is 0 °C. Because this weight rests on an area of 1 in.^{2}, the pressure exerted by the air column is the ratio, 14.7 lb/in.^{2}. This is roughly the normal, sea-level air pressure.

Our air column has less air in it when its bottom rests at a high altitude, like a mountain top, than when it rests on an ocean beach, so air pressure decreases with altitude. For example, at the top of Mount Everest, the highest mountain on earth, the pressure is about one-third what it is at sea level. It is the change in air pressure with altitude, particularly above 3000 m, that triggers a complex series of bodily changes that results in altitude sickness.

The pushing ability of air caused by its weight is the basis of a device to measure pressure, the mercury or Torricelli **barometer** (see Fig. 6.1), named after Evangelista Torricelli (1608 – 1647), an Italian scientist. It consists of a glass tube a little less than a meter long, sealed at one end, filled with mercury, and then inverted into a container of mercury. Some mercury immediately runs out of the tube, but no air can get in to fill the gap thus created at the top of the tube. Thus no air is in this gap, and any space without air or any gas is called a **vacuum.**[1] In other words, virtually nothing inside the tube at its top pushes down on the entrapped mercury. Thus, the outside air pressure on the mercury in the dish is unopposed, and this forces the remainder of the mercury to stay in the tube.

The height of the mercury column is about 760 mm at sea level, a height used to *define* a unit of pressure called the *standard atmosphere*. The **standard atmosphere, atm,** is the pressure exerted by a column of mercury 760 mm high at a temperature of 0 °C.

$$P = \frac{0.69 \text{ lb}}{1.3 \text{ in.}^2} = 5.3 \times 10^{-1} \text{ lb/in.}^2$$

$$P = \frac{0.69 \text{ lb}}{1.6 \times 10^{-5} \text{ in.}^2} = 4.3 \times 10^4 \text{ lb/in.}^2$$

■ By *outer space* we mean space beyond the earth's atmosphere where the atmosphere is so thin as to be almost nonexistent.

■ Altitude sickness is life-threatening and the only cure is supplemental oxygen-enriched air or, without a supply of this, *prompt* return to low altitude.

■ The height fluctuates with the temperature and weather.

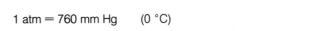

1 atm = 760 mm Hg   (0 °C)

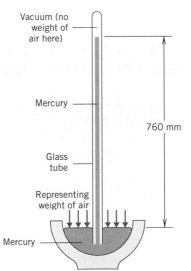

Vacuum (no weight of air here)

Mercury

760 mm

Glass tube

Representing weight of air

Mercury

FIGURE 6.1
The Torricelli barometer.

[1] The gap does not involve a *perfect* vacuum because this space contains a very small trace of vaporized mercury.

■ TV and radio weather reports in the United States use *inches* of mercury, not mm Hg, as the pressure unit.

1 atm = 29.9 in. Hg

A smaller unit, widely used by people in the life sciences, is called the **millimeter of mercury,** or **mm Hg** (pronounced *em em aitch gee*).

$$1 \text{ mm Hg} = \frac{1}{760} \text{ atm} \tag{6.1}$$

Many scientists are uncomfortable about using a length to express pressure, so an alternative name for the mm Hg, the **torr,** is widely used.

$$1 \text{ torr} = 1 \text{ mm Hg}$$

We will use the *mm Hg* unit, but you can see that it's easy to switch from mm Hg to torr.

The SI unit of pressure is called the *pascal.* One of the objectives of the SI is to have the units of all derived quantities, like pressure, be defined in terms of the units of the SI base quantities, like the kilogram for mass or the meter for length. *Force* and *area* are both derived quantities; the SI unit of force is called the *newton,* and the SI unit for area is the meter squared (m^2). Therefore the SI unit of pressure, the ratio of force to area, has the units of newtons/m^2, and is called the **pascal, Pa.** The *atmosphere,* in fact, is now *defined* in relationship to the pascal.

■ Remember that *derived quantities* are combinations of SI base quantities. Pressure is thus a derived quantity.

■ kPa = kilopascal
1 kPa = 1000 Pa

$$1 \text{ atm} = 101,325 \text{ Pa} \qquad \text{(exactly)}$$
$$1 \text{ atm} = 101.325 \text{ kPa} \qquad \text{(exactly)}$$

Thus it takes over 100,000 pascals to equal 1 atm, which tells us how extremely small the pascal is. As calculated in the margin,

$$1 \text{ mm Hg (1 torr)} = 133.322 \text{ Pa}.$$

■ $\dfrac{101,325 \text{ Pa}}{\text{atm}} \times \dfrac{1 \text{ atm}}{760 \text{ mm Hg}}$

   = 133.322 Pa/mm Hg

During your professional career you may see the pascal or an SI multiple like the kilopascal (kPa) used more and more, especially in formal scientific articles, including those in health care fields.

6.2

THE PRESSURE–VOLUME–TEMPERATURE RELATIONSHIPS FOR A FIXED AMOUNT OF GAS

The product of the pressure and volume of a gas divided by its temperature is a constant proportional to the number of moles of the gas.

Now that we have suitable physical units for describing gases, we can study how a fixed quantity of gas responds to changes in pressure, volume, or temperature.

The Volume of a Given Gas Sample Varies Inversely with the Pressure at Constant Temperature   When he trapped a gas sample in a device similar to that shown in Figure 6.2, English scientist Robert Boyle (1627–1691) discovered that an increase in gas pressure reduced the gas volume proportionately. If the pressure is doubled, the volume is cut in half, for example.

Whenever one quantity decreases in proportion to an increase in the other, we say that the quantities are *inversely proportional.* In time, other scientists found that all gases had this property, so we now have a law of nature called the **pressure–volume law** or **Boyle's law.**

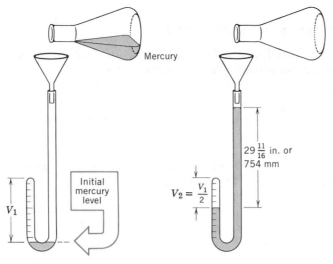

FIGURE 6.2

J-tube apparatus for pressure–volume data. On the left, the pressures in the two arms of the tubes are the same and equal, say, 754 mm Hg. A volume of gas, V_1, has been trapped in the shorter arm with the sealed end. On the right, enough mercury has been added to make the mercury column in the longer arm extend 754 mm above the top of the shorter mercury column. In other words, the pressure on the entrapped gas is now 754 + 754 mm Hg or twice the initial value. This has squeezed the volume of the gas to half its original size. Thus doubling the pressure halves the volume, as Robert Boyle discovered.

Boyle's Law (Pressure–Volume Law) The volume of a given amount of gas held at constant temperature varies inversely with the pressure.

$$V \propto \frac{1}{P} \qquad (T \text{ and mass are constant}) \qquad (6.2)$$

■ The symbol $\propto$ stands for *is proportional to*.

The proportionality sign, $\propto$, in Equation 6.2 can be replaced by introducing a proportionality constant, C.

$$V = \frac{1}{P} \times C$$

Rearranging gives

$$PV = C \qquad (6.3)$$

Let us now imagine that we have a fixed quantity of gas at constant temperature and that some change is made in the sample's volume or pressure. Using subscripts to distinguish the data, let the initial values of pressure and volume be P_1 and V_1 and the final values P_2 and V_2. Initially, before the change, Equation 6.3 tells us that

$$P_1 V_1 = C \qquad (6.4)$$

After the change,

$$P_2 V_2 = C \qquad (6.5)$$

The constant C, of course, *must be the same for both Equations 6.4 and 6.5.* Otherwise we could not call C a *constant.* Two things equal to the same thing, C in this instance, are equal to each other, so

$$P_1 V_1 = P_2 V_2 \qquad (6.6)$$

■ The same constant, C, applies. This is at the heart of Boyle's law.

This is probably the most useful way to express Boyle's law. When we know any three of the four values in Equation 6.6, we can calculate the fourth without the need to evaluate the Boyle's law constant, C.

151

EXAMPLE 6.1
Calculating with Boyle's Law

A sample of nitrogen occupies a volume at 25 °C of 5.65 L when the gas pressure is 740 mm Hg. If, at the same temperature, the pressure is changed to 760 mm Hg, what is the final volume?

ANALYSIS The problem is about the interaction of pressure with volume for a fixed amount of gas at constant temperature, all of the specifications of Boyle's law.

SOLUTION You will generally find gas law calculations easier if you assemble all pertinent data first.

$$V_1 = 5.65 \text{ L} \qquad P_1 = 740 \text{ mm Hg}$$
$$V_2 = ? \text{ L} \qquad P_2 = 760 \text{ mm Hg}$$

Now we apply Boyle's law using Equation 6.6, $P_1V_1 = P_2V_2$. After inserting the given data and rearranging, we have

$$V_2 = 5.65 \text{ L} \times \frac{740 \text{ mm Hg}}{760 \text{ mm Hg}} = 5.50 \text{ L}$$

CHECK Is the answer reasonable? Yes; 5.50 L is less than 5.65 L, and the *decrease* in volume is mandated by the *increase* in pressure (it's Boyle's law).

PRACTICE EXERCISE 1

If 2.5 L of a gas is at a pressure of 760 mm Hg, and its pressure changes to 730 mm Hg, what is the new volume (assuming a constant temperature)?

A Gas that Obeys the Gas Laws Exactly Is an *Ideal Gas* Most gases closely follow Boyle's law rather well at ordinary temperatures and pressures, like those in our calculations, but no gas obeys the law *exactly* over a wide range of pressures. A gas most poorly satisfies the Boyle's law equation when it is under high pressure (10 atm or more) or at a low temperature (below 200 K), or some combination of these. Under these conditions, most gases are close to changing to a liquid, and they depart from gas law behavior.

Although real gases do not *exactly* obey Boyle's law or any of the other gas laws, we can yet imagine a hypothetical gas that does. It's called an **ideal gas,** and is *defined* as one that obeys the gas laws exactly. A real gas increasingly shows ideal gas behavior as its pressure is decreased and its temperature is increased, changes that lower the tendency of the gas to change to its liquid state.

A Change in the Temperature of a Gas Sample *Proportionately* Changes Its Volume if the Pressure Is Held Constant Figure 6.3 shows how we could maintain a constant pressure on a gas sample while studying how a temperature change affects the gas volume. By raising and lowering a leveling bulb, the internal gas pressure can be kept equal to the external atmospheric pressure.

Plots of volumes versus temperatures for different amounts of the same gas under constant pressure are straight lines (see Figure 6.4). Thus, at constant pressure, the volume of each gas sample is directly proportional to the kelvin temperature. The gas involved in Figure 6.4 is one that happens to change to its liquid state at −100 °C but, because each plot is a straight line, it is easy to extrapolate (meaning, *reasonably extend*) the lines to temperatures below −100 °C. The dashed lines in Figure 6.4, therefore, are reasonable predictions of how the gas would behave if it never changed to a liquid.

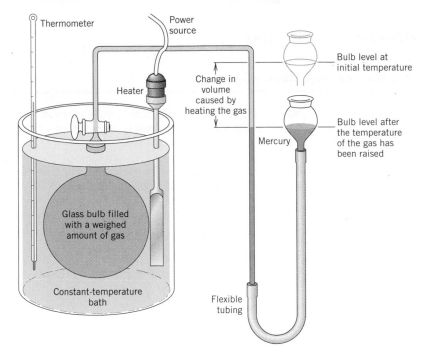

FIGURE 6.3
Obtaining temperature–volume data for a gas at constant pressure. The temperature of the gas can be changed by varying the temperature of the water bath.

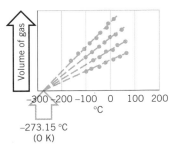

FIGURE 6.4
Plots of temperature–volume data obtained at constant pressure. For different masses of the same gas, the volume is directly proportional to the temperature.

Notice that all of the extrapolated lines point to -273.15 °C. The samples would all reach this temperature if their volumes could become zero. A zero volume is impossible for a *real* gas, of course. Yet this extrapolation tells us that -273.15 °C must theoretically be the coldest possible temperature. To imagine an even lower temperature would require that a gas be able to have a *negative* volume. So -273.15 °C is named **absolute zero** and is 0 K on the Kelvin temperature scale.

The **temperature–volume law,** named **Charles' law** after French scientist Jacques Charles (1746–1823), describes the way that volume and temperature interact for a gas at constant pressure.

Charles' Law (Temperature–Volume Law) The volume of a given amount of gas held at constant pressure is directly proportional to the Kelvin temperature.

$$V \propto T \qquad \text{(at constant } P \text{ and amount)} \qquad (6.7)$$

The proportionality given by 6.7 is changed to an equation by introducing a proportionality constant, C'. (The prime, $'$, indicates that this constant is not the same one used for Boyle's law.)

$$V = C'T \qquad (6.8)$$

Alternatively

$$C' = \frac{V}{T} \qquad (6.9)$$

By using subscripts as we did with Boyle's law, we can restate the relationship of Equation 6.9 as follows to make an evaluation of C' unnecessary.

$$\frac{V_1}{T_1} = \frac{V_2}{T_2} \qquad (6.10)$$

Equation 6.10 is true because each side equals the same constant, C'. Only by using the kelvin temperature scale can Charles' law be expressed in such simple terms, which is why we use kelvins in all calculations involving gases.

■ Using laser technology, scientists at Colorado's Joint Institute for Laboratory Astrophysics were able in 1991 to cool a sample of cesium vapor to within 10^{-6} degree of 0 K, the closest to absolute zero ever attained.

■ $\dfrac{V_1}{T_1} = C'$ and $\dfrac{V_2}{T_2} = C'$.
Hence $\dfrac{V_1}{T_1} = \dfrac{V_2}{T_2}$,
which rearranges to Equation 6.10.

EXAMPLE 6.2
Calculating with Charles' Law

Some of the anesthetics used in surgery are gases at 37 °C (body temperature). If 1.50 L of a gas is used at 20 °C, to what volume does the gas change when the temperature becomes 37 °C at the same pressure?

ANALYSIS The amount of gas and the pressure remain constant, so we know we are dealing with the temperature–volume law and can use Equation 6.10.

SOLUTION Let's collect the data, changing degrees Celsius to kelvins:

$$V_1 = 1.50 \text{ L} \qquad T_1 = 293 \text{ K } (20.0 + 273)$$
$$V_2 = ? \qquad T_2 = 310 \text{ K } (37.0 + 273)$$

■ The volumes can be in any units as long as they are the same units for both V_1 and V_2.

After putting these data into Equation 6.10 and rearranging we have

$$V_2 = 1.50 \text{ L} \times \frac{310 \text{ K}}{293 \text{ K}}$$

$$= 1.59 \text{ L}$$

The volume increases by 0.09 L, about 90 mL.

■ Although this is only a 6% change in volume, it is a factor that anesthesiologists have to consider.

CHECK The answer is reasonable because the volume must increase with the increase in temperature when the pressure is constant (Charles' law).

PRACTICE EXERCISE 2

A sample of cyclopropane, an anesthetic, with a volume of 575 mL at a temperature of 30 °C, is cooled to 15 °C at the same pressure. What is the new volume?

When the Temperature of a Gas in a Rigid Container Is Changed, the Pressure Changes *Proportionately* A rigid container might be an aerosol can idly tossed into a fire, contrary to the warning label. The sealed can is likely to explode from the increase in its internal pressure. Jacques Charles and Joseph Gay-Lussac (1778–1850), French scientists, discovered a relationship between gas temperature and pressure now called the **pressure–temperature law,** or **Gay-Lussac's law** that tells us how pressure and temperature interact.

Gay-Lussac's Law (Pressure–Temperature Law) The pressure of a given amount of gas held at constant volume is directly proportional to the Kelvin temperature.

$$P \propto T \qquad \text{(at constant } V \text{ and amount)} \tag{6.11}$$

The relationship of Equation 6.11 converts to Equation 6.12 in the same way that a similar equation for Charles' law was handled. (Do this as an exercise.)

$$\frac{P_1}{T_1} = \frac{P_2}{T_2} \tag{6.12}$$

Gay-Lussac's law calculations are similar to those for Charles' law. We are close to an important generalization about gases, however, so we won't take up specific examples.

The Combined Gas Law Says that the Ratio of *PV* to *T* Is a Constant In experimental work with gases, it is seldom necessary to maintain a *constant* value of *P*, or *V*, or *T* just to satisfy a specific condition of one of the individual gas laws just discussed. A more general law, the **combined gas law,** relates all three variables for a fixed amount of gas, and includes the preceding gas laws as special cases.

Combined Gas Law The volume of a given amount of a gas is proportional to the ratio of its Kelvin temperature and its pressure.

$$V \propto \frac{T}{P} \qquad (6.13)$$

Introducing a proportionality constant, C'', into Equation 6.13 and rearranging, we obtain

$$\frac{PV}{T} = C'' \qquad \text{(for a fixed amount of gas)} \qquad (6.14)$$

By using subscripts in the usual way to designate initial and final states, we can restate Equation 6.14 as follows.

$$\frac{P_1 V_1}{T_1} = \frac{P_2 V_2}{T_2} \qquad (6.15)$$

■ Because $(P_1 V_1)/T_1$ and $(P_2 V_2)/T_2$ equal the same constant, C'', they must equal each other.

Equation 6.15 is the most useful form of the combined gas law for calculations, because it bypasses the need for a value for C''. Although temperatures must be in kelvins, the pressures and volumes can be in any units we please provided we use them consistently.

Equation 6.15 includes the other gas laws as special cases. Boyle's law, for example, requires a constant temperature, so under this special circumstance, T_1 and T_2 cancel in Equation 6.15. The result is Equation 6.6, $P_1 V_1 = P_2 V_2$, used for Boyle's law calculations.

Under constant pressure, P_1 cancels P_2, so Equation 6.15 reduces to Equation 6.10, Charles' law. At constant volume, V_1 cancels V_2 in Equation 6.15 leaving Equation 6.12, Gay-Lussac's law. You can see why Equation 6.15 is called the *combined* gas law.

EXAMPLE 6.3
Using the Combined Gas Law

A 175-mL sample of argon, the gas in electric light bulbs, is at 25.0 °C and a pressure of 760 mm Hg. To change the pressure to 725 mm Hg and the volume to 200 mL, what change in temperature must occur if no change in the amount of gas is allowed?

■ Argon is one of the noble gases.

ANALYSIS The amount of argon is constant and *all* of the other three gas variables, P, V, and T, change. This is our clue that the combined gas law equation applies rather than any individual gas law.

SOLUTION Following our usual practice, we collect the data first.

$V_1 = 175$ mL $P_1 = 760$ torr $T_1 = 298$ K (25.0 + 273)
$V_2 = 200$ mL $P_2 = 725$ torr $T_2 = ?$

We next use these data in the combined gas law equation.

$$\frac{P_1 V_1}{T_1} = \frac{P_2 V_2}{T_2}$$

$$\frac{(760 \text{ torr})(175 \text{ mL})}{(298 \text{ K})} = \frac{(725 \text{ torr})(200 \text{ mL})}{(T_2)}$$

After rearranging we have

$$T_2 = \frac{(298\ K)(725\ torr)(200\ mL)}{(760\ torr)(175\ mL)}$$

$$T_2 = 325\ K,\ or\ 52°C\ (from\ 325 - 273)$$

CHECK The overall effect is that an *increase* in temperature is needed. Is this reasonable? Opposing forces are at work here, so we must be careful. The volume *increase* requires that the temperature go up (Charles' law), but the pressure *decrease* calls for a reduced temperature (Gay-Lussac's law). Notice, however, that the volume seems to be increased by a larger fraction (200/175) than the pressure is decreased (725/760). (It's not an easy call.) Therefore the volume increase requires a larger rise in temperature than can be offset by the pressure decrease.

PRACTICE EXERCISE 3

What will be the final pressure of a sample of oxygen with a volume of 775 mL at 745 torr and 25.0 °C if it is heated to 55.0 °C and given a final volume of 800 mL?

6.3

THE IDEAL GAS LAW

The ratio of *PV* to *nT* is a constant, the universal gas constant, *R*.

When Gases React, the Ratios of the *Volumes* that Combine Match the Mole Ratios Hydrogen gas reacts (explosively) with chlorine gas to give gaseous hydrogen chloride. The coefficients in the equation give the *mole* ratios; beneath the formulas stand the measured ratios by *volume,* and you can see that the sets of numbers — moles and volumes for the same substance — match.

■ Conditions of constant temperature and pressure are assumed throughout this discussion.

$$H_2(g) + Cl_2(g) \longrightarrow 2HCl(g)$$
$$1\ vol \qquad 1\ vol \qquad 2\ vol$$

The same kind of match occurs in the reaction of hydrogen with oxygen to give water, which is a gas above 100°C.

$$2H_2(g) + O_2(g) \longrightarrow 2H_2O(g)$$
$$2\ vol \qquad 1\ vol \qquad 2\ vol$$

The Match between Moles and Volumes of Reacting Gases Implies that Equal Volumes of Gas Have Equal Numbers of Moles Amedeo Avogadro, pondering the *whole number* relationships between the *volumes* of reacting gases, concluded that *equal volumes of gases (at the same temperature and pressure) must have identical numbers of molecules.* Today we call this **Avogadro's principle,** and knowing that "equal numbers of molecules" is the same as "equal numbers of *moles,*" we can express the principle as follows, where *n* = number of moles.

Avogadro's Principle When measured at the same temperature and pressure, equal volumes of gases contain equal numbers of moles.

$$V \propto n \qquad (at\ constant\ T\ and\ P)$$

TABLE 6.1
Molar Volumes of Some Gases at STP

Gas	Formula	Molar Volume (L)
Helium	He	22.398
Argon	Ar	22.401
Hydrogen	H_2	22.410
Nitrogen	N_2	22.413
Oxygen	O_2	22.414
Carbon dioxide	CO_2	22.414

The Same *Molar Volume* Applies to *All* Gases   Avogadro's principle could not be true unless the volume occupied by one mole of *any gas*—its *molar volume*—is the same for *all* gases (under the same pressure and temperature). For comparison purposes, scientists use the **standard conditions of temperature and pressure,** or **STP,** defined as 1 atm and 273.15 K (0 °C). The volume of one mole of any gas at STP is 22.4 L, or very close to this (see Table 6.1), so 22.4 L per mol at STP is called the **standard molar volume** of a gas.

The Combined Gas Law Becomes the *Ideal Gas Law* by Incorporating the Amount of Gas   The constant C'' in the combined gas law, $PV/T = C''$, is a constant only for a fixed quantity of gas. C'', in fact, is directly proportional to gas quantity. Using the mole unit, n, we can split C'' into the product of the number of moles, n, and still another constant:

$$\frac{PV}{T} = C'' = n \times \text{new constant}$$

The "new constant" is called the **universal gas constant** and is given the symbol R. We can now rewrite the combined gas law in a more general form called the **ideal gas law:** $PV/T = nR$. The equation is usually rearranged as follows, and is often called the **equation of state for an ideal gas.**

Ideal Gas Law (Equation of State for An Ideal Gas)

$$PV = nRT \qquad (6.16)$$

Equation 6.16 tells us that if we know values of *any* three of the four variables for a gas, P, V, n, and T, we can calculate the fourth. The equation also says that if any three of the four variables are fixed for a given gas, *the fourth can only have one value.* This is a remarkable fact about our world, and *it applies to all gases.* We can define the *state* of a given gas simply by specifying any three of the four variables. Nothing like the equation of state for an ideal gas exists for liquids and solids.

The value of the universal gas constant, R, is found by rearranging Equation 6.16 and inserting the values for the standard pressure, temperature, and molar volume. When $n = 1$ mol, $V = 22.4$ L, $P = 1.00$ atm, and $T = 273$ K,

$$R = \frac{PV}{nT} = \frac{(1.00 \text{ atm})(22.4 \text{ L})}{(1.00 \text{ mol})(273 \text{ K})}$$

$$= 0.0821 \frac{\text{atm L}}{\text{mol K}}$$

Or, arranging the units in their most commonly given order,

$$R = \textbf{0.0821 L atm/mol K}$$

The value of R obviously depends on the *units* used for pressure and volume. For example, if we calculate R using 760 mm Hg (the same as 1 atm) and 22.4×10^3 mL (the same as 22.4 L), we get

$$R = \frac{(760 \text{ mm Hg})(22.4 \times 10^3 \text{ mL})}{(1.00 \text{ mol})(273 \text{ K})}$$

$$= 6.24 \times 10^4 \text{ mL mm Hg/mol K}$$

EXAMPLE 6.4
Using the Ideal Gas Law

A sample of oxygen at 24.0 °C and 745 torr was found to have a volume of 455 mL. How many grams of O_2 were in the sample?

ANALYSIS In this question we are given pressure, volume, and temperature of a gas and then asked to determine the amount of gas. These quantities are related by the ideal gas law, so we need the equation $PV = nRT$. We can use the equation to calculate the number of moles of O_2, and then we can use the formula mass of O_2 as a tool to calculate the mass of O_2 in grams.

SOLUTION First, let's solve the ideal gas law for n.

$$n = \frac{PV}{RT}$$

To use the value of 0.0821 L atm/mol K for R, we must have all the data in matching units: V in liters, P in atmospheres, and T in kelvins. Gathering the data and making the necessary unit conversions, we have

■ Show that the identical answer, 0.0183 mol O_2, is obtained when the alternative value of R is used,

$$6.24 \times 10^4 \frac{\text{mL mm Hg}}{\text{mol K}}$$

$P = 0.980$ atm   From: $745 \text{ torr} \times \dfrac{1 \text{ atm}}{760 \text{ torr}}$

$V = 0.455$ L   From: 455 mL
$T = 297$ K   From: $24.0 + 273$

Substituting these values gives

$$n = \frac{(0.980 \text{ atm})(0.455 \text{ L})}{(0.0821 \text{ L atm/mol K})(297 \text{ K})}$$

$$= 0.0183 \text{ mol of } O_2$$

The formula mass of O_2 is 32.00. Therefore,

$$0.0183 \text{ mol } O_2 \times \frac{32.00 \text{ g } O_2}{1 \text{ mol } O_2} = 0.586 \text{ g } O_2$$

The sample of oxygen must have had a mass of 0.586 g.

PRACTICE EXERCISE 4

What volume in milliliters does a sample of nitrogen with a mass of 0.245 g occupy at 21.0 °C and 750 torr?

6.4

DALTON'S LAW OF PARTIAL PRESSURES

The total pressure of a mixture of gases is the sum of the partial pressures of the individual gases.

Many applications involving gases concern *mixtures* of nonreacting gases, like the air we breathe, and our question in this section is: How are the gas laws affected when we move from pure gases to mixtures? We are particularly interested in what effect the mixing of gases has on the overall pressure.

John Dalton discovered that each gas in a mixture contributes to the total pressure in proportion to its fraction, by volume, of the mixture. The individual contribution to the total pressure is called the **partial pressure** of the gas. It is the pressure that the gas would have if it were the only gas in the same container as the mixture and at the same temperature. The symbol for the partial pressure of gas a is P_a. The formula of a particular gas can be put in place of the subscript in P_a as in P_{O_2}, the partial pressure of oxygen. What Dalton found is now called **Dalton's law of partial pressures.**

■ You'll often see a partial pressure like P_{O_2} written as PO_2.

Dalton's Law of Partial Pressures The total pressure of a mixture of nonreacting gases is the sum of their individual partial pressures.

$$P_{total} = P_a + P_b + P_c + \cdots + \tag{6.17}$$

In dry CO_2-free air at STP, for example, P_{O_2} is 159 torr and P_{N_2} is 601 torr. The sum of these partial pressures is 760 torr or 1 atm. (Actually, about 1% of clean, dry air is taken up by other gases, mostly argon.)

■ For every 100.000 L of pure dry air, there is 0.934 L of argon, so air is nearly 1% in argon on a volume basis.

EXAMPLE 6.5
Using Dalton's Law of Partial Pressures

Suppose you want to fill a pressurized tank having a volume of 4.00 L with oxygen-enriched air for use in diving, and you want the tank to contain 50.0 g of O_2 and 150 g of N_2. What must be the total gas pressure at 25 °C?

ANALYSIS If we calculate the pressure that each gas would have if it were *alone* in the tank and then add the two partial pressures together, we'll have the answer. To find the individual partial pressures, we use the ideal gas law, $PV = nRT$, for a temperature of 25 °C or 298 K and a volume of 4.00 L. To use this law, we must first express the amounts of the gases in *moles*.

SOLUTION For moles of oxygen,

$$n = 50.0 \text{ g } O_2 \times \frac{1 \text{ mol } O_2}{32.0 \text{ g } O_2} = 1.56 \text{ mol } O_2$$

For moles of nitrogen,

$$n = 150 \text{ g } N_2 \times \frac{1 \text{ mol } N_2}{28.0 \text{ g } N_2} = 5.36 \text{ mol } N_2$$

Next, we calculate the partial pressures. For each, $P_a = (nRT)/V$, where V is the volume of the cylinder.

$$P_{O_2} = \frac{1.56 \text{ mol} \times 0.0821 \text{ L atm/K mol} \times 298 \text{ K}}{4.00 \text{ L}}$$
$$= 9.54 \text{ atm}$$

$$P_{N_2} = \frac{5.36 \text{ mol} \times 0.0821 \text{ L atm/K} \times 298 \text{ K}}{4.00 \text{ L}}$$
$$= 32.8 \text{ atm}$$

For the total pressure, using Dalton's law, we have

$$P_{total} = 9.54 \text{ atm} + 32.8 \text{ atm}$$
$$= 42.3 \text{ atm}$$

A relatively high pressure, 42.3 atm, is needed for the tank to hold only 50.0 g of O_2 and 150 g N_2.

PRACTICE EXERCISE 5

How many grams of oxygen are present at 25 °C in a 5.00-L tank of oxygen-enriched air under a total pressure of 30.0 atm when the only other gas is nitrogen at a partial pressure of 15.0 atm?

PRACTICE EXERCISE 6

At sea level and 0 °C and 760 mm Hg pressure, the partial pressure of nitrogen in clean dry air is 601 mm Hg. If oxygen is the only other component, what is its partial pressure?

PRACTICE EXERCISE 7

The atmospheric pressure atop Mount Everest (8.8 km) is 250 mm Hg and the partial pressure of nitrogen is 198 mm Hg. What is the partial pressure of oxygen, assuming no other gas is present? (The value is too low to force O_2 from the lungs into the bloodstream at a rate fast enough to sustain human activities for nearly all people.)

Vapors in Contact with Their Parent Liquids Are Treated as Gases We learned in Chapter 2 that liquids can change to their vapor states by *evaporation*. The air space in contact with any liquid therefore includes not only air but the vapor generated by this evaporation. The partial pressure exerted by the vapor in contact with its liquid in a closed container at a given temperature is called the **vapor pressure** of the liquid at that temperature.

Vapor pressure is measured using a closed container to prevent vapor from escaping (see Fig. 6.5). A device to measure pressure inside an enclosed space is called a **manometer.** In Figure 6.5, the manometer is the U-shaped tube, an example of an *open-end* manometer (meaning one end is open to the atmosphere). Both arms of the U contain mercury, initially at the same levels so the pressure inside the bulb equals that of the surrounding atmosphere. When a liquid is let into the bulb, its evaporation generates vapor and a vapor pressure, which pushes the mercury and changes the mercury levels in the manometer as shown in Figure 6.5. The difference in heights of the two mercury levels is the vapor pressure of the liquid.

As long as some liquid remains in contact with the vapor, *a liquid's vapor pressure depends only on the identity of the liquid and its temperature.* Different liquids have different vapor pressures (compared at the same temperature), and the vapor pressures of all liquids increase with temperature. Table 6.2, for example, gives the vapor pressure of water at different temperatures.

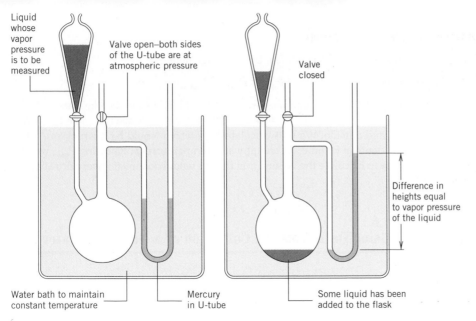

FIGURE 6.5
Measuring the vapor pressure of a liquid. A liquid is added to the glass bulb connected to an open-end manometer. On the left, just before the liquid is added, both mercury levels are the same in the manometer, so the pressure in the bulb equals the atmospheric pressure. On the right, the pressure exerted by the developing vapor builds up forcing one mercury level down and causing the other to rise. The difference in the heights of the mercury levels equals the vapor pressure of the liquid. (From J. E. Brady and J. R. Holum, *Chemistry. The Study of Matter and Its Changes,* 1993. John Wiley and Sons, Inc., New York. Used by permission.)

TABLE 6.2
Vapor Pressure of Water

Temperature (°C)	Vapor Pressure (mm Hg)	Temperature (°C)	Vapor Pressure (mm Hg)
18	15.5	32	35.7
20	17.5	34	39.9
22	19.8	36	44.6
24	22.4	37	47.1
26	25.2	38	49.7
28	28.3	40	55.3
30	31.8	50	92.5

Dalton's Law of Partial Pressures Applies to a Gas Collected Over Water   In the laboratory preparation of gases that do not react with water, a simple arrangement enables trapping the gas over water (see Fig. 6.6). The collected gas contains water vapor, as much as the vapor pressure of water permits at the water temperature. Because of the water vapor, the entrapped gas must be treated as a *mixture* of gases.

Usually we want to know how much *dry* gas is in a wet sample. This is calculated with the aid of Dalton's law, which lets us write

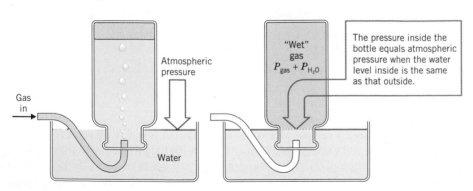

FIGURE 6.6
The collection of a gas over water produces a gas that contains water vapor. To collect the gas, a collecting bottle is first filled with water and then inverted over a basin of water. This water is displaced as the gas bubbles into the bottle. During this time, water vapor effectively saturates the gas and contributes its own partial pressure. (Adapted by permission from J. E. Brady and J. R. Holum, *Chemistry. The Study of Matter and Its Changes.* 1993, John Wiley and Sons, Inc.)

161

The partial pressure of the gas is then, by rearrangement,

$$P_{gas} = P_{total} - P_{water}$$

■ The *water* temperature is used (which might not be the same as the lab temperature) because it determines the vapor pressure of the water and thereby the contribution that water's vapor pressure makes to the total pressure.

The value of P_{water} at the water temperature, T, can be found in a table, like Table 6.2. P_{total} is read at the laboratory barometer, provided that the water levels are the same inside the gas collecting bottle and outside. We thus calculate P_{gas}, which would be the pressure if the gas were dry in the identical volume, V, and at water temperature T. Knowing P_{gas}, we use the combined gas law to calculate the volume that the gas would occupy if it were dry at any other values of P and T.

EXAMPLE 6.6
Calculation Using Data from the Collection of a Gas Over Water

Oxygen was collected in a 310-mL flask over water at 20 °C and a pressure of 738 mm Hg. (a) What is the partial pressure of the collected oxygen? (b) What would be the volume of the collected oxygen if it were dry at STP?

ANALYSIS The total pressure, 738 mm Hg, includes a contribution of the partial pressure of the water vapor, which (from Table 6.2) we find is 17.5 mm Hg at 20 °C. The difference, $[P_{total} - P_{H_2O}]$, will be the answer to part *a* (by Dalton's law).

Part *b* asks for the final volume, V, at STP. After answering part *a*, we will have values for P_1, V_1, and T_1, as well as values for P_2 and T_2. We're asked to calculate V_2. The combined gas law tells us how these values are all related.

SOLUTION First, we find the partial pressure of the oxygen, using Dalton's law.

$$P_{O_2} = P_{total} - P_{water}$$
$$= 738 \text{ torr} - 17.5 \text{ torr} = 721 \text{ torr}$$

Second, we assemble the data.

$P_1 = 721$ torr	$P_2 = 760$ torr (the standard atmosphere)
$V_1 = 310$ mL	$V_2 = ?$
$T_1 = (20 + 273) = 293$ K	$T_2 = 273$ K (standard temperature)

We use these in the combined gas law equation:

$$\frac{P_1 V_1}{T_1} = \frac{P_2 V_2}{T_2}$$

$$\frac{(721 \text{ torr})(310 \text{ mL})}{293 \text{ K}} = \frac{(760 \text{ torr})(V_2)}{273 \text{ K}}$$

Solving for V_2 gives us

$$V_2 = \frac{(721 \text{ torr})(310 \text{ mL})(273 \text{ K})}{(760 \text{ torr})(293 \text{ K})}$$

$$= 274 \text{ mL}$$

Thus, when the water vapor is removed from the gas sample, the dry oxygen will occupy a volume of 274 mL at STP.

PRACTICE EXERCISE 8

Suppose you prepared a sample of nitrogen and collected it over water at 15 °C at a total pressure of 745 torr and a volume of 310 mL. Find the partial pressure of the nitrogen and the volume it would occupy at STP. (The vapor pressure of water at 15 °C is 12.79 mm Hg.)

TABLE 6.3
The Composition of Air during Breathing

Gas	Partial Pressure (mm Hg)		
	Inhaled Air	Exhaled Air	Alveolar Air
Nitrogen	594.70	569	570
Oxygen	160.00	116	103
Carbon dioxide	0.30	28	40
Water vapor	5.00[a]	47	47
Totals	760.00	760	760

[a] A partial pressure of water vapor of 5.00 mm Hg corresponds to air with a relative humidity of about 20%, a condition in which air is holding 20% of the maximum amount of water vapor it is able to hold at the given temperature.

Water Vapor Is a Component of the Respiratory Gases Table 6.3 gives the values of the partial pressures of the gases in inhaled air, exhaled air, and alveolar air. The data show how the partial pressures of the various gases adjust to the changes in their immediate environments, but the total pressure stays constant. Notice particularly how the partial pressures of all gases but nitrogen significantly change. The partial pressure of water vapor, for example, increases to 47 mm Hg as air enters the body and encounters a more moist environment. The normal partial pressure of water at 37 °C (Table 6.2) is 47 mm Hg, so we know that exhaled and alveolar air are *saturated* in water vapor. Alveolar air is rich in carbon dioxide, a product of metabolism that needs to be exhaled. Alveolar air is poor in oxygen because oxygen is being rapidly absorbed out of the lungs and into blood capillaries. The differences between alveolar air and exhaled air stem largely from mixing.

■ Alveolar air is air inside *alveoli,* thin-walled air sacs enmeshed in beds of fine blood capillaries. We have roughly 300 million such air sacs in our lungs providing an enormous surface area for the exchange of O_2 and CO_2 between the alveolar air and the bloodstream.

6.5

THE KINETIC THEORY OF GASES

All the gas laws can be explained in terms of the model of an ideal gas described by the kinetic theory.

As the gas laws unfolded over the decades, more and more scientists asked the question: "What must gases really be like given the *experimental* facts about gases (the gas laws)?"

A Model of an Ideal Gas Led to Theoretical Conclusions that Matched the Gas Laws Scientists used the behavior of real gases to postulate the following features of an ideal gas.

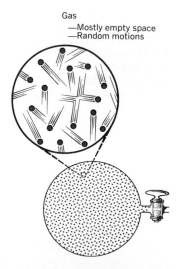

Gas
—Mostly empty space
—Random motions

Model of an Ideal Gas

1. The ideal gas consists of a large number of extremely tiny particles in a state of chaotic, utterly random motion.
2. The particles are perfectly hard, and when they collide they lose no energy because of friction.
3. The particles neither attract nor repel each other.
4. The particles move in accordance with the known laws of motion.

These postulates and the calculations based on them make up the **kinetic theory of gases.** The truth about gases, according to this theory, is that they consist of tiny particles in random motion.

■ The laws of motion are part of the science of physics, the science of motion and energy.

Theorists used the postulates and the equations for the laws of motion to see if they could derive the gas laws theoretically. They were splendidly successful. We won't develop the equations, but we will survey the results and see how the kinetic model explains the gas laws.

Gas Temperature Is Related to the Average Kinetic Energy of the Gas Particles

One of the outstanding results of the kinetic theory was an explanation of gas temperature. Using the model, theorists were able to show that *gas temperature is directly proportional to the average kinetic energy of the gas particles.* We'll speak of this energy as the average *molecular kinetic energy.* In other words, when we heat a gas and raise its temperature, we increase the average molecular kinetic energy of its particles. When we cool a gas, we reduce the average molecular kinetic energy.

■ The term "molecular" in "molecular kinetic energy" can be stretched to include the monatomic, noble gases.

What actually changes when the molecular kinetic energy changes, given that kinetic energy equals $(1/2)mv^2$? It cannot be the mass (m) of each particle, so the average *velocity* of the particles must be what changes. Heating a gas, therefore, increases the average molecular kinetic energy by increasing the average velocity of the gas molecules.

In the light of these ideas, it makes sense that if we cool a gas enough and slow its molecules to zero velocity, the molecular kinetic energy would decrease to zero. Because there's no such thing as *negative* kinetic energy, there ought to be a lower limit to which something can be cooled. As we know, this limit is $-273.15\ °C$, absolute zero.

The Kinetic Theory Explains Gas Pressure and Boyle's Law   Another result of the kinetic theory was an explanation of gas pressure. Theorists could show that the pressure exerted by a gas arises from the innumerable collisions per second that the particles make with each unit of area of the walls of the container. If we imagine, then, that the volume of the container is made less, the wall area battered by the gas particles is also less. More hits *per unit wall area* therefore occur after the volume has been reduced. Moreover, the reduction in volume does not affect the average speed at which the particles travel — only the temperature affects speed — and the particles don't have to travel as far to hit a wall. Thus reducing the volume of the gas increases the frequency with which its particles hit the walls without reducing the strengths of each tiny collision, so the gas pressure must increase (see Fig. 6.7).

■ Pressure is related to the ratio number of hits per second/area hit, so as the denominator (area hit) grows smaller, the ratio (pressure) *increases.*

When the theoretical calculations on which the preceding description is based were carried out, the result was identical with Boyle's pressure–volume law for gases. This kind of

FIGURE 6.7

The kinetic theory and the pressure–volume relationship (Boyle's law). The pressure of the gas is proportional to the frequency of collisions per unit area. When the gas volume is made smaller in part *b*, the frequency of the collisions per unit area of the container's walls increases. This is how the pressure increase occurs.

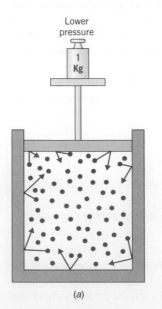
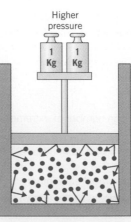

(a)   (b)

agreement and others like it give us confidence that the model of an ideal gas closely describes real gases, too.

The Kinetic Theory Explains Charles' Law If we make the gas molecules move with a higher average speed and energy (by heating the gas), the molecules will hit the container's walls more frequently and with greater energy. If we want to prevent the pressure from increasing — and remember that constant pressure is a condition of Charles' law — we have to let the volume expand. The theoretical calculations along these lines showed that gas volume at constant pressure is, in theory, proportional to the kelvin temperature, just as Charles and others had earlier shown by experiment.

The Kinetic Theory Also Explains Gay-Lussac's Law If we don't let the volume expand when we heat a confined gas, then the pressure of the gas must increase. The calculations based on the model of an ideal gas predicted the same relationship between P and T that had long been known from the pressure–temperature law.

■ This is the law behind the warning not to incinerate aerosol cans.

The Kinetic Theory Explains Gas Diffusion Another property of gases that the kinetic theory explains is the ability of a gas to diffuse or spread out throughout its entire container. Particles in random motion eventually find their way into all the space available. Almost everyone has experienced the diffusion of perfume, cologne, or after-shave fragrances throughout a room.

For us, perhaps the most useful single part of the kinetic theory of gases is the image of particles in rapid, utterly chaotic, random motion. The answer to the question, "What must gases really be like given the experimental truth of the gas laws?" is that real gases are very much like the ideal gas model. The kinetic theory also gives us something very fundamental for explaining some features of the liquid and solid states.

6.6

THE LIQUID STATE

Molecules in liquids experience attractions for each other but retain the freedom to move about.

There is no general pressure–volume law for liquids. The molecules in a liquid are quite tightly packed, so tightly that the volume of a liquid is barely changed even under pressures hundreds of times that of the atmosphere. The hydraulic brakes of a vehicle take advantage of this fact. When you push on the brake pedal, you exert a force on one end of the "plug" of liquid in the brake line. Because the brake fluid cannot be compressed, the force is "felt" instantly at the other end of the liquid by the mechanism that forces the brake shoe against the brake lining to slow the vehicle.

No general temperature–volume law exists for liquids either, although liquids do expand slightly when they are heated. The lack of large empty spaces between the molecules of a liquid, a condition sharply in contrast to a gas, explains why there are no laws similar to the gas laws that apply to all liquids. Let us now study the factors that keep a substance in its liquid state.

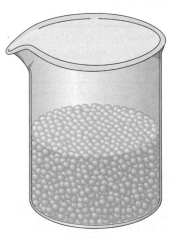

Liquid
Densely packed
Random motions

Substances with Polar Molecules Are More Likely to Be Liquids than Those with Nonpolar Molecules (When the Molecular Sizes Are Comparable) Recall (Section 5.8) that when the molecules of a substance are polar they have permanent $\delta+$ and $\delta-$ sites, and opposite charges attract, even when they are partial charges. The attraction between permanent $\delta+$ and $\delta-$ sites on separate polar molecules is called a **dipole–dipole attraction,** and it is one origin of the electrical force that draws molecules together and enables a substance to be in the liquid state. When the partial charges are very weak, however, this attractive force cannot outbalance the repelling force at work as the electron clouds of gas

■ We studied molecular polarity in Section 5.8, and we saw how strongly oppositely charged ions attract each other in Section 5.1.

■ Each gas has a particular temperature called its *critical temperature* above which no amount of pressure can liquify the gas. The critical temperature of water is 374.1 °C; for carbon dioxide it's 31 °C.

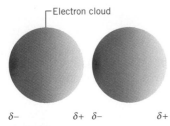

┌Electron cloud

$\delta-$ $\delta+$ $\delta-$ $\delta+$

FIGURE 6.8
London forces. Molecules can be temporarily polarized by coming close to each other, and the force of attraction between such temporary dipoles is called a London force.

■ Helium becomes a liquid only at about 4 degrees above absolute zero (4.18 K), and changing helium to a solid requires an extra 26 atm of pressure.

■

Noble Gas	Formula Mass	Boiling Point (°C)
He	4.00	−269
Ne	20.2	−246
Ar	39.9	−186
Kr	83.3	−152
Xe	131	−107

molecules draw near each other on a collision course. Gas molecules are too weakly polar in relationship to the energies of their collisions.

To change a gas into a liquid we must cool it, exert pressure on it, or carry out a combination of both. Chilling a gas slows its molecules down and so weakens the force of repulsion caused by colliding electron clouds. Exerting a pressure squeezes the gas molecules together, and when they are packed tightly enough, the gas changes over to the liquid state.

London Forces of Attraction Exist between Nonpolar Molecules   Even the most completely nonpolar substances, like the noble gases, are changed into liquids and solids by lowering the temperature enough. This behavior raises the question: How do electrical forces of attraction develop between *nonpolar* atoms or molecules? The answer lies in the nature of the electron cloud encompassing any atom or molecule.

As one nonpolar particle closely approaches another, their two somewhat "soft" electron clouds distort each other (see Fig. 6.8). ("Unlike charges repel.") The distortion gives to the approaching particles a *temporary* polarity; we say that polarization is *induced*. The induced polarization of molecules (or atoms) gives *temporary dipoles* even to molecules lacking permanent dipoles.

In a sample with billions and billions of particles without *permanent* dipoles, we can still easily imagine that there are innumerable, temporarily polarized molecules. Existing between these molecules, therefore, is a net electrical force of attraction, which is called the **London force** (after physicist Fritz London). At sufficiently low temperatures, London forces are strong enough to cause even such totally nonpolar substances as the noble gases to change from a gas to the liquid state or from the liquid to the solid state.

London forces are related to the polarization of electron clouds, so the larger the overall electron cloud per atom or molecule, the more induced polarization a particle can receive. Thus *substances with larger molecules or atoms generally have larger London forces than substances with smaller molecules or atoms.*

An excellent indication of the attractive forces between particles is simply the temperature required to make the substance boil, a temperature called the *boiling point.* When a liquid boils, its molecules are made to separate from each other. A high boiling point indicates that 0relatively large forces of attraction must be overcome before this separation can occur. Because greater London forces are possible with the larger electron clouds of bigger atoms or molecules, the rule of thumb for the boiling points of nonpolar substances is that *the higher the formula mass, the higher is the boiling point.* The boiling points of the noble gases, given in the margin, illustrate this rule. With sufficient molecular size, even nonpolar substances can be liquids and solids at room temperature and pressure. Examples are gasoline and paraffin wax, compounds made entirely of C and H atoms. The electronegativities of C and H are nearly the same (Fig. 5.8, page 135), so neither C—C nor C—H bonds are very polar. Hence, molecules with only these bonds are nearly nonpolar, yet London forces are great enough so that substances with such molecules can be liquids and solids under ordinary conditions.

6.7

VAPOR PRESSURE, DYNAMIC EQUILIBRIA, AND CHANGES OF STATE

When two opposing changes occur at the same rate, the system is in a state of dynamic equilibrium.

The Kinetic Theory Helps to Explain Vapor Pressure   The vapor pressures of several liquids are plotted against temperature in Figure 6.9. Notice that ether develops high vapor pressures even at low temperatures and such liquids are said to be **volatile;** they readily evaporate from open containers at room temperature. In contrast, propylene glycol (in Fig. 6.9) and salad oil have very low vapor pressures at room temperature and they are called

■ The "ether" in Figure 6.9 is diethyl ether, once widely used as an anesthetic.

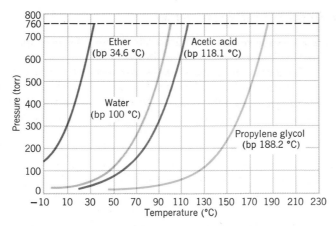

FIGURE 6.9
Equilibrium vapor pressure versus temperature. (Ether was once widely used as an anesthetic. Acetic acid is the sour component in vinegar. Propylene glycol is in several brands of antifreeze mixtures.)

nonvolatile liquids; they do not evaporate at room temperature. Water is more volatile than salad oil, but less volatile than ether.

Suppose we let some water enter the glass bulb of the device in Figure 6.5, page 161. Be sure to notice that the system in the figure is a *closed* system. Once in the bulb, the quantity of water is a constant, but the water can change between the liquid and vapor states. For a very brief moment we can imagine that no water molecules are in the air space above the sample, so the pressure inside and outside the bulb is initially the same, atmospheric pressure. This situation quickly changes, however, because water is somewhat volatile. Its molecules escape and move around in the space above the liquid. Such an escape is called **evaporation.** As more and more of the liquid water evaporates, the molecules entering the bulb's vapor space need more room and so force a change in the mercury level. In other words, water is exerting its vapor pressure.

Eventually, molecules of water vapor, suffering random collisions including some with the liquid surface, begin to return to the liquid state. The change of a vapor to its liquid form is called **condensation** (the verb is *to condense*). Water molecules are coming and going, some leaving the liquid for the vapor state and others returning. The two processes are said to *oppose* each other because one reverses the effect of the other. Eventually, for every molecule leaving the liquid state there is one entering it. The two opposing processes then occur at identical rates and so cancel each other's effect. After this, no further changes occur in the vapor pressure or in the actual masses of material in each of the two physical states. The system in the closed bulb has reached a steady state.

At Equilibrium in a Closed System, the Rates of Evaporation and Condensation Are Equal

When a dynamic process reaches a steady state in which opposing changes occur at identical rates and no further net change takes place, **dynamic equilibrium** exists for the system. We say "dynamic" because there is considerable coming and going; and we say "equilibrium" because there is no *net* change. An equilibrium is like a town with a stable population—births and deaths, departures and arrivals, but no net change in numbers. We have introduced the concept of a dynamic equilibrium with a simple *physical* system so that the associated terms can be defined with the fewest complications. Most of our applications, however, will concern *chemical* systems at equilibrium.

Dynamic Equilibria Are Described by a Special Vocabulary

To represent a dynamic equilibrium, we use an equation with double arrows. For convenience, we use the format of a chemical change even for one that is physical, and we may even refer to materials in two different physical states with the language of "reactant" and "product." Thus the equilibrium involving water is illustrated as follows.

$$\text{water}(l) \rightleftharpoons \text{water}(g) \tag{6.18}$$

This is an example of an *equilibrium expression* or an **equilibrium equation.** The change from left to right is the **forward reaction** or change, and the opposing change, from right to left, is

■ By **system** we simply mean something we have chosen to study, like the water–vapor equilibrium. Everything else in the universe would be called the **surroundings** to the system. Between the system and the surroundings stands a **boundary,** the walls of the container.

167

the **reverse reaction.** The double arrows ⇌ signify that equilibrium exists between the materials on both sides of the arrows.

Sometimes we include heat in an equilibrium expression as if it were also either a "reactant" or a "product." We know, for example, that a certain amount of heat of vaporization (discussed on page 41) is needed to change a liquid into its vapor. The same amount of heat is released when the vapor condenses. We can therefore place "heat" on the *left* in Equation 6.18 because it is "consumed" by the forward change (evaporation) and released by the reverse change (condensation).

$$\text{water}(l) + \text{heat} \rightleftharpoons \text{water}(g) \tag{6.19}$$

■ From *exo-* "out," *endo-* "in," and *-therm-* "heat."

Any change that consumes heat, like evaporation, is said to be **endothermic.** The opposite term is **exothermic,** used to describe a change that liberates heat. The condensation of steam to liquid water, or the condensation of any vapor, is an *exothermic* change.

■ *Spontaneous* here means "happening without the addition or removal of anything (or heat) to the system at equilibrium."

Dynamic Equilibria Can Be Shifted   Once an equilibrium is established, no net change occurs *spontaneously,* but this doesn't mean that we cannot cause a change. By including heat on the appropriate side of the equation, it's easy to predict how an equilibrium will *shift* in response to an increase or a decrease in temperature. If we add heat to the water in an equilibrium system, the rate of evaporation will increase. For a time, the system will not be in equilibrium because the forward rate is now greater than the reverse. We say that the addition of heat *upsets the equilibrium.*

Once we stop adding heat but yet maintain the temperature of the (closed) system at a constant but higher value, the rate at which vapor molecules return to the liquid will catch up. The two opposing rates of evaporation and condensation will become the same again. Both will be faster than at the lower temperature, but when both are *equally* fast there is no further net change. Once again, the system is in dynamic equilibrium.

At the higher temperature, of course, the vapor pressure of the liquid is higher because we have more of the gas in the same enclosed space. The vapor pressures plotted in Figure 6.9 are called the *equilibrium* vapor pressures; each value is the vapor pressure when the liquid and its vapor are in equilibrium at the particular temperature.

Generally, an equilibrium shifts in response to a disturbance, like the addition or removal of heat. A disturbance to an equilibrium is called a **stress,** which is anything that makes one of the two opposing changes faster than the other, at least for a while. Whichever change, forward or reverse, becomes faster is said to be *favored.* If the forward reaction becomes faster, we say that the equilibrium shifts to the right and at least some of the materials before the arrow are changed into materials after the arrow. Heating liquid water in equilibrium with its vapor places a stress on the equilibrium. In response, some liquid water evaporates, and more water vapor forms. On the other hand, an equilibrium shifts to the left when the reverse reaction is favored for any reason. Thus the removal of heat by cooling the water system slows the rate of evaporation and increases the rate of condensation. However, as more and more vapor condenses, the *rate* of this change slows down so, eventually, the rates of evaporation and condensation become equal, and once again there will be equilibrium. Less water mass will remain as a vapor and more will be in the liquid state (and the vapor pressure is less).

Shifts in Equilibria Can Be Predicted by Le Châtelier's Principle   Heat — its addition or removal — is just one stress that can affect an equilibrium, but it neatly illustrates a major principle about how nature works. The principle is called **Le Châtelier's principle,** after Henri Louis Le Châtelier (1850–1936), a French chemist.

> **Le Châtelier's Principle**   If a system in equilibrium is upset by a stress, the system shifts in whichever direction most directly absorbs the stress and restores equilibrium.

As anyone who has tried to cook foods in boiling water at a higher altitude knows, it takes longer than at sea level. Where the atmospheric pressure is lower, water boils at a lower temperature. The chemical reactions caused by cooking are all endothermic. Therefore they do not occur as rapidly at a temperature of, say, 95 °C (roughly the boiling point of water in Denver, Colorado) as they do at 100 °C. No matter how high you turn up the stove setting as you prepare a soft-boiled egg using boiling water, you cannot raise the temperature of boiling water above its value at your altitude. Turning up the stove boils the water away *faster*, but it does not raise its temperature.

Of course, you could use a special pan with a tight lid, such as a pressure cooker. Now the steam cannot escape and its pressure can build up until the safety valve is activated. This higher pressure means that the temperature of the boiling water in the pressure cooker is higher than in the open vessel, so the chemical reactions of cooking occur more rapidly.

These same principles are at work in steam sterilization equipment. To ensure that bacteria and viruses on surgical instruments are both quickly and completely destroyed, hospital workers place the instruments in the equivalent of a pressure cooker where the water and steam temperature can be raised well above the normal boiling point of water.

Provided that the applied stress hasn't been overwhelming, equilibrium will be restored. The rate of the unfavored reaction will eventually catch up to the opposing reaction. Both rates again become equal, and once more there is equilibrium. It's not the identical *system* that existed before the stress because the actual *quantities* of materials represented as reactants and products have changed, but there is once again equilibrium.

As we saw, the addition of heat to the water – water vapor system shifted the equilibrium to the right in favor of vapor. Le Châtelier's principle says that it *must* shift to the right because this is the only direction that absorbs the stress. The extra heat is used up as heat of vaporization and now exists as the molecular kinetic energy of the vapor molecules. Thus Le Châtelier's principle helps us predict the way a system at equilibrium *must* change under a given stress. Equilibria *always* change in whichever way absorbs an applied stress. It is as if the system "rolls with the punches."

■ The liquid – vapor equilibrium of ether would be overwhelmed if so much heat were added that *all* of the liquid changed to vapor.

A Liquid Boils When Its Vapor Pressure Equals the External Pressure Each liquid has a particular temperature at which its equilibrium vapor pressure exactly equals the external pressure of the atmosphere. At this point, the liquid's molecules can enter the vapor state not just at the surface *but everywhere throughout the liquid.* Bubbles of the vapor can now form *beneath* the surface, and they cause quite a commotion as they rise everywhere. This, of course, is the action called **boiling.** Each bubble is essentially pure vapor with a pressure equal to the opposing pressure of the atmosphere. Because the opposing pressure is no longer able to *squeeze* the bubbles back to liquid, and because the bubbles are much less dense than the liquid, they are free to rise to the surface.

The temperature at which a liquid's equilibrium vapor pressure equals 760 mm Hg (1 atm) is called the liquid's **normal boiling point.** Boiling, of course, can occur at other pressures. It happens at whatever temperature the vapor pressure equals the external pressure, but the associated temperature is not called the *normal* boiling point. For example, in Denver, Colorado, at an elevation of one mile where the atmospheric pressure is lower than it is at sea level, water boils at about 95 °C instead of 100 °C. The effect of this on cooking at high altitudes is described in Special Topic 6.1. On the summit of Mount Everest, water boils just under 70 °C!

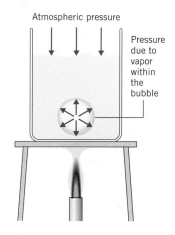

Atmospheric pressure

Pressure due to vapor within the bubble

■ On Mount Everest, the air pressure is roughly 250 mm Hg.

6.8

WATER AND THE HYDROGEN BOND

The physical properties of water are strongly affected by its polarity and the hydrogen bonds between its molecules.

Water is a particularly important liquid. Roughly 60% of the mass of an adult is water, and our bodies take in more water than all other materials combined. Water is the fluid in all cells, a heat-exchange agent, and the carrier in the bloodstream for the distribution of oxygen and all molecules from food, all hormones, minerals, and vitamins, and all disease-fighting agents. To understand many aspects of life at the molecular level, we must learn more about the high polarity of the water molecule and how this affects its properties.

■ Hydrides are compounds of hydrogen with another element.

Water's Boiling Point Is Unusually High   The boiling points of *similar* substances generally increase with formula mass, a rule of thumb illustrated by the noble gases, as we saw on page 166. The group IVA hydrides — methane (CH_4), silane (SiH_4), and germane (GeH_4) — also follow the rule. Their boiling points increase regularly with increasing formula mass (see Fig. 6.10). However, three simple hydrides with low formula masses, water, ammonia, and hydrogen fluoride, are each members of series displaying striking exceptions to the rule.

When the boiling points of the hydrides of the groups VA, VIA, and VIIA elements are plotted against their formula masses, the lowest members of series do not fit a straight line plot (see Fig. 6.10). Thus among the group VA hydrides, ammonia (NH_3) boils far higher than "it should." It is badly off the straight line on which the other group VA hydrides fall, those of phosphorus (PH_3), arsenic (AsH_3), and antimony (SbH_3). Similarly, hydrogen fluoride (HF), does not fit the plot of the boiling points of the hydrides of the group VIIA elements, the halogens.

Water departs most of all from the normal trend in the boiling points of the hydrides of its group, VIA. If the boiling point of H_2O fell on the same line as the boiling points of H_2S, H_2Se, and H_2Te, water "should" boil at about -100 °C. But it actually boils at $+100$ °C, 200 degrees higher.

The plots of Figure 6.10 suggest that forces of attraction are strong *between* molecules in liquid HF, H_2O, and NH_3, but not in methane (CH_4). These forces cannot be explained just as London forces, because the molecules of HF, H_2O, and NH_3 are too small. The molecules of these three hydrides, but not methane, must have unusually strong, *permanent* dipoles that allow relatively strong dipole–dipole attractions between their molecules.

Hydrogen Bonds Exist between Water Molecules   Oxygen is much more electronegative than hydrogen, so each of the two H—O bonds in H_2O are very polar with relatively large

FIGURE 6.10

Boiling points versus formula masses for the binary, nonmetal hydrides of the elements in groups IVA, VA, VIA, and VIIA.

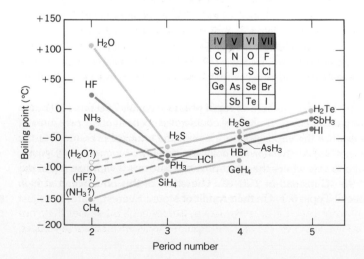

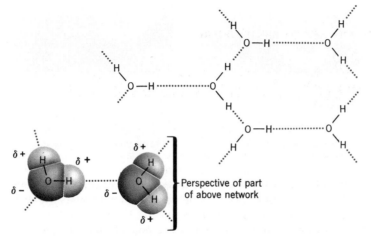

FIGURE 6.11
Hydrogen bonds (·······) in water.

Perspective of part of above network

values of $\delta+$ on H and of $\delta-$ on O. The individual bond polarities in the H_2O molecule do not cancel each other because the molecule is bent. The water molecule as a whole is thus very polar, so polar that chemists consider the dipole–dipole attraction between water molecules strong enough to be called a *bond*. It's not a covalent bond or an ionic bond, but a dipole–dipole attraction having a special name: *hydrogen bond*. It occurs in other systems besides water, but the sizes of $\delta+$ and $\delta-$ are large enough for hydrogen bonds only when H is attached to an atom of one of the three most electronegative elements: O, N, or F. The **hydrogen bond** is the force of attraction between the $\delta+$ on H when held by F, O, or N and the $\delta-$ on some other O, N, or F atom. The hydrogen bond explains why HF, H_2O, and NH_3 have boiling points that are "out of line" for their molecular sizes.

Here are most of the several possibilities where hydrogen bonds can exist. Only partial structures are shown, and a dotted line is used to represent the hydrogen bond. (The solid lines, of course, are covalent bonds.)

$$\overset{\delta+ \quad \delta-}{H-F} \cdots \overset{\delta+ \quad \delta-}{H-F} \qquad \overset{\delta+ \quad \delta-}{H-O} \cdots \overset{\delta+ \quad \delta-}{H-O} \qquad \overset{\delta+ \quad \delta-}{H-O} \cdots \overset{\delta+ \quad \delta-}{H-N}$$

$$\overset{\delta+ \quad \delta-}{H-N} \cdots \overset{\delta+ \quad \delta-}{H-O} \qquad \overset{\delta+ \quad \delta-}{H-N} \cdots \overset{\delta+ \quad \delta-}{H-N}$$

The hydrogen bond is a bridging bond between molecules, as illustrated for water in Figure 6.11. It is by no means as strong as a covalent bond (roughly only 5% as strong). Nevertheless, the hydrogen bond is strong enough to make a difference not just in water but in such important yet different substances as muscle proteins, cotton fibers, and the chemicals of genes, DNA. In fact, among the biochemicals where they occur, hydrogen bonds are more important structurally than any other bond—precisely because they are weak, not strong.

The Hydrogen Bond Is Weaker in Ammonia than in Water

The H—N bonds in ammonia aren't as polar as the H—O bonds in water because N is less electronegative than O. Therefore, the hydrogen bonds between molecules in NH_3 (ammonia) aren't as strong as those that exist in water. This is chiefly why ammonia doesn't have nearly as high a boiling point as that of water. Despite the lessened polarity of the H—N bond, and as suggested by the plots in Figure 6.10, hydrogen bonding nonetheless does occur in pure, liquid ammonia.

A Water Surface Acts like a Skin because of Hydrogen Bonding

All liquids possess a surface tension, but that of water is unusually high. **Surface tension** is a phenomenon in which the surface acts as though it were a thin, invisible, elastic membrane or skin. Water's surface tension explains why some bugs can skitter on the surface of a pond; why parlor magicians can set a steel needle afloat on water; and why water forms tight droplets and doesn't spread out on a waxy surface but does spread out on clean glass. Surface tension is also the reason why a lung collapses under certain conditions.

Surface tension exists in water because its molecules so attract each other that they jam together where water meets air (see Fig. 6.12). The dipole–dipole attractions (hydrogen

■ The direction of the polarity of the H—O bond is

$$\overset{\longrightarrow}{H-O}$$

■ In large flexible molecules (like those of proteins), attractions between $\delta+$ and $\delta-$ sites can occur between parts of the same molecule.

■ In a chain, the important link is the *weakest* link, not the strongest one.

■ H_2O, bp 100 °C
NH_3, bp −33.4 °C
CH_4, bp −161.5 °C

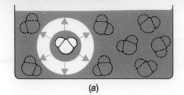
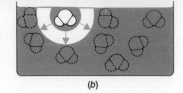

FIGURE 6.12
Surface tension. (*a*) In the interior of a sample of water, individual water molecules are attracted equally in all directions. (*b*) At the surface, nothing in the air counterbalances the downward pull that the surface molecules in water feel.

■ N₂ and O₂ molecules in air are nonpolar and so have no attraction for H_2O molecules.

bonds) that pull surface water molecules downward aren't counterbalanced by forces that pull the molecules upward. Thus a net downward pull causes the surface jam-up of water molecules responsible for surface tension.

On a greasy or waxed surface, water molecules gather together to form well-rounded beads (see Fig. 6.13). The molecules in greases and waxes are relatively nonpolar so the inward-pulling forces of attraction in the water bead aren't matched by outward-pulling forces from the nonpolar molecules in wax or grease. Thus water does not spread out on a greasy or waxy surface.

■ Glass consists mostly of silicon and oxygen. Si is less electronegative than H, so the Si—O bond is more polar than the H—O bond.

Water does spread out on a clean, grease-free glass surface (Fig. 6.14). Glass is rich in silicon-oxygen covalent bonds, so it is a very polar material. The $\delta+$ ends of H—O bonds in water are therefore attracted to the $\delta-$ charges on the oxygens in glass. Thus water molecules right at the glass surface find in the glass something to be more attracted to than the water molecules right behind them in the water. The water *cannot help* but spread out. For the same reason, in a glass graduated cylinder or in a glass pipet, the boundary between the surface of an aqueous solution and the air curves upward at the glass walls.

■ The curving surface is called the *meniscus.*

Surface-Active Agents Reduce the Surface Tension of Water

There are many substances, called **surface-active agents** or **surfactants,** that lower the surface tension of water. All soaps and detergents are surfactants, for example. Soapy water won't bead on glass. A magician has to be careful that there is no soap or detergent whatsoever in the water used for the trick with the floating needle.

At the molecular level of life, a particularly important situation involving surfactants occurs. The moist membrane of an air sac (an alveolus) in the lungs carries a natural surfactant secreted by the membrane itself. Without it, water molecules would be attracted so strongly to the membrane that the air sac would collapse. If enough air sacs did this, the entire lung would collapse. The membrane surfactant, however, prevents this and so protects the lungs. In some situations this surfactant is depleted or defective and the lungs do collapse. In milder cases, the alveoli shrink, making breathing difficult.

The hydrogen bond also helps us understand other properties of water such as the relatively high heats of fusion and vaporization of water noted in Section 2.4. An extra large input of heat per gram is necessary to melt or boil water because energy is needed to overcome the hydrogen bonding force of attraction between water molecules.

FIGURE 6.13
Water forms beads on a waxed or greasy surface. Nothing in the surface has enough polarity to attract water molecules to make the water spread out. The net inward pull created by water molecules at the surface creates the bead, because this shape minimizes the total area of the droplet.

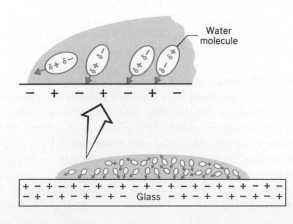

FIGURE 6.14
Water spreads out on a clean glass surface in response to strongly polar sites in the glass.

6.9

THE SOLID STATE

Forces of attraction between the particles in a solid are greater than in a liquid or a gas.

Solid (ionic)
Densely and orderly packed
Vibrations about fixed points

The Particles in a Solid Do Not Move Around   Not only are atoms, molecules, or ions tightly packed in a solid, they also have fixed positions and fixed neighboring particles. This does not mean that they are completely at rest; they jiggle and vibrate about their fixed positions. But in the solid the forces of attraction between particles are just too strong to permit any of the movement that occurs in liquids or gases.

A Solid – Liquid Equilibrium Exists at the Melting Point   As the temperature of a solid is increased, the molecular kinetic energy and the vibrations of the individual particles become more and more intense. Eventually neighboring particles bump each other strongly enough to overcome the forces of attraction between them. Now the solid passes over into the liquid state; it *melts*. If the temperature of the system is carefully controlled, the rate at which particles leave the solid state and move around as a liquid can be made equal to the rate at which they return and take up fixed positions in the solid again. In other words, at the right temperature, the following equilibrium will exist.

$$\text{solid} + \text{heat} \rightleftharpoons \text{liquid}$$

The *heat* in this equilibrium is the *heat of fusion* that we described on page 42. A mixture of ice pieces and water at 0 °C in a covered insulated cup is an common example of a system at equilibrium. (Heat slowly infiltrates from the surroundings, as you well know, so the equilibrium is not entirely stress-free, and eventually the ice melts.)

The temperature at which an equilibrium exists between the solid and the liquid states of a substance is called its **melting point**. The forward change in this equilibrium is endothermic; it consumes heat. So if we put a stress on the equilibrium by adding heat — by raising the temperature — the equilibrium will shift to the right. This shift uses up the stress, the added heat, and Le Châtelier's principle tells us that equilibria shift to absorb stresses. Of course, this shift means that the solid melts.

If we remove heat, if we cool the solid – liquid equilibrium, Le Châtelier's principle tells us that it must now shift to the left. More of the solid forms. Only a leftward shift can provide some heat (heat of fusion) to replace what is lost by cooling, and the equilibrium does what is necessary to replace heat being removed.

Solids also exert a vapor pressure, but it usually is very weak. Some solids, like mothballs or dry ice, readily change directly from the solid to the vapor state, a change called **sublimation**.

■ Heat of fusion is absorbed by a solid when it melts and the same heat is released when its liquid form solidifies again.

■ "Dry ice" is the solid form of carbon dioxide.

SUMMARY

Gas properties   The four important variables for describing the physical properties of gases are moles (n), temperature T (in kelvins), volume (V), and pressure (P). We express pressure — force per unit area — in atmospheres (atm), mm Hg, or torr. Other important physical quantities in the study of gases are partial pressures, standard pressure and temperature (STP = 273 K and 760 mm Hg), and the molar volume at STP (22.4 L).

Gas laws   All real gases obey, more or less, some important laws. Gas pressure is inversely proportional to volume (when n and T are fixed) — the pressure – volume law (Boyle's law). Gas volume is directly proportional to the Kelvin temperature (when n and P are fixed) — the temperature – volume law (Charles' law). The pressure of a gas is directly proportional to its Kelvin temperature (at fixed n and V) — Gay-Lussac's law. The combined gas law incorporates

Boyle's, Charles', and Gay-Lussac's law. Gas volume is directly proportional to the number of moles (when T and P are fixed). Equal volumes of gas (under identical T and P) have identical numbers of molecules (Avogadro's principle). Another gas law is that the total pressure of a mixture of gases equals the sum of their partial pressures (Dalton's law).

The ideal gas law incorporates all of the gas laws (except Dalton's): $PV = nRT$, where R, the universal gas constant, holds for all gases.

Kinetic theory If we imagine that an ideal gas consists of a huge number of very tiny, very hard particles in random, chaotic motion without attracting or repelling each other, the gas laws can be derived from the laws of motion. Out of this kinetic theory of gases came the insight that the Kelvin temperature of a gas is directly proportional to the average molecular kinetic energy of the gas particles.

Liquid state Liquids do not follow common laws as gases do, because essentially no space separates liquid particles from each other. In liquids, there are dipole–dipole attractions between particles originating from either permanent dipoles or temporary dipoles (London forces). Liquids can evaporate, and this "escaping tendency" gives rise to vapor pressure. Each liquid has a particular equilibrium vapor pressure that is a constant at each temperature, provided that care is taken to ensure a true dynamic equilibrium exists between the liquid and the vapor state.

When the liquid's vapor pressure equals the pressure of the atmosphere, the liquid boils. The normal boiling point is the temperature at which boiling occurs when the pressure of the atmosphere is 760 mm Hg (1 atm).

Water The higher electronegativity of oxygen over hydrogen and the angularity of the water molecule make it very polar, so polar that hydrogen bonds exist between its molecules. Hydrogen bonding helps to explain many of water's unusual thermal properties, such as its relatively high boiling point, its high heats of fusion and vaporization, and its high surface tension.

Hydrogen bonds When hydrogen is covalently bonded to atoms of any of the three most electronegative elements (O, N, or F), its partial positive charge is large enough to be attracted to the partial negative charge on an atom of O, N, or F on a nearby molecule. This force of attraction, although much weaker than a covalent bond, is large enough to have a special name, the hydrogen bond. (Never consider this bond to be a *covalent* bond within any molecule, neither the covalent bond in hydrogen itself, H_2, nor a covalent bond within any other molecule, like water.) The hydrogen bond is a (relatively weak) force of attraction between two dipoles, between the $\delta+$ on H in the polar bonds of the H—O, H—N, or H—F systems to the $\delta-$ of another O, N, or F.

Dynamic equilibria When the rates of two opposing changes, whether physical or chemical, are equal, the system is in dynamic equilibrium. If a stress, such as the addition or removal of heat, upsets an equilibrium, the equilibrium shifts in whichever direction absorbs the stress. If a change is exothermic, the reverse reaction is favored when more heat is added. If a change is endothermic, the forward reaction is favored by the addition of heat.

Solid state The particles in a solid vibrate about fixed equilibrium points, but if the solid is heated, these vibrations eventually become so violent that the particles enter the liquid state. The temperature at which liquid and solid are in equilibrium is the melting point.

REVIEW EXERCISES

The answers to Review Exercises whose numbers are in color are found in Appendix C. The answers to the other Review Exercises are found in the Study Guide that accompanies this book. The more challenging questions are marked with asterisks.

Pressure

6.1 When the word *gas* is used in chemistry, does it refer to a *kind* of matter or a *state* of matter? What are the three *kinds* of matter? What are the three *states* of matter?

6.2 Gases generally share some common features, unlike liquids and solids. What are they?

6.3 To describe the physical state of a gas we provide what information about it, besides its chemical identity?

6.4 The property of *gas diffusion* ensures that a given sample of a gas will behave in what way?

6.5 To what does the term *"weight"* of an object refer? (To review the concept of *mass*, discussed in Chapter 1, to what does *mass* refer?)

6.6 How is *pressure* defined?

6.7 Why isn't gas pressure (or the pressure of anything else) defined as *mass* per unit area?

6.8 What causes the atmosphere to have a pressure on the earth?

6.9 Why is the air pressure on top of a high mountain less than that at sea level?

6.10 In units of pounds, what is the approximate weight of a column of air pressing on a 1-in.^{2} area of the earth's surface at sea level?

6.11 Referring to Review Exercise 6.10, if the area on which the air column rests is doubled to 2 in.^{2}, what else doubles, the *pressure* or the *weight* of the column of air? Explain.

6.12 The diameter of the glass tube used to make a Torricelli barometer need not be the same from barometer to barometer; the measured pressures will be the same. How can this be?

6.13 What defines the *standard atmosphere*, the atmospheric pressure at sea level or something else? Explain.

*6.14 How high in *feet* will a column of water stand in a suitably constructed Torricelli barometer under standard atmosphere? (Assume that the water is in the liquid state.) At 0 °C, the density of mercury is 13.5955 g/mL and that of water is 0.99987 g/mL.

6.15 What is the relationship between the *torr* and the *mm Hg?*

6.16 What is the name of the SI unit of pressure? Is it larger or smaller than the mm Hg?

6.17 Which number is closer to the number of pascals needed to equal one standard atmosphere: 10 Pa, 100 Pa, 1×10^3 Pa, 1×10^5 Pa, or 1×10^8 Pa?

6.18 The atmospheric pressure on the summit of Mount Everest, the earth's highest mountain (8848 m, 29,029 ft), is roughly 250 mm Hg. What is this in atm? In torr? In kilopascals?

6.19 In an unpressurized aircraft, a pilot's upper limit on altitude (assuming the availability of oxygen-enriched air) is about 40,000 ft (13 km, 8 mi). The air pressure up there is roughly 0.20 atm. What is this in mm Hg? In torr? In kilopascals?

*6.20 When expressed in *inches* of mercury instead of mm Hg, the standard atmosphere is 29.92 in. Hg. This is the standard used by TV weather reporters in the United States. The lowest pressure ever recorded in the western hemisphere was 26.22 in. Hg (hurricane Gilbert, September 1988). What is this pressure in mm Hg?

6.21 What is the pressure in kilopascals (kPa) of each of the following? (They are, in order, the pressures exerted individually by the O_2, N_2, and CO_2 in typical inhaled air.)
(a) 160 mm Hg (b) 595 mm Hg (c) 0.380 mm Hg

The Individual Gas Laws

6.22 State the following laws both in words and in the forms of equations most useful for gas law calculations. By what other name is each law known?
(a) pressure–volume law
(b) pressure–temperature law
(c) volume–temperature law
(d) law of partial pressures

6.23 In each of the individual gas laws, one or more of the four variables involving gases are assumed to be held constant. Which are they for the following gas laws?
(a) Dalton's law (b) Gay-Lussac's law
(c) Boyle's law (d) Charles' law
(e) Avogadro's principle (f) Combined gas law

6.24 What did the data observed by Robert Boyle indicate was true about gases?

6.25 What happens to the volume of a gas sample if its pressure is tripled (at constant temperature)?

6.26 If, at constant temperature, the pressure on a sample of gas is changed from 745 mm Hg to 730 mm Hg, by what factor will the volume of the gas change, by 745/730 or by 730/745?

6.27 The pressure on a 2.56-L sample of nitrogen is changed from 750 mm Hg to 740 mm Hg while the temperature is left constant. Calculate the new volume of the gas in liters.

6.28 A gas sample under a pressure of 745 mm Hg has a volume of 16.4 L and a temperature of 25 °C. If the pressure on this sample is changed at 25 °C to 760 mm Hg, what is its new volume?

6.29 What must the pressure on a confined gas sample become if the volume of the sample, at constant temperature, is to change from 22.4 L at 760 mm Hg to 23.0 L?

6.30 At a constant temperature of 23.0 °C, the volume of a gas sample at 750 mm Hg pressure changed from 32.4 L to 30.4 L. (The *mass* of the sample did not change.) What else, therefore, must have changed and to what new value?

6.31 If 2.45 L of a gas at 25.0 °C is heated under constant pressure to 50.0 °C, what is the new volume of the gas?

6.32 A gas at 20.0 °C underwent an expansion at constant pressure to a final volume of 255 mL and a final temperature of 37.0 °C. What was the initial volume of the gas?

6.33 A heavy-walled steel cylinder with a volume of 25.0 L and containing oxygen at 25.0 °C and a pressure of 125 atm was in a building that caught fire. The temperature of the cylinder rose to 375 °C. If the cylinder did not burst, what was the final pressure of the oxygen?

6.34 A cylinder of oxygen-enriched air at a pressure of 15.0 atm and a temperature of 25.0 °C was backpacked to the summit of Mount Denali where the temperature was −35.0 °C. What was the internal pressure of the cylinder's contents under the new conditions? (Assume that the cold did not contract the cylinder's volume.)

Combined and Ideal Gas Laws

6.35 A sample of argon at a pressure of 740 mm Hg in a volume of 2.75 L was heated from 25.0 to 78.0 °C. The volume of the container expanded to 3.24 L. What was the final pressure in mm Hg of the gas?

6.36 What must be the final volume of a sample of neon (in L) if 2.55 L at 744 mm Hg and 24.0 °C is heated to 365 °C under conditions that let the pressure change to 760 mm Hg?

6.37 When 250 mL of nitrogen at 740 mm Hg and 19.0 °C is heated to 35.0 °C, the pressure changed to 760 mm Hg. What is the final volume in mL?

6.38 A sample of helium with a volume of 5.28 L, a pressure of 745 mm Hg, and a temperature of 25.0 °C expanded to a volume of 7.75 L with a pressure of 645 mm Hg. What final temperature (in °C) was needed to achieve this?

6.39 After a sample of oxygen with a volume of 550 mL was heated from 24.0 °C to 85.0 °C, its volume changed to 575 mL and its pressure became 770 mm Hg. What must have been its initial pressure in mm Hg?

6.40 A sample of neon with a volume of 650 mL and a pressure of 1.01 atm was compressed to a volume of 330 mL by changing the pressure to 1.75 atm. The compression caused the gas temperature to change to 27.0 °C. What was the initial temperature (in °C)?

6.41 What volume (in L) does 1.00 mol of N_2 occupy at 20.0 °C and 760 mm Hg?

6.42 A sample of 2.50 mol of O_2 at 20.0 °C occupies a volume of 12.5 L. Under what pressure (in atm) is this sample?

6.43 A steel cylinder with a volume of 12.5 L contains O_2 at a pressure of 127 atm at 25 °C. The tank contains how many moles and how many kilograms of O_2?

6.44 When the pressure in a certain gas cylinder with a volume of 5.00 L reaches 450 atm, the cylinder is likely to explode. Is it on the verge of exploding if it contains 44.5 mol of nitrogen at 25.0 °C? (Calculate the pressure in atm.)

6.45 A steel cylinder with a volume of 30.0 L contains nitrogen under a pressure of 138 atm and a temperature of 25 °C. The cylinder contains how many kilograms of nitrogen?

*6.46 A steel cylinder of oxygen with a volume of 15.0 L was available for medical purposes. The cylinder pressure decreased from 40.6 atm to 38.5 atm during which time the temperature remained at 24.0 °C. How many moles of oxygen had been removed?

6.47 What is meant by "STP," and what are the relevant values?

6.48 A container with a capacity of 750 mL held helium at a pressure of 740 mm Hg at a temperature of 25.0 °C.
(a) How many moles of helium were in the container?
(b) How many moles of H_2 could the container hold under identical conditions of pressure and temperature? (How can you tell without doing an additional calculation?)

6.49 How many millimoles of oxygen are needed to fill a container with a volume of 750 mL at a temperature of 25.0 °C under a pressure of 750 mm Hg (roughly, laboratory conditions)?

6.50 What size container (in mL) is needed to hold 12.5 mmol of O_2 at 25.0 °C and a pressure of 0.965 atm?

*6.51 Under suitable conditions, water can be broken down by an electric current into hydrogen and oxygen according to the following equation.

$$2H_2O \longrightarrow 2H_2 + O_2$$

In one experiment, 875 mL of a dry sample of one of the product gases was collected at 748 mm Hg and 23.0 °C.
(a) How many moles of the gas were in this sample?
(b) The sample was found to have a mass of 1.133 g. What is the formula mass of the gas in the sample? (Recall that formula mass is the ratio of grams to moles.)
(c) Which gas was it, hydrogen or oxygen?
(d) Based on stoichiometry and Avogadro's principle, how many moles of the other gas were also collected?
(e) How many grams of the other gas were collected?

*6.52 The following equation shows how ammonia can be made from nitrogen and hydrogen.

$$N_2(g) + 3H_2(g) \xrightarrow[\text{heat}]{\text{high pressure}} 2NH_3(g)$$

(a) If 6.00 mol of H_2 are consumed, how many moles of NH_3 are produced?
(b) If 250 L of nitrogen at 745 mm Hg and 25.0 °C are consumed, how many liters of ammonia can be made

when the measurements of V and T are made at 745 mm Hg and 25.0 °C?
(c) If 75.0 g of N_2 are consumed, how many grams of H_2 are required and how many grams of NH_3 can be made?

*6.53 Calcium carbonate decomposes as follows when it is strongly heated.

$$CaCO_3(s) \xrightarrow{\text{heat}} CaO(s) + CO_2(g)$$

In one experiment, 246 mL of CO_2 was collected at 740 mm Hg and 24.0 °C.
(a) How many moles of CO_2 formed?
(b) How many moles of $CaCO_3$ decomposed?
(c) How many grams of $CaCO_3$ decomposed?
(d) How many grams of CaO formed?

Dalton's Law of Partial Pressures

6.54 To what kind of a gas sample is the law of partial pressures relevant, and why is this law important for an understanding of the molecular basis of life?

6.55 What was it that John Dalton discovered about a mixture of gases?

6.56 The partial pressure of nitrogen in clean, dry air at 0 °C is 601 mm Hg when the total pressure is 760 mm Hg. The only other gas present to any appreciable extent is oxygen. If a sample of this air were sealed into a container at a pressure of 760 mm Hg and then all of the oxygen were removed, what would be the pressure exerted by the residual gas?

6.57 A gas mixture at 745 mm Hg consists of at least two gases, helium and argon. Their partial pressures are 350 mm Hg for helium and 375 mm Hg for argon. Is it likely that any other gas is also present? How can you tell?

*6.58 A sealed sample of helium at 760.0 mm Hg was saturated in water vapor. When the water vapor was completely removed from this sample, the pressure changed to 742.5 mm Hg. What was the temperature of the sample? (*Hint:* Use data in a table in this chapter for the vapor pressure of water.)

*6.59 A student was asked to prepare a sample of hydrogen gas by collecting it over water as it was produced by the reaction of zinc metal with hydrochloric acid.

$$Zn(s) + 2HCl(aq) \longrightarrow ZnCl_2(aq) + H_2(g)$$

The assignment was eventually to produce 250 mL of *dry* hydrogen as measured at STP from the wet sample produced at 740 mm Hg and 24.0 °C.
(a) What are the minimum numbers of moles and grams of Zn needed to make this much H_2?
(b) The bottle in which the *wet* hydrogen is to be collected must have what minimum volume in milliliters?

Kinetic Theory of Gases

6.60 Scientists asked, "What must gases be like for the gas laws to be true?" What was their answer?

6.61 What is true about an ideal gas that is not strictly true about any real gas?

6.62 Dalton's law of partial pressures implies that gas molecules from different gases actually leave each other alone in the mixture, both physically and chemically (except at moments of collisions, when they push each other around). Which one of the three postulates in the model of an ideal gas is based on Dalton's law?

6.63 How does the kinetic theory of gases explain the phenomenon of gas pressure?

6.64 How does the kinetic theory of gases account for Boyle's law (in general terms)?

6.65 Those working out the kinetic theory found that for 1 mol of an ideal gas, the product of pressure and volume is proportional to the average molecular kinetic energy of the ideal gas particles.
  (a) To which of the four physical quantities used to describe a gas is the product of pressure and volume for 1 mol of a gas also proportional, according to the universal gas law (which makes no mention of kinetic energy)?
  (b) If the product of P and V is proportional both to the average kinetic energy of the ideal gas particles and to the Kelvin temperature, what does this say about the relationship between the average kinetic energy and this temperature?

6.66 What happens to the motions of gaseous molecules at 0 K?

6.67 How does the kinetic theory explain (in general terms) the volume–temperature law?

6.68 The pressure–temperature law (Gay-Lussac's law) can be explained in terms of the kinetic theory in what way (in general terms)?

The Liquid State, Vapor Pressure, and Dynamic Equilibria

6.69 Why aren't there universal laws for the physical behavior of liquids (or solids) as there are for gases?

6.70 Explain how there can be forces of attraction between atoms like those of argon (a monatomic noble gas), or between molecules of N_2, when these particles have no permanent dipoles, as in H—Cl.

6.71 Which has the higher boiling point, octane (C_8H_{18}) or butane (C_4H_{10})? (How can we tell without a table of boiling points?)

6.72 How does the kinetic theory explain
  (a) How vapor pressure arises?
  (b) Why vapor pressure rises with increasing liquid temperature?

6.73 Dimethyl sulfoxide (DMSO) is a controversial pain-killing drug the use of which is permitted by only a few states. Its boiling point is 189 °C.
  (a) Is it more volatile or less volatile than water? Explain.
  (b) The structure of DMSO has a double bond between S and O. Which atom of the S=O system carries a permanent $\delta+$ and which a permanent $\delta-$? Explain.
  (c) What is the name of the chief force of attraction between DMSO molecules?

*6.74 Compare the following expressions.

  A Water + heat $\longrightarrow$ water vapor
  B Water + heat $\longleftarrow$ water vapor
  C Water + heat $\rightleftharpoons$ water vapor

Answer the following questions by using the letter A, B, or C to indicate which expression best represents the answer.
  (a) Which expression describes heat being liberated from the system?
  (b) Which expression describes the net formation of liquid water from water vapor?
  (c) Which expression describes a net endothermic change?
  (d) Which expression shows opposing changes?
  (e) Which expression, A or B, represents the forward change in expression C? Which represents the reverse change?
  (f) Expression C can be the correct description for the water–water vapor system at 1 atm only if the temperature is what?
  (g) What is true about the opposing changes in C?
  (h) What special term applies to expression C? (It does not represent a *reaction* or a permanent *change*, but what?)

6.75 When a liquid and its vapor are in dynamic equilibrium at a given temperature, the rates of what two changes are equal?

6.76 Explain why a liquid's boiling point decreases as the external pressure decreases.

6.77 What is Le Châtelier's principle?

*6.78 Consider the following equilibrium existing in a sealed container.

$$\text{alcohol}(l) + \text{heat} \rightleftharpoons \text{alcohol}(g)$$

  (a) What takes place when the reverse reaction occurs?
  (b) Which reaction, the forward or the reverse, is endothermic?
  (c) In which direction will the equilibrium shift if the system is in a sealed container and the pressure on the system is increased by some outside means not involving a change in the system temperature? Explain. (*Hint:* Which change, forward or reverse, is a volume-reducing change?)
  (d) In which direction will the equilibrium shift if alcohol vapor could be removed from the container? Explain.
  (e) If the system is cooled, will the equilibrium be affected? If so, in what way?

*6.79 Consider the following chemical equilibrium in which all the substances are gases.

$$N_2 + O_2 + \text{heat} \rightleftharpoons 2NO$$
Nitric oxide

  (a) Describe the chemical change for the forward reaction.
  (b) What is the chemical change when the reverse reaction occurs?
  (c) Which reaction, the forward or the reverse, is endothermic?
  (d) Which reaction, the forward or the reverse, will be favored by adding heat?

*6.80 Consider the following equilibrium that involves a common air pollutant, NO_2. Both substances are gases at room temperature.

$$2NO_2 \rightleftharpoons N_2O_4$$

When the temperature is reduced, this equilibrium shifts to the right.

(a) Which reaction, the forward or the reverse, is exothermic? Rewrite the equilibrium with "heat" positioned correctly.

(b) This reaction can be shifted in what direction by increasing the pressure on the mixture? Explain in terms of properties of gases and Le Châtelier's principle.

Water and the Hydrogen Bond

6.81 The hydrogen molecule, H—H, does not become involved in hydrogen bonding.
(a) What kind of bond occurs in a hydrogen molecule?
(b) Why can't this molecule become involved in hydrogen bonding?

6.82 The methane molecule, CH_4, does not become involved in hydrogen bonding. Why not?

6.83 Solid sodium hydroxide, NaOH, includes two kinds of chemical bonds. Which ones are they, and how do they differ?

6.84 Draw the structures of two water molecules. Write in $\delta+$ and $\delta-$ symbols where they belong. Then draw a correctly positioned dotted line between two molecules to symbolize a hydrogen bond.

6.85 Hydrogen bonds exist between two molecules of ammonia, NH_3. Draw the structures of two ammonia molecules. (You don't have to try to duplicate their tetrahedral geometry.) Put $\delta+$ and $\delta-$ signs where they should be located. Then draw a dotted line that correctly connects two points to represent a hydrogen bond.

*6.86 The hydrogen bond between two molecules of ammonia must be much weaker than the hydrogen bond between two molecules of water.
(a) How do boiling point data suggest this?
(b) What does this suggest about the relative sizes of the $\delta+$ and the $\delta-$ sites in molecules of water and ammonia?
(c) Why are the $\delta+$ and the $\delta-$ sites different in their relative amounts of fractional electric charge when we compare molecules of ammonia and water?

6.87 If it takes roughly 100 kcal/mol to break the covalent bond between O and H in H_2O, about how many kilocalories per mole are needed to break the hydrogen bonds in a sample of liquid water?

6.88 Explain in your own words how hydrogen bonding helps us understand each of the following.
(a) The high heats of fusion and vaporization of water
(b) The high surface tension of water

6.89 Explain in your own words why water forms tight beads on a waxy surface but spreads out on a clean glass surface.

6.90 What does a surfactant do to water's surface tension?

6.91 What common household materials are surfactants?

The Solid State

6.92 Describe the motions made by particles (e.g., ions or molecules) in a solid crystal.

6.93 What is the mechanism whereby heat causes a solid to melt?

6.94 What do we call the temperature at which a solid is in equilibrium with its liquid form?

6.95 At room temperature, nitrogen is a gas, water is a liquid, and sodium chloride is a solid. What do these facts tell us about the relative strengths of electrical forces of attraction in these substances?

6.96 Some solids, like dry ice (solid carbon dioxide), pass directly from their solid state to their vapor state when they absorb heat. This process is called *sublimation*. Write an equilibrium expression for the sublimation of solid CO_2. Include "heat" in the appropriate place.

6.97 The following are some common observations. Using the kinetic theory, explain how each occurs in terms of what molecules are doing.
(a) Moisture evaporates faster in a breeze than in still air.
(b) Ice melts much faster if it is crushed than if it is left in one large block.
(c) Even if hung out to dry in below-freezing weather, wet clothes will become completely dry even though they freeze first.

The Effect of Altitude on Cooking Times (Special Topic 6.1)

6.98 Cooking an egg involves heat-induced chemical reactions as well as some physical changes. Why does it take longer to prepare a soft-boiled egg in Denver than in New York City?

Additional Exercises

6.99 If 456 mL of oxygen at 912 mm Hg is allowed to expand at constant temperature until its pressure is 760 mm Hg, what will be the volume of the oxygen sample?

*6.100 A sample of carbon monoxide was prepared and collected over water at a temperature of 20.0 °C and a total pressure of 754 mm Hg. It occupied a volume of 268 mL. Calculate the partial pressure of the carbon monoxide in mm Hg as well as its dry volume (in mL) under a pressure of 1.00 atm at 20.0 °C.

*6.101 After the contents of a steel cylinder of pressurized oxygen with a volume of 18.5 L have been depleted until the pressure inside the cylinder is 1.00 atm, the cylinder is called "empty." How many moles of oxygen remain, however, if the temperature is 24.0 °C?

*6.102 Carbon dioxide can be made in the lab by the reaction of hydrochloric acid (HCl) with calcium carbonate according to the following equation.

$$CaCO_3(s) + 2HCl(aq) \longrightarrow CaCl_2(aq) + H_2O + CO_2(g)$$

How many grams of calcium carbonate and how many milliliters of 8.00 M HCl are needed to prepare 465 mL of dry CO_2 if it is collected at 20.0 °C and 745 torr?

SOLUTIONS AND COLLOIDS

7

White sunlight is actually a mixture of colors, the spectrum of the rainbow. We see brilliant sunsets because particles in air that are a little larger than molecules and ions — colloidal size smoke particles, for example — preferentially reflect and scatter colors at the blue end of the spectrum leaving those at the red end to reach our view. We'll learn about other properties of colloidal systems in this chapter, properties that are at work at the molecular level of life.

7.1

THE TYPES OF HOMOGENEOUS MIXTURES

The *sizes* of the particles intimately mixed with a solvent determine some physical properties of the mixtures.

Mixtures are either *homogeneous* or *heterogeneous,* but in either case they are always made of two or more nonreacting, pure substances occurring in *variable* proportions. **Homogeneous mixtures** are those in which the tiniest samples are everywhere identical in composition and properties. **Heterogeneous mixtures** are any that are not homogeneous. Most beverages are homogeneous mixtures, but orange juice with pulp is heterogeneous because a piece of the pulp itself does not have the same composition as the clear fluid. You can see, however, that the distinctions between homogeneous and heterogeneous depend altogether on what is meant by "tiniest sample." By this we normally mean a sample that can be manipulated, and that might be as small as a fraction of a microgram or microliter. Even a microgram quantity of a pure substance will have a billion billion formula units.

There are two kinds of relatively stable homogeneous mixtures: *solutions* and *colloidal dispersions.* A third kind, the *suspension,* is unstable, being homogeneous only when constantly stirred. The kinds of homogeneous mixtures differ fundamentally in the sizes of the particles involved, and the differences in size can alone cause interesting and important changes in properties.

In Solutions, the Dispersed Particles Are Smallest A **solution** is a homogeneous mixture in which the particles of both solvent and solutes have sizes of atoms, or ordinary ions and molecules. They have formula masses of no more than a few hundred and diameters in the range of 0.1 to 1 nm.

■ Unlike pure substances (elements and compounds), mixtures do not obey the law of definite proportions.

■ "Normally" is a weasel word. What we really want to do is retain a flexibility in how to apply the terms "homogeneous" and "heterogeneous."

■ 1 nm = 10^{-9} m = 1 nanometer.

TABLE 7.1
Solutions

Kinds	Common Examples
Gaseous Solutions	
Gas in a gas	Air
Liquid in a gas	(If droplets are present, a colloidal system)
Solid in a gas	(If particles are present, a colloidal system)
Liquid Solutions	
Gas in a liquid	Carbonated beverages (carbon dioxide in water)
Liquid in a liquid	Vinegar (acetic acid in water), gasoline
Solid in a liquid	Sugar in water, seawater
Solid Solutions	
Gas in a solid	Alloy of palladium and hydrogen[a]
Liquid in a solid	Toluene in rubber (e.g., rubber cement)
Solid in a solid	Carbon in iron (steel)[a]

[a] There is some doubt that this is a true solution.

TABLE 7.2
Colloidal Systems

Type	Dispersed Phase[a]	Dispersing Medium[b]	Common Examples
Foam	Gas	Liquid	Suds, whipped cream
Solid foam	Gas	Solid	Pumice, marshmallow
Liquid aerosol	Liquid	Gas	Mist, fog, clouds, certain air pollutants
Emulsion	Liquid	Liquid	Cream, mayonnaise, milk
Solid emulsion	Liquid	Solid	Butter, cheese
Smoke	Solid	Gas	Dust in smog
Sol	Solid	Liquid	Starch in water, jellies,[c] paints
Solid sol	Solid	Solid	Black diamonds, pearls, opals, alloys

[a] The colloidal particles constitute the dispersed phase.

[b] The continuous matter into which the colloidal particles are scattered is called the dispersing medium.

[c] Sols that adopt a semisolid, semirigid form (e.g., gelatin desserts, fruit jellies) are called **gels.**

We usually think of solutions as being liquids, but in principle the solvent can be in any state — solid, liquid, or gas — and so can the solute. Table 7.1 is a list of the several combinations that can form a solution. Solutions are generally transparent — you can see through them — but they often are colored. Solutes do not settle out of solutions under the influence of gravity, and they can't be separated from solutions by filter paper. The blood carries many substances in solution, like the ions of NaCl and the molecules of glucose ($C_6H_{12}O_6$).

In Colloidal Dispersions, the Particle Sizes Are Larger than in a Solution
A **colloidal dispersion** is a homogeneous mixture in which the dispersed particles have diameters in the range of 1 nm to 1000 nm and consist of very large clusters of ions or molecules.

Table 7.2 gives several examples of colloidal dispersions, and they include many familiar substances such as whipped cream, milk, dusty air, jellies, and pearls. The blood also carries many substances in colloidal dispersions, including a variety of proteins. The molecules of many proteins are so huge that their formula masses run into the hundreds of thousands and so they are called **macromolecules.** Some macromolecules are large enough to be of colloidal size and so form colloidal dispersions rather than solutions in water.

When colloidal dispersions are in a fluid state — liquid or gas — the dispersed particles, although large, are not large enough to be trapped by ordinary filter paper during filtration. They are large enough, however, to reflect and scatter light (Fig. 7.1). Light scattering by a

■ The prefix "macro-" signifies something enormous in size relative to the suffix, such as "molecule" in *macromolecule.*

FIGURE 7.1
Tyndall effect. The tube on the left contains a colloidal starch dispersion, and the tube on the right has a colloidal dispersion of Fe_2O_3 in water. The middle tube has a solution of Na_2CrO_4, a colored solute. The thin red laser light is partly scattered in the two colloidal dispersions so it can be seen, but it passes through the middle solution unchanged.

■ Solutions do not exhibit the Tyndall effect because the solute particles are too small.

colloidal dispersion is called the **Tyndall effect,** after British scientist John Tyndall (1820–1893). The effect is responsible for the milky, partly obscuring character of smog, or the way sunlight sometimes seems to stream through a forest canopy.

When a colloidal dispersion is a fluid, the large colloidally dispersed particles eventually settle out under the influence of gravity. The settling process can take time ranging from a few seconds to many decades, depending on the system. One of the factors that keeps the particles dispersed is their constant buffeting about by molecules of the solvent. Evidence for this buffeting can be seen by looking at the colloidal system under a good microscope. You can't actually see the colloidal particles, but you can see the light scintillations caused as they move erratically and unevenly about. This motion of colloidal particles is called the **Brownian movement.**

■ Robert Brown (1773–1858), an English botanist, first observed this phenomenon when he saw the trembling of particles inside pollen grains viewed with a microscope.

In the most stable colloidal systems a second stabilizing factor is at work. All of the dispersed particles bear like electrical charges. In living systems, it's common for colloidally dispersed proteins to be like this, for example. (Other dissolved species of opposite charge, such as small ions, balance the charges on the colloidal particles.) Because like-charged colloidal particles repel each other, they cannot coalesce to make particles large and heavy enough to settle out.

Emulsions are colloidal dispersions of two liquids in each other, like oil and vinegar in salad dressing. Emulsions usually are not stable; the oil soon separates from the aqueous layer. Sometimes, however, an emulsion can be stabilized by a third component called an *emulsifying agent.* For example, mayonnaise is stabilized by egg yolk whose protein molecules coat the microdroplets of olive oil or corn oil and prevent them from merging into drops large enough to rise to the surface.

TABLE 7.3

Characteristics of Three Homogeneous Mixtures: Solutions, Colloidal Dispersions, and Suspensions

Particle Sizes Become Larger →		
Solutions	**Colloidal Dispersions**	**Suspensions**
All particles are on the order of atoms, ions, or small molecules (0.1–1 nm)	Particles of at least one component are large clusters of atoms, ions, or small molecules, or are very large ions or molecules (1–1000 nm)	Particles of at least one component may be individually seen with a low-power microscope (over 1000 nm)
Most stable to gravity	Less stable to gravity	Unstable to gravity
Most homogeneous	Also homogeneous but borderline	Homogeneous only if well stirred
Transparent (but often colored)	Often translucent or opaque, but may be transparent	Often opaque, but may appear translucent
No Tyndall effect	Tyndall effect	Not applicable (suspension cannot be transparent)
No Brownian movement	Brownian movement	Particles separate unless system is stirred
Cannot be separated by filtration	Cannot be separated by filtration	Can be separated by filtration

Homogeneous   to   Heterogeneous →

Fluid Suspensions Must Be Stirred to Remain Homogeneous In **suspensions,** the dispersed or suspended particles are over 1000 nm in average diameter, they separate under the influence of gravity, and they are large enough to be trapped by filter paper. A fluid suspension such as clay in water has to be stirred constantly to keep it from separating. A suspension is thus always on the borderline between being homogeneous and heterogeneous. The blood, while it is moving, is a suspension, besides being a solution and a colloidal dispersion. Suspended in circulating blood are its red and white cells and its platelets.

See Table 7.3 for a summary of the chief features of solutions, colloidal dispersions, and suspensions.

7.2
AQUEOUS SOLUTIONS AND HOW THEY FORM

Water dissolves best those substances whose ions or molecules can strongly attract water molecules.

When crystals of a solid are placed into a potential solvent, the solvent molecules bombard the crystal surfaces. Figure 7.2 shows how such kinetic action would tend to dislodge Na^+ and Cl^- ions from a crystal of sodium chloride. The individual ions, however, have a very stable

■ We introduced the vocabulary of solutions in Section 3.4. You may want to review the meanings of *solute, solvent, unsaturated,* and *saturated* before continuing.

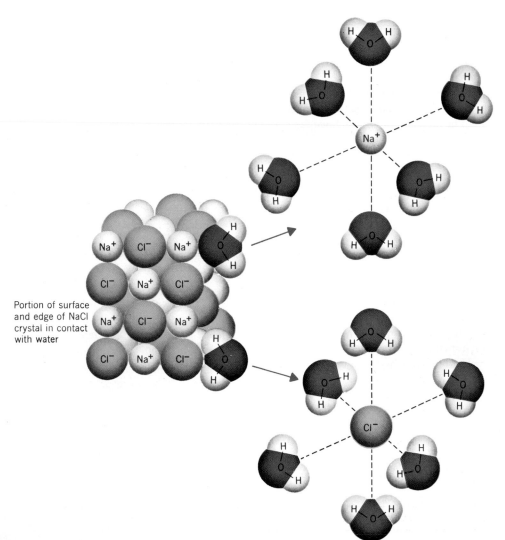

Portion of surface and edge of NaCl crystal in contact with water

FIGURE 7.2
The hydration of ions helps ionic substances dissolve in water.

environment in the crystal, where each is surrounded with oppositely charged ions as nearest neighbors. They simply will not leave this environment unless something else substitutes for it, because oppositely charge ions attract each other too strongly. Water, however, can provide a substitute environment.

Water Molecules Form Solvent Cages around Ions Water molecules are very polar and have sizable partial charges. The $\delta-$ sites on their oxygen atoms attract Na$^+$ ions, so once these ions are dislodged from a crystal of NaCl, they are surrounded by water molecules whose $\delta-$ ends point toward the positively charged ion (see Fig. 7.2). Similarly, water molecules also attract chloride ions and surround them (Fig. 7.2). The $\delta+$ sites of water molecules point toward the negatively charged Cl$^-$ ions. Both of the ions of NaCl thus become surrounded by "cages" of water molecules — *solvent cages.*

In its solvent cage, a chloride ion no longer has Na$^+$ ions as nearest neighbors, but several water molecules perform the same service, namely providing Cl$^-$ ions with an environment of opposite charge. In their solvent cages, sodium ions no longer have chloride ions as nearest neighbors; water molecules take their place. This phenomenon whereby water molecules are attracted to solute particles is called **hydration** and the resulting clusters are called **hydrated ions.** The force of attraction between an ion and one end of the dipole of a polar molecule is called an **ion – dipole attraction.**

Of all of the common solvents, only water has molecules both polar enough and small enough to form effective solvent cages around ions. Of course, in order for water to do this, water molecules must give up some of their attractions for each other. Only ionic substances or compounds made of very polar molecules can break up the hydrogen-bonded network between water molecules. Thus the formation of a solution isn't simply the separation of the solute particles from each other. It is also, to some extent, the separation of solvent molecules from each other.

When an ionic compound dissolves in water, we say that **dissociation** occurs because the oppositely charged ions separate — dissociate — from each other. Dissociation can be represented by an equation in which we take note of the physical states of the species.

$$NaCl(s) \xrightarrow{\text{dissociation}} Na^+(aq) + Cl^-(aq)$$

We should note here that chemists do not always include "(*aq*)" with the symbol of an ion dissolved in water. Often this designation is "understood." Usually, the context of the discussion makes very clear what is intended; whenever an ionic compound dissolves in water, its ions are *always* hydrated. We should also note here that the effect of forming a solvent cage about an ion is not only to aid in forming a solution. The cage also increases the effective size of the ion.

Polar Molecular Compounds Dissolve in Water, Too Water is able to form solvent cages about polar molecules (see Fig. 7.3), particularly when they have oxygen or nitrogen atoms with $\delta-$ charges on them (the more such atoms the better). Hydrogen bonds can then form from the $\delta-$ sites to the $\delta+$ sites on water molecules. Many molecular compounds have O—H groups attached by covalent bonds to carbon atoms. These are not hydroxide *ions* free

■ *Hydr-* is from the Greek *hydor,* water.

■ Usually, when both ions of an ionic compound carry charges of two or three units, the compound isn't very soluble in water. The ions find more stability by remaining in the crystal than they can replace by accepting solvent cages.

FIGURE 7.3
The hydration of a polar molecule helps polar molecular substances to dissolve in water.

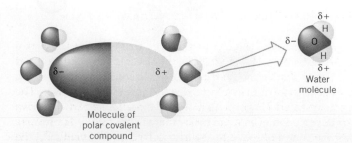

Molecule of polar covalent compound

Water molecule

to dissociate in water, but molecular groups held strongly to carbon. Methyl alcohol is the simplest example.

Methyl alcohol

Hydrogen bonds (•••) are
accepted and donated by
molecules with O—H groups.

As you can see, the waterlike O—H group in methyl alcohol not only can form a hydrogen bond from its O atom to $\delta+$ on a water molecule, it can also form a second hydrogen bond from the H atom of O—H to $\delta-$ of a water molecule. This enables methyl alcohol to be soluble in water in all proportions, and it also explains why ethane is insoluble. The ethane molecule has about the same size as that of methyl alcohol, but it has no partially charged sites on an O or an N atom. Ethane molecules thus cannot form hydrogen bonds to water.

Ethane

Dynamic Equilibrium Exists in a Saturated Solution

A **saturated solution** is one in which there is a dynamic equilibrium between the undissolved and the dissolved solute, which we represent as follows.

$$\text{solute}_{\text{undissolved}} \rightleftharpoons \text{solute}_{\text{dissolved}} \qquad (7.1)$$

In a saturated solution there is coming and going as solute particles leave the undissolved state and go into solution (the forward change) and others, at the same rate, leave the dissolved state and return to the undissolved condition (the reverse change). The *system,* here, is not homogeneous because both a solid and a liquid (the solution) are present. Hence, this equilibrium is called a *heterogeneous equilibrium.*

Each solid and most liquids have a limited solubility in water at a specific temperature. Hundreds of ionic compounds are only slightly soluble, and hundreds seemingly do not dissolve at all. In Section 3.4 we introduced such limits.

Dynamic equilibrium in a saturated solution

The Solubilities of Most Solids Increase with Temperature

Suppose we have a saturated solution of some solid in physical contact and in equilibrium with undissolved solid resting on the bottom of the container at, say, 20 °C. When we increase the temperature of this stable system, what will happen? Will more solid dissolve, or will some come out of solution? For most (but not all) solids, *more* will dissolve as the temperature is increased. The reason is that *most solids dissolve endothermically* when the solution into which they are dissolving is at or very near the point of being saturated. They require heat to dissolve into a solution already saturated. Thus, for solutes that dissolve endothermically, we can rewrite equilibrium expression 7.1 by introducing an energy term, the *heat of solution.*

$$\text{Solid}_{\text{undissolved}} + \text{solution} + \text{heat of solution} \rightleftharpoons \text{more concentrated solution}$$

Le Châtelier's principle applies to this equilibrium as it does to all. When we heat a saturated solution of anything that dissolves endothermically, the stress is absorbed by a shift of the equilibrium to the right. Only a shift in this direction absorbs the additional heat and so "absorbs" the stress. The rate of the forward change increases, so more solute dissolves. The solution becomes even more concentrated. However, as more and more solute particles move out into the solution to make the solution *even more concentrated,* the rate of the reverse change also increases, namely the return of solute to the undissolved state. Eventually, the two rates again become equal (although higher), and equilibrium is restored (assuming that undissolved solute is still present, of course).

The solubilities of some ionic compounds *decrease* with increasing temperature. Most examples are salts of the sulfate ion, SO_4^{2-}, with metal ions carrying $2+$ and $3+$ charges, but a few are metal hydroxides like calcium hydroxide, $Ca(OH)_2$.

The Ions in Some Crystals Are Hydrated   If we let water evaporate from an aqueous solution of any one of several substances, the dry-appearing crystalline residue contains intact water molecules. They are held within the crystals in *definite* proportions so these substances are true compounds; they obey the law of definite proportions. Such water-containing solids are called **hydrates.**

We write the formulas of hydrates in a special way to emphasize that intact water molecules are present. Thus the formula of the pentahydrate of copper(II) sulfate is written as $CuSO_4 \cdot 5H_2O$, where a raised dot separates the two parts of the formula. Table 7.4 lists a number of other common hydrates.

The water indicated by the formula of a hydrate is called the **water of hydration.** It usually can be driven away from the hydrate by heat, and when this is done the residue is sometimes called the **anhydrous form** of the compound. For example,

$$CuSO_4 \cdot 5H_2O(s) \xrightarrow{\text{heat}} CuSO_4(s) \quad + 5H_2O(g)$$

| Copper(II) sulfate pentahydrate (deep blue crystals) | Copper(II) sulfate (anhydrous form is white) | (as steam) |

Many anhydrous forms even as solids readily take up water and reform their hydrates. Plaster of paris, for example, although not completely anhydrous, contains relatively less

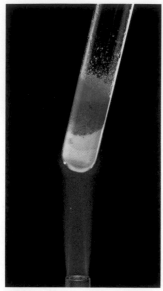

When a sample of blue $CuSO_4 \cdot 5H_2O$ is heated, water is expelled (see beads on the upper test tube wall) and anhydrous $CuSO_4$, which is white, forms.

TABLE 7.4
Some Common Hydrates

Formulas	Names	Decomposition Modes[a]	Uses
$(CaSO_4)_2 \cdot H_2O$	Calcium sulfate hemihydrate (plaster of paris)	$-H_2O$ (163)	Casts, molds
$CaSO_4 \cdot 2H_2O$	Calcium sulfate dihydrate (gypsum)	$-2H_2O$ (163)	Casts, molds, wallboard
$CuSO_4 \cdot 5H_2O$	Copper(II) sulfate pentahydrate (blue vitriol)	$-5H_2O$ (150)	Insecticide
$MgSO_4 \cdot 7H_2O$	Magnesium sulfate heptahydrate (epsom salt)	$-6H_2O$ (150)	Cathartic in medicine
		$-7H_2O$ (200)	Used in tanning and dyeing
$Na_2B_4O_7 \cdot 10H_2O$	Sodium tetraborate decahydrate (borax)	$-8H_2O$ (60) $-10H_2O$ (320)	Laundry
$Na_2CO_3 \cdot 10H_2O$	Sodium carbonate decahydrate (washing soda)	$-H_2O$ (33.5)	Water softener
$Na_2SO_4 \cdot 10H_2O$	Sodium sulfate decahydrate (glauber's salt)	$-10H_2O$ (100)	Cathartic
$Na_2S_2O_3 \cdot 5H_2O$	Sodium thiosulfate pentahydrate (photographer's hypo)	$-5H_2O$ (100)	Photographic developing

[a] Loss of water is indicated by the minus sign before the symbol, and the loss occurs at the temperature in °C that is given in parentheses.

water than gypsum. When we mix plaster of paris with water, it soon sets into a hard, crystalline mass according to the following reaction.

$$(CaSO_4)_2 \cdot H_2O + 3H_2O \longrightarrow 2CaSO_4 \cdot 2H_2O$$

Plaster of   Gypsum
paris

(Notice that the first 2 in gypsum's formula is a *coefficient* for the *entire* formula, including the $2H_2O$, so a total four H_2O molecules are represented in $2CaSO_4 \cdot 2H_2O$.)

Some compounds in their anhydrous forms are used as drying agents or desiccants. A **desiccant** is a substance that removes moisture from air by forming a hydrate. Any substance that can do this is said to be **hygroscopic.** Anhydrous calcium chloride, $CaCl_2$, is a common desiccant, and in humid air it draws enough water to form a liquid solution. Any substance this active as a desiccant is also said to be **deliquescent.** Thus calcium chloride is often used to dehumidify damp basements.

7.3
SOLUBILITIES OF GASES

Pressure, temperature, and sometimes the reactions of gases with water affect the solubility of a gas in an aqueous solution.

By some means or another, all living things must exchange gases with their environment. Our bodies, for example, take in oxygen from the air and expel carbon dioxide. The processes intimately involve aqueous systems, so to understand these vital matters at the molecular level of life we have to look at the factors that affect the solubilities of gases in water.

All Gases Are Less Soluble in Water at Higher Temperatures The solubilities of gases in water always decrease with increasing temperature, as the data in Table 3.1 (page 66) show. This is because the dissolving of gases in liquids, represented by the following equilibrium, is always an exothermic process.

$$gas_{undissolved} + solution \rightleftharpoons more\ concentrated\ solution + heat\ of\ solution$$

When heat is added, this equilibrium must shift to the left in favor of undissolved gas, in accordance with Le Châtelier's principle. It's the only way that the stress, heat, can be absorbed by the system. A shift to the left, of course, means that some of the gas leaves the solution. You have seen this often. Small bubbles appear in water soon after you start to heat it on the stove because dissolved air is leaving the water.

Gases Are More Soluble under Higher Partial Pressures Pressure is a factor that affects solubility only if the solute is a gas. As seen in Figure 7.4, the solubilities of oxygen and nitrogen, two typical gases, are directly proportional to the applied pressure. The equilibrium expression is

$$gas_{undissolved} + solvent \rightleftharpoons solution \qquad (7.2)$$

The equilibrium shifts to the right with increasing pressure because only such a change can absorb the volume-squeezing stress of extra pressure — yet another illustration of Le Châtelier's principle.

Similarly, if we reduce the pressure above a liquid that has a dissolved gas, we create a volume-expanding stress, and now equilibrium 7.2 shifts to the left. Dissolved gas now leaves the solution, something you have often observed when you have opened a can or bottle of a soft drink. If we apply both suction and heat to a solution of a gas, we very rapidly de-gas the solution.

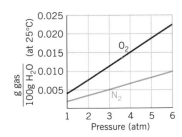

FIGURE 7.4
The solubilities of oxygen and nitrogen in water versus pressure.

William Henry (1775–1836) was the first to notice that gas solubility is directly proportional to gas pressure, so we now call this relationship **Henry's law** or the **pressure–solubility law.**

Henry's Law (Pressure–Solubility Law) The concentration of a gas in a liquid at any given temperature is directly proportional to the partial pressure of the gas on the solution.

Stated in the form of an equation, Henry's law says

$$C_g = k_g P_g$$

where C_g is the concentration of the gas, k_g is a constant of proportionality, and P_g is the partial pressure of the gas above the solution. The reference is to *partial* pressure because each gas in a mixture of gases, like air, dissolves individually according to its own partial pressure and its own value of k_g.

The value of k_g *for a given gas at a given temperature* is a constant, and it does not change with the partial pressure. This lets us change the Henry's law equation to a form easier to use in calculations, because it lets us avoid actually having to know the constant k_g. For a given gas and using subscripts 1 and 2 to refer to different partial pressures.

■ The two sides of 7.3 equal each other because they each equal the same Henry's law constant. This transformation is identical in type to what we did with Charles' law on page 153.

$$\frac{C_1}{P_1} = \frac{C_2}{P_2} \qquad \text{(at constant temperature)} \qquad (7.3)$$

EXAMPLE 7.1
Using Henry's Law

At 760 mm Hg and 20 °C, the solubility of oxygen in water is 4.30 mg O_2/100 g H_2O. When air is itself saturated with water, the partial pressure of oxygen is 156 mm Hg. How many milligrams of oxygen dissolve in 100 g of water when the water is saturated with air and is in equilibrium with air that is saturated with water vapor?

ANALYSIS This problem involves the solubility of a gas under different pressures, so Henry's law equation applies. Let's first collect the data to see what we have.

$$C_1 = 4.30 \text{ mg/100 g} \qquad C_2 = \,?$$
$$P_1 = 760 \text{ mm Hg} \qquad P_2 = 156 \text{ mm Hg}$$

SOLUTION Using Equation 7.3 gives us

$$\frac{C_1}{P_1} = \frac{C_2}{P_2}$$

$$\frac{4.30 \text{ mg/100 g}}{760 \text{ mm Hg}} = \frac{C_2}{156 \text{ mm Hg}}$$

When we solve this for C_2, we get

$$C_2 = 4.30 \text{ mg/100 g} \times \frac{156 \text{ mm Hg}}{760 \text{ mm Hg}}$$

$$= 0.883 \text{ mg/100 g}$$

PRACTICE EXERCISE 1

How many milligrams of nitrogen dissolve in 100 g of water when the water is saturated with air and is in equilibrium with air that is saturated with water vapor? The partial pressure of nitrogen in air that is itself saturated with water vapor is 586 mm Hg. The solubility of pure nitrogen in water at 760 mm Hg is 1.90 mg/100 g H_2O.

The pressure-solubility relationship for solutions of gases in water is particularly important to people exposed to possible decompression sickness (the bends), as discussed in Special Topic 7.1.

Water Reacts with Some Gases to Aid in Dissolving Them   A number of important gases, like carbon dioxide, sulfur dioxide, and ammonia, are far more soluble in water than are oxygen and nitrogen. At 20 °C, only 4.30 mg of oxygen dissolves in 100 g of water, but carbon dioxide is nearly 40 times as soluble, sulfur dioxide is nearly 2500 times as soluble, and ammonia is a whopping 12,000 times as soluble in water as oxygen.

There are two reasons for this. First, a physical fact: The molecules of these gases have $\delta+$ and $\delta-$ sites that can attract water molecules and be attracted by them. Figure 7.5, for example, shows how ammonia molecules can form hydrogen bonds to water molecules. Second, a chemical fact: A fraction of the gases actually reacts with water in solution. When water contains NH_3, for example, the following two equilibria exist, the first being physical and the second chemical.

$$NH_3(g) \rightleftharpoons NH_3(aq)$$

$$NH_3(aq) + H_2O \rightleftharpoons NH_4^+(aq) + OH^-(aq)$$
$$\quad\quad\quad\quad\quad\quad\quad\quad\quad \text{Ammonium} \quad \text{Hydroxide}$$
$$\quad\quad\quad\quad\quad\quad\quad\quad\quad \text{ion} \quad\quad \text{ion}$$

The forward reaction of the first equilibrium is aided by the hydrogen bonding that occurs between water and ammonia, but which cannot occur with gases like oxygen or nitrogen. The *chemical* nature of ammonia helps the forward reaction of the second equilibrium; NH_3 tends to react with water to produce the ions shown. At equilibrium, the molar concentrations of the ions are only a small fraction of all unreacted molecules of $NH_3(aq)$. However, the fact that the forward reaction occurs at all helps to draw $NH_3(aq)$ *out of the product side of the first equilibrium* and into solution by way of the second equilibrium's forward reaction.

We'll introduce a new term here often used in connection with equilibria, a term to describe which of the two sides of an equilibrium is *favored*. The **favored** side in an equilibrium has the *higher* concentration of the chemical species involved. In both of the ammonia equilibria, the favored side is the left; we say that the reactants are favored. Which side is favored for each solute has to be discovered by actual measurements.

Solubilities of Some Gases in Water at 20 °C in mg/100 g H_2O

O_2	4.3
CO_2	169
SO_2	10,600
NH_3	51,800

■ The nose readily knows that $NH_3(g)$ is present in the space above aqueous ammonia. Probably no other substance is more widely recognized *and correctly named* than ammonia.

■ In 1 M $NH_3(aq)$, only about 0.5% of the NH_3 molecules have reacted to form $NH_4^+(aq) + OH^-(aq)$.

189

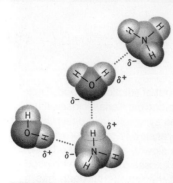

FIGURE 7.5
Hydrogen bonds (·····) between
molecules of ammonia and water
help to keep ammonia in solution.

■ *Respiration* means all the
activities that bring in and use
oxygen and get rid of carbon
dioxide.

■ Venous blood is the blood
returning to the lungs in veins.
Arterial blood is blood leaving
the lungs.

■ If you've ever heard of a
mountain road or path having a
"steep grade," you've
encountered the concept,
because "grade" means
"gradient," here a rapid change
in altitude over distance.

The two equilibria involving the formation of aqueous CO_2 are similar to those of ammonia. In both, the reactants are favored.

$$CO_2(g) \rightleftharpoons CO_2(aq)$$
$$CO_2(aq) + H_2O \rightleftharpoons H_2CO_3(aq)$$
Carbonic acid

Thus, to the (small) extent that $CO_2(aq)$ is removed by the second equilibrium and changed to H_2CO_3, the first equilibrium shifts to the right and so causes more $CO_2(g)$ to go into solution. Similarly, SO_2 is involved in two equilibria in water.

$$SO_2(g) \rightleftharpoons SO_2(aq)$$
$$SO_2(aq) + H_2O \rightleftharpoons H_2SO_3(aq)$$
Sulfurous acid

Associated with Any Solution of a Gas in Water Is a *Gas Tension* Sometimes in discussions of human or animal respiration, the term *gas tension* is used to describe the availability of a gas from some body fluid. **Gas tension** is the partial pressure of a gas over a solution with which it is in equilibrium. It is an indirect measure of how much gas is in solution, because the more there is in solution the more there will be of the gas above the solution exerting a partial pressure. Thus, a high gas tension means a high availability of the gas from the solution.

In venous blood, the oxygen tension is about 40 mm Hg. In the air inside the lungs, it is about 100 mm Hg. Gases always tend to diffuse from a higher to a lower pressure, so oxygen naturally tends to move from the lungs into the returning blood. This, of course, is the direction it must move if we are to live. Carbon dioxide, on the other hand, has a gas tension in venous blood of about 46 mm Hg, but in the air inside the lungs its gas tension is about 40 mm Hg. Thus this gas, a waste product, naturally tends to migrate from venous blood to the lungs, as it must, where it can then be discharged in exhaled air. The difference in gas tension between two regions of the same system is called a *pressure gradient*. **Gradient** is a general term used to describe the occurrence of a quantity, like pressure, that varies in size from one location to another within the system.

7.4

PERCENTAGE CONCENTRATION EXPRESSIONS

The number of grams of solute in 100 g of solution is the percentage concentration of the solution.

For working with the stoichiometries of reactions in solution, no better concentration expression exists than that of molarity, the concentration in moles per liter. Many chemical situations arise, however, in which there is little need to know exact stoichiometries, yet some idea of concentration is useful. Test-tube tests, for example, are somewhat casual about stoichiometry. A solution, called a **reagent,** is added drop by drop to another solution to see if some solute is present. It isn't always crucial to know the molarity of the reagent for such an experiment. Yet some information about its concentration is needed, if only to be able to prepare more of it. Here is where a number of other expressions for concentrations have been developed. The most important of these are percentage concentrations.

■ We now use *percent*
instead of *percentage* to
conform to common usage.

Weight/Weight Percent The **weight/weight percent (w/w%) concentration** of a solution is the number of grams of solute in 100 g of the solution. For example, a 10.0% (w/w) glucose solution has a concentration of 10.0 g of glucose in 100 g of solution. To make 100 g of this solution, you would mix 10.0 g of glucose with 90.0 g of the solvent for a total of 100 g.

190

EXAMPLE 7.2
Using Weight/Weight Percents

How many grams of 0.900% (w/w) NaCl solution contain 0.250 g of NaCl?

ANALYSIS The concentration term, 0.900% (w/w), translates into the units of 0.900 g NaCl/100 g NaCl solution, so it gives us the following two conversion factors.

$$\frac{0.900 \text{ g NaCl}}{100 \text{ g NaCl soln}} \quad \text{and} \quad \frac{100 \text{ g NaCl soln}}{0.900 \text{ g NaCl}}$$

These two ratios are just equivalent ways of understanding the concentration, and we should remind ourselves that any expression of a concentration in any units can be expressed as either of two ratios, as we have done here. If we multiply the given, 0.250 g NaCl, by the second factor, the final units will be g NaCl soln.

SOLUTION

$$0.250 \text{ g NaCl} \times \frac{100 \text{ g NaCl soln}}{0.900 \text{ g NaCl}} = 27.8 \text{ g NaCl soln}$$

Thus 27.8 g of 0.900% (w/w) NaCl contains 0.250 g of NaCl.

EXAMPLE 7.3
Preparing Weight/Weight Percent Solutions

A special kind of saline solution, called isotonic saline, is sometimes used in medicine. Its concentration is 0.90% NaCl (w/w). How would you prepare 750 g of such a solution?

ANALYSIS Once again, we have to translate the label on the bottle, 0.90% (w/w) NaCl, into conversion factors.

$$\frac{0.90 \text{ g NaCl}}{100 \text{ g NaCl soln}} \quad \text{and} \quad \frac{100 \text{ g NaCl soln}}{0.90 \text{ g NaCl}}$$

What we basically have to determine is the number of grams of NaCl that we must weigh out and dissolve in water to make the final mass equal to 750 g.

SOLUTION To find the number of grams of NaCl, we multiply the given mass of the NaCl solution by the first conversion factor.

$$750 \text{ g NaCl soln} \times \frac{0.90 \text{ g NaCl}}{100 \text{ g NaCl soln}} = 6.8 \text{ g NaCl} \quad \text{(rounded from 6.75)}$$

Thus if we dissolve 6.8 g NaCl in water and add enough water to make the final mass equal to 750 g, we can write the label to read 0.90% (w/w) NaCl.

PRACTICE EXERCISE 2

Sulfuric acid can be purchased from a chemical supply house as a solution that is 96.0% (w/w) H_2SO_4. How many grams of this solution contain 9.80 g of H_2SO_4 (or 0.100 mol)?

PRACTICE EXERCISE 3

How many grams of glucose and how many grams of water are needed to prepare 500 g of 0.250% (w/w) glucose?

Sometimes weight/weight percent solutions are prepared by diluting a more concentrated solution. The equation used to make the necessary calculation is similar to the one we derived in Section 3.6 for dilutions involving molar concentrations. We will simply give the equation here.

$$g_{\text{concd soln}} \times \text{percent (w/w)}_{\text{concd soln}} = g_{\text{dil soln}} \times \text{percent (w/w)}_{\text{dil soln}}$$

Volume/Volume Percent   A concentration expressed as a **volume/volume percent (v/v%)** gives us the number of volumes of one substance dissolved in 100 volumes of the mixture. This is often used for solutions of gases or for solutions of liquids. For example, the concentration of oxygen in air is 21% (v/v), which means that there are 21 volumes of oxygen in 100 volumes of air. The unit used for volume can be any unit, as long as we use the same unit for the solute as for the solution. The units cancel when we deal with true percentages. Calculations involving volume/volume percents involve the same kinds of steps we used for problems of weight/weight percents.

Weight/Volume "Percent"   Although this is a fairly common concentration expression, a weight/volume percent isn't a true percent because the units don't cancel. When a concentration is given as a **weight/volume percent (w/v%),** it means the number of grams of solute in 100 mL of the solution. Thus a concentration given as 0.90% (w/v) NaCl solution means

■ Often the units of g/100 mL will be given as g/dL because 100 mL = 1 dL.

0.90 g of NaCl/100 mL NaCl solution

Weight/volume percent problems are handled through conversion factors just as we did for weight/weight percent problems.

Milligram "Percent," Parts per Million, and Parts per Billion   You'll occasionally encounter some special concentration expressions that are handy when the solutions are very dilute. **Milligram percent** means the number of milligrams of solute in 100 mL of the solution.

For very dilute solutions, the concentration might be given in **parts per million (ppm),** which means the number of parts (in any unit) in a million parts (the same unit) of the solution. Parts per million might be interpreted as grams per million grams or pounds per million pounds. One ppm is analogous to one penny in a million pennies ($10,000) or 1 minute in a million minutes (about two years).

When water is the solvent, the volume in milliliters and the mass in grams of a dilute solution are numerically the same (at least to two significant figures), and it can be shown that 1 ppm is identical with 1 mg/L.

Parts per billion (ppb) similarly means parts per billion parts, such as grams per billion grams. This expression is used for extremely dilute systems. One ppb is like one penny in $10 million (a billion pennies), or like two drops of a liquid in a full, 33,000-gallon tank car.

Because of the many ways the term *percent* can be taken, there is a trend away from using it, which should be encouraged. Instead, the explicit units are given. Thus, instead of referring to a concentration of, say, 10.0% (w/v) KCl, the label or the report should read 10.0 g KCl/100 mL. When no units are given, just a percent, and until you find out otherwise, you have to interpret the percent as weight/weight if the solute is a solid when pure and as a volume/volume percent if the solute is a liquid when pure. Usually, however, the labels are clear and, as always, *labels should be read carefully.*

7.5

OSMOSIS AND DIALYSIS

The selective migration of ions and molecules through cell membranes is an important mechanism for getting nutrients inside cells and waste products out. Solutions generally have lower melting points and higher boiling points than their pure solvents. Aqueous solutions, for example, freeze not at 0 °C but at slightly lower temperatures. They also boil not at 100 °C but at slightly higher temperatures (both compared at 760 mm Hg). Interestingly, these effects of solutes depend not on what they are, but only on the ratio of solute to solvent particles. Properties of solutions or colloidal dispersions that depend only on the *number* of solute particles per unit volume of solvent, not on the chemical identities of their solutes, are called **colligative properties.** The depression of the freezing point and the elevation of the boiling point are two examples. The relative abilities of components of a solution to migrate through certain kinds of membranes — osmosis and dialysis — are also colligative properties.

■ From the Greek *kolligativ*, depending on number and not on nature.

The effects aren't very large unless the concentrations are very large. As we said, however, they are related to concentrations, not chemical identities. For example, if we prepare two solutions, one that has 1.0 mole of NaCl in 1000 g of water and the other with 1.0 mole of KBr in 1000 g of water, each freezes at −3.4 °C (and not at 0 °C), and each boils at 101 °C (at 760 mm Hg). Both the freezing and the boiling points are the same, despite the difference in solutes, because the ratios of the moles of ions to the moles of water in both solutions are identical. Otherwise, the identities of the solutes are immaterial.

■ The temperature of a mixture made from 33 g NaCl and 100 g ice is about −22 °C (−6 °F).

When the concentrations are very high, the effects can be quite large and important. The whole basis for the use of an antifreeze mixture in an automobile radiator is the large depression of the freezing point of the radiator fluid caused by the presence of the antifreeze. A 50/50 mixture (vol/vol) of almost any commercially available automobile antifreeze in water gives protection to about −40 °F.

Osmosis Is the Diffusion of Solvent Molecules through Membranes
Cells in living systems are enclosed by cell membranes. Generally, on both sides of them there are aqueous systems that contain substances both in solution and in colloidal dispersion. Materials and water have to be able to move through cell membranes in either direction so that nutrients can enter cells and wastes can leave. Two factors control these movements, *active transport* and *dialysis.*

Active transport is the active involvement of specialized materials (proteins) embedded in a cell membrane to propel ions and molecules through it. These substances accept and move ions and molecules by endothermic chemical reactions. We can do no more than mention active transport at this time.

■ The movements of Na^+ and K^+ ions into and out of cells is controlled by active transport mechanisms.

Dialysis, the other mechanism for getting substances through cell membranes, depends on a property of the membrane that separates two solutions of unequal particle concentration. Membranes of cells, for example, are **semipermeable.** This means that they can let some but not all kinds of molecules and ions pass through. Cellophane, for example, is a synthetic, semipermeable membrane. When it is in contact with an aqueous solution, only water molecules and other small molecules and ions can migrate through it. Molecules of colloidal size are stopped. Membranes with this selectivity are called **dialyzing membranes.** Cellophane is a dialyzing membrane. Evidently, it has ultrafine pores just large enough to let small particles through but too small for larger particles.

Some membranes have pores so small that only water molecules can get through them. A semipermeable membrane that is so selective that only the solvent molecules can get through is called an **osmotic membrane.** Because ions are hydrated, their effective sizes are apparently too large for the pores. No other molecules can pass either.

■ Essentially no purely osmotic membranes occur naturally.

This situation gives us a special case that is enough different from dialysis to have its own name — osmosis. **Osmosis** is the net migration of the solvent from a solution with the lower

FIGURE 7.6

Osmosis and osmotic pressure. (*a*) In the beaker, *A*, there is pure water and in the tube, *B*, there is a solution. An osmotic membrane closes the bottom of the tube. (*b*) A microscopic view at the osmotic membrane shows how solute particles interfere with the movements of water molecules from *B* to *A*, but not from *A* to *B*. (*c*) The level in *A* has fallen and that in *B* has risen because of osmosis. A back pressure would be needed, (*d*), to prevent osmosis, and the exact amount of pressure is the osmotic pressure of the solution in *B*.

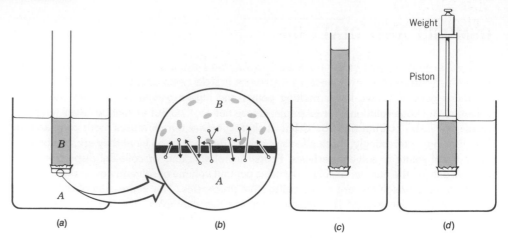

concentration of solute into the solution with the higher concentration through an osmotic membrane.

Figure 7.6 shows why there is a net flow in one direction in osmosis. It shows the special case in which pure water is on one side of the membrane. Water molecules can move in *both* directions, but when solute particles are on one side, they get in the way. Water molecules, therefore, are prevented from leaving as frequently from that side as they are able to come in from the opposite side where no solute particles interfere. Thus more water molecules move into the concentrated solution than leave it, and so this solution becomes increasingly diluted. Eventually, the column of water shown in Figure 7.6 will exert a high enough back pressure to prevent any further rise.

Sometimes students have a problem with remembering the direction of osmosis. Think of it this way. The net flow always makes the concentrated solution more dilute. If osmosis could continue long enough, the concentrated solution would become dilute enough and the other solution (by losing solvent) would become concentrated enough so that the two concentrations would become equal.

Osmotic Pressure Is a Measure of Concentration   The exact back pressure necessary to prevent osmosis is called the **osmotic pressure** of the solution, and its symbol is Π.

■ Π is the Greek capital letter pi.

The value of osmotic pressure is directly proportional to the molar concentration of the particles in the solution (at least at relatively low molarities), and the equation for osmotic pressure is almost identical to the ideal gas equation, $PV = nRT$. For osmotic pressure, Π,

$$\Pi V = nRT$$

If we rearrange terms, and remember that n/V = molarity (moles per liter), then

$$\Pi = \frac{n}{V} \times RT$$

or,

$$\Pi = MRT \tag{7.4}$$

This equation shows how osmotic pressure is directly proportional to the concentration. R, the gas constant, may be taken as 0.0821 L atm/mol K when the pressure in atm is sought or as 6.24×10^4 mL mm Hg/mol K when the pressure in mm Hg is sought. See page 157.

The osmotic pressure of a solution has to be understood not as something that the solution is actually exerting, like some hand pushing on a surface. Instead, osmotic pressure is a potential pressure, directly related to concentration, that can be realized only when an osmotic membrane separates the solution from pure water.

Even in relatively dilute solutions, the osmotic pressure can be very high, as the next worked example illustrates.

EXAMPLE 7.4
Calculating Osmotic Pressure

A dilute solution, 0.100 M sugar in water, is separated from pure water by an osmotic membrane. What is its osmotic pressure in mm Hg at a temperature of 25 °C or 298 K?

ANALYSIS The calculation of osmotic pressure from molarity and temperature data requires Equation 7.4. To obtain the pressure in mm Hg we can use $R = 6.24 \times 10^4$ mL mm Hg/mol K, remembering that M stands for mol solute/1000 mL solution.

SOLUTION

$$\Pi = MRT$$
$$= \frac{0.100 \text{ mol}}{1000 \text{ mL}} \times 6.24 \times 10^4 \frac{\text{mL mm Hg}}{\text{mol K}} \times 298 \text{ K}$$
$$= 1.86 \times 10^3 \text{ mm Hg}$$

The solution has an osmotic pressure of 1.86×10^3 mm Hg. This means that a solution with this concentration can support a column of mercury 1.86×10^3 mm high, over six feet.

PRACTICE EXERCISE 4

What is the osmotic pressure in mm Hg of a 0.900 M glucose solution at 25 °C?

If, instead of mercury, the column in Example 7.4 had been water (or the dilute solution), the column supported would have been 25.3 m (83.0 ft) high. Thus a relatively dilute solution, when separated from pure water, can be driven to a column height of *several dozen feet*. This phenomenon is one of the factors in the rise of sap in tall trees.

■ Water's density (1.00 g/mL) is much less than mercury's (13.6 g/mL), so the water column is 13.6 times higher.

The Ions of an Ionic Compound Individually Affect Osmotic Pressure Solute particles that cause osmotic pressure can be ions, molecules, or macromolecules. Just remember that osmotic pressure is a *colligative* property, so it depends only on the concentrations of the particles. Thus when the solute is an ionic compound such as sodium chloride, the concentration of particles — ions in this situation — is twice the molar concentration of the salt given by the label on the bottle. For example, 0.10 M NaCl has a concentration of $2 \times (0.10) = 0.20$ mole of all ions per liter, because NaCl breaks up into two ions for each formula unit that goes into solution. The osmotic pressure of 0.10 M NaCl is therefore twice as large as that of 0.10 M glucose, which does not break up into ions.

For an ionic compound like Na_2SO_4, for which three ions are released for each formula unit that dissolves — two Na^+ and one SO_4^{2-} — the concentration of particles in a 0.10 M solution is $3 \times (0.10) = 0.30$ mole of all ions per liter.

The labeled molarity of a solution thus does not reveal enough about a solution when we think about its osmotic pressure. To express the concentration of all osmotically active particles in the solution, scientists involved in the chemistry of health sometimes use a related concentration expression, called the solution's **osmolarity,** the molar concentration of all solute particles active in osmosis or dialysis. Thus 0.10 M NaCl has a molarity of 0.10 mol/L of NaCl but an osmolarity of 0.20 mol/L. The osmolarity of 0.10 M Na_2SO_4 is 0.30 mol/L. The concentration term in Equation 7.4, M, must refer to the osmolarity of the solution.

The abbreviation used in labeling osmolarities is **Osm.** Thus 0.0125 Osm means that the osmolarity of this solution is 0.0125 mol/L of all of the particles that contribute to the osmotic pressure.

PRACTICE EXERCISE 5

Assuming that any *ionic* solutes in this exercise break up completely into their constituent ions when they dissolve in water, what is the osmolarity of each solution?

(a)   0.010 *M* NH$_4$Cl (which ionizes as NH$_4^+$ and Cl$^-$)

(b)   0.005 *M* Na$_2$CO$_3$ (which ionizes as 2Na$^+$ and CO$_3^{2-}$)

(c)   0.100 *M* fructose (a sugar and a molecular substance)

(d)   A solution that contains both fructose and NaCl with concentrations of 0.050 *M* fructose and 0.050 *M* NaCl

Small Solute Particles Pass through Dialyzing Membranes   **Dialysis** is osmosis when the membrane is more permeable. In dialysis, not only water molecules but also ordinary-sized ions and molecules move through the membrane. A *dialyzing membrane* can be thought of as having larger pores than an osmotic membrane. Cell membranes are, in part, dialyzing membranes. However, cell membranes also have built into them clusters of protein molecules serving as "gates" and "portals" that, with the aid of energy-generating chemical reactions, exert selective control over the directions of migration of some small particles. Cell membranes, in other words, incorporate chemical systems that manage *active* transport.

Dialysis produces a net migration of water only if the fluid on one side of the dialyzing membrane has a higher concentration in colloidal substances than the other. Colloidal-sized particles are blocked by dialyzing membranes, so they get in the way of the movements of smaller particles through the membrane. The net flow of fluid in dialysis, as in osmosis, is from the side that has the lower concentration of colloidal substances to the side with the higher concentration. The effect is to make the concentrated solution more dilute.

The imbalance in concentration that is related to colloidally dispersed materials causes a **colloidal osmotic pressure,** which is similar to osmotic pressure in meaning. In the next section we present some situations involving life at the molecular level where osmotic pressure relationships are very critical and depend on the colloidal osmotic pressure of blood.

7.6

DIALYSIS AND THE BLOODSTREAM

When the osmotic pressure of blood varies too much, the result can be shock or damage to red blood cells.

The body tries to maintain the concentrations of all of the substances that circulate in blood within fairly narrow limits. Quite complicated mechanisms exist to excrete or retain solutes or to excrete or retain water. When they fail, the consequences can be life-threatening. In this section we will look briefly at two situations that arise when the osmotic pressure of blood changes too much.

The Brain Loses Blood Flowage in Shock   One feature of the shock syndrome is a dramatic increase in the permeability of the blood capillaries to colloidal-sized particles, particularly protein molecules. When these leave the blood, the colloidal osmotic pressure of blood decreases, which is another way of saying that the concentration of colloidal substances in blood decreases. The blood, in effect, becomes less concentrated. It is less able, therefore, to take up water from the spaces that surround the blood capillaries.

A sufficient decrease in the colloidal osmotic pressure of blood makes a net loss of water from blood possible. A loss of water means a loss of total blood volume. This makes it more difficult to bring nutrients to brain cells and carry wastes away. The brain functions much less

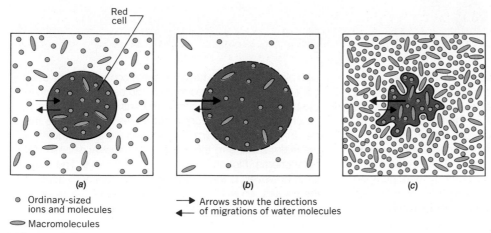

Red cell

(a)   (b)   (c)

○ Ordinary-sized
  ions and molecules

⬭ Macromolecules

→ Arrows show the directions
← of migrations of water molecules

FIGURE 7.7

Dialysis. (a) The red cell is in an isotonic environment. (b) Hemolysis is about to occur because the red cell is swollen by extra fluids brought into it from its hypotonic environment. (c) The red cell experiences crenation when it is in a hypertonic environment.

well, and the result to the nervous system is called shock. When a person goes into shock, one of the many problems — but one that lies close to the central cause — is that blood capillaries become temporarily more permeable to the loss of macromolecules from blood.

Red Blood Cells Hemolyze in Water Millions of red blood cells circulate in the bloodstream, and their membranes behave as dialyzing membranes. Within each red cell is an aqueous fluid with dissolved and colloidally dispersed substances (see Fig. 7.7a). Although the colloidal particles are too large to dialyze, they contribute to the colloidal osmotic pressure. They help, therefore, to determine the direction of dialysis through the red cell membrane.

When red cells are placed in pure water, the fluid inside the red cell is more concentrated than the surrounding liquid. Dialysis now occurs to bring fluid *into the red cell.* Enough fluid moves in to make the cell burst open, as seen in Figure 7.7b. The rupturing of red cells is called **hemolysis,** and we say that the cells hemolyze.

On the other hand, when red cells are put into a solution with an osmolarity greater than their own fluid, dialysis occurs in the opposite direction — out of the cell and into the solution. Now the cells lose fluid volume, and shrivel and shrink. This process is called **crenation** (Fig. 7.7c).

In some medical situations, body fluids need replacement or nutrients have to be given by intravenous drip. The osmolarity of the solution being added should match that of the fluid inside the red cells. Otherwise, hemolysis or crenation will occur.

Two solutions of equal osmolarity are called **isotonic solutions.** If one has a lower osmotic pressure than the other, the first is said to be *hypotonic* with respect to the second. A **hypotonic solution** has a lower osmolarity than the one to which it is compared. Red cells hemolyze if placed in a hypotonic environment such as pure water.

A **hypertonic solution** is one with a higher osmotic pressure than another. Thus 0.14 M NaCl is hypertonic with respect to 0.10 M NaCl. Red cells undergo crenation when they are in a hypertonic environment.

A 0.9% (w/w) NaCl solution, called **physiological saline solution,** is isotonic with respect to the fluid inside a red cell. Any solution to be added in any large quantity into the bloodstream has to be isotonic in this way.

All the topics we have studied in this and the preceding section are important factors in the operation of artificial kidney machines, which are discussed in Special Topic 7.2.

The kidneys cleanse the bloodstream of nitrogen waste products such as urea and other wastes. If the kidneys stop working efficiently or are removed, these wastes build up in the blood and threaten the life of the patient. The artificial kidney is one remedy for this.

The overall procedure is called hemodialysis—the dialysis of blood—and Figure 1 shows how it works. The bloodstream is diverted from the body and pumped through a long, coiled cellophane tube that serves as the dialyzing membrane. (The blood is kept from clotting by an anticlotting agent such as heparin.) A solution called the dialysate circulates outside of the cellophane tube. This dialysate is very carefully prepared not only to be isotonic with blood but also to have the same concentrations of all the essential substances that should be left in solution in the blood. When these concentrations match, the rate at which such solutes migrate out of the blood equals the rate at which they return. In this way several key equilibria are maintained, and there is no net removal of essential components. Figure 2 shows how this works. The dialysate, however, is kept very low in the concentrations of the wastes, so the rate at which they leave the blood is greater than the rate at which they can get back in. In this manner, hemodialysis slowly removes the wastes from the blood.

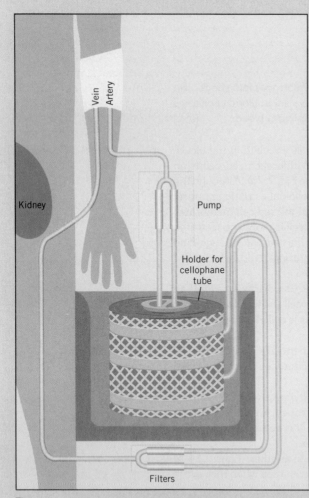

Figure 1

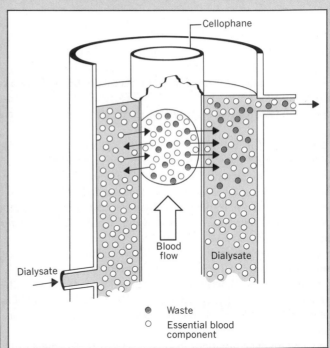

Figure 2

SUMMARY

Solutions Ions and molecules of ordinary size, if soluble in water at all, form solutions. These are homogeneous mixtures that neither gravity nor filtration can separate. The solubilities of most solids increase with temperature, because their dissolving is usually endothermic. (More energy is needed to break up the crystal than is recovered as the solvent cages form about the ions or molecules.)

Hydration The attraction of water molecules to ions or to polar molecules leads to a loose solvent cage that shields the ions or molecules from each other. This phenomenon is called hydration, and it helps to explain why some substances dissolve in water. Sometimes water of hydration is present in a crystalline material in a definite proportion to the rest of the formula unit, and such a substance is a hydrate. Heat converts most hydrates to their anhydrous forms. And some anhydrous forms serve as drying agents — desiccants.

Gas solubilities Gases dissolve in water exothermically, so the addition of heat to an aqueous solution of a gas drives the gas out of solution. The solubility of a gas is directly proportional to its partial pressure in the space above the solution (Henry's law). Some gases do more than mechanically dissolve in water; part of what dissolves forms hydrogen bonds with water and part reacts with water to form soluble species.

Percent concentration A variety of concentration expressions have been developed to provide ways to describe a concentration without going into molar concentrations. These include weight/weight percents, volume/volume percents, and hybrid descriptions that aren't true percentages — weight/volume percent, milligram percent, parts per million, and parts per billion.

Colloidal dispersions Large clusters of ions or molecules or macromolecules do not form true solutions but colloidal dispersions. These can reflect and scatter light (Tyndall effect), experience Brownian movement, and (in time) succumb to the force of gravity (if the medium is fluid). Protective colloids, such as emulsifying agents, sometimes stabilize these systems. If the dispersed particles grow to an average diameter of about 1000 nm, they slip over into the category of suspended matter; such systems must be stirred to maintain the suspension.

Osmosis and dialysis When a semipermeable membrane separates two solutions or dispersions of unequal osmolarities, a net flow occurs in the direction that, if continued, would produce solutions of identical osmolarities. When the membrane is osmotic, only the solvent can migrate, and the phenomenon is osmosis. The back pressure needed to prevent osmosis is called the osmotic pressure, and it's directly proportional to the concentration of all particles of solute that are osmotically active — ions, molecules, and macromolecules.

When macromolecules are present, their particular contribution to the osmotic pressure is called the colloidal osmotic pressure of a solution. It is this factor that operates when the membrane is a dialyzing membrane.

The permeability of blood capillaries changes temporarily when a person experiences shock, and macromolecules leave the blood. Their departure results in the loss of water, too, and the blood volume decreases.

Solutions of matched osmolarity are isotonic. Otherwise, one is hypertonic (more concentrated) with respect to the other, and the other is hypotonic (less concentrated). Only isotonic solutions, or those that are nearly so, should be administered in large quantities intravenously.

REVIEW EXERCISES

The answers to Review Exercises whose numbers are in color are found in Appendix C. The answers to the other Review Exercises are found in the Study Guide that accompanies this book. The more challenging questions are marked with asterisks.

Homogeneous Mixtures

7.1 The definition of *homogeneous* depends on how we define "smallest sample." Explain.

7.2 Chlorine gas is a mixture of the following molecules, where the left superscripts denote the mass numbers of individual chlorine isotopes.

$$^{35}Cl-^{35}Cl \qquad ^{35}Cl-^{37}Cl \qquad ^{37}Cl-^{37}Cl$$

Under what circumstances can we call this mixture *homogeneous?* Explain.

7.3 Particle *size* is one basis for distinguishing among solutions, colloidal dispersions, and (stirred) suspensions. Explain why size works for this purpose.

7.4 Why are suspensions usually not considered homogeneous?

7.5 Which of the three kinds of homogeneous mixtures
 (a) Can be separated into its components by filtration?
 (b) Exhibits the Tyndall effect?
 (c) Shows observable Brownian movement?
 (d) Has the smallest particles of all kinds?
 (e) Is likeliest to be the least stable at rest over time?

7.6 What kinds of colloidal dispersions are generally the most stable? Explain.

7.7 The blood is simultaneously a solution, a colloidal dispersion, and a suspension. Explain.

7.8 A colloidal dispersion gives the Tyndall effect, but a solution doesn't. Explain.

7.9 The observation of the Brownian movement was important historically in the development of the kinetic theory of gases. Suggest a reason.

7.10 What simple test could be used to tell if a clear, colorless solution contained substances in colloidal dispersion?

7.11 What is an emulsion? Give some examples.

7.12 What is a sol? Give some examples.

7.13 What is a gel? Give an example.

Aqueous Solutions

7.14 In a crystal of sodium chloride the chloride ions are surrounded by oppositely charged ions (Na^+) as nearest neighbors. What replaces this kind of electrical environment for chloride ions when sodium chloride dissolves in water?

7.15 When we say that a sodium ion in water is *hydrated*, what does this mean? (Make a drawing as part of your answer.)

7.16 Carbon tetrafluoride, CF_4, has the unusually polar C—F bonds, and its molecules are tetrahedral, like those of methane, CH_4. This substance does not dissolve in water. Why won't water let CF_4 molecules in?

7.17 We have to distinguish between *how fast* something dissolves in water and *how much* can dissolve to make a saturated solution. The speed with which we can dissolve a solid in water increases if we (a) crush the solid to a powder, (b) stir the mixture, or (c) heat the mixture. Use the kinetic theory as well as the concept of forward and reverse processes to explain these facts.

7.18 Assuming that solid potassium bromide is dissolved in water.
(a) Write the equation for the dissociation of this compound as its solution forms.
(b) If a saturated solution is prepared with excess, undissolved KBr present, the rates of what two changes are equal in this saturated solution? Write an equilibrium expression.

7.19 Suppose that you do not know and do not have access to a reference in which to look up the solubility of sodium nitrate, $NaNO_3$, in water at room temperature. Yet you need a solution that you know beyond doubt is saturated. How can you make such a saturated solution and know that it is saturated?

7.20 Ammonium chloride dissolves in water endothermically. Suppose that you have a saturated solution of this compound, that its temperature is 30 °C, and that undissolved solute is present. Write the equilibrium expression for this saturated solution, and use Le Châtelier's principle to predict what will happen if you cool the system to 20 °C.

Hydrates

7.21 Epsom salt is magnesium sulfate heptahydrate. Write the equation for the decomposition of this hydrate to its anhydrous form. (*Hepta-* denotes seven.)

7.22 Why are hydrates classified as compounds and not as wet mixtures?

7.23 When water is added to anhydrous sodium sulfate, the decahydrate of this compound forms. Write the equation. (*Deca-* denotes ten.)

7.24 Anhydrous calcium chloride is hygroscopic. What does this mean? Does this property make it useful as a desiccant?

7.25 Sodium hydroxide is sold in the form of small pellets about the size and shape of split peas. It is a very deliquescent substance. What can happen if you leave the cover off of a bottle of sodium hydroxide pellets?

*7.26 When 6.47 g of the hydrate of compound X was strongly heated to drive off all of the water of hydration, the residue, the anhydrous form of X, had a mass of 3.90 g. What number should y be in the formula of the hydrate, $X \cdot yH_2O$? The formula mass of X is 82.0.

*7.27 When all of the water of hydration was driven off of 4.32 g of a hydrate of compound Z, the residue, the anhydrous form, Z, had a mass of 3.09 g. What is the formula of the hydrate (using the symbol Z as part of it)? The formula mass of Z is 90.0.

Gas Solubilities

7.28 The solubility of methane, the chief component in bunsen burner gas, in water at 20 °C and 1.0 atm is 0.025 g/L. What will be its solubility at 1.4 atm?

7.29 At 20 °C the solubility of nitrogen in water is 0.0150 g/L when the partial pressure of the nitrogen is 580 mm Hg. What is its solubility when the partial pressure is raised to 740 mm Hg?

7.30 Explain why carbon dioxide is more soluble than nitrogen in water.

7.31 Explain why sulfur dioxide is much more soluble in water than oxygen.

7.32 Using Le Châtelier's principle, explain why the solubility of a gas in water should decrease with decreasing partial pressure of the gas.

7.33 If the gas tension of O_2 in blood is described as 80 mm Hg, what specifically does this mean?

7.34 If in one region of the body the gas tension of oxygen over blood is 79 mm Hg and in a second region it is 60 mm Hg, which region (the first or the second) has a higher concentration of oxygen in the blood itself?

Percent Concentrations

7.35 If a solution has a concentration of 0.915% (w/w) NaOH, what two conversion factors can we write based on this value?

7.36 A solution bears the label 1.42% (w/v) KCl. What two conversion factors can be written for this value?

7.37 The concentration of a pollutant in water is reported as 2.0 ppm. What is the concentration of this pollutant in units of mg/L?

7.38 A solution of rubbing alcohol in water is described as 30% (v/v). What two conversion factors are possible from this value?

7.39 How many grams of solute are needed to prepare each of the following solutions?
(a) 250 g of 0.900% (w/w) NaCl
(b) 500 g of 3.22% (w/w) $NaC_2H_3O_2$
(c) 125 g of 6.75% (w/w) NH_4Cl
(d) 250 g of 1.25% (w/w) Na_2CO_3

7.40 Calculate the number of grams of solute needed to make each of the following solutions.
(a) 250 g of 0.625% (w/w) NaI
(b) 125 g of 0.375% (w/w) NaBr
(c) 100 g of 1.00% (w/w) $C_6H_{12}O_6$ (glucose)
(d) 50.0 g of 8.50% (w/w) H_2SO_4

7.41 How many grams of solute have to be weighed out to make each of the following solutions?
(a) 250 mL of 12.5% (w/v) NaCl
(b) 500 mL of 1.50% (w/v) KBr
(c) 100 mL of 1.12% (w/v) $CaCl_2$
(d) 500 mL of 0.900% (w/v) NaCl

7.42 In order to prepare the following solutions, how many grams of solute are required?
(a) 125 mL of 6.00% (w/v) $Mg(NO_3)_2$
(b) 250 mL of 2.25% (w/v) NaBr
(c) 100 mL of 5.00% (w/v) KI
(d) 75.0 mL of 1.25% (w/v) $Ca(NO_3)_2$

7.43 How many milliliters of ethyl alcohol have to be used to make 500 mL of 20.0% (v/v) aqueous ethyl alcohol solution?

7.44 A sample of 750 mL of 6.25% (v/v) aqueous methyl alcohol contains how many milliliters of pure methyl alcohol?

*7.45 A chemical supply room has supplies of the following solutions: 4.00% (w/w) NaOH, 10.00% (w/w) Na_2CO_3, and 4.00% (w/v) glucose. If the densities of these solutions can be taken to be 1.00 g/mL, how many milliliters of the appropriate solution would you have to measure out to obtain the following quantities?
(a) 3.00 g of NaOH
(b) 0.325 g of Na_2CO_3
(c) 0.255 g of glucose
(d) 0.115 mol of NaOH
(e) 0.200 mol of glucose ($C_6H_{12}O_6$)

*7.46 The stockroom has the following solutions: 3.00% (w/w) KOH, 0.600% (w/w) HCl, and 1.00% (w/v) NaCl. Assuming that the densities of these solutions are all 1.00 g/mL, how many milliliters of the appropriate solution have to be measured out to obtain the following quantities of solutes?
(a) 0.100 g of KOH   (b) 0.200 g of HCl
(c) 0.115 mol of NaCl   (d) 0.125 mol of KOH

*7.47 A student needed 75.0 mL of 12.5% (w/w) aqueous sodium acetate, $NaC_2H_3O_2$. (The density of this solution is 1.05 g/mL.) Only the trihydrate of this compound, $NaC_2H_3O_2 \cdot 3H_2O$, was available, and the student knew that

the water of hydration would just become part of the solvent once the solution was made. How many grams of the trihydrate would have to be weighed out to prepare the needed solution?

*7.48 How many grams of $Na_2SO_4 \cdot 10H_2O$ have to be weighed out to prepare 250 mL of 5.00% (w/w) Na_2SO_4 in water? (The density of this solution is 1.09 g/mL.)

7.49 A student has to prepare 250 g of 2.50% (w/w) NaOH. The stock supply of NaOH is in the form of 10.0% (w/w) NaOH. How many grams of the stock solution have to be diluted to make the desired solution?

7.50 10.0% (w/w) HCl solution is available from the stockroom. How many grams of this solution have to be weighed out to prepare, by dilution, 500 g of 0.250% (w/w) HCl? If the density of the 10.0% solution is 1.05 g/mL, how many milliliters would provide the grams of the concentrated solution that are called for?

*7.51 Concentrated hydrochloric acid is available as 11.6 *M* HCl. The density of this solution is 1.18 g/mL.
(a) Calculate the percent (w/w) of HCl in this solution.
(b) How many milliliters of this concentrated acid have to be taken to prepare 250 g of a solution that is 15.0% (w/w) HCl?

*7.52 Commercial nitric acid comes in a concentration of 16.0 mol/L. The density of this solution is 1.42 g/mL.
(a) Calculate the percent (w/w) of nitric acid, HNO_3, in this solution.
(b) How many milliliters of the concentrated acid have to be taken to prepare 500 g of a solution that is 6.00% (w/w) HNO_3?

Osmosis and Dialysis

7.53 If a solution that contains 1.00 mol of sucrose in 1000 g of water freezes at -1.86 °C, what is the freezing point of a solution that contains 1.00 mol of glucose in 1000 g of water? (Both are compounds that do not break up into ions when they dissolve.)

7.54 A solution that contains 1.00 mol of glucose in 1000 g of water has a normal boiling point of 100.5 °C. Another solution that contains 1.00 mol of an unknown compound in 1000 g of water has a normal boiling point of 101.0 °C. What is the likeliest explanation for the higher boiling point of the second solution?

7.55 Explain in your own words and drawings how osmosis gives a net flow of water from pure water into a solution on the other side of an osmotic membrane.

7.56 In general terms, how does an osmotic membrane differ from a dialyzing membrane?

7.57 Explain in your own words why the osmotic pressure of a solution should depend only on the concentration of its solute particles and not on their chemical properties.

7.58 The equation for osmotic pressure (Eq. 7.4) shows that this pressure is directly proportional to the Kelvin temperature.

Use the kinetic model of molecules and ions in motion and other aspects of the general kinetic theory to explain why the osmotic pressure should increase with an increase in temperature.

7.59 Why is the osmolarity of 1.0 M NaCl not the same as its molarity?

*7.60 Which has the higher osmolarity, 0.10 M NaCl or 0.080 M Na_2SO_4? Explain.

*7.61 Which solution has the higher osmotic pressure, 5.0% (w/w) NaCl or 5.0% (w/w) KI? Both NaCl and KI break up in water in the same way—two ions per formula unit.

*7.62 Solution A consists of 0.60 mol of NaCl, 0.12 mol of $C_6H_{12}O_6$ (glucose, a molecular substance), and 0.055 mol of starch (a colloidal, macromolecular substance), all in 1000 g of water. Solution B is made of 0.60 mol of NaBr, 0.12 mol of $C_6H_{12}O_6$ (fructose, a molecular substance related to glucose), and 0.005 mol of starch all in 1000 g of water. Which solution, if either, has the higher osmotic pressure? Explain.

7.63 What is the osmotic pressure (in mm Hg) of a 0.0100 M solution in water of a molecular substance at 25 °C?

7.64 Calculate the osmotic pressure in mm Hg of a 0.0125 M solution in water at 20.0 °C of a compound that breaks up into two ions per formula unit when it dissolves.

7.65 What happens to red blood cells in crenation?

7.66 Physiological saline solution has a concentration of 0.90% (w/w) NaCl.
(a) Is a solution that is 1.1% (w/w) NaCl described as hypertonic or hypotonic with respect to physiological saline solution?
(b) What would happen, crenation or hemolysis, if a red blood cell were placed (1) in 0.5% (w/w) NaCl? (2) In 1.5% (w/w) NaCl?

7.67 Explain how the loss of macromolecules from the blood can lead to the increased loss of water from blood and a reduction in blood volume.

Decompression Sickness (Special Topic 7.1)

7.68 The solubilities of which gases increase in blood to cause decompression sickness? Why do they increase?

7.69 How does an increased solubility of a gas in blood cause a problem when the individual comes back to normal pressure?

7.70 How does a slow decompression reduce the possibility of decompression sickness?

7.71 What is the "rule of thumb" about the rate of decompression needed to avoid decompression sickness?

Hemodialysis (Special Topic 7.2)

7.72 What does *hemodialysis* mean?

7.73 During hemodialysis, what is the *dialysate*?

7.74 With respect to the following solutes in blood, what should be the concentration of the dialysate for effective hemodialysis, more or less concentrated or the same concentration? (a) Na^+ (b) Cl^- (c) urea

Additional Exercises

7.75 Write the equation for the dissociation of $Ca(NO_3)_2$ in water (using appropriate symbols for the *states* of the various species).

7.76 The solubility of a gas in water at 20 °C is 0.0176 g/L at 681 mm Hg. What is its solubility at 0.989 atm at 20 °C?

*7.77 The stockroom has a solution labeled 50.0% H_2SO_4. Its density at 20 °C is 1.40 g/mL. What is the molarity of the solution?

7.78 What is the osmotic pressure (in mm Hg) of a 0.0125 M solution of sugar in water at 20.0 °C?

*7.79 For rehydration therapy for cholera patients, the World Health Organization (WHO) uses an aqueous solution with the following concentrations: 3.5 g NaCl/L, 2.5 g $NaHCO_3$/L, 1.5 g KCl/L, and 20 g of glucose per liter. Assuming that the ionic compounds break up fully into their ions (Na^+, Cl^-, HCO_3^- and K^+), calculate the osmolarity of the solution. (Glucose is $C_6H_{12}O_6$.)

*7.80 Magnesium carbonate, a constituent of a limestone-like rock called dolomite, reacts with hydrochloric acid to give magnesium chloride, carbon dioxide, and water.
(a) Write the balanced equation for the reaction.
(b) How many milliliters of 15.0% hydrochloric acid (density = 1.073 g/mL) are necessary to react completely and exactly with 5.00 g of magnesium carbonate according to the balanced equation?
(c) How many milliliters of carbon dioxide would be produced if measured at 20.0 °C and 745 mm Hg?

ACIDS, BASES, AND IONIC COMPOUNDS

8

Sulfur and nitrogen oxide air pollutants that make rain acidic accelerate the decay of stone sculptures, such as this statue in France. We'll learn about acids in this chapter.

8.1

ELECTROLYTES

Solutions of ionic compounds in water readily conduct electricity.
All aqueous fluids of living systems—plants or animals—contain dissolved ions and molecules. Blood, for example, contains sodium and chloride ions at low concentrations, several other ions at even smaller concentrations, as well as molecules of glucose and other molecular compounds. To understand these fluids at their molecular level, therefore, requires a study of the chemical properties of ions.

■ Pure water conducts electricity very poorly.

One of the great differences made by the presence of ions in water is the ability of such solutions to conduct electricity. This is one reason why being on moist ground near trees during electrical storms is so dangerous. The sap of trees has dissolved ions, and tree roots extend outward some distance. When lightning strikes a tree, it discharges some of its electricity through the roots, and if you are on the ground nearby you could be seriously injured or killed. Yet, on the other hand, the ability of solutions with ions to conduct electricity also makes life-helping procedures possible. If you ever need an electrocardiogram, you will value this property of blood and fluids in the skin.

Solutes Can Release Ions in Water by Dissociation or by Ionization We learned in the previous chapter that when ionic compounds dissolve in water their ions *dissociate;* they separate from each other as the crystals of the ionic compound break up.

Many molecular compounds also generate ions in water but the process is *ionization,* not dissociation. **Ionization** is the formation of ions *by a chemical reaction* of a molecular compound with the solvent. Hydrogen chloride, for example, undergoes *ionization* as it dissolves in water. We can describe this reaction using Lewis structures as follows.

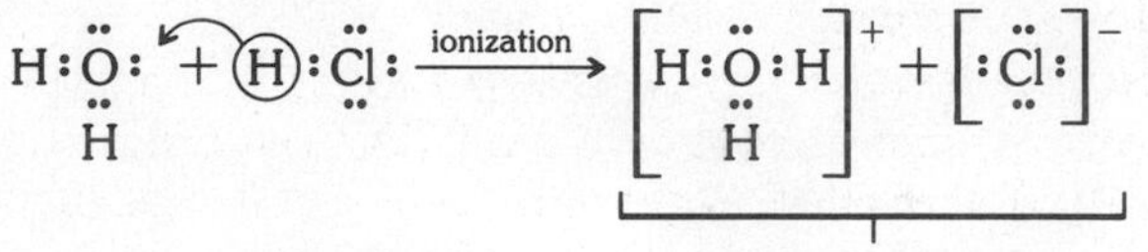

Hydrogen
chloride

Hydrochloric acid

■ For all practical purposes, no *molecules* of HCl remain unionized in dilute hydrochloric acid.

Pure hydrogen chloride, either as a gas or a liquid, contains no ions. Yet when HCl(*g*) dissolves in water, essentially 100% of its molecules react with water to give hydronium and chloride ions. It is this solution that is called *hydrochloric acid.* HBr(*g*) and HI(*g*) react the same way with water to give hydrobromic acid and hydriodic acid, respectively.

Ammonia is another compound that produces ions by reacting with water, but the extent of its ionization is quite small (about 0.5% in 1 *M* NH$_3$). Lewis structures help us see what happens.

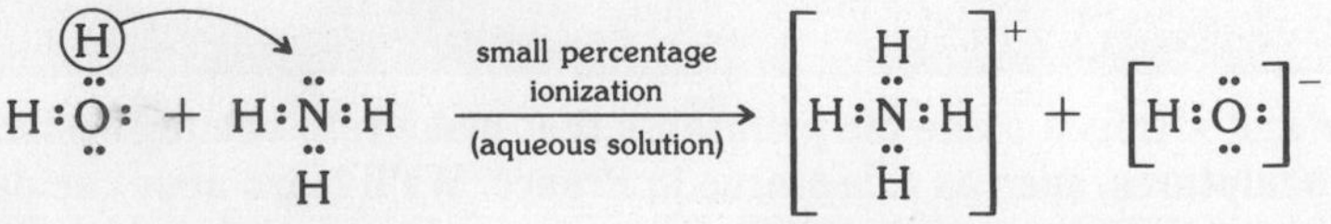

Ammonia

Ammonium ion

Hydroxide ion

As we learned in Section 7.3, this is really the forward reaction of an equilibrium:

$$NH_3(aq) + H_2O \rightleftharpoons NH_4^+(aq) + OH^-(aq)$$

Ions in Water Can Carry Electricity Electricity in metals is a flow of electrons. A complete circuit includes a battery or generator that forces the electrons to move. (The energy of the flow is called *electrical energy*.) If the circuit is broken, the flow stops. The break might be just air at the gap of an open switch, or it might be a space filled with some inert insulating fluid, like sulfur hexafluoride. Pure water is also a relatively good insulator. If we add ions to the water, however, the solution conducts electricity much more readily.

■ Electrical insulators prevent electricity from flowing.

The passage of electricity through a solution with dissolved ions is called **electrolysis**. A solute that enables a solution to conduct electricity is called an **electrolyte**. (Sometimes the solution itself is called the electrolyte.) The question now is, "How do electrolytes make electrolysis possible?"

Electrons do not move directly through a solution of electrolytes in the same way that they move through metals. Instead the dissolved ions move. Figure 8.1 shows a typical setup. The plates or wires that dip into the solution are called **electrodes**. The battery forces electrons to one electrode to make it electron-rich and negatively charged. In electrolysis, the negative electrode is called the **cathode**. The positive ions in solution—the *cations*—naturally are attracted to the cathode, because opposite charges attract.

The electrons that make the cathode electron-rich are "pumped" by the battery or generator from the other electrode, called the **anode,** which becomes electron-poor and positively charged. The negative ions or *anions* are naturally attracted to the anode.

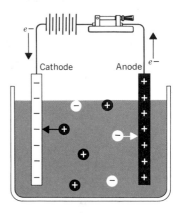

● Cation   ⊖ Anion

FIGURE 8.1
Electrolysis. Cations, positive ions, migrate to the cathode and remove electrons. Anions, negative ions, migrate to the anode and deposit electrons. The effect is a closed circuit.

Electrolysis Causes a Redox Reaction We will take a simple example of an electrolysis, that of aqueous copper(II) bromide, to show how a cation removes electrons at the cathode at the same instant that an anion delivers them to the anode. Copper(II) bromide dissociates in water as follows.

$$CuBr_2(s) \xrightarrow{\text{dissociation}} Cu^{2+}(aq) + 2Br^-(aq)$$

When an electric current is passed through the solution, the following reaction occurs.

$$Cu^{2+}(aq) + 2Br^-(aq) \xrightarrow{\text{electrolysis}} Cu(s) + Br_2(l)$$

The formation of copper, Cu(s), occurs at the cathode and bromine, $Br_2(l)$, forms at the anode (see Fig. 8.2). Let's see how this happens.

Cu^{2+} cations are attracted to the cathode where they pick electrons from the cathode's electron-rich surface and so are changed to Cu atoms. We can represent this by an equation in which electrons are shown as actual reactants:

$$Cu^{2+}(aq) + 2e^- \xrightarrow{\text{reduction}} Cu(s)$$

The oxidation number of copper thus changes from +2 to 0 and so becomes less positive. *Reduction,* therefore, is what happens to the Cu^{2+} ion. (*Reduction* here is clearly also a *gain of electrons,* the older definition of reduction.) Thus in electrolysis, reduction occurs at the cathode.

Br^- anions are attracted to the anode where they deposit electrons at the anode's electron-poor surface. Two Br^- ions give up electrons and one molecule of Br_2 forms. The oxidation number of bromine thus changes from -1 to 0 and so becomes more positive (less negative). *Oxidation* is happening to bromide ions. (And, by the older definition, there is clearly a loss of electrons as two Br^- change to Br_2.) The anode reaction is

$$2Br^-(aq) \xrightarrow{\text{oxidation}} Br_2(l) + 2e^-$$

Thus, oxidation occurs at an anode during electrolysis.

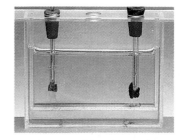

FIGURE 8.2
Electrolysis of $CuBr_2(aq)$. The solution is blue because of the copper (II) ion. The cathode, on the right, has a deposit of copper metal, and some has flaked off and fallen below it. The color around and below the anode, on the left, is brownish because Br_2 is forming.

■ The ions of some electrolytes, like KNO_3, give more complex reactions at the electrodes. The solute does not change but water breaks down to H_2 and O_2, instead.

During electrolysis we have something taking electrons from one electrode and something else putting them *simultaneously* on the other electrode. The effect is the same as if the electrons themselves were moving through the solution, but actually they move only through the wire.

Notice how the sum of the cathode and anode reactions gives the overall equation for the electrolysis.

■ Br_2 is somewhat soluble in water, so $Br_2(aq)$ could just as well be used in these equations as $Br_2(l)$.

$$\text{At the cathode:} \qquad Cu^+(aq) + 2e^- \xrightarrow{\text{reduction}} Cu(s)$$

$$\text{At the anode:} \qquad \underline{2Br^-(aq) \xrightarrow{\text{oxidation}} Br_2(l) + 2e^-}$$

$$\text{Sum:} \qquad Cu^{2+}(aq) + 2Br^-(aq) \xrightarrow{\text{electrolysis}} Cu(s) + Br_2(l)$$

The electrons cancel as this summation is made. They must cancel, of course, because we cannot have free electrons as actual reactants or products. But electrons can transfer. In electrolysis, they transfer from one dissolved species to the other through the wiring of the external circuit.

■ The electrolysis of anhydrous molten NaCl is the industrial synthesis of both sodium and chlorine.

$$2NaCl(l) \xrightarrow{\text{electrolysis}} 2Na(l) + Cl_2(g)$$

Molten Ionic Compounds Are Also Electrolytes   For electrolysis to happen, ions must be mobile. When ions are immobilized in the solid state, no electrolysis occurs. If the solid, ionic compound is heated until it melts, however, then the ions become mobile, and molten salts conduct electricity. Thus the term *electrolyte* can refer either to a solution of ions or to the pure, solid ionic compound. (The term does not apply to metals. Metals that conduct electricity are simply called *conductors*.)

Strong Electrolytes Give High Concentrations of Ions in Water   Electrolytes are not equally good at enabling the flow of electricity. A "good" electrolyte is a substance that even in low concentrations enables a strong electrical current to flow. It's a solute that can readily supply ions. Good electrolytes, in this sense, are called **strong electrolytes,** which means that in aqueous solutions essentially 100% of the formula units initially put into solution have dissociated or ionized. Sodium hydroxide and sodium chloride are strong electrolytes because they *dissociate* 100% in water. Hydrochloric acid is a strong electrolyte because its initial solute, hydrogen chloride, has *ionized* 100% in water.

■ In 1 M ammonia and 1 M acetic acid, the percentage ionization is less than 0.5%.

A **weak electrolyte** is a substance that generates ions in water only to a small percentage of its molar concentration. A typical example is aqueous ammonia. Thus 1 M aqueous ammonia is a poor conductor and a weak electrolyte because only a small percentage of dissolved NH_3 molecules react with water to give ions. Acetic acid, the acid that gives vinegar its tart taste, is also a weak electrolyte and a weak conductor.

Many substances are **nonelectrolytes,** whether they are in the liquid state or in solution. They do not conduct ordinary currents of electricity (e.g., household currents) at all. Although pure water will conduct a current under unusually high voltages, it is regarded as an example of a nonelectrolyte; under ordinary voltages it is a nonconductor. Ethyl alcohol and gasoline are others.

■ The *volt* is a unit of electrical force. Electricity at high voltage in a wire involves a powerful impelling force acting to move electrons.

We can summarize the relationships we have just studied as follows. Be sure to notice the emphasis on *percentage* ionization as the feature dominating these definitions.

■ Weak electrolytes are generally molecular compounds that give ions by ionization, not dissociation.

Strong Electrolyte   One that is strongly dissociated or ionized in water—a high percent ionization

Weak Electrolyte   One that is weakly ionized in water—a low percent ionization

Nonelectrolyte   One that does not dissociate or ionize in water—essentially zero percent ionization

---

8.2
ACIDS AND BASES AS ELECTROLYTES

Acids supply hydrogen ions, and bases neutralize hydrogen ions.
The three principal ion-producers in water are *acids, bases,* and *salts.* At the molecular level of life, the *acid–base status* of body fluids, part of the overall *electrolyte status,* is a matter of life and death. In this section we will learn about the major acids and bases and what it means for an aqueous solution to be acidic, basic, or neutral. We begin with another look at water, because some of our definitions are related to its self-ionization.

Traces of H_3O^+ and OH^- Ions Form from the Self-Ionization of Water   We said in the previous section that pure water is a nonconductor. Traces of ions, H_3O^+ and OH^-, are present, however, but not at concentrations high enough to conduct electricity at ordinary voltages. These ions come from the *self-ionization of water.* This self-ionization of water is actually the forward reaction in the following chemical equilibrium:

$$2H_2O \rightleftharpoons H_3O^+(aq) + OH^-(aq)$$

A transfer of H^+ occurs from one molecule of water to another to give two ions in a 1 : 1 mole ratio, the **hydronium ion,** H_3O^+, and the **hydroxide ion,** OH^- (see Fig. 8.3). The forward reaction is certainly not favored; at 25 °C the concentration of each product ion is only 1.0×10^{-7} mol/L. This means that out of a little over 1 billion water molecules, only 2 have changed into these ions at any one moment.

It is customary to use brackets, [ ], around a formula of a solute when we mean its concentration in the specific units of moles per liter. Thus, in pure water at 25 °C,

$$[H_3O^+] = [OH^-] = 1.0 \times 10^{-7} \text{ mol/L}$$

Although concentrations this low may seem too unimportant to mention, life itself hinges on holding the molar concentrations of H_3O^+ and OH^- ions in body fluids at about this level.

We have already learned that equilibria can be shifted, and shifts in water's self-ionization equilibrium are life-threatening. *Almost all of what we will be studying about acids and bases is essential to the study of how the body controls the equilibrium for the self-ionization of water and the acid–base status of body fluids.*

The Hydronium Ion Is Often Referred to as the Hydrogen Ion   The existence of ions in aqueous solutions of *electrolytes* was first proposed by Svante Arrhenius (1859 – 1927), a Swedish scientist. In the *Arrhenius theory of acids and bases,* all acids produce hydrogen ions in water. Arrhenius actually spoke of *hydrogen* ions, H^+, not hydronium ions, H_3O^+, but he had no way then of knowing that bare protons, which H^+ really represents, are *always* piggybacked onto something else in solution. Although protons can be *transferred* from one place to another, they have no independent existence as separate entities in solution any more than do electrons. H^+ is always held by an electron-pair bond to a water molecule or to something else. It is the ability of an acid to transfer H^+ that makes it an acid.

■ The phrase *acid–base balance* is sometimes used for *acid–base status.*

■ Minute changes in the concentrations of acids or bases can switch enzymes on or off, and enzymes are essential to almost all reactions in living systems.

■ $$\left[\begin{array}{c} H \colon \ddot{O} \colon H \\ H \end{array} \right]^+$$
Hydronium ion

$$\left[\colon \ddot{O} \colon H \right]^-$$
Hydroxide ion

■ In 1884, Arrhenius nearly lost his bid for a doctoral degree for proposing ions, so rash was the idea considered. But in 1903, the idea earned him a Nobel prize.

FIGURE 8.3
In the self-ionization of water, H^+ transfers from one water molecule to another.

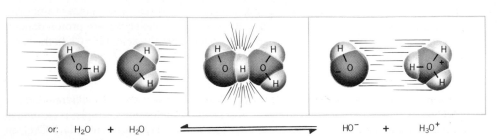

or: H_2O + H_2O ⇌ HO^- + H_3O^+

In a practical sense, Arrhenius's supposition about H^+ wasn't too wide of the mark. H^+ is so easily available from H_3O^+ that scientists today commonly use the terms *proton, hydrogen ion,* and *hydronium ion* interchangeably. We will use *hydrogen ion* as a convenient nickname for *hydronium ion* ourselves, and we will often employ the symbol $H^+(aq)$ as a simpler way of writing $H_3O^+(aq)$.

Acids Make the H^+ Level Exceed the OH^- Level and Bases Do the Opposite

When the molar concentrations of aqueous hydrogen ions and hydroxide ions are exactly equal, as they are in pure water at any temperature, the solution is called a **neutral solution.**

Acids make the molar concentration of hydrogen ion higher than that of hydroxide ion. Because acidic solutions all have the hydrogen ion, they possess many common properties. Acidic solutions, for example, turn blue litmus to a red color. Litmus is an example of an **acid–base indicator,** a compound whose color is different in acid than in base, so it can be used to tell if an aqueous solution is acidic or basic. Acidic solutions also have tart tastes, like solutions of acetic acid, citric acid, lactic acid, oxalic acid, and hydrochloric acid. But don't make a taste test without great care. Even dilute acids can corrode teeth.

Bases make the molar concentration of hydroxide ion greater than that of hydronium ion. Such solutions usually have a bitter taste and a soapy "feel," and they turn red litmus blue.

We can summarize the important conditions that define acidic, basic, and neutral solutions as follows.

Acidic solutions:	$[H^+] > [OH^-]$
Neutral solutions:	$[H^+] = [OH^-]$
Basic solutions:	$[H^+] < [OH^-]$

Brønsted Broadened the Concept of Acids and Bases

We said earlier that it is the ability of something to transfer H^+ that makes it an acid. We can also say that it is the ability of something to accept H^+ that makes it a base. Johannes Brønsted (1879–1947), a Danish chemist, is generally credited with the development of these definitions.

Brønsted Definitions of Acids and Bases
Acids are proton donors.
Bases are proton acceptors.

These definitions apply regardless of the solvent and even in the absence of any liquid solvent. Hydrogen chloride gas, for example, reacts with ammonia gas in a proton-transfer reaction.

$$HCl(g) + NH_3(g) \longrightarrow NH_4Cl(s)$$

The product is a crystalline solid that forms in a cloud of microcrystals when fumes of ammonia and hydrogen chloride intermingle (see Fig. 8.4). The H—Cl molecules are proton donors; NH_3 molecules are proton acceptors. When H^+ transfers from the acid, H—Cl, to the base, NH_3, NH_4^+ ions and Cl^- ions form, and no solvent is involved.

Acids and Bases Vary Widely in Strength

Acids and bases are quite different in their abilities to function as proton donors or proton acceptors. Water, for example, is extremely weak as both a donor and an acceptor, as we have just learned. Hydrogen chloride, $HCl(g)$, on the other hand, so readily donates H^+ that even such a weak acceptor as H_2O is able to take

---

Margin notes:

■ Paper impregnated with litmus dye is called litmus paper.

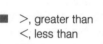

Acid	A Natural Source
Acetic acid	Vinegar
Citric acid	Lemons
Lactic acid	Sour milk
Oxalic acid	Rhubarb
Hydrochloric acid	Gastric juice

■ >, greater than
< , less than

■ Thomas Martin Lowry (1847–1936), an English chemist, proposed the same idea independently.

FIGURE 8.4
The reaction of $NH_3(g)$, from the bottle on the left, with $HCl(g)$, from the bottle on the right, produces a cloud of microcrystals of $NH_4Cl(s)$ by a neutralization reaction. These gaseous reactants are always present in the air spaces above concentrated solutions of aqueous ammonia and hydrochloric acid.

H^+ from $HCl(g)$. Essentially 100% of all hydrogen chloride molecules that dissolve in water react as follows.

$$HCl(g) + H_2O \longrightarrow H_3O^+(aq) + Cl^-(aq)$$

Because water is the solvent nearly always used in acid–base chemistry, *we normally define all strong acids with reference to their reactions with water.* If we use the general symbol HA for any acid, whether it is a gas, liquid, or solid, then all strong acids react as follows essentially 100%.

$$HA + H_2O \longrightarrow H_3O^+(aq) + A^-(aq)$$

■ Think of *A* in H*A* or *A*⁻ as standing for the anion of the acid.

We define a **strong acid** as one that is 100% ionized in this proton-donating reaction. The hydronium ion, itself, so readily donates a proton that we also apply the term *strong acid* to any aqueous solution of a strong acid.

A **weak acid** is one for which only a small percentage of its molecules ionize by reacting with water. You may have already noticed that the terms *strong* and *weak* are used alike with both acids and electrolytes. They refer to *percent* ionization. All strong acids are strong electrolytes. All weak acids are weak electrolytes. Let's now look at a few of the most common strong and weak acids.

Hydrochloric Acid and Nitric Acid Are Strong Monoprotic Acids

Hydrochloric acid is a **monoprotic acid** because the ratio of $[H_3O^+]$ ions to $[Cl^-]$ ions in an aqueous solution of HCl is 1 to 1. Other monoprotic acids are those that can be made by dissolving the other hydrogen halide gases in water, for example, hydrofluoric acid, $HF(aq)$, hydrobromic acid, $HBr(aq)$, and hydroiodic acid, $HI(aq)$. All but hydrofluoric acid are strong acids.[1]

■ Because F⁻ ion can tie up Ca²⁺ ion at nerve endings — ions essential for sending nerve signals — and because HF(*aq*) can carry F⁻ through the skin, HF causes excruciating pains and is a dangerous chemical.

PRACTICE EXERCISE 1

Depict the ionization of HBr and HI in water using Lewis structures such as we did for HCl in the previous section.

Nitric acid, $HNO_3(aq)$, is also a strong, monoprotic acid. We can represent its ionization as follows.

$$H_2O + HNO_3(aq) \longrightarrow H_3O^+(aq) + NO_3^-(aq)$$

Nitric acid (molecular formula)   Hydronium ion   Nitrate ion

Nitric acid in water

■

Nitric acid

Acetic Acid Is a Weak Monoprotic Acid

Acetic acid is a typical organic acid, and like virtually all organic acids, it is a weak acid. **Organic compounds** are the compounds of carbon other than its oxides, the cyanides, or those related to earth-like substances like limestone rocks and other carbonates. Compounds that are not organic compounds are called **inorganic compounds.**

The acetic acid molecule, $HC_2H_3O_2$, has four hydrogen atoms, but only one is attached to an oxygen atom. Only this one can transfer to a water molecule (see Fig. 8.5). The H—O bond in acetic acid is considerably stronger than the H—Cl bond in HCl(*g*), so acetic acid does not

■

$HC_2H_3O_2$
Acetic acid
(The H in color is the proton available in acid–base reactions.)

---

[1] Hydrofluoric acid, $HF(aq)$, is unusual because its ions, H_3O^+ and F^-, attract each other so strongly in solution that they behave as if they were not very free of each other. $HF(aq)$ is thus classified as a *weak acid.*

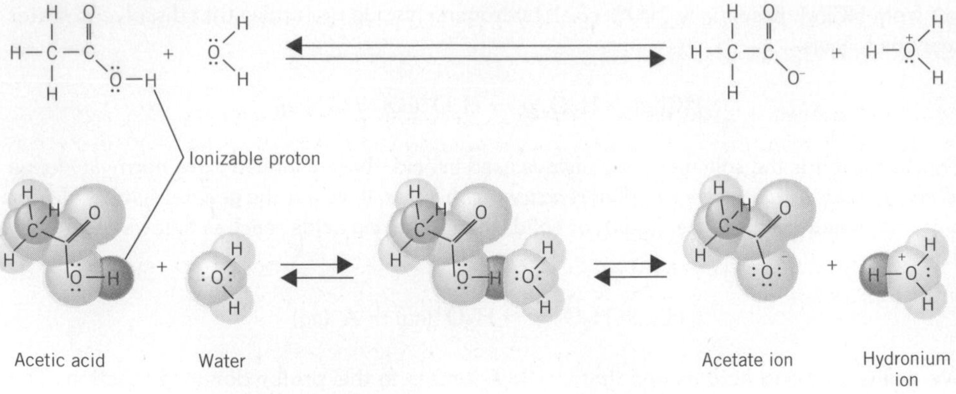

FIGURE 8.5
The ionization of acetic acid in water. The forward reaction is not favored, and acetic acid is a weak acid.

as readily as H—Cl transfer H^+ to H_2O. A chemical equilibrium is present in aqueous acetic acid, which can be represented as follows.

$$H_2O + HC_2H_3O_2 \rightleftharpoons H_3O^+(aq) + C_2H_3O_2^-(aq)$$

Acetic acid[2]   Acetate ion

In 0.1 M acetic acid, only 0.5% of all acetic acid molecules are at any moment ionized. But there is the coming and going characteristic of all dynamic equilibria.

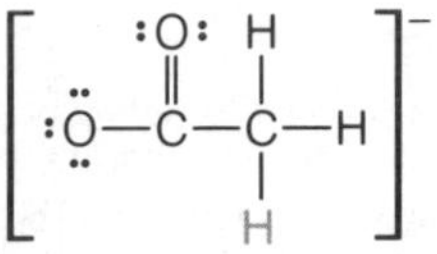

The C$=$O Group Makes Acetic Acid More Acidic than Water We might pause to ask here why acetic acid is an acid at all, why the proton of its H—O group more easily transfers than the proton of the H—O group in water. The difference is caused by the group that directly holds H—O in the acetic acid molecule, the system of the carbon–oxygen double bond, C$=$O. This is an electronegative group. Its electronegative oxygen atom makes it so, and it pulls some electron density away from the oxygen atom of the H—O group in acetic acid. This weakens the H—O bond, so the hydrogen more easily transfers to a water molecule than the hydrogen of a water molecule itself.

■ Sulfuric acid is the most widely used acid in industrial applications. Over 70 billion pounds (over 250 billion moles) are annually used in the United States.

Sulfuric Acid Sulfuric acid, H_2SO_4, is the only common, stable inorganic **diprotic acid,** one that is able to give up two hydronium ions per formula unit. The ionization of the first hydrogen ion is so easy that we do not write equilibrium arrows for the reaction; sulfuric acid is a strong acid.

$$H_2O + H_2SO_4 \longrightarrow H_3O^+(aq) + HSO_4^-(aq)$$

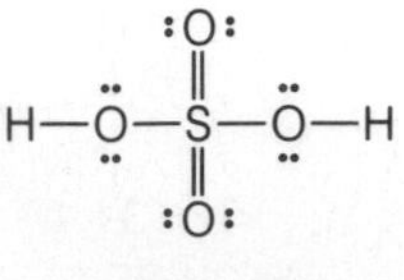

Sulfuric   Hydrogen
acid   sulfate ion

The two extra "lone oxygens" on sulfur in sulfuric acid help to make sulfuric acid an acid in a way similar to the effect of the lone oxygen in acetic acid.

■ A "lone oxygen" is one joined only to one other atom.

The Hydrogen Sulfate Ion Is an Acid The hydrogen sulfate ion is a monoprotic acid. The transfer of H^+ from it, however, requires a positively charged particle to leave one already oppositely charged. Although this is harder than the transfer of the first H^+ from H_2SO_4, it still happens. The following equilibrium exists, and the *reactants* are favored.

$$H_2O + HSO_4^-(aq) \rightleftharpoons H_3O^+(aq) + SO_4^{2-}(aq)$$

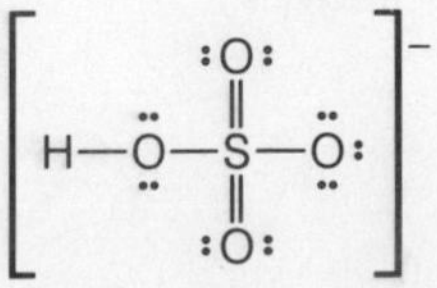

Hydrogen   Sulfate
sulfate ion   ion

[2] We will usually use $HC_2H_3O_2$ instead of the full structure as our symbol for acetic acid in this chapter and the next. Just remember that only one hydrogen is active in acid–base reactions, and that acetic acid is monoprotic.

[3] The double bonds from S to O and from P to O shown in the structures of the sulfuric acid and phosphoric acid systems violate the octet rule as we learned it. However, the valence shells of S and P, which are higher than level 2, can and often do hold more than 8 electrons. The double bonds shown in the margin structures reflect both this flexibility and the extra stabilization that additional bonds confer to the systems.

■ $C_2H_3O_2^-$
Acetate ion

■ Sulfuric acid

■ Hydrogen sulfate ion

In $0.10\ M\ HSO_4^-$ (present, for example, in $0.10\ M\ NaHSO_4$), the percentage ionization of the HSO_4^- ion is slightly less than 40%.[3]

Phosphoric Acid Is a Moderately Strong, Triprotic Acid

A **triprotic acid** is one that can potentially release three H^+ ions to proton acceptors, but each H^+ separates with greater difficulty than the previous one. Phosphoric acid, H_3PO_4, is the only common example of an inorganic triprotic acid. Like the ionization of sulfuric acid, that of phosphoric acid occurs in steps, each one more difficult than the previous. Even the first step does not occur to 100% of the phosphoric acid molecules. In $1\ M\ H_3PO_4$, less than a third are ionized, so we have to write equilibrium expressions for all steps. In all, the *reactants* are favored.

$$H_2O + H_3PO_4(aq) \rightleftharpoons H_3O^+(aq) + H_2PO_4^-(aq)$$

Phosphoric acid   Dihydrogen phosphate ion

$$H_2O + H_2PO_4^-(aq) \rightleftharpoons H_3O^+(aq) + HPO_4^{2-}(aq)$$

Monohydrogen phosphate ion

$$H_2O + HPO_4^{2-}(aq) \rightleftharpoons H_3O^+(aq) + PO_4^{3-}(aq)$$

Phosphate ion

The percentage ionization of H_3PO_4 is roughly 24% in a $0.10\ M$ solution of phosphoric acid. This percentage ionization is too low to let us call phosphoric acid a strong acid. Although phosphoric acid is a good conductor of electricity in water, it's classified as a *moderate acid*.

It is important to learn the names and formulas of the three ions available from phosphoric acid, because the phosphate ion system occurs widely in the body. Relatives of phosphoric acid — diphosphoric acid and triphosphoric acid — are important systems in metabolism.

Carbonic Acid Is Involved in Respiration

Carbonic acid, H_2CO_3, is a weak diprotic acid that is unusual because it is unstable. Its instability, however, is an important property when the body has to manage one of the respiratory gases, carbon dioxide. When carbon dioxide dissolves in water, a trace reacts with water to form carbonic acid, a weak acid.

$$CO_2(aq) + H_2O \rightleftharpoons H_2CO_3(aq)$$

Carbonic acid

Then a small fraction of the carbonic acid molecules ionizes, a fraction small enough to make carbonic acid a weak acid.

$$H_2CO_3(aq) + H_2O \rightleftharpoons H_3O^+ + HCO_3^-(aq)$$

Bicarbonate ion

The bicarbonate ion is itself a (very) weak acid. A strong base, like OH^-, can remove its proton, however.

$$HCO_3^-(aq) + OH^- \rightleftharpoons CO_3^{2-}(aq) + H_2O$$

Carbonate ion

The bicarbonate ion is the chief form in which waste carbon dioxide is carried from body tissues to the lungs.

Sulfate ion

Phosphoric acid

Diphosphoric acid

Triphosphoric acid

Carbonic acid

Bicarbonate ion

Carbonate ion

---

[3] The double bonds from S to O and from P to O shown in the structures of the sulfuric acid and phosphoric acid systems violate the octet rule as we learned it. However, the valence shells of S and P, which are higher than level 2, can and often do hold more than 8 electrons. The double bonds shown in the margin structures reflect both this flexibility and the extra stabilization that additional bonds confer to the systems.

TABLE 8.1
Common Acids[a]

Acid	Formula	Percent Ionization
Strong Acids		
Hydrochloric acid	HCl	Very high
Hydrobromic acid	HBr	Very high
Hydroiodic acid	HI	Very high
Nitric acid	HNO_3	Very high
Sulfuric acid[b]	H_2SO_4	Very high
Moderate Acids		
Phosphoric acid	H_3PO_4	27
Sulfurous acid[c]	H_2SO_3	20
Weak Acids		
Nitrous acid[c]	HNO_2	1.5
Acetic acid	$HC_2H_3O_2$	1.3
Carbonic acid[c]	H_2CO_3	0.2

[a] Data are for 0.1 M solutions of the acids in water at room temperature.

[b] *Concentrated* sulfuric acid (99%) is particularly dangerous not only because it is a strong acid but also because it is a powerful dehydrating agent. This action generates considerable heat at the reaction site, and at higher temperatures sulfuric acid becomes even more dangerous. Moreover, concentrated sulfuric acid is a thick, viscous liquid that does not wash away from skin or fabric very quickly.

[c] An unstable acid.

PRACTICE EXERCISE 2

■ Sulfurous acid is actually $SO_2 \cdot H_2O(aq)$, not $H_2SO_3(aq)$, but the latter formula is the more commonly used.

When sulfur dioxide dissolves in water, some of it reacts with the water to give a solution called sulfurous acid, traditionally written as H_2SO_3. Write the equilibrium equations for the successive ionizations of this weak acid.

■ You can know that lactic, acetic, oxalic, and citric acids are weak acids because they aren't on our list of strong acids.

Table 8.1 summarizes the common aqueous acids. You should memorize the names and formulas of all the strong and moderate acids on this list. We'll need this knowledge as we go along. There are very few of them, and once they are learned, you can be fairly certain that any unfamiliar acid you encounter will be a weak acid. It's easier to learn a few strong and moderate acids than several hundred weak acids.

Sodium Hydroxide Is the Most Common Strong Base   Table 8.2 lists the common bases. The **strong bases** are those that dissociate nearly 100% in water, and they furnish a strong proton-binding or proton-accepting species, like the hydroxide ion. Sodium hydroxide, NaOH, and potassium hydroxide, KOH, are examples of strong bases because they break up in water essentially 100% into the metal ion and the species that is the actual base or proton acceptor, OH^-.

■ Solid NaOH is very hygroscopic, so its container must be promptly and tightly reclosed each time some is taken.

$$NaOH(s) \xrightarrow{\text{water}} Na^+(aq) + OH^-(aq)$$
$$KOH(s) \xrightarrow{\text{water}} K^+(aq) + OH^-(aq)$$

Two other strong bases are the hydroxides of Group IIA metals. These are magnesium hydroxide, $Mg(OH)_2$, and calcium hydroxide, $Ca(OH)_2$. They also ionize essentially 100% in water, but as the data in Table 8.2 show, they are so insoluble in water that even saturated solutions provide only very dilute solutions of hydroxide ions.

$$Ca(OH)_2(s) \longrightarrow Ca^{2+}(aq) + 2OH^-(aq)$$
$$Mg(OH)_2(s) \longrightarrow Mg^{2+}(aq) + 2OH^-(aq)$$

$H_2O + SO_2 \longrightarrow H_2SO_3$

TABLE 8.2
Common Bases

Base	Formula	Solubility[a]	Percent Ionization
Strong Bases			
Sodium hydroxide	NaOH	109	>90 (0.1 M solution)
Potassium hydroxide	KOH	112	>90 (0.1 M solution)
Calcium hydroxide	Ca(OH)$_2$	0.165	100 (saturated solution)
Magnesium hydroxide	Mg(OH)$_2$	0.0009	100 (saturated solution)
Weak Base			
Ammonia, aqueous	NH$_3$	89.9	1.3 (18 °C)[b]

[a] Solubilities are in grams of solute per 100 g of water at 20 °C except where otherwise noted.

[b] The ionization referred to here is the equilibrium: $NH_3(aq) + H_2O \rightleftharpoons NH_4^+(aq) + OH^-(aq)$

Milk of magnesia, a slurry of Mg(OH)$_2$(s) in water, is a common remedy for acid indigestion. Too heavy and frequent doses, however, can upset the levels of $Mg^{2+}(aq)$ in various body fluids.

In sufficient concentration, the hydroxide ion causes a severe chemical burn, certainly enough to be very hazardous to the eyes. Both sodium and potassium hydroxides can be prepared in solutions concentrated enough to be dangerous chemicals. Calcium and magnesium hydroxides, however, are so insoluble that they not only pose no grave danger to the skin, they are used internally in home remedies. Calcium hydroxide is a component of one commercial antacid tablet. A slurry of magnesium hydroxide in water, called "milk of magnesia," is used as an antacid and a laxative.

A **weak base** is a poor proton acceptor, one that is unable to take protons from water molecules to any appreciable extent. Ammonia is the most common example. A solution of ammonia in water, called *aqueous ammonia,* does have some excess hydroxide ion, but only a small percent of ammonia molecules react to produce it, as we earlier learned. The following equilibrium is present, and the reactants are strongly favored:

■ You'll sometimes see aqueous ammonia called "ammonium hydroxide," but this is misleading. NH$_4$OH is unknown as a pure compound.

$$NH_3(aq) + H_2O \rightleftharpoons NH_4^+(aq) + OH^-(aq)$$

Ammonia   Ammonium
                                        ion

A dilute solution (about 5%) of ammonia in water is sold as household ammonia in supermarkets. It's a good cleaning agent, but watch out for its fumes.

The names and formulas of the bases in Table 8.2 should also be memorized.

All Salts Are Strong Electrolytes The third major family of ion-producing substances is the salts. We have to say something about them here — they'll be treated more fully in a later section — because salts are products of the reactions of acids and bases.

Salts are ionic compounds in which the positive ion is a metal ion or any other positive ion except H$^+$, and the negative ion is any except OH$^-$ or O^{2-}. All salts are crystalline ionic solids at room temperature. All are strong electrolytes. When salts dissolve in water, their ions dissociate essentially 100%. Even for very insoluble salts, what little of them that does dissolve becomes 100% dissociated, and many can be made in aqueous solution by the reaction of an acid with a base.

■ Some chemists classify metal oxides, compounds with the oxide ion (O^{2-}), as *salts.* However, those that dissolve in water *react* with it completely, and OH$^-$(aq) forms, not O^{2-}(aq).

8.3

THE CHEMICAL PROPERTIES OF AQUEOUS ACIDS AND BASES

Acids react with hydroxides, bicarbonates, carbonates, ammonia, and active metals.

In this section we will principally study the reactions of the hydronium ion, H$_3$O$^+$. We will see how a variety of substances are able to accept H$^+$ from this ion in reactions that neutralize acidic solutions. They include the hydroxide ion, the bicarbonate ion, the carbonate ion, and the ammonia molecule. All are Brønsted bases, of course, because all accept H^+. We'll also see that many metals are able to reduce two H^+ ions to H_2 and themselves be oxidized to metal ions. Some metals are so active that they can even abstract protons directly from molecules of water.

As we mentioned before, we'll use the symbol $H^+(aq)$ as a shorthand symbol for the hydronium ion in most of the equations. The anions of the acids have their own chemical reactions, of course, but we'll not be concerned about them here unless they are Brønsted bases whose reactions we have to study at this time.

There is a special kind of equation, the *net ionic equation,* that is particularly helpful when we concentrate on the chemical properties of ions. We have occasionally been using such equations, and now we will learn how to write them.

A Net Ionic Equation Omits "Spectator" Species

■ We say *molecular* equation even though some of the chemicals in the equation might be ionic compounds.

The conventional equation for a reaction is called a **molecular equation** because it shows all of the substances in the molecular or empirical formulas that we would need for planning an actual experiment. The molecular equation for the reaction of sodium carbonate decahydrate with hydrochloric acid, for example, is

$$Na_2CO_3 \cdot 10H_2O(s) + 2HCl(aq) \longrightarrow 2NaCl(aq) + CO_2(g) + 11H_2O$$

Similarly, the molecular equation for the reaction of hydrochloric acid with aqueous sodium hydroxide is

$$HCl(aq) + NaOH(aq) \longrightarrow NaCl(aq) + H_2O$$

However, as we now know, $HCl(aq)$ is really $H^+(aq)$ and $Cl^-(aq)$, and $NaOH(aq)$ is actually $Na^+(aq)$ and $OH^-(aq)$. We know this because we know the acid is a *strong* acid and the base is a *strong* base, so both must be essentially fully ionized in solution. We don't have nonionized molecules of HCl or NaOH in the solution. We also know that $NaCl(aq)$ is fully ionized because all salts are strong electrolytes, and the (aq) by its formula tells us that the NaCl is in solution. [An insoluble salt would have had (s) after its formula.] The fourth formula in the equation is that of water, a nonelectrolyte. Its molecules aren't separated into ions. (We ignore, of course, the self-ionization of water, because it occurs to an exceedingly low percentage.) We'll always assume that H_2O means $H_2O(l)$.

Using these facts we can expand the molecular equation into what is called the **ionic equation,** one that shows all of the dissolved species, whether ionic or molecular. We do this by replacing anything that we know is present as ions by the actual formulas of these ions. Thus the ionic equation for our example is

$$H^+(aq) + Cl^-(aq) + Na^+(aq) + OH^-(aq) \longrightarrow Na^+(aq) + Cl^-(aq) + H_2O$$

| These came from HCl(aq) | These came from NaOH(aq) | These came from NaCl(aq) | Not ionized |

The preparation of an ionic equation is actually just a scratch paper operation, because we next cancel all of the formulas that appear *identically* on opposite sides of the arrow. The foregoing ionic equation shows us, for example, that nothing happens either to $Na^+(aq)$ or to $Cl^-(aq)$. There is no reason, therefore, to let them remain in the equation when we just want to give full attention to the species that react or form. [$Na^+(aq)$ and $Cl^-(aq)$, of course, do serve one function; they give electrical neutrality to their respective compounds. Otherwise, they are nothing more than *spectator particles* in this reaction.] Formulas must be of the same physical state before they can be canceled. We could not, for example, cancel $HCl(g)$ by $HCl(aq)$, because their states are different.

We can cancel the spectator particles, whether they are ions or molecules, from the ionic equation. This leaves the **net ionic equation,** one that shows only the reacting species and the products they form. As you can see, this equation is a simple description of what happens when hydrochloric acid, *or any strong acid,* neutralizes sodium hydroxide, *or any hydroxide base,* in solution.

$$H^+(aq) + OH^-(aq) \longrightarrow H_2O$$

Net Ionic Equations Must Balance Both Electrically and Materially For a net ionic equation to be balanced, two conditions must be met: a material balance and an electrical balance. We have **material balance** when the numbers of atoms of each element, regardless of how they are chemically present, are the same on both sides of the arrow. We have **electrical balance** when the algebraic sum of the charges to the left of the arrow equals the sum of the charges to the right.

■ Sometimes we don't cancel, but only reduce in number. If an ionic equation, for example, has

$$\ldots + 4H_2O \longrightarrow \ldots + 2H_2O$$

we can simplify it to

$$\ldots + 2H_2O \longrightarrow \ldots$$

■ The material balance is what we need in any kind of balanced equation.

EXAMPLE 8.1
Writing a Net Ionic Equation

Sulfuric acid is the most important acid in industrial use, and sometimes it has to be neutralized by sodium hydroxide. The reaction can be carried out to produce sodium sulfate, $Na_2SO_4(aq)$, and water. Write the molecular, ionic, and net ionic equations.

ANALYSIS-1 The question calls for three answers, so we'll proceed in steps. First we deal with the molecular equation. The complete formulas of the reactants and products must first be assembled in the pattern of an equation, which is then balanced to give the molecular equation.

SOLUTION-1, THE MOLECULAR EQUATION The balanced molecular equation is

$$H_2SO_4(aq) + 2NaOH(aq) \longrightarrow Na_2SO_4(aq) + 2H_2O$$

ANALYSIS-2 Using our knowledge about which acids and bases are strong and which salts are fully ionized, we disassemble the molecular equation to display all of the ions present or provided. The following facts about the reactants and products are relevant to this exercise. We show H_2SO_4 as breaking up entirely into $2H^+$ and SO_4^{2-} because the given reaction is with a strong base, not with water, and SO_4^{2-}, not HSO_4^-, forms.

■ Sulfuric acid must be handled very carefully. See Table 8.1.

$H_2SO_4(aq)$ means $2H^+(aq) + SO_4^{2-}(aq)$	(H_2SO_4 gives up *two* H^+ ions to the other reactant.)
$2NaOH(aq)$ means $2Na^+(aq) + 2OH^-(aq)$	(This is a strong, fully ionized metal hydroxide.)
$Na_2SO_4(aq)$ means $2Na^+(aq) + SO_4^{2-}(aq)$	(This is a salt, and *aq* tells us that it is in solution; hence, it is fully ionized.)
$2H_2O$ means $2H_2O$	(No breaking up occurs with this non-electrolyte.)

SOLUTION-2, THE IONIC EQUATION The result of the analysis is the ionic equation.

$$2H^+(aq) + SO_4^{2-}(aq) + 2Na^+(aq) + 2OH^-(aq) \longrightarrow 2Na^+(aq) + SO_4^{2-}(aq) + 2H_2O$$

ANALYSIS-3 Finally, identical species in identical states on opposite sides of the arrow in the ionic equation are canceled.

$$2H^+(aq) + \cancel{SO_4^{2-}(aq)} + \cancel{2Na^+(aq)} + 2OH^-(aq) \longrightarrow \cancel{2Na^+(aq)} + \cancel{SO_4^{2-}(aq)} + 2H_2O$$

SOLUTION-3, THE NET IONIC EQUATION This leaves us with a net ionic equation.

$$2H^+(aq) + 2OH^-(aq) \longrightarrow 2H_2O$$

Note that we can divide all the coefficients by 2 and convert them to smaller whole numbers. Thus the final net ionic equation is

$$H^+(aq) + OH^-(aq) \longrightarrow H_2O$$

CHECK We have both a material and an electrical balance for each equation. ∎

PRACTICE EXERCISE 3

Write the molecular, the ionic, and the net ionic equations for the neutralization of nitric acid by potassium hydroxide. A water-soluble salt, $KNO_3(aq)$, and water form. ∎

Strong Acids React with Metal Hydroxides to Give Water and a Salt Example 8.1 and Practice Exercise 3 illustrate reactions of strong acids with metal hydroxides, and we saw in the net ionic equations that they are the reaction of a Brønsted acid, H^+, with a Brønsted base, OH^-. If we let M stand for any group IA metal, we can write the reactions of aqueous solutions of their hydroxides with a strong acid such as hydrochloric acid by the following general equation.

$$MOH(aq) + HCl(aq) \longrightarrow MCl(aq) + H_2O$$

■ The group IA hydroxides are LiOH, NaOH, KOH, RbOH, and CsOH.

The net ionic equation is

$$OH^-(aq) + H^+(aq) \longrightarrow H_2O$$

Only the group IA hydroxides are very soluble in water. Most of the others are relatively insoluble, but their solid forms can still neutralize strong acids. The net ionic equations for the reactions of the water-insoluble metal hydroxides with acid, therefore, reflect the insolubility of such hydroxides. The equations cannot show OH^- as a dissociated species.

$$M(OH)_2(s) + 2HCl(aq) \longrightarrow MCl_2(aq) + 2H_2O$$

The net ionic equation is

$$M(OH)_2(s) + 2H^+(aq) \longrightarrow M^{2+}(aq) + 2H_2O$$

Here M stands for any metal in group IIA (except beryllium).

PRACTICE EXERCISE 4

When milk of magnesia is used to neutralize hydrochloric acid (stomach acid), solid magnesium hydroxide in the suspension reacts with the acid. Write the molecular and net ionic equations for this reaction. ∎

Strong Acids React with Metal Bicarbonates to Give CO_2, H_2O, and a Salt All metal bicarbonates react the same way with strong, aqueous acids. They react to give carbon dioxide, water, and a salt. For example, sodium bicarbonate and hydrochloric acid react as follows.

$$HCl(aq) + NaHCO_3(aq) \longrightarrow CO_2(g) + H_2O + NaCl(aq)$$

Potassium bicarbonate and hydrobromic acid give a similar reaction.

$$HBr(aq) + KHCO_3(aq) \longrightarrow CO_2(g) + H_2O + KBr(aq)$$

$NaHCO_3(aq)$ and $HCl(aq)$ react to give $NaCl(aq)$, H_2O, and $CO_2(g)$, which can be seen bubbling out of this tube.

What is believed to form initially is not CO_2 and H_2O but H_2CO_3, carbonic acid. However, almost all of it promptly decomposes to CO_2 and H_2O, so the solution fizzes strongly as the reaction proceeds and CO_2 evolves.

Notice that the salt whose formula appears as a product in the molecular equation is always a combination of the cation of the bicarbonate (Na^+ or K^+ in our examples) and the anion of the acid (Cl^- or Br^- in our examples). Let's be sure we can write the formula of the salt that forms in these reactions before we continue.

EXAMPLE 8.2
Writing the Formula of the Salt That Forms When a Bicarbonate Reacts with an Acid

What salt forms when lithium bicarbonate reacts with nitric acid?

ANALYSIS   The name, lithium bicarbonate, gives us the names of the ions involved, the lithium ion (Li^+) and the bicarbonate ion (HCO_3^-). The anion of nitric acid is NO_3^-, so this is the anion that must be paired with Li^+ if a salt is to form. The electrical charges, $1+$ and $1-$, on these ions tell us that they *must* combine in a $1:1$ ratio.

SOLUTION   The salt's formula is $LiNO_3$.

PRACTICE EXERCISE 5

What is the formula of the salt that forms when potassium bicarbonate reacts with sulfuric acid? Assume the salt is a sulfate and not a hydrogen sulfate.

In writing net ionic equations of reactions between strong acids and metal bicarbonates, we'll treat all metal bicarbonates as dissociated in water to the metal ion and the bicarbonate ion. As we will see, the reaction of bicarbonates with acids is one in which the bicarbonate ion is a base, a proton acceptor.

■ Pure bicarbonate salts generally involve only the group IA metal ions.

EXAMPLE 8.3
Writing Equations for the Reactions of Bicarbonates with Strong Acids

What are the molecular and the net ionic equations for the reaction of potassium bicarbonate with hydroiodic acid?

ANALYSIS   The products are CO_2, H_2O, and a salt. Using what we learned in Example 8.2, the salt must be a combination of the potassium ion, K^+, and the iodide ion, I^-. The salt is KI.

SOLUTION   The balanced molecular equation is

$$KHCO_3(aq) + HI(aq) \longrightarrow CO_2(g) + H_2O + KI(aq)$$

To prepare the ionic equation, we analyze each of the formulas in the molecular equation to see which can be disassembled into ions.

$KHCO_3(aq)$ means $K^+(aq)$ and $HCO_3^-(aq)$	(As we were told.)
$HI(aq)$ means $H^+(aq) + I^-(aq)$	(Because this is a fully ionized acid.)
$KI(aq)$ means $K^+(aq) + I^-(aq)$	(Because we treat all water-soluble salts as fully ionized.)
$CO_2(g)$ and H_2O stay the same	(Neither is ionized.)

Solid sodium carbonate is being pumped by a highway snow blower onto concentrated nitric acid spilling from a ruptured tank car. (This accident occurred in April 1983, in a Denver, Colorado railyard.)

Using these facts, we expand the molecular equation into the ionic equation.

$$[K^+(aq) + HCO_3^-(aq)] + [H^+(aq) + I^-(aq)] \longrightarrow CO_2(g) + H_2O + [K^+(aq) + I^-(aq)]$$

The $K^+(aq)$ and the $I^-(aq)$ cancel from each side of the arrow. This leaves the following net ionic equation.

$$H^+(aq) + HCO_3^-(aq) \longrightarrow CO_2(g) + H_2O$$

CHECK   The equation is balanced both materially and electrically.

The equation produced by Example 8.3 is the same net ionic equation for the reactions of all metal bicarbonates with all strong, aqueous acids. Had we wanted to be a bit more exact and used $H_3O^+(aq)$ instead of $H^+(aq)$, the net ionic equation would have been

$$H_3O^+(aq) + HCO_3^-(aq) \longrightarrow CO_2(g) + 2H_2O$$

The only difference is in how H_2O becomes balanced. The essential chemistry has not changed.

Because this reaction destroys the hydrogen ions of the acid, it must also be called an acid neutralization. In fact, the familiar "bicarb" used as a home remedy for acid stomach is nothing more than sodium bicarbonate. Stomach acid is roughly 0.1 M HCl, and bicarbonate ion neutralizes this acid by the reaction we have just studied. An overdose of "bicarb" must be avoided because it can cause a medical emergency involving the respiratory gases. Another use of sodium bicarbonate is as an isotonic solution given intravenously to neutralize acid in the blood. For still another use, see Special Topic 8.1.

PRACTICE EXERCISE 6

Write the molecular, ionic, and net ionic equations for the reaction of sodium bicarbonate with sulfuric acid in which sodium sulfate, $Na_2SO_4(aq)$, is one of the products.

Strong Acids React with Carbonates to Give CO_2, H_2O, and a Salt   Carbonates react with hydrogen ions to give the same products as bicarbonates. Only the stoichiometry changes. The CO_3^{2-} ion is thus a base and, mole for mole, it neutralizes twice as much H^+ as the HCO_3^- ion, as we'll see in the next example.

EXAMPLE 8.4
Writing Equations for the Reactions of Metal Carbonates with Strong, Aqueous Acids

Sodium carbonate, Na_2CO_3, neutralizes nitric acid. Sodium nitrate is one product. Write the molecular, ionic, and net ionic equations for this reaction. Assume that the reaction occurs in an aqueous solution.

ANALYSIS   The pattern of the previous example repeats.

1.   Write the formulas into a conventional, molecular equation and balance it.
2.   Analyze the formulas in the equation to see how to disassemble them to prepare the ionic equation.
3.   Expand the molecular equation to the ionic equation.
4.   Cancel spectator ions and write the net ionic equation.

As you no doubt know, you can buy fruit-flavored tablets that dissolve in water to give a fizzy drink. Alka-Seltzer and similar tablets contain a solid acid, citric acid, and solid sodium bicarbonate, besides aspirin. The way that these tablets respond when they're dropped into water illustrates the importance of water as a solvent in the reactions of ions. In the crystalline materials, ions are not free to move, but as soon as these tablets hit the water and the ions become mobile, the ions start to react.

Citric acid is a triprotic acid, and we can represent it as H_3Cit, where Cit stands for the citrate ion, an ion with a charge of 3−. The hydrogen ions liberated by citric acid when it is in solution react with the bicarbonate ions that become free to move around when sodium bicarbonate dissolves. This reaction gives the CO_2 that fizzes out of solution as it forms.

$$H^+(aq) + HCO_3^-(aq) \longrightarrow CO_2(g) + H_2O$$

SOLUTION The molecular equation is

$$2HNO_3(aq) + Na_2CO_3(aq) \longrightarrow CO_2(g) + H_2O + 2NaNO_3(aq)$$

To disassemble species that exist as ions in solution, we recall the following.

$2HNO_3(aq)$ means $2H^+(aq) + 2NO_3^-(aq)$ (The acid is strong and fully ionized.)

$Na_2CO_3(aq)$ means $2Na^+(aq) + CO_3^{2-}(aq)$ (This is a salt and it is written with aq, so it is in solution and fully ionized.)

$2NaNO_3(aq)$ means $2Na^+(aq) + 2NO_3^-(aq)$ (This soluble salt is treated, like all salts dissolved in water, as fully ionized.)

$CO_2(g)$ and H_2O remain unchanged

With these facts, we prepare the ionic equation.

$$[2H^+(aq) + 2NO_3^-(aq)] + [2Na^+(aq) + CO_3^{2-}(aq)] \longrightarrow$$
$$CO_2(g) + H_2O + [2Na^+(aq) + 2NO_3^-(aq)]$$

We can cancel the $2Na^+(aq)$ and the $2NO_3^-(aq)$ from both sides of the equation, which leaves the following net ionic equation.

$$2H^+(aq) + CO_3^{2-}(aq) \longrightarrow CO_2(g) + H_2O$$

CHECK There is both an electrical and a material balance.

Notice in Example 8.4 that one carbonate can neutralize two hydrogen ions, twice as many as are neutralized by a bicarbonate ion. The net ionic equation obtained in Example 8.4,

$$2H^+(aq) + CO_3^{2-}(aq) \longrightarrow CO_2(g) + H_2O$$

is the same for the reactions of all of the carbonates of the group IA metals with all strong, aqueous acids.

PRACTICE EXERCISE 7

Write the molecular, ionic, and net ionic equations for the reaction of aqueous potassium carbonate, $K_2CO_3(aq)$, with sulfuric acid to give potassium sulfate, $K_2SO_4(aq)$, a water-soluble salt, and the other usual products.

■ Stalactites and stalagmites in limestone caverns are chiefly deposits of limestone.

Only the carbonates of group IA metal ions (as well as ammonium carbonate) are very soluble in water. Most other carbonates are water-insoluble compounds. Calcium carbonate, $CaCO_3$, for example, is the chief substance in limestone and marble. Despite its insolubility in water, calcium carbonate reacts readily with strong, aqueous acids (with their hydrogen ions, of course). The products are soluble in water, so the insoluble carbonates dissolve by this reaction.

$$CaCO_3(s) + 2HCl(aq) \longrightarrow CO_2(g) + H_2O + CaCl_2(aq)$$

For water-insoluble carbonates, we have to write their entire formulas in net ionic equations, so the net ionic equation for the above reaction is

$$CaCO_3(s) + 2H^+(aq) \longrightarrow CO_2(g) + H_2O + Ca^{2+}(aq)$$

■ The addition of a few drops of hydrochloric acid to a rock sample is a field test for carbonate rocks. A positive test is the fizzing of an odorless gas.

PRACTICE EXERCISE 8

Dolomite, a limestone-like rock, contains both calcium and magnesium carbonates. Magnesium carbonate is attacked by nitric acid. The salt that forms is water soluble. Write the molecular, ionic, and net ionic equations for this reaction.

Ammonia Neutralizes Strong Aqueous Acids An aqueous solution of ammonia is an excellent reagent for neutralizing acids. We learned in Section 5.5 how an unshared pair of electrons on nitrogen in ammonia can form a coordinate covalent bond to H^+ furnished by an acid. This makes the ammonium ion an effective Brønsted base. For example,

$$NH_3(aq) + HCl(aq) \longrightarrow NH_4Cl(aq)$$

The net ionic equation is

$$NH_3(aq) + H^+(aq) \longrightarrow NH_4^+(aq)$$

All ammonium salts are soluble in water, so they liberate NH_4^+ ions in aqueous solutions. Many biochemicals have ammonia-like molecules that also neutralize hydrogen ions.

PRACTICE EXERCISE 9

Write the molecular and net ionic equations for the reaction of aqueous ammonia with (a) HBr(aq) and (b) H_2SO_4(aq).

Active Metals React with Strong Acids to Give Hydrogen and a Salt Many metals are attacked more or less readily by the hydrogen ion in solution. The products are generally hydrogen gas and a salt made of the cation from the metal and the anion from the acid. Zinc, for example, reacts with hydrochloric acid as follows:

$$Zn(s) + 2HCl(aq) \longrightarrow H_2(g) + ZnCl_2(aq)$$

The net ionic equation is

$$Zn(s) + 2H^+(aq) \longrightarrow H_2(g) + Zn^{2+}(aq)$$

Aluminum is also attacked by acids. Its reaction with nitric acid, for example, can be written as follows:

$$2Al(s) + 6HNO_3(aq) \longrightarrow 2Al(NO_3)_3(aq) + 3H_2(g)$$

The net ionic equation is

$$2Al(s) + 6H^+(aq) \longrightarrow 2Al^{3+}(aq) + 3H_2(g)$$

PRACTICE EXERCISE 10

Write the molecular and the net ionic equations for the reaction of magnesium with hydrochloric acid.

Metals Form an Activity Series in Their Reactions with Acids

Metals differ greatly in their tendencies to react with aqueous hydrogen ions. When they do, atoms of the metal are oxidized because they lose electrons and become metal ions. The oxidizing agent is H^+. The electrons are transferred to H^+, taken from H_3O^+ ions (sometimes from H_2O), and these protons are reduced and made electrically neutral. Two H atoms combine and emerge as a molecule of hydrogen gas, H_2. Thus the metal is oxidized by H^+ to a cation and H^+ is reduced by the metal to H_2.

The group IA metals such as sodium and potassium include the most reactive metals of all. They not only reduce H^+ taken from hydronium ions, they also reduce H^+ taken from water molecules. No acid need be present. The following reaction of sodium metal with water is extremely violent, and it should never be attempted except by an experienced chemist working with safety equipment, including a fire extinguisher (see Fig. 8.6).

$$2Na(s) + 2H_2O \longrightarrow 2NaOH(aq) + H_2(g)$$

This reaction, violent in water, is even more violent in aqueous acids.

Gold, silver, and platinum, in contrast, are stable not only toward water but also toward hydronium ions. Lead and tin react very slowly with acids. Figure 8.7 shows how the reactivities of iron, zinc, and magnesium differ toward 1 M HCl.

The vast differences in the reactivities of metals toward acids make it possible to arrange the metals in an order of reactivity. The result is the **activity series** of the metals, given in Table 8.3. Atoms of any metal above hydrogen in the series can transfer electrons to H^+, either from H_2O or from H_3O^+, to form hydrogen gas, and the metal atoms change to ions. The farther a metal is above hydrogen, the more reactive it is toward acids. The metals below hydrogen in the activity series do not transfer electrons to H^+.

■ In Chapter 10 we will study redox reactions like these in a *quantitative* way. The intent here is simply to introduce the *idea* of an activity series of the metals as part of an introduction to the properties of acids.

■ In oxidation, an oxidation number becomes more positive. In reduction, an oxidation number becomes more negative. Oxidation numbers were introduced in Section 5.2.

FIGURE 8.6
The reaction of metallic sodium with water is violent. It produces hydrogen gas and enough heat to ignite the sodium metal. You can see it burning at the surface and sending out a shower of sparks.

FIGURE 8.7
Metals vary widely in their ease of oxidation. Iron is in the first tube, zinc in the second, and magnesium in the third, and all are exposed to HCl(aq) at the same molarity. All these metals can be oxidized to their metal ion states by H^+, and H^+ is reduced to hydrogen gas. Iron, the least readily oxidized of these metals, produces hardly any visible fizzing of hydrogen gas. Zinc, the next most easily oxidized of the three, reacts rather well. Magnesium reacts very vigorously and is the most easily oxidized of the three.

TABLE 8.3
The Activity Series of the Common Metals

Decreasing tendency to become ionic →	**Greatest tendency to become ionic**	Potassium Sodium	React violently with water
		Calcium	Reacts slowly with water
	React with hydrogen ions to liberate H_2	Magnesium Aluminum Zinc Chromium	React very slowly with water
		Iron Nickel Tin Lead	
		HYDROGEN	
	Do not react with hydrogen ions	Copper Mercury Silver Platinum	
	Least tendency to become ionic	Gold	

Strong, Moderate, and Weak Acids React at Different Rates with the Same Metal   The rate of the reaction of an acid with a metal depends on the acid as well as the metal. When compared at the same molar concentrations, strong acids react far more rapidly than weak acids (see Fig. 8.8). These differences reflect the differences in percentage ionizations, because the actual reaction, as we have said, is with the hydrogen ion. When the

FIGURE 8.8
Relative hydrogen ion concentrations and the reactivity of zinc. Zinc reacts with hydrogen ion to give zinc ion and hydrogen gas, which fizzes out of the test tubes. Three different acids, ranging from strong to moderate to weak, are used here in identical molar concentrations. *(Left)* The acid is HCl(*aq*), a strong, fully ionized acid. *(Center)* The acid is H_3PO_4(*aq*), a moderate, partly ionized acid. *(Right)* The acid is acetic acid, $HC_2H_3O_2$(*aq*), a weak, poorly ionized acid. Although the molarities of the acids are the same, the actual molar concentrations of their hydrogen ions are greatly different, being highest in HCl(*aq*), where the bubbles of hydrogen are evolving the most vigorously, next highest in H_3PO_4(*aq*), and lowest in $HC_2H_3O_2$(*aq*).

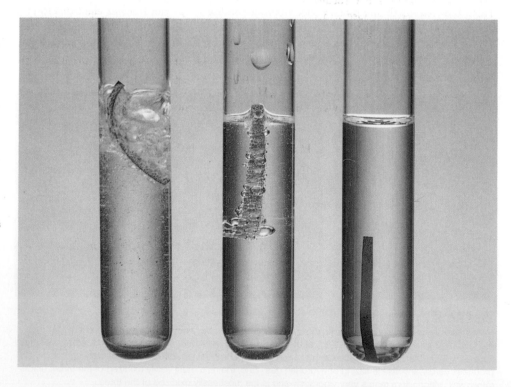

the molecule of the acid. In other words, *every weak acid has a moderately strong conjugate base.* If the acid is *very* weak, like H_2O, its conjugate base, OH^-, is a very strong base.

We can summarize these observations as rules of thumb for conjugate acid–base relationships. The last two rules are just "opposite sides of the same coin" of the first two.

Conjugate Acid–Base Relationships, Brønsted Concept

The stronger an acid is, the weaker is its conjugate base.
The weaker an acid is, the stronger is its conjugate base.

The stronger a base is, the weaker is its conjugate acid.
The weaker a base is, the stronger is its conjugate acid.

These rules will enable us to judge when to expect a base to be relatively strong or weak just by using our knowledge of the list of strong acids. Let's work an example to show how this list enables us to figure out if a particular species is strong or weak. We'll first review how to tell if an acid not studied before is strong or weak.

EXAMPLE 8.7
Deducing Whether an Ion or Molecule Is a Weak Brønsted Acid

Lactic acid is the acid responsible for the tart taste of sour milk. Is lactic acid a strong or a weak acid?

SOLUTION The list of strong acids does not include lactic acid. Hence, it is a weak acid. It's as simple as that (and we'd err very few times).

EXAMPLE 8.8
Deducing Whether an Ion or Molecule Is a Strong or a Weak Brønsted Base

Is the bromide ion a strong or a weak Brønsted base?

ANALYSIS When the question deals with a potential *base,* we have to find the answer in a roundabout fashion. We accept this because the alternative would be to memorize a rather extensive list of the stronger Brønsted bases. Here's how to go about it.

Pretend that the potential base actually functions as a base, so write the formula of its conjugate acid. If Br^- were to be a base, its conjugate acid would be HBr. In water, HBr is hydrobromic acid. Now comes the crucial question. Is hydrobromic acid a strong acid? *We have to know the list of strong acids,* and HBr is on this list. So we know that HBr easily gives up a proton. Therefore we know that what remains when the proton so readily leaves HBr, Br^-, has to be a poor proton binder.

SOLUTION Our answer is that Br^- is a weak Brønsted base.

EXAMPLE 8.9
Deducing Whether an Ion or a Molecule Is a Strong or a Weak Brønsted Base

Is the phosphate ion, PO_4^{3-}, a strong or a weak Brønsted base?

ANALYSIS Using the strategy described in Example 8.8, we pretend that this ion actually is a base, a proton acceptor. So we give it a proton, and write the result, the conjugate acid. The conjugate acid of PO_4^{3-} is HPO_4^{2-}. This Brønsted acid isn't on our list of strong acids, so we conclude it's a weak acid. This means that PO_4^{3-} is a good proton binder (holding the proton as HPO_4^{2-}).

SOLUTION A good proton binder is a relatively *strong* base, so the answer is that PO_4^{3-} is such a base.

PRACTICE EXERCISE 13

Classify the following particles as strong or as weak Brønsted acids.

(a) HSO_3^- **(b)** HCO_3^- **(c)** $H_2PO_4^-$

PRACTICE EXERCISE 14

Classify the following ions as strong or as weak Brønsted bases.

(a) I^- **(b)** NO_3^- **(c)** CN^- **(d)** NH_2^-

The Strongest Base We Can Have in Water Is OH^-

■ OH^- is the conjugate acid of O^{2-}.

If we try to dissolve a base stronger than the OH^- ion in water, it reacts with water, takes a proton, and changes to the conjugate acid. For example, the oxide ion, O^{2-}, which is the conjugate *base* of OH^-, is a much stronger base than OH^-. If we add it to water in the form of sodium oxide, the following reaction occurs (very exothermically), and none of the oxide ions supplied by Na_2O is in the solution. They have all changed to hydroxide ions by the following reaction.

■ Solid Na_2O or any other group IA oxide cannot be stored exposed to humid air.

$$Na_2O(s) + H_2O \longrightarrow 2NaOH(aq)$$
Sodium oxide   Sodium hydroxide

Thus, although sodium oxide is very soluble in water, its solution contains no oxide ions. It dissolves by reacting with water, and its oxide ions change to hydroxide ions. Most metal oxides that dissolve in water do so by reacting in this way. The ionic equation can be written:

$$O^{2-}(s) + H_2O \longrightarrow 2OH^-(aq)$$

Thus the strongest base that can exist in water is the hydroxide ion.

The Strongest Acid We Can Have in Water Is H_3O^+

If we try to dissolve in water any acid stronger than H_3O^+, it reacts with water to give hydronium ion. Hydrogen chloride, for example, is a stronger proton donor than H_3O^+. As we already know, when we bubble $HCl(g)$ into water, the following reaction occurs, a typical Brønsted acid–base reaction. We'll write it as an equilibrium, although the forward reaction occurs essentially 100%.

$$HCl(g) + H_2O \rightleftharpoons H_3O^+(aq) + Cl^-(aq)$$
Stronger   Stronger   Weaker   Weaker
acid   base   acid   base

Evidently, the hydronium ion holds the proton better than it is held by the Cl atom in $HCl(g)$.

In All Brønsted Acid–Base Equilibria, the Weaker Acid and Weaker Base Are Favored Now that we can make reasonable predictions of relative acid or base strengths, let's see how we can use this skill in predicting reactions. This will enable us to judge which side of an acid–base equilibrium is favored.

A logical consequence of our rules of thumb about acid–base strengths is that in all proton-transfer equilibria the side having the weaker acid and base will always be favored over the side with the stronger acid and base. We can condense this to another rule of thumb concerning what to expect when we mix chemicals that can participate in acid–base equilibria: *The stronger always gives way to the weaker in acid–base reactions.*

EXAMPLE 8.10
Predicting Which Substances Are Favored in an Acid–Base Equilibrium

If we add hydrochloric acid to an aqueous solution of sodium cyanide, NaCN, will HCN and NaCl form to any significant extent? (If they do, the evolving HCN, hydrogen cyanide, might kill anyone who mixes these substances. Hydrogen cyanide is a very dangerous poison.)

ANALYSIS Because we are dealing with HCl(aq), the reagent actually consists of $H_3O^+(aq)$ and $Cl^-(aq)$. Because NaCN is a salt in solution, we are dealing with $Na^+(aq)$ and $CN^-(aq)$. Sodium ions and chloride ions would be spectator ions. The question, therefore, is, Does the following reaction occur?

$$H_3O^+(aq) + CN^-(aq) \longrightarrow H_2O + HCN(aq)$$

To apply the rule "the stronger gives way to the weaker" in proton-transfer reactions, we first need to identify the acids and bases and then infer what is stronger and what is weaker. When we look for the conjugate pairs, we can see them as follows:

Conjugate pair

$$CN^-(aq) + H_3O^+(aq) \rightleftharpoons HCN(aq) + H_2O$$

Conjugate pair

SOLUTION Each pair must have an acid and each must have a base, so let's write in these labels (and, to reduce clutter, omit the lines that have served to connect conjugate pairs).

$$CN^-(aq) + H_3O^+(aq) \longrightarrow HCN(aq) + H_2O$$
Base   Acid   Acid   Base

Now we decide which of the two acids is stronger. We know that H_3O^+ is the strongest acid species we can have in water. (We also know that because HCN isn't on the list of strong acids, it must be relatively weak.) So we modify our labels with this new information.

$$CN^-(aq) + H_3O^+(aq) \longrightarrow HCN(aq) + H_2O$$
Base   Stronger   Weaker   Base
        acid   acid

Because the conjugate of a stronger acid must be a weaker base, and the conjugate of a weaker acid must be a stronger base, we can modify the remaining labels as follows.

$$CN^-(aq) + H_3O^+(aq) \longrightarrow HCN(aq) + H_2O$$
Stronger   Stronger   Weaker   Weaker
base   acid   acid   base

Finally, we can tell that the reaction must occur as written, because the stronger conjugates are always replaced by the weaker in acid–base reactions. If we write the equation as an equilibrium (as we should),

$$CN^-(aq) + H_3O^+(aq) \rightleftharpoons HCN(aq) + H_2O$$

Stronger   Stronger   Weaker   Weaker
base   acid   acid   base

then we must note that the products of the forward reaction are favored (the double arrows give no indication of this).

PRACTICE EXERCISE 15

When the meat preservative sodium nitrite, $NaNO_2$, enters the stomach and encounters the hydrochloric acid in gastric juice, can nitrous acid, HNO_2, be produced? Write the equilibrium expression for any net ionic interactions. State which are favored—the reactants or the products. (Nitrous acid is suspected of being a cause of cancer, but no evidence presently exists that it actually causes cancer in humans.)

Acids and Bases Can Be Organized in Their Order of Strengths

Table 8.4 lists several Brønsted acids and bases in their orders of increasing strength. Carbonic acid, for example, is a weaker acid than acetic acid so it stands below acetic acid in the table.

All acids above H_3O^+ in the column of acids in the table—those that we have learned are the strong acids—are essentially 100% ionized into the hydronium ion and the conjugate base in aqueous solutions. We also treat $HSO_4^-(aq)$ as a strong acid, although it is less strong than H_2SO_4. (Why?) $H_3PO_4(aq)$ is a moderate acid.

■ Perchloric acid in Table 8.4 was not on our earlier list of strong acids because it is less common.

Moving down the column, the next acids—acetic acid, carbonic acid, the dihydrogen phosphate ion, the ammonium ion, the bicarbonate ion, and the monohydrogen phosphate ion—are all weak acids (becoming progressively weaker as we move down the column).

The acids from water and below in Table 8.4 ionize in water to such a low percentage that, except in discussions of the Brønsted concept, we do not refer to them as proton donors.

Moving over to the column of bases, the conjugate bases of all strong acids in Table 8.4 are such weak bases that we almost never refer to them as bases. Neither water nor the sulfate ion are routinely called bases either (except in discussions of the Brønsted concept). The dihydrogen phosphate ion is a weak base, and as we move down the list through the acetate ion, the bicarbonate ion, the monohydrogen phosphate ion, ammonia, the carbonate ion, and the phosphate ion, the bases get stronger. This means that as we move down through this series, the products in the following equilibrium expression become more and more favored. (We let B^- represent any base except NH_3.)

$$B^-(aq) + H_2O \rightleftharpoons BH(aq) + OH^-(aq)$$

■ Sodium salts of all these strongly basic anions are known—$NaOCH_3$, $NaNH_2$, and NaH.

The ions below OH^- in the table—CH_3O^-, NH_2^-, O^{2-}, and H^-—react quantitatively with water to give their conjugate acids. As we said, no base stronger than OH^- can exist in water. In the reactions, for example, of NH_2^- and H^- with water, we don't normally even use equilibrium arrows; the reactions go to completion for all practical purposes.

$$NH_2^- + H_2O \longrightarrow NH_3(aq) + OH^-(aq)$$
$$H^- + H_2O \longrightarrow H_2(g) + OH^-(aq)$$

Our discussion of relative strengths of acids and bases has been qualitative because that is all that is needed for many uses. Sometimes, however, it is necessary to have numbers to describe these relative strengths, and we will describe such numbers in the next chapter.

TABLE 8.4
Relative Strengths of Some Brønsted Acids and Bases

Brønsted Acid		Brønsted Base	
Name	Formula	Name	Formula
Perchloric acid	$HClO_4$	Perchlorate ion	ClO_4^-
Hydrogen iodide	HI	Iodide ion	I^-
Hydrogen bromide	HBr	Bromide ion	Br^-
Sulfuric acid	H_2SO_4	Hydrogen sulfate ion	HSO_4^-
Hydrogen chloride	HCl	Chloride ion	Cl^-
Nitric acid	HNO_3	Nitrate ion	NO_3^-
HYDRONIUM ION	H_3O^+	WATER	H_2O
Hydrogen sulfate ion	HSO_4^-	Sulfate ion	SO_4^{2-}
Phosphoric acid	H_3PO_4	Dihydrogen phosphate ion	$H_2PO_4^-$
Acetic acid	$HC_2H_3O_2$	Acetate ion	$C_2H_3O_2^-$
Carbonic acid	H_2CO_3	Bicarbonate ion	HCO_3^-
Dihydrogen phosphate ion	$H_2PO_4^-$	Monohydrogen phosphate ion	HPO_4^{2-}
Ammonium ion	NH_4^+	Ammonia	NH_3
Bicarbonate ion	HCO_3^-	Carbonate ion	CO_3^{2-}
Monohydrogen phosphate ion	HPO_4^{2-}	Phosphate ion	PO_4^{3-}
WATER	H_2O	HYDROXIDE ION	OH^-
Methyl alcohol	CH_3OH	Methoxide ion	CH_3O^-
Ammonia	NH_3	Amide ion	NH_2^-
Hydroxide ion	OH^-	Oxide ion	O^{2-}
Hydrogen	H_2	Hydride ion	H^-

Increasing acid strength (↑)   *Increasing base strength* (↓)

(handwritten annotations: "weak acid", "weak")

The Ammonium Ion Is a Brønsted Acid

The ammonium ion occupies a special place in our study because many biochemicals, like proteins, have a molecular part that is very much like this ion. Although the ammonium ion is a weak acid (Table 8.4), it still is a Brønsted acid, and it can neutralize the hydroxide ion. When we add sodium hydroxide to a solution of ammonium chloride, the following reaction occurs.

$$NaOH(aq) + NH_4Cl(aq) \longrightarrow NH_3(aq) + H_2O + NaCl(aq)$$

The net ionic equation is

$$OH^-(aq) + NH_4^+(aq) \longrightarrow NH_3(aq) + H_2O$$

This reaction neutralizes the hydroxide ion, and it leaves a solution of the weaker base, NH_3. (If the initial solution is concentrated enough, the final solution has a strong odor of ammonia.)

In some medical emergencies, when the blood has become too alkaline or too basic, an isotonic solution of ammonium chloride is administered by intravenous drip. Its ammonium ions can neutralize some of the base in the blood and bring the acid–base status back to normal.

■ The *protonated amino group*, NH_3^+, is covalently bound to carbon in a large number of organic compounds, including proteins.

■ NH_3 is a much weaker base than OH^- so, when we compare their conjugate acids, we know that NH_4^+ is a stronger acid than H_2O.

8.5

SALTS

A very large number of ionic reactions can be predicted from a knowledge of the solubility rules of salts.

Salts, as we have said, are ionic compounds whose cations are any except H^+ and whose anions are any except OH^- or O^{2-}. All are crystalline solids at room temperature, because forces of attraction between ions in crystals are very strong.

A **simple salt** is one that is made of just *two* kinds of oppositely charged ions. Examples are $NaCl$, $MgBr_2$, and $CuSO_4$. *Mixed salts* are those that have three or more different ions. Alum, used in water purification, is an example: $K_2SO_4 \cdot Al_2(SO_4)_3 \cdot 24H_2O$. As the formula of alum illustrates, the salt family includes hydrates. Some salts of practical value are given in Table 8.5.

Formation of Salts   In the laboratory, salts are obtained whenever an acid is used in any of the following ways. We summarize and review these methods here and show their similarities.

$$\text{Acid} + \text{metal hydroxide} \longrightarrow \text{a salt} + H_2O$$
$$\text{Acid} + \text{metal bicarbonate} \longrightarrow \text{a salt} + H_2O + CO_2$$
$$\text{Acid} + \text{metal carbonate} \longrightarrow \text{a salt} + H_2O + CO_2$$
$$\text{Acid} + \text{metal} \longrightarrow \text{a salt} + H_2$$

If the salt is soluble in water, we have to evaporate the solution to dryness to isolate it.

Many Salts Are Insoluble in Water   Sometimes a salt precipitates as it forms instead of remaining in solution. To predict when to expect this, we use a small number of solubility rules for ionic compounds. We say that a compound is *soluble* in water if it can form a solution with a concentration of at least 3 to 5% (w/w). When a *counter ion* is referred to in the following rules,

TABLE 8.5
Some Salts and Their Uses

Formula and Name	Uses
$BaSO_4$ Barium sulfate	Used in the "barium cocktail" given prior to X-raying the gastrointestinal tract
$MgSO_4 \cdot 7H_2O$ Magnesium sulfate heptahydrate (epsom salt)	Purgative
$(CaSO_4)_2 \cdot H_2O$ Calcium sulfate hemihydrate (plaster of paris)	Plaster casts. Wall stucco. Wall plaster
$AgNO_3$ Silver nitrate	Antiseptic and germicide. Used in eyes of infants to prevent gonorrheal conjunctivitis. Photographic film
$NaHCO_3$ Sodium bicarbonate (baking soda)	Baking powders. Effervescent salts. Stomach antacid. Fire extinguishers
$Na_2CO_3 \cdot 10H_2O$ Sodium carbonate decahydrate (soda ash, sal soda, washing soda)	Water softener. Soap and glass manufacture
$NaCl$ Sodium chloride	Manufacture of chlorine, sodium hydroxide. Preparation of food
$NaNO_2$ Sodium nitrite	Meat preservative

it means the unnamed ion of the ionic compound. For example, in the lithium salt, LiCl, the counter ion is the chloride ion. In the hydroxide, $Ca(OH)_2$, the counter ion is Ca^{2+}.

Solubility Rules for Ionic Compounds in Water

1. All lithium, sodium, potassium, and ammonium salts are soluble regardless of the counter ion.

2. All nitrates and acetates are soluble, regardless of the counter ion.

3. All chlorides, bromides, and iodides are soluble, *except* when the counter ion is lead, silver, or mercury(I).

4. All sulfates are soluble *except* those of lead, calcium, strontium, mercury(I), and barium.

5. All hydroxides and metal oxides are insoluble *except* those of the group IA cations and those of calcium, strontium, and barium.

6. All phosphates, carbonates, sulfites, and sulfides are insoluble *except* those of the group IA cations and NH_4^+.

There are exceptions to the solubility rules, but we will seldom be wrong in applying the rules. One of the many applications of the rules is to predict possible reactions involving ionic compounds. By developing the skill of *predicting* reactions from a few facts, we sharply reduce the quantity of facts that should be memorized.

Salts Can Form by *Double Replacement* ("Exchange of Partners") Reactions In addition to the reactions we have already studied to make salts by acid–base neutralization, we can also make salts by a "change of partners" reaction called **double replacement.** For example, sodium carbonate and calcium chloride are both soluble in water. But if we mix aqueous solutions of the two, the following reaction occurs because CO_3^{2-} ion and Ca^{2+} ions cannot remain in solution in each others presence beyond extremely trace concentrations. Their combination, $CaCO_3$, is too insoluble.

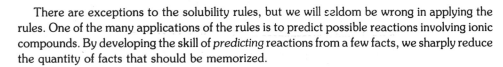

Combining CO_3^{2-} with Ca^{2+}

$$Na_2CO_3(aq) + CaCl_2(aq) \longrightarrow CaCO_3(s) + 2NaCl(aq)$$

Combining Na^+ with Cl^-

An ion from each salt combines with an ion from the other salt, which gives the informal name, "exchange of partners," to the reaction or, more formally, *double replacement.* The ionic equation for this reaction more clearly shows this exchange.

$$[2Na^+(aq) + CO_3^{2-}(aq)] + [Ca^{2+}(aq) + 2Cl^-(aq)] \longrightarrow CaCO_3(s) + [2Na^+(aq) + 2Cl^-(aq)]$$

The net ionic equation is

$$Ca^{2+}(aq) + CO_3^{2-}(aq) \longrightarrow CaCO_3(s)$$

The other product, NaCl, stays in solution in its dissociated form. You would have to filter off the precipitate of calcium carbonate and then evaporate the filtrate to dryness to obtain crystalline NaCl.

Not all combinations of solutes give double replacement reactions. If you mixed solutions of NaI and KCl, no combination of oppositely charged ions is insoluble, so the solution would just contain separated (and hydrated) ions of Na^+, K^+, I^-, and Cl^-.

■ Some references use the term **metathesis reaction** for double replacement reaction. (The term *double displacement* is used by some, but it is not the preferred term.)

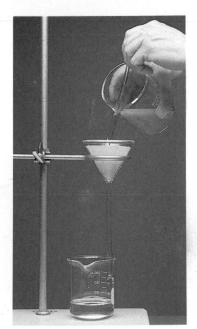

The precipitate in the beaker is being separated by *filtration*. It collects in a cone of filter paper in the funnel, and the *filtrate*—the clear solution—flows through into the beaker below.

EXAMPLE 8.11
Predicting Double Replacement Reactions of Salts

What happens if we mix aqueous solutions of sodium sulfate and barium nitrate?

ANALYSIS By the solubility rules, we know that both sodium sulfate and barium nitrate are soluble in water, so their solutions contain their separated ions. When we pour the two solutions together, four ions experience attractions and repulsions. Hence, we must examine each possible combination of oppositely charged ions to see which, if any, makes a water-insoluble salt. When we find one, then we can write an equation for the reaction that produces this salt. Here are the possible combinations when Ba^{2+}, NO_3^-, Na^+, and SO_4^{2-} ions intermingle in water.

$Ba^{2+} + 2NO_3^- \xrightarrow{?} Ba(NO_3)_2(s)$ This possibility is obviously out, because barium ions and nitrate ions do not precipitate together from water. ("All nitrates are soluble.")

$2Na^+ + SO_4^{2-} \xrightarrow{?} Na_2SO_4(s)$ This possibility is also out. ("All sodium salts are soluble.")

$Na^+ + NO_3^- \xrightarrow{?} NaNO_3(s)$ No. Again, "All sodium salts are soluble."

$Ba^{2+} + SO_4^{2-} \xrightarrow{?} BaSO_4(s)$ Yes. Barium sulfate, $BaSO_4$, is not in any of the categories of water-soluble salts, so we conclude that it is an insoluble salt (rule 4).

■ $BaSO_4$ is the white, insoluble substance in a flavored slurry given to patients as a "barium cocktail" before an X-ray is taken of the intestinal tract. The barium ion stops X-rays, so the tract is outlined on the film.

SOLUTION Because we predicted that $BaSO_4$ can form a precipitate, we can write a molecular equation, and we'll use some connector lines to show how partners exchange—how *double* replacement occurs.

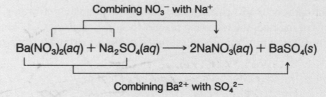

The net ionic equation, however, is a better way to describe what happens.

$$Ba^{2+}(aq) + SO_4^{2-}(aq) \longrightarrow BaSO_4(s)$$

The sodium and nitrate ions are only spectators. To obtain the solid barium sulfate, we would filter the mixture and collect this compound on the filter. If we also wanted the sodium nitrate, we would evaporate the clear filtrate to dryness.

PRACTICE EXERCISE 16

If solutions of sodium sulfide, Na_2S, and copper(II) nitrate, $Cu(NO_3)_2$, are mixed, what if anything will happen chemically? Write a molecular and a net ionic equation for any reaction.

One of the many uses of the solubility rules is to understand what it means for water to be called *hard water* and what it means to *soften* such water. These are discussed in Special Topic 8.2.

To Summarize, the Chief Reactions of Ions Are Those That Form Gases, Molecules in Solution, or Precipitates Our study of the reaction of sodium sulfate and barium nitrate in Example 8.11 illustrates the power of knowing just a few facts for the sake of predicting an enormous number of others with a high probability of success. The following should now be well learned.

HARD WATER

Groundwater that contains magnesium, calcium, or iron ions at a high enough level to interact with ordinary soap to form scum is called **hard water**. In **soft water** these "hardness ions"—Ca^{2+}, Mg^{2+}, Fe^{2+}, or Fe^3—are either absent or are present in extremely low concentrations. (The anions that most frequently accompany the hardness ions are SO_4^{2-}, Cl^-, and HCO_3^-.)

Hard water in which the principal anion is the bicarbonate ion is called **temporary hard water**. Hard water in which the chief negative ions are anything else is called **permanent hard water**. When temporary hard water is heated near its boiling point, as in hot boilers, steam pipes, and instrument sterilizers, the bicarbonate ion breaks down to the carbonate ion. And this ion forms insoluble precipitates with the hardness ions. Their carbonate salts form, come out of solution, and deposit as scaly material that can even clog the equipment (see Figure 1). The equations for these changes are as follows.

The breakdown of the bicarbonate ion:

$$2HCO_3^-(aq) \longrightarrow CO_3^{2-}(aq) + CO_2(g) + H_2O$$

The formation of the scaly precipitate (using the calcium ion to illustrate):

$$CO_3^{2-}(aq) + Ca^{2+}(aq) \longrightarrow CaCO_3(s)$$

Water Softening Removes the Hardness Ions Chemically Hard water can be softened in various ways. Most commonly, excess soap is used. Some scum does form, but then the extra soap does the cleansing work. To avoid the scum altogether, softening agents are added before the soap is used. One common water-softening chemical is sodium carbonate decahydrate, known as washing soda. Its carbonate ions take out the hardness ions as insoluble carbonates by the kind of reaction for which we wrote the previous net ionic equation.

Another home water-softening agent is household ammonia—5% (w/w) NH_3. We've already learned about the following equilibrium in such a solution.

$$NH_3(aq) + H_2O \rightleftharpoons NH_4^+(aq) + OH^-(aq)$$

A deposit of calcium carbonate has nearly closed this 2-in. hot-water pipe after just two years of service in northeastern New Jersey.

In other words, aqueous ammonia has some OH^- ions, and the hydroxides of the hardness ions are not soluble in water. Therefore, when aqueous ammonia is added to hard water, the following kind of reaction occurs (illustrated using the magnesium ion this time):

$$Mg^{2+}(aq) + 2OH^-(aq) \longrightarrow Mg(OH)_2(s)$$

As hydroxide ions are removed by this reaction, more are made available from the ammonia–water equilibrium. (A loss of OH^- ion from this equilibrium is a stress, and the equilibrium shifts to the right in response, as we'd predict using Le Châtelier's principle.)

Still another water-softening technique is to let the hard water trickle through zeolite, a naturally occurring porous substance that is rich in sodium ions. When the hard water is in contact with the zeolite, sodium ions go into the water and the hardness ions leave solution and attach themselves to the zeolite. Later, the hardness ions are flushed out by letting water that is very concentrated in sodium chloride trickle through the spent zeolite, and this restores the zeolite for reuse. Synthetic ion-exchange materials are also used to soften water by roughly the same principle.

Perhaps the most common strategy in areas where the water is quite hard is to use synthetic detergents instead of soap. Synthetic detergents do not form scums or precipitates with the hardness ions.

1. The solubility rules of ionic compounds (because then we can assume that all the other ionic compounds are insoluble).

2. The five strong acids in Table 8.1 (because then we can assume that all the other acids, including organic acids, are weak).

3. The first two strong bases of Table 8.2 (because then we can assume that all the other bases are either weak or are too insoluble in water to matter much).

To summarize, we expect ions to react with each other if any one of the following possibilities is predicted.

1. A gas forms that (mostly) leaves the solution. It could be
   (a) Hydrogen — from the action of acids on metals, or
   (b) Carbon dioxide — from acids reacting with carbonates or bicarbonates

2. An un-ionized, molecular compound forms that remains in solution. It could be
   (a) Water — from acid–base neutralizations or
   (b) A weak acid — by the action of H^+ on a strong Brønsted base, the conjugate base of any weak acid, or
   (c) Ammonia — by the reaction of OH^- with NH_4^+.

3. A precipitate forms — a water-insoluble salt or one of the water-insoluble hydroxides.

PRACTICE EXERCISE 17

When a solution of hydrochloric acid is mixed in the correct molar proportions with a solution of sodium acetate, $NaC_2H_3O_2$, essentially all of the hydronium ion concentration vanishes. What happens and why? Write the net ionic equation.

PRACTICE EXERCISE 18

What, if anything, happens chemically when each pair of solutions is mixed? Write net ionic equations for any reactions that occur.

(a) NaCl and $AgNO_3$ (b) Na_2CO_3 and HNO_3 (c) KBr and NaCl

The Common Ion Effect Shifts Solubility Equilibria to Favor Insoluble or Un-ionized Species   The solubility rules predict the solubility of an individual salt when it is the lone solute in solution. In nature and in living systems, however, solutions this simple seldom occur. Usually, two or more electrolytes are present, so it is important to consider any changes in the solubility of one compound that might be caused by the presence of another.

If the other solutes provide *different* ions entirely, then the solubility rules work just by themselves, at least in solutions that are relatively dilute in all dissolved species. But if some other solute contributes one ion that is *common* to the salt whose solubility we are studying, then the solubility of the latter is reduced. This reduction in solubility of one salt by the addition of a common ion is called the **common ion effect.** Let's see how it works.

Suppose that we have a *saturated* solution of sodium chloride. The following equilibrium exists.

$$NaCl(s) \rightleftharpoons Na^+(aq) + Cl^-(aq)$$

What happens if we now pour into this solution some concentrated hydrochloric acid, a fully ionized acid? By using *concentrated* HCl(aq) we can quickly increase the concentration of the chloride ion in the solution, and this ion is *common* to the original solute, NaCl. By increasing the concentration of Cl^-, we place a stress on the equilibrium. In accordance with Le Châtelier's principle, the equilibrium must shift to absorb this stress, and it has to shift to the left. Only by running the reverse reaction of the equilibrium can the system reduce the concentration of dissolved Cl^- and so reduce the stress. But this has to cause the precipitation of some solid sodium chloride, and this is exactly what happens (see Fig. 8.9). After the activity has quieted, we still have a saturated solution, but we also have more solid NaCl, and we have a lower concentration of dissolved Na^+ ions.

FIGURE 8.9
The common ion effect. At the start there is a saturated solution of sodium chloride (first frame). When concentrated hydrochloric acid is added (middle frame), a white precipitate of sodium chloride appears, grows in quantity, and begins to settle (last frame).

Special Topic 8.3 goes into a quantitative method for dealing with the solubilities of ionic substances and provides an application involving the molecular basis for the use of calcium channel blockers in the treatment of certain kinds of heart conditions. Special Topic 8.4 builds on Special Topic 8.3 to show how the earth manages atmospheric CO_2 levels and how this level is an important factor in global warming — the *greenhouse effect.*

Concentrations of Individual Ions in Solutions of Several Substances Are Often Given in Equivalents or Milliequivalents per Liter

Before we leave this introduction to salts, we must look at a concentration expression often used for their individual ions. It is based not on the moles of an ion per liter but on a quantity called an *equivalent* per liter. We'll see why soon.

One **equivalent** of an ion, abbreviated **eq**, is the number of grams of the ion that corresponds to Avogadro's number, one mole, of electrical charges. For example, when the charge is unity, either $1+$ or $1-$, it takes Avogadro's number of ions to have Avogadro's number of electrical charges. Thus 1 eq for ions such as Na^+, K^+, Cl^-, or Br^- is the same as the molar mass of each ion. The molar mass of Na^+ is 23.0 g Na^+/mol, so 1 eq of $Na^+ = 23.0$ g of Na^+. This much sodium ion, 23.0 g of Na^+, contributes Avogadro's number of positive charges. Thus the equivalent mass of the sodium ion is 23.0 g Na^+/eq.

When an ion has a double charge, either $2+$ or $2-$, then the mass of one equivalent equals the molar mass divided by 2. For example, 1 mol of CO_3^{2-} ion $= 60.0$ g of CO_3^{2-}, so 1 eq of CO_3^{2-} ion $= 30.0$ g of CO_3^{2-} ion. This much carbonate ion carries Avogadro's number of negative charge. The extension of this to ions of higher charges should now be obvious.

The Equivalent Mass of an Ion Is Its Formula Mass Divided by Its Charge

You can see this in Table 8.6, which gives equivalent masses for a number of ions.

The advantage of the concept of the equivalent is the simplicity of a 1 : 1 ratio. Regardless of the amounts of charges on the individual ions, when cations and anions are present either in an ionic crystal or in a solution, we can always be sure that for every equivalent of positive

TABLE 8.6
Equivalents of Ions

Ion	g/mol	g/eq
Na^+	23.0	23.0
K^+	39.1	39.1
Ca^{2+}	40.1	20.1
Mg^{2+}	24.3	12.2
Al^{3+}	27.0	9.00
Cl^-	35.5	35.5
HCO_3^-	61.0	61.0
CO_3^{2-}	60.0	30.0
SO_4^{2-}	96.1	48.1

charge there has to be one equivalent of negative charge. The condition of electrical neutrality in an ionic compound or a solution of ions is that

$$eq\ of\ cations = eq\ of\ anions$$

Or,

$$meq\ of\ cations = meq\ of\ anions$$

where the **milliequivalent,** or **meq,** is related to the equivalent by the relationship, 1000 meq = 1 eq.

The normal ranges of values of the concentrations of several components of blood are listed on the inside back cover where you will see that many are given in units of meq/L. There is also an application, described in Special Topic 8.5, in which the concentrations of anions in blood that are hard to determine directly can be estimated. When the level of such anions increases sufficiently, it can indicate a malfunction somewhere in the body.

SPECIAL TOPIC 8.3
SOLUBILITY PRODUCT CONSTANTS

For each sparingly soluble ionic compound, an important relationship exists for the molar concentrations of the ions. Silver bromide(AgBr), the light-sensitive material in photographic film, is an insoluble compound, for example. Yet a trace does dissolve to establish the following heterogeneous equilibrium with undissolved AgBr.

$$AgBr(s) \rightleftharpoons Ag^+(aq) + Br^-(aq)$$

One liter of water is able to dissolve only 0.13 mg of AgBr at 25 °C, which corresponds to a molar solubility of AgBr of only 7.1×10^{-7} mol/L (at 25 °C). For each mole of AgBr that dissolves and dissociates, one mole each of Ag^+ and Br^- ion is set free in solution. So if 7.1×10^{-7} mol of AgBr dissolves, we must obtain 7.1×10^{-7} mol of each ion. Recalling that the units of molar concentration of a species are indicated when we place brackets around the formula of the species, we can write

$$[Ag^+] = 7.1 \times 10^{-7}\ mol/L$$
$$[Br^-] = 7.1 \times 10^{-7}\ mol/L$$

At any given temperature and for any particular ionic compound present as the sole solute, the molar concentrations of its ions in a *saturated* solution can have only one value. Hence, the product of multiplying the molar concentrations of its ions in a saturated solution must also be a constant. Chemists calculate and record such **ion product constants,** symbolized as K_{sp}, as a way of storing solubility information in a form useful for a variety of purposes. For AgBr,

$$K_{sp}(AgBr) = [Ag^+][Br^-]$$

In a saturated solution (at 25 °C) where $[Ag^+] = [Br^-] = 7.1 \times 10^{-7}$ mol/L at equilibrium,

$$K_{sp}(AgBr) = (7.1 \times 10^{-7})(7.1 \times 10^{-7})$$
$$= 5.0 \times 10^{-13}$$

(Only the *numerical* parts of the molar concentrations of the ions are multiplied; the units are understood.)

For salts whose formulas have subscripts, like PbI_2, the ion product constant includes the subscripts as exponents. For a general salt, M_mX_x, the ion product constant has the form

$$K_{sp}(M_mX_x) = [M]^m[X]^x$$

Thus the *ion product constant* of a salt is the product of the molar concentrations of the ions in a saturated solution, each concentration raised to a power that equals the number of ions obtained from one formula unit of the salt. The ion product for PbI_2, for example, is $[Pb^{2+}][I^-]^2$. The solubility of PbI_2 at 25 °C is 1.25×10^{-3} mol/L. From the equation for the dissociation of PbI_2,

$$PbI_2(s) \longrightarrow Pb^{2+}(aq) + 2I^-(aq)$$

we can see that for every formula unit of PbI_2 that dissolves, the solution picks up one Pb^{2+} ion and *two* I^- ions. This lets us assemble the following concentration data for the ions in saturated PbI_2 (at 25 °C).

$$[Pb^{2+}] = 1.25 \times 10^{-3}\ mol/L$$
$$[I^-] = 2 \times 1.25 \times 10^{-3}\ mol/L = 2.50 \times 10^{-3}\ mol/L$$

Therefore,

$$K_{sp}(PbI_2) = (1.25 \times 10^{-3})(2.50 \times 10^{-3})^2$$
$$= 7.8 \times 10^{-9}$$

The accompanying table gives the values of several solubility product constants.

The smaller K_{sp} is, the less soluble is the substance. For example the K_{sp} of $Ca_3(PO_4)_2$ is only 1.3×10^{-32}. The only way in which K_{sp} for $Ca_3(PO_4)_2$ can be so small is that, at equilibrium in the saturated solution of this salt, the values of $[Ca^{2+}]$ and $[PO_4^{3-}]$ are themselves very small. This means that *very* little $Ca_3(PO_4)_2$ dissolves and very little Ca^{2+} ion can exist dissociated from the PO_4^{3-} ion.

 Managing Cellular Calcium Ion in the Presence of Phosphate Ion The facts just mentioned about $Ca_3(PO_4)_2$ present a problem at the molecular level of life. The fluid inside cells contains phosphate ion (or phosphate ion donors) at a concentration of about 10^{-3} mol/L. Calculations can show that this is much too high to allow enough Ca^{2+} to be present as such *and in solution* inside the cell to meet the cellular needs for this ion. Yet, calcium ion *must* be available for a variety of cellular functions. The body solves this problem by keeping Ca^{2+} largely *outside* the cell in the surrounding fluid and away from PO_4^{3-} (or its donors).

There are mechanisms that permit the flow of Ca^{2+} ions through *calcium channels* embedded in the cell membrane only when these ions are needed inside the cell. Once in, Ca^{2+} combines with a *receptor* protein so that Ca^{2+} ions are not precipitated as $Ca_3(PO_4)_2$. The combination of Ca^{2+} with this protein changes the protein and converts it into the form needed by the cell.

Ca^{2+} is particularly needed by heart muscle tissue to enable muscle contractions. When the contractions must be reduced in order to ease the burden on a stressed heart, medications called "calcium channel blockers" can be administered. They reduce the rate of flow of Ca^{2+} ions through the calcium channels and so retard the rate of contraction.

Solubility Product Constants (at 25 °C)

Salt	Solubility Equilibrium	K_{sp}
AgCl	$AgCl(s) \rightleftharpoons Ag^+(aq) + Cl^-(aq)$	1.8×10^{-10}
Hg_2Cl_2	$Hg_2Cl_2(s) \rightleftharpoons Hg_2^{2+} + 2Cl^-(aq)$	1.2×10^{-18}
$PbCl_2$	$PbCl_2(s) \rightleftharpoons Pb^{2+}(aq) = 2Cl^-(aq)$	1.7×10^{-5}
AgBr	$AgBr(s) \rightleftharpoons Ag^+(aq) + Br^-(aq)$	5.0×10^{-13}
AgI	$AgI(s) \rightleftharpoons Ag^+(aq) + I^-(aq)$	8.3×10^{-17}
$Mg(OH)_2$	$Mg(OH)_2(s) \rightleftharpoons Mg^{2+}(aq) + 2OH^-(aq)$	7.1×10^{-12}
$Ca(OH)_2$	$Ca(OH)_2(s) \rightleftharpoons Ca^{2+}(aq) + 2OH^-(aq)$	6.5×10^{-6}
$CaSO_4$	$CaSO_4(s) \rightleftharpoons Ca^{2+}(aq) + SO_4^{2-}(aq)$	2.4×10^{-5}
$BaSO_4$	$BaSO_4(s) \rightleftharpoons Ba^{2+}(aq) + SO_4^{2-}(aq)$	1.1×10^{-10}
$MgCO_3$	$MgCO_3(s) \rightleftharpoons Mg^{2+}(aq) + CO_3^{2-}(aq)$	3.5×10^{-8}
$CaCO_3$	$CaCO_3(s) \rightleftharpoons Ca^{2+}(aq) + CO_3^{2-}(aq)$	4.5×10^{-9}
$Mg_3(PO_4)_2$	$Mg_3(PO_4)_2(s) \rightleftharpoons 3Mg^{2+}(aq) + 2PO_4^{3-}(aq)$	6.3×10^{-26}
$Ca_3(PO_4)_2$	$Ca_3(PO_4)_2(s) \rightleftharpoons 3Ca^{2+}(aq) + 2PO_4^{3-}(aq)$	1.3×10^{-32}

SPECIAL TOPIC 8.4
THE EARTH'S MANAGEMENT OF CO_2 AND THE GREENHOUSE EFFECT

The earth constantly radiates energy to outer space. It *must* do so at exactly the same average rate as energy is received from the sun. Otherwise, the earth's average temperature will either increase or decrease drastically. The term **greenhouse effect** refers to the insulating effect of the earth's atmosphere caused chiefly by the presence of what are called the *greenhouse gases* — natural substances like carbon dioxide, methane, and dinitrogen monoxide (N_2O), and synthetic chemicals like the chlorofluorocarbons.

Greenhouse Gases as Insulators The greenhouse gases act as *necessary* insulators by first absorbing some of the energy radiating from the earth. Then the molecules of these gases, now more rich in energy, re-radiate it in all directions. Some comes back to the earth again with the net effect being the retention of some of the energy by the atmosphere. In the absence of this insulating action, the earth would chill. Because of the greenhouse effect, the earth's surface is currently an estimated 33 °C warmer than it would otherwise be.

If the levels of the greenhouse gases provided too much insulation over a long enough time, the earth could eventually become too hot for life. There is concern not about the presence of greenhouse gases — they are partly of natural origins and perform an essential service — but about too high a concentration of them. Today, when news services and publications speak of the "greenhouse effect," they refer not to the normal phenomenon described above but to a possible *accelerated* warming of the earth caused by increasing levels of the greenhouse gases.

The Greenhouse Effect and the Ocean Level No one knows with certainty what the effect of a small increase in average global temperature would mean. If warmer ocean water seeped under the ice shelves of the Antarctic, for example, could this cause the ice sheets to slip more rapidly into the oceans? If so, the additional water from the melting of all of this ice — both from the Antarctic and from Greenland — would cause the levels of the oceans to rise enough to inundate some of the world's coastal cities and

(Continued)

island nations in the Caribbean Sea and many in the Pacific Ocean.

The Greenhouse Effect and Climate Small changes in *average* air temperatures could have major changes in world climate patterns. Tropical storms and hurricanes would intensify and perhaps destroy the coral reefs that protect coastlines. During the last century, the average global air temperature rose by 0.4 °C. The average *annual* temperature of the earth rose more than this in 1990 alone, but there have always been yearly fluctuations. If the increase of the 1990s were to continue, however, huge ecosystems, like the Arctic ocean and nearby Arctic tundra as well as mangrove swamps and other coastal wetlands, would be under severe pressure. It is not that the earth has never before undergone large changes in overall climate with successful adaptations by most living species, but the greenhouse warming entails an unprecedented *rate* of change with which vulnerable species could not cope.

Carbon Dioxide as the Chief Greenhouse Gas The atmosphere receives each year an estimated 9×10^{15} mol of CO_2, largely from three sources: the respiration of plants and animals (over 93%), forest fires and the burning of other vegetable matter (2%), and the burning of fossil fuels — oil, coal, and natural gas (5%). Two natural "sinks" take away over 95% of the released CO_2, each contributing about equally. One is *photosynthesis,* the conversion of CO_2 and H_2O in plants with the aid of chlorophyll (the green pigment in plants) and energy from the sun. The other sink for CO_2 is its uptake by the oceans as it reacts with calcium (and magnesium) ion to form carbonate deposits. The total uptake leaves roughly 5% of the CO_2 in the air, a small but important net amount. The increased rate of burning of fossil fuels throughout the last several decades is believed to be chiefly responsible for this extra amount of atmospheric CO_2.

The additional carbon dioxide annually adds more to the increase in the concentration of the greenhouse gases than all of the others combined, about 55% of the total additions. See Figure 1. Studies of ocean sediment cores and tree rings indicate that the carbon dioxide level of the atmosphere fluctuated between 200 and 300 ppm until the end of the last century. What concerns scientists and government officials is the steady increase in the concentration of CO_2 in the atmosphere since the time when the burning of coal and oil became a major source of energy. Since about 1900, the average CO_2 level in the atmosphere has risen from 280 to 345 ppm, and some scientists see this going to 600 ppm in another 50 to 75 years, maybe in less time. This change alone could increase the average global temperature by 1.5 degrees. A phased reduction, worldwide, in CO_2 emissions was thus the subject of urgent international negotiations in the early 1990s.

Other Greenhouse Gases A group of synthetic compounds called the *chlorofluorocarbons* or CFCs contribute nearly 25% of the increase in the concentration of the greenhouse gases. As described in Special Topic 5.1, these gases threaten the stratospheric ozone shield. They also contribute to greenhouse warming (although not as much as once thought).

Methane — according to one estimate, about 15% of the load of greenhouse gases — originates from several sources, chiefly the action of bacteria in soil, in wetlands and rice paddies, and in the digestive tracts of animals. Oil and gas production releases some as well. There are several "sinks" for methane, but they do not remove it as rapidly as it is being generated. The atmosphere's concentration of methane has risen over the last two centuries from 750 ppb to about 1700 ppb.

Dinitrogen monoxide, N_2O, which contributes a little over 5% of the total greenhouse gas concentration, arises from the action of soil bacteria on the increased amounts of fertilizers being used worldwide.

Current Status as Estimated by Climatologists Has the "greenhouse effect" arrived? Is accelerated global warming now a certainty? These are urgently debated questions among climatologists. The United Nations Intergovernmental Panel on Climate Change (IPCC) is a group of scientists from several countries seeking answers. Its consensus — a "best guess" scenario — in the early 1990s was that the planet's average temperature will increase between 1.7 and 3.8 °C by the year 2100. The IPCC also predicted that the sea level would rise between 15 and 90 cm. These changes, said the IPCC report, are far beyond the changes that would otherwise occur if human activities were not a factor in world climate.

Atmospheric Haze and the Greenhouse Effect The problems of making long-range forecasts of changes in climate are huge, given our uncertain knowledge about how climate works. In its 1992 report, the IPCC said that as much as 50% of the greenhouse warming is presently being cancelled by the haze particles that blanket large areas of the world. These colloidal aerosol particles consist largely of microscopic droplets of sulfuric acid formed from sulfur oxides made by the combustion of sulfur-bearing fuels and vegetation. Some are also contributed

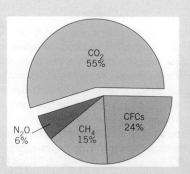

FIGURE 1

The contribution of each of the major greenhouse gases to the greenhouse effect as estimated for the period 1980–1990. ("Scientific Assessment of Climate Change," Intergovernmental Panel on Climate Change, United Nations publication.)

by volcanic eruptions. The most numerous of these aerosol particles are also the smallest and the most effective in reflecting incoming solar radiation back to space. This haze is not expected to compensate fully for the greenhouse warming; over a period of decades, the earth's temperature is expected to change upward.

SPECIAL TOPIC 8.5
ESTIMATING UNMEASURED ANION CONCENTRATION BY THE ANION GAP

The ions in blood that contribute the highest levels of concentration and charge are Na^+, Cl^-, and HCO_3^-. For example, the concentration of Na^+ normally is in the range of 135–145 meq/L; of Cl^-, 100–108 meq/L; and of HCO_3^-, 21–29 meq/L. In contrast, the levels of K^+, Ca^{2+}, and Mg^{2+} ions are on the order of only 2 to 5 meq/L each. The blood also carries varying concentrations of negatively charged ions of organic acids, such as the anions (the conjugate bases) of acetic acid, citric acid, and many others.

The levels of the organic anions tend to rise in several metabolic disturbances such as diabetes or kidney disease, but measuring these anions is difficult. The combined levels of the organic ions, however, can be estimated by calculating a quantity known as the **anion gap**, which can be defined by the following equation:

$$\text{anion gap} = \left(\frac{\text{meq of } Na^+}{L}\right) - \left(\frac{\text{meq of } Cl^-}{L} + \frac{\text{meq of } HCO_3^-}{L}\right)$$

To determine the anion gap in a patient's blood, a sample is analyzed for the concentrations in meq/L of Na^+, Cl^-, and HCO_3^-, which as we said are the most abundant ions and are also relatively easy to analyze. Then the concentration data are fed into this equation. For example, suppose that analyses found the following data: Na^+ = 137 meq/L; Cl^- = 100 meq/L; and HCO_3^- = 28 meq/L.

Then the anion gap is found by

$$\text{anion gap} = \left(\frac{137 \text{ meq}}{L}\right) - \left(\frac{100 \text{ meq}}{L} + \frac{28 \text{ meq}}{L}\right)$$
$$= 9 \text{ meq/L}$$

The normal range for the anion gap is 5 to 14 meq/L, so in our example the anion gap of 9 meq/L falls within the normal range. This 9 meq/L is accounted for by the presence of unmeasured ions of low concentration.

If metabolic disturbances cause the levels of organic anions to rise, the body must retain cations in the blood and excrete some of the more common anions such as Cl^- and HCO_3^- to maintain the absolute requirement that the blood be electrically neutral. In other words, negative organic ions tend to expel other negative ions but retain whatever positive ions are available. This is how the anion gap widens in metabolic disturbances that generate organic anions. The anion gap routinely rises above 14 meq/L in untreated diabetes.

You can see that by using rather easily measured data on the meq/L concentrations of Na^+, Cl^-, and HCO_3^-, and calculating the anion gap from these data, the clinical chemist can inform the health care professionals of any unusual buildups in anions which indicate possible disease. In severe exercise, the anion gap rises above normal, too, but it goes back down again in time. Thus an above normal anion gap has to be interpreted in the light of other facts.

SUMMARY

Ionization of water Trace concentrations of hydronium ions, H_3O^+, and hydroxide ions, OH^-, are always present in water. In neutral water, their molar concentrations are equal (and very low). In writing equations, we usually write H_3O^+ as H^+, calling the latter either the hydrogen ion or the proton. In explaining these reactions, however, we usually find it necessary to use the correct formula, H_3O^+.

Aqueous solutions of ions form either by the dissociation of ionic compounds as they dissolve or by the ionization of molecular substances as they react with water.

Electrolytes Solutes that are dissociated or ionized in water are electrolytes. Their solutions conduct electricity between a positively charged electrode, or anode, and a negatively charged electrode, or cathode. Cations that accept electrons from cathodes are reduced. Anions that deliver electrons to anodes are oxidized.

Chief ion producers Acids, bases, and salts are the common electrolytes. In the Brønsted concept, acids are chemical species that can donate hydrogen ions. Their aqueous solutions are also called acids. The five most common strong acids are hydrochloric, hydrobromic, hydroiodic, sulfuric, and nitric acid. All are monoprotic except sulfuric acid, which is diprotic. The chief acidic species in all is H_3O^+.

Bases are substances that accept hydrogen ions. Among the common bases are those that directly supply OH^- ion in water, like the hydroxides and oxides of sodium and potassium. Other common proton acceptors that readily take H^+ from H_3O^+ (but not from H_2O) are the carbonate and bicarbonate ions, the monohydrogen phosphate and phosphate ions, and ammonia.

Salts are ionic compounds involving any other ions but H^+, OH^-, or O^{2-}.

Strong and weak electrolytes; strong and weak acids and bases A strong electrolyte is one fully ionized or dissociated in solution, and all strong acids and strong bases are strong electrolytes. Salts in their molten states or in aqueous solutions are fully dissociated and are therefore all strong electrolytes. Remember that *strong* refers to percentage dissociation or ionization. Many salts are quite insoluble in water and so cannot supply a high concentration of ions. But what does dissolve of such salts is 100% dissociated.

Many molecular acids and bases ionize to a small percentage in water and so are weak electrolytes. Acetic acid and ammonia are examples. Many other molecular substances can be present in an aqueous system without being ionized and so are called nonelectrolytes. Pure water is a nonelectrolyte.

H_3O^+ is the strongest acid that can be present in water, and OH^- is the strongest base.

Conjugate acids and bases in the Brønsted concept All proton-transfer reactions can be expressed in terms of equilibria in which two acids and two bases appear. An acid and a base whose formulas differ only by one H^+ are a conjugate acid–base pair.

In the Brønsted concept, the terms *strong* and *weak* are enlarged. A strong acid is one that is a good proton donor. It readily gives up H^+. A strong acid has a weak conjugate base. A weak acid has a strong conjugate base. A strong base is one that strongly binds a proton, and a weak base is one that cannot hold H^+ very well. A strong base has a weak conjugate acid, and a weak base has a strong conjugate acid.

Reactions of aqueous acids The hydronium ion in strong aqueous acids reacts with:

Metal hydroxides, to give a salt and water

Metal carbonates, to give a salt, carbon dioxide, and water

Metal bicarbonates, to give a salt, carbon dioxide, and water

Metals, to give the salt of the metal and hydrogen gas

A solution of an acid is neutralized when any sufficiently strong proton-binding species is added in the correct mole proportion to make the concentration of hydrogen ion and hydroxide ion equal (and very small).

Carbonic acid and carbonates Carbonic acid, H_2CO_3, is both a weak acid and an unstable acid. When it is generated in water by the reaction of any stronger acid with a bicarbonate or a carbonate salt, virtually all of the carbonic acid decomposes to carbon dioxide and water, and most of the carbon dioxide fizzes out. The carbonate ion and the bicarbonate ion are both Brønsted bases, and the bicarbonate ion is involved in carrying waste carbon dioxide from cells, where it is made, to the lungs.

Ammonia and the ammonium ion Ammonia is a strong base toward H_3O^+ but a weak base toward H_2O. The ammonium ion is a strong acid toward OH^- but a weak acid toward H_2O. Ammonia can neutralize strong acids and the ammonium ion can neutralize strong bases.

Salts The chemical properties of salts in water are the properties of their individual ions. If the anion of the salt is the conjugate base of a weak acid, as HCO_3^- is the conjugate base of H_2CO_3, then the salt can neutralize strong acids. Thus bicarbonates, carbonates, acetates, and the salts of other organic acids supply Brønsted bases, namely their anions.

If the cation of the salt is the conjugate acid of a weak base, as NH_4^+ is the conjugate acid of NH_3, then the salt supplies a Brønsted acid in water.

Salts can be produced by any of the reactions of strong acids that were studied (and summarized, above) as well as by double replacement reactions. The solubility rules are guides for the prediction of their reactions. If a combination of oppositely charged ions can lead to an insoluble salt, an un-ionized species that stays in solution, or a gas, then the ions react.

If a different compound that can furnish an ion that is common to an ion of a salt already in solution is added to this solution, the solubility of the salt might be reduced enough to force it out of solution (common ion effect).

Equivalents of ions An equivalent (eq) of an ion is the number of grams of the ion that carry Avogadro's number of positive or negative charges. It is calculated by dividing the molar mass of the ion by the size of the charge it carries. The concentration of an ion in a dilute solution is often given in meq/L, where 1000 meq = 1 eq, and meq means milliequivalent.

REVIEW EXERCISES

The answers to Review Exercises whose numbers are in color are found in Appendix C. The answers to the other Review Exercises are found in the Study Guide that accompanies this book. The more challenging questions are marked with asterisks.

Electrolytes

8.1 What occurs chemically in the *ionization* reaction of HCl(*g*) when it dissolves in water? (Write the equation.) How does the *formation* of this solution differ from the way in which a solution of NaCl forms when crystals of NaCl are added to water?

8.2 K_2SO_3 is a crystalline solid and an *electrolyte*.
(a) What does *electrolyte* mean in this case?
(b) By what process, dissociation or ionization, does K_2SO_3 form its solution?

8.3 SO_3 is a colorless gas that dissolves in water to give a solution that conducts electricity. By what *process*—dissociation or ionization—does it form this solution? How can you tell?

8.4 The word electrolyte can be understood in two ways. What are they? Give examples.

8.5 To which electrode do anions migrate?

8.6 The cathode has what electrical charge, positive or negative?

8.7 An electrode that is positively charged attracts what kinds of ions, cations or anions?

8.8 Explain in your own words how the presence of cations and anions in water enables the system to conduct electricity.

8.9 When KOH(s) is dissolved in water, the solution is an excellent conductor of electricity, but when ethyl alcohol is dissolved in water, the solution won't conduct electricity at all. What does this behavior suggest about the structural natures of KOH and ethyl alcohol? (Notice that both appear to have OH groups in their formulas.)

$$
\begin{array}{c}
\text{H} \quad \text{H} \\
| \quad\; | \\
\text{H}-\text{C}-\text{C}-\text{O}-\text{H} \\
| \quad\; | \\
\text{H} \quad \text{H}
\end{array}
$$

Ethyl alcohol

8.10 When lithium nitrate dissolves in water, its crystals break up entirely into Li^+ and NO_3^- ions. Do we call this compound a *weak* or a *strong* electrolyte?

8.11 In the liquid state, tin(IV) chloride, $SnCl_4$, is a nonconductor. What does this suggest about the structural nature of this compound?

•8.12 Molten sodium chloride conducts electricity. At the cathode one of its ions is reduced, and at the anode the other ion is oxidized.
(a) Write an equation for the reaction at the cathode (using electrons as species in the reaction).
(b) Write an equation for the reaction at the anode (again, using electrons as species in the reaction).
(c) Write the overall reaction for the electrolysis.

8.13 What families of compounds are the principal sources of ions in aqueous solutions?

8.14 Review the differences between atoms and ions by answering the following questions.
(a) Are there any atoms that have more than one nucleus? If so, give an example.
(b) Are there any ions with more than one nucleus? If so, give an example.
(c) Are there any ions that are electrically neutral? If so, give an example.

(d) Are there any atoms that are electrically charged? If so, give an example.

Acids and Bases as Electrolytes

8.15 Write the equilibrium equation for the self-ionization of water, and label the ions that are present.

8.16 Tell whether each of the following solutions is acidic, basic, or neutral.
(a) $[H^+] = 7.9 \times 10^{-6}$ mol/L and $[OH^-] = 1.3 \times 10^{-9}$ mol/L
(b) $[H^+] = 1.0 \times 10^{-7}$ mol/L and $[OH^-] = 1.0 \times 10^{-7}$ mol/L
(c) $[H^+] = 8.6 \times 10^{-8}$ mol/L and $[OH^-] = 1.2 \times 10^{-7}$ mol/L

8.17 Salts are all crystalline solids at room temperature. Why do you suppose this is?

8.18 How did Arrhenius define an acid? A base?

8.19 What features do the common aqueous acids share?

8.20 In the context of acid–base discussions, what are two other names that we can use for "hydrogen ion"?

8.21 Acids have a set of common reactions, and so do bases, but not salts. Explain.

8.22 How does litmus paper work to tell whether a solution is acidic, basic, or neutral?

8.23 How did Brønsted define an acid? A base?

8.24 What are the names and the formulas of the aqueous solutions of the four hydrohalogen acids?

8.25 What is the difference between hydrochloric acid and hydrogen chloride?

8.26 In which species is the covalent bond to hydrogen stronger, in HI(g) or in $H_3O^+(aq)$? How do we know?

8.27 $HClO_4$ (perchloric acid) is a less common, strong acid. Represent its ionization in water by an equation.

8.28 Write the equation for the ionization of nitric acid in water.

8.29 Why is sulfuric acid with only 2 H in its formula called a diprotic acid but acetic acid, which has 4 H in its formula, is a monoprotic acid?

8.30 If we represent all diprotic acids by the symbol H_2A, write the equilibrium expressions for the two separate ionization steps.

8.31 Would the ionization of the second proton from a diprotic acid occur with greater ease or with greater difficulty than the ionization of the first proton? Explain.

8.32 Write the equations for the progressive ionizations of sulfuric acid. Include the names of the ions.

8.33 Write the equations for the progressive ionizations of phosphoric acid, including the names of the ions.

8.34 Which is the stronger acid in water, sulfurous acid or sulfuric acid? How can you tell?

$$
\begin{array}{cc}
\quad\;\; :\!\!\overset{..}{O}\!\!: & \quad\;\; :\!\!\overset{..}{O}\!\!: \\
\quad\;\; \| & \quad\;\; \| \\
\text{H}-\overset{..}{\underset{..}{O}}-\text{S}-\overset{..}{\underset{..}{O}}-\text{H} & \text{H}-\overset{..}{\underset{..}{O}}-\text{S}-\overset{..}{\underset{..}{O}}-\text{H} \\
& \quad\;\; \| \\
& \quad\;\; :\!\!\overset{..}{O}\!\!:
\end{array}
$$

Sulfurous acid Sulfuric acid

8.35 Compare the structures of chlorous acid, $HClO_2$, and chloric acid, $HClO_3$.

Chlorous acid   Chloric acid

Which is the stronger acid? How can you tell?

8.36 Write the equilibrium expression for the solution of carbon dioxide in water that produces some carbonic acid.

8.37 Write the equilibrium expressions for the successive steps in the ionization of carbonic acid.

8.38 KOH is a strong base and a strong electrolyte. What do these terms mean in connection with this compound?

8.39 Calcium hydroxide is only slightly soluble in water, and yet it is classified as a *strong* base. Explain.

8.40 Ammonia is very soluble in water, and yet it is called a weak base. Explain.

8.41 What are the names and formulas of two bases that are both strong and are capable of forming relatively concentrated solutions in water?

8.42 When carbon dioxide is bubbled into pure water to form a solution, it takes only time and the help of a little warming to drive essentially all of it out of solution again. When this gas is bubbled into aqueous sodium hydroxide, however, it is completely trapped by a chemical reaction. If we assume that the reaction involves CO_2 and NaOH in a mole ratio of one to one, what is the molecular equation for this trapping reaction?

8.43 What is meant by aqueous ammonia? Why don't we call it "ammonium hydroxide"?

8.44 Write the names and the formulas of the five strong acids that we have studied.

8.45 What are the four strong bases—both the names and formulas? Which are quite soluble in water?

Net Ionic Equations

8.46 Consider the following net ionic equation.

$$6H^+(aq) + Cu(s) + 3NO_3^-(aq) \longrightarrow$$
$$Cu^{2+}(aq) + 3NO_2(g) + 3H_2O$$

(a) Does it have material balance?
(b) Does it have electrical balance?
(c) Is it an *equation*?

°8.47 Complete and balance the following molecular equations, and then write the net ionic equations.
(a) $HNO_3(aq) + KOH(aq) \rightarrow$
(b) $HCl(aq) + NaHCO_3(aq) \rightarrow$
(c) $HBr(aq) + MgCO_3(s) \rightarrow$
(d) $HNO_3(aq) + KHCO_3(aq) \rightarrow$
(e) $HBr(aq) + NH_3(aq) \rightarrow$
(f) $HNO_3(aq) + Ca(OH)_2(s) \rightarrow$
(g) $HCl(aq) + Mg(s) \rightarrow$

°8.48 Complete and balance the following molecular equations, and then write the net ionic equations.
(a) $NaOH(aq) + H_2SO_4(aq) \rightarrow$
(b) $K_2CO_3(aq) + HNO_3(aq) \rightarrow$
(c) $NaHCO_3(aq) + HBr(aq) \rightarrow$
(d) $CaCO_3(s) + HI(aq) \rightarrow$
(e) $NH_3(aq) + HI(aq) \rightarrow$
(f) $Mg(OH)_2(s) + HBr(aq) \rightarrow$
(g) $Al(s) + HCl(aq) \rightarrow$

8.49 What are the net ionic equations for the following reactions of strong, aqueous acids? (Assume that all reactants and products are soluble in water.)
(a) With metal hydroxides
(b) With metal bicarbonates
(c) With metal carbonates
(d) With aqueous ammonia

8.50 Write net ionic equations for the reactions of all the water-insoluble group IIA carbonates, where you use $MCO_3(s)$ as their general formula, with nitric acid (chosen so that all the products are soluble in water).

8.51 If we let $M(OH)_2(s)$ represent the water-insoluble group IIA metal hydroxides, what is the general net ionic equation for all of their reactions with hydrochloric acid (chosen so that all the products are soluble in water)?

8.52 If we let $M(s)$ represent either calcium or magnesium metal, what net ionic equation represents the reaction of either with hydrochloric acid?

°8.53 Sodium and potassium in group IA are higher in the activity series than calcium and magnesium in group IIA.
(a) What does it mean to be higher in the activity series?
(b) If you check back to Figure 4.1*b* on page 81, you will see that sodium and potassium have lower ionization energies than calcium and magnesium. In what way does this fact correlate with their *higher* position in the activity series of the metals?

8.54 Zinc metal reacts more rapidly with which acid, 1 *M* hydrochloric acid or 1 *M* phosphoric acid?

8.55 How many moles of potassium bicarbonate can react quantitatively with 0.355 mol of HCl?

8.56 How many moles of sodium hydroxide can react quantitatively with 0.256 mol of H_2SO_4 (assuming that both H^+ in H_2SO_4 are neutralized)?

8.57 How many grams of sodium carbonate will neutralize 5.24 g of HCl?

8.58 How many grams of calcium carbonate react quantitatively with 7.98 g of HNO_3?

8.59 How many grams of sodium bicarbonate does it take to neutralize all the acid in 28.9 mL of 1.05 *M* H_2SO_4?

8.60 How many grams of potassium carbonate will neutralize all of the acid in 32.9 mL of 0.435 *M* HCl?

°8.61 How many milliliters of 0.165 *M* NaOH are needed to neutralize the acid in 28.6 mL of 0.212 *M* HNO_3?

°8.62 How many milliliters of 0.115 *M* KOH are needed to neutralize the acid in 14.6 mL of 0.161 *M* H_2SO_4?

°8.63 For an experiment that required 13.5 L of dry CO_2 gas (as

measured at 745 mm Hg and 24 °C), a student let 4.62 *M* HCl react with marble chips, $CaCO_3$.
  (a) Write the molecular and net ionic equations for this reaction.
  (b) How many grams of $CaCO_3$ and how many milliliters of the acid are needed?

*8.64 How many liters of dry CO_2 gas are generated (at 740 mm Hg and 25 °C) by the reaction of $Na_2CO_3(s)$ with 325 mL of 5.85 *M* HCl? Write the molecular and the net ionic equations for the reaction, and calculate how many grams of Na_2CO_3 are needed.

Strengths of Conjugate Brønsted Acids and Bases

8.65 What is the reason that OH^- is the strongest base we can have in water?

8.66 Why is H_3O^+ the strongest acid we can have in water?

8.67 What are the formulas of the conjugate acids of the following?
  (a) HSO_3^- (b) Br^- (c) H_2O (d) $C_2H_3O_2^-$

8.68 Write the formulas of the conjugate acids of the following.
  (a) HSO_4^- (b) HCO_3^- (c) I^- (d) NO_2^-

8.69 Write the formulas of the conjugate bases of the following.
  (a) NH_3 (b) HNO_2 (c) HSO_3^- (d) H_2SO_3

8.70 What are the conjugate bases of the following? Write their formulas.
  (a) H_2CO_3 (b) $H_2PO_4^-$ (c) NH_4^+ (d) OH^-

8.71 Which member of each pair is the stronger Brønsted base?
  (a) NH_3 or NH_2^- (b) OH^- or H_2O (c) HS^- or S^{2-}

8.72 Which member of each pair is the stronger Brønsted base?
  (a) Br^- or HCO_3^- (b) $H_2PO_4^-$ or HSO_4^-
  (c) NO_2^- or NO_3^-

8.73 Which member of each pair is the stronger Brønsted acid?
  (a) H_2CO_3 or HCl (b) H_2O or OH^-
  (c) HSO_4^- or HSO_3^-

8.74 Study each pair and decide which is the stronger Brønsted acid.
  (a) $H_2PO_4^-$ or HPO_4^{2-} (b) H_2SO_3 or HSO_3^-
  (c) NH_4^+ or NH_3

*8.75 If sodium phosphate and sodium hydrogen sulfate solutions are mixed in equimolar amounts of their solutes, the following ionic equilibrium is established.

$$HPO_4^{2-}(aq) + SO_4^{2-} \rightleftharpoons PO_4^{3-}(aq) + HSO_4^-(aq)$$

Which side is favored, the reactants or products? How can you tell?

*8.76 Aspirin is a weak acid. We can represent it as H(*Asp*), and it has a sodium salt that we can symbolize as Na(*Asp*). When the sodium salt of aspirin is given as a medication and it encounters gastric juice, which contains HCl(*aq*), the following ionic equilibrium is established (at least temporarily). Which side is favored, the reactants or products? How can you tell?

$$(Asp)^-(aq) + H_3O^+(aq) \rightleftharpoons H(Asp)(aq) + H_2O$$

*8.77 Suppose you are handed a test tube and told that it contains a concentrated solution of either ammonium chloride or potassium chloride. An aqueous solution of one of the substances that we have studied in this chapter could be added to the unknown solution as a test for deciding which of the two solutes is present. What is this test reagent, and what would you observe as a result of the test if the unknown contained ammonium chloride?

8.78 Complete and balance the following molecular equations.
  (a) $Na_2O(s) + H_2O \rightarrow$
  (b) $KNH_2(s) + H_2O \rightarrow$
  (c) $LiH(s) + H_2O \rightarrow$

Salts

*8.79 Write the names and formulas of three compounds that, by reacting with hydrochloric acid, give a solution of potassium chloride. Write the molecular equations for these reactions.

*8.80 Write the names and formulas of three compounds that will give a solution of sodium bromide when they react with hydrobromic acid. Write the molecular equations for these reactions.

8.81 Which of the following compounds are insoluble in water (as we have defined solubility)?
  (a) KOH (b) NH_4Cl (c) Hg_2Cl_2
  (d) $Mg_3(PO_4)_2$ (e) $NaBr$ (f) Li_2SO_4

8.82 Which of the following compounds are insoluble in water?
  (a) $(NH_4)_2SO_4$ (b) $NaNO_2$ (c) $LiBr$
  (d) $AgBr$ (e) $Ca_3(PO_4)_2$ (f) KNO_3

8.83 Identify the compounds that do not dissolve in water.
  (a) $BaCO_3$ (b) NH_4NO_3 (c) K_2CO_3
  (d) $PbCl_2$ (e) Na_2SO_4 (f) $LiC_2H_3O_2$

8.84 Which of the following compounds do not dissolve in water?
  (a) K_2CrO_4 (b) $AgBr$ (c) $FeCO_3$
  (d) $Na_2Cr_2O_7$ (e) Li_2CO_3 (f) NH_4I

*8.85 Assume you have separate solutions of each compound in the pairs below. Predict what happens chemically when the two solutions of a pair are poured together. If no reaction occurs, state so. If there is a reaction, write its net ionic equation.
  (a) KCl and $AgNO_3$ (b) KNO_3 and $MgCl_2$
  (c) $NaOH$ and H_2SO_4 (d) $Pb(NO_3)_2$ and $NaCl$
  (e) NH_4Cl and K_2SO_4 (f) Na_2S and $CuSO_4$
  (g) Na_2SO_4 and $Ba(NO_3)_2$ (h) $NaOH$ and HI
  (i) Na_2S and $NiCl_2$ (j) $AgNO_3$ and $NaBr$
  (k) $LiHCO_3$ and HI (l) $MgCl_2$ and KOH

*8.86 If you have separate solutions of each of the compounds given below and then mix the two of each pair together, what (if anything) happens chemically? If no reaction occurs, state so, but if there is a reaction write its net ionic equation.
  (a) H_2S and $Cu(NO_3)_2$ (b) $LiOH$ and HBr
  (c) Na_2SO_4 and $Ba(NO_3)_2$ (d) $Pb(C_2H_3O_2)_2$ and Na_2SO_4
  (e) $Ba(NO_3)_2$ and $NaCl$ (f) $KHCO_3$ and H_2SO_4
  (g) Na_2S and $Cd(NO_3)_2$ (h) $NaOH$ and HBr
  (i) $Hg(NO_3)_2$ and KCl (j) $NaHCO_3$ and HI
  (k) KBr and $NaCl$ (l) $Pb(NO_3)_2$ and Na_2CrO_4

*8.87 Soap is a mixture of the sodium salts of certain organic acids. One is sodium stearate, which we can represent as Na(Ste).

(a) Write the equilibrium expression for a saturated solution of this salt in water.

(b) What would happen to this equilibrium if a concentrated solution of sodium chloride were added to it?

(c) The NaCl solution need not be concentrated. Seawater is about 3% (w/w) NaCl, and soap doesn't work well when seawater is used. Suggest a reason.

Equivalents and Milliequivalents of Ions

8.88 The concentration of potassium ion in blood serum is normally in the range of 0.0035 to 0.0050 mol K^+/L. Express this range in units of milliequivalents of K^+ per liter.

8.89 The concentration of calcium ion in blood serum is normally in the range of 0.0042 to 0.0052 eq Ca^{2+}/L. Express this range in units of milliequivalents of Ca^{2+} per liter.

*8.90 The potassium ion level of blood serum normally does not exceed 0.196 g of K^+ per liter. How many milliequivalents of K^+ ion are in 0.196 g of K^+?

*8.91 The magnesium ion level in plasma normally does not exceed 0.0243 g of Mg^{2+}/L. How many milliequivalents of Mg^{2+} are in 0.0243 g of Mg^{2+}?

Ion Mobility and Ionic Reactions (Special Topic 8.1)

8.92 How can an acid, like citric acid, and a bicarbonate salt be stable in each other's presence since we know that acids and bicarbonates react to give an unstable acid?

8.93 What is the net ionic equation for the reaction between citric acid and sodium bicarbonate when something like an Alka-Seltzer tablet is dropped into water?

Hard Water (Special Topic 8.2)

8.94 What is hard water?

8.95 What are the formulas of the "hardness ions"?

8.96 What chemical property of these ions and of ordinary soap makes it difficult to use such soap in hard water?

8.97 What is temporary hard water? Why is it designated temporary?

8.98 What is permanent hard water?

8.99 What is meant by water softening?

8.100 Concerning washing soda as a water-softening agent,

(a) What is its molecular formula?

(b) What part of its formula is the active softening agent?

(c) What is the net ionic equation for its work in water where the hardness is caused by Ca^{2+}? By Mg^{2+}?

Solubility Product Constant (Special Topic 8.3)

8.101 Consider the equilibrium present in a saturated solution of the sparingly soluble salt, $PbSO_4$, when undissolved $PbSO_4$ is also present.

(a) Write an equation for the equilibrium.

(b) In terms of the molar solubilities of the ions of $PbSO_4$, what is the expression for the solubility product of this salt?

(c) Under what condition is the ion product of $PbSO_4$ equal to the solubility product constant of $PbSO_4$?

(d) The solubility of $PbSO_4$ in water at 25 °C is 7.9×10^{-4} mol/L. What is the value of K_{sp} for this compound?

Greenhouse Effect (Special Topic 8.4)

8.102 Give the names of the chief greenhouse gases and their main sources.

8.103 Explain briefly how the greenhouse effect works. What happens that tends to promote global warming?

8.104 What has largely been responsible for the steady increase in the atmospheric concentration of CO_2 during this century?

8.105 What are the two principal ways by which CO_2 is removed from the atmosphere?

8.106 What causes the haze that climatologists see as partially reversing the greenhouse effect? How does haze cause this reversal?

The Anion Gap (Special Topic 8.5)

8.107 A patient on a self-prescribed diet consisting essentially only of protein was found to have the following blood analyses after two weeks of the diet: Na^+, 174 meq/L; Cl^-, 135 meq/L; and HCO_3^-, 20 meq/L. Calculate the anion gap. Does it suggest a disturbance in metabolism?

Additional Exercises

*8.108 Consider the following compounds:

$$Ca_3(PO_4)_2 \quad CaHPO_4 \quad Ca(H_2PO_4)_2$$

(a) How can the *relative* molar solubilities of these compounds be *predicted* rather than looked up in a table?

(b) Which would have the highest molar solubility in water? (Write its formula.)

(c) Which has the lowest molar solubility in water? (Write its formula.)

*8.109 Compounds A and B are both white solids that dissolve in water. One is an ionic compound and the other is molecular. Discuss an experiment that could be conducted to find out which is molecular and mention any possible drawbacks to the kind of experiment you select. (How might the experiment give ambiguous results?)

*8.110 A white solid is either KNO_3 or K_2O. Describe an experiment that you could carry out using only test tubes and aqueous solutions that would tell which compound is present. Assume that the lab has whatever other chemicals you need.

*8.111 A white solid is either Na_2CO_3 or $NaHCO_3$. A sample of the solid with a mass of 0.144 g requires 15.0 mL of 0.114 M HCl to react with it fully until the exact point is reached when no more CO_2 forms. Which compound is it?

*8.112 A white solid was a mixture of K_2CO_3 and KNO_3. A 0.624-g sample of the mixture consumed 21.5 mL of 0.156 M HCl before CO_2 stopped evolving. How many grams of K_2CO_3 were in the mixture?

REACTION KINETICS AND CHEMICAL EQUILIBRIA. ACID–BASE EQUILIBRIA

9

If the water in Lake Michigan, seen here in a satellite photo, were 1 M NaOH (which it isn't, of course), the entire lake would contain no more than 50 g of hydrogen ion, H^+. Yet trace concentrations of this ion affect chemical equilibria vital to life. We'll learn how to describe such concentrations and how the body controls them in this chapter.

9.1

FACTORS THAT AFFECT REACTION RATES

The rate of a chemical reaction is affected by the nature of the reactants, their concentrations, the temperature, and sometimes by outside agents.

The field of chemistry that deals with the rates of chemical reactions is called **kinetics.** It gives us the factors that affect the rates of reactions and an understanding of how these factors work. We need this information to help us understand the relative rates of metabolic reactions at the molecular level of life.

The Chemical Nature of the Reactants Is the Most Important Factor That Affects Reaction Rates When you turn on the gas of a bunsen burner and ignite it, it's obvious that a chemical reaction is occurring very rapidly. The reaction is the oxidation of methane, CH_4.

$$CH_4(g) + 2O_2(g) \longrightarrow CO_2(g) + 2H_2O(g)$$

When iron rusts, a similar oxidation occurs.

$$4Fe(s) + 3O_2(g) \longrightarrow 2Fe_2O_3(s)$$

But you know that this reaction is not like the burning of methane; the rusting of iron is a slow reaction. Clearly, *what* reacts is the first factor to determine reaction rate.

The Physical States of the Reactants Affect Reaction Rates A *homogeneous reaction* is one in which all of the reactants are in the dissolved state either in a liquid solvent or in the gaseous state. In a *heterogeneous reaction,* at least one reactant is not dissolved in the same medium as the others. *Homogeneous reactions are faster,* provided that all other factors are equal — the reactants, the temperature, and the relative amounts. The rate of a heterogeneous reaction is limited by the total surface area of the undissolved reactant because the reaction can occur only at the surface.

One place where this factor impinges at the molecular level of life occurs with the fats and oils in our diets, like butter or salad oil. They are insoluble in water, but the digestive juices are dilute aqueous solutions. This could present a problem for our digesting fats and oils, because their globules can be chemically attacked largely only at their surfaces. The "trick" is to break the globules up into innumerable microglobules. Powerful surface-active agents secreted in the bile juice enable this emulsification of fats and oils. Each resulting tiny globule is in the colloidal state and so has a very small surface area, *but the total surface area of all of the globules is immense*. Digestive reactions affecting the dietary fats and oils can therefore occur quite rapidly.

■ If an egg-sized lump of coal is pulverized to the fineness of soot, the total surface area of all of the tiny particles adds up to two to three football fields.

The study of reaction kinetics is difficult for heterogeneous reactions because it is hard to control surface area. We will continue our study, therefore, with the kinetics of homogeneous reactions.

Concentration Is Another Factor That Affects Reaction Rates "Rate" is like "speed." You are familiar with the speed of travel and its most familiar units (at least in the U.S.), miles per hour. Thus speed or rate is always expressed as a ratio.

$$\text{Speed of travel} = \text{rate of vehicle motion} = \frac{\text{change in position}}{\text{time}} = \frac{\text{miles}}{\text{hour}}$$

Similarly, the rate of a chemical reaction is a ratio but, of course, the units are different.

$$\text{Rate of reaction} = \frac{\text{change in concentration}}{\text{time}} = \frac{\text{mol/L}}{\text{s}}$$

The "concentration" referred to in this equation is that of one of the species in the reaction, nearly always one of the reactants. If it's a reactant, then the rate of the reaction means the rate at which the molarity of the reactant decreases.

One of the important properties of chemical reactions is that their rates are not constant. They change continuously as the reaction proceeds until the reaction finally stops. We can best illustrate this by supposing the simplest possible reaction, the hypothetical conversion of substance X to Y.

$$X \longrightarrow Y$$

We can express the molar concentrations with the usual symbols, $[X]$ and $[Y]$. Initially, $[Y]$ is zero and $[X]$ is at its highest. As the reaction proceeds over time, however, the value of $[X]$ decreases and that of $[Y]$ increases. Figure 9.1 plots these changes in concentration with time for the hypothetical reaction. Early in the reaction, where the curves are most steep, the rates of disappearance of X and appearance of Y are rapid. Late into the reaction where the curves flatten out, further changes in concentrations take longer and longer, which is what we mean by saying that the *rate* of the reaction decreases with time.

In our hypothetical reaction, the rate at which $[X]$ decreases is identical to the rate at which $[Y]$ increases. In more complex reactions, a relationship this simple almost never is observed. Yet, what is observed experimentally is that the rate of a reaction is proportional to some mathematical combination of the molarities of the reactants, each molarity raised to some power. In a slightly more complex but still hypothetical reaction of the type

$$X + Y \longrightarrow Z$$

the rate of the reaction might turn out to be related to concentrations in any one of the ways suggested by varying values of the exponents in the following equation.

$$\text{rate} \propto [X]^x[Y]^y$$

The values of x and y are exponents *that must be discovered* by doing experiments that measure the actual rates at different concentrations. The point made by this discussion, the

■ The brackets, [ ], denote *moles per liter* concentrations.

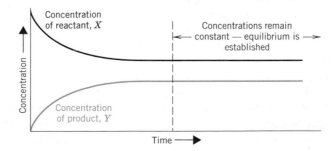

FIGURE 9.1
Changes in the concentrations of X and Y over time in the hypothetical reaction: $X \rightarrow Y$.

chief point that we must carry forward, is that *one of the factors that affects reaction rates is concentration*. In many reactions, for example, simply doubling the concentration of one reactant doubles the rate.

■ Recall that *metabolism* refers to the sum total of all of the reactions of the body.

Temperature Is Another Factor That Affects Reaction Rates   As a rough rule of thumb, an increase in the temperature of a reaction mixture by only 10 °C can be expected to double or triple the rate of a reaction. All other factors being equal, food spoils faster at room temperature than in the refrigerator. Metabolism also accelerates during a fever. Medical scientists tell us that an increase in body temperature of about 0.5 °C increases the metabolic rate so much that the oxygen requirement of the body increases by 7%. The heart responds to the increased demand for oxygen by speeding up its heartbeat, which puts a strain on the heart. Although a higher metabolic rate during a fever is part of the body's disease-fighting mechanism, a prolonged episode must be avoided.

■ With an increase in the heart beat, blood moves through the lungs faster, where it is oxygenated and where waste CO_2 is discharged.

"Outside Agents" Called *Catalysts* Also Accelerate Reactions   How the phenomenon was discovered no one knows, but many reactions go at higher rates in the presence of small amounts of what seems to be a nonreactant. At least, the substance causing the enhanced rate can be recovered unchanged at the end of the reaction. Outside agents that, in relatively trace concentrations, accelerate reactions without themselves being changed are called **catalysts**, and the phenomenon of rate acceleration by such a substance is called **catalysis.**

   Virtually all biochemical reactions *require* catalysts. In fact, the body's mainline mechanism for controlling metabolism is by controlling the availability of its cellular catalysts, called **enzymes.** Nearly all known enzymes are in a large family of biochemicals called *proteins*.

   You can easily observe the action of an enzyme if you have access to a drop of blood (or a slice of raw liver) and a dilute solution of hydrogen peroxide, which is sold in drugstores as a bleach and disinfectant. Hydrogen peroxide spontaneously (but slowly) decomposes as follows.

$$2H_2O_2 \longrightarrow 2H_2O + O_2$$

This reaction is so slow at room temperature that if you look at a sample of hydrogen peroxide, you won't notice any bubbling action. However, if you add a drop of blood or a tiny slice of liver to the hydrogen peroxide, an enzyme called *catalase* catalyzes this decomposition, and you'll see a vigorous evolution of oxygen. The frothing that you see if you ever use hydrogen peroxide to disinfect a wound is the same reaction. Hydrogen peroxide is toxic, and it forms in certain reactions of the metabolism of oxygen. Therefore catalase acts to detoxify hydrogen peroxide. The enzyme is not itself permanently changed by this work.

   Catalase makes a reaction occur much faster than it would at the same temperature in the absence of a catalyst. A catalyst can also cause a reaction to take place at a much lower temperature than otherwise. A classic illustration is the decomposition of potassium chlorate into potassium chloride and oxygen. Notice in the following equations how the temperature at which the reaction can occur varies according to the presence or absence of manganese dioxide, a catalyst for the decomposition.

■ Sometimes the special conditions for a reaction are written above or below the arrow in the equation.

Without MnO_2, the temperature has to be 420 °C
$$2KClO_3 + heat \xrightarrow{420\ °C} 2KCl + 3O_2$$

With MnO_2, the temperature can be only 270 °C
$$2KClO_3 + heat \xrightarrow[MnO_2]{270\ °C} 2KCl + 3O_2$$

The rates of the evolution of oxygen are approximately equal under the sets of conditions given here. But the catalyst permits it to happen at a much lower temperature.

   You can see why catalysts would be immensely important in industries that make large quantities of chemicals, like plastics or synthetic gasoline. Energy costs money, and catalysts in trace quantities lower this cost by reducing the energy needed. Catalysts thus make many products less expensive and indirectly extend the world's supplies of energy.

9.2

THE KINETIC THEORY AND CHEMICAL REACTIONS

Collision theory helps to explain the factors that affect reaction rates.
In chemical reactions, electrons and nuclei become reorganized relative to each other. The view provided by the kinetic theory, namely that particles in fluids (gases or liquids) are in constant random motion, helps us see how such reorganizations are made possible. The kinetic theory will help us understand how concentration and temperature are factors that affect reaction rates.

■ The original kinetic theory concerned only an *ideal gas,* but the idea of particles in motion applies to all fluids.

FIGURE 9.2
Steel wool, after being heated to redness in a flame, burns spectacularly when dropped into pure oxygen.

One of the Major Theories in Kinetics Is *Collision Theory*

For the particles of two reactants to change each other chemically, they have to collide. Only by a collision can the electrons and nuclei of the reactant particles be forced into the new arrangements of the products. This is the heart of a major theory concerning how reactions occur, **collision theory.** A given reaction will be faster the more frequently the reactant particles collide. We can define a *collision frequency* as the total number of collisions occurring between the reactant particles per unit of volume per second. The only way to increase the collision frequency *without changing the temperature* of the reacting mixture is to increase the concentrations of the reactant particles. It's like going from a stroll down a lonely country lane to an aisle of a very crowded store. An increase in the concentration of people in motion causes an increase in the "excuse-me" kind of bumps and collisions. If the molar concentration of one reactant is doubled, the frequency of all collisions must double because there are twice as many of its particles *in the same volume.* This is why reaction rate is a function of reactant molarities.

One of the spectacular results of increasing the concentration of a reactant can be observed by comparing the rate at which something burns in air with the rate of the same reaction in pure oxygen. Air is about 21% oxygen, which means that in every 100 liters of air there are 21 liters of oxygen (and virtually all of the rest is nitrogen). You can make steel wool glow and give off sparks if you direct a bunsen burner flame at it when it is in air. But if glowing steel wool is thrust into pure oxygen, it bursts into flame (see Fig. 9.2). The higher oxygen concentration makes the difference in rate.

Someone has estimated that if the air we breathe were 30% oxygen instead of 21%, no forest fire could ever be put out, and eventually all of the world's forests would disappear by uncontrollable fire. The higher concentration of oxygen would accelerate combustion too much.

■ Combustion means burning, a reaction with oxygen so rapid that heat, light, and some sound are rapidly released.

An Increase in Temperature Also Increases the Collision Frequency

The kinetic theory showed us that the temperature of a gas is proportional to the average molecular kinetic energy of the gas molecules. For a given gas, we saw that the increase in energy came as a result of an increase in molecular speeds, not molecular mass. The increase in speed must result in more frequent collisions, so this helps us understand why temperature is a factor in reaction rate.

The Minimum Collision Energy for a Reaction Is Called the *Energy of Activation*

Generally, very light tap-like collisions do not result in a reaction between the colliding particles. For each chemical reaction, there is a certain minimum collision energy that must develop before the new chemical bonds in the products can form. This minimum energy is called the reaction's **energy of activation** symbolized as E_{act}. Figure 9.3a illustrates what this means.

The vertical axis in Figure 9.3 represents changes in the *fraction* of all of the collisions that are occurring. The horizontal axis corresponds to values that the collision energy can have, ranging from zero on the left to very large values — approaching infinity — on the right. The reactant particles have a large range of speeds, ranging from very low values (even a zero

■ The *collision energy* is the sum of the kinetic energies of the colliding particles.

FIGURE 9.3

Energy of activation. (*a*) Only a small fraction of all collisions, represented by the ratio of areas, $A/(A + B)$, has enough energy for reaction. (*b*) This fraction greatly increases when the temperature of the reacting mixture is increased.

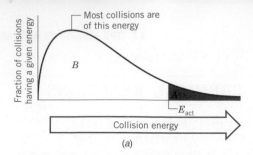

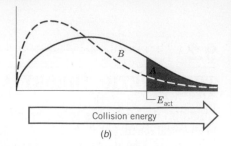

(*a*)　　　　　　　　　　　　(*b*)

■ All of the fractions represented by points on the curve add up to unity, the total number of all collisions.

value for some, for a moment) to very high speeds. Therefore some collisions will be mere taps, whereas others will be *extremely violent.*

Figure 9.3*a* shows that in a sample of reactant particles, some collisions will be such slight taps that virtually no distortions of electron clouds can occur. Little if any kinetic energy changes into potential energy. However, the fraction of all collisions that have zero collision energy is essentially zero. Then, to follow the curve to the right, as the collision energy increases, the fraction of the collisions having a particular energy also increases. We eventually reach a maximum value. Beyond it, collisions with increasingly higher energies become less and less likely. The fractions with very high energies decline and the curve moves back down to the baseline.

■ The *total* energy—kinetic plus potential—stays constant throughout the change, but it becomes apportioned differently. The molecular kinetic energy of the collision changes to the internal energies of the particles.

Eventually we reach a value of collision energy that provides the exact amount of energy to enable the electron–nuclei rearrangement, the chemical reaction, to occur. This particular collision energy is the energy of activation in the figure. When collisions have this much energy, or more, the reaction can take place. Collisions with energies less than the energy of activation cannot cause a chemical change. The colliding particles just bounce away chemically unchanged.

It isn't enough, of course, to have sufficient energy. The colliding particles must hit each other just right, much as the runners in a relay race have to pass the baton correctly regardless of how fast or slowly they are moving at this critical moment in the race.

■ A *successful* collision is simply one that produces products.

The **rate of a reaction** is the number of *successful* collisions that occur each second in each unit of volume of the reacting mixture. Generally, only a small fraction of the collisions have enough energy to be successful. This fraction is represented in Figure 9.3 by the ratio of the shaded area marked A to the total area under the curve, $(A + B)$. Thus this ratio is $A/(A + B)$.

You can see from the figure that if the energy of activation were very high, the shaded area on the right would be even smaller, so the fraction $A/(A + B)$ would be much smaller. A high energy of activation, in other words, means a small fraction of successful collisions and a slow rate of reaction.

On the other hand, a reaction with a very small energy of activation would have a large fraction of successful collisions and a high rate of reaction. In the extreme, the fraction might equal 1, which would mean that *every* collision would be successful no matter how low the energy of the collision. In practical terms, such a reaction would be extremely rapid—an explosion, essentially—because it would mean that simply mixing the reactants causes instantaneous change.

At the other extreme, the energy of activation could be so high that the fraction $A/(A + B)$ might be virtually zero. Now no reaction ever occurs, and the "reactants" are eternally stable in each other's presence.

This analysis tells us that the rate of a reaction depends greatly on its energy of activation. A high energy of activation means a slow rate, and a low energy of activation means a fast rate.

■ The energy of activation sometimes changes as the temperature changes, but usually by relatively little. It's generally safe to say as a rule of thumb that E_{act} is not affected by temperature under ordinary conditions.

Figure 9.3*b* shows another way in which temperature affects reaction rate. An increase in temperature not only increases the frequency of *all* collisions, it hugely increases the fraction of *successful* collisions. The calculations of the kinetic theory show that the curve of Figure 9.3*a* flattens as the temperature increases, and the curve's maximum shifts to the right. However, *the energy of activation stays put.* Consequently, the area under the curve to the right of E_{act}, namely area A, grows as the curve flattens, and you can see that $A/(A + B)$, therefore, increases. The increase in this ratio, of course, means a more rapid rate. We cannot go into the mathematical details, but they show that the influence of temperature on reaction rate is felt far more through its effect on the ratio $A/(A + B)$ than its enhancement of the total collision frequency.

Kinetic Energy Becomes Chemical Energy during Collisions between Reactant Particles Nature operates under the law of conservation of energy. When two moving particles are about to collide, each has a certain kinetic energy. One could imagine a collision in which *both* particles stop. (This happens all the time in highway accidents.) If they stop, their kinetic energies go to zero, because $KE = (1/2)mv^2$ and the value of v is now zero. Where did the kinetic energy go? Is it lost? If so, what of the law of conservation of energy?

Actually, the energy that existed as kinetic energy is not lost; it's transformed into the potential energy of distorted electron clouds. Relatively stable electron–nuclei arrangements are twisted temporarily into less stable arrangements that cannot last. They might, of course, twist back to the original electron–nuclei arrangements of the reactants. When this happens, and it often does, the effect is that the colliding particles simply bounce off each other. The potential energy in the temporary and unstable arrangement at the instant of collision reconverts to kinetic energy much as a bouncing ball can hit a sidewalk, briefly stop, then bounce away — still as a ball and not as some other substance. In other words, some collisions lead to no permanent change.

Following other, perhaps more violent collisions, reactant particles, during the brief moment of deformation, go through a rearrangement of their electrons and nuclei. As the system relaxes into a more permanent state, product particles form and we say that a *successful* collision has occurred. Thus the conversion of the kinetic energy of collision into the potential energy of distorted, unstable configurations makes a chemical reaction possible.

A Slow Reaction Could Still Be Very Exothermic We must make an important distinction now between a reaction's energy of activation and its *heat of reaction.* We'll use a *progress of reaction diagram* to do this (see Fig. 9.4). In such a diagram, the vertical axis gives *relative* values of the potential energies of the substances, either the reactants or the products, depending on which part of the plot gets our attention. The horizontal axis simply shows the direction of the chemical change.

We'll use the combustion of carbon and oxygen to illustrate how to follow a progress of reaction diagram. It's an exothermic reaction.

$$C + O_2 \longrightarrow CO_2 + \text{heat of reaction}$$

Begin in Figure 9.4 on the left at site A with the unchanged reactants, carbon and oxygen. We know that these are quite stable together at or near room temperature. Coal (mostly carbon), after all, can be stored in air (with its 21% oxygen). To initiate a reaction between carbon and oxygen, we have to heat them (ignite the system). Heat gives their particles higher kinetic energies, and more and more collisions become closer to being successful. In the diagram, we are moving up the curve from A, because the potential energy of the system is increasing. We are climbing an "energy hill."

If a collision provides sufficient energy of activation, we are at the top of the energy barrier at location B in Figure 9.4. The electrons and nuclei of the reactants can now rearrange to give molecules of carbon dioxide. Some of the potential energy in the complex of electrons and nuclei at the top of the energy hill now changes into the kinetic energy of newly forming molecules of carbon dioxide.

There is quite a drop in potential energy now as the reaction progresses to C in the diagram. Some of this potential energy goes to repay the cost of climbing the energy hill, but there is a net excess that is liberated as heat. This net energy difference between the reactants and the products is called the **heat of reaction.** The CO_2 molecules emerge from B going to C with greater average molecular kinetic energy and so the system becomes hotter. The heat of reaction in this chemical change comes from the conversion of some of the chemical energy in the electron–nuclei arrangements of carbon and oxygen into the kinetic energy of the molecules of CO_2.

As we know, once the reaction of carbon and oxygen starts, it continues spontaneously.

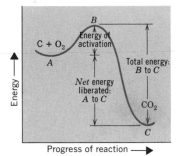

FIGURE 9.4

Progress of reaction diagram for the exothermic reaction of carbon with oxygen that produces carbon dioxide.

FIGURE 9.5

Because energies of activation, not heats of reaction, dominate reaction rates, it is possible to have (a) a fast reaction with a small heat of reaction or (b) a slow reaction with a large heat of reaction. We can tell that the rate in part a is faster than in part b because of its smaller E_{act}.

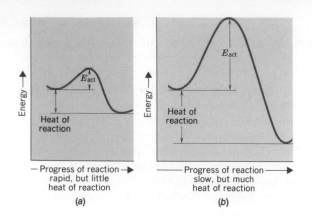

(a)

(b)

The reaction is exothermic, and some of the energy represented in Figure 9.4 in the change from B to C activates (ignites) still unchanged particles of the reactants.

One might think that a rapid exothermic reaction always means a high heat of reaction, but this is not necessarily so. The oxidation of 1 mol of iron, a very slow reaction,

$$4Fe + 3O_2 \longrightarrow 2Fe_2O_3$$

liberates over twice as much heat as the oxidation of one mole of carbon. Yet iron oxidizes very slowly, and carbon oxidizes so rapidly that we call it combustion. The reason lies in the energy barriers, the energies of activation. The energy barrier in the oxidation of iron is considerably higher than for the oxidation of carbon, so the rate is slower.

Figure 9.5 explains this in terms of two hypothetical reactions shown in progress of reaction diagrams. The reaction on the right has the higher heat of reaction but also a higher energy of activation. It is the slower reaction. The one on the left has the much lower energy of activation, so its rate will be much faster. Yet it gives off less heat. Thus there is really no simple relationship between how rapidly an exothermic reaction occurs and how large its heat of reaction is. In living systems, there are many highly exothermic changes that happen extremely slowly by themselves. Living systems have ways to switch slow reactions to higher rates, and the chemistry of heredity as well as the work of hormones have much to do with this.

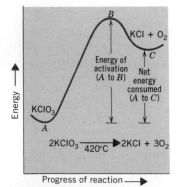

FIGURE 9.6

Progress of reaction diagram for the endothermic conversion of potassium chlorate into potassium chloride and oxygen.

No Net Release of Heat Occurs in Endothermic Reactions

Not all reactions liberate energy. Many won't occur without a continuous input of energy. The conversion of potassium chlorate into potassium chloride and oxygen, mentioned earlier, is an example. Now the heat of reaction has to be shown as if it were a reactant, because the reaction is endothermic.

$$2KClO_3 + \text{heat of reaction} \longrightarrow 2KCl + 3O_2$$

The energy relationships for this reaction are shown in the progress of reaction diagram of Figure 9.6. A good share of the energy of activation (A to B in this figure) is permanently retained by the product molecules as internal or potential energy. The net energy retained is represented in Figure 9.6 by the vertical distance between A and C.

In an endothermic reaction there is a net conversion of kinetic energy (supplied by the steady input of heat) into the potential or chemical energy of the products. Thus you can see that both exothermic and endothermic reactions have energies of activation. But in the exothermic reaction there is still a net release of energy, whereas in the endothermic reaction there is a net absorption of energy.

Catalysts Decrease Energies of Activation by Permitting Lower-Energy Paths

With or without a catalyst, the decomposition of potassium chlorate is endothermic, as we saw in Figure 9.6. Figure 9.7 shows the progress of reaction diagram for the same reaction except that the catalyst MnO_2 is present. It illustrates one of the major facts about the entire phenomenon of catalysis, *a catalyst does not change the heat of reaction.* What the catalyst somehow does is enable the reactants to change to products by a different pathway, by letting different lower-energy collisions or a different sequential set of lower-energy collisions lead to products. This is why the reaction happens faster. The overall energy barrier is reduced, so the fraction of all collisions that have enough energy is larger with the catalyst than without.

It's common (and permissible) to say that a catalyst lowers the energy of activation for a reaction, but this refers to the *overall* change. It does not mean that the overall reaction otherwise occurs by the *same* path. We'll see later, for example, that enzymes *temporarily* react with one (or more) reactant molecules, which then change to products on the enzyme's surface. The product molecules are then released as the enzyme (catalyst) reverts to its original state. We said earlier that a catalyst is not itself *permanently* changed in the reaction it catalyzes, but this does not rule out temporary changes. Taking a hypothetical reaction

$$X + Y \longrightarrow Z$$

suppose the *direct* reaction of X and Y has a virtually insurmountable energy of activation. X, however, might react readily with a catalyst to form something that holds a molecule of X, *but now with an altered electron cloud around it.* In this form, a collision with a molecule of Y might more easily (i.e., with a lower E_{act}) accomplish the electron–nuclei rearrangement necessary to form a molecule of Z. The series would be

$$X + \text{catalyst} \longrightarrow \text{catalyst} — X$$
$$\text{catalyst} — X + Y \longrightarrow \text{catalyst} — X — Y$$
$$\text{catalyst} — X — Y \longrightarrow \text{catalyst} + Z$$

Despite having three steps in the pathway from reactants to products, no step has such a high E_{act} that the overall rate cannot exceed that of the direct pathway.

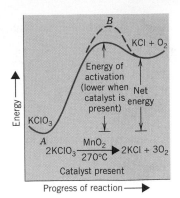

FIGURE 9.7
Progress of reaction diagram for the endothermic, catalyzed conversion of potassium chlorate into potassium chloride and oxygen. The dashed-line curve shows where the energy barrier was in the uncatalyzed reaction sketched in Figure 9.6. Notice that the net energy consumed, the heat of reaction, is identical to that of the uncatalyzed reaction, but the energy of activation is lower.

9.3

CHEMICAL EQUILIBRIA REVISITED

An equilibrium law exists for every chemical equilibrium.

One of the major facts about our world is that all chemical systems tend toward equilibrium. We have already used the concepts and vocabulary of equilibria, starting with the liquid–vapor equilibrium and vapor pressure and continuing with the heterogeneous equilibrium of a saturated solution of a slightly soluble salt. We'll pursue here a more quantitative treatment of *homogeneous* equilibria, those like acid–base equilibria in which all species are in the same physical phase, an *aqueous* solution.

In 1867, C. M. Guldburg and Peter Waage, two Norwegian scientists, discovered a relationship concerning the molar concentrations of the species in a chemical equilibrium that we now call the *equilibrium law* for the system. Let's illustrate this first with a general equation. If reactants A and B are in equilibrium with products C and D, according to the equilibrium equation (where a, b, c, and d are the coefficients):

$$aA + bB \rightleftharpoons cC + dD$$

then the **equilibrium law** for the system is

$$K_{eq} = \frac{[C]^c[D]^d}{[A]^a[B]^b} \tag{9.1}$$

The equilibrium law for any chemical equilibrium is thus an equation patterned after Equation 9.1, and there is a unique equation, a unique equilibrium law, for each and every chemical equilibrium. K_{eq}, called the **equilibrium constant** and calculated from measured molar concentrations, therefore, has a different value for each chemical equilibrium.

■ Equation 9.1 is often called the **law of mass action** (for historical reasons of no importance here), the ratio following the equals sign is the *mass action expression,* and the numerical value of this expression is called the *reaction quotient.*

Because equilibria can be shifted by increasing or decreasing the temperature, the value of K_{eq} depends on the temperature, and its value at 25 °C, for example, is not the same as at 30 °C.

The Size of K_{eq} Indicates the *Position of Equilibrium*

Notice in Equation 9.1 that the products appear in the numerator and the reactants in the denominator. By universal convention, the value of the equilibrium constant is always meant to correspond to this arrangement. This lets us always associate the size of K_{eq}, regardless of the reference table, with the *position* of the equilibrium.

The value of K_{eq} is small, less than 1, when the denominator in the equilibrium law is larger than the numerator. The denominator carries the reactant concentrations, so a larger denominator means that *reactants'* concentrations are greater than those of the products. A small value of K_{eq} means that the reactants are favored at equilibrium.

Conversely, a value of K_{eq} greater than 1 means that the *products* are favored, because their molarities appear in the numerator of Equation 9.1. We can summarize the relationships of K_{eq} to positions of equilibria as follows.

$K_{eq} < 1$, reactants are favored at equilibrium
$K_{eq} > 1$, products are favored at equilibrium

In all the equilibria we will study, those of weak acids and bases or the self-ionization of water, K_{eq} will be less than 1. In aqueous acetic acid, for example, we have the following equilibrium,

$$HC_2H_3O_2(aq) + H_2O \rightleftharpoons H_3O^+(aq) + C_2H_3O_2^-(aq)$$

Putting the molar concentrations of the products in the numerator (and noting that all chemical coefficients are 1), we find the equilibrium law with the known value of K_{eq} to be

$$K_{eq} = \frac{[H_3O^+][C_2H_3O_2^-]}{[H_2O][HC_2H_3O_2]} = 3.2 \times 10^{-7} \quad (25 °C)$$

We now at last have a number, K_{eq}, that *quantitatively* indicates how weak acetic acid is as an acid. K_{eq} is small for acetic acid, 3.2×10^{-7}, so the molarities of the product ions must be smaller than those of the un-ionized reactants. All that we have been able to say thus far about the strength of acetic acid is that it has a low percentage ionization, but this is different for different concentrations of the acid. K_{eq} is independent of concentration, however, so K_{eq} has the advantage over percentage ionization for comparing the relative strengths of acids.

K_{eq} Remains Constant Even When the Equilibrium Shifts

If we add sodium acetate to the equilibrium in aqueous acetic acid, we add the acetate ion. The common ion effect now operates — a special case of Le Châtelier's principle — and the equilibrium shifts to the left to use up as much of the added acetate ion as possible. This shift changes the values of every term in the equilibrium law. However, an increase in $[C_2H_3O_2^-]$ is offset by a decrease in $[H_3O^+]$ and comparable changes in the denominator as the equilibrium adjusts to the stress of added acetate ion. When the new equilibrium values of concentrations are measured and then inserted into the equilibrium law equation, the calculated K_{eq} is the same as before.

K_{eq} is a constant in the midst of other changes. Guldburg and Waage were the first to realize this important fact about chemical equilibria, and scientists have always paid attention to anything in nature that is a constant in the midst of change. No matter how we try to change the concentrations of individual species in the equilibrium and thereby make the equilibrium shift, the value of K_{eq} remains constant. This is why the word *law* is used in *equilibrium law;* it is the way nature behaves consistently.

■ Reference tables for K_{eq} values do not record the actual equilibrium law equations because they can always be written according to the convention.

■ When K_{eq} is greater than 10^2, we almost never express the equation as an equilibrium but use just a single arrow.

■ Other constants in the midst of change are the total energy of the universe and the masses of chemicals before and after a reaction. (Perhaps we might add death and taxes to the list.)

9.4

THE ION PRODUCT CONSTANT OF WATER

The ion product constant of water is a modified equilibrium law.
Certain kinds of ionic equilibria have equilibrium laws that can be simplified without any loss in meaning. The self-ionization of water is one example. It is one of nature's most important chemical equilibria.

$$H_2O \rightleftharpoons H^+(aq) + OH^-(aq)$$

The equilibrium law for this system is

$$K_{eq} = \frac{[H^+][OH^-]}{[H_2O]} \tag{9.2}$$

Equation 9.2 can be simplified with no loss of precision. The values of $[H^+]$ and $[OH^-]$ in pure water must be equal (one of each ion forms) and they are each 1.0×10^{-7} mol/L at 25 °C. These concentrations are so low that the formation of H^+ and OH^- ions from water molecules has no effect on the value of $[H_2O]$ in Equation 9.2, even if we round to seven significant figures. The value of $[H_2O]$ is a constant, for all practical purposes, in Equation 9.2.

In mathematics we learn that if we multiply one constant by another, we just get a new constant. So if we multiply Equation 9.2 on both sides by the constant value of $[H_2O]$, then do an obvious cancellation, we obtain a new expression and a new constant.

$$K_{eq} \times [H_2O] = \frac{[H^+][OH^-]}{[H_2O]} \times [H_2O] = \text{a new constant}$$

■ Values of $[H^+]$ and $[OH^-]$ this low are determined by measuring the ability of ultrapure water to conduct electricity.

■ $K_{eq} \times [H_2O]$ is one constant multiplied by another.

The new constant is called the **ion product constant of water,** and its symbol is K_w.

$$K_w = [H^+][OH^-] \tag{9.3}$$

Equation 9.3 is not a true equilibrium law, because it omits the reactant's term in the denominator, but it still behaves exactly like such a law. In accordance with the equilibrium law, no matter how we change $[H^+]$ by adding acid or base to a solution, the value of $[OH^-]$ in Equation 9.3 adjusts, and the *product* of the two terms remains equal to the constant, K_w. The only way to change K_w is to change the temperature, as data in the margin show.

At 25 °C (and to two significant figures),

$$[H^+] = 1.0 \times 10^{-7} \text{ mol/L}$$

and

$$[OH^-] = 1.0 \times 10^{-7} \text{ mol/L}$$

Therefore at 25 °C,

$$K_w = (1.0 \times 10^{-7})(1.0 \times 10^{-7})$$
$$= 1.0 \times 10^{-14}$$

In all our work, we will assume a temperature of 25 °C.

Knowing that $K_w = 1.0 \times 10^{-14}$, we can calculate the value of one of the two concentration terms, $[H^+]$ or $[OH^-]$, if we know the other.

■ If we add H^+, we shift

$$H_2O \rightleftharpoons H^+ + OH^-$$

to the left and so diminish $[OH^-]$.

■

K_w at Various Temperatures

Temperature (° C)	K_w
0	1.5×10^{-15}
10	3.0×10^{-15}
20	6.8×10^{-15}
25	1.0×10^{-14}
30	1.5×10^{-14}
40	3.0×10^{-14}
50	5.5×10^{-14}
60	9.5×10^{-14}

EXAMPLE 9.1
Using the Ion Product Constant of Water

The value of [H^+] of blood (when measured at 25 °C, not body temperature) is 4.5×10^{-8} mol/L. What is the value of [OH^-], and is the blood acidic, basic, or neutral?

ANALYSIS The values of [H^+] and [OH^-] in any aqueous solution are always related by the equation for K_w, so we use the value of [H^+] in this equation.

■ A review of exponents is in Appendix A.

SOLUTION

$$K_w = 1.0 \times 10^{-14} = (4.5 \times 10^{-8}) \times [OH^-]$$

$$[OH^-] = \frac{1.0 \times 10^{-14}}{4.5 \times 10^{-8}}$$

$$= 2.2 \times 10^{-7} \text{ mol/L}$$

Because the value of [H^+] is less than the value of [OH^-], the blood is basic.

PRACTICE EXERCISE 1

For each of the following values of [H^+], calculate the value of [OH^-] and state whether the solution is acidic, basic, or neutral.

(a) [H^+] = 4.0×10^{-9} mol/L
(b) [H^+] = 1.1×10^{-7} mol/L
(c) [H^+] = 9.4×10^{-8} mol/L

9.5

THE pH CONCEPT

Very low levels of H^+ are more easily described and compared in terms of pH values than as molar concentrations.

Our interest in acid–base balance at the molecular level of life is usually with *weak* acids and bases and with very low concentrations of H^+ or OH^-. We therefore encounter very small numbers quite frequently, numbers usually expressed as negative exponentials, like 10^{-7}. When we have to compare two such numbers very often to see which is larger, like those in Example 9.1 (4.5×10^{-8} versus 2.2×10^{-7}), we usually must look in *two* places in each number, the exponents of 10 and the numbers before the 10's. To make the comparisons of very small quantities easier, the Danish biochemist S. P. L. Sørenson (1868–1939), devised the concept of pH.

The pH of a Solution Is the Negative Logarithm of Its [H^+] There are two completely equivalent ways of defining pH.

$$[H^+] = 1 \times 10^{-pH} \tag{9.4}$$

$$pH = -\log [H^+] \tag{9.5}$$

Equation 9.4 tells us that the **pH** of a solution is the negative power (the p in pH) to which the number 10 must be raised to express the molar concentration of a solution's hydrogen ions (hence, the H in pH). Equation 9.5 is the result of taking the logarithms of both sides of Equation 9.4 and relocating the minus sign.[1]

In pure water at 25 °C, $[H^+] = 1.0 \times 10^{-7}$ mol/L. This value has the same form as the pH-defining Equation 9.4, so we can tell at a glance that the pH of pure water at 25 °C is 7.00. Thus a pH of 7.00 corresponds to a neutral solution at 25 °C.[2]

There are analogous equations for expressing low concentrations of OH^- in terms of the **pOH** of a solution.

$$[OH^-] = 1 \times 10^{-pOH} \tag{9.6}$$
$$pOH = -\log [OH^-] \tag{9.7}$$

Values of pOH are seldom used but when they are, a simple relationship between pH and pOH exists. If we insert the pH and the pOH expressions for $[H^+]$ and $[OH^-]$ into the equation for the ion product constant of water, Equation 9.3, we obtain

$$(1.0 \times 10^{-pH})(1.0 \times 10^{-pOH}) = 1.0 \times 10^{-14} \qquad \text{(at 25 °C)}$$

Recall that when we multiply numbers that involve exponents we *add* the exponents, so this equation means that

$$-pH + (-pOH) = -14$$

If we multiply both sides of this by -1, we get the following relationship between pH and pOH at 25 °C.

$$pH + pOH = 14.00 \qquad \text{(at 25 °C)} \tag{9.8}$$

Acidic Solutions Have pH Values Less than 7
Because pH occurs as a *negative* exponent in Equation 9.4, it takes a pH value that is less than 7.00 for a solution to be acidic, and a value more than 7.00 for it to be basic. In pH terms, then, we have the following definitions of acidic, basic, and neutral solutions when their temperatures are 25 °C.

At 25 °C,

Acidic solution	pH < 7.00
Neutral solution	pH = 7.00
Basic solution	pH > 7.00

$$\tag{9.9}$$

The pH values of several common substances are shown in Figure 9.8. Soft drinks, beer, and even milk are slightly acidic, as you can see, and sour pickles are sour for a now obvious reason.

Table 9.1 gives the correlations of pH, $[H^+]$, $[OH^-]$, and pOH values for the entire useful range of pH, 0 to 14. When the value of $[H^+]$ is 1 mol/L or higher, the pH concept is almost never used. The exponents would no longer be negative, so there would be none of the confusion that Sørenson addressed when he developed the pH concept.

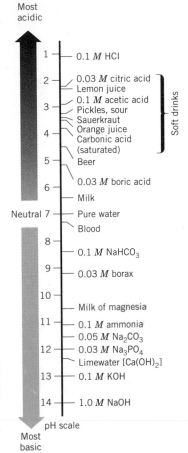

FIGURE 9.8
The pH scale and the pH values of several common substances.

---

[1] Appendix A has a unit on logarithms as well as directions for using a scientific calculator to work with equations like either 9.4 or 9.5 in solving pH/$[H^+]$ problems.

[2] A word about significant figures in logarithms. The 7 in 7.00 comes from the *exponent* in 1.0×10^{-7}, so it actually does nothing more than set off a decimal point when we rewrite the number as 0.00000010. Hence the 7 in the pH value of 7.00 can't be counted as a significant figure. A pH value of 7.00 therefore has just *two* significant figures, those that *follow* the decimal point, just as there are but two significant figures in the value of the molar concentration of $[H^+]$, 1.0×10^{-7} mol/L. To repeat, the number of significant figures in any value of pH is the number of figures that *follow* the decimal point.

TABLE 9.1
pH, [H$^+$], [OH$^-$], and pOH^a

pH	[H$^+$]	[OH$^-$]	pOH	
0	1	1×10^{-14}	14	
1	1×10^{-1}	1×10^{-13}	13	
2	1×10^{-2}	1×10^{-12}	12	
3	1×10^{-3}	1×10^{-11}	11	Acidic solutions
4	1×10^{-4}	1×10^{-10}	10	
5	1×10^{-5}	1×10^{-9}	9	
6	1×10^{-6}	1×10^{-8}	8	
7	1×10^{-7}	1×10^{-7}	7	Neutral solution
8	1×10^{-8}	1×10^{-6}	6	
9	1×10^{-9}	1×10^{-5}	5	
10	1×10^{-10}	1×10^{-4}	4	
11	1×10^{-11}	1×10^{-3}	3	Basic solutions
12	1×10^{-12}	1×10^{-2}	2	
13	1×10^{-13}	1×10^{-1}	1	
14	1×10^{-14}	1	0	

^{a} Concentrations are in mol/L at 25 °C.

Seemingly Small pH Changes Can Mean Large [H$^+$] Changes   One of the very deceptive features of the pH concept is that the actual hydrogen ion concentration changes greatly — by a factor of 10 — for each change of only one unit of pH. For example, if the pH of a solution is zero (meaning that [H$^+$] $= 1 \times 10^0$ mol/L $= 1$ mol/L), only 1 L of water is needed to contain 1 mol of H$^+$. When the pH is 1, however, then 10 L of water (about the size of an average wastebasket) is needed to hold 1 mol of H$^+$. At a pH of 5, it takes a large railroad tank car full of water to include just 1 mol of H$^+$. If the pH of the water flowing over Niagara Falls, New York, were 10 (which, of course, it isn't), an entire 1-hour flowage would be needed for 1 mol of H$^+$ to pass by. And at a pH of 14, the volume that would hold 1 mol of H$^+$ is about a quarter of the volume of Lake Erie, one of the Great Lakes. Seemingly small changes in pH numbers thus signify enormous changes in real concentrations of hydrogen ions.

■ In mass, 1 mol of H$^+$ has a mass of only 1.0 g, so 1×10^{-7} mol of H$^+$ weighs 0.1 microgram (0.1 μg).

pH Refers to [H$^+$], Not to Un-Ionized Acid Concentration   Another point about pH to be emphasized is that it refers to the molar concentration of *hydrogen ions*, not necessarily to the molar concentration of the solute contributing these ions. When the solute is a *weak* acid, only a small percentage of its molecules are ionized at equilibrium, so no simple correlation exists between the concentration of the weak acid and the pH of the solution. The pH of such a solution tells us about [H$^+$], not the molarity of the weak acid.

■ A solution at pH 4.56 has 10 times the concentration of H$^+$ as one at a pH of 5.56.

Only with dilute solutions of strong, 100% ionized acids is there a simple correlation between pH and the molarity of the acid. For example, each molecule of HCl that goes into solution ionizes to give one H$^+$ ion and one Cl$^-$ ion, because HCl is a strong acid. Therefore, for example, in a 0.010 M HCl solution, [H$^+$] $= 0.010$ mol H$^+$/L $= 1.0 \times 10^{-2}$ mol H$^+$/L. Because [H$^+$] $= 1.0 \times 10^{-2}$ mol/L, the pH is simply 2.00. Similarly, a solution that is 0.00010 M HNO$_3$, another strong monoprotic acid, has [H$^+$] $= 0.00010$ mol/L $= 1.0 \times 10^{-4}$ mol/L. So the pH of this solution is 4.00.

The correlation between pOH and the concentration of a strong base, like NaOH, is also simple. In 0.0010 M NaOH, for example, [OH$^-$] $= 0.0010$ mol/L $= 1.0 \times 10^{-3}$ mol/L, so the pOH is simply 3.00. Because pH + pOH $= 14.00$ at 25 °C, a pOH of 3.00 means a pH of 11.00.

In all these simple correlations, the numbers were picked to let 1.0 stand before the 10 in the exponential expression. We will work one example involving a strong acid for which the numbers do not have this relation, just to get used to using a scientific calculator for pH calculations.

EXAMPLE 9.2
Calculating pH from [H⁺]

Lakes in upper New York State, and some New England areas as well as in the Boundary Waters Canoe Area of northern Minnesota, are receiving rain-dissolved air pollutants, such as oxides of sulfur and nitrogen, that make the lake waters more acidic than normal. The water in one lake was found to have $[H^+] = 3.1 \times 10^{-5}$ mol/L. Calculate the pH and the pOH of the lake water.

ANALYSIS The defining equation for pH, Equation 9.5, gives the relationship between $[H^+]$ and pH.

SOLUTION

$$pH = -\log [H^+]$$
$$= -\log(3.1 \times 10^{-5})$$

Enter 3.1×10^{-5} into your calculator. If your calculator has the function keys, $\boxed{10^x}$ and $\boxed{\log}$, it almost certainly also has the keys, $\boxed{EXP}$ and $\boxed{+/-}$. (Your $\boxed{EXP}$ key might be labeled $\boxed{EE}$. Check your manual. If your calculator does not have these functions, you should buy one that does.) Remember, EXP means "times ten to the" as in "3.2 *times 10 to the* minus 5 power." And be doubly sure to remember to use the $\boxed{+/-}$ key to get a negative exponent from an entered positive number. To enter 3.1×10^{-5}, therefore, press the following keys.

$$\boxed{3}\boxed{.}\boxed{1}\boxed{exp}\boxed{+/-}\boxed{5}$$

The display screen should now look something like $3.1 - 05$. Now all you do is press the $\boxed{\log}$ key. The display should now read -4.508638306. The pH is the negative of this, so change the sign. Also round off to two significant figures, the number allowed by the value of $[H^+]$, 3.1×10^{-5}. The answer, therefore, is pH = 4.51.
The pOH is found from Equation 9.8.

$$pH + pOH = 14.00$$
$$4.51 + pOH = 14.00$$
$$pOH = 9.49$$

■ Rain made acidic by air pollutants is called **acid rain.**

PRACTICE EXERCISE 2

Calculate the pH and the pOH in each of the following solutions. (a) 0.025 *M* HCl, (b) 0.00025 *M* NaOH (hint: calculate pOH first, then the pH using Equation 9.8), (c) 0.00025 *M* Ba(OH)₂ (consider this to be 100% dissociated).

PRACTICE EXERCISE 3

A blood specimen was found to have $[H^+] = 7.3 \times 10^{-8}$ mol/L. Calculate its pH. Is it acidic, basic, or neutral?

Another calculation that sometimes has to be made is to find $[H^+]$ from the pH. We'll work an example to show how your calculator can handle this.

EXAMPLE 9.3
Calculating [H⁺] from pH

Because of acid rain, thousands of lakes in southern Norway no longer have game fish. The pH of the lake waters is below 5.50. What $[H^+]$ corresponds to a pH of 5.50?

■ Acid rain also harms forests and accelerates the corrosion of exposed objects made of metal, limestone, or marble.

ANALYSIS Equation 9.4 now becomes the best equation to use.

SOLUTION

$$[H^+] = 1 \times 10^{-pH}$$

We have to get 5.50 into the exponent as a negative number, so enter 5.50 into your scientific calculator *and then press the* $\boxed{+/-}$ *key.* You now have actually entered x for the $\boxed{10^x}$ key, so now press this key and the display will read 3.16227766^{-06}. This means 3.2×10^{-6}, after we round to the two significant figures allowed in the pH value of 5.50. Thus a pH of 5.50 means $[H^+] = 3.2 \times 10^{-6}$ mol/L. ∎

PRACTICE EXERCISE 4

Calculate the values of $[H^+]$ for solutions of the following pH and state if the solutions are acidic or basic. (a) 6.34, (b) 7.89

PRACTICE EXERCISE 5

A blood sample had a pH of 7.28. What is $[H^+]$ for this sample? 5.2×10^{-8}

PRACTICE EXERCISE 6

The pOH of a solution was 4.56. Calculate $[H^+]$.

The question now arises, "How are very small values of $[H^+]$ or their associated pH values measured in the lab?"

■ Phenolphthalein (fee-noll-THAY-lean).

Litmus Is One of Many Acid–Base Indicators The most common way to get a very rough idea of the pH of a solution is to use an acid–base indicator or a combination of them. A number of organic dyes are available for this purpose. Litmus, which we mentioned in the previous chapter, is one example; it is blue above a pH of about 8.5 and red below a pH of 4.5. Each indicator has its own pH range and set of colors, and Figure 9.9 gives a few examples. Phenolphthalein, for example, has a bright pink color at a pH above 10.0 and is colorless below a pH of 8.2. In the range of 8.2 to 10.0 phenolphthalein undergoes a gradual change from colorless to pale pink to deep pink. Bromothymol blue is pink above a pH of 7.6 and yellow below 6.0. Thus if you found that a solution turns phenolphthalein colorless (so the solution's pH is no higher than 8.2) but it makes bromothymol blue become blue (so the solution's pH is above 7.6), you would know that the pH of the solution is between 7.6 and 8.2.

Commercial test papers are available that are impregnated with several indicator dyes. Their containers carry a pH–color code, so you can match the color produced by a drop of solution to this code and so learn the pH of the solution (see Fig. 9.10).

When the solutions to be tested for pH are themselves highly colored, we can't use indicators. Moreover, we often need more than a rough idea of pH. For such situations there are available commercial pH meters (Fig. 9.11) that come equipped with specially designed electrodes that can be dipped into the solution to be tested. With a good pH meter, pH values can be read to the second decimal place. In addition, pH meters with microelectrodes are commercially available. These enable operating room specialists, for example, to measure the pH of very tiny samples of a body fluid of interest.

pH 8.2   pH 10.0
        Phenolphthalein

pH 6.0   pH 7.6
      Bromothymol blue

pH 3.2   pH 4.4
        Methyl orange

pH 9.4   pH 10.6
      Thymolphthalein

FIGURE 9.9
The colors of some common acid–
base indicators.

9.6

ACID IONIZATION CONSTANTS

The strengths of weak acids are described quantitatively by their acid ionization constants, K_a.

Weak acids vary widely in weakness. To compare the relative strengths of weak acids, we use a new kind of constant called an *acid ionization constant, K_a*. It is obtained by a simplification of the equilibrium law for the ionization of the weak acid in water. The objectives of this section are to know what K_a means, how to write the K_a expression for any weak acid, and how to use values of K_a to judge the relative weakness of an acid. Eventually, we need K_a to study systems called *buffers* that protect fluids at the molecular level of life from lethal changes in pH.

Acid Ionization Constants Are Very Large for Strong Acids The K_a values for strong acids are all so large that we ignore them. We assume that, for all practical purposes, strong acids are 100% ionized in dilute aqueous solutions.

Acid Ionization Constants Become Smaller As the Acid Becomes Weaker We will represent any weak acid by the symbol HA, where A denotes the species that separates from H^+ when HA ionizes. The weak acid might be electrically neutral like acetic acid, positively charged like the ammonium ion, or negatively charged like the bicarbonate ion. All these are proton donors, but they differ in acid strengths. If the acid is diprotic or triprotic, we use HA to consider the ionization of just one of its protons. The equilibrium equation for the ionization of HA in water is as follows. Notice particularly that one product is always H_3O^+ (or, later, H^+) and the other is always the conjugate base of the acid.

$$\underset{\substack{\text{Weak Brønsted} \\ \text{acid}}}{HA} + H_2O \rightleftharpoons \underset{\substack{\text{Conjugate} \\ \text{base}}}{H_3O^+ + A^-}$$

FIGURE 9.10
A pH test paper. A drop of the lemon juice in the beaker has been removed and put on the test paper, causing the orange-red color. The pH of the lemon juice is therefore between 2 and 3, according to the color code.

261

FIGURE 9.11
A pH meter

■ The concentrations in the brackets are the concentrations *after equilibrium has been established.*

■ Some references call it the *acid dissociation constant.*

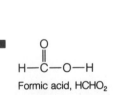

To illustrate, the equilibria for the Brønsted acids already mentioned are

$$HC_2H_3O_2 + H_2O \rightleftharpoons H_3O^+ + C_2H_3O_2^-$$
$$NH_4^+ + H_2O \rightleftharpoons H_3O^+ + NH_3$$
$$HCO_3^- + H_2O \rightleftharpoons H_3O^+ + CO_3^{2-}$$

The general form for the equilibrium law for these equilibria is

$$K_{eq} = \frac{[H_3O^+][A^-]}{[HA][H_2O]} \tag{9.10}$$

The Value of $[H_2O]$ in Equation 9.10 Is Essentially a Constant We can now simplify Equation 9.10 exactly as we simplified Equation 9.2 for the self-ionization of water on our way to the ion-product constant of water.

In dilute solutions of *weak* acids, the ionization equilibrium does not disturb the molar concentration of water even if we think in terms of several significant figures. Therefore the value of $[H_2O]$ in Equation 9.10 is essentially a constant. Let us multiply both sides of Equation 9.10 by this constant and then do an obvious cancellation.

$$K_{eq} \times [H_2O] = \frac{[H_3O^+][A^-]}{[HA][\cancel{H_2O}]} \times [\cancel{H_2O}] = \text{a new constant}$$

The new constant is called the **acid ionization constant,** and its symbol is K_a. If we switch from H_3O^+ to H^+, the equation for K_a is

$$K_a = \frac{[H^+][A^-]}{[HA]} \tag{9.11}$$

Equation 9.11 is the equilibrium law for all Brønsted acids that ionize as monoprotic acids.

Weak, Moderate, and Strong Acids Can Be Defined by K_a Values The K_a values of several acids are given in Table 9.2. The equilibrium equations and the K_a for each acid are figured out as needed.

EXAMPLE 9.4
Writing Equations for K_a for Acids

■

O
||
H—C—O—H

Formic acid, HCHO$_2$

O
||
H—C—O$^-$

Formate ion, CHO$_2^-$

What is the equation for K_a for formic acid, $HCHO_2$, the acid present in the stinging juices of ants?

ANALYSIS The equilibrium equation for the ionization of this acid must first be written, and then the K_a equation can be prepared.

SOLUTION Letting H^+ represent H_3O^+, the ionization of formic acid establishes the following equilibrium.

$$HCHO_2(aq) \rightleftharpoons H^+(aq) + CHO_2^-(aq)$$

Next, we write the equation for K_a. We have to remember that the products always appear in the numerator and the reactants in the denominator. The equation for K_a, therefore, is

$$K_a = \frac{[H^+][CHO_2^-]}{[HCHO_2]}$$

TABLE 9.2
K_a and pK_a Values for Acids at 25 °C^a

Name of Acid	Formula	K_a	pK_a
Perchloric acid	$HClO_4$	Large	
Hydroiodic acid	HI	Large	
Hydrobromic acid	HBr	Large	
Sulfuric acid	H_2SO_4	Large	
Hydrochloric acid	HCl	Large	
Nitric acid	HNO_3	Large	
HYDRONIUM ION	H_3O^+	55	
Hydrogen sulfate ion	HSO_4^-	1.0×10^{-2}	2.00
Phosphoric acid	H_3PO_4	7.1×10^{-3}	2.15
Citric acid	$H_3C_6H_5O_7$	7.1×10^{-4}	3.15
Ascorbic acid (vitamin C)	$H_2C_6H_6O_6$	7.9×10^{-5}	4.10
Acetic acid	$HC_2H_3O_2$	1.8×10^{-5}	4.74
Carbonic acid	H_2CO_3	4.5×10^{-7}	6.35
Dihydrogen phosphate ion	$H_2PO_4^-$	6.3×10^{-8}	7.20
Ammonium ion	NH_4^+	5.7×10^{-10}	9.24
Bicarbonate ion	HCO_3^-	4.7×10^{-11}	10.33
Monohydrogen phosphate ion	HPO_4^{2-}	4.5×10^{-13}	12.35
WATER	H_2O	1.8×10^{-16}	15.74
Hydroxide ion	OH^-	est 1×10^{-36}	36

^{a} For diprotic and triprotic acids, K_a is for the ionization of the first proton only. Data are rounded to two significant figures from the values given in E. H. Martell and R. M. Smith, *Critical Stability Constants,* Plenum Press, New York, 1974. For water, see R. Starkey, J. Norman, and M. Hintze, *J. Chem. Ed. 63* (1986), p. 473. For OH^- see R. J. Myers, *J. Chem. Ed. 63* (1986), p. 687, and references cited therein.

PRACTICE EXERCISE 7

Write the equilibrium equation and the K_a equation for acetic acid, $HC_2H_3O_2$.

PRACTICE EXERCISE 8

Write the equilibrium equation and the K_a equation for the bicarbonate ion.

PRACTICE EXERCISE 9

Write the equilibrium equation and the K_a equation for the ammonium ion.

Because the products, including H^+, are in the *numerator* of the general equation for K_a (Eq. 9.11), the value of K_a is small when the products are not favored. *The weaker the acid, the smaller is its K_a.* A strong acid generates a high percentage of H^+, so the K_a values of strong acids are high. *The stronger the acid, the greater is its K_a.*

PRACTICE EXERCISE 10

The K_a for HCN, hydrogen cyanide, is 6.2×10^{-10} and for ascorbic acid (vitamin C) it is 7.9×10^{-5}. Which is the stronger acid?

Acid ionization constants are used to classify acids as weak, moderate, or strong according to the following criteria.

$K_a < 10^{-3}$	**Weak acid**
$K_a = 1$ to 10^{-3}	**Moderate acid**
$K_a > 1$	**Strong acid**

K_a values are obtained by a calculation using the measured pH of a solution and the molarity of the acid. Special Topic 9.1 shows how. It's also possible to estimate the values of [H⁺] and the pH of a solution of a given weak acid using the K_a of the acid and its molar concentration. Special Topic 9.2 shows how this is done.

Most Transition Metal Cations and the Ammonium Ion Hydrolyze to Generate H_3O^+ Ions and Lower the pH of the Solution

An aqueous solution of ammonium chloride, NH_4Cl, turns blue litmus red, so [H⁺] > [OH⁻] in the solution. The extra hydrogen ions come from the forward reaction of the following equilibrium involving the NH_4^+ ion.

$$NH_4^+ + H_2O \rightleftharpoons NH_3(aq) + H_3O^+(aq)$$

This, of course, is nothing more than the ammonium ion acting as a weak acid. (Its K_a is 5.7×10^{-10}.) But NH_4^+ is not so weak that it cannot produce enough hydronium ions in water to turn blue litmus red.

The reaction of a cation with water to generate hydronium ion is called the **hydrolysis of the cation.** The lesson here is that certain *salts,* like ammonium chloride, can make a solution acidic even though their names do not have the word *acid* in them. It's a lesson we need to know, because of our interest in the acid–base status of a fluid at the molecular level of life. We have to be aware of any solute that can make a solution have a pH other than 7.00. Ammonium salts are such solutes.

Even the hydrated cations of most metals can make a solution test acidic. The aluminum ion in water, for example, exists largely as $[Al(H_2O)_6]^{3+}$. The high positive charge on the central metal ion in this hydrated ion attracts electron density from the H—O bonds of the H_2O molecules it holds. These bonds are thus weakened so $[Al(H_2O)_6]^{3+}$ can donate H⁺ to water as follows.

$$[Al(H_2O)_6]^{3+}(aq) + H_2O \longrightarrow [Al(H_2O)_5(OH)]^{2+} + H_3O^+(aq)$$

A proton transfers from a water molecule of the hydrated ion to a molecule of the surrounding solvent; H_3O^+ forms and the solution thereby is made acidic. A solution of 0.1 M $AlCl_3$ has a pH of about 3, for example, the same as 0.1 M acetic acid, but there is nothing about the formula, $AlCl_3$, to suggest this property.

We don't have K_a values for hydrated metal cations, but all cations with charges of 3 + and most with charges of 2 + can generate hydronium ions in water. Metal ions with such charges have relatively small ionic radii, so they have a relatively high *density* of charge, a high charge *per unit volume* as in the hydrated aluminum ion. A high positive charge density is able to act in a strongly electronegative way to weaken H—O bonds in the surrounding water molecules of their hydrated forms.

The only common metal ions that do *not* hydrolyze to give acidic solutions are those of groups IA and IIA (except Be^{2+} of IIA) — for example, Li⁺, Na⁺, and K⁺ and Mg^{2+}, Ca^{2+}, and Ba^{2+} do not hydrolyze. Evidently, except for Be^{2+}, the cations of groups IA and IIA do not have sufficiently high positive charge densities.

■ "Hydrolysis" is from the Greek *hydro,* water, and *lysis,* loosening or breaking — breaking or loosening by water.

CALCULATING K_a FROM pH

We will show here how the K_a of an acid can be calculated from the pH of a solution of a known molar concentration.

Formic acid, $HCHO_2$, is a monoprotic acid. The pH of a 0.100 M solution of formic acid is 2.38 at 25 °C. Our assignment is to calculate K_a for formic acid at this temperature.

We begin any equilibrium problem by writing the equation for the equilibrium. We need this equation here so that we can set up the expression for K_a.

$$HCHO_2 \rightleftharpoons H^+ + CHO_2^- \qquad K_a = \frac{[H^+][CHO_2^-]}{[HCHO_2]}$$

Always remember that the terms in brackets all refer to concentrations at *equilibrium,* not to initial concentrations. We also note that the equilibrium concentration of CHO_2^- must be the same as that of H^+ because they form in a 1 : 1 ratio by the ionization. So when we find $[H^+]$ from the pH we also find $[CHO_2^-]$. To find $[H^+]$ (and $[CHO_2^-]$), we use one of the defining equations for pH:

$$[H^+] = 10^{-pH} \text{ mol/L}$$
$$= 10^{-2.38} \text{ mol/L}$$
$$= 4.2 \times 10^{-3} \text{ mol/L}$$
$$= 0.0042 \text{ mol/L} \quad \text{(the form most useful next)}$$

This result means that

$$[CHO_2^-] = 0.0042 \text{ mol/L}$$

So we have two of the three equilibrium concentrations. We next have to find the value of $[HCHO_2]$ *at equilibrium* (which is not the same as the molarity of the solution).

The best strategy to keep both our data and our thinking straight is to prepare a *concentration table.* The column headings are the formulas of the species in the equilibrium, so we rewrite the equilibrium equation at the top of our table. Then we will prepare three rows of data. The first row is for *initial concentrations.* For this row we imagine the solution immediately after the solute has been added and before any ionization has occurred. No H^+ or CHO_2^- have yet formed. The second row will express all *changes* in concentration that are caused by the ionization. The concentration of H^+, for example, increases from 0 to 0.0042 mol/L, because we calculated this value from the pH earlier.

The third row of data in the concentration table will have the final, equilibrium concentrations. We get these by a simple algebraic addition of the data in the respective columns, as we will show. The data in the third row, all equilibrium data, are then plugged into the equation for K_a. (All concentrations in such a table, of course, are in moles per liter.) Here is how the table for this problem looks.

	$HCHO_2 \rightleftharpoons$	H^+	$+ CHO_2^-$
Initial concentrations (M)	0.100	0	0 (Note 1)
Changes in concentrations caused by the ionization (M)	-0.0042	$+0.0042$	$+0.0042$ (Note 2)
Final concentrations at equilibrium (M)	$(0.100 - 0.0042)$ = 0.096 (correctly rounded)	0.0042	0.0042

Note 1. The *initial* concentration of the acid is what the label tells us, 0.100 M; it ignores whatever ionization occurs. However, by ignoring this we have to set the *initial* values of $[H^+]$ and $[CHO_2^-]$ equal to zero. (The ultra-trace concentration of H^+ contributed by the self-ionization of water can be safely ignored. It is just too small, and we will ignore it in all calculations in this chapter.)

Note 2. These *changes* are the values we calculated from the pH. For every H^+ ion that forms by the ionization, there is one less molecule of the initial acid. The minus sign in -0.0042 in the column for $HCHO_2$ is used because the initial concentration of this species is *decreased* by the ionization.

The last row of data gives us the equilibrium concentrations that we now use to calculate K_a.

$$K_a = \frac{(4.2 \times 10^{-3})(4.2 \times 10^{-3})}{0.096}$$
$$= 1.8 \times 10^{-4}$$

Thus the acid ionization constant for formic acid is 1.8×10^{-4}.

PRACTICE EXERCISE A 0.0100 M solution of butyric acid has a pH of 3.40 at 20 °C. Calculate the K_a of butyric acid at 20 °C. (Use the symbols H*Bu* and *Bu*$^-$ for butyric acid and its conjugate base.) Butyric acid is the odorous compound in rancid butter. Answer, $K_a = 1.7 \times 10^{-5}$.

CALCULATING [H⁺] AND pH FROM K_a AND [ACID]

The molar concentration $[H^+]$ in a solution of a weak acid does not equal the molarity of the weak acid, because the weak acid is poorly ionized. If we want to know the value $[H^+]$ in such a solution, we have two choices. One is experimental. We measure the pH and convert it to $[H^+]$. The other is to calculate the pH (estimate it might be a better term) from the K_a of the acid and its molarity. We will see how such a calculation can be done in this Special Topic.

A sample of vinegar was found to be $0.100\ M$ acetic acid, $HC_2H_3O_2$. Our assignment is to calculate the values of $[H^+]$ and pH of this sample.

We can use the same general approach given in Special Topic 9.1 — writing the chemical equilibrium and then preparing a concentration table. Of course, we do not know $[H^+]$ (we are to find it), so we have to let x stand for it.

	$HC_2H_3O_2 \rightleftharpoons$	$H^+ +$	$C_2H_3O_2^-$
Initial concentrations (M)	0.100	0	0
Changes in concentrations caused by the ionization (M)	$-x$	$+x$	$+x$ (Note 1)
Final concentrations at equilibrium (M)	$(0.100 - x)$ ≈ 0.100 (Note 2)	x	x

Note 1. We let $+x$ stand for the increase in the concentration of both H^+ and $C_2H_3O_2^-$, and so the initial concentration of $HC_2H_3O_2$ is reduced by the same amount.

Note 2. Saying that $0.100 = (0.100 - x)$ is an important simplification. But we can do this with almost no loss in precision and it makes the calculation much easier. It is studied in detail, next.

The problem we would have without the simplification mentioned in Note 2 is as follows. If we use $(0.100 - x)$ for $[HC_2H_3O_2]$, we would soon have to solve for x in an equation with both an x^2 and an x term — a quadratic equation. In the present example, the quadratic equation would be obtained by making the following substitutions from our concentration table, above.

$$K_a = \frac{[H^+][C_2H_3O_2^-]}{[HC_2H_3O_2]} = \frac{(x)(x)}{(0.100 - x)} = 1.8 \times 10^{-5}$$

We would eventually have to solve for x in

$$x^2 + 1.8 \times 10^{-5}x - 1.8 \times 10^{-6} = 0$$

We can certainly solve for x, but the question here is not if this can be done, but whether it has to be done. Since K_a is very small, the calculated value of x is likely to be very small, too. If it is, then the value of $(0.100 - x)$ will likely be very nearly equal to 0.100, itself. Let's see if replacing $(0.100 - x)$ by x works.

We set $(0.100 - x) = 0.100$, so $[HC_2H_3O_2] = 0.100$ mol/L. From our table, $x = [H^+] = [C_2H_3O_2^-]$. Then we use these values in the equation for K_a for acetic acid.

$$K_a = \frac{[H^+][C_2H_3O_2^-]}{[HC_2H_3O_2]} = \frac{(x)(x)}{(0.100)} = 1.8 \times 10^{-5}$$
$$x^2 = (0.100)(1.8 \times 10^{-5})$$
$$= 1.8 \times 10^{-6}$$
$$x = 1.3 \times 10^{-3}$$

Because $x = [H^+]$,

$$[H^+] = 1.3 \times 10^{-3}\ \text{mol/L} = 0.0013\ \text{mol/L}$$

You should always check to see if this kind of approximation is valid. In this case, $(0.100 - 0.0013) = 0.0987$ (unrounded), which is almost identical to 0.100. The error is only 1%. The assumption was valid; the simplification worked and it made the calculation both easier *and just as valid.*

Having found that at equilibrium $[H^+] = 1.3 \times 10^{-3}$ mol L^{-1}, we next calculate the pH of this solution using the equation that defines pH.

$$pH = -\log [H^+]$$
$$= -\log (1.3 \times 10^{-3})$$
$$pH = 2.89$$

PRACTICE EXERCISE Nicotinic acid, $HC_2H_4NO_2$, is a B vitamin. It is also a weak acid with $K_a = 1.4 \times 10^{-5}$. What is the $[H^+]$ and the pH of a $0.010\ M$ solution of $HC_2H_4NO_2$? Answer, $[H^+] = 3.7 \times 10^{-4}$ mol/L; pH = 3.43.

9.7

BASE IONIZATION CONSTANTS

Base ionization constants let us compare the strengths of weak bases.
Strong bases, like sodium hydroxide, dissociate completely in water to release OH^- ions.

$$NaOH(s) \xrightarrow{\text{dissociation}} Na^+(aq) + OH^-(aq)$$

Other, even stronger bases, like the oxide ion in sodium oxide, react completely with water and generate OH^- ions.

$$Na_2O(s) + H_2O \rightarrow 2Na^+(aq) + 2OH^-(aq)$$

Weak bases, like ammonia or the bicarbonate ion, react incompletely with water, usually to a small percentage, to make some OH^-. An equilibrium is established in which the unchanged base is favored. Ammonia and the bicarbonate ion, for example, generate OH^- ions in water in the following equilibria.

$$NH_3(aq) + H_2O \rightleftharpoons NH_4^+(aq) + OH^-(aq)$$
$$HCO_3^-(aq) + H_2O \rightleftharpoons H_2CO_3(aq) + OH^-(aq)$$

■ The weak bases of greatest importance in our study are NH_3, HCO_3^-, CO_3^{2-}, HPO_4^{2-}, $H_2PO_4^-$, and any of the conjugate bases of the weak organic acids that we will encounter.

Enough forward reaction occurs to make $[OH^-] > [H^+]$ in the resulting solutions, so they test basic to litmus.

Weak bases vary considerably in their abilities to accept protons from water molecules and generate hydroxide ions. To compare these abilities, we use a special equilibrium constant called the *base ionization constant,* K_b. The equilibrium this refers to is always of the following type, where we represent any base by the symbol B, regardless of its electrical charge. In this equilibrium, one product is always OH^- and the other is always the conjugate acid of the base.

$$B(aq) + H_2O \rightleftharpoons BH^+(aq) + OH^-(aq)$$

Weak   Conjugate
base   acid

The **base ionization constant** for this equilibrium, K_b, is defined by the following equation. (Note how closely it parallels the definition of the acid ionization constant, K_a.)

$$K_b = \frac{[BH^+][OH^-]}{[B]} \qquad (9.12)$$

■ $[H_2O]$ is incorporated into K_b just as it was into K_a.

The K_b values for several bases are given in Table 9.3.

The Smaller the K_b, the Weaker the Base Equation 9.12 has the products of the base ionization equilibrium in the numerator. When their concentrations are small, therefore, the base is weak and so K_b has a small value. *The smaller the K_b, the weaker the base.* When the base is strong, then the products are in relatively high concentration, the numerator in Equation 9.12 now is larger, and the K_b value is higher. *The larger the K_b, the stronger the base.*

TABLE 9.3
K_b and pK_b Values for Bases at 25 °C[a]

Name of Base	Formula	K_b	pK_b
Oxide ion	O^{2-}	1×10^{22}	
HYDROXIDE ION	OH^-	55	
Phosphate ion	PO_4^{3-}	2.2×10^{-2}	1.66
Carbonate ion	CO_3^{2-}	2.1×10^{-4}	3.68
Ammonia	NH_3	1.8×10^{-5}	4.74
Monohydrogen phosphate ion	HPO_4^{2-}	1.6×10^{-7}	6.80
Bicarbonate ion	HCO_3^-	2.2×10^{-8}	7.65
Acetate ion	$C_2H_3O_2^-$	5.6×10^{-10}	9.26
Ascorbate ion	$HC_6H_6O_6^-$	1.3×10^{-10}	9.89
Citrate ion	$H_2C_6H_5O_7^-$	1.4×10^{-11}	10.85
Dihydrogen phosphate ion	$H_2PO_4^-$	1.4×10^{-12}	11.85
Sulfate ion	SO_4^{2-}	9.8×10^{-13}	12.01
WATER	H_2O	1.8×10^{-16}	15.74
Nitrate ion	NO_3^-	Very small	
Chloride ion	Cl^-	Very small	
Hydrogen sulfate ion	HSO_4^-	Very small	
Bromide ion	Br^-	Very small	
Iodide ion	I^-	Very small	
Perchlorate ion	ClO_4^-	Very small	

[a] K_b values (except for ammonia) were calculated from the K_a values obtained from the references cited for Table 9.2 and then rounded to two significant figures.

PRACTICE EXERCISE 11

The base ionization constant for CN^-, the cyanide ion, is 1.6×10^{-5} and that for the bicarbonate ion is 2.6×10^{-8}. Which is the stronger base? ∎

The interpretation of K_b values depends always on the ability to translate just the formula of the base both into its chemical equilibrium in water and into the specific equation for its K_b. The next example shows how this is done.

EXAMPLE 9.5
Writing Expressions for K_b for Brønsted Bases

The monohydrogen phosphate ion is a base. Write the equilibrium expression on which its K_b is based and then write the equation for K_b.

ANALYSIS To write the equilibrium expression for HPO_4^{2-} in water, we put HPO_4^{2-} and H_2O as the reactants, and OH^- and the conjugate acid of HPO_4^{2-} as the products.

SOLUTION We figure out the formula of the conjugate acid of any base by adding one H^+ to the formula of the base, remembering to adjust the charge correctly. So the conjugate acid of HPO_4^{2-} is $H_2PO_4^-$. Our equilibrium equation is

$$HPO_4^{2-}(aq) + H_2O \rightleftharpoons H_2PO_4^-(aq) + OH^-(aq)$$

Now we can write the equation for K_b, omitting H_2O and remembering that the products are always in the numerator. The answer, then, is

$$K_b = \frac{[H_2PO_4^-][OH^-]}{[HPO_4^{2-}]}$$

PRACTICE EXERCISE 12

Write the equilibrium equations and the equations for K_b for each of the following Brønsted bases.

(a) CO_3^{2-} (b) $C_2H_3O_2^-$ (acetate ion) (c) NH_3

Two kinds of calculations are possible with what has been studied thus far in this section, the calculation of a value of K_b from $[OH^-]$ or pOH and the calculation of equilibrium concentrations of $[H^+]$ or $[OH^-]$ from values of K_b and $[B]_{init}$. These calculations are very similar to those in Special Topics 9.1 and 9.2, and we will not pursue them further.

The Reaction of an Anion with Water to Produce OH⁻ Is Called the Hydrolysis of the Anion

Most of the Brønsted base anions with K_b values greater than 10^{-13} are available as their sodium salts, like Na_3PO_4, Na_2CO_3, NaCN, Na_2HPO_4, $NaHCO_3$, $NaC_2H_3O_2$, and NaH_2PO_4. Aqueous solutions of these salts test basic to litmus because their anions have reacted with water — it's called the **hydrolysis of anions** — to generate an excess of OH^- over H^+. The bicarbonate ion in aqueous $NaHCO_3$, for example, hydrolyzes to establish the following equilibrium in which OH^- ions are present in excess over H^+ ions.

$$HCO_3^-(aq) + H_2O \rightleftharpoons H_2CO_3(aq) + OH^-(aq)$$

Anions like Cl^-, Br^-, I^-, NO_3^-, and SO_4^{2-}, which are conjugate bases of strong acids, do not hydrolyze in this way. To summarize,

> Anions whose conjugate acids are weak acids hydrolyze in water and tend to make the solution basic.
>
> Anions of strong acids do not hydrolyze.

In the previous section we developed similar rules of thumb about cations, which we'll repeat here.

> Metal ions from group IA or IIA (except Be^{2+}) do not hydrolyze.
>
> Expect other metal ions as well as NH_4^+ to hydrolyze and generate H^+.

With these rules we can generally predict correctly whether a given salt will affect the pH of an aqueous solution. The exceptions would be salts where both cation and anion can hydrolyze, like $NH_4C_2H_3O_2$, ammonium acetate. These have to be taken on a case-by-case basis with the result hinging on the relative strengths of the cation as a proton producer and the base as a proton neutralizer. We will not work with such salts. But let's see how we can predict the hydrolysis of salts that respond to a simpler analysis.

■ One reason we are interested in any solute that can affect the pH of a solution is that enzyme action is very sensitive to pH.

EXAMPLE 9.6
Predicting How a Salt Affects the pH of Its Solution

Sodium phosphate, Na_3PO_4 ("trisodium phosphate"), is a strong cleaning agent for walls and floors. Is its aqueous solution acidic or basic?

ANALYSIS We must consider the *ions* that the salt releases in solution and decide what to expect of each.

SOLUTION Na_3PO_4 involves Na^+ and PO_4^{3-}. Na^+ does not hydrolyze, but PO_4^{3-} does. Its conjugate acid, HPO_4^{2-}, is not on our list of strong acids, so we can infer that it is a weak acid. We therefore expect PO_4^{3-} to be a relatively strong base and we expect it to hydrolyze as follows.

$$PO_4^{3-}(aq) + H_2O \rightleftharpoons HPO_4^{2-}(aq) + OH^-(aq)$$

This equilibrium generates some OH^- ions, so the solution will be basic. ∎

In working Example 9.6 we did not need a table of Brønsted bases to predict that PO_4^{3-} would hydrolyze. We used our knowledge of just a few facts — which acids are strong acids in water and which cations do not hydrolyze — to figure out what we needed. We will work another example to practice using the list of strong aqueous acids to decide whether a given salt can affect the pH of its solution.

EXAMPLE 9.7
Predicting whether a Salt Affects the Acidity of Its Solution

Chromium(III) nitrate, $Cr(NO_3)_3$, is soluble in water. Does this salt make its aqueous solution acidic or basic?

ANALYSIS We have to work with the individual *ions* made available by this salt and how they might react with water.

SOLUTION $Cr(NO_3)_3$ dissociates into $Cr^{3+}(aq)$ and three $NO_3^-(aq)$ ions. Because the nitrate ion is the conjugate base of a *strong* acid, HNO_3, we know that it is unable to react with water to generate hydroxide ions. The nitrate ion does not hydrolyze. The Cr^{3+} ion, however, is not from the metals of groups IA or IIA. Moreover, like the aluminum ion, it has a high positive charge, so we expect Cr^{3+} to hydrolyze and generate some hydrogen ion. Because of the Cr^{3+} ion, a solution of $Cr(NO_3)_3$ tests acidic. ∎

PRACTICE EXERCISE 13

Determine without the use of tables whether each ion can hydrolyze. If so, state whether it tends to make the solution acidic or basic.

(a) CO_3^{2-} **(b)** S^{2-} **(c)** HPO_4^{2-} **(d)** Fe^{3+} **(e)** NO_2^- **(f)** F^-

PRACTICE EXERCISE 14

Is a solution of potassium acetate, $KC_2H_3O_2$, acidic, basic, or neutral to litmus?

PRACTICE EXERCISE 15

Is a solution of copper(II) nitrate, $Cu(NO_3)_2$, acidic, basic, or neutral?

PRACTICE EXERCISE 16

Ammonium sulfate, $(NH_4)_2SO_4$, is a nitrogen fertilizer. Could the application of an aqueous solution of this fertilizer affect the pH of the soil? If so, will it increase or decrease the pH?

9.8

THE pK_a AND pK_b CONCEPTS

The negative logarithms of K_a and K_b, pK_a and pK_b, are useful in the same way as pH.

For the same reason that the pH concept was devised, analogous pK_a and pK_b expressions, based on K_a and K_b, have been defined. The **pK_a** is the negative logarithm of K_a, and the **pK_b** is the negative logarithm of K_b.

$$pK_a = -\log K_a \qquad (9.13)$$
$$pK_b = -\log K_b \qquad (9.14)$$

Note carefully that pK_a and pK_b are *negative* logarithms. Therefore, the generalizations we want to carry forward from Equations 9.13 and 9.14 are the following.

The larger the pK_a, the weaker the acid.
The larger the pK_b, the weaker the base.

EXAMPLE 9.8
Calculating pK_a from K_a

The K_a for acetic acid is 1.8×10^{-5} (at 25 °C). What is its pK_a at this temperature?

ANALYSIS   We use the defining equation for pK_a.

SOLUTION
$$pK_a = -\log K_a$$
$$= -\log (1.8 \times 10^{-5})$$
$$pK_a = 4.74 \qquad \text{(rounded as per footnote 2, page 257)}$$

PRACTICE EXERCISE 17

A base has a K_b value of $4.7 = 10^{-11}$. Calculate its pK_a value.

EXAMPLE 9.9
Using pK_a Values to Compare Strengths of Acids

Hypoiodous acid, HIO, has a pK_a of 10.6. The pK_a of hypobromous acid, HBrO, is 8.64. Which is the weaker acid?

ANALYSIS   Remember, the larger the pK_a is, the weaker is the acid.

SOLUTION   HIO has the larger pK_a, so it is the weaker acid.

PRACTICE EXERCISE 18

The pK_a of carbonic acid, H_2CO_3, is 6.35, and that of acetic acid is 4.76. Which is the stronger acid?

EXAMPLE 9.10
Calculating pK_b from K_b

The K_b of ammonia is 1.8×10^{-5} (at 25 °C). What is its pK_b?

ANALYSIS For this we simply use the equation that defines pK_b.

SOLUTION

$$pK_b = -\log K_b$$
$$= -\log (1.8 \times 10^{-5})$$
$$pK_b = 4.74 \quad \text{(rounded as per footnote 2)}$$

PRACTICE EXERCISE 19

The K_b value for the monohydrogen phosphate ion is 1.6×10^{-7}. Calculate its pK_b.

For a Conjugate Acid–Base Pair, the Product of K_a and K_b Is K_w A very simple relationship exists between K_a and K_b when we work with a conjugate acid–base pair.

$$K_aK_b = K_w \quad \text{(for a conjugate acid–base pair)} \quad (9.15)$$

To prove Equation 9.15, we substitute the expressions for K_a, K_b, and K_w into it and cancel what terms we can. In the equilibrium in a solution of a weak acid we have

$$HA \rightleftharpoons H^+ + A^- \quad \text{and} \quad K_a = \frac{[H^+][A^-]}{[HA]}$$

For a solution of the conjugate base of HA, which is A^- (put into solution as some salt, like NaA), we have

$$A^- + H_2O \rightleftharpoons HA + OH^- \quad \text{and} \quad K_b = \frac{[HA][OH^-]}{[A^-]}$$

We now multiply the expressions for K_a and K_b and cancel what we can.

$$K_a \times K_b = \frac{[H^+][A^-]}{[HA]} \times \frac{[HA][OH^-]}{[A^-]} = [H^+][OH^-]$$
$$= K_w \quad \text{(proving Equation 9.15)}$$

■ Because $K_a \times K_b = K_w$,

$$K_a = \frac{K_w}{K_b} \quad \text{and} \quad K_b = \frac{K_w}{K_a}$$

When We Know Either K_a or K_b for a Conjugate Acid–Base Pair, We Can Calculate the Other Equation 9.15 lets us calculate K_b when we know K_a for its conjugate acid or it lets us calculate K_a for an acid when we know K_b for its conjugate base. All the K_b values in Table 9.3, for example, were calculated from the values of K_a in Table 9.2.

EXAMPLE 9.11
Calculating K_b from K_a

What is the value of K_b at 25 °C for the conjugate base of hypochlorous acid, HOCl, whose K_a is 3.0×10^{-8}?

ANALYSIS At 25 °C, $K_w = 1.0 \times 10^{-14}$, so we simply substitute this value and the given value of K_a, 3.0×10^{-8}, into Equation 9.15.

SOLUTION

$$(3.0 \times 10^{-8}) \times K_b = 1.0 \times 10^{-14}$$

Solving for K_b gives

$$K_b = \frac{1.0 \times 10^{-14}}{3.0 \times 10^{-8}}$$
$$= 3.3 \times 10^{-7}$$

Thus the K_b for OCl⁻, the conjugate base of HOCl, is 3.3×10^{-7}.

PRACTICE EXERCISE 20

For NH_4^+, $K_a = 5.7 \times 10^{-10}$. What is the conjugate base of this weak acid, and what is its K_b?

For a Conjugate Acid–Base Pair, pK_a + pK_b = 14.00 (25 °C) The relationship among K_a, K_b, and K_w leads to a simple relationship between pK_a and pK_b for a conjugate acid–base pair. If we take the logarithms of both sides of Equation 9.15, we obtain

$$\log(K_a \times K_b) = \log K_w$$

Or

$$\log K_a + \log K_b = \log K_w$$

After multiplying both sides by -1, we get

$$(-\log K_a) + (-\log K_b) = (-\log K_w)$$

But the first two terms define pK_a and pK_b, respectively, so this equation is equivalent to writing

$$pK_a + pK_b = -(\log 1.0 \times 10^{-14}) \qquad \text{(at 25 °C)}$$

Because $\log 1.0 \times 10^{-14} = 14.00$, we have the following relationship between the pK_a and pK_b values for an acid and its conjugate base.

$$pK_a + pK_b = 14.00 \qquad (25\ °C) \tag{9.16}$$

■ One of the rules of logarithms is

$$\log(a \times b) = \log a + \log b$$

■ We could also say, pK_w = $-\log K_w$, so at 25 °C, pK_w = 14.00.

EXAMPLE 9.12
Finding pK_a from pK_b or pK_b from pK_a for Conjugate Acid–Base Pairs

The pK_a of acetic acid at 25 °C is 4.74. What is the pK_b of its conjugate base, the acetate ion, $C_2H_3O_2^-$?

ANALYSIS Equation 9.16 applies when the relationship involves an acid and its own conjugate base.

SOLUTION
$$pK_a + pK_b = 14.00$$
$$4.74 + pK_b = 14.00$$
$$pK_b = 14.00 - 4.74$$
$$= 9.26$$

Thus, the pK_b for the acetate ion is 9.26.

PRACTICE EXERCISE 21

■ Hydrocyanic acid, HCN(*aq*), is a solution of hydrogen cyanide, HCN(*g*), in water.

Hydrocyanic acid, HCN, is a very weak acid with a pK_a of 9.2 at 25 °C. Calculate the pK_b of its conjugate base, CN⁻. Write the equation for the chemical equilibrium in which CN⁻ acts as a Brønsted base in water, and then write the equation for K_b.

9.9

BUFFERS

The pH of a solution can be held relatively constant if it contains a buffer—a weak base and its conjugate acid.

It takes just a little strong acid or strong base to cause a large change in the pH of a solution. The addition of only one drop of concentrated hydrochloric acid to a liter of pure water drops the pH of the system from 7 to 4, a change of three units in pH but a 10^3 or 1000-fold change in acidity. If such a change occurred in the bloodstream, you would die. The pH of blood can't be allowed to change by more than 0.2 to 0.3 pH units from its normal pH of 7.35. Special Topic 9.3 tells of the threat posed by *acid rain* to living systems.

■ $\dfrac{10^{-4}}{10^{-7}} = 1000$

Acidosis and Alkalosis Are Life Threatening So critical is the maintenance of the pH of the blood that a special vocabulary exists to describe small shifts away from it. If the pH becomes lower, which means that the acidity of the blood is increasing, the condition is called **acidosis.** Acidosis is characteristic of untreated diabetes, emphysema, and other conditions. If the pH of the blood increases, which means that the blood is tending to become more basic, the condition is called **alkalosis.** An overdose of bicarbonate, exposure to the low partial pressure of oxygen at high altitudes, or prolonged hysteria can cause alkalosis.

In their more advanced stages, acidosis and alkalosis are medical emergencies because they interfere with the smooth working of *respiration*, the physical and chemical apparatus that brings oxygen in, uses it, and then removes waste carbon dioxide. We need much more background before we can really understand this complex subject, but important parts of the preparation are in this chapter. We will learn in general terms in this section, for example, how the body controls the pH of its fluids.

■ Emergency room personnel at hospitals must be extremely well versed in recognizing signs of acidosis or alkalosis.

Buffers Prevent Serious Changes in pH Certain combinations of solutes, called **buffers,** keep changes in pH to a minimum when strong acids or bases are added to an aqueous solution. One part of the buffer system can neutralize H⁺, and the other part can neutralize OH⁻. Fluids that contain buffers are said to be *buffered* against the changes in pH that H⁺ or OH⁻ ions otherwise cause. The blood and other body fluids include buffers, and much of the body's work in maintaining its acid–base status depends on buffers. Both acidosis and alkalosis are greatly restrained and sometimes totally prevented by buffers. Let's first see what buffers are and how they work, and then we can study them quantitatively.

■ Every form of life is very sensitive to slight changes of pH in internal fluids.

The Phosphate Buffer Is Important within Cells The principal buffer at work inside cells is called the **phosphate buffer**. It consists of the pair of ions, HPO_4^{2-} and $H_2PO_4^-$, the monohydrogen and the dihydrogen phosphate ions. Notice that $H_2PO_4^-$ is the conjugate acid of HPO_4^{2-}, so $H_2PO_4^-$ is the member of this pair that is better able to neutralize base. Any added OH^- is neutralized by $H_2PO_4^-$, and this keeps the pH from increasing.

$$H_2PO_4^-(aq) + OH^-(aq) \longrightarrow HPO_4^{2-}(aq) + H_2O$$

The proton acceptor or base in the phosphate buffer is the conjugate base of $H_2PO_4^-$, the HPO_4^{2-} ion. It can neutralize H^+ and so keep the pH from decreasing.

$$HPO_4^{2-}(aq) + H^+(aq) \longrightarrow H_2PO_4^-(aq)$$

The Carbonate Buffer Is Important in the Blood The principal buffer in blood is called the **carbonate buffer**. Traditionally, it has been described as the conjugate pair, H_2CO_3 and HCO_3^-, carbonic acid and the bicarbonate ion. Actually, the carbonic acid in blood is almost entirely in the form of $CO_2(aq)$. For every molecule of $H_2CO_3(aq)$, there are 400 molecules of $CO_2(aq)$. $CO_2(aq)$, however, is able to *directly* neutralize hydroxide ion by the following reaction.

$$CO_2(aq) + OH^-(aq) \longrightarrow HCO_3^-(aq) \qquad (9.17)$$

Actually (and importantly), this is really the forward step in an equilibrium.

$$CO_2(aq) + OH^-(aq) \rightleftharpoons HCO_3^-(aq) \qquad (9.18)$$

■ In blood, $[HCO_3^-]$ is normally about 24 mmol/L (about 1.5 g/L).

■ The concentration of CO_2 in blood is normally about 1.2 mmol/L (about 0.05 g/L).

■ The "total CO_2" concentration of blood is $[HCO_3^-] + [CO_2]$ and is normally about 26 mmol/L.

The equilibration of $CO_2(aq)$ and $OH^-(aq)$ with $HCO_3^-(aq)$ is catalyzed by what is one of the most rapidly acting enzymes in the body, *carbonic anhydrase*. Because of this enzyme's work, we can use $CO_2(aq)$ as a stand-in for $H_2CO_3(aq)$ in discussing the carbonate buffer, even though $CO_2(aq)$ is not an acid in the Brønsted sense.

The $CO_2(aq)$ of the blood's carbonate buffer can neutralize OH^-, as we just said, and thus prevent an increase in pH. If there should be some metabolic or respiratory problem that increases the blood's OH^- level, this OH^- is neutralized by $CO_2(aq)$, Equation 9.17, and alkalosis is prevented.

The bicarbonate ion is the base of the blood's carbonate buffer. If in a particular metabolic or respiratory situation the blood's level of H^+ ion increases, the H^+ is neutralized by HCO_3^- and acidosis is prevented.

$$HCO_3^-(aq) + H^+(aq) \longrightarrow CO_2(aq) + H_2O \qquad (9.19)$$

The Ability to Breathe Out CO_2 Is Essential to the Control of Acidosis When acid is neutralized by the carbonate buffer (Eq. 9.19), the blood's level of $CO_2(aq)$ increases. However, when the blood moves through the capillaries in the lungs, gaseous CO_2 is released from dissolved CO_2 and breathed out. Because the gas leaves, *we cannot write this change as an equilibrium.*

$$CO_2(aq) \longrightarrow CO_2(g) \qquad (9.20)$$

The loss of one molecule of $CO_2(g)$ by this change means that the H^+ ion neutralized by the buffer action (Eq. 9.19) is now *permanently* neutralized. All these steps are summarized in Figure 9.12, where you can see that the H^+ to be neutralized ends up in a molecule of water. The ability of this water molecule to form finally depends on the loss of the CO_2 molecule from the body.

Enormous quantities of sulfur dioxide are generated worldwide from the combustion of coal and oil, which contain relatively small quantities of sulfur compounds. This sulfur becomes oxidized to gaseous SO_2 as the fuels burn. Although the sulfur content of a fuel seldom exceeds 3%, often much less, the vast tonnages of fuels consumed worldwide annually release hundreds of millions of tons of SO_2 per year into the atmosphere. It is a major contributor to "acid rain."

Sulfur dioxide dissolves in water by forming hydrates, $SO_2 \cdot nH_2O$, where n varies with concentration, temperature, and pH. The hydrates are in equilibrium with some hydronium ion and hydrogen sulfite ion, HSO_3^-, whose presence has long been explained simply in terms of $H_2SO_3(aq)$, sulfurous acid. Actual molecules of this species — H_2SO_3 — have never been detected in or out of water, however. Nevertheless, for convenience in writing chemical equations, the formula H_2SO_3 is widely used for the solute in aqueous sulfur dioxide. It is the first ionization of sulfurous acid that generates virtually all of the hydrogen ion that this acid can produce in water.

$$H_2SO_3(aq) \rightleftharpoons$$
$$H^+(aq) + HSO_3^-(aq) \qquad pK_a = 1.92 \ (25 \ °C)$$

The relatively low value of pK_a makes sulfurous acid a moderate acid. Thus when rain washes gaseous SO_2 from the atmosphere, the rainwater is acidic. Moreover, both oxygen and ozone (O_3) in smog convert some SO_2 to SO_3, particularly in sunlight when fine dust is present. When SO_3 reacts with water, sulfuric acid, a strong acid, forms. It also contributes to the acidity of rain where air pollution occurs.

Nitrogen Dioxide Is Another Major Air Pollutant That Contributes to Acid Rain As you know, nitrogen and oxygen are very stable toward each other *at ordinary temperatures and pressures.* When fuels are burned in vehicles, however, high temperatures and higher pressures cause some reaction to occur between nitrogen and oxygen to give nitrogen monoxide.

$$N_2(g) + O_2(g) \longrightarrow 2NO(g)$$

When the vehicle exhaust leaves and encounters a much cooler atmosphere, nitrogen monoxide reacts with oxygen to give nitrogen dioxide.

$$NO_2(g) + O_2(g) \longrightarrow 2NO_2(g)$$

Nitrogen dioxide is responsible for the reddishness of smog. In water, NO_2 reacts to give two acids: HNO_3, a strong acid, and HNO_2, a weak acid.

$$2NO_2(g) + H_2O \longrightarrow HNO_3(aq) + HNO_2(aq)$$

Thus, oxides of both nitrogen and sulfur are chiefly responsible for **acid rain.** Rain as acidic as lemon juice (pH 2.1) was observed in 1964 in the northeastern section of the United States, and rain as acidic as vinegar (pH 2.4) fell at Pitlochry, Scotland, in 1974.

A Better Term for Acid Rain Is Acid Deposition Dry dust particles settling on buildings, metals, and soil also carry acidic materials adhering to their surfaces. Thus *acid deposition* refers to all means whereby acids enter the earth's system. Acid deposition is an acute problem in regions downwind from major users of sulfur-containing fuels. Southern parts of the Scandinavian peninsula receive acid deposition from Germany's Ruhr valley, the English Midlands, and countries of eastern Europe. Parts of southern Canada and the northern United States get acid deposi-

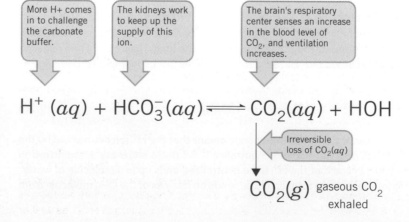

FIGURE 9.12
The irreversible neutralization of H^+ through the loss of CO_2 is one way the carbonate buffer system handles acidosis. It is the last step, the change of dissolved CO_2 into gaseous CO_2, which is exhaled, that draws all the equilibria to the right and makes H^+ "disappear" into H_2O.

tion from the great industrial belt curving from Boston to Chicago. Industrial areas of eastern Europe have released enormous loads of sulfur and nitrogen oxides into their atmospheres.

Acid deposition affects lakes, soil, vegetation, and building stone. It makes bodies of water too acidic for much aquatic life. Because CO_2 from the air is naturally present in water, the pH of fresh water in equilibrium with air is about 5.7. Below a pH of 5.5, the hatchlings of most game fish are unable to live. Dissolved acids leach calcium and magnesium ions from the soil and thus adversely affect vegetation. Severe damage has occurred to many mature forests in or near heavily industrialized nations.

Acid deposition is corrosive to exposed metals such as railroad rails, vehicles, and machinery, as well as to stone building materials (Figure 1). Limestone and marble are particularly sensitive, because they are chiefly calcium carbonate, and carbonates are dissolved by acids.

$$CaCO_3(s) + 2H^+(aq) \longrightarrow Ca^{2+}(aq) + CO_2(g) + H_2O$$

Several major cathedrals in Europe need constant repair because of the attack of deposited acids on their limestone and marble.

No Easy Alternatives Exist to Sulfur-Containing Fuels
Less reliance on sulfur-containing coal and oil might be thought to be a solution to the acid deposition problem. No doubt it would help considerably. The alternatives to coal and oil are either a drastic cutback in energy consumption made possible by a simpler less consuming lifestyle, or a switch to a heavier reliance on nuclear power, or new technology not yet tested on a huge scale. Nuclear power bears its own pollution ills and it presently is costlier in every way than power obtained from the burning of coal or oil. Meanwhile, as coal and oil are used, the removal of

most of the SO_2 from smokestack gases is possible. SO_2, for example, is absorbed by wet calcium hydroxide by the following reaction.

$$SO_2(g) + Ca(OH)_2(s) \longrightarrow CaSO_3(s) + H_2O$$

Not all SO_2 is removed, however, and given the enormous quantities of coal and oil burned worldwide, emissions of SO_2 still occur. In a *technological* sense, the problem is controllable. The remaining issues concerning emissions of the sulfur oxides are mostly personal, political, economic, and diplomatic.

FIGURE 1
Acid deposition increased the rate of decay of this gargoyl on Old City Hall in Munich, Germany.

The blood thus brings to bear *two* mechanisms that rapidly handle an influx of H^+. It neutralizes H^+ by the work of the carbonate buffer, and it uses a physical process, ventilation, to make this neutralization permanent.

Ventilation is the circulation of air into and out of the lungs. The brain has a site called the *respiratory center* that monitors the level of $CO_2(aq)$ in the blood. When this level increases, the brain instructs the breathing apparatus to breathe more rapidly and deeply, a response called **hyperventilation.** This response increases the rate at which CO_2 can be exhaled, and the permanent loss of CO_2 shifts all the carbonate equilibria in their acid-neutralizing directions.

One of many lessons we can draw from this discussion is that anything that interferes with the loss of CO_2 inhibits the neutralization of the H^+ ion and causes acidosis. One such interference is involuntary **hypoventilation,** the slow and shallow breathing found in people with emphysema. They are unable to breathe deeply enough. They struggle with acidosis as well, because they are unable to get rid of CO_2 and so are unable to neutralize H^+ as well as

■ A third mechanism involves the kidneys, which remove H^+ from blood and resupply HCO_3^-. But this work takes hours to days.

■ This retention of CO_2 because of hypoventilation is called "the retention of acid" because CO_2, one way or another, neutralizes OH^-.

they should. Any other cause of involuntary hypoventilation that renders the body unable to breathe out CO_2, like asthma, pneumonia, or overdoses of narcotics or barbiturates, also threatens the system with acidosis.

There is much more to say about the various ways in which metabolism and respiration interact with the buffer systems in the blood, and we will return to this topic in considerable detail in a later chapter, after we have learned more about metabolism and about the chemistry of the blood.

9.10

SOME QUANTITATIVE ASPECTS OF BUFFERS

Buffers help to keep a pH constant, but not necessarily at pH 7.
In this section we move our study of buffers to a more quantitative level. We'll first explore the relationship between the pH of a buffered solution and the concentrations and relative acid–base strengths of its components.

The pH Values of Buffered Solutions Can Be Calculated from Ionization Constants and Buffer Concentrations To make the discussion general, we assume that the buffer is made by dissolving some weak acid, HA, together with some of its sodium salt, NaA, in water. We use a group IA salt, like the sodium salt, because we generally want a fully soluble salt. Then it is very soluble, 100% ionized, and thus makes available the maximum amount of the conjugate base, A^-, of the weak acid.

The HA/A^- type of buffer system could involve any one of a number of weak acids of widely varying K_a values. So we cannot expect just one weak acid to work for the buffering of all ranges of pH. We therefore need an equation to tell us at what pH a specific HA/A^- buffer system will work.

Recall Equation 9.11, which defines K_a for a weak acid, HA:

$$K_a = \frac{[H^+][A^-]}{[HA]} \qquad \text{(This was Equation 9.11.)}$$

Because we want to know $[H^+]$ and then pH, let us rearrange this equation to give us an expression for $[H^+]$. (We can find pH after we find $[H^+]$.)

$$[H^+] = K_a \times \frac{[HA]}{[A^-]} \qquad (9.21)$$

All molar concentrations in Equations 9.11 and 9.21, remember, are the concentrations *at equilibrium* after the solution has been prepared. The terms [HA] and $[A^-]$ do not mean the *initial* concentrations of solutes used to prepare the solution. To keep this point before us in our discussion, we probably should write [HA] and $[A^-]$ as $[HA]_{eq}$ and $[A^-]_{eq}$.

Yet when we deal with buffers, we can safely make some very important simplifications. We actually can use *initial* values of molarities of the weak acid, HA, and of the anion, A^- (as supplied by the salt), to be the same as the equilibrium values. Initial values are easier to use because we get them directly from the moles of solutes used to make the buffer solution. Here is why these simplifications work.

Because HA is a weak acid, little would be ionized at equilibrium, even if it were the only solute. But when its anion, A^-, is also present (supplied by the salt), the ionization of HA is suppressed even more. The presence of A^- from the salt, in other words, acts as a stress on the following equilibrium and keeps it shifted to the left, in favor of un-ionized HA:

$$HA(aq)_{eq} \rightleftharpoons H^+(aq)_{eq} + A^-(aq)_{eq}$$

The result is that the value of $[HA(aq)]_{eq}$ is essentially identical to that of $[HA(aq)]_{initial}$, the value we know about from preparing the solution. Thus our first simplification is

$$[HA(aq)]_{eq} = [HA(aq)]_{\text{from initial}} = [\text{acid}]$$
$$\underset{\substack{\text{concentration}\\\text{of the acid}}}{}$$

We will let [acid] be our symbol for the *initial* molar concentration of the weak acid in the buffer.

Now let us see what we can do about simplifying $[A^-(aq)]_{eq}$. Not much $A^-(aq)_{eq}$ is supplied by the ionization of the weak acid. ("Weak" implies this.) Nearly 100% of the $A^-(aq)_{eq}$ is supplied, instead, by the salt, because the ionization of the weak acid is suppressed. Thus not much of anything changes the value of $[A^-(aq)]$ *as initially supplied by the salt*. We conclude that the value of $[A^-(aq)]_{eq}$ is essentially the same as $[A^-(aq)]_{\text{initial}}$. So our second simplification is

■ For all practical purposes, the *only* source of A^- is the salt, NaA, used to prepare the buffer.

$$[A^-(aq)]_{eq} = [A^-(aq)]_{\text{from initial}} = [\text{anion}]$$
$$\underset{\substack{\text{concentration}\\\text{of the salt}}}{}$$

We will let [anion] stand for the *initial* molar concentration of the other component of the buffer, the component supplied by the salt.

These relationships define new terms, [acid] and [anion], which we can now substitute into Equation 9.21 for [HA] and $[A^-]$, respectively. Always remember that [acid] refers to the *initial* concentration of the weak acid and [anion] refers to the *initial* concentration of the anion directly provided by the salt in the buffer system. So for a buffer system made of the HA/A^- pair, we have the following equation for $[H^+]$.

For HA/A^- buffers:

$$[H^+] = K_a \times \frac{[\text{acid}]}{[\text{anion}]} \qquad (9.22)$$

Let's see how we can use what we have learned to find $[H^+]$ and, from it, the pH of a buffer solution.

EXAMPLE 9.13
Calculating the pH of a Buffered Solution

To study the effect of a weakly acidic medium on the rate of growth of a species of bacteria, a biochemist prepared a buffer from the weak acid, acetic acid, and used sodium acetate as the source of the conjugate base, the acetate ion. The buffer solution was made so that it had concentrations of sodium acetate at 0.11 M $NaC_2H_3O_2$ and acetic acid at 0.090 M $HC_2H_3O_2$. What is the pH of this solution?

ANALYSIS Equation 9.11 defines K_a, an equation you should now know. It is the only equation you really need to remember to work buffer problems. The rearranged form, Equation 9.22, is derived from it. But we'll use 9.11 to emphasize how the simplifications work.

SOLUTION We note first, from Table 9.2, that K_a for acetic acid is 1.8×10^{-5}. We have

$$K_a = \frac{[H^+][A^-]}{[HA]} \qquad \text{(This is Equation 9.11.)}$$

Here is where we remember the crucial substitutions allowed when we are working with a buffer system — [anion] for $[A^-]$ and [acid] for [HA].

$$K_a = \frac{[H^+][A^-]}{[HA]} = \frac{[H^+][\text{anion}]}{[\text{acid}]}$$

Because [anion] = 0.11 mol/L, [acid] = 0.090 mol/L, and K_a = 1.8 × 10⁻⁵, and solving for [H⁺], we have

$$[H^+] = K_a \times \frac{[acid]}{[anion]} = 1.8 \times 10^{-5} \times \frac{0.090}{0.11} \text{ mol/L}$$

$$= 1.5 \times 10^{-5} \text{ mol/L}$$

Finally, we calculate the pH.

$$pH = -\log[H^+]$$
$$pH = -\log(1.5 \times 10^{-5})$$
$$= 4.82$$

Thus the pH of this buffer solution is 4.82.

PRACTICE EXERCISE 22

A buffer solution was prepared using 0.085 M formic acid, $HCHO_2$, and sodium formate, $NaCHO_2$, dissolved in the same solution at a concentration of 0.12 mol/L. Calculate the pH of this solution. For formic acid, K_a = 1.8 × 10⁻⁴.

The pH of a Buffer Solution Can Also Be Found by the Henderson-Hasselbalch Equation

We can convert Equation 9.22 into a form that includes pH instead of [H⁺]. We'll use the result, called the *Henderson-Hasselbalch equation,* to continue our study of the carbonate buffer in blood.

If we take the logarithm of both sides of Equation 9.22, and then multiply every resulting term by −1, we get

$$-\log[H^+] = -\log K_a - \log \frac{[acid]}{[anion]}$$

We can recognize expressions for pH and pK_a in this, so we can write

■ $-\log[H^+] = pH$
   $-\log K_a = pK_a$

If we note that

$$pH = pK_a - \log \frac{[acid]}{[anion]}$$

■ Another rule of logarithms:

$-\log a/b = +\log b/a$

$$-\log \frac{[acid]}{[anion]} = \log \frac{[anion]}{[acid]}$$

we obtain

$$pH = pK_a + \log \frac{[anion]}{[acid]} \tag{9.23}$$

Equation 9.23 is the **Henderson-Hasselbalch equation.**[3] Let's rework Example 9.13 using it.

---

[3] Some references give the Henderson-Hasselbalch equation as

$$pH = pK_a + \log \frac{[salt]}{[acid]}$$

In other words, [salt] is used instead of [anion], as though the two were always identical in value. They are identical only when the cation of the salt is of the form M^+ (e.g., Na⁺ or K⁺) so that each formula unit of the salt furnishes only *one* anion. But when the cation is of the form M^{2+} (e.g., Ca²⁺), then *two* anions are released by the dissociation of only one formula unit of the salt. With such salts the value of [anion] is twice the value of [salt]. It is safest to stick with the form of Equation 9.23.

EXAMPLE 9.14
Calculating the pH of a Buffered Solution

Calculate the pH of the buffer solution described in Example 9.13.

ANALYSIS This is a problem for which the Henderson-Hasselbalch equation may be used.

SOLUTION We note first that for acetic acid $K_a = 1.8 \times 10^{-5}$, so $pK_a = 4.74$. ($pK_a = -\log K_a$.) In the buffer of Example 9.13, [anion] = 0.11 mol/L and [acid] = 0.090 mol/L. Substituting these values into the Henderson-Hasselbalch equation, we get

$$pH = pK_a + \log \frac{(0.11)}{(0.090)}$$
$$= 4.74 + \log 1.2$$
$$= 4.74 + 0.079$$
$$pH = 4.82$$

PRACTICE EXERCISE 23

Calculate the pH of a buffered solution made up to be 0.016 M sodium acetate and 0.12 M acetic acid.

Buffers Hold a pH Steady, but Not Necessarily at pH 7 One important point about buffers is the distinction between keeping a solution at a particular pH and keeping it neutral — at a pH of 7. Although it is certainly possible to prepare a buffer to work at pH 7, buffers can be made that will work at any pH value throughout the pH scale.

The Ratio [Anion]/[Acid] Dominates the pH Once a Buffer Pair Is Selected The Henderson-Hasselbalch equation very clearly shows that two factors govern the pH of a buffered solution. The first is the pK_a of the weak acid in the buffer pair and the second is the ratio [anion]/[acid]. To decide what weak acid and salt to use in a buffer, we first decide the pH that we want to protect. Then we look for a weak acid with a pK_a as close to it as possible. We can see from Equation 9.23 that if we prepare a buffer with the ratio of [anion] to [acid] made equal to 1, then the pH of the buffered solution equals the pK_a of the weak acid.

■ When [anion] = [acid],

$$\log \frac{[\text{anion}]}{[\text{acid}]} = \log \frac{1}{1} = 0$$

So then pH = pK_a + 0.

Of course, when we work with the buffered solutions found in a living system, we have to take what nature gives us. And nature gives us carbonic acid [or the stand-in, $CO_2(aq)$] as the acid component of the chief blood buffer. This makes the second factor in the Henderson-Hasselbalch equation decisive for this buffer.

The second factor, as we said, is the *ratio* [anion]/[acid]. The pH of a buffer solution depends entirely on this once the acid component with its pK_a has been picked. Notice particularly that it isn't the absolute values of [anion] and of [acid] *but the ratio of these values* that determines the pH of the buffer. You could get a 1 : 1 ratio, for example, with [anion] and [acid] both equal to 0.50 mol/L or both equal to 0.25 mol/L or to 0.10 mol/L. The pH of the buffer would be the same as long as the acid has not been changed.

PRACTICE EXERCISE 24

How is the pH of a buffered solution related by an equation to the pK_a of the weak acid in the buffer when

(a) The ratio of [anion] to [acid] is 10 to 1?

(b) The ratio of [anion] to [acid] is 1 to 10?

Buffers Minimize but Do Not Completely Prevent pH Changes Let's compare what happens if we add a small amount of NaOH, a strong base, to pure water with what happens when we add the same amount to a buffer solution.

■ $[OH^-] = 1.0 \times 10^{-pOH}$

When we add 0.010 mol of NaOH to 1.0 L of pure water, the concentration of OH^- ion becomes 0.010 mol/L. Because 0.010 mol OH^-/L is the same as 1.0×10^{-2} mol OH^-/L, we can see that the pOH is 2 ($pOH = -\log[OH^-]$). This makes the pH 12 ($pH + pOH = 14.00$), so the addition of 0.010 mol of NaOH (only 0.40 g of NaOH) to a liter of pure water has changed the pH from 7 to 12. The concentration of OH^- ion has gone from 1.0×10^{-7} to 1.0×10^{-2} mol/L, a whopping 100,000-fold increase in $[OH^-]$. Now let's add 0.010 mol of NaOH to 1.0 L of a hypothetical buffer in which the weak acid component has $pK_a = 7.00$ and in which [anion] = [acid] = 0.10 mol/L. (Because [anion] = [acid], the pH of the buffer equals the specified pK_a or 7.00.) When we add 0.010 mol of OH^- to this solution, we neutralize 0.010 mol of the acid. This reduces the amount of acid by 0.010 mol, but it also increases the amount of anion by 0.010 mol. So after the addition of the sodium hydroxide, we have the following new concentrations of the buffer components.

$$[anion] = (0.10\ mol + 0.010\ mol)/L = 0.11\ mol/L$$
$$[acid] = (0.10\ mol - 0.010\ mol)/L = 0.090\ mol/L$$

By the Henderson–Hasselbalch equation, the new pH is

$$pH = 7.00 + \log \frac{(0.11)}{(0.090)}$$
$$= 7.09$$

When the buffer is present, in other words, the addition of 0.010 mol of NaOH causes a change in pH from 7.00 to 7.09, a change of only 0.09 unit. (It can be calculated that this amounts to a 1.25-fold increase in OH^- concentration, a vastly smaller change than the 100,000-fold increase.) Thus, although the buffer has not prevented some change in pH, it has surely kept the change very small.

PRACTICE EXERCISE 25

Suppose that in the illustration just concluded the hypothetical buffer had been more dilute, say, [anion] = [acid] = 0.050 mol/L. (The weak acid still has $pK_a = 7.00$.)

(a) What is the pH of this buffer?

(b) What is the pH of 1.0 L of this buffer after 0.010 mol of NaOH has been added?

(c) Compare the result in part (b) to the result of the previous illustration; there we began with 1.0 L in which [anion] = [acid] = 0.10 mol/L. Can we conclude that the *capacity* of a buffer solution to keep a pH change to a minimum depends on the *absolute values* of the concentrations of acid and anion or does it depend only on the *ratio* of these concentrations?

PRACTICE EXERCISE 26

Suppose that so much strong base is added to a buffer made of HA and A^-, initially at a ratio of one to one so that the ratio of [anion] to [acid] changes to 1000 to 1. The pK_a of HA is 4.75. What is the new pH of the solution? Can one conclude from this result that buffers have an *unlimited* capacity to hold a pH fairly constant?

Buffers Resist Changes in Their pH Values Even as They Are Diluted with Water
If you have ever donated blood you know that afterward you are asked to drink more fluids than usual to replace the lost volume. Fortunately, and for good reason, replacing the lost

blood volume by *water* rather than a buffer solution does not seriously affect the pH of your blood. The "good reason" is the relationship of Equation 9.23:

$$pH = pK_a + \log \frac{[anion]}{[acid]}$$

The value of [anion] is a *ratio* — mole of anion per liter of solution. Likewise, the value of [acid] is a *ratio* — mole of acid per liter of solution. If we change the volume of the solution *identically* for both anion and acid simply by adding water, the *ratio* of [anion] to [acid] in the equation must remain the same as before. The actual values of the concentrations, [anion] and [acid], do decrease because the blood is being diluted, but both concentrations decrease *by the same factor*. Thus the ratio of the two — [anion] to [acid] — is unchanged by the dilution and so the pH is unchanged. There is a limit to how much dilution of the blood one's system can stand, of course, but your thirst mechanism will switch off any desire to exceed the limit.

Dissolved CO₂ Must Be Factored into a Henderson-Hasselbalch Treatment of the Blood's Carbonate Buffer

The Henderson-Hasselbalch equation must be slightly modified so that it uses the formula of carbon dioxide, not the formula of the associated weak acid, H_2CO_3. We also make a small but significant change in one symbol: pK' for pK_a.

$$pH = pK' + \log \frac{[HCO_3^-(aq)]}{[CO_2(aq)]} \qquad (9.24)$$

We cannot employ the usual symbol, pK_a, because we are not working with a system at 25 °C but at 37 °C. Moreover, we're not working with H_2CO_3 for which K_a is 4.5×10^{-7} and pK_a is 6.35 (at 25 °C for the ionization of the first proton). To handle these different conditions, we use what is called an *apparent acid ionization constant*, symbolized as K', and a corresponding apparent pK'. The accepted value of pK' for the carbonate buffer under body conditions is 6.1, so we can rewrite Equation 9.24 as follows when we are dealing with the carbonate buffer in the body:

■ K_a values are usually assumed to be for 25 °C unless another temperature is specified, like body temperature.

$$pH = 6.1 + \log \frac{[HCO_3^-(aq)]}{[CO_2(aq)]} \qquad (9.25)$$

Under normal pH conditions in human arterial blood, $[HCO_3^-] = 24$ mmol/L, and $[CO_2] = 1.2$ mmol/L. The pH of human arterial blood is then calculated, as follows, to be 7.4.

$$pH = 6.1 + \log \frac{(24 \text{ mmol/L})}{(1.2 \text{ mmol/L})}$$

$$= 7.4$$

Statistically, the average human arterial blood in health is 7.35, so the calculated and the observed values agree well.

Normal Exhaling of CO₂, Hyperventilation, and Resupply of HCO₃⁻ by the Kidneys All Work to Combat Acidosis

Let's suppose that the blood is challenged with a sudden influx of acid in the equivalent of 10 mmol of HCl per liter of blood. This would neutralize 10 mmol/L of HCO_3^- and so reduce its concentration from 24 mmol/L to 14 mmol/L. The HCO_3^-, of course, changes almost entirely to $CO_2(aq)$, so 10 mmol/L of new $CO_2(aq)$ appears in the blood. If this new CO_2 could not be removed by breathing, the level of $CO_2(aq)$ would increase by 10 mmol/L, from 1.2 mmol/L to 11.2 mmol/L. This would be fatal because, by Equation 9.25, the resulting pH of the blood would be 6.2, far, far too low to permit life to continue.

$$pH = 6.1 + \log \frac{(14 \text{ mmol/L})}{(11.2 \text{ mmol/L})}$$

$$= 6.2$$

However, essentially all the $CO_2(aq)$ leaves the blood in the lungs and is breathed out as $CO_2(g)$. Although the level of HCO_3^- stays the same, at 14 mmol/L, the level of $CO_2(aq)$ quickly drops back to 1.2 mmol/L. So the pH quickly bounces back up to 7.2.

$$pH = 6.1 + \log \frac{(14 \text{ mmol/L})}{(1.2 \text{ mmol/L})}$$

$$= 7.2$$

■ A pH of 7.2 corresponds to fairly severe acidosis.

Of course, a blood pH of 7.2 is still too low for health, but it doesn't cause death.

There is another mechanism to provide further upward readjustment of the blood pH. The body, when healthy, responds quickly to a lowering of the blood pH by increasing the rate of breathing, by causing *hyperventilation*. This works to force even more CO_2 out of the blood and into the exhaled air. It's quite common for hyperventilation to pull the level of $CO_2(aq)$ from 1.2 mmol/L down to 0.70 mmol/L, sometimes a bit lower. This would bring the pH of the blood in our example — the sudden influx of 10 mmol/L of acid — back to 7.4.

$$pH = 6.1 + \log \frac{(14 \text{ mmol/L})}{(0.70 \text{ mmol/L})}$$

$$= 7.4$$

Thus two mechanisms have protected the system against the otherwise lethal assault of an influx of 10 mmol/L of strong acid. The buffer system has neutralized the acid, and the respiratory system has, by removing CO_2, readjusted the ratio of $[HCO_3^-]/[CO_2]$. Neither of these two mechanisms could save the situation alone.

The situation, however, is not back to normal in all respects. The system will continue to produce CO_2, which will increase the denominator in $[HCO_3^-]/[CO_2]$, and so the pH will gradually decline again. The system must, therefore, also increase the level of HCO_3^-, and this is done principally by the chemical work of the kidneys. The kidneys, as we have said, can manufacture "new" HCO_3^- to replenish the base of the blood's carbonate buffer, and as they do this the kidneys can also export H^+. We won't delve further into this here, but you can see that the management of the pH of the blood, so that respiration may proceed well, involves an intricate interplay of chemical and physiological events at the molecular level of life. The smooth, harmonious working of these events in the healthy body is one of the grandest, most beautiful aspects in all of nature. Moreover, if your intended career is a field of medicine, particularly in a primary care area like nursing, you should master the concepts of this section. The author has spoken with many experienced nurses and physicians who vigorously agree.

■ In battling severe acidosis, the pH of the urine can go as low as 4.

9.11

ACID–BASE TITRATIONS

At the end point of an acid–base titration, the number of moles of H^+ should match the number of moles of H^+ acceptor.

One of the very common kinds of chemical analysis is to determine the concentration of an acid or a base. The purpose is to find the molarity of some whole solute, like moles of acetic acid per liter of solution, not just to measure the pH of the solution. Thus we have to make some distinctions between the kinds of acid species present.

The pH of a Solution Refers to $[H^+]$, Not to $[HA]$ The pH of a solution tells us indirectly the *acidity* of a solution, what the concentration of hydronium ions is. It does not, however, disclose the **neutralizing capacity** of the solution — its capacity to neutralize a strong base. The 1 mol of acetic acid in 1 L of 1 M $HC_2H_3O_2$ can neutralize 1 mol of NaOH, yet this solution has an actual quantity of H_3O^+ of only about 0.004 mol, the result of acetic acid being a weak acid.

Always remember that acids are classified as *weak* or *strong* according to their abilities to transfer a proton to one particular and very weak base, H_2O. When sodium hydroxide is added to an acid, however, we are adding a very strong base, OH^-. This base can take H^+ not only from H_3O^+ but also from $HC_2H_3O_2$. Thus the neutralizing capacity of 1 M acetic acid is considerably greater than its concentration of hydronium ions.

The Titration of an Acid with a Base Gives Data from Which Concentrations Can Be Calculated   The procedure used to measure the total acid (or base) neutralizing capacity of a solution is called **titration**. It involves comparing the volume of a solution of unknown concentration to the volume of a *standard solution* that exactly neutralizes it. A **standard solution** is simply one whose concentration is accurately known.

The apparatus for titration is shown in Figure 9.13. When a titration is used for an acid–base analysis, a carefully measured volume of the solution of unknown acidity (or basicity) is placed in a beaker or a flask. A very small amount of an acid–base indicator, like phenolphthalein, is added. Then a *standard solution* of the neutralizing reagent is added through a stopcock, portion by portion, from a special tube called a *buret*, marked in 1-mL and 0.1-mL divisions (Fig. 9.13). This addition is continued until the end point is reached—a change in color, caused by the acid–base indicator that signals that the unknown has been exactly neutralized.

■ The careful measurement of the concentration of a standard solution is called **standardization.** We say that we standardize the solution.

End Points Ideally Occur at Equivalence Points   With a carefully selected acid–base indicator, the color change in an acid–base titration occurs when all the available hydrogen ions have reacted with all the available proton acceptors. This point in a titration is called the **equivalence point.**

A well-chosen indicator is one whose color at the equivalence point is the same as it would be in a solution made up of the *salt* that forms in the titration (and in the same concentration). When this salt has an ion that hydrolyzes, the equivalence point cannot be at pH 7.00. For example, when one mole of acetic acid has been exactly neutralized by one mole of sodium hydroxide, exactly one mole of sodium acetate has been made. Because the acetate ion hydrolyzes (but not the sodium ion), this salt gives a solution that is slightly basic to litmus, not a solution with a pH of 7.00. So it would be poor to pick an indicator that changes color over an acidic range. (Phenolphthalein works very well in this titration.) Whether or not the indicator has been well-chosen, the analyst has little choice but to stop the titration when the indicator's color changes. This stopping point is called the **end point** of the titration. In a well-run titration, of course, the end point and the equivalence point coincide.

■ To reach the equivalence point the moles of H^+ used must be the same as (must be *equivalent to*) the moles of proton acceptor present.

■ The pH of 1 M $NaC_2H_3O_2$ is about 9.4.

In Chapter 3 we studied how to do calculations involving volumes and concentrations of solutions with an emphasis on calculating volumes. Acid–base titrations, however, are usually done to determine concentrations, so we'll work through an example to see how molarities can be calculated from titration data.

EXAMPLE 9.15
Calculating Molarities from Titration Data

A student titrated 25.0 mL of sodium hydroxide solution with standard sulfuric acid. It took 13.4 mL of 0.0555 M H_2SO_4 to neutralize the sodium hydroxide in the solution. What was the molarity of the sodium hydroxide solution? The equation for the reaction is

$$2NaOH(aq) + H_2SO_4(aq) \longrightarrow Na_2SO_4(aq) + 2H_2O$$

ANALYSIS   Be sure to understand the goal first. We are to calculate the molarity of the NaOH, which means the ratio of the moles of NaOH to liters of NaOH solution. We were given (indirectly) the liters of the NaOH solution, 25.0 mL = 0.0250 L. To find the moles

FIGURE 9.13
The apparatus for titration. By manipulating the stopcock, the analyst controls the rate at which the solution in the buret is added to the flask below.

of NaOH we need two conversion factors, one involving the molarity of the acid, and the other involving the coefficients in the equation.

The molarity of the H_2SO_4 solution, 0.0555 M, gives us the following conversion factors.

$$\frac{0.0555 \text{ mol } H_2SO_4}{1000 \text{ mL } H_2SO_4 \text{ soln}} \quad \text{and} \quad \frac{1000 \text{ mL } H_2SO_4 \text{ soln}}{0.0555 \text{ mol } H_2SO_4}$$

The first of these will enable us to calculate the moles of H_2SO_4 in 13.4 mL of 0.0555 M H_2SO_4.

The balanced equation gives us the following conversion factors that relate moles of acid used to moles of NaOH required.

$$\frac{1 \text{ mol } H_2SO_4}{2 \text{ mol NaOH}} \quad \text{and} \quad \frac{2 \text{ mol NaOH}}{1 \text{ mol } H_2SO_4}$$

The second factor is the one we'll use to find the moles of NaOH that are equivalent to the moles of acid in 13.4 mL of 0.0555 M H_2SO_4.

SOLUTION   We first take the given volume of the acid and convert it into the number of moles of H_2SO_4 that were taken.

$$13.4 \text{ mL } H_2SO_4 \text{ soln} \times \frac{0.0555 \text{ mol } H_2SO_4}{1000 \text{ mL } H_2SO_4 \text{ soln}} = 7.44 \times 10^{-4} \text{ mol } H_2SO_4$$

The number of moles of NaOH that are equivalent to 7.44×10^{-4} mol H_2SO_4 *in the given reaction* is found by

$$7.44 \times 10^{-4} \text{ mol } H_2SO_4 \times \frac{2 \text{ mol NaOH}}{1 \text{ mol } H_2SO_4} = 1.49 \times 10^{-3} \text{ mol NaOH}$$

Thus 0.00149 mol of NaOH was present in 25.0 mL or 0.0250 L of NaOH solution. To find the molarity of the NaOH solution, we take the following ratio of moles to liters:

$$\frac{0.00149 \text{ mol NaOH}}{0.0250 \text{ L NaOH soln}} = 0.0596 \text{ } M \text{ NaOH}$$

Thus the concentration of the NaOH solution is 0.0596 M.

PRACTICE EXERCISE 27

If it takes 24.3 mL of 0.110 M HCl to neutralize 25.5 mL of freshly prepared sodium hydroxide solution, what is the molarity of the NaOH solution?

PRACTICE EXERCISE 28

If 20.0 mL of 0.125 M solution of NaOH exactly neutralized the sulfuric acid in 10.0 mL of H_2SO_4 solution, what was the molarity of the H_2SO_4 solution?

SUMMARY

Reaction kinetics   Virtually all chemical reactions have an energy of activation. Reactant particles more frequently surmount this barrier—the reaction happens faster—the more concentrated they are and the more readily their collisions have the proper combined total collision energy. Raising the temperature of a react- ing mixture increases the frequency of successful collisions (makes the rate of reaction faster). A catalyst, such as any enzyme, lowers the energy of activation without affecting the overall heat of reac- tion, so a catalyst also increases the rate of a reaction.

Equilibrium laws An equilibrium law exists for every chemical equilibrium. It is of the form

$$K_{eq} = \frac{[C]^c[D]^d}{[A]^a[B]^b}$$

when the equilibrium is of the type

$$aA + bB \rightleftharpoons cC + dD$$

The equilibrium constant, K_{eq}, which is different for each equilibrium, depends only on the temperature. If any stress other than a temperature change is placed on an equilibrium, such as the addition of a common ion, all concentrations adjust, but K_{eq} stays the same. A large value of K_{eq} means that the products are favored at equilibrium; a small value means that the reactants are favored.

Ion product constant of water, K_w The ion-product constant of water, K_w, equals $[H^+][OH^-]$. At 25 °C, $K_w = 1.0 \times 10^{-14}$. If acids or bases are added, the value of K_w stays the same but individual values of $[H^+]$ and $[OH^-]$ adjust.

The pH concept A simple way to express very low values of $[H^+]$ is by pH, where $[H^+] = 1 \times 10^{-pH}$, or $pH = -\log [H^+]$. When $pH < 7$, the solution is acidic. When $pH > 7$, the solution is basic. An analogous pOH concept exists. $pOH = -\log [OH^-]$, and at 25 °C, $pH + pOH = 14.00$.

To measure the pH of a solution we use indicators, dyes whose colors change over a narrow range of pH, or we use a pH meter.

Acid ionization constants A modified equilibrium law defining the acid ionization constant, K_a, exists for weak acids. Letting HA represent the acid,

$$K_a = \frac{[H^+][A^-]}{[HA]}$$

for the equilibrium,

$$HA \rightleftharpoons H^+ + A^-$$

For weak acids, $K_a < 10^{-3}$. For moderate acids K_a is roughly 1 to 10^{-3}, and for strong acids $K_a > 1$.

Besides weak acids, like HA, most metal ions in water generate hydrogen ions by the ionization of a water molecule attracted to the metal ion (e.g., in the hydrated metal ion). The exceptions that do not make a solution acidic are the group IA and IIA cations (below beryllium). Ammonium salts also give acidic solutions.

Base ionization constants Bases weaker than OH^- or O^{2-} occur mostly as anions, the conjugate bases of weak acids. Ammonia is also a base weaker than OH^-. A base, B, ionizes according to the equilibrium,

$$B + H_2O \rightleftharpoons BH^+ + OH^-$$

The base ionization constant is then

$$K_b = \frac{[BH^+][OH^-]}{[B]}$$

Strong bases have large values of K_b. Weak bases have small values of K_b.

The reaction of a basic anion with water is called the hydrolysis of the anion, and a salt of such an anion with any group IA or IIA metal (except beryllium) gives a basic solution in water. The anions that do not hydrolyze are the conjugate bases of strong acids, like Cl^-, Br^-, I^-, NO_3^-, and SO_4^{2-}.

The pK_a and pK_b concepts The negative logarithm of K_a is pK_a, and the negative logarithm of K_b is pK_b.

$$pK_a = -\log K_a$$
$$pK_b = -\log K_b$$

The weaker the acid the larger its pK_a. The weaker the base the larger its pK_b. $pK_a + pK_b = 14.00$ at 25 °C.

Buffers Solutions that contain something that can neutralize OH^- ion (such at a weak acid) and something else that can neutralize H^+ ion (such as the conjugate base of the same weak acid) are buffered against changes in pH when either additional base or acid is added. The phosphate buffer, which is present in the fluids inside cells of the body, consists of HPO_4^{2-} (to neutralize H^+) and $H_2PO_4^-$ (to neutralize OH^-).

The normal pH of blood is 7.35. A decrease in the pH of blood is called acidosis and an increase is called alkalosis. Either condition interferes with respiration, and extreme cases ($pH < 7$ or $pH > 8$) are lethal.

The carbonate buffer, which is the chief buffer in blood, consists of HCO_3^- (to neutralize H^+) and dissolved CO_2 (to neutralize OH^-). Dissolved CO_2, or CO_2 (aq), reacts directly with OH^-. In the lungs, CO_2 is expelled. The ability to exhale CO_2 is essential to the prevention of acidosis.

When metabolism or some deficiency in respiration produces or retains H^+ at a rate faster than the blood buffer can neutralize it, the lungs try to remove CO_2 at a faster rate (hyperventilation). Overall, for each molecule of CO_2 exhaled, one proton is neutralized.

The pH of a buffer solution consisting of a weak acid, HA, and its conjugate base, A^-, can be calculated from the defining equation for K_a. In such a buffer at equilibrium, for all practical purposes, $[HA]_{eq} = [acid]_{init}$ and $[A^-]_{eq} = [anion]_{init}$. So the defining equation for K_a can be modified to express $[H^+]$ as

$$[H^+] = K_a \times \frac{[acid]}{[anion]}$$

The defining equation for pH lets us convert this equation to the Henderson-Hasselbalch equation:

$$pH = pK_a + \log \frac{[anion]}{[acid]}$$

The two factors that determine at what pH a buffer works are the pK_a of the weak acid component and the *ratio* [anion]/[acid]. As this ratio varies from 10/1 to 1/1 to 1/10, the pH varies as $pK_a \pm 1$.

A buffer thus does not keep a solution necessarily at a pH of 7, but it holds a fairly constant pH when extra acid or base enter.

At body temperature Henderson-Hasselbalch equation takes the following working form in which $[CO_2(aq)]$ appears in the place of $[H_2CO_3]$.

$$pH = 6.1 + \log \frac{[HCO_3^-]}{[CO_2(aq)]}$$

If an acid neutralizes some HCO_3^-, the ratio in the log term decreases too much unless the lungs simultaneously breathe out the extra CO_2 produced. If the lungs are able to work, some hyperventilation helps this process, and the pH of the blood stays quite close to

7.35. Over a longer period, the supply of HCO_3^- is replenished by the work of the kidneys.

Acid–base titration The concentration of an acid or a base in water can be determined by titrating the unknown solution with a standard solution of what can neutralize it. The indicator is selected to have its color change occur at whatever pH the final solution would have were it made from the salt that forms by the neutralization. The unknown concentration can be calculated from the volumes of the acid and base used, the molarity of the standard solution, and the coefficients in the equation for the specific neutralization reaction.

REVIEW EXERCISES

The answers to Review Exercises whose numbers are in color are found in Appendix C. The answers to the other Review Exercises are found in the Study Guide that accompanies this book. The more challenging questions are marked with asterisks.

Collision Theory and the Factors That Affect Reaction Rates

9.1 What aspects of chemical reactions are studied in the field of *kinetics*?

9.2 The *rate of reaction* is defined in terms of a ratio.
(a) What is the ratio?
(b) What *units* are usually used to express the ratio?

9.3 What are the main factors that govern the rate of a reaction? What factor distinguishes a homogeneous from a heterogeneous reaction?

9.4 The rate of a given reaction is generally most rapid at the beginning of the reaction, rather than later.
(a) What factor affecting reaction rates is at work to account for this?
(b) How do we use the kinetic theory to explain this?

9.5 When an increase in the rate of a reaction has been caused by an increase in the concentration of one of the reactants, which of the following factors has been changed? (Identify the factors by letter.)
A The energy of activation
B The heat of reaction
C The frequency of collisions
D The frequency of successful collisions

9.6 How do we explain the rate-increasing effect of an increase in temperature?

9.7 As a rule of thumb, how much of a temperature increase doubles or triples the rates of most reactions?

9.8 Explain how an increase in body temperature can lead to a strain on the heart.

9.9 How can we increase the frequency of all collisions in a reacting mixture without raising the temperature?

9.10 In what way, if any, does a catalyst affect the following factors of a chemical reaction?

(a) The heat of reaction
(b) The energy of activation
(c) The frequency of collisions
(d) The frequency of successful collisions

9.11 What is the general name for the catalysts found in living systems?

9.12 Name the two catalysts mentioned in this chapter and describe what they do by means of equations.

9.13 A catalyst speeds up a reaction at a given temperature.
(a) In what other way can a catalyst affect the ease of the reaction?
(b) Why are catalysts strongly sought for large-scale industrial syntheses of compounds?

9.14 Collision theory holds that something must happen for a reaction to occur between two particles. What must occur and why?

9.15 Distinguish between *collision frequency* and *rate of reaction*.

9.16 In what *two* ways does an increase in temperature affect a reaction rate?

9.17 Consider the concepts of E_{act} and heat of reaction.
(a) Which of the two is associated with *rate*?
(b) Which of the two is associated with whether the reaction is endothermic or exothermic?
(c) How can a very exothermic reaction be very *slow*?
(d) If an energy of activation for a reaction is zero, what can be said to describe the reaction rate?
(e) A catalyst affects which factor more, E_{act} or heat of reaction?

9.18 In terms of what we visualize as happening when two molecules interact to form products, how do we explain the existence of an energy barrier to the reaction—an energy of activation?

9.19 Study the accompanying progress of reaction diagram for the conversion of carbon monoxide and oxygen to carbon dioxide, and then answer the questions. The equation for the reaction is

$$2CO(g) + O_2(g) \longrightarrow 2CO_2(g)$$

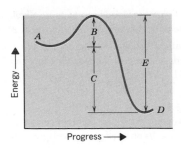

(a) What substance or substances occur at position A?

(b) What substance or substances occur at position D?

(c) Which letter labels the arrow that represents the heat of reaction?

(d) Which letter labels the arrow that stands for the energy of activation?

(e) Is this reaction endothermic or exothermic? How can you tell?

(f) Which letter labels the arrow that would correspond to the energy of activation if the reaction could go in reverse?

*9.20 Suppose that the following hypothetical reaction occurs.

$$A + B \longrightarrow C + D$$

Suppose further that this reaction is endothermic and that the energy of activation is numerically twice as large as the heat of reaction. Draw a progress of reaction diagram for this reaction, and draw and label arrows that correspond to the energy of activation and the heat of reaction.

9.21 The reaction of X and Y to form Z is exothermic. For every mole of Z produced, 10 kcal of heat are generated. The energy of activation is 3 kcal. Sketch the energy relationships on a progress of reaction diagram.

9.22 What did Guldburg and Waage discover about chemical equilibria?

*9.23 In setting up an equilibrium law for a system,

(a) What concentration units are assumed?

(b) How are the coefficients of the equation used?

(c) The terms for what substances, reactants or products, appear in the denominator?

(d) Why must the temperature be specified?

9.24 Write the equilibrium laws for the following.

(a) $NH_3(aq) + H_2O \rightleftharpoons NH_4^+(aq) + OH^-(aq)$

(b) $2H_2(g) + O_2(g) \rightleftharpoons 2H_2O(g)$

(c) $N_2(g) + 3H_2(g) \rightleftharpoons 2NH_3(g)$

9.25 The equilibrium in which ethylene reacts with water to give ethyl alcohol is

$$C_2H_4(g) + H_2O(g) \rightleftharpoons C_2H_5OH(g) \quad K_{eq} = 8.3 \times 10^3$$

Is the product favored? How can you tell?

Ion Product Constant of Water

9.26 Write the equation that defines K_w. How does it differ from the equilibrium law for water (written using H^+ and OH^-)?

9.27 What is the value of K_w at 25 °C?

9.28 The higher the temperature, the higher the value of K_w. Why should there be this trend?

9.29 At the temperature of the human body, 37 °C, the concentration of hydrogen ion in pure water is 1.56×10^{-7} mol/L. What is the value of K_w at 37 °C? Is this pure water acidic, basic, or neutral?

*9.30 "Heavy water" or deuterium oxide, D_2O, is used in nuclear power plants. It self-ionizes like water, and at 20 °C there is a concentration of D^+ ion of 3.0×10^{-8} mol/L. What is the value of K_w for heavy water at 20 °C?

pH and pOH

9.31 What equation defines pH in exponential terms? In log terms?

9.32 The average pH of saliva is about 6.8. Is saliva acidic, neutral, or basic?

9.33 The pH of pancreatic juice, a digestive juice of the intestinal tract, is in the range of 7 to 8. Is pancreatic juice acidic, basic, or neutral?

9.34 What is the pH of 0.001 M HCl(aq), assuming 100% ionization?

9.35 What is the pOH of 0.001 M NaOH(aq), assuming 100% ionization? What is the pH of this solution?

9.36 Explain why a pH of 7.00 corresponds to a neutral solution at 25 °C.

9.37 A certain brand of soft drink has a pH of 5.35. What is the concentration of hydrogen ion in moles per liter? Is the soft drink slightly acidic or basic?

9.38 The pH of a fruit juice was found to be 4.82. What is the value of $[H^+]$?

*9.39 A solution of a monoprotic acid was prepared with a molar concentration of 0.010 M. Its pH was found to be 2.00. Is the acid a strong or a weak acid? Explain.

*9.40 The pH of a solution of a monoprotic acid was found to be 5.72, whereas its molar concentration was 0.0010 M. Is this acid a strong or a weak acid? Explain.

9.41 When a soil sample was stirred with pure water, the pOH of the water changed to 6.24. Did the soil produce an acidic or a basic reaction with the water?

*9.42 A solution was prepared by dissolving 0.316 g of Ba(OH)$_2$ in a final volume of solution of 100 mL. Calculate the pOH and the pH of this solution, assuming that the Ba(OH)$_2$ is fully dissociated.

Acid Ionization Constants

9.43 Write the equilibrium equation and the equation for K_a for the ionization of nitrous acid, HNO_2.

9.44 Write the equilibrium equation and the equation for K_a for the ionization of the hydrogen sulfate ion, HSO_4^-.

9.45 Write the equilibrium equation and the equation for K_a for the ionization of ammonia as an *acid*.

9.46 The K_a for the hydrogen sulfite ion, HSO_3^-, is 6.6×10^{-8} and for barbituric acid is 9.9×10^{-5}. Which is the stronger acid?

9.47 The K_a for the ammonium ion is 5.7×10^{-10} and for the hydrogen sulfide ion, HS^-, is 1×10^{-19}. Which is the stronger acid?

Base Ionization Constants

9.48 Write the equilibrium equation and the equation for K_b for the formate ion, HCO_2^-, acting as a base.

9.49 Write the equilibrium equation and the equation for K_b for the nitrite ion, NO_2^-, acting as a base.

9.50 Write the equilibrium equation and the equation for K_b for the monohydrogen phosphate ion acting as a base.

9.51 Which is the stronger base, ammonia ($K_b = 1.8 \times 10^{-5}$) or the hypochlorite ion, OCl^- ($K_b = 3.3 \times 10^{-7}$)?

9.52 Explain why a solution of ammonium bromide tests slightly acidic.

9.53 Explain in your own words why a solution of sodium sulfide, Na_2S, is basic, not neutral.

*9.54 Predict whether each of the following solutions is acidic, basic, or neutral.
(a) KNO_3 (b) NH_4Br (c) $NaHCO_3$ (d) $FeCl_3$
(e) Li_2CO_3

*9.55 Predict whether each of the following solutions is acidic, neutral, or basic.
(a) Na_2SO_4 (b) K_2HPO_4 (c) K_3PO_4 (d) $Cr(NO_3)_3$
(e) $KC_2H_3O_2$

9.56 Aspirin is a weak, monoprotic acid for which $K_a = 3.3 \times 10^{-4}$. Does a solution of the sodium salt of aspirin test acidic, basic, or neutral?

9.57 The K_b of the hydrogen sulfide ion, HS^-, is 1.1×10^{-7}. Does a solution of sodium hydrogen sulfide, $NaSH$, test acidic, basic, or neutral?

pK_a and pK_b

9.58 Calculate the pK_a values of the following acids.
(a) HF ($K_a = 6.8 \times 10^{-4}$), hydrofluoric acid
(b) HOCl ($K_a = 3.0 \times 10^{-8}$), hypochlorous acid

9.59 Calculate the pK_b values of the following bases.
(a) NO_2^- ($K_b = 1.4 \times 10^{-11}$), nitrite ion
(b) CHO_2^- ($K_b = 5.6 \times 10^{-11}$), formate ion

9.60 What are the K_b and the pK_b values for the conjugate bases of the acids in Practice Exercise 9.58?

9.61 What are the K_a and the pK_a values for the conjugate acids of the bases given in Practice Exercise 9.59?

9.62 Acid X has a pK_a of 5.68 and acid Y has a pK_a of 6.25. Which is the stronger acid? Which has the stronger conjugate base?

9.63 Base M has a pK_b of 8.02 and base N has a pK_b of 5.67. Which is the stronger base? Which has the weaker conjugate acid?

Buffers

9.64 In the study of the molecular basis of life, why is the study of buffers important?

9.65 What is acidosis? What is alkalosis?

9.66 In very general terms, why are both acidosis and alkalosis serious?

9.67 Following surgery, a patient experienced persistent vomiting and the pH of his blood became 7.49. (Normally it is

7.35.) Has the blood become more alkaline or more acidic? Is the patient experiencing acidosis or alkalosis?

9.68 A patient brought to the emergency room following an overdose of aspirin was found to have a pH of 7.18 for the blood. (Normally the pH of blood is 7.35.) Has the blood become more acidic or more basic? Is the condition acidosis or alkalosis?

9.69 A patient entered the emergency room of a hospital after three weeks on a self-prescribed low-carbohydrate, high-fat diet and the regular use of the diuretic, acetazolamide. (A diuretic promotes the formation of urine and thus causes the loss of body fluid.) The pOH of the patient's blood was 6.82. What was the pH, and was the condition acidosis or alkalosis?

9.70 What does it mean when we say that a solution of pH 7.45 is *buffered* at this pH?

9.71 What chemical species constitute the chief buffer inside cells?

9.72 Write the net ionic equation that shows how the phosphate buffer neutralizes OH^-.

9.73 How does the phosphate buffer neutralize acid? Write the net ionic equation.

9.74 What two chemical species make up the chief buffer in blood?

9.75 Write the net ionic equations that show how the chief buffer system in the blood works to neutralize hydroxide ion and hydrogen ion.

9.76 Explain in your own words, using equations as needed, how the loss of a molecule of CO_2 at the lungs permanently neutralizes a hydrogen ion.

9.77 What is meant by *ventilation* in connection with respiration? What is hyperventilation? Hypoventilation?

9.78 What does the respiratory center in the brain instruct the lungs to do when the level of CO_2 in blood increases? Why?

9.79 Why does the hypoventilation of someone with emphysema lead to acidosis?

*9.80 In high-altitude sickness, the patient *involuntarily* over-breathes, and expels CO_2 from the body at a faster than normal rate. This results in an *increase* in the pH of the blood.
(a) Is this condition alkalosis or acidosis?
(b) Why should excessive loss of CO_2 result in an increase in the pH of the blood? (*Note:* Such a patient should be returned to lower elevations as soon as possible. It helps to rebreathe one's own air, as by breathing into a paper sack, because this helps the system retain CO_2.)

*9.81 A patient with untreated diabetes tends to hyperventilate. This is a natural response to what kind of change in the pH of the blood? Explain.

Buffer Calculations

9.82 When we use the defining equation for K_a, the concentrations in brackets refer to what condition, to the *initial* quantities of solutes used to prepare the solution or to the concentrations *after equilibrium* has been established?

9.83 Why is it, in using the defining equation for pK_a in working buffer problems, we can substitute initial concentrations of acid and conjugate base for equilibrium concentrations?

9.84 Calculate the pH of a buffer solution consisting of 0.21 M acetic acid and 0.26 M sodium acetate.

*9.85 What is the pH of a buffer solution made to be 0.19 M in dihydrogen phosphate ion and 0.22 M in its conjugate base, the monohydrogen phosphate ion?

*9.86 A 500-mL supply of a buffer solution is made of a weak acid, HA ($K_a = 6.21 \times 10^{-6}$), and its conjugate base, A^-, so that the solution contained 0.14 mol of the weak acid and 0.11 mol of the base.
(a) Calculate the pH of this solution.
(b) Calculate the pH of the solution after the addition of 0.020 mol of strong, monoprotic acid.
(c) Calculate the pH of a different sample of the original solution after the addition of 0.020 mol of NaOH.
(d) Calculate the pH that 500 mL of water would have had in parts (b) and (c) had no buffer been present.

9.87 Why is it better to use $[CO_2(aq)]$ instead of $[H_2CO_3(aq)]$ in the Henderson-Hasselbalch equation for the carbonate buffer in blood?

9.88 What is the apparent pK_a of the weak acid component of the carbonate buffer in blood?

9.89 Under normal pH conditions in human arterial blood, what values are usually assigned to the following? Be sure to include the correct units.
(a) $[HCO_3^-]$ (b) $[CO_2]$

*9.90 Suppose that normal human arterial blood is suddenly made to accept 11 mmol of HCl(aq) per liter of blood.
(a) What is the resulting pH of the blood if no CO_2 is allowed to escape?
(b) What is the resulting pH of the blood if 11 mmol/L of CO_2 can be quickly exhaled by normal processes?
(c) What further event will happen quickly to help bring the pH up still further toward normal?
(d) How does the body normally replace HCO_3^- lost by a battle with developing acidosis?

Acid–Base Titrations[4]

9.91 What does it mean to have a *standard* solution of, say, HCl(aq)?

9.92 When doing a titration, how does one know when the end point is reached?

9.93 What steps does an analyst take to ensure that the end point and the equivalence point in an acid–base titration occur together?

9.94 Give an example of a titration in which the equivalence point has a pH that is equal to 7.00. (Give a specific example of an acid and a base that, when titrated together, produce such a solution.)

9.95 Give a specific example of an acid and a base that, when titrated together, produce a solution with a pH greater than 7.00.

9.96 At the equivalence point in a titration the pH is less than 7.00. Give a specific example of an acid and a base that, when titrated together, would produce this result.

9.97 Individual aqueous solutions were prepared that contained the following substances. Calculate the molarity of each solution.
(a) 6.892 g of HCl in 500.0 mL of solution
(b) 8.090 g of HBr in 250.0 mL of solution

9.98 What is the molarity of each of the following solutions?
(a) 32.68 g of HI in 750.0 mL of solution
(b) 4.9048 g of H_2SO_4 in 250.0 mL of solution

*9.99 If 20.00 g of a monoprotic acid in 100.0 mL of solution gives a concentration of 0.5000 M, what is the formula mass of the acid?

*9.100 A solution with a concentration of 0.2500 M could be made by dissolving 4.000 g of a base in 250.0 mL of solution. What is the formula mass of this base?

*9.101 How many grams of each solute are needed to prepare the following solutions?
(a) 1000 mL of 0.2000 M HCl
(b) 750.0 mL of 0.1025 M HNO$_3$
(c) 500.0 mL of 0.01125 M H$_2$SO$_4$

*9.102 To prepare each of the following solutions would require how many grams of the solute in each case?
(a) 100.0 mL of 0.1000 M HI
(b) 750.0 mL of 0.2000 M H$_2$SO$_4$
(c) 500.0 mL of 1.125 M Na$_2$CO$_3$

*9.103 In standardizing a sodium carbonate solution, 22.48 mL of this solution was titrated to the end point with 19.82 mL of 0.1181 M HCl.
(a) Calculate the molarity of the sodium carbonate solution.
(b) How many grams of Na$_2$CO$_3$ does it contain per liter?

*9.104 A freshly prepared solution of sodium hydroxide was standardized with 0.1024 M H$_2$SO$_4$.
(a) If 19.46 mL of the base was neutralized by 21.28 mL of the acid, what was the molarity of the base?
(b) How many grams of NaOH were in each liter of this solution?

Finding K_a from pH (Special Topic 9.1)

*9.105 The pH of 0.10 M chloroacetic acid, HC$_2$H$_2$ClO$_2$ is 1.96. Calculate the K_a and the pK_a of this acid.

*9.106 At 60 °C, the pH of 0.010 M butyric acid is 2.98. What are the K_a and pK_a of this acid at this temperature?

Finding the pH of a Solution of a Weak Acid (Special Topic 9.2)

*9.107 Calculate the pH of a 0.030 M solution of *para*-aminobenzoic acid (PABA) at 25 °C. Its pK_a is 4.92 (25 °C). (Use the symbol H-*Paba* for the acid.) PABA is used in powerful sunscreening ointments.

*9.108 Calculate the pH of a 0.20 M solution of HF at 25 °C. Its K_a is 6.8×10^{-4} at this temperature.

[4] For all the calculations in Review Exercises that follow, round atomic masses to their *second* decimal places before adding them to find formula masses.

Acid Rain (Special Topic 9.3)

9.109 What are the chief gases that are responsible for acid rain and how do they get into the earth's atmosphere? (Write equations.) Why is *acid deposition* a better term?

9.110 Briefly describe some of the problems caused by acid rain in living systems and materials.

Additional Exercises

*9.111 Calculate the pH of 0.115 *M* HCl.

*9.112 Calculate the pH of 0.00045 *M* NaOH.

*9.113 Calculate the pH of a buffer solution prepared as 0.015 *M* sodium acetate and 0.10 *M* acetic acid.

*9.114 A buffer solution is prepared by dissolving 0.10 mol of solid sodium acetate in 1.0 L of water and as many moles of pure acetic acid as will make the final pH of the buffer equal to 5.00. Assuming that the amount of solutes added to the 1.0 L of water do not affect the total volume in any significant way, how many moles of acetic acid must be used to make this buffer?

*9.115 A solution of NH_4Cl and NH_3 in water of roughly equal molarities is a buffer system.
   (a) Which specific species in this buffer is able to neutralize OH^-? Write the net ionic equation.
   (b) Which specific species in this buffer is able to neutralize H^+? Write the net ionic equation.
   (c) The K_a of NH_4^+ is 5.7×10^{-10} (25 °C). Calculate the pH of a solution of this buffer in which $[NH_4Cl] = [NH_3]$.

OXIDATION–REDUCTION EQUILIBRIA

10

For those who are able to mix business with pleasure, the battery-powered laptop computer makes it easier, although it does seem unrealistic to suppose that much work is done on a pearl white dock jutting over an emerald green lagoon under a cobalt blue sky on Aruba in the Caribbean Sea. But that's not our business, which rather is the study in this chapter of the principles that enable a battery to work.

10.1
OXIDATION NUMBERS
10.2
BALANCING EQUATIONS OF REDOX REACTIONS

10.3
REDUCTION POTENTIALS
10.4
REDOX EQUILIBRIA

10.1

OXIDATION NUMBERS

A redox reaction is one in which oxidation numbers change.
At the molecular level of life, redox reactions are at the heart of an organism's use of oxygen. In redox reactions, oxidation numbers change, and electrons transfer. In acid–base reactions, protons transfer. Thus the transfer of two of nature's tiniest particles, the electron and proton, dominate two of the most important kinds of chemical events in nature: acid–base and redox reactions.

Just as in acid–base equilibria, so in redox equilibria: the extent to which products or reactants are favored varies widely. In this chapter one goal will be to learn how to assess the potential for a chemical species to remain in its reduced state when something else is present. Another goal is to use these potentials to predict how much a redox equilibrium favors the products. This study can help us appreciate better what drives biological oxidations when we study metabolism. In the major metabolic oxidations, electrons are passed along a series of enzymes to one uniquely able to transfer electrons to O_2 and reduce it to water. The final event, with H^+ supplied from the cell buffer system, is

$$4H^+ + O_2 + 4e^- \longrightarrow 2H_2O$$

Indirectly, the energy provided by this reaction is what ultimately drives virtually all of metabolism. The chief chemical function of many of the substances released into circulation by digestion is to supply electrons for this reduction of oxygen.

Because we define **redox reactions** as those in which oxidation numbers change, let's first review and extend our knowledge of how to assign these numbers, a skill begun in Section 5.2.

■ The chemical energy available by the reduction of oxygen supplies essentially all of the energy we use for metabolism.

■

H:F:

Hydrogen
fluoride

Oxidation Numbers Are the Charges Atoms in a Molecule Would Have if the Electrons of Each Bond Belonged to the More Electronegative Atoms The Lewis structure of a molecule of hydrogen fluoride is shown in the margin. Remember that the dots designate the valence electrons, one from H and seven from F (because F is in group VIIA). One of the two shared electrons of the covalent bond has been furnished by the H atom. If *both* shared electrons, however, were given *entirely* to the more electronegative atom, F, the charge on F would be 1 − and the charge on H would be 1 +. (Values of electronegativities are found on page 135.)

This is roughly how oxidation numbers are assigned. They're determined by the charge that results when the shared pair of electrons in each bond are all given to the more electronegative atom. They are assigned as though every compound were made entirely of ions, which obviously isn't true. Thus oxidation numbers do not necessarily correspond to real electrical charges, although sometimes they do, as we learned in Section 5.2. Oxidation numbers are assigned by using a set of rules. There aren't many, they are logical, and they let us assign oxidation numbers without having to remember a large number of electronegativities.

To distinguish an oxidation number from a real charge, we will reverse the number and the charge sign. For example, the charge on the sodium ion is written as 1 +, but its oxidation number is given as +1.

Rules for Assigning Oxidation Numbers

1. Atoms of any element not combined with atoms from another element have oxidation numbers of zero.
    Examples: The oxidation numbers of the atoms in N_2, Br_2, Cl_2, P_4, and S_8 are zero.

2. The oxidation numbers of monatomic ions equal their ionic charge.
    Examples: The oxidation numbers of Na^+ and K^+ are $+1$; of Ca^{2+}, Cu^{2+}, and Mg^{2+} are $+2$; and of Cl^- and Br^- are -1.

3. *In their compounds,* the oxidation number of any atom of the group IA elements is $+1$ (e.g., Na^+, K^+); of the group IIA elements is $+2$ (e.g., Ca^{2+}, Mg^{2+}); and the oxidation number of aluminum (group IIIA) is $+3$.

4. The oxidation number of any nonmetal *in its binary compounds with metals* equals the charge of the monatomic anion.
    Example: The oxidation number of Br in $CrBr_3$ is -1 because the monatomic ion of Br is the bromide ion, which has a charge of $1-$.

5. In compounds, the oxidation numbers of O, H and F are the following.
    O is almost always -2. (Exceptions occur in H_2O_2, hydrogen peroxide, and whenever the rules for H or F would be violated.)
    H is almost always $+1$. (The exceptions are in metal hydrides, binary compounds with metals like NaH in which H has the oxidation number -1.)
    F is always -1. (No exceptions. Fluorine is the most electronegative of all elements.)

6. The sum of the oxidation numbers of all the atoms in the formula of the atom, ion, or molecule must equal the overall charge given for the formula — the *sum rule*.

Now let's work some examples to show how the oxidation numbers of the atoms in some molecules or polyatomic ions can be assigned.

■ All monatomic ions, recall, have oxidation numbers numerically equal to their ionic charges.

■ The oxidation number of O in OF_2 is $+2$, not -2. The oxidation number of H in NaH, KH, or CaH_2 is -1, not $+1$.

EXAMPLE 10.1
Assigning Oxidation Numbers

Calomel, long used in medicine, has the formula Hg_2Cl_2. What are the oxidation numbers of the atoms in this compound?

ANALYSIS This is a binary compound of a nonmetal with a metal. The oxidation number of Cl must be -1 (rule 4). To determine the oxidation number of Hg, we use the sum rule, rule 6, which tells us that the sum of oxidation numbers must be zero.

SOLUTION Let x be the oxidation number of Hg.

For Hg_2Cl_2: Cl 2 atoms $\times (-1) = -2$
Hg 2 atoms $\times (x) = 2x$
Sum $= 0$

The value of x comes from the sum,

$$2x + (-2) = 0$$
$$2x = +2$$
$$x = +1$$

The oxidation number of Hg in Hg_2Cl_2 is thus $+1$.

■ Calomel has been used as a laxative, but if it's contaminated by $HgCl_2$, it contains a dangerous poison.

EXAMPLE 10.2
Assigning Oxidation Numbers

What is the oxidation number of carbon in ethane, C_2H_6?

ANALYSIS Both carbon and hydrogen are nonmetals. Of the two, we assign a number, $+1$, to hydrogen first because of rule 5. Then we calculate the oxidation number of C using the sum rule.

SOLUTION Letting x be the oxidation number of carbon, we have

$$\text{For } C_2H_6: \qquad \begin{aligned} \text{H 6 atoms} \times (+1) &= +6 \\ \text{C 2 atoms} \times (x) &= 2x \\ \text{Sum} &= 0 \end{aligned}$$

So

$$2x + 6 = 0$$
$$2x = -6$$
$$x = -3$$

The oxidation number of C in C_2H_6 is -3.

EXAMPLE 10.3
Assigning Oxidation Numbers

What are the oxidation numbers of the atoms in the nitrate ion, NO_3^-?

ANALYSIS Of the two nonmetals, N and O, O always takes priority over N (a consequence of rule 5) so its oxidation number is -2. We note that we have a charged ion, so the sum of oxidation numbers must equal this charge, -1. Using these data and the sum rule will give us the oxidation number of N.

SOLUTION Letting x equal the oxidation number of N, we have

$$\text{For } NO_3^-: \qquad \begin{aligned} \text{O 3 atoms} \times (-2) &= -6 \\ \text{N 1 atom} \times (x) &= x \\ \text{Sum} &= -1 \end{aligned}$$

So

$$x + (-6) = -1$$
$$x = +5$$

The oxidation number of N in NO_3^- is $+5$.

■ The $+5$ state is the highest oxidation state of nitrogen, and it makes NO_3^- a good oxidizing agent, particularly in acid.

PRACTICE EXERCISE 1

Assign the oxidation numbers to the atoms in the following.
(a) H_2S (b) SO_2 (c) H_2SO_3 (d) SO_3 (e) H_2SO_4 (f) SO_4^{2-}

PRACTICE EXERCISE 2

Assign the oxidation numbers to the atoms in the following.
(a) CH_4 (methane) (b) CH_3OH (methyl alcohol) (c) CH_2O (formaldehyde)
(d) $HCHO_2$ (formic acid) (e) H_2CO_3 (carbonic acid)

By Noting whether Oxidation Numbers Change, We Can Tell whether a Redox Reaction Has Occurred   The compounds in Practice Exercise 2 can be prepared by a series of redox reactions. If you worked this exercise you found that the oxidation number of carbon steadily becomes more positive in going from methane to carbonic acid. Thus the series represents a steady **oxidation** because this term means that an oxidation number becomes more positive or less negative. The opposite, **reduction,** corresponds to an oxidation number becoming less positive or more negative.

PRACTICE EXERCISE 3

Calculate the oxidation number of carbon in (a) C_2H_2 (acetylene), (b) C_2H_4 (ethylene), and (c) C_2H_6 (ethane). Does the series, from compounds (a) to (c), which can be carried out experimentally, represent oxidation or reduction (or neither)?

10.2

BALANCING EQUATIONS OF REDOX REACTIONS

An equation for a redox reaction can be balanced by separately balancing the equations of its half-reactions and then adding them.

The method that we will learn for balancing a redox reaction is called the *ion-electron method.* This procedure views a redox reaction as consisting of two parts, called **half-reactions,** one an oxidation and the other a reduction. Each half-reaction is separately balanced, first by atoms (the material balance) and then, using electrons, by net charge (the electrical balance). Only when both half-reactions are balanced in both atoms and charges are they combined to give the final, balanced redox reaction.

Using the ion-electron method sometimes causes a problem in balancing O or H atoms. However, whenever water is present, as it nearly always is, *molecules* of H_2O are used either to obtain O atoms or as places to put extra O atoms. This, of course, worsens the balance of H atoms, because we can't take O from H_2O without getting 2H. We solve this by using H^+ as either a reactant or a product, taking as many H^+ as needed to get the half-reaction equation to balance. There is no logical (or chemical) problem in doing this when the medium is (or becomes) acidic, so we'll learn how to balance equations for such systems first. Then we can build on that for situations in which the medium is basic.

We will first simply list the steps to balance a redox reaction in an acidic (or neutral) medium, and then go back and illustrate each one. *The steps must be taken in the order given.*

> **Balancing Redox Reactions by the Ion-Electron Method**
>
> **1.** Write a skeletal equation that shows only the ions or molecules involved in the reaction.
> **2.** Divide the skeletal equation into two half-reactions.
> **3.** Balance all atoms that are not H or O.
> **4.** Balance O by adding H_2O.
> **5.** Balance H by adding H^+ (not H or H_2 or H^-, but H^+).
> **6.** Balance the net charge by adding e^-. (Remember its minus sign.)
> **7.** Multiply entire half-reactions by simple whole numbers, as needed, to get the gain of e^- in one half-reaction to match the loss of e^- in the other. Then add the half-reactions.
> **8.** Cancel whatever is the same on both sides of the arrow.

Now let's see how to apply these rules in balancing the equation for the oxidation of methyl alcohol, CH_3OH, to formic acid, $HCHO_2$, using the dichromate ion, $Cr_2O_7{}^{2-}$, in an acidic medium. As this reaction proceeds, the chromium changes to Cr^{3+}.

■ Another method, the oxidation-number-change method, is trickier, in the view of many teachers.

■ Electrical balance means that the net charge to the left of the arrow equals the net charge to the right.

■

H
|
H—C—O—H
|
H

Methyl alcohol

■

$$H-\overset{\overset{O}{\|}}{C}-O-H$$

Formic acid

$$\left[\begin{array}{c} :\!\overset{..}{O}\!: \qquad :\!\overset{..}{O}\!: \\ \overset{..}{:}\overset{|}{O}-\overset{|}{Cr}-\overset{..}{O}-\overset{|}{Cr}-\overset{..}{O}\!: \\ :\!\overset{..}{O}\!: \qquad :\!\overset{..}{O}\!: \end{array}\right]^{2-}$$

$Cr_2O_7{}^{2-}$

Dichromate ion

Step 1. *Write a skeletal equation showing reactants and products as given.*

$$CH_3OH + Cr_2O_7{}^{2-} \longrightarrow HCHO_2 + Cr^{3+}$$

Step 2. *Divide the skeletal equation into two half-reactions. Except for H and O, the same elements must appear on both sides of each half-reaction.*

$$CH_3OH \longrightarrow HCHO_2$$
$$Cr_2O_7{}^{2-} \longrightarrow Cr^{3+}$$

Step 3. *Balance all atoms that are not H or O.*

$$CH_3OH \longrightarrow HCHO_2 \qquad \text{(No change, yet.)}$$
$$Cr_2O_7{}^{2-} \longrightarrow 2Cr^{3+} \qquad \text{(Balances Cr atoms.)}$$

Step 4. *Balance O by adding H_2O.*

$$CH_3OH + H_2O \longrightarrow HCHO_2 \qquad \text{(C's and O's balance.)}$$
$$Cr_2O_7{}^{2-} \longrightarrow 2Cr^{3+} + 7H_2O \qquad \text{(Cr's and O's balance.)}$$

■ To reduce clutter, the physical states, e.g., (*aq*), are not included until the last step, but remember that you can't cancel species that are not in the same state.

Step 5. *Balance H by adding H^+.*

$$CH_3OH + H_2O \longrightarrow HCHO_2 + 4H^+ \qquad \text{(All atoms now balance.)}$$
$$Cr_2O_7{}^{2-} + 14H^+ \longrightarrow 2Cr^{3+} + 7H_2O \qquad \text{(All atoms now balance.)}$$

Step 6. *Balance the net charge by adding e^-.* The first half-reaction after Step 5 has 0 charge on the left and +4 on the right, so we have to add $4e^-$ to the right side to get electrical balance. The second half-reaction has a net of +12 on the left side [(−2) + (+14)] and +6 on the right [2 × (+3)]. There is a net excess of +6 charge on the left. We therefore have to add $6e^-$ to the left side to make the net charge on the left side of the arrow equal to the net charge on the right side. Our half-reactions are now fully balanced.

$$CH_3OH + H_2O \longrightarrow HCHO_2 + 4H^+ + 4e^-$$
$$Cr_2O_7{}^{2-} + 14H^+ + 6e^- \longrightarrow 2Cr^{3+} + 7H_2O$$

Step 7. *Multiply half-reactions by whole numbers so that the electrons will cancel when the half-reactions are added.* When the electrons cancel from both sides of the equation, it means that the electrons gained equals the electrons lost. The number of electrons received by the oxidizing agent (as it becomes reduced) equals the number of electrons given up by the reducing agent (as it becomes oxidized). If we multiply every item in the first half-reaction by 3 we will have $3 \times (4e^-) = 12e^-$ shown as products. And if we multiply the second half-reaction by 2, we will have $2 \times (6e^-) = 12e^-$ shown as reactants. After this, when we add the two adjusted equations, the electrons will cancel.

$$3 \times [CH_3OH + H_2O \longrightarrow HCHO_2 + 4H^+ + 4e^-]$$
$$2 \times [Cr_2O_7{}^{2-} + 14H^+ + 6e^- \longrightarrow 2Cr^{3+} + 7H_2O]$$

Sum: $\quad 3CH_3OH + 2Cr_2O_7{}^{2-} + 3H_2O + 28H^+ + 12e^- \longrightarrow$
$$3HCHO_2 + 4Cr^{3+} + 12H^+ + 14H_2O + 12e^-$$

Step 8. *Cancel everything that can be canceled.* The 12 electrons on each side obviously cancel. But we can also get rid of some water molecules. There are 3 on the left and 14 on the right, so we can strike those on the left and change those on the right to 11. Thus

$$\ldots + 3H_2O + \ldots \longrightarrow \ldots + 14H_2O + \ldots$$

becomes

$$\ldots \longrightarrow \ldots + 11H_2O + \ldots$$

We can also cancel some H^+:

$$\ldots + 28H^+ \ldots \longrightarrow \ldots + 12H^+ \ldots$$

becomes

$$\ldots + 16H^+ \ldots \longrightarrow \ldots$$

This leaves, putting in the physical states,

$$3CH_3OH(aq) + 2Cr_2O_7{}^{2-}(aq) + 16H^+(aq) \longrightarrow$$
$$3HCHO_2(aq) + 4Cr^{3+}(aq) + 11H_2O$$

Check to see that both material and electrical balance exist.

PRACTICE EXERCISE 4

Balance the following equation, which occurs in an acidic medium.

$$Cu(s) + NO_3{}^-(aq) \longrightarrow Cu^{2+}(aq) + NO_2(g)$$

To Balance a Redox Equation when the Medium Is Basic, First Balance It for an Acidic Medium and Then Neutralize the Acid We really should not employ H^+ to balance H atoms when the medium is basic, but it turns out to be much simpler if we begin this way. Then, after all the steps have been done, we neutralize the H^+ by adding enough OH^- *to both sides of the equation* to neutralize any H^+. Acid–base neutralization is not a redox process, so this approach does not upset any redox balance. Then we recheck the equation to cancel out extra H_2O molecules.

If, for example, you had been asked to balance the following equation for the reaction carried out in base,

$$MnO_4{}^-(aq) + SO_3{}^{2-}(aq) \longrightarrow MnO_2(s) + SO_4{}^{2-}(aq)$$

you would have obtained the following by using Steps 1 through 8 for acidic solutions (omitting physical states for the moment).

$$2MnO_4{}^- + 3SO_3{}^{2-} + 2H^+ \longrightarrow 2MnO_2 + 3SO_4{}^{2-} + H_2O$$

Step 9. *Add as many OH^- as there are H^+ to both sides of the equation.* There are $2H^+$ on the left so we add $2OH^-$ to both sides. (It must be done to both sides so we do not upset either the material or the electrical balance.)

$$2OH^- + 2MnO_4{}^- + 3SO_3{}^{2-} + 2H^+ \longrightarrow 2MnO_2 + 3SO_4{}^{2-} + H_2O + 2OH^-$$

Step 10. *When they occur on the same side of the arrow, combine H^+ and OH^- into H_2O.* On the left side we have $2OH^-$ and $2H^+$, so we combine them into $2H_2O$.

$$2H_2O + 2MnO_4{}^- + 3SO_3{}^{2-} \longrightarrow 2MnO_2 + 3SO_4{}^{2-} + H_2O + 2OH^-$$

■ As this oxidation proceeds, the purple color of $MnO_4{}^-(aq)$ disappears and a brown, mud-like precipitate of $MnO_2(s)$ appears.

Step 11.   *Cancel H₂O molecules as possible.* We can cancel one H_2O molecule from each side, which leaves the final equation, with all physical states again in place, as

$$H_2O + 2MnO_4^-(aq) + 3SO_3^{2-}(aq) \longrightarrow 2MnO_2(s) + 3SO_4^{2-}(aq) + 2OH^-(aq)$$

PRACTICE EXERCISE 5

■

Ethyl alcohol

Balance the following equation for the oxidation of ethyl alcohol to the acetate ion by the permanganate ion in base.

$$C_2H_6O(aq) + MnO_4^-(aq) \longrightarrow C_2H_3O_2^-(aq) + MnO_2(s)$$

■

Notice that the half-reactions that we produced in the preceding discussion clearly showed which part of the redox reaction represented an electron loss (an oxidation) and which was an electron gain (a reduction).

Half-reactions written as reductions, those in which electrons always appear as reactants, can be assigned numbers whose signs and magnitudes measure the potential for the reaction to occur. We'll study this in the next section.

10.3

REDUCTION POTENTIALS

The more positive (or less negative) the reduction potential of a half-reaction, the greater its tendency to operate as a reduction.

As we saw in the previous chapter, *every acid-base equilibrium involves the exchange of* H^+ between a proton donor and acceptor. Each equilibrium, however, has its own potential for going to completion. We used K_a values for the equilibria of weak acids in water to compare certain of these potentials. Redox reactions also involve the exchange of something, electrons, but these are not *chemical* species like H^+ that can be described by an equilibrium concentration. Yet, like acid-base equilibria, redox equilibria have varying potentials for going to completion. So it would be nice to have a table of numbers, like the K_a values of weak acids, that indicate this potential. The dilemma is that no term like $[e^-]$, the molar concentration of electrons, can appear in any equilibrium law. The one "species" common to all redox equilibria, electrons, cannot be measured as we can measure $[H^+]$.

Half-Reactions, Written as Reductions, Have Different Potentials for Proceeding   To solve the dilemma just described, we use a feature of redox equilibria that we have already found useful, the half-reaction. Every redox reaction has two. Each involves electrons either as a reactant or as a product. It should be possible, therefore, to assign a number to each half-reaction that describes its relative potential to manage electrons. Then we could make a table of half-reactions, each with its own number (potential), and thereby know the relative potentials of all half-reactions to proceed to completion.

This strategy is a bit tricky, because we cannot run a half-reaction by itself. We cannot, therefore, *directly* measure anything that we could call a "potential for going to completion" for any half-reaction. To get around the problem, we pick one particular half-reaction to be the standard of comparison. Its potential for proceeding is arbitrarily assigned a value of zero. *The potentials of all other half-reactions are then made relative to this reference half-reaction.* We will not be able to study the experimental details of how to get these data, but let's see how the results work out.

To make comparisons as easy as possible when we finally construct a table, we write all half-reactions in the same pattern, just as we wrote acid equilibria in the same pattern. The pattern shows electrons as *reactants* in the *forward* reaction. *Thus the forward reactions in the table will all be reductions.* Each such half-reaction has a **reduction potential,** the potential, relative to the reference, for one of its chemical species to be reduced. A *high* positive value for a reduction potential will thus mean a half-reaction with some chemical that readily *consumes* electrons, because it is prone to being reduced. Half-reactions like this will have the more *positive* numbers in the table. Half-reactions with chemicals that easily give up electrons (because they are easily oxidized) will have lower reduction potentials and many will have negative values in the table.

The reference half-reaction against which all others are rated is the following equilibrium.

$$2H^+(aq) + 2e^- \rightleftharpoons H_2(g) \tag{10.1}$$

Notice that it conforms to the pattern; the forward reaction is a reduction and electrons are shown as *reactants.* When the solution is exactly 1 M in H^+, the temperature is 25 °C, and the pressure of the H_2 is exactly 1 atm, the reduction potential for equilibrium 10.1 is *defined* as 0.00 volt. Any reduction potential that corresponds to these arbitrary but standard conditions — concentrations of all chemical species of 1 M, a temperature of 25 °C, and pressures of any gaseous species at 1 atm — is called a **standard reduction potential,** $E°$.

As we indicated, the unit for a reduction potential is the **volt, V,** the SI unit for electrical potential. The symbol for reduction potential is E except when it refers to a standard reduction potential, when $E°$ is used. The volt is to the flow of electrons in a conductor roughly what pressure is to the flow of water in a conduit. We can think of the volt as the force that pushes an electrical current through a wire. This force, to distinguish it from other forces, is called the *electromotive force,* or the *emf,* of the electrical system.

Table 10.1 gives the standard reduction potentials for several common half-reactions. Those with positive $E°$ values all have greater tendencies to run as reductions than does the reference $2H^+/H_2$ half-reaction. We already know, for example, that Cl_2 has a strong tendency to change to Cl^- ions, which is a reduction. The half-reaction is

$$Cl_2(g) + 2e^- \rightleftharpoons 2Cl^-(aq) \qquad E° = +1.36 \text{ V}$$

For this half-reaction, $E° = +1.36$ V. From chemical knowledge, we know that fluorine (also in group VIIA) has an even more powerful tendency than chlorine to change to its negative ion, F^-, and be reduced. So its reduction potential should have a higher positive number than that of chlorine, and it does.

$$F_2(g) + 2e^- \rightleftharpoons 2F^-(aq) \qquad E° = +2.87 \text{ V}$$

Recall that F_2 is the most reactive of all elements and always gets an oxidation number of -1. It is also the most electronegative element. No wonder, therefore, that the reduction half-reaction for F_2 has the highest, most positive reduction potential of all, $E° = +2.87$ V. Thus the reduction potentials parallel, as they must, our knowledge of which systems are relatively easily reduced.

The half-reactions in Table 10.1 with negative $E°$ values are less able to run as reductions than the $2H^+/H_2$ half-reaction. They are, in fact, more prone to run as oxidations — that is, as the *reverse* of the half-reactions shown in the table. For example, we already know that sodium metal, like all group IA elements, has a powerful tendency to lose electrons — an oxidation — and change to sodium ions. The equation for the half-reaction representing this strong tendency for Na to be oxidized to Na^+ has to be shown with the electron as a *product:*

$$Na(s) \longrightarrow Na^+(aq) + e^- \qquad E° = +2.71 \text{ V}$$

■ We can always write a reduction half-reaction in reverse when we want an oxidation half-reaction.

This equation is opposite to the way it is given in Table 10.1. Thus, the $E°$ value for the half-reaction involving Na given in Table 10.1 has a negative sign, -2.71 V. The high *negative* value tells us that the reaction, as written in the table, has little tendency to be spontaneous.

■ Notice that Au^{3+}, the gold(III) cation, has the most powerful tendency of all metals to stay in its reduced form, Au.

TABLE 10.1
Standard Reduction Potentials (at 25 °C)

Half-Reaction	$E°$(volts)
$F_2(g) + 2e^- \rightleftharpoons 2F^-(aq)$	+2.87
$PbO_2(s) + SO_4^{2-}(aq) + 4H^+(aq) + 2e^- \rightleftharpoons PbSO_4(s) + 2H_2O$	+1.69
$MnO_4^-(aq) + 8H^+(aq) + 5e^- \rightleftharpoons Mn^{2+}(aq) + 4H_2O$	+1.49
$PbO_2(s) + 4H^+(aq) + 2e^- \rightleftharpoons Pb^{2+}(aq) + 2H_2O$	+1.46
$Au^{3+}(aq) + 3e^- \rightleftharpoons Au(s)$	+1.42
$Cl_2(g) + 2e^- \rightleftharpoons 2Cl^-(aq)$	+1.36
$O_2(g) + 4H^+(aq) + 4e^- \rightleftharpoons 2H_2O$	+1.23
$Br_2(aq) + 2e^- \rightleftharpoons 2Br^-(aq)$	+1.07
$NO_3^-(aq) + 4H^+(aq) + 3e^- \rightleftharpoons NO(g) + 2H_2O$	+0.96
$Ag^+(aq) + e^- \rightleftharpoons Ag(s)$	+0.80
$Fe^{3+}(aq) + e^- \rightleftharpoons Fe^{2+}(aq)$	+0.77
$I_2(s) + 2e^- \rightleftharpoons 2I^-(aq)$	+0.54
$Cu^{2+}(aq) + 2e^- \rightleftharpoons Cu(s)$	+0.34
$SO_4^{2-}(aq) + 4H^+(aq) + 2e^- \rightleftharpoons H_2SO_3(aq) + H_2O$	+0.17
$2H^+(aq) + 2e^- \rightleftharpoons H_2(g)$	0.00
$Pb^{2+}(aq) + 2e^- \rightleftharpoons Pb(s)$	−0.13
$Sn^{2+}(aq) + 2e^- \rightleftharpoons Sn(s)$	−0.14
$Ni^{2+}(aq) + 2e^- \rightleftharpoons Ni(s)$	−0.25
$Co^{2+}(aq) + 2e^- \rightleftharpoons Co(s)$	−0.28
$Cd^{2+}(aq) + 2e^- \rightleftharpoons Cd(s)$	−0.40
$Fe^{2+}(aq) + 2e^- \rightleftharpoons Fe(s)$	−0.44
$Cr^{3+}(aq) + 3e^- \rightleftharpoons Cr(s)$	−0.74
$Zn^{2+}(aq) + 2e^- \rightleftharpoons Zn(s)$	−0.76
$2H_2O + 2e^- \rightleftharpoons H_2(g) + 2OH^-(aq)$	−0.83
$Al^{3+}(aq) + 3e^- \rightleftharpoons Al(s)$	−1.66
$Mg^{2+}(aq) + 2e^- \rightleftharpoons Mg(s)$	−2.37
$Na^+(aq) + e^- \rightleftharpoons Na(s)$	−2.71
$Ca^{2+}(aq) + 2e^- \rightleftharpoons Ca(s)$	−2.76
$K^+(aq) + e^- \rightleftharpoons K(s)$	−2.92
$Li^+(aq) + e^- \rightleftharpoons Li(s)$	−3.05

■ Notice in Table 10.1 how the group IA and IIA elements are clustered at the bottom, which indicates how powerfully they tend to be in their oxidized forms.

Standard Reduction Potentials Enable Us to Predict Redox Reactions   We can now use the tabulated half-reactions to predict whether a given combination of such reactions, one a reduction as written and one an oxidation and so written in reverse, can make up a full redox reaction that proceeds spontaneously. There is a pattern to the way half-reactions cooperate in all spontaneous redox reactions. Let's first summarize it and then show how it works.

> **Rule for Combining Reduction Half-Reactions**   When two reduction half-reactions are combined into a full redox reaction the one with the more positive $E°$ always runs as written, as a reduction, and it forces the other, with the less positive $E°$, to run in reverse, as an oxidation.

To illustrate, suppose we came for the first time to the question, ''Will sodium react with chlorine?'' Standard reduction potentials of half-reactions can be used to answer a question like this. Let's take the relevant data from Table 10.1. The product, NaCl, will be shown not as a solid, NaCl(s), but in solution as separated, hydrated ions, because the data are presented this way.

$$Na^+(aq) + e^- \rightleftharpoons Na(s) \qquad E° = -2.71 \text{ V}$$
$$Cl_2(g) + 2e^- \rightleftharpoons 2Cl^-(aq) \qquad E° = +1.36 \text{ V}$$

The $E°$ for the reaction of chlorine is more positive than the $E°$ for the reaction of sodium. This tells us that the reaction involving chlorine will proceed as written, as a reduction, and that the reaction involving sodium is forced to proceed in the opposite direction, as an oxidation. So let's rewrite the two half-reactions to reflect these facts. We have to reverse the reaction for sodium. (We will now use only a forward arrow, not equilibrium arrows. We have already learned how powerfully this reaction proceeds.)

$$Na(s) \longrightarrow Na^+(aq) + e^- \qquad \text{(oxidation)}$$
$$Cl_2(g) + 2e^- \longrightarrow 2Cl^-(aq) \qquad \text{(reduction)}$$

To get the net reaction, we multiply the coefficients of the first half-reaction by 2. Then the electrons can cancel as we add the two.

$$2Na(s) \longrightarrow 2Na^+(aq) + 2e^-$$
$$\underline{Cl_2(g) + 2e^- \longrightarrow 2Cl^-(aq)}$$
$$\text{Sum:} \quad 2Na(s) + Cl_2(g) \longrightarrow 2Na^+(aq) + 2Cl^-(aq)$$

This is the equation for the reaction we would predict, using standard reduction potentials, between sodium and chlorine. We would not predict that Na^+ and Cl^- ions would *spontaneously* combine to give sodium metal and chlorine gas.

Before we turn to a less obvious example, we must issue a caution. Just because we can write an equation for a redox reaction that should (on paper) have a powerful tendency to take place doesn't mean that we can go into the lab and set up the reaction exactly as the equation is written. If we tried to mix sodium and chlorine so as to get *aqueous* sodium ions and chloride ions *directly*, we'd be in great trouble. We have already learned that sodium reacts powerfully with water, too, to give hydrogen and NaOH,

$$2Na(s) + 2H_2O \longrightarrow H_2(g) + 2NaOH(aq)$$

The earlier reaction of sodium with chlorine to give *aqueous* sodium and chloride ions would require *two* experimental steps. The first would let sodium and chlorine react in the absence of water to give crystalline NaCl (a dangerous reaction). Then we would dissolve the crystals in water. Thus, although the data in Table 10.1 can be used to indicate how powerful the driving force for a postulated redox reaction might be, we have to think about water itself as a possible reactant before we try the reaction.

Now let's study another illustration of the use of reduction potentials to predict a redox reaction.

EXAMPLE 10.4
Predicting Spontaneous Redox Reactions

If both Cl_2 and I_2 are present in a solution that contains Cl^- and I^- ions, what spontaneous reaction will occur? (The chlorine would have to be bubbled into the solution.)

ANALYSIS We notice that the conditions provide both the oxidized and reduced forms of chlorine and of iodine. This tells us we have a redox situation. There are two possible half-reactions (and both are in Table 10.1).

$$Cl_2(g) + 2e^- \rightleftharpoons 2Cl^-(aq) \qquad E° = +1.36 \text{ V}$$
$$I_2(s) + 2e^- \rightleftharpoons 2I^-(aq) \qquad E° = +0.54 \text{ V}$$

Thus the $Cl_2/2Cl^-$ system has the more positive reduction potential than the $I_2/2I^-$ system. This tells us that the $Cl_2/2Cl^-$ system must run as a reduction, which forces the $I_2/2I^-$ system to run in reverse, as an oxidation.

SOLUTION When a reaction occurs in the mixture, the half-reactions will be

$$Cl_2(g) + 2e^- \longrightarrow 2Cl^-(aq) \qquad \text{(reduction)}$$
$$2I^-(aq) \longrightarrow I_2(s) + 2e^- \qquad \text{(oxidation)}$$

The net redox reaction, therefore, is

$$Cl_2(g) + 2I^-(aq) \longrightarrow 2Cl^-(aq) + I_2(s) \qquad \text{(net reaction)}$$

This reaction is actually one method for preparing iodine, I_2. Chlorine gas is bubbled through a solution of sodium iodide, where it reacts by the equation we just figured out. The reverse reaction, which can be attempted by mixing iodine with aqueous sodium chloride, does not happen. Chlorine can oxidize iodide ion to iodine, but iodine cannot oxidize chloride ion to chlorine.

EXAMPLE 10.5
Predicting Spontaneous Redox Reactions

Predict what will happen if both lead and silver are placed in contact with a solution that contains both Cu^{2+} and Ag^+ ions.

ANALYSIS If anything happens, it will be a redox reaction because we are obviously seeing these two elements in two different oxidation states. So we can go to Table 10.1 for help. The relevant half-reactions from the table are

$$Cu^{2+}(aq) + 2e^- \rightleftharpoons Cu(s) \qquad E° = +0.34 \text{ V}$$
$$Ag^+(aq) + e^- \rightleftharpoons Ag(s) \qquad E° = +0.80 \text{ V}$$

The Ag^+/Ag system has the more positive reduction potential, so it will run as written, as a reduction. It will force the Cu^{2+}/Cu system to run in reverse, as an oxidation.

SOLUTION To balance the two half-reactions so that the electrons cancel as we add the two equations, we have to multiply the Ag^+/Ag half-reaction by 2. When the redox reaction occurs, the two half-reactions will be

$$2Ag^+(aq) + 2e^- \longrightarrow 2Ag(s) \qquad \text{(reduction)}$$
$$\underline{Cu(s) \longrightarrow Cu^{2+}(aq) + 2e^- \qquad \text{(oxidation)}}$$
$$2Ag^+(aq) + Cu(s) \longrightarrow 2Ag(s) + Cu^{2+}(aq) \qquad \text{(net reaction)}$$

This is a pretty reaction (Fig. 10.1). A coil of copper wire is immersed into a solution of silver nitrate, and crystals of silver metal begin to deposit on the coil as copper atoms change to ions and go into solution. Given time, the solution acquires the blue color of the newly formed $Cu^{2+}(aq)$ ion.

PRACTICE EXERCISE 6

Using the data in Table 10.1, predict the net ionic equation for what can happen if zinc and iron metal strips are placed in contact with a solution that contains $Fe^{2+}(aq)$ and $Zn^{2+}(aq)$ ions.

Often the question is not "What can happen?" but "Will a given reaction occur as written?" We'll show an example.

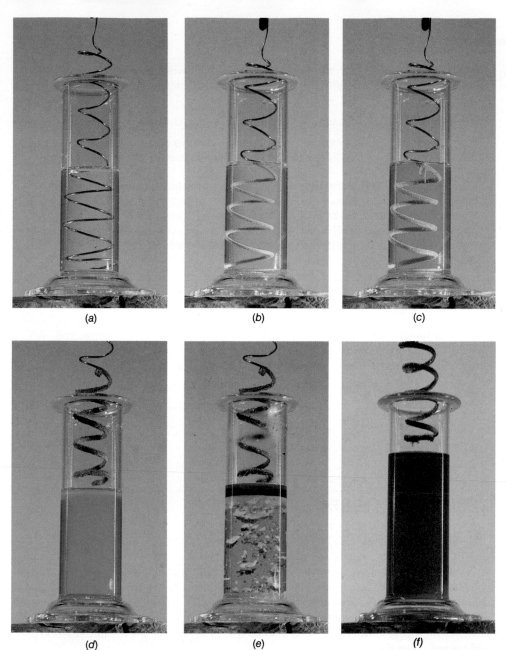

The redox reaction of copper with silver ion to give copper(II) ion and silver metal. As the reaction proceeds, silver deposits form, and the solution becomes ever more intensely blue because of the hydrated copper(II) ion. (a) A coil of copper has just been placed into the colorless solution of silver nitrate. (b) Crystals of silver have formed on the wire and the solution is now tinged with blue. (c) More silver has deposited on the wire, and the solution is a deeper blue. (d) The solution is a still deeper blue. The wire has been lifted from the solution to show the silver crystals deposited on it. (e) The wire has been lowered a short distance into the solution, and a layer of the solution becomes intensely blue as silver particles dislodge. (f) A small lower portion of the copper wire has dissolved completely, and the entire solution is deep blue.

(a)   (b)   (c)

(d)   (e)   (f)

EXAMPLE 10.6
Predicting whether a Redox Reaction Is Spontaneous

The following is a balanced equation of a redox reaction between copper metal and nitric acid. But does it occur?

$$3Cu(s) + 8H^+(aq) + 2NO_3^-(aq) \longrightarrow 3Cu^{2+}(aq) + 2NO(g) + 4H_2O$$

ANALYSIS We must first separate the half-reactions from this equation. Remember that except for H and O, other elements *must* appear on both sides of the arrows in their respective half-reactions. So one half-reaction involves copper, and the other involves nitrogen.

$$Cu(s) \longrightarrow Cu^{2+}(aq) + 2e^- \quad \text{(an oxidation)}$$
$$NO_3^-(aq) + 4H^+(aq) + 3e^- \longrightarrow NO(g) + 2H_2O \quad \text{(a reduction)}$$

■ Depending on the concentration of the nitric acid, either NO or NO$_2$ forms.

305

In Table 10.1, the $E°$ for the reduction half-reaction, which here involves NO_3^-, is +0.96 V. The oxidation half-reaction, which involves Cu, when reversed and written as a reduction is

$$Cu^{2+}(aq) + 2e^- \longrightarrow Cu(s) \qquad \text{(a reduction)}$$

It has a reduction potential of +0.34 V. The more positive of the reduction potentials, therefore, is for the NO_3^-/NO system, so its half-reaction *must* run as a reduction.

SOLUTION This is just what the NO_3^-/NO system does in the given redox reaction; it runs as a reduction. It forces the Cu^{2+}/Cu system to run as an oxidation, and that is what it actually does in the given redox reaction. So these two half-reactions proceed as they must when put together. The given redox reaction does, in fact, occur. Copper metal dissolves in nitric acid. The oxidizing agent provided by nitric acid is the nitrate ion, not the hydrogen ion.

PRACTICE EXERCISE 7

We saw in the previous example that copper metal should dissolve in nitric acid. Will it dissolve in hydrochloric acid? Will the following reaction occur? Because $Cl^-(aq)$ cannot accept electrons from Cu (or anything), we have to suppose that $H^+(aq)$ would.

$$Cu(s) + 2H^+(aq) \longrightarrow Cu^{2+}(aq) + H_2(g)$$

10.4

REDOX EQUILIBRIA

The equilibrium constant for a redox equilibrium can be calculated from the standard reduction potentials of its half-reactions.

In the previous section we learned that the half-reaction with the more positive reduction potential in a redox system forces the other half-reaction to run as an oxidation. Our task in this section is to use the standard reduction potentials to calculate a number that can be related to an equilibrium constant for a redox reaction. When K_{eq} is high, you will recall, the products are strongly favored, which is another way of saying that the forward reaction goes well toward completion.

The Spread between $E°$ for the Substance Reduced and the $E°$ for the Substance Oxidized Is a Measure of the Potential for a Redox Reaction If we can think of a *driving force* for a reaction, it would be some characteristic of the reactants and products that impels the former to change into the latter. With redox reactions, this characteristic might logically reside in the relative potentials for the oxidation and reduction half-reactions to proceed. It would seem logical that the more dissimilar two half-reactions are in their $E°$ values, the greater is the driving force for the redox reaction (see Fig. 10.2). The redox reaction of Figure 10.2a is

■ Another powerful driving force for reactions is to make stronger covalent bonds from weaker bonds.

$$Zn(s) + Cl_2(g) \longrightarrow Zn^{2+}(aq) + 2Cl^-(aq)$$

This is a combination of the following half-reactions

$$Zn(s) \longrightarrow Zn^{2+}(aq) + 2e^- \qquad \text{(oxidation)}$$
$$Cl_2(g) + 2e^- \longrightarrow 2Cl^-(aq) \qquad \text{(reduction)}$$

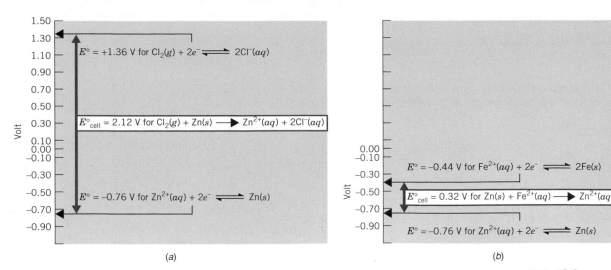

FIGURE 10.2
Standard cell potentials as the difference between $E°_{cell}$ for the substance reduced and the $E°_{cell}$ for the substance oxidized.
(a) For the reaction, $Zn(s) + Cl_2(g) \rightarrow Zn^{2+}(aq) + 2Cl^-(aq)$.
(b) For the reaction, $Zn(s) + Fe^{2+}(aq) \rightarrow Zn^{2+}(aq) + Fe(s)$.

To compare these in terms of their reduction potentials, we have to reverse the oxidation half-reaction. Then we have, together with $E°$ values:

$$Zn^{2+}(aq) + 2e^- \rightleftharpoons Zn(s) \qquad E° = -0.76 \text{ V}$$
$$Cl_2(g) + 2e^- \rightleftharpoons 2Cl^-(aq) \qquad E° = +1.36 \text{ V}$$

Obviously, the reduction of Cl_2 to $2Cl^-$ is able to force the other half-reaction to run in reverse; the $Cl_2/2Cl^-$ system has the more positive reduction potential. The spread between the two values of $E°$ is $(+1.36 \text{ V}) - (-0.76 \text{ V}) = 2.12$ V, as Figure 10.2a also shows.

Now let's consider a redox reaction with a smaller spread. If we take another reaction of zinc,

$$Zn(s) + Fe^{2+}(aq) \longrightarrow Zn^{2+}(aq) + Fe(s)$$

and break it down into its two half-reactions, we get

$$Zn(s) \longrightarrow Zn^{2+}(aq) + 2e^- \qquad \text{(an oxidation)}$$
$$Fe^{2+}(aq) + 2e^- \longrightarrow Fe(s) \qquad \text{(a reduction)}$$

For comparison, and to illustrate the spread in $E°$ values, we rewrite the oxidation as a reduction and, after inserting the $E°$ values, we have

$$Zn^{2+}(aq) + 2e^- \rightleftharpoons Zn(s) \qquad E° = -0.76 \text{ V}$$
$$Fe^{2+}(aq) + 2e^- \rightleftharpoons Fe(s) \qquad E° = -0.44 \text{ V}$$

The Fe/Fe^{2+} system has the more positive (less negative) $E°$, so it runs as the reduction, forcing the Zn/Zn^{2+} system to again run as the oxidation. The spread between the $E°$ values is $(-0.44 \text{ V}) - (-0.76 \text{ V}) = 0.32$ V, shown in Figure 10.2b, which is much smaller than the 2.12 V spread when zinc reacts with chlorine.

The "spread" we have been talking about has the technical name of standard cell potential. (Why the word *cell* is used will be explained shortly.) The **standard cell potential, $E°_{cell}$**, is the magnitude of the difference between the reduction potentials of the two half-cells of a redox reaction. The difference is always taken as follows.

$$E°_{cell} = \left(\begin{array}{c} \text{standard reduction} \\ \text{potential of} \\ \text{substance reduced} \end{array} \right) - \left(\begin{array}{c} \text{standard reduction} \\ \text{potential of} \\ \text{substance oxidized} \end{array} \right) \qquad (10.2)$$

307

A GALVANIC CELL

If silver and copper electrodes as well as an aqueous solution of silver nitrate and copper(II) nitrate were all present in the same vessel, a spontaneous redox reaction would occur:

$$Cu(s) + 2Ag^+(aq) \longrightarrow Cu^{2+}(aq) + 2Ag(s)$$

It would not give electricity, however, because the electrons would transfer directly from copper atoms to silver ions. For the transfer to go through wires, the chemicals involved in each of the two half-reactions of the redox system have to be in separate places and be kept from mixing. Figure 1 shows a simple way in which this could be done.

In the left beaker is a solution of $AgNO_3$ with a silver electrode, the cathode. In the right beaker is a solution of $Cu(NO_3)_2$ with a copper electrode, the anode. The electrons that copper atoms leave behind at the anode cannot get to the silver ions that are able to accept them except by going through the external circuit (with the switch closed). As these electrons reach the cathode, they attract silver ions and reduce them to silver atoms, which deposit on the cathode as silver metal.

The *salt bridge* shown in Figure 1 has an essential function. It is made of open-ended bent glass tubing and is filled with a solution of a neutral electrolyte, such as $KNO_3(aq)$. Tufts of glass wool or cotton stuck into the open ends keep the solution from spilling out. But, being soaked with the electrolyte, they do not prevent its K^+ ions or NO_3^- ions from moving in whichever directions events elsewhere in the system force them to move. And move they must. When a copper atom gives up two electrons at

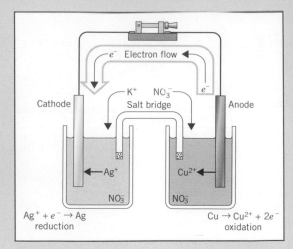

FIGURE 1

A galvanic cell. The "driving force" is the natural tendency for copper atoms to give electrons to silver ions, as shown in Figure 10.1. (From J. E. Brady and J. R. Holum, *Chemistry. The Study of Matter and Its Changes,* 1993. John Wiley & Sons, Inc. Used by permission.)

the anode and moves into the solution, the solution would, if nothing else happened, have an excess of positive charge. This is not permitted in nature. But two nitrate ions are attracted into the solution to balance the charge. Similarly, at the cathode, when two silver ions pick up the two electrons and become silver atoms, they would, should nothing else happen, leave the solution in the left beaker with an excess of negative charge. However, two K^+ ions are attracted from the salt bridge into the beaker and so keep the solution electrically neutral. The salt bridge thus allows for a closed circuit. If it were not present, no current could flow. It would be like opening the switch, which would also stop the redox reaction.

Thus we can rewrite the two redox reactions of Figure 10.2 so as to include their standard cell potentials.

$$Zn(s) + Cl_2(g) \longrightarrow Zn^{2+}(aq) + 2Cl^-(aq) \qquad E^\circ_{cell} = 2.12 \text{ V}$$
$$Zn(s) + Fe^{2+}(aq) \longrightarrow Zn^{2+}(aq) + Fe(s) \qquad E^\circ_{cell} = 0.32 \text{ V}$$

The first reaction, you can see, has a substantially greater standard cell potential, implying that it has a far greater tendency to go to completion as shown than the second reaction. Before we see how much greater it is, let's clear up the use of the word *cell* here.

A standard cell potential is actually a measure of the voltage of a battery made up to correspond to the redox reaction and its half-reactions. All that a battery does is arrange to have the electron transfer of a redox reaction occur through an external circuit. The electrons given up by the reducing agent do not go directly to the atoms or ions of the oxidizing agent but through the circuit instead. As the electrons move in the external circuit they do some work for us, like start a car, or trip a camera shutter. Special Topic 10.1 describes how the reaction of Cu with Ag^+ of Example 10.5 (and Figure 10.1) can be set up to make electricity flow in an external circuit.

■ Scientists call devices that use redox reactions to generate electricity *galvanic cells* after Luigi Galvani (1737–1798), an Italian scientist.

Unfortunately, we cannot delve more deeply into the applications of redox reactions to batteries, but we now can see why the term *cell* and the unit of *volt* came into use in this field of chemistry. Special Topic 10.2 describes the chemistry of the standard lead storage battery used in vehicles and the silver oxide battery used in cameras, wristwatches, and hand-held calculators.

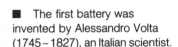

■ The first battery was invented by Alessandro Volta (1745–1827), an Italian scientist.

The connection between a standard cell potential and a real battery explains why the difference in Equation 10.2 is always taken this way, as the $E°$ of the reduced substance minus the $E°$ of the oxidized substance. The result of this calculation must always be a positive number (or zero). No battery has a negative voltage, and thus no $E°_{cell}$ can be negative. If you make a mistake and obtain a negative value, the redox reaction you are studying would actually run the opposite way from what you have written.

The Standard Cell Potential Is Proportional to the Log of the Equilibrium Constant for a Redox Equilibrium

We can now get back to the question "How far toward completion do redox equilibria go?" We have a hint in terms of the relative $E°_{cell}$ values, which we just learned to calculate. But these are not K_{eq} values.

The relationship between $E°_{cell}$ and K_{eq} is one that we will have to state without explaining how it was derived. (We would have to go much farther into the electricity of these reactions as well as into the laws of thermodynamics than we can.) Ultimately, we want only to demonstrate how even extremely small values of $E°_{cell}$ correspond to large values of K_{eq}. The equation relating $E°_{cell}$ to K_{eq} is

$$E°_{cell} = \frac{0.0592}{n} \log K_{eq} \qquad (10.3)$$

where n is the number of moles of electrons that transfer in the balanced redox equation. Let's use Equation 10.3 to calculate values of K_{eq} for the two redox reactions involving zinc that we studied earlier.

EXAMPLE 10.7
Calculating K_{eq} from $E°_{cell}$

What are the K_{eq} values for the following redox reactions?

(a) $Zn(s) + Cl_2(g) \longrightarrow Zn^{2+}(aq) + 2Cl^-(aq)$ $E°_{cell} = 2.12$ V
(b) $Zn(s) + Fe^{2+}(aq) \longrightarrow Zn^{2+}(aq) + Fe(s)$ $E°_{cell} = 0.32$ V

ANALYSIS Both equations involve *two*-electron transfers, so when Equation 10.3 is used, $n = 2$.

SOLUTION (a) Substituting $n = 2$ and the value of $E°_{cell}$ (2.12 V) into Equation 10.3 gives

$$2.12 = \frac{0.0592}{2} \log K_{eq}$$

Or,

$$\log K_{eq} = \frac{2.12 \times 2}{0.0592}$$
$$= 71.6$$

Therefore,

$$K_{eq} = 10^{71.6}$$
$$= 4 \times 10^{71} \text{ (!)}$$

(b) Again $n = 2$, but now $E°_{cell} = 0.32$ V. So

$$0.32 = \frac{0.0592}{2} \log K_{eq}$$

THE LEAD STORAGE BATTERY AND THE SILVER OXIDE BATTERY

Portable sources of electricity are popularly called batteries. They are packages of chemicals that can react, when the external electrical circuit is closed (the switch is "on"), and cause a flow of electrons through some conducting system. We'll describe two common batteries here.

The Lead Storage Battery We use the lead storage battery to start vehicles. The most common come in a 12-V size, but they are available in 6-, 24-, and 32-V sizes, too. To develop such relatively high voltages by *chemical* reactions requires that several individual cells, each with a cell potential of about 2 V, be combined.

Figure 1 sketches the working parts of a 6-V lead storage battery, which has three individual cells. The anode, where oxidation occurs, is made of several plates of lead metal. The cathode, where reduction happens, is made of slabs of compressed lead dioxide, PbO_2. The solution is aqueous sulfuric acid. When we draw electricity from the battery, the following half-cell reactions are allowed to happen spontaneously inside each cell.

Cathode
$$PbO_2(s) + 4H^+(aq) + SO_4^{2-}(aq) + 2e^- \longrightarrow PbSO_4(s)$$
$$E° = +1.69 \text{ V}$$
Anode
$$Pb(s) + SO_4^{2-}(aq) \longrightarrow PbSO_4(s) + 2e^-$$
$$E° = +0.36 \text{ V}$$

The net reaction within each cell is thus

$$PbO_2(s) + Pb(s) + 4H^+(aq) + SO_4^{2-}(aq) \longrightarrow$$
$$2PbSO_4(s) + 2H_2O \qquad E°_{cell} = +2.05 \text{ V}$$

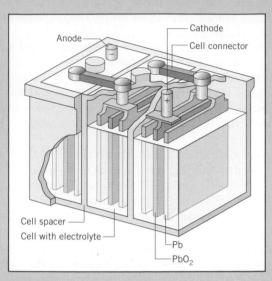

FIGURE 1
The 6-V lead storage battery consists of three cells each able to produce about 2 V.

When three such cells are combined—wired *in series* is the technical description—the total voltage is about three times the individual cell potential, or about 6 V.

As electricity is drawn, the net reaction proceeds from left to right. This reaction consumes sulfuric acid, because its sulfate ion ends up in solid lead sulfate and its hydrogen ions in molecules of water. As sulfuric acid is used up, the density of the aqueous phase in the battery diminishes.

Or,

$$\log K_{eq} = \frac{0.32 \times 2}{0.0592}$$
$$= 10.81$$

Therefore,

$$K_{eq} = 10^{10.81}$$
$$= 6.5 \times 10^{10} \text{ (Also !)}$$

Even this reaction with a small $E°_{cell}$ proceeds essentially to completion, for all practical purposes, before equilibrium is established. (We normally would not even write the double arrows, just a single forward arrow for the redox reaction.)

You can see by the examples just worked that it would take extremely small values of $E°_{cell}$ to have a redox equilibrium with K_{eq} between 1 and 10. Even then, we would say that the products are favored. Even if you found a value of $E°_{cell}$ equal to 0, K_{eq} would equal 1, since $10^0 = 1$. And we do not have values of $E°_{cell}$ less than 0 (i.e., negative numbers). Later, when

It's easy to measure this change in density by a battery hydrometer, Figure 2; service stations perform this routine operation to check on the status of a car's battery.

One property of the lead storage battery that continues to make it popular for vehicles is the ability to reverse the net discharge reaction. This more or less restores the cells to their original conditions, and after such a "recharge" the battery is ready to go again. The recharge is done by passing direct electrical current backward through the system. If done slowly, it works well.

The Silver Oxide Battery Figure 3 shows the inner parts of the silver oxide battery, which is very popular for use with calculators and automatic cameras. The half-reactions are the following.

Cathode
$$Ag_2O(s) + H_2O + 2e^- \longrightarrow 2Ag(s) + 2OH^-(aq)$$
Anode
$$Zn(s) + 2OH^-(aq) \longrightarrow Zn(OH)_2(s) + 2e^-$$

An alkaline, moist paste makes up the electrolyte medium. This cell is able to generate about 1.5 V.

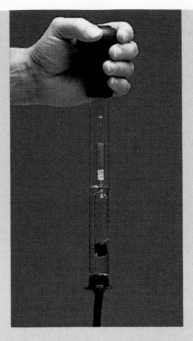

FIGURE 2
This battery hydrometer is used to monitor the charge of a lead storage battery. Battery acid is drawn up into the glass cylinder which contains a float. The float will sink further, the more dilute the acid is and, therefore, the more run-down the battery is.

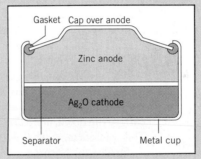

FIGURE 3
The silver oxide battery is a miniature cell capable of delivering 1.5 V.

we study some redox reactions that occur during the body's use of oxygen in cells, we'll better appreciate how powerfully these tend to go to completion.

In summary, when $E°_{cell} > 0$ for a redox reaction, the forward reaction has a strong tendency to proceed to completion.

PRACTICE EXERCISE 8

Calculate $E°_{cell}$ and K_{eq} for the following redox reaction, which is a commercial preparation of bromine from sodium bromide and chlorine gas.

$$Cl_2(g) + 2Br^-(aq) \longrightarrow 2Cl^-(aq) + Br_2$$

We have two cautions to issue before we continue. The first is a reminder that a "tendency to go to completion" is no more than that. It carries no indication of *how rapidly* the reaction goes. The rate depends on the energy of activation, and nothing in $E°$, $E°_{cell}$, or K_{eq} carries the slightest clue about energy of activation.

The second caution is that the actual cell potential of a redox reaction depends not just on the calculated value of $E°_{cell}$, the *standard* cell potential, but also on the concentrations of the chemicals in solution. We'll look very briefly at this next.

The Relationship between E_{cell}, E_{cell}°, and Concentrations Is Given by the Nernst Equation   We again must simply state an equation without justifying it as fully as might be desired. But our goals are very limited in this area. We want to point out that E_{cell} depends on concentration and that this fact is behind the way pH meters and similar devices work.

The effect of the concentrations of reactants and products on the cell potential of a redox reaction, E_{cell}, is given by an equation developed by Walter Nernst and, appropriately, now called the *Nernst equation*.

$$E_{cell} = E_{cell}^{\circ} - \frac{0.0592}{n} \log Q \qquad (10.4)$$

The symbol Q stands for the expression that K_{eq} equals in the equilibrium law, with the difference that the concentrations in brackets need not be *equilibrium* concentrations. When an equilibrium can be written as

$$aA + bB \rightleftharpoons cC + dD$$

then

$$Q = \frac{[C]^c[D]^d}{[A]^a[B]^b}$$

■ Q stands for *reaction quotient*.

Only when the bracket concentrations are equilibrium concentrations does $Q = K_{eq}$.

Now notice in Equation 10.4 the condition needed for E_{cell} to equal E_{cell}°, the standard cell potential. For this to be true, the value of $\log Q$ must be zero. This can happen only when Q itself equals 1. This is possible when all concentrations, both reactants and products, are exactly 1 mol/L. When the concentrations have other values, the cell potential, E_{cell}, generally does not equal E_{cell}°.

Notice also that when $Q = K_{eq}$, the last term in Equation 10.4 equals E_{cell}°, as we learned from Equation 10.3. Thus at equilibrium, $E_{cell} = E_{cell}^{\circ} - E_{cell}^{\circ} = 0$ V. In other words, a redox system that has reached equilibrium has no more potential left to do anything by itself. It's a battery run down, literally so when the redox system has been used to construct a real battery. This, of course, is what we mean by a chemical equilibrium—no further net change.

Finally, the Nernst equation suggests a way to measure the concentration of some species in a redox equilibrium. Actual cell potentials are easily measured by devices called *voltmeters*. If for a particular system there is a difference between the measured E_{cell} and the known value of E_{cell}°, such data can be plugged into the Nernst equation for the system to calculate a molarity of some species present and in the equation for Q. The pH meter works roughly this way. The meter is able to use a special electrode, sensitive to $[H^+]$, to measure E_{cell} for its special cell, to compare it to E_{cell}°, and to translate the data directly into a pH reading. A large number of similar electrodes, ones that are able to relate values of E_{cell} to the molarity of a particular species, even when many other solutes are in solution, have been invented. Many are used in the analysis of body fluids, like blood or urine, particularly in an emergency when very rapid results are critical.

SUMMARY

Oxidation numbers   A reaction in which oxidation numbers change is a redox reaction—oxidation–reduction. An atom whose oxidation number becomes more positive is said to be oxidized. One whose number is made less positive (more negative) is reduced. One event cannot occur without the other in a reaction.

With monatomic ions, oxidation numbers are the same as the real electrical charges on the ions. Thus the ions of all group IA elements have oxidation numbers of $+1$; those of group IIA, $+2$, and aluminum (group IIIA), $+3$. The monatomic ions of group VIIA elements have oxidation numbers of -1; those in group VIA, -2.

With molecules and polyatomic ions, oxidation numbers are the hypothetical charges that their atoms would have if both electrons

in electron-pair bonds were assigned to the more electronegative atom. Fluorine, the most electronegative element, always has an oxidation number of -1. That of hydrogen is always $+1$ except in metal hydrides. That of oxygen is always -2 except when this would lead to a violation of the rules for F or H. When an atom is not combined with an atom from a different element, its oxidation number is zero. The sum of all oxidation numbers in a formula unit must equal the charge on the unit.

Balancing redox reactions   The ion-electron method breaks the skeletal equation of a redox reaction into two half-reactions, one an oxidation and the other a reduction. Atoms that are not H or O are balanced first, then O is balanced using H_2O. H is next balanced using H^+. The necessary electrical balance is obtained by using or releasing electrons. After each half-reaction has been balanced, small whole-number multipliers may be needed for one or both so that when the half-reactions are added the electrons will cancel. (A final trimming of excess H^+ or H_2O may be needed.)

If the medium is basic, the redox reaction is first balanced pretending that the medium isn't basic. Then OH^- ions are added to both sides of the balanced equation in enough numbers to enable neutralization of any H^+.

Reduction potentials The tendency for a half-reaction to proceed as a reduction, relative to the reduction of H^+, is described by the standard reduction potential, $E°$. Its value corresponds to concentrations of dissolved species at 1 M and gases above the solution at 1 atm pressure. A half-reaction with a positive reduction potential represents a system more easily reduced than H^+. Those with negative reduction potentials are less easily reduced than H^+. The $H_2/2H^+$ system is assigned a reduction potential of 0.00 V when $[H^+]$ is exactly 1 M, the temperature is 25 °C, and the pressure of H_2 is 1 atm.

The most easily reduced species have the most positive reduction potentials. They are the strongest oxidizing agents. Fluorine, F_2, is an example. The most easily oxidized species have the most

negative reduction potentials. The metals in groups IA and IIA are examples.

When two half-reactions are combined into a full redox reaction, the half-reaction with the more positive $E°$ always runs as a reduction and forces the other to run as an oxidation.

Redox equilibria The spread between $E°$ for the substance reduced and $E°$ for the substance oxidized in a redox reaction is called the standard cell potential, $E°_{cell}$, for the reaction. It is always a positive number (or zero), and is calculated by the algebraic subtraction of $E°$ for the oxidized species from $E°$ for the reduced species.

$E°_{cell}$ is proportional to the log of the equilibrium constant for the redox reaction.

$$E°_{cell} = \frac{0.0592}{n} \log K_{eq}$$

Even very small values of $E°_{cell}$ correspond to sizable values of K_{eq}, so just about all redox reactions go entirely to completion for all practical purposes.

$E°_{cell}$ is for all concentrations at 1 M, the temperature at 25 °C, and any gases in the redox reaction at 1 atm of pressure. E_{cell}, the cell potential under any other conditions, is related to $E°_{cell}$ and concentrations by the Nernst equation. When all concentrations are the *equilibrium* concentrations, E_{cell} is zero, and no potential remains for further change. (It's like a run down battery.)

REVIEW EXERCISES

The answers to Review Exercises whose numbers are in color are found in Appendix C. The answers to the other Review Exercises are found in the Study Guide that accompanies this book. The more challenging questions are marked with asterisks.

Oxidation Numbers

10.1 Define the following terms. (These also review parts of Section 5.2.)
(a) oxidation (b) reduction (c) oxidizing agent
(d) reducing agent (e) redox reaction

10.2 What are the oxidation numbers of the atoms in the following species?
(a) NH_3 (b) NF_3 (c) NO_2 (d) NO_2^- (e) N_2O_5

10.3 What are the oxidation numbers of the atoms in the following?
(a) HIO_4 (b) HIO_3 (c) HIO_2 (d) HIO

10.4 Calculate the oxidation numbers of the atoms in the following.
(a) HS^- (b) HSO_3^- (c) HSO_4^-

10.5 What are the oxidation numbers of the atoms in each?
(a) MnO_2 (b) Mn_2O_3 (c) MnO_4^- (d) $MnCl_2$

Balancing Redox Reactions

10.6 When electrons appear as *products* in a balanced equation for a half-reaction, does the reaction represent oxidation or reduction? What kind of change in oxidation number has occurred, and is the species to which the change happens an oxidizing agent or a reducing agent?

*10.7** Balance the following half-reactions occurring in aqueous acid and tell whether it represents oxidation or reduction.
(a) $MnO_4^- \rightarrow Mn^{2+}$ (b) $Fe^{2+} \rightarrow Fe_2O_3$
(c) $H_2C_2O_4 \rightarrow CO_2$ (d) $NO_2 \rightarrow N_2$

*10.8** Balance the following half-reactions occurring in aqueous acid. Tell whether they represent oxidation or reduction.
(a) $CH_3OH \rightarrow CO_2$ (b) $CH_4 \rightarrow CO$
(c) $CH_3OH \rightarrow CH_2O$ (d) $CO \rightarrow C$

*10.9** Balance the following redox reactions occurring in aqueous acid.
(a) $I^- + HNO_2 \rightarrow I_2 + NO$
(b) $Sn + NO_3^- \rightarrow SnO_2 + NO$
(c) $Mg + SO_4^{2-} \rightarrow Mg^{2+} + SO_2$
(d) $PbO_2 + Cl^- \rightarrow PbCl_2 + Cl_2$

*10.10** Balance the following redox reactions, which take place in acid.
(a) $Ag + NO_3^- \rightarrow NO_2 + Ag^+$
(b) $C_2O_4^{2-} + HNO_2 \rightarrow CO_2 + NO$
(c) $MnO_4^- + HNO_2 \rightarrow Mn^{2+} + NO_3^-$
(d) $H_3PO_2 + Cr_2O_7^{2-} \rightarrow H_3PO_4 + Cr^{3+}$

*10.11** Balance the following redox reactions occurring in a basic solution.
(a) $MnO_4^- + CH_2O \rightarrow CO_2 + MnO_2$
(b) $CrO_4^{2-} + S^{2-} \rightarrow CrO_2^- + S$

*10.12** Balance the following redox reactions that occur in aqueous base.
(a) $MnO_4^- + SO_3^{2-} \rightarrow MnO_2 + SO_4^{2-}$
(b) $C_6H_5CH_3 + CrO_4^{2-} \rightarrow C_6H_5CO_2^- + CrO_2^-$

Reduction Potentials

10.13 Describe the reference half-reaction to which other standard reduction potentials are related. Use the appropriate chemical equation and describe the experimental conditions that must exist if a value of E_{cell} can be called E_{cell}°.

˚10.14 The Ca/Ca^{2+} system has a standard reduction potential of -2.76 V. To show what this means,
(a) Write the equation of the half reaction for this system.
(b) Does this half-cell reaction have a greater or lesser tendency to occur under the standard conditions than the $H_2/2H^+$ system?
(c) Which is the stronger oxidizing agent under the standard conditions, Ca^{2+} or H^+?

˚10.15 The $I_2/2I^-$ system has a standard reduction potential of $+0.54$ V.
(a) Write the equation of the half reaction for this system.
(b) Which has the greater tendency to occur as a reduction under the standard conditions, this system or the Cu/Cu^{2+} system, whose $E^{\circ} = +0.34$ V?
(c) Which is the stronger reducing agent under the standard conditions, I_2 or H_2?

10.16 Find the standard half-reactions for the halogens listed in Table 10.1.
(a) Arrange the elements in their order of increasing ability to function as oxidizing agents. Place the formula of the halogen with the weakest oxidizing power on the left in the series.
(b) Arrange the elements in the order in which their monatomic anions are able to function as reducing agents. Place the formula of the halide ion with the weakest reducing power on the left.

10.17 Given the large negative standard reduction potentials of the group IA and IIA metals, should one expect them to have low or high ionization energies? Explain.

10.18 Find the symbol of the metal ion with the highest standard reduction potential in Table 10.1. How does its position in the Table correlate with a well-known fact about the metal itself?

10.19 Compare Table 10.1 with Table 8.3. What is the basis for the activity series of the metals?

˚10.20 If O_2, I_2, I^-, H^+, and water are together, what spontaneous reaction will occur? Write the equation for the redox reaction.

˚10.21 If Cl_2, Cl^-, Fe^{3+}, and Fe^{2+} are brought together, what spontaneous chemical reaction can occur? Write the equation.

10.22 Using data in Table 10.1, could the following reaction occur spontaneously?

$$Cu(s) + 2H^+(aq) \longrightarrow Cu^{2+}(aq) + H_2(s)$$

˚10.23 Could the following reactions occur spontaneously? Use data in Table 10.1.
(a) $Fe^{2+}(aq) + Ni(s) \rightarrow Fe(s) + Ni^{2+}(aq)$
(b) $PbO_2(s) + 4H^+(aq) + 2F^-(aq) \rightarrow Pb^{2+}(aq) + F_2(g) + 2H_2O$

Redox Equilibria

10.24 Calculate the standard cell potential for the following reaction.

$$Fe(s) + Br_2(aq) \longrightarrow Fe^{2+}(aq) + 2Br^-(aq)$$

10.25 What is the standard cell potential for the following reaction?

$$Sn^{2+}(aq) + Ni(s) \longrightarrow Sn(s) + Ni^{2+}(aq)$$

10.26 Consider the redox reaction of Review Exercise 10.24 as an equilibrium and calculate its equilibrium constant.

10.27 Calculate the equilibrium constant for the reaction of Practice Exercise 10.25. Are the products favored at equilibrium?

10.28 Under what conditions does $E_{cell} = E_{cell}^{\circ}$?

10.29 Under what conditions does $E_{cell} = 0$?

Galvanic Cells (Special Topic 10.1)

10.30 Prepare a drawing of a galvanic cell that could use the following redox reaction to generate electricity.

$$Fe^{2+}(aq) + Mg(s) \longrightarrow Fe(s) + Mg^{2+}(aq)$$

Label the parts and show which half-reactions occur in each chamber. (The salt bridge could contain aqueous KNO_3.)

10.31 Explain why a salt bridge is necessary in the cell of Review Exercise 10.30.

10.32 What would be the standard cell potential for the cell of Practice Exercise 10.30?

Batteries (Special Topic 10.2)

10.33 What are the half-reactions in a lead storage battery?

10.34 What is the net reaction in the lead storage battery?

10.35 What happens as a lead storage battery runs down that enables the use of a battery hydrometer?

10.36 What are the half-reactions and the net equation for the silver oxide battery?

Additional Exercises

10.37 In each pair, choose the better oxidizing agent.
(a) Br_2 or I_2 (b) O_2 or F_2

˚10.38 Sodium iodate, $NaIO_3$, reacts with sodium sulfite, Na_2SO_3, according to the equation

$$3Na_2SO_3(aq) + NaIO_3 \longrightarrow NaI(aq) + 3Na_2SO_4(aq)$$

How many grams of Na_2SO_3 are needed to react with 10.0 g of $NaIO_3$?

˚10.39 Balance the following equations for reactions that occur in acid solution.
(a) $I^- + HSO_4^- \rightarrow I_2 + SO_2$
(b) $Mn^{2+} + BiO_3^- \rightarrow MnO_4^- + Bi^{3+}$

˚10.40 Balance the following equations for reactions that occur in basic solution.
(a) $Au + CN^- + O_2 \rightarrow Au(CN)_2^- + OH^-$
(b) $ClO_3^- + N_2H_4 \rightarrow NO + Cl^-$

RADIOACTIVITY AND NUCLEAR CHEMISTRY

When atomic particles collide at very high speed (center), other small particles are created that make strange and beautiful tracks in special devices. Scientists have used such tracks to learn more about atomic nuclei. Some nuclei are unstable, and in this chapter we'll learn how their properties can be used for beneficial purposes.

11.1

ATOMIC RADIATION

Unstable atomic nuclei eject high-energy radiation as they change to more stable nuclei.

Some atomic nuclei are unstable and the isotopes with such nuclei are **radioactive,** meaning that they emit high-energy radiation. Each radioactive isotope is called a **radionuclide,** and its radiation can cause grave harm to human life. When carefully used, however, their potential benefits outweigh their possible harm. In this chapter we will study the various kinds of radiation, how they can be dangerous, how they can be used wisely, and how they are measured.

Unstable Nuclei Undergo Radioactive Decay, Emit Different Kinds of Radiation, and Transmute to Nuclei of Different Elements Radioactivity was discovered in 1896 when a French physicist, A. H. Becquerel (1852 – 1908), happened to store some well-wrapped photographic plates in a drawer that contained samples of uranium ore. The film became fogged, meaning that when developed the picture was like a photograph of fog.

Becquerel might have blamed the accident on faulty film or careless handling, but a mysterious radiation called X rays had recently been discovered by a German scientist, Wilhelm Roentgen (1845 – 1923). X rays were known to be able to penetrate the packaging of unexposed film and ruin it. What fogged Becquerel's film was a natural radiation that resembled X rays. It was soon found to be emitted by any compound of uranium as well as by uranium metal itself.

Several years later, two British scientists, Ernest Rutherford (1871 – 1937) and Frederick Soddy (1877 – 1956), explained radioactivity in terms of events inside unstable atomic nuclei. Such a nucleus undergoes a transformation, called a *disintegration,* and the overall process is called **radioactive decay.** A decaying nucleus may eject a tiny particle into space, or it may emit a powerful radiation like an X ray but called a gamma ray, or the nucleus may do both.

The nuclei that remain after decay almost always are those of an entirely different element, so decay is usually accompanied by the **transmutation** of one isotope into another. The natural sources of radiation on our planet emit one or more of three kinds: *alpha radiation, beta radiation,* and *gamma radiation.* We receive another, called *cosmic radiation,* from the sun and outer space. (See Special Topic 11.1.)

Alpha Particles Are the Nuclei of Helium Atoms One natural atomic radiation is called **alpha radiation.** It consists of particles called **alpha particles** that move with a velocity almost one-tenth the velocity of light as they leave the atom. Alpha particles are clusters of two protons and two neutrons, so they are actually the nuclei of helium atoms (see Fig. 11.1). They are the largest of the decay particles and they have the greatest charge, so when alpha particles travel in air, they soon collide with molecules of N_2 or O_2 and lose their energy (and charge). Alpha particles cannot penetrate even thin cardboard or the outer layer of dead cells on the skin, but they can cause a severe burn to the skin.

■ Becquerel shared the 1903 Nobel prize in physics with Pierre and Marie Curie.

■ Roentgen won the 1901 Nobel prize in physics, the first to be awarded.

■ Nobel prizes in chemistry were awarded to both Rutherford (1908) and Soddy (1921).

SPECIAL TOPIC 11.1
COSMIC RAYS

Cosmic rays are streams of particles that enter our outer atmosphere from the sun and outer space. They consist mostly of high-energy protons plus some alpha and beta particles and the nuclei of the lower-formula-mass elements (up through number 26, iron).

Cosmic ray particles do not travel far when they enter the atmosphere because they quickly collide with the air's molecules and atoms. These collisions, however, generate all the subatomic particles, including neutrons and some others we haven't studied in this chapter. These secondary cosmic rays are what we are exposed to on the earth's surface. High-energy electromagnetic radiation is also produced, and this is the most dangerous component from a radiological health perspective.

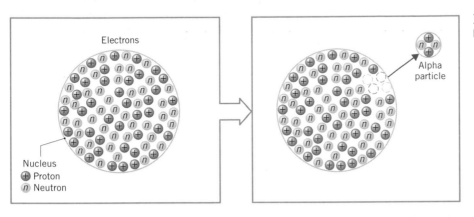

FIGURE 11.1
Emission of an alpha particle.

The most common isotope of uranium, uranium-238 or $^{238}_{92}U$, is an alpha emitter. When its nucleus ejects an alpha particle, it loses two protons, so the atomic number changes from 92 to 90. It also loses four units of mass number (two protons + two neutrons), so its mass number changes from 238 to 234. The result is that uranium-238 transmutes into an isotope of thorium, $^{234}_{90}Th$.

Beta Radiation Is a Stream of Electrons
Another natural radiation, **beta radiation,** consists of a stream of particles called **beta particles,** which are actually electrons. They are produced *within* the nucleus and then emitted (see Fig. 11.2). With less charge and a much smaller size, beta particles can penetrate matter, including air, more easily than alpha particles. Different sources emit beta particles with different energies, and those of lower energy are unable to penetrate the skin. Those of the highest energy can reach internal organs from outside the body.

As a nucleus emits a beta particle, a neutron changes into a proton (Fig. 11.2). Thus there is no loss in mass number, but the atomic number *increases* by 1 unit because of the new proton. For example, thorium-234, $^{234}_{90}Th$, is a beta emitter, and when it ejects a beta particle it changes to an isotope of protactinium, $^{234}_{91}Pa$.

■ Isotopes have special symbols that give the mass number as a left superscript and the atomic number as a left subscript. Thus in $^{238}_{92}U$, 238 is the mass number and 92 is the atomic number.

■ Losing one electron from the nucleus makes a proton out of a neutron, so no change in mass number occurs.

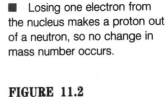

FIGURE 11.2
Emission of a beta particle.

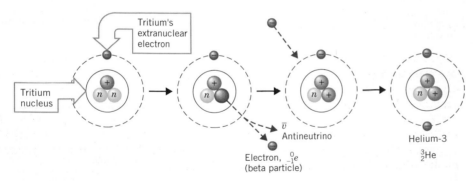

TABLE 11.1

Radiation from Naturally Occurring Radionuclides

Radiation	Composition	Mass Number	Electrical Charge	Symbols
Alpha	Helium nuclei	4	2+	$^4_2\text{He}, \alpha$
Beta	Electrons	0	1−	$^0_{-1}e, \beta$
Gamma	X-ray like	0	0	$^0_0\gamma$

Gamma Radiation Often Accompanies Other Radiation By emitting alpha or beta radiation, a radionuclide achieves greater stability. Atoms have different *nuclear* energy states just as they have different *electron* energy states. By ejecting small particles, unstable nuclei acquire a lower, more stable nuclear state. The energy lost by the nucleus is carried away by the moving particles, but often some photons of high-energy electromagnetic radiation are also emitted. This is called **gamma radiation,** and it is like X rays or ultraviolet rays, but with more energy. Like X rays, gamma radiation is quite penetrating and very dangerous. It easily travels through the *entire* body.

The composition and symbols of the three kinds of radiation studied thus far are summarized in Table 11.1.

Nuclear Equations Must Display Balances of Both Mass Numbers and Atomic Numbers In *chemical* reactions no changes in atomic nuclei occur. But nuclear reactions are nearly always accompanied by transmutations. Therefore **nuclear equations** are different from chemical equations in important ways. They must, in particular, describe the changes in atomic numbers, mass numbers, and identities of radionuclides.

In nuclear equations the alpha particle is symbolized as ^4_2He, and although it is positively charged, the charge is omitted from the symbol. The particle soon picks up electrons, anyway, taking them from the matter through which the particle moves. Thus the alpha particle becomes a neutral atom of helium. The beta particle has the symbol $^0_{-1}e$ because its mass number is 0 and its charge is 1 −. A photon of gamma radiation is symbolized simply by γ (or, sometimes, by $^0_0\gamma$). Because it is a photon, its mass number and charge are 0.

A nuclear equation is *balanced* when the sums of the mass numbers on either side of the arrow are equal and when the sums of the atomic numbers are equal. The alpha decay of uranium-238, for example, is represented by the following equation:

$$^{238}_{92}\text{U} \longrightarrow {}^{234}_{90}\text{Th} + {}^4_2\text{He}$$

Notice that the sums of the atomic numbers agree: $92 = 90 + 2$. The sums of the mass numbers also agree: $238 = 234 + 4$. The equation is balanced.

The beta decay of thorium-234, which also emits gamma radiation, is represented by the following nuclear equation.

$$^{234}_{90}\text{Th} \longrightarrow {}^{234}_{91}\text{Pa} + {}^0_{-1}e + {}^0_0\gamma$$

The sums of the atomic numbers agree: $90 = 91 + (-1)$. Likewise, the sums of the mass numbers agree; the equation is balanced.

■ It's proper to think of the electron as having an atomic number of −1.

EXAMPLE 11.1
Balancing Nuclear Equations

Cesium-137, $^{137}_{55}\text{Cs}$, is one of the radioactive wastes that form during fission in a nuclear power plant or an atomic bomb explosion. This radionuclide decays by emitting both beta and gamma radiation. Write the nuclear equation for this decay.

ANALYSIS We start with an incomplete equation using the given information. Then we figure out any additional data we need.

SOLUTION The incomplete nuclear equation is

$$^{137}_{55}Cs \longrightarrow \, ^{0}_{-1}e + \, ^{0}_{0}\gamma + \underline{\quad}\, \underline{\quad}$$

- mass number goes here
- atomic symbol goes here
- atomic number goes here

We have to figure out the atomic number, Z, first so that we can find the atomic symbol for the product in the table inside the front cover. To calculate Z we use the fact that the atomic number (55) on the left side of the equation must equal the sum of the atomic numbers on the right side.

$$55 = -1 + 0 + Z$$
$$Z = 56$$

In the periodic table we see that element 56 is Ba (barium), but which isotope of Ba? Recall that the sums of the mass numbers on either side of the equation must also be equal. Letting A equal the mass number of the barium isotope,

$$137 = 0 + 0 + A$$
$$A = 137$$

The balanced nuclear equation, therefore, is

$$^{137}_{55}Cs \longrightarrow \, ^{0}_{-1}e + \, ^{0}_{0}\gamma + \, ^{137}_{56}Ba$$

EXAMPLE 11.2
Balancing Nuclear Equations

Until the 1950s, radium-226 was widely used as a source of radiation for cancer treatment. It is an alpha emitter and a gamma emitter. Write the equation for its decay.

ANALYSIS We have to look up the atomic number of radium, which turns out to be 88, so the symbol we'll use for this radionuclide is $^{226}_{88}Ra$. When one of its atoms loses an alpha particle, $^{4}_{2}He$, it loses 4 units in mass number—from 226 to 222. And it loses 2 units in atomic number—from 88 to 86. Thus the new radionuclide has a mass number of 222 and an atomic number of 86. We have to look up the atomic symbol for element number 86, which turns out to be Rn, for radon.

SOLUTION Now we can assemble the nuclear equation.

$$^{226}_{88}Ra \longrightarrow \, ^{222}_{86}Rn + \, ^{4}_{2}He + \gamma$$

■ The radium used in cancer therapy was held in a thin, hollow gold or platinum needle to retain the alpha particles and all the decay products.

PRACTICE EXERCISE 1

Iodine-131 has long been used in treating cancer of the thyroid. This radionuclide emits beta and gamma rays. Write the nuclear equation for this decay.

PRACTICE EXERCISE 2

Plutonium-239 is a by-product of the operation of nuclear power plants. It can be isolated from used uranium fuel and made into fuel itself or into atomic bombs. A powerful alpha and gamma emitter, plutonium-239 is one of the most dangerous of all known substances. Write the equation for its decay.

A Short Half-Life Means a Rapid Decay

Some radionuclides are much more stable than others, and we use the concept of a half-life to describe the differences. The **half-life** of a radionuclide, symbolized as $t_{1/2}$, is the time it takes for half of the atoms in a sample of a single, pure isotope to decay. (The atoms that decay don't just vanish, of course. They change into different isotopes.) Table 11.2 gives the half-lives of several radionuclides.

The half-life of uranium-238 is 4.51×10^9 years, which means that an initial 100 g of this radionuclide would have only 50.0 g of uranium-238 left after 4.51×10^9 years. Strontium-90, a by-product of nuclear power plants, is a beta emitter with a half-life of 28.1 years. Figure 11.3 shows graphically how an initial supply of 40 g is reduced successively by units of one-half for each half-life period. At the end of seven half-life periods (196.7 years, from 7×28.1), only 0.3 g of strontium-90 remains in the sample.

The shorter the half-life, the larger the number of decay events per mole per second occurring in the isotope. Mole for mole, it's generally much safer to be near a sample that has a long half-life and thus decays very slowly than to be near one that has a short half-life and decays very rapidly. The potential danger of a radionuclide, however, is a function of more factors than its half-life, as we will soon learn.

■ An extremely long half-life is typical of the radionuclides that head a radioactive disintegration series.

A Succession of Decays Occurs in a Radioactive Disintegration Series

The decay of one radionuclide sometimes produces not a stable isotope but just another radionuclide. This might, in turn, decay to still another radionuclide, with the process repeating until a stable

TABLE 11.2

Typical Half-Life Periods

Element	Isotope	Half-Life	Radiation or Mode of Decay
Naturally Occurring Radionuclides			
Potassium	$^{40}_{19}K$	1.3×10^9 yr	Beta, gamma
Tellurium	$^{123}_{52}Te$	1.2×10^{13} yr	Electron capture
Neodymium	$^{144}_{60}Nd$	5×10^{15} yr	Alpha
Samarium	$^{149}_{62}Sm$	4×10^{14} yr	Alpha
Rhenium	$^{187}_{75}Re$	7×10^{10} yr	Beta
Radon	$^{222}_{86}Rn$	3.82 day	Alpha
Radium	$^{226}_{88}Ra$	1590 yr	Alpha, gamma
Thorium	$^{230}_{90}Th$	8×10^4 yr	Alpha, gamma
Uranium	$^{238}_{92}U$	4.51×10^9 yr	Alpha
Synthetic Radionuclides			
Tritium	$^{3}_{1}H$	12.26 yr	Beta
Oxygen	$^{15}_{8}O$	124 s	Positron
Phosphorus	$^{32}_{15}P$	14.3 day	Beta
Technetium	$^{99m}_{43}Tc$	6.02 h	Gamma
Iodine	$^{131}_{53}I$	8.07 day	Beta
Cesium	$^{137}_{55}Cs$	30 yr	Beta
Strontium	$^{90}_{38}Sr$	28.1 yr	Beta
Americium	$^{243}_{95}Am$	7.37×10^3 yr	Alpha

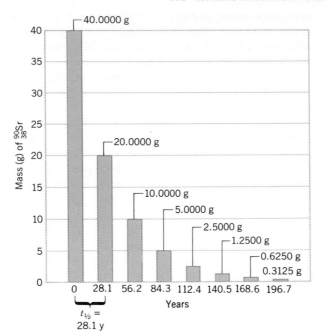

FIGURE 11.3
Each half-life period reduces the quantity of a radionuclide by a factor of 2. Shown here is the pattern for strontium-90, a radioactive pollutant with a half-life of 28.1 years.

nuclide is finally reached. There are three such series still active in nature, called **radioactive disintegration series,** and uranium-238 is at the head of one (see Fig. 11.4). This series ends in a stable isotope of lead.

11.2

IONIZING RADIATION—DANGERS AND PRECAUTIONS

Atomic radiation creates unstable ions and radicals in tissue, which can lead to cancer, mutations, tumors, or birth defects.
The undesired effects of radiation are both *acute* and *latent*. Acute effects can be burns and any of the symptoms that are a part of radiation sickness, which we will study in this section. Latent effects of radiation are those that do not show themselves until some time after exposure. Cancer, particularly leukemia, can be induced by radiation and is an example of a latent effect. Another example is the alteration of a gene.

Policy Makers Consider That No Safe *Threshold Exposure* to Radiation Exists
Any kind of radiation that penetrates the skin or enters the body on food or through the lungs is considered harmful, and *the damage can accumulate* over a lifetime. Even the ultraviolet

FIGURE 11.4
The uranium-238 radioactive disintegration series. The time given beneath the arrow is the half-life of the preceding isotope. (y = year, m = month, d = day, *min* = minute, and s = second).

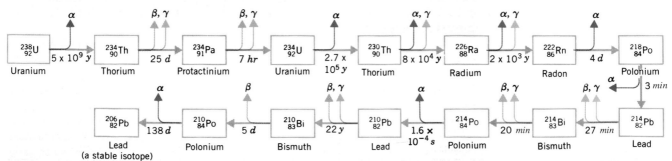

radiation in strong sunlight, which barely penetrates the skin, can alter the genetic molecules in skin cells so as to lead to skin cancer. No "tiny bit of exposure," no **threshold exposure** is considered to exist for ionizing radiation below which no harm is possible. Cells do have a significant capacity for the self-repair of radiation-caused damage, and some exposures carry very low risks. The threshold exposure, surprisingly, appears to be high for transmittable genetic defects. The threshold is much lower for a fetus than it is for men and nonpregnant women.

The latent effects of radiation correlate well with the accumulated dose regardless of how it is received, so medical personnel who work with radionuclides, X rays, or emitters of gamma rays wear devices that automatically record their exposure. The exposure data are periodically logged into a permanent record book, and if the maximum permissible dose is attained, the worker must be transferred. Let's now see how the radiation has its effects and what are some steps for self-protection.

Unstable Ions and Radicals Are Produced in Tissue by Radiation The different kinds of atomic radiation are dangerous because they can generate unstable, highly reactive particles as they travel through living tissues. For this reason, alpha and beta particles, as well as X rays and gamma rays, are called **ionizing radiations.** They can knock electrons from molecules as they strike them and so produce unstable polyatomic ions. Radiation, for example, can make ions from water molecules by the following reaction.

$$H—\ddot{O}—H \xrightarrow{\text{radiation}} \left[H—\dot{O}—H\right]^+ + _{-1}^{0}e$$

The new cation, $\left[H—\dot{O}—H\right]^+$, is unstable and one breakup path is

$$\left[H—\dot{O}—H\right]^+ \longrightarrow H^+ + :\dot{O}—H$$
Proton   Hydroxyl radical

A proton forms plus the hydroxyl radical, a kind of particle new to our study. It's a *neutral* particle with an unpaired electron and without an octet for its oxygen atom. Any particle with an unpaired electron is called a **radical** and, with few exceptions, radicals are very reactive species.

The new ions and radicals produced by ionizing radiation cause chemical reactions in the more stable substances around them, altering them in ways foreign to metabolism. If such chemical reactions happen in genes and chromosomes, the cell's genetic substances, subsequent reactions could lead to cancer, tumor growth, or a genetic mutation. If they happen in a sperm cell, an ovum, or a fetus, the result might be a birth defect. Relative to other tissues, the fetus is particularly sensitive to radiation. Prolonged and repeated exposures to *low* levels of radiation are more likely to induce these problems than bursts of high-level radiation. It depends on whether the injured cell is still able to duplicate itself by cell division. High-energy radiation bursts usually kill a cell outright, or at least render it reproductively dead. For this reason, high doses of radiation are used in cancer treatment. But low-level radiation that leaves a cell reproductively viable can alter the cell contents in ways that affect the way in which the cell is reproduced.

Cells do have a capacity for self-repair, as we said, but no exposure is entirely risk-free. The widespread and routine use of X rays for public health screenings has long been curtailed.

The Collection of Symptoms Caused by Ionizing Radiation Is Called Radiation Sickness Molecules of hereditary materials in the cell's chromosomes are the primary site of the most serious radiation damage. Damage to these leads to all other problems. The first symptoms of exposure to radiation, therefore, often occur in tissues whose cells divide most frequently, for example, the cells in bone marrow. These make white blood cells, so an early

■ Sometimes the term used is *free* radical.

■ Technical terms are
Carcinogen, a cancer causer
Tumorigen, a tumor causer
Mutagen, a mutation causer
Teratogen, a birth defect causer

■ No technology in any medical field is entirely risk-free. We take risks when we believe that the benefits outweigh them.

sign of radiation damage is a sharp decrease in the blood's white cell count. Cells in the intestinal tract also divide frequently, and even moderate exposure to X rays or gamma rays (as in cobalt-ray therapy for cancer) may produce intestinal disorders.

The set of symptoms caused by nonlethal exposures to atomic radiation or X rays is called **radiation sickness.** The symptoms include nausea, vomiting, a drop in the white cell count, diarrhea, dehydration, prostration, hemorrhaging, and the loss of hair. They often appear when sharp bursts of radiation are used to halt the spread of cancer.

Protection from Ionizing Radiation Is Achieved by Deploying Shields, Using Short Exposure Times, and Moving Away from the Source Shields have long been used for protection against ionizing radiation. (No doubt on a visit to the dentist you have had a lead apron placed over your chest before a dental X ray.) Alpha and beta rays are the easiest to stop as the data in Table 11.3 show. Gamma radiation and X rays are stopped effectively only by particularly dense substances. Lead, a very dense metal but still fairly inexpensive, is the most common material used to shield against gamma or X rays. But notice in the data in Table 11.3 that even 30 mm (3.0 cm, a little over an inch) of lead reduces the intensity of gamma radiation by only 10%. A vacuum is least effective, of course, and air isn't much better. Low-density materials such as cardboard, plastic, and aluminum are poor shielders, but concrete works well if it is thick (and it's much cheaper than lead). Thus by a careful choice of a shielding material, protection can be obtained.

Another strategy to minimize exposure when an ionizing radiation is used in medical diagnosis, such as in taking X rays, is to use fast film. With fast film, the *time* of exposure is kept as low as possible.

TABLE 11.3
Penetrating Abilities of Some Common Kinds of Radiation[a]

Type of Radiation	Common Sources	Approximate Energy When from These Sources	Approximate Depth of Penetration of Radiation into		
			Dry Air	Tissue	Lead
Alpha rays	Radium-226 Radon-222 Polonium-210	5 MeV	4 cm	0.05 mm	0
Beta rays	Tritium Strontium-90 Iodine-131 Carbon-14	0.01 to 0.2 MeV	0.3 to 6 cm[b]	0.06 to 4 mm[c]	0.005 to 0.3 mm
			Thickness to Reduce Initial Intensity by 10%		
Gamma rays	Cobalt-60 Cesium-137 Radium-226 decay products	1 MeV	400 cm	50 cm	30 cm
X Rays Diagnostic Therapeutic		Up to 90 keV Up to 250 keV	120 m 240 m	15 cm 30 cm	0.3 mm 1.5 mm

[a] Data from J. B. Little. *The New England Journal of Medicine*, vol. 275, pages 929–938, 1966.

[b] The range of beta particles in air is about 30 cm per MeV.

[c] The protective layer of skin is about 0.07 mm thick. To penetrate it, alpha particles need about 7.5 MeV of energy and beta particles about 0.07 MeV.

The least expensive self-protection step is to get as far from the radiation source as you can. Radiation, like light from a bulb, moves in straight lines spreading out in all the directions open to it from its source. From any point on the surface of the source, the radiation forms a cone of rays, so fewer rays can strike a unit of surface area the more distant the surface is from the source. The area of the base of such a cone increases with the square of the distance. Hence the radiation intensity I on a unit area diminishes with the square of the distance d from the source. This is the **inverse-square law** of radiation intensity.

> **Inverse-Square Law** The intensity of radiation is inversely proportional to the square of the distance from the source.
>
> $$\text{radiation intensity, } I \propto \frac{1}{d^2} \qquad (11.1)$$

This law holds strictly only in a vacuum, but it holds closely enough when the medium is air to make good estimates. If we move from one location, a, to another location, b, the variation of Equation 11.1 that we can use to compare the intensities, I_1 and I_2, at the two different places is given by the following equation.

$$\frac{I_1}{I_2} = \frac{d_2{}^2}{d_1{}^2} \qquad (11.2)$$

EXAMPLE 11.3
Using the Inverse Square Law of Radiation Intensity

At 1.0 m from a radioactive source the radiation intensity was measured as 30 units. If the operator moves away to a distance of 3.0 m, what will the radiation intensity be?

ANALYSIS Any problem involving changes of radiation intensity with distance calls for Equation 11.2.

SOLUTION As usual, it's a good idea to assemble the data in one place.

$$I_1 = 30 \text{ units} \qquad d_1 = 1.0 \text{ m}$$
$$I_2 = ? \qquad d_2 = 3.0 \text{ m}$$

Now we can use Equation 11.2:

$$\frac{30 \text{ units}}{I_2} = \frac{(3.0 \text{ m})^2}{(1.0 \text{ m})^2}$$

Solving for I_2, we get

$$I_2 = 30 \text{ units} \times \frac{1 \text{ m}^2}{9.0 \text{ m}^2}$$

$$= 3.3 \text{ units}$$

Thus tripling the distance cut the intensity almost by a factor of 10, from 30 to 3.3 units.

PRACTICE EXERCISE 3

If the intensity of radiation is 25 units at a distance of 10 m, what does the intensity become if you move to a distance of 0.50 m?

SPECIAL TOPIC 11.2
RADON IN THE ENVIRONMENT

Radon-222 is a naturally occurring radionuclide in the family of noble gases produced by the U-238 disintegration series. Chemically, it is as inert as the other noble gases, but radiologically it is a dangerous air pollutant. It is an alpha emitter and gamma emitter with a half-life of only four days. Produced in rocks and soil wherever uranium-238 is found, it migrates as a gas into the surrounding air. Basements not fully sealed act like fireplace chimneys to draw radon-222 into homes.

The first indication of how serious radon-222 pollution might be came when an engineer at a nuclear power plant in Pennsylvania set off radiation alarms just by his presence. The problem was traced to his home where the radiation level in the basement was 2700 picocuries per liter of air. In the average home basement, the level is just 1 picocurie/L. (The picocurie is 10^{-12} curie.) The engineer had carried radon-222 and its radioactive decay products on his clothing into his workplace.

As a result of the incident, geologists went looking for unusual concentrations of uranium-238 nearby. They found that the Reading Prong, a formation of bedrock that cuts across Pennsylvania, New Jersey, New York State, and up into the New England states, is relatively rich in U-238.

Radon-222 enters the lungs with breathing, and some decays within the lungs. Several decay products in the series after radon-222 are not gases, like polonium-218 ($t_{1/2}$ 3 min, alpha emitter), lead-214 ($t_{1/2}$ 27 min, beta and gamma emitter), and polonium-214 ($t_{1/2}$ 1.6×10^{-4} s, alpha and gamma emitter). Left in the lungs, these can cause cancer, and U.S. officials estimate that 7% to 8% of the country's annual deaths from lung cancer are caused by indoor radiation from radon-222.

The recommended upper limit on radon-222 concentration in home air is 4 picocuries/L. A conservative estimate puts the percentage of homes in the United States with levels above this at 1%.

PRACTICE EXERCISE 4

If you are receiving an intensity of 80 units of radiation at a distance of 6.0 m, to what distance would you have to move to reduce this intensity by half, to a value of 40 units?

No Escape from the Natural Background Radiation Is Possible Shielding materials and distance can never completely eliminate our exposure to ionizing radiation. We are constantly exposed to **background radiation,** the radiation given off, for example, by naturally occurring radionuclides, by radioactive pollutants, cosmic rays, and medical X rays. About 50 of the roughly 350 isotopes of all elements in nature are radioactive. Different natural radionuclides (e.g., potassium-40 and carbon-13) are in the food we eat, the water we drink, and the air we breathe. Radiation enters our bodies with every X ray taken of us. Radiation comes in cosmic-ray showers. On the average, the top 15 cm of soil on our planet has 1 g of radium per square mile. Thus radioactive materials are in the soils and rocks on which we walk and which we use to make building materials.

Radon, a *chemically* inert but radioactive gas and a product of the uranium-238 disintegration series, makes the largest contribution to our background radiation. See Special Topic 11.2. (We will discuss the problems of radioactive pollutants from the operation of nuclear reactors in Section 11.7.)

The background radiation varies widely from place to place, and only estimates (which vary widely with the estimator) are possible. Table 11.4 gives the average radiation dose equivalents from various sources for the U.S. population. We will discuss the millirem (mrem), the unit of *dose equivalent* used in this table, shortly. For comparison purposes, a dose of 500 rem (500,000 mrem) given to the individuals in a large population would cause the deaths of half of them in 30 days. In relation to this, the intensity of natural background radiation is very small. At higher altitudes, the intensity of background radiation is greater because incoming cosmic rays, which contribute to the background, have had less opportunity to be absorbed and destroyed by the earth's atmosphere.

■ People became aware of the radon problem only in the early 1980s.

TABLE 11.4
Average Radiation Doses Received Annually by the United States Population[a]

Type of Radiation	Dose (mrem)	Percent of Total
Natural Radiation — 295 mrem, 82%		
Radon[b]	200	55
Cosmic rays[c]	27	8
Rocks and soil	28	8
From inside the body	40	11
Artificial Radiation — 65 mrem, 18%		
Medical X rays[d]	39	11
Nuclear medicine	14	4
Consumer products	10	3
Others	2	<1
Total[e]	360	

[a] Data from "Ionizing Radiation Exposure of the Population of the United States." Report 93, 1987, National Council on Radiation Protection. These are averages. Individual exposures vary widely.

[b] See Special Topic 11.2

[c] Travelers in jet airplanes receive about 1 mrem per 1000 miles of travel.

[d] A normal chest X ray entails an exposure of 10–20 mrem.

[e] The federal standard for maximum safe occupational exposure in the United States is roughly 5000 mrem/yr.

11.3

UNITS TO DESCRIBE AND MEASURE RADIATION

Units have been devised to describe the activity of a radioactive sample, radiation exposure, and radiation dose.

A number of units exist for a variety of measurements of radiation, and each has been invented to serve in answering only a particular question. Note carefully what these questions are and the units will be easier to learn. The SI units that will be described have been adopted in the United States by the National Council on Radiation Protection and Measurements (NCRP Report 82, 1985). Most scientists, including those in health areas, now use them, although older units appear in the technical literature.

The *Becquerel* Describes How Active a Sample Is   The **becquerel, Bq,** is the SI unit of activity and is used to answer the question, "How *active* is a sample of a radionuclide?" In terms of the number of disintegrations per second or dps,

$$1 \text{ Bq} = 1 \text{ dps}$$

The *curie,* the older unit of activity, is named after Marie Sklodowska Curie (1867–1934), a Polish scientist, who discovered radium. One **curie, Ci,** is the number of radioactive disintegrations that occur per second in a 1.0-g sample of radium, 3.7×10^{10} dps,

$$1 \text{ Ci} = 3.7 \times 10^{10} \text{ dps}$$

or

$$1 \text{ Ci} = 3.7 \times 10^{10} \text{ Bq}$$

This is an intensely active rate, so fractions of the Ci, such as the millicurie (mCi, 10^{-3} Ci), the microcurie (μCi, 10^{-6} Ci), and the picocurie (pCi, 10^{-12} Ci), are often used. Mole for mole, when the half-life is short, the radionuclide has a high activity.

■ Marie Curie is one of two scientists to win a Nobel prize in two fields of science, a share of the physics prize in 1903 and the chemistry prize in 1911.

Exposure to X Rays or Gamma Rays Is Described in Terms of the Quantity of Ions They Produce in Dry Air The older unit of X ray or gamma ray exposure is called the **roentgen,** which is based on a specific total number of charges (2.1×10^9 units) created in 1 cm^3 of dry air. If members of a large population are exposed to 650 roentgens, half will die in one to four weeks. (The rest will have radiation sickness.) The roentgen serves to answer the question, "How *intense* is the exposure to X-ray or gamma-ray radiation?"

The SI has no special unit for exposure to X rays or gamma rays, only a symbol, **X**. **X** is the ratio of the total charge on the ions (of one sign) produced per 1 kilogram of dry air by the radiation.

■ *X* is a derived SI quantity, a ratio, and so is based on the SI units of the quantities in the ratio.

$$X = \frac{\text{coulombs of charge}}{\text{kg dry air}}$$

The *coulomb,* the SI unit of quantity of charge, is the same as the charge on 6.25×10^{18} electrons.

The *Gray* Describes the *Absorbed Dose* The SI unit of *absorbed dose* is called the *gray,* named after a British radiologist, Harold Gray. It is used to answer the question, "How much *energy* is *absorbed* by a unit mass of tissue or other materials?" The **gray, Gy,** corresponds to the absorption of 1 joule (J) of energy per kilogram of tissue. (Recall that 1 J = 4.184 cal.)

$$1 \text{ Gy} = 1 \text{ J/kg}$$

The older unit of absorbed dose is the **rad,** which is 1/100th of a gray.

$$1 \text{ rad} = 10^{-2} \text{ Gy}$$

■ *Rad* comes from **r**adiation **a**bsorbed **d**ose.

The roentgen and the rad are close enough in magnitude to be nearly equivalent from a health standpoint. Thus one roentgen of gamma radiation from a cobalt-60 source, often used in cancer treatment, equals 0.96 rad in muscle tissue and 0.92 rad in compact bone.

About 6 Gy or 600 rad of gamma radiation would be lethal to most people despite the fact that this corresponds to a very small quantity of energy. We have to remember that it is not the quantity of energy that matters so much as the formation of unstable radicals and ions caused by this energy. A 6-Gy dose delivered to water breaks up only one molecule in every 36 million, but the radicals thus produced begin a cascade of harmful reactions inside a cell.

The *Sievert,* the Unit of *Dose Equivalent,* Adjusts Absorbed Doses for Different Effects in Different Tissues A dose of 1 Gy of gamma radiation is not biologically the same as a dose of 1 Gy of beta radiation or of neutrons. Thus the gray does not serve as a good basis for comparison when working with biological effects. The **sievert (Sv)** is the SI unit that satisfies the need for a way to express dose equivalent that is additive for different kinds of radiation and different target tissues. If we let D stand for absorbed dose in grays, Q for "quality factor" (meaning relative *effectiveness* for causing harm in tissue), and N for any other modifying factors, then the dose equivalent, H, in sieverts is defined as follows.

$$H = DQN$$

In other words, we multiply the absorbed dose from some radiation, D, by a factor (Q) that takes into account biologically significant properties of the radiation, and by any other factor, N, bearing on the net effect. The quality factor, Q, for alpha radiation is 20, but is only 1.0 for beta and gamma radiation. The much larger size of the alpha particle accounts for its extra danger. Its size and charge enable it to strike molecules with considerable (although undesirable) efficiency.

The older unit of dose equivalent is called the **rem.** One rem of any given radiation is the dose that has, in a human being, the effect of one roentgen. The sievert is 100 times larger than the rem but is still a quantity small in terms of energy and yet significant in terms of danger. Even millirem quantities of radiation should be avoided, and when this isn't possible the workers must wear monitoring devices that allow the day-to-day exposures to be determined.

■ *Rem* comes from **r**oentgen **e**quivalent for **m**an.

The Electron-Volt Describes the Energy of X Rays or Gamma Rays Those who work with X rays or gamma radiation use an old unit of energy to describe the energies of these kinds of radiation. This unit is the **electron-volt, eV,** defined as follows.

$$1 \text{ eV} = 1.602 \times 10^{-19} \text{ J}$$

■ The higher the energy possessed by X rays or gamma rays, the more penetrating the radiation is.

FIGURE 11.5
The basic features of a gas-
ionization radiation detection tube
such as used in a Geiger-Müller
counter.

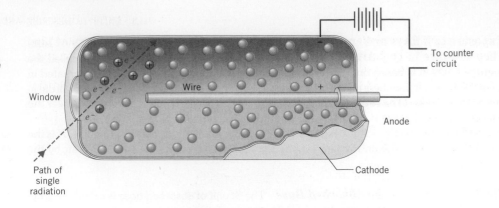

(Originally, the electron-volt was viewed as the energy an electron receives when it is accelerated by a voltage of 1 V.)

With an exponent of -19 in the equation defining the electron-volt, you might guess that it
is an extremely small amount of energy. Multiples of the electron-volt are therefore very
common, such as the kiloelectron-volt (1 keV $= 10^3$ eV) and the megaelectron-volt (MeV $=
10^6$ eV). X rays used for diagnosis are typically 100 keV or less. The chief gamma rays from
cobalt-60 used in cancer treatment have energies of 1.2 and 1.3 MeV. Beta radiation of
70 keV or more can penetrate the skin, but alpha particles (which are much larger than beta
particles) need energies of more than 7 MeV to do this. Alpha radiation from radium-226 has
an energy of 5 MeV. The cosmic radiation that enters our outer atmosphere has energies
ranging from 200 MeV to 200 GeV (1 GeV = 1 gigaelectron-volt $= 10^9$ eV).

**Film Dosimeters, Scintillation Counters, and Geiger Counters Measure Ionizing
Radiation**   A *dosimeter* is a device for measuring exposure. One common type is a film
badge that contains photographic film, which becomes fogged by radiations. The degree of
fogging, which is related to the exposure, can be measured.

Ionizing radiation also affects substances called phosphors, salts with traces of rare earth
metal ions that scintillate when struck by radiation. These scintillations — brief, spark-like
flashes of light — can be translated into doses of radiation received. Devices based on this
technology are called *scintillation counters.*

Evacuated tubes that are fitted with two electrodes and hold a gas at very low pressure are
used in devices such as the Geiger counter. This is a type of *ionization counter,* and Figure
11.5 shows how the tube itself is constructed. The tube, called a Geiger-Müller tube, is
especially useful in measuring the beta and gamma radiation that has enough energy to
penetrate the window. When this radiation enters the rarified gaseous atmosphere in the tube,
it creates ions that cause a brief pulse of electricity that the apparatus records as a "count."

11.4

SYNTHETIC RADIONUCLIDES

**Most radionuclides used in medicine are made by bombarding other atoms with
high-energy particles.**
Radioactive decay is nature's way of causing transmutations. Transmutations can also be
caused artificially by bombarding atoms with high-energy particles. Several hundred isotopes
that do not occur naturally have been made this way. Several have been used successfully in
medicine, both in diagnosis and in treatment.

Various bombarding particles can be used, including alpha particles, neutrons, and protons. The first artificial transmutation, observed by Rutherford, was the conversion of nitro-
gen-14 into oxygen-17 by alpha particle bombardment. Rutherford let alpha particles from a
naturally radioactive source travel through a tube that contained nitrogen-14. He soon

Linear accelerators produce
radiation for cancer treatment
in the range of 6 to 12 MeV.

Inside the screen of a color
TV tube there is a coating that
includes various phosphors;
these glow with different colors
when struck by the focused
electron beam in the tube.

These are gaseous
protons, true subatomic
particles, and not hydronium
ions.

detected that another radiation was being generated, one far more penetrating than the alpha radiation he used. He showed that it consisted of high-energy protons and that atoms of oxygen-17 now existed in the tube.

To explain his observations, Rutherford reasoned that alpha particles had plowed right through the electron clouds of nitrogen-14 atoms and buried themselves in their nuclei. The strange new nuclei, called *compound nuclei,* evidently had too much energy to exist for long. They were excited nuclei of fluorine-18, and to rid themselves of excess energy each ejected a proton, leaving behind an atom of oxygen-17. The equation is

$$\underset{\substack{\text{Alpha} \\ \text{particle}}}{^{4}_{2}\text{He}} + \underset{\substack{\text{Nitrogen} \\ \text{nucleus}}}{^{14}_{7}\text{N}} \longrightarrow \underset{\substack{\text{Fluorine} \\ \text{nucleus}}}{^{18}_{9}\text{F}^{*}} \longrightarrow \underset{\substack{\text{Oxygen} \\ \text{nucleus}}}{^{17}_{8}\text{O}} + \underset{\text{Proton}}{^{1}_{1}p}$$

(The asterisk by the symbol for fluorine-18 signifies that the particle is a high-energy, compound nucleus.) Oxygen-17 is a rare but nonradioactive isotope of oxygen. Usually, transmutations caused by bombardments produce *radioactive* isotopes of other elements.

Electrically charged bombarding particles, like the alpha particle and the proton, can be given greater velocity and therefore greater energy when attracted by opposite charge. Particle accelerators, devices that do this, include some of the multimillion-dollar hardware of atomic research. The interactions of their high-energy beams with selected targets have made possible the synthesis of dozens of new radionuclides.

Certain isotopes of uranium in atomic reactors eject neutrons, and although neutrons cannot be accelerated (they are electrically neutral), they have sufficient energy to serve as bombarding particles. Being neutral is an advantage, because the neutrons aren't repelled either by the electrons that surround an atom or by the nucleus. One of the very important applications of neutron bombardment is the synthesis of molybdenum-99 from molybdenum-98. With $^{0}_{1}n$ as the symbol for the neutron, the equation is

$$^{98}_{42}\text{Mo} + {}^{0}_{1}n \longrightarrow {}^{99}_{42}\text{Mo} + \gamma$$

As we will learn in Section 11.6, the decay of molybdenum-99 leads to one of the radionuclides most commonly used in medicine.

11.5

RADIATION TECHNOLOGY IN THE FOOD INDUSTRY

Food irradiated with controlled doses of X rays, gamma rays, or electron beams is less likely to spoil.
X rays, gamma rays, and particle beams have long been used in medicine as technologies for diagnosis and for cancer treatments. In treatments for cancer the aim is to kill the cells in cancer tissue. This same kind of aim, to kill cells that can subdivide, such as those of disease-causing microorganisms, is behind food irradiation technology.

Food Irradiation Inhibits, Inactivates, or Kills Molds and Bacteria
When food products are passed through a beam of gamma rays, X rays, or accelerated electrons, the effects depend on the energy of the beam. A low-dose beam — up to 100 kilorads — renders reproductively dead any insects that remain after harvest and inhibits the sprouting of potatoes and onions during storage. Such beams also inactivate trichinae (*Trichinella spiralis,* a nematode worm) in pork, the parasite that causes trichinosis.

Medium dosage beams of radiation — 100 to 1000 kilorads — significantly reduce the populations of salmonella bacteria in poultry, fish, and other meats. Such radiation also greatly extends the shelf lives of strawberries (see Fig. 11.6) and certain other fruits, which otherwise form molds quickly.

■ The nucleus of $^{18}_{9}\text{F}^{*}$ is *compounded* of the nuclei of $^{14}_{7}\text{N}$ and $^{4}_{2}\text{He}$, so is called a compound nucleus.

■ Remember that the energy of a moving object increases with the *square* of its velocity.

$$KE = (1/2)mv^2$$

■ The occurrence of trichinosis in the pork supply of the United States is small, and thorough cooking destroys it. (Pork is never served "rare.")

FIGURE 11.6

After 15 days of storage at 4 °C (38 °F), the unirradiated strawberries on the left became heavily covered by mold. Those on the right, however, had been protected by 200 kilorad of radiation.

High-dosage beams — 1000 to 10,000 kilorads — sterilize poultry, fish, and other meats. They also kill microorganisms and insects on seasonings and spices.

Food Irradiation Does Not Appear to Produce Any Unique Radiolytic Products

Ionizing radiation causes chemical reactions in foods or any insects or microorganisms in them. Since water is generally the most abundant substance present, the primary products result from the splitting of water into radicals and ions. Some recombine to form water and others combine to give hydrogen peroxide. Otherwise, the primary products of irradiation react with food molecules to give secondary products. Generally, however, these are the same as will be produced by cooking or baking, or by the subsequent digestion of foods.

■ In the body, hydrogen peroxide, H_2O_2, quickly breaks down to water and oxygen.

The irradiation of wheat and potatoes to control insects has been permitted in the United States for over twenty years, but no commercial operator is doing so. Some herbs and spices are now marketed after irradiation to reduce insects, bacteria, molds, and yeasts. The U.S. Food and Drug Administration (FDA) has approved the use of low dosage radiation to control trichinae in pork and to inhibit the spoilage of fruit and vegetables. None of these operations is as yet widespread, mostly because they are expensive.

Besides cost, customer confidence is another barrier to further use of irradiation. Any technology with the word *nuclear* or *radiation* in it makes people nervous. The legitimate question is, "Does radiation produce substances — radiolytic products — that would not be present once foods have been otherwise processed, cooked, and digested?" If so, "Are the radiolytic products harmful at the levels at which they are present?"

Specialists in food irradiation claim that their research has yet to turn up any unique radiolytic products. Even benzene, for example, is not uniquely a product of radiation. Repeated environmental exposure to benzene is known to increase a person's likelihood of contracting leukemia, and *high level* irradiation of certain foods gives traces of benzene. But small traces of this substance are naturally present in some nonirradiated foods, like boiled eggs, at much higher levels than produced by irradiation.

■ A half a pound of the botulinum toxin would be enough to kill all of the people on earth.

Still another safety issue concerns botulism, a particularly dangerous form of food poisoning caused by an odorless chemical, a toxin, produced by *Clostridium botulinum*. This bacterium is more resistant to radiation than the microorganisms that cause food spoilage. Organisms that spoil foods give them dreadful odors which warn people, but the botulinum toxin cannot be detected in this way. High-dose food irradiation could thus prevent the kind of food spoilage associated with odors, without necessarily destroying all of the botulinum bacillus.

Food irradiation technology has its strong proponents and opponents, and it is not yet clear how widely it will be used and accepted in the United States. Customers will in any event see labels to the effect that the food has been treated with radiation or that it has been made from such substances. This will give the customers the choice of buying or not buying the food. Irradiated food has been used for space missions because it minimizes the weight problems associated with other methods of food preservation.

11.6

RADIATION TECHNOLOGY IN MEDICINE

Both in diagnosis and in cancer treatment, ionizing radiation is used when its benefits are judged to outweigh its harm.

For medical uses, ionizing radiation either is supplied as X rays or electron beams generated by machinery or is emitted by selected radionuclides. In diagnostic work, radiation is used to locate a cancer or tumor or to assess the function of some organ, like the thyroid gland. In therapeutic work, radiation is used to kill cancerous cells or to inhibit their growth. Generally, therapeutic doses are of much higher energy than those used for diagnostic purposes. We will look here at the radioactive chemicals used in diagnosis, and discuss beam radiation technologies in Special Topics.

Both Chemical and Radiological Properties Are Important in the Selection of Radionuclides in Medicine   The *chemical* properties of radionuclides are identical to those of the stable isotopes of the same element. When radionuclides are selected for use in medicine, their chemistry, therefore, has to be considered. They must be *chemically* compatible with the living system when used for their ionizing radiation. Moreover, their *chemistry* is what guides them naturally to desired tissues. Iodine-127, for example, the only stable isotope of this element, is used chemically in the body by the thyroid gland to make the hormone thyroxin. This chemical property also will guide iodine-131, a beta emitter, to the thyroid gland, where its radiation can be used to assess thyroid function or to treat cancer of the thyroid gland, as we will see.

Minimizing Harm and Maximizing Benefit Guide the Selection of Radionuclides in Medicine   Exposing anyone to any radiation entails some risks, because prolonged exposure can produce cancer. No such exposure is permitted unless the expected benefit from finding and treating a dangerous disease is thought to be greater than the risk. To minimize the risks, the radiologist uses radionuclides that, as much as possible, have the following properties.

1.   The radionuclide should have a half-life that is short. (Then it will decay *during* the diagnosis when the decay gives some benefit, and as little as possible of the radionuclide will decay later, when the radiation is of no benefit.)

2.   The product of the decay of the radionuclide should have little if any radiation of its own. (Either the product should be a stable isotope or it should have a very long half-life.) It should also be quickly eliminated.

3.   The half-life of the radionuclide must be long enough for it to be prepared and administered to the patient.

4.   If the radionuclide is to be used for diagnosis, it should decay by penetrating radiation entirely, which means gamma radiation. (Nonpenetrating radiation, such as alpha and beta radiation, adds to the risk by causing internal damage without contributing to the detection of the radiation externally. For uses in *therapy*, as in cancer therapy, nonpenetrating radiation is preferred because a radionuclide well placed in cancerous tissue *should* cause damage to such tissue.)

5.   The diseased tissue should concentrate the radionuclide, giving a "hot spot" where the diseased area exists, or it should do the opposite and reject the radionuclide, making the diseased area a "cold spot" insofar as external detectors are concerned.

Let us now look briefly at a few of the more important radionuclides employed in medicine.

Technetium-99m Is the Most Widely Used Radionuclide in Medicine   Technetium-99m is a radionuclide produced by the decay of molybdenum-99. (The synthesis of molybdenum-99 was described at the end of Section 11.4.) See also Special Topic 11.3. You already

■ *Radiology:* the science of radioactive substances and of X rays.
Radiologist: a specialist in radiology who also usually has a medical degree.
Radiobiology: the science of the effects of radiation on living things.

TECHNETIUM-99*m* IN MEDICINE

To make a dilute solution of technetium-99*m* (in the form of TcO_4^-), a radiologist "milks a molybdenum cow" (see Figure 1). This device contains molybdenum-99 (in the form of MoO_4^{2-}) mixed with granules of alumina. (Alumina is one of the crystalline forms of Al_2O_3, and it is used here as a support for the molybdate ion.) The device is charged at a nuclear reactor facility by the neutron bombardment of the molybdenum-98 present (as MoO_4^{2-}) on the alumina granules, as described in Section 11.6. After neutron bombardment, the device is shipped to the hospital, and all the while the molybdenum-99 is decaying to technetium-99*m*.

Each morning, a member of the radiology staff lets a predetermined volume of isotonic salt solution trickle through the bed of granules. Some of the pertechnetate ions dissolve and are leached out. Then the solution so

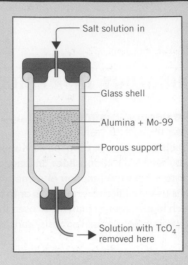

FIGURE 1
Molybdenum "cow."

obtained is used for the day's work. After several days, too little molybdenum-99 remains, so the device is shipped back to the reactor for recharging.

know that gamma radiation often accompanies the emission of other types of radiation, but with molybdenum-99 the gamma radiation comes after a pause. Molybdenum-99 first emits a beta particle:

$$_{42}^{99}Mo \longrightarrow _{43}^{99m}Tc + _{-1}^{0}e$$

The other decay product is a metastable form of technetium-99, hence the *m* in 99*m*. Metastable means poised to move toward greater stability. Technetium-99*m* decays by emitting gamma radiation with an energy of 143 keV:

$$_{43}^{99m}Tc \longrightarrow _{43}^{99}Tc + \gamma$$

Technetium-99*m* almost ideally fits the criteria for a radionuclide intended for diagnostic work. Its half-life is short, 6.02 h. Its decay product, technetium-99, has a very long half-life, 212,000 years, so it has too little activity to be of much concern. (Technetium-99 decays to a stable isotope of ruthenium, $_{44}^{99}Ru$.) The half-life of technetium-99*m*, although short, is still long enough to allow time to prepare it and administer it. It decays entirely by gamma radiation, which means that the maximum amount of the radiation gets to a detector to signal where the radionuclide is in the body. Finally, a variety of chemically combined forms of technetium-99*m* have been developed that permit either hot spots or cold spots to form.

One form of technetium-99*m* is the pertechnetate ion, TcO_4^-. It behaves in the body very much like a halide ion, so it tends to go where chloride ions, for example, go. It is eliminated by the kidneys, so it is used to assess kidney function. Other organs whose functions are also studied by technetium-99*m* are the liver, spleen, lungs, heart, brain, bones, and the thyroid gland.

Technetium-99*m* technology has received competition from the CT scan (Special Topic 11.4), the PET scan (Special Topic 11.5), and MRI imaging (Special Topic 11.6). But they are much more expensive, so the use of technetium-99*m* will continue for some time.

Iodine-131 and Iodine-123   The thyroid gland, the only user of iodine in the body, takes iodide ion and, as we said, makes the hormone thyroxin. When either an underactive or an overactive thyroid is suspected, one technique is to let the patient drink a glass of flavored water that contains some radioactive iodine as I^-. By placing radiation detection equipment

X RAYS AND CT SCANS

X rays are generated by bombarding a metal surface with high-energy electrons. These can penetrate the metal atom far enough to knock out one of its low-level orbital electrons, such as a 1s electron. This creates a "hole" in the electron configuration, and orbital electrons at higher levels begin to drop down. In other words, the creation of this "hole" leads to electrons changing their energy levels. The difference between two of the lower levels corresponds to the energy of an X ray, which is emitted.

The refinement of X-ray techniques and the development of powerful computers made possible the generation of a diagnostic technology called computerized tomography, or CT for short. A. M. Cormack (United States) and G. N. Hounsfield (England) shared the 1979 Nobel prize in medicine for their work in the development of this technology. The instrument includes a large array of care-fully positioned and focused X-ray generators. In the procedure called a CT scan, this array is rotated as a unit around the body or the head of the patient (see Figure 1). Extremely brief pulses of X rays are sent in from all angles across one cross section of the patient (see Figure 2).

The changes in the X rays that are caused by internal organs or by tumors are sent to a computer, which then processes the data and delivers a picture of the cross section. It's like getting a picture of the inside of a cherry pit without cutting open the cherry. The CT scan is widely used for locating tumors and cancers.

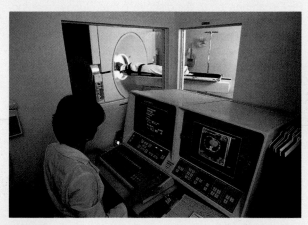

FIGURE 1
CT scan instrumentation

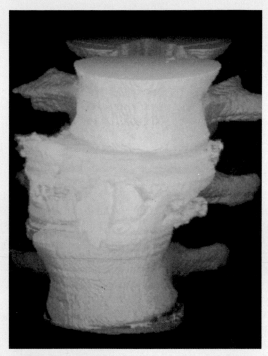

FIGURE 2
A three-dimensional image, based on 63 CT scans, of a section of the vertebrae of a man injured in a motorcycle accident. Both the compression and the twisting are clearly evident.

near the thyroid gland, the radiologist can tell how well this gland takes up iodide ion from circulation. For diagnostic purposes, iodine-123 is popular because it has a short half-life (13.3 h) and it emits only gamma radiation (159 keV). Iodine-131 also has a short half-life (8 days), but it emits both beta particles (600 keV) and gamma radiation (mostly 360 keV). Certain types of thyroid cancer have been treated with this radionuclide. The ability of this gland to concentrate iodide ion is so good that if a small whole-body dose of iodine-131 is given, nearly 1000 times this much dose concentrates in the thyroid.

Technetium-99m as the hydrated pertechnetate ion, TcO_4^-, has about the same radius and charge density as the hydrated iodide ion. Cells of the thyroid gland, therefore, do not distinguish between these two ions at the point where they move inside the cells. When the purpose is to get any kind of detectable radiation inside the thyroid gland in order to detect a tumor or cancer, then Tc-99m, because of its short half life (6.02 h), is preferred over any radionuclide of iodine.

SPECIAL TOPIC 11.5
POSITRON EMISSION TOMOGRAPHY — PET SCAN

A number of synthetic radionuclides emit positrons. These are particles that have the same small mass as an electron but carry one unit of *positive* charge. (They're sometimes called a positive electron.) A positron forms by the conversion of a proton into a neutron, as follows:

$$_1^1p \longrightarrow _0^1n + _1^0e$$

| Proton (in atom's nucleus) | Neutron (stays in nucleus) | Positron (is emitted) |

Positrons, when emitted, last for only a brief interval before they collide with an electron. The two particles annihilate each other, and in so doing their masses convert entirely into energy in the form of two photons of gamma radiation (511 keV). The gamma radiation formed in this way is called annihilation radiation.

$$_{-1}^0e + _1^0e \longrightarrow 2 _0^0\gamma$$

| Electron | Positron | Gamma radiation |

The two photons leave the collision site in almost exactly opposite directions.

To make a medically useful technology out of this property of the positron, a positron-emitting nuclide must be part of a molecule with a chemistry that will carry it into the particular tissue to be studied. Once the molecule gets in, the tissue now has a gamma radiator *on the inside.* Thus instead of X rays being sent through the body, as in a CT scan (Special Topic 11.4), the radiation originates right within the site being monitored. The overall procedure is called positron emission tomography, or PET for short.

Three positron emitters are often used: oxygen-15, nitrogen-13, and carbon-11. Glucose, for example, can be made in which one carbon atom is carbon-11 instead of the usual carbon-12. Glucose can cross the blood–brain barrier and get inside brain cells. If some part of the brain is

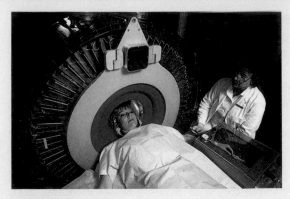

FIGURE 1
The patient is shown undergoing a brain scan performed by means of PET technology.

experiencing abnormal glucose metabolism, this will be reflected in the way in which positron-emitting glucose is handled, and gamma radiation detectors on the outside can pick up the differences (see Figure 1). The use of the PET scan has led to the discovery that glucose metabolism in the brain is altered in schizophrenia and manic depression. PET scanning technology is able to identify extremely small regions in the brain that are in early stages of breakdown and that CT and MRI techniques miss. A brain involved with Alzheimer's disease gives a PET scan different from a normal brain (see Figure 2). The PET scan is particularly useful in detecting abnormal brain function in infants, as in early stages of epilepsy.

PET technology is being used to study a number of neuropsychiatric disorders, including Parkinson's disease. When a drug labeled with carbon-11 is used, its molecules go to the parts of the brain which have nerve endings that release dopamine. A PET scan then discloses the dopamine-releasing potential of the patient. This potential becomes impaired in Parkinson's disease.

By labeling blood platelets with a positron emitter, scientists can follow the development of atherosclerosis in even the tiniest of human blood vessels. Blood flow in the heart can be monitored without having to insert a catheter.

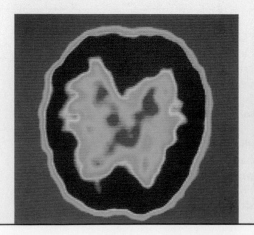
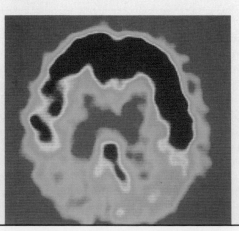

FIGURE 2
Left: Normal brain scan using PET technology. Right: PET scan of the brain of a patient with Alzheimer's disease

MAGNETIC RESONANCE IMAGING—MRI SCAN

The CT scan subjects a patient to large numbers of short bursts of X rays. The PET scan exposes the patient to gamma radiation that is generated on the inside. Thus both technologies carry the usual risks that attend ionizing radiation, and they are used when the potential benefits from correct diagnoses far outweigh such risks. The MRI imaging technology operates without these dangers. At least, none has been discovered thus far. The principal developer of the first hardware for MRI imaging was Raymond Damadian.

MRI stands for magnetic resonance imaging. Atomic nuclei that have odd numbers of protons and neutrons behave as though they were tiny magnets, hence the *magnetic* part of MRI. The nucleus of ordinary hydrogen (which has no neutrons and one proton) constitutes the most abundant nuclear magnet in living systems, because hydrogen atoms are parts of water molecules and all biochemicals. Nuclear magnets spin about an axis much as the earth spins about its axis.

When molecules with spinning nuclear magnets are in a strong magnetic field and are simultaneously bathed with properly tuned radiofrequency radiation (which is of very low energy), the nuclear magnets flip their spins. (This is the *resonance* part of MRI.) As they resonate, they emit electromagnetic energy that is biologically harmless, being of very low energy, and this energy is picked up by detectors. The data are fed into computers, which produce an image much like that of a CT or a PET scan, but without having subjected the patient to ultrahigh energy electromagnetic radiation such as X rays or gamma rays. The MRI images are actually sophisticated plots of the distributions of the spinning nuclear magnets, of hydrogen atoms, for example.

MRI scanning has proved to be especially useful for studying soft tissue, the sort of tissue least well studied by X rays. Different soft tissues have different population densities of water molecules or of fat molecules (which are loaded with hydrogen atoms). And tumors and cancerous tissue have their own water inventories. Calcium ions do not produce any signals to confuse MRI scanning, so bone, which is rich in calcium, is transparent to MRI.

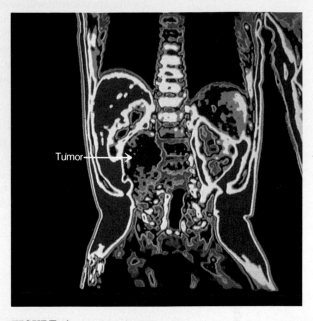

FIGURE 1

An MRI scan of a seven-month-old child revealed a malignant tumor pushing its way into the spinal canal. (The tumor was treated in time.)

MRI technology has developed during the 1980s as a method superior to the CT scan for diagnosing tumors at the rear and base of the skull and equal to the CT scan for finding other brain tumors. MRI is now the preferred technology for assessing problems in joints (particularly the knees) and in the spinal cord, such as ruptured (herniated, "slipped") disks. Patients with heart pacemakers or embedded shrapnel or surgical clips present problems to the use of MRI because of the powerful magnets used.

CT scans are still better than MRI for the early detection of hemorrhages in the brain, so CT is the method of choice for finding them in potential stroke victims. CT scans are also still preferred for detecting tumors in the kidneys, lungs, pancreas, and spleen (see Figure 1).

Cobalt-60 Provides Gamma Radiation above the Megaelectron-Volt Level Cobalt-60 emits beta radiation (315 keV) and gamma radiation (2.819 MeV), and it has a half-life of 5.3 years. Gram for gram, cobalt-60 has a source intensity more than 200 times that of pure radium, which until the 1950s had been widely used in treating cancer. The sample of cobalt-60 is placed in a lead container that is many centimeters thick and has an opening pointed toward the cancerous site. All beta radiation is shielded by a thin piece of aluminum.

■ Cobalt-60 therapy is being replaced by radiation from linear accelerators.

Linear Accelerators Make High-Energy X Rays Betatrons are linear accelerators that are tending to supplant cobalt-60 therapy for several reasons. These devices generate X rays for therapeutic uses with energies in the range of 6 to 12 MeV, which are much more powerful and so more penetrating than the gamma rays from cobalt-60. The output of the machines is stable; the output of cobalt-60 declines over time. The edges of the beams are sharply defined; those from cobalt-60 are less so. This means that X-ray pictures obtained as the therapy is applied are of higher quality. Linear accelerator machines are easily mounted so that they can rotate about a central point, which makes it easier to position the patient and plan the treatment.

Other Medically Useful Radionuclides Are Available for Special Purposes Indium-111 ($t_{1/2}$ = 2.8 days; gamma emitter, 173 and 247 keV) has been found to be a good labeler of blood platelets.

Gallium-67 ($t_{1/2}$ = 78 h; gamma emitter: 1.003 MeV) is used in the diagnosis of Hodgkin's disease, lymphomas, and bronchogenic carcinoma.

Phosphorus-32 ($t_{1/2}$ = 14.3 days; beta emitter: 1.71 MeV) in the form of the phosphate ion has been used to treat a form of leukemia, a cancer of the bone that affects white cells in the blood. Because the phosphate ion is part of the hydroxyapatite mineral in bone, this ion is a bone-seeker.

11.7

ATOMIC ENERGY AND RADIONUCLIDES

Nuclear power plants generate atomic wastes that must be kept from human contact for several centuries.

All operating nuclear power plants in the United States today use fission to generate heat. **Fission** is the disintegration of a large atomic nucleus into small fragments following neutron capture. It releases additional neutrons, radioactive isotopes, and enormous yields of heat. Unless the reactor is continuously cooled by a flowing coolant, usually water, the whole system will melt very quickly. The heat converts the coolant into a hot gas under high pressure, such as steam, which drives electric turbines (see Fig. 11.7). Thus some of the fission energy becomes electrical energy, and what isn't used for this is vented as heat to the atmosphere or into a moving stream of cooler water from a river or lake.

The uranium-235 isotope is the only naturally occurring radionuclide that spontaneously undergoes fission when it captures a slow-moving and relatively low-energy neutron. After

FIGURE 11.7

Pressurized water nuclear power station. Water in the primary coolant loop circulates around the reactor core and carries away the heat of fission. This water is sealed under pressure, so its temperature can rise well above its normal boiling point. The water delivers its heat to the water in the secondary loop, which then turns to high-pressure steam and drives the electrical turbines. Maximum efficiency is reached by having as large a temperature drop as possible between the inlet steam temperature in this loop and the outlet water temperature. (Drawing from WASH 1261, U. S. Atomic Energy Commission, 1973.)

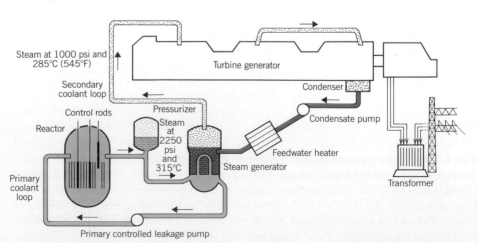

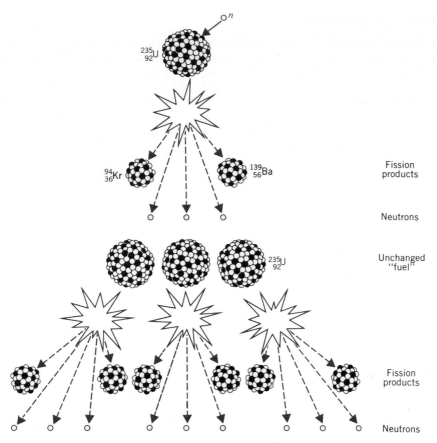

FIGURE 11.8
Fission is initiated when a nucleus of uranium-235 captures a neutron. The new nucleus splits apart, and more neutrons are released. These can initiate subsequent fission events, and unless this is prevented or at least tightly controlled, the whole mass of uranium-235 will detonate — as in the explosion of an atomic bomb. One method of control is to keep the mass of uranium-235 low, below what is called the critical mass. Now, enough neutrons escape before being captured by neighboring nuclei.

capturing such a neutron, the nucleus spontaneously splits apart. It can split in a number of ways that give different products. We'll give the equation for just one mode of splitting.

■ Neutrons moving too rapidly are poorly captured.

$$^{235}_{92}U + ^{1}_{0}n \xrightarrow[\text{capture}]{\text{neutron}} ^{236}_{92}U \xrightarrow{\text{fission}} ^{139}_{56}Ba + ^{94}_{36}Kr + 3^{1}_{0}n + \gamma + \text{heat}$$

More neutrons are released than are used up. Hence, one fission event produces particles that can initiate more than one new fission. In other words, a **nuclear chain reaction** can take place (Fig. 11.8) that would almost instantly envelop all U-235 atoms present if the system were designed to hold the ratio of neutrons produced to neutrons captured at 1 : 1, the *critical ratio.* When this ratio goes higher, becomes *supercritical,* and the concentration of U-235 is very high, an atomic explosion will occur. When, during a supercritical period, the U-235 concentration is low, no *atomic* explosion occurs. However, the heat generated cannot be removed rapidly enough to prevent the whole system from melting.

■ Plutonium-239, made in reactors, also can fission, and it is used as a nuclear fuel and in making atomic bombs.

A nuclear *reactor* is a device to enable the control of nuclear fission at the critical ratio so that heat is generated without running any risk of an atomic explosion and is removed rapidly enough to prevent a meltdown. The nuclear "fuel" is U-235 at a concentration of 3% to 5%. It is distributed among a large number of small-radius tubes (called cladding tubes) so that a significant fraction of neutrons can escape without causing fission. Control rods can be moved into or out of the spaces between these tubes. They are made of materials that can capture neutrons harmlessly. Circulating around and among the tubes is the coolant (water in nearly all reactors), which is heated by the atomic fission taking place.

■ The fuel is a mixture of powdered oxides of U-235 and U-238 baked into ceramic-like pellets.

■ The circulating water is also a good neutron moderator.

In *pressurized water reactors,* the water is kept under pressure so that it can be heated above its normal boiling point. As seen in Figure 11.7, this water is in a closed, primary loop engineered so as to heat the water in a secondary loop. The steam thus generated in the secondary loop drives the turbines, and this operation cools the fluid back to liquid water for recirculation.

Fission Products Are Potential Pollutants The new isotopes produced by fission are radioactive, and their decay leads to radioactive pollutants, which must be contained by the reactor and, later, by safe storage. They include strontium-90 (a bone-seeking element in the calcium family), iodine-131 (a thyroid gland seeker), and cesium-137 (a group IA radionuclide that goes wherever Na^+ or K^+ can go). The U.S. government has set limits to the release of each radioactive isotope into the air and into the cooling water of nuclear power plants. Plants that operate in compliance with these standards expose people living near them to an extra dose of no more than 5% of the dose they normally receive from background radiation.

Two Steam Explosions Ruptured a Large Atomic Reactor at Chernobyl Nuclear reactors are encased in huge chambers called containment vessels with thick walls made of steel-reinforced concrete. If a rupture occurs in the coolant lines that lead into and out of the reactor's core, and there is suddenly a loss of coolant around the core, an emergency backup coolant system is activated, and the reactor is shut down by means of the control rods. Between the time of the first loss of coolant and the successful operation of the backup and shutdown activities, the containment vessel is meant to retain any radioactivity.

A loss-of-coolant emergency occurred in 1979 at a nuclear power plant at Three Mile Island in Pennsylvania. The emergency systems worked, but it took nearly seven years of cleanup just to get the *undamaged* reactor back into operation. There was widespread public alarm over the accident.

The nightmare scenario of nuclear power, a breach in the reactor vessel, occurred on April 26, 1986, at the great atomic energy park at Chernobyl, Ukraine (then the Soviet Union). The Chernobyl reactors were not encased in massive containment structures. Through operator error, the emergency cooling system was not operational when a loss of coolant occurred in one reactor. Within 5 seconds, the reactor's power increased 1400-fold, and steam developed with enough pressure to blow off a cover plate weighing 1000 tons. The tops of all the some 1600 cladding tubes were also torn off. Each erupted like a shotgun, and 8 tons of radioactive, incandescent fuel and wastes were hurled into the night sky. The reactor held an estimated 1×10^{20} Bq (3×10^9 Ci) of radioactivity at the time of the accident, and about 3–4% of this was expelled. A radioactive plume rose 3200 feet into the atmosphere and started to drift on the prevailing winds. The human exposure to radioactive cesium-137 in the northern hemisphere from this accident is estimated to be 60% of the human exposure to this radionuclide from all previous atmospheric nuclear tests.

The explosions that occurred were *not* atom-bomb-like explosions. The concentration of fissionable U-235 in the fuel of atomic reactors is, as we said, far too low for this to happen under any circumstances. In an atomic bomb, essentially pure U-235 is used, and when a sufficient mass is suddenly brought together, fission multiplies much more rapidly than neutrons can escape. The explosions at Chernobyl were those of steam suddenly raised to ultrahigh pressure, which the reactor could not contain. Of grave concern to policy makers worldwide is the existence of a number of reactors like those at Chernobyl throughout the former states of the USSR and in eastern Europe.

Wastes from Nuclear Power Plants Must Be Kept Apart from Human Contact for a Thousand Years One of the most vexing problems of nuclear energy has been the permanent storage of long-lived radioactive wastes. Most are now in temporary storage at nuclear power plants. Because several waste radionuclides have very long half-lives, the wastes must be sequestered from all human contact for at least a thousand years, and scientists are seeking deep geologic formations out of all contact with mining operations or underground water supplies into which these wastes can be placed. The sites must be marked and continuously maintained so that archeologists several centuries hence will not unknowingly venture into them.

Wastes Can Enter Food Chains Atomic wastes that do escape, like $^{131}_{53}I$ from Chernobyl, enter food chains wherever they deposit. Milk from cows that have grazed on pastures contaminated by $^{131}_{53}I$ fallout carries this isotope into the human diet. We indicated earlier how

■ If the reactor core starts to melt, it cannot be shut down and will melt its way through the containment vessel.

■ At Chernobyl the graphite used as a moderator ignited and burned like any charcoal fire. Some extremely brave fire fighters managed to extinguish this fire. All but one died.

■ The entire reactor becomes a radioactive waste once the power plant's working lifetime is over.

effectively the human thyroid gland concentrates iodine, so those parts of Europe where the Chernobyl plume drifted were suddenly faced with hard decisions. One strategy used was to distribute sodium iodide to the populations with instructions about taking it in drinking water. By thus raising the level of nonradioactive iodide ion, the fraction of radioactive iodide ion taken up by the thyroid gland would be reduced. Despite these measures, scientists of Belarus (the former Soviet state just north of Ukraine) reported in 1992 that a great increase in cancer of the thyroid has occurred among Belarus children since 1990 (114 new cases from 1990 to 1992 as compared to 15 cases from 1987 to 1989). Another strategy following the Chernobyl accident was to forbid the consumption of meat products from animals that tested at unnaturally high radiation levels. (Two years after Chernobyl, for example, this affected roughly 300,000 sheep in northern Great Britain.) Estimates of human deaths traceable to Chernobyl radiation releases are hard to make and so they vary widely among various reports.

Alternatives to nuclear power pose several problems. Power plants that use petroleum or coal are huge emitters of carbon dioxide, a greenhouse gas. When coal or petroleum also contain sulfur impurities, the power plants emit sulfur dioxide, a contributor to acid rain. Moreover, what few people know is that deaths from the mining of coal (so far) greatly exceed deaths related to nuclear power, including the mining of nuclear fuels. Many people call for drastic reductions in the use of all currently exploited forms of energy — by the conservation of energy, by the use of wind energy and solar energy, and by other technologies.

■ Workers at Chernobyl received at least 400 rem of radiation dose, and 31 died. In a town nearby, 135,000 people were evacuated; they were receiving 1 rem/h the day after the explosion.

SUMMARY

Atomic radiation Radionuclides in nature emit alpha radiation (helium nuclei), beta radiation (electrons), and gamma radiation (high-energy X-ray-like radiation). This radioactive decay causes transmutation. The penetrating abilities of the different kinds of radiation are a function of the sizes of the particles, their charges, and the energies with which they are emitted. Gamma radiation, which has no associated mass or charge, is the most penetrating.

Each decay can be described by a nuclear equation in which mass numbers and atomic numbers on either side of the arrow must balance. To describe how stable a radionuclide is we use its half-life, and the shorter this is, the more radioactive is the radionuclide.

The decay of one radionuclide doesn't always produce a stable nuclide. Uranium-238 is at the head of a radioactive disintegration series that involves several intermediate radionuclides until a stable isotope of lead forms.

Ionizing radiation — dangers and precautions When radiation travels in matter, it creates unstable ions and radicals that have chemical properties dangerous to health. Intermittent exposure can lead to cancer, tumors, mutations, or birth defects. Intense exposures cause radiation sickness and death. Intense exposures focused on cancer tissue are used in cancer therapy. The use of distance, fast film, and dense shielding material are the best strategies to guard against the hazards of ionizing radiation. According to the inverse-square law, the intensity of radiation falls off with the square of the distance from the source. Complete protection, however, is not possible because of the natural background radiation that now includes traces of radioactive pollutants.

Units of radiation measurement To describe activity we use the SI unit of the becquerel, Bq, or the curie, Ci. To describe the intensity of exposure to X rays or gamma radiation, we use the SI unit X or the roentgen. The gray (Gy) or the older rad is used to describe the absorbed dose, how much energy has been absorbed by a unit mass of tissue (or other matter). To put the damage that different kinds of radiation can cause when they have the same values of rads (or grays) on a comparable and additive basis, we use the sievert or the rem. Finally, to describe the energy possessed by a radiation, we use the electron-volt. Diagnostic X rays are on the order of 100 keV. The radiation used in cancer treatment is in the low MeV range. To measure radiation there are devices such as film badges, scintillation counters, and ionization counters (Geiger-Müller tubes).

Synthetic radionuclides A number of synthetic radionuclides have been made by bombarding various isotopes with alpha radiation, neutrons, or accelerated protons. The target nucleus first accepts the mass, charge, and energy of the bombarding particle, and then it ejects something else to give the new nuclide.

Radiation technology in the food industry X rays, gamma rays, and accelerated electrons can be used to inhibit or to kill insects, molds, and bacteria on seeds, potatoes, fruit, and meats. High doses fully sterilize the products. Low doses inhibit bacterial or mold growth. Radiolytic products, those formed by the chemical reactions induced by the radiation, are generally the same as the substances found naturally in food either before cooking and digestion or after. The resistance of the botulinus bacillus to radiation is one problem with the technology.

Radionuclides in medicine For diagnostic uses, the radionuclide should have a short half-life (but not so short that it decays before any benefit can be obtained). It should decay by gamma radiation only, and it should be chemically compatible with the organ or tissue so that either a hot spot or a cold spot appears. Its decay products should be as stable as possible, and capable of being eliminated from the body.

Technetium-99m is almost ideal for diagnostic work, particularly for assessing the ability of an organ or tissue to function. Radionuclides of iodine ($^{123}_{53}$I and $^{131}_{53}$I) are used in diagnosing or treating thyroid conditions. Gallium-67, indium-111, and phosphorus-32 are a few of the many other radionuclides used in diagnosis. Cobalt-60 has powerful gamma radiation, and it is used in cancer treatment. Linear accelerators also provide high-energy (6 to 12 MeV) radiation for cancer therapy.

Atomic energy and radioactive pollutants The reactors of most nuclear power plants use U-235 as a fuel. When its atoms capture neutrons, they fission into smaller, usually radioactive atoms as neutrons are released that can cause additional fissions. The concentration of U-235 atoms is kept too small (3% to 5%) to make an atomic-bomb type of explosion possible at a nuclear reactor. Circulating water keeps the reactor cool enough not to melt. The heat generated by fission converts water into high-pressure steam that drives electrical turbines. A containment vessel surrounding the reactor is intended to prevent the escape of radioactive materials should a loss-of-coolant emergency arise. (The containment system worked at Three Mile Island.)

Radioactive by-products of fission, such as $^{131}_{53}$I, $^{90}_{38}$Sr, and $^{137}_{55}$Cs, cannot be allowed to enter the food supply and must be contained. Some fission products have such long half-lives that radioactive wastes must be kept from human contact for a thousand years.

REVIEW EXERCISES

The answers to Review Exercises whose numbers are in color are found in Appendix C. The answers to the other Review Exercises are found in the Study Guide that accompanies this book. The more challenging questions are marked with asterisks.

Radioactivity and the Kinds of Radiation

11.1 Distinguish between *radioactive decay* and *transmutation*.

11.2 What are the names and symbols used in nuclear equations for the three types of naturally occurring atomic radiation?

11.3 The emission of which naturally occurring atomic radiation would *not* be accompanied by transmutation if it were the sole radiation from a radionuclide?

11.4 In balancing *chemical* equations we seek both a material balance and a charge balance.
(a) How do we check for material balance in a nuclear equation?
(b) Why do we ignore the question of charge balance by not showing electrical charges on reactants or products?
(c) In what other sense, however, do we consider charge balance?

11.5 The energy of an alpha particle is often higher than that of beta or gamma rays. Why is it, then, the least penetrating of the kinds of radiation?

11.6 The loss of an alpha particle changes the radionuclide's mass number by how many units? Its atomic number by how many units?

11.7 Why does the loss of a beta particle not change the radionuclide's mass number but *increases* its atomic number?

11.8 How many neutrons are in the nucleus of $^{232}_{92}$U?

11.9 What is the one most appropriate term to apply to a *radionuclide*: element, isotope, compound, mixture, atom, ion, or molecule? Explain.

11.10 If electrons do not exist in the nucleus, how can one originate in a nucleus in beta decay?

Nuclear Equations

*11.11 Write the symbols of the missing particles in the following nuclear equations.
(a) $^{211}_{82}$Pb $\rightarrow$ $^{0}_{-1}e$ + _____   (b) $^{220}_{86}$Rn $\rightarrow$ $^{4}_{2}$He + _____
(c) $^{140}_{56}$Ba $\rightarrow$ $^{0}_{-1}e$ + _____

*11.12 Complete the following nuclear equations by writing the symbols of the missing particles.
(a) $^{149}_{62}$Sm $\rightarrow$ $^{4}_{2}$He + _____
(b) $^{245}_{96}$Cm $\rightarrow$ $^{4}_{2}$He + _____
(c) $^{22}_{9}$F $\rightarrow$ $^{0}_{-1}e$ + _____

*11.13 Write a balanced nuclear equation for each of the following changes.
(a) Alpha emission from neodymium-144
(b) Beta emission from potassium-40
(c) Beta emission from samarium-149
(d) Alpha and gamma emission from californium-251

*11.14 Give the nuclear equation for each of the following radioactive decays.
(a) Beta emission from rhenium-187
(b) Alpha and gamma emission from plutonium-242
(c) Beta emission from iodine-131
(d) Alpha emission from americium-243

Half-Lives

11.15 Lead-214 is in the uranium-238 disintegration series. Its half-life is 26.8 min. Explain in your own words what being in this series means and what *half-life* means.

11.16 Which would be more dangerous to be near, a radionuclide that has a short half-life and decays by alpha emission only, or a radionuclide that has the same half-life but decays by beta and gamma emission? Explain.

*11.17 A 12.00-ng sample of technetium-99m will still have how many nanograms of this radionuclide left after three half-life periods?

*11.18 If a patient is given 12.00 ng of iodine-123 (half-life 13.3 h), how many nanograms of this radionuclide remain after 12 half-life periods (about a week)?

Dangers of Ionizing Radiation

11.19 We have ions in every fluid of the body. Why, then, is ionizing radiation dangerous?

11.20 What is a chemical *radical,* and why is it chemically reactive?

11.21 Ionizing radiation is a teratogenic agent. What does this mean?

11.22 Cesium-137 is a carcinogen. What does this mean?

11.23 What two properties of ionizing radiation are exploited in strategies for providing radiation protection?

11.24 Ionizing radiation is said to have no exposure threshold. What does this mean?

11.25 How is it that the same agent, radiation from a radionuclide, can be used both to cause cancer and to cure it?

11.26 The inverse-square law tells us that if we double the distance from a radioactive source, we will reduce the radiation intensity received by a factor of what number?

11.27 What general property of radiation is behind the inverse square law?

11.28 List as many factors as you can that contribute to the background radiation.

11.29 Why does a trip in a high-altitude jet plane increase a person's exposure to background radiation?

11.30 A radiologist discovered that at a distance of 1.80 m from a radioactive source, the intensity of radiation was 140 millirad. How far should the radiologist move away to reduce the exposure to 2 millirad?

11.31 Using a Geiger-Müller counter, a radiologist found that in a 20-min period the dose from a radioactive source would measure 80 millirad at a distance of 10.0 m. How much dose would be received in the same time by moving to a distance of 2.00 m?

Units of Radiation Measurement

11.32 What SI unit is used to describe the activity of a radioactive sample?

11.33 A hospital purchased a sample of a radionuclide rated at 1.5 mCi. What does this rating mean?

11.34 What is the SI unit and the older unit that describe the intensity of an exposure to X rays?

11.35 What is the name and symbol of the SI unit used in describing how much energy a given mass of tissue receives from exposure to radiation? What is the name of the older, common unit?

11.36 How does the SI define the dose equivalent? What is meant by the *quality factor?* Which has the higher quality factor, alpha radiation or beta radiation? Explain.

11.37 Approximately how many rads would kill half of a large population within four weeks, assuming that each individual received this much? From a health protection standpoint, how do the roentgen and the rad compare in their potential danger?

11.38 We cannot add a 1-rad dose of gamma radiation to some organ to a 1-rad dose of neutron radiation to the same organ and say that the total biologically effective dose is 2 rads. Why not?

11.39 How is the problem implied by the previous review exercise resolved?

11.40 In units of mrem, what is the average natural background radiation received by the U.S. population?

11.41 What is the name of the energy unit used to describe the energy associated with an X ray or a gamma ray?

11.42 In the unit traditionally used (Review Exercise 11.41), how much energy is associated with diagnostic X rays?

11.43 Why should diagnostic radiation ideally be of much lower energy than radiation used in therapy, in cancer treatment, for example?

11.44 In general terms, how does a film badge dosimeter work?

11.45 In your own words, how does a Geiger-Müller counter work? Why doesn't it detect alpha radiation?

Synthetic Radionuclides

•11.46 When manganese-55 is bombarded by protons, the neutron is one product. What else is produced? Write a nuclear equation.

•11.47 To make indium-111 for diagnostic work, silver-109 is bombarded with alpha particles. What forms if the nucleus of silver-109 captures one alpha particle? Write the nuclear equation for this capture.

•11.48 The compound nucleus that forms when silver-109 captures an alpha particle (previous review exercise) decays directly to indium-111, plus *two* other identical particles. What are they? Write the nuclear equation for this decay.

•11.49 To make gallium-67 for diagnostic work, zinc-66 is bombarded with accelerated protons. When a nucleus of zinc-66 captures a proton, the nucleus of what isotope forms? Write the nuclear equation.

•11.50 When fluorine-19 is bombarded by alpha particles, both a neutron and a nucleus of sodium-22 form. Write the nuclear equation, including the compound nucleus that is the intermediate.

•11.51 When boron-10 is bombarded by alpha particles, nitrogen-13 forms and a neutron is released. Write the equation for this reaction.

•11.52 When nitrogen-14 is bombarded with deuterons, ^2_1H, oxygen-15 and a neutron form. Write the equation for this reaction.

•11.53 What bombarding particle could change aluminum-27 into phosphorus-32 and a proton? Write the equation.

•11.54 What bombarding particle can change sulfur-32 into phosphorus-32 and a proton? Write the equation.

Radiation Technology and the Food Industry

11.55 The unit commonly used to describe how much radiation has been given to a food product is the *kilorad,* krad. Using conversion factors supplied in Section 11.3, calculate how much energy is in 1.00 krad in units of J/kg.

11.56 What is meant by the term *radiolytic product?*

11.57 What is meant by the term *unique* radiolytic product?

11.58 Are any radiolytic products known cancer-causing substances? Are they also found in nonirradiated foods? Give an example.

11.59 What nonradiation processes produce the same compounds as food irradiation?

11.60 What potential hazard in food is the most difficult to remove by radiation?

Medical Applications of Radiation

11.61 Why is it desirable to use a radionuclide of short half-life in diagnostic work, when we know that even small samples of such isotopes can be very active?

11.62 We know that gamma radiation is the most penetrating of all natural radiation. Why, then, is a diagnostic radionuclide that emits only gamma radiation preferable to one that gives, say, only alpha radiation?

11.63 Why is iodine-123 better for diagnostic work than iodine-131?

11.64 Why did cobalt-60 replace radium for cancer treatment?

11.65 In what chemical form should phosphorus-32 be used to facilitate its seeking bone tissue? Explain.

Atomic Energy and Radionuclides

11.66 What is *fission*?

11.67 In general terms, how does fission differ from radioactive decay?

11.68 What fundamental aspect of fission makes it possible for it to proceed as a chain reaction?

11.69 Which naturally occurring radionuclide is able to undergo fission?

11.70 What is the concentration of the fissionable isotope in atomic reactors? What concentration is it in an atomic bomb?

11.71 The heat generated by fission in a power plant reactor is carried away in what way? And for what purpose?

11.72 What makes a loss of coolant an emergency?

11.73 What events are supposed to happen in a loss-of-coolant emergency?

11.74 What is a "containment vessel" at a nuclear power plant and what is its purpose?

11.75 Name three isotopes made in nuclear power plants that are particularly hazardous to health, and explain in what specific ways they endanger various parts of the body.

11.76 How can extra amounts of nonradioactive iodide ion in the diet provide some protection against radioactive iodide ion?

11.77 What fact about certain atomic wastes necessitates very long waste storage times?

Cosmic Rays (Special Topic 11.1)

11.78 Where do cosmic rays originate?

11.79 What is the chief particle found in primary cosmic rays?

11.80 What happens to cosmic rays as they enter the earth's atmosphere?

Radon in the Environment (Special Topic 11.2)

11.81 How is radon-222 produced in the environment?

11.82 What kinds of radiation does radon-222 emit?

11.83 Besides its own radiation, what other factors make radon-222 in the lungs particularly hazardous?

11.84 What upper limit on the level of radon-222 in the home is recommended?

Technetium-99*m* in Medicine (Special Topic 11.3)

11.85 What is done at a reactor facility to charge a molybdenum "cow"?

11.86 In what chemical form is Tc-99*m* prepared?

X Rays and CT Scans (Special Topic 11.4)

11.87 In general terms, how are X rays prepared?

11.88 How does the CT scanner differ from an ordinary X ray machine?

Positron Emission Tomography — The PET Scan (Special Topic 11.5)

11.89 Compare the positron and electron in terms of mass and charge.

11.90 Describe how a positron forms in a positron-emitting radionuclide.

11.91 What property makes the lifetime of a positron extremely short?

11.92 When a radiologist uses the PET scan, what radiation is converted into an X-ray-like picture?

11.93 In general terms, how can a positron-emitting radionuclide be gotten inside a tissue, like the brain?

Magnetic Resonance Imaging — The MRI Scan (Special Topic 11.6)

11.94 Why is the MRI less harmful a technique than the CT or PET scan?

11.95 How does MRI complement the use of X rays?

11.96 Why is bone transparent to the MRI?

Additional Exercises

11.97 Why is radiation sickness called an *acute* effect of radiation exposure? What is a particularly common *latent* effect of radiation exposure?

11.98 Explain how the control rods of a nuclear reactor control the rate at which energy is generated.

11.99 The hydroxyl radical is an electrically neutral species, yet it is dangerous in tissue. Explain.

*11.100 A few isotopes are known that have nuclei able to capture an electron from the atom's own electron energy levels. What changes, if any, in atomic number and mass number occur when a nucleus undergoes electron capture? Complete and balance the following nuclear equation.

$$_{23}^{50}V + _{-1}^{0}e \xrightarrow{\text{electron capture}}$$

ORGANIC CHEMISTRY. SATURATED HYDROCARBONS

12

Take a moment and marvel. All these people, and a few billion others, are individually unique "packages" of electrons, protons, and atomic nuclei (and actually the nuclei of only a handful of elements). In this chapter we begin the study of another major foundation of the molecular basis of life, the structures and properties of organic compounds.

12.1

ORGANIC AND INORGANIC COMPOUNDS

The major differences between organic and inorganic compounds stem from variations in composition, bond types, and molecular polarities.

Organic compounds are compounds made of carbon atoms covalently bonded to each other and to atoms of other nonmetals, like hydrogen, oxygen, nitrogen, sulfur, or the halogens. All other compounds are called **inorganic compounds** but even they include a few that contain carbon, like the carbonates, bicarbonates, cyanides, and the oxides of carbon.

In the popular press, "organic" (as in "organic foods") has come to mean "produced without the use of pesticides or synthetic fertilizers or hormones." We will use the traditional meaning. **Organic chemistry** is the study of the structures, properties, and syntheses of organic compounds. In this and the next few chapters we will study only those parts of organic chemistry that develop the principles or reactions needed to our study of the molecular basis of life.

Wöhler's Experiment Opened the Doors to the Laboratory Synthesis of Organic Compounds The word *organic* arose from an association with living organisms, because in the early days of organic chemistry all organic compounds were isolated from living systems or their remains. Until 1828, all efforts to synthesize organic from inorganic compounds had failed, and out of such repeated failures the **vital force theory** emerged. It stated that it is actually impossible to make organic compounds in glass vessels, that a special *vital force* said to be found only in living systems was essential.

In 1828, while trying to make a sample of crystalline ammonium cyanate, NH_4NCO, Friedrich Wöhler (1800–1882) boiled the water from an aqueous solution containing the ammonium ion and the cyanate ion, NCO^-. The white solid he obtained, however, was an unexpected compound, urea. Ammonium cyanate, then as now, is regarded as an inorganic compound, but urea is clearly a product of metabolism. Wöhler had succeeded in making the first organic compound in a glass vessel. The heat used for boiling evidently caused the following reaction.

■ *Vita-* is from a Latin root meaning "life."

■ Urea is the chief nitrogen waste in the urine of animals. It is also manufactured from ammonia for use as a commercial fertilizer. Whether made by animals or machinery, urea is urea and is "organic." Plants will accept no fertilizer than what they are used to.

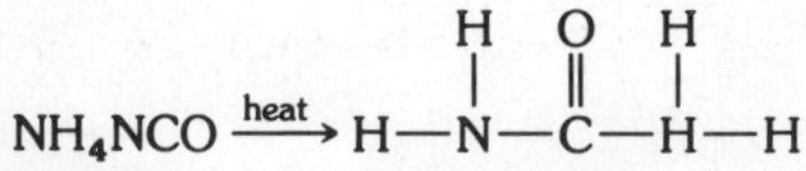

$$NH_4NCO \xrightarrow{\text{heat}} \underset{\text{Urea}}{H-\overset{\displaystyle H}{\underset{|}{N}}-\overset{\displaystyle O}{\underset{\|}{C}}-\overset{\displaystyle H}{\underset{|}{H}}-H}$$

Ammonium
cyanate

Following Wöhler's discovery, other organic compounds were made from inorganic substances, and the vital force theory was dead. Today, well over six million organic compounds are known, and all have been or could in principle be made from inorganic substances. The significance to human well-being of the development of organic chemistry as a science is incalculable. Although large numbers of useful organic substances occur in nature and are still obtained from nature, one cannot imagine today's world of synthetic fabrics, dyes, and plastics as well as most pharmaceuticals without the opening to synthetic organic compounds that Wöhler discovered. Among all scientific specialties, there are more organic chemists than any other single kind of chemist. The education of those entering biochemistry, medicinal chemistry, pharmaceutical chemistry, polymer chemistry, molecular biology, and the primary health care fields of nursing and doctoring includes at its core the study of organic chemistry. One specialist at The Johns Hopkins School of Medicine flatly states that we cannot say we *know* what a disease is until we know its *chemistry,* and organic chemical principles are at the heart of this knowledge. The study of these principles begins with the kinds of *bonds* that hold organic molecules together.

Covalent Bonds, Not Ionic Bonds, Dominate Organic Molecules   The overwhelming prevalence of nonmetal atoms in organic compounds means that their molecular structures are dominated by *covalent* bonds. In contrast, most inorganic compounds are *ionic.* As we will see, carbon–carbon and carbon–hydrogen bonds are the most prevalent in organic molecules. These bonds are essentially nonpolar, so organic compounds tend to be relatively nonpolar except when atoms of such electronegative elements as oxygen and nitrogen are present. These structural facts are behind several major properties of organic compounds, like melting and boiling points and solubility in water.

■ The covalent bond was studied in Section 5.3.

■ The concept of bond polarity was discussed in Section 5.8.

Most organic compounds have melting points and boiling points well below 400 °C, whereas most ionic compounds melt or boil far above this temperature. The reason is that the relatively nonpolar molecules in organic substances are unable to attract each other very strongly, in sharp contrast to the oppositely charged ions in ionic compounds. As we learned in Section 6.6, the *permanent* polarity of an organic molecule is only one factor that affects a boiling point or melting point. The *size* of the molecule is also a factor because the larger the size, the stronger are the London forces between molecules.

■ Factors that influence boiling points and melting points were described in Sections 6.6 and 6.9.

Weakly polar molecules are poorly hydrated, in contrast to ions, so most organic compounds are relatively insoluble in water, whereas many ionic compounds are soluble. This fact is particularly relevant at the molecular level of life where the central fluid is water. We must, therefore, be alert during our study of organic chemistry to any molecular features that increase solubility in water.

■ Organic *ions* tend to be very soluble in water.

12.2

SOME STRUCTURAL FEATURES OF ORGANIC COMPOUNDS

Organic molecules can have straight or branched chains; they can be open-chained or cyclic, saturated or unsaturated; and ring systems can be carbocyclic or heterocyclic.

The uniqueness of carbon among the elements is that its atoms can bond to each other successively many times and still form equally strong bonds to atoms of other nonmetals. A typical molecule in the familiar plastic polyethylene has hundreds of carbon atoms covalently joined in succession, and each carbon binds enough hydrogen atoms to fill out its full complement of four bonds.

Polyethylene (small segment of one molecule)

■ Only a short segment of a typical molecule of polyethylene is shown here.

The sequence of the heavier atoms, here the carbon atoms, is called the *skeleton* of the molecule, and it holds the hydrogen atoms. Many variations of heavier-atom skeletons occur, and we will look at them next.

Straight chain

Branched chain

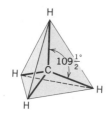

Carbon ring

Carbon Skeletons Can Be in Straight Chains or Branched Chains The carbon skeleton in the polyethylene molecule is described as a **straight chain.** *Straight* has a very limited and technical meaning here: the absence of carbon branches. This means that one carbon follows another, like the pearls in a single-strand necklace, with no additional carbons joined to the skeleton at intermediate points. Pentane illustrates a straight chain of five carbons. The 2-methylpentane molecule has a **branched chain,** a chain with at least one carbon atom joined to the skeleton between the ends of the main chain, like a charm hung on a bracelet.

Pentane
(straight chain)

2-Methylpentane
(branched chain)

Pentane skeleton

2-Methylpentane skeleton

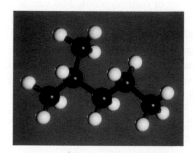

■

- The concept of an *electron cloud* was introduced in Section 4.4 (see also Section 5.7).

Some Features of Molecular Geometry Are Often "Understood" when Writing Structural Formulas When you compare the ball-and-stick models of pentane and 2-methylpentane with their structural formulas, be sure to notice that the written (or printed) structures disregard the correct bond angles at carbon. A carbon that has four single bonds has a tetrahedral geometry with bond angles of 109.5°. The ball-and-stick models faithfully show the correct angles at each carbon, but the printed symbols do not. The point here is that it is perfectly all right to let bond angles be "understood" unless there is some important reason to the contrary.

It is definitely not "all right" to forget about the *geometry* of molecules, however. When we enter the study of biochemistry, particularly the way that enzymes work, we will quickly learn that the geometry of a molecule is just as important as any other feature of its structure. Molecules generally take up whatever shape is permitted by their *bonds* and the *electron clouds* surrounding the individual parts of the molecules. Bonds hold atoms together, and whenever a carbon atom has four *single* bonds, the geometry at that point of the molecule will be tetrahedral. The electron clouds tend to push the parts of molecules away from each other and so influence the *overall* shape of the molecule. This "pushing" cannot succeed in splitting stable molecules apart, but it has a huge influence on how a molecule becomes twisted in shape, at least in molecules that have the flexibility permitted by "free rotation," studied next.

Free Rotation at Single Bonds Is Also "Understood" in Structural Formulas The molecules of pentane and 2-methylpentane are quite flexible, like a necklace. Figure 12.1 shows models of just a few of the many ways the carbon skeleton of the pentane molecule can be flexed. These twistings actually occur as pentane molecules collide with each other in the liquid or gaseous states, and they illustrate an important property of at least large segments of organic molecules, a property called **free rotation** about single bonds. Two clusters of atoms held by a single bond can rotate with respect to each other about the bond.

Free rotation is possible because the single bond is a *sigma bond* as described in Figure 12.2. The *strength* of a sigma bond lies in the degree of overlap of the hybrid orbitals that make up the molecular orbital. As long as the degree of such overlap is not affected too much while one group rotates about the bond, the rotation not only is allowed but it happens readily.

The differently twisted forms of pentane in Figure 12.1 are called *conformations,* and samples of liquid or gaseous pentane consisting entirely of molecules in only one conformation cannot be isolated. The physical and chemical properties of pentane, therefore, are the net results of the effects that the whole collection of conformations has on whatever physical agent or chemical reactant has been used to observe the property. Generally, however, one conformation is present in a relatively high concentration, the conformation that corresponds to the most stable arrangement of the electron clouds. The top conformation in Figure 12.1, for example, gets the electron clouds of the various parts of the pentane molecule as far apart as they can be within the molecule. With the forces of repulsion thus at a minimum in this conformation, the molecule has its lowest potential energy and so has its greatest stability. It's an example of *nature's preference for the lowest energy, most stable arrangements.* This preference is the basic principle behind all aspects of molecular shape within a given environment.

■ Because of free rotation, we have to be able to interpret zigzags. For example,

$$CH_3$$
$$|$$
$$CH_2CH_2CH_2$$
$$|$$
$$CH_3$$

is the same molecule as

$$CH_3CH_2CH_2CH_2CH_3$$

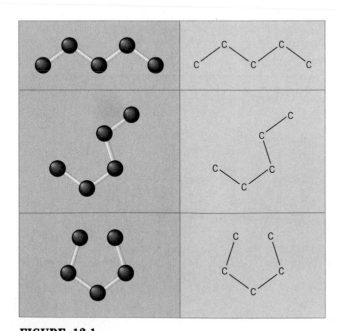

FIGURE 12.1
Free rotation at single bonds. Three of the innumerable conformations of the skeleton of the pentane molecule are shown here. (The hydrogens have been omitted.) Free rotation about single bonds easily converts one conformation into another.

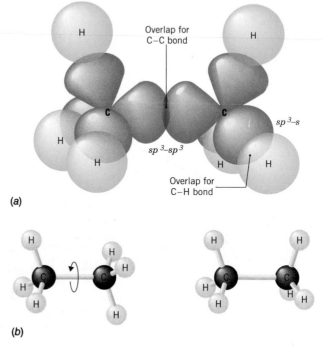

FIGURE 12.2
The sigma bonds in ethane. *(a)* Here are all the orbital overlappings that create the C—C and C—H bonds. *(b)* The sp^3 to sp^3 overlap that forms the C—C single bond, a sigma bond, is affected very little by a rotation of one CH_3 group relative to the other. Because such a rotation costs essentially no energy, it occurs easily.

Condensed Structural Formulas Reduce Clutter with Little Sacrifice in Information Just as we can leave some aspects of molecular geometry to the informed imagination, we can leave most of the bonds in a structural formula to it as well. Remembering that any carbon in a structural formula must have four bonds (or else carry some charge), we can group beside its symbol, C, all the hydrogen atoms held by it. If it holds three hydrogens, we can write CH_3 (or H_3C, but you don't see this as often). Just remember that these three hydrogen atoms are individually joined to the carbon. The structure of ethane illustrates this.

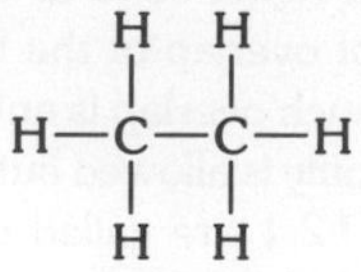

$$H-\underset{\underset{H}{|}}{\overset{\overset{H}{|}}{C}}-\underset{\underset{H}{|}}{\overset{\overset{H}{|}}{C}}-H \qquad\qquad CH_3-CH_3 \quad \text{or} \quad H_3C-CH_3$$

Ethane	Ethane
(expanded	(condensed
structure)	structures)

A carbon holding two hydrogen atoms can be represented as CH_2 or (seen less often) as H_2C. When a carbon holds just one hydrogen, we can write it as CH or HC.

The result of these simplifications is called a **condensed structure,** or simply the **structure.** These kinds of structures will be used almost exclusively in our continuing study.

EXAMPLE 12.1
Condensing a Full Structural Formula

Condense the structural formula for 2-methylpentane.

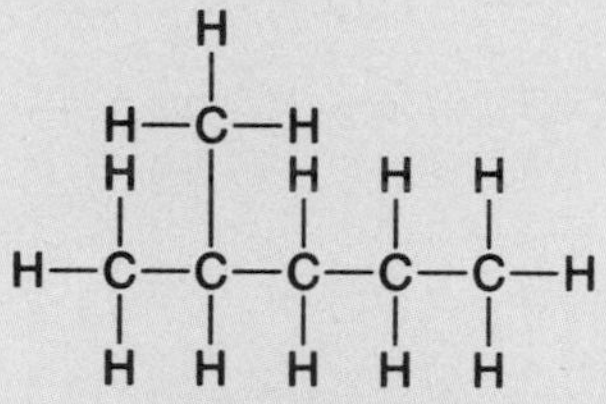

$$H-\underset{\underset{H}{|}}{\overset{\overset{H}{|}}{C}}-\underset{\underset{H}{|}}{\overset{\overset{\overset{\overset{H}{|}}{H-C-H}}{|}}{C}}-\underset{\underset{H}{|}}{\overset{\overset{H}{|}}{C}}-\underset{\underset{H}{|}}{\overset{\overset{H}{|}}{C}}-\underset{\underset{H}{|}}{\overset{\overset{H}{|}}{C}}-H$$

2-Methylpentane

ANALYSIS Each unit of 3 H attached to the same carbon becomes CH_3. Where 2 H are joined to the same carbon, write CH_2. A carbon holding only one H becomes CH.

SOLUTION

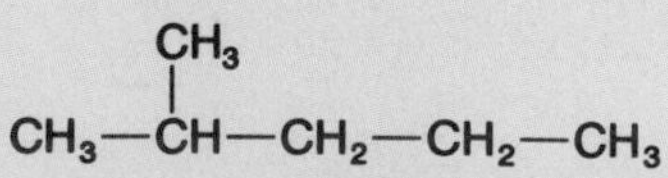

$$CH_3-\underset{\underset{CH_3}{|}}{CH}-CH_2-CH_2-CH_3$$

CHECK We formulate here a general check rule for all molecular structures. *Scan all bond connections to verify that the rules of covalence have been obeyed.* In the answer, each C has four bonds and each H one bond. When you find a violation, the answer cannot possibly be correct, so fix it.

PRACTICE EXERCISE 1

Condense the following expanded structural formulas.

Even Most Single Bonds Can Be "Understood" When a *single* bond appears on a *horizontal* line, we need not write a straight line to represent it; we can leave such single bonds to the imagination. We do not do this, however, for bonds that are not on a horizontal line. Thus we can write the structure of 2-methylpentane, Example 12.1, as follows. Notice that the vertically oriented bond is shown by a line but that all other single bonds are understood.

$$\begin{array}{c} CH_3 \\ | \\ CH_3CHCH_2CH_2CH_3 \end{array}$$

2-Methylpentane

PRACTICE EXERCISE 2

Rewrite the condensed structures that you drew for the answers to Practice Exercise 1 and let the appropriate carbon–carbon single bonds be left to the imagination.

PRACTICE EXERCISE 3

Just to be certain that you are comfortable with condensed structures, expand each of the following to make them full, expanded structures with no bonds left to the imagination.

(a) CH_3CH_3 (b) $\begin{array}{c} CH_3 \\ | \\ CH_3CHCHCH_3 \\ | \\ CH_3 \end{array}$ (c) $\begin{array}{c} CH_3 \\ | \\ CH_3CH_2CCH_2CH_2CH_3 \\ | \\ CH_3 \end{array}$

As indicated in the *check* part of Example 12.1, an important skill in using condensed structures is the ability to recognize errors. The most common is a violation of the rule that *every carbon atom in a structure that carries no electrical charge must have exactly four bonds, no more and no fewer.* Do the next Practice Exercise to test your skill in recognizing an error in structure.

PRACTICE EXERCISE 4

Which of the following structures cannot represent real compounds?

(a)
$$CH_3$$
$$|$$
$$CH_3CCH_3$$
$$|$$
$$CH_3$$

(b) $CH_3CH_2CHCH_3$
with CH_3 above the CH

(c)
$$CH_3 \quad CH_3$$
$$| \qquad |$$
$$CH_3CHCH_2CHCH_2CH_3$$
$$|$$
$$CH_3$$

When atoms other than carbon and hydrogen are present in a molecule, there is no major new problem in writing condensed structures. Remember that every oxygen or sulfur atom carrying no electrical charge must have two bonds, every nitrogen must have three, and every halogen atom must have just one.

Double and Triple Bonds Are Seldom Condensed

Another rule about condensed structures is that carbon–carbon double and triple bonds are never left to the imagination. Carbon–oxygen double bonds are sometimes condensed, as some of the following examples illustrate. Study them as illustrations of how to condense structures.

$$\begin{array}{cc} H & H \\ | & | \\ H-C-C-OH \\ | & | \\ H & H \end{array}$$ condenses to CH_3-CH_2-OH or to CH_3CH_2OH

Ethyl alcohol (in alcoholic drinks)

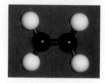

$$\begin{array}{cc} H & H \\ | & | \\ H-C=C-H \end{array}$$ condenses to $CH_2=CH_2$ or to $H_2C=CH_2$

Ethylene (raw material for making polyethylene)

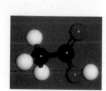

$$\begin{array}{cc} H & O \\ | & \| \\ H-C-C-OH \\ | \\ H \end{array}$$ condenses to $CH_3-\overset{\overset{O}{\|}}{C}-OH$ or to CH_3COH
and often to CH_3CO_2H or to CH_3COOH

Acetic acid (in vinegar)

$$\begin{array}{ccc} H & O & H \\ | & \| & | \\ H-C-C-C-H \\ | & & | \\ H & & H \end{array}$$ condenses to $CH_3-\overset{\overset{O}{\|}}{C}-CH_3$ or to CH_3CCH_3

Acetone (nail polish remover)

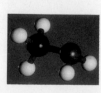

$$\begin{array}{cc} H & H \\ | & | \\ H-C-N-H \\ | \\ H \end{array}$$ condenses to CH_3-NH_2 or to CH_3NH_2

Methylamine (in decaying fish)

Parentheses Are Sometimes Used to Condense Structures Further Sometimes two or three identical groups that are attached to the same carbon are grouped inside a set of parentheses. For example,

$$
\begin{array}{c}
CH_3 \\
| \\
CH_3CHCH_2CH_3
\end{array}
\quad \text{can be written as} \quad (CH_3)_2CHCH_2CH_3
$$

$$
\begin{array}{c}
CH_3 \quad CH_3 \\
| \qquad | \\
CH_3CCH_2CHCH_3 \\
| \\
CH_3
\end{array}
\quad \text{can be written as} \quad (CH_3)_3CCH_2CH(CH_3)_2
$$

We will not do this often, but you will see it in many references and you should be aware of it.

Unsaturated Compounds Have Double or Triple Bonds If its molecules have only single bonds, the compound is called a **saturated compound.** When one or more double or triple bonds are present, the substance is said to be an **unsaturated compound.** Thus ethylene, acetic acid, and acetone, just shown, are all unsaturated compounds, but ethyl alcohol and methylamine are saturated.

Saturated describes any molecule whose atoms are directly holding as many other atoms as they can. Each carbon in ethylene, for example, is holding just three atoms, but in ethyl alcohol each is directly holding four. *Unsaturated* implies that something can be added, and we will see that unsaturated compounds can add certain substances, like hydrogen, to their double or triple bonds.

■ Molecules of edible oils, like olive oil or corn oil, have many double bonds and are described as polyunsaturated.

Many Organic Molecules Contain Rings of Carbon Atoms A carbon **ring** is an arrangement of three or more carbon atoms into a closed cycle. Molecules with this feature are called ring compounds or cyclic compounds. (Sometimes an all-carbon ring is described as *carbocyclic.*) Cyclohexane molecules, for example, have a ring of six carbon atoms. Cyclopropane, once an important anesthetic, is also a cyclic compound.

■ More than 2 billion pounds of cyclohexane are made annually in the United States, with over 90% being used to make nylon.

Cyclohexane   Cyclopropane

Cyclic compounds can have double bonds as in cyclohexene, but always remember that carbon–carbon double bonds are never left to the imagination. Rings, of course, can carry substituents, as in ethylcyclohexane. Not all the ring atoms have to be carbon atoms. They can be O, N, or S, too, and cyclic compounds with ring atoms other than C are called **heterocyclic compounds.** A simple example is tetrahydropyran.

Cyclohexene   Ethylcyclohexane   Tetrahydropyran

■ The ring system in tetrahydropyran—5 C plus 1 O in the ring—is widely present among molecules of carbohydrates.

The Rings of Cyclic Compounds Can Be Condensed to Simple Polygons Since we can leave to our imaginations so many structural features, a polygon, a many-sided figure, becomes a handy way to condense rings. A square, for example, can represent cyclobutane. The photograph of the ball-and-stick model of cyclobutane and its progressively more condensed structures show what is left to the imagination when just a square is used. At each corner, we have to understand that there is a CH_2 group. Each line in the square is a carbon–carbon single bond.

Three ways to represent the structure of cyclobutane

The model of methylcyclopentane and its progressively more condensed structures further show how a polygon can represent a ring.

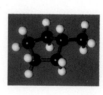

Three ways to represent the structure of methylcyclopentane

By convention, polygons like the hexagon for cyclohexane can be used to represent rings provided that we understand the following rules.

1. C occurs at each corner unless O or N (or another multivalent atom) is explicitly written at a corner.

2. A line connecting two corners is a covalent bond between adjacent ring atoms.

3. Remaining bonds, as required by the covalence of the atom at a corner, are understood to hold H atoms.

4. Double bonds are always explicitly shown.

We can illustrate these rules with the following cyclic compounds.

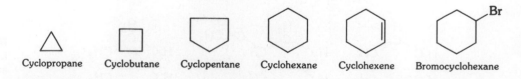

Cyclopropane   Cyclobutane   Cyclopentane   Cyclohexane   Cyclohexene   Bromocyclohexane

There is no theoretical upper limit to the size of a ring.

EXAMPLE 12.2
Understanding Condensed Structures of Ring Compounds

To make sure that you can read a condensed structure when it includes a polygon for a ring system, expand this structure, including its side chains.

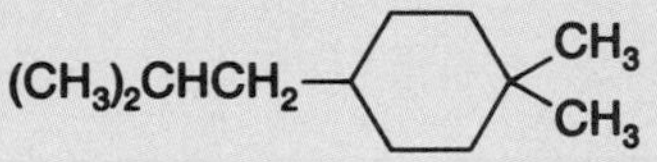

$(CH_3)_2CHCH_2$—

ANALYSIS Every carbon atom, every bond, and every H atom has to be shown explicitly.

SOLUTION

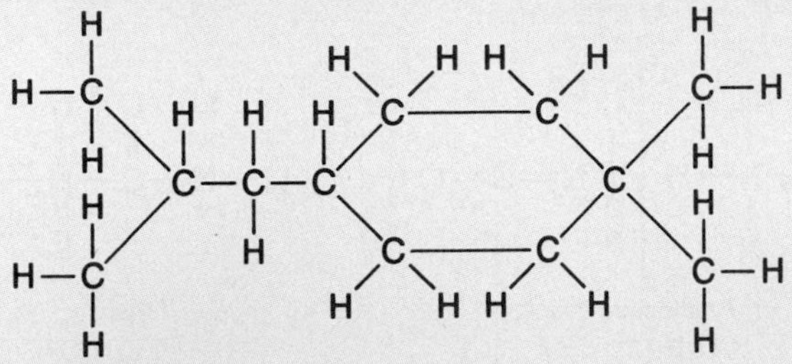

Note especially how we can tell the numbers of hydrogens that must be attached to a ring atom. We need as many as required to fill out a set of four bonds from each carbon. This example also shows a situation (on the right side of the ring) in which no bonding room is left for holding an H atom.

PRACTICE EXERCISE 5

Expand each of these two structures.

(a)   (b)

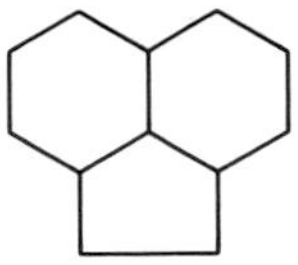

PRACTICE EXERCISE 6

Condense the following structure.

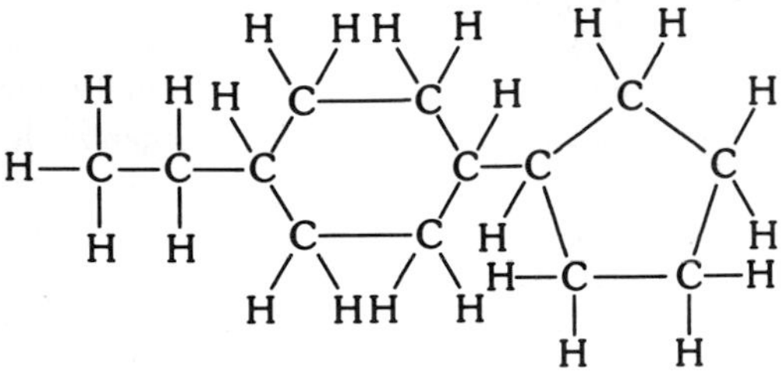

As you might expect, free rotation about the single bonds in a ring is not possible. Although there is a small amount of "flex" in rings, you'd have to break single bonds to get the kind of free rotation we observed with pentane (Fig. 12.1).

When we use polygons to represent saturated rings that have six or more ring atoms, we gloss over one feature of such molecules. The ring atoms that make up the rings of this size do not all lie in the same plane. We will postpone a study of what this fact implies.

12.3

ISOMERISM

Compounds can have identical molecular formulas but different structures. Ammonium cyanate and urea, the chemicals of Wöhler's important experiment, both have the molecular formula CH_4N_2O, but the atoms are organized differently:

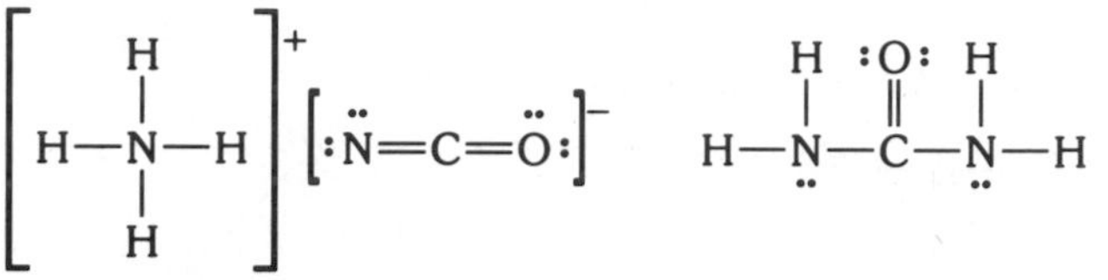

Ammonium cyanate CH_4N_2O — Urea CH_4N_2O

■ "Isomer" has Greek roots —*isos*, the same, and *meros*, parts; that is, "equal parts" (but put together differently).

Compounds that have identical molecular formulas but different structures are called **isomers** of each other, and the existence of isomers is a phenomenon called **isomerism.** Isomerism is one reason why there are so many organic compounds. There are several kinds of isomers, and we will consider just one type here, **constitutional isomers.** Constitutional isomers, once called *structural isomers,*[1] differ in the basic atom-to-atom connectivities. There are three constitutional isomers of C_5H_{12}, for example, pentane, 2-methylbutane, and 2,2-dimethylpropane.

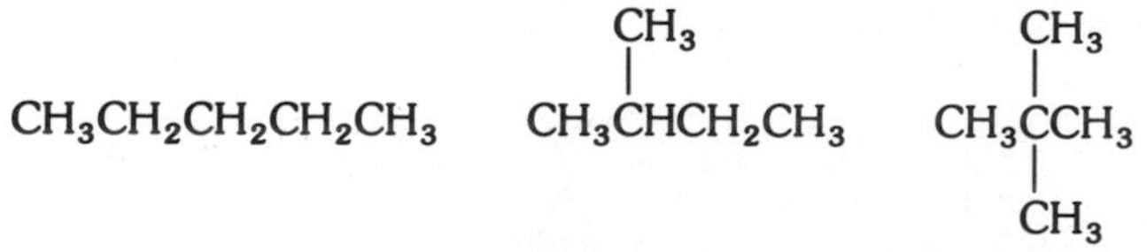

Pentane (*n*-pentane) — 2-Methylbutane (isopentane) — 2,2-Dimethylpropane (neopentane)

■ The names in parentheses are the common names of these compounds. The letter *n* stands for *normal,* meaning the straight-chain isomer. *Neo* signifies *new,* as in a new isomer.

The larger the number of carbon atoms per molecule, the larger the number of isomers. For example, C_8H_{18} has 18 isomers; $C_{10}H_{22}$ has 75, and $C_{20}H_{42}$ has 366,319. Someone has figured out that roughly 6.25×10^{13} isomers are possible for $C_{40}H_{82}$. (Very few have actually been prepared. It would take nearly 200 billion years to make each one at the rate of one per day.)

The isomers of pentane or of $C_{40}H_{82}$ have quite similar chemical properties because their molecules all have only C—C and C—H single bonds. Often, however, isomers have very different properties. There are two ways, for example, to organize the atoms in C_2H_6O into constitutional isomers, as seen near the top of Table 12.1. One isomer is ethyl alcohol and the other is dimethyl ether. They are radically different, as the data in Table 12.1 show. At room temperature, ethyl alcohol is a liquid and dimethyl ether is a gas. Ethyl alcohol reacts with sodium; dimethyl ether does not. It is quite common for isomers to have properties this

[1] The term "structural isomer" has fallen into disfavor because it is regarded as too broad, that it implies not only atom-to-atom connectivities but also geometrical differences. Another kind of isomerism, geometrical isomerism (Section 13.3), deals with the latter.

TABLE 12.1
Properties of Two Isomers: Ethyl Alcohol and Dimethyl Ether

Property	Ethyl Alcohol	Dimethyl Ether
Structure	CH_3CH_2OH	CH_3OCH_3
Boiling point	78.5 °C	−24 °C
Melting point	−117 °C	−138.5 °C
Density (25 °C)	0.79 g/mL (a liquid)	2.0 g/L (a gas)
Solubility in water	Soluble in all proportions	Slightly soluble

different. Therefore, we nearly always use *structural* rather than molecular formulas for organic compounds. Only structures let us see at a glance how the atoms in the molecules are organized. The ability to recognize two structures as identical molecules, or as isomers or something else, is very important.

EXAMPLE 12.3
Recognizing Isomers

Which pair of structures represents a pair of isomers?

1. CH_3—O—CH_2CH_3 and CH_3CH_2—O—CH_3

2. 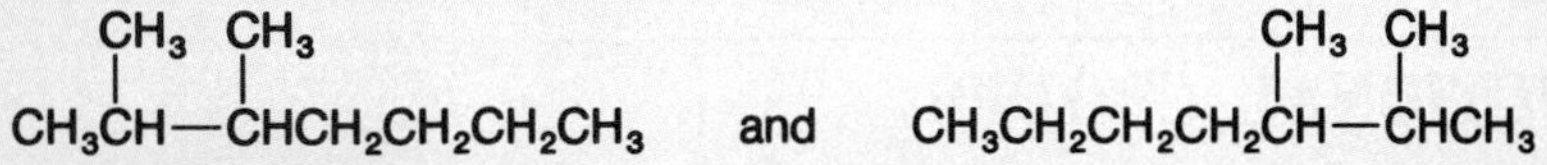 $CH_3\overset{\displaystyle CH_3}{\underset{|}{C}H}$—$\overset{\displaystyle CH_3}{\underset{|}{C}H}CH_2CH_2CH_2CH_3$ and $CH_3CH_2CH_2CH_2\overset{\displaystyle CH_3}{\underset{|}{C}H}$—$\overset{\displaystyle CH_3}{\underset{|}{C}H}CH_3$

3. $CH_3\overset{\displaystyle CH_3}{\underset{|}{C}H}CH_2CH_2CH_3$ and $CH_3CH_2\overset{\displaystyle CH_3}{\underset{|}{C}H}CH_2CH_3$

4. 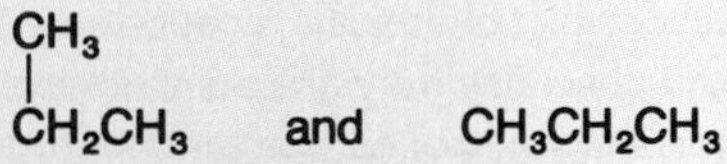 $\overset{\displaystyle CH_3}{\underset{|}{C}H_2CH_3}$ and $CH_3CH_2CH_3$

ANALYSIS Unless you spot a difference that rules out isomerism immediately, the *first step* is to see whether the molecular formulas are the same. If they aren't, the two structures are *not* isomers. If the molecular formulas of the two structures are identical, they might be identical or they might be isomers.

In this problem, the members of each pair share the same molecular formula.

Pair 1: C_3H_8O Pair 2: C_9H_{20} Pair 3: C_6H_{14} Pair 4: C_3H_8

Next, to see whether a particular pair represents isomers we try to find at least one structural difference. If we can't, the two structures are identical; they are just oriented differently on the page, or their chains are twisted into different conformations. Don't be fooled by an "east-to-west" versus a "west-to-east" type of difference. The difference must be *internal* within the structure. (Whether you face east or west you're the same person!)

SOLUTION Pair 1 are identical molecules; they're only oriented differently. (Imagine using a pancake turner to flip the one on the left, left to right; it would then be the structure on the right.)

Pair 2 is also an example of an east versus west difference in orientation. These two structures are identical. Their internal sequences—their atom-to-atom connectivities—are the same.

Pair 3 are isomers. In the first, a CH_3 group joins a five-carbon chain at the chain's second carbon, and in the second, this group is attached at the third carbon.

Pair 4 are identical. The two structures differ only in the conformations of their chains. Recall that free rotation about bonds allows us to imagine the straightening out of a continuous, open chain.

PRACTICE EXERCISE 7

Examine each pair to see whether the members are identical, are isomers, or are different in some other way.

(a) H—O—CH$_3$ and CH$_3$—O—H   *identical*

(b) CH$_3$—NH—CH$_3$ and CH$_3$—CH$_2$—NH$_2$ *isomers*

(c)

$$CH_2CH_3$$
$$|$$
CH$_2$CH$_2$CHCH$_3$ and

$$CH_3$$
$$|$$
CH$_3$CH$_2$CH$_2$CH$_2$CHCH$_3$ *differ*

(d) CH$_2$=CHCH$_2$CH$_3$ and CH$_3$CH=CHCH$_3$ *isomer*

(e)

$$\qquad\quad O$$
$$\qquad\quad \|$$
CH$_3$CH$_2$COH and

$$\qquad O$$
$$\qquad \|$$
HOCCH$_3$ *different*

12.4

FUNCTIONAL GROUPS

The study of organic chemistry is organized around functional groups.
Regions of molecules that have nonmetal atoms other than C and H or that have double or triple bonds are the specific sites in organic molecules that chemicals most often attack. These small structural units are called **functional groups,** because they are the chemically functioning parts of molecules. Sections of molecules consisting only of carbon and hydrogen and only single bonds are called the **nonfunctional groups.**

Each Functional Group Defines an Organic Family   Although over six million organic compounds are known, there are only a handful of functional groups, and each one serves to define a family of organic compounds. Our study of organic chemistry will be organized around just a few of these families, those outlined in Table 12.2. Let's see how the idea of a family will greatly simplify our study.

The *alcohols* are a major organic family of compounds. We have learned, for example, that ethyl alcohol is CH$_3$CH$_2$OH. Its molecules have the OH group attached to a chain of two carbons, but *chain length* is not what determines a compound's family. Chain length only bears on the *name* of that specific family member, as we'll soon see. The chain can be any length imaginable, and the substance will be in the alcohol family provided that somewhere on the chain there is an OH group attached to a carbon from which only single bonds extend. This functional group is called the *alcohol group;* it is very common in nature, being present in all carbohydrates and most proteins. Some examples of simple alcohols are

■ Isopropyl alcohol is commonly used as a rubbing alcohol.

$$\begin{array}{ccccc}
\quad | & & & & \\
-C-O-H & CH_3-OH & CH_3CH_2-OH & CH_3CH_2CH_2-OH & CH_3CHCH_3 \\
\quad | & & & & \quad\quad\quad | \\
 & & & & \quad\quad\quad OH
\end{array}$$

| Alcohol group | Methyl alcohol | Ethyl alcohol | Propyl alcohol | Isopropyl alcohol |

Because all these alcohols have the same functional group, they exhibit the same kinds of chemical reactions. When just one of these reactions is learned, it applies to all members of the family, literally to thousands of compounds. In fact, we will often summarize a particular reaction for an organic family by using a general family symbol. All alcohols, for example, can

TABLE 12.2
Some Important Families of Organic Compounds

Family	Characteristic Structural Feature[a]	Example
Hydrocarbons	Only C and H present **Families of Hydrocarbons:** Alkanes: only single bonds Alkenes: C=C Alkynes: C≡C Aromatic: benzene ring	 CH_3CH_3 $CH_2\!=\!CH_2$ $HC\!\equiv\!CH$
Alcohols	ROH	CH_3CH_2OH
Ethers	ROR′	CH_3OCH_3
Thioalcohols	RSH	CH_3SH
Disulfides	RS—SR	$CH_3S\!-\!SCH_3$
Aldehydes	$\overset{\displaystyle O}{\overset{\displaystyle \|}{RCH}}$	$\overset{\displaystyle O}{\overset{\displaystyle \|}{CH_3CH}}$
Ketones	$\overset{\displaystyle O}{\overset{\displaystyle \|}{RCR'}}$	$\overset{\displaystyle O}{\overset{\displaystyle \|}{CH_3CCH_3}}$
Carboxylic acids	$\overset{\displaystyle O}{\overset{\displaystyle \|}{RCOH}}$	$\overset{\displaystyle O}{\overset{\displaystyle \|}{CH_3COH}}$
Esters of carboxylic acids	$\overset{\displaystyle O}{\overset{\displaystyle \|}{RCOR'}}$	$\overset{\displaystyle O}{\overset{\displaystyle \|}{CH_3COCH_3}}$
Esters of phosphoric acid	$\overset{\displaystyle O}{\overset{\displaystyle \|}{ROPOH}}$ $\|$ OH	$\overset{\displaystyle O}{\overset{\displaystyle \|}{CH_3OPOH}}$ $\|$ OH
Esters of diphosphoric acid	$\overset{\displaystyle O\ \ O}{\overset{\displaystyle \|\ \ \|}{ROPOPOH}}$ $\|\ \ \|$ HO OH	$\overset{\displaystyle O\ \ O}{\overset{\displaystyle \|\ \ \|}{CH_3OPOPOH}}$ $\|\ \ \|$ HO OH
Esters of triphosphoric acid	$\overset{\displaystyle O\ \ O\ \ O}{\overset{\displaystyle \|\ \ \|\ \ \|}{ROPOPOP(OH)_2}}$ $\|\ \ \|$ HO OH	$\overset{\displaystyle O\ \ O\ \ O}{\overset{\displaystyle \|\ \ \|\ \ \|}{CH_3OPOPOP(OH)_2}}$ $\|\ \ \|$ HO OH
Amines	RNH_2, RNHR′, RNR′R″	CH_3NH_2 CH_3NHCH_3 $\overset{\displaystyle CH_3}{\overset{\displaystyle \|}{CH_3NCH_3}}$
Amides	$\overset{\displaystyle O\ \ \ R''(H)}{\overset{\displaystyle \|\ \ \ \ \|}{RC\!-\!NR'(H)}}$	$\overset{\displaystyle O}{\overset{\displaystyle \|}{CH_3CNH_2}}$

[a] R, R′, and R″ represent hydrocarbon groups—*alkyl groups*—defined in the text. R′(H) or R″(H) signifies that the substituent can be either a hydrocarbon group or hydrogen.

be symbolized by R—OH, where R stands for a carbon chain (or ring), one of whatever length or branchings or rings. All alcohols, for instance, react with sodium metal as follows:

$$2R\!-\!OH + 2Na \longrightarrow 2R\!-\!ONa + H_2$$

■ R is from the German word *Radikal*, which we translate here to mean *group* as in a group of atoms.

If we wanted to write the specific example of this reaction that involves, say, ethyl alcohol, all we have to do is replace R by CH_3CH_2.

$$2CH_3CH_2-OH + 2Na \longrightarrow 2CH_3CH_2-ONa + H_2$$

Notice that this reaction changes only the OH group of the alcohol. (Dimethyl ether, Table 12.1, which does not have the OH group, cannot give this reaction with sodium, as we learned in the previous section.)

Another organic family is that of the *carboxylic acids.* All their molecules have the *carboxyl group,* and this group makes all its compounds weak acids. We've often illustrated this with acetic acid.

■ The carboxyl group is present in all fatty acids, products of the digestion of the fats and oils in our diets.

$$\overset{\overset{\textstyle O}{\|}}{-C}-O-H \qquad RCO_2H \qquad \overset{\overset{\textstyle O}{\|}}{CH_3C}OH \quad \text{or} \quad CH_3CO_2H \quad \text{or} \quad CH_3COOH$$

Carboxyl group   Carboxylic acids   Acetic acid

We know that acetic acid neutralizes the hydroxide ion:

$$\overset{\overset{\textstyle O}{\|}}{CH_3C}OH + OH^- \longrightarrow \overset{\overset{\textstyle O}{\|}}{CH_3C}O^- + H_2O$$

Acetate ion

All molecules with the carboxyl group give the same reaction, so we can represent literally thousands of such reactions by just one simple equation:

$$\overset{\overset{\textstyle O}{\|}}{RC}OH + OH^- \longrightarrow \overset{\overset{\textstyle O}{\|}}{RC}O^- + H_2O$$

Carboxylate ion

The groups characteristic of both carboxylic acids and the carboxylate ions are present in all proteins and their building blocks, the amino acids.

The amino acids are examples of substances with more than one functional group in the same molecule. They have the carboxyl group as well as the amino group, a group that defines the *amines,* still another simple family.

■ The amino group is a proton acceptor, like ammonia.

$$NH_2 \qquad CH_3NH_2 \qquad RNH_2 \qquad NH_2CH_2CO_2H$$

Amino group   Methyl- amine   Simple amines   Glycine, the simplest amino acid

■ This icon will identify structural map signs:

Glycine, like all carboxylic acids, can neutralize hydroxide ion.

You can see how powerful a learning tool the functional group is. In the next few chapters, we will study just a few of the reactions of the most important functional groups found at the molecular level of life. *Learning these reactions will be like mastering a set of map signs.* You can read thousands of maps intelligently once you know their common signs and symbols. Similarly, we'll be able to "read" some of the chemical and physical properties of astonishingly complex molecules with the knowledge of the properties of a few functional groups.

The functional groups of a molecule generally occur in a setting of alkane-like groups. To be able to contrast the properties of functional groups with those of the alkane-like groups we turn our attention next to the alkanes themselves, the least reactive of the organic systems.

12.5

ALKANES AND CYCLOALKANES

Alkanes and cycloalkanes are saturated hydrocarbons.

Petroleum and natural gas are substances that consist almost entirely of a complex mixture of molecular compounds called hydrocarbons. Special Topic 12.1 describes the nature of petroleum and other chemical fuels in greater detail. **Hydrocarbons** are made from the atoms of just two elements, carbon and hydrogen, and the covalent bonds between the carbon atoms can be single, double, or triple. The carbon skeletons can be chains or rings. These possibilities define the various kinds of hydrocarbons, which are outlined in Figure 12.3.

The **alkanes,** whether open-chain or cyclic, are saturated hydrocarbons, those with only single bonds. Table 12.3 gives the first ten straight-chain members of this family. The **alkenes** are hydrocarbons with one or more carbon–carbon double bonds, whether the skeletons are chains or rings. Hydrocarbons with one or more carbon–carbon triple bonds are **alkynes.** We'll study the alkenes (and a little about the alkynes) in the next chapter. It is possible for one molecule to have both double and triple bonds, of course.

One important distinction in Figure 12.3 is between aliphatic and aromatic hydrocarbons. **Aliphatic compounds** of whatever family are any that have no benzene ring (or a similar system), and **aromatic compounds** are those with such rings. (We'll postpone the implications of the benzene ring to the next chapter.)

Hydrocarbons of All Types, Saturated or Not, Have Common Physical Properties Both the carbon-carbon and the carbon-hydrogen bonds are almost entirely nonpolar, so hydrocarbon molecules have very little if any overall polarity. Hydrocarbons of all types, for this reason, are insoluble in water, but they dissolve well in nonpolar solvents (like CCl_4). Indeed, many special mixtures of alkanes are themselves used as nonpolar solvents. Some people, for example, have used gasoline or lighter fluid — both are mixtures of alkanes — to remove tar spots or grease. (If you do, be sure to keep all flames away and work outside, never in a garage or other enclosed room.)

Hydrocarbons are not only insoluble in water but are generally less dense than water and so they will float on it. Thus using water to put out a hydrocarbon fire, like flaming gasoline, will only float the flames over a wider area. Nonflammable foams or CO_2 extinguishers must be used instead.

■ Because the bond angle at a triple bond is 180°, a ring has to be quite large to have a triple bond, and cycloalkynes are rare.

■ The first compounds found with benzene rings had pleasant odors and so were called *aromatic*. Aliphatic is from the Greek *aliphatos*, "fat-like."

■ In the right proportion in air, hydrocarbon vapors explode when ignited.

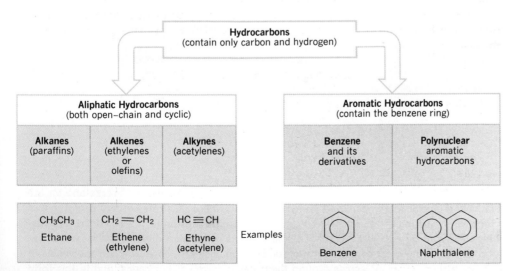

FIGURE 12.3
The several kinds of hydrocarbons. (The circles in the structures for benzene and naphthalene will be explained in the next chapter.)

The Fossil Fuels Long ago, nature used photosynthesis to transform solar energy into the chemical energy of ancient plants and then locked up this energy in the fossilized remains of these plants, the *fossil fuels* — principally petroleum, coal, and natural gas. These fuels are a legacy from the past now being consumed so rapidly that for the first time in history we are concerned about running out of them and about the impact on civilization if their disappearance occurs too suddenly for us to adapt.

Petroleum and Crude Oil The fossil fuels formed over a span of hundreds of thousands of years during the Carboniferous period in geologic history, roughly 280 to 345 million years ago. Vast areas of the continents, little more than monotonous plains, then basked in sunlight near sea level. In the oceans, countless tiny, photosynthesizing plants like the diatoms — tons of them per acre of ocean surface during the early spring — soaked up solar energy to power their fugitive lives. Then they died. According to one theory, the death of each such plant released a tiny droplet of oily material that eventually settled into the bottom muds. The muds grew in thickness and compacted, sometimes into a rock called shale, sometimes into limestone and sandstone deposits. Under the pressure and heat of compacting, the oily matter changed into petroleum, a mixture of crude oil, water, and natural gas. (*Petroleum* is from *petra,* "rock" and *oleum,* "oil.") In some parts of the world, the petroleum managed to move slowly through porous rock and collect into vast underground pools to form the great petroleum reserves of our planet. In other regions, this movement could not occur, and the oily substances remain to this day locked in enormous deposits of oil shales and oil sands.

The story is told of a gentleman in a western state who built a new home and made the fireplace out of a local shale. When he lit the first fire, both the fireplace and his house burned up! His shale was rich in oil; in the Green River region where Utah, Colorado, and Idaho meet, there is a shale formation estimated to contain 2000 billion barrels of *shale oil* (see Figure 1). (The United States uses roughly 5 billion barrels of oil per year.) When the oil shale is crushed and heated to about 260 °C, a substance es-

FIGURE 1
Oil shale in the Mahogony zone near Parachute, Colorado

sentially the same as petroleum is released. Rock qualifies as oil shale if it holds an average of 10 gallons of petroleum per ton of rock. The cost of wringing the petroleum from the rock is presently too high for commercial exploitation, even if the management of the associated environmental problems is excluded from the cost.

Near Lake Athabasca in the province of Alberta, Canada, in a land area about the size of Lake Michigan, there are huge reserves of a material much like crude oil but intermixed with sand, not shale (see Figure 2). Each ton of this *oil sand* holds about two-thirds of a barrel of oil, and the total deposit of oil sand is estimated to contain over 600 billion barrels of oil. This is equivalent to about half of the entire petroleum reserves of the world.

Coal On the marshy lands bordering the ancient oceans, lush vegetation flourished and died in a moist and sunny setting. The rate of decay of the remains of these plants, covered by stagnant, oxygen-poor, often acidic water, was slower than the rate at which the plants died. The slowly rotting mass accumulated to huge depths, became fibrous and turned to *peat,* a woody material used as fuel in some regions of the world.

Where peat layers became thick enough or were compressed by later deposits of sedimentary rock like limestone and sandstone, the peat changed into lignite ("brown coal"). Although lignite is over 40% water, it is

Hydrocarbon Solvents Illustrate the Like-Dissolves-Like Rule Grease and tar are relatively nonpolar materials, and their solubility in gasoline illustrates a very useful rule of thumb for predicting whether a solvent can dissolve some substance. It's called the **like-dissolves-like rule,** where "like" refers to a likeness in *polarity.* Polar solvents, such as water, are good for dissolving polar or ionic substances, like sugar or salt, because polar molecules or ions can attract water molecules around themselves, form solvent cages, and in this form

FIGURE 2
Syncrude's tar sands project in Alberta, Canada

Useful Substances from Crude Oil by Refining Crude oil is a complex mixture of organic compounds, but nearly all are hydrocarbons. Small but vexing amounts of sulfur-containing compounds are also present. The object of refining crude oil is to separate this mixture into products of varying uses. Refinery operations yield mixtures of compounds called *fractions* that boil over certain ranges of temperatures (see the accompanying table). Roughly 500 compounds occur among the fractions boiling up to 200 °C; about a third are alkanes, a third are cycloalkanes, and a third are aromatic hydrocarbons. You can see in the table where the chief fuels for transportation—gasoline, diesel oil, and jet fuel—originate.

The gasoline fraction of crude oil does not provide nearly enough of the world's needs for gasoline. One of the strategies at petroleum refineries to make more gasoline is to subject high boiling petroleum fractions to operations called *catalytic cracking* and *reforming*. In the presence of catalysts and when heated, large alkane molecules break up into smaller ones corresponding to lower boiling points that are in the range useful for gasoline engines.

still an important fuel. Many lignite deposits became thick enough for further compaction to occur, and most of the water was squeezed out. Thus bituminous coal ("soft coal") formed, which has less than 5% water but contains considerable quantities of volatile matter. (*Volatile* means "easily evaporated.") When still further compaction took place and nearly all of the volatile matter was forced out, anthracite ("hard coal") formed, which is over 95% carbon.

The energy content in the coal reserves of the world exceeds that of the known petroleum reserves by a wide margin. In estimates that have the available supplies of petroleum lasting less than a century, the coal reserves are considered to have a lifetime of up to three centuries. Coal contains very small quantities of compounds of mercury, sulfur, radioactive elements and other potential pollutants. Their concentrations are very small, as we said, but because of the enormous quantities of coal annually burned, mostly at electric power plants, air and water pollution problems are caused, which often lead to water pollution as well. The mercury pollution in many lakes north of the industrial regions of the United States stems in part from the mercury released by burning coal.

Natural Gas In both soft coal and petroleum, the most volatile hydrocarbons, methane and ethane, also accumulated. Natural gas is largely methane.

Principal Fractions from Petroleum

Boiling Point Range (in °C)	Molecular Size	Principal Uses
Below 20	$C_1–C_4$	Natural gas; heating and cooking fuel; raw materials for other chemicals
20–60	$C_5–C_6$	Petroleum "ether," a nonpolar solvent and cleaning fluid
60–100	$C_6–C_7$	Ligroin or light naphtha; nonpolar solvents and cleaning fluids
40–200	$C_5–C_{10}$	Gasoline
175–325	$C_{12}–C_{18}$	Kerosene; jet fuel; tractor fuel
250–400	C_{12} and higher	Gas oil; fuel oil; diesel oil
Nonvolatile liquids	C_{20} and up	Refined mineral oil; lubricating oil; grease (a blend of soap in oil)
Nonvolatile solids	C_{20} and up	Paraffin wax; asphalt; road tar; roofing tar

freely intermingle with water molecules. Nonpolar solvents, like gasoline, do not dissolve sugar or salt, because nonpolar solvent molecules cannot be attracted to polar molecules or ions and form solvent cages.

Because of their low polarity and small size, the hydrocarbons that have one to four carbons per molecule are generally gases at or near room temperature. As the molecular size increases, however, London forces between molecules become stronger. Therefore, the

■ Solvent cages as factors in the solubilities of ionic or polar molecular compounds were described in Section 7.2.

TABLE 12.3
Straight-Chain Alkanes

IUPAC Name	Carbons	Molecular Formula[a]	Structure	BP (°C)	MP (°C)	Density (g/mL, 20 °C)
Methane	1	CH_4	CH_4	−161.5	−182.5	—
Ethane	2	C_2H_6	CH_3CH_3	−88.6	−183.3	—
Propane	3	C_3H_8	$CH_3CH_2CH_3$	−42.1	−189.7	—
Butane	4	C_4H_{10}	$CH_3(CH_2)_2CH_3$	−0.5	−138.4	—
Pentane	5	C_5H_{12}	$CH_3(CH_2)_3CH_3$	36.1	−129.7	0.626
Hexane	6	C_6H_{14}	$CH_3(CH_2)_4CH_3$	68.7	−95.3	0.659
Heptane	7	C_7H_{16}	$CH_3(CH_2)_5CH_3$	98.4	−90.6	0.684
Octane	8	C_8H_{18}	$CH_3(CH_2)_6CH_3$	125.7	−56.8	0.703
Nonane	9	C_9H_{20}	$CH_3(CH_2)_7CH_3$	150.8	−53.5	0.718
Decane	10	$C_{10}H_{22}$	$CH_3(CH_2)_8CH_3$	174.1	−29.7	0.730

[a] The molecular formulas of the open-chain alkanes fit the general formula C_nH_{2n+2}, where n = the number of carbon atoms per molecule.

boiling points of the straight-chain alkanes increase with chain length. Hydrocarbons that have from 5 to about 16 carbon atoms per molecule are usually liquids at room temperature (see Table 12.3). When alkanes have approximately 18 or more carbon atoms per molecule, the London forces are strong enough to make the alkanes (waxy) solids at room temperature. Paraffin wax, for example, is a mixture of alkanes whose molecules have 20 or more carbon atoms.

■ Most candles are made from paraffin.

Our First Molecular "Map Sign," Hydrocarbon-Like Portions of Molecules Before moving on, let's pause to reflect on what we have done in relating physical properties to structural features. We have introduced the first molecular "map sign" in our study: *substances whose molecules are entirely or even mostly hydrocarbon-like are likely to be insoluble in water but soluble in nonpolar solvents.* When we see an unfamiliar structure, we can tell at a glance whether it is mostly hydrocarbon-like. If it is, we can predict with considerable confidence that the compound is not soluble in water. The structure of cholesterol illustrates our point.

■ The cholesterol structure shown here extends to open-chain systems the formalism of rings. Thus we use a point *where two lines meet* to denote one C and as many Hs as needed to fill out a covalence of 4 for the C atom. A line that terminates with nothing would denote CH_3. Thus, we could represent CH_3CH_2OH simply as

Cholesterol

You probably know that cholesterol can form solid deposits in blood capillaries, and even close them. The heart must then work harder to sustain the flow of blood, and under this stress a heart attack can occur. Notice, now, that virtually the entire cholesterol molecule is hydrocarbon-like. It has only one polar group, the OH or alcohol group, and this takes up too small a portion of the molecule to make cholesterol sufficiently polar to dissolve either in water or in blood (which is mostly water).

What cholesterol illustrates is that by learning one very general fact, one "map sign," we don't have to memorize a long list of separate (but similar) facts about an equally long list of separate compounds that occur at the molecular level of life. With what we have just learned, we can look at the structures of hundreds of complicated compounds and confidently predict particular properties such as the likelihood of their being soluble in water. You have gained a powerful tool, one that immensely reduces what otherwise might have to be memorized.

EXAMPLE 12.4
Predicting Physical Properties from Structures

Study the following two structures and tell which is the structure of the more water-soluble compound.

$$HO-CH_2-CH-CH-CH-CH-C-H$$
$$\qquad\qquad OH\quad OH\quad OH\quad OH \qquad \overset{O}{\|}$$

$$CH_3CH_2CH_2CH_2CH_2CH_2CH_2CH_2CH_2CH_2CH_2CH_2CH_2CH_2CH_2CH_2CH_2CO_2H$$

ANALYSIS The structure of the first compound has several polar OH groups but the second is almost entirely like an alkane.

SOLUTION The first should be (and is) much more soluble in water. The first compound is glucose (in one of its forms), the chief sugar in the bloodstream. The second is stearic acid, which does not dissolve in water. It forms when we digest the fats and oils (lipids) in the diet.

PRACTICE EXERCISE 8

Which of the following is more soluble in gasoline?

$$HOCH_2-CH-CH_2OH \qquad CH_3CH_2CHCH_2OH$$
$$\qquad\quad OH \qquad\qquad\qquad\qquad CH_3$$

Glycerol   2-Methyl-1-butanol

12.6

NAMING THE ALKANES AND CYCLOALKANES

An IUPAC name discloses the compound's family, the number of carbons in the parent chain, and the kinds and locations of substituents.

In chemistry, **nomenclature** refers to the rules used to name compounds. The International Union of Pure and Applied Chemistry (IUPAC) is the organization that now develops the rules of chemical nomenclature. All scientific societies in the world accept the rules, known as the **IUPAC rules.** They are so carefully constructed that only one name could be written for each compound, and only one structure could be drawn for each name.

The IUPAC names, unfortunately, are sometimes very long and difficult to write or pronounce. It's much easier to call table sugar *sucrose* than α-D-glucopyranosyl-β-D-fructofuranoside. This illustrates why shorter names, referred to as *common names,* are still widely used. We will want to learn some common names, too, and you will see that even they usually have some system to them.

■ "Nomenclature" is from the Latin *nomen,* "name," and *calare,* "to call." Wealthy Romans had slaves called *nomenclators* whose duty it was to remind their owners of the names of important people who approached them on the street.

IUPAC Rules for Naming the Alkanes

1. The name ending for all alkanes (and cycloalkanes) is -ane.

2. The *parent chain* is the longest continuous chain of carbons in the structure. For example, the branched-chain alkane

$$CH_3CH_2\overset{\overset{\displaystyle CH_3}{|}}{C}HCH_2CH_2CH_3$$

is regarded as being "made" from the following parent

$$CH_3CH_2CH_2CH_2CH_2CH_3$$

by replacing a hydrogen atom on the third carbon from the left with CH_3.

$$CH_3CH_2\overset{\overset{\displaystyle CH_3}{\underset{\displaystyle H}{\searrow}}}{C}HCH_2CH_2CH_3 \longrightarrow CH_3CH_2\overset{\overset{\displaystyle CH_3}{|}}{C}HCH_2CH_2CH_3$$

3. A prefix is attached to the name-ending, -ane, that specifies the number of carbon atoms *in the parent chain.* The prefixes through parent chain lengths of 10 carbons are as follows *and should be learned.* The names in Table 12.3 show their use.

■ We won't need to know the prefixes for the higher alkanes.

meth-	1 C	hex-	6 C
eth-	2 C	hept-	7 C
prop-	3 C	oct-	8 C
but-	4 C	non-	9 C
pent-	5 C	dec-	10 C

Because the parent chain of our example has six carbons, the parent chain is named hexane — *hex* for six carbons and *ane* for being in the alkane family. The alkane whose name we are devising is regarded as a derivative of this parent, *hexane.*

4. The carbon atoms of the parent chain are numbered starting from whichever end of the chain gives the location of the first branch the lower of two possible numbers. Thus the correct direction for numbering our example is from left to right.

$$\underset{\scriptstyle 1 \quad 2 \quad 3 \quad 4 \quad 5 \quad 6}{CH_3CH_2\overset{\overset{\displaystyle CH_3}{|}}{C}HCH_2CH_2CH_3}$$

(correct direction of numbering)

Had we numbered from right to left, the carbon holding the branch would have had a higher number, which is not allowed by the IUPAC rules for alkanes.

$$\underset{\scriptstyle 6 \quad 5 \qquad 4 \quad 3 \quad 2 \quad 1}{CH_3CH_2 \quad \overset{\overset{\displaystyle CH_3}{|}}{C}HCH_2CH_2CH_3}$$

(incorrect direction of numbering)

5. Name each branch attached to the parent chain. We must now pause and learn the names of some of the *alkyl groups,* groups with alkane-like branches.

The Alkyl Groups

Any branch that consists only of carbon and hydrogen and has only single bonds is called an **alkyl group,** and the names of all alkyl groups end in -*yl*. Think of an alkyl group as an alkane minus one H.

$$H-\underset{\underset{H}{|}}{\overset{\overset{H}{|}}{C}}-H \xrightarrow{\text{remove one H}} H-\underset{\underset{H}{|}}{\overset{\overset{H}{|}}{C}}- \qquad \text{or } CH_3-$$

Methane   **Methyl**

$$H-\underset{\underset{H}{|}}{\overset{\overset{H}{|}}{C}}-\underset{\underset{H}{|}}{\overset{\overset{H}{|}}{C}}-H \xrightarrow{\text{remove one H}} H-\underset{\underset{H}{|}}{\overset{\overset{H}{|}}{C}}-\underset{\underset{H}{|}}{\overset{\overset{H}{|}}{C}}- \qquad \text{or } CH_3CH_2-$$

Ethane   **Ethyl**

Two alkyl groups can be obtained from propane because the middle position in its chain of three is not equivalent to either of the end positions.

$$H-\underset{\underset{H}{|}}{\overset{\overset{H}{|}}{C}}-\underset{\underset{H}{|}}{\overset{\overset{H}{|}}{C}}-\underset{\underset{H}{|}}{\overset{\overset{H}{|}}{C}}-H \xrightarrow[\text{(from either end)}]{\text{remove one H}} H-\underset{\underset{H}{|}}{\overset{\overset{H}{|}}{C}}-\underset{\underset{H}{|}}{\overset{\overset{H}{|}}{C}}-\underset{\underset{H}{|}}{\overset{\overset{H}{|}}{C}}- \qquad \text{or } CH_3CH_2CH_2-$$

Propane   **Propyl**

$$H-\underset{\underset{H}{|}}{\overset{\overset{H}{|}}{C}}-\underset{\underset{H}{|}}{\overset{\overset{H}{|}}{C}}-\underset{\underset{H}{|}}{\overset{\overset{H}{|}}{C}}-H \xrightarrow[\text{(from middle C)}]{\text{remove one H}} H-\underset{\underset{H}{|}}{\overset{\overset{H}{|}}{C}}-\overset{\overset{H}{|}}{C}-\underset{\underset{H}{|}}{\overset{\overset{H}{|}}{C}}-H \qquad \text{or } CH_3CHCH_3$$

Propane   **Isopropyl**

Two alkyl groups can similarly be obtained from butane.

$$H-\underset{\underset{H}{|}}{\overset{\overset{H}{|}}{C}}-\underset{\underset{H}{|}}{\overset{\overset{H}{|}}{C}}-\underset{\underset{H}{|}}{\overset{\overset{H}{|}}{C}}-\underset{\underset{H}{|}}{\overset{\overset{H}{|}}{C}}-H \xrightarrow[\text{(from either end)}]{\text{remove one H}} H-\underset{\underset{H}{|}}{\overset{\overset{H}{|}}{C}}-\underset{\underset{H}{|}}{\overset{\overset{H}{|}}{C}}-\underset{\underset{H}{|}}{\overset{\overset{H}{|}}{C}}-\underset{\underset{H}{|}}{\overset{\overset{H}{|}}{C}}- \qquad \text{or } CH_3CH_2CH_2CH_2-$$

Butane   **Butyl**

$$H-\underset{\underset{H}{|}}{\overset{\overset{H}{|}}{C}}-\underset{\underset{H}{|}}{\overset{\overset{H}{|}}{C}}-\underset{\underset{H}{|}}{\overset{\overset{H}{|}}{C}}-\underset{\underset{H}{|}}{\overset{\overset{H}{|}}{C}}-H \xrightarrow[\substack{\text{(from either of} \\ \text{the two interior} \\ \text{C atoms)}}]{\text{remove one H}} H-\underset{\underset{H}{|}}{\overset{\overset{H}{|}}{C}}-\underset{\underset{H}{|}}{\overset{\overset{H}{|}}{C}}-\overset{\overset{H}{|}}{C}-\underset{\underset{H}{|}}{\overset{\overset{H}{|}}{C}}- \qquad \text{or } CH_3CH_2CHCH_3$$

Butane   *sec*-**Butyl**

The last alkyl group is called the *secondary* butyl group (abbreviated *sec*-butyl) because the open bonding site is at a **secondary carbon,** a carbon that is directly attached to just two other carbons. A **primary carbon** is one to which just one other carbon is directly attached. The open bonding site in the butyl group, for example, is at a primary carbon atom. A **tertiary carbon** is one that holds directly three other carbons. We will encounter a tertiary carbon in a group that we will soon study.

■ Primary carbons

$$CH_3-CH-CH_2-CH_3$$
$$\qquad\qquad |$$
$$\qquad\quad CH_3$$

Tertiary carbon   Secondary carbon

Butane is the smallest alkane to have an isomer. The common name of the isomer is isobutane, and we can derive two more alkyl groups from it.

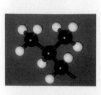

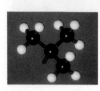

■ The *Study Guide* accompanying this book has exercises that provide drills in recognizing alkyl groups when they are positioned in different ways on the page.

Notice that the open bonding site in the *tertiary*-butyl group (abbreviated *t*-butyl) occurs at a tertiary carbon.

The names and structures of these alkyl groups must now be learned. If you have access to ball-and-stick models, make models of each of the parent alkanes and then remove hydrogen atoms to generate the open bonding sites and the alkyl groups.

The prefix *iso-* in the name of an alkyl group, such as in isopropyl or isobutyl, has a special meaning. It can be used to name any alkyl group that has the following general features.

$$CH_3 \diagdown CH(CH_2)_n - \qquad (n = 0, 1, 2, 3)$$
$$CH_3 \diagup$$

$n = 0$, isopropyl group
$= 1$, isobutyl group
$= 2$, isopentyl group
$= 3$, isohexyl group

■ Here is another way to condense a structure. Thus $CH_3(CH_2)_3CH_3$ represents $CH_3CH_2CH_2CH_2CH_3$.

Notice that each of these names has a word fragment (-*prop*-, -*but*-, and so forth) associated with a number of carbon atoms. When these word fragments are attached to *iso*, they specify the *total* number of carbons in the alkyl group. Thus the isopropyl group has three carbons (indicated by *prop*) and the isobutyl group has four carbons (indicated by *but*).

We can now continue with the IUPAC rules for naming alkanes.

6. Attach the name of the alkyl group to the name of the parent as a prefix. Place the location number of the group in front of the resulting name and separate the number from the name by a hyphen. Returning to our original example, its name is 3-methylhexane.

$$CH_3$$
$$|$$
$$CH_3CH_2CHCH_2CH_2CH_3$$

3-Methylhexane

7. When two or more groups are attached to the parent, name each and locate each with a number. The names of alkyl substituents are assembled in their alphabetical order. Always use *hyphens* to separate numbers from words. Here is an application.

$$\begin{array}{cc} CH_3CH_2 & CH_3 \\ | & | \end{array}$$
$$CH_3CH_2CH_2CHCH_2CHCH_3$$
$$\;\;\;7\;\;\;\;6\;\;\;\;5\;\;\;\;4\;\;\;3\;\;\;\;2\;\;\;1$$

4-Ethyl-2-methylheptane

8. When two or more substituents are identical, use such prefixes as di- (for 2), tri- (for 3), tetra- (for 4), and so forth; and specify the location number of *every* group. Always separate a number from another number in a name by a *comma*. For example,

$$\begin{array}{cc} CH_3 & CH_3 \\ | & | \end{array}$$
$$CH_3CHCH_2CHCH_2CH_3$$

Correct name: 2,4-dimethylhexane
Incorrect names: 2,4-methylhexane
3,5-dimethylhexane
2-methyl-4-methylhexane

9. When identical groups are on the *same* carbon, repeat the number locating this carbon in the name. For example,

$$\begin{array}{c} CH_3 \\ | \end{array}$$
$$CH_3CCH_2CH_2CH_3$$
$$\begin{array}{c} | \\ CH_3 \end{array}$$

Correct name: 2,2-dimethylpentane
Incorrect names: 2-dimethylpentane
2,2-methylpentane
4,4-dimethylpentane

These are not all of the IUPAC rules for alkanes, but they will handle all of our needs. Study the following examples of correctly named compounds. Be sure to notice that in choosing the parent chain we sometimes have to go around a corner as the chain zigzags on the page.

$$\begin{array}{c} CH_3 \\ | \\ CH_3-C-CH_3 \\ | \\ CH_2 \\ | \\ CH_3 \end{array}$$

2,2-Dimethylbutane
not 2-ethyl-2-methylpropane

$$\begin{array}{cc} H_3C & CH_3 \\ \diagdown & \diagup \end{array}$$
$$CH_3-CH_2\quad CH$$
$$\begin{array}{c} | \\ CH_2-CH-CH_3 \end{array}$$

2,3-Dimethylhexane
not 2-isopropylpentane

$$\begin{array}{c} CH_3 \\ | \\ CH_3-C-H \\ | \\ CH_3 \end{array}$$

2-Methylpropane
not 1,1-dimethylethane

$$\begin{array}{c} CH_3 \\ | \\ CH_3CH_2CH_2CHCH_2CHCH_3 \\ | \\ CH_3-CH-CH_3 \end{array}$$

4-Isopropyl-2-methylheptane
not 4-isopropyl-6-methylheptane

EXAMPLE 12.5
Using the IUPAC Rules to Name an Alkane

What is the IUPAC name for the following compound?

$$CH_3 \quad CH_2CH_2CH_2CH_3$$
$$CH_3CHCHCHCHCH_2CH_2CH_3$$
$$CH_3 \quad CH$$
$$CH_3 \quad CH_3$$

ANALYSIS   The compound is an alkane because it is a hydrocarbon with only single bonds. We must therefore use the IUPAC rules for alkanes.

$$\overset{6\quad 7\quad 8\quad 9}{}$$
$$CH_3 \quad CH_2CH_2CH_2CH_3$$
$$\overset{1\quad 2}{CH_3}CHCHCHCHCH_2CH_2CH_3$$
$$\overset{3}{CH_3} \quad \overset{4\ 5}{CH}$$
$$CH_3 \quad CH_3$$

SOLUTION   The ending to the name must be -ane. The next step is to find the longest chain even if we have to go around corners. This chain is nine carbons long, so the name of the parent alkane is nonane. We have to number the chain from left to right, as follows, in order to reach the first branch with the lower number.
At carbons 2 and 3 there are the one-carbon methyl groups. At carbon 4, there is a three-carbon isopropyl group (not the propyl group, because the bonding site is at the middle carbon of the three-carbon chain). At carbon 5, there is a three-carbon propyl group. (It has to be this particular propyl group because the bonding site is the end of the three carbon chain in the group.) Alphabetically, isopropyl comes before methyl which comes before propyl, so we must assemble these names as follows to make the final name. (Names of alkyl groups are alphabetized before any prefixes such as di- or tri- are affixed.)

4-Isopropyl-2,3-dimethyl-5-propylnonane

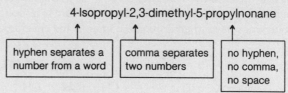

| hyphen separates a number from a word | comma separates two numbers | no hyphen, no comma, no space |

CHECK   The most common mistake students make is in the discovery of the longest chain. Check your answer to make sure that you have not erred in this. Then be sure you have numbered from the correct end.

PRACTICE EXERCISE 9

Write the IUPAC names of the following compounds.

(a)
$$CH_3CH_2$$
$$CHCH_3$$
$$CH_2CH_2$$
$$CH_3$$

(b)
$$CH_3 \quad CH_2CH_2CH_3$$
$$CHCHCH_2CH_3$$
$$CH_3CH$$
$$CH_3$$

(c)
$$CH_3 \quad CH_3 \quad CH_3$$
$$CH_3CH_2CHCHCHCH_2CHCH_3$$
$$CH_3CH_2$$

IUPAC Rules for Naming Cycloalkanes To name a cycloalkane, place the prefix *cyclo-* before the name of the straight-chain alkane that has the same number of carbon atoms as there are in the ring. This is illustrated in Table 12.4. When necessary, give numbers to the ring atoms by giving location 1 to a ring position that holds a substituent and numbering around the ring in whichever direction reaches the nearest substituent first. For example,

■ No number is needed when the ring has only one group. Thus,

1,2-Dimethylcyclohexane   1,2,4-Trimethylcyclohexane

is named methylcyclohexane, not 1-methylcyclohexane.

IUPAC Names of Substituents Other than Alkyl Groups When halogen atoms, or nitro or amino groups are joined to a carbon of a chain or ring, the following names are used for them in IUPAC nomenclature.

—F	fluoro	—I	iodo
—Cl	chloro	—NO_2	nitro
—Br	bromo	—NH_2	amino

For example, $CH_3CH_2CH_2CHCl_2$ is named 1,1-dichlorobutane in the IUPAC system. Nitroethane is $CH_3CH_2NO_2$.

■ No number is needed in *nitroethane* because the name is unambiguous as it stands.

TABLE 12.4
Some Cycloalkanes

IUPAC Name	Structure	BP (°C)	MP (°C)	Density (20 °C)
Cyclopropane		−33	−127	1.809 g/L (0 °C)
Cyclobutane		−13.1	−80	0.7038 g/L (0 °C)
Cyclopentane		49.3	−94.4	0.7460 g/mL
Cyclohexane		80.7	6.47	0.7781 g/mL
Cycloheptane		118.5	−12	0.8098 g/mL

PRACTICE EXERCISE 10

Write the condensed structures of the following compounds.

(a) 1-Bromo-2-nitropentane

(b) 5-Isopropyl-2,2,3,3,4,4-hexamethyloctane

(c) 5-*sec*-Butyl- 2,2-diiodo- 4-isopropyl-3-methylnonane

(d) 1-Bromo-1-chloro-2-methylpropane

(e) 5,5-Di-*sec*-butyldecane

PRACTICE EXERCISE 11

Examine the structure of part (e) of Practice Exercise 10. Underline each primary carbon, draw an arrow to each secondary carbon, and circle each tertiary carbon.

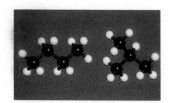

n-Butane and Isobutane.

Common Names In some references you might see the names of straight-chain alkanes with the prefix *n*-, as in *n*-butane, the common name of butane. It stands for *normal,* which is a way of designating that the straight-chain isomer is regarded as the *normal* isomer, as in the common names, *n*-pentane, *n*-hexane, and so forth. It is used only when isomers are possible. (You would never see *n*-propane printed as a name, for example, because there are no isomers of propane.)

Common Names of Alcohols, Amines, and Haloalkanes Employ the Names of the Alkyl Groups The following examples of some halogen derivatives of the alkanes, called *haloalkanes,* illustrate how common names are easily constructed. The IUPAC names are given for comparison.

■ The haloalkanes are examples of *organohalogen compounds.*

Structure	Common Name	IUPAC Name
CH_3Cl	Methyl chloride	Chloromethane
CH_3CH_2Br	Ethyl bromide	Bromoethane
$CH_3CH_2CH_2Br$	Propyl bromide	1-Bromopropane
CH_3CHCH_3 $\|$ Cl	Isopropyl chloride	2-Chloropropane
$CH_3CH_2CH_2CH_2Cl$	Butyl chloride	1-Chlorobutane
$CH_3CHCH_2CH_3$ $\|$ Br	*sec*-Butyl bromide	2-Bromobutane
CH_3 $\|$ $CH_3C{-}Br$ $\|$ CH_3	*t*-Butyl bromide	2-Bromo-2-methylpropane

Ethane

PRACTICE EXERCISE 12

Give the common names of the following compounds.

(a) $ClCH_2CH_3$ (b) $BrCH_2CH_2CH_2CH_3$ (c) CH_3CHCH_2Cl with CH_3 above (d) CH_3CCH_3 with CH_3 above and Br below

12.7

CHEMICAL PROPERTIES OF ALKANES

Alkanes can burn and can give substitution reactions with the halogens, but they undergo almost no other reaction.

The chemistry of the alkanes and cycloalkanes is quite simple. Very few chemicals react with them. This is why they are nicknamed the *paraffins* or the paraffinic hydrocarbons, after the Latin *parum affinis*, meaning "little affinity" or "little reactivity." The alkanes are not chemically attacked by water, by strong acids such as sulfuric or hydrochloric acid, by strong bases like sodium hydroxide, by active metals such as sodium, by strong oxidizing agents such as the permanganate ion or the dichromate ion, or by any of the reducing agents. Among the few reactions of alkanes are combustion and halogenation—reactions with F_2, Cl_2, Br_2 (but not I_2), the halogens.

■ Mineral oil is a safe laxative (when used with care) because it is a mixture of high-formula-mass alkanes that undergo no chemical reactions in the intestinal tract.

The Combustion of Alkanes or Any Hydrocarbon Gives CO₂ and H₂O Virtually all organic compounds burn, and the hydrocarbons are no exception. We burn mixtures of alkanes, for example, as fuel to obtain energy. Bunsen burner gas is mostly methane. Liquified propane is used as fuel in areas where gas lines have not been built. Gasoline, diesel fuel, jet fuel, and heating oil are all mixtures of hydrocarbons, mostly alkanes.

If enough oxygen is available, the sole products of the complete combustion of *any hydrocarbon,* not just alkanes, are carbon dioxide and water plus heat. The (unbalanced) equation, regardless of the kind of hydrocarbon is

$$\text{Hydrocarbon} + O_2 \longrightarrow CO_2 + H_2O + \text{heat}$$

To illustrate, using propane,

$$CH_3CH_2CH_3 + 5O_2 \longrightarrow 3CO_2 + 4H_2O + 531 \text{ kcal/mol propane}$$

The same products, carbon dioxide and water, are also obtained by the complete combustion of any organic compound that consists only of carbon, hydrogen, and oxygen (for example, the alcohols). If insufficient oxygen is present, some carbon monoxide forms.

The Chlorination of Alkanes Is a *Substitution Reaction* In the presence of ultraviolet radiation or at a high temperature, alkanes react with chlorine to give organochlorine compounds and hydrogen chloride. An atom of hydrogen in the alkane can be replaced by an atom of chlorine, and this kind of replacement of one atom or group by another is called a **substitution reaction.** For example,

$$CH_4 + Cl_2 \xrightarrow[\text{or heat}]{\text{ultraviolet light}} CH_3Cl + HCl$$
$$\text{Methyl chloride}$$

The reaction takes place by a chemical chain reaction such as was described in Special Topic 5.1 on page 120. We'll use Special Topic 12.2 to explain how the chlorination of an alkane, like methane, occurs.

The hydrogen atoms in methyl chloride can also be replaced by chlorine, so as methyl chloride starts to form, its molecules compete with those of still unreacted methane for the remaining chlorine. This is how some methylene chloride (dichloromethane) forms when the chlorination of methane is carried out. In fact, when 1 mol of CH_4 and 1 mol of Cl_2 are mixed

THE CHLORINATION OF METHANE— A FREE RADICAL CHAIN REACTION WITH COUNTERPARTS AT THE MOLECULAR LEVEL OF LIFE

Free Radicals In Special Topic 5.1 (page 120), "Ozone in the Stratosphere," we introduced the basic principles of a chemical chain reaction initiated by photons of ultraviolet light. We learned there that photons of the proper energy are absorbed by the electron pairs of covalent bonds and cause the bonds to break. This kind of bond breaking releases not ions but electrically neutral particles in which an atom lacks an octet and has an unpaired electron. Particles with unpaired electrons are called *radicals*, sometimes *free radicals* because they so often form with the freedom to move about. The breaking of the bond in the chlorine molecule, Cl_2, for example, occurs as follows.

$$:\overset{..}{\underset{..}{Cl}}:\overset{..}{\underset{..}{Cl}}: + \text{UV energy or heat} \longrightarrow :\overset{..}{\underset{..}{Cl}}\cdot + \cdot\overset{..}{\underset{..}{Cl}}:$$

Two chlorine atoms
(two free radicals)

Free Radical Chain Reaction in the Chlorination of Methane The reaction of methane with chlorine is a free radical reaction that begins by the breaking of the bond in Cl_2, as we just described. (It is called the *chain initiating* step.) Chlorine is in group VIIA of the periodic table, so each chlorine atom has but seven valence shell electrons, not an octet. Each Cl atom is thus unstable and is able to launch the first step of a chemical chain reaction. In the end, the Cl atom recovers its outer octet but within a molecule of H—Cl. The first step of the actual *chain* reaction part of the overall reaction is that of a ·Cl atom with methane to generate ·CH$_3$, a neutral particle called the methyl radical. (We'll often explicitly show only the unpaired electron of a radical or atom, not the full population of the valence shell.)

$$\cdot Cl + CH_4 \longrightarrow H{-}Cl + \cdot CH_3 \qquad (1)$$

If we use molecular models we can better imagine what happens when a chlorine atom strikes the H end of an H—C bond in methane hard enough.

The carbon atom in the methyl radical also lacks an octet; it has seven valence shell electrons, one unshared (and shown by the dot). When it collides hard enough with an unreacted molecule of Cl—Cl, the Cl—Cl bond breaks and the new C—Cl bond in methyl chloride forms.

This is the second step of the chain itself.

$$\cdot CH_3 + Cl_2 \longrightarrow CH_3Cl + \cdot Cl \qquad (2)$$

These two steps are called the *chain propagating* steps.

Notice that one product of reaction (2), ·Cl, is a necessary reactant for reaction (1). Thus chlorine atoms are generated not only by the action of heat or UV radiation, but also by reaction (2). This is why a relatively small amount of radiation can initiate chains that lead to the formation of a huge number of product molecules. The $\cdot Cl$ forming in reaction (2) appears in the midst of many still unreacted CH_4 molecules, so step (1) can occur again. This only makes another $\cdot CH_3$ radical, so a repeat of reaction (2) is set up. The sequence throughout chain propagation is thus (1) then (2), followed by (1) then (2), and so on.

Two chains are initiated by the breaking of the bond in one molecule of Cl_2 by heat or UV light because the chain-initiating event produces *two* $\cdot Cl$ atoms. Upwards of 8000 molecules of methyl chloride can be generated by one chlorine atom formed from the initiating action of only 1 photon of UV radiation. The chains continue until two radicals happen to find each other and join. Some *chain-termination* reactions are

$$2Cl\cdot \longrightarrow Cl_2$$
$$2CH_3\cdot \longrightarrow CH_3-CH_3$$
$$CH_3\cdot + Cl\cdot \longrightarrow CH_3Cl$$

Other chains are launched, however, as some terminate until one or both reactants are used up.

Multiple Chlorinations Because of the nature of the reaction, additional products are inevitable. Methyl chloride, CH_3Cl, forms while some unreacted Cl_2 remains, so a chlorine atom might collide with the H end of a $H-C$ bond in CH_3Cl instead of at a $H-C$ bond in another molecule of CH_4. This launches a new chain of reactions that converts CH_3Cl into CH_2Cl_2. The latter, of course, has its own $H-C$ bonds, so still other chains can be started that convert CH_2Cl_2 into $CHCl_3$. You can see that the latter also has a $H-C$ bond, so further chains can commence that lead to the formation of some CCl_4. These events occur more or less at random as the atoms and molecules whiz about in the gaseous state. Statistical probabilities largely govern what collisions occur but, as we mentioned in the text, a mixture of products is bound to form when chlorination is initiated in a mixture having a 1 : 1 mole ratio of Cl_2 to CH_4.

Cancer, Aging, and Free Radicals Ultraviolet radiation and the generation of free radicals also have roles in the development of skin cancer. Remember that nearly all free radicals are inherently unstable because they lack outer octets. Free radicals are thus rogue species and trigger unwanted events in cells, some leading to cancer.

Atomic radiation is dangerous partly because some of the particles it generates in cells are radicals. The chemi-

FIGURE 1
Prolonged exposure to sunlight contributes to deep wrinkles.

cal changes that occur during aging, when muscles lose their suppleness and flexibility, involve the formation of free radicals that form not so much from exposure to radiation or sunlight but by natural processes involving peroxides, compounds with such general formulas as $R-O-O-H$ and $R-O-O-R'$. Their $O-O$ bonds break rather easily to give free radicals of the $R-O\cdot$ type that lead to the crosslinking of protein molecules. Heavy smoking and excessive exposure to sunlight (for example, by overtanning) also contribute, by means of free radical chemistry, to the deep wrinkling of the skin (see Figure 1). Vitamins A and C are known to scavenge and destroy free radicals and so they seem to provide some protection.

WHY EQUATIONS FOR ORGANIC REACTIONS CANNOT ALWAYS BE BALANCED

Chemical reactions involve the breaking and reforming of bonds. Because organic molecules have several bonds of similar strengths, several products can sometimes form, even when we are trying to make just one of them. The substances we don't want are called the *by-products,* and the reactions that produce by-products are called *side reactions.* The reaction that produces the largest relative quantity of product is called the *main reaction.* Naturally, the chemist hopes this is the reaction that produces the desired product. In any case, most organic reactions produce a set of products, a mixture that must then be separated into the constituents. This often takes more time and effort than any other aspect of the synthesis.

As an example of a reaction that produces a mixture, we can use the chlorination of ethane. If we were to mix 1 mol of chlorine with 1 mol of ethane, naively trying to prepare 1 mol of ethyl chloride, we would obtain a mixture of mono-, di-, tri-, and possibly still more highly chlorinated molecules (plus HCl).

$$CH_3CH_3 + Cl_2 \xrightarrow{\text{ultraviolet light}} CH_3CH_2Cl + CH_3CHCl_2$$
$$+ ClCH_2CH_2Cl + CH_3CCl_3 + ClCH_2CHCl_2 + \text{etc.} + HCl$$

It would be foolish to write coefficients in front of any of the products or to change the coefficients in front of the reactants—which, as they stand, say that a 1:1 molar ratio was taken—in an effort to balance the equation. If the mixture were to be completely separated, then the molar percentages of the various products could be reported, but seldom is such a thorough separation performed when just one product is sought.

Most organic reactions pose this kind of balancing problem, so the equation written is the equation for the main reaction only. Often it is balanced, but for just one reason—to provide a basis for the selection of the relative numbers of moles of the reactants to be mixed at the start. In nearly all the reactions that we will study, side reactions occur, which we generally will ignore.

and allowed to react, several reactions eventually occur, and a mixture forms of four chlorinated methanes, hydrogen chloride, plus some unreacted methane. It isn't possible to write a balanced equation, but what we can do is represent the reaction by a flow of symbols.

■ Notice how the boiling points of the compounds (in parentheses) increase with formula mass.

$$CH_4 + Cl_2 \xrightarrow[\text{HCl}]{} CH_3Cl \xrightarrow[\text{HCl}]{Cl_2} CH_2Cl_2 \xrightarrow[\text{HCl}]{Cl_2} CHCl_3 \xrightarrow[\text{HCl}]{Cl_2} CCl_4$$

Methane (bp −162 °C)   Methyl chloride (bp −24 °C)   Methylene chloride (bp 40 °C)   Chloroform (bp 61 °C)   Carbon tetrachloride (bp 77 °C)

For a discussion of the use of unbalanced reaction sequences to represent organic reactions, see Special Topic 12.3.

Methylene chloride, chloroform, and carbon tetrachloride are examples of *organochlorine compounds,* and all are used as nonpolar solvents. Chloroform has been used as an anesthetic.

■ Chloroform (bp 61 °C) was an anesthetic better suited for use in the tropics than diethyl ether (bp 35 °C) because of its higher boiling point.

Bromine reacts with methane by the same kind of substitution as chlorine. Iodine does not react. Fluorine combines explosively with most organic compounds at room temperature, and complex mixtures form.

The higher alkanes can also be chlorinated. Ethyl chloride, plus more highly chlorinated products, form by the chlorination of ethane, and ethyl chloride (bp 12.5 °C) is used as a local anesthetic. When it is sprayed on the skin, it evaporates very rapidly, and this cools the area enough to prevent the transmission of pain signals during minor surgery at the site.

When propane is chlorinated, both propyl chloride and isopropyl chloride form in roughly equal amounts.

$$CH_3CH_2CH_3 + Cl_2 \xrightarrow{\text{ultraviolet light}} CH_3CH_2CH_2Cl + CH_3\overset{\displaystyle Cl}{\underset{\displaystyle |}{CH}}CH_3$$

Propane   Propyl chloride   Isopropyl chloride

Besides these products, some higher chlorinated compounds also form.

PRACTICE EXERCISE 13

(a) How many monochloro compounds of butane are possible? Give both their common and IUPAC names. (Consider only the monochloro compounds with the formula C_4H_9Cl.) (b) How many monochloro derivatives of isobutane are possible? Write both their common and IUPAC names.

SUMMARY

Organic and inorganic compounds Most organic compounds are molecular and the majority of inorganic compounds are ionic. Molecular and ionic compounds differ in composition, in types of bonds, and in several physical properties.

Structural features of organic molecules The ability of carbon atoms to join to each other many times in succession—in straight chains, in branched chains, as well as in rings—accounts in large measure for the existence of several million organic compounds. The skeletons of the rings can be made entirely of carbon atoms or there may be one or more other nonmetal atoms (heterocyclic compounds).

Full structural formulas of organic compounds are usually condensed by grouping the hydrogens attached to a carbon immediately next to this carbon; by letting single bonds on a horizontal line be understood; and by leaving bond angles and conformational possibilities to the informed imagination. Skeletons of rings are usually represented by simple polygons. Free rotation about single bonds is possible in open-chain compounds but not in rings.

Compounds without multiple bonds are saturated. Those with double or triple bonds are unsaturated. Carbon–carbon double or triple bonds are never "understood" in structures.

The families of organic compounds are organized around functional groups, parts of molecules at which most of the chemical reactions occur. Nonfunctional units can sometimes be given the general symbol R, as in ROH, the general symbol for all alcohols. These R groups are hydrocarbon-like groups.

Isomerism Differences in the conformations of carbon chains do not create new compounds, but differences in the organizations of parts do. Isomers are compounds with identical molecular formulas but different structures. Constitutional isomers make up one kind of isomer, those whose molecules have different atom-to-atom connectivities. Sometimes constitutional isomers are in the same fam-

ily, like butane and isobutane. Often they are not, like ethyl alcohol and dimethyl ether.

Hydrocarbons Hydrocarbons are compounds in which the only elements are carbon and hydrogen. The alkanes are saturated hydrocarbons; the alkenes and alkynes are unsaturated. The alkenes have at least one double bond. The alkynes have at least one triple bond. The aromatic hydrocarbons have a benzene ring, and the aliphatic hydrocarbons do not. Being nonpolar compounds, the hydrocarbons are all insoluble in water, and many mixtures of alkanes are common, nonpolar solvents. The rule like-dissolves-like lets us predict solubilities.

Nomenclature of alkanes In the IUPAC system, a compound's family is always indicated by a name ending, like *ane* for the alkanes. The number of carbons in the parent chain is indicated by a unique prefix for each number, like the prefix *but* to denote four carbons in *butane*. The locations of side chains or other kinds of atoms or groups are specified in the final name by numbers assigned to the carbons to which they are attached. The numbering of the parent chain is done in the direction that locates the first branch at the lower of two possible numbers.

Alkane-like substituents are called alkyl groups, and the names and formulas of those having from one to four carbon atoms must be learned. Common names are still popular, particularly when the IUPAC names are long and cumbersome.

Chemical properties of alkanes Alkanes and cycloalkanes are generally unreactive at room temperature toward concentrated acids and bases, toward oxidizing and reducing agents, toward even the most reactive metals, and toward water. They burn, giving off carbon dioxide and water, and in the presence of ultraviolet light (or at a high temperature) they undergo useful substitution reactions with chlorine and bromine.

REVIEW EXERCISES

The answers to Review Exercises whose numbers are in color are found in Appendix C. The answers to the other Review Exercises are found in the Study Guide that accompanies this book. The more challenging questions are marked with asterisks.

Organic and Inorganic Compounds

12.1 In terms of *bonding abilities,* what is unique about the element carbon? How does this contribute to the huge *number* of possible organic compounds?

12.2 With respect to the *synthesis* of organic compounds, what specifically was the problem that organic chemists faced prior to 1828? What scientific theory had been devised to meet this problem?

12.3 What was Wöhler's goal when he evaporated an aqueous solution of ammonium cyanate to dryness? What happened instead? With respect to *scientific theory* at the time, what specifically did Wöhler accomplish?

12.4 What kind of bond between atoms predominates among organic compounds?

12.5 How many single bonds are observed in neutral molecules at each of the following atoms?
(a) C   (b) O   (c) N   (d) H   (e) Cl

12.6 Which of the following compounds are inorganic?
(a) CH_3CH_2OH (b) CO_2 (c) $CHCl_3$ (d) $KHCO_3$
(e) Na_2CO_3

12.7 Are the majority of all compounds that dissolve in water ionic or molecular? Inorganic or organic?

12.8 Explain why very few organic compounds can conduct electricity either in an aqueous solution or as molten materials.

*12.9** Each compound described below is either ionic or molecular. State which it most likely is, and give one reason.
(a) A compound that melts at 281 °C, and burns in air.
(b) A compound that dissolves in water. When hydrochloric acid is added, the solution fizzes and an odorless, colorless gas is released, which can extinguish a burning flame.
(c) A compound that is a colorless gas at room temperature.
(d) A compound that melts at 824 °C and becomes white when heated.
(e) A compound that is a liquid and does not dissolve in water but does burn.
(f) A compound that is a liquid and does dissolve in water as well as burns.

Structural Features of Organic Molecules

12.10 One can write the structure of propane, a common heating gas, as follows.

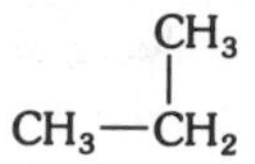

$$CH_3-CH_2 \quad \overset{CH_3}{|} \quad \text{Propane}$$

Are propane molecules properly described as straight chain or as branched chain, in the sense in which we use these terms? Explain.

12.11 Which of the following structures are possible, given the numbers of bonds that various atoms can form?
(a) $CH_3CH_2CH_2OCH_3$
(b) $CH_2CH_2CH_2CH_4$
(c) $CH_3{=}CHCH_2CH_3$
(d) $CH_3CH{=}CHCH_2CH_3$
(e) $NH_2CH_2CH_2CH_3$

*12.12** Write full (expanded) structures for each of the following *molecular* formulas. Remember how many covalent bonds

nonmetals have in molecules: C, 4; N, 3; O, 2; H, Cl, and Br, 1 each. In some structures you will have to use double or triple bonds. (*Hint:* A trial-and-error approach will have to be used.)
(a) CH_4O (b) CH_2Cl_2 (c) N_2H_4 (d) C_2H_6
(e) CH_2O (f) CH_2O_2 (g) NH_3O (h) C_2H_2
(i) $CHCl_3$ (j) HCN (k) C_2H_3N (l) CH_5N

12.13 Expand the following structure of nicotinamide, one of the B-vitamins.

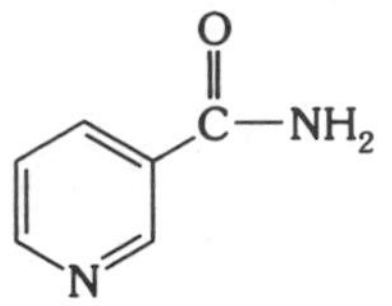

Nicotinamide

12.14 Expand the structure of thiamine, vitamin B_1. Notice that one nitrogen has four bonds so it has a positive charge.

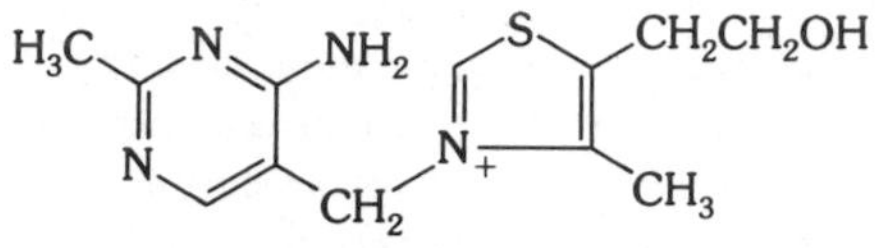

Thiamine

12.15 Write neat, condensed structures of the following.

(a)

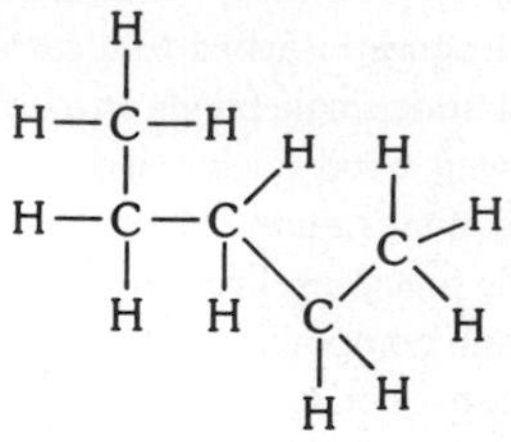

(b)

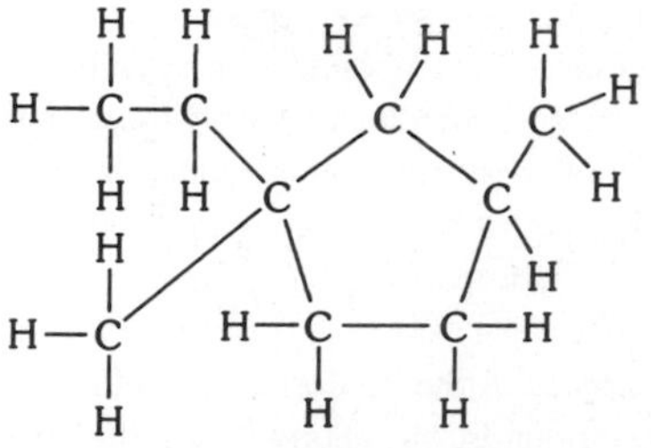

(c)

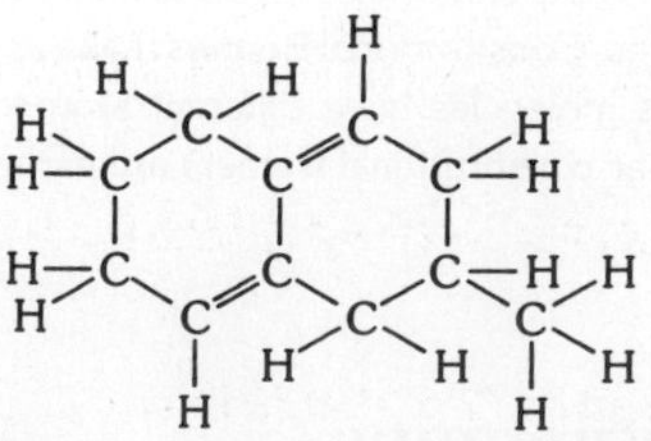

12.16 Why is the topic of molecular shape important at the molecular level of life?

12.17 What kinds of orbitals overlap when the C—C bond forms in ethane?

12.18 Free rotation can occur about single bonds (in open-chain structures) without breaking or weakening the bonds. Why is this possible?

12.19 Which of the following structures represent unsaturated compounds?

(a)

(b) O—O ring structure

(c) $CH_3CCH_2CH_3$ (with C=O)

(d) $CH_3C{\equiv}CCH_3$

12.20 Which compounds are saturated?

(a) cyclopentene structure

(b) $CH_3CH{=}NCH_3$

(c) C_4H_{10}

(d)

OH (cyclohexane)

(e) CO_2H / $OCCH_3$ structure

Aspirin

Cyclohexanol

*12.21 Decide whether the members of each pair are identical, are isomers, or are unrelated.

(a) CH_3 and $CH_3{-}OH$

    OH

(b) 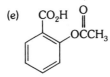 CH_3 / OH / CH / CH_3 and CH_3 / $CH{-}OH$ / CH_3

(c) CH_3CH_2SH and $CH_3CH_2CH_2SH$

(d) $CH_3CH{=}CH_2$ and $CH_2{-}CH_2$ / CH_2 *Isomer*

(e) $CH_3CCH_2CH_3$ and $CH_3CH_2CCH_3$ *Same*

(f) CH_3CHCH_3 and CH_3CH / CH_3 *Same*
    CH_3 CH_3

(g) $CH_3CH_2CH_2NH_2$ and $CH_3NHCH_2CH_3$ *Isomer*

Acid organic (h) CH_3CH_2COH and $HOCH_2CCH_3$ *Isomer Kiton*

Aldhyde (i) $HCOCH_2CH_3$ and *ACi* CH_3CH_2COH *Isomer*

(j) $HCOCH_2CH_2OH$ and $HOCH_2CH_2COH$ *Isomer*

Alcohol (k) $CH_3OCH_2CCH_3$ and *keton* $CH_3CH_2COCH_3$ *ACid*

Aster *Isomers* *Aster*

Ether

(l) $CH_3{-}CH{-}CH_2{-}CH_3$
   $CH_2{-}CH_2$ CH_3 CH_3
         $CH_2{-}C{-}CH$
           CH_3 CH_3

and

    CH_3 CH_3 CH_3
$CH_3{-}CH{-}CH_2{-}CH_2{-}CH_2{-}CH_2{-}C{-}CH{-}CH_3$
                          CH_3

(m) 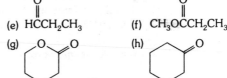 cyclohexanone-type ${=}O$ and methylcyclohexanone structure

(n) $HO{-}O{-}CH_3$ and $HO{-}O{-}CH_2{-}OCH_3$

Families of Organic Compounds

12.22 Name the family to which each compound belongs.

(a) $CH_3CH_2CH_3$

(b) $HOCH_2CH_2CH_3$

(c) cyclopentane with SH

(d) $CH_3C{\equiv}CH$

(e) $HCCH_2CH_3$ (with C=O)

(f) $CH_3OCCH_2CH_3$ (with C=O)

(g) lactone ring structure O / O

(h) cyclohexanone ${=}O$

(i) $CH_3CH_2CH_2NH_2$

(j) $CH_3OCH_2CH_3$

*12.23 Name the families to which the compounds in parts (a)–(m) of Practice Exercise 12.21 belong. (A few belong to more than one family.)

Physical Properties and Structure

12.24 Which compound must have the higher boiling point? Explain.

    $CH_3CH_2CH_3$ $CH_3CH_2CH_2CH_2CH_3$
       A   **B**

12.25 Which compound must be less soluble in gasoline? Explain.

  $ClCH_2CH_2CH_2CH_2CH_2Cl$ $HOCH_2CH_2CH_2CH_2CH_2OH$
        A   **B**

12.26 Suppose that you are handed two test tubes containing colorless liquids, and you are told that one contains pentane and the other holds hydrochloric acid. How can you use just water to tell which tube contains which compound without carrying out any chemical reaction?

12.27 Suppose that you are given two test tubes and are told that one holds methyl alcohol, CH_3OH, and the other hexane. How can water be used to tell these substances apart without carrying out any chemical reaction?

Nomenclature

12.28 There are five isomers of C_6H_{14}. Write their condensed structures and their IUPAC names.

12.29 Which of the isomers of hexane (Review Exercise 12.28) has the common name *n*-hexane? Write its structure.

12.30 Which of the hexane isomers (Review Exercise 12.28) has the common name isohexane? Write its structure.

*12.31 There are nine isomers of C_7H_{16}. Write their condensed structures and their IUPAC names.

12.32 Write the condensed structures of the isomers of heptane (Review Exercise 12.31) that have the following names.
(a) *n*-heptane (b) isoheptane

*12.33 Write the IUPAC names of the following compounds.

(a)

(b)

12.34 Write condensed structures for the following compounds.
(a) *n*-Butyl bromide (b) Isohexyl iodide
(c) *sec*-Butyl chloride (d) Isopropylcyclohexane

12.35 Write condensed structures for the following compounds.
(a) Propyl chloride (b) Isobutyl iodide
(c) *t*-Butyl bromide (d) Ethyl bromide

*12.36 The following are incorrect efforts at naming certain compounds. What are the most likely condensed structures and correct IUPAC names?
(a) 1,6-Dimethylcyclohexane (b) 2,4,5-Trimethylhexane
(c) 1-Chloro-*n*-butane (d) Isopropane

12.37 The following names cannot be the correct names, but it is still possible to write structures from them. What are the correct IUPAC names and the condensed structures?
(a) 1-Chloroisobutane (b) 2,4-Dichlorocyclopentane
(c) 2-Ethylbutane (d) 1,3-6-Trimethylcyclohexane

Reactions of Alkanes

12.38 Write the balanced equation for the complete combustion of heptane, a component of gasoline.

12.39 Gasohol is a mixture of ethyl alcohol, CH_3CH_2OH, in gasoline. Write the equation for the complete combustion of ethyl alcohol.

12.40 What are the formulas and common names of all the compounds that can be made from methane and chlorine?

12.41 There are two isomers of $C_2H_4Cl_2$. What are their structures and IUPAC names?

*12.42 When propane reacts with chlorine, in addition to the two isomeric monochloropropanes, some dichloropropanes also form. Write the structures and the IUPAC names for all these possible dichloropropanes.

Fossil Fuels (Special Topic 12.1)

12.43 What are the three principal fossil fuels being used today?

12.44 What is the difference between petroleum, crude oil, and natural gas?

12.45 In general terms, describe what happened to change peat into lignite, lignite into soft coal, and soft coal into hard coal.

12.46 What does *fraction* mean in connection with oil refining?

12.47 What kinds of compounds predominate in the crude oil fractions that boil below 200 °C?

12.48 How do petroleum refineries increase the supply of gasoline?

Free Radical Reactions (Special Topic 12.2)

12.49 Why is the chlorine atom called a free radical but the chlorine molecule is not?

12.50 The bromination of methane proceeds by a series of steps just like those of the chlorination of methane.
(a) Write the equation for the reaction that initiates chains.
(b) Write the two equations that constitute the chain reaction.
(c) Write two equations that illustrate how chains can be broken.

*12.51 The explosive gas-phase reaction of H_2 with Cl_2 that makes HCl is a free radical chain reaction.
(a) Write the balanced equation for the reaction.
(b) It is initiated by the breaking of the bond in Cl_2. Write the equation for this chain-initiating reaction as well as the equations for the chain reaction itself.

On Balancing Organic Reactions (Special Topic 12.3)

12.52 Explain in your own words why it is impossible to write a balanced equation for what actually happens when methane is chlorinated.

Additional Exercises

*12.53 When cyclohexane is chlorinated, how many monochloro derivatives are possible? Write the name or names and structure or structures.

*12.54 What reaction, if any, will cyclohexane give with each reactant at room temperature?
(a) Aqueous NaOH
(b) Concentrated sulfuric acid
(c) Iodine

12.55 What is the name of the butyl group with nine equivalent hydrogen atoms?

*12.56 If 7.46 g of cyclopentane is chlorinated, what is the maximum number of grams of chlorocyclopentane that can be obtained?

UNSATURATED HYDROCARBONS

13

People who climb mountains, such as this climber near the Half Dome of Yosemite National Park, routinely trust their lives to the strengths of synthetic fibers. The molecules of these synthetics are polymers, and the principles behind their formation are one topic in this chapter.

13.1

OCCURRENCE

The carbon–carbon double bond occurs in nature largely in compounds having other functional groups.

■ The carbon–carbon double bond is sometimes called the *ene* function.

Unsaturation in hydrocarbons occurs as double bonds, triple bonds, or benzene rings. Hydrocarbons with double bonds are called **alkenes** and the carbon–carbon double bond is the **alkene group.** The double bond is common at the molecular level of life, particularly in the edible fats and oils and in related compounds that make up most of a cell membrane. Special Topic 13.1 briefly describes a few of the large number of compounds with one or more alkene groups. This chapter is mostly about the properties of the alkene group *wherever it occurs,* whether in simple alkenes or in polyfunctional compounds (those having two or more functional groups).

The carbon–carbon triple bond occurs in hydrocarbons called **alkynes** but is uncommon in nature. We will use Special Topic 13.4 later in the chapter to discuss it briefly. The *benzene ring* (Section 13.7) and rings similar to it occur in proteins and nucleic acids, and it is widely present in pharmaceuticals like aspirin (structure on page 407).

Table 13.1 shows the structures and some physical properties of several alkenes. As with the alkanes, the first four alkenes are gases at room temperature, and all are much less dense than water. Like all hydrocarbons, alkenes are insoluble in water and soluble in nonpolar solvents for reasons we discussed in Section 12.1.

TABLE 13.1
Properties of Some 1-Alkenes

Name (IUPAC)	Structure	BP (°C)	MP (°C)	Density (g/mL)[a]
Ethene	$CH_2\!=\!CH_2$	−104	−169	—
Propene	$CH_2\!=\!CHCH_3$	−48	−185	—
1-Butene	$CH_2\!=\!CHCH_2CH_3$	−6	−185	—
1-Pentene	$CH_2\!=\!CHCH_2CH_2CH_3$	30	−165	0.641
1-Hexene	$CH_2\!=\!CHCH_2CH_2CH_2CH_3$	64	−140	0.673
1-Heptene	$CH_2\!=\!CHCH_2CH_2CH_2CH_2CH_3$	94	−119	0.697
1-Octene	$CH_2\!=\!CHCH_2CH_2CH_2CH_2CH_2CH_3$	121	−102	0.715
1-Nonene	$CH_2\!=\!CHCH_2CH_2CH_2CH_2CH_2CH_2CH_3$	147	−81	0.729
1-Decene	$CH_2\!=\!CHCH_2CH_2CH_2CH_2CH_2CH_2CH_2CH_3$	171	−66	0.741
Cyclopentene		44	−135	0.722
Cyclohexene		83	−104	0.811

[a] At 10 °C.

THE ALKENE DOUBLE BOND IN NATURE

Molecules with alkene groups are found among most of the major families of biochemically important substances. Cholecalciferol (vitamin D_3) and retinol (vitamin A_1), for example, are two of the vitamins that have several alkene groups. This group renders both vitamins sensitive to conditions that favor oxidation, such as prolonged standing in air, particularly if it is warm. Food sources of these vitamins, therefore, should not be overcooked. Because the double bond carries two electrons in its pi bond, it has more electron density than a single bond. The defining property of an oxidizing agent is its attraction to electron-rich, electron-donating sites. With increased electron density, the alkene group is thus open to attack by oxidizing agents.

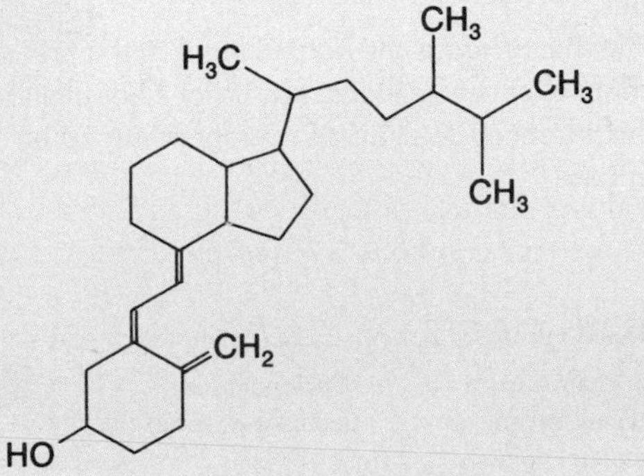

Cholecalciferol (Vitamin D_3)

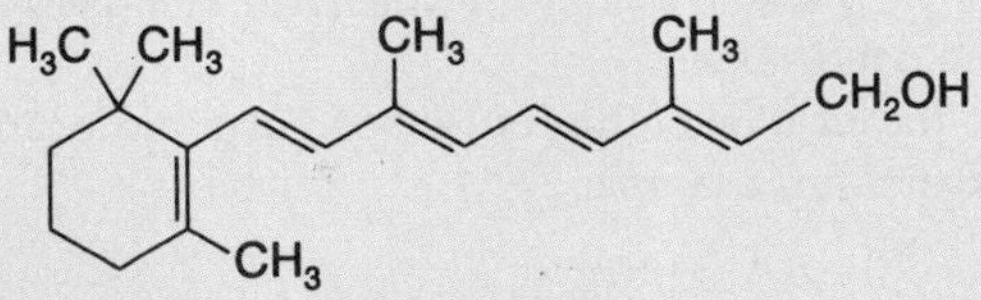

Retinol (Vitamin A_1)

β-Carotene is one of the bright yellow-orange compounds in carrots (Figure 1) and is called a *provitamin* because the body is able to convert it into vitamin A. You can see that β-carotene has eleven alkene groups and no other functional group. Compounds like this have the general name of *polyene*. By comparing the structures of β-carotene and retinol, you can see how the body can obtain the basic retinol system from β-carotene.

Several of the human sex hormones have alkene groups, including testosterone, the chief male sex hormone. It is classified as a *steroid* hormone because it has the special four-ring system shown in its structure and called the steroid system.

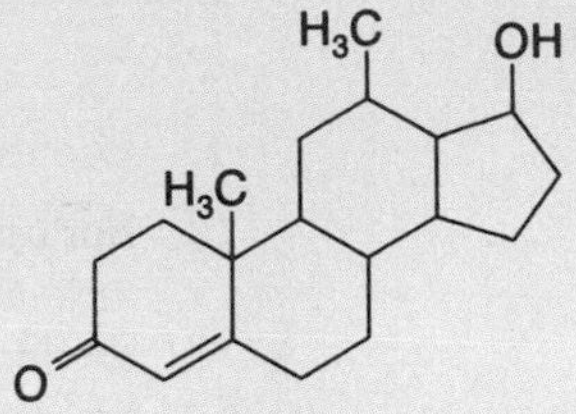

Testosterone (male sex hormone)

The edible fats and oils all have at least one alkene group per molecule, and the vegetable oils generally have two to four. Shown here is the structure of a molecule found in corn oil, a typical vegetable oil and a product widely used to make salad dressing. You can see why vegetable oils are described as *polyunsaturated;* their molecules have several unsaturated sites (specifically meaning alkene groups rather than the carbon–oxygen double bonds). In animal fats like butterfat and tallow (beef fat), the number of double bonds per molecule is smaller so they are more "saturated." Otherwise, animal fats are structurally like the vegetable oils.

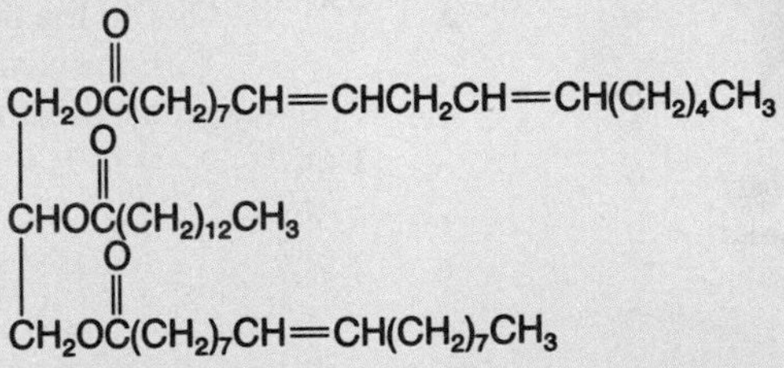

A typical molecule in a vegetable oil

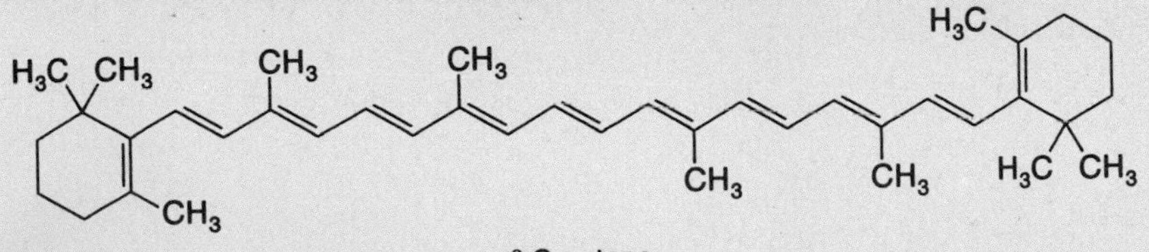

β-Carotene

13.2

NAMING THE ALKENES

The IUPAC names of alkenes end in -ene, and the double bond takes precedence over substituents in numbering the parent chain.

Consistent with the IUPAC rules for any family, the rules for the alkenes specify the *name ending,* how to pick out the *parent chain* or *parent ring,* how to *number the chain or ring,* and how to designate substituent groups. For alkenes and cycloalkenes, the IUPAC rules are as follows.

1. Use the ending *-ene* for all alkenes and cycloalkenes.

2. For open-chain alkenes, identify the parent chain as the longest sequence of carbons that *includes the double bond.* Name this chain as if it were that of an alkane and then change the *ane* ending to *ene.* This gives the basic *name* of the parent chain, except that the location of the double bond is yet to be specified. For cyclic alkenes, the ring is the parent in all situations we will encounter.

   For example, the longest chain *that includes the double bond* in the first structure has six carbons. There is a longer chain of seven carbons, but it does not include the double bond. In the second structure, the parent cycloalkane is cyclopentane, so by changing the ending to *ene* we have cyclopentene.

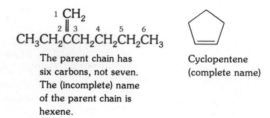

The parent chain has six carbons, not seven. The (incomplete) name of the parent chain is hexene.

Cyclopentene (complete name)

3. For open-chain alkenes, number the parent chain from whichever end gives the lower number to the first carbon of the double bond.

   This rule gives precedence to the location of the double bond over the location of the first substituent on the parent chain. For example,

$$\underset{\substack{5\ \ \ 4\ \ \ 3\ \ \ 2\ \ \ \ 1 \\ 1\ \ \ 2\ \ \ 3\ \ \ 4\ \ \ \ 5}}{\overset{\overset{\displaystyle CH_3}{|}}{CH_3CHCH_2CH\!=\!CH_2}}$$ The double bond is at position 1, not 4.

Not

4-Methyl-1-pentene (complete name)
Not: 2-methyl-4-pentene

4. For cycloalkenes, always give position 1 to one of the two carbons at the double bond. To decide which carbon gets this number, number the ring atoms from carbon 1 *through the double bond* in whichever direction reaches a substituent first.

   For example, the numbers inside the ring represent the correct numbering, not the numbers outside the ring.

3-Methylcyclohexene (complete name)
Not: 6-methylcyclohexene

5. To the name begun with rules 1 and 2, place the number that locates the first carbon of the double bond as a prefix, and separate this number from the name by a hyphen.

6. If substituents are on the parent chain or ring, complete the name obtained by rule 5 by placing the names and location numbers of the substituents as prefixes.

   Remember to separate numbers from numbers by commas, but use hyphens to connect a number to a word.

Several correctly named alkenes are shown next[1] with the common names of three given in parentheses. The ending *-ylene* characterizes the common names of open-chain alkenes.

$$CH_2\!=\!CH_2 \qquad CH_3CH\!=\!CH_2 \qquad \overset{\displaystyle CH_3}{\underset{}{CH_3C\!=\!CH_2}}$$

Ethene   Propene   2-Methylpropene
(ethylene)   (propylene)   (isobutylene)

■ The common name, ethylene, is also allowed by the IUPAC as the name for ethene.

$$\overset{\displaystyle CH_3}{CH_3CH_2CHCH_2CH}\!=\!\overset{\displaystyle CH_3}{CCH_3} \qquad CH_3CH_2CH_2C\!=\!CH_2 \quad \overset{\displaystyle CH_3}{CHCH_2CH_2CH_3}$$

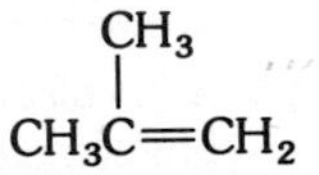

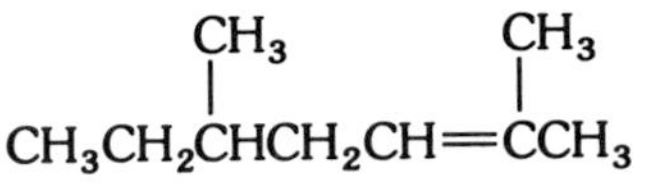

2,5-Dimethyl-2-heptene 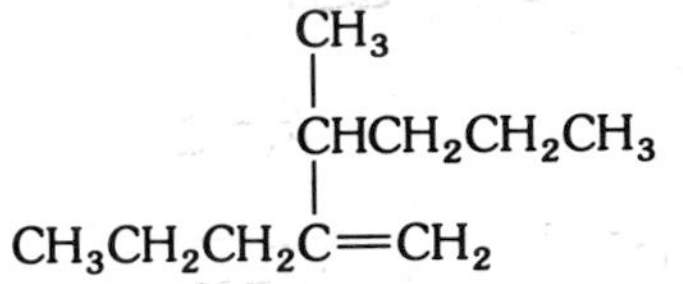   3-Methyl-2-propyl-1-hexene

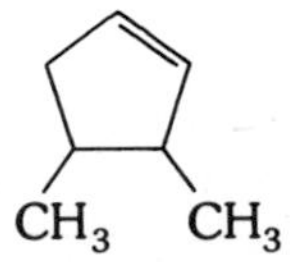

3,4-Dimethylcyclopentene
Not 4,5-dimethylcyclopentene

$$Cl(CH_2)_6CH\!=\!CH_2$$

8-Chloro-1-octene

No number is used to locate the double bond in 3,4-dimethylcyclopentene because, by rule 4, the double bond can only be at position 1.

EXAMPLE 13.1
Naming an Alkene

Write the name of the following alkene.

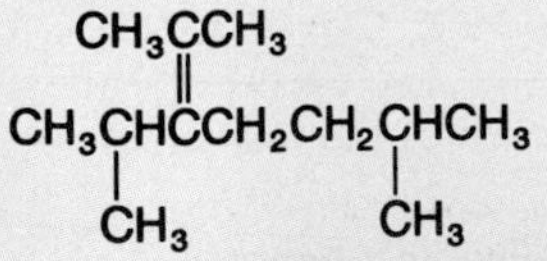

$$CH_3CCH_3$$
$$CH_3CHCCH_2CH_2CHCH_3$$
$$\overset{}{\underset{CH_3}{|}} \qquad \overset{}{\underset{CH_3}{|}}$$

ANALYSIS The longest chain that includes the double bond must be identified and numbered from whichever end gives the first carbon of the double bond the lower of two possible numbers. The parent is an *alkene* with a chain of seven carbons, a heptene. The following numbering of the parent chain gives the double bond position 2. (The alternative numbering, right to left, would have given the double bond position 5.)

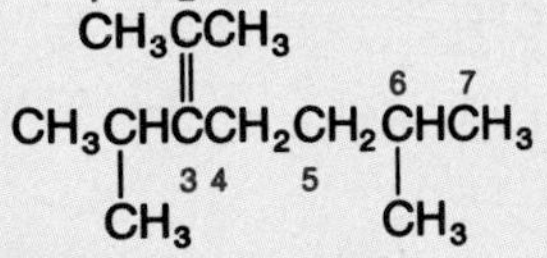

$$\overset{1\quad 2}{CH_3CCH_3}$$
$$\overset{}{\underset{\underset{CH_3}{|}\;\; 3\;4 \qquad 5 \qquad \underset{CH_3}{|}}{CH_3CHCCH_2CH_2CHCH_3}}\;^{6\;\;7}$$

[1] Propene is not named 1-propene because by Rule 3 there is no 2-propene. 2-Methylpropene is not named 2-methyl-1-propene because the 1 is not needed.

The parent alkene is thus 2-heptene. It holds two methyl groups (positions 2 and 6) and one isopropyl group (position 3). The names and location numbers are next assembled into the name.

SOLUTION   The correct name is

3-Isopropyl-2,6-dimethyl-2-heptene

A comma separates two numbers

Hyphens separate numbers from names

CHECK   The most common error that students make is *to fail to find the longest chain* that includes the double bond. The first check step, then, is to go back over the answer to see if there is a chain holding the alkene group that is longer than 7 carbons. Use a colored pen to draw an enclosure for this chain so that all substituents are outside. Then move in from either end of the chain, counting carbons, to see which starting point yields the lower number for the beginning of the double bond. All the other numbers must then fall into place. Another common error is failure to identify alkyl groups correctly, so double-check these.

PRACTICE EXERCISE 1

Write the IUPAC names for the following compounds.

(a) 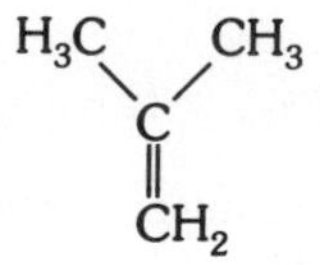
$$H_3C \quad CH_3$$
$$C$$
$$CH_2$$

(b) 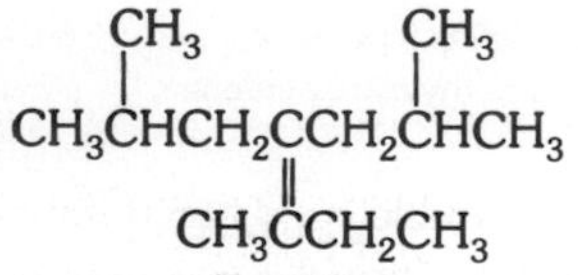
$$CH_3 \qquad CH_3$$
$$CH_3CHCH_2CCH_2CHCH_3$$
$$CH_3CCH_2CH_3$$

(c) $CH_3CH{=}CHCl$

(d) $BrCH_2CH{=}CH_2$

(e) 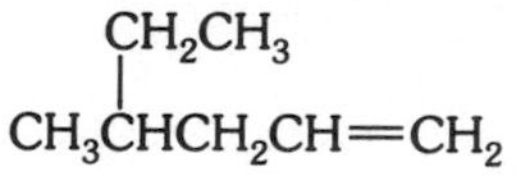
$$CH_2CH_3$$
$$CH_3CHCH_2CH{=}CH_2$$

(f) 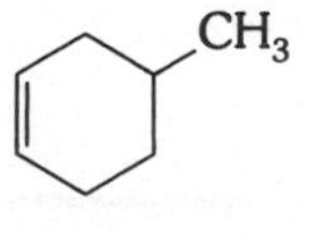
$$CH_3$$

PRACTICE EXERCISE 2

Write condensed structures for each of the following.

(a)   4-Methyl-2-pentene   (b)   3-Propyl-1-heptene

(c)   4-Chloro-3,3-dimethyl-1-butene   (d)   2,3-Dimethyl-2-butene

When a compound has two double bonds, it is named as a *diene* with two numbers in the name to specify the locations of the double bonds. For example,

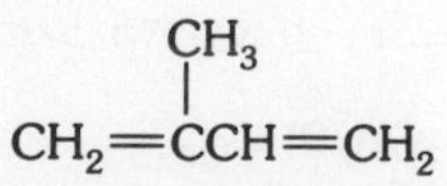

$$CH_3$$
$$CH_2{=}CCH{=}CH_2$$

2-Methyl-1,3-butadiene 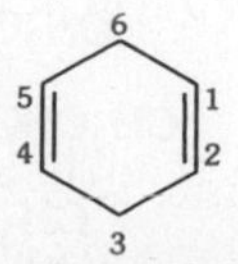   1,4-Cyclohexadiene

This pattern can be easily extended to *trienes, tetraenes,* and so forth.

13.3

GEOMETRIC ISOMERS

The alkenes and cycloalkanes can exhibit geometric isomerism because there is no free rotation at the double bond or in a ring.

The six atoms at a double bond, the two carbons and the four atoms attached to them, all lie in the same plane, as illustrated in Figure 13.1, which shows the simplest alkene. The bond angles are 120°.

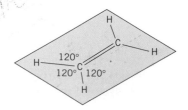

FIGURE 13.1
The geometry at a carbon–carbon double bond.

Some Alkene Isomers Have Identical Constitutions but Different Geometries

Alkenes can exist as isomers in three ways. They can have different carbon skeletons, as shown by 1-butene and 2-methylpropene. These two compounds are *constitutional isomers*. Two alkenes can also have identical carbon skeletons but differ in the locations of their double bonds, as in 1-butene and 2-butene. These two are also constitutional isomers because their H atoms are attached differently to the skeleton.

$$CH_2{=}CHCH_2CH_3 \qquad CH_2{=}\underset{\underset{CH_3}{|}}{\overset{\overset{CH_3}{|}}{C}}CH_3 \qquad CH_3CH{=}CHCH_3$$

1-Butene   2-Methylpropene   2-Butene

Some alkenes have identical constitutions *including the location of the double bond,* but differ only in the geometry at this bond, as seen in the two isomers with the constitution of 2-butene, $CH_3CH{=}CHCH_3$.

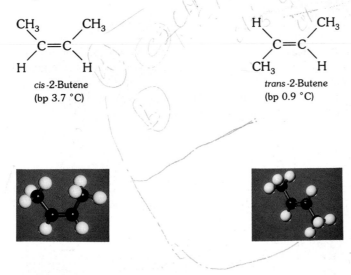

cis-2-Butene
(bp 3.7 °C)

trans-2-Butene
(bp 0.9 °C)

cis-2-Butene and *trans*-2-butene differ only in the *directions* taken by their end-of-chain methyl groups. Isomers that differ in geometry but have identical constitutions are called **geometric isomers,** and the phenomenon is called **geometric isomerism.**

Geometric Isomers Are Possible because There Is No Free Rotation at the Double Bond

Recall (Section 5.9) that one of the bonds at a double bond is a pi bond. It results from the side-to-side overlap of two unhybridized 2*p* orbitals, one on each carbon (see Fig. 13.2). The rotation of two groups joined by a double bond could happen only if the pi bond breaks. This ordinarily costs too much energy, making geometric isomers possible.

FIGURE 13.2
Overlapping 2p orbitals form the pi bond at a double bond.

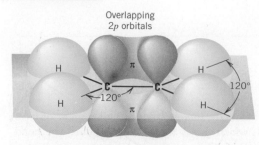

Overlapping 2p orbitals

■ The *side* of a double bond is not the same as the *end* of a double bond.

When two designated substituents are on the same side of the double bond, they are said to be *cis* to each other. When they are on opposite sides, they are *trans* to each other. (Sometimes geometric isomerism is called *cis–trans isomerism*.) The designations of cis and trans can be made parts of the names of geometric isomers, as the examples of *cis-* and *trans-*2-butene earlier showed.

When There Are Two Identical Groups at One End of a Double Bond, Geometric Isomers Are Not Possible If one end of a double bond has two *identical* groups, like two Hs or two methyls, then there is nothing for a group at the other end to be *uniquely* cis or trans to. 1-Butene, for example, has two H atoms at one end of its double bond, so the ethyl group at the other end cannot be positioned to give geometric isomers. We can *write* structures that might appear to be isomers:

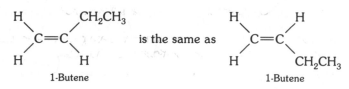

is the same as

1-Butene   1-Butene

1-Butene

However, they are actually identical. Simply flop the *whole* first structure over, top to bottom, to obtain the second. Whole-molecule flopping, of course, does not reorganize bonds into any new structure or geometry. Thus, there are no geometric isomers of 1-butene.

Geometric isomerism also occurs when the atoms involved at the ends of the double bond are halogen atoms or other groups, for example:

trans-1-Chloro-1-propene   *cis*-1-Chloro-1-propene

EXAMPLE 13.2
Writing the Structures of Cis and Trans Isomers

Write the structures of the cis and trans isomers, if any, of the following alkene.

$$CH_3CH=CHCH_2CH_3$$
2-Pentene

ANALYSIS Notice first that geometric isomerism is possible in 2-pentene. At *neither* end of the double bond are the two groups identical. At one end there are H and CH_3; at the other end, H and CH_2CH_3. Therefore, we must draw structures of the geometric isomers.

To show the geometry of each isomer correctly, we start by writing a carbon–carbon double bond without any attached groups *spreading the single bonds at the carbon atoms at angles of about 120°*.

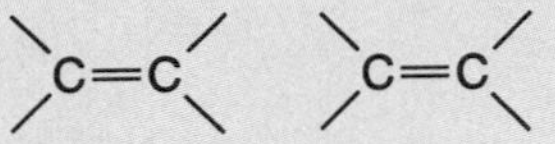

$$\diagdown \!\! C \!\!=\!\! C \!\! \diagdown \qquad \diagdown \!\! C \!\!=\!\! C \!\! \diagdown$$

Then we attach the two groups that are at one of the ends of the double bond. We attach them *identically* to make identical partial structures.

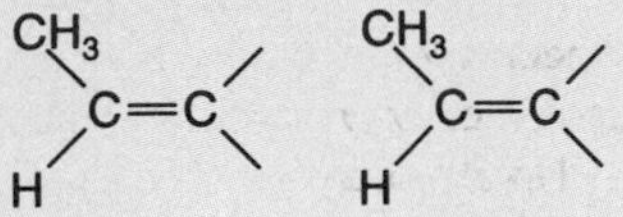

$$\underset{H}{\overset{CH_3}{\diagdown}} C \!\!=\!\! C \diagdown \qquad \underset{H}{\overset{CH_3}{\diagdown}} C \!\!=\!\! C \diagdown$$

Finally, at the other end of the double bond, we draw the other two groups, only this time be sure that they are switched in their relative positions.

SOLUTION   The geometric isomers of 2-pentene are

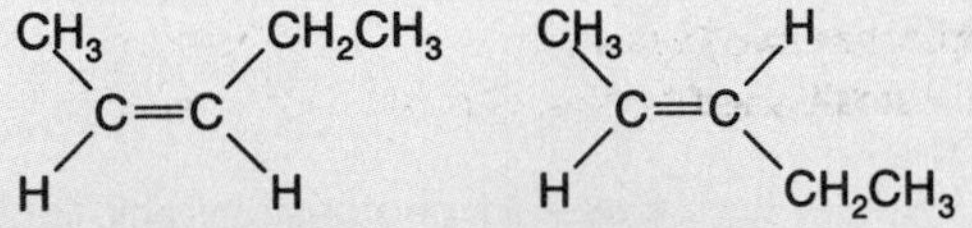

$$\underset{H}{\overset{CH_3}{\diagdown}} C \!\!=\!\! C \underset{H}{\overset{CH_2CH_3}{\diagup}} \qquad \underset{H}{\overset{CH_3}{\diagdown}} C \!\!=\!\! C \underset{CH_2CH_3}{\overset{H}{\diagup}}$$

$$\text{\textit{cis}-2-Pentene} \qquad\qquad \text{\textit{trans}-2-Pentene}$$

CHECK   Be sure to check whether the two structures are geometric *isomers* and not two identical structures that are merely flip-flopped on the page.

PRACTICE EXERCISE 3

Write the structures of the cis and trans isomers, if any, of the following compounds.

(a) $CH_3CH_2\underset{\underset{CH_3}{|}}{C}\!\!=\!\!CHCH_3$ (b) $ClCH\!\!=\!\!CHCl$ (c) $CH_3\underset{\underset{CH_3}{|}}{C}\!\!=\!\!CH_2$ (d) $Cl\underset{\underset{Cl}{|}}{C}\!\!=\!\!CHBr$

Cyclic Compounds Can Also Have Geometric Isomers

The double bond is not the only source of restricted rotation; the ring is another. For example, two geometric isomers of 1,2-dimethylcyclopropane are known, and neither can be twisted into the other without breaking the ring open. This costs too much energy to occur spontaneously even at quite high temperatures.

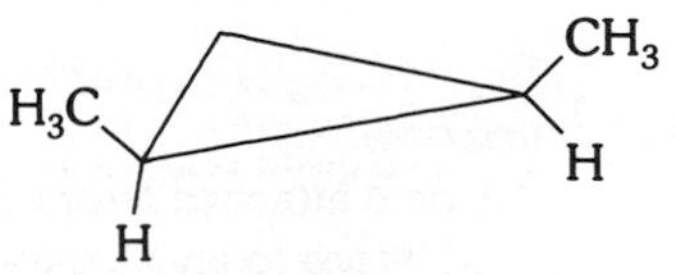

cis-1,2-Dimethyl-
cyclopropane
(bp 37 °C)

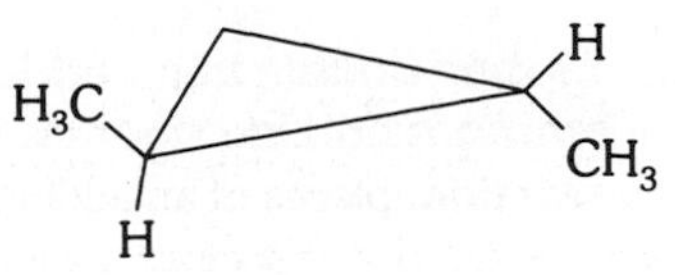

trans-1,2-Dimethyl-
cyclopropane
(bp 28 °C)

We'll see this kind of cis–trans isomerism in the many cyclic structures of carbohydrates; their *geometric* differences alone make most carbohydrates unusable in human nutrition. As we have said before, molecular geometry is as important in living processes as functional groups. For an interesting occurrence of cis–trans isomerism at the molecular level of life, see Special Topic 13.2.

The retina in the eye of a human being has two kinds of cells, rods and cones, that can convert absorbed light into signals which the brain perceives as vision. Rods equip us to see shades of gray; cones give us color vision. We know more about how rods work than cones, so our brief discussion is limited to what happens in the rods when they absorb light.

In Special Topic 5.1, page 120, light was described as a stream of tiny energy packets called *photons* having a wide range of energies. Photons of the higher energy regions of ultraviolet light have sufficient energy to break sigma bonds. Photons in visible light have less energy, but still enough to break the weaker pi bonds, particularly in molecules with repeating patterns of alternating double and single bonds. Just such a sequence of alternating double and single bonds occurs in a polyunsaturated compound called 11-*cis*-retinal, the primary absorber of visible light photons in the rods of our eyes. Notice that the chain emerges on the same side or the *cis* side of the double bond at carbons numbered 11 and 12 (according to the numbering used for this system). Otherwise the structures are the same.

If you compare the structures of 11-*cis*- and all-*trans*-retinal to the structure of vitamin A, Special Topic 13.1, you will notice a remarkable similarity. The body needs vitamin A to make the retinals.

Molecules of 11-*cis*-retinal must be joined to a protein called *opsin* in order for the absorption of light photons to be coupled to a nerve signal. The carbon–oxygen double bond in 11-*cis*-retinal makes the joining possible. If we represent everything except the carbon–oxygen double bond of 11-*cis*-retinal by R, then the coupling to opsin can be represented by the structure R—CH=N—{opsin}. This material is called *rhodopsin*.

11-*cis*-Retinal

All-*trans*-retinal

As we said, photons of visible light can break certain pi bonds. When this occurs at a double bond, there is still the sigma bond to hold the molecule together. With the pi bond broken, however, *free rotation now becomes possible*. This is what happens when the vast carbon network of 11-*cis*-retinal in rhodopsin accepts a light photon. One side of the molecule flips to a *trans* configuration at the 11,12 linkage to form all-*trans*-retinal. This change in geometry affects the properties of the opsin portion of rhodopsin, and a series of chemical changes swiftly occurs that sends a signal to the brain. An equally swift set of chemical changes occurs in the cone of the eye to restore rhodopsin to its photon-accepting geometry. Thus cis–trans isomerization is at the heart of vision.

13.4

ADDITION REACTIONS OF THE DOUBLE BOND

The carbon–carbon double bond adds H_2, Cl_2, Br_2, HX, H_2SO_4, and H_2O, and it is attacked by strong oxidizing agents, including ozone.

In an **addition reaction,** pieces of an adding molecule become attached to opposite ends of the double bond which then becomes a single bond. All additions to an alkene double bond thus have the following features, where $X-Y$ is the adding molecule:

$$\diagdown\!\!C=C\!\!\diagup + X-Y \longrightarrow -\overset{|}{\underset{X}{C}}-\overset{|}{\underset{Y}{C}}-$$

We'll study a few examples and then see how addition reactions take place.

Hydrogen Adds to a Double Bond and Saturates It In the presence of a powdered metal catalyst, like powdered nickel or platinum metal, hydrogen adds to double bonds. The reaction, sometimes called **hydrogenation,** converts an alkene to an alkane as follows.

$$\text{C}=\text{C} + \text{H}-\text{H} \xrightarrow{\text{Ni catalyst}} -\overset{|}{\underset{|}{\text{C}}}-\overset{|}{\underset{|}{\text{C}}}-$$
$$\qquad\qquad\qquad\qquad\qquad \text{H}\ \ \text{H}$$

Specific examples are

$$CH_2\!=\!CH_2 + H\!-\!H \xrightarrow{\text{Ni catalyst}} \underset{\underset{H}{|}}{CH_2}\!-\!\underset{\underset{H}{|}}{CH_2} \ \text{ or } \ CH_3CH_3$$

Ethene   Ethane

3-Methylcyclo-pentene   + H_2 $\xrightarrow{\text{Ni catalyst}}$ Methylcyclopentane   or

Notice that each alkane product has the same carbon skeleton as the alkene used to make it.

The Net Effect of Hydrogenation Occurs at the Molecular Level of Life Molecules of H_2 and powdered metal catalysts, of course, are unavailable in the body. Cells, however, have molecular carrier systems that deliver the pieces of H—H to carbon–carbon double bonds. One piece is $H:^-$, donated by a carrier enzyme to one end of the double bond. The other piece is H^+, either donated by the carrier or plucked from a proton donor of the surrounding buffer and given to the carbon that was at the other end of the double bond. The *net* effect is the addition of H_2 because $H:^-$ and H^+ together add up to one $H:H$ molecule.

■ $H:^-$ (the hydride ion) must be donated by the carrier *directly* to the acceptor and not through the solution. When exposed directly to water, $H:^-$ reacts vigorously:

$$H:^- + H-OH \longrightarrow$$
$$H-H + OH^-$$

■ We studied buffers and how they work in Section 9.9.

EXAMPLE 13.3
Writing the Structure of the Product of the Addition of Hydrogen to a Double Bond

Write the structure of the product of the following reaction.

$$\underset{\qquad\quad\overset{\overset{\displaystyle CH_3CH_3}{|\quad\ |}}{}}{CH_3CH_2C\!=\!CCH_2CH_2CH_3} + H_2 \xrightarrow{\text{Ni catalyst}} ?$$

ANALYSIS The only change occurs at the double bond, and all the rest of the structure goes through the reaction unchanged. *This is true of all of the addition reactions that we will study.* Therefore copy the structure of the alkene just as it is, except leave only a single bond where the double bond was. Then increase by one the number of hydrogens at each carbon of the original double bond.

SOLUTION The structure of the product can be written as

$$\underset{\quad\ \ \overset{\overset{\displaystyle H\ \ H}{|\ \ |}}{}}{\underset{\overset{\displaystyle CH_3CH_3}{|\quad|}}{CH_3CH_2C\!-\!CCH_2CH_2CH_3}} \ \text{ or } \ \underset{\qquad\ \overset{\displaystyle CH_3\ \ CH_3}{|\qquad\ |}}{CH_3CH_2CH\!-\!CHCH_2CH_2CH_3}$$

■ The addition of hydrogen is often called the *reduction* of the double bond.

CHECK Make sure that the product has the identical carbon skeleton as the starting material. Then see that *each* carbon of the original alkene group has one more H.

One major goal in these chapters is to learn some chemical properties of functional groups. We have just learned a chemical "map sign" for the carbon–carbon double bond, one of its important chemical properties. It can be made to add hydrogen, and when it does, it changes to a single bond as each of its carbon atoms picks up one hydrogen atom. This sentence states a chemical fact about the double bond that has to be learned. Learn it, however, by working illustrations involving specific alkenes. There are too many individual reactions to memorize, so use your memory work to learn the *kinds* of reactions and what they do *in general* to change a molecule. As you work each part of the following practice exercise, say to yourself the *general* fact, the chemical map sign about hydrogenation, each time you apply it to a specific case.

PRACTICE EXERCISE 4

Write the structures of the products, if any, of the following.

(a) $CH_3CH{=}CH_2 + H_2 \xrightarrow{\text{Ni catalyst}}$

(b) $CH_3CH_2CH_3 + H_2 \xrightarrow{\text{Ni catalyst}}$

(c) 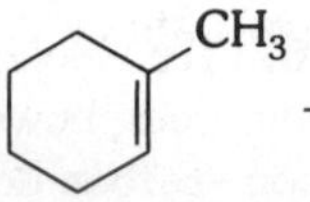 $+ H_2 \xrightarrow{\text{Ni catalyst}}$

(d) $CH_3(CH_2)_7CH{=}CH(CH_2)_7CO_2H + H_2 \xrightarrow{\text{Ni catalyst}}$

Chlorine and Bromine Also Add to Double Bonds Without any need for a special catalyst, both Cl_2 and Br_2 *rapidly* add to the carbon–carbon double bond. Iodine (I_2) does not add, and fluorine (F_2) reacts explosively with almost any organic compound to give a mixture of products.

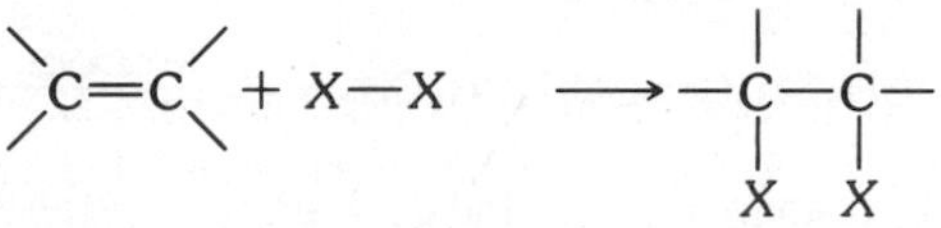

X = Cl or Br

■ Use a good fume hood and protective gloves when dispensing bromine.

Specific examples are

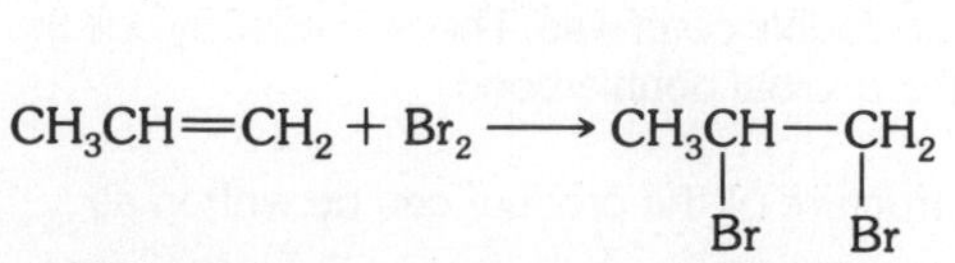

$CH_3CH{=}CH_2 + Br_2 \longrightarrow$ 1,2-Dibromopropane

Propene

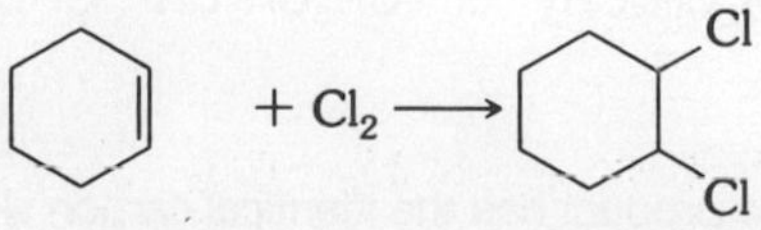

Cyclohexene   1,2-Dichlorocyclohexane

EXAMPLE 13.4
Writing the Structure of the Product of the Addition of Chlorine or Bromine to a Carbon–Carbon Double Bond

What compound forms in the following situation?

$$CH_3CH\!=\!CHCH_3 + Br_2 \longrightarrow ?$$

ANALYSIS This problem is very similar to that of the addition of hydrogen. We rewrite the alkene except that a single bond is left where the double bond was.

$$CH_3CH\!-\!CHCH_3 \quad \text{(Incomplete)}$$

Then we attach a bromine atom by a single bond to each carbon of the former double bond.

SOLUTION

$$\begin{array}{c} CH_3CH\!-\!CHCH_3 \\ \quad | \qquad | \\ \quad Br \quad\; Br \end{array}$$

2,3-Dibromobutane

CHECK Make sure that the carbon skeleton in the product is identical to that of the original alkene, that the double bond has been replaced by a single bond, and that each carbon of the double bond carries one of the pieces (Br in this example) of the adding molecule.

■ Bromine is dark brown, but the dibromoalkanes are virtually colorless. Therefore an unknown organic compound that rapidly decolorizes bromine quite likely has a carbon–carbon double bond.

PRACTICE EXERCISE 5

Complete the following equations by writing the structures of the products. If no reaction occurs under the conditions shown, write "no reaction."

(a)
$$\begin{array}{c} CH_3 \\ | \\ CH_3C\!=\!CH_2 + Br_2 \longrightarrow \end{array}$$
(b) $CH_3CH_2CH_2CH_3 + Cl_2 \longrightarrow$

(c) $CH_2\!=\!CHCH_2CH_3 + Cl_2 \rightarrow$
(d) $CH_3CH\!=\!CHCH_3 + H_2 \xrightarrow{\text{Ni catalyst}}$

Hydrogen Chloride, Hydrogen Bromide, and Sulfuric Acid Add Easily to Double Bonds

If gaseous hydrogen chloride or hydrogen bromide is bubbled into an alkene or if concentrated sulfuric acid is mixed with it, the following kind of reaction takes place. The pattern is the same in all these additions. We'll let H—G represent any of these reactants, where G stands for any electron-rich group that we will study, like Cl or OSO_3H.

$$\begin{array}{c} \diagdown \qquad \diagup \\ C\!=\!C \qquad + \; H\!-\!G \longrightarrow \\ \diagup \qquad \diagdown \end{array} \begin{array}{c} | \quad\; | \\ -\!C\!-\!C\!- \\ | \quad\; | \\ H \quad G \end{array} \quad (G = Cl, Br, \text{ or } OSO_3H)$$

An example:

$$CH_2\!=\!CH_2 + H\!-\!Cl \longrightarrow \begin{array}{c} CH_2\!-\!CH_2 \\ | \qquad | \\ H \qquad Cl \end{array} \quad \text{or} \quad CH_3CH_2Cl$$

Unsymmetrical Reactants Add Selectively to Unsymmetrical Double Bonds   We now have a small complication. H—G is not a symmetrical molecule, like H—H, Br—Br, or Cl—Cl. When symmetrical molecules add to a double bond, it doesn't matter which end of the double bond gets which half of the adding molecule. But it matters when H—G adds, *if the double bond is itself unsymmetrical.* An *unsymmetrical double bond* is one whose two carbon atoms hold unequal numbers of hydrogen atoms. For example, 1-butene, CH_2=$CHCH_2CH_3$, and propene, CH_3CH=CH_2, both have unsymmetrical double bonds. Each molecule has one carbon at the double bond with two Hs, but the other carbon has one H, an unequal number. The double bond in 2-butene, CH_3CH=$CHCH_3$, however, is symmetrical; one H is at each carbon.

When H—Cl(*g*) adds to propene we could imagine obtaining 1-chloropropane, 2-chloropropane, or a mixture of the two, perhaps 50 : 50. Let's see what actually happens. We have

$$CH_3CH\text{=}CH_2 + H\text{---}Cl \longrightarrow CH_3CH\text{---}CH_2 \quad \text{or} \quad CH_3CH_2CH_2Cl$$
$$\underset{\displaystyle H}{|} \quad \underset{\displaystyle Cl}{|}$$

1-Chloropropane
(Very little forms.)

or

$$CH_3CH\text{=}CH_2 + H\text{---}Cl \longrightarrow CH_3CH\text{---}CH_2 \quad \text{or} \quad CH_3CHCH_3$$
$$\underset{\displaystyle Cl}{|} \quad \underset{\displaystyle H}{|} \qquad\qquad \underset{\displaystyle Cl}{|}$$

2-Chloropropane
(The major product)

The actual product is largely 2-chloropropane, and very little of its isomer, 1-chloropropane, forms. In other words, the reactant, H—Cl, adds to the unsymmetrical double bond selectively.

Markovnikov's Rule Predicts Directions in Unsymmetrical Additions   Vladimer Markovnikov (1838–1904), a Russian chemist, was the first to notice that unsymmetrical alkenes add unsymmetrical reactants in one direction. Which direction can be predicted by **Markovnikov's rule.**[2]

■ "Them that has, gits" applies here, too.

> **Markovnikov's Rule**   When an unsymmetrical reactant of the type H—G adds to an unsymmetrical alkene, the carbon with the greater number of hydrogens gets one more H.

The following examples illustrate Markovnikov's rule in action.

$$CH_3C\text{=}CH_2 + H\text{---}Cl \longrightarrow CH_3CCH_3 \quad \left(Not\ CH_3CHCH_2Cl \right)$$
$$\underset{\displaystyle CH_3}{|} \qquad\qquad \overset{\displaystyle Cl}{\underset{\displaystyle CH_3}{|}} \qquad\qquad \underset{\displaystyle CH_3}{|}$$

2-Methylpropene   *t*-Butyl chloride

[2] When hydrogen bromide is used, it is important that no peroxides, compounds of the type R—O—O—H or R—O—O—R, be present. Traces of peroxides commonly form in organic liquids that are stored in contact with air for long periods. When peroxides are present, the addition of H—Br occurs in the direction opposite to that predicted by Markovnikov's rule. Peroxides catalyze the anti-Markovnikov addition *only* of HBr, not of HCl or H_2SO_4.

1-Methylcyclohexene + H—Cl ⟶ 1-Chloro-1-methylcyclohexane

$$\left(Not \right)$$

Concentrated Sulfuric Acid Actually Dissolves Alkenes

When an alkene is mixed with concentrated sulfuric acid, the hydrocarbon dissolves and heat evolves. How can two substances of such radically different polarities dissolve together? An alkane does not behave this way at all but merely forms a separate layer that floats on the sulfuric acid (see Fig. 13.3).

The alkene dissolves because it reacts by an addition reaction to form an alkyl hydrogen sulfate, which is very polar. For example,

$$CH_3CH{=}CH_2 + H{-}O{-}\overset{\overset{O}{\|}}{\underset{\underset{O}{\|}}{S}}{-}O{-}H \longrightarrow CH_3\overset{\overset{CH_3}{|}}{CH}{-}O{-}\overset{\overset{O}{\|}}{\underset{\underset{O}{\|}}{S}}{-}O{-}H$$

1-Propene   Sulfuric acid (H_2SO_4)   Isopropyl hydrogen sulfate

Molecules of the alkyl hydrogen sulfates are very polar, and as soon as they form, they move smoothly into the polar sulfuric acid layer and out of the nonpolar alkene layer. The heat generated by the reaction and the strongly acidic nature of the mixture causes side reactions that generate black by-products.

■ The sodium salts of long-chain alkyl hydrogen sulfates are detergents, for example, $CH_3(CH_2)_{11}OSO_3Na$.

Symmetrical Double Bonds Do Not Add H—G Reactants Selectively

Markovnikov's rule does not apply when the double bond is symmetrical. Reactants like H—G add in both of the two possible directions, and a mixture of isomers can form. For example,

$$CH_3CH_2CH{=}CHCH_3 + H{-}Br \longrightarrow CH_3CH_2\overset{\overset{}{|}}{\underset{\underset{Br}{|}}{C}}HCH_2CH_3 + CH_3CH_2CH_2\overset{\overset{}{|}}{\underset{\underset{Br}{|}}{C}}HCH_3$$

2-Pentene   3-Bromopentane   2-Bromopentane

(For a reminder of why we don't try to balance an equation such as this, check back to Special Topic 12.3.)

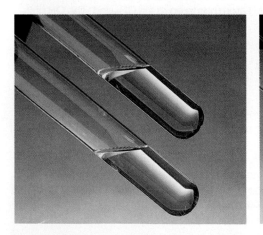

FIGURE 13.3

The effect of concentrated sulfuric acid on an alkane and an alkene. In the photo on the left, the alkane (top tube) and the alkene (bottom tube) are seen as clear, colorless liquids. The photo on the right shows the two systems soon after concentrated sulfuric acid has been added to each. The alkane floats unaffected on the acid (top tube), but the alkene (bottom tube) has already begun to react and form an alkyl hydrogen sulfate (together with darkly colored matter produced by side reactions). The alkyl hydrogen sulfate is soluble in the remaining concentrated sulfuric acid, and so all of the alkene will appear to dissolve. (The alkane is cyclohexane and the alkene is cyclohexene.)

Water Adds to Double Bonds to Give Alcohols Water adds to the carbon–carbon double bond provided that an acid catalyst (or the appropriate enzyme) is present. The product is an alcohol. Water alone, or aqueous bases have no effect on alkenes whatsoever.

$$\underset{\text{Alkene}}{\diagup\kern-0.5em\diagdown C=C\diagup\kern-0.5em\diagdown} + H-OH \xrightarrow[\text{heat}]{H^+} \underset{\underset{\displaystyle H\;\;OH}{|\;\;\;|}}{-\overset{|}{C}-\overset{|}{C}-} \atop \text{Alcohol}$$

Specific examples are

$$\underset{\text{Ethene}}{CH_2\!=\!CH_2} + H-OH \xrightarrow[\substack{240\ ^\circ C \\ \text{(closed vessel)}}]{10\%\ H_2SO_4} \underset{\text{Ethyl alcohol}}{CH_3CH_2OH}$$

$$\underset{\substack{|\\ CH_3 \\ \text{2-Methylpropene}}}{CH_3\overset{\displaystyle |}{C}\!=\!CH_2} + H-OH \xrightarrow[25\ ^\circ C]{10\%\ H_2SO_4} \underset{\substack{|\\ CH_3 \\ \textit{t-Butyl alcohol}}}{CH_3\overset{\displaystyle OH}{\underset{\displaystyle |}{\overset{\displaystyle |}{C}}}CH_3} \quad \left(\textit{Not } CH_3\overset{\displaystyle |}{\underset{\substack{|\\ CH_3}}{CH}}CH_2OH\right)$$

As you can see, Markovnikov's rule applies to this reaction, too. One H in H—OH goes to the carbon with the greater number of hydrogens, and the OH goes to the other carbon of the double bond.

Notice in the last example, and in all previous examples of addition reactions, that the carbon skeleton does not change. Although this is not always true, it will be in all the examples we will use as well as in all the Practice and Review Exercises.

EXAMPLE 13.5
Using Markovnikov's Rule

What product forms in the following situation?

$$CH_3CH\!=\!CH_2 + H-OH \xrightarrow[\text{heat}]{H^+}$$

ANALYSIS As in all of the addition reactions that we are studying, the carbon skeleton of the alkene can be copied over intact, except that a single bond is shown where the double bond was.

$$CH_3CH-CH_2 \qquad \text{(Incomplete)}$$

To decide which carbon of the original double bond gets the H atom from the water molecule, we use Markovnikov's rule. The H atom has to go to the CH_2 end. The OH unit from H—OH goes to the other carbon.

SOLUTION The product is

$$\underset{\substack{|\;\;\;|\\ OH\;\;H \\ \text{Isopropyl} \\ \text{alcohol}}}{CH_3CH-CH_2} \quad \text{or} \quad \underset{\substack{|\\ OH}}{CH_3CHCH_3}$$

CHECK At this stage be sure to see whether each carbon in the product has four bonds. If not, you can be certain that some mistake has been made. This is always a useful way to avoid at least some of the common mistakes made in solving a problem such as this. Then double-check that the carbon of the double bond initially having the greater number of Hs has been given one more.

PRACTICE EXERCISE 6

Write structures for the product(s), if any, that would form under the conditions shown. If no reaction occurs, write "no reaction."

(a) CH_2=$CHCH_2CH_3$ + HCl $\longrightarrow$

(b) $\underset{\underset{\textstyle CH_3}{\mid}}{CH_3C}$=$CH_2$ + HBr $\longrightarrow$

(c) CH_3CH=$\overset{\overset{\textstyle CH_3}{\mid}}{C}$-⬡ + H_2O $\xrightarrow[\text{heat}]{H^+}$

(d)   + H—OH $\xrightarrow[\text{heat}]{H^+}$

(e)   ⬡ + H—OH $\xrightarrow[\text{heat}]{H^+}$

Double Bonds Are Attacked by Many Oxidizing Agents

With two pairs of electrons, the double bond is more electron-rich than a single bond, so electron-seeking reagents attack it. These include oxidizing agents. Hot solutions of potassium permanganate ($KMnO_4$) and potassium dichromate ($K_2Cr_2O_7$), for example, vigorously oxidize molecules at carbon–carbon double bonds. The oxidations begin as addition reactions, but continue beyond to the point where alkene molecules are split apart at the double bond. We will not study any of the details or learn how to predict products. We only note the fact that the carbon–carbon double bond makes a molecule susceptible to attack by strong oxidizing agents. This is the fact to remember. The products of alkene oxidations can be ketones, carboxylic acids, carbon dioxide, or mixtures of these. Alkanes, in sharp contrast, are inert toward these oxidizing agents.

Ketones

Carboxylic acid

The permanganate ion is intensely purple in water, and as it oxidizes double bonds it changes to manganese dioxide, MnO_2, a brownish, sludge-like, insoluble solid (see Fig. 13.4). The dichromate ion is bright orange in water, and when it acts as an oxidizing agent, it changes to the bright green, hydrated chromium(III) ion, $Cr^{3+}(aq)$, as seen in Figure 13.5.

So far we have studied four tests for distinguishing an alkane and an alkene: the bromine test, the concentrated sulfuric acid test, and the use of aqueous permanganate or aqueous dichromate.

(a)

(b)

FIGURE 13.4
Permanganate oxidation.
(a) Crystals of potassium permanganate are so deeply purple that they appear almost black. The dilute $KMnO_4$ solution has a purple color. (b) After an oxidizable compound has been added to the $KMnO_4$ solution and the mixture has been heated to complete the oxidation of the compound, the solution has no color but now contains a precipitate of MnO_2.

FIGURE 13.5

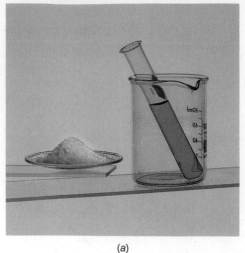

Dichromate oxidation (a) Crystals of sodium dichromate are bright orange, and a dilute $Na_2Cr_2O_7$ solution has the orange color of the hydrated dichromate ion. (b) After the dichromate solution has been mixed with some oxidizable compound and heated, the solution takes on the green color of the hydrated Cr^{3+} ion.

(a)   (b)

■ Ozone destroys any vegetation that has the green pigment chlorophyll, because chlorophyll contains double bonds.

Ozone Is One of the Most Powerful Oxidizing Agents Ozone, O_3, a pollutant in smog, is dangerous because it is a very powerful oxidizing agent that attacks biochemicals wherever they have carbon–carbon double bonds. Because such bonds occur in the molecules of all cell membranes, you can see that exposure to ozone must be kept very low. Even a concentration in air of only one part per million parts (1 ppm) warrants the declaration of a smog emergency condition. Special Topic 13.3 describes how incompletely burned hydrocarbons released in the exhaust from vehicle fuels contribute to the generation of ozone in smog.

Although ozone is dangerous to us where we live and breathe, it is vital to our well-being that ozone be in the stratosphere, the zone of the atmosphere in a band 10 to 25 miles above us. How stratospheric ozone benefits us and how its work is being threatened by air pollution were discussed in Special Topic 5.1 (page 120).

SPECIAL TOPIC 13.3
OZONE IN SMOG

We have noted elsewhere how dangerous ozone is to materials, including plants and lung tissue. The U.S. National Ambient[1] Air Quality Standard for ozone is a daily maximum one-hour average ozone concentration of only 0.12 ppm (120 ppb). Dozens of U.S. urban areas exceed this at least once per year. How does ozone originate in the air where we live?

The *direct* and only source of ozone in the lower atmosphere is the combination of oxygen atoms with oxygen molecules.

$$O + O_2 + M \longrightarrow O_3 + M \qquad (1)$$

[1] *Ambient* means "all surrounding, all encompassing."

M is any molecule, like N_2 or O_2, that can absorb some of the kinetic energy involved in the collision of O with O_2. The collision that leads to product occurs on the surface of M. (Following collisions occurring elsewhere, the ozone molecule breaks up instantly because it carries too much energy.) Thus the earlier question, "How does ozone originate?" becomes a new question, "How are oxygen *atoms* generated?" Once oxygen atoms are generated, there will be ozone. The chief source of oxygen atoms is the breakup of molecules of nitrogen dioxide, an air pollutant, a breakup made possible by solar energy.

$$NO_2 + solar\ energy \longrightarrow NO + O \qquad (2)$$

Oxides of Nitrogen The NO_2 for Equation (2) is made from NO, nitrogen monoxide. NO is produced inside vehicle engine cylinders where the direct combination of nitro-

The effect of smog on visibility. (a) A clear day. (b) A day of heavy smog.

gen and oxygen is possible because of the high temperature and pressure.

$$N_2 + O_2 \xrightarrow[\text{pressure}]{\text{high temperature}} 2NO \qquad (3)$$

As soon as newly made NO, now in the exhaust gas, hits the cooler outside air, it reacts further with oxygen to give nitrogen dioxide, NO_2. NO_2 gives smog its reddish-brown color (see Figure 1).

$$2NO + O_2 \longrightarrow 2NO_2 \qquad (4)$$

There is thus an ample supply of NO_2 in air made smoggy by heavy vehicle traffic.

Ultraviolet Energy As described in Special Topics 5.1 and 13.2, solar energy, which can break down NO_2 (Equation 2), consists of a stream of *photons*. Some have relatively high energies and so are in the ultraviolet region of light. Such photons give us no visual sensation, but their energies cause grave harm, like severe sunburn, eye burn, and even skin cancer.

Virtually all of the dangerous ultraviolet radiation coming from the sun is absorbed high in the stratosphere (Special Topic 5.1). However, some ultraviolet energy gets to where we live where it is able to generate ozone in smog-filled air.

An interesting feature of reaction 2 is that it produces nitrogen monoxide, NO, along with oxygen atoms. *Nitrogen monoxide is able to destroy ozone as follows.*

$$NO + O_3 \rightarrow NO_2 + O_2 \qquad (5)$$

Reactions 1, 2, and 5, when added together, give us *no net chemical effect!* (Try it.) How then does ozone develop in our atmosphere at all?

Reactions 1–5 proceed at their own rates under their own kinetic laws. These rates are unequal, so the reactions actually do not exactly cancel each other. The net effect is that a small, steady-state concentration of ozone develops even in clean air. The range of ozone concentration in clean air, however, is very low, between 20 and 50 ppb. In the more polluted urban areas, levels as high as 400 ppb commonly occur for brief periods of time each year.

Unburned Hydrocarbons and the Ozone in Smog Reaction 5 is the reaction that can make most ozone disappear, *but other substances are able to remove NO from reaction 5 before it is used to destroy ozone.* Such removal of NO other than by reaction 5 enables a buildup in the ozone level of the lower atmosphere.

Net destroyers of nitrogen monoxide are themselves present in vehicle exhaust — the unburned or partially oxidized hydrocarbons remaining from the incomplete combustion of the fuel. In sunlight and oxygen, some unburned hydrocarbon molecules are changed into organic derivatives of hydrogen peroxide called *peroxy radicals* usually symbolized as $R—O—O\cdot$, or simply RO_2. Peroxy radicals originate chiefly by the reaction of unburned hydrocarbons with hydroxyl radicals, HO. These arise by a number of mechanisms in polluted air, but we will not go into the details of how HO radicals form. Suffice it to say, HO radicals and hydrocarbons lead to ROO radicals, and these destroy NO molecules as follows.

$$ROO + NO \longrightarrow RO + NO_2$$

This reaction reduces the supply of ozone-destroying NO.

The development of ozone in smog thus depends on the formation of ROO radicals whose formation, in turn, depends on hydrocarbon emissions. Specialists generally agree that emissions of hydrocarbons must be significantly reduced before the ozone problem will be appreciably solved. Even the loss of hydrocarbons in the vapors from fuel tanks or at gas pumps has to be reduced.
[A reference: J. H. Seinfeld. "Urban Air Pollution: State of the Science." *Science,* February 10, 1989, page 745.]

13.5

HOW ADDITION REACTIONS OCCUR

The carbon–carbon double bond can accept a proton from an acid and change into a carbocation.

Because the double bond is somewhat electron-rich, it functions as a base, a proton acceptor, particularly when the proton donor is strong, like HCl, HBr, or $HOSO_3H$. When alkenes react with water, the *initial* step is the donation of a proton to the double bond from the acid catalyst, not from the water molecule. The water molecule is too weak a proton donor when the acceptor is a double bond.

■ In H—Cl, G is Cl. In H_2SO_4, G is OSO_3H. In dilute acid, the proton-donating species is H_3O^+, so G here is H_2O.

When an Alkene Accepts a Proton, a Reactive Carbocation Forms Using the symbol H—G to represent any proton donor, the reaction we will now explain is the following addition of HG to propene.

$$CH_3CH{=}CH_2 + H{-}G \longrightarrow CH_3\underset{\underset{G}{|}}{C}HCH_3 \qquad (not\ CH_3CH_2CH_2{-}G)$$

We can write the first step in this addition reaction as follows.

> What once were two electrons in a double bond are now the electrons of this bond.

$$CH_3CH{=}CH_2 + H{-}\ddot{\underset{..}{G}}: \longrightarrow CH_3\overset{+}{C}H{-}CH_2{-}H + :\ddot{\underset{..}{G}}:^-$$

Propene   Isopropyl carbocation

The curved arrows show how one of the two pairs of electrons of the double bond swings away from one carbon to form a bond to the proton.

One Carbon in a Carbocation Lacks an Octet The isopropyl cation that forms is an example of a **carbocation,** a positive ion in which carbon has a sextet not an octet of electrons. But notice that it is the *isopropyl* carbocation that forms, not the propyl carbocation, $CH_3CH_2CH_2^+$. We'll see why shortly, and when we do we'll understand Markovnikov's rule.

Carbocations are particularly unstable cations. All their reactions are geared to the recovery of an outer octet of electrons for carbon. The instant a carbocation forms, it strongly attracts any electron-rich species, and just such particles are produced when H—Cl, H—Br, or H—OSO_3H donate a proton to the double bond. When the proton leaves any of these acids, their conjugate base is released, which is Cl⁻, Br⁻, or $^-OSO_3H$ depending on which acid is used. We'll continue to generalize by using the symbol G^- for any of these conjugate bases.

The newly formed carbocation reacts with G^- in the next step of the addition.

$$CH_3\overset{+}{C}HCH_3 + :\ddot{\underset{..}{G}}:^- \longrightarrow CH_3\underset{\underset{:\ddot{G}:}{|}}{C}HCH_3$$

Isopropyl
carbocation

This restores the octet to carbon at the same time as it gives the product. As we noted earlier, the product is *not* $CH_3CH_2CH_2G$. Let us see why.

The More Stable Carbocation Preferentially Forms   The stability of a carbocation, although always very low, varies with the number of neighboring electron clouds surrounding its positively charged center. Alkyl groups provide larger overall electron clouds than H atoms. Therefore packing alkyl groups instead of H atoms around the carbon with the positive charge *helps to stabilize the carbocation.* When the positive charge is on a secondary carbon, as in the isopropyl carbocation, the electron clouds of *two* alkyl groups crowd around it. When the charge is on the end carbon of the propyl carbocation, the electron cloud of only one alkyl group is nearby.

$$\overset{+}{CH_3CHCH_3} \qquad CH_3CH_2CH_2{}^+$$

Isopropyl carbocation   Propyl carbocation

As a result, the isopropyl carbocation is more stable than the propyl carbocation. Or, to generalize, a secondary carbocation is more stable than a primary. By extension, a tertiary carbocation is more stable than a secondary. The order of stability of carbocations is thus

$$CH_3{}^+ \quad < \quad R{-}CH_2{}^+ \quad < \quad R{-}\overset{R}{\underset{}{CH^+}} \quad < \quad R{-}\overset{R}{\underset{R}{C^+}}$$

Methyl carbocation   Primary carbocation   Secondary carbocation   Tertiary carbocation

Increasing stability of carbocations

least stable   most stable

■ The R groups are alkyl groups, and they need not be identical.

This order of stability explains Markovnikov's rule. When the option of two different *kinds* of carbocations exists, as in the addition of H—G to propene, *the most stable carbocation always preferentially forms.* The double bond accepts H$^+$ from H—G so as to produce the more stable carbocation.

   When the two possible carbocations are of the *same* type, both secondary, for example, then there is no preference. Both possible carbocations actually form, both accept G$^-$, and a mixture of isomeric products is produced. (We'll leave an example to a practice exercise.)

The Water Molecule Is Too Weak a Proton Donor to Make a Carbocation from an Alkene   Water does not react with an alkene at all in the absence of an acid catalyst. The H$_2$O molecule is too weak a donor of H$^+$, as we said; H$_2$O holds its protons much too strongly. Thus when water does add to an alkene under acid catalysis, the first step in the mechanism of the reaction is a proton transfer *from the acid catalyst,* not from a water molecule. The water molecule has to wait for this to happen. The acid catalyst converts the double bond temporarily into an *electron-poor site,* a carbocation.

t-Butyl carbocation
(three alkyl groups on C)

The other possible but less stable carbocation, which does *not* form, is

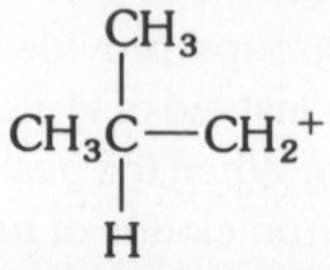

$$
\begin{array}{c}
CH_3 \\
| \\
CH_3C\!-\!CH_2{}^+ \\
| \\
H
\end{array}
$$

Isobutyl carbocation
(one alkyl group on C)

Once the *t*-butyl carbocation forms, what can it attract? What electron-rich particle is most abundantly available in water containing only a *trace* concentration (a *catalytic* amount) of acid? It is the electron-rich oxygen atom of the water molecule that is most likely to be attracted to the carbocation. The *t*-butyl carbocation, therefore, combines with a water molecule as follows to give what is essentially a hydronium ion in which one H has been replaced by a *t*-butyl group.

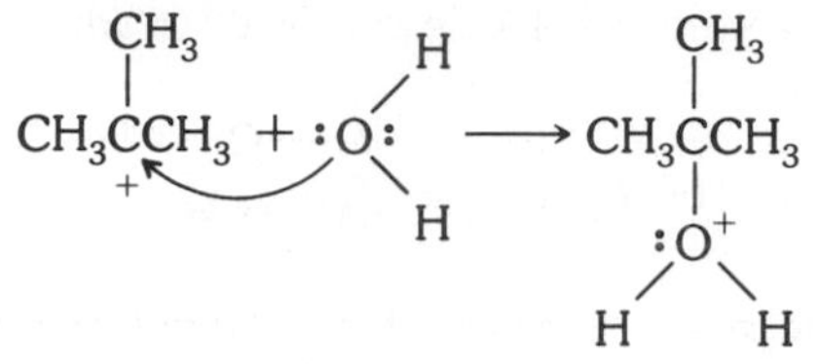

$$
\underset{+}{CH_3\overset{CH_3}{\overset{|}{C}}CH_3} + \;:\!\overset{H}{\underset{H}{\overset{|}{O}}}\!: \longrightarrow CH_3\overset{CH_3}{\overset{|}{C}}CH_3 \;\; \underset{H\quad H}{\overset{|}{\underset{}{:O^+}}}
$$

Now the positive charge is on oxygen, but this is acceptable in this instance because oxygen still has an outer octet.

In the last step, the catalyst, H_3O^+, is recovered by another proton transfer, this time from the *t*-butyl-substituted hydronium ion.

$$
CH_3\overset{CH_3}{\overset{|}{C}}CH_3 + \;:\!\overset{H}{\underset{H}{\overset{|}{O}}}\!: \longrightarrow CH_3\overset{CH_3}{\overset{|}{C}}CH_3 + H\!-\!\overset{H}{\underset{H}{\overset{|}{\underset{+}{O}}}}\!:
$$

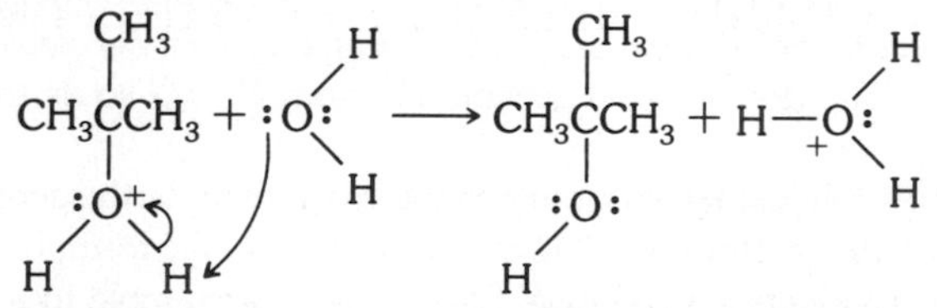

An alkyl
substituted
hydronium ion

t-Butyl
alcohol

Recovered
catalyst

PRACTICE EXERCISE 7

Write the condensed structures for the two carbocations that could conceivably form if a proton became attached to each of the following alkenes. Circle the carbocation that is preferred. If both are reasonable, state that they are. Then write the structures of the alkyl chlorides that would form by the addition of hydrogen chloride to each alkene.

(a) $CH_3CH_2CH\!=\!CH_2$ (b) $CH_3\overset{CH_3}{\overset{|}{C}}\!=\!CH_2$ (c) 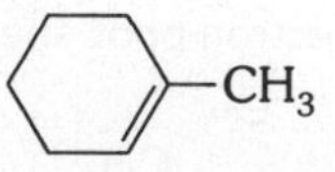 —CH_3

(d) $CH_3CH\!=\!CHCH_3$ (e) $CH_3CH\!=\!CHCH_2CH_3$

(f) The addition of water to 2-pentene (part e) gives a mixture of alcohols. What are their structures? Why is the formation of a mixture to be expected here but not when water adds to propene?

13.6

ADDITION POLYMERS

Hundreds to thousands of alkene molecules can join together to make one large molecule of a polymer.

Macromolecules Abound in Nature A **macromolecule** is simply a distinct molecule with a formula mass in thousands of daltons. Out of all of the many known macromolecular substances there are some, the *polymers,* with a unique structural feature. A **polymer** is a substance consisting of macromolecules *all of which have repeating structural units,* up to many thousands. Two of the carbohydrates that we will study — cellulose and one component of starch — are *polysaccharides* characterized by the following system, where Gl stands for a molecular unit made from a glucose molecule (the details of which we'll leave to Chapter 19).

- Macro signifies "huge" or "large scale." *Polymer* has Greek roots: *poly,* "many," and *meros,* "parts." (The *dalton* was defined on page 55.)

etc.—Gl—O—Gl—O—Gl—O—Gl—O—Gl—O—Gl—O—Gl—O—Gl—O—etc.

Section of a polysaccharide

- Starch molecules are a storage form for glucose in plants.

Another component of starch as well as the starch-like polymer glycogen, which we use to store glucose units, have similar sections but with many long branches similarly made of repeating Gl—O units.

- We store glucose units in the form of glycogen molecules principally in the liver, the kidneys, and in muscles.

  Proteins consist almost entirely (and often completely) of polymers called *polypeptides* whose molecules have the following features, showing only the molecular "backbone" and omitting the substituents or *side chains* appended to it. The backbone is like the chain of a charm bracelet, and such chains have *identical* links, but the bracelet acquires its uniqueness in the kinds of its charms (side chains) and the order in which they are hung.

$$\text{etc.}-\underset{|}{\overset{\overset{\displaystyle O}{\|}}{N}}HCHC-\underset{|}{\overset{\overset{\displaystyle O}{\|}}{N}}HCHC-\underset{|}{\overset{\overset{\displaystyle O}{\|}}{N}}HCHC-\underset{|}{\overset{\overset{\displaystyle O}{\|}}{N}}HCHC-\underset{|}{\overset{\overset{\displaystyle O}{\|}}{N}}HCHC-\text{etc.}$$

Section of a polypeptide (the vertical single bonds are sites where
side chains are attached, about 20 different kinds being the options)

Like charm bracelets, the backbones (chains) can be of varying lengths.

  The molecules of DNA, one of the kinds of nucleic acids, the chemicals of heredity, similarly have identical backbones, but they can be of different lengths and have different side chains (which we again omit).

$$\text{etc.}-\underset{\underset{\displaystyle O^-}{|}}{\overset{\overset{\displaystyle O}{\|}}{O}}PO-\text{pentose}-\underset{\underset{\displaystyle O^-}{|}}{\overset{\overset{\displaystyle O}{\|}}{O}}PO-\text{pentose}-\underset{\underset{\displaystyle O^-}{|}}{\overset{\overset{\displaystyle O}{\|}}{O}}PO-\text{pentose}-\underset{\underset{\displaystyle O^-}{|}}{\overset{\overset{\displaystyle O}{\|}}{O}}PO-\text{pentose}-\text{etc.}$$

Section of a DNA molecule with the vertical bonds being sites for the attachment of side chains, four kinds being used. ("Pentose" is a unit made from a
sugar called deoxyribose, which gives the D to DNA.)

  These examples are intended to illustrate just one feature of all polymers, the fact that they have *repeating structural units.* Thus the study of polymers focuses on the origins of these units and how they are joined together. Let's learn the rudiments of this chemistry by studying easier systems.

Polyethylene Is a Simple but Commercially Important Polymer Under a variety of conditions, many hundreds of ethylene molecules can reorganize their bonds, join together, and change into one large molecule.

■ Certain acids also catalyze this polymerization.

$$n\text{CH}_2\!\!=\!\!\text{CH}_2 \xrightarrow[\text{trace of O}_2]{\text{heat, pressure}} \left(\text{CH}_2\!-\!\text{CH}_2\right)_n \qquad (n = \text{a large number})$$

Ethylene   Polyethylene
                              (repeating unit)

The *repeating unit* in polyethylene is $\text{CH}_2\!-\!\text{CH}_2$, and one of these units after another is joined together into an *extremely long chain*. We see in the structure given a common way to write a polymer structure, by using parentheses to enclose the repeating unit.

Chain-branching reactions also occur during the formation of polyethylene, so the final product includes both straight and branched chain molecules. The molecules in a sample of a commercial polymer like polyethylene are never exact copies of each other. Their chain lengths vary, and the extent of branching varies, but it is still convenient to represent the polymer by showing its most characteristic repeating unit.

The starting material for making a polymer is a compound called a **monomer,** and the reaction in which a monomer changes to a polymer is called **polymerization.** Because alkenes are nicknamed *olefins,* the polymers of alkenes are usually called *polyolefins* (pol-y-**ol**-uh-fins).

Carbocations Are Intermediates in Some Polymerizations The catalysts used industrially to cause polymers to form vary widely. Some, for example, work by generating carbocation intermediates. When the catalyst is a proton donor, like our generalized acid H*G*, it can convert ethylene into the ethyl carbocation:

■ The negative ions, G^-, are also electron-rich but are present in minute traces compared to molecules of unreactive alkene.

$$G\!\!-\!\!\text{H} + \text{CH}_2\!\!=\!\!\text{CH}_2 \longrightarrow \text{H}\!-\!\text{CH}_2\!-\!\text{CH}_2{}^+ + G^-$$

Acid   Ethylene   Ethyl
catalyst   carbocation

When the only *abundant* electron-rich species around is unreacted alkene, the new carbocation attracts largely only the electron-rich carrier of an electron pair, *an alkene molecule.* Thus, the new carbocation restores an octet to its positively charged site. For example,

$$\text{CH}_3\!-\!\text{CH}_2{}^+ + \text{CH}_2\!\!=\!\!\text{CH}_2 \longrightarrow \text{CH}_3\!-\!\text{CH}_2\!-\!\text{CH}_2\!-\!\text{CH}_2{}^+$$

Ethylene

This reaction, of course, only creates a new and longer carbocation, one still surrounded mostly by unreacted molecules of alkene. So the new carbocation attracts still another molecule of alkene:

$$\text{CH}_3\!-\!\text{CH}_2\!-\!\text{CH}_2\!-\!\text{CH}_2{}^+ + \text{CH}_2\!\!=\!\!\text{CH}_2 \longrightarrow \text{CH}_3\!-\!\text{CH}_2\!-\!\text{CH}_2\!-\!\text{CH}_2\!-\!\text{CH}_2\!-\!\text{CH}_2{}^+$$

■ Polymerization is an example of a *chemical chain reaction* because the product of one step initiates the next step.

You can begin to see how this works. Yet another carbocation is produced, and it attracts another molecule of alkene. In this repetitive manner, the chain grows step by step, by one repeating unit of $\text{CH}_2\!-\!\text{CH}_2$ after another, until the positively charged site on a growing chain happens to pick up a stray anion, such as an anion left over from the catalyst. In this way, chains stop growing, some of different lengths than others. Actually, the catalyst should not be called a *catalyst,* because it is finally consumed and not regenerated. The term *promoter* is better than *catalyst* for polymerizations.

The Methyl Side Chains Occur Regularly in Polypropylene When propene (common name, propylene) polymerizes, the methyl groups appear regularly, *on alternate carbons* of the main chain.

$$n CH_2 = \underset{\underset{CH_3}{|}}{CH} \xrightarrow{\text{polymerization}}$$

$$\text{etc.} - CH_2 - \underset{\underset{CH_3}{|}}{CH} - CH_2 - \underset{\underset{CH_3}{|}}{CH} - CH_2 - \underset{\underset{CH_3}{|}}{CH} - CH_2 - \underset{\underset{CH_3}{|}}{CH} - \text{etc.} \quad \text{or} \quad \left(CH_2 - \underset{\underset{CH_3}{|}}{CH} \right)_n$$

Polypropylene   Polypropylene
(condensed structure)

Polymerizations generally take place in orderly fashions such as this because the reactive intermediates are governed by the same rules of stability that apply to simple carbocations. By the appropriate choice of the polymerization catalyst, however, the methyl groups in the polypropylene product can be made to line up all on the same side of the chain, or to alternate from side to side, or to project at random.

■ Each kind of polypropylene has its own commercial uses.

The Polyolefins Are Chemically Very Stable Because polyolefins, like polypropylene and polyethylene, are fundamentally alkanes, they have all the chemical inertness of alkanes. Polyolefins, therefore, are popular raw materials for making many items including containers that must be inert to food juices and to fluids used in medicine (see Fig. 13.6). Refrigerator boxes and bottles, containers for chemicals, sutures, catheters, various drains, and wrappings for aneurysms are commonly made of polyolefins. Polypropylene fibers are used to make indoor–outdoor carpeting and artificial turf because polypropylene is inert, won't mold, and is wear-resistant.

Substituted Alkenes Are Monomers for Important Polymers Monomers with carbon–carbon double bonds often carry other functional groups or halogen atoms. The resulting polymers are extremely important commercial substances, which we encounter often in our daily lives. Table 13.2 lists just a few examples (see Fig. 13.7).

Many dienes are used as monomers, too. Natural rubber is a polymer of a diene called isoprene (Table 13.2), and rubber is now made industrially.

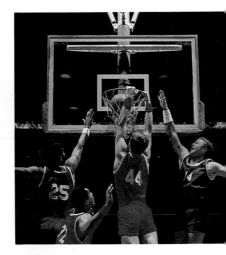

FIGURE 13.7
Lucite® and Plexiglas® are two trade names for the polymer of methyl methacrylate. The basketball backboard resists shattering when made of this polymer.

FIGURE 13.6
Polypropylene is used to make many items used in clinics and hospitals.

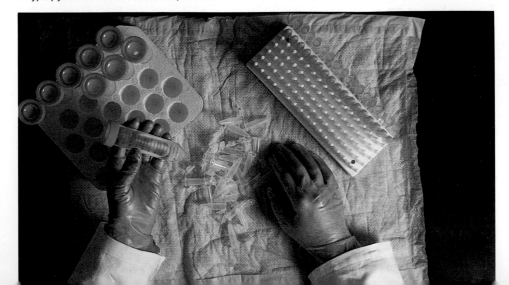

TABLE 13.2

Some Polymers of Substituted Alkenes[a]

Polymer	Monomer	Uses
Polyvinyl chloride (PVC)	$CH_2{=}CHCl$ Vinyl chloride	Insulation, credit cards, bottles, plastic pipe
Saran	$CH_2{=}CCl_2$ Vinylidene chloride and $CH_2{=}CHCl$ Vinyl chloride	Packaging film, fibers, tubing
Teflon	$F_2C{=}CF_2$ Tetrafluoroethylene	Nonstick surfaces, valves
Orlon	$CH_2{=}CH{-}C{\equiv}N$ Acrylonitrile	Fabrics
Polystyrene	⬡$-CH{=}CH_2$ Styrene	Foamed items, insulation
Lucite	$\overset{\displaystyle CH_3}{\underset{\displaystyle \text{Methyl methacrylate}}{CH_2{=}C{-}CO_2CH_3}}$	Windows, coatings, molded items
Natural Polymer		
Rubber	$\overset{\displaystyle CH_3}{CH_2{=}C{-}CH{=}CH_2}$ Isoprene	Tires, hoses, boots

[a] The common names rather than the IUPAC names of the monomers are given.

13.7

THE BENZENE RING AND AROMATIC PROPERTIES

The benzene ring undergoes substitution reactions instead of addition reactions despite a high degree of unsaturation.

The molecular formula of benzene is C_6H_6, which indicates considerable unsaturation.[3] Its ratio of hydrogen to carbon is much lower than in two simple, saturated hydrocarbons with six carbons, hexane (C_6H_{14}) and cyclohexane (C_6H_{12}). We should expect benzene, therefore, to be some kind of alkene, or alkyne, or a combination. Alkynes give addition reactions very similar to those of alkenes (Special Topic 13.4), and we might expect benzene to give addition reactions just as readily. There is one addition reaction that benzene does give; it adds hydrogen to give cyclohexane. Unlike the addition of hydrogen to an alkene, however, unusually rigorous conditions of pressure and temperature are necessary to make benzene add hydrogen.

$$C_6H_6 + 3H_2 \xrightarrow[\substack{\text{high pressure}\\ \text{and temperature}}]{\text{catalyst}} \bighexagon$$

Benzene   Cyclohexane
C_6H_{12}

[3] The molecular formulas of all saturated, open-chain alkanes fit the general formula, C_nH_{2n+2}. For cyclic alkanes and open-chain alkenes, the general formula is C_nH_{2n}. For open-chain alkynes, it's C_nH_{2n-2}. Thus benzene, C_6H_6, or C_nH_{2n-6}, is highly unsaturated.

REACTIONS OF ALKYNES

Because the triple bond has pi bonds like the double bond, alkynes give the same kinds of addition reactions as alkenes. Some examples are the following. Notice that the triple bond can add *two* molecules of a reactant. Usually, it is possible to control the reaction so that only one molecule adds.

$$CH_3C\equiv CH + H_2 \xrightarrow[\text{pressure}]{\text{Ni, heat,}} CH_3CH=CH_2 \xrightarrow{\text{more } H_2} CH_3CH_2CH_3$$

Propyne   Propene   Propane

$$CH_3C\equiv CH + HCl \longrightarrow \underset{\text{2-Chloropropene}}{CH_3\overset{\displaystyle Cl}{\overset{\displaystyle |}{C}}=CH_2} \xrightarrow{+HCl} \underset{\text{2,2-Dichloropropane}}{CH_3\overset{\displaystyle Cl}{\overset{\displaystyle |}{\underset{\displaystyle |}{\underset{\displaystyle Cl}{C}}}}CH_3}$$

Benzene's Typical Reactions Are Substitutions, Not Additions

Alkenes (and alkynes) readily *add* chlorine and bromine without a catalyst, but benzene needs a catalyst, and the reaction is not simple addition but substitution. The catalyst is generally an iron halide (or iron itself).

$$\underset{\text{Benzene}}{C_6H_6} + Cl_2 \xrightarrow[\text{FeCl}_3]{\text{Fe or}} \underset{\text{Chlorobenzene}}{C_6H_5Cl} + HCl$$

$$C_6H_6 + Br_2 \xrightarrow[\text{FeBr}_3]{\text{Fe or}} \underset{\text{Bromobenzene}}{C_6H_5Br} + HBr$$

Chlorobenzene

Benzene also reacts, *by substitution,* with sulfur trioxide dissolved in concentrated sulfuric acid. (Recall that alkenes react exothermically with concentrated sulfuric acid by *addition.*)

$$C_6H_6 + SO_3 \xrightarrow[\text{room temperature}]{\text{H}_2\text{SO}_4 \text{ (concd.)}} C_6H_5 - \overset{\displaystyle O}{\underset{\displaystyle O}{\overset{\displaystyle \|}{\underset{\displaystyle \|}{S}}}} - O - H$$

Benzenesulfonic acid

■ Benzenesulfonic acid is about as strong an acid as hydrochloric acid. It is a raw material for the synthesis of aspirin.

Benzene reacts with warm, concentrated nitric acid when it is dissolved in concentrated sulfuric acid. We will now represent nitric acid as $HO{-}NO_2$, instead of HNO_3, because it loses the HO group during the reaction.

$$C_6H_6 + \underset{\text{Nitric acid}}{HO - NO_2} \xrightarrow[\text{50--55 °C}]{\text{H}_2\text{SO}_4 \text{ (concd)}} \underset{\text{Nitrobenzene}}{C_6H_5 - NO_2} + H_2O$$

We have learned that alkenes (and alkynes) are readily oxidized by permanganate or dichromate ion, but benzene is utterly unaffected by these strong oxidizing agents even when boiled with them. (Ozone does attack benzene.)

In the light of all these chemical properties, whatever benzene is, it isn't an alkene or alkyne. Yet it surely is unsaturated. The problem of what benzene is wasn't satisfactorily solved in organic chemistry until the early 1930s, roughly a century after its molecular formula was known and half a century after its skeleton structure had been determined. Many of the reactions referred to above helped to establish the structure of benzene. Let's see how this was done.

The Six H Atoms in C_6H_6 Are Chemically Equivalent to Each Other

When benzene is used to make chlorobenzene (or any of the other products shown above), only *one* mono-substituted compound forms. Only one C_6H_5—Cl exists. This reminds us of what happens when ethane, CH_3CH_3, is chlorinated. Only one *mono*-substituted compound forms; only one CH_3CH_2Cl exists. It doesn't matter which H in CH_3CH_3 is replaced by Cl. The same is true for benzene. All six H atoms in C_6H_6 are equivalent to each other.

The Benzene Skeleton Has a Six-Membered Ring with a Hydrogen on Each Carbon

Cyclohexane forms when benzene is hydrogenated, so the six carbons of a benzene molecule must also form a six-membered ring. Because benzene's six Hs are chemically equivalent, it seems reasonable to put one H on each of the ring carbons. This gives a very symmetrical structure.

■ The older structure, **2**, is still widely used to represent benzene, although it is usually abbreviated further:

1
(Incomplete)

2
(Older structure for benzene)

The trouble with the incomplete structure identified as **1** is that each carbon has only three bonds, not four. To solve this, chemists for several decades simply wrote in three double bonds, as seen in structure **2**. As they well knew, the difficulty with **2** is that it says that benzene is a triene, a substance with three alkene groups per molecule.

Because the three double bonds indicated in structure **2** are misleading in a chemical sense, scientists often represent benzene simply by a hexagon with a circle inside, structure **3**.

■ Open-chain trienes, like 1,3,5-hexatriene, give addition reactions and are easily oxidized, like any ordinary alkene.

$$CH_2{=}CHCH{=}CHCH{=}CH_2$$
1,3,5-Hexatriene

3
Benzene

The ring system in **3**, the *benzene ring,* is planar. All its atoms lie in the same plane, and all the bond angles are 120°, as seen in the scale model of Figure 13.8.

Three Electron Pairs Are Delocalized in the Benzene Ring

The development of the chemical bonding theory of overlapping orbitals finally provided a satisfactory model for the structure of benzene (see Fig. 13.9). Part (*a*) of Figure 13.9 shows the sigma bond network. Each carbon atom holds three other atoms, not four, so each carbon is sp^2-hybridized and each has an unhybridized $2p$ orbital. The ring thus has six $2p$ orbitals, and their axes are coparallel and perpendicular to the plane of the benzene ring (see Fig. 13.9*b*).

The novel and important feature of benzene is that the six $2p$ orbitals overlap side to side *all around the ring.* They don't just pair off as in ethene and form three isolated double bonds. The result is a large, circular, double-doughnut shaped space above and below the plane, as seen in part (*c*) of the figure. Six electrons are in this space, and we can refer to them as the pi electrons of the benzene ring.

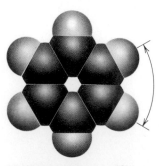

FIGURE 13.8
Scale model of a molecule of benzene. It shows the relative volumes of space occupied by the electron clouds of the atoms.

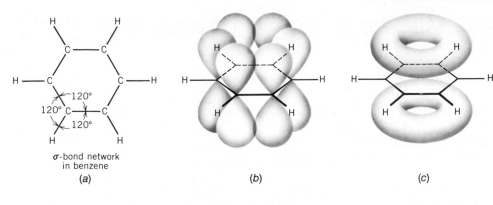

FIGURE 13.9
The molecular orbital model of benzene. (a) The sigma-bond framework. (b) The six 2p orbitals of the ring carbon atoms. (c) The double-doughnut shaped region formed by the side-by-side overlapping of the six 2p orbitals.

σ-bond network in benzene

(a)   (b)   (c)

A molecular orbital, like an atomic orbital, can hold no more than two electrons. The six pi electrons, therefore, are in three molecular orbitals within the double-doughnut shaped space, and they enjoy considerably more room and freedom of movement than if they were in isolated double bonds. The electrons are said to be *delocalized*. When electrons have more room, the system is more stable. The delocalization of the pi electrons, in fact, explains much of the unusual stability of benzene and its resistance to addition reactions. When a hydrogen atom held by a ring carbon is replaced by another group, the closed-circuit pi electron network is not broken up. But if an addition were to occur, the ring system would no longer be that of benzene but, instead, that of a cyclic diene. For example,

■ Electrons repel each other, so they are managed with greater overall stability in systems that allow them more room.

$$\text{benzene} + Cl_2 \xrightarrow[\text{(does not happen)}]{\times} \text{product}$$

It costs the system far more energy to react this way than to react by a substitution reaction, so the benzene ring strongly resists addition reactions. Just *how* chlorination does occur, however, involves a *temporary* rupture of the closed-circuit pi electron network as explained in Special Topic 13.5.

Aromatic Compounds Have Unsaturated Rings That Give Substitution Reactions, like Benzene

Any substances whose molecules have benzene rings and whose rings give substitution reactions instead of addition reactions are called **aromatic compounds.** The term is a holdover from the days when most of the known compounds of benzene actually had aromatic fragrances, but now the term does not refer to odor. Although oil of wintergreen and vanillin do have pleasant fragrances, aspirin does not. Yet all three have the benzene ring and all are classified as aromatic compounds.

Oil of wintergreen   Vanillin   Aspirin

1-Phenylpropane is an example of an aromatic compound with an aliphatic side chain. (In IUPAC nomenclature, the group C_6H_5, derived from benzene by removing one H atom, is called the **phenyl** group and rhyming with "kennel.") So stable is the benzene ring toward

■ $\bigcirc = C_6H_5—$

Phenyl group

HOW CHLORINE REACTS WITH BENZENE TO MAKE CHLOROBENZENE

Benzene and chlorine undergo essentially no reaction, even in the presence of ultraviolet radiation. However, when a trace of iron(III) chloride, $FeCl_3$, is present, chlorine changes benzene into chlorobenzene and HCl. The reaction that we'll explain in this Special Topic is

$$Cl_2 + C_6H_6 \xrightarrow{FeCl_3} HCl + C_6H_5Cl$$

To get a chlorine species reactive enough to break into benzene's pi electron network, we have to make the chlorine species temporarily even more reactive than it is in $Cl\cdot$ (Special Topic 12.2). This is the function of the $FeCl_3$ catalyst.

$FeCl_3$ is able to react with Cl_2 as follows to give $Cl^+FeCl_4^-$.

The species $Cl^+FeCl_4^-$ is not very stable, at least not enough to be stored, but it has a chlorine atom in a *positively* charged state, Cl^+. This is like $Cl\cdot$ minus its unpaired electron. In other words, Cl^+ has just *six* electrons in its outside level, even farther from an octet than $Cl\cdot$. Although Cl^+ is not a freely wandering species, the next step in the chlorination of benzene is envisioned as the attack on Cl^+ by benzene. To show what happens, we have to use one of the Lewis structures of benzene, and we will now show one of the H atoms attached to the benzene ring (leaving other five understood).

This unquestionably is an energy-expensive step, because it disrupts the uniquely stable pi electron network of the ring. However, the chlorine cation, Cl^+, recovers its octet, which is some compensation. Any remaining energy cost is repaid in the next step, because the pi network of the ring is restored and a *species more stable than Cl^+ is released*, namely H^+ *(which becomes joined to chlorine to give HCl)*. We'll use the $FeCl_4^-$ species made in the first step to provide the acceptor.

Thus the equivalent of Cl^- accepts H^+, the pi electron network of the ring is restored, chlorobenzene and HCl form, and *the catalyst is recovered to facilitate further reaction*. The mechanism of the substitution reaction thus involves *ions* and not the free radicals that we've seen in substitution reactions between alkanes and chlorine or bromine.

---

oxidizing agents that alkylbenzenes like 1-phenylpropane are attacked by hot permanganate at *the side chain* and not at the ring. Benzoic acid can be made by the oxidation of 1-phenyl-propane (to use an unbalanced equation).

Most of the side chain is destroyed, but the ring is not attacked.

Not all benzene derivatives have rings that resist oxidation as strongly as this. Rings, for example, that hold either the OH or the NH_2 groups are *very* readily oxidized. The unshared electrons on O and N in these groups are also somewhat delocalized into the ring network, which gives the rings more electron density than in benzene. Oxidizing agents, which seek electrons, are therefore able to react with C_6H_5OH and $C_6H_5NH_2$. We won't take this any further because our goal has been to learn *general* properties of the benzene ring. This ring is present in proteins, and benzene-like aromatic rings occur in the heterocyclic rings of all nucleic acids.

■ The liver has enzymes that put OH groups on benzene rings, making the products easier to break down into manageable wastes or to be made into substances needed by the body.

13.8

NAMING COMPOUNDS OF BENZENE

Common names dominate the nomenclature of simple derivatives of benzene. The names of several monosubstituted benzenes are straightforward. The substituent is indicated by a prefix to the word *benzene*. For example,

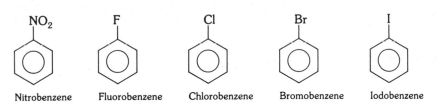

Nitrobenzene   Fluorobenzene   Chlorobenzene   Bromobenzene   Iodobenzene

■ All these compounds are oily liquids.

Other derivatives of benzene have common names that are always used.

Toluene   Phenol   Aniline   Benzoic acid   Benzaldehyde   Benzene-sulfonic acid

■ Phenol was the first antiseptic used by British surgeon Joseph Lister (1827 – 1912), the discoverer of antiseptic surgery.

***Ortho, Meta,* and *Para* Are Terms for 1,2-, 1,3-, and 1,4-Relationships**   When two or more groups are attached to the benzene ring, both what they are and where they are must be specified. One common way to indicate the relative locations of two groups in disubstituted benzenes is by the prefixes *ortho-, meta-,* and *para-,* which usually are abbreviated *o-, m-,* and *p-,* respectively. Two groups that are in a 1,2-relationship are *ortho* to each other, as in 1,2-dichlorobenzene, commonly called *o*-dichlorobenzene. A 1,3-relationship is designated *meta,* as in *m*-dichlorobenzene. In *p*-dichlorobenzene, the substituents are 1,4- or *para* relative to each other.

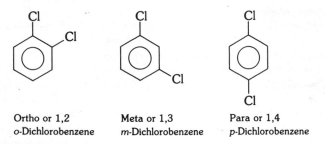

Ortho or 1,2   Meta or 1,3   Para or 1,4
o-Dichlorobenzene   *m*-Dichlorobenzene   *p*-Dichlorobenzene

A disubstituted benzene is usually named as a derivative not of benzene but of a monosubstituted benzene when the latter has a common name, like toluene or aniline. Then the *o-, m-,* or *p*- designations are used to specify relative positions of the two groups. For example,

p-Nitrotoluene

(*not* 4-nitro-1-
methylbenzene)

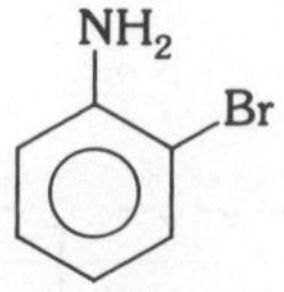

o-Bromoaniline

(*not* 2-bromo-1-
aminobenzene)

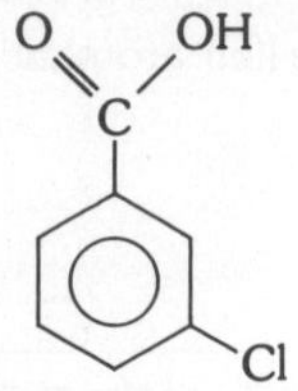

m-Chlorobenzoic acid

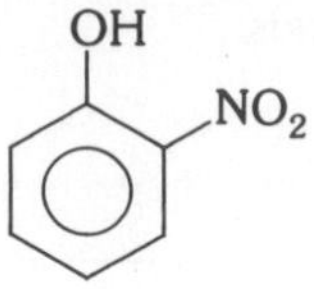

o-Nitrophenol

When we have trisubstituted benzenes (or higher), we cannot use the *ortho, meta,* or *para* designations with sufficient precision to make an unambiguous name. We must now use numbers assigned to ring positions in such a way as to use the lowest numbers possible. For example,

■ TNT is an important explosive. TNB is an even better explosive than TNT, but it is more expensive to make.

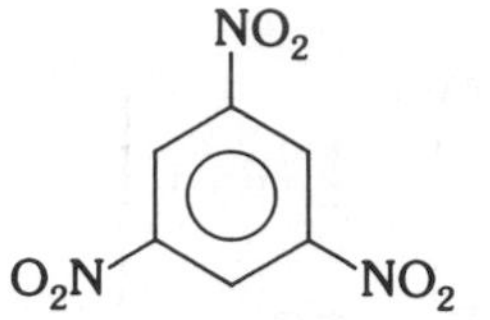

1,3,5-Trinitrobenzene, TNB

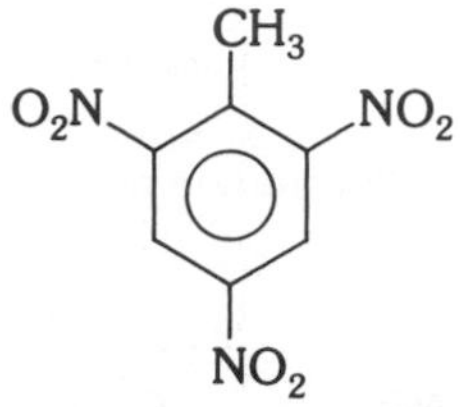

2,4,6-Trinitrotoluene, TNT

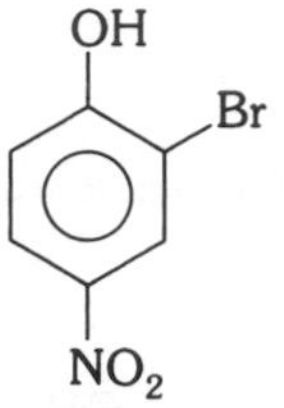

2-Bromo-4-nitrophenol

SUMMARY

Alkenes The lack of free rotation at a double bond makes geometric (cis–trans) isomers possible, but they exist only when the two groups are not identical at *either* end of the double bond. Cyclic compounds also exhibit cis–trans isomerism.

Alkenes and cycloalkenes are given IUPAC names by a set of rules very similar to those used to name their corresponding saturated forms. However, the double bond takes precedence both in selecting and in numbering the main chain (or ring). The first unsaturated carbon encountered in moving down the chain or around the ring through the double bond must have the lower number.

Addition reactions Several compounds add to the carbon–carbon double bond—H_2, Cl_2, Br_2, HCl, HBr, $HOSO_3H$, and H_2O (in the presence of an acid catalyst). The kinds of products that can be made are outlined in the accompanying chart of the reactions of alkenes. When both the alkene and the reactant are unsymmetrical, the addition proceeds according to Markovnikov's rule—the end of the double bond that already has the greater number of hydrogens gets one more. The double bond is vigorously attacked by strong oxidizing agents, like the permanganate ion (MnO_4^-), the dichromate ion ($Cr_2O_7^{2-}$), and ozone (O_3).

How additions occur The carbon–carbon double bond is a proton acceptor toward strong proton donors. The proton becomes at-

tached to one carbon at the double bond, using the pair of electrons in the pi bond to make the sigma bond to this hydrogen. The other carbon of the original double bond becomes positively charged. The result is a carbocation, which then accepts an electron-rich particle to complete the formation of the product.

Polymerization of alkenes The polymerization of an alkene is like an addition reaction. The alkene serves as the monomer, and one alkene molecule adds to another, and so on, until a long chain with a repeating unit forms—the polymer molecule.

Aromatic compounds Aromatic compounds contain rings having alternating double and single bonds, a closed-circuit pi-electron network commonly seen in the benzene ring. This ring is resistant to oxidation except when it holds OH or NH_2 as a substituent. When aromatic compounds undergo reactions at the benzene ring, substitutions rather than additions occur. In this way, the closed circuit pi-electron network of the ring remains unbroken. This network forms when six unhybridized 2*p* orbitals of the ring carbon atoms overlap side-by-side to form a double-doughnut shaped space found above and below the plane of the ring. There are three sublevels in this space, and the pi electrons enjoy considerable room and freedom of motion.

Chart of chemical properties

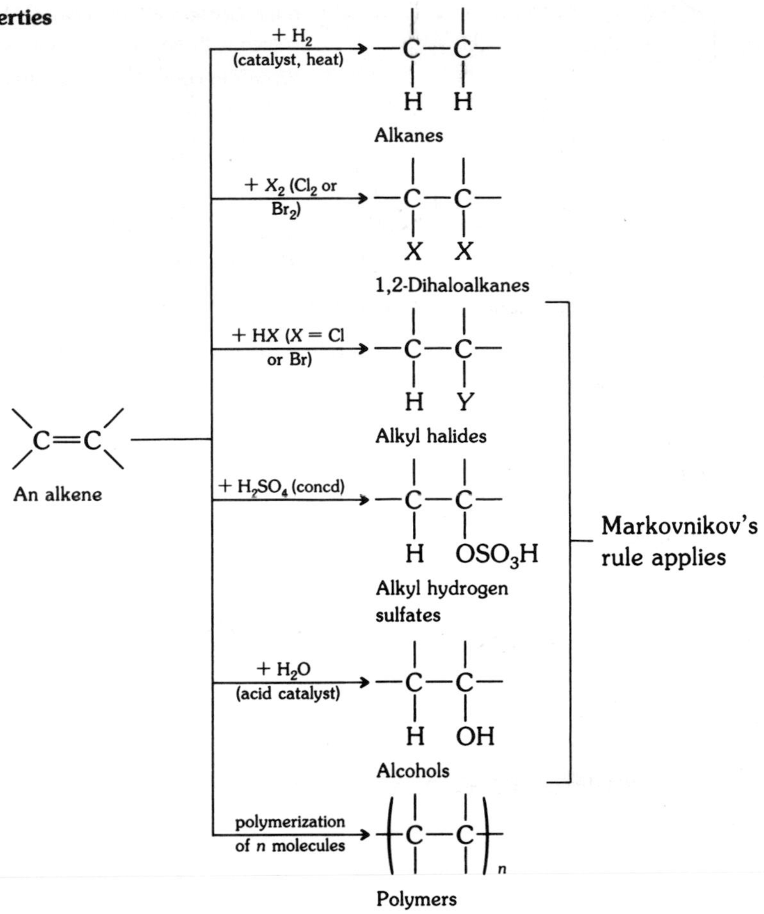

$+ H_2$ (catalyst, heat) → Alkanes

$+ X_2$ (Cl_2 or Br_2) → 1,2-Dihaloalkanes

$+ HX$ ($X = Cl$ or Br) → Alkyl halides

$+ H_2SO_4$ (concd) → Alkyl hydrogen sulfates

$+ H_2O$ (acid catalyst) → Alcohols

Markovnikov's rule applies

polymerization of n molecules → Polymers

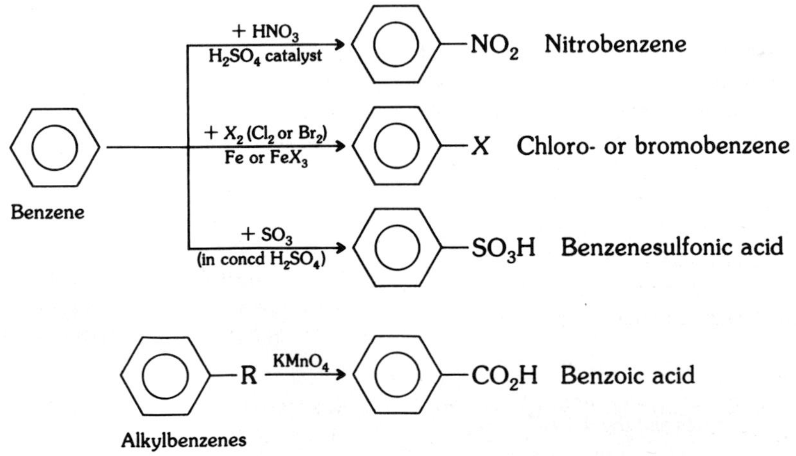

An alkene

$+ HNO_3$ H_2SO_4 catalyst → NO_2 Nitrobenzene

$+ X_2$ (Cl_2 or Br_2) Fe or FeX_3 → X Chloro- or bromobenzene

$+ SO_3$ (in concd H_2SO_4) → SO_3H Benzenesulfonic acid

Benzene

$-R$ $\xrightarrow{KMnO_4}$ $-CO_2H$ Benzoic acid

Alkylbenzenes

REVIEW EXERCISES

The answers to Review Exercises whose numbers are in color are found in Appendix C. The answers to the other Review Exercises are found in the Study Guide that accompanies this book. The more challenging questions are marked with asterisks.

Occurrence and Physical Properties

13.1 Examine the following structural formulas and *use their identifying letters* to answer the questions about them.

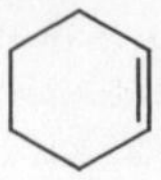

$CH_3CH\!=\!CH_2$  $CH_3CH_2C\!\equiv\!CH$

A   B   C   D

$CH_2\!=\!CHCH\!=\!CHCH_3$

E

(a) What are the letters of the compounds that are members of the *alkene* family?

(b) Which is a *saturated* compound?

(c) The prefix *pent-* would appear in the names of which compounds?

(d) Which compounds are relatively insoluble in water?

(e) The general formula C_nH_{2n} fits which compounds? Are they all alkenes?

(f) The general formula C_nH_{2n-2} fits which compounds? Are they all alkynes?

(g) In view of your observations regarding parts (e) and (f), how well do the general formulas C_nH_{2n} and C_nH_{2n-2} specify hydrocarbon families?

13.2 In view of the theme of this book, the molecular basis of life, what has justified our including the alkene group in our study?

Nomenclature

13.3 Write the condensed structures of the following compounds.

(a) Isobutylene

(b) Propylene

(c) *trans*-2-Hexene

(d) 3-Bromo-2-pentene

(e) 1,2-Dimethylcyclohexene

(f) 2,4-Dimethylcyclohexene

13.4 Write the IUPAC names of the following compounds.

(a) $CH_2\!=\!CH(CH_2)_5CH_3$

(b) $CH_3\overset{\displaystyle CH_3}{\underset{|}{C}}HCH\!=\!CHBr$

(c) $CH_3CH_2\overset{\displaystyle CH_3}{\underset{|}{C}}HCH_2\overset{\displaystyle CH_2}{\underset{|}{C}}CH_2CH_2CH_3$

(d) $CH\!=\!\overset{\displaystyle CH_3}{\underset{|}{C}}H\overset{\displaystyle CH_2CH_3}{\underset{|}{C}}CH_3$
$\qquad\qquad\;\underset{|}{CH_3}$

°13.5 Write the condensed structures and the IUPAC names for all the isomeric pentenes, C_5H_{10}. Include cis and trans isomers.

13.6 Write the condensed structures and the IUPAC names for all the isomeric methylcyclopentenes.

°13.7 Write the condensed structures and IUPAC names for all the isomeric butynes. The IUPAC rules for naming alkynes are identical with the rules for naming alkenes, except that the name ending is -*yne*, not -*ene*.

°13.8 Write the condensed structures and the IUPAC names for all the isomeric dimethylcyclopentenes. Include the cis and

trans isomers. Remember that all *six* atoms directly involved with an alkene system are in the same plane.

°13.9 Write the condensed structures and the names for all the open chain dienes with the molecular formula C_5H_8. (Note that there can be two double bonds from the same carbon.)

Cis–Trans Isomerism

13.10 What are the *names* of the molecular orbitals in which electron pairs reside between the two carbon atoms of an alkene group?

13.11 Given the options of the following types of level 2 atomic orbitals — s, p, sp, sp^2, and sp^3 — what are the *symbols* of the specific orbitals at a carbon atom that are used to make each of the two bonds to the other carbon of the alkene group?

13.12 Why is there a lack of free rotation at an alkene group, and why does this matter?

13.13 Which of the following pairs of structures represent identical compounds or isomers?

(a)

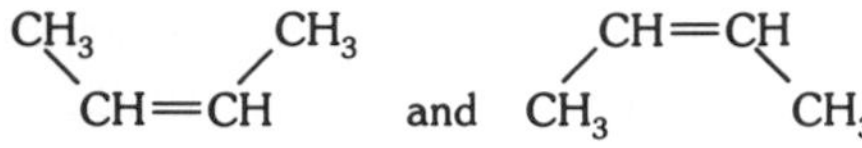

(b)

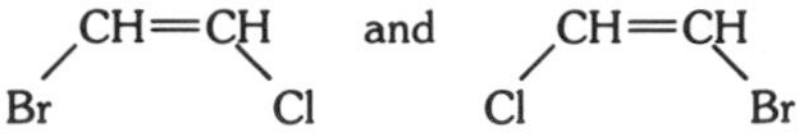

(c)

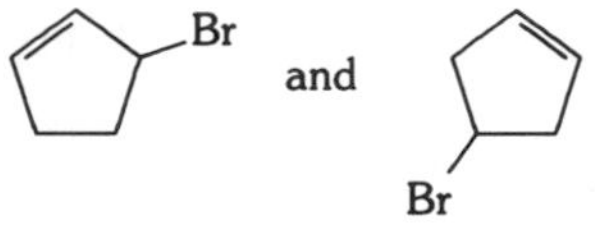

(d) $CH_3CH_2C\!=\!\overset{}{\underset{}{C}}HCH_3$ and $CH_3CH\!=\!CCH_2CH_3$
with CH_3 substituents

(e) $CH_3\overset{\displaystyle CH_3}{\underset{}{C}}\!=\!CHCHCH_3$ and $CHCHCH_3$ / CH_3CCH_3

13.14 Study the following structures to discover which are able to exhibit cis–trans isomerism. For those that do, write the structures of the cis and trans isomers.

(a) $CH_3CH\!=\!CHCH_2CH_3$

(b) $CH_3\overset{}{\underset{Cl}{C}}\!=\!\overset{}{\underset{Br}{C}}CH_2CH_3$

(c)

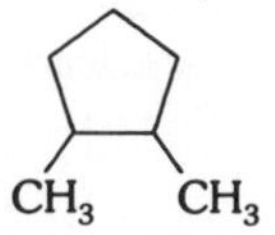

(d) $CH_3\overset{}{\underset{Cl}{C}}\!=\!CH\overset{\displaystyle CH_3}{\underset{}{C}}HCH_3$

13.15 Identify which of the following compounds can exist as cis and trans isomers and write the structures of these isomers.

(a) $FBrC=CHCl$

(b)

$$\begin{array}{c} H_3C \quad CH_3 \end{array}$$

(c) $CH_3CH=CHCH=CH_2$

Reactions of the Carbon–Carbon Double Bond

13.16 Write equations for the reactions of 2-methylpropene with the following reactants.
(a) Cold, concentrated sulfuric acid
(b) Hydrogen in the presence of a nickel catalyst
(c) Water in the presence of an acid catalyst
(d) Hydrogen chloride
(e) Hydrogen bromide
(f) Bromine

*13.17** The molecules of a hydrocarbon, A, with the molecular formula C_7H_{12} have the following structure, which is complete except for the location of one double bond. Compound A is *not* capable of existing as cis and trans isomers. However, the *product* of the hydrogenation of compound A is capable of existing as cis and trans isomers. Write the structure of A. (If more than one structure is possible, write them all.)

$$\begin{array}{c} \\ CH_3 \quad CH_3 \end{array}$$

(Incomplete
structure of A)

13.18 Write equations for the reactions of 1-methylcyclopentene with the reactants listed in Review Exercise 13.16.

13.19 Write equations for the reactions of 2-methyl-2-butene with the compounds given in Review Exercise 13.16.

13.20 Write equations for the reactions of 1-methylcyclohexene with the reactants listed in Review Exercise 13.16. (Do not attempt to predict whether cis or trans isomers form.)

13.21 Ethane is insoluble in concentrated sulfuric acid, but ethene dissolves readily. Write an equation to show how ethene is changed into a substance polar enough to dissolve in concentrated sulfuric acid, which, of course, is highly polar.

*13.22** Pentene, C_5H_{10}, has several isomers, and most but not all of them add bromine, dissolve in concentrated sulfuric acid, and react with potassium permanganate. Give the structure of at least one isomer of C_5H_{10} that gives *none* of these reactions.

*13.23** One of the raw materials for the synthesis of nylon, adipic acid, can be made from cyclohexene by oxidation using potassium permanganate. The balanced equation for the first step in which $K_2C_6H_8O_4$, the potassium salt of adipic acid, forms is as follows:

$$3C_6H_{10} + 8KMnO_4 \longrightarrow$$
$$3K_2C_6H_8O_4 + 8MnO_2 + 2KOH + 2H_2O$$

How many grams of potassium permanganate are needed for the oxidation of 11.2 g of cyclohexene, assuming that the reaction occurs exactly and entirely as written?

*13.24** Referring to Review Exercise 13.23, how many grams of the potassium salt of adipic acid can be made if 21.0 g of $KMnO_4$ are used in accordance with the equation given?

How Addition Reactions Occur

13.25 When 1-butene reacts with hydrogen chloride, the product is 2-chlorobutane, not 1-chlorobutane. Explain.

13.26 When 4-methylcyclohexene reacts with hydrogen chloride, the product consists of a mixture of roughly equal quantities of 3-chloro-1-methylcyclohexane and 4-chloro-1-methylcyclohexane. Explain why substantial proportions of *both* isomers form.

Polymerization

13.27 Rubber cement can be made by mixing some polymerized 2-methylpropene with a solvent such as toluene. When the solvent evaporates, a very tacky and sticky residue of the polymer (called polyisobutylene) remains, which soon hardens and becomes the glue. The structure is quite regular, like polypropylene. Write the structure of the polymer of 2-methylpropene in two ways.
(a) One that shows four repeating units, one after the other.
(b) The condensed structure.

13.28 Polyvinyl acetate is a soft adhesive that is modified (by reactions that we have yet to study) into a material (Butvar) that bonds two glass sheets together in safety glass. Thus when safety glass breaks, the broken pieces cannot fly around as easily. Using four vinyl acetate units, write part of the structure of a molecule of polyvinyl acetate. Also write its condensed structure. The structure of vinyl acetate is as follows.

$$\begin{array}{c} O \\ \parallel \\ O-CCH_3 \\ \mid \\ CH_2=CH \end{array}$$

Vinyl acetate

When it polymerizes, its units line up in a regular way as in the polymerization of propene.

13.29 Gasoline is mostly a mixture of alkanes. However, when a sample of gasoline is shaken with aqueous potassium permanganate, a brown precipitate of MnO_2 appears and the purple color of the permanganate ion disappears. What kind of hydrocarbon was evidently also in this sample?

13.30 If gasoline rests for months in the fuel line of some engine, the line slowly accumulates some sticky material. This can clog the line and also make the carburetor work poorly or not at all. In view of your answer to Review Exercise 13.29, what is likely to be happening, chemically?

Aromatic Properties

13.31 Dipentene has a very pleasant, lemon-like fragrance, but it is not classified as an aromatic compound. Why?

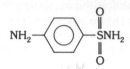

Dipentene

13.32 Sulfanilamide, one of the sulfa drugs, has no odor at all, but it is still classified as an aromatic compound. Explain.

Sulfanilamide

13.33 Write equations for the reactions, if any, of benzene with the following compounds.
(a) Sulfur trioxide (in concentrated sulfuric acid)
(b) Concentrated nitric acid (in concentrated sulfuric acid)
(c) Hot sodium hydroxide solution
(d) Hydrochloric acid
(e) Chlorine (alone)
(f) Hot potassium permanganate
(g) Bromine in the presence of $FeBr_3$

13.34 Write the structure of any compound that would react with hot potassium permanganate to give benzoic acid.

13.35 Phthalic acid is one of the raw materials for making a polymer that is used in automobile finishes.

Phthalic acid

What hydrocarbon with the formula C_8H_{10} could be changed to phthalic acid by the action of hot potassium permanganate? (Write its structure.)

The Bonds in Benzene

13.36 Describe in your own words, making your own drawings, the ways in which the bonds in benzene form from pure or hybrid atomic orbitals.

13.37 What is it about the delocalization of benzene's pi electrons that makes the benzene ring relatively stable?

13.38 Explain why benzene strongly resists addition reactions and gives substitution reactions instead.

Names of Aromatic Compounds

13.39 Write the condensed structure of each compound.
(a) toluene   (b) aniline
(c) phenol   (d) benzoic acid
(e) benzaldehyde (f) nitrobenzene

13.40 Give the condensed structures of the following compounds.
(a) p-nitrobenzoic acid
(b) m-bromotoluene
(c) o-chloronitrobenzene
(d) 2,4-dinitrophenol

Alkene Group in Nature (Special Topic 13.1)

13.41 What structural feature occurs in vitamins D and A that explains why these vitamins deteriorate somewhat when heated in air?

13.42 What structural difference exists between vegetable oils and animal fats?

Chemistry of Vision (Special Topic 13.2)

13.43 When cis–trans isomerization takes place, do groups at the double bond break off from the chain and exchange places?

13.44 How does a photon of the proper energy enable a cis isomer to change into a trans isomer?

13.45 The retina in the eye of an owl has only rods, not cones. What connection is there between this circumstance and the ability of an owl to see in very dim light? Can an owl see in a situation of total blackness?

Ozone in Smog (Special Topic 13.3)

13.46 What event in a vehicle engine launches the production of ozone in smog? (Write an equation.)

13.47 How is NO_2 formed in smog? (Write an equation.)

13.48 How is NO_2 involved in the production of ozone in smog? (Write equations.)

13.49 Why is ozone dangerous?

13.50 In areas with severe smog problems, air quality authorities seek to reduce emissions of hydrocarbons, even those arising from the use of power lawn mowers and similar machines. What is the connection between these uses of fuels and the ozone in smog?

Reactions of Alkynes (Special Topic 13.4)

13.51 Write the structures of the products that form when one mole of 2-butyne reacts with one mole of each of the following.
(a) H_2 (b) Cl_2 (c) Br_2

13.52 Write the structures of the products that form when one mole of 2-butyne reacts with *two* moles of each of the following.
(a) H_2 (b) Cl_2 (c) Br_2

How Benzene Reacts with Chlorine (Special Topic 13.5)

13.53 Why doesn't the benzene ring simply *add* chlorine molecules, like any alkene system?

13.54 Iron(III) bromide, $FeBr_3$, can be used as a catalyst for *chlorination* with little chance of any bromobenzene forming. Explain.

13.55 When bromine is mixed with benzene in the presence of iron(III) bromide as a catalyst, bromobenzene and HBr form. Write the mechanism; it is just like that of the chlorination of benzene.

Additional Exercises

13.56 Write the structures of the products to be expected in the following situations. If no reaction is to be expected, write "no reaction." To work this kind of exercise, you have to be able to do three things.

(1) *Classify* a specific organic reactant into its proper family. Do this first.

(2) *Recall* the short list of chemical facts about the family. (If there is no matchup between this list and the reactants and conditions specified by a given problem, assume that there is no reaction.)

(3) *Apply* the recalled chemical fact, which might be some "map sign" associated with a functional group, to the specific situation.

Study the next two examples before continuing.

EXAMPLE 13.6 Predicting Reactions

What is the product, if any, of the following?

$$CH_3CH_2CH_2CH_3 + H_2SO_4 \longrightarrow ?$$

Analysis: We note first that the organic reactant is an alkane, so we next turn to the list of chemical properties about all alkanes that we learned. With this family, of course, the list is very short. Except for combustion and halogenation, we have learned no reactions for alkanes, and we assume, therefore, that there aren't any others, not even with sulfuric acid. Hence, the answer is "no reaction."

EXAMPLE 13.7 Predicting Reactions

What is the product, if any, in the following situation?

$$CH_3CH=CHCH_3 + H_2O \xrightarrow{\text{acid catalyst}} ?$$

Analysis We first note that the organic reactant is an alkene, so we review our mental "file" of reactions of the carbon–carbon double bond.

1. Alkenes add hydrogen (in the presence of a metal catalyst and under heat and pressure) to form alkanes.

2. They add chlorine and bromine to give 1,2-dihaloalkanes.

3. They add hydrogen chloride and hydrogen bromide to give alkyl halides.

4. They add sulfuric acid to give alkyl hydrogen sulfates.

5. They add water in the presence of an acid catalyst to give alcohols.

6. They are attacked by strong oxidizing agents.

7. They polymerize.

These are the chief chemical facts, the principal "map signs," about the carbon–carbon double bond that we have studied, and we see that the list includes a reaction with water in the presence of an acid catalyst. We remember that in all addition reactions the double bond changes to a single bond and the pieces of the adding molecule end up on the carbons at ends of the double bond. We also have to remember Markovnikov's rule to tell us which pieces of the water molecule go to each carbon. However, in this specific example, Markovnikov's rule does not apply because the alkene is symmetrical.

Solution

iso butyl alcohol (handwritten)

$$\underset{\underset{OH}{|}}{CH_3CH_2CHCH_3}$$

Now work the following parts. (Remember that C_6H_6 stands for benzene and that C_6H_5 is the phenyl group.)

(a) $CH_3CH_2CH=CHCH_2CH_3 + H_2O \xrightarrow{\text{acid catalyst}}$

(b)
$$\underset{\underset{CH_3}{|}}{CH_3CHCH}=CH_2 + H_2 \xrightarrow{\text{Ni catalyst}}$$

(c) $C_6H_6 + Br_2 \xrightarrow{FeBr_3}$ *Bromo* (handwritten)

(d) $CH_3CH_2CH_2CH_2CH_2CH_3 + H_2SO_4(\text{concd}) \longrightarrow$

(e) (cyclopentane ring) $+ H_2O \xrightarrow{\text{acid catalyst}}$ *cyclo pentano* (handwritten)

(f) $CH_2=CHCH_2CH_2CH_3 + H_2 \xrightarrow{\text{Ni catalyst}}$

(g) (benzene ring)$-CH_2CH_3 + H_2SO_4(\text{concd}) \longrightarrow$

(h) $C_6H_5CH=CHC_6H_5 + H_2O \xrightarrow{\text{acid catalyst}}$

(i) (cyclopentane ring)$-CH_3 + H_2 \xrightarrow{\text{Ni catalyst}}$

(j) $CH_3CH_2CH(CH_3)_2 + O_2 \xrightarrow[\substack{\text{(Balance the} \\ \text{equation.)}}]{\substack{\text{complete} \\ \text{combustion}}}$

(k) $C_6H_6 + H_2O \xrightarrow{\text{acid catalyst}}$

13.57 Write the structures of the products in the following situations. If no reaction is to be expected, write "no reaction."

(a) $CH_2{=}CHCH_2CH_2CH{=}CH_2 + 2H_2 \xrightarrow{\text{Ni catalyst}}$

(b) [cyclopentene with CH_3] $+ H_2O \xrightarrow{\text{acid catalyst}}$ *1 methyl cyclopentanol*

(c) $\begin{array}{c} H_3C \quad CH_3 \\ | \quad\quad | \\ CH_3C{=}CCH_3 \end{array} + HBr \longrightarrow$ *2,3-dimethyl 2 bromobutane*

(d) [cyclohexyl-cyclopentyl] $+ H_2SO_4\text{(concd)} \longrightarrow$

(e) [cyclopentane with =CH_2] $+ Cl_2 \longrightarrow$ *[cyclobutane with CH₂Cl and Cl] cloromethy 1-cloro cyclopenten*

(f) [benzene ring with two CH_3 groups] $\xrightarrow[\text{(Balance the equation.)}]{\text{complete combustion}}$

(g) $C_6H_6 + Cl_2 \xrightarrow{\text{FeCl}_3}$

(h) $C_6H_6 + NaOH(aq) \longrightarrow$

(i) $\begin{array}{c} CH_3 \\ | \\ CH_3CH_2C{=}CHCH_3 \end{array} + HCl \longrightarrow$

(j) [cyclopentane] $+ NaOH(aq) \longrightarrow$

(k) $CH_3CH{=}CHCH_2CH{=}CH_2 + 2Cl_2 \longrightarrow$

(l) $C_6H_5CH{=}CHCH_3 + Br_2 \longrightarrow$

ALCOHOLS, PHENOLS, ETHERS, AND THIOALCOHOLS

14

The sugar in the juice of grapes, as from this vineyard in the Sonoma Valley of California, changes to the ethyl alcohol of wine during fermentation. Substances with alcohol groups occur widely at the molecular level of life and we'll learn several properties of alcohols in this chapter.

14.1

OCCURRENCE, TYPES, AND NAMES OF ALCOHOLS

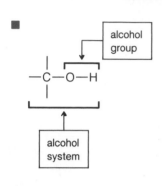

alcohol group

alcohol system

■ We'll soon learn about systems with the OH group that are not alcohols.

In the alcohol family, molecules have the OH group attached to a saturated carbon atom.

The alcohol system is one of the most widely occurring in nature. It is present in most of the major kinds of organic substances found in living things such as carbohydrates, proteins, and nucleic acids. The chemistry of carbohydrates, for example, is little more than the chemistry of alcohols (this chapter) and that of aldehydes and ketones (next chapter). Members of the alcohol family also include many common commercial products such as wood alcohol (methanol), rubbing alcohol (2-propanol), beverage alcohol (ethanol), and compounds in antifreezes (see Special Topic 14.1).

In **alcohols,** the OH group is covalently held by a *saturated* carbon atom, a carbon atom from which only *single* bonds extend. Only when present in this form is the OH group called the **alcohol group.** Alcohols with one OH group per molecule as the sole functional group are called the *simple alcohols* or **monohydric alcohols** (see Table 14.1). Molecules of **dihydric alcohols,** sometimes called **glycols,** have two OH groups: 1,2-ethanediol (ethylene glycol) is an example (Table 14.1). A **trihydric alcohol** is one whose molecules have three alcohol groups, such as 1,2,3-propanetriol (glycerol).

CH$_2$—CH$_2$
|　　|
OH　OH

1,2-Ethanediol
(ethylene glycol,
a dihydric alcohol)

CH$_2$—CH—CH$_2$
|　　|　　|
OH　OH　OH

1,2,3-Propanetriol
(glycerol, a trihydric
alcohol)

CH$_2$—CH—CH—CH—CH—CH
|　　|　　|　　|　　|　　‖
OH　OH　OH　OH　OH　O

Glucose, a sugar
(open form of molecule)

■ The following system, the 1,1-diol, is unstable; only rare examples are known.

OH
|
R—C—OH
|
R'

1,1-Diols

Many substances, particularly among the carbohydrates, have several OH groups per molecule. One important structural restriction on stable alcohols is that *almost no stable system is known in which one carbon holds two or more OH groups.* If such should form during a reaction, it breaks up. (More will be said about this in Chapter 15.)

The Alcohol Group Occurs as a 1°, 2°, or 3° System An alcohol is classified as primary (1°), secondary (2°), or tertiary (3°) according to the kind of carbon that holds the OH group. When the OH group is held by a 1° carbon, one that has only one carbon atom directly joined to it, the alcohol is a **primary alcohol.** In a **secondary alcohol,** the OH group is held by a 2° carbon. When the OH group is joined to a 3° carbon, the alcohol is a **tertiary alcohol.** We will see the usefulness of these subclasses when we learn that not all alcohols respond in the same way to oxidizing agents.

■ Pronounce 1° as *primary,* 2° as *secondary,* and 3° as *tertiary.*

■ The R— groups in 2° and 3° alcohols don't have to be the same.

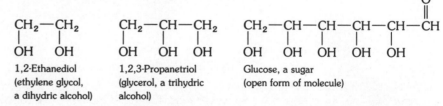

R—CH$_2$—OH

Primary
alcohol

R—CH—OH
　　|
　　R'

Secondary
alcohol

　　R'
　　|
R—C—OH
　　|
　　R''

Tertiary
alcohol

TABLE 14.1
Some Common Alcohols

Name[a]	Structure	BP (°C)
Methanol	CH_3OH	65
Ethanol	CH_3CH_2OH	78.5
1-Propanol	$CH_3CH_2CH_2OH$	97
2-Propanol	CH_3CHCH_3 $\quad\ \ \vert$ $\quad\ \ OH$	82
1-Butanol	$CH_3CH_2CH_2CH_2OH$	117
2-Butanol (*sec*-butyl alcohol)	$CH_3CH_2CHCH_3$ $\qquad\quad \vert$ $\qquad\quad OH$	100
2-Methyl-1-propanol (isobutyl alcohol)	$\qquad\ CH_3$ $\qquad\ \ \vert$ CH_3CHCH_2OH	108
2-Methyl-2-propanol (*t*-butyl alcohol)	$\qquad CH_3$ $\qquad\ \vert$ CH_3COH $\qquad\ \vert$ $\qquad CH_3$	83
1,2-Ethanediol (ethylene glycol)	$CH_2\!-\!CH_2$ $\ \vert\qquad\ \vert$ $OH\quad OH$	197
1,2-Propanediol (propylene glycol)	$CH_3\!-\!CH\!-\!CH_2$ $\qquad\ \vert\qquad\ \vert$ $\qquad OH\quad OH$	189
1,2,3-Propanetriol (glycerol)	$CH_2\!-\!CH\!-\!CH_2$ $\ \vert\qquad\ \vert\qquad\ \vert$ $OH\quad OH\quad OH$	290

[a] The IUPAC names with the common names in parentheses.

PRACTICE EXERCISE 1

Classify each of the following as monohydric or dihydric. For each found to be monohydric, classify it further as 1°, 2°, or 3°. If the structure is too unstable to exist, state so.

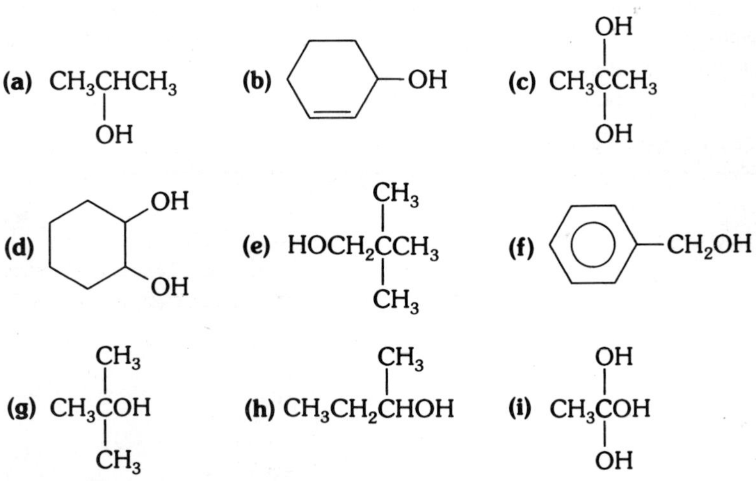

(a) CH_3CHCH_3
$\qquad\ \ \vert$
$\qquad\ \ OH$

(b) —OH

(c) $\quad\ \ OH$
$\qquad\ \ \vert$
CH_3CCH_3
$\qquad\ \ \vert$
$\qquad\ \ OH$

(d) —OH / OH

(e) $\qquad\quad CH_3$
$\qquad\qquad \vert$
$HOCH_2CCH_3$
$\qquad\qquad \vert$
$\qquad\qquad CH_3$

(f) —CH_2OH

(g) $\quad\ CH_3$
$\qquad\ \vert$
CH_3COH
$\qquad\ \vert$
$\quad\ CH_3$

(h) $\qquad\qquad\ CH_3$
$\qquad\qquad\quad \vert$
CH_3CH_2CHOH

(i) $\quad\ \ OH$
$\qquad\ \vert$
CH_3COH
$\qquad\ \vert$
$\qquad\ OH$

ALCOHOLS IN OUR DAILY LIVES

Methanol (methyl alcohol, wood alcohol) When taken internally in enough quantity, methanol causes either blindness or death. In industry, it is used as the raw material for making formaldehyde (which is used to make polymers), as a solvent, and as a denaturant (poison) for ethanol. It is also used as the fuel in "canned heat" (e.g., Sterno), as well as in burners for fondue pots.

Most methanol is made by the reaction of carbon monoxide with hydrogen under high pressure and temperature:

$$2H_2 + CO \xrightarrow[\substack{350-400\ °C \\ ZnO/Cr_2O_3\ catalyst}]{3000\ lb/in.^2} CH_3OH$$

Ethanol (ethyl alcohol, grain alcohol) Some ethanol is made by the fermentation of sugars, but most is synthesized by the addition of water to ethene in the presence of a catalyst. A 70% (v/v) solution of ethanol in water is used as a disinfectant.

In industry, ethanol is used as a solvent and to prepare pharmaceuticals, perfumes, lotions, and rubbing compounds. For these purposes, the ethanol is adulterated by poisons that are very difficult to remove so that the alcohol cannot be sold or used as a beverage. (Nearly all countries derive revenue by taxing potable, i.e., drinkable, alcohol.)

Ethanol, a toxic substance responsible for widespread human misery, is absorbed directly into the bloodstream along any part of the intestinal tract. Enzymes in the liver work to detoxify it. One enzyme takes it to the aldehyde stage, producing acetaldehyde (CH_3CHO), which is chiefly responsible for liver fibrosis. In time, the acetaldehyde is oxidized to the acetate ion, $CH_3CO_2^-$, which can be metabolized normally. Fetal alcohol syndrome appears to be caused directly by excessive ethyl alcohol itself, not by its oxidation products. This syndrome leads to babies with impaired mental abilities and many other disorders.

2-Propanol (isopropyl alcohol) 2-Propanol is a common substitute for ethanol for giving back rubs. In solutions with concentrations from 50% to 99% (v/v), 2-propanol is used as a disinfectant. It is twice as toxic as ethanol.

1,2-Ethanediol (ethylene glycol), **and 1,2-Propanediol** (propylene glycol) Ethylene glycol and propylene glycol are the chief components in permanent-type antifreezes. Their great solubility in water and their very high boiling points make them ideal for this purpose. An aqueous solution that is roughly 50% (v/v) in either glycol does not freeze until about $-40\ °C\ (-40\ °F)$. Ethylene glycol has a sweet taste, but it is highly toxic. The lethal dose for adults is about 100 mL. Propylene glycol, on the other hand, is far less toxic (possibly because the anions of its oxidation products, pyruvic acid and acetic acid, are normally formed in metabolism).

1,2,3-Propanetriol (glycerol, glycerin) Glycerol, a colorless, syrupy liquid with a sweet taste, is freely soluble in water and insoluble in nonpolar solvents. It is one product of the digestion of the fats and oils in our diets. Because it has three OH groups per molecule, each capable of hydrogen bonding to water molecules, glycerol can draw moisture from humid air. It is sometimes used as a food additive to help keep foods moist.

An oily compound made from glycerol and nitric acid, **nitroglycerin,** is a powerful explosive. When pure, it detonates from concussion. Interestingly, although nitroglycerin is toxic, it has a place in medicine.

$$O_2NOCH_2CHCH_2ONO_2$$
$$|$$
$$ONO_2$$

Nitroglycerin
(1,2,3-glyceryl trinitrate)

People who have periodic attacks of intense pain (angina pectoris), centered in heart muscle because of vasoconstriction (constriction of the blood vessels), are able to administer to themselves carefully controlled amounts of a vasodilator (a dilator of blood vessels). Nitroglycerin is commonly used for this purpose.

Sugars All carbohydrates consist of polyhydroxy compounds, and we will study them in Chapter 19.

Not All Compounds with the OH Group Are Alcohols   As you can see by the following structures, the OH group also occurs in the family of the phenols, where it is attached to a benzene ring, and in the family of the carboxylic acids, where it is attached to a carbon that has a double bond to oxygen.

■ Phenol is only the simplest member of the family of *phenols,* compounds with OH groups attached to benzene rings that may carry any number and kind of other ring substituents.

alcohol group

alcohol system

Phenol   Carboxylic acid   Enol system (unstable)

When the OH is attached to an alkene group, the system is called an *enol* (ene + ol), but it is unstable. Most of our attention in this chapter will be given to simple alcohols but with some study of phenols. We will study the carboxylic acids in a later chapter.

PRACTICE EXERCISE 2

Classify the following as alcohols, phenols, or carboxylic acids.

(a) CH_3—⬡—CH_2OH *alcohol*

(b) CH_3—⬡—OH *phenol*

(c) CH_3—⬡—$\overset{\displaystyle O}{\overset{\|}{C}}OH$ *carboxylic*

(d) CH_2=$CHOH$ *alcohols*

(e) $CH_3CH_2CH_2CH_2OH$ *alcohols*

(f) ⬡—OH *alcohol*

Common Names of Alcohols Are Popular When Their Alkyl Groups Are Easily Named   Simple alcohols have common names devised by writing the word *alcohol* after the name of the alkyl group holding the OH group. For example,

CH_3OH	CH_3CH_2OH	$CH_3\underset{OH}{CH}CH_3$	$CH_3\underset{CH_3}{\overset{CH_3}{CH}}CH_2OH$	$CH_3\underset{CH_3}{\overset{CH_3}{C}}OH$
Methyl alcohol	Ethyl alcohol	Isopropyl alcohol	Isobutyl alcohol	t-Butyl alcohol

IUPAC Names of Alcohols End in *-ol*   The IUPAC rules for naming alcohols are similar to those for naming alkanes. The full name of the alcohol is based on the idea of a *parent alcohol* that has substituents on its carbon chain. The rules are as follows.

1.   Determine the parent alcohol by selecting the longest chain of carbons *that includes the carbon atom to which the OH group is attached.* Name the parent alcohol by changing the

name ending of the alkane that corresponds to this chain from *-e* to *-ol*. Examples are

$$CH_3OH \qquad CH_3CH_2CH_2OH \qquad \overset{\displaystyle CH_3}{\underset{\displaystyle |}{CH_3CHCH_2CH_2OH}}$$

Methanol   Propanol   A substituted butanol

(complete name;   (incomplete name;   (incomplete name;

parent alkane   parent alkane   parent alkane

is methane)   is propane)   is butane)

■ No number is needed to specify the location of the OH group in the names methanol and ethanol.

2. **Number the parent chain from whichever end gives the carbon that holds the OH group the lower number.** Be sure to notice this departure from the IUPAC rules for numbering chains for alkanes; in the rules for alkanes the location of the first branch determines the direction of numbering. With alcohols, *the location of the OH group takes precedence over alkyl groups or halogen atoms in deciding the direction of numbering.* Examples are

$$\underset{\displaystyle 4 \quad 3 \quad 2 \quad 1}{\overset{\displaystyle CH_3}{\underset{\displaystyle |}{CH_3CHCH_2CH_2OH}}} \qquad \underset{\displaystyle 7 \ 6 \ 5 \quad 4 \quad 3 \ 2 \ 1}{CH_3CCH_2CH_2CHCH_2CH_3}$$

3-Methyl-1-butanol   6,6-Dimethyl-3-heptanol

Not 2-methyl-4-butanol   *Not* 2,2-dimethyl-5-heptanol

3. Write the number that locates the OH group in front of the name of the parent, and separate this number from the name of the parent by a hyphen. Then, as prefixes to what you have just written, assemble the names of the substituents and their location numbers. Use commas and hyphens in the usual way. For illustrations, see the examples above that follow rule 2.

4. When two or more OH groups are present, use name endings such as *-diol* (for two OH groups), *-triol,* and so forth. Immediately in front of the name of the *parent portion* of the alcohol, write the two, three, or more numbers that show the locations of the OH groups. For example,

$$\underset{\displaystyle OH}{\overset{\displaystyle CH_3}{CH_3CCH_2OH}} \qquad \underset{\displaystyle HO \quad OH}{CH_3CHCHCH_2OH}$$

2-Methyl-1,2-propanediol   1,2,3-Butanetriol

5. If no parent alcohol name is possible or convenient, then the OH group can be treated as another substituent, and it is named *hydroxy.* For example,

$$HO-\!\!\!\bigcirc\!\!\!-\overset{\displaystyle O}{\overset{\displaystyle \|}{C}}OH$$

4-Hydroxybenzoic acid

EXAMPLE 14.1
Using the IUPAC Rules to Name an Alcohol

What is the name of the following compound?

$$\underset{\displaystyle CH_2CH_2CH_3}{CH_3CH_2CH_2CH - CCH_2OH} \quad \overset{\displaystyle CH_3CH_2 \quad CH_3}{}$$

ANALYSIS The compound is in the alcohol family, so the ending to its IUPAC name is *-ol*. The IUPAC rules for alcohols require the parent chain to be the longest carbon chain *that includes the carbon atom to which the OH group is attached.* This is a six-carbon chain in the structure. The IUPAC rules then require us to number the chain from whichever of its ends lets the carbon bearing the OH group have the lower number. Thus

$$CH_3CH_2CH_2\underset{6\quad 5\quad 4\quad 3}{CH}-\underset{2\ |\ 1}{\overset{\overset{\displaystyle CH_3}{|}}{C}}CH_2OH$$

with CH_3CH_2 on carbon 3 and $CH_2CH_2CH_3$ on carbon 2.

The final name must end in -1-hexanol. Carbon 2 holds both a methyl and a propyl group, and carbon 3 has an ethyl group.

SOLUTION The final name, arranging the alkyl groups alphabetically, is

3-Ethyl-2-methyl-2-propyl-1-hexanol

■ The longest carbon chain in the molecule has eight carbon atoms, but it doesn't include the OH group.

PRACTICE EXERCISE 3

Name the following compounds by the IUPAC system.

(a) $CH_3\overset{\overset{\displaystyle CH_3}{|}}{C}HCH_2CH_2CH_2OH$

(b) $CH_3\overset{\overset{\displaystyle CH_3}{|}}{\underset{\underset{\displaystyle CH_3}{|}}{C}}OH$

(c) $CH_3CH_2\overset{\overset{\displaystyle CH_3}{|}}{\underset{\underset{\displaystyle CH_2CH_2CH_3}{|}}{C}}CH_2OH$

(d) $HOCH_2\overset{\overset{\displaystyle CH_3}{|}}{C}HCH_2OH$

14.2
PHYSICAL PROPERTIES OF ALCOHOLS

Hydrogen bonding dominates the physical properties of alcohols.
The OH group is quite polar, and it can both <u>donate</u> and <u>accept</u> hydrogen bonds. Hydrogen bonding gives alcohols much <u>higher</u> boiling points and much <u>greater</u> solubilities in water than hydrocarbons of the same formula mass.

The effect of the OH group on water solubility is particularly important at the molecular level of life, because some substances in cells must be in solution and others must not. Moreover, by using special enzymes in the liver, the body attaches OH groups to the molecules of many toxic substances to make them more soluble and thus more easily carried in the blood and eliminated in the urine.

Hydrogen Bonds between Molecules Raise Boiling Points Table 14.2 compares the boiling points of some alcohols to those of alkanes with comparable formula masses. The differences in boiling points are caused by hydrogen bonding made possible by the considerable difference in the electronegativities of O and H (page 171). The hydrogen bond, remember, is a force of attraction between opposite *partial* charges, such as the $\delta+$ charge on H in the OH group and the $\delta-$ charge on the O of another group. No such partial charges exist in molecules of alkanes because C and H have nearly identical <u>electronegativities</u>. An individual

■ The *donor* OH group has the H from which the H-bond ($\cdots$) extends to the $\delta-$ site on the *acceptor*,

■ H-bond *acceptor*

$$\begin{array}{c} H \\ \diagdown \delta- \\ O-R \\ \vdots \\ H\ \delta+ \\ R-O \end{array}$$

H-bond *donor*

FIGURE 14.1
Hydrogen bonding in alcohols (*a*) and in water (*b*).

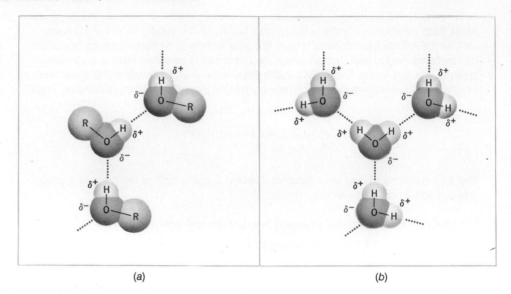

(a)   (b)

alcohol molecule is attracted to two neighboring molecules by two hydrogen bonds (see Fig. 14.1*a*). Water molecules have three hydrogen bonds between them (see Fig. 14.1*b*). So water, despite having a lower formula mass than methyl alcohol, has a higher boiling point.

Ethylene glycol (Table 14.2), a dihydric alcohol, has four hydrogen bonds between molecules, two from each OH group, and it boils nearly 200 °C higher than butane and 100 °C higher than water. You can see by these data how significantly the OH group provides forces of attraction between molecules.

The OH Group Makes Compounds More Soluble in Water   Methane, like all hydrocarbons, is insoluble in water.[1] Methyl alcohol, in contrast, dissolves in water in all proportions. The difference is caused by the OH group. Hydrogen bonds can be formed between methyl alcohol molecules and water molecules, and this enables methyl alcohol molecules to slip into the hydrogen-bonding network in water (see Fig. 14.2). The CH_3 group in CH_3OH is too small to interfere.

■ Only three elements, F, O, and N, have atoms that are electronegative enough to participate significantly in hydrogen bonds.

TABLE 14.2
The Influence of the Alcohol Group on Boiling Points

Name	Structure	Formula Mass	BP (°C)	Difference in BP
Ethane	CH_3CH_3	30	−89	
				154
Methyl alcohol	CH_3OH	32	65	
Propane	$CH_3CH_2CH_3$	44	−42	
				120
Ethyl alcohol	CH_3CH_2OH	46	78	
Butane	$CH_3CH_2CH_2CH_3$	58	0	
				197
Ethylene glycol	$HOCH_2CH_2OH$	62	197	

[1] Nothing is *totally* insoluble in water. The size of Avogadro's number (6.02×10^{23}) is so great that even if something can form a solution with a molarity of only, say, 1×10^{-14} mol/L — immeasurably small — there are still roughly a little over a billion (10^9) molecules of solute in each liter of solution!

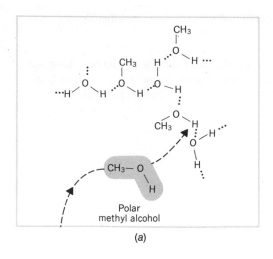

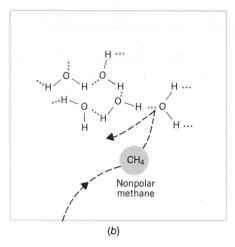

FIGURE 14.2
How a short-chain alcohol dissolves in water. (*a*) The alcohol molecule can take the place of a water molecule in the hydrogen-bonding network of water. (*b*) An alkane molecule cannot break into the hydrogen-bonding network in water, so the alkane cannot dissolve.

As the size of the R group in a monohydric alcohol molecule increases, however, alcohol molecules become more and more alkane-like. A long R group interferes with the alcohol's ability to dissolve in water. In 1-decanol, for example,

$$CH_3CH_2CH_2CH_2CH_2CH_2CH_2CH_2CH_2CH_2OH$$
1-Decanol

the small, water-like OH group is overwhelmed by the long hydrocarbon chain. Water molecules have no attraction for this part of the 1-decanol molecule. The flexings and twistings of its long chain interfere too much with the hydrogen-bonding networks in water, and water molecules will not separate to let 1-decanol molecules into solution. This alcohol, and most with five or more carbons, are thus insoluble in water. However, they do dissolve in such nonpolar solvents as diethyl ether, gasoline, and petroleum ether.

PRACTICE EXERCISE 4

1,2-Propanediol (propylene glycol) and 1-butanol have similar formula masses. In which of these two compounds is hydrogen bonding between molecules more extensive and stronger? How do the data in Table 14.1 support the answer? Which would be more soluble in water? ∎

14.3

CHEMICAL PROPERTIES OF ALCOHOLS

The loss of water (dehydration) and the loss of hydrogen (oxidation) are two very important reactions of alcohols.
Alcohols react with both inorganic and organic compounds, but we will study only the inorganic reactants in this chapter. Before continuing, however, we need to point out some properties that the alcohols do *not* have, but which might be assumed because of their OH groups.

Despite the OH Group, Alcohols Are Not Strong Bases We can think of alcohols as mono-alkyl derivatives of water and, like water, they are extremely weak proton donors (acids) and proton acceptors (bases). Alcohol molecules do not ionize to give either OH⁻ ions or H⁺ ions. Water, as you know, only *very* slightly ionizes; at room temperature, not much more than about 0.0000001% of its molecules are ionized. *The percentage ionization of any*

■ When an alcohol dissolves in water, it doesn't raise or lower the pH.

alcohol group in water is even smaller than that of water. A solution of methyl alcohol (or any alcohol) in water thus has a neutral pH.

Alcohols Can Be Dehydrated to Alkenes   In laboratory vessels, the action of heat and a strong acid catalyst causes the dehydration of an alcohol to an alkene. A water molecule splits out and a carbon-carbon double bond emerges. The pieces of the water molecule, one H and one OH, come from *adjacent* carbons.

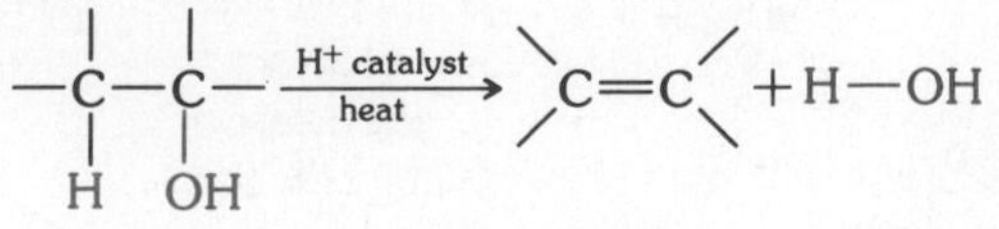

$$-\underset{\underset{H}{|}}{C}-\underset{\underset{OH}{|}}{C}- \xrightarrow[\text{heat}]{\text{H}^+ \text{ catalyst}} \diagdown C = C \diagup + H-OH$$

$$\qquad\text{Alcohol}\qquad\qquad\qquad\text{Alkene}$$

Specific examples are as follows.

$$CH_3CH_2OH \xrightarrow[\text{heat}]{\text{H}^+ \text{ catalyst}} CH_2=CH_2 + H_2O$$

$$\qquad\text{Ethanol}\qquad\qquad\qquad\text{Ethene}$$

$$CH_3CH_2\underset{\underset{OH}{|}}{C}HCH_3 \xrightarrow[\text{heat}]{\text{H}^+ \text{ catalyst}} CH_3CH=CHCH_3 + CH_3CH_2CH=CH_2 + H_2O$$

$$\qquad\text{2-Butanol}\qquad\qquad\quad\text{2-Butene}\qquad\qquad\text{1-Butene}$$
$$\qquad\qquad\qquad\qquad\qquad\quad\text{(chief product)}\qquad\text{(minor product)}$$

Special Topic 14.2 explains how the acid works as the catalyst for these reactions.

When It Is Possible for Two Alkenes to Be Made, the More Highly Branched Alkene Forms   Water can split out in two ways from 2-butanol (as shown above). Molecules of 2-butanol have H atoms on *two* carbons adjacent to the carbon holding the departing OH group. Two alkenes are therefore possible, 2-butene and 1-butene, and some of each forms. When options like this exist, however, one alkene predominates, generally *the more highly branched alkene*. This is the alkene with the greatest number of alkyl groups attached to the double bond. 2-Butene is more branched than 1-butene because it has two alkyl groups at the double bond (two methyl groups) and 1-butene has only one (an ethyl group).

In acid-catalyzed dehydrations of alcohols, the more branched alkene predominates because it is the more stable. Information about the relative stabilities of isomeric alkenes comes from combustion experiments. For example, both 1-butene and 2-butene are C_4H_8 and the combustion of either occurs as follows.

$$C_4H_8(g) + 6O_2(g) \longrightarrow 4CO_2(g) + 4H_2O + \text{heat of combustion}$$

However, slightly less heat per mole is released when 2-butene is burned, despite the fact that it has the same number of carbons and hydrogens as 1-butene. The only way by which 2-butene can release less energy than 1-butene is *to have less energy initially*. Having less energy always means being more stable, so 2-butene must be slightly more stable than 1-butene. None of this explains *why* 2-butene is more stable. (The explanation is complex.) The release of less energy by the combustion of 2-butene compared to 1-butene is mentioned only to give one piece of experimental support for the generalization: *the more branched alkene is the more stable*.

The complication of two possible products of dehydration is not a problem when the reaction happens to an alcohol system in the body. Such dehydrations are enzyme-catalyzed, not acid-catalyzed, and enzymes direct reactions in very specific ways. It is even possible for the less stable double bond to be produced when enzymes catalyze the reaction.

■ $CH_3-CH=CH-CH_3$

2-Butene
(two alkyl groups)

$CH_2=CH-CH_2CH_3$

1-Butene
(one alkyl group)

■ Enzymes are exceedingly selective in what they do and how they control reactions.

HOW ACIDS CATALYZE THE DEHYDRATION OF ALCOHOLS

When a strong acid is added to an alcohol, the first chemical event is the ionization of the acid. The reaction is exactly analogous to the ionization of a strong acid when it is added to water; a proton transfers from the acid to the oxygen atom of a solvent molecule. Because sulfuric acid is often used as the catalyst in the dehydration of an alcohol, we will use it to illustrate these reactions. Its ionization in water establishes the following equilibrium, with the products very strongly favored.

$$HO_3SO-H + :\overset{H}{\underset{H}{O}:} \rightleftharpoons HO_3SO^- + H-\overset{+}{\underset{H}{O}:}H$$

Sulfuric acid,	Water	Hydrogen	Hydronium
H_2SO_4		sulfate ion	ion

The same ionization happens in ethyl alcohol. (Be sure to notice the similarity to the previous reaction.)

$$HO_3SO-H + :\overset{CH_2CH_3}{\underset{H}{O}:} \rightleftharpoons$$

$$HO_3SO^- + H-\overset{+}{\underset{H}{O}}:{}^{CH_2CH_3}$$

Protonated form
of ethyl alcohol

All the covalent bonds to the oxygen atoms in either the hydronium ion or in the protonated form of the alcohol are weak, including, in the latter, *the covalent bond to carbon.* This bond is weakened through the action of the catalyst. As ions and molecules bump into one another, some of the ions of the protonated form of the alcohol break up, as we can illustrate using the protonated form of ethyl alcohol.

$$H-\overset{+}{\underset{H}{O}:}{}^{CH_2CH_3} \rightleftharpoons CH_3CH_2{}^+ + :\overset{H}{\underset{H}{O}}$$

Protonated form	Ethyl
of ethyl alcohol	carbocation

All carbocations are unstable, because a carbon atom does not handle anything less than an octet very well. The octet for carbon in the ethyl carbocation is restored when a proton from the carbon adjacent to the site of the positive charge transfers to a proton acceptor. As this proton transfers, the electron pair of its covalent bond to carbon pivots in to form the second bond of the emerging double bond. It is a smooth, synchronous operation. One acceptor for the proton is the hydrogen sulfate ion, and its acceptance of a proton restores the catalyst, as follows.

$$HO_3SO^- + H-CH_2-CH_2{}^+ \longrightarrow$$

Ethyl
carbocation

$$CH_2{=}CH_2 + HO_3SO-H$$

Once water starts to appear as another product, its molecules can also accept the proton from the ethyl carbocation (to give H_3O^+), but this is also equivalent to the recovery of the catalyst.

EXAMPLE 14.2
Writing the Structure of the Alkene That Forms When an Alcohol Undergoes Dehydration

What is the product of the dehydration of isobutyl alcohol?

$$\overset{\displaystyle CH_3}{\underset{}{CH_3CHCH_2OH}}$$

Isobutyl alcohol

ANALYSIS Predicting the product of the dehydration of this alcohol involves rewriting the structure of the alcohol but leaving off the OH group and one H from a carbon adjacent to the carbon that holds the OH group. A double bond is then written between these two carbon atoms.

SOLUTION

$$CH_3CC{=}CH_2$$

with CH_3 above the central carbon

2-Methylpropene

■ When some alcohols undergo dehydration, their carbon skeletons rearrange, but we won't study any examples of this.

CHECK Ask the following questions of the answer. Is it an *alkene*? (Alcohol dehydrations give alkenes.) Is its carbon skeleton the same as in the starting material? (Changes in the carbon skeleton do not occur in all of the examples that we will study.) Does each carbon have four bonds from it? (The rules of covalence must be obeyed.) By answering ''yes'' to these questions you have made a thorough check. ■

PRACTICE EXERCISE 5

Write the structures of the alkenes that can be made by the dehydration of the following alcohols.

(a) $CH_3CH_2CH_2OH$ **(b)** CH_3CHCH_3 with OH below **(c)** CH_3COH with CH_3 above and CH_3 below **(d)** cyclohexane ring—OH

1° and 2° Alcohols Are Dehydrogenated by Strong Oxidizing Agents We learned in an earlier chapter that an oxidation number becomes more positive for a species being oxidized. To use this definition to recognize whether something has been oxidized, however, requires the calculation of oxidation numbers. Organic chemists often use a shortcut summarized by the following two rules of thumb. With organic compounds,

1. An *oxidation* is the loss of H or the gain of O by a molecule.
2. A *reduction* is the loss of O or the gain of H by a molecule.

■ In the body, the enzyme *alcohol dehydrogenase* (ADH) oxidizes methyl alcohol (wood alcohol) to formaldehyde, $CH_2{=}O$, which has toxic effects and can cause blindness and death.

■ The term *in vitro* means ''in a glass vessel,'' that is, carried out in laboratory glassware. The term *in vivo* means in a living cell.

The oxidation of an alcohol is an example of the loss of H. Sometimes, therefore, the reaction is called *dehydrogenation*. Enzymes that catalyze such reactions in living systems are called *dehydrogenases*.

In studying dehydrogenation or oxidation, we are particularly interested in what happens to the organic molecule being oxidized. What does it change into? To serve this limited interest, we make two further simplifications. We largely use unbalanced ''reaction sequences,'' not balanced equations, and we'll use the symbol (O) *for any oxidizing agent that can bring about the oxidation given by a reaction sequence.*

One common strong oxidizing agent used *in vitro* is potassium permanganate, $KMnO_4$, in which the actual oxidizing agent is the deeply purple colored permanganate ion, MnO_4^-. Another common strong oxidizing agent used to oxidize alcohols is sodium dichromate, $Na_2Cr_2O_7$, which furnishes the brightly orange colored dichromate ion, $Cr_2O_7^{2-}$.

When an oxidizing agent causes an alcohol molecule to lose hydrogen, molecular hydrogen (H_2) does not itself form either *in vitro* or *in vivo*. Instead, the hydrogen atoms end up in a water molecule whose oxygen atom comes from the oxidizing agent. One H comes from the OH group of the alcohol, and the other H comes from the carbon that has been holding the OH group. Left behind in the organic molecule is a carbon–oxygen double bond, a carbonyl group. Study the following schematic carefully to learn where precisely the two H atoms originate.

What makes the reaction an *oxidation* of the alcohol group is specifically the *loss of the electron pair* of the C—H bond in the alcohol system. In the schematic above, we see this pair first on H:$^-$ and then incorporated into the water molecule. The H:$^-$ ion (the hydride ion) never becomes free in the aqueous medium. We show it in brackets above only to help you track the oxidation. In the end, a water molecule is exactly where the electron pair of the C—H bond of the alcohol system finally lodges when the alcohol is oxidized *in vivo* by a series of reactions called the *respiratory chain* (to be studied in Chapter 25). When (O) is MnO_4^- or $Cr_2O_7^{2-}$, however, the electrons of the oxidized alcohol group are accepted by these species to change the oxidation states of the central atoms. Thus the oxidation state of Mn is reduced from $+7$ in MnO_4^- to $+4$ in MnO_2. When (O) is $Cr_2O_7^{2-}$, the electrons end up in reducing the oxidation state of Cr from $+6$ in $Cr_2O_7^{2-}$ to $+3$ in Cr^{3+}. By *accepting* the electrons, MnO_4^- and $Cr_2O_7^{2-}$ are reduced.

■ The powerfully basic hydride ion, if let loose, would react with water to give the hydroxide ion and hydrogen gas.

$$H:^- + H_2O \rightarrow H-H + OH^-$$

3° Alcohols Cannot Be Dehydrogenated Only 1° and 2° alcohols can be oxidized by the loss of 2 H atoms. Molecules of 3° alcohols do not have an H atom on the carbon that holds the OH group, so 3° alcohols cannot be oxidized by dehydrogenation.

In Aqueous Systems, Strong Oxidizing Agents Are Used to Change the 1° Alcohol Group to a Carboxyl Group The organic product of the oxidation of either a 1° or a 2° alcohol has a carbon–oxygen double bond. Each subclass of alcohol, however, is oxidized to a different organic family. A 1° alcohol is oxidized first to an aldehyde, but because *aldehydes are more easily oxidized than 1° alcohols,* particularly in aqueous media, it is seldom practical to use $MnO_4^-(aq)$ or $Cr_2O_7^{2-}(aq)$ to prepare aldehydes. As soon as aldehyde groups appear in the presence of these ions, the aldehyde groups begin to use up oxidizing agent and change to carboxyl groups. In the presence of sufficient oxidizing agent, the successive stages in the oxidation of a 1° alcohol group are

■

$$RCH_2OH \xrightarrow[\text{$Cr_2O_7^{2-}(aq)$}]{\text{(O)} \atop \text{$MnO_4^-(aq)$ or}} RCH \xrightarrow{\text{more (O)}} RCOH$$

1° Alcohol   Aldehyde   Carboxylic acid

In the lab, therefore, when either aqueous permanganate or dichromate is used, the oxidation of a 1° alcohol is generally carried out with enough oxidizing agent to take the oxidation of the alcohol all the way to its corresponding carboxylic acid. Carboxylic acids strongly resist further oxidation. The dichromate oxidation of 1-propanol to propanoic acid, for example, occurs by the following equation.

$$3CH_3CH_2CH_2OH + 2Cr_2O_7^{2-} + 16H^+ \longrightarrow 3CH_3CH_2CO_2H + 4Cr^{3+} + 11H_2O$$

1-Propanol   Propanoic acid

Enzyme-Catalyzed Oxidations Convert a 1° Alcohol Group to an Aldehyde Group The problem of halting the oxidation of a 1° alcohol at the aldehyde stage does not occur in body cells, because *different enzymes are required for each oxidation step.* One

■ Certain vitamins in the diet, like riboflavin and niacin, provide molecular acceptor units for H:⁻ in dehydrogenase enzymes.

enzyme handles the oxidation of a 1° alcohol group to the corresponding aldehyde. A different enzyme is needed to take an aldehyde group to the next oxidation stage, and this enzyme is generally not in the same place where the aldehyde is made. In general, remembering that (O) stands for an oxidizing agent capable of accomplishing the given oxidation, the reaction is

$$RCH_2OH \xrightarrow[\text{(enzyme-catalyzed)}]{\text{(O)}} \underset{\text{Aldehyde}}{R\overset{\displaystyle O}{\overset{\displaystyle \|}{C}}H} + H_2O$$

$$\underset{1° \text{ Alcohol}}{}$$

■ A dehydrogenase changes (temporarily) to its reduced form when it accepts H:⁻.

Hydrogen is removed by the transfer of the *pieces* of H—H. An oxidizing enzyme, a dehydrogenase, accepts H:⁻ from the CH unit that holds the OH group, and a proton, H⁺, slips away from the O atom of the OH group. With some enzymes, H⁺ is simply neutralized by the buffer system of the cell fluid. Other enzymes accept both H:⁻ and H⁺. We'll leave details to later chapters.

Because our chief interest lies in what can occur in living systems, we will not be further concerned about *in vitro* oxidations using strong, aqueous oxidizing agents except as they may be used in lab tests. What interests us almost entirely is what aldehyde can be made from a 1° alcohol *in vivo* and what carboxylic acid can eventually be formed from a 1° alcohol.

EXAMPLE 14.3
Writing the Structure of the Product of the Oxidation of a Primary Alcohol

What aldehyde and what carboxylic acid could be made by the oxidation of ethyl alcohol?

ANALYSIS The CH_2OH group is changed to $CH{=}O$ when a 1° alcohol is oxidized to an aldehyde. Anything attached to the carbon of the CH_2OH group, like the CH_3 group in our example, is retained in the final structure of the aldehyde. When the aldehyde is further oxidized to a carboxylic acid, the $CH{=}O$ group is changed to CO_2H.

SOLUTION The aldehyde corresponding to CH_3CH_2OH is $CH_3CH{=}O$, ethanal. The carboxylic acid is CH_3CO_2H, acetic acid.

CHECK Is the first product an *aldehyde*? (Does it have the $CH{=}O$ group?) Is the second product a *carboxylic acid*? (Does it have the CO_2H group, sometimes written COOH?) The oxidation of a 1° alcohol first gives an aldehyde and then, with more oxidizing agent, the carboxylic acid. Are the carbon skeletons of the two products identical with that of the 1° alcohol? Does each carbon have four bonds and each oxygen two? By answering "yes" to these questions you have made a thorough check.

PRACTICE EXERCISE 6

Write the structures of the aldehydes and carboxylic acids that can be made by the oxidation of the following alcohols.

(a) $\underset{\displaystyle \overset{\textstyle |}{CH_3}}{CH_3CHCH_2OH}$ (b) ⬡—CH_2OH

Secondary Alcohols Are Oxidized to Ketones Ketones strongly resist further oxidation, so they are easily made by the oxidation of 2° alcohols using strong oxidizing agents like MnO_4^- or $Cr_2O_7^{2-}$. *In vivo*, dehydrogenases accomplish the identical overall reaction, the dehydrogenation of a 2° alcohol group to a keto group. In general, for 2° alcohols,

$$\underset{\text{2° alcohol}}{\overset{\overset{\displaystyle OH}{|}}{RCHR'}} + (O) \longrightarrow \underset{\text{Ketone}}{\overset{\overset{\displaystyle O}{\|}}{RCR'}} + H_2O$$

Specific *in vitro* examples are

$$\underset{\text{2-Butanol}}{\overset{\overset{\displaystyle OH}{|}}{CH_3CHCH_2CH_3}} \xrightarrow{Cr_2O_7^{2-},\ H^+} \underset{\text{2-Butanone}}{\overset{\overset{\displaystyle O}{\|}}{CH_3CCH_2CH_3}}$$

Cyclohexanol $\xrightarrow{Cr_2O_7^{2-},\ H^+}$ Cyclohexanone

EXAMPLE 14.4
Writing the Structure of the Product of the Oxidation of a Secondary Alcohol

What ketone forms when 2-propanol is oxidized?

ANALYSIS 2-Propanol is a 2° alcohol, and every 2° alcohol has a CHOH group. The oxidation of a 2° alcohol strips the 2 H atoms from CHOH, creates a double bond between C and O, and so changes CHOH to C=O, a keto group. (The two H atoms emerge in a molecule of water.) *The fundamental skeleton of all of the heavy atoms in the 2° alcohol, like C and O, remains intact.*

SOLUTION

$$\underset{\text{2-Propanol}}{\overset{\overset{\displaystyle OH}{|}}{CH_3CHCH_3}} \text{ is oxidized to } \underset{\text{Propanone}}{\overset{\overset{\displaystyle O}{\|}}{CH_3CCH_3}} \text{ plus } H_2O$$

■ The name chemists commonly use for propanone is *acetone*.

CHECK Is the product a *ketone*? (°2 Alcohols are oxidized to ketones.) Does the product have the same carbon skeleton as the starting material? Does each carbon have four bonds and each oxygen atom two?

PRACTICE EXERCISE 7

Write the structures of the ketones that can be made by the oxidation of the following alcohols.

(a) $\overset{\overset{\displaystyle OH}{|}}{CH_3CHCH_2CH_3}$ (b) —$\overset{\overset{\displaystyle OH}{|}}{CHCH_3}$ (c)

PRACTICE EXERCISE 8

What are the products of the oxidation of the following alcohols? If the alcohol is a 1° alcohol, show the structures of both the aldehyde and the carboxylic acid that could be made, depending on the conditions. If the alcohol cannot be oxidized, write "no reaction."

$$
\begin{array}{cccc}
& \overset{\text{OH}}{|} & \overset{\text{CH}_3}{|} & \overset{\text{CH}_3}{|} & \overset{\text{OH}}{|} \\
\textbf{(a)}\ \ \text{CH}_3\text{CHCH}_2\text{CH}_2\text{OH} & \textbf{(b)}\ \ \text{HOCCH}_3 & \textbf{(c)}\ \ \text{CH}_3\text{CCH}_2\text{OH} & \textbf{(d)}\ \ \text{CH}_3\text{CHCHCH}_3 \\
& \underset{\text{CH}_3}{|} & \underset{\text{CH}_3}{|} & \underset{\text{CH}_3}{|}
\end{array}
$$

The foregoing discussion of the chemical properties of alcohols leaves us with the following structural "map signs."

1. Alcohols can be dehydrated to alkenes, with the most branched alkene generally forming *in vitro*.

2. 1° Alcohols are oxidized *in vivo* to aldehydes; *in vitro* to carboxylic acids.

3. 2° Alcohols are oxidized both *in vivo* and *in vitro* to ketones.

14.4

PHENOLS

Phenols are weak acids that can neutralize sodium hydroxide, and their benzene rings are easily oxidized.

For a compound to be classified as a *phenol,* its molecules must have at least one OH group directly attached to a benzene ring. The simplest member of this family also carries the name phenol, and it is a raw material for making aspirin. Phenols are widespread in nature and in commerce (see Special Topic 14.3).

■ For phenol itself, $K_a = 1.0 \times 10^{-10}$ (25 °C), which can be compared to $K_a = 1.8 \times 10^{-5}$ for acetic acid.

Phenols Are Weak Acids In sharp contrast to alcohols, phenols are acidic, but they are *weak* acids, as the K_a values in the margin show. Phenol is itself a much weaker acid than acetic acid. Yet phenols in general are strong enough acids to neutralize OH^-. For example,

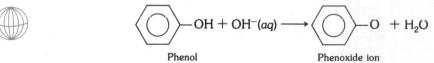

Phenol   Phenoxide ion

Understanding why OH^- can take H^+ from a phenol but not from an alcohol comes down to considering what makes the *anion* of a phenol, a phenoxide ion ($C_6H_5O^-$) more stable than the anion from an alcohol (RO^-). Both anions appear to have one negative charge on oxygen. What the structures do not reveal, however, is that the negative charge on the phenoxide ion is somewhat spread out into the pi-electron network of the benzene ring. *The surest way to stabilize negative charge is to spread it out,* but the negative charge on the RO^- ion cannot spread out as it does in the phenoxide ion. The alcohol's anion, RO^-, therefore, is less stable and it forms to a lesser extent from ROH than does the phenoxide ion from a phenol. Therefore, ROH is an *extremely poor* proton donor compared to a phenol.

The Ring in a Phenol Is Easily Oxidized Some of the electron density on the oxygen in phenol also spreads out over the pi-electron network of the ring. Oxidizing agents are electron seekers, so this makes the benzene ring in a phenol more easily oxidized than the ring in benzene. Oxidations of phenols *in vitro,* however, produce complex mixtures that include some highly colored compounds. Even phenol crystals left exposed to air slowly turn dark because the oxygen in the air can attack phenol. *In vivo,* special hydroxylating enzymes are able to steer the oxidations of the benzene rings in certain phenols to specific products (as we will learn in Section 28.2).

Phenols Are Not Dehydrated Unlike alcohols, phenols cannot be dehydrated. Dehydration would put a *triple* bond into a six-membered ring, and this ring is too small to accommodate the linear geometry at a triple bond.

14.5

ETHERS

The ethers are almost as chemically unreactive as the alkanes.

Ethers are compounds in whose molecules two organic groups are joined to the same oxygen atom, and their family structure is R—O—R′. The carbon joined to the bridging oxygen atom cannot be a carbonyl carbon, the one in C=O. Thus the first three compounds given below are ethers, but methyl acetate is in the ester family, not the ether family, because the bridging oxygen is attached to a carbonyl carbon. (We will study esters in a later chapter.)

$CH_3CH_2-O-CH_2CH_3$

Diethyl ether

CH_3-O-⬡

Methyl phenyl ether

⬡$-O-$⬡

Diphenyl ether

$CH_3\overset{\displaystyle O}{\overset{\|}{C}}-O-CH_3$

Methyl acetate (an *ester,* not an ether)

TABLE 14.3
Some Ethers

Common Name	Structure	BP (°C)
Dimethyl ether	CH_3OCH_3	−23
Methyl ethyl ether	$CH_3OCH_2CH_3$	11
Methyl t-butyl ether	$CH_3OC(CH_3)_3$	55.2
Diethyl ether	$CH_3CH_2OCH_2CH_3$	34.5
Dipropyl ether	$CH_3CH_2CH_2OCH_2CH_2CH_3$	91
Methyl phenyl ether	$CH_3OC_6H_5$	155
Diphenyl ether	$C_6H_5OC_6H_5$	259
Divinyl ether	$CH_2{=}CHOCH{=}CH_2$	29

■

Ether

Compound	BP (°C)	Solubility in water
Pentane	36	0.036 g/dL[a]
Diethyl ether	35	8.4 g/dL[a]
1-Butanol	118	11 g/dL[b]

[a] At 15 °C   [b] At 25 °C

■ Hydroperoxides are R—O—O—H and peroxides are R—O—O—R'.

Table 14.3 gives some examples of ethers, and three are described in Special Topic 14.4.

The common names of ethers are made by naming the groups attached to the oxygen and adding the word *ether*, as illustrated in Table 14.3.

Ethers Cannot Donate Hydrogen Bonds   Because the ether group cannot donate hydrogen bonds, the boiling points of simple ethers are more like those of the alkanes of comparable formula masses than those of the alcohols. The oxygen of the ether group can accept hydrogen bonds, however, so ethers are more soluble in water than alkanes. For example, both 1-butanol and its isomer, diethyl ether, dissolve in water to a much greater extent than pentane, which can neither accept nor donate hydrogen bonds.

Ethers Have Few Chemical Reactions   At room temperature, ethers do not react with strong acids, bases, or with strong oxidizing or reducing agents. On long standing in the presence of air, however, liquid ethers slowly react with oxygen, and compounds called *hydroperoxides* and *peroxides* gradually form. These compounds, when concentrated, can explode, so chemists are wary of using aged ether supplies. Like all organic compounds, ethers burn. We will learn no other reactions of ethers, but we must be able to recognize the ether group and to remember that it is not very reactive toward anything, particularly in the environment within the body.

Ethers Can Be Prepared from Alcohols   We learned earlier in this chapter that alcohols can be dehydrated by the action of heat and an acid catalyst to give alkenes. The precise temperature that works best has to be discovered experimentally for each alcohol, because if the temperature is not set correctly, a different pathway for dehydration can occur. Water can split out *between* two alcohol molecules rather than from within one alcohol molecule, and the product is an ether. In general,

■ Earlier we learned that concentrated H_2SO_4 acts on ethanol to give ethene when the temperature is higher (170 °C).

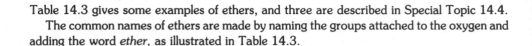

$$R{-}O{-}H + H{-}O{-}R \xrightarrow{\text{acid catalyst}} R{-}O{-}R + H_2O$$

Two alcohol molecules   Ether

A specific example is

$$2CH_3CH_2OH \xrightarrow[140\ °C]{H_2SO_4} CH_3CH_2{-}O{-}CH_2CH_3 + H_2O$$

Ethyl alcohol   Diethyl ether

The dehydration that produces an ether usually requires a lower temperature than that which gives an alkene, but we won't need the details. Our interest is simply in the *possibility* of making an ether from an alcohol as well as the structure of the ether that can be made.

ETHERS IN MEDICINE

Diethyl Ether ("ether") Diethyl ether is a colorless, volatile liquid with a pungent, somewhat irritating odor; it was once widely used as an anesthetic. It acts as a depressant for the central nervous system and a mild stimulant for the sympathetic system. It exerts an effect on nearly all tissues of the body.

Because mixtures of ether and air in the right proportions can explode, anesthesiologists avoid using it as an anesthetic whenever possible.

Divinyl Ether (vinethene) Divinyl ether is another anesthetic, but it also forms an explosive mixture in air. Its anesthetic action is more rapid than that of diethyl ether.

Methyl *t*-Butyl Ether The standard treatment for stones in the gallbladder or the gallbladder duct—"gallstones" —has been surgical removal of the gallbladder. The chief constituent of gallstones is cholesterol, which is mostly hydrocarbon-like. In the 1980s, methyl *t*-butyl ether, a relatively nontoxic solvent for cholesterol, was discovered by Mayo Clinic scientists to be an effective, nonsurgical agent for removing gallstones. Methyl *t*-butyl ether dissolves the stones without causing serious side effects, and it is less toxic than diethyl ether. It can be inserted into the gallbladder by a catheter, it remains a liquid at body temperature, and it works in just a few hours with no evidence of disagreeable side effects.

Dioxin Dioxin is the name for a family of compounds present as trace impurities in certain weed killers. There are actually 75 chlorine-containing dioxins, but one member, TCDD, is regarded as the most dangerous.

TCDD
(2,3,7,8-tetrachlorodibenzo-*p*-dioxin)

The TCDD molecule has two ether groups that bridge two parallel benzene rings each holding two chlorine atoms.

In male guinea pigs, TCDD is extremely toxic, more so than strychnine, the nerve gases, and cyanide. In male or female hamsters, however, the lethal dose is nearly 2000 times greater than for male guinea pigs, demonstrating that toxicity can be very species sensitive. In humans, TCDD appears to be even less toxic. TCDD does cause a form of acne (chloracne) in humans as well as digestive distress, pains in the joints, and psychiatric effects. TCDD is teratogenic (causer of birth defects) and carcinogenic (causer of cancer) in experimental animals. Whether it is carcinogenic or teratogenic in humans is the serious public health issue (and is hotly debated).

TCDD and other dioxins are very stable to heat, and they form in incinerators whenever chlorinated compounds are part of the wastes being burned. The waste gases from such incinerations contain levels of dioxins ranging from several parts per trillion to many parts per million. In the air, the dioxins are degraded by sunlight in a matter of days when moisture is present. TCDD is only slowly broken down in soil, however. None of the dioxins has any known value to humans; they offer nothing but risks. Some believe the risks are so small that municipal incinerators should not be stalled by them; others strongly disagree. [A reference: F. H. Tschirley, "Dioxin," *Scientific American,* February 1986, page 29.]

EXAMPLE 14.5
Writing the Structure of an Ether That Can Form from an Alcohol

If the conditions are right, 1-butanol can be converted to an ether. What is the structure of this ether?

ANALYSIS What this question asks is to complete the following equation:

$$CH_3CH_2CH_2CH_2OH \xrightarrow[\text{heat}]{\text{acid catalyst}} ?$$

As usual, the structure of the starting material gives us most of the answer. We know that the ether must get its organic groups from the alcohol, so to write the structure of the ether we write one O atom with two bonds from it.

—O—

Then we attach the alkyl group of the alcohol, one to each of the bonds.

SOLUTION   The structure of the ether is

$$CH_3CH_2CH_2CH_2—O—CH_2CH_2CH_2CH_3$$

Of course, we need *two* molecules of the alcohol to make one molecule of this ether, but always remember, *we balance an equation after we have written the correct formulas for reactants and products.* Remembering that water is the other product, we have as the balanced equation

$$2CH_3CH_2CH_2CH_2OH \xrightarrow[\text{heat}]{\text{acid catalyst}} CH_3CH_2CH_2CH_2—O—CH_2CH_2CH_2CH_3 + H_2O$$
1-Butanol   Dibutyl ether

CHECK   Is the product truly an *ether?* (Is it of the form R—O—R?) Are the R groups identical to the R group of the parent alcohol? (In this example are they *butyl* groups?) Does each carbon atom have four bonds and each oxygen two?

PRACTICE EXERCISE 9

Write the structures of the ethers to which the following alcohols can be converted.

(a) CH_3OH **(b)** $CH_3CH_2CH_2OH$ **(c)** ⬡—OH

14.6

THIOALCOHOLS AND DISULFIDES

Both the SH group, an easily oxidized group, and the S—S system, an easily reduced group, are important groups in proteins.

Alcohols, R—O—H, can be viewed as alkyl derivatives of water, H—O—H. Similar derivatives of hydrogen sulfide, H—S—H, are also known and are in the family called the **thioalcohols** or the **mercaptans.**

R—S—H	R—S—R′	R—S—S—R′
Thioalcohols (mercaptans)	Thioethers	Disulfides

■ *Mercaptan* is a contraction of *mercury-capturer.* Compounds with SH groups form precipitates with mercury ions.

The SH group is variously called the *thiol group,* the *mercaptan group,* or the *sulfhydryl group.*

Table 14.4 gives the IUPAC names and structures of a few thioalcohols. (We will not study the IUPAC nomenclature as a separate topic because it is straightforward, and because our interest in thioalcohols is limited just to one reaction.) Some very important properties of proteins depend on the presence of the SH group located on one of the building blocks of proteins, the amino acid called cysteine (Table 14.4).

■ Lower-formula-mass thioalcohols are present in and are responsible for the considerable respect usually given to skunks.

Dialkyl derivatives of water, the ethers (R—O—R′), have their sulfur counterparts, too, the thioethers. (You can see that the prefix *thio-* indicates the replacement of an oxygen atom by a sulfur atom.) Although proteins also have the thioether system, our studies will not require that we learn anything of the chemistry of this group.

Thioalcohols Are Oxidized to Disulfides   The one reaction of thioalcohols that will be important in our study of proteins is oxidation. Thioalcohols can be oxidized to **disulfides,** compounds whose molecules have two sulfur atoms joined by a covalent bond, R—S—S—R′. In general,

TABLE 14.4
Some Thioalcohols

Name	Structure	BP (°C)
Methanethiol	CH_3SH	6
Ethanethiol	CH_3CH_2SH	36
1-Propanethiol	$CH_3CH_2CH_2SH$	68
1-Butanethiol	$CH_3CH_2CH_2CH_2SH$	98
Cysteine (a monomer for proteins)	$^+NH_3CHCO_2^-$ $\mid$ CH_2SH	(Solid)

$$R—S—H + H—S—R + (O) \longrightarrow R—S—S—R + H_2O$$

Two molecules of   One molecule
a thioalcohol   of a disulfide

A specific example is

$$2CH_3SH + (O) \longrightarrow CH_3—S—S—CH_3 + H_2O$$

Methanethiol   Dimethyl disulfide

EXAMPLE 14.6
Writing the Product of the Oxidation of a Thioalcohol

What is the product of the oxidation of ethanethiol?

ANALYSIS Because the oxidation of the SH group generates the —S—S— group, begin simply by writing this group down.

$$—S—S—$$

Then attach the alkyl group from the thioalcohol, one on each sulfur atom.

SOLUTION Ethanethiol furnishes ethyl groups, so attach one of each of these groups to the S atoms.

$$CH_3CH_2—S—S—CH_2CH_3$$
Diethyl disulfide

If the problem had called for an equation, we would have had to use the coefficient of 2 for the ethanethiol.

$$2CH_3CH_2SH + (O) \longrightarrow CH_3CH_2—S—S—CH_2CH_3 + H_2O$$

However, this coefficient isn't necessary when all we are asked to do is to write the structure of the disulfide that can be made by the oxidation of ethanethiol. Always remember that balancing an equation comes *after* you have written the correct formulas for the reactants and products.

CHECK Is the product a *disulfide*? (Thiols are oxidized to disulfides.) Are the groups attached to the S atoms the same as present in the reactant? Does each carbon have four bonds and each sulfur two?

Disulfides Are Reduced to Thioalcohols The sulfur–sulfur bond in disulfides is very easily reduced, and the products are molecules of the thioalcohols from which the disulfide could be made. We will use the symbol (H) to represent any reducing agent that can do the task, just as we used (O) for an oxidizing agent. In general,

$$R\!-\!S\!-\!S\!-\!R + 2(H) \longrightarrow R\!-\!S\!-\!H + H\!-\!S\!-\!R$$

One molecule   Two molecules of a thioalcohol
of disulfide

A specific example is

$$CH_3CH_2\!-\!S\!-\!S\!-\!CH_2CH_2CH_3 + 2(H) \longrightarrow$$

$$CH_3CH_2\!-\!S\!-\!H + H\!-\!S\!-\!CH_2CH_2CH_3$$

PRACTICE EXERCISE 10

Complete the following equations by writing the structures of the products that form. If no reaction occurs, write "no reaction."

(a) $CH_3SSCH_3 + (H) \rightarrow$?   **(b)** $CH_3CHCH_3 + (O) \rightarrow$?
$$\qquad\qquad\qquad\qquad\qquad\qquad\qquad\qquad |$$
$$\qquad\qquad\qquad\qquad\qquad\qquad\qquad\; SH$$

(c) ⟨ring⟩$\genfrac{}{}{0pt}{}{S}{S}$ + (H) ⟶ ?   **(d)** ⟨ring⟩—SH + (O) ⟶ ?

SUMMARY

Alcohols The alcohol system has an OH group attached to a saturated carbon. The IUPAC names of simple alcohols end in *-ol,* and their chains are numbered to give precedence to the location of the OH group. The common names have the word *alcohol* following the name of the alkyl group.

Alcohol molecules hydrogen-bond to each other and to water molecules. By the action of heat and an acid catalyst, alcohols can be dehydrated internally to give carbon–carbon double bonds or externally to give ethers. Primary alcohols can be oxidized to aldehydes and ultimately to carboxylic acids. Secondary alcohols can be oxidized to ketones. Tertiary alcohols cannot be oxidized (without breaking up the carbon chain). The OH group in alcohols does not function well as either an acid or a base.

Phenols When the OH group is attached to a benzene ring, the system is the phenol system, and it is now acidic enough to neutralize strong bases. In addition, the ring is vulnerable to oxidizing agents.

Ethers The ether system, R—O—R', does not react at room temperature or body temperature with strong acids, bases, oxidizing agents, or reducing agents. It can accept hydrogen bonds but cannot donate them.

Thioalcohols The thioalcohols or mercaptans, R—S—H, are easily oxidized to disulfides, R—S—S—R. Disulfides are reduced to the original thioalcohols.

Reactions studied Without attempting to present balanced equations, or even all of the inorganic products, we can summarize the reactions studied in this chapter as follows. (We omit ethers.)

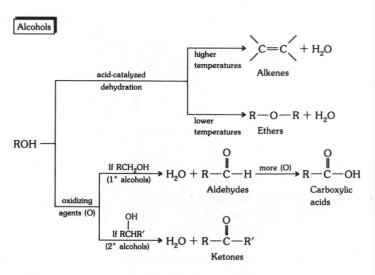

Alcohols

Phenols

Thioalcohols and Disulfides

REVIEW EXERCISES

The answers to Review Exercises whose numbers are in color are found in Appendix C. The answers to the other Review Exercises are found in the Study Guide that accompanies this book. The more challenging questions are marked with asterisks.

Functional Groups

14.1 Name the functional groups identified by the numbers in the structure of cortisone, a drug used in treating certain forms of arthritis. If a group is an alcohol, state if it is a 1°, 2°, or 3° alcohol.

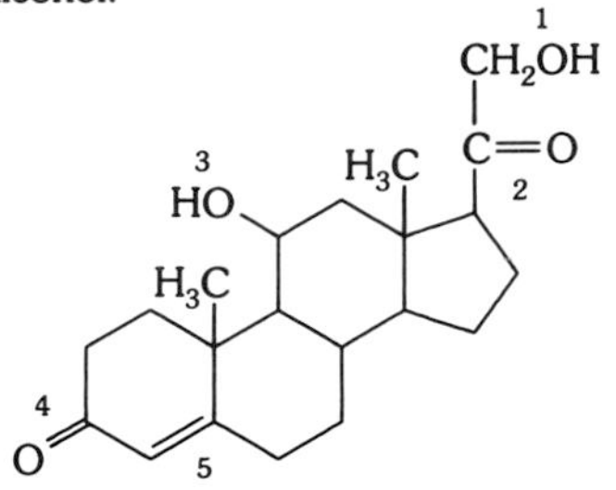

Cortisone

14.2 Give the names of the functional groups identified by the numbers in prostaglandin E_1, one of a family of compounds that are smooth muscle stimulants.

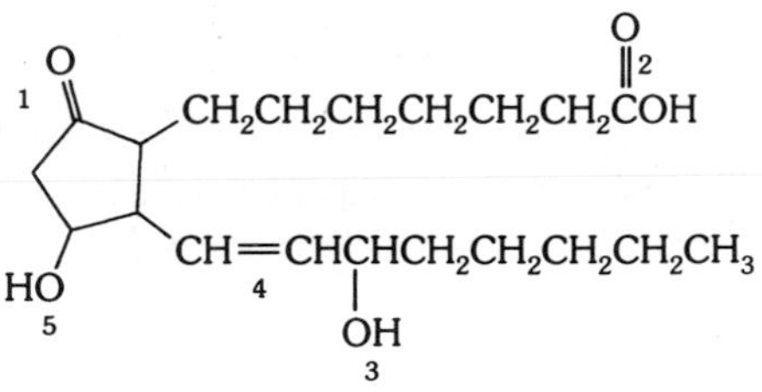

Prostaglandin E_1

Structures and Names

14.3 Write the structure of each compound.
(a) isobutyl alcohol (b) isopropyl alcohol
(c) propyl alcohol (d) glycerol

14.4 Write the structures of the following compounds.
(a) methyl alcohol (b) *t*-butyl alcohol
(c) ethyl alcohol (d) butyl alcohol

14.5 What are the common names of the following compounds?

(a) $CH_3CH_2CH_2OH$ (b) $HOCH_2CH_2CH_2CH_3$

(c) $HOCCH_3$ with CH_3 above and CH_3 below

(d) $HOCH_2CH$ with CH_3 and CH_3

14.6 What is the structure and the IUPAC name of the simplest, *stable* dihydric alcohol?

14.7 Give the structure and the IUPAC name of the simplest, *stable* trihydric alcohol?

14.8 Give the IUPAC names for the compounds listed in Review Exercise 14.3.

14.9 Give the IUPAC names for the compounds listed in Review Exercise 14.4.

14.10 Give the IUPAC names for the compounds listed in Review Exercise 14.5.

14.11 Write the IUPAC name of the following compound.

$$CH_3CH_2CHCH_2CH_2CH_3$$
$$|$$
$$CH_2OH$$

Physical Properties

14.12 When ethyl alcohol dissolves in water, its molecules slip into the hydrogen-bonding network in water. Draw a figure that illustrates this. Use dotted lines for hydrogen bonds, and place the $\delta+$ and $\delta-$ symbols where they belong.

*14.13 Arrange the following compounds in their order of increasing boiling points. Place the letter symbol of the compound that has the lowest boiling point on the left end of the series, and arrange the remaining letters in the correct order.

$CH_3CH_2CH_2OH$ CH_3CH_3 $HOCH_2CH_2OH$ $CH_3CH_2OCH_3$
   A   B   C   D

$$\underline{\qquad < \qquad < \qquad < \qquad}$$
Lowest bp   Highest bp

Chemical Properties of Alcohols

14.14 Write the structures of the alkenes that form when the following alcohols undergo acid-catalyzed dehydration. Where more than one alkene is possible, identify which most likely forms in the greatest relative amount.

(a) CH_3CHCH_2OH (b) $CH_3CHCH_2CH_3$
$\quad\quad |$ $\quad\quad\quad |$
$\quad\quad CH_3$ $\quad\quad\quad OH$

(c) 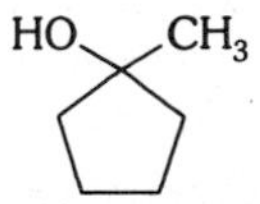 (d)

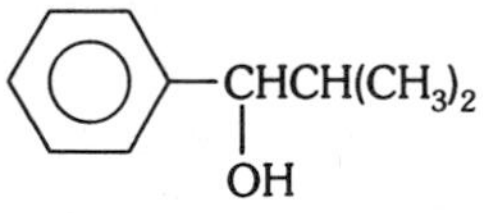

(e) $CH_3CH_2CH_2CCH_3$ with CH_3 above and OH below (f)

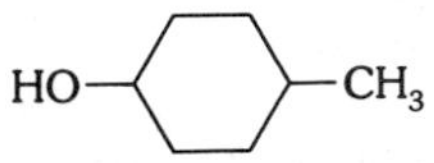

14.15 Write the structures of the products of the oxidation of the alcohols given in Review Exercise 14.14. If the alcohol is a 1° alcohol, give the structures of both the aldehyde and the carboxylic acid that could be made by varying the quantities of the oxidizing agent.

*14.16 Write the structures of any alcohols that could be dehydrated to give each of the following alkenes. In some instances, more than one alcohol would work.

(a) $CH_2{=}CHCH_3$　(b) [structure: cyclopentene]

(c) $CH_3\overset{\overset{\displaystyle CH_2}{\|}}{C}CH_3$　(d) [structure: cyclohexene ring with $-CH_3$]

*14.17 Write the structure of any alcohol that could be used to prepare each of the following compounds by an oxidation.

(a) $HO\overset{\overset{\displaystyle O}{\|}}{C}CH_2CH_2CH_3$　(b) $CH_3CH_2CH_2\overset{\overset{\displaystyle O}{\|}}{C}CH(CH_3)_2$

(c) $H{-}\overset{\overset{\displaystyle O}{\|}}{C}$[cyclopentane ring]　(d) [benzene ring]$-\overset{\overset{\displaystyle O}{\|}}{C}OH$

Phenols

14.18 Write the structures of the three isomeric monochlorophenols and give their names (using the *o-*, *m-*, and *p*-designations).

*14.19 What is one difference in *chemical* properties between **A** and **B**?

[structures of A and B: fused bicyclic ring systems with OH groups]

A　　　　**B**

14.20 A compound was either **A** or **B**.

[structures: A is benzene ring with Br and OH; B is cyclohexane ring with $-CH_2OH$]

A　　　　**B**

The compound dissolved in aqueous sodium hydroxide but not in water. Which compound was it? How can you tell? (Write an equation.)

Ethers

14.21 Write the structures of the ethers that can be made from the following alcohols.

(a) $HOCH_2CH_3$　(b) $CH_3\underset{\underset{\displaystyle OH}{|}}{CH}CH_3$

(c) $CH_3OCH_2CH_2OH$　(d) [cyclopentane ring]$-CH_2OH$

*14.22 Write the structures of the alcohols that could serve as the starting materials to prepare each of the following ethers.

(a) $CH_3CH_2CH_2OCH_2CH_2CH_3$

(b) $CH_3\underset{\underset{\displaystyle CH_3}{|}}{CH}CH_2OCH_2\underset{\underset{\displaystyle CH_3}{|}}{CH}CH_3$

(c) [cyclohexane ring]$-O-$[cyclohexane ring]

(d) [cyclopentane ring]$-O-$[cyclopentane ring]

*14.23 Suppose that a mixture of 0.50 mol of ethanol, 0.50 mol of methanol, and a catalytic amount of sulfuric acid is heated under conditions that favor only ether formation. What organic products will be obtained? Write their structures.

14.24 What happens chemically when the following compound is heated with aqueous sodium hydroxide?

$$CH_3CH_2CH_2OCH_2CH_2CH_3$$

Thioalcohols and Disulfides

14.25 We did not study rules for naming thioalcohols, but the patterns of the names in Table 14.4 make these rules obvious. Write the structures of the following compounds.
(a) diethyl disulfide　(b) 1,2-propanedithiol
(c) isopropyl mercaptan　(d) 1-propanethiol

14.26 Complete the following reaction sequences by writing the structures of the organic products that form.

(a) $CH_3CH_2CH_2SH + (O) \longrightarrow$
(b) $(CH_3)_2CHCH_2{-}S{-}S{-}CH_2CH(CH_3)_2 + (H) \longrightarrow$

(c) [cyclopentane ring with $S{-}S$] $+ (H) \longrightarrow$

(d) $HSCH(CH_3)_2 + (O) \longrightarrow$

14.27 Ethanol, methanethiol, and propane have nearly the same formula masses, but ethanol boils at 78 °C, methanethiol at 6 °C, and propane at -42 °C. What do the boiling points suggest about the possibility of hydrogen bonding in the thioalcohol family? Does hydrogen bonding occur at all? Are the hydrogen bonds as strong as those in the alcohol family?

Alcohols In Our Daily Lives (Special Topic 14.1)

14.28 Give the name of the alcohol used in each of the following ways.
(a) As the alcohol in beverages
(b) As a rubbing alcohol (Name two.)
(c) As a moisturizer in some food products
(d) As a fuel in "canned heat"
(e) As a permanent antifreeze (Name two.)
(f) To manufacture a vasodilator

14.29 Give the name of the alcohol made by
   (a) The digestion of fats or oils in the diet
   (b) The fermentation of sugars
   (c) The hydrogenation of carbon monoxide

Acid-Catalyzed Dehydration of Alcohols (Special Topic 14.2)

14.30 Write an equation for the equilibrium that forms when sulfuric acid is dissolved in cyclopentanol but before any further steps in the dehydration of this alcohol occur.

14.31 What is the structure of the carbocation that can form following the reaction described in Review Exercise 14.42?

14.32 When the cyclopentyl carbocation changes into cyclopentene, the carbocation must lose something. What is the formula of the species it must lose as the double bond forms? What is the name of the likeliest acceptor of this species when the surrounding medium contains mostly cyclopentanol and catalytic amounts of sulfuric acid?

14.33 Explain (briefly) why carbocations are unstable in water but something like the sodium ion is stable.

*__14.34__ Cyclopentene can be made to *add* a water molecule to give cyclopentanol when the medium is *dilute* sulfuric acid. The steps in the mechanism are the exact reverse of the kinds of steps for the dehydration of cyclopentanol to cyclopentene. Write the steps in the mechanism of the acid-catalyzed addition of water to cyclopentene.

Phenols in Our Daily Lives (Special Topic 14.3)

14.35 What is the name of a member of the phenol family that is used in or involved in each of the following ways?
   (a) A flavoring agent
   (b) Lister's original antiseptic
   (c) A substitute for cloves
   (d) An irritant in poison ivy

14.36 Both BHA and BHT interfere with the air oxidation of food materials. What chemical property do these food additives have that accounts for this?

14.37 Lister's original antiseptic is no longer used. Why?

Ethers in Medicine (Special Topic 14.4)

14.38 What physical property (aside from its anesthetic quality) made diethyl ether workable as an anesthetic?

14.39 What chemical property of diethyl ether makes it undesirable as an anesthetic, a property having nothing to do with its ability to induce the anesthetic state?

14.40 What is it about the structure of methyl *t*-butyl ether that accounts for its ability to dissolve cholesterol (whose structure may be looked up using the index)?

14.41 TCDD is a member of what family of pollutants?

14.42 What is currently an entry for trace quantities of TCDD into the environment?

Additional Exercises

*__14.43__ Examine each of the following sets of reactants and conditions and decide if a reaction occurs. If one does, write the structures of the organic products. If no reaction occurs, write "no reaction."

Some of the parts involve *alcohols* and their reactions. If the reaction is an oxidation of a 1° alcohol, write the structure of the *aldehyde* that can form, not the carboxylic acid.

When an alcohol is in the presence of an acid catalyst, we have learned that the alcohol might be dehydrated to an *alkene* or to an *ether*. To differentiate between these, use the following guide. When the alcohol structure has no coefficient, then write the structure of the alkene that can form. When the alcohol structure has a coefficient of 2, then give the structure of the ether that is possible. (This violates our rule that balancing an equation is the *last* step in writing an equation, but we need a signal here to tell what kind of reaction is intended.)

(a) [cyclopentane]—OH $\xrightarrow[\text{heat}]{H_2SO_4}$ *Ether*

(b) $2HOCH_2CH_3$ $\xrightarrow[\text{heat}]{H_2SO_4}$ *diether*

(c) $CH_3\overset{\overset{\displaystyle OH}{|}}{C}HCH_2CH_3 + (O) \longrightarrow$ *2-butanol*

(d) [cyclopentene ring]—$CH_3 + H_2$ $\xrightarrow{\text{Ni catalyst}}$

(e) $CH_3CH_2OH + NaOH(aq) \longrightarrow$

(f) $CH_3\overset{\overset{\displaystyle OH}{|}}{\underset{\underset{\displaystyle CH_3}{|}}{C}}CH_2CH_2CH_3 + (O) \longrightarrow$ *no react*

(g) $CH_3CH_2\overset{\overset{\displaystyle}{}}{\underset{\underset{\displaystyle CH_3}{|}}{C}}HCH_3$ $\xrightarrow[\text{heat}]{H_2SO_4}$ *no reaction*

(h) $CH_3CH=\overset{\overset{\displaystyle}{}}{\underset{\underset{\displaystyle CH_3}{|}}{C}}CH_3 + H_2$ $\xrightarrow{\text{Ni catalyst}}$

(i) [cyclohexene]—$CH_3 + H_2O$ $\xrightarrow[\text{heat}]{H_2SO_4}$ *addition*

(j) HO—[cyclopentane ring]—$CH_3 + (O) \longrightarrow$ *2° alcohol* *can make ketone*

*14.44 Write the structure of the principal organic product that would be expected in the following situations. Follow the directions given for Review Exercise 14.40. If no reaction occurs, write "no reaction."

(a) $CH_3CH_2CHCH_3$ + NaOH(aq) $\longrightarrow$
    |
    CH_2CH_3

(b) $CH_3CH{=}CCH_3$ + HCl(g) $\longrightarrow$ *Additional*
    |
    CH_3

(c) $CH_3CHCHCH_3$ + (O) $\longrightarrow$ *Keto*
    | |
    CH_3 OH

(d) $2CH_3CHOH$ $\xrightarrow{\text{H}_2\text{SO}_4 \text{ heat}}$ *Ether*
    |
    CH_3

(e) $HOCH_2CH_2CH_3$ + (O) $\longrightarrow$ *alcohol — Aldehyde*

(f) H₃C⟨ ⟩CH₃ + H₂O $\longrightarrow$ *No reaction*

(g) ⟨◯⟩ + NaOH(aq) $\longrightarrow$ *No reaction*

(h) CH_3–⟨ ⟩–OH + (O) $\longrightarrow$ *〉=O*

(i) CH_3CHCH_3 $\xrightarrow{\text{H}_2\text{SO}_4 \text{ heat}}$
    |
    OH

(j) $CH_3CHCH_2OCH_3$ + (O) $\longrightarrow$
    |
    HO

*14.45 2-Propanol (C_3H_8O) can be oxidized to acetone (C_3H_6O) by potassium permanganate according to the following equation.

$$3C_3H_8O + 2KMnO_4 \longrightarrow 3C_3H_6O + 2MnO_2 + 2KOH + 2H_2O$$

(a) How many moles of acetone can be prepared from 2.50 mol of 2-propanol?

(b) How many moles of potassium permanganate are needed to oxidize 0.180 mol of 2-propanol?

(c) A student began with 12.6 g of 2-propanol. What is the minimum number of grams of potassium permanganate needed for this oxidation?

(d) Referring to part (c), what is the maximum number of grams of acetone that could be made? How many grams of MnO_2 are also produced?

*14.46 1-Propanol can be oxidized to propanoic acid.

$$CH_3CH_2CH_2OH + Cr_2O_7^{2-} \longrightarrow CH_3CH_2CO_2H + Cr^{3+}$$

(a) Write the *net ionic equation* using the ion-electron method (Section 10.2) for balancing the equation. Assume that the medium is acidic.

(b) Transform the equation of part (a) into a *molecular equation* assuming that the potassium salt of the dichromate ion is used and that the aqueous acid is HCl(aq).

(c) For each mole of 1-propanol used, how many moles (in theory) of potassium dichromate are needed according to the molecular equation?

(d) The stockroom carries potassium dichromate only as its dihydrate. Write the formula of the dihydrate. How many moles of the dihydrate are needed for each mole of 1-propanol used, according to the equation?

(e) What is the maximum number of grams of propanoic acid that could be obtained if the reaction is carried out starting with 12.4 g of 1-propanol?

(f) What is the minimum number of grams of potassium dichromate dihydrate that would be needed to use up 14.5 g of 1-propanol according to the balanced equation?

*14.47 Write the balanced net ionic equation for the oxidation of cyclopentanol to its corresponding ketone using sodium dichromate in aqueous sulfuric acid. The dichromate ion is changed to the chromium(III) cation. Assume that because of side reactions and losses during the purification of the product ketone, only 50.0% of the theoretically possible ketone can actually be isolated. If you are assigned the task of preparing 10.0 g of the ketone under these limitations, what is the minimum number of grams of cyclopentanol that you must use at the beginning? What is the minimum number of grams of sodium dichromate that you must also use?

ALDEHYDES AND KETONES

15

Architects use mirrored building materials to create spectacular effects. A simple chemical reaction, introduced in this chapter, creates silvered mirrors when one reactant is an aldehyde.

15.1

STRUCTURES AND PHYSICAL PROPERTIES OF ALDEHYDES AND KETONES

Molecules of both aldehydes and ketones contain the carbonyl group.
A knowledge of some of the properties of aldehydes and ketones is essential to understanding carbohydrates. All simple sugars are either polyhydroxy aldehydes or polyhydroxy ketones, as illustrated in the structures of glucose and fructose.

■ We show here only one form of each of the glucose and fructose molecules.

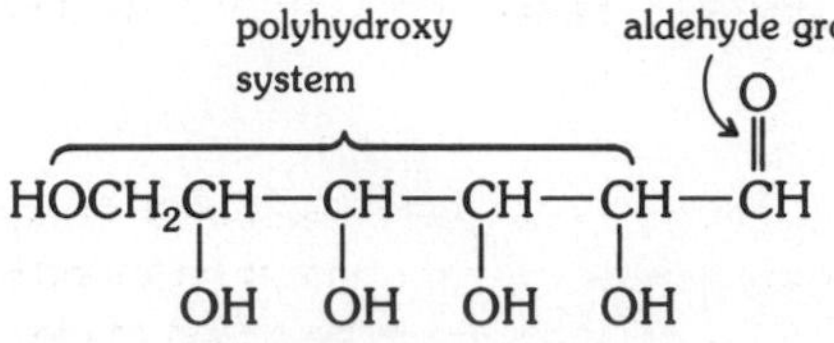

Glucose (open-chain form) 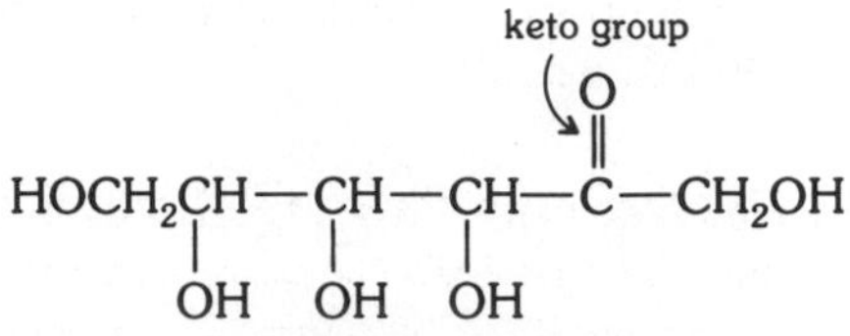   Fructose (open-chain form)

Many intermediates in metabolism are also aldehydes or ketones, and Special Topic 15.1 describes just a few.

Aldehydes and Ketones Have the *Carbonyl Group* Both aldehydes and ketones contain the carbon–oxygen double bond, which is called the **carbonyl group** (pronounced car-bon-EEL). The nature of this double bond is discussed in Special Topic 15.2. Like the alkene group, the carbon–oxygen double bond consists of one sigma bond and one pi bond.

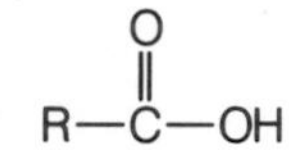

Carboxylic acids

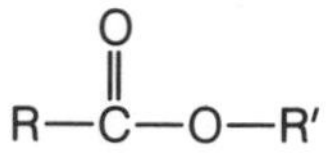

Esters

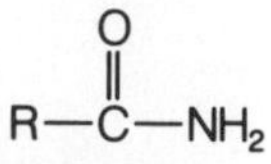

Amides

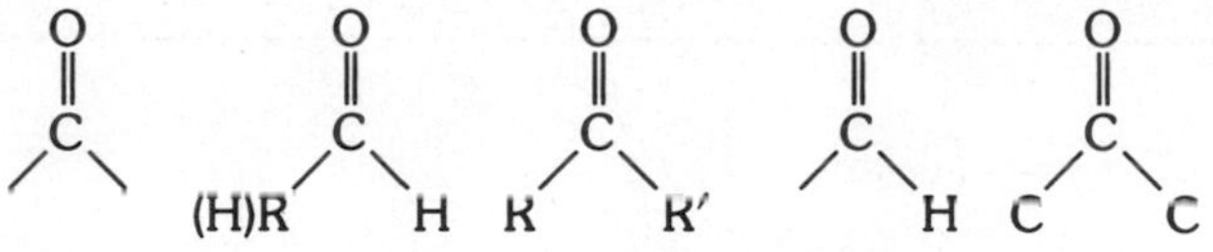

Carbonyl group   Aldehydes   Ketones   Aldehyde group   Ketone system

The carbonyl group also occurs in carboxylic acids, and we will study their salts, esters, anhydrides, and amides in the two chapters following this one.

***Aldehydes* Have the CH=O Group** For a compound with a carbonyl group to be called an **aldehyde,** the carbon atom of C=O *must have at least one H atom attached to it.* The other single bond from this carbon *must* be a bond to C (or to a second H), but not a bond to O, N, or S. Thus all aldehydes have the CH=O group, called the **aldehyde group.** (You will often see CH=O written as CHO as in the general formula for aldehydes, RCHO.) The simplest aldehyde, HCH=O, methanal (formaldehyde), has two hydrogens on the carbonyl carbon atom. Table 15.1 shows some other relatively simple aldehydes. They have very distinctive, unpleasant odors partly because almost any aldehyde exposed to air also has some of its

444

TABLE 15.1
Aldehydes

Name	Structure	Formula Mass	BP (°C)	Solubility in Water
Methanal	$CH_2{=}O$	30.0	−21	Very soluble
Ethanal	$CH_3CH{=}O$	44.0	21	Very soluble
Propanal	$CH_3CH_2CH{=}O$	58.1	49	16 g/dL (25 °C)
Butanal	$CH_3CH_2CH_2CH{=}O$	72.1	76	4 g/dL
Benzaldehyde	$C_6H_5CH{=}O$	106.0	178	0.3 g/dL
Vanillin	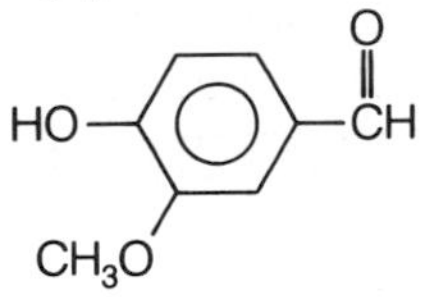	152.1	285	1 g/dL

corresponding carboxylic acid. (The aldehyde group slowly oxidizes in air.) The carboxylic acids of lower formula mass, to most people, have some of the most disagreeable odors of all organic compounds.

In *Ketones,* the Carbonyl Carbon Is Joined Directly to Two Carbons For a compound with a carbonyl group to be called a **ketone,** its carbonyl group *must* be joined on *both* sides by bonds to carbon atoms. Only then can the carbonyl group be called the **keto group.** Several ketones are listed in Table 15.2. Sometimes the structure of a ketone is condensed to RCOR′. Ketones also have distinctive but generally not disagreeable odors.

TABLE 15.2
Ketones

Name	Structure	Formula Mass	BP (°C)	Solubility in Water
Acetone	$CH_3\overset{O}{\overset{\|}{C}}CH_3$	58.1	57	Very soluble
Butanone	$CH_3\overset{O}{\overset{\|}{C}}CH_2CH_3$	72.1	80	33 g/dL (25 °C)
2-Pentanone	$CH_3\overset{O}{\overset{\|}{C}}CH_2CH_2CH_3$	86.1	102	6 g/dL *limited*
3-Pentanone	$CH_3CH_2\overset{O}{\overset{\|}{C}}CH_2CH_3$	86.1	102	5 g/dL
Cyclopentanone	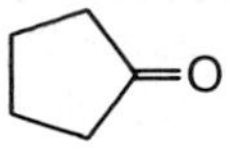	84.1	129	Slightly soluble
Cyclohexanone	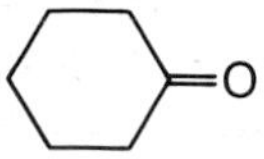	98.1	156	Slightly soluble

SOME IMPORTANT ALDEHYDES AND KETONES

Formaldehyde Pure formaldehyde is a gas at room temperature, and it has a very irritating and distinctive odor. It is quite soluble in water, so it is commonly marketed as a solution called formalin (37%) to which some methanol has been added. In this and more dilute forms, formaldehyde was once commonly used as a disinfectant and as a preservative for biological specimens. (Concern over formaldehyde's potential hazard to health has caused these uses to decline.) Most formaldehyde today is used to make various plastics such as Bakelite.

Acetone Acetone is valued as a solvent. Not only does it dissolve a wide variety of organic compounds, but it is also miscible with water in all proportions. Nail polish remover is generally acetone. Should you ever use "superglue," it would be a good idea to have some acetone (nail polish remover) handy because superglue can stick your fingers together so tightly that it takes a solvent such as acetone to get them unstuck.

Acetone is a minor by-product of metabolism, but in some situations (e.g., untreated diabetes) enough is produced to give the breath the odor of acetone.

Some Aldehydes and Ketones in Metabolism The aldehyde group and the keto group occur in many compounds at the molecular level of life. The following are just a few examples of substances with the aldehyde group.

$$HCCHCH_2OPO_3^{2-}$$
$$\overset{|}{OH}$$

Glyceraldehyde-3-phosphate, an intermediate in the metabolism of glucose

Pyridoxal, one of the vitamins (B$_6$)

Just a few of the many substances at the molecular level of life that contain the keto group are the following.

$$CH_3CCO_2^-$$

Pyruvate ion, a product of the metabolism of glucose and fructose

$$CH_3CCH_2CO_2^-$$

Acetoacetate ion, a product of the metabolism of long-chain carboxylic acids

$$HOCH_2CCH_2OPO_3^{2-}$$

Dihydroxyacetone phosphate, an intermediate in the metabolism of glucose and fructose

Estrone, a female sex hormone

Polarization of the carbonyl group

The Carbonyl Group Is Planar, Unsaturated, and Moderately Polar Because the carbonyl group is a *double bond,* it should be no surprise that the group can undergo addition reactions. Whether such reactions produce *stable* products, however, has to be studied on a case-by-case basis. The *direction* of the addition of an *unsymmetrical* reactant to the C=O group is governed by the group's permanent polarity. We'll be studying only one such reactant, ROH. The unsymmetrical reactant's more positive unit (H in ROH) always goes to the O of C=O because oxygen is more electronegative than carbon and so has a permanent $\delta-$ charge. The C of C=O has a permanent $\delta+$ charge and so can accept only the more electron-rich unit of the unsymmetrical reactant (the RO in ROH).

THE NATURE OF THE CARBON–OXYGEN DOUBLE BOND

We have learned that the carbon–carbon double bond consists of one sigma bond and one pi bond. The carbon–oxygen double bond is exactly like this, except that an oxygen atom has replaced a carbon atom. Both the carbon atom and the oxygen atom of the carbonyl group are sp^2 hybridized, and in methanal the carbon–oxygen sigma bond forms by the overlap of two such hybrid orbitals. The pi bond results from the side-to-side overlap of two unhybridized p orbitals (see Figure 1).

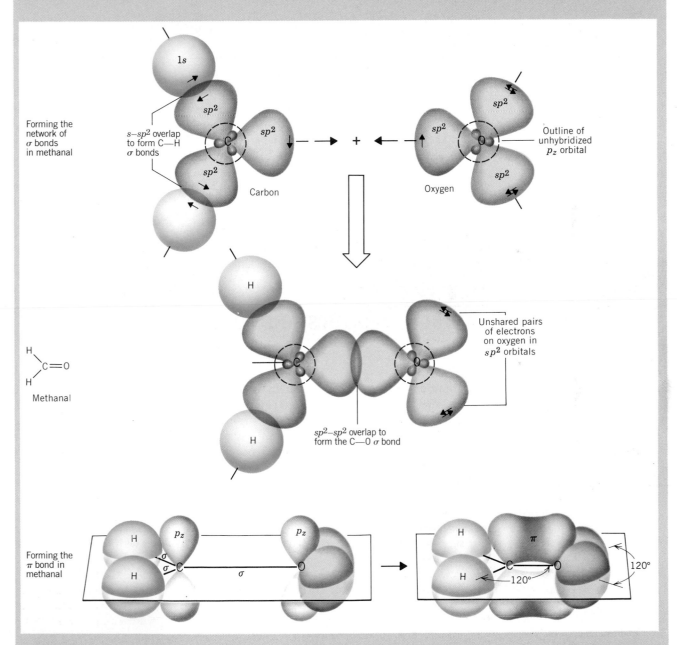

FIGURE 1
The formation of the bonds of the carbonyl group in formaldehyde. The C=O group has one sigma and one pi bond.

TABLE 15.3
Boiling Point versus Structure

Name	Structure	Formula Mass	BP (°C)
Butane	$CH_3CH_2CH_2CH_3$	58.2	0
Propanal	$CH_3CH_2CH{=}O$	58.1	49
Acetone	$CH_3\overset{\displaystyle O}{\overset{\|}{C}}CH_3$	58.1	57
1-Propanol	$CH_3CH_2CH_2OH$	60.1	98
1,2-Ethanediol (ethylene glycol)	$HOCH_2CH_2OH$	62.1	198

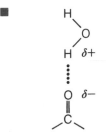

Hydrogen bond (•••••) between a water molecule and a carbonyl group

Because of the polarity of the C=O group, aldehydes and ketones are (moderately) polar compounds. Their molecules are attracted to each other, but not as strongly as they would be if they had OH groups instead and so could participate in hydrogen bonding. Evidence for the polarity of the carbonyl group can be seen in physical properties, like boiling points and solubilities in water. When comparing substances of nearly the same formula masses but from different families, we find that aldehydes and ketones boil higher than alkanes, but lower than alcohols (see Table 15.3). The lack of the OH group means that aldehydes and ketones cannot donate hydrogen bonds, only accept them. This is sufficient to make the low-formula-mass aldehydes and ketones relatively soluble in water, but as the total carbon content increases, their solubility decreases (see Tables 15.1 and 15.2).

15.2

NAMING ALDEHYDES AND KETONES

The IUPAC name ending for aldehydes is -al and for ketones is -one.

The Common Names of Simple Aldehydes Are Derived from Those of Carboxylic Acids What is easy about the *common* names of aldehydes is that they all (well, nearly all) end in *-aldehyde*. The prefixes to this are the same as found in the common names of the carboxylic acids to which the aldehydes are easily oxidized. We will, therefore, study the common names of these two families here in one place. (Common names are actually more often used than the IUPAC names.)

The simple carboxylic acids have been known for centuries, and their common names are based on natural sources of the acids. Formic acid, for example, is present in the stinging fluid of ants, and the Latin root for ants is *formica*. So this one-carbon acid is called formic acid. The prefix in formic acid is *form-*, so the one-carbon aldehyde is called *formaldehyde*. Here are the four simplest carboxylic acids and their common names together with the structures and names of their corresponding aldehydes.

■ Formic acid also appears to have an aldehyde group, but its second bond from C is to another O, not to H (or C), and it is classified as a carboxylic acid.

■ L. *acetum,* vinegar

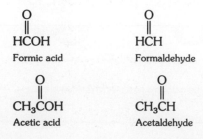

$$CH_3CH_2\overset{\displaystyle O}{\overset{\displaystyle \|}{C}}OH$$
Propionic acid

$$CH_3CH_2\overset{\displaystyle O}{\overset{\displaystyle \|}{C}}H$$
Propionaldehyde

- Gr. *proto*, first, and *pion*, fat

$$CH_3CH_2CH_2\overset{\displaystyle O}{\overset{\displaystyle \|}{C}}OH$$
Butyric acid

$$CH_3CH_2CH_2\overset{\displaystyle O}{\overset{\displaystyle \|}{C}}H$$
Butyraldehyde

- L. *butyrum*, butter

In the aromatic series we have the following examples.

Benzoic acid   Benzaldehyde

The IUPAC Names of Aldehydes End in -*al* As with the IUPAC names of the alcohols, those of the aldehydes are based on the idea of a *parent aldehyde*. Here are the rules.

1. Select as the parent aldehyde the longest chain *that includes the carbon atom of the aldehyde group.*

    The parent aldehyde in the following structure is a five-carbon aldehyde.

$$CH_3CH_2\underset{\displaystyle \underset{\displaystyle CH_3CH_2CH_2}{|}}{CH}\overset{\displaystyle O}{\overset{\displaystyle \|}{C}}H$$

    There is a longer chain in the structure, one of six carbons, but it doesn't include the carbon of the aldehyde group, so this longer chain may not be used in selecting and naming the parent.

2. Name the parent by changing the *-e* ending of the corresponding alkane to *-al*.

    In the example shown with rule 1, the alkane that corresponds to the correct chain is pentane, so the name of the parent aldehyde in this structure is *pentanal.*

3. Number the chain to give the carbon atom of the carbonyl group number 1.

    Precedence is accorded the aldehyde group. Regardless of where other substituents occur on the parent chain, they have to take whatever numbers they receive following the assignment of 1 to the carbonyl carbon atom.

4. Assemble the rest of the name in the same way that was used in naming alcohols, except do not include "1" to specify the location of the aldehyde group.

    The carbonyl carbon cannot have any other number but 1, so we do not include this number. Thus using the example begun under rule 1, we have

$$CH_3CH_2\underset{\displaystyle \underset{\displaystyle \underset{5\ \ \ 4\ \ \ \ 3}{CH_3CH_2CH_2}}{|}}{\overset{2}{CH}}\overset{\displaystyle O}{\overset{\displaystyle \|}{\underset{1}{C}}}H$$

2-Ethylpentanal
Not: 2-ethyl-1-pentanal

EXAMPLE 15.1
Writing the IUPAC Name of an Aldehyde

What is the IUPAC name of the following compound?

$$CH{=}O$$
$$BrCH_2CH_2CHCHCH_2CH_3$$
$$CH_3$$

ANALYSIS First, we identify the parent aldehyde. The longest chain that includes the carbon atom of the aldehyde group has five carbons, so the name of the parent aldehyde is *pentanal.* Next, we number the chain beginning with the carbon atom of the aldehyde group.

$$\overset{1}{CH}{=}O$$
$$\overset{5}{Br}CH_2\overset{4}{CH}_2\overset{3}{CH}\overset{}{CH}\overset{}{CH}_2CH_3$$
$$\overset{2}{CH_3}$$

At position 2 there is an ethyl group; at 3, a methyl group; and at 5, a bromo group.

SOLUTION   We organize the names of the groups alphabetically. The name is

5-bromo-2-ethyl-3-methylpentanal

CHECK Remember that the most common error is in failing to find the *longest chain that includes the carbonyl group.* Whenever you check your work in writing a name from a structure, test every conceivable way of finding the parent chain.

PRACTICE EXERCISE 1

Write the IUPAC names of the following aldehydes.

(a) $CH_3CHCH{=}O$
CH_3

(b) $CH_3CHCH_2CH{=}O$
Br

(c) $CH_3CHCH_2CCH_2CHCH{=}O$
$CH_3 \quad CH_3$
$CH_3CH_2 \quad CH_3$

PRACTICE EXERCISE 2

What is wrong with the name 2-*isopropylpropanal?*

IUPAC Names of Ketones End in *-one*   The IUPAC rules for naming ketones are identical to those for the aldehydes, except for two obvious changes. The name of the parent ketone must end in *-one* (not *-al*) and the keto group must be located by a number. In numbering the chain, the location of the keto group takes precedence, not the locations of substituents.

■ Pronounce *-one* as *own.*

EXAMPLE 15.2
Writing the IUPAC Name of a Ketone

What is the IUPAC name of the following ketone?

$$\underset{CH_3CH_2CH_2}{\overset{CH_3\quad CH_3\ O}{CH_3CH-C-CCH_3}}$$

3-isopropyl-3-meth... ...cyclohexanone

ANALYSIS There are two chains that have six carbon atoms, but we have to use the one that has the carbon atom of the carbonyl group. We number this chain to give the location of the carbonyl group the lower number.

$$\underset{\underset{6\ \ 5\ \ 4}{CH_3CH_2CH_2}}{\overset{CH_3\quad CH_3\ \ O}{CH_3CH-\underset{3}{C}-\underset{2\ 1}{CCH_3}}}$$

The parent ketone is therefore 2-hexanone. At carbon 3 its chain has a methyl group plus an isopropyl group.

SOLUTION What remains is to assemble these names into the complete name of the ketone:

$$3\text{-isopropyl-3-methyl-2-hexanone}$$

CHECK Another frequent error is to give incorrect names to alkyl groups. Once you are certain of the parent chain and the direction of numbering, double-check the alkyl groups. If an alkyl group has three carbons *it must be one of the propyl groups* because *prop* goes with three carbons. But which one? When the bonding site is from the middle carbon, it is *isopropyl*.

PRACTICE EXERCISE 3

Write the IUPAC names for the following ketones.

(a) $\underset{}{\overset{O}{CH_3CH_2CCH_3}}$ (b) $\underset{}{\overset{CH_3}{\underset{}{CH_3CHCH_2CH_2CH_2CCH_3}}}$ (c) [structure of cyclohexanone with CH₃]

The Simplest Ketone Is Usually Called Acetone, Not Propanone
Quite often the simpler ketones are given common names that are made by naming the two alkyl groups attached to the carbon atom of the carbonyl group and then following these names by the word *ketone*. For example,

$\underset{\text{Methyl ethyl ketone}}{\overset{O}{CH_3CH_2CCH_3}}$ $\underset{\text{Diethyl ketone}}{\overset{O}{CH_3CH_2CCH_2CH_3}}$ $\underset{\substack{\text{(Dimethyl ketone)}\\\text{Acetone}}}{\overset{O}{CH_3CCH_3}}$

■ The name *acetone* stems from the fact that this ket*one* can be made by heating the calcium salt of *acetic* acid.

As we noted, the name *acetone* is almost always used for dimethyl ketone or propanone.

PRACTICE EXERCISE 4

Write the structures of the following ketones.

(a) ethyl isopropyl ketone **(b)** methyl phenyl ketone

(c) dipropyl ketone **(d)** di-*t*-butyl ketone

15.3

THE OXIDATION OF ALDEHYDES AND KETONES

The aldehyde group is easily oxidized to the carboxylic acid group, but the keto group is difficult to oxidize.

We learned in Chapter 14 that the oxidation of a 1° alcohol to an aldehyde requires special reagents, because aldehydes are themselves easily oxidized. We also learned that much less care is needed to oxidize a 2° alcohol to a ketone, because ketones resist further oxidation.

$$\underset{\text{1° Alcohol}}{RCH_2OH} + \underset{\substack{\text{Oxidizing} \\ \text{agent}}}{(O)} \longrightarrow \underset{\text{Aldehyde}}{R\overset{\displaystyle O}{\overset{\|}{C}}H} + H_2O$$

$$\underset{\text{2° Alcohol}}{RCHR'} + (O) \longrightarrow \underset{\text{Ketone}}{R\overset{\displaystyle O}{\overset{\|}{C}}R'} + H_2O$$

The ease with which the aldehyde group is oxidized by even mild reactants has led to some simple test tube tests for aldehydes.

The *Tollens' Test* Produces a Silver Mirror One very mild oxidizing agent, called **Tollens' reagent,** consists of an alkaline solution of the silver ion in combination with two ammonia molecules, $[Ag(NH_3)_2]^+$. This species in Tollens' reagent oxidizes the aldehyde group to a carboxyl group (rather, to its anion form). The silver ion is reduced to metallic silver. In general,

$$RCH{=}O(aq) + 2[Ag(NH_3)_2]^+(aq) + 3OH^-(aq) \longrightarrow$$
$$RCO_2^-(aq) + 2Ag(s) + 2H_2O + 4NH_3(aq)$$

When this reaction occurs in a thoroughly clean, grease-free test tube, the silver plates to the glass as a beautiful mirror. In fact, this is one technique used to manufacture mirrors. A positive Tollens' test is actually the formation of metallic silver *in any form,* as a mirror in a clean test tube or as a grayish precipitate otherwise. The appearance of silver is dramatic evidence that a reaction occurs, so Tollens' reagent provides a simple test, called **Tollens' test,** to tell whether an unknown compound is an aldehyde or a ketone.

Benedict's Test **Produces a Brick-Red Precipitate** **Benedict's reagent,** another mild oxidizing agent, consists of a basic solution of the copper(II) ion and the citrate ion, the anion of citric acid, which causes the tart taste of citrus fruits. The medium has to be slightly basic in order for an aldehyde group to be oxidized by the reagent. However, Cu^{2+} normally forms an

■ The Tollens' test is sometimes called the *silver mirror test.* Glucose gives this test.

FIGURE 15.1
Benedict's solution is a slightly alkaline solution of citrate ion and Cu^{2+} ion, which makes the solution in the test tube at the rear intensely blue. In a positive Benedict's test, a brick red slurry of Cu_2O forms, seen here in the other test tube before settling.

extremely insoluble precipitate of CuO in a basic environment. The citrate ion prevents this by a mechanism to be studied shortly.

When an easily oxidized compound like glucose is added to a test tube that contains some Benedict's reagent, and the solution is warmed, Cu^{2+} ions are reduced to Cu^+ ions. The citrate ion is unable to keep the Cu^+ ion in the dissolved state in the basic medium. The newly formed Cu^+ ions are instantly changed by the base (OH^-) into a precipitate of copper(I) oxide, Cu_2O. We'll write the equation using the general structure of an aldehyde, $RCH{=}O$, but *simple* aldehydes (implied by this structure) give complex results.[1]

$$RCH{=}O(aq) + 2Cu[citrate]^{2+}(aq) + 2OH^-(aq) \longrightarrow RCO_2^-(aq) + Cu_2O(s) + 3H_2O$$

The Benedict's reagent has a brilliant blue color caused by the copper(II) ion, but Cu_2O has a brick red color (see Fig. 15.1). Therefore the visible evidence of a positive **Benedict's test** is the disappearance of a blue color and the appearance of a reddish precipitate.

Simple aldehydes, as we said, do not give the test as well as aldehydes with neighboring oxygens. Three systems, one not even an aldehyde, and all of which occur among various carbohydrates, give positive Benedict's tests:

■
$$\begin{array}{c} CH_2CO_2^- \\ | \\ HOCCO_2^- \\ | \\ CH_2CO_2^- \end{array}$$
Citrate ion

$$\begin{array}{ccc}
\overset{\displaystyle O}{\overset{\|}{RCHCH}} & \overset{\displaystyle O\ \ O}{\overset{\|\ \ \|}{RC{-}CH}} & \overset{\displaystyle O}{\overset{\|}{RCHCR'}} \\
| & & | \\
OH & & OH \\
\alpha\text{-Hydroxy aldehyde} & \alpha\text{-Keto aldehyde} & \alpha\text{-Hydroxy ketone} \\
\text{(present in glucose)} & & \text{(present in fructose)}
\end{array}$$

■ A carbon atom immediately adjacent to a carbonyl group is often called an alpha (α) carbon:

$$\begin{array}{c} O \\ | \ \ \| \\ {-}C{-}C{-} \\ | \\ \end{array}$$
Alpha carbon

Benedict's Test Has Been a Common Method for Detecting Glucose in Urine

In certain conditions, like diabetes, the body cannot prevent some of the excess glucose in the blood from being present in the urine, so testing the urine for its glucose concentration has long been used in medical diagnosis. Clinitest tablets, a convenient solid form of Benedict's reagent, contain all the needed reactants in their solid forms. To test for glucose, a few drops

---

[1] Although simple aldehydes give a reaction with the Benedict's reagent that reduces the blue color, the gummy solid that forms is not Cu_2O. Even the equation that we write here for carbohydrates is an oversimplification because in a *warm* alkaline medium, which is involved in the Benedict's test, carbohydrates undergo complex reactions. Yet, Cu_2O does form when carbohydrates are tested with Benedict's reagent. See R. Daniels, C. C. Rush, and L. Bauer, *Journal of Chemical Education*, Vol. 37, page 205 (1960).

■ Other tests for glucose are based on enzyme-catalyzed reactions, which we will learn more about later.

■ Tollens' reagent must be freshly made, because it deteriorates on standing.

When an aqueous hydroxide is added to aqueous silver nitrate, a tan, mud-like precipitate of silver oxide forms.

■ Hemoglobin exists inside erythrocytes — red blood cells.

of urine are mixed with a tablet and, as the tablet dissolves, the heat needed for the test is generated. The color that develops is compared with a color code on a chart provided with the tablets. Specialists in the control of diabetes, however, prefer to monitor the carbohydrate status of a diabetic person by determining the glucose in the *blood* instead of in the urine. Not all patients, however, can or are willing to manage blood tests, particularly when they are needed frequently.

Complex Ions Are Present in Tollens' and Benedict's Reagents

The reagents for the Tollens' and Benedict's tests involve a species new to our study, one of great importance at the molecular level of life, the *complex ion.* Tollens' reagent is prepared by adding sodium hydroxide to dilute silver nitrate. This causes the very insoluble silver oxide, Ag_2O, to precipitate. Undissolved Ag_2O would be unable to give the Tollen's test. When dilute ammonia is added next, however, its molecules are able to pull silver ions out of the solid silver oxide by forming a soluble complex ion, called the *silver diammine ion*, $[Ag(NH_3)_2]^+$. Thus the silver oxide dissolves.

A **complex ion** — often simply called a **complex** — consists of a metal ion that has strongly attracted a definite number of **ligands,** species that are either negative ions or neutral but electron-rich molecules, like ammonia. Examples of ligands include any of the halide ions (F^-, Cl^-, Br^-, and I^-), the cyanide ion (CN^-), the hydroxide ion, and many anions of organic acids (like the citrate ion). Among the common, electrically neutral ligands are water and ammonia. Other ligands are organic molecules with *amino groups,* NH_2. The most common metal ions that can form complex ions are those of the transition metal elements in the periodic table, like Ag^+ and Cu^{2+}. Two complex ions of Cu^{2+} and neutral ligands are $Cu(H_2O)_4^{2+}$ and $Cu(NH_3)_4^{2+}$, which are both blue but with strikingly different intensities of color (see Fig. 15.2). The citrate ion is a negatively charged ligand that forms the (blue) complex with Cu^{2+} in Benedict's reagent.

Many Complex Ions Are Important at the Molecular Level of Life

Uncomplexed transition metal ions are, in general, insoluble when the pH is greater than 7; they precipitate as their hydroxides or oxides. Thus virtually all of the trace transition metal ions required in nutrition, like Cu^{2+}, Co^{2+}, Fe^{2+}, and several others, can exist in the slightly alkaline fluids of the cell or in blood only as complex ions. Many electron-rich organic ligands, however, are able to form water-soluble complexes with such metal ions and so allow them to be in solution even at pHs greater than 7. The iron(II) ion, for example, is insoluble in base, but in blood (pH 7.35) it occurs in a complex ion called *heme,* the red-colored species in hemoglobin and the oxygen carrier in blood.

Inside cells, the phosphate ion level is sufficiently high to form insoluble phosphates with calcium ion, but this must be prevented for many reasons, not least of which is that cells would mineralize and die. Cells prevent this by forming soluble complex ions between Ca^{2+} and a variety of electron-rich molecular units on protein molecules. Complex ions thus have absolutely vital functions at the molecular level of life.

FIGURE 15.2

The hydrated copper (II) ion, $Cu(H_2O)_4^{2+}$ *(left)* gives a bright blue color to its aqueous solution. At the same molar concentration the ammoniated copper (II) ion, $Cu(NH_3)_4^{2+}$ *(right)* causes a much deeper blue.

15.4

THE REDUCTION OF ALDEHYDES AND KETONES

Aldehydes and ketones are reduced to alcohols when hydrogen adds to their carbonyl groups.

Aldehydes are reduced to 1° alcohols and ketones are reduced to 2° alcohols by a variety of conditions. We will study two methods, the direct addition of hydrogen and reduction by hydride ion transfer. Either method can be called the *hydrogenation* or the *reduction* of an aldehyde or ketone.

Aldehyde and Keto Groups Can Be Reduced to Alcohol Groups   Under heat and pressure and in the presence of a finely divided metal catalyst, the carbonyl groups of aldehydes and ketones add hydrogen.

$$\overset{\displaystyle O}{\underset{\text{Aldehyde}}{\| \atop RCH}} + H_2 \xrightarrow[\text{heat, pressure}]{Ni} \underset{\text{1° Alcohol}}{RCH_2OH}$$

$$\overset{\displaystyle O}{\underset{\text{Ketone}}{\| \atop RCR'}} + H_2 \xrightarrow[\text{heat, pressure}]{Ni} \underset{\text{2° Alcohol}}{\overset{\displaystyle OH}{\underset{}{| \atop RCHR'}}}$$

The experimental *conditions* for these catalytic hydrogenations are impossible in living systems, of course, but they do show the *net effect* of the reductions of aldehyde and keto groups that is also accomplished in living cells. *The aldehyde group is reduced to a 1° alcohol and the keto group to a 2° alcohol.*

The Aldehyde or Keto Group Is Reduced by Acceptance of the Hydride Ion   The hydride ion, $H\!:^-$, is a powerful reducing agent. As we have commented before, however, $H\!:^-$ is also an extremely powerful base. *The hydride ion cannot exist as an independent species in the aqueous medium of cells.* In its free form, $H\!:^-$, it reacts vigorously with water to give hydrogen gas, leaving the relatively much weaker base, OH^-.

■ In the lab, $H\!:^-$ can be supplied by $LiAlH_4$ or $NaBH_4$ for these reductions, but water must be rigorously excluded from the reaction mixture.

$$H\!:^- + H\!-\!OH \longrightarrow H\!-\!H + OH^-$$

If we add sodium hydride, NaH, to water, for example, the following reaction occurs, and a caustic solution containing sodium hydroxide, lye, forms.

$$NaH(s) + H_2O \longrightarrow H_2(g) + NaOH(aq)$$

The only way hydride ion can be supplied in living systems, therefore, is by a hydride-ion donor that transfers $H\!:^-$ *directly* to the hydride-ion acceptor.

The carbonyl group is an excellent acceptor of hydride ion. We'll represent an organic donor of hydride ion by the symbol *Mtb*:H, where *Mtb* refers to a *metabolite*, a chemical intermediate in metabolism. When an aldehyde or ketone group accepts a hydride ion, the following reaction occurs.

■ *Mtb*:H in the body is often made from a B vitamin unit incorporated into an enzyme.

$$\underset{\substack{\text{Hydride}\\\text{donor}}}{Mtb\!:\!H} + \underset{\substack{\text{Aldehyde}\\\text{or ketone}}}{\overset{}{C}\!=\!\overset{..}{\underset{..}{O}}\!:} \longrightarrow \underset{\substack{\text{Anion of}\\\text{an alcohol}}}{H\!-\!\overset{|}{\underset{|}{C}}\!-\!\overset{..}{\underset{..}{O}}\!:^-\!Mtb^+}$$

The anion of an alcohol is a stronger proton acceptor than a hydroxide ion. So in the instant when the newly formed anion emerges, it takes a proton either from a water molecule or from some other proton donor in the surrounding buffer system. Thus the final organic product is an alcohol.

$$H-\overset{|}{\underset{|}{C}}-\overset{..}{\underset{..}{O}}:^- + H \overset{..}{O}H \longrightarrow H-\overset{|}{\underset{|}{C}}-\overset{..}{\underset{..}{O}}-H + :\overset{..}{\underset{..}{O}}H^-$$

Anion of an alcohol   Alcohol

■ Remember, a gain of electrons is reduction because it makes oxidation numbers less positive.

As we have already noted, another name for *hydrogenation* is *reduction,* and when a carbonyl carbon atom accepts the pair of electrons carried by the hydride ion, it *gains* this pair and so is reduced.

One of the many examples in cells of reduction by the donation of hydride ion is the reduction of the keto group in the pyruvate ion to form the lactate ion, a step in the metabolism of glucose.

■ NAD+ is a structural unit in several enzymes and is made from a B vitamin. NAD:H, usually written NADH, is the reduced form of NAD$^+$. We explicitly use NAD:H here to emphasize that the species is a donor of H:$^-$. Later, we'll generally write NADH.

$$\underset{\text{Pyruvate ion}}{CH_3\overset{:O:}{\overset{||}{C}}CO_2^-} + \underset{\substack{\text{Hydride}\\\text{ion donor}}}{NAD\!:\!H} \longrightarrow \underset{}{CH_3\overset{:\overset{..}{O}:^-}{\overset{|}{C}}CO_2^-} + \underset{\substack{\text{Oxidized form of}\\\text{hydride ion donor}}}{NAD^+}$$

$$\underset{\substack{\text{(rapid}\\\text{reaction)}}}{HO-H}$$

$$\longrightarrow \underset{\text{Lactate ion}}{CH_3\overset{OH}{\overset{|}{C}}HCO_2^-} + HO^-$$

In this sequence, NAD$^+$ stands for *nicotinamide adenine dinucleotide,* a species that we will discuss further in Section 22.1. Right now, all we need to know about NAD$^+$ is that its reduced form, NAD:H, is a good donor of the hydride ion.

With respect to the reductions of aldehydes and ketones, our needs center on what these reactions produce *in vivo* rather than *in vitro.* We'll study, therefore, only the *net effects* of such reactions and not the specific reagents and conditions commonly used *in vitro.* What we need to learn, therefore, is only how write the products when we know the reactants.

EXAMPLE 15.3
Writing the Structure of the Product of the Reduction of an Aldehyde or Ketone

What is the product of the reduction of propanal?

ANALYSIS All the action is at the carbonyl group. It changes to an alcohol group. Therefore all we have to do is copy over the structure of the given compound, change the double bond to a single bond, and supply the two hydrogen atoms—one to the oxygen atom of the original carbonyl group and one to the carbon atom.

SOLUTION The product of the reduction of propanal is 1-propanol.

$$\underset{\text{Propanal}}{CH_3CH_2\overset{O}{\overset{||}{C}}H} \xrightarrow{\text{reduction}} \underset{\text{1-Propanol}}{CH_3CH_2CH_2\overset{OH}{\overset{|}{}}}$$

> **CHECK** One common error is to change the structural skeleton, so be sure to check
> that the *sequence* of all of the atoms heavier than H, like C and O, *has not changed.*
> Another common error is to write a structure which includes a violation of the cova-
> lences of the heavy atoms — 4 for C and 2 for O in neutral species. So go down the chain
> in the answer, atom by atom, to see that each carbon has four bonds and each oxygen
> has two. If you find an error, fix it. ∎

PRACTICE EXERCISE 5

Write the structures of the products that form when the following aldehydes and ketones are
reduced.

             O   O

                 ‖   ‖

(a) $CH_3CH_2CCH_3$ **(b)** CH_3CHCH_2CH **(c)** [cyclohexanone ring] $=O$

                                       CH_3

15.5

THE REACTIONS OF ALDEHYDES AND KETONES WITH ALCOHOLS

1,1-Diethers — acetals or ketals — form when aldehydes or ketones react with alcohols in the presence of an acid or enzyme catalyst.
This section is background to the study of carbohydrates whose molecules have the functional
groups introduced here. We will study the simplest possible examples of these groups now so
that carbohydrate structures will be easier to understand.

Alcohols Add to the Carbonyl Groups of Aldehydes and Ketones When a solution of
an aldehyde in an alcohol is prepared, molecules of the alcohol add to molecules of the
aldehyde and the following equilibrium mixture forms.

$$:\!\overset{\displaystyle :O:}{\underset{\displaystyle R'CH}{\|}} \ + \ \overset{\displaystyle H}{\underset{\displaystyle \ddot{O}R}{}} \ \rightleftharpoons \ \overset{\displaystyle :\ddot{O}H}{\underset{\displaystyle R'CH\ddot{O}R}{|}}$$

      Aldehyde   Alcohol   Hemiacetal

The product, a **hemiacetal,** has molecules that always have a carbon atom holding both an
OH group and an OR group. When these two groups are this close to each other, they so
modify each other's properties that it's useful to place the whole system into its own separate
family rather than view the system as only a alcohol–ether combination. For example,
hemiacetals, when formed by indirect means, very readily break down to aldehydes and
alcohols. Ordinary ethers, we learned, strongly resist reactions that break their molecules.
The hemiacetal system occurs among carbohydrates where, however, it is relatively stable. In
one form, glucose molecules exist as cyclic hemiacetals (Chapter 19).

■ The hemiacetal system:

        O—H

   C—C—H

        O—R

This originally was
the carbon atom of an
aldehyde group.

■ The hemiketal system:

```
     O—H
     |
C —  C — C
     |
     O—R
```

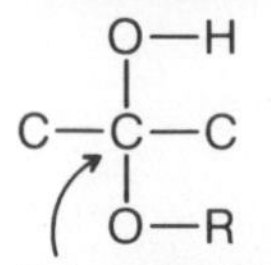

This originally was
the carbon atom of an
keto group.

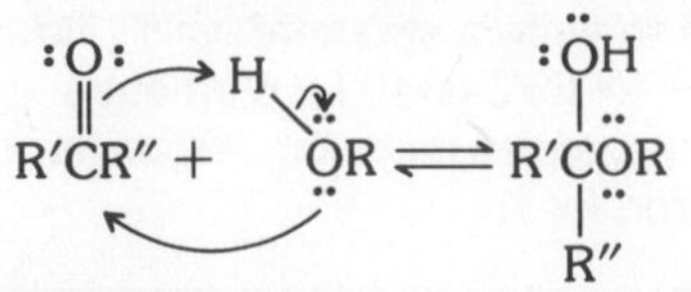

When a ketone is dissolved in an alcohol, a similar reaction occurs to give an equilibrium in which the product is called a **hemiketal** to signify its origin from a ketone.

$$:O: \quad\quad H \quad\quad :OH$$
$$\| \quad\rightarrow\quad \diagdown \quad\quad |$$
$$R'CR'' \;+\; \;OR \;\rightleftharpoons\; R'COR$$
$$\quad\quad\quad\quad\quad\quad\quad |$$
$$\quad\quad\quad\quad\quad\quad\quad R''$$

Ketone　　Alcohol　　Hemiketal

The position of equilibrium overwhelmingly favors the reactants, the ketone and alcohol, so hemiketals are even less stable than hemiacetals. The hemiketal system, however, does occur among carbohydrates. One form of fructose, for example, is a cyclic hemiketal (Chapter 19). The continuation of our study of these systems will deal almost entirely with hemiacetals because the extension of the principles to hemiketals is straightforward.

The Polarity of the Carbonyl Group Causes the Alcohol to Add to an Aldehyde in Only One Direction The —OR part of the alcohol molecule has $\delta-$ on O and the carbonyl group of the aldehyde has $\delta+$ on C. Because unlike charges attract, the alcohol's —OR unit *always* ends up attached to the carbon atom of the original carbonyl group, and the H atom of the alcohol always goes to the carbonyl oxygen. Thus the *direction* of this addition reaction is determined by the polarity of the carbonyl group.

　　Except among carbohydrates, the hemiacetal system is almost always too unstable to exist in a pure compound. If we try to isolate and purify an ordinary hemiacetal, it breaks back down, and only the original aldehyde and alcohol are obtained. Hemiacetals generally exist only in the equilibrium that includes their parent aldehydes and alcohols. Despite this, we're still interested in the hemiacetal system for two reasons. Not only is it relatively stable among carbohydrates, it is an intermediate in the formation of "1,1-diethers" or acetals, which are stable enough to be isolated. The acetal system is also common among carbohydrates.

　　The relative ease with which hemiacetals break back down means that the hemiacetal system is a site of structural weakness, even among carbohydrates. For this reason, we have to learn how to recognize the system when it occurs in a structure.

EXAMPLE 15.4
Identifying the Hemiacetal System

■ The ring system of **3** also occurs in glucose.

Which of the following structures has the hemiacetal system? Draw an arrow pointing to any carbon atoms that were initially the carbon atoms of aldehyde groups.

CH_3—O—CH_2—CH_2—OH　　CH_3—O—CH_2—OH

1　　　　　**2**　　　　　**3**

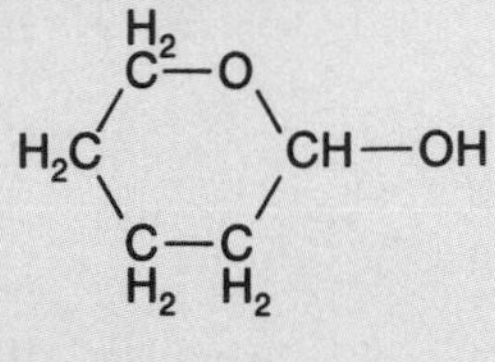

ANALYSIS To have the hemiacetal system, the molecule must have a carbon to which are attached one OH group and one —O—C unit.

SOLUTION In structure **1** there is an OH group and an —O—C unit, but they are not joined to the *same* carbon. Therefore **1** is not a hemiacetal. It has only an ordinary ether group plus an alcohol group.

　　In structure **2**, the OH and the —O—C are joined to the same carbon, so **2** is a hemiacetal. Similarly, in structure **3**, the carbon on the far right corner of the ring holds both an OH group and —O—C unit, and **3** is also a hemiacetal, a cyclic hemiacetal. Structure **3** shows the way in which the hemiacetal system occurs in many carbohydrates, as a cyclic hemiacetal. The asterisks (*) in the following structures identify carbon atoms that were originally carbonyl carbons.

$$CH_3 \text{—} O \text{—} \overset{*}{C}H_2 \text{—} OH$$

2   **3**

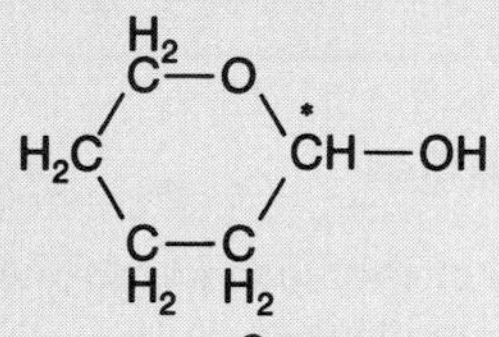

CHECK Make sure that any structure identified as a hemiacetal has at least one carbon attached to *two* O atoms by single bonds, that *one* of the O atoms is part of the OH group, and that the other is joined by its second single bond to C.

PRACTICE EXERCISE 6

Identify the hemiacetals or hemiketals among the following structures, and draw arrows that point to the carbon atoms that initially were part of the carbonyl groups of parent aldehydes (or ketones).

(a) 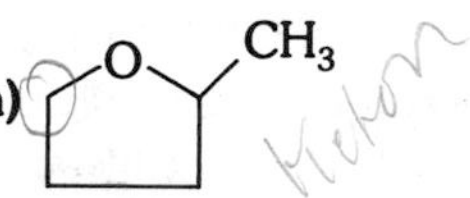   (b) 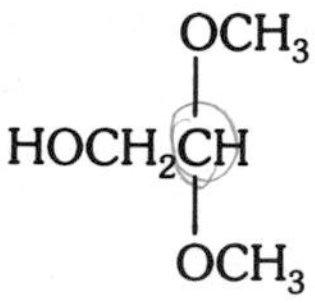 $HOCH_2CH$ with OCH_3 above and OCH_3 below

(c) $HOCH_2OCH_2CH_3$ (d) 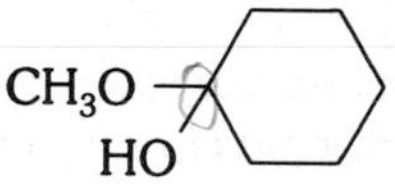 CH_3O and HO attached to a ring

Another skill that will be useful in our study of carbohydrates is the ability to write the structure of a hemiacetal that could be made from a given aldehyde and alcohol.

EXAMPLE 15.5
Writing the Structure of a Hemiacetal Given Its Parent Aldehyde and Alcohol

Write the structure of the hemiacetal that is present at equilibrium in a solution of propanal in ethanol.

ANALYSIS A hemiacetal must have a carbon atom to which both an OH group and an OR group are attached. This carbon atom *is provided by the aldehyde.* The structure of the alcohol, CH_3CH_2OH, tells us that the R group in OR of the hemiacetal is CH_2CH_3.

SOLUTION The structure of the hemiacetal formed from propanal and ethyl alcohol is

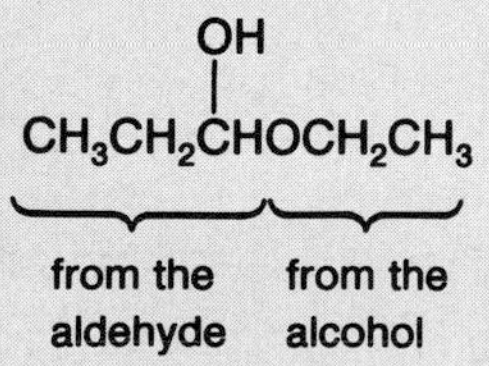

$$\overset{\displaystyle OH}{\underset{}{CH_3CH_2\overset{|}{C}HOCH_2CH_3}}$$

from the   from the
aldehyde   alcohol

CHECK Find the carbon holding *two* O atoms and check the other two atoms or groups that it also holds *against the original aldehyde* (or ketone). These two atoms or groups —here, H and CH_3CH_2—must match those of the aldehyde (or ketone). Finally check what else the two O atoms are holding; one must hold H (to make it an OH group) and the other must hold the alkyl group from the original alcohol. ∎

PRACTICE EXERCISE 7

Write the structures of the hemiacetals that are present in the equilibria that involve the following pairs of compounds.

(a)   ethanal and methanol   **(b)**   butanal and ethanol

(c)   benzaldehyde and 1-propanol   **(d)**   methanal and methanol

Still another skill that will be useful in studying carbohydrates is the ability to write the structures of the aldehyde and alcohol that are liberated by the breakdown of a hemiacetal.

EXAMPLE 15.6
Writing the Breakdown Products of a Hemiacetal

What aldehyde and alcohol form when the following hemiacetal breaks down?

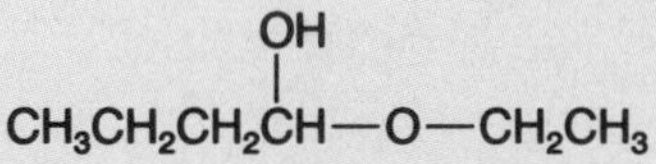

$$\underset{\displaystyle }{CH_3CH_2CH_2\overset{\displaystyle \overset{OH}{|}}{C}H-O-CH_2CH_3}$$

ANALYSIS The key step to solving this problem lies in analyzing the given structure. First, pick out the carbon atom of the original carbonyl group; it's the one holding *both* an OH group and an OR unit. Anything else this carbon holds—usually one H and one hydrocarbon group—completes what we need in order to write the structure of the aldehyde. The R group of OR is the hydrocarbon group of the original alcohol. Our analysis thus gives us:

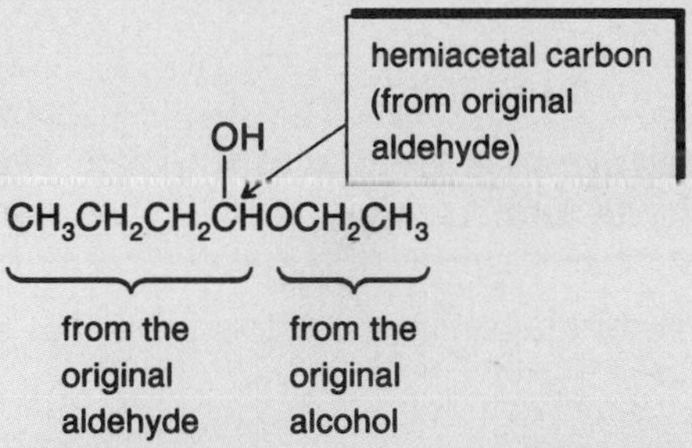

The original aldehyde is thus the unbranched four-carbon aldehyde, butanal, and the original alcohol is seen to be the two-carbon alcohol, ethanol.

SOLUTION The products of the breakdown of the given hemiacetal are

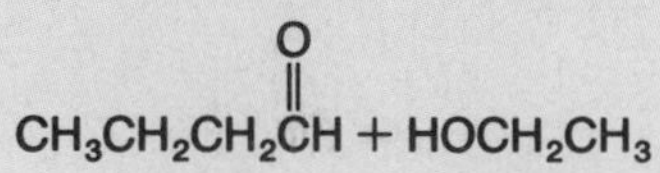

$$\underset{\text{Butanal}}{CH_3CH_2CH_2\overset{\displaystyle \overset{O}{\|}}{C}H} + \underset{\text{Ethanol}}{HOCH_2CH_3}$$

CHECK It is essential to notice that *only one bond* is affected when we disassemble the initial hemiacetal, *the C—O bond of the hemiacetal carbon,* not any other C—O bond, not a C—C bond, and not a C—H bond. Failure to learn this is the most common error that students make in working problems like this. Examine closely, therefore, your answer to see that the original molecule is ruptured *only* at the C—O bond at the hemiacetal carbon. *Break no other bond.*

PRACTICE EXERCISE 8

Write the structures of the breakdown products of the following hemiacetals.

       OH   OH
       | |
(a) $CH_3CH_2CHOCH_3$ **(b)** $CH_3CH_2OCHCH_2CH_3$

Acetals and Ketals Form When Alcohols React Further with Hemiacetals and Hemiketals Hemiacetals and hemiketals are special kinds of alcohols, and they resemble alcohols in one important property. They can undergo a reaction that looks like the formation of an ether. A ordinary ether does not form, however, but a special kind, a 1,1-diether called an **acetal** or a **ketal**. The overall change that leads to an acetal is as follows.

        OH   OR
        | |
R′CHOR + H—OR $\xrightarrow{\text{acid catalyst}}$ R′CHOR + H_2O
  Hemiacetal   Acetal

Hemiketals give the identical kind of reaction, but the products are called *ketals*. Unlike hemiacetals and hemiketals, both acetals and ketals are stable compounds that can be isolated and stored.

The difference between the formation of an acetal and an ordinary ether is that *acetals form and break more readily*. As a rule, when two functional groups are very close to each other in a molecule, each modifies the properties of the other in some way. Here, the OR group makes the OH group attached to the same carbon much more reactive toward the splitting out of water with an alcohol.

In a structural sense, an acetal is a 1,1-diether, but "1,1" does not refer to the numbering of the chain and the "ether" part of 1,1-diether does not connote "resistance to breaking up." The "1,1" means only that the two OR groups come to the *same* carbon. In this sense, ketals are also 1,1-diethers. The molecules of many carbohydrates, like sucrose (table sugar), lactose (milk sugar), and starch also have the 1,1-diether system, which is why we study acetals and ketals.

Ordinary ethers, R—O—R, do not break up in dilute acid or base, but *acetals and ketals are stable only if they are kept out of contact with aqueous acids*. In water, acids (or enzymes) catalyze the hydrolysis of acetals and ketals to their parent alcohols and aldehydes (or ketones). In aqueous *base*, however, the acetal (ketal) system is stable.

Hydrolysis is the only chemical reaction of acetals and ketals that we need to study; it is the reaction by which carbohydrates are digested. Before we study this reaction further, we must be sure that we can recognize the acetal or ketal system when it occurs in a structure.

■ The acetal system:

          O—R
          |
    C—C—H
        |
        O—R

This originally was the carbon atom of an aldehyde group.

■ The ketal system:

          O—R
          |
    C—C—C
        |
        O—R

This originally was the carbon atom of an keto group.

EXAMPLE 15.7
Recognizing the Acetal or Ketal Systems and Identifying Which Carbon Atoms Came from Parent Carbonyl Carbons

Examine each structure to see whether it is an acetal or a ketal. If it is either, identify the carbon atom furnished by the carbonyl carbon of the parent aldehyde or ketone.

$$CH_3OCH_2OCH_3 \quad CH_3OCHCH_3 \quad$$

$$\underset{4}{} \quad \underset{5}{\overset{OCH_3}{|}} \quad \underset{6}{}$$

ANALYSIS We have to find one carbon that holds two OR types of groups. The two *must* join to the *same* carbon.

SOLUTION Structure **4** has such a carbon in its central CH_2 unit. This carbon was initially the carbonyl carbon atom of an aldehyde (methanal), because it also holds at least one H atom. Structure **5** similarly has such a carbon, in the CH unit, a carbon that also came from an aldehyde group (because it has at least one H atom). In structure **6**, we can also find a carbon that holds two O—C bonds. This carbon lacks an H atom, however, so it must have come from the carbonyl group of a ketone system.

Initially a ketone carbonyl carbon atom

PRACTICE EXERCISE 9

Which of the following two compounds, if either, is an acetal or a ketal? Identify the carbon atom that came originally from the carbonyl group of a parent aldehyde or ketone.

(a) $CH_3OCH_2CH_2OCH_3$ **(b)** $CH_3CH_2OCOCH_2CH_3$ with CH_3 substituents

Because acetals and ketals can be hydrolyzed, we have to be able to examine the structures of such compounds and write the structures of their parent alcohols and aldehydes (or ketones), the hydrolysis products. A worked example shows how this can be done.

EXAMPLE 15.8
Writing the Structures of the Products of the Hydrolysis of Acetals or Ketals

What are the products of the following reaction?

$$CH_3CHOCH_3 + H_2O \xrightarrow{\text{acid catalyst}} ?$$

ANALYSIS The best way to proceed is to find the carbon atom in the structure that holds *two* oxygen atoms. This carbon is the carbonyl carbon atom of the parent aldehyde (or ketone). Break both of its bonds to these oxygen atoms. *Do not break any other bonds.* Then at the carbon that once held two oxygens, make a carbonyl group. The other groups, those of the OR type (here, OCH$_3$) become alcohols.

SOLUTION The final products of the hydrolysis of the given acetal are ethanal and methanol.

$$CH_3\overset{\underset{\displaystyle |}{OCH_3}}{C}HOCH_3 + H_2O \xrightarrow{\text{acid catalyst}} CH_3\overset{\displaystyle O}{\overset{\displaystyle \|}{C}}H + 2HOCH_3$$

Initially a carbonyl carbon

PRACTICE EXERCISE 10

Write the structures of the aldehydes (or ketones) and the alcohols that are obtained by hydrolyzing the following compounds. If they do not hydrolyze like acetals or ketals, write "no reaction."

back

(a) $CH_3OCH_2OCH_3$ (b) $CH_3OCH_2CH_2OCH_2CH_3$ (c) $CH_3\overset{\underset{\displaystyle |}{CH_3}}{\underset{\displaystyle |}{C}}\overset{\overset{\displaystyle H_3C}{|}}{}HCOCH_3$ with OCH$_3$

Other Reactions Involving Aldehydes and Ketones

Aldehydes and ketones are able to enter into reactions that create larger molecules from smaller molecules by forming new carbon–carbon bonds. The *aldol condensation* is one such reaction. The simplest illustration is the reaction of ethanal with itself in the presence of dilute sodium hydroxide or an enzyme (called *aldolase*).

New carbon–carbon bond

$$CH_3\overset{\displaystyle O}{\overset{\displaystyle \|}{C}}H + CH_2\overset{\displaystyle O}{\overset{\displaystyle \|}{C}}H \underset{\text{enzyme}}{\overset{\text{base or}}{\rightleftharpoons}} CH_3\overset{\underset{\displaystyle |}{OH}}{C}H-CH_2\overset{\displaystyle O}{\overset{\displaystyle \|}{C}}H$$

Ethanal Ethanal, second molecule 3-Hydroxybutanal "aldol"

It is little more than the addition of one aldehyde molecule to the carbonyl double bond of another, but it is a *reversible* reaction, as the equilibrium arrows indicate. The forward reaction makes a new carbon–carbon single bond; the reverse reaction, called the *reverse aldol,* breaks this same bond.

An aldol condensation occurs in those cells of ours where glucose molecules are made from smaller molecules by a series of steps called *gluconeogenesis* (Section 26.4). The breakdown of glucose occurs by a different series of reactions called *glycolysis* (Section 26.3), and one step is a reverse aldol. Because these are relatively complex series, we'll delay further study of the aldol condensation until the place where it most directly is used, Chapter 26.

■ The mechanism of the aldol condensation is given in Special Topic 26.3. It could easily be brought forward for study here.

SUMMARY

Naming aldehydes and ketones The IUPAC names of aldehydes and ketones are based on a parent compound, one with the longest chain that includes the carbonyl group. The names of aldehydes end in -al and of ketones in -one, and the chains are numbered so as to give the carbonyl carbons the lower of two possible numbers.

Physical properties of aldehydes and ketones The carbonyl group confers moderate polarity, which gives aldehydes and ketones higher boiling points and solubilities in water than hydrocarbons but lower boiling points and solubilities in water than alcohols (that have comparable formula masses).

Ease of oxidation of aldehydes Aldehydes are easily oxidized to carboxylic acids, but ketones resist oxidation. Aldehydes give a positive Tollens' test and ketones do not. α-Hydroxy aldehydes and ketones give the Benedict's test.

Complex ions The reagents for the Tollens' and Benedict's tests contain complex ions, the silver diammine complex in Tollens' reagent and the copper(II) citrate complex ion in Benedict's test. In the first, Ag^+, the central metal ion, holds two molecules of the ligand, NH_3. The ligand for the Cu^{2+} ion in Benedict's reagent is the negative ion of citric acid. Complex ions help to keep transition metal ions in solution, even in base.

Hemiacetals and hemiketals When an aldehyde or a ketone is dissolved in an alcohol, some of the alcohol adds to the carbonyl group of the aldehyde or ketone. The equilibrium thus formed includes molecules of a hemiacetal (or hemiketal). The chart at the end of this summary outlines the chemical properties of the aldehydes and ketones we have studied.

Hemiacetals and hemiketals are usually unstable compounds that exist only in an equilibrium involving the parent carbonyl compound and the parent alcohol (which generally is the solvent). Hemiacetals and hemiketals readily break back down to their parent carbonyl compounds and alcohols. When an acid catalyst is added to the equilibrium, a hemiacetal or hemiketal reacts with more alcohol to form an acetal or ketal.

Acetals and ketals Acetals and ketals are 1,1-diethers that are stable in aqueous base or in water, but not in aqueous acid. Acids

catalyze the hydrolysis of acetals and ketals, and the final products are the parent aldehydes (or ketones) and alcohols.

Summary of reactions (We omit the aldol condensation here.)

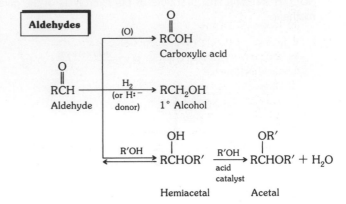

REVIEW EXERCISES

The answers to Review Exercises whose numbers are in color are found in Appendix C. The answers to the other Review Exercises are found in the Study Guide that accompanies this book. The more challenging questions are marked with asterisks.

Names and Structures

15.1 Give the names of the functional groups present in the following structural formulas.

(a) $CH_3CH_2\overset{O}{\underset{\|}{C}}CH_2CH_3$ (b) $H\overset{O}{\underset{\|}{C}}CH_2\overset{CH_3}{\underset{|}{C}}HCH_3$ (c)

(d) cyclopentane—CO_2H (e) benzene—CHO (f)

15.2 To display the *structural* differences among aldehydes, ketones, carboxylic acids, and esters, write the structure of one example of each using three carbons per molecule.

15.3 Write the structure of each of the following compounds.
(a) 2-methylbutanal
(b) 2,3-dichlorocyclohexanone
(c) 1-phenyl-1-ethanone
(d) diisobutyl ketone
(e) hexane-2,5-dione

15.4 What are the structures of the following compounds?
(a) 2-cyclohexylcyclopentanone
(b) 2-methylhexanal
(c) di-*sec*-butyl ketone
(d) 1,3-diphenyl-2-propanone
(e) 1,3,5-cyclohexanetrione

15.5 Although we can write structures that correspond to the following names, when we do, we find that the names aren't proper. How should these compounds be named in the IUPAC system?
(a) 6-methylcyclohexanone
(b) 1-hydroxy-1-propanone (Give common name.)
(c) 1-methylbutanal
(d) 2-methylethanal
(e) 2-propylpropanal

15.6 The following names can be used to write structures, but the names turn out to be improper. What should be their IUPAC names?
(a) 2-*sec*-butylbutanal
(b) 1-phenylethanal
(c) 4,5-dimethylcyclopentanone
(d) 1-hydroxyethanal (Give common name.)
(e) 1-butanone

15.7 If the IUPAC name of **A** is 2-ketopropanal, then what is the IUPAC name of compound **B**?

$$
\underset{\textbf{A}}{CH_3\overset{O}{\overset{\|}{C}}-\overset{O}{\overset{\|}{C}}H} \qquad \underset{\textbf{B}}{CH_3\overset{O}{\overset{\|}{C}}CH_2CH_2CH_2\overset{O}{\overset{\|}{C}}H}
$$

15.8 We can name compound **C** as 3-methylformylcyclopentane. Taking a cue from this, how might compound **D** be named?

C

D

15.9 Write the IUPAC names of the following compounds.

(a) $CH_3\underset{\underset{CH_3}{|}}{CH}CH_2CHO$

(b) $CH_3\underset{\underset{CH_3CH_2}{|}}{CH}CH_2CHO$

(c) $CH_3\underset{\underset{CH_3}{|}}{CH}CH_2\underset{\underset{CH_2CH_3}{|}}{CH}CH_2CHO$

(d) $CH_3\underset{\underset{CH_3}{|}}{CH}CH_2\overset{O}{\overset{\|}{C}}CH_3$

(e)

$CH_3CH_2CH_2$

*15.10 What are the IUPAC names of the following compounds?

(a)

$$CH_3CH_2\underset{\underset{CH_2}{|}}{\overset{\overset{CH_3}{|}}{CH}}$$
$$CH_3\underset{\underset{CH_3}{|}}{CH}CH_2\underset{\overset{\|}{O}}{CH}CCH_3$$

(b) $(CH_3)_3CCH\underset{\underset{CH_2CHCH_3}{\underset{\underset{CH_2CH_3}{|}}{|}}}{}CH_2CH_2CHO$

(c)

$-CH_2CH_2\overset{O}{\overset{\|}{C}}CH_3$

(d)

(e)

15.11 If the common name of $CH_3CH_2CH_2CH_2CO_2H$ is valeric acid, then what is the most likely common name of the following compound?

$$CH_3CH_2CH_2CH_2CHO$$

15.12 If the common name of **E** is glyceraldehyde, what is the most likely common name of **F**?

$$\underset{\underset{OH}{|}}{HOCH_2CHCHO} \qquad \underset{\underset{OH}{|}}{HOCH_2CHCO_2H}$$
$$\textbf{E} \qquad\qquad \textbf{F}$$

Physical Properties of Aldehydes and Ketones

15.13 Arrange the following compounds in their order of increasing boiling points. Do this by placing the letters that identify them in the correct order, starting with the lowest-boiling compound and moving in order to the highest-boiling compound. (They have about the same formula masses.)

A   B   C   D

15.14 Arrange the following compounds in their order of increasing boiling points. Do this by placing the letters that identify them in the correct order, starting with the lowest-boiling compound on the left in the series and moving to the highest-boiling compound. (They all have about the same formula mass.)

CH_3
|
$HOCH_2CH_2CH_2CHOH$ $CH_3CH_2CH_2CH_2CCH_3$
A   B

CH_3 CH_3
| |
$CH_3CH_2CH_2CH_2CHCH_3$ $HOCH_2CH_2CH_2CHCH_3$
C   D

15.15 Reexamine the compounds of Review Exercise 15.13, and arrange them in their order of increasing solubility in water.

15.16 Arrange the compounds of Review Exercise 15.14 in their order of increasing solubility in water.

15.17 Draw the structure of a water molecule and a molecule of propanal and align them on the page to show how the propanal molecule can accept a hydrogen bond from the water molecule. Use a dotted line to represent this hydrogen bond and place $\delta+$ and $\delta-$ symbols where they are appropriate.

15.18 Draw the structures of molecules of methanol and acetone, and align them on the page to show how a hydrogen bond (which you are to indicate by a dotted line) can exist between the two. Place $\delta+$ and $\delta-$ symbols where they are appropriate.

Oxidation of Alcohols and Aldehydes

15.19 What are the structures and the IUPAC names of the aldehydes and ketones to which the following compounds can be oxidized?

(a)
CH_3
|
CH_3CH_2CHOH

(b)
CH_3
|
$CH_3CHCH_2CH_2CH_2OH$

(c)

(d) $C_6H_5CH_2CHCH_2OH$
                    |
                    CH_3

15.20 Examine each of the following compounds to see whether it can be oxidized to an aldehyde or to a ketone. If it can,

write the structure of the aldehyde or ketone. If the aldehyde or ketone is *aliphatic*, write its IUPAC name.

(a) CH_3
       |
$HOCHCH_2CH_3$

(b) CH_3
       |
$HOCH_2CHCH_3$

(c)

(d)   O
            ||
CH_3CH_2COH

(e) CH_3
            |
$CH_3OCH_2CH_2CHOH$

(f) $HOCH_2$——CO_2H

*15.21 An unknown compound, C_3H_6O, reacted with permanganate ion to give $C_3H_6O_2$, and the same unknown also gave a positive Tollens' test. Write the structures of C_3H_6O and $C_3H_6O_2$.

*15.22 An unknown compound, $C_3H_6O_2$, could be oxidized easily by permanganate ion to $C_3H_4O_3$, and it gave a positive Benedict's test. Write structures for $C_3H_6O_2$ and $C_3H_4O_3$.

*15.23 Which of the following compounds can be expected to give a positive Benedict's test? All are intermediates in metabolism.

(a) $HOCH_2CCH_2OH$ (b) $HOCH_2CHCHO$
         || |
         O   OH

(c) $CH_3CCH_2CO_2H$ (d) $CH_3CHCH_2CO_2H$
         || |
         O   OH

*15.24 Which of the following compounds give a positive Benedict's test? (Most are intermediates in metabolism.)

(a) CH_3CCO_2H (b) CH_3CCHO
         || ||
         O   O

(c) $HOCH_2CH_2CCH_3$ (d) $HOCH_2CHCHCHO$
         || | |
         O   HO OH

15.25 Concerning complex ions,
(a) The cations of what kinds of elements are usually involved?
(b) What *kinds* of particles commonly are ligands?
(c) Give the *formulas* of three examples of negatively charged ligands that come from the same family in the periodic table.
(d) Give the *formulas* of two electrically neutral ligands that form complex ions with Cu^{2+}. Write the formulas of these complex ions.

15.26 Concerning complex ions in test reagents,
(a) What is the *formula* of the complex ion in Tollens' reagent and why is it important that Ag^+ be so complexed?
(b) What is the *function* of the citrate ion in Benedict's reagent?

15.27 What is the *formula* of the precipitate that forms in a positive Benedict's test?

15.28 Clinitest tablets are used for what?

15.29 What is one practical commercial application of the Tollens' reagent system?

*15.30 Neither the lactate ion nor the pyruvate ion gives a positive Tollens' test. When the body metabolizes the lactate ion, it oxidizes it to the pyruvate ion, $C_3H_3O_3^-$. Using only these facts, write the structure of the lactate ion.

15.31 One of the steps in the metabolism of fats and oils in the diet is the oxidation of the following compound:

$$\underset{\overset{|}{CH_3\overset{\overset{OH}{|}}{C}HCH_2CO_2^-}}{}$$

Write the structure of the product of this oxidation.

*15.32 One of the important series of reactions in metabolism is called the *citric acid cycle*. Structures **A** and **B** are of compounds (actually, anions) that participate in this cycle. One of them, isocitric acid, is oxidized to a ketone. The other is not. Which one is isocitric acid, **A** or **B**? Write the structure of its corresponding ketone.

$$\underset{\textbf{A}}{\overset{CH_2CO_2^-}{\underset{CH_2CO_2^-}{HO-CCO_2^-}}} \qquad \underset{\textbf{B}}{\overset{HO-CHCO_2^-}{\underset{CH_2CO_2^-}{CHCO_2^-}}}$$

Reduction of Aldehydes and Ketones

15.33 The hydride ion, as we learned, reacts as follows with water:

$$H:^- + H-OH \longrightarrow H-H + OH^-$$

The hydride ion reacts in a similar way with CH_3CH_2OH. Write the net ionic equation for this reaction.

15.34 Consider the reaction that occurs when a hydride ion transfers from its donor (which we can write as *Mtb*:H) to ethanal.
(a) Write the structure of the organic anion that forms when the hydride ion is transferred to ethanal.
(b) What is the net ionic equation of the reaction of water with the anion formed in part a?
(c) What is the IUPAC name of the organic product of the reaction of part b?

15.35 If a donor of a hydride ion (*Mtb*:H) transfers it to a molecule of acetone,
(a) What is the structure of the organic ion that forms?
(b) What happens to the organic ion (formed in part a) in the presence of water? (Write a net ionic equation.)
(c) What is the IUPAC name of the organic product of the reaction of part b?

*15.36 The metabolism of aspartic acid, an amino acid, occurs by a series of steps. A portion of this series is indicated below, where NAD:H is a reducing agent that becomes NAD^+ as it transfers hydride ion.

$$\underset{\text{Aspartate ion}}{\overset{\overset{+}{H_3NCHCO_2^-}}{\underset{CH_2CO_2^-}{|}}} \xrightarrow{\text{two steps}} \overset{\overset{+}{H_3NCHCO_2^-}}{\underset{CH_2CHO}{|}} \xrightarrow{NAD:H}$$

$$NAD^+ + \overset{+}{H_3NCHCO_2^-}$$

$$\underset{\textbf{A}}{\overset{?}{\bigcirc}} \quad \underset{H_2O}{\overset{}{\longrightarrow}} \textbf{B} + OH^-$$

Complete the structure of **A**, and write the structure of **B**.

*15.37 One of the steps the body uses to make long-chain carboxylic acids is a reaction similar to the following reaction.

$$CH_3\overset{\overset{O}{\|}}{C}CH_2\overset{\overset{O}{\|}}{C}S-\boxed{enzyme} + NAD:H \longrightarrow$$

$$?-CH_2\overset{\overset{O}{\|}}{C}S-\boxed{enzyme} + NAD^+$$

$$\underset{\textbf{A}}{} \quad \underset{H_2O}{\overset{}{\longrightarrow}} \textbf{B} + OH^-$$

Complete the structure of **A** and write the structure of **B**.

15.38 Write the structures of the aldehydes or ketones that could be used to make the following compounds by reduction (hydrogenation).

(a) $CH_3\overset{\overset{OH}{|}}{C}HCH_2CH_3$

(b) $HOCH_2\overset{\overset{CH_3}{|}}{C}HCH_2CH_3$

(c) $\bigcirc-OH$

(d) $CH_3-\bigcirc-CH_2OH$

15.39 Either an aldehyde or a ketone could be used to make each of the following compounds by hydrogenation. Write the structure of the aldehyde or ketone suitable in each part.

(a) $HOCH_2CH_2OCH_3$

(b) CH_3OCHCH_2OH
$\quad\quad\; \overset{|}{CH_3}$

(c) $CH_3\overset{\overset{OH}{|}}{C}HCH_2\overset{\overset{OH}{|}}{C}CH_3$
$\quad\quad\quad\quad\quad\quad \overset{|}{CH_3}$

(d) $CH_3CH_2O-\bigcirc-\overset{\overset{OH}{|}}{C}HCH_3$

Hemiacetals and Acetals. Hemiketals and Ketals

15.40 Examine each structure and decide whether it represents a hemiacetal, hemiketal, acetal, ketal, or something else.

(a) CH_3OCHOH with CH_3 substituent

(b) CH_3CHOCH_3 with OCH_3 substituent

(c) $CH_3OCHCH_2OCH_3$ with CH_3 substituent

(d) CH_3OCOCH_3 with CH_3 substituents (two)

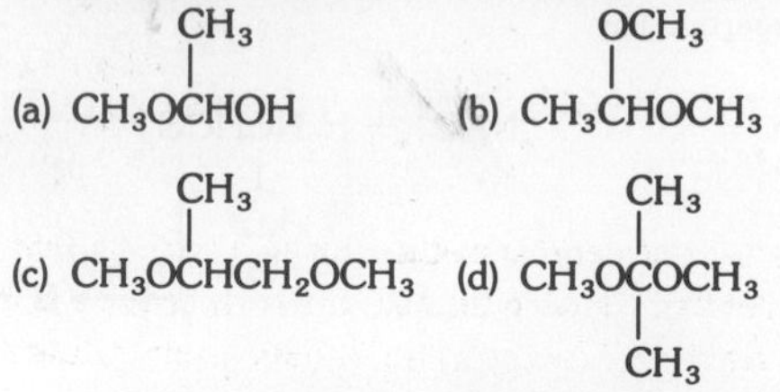

15.41 Examine each structure and decide whether it represents a hemiacetal, hemiketal, acetal, ketal, or something else.

(a) $HOCH_2CH_2CHOCH_3$ with OCH_2CH_3 substituent

(b) [six-membered ring with O and OH]

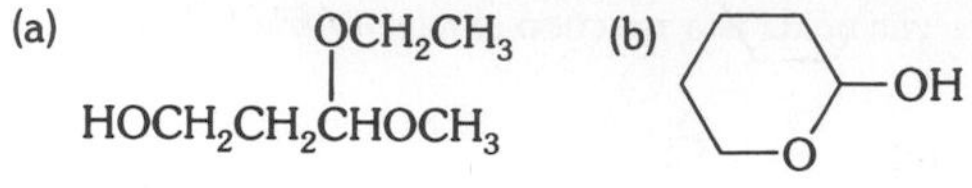

(c) $CH_3CH_2OCH_2OH$

(d) $HOCH_2CH_2OCH_2CH_2CH_3$

15.42 Write the structures of the hemiacetals and the acetals that can form between propanal and the following two alcohols. (a) methanol (b) ethanol

15.43 What are the structures of the hemiketals and the ketals that can form between acetone and these two alcohols? (a) methanol (b) ethanol

*15.44** Write the structure of the hydroxyaldehyde (a compound having both the alcohol group and the aldehyde group in the same molecule) from which the following hemiacetal forms in a ring-closing reaction. (You may leave the chain of the open-chain compound somewhat coiled.)

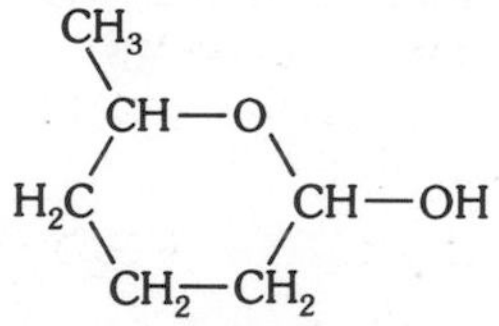

*15.45** One form in which a glucose molecule exists is given by the following structure. (*Note:* The atoms and groups that are attached to the carbon atoms of the six-membered ring must be seen as projecting *above* or *below* the ring.)

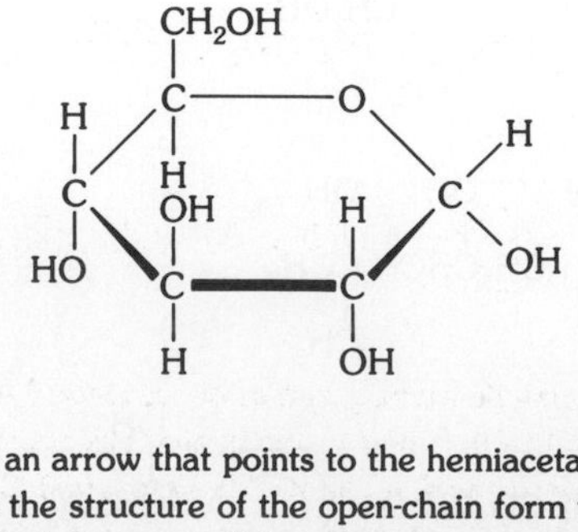

(a) Draw an arrow that points to the hemiacetal carbon.
(b) Write the structure of the open-chain form that has a free aldehyde group. (You may leave the chain coiled.)

*15.46** Write the structure of a hydroxy ketone (a molecule that has both the OH group and the keto group) from which the following hemiketal forms in a ring-closing reaction. (You may leave the chain of the open-chain compound somewhat coiled.)

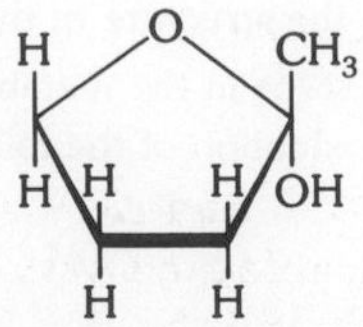

*15.47** Fructose occurs together with glucose in honey, and it is sweeter to the taste than table sugar. One form in which a fructose molecule can exist is given by the following structure.

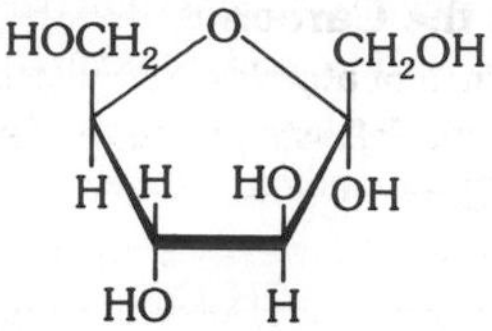

(a) Draw an arrow to the carbon of the hemiketal system that came initially from the carbon atom of a keto group.
(b) In water, fructose exists in an equilibrium with an open-chain form of the given structure. This form has a keto group in the same molecule as five OH groups. Draw the structure of this open-chain form (leaving the chain coiled somewhat as it was in the structure that was given).

15.48 The digestion of some carbohydrates is simply their hydrolysis catalyzed by enzymes. Acids catalyze the same kind of hydrolysis of acetals and ketals. Write the structures of the products, if any, that form by the action of water and an acid catalyst on the following compounds.

(a) $CH_3OCHOCH_3$ with CH_2CH_3 substituent

(b) $CH_3OCH_2CHOCH_3$ with CH_3 substituent

(c) $CH_3CH_2OCOCH_2CH_3$ with CH_3 substituents (two)

(d) [cyclopentane ring with two OCH_3 groups]

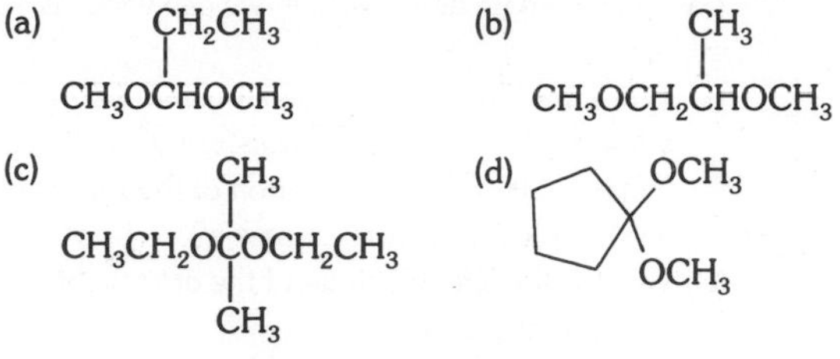

15.49 What are the structures of the products, if any, that form by the acid-catalyzed reaction of water with the following compounds?

(a) $CH_3CH_2CHOCH_2CH_3$ with $OCH(CH_3)_2$ substituent

(b) $CH_3OCH_2CHOCH_2CH_3$ with $CH(CH_3)_2$ substituent

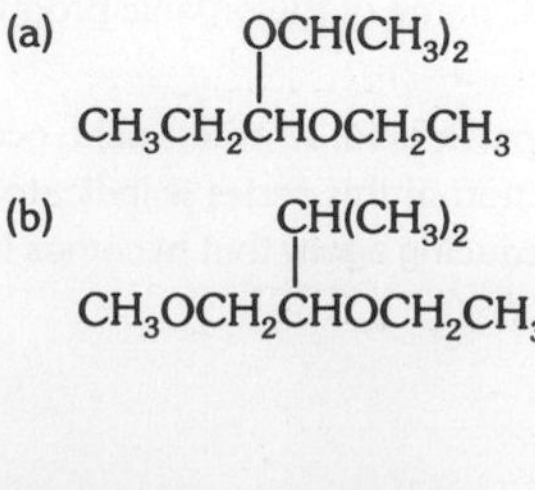

(c)

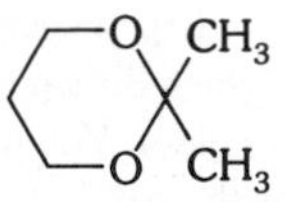

(d)

$$CH_3OCH_2CHOCH_3$$

(with OCH_2CH_3)

(i)

CHO $\xrightarrow[\text{(e.g., } Cr_2O_7{}^{2-}, H^+)]{\text{(O)}}$ *acid*

(j)

$CH_3 \xrightarrow[\text{(e.g., } MnO_4{}^-, OH^-)]{\text{(O)}}$ *no reaction*

Important Aldehydes and Ketones (Special Topic 15.1)

15.50 Give the name of a specific aldehyde or ketone described in Special Topic 15.1 that is
(a) a female sex hormone
(b) a good nail polish remover
(c) a preservative

The Bonds in the Carbonyl Group (Special Topic 15.2)

15.51 What kinds of atomic orbitals (pure or hybridized) overlap to form the following bonds in formaldehyde?
(a) the C—H bonds
(b) the sigma bond in the carbonyl group
(c) the pi bond in the carbonyl group

15.52 What are the bond angles in formaldehyde? Do all its atoms lie in the same plane?

Additional Exercises

*15.53 Complete the following reaction sequences by writing the structures of the organic products that form. If no reaction occurs, write "no reaction." (Reviewed here too are some reactions of earlier chapters.)

(a) *aldehyd*   *isobutyl*
$$CH_3CHCHO + H_2 \xrightarrow[\text{heat, pressure}]{\text{Ni catalyst}}$$ *alcohol*
(with CH_3, circled R)

(b) OH
$$(CH_3)_2CHCHCHCH_3 \xrightarrow[\text{(e.g., } Cr_2O_7{}^{2-}, H^+)]{\text{(O)}}$$ *2-alcohol*

(c) 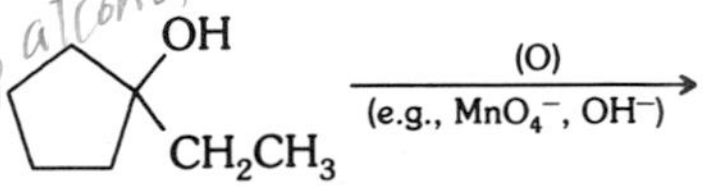 *tertiary alcohol*   OH
$\xrightarrow[\text{(e.g., } MnO_4{}^-, OH^-)]{\text{(O)}}$ *no reaction*
(with CH_2CH_3)

(d) $CH_3CH{=}CHCH_2CH_3 + H_2 \xrightarrow{\text{Ni catalyst}}$ *saturate*

(e) $CH_3OH + CH_3CH_2CHO \rightleftharpoons$ *Hemiacetal*

(f) $CH_3CHO + Mtb{:}H \xrightarrow[\text{by } H^+]{\text{(followed}}$ *alcohol*
(where *Mtb* :H is a metabolite able to donate hydride ion)

(g) $CH_3CHO + 2CH_3CH_2OH \xrightarrow[\text{catalyst}]{\text{acid}}$ *not make it*

(h) OCH_3
$$CH_3CHOCH_3 + H_2O \xrightarrow[\text{catalyst}]{\text{acid}}$$
hydrolysis

*15.54 Write the structures of the organic products that form in each of the following situations. If no reaction occurs, write "no reaction." (Some of the situations constitute a review of reactions in earlier chapters.)

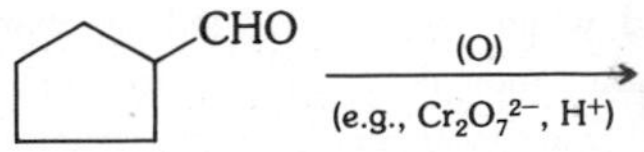

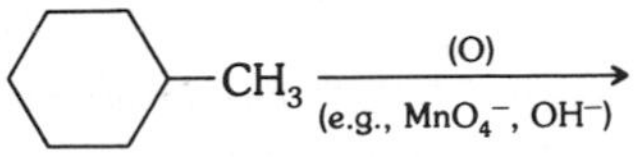

(a) 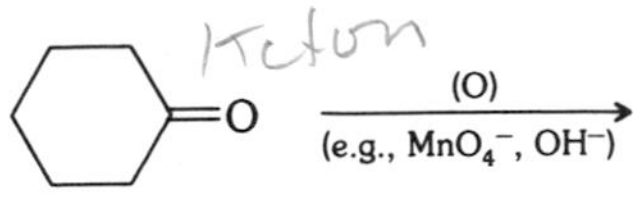 *ketone*
$=O \xrightarrow[\text{(e.g., } MnO_4{}^-, OH^-)]{\text{(O)}}$ *No reaction*

(b) CH_3
$$CH_3CH_2COH \xrightarrow[\text{(e.g., } MnO_4{}^-, OH^-)]{\text{(O)}}$$ *no reaction*
(with CH_3CH_2)

(c) $CH_3CH_2CH_2OH + CH_3CHO \rightleftharpoons$ *Hemiacetal*

(d) 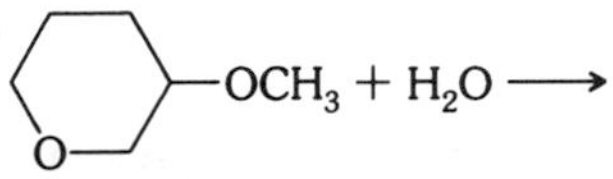
$-OCH_3 + H_2O \longrightarrow$ *hydrolysis*

(e) OH
$$CH_3COCH_2CH_3 \rightleftharpoons$$ *hydrolysis*
(with CH_3)

(f) $CH_3OCH_2CH_2CHO + Mtb{:}H \xrightarrow[\text{by } H^+]{\text{(followed}}$ *alcohol*
(where *Mtb* :H is a metabolite able to donate hydride ion)

(g) OCH_2CH_3
$$CH_3CH_2CCH_3 + H_2O \xrightarrow[\text{catalyst}]{\text{acid}}$$
(with OCH_2CH_3)

(h) 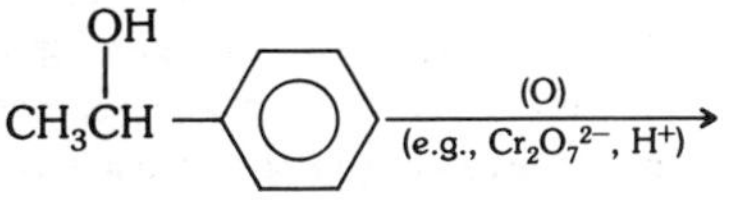 OH
$$CH_3CH \xrightarrow[\text{(e.g., } Cr_2O_7{}^{2-}, H^+)]{\text{(O)}}$$ *Ketone +*   *H₂O*
(attached to a ring)

(i) 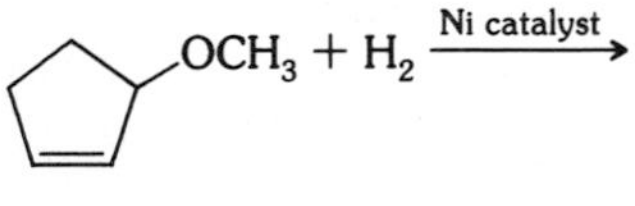
$-OCH_3 + H_2 \xrightarrow{\text{Ni catalyst}}$ *saturated*

(j) 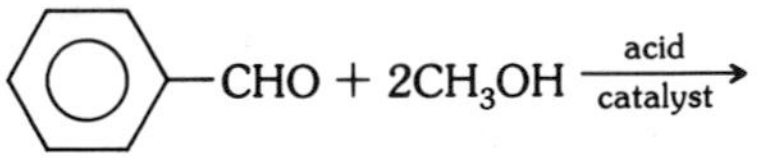
$-CHO + 2CH_3OH \xrightarrow[\text{catalyst}]{\text{acid}}$ *acetal*

*15.55 Catalytic hydrogenation of compound **A** (C_3H_6O) gave **B** (C_3H_8O). When **B** was heated strongly in the presence of sulfuric acid, it changed to compound **C** (C_3H_6). The acid-catalyzed addition of water to **C** gave compound **D**

(C_3H_8O); and when **D** was oxidized, it changed to **E** (C_3H_6O). Compounds **A** and **E** are isomers, and compounds **B** and **D** are isomers. Write the structures of compounds **A** through **E**.

•15.56 When compound **F** ($C_4H_{10}O$) was gently oxidized, it changed to compound **G** (C_4H_8O), but vigorous oxidation changed **F** (or **G**) to compound **H** ($C_4H_8O_2$). Action of hot sulfuric acid on **F** changed it to compound **I** (C_4H_8). The addition of water to **I** (in the presence of an acid catalyst) gave compound **J** ($C_4H_{10}O$), a compound that could not be oxidized. Compounds **F** and **J** are isomers. Write the structures of compounds **F** through **J**.

•15.57 A student was assigned the preparation of the dimethyl acetal of butanal for which the equation is

$$CH_3CH_2CH_2CH{=}O + 2CH_3OH \xrightarrow[\text{catalyst}]{\text{acid}}$$
$$CH_3CH_2CH_2CH(OCH_3)_2 + H_2O$$

A solution of 12.5 g of butanal in 50.0 mL of methanol was prepared for accomplishing this reaction. The density of methanol is 0.787 g/mL.

(a) How many moles of butanal were taken?
(b) How many moles of methanol were used?
(c) Was sufficient methanol taken? (Calculate the minimum number of grams of methanol that would be required.)
(d) How many grams of water would be obtained?
(e) Offer a reason for using an excess quantity of methanol.

•15.58 In Review Exercise 13.28 (page 413), polyvinyl acetate was described as a polymer from which Butvar is made and that Butvar is used to make safety glass. You may wish to refer to your answer to 13.28 for the parts of this Review Exercise.

(a) Prepare the structure of a segment of the polyvinyl acetate molecule consisting of *two* repeating units.
(b) Polyvinyl acetate can be converted into the corresponding *polyvinyl alcohol* by the replacement of all of the CH_3CO groups attached to the oxygens of polyvinyl acetate's chain by hydrogen atoms. This leaves a very long alkane chain with OH groups on every other carbon atom. Write the structure of a segment of polyvinyl alcohol consisting of *two* repeating units.
(c) Write the structure of the *monomer* of polyvinyl alcohol. This monomer does not exist (explaining why its polymer must be made indirectly). Is the monomer properly called an *alcohol* by our definition? What kind of an "alcohol" is it? Is this kind of "alcohol" stable? Does polyvinyl alcohol have OH groups that are properly called alcohol groups? Explain.
(d) Butvar is made by the combination of butanal with polyvinyl alcohol. *Cyclic* acetal systems form in which every other carbon of the main alkane chain is part of the ring system of these cyclic acetals. The generic name of this new polymer is polyvinyl butyral. Write the structure of a segment of this polymer that includes one cyclic acetal system.

CARBOXYLIC ACIDS AND ESTERS

16

Strong, light-weight fabrics made of synthetic polymers have transformed virtually all recreational activities, such as extended backpacking and tenting ventures into remote wilderness areas. Dacron, described in this chapter, is one of many polyesters.

16.1

OCCURRENCE, NAMES, AND PHYSICAL PROPERTIES OF CARBOXYLIC ACIDS

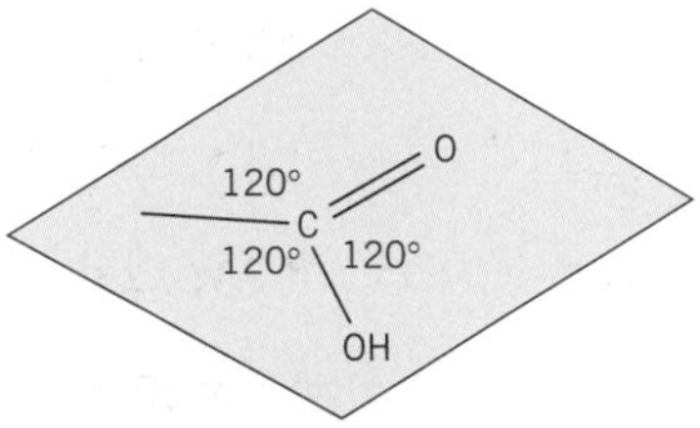

The carboxyl group has a planar geometry at the carbonyl group.

The carboxylic acids are polar compounds whose molecules form strong hydrogen bonds to each other.

The two principal types of organic acids are the carboxylic acids and the sulfonic acids. Sulfonic acids are much less common than carboxylic acids, and we will not study them.

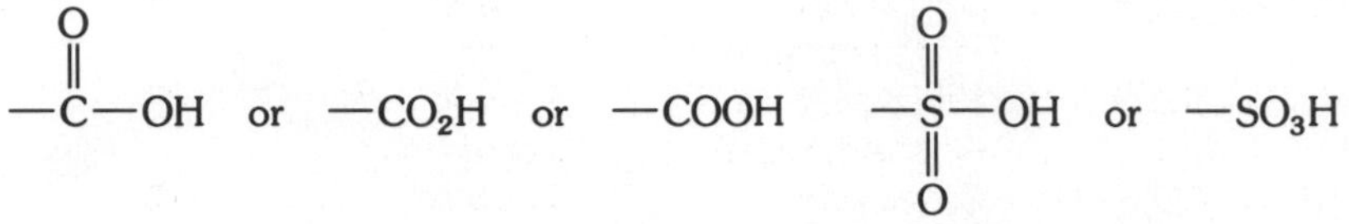

Carboxyl group

Sulfonic acid group

■ "Carboxyl" comes from *carb*onyl + hydr*oxyl*.

In **carboxylic acids**, the carbonyl carbon holds a hydroxyl group and either another carbon atom or a hydrogen atom. A number of specific examples are given in Table 16.1, and information in the table shows how widely the carboxyl group occurs in nature. The acids with long, straight, alkane-like chains are often called the **fatty acids**, because they are products of the hydrolysis of butterfat, olive oil, and similar substances in the diet.

Formic acid, the simplest acid, has a sharp, irritating odor and is responsible for the sting of nettle plants and certain ants. The next acid in Table 16.1, acetic acid, gives tartness to vinegar, where its concentration is 4% to 5%. Butyric acid causes the odor of rancid butter. Valeric acid gets its name from the Latin *valerum*, meaning to be strong. What is strong about valeric acid is its odor. Other acids with vile odors are caproic, caprylic, and capric acids, which get their names from the Latin *caper*, meaning "goat," a reference to their odor.

Some acids are *dicarboxylic acids* with two carboxyl groups per molecule. Oxalic acid, the simplest example, gives the sour taste to rhubarb. Citric acid, a *tricarboxylic acid*, causes the tartness of citrus fruits. Lactic acid, which has both a carboxyl group and a 2° alcohol group, gives the tart taste to sour milk.

■ The lactate ion is produced in muscles engaged in strenuous exercise.

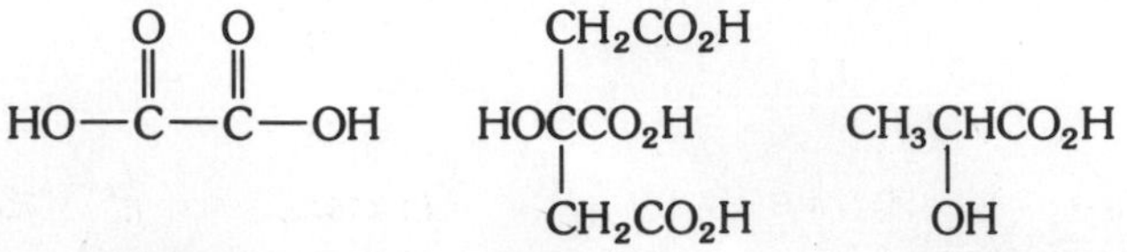

Oxalic acid
(in rhubarb)

Citric acid
(in lemon juice)

Lactic acid
(in sour milk)

All carboxylic acids are weak Brønsted acids, and they exist in the form of their anions both in basic solutions and in their salts. It is largely as their anions that they occur in living cells and body fluids.

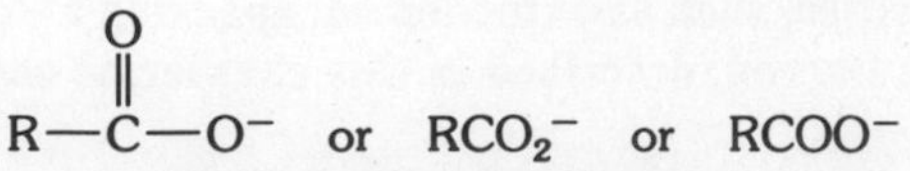

Symbols of anions of carboxylic acids

TABLE 16.1
Carboxylic Acids

(handwritten: all of them weak acides)

n^a Structure	Name^b	Origin of Name	MP (°C)	BP (°C)	Solubility in Water^c	K_a (25 °C)
1 HCO_2H	Formic acid (methanoic acid)	L. *formica*, ant	8	101	∞	1.8×10^{-4} (20°C)
2 CH_3CO_2H	Acetic acid (ethanoic acid)	L. *acetum*, vinegar	17	118	∞	1.8×10^{-5}
3 $CH_3CH_2CO_2H$	Propionic acid (propanoic acid)	L. *proto, pion*, first, fat	−21	141	∞	1.3×10^{-5}
4 $CH_3(CH_2)_2CO_2H$	Butyric acid (butanoic acid)	L. *butyrum*, butter	−6	164	∞	1.5×10^{-5}
5 $CH_3(CH_2)_3CO_2H$	Valeric acid (pentanoic acid)	L. *valere*, valerian root	−35	186	4.97	1.5×10^{-5}
6 $CH_3(CH_2)_4CO_2H$	Caproic acid (hexanoic acid)	L. *caper*, goat	−3	205	1.08	1.3×10^{-5}
7 $CH_3(CH_2)_5CO_2H$	Enanthic acid (heptanoic acid)	Gr. *oenanthe*, vine blossom	−9	223	0.26	1.3×10^{-5}
8 $CH_3(CH_2)_6CO_2H$	Caprylic acid (octanoic acid)	L. *caper*, goat	16	238	0.07	1.3×10^{-5}
9 $CH_3(CH_2)_7CO_2H$	Pelargonic acid (nonanoic acid)	Pelargonium, geranium	15	254	0.03	1.1×10^{-5}
10 $CH_3(CH_2)_8CO_2H$	Capric acid (decanoic acid)	L. *caper*, goat	32	270	0.015	1.4×10^{-5}
12 $CH_3(CH_2)_{10}CO_2H$	Lauric acid (dodecanoic acid)	Laurel	44	—	0.006	—
14 $CH_3(CH_2)_{12}CO_2H$	Myristic acid (tetradecanoic acid)	*Myristica*, nutmeg	54	—	0.002	—
16 $CH_3(CH_2)_{14}CO_2H$	Palmitic acid (hexadecanoic acid)	Palm oil	63	—	0.0007	—
18 $CH_3(CH_2)_{16}CO_2H$	Stearic acid (octadecanoic acid)	Gr. *stear*, solid	70	—	0.0003	—

Miscellaneous Carboxylic Acids

$C_6H_5CO_2H$	Benzoic acid	Gum benzoin	122	249	0.34 (25 °C)	6.5×10^{-5}
$C_6H_5CH{=}CHCO_2H$	Cinnamic acid (trans isomer)	Cinnamon	132	—	0.04	3.7×10^{-5}
$CH_2{=}CHCO_2H$	Acrylic acid	L. *acer*, sharp	13	141	Soluble	5.6×10^{-5}
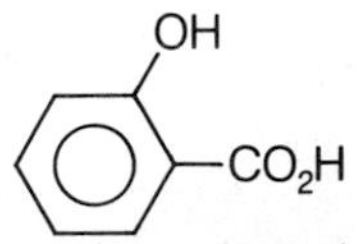 (salicylic acid structure, OH and CO_2H on benzene ring)	Salicylic acid	L. *salix*, willow	159	211	0.22 (25 °C)	1.1×10^{-3} (19 °C)

^{a} n = total number of carbon atoms per molecule

^{b} In parentheses below each common name is the IUPAC name.

^{c} In grams of acid per 100 g H_2O at 20 °C except where noted otherwise.

(handwritten notes: "fats animal" near stearic acid; "yugort milk" and "6, 8, 10" at bottom)

IUPAC Names of Carboxylic Acids end in *-oic Acid* In the IUPAC rules for naming carboxylic acids, the *parent acid* is that of the longest chain that includes the carboxyl group. To name the parent acid, change the ending of the name of the alkane having the same number of carbons (the parent alkane) from *-e* to *-oic acid*. To number the chain of the parent acid, *always* begin with the carbon atom of the carbonyl group, giving it position 1. The names in parentheses in Table 16.1 are IUPAC names.

■ Use the name of the *acid*, not the name of the alkane, to devise the name of the anion of the acid.

To name anions of carboxylic acids, the *carboxylate ions,* change the ending of the name of the parent *acid* from *-ic* to *-ate,* and omit the word *acid*. This rule applies both to the IUPAC names and to the common names. For example,

■ Pronounce "oate" as "oh-ate."

Methanoic acid
(formic acid)

Methanoate ion
(formate ion)

Sodium methanoate
(sodium formate)

EXAMPLE 16.1
Naming a Carboxylic Acid and Its Anion

The following carboxylic acid has the common name of isovaleric acid. What is the IUPAC name of this acid and its sodium salt?

$$CH_3CHCH_2COH$$

ANALYSIS The longest chain that includes the carboxyl group has four carbon atoms, so the parent acid is named by changing the name *butane,* the parent alkane, to *butanoic acid.* Next, we have to number the chain, starting with the carboxyl group's carbon.

$$\underset{4\ \ \ 3\ \ 2\ \ \ 1}{CH_3CHCH_2COH}$$

SOLUTION The methyl group is at position 3, so the IUPAC name of this acid is 3-methylbutanoic acid.

To name its anion, we drop *-ic acid* from the name of the acid and add *-ate.* Therefore the name of the anion is 3-methylbutanoate, and the name of the sodium salt of this acid is sodium 3-methylbutanoate. (The common name is sodium isovalerate.)

PRACTICE EXERCISE 1

What are the IUPAC names of the following compounds?

(a) CH_3CCO_2H

(b) $CH_3CH_2CH_2CCH_2CHCH_2CO_2H$

(c) CH_3CO_2Na

(d) $CH_3CH_2CHCH_2CHCH_2CO_2H$

PRACTICE EXERCISE 2

If the IUPAC name of $HO_2CCH_2CO_2H$ is propanedioic acid (and not 1,3-propanedioic acid), what must be the IUPAC name for $HO_2CCH_2CH_2CH_2CO_2H$?

PRACTICE EXERCISE 3

If the IUPAC name of $CH_3CH{=}CHCH_2CH_2CO_2H$ is 4-hexenoic acid, what must be the IUPAC name of the following acid? Its common name is *oleic acid,* and it is one of the products of the hydrolysis of almost any edible vegetable oil or animal fat. (The name of the straight-chain alkane with 18 carbon atoms is octadecane.)

$$CH_3(CH_2)_7CH{=}CH(CH_2)_7CO_2H$$

Oleic acid (common name)

■ Oleic acid is actually the *cis* isomer. The name of the *trans* isomer is elaidic acid.

Carboxylic Acid Molecules Hydrogen-Bond to Each Other Carboxylic acids have higher boiling points than alcohols of comparable formula masses, because molecules of carboxylic acids form hydrogen-bonded pairs:

Hydrogen bonds (•••) hold two
molecules of a carboxylic acid together.

This makes the *effective* formula mass of a carboxylic acid much higher than its calculated formula mass, and therefore the boiling point is higher.

The lower-formula-mass carboxylic acids ($C_1 - C_4$) are soluble in water largely because the carboxyl group has *two* oxygen atoms that can accept hydrogen bonds from water molecules. In addition, the carboxyl group has the OH group that can donate hydrogen bonds.

■ Remember, we're interested in how structure affects solubility in water because water is the fluid medium in the body.

16.2

THE ACIDITY OF CARBOXYLIC ACIDS

The carboxylic acids are weak Brønsted acids toward water but strong Brønsted acids toward the hydroxide ion.

Aqueous solutions of carboxylic acids contain the following species in equilibrium:

$$K_a = \frac{[RCO_2^-][H^+]}{[RCO_2H]}$$

| Carboxylic acid (weaker acid) | Water (weaker base) | Carboxylate ion (stronger base) | Hydronium ion (stronger acid) |

The K_a values of several carboxylic acids are given in Table 16.1, and you can see that most are on the order of 10^{-5}. Thus the carboxylic acids are weak acids toward water, and their percentage ionizations are low. For example, in a 1 M solution at room temperature, acetic acid is ionized only to about 0.5%.

Carboxylic Acids Are Stronger Acids than Phenols or Alcohols

Carboxylic acids ($K_a \approx 10^{-5}$) are several billion times stronger acids than alcohols ($K_a \approx 10^{-16}$), and roughly 100,000 times stronger acids than phenols ($K_a \approx 10^{-10}$). The greater acidity of carboxylic acids reflects the greater relative stability of carboxylate anions compared to the anions of alcohols or phenols. In the carboxylate ion, the negative charge is adjacent to the strongly electronegative carbonyl group. This helps to stabilizes the anion of the acid by helping to spread the negative charge over a wider area than is possible in the anions of alcohols or phenols. A more stable ion means one easier to form.

■ Phenols are strong enough acids to neutralize OH^-, a strong base, but are not strong enough acids to neutralize HCO_3^-, a weak base.

Carboxylic Acids Are Neutralized by Strong Bases

The hydroxide ion, the carbonate ion, and the bicarbonate ion are bases strong enough to neutralize carboxylic acids. This reaction is important at the molecular level of life, because the carboxylic acids we normally produce by metabolism must be neutralized. Otherwise, the pH of body fluids, such as the blood, would fall too low to sustain life.

With hydroxide ion, the reaction is as follows.

$$RCO_2H + OH^- \longrightarrow RCO_2^- + H—OH$$

Stronger   Stronger   Weaker   Weaker
acid   base   base   acid

■ The negative charges on the anions are balanced by the presence of some cation, the Na^+ or K^+ ion, for example. However, we'll work largely with net ionic equations.

With bicarbonate ion, the chief base in the buffer systems of the blood, the following reaction occurs.

$$RCO_2H + HCO_3^- \longrightarrow RCO_2^- + H_2O + CO_2$$

Stronger   Stronger   Weaker   Weaker
acid   base   base   acid

Some specific examples are as follows.

$$CH_3CO_2H + OH^- \longrightarrow CH_3CO_2^- + H_2O$$
Acetic acid   Acetate ion

■ The stearate ion is one of several organic ions in soap. The ion is so large that it might be better to say that it is "dispersed" in water rather than "soluble."

$$CH_3(CH_2)_{16}CO_2H + OH^- \longrightarrow CH_3(CH_2)_{16}CO_2^- + H_2O$$
Stearic acid   Stearate ion
(insoluble in water)   (soluble in water)

$$C_6H_5CO_2H + HCO_3^- \longrightarrow C_6H_5CO_2^- + H_2O + CO_2$$
Benzoic acid   Benzoate ion
(insoluble in water)   (soluble in water)

PRACTICE EXERCISE 4

Write the structures of the carboxylate anions that form when the following carboxylic acids are neutralized.

(a) $CH_3CH_2CO_2H$ **(b)** $CH_3O\!-\!\!\bigcirc\!\!-\!CO_2H$ **(c)** $CH_3CH{=}CHCO_2H$

Carboxylate Ions Are More Soluble in Water than Their Parent Acids

The purified salts that combine carboxylate ions and metal ions are genuine salts, assemblies of oppositely charged ions, so all are solids at room temperature. Table 16.2 gives a few examples of sodium salts. Like all sodium salts, they are soluble in water but completely insoluble in such nonpolar solvents as ether or gasoline. Several are used as decay inhibitors in foods, as described in Special Topic 16.1.

SOME IMPORTANT CARBOXYLIC ACIDS AND SALTS

Acetic Acid Most people experience acetic acid directly in the form of its dilute solution in water, which is called vinegar. Because blood is slightly alkaline, acetic acid circulates as the acetate ion, and we will meet this species many times when we study metabolic pathways. The acetate ion, in fact, is one of the major intermediates in the metabolism of carbohydrates, lipids, and proteins.

Acetic acid is also an important industrial chemical, and more than 3 billion pounds (23 billion moles) are manufactured each year in the United States. Acetate rayon is just one consumer product made using it (see Figure 1).

Propanoic Acid and Its Salts Propanoic acid occurs naturally in Swiss cheese in a concentration that can be as high as 1%. Its sodium and calcium salts are food additives used in baked goods and processed cheese to retard the formation of molds or the growth of bacteria. (On ingredient labels, these salts are listed under their common names, sodium or calcium propionate.)

Sorbic Acid and the Sorbates Sorbic acid, or 2,4-hexanedienoic acid, $CH_3CH{=}CHCH{=}CHCO_2H$, and its sodium or potassium salts are added in trace concentrations to a variety of foods to inhibit the growth of molds and yeasts. The sorbates often appear on ingredient lists for fruit juices, fresh fruits, wines, soft drinks, sauerkraut and other pickled products, and some meat and fish products. For food products that usually are wrapped, such as cheese and dried fruits, solutions of sorbate salts are sometimes sprayed onto the wrappers.

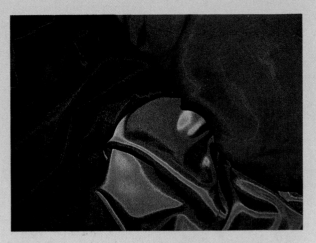

FIGURE 1
Acetate rayon is a lustrous fabric made from cellulose and acetic acid.

Sodium Benzoate Traces of sodium benzoate inhibit molds and yeasts in products that normally have pH values below 4.5 or 4.0. (The sorbates work better at slightly higher pH values—up to 6.5.) You'll see sodium benzoate on ingredient lists for beverages, syrups, jams and jellies, pickles, salted margarine, fruit salads, and pie fillings. Its concentration is low—0.05 to 0.10%—and neither benzoic acid nor the benzoate ion accumulates in the body.

TABLE 16.2
Some Sodium Salts of Carboxylic Acids

Common Name[a]	Structure	MP (°C)	Solubility[a] Water	Solubility[a] Ether
Sodium formate (sodium methanoate)	HCO_2Na	253	Soluble	Insoluble
Sodium acetate (sodium ethanoate)	CH_3CO_2Na	323	Soluble	Insoluble
Sodium propionate (sodium propanoate)	$CH_3CH_2CO_2Na$	—	Soluble	Insoluble
Sodium benzoate	$C_6H_5CO_2Na$	—	66 g/100 mL	Insoluble
Sodium salicylate	(benzene ring with OH and $-CO_2Na$)		111 g/100 mL	Insoluble

[a] At 20 °C.

477

Carboxylate Ions Are Good Proton Acceptors or Bases Because carboxylate ions are the anions of *weak* acids, they themselves must be relatively good bases, especially toward the hydronium ion, a strong proton donor. At room temperature, the following neutralization of a strong acid by a carboxylate ion occurs virtually instantaneously. It is the most important reaction of the carboxylate ion that we will study, because it makes the carboxylate ion group a neutralizer of excess acid at the molecular level of life.

For example,

Carboxylate ion	Hydronium ion	Carboxylic acid	Water
(stronger base)	(stronger acid)	(weaker acid)	(weaker base)

$$C_6H_5CO_2^- + H_3O^+ \longrightarrow C_6H_5CO_2H + H_2O$$

Benzoate ion
(soluble in water)

Benzoic acid
(insoluble in water)

$$CH_3(CH_2)_{16}CO_2^- + H_3O^+ \longrightarrow CH_3(CH_2)_{16}CO_2H + H_2O$$

Stearate ion
(soluble in water)

Stearic acid
(insoluble in water)

The Carboxylic Acid Group Is a Solubility "Switch" In the reactions just studied, we noted the solubilities in water of several species to draw attention to a very important property given to a molecule by the carboxylic acid group. This group can be used to switch on or off the solubility in water of any substance that contains it. When we try to increase the pH of a solution — by adding a strong base — a water-insoluble carboxylic acid almost instantly dissolves, because it changes to its carboxylate ion. Similarly, when we try to decrease the pH of a solution — by adding a strong acid — a water-soluble carboxylate anion instantly changes to its much less soluble, free carboxylic acid form. In other words, by suitably adjusting the pH of an aqueous solution, we can make a substance with a carboxyl group more soluble or less soluble in water.

PRACTICE EXERCISE 5

Write the structures of the organic products of the reactions of the following compounds with dilute hydrochloric acid at room temperature.

(a) CH_3O—⟨O⟩—$CO_2^-K^+$ **(b)** $CH_3CH_2CO_2^-Li^+$

(c) $(CH_3CH{=}CHCO_2^-)_2Ca^{2+}$

Carboxylic Acids Strongly Resist Oxidation Simple carboxylic acids (those with no other functional groups) or their anions are the stable end products of the oxidations of 1° alcohols and aldehydes, as we have learned. Not even hot solutions of permanganate or dichromate ion break down the carboxyl group. (The carboxylic acids will burn, of course, to give carbon dioxide and water.) The same end products, CO_2 and H_2O, form when carboxylic acids are completely metabolized in the body, but an elaborate, multistep, oxidative process is required.

16.3

THE CONVERSION OF CARBOXYLIC ACIDS TO ESTERS

Carboxylic acids can be used directly or indirectly to make esters from alcohols.
The carboxylic acids are the parent compounds for several families that collectively are called **acid derivatives.** These include the **acid chlorides,** the **anhydrides** (both common and mixed anhydrides), the **esters,** and the **amides.** They are called acid *derivatives* because they can be made from the acids and they can be hydrolyzed back to the acids.

■ We'll learn how to name esters soon, but we will not develop the rules for naming acid chlorides or acid anhydrides.

$$\underset{\substack{\text{Acid}\\\text{anhydrides}}}{R-\overset{\overset{\displaystyle O}{\|}}{C}-O-\overset{\overset{\displaystyle O}{\|}}{C}-R} \qquad \underset{\substack{\text{Mixed anhydrides}\\\text{with phosphoric acid}}}{R-\overset{\overset{\displaystyle O}{\|}}{C}-O-\underset{\underset{\displaystyle OH}{|}}{\overset{\overset{\displaystyle O}{\|}}{P}}-OH} \qquad \underset{\text{Esters}}{R-\overset{\overset{\displaystyle O}{\|}}{C}-O-R'} \qquad \underset{\text{Amides}}{R-\overset{\overset{\displaystyle O}{\|}}{C}-NH_2} \qquad \underset{\substack{\text{Acid}\\\text{chlorides}}}{R-\overset{\overset{\displaystyle O}{\|}}{C}-Cl}$$

The molecules of all of these compounds possess the **acyl group,** which is simply a carboxylic acid minus its OH group.

$$\overset{\displaystyle \not\!\!/\!\!/}{\underset{\text{Acyl group}}{R-\overset{\overset{\displaystyle O}{\|}}{C}-}} \qquad \text{For example:} \quad \underset{\text{Acetyl group}}{CH_3-\overset{\overset{\displaystyle O}{\|}}{C}-} \qquad \underset{\text{Benzoyl group}}{C_6H_5-\overset{\overset{\displaystyle O}{\|}}{C}-}$$

The reactions by which the acid derivatives are made as well as the reactions of the derivatives themselves generally occur as **acyl group transfer reactions.** The ability to transfer an acyl group, however, varies widely among the acid derivatives, as we will see.

Acid Chlorides Are the Most Reactive of Acid Derivatives

Acid chlorides react readily and exothermically with water to give the parent acids and hydrogen chloride (which, because of the *excess* water, forms as hydrochloric acid). In this reaction an acyl group transfers from Cl to OH. We call the chloride ion a *leaving group.*

$$\underset{\text{Acid chloride}}{R-\overset{\overset{\displaystyle O}{\|}}{C}-Cl} + H-O-H \longrightarrow \underset{\text{Carboxylic acid}}{R-\overset{\overset{\displaystyle O}{\|}}{C}-OH} + \underset{\text{Hydrochloric acid}}{\left\lfloor H^+(aq) + Cl^-(aq) \right\rfloor}$$

Acid chlorides also react vigorously with alcohols to give esters. In this reaction an acyl group transfers to an alcohol unit.

$$\underset{\text{Acid chloride}}{R-\overset{\overset{\displaystyle O}{\|}}{C}-Cl} + \underset{\text{Alcohol}}{H-O-R'} \longrightarrow \underset{\text{Ester}}{R-\overset{\overset{\displaystyle O}{\|}}{C}-O-R'} + HCl$$

For example,

$$\underset{\text{Acetyl chloride}}{CH_3-\overset{\overset{\displaystyle O}{\|}}{C}-Cl} + \underset{\text{Ethyl alcohol}}{H-O-CH_2CH_3} \longrightarrow \underset{\text{Ethyl acetate}}{CH_3-\overset{\overset{\displaystyle O}{\|}}{C}-O-CH_2CH_3} + HCl$$

This is an example of an **esterification** reaction, the synthesis of an ester. We say that both the alcohol and carboxylic acid are *esterified.*

Acid Anhydrides Are Good Acyl Transfer Reactants When a carboxylic acid anhydride reacts with an alcohol, an acyl group transfers from a carboxylate group to OR. This reaction occurs with roughly the same ease as the reaction of an acid chloride with an alcohol. Acid anhydrides react with alcohols as follows:

■ Either of the two acyl groups of the acid anhydride could be transferred. We've picked one.

$$R-\overset{\overset{\displaystyle O}{\|}}{C}-O-\overset{\overset{\displaystyle O}{\|}}{C}-R + H-O-R' \xrightarrow[\text{heat}]{H^+} R-\overset{\overset{\displaystyle O}{\|}}{C}-O-R' + H-O-\overset{\overset{\displaystyle O}{\|}}{C}-R$$

Acid anhydride　　　　Alcohol　　　　Ester　　　　Carboxylic acid

Phenols, like alcohols, can be also esterified by acid anhydrides. When the acid anhydride is acetic anhydride and the phenol group is in salicylic acid, one product is aspirin.

■ Both salicylic acid and acetic anhydride are common, readily available organic chemicals.

Salicylic acid　　　Acetic anhydride　　　Acetylsalicylic acid (aspirin)　　　Acetic acid

Direct Esterification of Acids Is Another Synthesis of Esters When a solution of a carboxylic acid in an alcohol is heated in the presence of a strong acid catalyst, the following species become involved in an equilibrium.

$$R-\overset{\overset{\displaystyle O}{\|}}{C}-O-H + H-O-R' \xrightleftharpoons{H^+} R-\overset{\overset{\displaystyle O}{\|}}{C}-O-R' + H-O-H$$

Carboxylic acid　　　Alcohol　　　Ester

When the alcohol is in *excess*, the equilibrium shifts so much to the right (in accordance with Le Châtelier's principle) that this reaction is a good method for making an ester. This synthesis of an ester is called *direct esterification*. Some specific examples are as follows.

$$CH_3-\overset{\overset{\displaystyle O}{\|}}{C}-O-H + H-O-CH_2CH_3 \xrightarrow{H^+} CH_3-\overset{\overset{\displaystyle O}{\|}}{C}-O-CH_2CH_3 + H_2O$$

Acetic acid　　　Ethyl alcohol (large excess)　　　Ethyl acetate

Salicylic acid + Methyl alcohol (large excess) $\xrightarrow[\text{heat}]{H^+}$ Methyl salicylate + H_2O

Notice that salicylic acid has *two* groups that can form an ester. Esterification of its phenolic OH group by acetic anhydride gives aspirin; esterification of its carboxyl group by methyl alcohol gives methyl salicylate.

EXAMPLE 16.2
Writing the Structure of a Product of Direct Esterification

What is the ester that can be made from benzoic acid and methyl alcohol?

ANALYSIS We need the *structures* of the starting materials, and in an esterification it is sometimes helpful to let the OH groups "face" each other.

$$\underset{\text{Benzoic acid}}{C_6H_5-\overset{\displaystyle O}{\overset{\|}{C}}-OH} + \underset{\text{Methyl alcohol}}{H-O-CH_3}$$

We know that a molecule of water splits out between the acid and the alcohol during esterification. For the pieces of H_2O, we take the OH from the acid and the H from the alcohol. This operation leaves the fragments:

$$C_6H_5-\overset{\displaystyle O}{\overset{\|}{C}}- \quad \text{and} \quad -O-CH_3$$

All that remains is to join these fragments.

SOLUTION The ester that forms is methyl benzoate.

$$C_6H_5-\overset{\displaystyle O}{\overset{\|}{C}}-O-CH_3$$

■ Although it is not important in predicting correct structures of products, always erase the OH group from the carboxylic acid, not the alcohol. This will make it easier to learn a reaction coming up in the next chapter.

PRACTICE EXERCISE 6

Write the structures of the esters that form by the direct esterification of acetic acid by the following alcohols.
(a) methyl alcohol (b) propyl alcohol (c) isopropyl alcohol

PRACTICE EXERCISE 7

Write the structures of the esters that can be made by the direct esterification of ethyl alcohol by the following acids.
(a) formic acid (b) propionic acid (c) benzoic acid

The Acid Catalyst Helps Direct Esterification Without an acid catalyst, direct esterification would proceed very slowly. The acid catalyst, by donating H^+ to the O atom of the carbonyl group of the carboxylic acid, makes the C atom of the carbonyl group much more positive in charge.

Acid catalyst   Protonated form of
                  the carboxylic acid

In the protonated form of the carboxyl group, the original carbonyl carbon atom is now much more attractive to the O atom of the alcohol. The alcohol molecule now attacks, and there follows a shift of a proton from one O atom to another.

After the proton transfer occurs, the system now has a very stable leaving group built into it, the water molecule, which leaves.

Finally, a proton transfers to an acceptor in the medium (thus freeing the proton catalyst for more chemical work) and the ester molecule emerges.

Every step involves an equilibrium. We have gone into detail about the mechanism of esterification to show what might seem to be very mysterious is not mysterious at all. Each step in direct esterification involves simple principles of proton transfers and the attractions of unlike charges. What can drive all of the equilibria to the right in favor of the ester *in vitro* is simply an excess of the alcohol. *All* of the equilibria shift to the right in accordance with Le Châtelier's principle.

■ The molecules of fats and oils — triacylglycerols — have three ester groups per molecule.

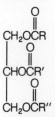

Triacylglycerol
(general structural
features)

16.4

OCCURRENCE, NAMES, AND PHYSICAL PROPERTIES OF ESTERS

Esters can be broken apart by reactions with water and aqueous base.
The functional group of an ester is the central structural feature of all the edible fats and oils as well as a number of constituents of body cells. Be sure that you can recognize this group and can pick out what we will call the *ester linkage,* the single bond between the carbonyl carbon atom and the oxygen atom that holds the ester's alkyl group. This linkage is where an ester breaks apart when it reacts with water.

$$\underset{\substack{\text{Two general formulas for}\\\text{esters}}}{(H)R-\overset{\displaystyle O}{\overset{\|}{C}}-O-R' \quad \text{or} \quad (H)RCO_2R'}$$

Ester linkage

$$\underset{\substack{\text{Ester group (carbonyl-}\\\text{oxygen-carbon system)}}}{-\overset{\displaystyle O}{\overset{\|}{C}}-O-\overset{\displaystyle |}{\underset{\displaystyle |}{C}}-}$$

Table 16.3 lists several common esters.

TABLE 16.3
Esters of Carboxylic Acids

Name[a]	Structure	MP (°C)	BP (°C)	Solubility in Water[b]
Ethyl Esters of Straight-Chain Carboxylic Acids, $RCO_2C_2H_5$				
Ethyl formate (ethyl methanoate)	$HCO_2C_2H_5$	−79	54	Soluble
Ethyl acetate (ethyl ethanoate)	$CH_3CO_2C_2H_5$	−82	77	7.35[c]
Ethyl propionate (ethyl propanoate)	$CH_3CH_2CO_2C_2H_5$	−73	99	1.75
Ethyl butyrate (ethyl butanoate)	$CH_3(CH_2)_2CO_2C_2H_5$	−93	120	0.51
Ethyl valerate (ethyl pentanoate)	$CH_3(CH_2)_3CO_2C_2H_5$	−91	145	0.22
Ethyl caproate (ethyl hexanoate)	$CH_3(CH_2)_4CO_2C_2H_5$	−68	168	0.063
Ethyl enanthate (ethyl heptanoate)	$CH_3(CH_2)_5CO_2C_2H_5$	−66	189	0.030
Ethyl caprylate (ethyl octanoate)	$CH_3(CH_2)_6CO_2C_2H_5$	−43	208	0.007
Ethyl pelargonate (ethyl nonanoate)	$CH_3(CH_2)_7CO_2C_2H_5$	−45	222	0.003
Ethyl caproate (ethyl decanoate)	$CH_3(CH_2)_8CO_2C_2H_5$	−20	245	0.0015
Esters of Acetic Acid, CH_3CO_2R				
Methyl acetate	$CH_3CO_2CH_3$	−99	57	24.4
Ethyl acetate	$CH_3CO_2CH_2CH_3$	−82	77	7.39[c]
Propyl acetate	$CH_3CO_2CH_2CH_2CH_3$	−93	102	1.89
Butyl acetate	$CH_3CO_2CH_2CH_2CH_2CH_3$	−78	125	1.0[d]
Miscellaneous Esters				
Methyl acrylate	$CH_2{=}CHCO_2CH_3$	—	80	5.2
Methyl benzoate	⬡—CO_2CH_3	−12	199	Insoluble
Natural waxes	$CH_3(CH_2)_nCO_2(CH_2)_nCH_3$	$n = 23{-}33$, carnauba wax $n = 25{-}27$, beeswax $n = 14{-}15$, spermaceti		

[a] Common names; IUPAC names are in parentheses.

[b] In grams of ester per 100 g H_2O at 20 °C (unless otherwise specified).

[c] At 25 °C.

[d] At 22 °C.

SOME IMPORTANT ESTERS

Esters of *p*-Hydroxybenzoic Acid — the Parabens Several alkyl esters of *p*-hydroxybenzoic acid—referred to as *parabens* on ingredient labels—are used to inhibit molds and yeasts in cosmetics, pharmaceuticals, and food.

Salicylates Certain esters and salts of salicylic acid are analgesics, the pain suppressants; and antipyretics, the fever reducers. The parent acid, salicylic acid, is itself too irritating to the stomach for these uses, but sodium salicylate and acetylsalicylic acid (aspirin) are commonly used. Methyl salicylate, a pleasant-smelling oil, is used in liniments, for it readily migrates through the skin.

Sodium salicylate

Acetylsalicylic acid (aspirin)

Methyl salicylate (oil of wintergreen)

FIGURE 2
The knitted tubing for this aortic heart valve in an operative site is made of Dacron fibers.

Dacron Dacron, a polyester of exceptional strength, is widely used to make fabrics and film backing for recording tapes. (Actually, the name *Dacron* applies just to the fiber form of this polyester. When it is cast as a thin film, its name is *Mylar*.) Dacron fabrics have been used to repair or replace segments of blood vessels (see Figure 1).

The formation of Dacron and many other polyfunctional polymers starts with two difunctional monomers, *aAa* and *bBb*. Their functional groups are able to react with each other to split out a small molecule, *ab*. The monomer fragments, *A* and *B,* join end to end to make a

very long, polymer molecule. In principle, the polymerization can be represented as follows:

$$aAa + bBb + aAa + bBb + aAa + bBb + \ldots \text{ etc.} \longrightarrow$$
$$-A-B-A-B-A-B-\ldots \text{ etc.} + n(ab)$$

A copolymer

Because two monomers are used, the reaction is called *copolymerization.*

One monomer used to make Dacron is ethylene glycol, which has two alcohol OH groups. The other monomer is dimethyl terephthalate, which has two methyl ester groups. The copolymerization of these two monomers depends on a reaction of esters with alcohols that we will not study; the *ab* molecule that splits out is methyl alcohol. The copolymerization proceeds as follows.

Ethylene glycol

Dimethyl terephthalate

(Repeating unit)
Dacron/Mylar

TABLE 16.4
Fragrances of Some Esters

Name	Structure	Fragrance
Ethyl formate	$HCO_2CH_2CH_3$	Rum
Isobutyl formate	$HCO_2CH_2CH(CH_3)_2$	Raspberries
Pentyl acetate	$CH_3CO_2CH_2CH_2CH_2CH_2CH_3$	Bananas
Isopentyl acetate	$CH_3CO_2CH_2CH_2CH(CH_3)_2$	Pears
Octyl acetate	$CH_3CO_2(CH_2)_7CH_3$	Oranges
Ethyl butyrate	$CH_3CH_2CH_2CO_2CH_2CH_3$	Pineapples
Pentyl butyrate	$CH_3CH_2CH_2CO_2(CH_2)_4CH_3$	Apricots
Methyl salicylate	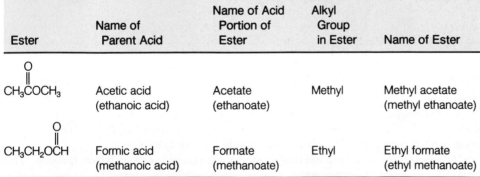	Oil of wintergreen

One interesting feature about acids and their esters is that the low-formula-mass acids have vile odors, but their esters have some of the most pleasant fragrances in all of nature (see Table 16.4). Special Topic 16.2 describes some important esters in more detail.

The Acid Portions of Esters and Carboxylate Ions Have Identical Names

Common and IUPAC names of esters are devised in the same way. For the moment, simply ignore the R′ group of an ester, the part supplied by the alcohol, and focus on the acid portion. Pretend you are naming the *anion* of the acid. Remember that in both the common and the IUPAC names for this anion, the *-ic* ending of the name of the parent acid is changed to *-ate*. Thus salts of acetic acid (ethanoic acid) are called acetate salts (common name) or ethanoate salts (IUPAC). Similarly, esters of this acid are called acetate esters (common) or ethanoate esters (IUPAC).

Once you have the name of the acid portion of the ester, simply write the name of the alkyl group in the ester's alcohol portion in front of this name (as a separate word). Here are some examples that show the pattern. (The IUPAC names are in parentheses.)

■ Acid portion (the acyl group)

$$R-\overset{\overset{\textstyle O}{\|}}{C}-O-R'$$

Alcohol portion of the ester

Ester	Name of Parent Acid	Name of Acid Portion of Ester	Alkyl Group in Ester	Name of Ester
$CH_3\overset{\overset{\textstyle O}{\|}}{C}OCH_3$	Acetic acid (ethanoic acid)	Acetate (ethanoate)	Methyl	Methyl acetate (methyl ethanoate)
$CH_3CH_2\overset{\overset{\textstyle O}{\|}}{O}CH$	Formic acid (methanoic acid)	Formate (methanoate)	Ethyl	Ethyl formate (ethyl methanoate)

EXAMPLE 16.3
Writing IUPAC Names for Esters

What is the IUPAC name for the following ester?

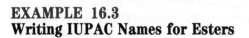

$$CH_3CH_2CH_2CH_2CH_2\overset{\overset{\textstyle O}{\|}}{C}-O-\overset{\overset{\textstyle CH_3}{|}}{C}HCH_3$$

ANALYSIS To devise the name of any ester we start with naming its parent carboxylic acid and then change -ic acid to -ate. The first step toward the ester's name, therefore, is to determine which part of the ester is contributed by a carboxylic acid. It is the acyl group in the molecule, that section of the molecule with the carbonyl group and anything joined to it by a carbon–carbon bond, not a carbon–oxygen bond.[1] Thus, the given ester has a straight-chain, six-carbon acyl group, so the parent acid is hexanoic acid.

The non-acyl part of the ester came from a parent alcohol. In the given ester, the non-acyl group is isopropyl.

SOLUTION We change the -ic acid ending of hexanoic acid to -ate, so the ester is a hexanoate ester. We add the name isopropyl as a separate word. The ester's name is

Isopropyl hexanoate

CHECK Compare the parts of the name with the original structure. The R group attached to O is a three-carbon, branched group named isopropyl. The rest of the ester unit has six carbon atoms and so the prefix hex- must be in its name.

PRACTICE EXERCISE 8

Write the IUPAC names of the following esters.

$$\text{(a)} \quad CH_3CH_2\overset{\displaystyle O}{\overset{\|}{C}}OCH_3 \qquad \text{(b)} \quad CH_3CH_2\overset{\displaystyle CH_3}{\underset{}{C}}HCH_2\overset{\displaystyle O}{\overset{\|}{C}}OCH_2CH_2CH_3$$

PRACTICE EXERCISE 9

Using the patterns developed, write the common names of the following esters.

$$\text{(a)} \quad CH_3CO_2\overset{\displaystyle CH_3}{\underset{\displaystyle CH_3}{C}}CH_3 \qquad \text{(b)} \quad CH_3CH_2CH_2CO_2CH_2CH_3$$

The Ester Group Is Polar but It Cannot Donate Hydrogen Bonds The inability of the ester group to donate hydrogen bonds affects the boiling points of esters. Thus esters of the lower-formula-mass alcohols, like methyl and ethyl alcohol, have lower boiling points than their parent acids. For example, although methyl acetate has a higher formula mass than acetic acid, it boils at 57 °C, whereas acetic acid boils at 118 °C.

The ester group, because it has oxygen atoms, can accept hydrogen bonds, however. This allows the lower-formula-mass esters to be relatively soluble in water.

Although the ester group is itself polar, the overall polarity of ester molecules is relatively low. Esters, therefore, are generally soluble in nonpolar solvents. In Chapter 20, we'll learn how molecules with three ester groups — the lipids (e.g., butterfat or olive oil) — are relatively only weakly polar and so are insoluble in water but soluble in solvents like ether and carbon tetrachloride.

---

[1] A bond from the carbonyl carbon to hydrogen occurs in formate esters. Esters of formic acid have no carbon attached to the carbonyl carbon.

16.5

SOME REACTIONS OF ESTERS

Ester molecules are broken apart by water in the presence of either acids or bases.

The reaction of esters with water is very slow unless some catalyst or promoter is present. Strong acids as well as special enzymes are good catalysts, and strong bases promote the reaction while becoming neutralized. Ester hydrolysis, catalyzed by enzymes, is the chemistry of the digestion of fats and oils.

Esters Hydrolyze to Their Parent Acids and Alcohols   An ester reacts with water to give the carboxylic acid and the alcohol from which the ester could be made. This reaction is called the *hydrolysis of an ester,* and a strong acid catalyst is generally used. (In the body, an enzyme acts as the catalyst.)

In general,

$$
\underset{\text{Ester}}{R-\overset{\overset{\text{O}}{\|}}{C}-O-R'} + H-O-H \xrightarrow[\text{heat}]{H^+} \underset{\text{Carboxylic acid}}{R-\overset{\overset{\text{O}}{\|}}{C}-O-H} + \underset{\text{Alcohol}}{H-O-R'}
$$

Specific examples are

$$
\underset{\text{Ethyl acetate}}{CH_3-\overset{\overset{\text{O}}{\|}}{C}-O-CH_2CH_3} + H_2O \xrightarrow{H^+} \underset{\text{Acetic acid}}{CH_3-\overset{\overset{\text{O}}{\|}}{C}-O-H} + \underset{\text{Ethyl alcohol}}{H-O-CH_2CH_3}
$$

$$
\underset{\text{Methyl benzoate}}{CH_3-O-\overset{\overset{\text{O}}{\|}}{C}-\bigcirc} + H_2O \xrightarrow[\text{heat}]{H^+} \underset{\text{Methyl alcohol}}{CH_3O-H} + \underset{\text{Benzoic acid}}{H-O-\overset{\overset{\text{O}}{\|}}{C}-\bigcirc}
$$

To avoid a mistake that students often make, notice that the *only* bond to break in ester hydrolysis is the one that joins the carbonyl group to the oxygen atom, the "ester bond." Notice also that the products are always the "parents" of the ester and that the names of these parents are strongly implied in the name of the ester itself. Thus methyl benzoate hydrolyzes to *methyl* alcohol and *benzoic* acid. Let's now work an example.

EXAMPLE 16.4
Predicting the Products of an Ester Hydrolysis

What are the products of the hydrolysis of the following ester?

$$
CH_3\overset{\overset{\text{O}}{\|}}{C}-O-CH_2CH_2CH_3
$$

ANALYSIS Finding the ester bond, the carbonyl-to-oxygen bond, is the crucial step because this is the bond that is broken when an ester hydrolyzes. It doesn't matter in which direction this bond happens to point on the page, it is the *only* bond that breaks.

$$\text{CH}_3\overset{\displaystyle O}{\overset{\|}{\text{C}}}\text{—O—CH}_2\text{CH}_2\text{CH}_3 \quad \text{or} \quad \text{CH}_3\text{CH}_2\text{CH}_2\text{—O—}\overset{\displaystyle O}{\overset{\|}{\text{C}}}\text{CH}_3$$

These are identical compounds

Carbonyl-to-oxygen bond, the ester bond

Break the carbonyl–oxygen bond. Erase it and separate the fragments. If the ester were written as follows:

$$\text{CH}_3\overset{\displaystyle O}{\overset{\|}{\text{C}}}\text{—O—CH}_2\text{CH}_2\text{CH}_3 \dashrightarrow \text{CH}_3\overset{\displaystyle O}{\overset{\|}{\text{C}}} + \text{O—CH}_2\text{CH}_2\text{CH}_3$$

On the other hand, if the ester's structure were written in the opposite direction:

$$\text{CH}_3\text{CH}_2\text{CH}_2\text{—O—}\overset{\displaystyle O}{\overset{\|}{\text{C}}}\text{CH}_3 \dashrightarrow \text{CH}_3\text{CH}_2\text{CH}_2\text{—O} + \overset{\displaystyle O}{\overset{\|}{\text{C}}}\text{CH}_3$$

Either way gives the same results. Next we attach the pieces of the water molecule to make the "parents" of the ester. We attach OH to the carbonyl carbon and we put H on the oxygen atom of the other fragment. The products therefore are propyl alcohol and acetic acid.

$$\text{CH}_3\text{CH}_2\text{CH}_2\text{OH} + \text{HO}\overset{\displaystyle O}{\overset{\|}{\text{C}}}\text{CH}_3$$

CHECK Reexamine the structure of the ester. Its alcohol portion, the R group on O, has three carbons in a straight chain, so the alcohol we have written is correct. The acid portion has two carbons, so the acid we wrote is correct. (Double-check each C and O for the correct number of bonds.)

PRACTICE EXERCISE 10

Write the structures of the products of the hydrolysis of the following esters.

(a) $\text{CH}_3\overset{\displaystyle O}{\overset{\|}{\text{O}}}\text{CCH}_3$ (b) $\text{CH}_3\text{CH}_2\overset{\displaystyle O}{\overset{\|}{\text{C}}}\text{—O—}\overset{\displaystyle \text{CH}_3}{\underset{}{\text{CHCH}_3}}$ (c) $\overset{\displaystyle \text{CH}_3}{\text{CH}_3\text{CH}}\text{—}\overset{\displaystyle O}{\overset{\|}{\text{C}}}\text{—OCH}_2\text{CH}_2\text{CH}_3$

Ester hydrolysis is the reverse of direct ester formation. Both involve the identical species in a chemical equilibrium, the equilibrium that we studied in the previous section for direct esterification. So if we were to take an ester and water in a 1 : 1 mole ratio, not all of the ester molecules would change into molecules of the parent acid and alcohol. Some of the ester molecules and some of the water molecules would still be unchanged. When we want to make sure that all of the ester is hydrolyzed, we use a large excess of water. In accordance with Le Châtelier's principle, this *excess of one reactant* shifts the equilibrium in favor of making the products of hydrolysis.

Esters Are Saponified by Bases If an aqueous base instead of a strong acid is used to promote the breakup of an ester, the products are the salt of the parent acid and the parent alcohol. The reaction is called **saponification,** and it requires a full mole of base (not just a catalytic trace) for each mole of ester bonds. The base *promotes* the reaction but, unlike a true catalyst, it is permanently changed (neutralized). No equilibrium forms, because one product, the *anion* of the parent acid, cannot be converted into an ester by a direct reaction with alcohols. In the lab, the base often used is sodium hydroxide. We'll write net ionic equations, but remember that wherever you see an anion, there is always some cation also around (the Na^+ ion when NaOH is used). In general:

■ L. *sapo,* soap, and *onis,* to make. Ordinary soap is made by the saponification of the ester groups in fats and oils.

$$R\overset{\displaystyle O}{\overset{\|}{-C}}-O-R' + OH^-(aq) \xrightarrow{heat} R\overset{\displaystyle O}{\overset{\|}{-C}}-O^- + H-O-R'$$

Ester   Carboxylate   Alcohol
                                  anion

Specific examples are as follows. (Assume that OH^- comes from an aqueous solution of NaOH or KOH.)

$$CH_3\overset{\displaystyle O}{\overset{\|}{-C}}-O-CH_2CH_3 + OH^-(aq) \xrightarrow{heat} CH_3\overset{\displaystyle O}{\overset{\|}{-C}}-O^- + H-O-CH_2CH_3$$

Ethyl acetate   Acetate ion   Ethyl alcohol

$$C_6H_5\overset{\displaystyle O}{\overset{\|}{-C}}-O-CH_3 + OH^-(aq) \xrightarrow{heat} C_6H_5\overset{\displaystyle O}{\overset{\|}{-C}}-O^- + H-O-CH_3$$

Methyl benzoate   Benzoate ion
Methyl alcohol

Saponification occurs approximately as follows.

■ The steps occur much more smoothly than implied by the way the mechanism has been written.

In step (1), the hydroxide ion is attracted to the $\delta+$ site on the carbonyl group's carbon atom, and the double bond breaks open to a single bond. In the step (2), the double bond re-forms as $^-OR'$ is expelled. This species is an even stronger base than OH^-, and in step (3) it takes a proton where indicated. The end products are the anion of the parent acid and the parent alcohol. The steps are not reversible because the $\delta-$ site on the O atom of the alcohol has no attraction for the carboxylate ion; like charges repel.

EXAMPLE 16.5
Writing the Structures of the Products of Saponification

What are the products of the saponification of the following ester?

$$CH_3CH_2\overset{\displaystyle O}{\overset{\displaystyle \|}{C}}-O-CH_3$$

ANALYSIS Saponification is very similar to ester hydrolysis. The ester bond is broken. *Break only this bond.* Separate the fragments:

$$CH_3CH_2\overset{\displaystyle O}{\overset{\displaystyle \|}{C}}-O-CH_3 \dashrightarrow CH_3CH_2\overset{\displaystyle O}{\overset{\displaystyle \|}{C}} + -O-CH_3$$

Now change the fragment that has the carbonyl group into the *anion* of a carboxylic acid. Do this by attaching O^- to the carbonyl carbon atom. Then attach an H atom to the oxygen atom of the other fragment to make the alcohol molecule.

SOLUTION The products are

$$CH_3CH_2\overset{\displaystyle O}{\overset{\displaystyle \|}{C}}-O^- + H-O-CH_3$$

PRACTICE EXERCISE 11

Write the structures of the products of the saponification of the following esters.

(a) ⬡—O—C(=O)—CH₃   **(b)** CH₃—O—C(=O)—⬡—O—CH₃

Other Reactions of Esters Esters participate in a reaction called the *Claisen ester condensation* that generates considerably larger molecules from smaller ones. We'll take this somewhat complicated reaction up in Chapter 25, closer to where it is needed in an application involving metabolism.

16.6

ORGANOPHOSPHATE ESTERS AND ANHYDRIDES

Some of the most widely distributed kinds of esters and anhydrides in living organisms are the anions of esters of phosphoric acid, diphosphoric acid, and triphosphoric acid.

Phosphoric acid appears in several forms and anions in the body, but the three fundamental parents of all these forms are phosphoric acid, diphosphoric acid, and triphosphoric acid.

Phosphoric acid   Diphosphoric acid   Triphosphoric acid

These are all polyprotic acids, but at the slightly alkaline pHs of body fluids, they cannot exist as free acids. They occur, instead, as a mixture of negative ions. The net charge on each ion and the relative amounts of the ions are functions of the pH of the medium.

Esters of Alcohols and Phosphoric Acid Are Monophosphate Esters

If you look closely at the structure of phosphoric acid, you can see that part of it resembles a carboxyl group.

Part of a phosphoric acid molecule   Part of a carboxylic acid molecule

It isn't surprising therefore that esters of phosphoric acid exist and that they are structurally similar to esters of carboxylic acids.

Part of a phosphate ester   Part of a carboxylate ester

One large difference between a phosphate ester and a carboxylate ester is that a phosphate ester is still a diprotic acid. Its molecules still carry two proton-donating OH groups. Therefore, depending on the pH of the medium, a phosphate ester can exist in any one of three forms, and usually there is an equilibrium mixture of all three.

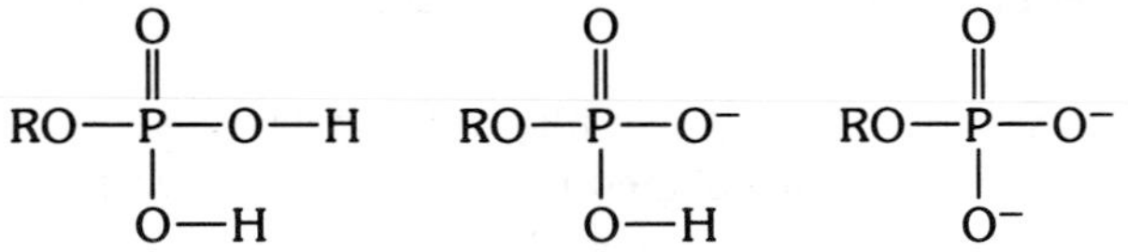

Phosphate ester (as a diprotic acid)

Phosphate ester (as a singly ionized species)

Phosphate ester (as a doubly ionized species)

Favored at low pH

Favored at pH values just below 7

Favored at pH values above 7

At the pH of most body fluids (just slightly more than 7), phosphate esters exist mostly as the doubly ionized species — as the di-negative ion. All forms, however, are generally soluble in water, and one reason that the body converts so many substances into their phosphate esters may be to improve their solubilities in water.

Alcohols and Diphosphoric Acid Form Diphosphate Esters

A diphosphate ester actually has three functional groups: the phosphate ester group, the proton-donating OH groups, and the phosphoric anhydride system.

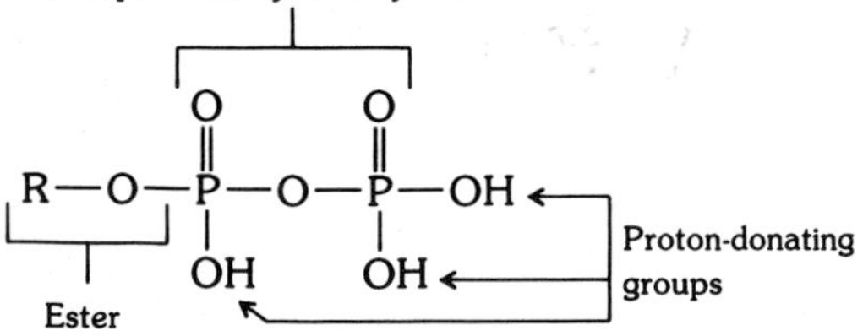

Phosphoric anhydride system

Proton-donating groups

Ester group

Diphosphate ester

■ Each of the two OH groups in a phosphate ester can be converted into an ester. Nucleic acid molecules, for example, have the phosphate diester system.

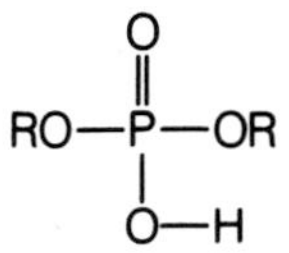

A phosphate diester

Notice the similarity of part of the structure of this diphosphate ester to that of an anhydride of a carboxylic acid:

O O	O O
‖ ‖	‖ ‖
—P—O—P—	—C—O—C—
| |	
Part of the diphosphate system	Part of a carboxylic acid anhydride system
The phosphoric anhydride group	The carboxylic anhydride group

One of the many diphosphate esters in the body is called adenosine diphosphate, or ADP. We will show its structure as its triply charged anion, because it exists largely in this fully ionized form at the pH of most body fluids.

Adenosine diphosphate, ADP

(fully ionized form)

 The Phosphoric Anhydride Group Is a Major Storehouse of Chemical Energy in Living Systems ADP can be hydrolyzed to adenosine and two phosphate ions, or to adenosine monophosphate and one phosphate ion. Although either hydrolysis is *very* slow in the absence of an enzyme, the breaking up of the phosphoric anhydride group generates considerable energy.

ADP can also react with alcohols. This reaction resembles hydrolysis because it breaks up the phosphoric anhydride group. This group in ADP and similar compounds (like ATP, below) turns out to be the chief means for storing chemical energy in cells. The phosphoric anhydride system is so important in this way that we should learn how it holds its chemical energy.

The source of the internal energy in the triply charged anion of ADP is the tension up and down the anhydride chain. The central chain bears oxygen atoms with full negative charges, and these charges repel each other. The internal repulsion primes the phosphoric anhydride system for breaking apart exothermically when it is attacked by a suitable reactant.

Molecules with alcohol groups are examples of such reactants in the body, but these reactions require enzymes for catalysis. Without a catalyst, the negatively charged ADP anion actually *repels* electron-rich species. Yet if an alcohol group is to attack the phosphoric anhydride system and break it apart, the oxygen atom of the alcohol group must be able to strike a phosphorus atom. We can visualize this attack as follows.

O   O   O   O

RO—P—O—P—O⁻ ⟶ RO—P—O⁻ + R′—O—P—O⁻

R′   H ---→ (proton to be buffered)

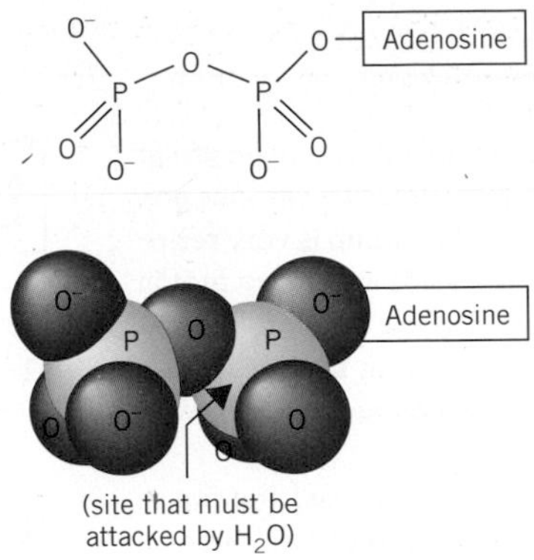

(site that must be
attacked by H_2O)

FIGURE 16.1
The oxygen atoms in the phosphoric anhydride system of ADP screen the phosphorus atoms. The negative charges on these oxygen atoms deflect incoming, electron-rich particles such as molecules of an alcohol or of water. Therefore this kind of anhydride system reacts very slowly with these reactants, unless a special catalyst such as an enzyme is also present.

As seen in Figure 16.1, however, the phosphorus atoms in the chain are buried within a clutch of negatively charged oxygen atoms that repel the alcohol molecule. Thus the internal tension cannot be relieved by this kind of reaction *unless an enzyme for the reaction is present.* You have probably already realized that *the body exerts control over energy-releasing reactions of diphosphates by its control of the enzymes for these reactions.*

Water could make the same kind of exothermic attack on a diphosphate ester as an alcohol, but *the body has no enzymes inside cells that catalyze this reaction.* Hence, energy-rich diphosphates can exist in cells despite the abundance of water.

Alcohols and Triphosphoric Acid Form Triphosphate Esters Adenosine triphosphate or ATP is the most common and widely occurring member of a small family of energy-rich triphosphate esters. Because the triphosphates have two phosphoric anhydride systems in each molecule, on a mole-for-mole basis the triphosphates are among the most energy-rich substances in the body.

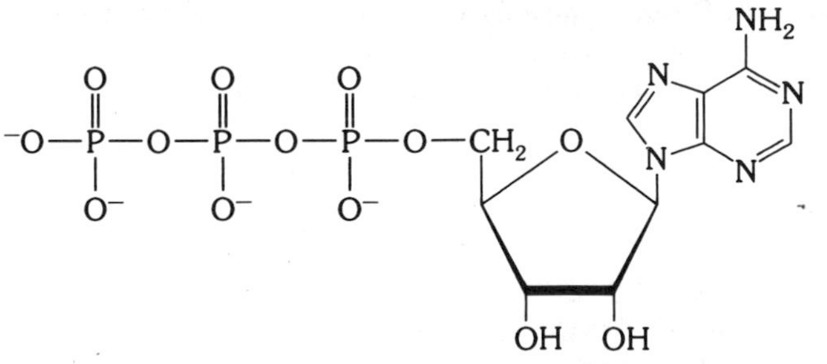

Adenosine triphosphate, ATP
(fully ionized form)

Triphosphates are much more widely used in cells as sources of energy than the diphosphates. The overall reaction for the contraction of a muscle, for example, can be written as follows. We now introduce the symbol P_i to stand for the set of inorganic phosphate ions, mostly $H_2PO_4^-$ and HPO_4^{2-}, produced in the breakup of ATP and present at equilibrium at body pH.

$$\text{Relaxed muscle} + \text{ATP} \xrightarrow{enzyme} \text{contracted muscle} + \text{ADP} + P_i$$

Muscular work requires ATP, and if the body's supply of ATP were used up *with no way to remake it,* we'd soon be helpless.

The resynthesis of ATP from ADP and P_i is one of the major uses of the chemical energy in the food we eat. What we have learned here is *how* these tri- and diphosphates can be energy-rich and, at the same time, not be destroyed by uncatalyzed reactions in the body. Later, we will study more details of how a cell makes and uses these phosphate systems.

SUMMARY

Acids and their salts The carboxyl group, CO_2H, is a polar group that confers moderate water solubility to a molecule without preventing its solubility in nonpolar solvents. This group is very resistant to oxidation and reduction. Carboxylic acids are strong proton donors toward hydroxide ions, whereas alcohols are not. Toward water, carboxylic acids are weak acids. Therefore their conjugate bases, the carboxylate anions, are good proton acceptors toward the hydronium ions of strong acids.

Salts of carboxylic acids are ionic compounds, and the potassium or sodium salts are very soluble in water. Hence, the carboxyl group is one of nature's important "solubility switches." An insoluble acid becomes soluble in base, but it is thrown out of solution again by the addition of acid.

The derivatives of acids studied in this chapter — acid chlorides, anhydrides, and esters — can be made from the acids and are converted back to the acids by reacting with water. We can organize the reactions we have studied for the carboxylic acids and esters as follows.

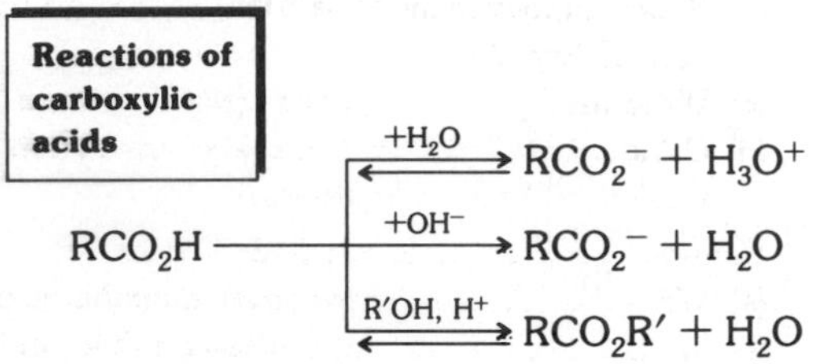

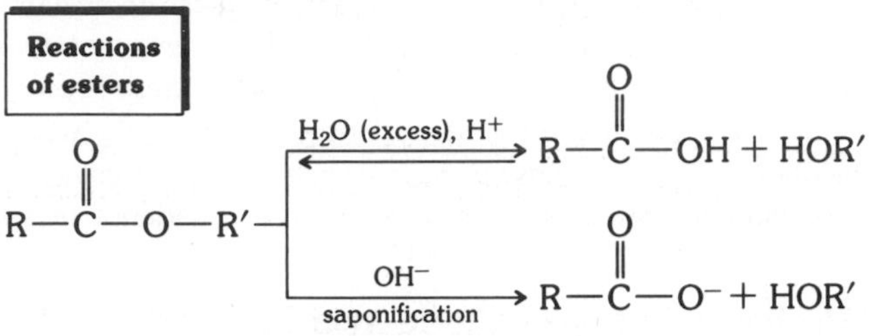

Reactions of carboxylic acids

$$RCO_2H \xrightarrow{\substack{+H_2O}} RCO_2^- + H_3O^+$$
$$\xrightarrow{\substack{+OH^-}} RCO_2^- + H_2O$$
$$\xrightarrow{\substack{R'OH,\ H^+}} RCO_2R' + H_2O$$

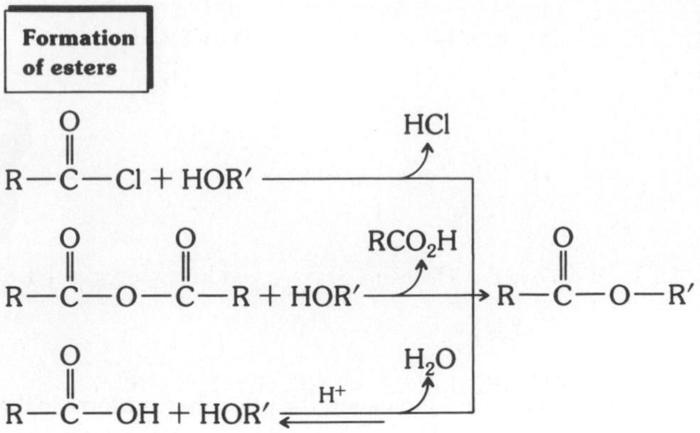

Formation of esters

$$R-\overset{O}{\overset{\|}{C}}-Cl + HOR' \longrightarrow$$
$$R-\overset{O}{\overset{\|}{C}}-O-\overset{O}{\overset{\|}{C}}-R + HOR' \longrightarrow$$
$$R-\overset{O}{\overset{\|}{C}}-OH + HOR' \underset{H^+}{\longleftarrow}$$

(products: HCl, RCO_2H, H_2O leading to $R-\overset{O}{\overset{\|}{C}}-O-R'$)

Reactions of esters

$$R-\overset{O}{\overset{\|}{C}}-O-R' \xrightarrow{\substack{H_2O\ (excess),\ H^+}} R-\overset{O}{\overset{\|}{C}}-OH + HOR'$$
$$\xrightarrow[\text{saponification}]{\substack{OH^-}} R-\overset{O}{\overset{\|}{C}}-O^- + HOR'$$

Esters and anhydrides of the phosphoric acid system Esters of phosphoric acid, diphosphoric acid, and triphosphoric acid occur in living systems largely as anions, because these esters are also polyprotic acids. In addition, esters of diphosphoric and triphosphoric acid are phosphoric anhydrides. These anhydrides are energy-rich compounds. Their reactions with water or alcohols are very exothermic, but the reactions are also very slow unless a catalyst (an enzyme) is present.

REVIEW EXERCISES

The answers to Review Exercises whose numbers are in color are found in Appendix C. The answers to the other Review Exercises are found in the Study Guide that accompanies this book. The more challenging questions are marked with asterisks.

Structures and Names of Carboxylic Acids and Their Salts

16.1 Of the following structures, which has the following functional group?
(a) alcohol   (b) carboxyl group   (c) ketone   (d) enol

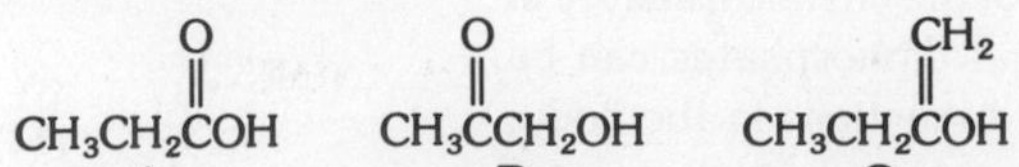

$$\underset{\textbf{A}}{CH_3CH_2\overset{O}{\overset{\|}{C}}OH} \qquad \underset{\textbf{B}}{CH_3\overset{O}{\overset{\|}{C}}CH_2OH} \qquad \underset{\textbf{C}}{CH_3CH_2\overset{CH_2}{\overset{\|}{C}}OH}$$

16.2 The carboxylic acids obtained by the hydrolysis of fats and oils in the diet have what general name?

16.3 What is the common name of the acid in vinegar? In sour milk?

16.4 Write the structures of the following substances.
(a) propionic acid   (b) benzoic acid
(c) acetic acid   (d) formic acid

16.5 What are the structures of the following?
(a) 2,2-dimethylbutanoate ion
(b) 4-chloro-2-methylheptanoic acid
(c) butanedioic acid
(d) benzoate ion

16.6 Write the IUPAC names of the following compounds.

(a) CH_3CHCO_2H with CH_3 above

(b) HO_2CCCH_3 with CH_2CH_3 above and CH_3 below   *cyl*

3-chloro propanoic

(c) $CH_3CH_2CHCH_2CO_2Na$ with Cl below   (d) 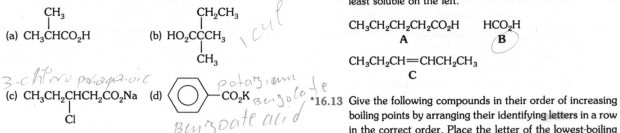 —CO_2K potassium benzolate

Benzoate acid

16.7 What are the IUPAC names of the following compounds?

(a) $CH_3CHCH_2CHCO_2H$ with CH_3 above and CH_3 below

(b) $HO_2CCH_2CHCH_3$ with CH_3CHCH_3 below

(c) $CH_3(CH_2)_8CO_2Na$

(d) CH_3CHBr with CO_2K below

16.8 One of the compounds whose concentration in blood increases in unchecked diabetes has the following structure.

$$CH_3CHCH_2CO_2H$$ with OH above

If its IUPAC name is 3-hydroxybutanoic acid and its common name is β-hydroxybutyric acid, what are the IUPAC and common names for its sodium salt?

*16.9 The citric acid cycle is one of the major metabolic sequences of reactions in the body. One of the acids in this series of reactions is commonly called fumaric acid, which has the following structure.

$$HO_2C\text{—}C\text{=}C\text{—}CO_2H$$ (with H on each carbon, trans arrangement)

Which of the following names is its correct IUPAC name?
(a) *trans*-dibutenoic acid
(b) *trans*-butenedioic acid
(c) *trans*-dibutenedioic acid
(d) *cis*-ethenedicarboxylic acid

Physical Properties of Carboxylic Acids

16.10 Draw a figure that shows how two acetic acid molecules can pair in a hydrogen-bonded form.

*16.11 The hydrogen bond system in formic acid includes an array of molecules, one after the other, each carbonyl oxygen of one molecule attracted to the HO group of the next molecule in line. Represent this linear array of hydrogen-bonded molecules of formic acid by a drawing.

16.12 Give the following compounds in their order of increasing solubility in water. Do this by arranging their identifying

letters in a row in the correct order, placing the letter of the least soluble on the left.

$CH_3CH_2CH_2CH_2CO_2H$ HCO_2H
    A   **B**

$CH_3CH_2CH\text{=}CHCH_2CH_3$
    C

*16.13 Give the following compounds in their order of increasing boiling points by arranging their identifying letters in a row in the correct order. Place the letter of the lowest-boiling compound on the left.   → Di Carboxylic acid

$HO_2CCH_2CH_2CO_2H$ $CH_3CH_2CH_2CH_3$
    A   **B**

CH_3CH_2OH CH_3CO_2H the least
    C   **D**

Carboxylic Acids as Weak Acids

16.14 Concerning an aqueous solution of acetic acid:
(a) Write the equation for the equilibrium that is present.
(b) In what direction will this equilibrium shift, toward acetic acid or toward the acetate ion, if hydrochloric acid is added? Explain.
(c) Write the K_a equation for acetic acid.
(d) Using data in Table 16.1, is acetic acid a stronger or a weaker acid than salicylic acid?

16.15 Consider a solution of formic acid in water.
(a) Write the equation for the equilibrium that is present.
(b) In what direction will this equilibrium shift, toward formic acid or toward the formate ion, if sodium hydroxide is added? Explain.
(c) Write the K_a equation for formic acid.
(d) Using data in Table 16.1, is formic acid a stronger or a weaker acid than benzoic acid?

16.16 Arrange the following compounds in their order of increasing acidity by placing their identifying letters in a row in the correct sequence. (Place the letter of the least acidic compound on the left.)

CH_3CH_2OH H_2SO_4 CH_3—⬡—OH CH_3CO_2H
    A   **B**   **C**   **D**

*16.17 Give the order of increasing acidity of the following compounds. Arrange their identifying letters in the order that corresponds to their acidity, with the letter of the least acidic compound on the left.

CH_3—⬡—OH CH_3—⬡—CH_2OH
      A   **B**

CH_3—⬡—CO_2H HNO_3
      C   **D**

16.18 Write the net ionic equation for the complete reaction, *if any*, of aqueous sodium hydroxide with each of the compounds in Review Exercise 16.17 at room temperature.

16.19 What are the net ionic equations for the reactions of the following compounds with aqueous potassium hydroxide at room temperature?

(a) $HO_2CCH_2CH_2CO_2H$ (b) $HOCH_2CH_2CH_2CO_2H$

(c) $\overset{O}{\overset{\|}{H}CCH_2CH_2CH_2CO_2H}$ (d) $O=\bigcirc-CO_2H$

Salts of Carboxylic Acids

16.20 Which compound, **A** or **B,** is more soluble in ether? Explain.

$CH_3(CH_2)_6CO_2Na$ $CH_3(CH_2)_6CO_2H$
    A   **B**

16.21 Which compound, **A** or **B,** is more soluble in water? Explain.

more soluble

$CH_3CH_2-\bigcirc-ONa$ $CH_3CH_2-\bigcirc-OH$
       A   **B**

•16.22 Suppose that you add 0.1 mol of hydrochloric acid to an aqueous solution that contains 0.1 mol of the compound given in each of the following parts. If any reaction occurs rapidly at room temperature, write its net ionic equation.
(a) $CH_3CH_2CO_2^-$
(b) $^-O_2CCH_2CH_2CH_2CO_2^-$
(c) NH_3

•16.23 Suppose that you have each of the following compounds in a solution in water. What reaction, if any, will occur rapidly at room temperature if an equimolar quantity of hydrochloric acid is added? Write net ionic equations.

(a) $HOCH_2CH_2CO_2^-$
(b) $HOCH_2CH_2CO_2H$

(c) $\bigcirc-O^-$

Esterification and Reactivity

16.24 What are the structures of the reactants that are needed to make ethyl propanoate from ethanol and each of the following kinds of starting materials?
(a) an acid chloride
(b) a carboxylic acid anhydride
(c) by direct esterification

16.25 In order to prepare methyl benzoate, what are the structures of the reactants needed for each kind of approach?

(a) by direct esterification
(b) from an acid chloride
(c) from an acid anhydride

•16.26 The reaction of ethyl alcohol with acetyl chloride is rapid.
(a) What is the structure of the organic product?
(b) How might the relatively high speed of the reaction be explained?

•16.27 Methyl alcohol reacts rapidly with acetic anhydride to give methyl acetate and acetic acid. How might the very rapid rate of the reaction be explained?

16.28 What are the structures of the products of the esterification by ethyl alcohol of each compound?
(a) propionic acid
(b) 2-methylpropanoic acid
(c) *p*-nitrobenzoic acid
(d) terephthalic acid (Show the esterification of both of the carboxyl groups.)

$HO_2C-\bigcirc-CO_2H$

Terephthalic acid

16.29 When ethanoic acid is esterified by each of the following compounds, what are the structures of the esters that form?
(a) methanol
(b) 2-methyl-1-propanol
(c) phenol
(d) $HOCH_2CH_2OH$ (1,2-ethanediol) (Show the esterification of both alcohol groups.)

•16.30 Explain by means of equations how H^+ works as a catalyst in the direct esterification of acetic acid by ethyl alcohol.

•16.31 Water does not react readily with methyl acetate to form methyl alcohol and acetic acid. The hydroxide ion, on the other hand, more readily attacks methyl acetate (to form methyl alcohol and the acetate ion). What might be the reason for the higher reactivity of the OH^- ion over H_2O toward methyl acetate?

16.32 Suppose that a way could be found to remove H_2O as rapidly as it is produced in direct esterification. What would this do to the equilibrium in this reaction, shift it to the right (favoring the ester) or to the left (favoring the carboxylic acid and the alcohol)? Explain.

•16.33 How do we explain the fact that esters react much more slowly with water than acid chlorides do?

Structures and Physical Properties of Esters

16.34 Write the structures of the following compounds.
(a) ethyl formate (b) ethyl *p*-chlorobenzoate

16.35 What are the structures of the following compounds?
(a) *t*-butyl propanoate (b) isopropyl 2-methylbutanoate

•16.36 Arrange the following compounds in their order of increasing boiling points. Do this by placing their identifying letters in a row, starting with the lowest-boiling compound on the left.

$$CH_3CO_2CH_2CH_2CH_3 \qquad CH_3CH_2CH_2CH_2CO_2H$$
$$\textbf{A} \qquad\qquad\qquad \textbf{B}$$

$$CH_3CH_2CH_2CH_2CH_3 \quad HO_2CCH_2CH_2CH_2CH_2CH_2CH_3$$
$$\textbf{C} \qquad\qquad\qquad\qquad \textbf{D}$$

*16.37 Arrange the following compounds in their order of increasing solubilities in water by placing their identifying letters in the correct sequence, beginning with the least soluble on the left.

$$CH_3CH_2CH_2CH_2CO_2Na \quad CH_3CH_2CH_2CH_2CO_2CH_2CH_3$$
$$\textbf{A} \qquad\qquad\qquad\qquad \textbf{B}$$

$$CH_3CH_2CH_2CH_2CO_2H \qquad CH_3CH_2CH_2CH_2CH_3$$
$$\textbf{C} \qquad\qquad\qquad\qquad \textbf{D}$$

Reactions of Esters

16.38 Write the equation for the acid-catalyzed hydrolysis of each compound. If no reaction occurs, write "no reaction."

(a) (b)

(c) (d) $CH_3CH_2OCH_2CH_2CH_3$

16.39 What are the equations for the acid-catalyzed hydrolyses of the following compounds? If no reaction occurs, write "no reaction."

(a)

(b)

(c)

(d)

16.40 The metabolism of fats and oils involves the complete hydrolysis of molecules such as the following. What are the structures of its hydrolysis products?

*16.41 Cyclic esters are known compounds. What is the structure of the product when the following compound is hydrolyzed?

16.42 What are the structures of the organic products of the saponification of the compounds in Review Exercise 16.38 by aqueous NaOH?

16.43 What forms, if anything, when the compounds of Review Exercise 16.39 are subjected to saponification by aqueous KOH? Write the structures of the organic products.

16.44 What are the products of the saponification of the compound given in Review Exercise 16.40? (Assume that aqueous NaOH is used.)

16.45 Write the structure of the organic ion that forms when the compound of Review Exercise 16.41 is saponified.

*16.46 A pharmaceutical chemist needed to prepare the ethyl ester of an extremely expensive and rare carboxylic acid in order to test this form of the drug for its side effects. Direct esterification had to be used. How could the conversion of all the acid to its ethyl ester be maximized? Use RCO_2H as a symbol for the acid in any equations you write.

*16.47 Write the steps in the mechanism of the acid-catalyzed hydrolysis of methyl acetate. (Remember, this is the exact reverse of the acid-catalyzed, direct esterification of acetic acid by methyl alcohol.)

Phosphate Esters and Anhydrides

16.48 Write the structures of the following compounds.
(a) monomethyl phosphate
(b) monoethyl diphosphate
(c) monopropyl triphosphate

16.49 State one apparent advantage to the body of its converting many compounds into phosphate esters.

16.50 What part of the structure of ATP is particularly responsible for its being described as an *energy-rich* compound? Explain.

16.51 Why is ATP more difficult to hydrolyze than acetyl chloride?

Common Acids and Salts (Special Topic 16.1)

16.52 Name a compound that is
(a) Used to manufacture a kind of rayon.
(b) Present in vinegar.
(c) A food additive put into wrappers of cheese.

Common Esters (Special Topic 16.2)

16.53 Esters of *p*-hydroxybenzoic acid are referred to by what common name? How are these esters used in commerce?

16.54 What is meant by a *copolymer?*

16.55 What copolymer has been used in surgical grafts?

16.56 Salicylates are described as analgesics and antipyretics. What do these terms mean?

16.57 Why is salicylic acid, the parent of the salicylates, structurally modified for medicinal uses?

16.58 Concerning salicylic acid,
   (a) What two functional groups does it have?
   (b) Which functional group is esterified in acetylsalicylic acid?
   (c) Which group is esterified in methyl salicylate?

Additional Exercises

16.59 Consider compounds **A** and **B**.

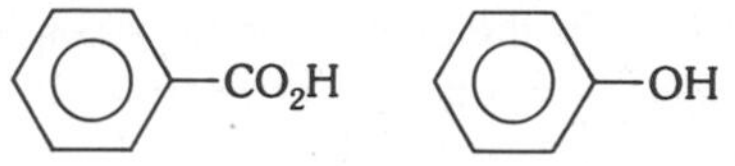

   A   **B**

   (a) Which is the stronger acid?
   (b) Write the structure of the conjugate base of each compound.
   (c) Which is the stronger conjugate base of the two you wrote for part (b)?

*16.60 To prepare a sample of oil of Niobe, methyl benzoate, a student heated a solution of 5.64 g of benzoic acid in 25.0 mL of methyl alcohol in the presence of a small amount of sulfuric acid as a catalyst.
   (a) What is the maximum number of grams of methyl benzoate obtainable from the mass of benzoic acid used?
   (b) What is the minimum number of grams and milliliters of methyl alcohol needed for the complete conversion of benzoic acid to methyl benzoate? (The density of methyl alcohol is 0.787 g/mL.)
   (c) What advantage is there in using an excess of methyl alcohol? (What principle is involved?)

*16.61 Complete the following reaction sequences by writing the structures of the organic products. If no reaction occurs, state so. (These constitute a review of this and earlier chapters on organic chemistry.)

   (a) $CH_3CHCO_2H + NaOH(aq) \longrightarrow$
        |
        CH_3

   (b) $CH_3CH_2COCH_3 + H_2O \xrightleftharpoons[\text{catalyst}]{\text{acid}}$

   (c) $CH_3CHO \xrightarrow{K_2Cr_2O_7(aq)}$

   (d) 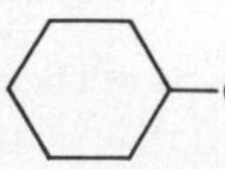 ⬡—OH $\xrightarrow[\text{heat}]{H_2SO_4}$

   (e) $CH_3CHCH_2COCH_3 + NaOH(aq) \longrightarrow$
        |
        CH_3

   (f) $CH_3CHCHCH_3 \xrightarrow{KMnO_4(aq)}$
        | |
        H_3C OH

(g) $CH_3CH_2CCl + CH_3CH_2OH \longrightarrow$

(h) $CH_3CH_2CHOCH_3 + H_2O \xrightarrow{\text{acid}}$ (OCH$_3$ substituent) $\xrightarrow[\text{catalyst}]{\text{acid}}$

(i) 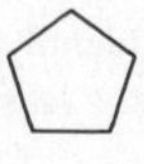 ⬠ $+ H_2SO_4 \longrightarrow$

(j) $CH_3CH_2OH + $ ⬡$-CO_2H \xrightleftharpoons[\text{catalyst}]{\text{acid}}$

(k) $H_2C=CHCH_2CH_3 + HCl(g) \longrightarrow$

(l) $CH_3CH_2OCH_2CH_2CCH_3 + H_2O \longrightarrow$

*16.62 Write the structures of the organic products, if any, that form in the following situations. If no reaction occurs, state so. (Some of these constitute a review of the reactions of earlier chapters.)

(a) 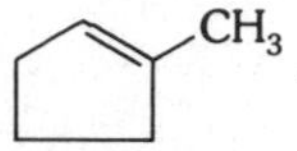 (cyclopentene with CH$_3$) $+ H_2 \xrightarrow{\text{Ni or Pt}}$

(b) $CH_3COCCH_3 + HOCH_2CH_2CH_3 \longrightarrow$

(c) $CH_3CHCH_2CO_2^- + HCl(aq) \longrightarrow$
        |
        CH_3

(d) 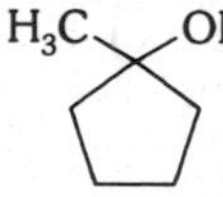 (cyclopentane with H$_3$C and OH) $\xrightarrow{KMnO_4(aq)}$

(e) $CH_3CH_2OCCH_2CH_2COCH_3 + NaOH(aq) \longrightarrow$ (excess)

(f) 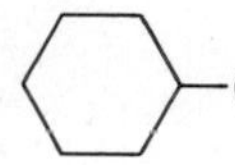 ⬡$-CO_2H + CH_3OH \xrightleftharpoons[\text{catalyst}]{\text{acid}}$

(g) $CH_3CH_2OCCH_2CH_2OCCH_2CH_3 + NaOH(aq) \longrightarrow$ (excess)

(h) 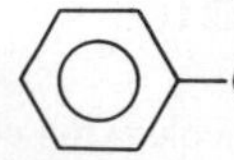 ⬡$-CHOCH_3 + H_2O \xrightarrow[\text{catalyst}]{\text{acid}}$ (OCH$_3$ substituent)

(i) $HO_2CCH_2CH_2CH_2CH_3 + NaOH(aq) \longrightarrow$

(j) $CH_3CH_2CH_2OCCH_2CH_2OCCH_3 + H_2O \xrightleftharpoons[\text{catalyst}]{\text{acid}}$ (excess)

(k) HO_2C-⬡$-CO_2H + CH_3OH \xrightleftharpoons[\text{catalyst}]{\text{acid}}$ (excess)

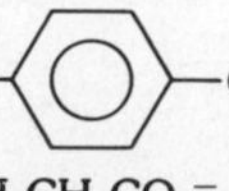

(l) $CH_3OCH_2CH_2CO_2^- + HCl(aq) \longrightarrow$

AMINES AND AMIDES

17

One of the very first uses of nylon was to make parachutes for the U. S. Army Airforce in World War II, when silk became unavailable. Now sky diving is a sport enjoyed by many who trust their lives to nylon, a polyamide described in this chapter.

17.1
OCCURRENCE, NAMES, AND PHYSICAL
PROPERTIES OF AMINES
17.2
CHEMICAL PROPERTIES OF AMINES

17.3
AMIDES OF CARBOXYLIC ACIDS

17.1

OCCURRENCE, NAMES, AND PHYSICAL PROPERTIES OF AMINES

The amino group, NH_2, has some of the properties of ammonia, including the ability to be involved in hydrogen bonding.

Both the amino group and its protonated form occur in living things in proteins, enzymes, and nucleic acids (the chemicals that carry our genes). When a carbonyl group is attached to nitrogen, the properties change sufficiently to make it convenient to create a different chemical family, the amides. The amide system also occurs widely in living things.

$$-NH_2 \qquad -NH_3^+ \qquad \overset{\displaystyle O}{\underset{|}{\overset{\|}{-C}}}-\overset{|}{\underset{|}{N}}-$$

Amino   Protonated   Amide system
group   amino group

Amines Are Ammonia-like Compounds   The **amines** are organic relatives of ammonia in which one, two, or all three of the hydrogen atoms on an ammonia molecule have been replaced by a hydrocarbon group. Some examples are

■ These are the common names, not the IUPAC names.

$$CH_3NH_2 \qquad CH_3NHCH_3 \qquad CH_3\overset{\displaystyle CH_3}{\underset{|}{N}}CH_3 \qquad CH_3NHCH_2CH_3$$

Methylamine   Dimethylamine   Trimethylamine   Methylethylamine

Several amines are listed in Table 17.1. All these are classified as *amines,* and all are basic, like ammonia.

It's quite important to realize that for a compound to be an amine, not only must its molecules have a nitrogen with three bonds, but also none of these bonds can be to a carbonyl group. If such a system is present — a carbonyl–nitrogen bond — the substance is an **amide.** Thus the structure given by **1** is an amide, not an amine. However, the structure given by **2** is not an amide, because there is no carbonyl–nitrogen bond. Instead, **2** has two functional groups, a keto group and an amino group. The chemical difference is that amines are basic and amides are not. Another difference is that the carbon–nitrogen bond in amines can't be broken by water, but the carbonyl–nitrogen bond in amides can, provided an appropriate catalyst is present.

■ The carbonyl–nitrogen bond occurs in protein molecules, where it is called the **peptide bond.**

$$R-\overset{\displaystyle O}{\overset{\|}{C}}-NH_2 \qquad\qquad R-\overset{\displaystyle O}{\overset{\|}{C}}-CH_2-NH_2$$

        1   **2**

If one or more of the groups attached directly to nitrogen in an amine is a benzene ring, then the amine is an *aromatic* amine. Otherwise, it is classified as an *aliphatic* amine. Thus benzylamine is an aliphatic amine and aniline, *N*-methylaniline, and *N,N*-dimethylaniline are all aromatic amines.

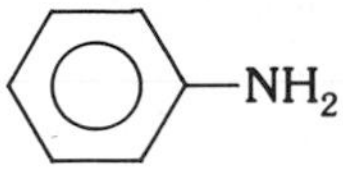

Aniline	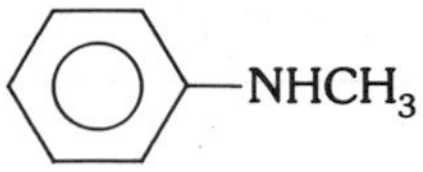 *N*-Methylaniline	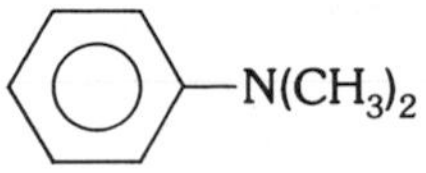 *N,N*-Dimethylaniline	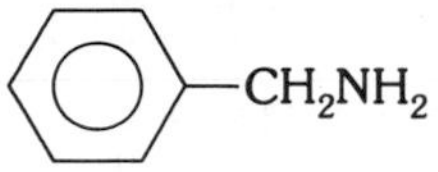 Benzylamine

The Common Names of Amines Usually End in -*amine*   The common names of the simple, aliphatic amines are made by writing the names of the alkyl groups attached to nitrogen in front of the word *amine* (and leaving no space). We have already seen how this works. Here are three more examples.

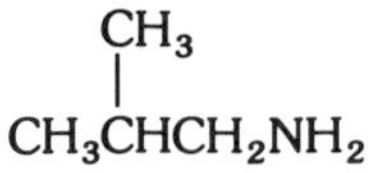

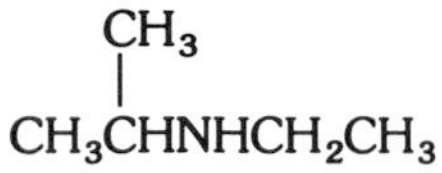

In complex systems, the names of some amines use the name *amino* as a substituent name for the NH_2 group. Thus isobutylamine can be named 1-amino-2-methylpropane. We will not develop IUPAC names for amines.

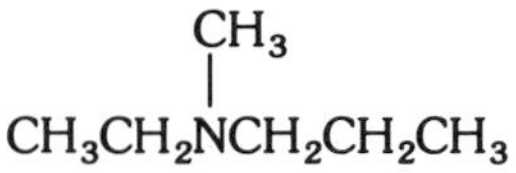

PRACTICE EXERCISE 1

Give common names for the following compounds.

(a) $(CH_3)_2NCH(CH_3)_2$ **(b)** 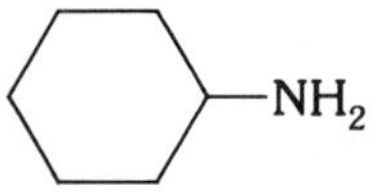 —NH_2 **(c)** $(CH_3)_2CHCH_2NHC(CH_3)_3$

PRACTICE EXERCISE 2

Write the structures of the following compounds.

(a) *t*-butyl-*sec*-butylamine   **(b)** *p*-nitroaniline

(c) *p*-aminobenzoic acid (the PABA of sun-screening lotions)

Heterocyclic Amines Have N as a Ring Atom   Both proteins and nucleic acids are rich in nitrogen-containing heterocyclic rings. For example, the amino acid proline has a saturated ring that includes a nitrogen atom. Unsaturated rings are present in other amino acids such as tryptophan.

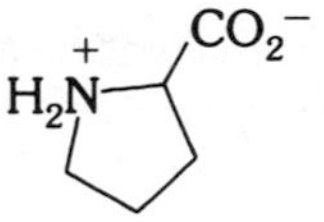

Proline

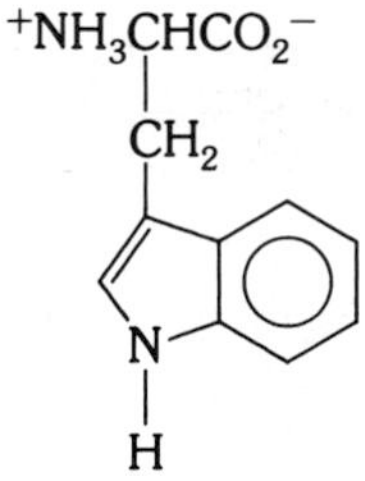

Tryptophan

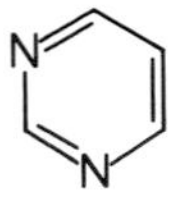

Pyrimidine

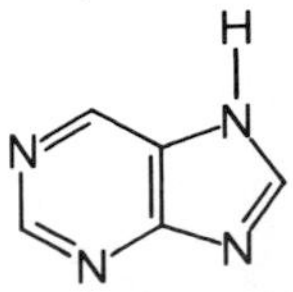

Purine

In molecules of nucleic acids, heterocyclic amine systems generally involve either the pyrimidine ring or the purine ring system.

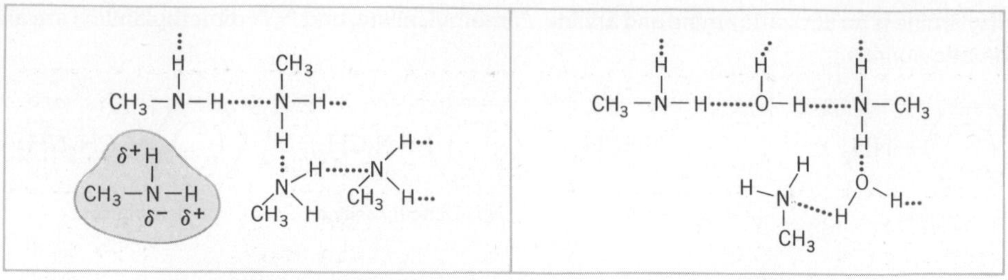

FIGURE 17.1
Hydrogen bonds in (a) amines and in (b) aqueous solutions of amines.

Compound	Formula Mass	BP (°C)
CH_3CH_3	30	−89
CH_3NH_2	31	−6
CH_3OH	32	65

■ The processes in the body that lead to the sensations of odor or taste begin with *chemical* reactions.

N—H Groups in Amines Are Involved in Hydrogen Bonding We have sometimes used boiling point data to tell us something about forces between molecules. The data in the margin tell us, for example, that when compounds of similar formula mass are compared, the boiling points of amines are higher than those of alkanes but lower than those of alcohols. This suggests that the forces of attraction between molecules are stronger in amines than in alkanes, but they are weaker in amines than in alcohols.

We can understand these trends in terms of hydrogen bonds. When a hydrogen atom is bound to oxygen or nitrogen but not to carbon, the system can donate and accept hydrogen bonds. Nitrogen, however, has a lower electronegativity than oxygen, so the polarity of the N—H bond in amines is weaker than the polarity of the O—H bond in alcohols. *The N—H system, therefore, develops weaker hydrogen bonds than the O—H system.* As a result, amine molecules can't attract each other as strongly as alcohol molecules, so amines boil lower than alcohols (of comparable formula masses). But amine molecules, nevertheless, do develop some hydrogen bonds (Fig. 17.1), which alkane molecules cannot do, so amines boil higher than alkanes (of comparable formula masses). Hydrogen bonding also helps amines to be much more soluble in water than alkanes, as Figure 17.1 also depicts.

Although hydrogen bonding between molecules of amines is weaker than hydrogen bonding between molecules of alcohols, it has a very important function at the molecular level of life. Among the molecules of proteins and nucleic acids, for example, the hydrogen bond stabilizes their special shapes, without which the substances lose their abilities to carry out their biological functions.

Odor isn't actually a *physical* property, but we should note that the amines with lower formula masses smell very much like ammonia. The odors of amines become very "fishy" at slightly higher formula masses.

17.2

CHEMICAL PROPERTIES OF AMINES

The amino group is a proton acceptor, and the protonated amino group is a proton donor.

We will examine two chemical properties of amines that will be particularly important to our study of biochemicals: the basicity of amines, in this section, and their conversion to amides, in the next.

Aliphatic Amines Are About as Basic as Ammonia When ammonia dissolves in water, the following equilibrium becomes established and a small concentration of hydroxide ion forms together with the ammonium ion.

$$NH_3 + H_2O \rightleftharpoons NH_4^+ + OH^-$$

Ammonia   Ammonium
                   ion

TABLE 17.1
Amines

Common Name	Structure	BP (°C)	Solubility in Water (20 °C)	K_b (25 °C)
Methylamine	CH_3NH_2	−6	very soluble	4.4×10^{-4}
Dimethylamine	$(CH_3)_2NH$	8	very soluble	5.3×10^{-4}
Trimethylamine	$(CH_3)_3N$	3	very soluble	0.5×10^{-4}
Ethylamine	$CH_3CH_2NH_2$	17	very soluble	5.6×10^{-4}
Diethylamine	$(CH_3CH_2)_2NH$	55	very soluble	9.6×10^{-4}
Triethylamine	$(CH_3CH_2)_3N$	89	14 g/dL	5.7×10^{-4}
Propylamine	$CH_3CH_2CH_2NH_2$	49	very soluble	4.7×10^{-4}
Aniline	$C_6H_5NH_2$	184	4 g/dL	3.8×10^{-10}

A very similar equilibrium forms when an amine dissolves in water. All that is different is that an alkyl group has replaced one of the H atoms of NH_3 (or NH_4^+).

$$R—NH_2 + H_2O \rightleftharpoons R—NH_3^+ + OH^-$$

Amine　　　　　　　Protonated
　　　　　　　　　　amine

How much the products are favored in this equilibrium is expressed by the value of the **base ionization constant, K_b.**

$$K_b = \frac{[RNH_3^+][OH^-]}{[RNH_2]}$$

■ Remember, the brackets [ ] denote the moles-per-liter concentration of the compound or ion that the brackets enclose.

Table 17.1 gives the K_b values for several amines. The K_b of ammonia is 1.8×10^{-5}, so you can see that most amines have K_b values slightly greater than that of ammonia. Remember, the larger the K_b value, the *stronger* is the base, because a larger value means that the terms in the numerator, including [OH⁻], have to be larger. Thus the aliphatic amines are generally slightly stronger bases than ammonia. Like ammonia, water-soluble amines cause the hydroxide ion concentration of the medium to become greater than the hydrogen ion concentration, and the aqueous solutions of amines are basic (and turn red litmus blue). Compounds with aliphatic amino groups, in other words, tend to increase the pH of an aqueous solution.

Like ammonia, compounds with amino groups can also neutralize hydronium ions. The following acid–base neutralization occurs rapidly and essentially completely at room temperature.

Amine　　　　Hydronium　　Protonated amine
(or ammonia,　ion　　　　　(or the ammonium
when R = H)　　　　　　　ion when R = H)

For example, hydrochloric acid is neutralized by methylamine as follows.

$$CH_3NH_2 + H_3O^+ + Cl^- \longrightarrow CH_3NH_3^+Cl^- + H_2O$$

Methylamine　　Hydrochloric　　Methylammonium
　　　　　　　acid　　　　　chloride

■ Tetraalkylammonium ions:

$$R-\overset{\underset{\displaystyle R}{|}}{\overset{\displaystyle R}{\underset{+}{N}}}-R$$

are also known species, but such cations can't be basic because they have no unshared pair of electrons on nitrogen.

It doesn't matter if the nitrogen atom in an amine bears one, two, or three hydrocarbon groups. The amine can still neutralize strong acids, because the reaction involves just the unshared pair of electrons on the nitrogen, not any of the bonds to the other groups.

The previously unshared electron pair on N now holds the H atom to N.

$$CH_3-\overset{\underset{\displaystyle H}{|}}{N}: \quad + H-\overset{\underset{\displaystyle H}{|}}{\overset{+}{O}}: \quad \longrightarrow \quad CH_3-\overset{\underset{\displaystyle H}{|}}{\overset{+}{N}}-H \; + \; :\overset{\underset{\displaystyle H}{|}}{O}:$$

Dimethyl-   Hydronium   Dimethylamine   ion   ammonium ion

EXAMPLE 17.1
Writing the Structure of the Product when an Amine Is Neutralized by a Strong Acid

What organic cation forms when hydrochloric acid (or any strong acid) neutralizes each of the following amines?

(a) $CH_3CH_2NH_2$ (b) $CH_3CH_2\overset{\underset{\displaystyle CH_3}{|}}{N}CH_3$ (c) ⬡$N-CH_3$

ANALYSIS The nitrogen atom of any amine group can accept H^+, so all that we do is increase the number of H atoms attached to the nitrogen atom by one, and then write a positive sign to show the charge.

SOLUTION The protonated forms of the given amines are

(a) $CH_3CH_2NH_3^+$ (b) $CH_3CH_2\overset{\underset{\displaystyle CH_3}{|}}{\overset{+}{N}}HCH_3$ (c) ⬡$\overset{\underset{\displaystyle CH_3}{|}}{\overset{\displaystyle \overset{H}{|}}{\overset{+}{N}}}$

PRACTICE EXERCISE 3

What are the structures of the cations that form when the following amines react completely with hydrochloric acid?

(a) aniline   **(b)** trimethylamine   **(c)** $NH_2CH_2CH_2NH_2$

Protonated Amines Can Neutralize Strong Bases A combination of a protonated amine and an anion make up an organic salt called an **amine salt**. Table 17.2 gives some examples and, like all salts, amine salts are crystalline solids at room temperature. In addition,

TABLE 17.2
Amine Salts

Name	Structure	MP (°C)
Methylammonium chloride	$CH_3NH_3{}^+Cl^-$	232
Dimethylammonium chloride	$(CH_3)_2NH_2{}^+Cl^-$	171
Dimethylammonium bromide	$(CH_3)_2NH_2{}^+Br^-$	134
Dimethylammonium iodide	$(CH_3)_2NH_2{}^+I^-$	155
Tetramethylammonium hydroxide[a]	$(CH_3)_4N^+OH^-$	130–135 (decomposes)

[a] This compound is as strong a base as NaOH because its OH^- ion fully dissociates in water.

like the salts of the ammonium ion, nearly all amine salts of strong acids are soluble in water *even when the parent amine is not.* Amine salts are much more soluble in water than amines because the *full* charges carried by the ions of an amine salt can be much better hydrated by water molecules than the amine itself, where only the small, partial charges of polar bonds occur.

Protonated amine cations also neutralize the hydroxide ion and revert to amines in the following manner (where we show only skeletal structures).

■ Some amine salts are internal salts, like all of the amino acids, the building blocks of proteins.

$$^+NH_3-CH-CO_2{}^-$$
$$|$$
$$R$$

General formula of all amino acids. R = an organic group but not always a simple alkyl group

Protonated amine   Hydroxide ion   Amine

For example,

$$CH_3NH_3{}^+ + OH^- \longrightarrow CH_3NH_2 + H_2O$$

$$CH_3CH_2\overset{+}{N}H_2CH_3 + OH^- \longrightarrow CH_3CH_2NHCH_3 + H_2O$$

The Amino Group Is a Solubility Switch
We have just learned that putting a proton on an amino group and taking it off are easily done at room temperature simply by adding an acid and then a base. We have also learned that the protonated amine is more soluble in water than the amine. This makes the amino group an excellent "solubility switch."

The solubility of an amine can be switched on simply by adding enough strong acid to protonate it. Triethylamine, for example, is insoluble in water, but we can switch on its solubility by adding a strong acid, like hydrochloric acid. The amine dissolves as its protonated ionic form is produced.

■ The other important solubility switch that we have studied involves the carboxylic acid group:

$$RCO_2H$$
(less soluble)

H^+ OH^-

$$RCO_2{}^-$$
(more soluble)

$$(CH_3CH_2)_3N\colon + HCl(aq) \longrightarrow (CH_3CH_2)_3\overset{+}{N}HCl^-(aq)$$

Triethylamine (water insoluble)   Hydrochloric acid   Triethylammonium chloride (water soluble)

We can just as quickly and easily bring the amine back out of solution by adding a strong base, like the hydroxide ion. It takes the proton off the protonated amine and gives the less soluble form.

$$(CH_3CH_2)_3\overset{+}{N}HCl^-(aq) + OH^- \longrightarrow (CH_3CH_2)_3N\!:\; + H_2O$$

Triethylammonium chloride (water soluble)   (supplied by NaOH, for example)   Triethylamine (water insoluble)

The significance of this "switching" relationship is that the solubilities of complex compounds that have the amine function can be changed almost instantly simply by adjusting the pH of the medium.

One application of this property involves medicinals. A number of amines obtained from the bark, roots, leaves, flowers, or fruits of various plants are useful drugs. These naturally occurring, acid-neutralizing, physiologically active amines are called **alkaloids,** and morphine, codeine, and quinine are just three examples.

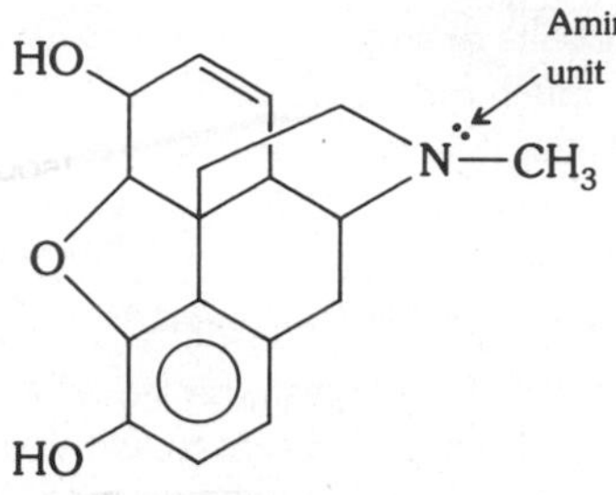

Morphine

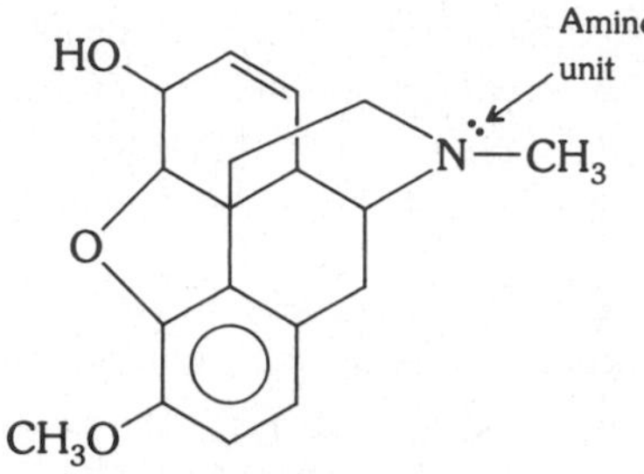

Codeine

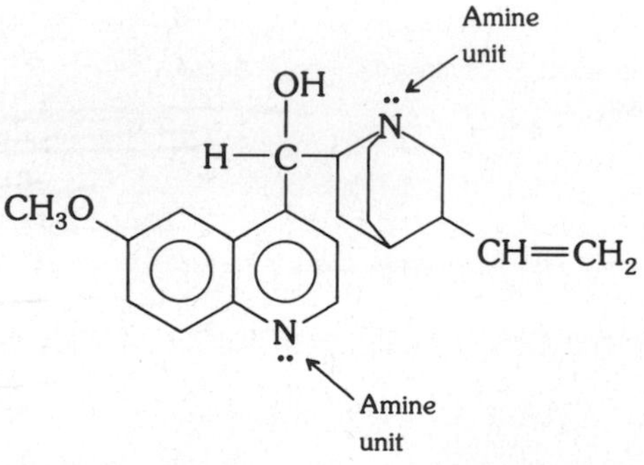

Quinine

To make it easier to administer alkaloidal drugs in the dissolved state, we often prepare them as their water-soluble amine salts. Morphine, for example, a potent sedative and painkiller, is often given as morphine sulfate, the salt of morphine and sulfuric acid. Quinine, an antimalarial drug, is available as quinine sulfate. Codeine, sometimes used in cough medicines, is often present as codeine phosphate. Special Topic 17.1 tells about a few other physiologically active amines, most of which are also prepared as their amine salts.

EXAMPLE 17.2
Writing the Structure of the Product of Deprotonating the Cation of an Amine Salt

The protonated form of amphetamine is shown below. What is the structure of the product of its reaction with OH⁻?

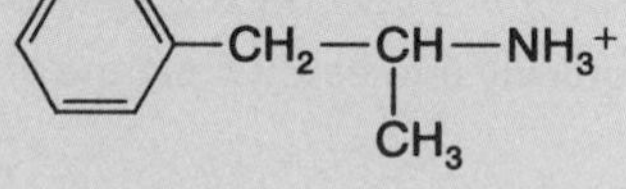

SOME PHYSIOLOGICALLY ACTIVE AMINES

Epinephrine and norepinephrine are two of the many hormones in our bodies. We will study the nature of hormones in a later chapter, but we can use a definition here. **Hormones** are compounds the body makes in special glands to serve as chemical messengers. In response to a stimulus somewhat unique for each hormone, such as fright, food odor, sugar ingestion, and others, the gland secretes its hormone into circulation in the bloodstream. The hormone then moves to some organ or tissue where it activates a particular metabolic series of reactions that constitute the biochemical response to the initial stimulus. Maybe you have heard the expression, "I need to get my adrenalin flowing." Adrenalin — or epinephrine, its technical name — is made by the adrenal gland. If you ever experience a sudden fright, a trace amount of epinephrine immediately flows and the results include a strengthened heartbeat, an increase in blood pressure, and a release of glucose into circulation from storage — all of which get the body ready to respond to the threat.

Norepinephrine has similar effects, and because these two hormones are secreted by the adrenal gland, they are called **adrenergic agents.**

Epinephrine

Norepinephrine

β-Phenylethanolamine

Several useful drugs mimic epinephrine and norepinephrine, and all are classified as *adrenergic drugs.* Most of them, like epinephrine and norepinephrine, are related structurally to β-phenylethanolamine (which is not a formal name, obviously). In nearly all their uses, these drugs are prepared as dilute solutions of their amine acid salts. Several of the β-phenylethanolamine drugs are even more structurally like epinephrine and norepinephrine, because they have the structural features of 1,2-dihydroxybenzene, commonly called *catechol.* The catechol-like adrenergic drugs are called the **catecholamines.** Synthetic epinephrine (an agent in Primatene Mist), ethylnorepinephrine, and isoproterenol are examples.

Ethylnorepinephrine
(Used against asthma
in children)

Isoproterenol (Used in treating
emphysema and asthma)

The β-phenylethylamines are another family of physiologically active amines. For example, dopamine (which is also a catecholamine) is the compound the body uses to make norepinephrine. Its synthetic form is used to treat shock associated with severe congestive heart failure.

The amphetamines are a family of β-phenylethylamines that include Dexedrine ("speed") and Methedrine ("crystal," "meth"). The amphetamines can be legally prescribed as stimulants and antidepressants, and sometimes they are prescribed for weight-control programs. However, millions of these "pep pills" or "uppers" are sold illegally, and this use of amphetamines constitutes a serious drug abuse problem. The dangers of overuse include suicide, belligerence and hostility, paranoia, and hallucinations.

Dopamine

Dexedrine

Methedrine

In a later chapter we will discuss the mechanisms by which these drugs and the naturally occurring hormones work.

■ Amphetamine sulfate (also known as benzedrine sulfate) is the form in which this drug is administered. It can be prescribed as an anorexigenic agent—one that reduces the appetite.

ANALYSIS Because OH^- removes just one H^+ from the nitrogen atom of a protonated amine's cation, all we have to do is reduce the number of H atoms on the nitrogen by one and cancel the positive charge.

SOLUTION Amphetamine is

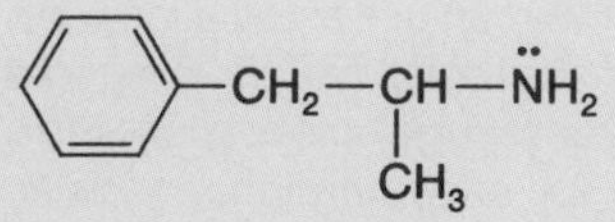

Amphetamine

(The structure of amphetamine, given here, appears to be identical with that of Dexedrine shown in Special Topic 17.1. However, there is an important difference that we will explore in the next chapter.)

PRACTICE EXERCISE 4

Write the structures of the products after the following protonated amines have reacted with OH^- in a 1 : 1 mole ratio.

(a)

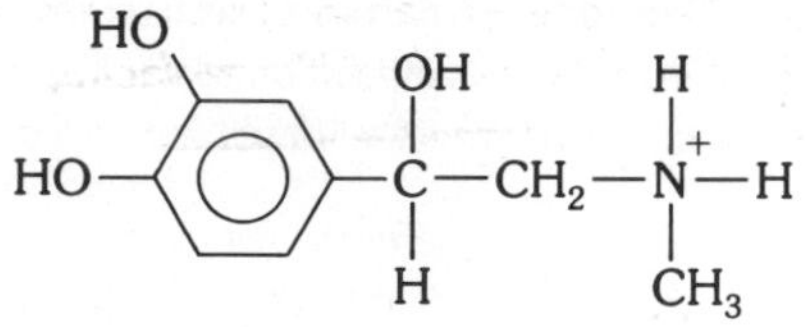

Epinephrine (adrenaline), a hormone given here in its protonated form. As the chloride salt in a 0.1% solution, it is injected in some cardiac failure emergencies. (See also Special Topic 17.1.)

(b)

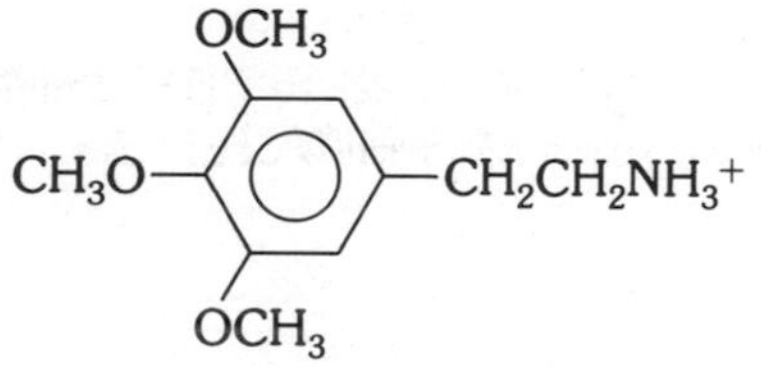

■ *Hallucinogens* are drugs that cause illusions of time and place, make unreal experiences or things seem real, and distort the qualities of things.

Mescaline, a mind-altering hallucinogen shown here in its protonated form. It is isolated from the mescal button, a growth on top of the peyote cactus. Indians in the southwestern United States have used it in religious ceremonies.

17.3

AMIDES OF CARBOXYLIC ACIDS

Amides are neutral nitrogen compounds that can be hydrolyzed to carboxylic acids and ammonia (or amines).
The carbonyl–nitrogen bond is sometimes called the **amide bond,** because it is the bond that forms when amides are made, and it is the bond that breaks when amides are hydrolyzed. As

the following general structures show, an amide can be derived either from ammonia or from amines. Those derived from ammonia itself are often referred to as *simple* amides.

$$
\begin{array}{cccc}
\underset{\substack{\| \\ R-C-NH_2 \\ \text{Amides of} \\ \text{ammonia} \\ \text{(simple amides)}}}{O} &
\underset{\substack{\| \\ R-C-NHR' \\}}{O} &
\underset{\substack{\| \\ R-C-NR' \\}}{O} \overset{R''}{\underset{}{|}} &
\underset{\substack{\| \\ -C-N- \\ \text{Amide group}}}{O}\!\!\!\!\overset{\text{Amide bond}}{\diagup}
\end{array}
$$

Amides of amines

We study the amide system because all proteins are essentially polyamides, polymers whose molecules have regularly spaced amide bonds. Nylon is a synthetic polymer with repeating amide groups (see Special Topic 17.2).

Table 17.3 lists several low-formula-mass amides. Their molecules are quite polar, and when they have an H atom bonded to N, they can both donate and accept hydrogen bonds. These forces add up so much in simple amides that all except methanamide are solids at room temperature. Simple amides have considerably higher boiling points than alkanes, alcohols, or even carboxylic acids of comparable formula mass, as the data in the margin show. When we study proteins, we'll see how hydrogen bonding is involved in stabilizing the shapes of protein molecules, shapes that are as important to the functions of proteins as anything else about their structures.

Compound	Formula mass	BP (°C)
$CH_3CH_2CH_2CH_3$	58	−42
$CH_3CH_2CH_2OH$	60	97
CH_3CO_2H	60	118
CH_3CONH_2	59	222

The Names of Simple Amides End in -*amide*   The common names of simple amides are made by replacing *-ic acid* by *-amide,* and their IUPAC names are devised by replacing *-oic acid* by *-amide.* In the following examples, notice how we can condense the structure of the amide group.

$$
\underset{\substack{\text{Acetamide (common name)} \\ \text{Ethanamide (IUPAC name)}}}{\overset{\displaystyle O}{\overset{\|}{CH_3CNH_2}} \;\; \text{or} \;\; CH_3CONH_2}
\qquad
\underset{\substack{\text{Butyramide (common name)} \\ \text{Butanamide (IUPAC name)}}}{\overset{\displaystyle O}{\overset{\|}{CH_3CH_2CH_2CNH_2}} \;\; \text{or} \;\; CH_3CH_2CH_2CONH_2}
$$

The simplest aromatic amide is called benzamide, $C_6H_5CONH_2$, where C_6H_5 signifies the phenyl group. We'll not need to know the rules for naming other kinds of amides.

$$C_6H_5 = \langle \bigcirc \rangle -$$

TABLE 17.3
Amides of Carboxylic Acids

IUPAC Name	Structure	MP (°C)
Methanamide	$HCONH_2$	3
N-Methylmethanamide	$HCONHCH_3$	−5
N,N-Dimethylmethanamide	$HCON(CH_3)_2$	−61
Ethanamide	CH_3CONH_2	82
N-Methylethanamide	$CH_3CONHCH_3$	28
N,N-Dimethylethanamide	$CH_3CON(CH_3)_2$	−20
Propanamide	$CH_3CH_2CONH_2$	79
Butanamide	$CH_3CH_2CH_2CONH_2$	115
Pentanamide	$CH_3CH_2CH_2CH_2CONH_2$	106
Hexanamide	$CH_3CH_2CH_2CH_2CH_2CONH_2$	100
Benzamide	$C_6H_5CONH_2$	133

The term *nylon* is a coined name that applies to any synthetic, long-chain, fiber-forming polymer with repeating amide linkages. One of the most common members of the nylon family, nylon-66, is made from 1,6-hexanediamine and hexanedioic acid.

$$NH_2CH_2CH_2CH_2CH_2CH_2CH_2NH_2$$

1,6-Hexanediamine

$$\underset{\text{Hexanedioic acid}}{HOCCH_2CH_2CH_2CH_2COH}$$

etc.$-[\overset{O}{\overset{\|}{C}}(CH_2)_4\overset{O}{\overset{\|}{C}}NH(CH_2)_6NH]_n-$ etc.

Repeating unit in nylon-66

(The "66" means that each monomer has six carbon atoms.) To be useful as a fiber-forming polymer, each nylon-66 molecule should contain from 50 to 90 of each of the monomer units. Shorter molecules form weak or brittle fibers.

When molten nylon resin is being drawn into fibers, newly emerging strands are caught up on drums and stretched as they cool. Under this tension, the long polymer molecules within the fiber line up side by side, overlapping each other, to give a finished fiber of unusual strength and beauty (see Figure 1). Part of nylon's strength comes from the innumerable hydrogen bonds that extend between the polymer molecules and that involve their many regularly spaced amide groups.

Nylon is more resistant to combustion than wool, rayon, cotton, or silk, and it is as immune to insect attack

FIGURE 1
This woman's life depends on the strength of nylon when she is parasailing.

as fiberglass. Molds and fungi do not attack nylon molecules either. In medicine, nylon is used in specialized tubing, and as velour for blood contact surfaces. Nylon sutures were the first synthetic sutures and are still commonly used.

PRACTICE EXERCISE 5

Write the IUPAC names of the following amides.

(a) $CH_3CH_2\underset{\underset{CH_3}{|}}{CH}CH_2CH_2\overset{O}{\overset{\|}{C}}NH_2$ (b) $CH_3CH_2\underset{\underset{CH_3CH_2}{|}}{CH}\overset{O}{\overset{\|}{C}}NH_2$

 Unlike the Amines, Amides Are Not Proton Acceptors One reason for creating a separate family for the amides apart from the amines is that, unlike amines, amides are not proton acceptors or bases. They're not proton donors or acids either. *Amides are neutral in an*

acid – base sense. The amide group, in other words, does not affect the pH of an aqueous system.

The electronegative carbonyl group on the nitrogen atom causes the acid – base neutrality of amides. Although both an amide and an amine have an unshared pair of electrons on nitrogen, in the amide this pair is drawn back so tightly by the electron-withdrawing ability of the carbonyl group that the electron pair shown on the nitrogen atom of the amide cannot actually accept and hold a proton.

■ The oxygen atom of the carbonyl group is what makes the whole group electronegative.

Amides Are Made from Amines by Acyl Group Transfer Reactions

Amides can be made from amines just as esters can be made from alcohols. Either acid chlorides or acid anhydrides react smoothly with ammonia or amines to give amides. (The amine, of course, must have at least one hydrogen atom on nitrogen, because one hydrogen has to be replaced as the amide forms.) We can illustrate these reactions using ammonia.

$$R-\overset{\overset{\displaystyle O}{\|}}{C}-Cl + 2NH_3 \longrightarrow R-\overset{\overset{\displaystyle O}{\|}}{C}-NH_2 + NH_4^+Cl^-$$

Acid chloride   Amide   Ammonium chloride

$$R-\overset{\overset{\displaystyle O}{\|}}{C}-O-\overset{\overset{\displaystyle O}{\|}}{C}-R + 2NH_3 \longrightarrow R-\overset{\overset{\displaystyle O}{\|}}{C}-NH_2 + NH_4^{+-}O-\overset{\overset{\displaystyle O}{\|}}{C}-R$$

Acid anhydride   Amide   Ammonium salt of the carboxylic acid

■ When heated strongly, the ammonium salts of carboxylic acids change to the corresponding simple amides as water is expelled.

These reactions are further examples of *acyl group transfer reactions.* The acyl group in the acid chloride, for example, transfers from the Cl atom to the N atom of the amine (or ammonia). An acyl group can transfer from an acid anhydride to N, also. In both of these reactions, a stable, weakly basic leaving group, Cl^- or RCO_2^-, is released by the transferring acyl group. Direct acyl transfer from a carboxylic acid, however, is more difficult because the leaving group is the less stable strong base, OH^-.

In the body, other kinds of acyl carrier molecules serve instead of ordinary acid chlorides and anhydrides as sources of the acyl group. When proteins are made from amino acids, for example, the acyl portions of amino acids — they are called *aminoacyl units* — are held by carrier molecules.

$$NH_2-\underset{\underset{\displaystyle R}{|}}{CH}-\overset{\overset{\displaystyle O}{\|}}{C}-\boxed{\begin{array}{c}\text{Carrier}\\\text{molecule}\end{array}}$$

Aminoacyl unit

■ R is some organic group, but not necessarily an alkyl group.

When a cell makes an amide bond, it transfers an aminoacyl group from its carrier molecule to the nitrogen atom of the amino group. The carrier molecule is released to be reused.

$$NH_2-\underset{\underset{\displaystyle R}{|}}{CH}-\overset{\overset{\displaystyle O}{\|}}{C}-\boxed{\begin{array}{c}\text{Carrier}\\\text{molecule}\end{array}} + NH_2-R' \xrightarrow{\overset{\text{Aminoacyl}}{\text{group transfer}}}$$

$$NH_2-\underset{\underset{\displaystyle R}{|}}{CH}-\overset{\overset{\displaystyle O}{\|}}{C}-NH-R' + \boxed{\begin{array}{c}\text{Carrier}\\\text{molecule}\end{array}} + H^+$$

■ The molecule given here as NH_2—R can be another aminoacyl group that is bound to another carrier molecule.

This is the aspect of making amides — aminoacyl transfers — that is of greatest interest as we prepare for our upcoming study of biochemistry. The skill we'll need is to figure out the structure of the amide that can be made from ammonia (or some amine) and a carboxylic acid regardless of the exact nature of the acyl transfer agent. We'll practice this in the next worked example.

EXAMPLE 17.3
Writing the Structure of an Amide That Can Be Made from the Acyl Group of an Acid and an Amine

What amide can be made from the following two substances, assuming that a suitable acyl group transfer process is available?

$$
\underset{\substack{\displaystyle \text{O} \\ \|}}{CH_3CH_2COH} \quad \text{and} \quad \underset{\substack{\displaystyle CH_3 \\ |}}{CH_3CHNH_2}
$$

ANALYSIS The amide system must be part of the structure we seek, so the best way to proceed is to write the skeleton of the amide system and then build on it. It doesn't matter how we orient this skeleton, left-to-right or right-to-left, as we'll demonstrate by showing both approaches.

$$
\underset{\substack{\displaystyle \text{O} \\ \|}}{-C-N-} \quad \text{or} \quad \underset{\substack{\displaystyle \text{O} \\ \|}}{-N-C-} \quad \text{(Incomplete)}
$$

Then we look at the acid to see what else must be on the carbon atom of this skeleton. It's an ethyl group, so we write it in:

$$
\underset{\substack{\displaystyle \text{O} \\ \|}}{CH_3CH_2-C-N-} \quad \text{or} \quad \underset{\substack{\displaystyle \text{O} \\ \|}}{-N-C-CH_2CH_3} \quad \text{(Incomplete)}
$$

Then we look at the amine to see what group(s) it carries. It has an isopropyl group, so we attach it to the N atom. (If there *had been two* organic groups on N in the amine, we would attach both, of course.)

$$
\underset{\substack{\displaystyle \text{O} \quad CH_3 \\ \| \quad\quad |}}{CH_3CH_2-C-N-CHCH_3} \quad \text{or} \quad \underset{\substack{\displaystyle CH_3 \quad \text{O} \\ | \quad\quad \|}}{CH_3CH-N-C-CH_2CH_3} \quad \text{(Incomplete)}
$$

Finally, of the two H atoms on N in the amine, one survives, and our last step is to write it in. (Recall that N needs three bonds in a neutral species.)

SOLUTION The final answer is

$$
\underset{\substack{\displaystyle \text{O} \quad \text{H} \quad CH_3 \\ \| \quad | \quad\quad |}}{CH_3CH_2-C-N-CHCH_3} \quad \text{or} \quad \underset{\substack{\displaystyle CH_3 \quad \text{H} \quad \text{O} \\ | \quad\quad | \quad \|}}{CH_3CH-N-C-CH_2CH_3}
$$

These structures, of course, are identical.

PRACTICE EXERCISE 6

What amides, if any, could be made by suitable acyl group transfer reactions from the following pairs of compounds?

(a) CH_3NH_2 and $\underset{\substack{\displaystyle | \\ CH_3}}{CH_3CHCO_2H}$ (b) $NH_2C_6H_5$ and CH_3CO_2H

(c) $\underset{\substack{\displaystyle \text{O} \\ \|}}{CH_3CCH_2NH_2}$ and CH_3NH_2 (d) CH_3CO_2H and $\underset{\substack{\displaystyle CH_3 \\ |}}{CH_3NCH_3}$

Amides Are Hydrolyzed to Their Parent Amines and Acids The only reaction of amides that we will study is their hydrolysis, a reaction in which the amide bond breaks and we obtain the amide's parent acid and amine (or ammonia). The hydrolysis of an amide does not occur easily. *In vitro,* either acids or bases are needed to promote the reaction. *In vivo,* enzymes catalyze amide hydrolysis, and this reaction is all that is involved in the overall chemistry of the digestion of proteins.

■ The digestive tract provides several protein-digesting enzymes called *proteases.*

In vitro, when an acid promotes the hydrolysis of an amide, one of the products, the amine, neutralizes the acid. (This is why we don't say that the acid *catalyzes* the hydrolysis. Catalysts, by definition, are reaction promoters that are not used up.) Thus instead of obtaining the amine itself, we get the salt of the amine. For example,

$$\underset{\substack{\| \\ O}}{R-C}-NH-CH_3 + H-OH + HCl(aq) \longrightarrow \underset{\substack{\| \\ O}}{R-C}-OH + H-\overset{+}{\underset{H}{N}}-CH_3\ Cl^-$$

On the other hand, if we use a base to promote amide hydrolysis, then the carboxylic acid that forms neutralizes the base, and we get the salt of the carboxylic acid. For example,

$$\underset{\substack{\| \\ O}}{R-C}-NH-CH_3 + NaOH(aq) \longrightarrow \underset{\substack{\| \\ O}}{R-C}-O^-Na^+ + H-\overset{H}{\underset{}{N}}-CH_3$$

When enzymes catalyze this hydrolysis, they are not used up by the reaction. Because our applications of this reaction concern *in vivo* situations at the molecular level of life, we'll write amide hydrolysis as a simple reaction with water to give the free carboxylic acid and the free amine. Here are some examples. How the acid and the amine actually emerge depends on the pH of the medium and the buffers present.

$$\underset{\substack{\| \\ O}}{R-C}-NH_2 + H_2O \xrightarrow{enzyme} \underset{\substack{\| \\ O}}{R-C}-OH + NH_3$$

$$\underset{\substack{\| \\ O}}{R-C}-NHR' + H_2O \xrightarrow{enzyme} \underset{\substack{\| \\ O}}{R-C}-OH + NH_2-R'$$

$$\underset{\substack{\| \\ O}}{R-C}-\underset{\substack{| \\ R''}}{N}R' + H_2O \xrightarrow{enzyme} \underset{\substack{\| \\ O}}{R-C}-OH + H-\underset{\substack{| \\ R''}}{N}-R'$$

EXAMPLE 17.4
Writing the Products of the Hydrolysis of an Amide

Acetophenetidin (phenacetin) was once used in some brands of headache remedies. (APC tablets, for example, consisted of aspirin, phenacetin, and caffeine.)

$$CH_3CH_2-O-\underset{}{\bigcirc}-NH-\underset{\substack{\| \\ O}}{C}-CH_3$$

Acetophenetidin

If this compound is an amide, what are the products of its hydrolysis?

ANALYSIS Acetophenetidin does have the amide bond, NH to carbonyl, so it can be hydrolyzed. (The functional group on the left side of this structure is an *ether,* and *ethers do not react with water.*) Because the amide bond breaks when an amide is hydrolyzed, simply erase this bond from the structure and separate the parts. *Do not break any other bond.*

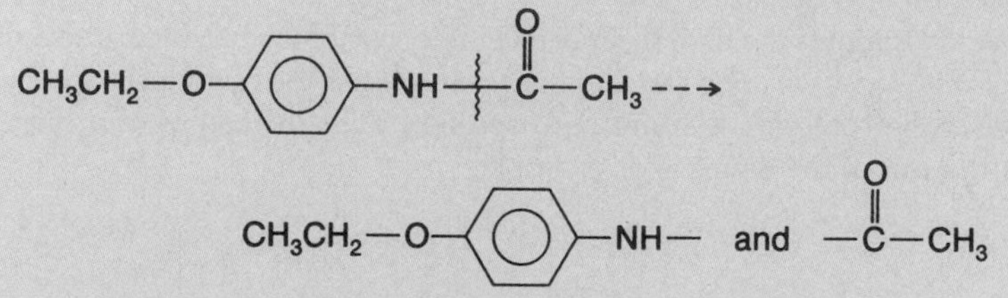

We know that the hydrolysis uses HO—H to give a *carboxylic acid* and an *amine,* so we put a HO group on the carbonyl group we put H on the nitrogen of the other fragment.

SOLUTION The products of the hydrolysis of acetophenetidin are

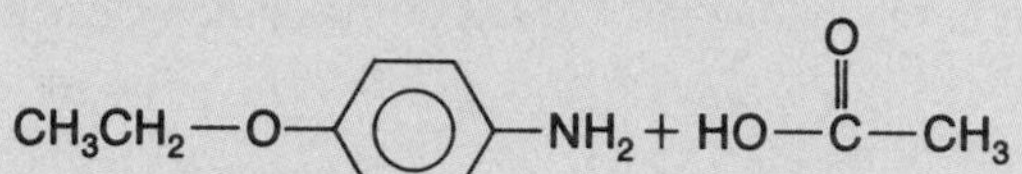

PRACTICE EXERCISE 7

For all compounds in the following list that are amides, write the products of their hydrolysis.

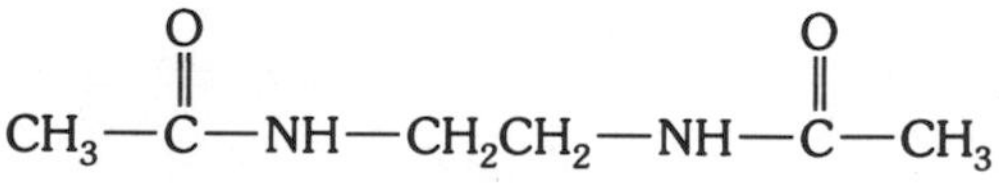

PRACTICE EXERCISE 8

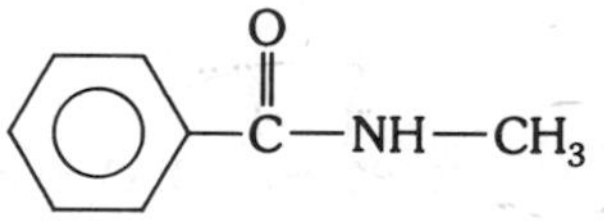

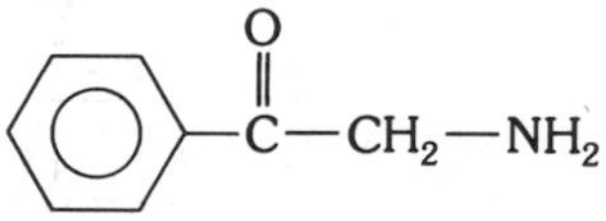

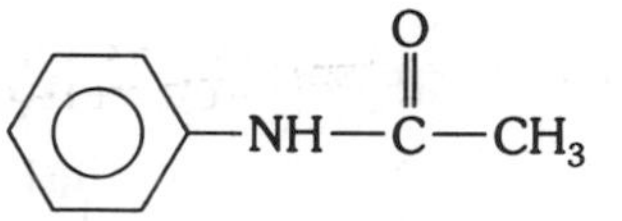

The following structure illustrates some of the features of protein molecules. What are the products of the complete, enzyme-catalyzed hydrolysis (the digestion) of this substance? (A typical protein would hydrolyze to give several hundred and up to several thousand of the kinds of small molecules produced by hydrolysis in this example.)

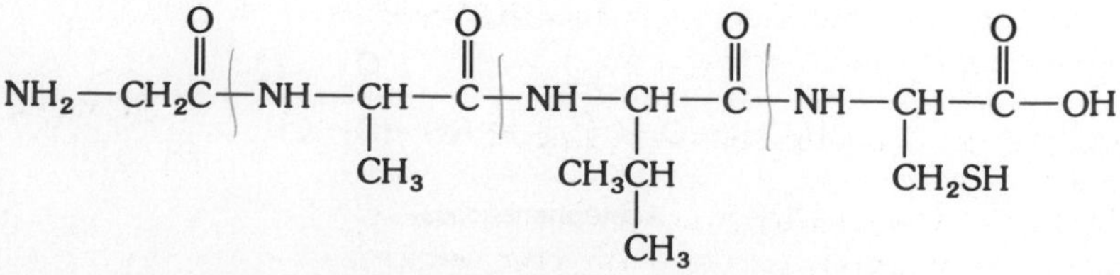

SUMMARY

Amines and protonated amines When one, two, or three of the hydrogen atoms in ammonia are replaced by an organic group (other than a carbonyl group), the resulting compound is an amine. The nitrogen atom can be part of a ring, as in heterocyclic amines. Like ammonia, the amines are weak bases, and all can form salts with strong acids. The cations in these salts are protonated amines.

Amine salts are far more soluble in water than their parent amines. Protonated amines are easily deprotonated by any strong base to give back the original and usually far less soluble amine. Thus any compound with the amine function has a "solubility switch," because its solubility in an aqueous system can be turned on by adding acid (to form the amine salt) and turned off again by adding base (to recover the amine).

Amides The carbonyl–nitrogen bond, the amide bond, can be formed by letting an amine or ammonia react with anything that can transfer an acyl group (e.g., an acid chloride or an acid anhydride). Amides are neither basic nor acidic, but are neutral compounds. Amides can be made to react with water to give back their parent acids and amines. The accompanying chart summarizes the reactions studied in this chapter.

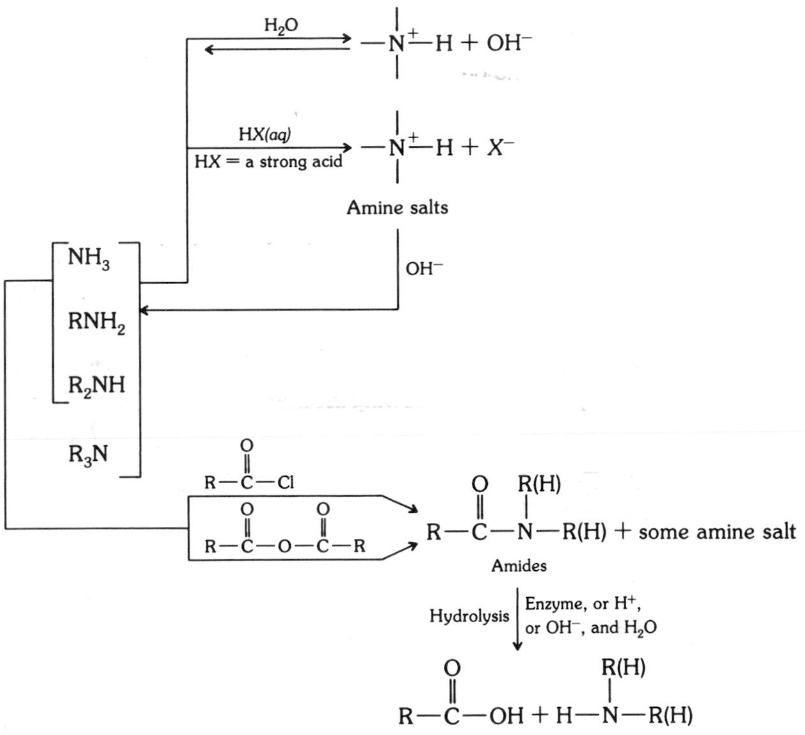

REVIEW EXERCISES

The answers to Review Exercises whose numbers are in color are found in Appendix C. The answers to the other Review Exercises are found in the Study Guide that accompanies this book. The more challenging questions are marked with asterisks.

Structures of Amines and Amides — Review of Functional Groups

17.1 Classify the following as aliphatic, aromatic, or heterocyclic amines or amides, and name any other functional groups, too.

(a) $CH_3OCH_2CNH_2$ (with C=O)

(b) $CH_3OCCH_2NH_2$ (with C=O)

(c) (ring)—N—CCH₃ (with C=O)

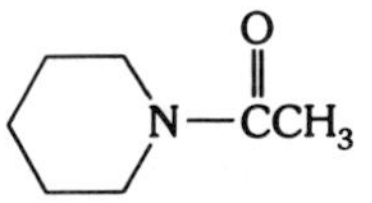

(d) 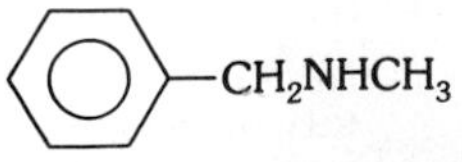 (benzene ring)—CH₂NHCH₃

17.2 Classify each of the following as aliphatic, aromatic, or heterocyclic amines or amides. Name any other functional groups that are present.

(a) 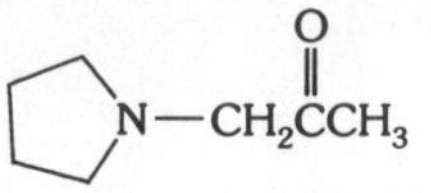 [structure: pyrrolidine N—CH_2CCH_3 with C=O]

(b) [structure: pyrrolidine N—CCH_2CH_3 with C=O]

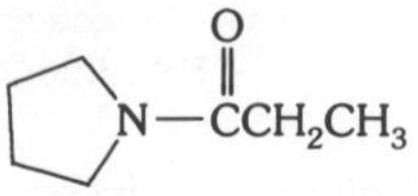

(c) [structure: ring with NH and C=O]

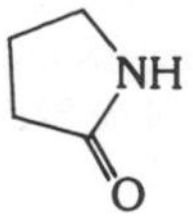

(d) 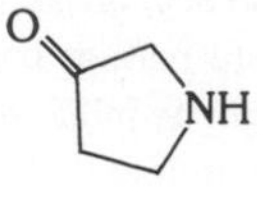 [structure: ring with O=C and NH]

·17.3 The following compounds are all very active physiological agents. Name the numbered functional groups that are present in each.

(a) 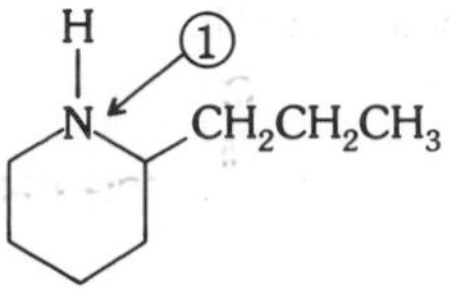 [structure]
H ①
N— $CH_2CH_2CH_3$

Coniine, the poison in the extract of hemlock that was used to execute the Greek philosopher Socrates

(b) 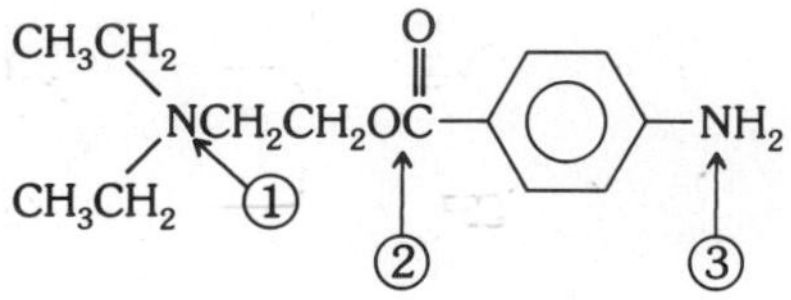 CH_3CH_2
        NCH_2CH_2OC ② — NH_2 ③
CH_3CH_2 ①

Novocaine, a local anesthetic

(c) 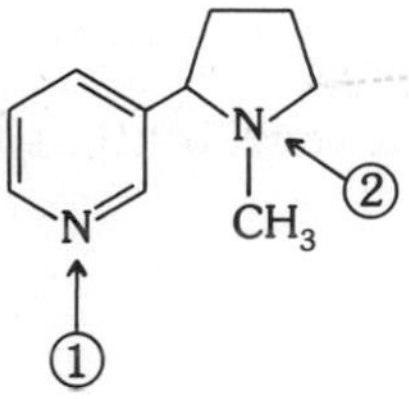 [structure]
N
CH_3 ②
N ①

Nicotine, a poison in tobacco leaves

(d) 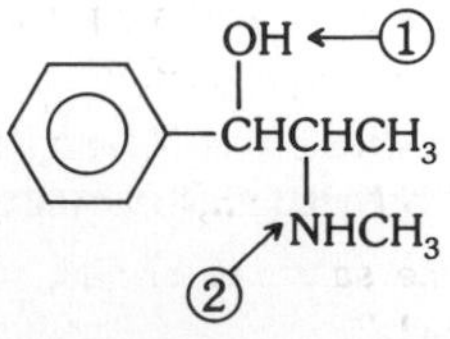 [structure]
OH ← ①
—CHCHCH_3
↑NHCH_3
②

Ephedrine, a bronchodilator

·17.4 Some extremely potent, physiologically active compounds are in the following list. Name the functional groups that they have.

(a) 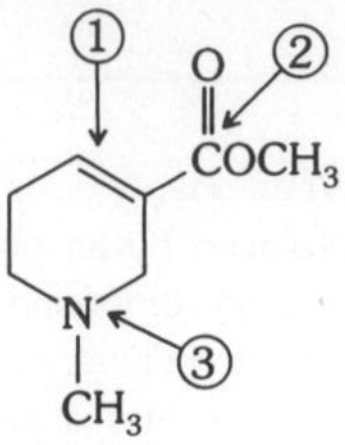 [structure]
① ② O
COCH_3
N
CH_3 ③

Arecoline, the most active component in the nut of the betel palm. This nut is chewed daily as a narcotic by millions of inhabitants of parts of Asia and the Pacific islands.

(b) 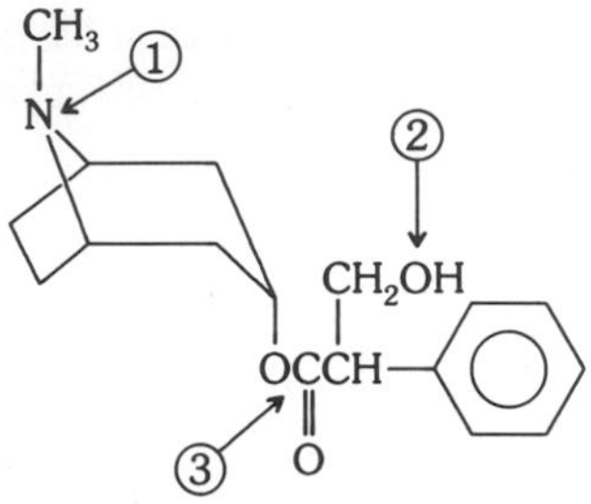 [structure]
CH_3 ①
N
② CH_2OH
OCCH— ③ O

Hyoscyamine, a constituent of the seeds and leaves of henbane, and a smooth muscle relaxant. (A similar form is called atropine, a drug used to counteract nerve poisons.)

(c) 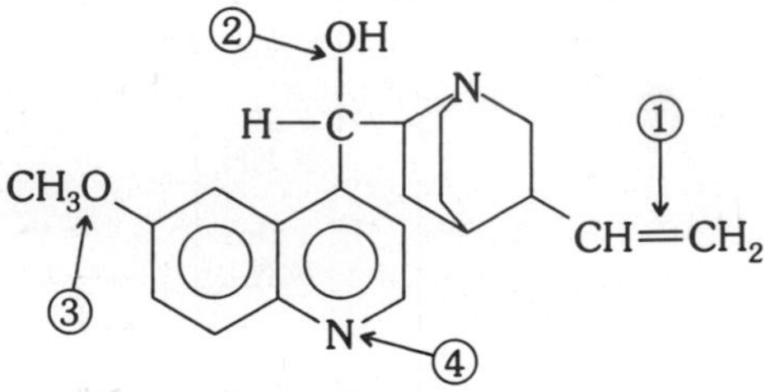 [structure]
② OH
H—C N ①
CH_3O
③ N CH=CH_2
④

Quinine, a constituent of the bark of the chinchona tree in South America and used to treat malaria

(d) 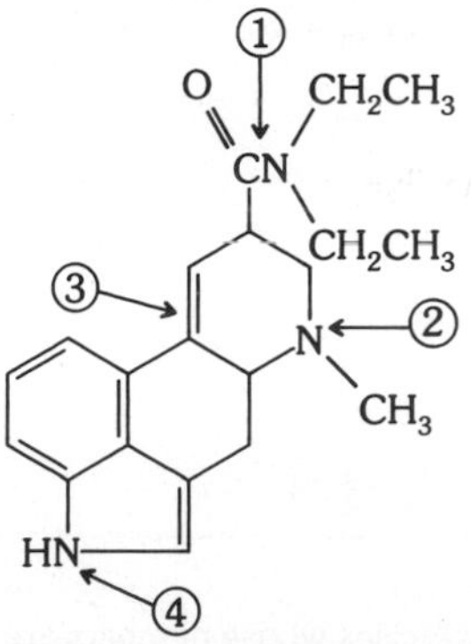 [structure]
① CH_2CH_3
O
CN
CH_2CH_3
③ N ②
CH_3
HN
④

Lysergic acid diethylamide (LSD), a constituent of diseased rye grain and a notorious hallucinogen

17.5 Give the common names of the following compounds or ions.

CH_3
(a) $CH_3CH_2CH_2NHCHCH_3$

(b) $CH_3CH_2CH_2\overset{\overset{\displaystyle CH_3}{|}}{N}CH_2CH_3$

(c) 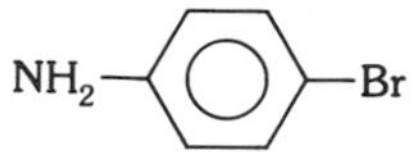 NH_2—⬡—Br

(d) $CH_3CH_2CH_2NHCH_2\overset{\overset{\displaystyle CH_3}{|}}{C}H_2$

17.6 What are the common names of the following compounds?

(a) $(CH_3)_3NH^+Cl^-$

(b) 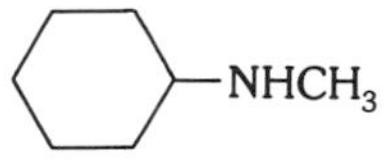 ⬡—$NHCH_3$

(c) 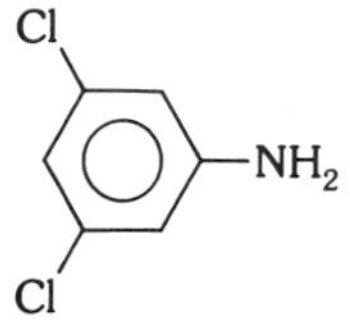 Cl-substituted benzene—NH_2

(d) $[(CH_3)_2CH]_3N$

Chemical Properties of Amines and Amine Salts

17.7 Complete the following reaction sequences by writing the structures of the organic products. If no reaction occurs, write "no reaction."
(a) $CH_3CH_2CH_2NH_2 + HCl(aq) \longrightarrow$
(b) $CH_3CH_2CH_2NH_3{}^+Cl^- + NaOH(aq) \longrightarrow$
(c) $CH_3CH_2CH_2NH_3{}^+Cl^- + HCl(aq) \longrightarrow$
(d) $CH_3CH_2CH_2NH_2 + NaOH(aq) \longrightarrow$

17.8 Write the structures of the organic products that form in each situation. Assume that all reactions occur at room temperature. (Some of the named compounds are described in Review Exercises 17.3 and 17.4.) If no reaction occurs, write "no reaction."

(a) 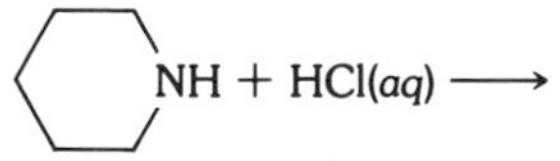 ⬡$NH + HCl(aq) \longrightarrow$

(b) 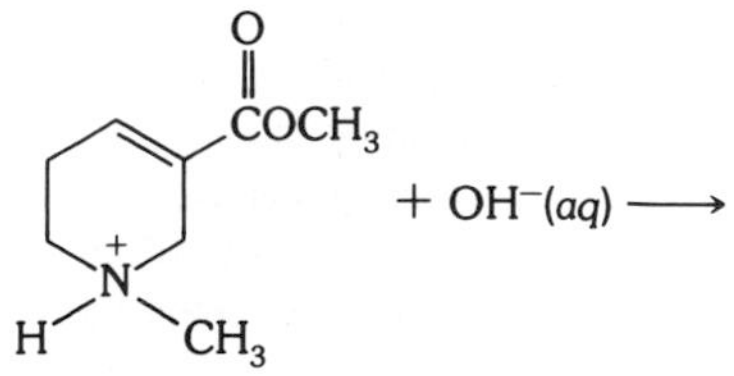 $+ OH^-(aq) \longrightarrow$

Protonated form
of arecoline

(c) 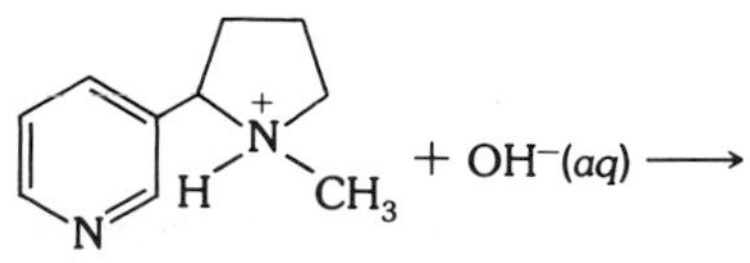 $+ OH^-(aq) \longrightarrow$

Protonated form
of nicotine

(d) 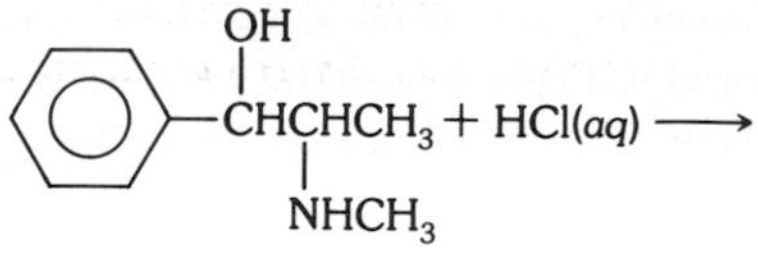 $+ HCl(aq) \longrightarrow$

Ephedrine

(e) 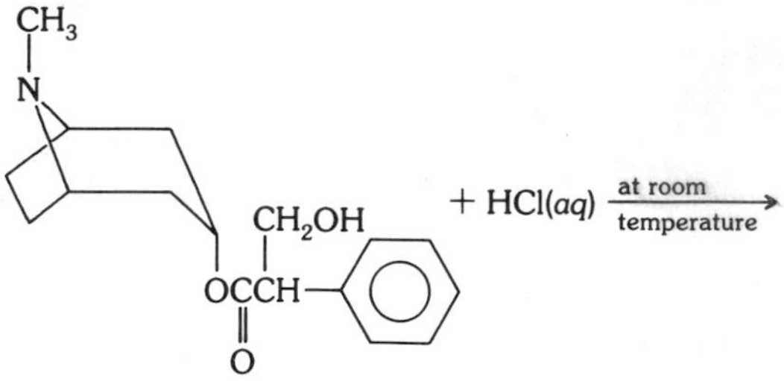 $+ HCl(aq) \xrightarrow{\text{at room} \atop \text{temperature}}$

Hyoscyamine

17.9 Which is the stronger base, **A** or **B**? Explain.

$$\underset{\textbf{A}}{NH_2CH_2\overset{\overset{\displaystyle O}{\|}}{C}CH_3} \qquad \underset{\textbf{B}}{CH_3\overset{\overset{\displaystyle O}{\|}}{C}NHCH_2CH_3}$$

17.10 Which is the stronger proton acceptor, **A** or **B**? Explain.

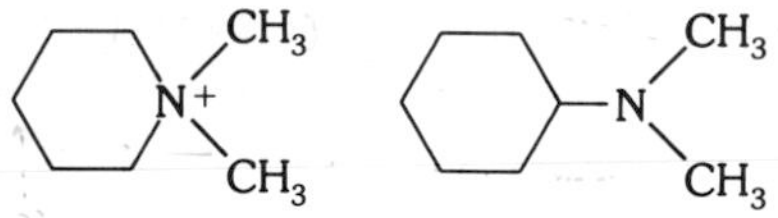

Names and Structures of Amides

17.11 What are the IUPAC names of the following compounds?

(a) $CH_3CH_2CH_2\overset{\overset{\displaystyle O}{\|}}{C}NH_2$

(b) $CH_3\overset{\overset{\displaystyle CH_3}{|}}{C}HCH_2\overset{\overset{\displaystyle O}{\|}}{C}NH_2$

17.12 If the common name of hexanoic acid is caproic acid, what is the common name of its simple amide?

17.13 If $C_6H_5CONHCH_3$ is the structure of *N*-methylbenzamide, what is the structure of *N,N*-dimethylbenzamide?

17.14 If ethanediamide has the structure shown, what is the structure of butanediamide?

$$NH_2\overset{\overset{\displaystyle O}{\|}}{C}-\overset{\overset{\displaystyle O}{\|}}{C}NH_2$$

Ethanediamide

•17.15 What is the structure of lysergic acid? The structure of its *N,N*-diethylamide was given in Review Exercise 17.4, part (d).

17.16 What is the structure of the amide that can form between acetic acid and ephedrine? (The structure of ephedrine is given in part (d) of Review Exercise 17.3).

Synthesis of Amides

17.17 Write the equations for two ways to make acetamide using ammonia as one reactant.

17.18 What are two different ways to make *N*-methylacetamide if methylamine is one reactant? Write the equations.

*17.19 Examine the following acyl group transfer reaction.

$$NH_2CH_2\overset{O}{\underset{\|}{C}}NH\overset{CH_3}{\underset{|}{CH}}\overset{O}{\underset{\|}{C}} - \boxed{\text{Carrier molecule}}_1 + NH_2\overset{O}{\underset{\|}{C}H}\overset{O}{\underset{\|}{C}} - \boxed{\text{Carrier molecule}}_2$$
$$\underset{CH_2C_6H_5}{}$$

$$NH_2CH_2\overset{O}{\underset{\|}{C}}NH\overset{CH_3}{\underset{|}{CH}}\overset{O}{\underset{\|}{C}}NH\overset{CH_2C_6H_5}{\underset{|}{CH}}\overset{O}{\underset{\|}{C}} - \boxed{\text{Carrier molecule}}_2 + \boxed{\text{Carrier molecule}}_1$$

(a) Which specific acyl group transferred? (Write its structure.)

(b) How many amide bonds are showing (or implied) in the product?

*17.20 If the following anhydride is mixed with ammonia, what possible organic products that are not salts can form? Write their structures.

$$CH_3CH_2\overset{O}{\underset{\|}{C}}O\overset{O}{\underset{\|}{C}}CH_3 \quad + NH_3$$

Reactions of Amides

17.21 What are the products of the hydrolysis of the following compounds? (If no hydrolysis occurs, state so.)

(a) $CH_3CH_2NH\overset{O}{\underset{\|}{C}}CH_3$

(b) $CH_3\overset{CH_3}{\underset{|}{CH}}NH\overset{O}{\underset{\|}{C}}CH_2CH_3$

(c) $CH_3NH\overset{O}{\underset{\|}{C}}-\overset{CH_3}{\underset{|}{CH}}CH_3$

(d) $CH_3\overset{O}{\underset{\|}{C}}CH_2NHCH_3$

17.22 Write the structures of the products of the hydrolysis of the following compounds. If no reaction occurs, state so. If more than one bond is subject to hydrolysis, be sure to hydrolyze all of them.

(a) $NH_2\overset{CH_3}{\underset{|}{CH}}CH_2\overset{O}{\underset{\|}{C}}NHCH_2\overset{O}{\underset{\|}{C}}OH$

(b) $NH_2\overset{O}{\underset{\|}{C}}CH_2\overset{H_3C}{\underset{|}{CH}}\overset{O}{\underset{\|}{C}}NH_2$

(c)
$$\overset{O}{\underset{\|}{}} \\ \boxed{}NH \\ \underset{CH_3}{}$$

(d) $NH_2\overset{O}{\underset{\|}{C}}NH_2$

Physiologically Active Amines (Special Topic 17.1)

17.23 What are hormones and, in very broad terms, what is their function?

17.24 Hormones secreted by the adrenal gland are called what kinds of agents?

17.25 Name two hormones secreted by the adrenal gland.

17.26 Drugs that tend to mimic the two hormones secreted by the adrenal gland are called what kinds of drugs?

17.27 To be a *catecholamine* as well as a *β*-phenylethanolamine, a compound must have what structural features?

17.28 Is dopamine, a *β*-phenylethylamine, also a catecholamine?

17.29 In what general family of the physiologically active amines are the amphetamines found?

17.30 What is the chemical name of each?
(a) "Speed"   (b) "Uppers"

Nylon (Special Topic 17.2)

17.31 What functional group is present in nylon-66?

17.32 The strength of a nylon fiber is attributed in part to what relatively weak bond.

Additional Exercises (A Review of Organic Reactions

*17.33 What are all the functional groups we have studied that can be changed by each of the following reactants? Write the equations for the reactions, using general symbols such as ROH or RCO_2H and so forth to illustrate these reactions, and name the organic families to which the reactants and products belong.
(a) Water, either with an acid or an enzyme catalyst.
(b) Hydrogen (or a hydride ion donor) and any needed catalysts and special conditions.
(c) An oxidizing agent represented by (O), such as $Cr_2O_7{}^{2-}$ or $MnO_4{}^-$, but not ozone and not oxygen as used in combustion.

17.34 We have described three functional groups that typify those involved in the chemistry of the digestion of carbohydrates, fats and oils, and proteins. What are the names of these groups and to which type of food does each belong?

*17.35 A student performed an experiment that hydrolyzed 1.65 g of benzamide.
(a) What is the maximum number of grams of benzoic acid that could be obtained?

(b) How many milliliters of 0.482 M HCl would be needed to convert all of the ammonia that can form from the hydrolysis of the sample of benzamide into ammonium chloride?

*17.36 Write the structures of the organic products that would form in the following situations. If no reaction occurs, state so. These constitute a review of nearly all the organic reactions we have studied, beginning with Chapter 12.

(a) $CH_3COCH_3 + H_2O \xrightarrow[\text{catalyst}]{\text{acid}}$

(b) $CH_3CCH_2CH_3 + H_2 \xrightarrow[\text{pressure}]{\text{Ni or Pt}}$

(c) $CH_3CH_2CH_2CH_2CH_3 + MnO_4^-(aq) \longrightarrow$

(d) $CH_3CH_2CCl + NH_3$ (excess) $\longrightarrow$

(e) ⬡—$CH_2CH_3 + Cr_2O_7^{2-}(aq) \longrightarrow$

(f) $CH_3CHCH_2CH_3 + NaOH(aq) \longrightarrow$ (with CH_3 substituent)

(g) $CH_3CHOCH_2CH_3 + H_2O \xrightarrow[\text{catalyst}]{\text{acid}}$ (with OCH_2CH_3 substituent)

(h) $CH_3CH_2CH + Cr_2O_7^{2-}(aq) \longrightarrow$

(i) $CH_3CH_2CCH_3 + Cr_2O_7^{2-}(aq) \longrightarrow$

(j) $CH_3OH + CH_3CH_2COH \xrightleftharpoons{H^+}$

(k) $CH_3CH_2CNH_2 + NaOH(aq) \xrightarrow{\text{heat}}$

(l) $CH_3CH_2CH + 2CH_3OH \xrightarrow[\text{catalyst}]{\text{acid}}$

(m) $CH_3CH_2SH + (O) \longrightarrow$

(n) $C_6H_5COCH_2CHCH_3 + NaOH(aq) \longrightarrow$ (with CH_3 substituent)

(o) $NH_2CH_2CH_2CHCH_3 + HCl(aq) \longrightarrow$ (with CH_3 substituent)

(p) $CH_3CH{=}CHCH_2OCH_3 + H_2 \xrightarrow{\text{Ni or Pt}}$

*17.37 What are the structures of the organic products that form in the following situations? (If there is no reaction, state so.) These reactions review most of the chemical properties of functional groups we have studied, beginning with Chapter 12.

(a) $HOCCH_2CH_2CH_3 + NaOH(aq) \longrightarrow$ *sodium butanoid ion*

(b) ⬠O + NaOH(aq) $\longrightarrow$ *unknown*

(c) $CH_3CH_2OCH_2CH_2CH + MnO_4^-(aq) \longrightarrow$ *Acid*

(d) $NH_2CH_2CH_2NH_2 + HCl(aq) \longrightarrow$ (excess) *2 Amin*

(e) ⬡—$CCH_2CH_3 + H_2 \xrightarrow[\text{pressure}]{\text{Ni or Pt}}$ *Keton ethyle di Alcohol*

(f) $CH_3(CH_2)_5CH_3 + Cr_2O_7^{2-}(aq) \longrightarrow$ *no react*

(g) CH_3—⬡—$CH + 2CH_3OH \xrightarrow[\text{catalyst}]{\text{acid}}$ *Hemiacital*

(h) CH_3—⬡—$COCH_2CHCH_3 + NaOH(aq) \longrightarrow$ (with CH_3 substituent) *Ester*

(i) $CH_3CHCH_2CH_3 + MnO_4^-(aq) \longrightarrow$ (with OH substituent) *butano keton*

(j) $CH_3COCH_2CH_3 + H_2O \xrightarrow[\text{catalyst}]{\text{acid}}$ (with OCH_2CH_3 and CH_3 substituents) *Ketal aldehyd + alcoho*

(k) $CH_3CH_2OC(CH_2)_3COCH_2CH_3 + H_2O \xrightarrow[\text{catalyst}]{\text{acid}}$ *Easter no reaction mau hydrol*

(l) $CH_3OCHOCH_3 + H_2O \xrightarrow[\text{catalyst}]{\text{acid}}$ (with CH_2CH_3 substituent) *Hemi acetal aldehyde + alcoh*

(m) $CH_3COCCH_3 + NH_3$ (excess) $\longrightarrow$ *base Amide*

(n) $CH_3SSCH_3 \xrightarrow{\text{reduction (2H)}}$ *methy disulfide*

(o) $CH_3(CH_2)_5CO_2H + CH_3OH \xrightleftharpoons{H^+}$ *easter*

(p) $CH_3NHCCH_2CH_2NHCCH_2CH_3 + H_2O \xrightarrow{\text{enzyme}}$ (excess) *amibt*

17.38 Write the IUPAC names of the following compounds.
(a) $CH_3CH_2CH_2CO_2CH_3$ ~~Easter~~—Butanote
(b) $(CH_3)_2CHCH_2Br$ 1-Bromo-2 methyl propane
(c) $O=CHCH_2CH(CH_3)_2$ Butanale - 3 methyl
(d) $CH_3CH_2CH=CHCH_3$ 2-p entene
(e) $(CH_3)_3CCH_2CH(CH_3)_2$ Dimethyl
(f) $CH_3CH_2CH_2COCH(CH_3)_2$
(g) $(CH_3)_3COH$
(h) $HO_2C(CH_2)_3CH_3$
(i) CH_3SH
(j) CH_3CO_2Na

17.39 Write the common names of the following compounds.
(a) $C_6H_5CO_2Na$
(b) $NH_2CH_2CH_2CH_3$
(c) $C_6H_5NH_2$
(d) $HO_2CCH_2CH_3$
(e) $O=CHCH_2CH_2CH_3$
(f) $HOCH_2CH(CH_3)_2$
(g) HOC_6H_5
(h) $CH_3CH_2OCH_2CH_3$

(i) $CH_3CH_2CH_2CO_2CH_2CH_3$
(j) CH_3CONH_2
(k) CH_3COCH_3

17.40 Identify by letter which of the following compounds would be more soluble in water at a pH of 12 than at a pH of 7. Explain.

$$CH_3(CH_2)_6CO_2CH_3 \quad CH_3(CH_2)_6CO_2H$$
$$\textbf{A} \qquad\qquad \textbf{B}$$
$$CH_3(CH_2)_6CH_2NH_2$$
$$\textbf{C}$$

17.41 Identify by letter which of the following compounds would be more soluble in water at a pH of 2 than at a pH of 7. Explain.

$$C_6H_5CH_2CH_2CONH_2 \quad C_6H_5CH_2COCH_2NH_2$$
$$\textbf{A} \qquad\qquad \textbf{B}$$
$$C_6H_5CH_2CH_2CO_2H$$
$$\textbf{C}$$

e = $CH_3 - \overset{CH_3}{\underset{CH_3}{\overset{|}{C}}} - CH_2 \overset{CH_3}{\underset{CH_3}{\overset{|}{C}H}}$

2,2,4-trimethyl pentane

f = $\underset{6}{CH_3} \underset{5}{CH_2} - \underset{4}{CH_2} - \overset{O}{\overset{||}{\underset{3}{C}}} - \underset{2}{CH} \overset{CH_3}{\underset{CH_3}{|}}$ = 2-methyl-3-hexanone

STEREOISOMERISM

Light that reflects from a flat surface is rich in polarized light, which causes glare. A polarizing lens, however, removes polarized light and reduces glare, making this photo of a stream near Sedona, Arizona, very pleasant.

18.1
TYPES OF ISOMERISM

18.2
MOLECULAR CHIRALITY

18.3
OPTICAL ACTIVITY

18.1

TYPES OF ISOMERISM

Constitutional isomers and stereoisomers are the two broad classes of isomers.
Compounds that have identical molecular formulas can be different in two general ways, as *constitutional isomers* or as *stereoisomers*.

■ Constitutional isomers are often called *structural isomers,* an older name now being replaced.

Constitutional Isomers Differ in Molecular Frameworks Butane and isobutane, both C_4H_{10}, have different chains. Ethanol and dimethyl ether, both C_2H_6O, have different functional groups. Each pair illustrates **constitutional isomerism** because their members differ in the ways by which the atoms are joined to each other. The molecules of **constitutional isomers** have different atom-to-atom sequences.

$$CH_3CH_2CH_2CH_3 \qquad CH_3\overset{\overset{\textstyle CH_3}{|}}{C}HCH_3 \qquad CH_3CH_2OH \qquad CH_3OCH_3$$

Butane   2-Methylpropane   Ethanol   Dimethyl ether
                    (isobutane)

Stereoisomers Differ Only in Geometry *cis*-2-Butene and *trans*-2-butene illustrate **stereoisomerism** because they have the same constitutions — identical molecular formulas, functional groups, and heavy-atom skeletons — but display different geometries.

$$\underset{\textit{cis}\text{-2-Butene}}{\overset{\textstyle H_3C \qquad\qquad CH_3}{\underset{\textstyle H \qquad\qquad H}{C=C}}} \qquad\qquad \underset{\textit{trans}\text{-2-Butene}}{\overset{\textstyle H_3C \qquad\qquad H}{\underset{\textstyle H \qquad\qquad CH_3}{C=C}}}$$

■ STER-ee-oh-EYE-som-ers

Stereoisomers have identical constitutions but different geometries. Thus *cis*- and *trans*-2-butene are identical in sharing the same constitution, $CH_3CH=CHCH_3$, but they differ in geometry. The lack of free rotation at the double bond ensures that one isomer cannot easily switch to the other.

■ Recall that the four atoms attached directly to the carbon atoms of the double bond all lie in the same plane.

There are two broad families of stereoisomers, *diastereomers* (illustrated by *cis*- and *trans*-2-butene) and *enantiomers*. This chapter is mostly about the latter, but to define them requires more background. Enantiomers come about as close to being identical as your left and right hands, yet they display dramatic differences in chemistry at the molecular level of life. One such difference is illustrated by the bittersweet story of asparagine.

■ dye-a-STER-ee-o-mers
en-AN-tee-o-mers

■ Asparagine is a building block for making proteins in the body.

Asparagine is a white solid that was first isolated in 1806 from the juice of asparagus. The asparagine obtained from this source has a bitter taste. Its molecular formula is $C_4H_8N_2O_3$, and its structure is given by structure **1**.

$$\underset{\textstyle 1 \text{ Asparagine}}{NH_2\overset{\overset{\textstyle O}{\|}}{C}CH_2\underset{\underset{\textstyle NH_2}{|}}{C}HCO_2H}$$

In 1886, a chemist isolated from sprouting vetch a white substance with the same molecular formula *and constitution,* but it had a sweet taste. To have names for these, the one isolated from asparagus is now called L-asparagine, and the one from vetch sprouts is called D-asparagine.

Here were two substances seemingly answering to the same structure despite what is almost a dogma in chemistry, the principle of *one-substance – one-structure.* If two samples of matter have identical physical and chemical properties, then they must be identical at the level of their individual formula units. If two samples differ in even one way in their fundamental properties, then there has to be at least one difference in the way that the atoms are put together into their molecules.

Taste is a chemical sense, so the two samples of asparagine do have one difference in chemical property. Under the one-substance – one-structure rule, therefore, the molecules of D- and L-asparagine must be structurally different in at least one way. This difference arises from a peculiar lack of symmetry in their molecules that makes possible two different relative configurations of their molecular parts.

■ Vetch is a member of a genus of herbs, some of which are useful as fodder for cattle.

■ The letters D and L will acquire more specific meaning in the next chapter. Consider them to be only labels now.

18.2

MOLECULAR CHIRALITY

The molecules of many substances have a handedness like that of the left and the right hands.

Two partial ball-and-stick molecular models of asparagine are shown in Figure 18.1a. Examine each model to make sure that it faithfully represents asparagine, structure **1**. To make the study of these models easier, we have simplified them as shown in Figure 18.1b. Notice again how alike the two molecular structures are. Notice particularly that the same four groups are attached to a central carbon atom, and that there is no cis – trans kind of difference between them. Yet the fact remains that one represents a bitter-tasting compound, and the other is of a sweet-tasting compound. In some way these two similar structures have to be different, and the difference isn't something that can be removed by rotating groups around single bonds.

Two Materials Whose Molecules Can Be Superimposed Are Samples of the Same Compound
To understand how the two asparagine structures are different, we first have to learn the ultimate test for deciding whether two structures are the same. *For two structures to be identical, it must be possible to superimpose them.* To superimpose two structures means to do a manipulation that you can carry to completion only in your mind, but the use of molecular models of the structures is a great help. **Superimposition** is the mental blending of one molecular model with another so that the two would coincide in *every* atom and bond simultaneously if the operation could actually be completed. (When working with molecular models, it's fair to twist parts about single bonds to find conformations that can be superimposed, but it's not legal to break any bonds.) Superimposition is illustrated in the lower left-hand side of Figure 18.1c, where two *identical* models of *one* of the asparagines are used.

As we said, the fundamental criterion of identity of two structures is that they pass the test of superimposition, a test that we will see is failed by the two *different* asparagines of Figure 18.1b. In Figure 18.1c, the model on the left in Figure 18.1b has been turned counterclockwise by 120° around the vertical bond from C to the group **g** and then placed as the object in front of the mirror. Now look at the reflection of this model in the mirror. If you made an exact molecular model of its reflection, the new model would be identical to that of the *other* asparagine, the one on the right in Figure 18.1b. This is how nearly identical the two asparagines are. In the lower right of part Figure 18.1c, you can see that these two models, the one in front of the mirror as the object and the model of its reflection, do not superimpose.

■ In some references you'll see the word *superposition* used for *superimposition.*

FIGURE 18.1

The two stereoisomers of asparagine. (*a*) Two ways of joining the four groups to the carbon marked by the asterisk. (*b*) Simplified representations of the models in part *a*. (*c*) What is in front of the mirror is identical with the model on the left in part *b*. What is seen in the mirror as the image is identical with the model on the right in part *b*. You'll mentally have to spin the mirror image 120° counterclockwise about the bond from C to **g** to see that they are the same. The object and its mirror image do not superimpose, so they can't be identical. Instead, they are enantiomers.

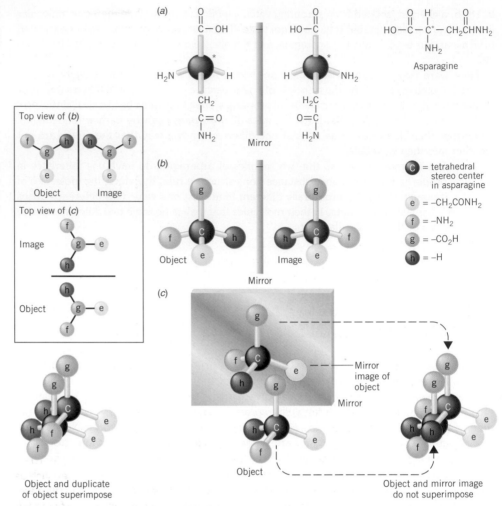

- ■ **Constitutional isomers:** Identical molecular formulas but different arrangements of atoms.
Stereoisomers: Identical *constitutions* but different geometries.
Enantiomers: Pairs of *stereoisomers* whose molecules are related as object to mirror image but cannot be superimposed.
Diastereomers: *Stereoisomers* that are not enantiomers.

The Molecules of Enantiomers Are Related as an Object to a Mirror Image but Cannot Be Superimposed As we said, the two different asparagine molecules are so alike that the molecule of one is the mirror image of the other. Yet the model and its mirror image do not superimpose. Pairs of stereoisomers whose molecules are related as object to mirror image that cannot be superimposed are called **enantiomers.**

Always remember two general facts about isomers of any kind. They are truly *different substances,* different compounds, but they share the same molecular formula while differing in the arrangements of their atoms. Enantiomers are just special kinds of stereoisomers. All other isomers that qualify as *stereoisomers* are called **diastereomers.** Diastereomers are stereoisomers that are not enantiomers. Thus all purely geometric (cis–trans) isomers are diastereomers but other examples also exist, as we'll see in Special Topic 18.1 later in this chapter. Figure 18.2 sorts out the kinds of isomers we have studied.

Molecules of a Pair of Enantiomers Have Opposite Configurations Lack of free rotation is not the cause of the asparagine enantiomers. Their molecules, instead, are described as having *opposite configurations.* To show what this means, we have repositioned their abbreviated molecular models in Figure 18.3. (Imagine a mirror standing between the two and perpendicular to the page to see that they are related as object to mirror image.)

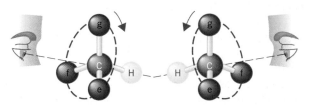

FIGURE 18.2
The relationships of various kinds of isomers.

You're now going to let your *eyes* make a special scanning trip around each molecule. Imagine that the bond from C to H in each model is the steering column of the steering wheel of a car. Then imagine that the remaining three groups — **e**, **f**, and **g** — are distributed around the steering wheel itself. Now move your eyes from **g** to **e** and then to **f**. When you do this with the asparagine model on the left in Figure 18.3, your eyes move clockwise. But to make the identical trip — **g** to **e** to **f** — in the model on the right, your eyes move counterclockwise. These clockwise versus counterclockwise arrangements of identical parts around the same central axis are what having *opposite configuration* means. The four groups on the central carbon in one asparagine are the same four groups as in the other, but they are configured oppositely in space.

Molecules of a Pair of Enantiomers Have Opposite Chirality

We have to remind ourselves now that any object has a mirror image. Spheres, cubes, broom handles, water glasses, and so on, can all be reflected in a mirror. It's only when the model of the object and the model of the mirror image cannot be superimposed that we call the two enantiomers.

Your two hands are like a pair of enantiomers, if you disregard small differences such as wrinkles, scars, rings, and fingerprints. Place your left hand in front of a mirror near its edge. Place your other hand just off the edge of the mirror and notice that the reflection of your left hand in the mirror is the same as your right hand (disregarding, as we said, the small differences). Your right hand is the mirror image reflection of your left hand.

Next, try to superimpose the two hands. Because the mirror image of your left hand is your real right hand, use your two hands to see whether they superimpose. If you put them palm to palm it seems as though all the fingers do superimpose. But remember, you have to carry this blending through to completion (in your mind), and when you do, the palm of one hand comes out on the back side of the other. The palms won't superimpose when the fingers seem to. And if you try to get the palms to come out right, then the fingers come out all wrong. Left and right hands, although related as object to mirror image, don't superimpose. They are related as enantiomers.

Notice, now, how the two hands have opposite configurations. Look at them down the same axis, as we did with the asparagine models in Figure 18.3, say, down the axis from palm to backside. This means that both palms will face you. To make the trip from thumb to little

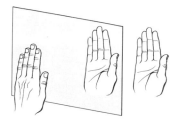

The image of the left hand is just like the right hand in the relative orientations of the fingers, thumb, and palm.

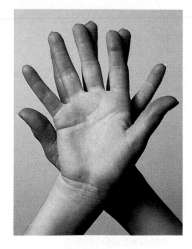

The two hands are related as an object to its mirror image, but they can't be superimposed.

FIGURE 18.3
When the two stereoisomers of asparagine are viewed down the same axis, the C—H bond axis, the remaining three groups at the central carbon atom have opposite configurations.

finger, touching every other finger on the way, you have to scan in one direction for one hand and in exactly the opposite direction for the other. Thus the two hands have opposite configurations.

The little experiment with the hands has been used for decades to teach about the configurational differences of enantiomers, and this is why we say that molecules of different enantiomers have different *handedness*. The technical term for this configurational property is **chirality** (from the Greek word for "hand," which is *cheir*). We say that *the molecules of enantiomers have opposite chirality,* meaning opposite handedness. We also say that the molecules of any given enantiomer are **chiral**—they possess handedness or chirality.

The opposite of chiral is **achiral.** An achiral molecule is one that is symmetrical enough so that its model and the model of its mirror image do superimpose. The methane molecule is achiral, for example. Some examples of larger achiral objects include a cube, a sphere, a broom handle, and a water glass.

Chirality Can Make Enormous Differences at the Molecular Level of Life

One asparagine enantiomer tastes sweet and the other tastes bitter—a large (although a somewhat trivial) difference that chirality can make. The details are not fully known, but the taste mechanism probably begins with a chemical reaction that is catalyzed by an enzyme. The enzymes involved, *like all enzymes,* themselves consist of very large, chiral molecules with molecular surfaces that are different for each enzyme.

The substance whose reaction an enzyme catalyzes is called the **substrate** for that enzyme. An enzyme works by letting a molecule of the substrate come and temporarily fit into the contours on the enzyme's surface. This idea of fitting can be illustrated by a return to our hands, only now we'll add gloves. Gloves, like hands, are chiral, and a glove fits well only to its matching hand. The left hand fits well into the left glove, not the right glove. Now suppose that an enzyme responsible ultimately for the sensation of a sweet taste is like a glove for the right hand. This means that only the substrate molecules that have the matching handedness can interact with this enzyme. Substrate molecules of the opposite handedness can't fit to this enzyme.

We can now shift back from this analogy of hands and gloves to chiral molecules with the aid of Figure 18.4. To make it easier, we have used simple geometric forms to create two enantiomers, and indentations that match these forms are part of the enzyme surface. One enantiomer can fit to the enzyme surface, but no matter how you turn the model of the other enantiomer you can't get it to fit to the same enzyme surface.

There is, evidently, a different enzyme whose surface chirality matches the other asparagine enantiomer that lets us know that this other enantiomer has a different taste. The

A left-hand glove does not fit the right hand.

FIGURE 18.4
Because an enzyme is chiral, it can accept substrate molecules of only one of a pair of enantiomers. To illustrate this difference, we have used simple geometric forms. On the left, the enzyme can accept as a substrate the molecule of one enantiomer. On the right, the same enzyme can't accept a molecule of the other enantiomer, because the shapes don't match.

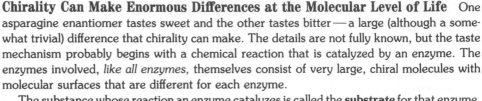
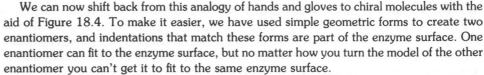
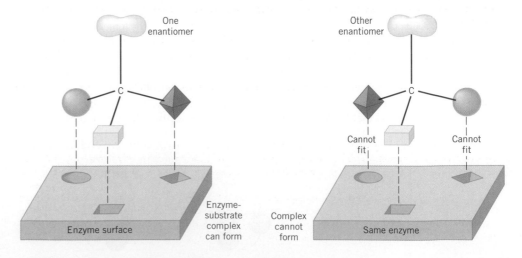

phenomenon of chirality is absolutely central to this difference. And the different chemical properties that relate to the taste of asparagine are illustrated in countless ways at the molecular level of life. We'll see time after time that *differences in geometry and configuration are as important as functional groups to the chemical reactions of life.*

Chirality Does Not Affect Reactions with Achiral Compounds If the enzyme molecule were not chiral, it could not discriminate between enantiomers. In fact, *enantiomers have identical chemical properties toward all reactants whose molecules are achiral* — reactants such as H_2O, NaOH, HCl, Cl_2, NH_3, and H_2. A broom handle, which isn't chiral, is an analogy. It fits just as easily to the left hand as to the right. It can't discriminate between the hands. In like manner, *the molecules of an achiral reactant cannot tell the difference between molecules of enantiomers.* To summarize, enantiomers react differently toward reactants whose molecules are chiral but identically toward achiral reactants.

Two Enantiomers Have Identical Physical Properties, with One Exception Two enantiomers have identical melting points, boiling points, densities, and solubilities in common (i.e., achiral) solvents like water, diethyl ether, or ethanol. This *must* be so, because the molecules of two enantiomers *must* have identical molecular polarities. They have, after all, identical intranuclear distances and bond angles.

Returning once again to our analogy with the hands, you can see that the distance, say, from the tip of the thumb to the tip of the little finger is the same in both hands. You can pick any other such intrahand distance that you please, and it is the same in both hands. Similarly, the angle between any two fingers is the same in both hands when the hands are spread identically. In like manner, if we pick any distance between atoms or any bond angle in one asparagine enantiomer, we will find it to be the same in the other. When all these distances and angles are identical in the two enantiomers, their molecules as a whole must have identical polarities. This is responsible for the identical physical properties that we mentioned. There is one difference in physical property, however, which we will study in Section 18.3.

Molecules with One Carbon Holding Four Different Groups Are Chiral It is important that we be able to recognize when a potential substrate is chiral, because molecules of opposite chirality have such different chemical properties at the molecular level of life. How, then, can we look at a structure and tell whether its molecules are chiral without making molecular models of both object and mirror image?

In all examples of chiral molecules that we'll encounter, their molecules always have at least one carbon to which *four different groups* are attached. A carbon that holds four different atoms or groups is called a **tetrahedral stereocenter.** A *stereocenter* is an atom in a molecule whose attached groups can be arranged in different configurations to give different stereo-isomers. A *tetrahedral* stereocenter is simply a stereocenter having four single bonds arranged tetrahedrally.[1]

The asparagine molecule has one (and just one) tetrahedral stereocenter, and when a molecule has only one such center, we can be absolutely certain that the molecule as a whole is itself chiral. Here, then, is one way to predict if a substance consists of chiral molecules; we look for a tetrahedral stereocenter. If we find *one,* we can be certain that the molecules are chiral.[2]

■ An older (but still widely used) name for *tetrahedral stereocenter* is **chiral atom.** When the atom is C, the term is *chiral carbon.*

---

[1] Alkenes that are cis and trans isomers of each other (e.g., the geometric isomers of 2-butene) have two stereocenters, two carbon atoms each holding *three* atoms or groups that, when differently arranged, give the two stereoisomers — the cis and the trans. The two carbon atoms at the double bond are thus not tetrahedral stereo-centers but *trigonal* stereocenters.

[2] If we find more than one tetrahedral stereocenter, we have to be careful. The complication is treated in Special Topic 18.1. Probably 99.99% of all examples of substances that have two or more tetrahedral stereocenters also have chiral molecules. A few exceptions, however, exist where the molecule is achiral despite having tetrahedral stereocenters. These are the *meso compounds* discussed in Special Topic 18.1.

MESO COMPOUNDS AND DIASTEREOMERS

Tartaric acid is a normal constituent of grapes. During the fermentation of grape juices into wine, the monopotassium salt separates as an insoluble substance called "tartar."

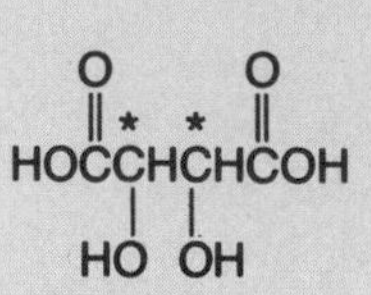

$$\underset{\text{Tartaric acid}}{\overset{\displaystyle \overset{O}{\underset{*}{\|}} \qquad \overset{O}{\underset{*}{\|}}}{HOCCHCHCOH}}$$

$$\underset{}{\overset{}{\qquad\; HO\;\; OH}}$$

Tartaric acid

Tartaric acid molecules have two tetrahedral stereocenters. They are *identical* because the sets of four atoms or groups attached to each are identical; both centers hold HO, H, CO_2H, and $CH(OH)CO_2H$. When we prepare perspective drawings of the structures of all of the possible stereoisomers that have the same constitutions as tartaric acid, we obtain those shown below, where solid wedges denote bonds coming forward and dashed-line wedges mean bonds going rearward. You can see that the first two structures, those of D- and L-tartaric acid, are related as an object to its mirror image and that by no allowed manipulation can the two be superimposed. The two are thus *enantiomers*. Notice that they have identical melting points and identical degrees but opposite signs of specific rotation.

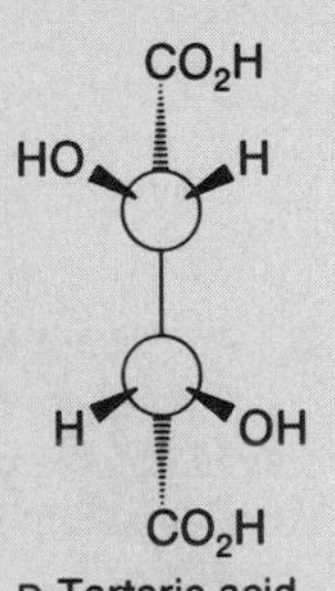

D-Tartaric acid
$[\alpha]_D^{20} -11.98°$
mp 170°C

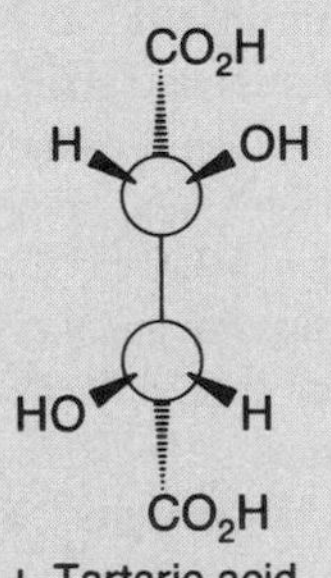

L-Tartaric acid
$[\alpha]_D^{20} +11.98°$
mp 170°C

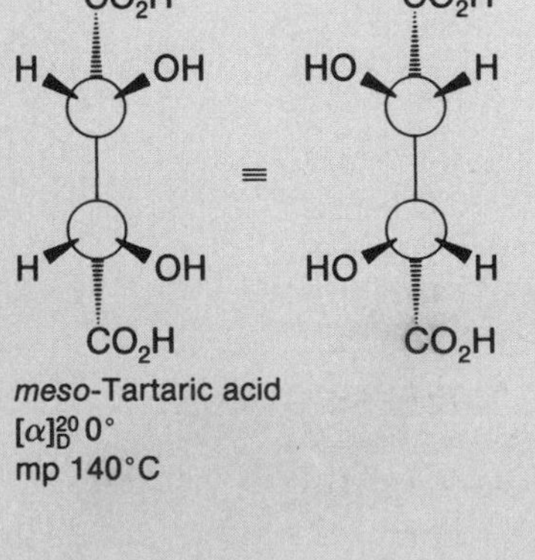

meso-Tartaric acid
$[\alpha]_D^{20} 0°$
mp 140°C

Only one of the remaining two perspective structures has been given a name—*meso*-tartaric acid. Although its mirror image structure has been drawn to make a point, the object and mirror image are actually superimposable. If you rotate the mirror image of *meso*-tartaric acid by 180° in the plane of the paper and around an axis perpendicular to the plane (piercing the paper at the bond *between* the two carbons), you soon realize that the two structures are identical (and so superimposable). Despite having two tetrahedral stereocenters, the *meso*-tartaric acid molecule is achiral. Therefore *meso*-tartaric acid is optically inactive; it has zero specific rotation. Optical activity *requires* chirality, and *meso*-tartaric acid doesn't have it. According to our definitions, because *meso*-tartaric acid is a nonenantiomeric isomer of a pair of enantiomers, it is a *diastereomer* of each enantiomer.

meso-Tartaric is one example of a common occurrence and, historically, it gave part of its name to the kind of diastereomer it exemplifies. An achiral diastereomer of a set of stereoisomers that includes some that are chiral is called the **meso** isomer. There are thus only three stereoisomers of tartaric acid. The equation, 2^n = number of stereoisomers, cannot be applied to tartaric acid because it works only when the *n* tetrahedral stereocenters are *different*, when the sets of four atoms or groups at the center differ in at least one way. No general equation exists for calculating the number of stereoisomers when two (or more) stereocenters are identical.

In Example 18.2, the structures of the four stereoisomers of threonine were shown. (Two bear the name "threonine" and two have the name "allothreonine.") Their molecules also have two tetrahedral stereocenters, but the centers are not identical. Hence the correct number of stereoisomers is predicted by the equation we have used. Each of the threonine enantiomers is a diastereomer of each of the allothreonine enantiomers.

Diastereomers do not have identical physical or chemical properties (although the chemical properties will be very similar to all achiral reactants). You can see in the threonine—allothreonine system how melting points and specific rotations for diastereomers are different. The set of intramolecular distances and bond angles in one diastereomer isn't exactly duplicated in any other diastereomer. Hence, molecules of one diastereomer should be expected to have at least slightly different polarities than those of any other diastereomer in the set. Such differences cannot help but cause differences in physical properties.

EXAMPLE 18.1
Identifying Tetrahedral Stereocenters

Amphetamine exists as a pair of enantiomers. One of them has its own name — Dexedrine. Find the tetrahedral stereocenter in amphetamine, and list the four groups attached to it.

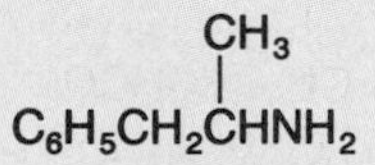

$$CH_3$$
$$C_6H_5CH_2CHNH_2$$
Amphetamine

ANALYSIS A tetrahedral stereocenter has *four different* attached atoms or groups.

SOLUTION Amphetamine has one tetrahedral stereocenter labeled with an asterisk:

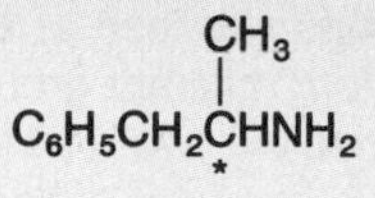

$$CH_3$$
$$C_6H_5CH_2\overset{*}{C}HNH_2$$

The four groups are $C_6H_5CH_2$, H, CH_3, and NH_2.

■ Amphetamine, a stimulant, is a controlled substance in the United States.

PRACTICE EXERCISE 1

Place an asterisk next to each tetrahedral stereocenter in the following structures.

(a)

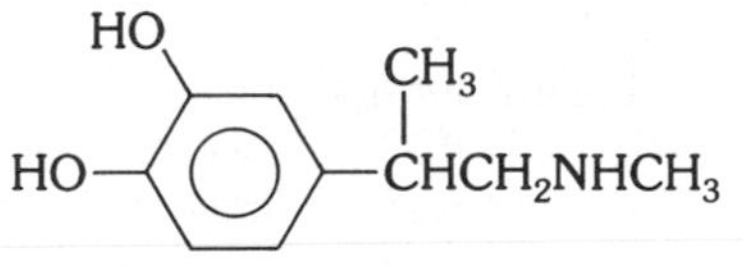

HO

CH_3

HO—⟨ ⟩—$CHCH_2NHCH_3$

Epinephrine, a hormone (See Special Topic 17.1.)

(b) 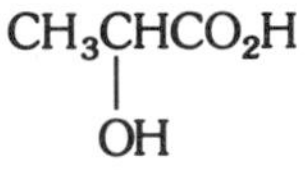 CH_3CHCO_2H
$\quad\quad\;|$
$\quad\quad\,OH$

Lactic acid, the sour constituent in sour milk

(c) 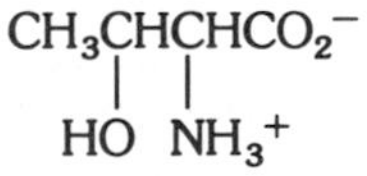 $CH_3CHCHCO_2^-$
$\quad\quad\;|\;\;|$
$\quad\;\;HO\;\;NH_3^+$

Threonine, one of the amino acid building blocks of proteins

(d) 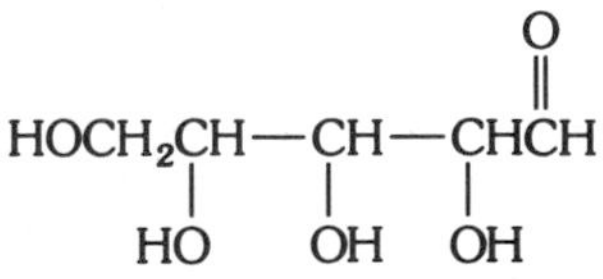
$$\qquad\qquad\qquad\overset{\textstyle O}{\overset{\|}{}}$$
$HOCH_2CH—CH—CHCH$
$\qquad\;|\quad\;\;|\quad\;\;|$
$\qquad HO\quad OH\quad OH$

Ribose, a sugar unit in one of the two kinds of nucleic acids (ribonucleic acid or RNA)

A Molecule with *n* Different Tetrahedral Stereocenters Has 2^n Stereoisomers

When a molecule has two or more tetrahedral stereocenters, as in parts (c) and (d) of Practice Exercise 1, then it becomes useful to judge whether these centers are *different*. When used in this context, *different* means that the sets of four atoms or groups at the various tetrahedral stereocenters have at least one difference. Two tetrahedral stereocenters are said to be *different* if the set of four groups at one is not duplicated by the set at the other. Whenever the tetrahedral stereocenters in a molecule are different in this sense — as they were in parts (c) and (d) of Practice Exercise 1 — then the substance can exist in the forms of 2^n stereoisomers, where *n* is the number of different tetrahedral stereocenters. These 2^n stereoisomers occur as half as many *pairs* of enantiomers. We'll see this illustrated in the next example.

EXAMPLE 18.2
Judging Whether Tetrahedral Stereocenters Are Different and Calculating the Number of Stereoisomers

The threonine molecule, part (c) of Practice Exercise 1, has two tetrahedral stereo-centers, labeled here by asterisks.

$$CH_3\overset{*}{C}H\overset{*}{C}HCO_2^-$$
$$\underset{HO}{|} \quad \underset{NH_3^+}{|}$$

Threonine

Are these tetrahedral stereocenters different? If so, how many stereoisomers of threonine are there?

ANALYSIS To compare the groups attached at each tetrahedral stereocenter, we should make a list of the sets of four different groups and compare them. If the lists aren't identical in every respect, then the two tetrahedral stereocenters are different.

At one tetrahedral stereocenter:

CH₃   H

HO   CHCO₂⁻
        |
        NH₃⁺

At the other tetrahedral stereocenter:

CH₃CH   H
    |
    HO

CO₂⁻   NH₃⁺

SOLUTION The sets are obviously different, so $n = 2$ is the number of different tetra-hedral stereocenters in a threonine molecule. Therefore, $2^n = 2^2 = 4$, the number of stereoisomers of threonine. These occur as half of 4 or 2 pairs of enantiomers. The complete set has the following structures. One pair of enantiomers is on the left. Just imagine that the mirror is between them and is perpendicular to the page. The other pair of enantiomers is on the right.

■ Only L-threonine works as a building block for making proteins in the body. There is no enzyme that can accept any of the other optical isomers as substrates.

■ The labels D and L will be explained in the next chapter, as we said. The meaning of the experimental values given for the symbol $[\alpha]_D^{26}$ is explained in the next section.

D-Threonine
$[\alpha]_D^{26} +28.3°$

L-Threonine
$[\alpha]_D^{26} -28.3°$

D-Allothreonine
$[\alpha]_D^{26} -9.6°$

L-Allothreonine
$[\alpha]_D^{26} +9.6°$

PRACTICE EXERCISE 2

Examine the structure of ribose that was given in part (d) of Practice Exercise 1. (a) How many different tetrahedral stereocenters does it have? (b) How many stereoisomers are there of this structure? (Only one is actually the ribose that can be used by the body.) (c) How many pairs of enantiomers correspond to this structure?

PRACTICE EXERCISE 3

Write the structure of 2,3-butanediol and place an asterisk by each carbon that is a tetrahedral stereocenter. Are they *different* tetrahedral stereocenters?

18.3

OPTICAL ACTIVITY

The members of a pair of enantiomers affect polarized light in equal and opposite ways when compared under identical conditions.

We mentioned earlier that the two members of a pair of enantiomers differ in one physical property, and to describe it we first have to learn something about polarized light.

The Electromagnetic Oscillations of Polarized Light Are All in the Same Plane

Light is electromagnetic radiation in which the intensities of the electric and magnetic fields set up by the light source oscillate in a regular way. In ordinary light, these oscillations occur equally in all directions about the line that defines the path of the light ray.

Certain materials, such as the polarizing film in the lenses of Polaroid sunglasses, affect ordinary light in a special way. Polarizing film interacts with the oscillating electrical field of any light passing through it to make this field oscillate *in just one plane.* The light that emerges is now **plane-polarized light** (see Figures 18.5 and 18.6*a*).

If we look at some object through polarizing film and then place a second film in front of the first, we can rotate one film until the object can no longer be seen (see Figures 18.6*b* and 18.7). If we now rotate one film by 90°, we'll see the object at maximum brightness again. The first film seems to act as a lattice fence, forcing any light that goes through it to vibrate only in the direction allowed by the long spaces between the slats. This light then moves on to the molecular slats of the second film. If the second film's slats are perpendicular to those of the first, the light has no freedom to oscillate, and it cannot get through the second film. At intermediate angles, fractional amounts of light can go through the second film. Only when the slats of both films are *parallel* to each other can the light leaving the first film slip easily through the second film with its maximum intensity.

■ You can try this out using two Polaroid sunglass lenses. The lenses of these glasses reduce glare by cutting out the plane-polarized light produced when sunlight reflects from a plane surface such as a road, a snowfield, or a lake.

An Enantiomer Can Rotate the Plane of Plane-Polarized Light

When a solution of D-asparagine in water is placed in the path of plane-polarized light, the *plane* of polarization is

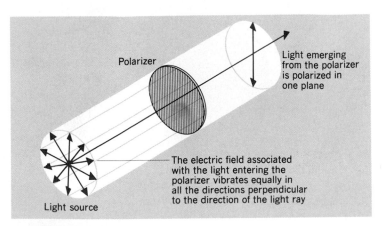

Polarizer

Light emerging from the polarizer is polarized in one plane

The electric field associated with the light entering the polarizer vibrates equally in all the directions perpendicular to the direction of the light ray

Light source

FIGURE 18.5
When light passes through polarizing film, it becomes polarized light.

FIGURE 18.6

The principal working parts of a polarimeter and how optical rotation can be measured.

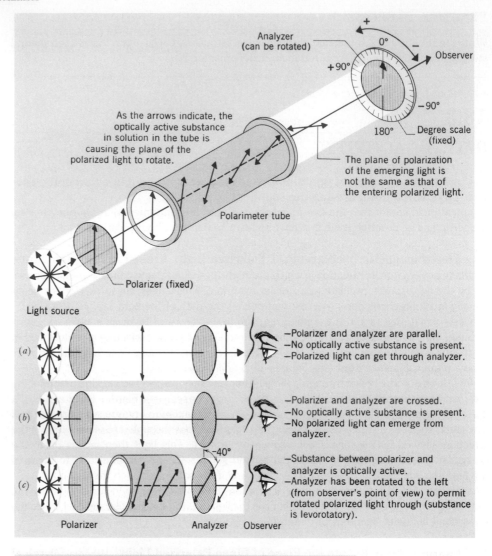

As the arrows indicate, the optically active substance in solution in the tube is causing the plane of the polarized light to rotate.

Analyzer (can be rotated)

Observer

Degree scale (fixed)

The plane of polarization of the emerging light is not the same as that of the entering polarized light.

Polarimeter tube

Polarizer (fixed)

Light source

(a)
—Polarizer and analyzer are parallel.
—No optically active substance is present.
—Polarized light can get through analyzer.

(b)
—Polarizer and analyzer are crossed.
—No optically active substance is present.
—No polarized light can emerge from analyzer.

(c)
—Substance between polarizer and analyzer is optically active.
—Analyzer has been rotated to the left (from observer's point of view) to permit rotated polarized light through (substance is levorotatory).

Polarizer   Analyzer   Observer

FIGURE 18.7

When two polarizing films are "crossed," no light can get through to the observer.

twisted or rotated. Any substance that can rotate the plane of plane-polarized light is said to be **optically active.** We have not actually explained *how* this phenomenon happens — we are unable to do so. We have only reported that it does take place. Quite often the members of a set of stereoisomers in which any or all are optically active are called **optical isomers.** Thus the two asparagines are optical isomers. All of the threonine stereoisomers of Example 18.2 are all likewise optical isomers of each other.

The *Polarimeter* Measures the Degree of Optical Activity

The instrument used to detect and measure optical activity is called a **polarimeter** (see Fig. 18.6). Its principal working parts consist of a *polarizer* for the light beam, a tube for holding solutions in the path of the polarized light, an *analyzer* (actually, just another polarizing device), and a circular scale for measuring the number of degrees of rotation. When the "slats" of the polarizer and the analyzer are parallel and the tube contains no optically active material, the polarized light emerges from the analyzer with maximum intensity. Let's assume that we start with this parallel orientation of polarizer and analyzer.

When a solution of one pure enantiomer is placed in the light path, the plane-polarized light encounters molecules of just one chirality. They cause the plane of oscillation of the polarized light to be rotated. The plane of oscillation of the polarized light that leaves the solution is now no longer parallel with the analyzer (see Fig. 18.6c). Consequently, not as much light gets through the analyzer, so the observed light intensity is now less than the original maximum. To restore the original intensity, the operator can rotate the analyzer to the right or to the left a definite number of degrees until the analyzer is once again parallel *with the light that emerges from the tube.*[3]

The operator, looking *toward* the light source, might find that rotating the analyzer to the right (clockwise) restores the original light intensity with fewer degrees of rotation than rotating the analyzer to the left (counterclockwise). When such a rightward rotation works, the degrees are recorded as positive, and the optically active substance is said to be **dextrorotatory.** If the fewer degrees of rotation are found by a leftward rotation, then the degrees are recorded as negative and the substance is said to be **levorotatory.** In Figure 18.6c, the reading is $\alpha = -40°$, where α (including the plus or minus sign) stands for the observed **optical rotation,** the observed number of degrees of rotation caused by the solution.

■ Latin, *dextro,* right, and *levo,* left.

The value of α varies both with the temperature of the solution and the frequency of the light used, but not in any simple direct way. Consequently, when α is recorded, both the temperature and the light frequency must also be recorded.

The *Specific Rotation* of an Optically Active Compound Is One of Its Physical Constants, Like Its Density, Boiling Point, or Melting Point

The observed rotation is related in simple ways to the concentration of the solution and the length of the tube.

$$\alpha \propto \text{concentration}$$
$$\alpha \propto \text{path length}$$

What both of these proportionalities are really saying is that the degree of rotation of the plane-polarized light is a function of the *population* of the chiral molecules. Either by increasing the concentration or by making the tube longer, we can force the polarized light to be in greater contact with chiral molecules.

Because α is *directly* proportional both to concentration (c) and to path length (l), α is proportional to the products of these two.

■ This is a rare use of the decimeter unit in chemistry.

1 decimeter (dm)
  = 10 centimeters (cm)

$$\alpha \propto cl \qquad (18.1)$$

The unit used for length, *l*, is the decimeter.

---

[3] This description of the measurement emphasizes only the essential principle involved. The operator has other options that yield identical results.

We can convert expression 18.1 to an equation by inserting a constant of proportionality, which is given the symbol $[\alpha]$ and the name **specific rotation.**

$$\alpha = [\alpha]cl \tag{18.2}$$

- Be sure to notice that the units of *c* are g/mL, not g/100 mL. If an optically active pure liquid is in the polarimeter tube, then the units of its concentration equal those of its density.

The unit traditionally used for concentration when reporting specific rotations is g/mL. As we said earlier, both the temperature (*t*) and the light frequency (λ) must be reported with any value of specific rotation. By rearranging Equation 18.2 and placing symbols for *t* and λ by the closing bracket, we obtain the usual form of the definition of specific rotation.

$$[\alpha]_\lambda^t = \frac{\alpha}{cl} \tag{18.3}$$

- L-Asparagine is the more common form.

Quite often polarimeters are used with the intense yellow light of a sodium vapor lamp like those that illuminate the streets in some cities. The symbol for this light is D, so when we write that $[\alpha]_D^{20} = +5.42°$ for D-asparagine, we mean that the temperature of the solution was 20 °C and that a sodium vapor lamp was used. For L-asparagine, $[\alpha]_D^{20} = -5.42°$. Thus we see that the *only* physical difference between the two enantiomers of asparagine is the *direction* of the rotation of the plane of plane polarized light. The numbers of degrees are identical; only the signs of rotation are opposite. Any pair of enantiomers is like this. All physical properties of a pair of enantiomers are the same — densities, boiling points, and melting points, for example — but the *signs* of the numerically identical degrees of rotation are opposite.

Specific Rotation Provides an Analytical Tool   The specific rotation of an optically active compound is an important physical constant, comparable to its melting point, boiling point, or density. If we know the value of $[\alpha]_\lambda^t$ for a compound, it's easy to see from Equation 18.3 that we have a way to determine the concentration of a solution. We measure the observed rotation, α, for the solution when it is in a tube of known path length, *l*, and use these data together with the specific rotation to calculate the concentration, *c*.

Sometimes the measurement of optical activity is used to identify a substance. A measurement of the observed rotation, α, of a solution of known concentration, *c*, in a tube of known path length, *l*, is made and the specific rotation is calculated. The calculated value is then compared to a table of specific rotations to see what matches.

A 50 : 50 Mixture of Enantiomers Is Optically Inactive   It is important to realize that *optical activity* and *optical isomerism* are not the same. *Optical activity* refers to a phenomenon observable with a special instrument, and what we see is the number of degrees of rotation of the solution. *Optical isomerism* is part of our explanation of optical activity. We *infer* optical isomerism from the observation of optical activity.

What happens if we mix two enantiomers together in a 1 : 1 ratio? Now as the plane-polarized light travels through the tube it encounters some molecules forcing its plane to twist to the right but it meets an identical number forcing its plane to twist just as much to the left. *The result is no net change to the plane of vibration of the plane-polarized light.* The operator of the polarimeter would be bound by the definitions that we have introduced to report that the substance in the tube is *optically inactive.* Any 50 : 50 mixture of enantiomers is optically inactive and is called a **racemic mixture.** Thus a substance, like a racemic mixture, can be made entirely of chiral molecules and yet be optically inactive.

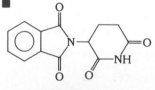

Thalidomide

- Thalidomide was never approved for use in the United States because Dr. Frances Oldham Kelsey of the U.S. Food and Drug Administration insisted *before* birth defects appeared in Europe that the testing of thalidomide had not been thorough enough.

Enantiomers Normally Have Large Differences in Biological Properties   It's both interesting and significant that if a racemic mixture of asparagine enantiomers is given in the diet, the body can use only one. This is generally true about pairs of enantiomers. If the body uses one enantiomer, it cannot use the other, which may even be a very dangerous substance. Thalidomide, for example, can exist as a pair of enantiomers. (Can you spot the tetrahedral stereocenter?) It was once widely prescribed in Europe as a sedative-tranquilizer, particularly for pregnant women. Tragically, the prescribed drug was the racemic mixture and only one enantiomer gives the desired effect. The other enantiomer disrupts fetal development during the first 12 weeks of pregnancy causing phocomelia — seal or flipperlike arms — and often abnormalities of the digestive tract, the eyes, and the ears. Between 1959 and 1962, from 2000 to 3000 babies were born in Germany with such thalidomide-caused problems. The drug was withdrawn from the market but not before many tragic births had occurred. The episode furnishes a dramatic example of how great the differences can be between enantiomers, even though their molecules are so nearly identical as to be related as object and mirror image.

■ In 1993, thalidomide was found to reduce the rate of activation of HIV-1, the virus that causes AIDS.

SUMMARY

Stereoisomerism Stereoisomers are isomers whose molecules have identical constitutions but different geometries. The two kinds of stereoisomers are enantiomers and diastereomers. Enantiomers are pairs of stereoisomers whose molecules are mirror images but they do not superimpose. Diastereomers are stereoisomers that are not related as enantiomers. Almost always, an enantiomer molecule has a tetrahedral stereocenter, which is usually a carbon atom holding four different atoms or groups. If a molecule has n *different* tetrahedral stereocenters, then the number of stereoisomers is 2^n, and these occur as half as many pairs of enantiomers. A 50:50 mixture of enantiomers, a racemic mixture, is optically inactive.

Optical activity Optical activity is a natural phenomenon detected by means of a polarimeter. A substance is optically active if polarized light that passes through a substance or its solution undergoes a rotation in its plane of polarization.

Specific rotation The specific rotation of an optically active substance is what its observed rotation is at one unit of concentration (1 g/mL) in one unit of path length (1 dm). It varies, but not in a simple way, with the temperature and the wavelength of the light used. Values of specific rotation can be used to determine concentrations of optically active substances.

Properties of enantiomers Enantiomers are identical in *every* physical property except the signs of their specific rotation. They are also identical in every chemical respect provided that the molecules or ions of the reactant are achiral. When the reactant particles are chiral, then one enantiomer reacts differently with the reactant than the other enantiomer, a phenomenon always observed when an enzyme is acting as a (temporary) reactant.

REVIEW EXERCISES

The answers to Review Exercises whose numbers are in color are found in Appendix C. The answers to the other Review Exercises are found in the Study Guide that accompanies this book. The more challenging questions are marked with asterisks.

Structural Isomers and Stereoisomers

18.1 What specifically must be true about two compounds before they can be called constitutional isomers?

18.2 What are the structures of the *simplest* alcohols that can exhibit constitutional isomerism?

18.3 What must be true about two compounds before they can be called stereoisomers?

18.4 There are two kinds of stereoisomers. What are their names and how is each kind defined?

18.5 Classify the following pairs of structures as constitutional isomers or as stereoisomers.

(a)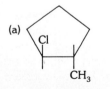

(b)

18.6 What is the "one-substance–one-structure" principle?

Optical Isomers

18.7 The general structure of several stereoisomers that include glucose is that of 2,3,4,5,6-pentahydroxyhexanal:

$$HOCH_2-CH-CH-CH-CH-CH=O$$
$$\quad\quad\ \ | \quad\ | \quad\ | \quad\ |$$
$$\quad\quad\ \ OH\ \ OH\ \ OH\ \ OH$$

(a) Place an asterisk by each tetrahedral stereocenter.
(b) How many of these tetrahedral stereocenters qualify as *different* tetrahedral stereocenters?
(c) How many stereoisomers of this compound are possible?
(d) These optical isomers occur as how many *pairs* of enantiomers?

18.8 One of the approximately 20 building blocks of protein molecules is glycine: $^+NH_3CH_2CO_2^-$. Does glycine have stereoisomers? How can you tell?

18.9 One of the important intermediate substances in the body's energy-producing metabolism is an anion of the following acid, citric acid.

$$CH_2CO_2H$$
$$HOCCO_2H$$
$$CH_2CO_2H$$

(a) Does citric acid have stereoisomers? How can you tell?

(b) One of the possible monomethyl esters of citric acid has chiral molecules. Write its structure. Place an asterisk by its tetrahedral stereocenter.

Properties of Enantiomers

18.10 The melting point of (−)-cholesterol is 148.5 °C. What is the melting point of (+)-cholesterol? How can we know what this melting point *must* be without actually making the measurement?

18.11 Explain why enantiomers should have identical physical properties (except for the sign of specific rotation).

18.12 Explain why enantiomers cannot have different chemical properties toward reactants whose molecules or ions are achiral. (Use an analogy if you wish.)

18.13 Explain why enantiomers have different chemical properties toward reactants whose molecules or ions are chiral. (Use an analogy if you wish.)

Specific Rotation

*18.14 Pantothenic acid was once called vitamin B_3.

$$H_3C \quad O$$
$$HOCH_2C—CHCNHCH_2CH_2CH_2CO_2H$$
$$H_3C \quad OH$$

Panthothenic acid

Only its dextrorotatory form can be used by the body, and its specific rotation is $[\alpha]_D^{25} = +37.5°$. In a tube 1.00 dm long and at a concentration of 1.00 g/100 mL, what is the *observed* optical rotation of the levorotatory enantiomer?

18.15 Ascorbic acid is also known as vitamin C. Only its dextrorotatory enantiomer can be used by the body.

$$HO \quad OH$$
$$HOCH_2CH \qquad =O$$
$$OH$$

Ascorbic acid
(vitamin C)

Its specific rotation is $[\alpha]_D^{25} = +21°$. What is the specific rotation of (−)-ascorbic acid?

18.16 Explain why the observed rotation is proportional to the concentration of the optically active compound and to the length of the tube through which the polarized light travels.

18.17 What becomes of the optical activity of a substance if it consists of a mixture of equal numbers of moles of its two enantiomers? Explain.

*18.18 A solution of sucrose (table sugar) in water at 25 °C in a tube that is 10.0 cm long gives an observed rotation of +2.00°. The specific rotation of sucrose in water at this temperature and the same wavelength of light is +66.4°. What is the concentration of sucrose in g/100 mL?

18.19 Quinine sulfate, an antimalarial drug, has a specific rotation in water at 17 °C of −214°. If a solution of quinine sulfate in this solvent in a 1.00-dm tube and under the same conditions of temperature and wavelength has an observed rotation of −10.4°, what is its concentration in g/100 mL?

*18.20 Strychnine and brucine are structurally similar compounds that have extremely bitter tastes, and both are very poisonous. The specific rotation in chloroform at 20 °C of brucine is −127° and that of strychnine under identical conditions is −139°. If a solution of one of these in chloroform at a concentration of 1.68 g/100 mL in a tube 1.00 dm long gave an observed rotation of −2.34°, which of the two compounds was it? Do the calculation.

18.21 Corticosterone and cortisone are two substances used to treat arthritis. Under identical conditions of solvent and temperature, the specific rotation of corticosterone is +223° and that of cortisone is +209°. If a solution of one of these at a concentration of 1.48 g/100 mL and in a tube 1 dm long has an observed rotation of +3.10°, which compound was in the solution? Do the calculation.

Other Optical Isomers (Special Topic 18.1)

18.22 Examine the structures of D-threonine and D-allothreonine as given in Example 18.2.

(a) Why don't these qualify as enantiomers?

(b) Why are they described as being members of the *same set* of optical isomers?

(c) Why aren't they called geometric isomers?

(d) What kind of stereoisomers are they?

18.23 Two models of methane, related as object to mirror image, superimpose. Why isn't methane called a meso compound?

18.24 Although *meso*-tartaric acid is optically inactive, it is described as an *optical isomer* of D- or L-tartaric acid. Why?

Additional Exercises

18.25 Explain how a substance could consist entirely of chiral molecules and yet be optically inactive.

18.26 What is the IUPAC name of each of the following?

(a) The monohydric alcohol of lowest formula mass whose molecules are chiral.

(b) The alkane of lowest formula mass whose molecules are chiral.

*18.27 There are *four* stereoisomers of the following compound. Explain.

$$CH_3CH=CHCHCH_3$$
$$OH$$

CARBOHYDRATES

19

In what a wonderful, roundabout way do we tap into solar energy! The chemical energy in the milk we drink comes through cows that feed on grass made possible by photosynthesis, one of the Special Topics in this chapter.

19.1

BIOCHEMISTRY — AN OVERVIEW

Building materials, information, and energy and are basic essentials for life.
Biochemistry is the systematic study of the chemicals of living systems, their organization, and the principles of their participation in the processes of life.

The Cell Is the Smallest Unit That Lives The molecules of living systems are lifeless, yet life has a molecular basis. Whether studied in cells or when isolated from them, the chemicals at the foundation of life obey all of the known laws of chemistry and physics. Yet, in isolation, not one compound of a cell has life. The intricate *organization* of compounds in a cell is as important to life as the chemicals themselves. Thus the cell is the smallest unit of matter that lives and that, in the proper environment, can make a new cell like itself.

The Life of a Cell Requires Materials, Information, and Energy Our purpose in the remainder of this book is to study the molecular basis of meeting the three basic needs of a living system, its needs for materials, information, and energy. Without the daily satisfaction of these, life at any of the many loftier levels, like creativity, relationships, and love, would be severely constrained. Most of our focus will be on the molecular basis of life in the human body.

We will begin in the next three chapters to study the organic materials of life, starting with the three main classes of foodstuffs: carbohydrates, lipids, and proteins. We use their molecules to build and run our bodies and to try to stay in some state of repair. Plants rely heavily on carbohydrates for cell walls, and animals obtain considerable energy from carbohydrates made by plants. Lipids (fats and oils) serve many purposes. They are used as materials for cell membranes, and as sources of chemical energy. Proteins are particularly important in both the structures and functions of cells, whether of plants or of animals. Because of the central catalytic role of proteins in regulating chemical events in cells, we will immediately follow our study of proteins with an examination of a particular family of proteins, the enzymes. We've mentioned them often as the special catalysts in living systems.

The Circulatory System Delivers Needed Compounds and Carries Wastes Away
Few of the substances in the diet are in forms directly usable by our bodies. Carbohydrates, lipids, and proteins must be broken down (hydrolyzed) to much smaller molecules. This work is done by the enzymes of our digestive juices, and the chemical reactions of digestion will be a part of our study. The small molecular products of digestion are delivered into circulation via the bloodstream, the vital conduit on which all tissues depend for raw materials and oxygen, for chemical signals such as hormones, for disease-fighting agents, and for the removal of wastes. One of the special emphases, already begun in our study, is the molecular basis for using oxygen and releasing carbon dioxide during metabolism.

Every Cell Has an Information System Enzymes, hormones, and neurotransmitters are components of the intricate information system in an organism. Without information — plans or blueprints — materials and energy could combine to produce only rubble and rubbish.

■ Cornstarch, potato starch, table sugar, and cotton are all carbohydrates.

■ Butter, lard, margarine, and corn oil are all examples of lipids known as fats and oils.

■ The reactions of *digestion* process food molecules into (usually) smaller molecules which then enter circulation and later participate in *metabolism.*

■ Neurotransmitters carry chemical signals from one nerve cell to the next.

Monkeys swinging hammers would only reduce a stack of lumber to splinters. Carpenters, using the same materials and expending no more raw energy, can build a building, because they possess information in the form of plans and experience.

Although enzymes are elements in the cell's information system, enzymes do not *originate* the blueprints. They only help to carry them out. The blueprint for any one member of a species is encoded in the molecular structures of its nucleic acids. These compounds are able to direct the synthesis of a cell's enzymes. Hormones and neurotransmitters, other elements of cellular information, depend on the presence of the right enzymes for their own existence. Thus a study of the enzyme-makers must be included in any study of the molecular basis of life. The study of nucleic acids will help us to see how different species can take essentially the same raw materials and energy, synthesize their enzymes, and thus lay the basis for making everything else needed.

■ The field of *molecular biology* deals with the work of nucleic acids, their synthesis, and uses.

There is both contentment and chemical energy in a peanut butter sandwich.

Some Materials Are Used Mainly for Their Chemical Energy The molecular basis of energy for life is another broad topic of our study. One of the kinds of questions that we will address is "How can one get the energy for running, skipping, and laughing out of a sandwich?" As we study biochemical energetics and its enzymes and metabolic pathways, we'll have numerous occasions to peer deeply into the molecular basis of some disorders and diseases.

To supply materials for any use—parts, information, or energy—each organism has basic nutritional needs. These include not just organic materials, but also minerals, water, and oxygen. Thus, after learning about the materials of life and how they are processed and used, we will close our study of the molecular basis of life with a broad survey of nutritional needs.

We Launch a New Beginning with the Study of Carbohydrates We have an exciting trip ahead. In the preceding chapters we have slowly and carefully built a solid foundation of chemical principles. It's been like a mountain-climbing trip where the route for a large part of the trek is through country with few grand vistas, and yet with a beauty of its own. Now we're moving to elevations where the vistas begin to open. It's like a new beginning, and we start it with a study of the first of the three chief classes of food materials, the carbohydrates.

■ When people *understand* how much health depends on the timely coming together of all of the proper substances, they become more interested in good nutrition.

19.2
INTRODUCTION TO THE MONOSACCHARIDES

The monosaccharides—polyhydroxyaldehydes and ketones—are carbohydrates that are not converted into substances with smaller molecules by hydrolysis.

Carbohydrates are aldehydes and ketones with many OH groups, or substances that form these when hydrolyzed. They include the simple sugars, like glucose, as well as table sugar, starch, and cellulose. Carbohydrates are the primary products of **photosynthesis,** the complex series of reactions in plants by which CO_2, H_2O, and minerals are converted to plant chemicals and oxygen using the solar energy absorbed by the green pigment, chlorophyll. Special Topic 19.1 discusses photosynthesis further.

The Simple Sugars Do Not React with Water The carbohydrates that cannot be hydrolyzed are called the **monosaccharides** or **simple sugars.** Their empirical formula is $(CH_2O)_n$. Those with aldehyde groups are called **aldoses** and those with keto groups are **ketoses.** Like these terms, the names of virtually all carbohydrates end in *-ose.*

Whether they are aldoses or ketoses, monosaccharides with three carbons are trioses, those with four are tetroses, and this pattern continues with pentoses, hexoses, and higher

■ The oxidized and reduced forms of polyhydroxy aldehydes and ketones as well as certain amino derivatives are also in the family of carbohydrates.

The energy released when a piece of wood burns came originally from the sun. The wood, of course, isn't just bottled sunlight. It's a complex, highly organized mixture of compounds, mostly organic. The solar energy needed to make these compounds is temporarily stored in wood in the form of distinctive arrangements of electrons and nuclei that characterize energy-rich molecules.

They are made from very simple, energy-poor substances such as carbon dioxide, water, and soil minerals. In the living world only plants have the ability to use solar energy to convert energy-poor substances into complex, energy-rich, organic compounds. The overall process by which plants do this is called **photosynthesis.**

The simplest statement of photosynthesis in equation form is

$$nCO_2 + nH_2O + \frac{solar}{energy} \xrightarrow[\text{plant enzymes}]{\text{chlorophyll}} (CH_2O)_n + nO_2$$

To make glucose, a hexose, n must equal 6. The symbol (CH_2O) stands for a molecular unit in carbohydrates, but plants can use the energy of carbohydrates (which came from the sun) to make other substances as well—proteins, lipids, and many others. In the final analysis, the synthesis of all the materials in our bodies consumes solar energy, and all our activities that use energy ultimately depend on a steady flow of solar energy through plants to the plant materials we eat. The meat and dairy products in our diets also depend on the consumption of plants by animals.

Chlorophyll is the green pigment in the solar-absorbing systems of plants, usually their leaves. Chlorophyll mole-cules absorb solar energy and, in their energized states, trigger the subsequent reactions leading to carbohydrates. A large number of steps and several enzymes are involved. The rate of photosynthesis increases as the air temperature increases and as the concentration of CO_2 in air increases.

Notice that another product of photosynthesis is oxygen, and this process continuously regenerates the world's oxygen supply. Roughly 400 billion tons of oxygen are set free by photosynthesis each year, and about 200 billion tons of carbon (as CO_2) is converted into compounds in plants. Of all of this activity, only about 10 to 20% occurs in land plants. The rest is done by tiny phytoplankton and algae in the earth's oceans. In principle it would be possible to dump so much poison into the oceans that the cycle of photosynthesis would be gravely affected. It is quite clear that the nations of the world must see that this does not happen.

When plants die and decay, their carbon atoms end up eventually in carbon dioxide again, and the reactions of decay consume oxygen. The combustion of fuels such as petroleum, coal, and wood also uses oxygen. And animals consume oxygen during respiration. Thus there exists a grand cycle in nature in which atoms of carbon, hydrogen, and oxygen move from CO_2 and H_2O into complex forms plus molecular O_2. The latter then interact in various ways to regenerate CO_2 and H_2O.

Someone has estimated that all the oxygen in the earth's atmosphere is renewed by this cycle once in about 20 centuries, and that all the CO_2 in the atmosphere and the earth's waters goes through this cycle every three centuries.

sugars as well. Two trioses, glyceraldehyde and dihydroxyacetone, occur in metabolism, and two pentoses, ribose and 2-deoxyribose, are essential to the nucleic acids.

■ *Deoxy* means lacking an oxygen where one normally is.

Glyceraldehyde   Dihydroxyacetone   Ribose   2-Deoxyribose

Glucose is a hexose, $(CH_2O)_6$ or $C_6H_{12}O_6$, with an aldehyde group, so it is also called an **aldohexose.** (We're interested only in terms here; structures will come soon.) Galactose is also an aldohexose, a stereoisomer of glucose. Fructose, also $(CH_2O)_6$ or $C_6H_{12}O_6$, has a keto group, so it's a **ketohexose.** You can see how parts of words can be combined into one very descriptive term. Glucose, galactose, and fructose are the nutritionally important monosaccharides.

Disaccharides and Polysaccharides Make up the Other Families of Carbohydrates

The monosaccharides are the monomer units of di- and polysaccharides. **Disaccharides** are carbohydrates that can be hydrolyzed to two monosaccharides. Sucrose, maltose, and lactose are common examples of disaccharides. Starch and cellulose are called **polysaccharides,** because when one of their molecules reacts with water, it gives hundreds of monosaccharide molecules.

■ *Oligosaccharide* molecules yield from three to a few dozen monosaccharide molecules when they are hydrolyzed.

All Monosaccharides Are Reducing Carbohydrates

Carbohydrates are sometimes described by their abilities to react with Tollens' and Benedict's reagents (pages 452 and 453). Something is reduced in these tests (e.g., Ag^+ or Cu^{2+}), so carbohydrates that give these tests are called **reducing carbohydrates.** All monosaccharides and nearly all disaccharides are reducing carbohydrates. Sucrose (table sugar) is not, and neither are the polysaccharides. We'll see why shortly.

■ Glucose is also called corn sugar, because it can be made by the hydrolysis of cornstarch.

Glucose Is Nature's Most Widely Used Organic Monomer

If we count all its combined forms, (+)-glucose is perhaps the most abundant organic species on earth. It's the building block for molecules of cellulose, a polysaccharide that makes up about 10% of all the tree leaves of the world (on a dry mass basis), about 50% of the woody parts of plants, and nearly 100% of cotton. Glucose is also the monomer for starch, a polysaccharide in many of our foods, particularly grains and tubers. Glucose and fructose are the major components of honey. Glucose is also commonly found in plant juices. Because it is by far the most common carbohydrate in blood, glucose is often called **blood sugar,** although this term strictly applies to the mixture of all the carbohydrates in blood.

■ Massachusetts General Hospital regards a concentration of glucose in the blood of 70–100 mg/100 mL (3.9–6.1 mmol/L) to be the "normal" range for a healthy adult who has not eaten for a few hours.

One Form of Glucose Is a Pentahydroxy Aldehyde

Simple alcohols, ROH, can form acetate esters, CH_3CO_2R. Glucose forms a pentaacetate, so five of the six oxygens in $C_6H_{12}O_6$ are in alcohol groups. The sixth oxygen is in an aldehyde group because glucose is easily oxidized to a C_6 monocarboxylic acid by reagents, like Tollens' reagent, that do not oxidize alcohol groups.

Under strong, forcing conditions, glucose can be reduced to a straight-chain derivative of hexane, so the six carbons in glucose must be in a straight chain. The five OH groups must be strung out, one on each of five carbons, because 1,1-diols are not stable. These data support the conclusion that glucose is 2,3,4,5,6-pentahydroxyhexanal. In fact, all aldohexoses are optical isomers of each other and so all have the same basic skeleton:

■ A 1,1-diol consists of the following system:

$$\begin{array}{c} OH \\ | \\ -C-OH \\ | \end{array}$$

Basic structure of all aldohexoses, including glucose

Carbons 2, 3, 4, and 5 in the glucose chain are all tetrahedral stereocenters. Each center has a *unique* set of four different groups, so the centers are all different. We learned in the previous chapter that the number of stereoisomers of a compound whose molecules have n different tetrahedral stereocenters is 2^n. In 2,3,4,5,6-pentahydroxyhexanal, $n = 4$, so there must be $2^4 = 16$ stereoisomers, or eight pairs of enantiomers. (+)-Glucose is one of these 16; galactose is another. None of the remaining 14 stereoisomers is nutritionally important. Clearly, to know what glucose really is, we must look more closely at the stereoisomers of glucose.

19.3

D- AND L-FAMILIES OF MONOSACCHARIDES

The tetrahedral stereocenter farthest from the carbonyl group of all naturally occurring monosaccharides has the same configuration as (+)-glyceraldehyde, which puts these monosaccharides in the D-family.
In this section we will study the question of the actual orientations or configurations of the tetrahedral stereocenters in the monosaccharides.

All Naturally Occurring Monosaccharides Belong to the Same Optical Family
To simplify this study, we'll retreat from the complexities of (+)-glucose and go back to the simplest aldose, glyceraldehyde. The structures of the two enantiomers of glyceraldehyde are shown in Figure 19.1. Both are known, and the enantiomer labeled D-(+)-glyceraldehyde actually has the absolute configuration shown. **Absolute configuration** refers to the actual arrangement in space about each stereocenter in a molecule. When we know the absolute configuration of (+)-glyceraldehyde, we also know that of (−)-glyceraldehyde, because its molecules *must* be the mirror image of the molecules of (+)-glyceraldehyde. (See also Figure 19.1.)

Chemists have used the absolute configurations of the enantiomers of glyceraldehyde to devise configurational or optical families for the rest of the carbohydrates. Any compound that has a configuration like that of (+)-glyceraldehyde and can be related to it by known reactions is said to be in the **D-family.** For example, (−)-glyceric acid is in the D-family because it can be made from D-(+)-glyceraldehyde by an oxidation that doesn't disturb any of the four bonds to the stereocenter, as the following equation shows.

■ When we use the term "stereocenter" we'll always mean "tetrahedral stereocenter." (The older term for such a center at a C atom is **chiral carbon.**)

■ As long as no bond to the stereocenter is disturbed in the reaction, no change in configuration can possibly occur.

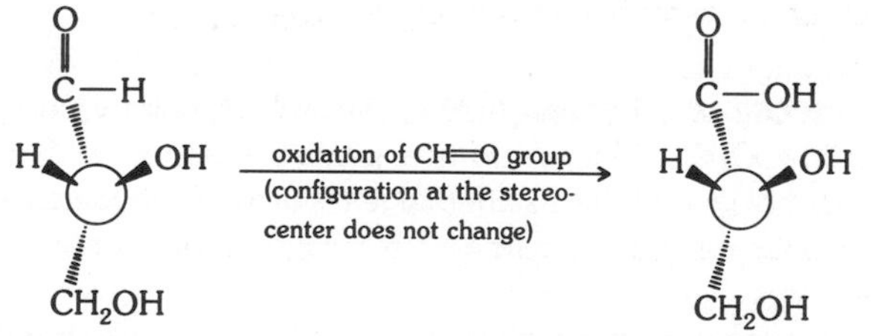

oxidation of CH=O group
(configuration at the stereocenter does not change)

D-(+)-Glyceraldehyde   D-(−)-Glyceric acid

When the molecules of a compound are the mirror images of an enantiomer in the D-family, the compound is in the **L-family.**

The letters D and L are only family names. *They have nothing to do with actual signs of the values of their specific rotations, [α].* No way exists, in fact, to tell from the *sign* of specific rotation whether a compound is in the D- or L-family. These letters signify something about absolute configuration only. Later we'll see that **D**-glucose is dextrorotatory but **D**-fructose is levorotatory.

FIGURE 19.1
The absolute configurations of the enantiomers of glyceraldehyde.

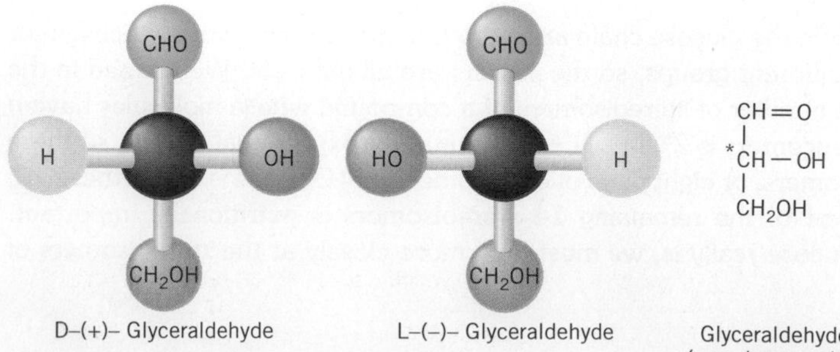

D–(+)– Glyceraldehyde   L–(−)– Glyceraldehyde   Glyceraldehyde (one stereocenter)

Fischer Projection Formulas Simplify Absolute Configurations When a molecule has several stereocenters, it becomes quite difficult for most people to make a perspective, three-dimensional drawing of an absolute configuration. Emil Fischer, a chemist who unraveled most of the carbohydrate structures, devised a way around this, and his structural representations are called *Fischer projection formulas.* To make them, we follow a set of rules that let us project onto a plane surface the three-dimensional configuration of each stereocenter in a molecule.

■ Emil Fischer (1852–1919), a German chemist, won the second Nobel prize in chemistry in 1902.

Rules for Writing Fischer Projection Formulas

1. Visualize the molecule with its main carbon chain vertical and with the bonds that hold the chain together projecting to the rear at each stereocenter. *Carbon-1 is at the top.*

2. Mentally flatten the structure, stereocenter by stereocenter, onto a plane surface. See Figures 19.2 and 19.3.

3. In the projected structure, represent each stereocenter either as the intersection of two lines or conventionally as C.

4. The horizontal lines at a stereocenter actually represent bonds that project *forward,* out of the plane of the paper.

5. The vertical lines at a stereocenter actually represent bonds that project *rearward,* behind the plane.

A Fischer projection formula can have more than one intersection of lines, each representing a stereocenter, as seen in Figure 19.3. Always remember that at each stereocenter, a horizontal line is a bond coming toward you and a vertical line is a bond going away from you.

Once we have one plane projection structure, it's easy to draw the mirror image, as we saw in Figures 19.2 and 19.3. We can easily test for superimposition, too, provided we strictly heed one important additional rule. We may never (mentally) lift a Fischer projection formula out of the plane of the paper. We may only slide it and rotate it within the plane and a rotation must be by 180°, not 90°. This rule is necessary because if we turn a Fischer projection formula out of the plane and over or rotate it by only 90°, we actually make groups that project in one direction project oppositely, but the operation that we do on the paper will not show this reversal.

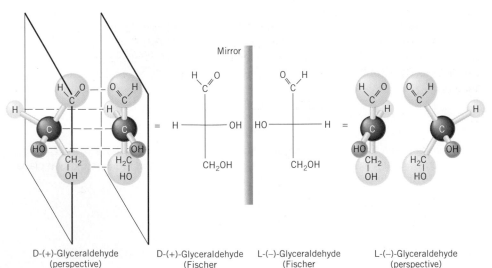

FIGURE 19.2
The relationships of the perspective (three-dimensional) drawings of D-(+)-glyceraldehyde and L-(−)-glyceraldehyde to their corresponding Fischer projection formulas.

D-(+)-Glyceraldehyde (perspective)   D-(+)-Glyceraldehyde (Fischer projection)   L-(−)-Glyceraldehyde (Fischer projection)   L-(−)-Glyceraldehyde (perspective)

FIGURE 19.3

The four aldotetroses in their perspective and Fischer projection formulas. There are two different stereocenters, so there are $2^2 = 4$ stereoisomers that occur as two pairs of enantiomers, those of D- and L-erythrose and those of D- and L-threose.

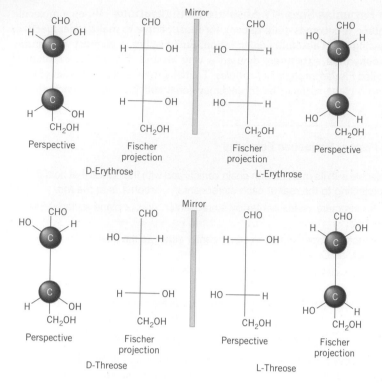

EXAMPLE 19.1
Writing Fischer Projection Formulas

Write the Fischer projection formulas for the stereoisomers of glyceric acid.

$$HOCH_2\overset{|}{\underset{OH}{C}}HCO_2H$$

Glyceric acid

ANALYSIS There are three carbons in the chain, but only one is joined to four different atoms or groups. Only the center carbon is a stereocenter. Therefore in our equation for calculating the number of stereoisomers, $n = 1$, so $2^n = 2$ and glyceric acid has two enantiomers. We represent its lone stereocenter by the intersection of two perpendicular lines, and we make two of these, one for each enantiomer:

$$+ \quad + \quad \text{(Incomplete)}$$

Then we attach the other two carbons. According to the rules, we have to put C-1, the carbon with the carbonyl group in CO_2H, at the top. This then requires that we place CH_2OH at the lower end of the vertical line.

$$\begin{array}{cc} CO_2H & CO_2H \\ + & + \\ CH_2OH & CH_2OH \end{array} \quad \text{(Incomplete)}$$

SOLUTION We know that the OH group at C-2 can be either on the right or the left, so we finish the Fischer projection formulas:

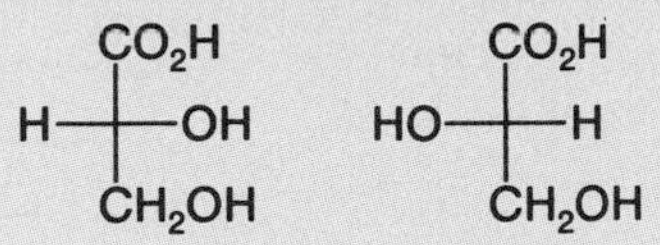

<pre>
        CO₂H   CO₂H
   H ——— OH   HO ——— H   (Complete)
       CH₂OH   CH₂OH
</pre>

These are the two enantiomers of glyceric acid. The one on the left is D-glyceric acid, and the other is L-glyceric acid.

■ D-Glyceric acid is levorotatory but its salts are dextrorotatory, which illustrates again that the sign of a specific rotation cannot be deduced from the D- or L-family membership.

PRACTICE EXERCISE 1

Fischer projection formulas that correspond to the various optical isomers of tartaric acid are given below. (a) Which are identical? (b) Which are related as enantiomers? (c) One is a *meso* compound (Special Topic 18.1). Which is it?

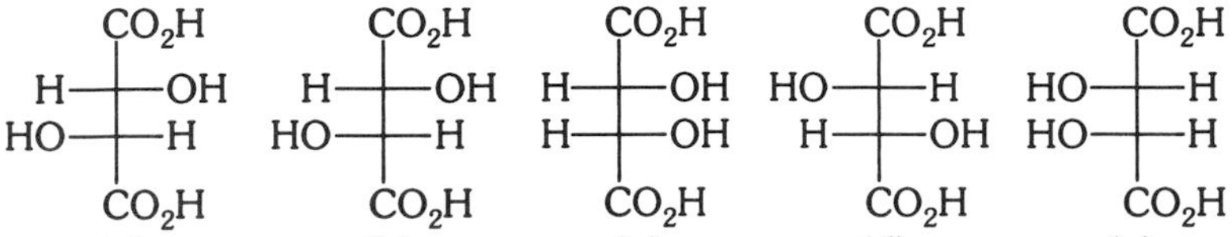

<pre>
      CO₂H   CO₂H   CO₂H   CO₂H   CO₂H
  H ——— OH   H ——— OH   H ——— OH   HO ——— H   HO ——— H
 HO ——— H   HO ——— H   H ——— OH   H ——— OH   HO ——— H
      CO₂H   CO₂H   CO₂H   CO₂H   CO₂H
       (a)   (b)   (c)   (d)   (e)
</pre>

PRACTICE EXERCISE 2

Write Fischer projection formulas for the stereoisomers of the following compound.

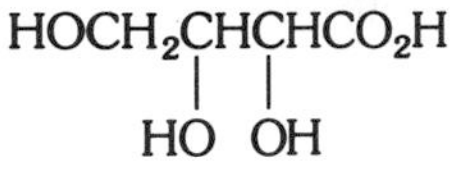

<pre>
   HOCH₂CHCHCO₂H
         | |
        HO OH
</pre>

All Naturally Occurring Carbohydrates Are in the D-Family

Monosaccharides are assigned to the D-family or the L-family according to the projection of the OH group *at the stereocenter farthest from the carbonyl group.* The compound is in the D-family when this OH group projects to the right *in a Fischer projection formula* oriented so that the carbonyl group is at or near the top. When this OH group projects to the left, the substance is in the L-family. We have already illustrated these rules by D- and L-glyceraldehyde (Fig. 19.2), and by the enantiomers of threose and erythrose (Fig. 19.3). It doesn't matter how the other OH groups at the other stereocenters project. Membership in the D- or L-family is determined solely by the projection of the OH on the stereocenter farthest from the carbonyl carbon in a properly drawn Fischer projection structure.

Because the nutritionally important carbohydrates are all in the D-family, throughout the rest of this book we'll assume that the D-family is meant whenever the family membership of a carbohydrate isn't given.

Figure 19.4 gives the Fischer projection formulas of all the aldoses in the D-family from the aldotriose through the aldohexoses. There are eight D-aldohexoses. The enantiomers of these constitute eight L-aldohexoses (not shown). In all, therefore, there are 16 optical isomers of the aldohexoses, as we calculated earlier. Figure 19.5 gives the Fischer projection formulas of several important ketoses. The ketotriose dihydroxyacetone, the two ketopentoses ribulose and xylulose, and the ketohexose fructose are the biologically important ketoses.

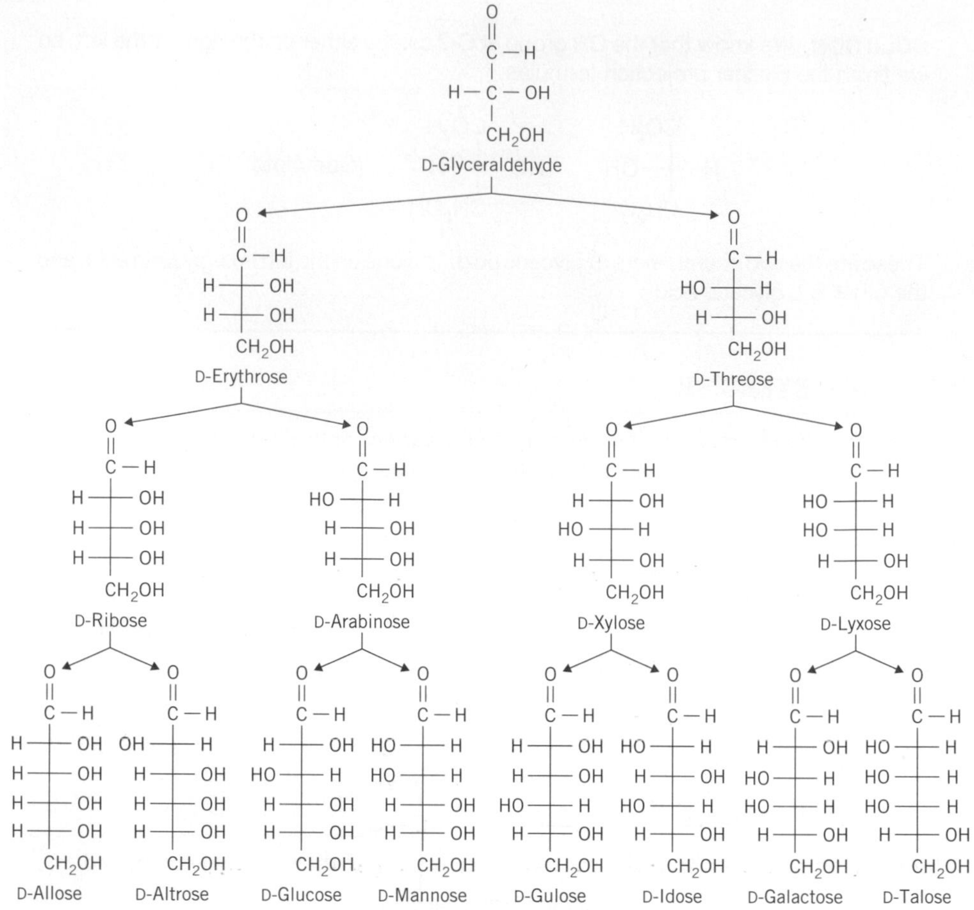

FIGURE 19.4

The D-family of the aldoses through the aldohexoses. Notice that in all of them, the OH group on the stereocenter that is farthest from the carbonyl group projects to the right. Each pair of arrows points to a pair of aldoses whose configurations are identical except at C-2.

19.4

CYCLIC FORMS OF THE MONOSACCHARIDES

The internal addition of an OH group to the double bond of the aldehyde or keto group makes possible two cyclic forms of monosaccharides.

Fresh Glucose Solutions Gradually Change in Optical Rotation Although glucose is optically active, when we try to measure its optical activity, it behaves in a very strange way. A *freshly prepared* solution of (+)-glucose has a specific rotation of $[\alpha]_D^{20} = +113°$. As this solution ages, however, its specific rotation slowly changes until it stabilizes at a value of $+52°$. We'll call a glucose solution with this specific rotation an "aged glucose solution."

■ The "aging" of a glucose solution occurs almost instantly when hydroxide ion is present.

By a special method of recovery (which we'll not discuss), we can recover crystalline (+)-glucose from the aged solution, the same glucose in every respect as before. A *freshly prepared* solution of this recovered glucose again shows a specific rotation of $+113°$. This new solution ages in the identical way until its specific rotation stabilizes at $+52°$. This cycle can be repeated as often as we please.

It is possible by another method of recovering glucose from an aged solution to obtain a slightly different crystalline compound. Its freshly prepared solution has a specific rotation of $+19°$, but it also changes with time to $+52°$, the same value that was observed for the other aged solution. Using this second method to recover glucose, we can repeat this cycle as often as we please, too.

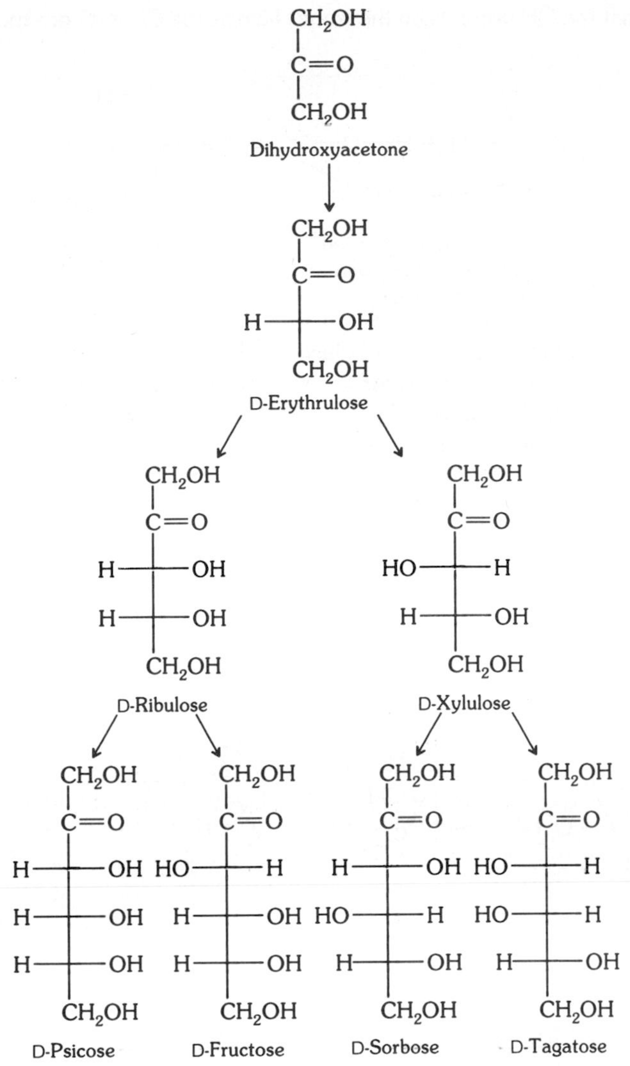

CH₂OH
|
C=O
|
CH₂OH

Dihydroxyacetone

↓

CH₂OH
|
C=O
|
H——OH
|
CH₂OH

D-Erythrulose

CH₂OH
|
C=O
|
H——OH
|
H——OH
|
CH₂OH

D-Ribulose

CH₂OH
|
C=O
|
HO——H
|
H——OH
|
CH₂OH

D-Xylulose

CH₂OH
|
C=O
|
H——OH
|
H——OH
|
H——OH
|
CH₂OH

D-Psicose

CH₂OH
|
C=O
|
HO——H
|
H——OH
|
H——OH
|
CH₂OH

D-Fructose

CH₂OH
|
C=O
|
H——OH
|
HO——H
|
H——OH
|
CH₂OH

D-Sorbose

CH₂OH
|
C=O
|
HO——H
|
HO——H
|
H——OH
|
CH₂OH

D-Tagatose

FIGURE 19.5

The D-ketoses having three to six carbon atoms. Notice that in all of them, the OH group on the stereocenter that is farthest from the carbonyl group projects to the right. Each pair of arrows points to a pair of ketoses whose configurations are identical except at C-3.

To summarize, from the original aged solution we can use one method to recover the solute and get back glucose with a specific rotation of $+113°$. With the second recovery technique, we get a glucose with a specific rotation of $+19°$. We can interconvert these forms through the aged solution as often as we please. This change over time of the optical rotation of an optically active substance, one that can be recovered from an aged solution without any other apparent change, is called **mutarotation.** All the hexoses and most of the disaccharides mutarotate. Let us now see what is behind it.

Glucose Molecules Exist Mostly in Cyclic Forms Built into the *same* molecule of 2,3,4,5,6-pentahydroxyhexanal are the two functional groups *needed* to make a hemiacetal, the OH group and the CH=O group. Recall that hemiacetal formation is represented as follows:

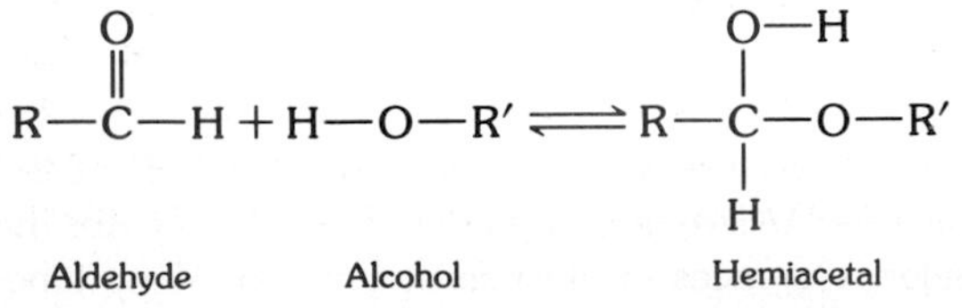

$$
\underset{\text{Aldehyde}}{R-\overset{\overset{\displaystyle O}{\|}}{C}-H} + \underset{\text{Alcohol}}{H-O-R'} \rightleftharpoons \underset{\text{Hemiacetal}}{R-\overset{\overset{\displaystyle O-H}{|}}{\underset{\underset{\displaystyle H}{|}}{C}}-O-R'}
$$

Suppose now that the OH group is on the *same chain* as the CH═O group. We would have something like

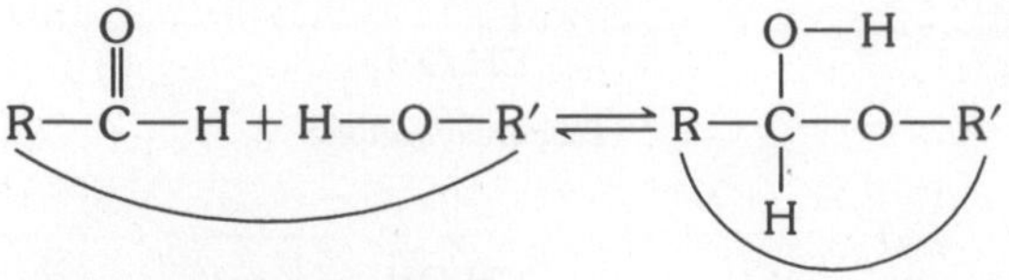

R and R′ would be joined
if CH═O and HO were in
the *same* molecule

A cyclic hemiacetal

This is what happens to the open form of glucose. The C-5 OH group adds to the aldehyde group. The open structure has to coil for this to happen, as shown in structure **2**, below. It undergoes ring closure, and a new OH group, the hemiacetal OH, appears at C-1. This C-1 OH, however, can emerge on one side of the ring or the other. It depends on the way the O atom of the C═O group points just before ring closure.

If, at the moment of ring closure, the C-1 OH comes out on the side of the ring opposite to the CH$_2$OH unit (involving C-6), one cyclic form of glucose emerges. It is the alpha form, **1**, called α-glucose. If the new C-1 OH group comes out on the same side of the ring as the CH$_2$OH unit, the beta form of glucose, called β-glucose, **3**, forms.

■ The H at C-2 and C-5 and the OH at C-3 do not stick *inside* the ring. They stick *above or below the plane* of the ring.

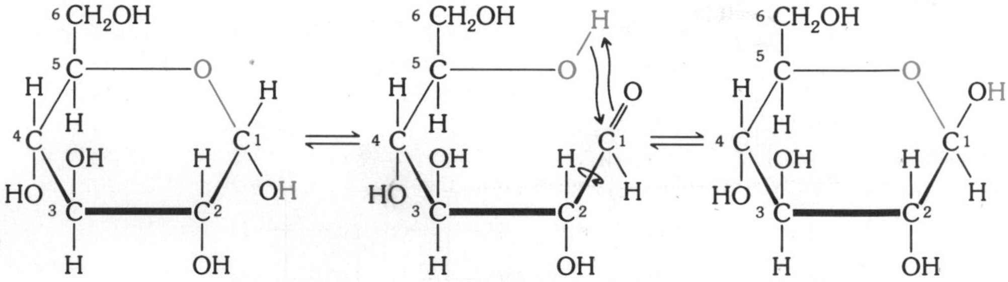

1
α-Glucose

2
Open form of glucose

3
β-Glucose

The Six-Membered Rings of Glucose Are Actually Not Flat

■ The isomeric, cyclic forms of any given carbohydrate that differ *only* in the configuration of the hemiacetal (or hemiketal) carbon are called **anomers.** Thus **4** (= **1**) and **6** (= **3**) are anomers.

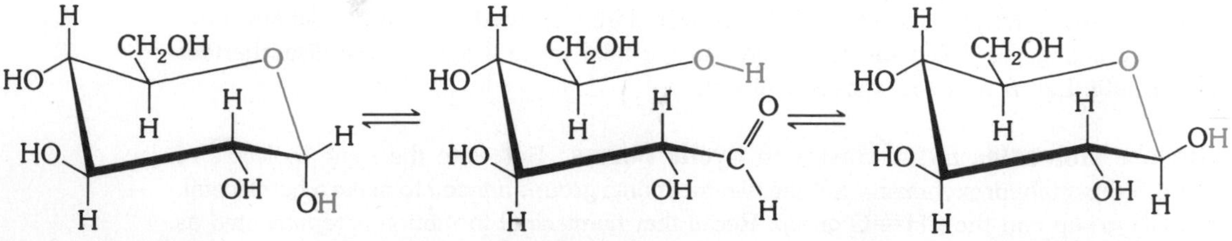

The carbon atoms in glucose are all tetrahedral, so the bonds from them normally are at angles of 109.5°. The two bonds from the O atom in the ring are also close to this. A *flat* hexagon ring, however, would have internal angles of 120°. A *saturated* six-membered ring, therefore, cannot be flat as indicated by structures **1** or **3**. Instead, such rings are nonplanar so that normal bond angles are possible. We show next the three forms of glucose as they are known to exist in nonplanar conformations called *chair forms.*

4 (= **1**)
α-Glucose

5 (= **2**)
Open form of glucose

6 (= **3**)
β-Glucose

The nonplanar forms of glucose

Special Topic 19.2 discusses these conformations in more detail and indicates why they are preferred. Having called this to your attention, we will often use the planar designations anyway, because major references in biochemistry do so. Many times we will show both forms, however.

In Aged Glucose Solutions, All Three Forms of Glucose Exist in Equilibrium

α-Glucose is the form of the glucose molecules when a freshly prepared aqueous solution has a specific rotation of $+113°$. Hemiacetals are unstable, however, and glucose in its cyclic forms is a hemiacetal. First one molecule of **1** and then another opens up to give form **2**. As soon as molecules of **2** appear, they can and do reclose. Figure 19.6 shows how free rotation about the C-1 to C-2 bond can reposition the aldehyde group, so that either one side or the other side of the carbonyl system faces the C-5 OH group at the moment of ring closure.

To keep the two glucose forms straight, use the CH_2OH group and the ring O atom as points of reference. Notice that in both of the cyclic forms of glucose the CH_2OH unit sticks upward from the plane *when the ring is drawn with its oxygen in the upper right-hand corner*. When we use these specific orientations, we are certain to be drawing a member of the **D**-family of the aldohexoses. Now notice in structure **3** that the OH group at C-1 projects *upward* in β-glucose and is on the same side of the ring as the CH_2OH group to the left of the ring O atom. In α-glucose, structure **1**, you can see that the OH at C-1 projects *downward* on the opposite side of the ring from the CH_2OH group. We should now use the names β-**D**-glucose and α-**D**-glucose for **3** and **1** but, as we have said, we'll always mean the **D**-family (unless something else is stated). The projections of all the other OH groups in both structures **1** and **3** are identical. If we changed any of their orientations, we would have the structure of a molecule that isn't any form of glucose, but one of its stereoisomers instead.

■ This arrangement of CH_2OH relative to the ring O atom also ensures that the optical configurations at all stereocenters of natural glucose are correctly displayed in our structures.

Ring Opening and Ring Closing Occur during Mutarotation   The two cyclic forms of glucose differ only in the orientation of the OH group at C-1. With this in mind, let's review what happens during mutarotation. As we said, when α-glucose is freshly dissolved in water, it has a specific rotation of $+113°$. But its molecules open and close, because the hemiacetal system easily breaks apart and reforms. Some of the newly formed open-chain molecules reclose as α-glucose, and some as β-glucose. These events take place during mutarotation, whether we start with α- or β-glucose, until one grand, dynamic equilibrium involving all three

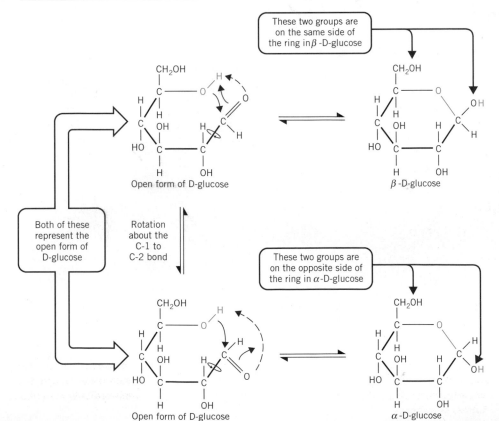

FIGURE 19.6
The α- and β-forms of D-glucose arise from the same intermediate, the open-chain form. Depending on how the aldehyde group, CH=O, is pointing when the ring closes, one ring form or the other results.

THE BOAT AND CHAIR FORMS OF SATURATED, SIX-MEMBERED RINGS

The inside angle of a regular hexagon is 120°, so the six atoms of a saturated six-membered ring cannot lie in the same plane and also have the tetrahedral bond angles of 109.5°. The cyclohexane ring resolves this by twisting into a nonplanar shape called the **chair form.**

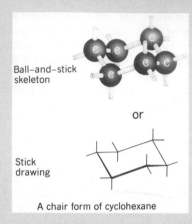

Ball–and–stick skeleton

or

Stick drawing

A chair form of cyclohexane

Even when one CH_2 unit of this ring is replaced by an oxygen atom, as in the rings of the glucose forms, the same kind of chair conformation predominates.

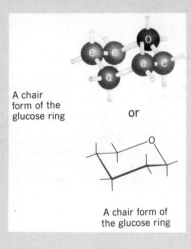

A chair form of the glucose ring

or

A chair form of the glucose ring

The **boat form** is another conformation of the ring that permits normal bond angles, and the ring has enough flexibility to be able to twist from the chair to the boat form. As the drawings show, if you twist one end of the chair form upward, you get the boat, and if you twist the opposite end downward you get an alternative chair form. Thus there are *two* chair forms and one boat. All three have normal bond angles.

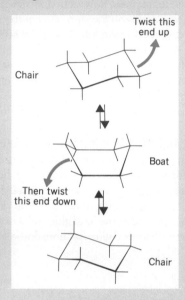

Twist this end up

Chair

Boat

Then twist this end down

Chair

The two chair forms of cyclohexane are equally stable, but the boat form is less stable than a chair. In the boat form, as you can see in the scale model, the electron clouds by the hydrogen atoms are closer to one another, particularly the two hydrogens at opposite ends, marked *a*, one at the "prow" and one at the "stern." They nudge each other in the boat form, but are as far from each other as possible in the chair form.

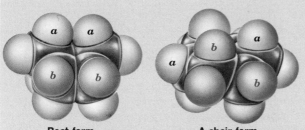

Boat form   A chair form
Cyclohexane (scale model)

Similarly, the electron clouds marked *b* are closer to each other in the boat than in the chair form. Electron clouds repel each other, so the boat form is less stable than the chair form.

When the six-membered ring holds substituents, as it does among the carbohydrates, the alternative chair forms are no longer equivalent. We must, therefore, have labels to distinguish the two ways that bonds and substituents can be oriented. Positions around the perimeter are called *equatorial positions* because they are located roughly on the ring's equator. They are indicated by the black bonds and their attached H atoms in structure **A**. The positions that project above or below the average plane parallel to the axis through this plane are called *axial substituents,* indicated in color in **A**.

When chair form **A** twists into its alternative chair form, **B**, every equatorial position changes to axial and every axial to equatorial.

In an actual sample of cyclohexane, the two chair forms exist in equilibrium, and they constantly flip-flop back and forth. In a sense, the flat hexagonal structure that we usually draw for the ring of cyclohexane is an average of these two forms, and in most situations we can ignore the true bond angles of the six-membered ring.

The equivalency of the two chair forms vanishes, as we said, when the ring bears substituents. The electron clouds of axial substituents nudge one another more than do those of equatorial substituents. Thus equatorial orientations are more stable. As a rule, therefore, saturated six-membered rings take up whichever chair form puts the maximum number of bulky substituents in equatorial positions. This important fact dominates the conformations of the ring forms of the aldohexoses. The beta form of glucose, for example, is able to have every ring substituent oriented equatorially, the only aldohexose in which this is possible. In its alternative chair form, however, all substituents would be oriented axially, and this form does not occur.

β-Glucose
(more stable;
all substituents
are equatorial)

β-Glucose
(less stable;
all substituents
are axial)

Since most people find it easier to draw flat rings for the aldohexoses, and most references in biochemistry use them widely, we will generally use them also. They correctly show *relative* projections of groups on a ring — the up or down orientations — but their bond angles are not correct.

A

B

FIGURE 19.7

How to draw the cyclic forms of D-glucose in a highly condensed way.

1. First write a six-membered ring with an oxygen in the upper right-hand corner.

2. Next "anchor" the terminal CH$_2$OH unit on the carbon to the left of the oxygen. (Let all the Hs attached to ring carbons be "understood.")

   CH$_2$OH   or condense to   OH

3. Continue in a *counterclockwise* way around the ring, placing the OHs first down, then up, then down.

   CH$_2$OH   or condense to   OH

4. Finally, at the last site on the trip, how the last OH is positioned depends on whether the alpha or the beta form is to be written. The alpha is "down," the beta "up."

   β-D-Glucose   or condense to   β-D-Glucose   or   α-D-Glucose

   If this detail is immaterial, or if the equilibrium mixture is intended, the structure may be written as   or condense to   α or β

forms of glucose is established. The identical equilibrium develops from either cyclic form of glucose, because we obtain an aged solution that has the identical specific rotation, +52°. We can express the equilibrium in words as follows.

$$\alpha\text{-Glucose} \rightleftharpoons \text{open-chain form of glucose} \rightleftharpoons \beta\text{-glucose}$$

At equilibrium, the solute molecules are 36% α-glucose, 64% β-glucose, and scarcely a trace (<0.05%) of the open-chain form.

The method of recovering solid glucose from an aged solution determines the form of glucose in the crystals. One recovery method succeeds in getting just α-glucose molecules to start the formation of crystals. The molecules of β-glucose can't fit to this crystal and help it grow, so they remain in solution. But the loss of molecules of α-glucose from the equilibrium puts a stress on it, and the equilibrium shifts (in accordance with Le Châtelier's principle) to replace the lost molecules. In this way, all the β-glucose molecules eventually get changed to α-glucose molecules and nestle into the growing crystals.

The other method of crystallization succeeds in getting crystals started from just the β-glucose molecules, so eventually all the α-glucose molecules get switched over to the β-form and join the growing crystals.

Molecules of the open-chain form never crystallize. They occur only in the solution. Of course, they are the ones attacked by Tollens' or Benedict's reagents and, as they are removed, they are replaced by a steady shifting of the equilibrium from closed forms to the open form. This is why glucose gives the chemical properties of a pentahydroxy aldehyde despite the fact that glucose is in either one cyclic form or the other — entirely in the solid state and almost entirely in solution. This is also why it is acceptable to define a monosaccharide as an aldehyde (or ketone) with multiple OH groups rather than as a cyclic hemiacetal.

Before continuing, you should now pause to learn how to write the cyclic forms of glucose (using flat rings). Figure 19.7 outlines the steps in mastering this that have worked well for many students. Although six-membered rings are not *flat*, as we said, the projections of the groups on the ring, relative to the CH$_2$OH unit, are faithfully shown by these kinds of drawings.

■ All forms of glucose are used by living systems.

■ This is another example of Le Châtelier's principle in action.

Galactose Is a Stereoisomer of Glucose   Galactose is an aldohexose that occurs in nature mostly as a structural unit in larger molecules such as the disaccharide lactose. It is also a sugar in peas. Galactose differs from glucose only in the orientation of the C-4 OH group. Like glucose, it is a reducing sugar, it mutarotates, and it exists in solution in three forms, α, β, and open.

■ Carbohydrates that differ *only* in the orientation of the OH at *only one* stereocenter other than the hemiacetal or hemiketal carbon are called **epimers** of each other.

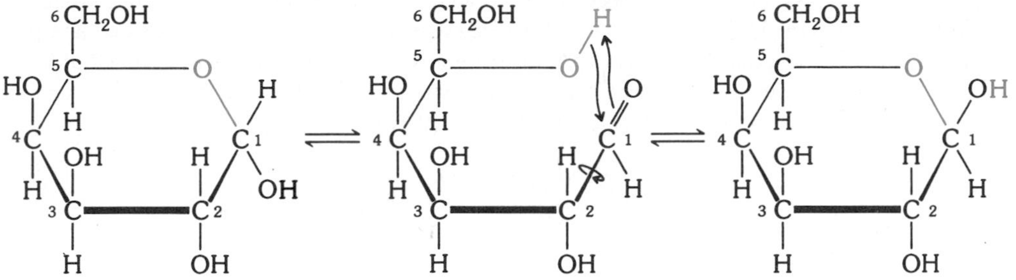

7
α-Galactose

8
Open form of galactose

9
β-Galactose

■ D-Glucose and D-galactose are *epimers* at C-4.

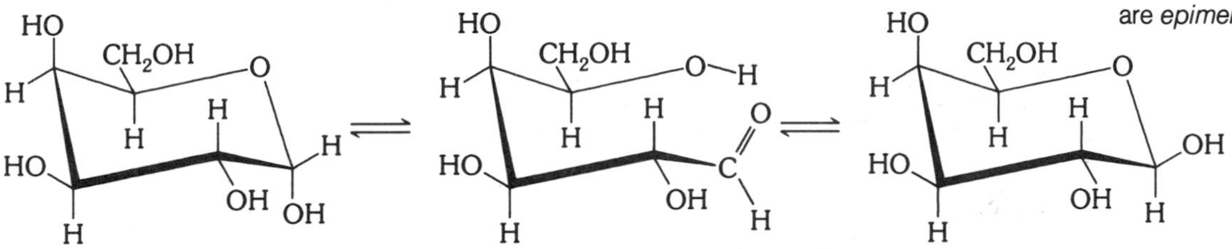

10 (= 7)
α-Galactose

11 (= 8)
Open form of galactose

12 (= 9)
β-Galactose

Fructose Occurs in a Five-Membered, Cyclic Hemiketal Form   Fructose, the most important ketohexose, is found with glucose and sucrose in honey and in fruit juices. Fructose exists in more than one form, including cyclic hemiketals. The hemiketal carbon is C-2.

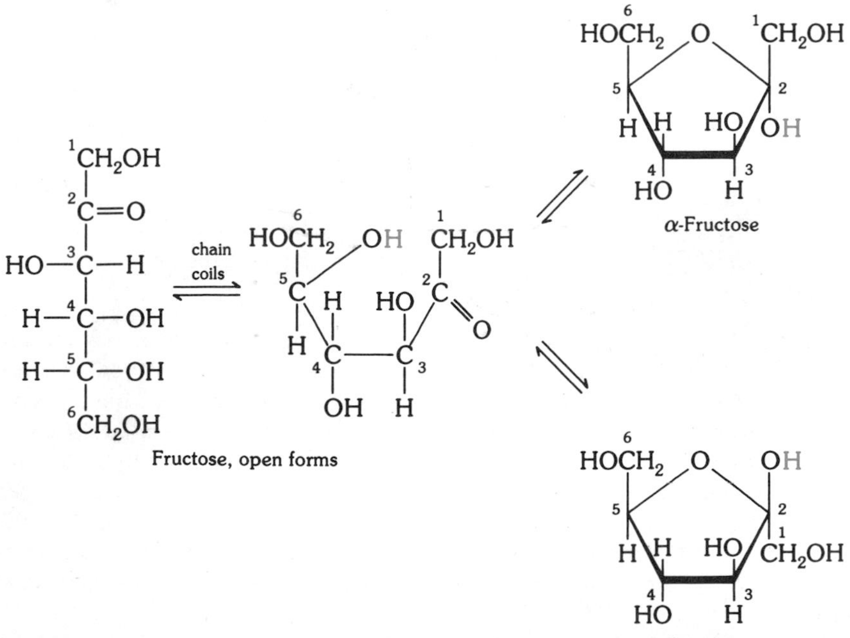

Fructose, open forms

α-Fructose

β-Fructose

■ The internal angle of a pentagon (108°) is very close to the tetrahedral angle, so the five-membered ring is nearly flat.

■ An old name for fructose is *levulose,* after its levorotatory power.

■ Mono- and diphosphoric acid esters of fructose and glucose are important compounds in metabolism.

D-Fructose is strongly levorotatory with a specific rotation of $[\alpha]_D^{20} = -92.4°$. Fructose is a reducing sugar, not because it's a ketone (simple ketones do not easily oxidize) but because its molecules have the α-hydroxyketone system mentioned on page 453 as a system that gives the Benedict's test.

Ribose and 2-Deoxyribose Are Aldopentoses Important to Nucleic Acid Structure Both ribose and 2-deoxyribose, as we noted earlier, are building blocks of nucleic acids, and each of these aldopentoses can exist in three forms, two cyclic hemiacetals and an open form. We show just one form of each. A *deoxycarbohydrate* is one with a molecule that lacks an OH group where normally such a group is expected. Thus 2-deoxyribose is the same as ribose except that there is no OH group at C-2, just two Hs instead.

■ Ribose gives the R to RNA, a nucleic acid. Deoxyribose gives the D to DNA, another nucleic acid and the chemical of genes.

β-Ribose β-2-Deoxyribose

Ribose furnishes the sugar units to molecules of ribonucleic acid or RNA. Deoxyribose provides sugar units to the molecules of deoxyribonucleic acid, DNA.

19.5

DISACCHARIDES

The disaccharides are glycosides (sugar acetals) that can be hydrolyzed to monosaccharides.

The aldoses, as we have just seen, are hemiacetals. Like all hemiacetals, they react with alcohols in the presence of a catalyst to give acetals.

The Sugar Acetals Are Called Glycosides Like all hemiacetals, those of the cyclic forms of the monosaccharides can also form acetals. Methanol, for example, can be made to react with the cyclic hemiacetal system in glucose to give either an α- or a β-acetal, depending on how the new OCH_3 group becomes oriented at the ring. If it's on the same side of the ring as our reference CH_2OH group, then we have the β-form. If it is on the opposite side, then we have the α-form.

An alpha glucoside   A beta glucoside

Alternatively,

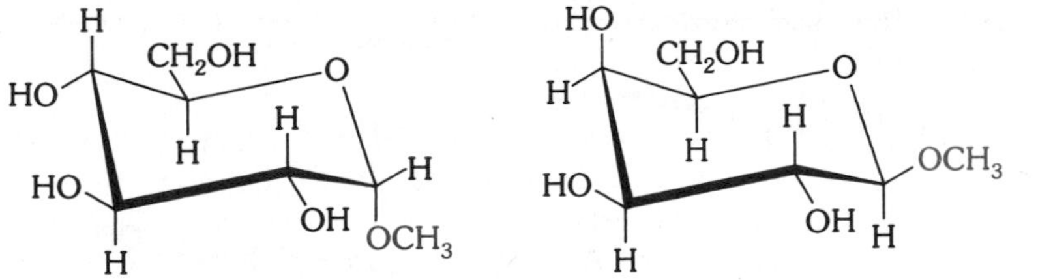

Methyl α-D-glucoside
$[\alpha]_D^{25} = +158°$

Methyl β-D-glucoside
$[\alpha]_D^{25} = -33°$

All sugar acetals have the general name of **glycosides.** To name the glycoside of a specific sugar, the -ose in the name of the sugar is replaced by -oside. Thus a glycoside made from glucose is called a *glucoside.* One made from galactose is a *galactoside.*

The sugar acetals or glycosides made from simple alcohols are stable enough to isolate, but they are readily hydrolyzed when an acid catalyst (or the appropriate enzyme) is present. The glycosides shown above do not mutarotate and do not give positive Benedict's tests because their rings cannot open to expose an aldehyde group.

The Disaccharides Are Glycosides That Use a Second Sugar as the Alcohol All the disaccharides are glycosides made from the cyclic hemiacetal unit of one sugar and one of the alcohol groups of another. *An acetal oxygen "bridge" thus links two monosaccharide units in disaccharides.* This acetal unit, like all acetals, reacts readily with water in the presence of an acid or enzyme catalyst, and this hydrolysis frees the original monosaccharide molecules.

The three nutritionally important disaccharides are maltose, lactose, and sucrose. All are in the D-family. We'll first show their relationships to monosaccharides by word equations, and then we'll look more closely, but briefly, at their structures.

$$\text{Maltose} + H_2O \xrightarrow[\text{enzyme (maltase)}]{H^+ \text{ or}} \text{glucose} + \text{glucose}$$

$$\text{Lactose} + H_2O \xrightarrow[\text{enzyme (lactase)}]{H^+ \text{ or}} \text{glucose} + \text{galactose}$$

$$\text{Sucrose} + H_2O \xrightarrow[\text{enzyme (sucrase)}]{H^+ \text{ or}} \text{glucose} + \text{fructose}$$

■ You can see why glucose is of such central interest to carbohydrate chemists.

Maltose Is Made from Two Glucose Units Maltose or malt sugar does not occur widely as such in nature, although it is present in germinating grain. It occurs in corn syrup, which is made from cornstarch, and it forms from the partial hydrolysis of starch. As the first equation indicated, maltose is made from two glucose units. They are joined by an acetal oxygen bridge that in carbohydrate chemistry is called a **glycosidic link.**

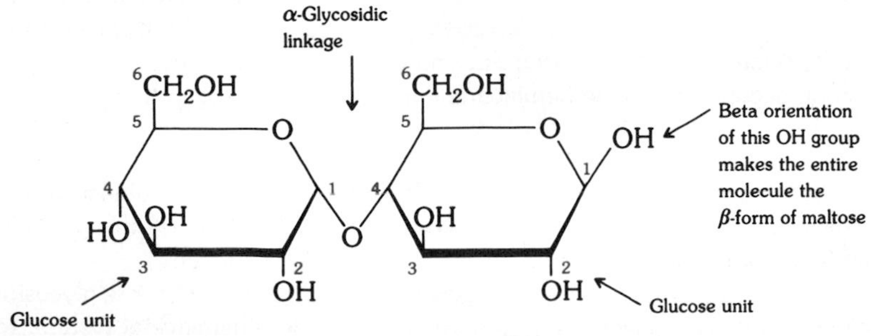

α-Glycosidic linkage

Beta orientation of this OH group makes the entire molecule the β-form of maltose

Glucose unit

Glucose unit

β-Maltose

■ If the OH group at C-1 on the far-right glucose unit projected downward instead of upward, the structure would be that of α-D-maltose instead of β-D-maltose.

Alternatively,

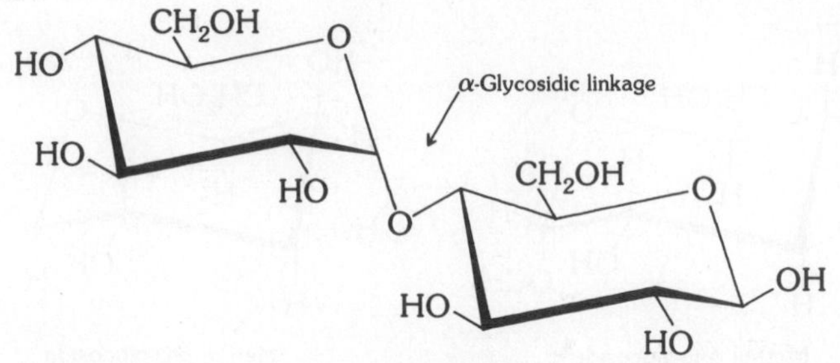

α-Glycosidic linkage

CH₂OH

β-Maltose

In maltose, the bridging oxygen of the glycosidic link joins the C-1 position of the glucose unit that serves as the hemiacetal partner, the unit on the left, to the C-4 of the glucose unit that is the alcohol partner, on the right in the structure. Such a glycosidic link is designated as $(1 \rightarrow 4)$.

The bond to the bridging oxygen from C-1 (on the left) points in the *alpha* direction, so the glycosidic link is more fully described as $\alpha(1 \rightarrow 4)$. Had this link pointed in the beta direction, it would have been described as $\beta(1 \rightarrow 4)$. But then the disaccharide would not have been maltose but a different disaccharide, cellobiose.

The purely geometric difference between $\alpha(1 \rightarrow 4)$ and $\beta(1 \rightarrow 4)$ that marks the difference between maltose and cellobiose may seem to be a trifle, but the difference to us is that we can digest maltose but not cellobiose. Just this difference in *geometry* bars humans from an enormous potential food source, for nature could supply much of it from the polysaccharide cellulose. We have an enzyme, maltase, that catalyzes the digestion (the hydrolysis) of maltose. We have no enzyme for cellobiose (or, for that matter, cellulose), although some organisms do.

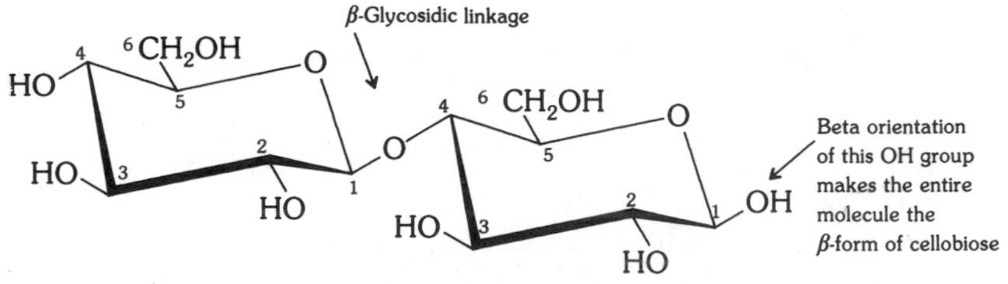

β-Cellobiose

Maltose Retains a Hemiacetal Unit, so It Is a Reducing Sugar and Mutarotates

The glucose unit on the 4-side of an $\alpha(1 \rightarrow 4)$ glycosidic link in maltose still has a hemiacetal group. This part of maltose, therefore, can open and close, so maltose can exist in three forms, α-, β-, and the open form. Maltose therefore mutarotates and is a reducing sugar. The ring opening action occurs only at the hemiacetal part, not at the oxygen bridge.

Lactose Links Galactose by a $\beta(1 \rightarrow 4)$ Bridge to Glucose

Lactose or milk sugar occurs in the milk of mammals — 4 to 6% in cow's milk and 5 to 8% in human milk. It is also a by-product in the manufacture of cheese.

Lactose is a galactoside. From C-1 of its galactose unit there is a $\beta(1 \rightarrow 4)$ glycosidic link to C-4 of a glucose unit. The glucose unit therefore still has a free hemiacetal system, so lactose mutarotates and is a reducing sugar.

Oxygen bridge

CH₂OH

CH₂OH

HO

OH

OH

OH

OH Glucose
unit

Galactose
unit

β-Lactose

Alternatively,

HO

4 ⁶CH₂OH

5

β-Glycosidic linkage

2

HO 3

1

OH

O

4 ⁶CH₂OH

5

O

2

HO 3

1 OH

HO

Beta orientation
of this OH group
makes the entire
molecule the
β-form of lactose

β-Lactose

Sucrose Links a Glucose Unit to a Fructose Unit   Sucrose, our familiar table sugar, is
obtained from sugar cane or from sugar beets. Its structure links a glucose to a fructose unit by
an oxygen bridge in such a way that *no hemiacetal or hemiketal group remains.* Neither ring in
sucrose, therefore, can open and close spontaneously in water. Hence, sucrose neither
mutarotates nor gives positive tests with Tollens' or Benedict's reagents. It's our only common nonreducing disaccharide.

■ Beet sugar and cane sugar
are identical compounds,
sucrose.

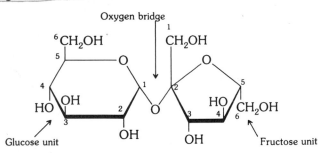

Oxygen bridge

1

⁶CH₂OH

5

CH₂OH

4

HO OH

2

3

OH

O

1

O

2

O

HO

5

CH₂OH

3

4 6

OH

Glucose unit

Fructose unit

Sucrose

   The 50 : 50 mixture of glucose and fructose that forms when sucrose is hydrolyzed is called
invert sugar, and it makes up the bulk of the carbohydrate in honey. (The sign of specific
rotation inverts from + to − when sucrose, $[\alpha]_D^{20}$ +66.5°, changes to invert sugar, $[\alpha]_D^{20}$
−19.9°, and this inversion of the sign is the origin of the term *invert* sugar.)

19.6

POLYSACCHARIDES

Starch, glycogen, and cellulose are all polyglucosides.
In this section we will study the structures and some of the properties of three polymers of
glucose — starch, glycogen, and cellulose.

■ Up to a million glucose units per molecule have been found in some amylose samples, making amylose one of nature's largest molecules.

(a)

(b)

(c)

The iodine test for starch. (a) The starch dispersion is so dilute it does not appear cloudy. (b) A few drops of iodine reagent cause a purple color to develop, which becomes quite intense. (c) The development of a purple color when a dilute iodine solution is added to a starch dispersion is a positive starch iodine test.

Plants Store the Chemical Energy of Glucose in Starch Molecules   When glucose is made by photosynthesis, solar energy becomes stored as chemical energy, which the plants can use for chemical work. Free glucose, however, is very soluble in water. A plant, therefore, would have to retain considerable water if its cells had to hold free glucose molecules in solution. Otherwise, the concentration of the cell fluid would be too high, thereby causing osmotic pressure problems (see Section 7.5) and upsetting proper movements of plant fluids. This problem of storing glucose without too much water is avoided by the conversion of glucose to its much less soluble polymer, starch. It is particularly abundant in plant seeds and tubers, where its energy is used for sprouting and growth. Animals that include plants in their diets also take advantage of the chemical energy in starch.

Starch is a mixture of two kinds of polymers of α-glucose, *amylose* and *amylopectin*. In amylose, the glucose units are joined by a linear succession of $\alpha(1 \rightarrow 4)$ glycosidic links, as seen in Figure 19.8. The lengths of the amylose "chains" vary within the same sample, but over 1000 glucose units occur per amylose molecule. Formula masses ranging from 150,000 to 600,000 have been measured. The long amylose molecules coil into spiral-like helices, which tuck a significant fraction of the OH groups inside and away from contact with water. Thus amylose is only slightly soluble in water.

Amylopectin molecules have both $\alpha(1 \rightarrow 4)$ and $\alpha(1 \rightarrow 6)$ glycosidic links, as seen in Figure 19.9. The $\alpha(1 \rightarrow 6)$ bridges link the C-1 ends of linear amylose-type units to C-6 positions of glucose units in other long amylose chains, as seen in Figure 19.9. There are hundreds of such links per molecule, so amylopectin is heavily branched, and the branches prevent any coiling of the polymer. This leaves many more OH groups exposed to water than in amylose, so amylopectin tends to be somewhat more soluble in water than is amylose. However, neither dissolves well. The "solution" is actually a colloidal dispersion, because it gives the Tyndall effect (page 182).

Natural starches are about 10 to 20% amylose and 80 to 90% amylopectin. Neither is a reducing carbohydrate and neither gives a positive Tollens' or Benedict's test. One unique test that starch does give is called the **iodine test,** and it can detect extremely minute traces of starch.[1] When a drop of iodine reagent is added to starch, an intensely purple color develops as the iodine molecules become trapped within the vast network of starch molecules. In a starch sample undergoing hydrolysis, this network gradually breaks up so the system slowly loses its ability to give the iodine test.

FIGURE 19.8
Amylose—partial structure.

$n > 1000$

[1] The starch-iodine reagent is made by dissolving iodine, I_2, in aqueous potassium iodide, KI. Iodine by itself is very insoluble in water, but iodine molecules combine with iodide ions to form the triiodide ion, I_3^-. Molecular iodine is readily available from this ion if some reactant is able to react with it.

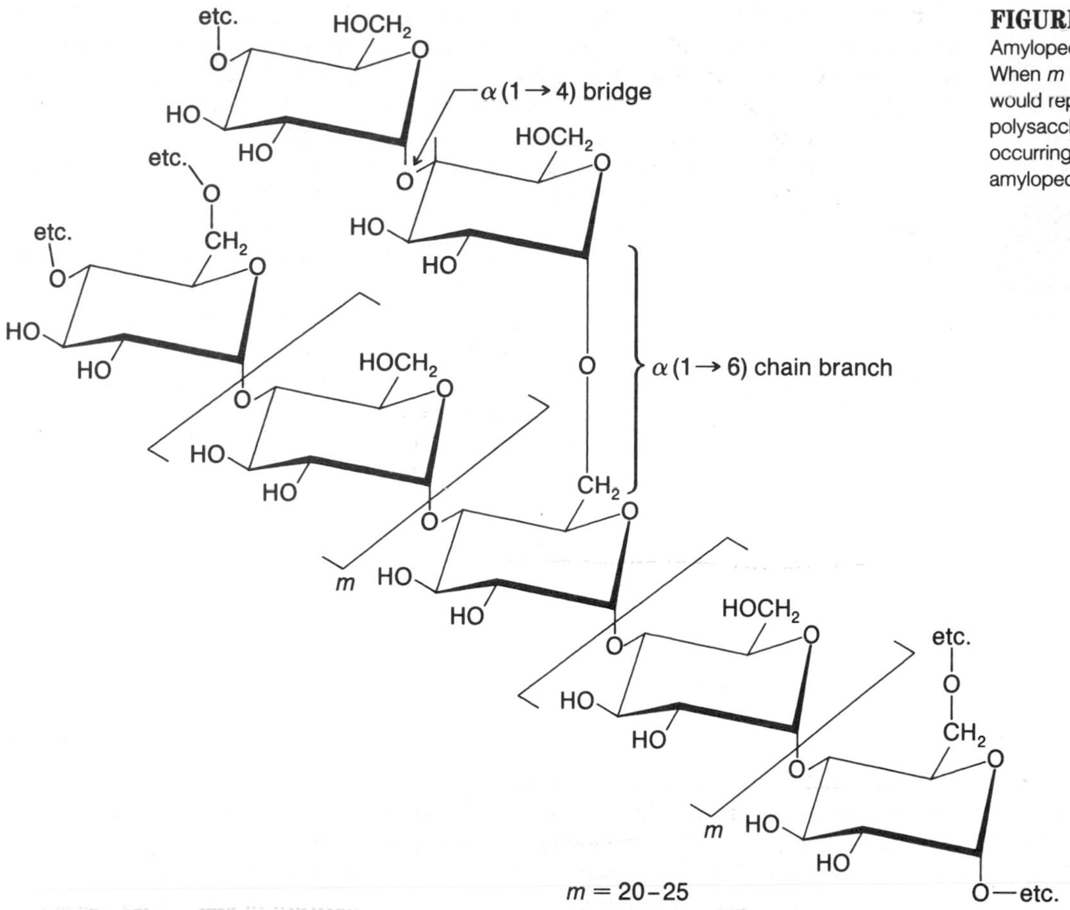
etc.
HOCH2
HO
HO
etc.
O
α(1→4) bridge
HOCH2
O
CH2
etc.
HO
HO
HOCH2
O
α(1→6) chain branch
CH2
m
HOCH2
etc.
CH2
m
m = 20–25
O—etc.

FIGURE 19.9
Amylopectin—partial structure. When $m = 6–12$, the structure would represent glycogen, a polysaccharide with branches occurring more frequently than in amylopectin.

The glycosidic links in starch are easily hydrolyzed in the presence of acids or the appropriate enzymes, which humans have. Thus the complete digestion of starch gives us only glucose. The partial hydrolysis of starch produces smaller polymer molecules that make up a substance known as *dextrin,* which has been used to manufacture mucilage and paste.

■ So-called *soluble* starch is partially hydrolyzed starch, and its smaller molecules more easily dissolve in water.

Glycogen Is the Storage Form of Glucose in Animals We and many animals use plant starch for food. Digestion hydrolyzes starch, and what glucose our bodies cannot use right away is changed into an amylopectin-like polymer called *glycogen*. In this form, we can store the chemical energy of glucose units. Normally we don't excrete any excess glucose. If we eat enough to replenish glycogen reserves in various tissues, any additional glucose is converted to fat (to the satisfaction of a huge weight-watcher industry).

■ Glycogen is sometimes called animal starch.

Glycogen molecules are essentially like those of amylopectin, perhaps even more branched. When the values of m in the structure of amylopectin (Fig. 19.9) are in the range of 6 to 12, the structure would be that of glycogen. The formula masses of various samples of glycogen have been reported in the range of 300,000 to 100,000,000, which correspond roughly to 1700 to 600,000 glucose units per molecule. We store glucose as glycogen principally in the liver and in muscle tissue.

Cellulose Is a Polymer of β-Glucose Much of the glucose a plant makes by photosynthesis goes to make cellulose and other substances that it needs to build its cell walls and its rigid fibers. Cellulose is thus a major component of the food fiber in our diets. Cellulose, unlike starch or glycogen, has a geometry that allows its molecules to line up side by side, overlap each other, and twist into fibers.

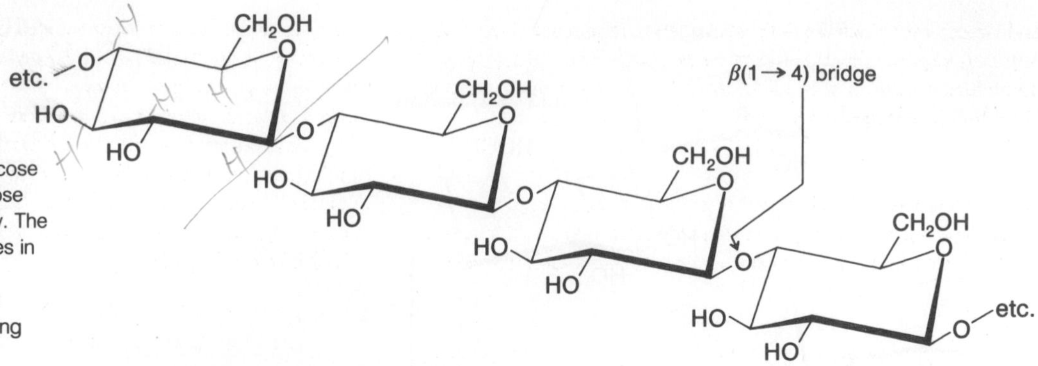

FIGURE 19.10

Cellulose—partial structure. In cotton, this polymer of β-D-glucose has from 2000 to 26,000 glucose units, depending on the variety. The strength of a cotton fiber comes in part from the thousands of hydrogen bonds that can exist between parallel and overlapping cellulose molecules.

- Cellobiose is to cellulose what maltose is to starch.

The huge geometric difference that allows cellulose to form fibers but not amylose is the orientation of the oxygen bridge. Cellulose is a polymer of the beta form of glucose. All the oxygen bridges are $\beta(1 \rightarrow 4)$. See Figure 19.10. Cellulose molecules, moreover, have no branches corresponding to the $\alpha(1 \rightarrow 6)$ branches in amylopectin. All the substituents in the rings in cellulose project in the most stable directions (the equatorial directions as discussed in Special Topic 19.2). The cellulose molecule is thus quite ribbon-like, so it's easy for neighboring molecules to nestle to each other where hydrogen bonds between molecules stabilize the aggregations. With twistings of these collections, cellulose fibers of considerable strength are possible.

- The adults in some ethnic human groups lack the digestive enzyme lactase that catalyzes hydrolysis of the β-glycosidic link between galactose and glucose units in lactose.

As we have noted, humans have no enzyme that can catalyze the hydrolysis of a beta-glycosidic link in cellulose, so none of the huge supply of cellulose in the world, or the cellobiose that could be made from it, is nutritionally useful to us. Many bacteria have this enzyme, however, and some strains dwell in the stomachs of cattle and other animals. Bacterial action converts cellulose in hay and other animal feed into small molecules that the larger animals can then use. Fungi and termites also can hydrolyze cellulose, enabling them to cause the decay of woody debris.

- If you spill *concentrated* acid on your clothes or skin, flush the area with water immediately.

The oxygen bridges in the cellulose of cotton fabrics, being acetal systems, are hydrolyzed when a trace of acid catalyst is present. Perhaps you have discovered this the morning after you spilled some acid on your jeans in lab. (If you know that you have spilled a small amount of dilute acid on jeans, put a small spatulaful of sodium bicarbonate or sodium carbonate on a towel, make a paste, and daub the spot with it. A towel moistened with dilute ammonia also works, but watch out for the ammonia odor. You might be able to save the fabric if you act quickly.)

SUMMARY

Carbohydrates Carbohydrates are aldehydes or ketones with multiple OH groups or are glycosides of these. Those that can't be hydrolyzed are the monosaccharides, which in pure forms exist as cyclic hemiacetals or cyclic hemiketals that can mutarotate and that are reducing sugars.

D- and L families of carbohydrates Fischer projection structures of open-chain forms of monosaccharides are made according to a set of rules. The carbon chain is positioned vertically with any carbonyl group as close to the top as possible. At each stereocenter, this chain projects toward the back. Any groups on bonds that appear horizontal project forward. If the OH group of a carbohydrate that is farthest from the carbonyl group projects to the right in a Fischer projection structure, the carbohydrate is in the D-family. If this OH group projects to the left, the substance is in the L-family.

Monosaccharides The three nutritionally important monosaccharides are glucose, galactose, and fructose—all in the D-family. Glucose is the chief carbohydrate in blood. Galactose, which differs from glucose only in the orientation of the OH at C-4, is obtained (together with glucose) from the hydrolysis of lactose. Fructose, a reducing ketohexose, differs from glucose only in the location of the carbonyl group. It's at C-2 in fructose and at C-1 in glucose.

Disaccharides The disaccharides are glycosides whose molecules hydrolyze into two monosaccharide molecules when they react with water. Maltose is made of two glucose units joined by an $\alpha(1 \rightarrow 4)$ glycosidic link. In a molecule of lactose (milk sugar)—a galactoside—a galactose unit joins a glucose unit by a $\beta(1 \rightarrow 4)$ oxygen bridge. In sucrose (cane or beet sugar), there is an oxygen bridge from C-1 of a glucose unit to C-2 of a fructose unit. Both maltose and lactose retain hemiacetal systems, so both mutarotate and are reducing sugars. They also exist in α- and β-forms. Sucrose is a nonreducing disaccharide. The digestion of these disaccharides gives their monosaccharide units.

Polysaccharides Three important polysaccharides of glucose are starch (a plant product), glycogen (an animal product), and cellulose (a plant fiber). In molecules of each, $(1 \rightarrow 4)$ glycosidic links occur.

They're alpha bridges in starch and glycogen and beta bridges in cellulose. In the molecules of the amylopectin portion of starch as well as in glycogen, numerous $\alpha(1 \rightarrow 6)$ bridges also occur. No polysaccharide gives a positive test with Tollens' or Benedict's reagents. Starch gives the iodine test. As starch is hydrolyzed its molecules successively break down to dextrins, maltose, and finally glucose. Humans have enzymes that catalyze the hydrolysis of $\alpha(1 \rightarrow 4)$ and $\alpha(1 \rightarrow 6)$ glycosidic links, but not the $\beta(1 \rightarrow 4)$ glycosidic links of cellulose.

REVIEW EXERCISES

The answers to Review Exercises whose numbers are in color are found in Appendix C. The answers to the other Review Exercises are found in the Study Guide that accompanies this book. The more challenging questions are marked with asterisks.

Biochemistry

19.1 Substances in the diet must provide raw materials for what three essentials for life?

19.2 What are the three broad classes of foods?

19.3 What kind of compound carries the genetic "blueprints" of a cell?

19.4 What is as important to the life of a cell as the chemicals that make it up or that it receives?

Carbohydrate Terminology

19.5 Examine the following structures and identify by letter(s) which structure(s) fit each of the labels. If a particular label is not illustrated by any structure, state so.

```
  CH=O          CH2OH
   |             |
  CHOH   CH=O   C=O    CH=O
   |      |      |      |
  CHOH   CHOH   CHOH   CHOH
   |      |      |      |
  CHOH   CHOH   CHOH   CHOH
   |      |      |      |
  CHOH   CHOH   CHOH   CH2
   |      |      |      |
  CH2OH  CH2OH  CH2OH  CH2OH
   A      B      C      D
```

(a) ketose (b) deoxy sugar
(c) aldohexose (d) aldopentose

19.6 Write the structure (open-chain form) that illustrates
(a) any ketopentose
(b) any aldotetrose

19.7 What is the structure and the common name of the simplest aldose?

19.8 What is the structure and the common name of the simplest ketose?

19.9 A sample of 0.0001 mol of a carbohydrate reacted with water in the presence of a catalyst and 1 mol of glucose was produced. Classify this carbohydrate as a mono-, di-, or polysaccharide.

19.10 An unknown carbohydrate gives a positive Benedict's test. Classify it as a reducing or a nonreducing carbohydrate.

19.11 An unknown carbohydrate, **A**, gives the following reaction.

$$A + H_2O \xrightarrow{H^+ \text{ catalyst}} \text{galactose} + \text{glucose} + \text{xylose}$$

(a) What should be the coefficient of H_2O to balance this equation?
(b) How is **A** classified, as a mono-, di-, or trisaccharide?

*19.12 A student in an advanced lab was assigned the task of determining the structure of a carbohydrate. The empirical formula was found to be CH_2O and the compound had a molecular mass of 150. When it was allowed to react with as much acetic anhydride as it could, it was changed to $C_{13}H_{18}O_8$. Very gentle oxidation (Tollens' reagent) changed the carbohydrate into a monocarboxylic acid, $C_5H_{10}O_6$. Vigorous reduction yielded pentane. Write a structure (open-chain) for the carbohydrate that is consistent with these observations.

*19.13 An unknown carbohydrate could be reduced by a series of steps to butane. It gave a positive Benedict's test, and its molecules had just one stereocenter. What is a structure consistent with these facts?

19.14 A student proposed the following two structures as the likeliest candidate structures for a carbohydrate being studied.

```
                                     O
                                     ||
HOCH—CH—CH—CH—CH—CH
     |   |   |   |   |
     OH  OH  OH  OH  OH
                  A
```

```
                                   O
                                   ||
CH2—CH—CH—CH—CH—CH
 |    |   |   |   |
 OH   OH  OH  OH  OH
                 B
```

One of these structures is highly unlikely. Which one, and why?

19.15 What is the name of the most abundant carbohydrate in blood?

Absolute Configurations

19.16 Consider the monomethyl ether of glyceraldehyde

$$CH_3OCH_2\overset{\underset{|}{OH}}{C}H\overset{\overset{O}{\|}}{C}H$$

(a) How many stereoisomers are possible for this compound?
(b) Draw the Fischer projection structures of the stereoisomers according to the conventions used for carbohydrates.
(c) Correctly label each structure as **D** or **L** according to the conventions used for carbohydrates.

19.17 The simplest ketotriose is never drawn in the form of a Fischer projection structure. Why not?

19.18 Write the Fischer projection formula of **L**-glucose. (Refer to Figure 19.4.)

19.19 What is the Fischer projection formula of **D**-2-deoxyribose?

19.20 Suppose that the aldehyde group of **D**-glyceraldehyde is oxidized to a carboxylic acid group, and that this is then converted to a methyl ester under conditions that do not touch any of the four bonds to C-2. What is the Fischer projection formula of this methyl ester? To what family, **D** or **L**, does it belong? Explain.

$$
\begin{array}{c}
CH_2OH \\
| \\
C=O \\
HO——H \\
H——HO \\
HO——H \\
| \\
CH_2OH
\end{array}
$$

19.21 Sorbose has the structure given below. It is made by the fermentation of sorbitol, and hundreds of tons of sorbose are used each year to make vitamin C.

(a) In what configurational family, **D** or **L**, is sorbose?
(b) Write the Fischer projection structure of the enantiomer of sorbose.
(c) Write the Fischer projection structure of **D**-3-deoxysorbose.

19.22 Sorbitol, $C_6H_{14}O_6$, is found in the juices of many fruits and berries (e.g., pears, apples, cherries, and plums). It can be made by the addition of hydrogen to **D**-glucose:

$$C_6H_{12}O_6 + H_2 \xrightarrow[\text{heat and pressure}]{\text{Ni}} C_6H_{14}O_6$$

D-Glucose   **D**-Sorbitol

Write the Fischer projection structure of **D**-sorbitol.

19.23 The magnesium salt of **D**-gluconic acid (Glucomag) is used as an antispasmodic and to treat dysmenorrhea. **D**-Gluconic acid, $C_6H_{12}O_7$, forms by the mild oxidation of **D**-glucose. What is the Fischer projection formula of **D**-gluconic acid?

Cyclic Forms of Carbohydrates

*19.24 Consider the following cyclic hemiacetal.

$$
\begin{array}{c}
CH_2OH \\
H\,C——O\quad H \\
C\,H\quad H\,C \\
HO\,C—C\quad OH \\
OH\ OH
\end{array}
$$

(a) What is the structure of its open form? (Write the open form with its chain coiled in the same way it is coiled in the closed form, above.)
(b) At which specific carbon (by number) does this compound differ from naturally occurring glucose? (The hemiacetal carbon has position 1 in the ring, and the ring is numbered clockwise from it.)
(c) Is the compound in the **D** or the **L**-family? (How can you tell without writing a Fischer projection structure?)
(d) With the aid of your answer to part (b) and Figure 19.4, what is the name of this compound?
(e) Is this compound an *anomer* or an *epimer* of **D**-glucose?

19.25 Examine the following structure. If you judge that it is either a cyclic hemiketal or a cyclic hemiacetal, write the structure of the open-chain form (coiled in like manner as the chain of the ring).

$$
\begin{array}{c}
HOCH_2\quad O\quad CH_2OH \\
H\ H\quad H\ OH \\
HO\quad HO
\end{array}
$$

*19.26 Mannose mutarotates like glucose. Mannose is identical with glucose except that in the cyclic structures the OH at C-2 in mannose projects on the same side of the ring as the CH_2OH group. Write the structures of the three forms of mannose that are in equilibrium after mutarotation gives a steady value of specific rotation. Identify which corresponds to α-mannose and which to β-mannose.

19.27 Allose is identical with glucose except that in its cyclic forms the OH group at C-3 projects on the opposite side of the ring from the CH_2OH group. Allose mutarotates like glucose. Write the structures of the three forms of allose that are in equilibrium after mutarotation gives a final value of specific rotation. Which structures are α- and β-allose?

19.28 If less than 0.05% of all galactose molecules are in their open-chain form at equilibrium in water, how can galactose give a strong, positive Tollens' test, a test good for the aldehyde group?

19.29 At equilibrium, after mutarotation, a glucose solution consists of 36% α-glucose and 64% β-glucose (and just a trace of the open form). Suppose that in some enzyme-catalyzed process the beta form is removed from this equilibrium. What becomes of the other forms of glucose?

19.30 Study the cyclic form of β-ribose on page 554 again. If its designation as β-ribose signifies a particular relationship between the CH_2OH group at C-4 and the OH group at C-1, what must the cyclic formula of α-ribose be?

19.31 Is the following structure that of α-fructose, β-fructose, or something else? Explain.

19.32 Write the cyclic structure of α-3-deoxyribose and draw an arrow that points to its hemiacetal carbon.

19.33 Could 4-deoxyribose exist as a cyclic hemiacetal with a five-membered ring (one of whose atoms is O)? Explain.

*19.34 With the aid of Figure 19.5 and the cyclic forms of D-fructose given in this chapter, write the structures of the cyclic forms of D-sorbose and correctly label your structures as α- or β-forms.

Glycosides

19.35 Using cyclic structures, write the structures of ethyl α-glucoside and ethyl β-glucoside. Are these two enantiomers or are they some other kind of stereoisomers?

19.36 What are the structures of methyl α-galactoside and methyl β-galactoside? Could these be described as cis—trans isomers? Explain.

Disaccharides

19.37 What are the names of the three nutritionally important disaccharides?

19.38 What is invert sugar?

19.39 Why isn't sucrose a reducing sugar?

*19.40 Examine the following structure and answer the questions about it.

(a) Does it have a hemiacetal system? Where? (Draw an arrow to it or circle it.)

(b) Does it have an acetal system? Where? (Circle it.)

(c) By what specific symbolism would the oxygen bridge be described? As an example of the kind of symbolism meant, recall that a bridge might be described as α(1 → 6).

(d) Does this substance give a positive Benedict's test? Explain.

(e) In what specific structural way does it differ from maltose?

(f) What are the names of the products of the acid-catalyzed hydrolysis of this compound.

19.41 Trehalose is a disaccharide found in young mushrooms and yeast, and it is the chief carbohydrate in the hemolymph of certain insects. On the basis of its structural features, answer the following questions.

(a) Is trehalose a reducing sugar? Explain.

(b) Can trehalose mutarotate? Explain.

(c) Identify, by name only, the products of the hydrolysis of trehalose.

*19.42 Maltose has a hemiacetal system. Write the structure of maltose in which this group has changed to the open form.

*19.43 When lactose undergoes mutarotation, one of its rings opens up. Write this open form of lactose.

Polysaccharides

19.44 Name the polysaccharides that give only D-glucose when they are completely hydrolyzed.

19.45 What is the main structural difference between amylose and cellulose?

19.46 How are amylose and amylopectin alike structurally?

19.47 How are amylose and amylopectin different structurally?

19.48 Why can't humans digest cellulose?

19.49 What is the iodine test? Describe the reagent and state what it is used to test for and what is seen in a positive test.

19.50 How do amylopectin and glycogen compare structurally?

19.51 How does the body use glycogen?

Photosynthesis (Special Topic 19.1)

19.52 The energy available in glucose originated in the sun. Explain in general terms how this happened.

19.53 Write the simple, overall equation for photosynthesis.

19.54 What is the name and color of the energy-absorbing pigment in plants?

19.55 In what general region of the planet earth is most of the photosynthesis carried out? By what organisms?

19.56 Describe in general terms the oxygen cycle of our planet including the function of photosynthesis in it.

Boat and Chair Forms of Saturated Six-Membered Rings (Special Topic 19.2)

Start by simply tracing the structures:, then practice drawing the skeletons of the two chair forms of six-membered rings.

19.57 Why are chair forms of saturated, six-membered rings more stable than boat forms?

19.58 Why are equatorial positions for groups attached to a six-membered ring more stable than axial?

*19.59 Draw the structures of the following.

(a) A chair form of cyclohexane. Label the bonds that can hold substituents as being axial (*a*) or equatorial (*e*).

(b) The least stable structure of *trans*-1,2-dimethylcyclohexane.

(c) The most stable structure of *trans*-1,2-dimethylcyclohexane.

(d) The most stable form of D-glucose.

(e) The most stable form of D-allose. (Hint: Refer to the Fischer projection structure of D-allose in Figure 19.4 and note where it is the same as the Fischer projection structure of D-glucose and where it differs.)

Additional Exercises

*19.60 A freshly prepared aqueous solution of sucrose gives a negative Tollens' test (as expected), but when the solution has stood at room temperature for about a week it gives this test. Explain.

*19.61 A freshly prepared solution (actually a dispersion) of starch in water gives a positive iodine test. If this solution is warmed with a trace of human saliva, however, the ability of the solution to give this test gradually disappears. How might this observation be explained?

*19.62 How many atoms (C and O) are there in the largest size ring possible for a cyclic hemiacetal form of an aldotetrose?

*19.63 Glyceraldehyde does not form a cyclic hemiacetal. Offer an explanation for this fact.

LIPIDS

Peanuts, olives, and corn. Humble fare, yet delightful to eat. That oils can be pressed from them was discovered ages ago, and such oils, members of the lipid family studied in this chapter, are the polyunsaturated oils widely recommended over saturated fats for heart-friendly diets.

20.1
WHAT LIPIDS ARE
20.2
CHEMICAL PROPERTIES OF TRIACYLGLYCEROLS
20.3
PHOSPHOLIPIDS

20.4
STEROIDS
20.5
CELL MEMBRANES — THEIR LIPID COMPONENTS

20.1

WHAT LIPIDS ARE

The lipids include the edible fats and oils whose molecules consist of esters of long-chain fatty acids and glycerol.

■ "Lipid" is from the Greek *lipos,* fat.

When undecomposed plant or animal material is crushed and ground with a nonpolar solvent such as ether, whatever dissolves is classified as a **lipid.** This operation catches a large variety of relatively nonpolar substances, and all are lipids. Thus this broad family is defined not by one common structure but by the technique used to isolate its members, *solvent extraction.* Among the many substances that won't dissolve in nonpolar solvents are carbohydrates, proteins, other very polar organic substances, inorganic salts, and water. The chart in Figure 20.1 outlines the many kinds of lipids.

■ Extraction means to shake or stir a mixture with a solvent that dissolves just part of the mixture.

Lipids Are Broadly Subdivided According to the Presence of Hydrolyzable Groups One of the major classes of lipids, the **hydrolyzable lipids,** consists of compounds with one or more groups that can be hydrolyzed. In nearly all examples, these are *ester groups.* A number of families are in the lipid group, including the neutral fats, the waxes, the phospholipids, and the glycolipids. The *neutral fats* include such familiar food products as butterfat, lard (pork fat), tallow (beef fat), olive oil, corn oil, and peanut oil. Thus some neutral fats are solids and others are liquids at room temperature. The solid neutral fats are generally from animals and so are called the *animal fats.* The liquid neutral fats are from plants and are called the *vegetable oils.* The fats and oils are the high-calorie components of the diet. The conversion factor used by the National Academy of Sciences is 9.0 kcal/g for food fat, which can be compared to 4.0 kcal/g for proteins and carbohydrates.

■ What is *neutral* about the neutral fats is the absence of electrical charges on their molecules.

■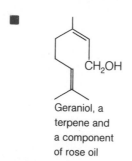

Geraniol, a
terpene and
a component
of rose oil

The **nonhydrolyzable lipids** lack groups that can be hydrolyzed. These include the steroids such as cholesterol and many sex hormones (Section 20.4). Many plants produce another group of nonhydrolyzable lipids called the *terpenes,* which often are responsible for the very pleasant odors of plant oils. Oil of rose, for example, is 40–60% geraniol, a terpene alcohol with a sizeable hydrocarbon unit.

The *Fatty Acids* Are Mostly Long-Chain, Unbranched Monocarboxylic Acids
When hydrolyzable lipids react with water, among the products are carboxylic acids (or their anions) that are dubbed the **fatty acids.** The fatty acids obtained from the lipids of most plants and animals share the following features.

Model of stearic acid, a saturated fatty acid.

Structural Features of the Common Fatty Acids

1. They are usually *mono*carboxylic acids, RCO_2H.
2. The R group is usually a long *unbranched* chain.
3. The number of carbon atoms is almost always *even.*
4. The R group can be saturated, or it can have one or more double bonds, which are cis.

FIGURE 20.1
Lipid families.

```
                              ┌─────────┐
                              │ LIPIDS  │
                              └────┬────┘
              ┌────────────────────┴────────────────────┐
      ┌───────────────┐                          ┌───────────────┐
      │ Hydrolyzable  │                          │ Nonhydrolyzable│
      │    lipids     │                          │    lipids     │
      └───────┬───────┘                          └───────┬───────┘
   (Only C, H, O)    (C, H, O, P, N)
┌─────┬──────────┐  ┌──────────┬──────────┐  ┌────────┬─────────┬────────┐
```

| Waxes | Triacylglycerols | | Esters of glycerol | | Esters of sphingosine | | Steroids | Terpenes | Others |

| | Animal fats | Vegetable oils | Phospha-tides | Plasmal-ogens | Sphingo-myelins | Cerebro-sides |

Thus just two functional groups are present in the fatty acids, both often in the same molecule — the alkene double bond and the carboxyl group.

The most abundant *saturated* fatty acids are palmitic acid, $CH_3(CH_2)_{14}CO_2H$, and stearic acid, $CH_3(CH_2)_{16}CO_2H$, which have 16 and 18 carbons, respectively. Refer back to Table 16.1 for the other saturated fatty acids obtainable from lipids — the acids with more carbon atoms than acetic acid but with *even* numbers of carbons, like butanoic, hexanoic, octanoic, and decanoic acids. Fatty acids with fewer than 16 carbons, however, are relatively rare in nature.

The *unsaturated* fatty acids most commonly obtained from lipids are listed in Table 20.1 and include palmitoleic acid (C_{16}) and the C_{18} acids, oleic, linoleic, and linolenic acids. The double bonds in the unsaturated fatty acids of Table 20.1 are cis. Oleic acid is the most abundant and most widely distributed fatty acid in nature.

The presence of cis alkene groups causes kinks in the long hydrocarbon groups of the unsaturated fatty acids, which affects their melting points, as you can see in Table 20.1. As more alkene groups per molecule are present, the melting points of the fatty acids decrease. The structural kinks inhibit the closeness of the packing of molecules in crystals, and such closeness is required for stronger forces of attraction between molecules that cause higher melting points. The relationship between double bonds per molecule and melting point carries over to the neutral fats. The animal fats happen to have fewer alkene groups per molecule than the vegetable oils and so the animal fats are likely to be solids at room temperature and the vegetable oils tend to be liquids.

Model of linoleic acid.

PRACTICE EXERCISE 1

To visualize how a cis double bond introduces a kink into a molecule, write the structure of oleic acid in a way that correctly shows the cis geometry of the alkene group. (Without the double bond, the entire sidechain can stretch out into a perfect zigzag conformation, as in stearic acid. This makes it easy for two sidechains to nestle very close to each other.)

TABLE 20.1
Common Unsaturated Fatty Acids

Name	Double Bonds	Total Carbons	Structure	MP (°C)
Palmitoleic acid	1	16	$CH_3(CH_2)_5CH{=}CH(CH_2)_7CO_2H$	32
Oleic acid	1	18	$CH_3(CH_2)_7CH{=}CH(CH_2)_7CO_2H$	4
Linoleic acid	2	18	$CH_3(CH_2)_4CH{=}CHCH_2CH{=}CH(CH_2)_7CO_2H$	−5
Linolenic acid	3	18	$CH_3CH_2CH{=}CHCH_2CH{=}CHCH_2CH{=}CH(CH_2)_7CO_2H$	−11
Arachidonic acid	4	20	$CH_3(CH_2)_4CH{=}CHCH_2CH{=}CHCH_2CH{=}CHCH_2CH{=}CH(CH_2)_3CO_2H$	−50

The prostaglandins were discovered in the mid-1930s by a Swedish scientist, Ulf von Euler (Nobel prize, 1970), but they didn't arouse much interest in medical circles until the late 1960s, largely through the work of Sune Bergstrom. It became apparent that these compounds, which occur widely in the body, affect a large number of processes. Their general name comes from an organ, the prostate gland, from which they were first obtained. About 20 are known, and they occur in four major subclasses designated as PGA, PGB, PGE, and PGF. (A subscript is generally placed after the third letter to designate the number of alkene double bonds that occur outside of the five-membered ring.) The structures of some typical examples are shown here.

Arachidonic acid

Several steps
(inhibited by aspirin)

CO_2H

$PGF_{2\alpha}$

PGA₁

PGB₁

PGE₁

Prostaglandins are made from twenty-carbon fatty acids, like arachidonic acid. By coiling a molecule of this acid, as shown, you can see how its structure needs only a ring closure (suggested by the dashed arrow) and three more oxygen atoms to become PGF_2. The oxygen atoms are all provided by molecular oxygen itself.

Prostaglandins as Chemical Messengers The prostaglandins are like hormones in many ways, except that they do not act globally, that is, over the entire body. They do their work within the cells where they are made or in nearby cells, so they are sometimes called *local hormones.* This is perhaps why the prostaglandins have such varied functions; they occur and express their roles in such varied tissues. They work together with hormones to modify the chemical messages that hormones bring to cells. In some cells, the prostaglandins inhibit enzymes and in others they activate them. In some organs, the prostaglandins help to regulate the flow of blood within them. In others, they affect the transmission of nerve impulses.

Some prostaglandins enhance inflammation in a tissue, and it is interesting that aspirin, an inflammation reducer, does exactly the opposite. This effect is caused by aspirin's ability to inhibit the work of an enzyme needed in the synthesis of prostaglandins.

Prostaglandins as Pharmaceuticals In experiments that use prostaglandins as pharmaceuticals, they have been found to have an astonishing variety of effects. One prostaglandin induces labor at the end of a pregnancy. Another stops the flow of gastric juice while the body heals an ulcer. Other possible uses are to treat high blood pressure, rheumatoid arthritis, asthma, nasal congestion, and certain viral diseases.

THE OMEGA-3 FATTY ACIDS AND HEART DISEASE

Omega-3 refers to the location of a double bond third in from the far end of a long-chain fatty acid, particularly those with 18, 20, and 22 carbons. Just as omega, ω, is the last letter in the Greek alphabet, so the omega position in a fatty acid is the one farthest from the carboxyl group. Arachidonic acid (Table 20.1) is an omega-6 C-20 fatty acid because it has a double bond at the sixth carbon from the omega end. Linolenic acid (Table 20.1) is an omega-3 fatty acid. Two other omega-3 acids are considered by some to be important in metabolism:

$$CH_3CH_2CH{=}CHCH_2(CH{=}CHCH_2)_4(CH_2)_2CO_2H$$
ω-3-Eicosapentaenoic acid

$$CH_3CH_2CH{=}CHCH_2(CH{=}CHCH_2)_5CH_2CO_2H$$
ω-3-Docosahexaenoic acid

The basis of the interest in these is that Eskimos have low incidences of heart disease despite relatively high cholesterol levels in their diets (from fish oils and fish liver). Some scientists believe that the high level of the omega-3 fatty acids in marine oils provides protection against disease.

The properties of the fatty acids are those to be expected of compounds with carboxyl groups, double bonds (where present), and long hydrocarbon chains. Thus they are insoluble in water and soluble in nonpolar solvents. The fatty acids are neutralized by bases to form salts, and they can be esterified by reacting with alcohols (see Section 16.5). Fatty acids with alkene groups react with bromine or hydrogen in the presence of a catalyst.

The *prostaglandins* are an unusual family of fatty acids with 20 carbons, five-membered rings, and a wide variety of effects in the body. See Special Topic 20.1.

In the late 1980s, one small group of fatty acids, the omega-3 fatty acids, appeared in scientific debates about the value of fish or marine oils in the diet. Special Topic 20.2 describes them further.

■ Omega (ω) designation

CH_3 ω-1
|
CH_2 ω-2
|
CH ω-3
‖
CH
|
(remainder of fatty acid)

The Triacylglycerols (or Triglycerides) Are Triesters of Glycerol and Fatty Acids

The molecules of the most abundant lipids are the **triacylglycerols** or the **triglycerides.** They are triesters between glycerol and three fatty acids.

Components of triacylglycerols (neutral fats and oils)

As you can see, there are no (+) or (−) charges on triacylglycerol molecules and so they are unlike the more complex hydrolyzable lipids to be studied in Section 20.3. The triacylglycerols include lard (pork fat), tallow (beef fat), butterfat — all animal fats — as well as such plant oils (vegetable oils) as olive oil, cottonseed oil, corn oil, peanut oil, soybean oil, coconut oil, and linseed oil.

In a particular fat or oil, certain fatty acids predominate, others either are absent or are present in trace amounts, and virtually all of the molecules are triacylglycerols. Data on the fatty acid compositions of several fats and oils are listed in Table 20.2. Oleic acid (C_{18}; one alkene group) is very common among both the fats and oils. Notice particularly, however, that

■ The name *triglycerides* is common in the older scientific literature on triacylglycerols.

TABLE 20.2
Fatty Acids Obtained from Neutral Fats and Oils

Type of Lipid	Fat or Oil	Average Composition of Fatty Acids (%)					
		Myristic Acid	Palmitic Acid	Stearic Acid	Oleic Acid	Linoleic Acid	Others
Animal fats	Butter	8–15	25–29	9–12	8–33	2–4	a
	Lard	1–2	25–30	12–18	48–60	6–12	b
	Beef tallow	2–5	24–34	15–30	35–45	1–3	b
Vegetable oils	Olive	0–1	5–15	1–4	67–84	8–12	
	Peanut	—	7–12	2–6	30–60	20–38	
	Corn	1–2	7–11	3–4	25–35	50–60	
	Cottonseed	1–2	6–10	2–4	20–30	50–58	
	Soybean	1–2	6–10	2–4	20–30	50–58	c
	Linseed	—	4–7	2–4	14–30	14–25	d
Marine oils	Whale	5–10	10–20	2–5	33–40	—	e
	Fish	6–8	10–25	1–3	—	—	e

a Also, 3–4% butyric acid, 1–2% caprylic acid, 2–3% capric acid, 2–5% lauric acid.

b Also, linolenic acid, 1%. c Also, linolenic acid, 5–10%. d Also, linolenic acid, 45–60%.

e Large percentages of other highly unsaturated fatty acids (see Special Topic 20.2).

■ An acyl group has the general structure:

$$\underset{RC-}{\overset{\displaystyle \overset{O}{\|}}{}}$$

the vegetable oils tend to incorporate more of the acyl groups of the unsaturated fatty acids, like those of oleic and linoleic acid, than do the animal fats. Thus vegetable oils have more double bonds per molecule and so are often described as *polyunsaturated*. The saturated fatty acyl units of palmitic and stearic acids are far more common in animal fats, which are thus sometimes called the *saturated fats*.

The three acyl units in the more common triacylglycerol molecules present in a given fat or oil are contributed by two or three *different* fatty acids. Fats and oils are thus mixtures of different molecules that share common structural features. Although we cannot give, for example, *one* structure for cottonseed oil, we can describe what is probably a fairly typical molecule, like that of structure **1**. Notice that the middle carbon of the glycerol unit in **1** is a stereocenter; it holds four different atoms or groups.

Model of structure **1**.

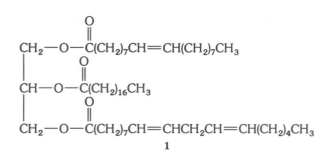

$$\begin{aligned}
&CH_2-O-\overset{\overset{\displaystyle O}{\|}}{C}(CH_2)_7CH=CH(CH_2)_7CH_3\\
&CH-O-\overset{\overset{\displaystyle O}{\|}}{C}(CH_2)_{16}CH_3\\
&CH_2-O-\overset{\overset{\displaystyle O}{\|}}{C}(CH_2)_7CH=CHCH_2CH=CH(CH_2)_4CH_3
\end{aligned}$$

1

Plant Waxes Are Simple Esters with Long Hydrocarbon Chains The waxes occur as protective coatings on fruit and leaves as well as on fur, feathers, and skin. Nearly all **waxes** are esters of long-chain monohydric alcohols and long-chain monocarboxylic acids in both of which there is an *even* number of carbons. As many as 26 to 34 carbon atoms can be incorporated into *each* of the alcohol and the acid units, which makes the waxes almost totally hydrocarbon-like.

$$R-O-\overset{\overset{\displaystyle O}{\|}}{C}-R'$$

Alcohol unit	Fatty acyl unit

Components of waxes

Any particular wax, like beeswax, consists of a mixture of similar compounds that share the kind of structure shown above. In molecules of lanolin (wool fat), however, the alcohol portion is contributed by steroid alcohols, which have large ring systems that we'll study in Section 20.4. Waxes exist in sebum, a secretion of human skin that helps to keep the skin supple.

■ Lanolin is used to make cosmetic skin lotions.

PRACTICE EXERCISE 2

One particular ester in beeswax can be hydrolyzed to give a straight-chain primary alcohol with 26 carbons and a straight-chain carboxylic acid with 28 carbons. Write the structure of this ester.

20.2

CHEMICAL PROPERTIES OF TRIACYLGLYCEROLS

Triacylglycerols can be hydrolyzed, saponified, and hydrogenated.

Triacylglycerols Can Be Hydrolyzed When we need the chemical energy of the triacylglycerols stored in our fat tissue, a special enzyme (a lipase) catalyzes their complete hydrolysis. The fatty acid molecules, bound to proteins (albumins) in the blood, are then sent to the liver. In general,

■ The solubility of free fatty acids in water is extremely low, only about 10^{-6} mol/L, so to be transported in blood fatty acids must be carried on protein molecules.

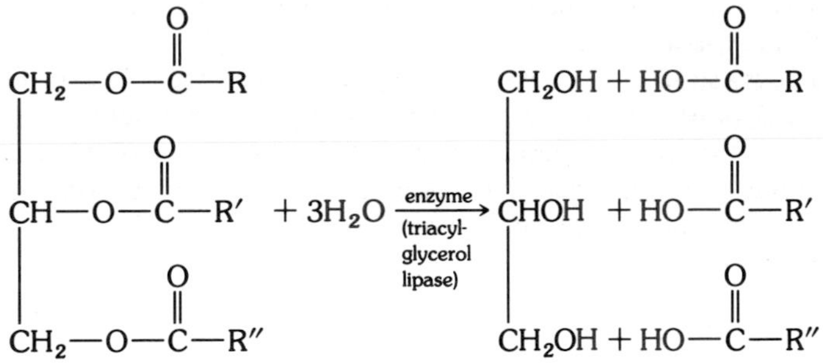

Triacylglycerol   Glycerol   Fatty acids

A specific example is

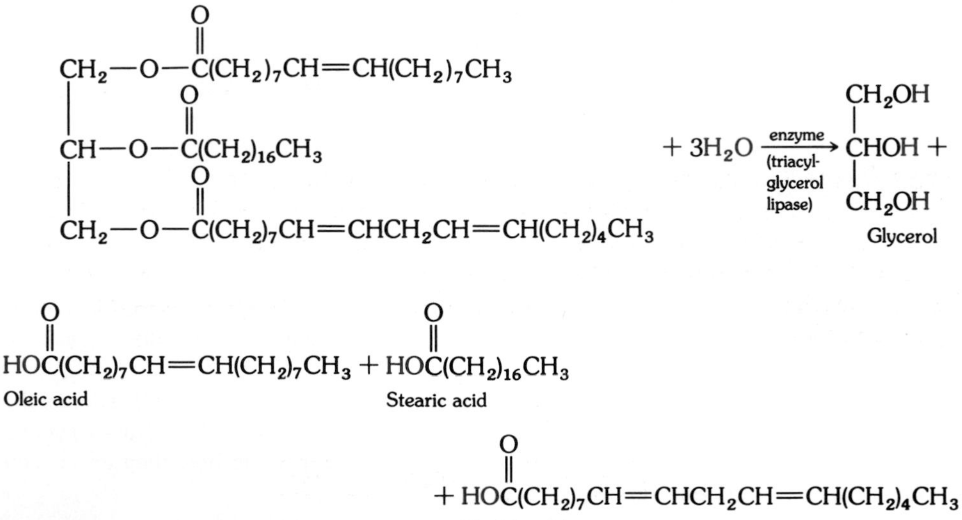

Glycerol

Oleic acid   Stearic acid

Linoleic acid

SPECIAL TOPIC 20.3
HOW DETERGENTS WORK

Soap Water is a very poor cleansing agent because it can't penetrate greasy substances, the "glues" that bind soil to skin and fabrics. When just a little soap is present, however, water cleans very well, especially warm water. Soap is a simple chemical, a mixture of the sodium or potassium salts of the long-chain fatty acids obtained by the saponification of fats or oils.

Detergents Soap is just one kind of detergent. All detergents are surface-active agents that lower the surface tension of water. All consist of ions or molecules that have long hydrocarbon portions plus ionic or very polar sections at one end. The accompanying structures illustrate these features and show the varieties of detergents that are available.

Although soap is manufactured, it is not called a synthetic detergent. This term is limited to detergents that are not soap, that is, not the salts of naturally occurring fatty acids obtained by the saponification of lipids. Most synthetic detergents are salts of sulfonic acids, but others have different kinds of ionic or polar sites. The great advantage of synthetic detergents is that they work in hard water and are not precipitated by the hardness ions — Mg^{2+}, Ca^{2+}, and the two ions of iron. These ions form messy precipitates ("bathtub ring") with the anions of the fatty acids present in soap. The anions of synthetic detergents do not form such precipitates.

Figure 1 shows how detergents work. In Figure 1a we see the hydrocarbon tails of the detergent work their way into the hydrocarbon environment of the grease layer. ("Like dissolves like" is the principle at work here.) The ionic heads stay in the water phase, and the grease layer becomes pincushioned with electrically charged sites. In Figure 1b we see the grease layer breaking up, aided with some agitation or scrubbing. Figure 1c shows a magnified view of grease globules studded with ionic groups and,

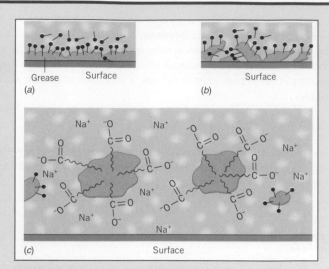

(a) Grease Surface   (b) Surface

(c) Surface

FIGURE 1

being like-charged, these globules repel each other. They also tend to dissolve in water, so they are ready to be washed down the drain.

$$CH_3(CH_2)_{14}CO_2^-Na^+$$

Soap—an anionic detergent

$$CH_3(CH_2)_{13}OSO_3^-Na^+$$

A sodium alkyl sulfate—an anionic detergent

$$CH_3(CH_2)_8-\text{benzene ring}-SO_3^-Na^+$$

A sodium alkylbenzenesulfonate—an anionic detergent

$$CH_3(CH_2)_{11}\overset{+}{N}(CH_3)_3Cl^-$$

A triethylalkylammonium ion—a cationic detergent

$$CH_3(CH_2)_8O(CH_2CH_2O)_nH$$

A nonionic detergent

When we *digest* triacylglycerols, hydrolysis is not complete. The digestive enzyme (pancreatic lipase) takes the hydrolysis only to monoacylglycerols, fatty acids, and some diacylglycerols.

Soaps Are Made by the Saponification of Triacylglycerols
The saponification of the ester links in triacylglycerols by the action of a strong base (e.g., NaOH or KOH) gives glycerol and a mixture of the salts of fatty acids. These salts are soaps, and how they exert their detergent action is described in Special Topic 20.3. In general,

$$\begin{array}{l}
\underset{\displaystyle \text{Triacylglycerol}}{
\begin{array}{l}
\mathrm{CH_2-O-\overset{\overset{\displaystyle O}{\|}}{C}-R} \\[1em]
\mathrm{CH-O-\overset{\overset{\displaystyle O}{\|}}{C}-R'} + 3\mathrm{NaOH}(aq) \\[1em]
\mathrm{CH_2-O-\overset{\overset{\displaystyle O}{\|}}{C}-R''}
\end{array}}
\xrightarrow{\text{heat}}
\underset{\displaystyle \begin{array}{l}\text{Glycerol} \quad \text{Mixture of sodium salts}\\ \qquad\qquad\text{of fatty acids (soap)}\end{array}}{
\begin{array}{l}
\mathrm{CH_2OH} + \mathrm{Na^+\ ^-O-\overset{\overset{\displaystyle O}{\|}}{C}-R} \\[1em]
\mathrm{CHOH} + \mathrm{Na^+\ ^-O-\overset{\overset{\displaystyle O}{\|}}{C}-R'} \\[1em]
\mathrm{CH_2OH} + \mathrm{Na^+\ ^-O-\overset{\overset{\displaystyle O}{\|}}{C}-R''}
\end{array}}
\end{array}$$

PRACTICE EXERCISE 3

Write a balanced equation for the saponification of **1** with sodium hydroxide.

The Hydrogenation of Vegetable Oils Gives Solid Shortenings and Margarine

When hydrogen is made to add to some of the double bonds in vegetable oils, the oils become like animal fats, both physically and structurally. One very practical consequence of such partial hydrogenation is that the oils change from being liquids to solids at room temperature. Many people prefer solid, lard-like shortening for cooking, instead of a liquid oil. Therefore the manufacturers of such "hydrogenated vegetable oils" as Crisco and Spry take inexpensive, readily available vegetable oils, like corn oil and cottonseed oil, and catalytically add hydrogen to some (not all) of the alkene groups in their molecules. We say that the double bonds become *saturated*. Unlike natural lard, the vegetable shortenings have no cholesterol.

■ Except for the absence of cholesterol, hydrogenated vegetable oils are chemically and nutritionally identical to animal fats.

PRACTICE EXERCISE 4

Write the balanced equation for the complete hydrogenation of the alkene links in structure **1**.

The chief lipid material in margarine is produced from vegetable oils in the same way. The hydrogenation is done with special care so that the final product can melt on the tongue, one property that makes butterfat so pleasant. (If all the alkene groups in a vegetable oil were hydrogenated, instead of just some of them, the product would be just like beef or mutton fat, relatively hard materials that would not melt on the tongue.)

The popular brands of peanut butter, those with peanut oils that do not separate, are made by the partial hydrogenation of the oil in real peanut butter. The peanut oil changes to a solid, hydrogenated form at room temperature and therefore it cannot separate.

20.3

PHOSPHOLIPIDS

Phospholipid molecules have very polar or ionic sites in addition to long hydrocarbon chains.

Phospholipids are esters of either glycerol or sphingosine, which is a long-chain, dihydric amino alcohol with one double bond.

$$\mathrm{CH_3(CH_2)_{12}CH=CHCHCHCH_2OH}$$
$$\qquad\qquad\qquad\underset{\text{HO}}{|}\ \underset{\text{NH}_2}{|}$$

Sphingosine

The phospholipids all have very polar but small molecular parts that are extremely important in the formation of cell membranes. We will survey their structures largely to demonstrate how they are both polar and hydrocarbon-like.

The Glycerophospholipids (Phosphoglycerides) Have Phosphate Units plus Two Acyl Units

The **glycerophospholipids** occur in two broad types, the *phosphatides* and the *plasmalogens*. Both are based on glycerol esters. Molecules of the **phosphatides** have two ester bonds from glycerol to fatty acids plus one ester bond to phosphoric acid. The phosphoric acid unit, in turn, is joined by a phosphate ester link to a small alcohol molecule. Without this link, the compound is called *phosphatidic acid*.

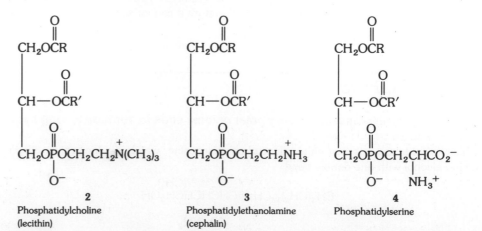

Phosphatidic acid components   Phosphatidic acid   Phosphatide components

Three particularly important phosphatides are esters between phosphatidic acid and either choline, ethanolamine, or serine, forming, respectively, phosphatidylcholine (lecithin), **2**, phosphatidylethanolamine (cephalin), **3**, and phosphatidylserine, **4**.

$$\overset{+}{HOCH_2CH_2N(CH_3)_3} \qquad HOCH_2CH_2NH_2 \qquad HOCH_2\underset{\underset{NH_3^+}{|}}{CH}CO_2^-$$

Choline   Ethanolamine   Serine

As the structures of **2**, **3**, and **4** given below show, one part of each phosphatide molecule is very polar because it carries full electrical charges. These charges are partly responsible for the phosphatides being somewhat more soluble in water than triacylglycerols. The remainder of a phosphatide molecule is nonpolar and hydrocarbon-like, so phosphatides can be extracted from animal matter by relatively nonpolar solvents.

A typical phosphatide.

■ *Cephalin* is from the Greek *kephale*, head. Cephalin is found in brain tissue.

$$
\begin{array}{ccc}
\overset{O}{\overset{\|}{CH_2OCR}} & \overset{O}{\overset{\|}{CH_2OCR}} & \overset{O}{\overset{\|}{CH_2OCR}} \\[2pt]
\overset{O}{\overset{\|}{CH-OCR'}} & \overset{O}{\overset{\|}{CH-OCR'}} & \overset{O}{\overset{\|}{CH-OCR'}} \\[2pt]
\underset{O^-}{\overset{O}{\overset{\|}{CH_2OPOCH_2CH_2\overset{+}{N}(CH_3)_3}}} & \underset{O^-}{\overset{O}{\overset{\|}{CH_2OPOCH_2CH_2\overset{+}{N}H_3}}} & \underset{O^- \ \ NH_3^+}{\overset{O}{\overset{\|}{CH_2OPOCH_2CHCO_2^-}}} \\[2pt]
\mathbf{2} & \mathbf{3} & \mathbf{4}
\end{array}
$$

Phosphatidylcholine (lecithin)   Phosphatidylethanolamine (cephalin)   Phosphatidylserine

These three are the most common hydrolyzable lipids used to make animal cell membranes.

When pure, lecithin is a clear, waxy solid that is very hygroscopic. In air, it is quickly attacked by oxygen, which makes it turn brown in a few minutes. Lecithin is a powerful emulsifying agent for triacylglycerols, and this is why egg yolks, which contain lecithin, are used to make the emulsions found in mayonnaise, ice cream, candies, and cake dough.

■ *Lecithin* is from the Greek *lekitos,* egg yolk—a rich source of this phospholipid.

The Plasmalogens Have Both Ether and Ester Groups

The **plasmalogens,** as we said, make up another family of glycerophospholipids, and they occur widely in the membranes of both nerve and muscle cells. They differ from the phosphatides by the presence of an unsaturated *ether* group instead of an acyl group at one end of the glycerol unit.

```
G ┌─ Fatty alcohol unit
l │
y │
c │
e ├─ Fatty acid unit
r │
o │
l └─ Phosphate unit ─── Amino alcohol unit
```

Plasmalogen components

Ether group

$$CH_2OCH=CHR$$
$$|$$
$$CH-OCR'$$ (with $C=O$)
$$|$$
$$CH_2OPOCH_2CH_2-\overset{+}{N}(CH_3)_3$$ (with $P=O$)
$$O^-\quad \text{or} -\overset{+}{N}H_3$$

Plasmalogens

Plasmalogen molecules, like phosphatides, also carry electrically charged positions as well as long hydrocarbon chains.

The Sphingolipids Are Based on Sphingosine, Not Glycerol

The two types of sphingosine-based lipids or **sphingolipids** are the *sphingomyelins* and the *cerebrosides,* and they are also important constituents of cell membranes, particularly those of nerve cells. The sphingomyelins are phosphate diesters of sphingosine. Their acyl units occur as acylamido parts, and they come from unusual fatty acids that are not found in neutral fats. Like the molecules of the phosphatides, those of the sphingolipids have two nonpolar tails.

The cerebrosides are not phospholipids. Instead they are **glycolipids,** lipids with a sugar unit (i.e., galactose or glucose) and not a phosphate ester system. The sugar unit, with its many OH groups, provides a strongly polar site, and it is usually a D-galactose or a D-glucose unit, or an amino derivative of these.

A sphingomyelin

```
S ┌─ OH
p │
h │
i │
n │
g ├─ Fatty acylamido unit
o │
s │
i │
n └─ Phosphate—alcohol
e      or
       Glycoside unit
```

Sphingolipid components

$$CH_3$$
$$(CH_2)_{12}$$
$$CH$$
$$\|$$
$$HC$$
$$CHOH$$
$$CHNHCR$$ (with $C=O$)
$$CH_2OPOCH_2CH_2\overset{+}{N}(CH_3)_3$$ (with $P=O$)
$$O^-$$

Sphingomyelins

β-D-galactose unit

(galactose ring structure with HO, CH_2OH, HO, OH, H, O linkages)

Cerebrosides

$$CH_3$$
$$(CH_2)_{12}$$
$$CH$$
$$\|$$
$$HC$$
$$CHOH$$
$$CHNHCR$$ (with $C=O$)
$$CH_2$$

The cerebrosides are particularly prevalent in the membranes of brain cells.

20.4

STEROIDS

Cholesterol and other steroids are nonhydrolyzable lipids.

Steroids are high-formula-mass aliphatic compounds whose molecules include a characteristic four-ring feature called the steroid nucleus. It consists of three six-membered rings and one five-membered ring, as seen in structure **5**. Several steroids are very active, physiologically.

■ Steroid alcohols are called *sterols*.

Cholesterol

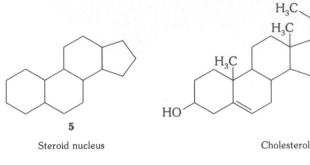

5

Steroid nucleus

Cholesterol

Table 20.3 lists several steroids and their functions, and you can see the diverse roles they have in the body.

Cholesterol Molecules Are Components of Cell Membranes Cholesterol is an unsaturated steroid alcohol that makes up a significant part of the membranes of the cells of animals. The membrane of a human red blood cell (erythrocyte), for example, has about 25% cholesterol by mass, and so is particularly rich in cholesterol. Cholesterol is the body's raw material for making bile salts and steroid hormones, including the sex hormones listed in Table 20.3. Little cholesterol is found in plants, but they have compounds with similar structures.

■ Cholesterol is the chief constituent in gallstones.

Cholesterol enters the body via the diet, but up to 800 mg per day can normally be synthesized in the liver from two-carbon acetate units. Cholesterol made in the liver is used to make the bile salts like sodium cholate (see Table 20.3) or is converted to esters of cholesterol.

Sodium cholate
(a bile salt)

Cholesteryl ester
(R is long-chain)

Bile salts are secreted into the intestinal tract where they function as powerful surface active agents (detergents). They aid both in the digestion of dietary lipids and also in the absorption of the fat-soluble vitamins and the fatty acids from the digestive tract (eventually) into circulation (see Section 6.8). Cholesterol is put into circulation in the bloodstream as components of *very-low-density lipoproteins* (or VLDL). Esters of cholesterol occur in higher density complexes. The relationships between cholesterol, esters of cholesterol, the various kinds of lipoproteins, and the risk of heart disease will be discussed when we study the metabolism of lipids. We need to know more about proteins, genes, and enzymes before we can discuss this topic.

■ *Lipoprotein* molecules are combinations of lipids and proteins.

TABLE 20.3
Important Steroids

Vitamin D₃ Precursor

Irradiation of this derivative of cholesterol by ultraviolet light opens one of the rings to produce vitamin D_3. Meat products are sources of this compound.

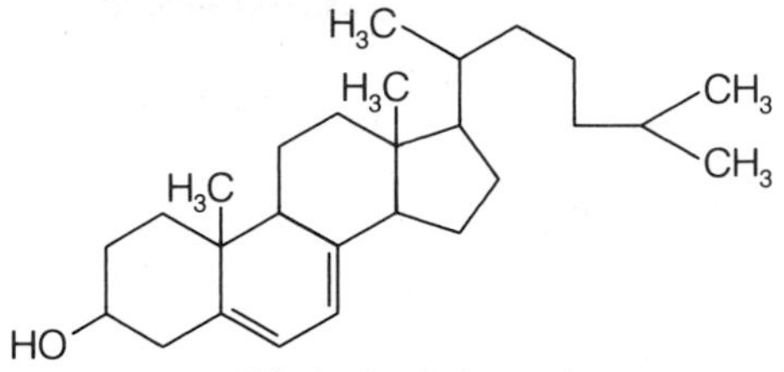

7-Dehydrocholesterol

Ultraviolet radiation

Vitamin D_3 is an antirachitic factor. Its absence leads to rickets, an infant and childhood disease characterized by faulty deposition of calcium phosphate and poor bone growth.

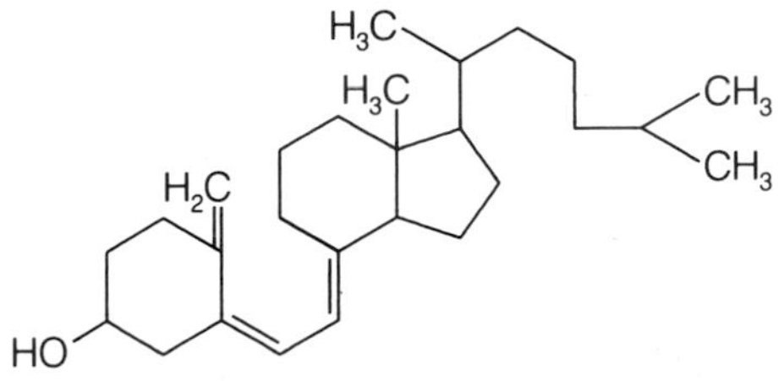

Vitamin D_3

Adrenocortical Hormone

Cortisol is one of the 28 hormones secreted by the cortex of the adrenal gland. Cortisone, very similar to cortisol, is another such hormone. When cortisone is used to treat arthritis, the body changes much of it to cortisol by reducing a keto group to the 2° alcohol group that you see in the structure of cortisol.

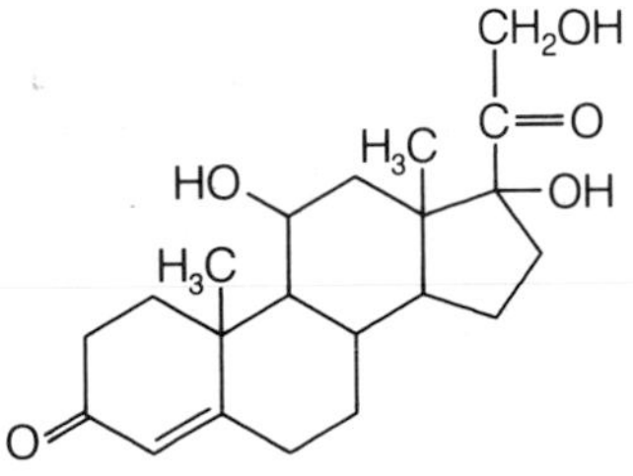

Cortisol

Sex Hormones

Estradiol is a human estrogenic hormone.

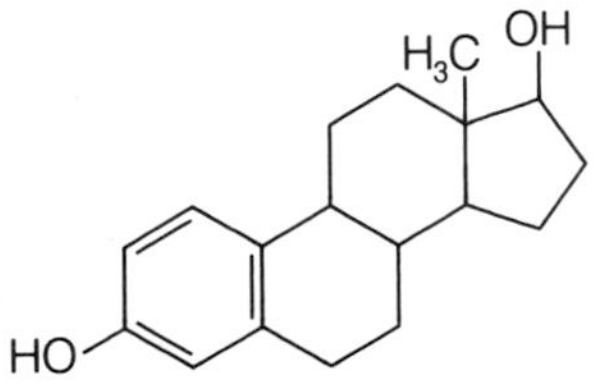

Estradiol

Progesterone, a human pregnancy hormone, is secreted by the corpus luteum.

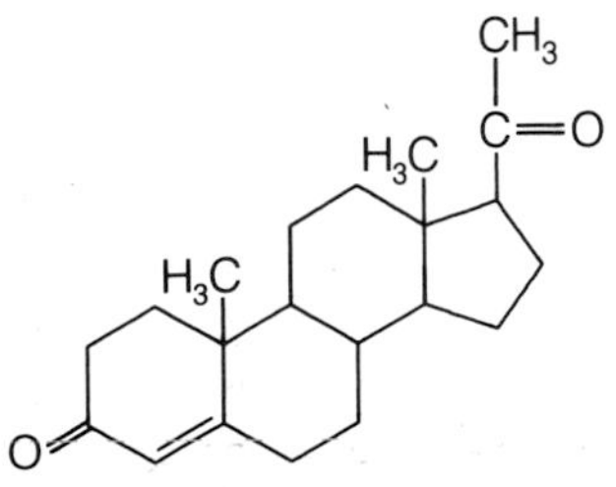

Progesterone

(Continued)

(Continued)

Testosterone, a male sex hormone, regulates the development of reproductive organs and secondary sex characteristics.

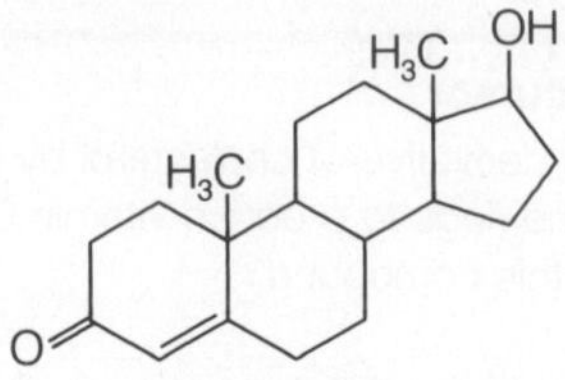

Testosterone

Androsterone is another male sex hormone.

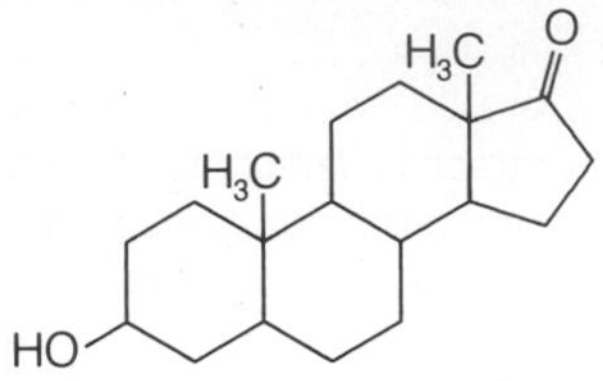

Androsterone

Synthetic Hormones in Fertility Control

Most oral contraceptive pills contain one or two synthetic, hormonelike compounds. (Synthetics must be used because the real hormones are broken down in the body.)

Synthetic Estrogens

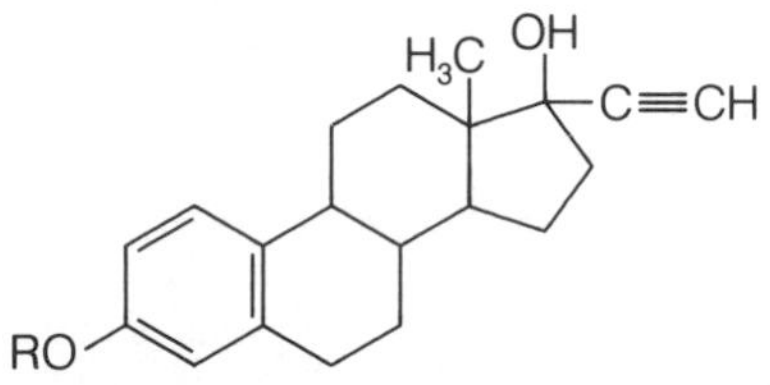

R=H, ethynylestradiol
R=CH$_3$, mestranol

Synthetic Progestin

The most widely used pills have a combination of an estrogen (20 to 100 μg if mestranol and 20 to 50 μg if ethinylestradiol) plus a progestin (0.35 to 2.5 mg depending on the compound). A relatively new birth control technology, one that prevents implantation of a fertilized ovum in the uterus, has been developed. The compound is an antiprogesterone called mifepristone, or RU 486, and how it works is described in Special Topic 21.2 (page 605).

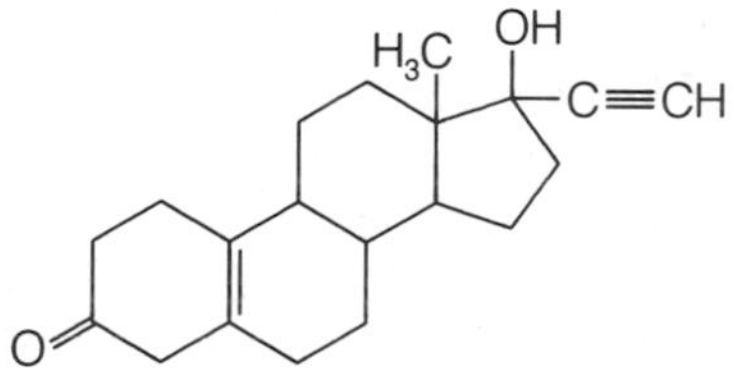

Norethynodrel

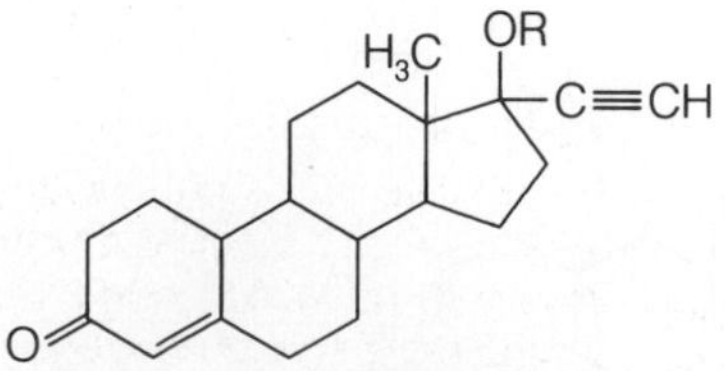

R=H, norethindrone
R=COCH$_3$, norethindrone acetate

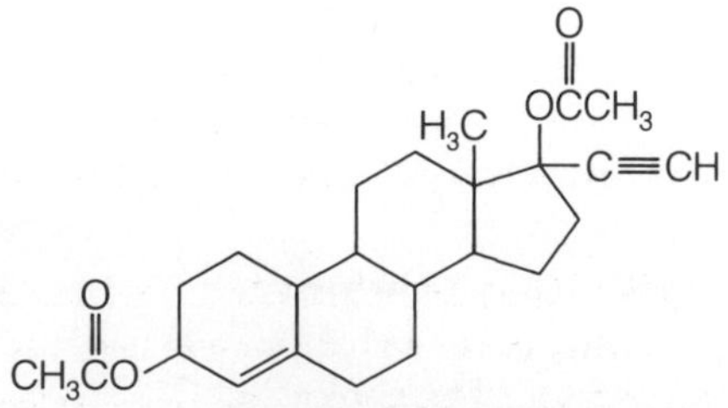

Ethynodiol diacetate

20.5

CELL MEMBRANES—THEIR LIPID COMPONENTS

Cell membranes consist mainly of a lipid bilayer plus molecules of proteins and cholesterol.

Cell membranes are made of both lipids and proteins. The lipid components are what keep a cell's insides within and its outsides without. The protein components perform services such as accepting hormone molecules and relaying them or their "messages" inside; for providing passages — closable molecular channels — for small ions and molecules; and for acting as "pumps" to move solutes across a cell membrane. We can only introduce the general features of animal cell membranes in this section. The services performed by the membrane proteins will be described in more detail after we have studied proteins and enzymes.

Both Hydrophilic and Hydrophobic Groups Are Necessary for Cell Membranes

The principal lipids of animal cell membranes are not triacylglycerols but more complex lipids, like the phospholipids and glycolipids, as well as cholesterol. The molecules of the hydrolyzable lipids in a membrane possess a part that is either very polar or fully ionic plus two nonpolar "tails." The polar or ionic sites are called **hydrophilic groups,** because they are able to attract water molecules. Hydrophilic groups force molecules to take up positions in membranes in such a way that the groups can be in contact with the water in body fluids both inside and outside the cell. In phospholipids, the hydrophilic groups are the phosphate diester units, which have ionic sites. In a glycolipid, the sugar unit with its many OH groups is the hydrophilic group.

The nonpolar, hydrocarbon sections of membrane lipids are called **hydrophobic groups** because they are water-avoiding. Hydrophobic groups tend to force molecules to become positioned *within* the membrane so that the groups are out of contact with water as much as possible. Substances like the phospholipids or glycolipids, with both hydrophilic and hydrophobic groups, are called **amphipathic compounds.** Soaps and detergents are also examples of amphipathic compounds.

The molecules of amphipathic compounds, when mixed in the right proportion with water, spontaneously become grouped into *micelles.* A **micelle** is a globular aggregation in which hydrophobic contacts are minimized and hydrophilic interactions are maximized. Figure 20.2 shows how a micelle forms when the amphipathic molecules have a *single* hydrocarbon "tail," like a detergent or soap molecule.

In the Lipid Bilayer of Cell Membranes, Hydrophobic Groups Intermingle between the Membrane Surfaces

As we said, molecules of phospholipids and glycolipids have two hydrocarbon "tails." These force a micelle made of such lipids to take up an extended disklike shape (see Fig. 20.3). Further extension of the shape shown in Figure 20.3 would produce two rows of molecules or a **lipid bilayer** arrangement, a sheet like array that consists of two layers of lipid molecules aligned side by side. This is the basic architecture of an animal cell membrane (see Fig. 20.4). The hydrophobic "tails" of the lipid molecules intermingle in the center of the bilayer away from water molecules. In a sense, these "tails" dissolve in each other, following the "like-dissolves-like" rule. The hydrophilic "heads" stick out into the aqueous phase in contact with water. These water-avoiding and water-attracting properties, not covalent bonds, are the major "forces" that stabilize the membrane.

Cholesterol Molecules Also Help to Stabilize Membranes

Cholesterol molecules are somewhat long and flat. In the lipid bilayer, they occur with their long axes lined up side by side with the hydrocarbon chains of the other lipids. The cholesterol OH groups are hydrogen-bonded to O atoms of ester groups in the membrane lipid molecules. Because the cholesterol units are relatively rigid, much more so than the fatty acid chains, cholesterol molecules help to keep a membrane from being too fluid like.

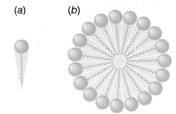

(a)   *(b)*

FIGURE 20.2
Detergent micelle. (a) Space-filling requirements of an amphipathic detergent or soap molecule with one hydrophobic tail (wavy line) and a hydrophilic head (blue sphere). (b) Micelle in water. The hydrophobic tails gather together as the hydrophilic heads have maximum exposure to the aqueous medium.

■ *Hydrophilic* — from the Greek *hydor*, water, and *philos*, loving.
Hydrophobic — from the Greek *phobikos*, hating.

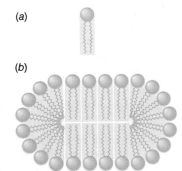

(a)

(b)

FIGURE 20.3
Phosphoglyceride micelle in water. (a) Space-filling requirements for an amphipathic phosphoglyceride, which has two hydrophobic tails (wavy lines) and a hydrophilic head (blue sphere). (b) A disklike micelle whose "wall" for the most part (top and bottom segments) is a lipid bilayer.

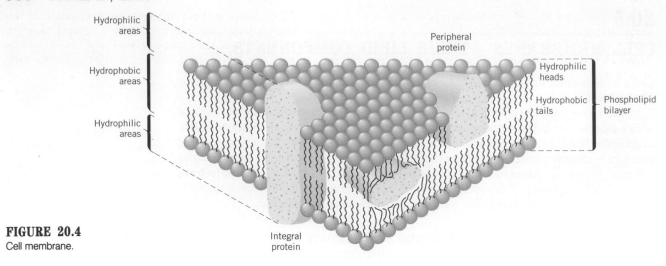

FIGURE 20.4
Cell membrane.

The Lipid Bilayer Is Self-Sealing If a pin were stuck through a cell membrane and then pulled out, the lipid layer would close back spontaneously. This flexibility is allowed because, as we said, no covalent bonds hold neighboring lipid molecules to each other. Only the net forces of attraction that we imply when we use the terms *hydrophobic* and *hydrophilic* are at work. Yet the bilayer is strong enough to hold a cell together, and it is flexible enough to let things in and out. Water molecules move back and forth easily through the membrane, but other molecules and ions are vastly less free to move. Their migrations depend on the protein components of the membrane, also indicated in Figure 20.4. We must, therefore, postpone further discussion of membranes until we have learned more about proteins, particularly such families of proteins as enzymes and receptors.

SUMMARY

Lipids Lipids are ether-extractable substances in animals and plants, and they include hydrolyzable esters and nonhydrolyzable compounds. The esters are generally of glycerol or sphingosine with their acyl portions contributed by long-chain carboxylic acids called fatty acids. The fatty acids obtained from lipids by hydrolysis generally have long chains of even numbers of carbons, seldom are branched, and often have one or more alkene groups. The alkene groups are cis. Because molecules of all lipids are mostly hydrocarbon-like, lipids are soluble in nonpolar solvents but not in water.

Triacylglycerols Molecules of neutral fats, those without electrically charged sites or sites that are similarly polar, are esters of glycerol and a variety of fatty acids, both saturated and unsaturated. Vegetable oils have more double bonds per molecule than animal fats. The triacylglycerols can be hydrogenated, hydrolyzed, and saponified.

Waxes Molecules of the waxy coatings on leaves and fruit, or in beeswax or sebum, are simple esters between long-chain alcohols and fatty acids.

Glycerophospholipids Molecules of the glycerophospholipids are esters both of glycerol and of phosphoric acid. A second ester bond

from the phosphate unit goes to a small alcohol molecule that can also have a positively charged group. Thus this part of a glycerophospholipid is strongly hydrophilic. The two types of glycerophospholipids are the phosphatides and the plasmalogens. Both are vital to animal cell membranes. In phosphatide molecules there are two fatty acyl ester units besides the phosphate system. In plasmalogens, there is one fatty acyl unit and a long-chain, unsaturated ether unit in addition to the phosphate system.

Sphingomyelins Sphingomyelins are esters of sphingosine, a dihydric amino alcohol. They also have a strongly hydrophilic phosphate system.

Glycolipids Also sphingosine-based, the glycolipids use a monosaccharide instead of the phosphate-to-small-alcohol unit to provide the hydrophilic section. Otherwise, they resemble the sphingomyelins.

Steroids Steroids are nonhydrolyzable lipids with the steroid nucleus of four fused rings (three being C-6 rings and one a C-5 ring). Several steroids are sex hormones, and oral fertility control drugs mimic their structure and functions. Cholesterol, the raw material used by the body to make bile salts and other steroids, is manufactured in the liver. Cholesterol is carried in circulation as cholesteryl esters in lipoprotein complexes. Cholesterol molecules are essential components of animal cell membranes.

Animal cell membranes A double layer of phospholipids or glycolipids plus cholesterol and assemblies of protein molecules make up the lipid bilayer part of a cell membrane. The hydrophobic tails of the amphipathic lipids intermingle within the bilayer, away from the aqueous phase. The hydrophilic heads are in contact with the aqueous medium. Cholesterol molecules help to stiffen the membrane.

REVIEW EXERCISES

The answers to Review Exercises whose numbers are in color are found in Appendix C. The answers to the other Review Exercises are found in the Study Guide that accompanies this book. The more challenging questions are marked with asterisks.

Lipids in General

20.1 Crude oil is soluble in ether, yet it isn't classified as a lipid. Explain.

20.2 Cholesterol has no ester group, yet we classify it as a lipid. Why?

20.3 Ethyl acetate has an ester group, but it isn't classified as a lipid. Explain.

20.4 What are the criteria for deciding if a substance is a lipid?

Fatty Acids

20.5 What are the structures and the names of the two most abundant *saturated* fatty acids?

20.6 Write the structures and names of the *unsaturated* fatty acids that have 18 carbons each and that have no more than three double bonds. Show the correct geometry at each double bond.

20.7 Write the equations for the reactions of palmitic acid with
(a) NaOH(*aq*)
(b) CH_3OH (when heated in the presence of acid)

*20.8 What are the equations for the reactions of oleic acid with each substance?
(a) Br_2
(b) KOH(*aq*)
(c) H_2 (in the presence of a catalyst)
(d) CH_3CH_2OH (heated with an acid catalyst)

20.9 Which of the following acids, **A** or **B**, is more likely to be obtained by the hydrolysis of a lipid? Explain.

$$CH_3(CH_2)_{12}CO_2H$$
A

$$CH_3CH(CH_2)_{11}CO_2H$$ with CH_3 branch
B

20.10 Without writing structures, state what kinds of chemicals the prostaglandins are.

Triacylglycerols

20.11 Write the structure of a triacylglycerol that involves linolenic acid, linoleic acid, and palmitic acid, besides glycerol.

20.12 What is the structure of a triacylglycerol made from glycerol, stearic acid, oleic acid, and palmitic acid?

20.13 Write the structures of all the products that would form from the complete hydrolysis of the following lipid. (Show the free carboxylic acids, not their anions.)

$$CH_2OC(CH_2)_7CH=CHCH_2CH=CH(CH_2)_4CH_3$$
(with O double bond)
$$CHOC(CH_2)_{12}CH_3$$
(with O double bond)
$$CH_2OC(CH_2)_7CH=CH(CH_2)_7CH_3$$
(with O double bond)

20.14 Write the structures of the products that are produced by the saponification (by NaOH) of the triacylglycerol whose structure was given in Review Exercise 20.13.

*20.15 The hydrolysis of a lipid produced glycerol, lauric acid, linoleic acid, and oleic acid in equimolar amounts. Write a structure that is consistent with these results. Is there more than one structure that can be written? Explain.

*20.16 The hydrolysis of 1 mol of a lipid gives 1 mol each of glycerol and oleic acid and 2 mol of lauric acid. The lipid molecule is chiral. Write its structure. Is more than one structure (constitution) possible? Explain.

20.17 What is the structural difference between the triacylglycerols of the animal fats and the vegetable oils?

20.18 Products such as corn oil are advertised as being "polyunsaturated." What does this mean in terms of the structures of the molecules that are present? Corn oil is "more polyunsaturated" than what?

20.19 What chemical reaction is used to make margarine?

20.20 Lard and butter are chemically almost the same substances, so what is it about butter that makes it so much more desirable a spread for bread than, say, lard or tallow?

Waxes

20.21 One component of beeswax has the formula $C_{36}H_{72}O_2$. When it is hydrolyzed, it gives $C_{18}H_{36}O_2$ and $C_{18}H_{38}O$. Write the most likely structure of this compound.

*20.22 When all the waxes from the leaves of a certain shrub are separated, one has the formula of $C_{60}H_{120}O_2$. Its structure is **A**, **B**, or **C**. Which is it most likely to be? Explain why the others can be ruled out.

$$CH_3(CH_2)_{56}CO_2CH_2CH_3$$
A

$$CH_3(CH_2)_{29}CO_2(CH_2)_{28}CH_3$$
B

$$CH_3(CH_2)_{28}CO_2(CH_2)_{29}CH_3$$
C

Phospholipids

20.23 Why are the phosphatides and plasmalogens both called glycerophospholipids?

20.24 What site in a glycerophospholipid carries a negative charge? What atom carries a positive charge?

20.25 In general terms, how are the sphingomyelins and cerebrosides structurally alike? How are they structurally different?

20.26 What structural unit provides the most polar groups in a molecule of a glycolipid? (Name it.)

20.27 Phospholipids are not classified as neutral fats. Explain.

20.28 Phospholipids are common in what part of a cell?

20.29 What are the names of the two types of sphingosine-based lipids?

20.30 Are the sugar units that are incorporated into the cerebrosides bound by glycosidic links or by ordinary ether links? How can one tell? Which kind of link is more easily hydrolyzed (assuming an acid catalyst)?

*20.31 The complete hydrolysis of 1 mol of a phospholipid gave 1 mol each of the following compounds: glycerol, linolenic acid, oleic acid, phosphoric acid, and the cation, $HOCH_2CH_2N(CH_3)_3^+$.
   (a) Write a structure of this phospholipid that is consistent with the information given.
   (b) Is the substance a glycerophospholipid or a sphingolipid? Explain.
   (c) Are its molecules chiral or not? How can you tell?
   (d) Is it an example of a lecithin or a cephalin? How can you tell?

*20.32 When 1 mol of a certain phospholipid was hydrolyzed, there was obtained 1 mol each of lauric acid, oleic acid, phosphoric acid, glycerol, and $HOCH_2CH_2NH_2$.
   (a) What is a possible structure for this phospholipid?
   (b) Is it a sphingolipid or a phosphoglyceride? Explain.
   (c) Can its molecules exist as enantiomers or not? Explain.
   (d) Is it a cephalin or a lecithin? Explain.

Steroids

20.33 What is the name of the steroid that occurs as a detergent in our bodies?

20.34 What is the name of a vitamin that is made in our bodies from a dietary steroid by the action of sunlight on the skin?

20.35 Give the names of three steroidal sex hormones.

20.36 What is the name of a steroid that is part of the cell membranes in animal tissues?

20.37 What is the raw material used by the body to make bile salts? How does the body use the bile salts?

20.38 How does the body carry cholesterol in circulation in the bloodstream?

Cell Membranes

20.39 Describe in your own words what is meant by the *lipid bilayer* structure of cell membranes.

20.40 How do the hydrophobic parts of phospholipid molecules avoid water in a lipid bilayer?

20.41 Besides lipids, what kinds of substances are present in a cell membrane?

20.42 What kinds of forces hold a cell membrane together?

20.43 What functions do the proteins of a cell membrane serve?

The Prostaglandins (Special Topic 20.1)

20.44 Name the fatty acid that is used to make the prostaglandins.

20.45 What effect is aspirin believed to have on prostaglandins, and how is this related to aspirin's medicinal value?

The Omega-3 Fatty Acids and Heart Disease (Special Topic 20.2)

20.46 What is it about the structure of linolenic acid that lets us call it an omega-3 acid?

20.47 What source of the omega-3 acids is relatively rich in the C-20 and C-22 acids?

20.48 Why have the omega-3 acids aroused the interests of people in nutrition and in medicine?

Detergent Action (Special Topic 20.3)

20.49 Which is the more general term, soap or detergent? Explain.

20.50 What kind of chemical is soap?

20.51 For household laundry work, which product is generally preferred, a synthetic detergent or soap? Why?

20.52 Why are soap and sodium alkyl sulfates called *anionic* detergents?

20.53 Explain in your own words how a detergent can loosen oils and greases from fabrics.

Additional Exercises

20.54 Examine the following structure.

$$CH_2OC(CH_2)_7CH{=}CHCH_2CH{=}CH(CH_2)_5CH_3$$
$$\overset{O}{\overset{\|}{\underset{}{}}}$$
$$CHOC(CH_2)_{11}CH_3$$
$$CH_2OC(CH_2)_7CH{=}CH(CH_2)_8CH_3$$

   (a) Is it a triester of glycerol?
   (b) Does it have hydrophobic groups?
   (c) What are its hydrophilic functional groups?
   (d) What would form if all of its alkene groups were hydrogenated?
   (e) Is this molecule likely to be found among naturally occurring triacylglycerols? Explain.

*20.55 Examine the following structure and answer the questions that follow.

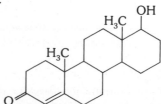

   (a) Is this compound amphipathic? Explain.
   (b) Is it a member of the steroid family? How can one tell?

PROTEINS

21

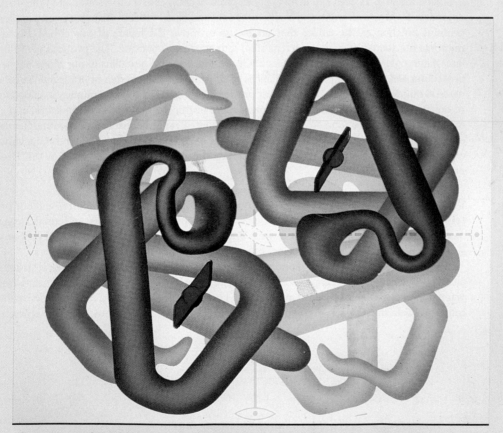

A representation of the hemoglobin molecule by artist Irving Geis. The flat red units are oxygen-holding molecules of heme, which are integral parts of hemoglobin (and which give the color to blood). Hemoglobin illustrates all four levels of complexity in protein structures, the principal topic of this chapter.

21.1

AMINO ACIDS. THE BUILDING BLOCKS OF PROTEINS

Living things select from among the molecules of about twenty α-amino acids to make the polypeptides in proteins.

Proteins, found in all cells and in virtually all parts of cells, constitute about half of the body's dry mass. They give strength and elasticity to skin and blood vessels. As muscles and tendons, proteins function as the cables that enable us to move the levers of our bones. Proteins reinforce our teeth and bones much as steel rods reinforce concrete. The proteins of antibodies, of hemoglobin, and of the various kinds of albumins and globulins in our blood serve as protectors and as the long-distance haulers of substances, like oxygen or lipids, which otherwise do not dissolve well in blood. Other proteins form parts of the communications network of our nervous system. Nearly all enzymes, some hormones and neurotransmitters, and cell membrane receptors are proteins that direct and control repair, construction, communication, and energy conversion in the body. No other class of compounds is involved in such a variety of functions, all essential to life. They deserve the name *protein,* taken from the Greek *proteios,* "of the first rank."

■ The chief elements in proteins are C, H, O, N, and S.

■ With few exceptions, the amino acids not in the "standard" set consist of modifications of "standard" amino acid molecules made *after* a polypeptide has been put together.

Polypeptides Are Made from α-Amino Acids The dominant structural units of **proteins** are high-formula-mass polymers called **polypeptides.** Metal ions and small organic molecules or ions are often present as well. The relationship of these parts to whole proteins is shown in Figure 21.1. Many proteins, however, are made entirely of polypeptides.

The monomer units for polypeptides are **α-amino acids,** which have the general structure given by **1.** Twenty such compounds make up the "standard" set (see Table 21.1), and in any given polypeptide some are used many times. Hundreds of **amino acid residues,** sometimes called **peptide units,** each derived from one or another of the various α-amino acids, are

FIGURE 21.1
Components of proteins. Some proteins consist exclusively of polypeptide molecules, but most also have nonpolypeptide units such as small organic molecules or metal ions, or both.

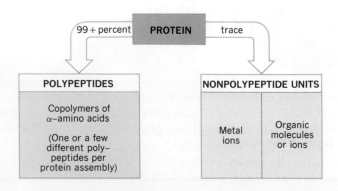

joined together in a single polypeptide molecule. Before we can study how polypeptides are put together, however, we must learn more about their α-amino acid building blocks.

$$\alpha\text{-position} \longrightarrow \quad \overset{\displaystyle \overset{O}{\|}}{^+NH_3CHCO^-} \qquad \qquad \overset{\displaystyle \overset{O}{\|}}{-NHCHC-}$$
$$\underset{R}{|} \qquad\qquad\qquad\qquad \underset{R}{|}$$

α-Amino acids, general structural features

Amino acid residue (peptide unit)

1

The same set of 20 standard amino acids (Table 21.1) is used by all species of plants and animals. A few others are present in certain tissues as well as in some strains of bacteria.

The α-Amino Acids Are Based on a Dipolar Form of α-Amino Acetic Acid but Have Different Sidechains at the α-Position

As you can see in Table 21.1, the amino acids differ in their **R** groups, called **sidechains,** located at the α-position in **1**.

In the solid state, amino acids exist entirely in the form shown by **1**, which is called a **dipolar ion** or a **zwitterion.** It is an electrically neutral particle but has a positive and a negative charge on different sites. Because dipolar ions are exceedingly polar, amino acids, like salts, have melting points that are considerably higher than those of most molecular compounds, and they tend to be much more soluble in water than in nonpolar solvents.

Structure **1** is actually an *internally neutralized molecule.* We can imagine that **1** started out with a regular amino group, NH_2, and an ordinary carboxyl group, CO_2H. But then the amino group, a proton acceptor, took a proton from the carboxyl group, a proton donor, to give the dipolar ionic form, **1**. Of course, **1** has its own (weaker) proton donating group, NH_3^+, and its own (also weaker) proton-accepting group, CO_2^-, so these dipolar ions can neutralize acids or bases of sufficient strength, like H_3O^+ and OH^-. Because polypeptides generally have at least one base-neutralizing NH_3^+ group and one acid-neutralizing CO_2^- group, *proteins are able to serve as buffers.*

For amino acids to exist in water as dipolar ions, **1**, the pH has to be about 6 to 7. If we make the pH much lower (more acidic) or much higher (more basic), the form of the amino acid changes. If we add enough strong acid like HCl*(aq),* for example, to a solution of an amino acid to lower the pH to about 1, the CO_2^- groups take on protons. They change to CO_2H groups shown in structure **2**. Now the amino acid molecules are positively charged (cations) and can migrate to a negatively charged electrode, the cathode, in an electrolysis experiment.

■ Hereafter, when we say "amino acid," we'll mean α-amino acid.

■ α-Amino acids melt around 300 °C but their simple esters, which cannot be dipolar ions, generally melt around 100 °C.

$$\overset{\displaystyle \overset{O}{\|}}{NH_2CHCOCH_3}$$
$$\underset{R}{|}$$

Methyl ester of α-amino acid

■ In an electrolysis experiment (page 205), the negatively charged electrode is called the *cathode,* so positively charged ions *(cations)* are attracted to it. Similarly, the positively charged electrode, the *anode,* attracts *anions,* ions bearing negative charge.

$$^+NH_3CHCO_2^-$$
$$\underset{R}{|}$$
1

OH^- ⇌ H^+ (left) OH^- ⇌ H^+ (right)

$$^+NH_3CHCO_2H \qquad\qquad\qquad NH_2CHCO_2^-$$
$$\underset{R}{|} \qquad\qquad\qquad\qquad \underset{R}{|}$$
2 $\qquad\qquad\qquad\qquad\qquad$ **3**

If we add enough strong base to an amino acid solution to raise the pH to about 11, then the amino acid molecules transfer protons from NH_3^+ groups to OH^- ions, which changes NH_3^+ groups to NH_2 groups as seen in structure **3**. Because **3** is a negatively charged ion (an anion), it can migrate to the positively charged electrode in an electrolysis apparatus (the anode).

TABLE 21.1

Amino Acids: $^{+}NH_3CHCO_2^{-}$
                        |
                        R

Type	Sidechain, R	Name	Symbol 3-Letter	Symbol 1-Letter	pl
Side chain is nonpolar	—H	Glycine	Gly	G	5.97
	—CH_3	Alanine	Ala	A	6.00
	—$CH(CH_3)_2$	Valine	Val	V	5.96
	—$CH_2CH(CH_3)_2$	Leucine	Leu	L	5.98
	—$CHCH_2CH_3$ (CH_3)	Isoleucine	Ile	I	6.02
	—$CH_2C_6H_5$	Phenylalanine	Phe	F	5.48
	—CH_2 (indole ring)	Tryptophan	Trp	W	5.89
	(complete structure) Proline ring	Proline	Pro	P	6.30
Side chain has a hydroxyl group	—CH_2OH	Serine	Ser	S	5.68
	—$CHOH$ (CH_3)	Threonine	Thr	T	5.64
	—CH_2—C₆H₄—OH	Tyrosine	Tyr	Y	5.66
Side chain has a carboxyl group (or an amide group)	—CH_2CO_2H	Aspartic acid	Asp	D	2.77
	—$CH_2CH_2CO_2H$	Glutamic acid	Glu	E	3.22
	—CH_2CONH_2	Asparagine	Asn	N	5.41
	—$CH_2CH_2CONH_2$	Glutamine	Gln	Q	5.65
Side chain has a basic amino group	—$CH_2CH_2CH_2CH_2NH_2$	Lysine	Lys	K	9.74
	—$CH_2CH_2CH_2NHCNH_2$ (=NH)	Arginine	Arg	R	10.76
	—CH_2—(imidazole ring)	Histidine	His	H	7.59
Side chain contains sulfur	—CH_2SH	Cysteine	Cys	C	5.07
	—$CH_2CH_2SCH_3$	Methionine	Met	M	5.74

A pH Exists for Each Amino Acid, Its Isoelectric Point, at Which No Net Migration in an Electric Field Occurs In an aqueous solution of an amino acid, a dynamic equilibrium exists between **1, 2,** and **3**. If now a current is passed between electrodes dipping into such a solution, cations of form **2** migrate to the cathode. Anions of form **3** migrate to the anode. Neutral molecules, **1**, migrate nowhere. A molecule with *equal* numbers of positive and negative charges, like **1**, is said to be an **isoelectric molecule,** and it cannot migrate in an electric field.

Because the equilibrium is *dynamic,* a migrating cation like **2** could flip a proton to some acceptor, become **1** and isoelectric (neutral), and stop moving to the cathode. In another instant, it (now **1**) could shed another proton, become **3**, and so made to turn around and head for the anode. Similarly, an anion on its way to the anode might pick up a proton, become neutral and also stop dead. Then it might take another proton, become **2**, and get turned around. In the meantime, an isoelectric molecule, **1**, might either donate or accept a proton, become electrically charged, and start its own migration. The question is, what overall *net* migration occurs and how is this net effect influenced by the pH of the solution?

Although much coming and going occurs in an amino acid solution, the net molar concentrations of the species stay the same at equilibrium. If either **2** or **3** is in any molar excess because of the pH, then some *net* migration occurs toward one electrode or the other. When the net molar concentration of **2**, for example, is greater than that of **3**, some statistical net movement to the cathode occurs.

Remember, however, that these equilibria can be shifted by adding acid or base. By carefully adjusting the pH, in fact, we can so finely tune the concentrations at equilibrium that *no net migration occurs.* At the right pH, the rates of proton exchange are such that each unit that is not **1** spends an equal amount of time as **2** and as **3**. (And the concentrations of **2** and **3** are very low.) Thus, any net migration to one electrode is blocked.

■ These shifts of H^+ ions illustrate Le Châtelier's principle at work.

The pH at which no net migration of an amino acid can occur in an electric field is called the **isoelectric point** of the amino acid, and the symbol of this pH value is **pI** (see Table 21.1, last column). Now let's see what the concepts of isoelectric molecules and pI values have to do with proteins.

Proteins, Like Amino Acids, Have Isoelectric Points As we will soon see, all proteins have NH_3^+ or CO_2^- groups or can acquire them by a change in the pH of the surrounding medium. Whole protein molecules, therefore, can also be isoelectric at the right pH. *Each protein thus has its own isoelectric point.* Now think of what can happen if the pH is changed from the pI value. The entire electrical condition of a huge protein molecule can be made either cationic or anionic almost instantly, at room temperature, by adding strong acid or base—by changing the pH of the medium.

Such changes in the electrical charge of a protein have serious consequences at the molecular level of life. Being electrically charged can dramatically affect chemical reactions, for example, or greatly alter protein solubility. If proteins are to serve their biological purposes, some must not be allowed to go into solution and others must not be permitted to precipitate. We'll return to this concept in this and later chapters, but the discussion focuses our attention again on how important it is that an organism control the pH values of its fluids.

■ Casein, the protein in milk, precipitates when milk turns sour because a change in pH causes the casein molecules to become isoelectric.

We will next survey the types of side chains in amino acids and how they affect the properties of polypeptides and proteins. These properties include how a polypeptide molecule will spontaneously fold and twist into its distinctive and vital final shape. You should memorize the structures of a minimum of five amino acids that illustrate the types we are about to study; glycine, alanine, cysteine, lysine, and glutamic acid are suggested. How to use Table 21.1 to write their structures is described in the following example.

EXAMPLE 21.1
Writing the Structure of an Amino Acid

What is the structure of cysteine?

■ The amino acid proline is the only one of the standard 20 in which the α-amino group is itself joined to one end of the sidechain.

$$H_2\overset{+}{N} \quad CO_2^-$$

Proline

ANALYSIS Regard any α-amino acid as having two structural features, a unit common to *all* α-amino acids plus a unique side chain at the α-position. So write the common unit first and then add the side chain.

SOLUTION   The common unit is

$$\overset{+}{N}H_3CHCO_2^-$$

Next, either look up or recall from memory the side chain for the particular amino acid. For cysteine, this is CH_2SH, so simply attach this group to the α-carbon. Cysteine is

$$\overset{+}{N}H_3CHCO_2^-$$
$$|$$
$$CH_2SH$$

PRACTICE EXERCISE 1

Write the structures of the dipolar ionic forms of glycine, alanine, lysine, and glutamic acid.

Several Amino Acids Have Hydrophobic Sidechains   The first amino acids in Table 21.1, including alanine, have essentially nonpolar, hydrophobic side chains. When a long polypeptide molecule folds into its distinctive shape, these hydrophobic groups tend to be folded next to each other as much as possible rather than next to highly polar groups or to water molecules in the solution. This water avoidance by nonpolar side chains is called the **hydrophobic interaction** of the side chains and two factors are at work. One is the strong tendency of water molecules to form hydrogen bonds between each other and thus "reject" the presence of molecules or groups that cannot themselves "offer" hydrogen-bonding capabilities to water. The water molecules, in a sense, club together forcing nonpolar groups to stay away and be by themselves. The other factor in hydrophobic interactions is that of *London forces* of attraction, which we discussed on page 166. These are attractive forces between nonpolar groups made possible by temporary dipoles.

Some Amino Acids Have Hydrophilic OH Groups on Their Sidechains   The second set of amino acids in Table 21.1 consists of those whose sidechains carry alcohol or phenol OH groups, which are polar and hydrophilic. They can donate and accept hydrogen bonds. As a long polypeptide chain folds into its final shape, these side chains tend to stick out into the surrounding aqueous phase to which they are attracted by hydrogen bonds.

■ Aspartame, a popular artificial sweetener, has an aspartic acid residue.

$$\overset{+}{N}H_3\overset{O}{\overset{\|}{C}HC}-\overset{O}{\overset{\|}{N}HCHCOCH_3} \quad \begin{array}{l}\text{Methyl}\\\text{ester}\end{array}$$
$$| \qquad\qquad |$$
$$CH_2CO_2^- \quad CH_2C_6H_5$$

Aspartic acid   Phenylalanine
residue   residue
Aspartame
(NutraSweet ®)

Two Amino Acids Have Carboxyl Groups on Their Sidechains   The sidechains of aspartic and glutamic acid carry proton donating CO_2H groups. Because body fluids are generally slightly basic, the protons available from side chain CO_2H groups have been neutralized, so these groups actually occur mostly as CO_2^- groups on protein molecules in body fluids. The aqueous medium would have to be made quite acidic to prevent this and force protons back onto the sidechain CO_2^- groups. This is why the pI values of aspartic and glutamic acid are lower (more acidic) than the pI values of amino acids with nonpolar side chains.

Aspartic acid and glutamic acid often occur as asparagine and glutamine in which their side chain CO_2H groups have become amide groups, $CONH_2$, instead. These are also polar, hydrophilic groups, *but they are not electrically charged.* They are neither proton donors nor proton acceptors, so the pI values of asparagine and glutamine are higher than those of aspartic or glutamic acids.

PRACTICE EXERCISE 2

Write the structure of aspartic acid (in the manner of **1**) with the sidechain carboxyl (a) in its carboxylate form and (b) in its amide form.

Lysine, Arginine, and Histidine Have Basic Groups on Their Side Chains The extra NH_2 group on lysine makes its side chain basic and hydrophilic. Remember that an amine in water makes the pH of the solution greater (more basic) than 7 because of the OH^- ion generated in the following equilibrium.

$$RNH_2(aq) + H_2O \rightleftharpoons RNH_3^+(aq) + OH^-(aq)$$

The addition of OH^- to this equilibrium shifts the equilibrium to the left in accordance with Le Châtelier's principle. (The extra base pulls H^+ from the NH_3^+ group.) A solution of lysine has to be made basic, therefore, to prevent its side chain NH_2 group from existing in the protonated form, NH_3^+, and so affect the net charge on the lysine molecule. This is why the pI value of lysine, 9.47, is relatively high. Arginine and histidine have similarly basic side chains.

PRACTICE EXERCISE 3

Write the structure of arginine in the manner of **1**, but with its sidechain amino group in its protonated form. (Put the extra proton on the $=NH$ unit, not the NH_2 unit of the side chain.)

PRACTICE EXERCISE 4

Is the side chain of the following amino acid hydrophilic or hydrophobic? Does the side chain have an acidic, basic, or neutral group?

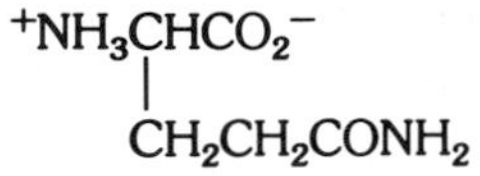

$$\overset{+}{N}H_3CHCO_2^-$$
$$|$$
$$CH_2CH_2CONH_2$$

Cysteine and Methionine Have Sulfur-Containing Sidechains The side chain in cysteine has an SH group. As we studied in Section 14.6, molecules with this group are easily oxidized to disulfide systems, and disulfides are easily reduced to SH groups:

$$2RSH \underset{\text{(reduction)}}{\overset{\text{(oxidation)}}{\rightleftharpoons}} RS{-}SR + H_2O$$

Cysteine and its oxidized form, cystine, are interconvertible by oxidation and reduction, a property of far-reaching importance in some proteins.

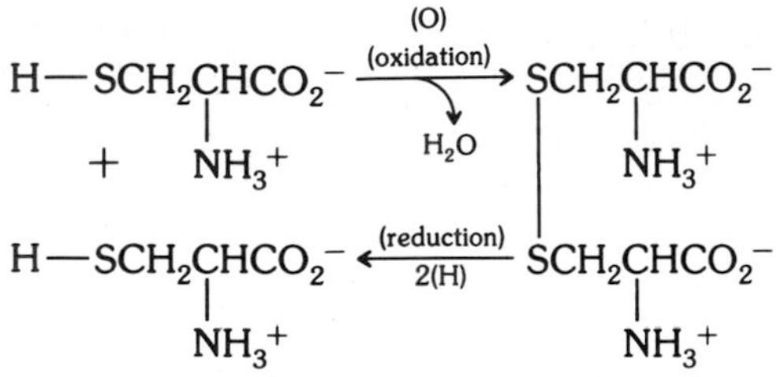

Cysteine
(two molecules)

Cystine

The **disulfide link** contributed by cystine is especially prevalent in the proteins that have a protective function, such as those in hair, fingernails, and the shells of certain crustaceans.

■ Some D-amino acid residues occur in bacterial cell membranes, which helps these disease-causing agents to survive in higher animals, where the enzymes for attacking polypeptides work only with L-forms.

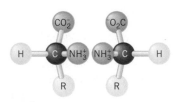

FIGURE 21.2
The two possible enantiomers of α-amino acids whose molecules have just one stereocenter. The absolute configuration on the right, which is in the L-family, represents virtually all the naturally occurring α-amino acids.

The α-Position in All Amino Acids except Glycine Is a Stereocenter All the amino acids except glycine consist of chiral molecules and can exist as pairs of enantiomers. For each possible pair of enantiomers, however, nature supplies just one of the two (with a few rare exceptions). All the naturally occurring amino acids, moreover, belong to the same optical family, the L-family. What this means is illustrated in Figure 21.2. It also means that all the proteins in our bodies, including all enzymes, are made from L-amino acids and are all chiral.

21.2

OVERVIEW OF PROTEIN STRUCTURE

Protein molecules can have four levels of complexity.
Protein structures are more complicated by far than those of carbohydrates or lipids, and every aspect of their structures is vital at the molecular level of life. We'll begin, therefore, with a broad overview of protein architecture.

Protein Structure Involves Four Features There are four possible levels of complexity in the structures of proteins. Disarray at any level almost always renders the protein biologically useless. A protein having the structure and overall shape that it normally possesses in a living system and that permit it to function biologically is called a **native protein.** The same protein when made to lose its molecular shape even while retaining its original molecular constitution (covalent structure), is a **denatured protein.** (We'll return to denaturing agents in Section 21.7.)

The first and most fundamental level of protein structure, the **primary structure,** concerns only the *sequence of amino acid residues* in the polypeptide(s) of the protein. However, it is this sequence that ultimately determines the three-dimensional structure of a protein and therefore determines how a protein can function.

The next level of protein complexity, the **secondary structure,** also concerns just individual polypeptides. It entails noncovalent forces, particularly the hydrogen bond, and it consists of the particular way in which a long polypeptide strand has coiled or in which strands have intertwined or lined up side to side.

The **tertiary structure** of a polypeptide concerns the further coiling, bending, kinking, or twisting of secondary structures. If you've ever played with a coiled door spring, you know that the coil (secondary structure) can be bent and twisted (tertiary structure). By and large, noncovalent forces such as hydrogen bonds and hydrophobic interactions stabilize these shapes. When polypeptides have disulfide bonds, they form from SH groups *after* the polypeptide has been synthesized in the cell, so the S—S covalent bond is usually classified as a feature of tertiary structure, not primary structure. In many polypeptides, attractions and repulsions between electrically charged sites on side chains are also involved in determining the overall shape of a polypeptide.

Finally, we deal with those proteins having still another complexity, **quaternary structure.** A protein's quaternary structure forms by a coming together of two or more polypeptides, often with other relatively small molecules or ions that aggregate in a precise manner to form one grand whole. Each polypeptide unit—now actually a *subunit* of the protein—has all previous levels of structure.

21.3

PRIMARY STRUCTURES OF PROTEINS

The backbones of all polypeptides of all plants and animals have a repeating series of N—C—C(=O) units.

The Peptide Bond (Amide Bond) Joins Amino Acid Residues Together in a Polypeptide

The **peptide bond** is the covalent bond that forms when amino acids are put together in a cell to make a polypeptide. It's nothing more than an amide system, carbonyl-to-nitrogen. To illustrate it and to show how polypeptides acquire their primary structure, we will begin by simply putting just two amino acids together to form a *dipeptide.*

Suppose that glycine acts at its carboxyl end and alanine acts at its amino end such that, by a series of steps (not given in detail), a molecule of water splits out and a carbonyl-to-nitrogen bond, a peptide bond, is created.

■ *Simple* amides have the following structure (see also Section 17.3).

$$R-\overset{\overset{\displaystyle O}{\|}}{C}-NH_2 \quad \boxed{\text{Amide bond}}$$

Glycine (Gly) Alanine (Ala) → Glycylalanine (Gly-Ala) **4**

Of course, there is no reason why we could not picture the roles reversed so that alanine acts at its carboxyl end and glycine at its amino end. This results in a different dipeptide (but an isomer of the first).

■ How a cell causes a peptide bond to form involves a cell's nucleic acids and is described in Chapter 24.

Alanine (Ala) Glycine (Gly) → Alanylglycine (Ala-Gly) **5**

The product of the union of any two amino acid residues by a peptide bond is called a **dipeptide,** and all dipeptides have the following features:

Dipeptide

■ The *di* in *di*peptide signifies that *two* amino acid residues are present and not the number of peptide bonds.

Structures **4** and **5** differ only in the sequence in which the sidechains, H and CH_3, occur on α-carbons. This is fundamentally how polypeptides also differ — in their sequences of sidechains.

EXAMPLE 21.2
Writing the Structure of a Dipeptide

What are the two possible dipeptides that can be put together from alanine and cysteine?

ANALYSIS Both dipeptides must have the same backbone, so we write two of these first. We follow the convention that such backbones are always written in the N to

C—left to right—direction, but this is only a convention:

$$\overset{O}{\underset{|}{^+NH_3CHC}}-\overset{O}{\underset{|}{NHCHCO^-}} \quad \text{and} \quad \overset{O}{\underset{|}{^+NH_3CHC}}-\overset{O}{\underset{|}{NHCHCO^-}}$$

Then either from memory or by using a table, we know that the two side chains are CH_3 for alanine and CH_2SH for cysteine. We simply attach these in their two possible orders to make the finished structures.

SOLUTION

$$\overset{O}{\underset{CH_3}{^+NH_3CHC}}-\overset{O}{\underset{CH_2SH}{NHCHCO^-}} \quad \text{and} \quad \overset{O}{\underset{CH_2SH}{^+NH_3CHC}}-\overset{O}{\underset{CH_3}{NHCHCO^-}}$$

Ala-Cys Cys-Ala

It would be worthwhile at this time simply to memorize the easy repeating sequence in a dipeptide, because it carries forward to higher peptides:

nitrogen – carbon – carbonyl – nitrogen – carbon – carbonyl

(Remember that the "carbon" in "nitrogen-carbon-carbonyl" is the *alpha* carbon. Remember also that the direction—left-to-right means nitrogen-carbon-carbonyl—is conventional, not a law of nature.)

PRACTICE EXERCISE 5

Write the structures of the two dipeptides that can be made from alanine and glutamic acid.

Three-Letter Symbols for Amino Acid Residues Simplify the Writing of Polypeptide Structures Each amino acid has been assigned a three-letter symbol, given in the third from the last column in Table 21.1. To use three-letter symbols to write a polypeptide structure, we have to follow certain rules. The convention is that a series of three-letter symbols, each separated by a hyphen, represents a polypeptide structure, provided that the first symbol (reading left to right) is the free amino end, NH_3^+, and the last symbol has the free carboxylate end, CO_2^-. The structure of the dipeptide **4**, for example, can be rewritten as Gly-Ala, and its isomer **5** as Ala-Gly. In both, the backbones are identical. In some applications, it is more convenient to use single-letter symbols, which are given in the next to the last column of Table 21.1.

Dipeptides still have NH_3^+ and CO_2^- groups, so a third amino acid can react at either end. In general,

■ Biochemists find it easier to use single letter symbols for the amino acid residues when they want to compare the amino acid sequences in several similar polypeptides.

$$\overset{O}{\underset{R^1}{^+NH_3CHC}}-\overset{O}{\underset{R^2}{NHCHC}}-O^- + H-\overset{H^+}{\underset{|_H}{N}}\overset{O}{\underset{R^3}{CHCO^-}} \xrightarrow{\text{(several steps)}} \overset{O}{\underset{R^1}{^+NH_3CHC}}-\overset{O}{\underset{R^2}{NHCHC}}-\overset{O}{\underset{R^3}{NHCHCO^-}} + H_2O$$

Peptide bonds

A tripeptide

If we start with glycine, alanine, and phenylalanine, the tripeptide, Gly-Ala-Phe, would be only one of six possible tripeptides involving these three different amino acids. The set of all possible sequences for a tripeptide made from Gly, Ala, and Phe is as follows:

Gly-Ala-Phe   Ala-Gly-Phe   Phe-Gly-Ala

Gly-Phe-Ala   Ala-Phe-Gly   Phe-Ala-Gly

Each of these tripeptides still has groups at each end, NH_3^+ and CO_2^-, that can interact with still another amino acid to make a tetrapeptide. And this product would still have the end groups from which the chain could be extended even further. You can see how a repetition of this pattern many hundreds of times can produce a long polymer, a polypeptide.

■ A polypeptide is actually a *copolymer*, because the monomer units are not identical.

The Sequence of Sidechains on the Repeating N—C—C(=O) Backbone Is the *Primary* Structure of All Polypeptides

All polypeptides have the following skeleton in common. They differ in length (n) and in the kinds and sequences of side chains.

Polypeptide "backbone" (n can equal several thousand)

■ Notice, for later reference, the designations *N-terminal unit* and *C-terminal unit* for the residues with the free α-NH_3^+ and the free α-CO_2^- groups, respectively.

The peptide bond, as we said, is the chief covalent bond that holds amino acid residues together in polypeptides. The disulfide bond is the only other covalent bond that affects the amino acid residues. As we said, the disulfide bond generally becomes established *after* polypeptide molecules have been put together, so it usually is called a tertiary structural feature.

As the number of amino acid residues increases, some used several times, the number of possible polypeptides increases rapidly. For example, if 20 different amino acids are incorporated, each used only once, there are 2.4×10^{18} possible isomeric polypeptides! (And a polypeptide with only 20 amino acid residues in its molecule is a very small polypeptide.) The maximum number of ways by which 20 *different* amino acid residues can be joined is 20^{100} or about 1.3×10^{130}. The estimated total number of atoms of all kinds and combinations everywhere in the entire universe is only(!) 9×10^{78}. You can see that the statistical possibilities for having many *different* polypeptides exceeds by far the available supply of atoms and makes possible the huge number of unduplicated small and large variations not only between species but among individuals within a species who have ever lived or ever will live.

The *Peptide Group* Is Trans-planar

Rotation about the peptide bond is not free despite its appearing to be a single bond. The unshared electron pair on the N atom of the NH group interacts with the C and O atoms of the carbonyl group to give the C—N peptide bond enough of the character of a double bond to prevent free rotation about it. Thus, in what we can call the *peptide group* of atoms — from one α-carbon through the carbonyl–nitrogen unit to the next α-carbon — the four atoms lie in the same plane (see Fig. 21.3). The H on N of the NH group and the O on C of the carbonyl group are held trans to each other, on opposite sides of the backbone. The rigidity of the peptide group places some constraints on the flexibility of the polypeptide chain and so affects its overall shape.

FIGURE 21.3

The trans nature of the *peptide group* showing how one C_α atom, the carbonyl carbon atom, the nitrogen atom, and the next C_α atom in the polypeptide chain are all in the same plane. The H in NH is trans to the O in C=O.

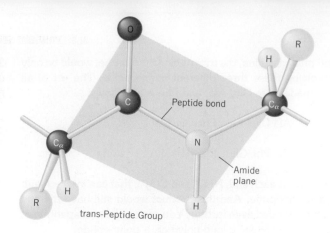

trans-Peptide Group

21.4

SECONDARY STRUCTURES OF PROTEINS

The α-helix and the β-pleated sheet are two important kinds of secondary protein structures.

Once a cell puts together a polypeptide, largely noncovalent forces, like the hydrogen bond and hydrophobic interactions, help to determine how a polypeptide twists into a particular native shape. The hydrophobic interactions are largely what "drive" the formation of an overall shape, and hydrogen bonds help to stabilize it.

The α-Helix Is a Major Secondary Structure of Polypeptides One of the most common configurations is the α-*helix*, a coiled configuration of a polypeptide strand discovered by Linus Pauling and R. B. Corey (see Fig. 21.4). In the α-**helix,** the polypeptide backbone coils as a right-handed screw, which permits all of its side chains to stick to the outside of the coil.

Hydrogen Bonds Stabilize α-Helices In α-helices, hydrogen bonds extend from the H atoms of polar NH units in peptide groups to oxygen atoms of polar carbonyl units four residues farther along the backbone. Individually, single hydrogen bonds are weak forces of attraction, but they add up much like the individual forces that hold a zipper strongly shut. Generally, in very long polypeptides, only *segments* of the molecules, not entire lengths, are in an α-helix configuration. Coils that are about 11 residues long are common, but as many as 53 residues have been observed in the α-helix portion of one polypeptide molecule. Uncoiled portions of a polypeptide strand or a segment having a pleated sheet array (discussed below) occur between α-helix segments.

A Left-Handed Helix Characterizes the Individual Polypeptide Strands in Collagen The collagens, the most abundant proteins in vertebrates, are a family of extracellular proteins that give strength to bone, teeth, cartilage, tendons, skin, blood vessels, and certain ligaments. Glycine contributes a third of the amino acid residues in collagen's polypeptide subunits. Another 15–30% of the residues are contributed by proline and 4-hydroxyproline, an amino acid not on the list of 20 standard amino acids. In addition, there are residues from 3-hydroxyproline and 5-hydroxylysine, also not on the list of 20.

■ The turns of a right-handed helix or screw are in the same direction taken as the fingers of your right hand curl when your thumb points along the axis of the helix in the direction in which the helix advances. Wood screws have right-handed helices.

■ | Hydrogen bond |

$$-N-H\cdots O=C\diagdown$$

■ The protein in the cornea of the eye is also a member of the collagen family.

4-Hydroxyproline   3-Hydroxyproline   5-Hydroxylysine

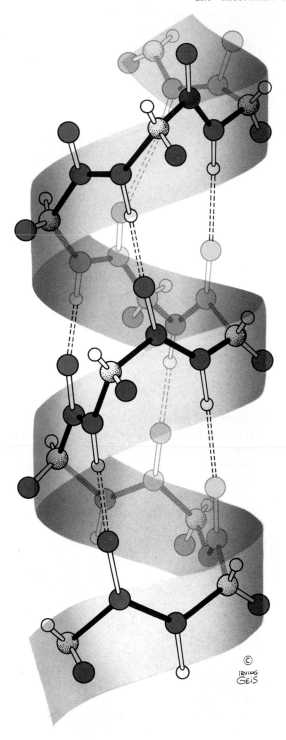

FIGURE 21.4
The α-helix. The polypeptide backbone follows the spiraling ribbon as a right-handed helix. The oxygen atoms of carbonyl groups are in red; nitrogen atoms of the peptide NH group are in dark blue with the H atoms of NH in white. Side chain groups, R, are represented here only by simple spheres in purple. Note how they project to the *outside* of the ribbon. Dashed lines show how hydrogen bonds extend from each NH group's H atom to a carbonyl group's O atom four residues along the backbone.

In some types of collagen, monosaccharide molecules are incorporated (as glycosides of side chain OH groups).

The ring systems in the molecules of proline and its hydroxylated relatives limit strand flexibility. They restrict the coiling of collagen polypeptides to *left-handed* helices. There is a further level of structure to collagen, tertiary structure, which we'll consider shortly.

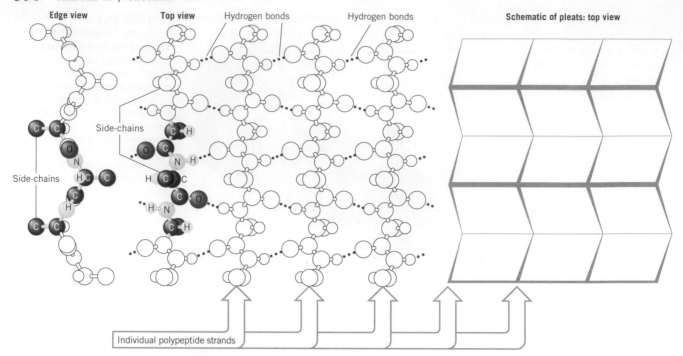

FIGURE 21.5
The β-pleated sheet.

Vitamin C Is Essential to the Synthesis of Collagen The hydroxylated derivatives of lysine and proline are made with the help of an enzyme that requires ascorbic acid (vitamin C). The reactions that put OH groups on proline rings (or lysine side chains) occur *after* the initial polypeptide is made. Thus vitamin C is essential to growing children for the formation of strong bones and teeth as well as all other tissues that rely on collagen. When an adult's diet is deficient in ascorbic acid, wounds do not heal well, blood vessels become fragile, and an overall vitamin-deficiency condition called *scurvy* results.

■ A collagen fibril only 1 mm in diameter can hold a suspended mass as large as 10 kg (22 lb).

The β-Pleated Sheet Is a Side-by-Side Array of Polypeptide Units Pauling and Corey also discovered that adjacent segments in some polypeptides line up side by side, to form a sheet-like array that is somewhat pleated and called the **β-pleated sheet** (see Fig. 21.5). The side chains project above and below the surface of the sheet. This is another kind of secondary structure in which hydrogen bonds hold things together. As few as two segments of

FIGURE 21.6
Segments of polypeptide strands can become adjacent to and parallel with each other in more than one way. (*a*) A hairpin loop brings the next segment into an antiparallel arrangement. (*b*) A back-and-over loop allows the next segment to have a parallel alignment. (*c*) A left-handed crossover loop also permits a parallel alignment. The loops are made of segments of the polypeptide chain.

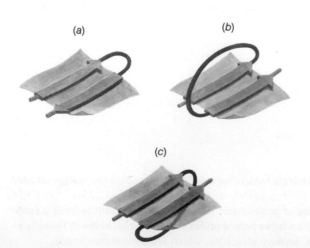

(a) (b)

(c)

a polypeptide strand and as many as 15 have been found in the same pleated sheet, with each strand ranging from 6 to 15 residues long.

When a hairpin turn carries the polypeptide chain from one segment of a pleated sheet to the next, the strands run in opposite directions. Other kinds of turns are possible, however, but are less common (see Fig. 21.6). The pleated sheet is a feature in portions of the polypeptide strands of many proteins and is the dominant feature in fibroin, silk protein (seen in Fig. 21.5).

■ Linus Pauling won the 1954 Nobel prize in chemistry for discovering the α-helix and β-sheet configurations.

21.5

TERTIARY STRUCTURES OF PROTEINS

Tertiary structures are the results of folding, bending, and twisting of secondary structures.
Once primary and secondary structures are in place, the final shaping of a protein occurs. All these activities happen spontaneously in cells, sometimes in a matter of seconds after the polypeptide molecule has been made and sometimes it takes several minutes. The "rules" followed by the polypeptide to give these shapes are still not fully known.

Tertiary Protein Structure Involves the Folding and Bending of Secondary Structure When α-helices take shape, their side chains tend to project outward where, in an aqueous medium, they can be in contact with water molecules. Even in water-soluble proteins, however, as many as 40% of the sidechains are hydrophobic. Because such groups cannot break up the hydrogen-bonding networks among the water molecules, an entire α-helix or β-sheet undergoes further twisting and folding until the hydrophobic groups, as much as possible, are tucked to the inside, away from the water, and the hydrophilic groups stay exposed to the water. Thus the final shape of the polypeptide, its **tertiary structure,** emerges in response to simple molecular forces set up by the water-avoiding and the water-attracting properties of the side chains. In fact, these hydrophobic interactions on which *tertiary* structure depends sometimes determine the best secondary structure for the polypeptide.

Disulfide Bonds Can Give Loops in Polypeptides or Join Two Strands Together
Polypeptides that are to have disulfide bonds receive them by the oxidation of SH groups during the development of tertiary structure. If the SH group on the side chain of cysteine appears on two neighboring polypeptide molecules, then mild oxidation is all it takes to link the two molecules by a disulfide bond. This cross-linking, facilitated by proper enzymes, can also occur between parts of the same polypeptide molecule, in which case a closed loop results.

■ The letter symbols for cystine are

$$\begin{array}{ccc} \text{Cys} & & \text{C} \\ | & \text{and} & | \\ \text{Cys} & & \text{C} \end{array}$$

Ionic Bonds Also Stabilize Tertiary Structures Another force that can stabilize a tertiary structure is the attraction between a full positive and a full negative charge, each occurring on a particular side chain. At the pH of body fluids, the side chains of both aspartic acid and glutamic acid carry CO_2^- groups. The side chains of lysine and arginine carry NH_3^+ groups. These oppositely charged groups can attract each other, like the attraction of oppositely charged ions in an ionic crystal. The attraction is an *electrostatic attraction,* and is sometimes called a **salt bridge.**

Hydrophobic Interactions Significantly Affect Polypeptide Shape In the tertiary structure of myoglobin, the oxygen-holding protein in muscle tissue, about 75% of the single polypeptide molecule consists of α-helix segments (see Fig. 21.7a). Virtually all of the hydrophobic groups of myoglobin are folded inside where they avoid water molecules as

FIGURE 21.7
(a) Myoglobin (sperm whale), a polypeptide with 153 amino acid residues. The tube-like forms outline the eight segments that are in an α-helix. The flat, purple structure is the heme unit and the red circle is an oxygen molecule. Only the atoms that make up the backbone of the chain are indicated (by circles). The side chains have been omitted (b) Heme molecule with its Fe²⁺ ion.

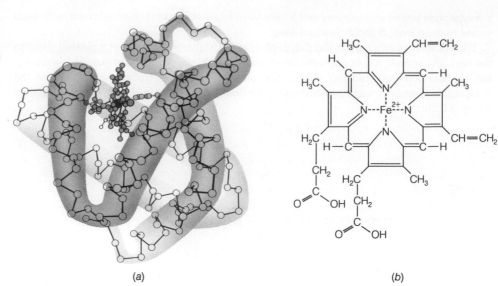

(a)   (b)

much as possible, and its hydrophilic groups are on the outside. A nonprotein molecule, heme, completes the native structure of myoglobin (Fig. 21.7b).

Polypeptides Often Incorporate Prosthetic Groups into Their Tertiary Structures

A nonprotein, organic compound that associates with a polypeptide, like heme in myoglobin, is called a **prosthetic group.** It is often the focus of the protein's biological purpose. Heme is the actual oxygen holder in myoglobin. Heme also serves the same function in **hemoglobin,** the oxygen carrier in blood. The heme molecule is held in the folded globin molecule by electrostatic attractions between two electrically charged side chains and the Fe²⁺ ion in heme.

■ "Prosthetic" is from the Greek *prosthesis,* an addition.

β-Sheets Are Often Twisted as Well as Pleated

Many polypeptides incorporate both helices and sheets within the same structure. In Figure 21.8, we see an artist's representation of two views of the molecule of a 247-residue enzyme, triose phosphate isomerase. Each broad, flat arrow is a segment of the chain that is in a β-sheet arrangement with a neighboring segment, also shown as a broad arrow. There are eight segments of the strand in the β-sheet regions of this enzyme, and the entire sheet is itself twisted to form a barrel-like configuration (called a β-barrel). α-Helix units occur between the strands involved in the β-sheet.

FIGURE 21.8
The molecule of the enzyme triose phosphate isomerase illustrates how both α-helix segments and β-sheet arrays can be incorporated together. The β-sheet consists of 8 parallel segments, but as indicated by the twists of the arrows, the sheet is itself twisted. The top view (a) and the side view (b) both indicate how this twist creates a barrel-like cylindrical structure.

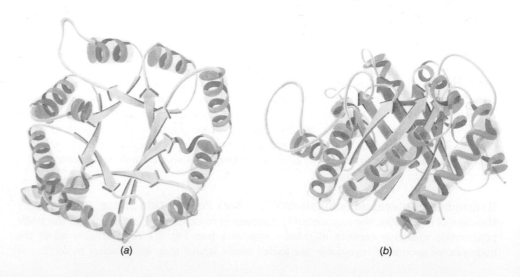

(a)   (b)

21.6

QUATERNARY STRUCTURES OF PROTEINS

For many proteins, the native form emerges only as two or more polypeptides assemble into a quaternary structure.

Proteins, like myoglobin, have finished shapes at the tertiary level. They are made up of single polypeptide molecules, sometimes with prosthetic groups. Many proteins, however, are aggregations of two or more polypeptides, and these aggregations constitute **quaternary structures.** One molecule of the enzyme phosphorylase, for example, consists of two tightly aggregated molecules of the same polypeptide. If the two become separated, the enzyme can no longer function. Individual molecules of polypeptides that make up an intact protein molecule are called the protein's *subunits*.

Hemoglobin has four subunits, two of one kind (designated α-subunits) and two of another (called the β-subunits), each subunit supporting a heme molecule (see Fig. 21.9). A combination of hydrophobic and electrostatic interactions as well as hydrogen bonds hold the subunits together. These forces do not work unless each subunit has the appropriate primary, secondary, and tertiary structural features. If even one amino acid residue is wrong, the results can be very serious, as in the example of sickle-cell anemia, described in Special Topic 21.1.

FIGURE 21.9

Hemoglobin. Four polypeptide chains, each with one heme molecule represented here by the colored, flat plates that contain spheres (Fe^{2+} ions), are nestled together. Only the atoms of the backbones are shown (by numbered circles). The central cavity, indicated by the double-headed arrow, has enough room to hold an organic anion not shown here, 2,3-bisphosphoglycerate (BPG), until hemoglobin starts to load up with oxygen molecules.

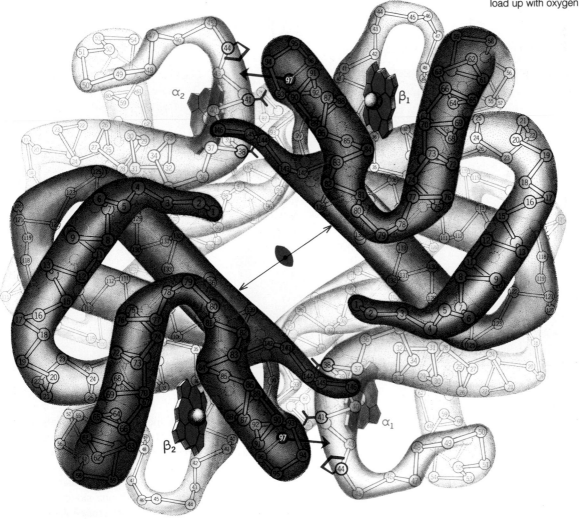

The decisive importance of the primary structure to all other structural features of a polypeptide or its associated protein is illustrated by the grim story of sickle-cell anemia. This inherited disease is widespread among those whose roots are in equatorial regions of central and western Africa.

In its mild form, where only one parent carried the genetic trait, the symptoms of sickle-cell anemia are seldom noticed except when the environment has a low partial pressure of oxygen, as at high altitudes. In the severe form, when both parents carried the trait, the infant usually dies by the age of 2 unless treatment is begun early. The problem is *an impairment in blood circulation traceable to the altered shape of hemoglobin in sickle-cell anemia.* The altered shape is particularly a problem after the hemoglobin has delivered oxygen and is on its way back to the heart and lungs for more.

The fault at the molecular level lies in a β-subunit of hemoglobin. One of the amino acid residues should be glutamic acid, but is valine instead. Thus instead of a side-chain CO_2^- group, which is electrically charged and hydrophilic, there is an isopropyl side chain, which is neutral and hydrophobic. Normal hemoglobin, symbolized as HbA, and sickle-cell hemoglobin, HbS, therefore have different patterns of electrical charges. Both have about the same solubility in well-oxygenated blood, but oxygen-free molecules of HbS clump together inside red cells and precipitate. This deforms the cells into a telltale sickle shape (see Figure 1). The distorted cells are harder to pump through capillaries, where the cells often create plugs. Sometimes the red cells split open. Any of these events place a strain on the heart. The error in one sidechain seems insignificant, but it is far from small in human terms.

The sickle-cell trait offers some resistance to malaria, which almost certainly explains why the trait survives largely where this tropical disease is most common. Normally, the mosquito-borne parasite that causes malaria resides within a red blood cell. However, the parasite cannot survive very long inside a sickled cell. The parasite has a high need for potassium ion, but the membrane of a sickled cell allows too much potassium ion to get through and escape. Thus people with the sickle-cell trait are statistically more likely than individuals unprotected from malaria to live long enough to bear children and so pass the trait to their offspring.

FIGURE 1   Electron micrographs of a normal red blood cell *(left)*, and a sickle cell *(right)*.

Covalent Cross-Linkages Occur in Collagen   The polypeptide units in collagen, each with about 1000 amino acid residues and each in a left-handed helix, assemble in units of three molecules each. The three left-handed helices wrap around each other in a relatively open *right*-handed helix of helices to form the **triple helix,** cable-like system called *tropocollagen* (see Fig. 21.10). Between the polypeptide strands of the triple helix *covalently bonded* molecular bridges are erected by a series of reactions that cause lysine sidechains to link together. The details are beyond the scope of our study, but *covalent* cross-links are better

(a)   (b)

FIGURE 21.10
The triple helix of collagen. (*a*)
Schematic drawing. (*b*) Electron
micrograph of collagen fibrils from
the skin. A fibril is an orderly
aggregation of collagen molecules
aligned side by side by overlapping
each other in a regularly repeating
manner that produces the banded
appearance. (Micrograph courtesy
of Jerome Gross, Massachusetts
General Hospital.)

able than hydrogen bonds or hydrophobic interactions to resist forces that would work to undo the tertiary structure of collagen. With aging, additional covalent cross-links develop between collagen strands, leading to less muscular flexibility and agility.

■ Meat from old animals is tougher because of their more highly cross-linked collagen.

A microfiber or *fibril* of collagen forms when individual tropocollagen cables overlap lengthwise. The mineral deposits in bones and teeth become tied into the protein at the gaps between the heads of tropocollagen molecules and the tails of others.

21.7

COMMON PROPERTIES OF PROTEINS

Even small changes in the pH of a solution can affect a protein's solubility and its physiological properties.

Although proteins come in many diverse biological types, they generally have similar chemical properties toward ordinary substances because they have similar functional groups.

Protein Digestion Is Hydrolysis The digestion of a protein is nothing more than the hydrolysis of its peptide bonds to give a mixture of amino acids. Different digestive *enzymes* —all in the family of *proteases*—handle the cleavage of peptide bonds according to the nature of the sidechains nearby. The hydrolysis of a tripeptide illustrates digestion.

$$^+NH_3CH_2C\overset{O}{\overset{\|}{\text{—}}}NHCHC\overset{O}{\overset{\|}{\text{—}}}NHCHCO^- + 2H_2O \xrightarrow[\text{(enzyme catalyzed)}]{\text{digestion}} {}^+NH_3CH_2CO^- + {}^+NH_3CHCO^- + {}^+NH_3CHCO^-$$

Peptide bonds are hydrolyzed

CH_3 $CH_2C_6H_5$ CH_3 $CH_2C_6H_5$

Glycylalanylphenylalanine (Gly-Ala-Phe)   Glycine (Gly)   Alanine (Ala)   Phenylalanine (Phe)

Protein *Denaturation* Is the Loss of Protein Shape   Peptide bonds are not hydrolyzed when a protein is denatured. All that has to happen is some disruption of secondary or higher structural features. **Denaturation** is the disorganization of the overall molecular shape of a native protein. It can occur as an unfolding or uncoiling of helices, or as the separation of subunits. Because native proteins have their overall shapes in an aqueous environment where water molecules are intimately involved with hydrophobic interactions, even the removal of water can cause the denaturation of many proteins.

Usually, denaturation is accompanied by a major loss of solubility in water. When egg white is whipped or is heated, for example, as when you cook an egg, the albumin molecules unfold and become entangled among themselves. The system no longer blends with water — it's insoluble — and it no longer allows light to pass through.

Table 21.2 has a list of several reagents or physical forces that cause denaturation, together with brief explanations of how they work. How effectively a given denaturing agent is depends on the kind of protein. The proteins of hair and skin and of fur or feathers quite strongly resist denaturation because they are rich in disulfide links.

In recent years, several proteins have been discovered that, after denaturation, can be *renatured*. Often the denaturation of such proteins results by the cleavage of disulfide links. When an enzyme that handles the oxidation of SH groups back to S—S groups is presented to such denatured proteins, they are restored to their native forms. The original S—S linkages are faithfully remade.

Protein Solubility Depends Greatly on pH   Because some sidechains as well as the end groups of polypeptides bear electrical charges, an entire polypeptide molecule can bear a net charge. Because these charged groups are either proton donors or proton acceptors, the net

TABLE 21.2
Denaturing Agents for Proteins

Denaturing Agent	How the Agent May Operate
Heat	Disrupts hydrophobic interactions and hydrogen bonds by making molecules vibrate too violently. Produces coagulation, as in the frying of an egg.
Microwave radiation	Causes violent vibrations of molecules that disrupt hydrogen bonds and hydrophobic interactions.
Ultraviolet radiation	Probably operates much like the action of heat (e.g., sunburning).
Violent whipping or shaking	Causes molecules in globular shapes to extend to longer lengths and then entangle (e.g., beating egg white into meringue).
Soaps	Probably affect hydrogen bonds and salt bridges.
Organic solvents (e.g., ethanol, acetone, 2-propanol)	May interfere with hydrogen bonds because these solvents can also form hydrogen bonds or can disrupt hydrophobic interactions. Quickly denature proteins in bacteria, killing them (e.g., the disinfectant action of 70% ethanol).
Strong acids and bases	Disrupt hydrogen bonds and salt bridges. Prolonged action leads to actual hydrolysis of peptide bonds.
Salts of heavy metals (e.g., salts of Hg^{2+}, Ag^+, Pb^{2+})	Cations combine with SH groups and form precipitates. (These salts are all poisons.)

charge is easily changed by changing the pH. For example, CO_2^- groups become electrically neutral CO_2H groups when they pick up protons as a strong acid is added.

Suppose that the net charge on a polypeptide is $1-$, and that one extra CO_2^- is responsible for it. When acid is added, we might imagine the following change, where the elongated shape is the polypeptide system.

$$ \text{(NH}_3^+ \quad CO_2^- \quad CO_2^- \text{)} + H_3O^+ \longrightarrow \text{(NH}_3^+ \quad CO_2^- \quad CO_2H \text{)} + H_2O $$

net charge: $1-$ net charge: 0

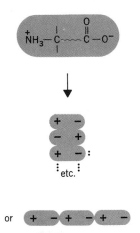

FIGURE 21.11
Several isoelectric protein molecules (top) can aggregate into very large clusters that no longer dissolve in water.

Now the product polypeptide is isoelectric and neutral. On the other hand, a polypeptide might have a net charge of $1+$, caused by an excess of one NH_3^+ group. The addition of OH^- can cause the polypeptide to become isoelectric and neutral.

Like each amino acid, each protein has a characteristic pH, its isoelectric point, at which its net charge is zero and at which it cannot migrate in an electric field. One major significance of this is that polypeptide molecules that are neutral can aggregate and clump together to become particles of enormous size that simply drop entirely out of solution (see Fig. 21.11). *A protein is least soluble in water when the pH equals the protein's isoelectric point.* Therefore, whenever a protein must be *in solution* to work, as is true for many enzymes, the pH of the medium must be kept away from the protein's isoelectric point. Buffers in body fluids have the task of ensuring this.

An example of the effect of pH on solubility, mentioned earlier, is given by casein, milk protein, whose pI value is 4.7. As milk turns sour, the pH drops from its normal value of $6.3-6.6$ to 4.7, and more and more casein molecules become isoelectric, denature, clump together, and separate as curds. As long as the pH of milk is something *other than* the pI for casein, the protein remains colloidally dispersed.

21.8

CELL MEMBRANES REVISITED—GLYCOPROTEIN COMPONENTS

Glycoproteins provide "recognition sites" on the surfaces of cell membranes. In Section 20.5 we introduced the general features of cell membranes, giving particular attention to their lipid components. With the knowledge about protein structure gained in this chapter and about carbohydrates learned in Chapter 19, we can take a second, deeper look at membranes. We'll also learn how glycoproteins are involved in giving shock absorbancy to the cartilage that cushions bone joints.

Membrane Proteins Help to Maintain Concentration Gradients If the cell membrane were an ordinary dialyzing membrane, any kind of small molecule or ion could move freely back and forth between the interior of the cell and any surrounding fluid. The working of cells, however, demands that only some things be let in from this fluid and that others be let out. This means that many concentration *gradients* occur between the cell interior and whatever fluid is outside. A **gradient** is an unevenness in the value of some physical property throughout a system. A *concentration gradient* exists in a solution, for example, when one region of the solution has a higher concentration of solute than another, such as the variations in sugar concentration in unstirred coffee just after you add sugar. Gradients are generally unstable systems compared to systems with the same components but more thoroughly mixed up. Because of the random motions of ions and molecules in liquids and gases, the natural

Ion	Concentration (mmol/L)	
	Plasma	Cells
Na^+	135–145	10
K^+	3.5–5.0	125

tendency is for gradients eventually to disappear in such media and for solute concentrations to become uniform. Yet living cells maintain a number of gradients.

As the data in the table in the margin show, both sodium ions and potassium ions have quite different concentrations in the fluids on the inside of a cell as compared to fluids on the outside. Thus between the inside and the outside of a cell there is a considerable concentration gradient for both of these ions. *This gradient must be maintained against nature's spontaneous tendency to remove concentration gradients.*

Here is where some of the proteins in cell membranes carry out a vital function. One kind of assembly of membrane protein molecules can move sodium ions against their gradient. When too many sodium ions move to the inside of a cell, they are "pumped" back out to the external fluid by a special molecular machinery called the *sodium–potassium pump.* The same pump can move potassium ions back inside a cell. This movement of any solute *against* its concentration gradient requires chemical energy and is an example of **active transport.** Other reactions in cells supply the chemical energy that lets it work.

Gap Junctions Enable Substances to Move Directly from One Cell to Another In the cells of most tissues of multicelled organisms, membrane proteins provide a route for the *direct* movements of ions and molecules from one cell to another. These routes are through **gap junctions,** tubules fashioned from membrane proteins that "rivet" cells together (see Fig. 21.12). So many of such junctions occur in some tissues that the entire tissue is interconnected from within. Gap junctions in bone tissue, for example, enable bone cells at some distance from capillaries to receive nourishment and to remove wastes. Heart muscle is able to contract *synchronously* because gap junctions allow ions to move easily between cells. The gaps are large enough to allow certain ions, like Ca^{2+}, and certain relatively small molecules (up to formula masses of about 1200) to pass, but are not large enough for macromolecules like proteins and nucleic acids to get through.

Calcium ion appears to control the diameters of gap junctions. When the concentration of this ion is very low ($< 10^{-7}$ M), the channels are fully open. As $[Ca^{2+}]$ increases, the gaps close, and they become completely shut when $[Ca^{2+}]$ reaches about 5×10^{-5} M. One consequence of this control is that if part of an interconnected mass of cells is injured, the closure of gaps limits the damage that might otherwise happen.

Some Proteins in Cell Membranes Are Receptors for Hormones and Neurotransmitters A **receptor molecule** is one whose unique shape enables it to fit only to the molecule of a compound that it is supposed to receive, its *substrate.* It is thus able to "recognize" the molecules of just one compound from among the hundreds whose molecules bump against it. This is roughly how specific hormones are able to find and stop only at the

FIGURE 21.12

Gap junctions. A protein-fashioned channel (in brown) between two cells enables some small particles to pass directly from one cell to another. The blue spheres each with two wavy tails are phospholipid molecules of the cell membranes.

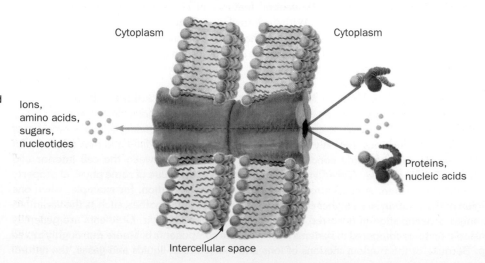

Cytoplasm

Cytoplasm

Ions, amino acids, sugars, nucleotides

Proteins, nucleic acids

Intercellular space

MIFEPRISTONE (RU 486)—RECEPTOR BINDING OF A SYNTHETIC ANTIPREGNANCY COMPOUND

When released from an ovarian follicle, the natural female hormone, progesterone, acts to prepare the system for pregnancy both by inhibiting further production of ova (egg cells) and by preparing the uterus for the implantation of the fertilized ovum. The action of progesterone involves the binding of its molecules to protein receptors within cells of the lining of the uterus (the endometrium).

Antiprogestins (Table 20.3) mimic the work of progesterone in that they cause a pseudopregnant state ("false pregnancy state") and so suppress the production of ova. Fertilization cannot occur without an ovum, of course, and so the synthetic progestin-containing medications are birth control pills.

Progesterone

Mifepristone (RU 486)

RU 486 (RU for Roussel-Uclaf, a French pharmaceutical company) was first prepared in 1980. It was soon discovered to be a strong binder to the progesterone binding sites on the receptor protein for progesterone. This action blocks the normal action of progesterone, and RU 486 is thus an *antiprogesterone.* If RU 486 is taken during the five to six day postcoital period (the period immediately following intercourse), its blocking action prevents pregnancy by suppressing the implantation of a fertilized ovum in the uterus. If used within 72 hours after unprotected intercourse, its failure rate is very low. It can thus be used as a "morning after" pill.

RU 486, followed by the use of two prostaglandin-like compounds, also induces abortion. Thus RU 486 is also an abortifacient (an abortion inducing agent). Under medical supervision, the use of RU 486 for this purpose has a 96% success record. The failures include continued pregnancy, only partial expulsion of the fetus, and the need for procedures to stem uterine bleeding.

That RU 486 "prevents pregnancy" is a controversial statement, because some view the onset of pregnancy as occurring at the moment of fertilization. Others regard pregnancy as not starting until the fertilized ovum has become implanted in the uterus. The controversy thus involves the question, "When does *pregnancy* begin?" and the not identical question, "When does *human life* begin?" Around these questions have surged some of the stormiest waters of the prolife–prochoice controversy.

---

cells where they are meant to stop, by being able to fit to unique receptors. An antiprogesterone birth control agent, RU 486, acts as a hormone-mimic, binding to a receptor protein (see Special Topic 21.2).

This fitting to a membrane protein also helps neurotransmitter molecules to become quickly attached to the right spot on the membrane of the next nerve cell after moving from one nerve cell to the next across the very narrow gap between them. Once a receptor molecule in a cell membrane accepts its unique substrate, further biochemical changes occur. An enzyme or a gene in the cell might be activated, for example.

■ Neurotransmitters are organic molecules that help carry nerve signals from the end of one nerve cell to the beginning of the next.

Membrane Proteins Are Glycoproteins Essentially all cells are "sugar-coated." The sugars are not the ordinary sugars of nutrition but are related to monosaccharides and *oligosaccharides.* The latter are carbohydrates that can be hydrolyzed to three or more — up to a few dozen — monosaccharide molecules. Some of the membrane carbohydrate units are

■ We learned in Section 20.3 that cerebrosides are glycolipids.

covalently joined to lipids — the **glycolipids** — of the membrane. Other membrane carbohydrates are bound to proteins, forming the **glycoproteins** of membranes (see Fig. 21.13). Most proteins are glycoproteins; several thousand have been identified and the list is growing.

The oligosaccharides of glycoproteins generally contain nitrogen or sulfur as additional elements. Nitrogen is present as an amine or an amide group. Sulfur occurs in SO_3^- groups joined to O atoms of carbohydrate systems or to N atoms of amino sugars. Amino sugars are those in whose molecules an OH group is replaced by an NH_2 group. Perhaps the most common amino sugar built into glycoproteins is D-glucosamine in the form of its *N*-acetyl derivative — *N*-acetyl-D-glucosamine. It and systems like it are usually joined to a polypeptide at an asparagine residue by what is called an *N-link*.

■ The *N* signifies that the acetyl group is attached to *nitrogen*.

β-D-Glucosamine

N-Acetyl-β-D-glucosamine

N-Acetyl-β-D-glucosamine unit N-linked to a polypeptide unit

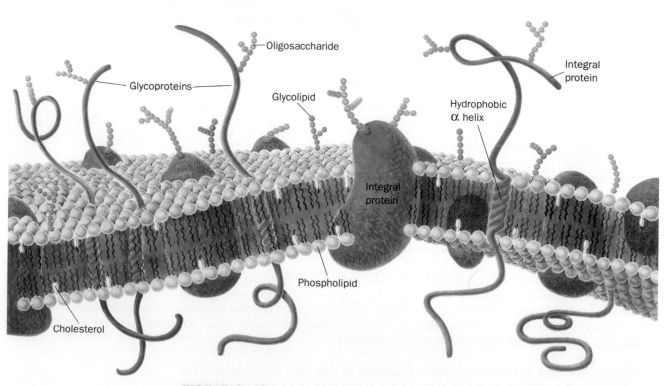

FIGURE 21.13

Glycoproteins as structural units in a cell membrane. The blue spheres each with two wavy tails are phospholipid molecules. Shown in yellow are cholesterol molecules. Chains of green beads represent glycolipids. Chains of yellow beads attached to polypeptides and proteins (in orange) are oligosaccharide units.

TABLE 21.3
Glycosaminoglycans—Their Repeating Disaccharide Units[a]

Hyaluronate monomer system	A component of ground substance particularly in connective tissue, in fluid that lubricates joints (synovial fluid), and the vitreous humor of the eye. Depending on the location, from 250 to 25,000 of these disaccharide units are joined into the polymer. (Note the N-acetyl-D-glucosamine unit on the right.)
Chondroitin-4-sulfate monomer system	A major component of cartilage and other connective tissues (after the Greek *chondros,* cartilage). There is also a 6-sulfate relative of this system. (Note that the unit on the right is derived from D-galactose, not D-glucose.)
Heparin monomer system	Occurs not in connective tissue, like the above, but in mast cells, cells that line the walls of arteries, particularly in the lungs, liver, and skin. Heparin inhibits the formation of blood clots, and its release from mast cells when an injury occurs prevents clotting from going too far. Heparin is widely used postsurgically to control clotting.

[a] In order to show the ring structures of the individual sugar units in their conventional array, distortions of connecting bonds must sometimes be tolerated in such structures.

One General Kind of Glycoprotein Gives Resiliency to Cartilage
Table 21.3 gives the structures and chief uses of three of the several members of carbohydrate polymers known as the *glycosaminoglycans*[1] or *mucopolysaccharides.* Two of the glycosaminoglycans form the gel-like material called **ground substance** found in cartilage and other extracellular spaces and in which fibers of collagen and another fibrous protein called *elastin* are embedded. The fibers give the *tensile strength* to cartilage, the strength to withstand stretching tension without breaking. The ground substance holding the fibers gives *flexibility* and shock absorbancy to cartilage for reasons described next.

With molecules having so many polar groups with hydrogen bonding ability, it isn't surprising that glycosaminoglycan molecules are "sticky," and that the substances are thick, slimy, viscous materials resembling mucus. It also is not surprising that their molecules can attract and hold large numbers of water molecules of hydration. This is what gives ground

---

[1] In the term glycosaminoglycan, "glycos-" is from glycose, the generic name for monosaccharides; "glycan" is the generic name of all polysaccharides. The "gly-" part of both glycose and glycan is replaced by the prefix of a specific monosaccharide when naming specific polysaccharides. Thus starch is a *glucan* because it is a polymer of glucose.

substance the spongy, resilient nature needed for the cartilage in bone joints. As cartilage is squeezed during bumps and jolts, water is forced out. When the pressure is released, water rushes back in. In fact, this "tidal flow" of water is what carries metabolic wastes away and brings in nutrients to cartilage tissue, a tissue that lacks blood vessels. The movement of water caused by flexing the joints makes it easy to understand why the cartilage in joints becomes somewhat fragile, even brittle, during long periods of no exercise.

Multifunctional Monosaccharide Units Have Numerous Ways of Combining into Oligosaccharides   Glycoproteins include a large subfamily of N-linked oligosaccharides whose structures include several monosaccharides in both "straight chain" and "branched chain" connections. Many of the linkages between monosaccharide units involve oxygen atoms at ring positions not used for linking purposes by the nutritionally important disaccharides or polysaccharides that we studied in Chapter 19. Table 21.3, in fact, illustrates $(1 \rightarrow 3)$ linkages in the first two structures. In Chapter 19 we saw mostly $(1 \rightarrow 4)$ linkages, but $(1 \rightarrow 6)$ linkages are also present in amylopectin and glycogen. The possibilities for a number of $(1 \rightarrow n)$ linkages having varying geometries give living systems far more options for joining monosaccharide units than for joining amino acid units. Two identical amino acids, for example, can be joined in only one way by a peptide bond, but two identical monosaccharides can be linked to form 11 different disaccharides. Someone has calculated that only four *different* monosaccharide units can be linked in over 35,000 unique tetrasaccharides. *Thus an almost unlimited variety of structurally different oligosaccharides is available for making glycoproteins.* This momentous fact at the molecular level of life is behind the almost incredible spectrum of biological properties observed throughout the living world. Different oligosaccharides, N-linked to polypeptides in a cell membrane, make possible the unique abilities of the membranes not only of species but also of individuals within species to interact discriminately among substances cruising near cells. The existence of the ABO blood groups; the inability of the sperm of one species to fertilize the egg of any other; the action of bacteria normally at just one tissue and not at any others; the existence of bacterial infections in animals that do not affect humans (and vice versa); the ability of hormones to be snared only by their own "target" tissues; these and many other properties can be traced to the occurrence of so many different oligosaccharides residing on the surfaces of cell membranes.

The Linkage of Oligosaccharides to Proteins Largely Occurs at Protein Surfaces, Not in Their Interiors   The N-links of oligosaccharides to polypeptides occur most often where the polypeptide strand is following a bend between segments of secondary structure, like β-sheets or α-helices. The oligosaccharide units, therefore, have little if any direct effect on tertiary structure but project, instead, from protein surfaces. This is why we could say near the start of this section that cell membranes are "sugar coated." The "sugar" consists of oligosaccharide units extending into the surrounding spaces from the glycoproteins that make up parts of cell membranes. Oligosaccharide units contribute to the adhesion between cells, and they have critical functions in all of the activities that depend on the "recognition" of hormones and neurotransmitters by a cell. In the next chapter we'll learn about the mechanism of recognition at the molecular level of life.

21.9

CLASSES OF PROTEINS

Three criteria for classifying proteins are their solubility in aqueous systems, their compositions, and their biological functions.

We began this chapter with hints about the wide diversities of the kinds and uses of proteins. Now that we know about their structures, we can better understand how so many types of proteins with so many functions are possible. The following three major classifications of proteins and several examples give substance to the chapter's introduction.

■ In carbohydrate chemistry, amylose and cellulose are examples of "straight chain" polysaccharides; amylopectin and glycogen are "branched chain."

■ The varying glycoproteins that provide this uniqueness are called **glycoforms.**

■ The oligosaccharide units also are involved in the actions of toxins, viruses, and bacteria.

Proteins Can Be Classified According to Solubility When proteins are classified by their solubilities, two families are the **fibrous proteins** and the **globular proteins.**

Fibrous Proteins

1. **Collagens** occur in bone, teeth, tendons, skin, blood capillaries, cartilage, and some ligaments. When such tissue is boiled with water, the portion of its collagen that dissolves is called *gelatin*.

2. **Elastins,** which have elastic, rubber-like qualities, are also in cartilage and are found in stretchable ligaments, the walls of large blood vessels like the aorta, the lungs, and the necks of grazing animals. Elastin, like collagen, is rich in glycine residues and proline, but not in hydroxyproline. Elastin chains are cross-linked by covalently bonded units that are largely responsible for elastin's elasticity.

3. **Keratins** occur in hair, wool, animal hooves, horns, nails, porcupine quills, and feathers. The keratins are rich in disulfide links, which contribute to the unusual stabilities of these proteins to environmental stresses.

4. **Myosins** are the proteins in contractile muscle.

5. **Fibrin** is the protein of a blood clot. During clotting, fibrin forms from its precursor, fibrinogen, by an exceedingly complex series of reactions.

■ When meat is cooked, some of its collagen changes to gelatin, which makes the meat easier to digest.

■ Elastin is not changed to gelatin by hot water.

Globular Proteins Globular proteins are soluble in water or in water that contains salts.

1. **Albumins** are present in egg white and in blood. In the blood, the albumins are buffers, transporters of water-insoluble molecules of lipids or fatty acids, and carriers of metal ions, like Cu^{2+} ions, that are insoluble in aqueous media at pH values higher than 7.

2. **Globulins** include antibodies, factors of the body's defenses against diseases. In addition, enzymes, many transport proteins, and receptor proteins are globulins.

Proteins Can Be Classified According to Biological Function Perhaps no other system more clearly dramatizes the importance of proteins than classifying them by their biological function.

1. Enzymes. The biological catalysts.

2. Contractile muscle. With stationary filaments, myosin, and moving filaments, actin.

3. Hormones. Such as growth hormone, insulin, and others.

4. Neurotransmitters. Such as the enkephalins and endorphins.

5. Storage proteins. Those that store nutrients that the organism will need such as seed proteins in grains, casein in milk, ovalbumin in egg white, and ferritin, the iron-storing protein in human spleen.

6. Transport proteins. Those that carry things from one place to another. Hemoglobin and the serum albumins are examples already mentioned. Ceruloplasmin is a copper-carrying protein.

7. Structural proteins. Proteins that hold a body structure together, such as collagen, elastin, keratin, and glycoproteins in cell membranes.

8. Protective proteins. Those that help the body to defend itself. Examples are the antibodies and fibrinogen.

9. Toxins. Poisonous proteins. Examples are snake venom, diphtheria toxin, and clostridium botulinum toxin (a toxic substance that causes some types of food poisoning).

SUMMARY

Amino acids About 20 α-amino acids supply the amino acid residues that make up a polypeptide. The molecules of all but one (glycine) are chiral and in the L-family. In the solid state or in water at a pH of roughly 6 to 7, amino acids exist as dipolar ions or zwitterions. Isoelectric points are the pH values of solutions in which amino acid (or protein) molecules are isoelectric. For amino acids without acidic or basic side chains, the pI values are in the range of 6 to 7. Amino acids with CO_2H groups on sidechains have lower pI values. Those with basic sidechains have higher pI values. Several amino acids have hydrophobic sidechains, but the sidechains in others are strongly hydrophilic. The SH group of cysteine opens the possibility of disulfide cross-links between or within polypeptide units.

Polypeptides Amino acid residues are held together by peptide (amide) bonds, so the repeating unit in polypeptides is —NH—CH—CO—. Each amino acid residue has its own sidechain. This repeating system with a unique sequence of sidechains constitutes the primary structure of a polypeptide.

Once the primary structure is fashioned, the polypeptide coils and folds into higher features — secondary and tertiary — that are stabilized largely by hydrophobic interactions and hydrogen bonds. The most prominent secondary structures are the α-helix — a right-handed helix — and the β-pleated sheet. Individual polypeptides in collagen, which has an abundance of glycine, proline, and hydroxylated proline residues, are in a left-handed helix. Disulfide bonds form from SH groups on cysteine residues as many proteins assume their tertiary structure.

Proteins Many proteins consist just of one kind of polypeptide. Many others have nonprotein, organic groups — prosthetic groups — or metal ions. And still other proteins — those with quaternary structure — involve two or more polypeptides whose molecules aggregate in definite ways, stabilized by hydrophobic interactions, hydrogen bonds, and salt bridges. Thus the terms *protein* and *polypeptide* are not synonyms, although for some specific proteins they turn out to be.

Because of their higher levels of structure, proteins can be denatured by agents that do nothing to peptide bonds. A few denatured proteins can be renatured, but this is uncommon. The acidic and basic sidechains of polypeptides affect protein solubility, and when a protein is in a medium whose pH equals the protein's isoelectric point, the substance is least soluble. The amide bonds (peptide bonds) of proteins are hydrolyzed during digestion.

Membrane proteins — glycoproteins Incorporated into the lipid bilayer membranes of cells are proteins (and lipids) with attached oligosaccharide units of widely varying structure. Some of the proteins of a cell membrane provide conduits by which active transport processes can maintain concentration gradients. Other proteins provide gap junctions for direct movements, cell-to-cell, of certain dissolved species. The oligosaccharides of the membrane proteins stick out away from the membrane surface. They do not appear to affect the overall shapes of their attached polypeptides, and they serve as cell-recognition features for molecules moving near the cell. Certain oligosaccharides, the glycosaminoglycans, make up ground substance, which gives elasticity and shock absorbancy to cartilage.

REVIEW EXERCISES

The answers to Review Exercises whose numbers are in color are found in Appendix C. The answers to the other Review Exercises are found in the Study Guide that accompanies this book. The more challenging questions are marked with asterisks.

Amino Acids

21.1 One of the following structures is *not* of an amino acid on the list of standard 20. Which one is not on the list? How can you tell without looking at Table 21.1?

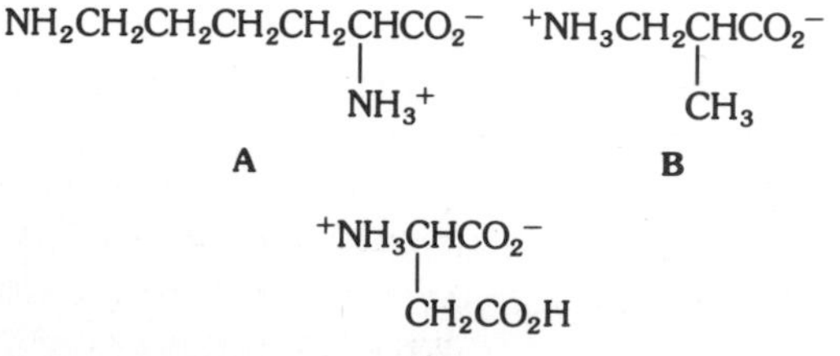

$$NH_2CH_2CH_2CH_2CH_2CHCO_2^-$$
$$| \atop NH_3^+$$
A

$$^+NH_3CH_2CHCO_2^-$$
$$| \atop CH_3$$
B

$$^+NH_3CHCO_2^-$$
$$| \atop CH_2CO_2H$$
C

21.2 The following amino acid is on the standard list of 20.

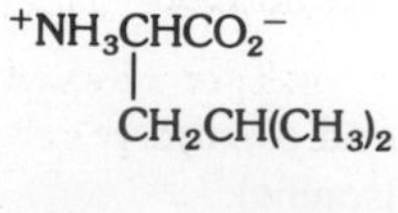

$$^+NH_3CHCO_2^-$$
$$| \atop CH_2CH(CH_3)_2$$

(a) What part of its structure would be its amino acid *residue* in the structure of a polypeptide? (Write the structure of this residue.)

(b) With the aid of Table 21.1, write the name and the three-letter symbol of this amino acid.

(c) Is its sidechain hydrophobic or hydrophilic?

21.3 What structure will nearly all the molecules of glycine have at a pH of about 1?

21.4 What structure will most of the molecules of alanine have at a pH of about 12?

21.5 Pure alanine does not melt, but at 290 °C it begins to char and decompose. However, the ethyl ester of alanine, which has a free NH_2 group, has a low melting point, 87 °C. Write the structure of this ethyl ester, and explain this large difference in melting point.

*21.6 The ethyl ester of alanine (Review Exercise 21.5) is a much stronger base — more like ammonia — than alanine. Explain this.

21.7 Which of the following amino acids has the more hydrophilic side chain? Explain.

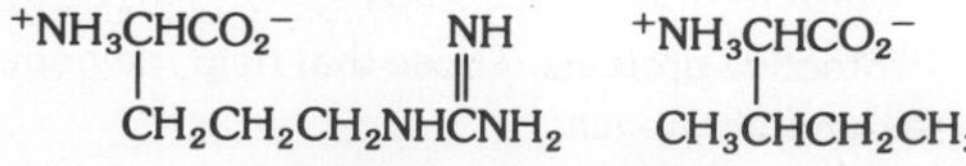

$$^+NH_3CHCO_2^-$$
$$| \atop CH_2CH_2CH_2NHCNH_2 \atop \| \atop NH$$
A

$$^+NH_3CHCO_2^-$$
$$| \atop CH_3CHCH_2CH_3$$
B

21.8 Which of the following amino acids has the more hydrophobic side chain? Explain.

$$^+NH_3CHCO_2^- \qquad ^+NH_3CHCO_2^-$$
$$\underset{\textstyle CH_2OH}{|} \qquad \underset{\textstyle CH_2C_6H_5}{|}$$
$$\textbf{A} \qquad\qquad \textbf{B}$$

21.9 Glutamic acid can exist in the following form.

$$NH_2CHCO_2^-$$
$$\underset{\textstyle CH_2CH_2CO_2^-}{|}$$

(a) Would this form predominate at a pH of 2 or a pH of 10? Explain.

(b) To which electrode, the anode or the cathode — or to neither — would aspartic acid in this form migrate in an electric field?

21.10 When it is said that a substance is poorly soluble in water because of a *hydrophobic interaction*, what does "hydrophobic interaction" mean?

21.11 What kind of a reactant is required to convert cysteine into cystine: an acid, base, oxidizing agent, or reducing agent?

*21.12 Write two equilibrium equations that show how glycine, in its isoelectric form, can serve as a buffer.

21.13 Complete the following Fischer projection formula to show correctly the absolute configuration of L-serine.

$$CO_2^-$$
$$+$$
$$CH_2OH$$

Polypeptides

*21.14 Each of the following structures has an amide linkage. Each can be hydrolyzed to glycine and lysine. The amide linkage in one of the two structures, however, cannot properly be called a *peptide bond*. This is true of which structure? Why?

$$^+NH_3CH(CH_2)_4NHCCH_2 \qquad ^+NH_3CH_2CNHCHCO_2^-$$

(with carbonyl O groups, CO$_2^-$, NH$_3^+$, (CH$_2$)$_4$NH$_2$ substituents)

$$\textbf{A} \qquad\qquad\qquad \textbf{B}$$

21.15 Write both the conventional and the condensed structure (three-letter symbols) of the dipeptides that can be made from lysine and cysteine.

21.16 What are the condensed structures of the dipeptides that can be made from glycine and glutamic acid? (Do not use the three-letter symbols.)

21.17 Using three-letter symbols, write the structures of all of the tripeptides that can be made from lysine, glutamic acid, and cysteine.

21.18 Write the structures in three-letter symbols of all of the tripeptides that can be made from glycine, cysteine, and alanine.

21.19 What is the conventional structure of Val-Ile-Phe?

*21.20 Write the conventional structure for Val-Phe-Ala-Gly-Leu.

*21.21 Write the conventional structure for Asp-Lys-Glu-Thr-Tyr.

*21.22 Compare the side chains in the pentapeptide of Review Exercise 21.20 (call it **A**) with those in the following, which we can call **B**

Lys-Glu-Asp-Thr-Ser

(a) Which of the two, **A** or **B**, is the more hydrocarbon-like?

(b) Which is probably more soluble in water? Explain.

*21.23 Compare the sidechains in the pentapeptide of Review Exercise 21.21, which we'll label **C**, with those in Phe-Leu-Gly-Ala-Val, which we can label **D**. Which of the two would tend to be less soluble in water? Explain.

*21.24 If the tripeptide Gly-Cys-Ala were subjected to mild oxidizing conditions, what would form? Write the structure of the product using three-letter symbols.

21.25 What is meant by a *peptide group?* Describe its geometry.

21.26 What atoms or groups are *trans* to each other in a trans-planar peptide group?

Higher Levels of Protein Structure

21.27 Which *level* of polypeptide complexity concerns the molecular "backbone" and the sequence of side chains?

21.28 What is meant by *native* protein?

21.29 To what level of protein complexity is the disulfide bond normally assigned?

21.30 The disulfide bond is a *covalent* bond. Why isn't it assigned to the primary level of polypeptide structure?

21.31 An enzyme consists of two polypeptide chains associated together in a unique manner. To what level of protein structure is this detail assigned?

21.32 Does the trans-planar nature of the peptide group enlarge or reduce the *range of geometrical options* available to a polypeptide?

21.33 Describe the specific geometrical features of an α-helix structure. What force of attraction stabilizes it? Between what two kinds of sites in the α-helix does this force operate? How do the sidechains become positioned in the α-helix?

21.34 Give a brief description of the secondary structure of an individual polypeptide strand in collagen.

21.35 What function does ascorbic acid (vitamin C) perform in the formation of strong bones?

21.36 Describe the structure and geometry of tropocollagen. How is tropocollagen made into a collagen fibril?

21.37 Bridges between the polypeptide strands in collagen have what principal feature: a hydrophobic interaction, an electrostatic attraction (salt bridge), a disulfide system, or some other kind of covalent linkage?

21.38 What specific force of attraction stabilizes a β-pleated sheet? Where do the sidechains take up positions?

21.39 Does an α-helix or a β-sheet describe the *entire* secondary structure of a polypeptide? If not, how do these features occur?

21.40 What factors affect the bending and folding of α-helices in the presence of an aqueous medium?

21.41 What is meant by a salt bridge?

21.42 When is the disulfide bond normally put into place during the formation of a protein?

21.43 In what way does hemoglobin represent a protein with quaternary structure (in general terms only)?

21.44 How do myoglobin and hemoglobin compare (in general terms only)?
(a) Structurally—at the quaternary level
(b) Where they are found in the body
(c) In terms of their prosthetic group(s)
(d) In terms of their functions in the body

Properties of Proteins

*21.45 What products form when the following polypeptide is completely digested?

$$^+NH_3CHCONHCHCONHCHCONHCHCONHCH_2CO_2^-$$
$$\quad\ |\qquad\quad |\qquad\quad\ \ |\qquad\qquad |$$
$$\ CH_2OH\quad CH_3\qquad CH\qquad (CH_2)_4NH_2$$
$$\qquad\qquad\qquad\quad\ \ /\ \ \backslash$$
$$\qquad\qquad\qquad H_3C\quad CH_3$$

21.46 Explain why a protein is least soluble in an aqueous medium that has a pH equal to the protein's pI value.

21.47 What is the difference between the *digestion* and the *denaturation* of a protein?

21.48 Some proteins can be denatured by a reducing agent but then completely renatured by a mild oxidizing agent. What functional groups are involved?

Cell Membranes

21.49 What is meant by "gradient" in the term *concentration gradient?*

21.50 Which has the higher level of Na^+, plasma or cell fluid?

21.51 Does cell fluid or plasma have the higher level of K^+?

21.52 In which fluid, plasma or cell fluid, would the level of sodium ion increase if the sodium ion gradient could not be maintained?

21.53 What does the sodium–potassium pump do?

21.54 What does "active" refer to in the term *active transport?*

21.55 What is a *gap junction* and what services does it perform?

21.56 The concentration of what species appears to control the size of the opening of a gap junction?

Glycoproteins and Cell Membranes

21.57 What does the prefix "glyco-" refer to in *glycoprotein?*

21.58 In general terms only, in what structural way does an oligosaccharide differ from a mono- or a disaccharide? From a polysaccharide?

21.59 The term "glycan" refers to what?

21.60 A "D-glucosaminoglucan" would be made of what monomer?

21.61 What is *ground substance?*

21.62 What kinds of substances provide tensile strength to carti-

lage and what substance gives cartilage its resiliency and shock-absorbing properties?

21.63 What role does the hydrogen bond play in the ability of cartilage tissue to carry out its functions?

21.64 What structural fact about monosaccharides (including the amino sugars) makes possible the huge variety of possible oligosaccharides?

Types of Proteins

21.65 What experimental criterion distinguishes between fibrous and globular proteins?

21.66 What is the relationship between collagen and gelatin?

21.67 How are collagen and elastin alike? How are they different?

21.68 What experimental criterion distinguishes between the albumins and the globulins?

21.69 What is fibrin and how is it related to fibrinogen?

21.70 What general name can be given to a protein that carries a carbohydrate molecule?

Sickle-Cell Anemia (Special Topic 21.1)

21.71 What is the primary *structural* fault in the hemoglobin of sickle-cell anemia?

21.72 What happens in blood cells in sickle-cell anemia that causes their shapes to become distorted?

21.73 What problems are caused by the distorted shapes of the red cells?

Mifepristone (RU 486) (Special Topic 21.2)

21.74 What is meant by a "receptor protein?"

21.75 In general terms only, how does RU 486 work in the post-coital period?

Additional Exercises

21.76 When an oligosaccharide unit is cleaved from its glycoprotein, the overall *shape* of the protein section is largely unchanged. Explain.

21.77 Write the structure of a pentapeptide that would hydrolyze to give only alanine.

*21.78 Consider the following structure.

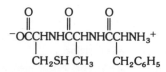

(a) If a polypeptide were *partially* hydrolyzed, could a molecule of this structure possibly form in theory? Explain.
(b) What is the three-letter symbol of the N-terminal residue?
(c) How would the structure of this compound be represented using the three-letter symbols and following the rules for writing such a structure?
(d) Would a mild reducing agent have any affect on this compound? If so, write the structure of the product.

ENZYMES, HORMONES, AND NEUROTRANSMITTERS

22

In the most tense situations requiring the keenest reflexes, many of the body's hormones participate in assuring the needed energy supply. You can be sure that the adrenaline is flowing in this surfer, catching the curl off the northern coast of Oahu, Hawaii.

22.1

ENZYMES

Enzymes are biological catalysts whose activities often depend on cofactors made from B vitamins or metal ions.

Virtually all enzymes are proteins. A few have been discovered that are made of RNA, but they are exceptions. In this chapter we will only study enzymes that are proteins.

■ All enzyme molecules and most substrate molecules are chiral; they have handedness.

Enzymes Are Very Specific for the Substrates They Accept Enzyme specificity is one property that a theory of enzyme action has to explain. *In vivo*, a given enzyme acts on just one substrate. The digestive enzymes are even specific about which part of a substrate they hydrolyze. Some digestive enzymes specialize, for example, in hydrolyzing peptide bonds adjacent to particular sidechains on amino acid residues as the enzyme works to split long polypeptides into shorter molecules. Other digestive enzymes work exclusively to cleave C-terminal amino acid residues and still others to split off N-terminal residues. The starch hydrolyzing enzyme that catalyzes the hydrolysis of $\alpha(1 \rightarrow 4)$ oxygen bridges in starch does not work on the $\alpha(1 \rightarrow 6)$ bridges. Enzymes, as we said, are unusually specific *in vivo*.

■ Many enzymes have been isolated for use in catalyzing chemical reactions *in vitro*.

Enzymes Work Best over Relatively Narrow Ranges of pH and Temperature Because nearly all enzymes are proteins, they are constrained by the general properties of proteins. The electrical charges on proteins, for example, vary and are sometimes differently distributed depending on the availability of acids or bases in the surrounding media, as we learned in Section 21.7. Changes in electric charge distribution can even affect the overall shape of a protein. We'll soon see that this explains why enzymes are active only in the pH range normal to their environment in the body. Pepsin, the gastric protein-digesting enzyme in the acidic medium of the stomach, is most active at a pH of about 2. Fumarase, an enzyme involved within cells in metabolizing acetyl units (from the breakdown of sugar or fatty acids), works best at a pH of just below 7 (see Fig. 22.1).

FIGURE 22.1
The effect of pH on the rate of an enzyme-catalyzed reaction, the enzyme being fumarase.

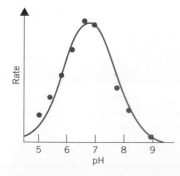

We learned in Section 9.1 that the rates of many reactions slow as the temperature of the reacting mixture decreases and that rates increase with increasing temperature. The rate-slowing effect is also observed with enzyme catalysis. However, because enzymes are proteins, the opposite effect is not observed; the rates of enzyme-catalyzed reactions do *not* increase steadily with increasing temperature. Proteins are denatured by heat. Thus the efficiency of an enzyme will reach a peak over a small range of temperatures.

Enzymes Display Remarkable Rate Enhancements Enzymes, like all catalysts, affect the *rates* of reactions by providing a reaction pathway with a lower energy of activation than the uncatalyzed reaction can take (see Sections 9.1–9.2). Even small reductions of energy barriers can cause spectacular increases in rates. The enzyme carbonic anhydrase (CA) is an example. It catalyzes the interconversion of bicarbonate ion and protons with carbon dioxide and water:

$$CO_2 + H_2O \underset{\text{carbonic anhydrase}}{\overset{\text{carbonic anhydrase}}{\rightleftharpoons}} HCO_3^- + H^+$$

In actively metabolizing cells, where the supply of CO_2 is relatively high, this equilibrium shifts to the right. In blood that circulates into the lungs, where exhaling keeps the supply of CO_2 low, this equilibrium must shift to the left, *and the same enzyme participates in this change.* Each molecule of carbonic anhydrase aids in the conversion of 600,000 molecules of CO_2 *each second!* This is ten million times faster than the uncatalyzed reaction, which makes the speed of action of carbonic anhydrase among the highest of all known enzymes.

■ The equilibrium *must* shift to the right to make HCO_3^- and H^+ when the supply of CO_2 is high — a consequence of Le Châtelier's principle.

Enzymes Get Equilibria Established Extremely Rapidly In a chemical equilibrium, like that involving carbonic anhydrase, it is important to remind ourselves that the catalyst speeds up *equilibration.* It accelerates *both* the forward and the reverse reactions. Whether the equilibrium shifts to the right or to the left doesn't depend on the catalyst at all. It depends strictly on the inherent equilibrium constants; on the relative concentrations of reactants and products; on whether other reactions feed substances into the equilibrium or continuously remove them; and on the temperature. All the catalyst does is to make whatever shift is mandated by these conditions occur very rapidly.

Most Enzymes Consist of Polypeptides plus Cofactors The molecules of most enzymes include a nonpolypeptide component called a **cofactor.** The polypeptide is called the **apoenzyme,** but without the cofactor there is no enzymatic activity.

The cofactor of some enzymes is simply a metal ion. Zn^{2+} is the metal ion in carbonic anhydrase, for example. Fe^{2+} occurs in the cytochromes, a family of enzymes involved in biological oxidations. Most of the trace metal ions of nutrition are enzyme cofactors.

In other enzymes the cofactor is an organic molecule or ion called a **coenzyme.** Some enzymes have both a coenzyme and a metal ion cofactor.

B Vitamins Are Used to Make Coenzymes Thiamine diphosphate, a coenzyme with structure **1**, is a diphosphate ester of thiamine, a B vitamin, shown here in its fully ionized form.

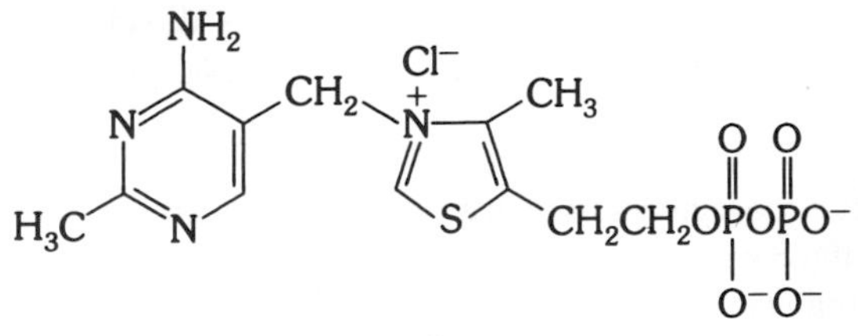

1

Thiamine diphosphate

■ When the diet is deficient in thiamine, often called vitamin B_1, a disease called *beriberi* results.

When pure, this coenzyme is a triprotic acid, but at the pH of body fluids it is ionized approximately as shown.

■ Nicotinamide's other name is *niacin*. A deficiency of this vitamin leads to *pellegra*.

Nicotinamide, another B vitamin, is part of the structure of nicotinamide adenine dinucleotide, **2a**, another important coenzyme. Mercifully, its long name is usually shortened to NAD^+ (or, sometimes, just NAD). The bottom half of the NAD^+ molecule is from adenosine monophosphate, AMP, a phosphate ester described on page 244. The upper half of NAD^+ is almost like its lower portion except that a molecule of nicotinamide has replaced the two-ring heterocyclic unit below it.

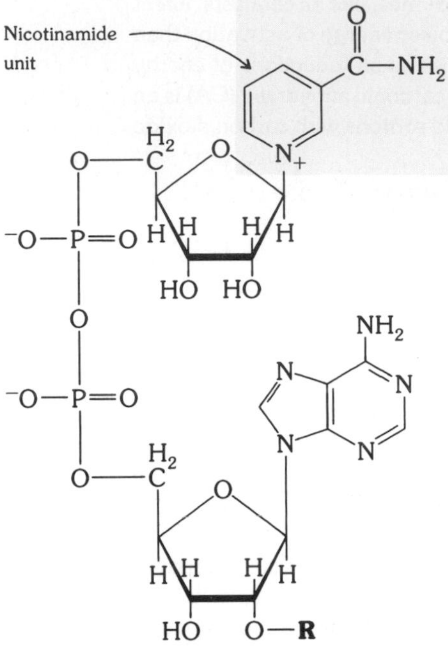

2

a NAD^+ **R**=H
b $NADP^+$ **R**=OPO_3^{2-}

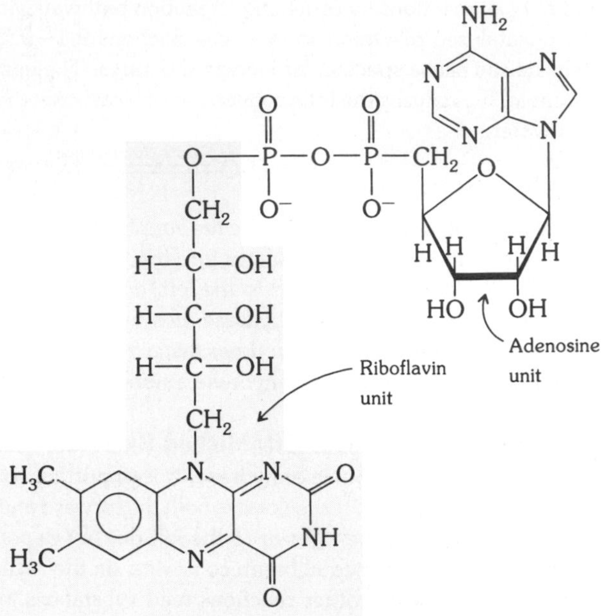

3

FAD

Nicotinamide occurs in yet another important coenzyme, nicotinamide adenine dinucleotide phosphate, **2b**, a phosphate ester of NAD^+. Its name is usually shortened to $NADP^+$ (or, sometimes, just NADP). Both NAD^+ and $NADP^+$ are coenzymes in major biological oxidation – reduction reactions.

■ The P in $NADP^+$ refers to the extra phosphate ester unit.

Quite often equations that involve enzymes with recognized coenzymes are written with the symbol of the coenzyme as a reactant or as a product. NAD^+, for example, is the cofactor for the enzyme that catalyzes the body's oxidation of ethyl alcohol to acetaldehyde. It serves, in fact, as the actual acceptor of the hydride ion, $H:^-$, given up by ethyl alcohol. The details are reserved to Special Topic 22.1, but the overall equation is written simply as follows.

$$CH_3CH_2OH + NAD^+ \longrightarrow CH_3CH{=}O + NAD:H + H^+$$

Ethanol   Ethanal   Reduced   Hydrogen
                                                 form of NAD^+ ion (buffered)

When the symbol of a coenzyme is used in an equation, remember that it stands for the entire enzyme that bears the coenzyme. In this reaction, the NAD^+ unit in the enzyme accepts $H:^-$ from the alcohol, and we can write this part of the reaction by the following equation.

$$NAD^+ + H:^- \longrightarrow NAD:H$$

By accepting the *pair of electrons* in $H:^-$, NAD^+ is reduced, and $NAD:H$ (usually written as NADH) is called the *reduced form* of NAD^+. $NADP^+$ can also accept hydride ion, and its reduced form is written as NADPH.

HOW NAD⁺ AND FAD (OR FMN) PARTICIPATE IN ELECTRON TRANSFER

In the nicotinamide unit of both NAD^+ and $NADP^+$ there is a positive charge on the nitrogen atom, an atom that is also part of an aromatic ring. Despite how rich this ring is in electrons, the positive charge on N makes the ring an electron acceptor. Without the charge, the ring would otherwise be very similar to the ring in benzene, and it would strongly resist any reaction that disrupted its closed-circuit electron system. The positive charge on N, however, places the ring of the nicotinamide unit on an energy teeter-totter that the cell can tip either way without too much energy cost.

As seen in the following equations, when the donor of an electron pair, such as an oxidizable alcohol, arrives at the enzyme, the donor can transfer an electron pair (as $H:^-$) to the ring. Although the ring will temporarily lose its aromatic character, the positive charge of the ring nitrogen atom encourages this transfer of negative charge. In the next step of this metabolic pathway, the pair of electrons, still as $H:^-$, transfers to a riboflavin unit, also shown in the equations below. As this occurs, the ring of the nicotinamide unit recovers its stable, benzene-like nature.

Eventually, the pair of electrons winds up on an oxygen atom when, in the last step of this long metabolic pathway, oxygen that has been supplied by the air we breathe is reduced to water. We'll return to this in Chapter 25.

In the first reaction sequence, the alcohol is oxidized to an aldehyde as NAD^+ is reduced to NADH.

In the second reaction, NADH transfers $H:^-$ to FMN (or FAD). The reaction thus reoxidizes NADH to NAD^+ as FMN is reduced to $FMNH_2$ (or as FAD is reduced to $FADH_2$).

We learned earlier that we may not call something a catalyst unless it undergoes no *permanent* change, but the foregoing examples seem to contradict this. In the body, however, a reaction that alters an enzyme is followed at once by one that regenerates the enzyme. The NADH produced by the oxidation of ethyl alcohol, for example, is recovered in the next step. One of the possible enzymes for the next step has the cofactor FAD (for flavin adenine dinucleotide), **3**. FAD incorporates still another B vitamin, riboflavin. FAD can accept $H : ^-$ from NAD$:$H, for example, change to FADH$_2$ (the second H is H$^+$ from the buffer), and so regenerate NAD$^+$. This is also shown in Special Topic 22.1, but the overall reaction is

■ Riboflavin is vitamin B$_2$.

$$NADH + FAD + H^+ \longrightarrow NAD^+ + FADH_2$$

The FAD-containing enzyme is, of course, now in its reduced form, FADH$_2$. FADH$_2$ passes on its load of hydrogen and electrons in yet another step and so is reoxidized and restored to FAD. The steps continue, but we'll stop here. The main points are that B vitamins are key parts of coenzymes, and that the catalytic activities of the associated enzymes directly involve the parts of the molecules contributed by these vitamins.

Flavin mononucleotide or FMN is a near relative of FAD that contains riboflavin. FMN is FAD minus the adenosine unit. The reduced form of FMN is FMNH$_2$, and FMN is also involved in biological oxidations.

Enzymes Are Named after Their Substrates or Reaction Types

■ Whenever we see *-ase* as a suffix in the name of any substance or type of reaction, the word is the name of an enzyme.

Nearly all enzymes have names that end in *-ase*. The prefix is either from the name of the substrate or from the kind of reaction. For example a **hydrolase** catalyzes hydrolysis reactions. An *esterase* is a hydrolase that aids the hydrolysis of esters. A *lipase* works on the hydrolysis of lipids. A *peptidase* or a *protease* catalyzes the hydrolysis of peptide bonds.

■ The digestive enzymes *trypsin, chymotrypsin,* and *pepsin,* all peptidases, have old (nonsystematic) names that do not end in *-ase*.

An **oxidoreductase** handles a redox equilibrium. Sometimes an oxidoreductase is called an *oxidase* when the favored reaction is an oxidation and a *reductase* when the reaction is a reduction. A **transferase** catalyzes the transfer of a group from one molecule to another, and a *kinase* is a special transferase that handles phosphate groups. Other broad categories of enzymes are the **lyases,** which catalyze elimination reactions that form double bonds; **isomerases,** which cause the conversion of a compound into an isomer; and **ligases,** which cause the formation of bonds at the expense of chemical energy in triphosphates, like ATP.

■ ATP is adenosine triphosphate, an important, high-energy triphosphate ester (see page 493).

An International Enzyme Commission has developed a system of classifying and naming enzymes that places considerable chemical information into the enzyme's name. The names of the principal reactants, separated by a colon, are written first and then the name of the kind of reaction is written as a prefix to *-ase*. For example, in moving from left to right in the following equilibrium, an amino group transfers from the glutamate ion to the pyruvate ion:

■ This reaction, incidentally, is an example of how the body can make an amino acid — here, alanine — from other substances.

$$^-O_2CCH_2CH_2\underset{\underset{NH_3^+}{|}}{CH}CO_2^- + CH_3\overset{\overset{O}{\|}}{C}CO_2^- \rightleftharpoons {}^-O_2CH_2CH_2\overset{\overset{O}{\|}}{C}CO_2^- + CH_3\underset{\underset{NH_3^+}{|}}{CH}CO_2^-$$

Glutamate ion   Pyruvate ion   α-Ketoglutarate ion   Alanine

The systematic name for the enzyme is *glutamate : pyruvate aminotransferase*. In all but formal publications, such a cumbersome (but unambiguous) name is seldom used. This enzyme, for example, is often referred to simply as GPT (after the older name, glutamate : pyruvate transaminase) and sometimes as alanine transaminase or ALT. We'll largely stick with common names of enzymes.

PRACTICE EXERCISE 1

What is the most likely substrate for each of the following enzymes?

(a) sucrase **(b)** glucosidase **(c)** protease **(d)** esterase

Enzymes Often Occur as a Family of Similar Compounds Called *Isoenzymes* with Identical Functions Identical reactions are often catalyzed by enzymes with identical cofactors but slightly different apoenzymes. These variations are called **isoenzymes** or **isozymes**.

Creatine kinase or CK, for example, consists of two polypeptide chains labeled *M* (for skeletal muscle) and *B* (for brain). It occurs as three isoenzymes. All catalyze the transfer of a phosphate group in the following equilibrium:

$$NH_2CNCH_2CO_2^- + ATP \underset{\text{creatine kinase}}{\rightleftharpoons} {}^-OPONHCNCH_2CO_2^- + ADP$$

Creatine   Creatine phosphate

■ Here "iso-" signifies the same catalytic function, not identical molecular formulas.

■ When supplies of ATP are low and those of ADP are therefore high, the *reverse* of this equilibrium becomes a major path for making more ATP in muscle cells.

One CK isoenzyme, called CK(*MM*), has two *M* units and occurs in skeletal muscle. Another, CK(*BB*), has two *B* units and occurs in brain tissue. The third, CK(*MB*), has one *M* and one *B* polypeptide, and it is present almost exclusively in heart muscle, where it accounts for 15% to 20% of the total CK activity. The rest is contributed by CK(*MM*).

We have given this much detail about creatine kinase because this and similar enzymes have an extraordinarily important function in clinical analysis and diagnosis, as we'll see later in the chapter.

22.2

THE ENZYME–SUBSTRATE COMPLEX

The chirality and flexibility of an enzyme and the sidechains of its amino acid residues allow only the enzyme's substrate to fit to it and to become activated for a reaction.

When an enzyme catalyzes a reaction of a substrate, molecules of each must momentarily fit to each other. This temporary combination is called an **enzyme–substrate complex.** It is part of a series of chemical equilibria that carry the substrate through a number of changes until the products of the overall reaction form.

$$E + S \rightleftharpoons E{-}S \rightleftharpoons E{-}S^* \rightleftharpoons E{-}P \rightleftharpoons E + P$$

Enzyme   Substrate   Enzyme– substrate complex   Substrate– activated $E{-}S$ complex   Enzyme– product complex   Enzyme (recovered)   Product

The first equilibrium is the binding of the enzyme to the substrate like the fitting of a key (the substrate molecule) to a tumbler lock (the enzyme), so this theory is often called the **lock-and-key theory** of enzyme action. Shaped pieces that fit together are said to have *complementary shapes;* or we say that there is *complementarity* between the two shapes (see Fig. 22.2).

For an enzyme–substrate complex to form, there must be two kinds of complementarity. The first is what we have already implied — *geometrical complementarity:* a square peg fits a square hole better than to a round hole, for example. The other is *physical complementarity,* which concerns factors other than shape — hydrophobic interactions, hydrogen bonds, and electrical charges of *opposite* nature nestling *nearest* each other as the complex forms.

The chiral natures of enzymes and substrates place severe constraints on lock-and-key fitting and contribute to the high degree of specificity of enzymes. One illustration of the significance of chirality to fitting was observed in the early 1990s with a certain protease, HIV protease. The natural form of the protease, like all enzymes, is made from amino acids that are

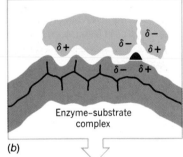

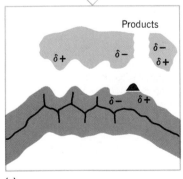

(c)

FIGURE 22.2
The lock-and-key model for enzyme action. (*a*) The enzyme and its substrate fit together to form an enzyme–substrate complex. (*b*) A reaction, such as the breaking of a chemical bond, occurs. (*c*) The product molecules separate from the enzyme.

FIGURE 22.3

Induced fit theory. (*a*) A molecule of an enzyme, hexokinase, has a gap into which a molecule of its substrate, glucose, can fit. (*b*) The entry of the glucose molecule induces a change in the shape of the enzyme molecule, which now surrounds the substrate entirely. (Courtesy of T. A. Steitz, Department of Molecular Biophysics and Biochemistry, Yale University, New Haven, CT.)

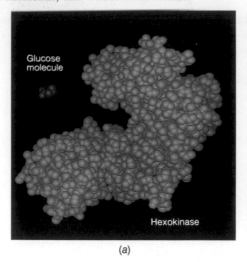 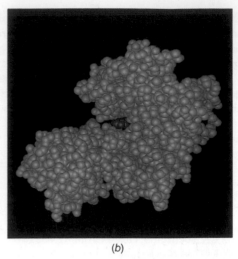

(*a*)   (*b*)

■ Hormone uptake only by specific cells also depends on a flexible lock-and-key kind of recognition.

all in the L-family. Likewise, the natural substrate protein for HIV protease is made of all L-family amino acids. Both the protease and the substrate protein made from all D-amino acids have been synthesized in the lab. The all-D enzyme worked only with the all-D substrate; the all-L (natural) enzyme cleaved only the all-L substrate. Another example of the significance of chirality to complementarity is the ability of trypsin, a digestive protease, to affect only substrates made of L-amino acids. The enzymes involved in the metabolism of glucose are similarly effective only with D-glucose, not L-glucose units.

To get the substrate to fit to the enzyme depends on some flexibility in the enzyme molecule much as a lock flexibly adapts as the key is inserted. As the substrate molecule nestles onto the enzyme, the molecular groups of the substrate induce the enzyme molecule to adjust its shape to achieve the best fit (see Fig. 22.3). The initial contact with substrate and enzyme may cause changes in tertiary structure in the polypeptide of the protein. Such changes, which induce stress in the polypeptide, force the enzyme to modify its shape further. This **induced fit** model of how enzymes work describes the true nature of what is traditionally called the "lock-and-key" interaction.

Proteins, as we have indicated, consist of huge molecules, and not *all* parts of their molecules are ever *directly* involved in catalysis. Some of the enzyme's amino acid residues have side chains (or groups of side chains) with shapes and polar sites complementary to the substrate and so are called the enzyme's *binding sites*. Other groups on the enzyme, called *catalytic sites*, handle the actual catalytic work in the complex. As stated in Section 22.1, catalytic sites are often supplied by molecules of coenzymes, and these must be bound to the apoenzyme to become integral parts of whole enzyme.

The Activation of the Substrate Changes It into Its Transition State

The fit achieved by the enzyme and the substrate molecules in the $E-S$ complex is not perfect. However, the intermolecular forces that cause the complex to *begin* its formation in the first place now continue to work. These forces distort and stretch chemical bonds in the substrate to improve the fit. The chemical energy for this distortion is generally provided by the *gain in overall stability achieved in the complex*. The result of such changes is the conversion of the initial enzyme–substrate complex, $E-S$, into a substrate-activated complex, $E-S^*$. In $E-S^*$, the fit between enzyme and substrate is as good as it can be, and the substrate molecule has reached a unique condition of both shape and internal energy called its *transition state*. The perfecting of the enzyme–substrate fit as the transition state *forms*, rather than the initial, somewhat imperfect fitting of enzyme to substrate, largely accounts for the high catalytic power of an enzyme.

Whether $E-S^*$ collapses to return to the reactants or to proceed to the products depends on how much reactant and product concentrations are building up or declining, all in accordance with Le Châtelier's principle. If product forms (Fig. 22.2*b*), we might suppose that its molecule has a different distribution of electrical charges than that of the reactant molecule (see Fig. 22.2*c*). The enzyme–product complex, $E–P$, does not hold together, and the product molecule P slips off. The enzyme is then ready to receive another substrate molecule.

This has been a very broad and vastly simplified view of enzyme catalyses intended almost entirely to fortify one point: theories of how enzymes work start with the idea of induced fitting based on both geometrical and physical complementarity. An increasing number of detailed mechanisms that explain how specific enzymes or teams of enzyme work is accumulating. For an application of the principle of complementarity to antibodies, antigens, and the ABO blood groups, see Special Topic 22.2.

22.3

KINETICS OF SIMPLE ENZYME–SUBSTRATE INTERACTIONS

At high substrate concentrations, an enzyme's rate enhancement levels off.
We know that reaction rates are sensitive to the *concentrations* of reactants. In many reactions between two species, if you double the initial concentration of one species, holding the other's constant, the rate doubles. Many enzyme-catalyzed reactions are like this if we treat their enzymes as actual (although temporary) reactants. At some fixed initial enzyme concentration, $[E_o]$, doubling the concentration of the substrate, $[S]$, doubles the reaction rate, V. What is significant about enzyme-catalyzed reactions is that this rate enhancement cannot be indefinitely extended to higher and higher initial substrate concentrations. Eventually, at some higher initial substrate concentration, no further rate acceleration occurs. The rate levels off.

We can see what this means with the aid of Figure 22.4, a plot of initial rates versus initial values of $[S]$. Imagine a series of experiments in all of which the molar concentration of the enzyme, $[E_o]$, is the same. We assume a simple reaction, meaning one for which the enzyme is able to handle only *one* substrate molecule at a time. We will vary only the initial concentration of the substrate, $[S]$, from experiment to experiment. As we said, in most ordinary reactions, the initial rate would double each time we doubled the initial concentration of one reactant.

In our series of experiments, we do observe something like this, but only in those trials that have *low* initial values of $[S]$, as in part *A* of the plot in Figure 22.4. Here, a small increase in $[S]$ does cause a proportionate increase in initial rate. The curve rises steadily.

In succeeding experiments, however, at higher and higher initial values of $[S]$, the initial rates respond less and less until, in part *B* of the plot, the initial rates are constant *regardless of the value of* $[S]$. The reason for the leveling off is that we now have enough substrate molecules to saturate all of the active sites of all the enzyme molecules. Any additional substrate molecules have to wait their turns, so to speak.

The relationship between V, $[E_o]$, and $[S]$ for the kind of reaction just described is given by the following equation.

$$V = \frac{k[E_o][S]}{K_M + [S]} \tag{22.1}$$

The symbol k stands simply for a proportionality constant. The symbol K_M is another constant, called the *Michaelis constant,* and it has a particular value for a specific enzyme-catalyzed reaction. Notice what happens when $[S]$ has very small values, approaching zero. At very low values of $[S]$ the denominator becomes *essentially identical with* K_M, and Equation 22.1 becomes

$$V = \frac{k}{K_M} [E_o][S] \tag{22.2}$$

The ratio, k/K_M is a constant, being a ratio of other constants. Equation 22.2, in other words, says that the velocity of the reaction is directly proportional to $[S]$ at *low* values of $[S]$. "Directly proportional" translates into a *straight line* or *linear* plot of initial rate versus $[S]$,

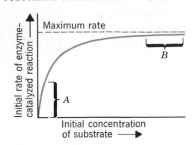

FIGURE 22.4
Initial rates of an enzyme-catalyzed reaction plotted versus the initial concentrations of substrates when the concentration of the enzyme is fixed in each experiment. The sections labeled *A* and *B* are discussed in the text.

■ Many enzymes consist of two or more polypeptide subunits *each of which has binding and catalytic sites and all of which become involved in the overall reaction.*

■ In the early part of this century, biochemists Leonor Michaelis and Maude Menten were pioneer scientists in the field of enzyme kinetics.

MOLECULAR COMPLEMENTARITY AND IMMUNITY, AIDS, EVEN THE ABO BLOOD GROUPS

The *immune system,* as large and complex as the nervous system, is the body's array of defenses against *pathogens* — disease-causing microorganisms and viruses. We cannot in a brief Special Topic do justice to the immune system, of course, but we can take note of some of the ways in which it shares basic operating principles with the enzyme–substrate reaction. The concept of the fitting of a substrate to an enzyme by means of geometrical and physical complementarity is also at the molecular base of the body's immune system as well as the existence of blood type groups.

When a pathogen has penetrated the first line of defense, the physical barriers of skin and mucous membranes, white blood cells known as *lymphocytes* go into action. They all begin life in bone marrow, but not all mature there. Two kinds of immunity involving two kinds of lymphocytes are recognized. One is *cellular immunity,* and it is handled by *T lymphocytes* or *T-cells* (after thymus tissue, where T-cells mature). Cellular immunity handles viruses that have gotten inside cells, as well as parasites, fungi, and foreign tissue.

AIDS The human immunodeficiency virus (HIV) is able to destroy certain kinds of T-cells, the *helper T-cells.* This renders the immune system deficient in its ability to handle infections and results in AIDS, acquired immune deficiency syndrome. Relatively non-lethal diseases normally handled routinely by the body thus become lethal in AIDS victims.

Antigen–Antibody Reaction The second kind of immunity is *humoral immunity* (after an old word for fluid, *humor*). Humoral immunity is the responsibility of the *B lymphocytes* or *B-cells* (because they mature in *bone marrow*). B-cells act mostly against bacterial infections but also against those workings of viral infections that

occur outside of cells. We'll limit the continuing discussion to the work of B-cells.

B-cells carry and manufacturer *antibodies,* glycoproteins that are able, by an interaction like that between substrate and enzyme, to attract and take antigens out of circulation and defeat the spread of the pathogen. An *antigen* is any molecular species or any pathogen that induces the immune system to make antibodies as well as gives the immune system a molecular-cellular memory for the antigen. Thus, at a later invasion of the same antigen, the immune system is poised for a far more rapid defensive response than it initially had. A *vaccine* is able to start the initial defensive response leading to the molecular memory for the antigen without causing the disease itself.

Figure 1 represents an antibody that is *dipolar;* it has two cross-linking molecular groups. The antigen in Figure 1 is represented by a unit that can become bound to at least three antibody-binding sites. *The antibody protein is specific for just one antigen.* As you can see, the interaction of antibody with its antigen essentially "polymerizes" the entire system into one vast "copolymer." The product is now in a far less soluble form and it bears molecular markings that are recognized by other white cells (phagocytes) that engulf the "polymer" and destroy it. There are some antigen–antibody complexes that are destroyed by a series of interacting proteins called the *complement system.* By tying up the antigens, the antibodies prevent the spread of the infection and thus allow the system the time needed to destroy the antigens.

The ABO Blood Groups The surfaces of red blood cells carry projecting oligosaccharide units of glycolipids. There are differences, however, among individuals in the structures of these sugar residues. One of the consequences of these differences is the existence of blood group systems, one being the *ABO system.* You might

which is approximately what we see in Figure 22.4. At low initial values of [S], the initial rates lie nearly on a straight line. The curve bends, as you can see, at higher values of [S]. As [S] becomes high enough, the K_M term in the denominator in Equation 22.1 is overwhelmed by the [S] term. Eventually, at sufficiently high values of [S], the [S] term in the *numerator* can be canceled by the entire denominator (which is now almost entirely contributed by [S] anyway), leaving the following simple expression as the equation for the velocity, where V_m means the *maximum* velocity.

$$V_m = k[E_o] \tag{22.3}$$

But $[E_o]$ is a *constant,* the concentration of the enzyme (in any form, free or combined in

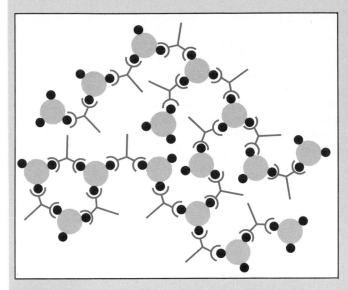

FIGURE 1

Molecules of the antibody, in green, cross-link with antigen particles, in red, to form a mass resembling a huge copolymer.

have type A, type B, or type O blood. Some have a combination type, AB blood.

If you are of type A and by a transfusion are given blood from a type B person, the red cells from the type B blood will clump together (agglutinate) likely causing a blockage of blood capillaries that could be fatal. Thus your type A blood is able to "see" something in type B blood as a foreign material. Type A blood contains in the serum portion (the liquid minus the cells) an antibody against type B blood. What specifically is the antigen in the type B blood is the molecular unit at the tip of an oligosaccharide joined to a glycolipid of the (type B) red cell. This is why each kind of red cell is described as carrying an *antigen,* one of three types, A, B, and H.

If you are of type A, your serum includes anti-B antibodies. People with type B blood have anti-A antibodies. Those with AB blood have neither anti-A nor anti-B antibodies. AB blood type people are able to accept transfusions from people of any blood type. However, in all but emergency situations, transfusions are normally done using type AB blood. Type O people carry the H "antigen" on their red cells, and their blood has *both* anti-A and anti-B antibodies. Type O people, therefore, can *receive* transfusions only from individuals with type O blood. At the same time, type O people can *give* transfusions to all types — they are universal donors — because the antigen on the red cells in type O people actually has no "enemies" in other types of blood, no antibodies that can attack and agglutinate type O red cells when they are transfused into people of other blood types.

The H antigen in type O blood is given the name antigen because it is the precursor to the A and B antigens of other blood types. Type A individuals make type A antigen by adding an *N*-acetylgalactosamine residue to the tip of a glycolipid on the red cell. Type B people make type B antigen by adding a galactose residue to the same glycolipid. Type O individuals simply lack the enzymes needed for these transformations. The differences among these enzymes are thought to involve single amino acid residue substitutions in the enzymes' polypeptides. Table 1 summarizes donor–acceptor relationships for the blood types.

Table 1   Blood Type Acceptor-Donor Options

If Your Blood Type Is	You Can Accept Blood from One of This Type	You Can Donate Blood to One of This Type
O	O	O, A, B, AB
A	A or O	A or AB
B	B or O	B or AB
AB	AB, A, B, O	AB

complex) and k is, of course, also a constant. So the maximum velocity, V_m, *must* be a constant. Thus at high values of [S], the rate of the enzyme catalyzed reaction must level off at a maximum, as you can see it does in Figure 22.4. If we substitute the expression from Equation 22.3 into Equation 22.1, we obtain what is called the *Michaelis-Menten equation* for the rate of a simple enzyme catalyzed reaction.

$$V = \frac{V_m[S]}{K_M + [S]} \tag{22.4}$$

The reason for a *constant* rate at high values of [S] is that all enzyme molecules are now saturated with substrate and are in the form of the complex $E - S$. Further catalysis depends on

■ The $E - S$ complex is sometimes called the Michaelis complex to honor the work of Leonor Michaelis.

the freeing of enzyme from enzyme – product complexes. At one specific value of $[S]$, $[S] = K_M$, Equation 22.4 reduces to give

$$V = \frac{1}{2} V_m \qquad (22.5)$$

This equation gives meaning to the Michaelis constant, K_M; K_M is the substrate concentration at which the reaction rate is one-half of the maximum rate.

Our discussion involving Equations 22.1 – 22.5 concerned the response of rate to concentrations of enzyme and substrate when the enzyme carries only one active catalytic site. Let's now turn our attention to enzymes with more than one site; they offer the system numerous pathways for the regulation of enzyme action.

22.4

THE REGULATION OF ENZYMES

Enzymes are switched on and off by initiators, effectors, inhibitors, genes, poisons, hormones, and neurotransmitters.
A cell cannot be allowed to do everything at once. Some of its possible reactions have to be shut down while others occur. One way to keep a reaction switched off is to prevent its enzyme from being made in the first place. Another but harmful way is to deny the cell the amino acids, vitamins, and minerals that it needs to make enzymes. Hormones and neurotransmitters are natural regulators of enzymes, and we'll learn about them in the next section. In this section we study several other means to control enzymes.

■ The prevention of enzyme synthesis often involves the regulation of genes and their work of directing the synthesis of polypeptides, topics for Chapter 24.

Some Enzymes Display an Initial Resistance to the Formation of an Enzyme – Substrate Complex
One of the very significant features of many enzymes is that the initial sharp rise of the curve in Figure 22.4 does not occur when the concentration of the substrate, $[S]$, is low. It's as though the enzyme is inactive at low substrate concentrations. For these enzymes, the plots of initial values of $[S]$ versus initial rates look more like the curve in Figure 22.5. The plot has a lazy "s" shape, so it's called a *sigmoid plot* (after *sigma*, Greek for S). The rate increases very slowly with initial substrate concentration, then the rate takes off in the normal response of rate to concentration, and finally the rate levels off in the usual way.

Sigmoid rate plots are found among enzymes that remain inactive *until a sufficient concentration of substrate forces them into active forms.* What the curve suggests is something about *enzyme activation,* as we'll learn next.

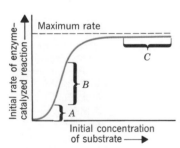

FIGURE 22.5
Initial rates of enzyme-catalyzed reactions plotted versus initial values of substrate concentrations, at fixed enzyme concentration, when an allosteric effect is observed. Sections labeled *A* and *B* of the curve — a sigmoid curve — are discussed in the text.

The Catalytic Sites in Some Enzymes Are Given Active Shapes by Reactions at Other Sites
Enzymes with sigmoid rate curves (Fig. 22.5) consist of two or more polypeptide units *each* with a catalytic site that normally is in an *inactive* configuration, even in the presence of some substrate. Inactive catalytic sites, however, are eventually activated by the substrate, but only when the initial concentration of the substrate is high enough.

Let's suppose, for simplicity, that our enzyme is made of just two polypeptide chains. It has two catalytic sites, one on each polypeptide. We'll represent each site by a geometric shape, as shown in Figure 22.6. We have to suppose that the shape of each site is not quite complementary to the substrate, that the substrate must itself induce the correct fit, because the enzyme's response is sluggish at low substrate concentration. We're in region *A* of the sigmoid rate curve (Fig. 22.5), the region of the slower than normal rate.

When a substrate molecule induces a conformational change in the polypeptide unit to enable it to fit to one catalytic site, it simultaneously induces the other polypeptide to adopt a new shape *causing the second catalytic site to become active.* Now the same enzyme molecule can accept the second substrate molecule *much more easily than the first.* The enzyme is now being used at maximum efficiency, and the rate of reaction takes off, putting the system into part *B* of Figure 22.5.

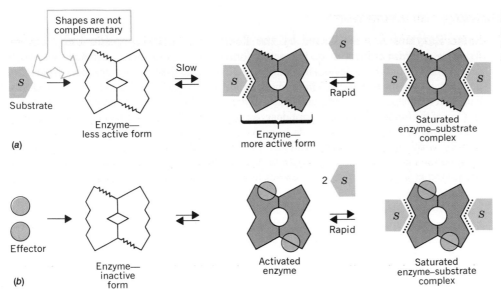

Shapes are not complementary

S Substrate

Slow

Enzyme— less active form

S

Rapid

Enzyme— more active form

S *S*

Saturated enzyme–substrate complex

(*a*)

Effector

Enzyme— inactive form

Rapid

2 *S*

Activated enzyme

S *S*

Saturated enzyme–substrate complex

(*b*)

This phenomenon in which one active site is activated by an event that occurs elsewhere on the enzyme is called **allosteric activation.** The enzyme's subunits cooperate with each other to cause full activation, but this doesn't happen unless the level of substrate concentration has climbed high enough to start the process. *The activity of an enzyme subject to allosteric activation is thus regulated by how much its services are needed.* The "need," of course, increases as the concentration of substrate increases, because something has to be done about the increasing level of substrate.

In the next chapter, we'll see how a similar allosteric effect is caused by oxygen when it interacts with hemoglobin, which is not an enzyme, and how this enables hemoglobin to operate at 100% efficiency, or very nearly so.

■ *Allo-*, other; *-steric*, space — the other space or the other site.

Allosteric Activation Can Be Caused by Effectors Instead of Substrates The catalytic sites of some enzymes are activated by substances called **effectors** that are not substrates. When their molecules bind allosterically to the enzyme, meaning at a location distinct from the catalytic site, they force a configurational change that activates the enzyme (see Figure 22.6*b* where circles represent the effector molecules). The effector might, for example, be a molecule whose own metabolism *needs the products* made by the enzyme it activates.

Nerve Signals Can Indirectly Tell an Effector to Work Two of the very important effectors are *calmodulin,* a protein found in most cells, and *troponin,* present in muscle cells. Neither works as an effector, however, until it has itself been activated. The activator is calcium ion, and cells control their calcium ion levels by active transport mechanisms mediated by nerve signals.

Normally, the level of calcium ion that moves freely in solution in the cytosol is only about 10^{-7} mol/L. It must be kept extremely low, because the cytosol contains phosphate ion which forms an insoluble salt with Ca^{2+}. The level of Ca^{2+} just outside the cell is about 10^{-3} mol/L, very considerably higher. Despite the concentration gradient, which nature would normally erase by diffusion, the calcium ions stay outside until something changes the cell's permeability to them. Nerve signals do this.

The overall sequence is roughly as follows. A nerve signal opens protein channels in the cell membrane for Ca^{2+} ions and they enter the cell, where they bind to calmodulin or troponin. The effector is thus activated and it then activates an enzyme to cause some chemical work or, in muscles, to cause muscle contraction. When the signal is over, the channels close, and Ca^{2+} ions are pumped back out through other portals. The effector is thus inactivated. This mechanism thus connects nerve signals to specific chemical activities in cells.

■ The cytosol is the *solution* in the cytoplasm and does not include the organelles in the cytoplasm.

■ Other kinds of proproteins are known, like *proinsulin,* the precursor to the hormone *insulin,* a blood sugar regulator. The conversion of proinsulin to insulin entails the removal of a 33-residue polypeptide unit.

Some Enzymes Are Activated by the Removal of a Polypeptide Unit

Several digestive enzymes are first made in inactive forms called **zymogens** or **proenzymes.** Their polypeptide strands have several more amino acid residues than the enzyme, and these extra units cover over the active site. Then, when the active enzyme is needed, a complex process is launched that clips off the extra units and, by exposing the active site, activates the enzyme.

One of the functions of enteropeptidase, a compound released in the upper intestine when food moves in from the stomach, is to convert the zymogen trypsinogen into the enzyme trypsin, which helps to digest proteins. Trypsin is activated by the deletion of a small polypeptide unit in trypsinogen. When no food is present, there is no need for trypsin, but when food enters, enteropeptidase comes in as well, and then trypsin is activated.

■ Kinases are the enzymes for this phosphorylation, and they must themselves be activated (usually by Ca^{2+}) only when needed.

Phosphorylation Activates Some Enzymes

Glycogen phosphorylase, the enzyme that catalyzes the hydrolysis of glucose units from glycogen, is made in an inactive state. However, when a serine side chain (CH_2OH) is changed to its phosphate ester, $CH_2OPO_3^{2-}$, the enzyme is activated.

■ Inhibitors are *negative effectors.*

Inhibitors Can Keep Enzymes Switched Off until They Are Needed

Some substances, called **inhibitors,** bind reversibly to the enzyme and prevent it from working. In **allosteric inhibition,** molecules of the inhibitor bind to the enzyme somewhere other than the active site. This affects the shape of the active site or a binding site, and the enzyme–substrate complex cannot form (see Fig. 22.7a). Then if some reaction changes the inhibitor so that it no longer sticks to the enzyme, the catalyst becomes active.

In **competitive inhibition,** the inhibitor is a nonsubstrate molecule with a shape similar enough to that of the true substrate *that it can compete with the substrate for attachment to the active site.* When the inhibitor molecules lock to the enzyme's active sites, but do not undergo a reaction and leave, then the enzyme has become useless to the true substrate (see Fig. 22.7b).

A competitive inhibitor doesn't have to be a product of the enzyme's own work. It can be something else the cell makes, or it could be the molecules of a medication. What its molecules must do is *resemble* those of the normal substrate enough to bind to the active site of the enzyme.

Sometimes an inhibitor is the *product* of the reaction being catalyzed or one of the products produced later in a series of connected reactions. Now we have **feedback inhibition.** As the level of such a product increases, its molecules "feed back" with increasing success as inhibitors to one of the enzymes in the series of reactions that helped to make the product. A *feedback inhibitor works as an allosteric, noncompetitive inhibitor.* The amino acid isoleucine, for example, is a feedback inhibitor. It is made from another amino acid, threonine, by a series

FIGURE 22.7

Enzyme inhibition. (*a*) Allosteric inhibition. (*b*) Competitive inhibition.

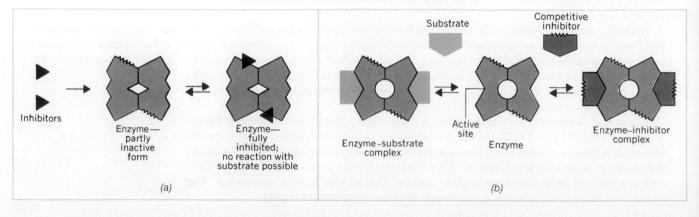

Inhibitors

Enzyme— partly inactive form

Enzyme— fully inhibited; no reaction with substrate possible

(a)

Substrate

Competitive inhibitor

Enzyme-substrate complex

Active site

Enzyme

Enzyme–inhibitor complex

(b)

of steps, each with its own enzyme. As more and more isoleucine is made, its molecules more and more inhibit the enzyme involved with threonine in the first step, E_1, of the series:

Threonine   Isoleucine

■ Both threonine and isoleucine have two tetrahedral stereocenters. Can you spot them, and can you tell how many stereoisomers there are of each?

The beautiful feature of feedback inhibition is that the system for making a product shuts down automatically when enough product is made. Then, as the cell consumes this product, it eventually uses even product molecules that have been serving as inhibitors. The result is that when the product concentration has dropped very low, the enzyme needed to make more is released from its bondage.

Feedback inhibition is very common in nature. It helps to maintain a condition of **homeostasis** in which disturbances to systems by stimuli are minimized, because the stimulus is able to start a series of events that restore the system to the original state. Body temperature is a condition maintained by homeostatic mechanisms. The body does this so well that even small changes in temperature tell us that something is wrong.

A familiar homeostatic mechanism in the home is the work of a furnace controlled by a thermostat. When the room becomes hot enough (the desired "product"), the thermostat trips and the furnace shuts off. In time, the temperature drops, the thermostat trips back, and the furnace restarts.

Inhibition by Antibiotics A broad family of compounds called **antimetabolites** includes some made by bacteria and fungi and called **antibiotics**. Antibiotics are substances that inhibit or prevent the normal metabolism of a disease-causing bacterial system. Some antibiotics work by inhibiting an enzyme that the bacterium needs for its own growth. Both the sulfa drugs and penicillin work in this way.

■ An antimetabolite is called an *antibiotic* when it is the product of the growth of a fungus or a natural strain of bacteria.

Poisons Often Cause Irreversible Enzyme Inhibition The most dangerous **poisons** are effective even at very low concentrations because they are powerful inhibitors of enzymes, themselves not requiring high concentrations. By inhibiting an enzyme, "a little goes a long way." The cyanide ion, for example, forms a strong complex with one of the metal ion cofactors in an enzyme needed for our use of oxygen.

Enzymes that have SH groups are denatured and deactivated by such poisonous heavy metal ions as Hg^{2+}, Pb^{2+}, Cu^{2+}, and Ag^+.

Nerve gases and their weaker cousins, the organophosphate insecticides, inactivate enzymes of the nervous system.

PRACTICE EXERCISE 2

The following overall change is accomplished by a series of steps, each with its own enzyme.

1,3-Bisphosphoglycerate (1,3-BPG)   2,3-Bisphosphoglycerate (2,3-BPG)

One of the enzymes in this series is inhibited by 2,3-BPG. What kind of control is exerted by 2,3-BPG on this series? (Name it.)

22.5

ENZYMES IN MEDICINE

The specificity of the enzyme for its substrate and the slight differences in properties of isoenzymes provide several unusually sensitive methods of medical diagnosis.

Enzymes that normally work only inside cells are not found in the blood except at extremely low concentrations. When cells are diseased or injured, however, their enzymes spill into the bloodstream. Much can be learned about the disease or injury by detecting such enzymes and measuring their levels.

Enzyme Assays of Blood Use Substrates as Chemical "Tweezers"   Despite the enormous complexity of blood and the very low levels of enzymes in it, enzyme assays are relatively easy to carry out. The substrate for the enzyme is used to find its own enzyme, and the specificity of the enzyme–substrate system ensures that it will find nothing else. If no enzyme is present to match the substrate, nothing happens. Otherwise, the extent of the reaction of the substrate measures the concentration of the enzyme. In this section, we learn about some examples of this medical technology.

Viral Hepatitis Is Detected by the Appearance of GPT and GOT in Blood   Heart, muscle, kidney, and liver tissue all contain the enzyme glutamate : pyruvate aminotransferase or GPT, which we introduced earlier. The liver, however, has about three times as much GPT as any other tissue, so the appearance of GPT in the blood generally indicates liver damage or a virus infection of the liver, such as viral hepatitis.

■ GPT is the transaminase introduced on page 618.

The level of another enzyme, glutamate : oxaloacetate aminotransferase or GOT, also increases in viral hepatitis, but the GPT level goes much higher than the GOT level. The ratio of GPT to GOT in the serum of someone with viral hepatitis is typically 1.6, compared with a level of 0.7 to 0.8 in healthy individuals. (Notice that we speak here of *ratios*, not absolute amounts, which are normally very low.)

Heart Attacks Cause Increased Levels of Three Enzymes in Blood Serum   A *myocardial infarction* (MI) is the withering of a portion of the heart muscle following some blockage of the blood vessels that supply it with oxygen and nutrients. Such blockage can be caused by deposits, by hardening, or by a clot. If the patient survives, the withered muscle becomes scar tissue, and the outlook for a reasonably active life is generally good, particularly

■ The popular term for this set of events is *heart attack*.

FIGURE 22.8
The concentrations of three enzymes in blood serum increase after a myocardial infarction. Here CK is creatine kinase; GOT is glutamate : oxaloacetate aminotransferase; and LD is lactate dehydrogenase.

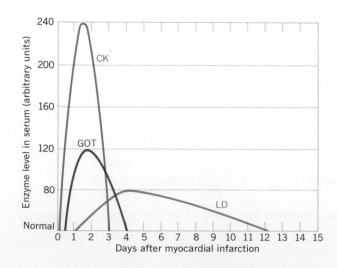

TABLE 22.1
Clinical Report: Myocardial Infarction

DAY I	DAY II	DAY III
DATE *5-25*	DATE *5-26*	DATE *5-27*
CK *51*	CK *552*	CK *399*
CK (MB) *Negative*	CK (MB) *Moderate Positive*	CK (MB) *Weak Positive*
GOT *19*	GOT *91*	GOT *117*
LD *88*	LD *151*	LD *247*
LD ISOENZYME	LD ISOENZYME	LD ISOENZYME
% of Total LD activity	% of Total LD activity	% of Total LD activity
LD_1 *28.3* %	LD_1 *33.8* %	LD_1 *37.9* %
LD_2 *32.9* %	LD_2 *32.3* %	LD_2 *32.8* %
LD_3 *19.7* %	LD_3 *16.3* %	LD_3 *15.7* %
LD_4 *11.6* %	LD_4 *9.1* %	LD_4 *7.6* %
LD_5 *7.5* %	LD_5 *8.5* %	LD_5 *6.0* %
CB	*C.B*	*B.L.*

Normal Range		LD Isoenzymes Normals	
CK	Male 5–75 mU/ml	LD_1	14–29%
	Female 5–55 mU/ml	LD_2	29–40%
		LD_3	18–28%
GOT	5–20 mU/ml	LD_4	7–17%
		LD_5	3–16%
LD	30–100 mU/ml		

One International Unit (U) of activity is the reaction under standard conditions of one micromole per minute of a particular substrate used in the test. (Data courtesy of Dr. Gary Hemphill, Clinical Laboratories, Metropolitan Medical Center, Minneapolis, Minn.)

if treatment is started promptly. A diagnosis of an infarction of exceptionally high reliability can be made by the analysis of the serum for several enzymes and isoenzymes.

When a myocardial infarction occurs, the serum levels of three enzymes normally confined inside heart muscle cells begin to rise. See Figure 22.8. These enzymes are CK (page 619), GOT (just described) and LD. LD stands for lactate dehydrogenase, which catalyzes the formation of the oxidation–reduction equilibrium between lactate and pyruvate.

$$\underset{\text{Pyruvate}}{CH_3\overset{O}{\overset{\|}{C}}CO_2^-} + NAD:H + H^+ \underset{\text{(an NAD}^+\text{ enzyme)}}{\overset{\text{lactate dehydrogenase (LD)}}{\rightleftharpoons}} \underset{\text{Lactate}}{CH_3\overset{OH}{\overset{|}{C}}HCO_2^-} + NAD^+$$

A clinical report from a patient with a myocardial infarction is shown in Table 22.1. You can see how sharply the levels of these three enzymes rose between day I and day II. The line

FIGURE 22.9

The lactate dehydrogenase isoenzymes. (*a*) The normal pattern of the relative concentrations of the five isoenzymes. (*b*) The pattern after a myocardial infarction. Notice the reversal in relative concentration between LD_1 and LD_2. This is the LD_1–LD_2 flip.

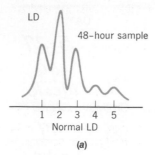
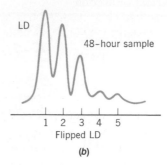

(*a*) Normal LD

(*b*) Flipped LD

of data labeled "CK(MB)" and the data in the columns headed by "LD ISOENZYME" provided the clinching evidence for the infarction. As we learned on page 619, the CK enzyme occurs as three isoenzymes, CK(*MM*), CK(*BB*), and CK(*MB*). The technique that uses a chemical substrate to determine the serum CK level can't tell these isoenzymes apart, because all three catalyze the reaction with the substrate.

To be sure that the increase in serum CK level is caused by injury to the *heart* tissue, the clinical chemist has to use a special technique to analyze specifically for the CK(*MB*) isoenzyme common to heart muscle tissue. As you can see in the chart, the patient's serum CK(*MB*) level did increase.

As additional confirmation of an infarction, the serum LD fraction is further separated by a special method into the five LD isoenzymes, and each is individually analyzed. The *relative* serum concentrations of these five differ distinctively between a healthy person (Fig. 22.9*a*) and one who has suffered an infarction (Fig. 22.9*b*). Of particular importance are the relative levels of the LD_1 and the LD_2 isoenzymes.

Normally the LD_1 level is less than that of the LD_2, but following a myocardial infarction what is called an "LD_1–LD_2 flip" occurs. The relative concentrations of LD_1 and LD_2 become reversed and the level of LD_1 rises higher than that of LD_2. When both the CK(*MB*) band and the LD_1–LD_2 flip occur, the diagnosis of a myocardial infarction is essentially 100% certain.

The Blood Glucose Level Can Be Determined Enzymatically The regular determination of the level of glucose in blood is important to people with diabetes, because when this level is poorly managed, several complications can occur. One commercially available test uses a combination of chemicals, including enzymes, that react with blood glucose to generate a dye. The intensity of the resulting dye is proportional to the blood glucose level.

In the CHEMSTRIP MatchMaker device, the blood glucose level is measured by the intensity of a dye produced enzymatically and converted into mg of glucose per deciliter (100 mL) of blood.

Similar enzyme-based tests are available to measure the serum levels of urea, triacylglycerols, bilirubin (a breakdown product of hemoglobin), and other compounds.

Enzymes Can Be Immobilized on Solid Supports In some applications, enzymes are physically immobilized onto the surfaces of extremely tiny, inert plastic beads, which makes it easier to separate the products from the enzymes. Immobilized enzymes last longer, are less sensitive to temperature, and are less vulnerable to oxygen. Immobilized enzymes, for example, are used in filtering systems to remove bacteria and viruses from air and water.

The enzyme *heparinase* is immobilized onto plastic beads that are then used to catalyze the breakdown of the heparin (page 607) added to blood sent through a hemodialysis machine. The added heparin inhibits the clotting of the blood outside the body, but it has to be removed before the blood goes back into the body. The heparinase catalyzes this removal.

■ The level of urea in the blood is called the BUN level (for blood urea nitrogen). A high BUN level indicates a kidney disorder.

Enzymes in Electrode Tips Make Possible Several Serum Assays Clinical chemists have a variety of electrodes made with immobilized enzymes for analyses that resemble the use of pH electrodes to measure pH. The specificity of the enzyme, immobilized on the electrode's tip, is a key factor in this technology. The level of urea in blood, for example, can be measured as a function of the concentration of the ammonium ions produced when urea is hydrolyzed. The tip of the urea electrode is lightly coated with a polymer that immobilizes urease, the enzyme that catalyzes the hydrolysis of urea.

$$NH_2\!-\!\overset{\displaystyle O}{\overset{\displaystyle \|}{C}}\!-\!NH_2 + H_2O + 2H^+ \xrightarrow{\text{urease}} 2NH_4^+ + CO_2$$

When this electrode is dipped into blood serum containing urea, urea and water migrate into the polymer, the urea is promptly hydrolyzed (thanks to the urease), and the electrode helps to register the appearance of the newly formed ammonium ion. It is relatively easy to correlate the response of the electrode to the concentration of the urea.

Enzyme electrodes for the determination of glucose, uric acid, tyrosine, lactic acid, acetylcholine, cholesterol, and other substances have been developed.

The removal of pollutants from water supplies is receiving the attention of specialists in immobilized enzymes. Both the nitrate and nitrite ions are pollutants that get into the ground water by the excessive use of nitrogen fertilizers. One problem with the nitrite ion is that it alters the hemoglobin into forms that cannot carry oxygen, a problem particularly dangerous to infants. Enzymes that reduce nitrate and nitrite ions to nitrogen have been successfully immobilized onto an electrode system that works effectively to remove these ions from polluted water.

■ The nitrate ion is the more common of the two, but NO_3^- is converted to NO_2^- in the body. A level of NO_3^- in water exceeding 50 mg/L is unacceptable for drinking water.

A Natural Blood-Clot-Dissolving Enzyme Can Be Activated by Other Enzymes

Three of the enzymes available to aid in dissolving blood clots that cause myocardial infarctions are streptokinase, tissue plasminogen activator (TPA), and a modified streptokinase called APSAC (for acylated plasminogen–streptokinase-activator complex).

If you cut yourself, there is set in motion a huge cascade of enzyme-catalyzed reactions that bring about the formation of fibrin from a circulating polypeptide, fibrinogen. The long, stringy fibrin molecules form a brush-heap mat that entraps water and puts a seal, a blood clot, on the cut. After the wound heals, the clot must be dissolved. No part of a blood clot must break loose and circulate to the heart, because it will be stopped by tiny capillaries in heart muscle tissue. Such a blockage is one cause of a myocardial infarction. Clots in the lungs are also very serious.

To dissolve the fibrin of a clot, the body normally converts a zymogen, *plasminogen,* into the enzyme *plasmin.* We described the chemistry of zymogen activation in the previous section. Plasminogen actually is absorbed out of circulation by the fibrin as the clot forms. Its eventual activation, therefore, occurs exactly where its active form, plasmin, is needed. Plasmin, a protease, then catalyzes the hydrolysis of fibrin, and the clot "dissolves."

One of the interesting facts about plasminogen is that it becomes more susceptible to activation when bound to fibrin than when it is simply in circulation. In time, a circulating *tissue plasminogen activator* does what its name implies. It catalyzes the conversion of plasminogen to plasmin at the site of the blood clot.

Therapy for a myocardial infarction is intended to open blocked capillaries as rapidly as possible before the oxygen starvation of surrounding heart muscle tissue spreads the damage too widely. *Plasminogen activation therapy* has became the most commonly used method for treating clot-related infarctions. Obviously, the sooner this therapy is applied following a myocardial infarction, the better the chances that long-term heart muscle damage will be slight.

Plasminogen activation therapy involves introducing into circulation one of the plasminogen-activating enzymes. Among the most commonly used are streptokinase (with aspirin) and tissue plasminogen activator (TPA). When therapy is started within the first four hours of a myocardial infarction, TPA therapy appears to have a small edge in effectiveness. There is a huge difference in cost, however; in the early 1990s, TPA cost about $2200 per dose, but the cost of streptokinase was about $200 per dose.

■ Tissue plasminogen activator made by genetic engineering (Section 24.6) is referred to as *recombinant tissue plasminogen activator* and symbolized as rt-PA.

■ About 150,000 people a year die from clots that start in their lungs.

■ This therapy is called *thrombolytic therapy* because it *lyses* (breaks down) *thrombi* (blood clots).

22.6

CHEMICAL COMMUNICATION—AN INTRODUCTION

Hormones and neurotransmitters are the chief methods by which cells communicate with one another.

Like any complex organism, the body is made up of highly specialized parts. Information, therefore, must flow among the parts to maintain a well-coordinated system. This flow is handled by chemical messengers, hormones or neurotransmitters, sent in response to a variety of signals.

■ Greek *hormon,* arousing.

Hormones are compounds made in specialized organs, the endocrine glands, secreted into the bloodstream, and usually sent some distance away where they launch responses in their particular **target tissue** or **target cells.** The distance might be as close as another neighboring tissue or as far away as 15 to 20 cm. The signal for releasing a hormone might be something conveyed by one of our senses, such as light or an odor, or it might be a stress, or a variation in the level of a particular substance in the blood or in another fluid. Insulin, for example, is released when the level of glucose in blood increases.

Neurotransmitters are chemicals made in nerve cells, called *neurons,* and sent to the next nerve cells. Thus, the distinctions between hormones and neurotransmitters concern differences in how far they go to exert their action. How these two kinds of messengers cause what they do bears many close similarities, however, as we will see.

■ We can imagine the unique oligosaccharide unit of one of the membrane's glycoproteins being like a fishing line dangling a unique lure at its tip that is attractive only to one kind of hormone or neurotransmitter molecule (refer back to Figure 21.13, page 606).

Target Cell Receptors Identify Chemical Messengers At a target cell, a hormone or neurotransmitter delivers its messages by binding to a cell **receptor.** Each receptor, a unique glycoprotein, has molecules so structured that it can accept just the messenger intended for it. It's another example of a lock-and-key mechanism at work to make interactions very specific. Sometimes molecules of toxic substances, virus particles, and even dangerous bacteria "recognize" a glycoprotein on the cells of one particular tissue and cause wholly unwanted changes to occur. Other receptor-like proteins recognize complementary molecular units on neighboring cells, lock to them, and so bind cells together. Such binding helps to control cell division. In cancerous tissue, cell-to-cell binding is weakened and cancer cells more easily proliferate and even break off, enter into circulation in blood or lymph, and so spread the cancer. Receptor-like glycoproteins on sperm cells are able to lock to molecular units on only the ova of the same species, thanks to a lock-and-key mechanism. Antibodies "recognize" antigens in the body's immune response by a lock-and-key mechanism. As evidence that such responses involve cellular *surfaces,* sometimes even killed bacteria or deactivated virus particles cause the body's immune system to develop antibodies. Once initiated, the body retains immunity for a long time.

■ When immunity is not life-lasting, as with the influenza virus, it is because the disease-causing material undergoes mutations.

Chemical Messengers Enter Cells by Four Major Mechanisms The formation of a receptor–messenger complex changes the receptor structure, so that now it is activated to do something. It might be to activate a gene, or an enzyme, or to alter the permeability of a cell membrane so that certain ions or small molecules can move across it. In neurons, the activation of a receptor sends the nerve signal on. We'll consider specific examples later. We offer only a broad overview here.

Figure 22.10 outlines the principal ways by which signals enter cells. Some hormones, once "recognized" by the target cell, move directly through the cell membrane, enter the cytosol, and find a receptor inside the cell. Steroid hormones work in this way. See $\boxed{1}$ in Figure 22.10. They bind to receptors close to or inside the cell nucleus, where they induce changes in the way the cell uses DNA.

Polypeptide hormones cannot migrate directly through cell membranes, so they bind to receptors that are integral parts of the membrane. See $\boxed{2}$ in Figure 22.10. Insulin and growth hormone work in this way.

Neurotransmitters also bind to membrane-bound receptors, and this opens channels through the membrane for metal ions, $\boxed{3}$. We have already seen how movements of the

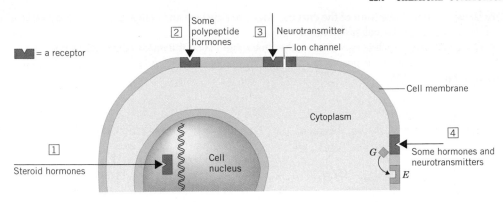

FIGURE 22.10
The ways by which hormones get chemical messages into cells

calcium ion can affect calmodulin or troponin and so activate a series of enzyme-catalyzed reactions.

Some receptors, in accepting neurotransmitters, hormones, or even light photons activate a small polypeptide of a *G-protein* complex, [4] in Figure 22.10. This leads to the activation of an enzyme (*E* in Figure 22.10). Remarkably, a large variety of cells share just a few mechanisms for taking advantage of the action of the G-protein. We'll study two, the cyclic AMP and the inositol phosphate cascades.

The Formation and Hydrolysis of Cyclic AMP Is a Major Mechanism by Which Many Cells Pass On Messages
Cyclic nucleotides, particularly 3′,5′-cyclic AMP, are important secondary chemical messengers. How cyclic AMP works is sketched in Figure 22.11.

At the top of the figure we see a hormone—it could just as well be a neurotransmitter—that can combine with a receptor molecule at the surface of a cell. Because of the requirements of fitting substrate and receptor molecules together, the hormone bypasses all cells that do not have a matching receptor. The complex that forms between hormone and receptor activates the G-protein molecule bound on the cytosol side of the lipid bilayer. The G-protein finds a

■ The G in G-protein stands for *guanyl nucleotide binding protein.* Of its three small polypeptide subunits, only one actually migrates to activate adenylate cyclase.

■ *Cyclic* refers to the *extra* ring of the phosphate diester system.

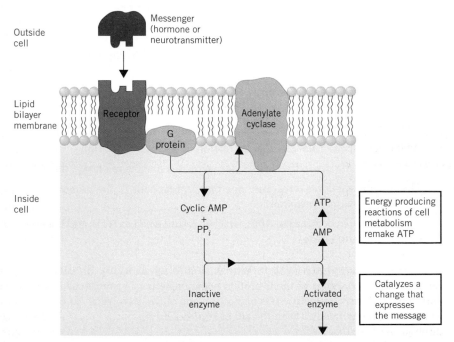

FIGURE 22.11
The activation of the enzyme adenylate cyclase by a hormone (or a neurotransmitter). The hormone–receptor complex releases a unit of the G-protein, which activates this enzyme. It then catalyzes the formation of cyclic AMP, which, in turn, activates an enzyme inside the cell.

molecule of an inactive form of the enzyme adenylate cyclase and activates it. This enzyme is an integral part of the cell membrane.

Once adenylate cyclase is activated, the "message" is on the inside of the cell membrane, because now the enzyme promptly catalyzes the conversion of ATP into cyclic AMP and diphosphate ion, PP_i.

■ The complete structure of ATP is shown on page 493.

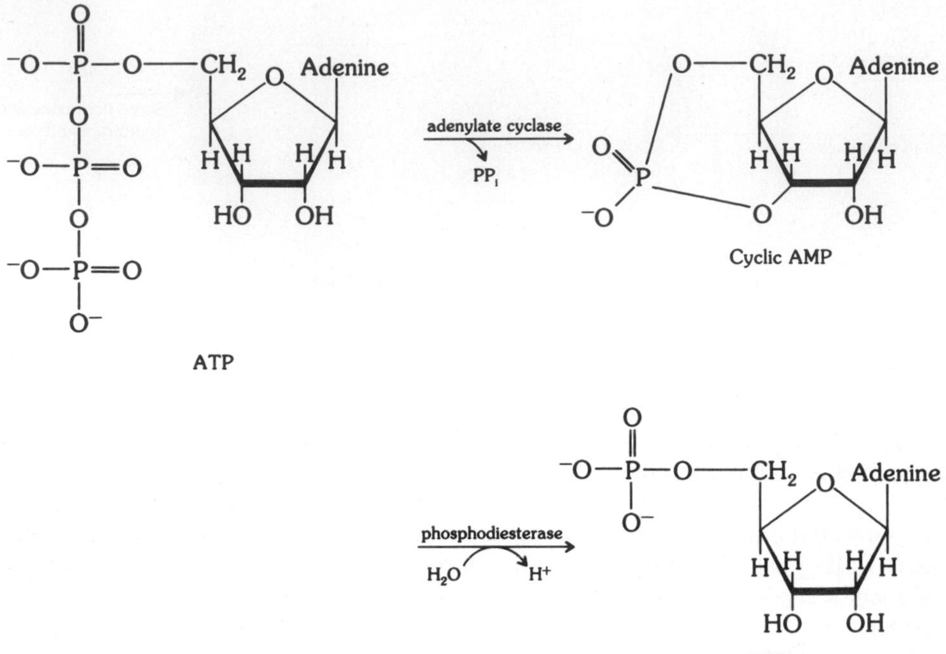

ATP

Cyclic AMP

AMP

■ E. W. Sutherland, Jr., an American scientist, won the 1971 Nobel prize in physiology and medicine for his work on cyclic AMP.

The newly formed cyclic AMP now activates an enzyme, which, in turn, catalyzes a reaction. This last event is what the original message was all about.

Finally, an enzyme called *phosphodiesterase* catalyzes the hydrolysis of cyclic AMP to AMP, and this shuts off the cycle. Energy-producing reactions in the cell will now remake ATP from the AMP.

Let's summarize the steps in this remarkable chemical cascade.

1. A signal releases the hormone or neurotransmitter.
2. It travels to its target cell, next door for a neurotransmitter but some farther distance away for a hormone.
3. The primary messenger molecule finds its target cell by a lock-and-key mechanism and binds to a receptor, which alters a polypeptide unit of the G-protein.

■ Hormones that work through the cyclic AMP cascade that we will encounter later include epinephrine, glucagon, norepinephrine, and vasopressin.

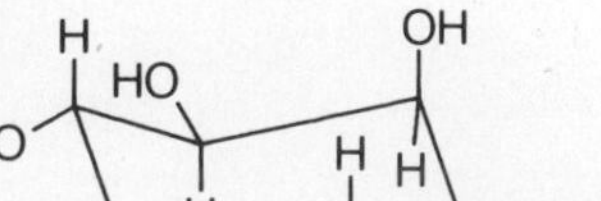

4. The altered G-protein activates the enzyme adenylate cyclase.
5. Adenylate cyclase catalyzes the conversion of ATP to cyclic AMP.
6. Cyclic AMP, the secondary messenger, activates an enzyme inside the cell.
7. The enzyme catalyzes a reaction, one that corresponds to the primary message of the hormone or neurotransmitter.
8. Cyclic AMP is hydrolyzed to AMP, which is reconverted to ATP, and the system returns to the preexcited state.

Inositol

The Inositol Phosphate System Works in a Way Roughly Similar to the Cyclic AMP System

Another of the G-protein communications systems used by cells involves phosphate and trisphosphate esters of inositol, one of the stereoisomers of hexahydroxycyclohexane. We can follow it with the aid of Figure 22.12.

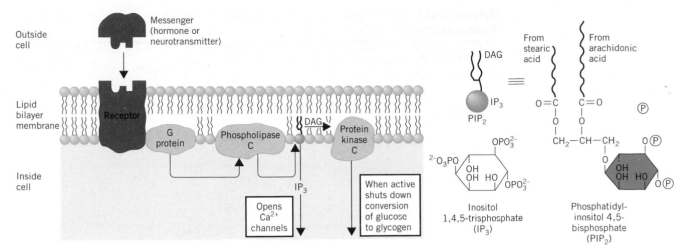

FIGURE 22.12
The inositol phosphate system that takes a message from a hormone or neurotransmitter and activates two enzymes to work cooperatively.

The primary messenger binds to the receptor, the G-protein is altered, and it activates an enzyme called phospholipase C. The pattern so far, as you can see, is quite similar to the cyclic AMP system. Phospholipase C now catalyzes the hydrolysis of a phosphate ester bond in a membrane-bound phospholipid, which we'll call PIP_2 (because its full name makes eyes glaze over). The products are two fragments *that are both messengers*. One we'll call IP_3 (an inositol trisphosphate) and the other DAG (a diacylglycerol).

DAG, being mostly hydrocarbon-like, stays in the lipid bilayer of the cell membrane where it activates a membrane-bound enzyme, protein kinase C. As we'll study under the metabolism of glucose, this enzyme aids the cell in developing a higher glucose level, which might be needed for energy.

The other fragment, IP_3, causes the rapid release of Ca^{2+} ion from intracellular storage systems. This ion might, for example, activate the contraction of a muscle. Of course, this is work requiring energy, so you can see that the two fragments, DAG and IP_3, work together. IP_3 tells the cell to do some work, DAG sees to the energy supply, and both are responses to one signal at the cell receptor.

Interestingly, malfunctions in the inositol phosphate system may be responsible for manic-depressive illness, one of the major psychiatric disorders. Lithium ion (as lithium carbonate) is used to treat this condition, and its action is evidently to shut down the inositol phosphate network in affected cells.

The inositol phosphate system may also be a target for the action of chemicals, including errant genes, that cause tumors and cancer. Protein kinase C has a function in cell division and in the control of the proliferation of cells, so chemicals that interact with this enzyme in the wrong way can affect cell division.

■ There are several variations of the G-protein.

■ The phosphoinositol cascade mediates the following activities:

glycogenolysis in the liver
insulin secretion
smooth muscle contraction
platelet aggregation

■ The cholera toxin is able to lock the G-protein into its active form, so adenylate cyclase cannot be shut off. This stimulates so much active transport of Na^+ ion into the gut, along with water, that massive diarrhea kills the victim.

22.7

HORMONES AND NEUROTRANSMITTERS

Interventions in the work of hormones and neurotransmitters are the bases of the action of a number of drugs, both licit and illicit.

Structurally, Hormones Come in Four Broad Types
The principal endocrine glands of the human body and the major hormones they secrete are too numerous to catalog in detail. It is impossible to do justice to a subject as vast as hormones in one section of one chapter, so

TABLE 22.2
Neurotransmitters

Monoamines	
Acetylchloline	$(CH_3)_3\overset{+}{N}CH_2CH_2O\overset{\overset{O}{\Vert}}{C}CH_3$
Dopamine	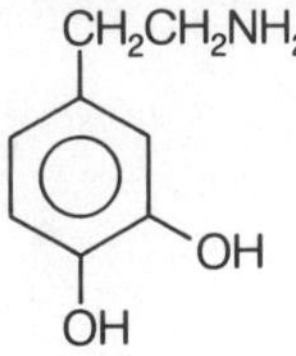
Norepinephrine	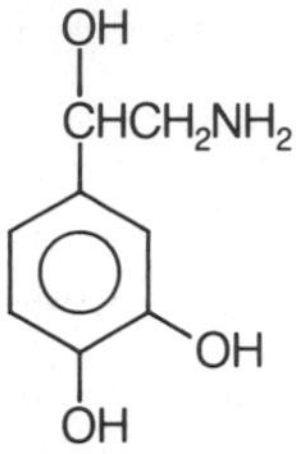
Serotonin	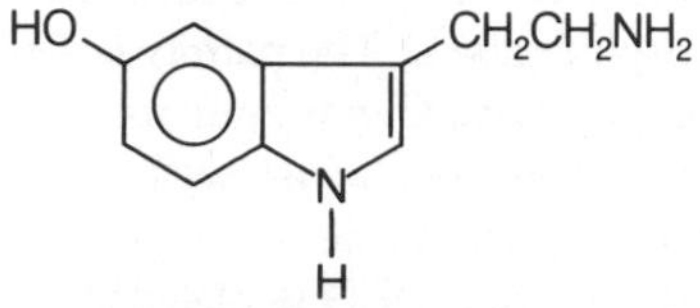
Amino Acids	
Glycine	$^{+}NH_3CH_2CO_2^-$
γ-Aminobutyric acid (GABA)	$^{+}NH_3CH_2CH_2CH_2CO_2^-$
Glutamic acid	$^{+}NH_3\underset{\vert}{C}HCO_2^-$ $CH_2CH_2CO_2H$
Neuropeptides	
Met-Enkephalin	Tyr-Gly-Gly-Phe-Met
Leu-Enkephalin	Tyr-Gly-Gly-Phe-Leu
β-Endorphin	Tyr-Gly-Gly-Phe-Met-Thr-Ser-Glu-Lys-Ser Gln-Thr-Pro-Leu-Val-Thr-Leu-Phe-Lys-Asn Ala-Ile-Val-Lys-Asn-Ala-His-Lys-Gly-Gln
Substance P	Arg-Pro-Lys-Pro-Gln-Gln-Phe-Phe-Gly-Leu-Met-NH_2
Angiotensin II	Asp-Arg-Val-Tyr-Ile-His-Pro-Phe-NH_2
Somatostatin	Ala-Gly-Cys-Lys-Asn-Phe Phe Cys Trp Ser-Thr-Phe-Thr

what follows is a very broad sketch of a few chemical aspects of hormone action. In later chapters we will mention specific hormones where they are particularly relevant to the metabolic activity being studied. Let's first consider some features of hormone molecules. They come in four general types.

Some hormones, the steroid hormones, are made from cholesterol and so have largely hydrocarbon-like molecules. This feature enables them to slip easily through the lipid bilayers of their target cells. Inside they find their final receptors, and the hormone – receptor complexes move to DNA molecules where they bind and affect the transcriptions of genetic messages. The sex hormones like estradiol, progesterone, and testosterone work in this way.

Many growth factors as well as insulin, oxytocin, and thyroid-stimulating hormone consist of polypeptides or proteins. These are able to alter the permeabilities of their target cells to the migrations of small molecules. The growth factors, for example, help get amino acids inside cells where they are needed for growth. Insulin helps to get glucose inside its target cells. We'll have much more to say about insulin in a later chapter. Either the absence of insulin or the absence (or inactivity) of insulin receptors results in the disease diabetes mellitus ("diabetes"). Several neurotransmitters are also polypeptides, and they alter the permeability of a neuron membrane to Ca^{2+} and Na^+. The cross-membrane movements of these ions are involved in the electrical signal that flows down a neuron.

The prostaglandins (Special Topic 20.1, page 568) are now classified as hormones, as *local hormones* because they work where they are made. The *eicosanoids* is the technical name for this family of compounds.

■ From *eicosane,* a C-20 hydrocarbon.

Finally, a number of hormones are relatively simple amino compounds made from amino acids. These include epinephrine (page 507) and thyroxine. Some of these are also neurotransmitters.

Neurotransmitters Move across the Narrow Synaptic Gap from One Neuron to the Next
A partial list of neurotransmitters is given in Table 22.2. Some are nothing more than simple amino acids. Others are β-phenylethylamines or catecholamines (Special Topic 17.1, page 507), and many are polypeptides.

Each nerve cell has a fiber-like part called an *axon* that reaches to the face of the next neuron or to one of its filament-like extensions called *dendrites.* A nerve impulse consists of a traveling wave of electrical charge that sweeps down the axon as small ions migrate at different rates between the inside and the outside of the neuron. The problem is how to get this impulse launched into the next neuron so that it can continue along the length of the nerve fiber. This is solved by a *chemical* communication from one neuron to the next. Neurotransmitters are the chemicals involved, and they are made from amino acids within the neuron and stored in sacs, called *vesicles,* located near the ends of the axons.

■ The traveling wave of electrical charge moves rapidly, but still not as rapidly as electricity moves in electrical wires.

Between the terminal of an axon and the end of the next neuron, there is a very narrow, fluid-filled gap called the *synapse.* Neurotransmitters move across the synapse when the electrical wave causes them to be released from their vesicles.

When neurotransmitter molecules lock into their receptors on the other side of the synapse, adenylate cyclase (or phospholipase C) is activated. (We'll use the adenylate cyclase system to illustrate the process.) See Figure 22.13. Now the formation of cyclic AMP is catalyzed, and newly formed cyclic AMP initiates whatever change is programmed by the chemicals in the target neuron. An enzyme then deactivates adenylate cyclase by catalyzing the release of the neurotransmitter molecule.

If the newly released neurotransmitter were a hormone, it would be swept away in the bloodstream, but it's not. It's still a neurotransmitter in the synapse, so unless the system wants it to act again, it must be removed or deactivated. A number of options are open, depending on the neurotransmitter.

FIGURE 22.13

In neurotransmission, the neurotransmitter molecules, released from vesicles of the presynaptic neuron, travel across the synapse. At the postsynaptic neuron, they find their receptors, and the cyclic AMP system similar to that shown in Figure 22.11 becomes activated.

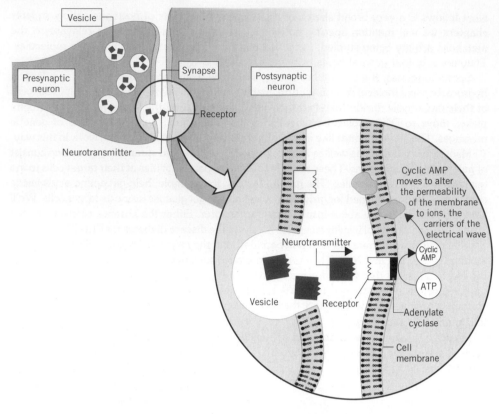

The Neurotransmitter Acetylcholine Is Swiftly Hydrolyzed

One method used to remove the neurotransmitter is to break it up by a chemical reaction. Acetylcholine, for example, a neurotransmitter in the autonomic nervous system or ANS, is catalytically hydrolyzed to choline and acetic acid. The enzyme is choline acetyltransferase.

■ The ANS nerves handle the signals that run the organs that have to work autonomously (without conscious effort), such as the heart and the lungs.

$$(CH_3)_3\overset{+}{N}CH_2CH_2O\overset{O}{\overset{\|}{C}}CH_3 + H_2O \underset{\text{acetyltransferase}}{\overset{\text{choline}}{\rightleftharpoons}} (CH_3)_3\overset{+}{N}CH_2CH_2OH + HO\overset{O}{\overset{\|}{C}}CH_3$$

Acetylcholine   Choline   Acetic acid

Within two milliseconds (2×10^{-3} s) of the release of acetylcholine in the synapse, all its molecules are broken down. The synapse is now cleared for a fresh release of acetylcholine from the presynaptic neuron if the signal for its release continues. If the signal does not come, then the action is shut down.

The Botulinum Toxin Prevents a Neuron from Making Acetylcholine

■ The ANS nerves that use acetylcholine are called the *cholinergic nerves.*

The botulinum toxin is an extremely powerful toxic agent made by the food-poisoning botulinum bacterium and works by preventing the *synthesis* of acetylcholine. Without this neurotransmitter, the cholinergic nerves of the ANS can't work.

Nerve Poisons Deactivate Choline Acetyltransferase

The nerve gases act by deactivating the enzyme needed to break acetylcholine down after it has done its work. The absence of this enzyme means, therefore, that the signal transmitted by acetylcholine can't be turned off. It continues unabated until the heart fails, usually in a minute or two.

An antidote for nerve gas poisoning, atropine, works by blocking the receptor protein for acetylcholine, so despite the continuous presence of this neurotransmitter, it isn't able to complete the signal-sending work. This tones the system down, and other processes slowly restore the system to normal. (Given the extreme speed with which nerve gases work, atropine must obviously be used very promptly.)

Some organophosphate insecticides are mild nerve poisons and work in the same way. Other blockers of the receptor protein for acetylcholine are some local anesthetics like nupercaine, procaine, and tetracaine.

Drugs that block the action of a neurotransmitter are called **antagonists** to the neurotransmitter. The neurotransmitter itself is sometimes referred to as an **agonist**.

Some Neurotransmitters Are Reabsorbed by the Presynaptic Neuron   Norepinephrine, another neurotransmitter, is deactivated by being reabsorbed by the neuron that released it, where it is then degraded. (Some is also deactivated right within the synapse.) The degradation of norepinephrine is catalyzed by enzymes called the **monoamine oxidases** or **MAO**.

■ The nerves that use norepinephrine are called the *adrenergic nerves* (after an earlier name for norepinephrine, noradrenaline).

Drugs that Inactivate the Monoamine Oxidases Are Used to Treat Depression   One place where norepinephrine works is in the brain stem, where mood regulation is centered. If for any reason the monoamine oxidases are inactivated, then an excess of norepinephrine builds up in brain stem cells, and some spills back into the synapse and sends signals on. In some mental states, like depression, an abnormally low level of norepinephrine develops, so now one would want to inactivate the monoamine oxidases. This would leave what norepinephrine there is to carry on its work. Thus, some of the antidepressant drugs, such as iproniazid, work by inhibiting the monoamine oxidases.

Other antidepressants, like amitriptyline (e.g., Elavil) and imipramine (e.g., Tofranil), inhibit the reabsorption of norepinephrine by the presynaptic neuron. Without this reabsorption into the degradative hands of the monoamine oxidases, the level of norepinephrine and its signal-sending work stay high.

Iproniazid

Amitriptyline (Elavil)

Imipramine (Tofranil)

Norepinephrine is both a hormone and a neurotransmitter. The adrenal medulla secretes it into the bloodstream in emergencies when it must be made available to all the nerve tissues that use it.

Dopamine Excesses Occur in Schizophrenia   Dopamine, like norepinephrine, is also a monoamine neurotransmitter. It occurs in neurons of the midbrain that are involved with feelings of pleasure and arousal as well as with the control of certain movements.

In schizophrenia the neurons that use dopamine are overstimulated, because either the releasing mechanism or the receptor mechanism is overactive. Drugs commonly used to treat schizophrenia such as chlorpromazine (e.g., Thorazine) and haloperidol (Haldol) bind to dopamine receptors and thus inhibit its signal-sending work.

Chloropromazine

Amphetamine Abuse Causes Schizophrenia-like Symptoms   Stimulants like the amphetamines (cf. Special Topic 17.1, page 507) work by triggering the release of dopamine into the arousal and pleasure centers of the brain. The effect is therefore a "high." But it's easy to abuse the amphetamines. When this occurs, the same kind of overstimulation associated with schizophrenia results, with such resulting symptoms as delusions of persecution, hallucinations, and other disturbances of the thought processes.

Haloperidol

$^+NH_3CHCO_2^-$

(L-DOPA structure)

L-DOPA

(MPTP and MPP$^+$ structures)

MPTP   MPP$^+$

$^+NH_3CH_2CH_2CH_2CO_2^-$

GABA

Dopamine-Releasing Neurons Have Degenerated in Parkinson's Disease When the dopamine-using neurons in the brain degenerate, as in Parkinson's disease, an extra supply of dopamine itself is then needed to compensate. This is why a compound called L-DOPA (*levorotatory dihydroxyphenylalanine*) is used. The neurons that still work can use it to make extra dopamine.

In the mid-1980s, by an accidental discovery, it was found that a contaminant in street heroin, called MPTP for short, rapidly destroys the same cells that degenerate in Parkinson's disease. The active agent is actually a metabolic breakdown product called MPP$^+$. The cells affected by Parkinson's disease apparently do not die at once but first go into a dormant state. The rejuvenation of such cells has long been sought. In the early 1990s, scientists, working with experimental rats, found that a substance called *brain-derived neurotrophic factor* or BDNF protects the dormant cells, stimulates them into recovery, and protects susceptible cells.

In 1993, another neurotrophic factor, GDNF (for glial cell line-derived neurotrophic factor) was also found to be able to promote the survival of neurons that release dopamine. Another line of research into treating Parkinson's disease involves transplants of healthy fetal brain tissue. Preliminary results of the use of this technique on human patients, made public in 1992, were promising.

GABA Inhibits Nerve Signals The normal function of some neurotransmitters is to *inhibit* signals instead of to initiate them. Gamma-aminobutyric acid (GABA) is an example, and as many as a third of the synapses in the brain have GABA available.

The inhibiting work of GABA can be made even greater by mild tranquilizers such as diazepam (e.g., Valium) and chlordiazepoxide hydrochloride (Librium), as well as by ethanol. The augmented inhibition of signals reduces anxiety, affects judgment, and induces sleep. Of course, you've probably heard of the widespread abuse of Valium and Librium, to say nothing of alcohol.

(Diazepam and Chlordiazepoxide hydrochloride structures)

Diazepam (Valium)   Chlordiazepoxide hydrochloride (Librium)

■ Greek, *chorea*, dance.

GABA Is Deficient in Huntington's Chorea The victims of Huntington's chorea, a hereditary neurological disorder, suffer from speech disturbances, irregular movements, and a steady mental deterioration, all related to a deficiency in GABA. Unhappily, GABA can't be administered in this disease, because it can't move out of circulation and into the regions of the brain where it works.

■ *En-* or *end-*, within; *kephale*, brain; *-orph-*, from morphine.

Several Polypeptides Act as Painkilling Neurotransmitters As we said, some neurotransmitters are relatively small polypeptides. See Table 22.2. One type includes the *enkephalins;* another consists of the *endorphins*. Both types are powerful pain inhibitors. One compound, dynorphin, is the most potent painkiller yet discovered, being 200 times stronger than morphine, an opium alkaloid that is widely used to relieve severe pain. Sites in the brain that strongly bind molecules of morphine also bind those of the enkephalins, so these natural painkillers are now often referred to as the body's natural opiates.

The Enkephalins Inhibit the Release of Substance P   Substance P is a pain-signaling, polypeptide neurotransmitter. According to one theory, when a pain-transmitting neuron is activated, it releases substance P into the synapse. However, butting against such neurons are other neurons that can release enkephalins. And these, when released, inhibit the work of substance P. In this way, the intensity of the pain signal is toned down. The action of enkephalin might explain the delay of pain that sometimes occurs during an emergency when the brain and the body must continue to function to escape the emergency.

Substance P might be involved in the link between the nervous system and the body's immune system. It is known that some forms of arthritis flare up under stress, and stress deeply involves the nervous system. But arthritis is generally regarded as chiefly a disease of the immune system. In tests on rats with arthritis, flare-ups of the arthritis could be induced by injections of substance P. What interests scientists about this is the possibility of controlling the severity of arthritis by somehow diminishing levels of substance P in the affected joints.

One of the interesting developments in connection with the endorphins is that acupuncture, a pain-alleviating procedure developed in China many centuries ago, might work by stimulating the production and release of endorphins.

Many Neurotransmitters Exert More than One Effect   Several neurotransmitters can be received by more than one kind of receptor. For example, at least three types of receptors for the opiates have been identified thus far. Such receptor multiplicity may explain how some neurotransmitters have multiple effects. Thus not only do opiates reduce pain, but they affect emotions, they induce sleep, and they affect the appetite, with each of the opiate receptors handling a different one of these functions.

Calcium Channel Blockers Are Drugs that Reduce the Vigor of Heart Muscle Contractions   As we have often seen, calcium ions are major secondary chemical messengers, and neurotransmitters are able to open channels for calcium ions through cell membranes. Heart muscle tissue receives such signals at a rate that paces the heart as its muscles contract and relax during the heartbeat. Calcium ions are what finally deliver the message to contract. Then the cell pumps them back out and the muscle relaxes until another cycle starts.

Drugs like nifedipine (Procardia, Adalat), diltiazem (Cardizem), and verapamil (Isoptin) find calcium ion channels in heart muscle and block them. Not all are blocked, of course, so the effect is to reduce the migrations of Ca^{2+} through cell membranes. These calcium channel blockers (also called calcium antagonists or slow channel blocking agents) thus make each heart muscle contraction less vigorous. This reduces the risk of heart attacks in people known to be at risk, like those who experience angina pectoris and cardiac arrhythmias (heartbeat irregularities).

Nitric Oxide Is a Retrograde Messenger that Might Be Involved in the Storage of Memory   Learning involves putting experiences, information, and data into memory in such a way that they can be retrieved and applied in new situations. If life has a molecular basis, then there must be a molecular basis to memory, too. What this basis is has long intrigued neuroscientists. According to one broad model of how some types of learning take place, a nerve cell that receives a signal associated with something to be remembered manufactures a *retrograde messenger*. This moves back to the signal-sending cell and strengthens the connection between the two cells. Each additional time that the same signal is sent, the signal is further strengthened.

One of the candidates for retrograde messenger is nitric oxide, NO. Its tiny molecules are able to slip easily through cell membranes, which is an essential property for any retrograde messenger. It is known that cultured brain cells can make nitric oxide when certain receptors are stimulated, but when a binder of NO is present, experimental rats are unable to learn certain spatial tasks. The discovery of a neurotransmitter role for an otherwise noxious gas, NO, means that not all chemical messenger work is done by a common mechanism. NO is a gas and can diffuse, as we said, from cell to cell relatively easily in several directions. The

■ "Retrograde" means moving or directed backward.

■ Abnormal NO systems have been found in some people with hypertension and in others with high blood cholesterol levels.

"classical" neurotransmitters must be stored and then released on signal. Next, they must find a receptor in a membrane in order to influence events within the cell. NO is able to slip directly through a membrane and so needs no membrane-bound receptor. (One enzyme that NO is known to stimulate directly is guanylate cyclase, an enzyme similar to adenylate cyclase.)

Carbon Monoxide Is Probably a Retrograde Messenger In the early 1990s, evidence was found that carbon monoxide, another noxious gas, might also be a brain messenger and possibly involved in a long-term learning-linked process. (One is certainly tempted to say, "Of all things!") In relatively large concentrations when carried into the body in inhaled air, CO binds so strongly to heme units in hemoglobin that the latter cannot carry oxygen, which causes death. Yet in small concentrations manufactured within the brain, CO activates an enzyme involved in learning. This is an area of rapidly moving research driven in part by intense curiosity about *how* we remember.

What we have done in this section is look at some *molecular* connections between conditions of the nervous system and particular chemical substances. This whole field is one of the most rapidly moving areas of scientific investigation today, and during the next several years we may expect to see a number of dramatic advances both in our understanding of what is happening and in the strategies of treating both mental and heart diseases.

■ Solomon Snyder (The Johns Hopkins University) has been a pioneer in studying how the brain works, biochemically.

SUMMARY

Enzymes Enzymes are the catalysts in cells. Some consist wholly of one or more polypeptides, and other enzymes include a cofactor besides the polypeptides. The cofactor can be an organic coenzyme, a metal ion, or both. Some coenzymes are phosphate esters of B vitamins and, in these examples, the vitamin unit usually furnishes the enzyme's active site. Because they are mostly polypeptide in nature, enzymes are vulnerable to all of the conditions that denature proteins. The name of an enzyme, which almost always ends in *-ase,* usually discloses either the identity of its substrate or the kind of reaction it catalyzes.

Some enzymes occur as small families called isoenzymes in which the polypeptide components vary slightly from tissue to tissue in the body. An enzyme is very specific both in the kind of reaction it catalyzes and in its substrate. Enzymes make possible reaction rates that are substantially higher than the rates of uncatalyzed reactions.

Induced fitting When an enzyme – substrate complex forms, the active site is brought together with the part of the substrate that is to react. Binding sites on the enzyme guide the substrate molecule in and induce a change in the conformation of the enzyme molecule to produce the best fit of the substrate. The recognition of the enzyme by the substrate occurs as a flexible, lock-and-key model that involves complementary shapes and electrical charges.

Enzyme kinetics An enzyme is a reactant in the initial phase of the reaction as it functions catalytically. At a sufficiently high initial substrate concentration, the active sites on all the enzyme molecules become saturated by substrate molecules, so at still higher initial substrate concentrations, there is little further increase in the initial rate. The rate behavior of enzyme-catalyzed reactions for which the enzyme has one catalytic site is described by the Michaelis-Menten equation. At low substrate concentration, the equation produces a straight-line plot of rate versus substrate concentration. At high substrate concentration, the equation shows that the rate must reach a maximum value.

Some enzymes seem to respond sluggishly to small increases in substrate concentration when the latter is very low. These display a sigmoid rate curve, and they require activation by substrate molecules before they produce their dramatic rate enhancements.

Regulation of enzymes Enzymes that display a sigmoid rate curve can be activated allosterically either by their own substrates or by effectors. Some enzymes are activated by genes. Some enzymes that are parts of the membranes of cells (or small bodies within cells) are activated by the interaction between a hormone or a neurotransmitter and its receptor protein. The work of many of these is to cause changes in the calcium ion level in a cell.

Other enzymes, such as certain digestive enzymes, exist as zymogens (proenzymes) and are activated when some agent acts to remove a small part that blocks the active site. The kinase enzymes are activated by being phosphorylated.

Enzymes can be inhibited by a non-product inhibitor or by competitive feedback that involves a product of the enzyme's action. Some inhibitors act allosterically; they bind to the enzyme at some location other than the catalytic site. Some of the most dangerous poisons bind to active sites and irreversibly block the work of an enzyme, or they carry enzymes out of solution by a denaturant action. Many antibiotics and other antimetabolites work by inhibiting enzymes in pathogenic bacteria.

Medical uses of enzymes The serum levels of many enzymes rise when the tissues or organs that hold these enzymes are injured or diseased. By monitoring these serum levels, and by looking for certain isoenzymes, we can diagnose diseases — for example, viral hepatitis and myocardial infarctions.

When a blood clot threatens or causes a heart attack, any one of three enzymes — streptokinase, ASPAC, or recombinant tissue plasminogen activator (TPA) — can be used to initiate the hydrolysis of the fibrin of the clot.

Enzymes are also used in analytical systems that measure concentrations of substrates, such as in tests for glucose. In some analytical systems, enzymes are immobilized on electrodes where they catalyze a reaction that produces a product; the electrode then senses and measures this product.

Chemical communication with hormones and neurotransmitters The carbohydrate tails of membrane glycoproteins project away from the membrane surface and provide recognition sites for chemical messengers (and other particles). The messenger molecule and the receptor protein form a complex. One common response to the formation of the complex is the release of a G-protein. In the adenylate cyclase cascade, this activates adenylate cyclase, which then triggers the formation of cyclic AMP. In turn, cyclic AMP sets off other events, such as the activation of an enzyme that catalyzes a reaction, one that is ultimately what the "signal" of the neurotransmitter was all about.

When the release of a G-protein is followed by the inositol phosphate cascade, the G-protein molecule activates phospholipase C, which breaks up a phospholipid in the membrane into two enzyme activators, PIP_2 and DAG. PIP_2 initiates the rapid release of Ca^{2+} from cytoplasm stores to cause muscle contraction, and DAG helps to keep the cell's glucose level high so that its metabolism can supply the energy.

Hormones Endocrine glands secrete hormones, and these primary chemical messengers travel to their target cells in the blood, where they activate a gene, or an enzyme, or affect the permeability of a cell membrane. They recognize their own target cells by binding to specific receptor proteins.

The steroid hormones can move into a cell to its nucleus and there find a receptor. Polypeptide hormones bind to membrane-bound receptors to initiate their action.

Neurotransmitters In response to an electrical signal, vesicles in an axon release a neurotransmitter that moves across the synapse. Its molecules bind to a receptor protein on the next neuron, and then the pattern is much like that of hormones. The result, however, is to open channels through the cell membrane for the migration of ions.

Neurotransmitters include amino acids, monoamines, and polypeptides. Some neurotransmitters *activate* some response in the next neuron, whereas others *deactivate* some activity. A number of medications work by interfering with neurotransmitters or with the opening of calcium ion channels.

REVIEW EXERCISES

The answers to Review Exercises whose numbers are in color are found in Appendix C. The answers to the other Review Exercises are found in the Study Guide that accompanies this book. The more challenging questions are marked with asterisks.

Nature of Enzymes
22.1 What are (a) the function and (b) the composition, in general terms only, of an enzyme?

22.2 To what does *specificity* refer in enzyme chemistry?

22.3 Define and distinguish among the following terms.
(a) apoenzyme (b) cofactor (c) coenzyme

22.4 Write the equation for the equilibrium catalyzed by carbonic anhydrase. What is particularly remarkable about the enzyme?

22.5 What in general does an enzyme do to an equilibrium?

Coenzymes
22.6 What B vitamin is involved in the NAD^+/NADH system?

22.7 The active part of either FAD or FMN is furnished by which vitamin?

22.8 Complete and balance the following equation.

$$\underset{CH_3CHCH_3}{\overset{OH}{|}} + NAD^+ \longrightarrow \underset{CH_3CCH_3}{\overset{O}{\|}} + \underline{\quad} + \underline{\quad}$$

22.9 Complete and balance the following equation.

$$\underline{\quad} + NADH + FAD \longrightarrow NAD^+ + \underline{\quad}$$

22.10 In what structural way do NAD^+ and $NADP^+$ differ? What formula can be used for the reduced form of $NADP^+$?

Kinds of Enzymes
22.11 What *kind* of reaction does each of the following enzymes catalyze?
(a) an oxidase (b) transmethylase
(c) hydrolase (d) oxidoreductase

22.12 What is the difference between lactose and lactase?

22.13 What is the difference between a hydrolase and hydrolysis?

22.14 What are isoenzymes (in general terms)?

22.15 What are the three isoenzymes of creatine kinase? Give their symbols and state where they are principally found.

Theory of How Enzymes Work
22.16 What name is given to the part of an enzyme where the catalytic work is carried out?

22.17 How is enzyme specificity explained?

22.18 What is the induced-fit theory?

22.19 The Michaelis-Menten equation applies to an enzyme having how many catalytic sites?

*22.20 Which relationship, $V \propto [S]$ or $V \propto [E_o]$, applies under each of the following conditions?
(a) A relatively high concentration of substrate.
(b) A relatively low concentration of substrate.

22.21 When the velocity of an enzyme-catalyzed reaction obeying the Michaelis-Menten equation equals half of the maximum velocity, what does the Michaelis constant compare to?

Enzyme Activation and Inhibition

22.22 What does *allosteric* mean?

22.23 If the plot of initial reaction rate versus initial substrate concentration at constant [E] has a sigmoid shape, what does this signify about the active site(s) on the enzyme?

22.24 How does a substrate molecule activate an enzyme whose rate curve is sigmoid?

22.25 How does an effector differ from a substrate in causing allosteric activation?

22.26 What are the names of two important effectors? Which one is used in muscle cells?

22.27 What are the approximate concentrations of calcium ion in the cytosol and the fluid just outside a cell? Why doesn't simple diffusion wipe out this concentration gradient?

22.28 Why must the concentration of Ca^{2+} be so low in the cytosol?

22.29 Does the concentration of Ca^{2+} in the cytosol measure the amount of Ca^{2+} in the whole cytoplasm? Explain.

22.30 What does Ca^{2+} do to calmodulin or troponin?

22.31 When Ca^{2+} combines with troponin, what happens with respect to other proteins in the cell? What then happens to Ca^{2+}?

22.32 What is the relationship of a zymogen to its corresponding enzyme? Give an example of an enzyme that has a zymogen.

22.33 What is enteropeptidase and what does it do?

22.34 What is plasmin and in what form does it normally circulate in the blood?

22.35 What role does a phosphorylation reaction have in connection with some enzymes?

22.36 How does competitive inhibition of an enzyme work?

22.37 Feedback inhibition of an enzyme works in what way?

22.38 Why is feedback inhibition an example of a homeostatic mechanism?

*22.39 How do competitive inhibition and allosteric inhibition differ?

22.40 How do the following poisons work?
(a) CN^- (b) Hg^{2+}
(c) nerve gases or organophosphate insecticides

22.41 What are antimetabolites, and how are they related to antibiotics?

22.42 In broad terms, how does penicillin work?

Enzymes in Medicine

22.43 If an enzyme such as CK or LD is normally absent from blood, how can a *serum* analysis for either tell anything? (Answer in general terms.)

22.44 What is the significance of CK(*MB*) in serum in trying to find out whether a person has had a heart attack and not just some painful injury in the chest region?

22.45 What CK isoenzyme would increase if the injury in Review Exercise 22.44 were to skeletal muscle?

22.46 What is the LD_1-LD_2 flip, and how is it used in diagnosis?

22.47 How is immobilized urease used?

22.48 How is immobilized heparinase used?

22.49 Describe an example of an immobilized enzyme on an electrode tip and the function it serves.

22.50 What three enzymes are available to help dissolve a blood clot? Which one occurs in human blood, and how is it obtained for therapeutic uses?

22.51 What substance makes up most of a blood clot, and what happens to it when TPA works?

Chemical Communication

22.52 What are the names of the sites of the synthesis of (a) hormones and (b) neurotransmitters?

22.53 What do the lock-and-key and induced fit concepts have to do with the work of hormones and neurotransmitters?

22.54 In what general ways do hormones and neurotransmitters resemble each other?

22.55 What general name is given to the substance on a target cell that recognizes a hormone or neurotransmitter?

22.56 Name the two kinds of "enzyme cascades" studied in this chapter.

22.57 Name the small polypeptide that both systems (Review Exercise 22.56) use to activate something inside the cell.

22.58 What function does adenylate cyclase have in the work of at least some hormones?

22.59 How is cyclic AMP involved in the work of some hormones and neurotransmitters?

22.60 After cyclic AMP has caused the activation of an enzyme inside a cell, what happens to the cyclic AMP that stops its action until more is made?

*22.61 In what ways does the inositol "cascade" resemble the cyclic AMP cascade?

22.62 When the G-protein of the inositol cascade completes its work, it has produced *two* new messengers. In general terms, what does each one do next?

Hormones

22.63 What are the four broad types of hormones?

22.64 What structural fact about the steroid hormones makes it easy for them to get through a cell membrane?

22.65 In each case, what substance (or kind of substance) can enter a target cell more readily following the action of the hormone?

(a) insulin (b) growth hormone (c) a neurotransmitter

22.66 Why are the prostaglandins called *local hormones?*

Neurotransmitters

22.67 What happens to acetylcholine after it has worked as a neurotransmitter? What is the name of the enzyme that catalyzes this change? In chemical terms, what does a nerve gas poison do?

22.68 How does atropine counter nerve gas poisoning?

22.69 How does a local anesthetic such as procaine affect the functioning of acetylcholine as a neurotransmitter?

22.70 How does the botulinum toxin work?

22.71 What, in general terms, are the monoamine oxidases, and in what way are they important?

22.72 What does iproniazid do chemically in the neuron-signaling that is carried out by norepinephrine?

22.73 In general terms, how do antidepressants such as amitriptyline or imipramine work?

22.74 Which neurotransmitter is also a hormone, and what is the significance of this dual character to the body?

22.75 The overactivity of which neurotransmitter is thought to be one biochemical problem in schizophrenia?

22.76 How do the schizophrenia-control drugs chlorpromazine and haloperidol work?

22.77 How can the amphetamines, when abused, give schizophrenia-like symptoms?

22.78 How does L-DOPA work in treating Parkinson's disease?

22.79 How does a neurotrophic factor like BDNF work in treating Parkinson's disease?

22.80 Which common neurotransmitter in the brain is a signal inhibitor? How do such tranquilizers as Valium and Librium affect it?

22.81 Why is enkephalin called one of the body's own opiates? How does it appear to work?

22.82 What does substance P do?

22.83 How do the calcium channel blockers reduce the risk of a heart attack?

22.84 What is meant by the term *retrograde messenger?*

22.85 What physical property of nitric oxide or carbon monoxide enables them to act in a retrograde manner?

Work of NAD⁺, FMN, or FAD (Special Topic 22.1)

22.86 When NAD^+ is involved in the oxidation of an alcohol, what small pieces leave the alcohol molecule? Where does each go?

22.87 When FMN or FAD is involved in the oxidation of an alcohol, what small pieces leave the alcohol molecule? Where does each go?

22.88 What specifically makes the NAD^+ unit an attractor of $H\!:^-$?

22.89 What molecule finally gets the electrons of $H\!:^-$ at the end of the entire metabolic pathway?

Complementarity and Immune Responses (Special Topic 22.2)

22.90 What is a pathogen? (In general terms.)

22.91 What two kinds of immunity are recognized and how do they differ?

22.92 The HIV particle attacks what kind of immunity and in what way?

22.93 What is an antibody? An antigen?

22.94 At the molecular level, what aspects of molecular structure explain the high specificity of the immune response?

22.95 In what kind of immunity are the B-cells operative, and what kind of substance do B-cells eventually make to counter an alien material?

22.96 Why is a glycolipid on a red blood cell of an A-type individual referred to as an antigen and not an antibody?

22.97 At the molecular level involving their red blood cells, in what specific ways do the A, B, and O types differ?

22.98 What is present in the blood of an A-type person that makes receiving blood from a B-type dangerous?

22.99 Why is it that O-types can donate blood to people of any type but can receive blood only from other O-types?

Additional Exercises

22.100 How does the plot of initial rate versus initial [S] at constant [E] look when (a) an allosteric effect is occurring and (b) no allosteric effect is observed? (Draw pictures.)

22.101 If you drink enough methanol, you will become blind or die. One strategy to counteract methanol poisoning is to give the victim a nearly intoxicating drink of dilute ethanol. As the ethanol floods the same enzyme that attacks the methanol, the methanol gets a lessened opportunity to react and it is slowly and relatively harmlessly excreted. Otherwise, it is oxidized to formaldehyde, the actual poison from an overdose of methanol:

$$CH_3OH \xrightarrow{\text{dehydrogenase}} CH_2O$$
Methanol　　　　　　　Formaldehyde

What kind of enzyme inhibition might ethanol be achieving here? (Name it.)

22.102 Truffles are edible, potato-shaped fungi that grow underground in certain parts of France, and are highly prized by gourmet cooks and gourmands. Pigs are used to locate truffles buried as much as one meter below the surface because truffles carry traces of a steroid, androsten-16-en-3-ol, which is a powerful sex attractant for pigs. A sex attractant is a species-specific chemical compound made and released in trace amounts by a female member of a species toward which a male member experiences a powerful sexual response. A male pig, of course, does not initially know that it is a truffle emitting the attractant. (It appears that all sex attractants for humans are *nonchemical* being better understood, perhaps, as public relations activities.)

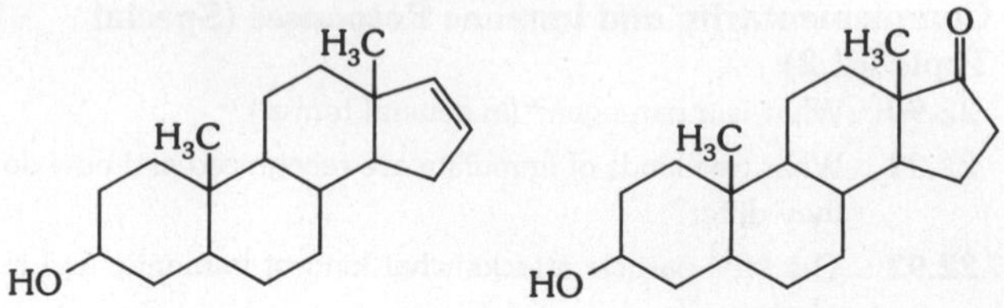

Androsten-16-en-3-ol

Androsterone

Notice the structural similarity between androsten-16-en-3-ol and a human sex hormone, androsterone.

(a) In terms of what general theory would we explain how androsten-16-en-3-ol has a particular specificity in pigs and not in humans but androsterone has a specificity in humans and not in pigs?

(b) In order to convert androsten-16-en-3-ol into androsterone *in vitro*, what specific *series* of changes in functional groups would have to be carried out. Your answer would begin with something like "First, we have to change such and such a group into. . . . . Then this new group would have to be changed . . ." (All needed changes involve one-step reactions we have studied.)

*22.103 The structure of isoleucine is given on page 627. Draw the Fischer projection structures of its stereoisomers. (See Special Topic 18.1.)

*22.104 Referring to the equation given on page 627 for the conversion of threonine to isoleucine, what is the maximum number of milligrams of isoleucine that could be made from 150.0 mg of threonine?

EXTRACELLULAR FLUIDS OF THE BODY

23

High-altitude work, such as climbing along the southeast buttress of Fitzroy in Patagonia, places intense demands on the buffer system of the blood. The tendency is to overbreathe and so remove carbon dioxide too rapidly from the body. This leads to alkalosis and possibly fatal high-altitude sickness. The acid-base status of the blood is the major topic of this chapter.

23.1

DIGESTIVE JUICES

The end products of the digestion of the nutritionally important carbohydrates, neutral fats and oils, and proteins are monosaccharides, fatty acids, monoacylglycerols, and amino acids.

Life engages two environments, the outside environment we commonly think of, and the internal environment, which we usually take for granted. When healthy, our bodies have nearly perfect control over the internal environment, and so we are able to handle large changes outside more or less well: large temperature fluctuations, chilling winds, stifling humidity, and a fluctuating atmospheric tide of dust and pollutants.

■ The fluids of the internal environment make up about 20% of the mass of the body.

Cells Exist in Contact with Interstitial Fluids and Blood The **internal environment** consists of all the **extracellular fluids,** those that aren't actually inside cells. About three-quarters consists of the **interstitial fluid,** the fluid in the spaces or interstices *between* cells. The blood makes up nearly all the rest. The lymph, cerebrospinal fluid, digestive juices, and synovial fluids (lubricants of joints) are also extracellular fluids.

The chemistry occurring inside cells, in the *intracellular fluid,* has been and will continue to be a major topic of our study. Here we will focus on two of the extracellular fluids, the digestive juices and the blood.

The Digestive Tract Is a Convoluted Tube Running through the Body with Access to Several Solutions of Hormones and Hydrolases The principal parts of the digestive tract are given in Figure 23.1. The **digestive juices** are dilute solutions of electrolytes and hydrolytic enzymes (or zymogens) either in the cells lining the intestinal tract or in solutions that enter the tract from various organs.

The intestinal tract also includes specialized cells that manufacture and release hormones in response to various signals — the presence of arriving nutrients, or the distension of the wall of the tract, or a change in pH. Some hormones are secreted into the tract itself, and others enter the bloodstream. Taken as a whole, the digestive tract is itself a vast endocrine gland, the largest we have. The lower part of the stomach releases *gastrin,* which helps to stimulate the release of gastric juice. Cells in the upper intestinal tract release the hormone *cholecystokinin.* It acts to modulate the release of material from the stomach into the intestinal tract, which means that digestion proceeds at a sufficiently slow rate to be complete. Cholecystokinin also tells the gallbladder to release bile, and it makes the pancreas release pancreatic enzymes. *Secretin* is another hormone of the upper intestinal tract, and it makes the pancreas release bicarbonate ion for neutralizing incoming gastric acid.

■ Dextrins are partial breakdown products of amylopectin, a component of starch.

Saliva Provides α-Amylase, a Starch-Splitting Enzyme The flow of **saliva** is stimulated by the sight, smell, taste, and even the thought of food. Besides water (99.5%), saliva includes a food lubricant called *mucin* (a glycoprotein) and an enzyme, α-amylase. This enzyme catalyzes the partial hydrolysis of starch to dextrins and maltose, and it works best at the pH of saliva, 5.8 – 7.1. Proteins and lipids pass through the mouth essentially unchanged.

648

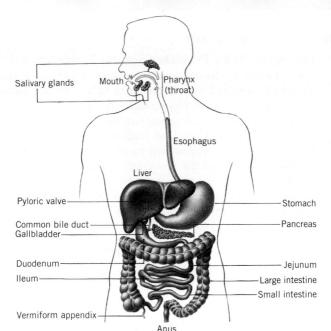

FIGURE 23.1
Organs of the digestive tract

Salivary glands

Mouth

Pharynx (throat)

Esophagus

Liver

Pyloric valve

Common bile duct

Gallbladder

Duodenum

Ileum

Vermiform appendix

Anus

Stomach

Pancreas

Jejunum

Large intestine

Small intestine

Gastric Juice Starts the Digestion of Proteins with Pepsin When food is on the way to the stomach, the neurotransmitter acetylcholine is produced in gastric cells and histamine is manufactured. When food arrives in the stomach and distends it, the cells in the gastric lining release gastrin. These three compounds, gastrin, acetylcholine, and histamine, stimulate the release of the fluids that combine to give **gastric juice.** One kind of gastric cell secretes mucin, which coats the stomach to protect it against its own digestive enzymes and its acid. Mucin is continuously produced and only slowly digested. If for any reason its protection of the stomach is hindered, part of the stomach itself could be digested, and this would lead to an ulcer.

Another kind of gastric cell (the parietal cells) secretes hydrogen ion at a concentration of roughly 0.15 mol/L (pH 0.8 or over a million times more acidic than blood). A $K^+ - H^+$ ion "pump" (using ATP chemical energy) takes K^+ ion out of stomach fluids and puts H^+ into the fluids. The K^+ is sent back out along with Cl^- ion. Because chloride ion is the chief anion in gastric juice, the overall effect is the secretion of hydrochloric acid. Histamine is the specific stimulator of the $K^+ - H^+$ ion pump. One of the most common and successful medications for the treatment of stomach ulcers is cimetidine (Tagamet®), and it acts to competitively inhibit the histamine receptor of the $K^+ - H^+$ ion pump. By preventing histamine binding, cimetidine shuts down the $K^+ - H^+$ ion pump, thus preventing the secretion of acid and so allowing the stomach to heal the ulcer.

The stomach acid coagulates proteins and activates a protease. Protein coagulation retains the protein in the stomach longer for exposure to the protease, *pepsin.*

Gastric cells also secrete the zymogen, *pepsinogen.* Pepsinogen is changed into pepsin by the action of hydrochloric acid and traces of pepsin. The optimum pH of pepsin is about 2, reflecting how relatively acidic stomach fluid becomes as it mixes with incoming food. Pepsin catalyzes the only important digestive work in the stomach, the hydrolysis of some of the peptide bonds of proteins to make shorter polypeptides.

Adult gastric juice also has a lipase, but it does not start its work until it arrives in the higher pH medium of the upper intestinal tract.

The gastric juice of infants is less acidic than the adult's. To compensate for the protein-coagulating work normally done by the acid, infant gastric juice contains rennin, a powerful protein coagulator. Because the pH of an infant's gastric juice is higher than that in the adult, its lipase gets an early start on lipid digestion.

The churning and digesting activities in the stomach produce a liquid mixture called *chyme.* This is released in portions through the pyloric valve into the duodenum, the first 12 inches of the upper intestinal tract.

■ Histamine is made by the loss of the CO_2H group from the amino acid histidine.

Histamine

Cimetidine

■ A *protease* is an enzyme that catalyzes the digestion of proteins.

649

Pancreatic Juice Furnishes Several Zymogens and Enzymes   As soon as chyme appears in the duodenum, hormones (cholecystokinin and secretin) are released that circulate to the pancreas and induce this organ to release two juices. One is almost entirely dilute sodium bicarbonate, which neutralizes the acid in chyme. The other juice is the one usually called **pancreatic juice.** It carries enzymes or zymogens that become involved in the digestion of practically everything in food. It contributes an α-amylase similar to that present in saliva, a *lipase, nucleases,* and zymogens for protein-digesting enzymes.

The conversion of the proteolytic zymogens to active enzymes begins with a "master switch" enzyme called *enteropeptidase,* which we mentioned in Section 22.4. It is released from cells that line the duodenum when chyme arrives, and it then catalyzes the formation of trypsin from its zymogen, trypsinogen.

- The nucleases include ribonuclease (RNase) and deoxyribonuclease (DNase).

- Enteropeptidase used to be called enterokinase.

$$\text{Trypsinogen} \xrightarrow{\text{enteropeptidase}} \text{trypsin}$$

Trypsin then catalyzes the change of the other zymogens into their active enzymes.

- These proteases must exist as zymogens first or they will catalyze the self-digestion of the pancreas, which does happen in acute pancreatitis.

$$\text{Procarboxypeptidase} \xrightarrow{\text{trypsin}} \text{carboxypeptidase}$$
$$\text{Chymotrypsinogen} \xrightarrow{\text{trypsin}} \text{chymotrypsin}$$
$$\text{Proelastase} \xrightarrow{\text{trypsin}} \text{elastase}$$

Trypsin, chymotrypsin, and *elastase* catalyze the hydrolysis of large polypeptides to smaller ones. *Carboxypeptidase,* working in from C-terminal ends of small polypeptides, carries the action further to amino acids and di- or tripeptides.

Bile Salts Are Powerful Surfactants Necessary to Manage Dietary Triacylglycerols and Fat-Soluble Vitamins   In order to digest triacylglycerols, the lipase in pancreatic juice needs the help of the powerful detergents in bile, called the *bile salts.* These help to emulsify water-insoluble fatty materials and so greatly increase the exposure of lipids to water and lipase. The digestion of triacylglycerols gives the anions of fatty acids and monoacylglycerols (plus some diacylglycerols).

- The structure of a typical bile salt was given on page 576.

Bile is a juice that enters the duodenum from the gallbladder. Its secretion is stimulated by a hormone released when chyme contains fatty material. Bile is also an avenue of excretion, because it can carry cholesterol and breakdown products of hemoglobin. These and further breakdown products constitute the bile pigments, which give color to feces.

The bile salts also assist in the absorption of the fat-soluble vitamins (A, D, E, and K) from the digestive tract into the blood. This work reabsorbs some bile pigments, some of which eventually leave the body via the urine. Thus the bile pigments are responsible for the color of both feces and urine.

Cells of the Intestines Carry Several Digestive Enzymes   The term **intestinal juice** embraces not only a secretion but also the enzyme-rich fluids found inside certain kinds of cells that line the duodenum and jejunum. The secretion of some of these cells delivers an amylase and enteropeptidase, which we just described.

- These intestinal cells last only about two days before they self-digest. They are constantly being replaced.

The other enzymes in this region work within their cells as digestible compounds are already being absorbed. An *aminopeptidase,* working inward from N-terminal ends of small polypeptides, hydrolyzes them to amino acids. The hydrolase enzymes *sucrase, lactase,* and *maltase* handle the digestion of disaccharides: sucrose to fructose and glucose; lactose to galactose and glucose; and maltose to glucose. An intestinal lipase and enzymes for the hydrolysis of nucleic acids are also present.

As the anions of fatty acids and mono- and diacylglycerols migrate into the cells of the duodenal lining, they are reconstituted into triacylglycerols, which are taken up by the lymph system and then delivered to the blood in special complexes of lipids and proteins called chylomicrons. (We'll study what happens to them and to dietary cholesterol in Chapter 27.)

Some Vitamins and Essential Amino Acids Are Made in the Large Intestine   No digestive functions are performed in the large intestine. Microorganisms in residence there, however, make vitamins K and B, plus some essential amino acids. These are absorbed by the body, but their contribution to overall nutrition in humans is not large.

Water and sodium chloride are reabsorbed from the ileum, and undigested matter (including fiber), microorganisms, and some water make up the feces.

23.2

BLOOD AND THE ABSORPTION OF NUTRIENTS BY CELLS

The balance between the blood's pumping pressure and its colloidal osmotic pressure tips at capillary loops.
The circulatory system, Figure 23.2, is one of our two main lines of chemical communication between the external and internal environments. All of the veins and arteries together are called the **vascular compartment**. The **cardiovascular compartment** includes this plus the heart.

■ The nervous system with its neurotransmitters is the other line of communication.

The Blood Moves Nutrients, Oxygen, Messengers, Wastes, and Disease Fighters throughout the Body   Blood in the pulmonary branches moves through the lungs where waste carbon dioxide is exchanged for oxygen. The oxygenated blood then moves to the rest of the system via the systemic branches.

■ About 8% of the body's mass is blood. In the adult, the blood volume is 5 to 6 L.

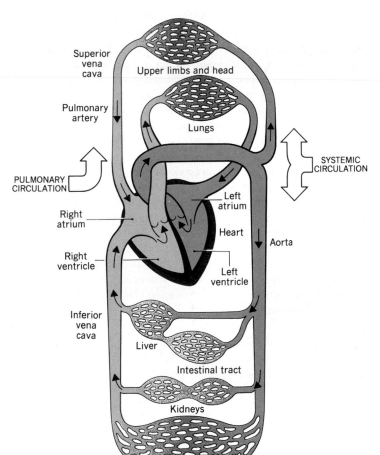

FIGURE 23.2
Human circulatory system. When oxygen-depleted venous blood (blue areas) returns to the heart, it is pumped into the capillary beds of the alveoli in the lungs to reload oxygenated blood (red areas) is dioxide. Then the freshly oxygenated blood (dark areas) is distributed by the arteries throughout the body, including the heart muscle.

FIGURE 23.3
Major components of blood

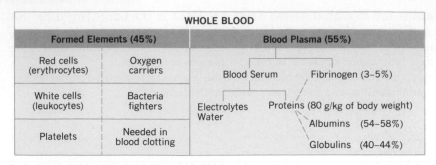

WHOLE BLOOD				
Formed Elements (45%)		**Blood Plasma (55%)**		
Red cells (erythrocytes)	Oxygen carriers	Blood Serum — Fibrinogen (3–5%)		
White cells (leukocytes)	Bacteria fighters	Electrolytes Water — Proteins (80 g/kg of body weight)		
Platelets	Needed in blood clotting	Albumins (54–58%) Globulins (40–44%)		

At the intestinal tract, the blood picks up the products of digestion. Most of these are immediately monitored at the liver and many alien chemicals are modified for elimination. In the kidneys, the blood replenishes its buffer supplies and eliminates waste nitrogen, mostly as urea. The pH of blood and its electrolyte levels depend largely on the kidneys. At endocrine glands, the blood picks up hormones whose secretions are often in response to something present in the blood.

White cells in blood provide protection against infection; red cells or **erythrocytes** carry oxygen and waste bicarbonate ion; and platelets are needed for blood clotting and other purposes. The blood also carries several zymogens that participate in the blood-clotting mechanism.

The Proteins in Blood Are Vital to Its Colloidal Osmotic Pressure   The principal types of substances in whole blood are summarized in Figure 23.3. Among the proteins, the **albumins** help carry hydrophobic molecules, like fatty acids and steroid hormones, and they contribute 75–80% of the osmotic effect of the blood. Some **globulins** carry ions (e.g., Fe^{2+} and Cu^{2+}) that otherwise are insoluble when the pH is greater than 7. Some globulins are antibodies that help to protect against infectious disease. **Fibrinogen** is converted to an insoluble form, **fibrin,** when a blood clot forms, as we discussed in Section 22.5.

■ About a quarter of the plasma proteins are replaced each day.

Figure 23.4 gives the levels of solutes, including proteins, in various components of the major body fluids. The higher level of protein in blood is the principal reason why blood has a higher osmotic pressure than interstitial fluid.[1] The *total* osmotic pressure of blood is caused, of course, by all the dissolved and colloidally dispersed solutes. The small ions and molecules, however, can move back and forth between the blood and the interstitial compartment. The large protein molecules can't do this, so it is their presence that gives to blood a higher effective osmotic pressure than interstitial fluid. The margin of difference is the blood's colloidal osmotic pressure.

■ The osmolarity of plasma is about 290 mOsm/L.

As a consequence of the higher osmotic pressure of blood, water tends to flow into the blood from the interstitial compartment. This cannot be allowed to happen everywhere and continually, however, or the interstitial spaces and then the cells would eventually become too dehydrated to maintain life. Before we see how this problem is managed, we need to survey some of the **electrolytes** in blood.

The Chief Ions in Blood Are Na^+, Cl^-, and HCO_3^-   Figure 23.4 gives the levels of the electrolytes in body fluids. The sodium ion is the chief cation in both the blood and the interstitial fluid and the potassium ion is the major cation inside cells. A sodium–potassium pump, a special ATP-run protein complex, maintains these gradients. Both ions are needed to maintain osmotic pressure relationships, and both are a part of the regulatory system for

---

[1] As a reminder and a useful memory aid, high solute concentration means high osmotic pressure; and solvent flows in osmosis or dialysis from a region where the solute is dilute to the region where it is concentrated. The "goal" of this flow is to even out the concentrations everywhere.

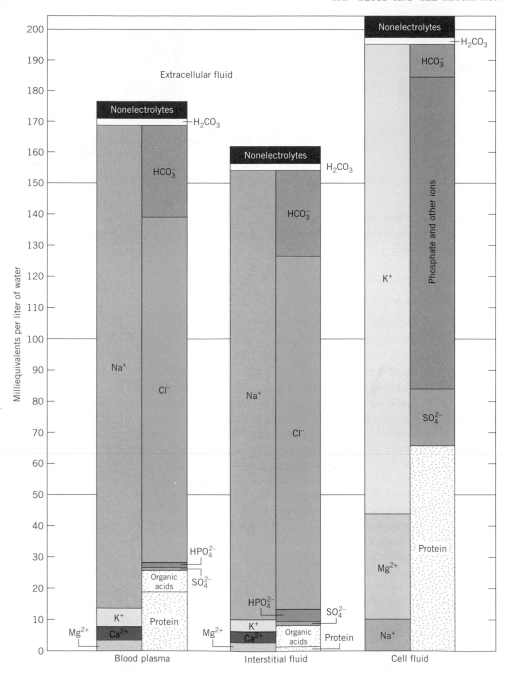

FIGURE 23.4
Electrolyte composition of body fluids. (Adapted by permission from J. L. Gamble, *Chemical Anatomy, Physiology and Pathology of Extracellular Fluids,* 6th ed. Harvard University Press, Cambridge, MA, 1954.)

acid–base balance. Both ions are also needed for the smooth working of the muscles and the nervous system.

Changes in the concentrations of sodium and potassium ion in blood can lead to serious medical emergencies, so a special vocabulary has been developed to describe various situations. We will see here how a technical vocabulary can be built on a few word parts, and we will use some of these word parts in later chapters, too. For example, we use *-emia* to signify "in the blood." *Hypo-* indicates a condition of a low concentration of something, and *hyper-* is the opposite, a condition of a high concentration of something. We can specify this something by a

■ Another word part is *-uria,* of the urine. Thus *glucosuria* means glucose in the urine.

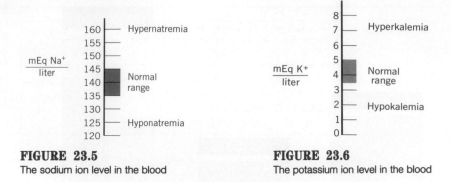

FIGURE 23.5
The sodium ion level in the blood

FIGURE 23.6
The potassium ion level in the blood

word part, too. Thus -nat- signifies sodium (from the Latin *natrium* for sodium), and -kal- designates potassium (from the Latin *kalium* for potassium). Putting these together gives us the following terms.

Hyponatremia: low level of sodium ion in blood

Hypokalemia: low level of potassium ion in blood

Hypernatremia: high level of sodium ion in blood

Hyperkalemia: high level of potassium ion in blood

Na⁺ and K⁺ Levels Are Regulated in Tandem Figure 23.5 shows the normal range for the sodium ion level in blood. Figure 23.6 does the same thing for the potassium ion level. Table 23.1 gives the normal ranges for these cations in various units.

■ You'll see it called the "sodium level," not sodium *ion* level, in nearly all references, but sodium *ion* level is always intended.

If our kidneys cannot make urine or if we drink water faster than the kidneys can handle it, the sodium level of the blood decreases and we will display the signs of hyponatremia: flushed skin, fever, and a dry tongue (and a noticeable decrease in urine output).

Hypernatremia occurs from excessive losses of water under circumstances in which sodium ions are not lost, as in diarrhea, diabetes, and even in some high-protein diets.

Because the level of potassium ion in blood is so low, small changes are particularly dangerous. Severe hyperkalemia leads to death by heart failure. This danger exists in victims with crushing injuries, severe burns, or heart attacks — anything that breaks cells open so that they spill their K⁺ ions into general circulation.

At the other extreme, severe hypokalemia, caused by any unusual losses of body fluids including prolonged, excessive sweating, can lead to death by heart failure. Both body fluids *and electrolytes* have to be replaced during severe exercise.

■ Experienced backpackers use salt tablets that contain not only NaCl but also some KCl to supply K⁺. Beverages like Gatorade® similarly resupply the body with electrolytes.

The serum levels of sodium ions and potassium ions are regulated in tandem by the kidneys. If the intake of Na⁺ is high, there will be a loss of K⁺ from the body. If the intake of K⁺ is high, there will be a loss of Na⁺ from the body. It's the total positive charge that must be maintained so as to balance the total negative charge. One reason we can't tolerate seawater (3% NaCl) and will die if we drink it is that it upsets the sodium–potassium balance in the body.

TABLE 23.1
Sodium and Potassium Ions in the Human Body

Area	Na⁺	K⁺
Total body	2700–3000 meq	3200 meq
Plasma level	135–145 meq/L	3.5–5.0 meq/L
Intracellular level	10 meq/L	125 meq/L
Mass of 1 meq	23.0 mg	39.1 mg

TABLE 23.2
Calcium and Magnesium Ions in the Human Body

Area	Ca^{2+}	Mg^{2+}
Total body	6×10^4 meq (1.2 kg)	1000 meq (24 g)
Plasma level	4.2–5.2 meq/L	1.5–2.0 meq/L
Intracellular level	a	35 meq/L
Mass of 1 meq	20.0 mg	12.2 mg

a Free in solution in the cytosol, about 10^{-7} mol/L.

Mg^{2+} Is Second to K^+ as a Cation Inside Cells

The normal ranges for the levels of calcium ion and magnesium ion in the body are given in Table 23.2. The level of magnesium ion in the blood is even lower than that of K^+ (Fig. 23.4), so small variations can mean large trouble. Hypomagnesemia, for example, is observed when the kidneys aren't working properly, or in alcoholism, or in untreated diabetes. Some of its signs are muscle weakness, insomnia, and cramps in the legs or the feet. Injections of isotonic magnesium sulfate solution are sometimes used to restore Mg^{2+} to the serum.

On the opposite side, hypermagnesemia can lead to cardiac arrest, and it can be brought on by the overuse of magnesium-based antacids such as milk of magnesia, $Mg(OH)_2$.

Nearly All the Body's Ca^{2+} Is in Bones and Teeth

The Ca^{2+} not in bones and teeth is absolutely vital, as we explained in the previous chapter. It is an important secondary chemical messenger involved in activating enzymes and in muscle contraction.

Hypocalcemia can be brought on by vitamin D deficiency, the overuse of laxatives, an impaired activity of the thyroid gland, and even by hypomagnesemia. Hypercalcemia, on the other hand, is caused by the opposite conditions: an overdose of vitamin D, the overuse of calcium ion-based antacids, or an overactive thyroid. In severe hypercalcemia, the heart functions poorly.

■ It's common among the elderly to suffer the loss of Ca^{2+} from bones, a condition called *osteoporosis*.

Cl^- and HCO_3^- Provide Almost All the Negative Charge to Balance the Cationic Charges in Blood

The chloride ion in blood helps to maintain osmotic pressure relationships, the acid–base balance, and the distribution of water in the body. It has a function in oxygen transport. The bicarbonate ion is the chief acid-neutralizing buffer in blood and the principal form in which waste CO_2 is carried.

In hypochloremia, there is an excessive loss of Cl^- from the blood. For every Cl^- lost, the blood either must lose one (+) charge, like Na^+, or retain one extra (−) charge on some other anion. Electrical neutrality dictates these simple facts. The chief ion retained is HCO_3^-. Because this ion tends to raise the pH of a fluid, *a condition of hypochloremia can cause the pH of blood to increase — alkalosis.*

By the same token, hyperchloremia, a rise in the level of Cl^-, could mean that HCO_3^- has to be dumped, which would mean a loss of base. Thus *hyperchloremia can cause a decrease in blood pH — acidosis.*

■ In blood, the normal range for $[Cl^-]$ is 100 to 106 meq/L (1 meq Cl^- = 35.5 mg).

Fluids That Leave the Blood Must Return in Identical Volume

The blood vessels undergo extensive branching until the narrowest tubes called the capillaries are reached. Blood enters a capillary loop (Fig. 23.7) as arterial blood, but it leaves on the other side of the loop as venous blood.

During the switch, fluids and nutrients leave the blood and move into the interstitial fluids and into the tissue cells themselves. *In the same volume* the fluids must return to the blood, but now they must carry the wastes of metabolism. The rate of this diffusion of fluids throughout the body is sizable, about 25 to 30 L/s. Some fluids return to circulation by way of the lymph ducts, which are thin-walled, closed-end capillaries that bed in soft tissue.

■ The lymph system makes antibodies and it has white cells that help defend the body against infectious diseases.

FIGURE 23.7

The exchange of nutrients and wastes at capillaries. As indicated at the top, on the arterial side of a capillary loop the blood pressure counteracts the pressure from dialysis and osmosis, and fluids are forced to leave the bloodstream. On the venous side of the loop, the blood pressure has decreased below that of dialysis and osmosis, so fluids flow back into the bloodstream. On the top right and the bottom is shown how a normal red cell distorts as it squeezes through a capillary loop. Red cells in sickle-cell anemia do not pass through as smoothly. The bottom drawing also shows how some fluids enter the lymph system.

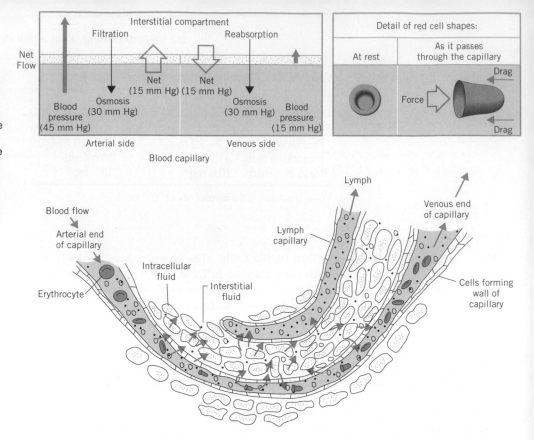

Blood Pressure Overcomes Osmotic Pressure on the Arterial Side of a Capillary Loop   Interstitial fluids have a natural tendency to dialyze into the bloodstream because the blood has the higher colloidal osmotic pressure. On the arterial side of a capillary loop, however, the blood pressure is sufficiently high to overcome this. Water and dissolved solutes are forced, instead, from the blood *into* the interstitial space. While they are there, exchanges of chemicals occur. Nutrients move into cells, and cells get rid of wastes.

Osmotic Pressure Overcomes Blood Pressure on the Venous Side of a Capillary Loop   As blood emerges from the thin constriction of a capillary into the venous side, its pressure drops. Now it is too low to prevent the natural diffusion of fluids back into the bloodstream. By this time, of course, the fluids are carrying waste products. These relationships are illustrated in Figure 23.7. It shows how the colloidal osmotic pressure contributed by the dispersed macromolecules in blood, particularly the albumins, make the difference in determining the direction of diffusion.

Blood Loses Albumins in the Shock Syndrome   When the capillaries become more permeable to blood proteins, as they do in such trauma as sudden severe injuries, major surgery, and extensive burns, the proteins migrate out of the blood. Unfortunately, this protein loss also means the loss of the colloidal osmotic pressure that helps fluids to return from the tissue areas to the bloodstream. As a result, the total volume of circulating blood drops quickly, and this drastically reduces the blood's ability to carry oxygen and to remove carbon dioxide. The drop in blood volume and the resulting loss of oxygen supply to the brain send the victim into traumatic **shock.**

■   The prompt restoration of blood volume is mandatory in the treatment of shock.

Blood Also Loses Proteins in Kidney Disease and Starvation   Sometimes the proteins in the blood are lost at malfunctioning kidneys. The effect, although gradual, is a slow but unremitting drop in the blood's colloidal osmotic pressure. Fluids accumulate in the interstitial regions. Because this takes place more slowly and water continues to be ingested, there is no sudden drop in blood volume as in shock. The victim appears puffy and water-logged, a condition called **edema.**

Edema can also appear at one stage of starvation, when the body has metabolized its circulating proteins to make up for the absence of dietary proteins.

Any obstruction in the veins can also cause edema, as in varicose veins and certain forms of cancer. Now it is the venous blood pressure that rises, creating a back pressure that reduces the rate at which fluids can return to circulation from the tissue areas. The localized swelling that results from a blow is a temporary form of edema caused by injuries to the capillaries.

■ Greek, *oidema,* swelling.

23.3

BLOOD AND THE EXCHANGE OF RESPIRATORY GASES

The binding of oxygen to hemoglobin is allosteric, and it is affected by the pH, pCO$_2$, and the pO$_2$ of the blood.

The carrier of oxygen in blood is **hemoglobin,** a complex protein found inside erythrocytes. It consists of four subunits, each with one molecule of heme, an organic group that holds an iron(II) ion, the actual oxygen-binding unit. Two of the subunits are identical and have the symbol α. The other two are also identical and have the symbol β.

When hemoglobin is oxygen free, it is called *deoxyhemoglobin;* in this state the molecule has a cavity in which a small organic anion nestles. This is the **2,3-bisphosphoglycerate** ion, called simply **BPG,** and it has an important function in oxygen transport.[2]

■ Each red cell carries about 2.8×10^8 molecules of hemoglobin.

■
$$^-O_3POCH_2CHCO_2^-$$
$$|$$
$$OPO_3^{2-}$$
BPG (2,3-bisphosphoglycerate)

The Subunits of Hemoglobin Cooperate in Its Oxygenation We define the **oxygen affinity** of the blood as the percent to which the blood has all of its hemoglobin molecules saturated with oxygen. A fully laden hemoglobin molecule carries four oxygen molecules and is called **oxyhemoglobin.** For the maximum efficiency in moving oxygen from the lungs to tissues that need it, all hemoglobin molecules should leave the lungs in the form of fully loaded oxyhemoglobin. Let's see what factors ensure this.

First, the partial pressure of oxygen is higher in the lungs than anywhere else in the body; pO$_2$ is 100 mm Hg in freshly inhaled air in alveoli and only about 40 mm Hg in oxygen-depleted tissues. Because of this partial pressure gradient, oxygen naturally migrates from the lungs into the bloodstream. It's as though the higher partial pressure *pushes* oxygen into the blood. Second, newly arrived oxygen actually reacts with hemoglobin, binding to its Fe^{2+} ions to form oxyhemoglobin, so this tends to *pull* oxygen into the blood.

An allosteric effect also helps to load hemoglobin with oxygen. Figure 23.8 shows a plot of oxygen affinity versus the oxygen partial pressure. It has the sigmoid shape that in Chapter 22

■ The structure in Figure 21.9, page 599, is actually deoxyhemoglobin, and the central cavity for BPG was pointed out.

■ We first learned about the allosteric effect on page 625.

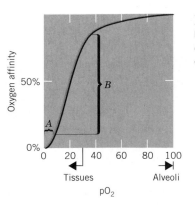

FIGURE 23.8
Hemoglobin–oxygen dissociation curve. Regions *A* and *B* are discussed in the text.

[2] The name 2,3-diphosphoglycerate (DPG) has been supplanted by 2,3-bisphosphoglycerate (BPG). "Diphospho-" signifies a diphosphate ester, an ester of diphosphoric acid (page 491). "Bisphospho" signifies two ("bis") monophosphate ester groups.

we learned to associate with the allosteric effect among enzymes. At low values of pO_2, in region A of the plot, the ability of the blood to take up oxygen rises only slowly with increases in pO_2. But eventually the oxygen affinity takes off, and rises very steeply with still more increases in pO_2, in region B of the plot. Eventually, the oxygen affinity starts to level off. It almost seems that a small "molecular dam" thwarts the efforts of oxygen molecules to be joined to hemoglobin at low partial pressures of oxygen.

■ This natural resistance is needed in working cells where we want no restrictions on the deoxygenation of blood.

What is thought to be happening is as follows. We'll represent deoxyhemoglobin by structure **1**, below. Each circle in **1** is a polypeptide subunit with its heme unit but without oxygen. The oval figure centered within the structure represents one BPG anion. When the first O_2 molecule manages to bind ($1 \rightarrow 2$), it induces a change in the shape of the affected subunit, which we have represented as a change from a circle to a square:

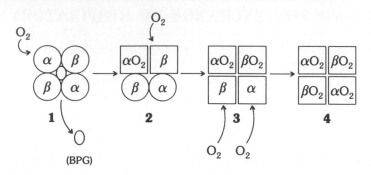

■ Carbon monoxide binds 150 to 200 times more strongly to hemoglobin than does oxygen and thus prevents the oxygenation of hemoglobin and causes internal suffocation.

This change tends to squeeze out the BPG unit, and this "breaks the dam." The next oxygen molecule enters more readily ($2 \rightarrow 3$). As its affected subunit changes its shape, the remaining two subunits allosterically change their shapes and become very receptive to the third and fourth molecules of oxygen. These two flood into this hemoglobin molecule ($3 \rightarrow 4$) far more readily than either would bind to a completely deoxygenated hemoglobin molecule. Thus if a hemoglobin molecule accepts just one oxygen molecule, it's almost certain that it will soon accept three more to become fully oxygenated. Few if any partially oxygenated hemoglobin molecules leave the lungs.

For the sake of the remaining discussion, we'll simplify what we have just described by letting the symbol HHb represent an entire hemoglobin molecule. The first H in HHb stands for a potential hydrogen ion, and we overlook the fact that more than one is actually present in hemoglobin. (We're now also overlooking the fact that *four* molecules of O_2 bind to one of hemoglobin.) With this in mind, we can represent the oxygenation of hemoglobin as the *forward* reaction in the following equilibrium where oxyhemoglobin is represented as the anion, HbO_2^-:

■ About 20% of a smoker's hemoglobin is more or less permanently tied up by carbon monoxide.

$$\underset{\text{Hemoglobin}}{HHb} \; + \; O_2 \rightleftharpoons \underset{\text{Oxyhemoglobin}}{HbO_2^-} \; + \; H^+ \qquad (23.1)$$

Two facts indicated by this equilibrium are that HHb is a weak acid and that it becomes a stronger acid as it becomes oxygenated. Being stronger, it is produced in its ionized state as HbO_2^- and H^+. *The presence of H^+ in equilibrium 23.1 means that the equilibrium can be shifted one way or another simply by changing the pH,* a fact of enormous importance at the molecular level of life, as we'll see.

To understand the oxygenation of hemoglobin, we have to see how various stresses shift equilibrium 23.1 to the *right* in the lungs. One stress, as we've already noted, is the relatively high value of pO_2 (100 mm Hg) in the alveoli. This stress acts on the left side of 23.1, so it helps to shift the equilibrium to the right.

$$\underset{\substack{\text{In the} \\ \text{red cell}}}{HHb} + \underset{\substack{\text{From} \\ \text{air}}}{O_2} \longrightarrow \underset{\substack{\text{In the} \\ \text{red cell}}}{HbO_2^-} + \underset{\substack{\text{In the} \\ \text{red cell}}}{H^+}$$

Another stress that isn't evident from equation 23.1 is the removal of H^+ *as it forms*. The red cell in the lungs is carrying waste HCO_3^-. The newly forming H^+ ions are therefore promptly equilibrated with HCO_3^- and CO_2 by the help of carbonic anhydrase. (We write only the forward reaction of this equilibrium here because this is how the equilibrium shifts when both H^+ and HCO_3^- are high.)

$$H^+ + HCO_3^- \xrightarrow{\text{carbonic anhydrase}} CO_2 + H_2O \qquad (23.2)$$

In the   In the   In the
red cell   red cell   red cell

This switch from the appearance of H^+ as a product to its disappearance as a reactant is called the **isohydric shift.**

We see here one of the beautiful examples of coordinated activity in the body, the coupling of the uptake of oxygen to the release of the carbon dioxide to be exhaled. The neutralization of the H^+ ions produced from the uptake of O_2 produces CO_2. The loss of CO_2 from the red cell by exhaling pulls this and all previous equilibria to the right in the lungs.

$$CO_2 \xrightarrow{\text{exhaling}} CO_2$$

In the   In exhaled
red cell   air

Thus the uptake of O_2 as hemoglobin oxygenates, which pushes the equilibria to the right, simultaneously produces a chemical (CO_2) whose loss pulls the same equilibria to the right.

Let's now see how this waste is picked up at cells that have produced it and is carried to the lungs; and let's also see how this works cooperatively with the *release* of oxygen at cells that need it.

The Hemoglobin Subunits and BPG Cooperate in the Deoxygenation of Oxyhemoglobin in Metabolizing Tissues

Consider, now, a tissue that has done some chemical work, used up some oxygen, and made some waste carbon dioxide. When a fully oxygenated red cell arrives in such a tissue, some of the events we just described reverse themselves.

We can think of this reversal as beginning with the diffusion of waste CO_2 from the tissue into the blood. An impetus for this diffusion is the higher pCO_2 (50 mm Hg) in active tissue versus its value in blood (40 mm Hg). Once the CO_2 arrives in the blood, it moves inside a red cell where it encounters carbonic anhydrase. Equation 23.2 is therefore run in reverse. It was part of the equilibrium managed by carbonic anhydrase, and it shifts to the left as more and more CO_2 arrives. In other words, the equilibrium now shifts to *make* HCO_3^-.

$$H_2O + CO_2 \xrightarrow{\text{carbonic anhydrase}} HCO_3^- + H^+ \qquad (23.2\text{—reversed})$$

From the   In the
working   red cell
tissue

Of course, this generates hydrogen ions, and if you'll look back to Equation 23.1, which is also part of an equilibrium, you will see that an increase in the level of H^+ (caused by the influx of waste CO_2) can only make Equation 23.1 run in reverse.

$$H^+ + HbO_2^- \longrightarrow HHb + O_2 \qquad (23.1\text{—reversed})$$

Just made   In the   In the   Will diffuse
in the red   red cell;   red cell   into the
cells at   just ar-   tissue needing
working   rived at   oxygen
tissue   tissue

This reaction, another isohydric shift, not only neutralizes the acid generated by the arrival of waste CO_2, but also helps to force oxyhemoglobin to give up its oxygen. Notice the coopera-

---

■ The stimulation of HHb to bind O_2 caused by the removal of H^+ is called the **Bohr effect** after Christian Bohr, a Danish scientist (and the father of nuclear physicist Niels Bohr).

■ As we learned in the last chapter, carbonic anhydrase is one of the body's fastest working enzymes, and it has to be fast because the red cells are always on the move.

■ CO_2 molecules diffuse in body fluids 30 times more easily than O_2 molecules, so the partial pressure gradient for CO_2 need not be as steep as that for O_2.

tion. The tissue that needs oxygen has made CO_2 and, hence, it has indirectly made the H^+ that is required to release this needed oxygen from newly arrived HbO_2^-.

The deoxygenation of HbO_2^- is also aided by the BPG anions that were pushed out when HbO_2^- formed. These anions are still inside the red cell, and as soon as an O_2 molecule leaves oxyhemoglobin, a BPG anion starts to move back in. The changes in shapes of the hemoglobin subunits now operate in reverse, and all oxygen molecules smoothly leave. It's all or nothing again, and the efficiency of the unloading of oxygen is so high that if one O_2 molecule leaves, the other three follow essentially at once. BPG helps this to happen. Partially deoxygenated hemoglobin units do not slip through and go back to the lungs.

■ The high negative charge on BPG keeps it from diffusing through the red cell's membrane.

BPG and Hemoglobin Levels Are Higher in People Living at High Altitudes

It's interesting that those who live and work at high altitudes, such as the populations in Nepal in the Himalayan Mountains or the people in the Andes Mountains in Bolivia, have 20% higher levels of BPG in their blood and more red blood cells than those who live at sea level.

The extra red cells give them more hemoglobin to help them carry more oxygen per milliliter of blood, and the extra BPG increases the efficiencies of both loading and unloading oxygen. When lowlanders take trips to high altitudes, their bodies start to build more red cells and to make more BPG so that they can function better where the partial pressure of the atmospheric oxygen is lower. Those who patiently wait during the few days that it takes for these events to occur before they set off on strenuous backpacking expeditions are less likely to suffer high altitude sickness, a condition that can cause death.

■ No conditioning at a low altitude can get the cardiovascular system ready for a low pO_2 at a high altitude.

To summarize the chemical reactions we have just studied, we can write the following equations. The cancel lines show how we can arrive at the overall net results.

Oxygenation:
(These reactions occur in the lungs.)

$$\text{HHb} + O_2 \longrightarrow HbO_2^- + \cancel{H^+}$$

In the red cell | From air | Will go in red cell to tissue | In the red cell

■ CA is carbonic anhydrase.

$$\cancel{H^+} + HCO_3^- \xrightarrow{\text{CA}} CO_2 + H_2O$$

Just made | In red cell (but from tissues) | In red cell (but will be exhaled)

■ **Net effect of oxygenating hemoglobin:**

$$\textbf{HHb} + \textbf{O}_2 + \textbf{HCO}_3^- \longrightarrow \textbf{HbO}_2^- + \textbf{CO}_2 + \textbf{H}_2\textbf{O} \qquad (23.3)$$

In the red cell | From air | In red cell (but from tissues) | Will go in red cell to tissue | Will leave the lungs in exhaled air

Deoxygenation:
(These reactions occur wherever tissues are low in oxygen.)

$$CO_2 + H_2O \xrightarrow{\text{CA}} HCO_3^- + \cancel{H^+}$$

Waste from tissues | In red cell (in blood still within tissue but will go to the lungs) | In the red cell

$$\cancel{H^+} + HbO_2^- \longrightarrow \text{HHb} + O_2$$

Just made | In red cell (in blood within tissues) | In red cell (will return to the lungs) | Goes into tissue needing it

■ **Net effect of deoxygenating oxyhemoglobin:**

$$\textbf{CO}_2 + \textbf{H}_2\textbf{O} + \textbf{HbO}_2^- \longrightarrow \textbf{HHb} + \textbf{HCO}_3^- + \textbf{O}_2 \qquad (23.4)$$

Waste from tissues | In red cell (in blood with tissues) | In red cell (will go to the lungs) | Goes in blood to the lungs | Goes into tissue

These summarizing equations omit two features. They do not show the importance of BPG and of the concentration of H^+ in both the loading and the unloading of oxygen. Concerning the concentration of H^+, Figure 23.9 shows the plots of oxygen affinity versus pO_2 at two different values of pH, one relatively low (pH 7.2) compared with the normal value of 7.35, and the other relatively high (pH 7.6). You may recall that the pO_2 in the vicinity of oxygen-starved cells is around 30 to 40 mm Hg. Notice in Figure 23.9 that in this range of the partial pressure of oxygen, the blood's ability to hold oxygen is much less at the lower pH of 7.2 than it is at a pH of 7.6. In actively metabolizing cells, there is a localized drop in pH caused chiefly by the presence of the CO_2 that these cells have made. This drop in pH caused by CO_2 cannot help but to assist in the deoxygenation of HbO_2^-.

Thus precisely where O_2 should be unloaded there is a chemical signal (a lower pH) that makes it happen. It's an altogether beautiful example of how a set of interrelated chemical equilibria shift in just the directions that are required for health and life. Figure 23.9 will also help us understand in what ways both acidosis and alkalosis are serious threats.

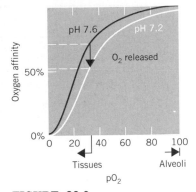

FIGURE 23.9
Hemoglobin–oxygen dissociation curves at two different values of the pH of blood

Some CO_2 Is Carried to the Lungs on Hemoglobin

There is another chemical reaction involving waste carbon dioxide that we must now mention. Not all the CO_2 made by metabolizing cells winds up as HCO_3^-. Some CO_2 reacts with the hemoglobin that has just been freed by deoxygenation:

$$CO_2 + HHb \longrightarrow Hb-CO_2^- + H^+ \qquad (23.5)$$

Waste	Just released	Carbamino-
from	by deoxygen-	hemoglobin
tissue	ation of HbO_2^-	(in red cells)

This is actually the forward reaction of an equilibrium. The product, $Hb-CO_2^-$, is called **carbaminohemoglobin,** and it is one form in which some of the waste CO_2 travels in the blood back to the lungs.

Notice in Equation 23.5 that this reaction also produces H^+, just as did the reaction of water and waste CO_2. Thus, whether waste CO_2 is changed to HCO_3^- and H^+ or to $Hb-CO_2^-$ and H^+, either fate helps to generate hydrogen ions that are needed to react with HbO_2^- and make it unload O_2.

When the red cell reaches the lungs, where H^+ will now be *generated* by

$$HHb + O_2 \longrightarrow HbO_2^- + H^+ \qquad (23.1, \text{again})$$

the reaction of Equation 23.5 will be forced to shift into reverse, because the added H^+ provides this kind of stress. Of course, this releases the CO_2 where it can be exhaled, and it gets hemoglobin ready to take on more oxygen.

Chloride Ion Is Also Needed to Deoxygenate Hemoglobin

Still another factor we have not mentioned thus far and that helps to unload oxygen from oxyhemoglobin is the chloride ion. Hemoglobin, HHb, binds chloride ion, and it binds it better than does oxyhemoglobin. The following equilibrium exists along with all the others.

$$HHb(Cl^-) + O_2 \longrightarrow HbO_2^- + Cl^- + H^+ \qquad (23.6)$$

Therefore to *unload* oxygen (do the reverse of 23.6), chloride ion must be available *inside* the red cell. The red cell obtains it by a mechanism called the *chloride shift*. Let's see how it occurs.

As chloride ions are drawn into the red cell to react with some of the HHb that is released in active tissue, negative ions have to move out so that there is electrical charge balance. The negative ions that leave are the newly forming bicarbonate ions. From 60% to 90% of all waste CO_2 returns to the lungs as the bicarbonate ion, but most of this makes the trip *outside* red cells.

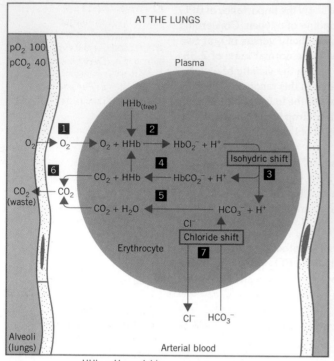

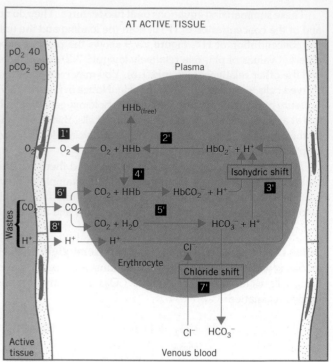

HHb = Hemoglobin
HbO$_2^-$ = oxyhemoglobin [actually, Hb(O$_2$)$_4$]

HbCO$_2^-$ = Carbaminohemoglobin [actually, Hb(CO$_2$)$_4$$^{n-}$
where $n-$ = 1– to 4–]

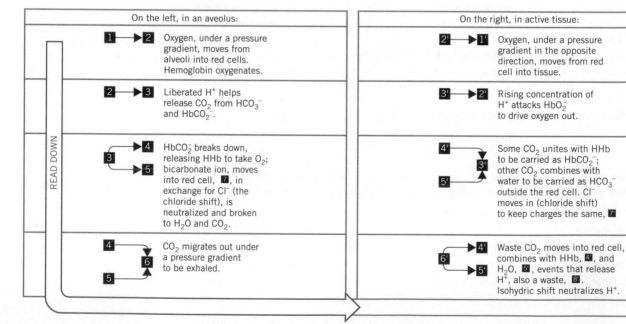

FIGURE 23.10
Oxygen and carbon dioxide exchange in the blood

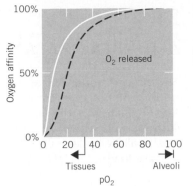

FIGURE 23.11
The myoglobin–oxygen dissociation curve is a solid line. The dashed line is the curve for hemoglobin. Over the entire range of pH in tissues that might need oxygen, the oxygen affinity of myoglobin is greater than that of hemoglobin.

For every chloride ion that enters the red cell, one HCO_3^- ion leaves, and this switch is called the **chloride shift.** When the red cell gets back to the lungs, the various new chemical stresses make all the equilibria, including the chloride shift, run in reverse.

If you're bewildered by all these equilibria and how they are made to shift in the correct directions, you're almost certainly not alone among your classmates. This isn't easy material, but it is so much at the heart of so many aspects of health and one's ability to have an active life that the effort to master it is very worthwhile. As you make this effort, get the key equilibria down and memorized. For each one ask: What are the stresses that can make it shift, and where in the body is each stress important, in the lungs or at actively metabolizing cells?

The key stresses are the following: the relative partial pressures of O_2, the relative partial pressures of CO_2, and the changes in the levels of H^+ caused by the influx of CO_2 or by its loss by means of exhaling.

After you have studied the various equilibria from the stress point of view, and you can write all the equilibria and discuss the influence of various stresses, then you might find Figure 23.10 a useful way to review. Follow the direction of the large U-shaped arrow that curves around the legend, and use the boxed numbers in the legend to follow the events in the figure. Notice particularly that the reactions that occur in the red cell when it is in metabolizing tissue are just the reverse of those that happen when the cell is in the lungs.

Myoglobin Binds O_2 More Strongly than Hemoglobin Myoglobin is a heme-containing protein in red muscle tissue such as heart muscle. Its function is to bind and store oxygen for the needs of such tissue. Unlike hemoglobin, myoglobin has only one polypeptide unit and only one heme unit per molecule. Moreover, there is no allosteric effect when it binds oxygen, as the shape of the myoglobin–oxygen dissociation curve indicates (see Fig. 23.11).

Myoglobin's oxygen affinity is greater than that of hemoglobin, especially in the range of pO_2 associated with actively metabolizing tissues. Consequently, *myoglobin is able to take oxygen from oxyhemoglobin:*

$$HbO_2^- + HMy \longrightarrow HHb + MyO_2^-$$

This ability is vital to heart muscle which, as much as the brain, must have an assuredly continuous supply of oxygen. When oxymyoglobin, MyO_2^-, gives up its oxygen for the cell's needs, it can at once get a fresh supply from the circulating blood. Not only does this cell now have CO_2 and H^+ available to deoxygenate HbO_2^-, it also has the superior oxygen affinity of its own myoglobin to draw more O_2 into the cell.

Fetal Hemoglobin Binds Oxygen More Strongly than Adult Hemoglobin The hemoglobin in a fetus is slightly different from that of an adult, and it has a higher oxygen affinity than adult hemoglobin. This helps to ensure that the fetus successfully pulls oxygen from the mother's oxyhemoglobin to satisfy its own needs.

■ For those entering careers in nursing and respiratory therapy, there is no other single topic of such career-lasting importance than the chemistry of respiration and its associated electrolyte balance.

■ For those who simply want to understand the respiratory demands (and limitations) of active sports, including skiing at altitude, the chemistry of respiration is the key.

■ HMy = myoglobin

23.4

ACID–BASE BALANCE OF THE BLOOD

The proper treatment of acidosis or of alkalosis depends on knowing if the underlying cause is a metabolic or a respiratory disorder.

Acid–base balance in the blood exists when the pH of blood is in the range of 7.35 to 7.45. A decrease in pH, acidosis, or an increase, alkalosis, is serious and requires prompt attention, because all the equilibria that involve H^+ in the oxygenation or the deoxygenation of blood are sensitive to pH. If the pH falls below 6.8 or rises above 7.8, life is not possible.

■ Acidosis is sometimes called *acidemia,* and alkalosis is called *alkalemia.*

Disturbances in Either Metabolism or Respiration Can Upset the Blood's Acid–Base Balance In general, acidosis results from either the retention of acid or the loss of base by the body, and these can be induced by disturbances in either metabolism or respiration. Similarly, alkalosis results either from the loss of acid or from the retention of base, and some disorder in either metabolism or respiration can be the underlying cause.

A malfunction in respiration can be caused by any kind of injury to the *respiratory centers.* These are units in the brain that sense changes in the pH and pCO_2 of the blood and instruct the lungs to breathe either more rapidly or more slowly. Another cause of a malfunction in respiration is any kind of injury or disease of the lungs.

What we'll do in this section is study four situations, metabolic and respiratory acidosis as well as metabolic and respiratory alkalosis. We will learn how the values of pH, pCO_2, and serum $[HCO_3^-]$ change in each situation. (Normal values are given in the margin.)

■ Normal values (arterial blood):

pH = 7.35 − 7.45
pCO_2 = 35–45 mm Hg
$[HCO_3^-]$ = 19–24 meq/L
(1.0 meq HCO_3^- = 61 mg HCO_3^-)

Metabolic Acidosis Receives a Respiratory Compensation, Hyperventilation In **metabolic acidosis,** the lungs and the respiratory centers are working, and the problem is metabolic. Acids are being produced faster than they are neutralized, or they are being exported too slowly.

Excessive loss of base, such as from severe diarrhea, can also result in metabolic acidosis. (In diarrhea, the alkaline fluids of the duodenum leave the body, and as base migrates to replace them, there will be a depletion of base somewhere else, such as in the blood, at least for a period of time.)

As the pH of the blood falls and the molar concentration of H^+ rises, there are parallel *but momentary* increases in the values of pCO_2 and $[HCO_3^-]$. The value of pCO_2 starts to increase because the carbonate buffer, working hard to neutralize the extra H^+, manufactures CO_2:

$$H^+ \; + \; HCO_3^- \longrightarrow H_2O + CO_2$$
Produced
by acidosis

The kidneys work harder during this situation to try to keep up the supply of HCO_3^-.

The chief compensation for metabolic acidosis, however, involves the respiratory system. The respiratory centers, which are sensitive to changes in pCO_2, instruct the lungs to blow the CO_2 out of the body. The lungs, in other words, hyperventilate. As the equation given just above shows, the loss of each molecule of CO_2 means a net neutralization of one H^+ ion.

Hyperventilation, however, is overdone. So much CO_2 is blown out that pCO_2 actually decreases; a low arterial pCO_2 is called **hypocapnia.** Thus as the blood pH decreases, so too do the values of pCO_2 (from hyperventilation) and $[HCO_3^-]$ (from the reaction with H^+). We can summarize the range of a number of clinical situations that involve metabolic acidosis as follows.

Clinical Situations of Metabolic Acidosis

Lab Results:	pH↓ (7.20); pCO_2↓ (30 mm Hg); [HCO_3^-]↓ (12 meq/L)
Typical Patient:	An adult male comes to the clinic with a severe infection. He does not know that he has diabetes.
Range of Causes:	Diabetes mellitus; severe diarrhea (with loss of HCO_3^-); kidney failure (to export H^+ or to make HCO_3^-); prolonged starvation; severe infection; aspirin overdose, alcohol poisoning.
Symptoms:	Hyperventilation (because the respiratory centers have told the lungs to remove excess CO_2 from the blood); increased urine output (to remove H^+ from the blood); thirst (to replace water lost as urine); drowsiness; headache; restlessness; disorientation.
Treatment:	If the kidneys function, use isotonic HCO_3^- intravenously to restore HCO_3^- level, thereby neutralizing H^+ and raising pCO_2. In addition, restore water. In diabetes, use insulin therapy. If the kidneys do not function, hemodialysis must be tried.

■ (↓) means a decrease from normal and (↑) means an increase. Some typical values are in parentheses. Note that some changes do not necessarily bring values outside the normal ranges.

Respiratory Acidosis Is Compensated by a Metabolic Response In **respiratory acidosis,** either the respiratory centers or the lungs have failed, and the lungs are hypoventilating *because they cannot help it.* The blood now cannot help but retain CO_2. An increase in arterial pCO_2 is called **hypercapnia.** The retention of CO_2 functionally means the retention of acid, because CO_2 can neutralize base by the following equation.

$$HO^-(aq) + CO_2(aq) \longrightarrow HCO_3^-(aq) \qquad (23.7)$$

The decrease in the level of base lowers the pH of the blood and gives rise to respiratory acidosis.

Clinical Situations of Respiratory Acidosis

Lab Results:	pH↓ (7.21); pCO_2↑ (70 mm Hg); [HCO_3^-]↑ (27 meq/L)
Typical Patient:	Chain smoker with emphysema or anyone with chronic obstructive pulmonary disease.
Range of Causes:	Emphysema, severe pneumonia, asthma, anterior poliomyelitis, or any cause of shallow breathing such as an overdose of narcotics, barbiturates, or general anesthesia; severe head injury.
Symptoms:	Shallow breathing (which is involuntary)
Treatment:	Underlying problem must be treated; possibly intravenous sodium bicarbonate; possibly hemodialysis.

The body responds metabolically as best it can to respiratory acidosis by using HCO_3^- to neutralize the acid, by making more HCO_3^- in the kidneys, and by exporting H^+ via the urine.

Metabolic Alkalosis Also Receives a Respiratory Compensation, Hypoventilation In **metabolic alkalosis,** the system has lost acid, or it has retained base (HCO_3^-), or it has been given an overdose of base (e.g., antacids). Metabolic alkalosis can also be caused by a kidney-associated decrease in the serum levels of K^+ or Cl^-. The loss of these ions means the retention of Na^+ and HCO_3^- ions, because these work in tandem and oppositely. The loss of acid could be from prolonged vomiting, which removes the gastric acid. This is followed by an effort to borrow serum H^+ to replace it, and the pH of the blood increases.

Whatever the cause, the respiratory centers sense an increase in the level of base in the blood (as the level of acid drops), and they instruct the lungs to retain the most readily available neutralizer of base it has, namely CO_2, which removes OH^- according to equation 23.7. To help retain CO_2 to neutralize base, the lungs hypoventilate. Thus metabolic alkalosis leads to hypoventilation.

■ Compensation by hypoventilation is obviously limited by the fundamental need of the body for some oxygen.

Notice carefully that hypoventilation alone cannot be used to tell whether the patient has metabolic *alkalosis* or respiratory *acidosis*. Either condition means hypoventilation. But one condition, respiratory acidosis, could be treated by intravenous sodium bicarbonate, a base. This would aggravate metabolic alkalosis.

You can see that the lab data on pH, pCO_2, and $[HCO_3^-]$ must be obtained to determine which condition is actually present. Otherwise, the treatment used could be just the opposite of what should be done.

■ An overdose of "bicarb" (NaHCO₃), from a too aggressive use of this home remedy for "heartburn," can cause metabolic alkalosis.

Clinical Situations of Metabolic Alkalosis

Lab Results:	pH↑ (7.53); pCO_2↑ (56 mm Hg); $[HCO_3^-]$↑↑ (45 meq/L)
Typical Patient:	Postsurgery patient with persistent vomiting.
Range of Causes:	Prolonged loss of stomach contents (vomiting or nasogastric suction); overdose of bicarbonate or of medications for stomach ulcers; severe exercise, or stress, or kidney disease (with loss of K^+ and Cl^-); overuse of a diuretic.
Symptoms:	Hypoventilation (to retain CO_2); numbness, headache, tingling; possibly convulsions.
Treatment:	Isotonic ammonium chloride (a mild acid), intraveneously with great care; replace K^+ loss.

■ Ammonium ion acts as a neutralizer as follows:

$$NH_4^+ + OH^- \longrightarrow NH_3 + H_2O$$

■ Tissue that gets too little O_2 is in a state of *hypoxia*. If it gets none at all, it is in a state of *anoxia*.

Respiratory Alkalosis Is Compensated Metabolically by a Reduced Bicarbonate Level In **respiratory alkalosis,** the body has lost acid usually by some involuntary hyperventilation such as hysterics, prolonged crying, or overbreathing at high altitudes, or by the mismanagement of a respirator. The respiratory centers have lost control, and the body expels CO_2 too rapidly. The loss of CO_2 means the loss of a base neutralizer from the blood. Hence, the level of base rises; the pH increases. To compensate, the kidneys excrete base, HCO_3^-, so the serum level of HCO_3^- decreases.

Extreme respiratory alkalosis can occur to high mountain climbers, like climbers of Mount Everest (8848 m, 29,030 ft). At its summit, the barometric pressure is 253 mm Hg and the pO_2 of the air is only 43 mm Hg (as compared to 149 mm Hg at sea level). Hyperventilation brings a climber's arterial pCO_2 down to only 7.5 mm Hg (compared to a normal of 40 mm Hg) and the blood pH is above 7.7!

Clinical Situations of Respiratory Alkalosis

Lab Results:	pH↑ (7.56); pCO_2↓ (23 mm Hg); $[HCO_3^-]$↓ (20 meq/L)
Typical Patient:	Someone nearing surgery and experiencing anxiety.
Range of Causes:	Prolonged crying; rapid breathing at high altitudes; hysterics; fever; disease of the central nervous system; improper management of a respirator.
Symptoms:	Hyperventilation (that can't be helped); numbness, headache, tingling. Convulsions may occur.
Treatment:	Rebreathe one's own exhaled air (by breathing into a sack); administer carbon dioxide; treat underlying causes.

Take careful notice that hyperventilation alone cannot be used to tell what the condition is. Either metabolic *acidosis* or respiratory *alkalosis* is accompanied by hyperventilation, but the treatments are opposite in nature.

■ People working in emergency care situations get the requisite lab data rapidly, and they must be able to interpret the data on the spot.

Combinations of Primary Acid–Base Disorders Are Possible We have just surveyed the *four primary acid–base disorders.* Combinations of these are often seen, and healthcare professionals have to be alert to the ways in which the lab data vary in such combinations. Someone with diabetes, for example, might also suffer from an obstructive pulmonary disease. Diabetes causes metabolic acidosis and a *decrease* in $[HCO_3^-]$. The pulmonary disease causes respiratory acidosis with an *increase* in $[HCO_3^-]$. In combination, then, the lab data on bicarbonate level will not be in the expected pattern for either. We will not carry the study of such complications further. We mention them only to let you know that they exist. There are standard ways to recognize them.[3]

23.5

BLOOD AND THE FUNCTIONS OF THE KIDNEYS

Both filtration and chemical reactions in the kidneys help to regulate the electrolyte balance of the blood.

Diuresis is the formation of urine in the kidneys, and it is an integral part of the body's control of the electrolyte and buffer levels in blood.

Urea Is the Chief Nitrogen Waste Exported in the Urine

Huge quantities of fluids leave the blood by diffusion each day at the hundreds of thousands of filtering units in each kidney. Substances in solution but not those in colloidal dispersions (e.g., proteins) leave in these fluids. Then active transport processes in kidney cells pull all of any escaped glucose, any amino acids, and most of the fluids and electrolytes back into the blood. Most of the wastes are left in the urine that is being made.

■ The net urine production is 0.6 to 2.5 L/day.

Urea is the chief nitrogen waste (30 g/day), but creatinine (1 to 2 g/day), uric acid (0.7 g/day), and ammonia (0.5 g/day) are also excreted with the urine. If the kidneys are injured or diseased and cannot function, wastes build up in the blood, which leads to a condition known as *uremic poisoning.*

■ *Ur-,* of the urine; *-emia,* of the blood. *Uremia* means substances of the urine present in the blood.

The Hormone Vasopressin Helps Control Water Loss

A nonapeptide hormone, **vasopressin,** instructs the kidneys to retain or excrete water and thus helps to regulate the overall levels of solutes in blood. The hypophysis, where vasopressin is made, releases it when the osmotic pressure of blood increases by as little as 2%. At the kidneys, vasopressin promotes the reabsorption of water, and therefore it is often called the *antidiuretic hormone* or ADH.

■ The monitors in the hypophysis of the blood's osmotic pressure are called *osmoreceptors.*

An osmotic pressure that is higher than normal (hypertonicity) means a higher concentration of solutes and colloids in blood. The released vasopressin therefore helps the blood to retain water and thus keeps the solute levels from going still higher. In the meantime, the thirst mechanism is stimulated to bring in water to dilute the blood.

Conversely, if the osmotic pressure of blood decreases (becomes hypotonic) by as little as 2%, the hypophysis retains vasopressin. None reaches the kidneys, so the water that has left the bloodstream at the glomeruli does not return as much. Remember that a low osmotic pressure means a low concentration of solutes, so the absence of vasopressin at the kidneys when the blood is hypotonic lets urine form. This reduces the amount of water in the blood and thereby raises the concentrations of its dissolved matter. You can see that with the help of vasopressin a normal individual can vary the intake of water widely and yet preserve a stable, overall concentration of substances in blood.

■ In *diabetes insipidus,* vasopressin secretion is blocked, and unchecked diuresis can make from 5 to 12 L of urine a day.

■ The inositol phosphate system (Figure 22.12, page 635) mediates the chemical signals of vasopressin.

The Hormone Aldosterone Helps the Blood Retain Sodium Ion

The adrenal cortex makes **aldosterone,** a hormone that works to stabilize the sodium ion level of the blood. This steroid hormone is secreted if the blood's sodium ion level drops. When aldosterone arrives at the kidneys, it initiates reactions that make sodium ions that have left the blood return again. Of course, to keep things isotonic in the blood, the return of sodium ions also requires the return of water.

---

[3] See, for example, H. Valtin and F. J. Gennari, *Acid–Base Disorders, Basic Concepts and Clinical Management,* 1987. Little, Brown and Company, Boston.

Conversely, if the sodium ion level of the blood increases, then aldosterone is not secreted, and sodium ions that have migrated out of the blood at a filtering unit are permitted to stay out. They remain in the urine being made, together with some extra water.

The Kidneys Make HCO₃⁻ for the Blood's Buffer System

The Kidneys Make HCO_3^- for the Blood's Buffer System We have seen that breathing is the body's most direct means of controlling acid as it removes or retains CO_2. The kidneys are the body's means of controlling base, as they make or remove HCO_3^-.

The kidneys also adjust the blood's levels of HPO_4^{2-} and $H_2PO_4^-$ the anions of the phosphate buffer. Moreover, when acidosis develops, the kidneys can put H^+ ions into the urine. Some neutralization of these ions by HPO_4^{2-} and by NH_3 takes place, but the urine becomes definitely more acidic as acidosis continues, as we've mentioned before.

Figure 23.12 shows the various reactions that take place in the kidneys, particularly during acidosis. (The numbers in the following boxes refer to this figure.) The breakdown of metabolites, $\boxed{1}$, makes carbon dioxide, which enters the equilibrium whose formation is catalyzed by carbonic anhydrase, $\boxed{2}$. The ionization of carbonic acid, $\boxed{3}$, makes both bicarbonate ion and hydrogen ion. The bicarbonate ion goes into the bloodstream, $\boxed{4}$, but the hydrogen ion is put into the tubule, $\boxed{5}$, where urine is accumulating. This urine already contains sodium ions and monohydrogen phosphate ions, but to make step $\boxed{5}$ possible, *some* positive ion has to go with the HCO_3^-. Otherwise, there would be no net electrical balance. The kidneys have the ability to select Na^+ to go with HCO_3^- at $\boxed{4}$. The kidneys can make Na^+ travel one way and H^+ the other. Newly arrived H^+ can be buffered by HPO_4^{2-} in the developing urine, $\boxed{6}$. Moreover, the kidneys have an ability not generally found in other tissues to synthesize ammonia and use it to neutralize H^+, $\boxed{7}$. Thus the ammonium ion also appears in the urine.

The Kidneys Excrete Organic Anions

The Kidneys Excrete Organic Anions When acidosis has a metabolic origin, the serum level of the anions of organic acids increases. Organic acids are made at accelerated rates in metabolic disorders, like diabetes or starvation, and are the chief cause of the pH change in metabolic acidosis. The base in the blood buffer has to neutralize them.

The kidneys let organic anions stay in the urine, but only by letting increasing quantities of water stay, too. There is a limit to how concentrated the urine can become, so as solutes stay

■ Urine taken after several hours of fasting normally has a pH of 5.5 to 6.5.

■ In severe acidosis, the pH of urine can go as low as 4.

FIGURE 23.12
Acidification of the urine. The numbers refer to the text discussion.

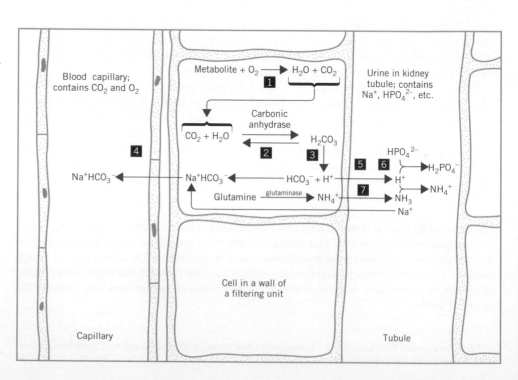

in the urine, water must also stay. Someone with metabolic acidosis, therefore, can experience a general dehydration as the system borrows water from other fluids to make urine. The thirst mechanism normally brings in replacement water, so the individual drinks copious amounts of fluids.

The Kidneys Can Export HCO_3^- In alkalosis, the kidneys can put bicarbonate ion into the urine, and it no longer uses HPO_4^{2-} to neutralize H^+. Both actions raise the pH of the urine, and in severe alkalosis it can go over a pH of 8.

The Kidneys Also Help to Regulate Blood Pressure If the blood pressure drops, as in hemorrhaging, the kidneys secrete a trace of renin into the blood. Renin is an enzyme that acts on one of the zymogens in blood, angiotensinogen, to convert it to the enzyme, angiotensin I. This, in turn, helps to convert still another protein in blood to angiotensin II, a neurotransmitter.

Angiotensin II is the most potent vasoconstrictor known. When it makes blood capillaries constrict, the heart has to work harder, and this makes the blood pressure increase. This helps to ensure that some semblance of proper filtration continues at the kidneys.

Angiotensin II also triggers the release of aldosterone, which we've already learned helps the blood to retain water. This is important because the maintenance of the overall blood volume is needed to sustain a proper blood pressure.

SUMMARY

Digestion The release of digestive juices is under the control of nerve signals and such digestive hormones as gastrin, cholecystokinin, and secretin. α-Amylase in saliva begins the digestion of starch. Pepsin in gastric juice starts the digestion of proteins. In the duodenum, trypsinogen (from the pancreas) is activated by enteropeptidase (from the intestinal juice) and becomes trypsin, which helps to digest proteins. Enteropeptidase also activates chymotrypsin (from chymotrypsinogen) and carboxypeptidase (from procarboxypeptidase). These also help to digest proteins. The pancreas supplies an important lipase, which, with the help of the bile salts, catalyzes the digestion of hydrolyzable lipids. The bile salts also aid in the absorption of the fat-soluble vitamins, A, D, E, and K.

Intestinal juice supplies enzymes for the digestion of disaccharides, nucleic acids, small polypeptides, and lipids.

The products of the digestion of proteins are amino acids; of carbohydrates: glucose, fructose, and galactose; and of the triacylglycerols: anions of fatty acids and monoacylglycerols (and some diacylglycerols). Complex lipids and nucleic acids are also hydrolyzed.

Blood Proteins in blood give it a colloidal osmotic pressure that assists in the exchange of nutrients at capillary loops. Albumins are carriers for hydrophobic molecules and serum-soluble metallic ions. Gamma globulin helps to defend the body against bacterial infections. Fibrinogen is the precursor of fibrin, the protein of a blood clot.

Among the electrolytes, anions of carbonic and phosphoric acid are involved in buffers, and all ions are involved in regulating the osmotic pressure of the blood. The chief cation in blood is Na^+, and the chief cation inside cells is K^+. The balances between Na^+ and K^+ as well as between Ca^{2+} and Mg^{2+} are tightly regulated. Ca^{2+} is vital to the operation of muscles, including heart muscle. Mg^{2+} is involved in a number of enzyme systems.

The blood transports oxygen and products of digestion to all tissues. It carries nitrogen wastes to the kidneys. It unloads cholesterol and heme breakdown products at the gallbladder. And it transports hormones to their target cells. Lymph, another fluid, helps to return some substances to the blood from tissues.

Sudden failure to retain the protein in blood leads to an equally sudden loss in blood volume and a condition of shock. Slower losses of protein, as in kidney disease or starvation, lead to edema.

Respiration The relatively high pO_2 in the lungs helps to force O_2 into HHb. This creates HbO_2^- and H^+. In an isohydric shift, the H^+ is neutralized by HCO_3^-, which is returning from working tissues that make CO_2. The resulting CO_2 leaves during exhaling. Some of the H^+ also converts $HbCO_2^-$ to HHb and CO_2.

In deoxygenating HbO_2^- at cells that need oxygen, the influx of CO_2 makes HCO_3^- and H^+. The H^+ then moves (isohydric shift) to HbO_2^- and breaks it down to HHb and O_2. Both Cl^- and BPG help to ease the last of the four O_2 molecules out of oxyhemoglobin. The low oxygen affinity of blood in tissues where pCO_2 is high helps in the release of oxygen also.

In red muscle tissue, myoglobin's superior oxygen affinity ensures that such tissue can obtain oxygen from the deoxygenation of oxyhemoglobin. Fetal hemoglobin also has an oxygen affinity superior to that of adult hemoglobin.

Acid–base balance The body uses the bicarbonate ion of the carbonate buffer to inhibit acidosis by irreversibly removing H^+ when the lungs release CO_2. HCO_3^- is replaced by the kidneys, which can also put excess H^+ into the urine. Dissolved CO_2 in the blood's carbonate buffer works to control alkalosis by neutralizing OH^-. Metabolic acidosis, with hyperventilation, and metabolic alkalosis, with hypoventilation, arise from dysfunctions in metabolism. Respiratory acidosis, with hypoventilation, and metabolic alkalosis, with hyperventilation, occur when the respiratory centers or the lungs are not working.

Diuresis The kidneys, with the help of hormones and changes in blood pressure, blood osmotic pressure, and concentrations of ions,

monitor and control the concentrations of solutes in blood. Vasopressin tells the kidneys to keep water in the bloodstream. Aldosterone tells the kidneys to keep sodium ion (and therefore water also) in the bloodstream. A drop in blood pressure tells the kidneys to release renin, which activates a vasoconstrictor, and aldosterone, which helps to raise blood pressure and retain water. In acidosis, the kidneys transfer H^+ to the urine and replace some of the HCO_3^- lost from the blood. In alkalosis the kidneys put some HCO_3^- into urine.

REVIEW EXERCISES

The answers to Review Exercises whose numbers are in color are found in Appendix C. The answers to the other Review Exercises are found in the Study Guide that accompanies this book. The more challenging questions are marked with asterisks.

Digestion

23.1 What are the names of the two chief extracellular fluids?

23.2 Name the fluids that have digestive enzymes or digestive zymogens.

23.3 Describe how the flow of the acid component of gastric juice is controlled.

23.4 Cimetidine is described as a competitive enzyme inhibitor. What does this mean? How does cimetidine work to aid in the healing of an ulcer?

23.5 What is the result of the work of cholecystokinin? Of secretin?

23.6 What enzymes or zymogens are there, if any, in each of the following?
(a) saliva   (b) gastric juice
(c) pancreatic juice (d) bile
(e) intestinal juice

23.7 Name the enzymes and the digestive juices that supply them (or their zymogens) that catalyze the digestion of each of the following.
(a) large polypeptides (b) triacylglycerols
(c) amylose   (d) sucrose
(e) di- and tripeptides (f) nucleic acids

23.8 What are the end products of the digestion of each of the following?
(a) proteins
(b) carbohydrates
(c) triacylglycerols

*23.9 What functional groups are hydrolyzed when each of the substances in Review Exercise 23.8 is digested? (Refer back to earlier chapters if necessary.)

23.10 In what way does enteropeptidase function as a "master switch" in digestion?

23.11 What would happen if the pancreatic zymogens were activated within the pancreas?

23.12 What services do the bile salts render in digestion?

23.13 What does mucin do (a) for food in the mouth and (b) for the stomach?

23.14 What is the catalyst for each of the following reactions?
(a) pepsinogen $\rightarrow$ pepsin
(b) trypsinogen $\rightarrow$ trypsin
(c) chymotrypsinogen $\rightarrow$ chymotrypsin
(d) procarboxypeptidase $\rightarrow$ carboxypeptidase
(e) proelastase $\rightarrow$ elastase

23.15 Rennin does what for an infant?

23.16 Why is gastric lipase unimportant to digestive processes in the adult stomach but useful in the infant stomach?

23.17 In terms of where they work, what is different about intestinal juice compared to pancreatic juice?

23.18 What secretion neutralizes chyme, and why is this work important?

23.19 What happens to the molecules of the monoacylglycerols and fatty acids that form from digestion?

23.20 In a patient with a severe obstruction of the bile duct the feces appear clay-colored. Explain why the color is light.

23.21 The cholesterol in the diet undergoes no reactions of digestion. Explain.

Substances in Blood

23.22 In terms of their general composition, what is the greatest difference between blood plasma and interstitial fluid?

23.23 What is the largest contributor to the net osmotic pressure of the blood as compared to the interstitial fluid?

23.24 What is fibrinogen? Fibrin?

23.25 What services are performed by albumins in blood?

23.26 In what two different regions are Na^+ and K^+ ions mostly found? What are the chief functions of these ions?

23.27 In hypernatremia, the sodium ion level of blood is above what value?

23.28 The sodium ion level of blood is below what value in hyponatremia?

23.29 What causes the hyperkalemia in crushing injuries?

23.30 Above what level is the blood described as hyperkalemic?

23.31 Excessive drinking of water tends to cause what condition, hyponatremia or hypernatremia?

23.32 The overuse of milk of magnesia can lead to what condition that involves Mg^{2+}?

23.33 Inside cells, what is a function that Mg^{2+} serves?

23.34 Where is most of the calcium ion in the body?

23.35 What does Ca^{2+} do in cells?

23.36 What condition is brought on by an overdose of vitamin D, hypercalcemia or hypocalcemia?

23.37 Injections of magnesium sulfate would be used to correct which condition, hypomagnesemia or hypermagnesemia?

23.38 What is the principal anion in both the blood and the interstitial fluid?

23.39 What is the normal range of concentration of Cl^- in blood? Explain how hypochloremia leads to alkalosis.

Exchange of Nutrients at Capillary Loops

23.40 What two opposing forces are at work on the arterial side of a capillary loop? What is the net result of these forces, and what does the net force do?

23.41 On the venous side of a capillary loop there are two opposing forces. What are they, what is the net result, and what does this cause?

23.42 Explain how a sudden change in the permeability of the capillaries can lead to shock.

23.43 Explain how each of the following conditions leads to edema.
(a) kidney disease   (b) starvation
(c) a mechanical blow

Exchange of Respiratory Gases

23.44 What are the respiratory gases?

23.45 What compound is the chief carrier of oxygen to actively metabolizing tissues?

23.46 The binding of oxygen to hemoglobin is said to be *allosteric*. What does this mean, and why is it important?

*23.47 Write the equilibrium expression for the oxygenation of hemoglobin. In what direction does this equilibrium shift when:
(a) The pH decreases?
(b) The pO_2 decreases?
(c) The red cell is in the lungs?
(d) The red cell is in a capillary loop of an actively metabolizing tissue?
(e) CO_2 comes into the red cell?
(f) HCO_3^- ions flood into the red cell?

*23.48 Using chemical equations, describe the isohydric shift when a red cell is (a) in actively metabolizing tissues and (b) in the lungs.

*23.49 In what two ways does the oxygenation of hemoglobin in red cells in alveoli help to release CO_2?

*23.50 In what way does waste CO_2 at active tissues help to release oxygen from the red cell?

*23.51 In what way does extra H^+ at active tissue help release oxygen from the red cell?

23.52 Where is carbonic anhydrase found in the blood, and what function does it have in the management of the respiratory gases in (a) an alveolus and (b) actively metabolizing tissues?

23.53 How does BPG help in the process of oxygenating hemoglobin?

23.54 In what way is BPG involved in helping to deoxygenate HbO_2^-?

23.55 What are some changes involving the blood that occur when the body remains at a high altitude for a period of time, and how do these changes help the individual?

23.56 How would the net equations for the oxygenation and the deoxygenation of blood be changed to include the function of BPG?

23.57 What are the two main forms in which waste CO_2 moves to the lungs?

23.58 How is oxygen affinity affected by pCO_2, and how is this beneficial?

23.59 What is the chloride shift and how does it aid in the exchange of respiratory gases?

23.60 In what way is the superior oxygen affinity of myoglobin over hemoglobin important?

23.61 Fetal hemoglobin has a higher oxygen affinity than adult hemoglobin. Why is this important to the fetus?

Acid–Base Balance of the Blood

23.62 Construct a table using arrows ($\uparrow$) or ($\downarrow$) and typical lab data that summarize the changes observed in respiratory and metabolic acidosis and alkalosis. The column headings should be as follows:

Condition	pH	pCO_2	$[HCO_3^-]$

*23.63 With respect to the *directions* of the changes in the values of pH, pCO_2, and $[HCO_3^-]$ in both respiratory acidosis and metabolic acidosis, in what way are the two types of acidosis the same? In what way are they different?

23.64 Hyperventilation is observed in what two conditions that relate to the acid–base balance of the blood? In one, giving carbon dioxide is sometimes used, and in the other, giving isotonic HCO_3^- can be a form of treatment. Which treatment goes with which condition and why?

23.65 In what two conditions that relate to the acid–base balance of the blood is hypoventilation observed? Isotonic ammonium chloride or isotonic sodium bicarbonate are possible treatments. Which treatment goes with which condition, and how do they work?

23.66 In which condition relating to acid–base balance does hyperventilation have a beneficial effect? Explain.

*23.67 Hyperventilation is part of the *cause* of the problem in which condition relating to the acid–base balance of the blood?

*23.68 Hypoventilation is the body's way of helping itself in which condition that relates to the acid–base balance of the blood?

*23.69 In which condition that concerns the acid–base balance of the blood is hypoventilation part of the *problem* rather than the cure?

23.70 How can a general dehydration develop in metabolic acidosis?

*23.71 Which condition, metabolic or respiratory acidosis or alkalosis, results from each of the following situations?
(a) hysterics
(b) overdose of bicarbonate
(c) emphysema
(d) narcotic overdose
(e) diabetes
(f) overbreathing at a high altitude
(g) severe diarrhea
(h) prolonged vomiting
(i) cardiopulmonary disease
(j) barbiturate overdose

*23.72 Referring to Review Exercise 23.71, which is happening in each situation, hyperventilation or hypoventilation?

23.73 Why does hyperventilation in hysterics cause alkalosis?

23.74 Explain how emphysema leads to acidosis.

23.75 Prolonged vomiting leads to alkalosis. Explain.

23.76 Uncontrolled diarrhea can cause acidosis. Explain.

23.77 Respiratory alkalosis causes hypocapnia or hypercapnia?

23.78 For each 1 °C above normal human body temperature, the rate of CO_2 production increases by 13%. If the rate of breathing does not increase, what results — hypocapnia or hypercapnia?

Blood Chemistry and the Kidneys

23.79 If the osmotic pressure of the blood has increased, what, in general terms, has changed to cause this?

23.80 How does the body respond to an increase in the osmotic pressure of the blood?

23.81 If the sodium ion level of the blood falls, how does the body respond?

23.82 What is the response of the kidneys to a decrease in blood pressure?

23.83 Alcohol in the blood suppresses the secretion of vasopressin. How does this affect diuresis?

23.84 In what ways do the kidneys help to reduce acidosis?

Additional Exercises

23.85 Monoacylglycerols are able to migrate through membranes of the cells of the intestinal tract that absorb them. Glycerol, however, is unable to accomplish this movement. How might we explain these relative abilities to migrate through a cell membrane?

*23.86 When the gallbladder is surgically removed, lipids of low formula mass are the only kinds that can be easily digested. Explain.

*23.87 It has been reported that some long-distance Olympic runners have trained at high altitudes and then had some of their blood withdrawn and frozen. Days or weeks after returning to lower altitudes and just prior to a long race, they have used some of this blood to replace an equal volume of what they are carrying. This is supposed to help them in the race. How would it work?

*23.88 Aquatic diving animals are known to have much larger concentrations of myoglobin in their red muscle tissue than humans. How is this important to their lives?

NUCLEIC ACIDS

24

Only a hippopotamus can have a hippopotamus baby (which is a good thing when you think about it). The molecular basis for such good things is presented in this chapter.

24.1

HEREDITY AND THE CELL

The chemicals associated with living things are organized in structural units called cells, and the instructions for making these chemicals are carried on molecules of nucleic acids.

We have learned that nearly every reaction in a living organism requires its special catalyst, and that these catalysts are called enzymes. The set of enzymes in one organism is not exactly the same as in another, although many enzymes in different species are quite similar. A major chemical requirement of reproduction is that each organism transmits to its offspring the capacity to possess the set of enzymes that are unique to the organism.

Nucleic Acids Carry Instructions for Making Enzymes In reproduction, an organism does not duplicate the enzymes themselves and then pass them on directly to its offspring. Instead, it sends on the *instructions* for making its unique enzymes from amino acids. It does this by duplicating the compounds of a different family, the *nucleic acids*. These then direct the synthesis of enzymes (and all other proteins) in the offspring. Certain nucleic acids called **ribozymes** are themselves also able to function as enzymes. Our purpose in this chapter is to study how nucleic acids carry genetic instructions and act on them, but we have to learn something about the biological context first.

Every Cell Carries Nucleic Acids The cell is the smallest unit of an organism that has life and can duplicate itself in the organism. Cells of different systems vary widely in shape and size, but they generally have similar parts. Figure 24.1 outlines the major features of an animal cell. The cell boundary is the cell membrane, which we studied in Sections 20.5 and 21.6. Everything enclosed by it is called *protoplasm,* which contains several discrete particles or cellular bodies. Prominent among them are the **mitochondria,** which are tiny *organelles* (subcellular bodies) where adenosine triphosphate (ATP) is made for the cell's needs for chemical energy.

The part of the cell outside of its nucleus is called the *cytoplasm.* (The liquid portion itself is called the *cytosol,* as we have learned.) One kind of particle in the cytoplasm is called a **ribosome,** which consists mostly of nucleic acids and polypeptides. Ribosomes have essential functions in the synthesis of polypeptides, including the polypeptides of enzymes. Polypeptide synthesis, however, is under the primary control of nucleic acids *inside* the cell's nucleus. Nucleic acids thus occur in both the cytoplasm and nucleus.

Genes Are the Fundamental Units of Heredity The nucleus has its own membrane, and inside is a web-like network of protein. The nucleus also contains twisted and intertwined filaments of nucleoprotein called *chromatin.* Chromatin is like a strand of pearls, each pearl made of proteins called *histones* around which are tightly coiled one of the kinds of nucleic acid called DNA for short. DNA also links the "pearls." Each "pearl" is called a *nucleosome.* We'll study the structure of DNA in the next section, but DNA molecules have portions that

■ T. R. Cech and S. Altman shared the 1989 Nobel prize in chemistry for discovering ribozymes. It came as a huge surprise in the 1980s to learn that the body uses anything other than proteins as enzyme catalysts.

■ The histones are not just spools for wrapping DNA strands; they contribute to the regulation of DNA activity.

674

FIGURE 24.1
Model of a generalized animal cell. Although cells differ greatly from tissue to tissue, most have the features shown here. ER stands for endoplasmic reticulum. (From G. C. Stephens and B. B. North, *Biology, A Contemporary Perspective*, 1974. John Wiley and Sons. Used by permission.)

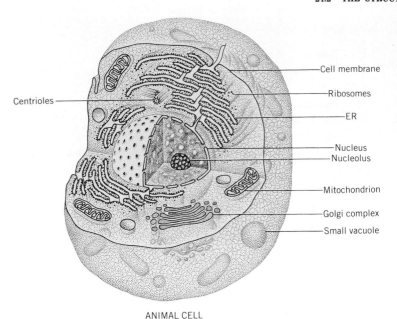

Centrioles

Cell membrane

Ribosomes

ER

Nucleus
Nucleolus

Mitochondrion

Golgi complex

Small vacuole

ANIMAL CELL

constitute individual **genes,** the fundamental units of heredity. A single gene, for example, contains the information for making a single enzyme, so unique genes translate into unique enzymes. A major goal of this chapter is to learn how the chemical structure of a gene enables it to function in this way.

Prior to Cell Division, Genes Replicate When cell division begins, the chromatin strands thicken and become rod-like bodies that accept staining agents and so can be seen under a microscope. These discrete bodies are called **chromosomes.** The thickening of chromatin into chromosomes is caused by the synthesis of new DNA and histones. The new chromosomes, including their DNA, are exact copies of the old, if all goes well as it does to a remarkable extent. This reproductive duplication of DNA is called **replication,** so by the replication of DNA, the genetic message of the first cell is passed to each of the two new cells.

When two germ cells, a sperm and an ovum, unite at conception to form the single cell from which the entire organism will grow, DNA from both germ cells combine. Genetic characteristics of both parents thus are passed on to the offspring. Every cell made from the first cell has the entire set of genes, but obviously most genes in the older organism are turned off most of the time. Genes that are behind the formation of fingernails, for example, must not operate in heart muscle cells! Another goal of our study is to learn how (some) genes might be regulated.

■ There are 23 matched pairs of chromosomes in the human cell for a total of 46 chromosomes.

■ Every human cell has between 50,000 and 100,000 genes.

24.2

THE STRUCTURE OF NUCLEIC ACIDS

Genetic information is carried by the sidechains of a twisted, double-stranded polymer called the DNA double helix.
The nucleic acids that store and direct the transmission of genetic information are polymers nicknamed DNA and RNA. **DNA** is **deoxyribonucleic acid** and **RNA** is **ribonucleic acid.** The monomer molecules for the nucleic acids are called **nucleotides.**

Nucleotides Are Monophosphate Esters of Pentoses to Which Heterocyclic Bases Are Joined Unlike the monomers of proteins, the nucleotides can be further hydrolyzed. As outlined in Figure 24.2, the hydrolysis of a representative mixture of nucleotides produces

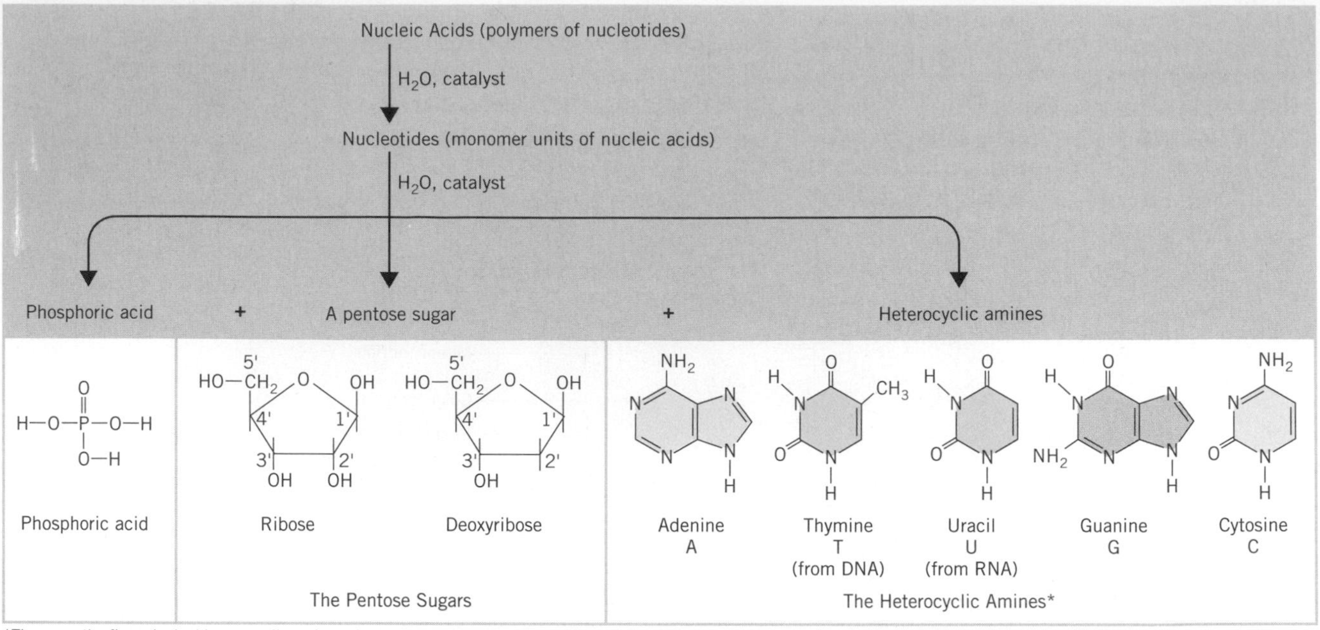

*These are the five principal heterocyclic amines obtainable from nucleic acids. Others, not shown, are known to be present. Although they differ slightly in structure, they are informationally equivalent to one or another of the five shown here.

FIGURE 24.2

The hydrolysis products of nucleic acids

■ Ring carbon number 1′ is the carbon of the aldehyde group when the pentose ring is open.

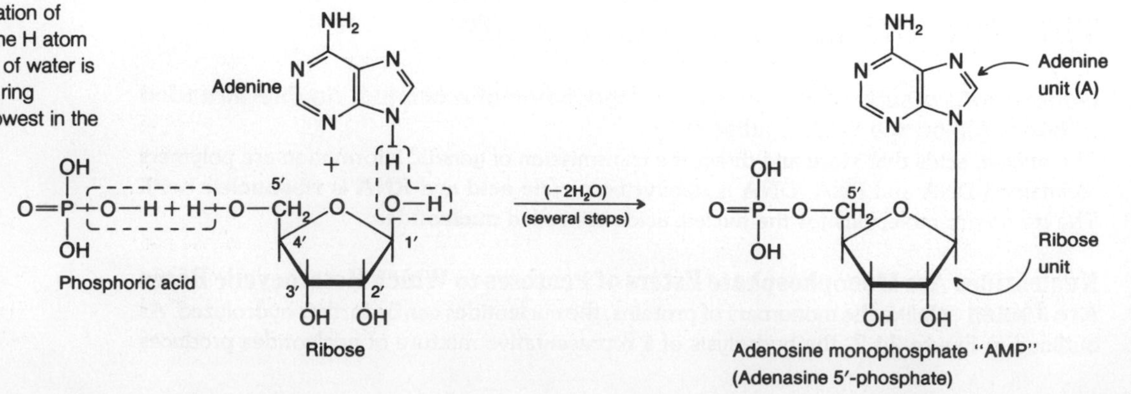

FIGURE 24.3

A typical nucleotide, AMP, and the smaller units from which it is assembled. The phosphate ester group forms by the splitting out of water between an OH group of phosphoric acid and an H atom at the 5′ OH group of the ribose unit. The splitting out of water between the 1′ OH group of the ribose unit and the H atom on a ring nitrogen of adenine (A) joins adenine to the sugar unit. Similar structures could be drawn with the other bases of Figure 24.2. In the formation of each such nucleotide, the H atom used in this splitting out of water is the one attached to the ring nitrogen drawn as the lowest in the structure of the base.

three kinds of products: inorganic phosphate, a pentose sugar, and a group of heterocyclic amines called the **bases,** which have single-letter symbols:

Bases from DNA		Bases from RNA	
Adenine	A	Adenine	A
Thymine	T	Uracil	U
Guanine	G	Guanine	G
Cytosine	C	Cytosine	C

Three bases are thus common to both DNA and RNA, and one base is different. The sugar unit is also different. It is a pentose in both, but the hydrolysis of RNA gives ribose — hence the R in RNA — and that of DNA gives deoxyribose — which lends the D to DNA.

In the structure of a typical nucleotide of RNA, adenosine monophosphate or AMP, the base adenine (A) is joined to the hemiacetal carbon (C-1′) of the pentose; the C-5′ OH group of the pentose has been changed into a ester of phosphoric acid (see Fig. 24.3). (The prime, as in 5′, refers to the numbering of the pentose ring; unprimed numbers are used to number positions on the rings of the bases.) All nucleotides are monophosphate esters with structures

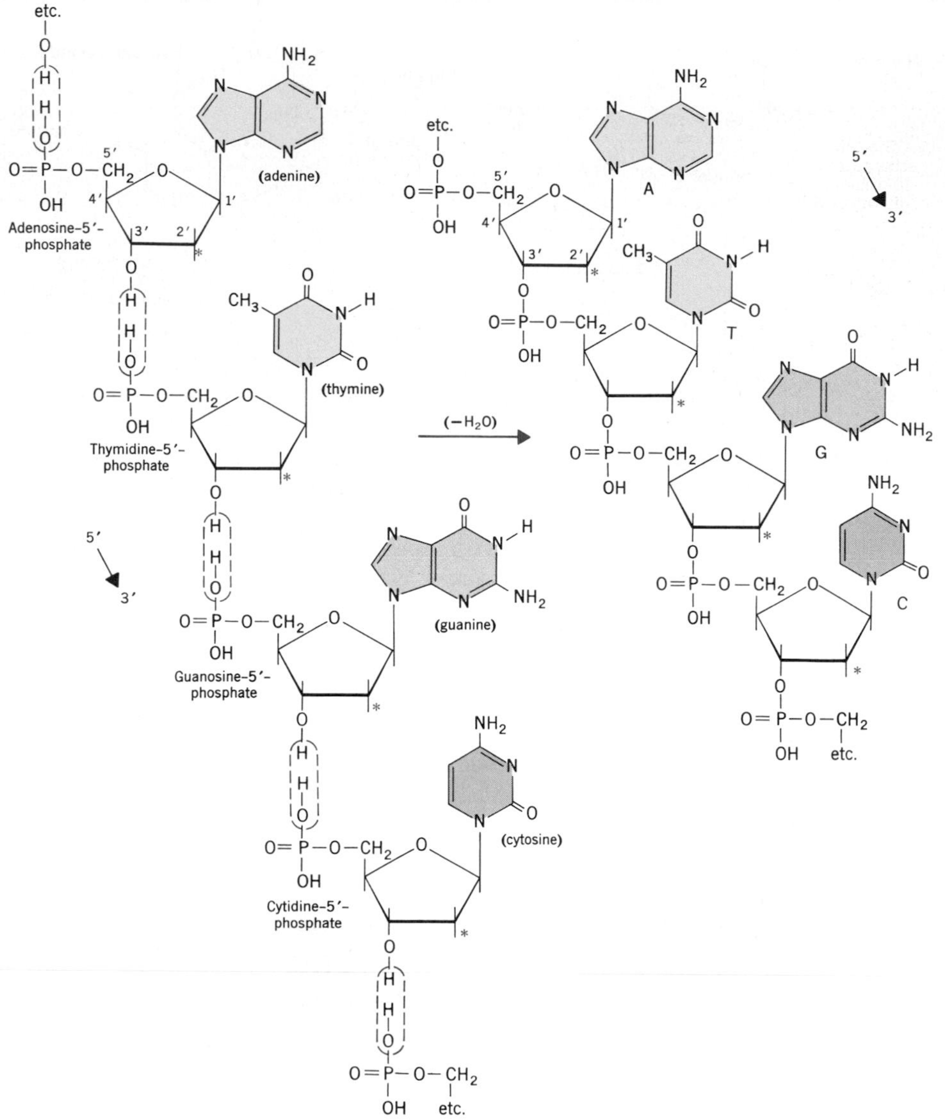

FIGURE 24.4

The relationship of a nucleic acid chain to its nucleotide monomers. On the right is a short section of a DNA strand. On the left are the nucleotide monomers from which it is made (after many steps). The colored asterisks by the pentose units identify the 2′ positions of these rings where there would be another OH group if the nucleic acid were RNA (assuming that uracil also replaced thymine).

like that of adenosine monophosphate. In making its nucleotides, the cell splits out two molecules of water as indicated in Figure 24.3 (but by *several* steps that we won't discuss).

When the OH at C-2′ is replaced by H, and the pentose is deoxyribose, the resulting nucleotide is one for DNA. The pattern for all nucleotides, therefore, is as follows.

$$\text{HO}-\text{phosphate}-\overset{\displaystyle \text{base} \atop |}{\text{pentose}}-\text{OH}$$

The Bases Project from the Backbones of Nucleic Acids Figure 24.4 shows the general scheme of how nucleotides are linked in nucleic acids. Many steps and many enzymes are required but, in the overall result, water splits out between an OH group at a C-3′ ring position of one nucleotide and the phosphate unit of the next nucleotide. The result of the splitting out of water is a *phosphodiester* system. When this linking is repeated thousands of times, a nucleic acid is the product. A nucleic acid is thus a copolymer for which there are four monomers. If the sequence at the top end of the chain in Figure 24.4 did not continue, the top would be regarded as the *starting* point of the polymer chain. It's called the 5′ end after the number of the first position encountered on a pentose ring. At the opposite end, the chain ends at a C-3′ OH group. Thus the chain in Figure 24.4 is said to be running, top to bottom

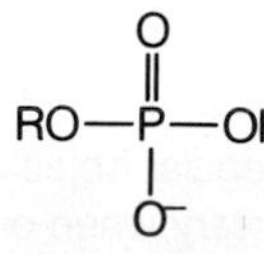

■ Some 20 enzymes are required, the chief being DNA polymerase, discovered in 1958 by Arthur Kornberg (Nobel prize, 1959).

■

Phosphodiester
(anion form)

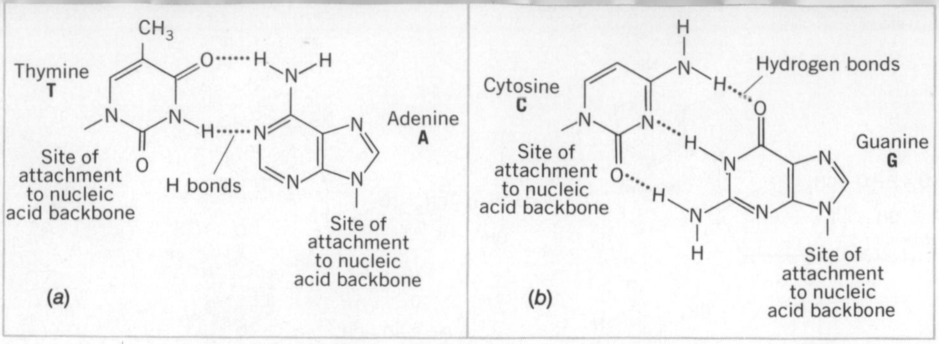

FIGURE 24.5

Hydrogen bonding between base pairs. (*a*) Thymine (T) and adenine (A) form one base pair between which are two hydrogen bonds. (*b*) Cytosine (C) and guanine (G) form another base pair between which are three hydrogen bonds. Adenine can also base-pair to uracil (U).

(beginning to end), in the $5' \rightarrow 3'$ direction. The bases are in the sequence of A to T to G to C, $5' \rightarrow 3'$, so the *structural formula* of the portion of the DNA polymer in Figure 24.4 can be highly condensed simply as ATGC.

What are shown in Figure 24.4 as free OH groups attached to the P atoms of the chain are moderately strong proton-donating groups, similar to the OH groups in phosphoric acid. The pH of cellular fluid is slightly higher (more basic) than 7, however, so these acidic groups have given up protons (which are neutralized by the buffer), and nucleic acids exist *in vivo* as multiply charged anions.

The repeating pattern for the backbone of any nucleic acid, DNA or RNA, thus shows alternating phosphate and pentose units, and projecting from each pentose is one of the bases. The backbone holds the system together, and the bases — their selection and sequence — carry the genetic information. *The distinctiveness of any one nucleic acid is in the sequence of its side chain bases.* There are 24 different sequences possible for just the four different bases of DNA.

ATGC	TAGC	GCAT	CATG
ATCG	TACG	GCTA	CAGT
AGTC	TGAC	GATC	CTAG
AGCT	TGCA	GACT	CTGA
ACTG	TCAG	GTAC	CGAT
ACGT	TCGA	GTCA	CGTA

Many genes involve thousands of bases (actually thousands of *pairs* of bases, as we will see), each base obviously used many times, and you can begin to see how the uniqueness of a given gene rests altogether on the sequences of the DNA bases.

Pairs of Bases Are Attracted to Each Other by Hydrogen Bonds The sidechain bases have functional groups so arranged geometrically that they fit to each other in pairs by means of hydrogen bonds, a phenomenon called **base pairing** (see Fig. 24.5). *The locations and geometries of the functional groups of the bases permit only certain base pairs to exist.* In DNA, G and C always form a pair, and A and T form another pair. In RNA, G and C always pair, and U and A always pair. Neither G nor C ever pairs with A, T, or U.

DNA Occurs as Paired Strands Twisted into a Double Helix In 1953, Francis Crick of England and James Watson of the United States proposed a structure for DNA, the **DNA double helix,** that made possible an understanding of how heredity works at the molecular level. Using X-ray data obtained by Rosalind Franklin, they deduced that two complementary DNA molecules line up side by side to form a double strand that then twists into a right-handed helix. The strands will not separate except by unwinding the helix. DNA in its double helix form is called **duplex DNA.**

The idea of complementary strands is a key feature of the double helix just as it was in our study of enzymes and substrates or receptors and hormones (or neurotransmitters). In DNA chemistry, *strand complementarity* also refers to the kind of fitting of one whole strand to the

■ Crick and Watson shared the 1962 Nobel prize in medicine and physiology with Maurice Wilkins.

■ Two irregular objects are *complementary* when one fits to the other, as your right hand would fit to its impression in clay.

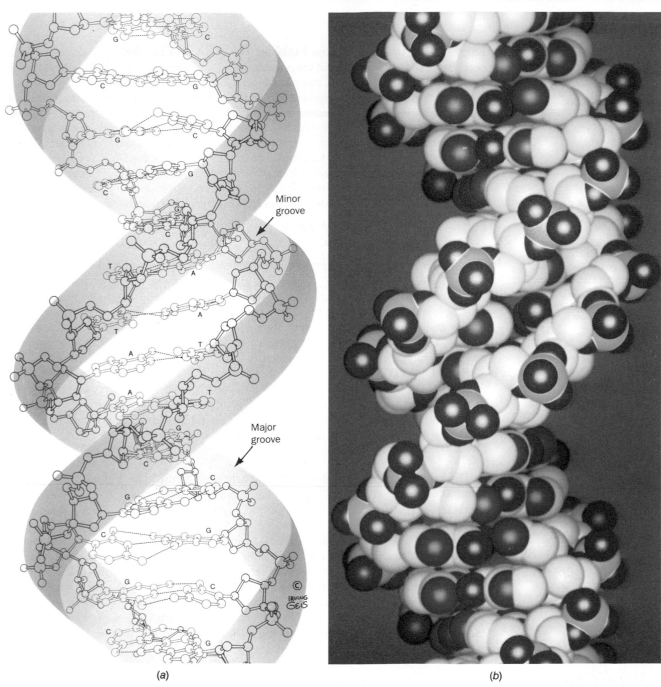

(a)

(b)

FIGURE 24.6

The structure of duplex DNA — the native form (B-DNA) — seen here in (a) a ball-and-stick drawing and (b) a computer-generated space-filling model. (Drawing copyrighted © by Irving Geis. Computer graphics courtesy of Robert Stodola, Fox Chase Cancer Center.)

Minor groove

Major groove

other. This occurs by base pairing *between the strands,* an idea suggested by two significant 1 : 1 mole ratios discovered earlier by Erwin Chargaff. He had found that A and T are always present in a 1 : 1 ratio in DNA *in all species,* and that G and C also occur in a 1 : 1 mole ratio in all species studied. As we have seen, A pairs to T and G pairs to C, so each pair *must* be in a 1 : 1 ratio. Crick and Watson made sense of the simple 1 : 1 ratios by proposing that A pairs to T *between two complementary strands,* and that G pairs to C also between the strands.

Whenever adenine (A) is attached to one strand, then thymine (T) is attached opposite it on the complementary strand. And whenever guanine (G) projects from one strand, then cytosine (C) is opposite it on the complementary strand. A molecular model of the DNA double helix discovered by Crick and Watson is shown in Figure 24.6. The system resembles a spiral staircase in which the steps, which are nearly perpendicular to the long axis of the spiral, consist of the base pairs.

■ The mole ratios of (A + T) to (C + G) vary between species.

679

Largely Hydrophobic Forces Stabilize the DNA Double Helix Hydrogen bonds occur between the base pairs, but these bonds are no longer regarded as the primary source of the stability of the double helix structure. Hydrogen bonds determine which bases can pair, but *hydrophobic interactions stabilize duplex DNA.* The rings themselves of the bases are planar and unsaturated, like benzene rings, and relatively nonpolar despite the ring N atoms. In an aqueous medium there is a natural tendency for these rings to stack closely together because they cannot offer much hydrogen-bonding alternatives to water molecules. Water excludes them, so they stack together, which is exactly what the DNA duplex structure portrays. Around the edge of each spiraling backbone project the negatively charged, anionic sites of the phosphate units, giving these strongly hydrophilic groups full access to the cellular fluid. What may not be immediately obvious from Figure 24.6 is that the two DNA strands run in opposite directions.

- The *in vivo* form of DNA is sometimes called B-DNA. Other forms are A-DNA and Z-DNA. They differ in coiling geometries.

Duplex DNA Has a Major and a Minor Groove The structure of duplex DNA in Figure 24.6 is the *in vivo* form of DNA. Notice that it has two grooves, one major and one minor. These are binding sites for molecules of proteins involved in several gene functions, including gene repression and gene activation. They are also sites for the initial molecular interactions of certain drugs, antibiotics, carcinogens (cancer-causing agents), and poisons. Polypeptides with net positive charges would be attracted to these spiraling duplex DNA grooves, and complementary fitting to the DNA becomes a factor in protein–DNA interactions.

Figure 24.6 shows only a short segment of duplex DNA, a sequence too short to show that duplex DNA is further twisted and coiled into superhelices. The looping and coiling are necessary to fit the cell's DNA into its nucleus. A typical human cell nucleus, for example, is only about 10^{-7} m across, but if all of its DNA double helices were stretched out, they would measure a little over 1 m, end to end.

- There are an estimated 3 billion base pairs in one nucleus of a human cell but only about 5% are a part of genes. The functions of the remaining 95% are not yet fully known.

In DNA Replication, the Bases on Each Strand Guide the Formation of New Complementary Strands When DNA replicates, the cell makes an exact complementary strand for each of the two original strands, so two identical double helices are made. The built-in guarantee that each new strand will be the complement of one of the old is the requirement that A pairs to T and C to G. A very general picture that shows how this works is in Figure 24.7. Realize that this figure explains only one of the many aspects of replication, how base pairing ensures two new complementary strands. Many of the details of how replication occurs without everything becoming hopelessly tangled in the nucleus are understood, but we must leave them to more advanced references. One feature of DNA replication correctly shown by Figure 24.7 is that the synthesis of replica DNA strands occurs as the parent strand unwinds. The formation of the replicas does not await the complete unwinding of the parent before the replication commences.

In Higher Organisms, Sections of DNA Molecules Called *Exons* Collectively Carry the Message of One Gene Although a single gene can itself be thought of as a continuous sequence of nucleotide units, each with its sidechain, *the gene does not occur as a continuous sequence within a DNA system of chromatin.* In their locations in chromatin, virtually all individual genes are divided or split. This means that a gene is made up of *sections,* called **exons** of a DNA chain. Interrupting the exons and separating them are other, usually longer sections of the DNA molecule called **introns.** The gene for the β-subunit of human hemoglobin, for example, consists of 990 bases, but two intron units, 120 and 550 bases long, interrupt the gene, as illustrated in Figure 24.8. Some introns are as short as 50 base pairs in length and others as long as 200,000 base pairs. Introns make up an average of roughly 80% of a human structural gene (that part of a gene that does not include regulator and control sites).

- *Exon* refers to the part that is *expressed* and *intron* to the segments that *int*errupt the exons.

FIGURE 24.7
The replication of DNA

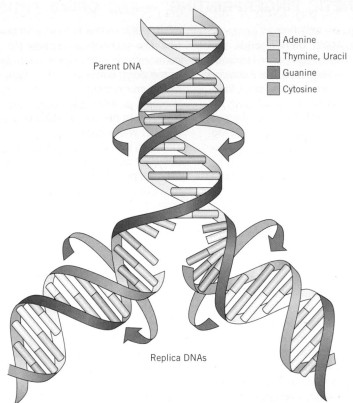

Parent DNA

☐ Adenine
☐ Thymine, Uracil
☐ Guanine
☐ Cytosine

Replica DNAs

A single gene can have as many as 50 intron segments, but not all DNA molecules in higher organisms have introns. A few human genes do exist with completely continuous sequences of bases. Just what purposes are served by the introns is currently under considerable speculation. They may serve a protective role or they may have a regulatory function.

We'll see later in this chapter how the exons of one gene manage to get the gene's message together. For the present, we can consider that a **gene** is a particular section of a DNA strand minus all the introns in this section. A gene, in other words, is a specific series of bases strung in a definite sequence along a DNA backbone.

Gene Sequencing Has Been Automated Chemists have developed automatic gene sequencing instruments that determine the sequences of bases in individual genes. The genes of one human being have to be similar, of course, to those of another for the two to be members of the same species. However, every individual human being that has ever lived or ever will live, except for identical twins, has genes unique in some ways. This uniqueness of individuals is so analogous to the uniqueness of fingerprints that genetic "fingerprints" often are introduced as evidence in certain criminal trials (see Special Topic 24.1).

■ There are no introns in human genes that code for histones or for α-interferon.

■ American chemists Richard Roberts and Phillip Sharp shared the 1993 Nobel prize in physiology and medicine for their independent discovery in 1977 of "split genes."

Bases   240   120   500   550   250

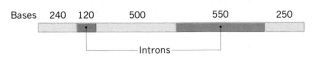

Introns

FIGURE 24.8
The gene for the β-subunit of hemoglobin is a split gene with two long intron sections.

DNA TYPING—"GENETIC FINGERPRINTING"—AND CRIME PROSECUTION

In the entire set of human genes, the human *genome,* there are many regions consisting of nucleotide sequences repeated in tandem. These regions, called *minisatellites,* all have a common core sequence, but the *number* of repeated sequences in the minisatellites varies from individual to individual. These variations, which can be measured, are so considerable between individuals that they are the basis of a major new technique in discovering the truth concerning a suspect in certain crimes, particularly rape.

Suppose, for example, a sample of semen can be obtained from the sperm left by a rapist. The DNA is "amplified" or cloned by means of the *polymerase chain reaction* to convert it into a sample large enough for the analysis. To apply the polymerase chain reaction, the DNA is first denatured by warming the sample, which causes the strands of duplex DNA to unwind and separate. (Denatured DNA renatures itself spontaneously if the temperature of the system is carefully controlled over a sufficient length of time.) A sample of the enzyme DNA polymerase is then added together with specially prepared, short DNA "primer" strands, and the mixture is incubated. During this period, cloning of the DNA occurs and new but identical copies of duplex DNA molecules form.

The sample of cloned DNA is next hydrolyzed with the use of a specific enzyme, called a *restriction enzyme.* The enzyme is able to catalyze the breakage (hydrolysis) of a DNA strand *only at specific sites,* which releases the core segments described earlier. The fragments of DNA can then be separated and made to bind (by base pairing) to short, radioactively labeled, specially made DNA that has been designed to be able to bind to core segments. Now the labeled DNA fragments are separated by a special technique so that the fragments occur as thin, parallel deposits spread out on a film. Their locations on the film can be detected by the ability of their atomic radiations to affect photographic film. The radiations darken the film only in characteristic locations and produce a series of 30 to 40 dark bands, roughly analogous to the bar codes on groceries that are used for pricing at checkout counters. Each individual has a unique genetic "bar code." Because

every cell in the body has the entire genome, a single hair of a rape suspect can provide the same fundamental genome as comes from a sample of semen or blood. Thanks to the polymerase chain reaction, enough DNA material can be made from the tiniest of samples to measure the person's genetic "bar code" and compare it to that obtained from the semen sample. Figure 1 shows how one suspect was trapped.

When the two "bar codes" match and there has been sufficient care in making sure that the samples are from *different* people and that the cloning work has been carefully done, the jury has evidence considered by nearly all scientists to be as powerful as fingerprints for a conviction. If "bar codes" do not match, the district attorney looks for another suspect. DNA typing evidence is thus as powerful an ally of the innocent as it is an enemy of the guilty. If a crime site specimen is old or has been subjected to harsh environmental conditions and has deteriorated, it will either give a true test or none at all. It does not give a false test that will lead to the conviction of an innocent person.

The use of DNA typing is going through rigorous court challenges. Does every individual actually have a completely unique genome? (The DNA "fingerprints" of identical twins are the same.) Is the evidence from DNA typing truly like fingerprint evidence? (Members of a South American tribe founded by a few individuals and now excessively inbred have very similar DNA fingerprints.) What are the statistical probabilities that two DNA samples from *different* people will be different? Enough doubts have been raised over these questions in the minds of some judges that (as of early 1993) the court systems of Massachusetts, California, and Guam will not admit DNA "fingerprints" as evidence unless or until these uncertainties are resolved. The use of real fingerprints went through similar challenges. During the period of the challenges to DNA typing, recognizing that the technique is an extraordinarily powerful tool to establish the truth, scientists have been at work to remove the doubts, improve the technology, and establish the reliability of every procedural step from the finding of raw evidence to the court appearance.

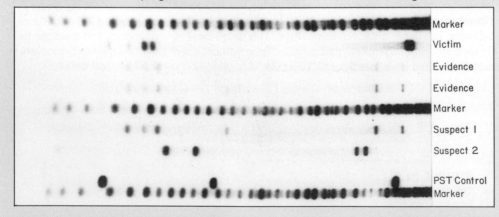

FIGURE 1

DNA fingerprinting. The banding pattern of the DNA of suspect 1 matches that of the evidence. Neither the rape victim's DNA nor that of suspect 2 does.

24.3

RIBONUCLEIC ACIDS

Triplets of bases on messenger RNA correlate with individual amino acids through the *genetic code.*

The general scheme that relates DNA to polypeptides is illustrated in Figure 24.9. We'll refer to it often but first we need to learn more about RNA and its various types, particularly the four that participate in expressing genes in higher organisms.

Ribosomal RNA (rRNA) Is the Most Abundant RNA We have already mentioned the small particles in the cytoplasm, called ribosomes (Fig. 24.1). Each forms from two subunits, as shown in Figure 24.9, which come together to form a complex with messenger RNA, another type that we'll study soon.

Ribosomes contain proteins and a ribonucleic acid called **ribosomal RNA,** abbreviated **rRNA.** Except in a few viruses, rRNA is single-stranded, but its molecules often have hairpin loops in which base pairing occurs. The rRNA of a ribosome is itself made from longer versions by losing some of its chain pieces. This loss is catalyzed by a region of the very RNA being groomed, so such RNA regions are actually working as enzymes, the ribozymes we mentioned on page 674.

Ribosomes are the sites of polypeptide synthesis. However, the ribosome does not know how to arrange the amino acid residues of the polypeptide in the right order. The rRNA of the

■ *Molecular biology* is the hybrid science that encompasses all of the chemistry of nucleic acids, genes, chromosomes, and genetic expressions in protein syntheses.

■ Over 50 kinds of polypeptides exist in one ribosome.

■ Other ribozymes are now known that catalyze reactions of substrates other than themselves.

FIGURE 24.9
The relationships of nuclear DNA to the various RNAs and to the synthesis of polypeptides

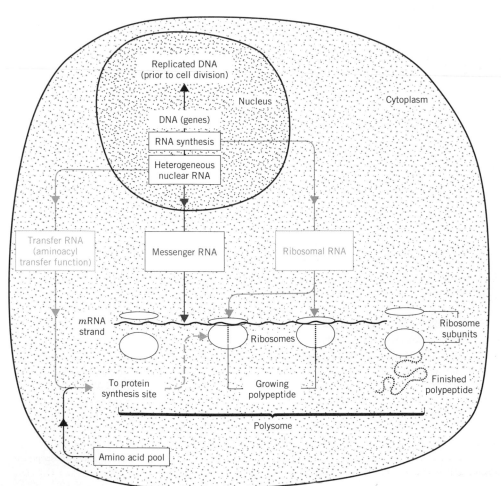

FIGURE 24.10

DNA-directed synthesis of hnRNA in the nucleus of a cell in a higher organism. The shaded oval on the left represents a complex of enzymes that catalyze this step. (Notice that the direction of the hnRNA strand is opposite to that of the DNA strand.)

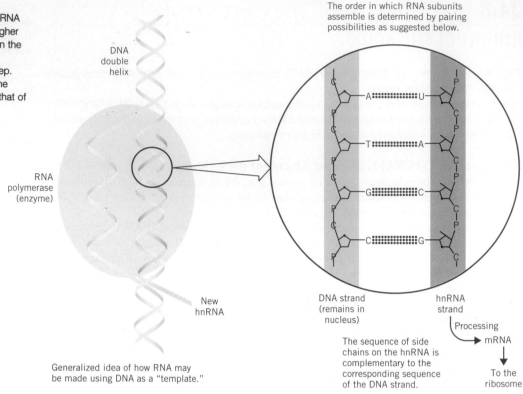

The order in which RNA subunits assemble is determined by pairing possibilities as suggested below.

DNA double helix

RNA polymerase (enzyme)

New hnRNA

Generalized idea of how RNA may be made using DNA as a "template."

DNA strand (remains in nucleus)

hnRNA strand

Processing → mRNA

To the ribosome

The sequence of side chains on the hnRNA is complementary to the corresponding sequence of the DNA strand.

---

ribosome cannot itself direct this aspect of putting a polypeptide together. Another kind of RNA called *messenger RNA* is the director of polypeptide synthesis, and messenger RNA is made from another kind of RNA that we will study.

Heterogeneous Nuclear RNA (hnRNA) Is Complementary to DNA — Exons and Introns When a cell uses a gene to direct the synthesis of a polypeptide, its first step is to use the single DNA strand bearing the polypeptide's gene to guide the assemblage of a complementary molecule of RNA, called **heterogeneous nuclear RNA,** abbreviated **hnRNA.**[1] Figure 24.10 gives a very broad picture of how sidechain bases on DNA guide the nucleotides with sidechain bases for hnRNA into the correct sequence during the assembly of hnRNA. Uracil (U) is now used instead of thymine (T), so when a DNA strand has an adenine (A) side chain, then uracil not thymine takes the position opposite it on the complementary RNA. The enzyme involved is *RNA polymerase.*

The next general step in gene-directed polypeptide synthesis is the processing of hnRNA to the next kind of RNA that we must study, messenger RNA or mRNA.

■ Part of the grooming of hnRNA molecules installs at their 3′ ends a long poly-A tail (about 200 adenosine units long), and at the other end a nucleotide triphosphate "cap."

Each Messenger RNA (mRNA) Is Complementary to One Gene Molecules of hnRNA have large sections complementary to the introns of the DNA, and these sections must be deleted. Special enzymes catalyze reactions that snip these pieces out of hnRNA and splice together just the units corresponding to the exons of the divided gene (see Fig. 24.11). The result is a much shorter RNA molecule called **messenger RNA,** or **mRNA.**

In mRNA we have a sequence of bases complementary just to the gene's exons, so this mRNA now carries the unsplit genetic message. We have now moved the genetic message

■ A small family of nuclear ribonucleoproteins, snRNPs (or "snurps"), helps this resplicing process.

---

[1] Heterogeneous nuclear RNA has been called *primary transcript RNA (ptRNA),* or simply *primary transcript,* and sometimes *pre-messenger RNA (pre-mRNA).*

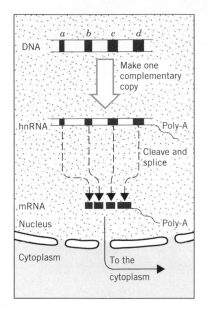

FIGURE 24.11
The RNA made directly at a DNA strand is hnRNA. Only the segments made at sites *a, b, c,* and *d* — the exons — are needed to carry the genetic message to the cytoplasm. The hnRNA is processed, therefore, and its segments that matched the introns of the gene are snipped out. Then the segments that matched the exons are rejoined to make the mRNA strand.

from a split gene on DNA to a molecule of mRNA, and the name for this overall process is **transcription.**

Triplets of Bases on mRNA Are Genetic Codons Each group of three adjacent bases on a molecule of mRNA constitutes a unit of genetic information called a **codon** (taken from the word *code*). Thus it is a sequence of *codons* on the mRNA backbone more than a sequence of individual bases that now carries the genetic message. (We'll explain shortly why *three* bases per codon are necessary.)

Once they are made, the mRNAs move from the nucleus to the cytoplasm where they attach ribosomes. Many ribosomes can be strung like beads along one mRNA chain, and such a collection is called a *polysome* (short for *polyribosome*).

Ribosomes are traveling packages of enzymes intimately associated with rRNA. Each ribosome moves along its mRNA chain while the codons on the mRNA guide the synthesis of a polypeptide. To complete this system, we need a way to bring individual amino acids to the polysome's polypeptide assembly sites. For this, the cell uses still another type of RNA.

■ F. Jacob and J. Monod (Nobel prizes, 1965) conceived the idea that a messenger RNA must exist.

Transfer RNA (tRNA) Molecules Can Recognize Both Codons and Amino Acids
The substances that carry aminoacyl units to mRNA *in the right order for a particular polypeptide* are a collection of similar compounds called **transfer RNA or tRNA.** Their molecules are small, each typically having only 75 nucleotides. As seen in Figure 24.12, they are single-stranded but with hairpin loops.

We're now dealing with the molecular basis of *information,* so we can use language analogies. On a human level, we use language to convey information, and language involves words built from a common alphabet. We are aware that many languages exist among human societies and that the world knows several alphabets. To communicate between languages, we have to translate. The same need for translation occurs at the level of genes and polypeptides. tRNA is the master translator in cells.

tRNA is able to work with two "languages," the nucleic acid or genetic and the polypeptide. The nucleic acid language is expressed in an alphabet of 4 letters, the four bases, A, T (or U), G, and C. The polypeptide language has an alphabet of 20 letters, the side chains on the 20 amino acids. To translate from a 4-letter language to a 20-letter language requires that the 4 nucleic acid letters be used in groups of a minimum of 3 letters. Then there will be enough combinations of letters for the larger alphabet of the amino acids. Thus there can be at least one nucleic acid "word," built of 3 letters, for each of the 20 amino acids. This is exactly how

NH_2CHC- with O double bonded and R below

Aminoacyl unit

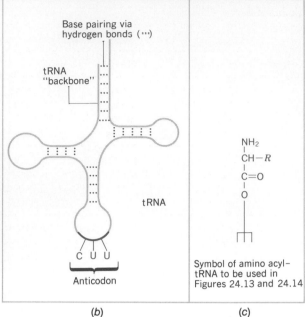

(a)   (b)   (c)

FIGURE 24.12

Transfer RNA (tRNA). (a) The tRNA for phenylalanine. Its anticodon occurs at the tip of the base, and the place where the phenylalanyl residue can be attached is at the upper left point. (b) Highly schematic representation of the model to highlight the occurrence of double-stranded regions. (c) Symbol of the aminoacyl–tRNA unit that will be used in succeeding figures. (Molecular model courtesy of Academic Press/Molecular Design Incorporated.)

tRNA is structured for its work of translating between the nucleic acid and the polypeptide languages. It is able to connect the three-letter codon "words" aligned along an mRNA backbone to a matching alignment of sidechains of individual amino acids in a polypeptide being made.

One part of a tRNA molecule can recognize a codon because it carries a triplet of bases complementary to the codon. This triplet on tRNA is called an **anticodon.** There are many individual kinds of tRNA, each carrying a particular anticodon. In the tRNA molecule in Figure 24.12, the triplet CUU is its anticodon.

Another functional group of each tRNA molecule, an OH group at a terminal ribose unit, can attach a particular aminoacyl unit (by an ester bond). We can use the symbol tRNA-aa for this new compound, where we use "aa" for the aminoacyl group. The anticodon on each tRNA-aa molecule is unique. A given tRNA-aa molecule, therefore, can be brought into alignment only with one codon of mRNA at a polysome. This ensures that the attached aa unit is brought into place in the correct sequence.

To be able to work with the polypeptide language, each tRNA is able to recognize the amino acid that "belongs" to it. Complementary fitting is at work along with special enzymes.

As each tRNA-aa molecule docks at its unique location at the mRNA chain, the growing polypeptide chain tethered "next door" is transferred to it. You can see that a unique series of codons *can allow the polypeptide chain to grow only with an equally unique sequence of amino acid residues.* The pairing of the triplets of bases between the codons and anticodons can permit only one sequence.

The Genetic Code Is the Correlation between Codons and Amino Acids Table 24.1 displays the known correlations of codons with amino acids, the **genetic code.** Most amino acids are associated with more than one codon, which apparently minimizes the harmful effects of genetic mutations. (These, in molecular terms, are small changes in the structures of genes.) Phenylalanine, for example, is coded either by UUU or by UUC. Alanine is coded by any one of the four triplets: GCU, GCC, GCA, or GCG. Only two amino acids go with single codons, tryptophan (Trp) and methionine (Met).

■ UUU and UUC are called *synonyms* for Phe.

686

is now a *di*peptidyl – tRNA unit over the P site. The third aminoacyl – tRNA now finds its anticodon-to-codon matching at the third codon of the mRNA strand, which is over the recently vacated A site.

Elongation now occurs; the dipeptide unit transfers to the amino group of the third amino acid. Another peptide bond forms. And a tripeptidyl system has been made.

The cycle of steps can now occur again, starting with a movement of the mRNA chain relative to the ribosome that shifts this tripeptidyl – tRNA and positions it over the P site. The fourth amino acid residue is carried to the mRNA; the tripeptidyl unit transfers to it to make a tetrapeptidyl unit, and so forth. This cycle of steps continues until a special chain-terminating codon is reached.

■ Polypeptide synthesis can occur at several ribosomes moving along the mRNA strand at the same time.

3. **Termination of polypeptide synthesis**   Once a ribosome has moved down to one of the chain-terminating codons of the mRNA strand (UAA, UAG, or UGA), the polypeptide synthesis is complete, and the polypeptide is released. The ribosome can be reused, and the polypeptide acquires its higher levels of structure. The folding and aligning of polypeptides into these higher structural levels are often directed by folding catalysts, enzymes that guide the polypeptides toward the correct overall aggregations and geometries. Because they have a chaperone function, these enzymes are actually called *chaperonines*.

■ In mammals, it takes only about one second to move each amino acid residue into place in a growing polypeptide.

A *Repressor* Can Keep a Gene Switched Off until an *Inducer* Acts

As we have said, every cell in the body carries the entire set of genes. In any given tissue, therefore, most must be permanently switched off. It wouldn't do, for example, for a cell to be making a protein-digesting enzyme that would then catalyze the destruction of its own proteins.

Because so many steps occur between the divided gene and the finished polypeptide, there are a large number of points at which the cell can control the overall process. We briefly discuss only one way this is done, one classic example, and we have to rely heavily on Figure 24.15. It comes from discoveries involving a bacterium called *Escherichia coli*, or *E. coli*, a one-celled organism found in our intestinal tracts.

E. coli are able to obtain all the carbon atoms they need from the metabolism of lactose, milk sugar. They must first hydrolyze this disaccharide to galactose and glucose, and the enzyme β-galactosidase is essential for this reaction. The enzyme, of course, is not needed until lactose is available, so the gene for the enzyme is switched off until lactose molecules arrive. We must leave some details to more advanced treatments. Our aim is only to illustrate in broad outline one of the most important kinds of activities at the molecular level of life, the control of gene expression. What we describe here is an *inducible gene* and just one strategy for the control of genes.

The DNA segment responsible for the structure of β-galactosidase is in a region of the DNA strand called the *structural gene* (see Fig. 24.15). Next to it, on the same DNA strand, is a segment of DNA to which a **repressor** molecule can bind. Because this DNA unit is part of the switch for the operation of the structural gene, it is named the *operator site*. Immediately next to it, down the same DNA chain, is a small segment called the *promoter site*. It holds in readiness the enzyme, *DNA polymerase*, needed to transcribe the structural gene. But this enzyme cannot work until the operator site is switched on. The structural gene plus the repressor and operator sites, taken together, make up a unit called the *lac operon*.

The repressor is a polypeptide made at the direction of its own gene, a *regulator gene*, located just above the promoter site in Figure 24.15. When repressor molecules are present, they bind to the DNA of the operator site. *This binding of the repressor is what prevents gene transcription and translation.*

Now suppose that some molecules of lactose appear. The enzyme made with the aid of the structural gene is now needed. The first of the lactose molecules to arrive are altered slightly (by steps we will not discuss) into **inducer** molecules. These have shapes that enable them to fit in some way to the repressor molecule at the operator site. As they attach to the repressor, the shape of the repressor becomes changed so that it no longer can stick to the operator gene, *and it drops off.* DNA polymerase is now released to work with the structural gene to help to make the *mRNA* needed for the manufacture of β-galactosidase.

FIGURE 24.15

Repression and induction of the enzyme β-galactosidase in *E. coli*. The inducer is lactose, whose hydrolysis requires this enzyme. The first lactose molecules to arrive bind to and remove the repressor from the operator gene. Now the structural gene is free to direct the synthesis of mRNA coded to make the enzyme. The repressor is made at the direction of another gene *(top)*, the regulator gene.

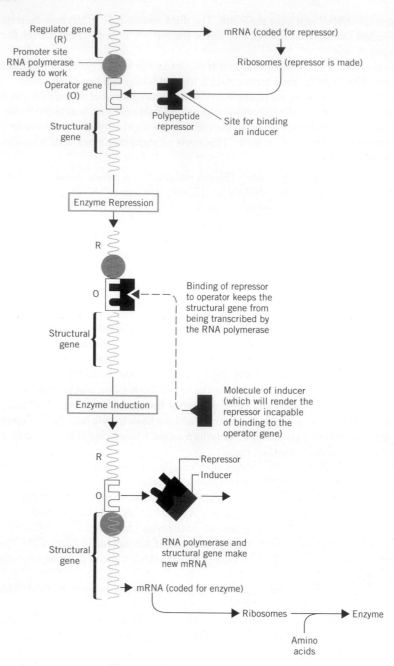

The overall process we have described is called **enzyme induction**, and β-galactosidase is one of many *inducible enzymes*. Don't let the beauty of it be smothered by the details. If no substrate (lactose in this example) is in the cell, an enzyme isn't needed, and the genetic machinery for making it stays switched off. Only when the enzyme is needed, signaled by the arrival of its substrate, is it manufactured. Many human enzymes no doubt work in the same way. Without thousands of events at the molecular level of life happening *automatically*, that is, without human thought, we couldn't imagine minds free enough for any higher thoughts.

Many Antibiotics Kill Bacteria by Interfering with Genetic Translation or Transmission Bacteria, like us, have to manufacture polypeptides to stay alive and multiply. Inhibiting the synthesis of bacterial polypeptides at any one of several steps in the overall process will kill the bacteria. Streptomycin, for example, inhibits the initiation of polypeptide synthesis. Chloramphenicol inhibits the ability to transfer newly arrived aminoacyl units to the elongating strand. The tetracyclines inhibit the binding of tRNA-aa units when they arrive at the ribosome. Actinomycin binds tightly to DNA. Erythromycin, puromycin, and cycloheximide interfere with elongation.

■ These inhibiting activities render the *enzymes* for the various steps inactive.

X Rays and Atomic Radiations Can Damage Genes   Atomic radiations, particularly X rays and gamma rays, go right through soft tissue where they create unstable ions and radicals (particles with unpaired electrons). New covalent bonds can form from radicals. Even side-chain bases might be annealed together. If such events were to happen to DNA molecules, the polypeptide eventually made at their direction might be faulty. The DNA made by replication might then be seriously altered. If the initial damage to the DNA is severe enough, replication won't be possible and the cell involved is reproductively dead. This, in fact, is the *intent* when massive radiation doses are used in cancer therapy — to kill cancer cells.

Chemicals that are used in cancer therapy often mimic radiations by interfering with the genetic apparatus of cancer cells. Such chemicals are called **radiomimetic substances.**

24.5

VIRUSES

Viruses take over the genetic machinery of host cells to make more viruses. They are at the borderline of living systems and are generally regarded only as unique packages of dead chemicals, except when they get inside their host cells. They then seem to be living things, because they reproduce. They are a family of materials called **viruses.**

Viruses Consist of Nucleic Acids and Proteins   Viruses are agents of infection made of nucleic acid molecules surrounded by overcoats of protein molecules. Unlike a cell, a virus has either RNA or DNA, but not both. Viruses must use host cells to reproduce because they can neither synthesize polypeptides nor generate their own energy for metabolism. The simplest virus has only four genes and the most complex has about 250.

■ Viruses are intracellular parasites.

Each kind of virus has something on its surface that is complementary to something on the surface of the host cell. The glycoproteins of membranes are involved. Thus each virus has one particular kind of host cell and does not attack all kinds. At least, this theory helps us understand how viruses are so unusually selective. A virus that attacks, for example, the nerve cells in the spinal cord has no effect on heart muscle cells. A large number of viruses exist that do not affect any kind of human cell. Many viruses attack only plants.

■ A complete virus particle *outside* its host cell is called a *virion.*

The protein overcoat of some viruses includes an enzyme that catalyzes the breakdown of the cell membrane of the host cell. When such a virus particle sticks to the surface of a host cell, its overcoat catalyzes the opening of a hole into the cell. Then the viral nucleic acid squirts into the cell, or the whole virus might move in. Each virus that works this way evidently has its own unique membrane-dissolving enzyme.

Once a virus particle or the parts of one gets inside its host cell one of two possible fates awaits it. It might become turned off and change into a *silent gene;* or it might take over the genetic machinery of the cell and reproduce so much of itself that it bursts the host cell walls. The new virus particles that spill out then infect neighboring host cells, and in this way the infection spreads. A virus that has become a silent gene might later be activated. Some cancer-causing agents, including ultraviolet light, may initiate cancer by this mechanism.

Most viruses contain RNA, not DNA, so the manufacture of more of their RNA must somehow be managed without the direction of DNA.

RNA Viruses Either Carry or Make Enzymes for Synthesizing More RNA   RNA viruses have to solve a major problem if they are to infect a host cell. Host cells normally (in health) have no enzyme that can direct the synthesis of a copy of an RNA molecule *from the* *instructions of another RNA.* In healthy host cells, copies of RNA molecules are made by the direction of DNA, not RNA, like the synthesis of hnRNA directed by DNA.

Two basic solutions to this problem occur involving two different enzymes. One is called *RNA replicase,* and it can catalyze the manufacture of RNA *from the directions encoded on RNA.* Some viruses carry this enzyme. Others direct the host cell to synthesize RNA replicase. Either way, once RNA replicase is inside the host cell, it handles the manufacture of the *mRNA* needed to make more viral RNA and protein, so new virus particles can form.

The second way by which RNA directs the synthesis of RNA occurs with viruses that carry a DNA polymerase enzyme. This enzyme, called *reverse transcriptase,* directs the synthesis of viral DNA, which subsequently is used to code for more viral RNA. Reverse transcriptase can use RNA information to make DNA. This is unusual because it's normally the other way around; DNA information is used to make RNA.

■ David Baltimore and Howard Temin shared the 1975 Nobel prize in physiology and medicine (together with Renato Dulbecco) for the discovery of reverse transcriptase.

Four Basic Strategies Are Used by RNA Viruses to Make More RNA

Figure 24.16 outlines the strategies used by four kinds of RNA viruses to manufacture more of their own RNA. The (+) and (−) signs denote single-stranded RNA molecules of opposite complementarity. A (±) sign denotes a double-stranded nucleic acid. By convention, the messenger RNA that the virus must make is designated (+)*mRNA.* Refer to this figure as we briefly discuss four kinds of RNA viruses.

The polio virus contains a single-stranded RNA molecule. Inside its host cell, it functions as a messenger RNA at the host's ribosomes for the synthesis of both overcoat proteins and molecules of RNA replicase. This enzyme then synthesizes (−)RNA molecules which, in turn, direct the synthesis of the (+)mRNA that now takes over the host cell's genetic machinery.

The rabies virus has (−)RNA molecules, but they are not messengers in their host cells. This virus carries its own RNA replicase into the host cells where it directs the synthesis of (+)RNA. This is then used to make the new (−)RNA molecules required for additional virus particles.

■ In the 1919 worldwide influenza epidemic, 20 million people died.

The influenza virus also has (−)RNA, each molecule bearing 10 genes. During infection, segments of the RNA exist that can become resorted as intact, new (−)RNA forms. In this way the influenza virus is able to change into new strains that make the job of immunizing large populations against influenza for long periods particularly difficult.

■ *Reo* in reovirus is from **r**espiratory, **e**nteric **o**rphan—a virus in search of a disease.

The reovirus of Figure 24.16 is present in the intestinal and respiratory tracts of mammals without causing disease. Its RNA is double-stranded RNA, or (±)RNA, and it carries its own RNA polymerase. It can use this on both (+) and (−) strands to make more viral (±)RNA.

The retroviruses (Fig. 24.16) form a family of viruses with (+)RNA that cannot make more RNA without first making double-stranded DNA. To do this, retroviruses carry reverse transcriptase. Thus in retroviruses, the flow of information goes from RNA to DNA. (This explains the *retro-* prefix; it suggests a reversal or retrograde action.) Reverse transcriptase uses retroviral (+)RNA to direct the formation of (−)DNA, which directs the formation of (±)DNA. This DNA is incorporated into the host cell's collection of genes and then directs the synthesis of the (+)mRNA needed to make more retrovirus particles.

Cancer-Causing Viruses Transform Normal Genes in Host Cells

The retroviruses include the only known cancer-causing RNA viruses, technically termed the *oncogenic RNA viruses.* (Several DNA-based viruses also cause cancer.) They transform host cells so that they

■ *Oncogenic* means cancer-inducing.

FIGURE 24.16
Overall strategies used by RNA viruses to make their messenger RNA

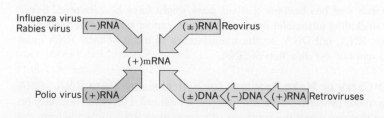

INTERFERON

The term *interferon* refers to a family of similar polypeptides that are chiefly characterized by an ability to inhibit viruses. There are at least three types in humans, designated as α-, β-, and γ-interferon. The interferons are glycoproteins that have about 150 amino acid residues per molecule. α-Interferon is from white blood cells; β-interferon is from certain connective tissue (fibroblasts); and γ-interferon is from cells of the immune system (lymphocytes).

Viruses are potent stimuli for the induction of interferon synthesis in humans. When an invading virus particle first encounters a white blood cell, particularly the type made in lymph tissue, the mechanism in the cell for making the mRNA coded for interferon is switched on. In a short time, the cell is manufacturing and releasing interferon, which then acts as a signal to other cells to make interferon, too. By binding to its host cell, α-interferon directly activates an enzyme in the cell called tyrosine kinase. This enzyme then brings together three polypeptide subunits and so activates a transcription factor that enters the cell nucleus where it completes its work. Circulating cells called killer cells are activated to attack and destroy virus particles. In this way, interferon works to inhibit a viral infection.

The discovery of interferon in 1957 came about when two scientists, Alick Isaacs and Jean Lindenmann, wondered why victims of one viral disease never seemed to come down with a second viral disease at the same time. It seemed to them that something in the first viral attack triggered a mechanism that provided protection against an attack by a different virus. The search for this "something" led to the discovery of interferon.

In the earliest clinical trials, interferon provided relief and sometimes a cure to a few patients who had rapidly acting cancer, such as osteogenic sarcoma (bone cancer), multiple myeloma, melanoma (a form of skin cancer), breast cancer, and some forms of leukemia and lymphoma. However, it takes nearly 25,000 pints of blood to make 100 mg of interferon, at a cost of a few billion dollars, so large-scale clinical tests really depend on interferon made by recombinant DNA technology.

The initial excitement that interferon would be the cure-all, free of side effects, for any kind of viral disease or any kind of cancer has faded. Yet tests with experimental animals continue to be promising, and several medical groups and biotechnology companies are continuing their testing efforts.

In 1988, the U.S. Food and Drug Administration approved the use of α-interferon for the treatment of genital warts. In a clinical trial, it eliminated these warts in over 40% of the patients, and in another 24% the warts were reduced in size. About 8 million Americans suffer from this sexually transmitted virus, which has been linked to cervical cancer. Other treatments were usually ineffective. Pregnant women were advised against this treatment because it could induce abortions.

---

grow chaotically and continuously. They do this by changing normal genes in the host cell to *oncogenes*, genes that henceforth are able to continue the cancerous growth.

The Host Cell of the AIDS Virus Is Part of the Human Immune System The acquired immunodeficiency syndrome, AIDS, is also caused by a retrovirus, the human immunodeficiency virus or HIV. One reason why this virus is so dangerous is that its host cell, the T4 lymphocyte, is a vital part of the human immune system, as we mentioned in Special Topic 22.2, page 622. By destroying T4 lymphocytes, HIV exposes the body to other infectious diseases, like pneumonia, or to certain rare types of cancer. One of the strategies being used to retard the development of AIDS, if not cure it in those with this syndrome, is to offer the HIV virus a nucleotide that can bind to its *reverse transcriptase* but, once bound, inhibit the further work of this enzyme. Thus AZT, a nucleotide with a modified sugar unit, has been a commonly used nucleotide medication against AIDS. Used by itself, however, and not as part of a combination of drugs, AZT does not delay the onset of AIDS in patients known to be infected with the HIV virus.

AZT

Some Viral Infections Produce Interferons, Which Fight Further Infection One of the many features of the body's defense against some viral infections is a small family of polypeptides called the *interferons*, described further in Special Topic 24.2. Supplies of interferons are now available by genetic engineering, which we will study next.

24.6

RECOMBINANT DNA TECHNOLOGY AND GENETIC ENGINEERING

Single-celled organisms can be made to manufacture the proteins of higher organisms.

Human insulin, human growth hormone, and human interferons are now being manufactured by a technology that involves the production of *recombinant DNA*. With the aid of Figure 24.17, we'll learn how this technology works. It represents one of the important advances in scientific technology of this century. It has permitted the *cloning,* the synthesis of identical copies, of a number of genes. The use of recombinant DNA to make genes and the products of such genes is called **genetic engineering.**

■ The term *cloning* is used for the operation that places new genetic material into a cell where it becomes a part of the cell's gene pool. The new cells that follow this operation are called *clones.*

Genes Alien to Bacteria Can Be Inserted into Bacterial Plasmids   Bacteria generally make polypeptides using the same genetic code as humans. There are some differences in the machinery, however. An *E. coli* bacterium, for example, has DNA not only in its single chromosome but also in large, circular, supercoiled DNA molecules called **plasmids.** Each plasmid carries just a few genes, but several copies of a plasmid can exist in one bacterial cell. Each plasmid can replicate independently of the chromosome.

The plasmids of *E. coli* can be removed and given new DNA material, such as a new gene, with base triplets for directing the synthesis of a particular polypeptide. It can be a gene completely alien to the bacteria, like the subunits of human insulin, or human growth hormone, or human interferon. The DNA of the plasmids is snipped open by special enzymes called *restriction enzymes* absorbed from the surrounding medium. This medium can also contain naked DNA molecules, such as those of the gene to be cloned. Then, with the aid of a DNA-knitting enzyme called *DNA ligase,* the new DNA combines with the open ends of the plasmid. This recloses the plasmid loops. The DNA of these altered plasmids is called **recombinant DNA.** The altered plasmids are then allowed to be reabsorbed by bacterial cells.

The remarkable feature of bacteria with recombinant DNA is that when they multiply, the plasmids in the offspring also have this new DNA. When these multiply, still more altered plasmids are made.

Between their cell divisions, the bacteria manufacture the proteins for which they are genetically programmed, including the proteins specified by the recombinant DNA. In this way, bacteria can be tricked into making the *human* proteins we have mentioned. The technology isn't limited to bacteria; yeast cells work, too.

People who rely on the insulin of animals, like cows, pigs, or sheep, sometimes experience allergic responses. They are also vulnerable to the availability of the pancreases of these animals, because they once were the *only* sources of insulin. The ability to make *human* insulin, therefore, has been a welcome development for diabetics.

If an interferon should prove effective in treating various types of viruses, hepatitis, and possibly even certain kinds of cancer, the ability to make this unusually rare and costly substance in large quantities at low cost by recombinant DNA technology will truly be a major technological advance. In fact, just the clinical tests alone would be too costly without the availability of manufactured interferon.

Recombinant DNA technology has interacted with archeology in an interesting way. In the early 1980s scientists were able to remove segments of DNA molecules from an Egyptian mummy and from an extinct horse-like animal (a quagga) and reproduce these segments using bacteria. No segment was long enough to include a complete gene, but the technique no doubt will be developed as another tool for the study of evolutionary history. To amplify faithfully (clone) the amount of "archeologic" DNA from a tiny sample, the *polymerase chain reaction* is used (see Special Topic 24.1).

■ Kary B. Mullis of the United States was a cowinner of the 1993 Nobel Prize in chemistry for discovering the polymerase chain reaction. He shared the prize with Michael Smith of Canada, who discovered how to make specific changes in the molecular coding of DNA so that custom-made proteins can be synthesized by living organisms.

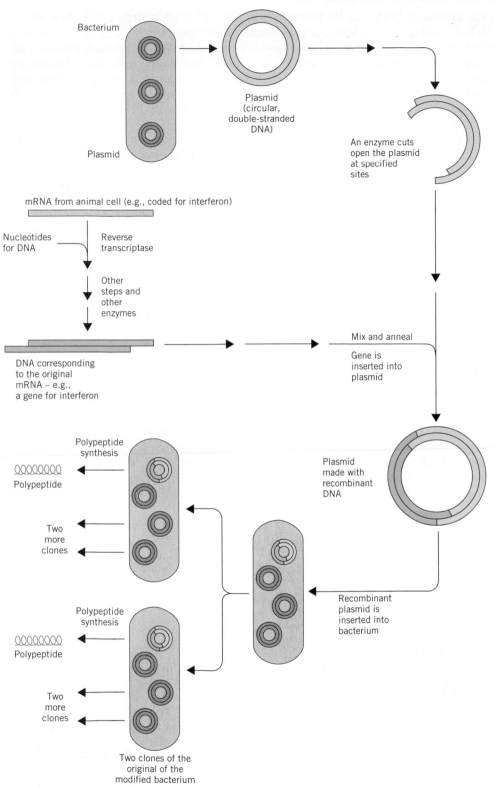

FIGURE 24.17
Recombinant DNA is made by inserting a DNA strand, coded for some protein not made by the bacteria, into the circular DNA of the bacterial plasmid.

Bacterium

Plasmid

Plasmid
(circular,
double-stranded
DNA)

An enzyme cuts
open the plasmid
at specified
sites

mRNA from animal cell (e.g., coded for interferon)

Nucleotides
for DNA

Reverse
transcriptase

Other
steps and
other
enzymes

DNA corresponding
to the original
mRNA – e.g.,
a gene for interferon

Mix and anneal

Gene is
inserted into
plasmid

Plasmid
made with
recombinant
DNA

Polypeptide
synthesis

Polypeptide

Two
more
clones

Recombinant
plasmid is
inserted into
bacterium

Polypeptide
synthesis

Polypeptide

Two
more
clones

Two clones of the
original of the
modified bacterium

Recombinant DNA Can Be Inserted into Cells of Higher Organisms Sometimes, before it will function, a desired polypeptide has to be "groomed" by a cell *after* it has been made by genetic translation. It might have to be attached to a carbohydrate molecule, for example. Bacteria lack the enzymes for such grooming work, so cells of higher organisms are used. When these cells are large enough, the new DNA can be inserted directly into them using glass pipets of extremely small diameters ($0.1~\mu$m). Although only a small fraction of such inserted DNA becomes taken up into the cell's chromosomes, it can be enough when amplified by successive cell divisions.

Because viruses are able to get inside cells, some have been used to carry new DNA along. Retroviruses can be customized for this purpose, for example (Section 24.7).

Genetic Engineering Offers Major Advances in Medicine One of the hopes of genetic engineering research is to have ways to correct genetic faults. As we will study in the next section, a number of undesirable conditions are caused by flawed or absent genes. Dwarfism, for example, is caused by a lack of growth hormone, a relatively small polypeptide. In experiments with mice, genetic engineering has successfully introduced growth hormone into mice, with dramatic effects on the mouse size.

In another medical application of genetic engineering, the smallpox vaccine is being remodeled to provide altered forms that might give immunity to many other diseases, ranging from malaria to influenza.

■ This hormone, called the atrial natriuretic factor (ANF), was discovered in the mid 1980s, but had been long-suspected.

A polypeptide hormone made by the heart, which reduces blood pressure, can be manufactured by genetic engineering and used by victims of high blood pressure. As we mentioned in Section 22.5, the clot-dissolving enzyme called tissue plasminogen activator has been genetically engineered for use in reducing the damage to heart tissue following a sudden heart attack.

The alteration of a sticky protein that mussels use to cling to underwater surfaces is being studied to find an adhesive that can be used after surgery.

The list of potential applications of genetic engineering to health problems grows yearly. It seems likely that kidney dialysis patients will need fewer blood transfusions if a blood-cell-producing substance, erythropoietin, can be made by this technology. Hemophiliacs who lack a blood-clotting factor may have it available. The synthesis of a number of drugs by genetic engineering is being studied, some that might be used against cancer.

Agriculture Is Affected by Genetic Engineering In agriculture, genetic engineering is being tried for developing pest-resistant strains of plants, even plants that manufacture their own fertilizer. Gene manipulations have developed cows that yield much larger quantities of milk. Pigs with leaner meat have been produced. The list of similar applications is certain to grow.

The list of problems for society will also grow. In an age of large milk surpluses, some ask, who needs cows that can make more? What of the dairy farmers with huge investments already? These and questions like them are serious enough that if you follow newspapers and news magazines, you no doubt will read of the debates. There has never been a technology yet that had no associated problems. The extent to which these problems are accepted depends on human values expressed through laws and the acceptance of laws.

24.7

HEREDITARY DISEASES AND GENETIC ENGINEERING

About 4000 inherited disorders in humans are caused directly or indirectly by flawed genes.

In Cystic Fibrosis, a Defective Gene Makes a Defective Membrane Protein The victims of cystic fibrosis overproduce a thick mucus in the lungs and the digestive tract, which clogs these systems and which often leads to death in children. About one person in 20 carries the defective gene associated with this disease, and it hits about one in every 1000 newborns. The gene that is defective in cystic fibrosis normally directs the synthesis of a transmembrane protein that lets chloride ion pass out of a cell. For those with the defective gene, this passage of chloride ion is impaired. When Cl⁻ passes unimpaired out of a cell *it also takes water along*, probably for osmotic pressure reasons. When water is kept in diminished supply outside of the cells of the lungs and airways, however, the mucus thickens and does not flow properly. The thickened mucus makes breathing more difficult and provides a breeding ground for the bacteria that cause a certain form of pneumonia.

■ The membrane protein is called the CFTR protein after cystic fibrosis, transmembrane conductance regulator.

Gene Therapy for Cystic Fibrosis Is Being Tried In late 1992, teams of scientists received the permission of a Federal panel to attempt *gene therapy* for cystic fibrosis. The general plan is to take advantage of the ability of retroviruses (page 694) to insert DNA into a chromosome of the host cell. If the piggybacked DNA is a replacement for the DNA of a defective gene, the host cell will thereby acquire the DNA needed to counter the inherited defect. The *surface* of the altered retrovirus particle is left the same; it still has the original proteins by means of which the particle "recognizes" and sticks to the host cell. Only the retroviral particle interior is altered so that the virus particles cannot manufacture more of themselves in the host and so cause serious infection. Instead, the altered virus particles must launch the desired genetic correction based on the correct DNA. One of the common cold viruses is the designated delivery vehicle for gene therapy against cystic fibrosis. The material is simply squirted into the airways. If corrected genes become installed in only 10 percent of the airway cells, it is thought that the defects in Cl⁻ and water flows will be rendered nonlethal.

Gene Therapy Has Worked against Enzyme Deficiencies The very first effort to cure a human disease using gene therapy began in 1990. It involved two little girls, ages 4 and 9, born with a defective ADA gene, which left them with almost no natural immunity. The children were injected with white blood cells that had been given the correct gene. Two years later, the girls had functioning immune systems and instead of having to lead very isolated lives were in public school. Sickle-cell anemia is another disease caused by a defective gene, as we described in Special Topic 21.1.

■ The ADA gene makes the enzyme *adenosine deaminase*, which is vital to the immune system.

Albinism Is a Genetic Defect Albinism, the absence of pigments in the skin and the irises of the eyes, is caused by a defect in a gene that directs the synthesis of an enzyme that is needed to make these pigments. The pigments absorb the ultraviolet rays in sunlight, radiation that can induce cancer, so victims of albinism are more susceptible to skin cancer.

PKU Is a Genetic Disease Treated Nutritionally Phenylketonuria, or PKU disease, is a brain-damaging genetic disease in which abnormally high levels of the phenylketo acid called phenylpyruvic acid occur in the blood. This condition causes permanent brain damage in the newborn. Because of a defective gene, an enzyme needed to handle phenylalanine is not made, and this amino acid is increasingly converted to phenylpyruvic acid.

$$C_6H_5CH_2\overset{\overset{\displaystyle O}{\|}}{C}CO_2H \qquad C_6H_5CH_2\underset{\underset{\displaystyle NH_3^+}{|}}{C}HCO_2^-$$

Phenylpyruvic acid   Phenylalanine

PKU can be detected by a simple blood test within four or five days after birth. If the infant's diet is kept very low in phenylalanine, it can survive the critical danger period and experience no brain damage. The infant's diet should include no aspartame, a low-calorie sweetener

■ There is enough phenylalanine in just one slice of bread to be potentially dangerous to a PKU infant.

(contained in NutraSweet®), because it is hydrolyzed by the digestive processes to give phenylalanine.

$$^+NH_3CHC-NHCHCOCH_3 + 2H_2O \longrightarrow {}^+NH_3CHCO_2^- + {}^+NH_3CHCO_2^- + CH_3OH$$

O || O ||			
CH$_2$	CH$_2$	CH$_2$	CH$_2$
CO$_2^-$	C$_6$H$_5$	CO$_2^-$	C$_6$H$_5$
Aspartame		Aspartic acid	Phenylalanine	Methanol

Maintaining a low-phenylalanine diet, of course, is not the easiest and best solution, so genetic engineers are working to correct the fundamental gene defect.

The *Human Genome Project* Aims to Map All Human Genes and Determine Their DNA Sequences The entire complement of genetic information of a species is called its **genome.** The *Human Genome Project,* formally launched in 1990, intends to discover the "map" of every human chromosome and to determine the sequence of bases in every human gene. It's perhaps the largest single project ever attempted in molecular biology, and its total cost may come within range of (but will probably be less than) what it cost to put an astronaut on the moon. If the genome is a library, if the chromosomes are book sections, and if the genes are individual books, then the Human Genome Project's goal in *mapping* chromosomes is to create the card catalog. "In which section of the library is the gene for cystic fibrosis located?" Answer: on chromosome 7. "In which particular part (shelf) of the chromosome 7 is the gene responsible for cystic fibrosis located?" Answer: in region q31. Now comes the *sequencing* question; it's not the same as the mapping question. "What is the sequence of bases in the cystic fibrosis gene?" The answer is known, but we can't give it here; the gene involves over 6000 bases. "What's wrong with this gene to make it cause cystic fibrosis?" For about 70 percent of cystic fibrosis mutations, there is a deletion (an absence) of three bases in exon number 10. So much human suffering from so small a molecular defect!

■ In early 1993, the location of the gene for Huntington's disease, a deadly neurological disorder (page 640) was found near the tip of chromosome 4.

SUMMARY

Hereditary information The genetic apparatus of a cell is mostly in its nucleus and consists of chromatin, a complex of DNA and proteins (histones). Strands of DNA, a polymer, carry segments that are individual genes. Chromatin replicates prior to cell division, and the duplicates segregate as the cell divides. Each new cell thereby inherits exact copies of the chromatin of the parent cell. If copying errors are made, the daughter cells (if they form at all) are mutants. They may be reproductively dead—incapable of themselves dividing—or they may transmit the mutant character to succeeding cells. The expression of this might be as a cancer, a tumor, or a birth defect. Atomic radiations, particularly X rays and gamma rays, are potent mutagens, but many chemicals, those that are radiomimetic, mimic these rays.

DNA Complete hydrolysis of DNA gives phosphoric acid, deoxyribose, and a set of four heterocyclic amines, the bases adenine (A), thymine (T), guanine (G), and cytosine (C). The molecular backbone of the DNA polymer is a series of deoxyribose units joined by phosphodiester groups. Attached to each deoxyribose is one of the four bases. The order in which triplets of bases occur is the cell's way of storing genetic information.

In higher organisms, a gene consists of successive groups of triplets, the exons, separated by introns. Thus the gene is a split system, not a continuous series of nucleotide units. DNA exists in cell nuclei as duplex DNA, double helices held to this geometry largely by hydrophobic forces. The base A always pairs with the base T and C always pairs with G. Using this structure and the faithfulness of base pairing, Crick and Watson explained the accuracy of replication. After replication, each new double helix has one of the parent DNA strands and one new, complementary strand.

RNA RNA is similar to DNA except that in RNA ribose replaces deoxyribose and uracil (U) replaces thymine (T). Four main types of RNA are involved in polypeptide synthesis. One is rRNA, which is in ribosomes. A ribosome contains both rRNA and proteins that have enzyme activity needed during polypeptide synthesis.

mRNA is the carrier of the genetic message from the nucleus to the site where a polypeptide is assembled. mRNA results from a chemical processing of the longer RNA strand, hnRNA, which is made directly under the supervision of DNA.

tRNA molecules are the smallest RNAs to participate directly in polypeptide synthesis. Their function is to convey aminoacyl units to the polypeptide assembly site. tRNAs recognize where they are to go by base pairing between an anticodon on the tRNA molecule and its complementary codon on mRNA. Both codon and anticodon consist of a triplet of bases.

Polypeptide synthesis Genetic information is first transcribed when DNA directs the synthesis of mRNA. Each base triplet on the exons of DNA specifies a codon on mRNA. The mRNA moves to the cytoplasm to form an elongation complex with subunits of a ribosome and the first and second tRNA-aa unit to become part of the developing polypeptide.

The ribosome then rolls down the mRNA as tRNA-aa units come to the mRNA codons during the moment when the latter are aligned over the proper enzyme site of a ribosome. Elongation of the polypeptide then proceeds to the end of the mRNA strand or to a chain-terminating codon. After chain termination, the polypeptide strand leaves, and it may be further modified to give it its final N-terminal amino acid residue. Chaperonine enzymes guide the polypeptides into their final geometries.

The whole operation can be controlled by a feedback mechanism in which an inducer molecule removes a repressor of the gene, thus letting the gene work.

Several antibiotics inhibit bacterial polypeptide synthesis, which causes the bacteria to die.

Viruses Viruses are packages of DNA or RNA encapsulated by protein. Once they get inside their host cell, virus particles take over the cell's genetic machinery, make enough new virus particles to burst the cell, and then repeat this in neighboring cells. Viruses are implicated in human cancer. Some viruses make new RNA under the direction of existing RNA, using RNA replicase. Others, the retroviruses, use RNA and reverse transcriptase to make DNA, which then directs the synthesis of new RNA.

Recombinant DNA Recombinant DNA is DNA made from bacterial plasmids and DNA obtained from another source and encoded to direct the synthesis of some desired polypeptide. The altered plasmids are reintroduced into the bacteria, where they become machinery for synthesizing the polypeptide (e.g., human insulin, or growth hormone, or interferon). Yeast cells can be used instead of bacteria for this technology.

Cells of higher organisms are also used as sites for inserting new DNA. In this kind of genetic engineering, microsyringes or tailored viruses have been used to insert the DNA. Many applications in medicine and agriculture exist.

Gene therapy Hereditary diseases stem from defects in DNA molecules that either prevent the synthesis of necessary enzymes or that make the enzymes in forms that won't work. The identification of the chromosomes bearing the defective genes has enabled the analyses of the genes themselves. In gene therapy, it is hoped that healthy genes can be substituted for those that are defective. Gene therapy for cystic fibrosis entails using a modified retrovirus to convey correct DNA material into the host cells without causing a viral infection.

REVIEW EXERCISES

The answers to Review Exercises whose numbers are in color are found in Appendix C. The answers to the other Review Exercises are found in the Study Guide that accompanies this book. The more challenging questions are marked with asterisks.

The Cell

24.1 What is the term used for each of the following?
(a) the *liquid* inside the cell but outside the nucleus
(b) the entire contents of the cell
(c) the region of the cell outside the nucleus
(d) the particle at which polypeptide synthesis occurs
(e) the particle in which ATP is synthesized
(f) the name of the chemical that makes up a gene
(g) the nucleoprotein material inside a cell nucleus
(h) the protein around which DNA strands are wound
(i) the fundamental unit of heredity

24.2 What is the relationship between a chromosome and chromatin?

24.3 The duplication of a gene occurs in what part of the cell?

24.4 In a broad, overall sense, what happens when DNA replicates?

Structural Features of Nucleic Acids

24.5 What is the general name for the chemicals that are most intimately involved in the storage and the transmission of genetic information?

24.6 The monomer units for the nucleic acids have what *general* name?

24.7 What are the names of the two sugars produced by the complete hydrolysis of all the nucleic acids in a cell?

24.8 What are the names and symbols of the four bases that are liberated by the complete hydrolysis of (a) DNA and (b) RNA?

24.9 How are all DNA molecules structurally alike?

24.10 How do different DNAs differ structurally?

24.11 How are all RNA molecules structurally alike?

24.12 What are the principal structural differences between DNA and RNA?

24.13 When DNA is hydrolyzed, the ratios of A to T and of G to C are each very close to 1 : 1, *regardless of the species investigated*. Explain.

24.14 What does base pairing mean, in general terms?

24.15 What is the chief stabilizing factor for the geometrical form taken by duplex DNA?

*24.16 If the AGGCTGA sequence appeared on a DNA strand, what would be the sequence on the DNA strand opposite it in a double helix?

24.17 The *accuracy* of replication is assured by the operation of what factors?

24.18 What is the relationship between a single molecule of single-stranded DNA and a single gene?

24.19 Suppose that a certain DNA strand has the following groups of nucleotides, where each lowercase letter represents a group several nucleotides long.

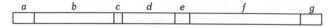

Which sections are likelier to be the introns? Why?

24.20 In general terms only, what particular contribution does a gene make to the structure of a polypeptide?

Ribonucleic Acids

24.21 What is the general composition of a ribosome, and what function does this particle have?

24.22 What is hnRNA, and what role does it have?

24.23 To which kind of RNA does the term "primary transcript" refer?

24.24 What is a codon, and what kind of nucleic acid is a continuous, uninterrupted series of codons?

24.25 What is an anticodon, and on what kind of RNA is it found?

24.26 Which triplet, ATA or CGC, cannot be a codon? Explain.

24.27 Which amino acids are specified by the following codons?
(a) UUU (b) UCC (c) ACA (d) GAU

*24.28 What are the anticodons for the codons of Review Exercise 24.27?

*24.29 Suppose that sections x, y, and z of the following hypothetical DNA strand are the exons of one gene.

3′				
AAA	GAA TAT CTC	AGG	GGT	TGT CTA
x		y		z

(with 5′ at the top right)

What is the structure of each of the following substances made under its direction?
(a) the hnRNA
(b) the mRNA
(c) the tripeptide that is made using the given genetic information (Use the three-letter symbol format for the tripeptide structure, referring as needed to Table 21.1 for these symbols.)

Polypeptide Synthesis

24.30 Use the identifying letters to arrange the following symbols or terms in the correct order in which they are synthesized in going from a gene to an enzyme. (Place the letter of the first material to be involved on the left.)

hnRNA	duplex DNA	polypeptide	mRNA
A	B	C	D
	< < <		
earliest to be involved			last to appear

24.31 What is meant by translation, as used in this chapter? And what is meant by transcription?

*24.32 To make the pentapeptide, Met-Ala-Trp-Ser-Tyr,
(a) What do the sequences of bases on the mRNA strand have to be?
(b) What is the anticodon on the first tRNA to move into place?

24.33 In general terms, explain how the overlapping of triplets on mRNA is avoided.

24.34 The discussion about the *lac operon* was included to illustrate what aspect of the chemistry of polypeptide synthesis?

24.35 Not all of the DNA of the *lac operon* codes for the synthesis of β-galactosidase. What is the name given to the part that is so coded? What names are given to other segments of the DNA of the *lac operon* and what is their function?

24.36 In general terms, what is a repressor, and what does it do?

24.37 What does an inducer do (in general terms)?

24.38 How do some of the antibiotics work at the molecular level?

24.39 The genetic code is the key to translating between what two "languages"?

24.40 What is meant by the statement that the genetic code is universal? And is the code strictly universal?

24.41 In general terms, how do X rays cause cancer?

24.42 How do X rays and gamma rays work in cancer therapy?

24.43 What is a radiomimetic substance?

Viruses

24.44 What is a virus made of?

24.45 In general terms, how does a virus discriminate among all possible host cells and "find" just one kind of host cell?

24.46 In general terms, once a virus particle has joined to the membrane of its host cell, what must occur next if the viral infection is to advance?

24.47 In general terms, in the systems having it, what does RNA replicase do? What is true about a normal host cell that requires a virus to have RNA replicase?

24.48 In general terms and in connection with the work of (some) viruses, what does reverse transcriptase do?

24.49 Which kind of mRNA enables a virus to use the host cell to make more virus particles, (−)mRNA or (+)mRNA?

24.50 If the mRNA in a given virus is (—)mRNA, what must the virus–host cell system accomplish if the virus is to multiply?

24.51 If the mRNA in a given virus is (—)mRNA, what enzyme is carried into the host cell by the virus?

24.52 What does the prefix *retro-* signify in retrovirus?

24.53 What is meant by a *silent gene* and where does it come from?

24.54 Some viruses are called *oncogenic RNA viruses*. What does "oncogenic" mean here?

24.55 What is the full name of the HIV system?

24.56 What is the host cell of HIV and how does this fact make AIDS so dangerous?

24.57 What is the theory concerning the action of AZT?

Recombinant DNA

24.58 What is a plasmid, and what is it made of?

24.59 What is the name of the enzyme that can snip open the DNA of a plasmid?

24.60 Recombinant DNA is made from the DNA of two different kinds of sources. What are they?

24.61 Recombinant DNA technology is carried out to accomplish the synthesis of what kind of substance (in general terms)?

24.62 What does "genetic engineering" refer to?

Hereditary Diseases

24.63 At the molecular level of life, what kind of defect is the fundamental cause of a hereditary disease?

24.64 The defective gene in cystic fibrosis leads to an impairment of what specific activity of the cells of the affected tissues? Why does this activity result in a problem with mucus?

24.65 Only in the broadest terms, what is gene therapy mean to accomplish?

24.66 A retrovirus instead of some other kind of virus is being used in gene therapy for cystic fibrosis. Why?

24.67 In altering the retrovirus used in gene therapy for cystic fibrosis, only the interior of the virus is changed. Why is the surface of the virus particle left alone?

24.68 What is the molecular defect in PKU, and how does it cause the problems of the victims? How is it treated?

DNA Typing (Special Topic 24.1)

24.69 What fact about cells makes it possible to use cells from any part of the body of a suspect for DNA fingerprinting in a rape case for which a semen sample has been obtained?

24.70 What is meant by the *polymerase chain reaction,* and why is it used as part of the technology of DNA typing?

24.71 A restriction enzyme separates DNA molecules into pieces given what general name? How are they used to give the genetic "bar code"?

Interferon (Special Topic 24.2)

24.72 What kinds of compounds are the interferons and what are they able to do that protects us from disease?

24.73 What scientific observation led scientists to look for something that would do what the interferons do?

24.74 What kinds of infectious agents stimulate the production of interferons?

24.75 How are interferons made for clinical trials?

Additional Exercises

24.76 Compare the structures of uracil and thymine (Fig. 24.2).
   (a) How do the structures differ?
   (b) Can the structural difference affect the hydrogen-bonding capabilities of these two bases?
   (c) How does the known behavior of uracil and thymine toward adenine bear on the answer to part (b)?

24.77 Write the structures of nucleotides that involve deoxyribose and each of the following bases.
   (a) adenine   (b) cytosine

*24.78 Consider the following compound.

   (a) Is it a mononucleotide, a dinucleotide, or a higher nucleotide (an oligonucleotide)? How can you tell?
   (b) Could it be obtained by the partial hydrolysis of DNA or RNA? How can you tell?
   (c) Where is the 5′ end, at the bottom or top of the structure as written?
   (d) In terms of the single-letter-symbols for bases, how is the structure of this compound written?

24.79 Consider the structure of AZT found on page 695.

(a) What is the name of its side chain base?

(b) The side chain base of AZT could form a hydrogen-bonded pair to which other side chain base among those that occur among the nucleic acids?

(c) Could AZT become hydrogen-bonded to side chains of *both* DNA and RNA? Or to only one of these two? (If so, which one?)

(d) If the N_3 group on the AZT molecule were replaced by an OH group, would the resulting molecule be a nucleotide? If so, would it be a nucleotide of RNA, DNA, both, or neither? If neither, what would have to be done to make the product of replacing N_3 by OH into a nucleotide?

BIOCHEMICAL ENERGETICS

25

When Gail Devers (USA) won a gold medal for the 100-m dash in the 1992 Olympic Games, her body mobilized an incredible burst of energy from chemicals, like glucose. In this chapter we'll study how the body handles such transformations at the molecular level.

25.1
ENERGY FOR LIVING—AN OVERVIEW
25.2
THE CITRIC ACID CYCLE

25.3
THE RESPIRATORY CHAIN

25.1

ENERGY FOR LIVING–AN OVERVIEW

High-energy phosphates, such as ATP, are the body's means of trapping the energy from the oxidation of the products of digestion.

We cannot use solar energy directly, like plants. We cannot use steam energy, like a coal-fired electrical power plant. We need *chemical* energy for living. We obtain it from food, and we use it to make high-energy molecules. These then drive the chemical "engines" behind muscular work, signal sending, and chemical manufacture in tissue. Our principal source of chemical energy is the **catabolism** (i.e., the breaking down) of carbohydrates and fatty acids, although we can also use most of the amino acids obtained from proteins for energy.

> ■ *Catabolism* is from the Greek, *cata-*, down; *ballein,* to throw or cast.

Heats of Combustion Disclose the Energy Available from Catabolism The energy from the *combustion* of 1 mol of glucose to CO_2 and H_2O is exactly the same as the energy available from any other method to convert glucose to the same products, 673 kcal/mol (2.82×10^3 kJ/mol). The difference is that our cells don't *burn* glucose or they would receive their energy solely as heat. We oxidize glucose by a number of small steps, some of which make other high-energy compounds. Thus some of the energy available from changing glucose to CO_2 and H_2O is used to run chemical reactions that make high-energy compounds, and the rest of the energy of catabolism is released as heat.

If we were to burn 1 mol of palmitic acid, $CH_3(CH_2)_{14}CO_2H$, a typical fatty acid, we could obtain 2400 kcal (1.00×10^4 kJ) of energy. In the body, we can also change this compound to CO_2 and H_2O, and we obtain the same energy per mole. However, as with glucose, not all this energy is released as heat. The catabolism of palmitic acid occurs by a number of steps some of which make other high-energy compounds, like the high-energy phosphates.

Each Organophosphate Has a Potential for Transferring Its Phosphate Unit to Another Molecule In Section 16.6 we first learned about the existence of energy-rich triphosphate esters. These and many similar high-energy compounds are involved in the mobilization of chemical energy in the body. Table 25.1 gives the names and structures of the principal phosphates we'll encounter. They are arranged in the order of their **phosphate group transfer potentials,** their relative abilities to transfer a phosphate group to an acceptor, as in the following equation:[1]

> ■ The bond in color denotes the P—O bond that breaks in an energy-releasing reaction of a high-energy phosphate.

$$R\text{—}O\text{—}PO_3{}^{2-} + R'\text{—}O\text{—}H \longrightarrow R\text{—}O\text{—}H + R'\text{—}O\text{—}PO_3{}^{2-} \qquad (25.1)$$

Higher   Lower
potential   potential

The numbers in the last column of Table 25.1 are measures of the relative potentials that the compounds have to transfer their phosphate groups to water under standard conditions. (We don't need to know anything about these conditions, just that they supply a common reference.) The numbers are given as *negative* values, the convention used when energy is *released* by (subtracted from) the molecule during its reaction.

We can use the positions of compounds in Table 25.1 to predict whether a particular transfer is possible. A compound with a higher (more negative) phosphate group transfer

---

[1] We will usually write the phosphate unit as $O\text{—}PO_3{}^{2-}$, but its state of ionization varies with the pH of the solution. At physiological pH, the unit is mostly in the singly and doubly ionized forms, $O\text{—}PO_3H^-$ and $O\text{—}PO_3{}^{2-}$.

TABLE 25.1
Some Organophosphates in Metabolism

Organophosphate	Structure	Phosphate Group Transfer Potential
Phosphoenolpyruvate		-14.8
1,3-Bisphosphoglycerate		-11.8
Creatine phosphate		-10.3
Acetyl phosphate		-10.1
Adenosine triphosphate, ATP		-7.3[a]
Glucose-1-phosphate		-5.0
Fructose-6-phosphate		-3.8
Glucose-6-phosphate		-3.3
Glycerol-1-phosphate		-2.2

[a] This value applies whether ADP or AMP forms.

potential can, in principal, always be used to make one with a lower (less negative) potential, assuming that the right enzyme and any other needed reactants are available.

PRACTICE EXERCISE 1

Using Table 25.1, tell whether each reaction can occur.

(a) ATP + glycerol-3-phosphate → ADP + glycerol-1,3-diphosphate

(b) ATP + glucose → ADP + glucose-1-phosphate

(c) Glucose-1-phosphate + creatine → glucose + creatine phosphate

Adenosine Triphosphate (ATP) Is the Body's Chief Energy Broker One compound in Table 25.1, **adenosine triphosphate** or **ATP,** is so important that we must review what we learned about it in Section 16.6. From the lowest to the highest forms of life, ATP is universally used as the principal carrier of energy for living functions. It is the chief means used by the body to trap energy available by oxidations.

We saw in Section 16.6 our first example of how ATP can be used — to power muscle contraction. This activity can be simply written as a chemical reaction of ATP with the proteins in relaxed muscle (or with something within the proteins):

$$\text{``Relaxed'' muscle} + \text{ATP} \longrightarrow \text{``contracted'' muscle} + \text{ADP} + \text{P}_i$$

This reaction produces changes in tertiary structures of muscle proteins that cause the fibers made from these proteins to contract. Simultaneously, ATP changes to **ADP,** which is **adenosine diphosphate,** plus inorganic phosphate, P_i.

ATP has two phosphate bonds either of whose rupture can release much useful energy. Occasionally the second bond (see Table 25.1) is broken in some transfer of chemical energy, and the products then are **AMP, adenosine monophosphate,** and the inorganic diphosphate ion, PP_i.

Occasionally triphosphates other than ATP are the carriers of energy. Guanosine triphosphate, GTP, is an example that we'll encounter later in this chapter.

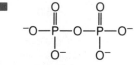

Diphosphate ion, PP_i

Adenosine diphosphate, ADP
(fully ionized form)

Adenosine monophosphate, AMP
(fully ionized form)

Guanosine triphosphate, GTP
(fully ionized form)

By convention, ATP and the phosphates higher than ATP in Table 25.1 are called **high-energy phosphates.** Notice that ATP is not at the head of the list. This means that ATP can be made from those above it in Table 25.1 by their transfer of phosphate to ADP. Then the ATP can, in turn, make any of the phosphates lower on the list. This intermediate position of the ATP and ADP is therefore one reason why this system is so useful as an energy broker. ADP can *accept* chemical energy from the phosphates with higher (more negative) potential and be changed itself to a high-energy phosphate (ATP). Then the newly made ATP can pass the chemical energy on by phosphorylations that make compounds lower on the list.

The Resynthesis of ATP Is a Major Goal of Catabolism

Almost any energy-demanding activity of the body consumes ATP. The adult human *at rest* consumes about 40 kg (about 80 mol) of ATP per day yet, at any one instant, there is less than 50 g (0.1 mol) of ATP in existence in the body. When we engage in exercise, our rate of consumption of ATP can go as high as 0.5 kg per minute! Obviously, the rapid resynthesis of ATP is one of the highest priority activities of the body, and it occurs continuously. Virtually all biochemical energetics come down to the synthesis and uses of this compound.

■ Between the completely resting and the vigorously exercising states, the rate of ATP consumption can vary by a factor of 100.

When one of the compounds in Table 25.1 *above* ATP is first made by catabolism, and then its phosphate group is made to transfer to ADP to give ATP, the overall ATP synthesis is called **substrate phosphorylation** (where the substrate is ADP). The transfer is direct from an organic phosphate donor to ADP. In substrate phosphorylation, in other words, the phosphate group doesn't come directly from a phosphate ion in solution. Nonetheless, it is possible for inorganic phosphate to be joined *directly* to ADP. The direct use of phosphate ion happens in another kind of phosphorylation, **oxidative phosphorylation,** done by a series of reactions called the **respiratory chain,** which we'll study soon.

Creatine Phosphate Phosphorylates ADP in Muscles

As you saw in Table 25.1, not all high-energy phosphates are triphosphates. Creatine phosphate, for example, is as important to muscle contraction as ATP. The ATP that is actually present in rested muscle can sustain muscle activity for only a fraction of a second. To provide for the *immediate* regeneration of ATP, muscle tissue makes and stores creatine phosphate (phosphocreatine) during periods of rest. Then, as soon as some ATP is used, and ADP plus P_i is made, creatine phosphate regenerates ATP.

■ The enzyme is *creatine kinase,* the CK enzyme we studied in Section 22.1.

What actually happens is that an increase in the supply of ADP shifts the following equilibrium to the right to raise the concentration of ATP. The high value of K_{eq} tells us that the forward reaction is favored.

$$\text{Creatine phosphate} + \text{ADP} \underset{}{\overset{\text{creatine kinase, CK}}{\rightleftharpoons}} \text{creatine} + \text{ATP} \qquad K_{eq} = 162$$

Then, during periods of rest, when the ATP level is substantially raised by other reactions, this equilibrium shifts back to the left to resupply the creatine phosphate reserves.

Although muscle tissue has three to four times as much creatine phosphate as ATP, even this reserve cannot supply the high-energy phosphate needs of muscle for more than a 100-m to 200-m sprint. For a long sustained period of work, the body uses other methods to make ATP. We'll now take an overview of all of the routes to ATP, and then we will look closely at individual pathways.

All Bioenergetic Pathways Converge on the Citric Acid Cycle and the Respiratory Chain

Our first interest is what *initiates* ATP synthesis. This process is under feedback control, and if the supply of ATP is high, no more needs to be made. Only as ATP is used up is first one mechanism and then another thrown into action. Generally, *it's an increase in the concentration of ADP that triggers the synthesis of ATP.*

■ When the level of ATP drops and the levels of ADP + P_i rise, the rate of breathing is also accelerated.

Figure 25.1 is a broad outline of the metabolic pathways that can generate ATP. Think of the last one shown in the figure, the respiratory chain, as being at the bottom of a tub, nearest the drain through which ATP will leave for some use. As with water that leaves a tub, the first to leave is at the plug. This activates the motions of water at higher levels.

Carbohydrate Lipid Protein

GLYCOLYSIS
(several
steps)

**β-OXIDATION
PATHWAY**
(several
steps)

Various
pathways

ATP ← Lactate → ATP → ATP
Pyruvate

Acetyl units
in
O
‖
CH_3C-SCoA

CITRIC ACID
CYCLE

(several steps)

→ ATP

RESPIRATORY
CHAIN → ATP
(several steps)

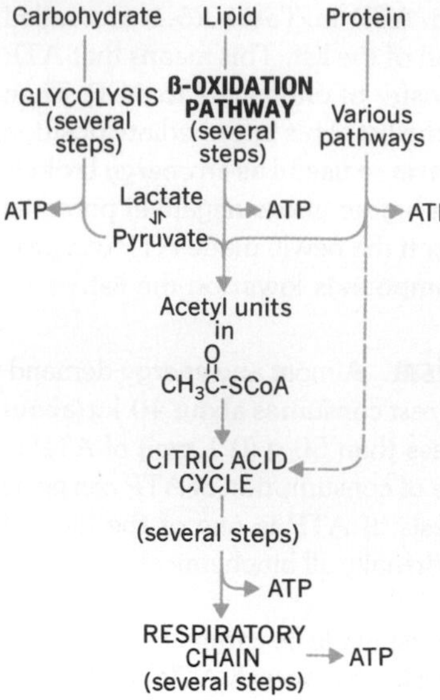

■ Hans Krebs won a share of the 1953 Nobel prize in medicine and physiology for his work on the citric acid cycle.

By analogy, once the respiratory chain is launched, the next pathway to be thrown into action is the one second from the bottom in Figure 25.1, the **citric acid cycle.** *The chief purpose of the citric acid cycle is to supply the chemical needs of the respiratory chain.*

The citric acid cycle also requires a "fuel," and this need is filled by an acetyl derivative of an enzyme cofactor called coenzyme A. **Acetyl coenzyme A,** or acetyl CoA, is the "fuel" for the citric acid cycle. *The catabolism of molecules from all three major foods — carbohydrates, lipids, and proteins — can produce acetyl coenzyme A.*

Fatty acids are a major source of acetyl CoA, particularly during periods of sustained activities. A series of reactions called the **β-oxidation pathway** breaks fatty acids into acetyl units. We will study this pathway in Chapter 27.

Most amino acids can also be catabolized to acetyl units or to intermediates of the citric acid cycle itself, and we'll survey these reactions in Chapter 28.

Starting with glucose, an important pathway called **glycolysis** breaks glucose units down to the pyruvate ion and also makes some ATP by substrate phosphorylation. Then a short pathway converts pyruvate to acetyl CoA from which more ATP is made by oxidative phosphorylation.

Glycolysis Can Make ATP When a Cell Is Deficient in Oxygen One important aspect of glycolysis is that it makes some ATP by substrate phosphorylation *under low oxygen conditions* and independently of the respiratory chain. Even when a cell is temporarily starved for oxygen, glycolysis is able to make some ATP for a while. Glycolysis is thus a backup source of ATP for cells (temporarily) running low on oxygen.

When glycolysis has to run with insufficient oxygen, then its end product is the lactate ion, not the pyruvate ion. Once the cell obtains sufficient oxygen, however, lactate is converted to pyruvate.

■ The pyruvate ⇌ lactate equilibrium is discussed further in Chapter 27.

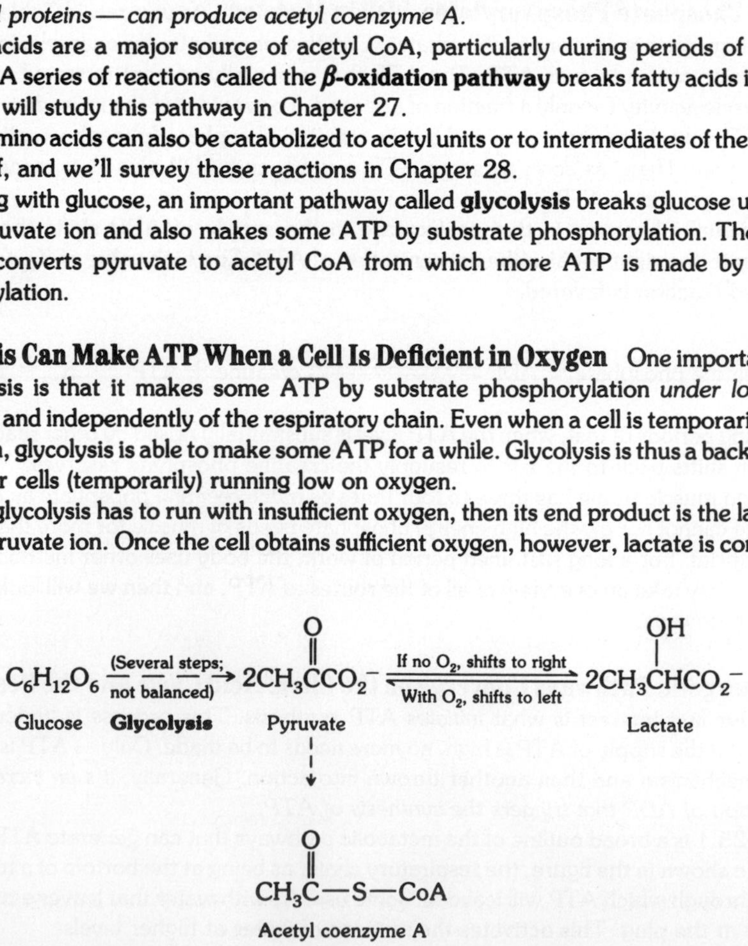

$$C_6H_{12}O_6 \xrightarrow[\text{not balanced}]{\text{(Several steps;}} 2CH_3CCO_2^- \underset{\text{With } O_2, \text{ shifts to left}}{\overset{\text{If no } O_2, \text{ shifts to right}}{\rightleftharpoons}} 2CH_3CHCO_2^-$$

Glucose   **Glycolysis**   Pyruvate   Lactate

O
‖
CH_3C—S—CoA

Acetyl coenzyme A

Thus glycolysis can end either with lactate or with pyruvate, depending on the oxygen supply. We will study details of glycolysis in the next chapter.

The full sequence of oxygen-consuming reactions from glucose to pyruvate ions to acetyl CoA and on through the respiratory chain is sometimes called the **aerobic sequence** of glucose catabolism. That part of the sequence that runs without oxygen and from glucose only as far as lactate ions is called the **anaerobic sequence** of glucose catabolism.

■ *Aerobic* signifies the use of air. *Anaerobic,* stemming from "not air," means in the absence of the use of oxygen.

Our interest in this chapter is in the citric acid cycle and the respiratory chain, pathways that can accept breakdown products from *any* food. Their reaction pathways converge on the citric acid cycle, which we will take up next.

25.2

THE CITRIC ACID CYCLE

Acetyl CoA is used to make the citrate ion, which is then broken down, bit by bit, to CO_2 as units of ($H:^- + H^+$) are sent into the respiratory chain.

Figure 25.2 gives the reactions of the citric acid cycle, a series of reactions that break down acetyl groups.[2] The two carbon atoms of this group end up in molecules of CO_2, and the hydrogen of the acetyl group is fed into the respiratory chain by means of transfers of $H:^-$ and H^+. In the end, these pieces of $H:H$ are irreversibly oxidized by oxygen (from respiration) to H_2O, a reaction of the respiratory chain that occurs in the innermost part of a mitochondrion.

■ The mitochondrion is an organelle (small body) in the cytoplasm (as we illustrated in Figure 24.1, page 675).

Coenzyme A Is the Common Carrier of Acetyl Units Before an acetyl group can enter the citric acid cycle, it must be made and attached to coenzyme A, which we represent as CoASH. In acetyl CoA, the acetyl residue replaces the H on HS of CoASH.

■ Pantothenic acid is another of the B vitamins.

Coenzyme A (CoASH)

As we said, amino acid residues, fatty acids, and glucose are all sources of acetyl groups for acetyl coenzyme A. When glucose is the source, it is catabolized by many steps to pyruvate ion.

The conversion of pyruvate ion to acetyl coenzyme A involves both a decarboxylation and an oxidation. The overall equation for this complicated and irreversible change is as follows.

[2] You should be aware that the citric acid cycle goes by two other names as well: the **tricarboxylic acid cycle** and **Krebs' cycle.** You might encounter any of these names in other references.

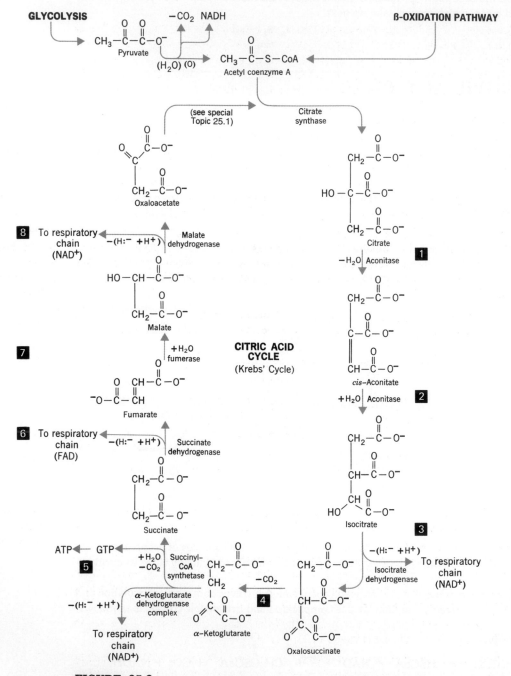

FIGURE 25.2

The citric acid cycle. The boxed numbers refer to the text discussion. The names of the enzymes for each step are given by the arrows.

We leave the details to more advanced treatments, but there are five actual steps catalyzed by a large, multienzyme complex called *pyruvate dehydrogenase.* Three B vitamins — thiamine, nicotinic acid (niacin), and riboflavin — are among the compounds needed to make the coenzymes for the complex.

Acetyl coenzyme A is a thioester, which means an ester in which an oxygen atom has been replaced by a sulfur atom. A thioester is far more reactive in transferring an acyl group than an ordinary ester. Thus acetyl CoA is a particularly active transfer agent for the acetyl group, and just such a transfer of an acetyl group launches one turn of the citric acid cycle.

The Citric Acid Cycle Dismantles Acetyl Groups, Sending Hydrogen to the Respiratory Chain and CO_2 to Waste Disposal

The enzymes for the citric acid cycle occur in mitochondria in close proximity to the enzymes of the respiratory chain. In the first step of the citric acid cycle, the acetyl group of acetyl CoA transfers to oxaloacetate ion, an ion that has two carboxylate groups and a keto group. The enzyme is *citrate synthase.* The acetyl unit of acetyl CoA adds across the keto group in a type of reaction that we have not studied before. For background to this reaction see Special Topic 25.1.

The product is the citrate ion. Now begins a series of reactions by which the citrate ion is degraded bit by bit until another oxaloacetate ion is regenerated. The numbers of the following steps match those in Figure 25.2.

1. Citrate is dehydrated to give the double bond of *cis*-aconitate. This is the dehydration of an alcohol. The enzyme is *aconitase.*

2. Water adds to the double bond of *cis*-aconitate to give an isomer of citrate called isocitrate; aconitase also catalyzes this step. Thus the net effect of steps 1 and 2 is to switch the alcohol group in citrate to a different carbon atom. This changes the alcohol from being tertiary to one that is secondary, from an alcohol that cannot be oxidized to one that can.

3. The secondary alcohol group of isocitrate is dehydrogenated (oxidized) by *isocitrate dehydrogenase* to give the keto group of oxalosuccinate. The alcohol system gives up both $H\!:^-$ and H^+.

An NAD^+ coenzyme accepts the $H\!:^-$ unit; the released proton (H^+) is transferred across a membrane within the mitochondrion. The NAD^+ thereby changes to $NAD\!:\!H$ (usually written simply as NADH).

What happens to the electron pair in $NAD\!:\!H$ and to the H^+ is described in Section 25.3. Overall, the fates of NADH and H^+ are intimately tied to the synthesis of three molecules of ATP, so the step we have here described is the first delivery of chemical energy from the citric acid cycle to a triphosphate.

4. Oxalosuccinate, still bound to isocitrate dehydrogenase, loses a carboxyl group — it decarboxylates — to give α-ketoglutarate.

5. α-Ketoglutarate now undergoes a very complicated series of reactions, all catalyzed by one team of enzymes called *α-ketoglutarate dehydrogenase,* a complex enzyme system resembling pyruvate dehydrogenase mentioned earlier. A carboxyl group is lost and another dehydrogenation occurs. The same kind of fate occurs to $H\!:^-$ and H^+ here as happened in step 3, so three more ATPs can now be made by the respiratory chain.

■
$$\underset{\text{Ester}}{\overset{\displaystyle O}{\overset{\displaystyle \|}{RCOR'}}}$$

$$\underset{\text{Thioester}}{\overset{\displaystyle O}{\overset{\displaystyle \|}{RCSR'}}}$$

■
$$\underset{\text{Oxaloacetate ion}}{\overset{\displaystyle O}{\overset{\displaystyle \|}{^-O_2CCH_2CCO_2^-}}}$$

■ At physiological pH, the acids in the cycle exist largely as their anions.

■ The fluoroacetate ion, $FCH_2CO_2^-$, one of the most toxic of small-molecule poisons, inhibits aconitase.

■ We studied how $H\!:^-$ transfers to NAD^+ in Section 22.1 and Special Topic 22.1.

■ $^-O_2CCH_2CH_2CO_2^-$
Succinate ion

■ $\begin{array}{c} HCCO_2^- \\ \| \\ ^-O_2CCH \end{array}$
Fumarate ion
(trans geometry)

■ $\begin{array}{c} OH \\ | \\ ^-O_2CCH_2CHCO_2^- \end{array}$
Malate ion

The product (not shown in Fig. 25.2) is the coenzyme A derivative of succinic acid. It is converted by the action of the enzyme *succinyl CoA synthetase* to the succinate ion, which is shown in Figure 25.2. The reaction also generates guanosine triphosphate, GTP, from its diphosphate. GTP is another high-energy triphosphate, which we mentioned on page 708 as being similar to ATP. GTP is able to phosphorylate ADP to make one ATP. Thus the citric acid cycle includes one substrate phosphorylation.

6. Succinate donates hydrogen to FAD (not to NAD$^+$) in a reaction catalyzed by *fumarate dehydrogenase,* and the fumarate ion forms. Both H $:^-$ and H$^+$ are accepted by FAD, which becomes FADH$_2$. FADH$_2$ also intersects with the respiratory chain, and its further involvement in this chain produces two ATPs (not three, which are possible from NADH).

7. Fumarate adds water to its double bond, and malate forms. The enzyme is *fumarase.* Note the 2° alcohol group in malate.

8. Malate is oxidized by *malate dehydrogenase* to oxaloacetate. The 2° alcohol group of malate, in a reaction resembling step 3, gives up H $:^-$ to NAD$^+$ to form NADH. The chemical energy in NADH will be used to make three more ATP molecules by the respiratory chain.

One turn of the citric acid cycle is now complete and one molecule of the carrier, oxaloacetate, has been remade to enable another turn of the cycle. It can accept another acetyl group from acetyl coenzyme A.

SPECIAL TOPIC 25.1
CONDENSING ESTERS—A MAJOR C—C BOND-MAKING REACTION

The reaction by which an acetyl group enters the citric acid cycle is just one example of many reactions of two carbonyl compounds in which an acyl group of one becomes joined to an alpha position of another. The product is a much larger molecule. We'll first explain how this happens under laboratory conditions with a very simple system, two molecules of the same ester, ethyl acetate. Organic chemists call the reaction the *Claisen ester condensation.*

The Alpha Hydrogen of a Carbonyl Compound Is "Mobile" The acid ionization constant of an alkane is estimated to be about 10^{-40}; clearly, no one would call an alkane an acid! The ionization constant of the H atom attached to the α-position of the CH$_3$CO unit in ethyl acetate is about 10^{-18}, making it a billion trillion times as strong an acid (but still nothing we'd call an acid in water).

The reason for this greater acidity is the presence of the two nearby electronegative oxygen atoms. This greater acidity (mobility) of a hydrogen on a carbon alpha

to oxygens is all we need to understand how the Claisen condensation and similar reactions in the body can occur.

With ethyl acetate, the reaction is as follows, where $B:^-$ is a powerful base, so powerful that the reaction cannot be run in water. (Usually it is run in ethyl alcohol.) In living systems, reactions like this use enzymes to handle necessary proton exchanges.

$$\underset{O}{\overset{\overset{O}{\|}}{CH_3C}}-OCH_2CH_3 + \underset{O}{\overset{\overset{O}{\|}}{CH_3C}}-OCH_2CH_3 \xrightarrow{B:^-}$$

$$\underset{O}{\overset{\overset{O}{\|}}{CH_3C}}-CH_2\overset{\overset{O}{\|}}{C}-OCH_2CH_3 + HOCH_2CH_3$$
Ethyl acetoacetate

Notice that an acetyl group (in red) has been joined to the alpha carbon of the second ester molecule. A new C—C bond has been made. Let's see how a strong base handles this reaction.

Step 1 The base takes an H+ from the alpha H—C position of one ethyl acetate.

$$B:^{-} + H-CH_2\overset{\overset{O}{\|}}{C}-OCH_2CH_3 \longrightarrow$$

Ethyl acetate

$$^{-}:CH_2\overset{\overset{O}{\|}}{C}-OCH_2CH_3 + B-H$$

Anion of ethyl acetate

Step 2 The new anion attacks the carbonyl carbon of another ester molecule. (This carbon has a partial positive charge on it.)

$$CH_3\overset{\overset{\ddot{O}:}{\|}}{C}\underset{OCH_2CH_3}{} + {}^{-}:CH_2\overset{\overset{O}{\|}}{C}-OCH_2CH_3 \longrightarrow$$

$$CH_3\overset{\overset{:\ddot{O}:^{-}}{|}}{C}-CH_2\overset{\overset{O}{\|}}{C}-OCH_2CH_3$$
$$\underset{OCH_2CH_3}{}$$

Step 3 The product of this reaction expels $CH_3CH_2O^{-}$, which takes H+ from $B:H$ (and the base is thus regenerated).

$$CH_3\overset{\overset{:\ddot{O}:^{-}}{|}}{C}-CH_2\overset{\overset{O}{\|}}{C}-OCH_2CH_3 \longrightarrow$$
$$\underset{OCH_2CH_3}{}$$
$$B:H$$

$$CH_3\overset{\overset{O}{\|}}{C}-CH_2\overset{\overset{O}{\|}}{C}-OCH_2CH_3 + HOCH_2CH_3 + B:^{-}$$

Ethyl acetoacetate

This overall reaction is formally very similar to the initial reactions the body uses for several biosyntheses: to make long hydrocarbon chains from acetate units, and to make cholesterol and the sex hormones. So we'll meet this kind of reaction in later chapters.

Citrate Synthase Manages a Claisen-like Condensation of Acetyl CoA with Oxaloacetate If we take acetyl CoA and oxaloacetate through the same steps, we have the following in which the services of a base, still represented by $B:^{-}$, are provided by the enzyme citrate synthase. Realize that what follows is a considerable simplification of a series of steps in which the way charges are managed by the enzyme are not indicated.

Step 1 Acetyl CoA gives up a proton.

$$B:^{-} + H-CH_2\overset{\overset{O}{\|}}{C}SCoA \longrightarrow BH + {}^{-}:CH_2\overset{\overset{O}{\|}}{C}SCoA$$

Acetyl CoA

Step 2 The new anion attacks the keto group in oxaloacetate. The new C—C bond forms.

$$\overset{\overset{\ddot{O}:}{\|}}{C}-CO_2^{-} + {}^{-}:CH_2\overset{\overset{O}{\|}}{C}SCoA \longrightarrow {}^{-}:\overset{\overset{CH_2\overset{\overset{O}{\|}}{C}SCoA}{|}}{\underset{CH_2CO_2^{-}}{\ddot{O}}}-C-CO_2^{-}$$
$$\underset{CH_2CO_2^{-}}{}$$

Step 3 The CoA–S group is hydrolyzed, and a proton is donated so that the 3° alcohol group can form. (The CoA–S unit is similar to an ordinary ester. It is a thioester, and these hydrolyze more readily than ordinary esters. In fact this hydrolysis is part of the driving force of the overall reaction.)

$$B-H + {}^{-}:\overset{\overset{CH_2\overset{\overset{O}{\|}}{C}SCoA}{|}}{\underset{CH_2CO_2^{-}}{\ddot{O}}}-C-CO_2^{-} + H_2O \longrightarrow$$

$$B:^{-} + H-\overset{\overset{CH_2\overset{\overset{O}{\|}}{C}O^{-}}{|}}{\underset{CH_2CO_2^{-}}{\ddot{O}}}-C-CO_2^{-} + H-SCoA + H^{+}$$

Citrate ion

Thus a Claisen-like condensation launches the acetyl group of acetyl CoA into the citric acid cycle.

25.3

THE RESPIRATORY CHAIN

The flow of electrons to oxygen in the respiratory chain creates a proton gradient in mitochondria that drives the synthesis of ATP.

The term *respiration* refers to more than just breathing. It includes the chemical reactions that use oxygen in cells. We are now ready to learn specifically how oxygen is reduced to water and how the chemical energy of this event helps to drives the entire respiratory chain. The most basic statement we can write for what happens when water forms emphasizes how electrons and protons combine with an oxygen atom:

$$(\overset{..}{:}) \;+\; 2H^+ \;+\; \cdot\overset{..}{O}\cdot \;\longrightarrow\; H—\overset{..}{\underset{..}{O}}—H \;+\; energy$$

| Pair of electrons | Pair of protons | Atom of oxygen | Molecule of water |

■ The gain of e^- or of e^- carriers such as $H{:}^-$ is *reduction;* the loss of e^- or of $H{:}^-$ is *oxidation.*

Remember that when any particle, like the oxygen atom, gains electrons it is reduced, and the donor of electrons is oxidized. The electrons and protons for the reduction of oxygen come mostly from intermediates in the citric acid cycle. As you saw in the discussion of steps 3, 5, and 6 of this cycle, the electrons of C—H bonds are used, not O—H bonds, and the hydride ion $H{:}^-$ is often (but not always) the vehicle for carrying the electrons from one species to another.

The Mitochondrion Is the Cell's "Powerhouse"   The respiratory chain is one long series of oxidation–reduction reactions. The flow of electrons from the initial donor is down an energy hill all the way to oxygen, and it is irreversible. As this flow occurs, other complex events take place to make ATP from ADP and P_i.

The cell's principal site of respiratory chain activity and ATP synthesis, a *mitochondrion,* is often dubbed the powerhouse of the cell (see Fig. 25.3). Some tissues have thousands of mitochondria in the cytoplasm of a single cell. A mitochondrion has two important mem-

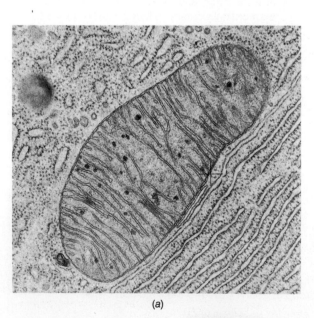

(a)

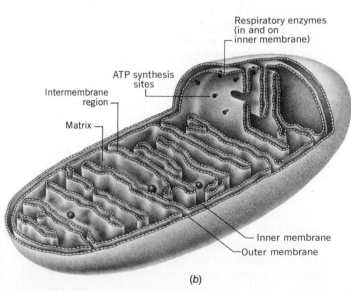

Respiratory enzymes
(in and on
inner membrane)

ATP synthesis
sites

Intermembrane
region

Matrix

Inner membrane

Outer membrane

(b)

FIGURE 25.3

A mitochondrion. (a) Electron micrograph ($\times 53{,}000$) of a mitochondrion in a pancreas cell of a bat. (b) Perspective showing the interior. The respiratory enzymes are incorporated into the inner membrane. On the matrix side of this membrane are enzymes that catalyze the synthesis of ATP. (Micrograph courtesy of Dr. Keith R. Porter.)

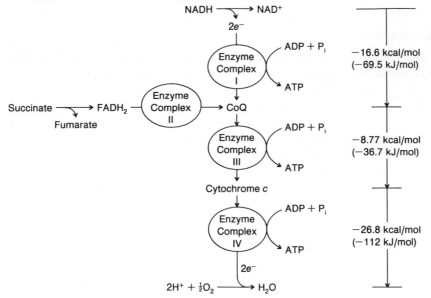

FIGURE 25.4
The enzyme complexes of the respiratory chain and the energy drop from NAD$^+$ to reduced oxygen (water). The figures for energy refer to the portion of the energy change that is not unavoidably dissipated as heat. The symbols for the enzymes are as follows.

NADH = enzyme with reduced nicotinamide cofactor

FADH$_2$ = enzyme with reduced riboflavin cofactor

CoQ = enzyme with coenzyme Q cofactor

branes, one outer and one inner. The inner membrane is very convoluted and has a surface area many times that of the outer membrane. The space between the two membranes is called the *intermembrane space*. The space deep inside the inner membrane is filled with a gel-like material called the *matrix*. It's less than 50% water and rich in soluble enzymes, cofactors, inorganic ions, and substrates.

■ A cell in the flight muscle of a wasp has about a million mitochondria.

The Chief Agents of Electron Transfer Are the Respiratory Enzymes
Built into the inner membrane itself are numerous immobile enzyme clusters designated as enzyme complexes I, II, III, and IV. Also existing as parts of the inner membrane are enzymes capable of moving from one complex to another and named coenzyme Q (CoQ) and cytochrome c. Taken together, these enzymes are called the **respiratory enzymes,** and they operate the respiratory chain.

The Operation of the Respiratory Chain Moves H$^+$ Ions from the Matrix to the Intermembrane Space
One very important property of the inner membrane is that *only at particular channels is it permeable to chemical species, particularly ions like* H$^+$. The operation of the respiratory chain transfers H$^+$ ions from the matrix to the intermembrane compartment. *This establishes an unstable proton gradient between the intermembrane region (more* H$^+$) *and the matrix (less* H$^+$). Eventually the fluid of the intermembrane space will have a pH 1.4 units lower (more acidic) than the matrix.

Remember, given the random motions of species in any fluid, that nature eventually destroys a gradient. However, the gradient-destroying flow of H$^+$ ion back into the matrix is permitted only at certain inner-membrane channels. The flowing protons, therefore, can return to the matrix only at certain locations, *places where the returning protons activate a complex enzyme system that makes and releases ATP.* This is basically how the cell connects activities of the respiratory chain with the synthesis of ATP. An outline of this understanding was first formulated by English scientist Peter Mitchell, and it is called the **chemiosmotic theory.**

■ Peter Mitchell received the 1978 Nobel prize in chemistry. His proposal was made in 1961 and inspired a flurry of controversy among scientists before intensive research led to its acceptance.

One Entry to the Respiratory Chain Is the Synthesis of NADH
If we let MH_2 represent any metabolite that can donate H:$^-$ to NAD$^+$, we can write the following equation.

$$MH_2 + NAD^+ \longrightarrow M + NADH + H^+$$

By this reaction, the electron pair has moved from MH_2 to NADH, which next interacts with enzyme complex I.

Figure 25.4 gives the sequence by which the respiratory enzyme complexes interact and shows the reaction-favoring releases of chemical energy that impel the overall sequence.

■ Think of MH_2 as M:H, which becomes M when H:$^-$ and H$^+$ leave.

Because these are all redox reactions, they have their own cell potentials, which are discussed in Special Topic 25.2.

Enzyme Complex I of the Respiratory Chain Oxidizes NADH and Reduces CoQ

As indicated in Figure 25.4, enzyme complex I first accepts a pair of electrons (carried by $H:^-$) from NADH. The electron flow has now started toward oxygen.

■ We studied the FMN enzyme in Section 22.1.

The acceptor of $H:^-$ in complex I is FMN, which carries the B vitamin riboflavin as part of its coenzyme. The hydride unit, $H:^-$, in NADH transfers to FMN according to the following equation.

$$NADH + H^+ + FMN \longrightarrow FMNH_2 + NAD^+$$

■ Think of NADH as $NAD:H$ in which the electron pair dots are those that move along the respiratory chain.

This restores the NAD^+ enzyme, and moves the electron pair one more step along the respiratory chain.

What happens next is the transfer of just a pair of electrons, not $H:^-$, from $FMNH_2$ while the hydrogen *nucleus* of $H:^-$ leaves the carrier as H^+. *This transfer induces conformational changes in an inner membrane protein that cause protons to transfer from the matrix into the intermembrane region.* Thus the buildup of the H^+ gradient across the inner mitochondrial membrane begins. The two electrons are accepted by an iron-sulfur protein, which we'll symbolize by FeS–P. The iron in FeS–P occurs as Fe^{3+}. By accepting one electron, it is reduced to the Fe^{2+} state, so we need *two* units of FeS–P to handle the pair of electrons now about to leave $FMNH_2$. This step can be written as

■ There are actually three types of iron–sulfur proteins having different atom ratios of iron to sulfur. We do not indicate the actual net charge; regardless of the type, the oxidized and reduced forms all differ by a net charge of 1+.

$$FMNH_2 + 2FeS-P \longrightarrow FMN + 2FeS-P\cdot + 2H^+$$

where we use FeS–P· to represent the reduced form of the iron–sulfur protein in which Fe^{2+} occurs.

Before we go on, we will introduce a way to display these reactions that helps to emphasize the restoration of each enzyme to its original condition. For example, the last three reactions we have studied can be represented as follows:

Will promote the migration of protons from the matrix to the intermembrane space.

Reading from left to right, at the point where the first pair of curved arrows touch, we see that as MH_2 changes to M, it passes $H:^-$ to NAD^+ to change it to NADH. H^+ is also released. The touching of the second pair of curved arrows tells us that as NADH and H^+ are changed back to NAD^+ the system passes one unit of $H:^-$ plus H^+ across to FMN to change it to $FMNH_2$. At the third pair of curved arrows, the display shows $FMNH_2$ changing back to FMN as it passes two electrons to two FeS–P molecules and also as it expels two protons (H^+).

Between three and four protons, on the average, are moved from the matrix into the intermembrane space by each pair of H^+ ions released by respiratory chain oxidation. From which *specific* chemical species the protons actually migrate is still being studied.

The last work of enzyme complex I is to pass electrons from the reduced iron–sulfur complex to coenzyme Q. We will now drastically reduce the attention we give to specific detail.

Respiratory Enzyme Complexes III and IV Continue the Respiratory Chain   As you have seen, the respiratory chain is quite complicated, and from this point on it becomes even more so. What follows next are further transfers of electrons through enzyme complexes III and IV, which involve a series of individual enzymes called the *cytochromes*. These are designated, in the order in which they participate, as cytochromes b, c_1, and c, which together with an iron–sulfur protein make up enzyme complex III. Cytochromes a and a_3 are parts of enzyme complex IV, often called *cytochrome c oxidase*.

■ *Cyto-*, cell; *-chrome*, pigment. The cytochromes are colored substances.

Complex IV is the enzyme complex that catalyzes the reduction of oxygen to water that we described earlier. It carries a copper ion that alternates between the Cu^{2+} and the Cu^+ states.

Figure 25.5 is a diagram of the electron movement through complexes I, III, and IV. The net result of the operation of this, the main branch of the respiratory chain, starting with MH_2 and NAD^+, can be expressed by the following overall equation.

$$MH_2 + nH^+ + \tfrac{1}{2}O_2 \xrightarrow[\text{(enzyme complexes I, III, and IV)}]{\text{respiratory chain}} M + H_2O + nH^+$$

From inside the inner membrane

These protons help make the proton gradient across the inner membrane of the mitochondrion.

Complex II, indicated in Figure 25.4 (but not in Fig. 25.5), is just another way to enter the respiratory chain at coenzyme Q. The succinate ion of step 6 of the citric acid cycle is one substrate for complex II.

■ Complex II includes the riboflavin-based coenzyme, FAD.

Two Gradients Are Established, a Proton and a Plus Charge Gradient   The operation of the respiratory chain pumps protons from the mitochondrial matrix to the intermembrane region *without also putting the equivalent of negatively charged ions there, too.* Two gradients are thus set up, a gradient of H^+ ions and a gradient of positive charge. The positive charge gradient could be erased either by the migration of negative ions to the intermembrane region *or by the migration of any kind of positive ion from this region into the matrix.* Calcium ions, for example, might move, *and precisely such movements of Ca^{2+} ions are involved with nerve signals and muscular work.* Thus the chemiosmotic theory helps us understand events other than ATP synthesis, like electrical signal sending in nerve tissue.

■ When a gradient of electrical charge is established by chemical species, it is called an *electrochemical gradient*.

■ Because chemical reactions create the gradients, we have the *chemi-* part of the term *chemiosmotic*.

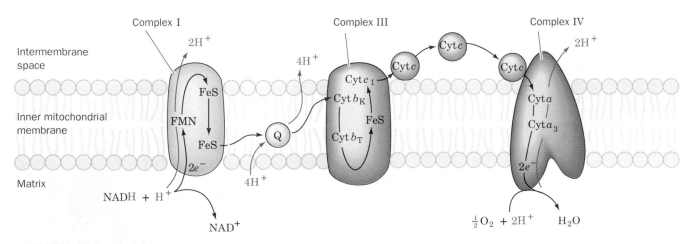

FIGURE 25.5
The electron flow (black) and the proton transfers (red) from NADH through enzyme complexes I, III, and IV of the respiratory chain. Q represents coenzyme Q, CoQ. Cyt c is cytochrome c. Complex II is not shown, but it transfers electrons from succinate ion to coenzyme Q. (From Voet and Voet, *Biochemistry*, John Wiley & Sons, Inc., 1990. Used by permission.)

The Proton Channels of the Inner Mitochondrial Membrane Are Part of an Enzyme for Making ATP from ADP and P_i

Embedded in the inner mitochondrial membrane are complexes of proteins that form a tube through it. At least one tube exists for every unit of respiratory enzymes. The tube itself is called the F_0 component of a complex enzyme called the **proton-pumping ATPase.** On the matrix side of the inner membrane occurs the other component, called the F_1 component, at which the proton channel ends. In electron micrographs of the inner membrane, the F_1 components project like lollipops from the matrix-side surface (see Fig. 25.6). The F_1 components consist of eight polypeptides one of which is the proton gate and three of which are the actual sites where ADP and P_i are put together to make ATP.

In a theory proposed by Paul Boyer, the linkage between the flow of protons through F_0 and the synthesis and release of ATP from F_1 is made as follows by a series of steps involving allosteric interactions (see Fig. 25.7). Each of the ATP-synthesizing F_1 subunits is capable of existing in any one of three configurations, open (O), loosely closed (L), and tightly closed (T). The O configuration has little ability to bind substrates, whether ADP plus P_i or ATP. The L configuration can bind ADP $+ P_i$ *but can do nothing catalytically to change these particles to ATP until protons arrive to cause a change in configuration of the protein shape from the L to the T state.* Now the formation of ATP is catalyzed, *but it is tightly held to the T-shaped subunit.* It cannot be released until the imbalance in proton concentration — the proton gradient — builds up sufficiently to drive protons through the F_0 channel and into F_1. The arriving protons cause conformational changes in all of the F_1 subunits. The T subunit with its bound ATP changes into an O subunit, which permits the ATP to drop off. The L subunit (with loosely bound ADP plus P_i) changes to the T form as a result of which another ATP is put together to await release by further surges of protons.

■ The proton-pumping ATPase is also called the *proton-translocating ATPase* as well as the $F_0F_1ATPase$.

FIGURE 25.6
Proton-pumping ATPase. (a) Electron micrograph of an intact inner mitochondrial membrane showing the "lollipop" projections of the enzyme on the matrix side. [From Parsons, D. F., *Science* **140**, 985 (1963). Copyright © 1963 American Association for the Advancement of Science. Used by permission.] (b) Interpretive drawing of the mitochondrion with the same projections shown in part a. (c) Interpretive drawing of one proton-pumping ATPase complex. The unit labeled δ is the proton gate between the F_0 proton channel (beneath) and the other subunits of the F_1 complex (above). The three subunits labeled β are ATP synthesis sites. (Adapted with permission from Voet and Voet, *Biochemistry,* John Wiley & Sons, Inc., 1990. Used by permission.)

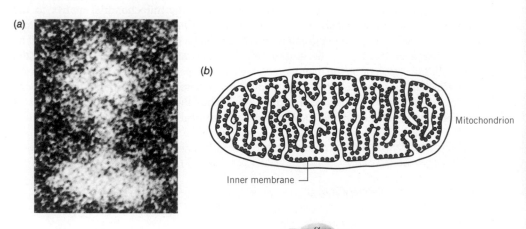

(a)

(b)

Mitochondrion

Inner membrane

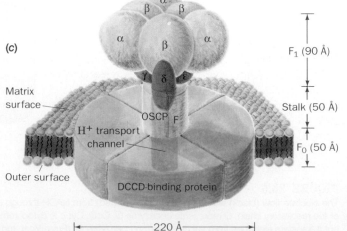

(c)

Matrix surface

OSCP

H⁺ transport channel

Outer surface

DCCD-binding protein

F_1 (90 Å)

Stalk (50 Å)

F_0 (50 Å)

220 Å

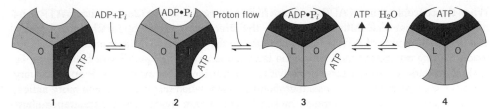

FIGURE 25.7
The synthesis and release of ATP from subunits of the F_0 portion of the proton-pumping ATPase complex. Each of the three β-subunits shown in Figure 25.6 is capable of being in the O, L, or T configuration discussed in the text. In set **1**, an ATP molecule, previously made, is tightly held. The system picks up ADP plus P_i, which become loosely held, as the system changes to **2**. Now, **2** to **3**, we imagine that protons in the gradient have entered the complex via F_0 to cause conformational changes that let ATP drop away, get ADP plus P_i ready to become ATP (**3** to **4**), and let an old O site change to an L site in readiness to accept ADP and P_i. (Adapted with permission from Voet and Voet, *Biochemistry*, John Wiley & Sons, Inc., 1990. Used by permission.)

It is because a series of *oxidations* creates the proton gradient that makes this ATP synthesis possible that the overall synthesis is called **oxidative phosphorylation,** or sometimes *respiratory chain phosphorylation*. Each molecule of NADH that enters the chain at enzyme complex I can lead to a maximum of three ATPs. Each $FADH_2$ that enters at complex II can cause just two ATP molecules to form. Let's now go back over the major aspects of oxidative phosphorylation

---

Oxidative Phosphorylation

1. The synthesis of ATP occurs at an enzyme located on the matrix side of the inner mitochondrial membrane.
2. ATP synthesis is driven by a flow of protons that occurs from the intermembrane side to the matrix side of the inner membrane.
3. The flow of protons is through special channels in the membrane and down a concentration gradient of protons that exists across the inner membrane.
4. The energy that creates the proton gradient and the (+) charge gradient is provided by the flow of electrons between the respiratory chain enzymes that make up integral packages of the inner membrane.
5. The inner mitochondrial membrane is a closed envelope except for the special channels for the flow of protons and for special transport systems that let needed solutes move into or out of the innermost mitochondrial compartment.

■ This migration through a semipermeable membrane explains the *osmotic* part of the term *chemiosmotic*.

A Transport Protein in the Mitochondrial Membrane Moves ADP Inside as It Carries ATP Out Both ATP and ADP are highly charged species and so could not easily get through a lipid bilayer membrane. A transport protein of an inner mitochondrial membrane "pump" called *ATP–ADP translocase* solves this problem. The migration of newly made ATP to the intermembrane region and thence to the cytosol outside is coupled to the movement of ADP into the matrix. The expression "is coupled" means that the membrane pump that moves ADP in one direction (in) simultaneously moves ATP in the other (out).

The Rate of ATP Resynthesis Is Sensitive to the Molar Concentrations of ADP and ATP With considerable simplification of a complex phenomenon, it is as if ADP, P_i, and ATP are involved in the following equilibrium.

$$ADP + P_i \rightleftharpoons ATP \tag{25.2}$$

When the concentrations of ADP and P_i are high, a stress thereby exists that causes ATP to be made; equilibrium 25.2 shifts to the right — the direction predicted by Le Châtelier's principle. When the supply of ATP is relatively high, on the other hand, little more is (or need be) made. During our quiet times, when the least amounts of ADP and P_i are being made by ATP-consuming, energy-demanding activities, the machinery for making ATP operates very slowly — enough to maintain basal metabolism. Then, when our lives become more active, the rate of use of ATP accelerates, and ADP plus P_i are made. Their appearance shifts equilibrium 25.2 to the right, thus using them up. A question not yet fully resolved is the *location* in the cell of these shifts. Is the rate of ATP synthesis controlled by changes in concentrations in the cytosol or in the mitochondrial matrix? Or is the rate controlled at the ATP–ADP translocase protein of the mitrochondrial membranes? In any case, the rate depends on the supply of ADP, and ATP is made no more rapidly than actually needed.

Some Antibiotics and Poisons Inhibit Oxidative Phosphorylation One of the barbiturates, amytal sodium, and the powerful insecticide, rotenone, block the respiratory chain by interfering at enzyme complex I. The antibiotic antimycin A stops the chain at complex III. The cyanide ion blocks the chain at its very end, at cytochrome a_3 in complex IV. These agents, by inhibiting the chain, thus work to inhibit the creation of the proton gradient and in this manner inhibit ATP synthesis.

■ Rotenone is a naturally occurring insecticide.

■ Remember that the respiratory chain also occurs in bacteria, so antibiotics that interfere with the chain can kill bacteria.

Other substances interfere with the *use* of the gradient. They do not prevent the respiratory chain from operating, but they cancel the effect of the gradient. They *uncouple* the chain from ATP synthesis. For example, 2,4-dinitrophenol increases the permeability of the inner membrane to protons so that they can use random routes to reenter the matrix. When the gradient is destroyed without the use of the F_0 proton channel, little if any ATP synthesis can occur. The chemical energy released by the operation of the chain is then converted to heat.

■ A natural and healthy mechanism for uncoupling the chain from ATP synthesis exists in the heat-generating, brown fat tissue that cold-adapted animals have (and which will be discussed further in Special Topic 27.2, page 753.)

Oxidative Phosphorylation Makes a Maximum of Twelve ATP Molecules Using Intermediates of the Citric Acid Cycle The data in Table 25.2 show where ATP production is generated as the citric acid cycle runs. If, to the twelve ATPs made this way, we add the three ATPs made possible by the conversion of pyruvate to acetyl CoA, then fifteen ATPs can be made by the degradation of one pyruvate ion.

Keep in mind that pyruvate isn't the only raw material either for acetyl coenzyme A or for other metabolites (MH_2) that can fuel the respiratory chain. Also remember that the energy of the respiratory chain can be used to run other operations besides making ATP — for example, the operation of nerves. Thus Table 25.2 has to be viewed as giving the upper limits to ATP production from pyruvate by the respiratory chain.

TABLE 25.2

ATP Production by Oxidative Phosphorylation

Step	Receiver of ($H{:}^- + H^+$) in the Respiratory Chain	Molecules of ATP Formed
Isocitrate → α-ketoglutarate	NAD^+	3
α-Ketoglutarate → succinyl CoA	NAD^+	3
Succinyl CoA → succinate (via GTP)	—	1
Succinate → fumarate	FAD	2
Malate → oxaloacetate	NAD^+	3
Total ATP via citric acid cycle		12
Pyruvate → acetyl CoA	NAD^+	3
Total ATP from pyruvate via the citric acid cycle		15

REDOX POTENTIALS AND BIOLOGICAL OXIDATIONS

Each step in the respiratory chain has its own potential for proceeding to completion. The first step, for example, in which NAD^+ accepts $H:^-$ from some donor is a redox reaction. We'll suppose that the donor is something simple, like ethyl alcohol. In Special Topic 22.1 we showed how NAD^+ could accept $H:^-$ from this alcohol and be reduced to NADH as the alcohol became oxidized to ethanal:

$$CH_3CH_2OH + NAD^+ \longrightarrow CH_3CH{=}O + NADH + H^+$$
Ethanol   Ethanal

We'll now use what we studied in Chapter 10 to assess the extent to which this reaction spontaneously runs from left to right.

In Chapter 10 we learned how we can separate a redox reaction into two half-reactions, each with its own reduction potential. The half-reaction with the more positive reduction potential is able to force the other half-reaction to run in reverse, as an oxidation.

Recall that we begin with the half-reactions *both* written as reductions. For the oxidation of ethanol by NAD^+ (or the reduction of NAD^+ by ethanol) we have the following two half-cell reactions with their reduction potentials.

$$NAD^+ + H^+ + 2e^- \longrightarrow NADH$$
$$E^{\circ\prime} = -0.32 \text{ V} \quad (1)$$
$$CH_3CH{=}O + 2H^+ + 2e^- \longrightarrow CH_3CH_2OH$$
$$E^{\circ\prime} = -0.20 \text{ V} \quad (2)$$

The Standard State Is Different When Working with Cellular Reactions We now have a prime (′) added to E° to indicate a change in the definition of a standard state. Earlier, we defined a standard state as involving all chemicals at concentrations of $1 M$. In living systems this is unrealistic, particularly for H^+. The molarity of H^+ in cells is closer to 1×10^{-7} mol/L. In using reduction potentials for living systems, therefore, scientists define the standard state as one in which $[H^+] = 1.0 \times 10^{-7} M$ with a temperature of 25 °C.

The Reduction of Ethanal, Not the Oxidation of Ethanol, Is Favored Getting back to the two half-reactions, we see that the reduction *of the aldehyde by Equation 2 has the more positive (less negative) reduction potential.* This half-reaction, therefore, is able to force the reaction of Equation 1 to run in the opposite direction, as an oxidation. So we reverse Equation 1 and then combine it with Equation 2 as follows.

$$NADH \longrightarrow NAD^+ + H^+ + 2e^- \quad \text{(1, reversed)}$$
$$CH_3CHO + 2H^+ + 2e^- \longrightarrow CH_3CH_2OH \quad (2)$$
$$\overline{CH_3CHO + NADH + H^+ \longrightarrow CH_3CH_2OH + NAD^+} \quad (3)$$

The cell potential for this (by Equation 10.2, page 281) is

$$E^{\circ\prime}_{cell} = (-0.20 \text{ V}) - (-0.32 \text{ V})$$
$$= 0.12 \text{ V}$$

We learned in Chapter 10 that even very small cell potentials correspond to huge equilibrium constants, and 0.12 V means that if we treat the reduction of ethanal as an equilibrium *under the original definitions of standard conditions*, K_{eq} is roughly 10^5! How in the world, we might ask, can ethanol be *oxidized* by NAD^+, if the reverse reaction (3) is so overwhelmingly favored? There are two pieces to the explanation.

First, we're not under standard conditions. The molarity of H^+ is not $1 M$ but $1 \times 10^{-7} M$. We'd have to use the Nernst equation (10.4, page 312) to calculate the cell potential under nonstandard conditions, and for this we would need data on the concentrations of all species at equilibrium. We do not have these data. But the second piece in the explanation makes this lack unimportant.

NAD^+ Does Not Act Alone, So the Oxidation of Ethanol Is Favored The second reason why NAD^+ can participate in the *oxidation* of ethanol and be reduced to NADH is that NAD^+ *is the cofactor for only the first enzyme in a series leading to the final step, the reduction of oxygen.* NAD^+ does not have to work alone. The whole respiratory chain pulls electrons from ethanol toward oxygen. Thus it is the reduction of *oxygen*, not the reduction of NAD^+ that we should use in setting up the half-reactions leading to the cell reaction. With this in mind, let's go back and redo the oxidation of ethanol. The two half-reactions, written as reductions and with their reduction potentials are

$$CH_3CH{=}O + 2H^+ + 2e^- \longrightarrow CH_3CH_2OH$$
$$E^{\circ\prime} = -0.20 \text{ V} \quad (2)$$
$$(\tfrac{1}{2})O_2 + 2H^+ + 2e^- \longrightarrow H_2O$$
$$E^{\circ\prime} = +0.82 \text{ V} \quad (4)$$

Now it's the oxygen reduction that has the more positive reduction potential and it is well able to force the half-reaction of Equation 2 to run as an oxidation. Reversing Equation 2 and rewriting the two equations, we have:

$$CH_3CH_2OH \longrightarrow CH_3CH{=}O + H^+ + 2e^- \quad \text{(2, reversed)}$$
$$(\tfrac{1}{2})O_2 + 2H^+ + 2e^- \longrightarrow H_2O \quad (4)$$
$$\overline{CH_3CH_2OH + (\tfrac{1}{2})O_2 \longrightarrow CH_3CH{=}O + H_2O}$$

The cell potential for this net equation is

$$E^{\circ\prime}_{cell} = (+0.82 \text{ V}) - (-0.20 \text{ V})$$
$$= 1.02 \text{ V}$$

(Continued)

This cell potential indicates an overwhelming driving force for the oxidation of ethanol to ethanal as oxygen is reduced to water.

The Reduction of Oxygen Has a Very Favorable Reduction Potential The same reduction of oxygen to water is the largest driving force in biological oxidations, too. Figure 25.4 shows the "energy cascade" along the respiratory chain until this last step, the largest energy drop, is accomplished.

We have illustrated that when the body causes an overall change by a series of small steps, like the oxidation of ethanol, one (or more) steps can be energetically quite unfavorable if at least one step (particularly the last step) is very favorable.

Small Steps Enable a Long Sequence to Trap Energy in ATP We have also, very indirectly, suggested something else about biological oxidations. You would think that the oxidation of an alcohol group in a cell, so overwhelmingly favored, would go extremely rapidly, almost like a combustion. So we get here a hint of the reason why the cell uses many steps.

Several steps provide two needs, the need for the control of overall rate being only one. The other need is to use at least some steps to run energy-consuming reactions. Direct combustion cannot provide such steps. It can produce only heat, which would be fine if we were steam engines. Instead, some of the energy of the oxidation of an alcohol group is used to create the proton gradient and the charge gradient, discussed in Section 25.3, that is so vital to the synthesis of ATP.

SUMMARY

High-energy compounds Organophosphates whose phosphate group transfer potentials equal or are higher than that of ATP are classified as high-energy phosphates. A lower energy phosphate can be made by phosphate transfer from a higher energy phosphate, a process called substrate phosphorylation.

Citric acid cycle Acetyl groups from acetyl coenzyme A are joined to a four-carbon carrier, oxaloacetate, to make citrate. This six-carbon salt of a tricarboxylic acid then is degraded bit by bit as pieces of hydrogen, $(H:^- + H^+)$, are fed to the respiratory chain. Each acetyl unit leads to a maximum of twelve ATP molecules.

Fatty acids and glucose are important suppliers of acetyl groups for the citric acid cycle. In aerobic glycolysis, glucose units are broken to pyruvate units. The oxidative decarboxylation of pyruvate leads to the synthesis of one NADH (which eventually leads to the formation of three ATPs), and it supplies acetyl units to the citric acid cycle from which twelve more ATPs are made.

Respiratory chain A series of electron transfer enzymes called the respiratory enzymes occur together as groups called respiratory assemblies in the inner membranes of mitochondria. These enzymes process NADH or $FADH_2$, made by reduction reactions from metabolites (MH_2) and NAD^+ or FAD. NADH or $FADH_2$ then pass on electrons until cytochrome oxidase uses them (together with H^+) to reduce oxygen to water.

Oxidative phosphorylation As electrons flow from NADH to oxygen in the respiratory chain, protons are released. They cause conformational changes in proteins imbedded in the inner membrane of a mitochondrion. These changes cause protons to move from the mitochondrial matrix to the intermembrane side of the inner membrane, making a proton gradient and a positive charge gradient across the inner membrane. As protons flow back at the allowed F_0 conduits of the proton-pumping ATPase complex of this membrane, ATP is made and released by the F_1 complex of this enzyme. In some systems, other kinds of positive ions migrate, as in the operation of nerves. Various drugs and antibiotics can block the respiratory chain. Other chemicals are able to uncouple the work of the respiratory chain from the synthesis of ATP so that the operation of the chain converts chemical energy to heat. The rate at which ATP is made increases as the concentration of ADP increases.

REVIEW EXERCISES

The answers to Review Exercises whose numbers are in color are found in Appendix C. The answers to the other Review Exercises are found in the Study Guide that accompanies this book. The more challenging questions are marked with asterisks.

Energy Sources

25.1 What products of the digestion of carbohydrates, triacylglycerols, and the polypeptides can be used as sources of biochemical energy to make ATP?

25.2 The complete catabolism of glucose gives what products?

25.3 The identical products form when glucose is burned in open air as when it is fully catabolized in the body. How, then, do these two processes differ in their overall accomplishments?

25.4 What are the end products of the complete catabolism of fatty acids?

High-Energy Phosphates

25.5 Complete the following structure of ATP.

$$\text{Adenosine}-\text{O}-\overset{\overset{\displaystyle O}{\|}}{\underset{\underset{\displaystyle O^-}{|}}{P}}-$$

25.6 Write the structures of ADP and of AMP in the manner started by Review Exercise 25.5.

25.7 At physiological pH, what does the term *inorganic phosphate* stand for? (Give formulas and names.)

25.8 Phosphate X has a phosphate group transfer potential of -13 kcal/mol. What does this quantity refer to?

25.9 Why are creatine phosphate and ATP called high-energy phosphates but glycerol-3-phosphate is not?

*25.10 If the organophosphate of M has a *lower* (less negative) phosphate group transfer potential than that of N, in which direction does the following reaction tend to go spontaneously, to the left or to the right?

$$M-OPO_3{}^{2-} + N \overset{?}{\rightleftharpoons} N-OPO_3{}^{2-} + M$$

25.11 Using data in Table 25.1, tell (yes or no) whether ATP readily transfers a phosphate group to each of the following possible compounds. (Assume, of course, that the right enzyme is available.)
(a) glycerol (b) fructose (c) creatine (d) acetic acid

25.12 All of the possible phosphate transfers in Review Exercise 25.11 are classified as *substrate* phosphorylations. What does this mean?

25.13 What is the function of creatine phosphate in muscle tissue?

25.14 Whether or not creatine phosphate is used in muscle tissue is under *feedback control*. Explain.

Overview of Metabolic Pathways

25.15 In the general area of biochemical energetics, what is the purpose of each of the following pathways?
(a) respiratory chain (b) anaerobic glycolysis
(c) citric acid cycle (d) fatty acid cycle

25.16 What prompts the respiratory chain to go into operation?

25.17 Which tends to increase the rate of breathing, an increase in the body's supply of ATP or an increase in its supply of ADP? Explain.

25.18 In general terms, the intermediates that send electrons down the respiratory chain come from what metabolic pathway that consumes acetyl groups?

25.19 Arrange the following sets of terms in sequence in the order in which they occur or take place. Place the identifying letter of the first sequence of a set to occur on the left of the row of letters.
(a) Citric acid cycle   pyruvate   acetyl CoA
      A   B   C
   respiratory chain   glycolysis
        D   E
(b) Citric acid cycle β-oxidation   acetyl CoA
      A   B   C
   respiratory chain
        D

25.20 The *aerobic sequence* begins with what metabolic pathway and ends with which pathway?

25.21 The β-oxidation pathway occurs to what kind of compound?

25.22 The anaerobic sequence begins and ends with what compounds?

25.23 What does it mean for the cell that anaerobic glycolysis is a "backup?"

Citric Acid Cycle

25.24 What makes the citric acid cycle start up?

25.25 What chemical unit is degraded by the citric acid cycle? Give its name and structure.

25.26 How many times is a secondary alcohol group oxidized in the citric acid cycle?

25.27 Water adds to a carbon–carbon double bond how many times in one turn of the citric acid cycle?

*25.28 The enzyme for the conversion of isocitrate to oxalosuccinate (Fig. 25.2) is stimulated by one of these two substances, ATP or ADP. Which one is the more likely activator? Explain.

25.29 What is the maximum number of ATP molecules that can be made from the use of respiratory chain phosphorylation to break down each of the following?
(a) pyruvate
(b) an acetyl group in acetyl CoA

*25.30 Glutamic acid, one of the amino acids, can be converted to α-ketoglutarate (Fig. 25.2). How many ATP molecules can be made from the entry of α-ketoglutarate into the citric acid cycle?

Respiratory Chain

25.31 Is the following (unbalanced) change a reduction or an oxidation? How can you tell?

$$\underset{\displaystyle CH_3CHCH_2CO_2{}^-}{\overset{\displaystyle \overset{OH}{|}}{}} \longrightarrow \underset{\displaystyle CH_3CCH_2CO_2{}^-}{\overset{\displaystyle \overset{O}{\|}}{}}$$

25.32 Which kind of enzyme would be more likely to cause the change given in Review Exercise 25.31, an enzyme like aconitase or an enzyme like isocitrate dehydrogenase? Explain.

25.33 What is missing in the following basic expression for what must happen in the respiratory chain?

$$\tfrac{1}{2}O_2 + 2H^+ \longrightarrow H_2O$$

25.34 What general name is given to the set of enzymes involved in electron transport?

*25.35 Write the following display in the normal form of a chemical equation.

$$
\begin{array}{c}
\overset{\text{OH}}{\underset{|}{}} \\
CH_3CHCO_2^- \qquad\qquad NAD^+ \\
\\
\overset{O}{\underset{\parallel}{}} \\
CH_3CCO_2^- \qquad\qquad NAD \colon H + H^+
\end{array}
$$

(a) Which specific species is oxidized? (Write its structure.)
(b) Which species is reduced?

*25.36 Write the following equation in the form of a display like that shown in Review Question 25.35.

$$^-O_2CCH_2CH_2CO_2^- + FAD \longrightarrow$$
$$^-O_2CCH{=}CHCO_2^- + FADH_2$$

25.37 Arrange the following in the order in which they receive and pass on electrons.
FMN   NAD$^+$ CoQ   FeS–P

25.38 What does respiratory enzyme complex IV do?

25.39 What is FAD, and where is it involved in the respiratory chain?

25.40 Across which cellular membrane does the respiratory chain establish a gradient of H$^+$ ions? On which side of this membrane is the value of the pH lower?

25.41 According to the chemiosmotic theory, the flow of what particles most directly leads to the synthesis of ATP?

25.42 If the inner mitochondrial membrane is broken, the respiratory chain can still operate, but the phosphorylation of ADP that normally results stops. Explain this in general terms.

25.43 Complete and balance the following equation. (Use 1/2 as the coefficient of oxygen as shown.)

$$MH_2 + H^+ \;+\; \tfrac{1}{2}O_2 \longrightarrow$$
From the
matrix

25.44 What is the difference between substrate and oxidative phosphorylation?

25.45 Besides a gradient of H$^+$ ions, what other gradient exists in mitochondria as a result of the operation of the respiratory chain? In terms of helping to explain how cations other than H$^+$ move across a membrane, of what significance is this gradient?

25.46 Briefly describe the theory presented in this chapter that explains how a flow of protons across the inner mitochondrial membrane initiates the synthesis of ATP from ADP and P$_i$.

Ester Condensations (Special Topic 25.1)

25.47 Which would be more acidic, acetone or propane? Explain.

25.48 Suppose that ethyl propanoate went through the steps of the Claisen condensation. Write the structure of the final product.

Redox Potentials (Special Topic 25.2)

25.49 What are *standard conditions* for reduction potentials when working with biological oxidations, and why are they different from the regular standard conditions?

25.50 Write the half-cell reaction for the reduction of acetate ion to ethanal. Its value of E'° is -0.60 V.

25.51 Using the data and equation of Review Exercise 25.50 plus such data as needed from Special Topic 25.2, what would be the cell reaction if the acetate–ethanal system were coupled with the NAD$^+$–NADH system? Calculate the cell potential. What is the spontaneous reaction, the oxidation of ethanal or the reduction of acetate ion according to your answers here?

25.52 Repeat Review Exercise 25.51, only this time couple the acetate–ethanal system to the $(1/2)O_2$–H$_2$O system.

25.53 What is one reason why one step in a series can be quite unfavorable and yet made to happen?

25.54 What is one reason why the body uses so many steps to bring about an overall oxidation of a metabolite?

Additional Exercises

*25.55 The conversion of pyruvate to an acetyl unit is both an oxidation and a decarboxylation.
(a) If *only* decarboxylation occurred, what would form from pyruvate? Write the structure of the other product in the following.

$$
\overset{O}{\underset{\parallel}{}}\\
CH_3CCO_2^- + H^+ \longrightarrow \underline{\hspace{2cm}} + O{=}C{=}O
$$

(b) If this product is oxidized, what is the name and the structure of the product of such oxidation?
(c) Referring to Figure 25.2, which specific compound undergoes an oxidative decarboxylation similar to that of pyruvate? (Give its name.)

*25.56 One cofactor in the enzyme assembly that catalyzes the oxidative decarboxylation of pyruvate requires thiamine, one of the B vitamins. Therefore in beriberi, the deficiency disease for this vitamin, the level of what substance can be expected to rise in blood serum (and for which an analysis can be made as part of the diagnosis of beriberi)?

*25.57 Consider the oxidation of one acetyl group via the citric acid cycle to two molecules of carbon dioxide.
(a) This overall change requires that how many *pairs* of electrons be transferred to the enzymes of the respiratory chain?
(b) In the transfer of these pairs of electrons, what else transfers?
(c) What is a word common to the names of the enzymes that catalyze these electron transfers?

METABOLISM OF CARBOHYDRATES

No doubt the act of mouth-watering has a molecular basis, but all we'll attempt to understand in this chapter is the metabolism of sugar and other carbohydrates.

26.1
GLYCOGEN METABOLISM
26.2
GLUCOSE TOLERANCE

26.3
THE CATABOLISM OF GLUCOSE
26.4
GLUCONEOGENESIS

26.1

GLYCOGEN METABOLISM

Much of the body's control over the blood sugar level is handled by its regulation of the synthesis and breakdown of glycogen.

The digestion of the starch and disaccharides in the diet gives glucose, fructose, and galactose, but their catabolic pathways all converge very quickly to that of glucose itself. Galactose, for example, is changed by a few steps in the liver to glucose-1-phosphate. Fructose is changed to a compound that occurs early in glycolysis. For these reasons, this chapter concentrates on the metabolism of glucose, but we'll begin with information about the glucose in circulation. The concentration of circulating glucose, as we soon will learn, is affected by several factors and demands involved in its distribution, storage, and use. We will go into details about the catabolism of glucose both by glycolysis and by the pentose phosphate pathway. Finally, we'll learn how glucose is synthesized in the body from smaller molecules.

A Special Vocabulary Exists to Describe Variations in the *Blood Sugar Level*

The concentration of monosaccharides in whole blood, expressed in milligrams per deciliter (mg/dL), is called the **blood sugar level.** This is very nearly the same as the glucose level, because glucose is overwhelmingly the major monosaccharide. When determined after several hours of fasting, the plasma sugar level, called the **normal fasting level,** is 70 to 110 mg/dL (3.9 to 6.1 mmol/L).[1] In a condition of **hypoglycemia,** the plasma sugar level is *below* normal, and in **hyperglycemia** it is *above* the normal fasting level. (Sometimes you will see the term *normoglycemia* used for levels in the normal range.)

When the blood sugar level becomes too hyperglycemic (becomes too high), the kidneys are unable to put back into the blood all of the glucose that leaves them. Glucose then appears in the urine, a condition called **glucosuria.** The blood sugar level above which this happens is called the **renal threshold** for glucose, and it is in the range of roughly 160 to 180 mg/dL, and higher in some individuals.

Excess Blood Glucose Normally Is Withdrawn from Circulation
When there is more than enough glucose in circulation to meet energy needs, the body does not eliminate the excess glucose but conserves its chemical energy. There are two ways to do this. One is to convert glucose to fat, and we'll study how this is done in the next chapter. The other is to synthesize glycogen, which we'll discuss here.

Liver and muscle cells can convert glucose to glycogen by a series of steps called **glycogenesis** ("glycogen creation"). The liver holds 70 to 110 g of glycogen, and the muscles, taken as a whole, contain 170 to 250 g. When muscles need glucose, they take it back out of glycogen. When the blood needs glucose because the blood sugar level has dropped too much, the liver hydrolyzes as much of its glycogen reserves as needed and then puts the glucose into circulation. The overall series of reactions in either tissue that hydrolyzes glycogen is called **glycogenolysis** (lysis or hydrolysis of glycogen), a process controlled by several hormones.

Epinephrine Launches a Multiple-Enzyme Glycogenolysis Cascade
When muscular work is begun, the adrenal medulla secretes the hormone **epinephrine.** Epinephrine (as

■ mg/dL = milligrams per deciliter, where 1 dL = 100 mL.

■ *-glyc-*, sugar
-emia, in blood
hypo-, under, below
hyper-, above, over
renal, of the kidneys
-uria, in urine

■ Glycogen (page 559), recall, is a polymer of glucose with both $\alpha(1 \rightarrow 4)$ and $\alpha(1 \rightarrow 6)$ bridges and resembling amylopectin, a component of starch.

[1] The normal reference laboratory values or "normals" in this text are those used by the Massachusetts General Hospital and published in *The New England Journal of Medicine*, **327,** page 718 (Sept. 3, 1992).

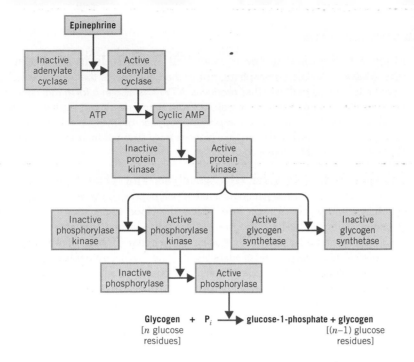

FIGURE 26.1
The epinephrine "cascade"

Glycogen + P$_i$ ⟶ glucose-1-phosphate + glycogen
[*n* glucose
residues]   [(*n*−1) glucose
residues]

well as a close relative, norepinephrine) activates glycogenolysis by the steps outlined in Figure 26.1. At its target cells, epinephrine activates the *enzyme adenylate cyclase,* which catalyzes the conversion of some ATP to cyclic AMP as we studied in Section 22.6.

Cyclic AMP then activates another enzyme, and this still another enzyme, and so on in a cascade of events (see Fig. 26.1). Each molecule of epinephrine, by activating adenylate cyclase, triggers the formation of dozens of molecules of cyclic AMP. *Each* of these activates a succeeding enzyme, and so on until the final enzyme appears not only suddenly but in a relatively large quantity. Thus one epinephrine molecule triggers the rapid mobilization of thousands of glucose units, and these are now ready to supply energy that the body needs.

While the epinephrine "cascade" proceeds to release glucose from glycogen, the affected tissue simultaneously shuts down a team of enzymes called *glycogen synthetase* that otherwise would do the opposite, change glucose to glycogen. This is done at the step in Figure 26.1 where the enzyme called *active protein kinase* catalyzes the phosphorylation of *two* enzymes. One is inactive phosphorylase kinase and the other is active glycogen synthetase. The first action continues the epinephrine cascade, until glucose units are obtained from glycogen. The second action, the phosphorylation of glycogen synthetase, *shuts down an enzyme that would help to remake glycogen.* The potential competition, therefore, cannot develop. It's a remarkable aspect of this system.

The end product of glycogenolysis isn't actually glucose but glucose-1-phosphate. Cells that can do glycogenolysis also have an enzyme called *phosphoglucomutase,* which catalyzes the conversion of glucose-1-phosphate to its isomer, glucose-6-phosphate:

■ Epinephrine (adrenaline) and norepinephrine (noradrenaline) — see Special Topic 17.1, page 507 — are called the "fight or flight" hormones because the brain causes their release in sudden emergencies.

■ An estimated 30,000 molecules of glucose are released from glycogen for each molecule of epinephrine that initiates glycogenolysis.

Glucose-1-phosphate ⇌ [phosphoglucomutase] ⇌ Glucose-6-phosphate

Glucose Is Trapped in the Muscle Cell When It Is in the Form of Glucose-6-Phosphate Glucose-6-phosphate, rather than glucose, is the form in which a glucose unit must be to enter a pathway that produces ATP. *It is also in a form that cannot migrate out of muscle cells.* It cannot be lost, therefore, from tissue needing it during exercise. Thus glycogenolysis in muscle tissue is an important supplier of energy. When the supply of muscle glycogen is low, muscle cells can take glucose from circulation, *trap it as glucose-6-phosphate,* and then convert this to glycogen.

■ The glucagon molecule has 29 amino acid residues.

Glucagon Activates Liver Glycogenolysis and Thus Affects the Blood Sugar Level The α-cells of the pancreas make a polypeptide hormone, **glucagon,** which helps to maintain a normal blood sugar level. When the blood sugar level drops, the α-cells release glucagon. Its target tissue is the liver, where it is a strong activator of glycogenolysis.

Glucagon works by a cascade process very similar to that initiated by epinephrine. Like epinephrine, glucagon activates adenylate cyclase. Unlike epinephrine, glucagon also inhibits glycolysis, so this action helps to keep the supply of glucose up. Glucagon, also unlike epinephrine, does not cause an increase in blood pressure or pulse rate, and it is longer acting than epinephrine.

Liver Cells Can Release Glucose to Circulation Glucose units released from liver glycogen as glucose-6-phosphate are converted to glucose by the enzyme *glucose-6-phosphatase* that the liver has but is not present in the muscles. This enzyme catalyzes the hydrolysis of glucose-6-phosphate to glucose and inorganic phosphate.

■ The letter P is often used to represent the whole phosphate group in the structures of phosphate ester intermediates in metabolism.

$$\text{Glucose-6-P} + H_2O \xrightarrow{\text{glucose-6-phosphatase}} \text{glucose} + P_i$$

Glucose can now leave the liver and so help raise the blood sugar level. During periods of fasting, therefore, the overall process in the liver from glucose-1-phosphate to glucose-6-phosphate to glucose is a major supplier of glucose for the blood. Glucagon, which triggers this, is thus an important regulator of the blood sugar level. The brain depends on the liver during fasting to maintain its favorite source of chemical energy, circulating glucose. (We are beginning to see in chemical terms how vital the liver is to the performance of other organs.) When circulating glucose is taken up by a brain cell, it is promptly trapped in the cell by being converted to glucose-6-phosphate.

Several hereditary diseases involve the glucose–glycogen interconversion, and some are discussed in Special Topic 26.1.

Human Growth Hormone Stimulates the Release of Glucagon Growth requires energy, so the action of glucagon that helps to supply a source of energy — glucose — aids in the work of the human growth hormone. In some situations, such as a disfiguring condition known as acromegaly, there is an excessive secretion of human growth hormone that promotes too high a level of glucose in the blood. This is undesirable because a prolonged state of hyperglycemia from any cause can lead to some of the same blood-capillary related complications observed when diabetes is poorly controlled.

■ Acromegaly is sometimes called *giantism* because the bone structures of victims are enlarged.

Insulin Strongly Lowers the Blood Sugar Level The β-cells of the pancreas make and release **insulin,** a polypeptide hormone. Its release is stimulated by an increase in the blood sugar level, such as normally occurs after a carbohydrate-rich meal. As insulin moves into action, it finds its receptors at the cell membranes of muscle and adipose tissue. The insulin–receptor complexes make it possible for glucose molecules to move easily into the affected cells, which lowers the blood sugar level.

■ Adipose tissue is the fatty tissue that surrounds internal organs.

Not all cells depend on insulin to take up glucose. Brain cells, red blood cells, and cells in the kidneys, the intestinal tract, and the lenses of the eyes take up glucose directly.

In one form of diabetes mellitus, type I diabetes, the β-cells have been destroyed, and the pancreas is unable to release insulin. Such individuals must receive insulin, usually by intravenous injection. If more insulin is put into circulation than needed for the management of the blood sugar level, this level falls too low and *insulin shock* results.

■ Life-saving first aid for someone in insulin shock is sugared fruit juice or candy to counter the hypoglycemia.

GLYCOGEN STORAGE DISEASES

A number of inherited diseases involve the storage of glycogen. For example, in **Von Gierke's disease,** the liver lacks the enzyme glucose-6-phosphatase, which catalyzes the hydrolysis of glucose-6-phosphate. Unless this hydrolysis occurs, glucose units cannot leave the liver. They remain as glycogen in such quantities that the liver becomes very large. At the same time, the blood sugar level falls, the catabolism of glucose accelerates, and the liver releases more and more pyruvate and lactate.

In **Cori's disease,** the liver lacks an enzyme needed to catalyze the hydrolysis of 1,6-glycosidic bonds, the bonds that give rise to the many branches of a glycogen molecule. Without this enzyme, only a partial utilization of the

glucose in glycogen is possible. The clinical symptoms resemble those of Von Gierke's disease, but they are less severe.

In **McArdle's disease,** the muscles lack phosphorylase, the enzyme needed to obtain glucose-1-phosphate from glycogen (see Fig. 26.6). Although the individual is not capable of much physical activity, physical development is otherwise relatively normal.

In **Andersen's disease,** both the liver and the spleen lack the enzyme for putting together the branches in glycogen. Liver failure from cirrhosis usually causes death by age 2.

Hypoglycemia Can Make You Faint Your brain relies almost entirely on glucose for its chemical energy, so if hypoglycemia develops rapidly, you can become dizzy and may even faint. The brain consumes about 120 g/day of glucose, and a quick onset of hypoglycemia starves the brain cells. They do have the ability to switch over to other nutrients, but brain cells can't do this very rapidly.

Somatostatin Inhibits Glucagon and Slows the Release of Insulin The hypothalamus, a specific region in the brain, makes **somatostatin,** another hormone that participates in the regulation of the blood sugar level. When the β-cells of the pancreas secrete insulin, which helps to *lower* the blood sugar level, the α-cells should not at the same time release glucagon, which helps to *raise* this level. Somatostatin acts at the pancreas to inhibit the release of glucagon as well as to slow down the release of insulin. It thus helps to prevent a wild swing in the blood sugar level that insulin alone might cause.

Persistent Hyperglycemia Indicates Diabetes Whenever hyperglycemia develops and tends to persist, something is wrong with the mechanisms for withdrawing glucose from circulation. In an individual with diabetes under poor control and so with sustained hyperglycemia, some blood glucose combines with hemoglobin to give glycohemoglobin (or glycosylated hemoglobin). The measurement of the level of this substance has become the best way to determine the average blood sugar level of an individual, better even than direct measurements of blood glucose. The level of glycohemoglobin doesn't fluctuate as widely and quickly as the blood sugar level. Let's now examine in more detail the body's ability to manage a somewhat steady concentration of blood glucose.

■ Diabetes is a common cause of hyperglycemia, but there are other possible causes.

26.2

GLUCOSE TOLERANCE

The ability of the body to tolerate swings in the blood sugar level is essential to health.

Your **glucose tolerance** is the ability of your body to manage its blood sugar level within the normal range. We'll take an overview here of the many factors that contribute to glucose tolerance.

■ Carl Cori and Gerti Cori shared the 1947 Nobel prize in physiology and medicine.

The Cori Cycle Describes the Distributions and Uses of Glucose

The strategies used by the body to maintain its blood sugar level within the normal range form a cycle of events called the **Cori cycle,** outlined in Figure 26.2.

At the bottom of the figure we see glucose as it enters the bloodstream from the intestinal tract. Its molecules either stay in circulation or are soon removed by various tissues. Two are shown in the figure, those of muscle and liver. Muscle cells can trap glucose molecules and use them either to make ATP by glycolysis or to replenish the muscle's glycogen reserves. Liver cells can similarly trap glucose, and the liver is able to release glucose back into the bloodstream when the blood sugar level must be raised.

When glucose is used in glycolysis, the end product is either pyruvate or lactate, depending on the oxygen supply, as we learned in the previous chapter. Either pyruvate or lactate can be used to make more ATP by means of the citric acid cycle and the respiratory chain. However, when extensive anaerobic glycolysis is carried out in a tissue, then the lactate level increases considerably.

■ *Neo,* new; *-neogenesis,* new creation:, *gluconeogenesis,* the synthesis of new glucose.

Because lactate still has C—H bonds, it continues to have useful chemical energy, so instead of simply excreting excess lactate at the kidneys, the body recycles it. It converts a fraction of it — about five-sixths — to glucose. This synthesis of glucose from smaller molecules is called **gluconeogenesis,** and it requires ATP energy. The remaining one-sixth of the lactate is catabolized to make the ATP needed for gluconeogenesis. (We go into more details about gluconeogenesis later in this chapter.) The recycling of lactate completes the Cori cycle.

We can see from all these processes that many factors affect the blood sugar level. Some tend to increase it and some do the opposite. Figure 26.3 summarizes them in a different kind of display.

Glucose Tolerance Can Be Measured by the Glucose Tolerance Test

In the **glucose tolerance test,** the individual is given a drink that contains glucose, generally 75 g for an adult and 1.75 g per kilogram of body weight for children, and then the blood sugar level is checked at regular intervals.

Figure 26.4 gives typical plots of this level versus time. The lower curve is that of a person with normal glucose tolerance, and the upper curve is of one whose glucose tolerance is typical of a person with diabetes. In both, the blood sugar level initially increases sharply. The healthy person, however, soon manages the high level and brings it back down with the help of a

FIGURE 26.2
The Cori cycle

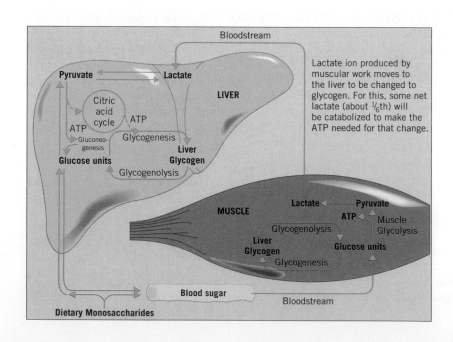

FIGURE 26.3
Factors that affect the blood sugar level

Blood sugar level; concentration of glucose in mg/dL

160 — Renal threshold via kidneys → Glucose is transferred to the urine (glucosuria)

Hyperglycemia ——→ Fatty acid, cholesterol, body fats are made

110 — Glycogenesis ——→ Muscle and liver glycogen form

Normal fasting level ——→ Glycolysis

←—— Absorption of dietary glucose, gluconeogenesis, or glycogenolysis puts glucose in circulation

70 —

Hypoglycemia ——→ Hyperinsulinism Fasting Starvation

normal flow of insulin and somatostatin. In the diabetic, the level comes down only very slowly and remains essentially in the hyperglycemic range throughout.

Notice that in the normal individual, the blood sugar level can sometimes drop to a mildly hypoglycemic level. Such hypoglycemia is possible also in someone who has eaten a carbohydrate-rich breakfast. With glucose pouring into the bloodstream, the release of a bit more insulin than needed can occur. This leads to the overwithdrawal of glucose from circulation, and midmorning brings dizziness and sometimes even fainting (or falling asleep in class). The prevention isn't more sugared doughnuts but a balanced breakfast.

■ Unhappily, most people with midmorning sag go into another round of sugared coffee and sugared rolls. The glucose gives a short lift, but then an oversupply of insulin restores the mild hypoglycemia of the sag.

Glucose Tolerance Is Poor in Those with Diabetes   The subject of glucose tolerance is nowhere of more concern than in connection with diabetes. **Diabetes** is defined clinically as a disease in which the blood sugar level persists in being much higher than warranted by the dietary and nutritional status of the individual. Invariably, a person with untreated diabetes has glucosuria, and the discovery of this condition often triggers the clinical investigations that are necessary to diagnose diabetes.

As discussed in Special Topic 26.2, there are two broad kinds of diabetes, type I and type II. Type I diabetics are unable to manufacture insulin (at least not enough), and they need daily insulin therapy to manage their blood sugar levels. Maintaining a relatively even and normal blood sugar level is the best single strategy that such diabetics have for the prevention of some of the vascular and neural problems that can complicate their health later. Most type II diabetics are able to manage their blood sugar levels by a good diet, weight control, exercise, and sometimes the use of oral medications.

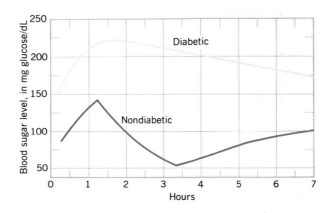

FIGURE 26.4
Glucose tolerance curves

DIABETES MELLITUS

The name for this disorder is from the Greek *diabetes*, to pass through a siphon, and *mellitus*, honey-sweet— meaning to pass urine that contains sugar. We'll call it diabetes for short. In severe, untreated diabetes, the victim's body wastes away despite efforts to satisfy a powerful thirst and hunger. To the ancients, it seemed as if the body were dissolving from within.

Between 1 and 6% of the United States population has diabetes, and almost as many others are believed to have this disease. It ranks third behind heart disease and cancer as a cause of death.

Between 10 and 25% of all cases of diabetes are of the severe, insulin-dependent variety in which the β-cells of the pancreas are unable to make and secrete insulin. This is **type I diabetes,** and insulin therapy is essential. It is also called **insulin-dependent diabetes mellitus** or **IDDM.** Most victims contract IDDM before the age of 40, often as adolescents, so IDDM has sometimes been called *juvenile-onset diabetes.*

The rest of all those with diabetes have a form called **type II diabetes,** or **non-insulin-dependent diabetes mellitus, NIDDM.** Most victims are able to manage their blood sugar levels by diet and exercise alone, without insulin injections. Their problem is not the lack of insulin but with a breakdown in the machinery for taking advantage of it at insulin's target cells. Most who contract NIDDM do so when they are over 40, so NIDDM has been called *adult-onset diabetes.*

Type I Diabetes Develops in Six Stages D. S. Eisenbarth, a diabetes specialist, divides the onset of type I diabetes into six stages. We'll review them as background for illustrations of equilibrium chemistry and factors that shift equilibria.

Stage 1 is thought to be an existing genetic defect most likely involving more than one gene. However, sets of identical twins are known in which only one twin becomes diabetic. In some way, the genetic problem must be related to a problem of the immune system.

Stage 2 is a triggering incident, like a viral infection. The mumps virus, for example, causes diabetes in some. Usually, the onset of virus-caused type I diabetes occurs slowly over a few years.

Stage 3 is the appearance in the blood of certain antibodies. (*Antibodies* are substances made by the immune system to counteract the effects of invading substances called *antigens* that are alien to the body.) Substances on the membranes of the pancreatic β-cells have been altered (perhaps by the virus) so that the body's immune system sees β-cells as foreign antigens and so makes antibodies against them. Type I diabetes is thus an autoimmune disease, *auto* because the body's immune system fails to recognize the proteins of its own body and sets out to destroy them and, therefore, itself.

Stage 4 is a period during which the pancreas loses its ability to secrete insulin. Stage 5 is diabetes and persistent hyperglycemia. Most of the pancreatic β-cells have disappeared. Stage 6 is the period following complete destruction of β-cells.

Immune-Suppressant Therapy Works if the Problem Is Caught in Its Early Stages Cyclosporine is an agent used to suppress the rejection of transplanted organs, like kidney transplants. When used in the early stages of the onset of type I diabetes, cyclosporine prevents insulin dependence in a significant fraction of individuals tested.

Insulin Receptors Are a Problem in Type II Diabetes The onset of type II diabetes is much slower than that for type I, and obesity, lack of exercise, and a sugar-rich diet are factors. Some scientists believe that sugar evokes such a continuous presence of insulin that the insulin-receptor proteins of target cells literally wear out faster than they can be replaced. When the weight is reduced, particularly in connection with physical exercise, the relative numbers of receptor proteins rebound.

Under the incessant demand to produce and secrete insulin, the β-cells can eventually give out. In a veritable epidemic of NIDDM, over 60% of the older adult population of the Pacific islet of Nauru has become diabetic following a change in lifestyle from active fishing and farming to a very sedentary life. (The discovery of huge phosphate reserves on Nauru Island made the Nauruans one of the world's wealthiest peoples.)

The conventional wisdom that places insulin and its receptors at the heart of the NIDDM problem has been recently challenged by the discovery that the β-cells secrete not only insulin but also another protein called *amylin*. Is it the relatively high level of amylin that suppresses the uptake of glucose by target cells? If so, would an amylin control drug be the answer to NIDDM? Needless to say, much research is in progress.

Glucosylation of Proteins May Cause the Long-Term Complications of Diabetes The immediate complications of IDDM are an elevated blood sugar level, metabolic acidosis, and eventual death from coma and uremic poisoning. Insulin therapy corrects these immediate problems, but it deals less well with the longer-term complications.

The continuous presence of a high level of blood glucose in both IDDM and NIDDM shifts certain chemical equilibria in favor of glucosylated compounds. The aldehyde group of the open form of glucose, for example, can react with amino groups, like those on side chains of lysine residues, to form products called *Schiff bases*.

$$-CH{=}O + H_2N{-} \rightleftharpoons -CH{=}N{-} + H_2O$$

| Aldehyde | Amino | Schiff base |
| group | group | system |

Hemoglobin, for example, gives this reaction, and a high level of glucose shifts this equilibrium to the right. The level of glucosylated hemoglobin thus increases. When the glucose level is brought down and kept within a normal range, the Schiff base level also declines.

The problem with the Schiff bases in the long term is that they undergo molecular rearrangements to more permanent products, called *Amadori compounds,* in which the C=N double bond has migrated to C=C positions. *After a time, the formation of the Amadori compounds is not reversible.*

When these reactions occur in the basement membrane of blood capillaries, they swell and thicken; the condition is called *microangiopathy*. (The basement membrane is the protein support structure that encases the single layer of cells of a capillary.) Microangiopathy is believed to lead to the other complications, most of which involve the vascular system or the neural networks: kidney problems, gangrene of the lower limbs, and blindness. Diabetes is the leading cause of new cases of blindness in the United States, and it is the second most common cause of blindness, overall.

Blindness from Diabetes May Also Reflect the Reduction of Glucose to Sorbitol Glucose is reduced by the enzyme *aldose reductase* to sorbitol. It's a minor reaction in cells of the lens of the eye, but *an abundance of glucose shifts equilibria in favor of too much sorbitol*. Sorbitol, unlike glucose, tends to be trapped in lens cells, and as the sorbitol concentration rises so does the osmotic pressure in the fluid. *This draws water into the lens cells, which generates pressure and leads to cataracts.*

Diet Control Is Mandatory The best single treatment of diabetes is any effort that keeps a strict control on the blood sugar level to avoid the episodes of upward surges followed by precipitous declines.

A Number of Technologies Are in Use or Are Being Tested People with IDDM gauge their insulin needs by blood tests for blood sugar levels. We mentioned an enzyme-based test in Section 22.5, but it requires a puncture to obtain a drop of blood. Another technology in a testing stage (Futrex Inc., Maryland) uses a hand-held, battery-driven source of infrared rays, which are focused onto the skin at the wrist or a fingertip. The meter converts the amount of light absorbed to a blood glucose level. With some diabetics, insulin pumps can be implanted much like heart pacemakers. These monitor the blood sugar level and release insulin according to the need. The use of insulin nasal sprays immediately before a meal is another approach being tested. Several groups of scientists are working on insulin pills, a technology made difficult by the fact that insulin, like any protein, is digested.

When stripped of neighboring cells, β-cells can be transplanted. They need not even be inserted into the receiver's pancreas, and they start to make insulin in a few weeks. Apparently the cells *adjacent* to the β-cells are responsible for inducing the body's immune-centered rejection process, so without such cells the β-cells are more safely transplanted.

Human β-cells work best, of course, but those from pigs and cows also appear to be usable provided that they are encapsulated in very small plastic spheres. These spheres have microscopic holes large enough to let insulin molecules escape but not large enough to let antibodies inside. This technique has cured type I diabetes in experimental animals. A test of this technique in a human was begun in 1993, and early indications were very promising.

26.3

THE CATABOLISM OF GLUCOSE

Glycolysis and the pentose phosphate sequence are the chief catabolic pathways open to glucose.

Glycolysis, as we noted in the previous chapter, is a series of reactions that change glucose to pyruvate or to lactate while a small but important amount of ATP is made. Other monosaccharides eventually enter the same glycolysis pathway as glucose, as seen in Figure 26.5, so when we study glycolysis we cover most of monosaccharide catabolism.

Anaerobic Glycolysis Ends in Lactate When a cell receives oxygen at a rate slower than needed, glycolysis can still operate, but it ends in lactate, not pyruvate. The overall equation for this *anaerobic glycolysis,* or the **anaerobic sequence,** is

$$C_6H_{12}O_6 + 2ADP + 2P_i \xrightarrow[\text{glycolysis}]{\text{anaerobic}} 2CH_3\overset{\overset{\displaystyle OH}{|}}{C}HCO_2^- + 2H^+ + 2ATP$$

Glucose   Lactate

Except during extensive exercise, glycolysis is operated with sufficient oxygen, and it is aerobic. Its overall equation is

$$C_6H_{12}O_6 + 2ADP + 2P_i + 2NAD^+ \longrightarrow 2CH_3\overset{\overset{\displaystyle O}{||}}{C}CO_2^- + 2ATP + 2NADH + 2H^+ + 2H_2O$$

Glucose   Pyruvate

The NADH produced by aerobic glycolysis is involved with the respiratory chain, so more ATP is made by using its $H:^-$ as NAD^+ is regenerated.

FIGURE 26.5
Convergence of the pathways in the metabolism of dietary carbohydrates

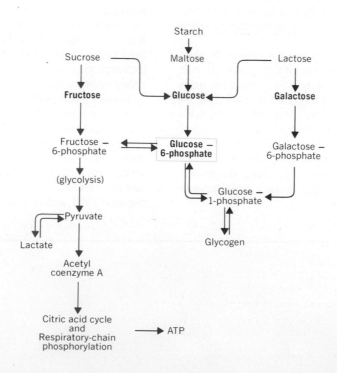

Glycolysis Begins with ATP Consumption but Then Generates More Figure 26.6 outlines the steps to pyruvate (or lactate) that can begin with either glucose or glycogen. The steps that lead to fructose-1,6-bisphosphate are actually up an energy hill, because they consume ATP. But this is like pushing a sled or bike up the short backside of a long hill, because the investment in energy is more than repaid by the long, downhill slide to lactate and more ATP. When glycogen is the source of glucose for glycolysis, the initial investment in ATP is slightly smaller.

■ The numbered steps in the discussion of glycolysis, below, refer to Figure 26.6.

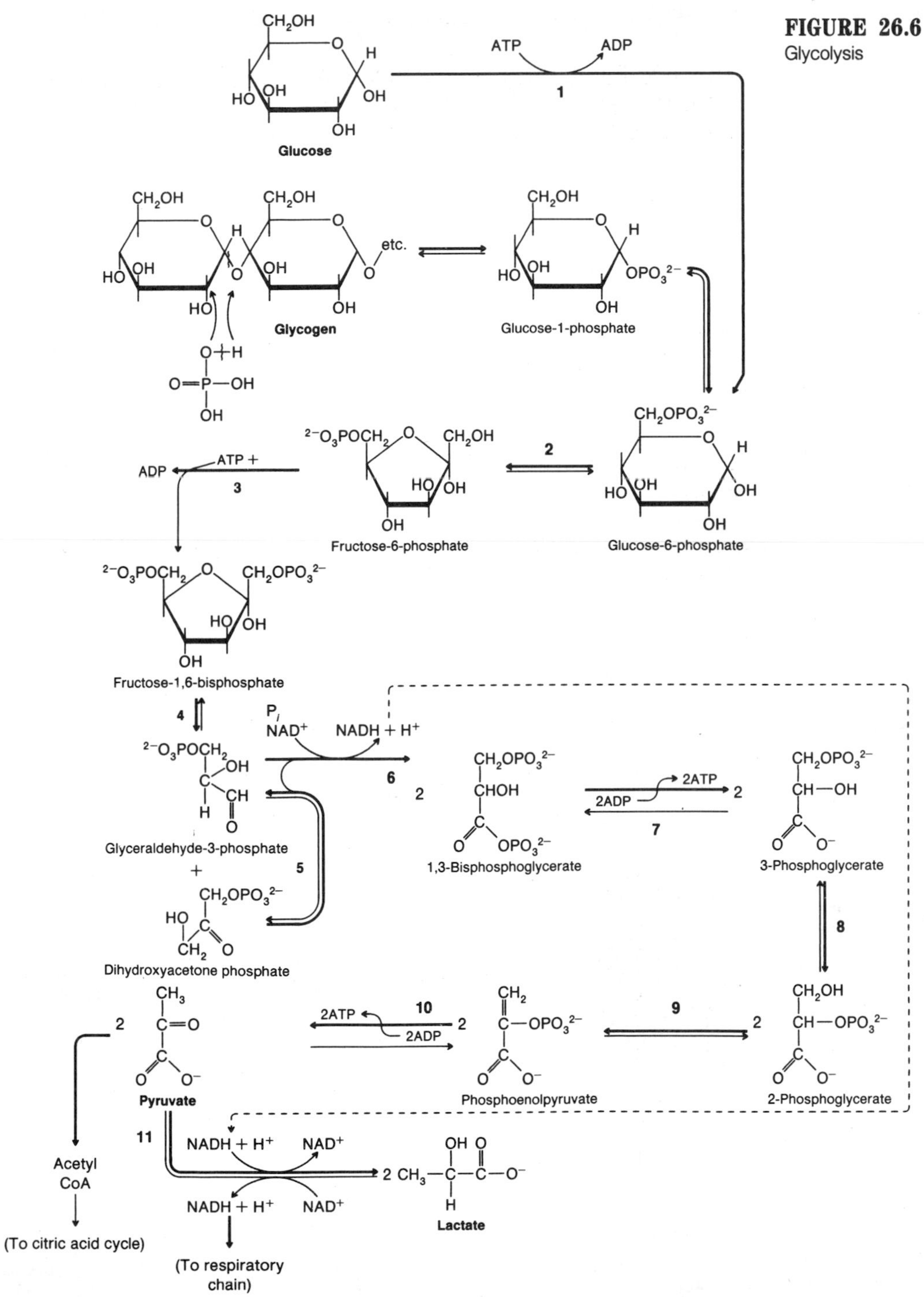

FIGURE 26.6
Glycolysis

■ A *kinase* is an enzyme that handles transfers of phosphate units between ATP and some other substrate.

1. Glucose is phosphorylated by ATP under catalysis by *hexokinase* to give glucose-6-phosphate.

2. Glucose-6-phosphate changes to its isomer, fructose-6-phosphate. The enzyme is *phosphoglucose isomerase*. This may seem to be a major structural change, but it involves little more than some shifts of bonds and hydrogens, as the arrows in the following sequence show:

■ These equilibria shift constantly to the right as long as later reactions continuously remove products as they form.

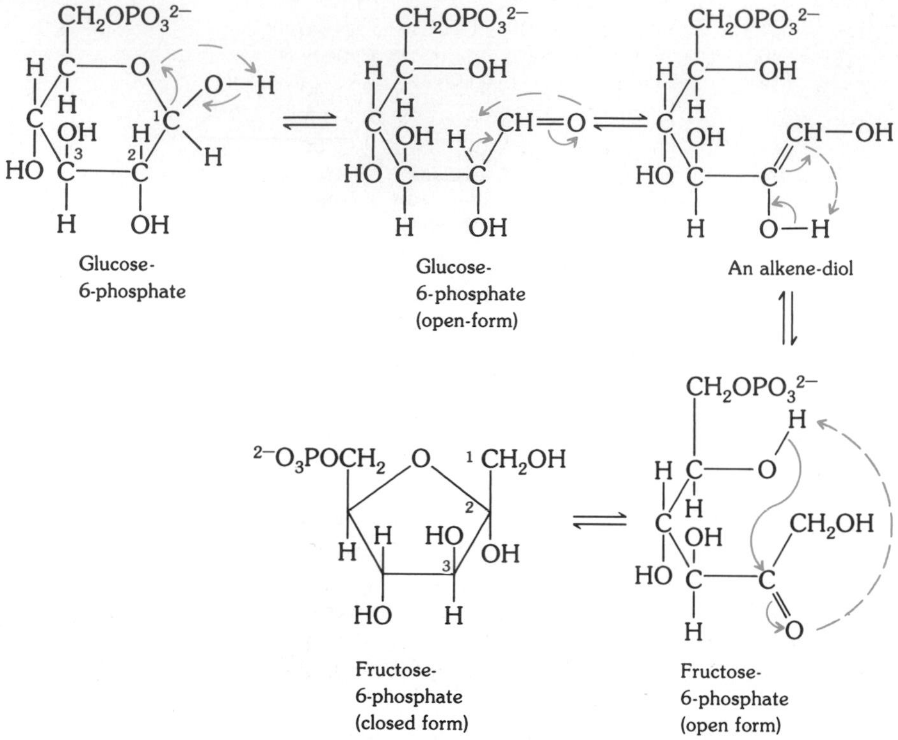

Glucose-6-phosphate

Glucose-6-phosphate (open-form)

An alkene-diol

Fructose-6-phosphate (closed form)

Fructose-6-phosphate (open form)

■ All kinases require Mg^{2+} as a cofactor.

3. ATP phosphorylates fructose-6-phosphate to make fructose-1,6-bisphosphate. The enzyme is *phosphofructokinase*. This step is essentially irreversible, and it ends the energy-using phase of glycolysis.

4. Fructose-1,6-bisphosphate breaks apart into two triose monophosphates. The reaction is catalyzed by *aldolase*. *This is actually a reverse aldol condensation,* a reaction whose study we postponed from Chapter 15. The regular aldol condensation is discussed in Special Topic 26.3. If you study this Special Topic now, it will be easier to see how we can visualize step 4 in glycolysis, as follows, in terms of a few simple and reasonable shifts of electrons and protons.

■ Side chain groups on certain proton-donating or proton-accepting amino acid residues of the enzyme participate in shuttling protons around.

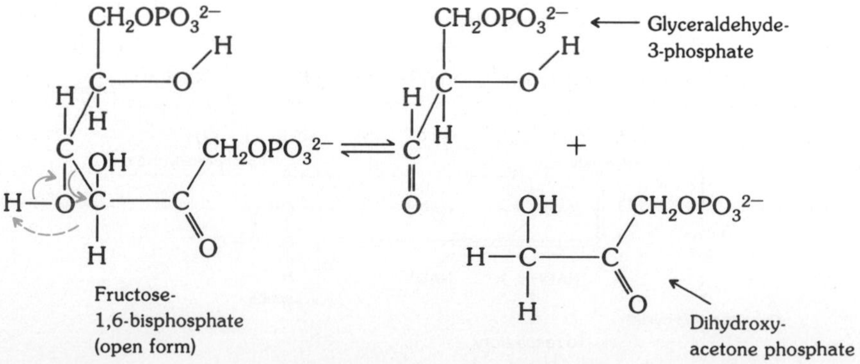

Fructose-1,6-bisphosphate (open form)

Glyceraldehyde-3-phosphate

Dihydroxy-acetone phosphate

THE ALDOL CONDENSATION AND REVERSE ALDOL CONDENSATION

Reactions that make new carbon–carbon bonds or break them have special places in both organic chemistry and biological chemistry. In this Special Topic we will examine the *aldol condensation,* one that joins aldehydes or ketones together and that can be reversed. The term *aldol* signifies that the product has both an aldehyde group and an alcohol group. Step 4 of glycolysis, catalyzed by aldolase, is a reverse of this reaction. An example of the forward running of an aldol condensation occurs in gluconeogenesis. Let's see how the aldol condensation occurs in the molecule-building or forward direction. Each step is reversible, however, so we'll write each step as an equilibrium. We'll also show the reaction using a aldehyde, but the same principles apply to the condensation of ketones or to a mixed reaction involving an aldehyde and a ketone.

The reaction whose mechanism we will examine is

(Two molecules of the same aldehyde)

An aldol (a β-hydroxy aldehyde)

Notice that the reaction is simply the addition of one molecule of the aldehyde (the second) to another molecule (the first). *It always involves the H atom of an alpha position of the adding molecule, never a hydrogen attached at any other position on the chain.* By the same token, *the reverse aldol condensation requires that the OH group be beta to the carbonyl group.*

Step 1 Proton transfer from the aldehyde to a proton acceptor, B:⁻. (*In vivo,* sidechains of amino acid residues, like the NH₂ group of lysine, serve as carriers and transfer agents in proton shuttles.)

(Aldehyde with one α-hydrogen)

We see here why only a hydrogen attached to the *alpha* position of the aldehyde can be involved. It's the only H atom on the chain with any acidity whatsoever, and it is only *very* weakly acidic. That it has any acidity at all is caused by the nearby carbonyl group, an electronegative group that can attract electron density in the anion of the aldehyde and so help to stabilize it.

Step 2 The anion of the aldehyde is attracted to the $\delta+$ charge on the permanently polarized carbonyl carbon of the intact aldehyde molecule. The new carbon–carbon bond forms in this step. The product is the anion of the aldol.

Step 3 The anion of the aldol is the conjugate base of an alcohol and so is a very strong base. It recovers a proton in the last step.

Aldol product

Ketones are able to engage in the same kind of reaction. All that would change in the mechanism given here is that instead of H on the carbonyl carbon there would be another R group.

In the reverse of an aldol condensation, each step is run in reverse, in turn. Thus the aldol product would first transfer H from its OH group to an acceptor to give the anion of the aldol. Next, the carbon–carbon bond would *break* as the two smaller fragments form. *It is precisely this kind of reaction that occurs in step 4 of glycolysis.* If you find the HO group that is *beta* to the keto group of fructose-1,6-bisphosphate (position 4 of the original ring), you will see how the reverse aldol condensation breaks the bisphosphate into two half-size fragments.

In gluconeogenesis (Fig. 26.7), you can see that glyceraldehyde-3-phosphate and dihydroxyacetone phosphate are joined by an aldol condensation to give fructose-1,6-bisphosphate.

5. Dihydroxyacetone phosphate, in the presence of *triose phosphate isomerase,* changes to its isomer, glyceraldehyde-3-phosphate.

Dihydroxy-
acetone phosphate

An alkene-diol

Glyceraldehyde-
3-phosphate

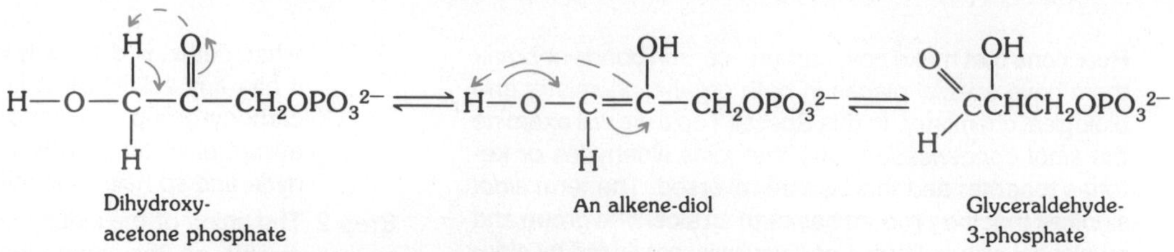

This change ensures that all the chemical energy in glucose will be obtained, because the main path of glycolysis continues with glyceraldehyde-3-phosphate. All dihydroxyacetone shuttles through glyceraldehyde-3-phosphate.

■ The continuous removal of glyceraldehyde-3-phosphate by its subsequent reaction shifts the dihydroxyacetone phosphate equilibrium to the right.

6. Glyceraldehyde-3-phosphate is simultaneously oxidized and phosphorylated. The enzyme, glyceraldehyde-3-phosphate dehydrogenase, has NAD^+ as a cofactor, and an SH group on the enzyme participates. Inorganic phosphate is the source of the new phosphate group. We can visualize how it happens as follows.

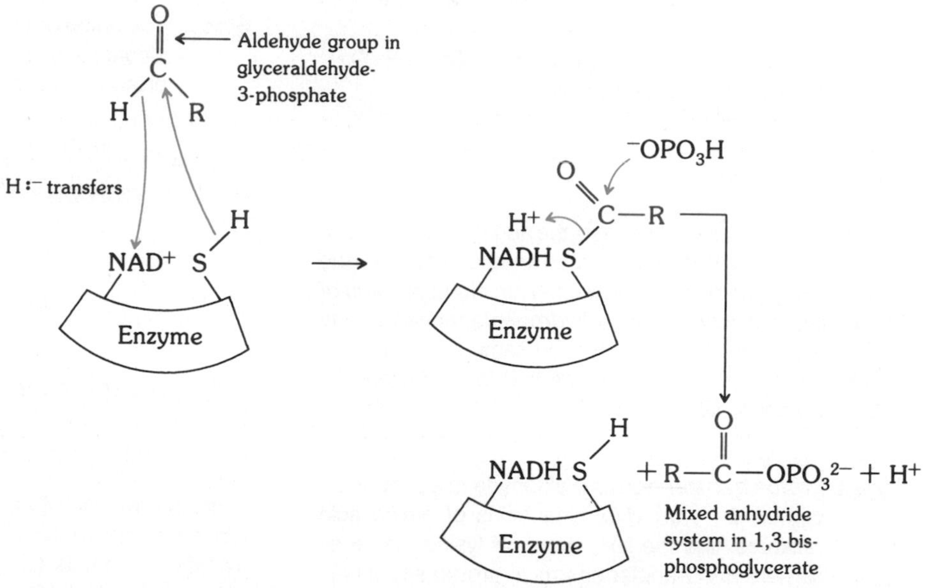

7. 1,3-Bisphosphoglycerate has a higher phosphate group transfer potential than ATP. With the help of *phosphoglycerate kinase,* it transfers a phosphate to ADP, so this gives us back the original investment of ATP. (Remember that each glucose molecule with six carbons is processed through *two* three-carbon molecules, so *two* ATPs are made here for each original single glucose molecule.)

8. The phosphate group in 3-phosphoglycerate shifts to the 2-position, catalyzed by *phosphoglycerate mutase.*

9. The dehydration of 2-phosphoglycerate to phosphoenolpyruvate[2] is catalyzed by *enolase.* This step has been compared to cocking a huge bioenergetic gun. A simple,

[2] When a carbon atom of an alkene group holds an OH group, the compound is called an *enol* ("ene" + "-ol"). Enols are unstable alcohols that spontaneously rearrange into carbonyl compounds:

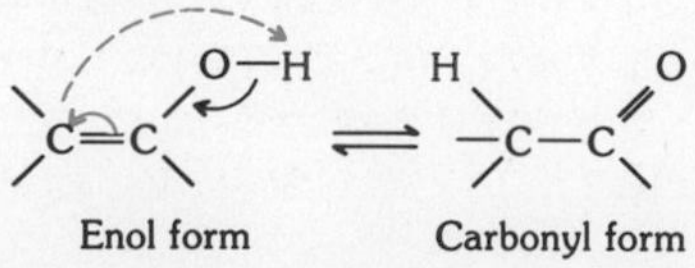

Enol form   Carbonyl form

low-energy reaction, dehydration of an alcohol group, converts a low-energy phosphate into the highest energy phosphate in all of metabolism, the phosphate ester of the enol form of pyruvate, phosphoenolpyruvate.

10. More ATP is made as the phosphate group in phosphoenolpyruvate transfers to ADP. The enzyme is *pyruvate kinase*. The enol form of pyruvate that is left behind promptly and irreversibly rearranges to the keto form of the pyruvate ion. The instability of enols (see footnote 2) is the driving force for this entire step.

11. If the mitochondrion is running aerobically when pyruvate is made, the pyruvate ion changes to acetyl coenzyme A, from which an acetyl group can enter the citric acid cycle (or be used in another way).

Anaerobic Glycolysis Provides a Way to Restore an Enzyme Vital to Continued Glycolysis in the Absence of Oxygen The enzyme at step 6 now has its coenzyme, NAD^+, in its reduced form, NADH. *As long as $H:^-$ is on an NADH unit, the enzyme lacks NAD^+ and so is "plugged." Further glycolysis, therefore, is blocked at step 6.* The block is continuously removed when the cell has sufficient oxygen because NADH simply gives its hydride unit to the respiratory chain, NAD^+ reappears, and glycolysis can run again. When the cell is deficient in oxygen, however, perhaps because of excessive work, glycolysis would shut down quickly without an alternative mechanism for changing NADH back to NAD^+. Exactly such an alternative is the conversion of pyruvate to lactate.

When oxygen isn't available, the $H:^-$ in NADH is unloaded onto the keto group of pyruvate, which thereby changes to the $2°$ alcohol group in lactate. *This is why lactate is the end product of anaerobic glycolysis.* Lactate serves to store $H:^-$ made at step 6 until the cell once again becomes aerobic. The overall change from glucose to lactate by anaerobic glycolysis can be represented by the following equation. (Note the generation of acid.)

$$\text{Glucose} + 2ADP + 2P_i \longrightarrow 2 \text{ lactate} + 2ATP + 2H^+$$

The importance of anaerobic glycolysis is that the cell can continue to make some ATP even when insufficient oxygen is available to run the respiratory chain. Of course, there are limits. The longer the cell operates anaerobically and the more that lactate accumulates, the more the cell runs an **oxygen debt.** Eventually the system has to slow down to let respiration bring back oxygen to metabolize lactate.

Excessive Exercise Causes Lactic Acid Acidosis By describing the production of lactate rather than lactic acid, we have obscured the generation of acid during glycolysis (but did note it in the equation, above). Extensive physical exercise that forces tissues to operate anaerobically can overtax the buffer. A form of metabolic acidosis called *lactic acid acidosis* is the result. When the pH of fluids in muscle tissue decreases, the muscles become tired and sore. The respiratory response to *metabolic acidosis* is hyperventilation, which blows out carbon dioxide and thus helps to remove acid from the body. During the cool down period following exercise, the body reestablishes the acid–base balance of the blood, and excess lactate ion is shuttled into the Cori cycle.

The *Pentose Phosphate Pathway* Makes NADPH The biosyntheses of some substances in the body require a reducing agent. Fatty acids, for example, are almost entirely alkane-like, and alkanes are the most reduced types of organic compounds. The reducing agent used in the biosynthesis of fatty acids is NADPH, the reduced form of $NADP^+$.

The body's principal route to NADPH is the **pentose phosphate pathway** of glucose catabolism. This complicated series of reactions (which we'll not study in detail) is very active in adipose tissue, where fatty acid synthesis occurs. Skeletal muscles have very little activity in this pathway.

■ We studied the oxidative decarboxylation of pyruvate to acetyl coenzyme A, catalyzed by pyruvate dehydrogenase, in Section 25.2.

■ No reaction in the body is truly complete until its enzyme is fully restored.

■ Hunters know that meat from animals run to exhaustion is sour.

■ $NADP^+$ is a phosphate derivative of NAD^+.

■ The pentose phosphate pathway also goes by the names *hexose monophosphate shunt* and *phosphogluconate pathway*.

The oxidative reactions in the pentose phosphate pathway convert a hexose phosphate into a pentose phosphate, hence the name of the series. There are two broad sequences, one oxidative and the other nonoxidative. The overall equation for the oxidative sequence is

$$\text{Glucose-6-phosphate} + 2\text{NADP}^+ + \text{H}_2\text{O} \longrightarrow$$
$$\text{ribose-5-phosphate} + 2\text{NADPH} + 2\text{H}^+ + \text{CO}_2$$

The ribose-5-phosphate can now be used to make the pentose systems in the nucleic acids, if they are needed. If not, ribose-5-phosphate undergoes a series of isomerizations and group transfers that make up the nonoxidative phase of the pentose phosphate pathway. These reactions have the net effect of converting three pentose units into two hexose units (glucose) and one triose unit (glyceraldehyde).

The hexose units can be catabolized by glycolysis, or they can be recycled into the oxidative series of the pentose phosphate pathway. Glyceraldehyde, as we'll soon see, can be converted to glucose and recycled as well. Thus glycolysis, the pentose phosphate reactions, and the resynthesis of glucose all interconnect, and specific bodily needs of the moment determine which pathway is operated.

Overall, with pentose recycled, the balanced equation for the complete oxidation of one glucose molecule via the pentose phosphate pathway is as follows.

$$6 \text{ Glucose-6-phosphate} + 12\text{NADP}^+ + 6\text{H}_2\text{O} \xrightarrow{\text{pentose phosphate pathway}}$$
$$5 \text{ glucose-6-phosphate} + 6\text{CO}_2 + 12\text{NADPH} + 12\text{H}^+ + \text{P}_i$$

26.4

GLUCONEOGENESIS

Some of the steps in gluconeogenesis are the reverse of steps in glycolysis. The overall scheme of gluconeogenesis, by which glucose is made from smaller molecules, is given in Figure 26.7. Excess lactate can be used as a starting material, as we have already mentioned. Just as important, several amino acids can be degraded to molecules that can be used to make glucose, too.

■ Most of our glucose needs are met by gluconeogenesis during periods of fasting.

You'll recall that the brain normally uses circulating glucose for energy. Therefore the ability of the body to manufacture glucose from noncarbohydrate sources such as amino acids — even those from the body itself — is an important backup during times when glucose either isn't in the diet (starvation) or cannot be effectively used (untreated diabetes).

Gluconeogenesis Is Not the Exact Reverse of Glycolysis There are three steps in glycolysis that cannot be directly reversed: steps 1, 3, and 10 of glycolysis (see again Figure 26.6). However, the liver and the kidneys have special enzymes that create bypasses.

■ The enzyme for this step requires the vitamin biotin.

In the bypass that gets back and around step 10 of glycolysis, the synthesis of phosphoenolpyruvate from pyruvate, carbon dioxide is used as a reactant; ATP energy is used to drive the steeply uphill reaction; and *pyruvate carboxylase* is the enzyme. Pyruvate is changed to oxaloacetate:

$$\underset{\text{Pyruvate}}{\text{CH}_3\overset{\overset{\text{O}}{\|}}{\text{C}}\text{CO}_2^-} + \text{HCO}_3^- + \text{ATP} \xrightarrow[\text{pyruvate carboxylase}]{\text{(in mitochondria)}} \underset{\text{Oxaloacetate}}{^-\text{O}_2\text{CCH}_2\overset{\overset{\text{O}}{\|}}{\text{C}}\text{CO}_2^-} + \text{ADP} + \text{P}_i$$

This reaction occurs in mitochondria, but the *use* of oxaloacetate in gluconeogenesis occurs in the cytosol. *By segregating anabolism reactions from their opposite, catabolism, the body is better able to control which pathways to operate.* After a complex shuttle mechanism gets

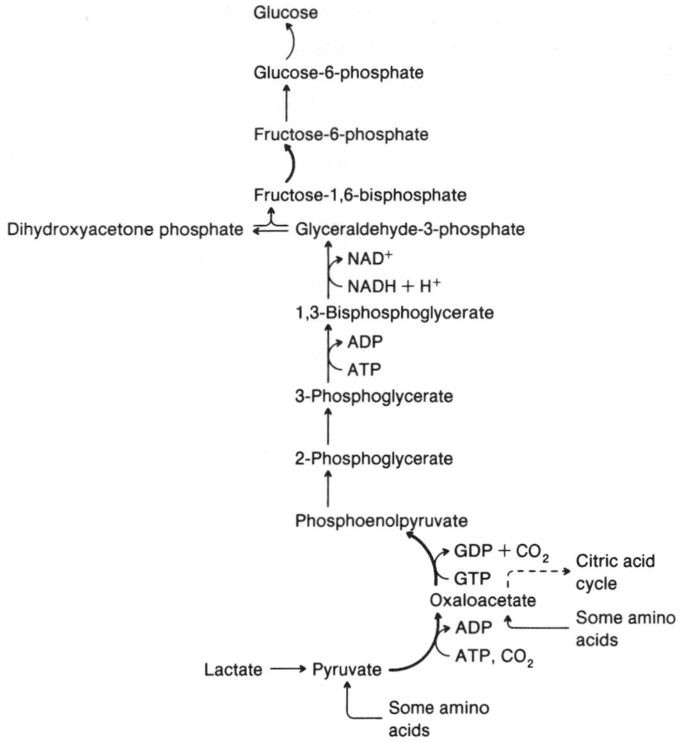

Glucose

Glucose-6-phosphate

Fructose-6-phosphate

Fructose-1,6-bisphosphate

Dihydroxyacetone phosphate ⇌ Glyceraldehyde-3-phosphate

NAD⁺
NADH + H⁺

1,3-Bisphosphoglycerate

ADP
ATP

3-Phosphoglycerate

2-Phosphoglycerate

Phosphoenolpyruvate

GDP + CO₂
GTP ----→ Citric acid cycle

Oxaloacetate

Some amino acids

ADP
ATP, CO₂

Lactate ⟶ Pyruvate

Some amino acids

FIGURE 26.7

Gluconeogenesis. The straight arrows signify steps that are the reverse of corresponding steps in glycolysis. The heavy, curved arrows denote steps that are unique to gluconeogenesis.

oxaloacetate through the mitochondrial membranes, it reacts in the cytosol with another triphosphate, guanosine triphosphate (GTP), as carbon dioxide splits out and bonds rearrange:

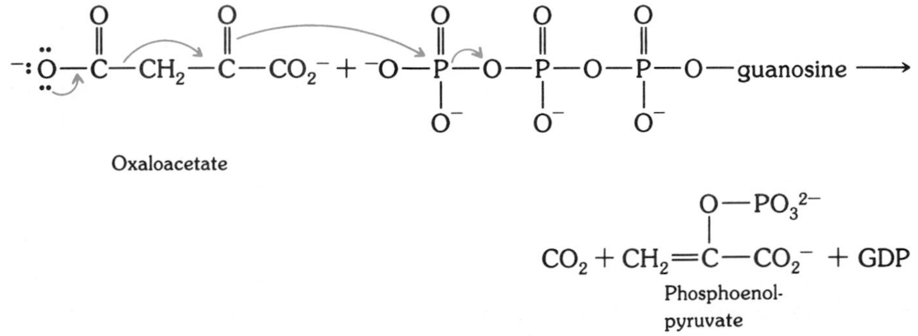

Oxaloacetate

$$CO_2 + CH_2{=}\overset{\displaystyle O-PO_3{}^{2-}}{\underset{\displaystyle \;}{C}}-CO_2{}^- + GDP$$

Phosphoenol-
pyruvate

Because each glucose to be made requires two pyruvates, and because each pyruvate uses two high-energy phosphates in gluconeogenesis, this bypass costs the equivalent of four ATPs (2ATP + 2GTP) per glucose molecule to be made.

At the reversal of step 7 (see again Fig. 26.6), two more ATPs are used per molecule of glucose made. Thus a total of six ATPs are needed to make one glucose molecule by gluconeogenesis that starts from pyruvate. This may be compared with two ATPs that are produced by anaerobic glycolysis (that begins with glucose).

The bypasses to the reverses of steps 3 and 1 of glycolysis require only specific enzymes that catalyze the hydrolysis of phosphate ester groups, not high-energy boosters. Such enzymes are integral parts of the enzyme team for gluconeogenesis.

The other steps in gluconeogenesis are run as reverse shifts in equilibria that occur in the opposite direction in glycolysis.

SUMMARY

Glycogen metabolism The regulation of glycogenesis and glycogenolysis is a part of the machinery for glucose tolerance in the body. Hyperglycemia stimulates the secretion of insulin and somatostatin, and insulin helps cells of adipose tissue to take glucose from the blood. Somatostatin helps to suppress the release of glucagon (which otherwise stimulates glycogenolysis and leads to an increase in the blood sugar level).

When glucose is abundant, the body either replenishes its glycogen reserves or makes fat. In muscular work, epinephrine stimulates a cascade of enzyme activations that begins with the activation of adenylate cyclase and ends with the release of many glucose molecules from glycogen.

When glucose is in short supply, the body makes its own by gluconeogenesis from noncarbohydrate molecules, including several amino acids. In diabetes, some cells that are starved for glucose make their own, also. Such cells are unable to obtain glucose from circulation, so the blood sugar level is hyperglycemic to a glucosuric level. The glucose tolerance test is used to see how well the body handles an overload of glucose. In the management of diabetes, the maintenance of a reasonably steady blood sugar level in the normal range is vital. The measurement of the glycohemoglobin in circulation serves to monitor how well the normal range is kept over a long period of time.

Following strenuous exercise, when lactate is plentiful, the liver makes glucose from lactate. The many pathways that involve glycogen and glucose form a cycle of events called the Cori cycle.

Glycolysis Under anaerobic conditions, glucose can be catabolized to lactate ion. (Galactose and fructose enter this pathway, too.) When lactate is used to store H:$^-$, one of the enzymes in glycolysis can be regenerated in the absence of oxygen, and glycolysis can be run to make some ATP without the involvement of the respiratory chain. Then, when the cell is aerobic again, the H:$^-$ that this enzyme must shed to work again is given directly into the respiratory chain, and pyruvate instead of lactate becomes the end product of glycolysis.

Pentose phosphate pathway The body's need for NADPH to make fatty acids is met by catabolizing glucose through the pentose phosphate pathway.

Gluconeogenesis Most of the steps in gluconeogenesis are simply the reverse of steps in glycolysis, but there are a few that require rather elaborate bypasses. Special teams of enzymes and supplies of high-energy phosphates are used for these. Many amino acids can be used to make glucose by gluconeogenesis.

REVIEW EXERCISES

The answers to Review Exercises whose numbers are in color are found in Appendix C. The answers to the other Review Exercises are found in the Study Guide that accompanies this book. The more challenging questions are marked with asterisks.

Blood Sugar

26.1 What are the end products of the complete digestion of the carbohydrates in the diet?

26.2 Why can we treat the catabolism of carbohydrates as almost entirely that of glucose?

26.3 What is meant by *blood sugar level*? By *normal fasting level*?

26.4 What is the range of concentrations in mg/dL for the normal fasting level of whole blood?

*26.5 A level of 5.5 mmol/L for glucose in the blood corresponds to how many milligrams of glucose per deciliter?

26.6 What characterizes the following conditions?
(a) glucosuria   (b) hypoglycemia
(c) hyperglycemia   (d) glycogenolysis
(e) renal threshold (f) glycogenesis

26.7 What is gluconeogenesis? Which condition, hypoglycemia or hyperglycemia, would activate gluconeogenesis?

26.8 Explain how severe hypoglycemia can lead to disorders of the central nervous system.

Hormones and the Blood Sugar Level

26.9 When epinephrine is secreted, what soon happens to the blood sugar level?

26.10 At which one tissue is epinephrine the most effective?

26.11 What does epinephrine activate at its target cell?

26.12 Arrange the following in the correct order in which they work in epinephrine-initiated glycogenolysis. Place the identifying letters in their correct order, left to right, in the series.

phosphorylase kinase	phosphorylase	cyclic AMP
A	**B**	**C**

adenylate cyclase	protein kinase
D	**E**

26.13 One epinephrine molecule triggers the ultimate formation of roughly how many glucose units, 10^2, 10^3, or 10^4?

26.14 What might be the result if phosphorylase and glycogen synthetase were both activated at the same time?

26.15 What switches glycogen synthetase off when glycogenolysis is activated?

26.16 What is the end product of glycogenolysis, and what does phosphoglucomutase do to it?

26.17 Why can liver glycogen but not muscle glycogen be used to resupply blood sugar?

26.18 What is glucagon, what does it do, and what is its chief target tissue?

26.19 Which is probably better at increasing the blood sugar level, glucagon or epinephrine? Explain.

26.20 How does human growth hormone manage to promote the supply of the energy needed for growth?

26.21 What is insulin, where is it released, and what is its chief target tissue?

26.22 What triggers the release of insulin into circulation?

*26.23 If brain cells are not insulin-dependent cells, how can too much insulin cause insulin shock?

26.24 What is somatostatin, where is it released, and what kind of effect does it have on the pancreas?

Glucose Tolerance and the Cori Cycle

26.25 What is meant by *glucose tolerance?*

26.26 How is glucose trapped in muscle cells and what happens to it?

26.27 What are the main steps in the Cori cycle?

26.28 In general terms, what happens to excess lactate produced in muscles during exercise?

26.29 What is the purpose of the glucose tolerance test?

26.30 Describe what happens when each of the following persons takes a glucose tolerance test.
(a) a nondiabetic individual
(b) a diabetic individual

26.31 Describe a circumstance in which hyperglycemia might arise in a nondiabetic individual.

Catabolism of Glucose

26.32 Fill in the missing substances and balance the following incomplete equation:

$$C_6H_{12}O_6 + 2\ ADP + \underline{\hspace{1cm}} \longrightarrow$$
Glucose

$$C_3H_5O_3^- + 2H^+ + \underline{\hspace{1cm}}$$
Lactate

26.33 What particular significance does glycolysis have when a tissue is running an oxygen debt?

26.34 Why is the rearrangement of dihydroxyacetone phosphate into glyceraldehyde-3-phosphate important?

26.35 What happens to pyruvate (a) under aerobic conditions and (b) under anaerobic conditions?

26.36 What happens to lactate when an oxygen debt is repaid?

*26.37 What is the maximum number of ATPs that can be made by the complete catabolism of (a) one molecule of glucose and (b) one glucose residue in glycogen?

26.38 The pentose phosphate pathway uses $NADP^+$, not NAD^+. What forms *from* $NADP^+$, and how does the body use what forms (in general terms)?

Gluconeogenesis

26.39 In a period of prolonged fasting or starvation, what does the system do to try to maintain its blood sugar level?

*26.40 Amino acids are not excreted, and they are not stored in the same way that glucose residues are stored in a polysaccharide. What probably happens to the excess amino acids in a high-protein diet of an individual who does not exercise much?

26.41 The amino groups of amino acids can be replaced by keto groups. Which amino acids could give the following keto acids that participate in carbohydrate metabolism?
(a) pyruvic acid (b) oxaloacetic acid

*26.42 The amino acids glutamic acid, arginine, histidine, and proline can all be catabolized to α-ketoglutarate. What metabolic cycle in the body makes it possible for α-ketoglutarate to be used eventually in gluconeogenesis? Explain.

*26.43 The carbon atoms of succinyl CoA can wind up in glucose molecules. What metabolic pathway in the body enables succinyl CoA to connect to gluconeogenesis?

Glycogen Storage Diseases (Special Topic 26.1)

26.44 For each of the following diseases, name the defective enzyme, and state the biochemical and physiological consequences.
(a) Von Gierke's disease (b) Cori's disease
(c) McArdle's disease (d) Andersen's disease

Diabetes Mellitus (Special Topic 26.2)

26.45 What is the biochemical distinction between type I and type II diabetes?

26.46 Juvenile-onset diabetes is usually which type?

26.47 Adult-onset diabetes is usually which type?

26.48 Briefly state the six stages in the onset of type I diabetes.

26.49 Viruses that cause diabetes attack which target cells?

26.50 Type I diabetes is an autoimmune disease. What does this mean?

26.51 What are some explanations for the lack of glucose uptake in NIDDM?

26.52 Sustained hyperglycemia causes damage to which specific tissue, causing damage that might be responsible for other complications?

26.53 How is glucose involved in the formation of a Schiff base? What other kinds of compounds react with glucose in this way?

26.54 When the glucose level in blood drops, what happens to the level of glucosylated hemoglobin? Why?

26.55 What happens to the Schiff bases involving glucose if given enough time? Why is this serious?

26.56 Describe a theory that explains how the hydrogenation of glucose might contribute to blindness.

Aldol Condensation (Special Topic 26.3)

26.57 Write the products that would form in an aldol condensation starting with (a) ethanal, (b) propanal, and (c) acetone.

26.58 What products would form if the following compound underwent a reverse aldol condensation?

$$C_6H_5CH_2\overset{\overset{\displaystyle OH}{|}}{C}H-\overset{\overset{\displaystyle O}{\|}}{\underset{\underset{\displaystyle C_6H_5}{|}}{C}}HCH$$

Additional Exercises

*26.59 Suppose that a sample of glucose is made using some atoms of carbon-13 in place of the common isotope, carbon-12. Suppose further that this is fed to a healthy, adult volunteer and that all of it is taken up by the muscles.

(a) Will some of the original molecules be able to go back out into circulation? Explain.

(b) Can we expect any carbon-13 compounds to end up in the liver? Explain.

(c) Can we ever expect to see carbon-13 labeled glucose molecules in circulation again? Explain.

*26.60 Referring to data in the previous chapter, (a) how much ATP can be made from the chemical energy in one pyruvate? (b) The conversion of one lactate to one pyruvate transfers one H:$^-$ to NAD$^+$. How many ATP molecules can be made just from the operation of this step? (c) If all the possible ATP that can be obtained from the complete catabolism of lactate were made available for gluconeogenesis, how many molecules of glucose could be made? (Assume that ATP can substitute for GTP.)

METABOLISM OF LIPIDS 27

For an activity lasting much longer than a sprint, an athlete must draw on the chemical energy of lipids, the major topic of this chapter. This kayaker is probably thinking about other things, like survival, while negotiating Rainbow Falls on the Tuolumne River in California.

27.1

ABSORPTION AND DISTRIBUTION OF LIPIDS

Several lipoprotein complexes in the blood transport triacylglycerols, fatty acids, cholesterol, and other lipids from tissue to tissue.

The digestion of triacylglycerols, as we learned in Chapter 23, produces a mixture of the anions of long-chain fatty acids and monoacylglycerols (plus some diacylglycerols). As these move into the cells of the intestinal membrane, they become reconstituted into triacylglycerols. These plus the cholesterol in the diet make up most of the *exogenous lipid* material eventually delivered into circulation.

■ *Exogenous* means "from the outside" (e.g., from the diet).

Chylomicrons Carry Dietary Lipids and Cholesterol Lipids are virtually insoluble in water and, together with proteins, they are transported in blood in tiny particles called **lipoprotein complexes.** There are several kinds, each with its own function, and they are classified according to their densities (see Table 27.1), which range from 0.95 g/cm^3 to 1.21 g/cm^3.

The lipoprotein complexes with the lowest density are called *chylomicrons.* They are put together from exogenous lipid material but include 2% protein or less. After their assembly within the cells of the intestinal membrane, chylomicrons are delivered to the lymph, which carries them to the bloodstream. When they enter the capillaries embedded in muscle and adipose tissue (fat tissue), chylomicrons encounter binding sites and are held up. An enzyme, *lipoprotein lipase,* now catalyzes the hydrolysis of the chylomicrons' triacylglycerols to fatty

■ The solubility in water of a long-chain fatty acid is typically less than 10^{-6} mol/L, but that of a complex of the fatty acid with a plasma albumin is about 2×10^{-3} mol/L.

TABLE 27.1
Composition of Plasma Lipoproteins

Type of Complex[a]	Chief Constituents (in order of amount)	Density (in g/cm³)
Chylomicrons and chylomicron remnants	Triacylglycerols and cholesterol from the diet	<0.95
VLDL	Triacylglycerols from the liver, cholesteryl esters, cholesterol	0.95–1.006
IDL	Cholesteryl esters, cholesterol, triacyl glycerols (trace)	1.066–1.019
LDL	Cholesteryl esters, cholesterol, triacyl glycerols (trace)	1.019–1.063
HDL	Cholesteryl esters, cholesterol	1.063–1.210

[a] VLDL = very-low-density lipoprotein complex, IDL = intermediate-density lipoprotein complex, LDL = low-density lipoprotein complex, HDL = high-density lipoprotein complex.

Data from M. S. Brown and J. L. Goldstein in *Harrison's Principles of Internal Medicine,* 11th ed (1987), page 1651. Edited by E. Braunwald, K. J. Isselbacher, R. G. Petersdorf, J. D. Wilson, J. B. Martin, and A. S. Fauci.

acids and monoacylglycerols, which are promptly absorbed by the nearby tissue. As the hydrolysis occurs, the chylomicrons shrink to *chylomicron remnants,* which still contain dietary cholesterol. The remnants break loose from the capillary surface, and circulate to the liver where they are absorbed (see Fig. 27.1). *The overall functions of the chylomicrons are thus to deliver exogenous fatty acids to muscle and adipose tissue and to carry dietary cholesterol to the liver.*

■ *Chyle* is the lymph from the intestines. (Lymph was discussed on page 655.)

Lipoprotein Complexes Transport Endogenous Lipids   The liver is a site where breakdown products from carbohydrates are used to make fatty acids and, from them, triacylglycerols. Lipids made in the liver are called *endogenous* lipids ("generated from within"). Three similar lipoprotein complexes are used to transport endogenous lipids in the bloodstream: *very-low-density lipoproteins (VLDL), intermediate-density lipoproteins (IDL),* and *low-density lipoproteins (LDL).* These three complexes also carry cholesterol. Another complex, *high-density lipoprotein (HDL),* carries back to the liver the cholesterol released from tissues and no longer needed by them.

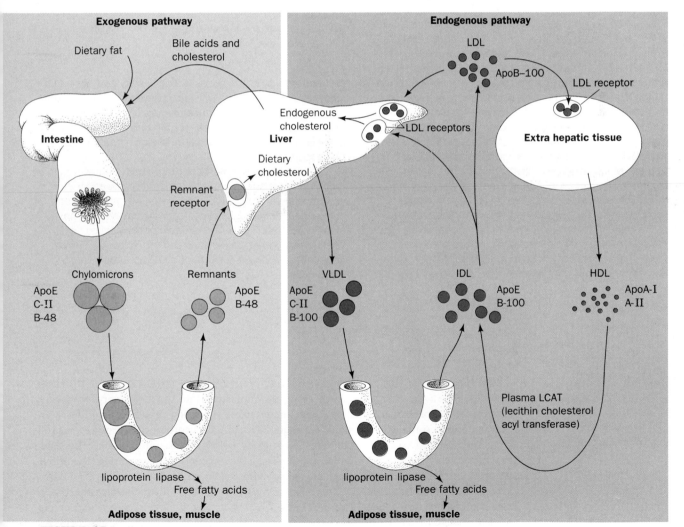

FIGURE 27.1
The transport of cholesterol and triacylglycerols by lipoprotein complexes. The abbreviations ApoE, C-II, B-48, and so forth refer to specific proteins involved with the lipoprotein complexes. [After M. S. Brown and J. L. Goldstein *in* E. Brunwald, K. J. Isselbacher, R. G. Petersdorf, J. B. Martin, and A. S. Fauci, Eds., *Harrison's Principles of Internal Medicine* (11th edition), p. 1652, McGraw-Hill (1987).]

SPECIAL TOPIC 27.1
"GOOD" AND "BAD" CHOLESTEROL AND LIPOPROTEIN RECEPTORS

The receptor proteins for LDL, whether on the liver or on extrahepatic tissue, have a crucial function. If they are reduced in number or are absent, there is little ability to remove cholesterol from circulation. The level of cholesterol in the blood, therefore, becomes too high. The result is *atherosclerosis,* a disease in which several substances, including collagen, elastic fibers, and triacylglycerols, but chiefly cholesterol and its esters, form plaques in the arterial wall (see Figure 1). Such plaques are the chief cause of heart attacks.

Inadequate LDL Receptors Cause Serum Levels of LDL, the "Bad Cholesterol," To Increase The ranges of serum cholesterol concentrations regarded as "desirable," "borderline high," and "risk" are given in Table 1. Some people have a genetic defect that bears specifically on the LDL receptors at the liver and causes elevated levels of LDL cholesterol. Two genes are involved. Those who carry two mutant genes have *familial hypercholesterolemia,* a genetically caused high level of cholesterol in the blood—3 to 5 times higher than average. Even on a zero-cholesterol diet, the victims have very high cholesterol levels. Their cholesterol along with other materials slowly comes out of the blood at valves and other sites, reduces the dimensions of the blood capillaries, and thus restricts blood flowage. Atherosclerosis has set in. Because elevated levels of LDL cholesterol are so often involved, LDL cholesterol is often called "bad cholesterol." Because the HDL complexes perform the desirable service of helping to carry cholesterol back to the liver, HDL cholesterol is sometimes called the "good cholesterol."

In atherosclerosis, the heart must work harder. Eventually arteries and capillaries in the heart itself become

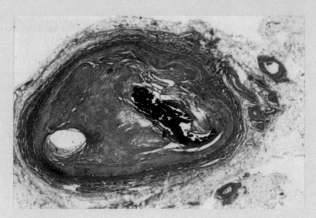

FIGURE 1
The buildup of plaque in this artery in the heart has almost closed it.

reduced so much in cross-sectional size that they no longer are able to bring sufficient oxygen to heart tissue. The plaques sometimes cause clots to form in capillaries of the heart and so block all oxygen delivery to affected heart tissue. Now a myocardial infarction ("heart attack") has occurred.

Victims of familial hypercholesterolemia generally have their first heart attacks as children and are dead by their early twenties. People with one defective gene and one normal gene for the LDL receptor proteins generally have blood cholesterol levels that are two or three times higher than normal. Although they number only about 0.5% of all adults, they account for 5% of all heart attacks among those younger than 60.

■ American scientists J. L. Goldstein and M. S. Brown shared the 1985 Nobel prize in medicine for their work on LDL receptor proteins and how they help to control blood cholesterol levels.

■ Cholesterol ($d = 1.05$ g/cm³) is more dense than triacylglycerols (density of about 0.9 g/cm³).

VLDL Changes to IDL and Then to LDL as VLDL Moves from the Liver to Other Tissues The liver packages triacylglycerols and cholesterol into VLDL, the very-low-density lipoprotein complexes, and releases them into circulation. During circulation, a VLDL complex undergoes somewhat continuous changes as its triacylglycerols are hydrolyzed in capillaries and the hydrolysis products are taken up by adipose tissue and muscles. By these processes, which strongly resemble the changes that occur to chylomicrons, the VLDL particles change to IDL—intermediate-density lipoprotein complexes (see Fig. 27.1). The loss of the lower-density triacylglycerols leaves a particle richer in the higher-density components, so the net density increases (see Table 27.1). With continued loss of triacylglycerols, the IDL change to low-density lipoprotein complexes, LDL (Fig. 27.1). While these activities take place, the VLDL lose some of their protein, and their cholesterol becomes largely esterified by fatty acids and changed to cholesteryl esters. Some LDL is reabsorbed by the liver, but *the main purpose of LDL is to deliver cholesterol to extrahepatic tissue* (tissue other

One Kind of Lipoprotein in HDL Is Undesirable It isn't only the *relative concentration* of HDL that confers protection against atherosclerosis; the *composition* of the HDL is also a factor. HDL is made of about equal parts of lipid and protein. The two most abundant proteins are apolipoprotein A-I and A-II (or apoA-I and apoA-II). When *both* are found in the same HDL particles, the HDL is less "good" than when only apoA-I is present. It is the apoA-I that confers on the HDL its ability to protect against atherosclerosis, but this ability deteriorates when apoA-II is also present (as shown in 1993 by experiments in mice). *Thus the ratio of HDL to LDL does not alone provide information concerning an individual's status regarding atherosclerosis. The composition of the HDL lipoproteins must also be considered.* Just how awaits further research.

Cholesterol Levels Can Be Reduced High blood cholesterol levels occur in many people besides those with genes for hypercholesterolemia, even in people with normal genes for the receptor proteins. The causes of their high cholesterol levels have not been fully unraveled. Smoking, obesity, and lack of exercise are known to contribute to the cholesterol problem. High-cholesterol foods also appear to be factors. There is some evidence that as the liver receives more and more cholesterol from the diet it loses more and more of its receptor proteins. This forces more and more cholesterol to linger in circulation. Some people, however, are able to sustain high-cholesterol and high-lipid diets with no ill effects. Eggs are 0.5% cholesterol (all in the yolk), so egg-less diets are commonly recommended for people with high cholesterol. Yet, one 88-year-old man had eaten 25 eggs a day for at least 15 years without having elevated blood cholesterol levels!

Changing from a typically American high-fat diet to a low-fat diet lowered the serum cholesterol levels of several dozen adults with borderline high levels by an average of only 5 percent. In the same study, reported in 1993, lovastatin brought about a 27 percent reduction in serum cholesterol. Thus the effect of diet on serum cholesterol levels among adults who are healthy, except for borderline high cholesterol levels, is relatively small. Serum cholesterol, of course, is not the only risk factor for heart disease, and the 1993 study in no way endorses a high-fat diet.

TABLE 1
Serum Levels of Cholesterol in All Forms

Form	Desirable	Borderline High	Risk
Total cholesterol	<200 mg/dL (<5.18 mmol/L)	200–239 mg/dL (5.18–6.19 mmol/L)	>239 mg/dL (>6.20 mmol/L)
LDL	<130 mg/dL	130–159 mg/dL	>160 mg/dL
HDL	<35 mg/dL (<0.91 mmol/L)		

Source: *New England Journal of Medicine*, Sept. 3, 1992, page 718.

than liver tissue) *to be used to make cell membranes and, in specialized tissues, steroid hormones.*

HDL Transports Cholesterol from Extrahepatic Tissue to the Liver

High-density lipoprotein complexes — HDL — are cholesterol scavengers. When cells of extrahepatic tissue break down for any reason, their cholesterol molecules are picked up by HDL particles and changed to cholesteryl esters. En route to the liver, HDL undergoes some changes of its own. There is evidence that HDL can transfer cholesteryl esters to VLDL. By one means or another, some of the HDL becomes more like LDL before entering liver cells by means of LDL receptors (Fig. 27.1). Some evidence also exists for specific HDL receptors on liver cells. When receptors for cholesterol-bearing lipoprotein complexes are absent, defective, or in too few numbers, either at the liver or at extrahepatic tissue, there are serious consequences, which are discussed in Special Topic 27.1.

27.2

STORAGE AND MOBILIZATION OF LIPIDS

The high energy density of stored triacylglycerol makes it a choice form for storing energy.

The energy stored per gram of tissue or solution is called the *energy density* of the material. Isotonic glucose solution, for example, carries only about 0.2 kcal per gram. When the glucose is changed into glycogen, however, *and is no longer in solution,* there is little associated water. Now we can get more energy into storage in 1 g; the energy density of wet glycogen is about 1.7 kcal/g. Triacylglycerol, in sharp contrast, has an energy density of about 7.7 kcal/g.

A 70-kg adult male has about 12 kg of triacylglycerol in storage. If he had to exist on no food, just water and a vitamin−mineral supplement, and if he needed 2500 kcal/day, this fat would supply his caloric needs for 43 days. Of course, during this time the body proteins would also be wasting away, and metabolic acidosis (page 664) would be a problem of growing urgency.

Adipose Tissue Is the Principal Lipid Storage Depot The chief depot for the storage of fatty acids is adipose tissue, a very metabolically active tissue. There are two kinds, brown and white. Both types are associated with internal organs, where they cushion the organs against mechanical bumps and shocks, and insulate them from swings in temperature. For a discussion of how brown adipose tissue uses the respiratory chain to generate heat, not ATP, see Special Topic 27.2. The discussion that continues concerns the metabolic activities of white adipose tissue. This tissue stores energy as triacylglycerols chiefly on behalf of the energy budgets of other tissues.

■ Because lipids are water-insoluble, they attract the least amount of associated water in storage.

■ These data are for information; they're certainly not recommendations!

■ The relatively high concentration of mitochondria, which hold iron-containing enzymes (cytochromes), causes the color of brown fat tissue.

FIGURE 27.2

Pathways for the mobilization of energy reserves in the triacylglycerols of adipose tissue

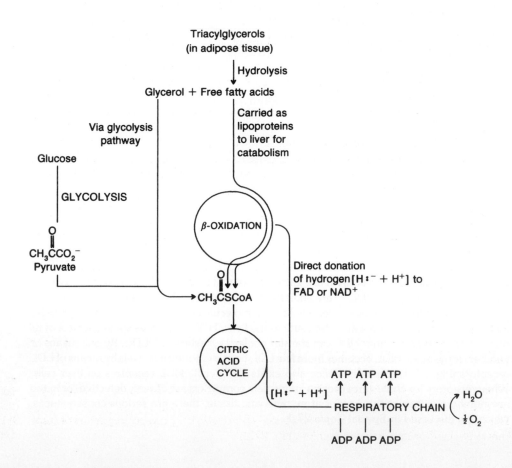

SPECIAL TOPIC 27.2
BROWN FAT AND THERMOGENESIS

Both kinds of adipose tissue, white and brown, store triacylglycerols, but white adipose tissue does not metabolize them except to break them down to free fatty acids and glycerol. The fatty acids from white adipose tissue are then exported to other tissues for catabolism.

Respiratory Chain Oxidation and ATP Synthesis Are Uncoupled in Brown Adipose Tissue The triacylglycerols in brown adipose tissue are catabolized within this tissue for little other use than to generate heat. This is possible because cells of brown adipose tissue can switch off the capability of the inner mitochondrial membranes to accept and hold a proton gradient. Recall that the respiratory chain normally establishes a proton gradient across the inner mitochondrial membrane, and that as protons flow back through selected channels, they trigger the synthesis of ATP from ADP and P_i.

When the respiratory chain runs but cannot set up the proton gradient, the chemical energy released by the chain emerges only as heat. Such generation of heat is called *thermogenesis.* Two stimuli, both mediated by the neurotransmitter norepinephrine, trigger this heat-generating activity, exposure to cold (and the resulting shivering) and the ingestion of food.

The advantage to the body of thermogenesis induced by a cold outside temperature is that the body can oxidize its own fat to help keep itself warm.

The advantage of food-induced thermogenesis is that the individual is protected from getting fat by too much eating in relationship to exercise. Thermogenesis in brown adipose tissue removes fat by catabolism, rather than by exercise, as new calories (in the food) are imported.

A Body in Dietary Balance Uses Fatty Acids as Well as Glucose for Energy Some tissues, like heart muscle and the renal cortex, use breakdown products of fatty acids in preference to glucose for energy. Skeletal muscles can use both fatty acids and short molecules made from them for energy. In fact, most of the energy needs of resting muscle tissue are met by intermediates of fatty acid catabolism, not from glucose. Given enough time for adjustment, during starvation, for example, even brain cells can obtain some energy from fatty acid breakdown products.

Fatty acids, consequently, come and go from adipose tissue, and the balance between this tissue's receiving or releasing them is struck by the energy requirements elsewhere. In extreme distress, when either glucose is in very low supply (as in starvation) or what is available can't be used (as in diabetes), the body must turn almost entirely to its fatty acids for energy.

Figure 27.2 outlines the many steps involved in tapping the lipid reserves for their energy. (It also reminds us of the connection between glucose and ATP.) Triacylglycerol molecules in adipose tissue are first hydrolyzed to free fatty acids and glycerol. The cellular lipase needed for this is activated by a process involving cyclic AMP and the hormone epinephrine. The fatty acids are carried as albumin complexes to the liver, the chief site for their catabolism.

■ Epinephrine is also involved in glucose metabolism.

Insulin suppresses the cellular lipase that releases fatty acids from adipose tissue. Thus when insulin is in circulation, and its presence is linked to the presence of blood glucose, the fatty acids are less required for energy.

The glycerol that is produced when the fatty acids are released is changed to dihydroxyacetone phosphate, and it enters the pathway of glycolysis.

27.3

OXIDATION OF FATTY ACIDS

Acetyl groups are produced by the β-oxidation of fatty acids and are fed into the citric acid cycle and respiratory chain.

The degradation of fatty acids takes place inside mitochondria by a repeating series of steps known as the **β-oxidation pathway** (see Fig. 27.3).

■ The β-oxidation pathway has also been called the *fatty acid cycle,* but it isn't exactly a true *cycle,* like the citric acid cycle.

753

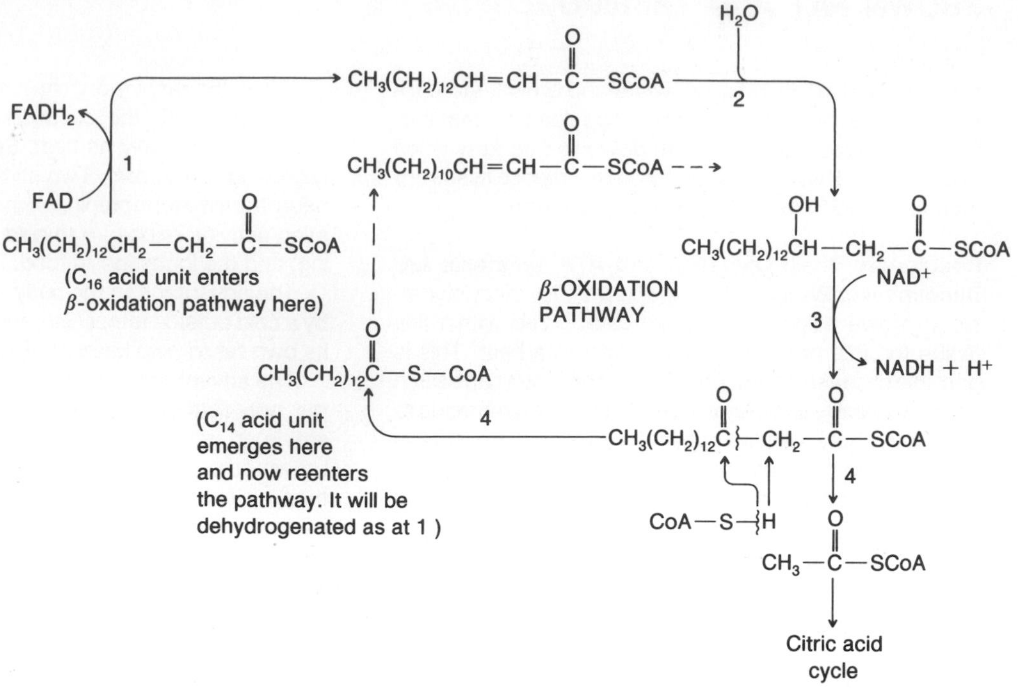

FIGURE 27.3`
The β-oxidation pathway. The numbers refer to the numbered steps discussed in the text.

A Fatty Acid Is First Joined to Coenzyme A

The β-oxidation pathway occurs in the mitochondrial matrix, and before a fatty acid can enter this pathway, it has to be joined to coenzyme A. It costs one ATP to do this, but now the fatty acyl unit is activated. The ATP itself breaks down to AMP and PP_i. *The subsequent hydrolysis of the diphosphate is the driving force for the overall change,* and this means that the actual cost in high-energy phosphate to commence the β-oxidation pathway is *two* high-energy phosphate bonds.

Getting a fatty acid inside the matrix of a mitochondrion from the cytosol begins with the formation of fatty acyl CoA in the cytosol. From this form, the fatty acyl unit is passed over to a protein of the mitochondrial membrane, migrates through, and is finally passed to coenzyme A inside. It's complicated, but the cell has now isolated the fatty acyl unit for *catabolism;* the steps by which a fatty acyl unit are made (its anabolism) occur in the cytosol. As we have said, nature segregates anabolism from catabolism and thereby is able to control both more easily.

Fatty Acyl CoA Is Catabolized by Two Carbons at a Time

The repeating sequence of the β-oxidation pathway consists of four steps as the result of which one molecule of $FADH_2$, one of NADH, and one of acetyl coenzyme A are made in addition to a fatty acyl unit with two less carbons.

■ Franz Knoop directed much of the research on the fatty acid cycle, so this pathway is sometimes called *Knoop oxidation.*

The now shortened fatty acyl unit is carried again through the four steps, and the process is repeated until no more two-carbon acetyl units can be made. The $FADH_2$ and the NADH fuel the respiratory chain. The acetyl groups pass into the citric acid cycle, or they enter the general pool of acetyl coenzyme A that the body draws from to make other substances (e.g., cholesterol). Let's now look at the four steps in greater detail. The numbers that follow refer to Figure 27.3.

1. The first step is dehydrogenation. FAD accepts $(H:^- + H^+)$ from the α- and the β-carbons of the fatty acyl unit of palmityl coenzyme A.

$$CH_3(CH_2)_{12}CH_2-CH_2\overset{\overset{\displaystyle O}{\|}}{C}SCoA + FAD \xrightarrow{\boxed{1}}$$

Palmityl coenzyme A

$$CH_3(CH_2)_{12}CH=CH\overset{\overset{\displaystyle O}{\|}}{C}SCoA + FADH_2 \longrightarrow [H\!:^- + H^+] \longrightarrow$$

An α,β-unsaturated
acyl derivative of
coenzyme A

FAD

Transfers to
respiratory
chain

■ The enzyme is *acyl-CoA
dehydrogenase.*

FADH$_2$ interacts with the respiratory chain at enzyme complex II (page 717).

2. The second step is hydration. Water adds to the alkene double bond formed by the previous step, and a 2° alcohol group forms.

■ The enzyme is *enoyl-CoA
hydratase.*

$$CH_3(CH_2)_{12}CH=CH\overset{\overset{\displaystyle O}{\|}}{C}SCoA + H_2O \xrightarrow{\boxed{2}} CH_3(CH_2)_{12}\overset{\overset{\displaystyle OH}{|}}{C}H CH_2\overset{\overset{\displaystyle O}{\|}}{C}SCoA$$

A β-hydroxyacyl derivative
of coenzyme A

3. The third step is another dehydrogenation—a loss of (H$:^-$ + H$^+$). This oxidizes the secondary alcohol to a keto group. Notice that these steps end in the oxidation of the β-position of the original fatty acyl group to a keto group, which is why the pathway is called *beta* oxidation.

■ *3-L-Hydroxyacyl-CoA
dehydrogenase* is the enzyme
for this oxidation. One form of
this enzyme has been found to
be deficient in about 10% of
infants who died of sudden
infant death syndrome (SIDS),
so without the enzyme there
may be a fatal imbalance
between glucose and fatty acid
metabolism.

$$CH_3(CH_2)_{12}\overset{\overset{\displaystyle OH}{|}}{C}H CH_2\overset{\overset{\displaystyle O}{\|}}{C}SCoA + NAD^+ \xrightarrow{\boxed{3}} CH_3(CH_2)_{12}\overset{\overset{\displaystyle O}{\|}}{C}CH_2\overset{\overset{\displaystyle O}{\|}}{C}SCoA + \underline{NADH + H^+}$$

A β-keto acyl derivative
of coenzyme A

NAD$^+$

Transfers to
respiratory chain

$\longleftarrow$ [H$:^-$+H$^+$]

4. The fourth step breaks the bond between the α-carbon and the β-carbon. This bond has been weakened by the stepwise oxidation of the β-carbon, and now it breaks to release one unit of acetyl coenzyme A. (This bond-breaking reaction is little more than a *reverse* Claisen condensation; we studied the Claisen reaction in Special Topic 25.1, page 714)

$$CH_3(CH_2)_{12}\overset{\overset{\displaystyle O}{\|}}{C}-CH_2\overset{\overset{\displaystyle O}{\|}}{C}SCoA \xrightarrow{\boxed{4}} CH_3(CH_2)_{12}\overset{\overset{\displaystyle O}{\|}}{C}SCoA + CH_3\overset{\overset{\displaystyle O}{\|}}{C}SCoA$$

CoA—S—H

Myristyl coenzyme A   Acetyl coenzyme A

Transfers to
citric acid cycle

12ATP $\longleftarrow$ via respiratory chain

■ The enzyme is *thiolase* (or
β-ketoacyl-CoA thiolase).

The remaining acyl unit, the original shortened by two carbons, now goes through the cycle of steps again: dehydrogenation, hydration, dehydrogenation, and cleavage. After seven such cycles, one molecule of palmityl coenzyme A is broken into eight molecules of acetyl coenzyme A.

One Palmityl Unit Yields 129 ATP Molecules
Table 27.2 shows how the maximum yield of ATP from the oxidation of one unit of palmityl coenzyme A adds up to 131 ATPs. The

TABLE 27.2
Maximum Yield of ATP from Palmityl CoA

Intermediates Produced by Seven Turns of the Fatty Acid Cycle	ATP Yield per Intermediate	Total ATP Yield
7 FADH$_2$	2	14
7 NADH	3	21
8 Acetyl Coenzyme A	12	96
		131 ATP
Deduct two high-energy phosphate bonds for activating the acyl unit		-2
Net ATP yield per palmityl unit		129 ATP

net from palmitic acid is two ATPs fewer, or 129 ATPs, because the activation of the palmityl unit — joining it to CoA — requires this initial investment, as we mentioned earlier.

27.4

BIOSYNTHESIS OF FATTY ACIDS

Acetyl CoA molecules that are not needed to make ATP can be made into fatty acids.

Acetyl CoA stands at a major metabolic crossroad. It can be made from any monosaccharide in the diet, from virtually all amino acids, and from fatty acids. Once made, it can be shunted into the citric acid cycle where its chemical energy can be used to make ATP; or its acetyl group can be made into other compounds that the body needs. In this section we'll see how acetyl CoA can be made into long-chain fatty acids.

Fatty Acid Synthesis Begins with the Activation of Acetyl CoA Whenever acetyl CoA molecules are made within mitochondria but aren't needed for the citric acid cycle and respiratory chain, they are exported to the cytosol. The enzymes for the synthesis of fatty acids are found there, not within the mitochondria, illustrating again the general rule that the body segregates its sequences of catabolism from those of anabolism.

As might be expected, because fatty acid synthesis is in the direction of climbing an energy hill, the cell has to invest some energy of ATP to make fatty acids from smaller molecules. The first payment occurs in the first step in which the bicarbonate ion reacts with acetyl CoA to form malonyl CoA.

■ The enzyme for this step, *acetyl CoA carboxylase*, requires the vitamin biotin.

$$CH_3\overset{O}{\underset{\|}{C}}SCoA + HCO_3^- + ATP \longrightarrow {}^-O\overset{O}{\underset{\|}{C}}CH_2\overset{O}{\underset{\|}{C}}SCoA + 2H^+ + ADP + P_i$$

Acetyl
coenzyme A

Malonyl
coenzyme A

The extra carbonyl group in malonyl CoA activates the CH$_2$ unit, now between two carbonyl groups, for fatty acid synthesis.

The Growing Fatty Acyl Unit Is Moved from Enzyme to Enzyme by a Construction Boom Molecular Unit The enzyme that now builds a long-chain fatty acid is actually a huge complex of seven enzymes called *fatty acid synthase* (see Fig. 27.4). In the center of this

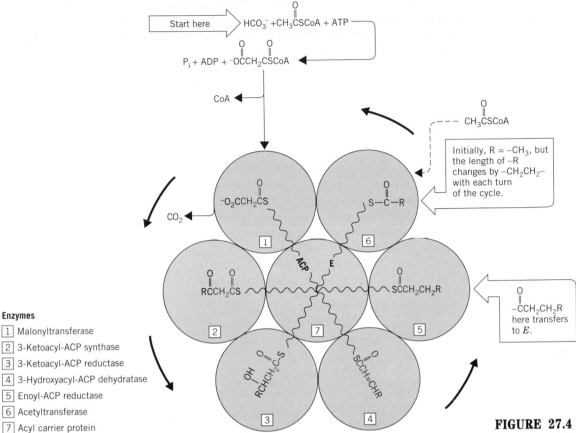

FIGURE 27.4

The synthesis of fatty acids. At the top, an acetyl group is activated and joined as a malonyl unit to an arm of the acyl carrier protein, ACP. Another acetyl group transfers from acetyl CoA to site E. In a second transfer, this acetyl group is then joined to the malonyl unit as CO_2 splits back out. This gives a β-ketoacyl system whose keto group is reduced to CH_2 by the next series of steps. One turn of the "cycle" adds a CH_2CH_2 unit to the growing acyl chain.

complex is a molecular unit long enough to serve as a swinging arm carrier. It's called the *acyl carrier protein*, or ACP, and like the boom of a construction crane, this arm swings from enzyme to enzyme in the synthase complex. Thus the arm brings what it carries over first one enzyme and then another, and at each stop a reaction is catalyzed that contributes to chain lengthening. Let's see how it works. The enzymes are represented as colored circles in Figure 27.4 and identified by numbered boxes used in the following discussion. The *names* of the enzymes involved are given in Figure 27.4.

The malonyl unit in malonyl CoA, just made, transfers to the swinging arm of ACP, placing the product over enzyme $\boxed{1}$:

$$\overset{O}{\overset{\|}{-OCCH_2}}\overset{O}{\overset{\|}{CS}}\!-CoA + ACP \longrightarrow \overset{O}{\overset{\|}{-OCCH_2}}\overset{O}{\overset{\|}{CS}}\!-ACP + CoA$$

Malonyl   Malonyl ACP
coenzyme A

In the meantime, a similar reaction occurs over enzyme $\boxed{6}$ to another molecule of acetyl CoA at a different unit of the synthase, a unit that we'll call simply E.

$$\overset{O}{\overset{\|}{CH_3CS}}\!-CoA + E \longrightarrow \overset{O}{\overset{\|}{CH_3CS}}\!-E + CoA$$

Acetyl E

■ Glucagon, epinephrine, and cyclic AMP—all stimulators of the use of glucose to make ATP—depress the synthesis of fatty acids in the liver. Insulin, however, promotes it.

Next, enzyme $\boxed{1}$ catalyzes the transfer of the acetyl group of acetyl E to the malonyl group of malonyl ACP as carbon dioxide, the initial activator, is ejected. The loss of CO_2, in fact, is the driving force for this transfer reaction, a driving force initially put in place by energy from ATP. A four-carbon derivative of ACP, acetoacetyl ACP, forms. It's over enzyme $\boxed{2}$. The E unit is vacated.

$$\underset{\text{Acetyl } E}{CH_3\overset{O}{\overset{\|}{C}}S-E} + \underset{\text{Malonyl ACP}}{^-O\overset{O}{\overset{\|}{C}}CH_2\overset{O}{\overset{\|}{C}}S-ACP} \longrightarrow \underset{\text{Acetoacetyl ACP}}{CH_3\overset{O}{\overset{\|}{C}}CH_2\overset{O}{\overset{\|}{C}}S-ACP} + CO_2 + E$$

The ketone group in acetoacetyl ACP is next reduced by enzyme $\boxed{3}$ to a secondary alcohol and passed to enzyme $\boxed{4}$. Then this alcohol is dehydrated by enzyme $\boxed{4}$ to introduce a double bond, putting the unit over enzyme $\boxed{5}$. The double bond is next reduced over enzyme $\boxed{5}$ to give butyryl ACP. The overall effect of these steps is to reduce the keto group to CH_2. Notice that NADPH, the reducing agent manufactured by the pentose phosphate pathway of glucose catabolism, is used here, not NADH.

$$\underset{\text{Acetoacetyl ACP}}{CH_3\overset{O}{\overset{\|}{C}}CH_2\overset{O}{\overset{\|}{C}}S-ACP} \xrightarrow[\substack{\text{(reduction of the keto} \\ \text{group by enzyme 3)}}]{NADPH + H^+ \qquad NADP^+} \underset{}{CH_3\overset{OH}{\overset{|}{C}}HCH_2\overset{O}{\overset{\|}{C}}S-ACP}$$

$$\Bigg\downarrow \substack{\text{(dehydration} \\ \text{by enzyme 4)}} \searrow H_2O$$

$$\underset{\text{Butyryl ACP}}{CH_3CH_2CH_2\overset{O}{\overset{\|}{C}}S-ACP} \xleftarrow[\substack{\text{(reduction of the double} \\ \text{bond at enzyme 5)}}]{NADP^+ \qquad NADPH + H^+} \underset{}{CH_3CH=CH\overset{O}{\overset{\|}{C}}S-ACP}$$

The butyryl group is now transferred to the *vacant* E unit of the synthase, the unit that initially held an acetyl group. Butyryl E instead of acetyl E is now over enzyme $\boxed{6}$. This ends one complete cycle of the β-oxidation pathway. To recapitulate, we have gone from two two-carbon acetyl units to one four-carbon butyryl unit.

The steps now repeat as indicated in Figure 27.4. A new malonyl unit is joined to the ACP. Then the newly made *butyryl* group is made to transfer to the malonyl unit as CO_2 is again ejected. This elongates the fatty acyl chain to six carbons in length, positions it on the swinging arm, and gets it ready for the several-step reduction of the keto group to CH_2. The swinging arm mechanism and the enzymes of the synthase complex go to work until the chain is that of the six-carbon acyl group, the hexanoyl group.

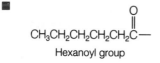

Hexanoyl group

In the next repetition of this series, the six-carbon acyl group will be elongated to an eight-carbon group, and the process will repeat until the chain is sixteen carbons long. Overall, the net equation for the synthesis of the palmitate ion from acetyl CoA is

■ Because the symbols ATP, ADP, and P_i are given without their electrical charges, we can't provide an electrical balance to equations such as this.

$$8CH_3\overset{O}{\overset{\|}{C}}SCoA + 7ATP + 14NADPH \longrightarrow$$
$$\underset{\text{Palmitate ion}}{CH_3(CH_2)_{14}CO_2^-} + 7ADP + 7P_i + 8CoA + 14NADP^+ + 6H_2O$$

■ The pentose phosphate pathway for the catabolism of glucose (page 741) is the body's chief supplier of NADPH.

The process creates a heavy demand for NADPH — 14 NADPH to make one palmitate ion. If acids with chains longer than the chain in the palmitate ion are needed, or acids with double bonds are to be made, additional steps or different pathways using different enzymes are taken.

27.5

BIOSYNTHESIS OF CHOLESTEROL

Excessive cholesterol can inhibit the formation of a key enzyme required in the multistep synthesis of cholesterol.

In addition to serving as a raw material for making fatty acids, acetyl CoA can be used to make the steroid nucleus. Cholesterol, an alcohol with this nucleus, is the end product of a long, multistep process, and is used by the body to make various bile salts and sex hormones and to be incorporated into cell membranes. In mammals, about 80% to 95% of all cholesterol synthesis takes place in cells of the liver and the intestines. We won't go into all the details, but we will go far enough to learn more about how the body normally controls the process. If sufficient cholesterol is provided by the diet, then the body's synthesis should be shut down. Let's see how this is done.

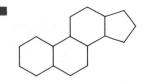

Steroid nucleus

Cholesterol

Cholesterol Is Made from Acetyl Units When the level of acetyl CoA builds up in the liver, the following equilibrium shifts to the right (in accordance with Le Châtelier's principle):

$$2CH_3\overset{O}{\overset{\|}{C}}SCoA \rightleftharpoons CH_3\overset{O}{\overset{\|}{C}}CH_2\overset{O}{\overset{\|}{C}}SCoA + CoASH$$

Acetyl CoA   Acetoacetyl CoA

■ The forward reaction is an example of the Claisen ester condensation (Special Topic 25.1, page 714), the joining of an acyl group of one ester molecule to the α-position of a second ester molecule.

When cholesterol synthesis is switched on, then acetoacetyl CoA combines with another acetyl CoA:

$$CH_3\overset{O}{\overset{\|}{C}}CH_2\overset{O}{\overset{\|}{C}}{-}SCoA + CH_3\overset{O}{\overset{\|}{C}}SCoA \xrightarrow[\text{synthase}]{\text{HMG-CoA}} {}^-O\overset{O}{\overset{\|}{C}}CH_2\overset{OH}{\overset{|}{C}}CH_2\overset{O}{\overset{\|}{C}}SCoA + CoASH \quad (26.1)$$
$$\underset{CH_3}{}$$

HMG-CoA
(β-hydroxy-β-methyl-
glutaryl CoA)

■ The reaction is the addition of one carbonyl compound (acetyl CoA) to the keto group of another, so it strongly resembles the aldol condensation (Special Topic 26.3, page 739).

The Reduction of HMG–CoA Commits the Cell to the Complete Cholesterol Synthesis Both a reduction and a hydrolysis occur in the next step, which is a complex change catalyzed by *HMG–CoA reductase*, a key enzyme.

$$HMG{-}CoA + 2NADPH + 2H^+ \xrightarrow[\text{reductase}]{\text{HMG-CoA}}$$

$$HOCH_2CH_2\overset{OH}{\overset{|}{\underset{|}{C}}}CH_2CO_2^- + 2NADP^+ + CoA{-}SH$$
$$\underset{CH_3}{}$$

Mevalonate

Mevalonate is next carried through a long series of reactions until cholesterol is made. As we said, we'll not take it that far, but consider, instead, how cholesterol synthesis is controlled.

Cholesterol Is a Natural Inhibitor of HMG-CoA Reductase The control of HMG–CoA reductase is the major factor in the overall control of the biosynthesis of cholesterol. Cholesterol itself is one inhibitor, and it works by inhibiting both the *synthesis* of the enzyme and the enzymatic activity of any of its existing molecules. In the presence of cholesterol, the

■ Fifteen Nobel prizes have gone to scientists who devoted the better parts of their careers to various aspects of cholesterol and its uses in the body.

■ On a low-cholesterol diet, an adult makes about 800 mg of cholesterol per day.

enzyme isn't totally deactivated. There is just *less* of it free to do catalytic work. Thus if the diet is relatively rich in cholesterol, the body tends to make less of it. If the diet is very low in cholesterol, the body makes more.

Lovastatin and Compactin Lower Blood Cholesterol Levels   A drug that lowers the cholesterol level, lovastatin (Mevacor), was approved in 1987 and has proved to be extraordinarily effective in suppressing the body's natural synthesis of cholesterol. It works partly as a competitive inhibitor of the enzyme HMG–CoA reductase, which is a key choke point in cholesterol synthesis. You can see from the resemblances of the highlighted parts of the structures below how lovastatin might compete with mevalonate for a position on the enzyme.

■ Individuals with severe hypercholesterolemia who take lovastatin show a decrease in serum cholesterol of 30%. When lovastatin and compactin are used in combination, the decrease is 50% to 60%.

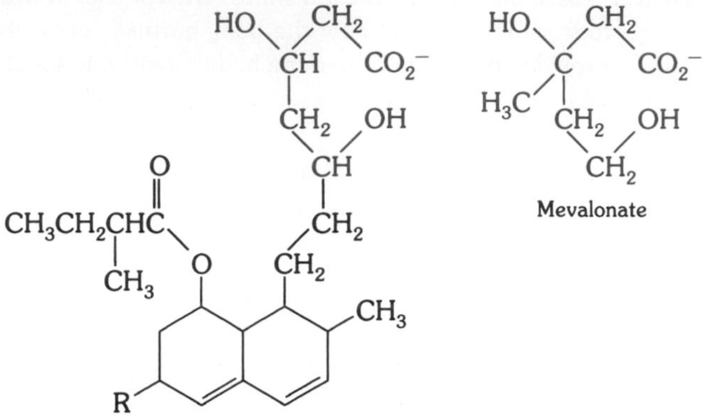

R = H   Compactin
R = CH$_3$ Lovastatin

Mevalonate

Lovastatin also increases the synthesis of the mRNA responsible for making the liver LDL receptors. With more of these, the liver is better able to reabsorb cholesterol, recycle it, or export it in the bile. In some people, lovastatin also enhances the levels of the HDL units that carry cholesterol back to the liver for removal.

Carbohydrate and Lipid Metabolisms Are Intertwined   Figure 27.5 provides a summary of much of what we have covered in this and the previous chapter about the chief uses of acetyl CoA and its relationship to carbohydrate and lipid catabolism. One point emphasized by this figure is that triacylglycerols can be made from any of the three dietary components: carbohydrates, lipids, and proteins.

27.6

KETOACIDOSIS

An acceleration of the fatty acid cycle tips some equilibria in a direction that leads to ketoacidosis.

Cells of certain tissues have to engage in gluconeogenesis in two serious conditions, starvation and uncontrolled diabetes mellitus. In starvation, the blood sugar level drops because of nutritional deficiencies, so the body (principally the liver) tries to compensate by making glucose. The consequences are fatal unless the underlying causes are treated.

The Level of Acetyl CoA Increases When Gluconeogenesis Is Accelerated   If you look back to Figure 26.7 (page 743), you will see that gluconeogenesis consumes oxaloacetate, the carrier of acetyl units in the citric acid cycle (Fig. 25.2, page 712). When

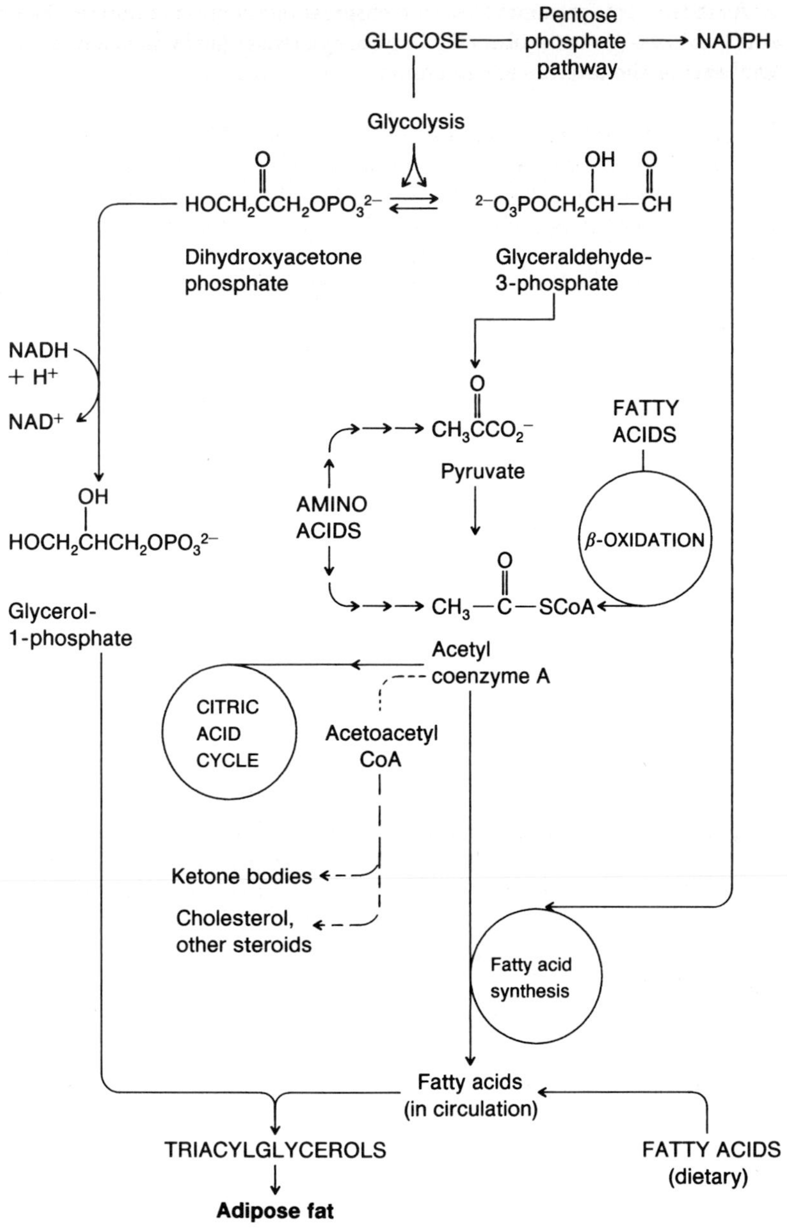

FIGURE 27.5
Principal sources of triacylglycerols for adipose tissue and the chief uses of acetyl CoA

oxaloacetate is diverted from the citric acid cycle, acetyl coenzyme A cannot put its acetyl group into the cycle. Yet acetyl coenzyme A continues to be made by the fatty acid cycle, *so acetyl CoA levels build up.*

As the supply of acetyl CoA increases in the liver, the following equilibrium shifts to the right to make acetoacetyl CoA.

$$2CH_3\overset{O}{\overset{\|}{C}}SCoA \rightleftharpoons CH_3\overset{O}{\overset{\|}{C}}CH_2\overset{O}{\overset{\|}{C}}SCoA + CoASH$$

Acetyl CoA   Acetoacetyl CoA

■ Again we see applications of Le Châtelier's principle; the increase in the level of acetyl CoA is the stress and the equilibrium, by shifting to the right, relieves the stress.

As the level of acetoacetyl CoA increases, equilibrium 26.1 shifts to the right to make more HMG–SCoA. As the level of HMG–SCoA increases (and little if any is being diverted to the synthesis of cholesterol), a liver enzyme splits it to acetoacetate ion and coenzyme A:

$$HMG-SCoA \longrightarrow \underset{\text{Acetoacetate}}{CH_3\overset{O}{\overset{||}{C}}CH_2\overset{O}{\overset{||}{C}}O^-} + CoASH$$

The net effect of these steps, starting from acetyl CoA, is the following:

$$2CH_3\overset{O}{\overset{||}{C}}SCoA + H_2O \longrightarrow \underset{\text{Acetoacetate}}{CH_3\overset{O}{\overset{||}{C}}CH_2\overset{O}{\overset{||}{C}}O^-} + 2CoASH + H^+$$

Notice the hydrogen ion. It makes the situation dangerous, and an increased synthesis of "new" glucose was the cause. As we said, conditions of starvation or untreated diabetes mellitus can be instigators of accelerated gluconeogenesis. Figure 27.6 outlines the chain of events that occurs in untreated diabetes, and you may wish to refer to this figure as you continue with the following discussion.

Accelerated Acetoacetate Production Leads to Acidosis   The acid produced by the formation of acetoacetate must be neutralized by the buffer. Under an increasingly rapid production of acetoacetate and hydrogen ion, the blood buffer slowly loses ground. A condition of *acidosis* sets in. It is *metabolic* acidosis, because the cause lies in a disorder of metabolism. Because the chief species responsible for this acidosis has a keto group, the condition is often called **ketoacidosis.**

Blood Levels of the Ketone Bodies Increase in Starvation and Diabetes   The *acetoacetate* ion is called one of the **ketone bodies.** The two others are *acetone* and the *β-hydroxybutyrate* ion. Both are produced from the acetoacetate ion. Acetone arises from acetoacetate by the loss of the carboxyl group:

$$\underset{\text{Acetoacetate}}{CH_3\overset{O}{\overset{||}{C}}CH_2\overset{O}{\overset{||}{C}}O^-} + H_2O \longrightarrow \underset{\text{Acetone}}{CH_3\overset{O}{\overset{||}{C}}CH_3} + HCO_3^-$$

■ *β*-Hydroxybutyrate is a *ketone* body not because it has a keto group but because it is made from and is found together with a species that does.

β-Hydroxybutyrate is produced when the keto group of acetoacetate is reduced by NADH:

$$\underset{\text{Acetoacetate}}{CH_3\overset{O}{\overset{||}{C}}CH_2\overset{O}{\overset{||}{C}}O^-} + NADH + H^+ \longrightarrow \underset{\text{β-Hydroxybutyrate}}{CH_3\overset{OH}{\overset{|}{C}}HCH_2\overset{O}{\overset{||}{C}}O^-} + NAD^+$$

The ketone bodies enter general circulation. Because acetone is volatile, most of it leaves the body via the lungs, and individuals with severe ketoacidosis have "acetone breath," the noticeable odor of acetone on the breath.

■ The vapor pressure of acetone at body temperature is nearly 400 mm Hg (about 0.53 atm), so it readily evaporates from the blood in the lungs.

Acetoacetate and *β*-hydroxybutyrate can be used in skeletal muscles to make ATP. Heart muscle uses these two for energy in preference to glucose. Even the brain, given time, can adapt to using these ions for energy when the blood sugar level drops in starvation or prolonged fasting. The ketone bodies are not in themselves abnormal constituents of blood. Only when they are produced at a rate faster than the blood buffer can handle them are they a problem.

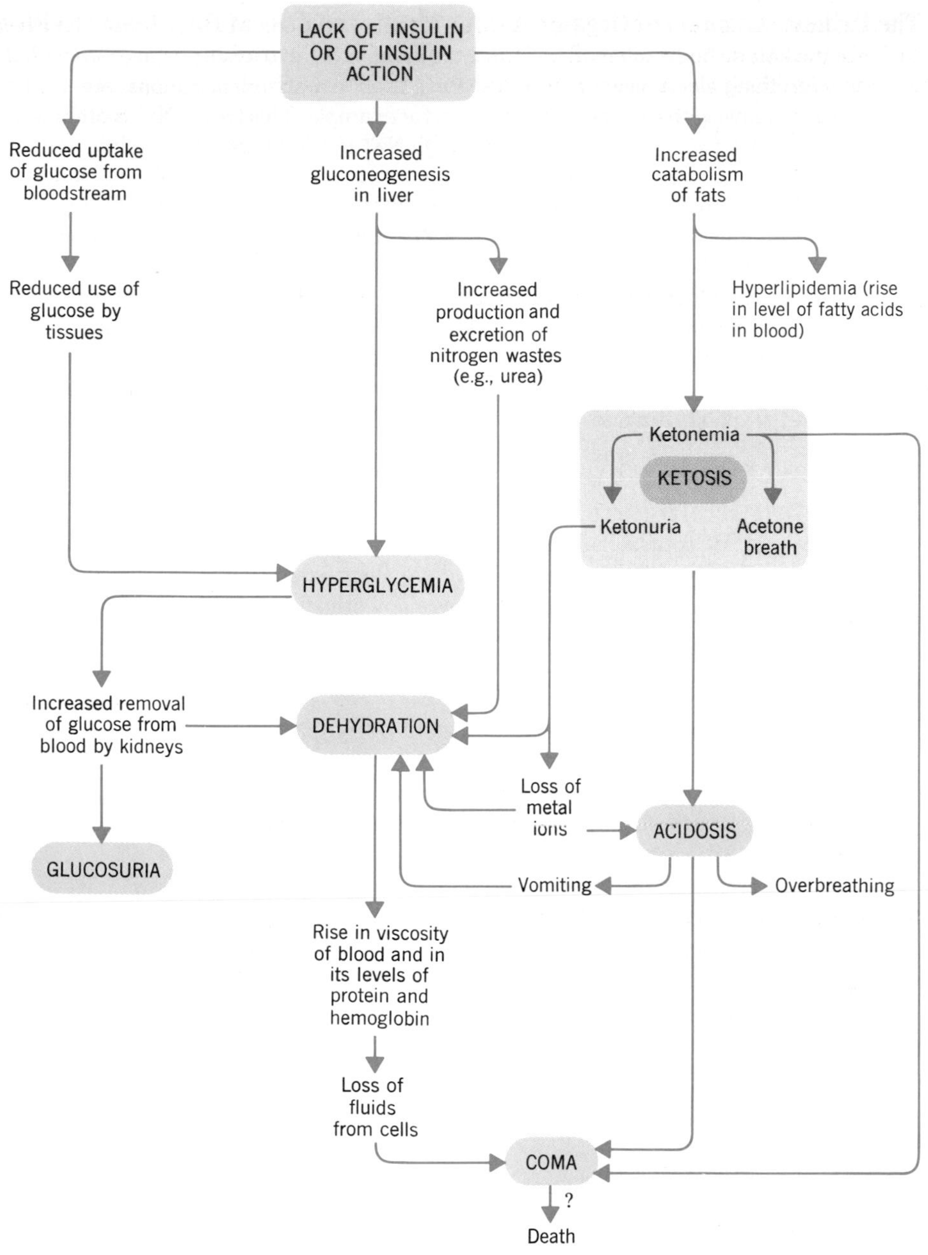

FIGURE 27.6
The principal sequence of events in untreated diabetes

The Conditions of Ketonemia, Ketonuria, and "Acetone Breath" Collectively Constitute Ketosis

Normally, the levels of acetoacetate and β-hydroxybutyrate in the blood are, respectively, 2 μmol/dL and 4 μmol/dL. In prolonged, undetected, and untreated diabetes, these values can increase as much as 200-fold. The condition of excessive levels of ketone bodies in the blood is called **ketonemia.**

As ketonemia becomes more and more advanced, the ketone bodies begin to appear in the urine, a condition called **ketonuria.** When there is a combination of ketonemia, ketonuria, and acetone breath, the overall state is called **ketosis.** The individual will be described as *ketotic.* As unchecked ketosis becomes more severe, the associated ketoacidosis worsens and the pH of the blood continues its fatal descent.

■ 1 μmol = 1 micromole = 10^{-6} mol.

Condition	$[HCO_3^-]_{blood}$ in mmol/L
Normal	22–30
Mild acidosis	16–20
Moderate acidosis	10–16
Severe acidosis	<10

■ **Polyuria** is the technical name for the overproduction of urine.

The Urinary Removal of Organic Anions Means the Loss of Base from the Blood

To leave the ketone body anions in the urine, the kidneys have to leave positive ions with them to keep everything electrically neutral. Na^+ ions, the most abundant cations, are used. One Na^+ ion has to leave with each acetoacetate ion, for example. This loss of Na^+ is often referred to as the "loss of base" from the blood, although Na^+ is not a base. But the loss of one Na^+ stems from the appearance of one acetoacetate ion *plus one* H^+ *ion* that the blood had to neutralize. Thus each Na^+ that leaves the body corresponds to the loss of one HCO_3^- ion, the true base, consumed in neutralizing one H^+. Hence, the loss of Na^+ is taken as an indicator of the loss of this true base.

Another way to understand the urinary loss of Na^+ as the loss of base from the blood is that a Na^+ ion has to accompany a bicarbonate ion when it goes from the kidneys into the blood. The kidneys manufacture HCO_3^- ions normally in order to replenish the blood buffer system. The greater the number of Na^+ ions that have to be left in the *urine* in order to clear ketone bodies from the blood, the less the amount of true base, HCO_3^-, that can be put into the *blood*.

Diuresis Must Accelerate to Handle Ketosis The solutes that are leaving the body in the urine cannot, of course, be allowed to make the urine too concentrated. Otherwise, osmotic pressure balances are upset. Therefore increasing quantities of water must be excreted. To satisfy this need, the individual has a powerful thirst. Other wastes, such as urea, are also being produced at higher than normal rates, because amino acids are being sacrificed in gluconeogenesis. These wastes add to the demand for water to make urine.

Internal Water Shortages in Ketosis Spell Dehydration of Critical Tissues If, during a state of ketosis, insufficient water is drunk, then water is simply taken from extracellular fluids. The blood volume therefore tends to drop, and the blood becomes more concentrated. It also thickens and becomes more viscous, which makes the delivery of blood more difficult.

Because the brain has the highest priority for blood flow, some of this flow is diverted from the kidneys to try to ensure that the brain gets what it needs. This only worsens the situation in the kidneys, and they have an increasingly difficult time clearing wastes. As the water shortage worsens, some water is borrowed from the intracellular supply. This, in addition to a combination of other developments, leads to coma and eventually death.

SUMMARY

Lipid absorption and distribution As fatty acids and monoacylglycerol move out of the digestive tract they become reconstituted as triacylglycerols. These, in addition to cholesterol and proteins, are packaged as chylomicrons. As chylomicrons move in the bloodstream, they unload some of their triacylglycerols and change to chylomicron remnants, which are taken up by the liver. The liver organizes cholesterol, triacylglycerols available to it, and proteins into VLDL and then sends these very-low-density lipoprotein complexes into circulation. Triacylglycerols are again unloaded where they are needed, and the VLDL become more dense, changing into intermediate-density complexes, IDL. Much of these are reabsorbed by the liver. Those that aren't become more dense and change over to low-density lipoprotein complexes, LDL. Most of the LDL finds its way to extrahepatic tissues to deliver cholesterol for cell membranes. Endocrine glands need cholesterol to make steroid hormones. LDL can be reabsorbed by the liver to recycle its lipids. Cholesterol not needed in extrahepatic tissue is carried back to the liver by high-density lipoprotein complexes, HDL. The liver can excrete excess cholesterol via the bile, or it makes bile salts.

Storage and mobilization of lipids The favorable energy density of triacylglycerol means that more energy is stored per gram of this material than can be stored by any other chemical system. The adipose tissue is the principal storage site, and fatty material comes and goes from this tissue according to the energy budget of the body. When the energy of fatty acids is needed, they are liberated from triacylglycerols and carried to the liver, the chief site of fatty acid catabolism.

Catabolism of fatty acids Fatty acyl groups, after being pinned to coenzyme A inside mitochondria, are catabolized by the β-oxidation pathway. By a succession of four steps — dehydrogenation, hydration of a double bond, oxidation of the resulting alcohol, and cleavage of the bond from the α- to the β-carbon — one cycle of the β-oxidation pathway removes one two-carbon acetyl group. The pathway then repeats another cycle as the shortened fatty acyl group continues to be degraded. Each cycle of reactions produces one $FADH_2$ and one NADH, which pass $H:^-$ to the respiratory chain for the synthesis of ATP. Each cycle of β-oxidation also sends one acetyl group into the citric acid cycle which, via the respiratory chain, leads to several more ATPs. The net ATP production is 129 ATPs per palmityl residue.

Biosynthesis of fatty acids Fatty acids can be made by a repetitive cycle of steps. It begins by building one butyryl group from two acetyl groups. The four-carbon butyryl group is attached to an acyl carrier protein that acts as a swinging arm on the enzyme complex. This arm moves the growing fatty acyl unit first over one enzyme and then another as additional two-carbon units are added. The process consumes ATP and NADPH.

Biosynthesis of cholesterol Cholesterol is made from acetyl groups by a long series of reactions. The synthesis of one of the enzymes is inhibited by excess cholesterol, which gives the system a mechanism for keeping its own cholesterol synthesis under control.

Ketoacidosis Acetoacetate, β-hydroxybutyrate, and acetone build up in the blood — ketonemia — in starvation or in diabetes. The first two are normal sources of energy in some tissues. When they are made faster than they can be metabolized, however, there is an accompanying increased loss of bicarbonate ion from the blood's carbonate buffer. This leads to a form of metabolic acidosis called ketoacidosis.

The kidneys try to leave the anionic ketone bodies in the urine, but this requires both Na^+ (for electrical neutrality) and water (for osmotic pressure balances). The excessive loss of Na^+ in the urine is interpreted as a loss of "base." With Na^+ leaving the body in the urine, less is available to accompany replacement HCO_3^-, made by the kidneys, when this base should be going into the blood. Under developing ketoacidosis, the kidneys have extra nitrogen wastes and, in diabetes, extra glucose that need to leave via the kidneys in the urine. This also demands an increased volume of water. Unless this is brought in by the thirst mechanism, it has to be sought from within. However, the brain has first call on blood flowage, so the kidneys suffer more. Eventually, if these events continue unchecked, the victim goes into a coma and dies.

REVIEW EXERCISES

The answers to Review Exercises whose numbers are in color are found in Appendix C. The answers to the other Review Exercises are found in the Study Guide that accompanies this book. The more challenging questions are marked with asterisks.

Absorption and Distribution of Lipids

27.1 What are the end products of the digestion of triacylglycerols?

27.2 What happens to the products of the digestion of triacylglycerols as they migrate out of the intestinal tract?

27.3 What are chylomicrons and what is their function?

27.4 What happens to chylomicrons as they move through capillaries of, say, adipose tissue?

27.5 What happens to chylomicron remnants when they reach the liver?

27.6 What are the two chief sources of cholesterol that the liver exports?

27.7 What do the following symbols stand for?
(a) VLDL   (b) IDL
(c) LDL   (d) HDL

27.8 The loss of what kind of substance from the VLDL converts them into IDL?

27.9 What happens to cause the increase in density between VLDL and LDL?

27.10 What tissue can reabsorb IDL complexes?

27.11 What is the chief constitutent of LDL?

27.12 In extrahepatic tissue, what two general uses await delivered cholesterol?

27.13 If the liver lacks the key receptor proteins, which specific lipoprotein complexes can't be reabsorbed?

27.14 Explain the relationship between the liver's receptor proteins for lipoprotein complexes and the control of the cholesterol level of the blood.

27.15 What is the chief job of the HDL?

*27.16 Why does HDL have a higher density than chylomicrons?

Storage and Mobilization of Lipids

27.17 With reference to the storage of chemical energy in the body, what is meant by *energy density*?

27.18 Arrange the following in their order of increasing quantity of energy that they store per gram.

Wet glycogen	Adipose lipids	Isotonic glucose
1	2	3

27.19 Briefly describe two conditions in which the body would have to turn to the fatty acids for energy.

27.20 How does insulin suppress the mobilization of fatty acids from adipose tissue?

27.21 Arrange the following processes in the order in which they occur when the energy in storage in triacylglycerols is mobilized.

β-Oxidation	Oxidative phosphorylation	Citric acid cycle
A	B	C

Lipoprotein formation	Lipolysis in adipose tissue
D	E

27.22 What specific function does β-oxidation have in obtaining energy from fatty acids?

27.23 What specific function does the citric acid cycle have in the use of fatty acids for energy?

*27.24 Name two hormones that activate the lipase in adipose tissue. Referring to the previous chapter, what does the presence of these hormones do for the blood sugar level?

*27.25 Explain how an increase in the blood sugar level indirectly inhibits the mobilization of energy from adipose tissue.

27.26 When lipolysis occurs in adipose tissue, what happens to the glycerol?

Catabolism of Fatty Acids

27.27 How are long-chain fatty acids activated for β-oxidation?

*27.28 Write the equations for the four steps of β-oxidation as it operates on butyryl CoA. How many more cycles are possible after this one?

27.29 How is the FAD enzyme recovered from its reduced form, $FADH_2$, when β-oxidation operates?

27.30 How is the reduced form of the NAD^+ enzyme that is used in β-oxidation restored to its oxidized form?

27.31 Why is fatty acid catabolism called *beta* oxidation?

Biosynthesis of Fatty Acids

27.32 Where are the principal sites for each activity in a liver cell?
(a) fatty acid catabolism
(b) fatty acid synthesis

*27.33 Outline the steps that make butyryl ACP out of acetyl CoA.

27.34 What metabolic pathway in the body is the chief supplier of NADPH for fatty acid synthesis?

Biosynthesis of Cholesterol

27.35 The enzyme for the formation of which intermediate in cholesterol synthesis is the major control point in this pathway?

27.36 How does cholesterol itself work to inhibit the activity of the enzyme referred to in Review Exercise 27.35?

Ketoacidosis

27.37 What species is diverted from the citric acid cycle to gluconeogenesis?

*27.38 Why does the diversion of Review Exercise 27.37 lead to an increase in the level of acetyl CoA?

27.39 Two molecules of acetyl CoA can combine to give the coenzyme A derivative of what keto acid? Give its structure.

*27.40 In two steps, the compound of Review Exercise 27.39 gives one unit of a ketone body and one other significant species (besides recovered CoA). What is it? Why is it a problem?

27.41 Give the names and structures of the ketone bodies.

27.42 What is ketonemia?

27.43 What is ketonuria?

27.44 What is meant by acetone breath?

27.45 Ketosis consists of what collection of conditions?

27.46 What is ketoacidosis? What form of acidosis is it, metabolic or respiratory?

27.47 The formation of which particular compound most lowers the supply of HCO_3^- in ketoacidosis?

27.48 What are the reasons for the increase in the volume of urine that is excreted by someone with untreated type I diabetes?

*27.49 If the ketone bodies (other than acetone) can normally be used by heart and skeletal muscle, what makes them dangerous in starvation or in diabetes?

*27.50 Why does the rate of urea production increase in untreated, type I diabetes?

27.51 When a physician refers to the loss of Na^+ as the loss of *base*, what is actually meant?

Good and Bad Cholesterol (Special Topic 27.1)

27.52 What is atherosclerosis?

27.53 What is familial hypercholesterolemia and how do people get it?

27.54 Which of the lipoprotein complexes is sometimes called "bad cholesterol?" Explain why it is so designated.

27.55 Which lipoprotein complex carries "good cholesterol?"

27.56 Some scientists believe that a relatively high level of HDL is associated with protection against atherosclerosis, and that one's risk of having heart disease declines when the level of HDL is raised by exercise and losing weight. Why would a low level of HDL tend to be associated with heart disease?

27.57 Which substance is believed to be associated with the *most common* inherited risk factor for heart attack?

Brown Adipose Tissue (Special Topic 27.2)

27.58 With respect to the catabolism of fatty acids, how do white and brown adipose tissue differ?

27.59 In the ordinary operation of the respiratory chain, the chain is coupled to the synthesis of ATP because of what condition concerning mitochondria?

27.60 What happens to the energy released by the respiratory chain when it isn't used to make ATP?

27.61 What is meant by *thermogenesis?*

Additional Exercises

*27.62 Complete the following equations for one cycle of β-oxidation by which a six-carbon fatty acyl group is catabolized.

(a) $CH_3CH_2CH_2CH_2CH_2\overset{\overset{\displaystyle O}{\|}}{C}SCoA + FAD \rightarrow$

_____ + _____

(b) _____ + H_2O → _____

(c) _____ + NAD^+ → _____ + $NADH + H^+$

(d) _____ + CoASH → _____ + _____

*27.63 Myristic acid, $CH_3(CH_2)_{12}CO_2H$, can be catabolized by β-oxidation just like palmitic acid.
(a) How many units of acetyl CoA can be made from it?
(b) In producing this much acetyl CoA, how many times does $FADH_2$ form and then deliver its hydrogen to the respiratory chain?
(c) Referring again to part (a), how many times does NADH form as acetyl CoA is produced and then delivers its hydrogen to the respiratory chain?
(d) Calculate the maximum net number of molecules of ATP that can be made by means of the catabolism of one molecule of myristic acid. (Table 27.2 provides clues about this calculation.)

METABOLISM OF NITROGEN COMPOUNDS

28

This fellow needs *meat*, and no arguments, please. Carnivores rely on weaker, slower, or less alert animals to supply them their needed amino acids. How amino acids are broken down and used by the body is studied in this chapter.

28.1
SYNTHESIS OF AMINO ACIDS IN THE BODY
28.2
CATABOLISM OF AMINO ACIDS

28.3
FORMATION OF UREA
28.4
CATABOLISM OF OTHER NITROGEN COMPOUNDS

28.1

SYNTHESIS OF AMINO ACIDS IN THE BODY

The body can manufacture a number of amino acids from intermediates that appear in the catabolism of nonprotein substances.

Amino acids, the end products of protein digestion, are rapidly transported across the walls of the small intestine. Some very small, simple peptides can also be absorbed. Once amino acids enter circulation, they become part of what is called the **nitrogen pool.**

■ The polypeptides in proteins that serve as enzymes have a particularly rapid turnover.

The *Nitrogen Pool* Consists of All Nitrogenous Compounds Anywhere in the Body Figure 28.1 illustrates the various compartments of the nitrogen pool and how they are interrelated. Amino acids enter the nitrogen pool not only as digestive products but also as products of the breakdown of proteins in body fluids and tissues. These are undergoing constant turnover, fairly rapidly among the liver proteins and those in the blood and quite slowly among muscle proteins.

As indicated in Figure 28.1, individual amino acids can be used in any one of the following ways, depending on the body's needs of the moment.

1. the synthesis of new or replacement proteins

2. the synthesis of such nonprotein nitrogen compounds as heme, creatine, nucleic acids, and certain hormones and neurotransmitters

■ In both starvation and diabetes the body draws from its amino acid pool to make glucose.

3. the production of ATP or of glycogen and fatty acids, substances with the potential for making ATP

4. the synthesis of any needed nonessential amino acids

The Body Uses Both Essential and Nonessential Amino Acids to Make Proteins We do not have to take in *all* of the 20 amino acids in the diet; we are able to make roughly half of them ourselves from other substances in food. Those that *must* be in the diet are called the

FIGURE 28.1
The nitrogen pool

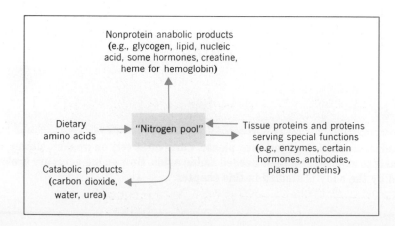

essential amino acids. The others are called the *nonessential amino acids*. Be sure to remember that "nonessential" in this context refers only to a *dietary* need because, one way or another, the body must have on hand all of the amino acids whenever it needs to make polypeptides. We'll broadly study how the body makes nonessential amino acids next. Figure 28.2 gives an overview to which we'll refer as we go along.

The Reductive Amination of α-Ketoglutarate to Give Glutamate Makes the Ammonium Ion a Source of Nitrogen for Nonessential Amino Acids

Many of the syntheses of nonessential amino acids outlined in Figure 28.2 depend on the availability of the glutamate ion. It is made from α-ketoglutarate by a reaction called **reductive amination,** in which the ammonium ion is the source of nitrogen and NADPH is the reducing agent. (In some cells, NADH enzymes can also work.) We need a reducing agent because the keto group (of

■ This reaction is the body's principal means for incorporating inorganic nitrogen (as NH_3) into organic compounds.

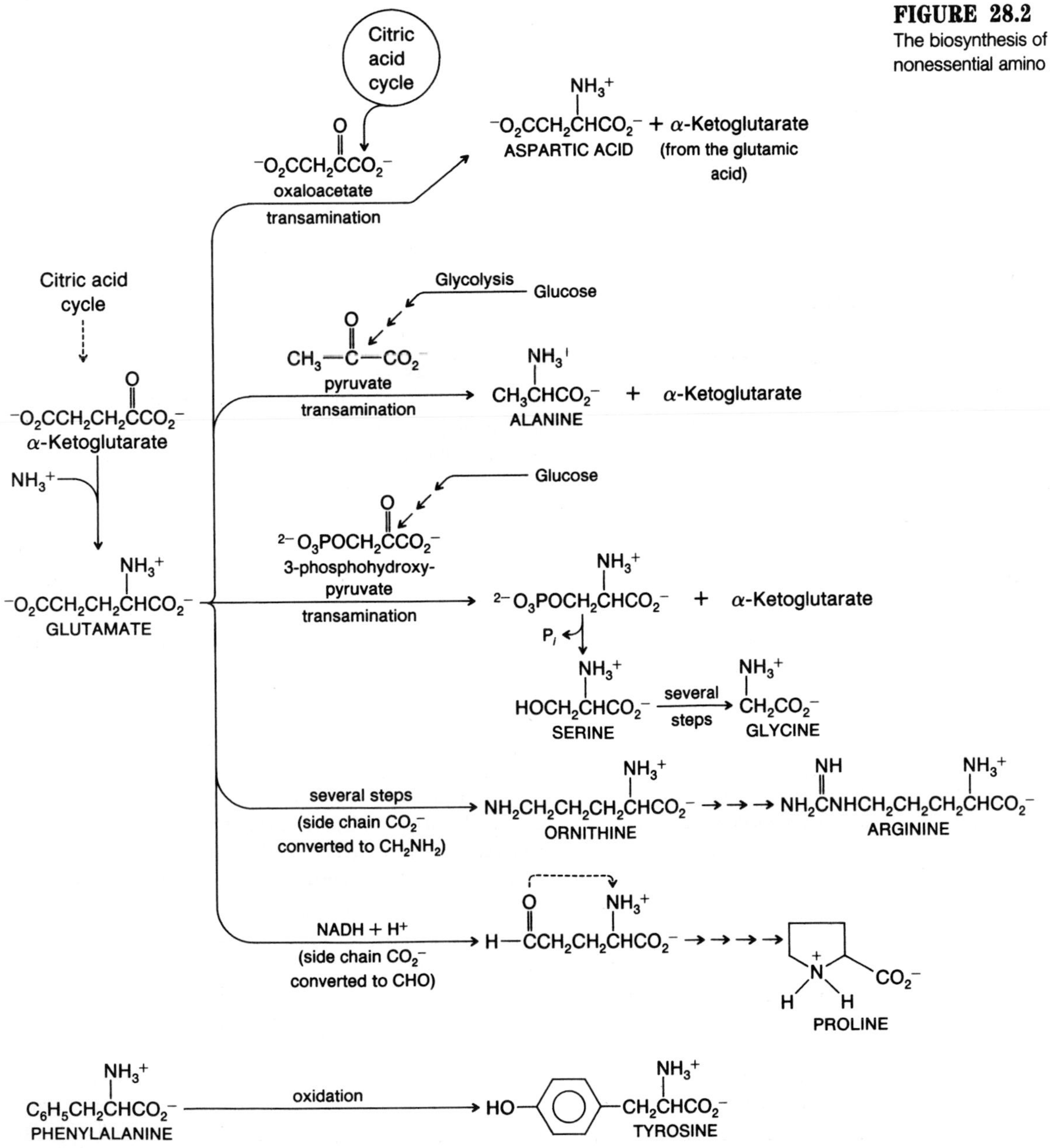

FIGURE 28.2
The biosynthesis of some nonessential amino acids

α-ketoglutarate) is in a higher oxidation state than the $CHNH_2$ unit in glutamate, the product. The overall result is

- The enzyme is *glutamate dehydrogenase.*

$$^-O_2CCH_2CH_2\overset{O}{\overset{\|}{C}}CO_2^- + NH_4^+ + NADPH + H^+ \rightleftharpoons$$
α-Ketoglutarate

$$^-O_2CCH_2CH_2\overset{NH_3^+}{\overset{|}{C}H}CO_2^- + NADP^+ + H_2O$$
Glutamate

- The coenzymes for aminotransferases are compounds that collectively are called vitamin B_6.

Glutamate Supplies the Amino Group to Make Many Nonessential Amino Acids

Special aminotransferase enzymes (sometimes called transaminases) are able to catalyze the transferral of an amino group from glutamate to the keto carbonyl group of a keto acid. The reaction is called **transamination.** The required ketone compounds can come from the catabolism of glucose, as you can see in Figure 28.2. We'll illustrate the general case of what happens in a transamination as follows.

$$R\overset{O}{\overset{\|}{C}}CO_2^- + {}^-O_2CCH_2CH_2\overset{NH_3^+}{\overset{|}{C}H}CO_2^- \rightleftharpoons R\overset{NH_3^+}{\overset{|}{C}H}CO_2^- + {}^-O_2CCH_2CH_2\overset{O}{\overset{\|}{C}}CO_2^-$$
An α-keto acid   Glutamate   An α-amino acid α-Ketoglutarate

28.2

CATABOLISM OF AMINO ACIDS

The breakdown products of amino acid catabolism eventually enter the pathways of catabolism of either carbohydrates or lipids.

There is no special mechanism for the storage of amino acids analogous to the storage system for glucose (glycogen) or for fatty acids (fat in adipose tissue). Amino acids in the nitrogen pool that aren't needed to make other amino acids or other nitrogen compounds are soon catabolized. Most of this work is done in the liver (indicating once again how important this organ is).

Amino Acids, When Stripped of Amino Groups, Can Be Used to Make Virtually Anything Else the Body Needs

$$NH_2\overset{O}{\overset{\|}{C}}NH_2$$
Urea

The ultimate end products of the complete catabolism of amino acids are urea, carbon dioxide, and water. On the way to these compounds, however, several intermediates form that can enter other pathways. In fact, all the pathways for the use of carbohydrates, lipids, and proteins are interconnected in one way or another (see Fig. 28.3).

Notice in Figure 28.3 the central importance of two small molecules, acetyl CoA and pyruvate. Notice also that there is no route from acetyl CoA to pyruvate, which means that (in all animals, at least) *glucose cannot be made from fatty acids.* (We'll return to this shortly.)

We won't study in detail how each amino acid is catabolized, because each requires its own particular scheme, usually quite complicated. Early in each pathway, however, the amino acid gets rid of its nitrogen, which is shuttled into the synthesis of urea. The nonnitrogen fragment then eventually enters a pathway we have already studied. There are three kinds of reactions, besides transamination, that occur often: *oxidative deamination, direct deamination,* and *decarboxylation.* We'll study these and how they apply to certain selected amino acids.

Amino Groups Are Shuttled through the α-Ketoglutarate–Glutamate Switch toward Urea Synthesis

One of the steps in the catabolic process that removes amino groups of amino acids is **oxidative deamination,** the name given to the reverse of reductive amination, studied on page 769. In the display that follows, the step at the first pair of curved

FIGURE 28.3
Interrelationships of major metabolic pathways

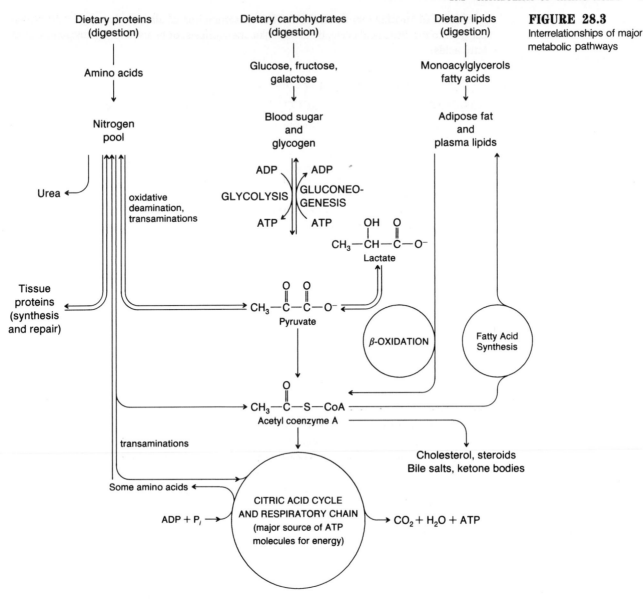

arrows is a transamination. Oxidative deamination occurs at the second pair, and the arrowheads point, left to right, in the direction of catabolism.

■ Most amino acids are deaminated by this route.

Notice that the nitrogen of the amino acid on the upper left ends up in urea, and that the α-ketoglutarate–glutamate pair provides a switching mechanism to convey this nitrogen in the right direction for catabolism.

Alanine Catabolizes to Pyruvate   The transamination of alanine gives pyruvate, which can go into the citric acid cycle, be used in gluconeogenesis, or be used for the biosynthesis of fatty acids.

■ This display shows how excess alanine from the diet, alanine not needed that day to make proteins or other nitrogen compounds, can be used to make glucose, glycogen, or fatty acids according to other needs.

$$\underset{\text{Alanine}}{CH_3\overset{\overset{\displaystyle NH_3^+}{|}}{C}HCO_2^-} \xrightarrow{\text{transamination}} \underset{\text{Pyruvate}}{CH_3\overset{\overset{\displaystyle O}{\|}}{C}CO_2^-} \longrightarrow \text{Acetyl CoA} \longrightarrow \text{citric acid cycle}$$

gluconeogenesis

fatty acid synthesis

Aspartic Acid Catabolizes to Oxaloacetate   The transamination of aspartic acid gives oxaloacetate, an intermediate in both gluconeogenesis and the citric acid cycle.

$$\underset{\text{Aspartic acid}}{^-O_2CH_2\overset{\overset{\displaystyle NH_3^+}{|}}{C}HCO_2^-} \xrightarrow{\text{transamination}} \underset{\text{Oxaloacetate}}{^-O_2CCH_2\overset{\overset{\displaystyle O}{\|}}{C}CO_2^-} \longrightarrow \text{citric acid cycle}$$

gluconeogenesis

Direct Deamination Removes Amino Groups without Oxidation   Two amino acids, serine and threonine, have OH groups which make possible a nonoxidative loss of NH_3, called **direct deamination.** They are able to undergo the simultaneous loss of water and ammonia because their OH groups are strategically located on the carbon adjacent to the one that holds an amino group. Here's how direct deamination happens with serine.

■ During prolonged fasting or starvation, the body degrades its own proteins to make the brain's favorite source of energy, glucose, via gluconeogenesis.

$$\underset{\text{Serine}}{HOCH_2\overset{\overset{\displaystyle NH_3^+}{|}}{C}HCO_2^-} \xrightarrow[\underset{H_2O}{}]{\text{(dehydration of alcohol)}} CH_2=\overset{\overset{\displaystyle NH_2^+}{|}}{C}CO_2^- \longrightarrow \underset{\text{An imine}}{CH_3\overset{\overset{\displaystyle NH}{\|}}{C}CO_2^- + H^+}$$

$$\underset{\text{Pyruvate}}{NH_3 + CH_3\overset{\overset{\displaystyle O}{\|}}{C}CO_2^-} \xleftarrow[\text{the imine group)}]{\text{(hydrolysis of}} \quad H_2O$$

The first step is the dehydration of the alcohol system of serine to give an unsaturated amine. This spontaneously rearranges into an imine, a compound with a carbon−nitrogen double bond. Water can add to this double bond, but the product spontaneously breaks up, so the net effect is the hydrolysis of the imine group to a keto group (in pyruvate) and ammonia. Thus serine breaks down to pyruvate, which, as we now well know, can send an acetyl group into the citric acid cycle, or can contribute an acetyl group to fatty acid synthesis, or can be used to make glucose.

■ Imine groups easily hydrolyze because they can add water and then split out ammonia.

Sustained Gluconeogenesis Necessarily Consumes Body Proteins   Figure 28.4 broadly outlines how the reactions just surveyed are involved in the catabolism of several amino acids. Notice particularly that oxaloacetate occurs in two places, as an intermediate in the citric acid cycle and as the product of the oxidative deamination of aspartate. The oxaloacetate made from aspartate has two options: to be used to make ATP or to make

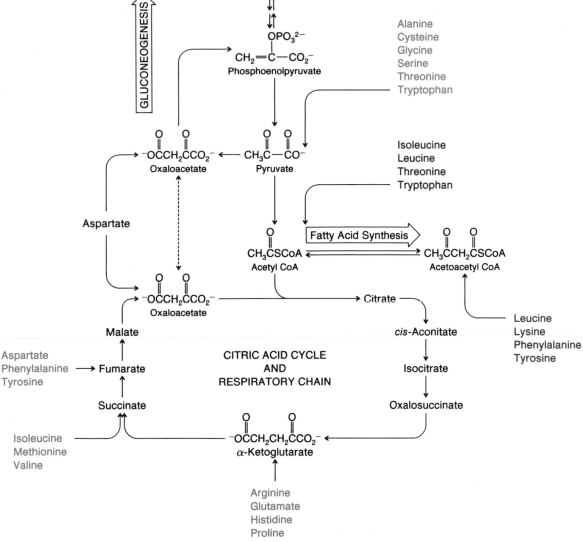

FIGURE 28.4
The catabolism of some amino acids. Glucogenic amino acids are shown in blue and ketogenic amino acids are given in red.

glucose by gluconeogenesis. Oxaloacetate thus connects the catabolism of amino acids to carbohydrate synthesis.

Because of the occurrence of oxaloacetate in the citric acid cycle, and because its carbons can originate in fatty acids (by way of acetyl coenzyme A), we might think that fatty acids could also be used to make glucose. Not so, at least *not for a net gain of glucose.* The removal of oxaloacetate from the citric acid cycle for gluconeogenesis means the removal of what carries acetyl units (of any origin) into the cycle. This leads to a backup of acetyl CoA (as we learned in Section 27.6) and a buildup of the ketone bodies.

For a *net gain* of glucose molecules via gluconeogenesis, the oxaloacetate of the citric acid cycle cannot be counted as available. *Only oxaloacetate made from amino acids can give a net gain of glucose this way.* This is why gluconeogenesis under conditions of starvation or diabetes necessarily breaks down body proteins. It needs some of their amino acids to make oxaloacetate to be able to make glucose.

The amino acids indicated in Figure 28.4 as sources of carbons for the synthesis of glucose are called the *glucogenic amino acids.* These are any that can be degraded to pyruvate, or to such citric acid cycle intermediates as α-ketoglutarate, succinate, fumarate, or oxaloacetate. Amino acids that can be degraded to acetyl CoA or acetoacetate can be used to make fatty acids and so are called the *ketogenic amino acids.* (Only leucine and lysine are exclusively ketogenic.) A few amino acids are both glucogenic and ketogenic.

The Decarboxylation of Some Amino Acids Leads to Neurotransmitters Some special enzymes can split out just the carboxyl groups from amino acids and so make amines. The reaction, called **decarboxylation,** is used to make some neurotransmitters and hormones. Dopamine, norepinephrine, and epinephrine are all made by steps that begin with the decarboxylation of dihydroxyphenylalanine, which the body makes from the amino acid tyrosine.

■ The symbol (O) signifies an oxidation step.

28.3

FORMATION OF UREA

Ammonia and amino groups are converted into urea by a complex cycle of reactions called the urea cycle.

Urea, as we have learned, is the chief nitrogen waste made by the body. Most of its nitrogen indirectly comes from amino acids, but some comes (also indirectly) from two of the side chain bases of nucleic acids.

Oxaloacetate Shuttles Nitrogen from Glutamate to Aspartate and Then into the Synthesis of Urea There are two direct sources for the nitrogen atoms in urea. One is the ammonium ion produced by the oxidative deamination of glutamate (page 771).

■ Recall that the amino group of glutamate can come from any other amino acid by way of another shuttle.

The other nitrogen atom in urea comes from a specific amino acid, aspartate. However, because aspartate can be made from glutamate by a transamination, you can see that

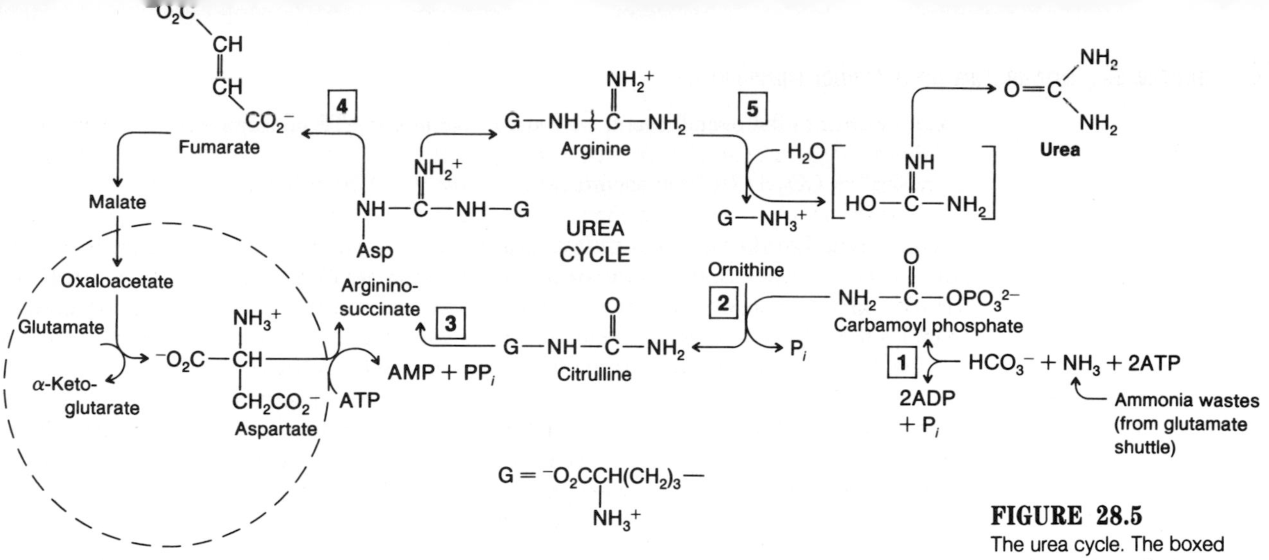

O_2C
CH
‖
CH
|
CO_2^- [4]

Fumarate

$G-NH \! + \! \overset{NH_2^+}{\underset{\|}{C}} \! -NH_2$ Arginine [5]

H_2O [$\overset{NH}{\underset{\|}{HO-C-NH_2}}$]

$O=C\overset{NH_2}{\underset{NH_2}{\big\langle}}$ **Urea**

Malate

$\overset{NH_2^+}{\underset{\|}{NH-C-NH-G}}$
|
Asp

$G-NH_3^+$

UREA CYCLE

Oxaloacetate

Glutamate

Arginino-succinate [3]

Ornithine [2]

$NH_2-\overset{O}{\underset{\|}{C}}-OPO_3^{2-}$
Carbamoyl phosphate

$\overset{NH_3^+}{\underset{|}{-O_2C-CH}}$
|
$CH_2CO_2^-$ ATP
Aspartate

α-Keto-glutarate

$AMP + PP_i$

$G-NH-\overset{O}{\underset{\|}{C}}-NH_2$ → P_i
Citrulline

[1] $\{ HCO_3^- + NH_3 + 2ATP$

$2ADP + P_i$

Ammonia wastes (from glutamate shuttle)

$G = {}^-O_2CCH(CH_2)_3-$
|
NH_3^+

FIGURE 28.5
The urea cycle. The boxed numbers refer to the text discussion. The dashed-line circle is the aspartate–oxaloacetate shuttle also discussed in the text.

glutamate is close to being the direct source of both nitrogens in urea. Here is the shuttle from glutamate to aspartate just described.

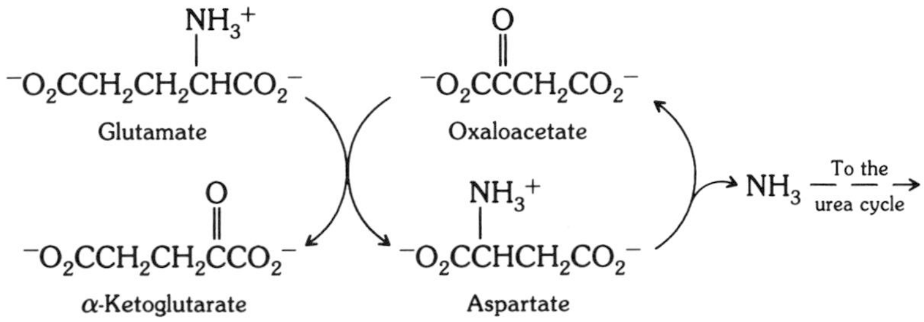

$\overset{NH_3^+}{\underset{|}{{}^-O_2CCH_2CH_2CHCO_2^-}}$
Glutamate

$\overset{O}{\underset{\|}{{}^-O_2CCCH_2CO_2^-}}$
Oxaloacetate

$\overset{O}{\underset{\|}{{}^-O_2CCH_2CH_2CCO_2^-}}$
α-Ketoglutarate

$\overset{NH_3^+}{\underset{|}{{}^-O_2CCHCH_2CO_2^-}}$
Aspartate

$NH_3 \xrightarrow[\text{urea cycle}]{\text{To the}}$

The Urea Cycle Is the Only Way the Body Has to Make Urea

A series of reactions called the **urea cycle** manufactures urea from ammonium ion, carbon dioxide, and aspartate. Figure 28.5 displays its steps, and the boxed numbers in this figure refer to the following discussion.

■ The urea cycle is sometimes called the *Krebs ornithine cycle*.

1. Ammonia, with the help of ATP, reacts with CO_2 to form carbamoyl phosphate, a high-energy phosphate. In a sense, this is an activation of ammonia that launches it into the next step that takes it into the cycle.

■ This step occurs insides a mitochondrion.

2. The carbamoyl group transfers to the carrier unit, ornithine, as P_i is ejected. This consumes energy from high-energy phosphate. Citrulline forms.

■ Steps 2 through 5 occur in the cell's cytosol.

3. Citrulline condenses with the alpha amino group of aspartate to give argininosuccinate.

4. Fumarate forms from the original aspartate as the amino group stays with the arginine that emerges. Fumarate is an intermediate in the citric acid cycle. By a transamination that involves glutamate, it is reconverted to aspartate.

5. Arginine is hydrolyzed. Urea forms and ornithine is regenerated to start another turn of the cycle.

The overall result of the urea cycle is given by the following equation (which, as Figure 28.5 makes clear, is extremely simplified):

$$2NH_3 + H_2CO_3 \longrightarrow NH_2\overset{O}{\underset{\|}{C}}NH_2 + 2H_2O$$
Urea

To do justice to the overall event, we must factor in the ATP consumption, as follows:

$$NH_4^+ + CO_2 + 3ATP + aspartate \longrightarrow urea + 2ADP + fumarate + 4P_i + AMP$$

At Elevated Levels, the Ammonium Ion Is Toxic If an infant is born without any one of the enzymes needed for the five steps of the urea cycle, it will die soon after birth. It will be unable, on its own, to clear ammonia from its blood. (Prior to birth, the mother's metabolism handles this.) Ammonia and the ammonium ion are toxic at sufficiently high levels.

Some inherited genetic defects produce enzymes for this cycle that have reduced activity. Such individuals have a condition called **hyperammonemia,** an elevated level of NH_3 in the blood. Infants that have this genetic defect improve on low-protein diets. If the level is not high enough to cause death, it can be expected to cause mental retardation. Periodic but unremembered episodes of bizarre behavior by one man — babbling, pacing, crying, glassy eyes — were not understood until it was found that he had a rare deficiency of the enzyme for step 2 of the urea cycle (ornithine transcarboxylase).

■ There are many genetic defects involving enzymes for the body's utilization of amino acids.

28.4

CATABOLISM OF OTHER NITROGEN COMPOUNDS

Uric acid and the bile pigments are other end products of the catabolism of nitrogen compounds.

The nitrogen of the purine bases of nucleic acids, adenine (A) and guanine (G), is excreted as uric acid, which also has the purine nucleus. After studying how uric acid forms, we'll see how defects in this pathway can lead to gout or to a particularly difficult disease of children, the Lesch-Nyhan syndrome.

The Catabolism of AMP Gives the Urate Ion The numbered steps in Figure 28.6 are discussed next to show how the adenine unit of AMP can be used to make uric acid.

1. A transamination removes the amino group of the adenine side chain in adenosine monophosphate, AMP.

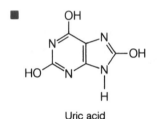

Uric acid

Purine

FIGURE 28.6
The catabolism of adenosine monophosphate, AMP. The boxed numbers refer to the discussion in the text.

AMP (Adenosine monophosphate) — Inosine — Hypoxanthine + Ribose—P — (recycle) Nucleotides — Xanthine — Uric acid (Phenolic form, Keto form)

2. Ribose phosphate is removed and will enter the pentose phosphate pathway of carbohydrate catabolism. The product is hypoxanthine.

3. An oxidation produces xanthine. (The steps from guanine lead to xanthine, too.)

4. Another oxidation produces the keto form of uric acid, which exists partly in the form of a phenol. Actually, it's the salt of uric acid that forms, sodium urate, because the acid is neutralized by base in the buffer system.

Overproduction of the Urate Ion Causes Gout In the disease known as *gout,* the rate of formation of sodium urate is more rapid than its rate of elimination. Crystals of this salt precipitate in joints where they cause painful inflammations and lead to a form of arthritis. Kidney stones may form as this salt comes out of solution in this organ.

Just why the formation of sodium urate accelerates isn't well understood, but genetic factors are involved. Normally, some of the hypoxanthine made in step 2 is recycled back to nucleotide bases that are needed to make nucleic acids or high-energy phosphates. Some individuals with gout are known to have a partial deficiency of the enzyme system required for this recycling of hypoxanthine. Hence, most if not all of their hypoxanthine ends up as more sodium urate than normal.

■ Ethyl alcohol accelerates the synthesis of urate ion by increasing the rate of catabolism of adenosine monophosphate.

The Absence of One Enzyme Needed for Hypoxanthine Recycle Leads to Self-mutilating Behavior in Infants In Lesch-Nyhan syndrome, the enzyme for recycling hypoxanthine is totally lacking. The result is both bizarre and traumatic. Infants with this syndrome develop compulsive, self-destructive behavior at age 2 or 3. Unless their hands are wrapped in cloth, they will bite themselves to the point of mutilation. They act with dangerous aggression toward others. Some become spastic and mentally retarded. Kidney stones develop early, and gout comes later.

■ The lack of one enzyme usually means great personal and family suffering.

The Catabolism of Heme Produces the Bile Pigments Erythrocytes have life spans of only about 120 days. Eventually they split open. Their hemoglobin spills out and then is degraded. Its breakdown products are eliminated via the feces and, to some extent, in the urine. In fact, the characteristic colors of feces and urine are caused by partially degraded heme molecules called the **bile pigments.**

The degradation of heme begins before the globin portion breaks away. The heme molecule partly opens up to give a system that has a chain of four small rings called pyrrole rings. (This is why the bile pigments are sometimes called the *tetrapyrrole pigments.*)

■

Pyrrole skeleton

Carbon skeleton of the bile pigments

The rings have varying numbers of double bonds according to the state of oxidation of the pigment.

The slightly broken hemoglobin molecule, now called verdohemoglobin, then splits into globin, iron(II) ion, and a greenish pigment called **biliverdin** (Latin *bilis,* bile, + *virdus,* green). Globin enters the nitrogen pool. Iron is conserved in a storage protein called ferritin and is reused. Biliverdin is changed in the liver to a reddish-orange pigment called **bilirubin** (Latin *bilis,* bile + *rubin,* red). Bilirubin is not only made by the liver but is also removed from circulation by the liver, which transfers it to the bile. In this fluid it finally enters the intestinal tract.

The pathway from hemoglobin to bilirubin after the rupture of an erythrocyte, as well as the fate of bilirubin, is shown in Figure 28.7.

Bilirubin is the principal bile pigment in humans. In the intestinal tract, bacterial enzymes convert bilirubin to a colorless substance called *mesobilirubinogen.* This is further processed to form a substance known as **bilinogen,** which usually goes by other names that describe

Jaundice (French, *jaune,* yellow) is a condition that is symptomatic of a malfunction somewhere along the pathway of heme metabolism. If bile pigments accumulate in the plasma in concentrations high enough to impart a yellowish coloration to the skin, the condition of *jaundice* is said to exist. Jaundice can result from one of three kinds of malfunctions.

Hemolytic jaundice results when hemolysis takes place at an abnormally fast rate. Bile pigments, particularly bilirubin, form faster than the liver can clear them. Hepatic diseases such as infectious hepatitis and cirrhosis sometimes prevent the liver from removing bilirubin from circulation. The stools are usually clay-colored, because the pyrrole pigments do not reach the intestinal tract.

Obstructions of bile ducts can prevent release of bile into the intestinal tract, and the tetrapyrrole pigments in bile cannot be eliminated. Under these circumstances, they tend to reenter general circulation. The kidneys remove large amounts of bilirubin, but the stools are usually clay-colored. As the liver works harder and harder to handle its task of removing excess bilirubin, it can weaken and become permanently damaged.

differences in destination rather than structure. Thus bilinogen that leaves the body in the feces is called *stercobilinogen* (Latin, *stercus,* dung). Some bilinogen is reabsorbed via the bloodstream, comes to the liver, and finally leaves the body in the urine. Now it is called *urobilinogen.* Some bilinogen is reoxidized to give **bilin,** a brownish pigment. Depending on its destination, bilin is called *stercobilin* or *urobilin.*

Special Topic 28.1 describes how the bile pigments are involved in jaundice.

FIGURE 28.7
The formation and the elimination of the products of the catabolism of hemoglobin.

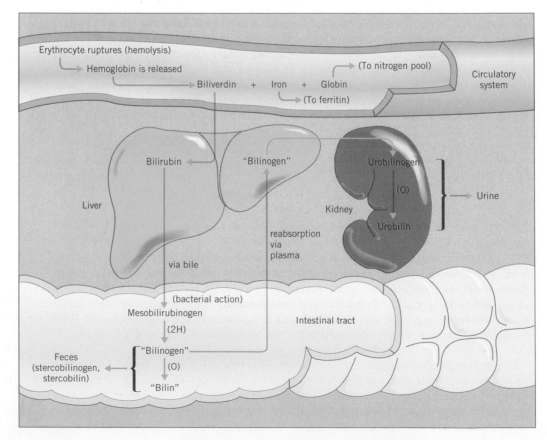

SUMMARY

Amino acid distribution The nitrogen pool receives amino acids from the diet, from the breakdown of proteins in body fluids or tissues, and from any synthesis of nonessential amino acids that occurs. Amino acids are used to build and repair tissue, replace proteins of body fluids, make nonprotein nitrogen compounds, provide chemical energy if needed, and supply molecular parts for gluconeogenesis or fatty acid synthesis.

Amino acid metabolism By reactions of transamination, oxidative deamination, direct deamination, and decarboxylation, the α-amino acids shuttle amino groups between themselves and intermediates of the citric acid cycle, or the synthesis of urea and nonprotein nitrogen compounds.

Deaminated amino acids eventually become acetyl CoA, ace-toacetyl CoA, pyruvate, or an intermediate in the citric acid cycle. The skeletons of most amino acids can be used to make glucose, or fatty acids, or the ketone bodies. Their nitrogen atoms become part of urea.

Metabolism of other nitrogen compounds The nitrogen atoms in some of the sidechain bases of nucleic acids end up in urea and those of the others are excreted as sodium urate. Urea is made by a complex cycle of reactions — the urea cycle.

Heme is catabolized to bile pigments and its iron is reused. The pigments — first, biliverdin (green), then bilirubin (red), then meso-bilirubinogen (colorless), and finally bilinogen and bilin (brown) — become stercobilin or stercobilinogen, urobilin or urobilinogen, depending on the route of elimination.

REVIEW EXERCISES

The answers to Review Exercises whose numbers are in color are found in Appendix C. The answers to the other Review Exercises are found in the Study Guide that accompanies this book. The more challenging questions are marked with asterisks.

Nitrogen Pool

28.1 What is the nitrogen pool?

28.2 What are four ways in which amino acids are used in the body?

28.3 When the body retains more nitrogen than it excretes in all forms, the system is said to be on a *positive nitrogen balance*. Would this state characterize infancy or old age?

28.4 What happens to amino acids that are obtained in the diet but aren't needed to make any nitrogeneous compounds?

Biosynthesis of Amino Acids

28.5 Which amino acid is the major supplier of amino groups for the synthesis of the nonessential amino acids?

28.6 Write the equation for the reductive amination that produces glutamate. (Use NADPH as the reducing agent.)

*28.7 Write the structure of the keto acid that forms when phenylalanine undergoes transamination with α-ketoglutarate.

*28.8 When valine and α-ketoglutarate undergo transamination, what new keto acid forms? Write its structure.

The Catabolism of Amino Acids

28.9 Cysteine is a glucogenic amino acid. What does this mean?

28.10 Lysine is exclusively a ketogenic amino acid. What does this mean?

28.11 An amino acid that can generate acetyl CoA without going through pyruvate is glucogenic or ketogenic?

*28.12 By means of two successive equations, one a transamination and the other an oxidative deamination, write the reactions that illustrate how the amino group of alanine can be removed as NH_4^+.

*28.13 Arrange the following compounds in the order in which they would be produced if the carbon skeleton of alanine were to appear in one of the ketone bodies.

Pyruvate	Acetoacetyl CoA	Acetoacetate
1	2	3

Alanine	Acetyl CoA
4	5

*28.14 In what order would the following compounds appear if some of the carbon atoms in glutamate were to become part of glycogen?

Oxaloacetate	α-Ketoglutarate	Glucose
1	2	3

Glycogen	Glutamate
4	5

28.15 Can any of the carbon atoms of glucose become part of alanine? If so, explain (in general terms).

28.16 Write the structure of the keto acid that forms by the direct deamination of threonine.

*28.17 When tyrosine undergoes decarboxylation, what forms? Write its structure.

*28.18 Write the structure of the product of the decarboxylation of tryptophan.

28.19 In the conditions of starvation or diabetes, what can the amino acids be used for?

The Formation of Urea

28.20 In the biosynthesis of urea, what are the sources of (a) the two NH_2 groups and (b) the $C{=}O$ group?

28.21 What is hyperammonemia and, in general terms, how does it arise and how can it be handled in infants?

28.22 What is the overall equation for the synthesis of urea?

28.23 Would a deficiency in ornithine transcarbamylase (the enzyme for step 2 of the urea cycle, Figure 28.5) cause hypoammonemia or hyperammonemia? Explain.

The Catabolism of Nonprotein Nitrogen Compounds

28.24 What compounds are catabolized to make uric acid?

28.25 What product of catabolism accumulates in the joints in gout?

*28.26 Arrange the names of the following substances in the order in which they appear during the catabolism of heme.

Biliverdin	Heme	Hemoglobin	Bilirubin
1	2	3	4

Mesobilirubinogen	Bilin	Bilinogen
5	6	7

Tetrapyrrole Pigments and Jaundice (Special Topic 28.1)

28.27 Briefly describe what jaundice does to the body.

28.28 Describe how the following kinds of jaundice arise.
(a) hemolytic jaundice
(b) the jaundice of hepatic diseases

28.29 Why should an obstruction of the bile ducts cause jaundice?

Additional Exercises

*28.30 From a study of the figures in this chapter, can the carbon atoms of serine become a part of a molecule of palmitic acid? If so, write the names of the compounds, beginning with serine, in the sequence to the start of fatty acid synthesis.

*28.31 The carbon atoms of a ketogenic amino acid eventually are present among intermediates in the citric acid cycle, including oxaloacetate, a starting material for gluconeogenesis. Why, then, aren't all ketogenic amino acids also glucogenic?

NUTRITION

Specialists in nutrition counsel us to have a variety of foods in our diets, and from both land and sea a bountiful nature obliges. This chapter describes the kinds and amounts of nutrients we must have for maximum health.

29.1

GENERAL NUTRITIONAL REQUIREMENTS

The intent of the recommended dietary allowances (RDAs) is to help meal planners provide for the nutritional needs of practically all healthy people.

■ Two of the great scientific triumphs of the nineteenth century were the germ theory of disease and the birth of the science of nutrition.

Nutrition can be defined in both technical and personal terms. Technically, nutrition is a field of science that investigates the identities, the quantities, and the sources of substances, called *nutrients,* that are needed for health.

In personal terms, we speak of our own nutrition, the sum of the foods taken in the proper proportions at the best moments as we try to maintain a state of well-being and avoid diet-related diseases and infirmities.

The science of **dietetics** is the application of the findings of the science of nutrition to feeding individual humans, whether they are ill or well.

This chapter is mostly about some of the major findings of the science of nutrition, a field that is so huge and complex that we can do little more than introduce some of its most important features and terms.

TABLE 29.1

Recommended Daily Dietary Allowances[a] of the Food and Nutrition Board, National Academy of Sciences National Research Council, Revised 1989

									Fat-Soluble Vitamins			
	Age	Weight		Height			Protein	A	D	E	K
Persons	(years)	(kg)	(lb)	(cm)	(in.)		(g)	(μg)[b]	(μg)[c]	(mg)[d]	(μg)
Infants	0.0–0.5	6	13	60	24		13	375	7.5	3	5
	0.5–1.0	9	20	71	28		14	375	10	4	10
Children	1–3	13	29	90	35		16	400	10	6	15
	4–6	20	44	112	44		24	500	10	7	20
	7–10	28	62	132	52		28	700	10	7	30
Males	11–14	45	99	157	62		45	1000	10	10	45
	15–18	66	145	176	69		59	1000	10	10	65
	19–24	72	160	177	70		58	1000	10	10	70
	25–50	79	174	176	70		63	1000	5	10	80
	51+	77	170	173	68		63	1000	5	10	80
Females	11–14	46	101	157	62		46	800	10	8	45
	15–18	55	120	163	64		44	800	10	8	55
	19–24	58	128	164	65		46	800	10	8	65
	25–50	63	138	163	64		50	800	5	8	65
	51+	65	143	160	63		50	800	5	8	65
Pregnant							60	800	10	10	65
Lactating	First 6 months						65	1300	10	12	65
	Second 6 months						62	1200	10	11	65

[a] The allowances are intended to provide for individual variations among most normal persons as they live in the United States under usual environmental stresses. Diets should be based on a variety of common foods in order to provide other nutrients for which human requirements have been less well defined.

[b] Retinol equivalents. 1 retinol equivalent = 1 μg retinol or 6 μg β-carotene.

Nutrients **Are Chemicals Required by Healthy Metabolism, and** *Foods* **Are Substances That Supply Them** When James Lind discovered in 1852 that oranges, lemons, and limes could cure scurvy, and K. Takadi in the 1880s found that a proper diet could cure beriberi, the science of nutrition was on its way. Good personal nutrition prevents a number of specific diseases and it cures several. It provides the correct nutrients for whatever happens to be the metabolic or health status of the body.

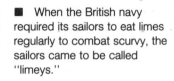 When the British navy required its sailors to eat limes regularly to combat scurvy, the sailors came to be called "limeys."

The **nutrients** are any chemical substances that take part in any nourishing, health-supporting, or health-promoting metabolic activity. Carbohydrates, lipids, proteins, vitamins, minerals, and trace elements are nutrients. Oxygen and water are usually not called nutrients, although they formally qualify under the definition. Materials that supply one or more nutrients are called **foods.**

The *Recommended Dietary Allowances* **Describe What Nutrients Are Adequate for Healthy People** For several decades, the Food and Nutrition Board of the National Research Council of the National Academy of Sciences has published what are known as the **recommended dietary allowances,** or the **RDAs.** In the judgment of this Board, these allowances are the intake levels of essential nutrients that are "adequate to meet the known nutritional needs of practically all healthy persons." The values of the RDAs, as revised in 1989, are given in Table 29.1.

A number of points and qualifications about the RDAs must be emphasized.

1. *The RDAs are not the same as the U.S. Recommended Daily Allowances (USRDAs).* The USRDAs are set by the U. S. Food and Drug Administration, based on the RDAs, as standards for nutritional information on food labels.

2. *The RDAs are not the same as the Minimum Daily Requirements (the MDRs) for any one individual.*

Water-Soluble Vitamins							Minerals						
Ascorbic acid (mg)	Folate (μg)	Niacin[e] (mg)	Riboflavin (mg)	Thiamine (mg)	Vitamin B_6 (mg)	Vitamin B_{12} (mg)	Calcium (mg)	Phosphorus (mg)	Iodine (μg)	Iron (mg)	Magnesium (mg)	Selenium (μg)	Zinc (mg)
30	25	5	0.4	0.3	0.3	0.3	400	300	40	6	40	10	5
35	35	6	0.5	0.4	0.6	0.5	600	600	50	10	60	15	5
40	50	9	0.7	0.7	1.0	0.7	800	800	70	10	80	20	10
45	75	12	0.9	0.9	1.1	1.0	800	800	90	10	120	20	10
45	100	13	1.0	1.0	1.4	1.4	800	800	120	10	170	30	10
50	150	17	1.5	1.3	1.7	2.0	1200	1200	150	12	270	40	15
60	200	20	1.8	1.5	2.0	2.0	1200	1200	150	12	400	55	15
60	200	19	1.7	1.5	2.0	2.0	1200	1200	150	10	350	70	15
60	200	19	1.7	1.5	2.0	2.0	800	800	150	10	350	70	15
60	200	15	1.4	1.2	2.0	2.0	800	800	150	10	350	70	15
50	150	15	1.3	1.1	1.4	2.0	1200	1200	150	15	280	45	12
60	180	15	1.3	1.1	1.5	2.0	1200	1200	150	15	300	50	12
60	180	15	1.3	1.1	1.6	2.0	1200	1200	150	15	280	55	12
60	180	15	1.3	1.0	1.6	2.0	800	800	150	15	280	55	12
60	180	13	1.2	1.1	1.6	2.0	800	800	150	10	280	55	12
70	400	17	1.6	1.5	2.2	2.2	1200	1200	175	30	320	65	15
95	280	20	1.9	1.6	2.1	2.6	1200	1200	200	15	355	75	19
90	280	20	1.7	1.6	2.1	2.6	1200	1200	200	15	340	75	16

[c] As cholecalciferol. 10 μg cholecalciferol = 400 IU of vitamin D.

[d] α-Tocopherol equivalents. 1 mg D-α-tocopherol = 1 α-tocopherol equivalent.

[e] 1 niacin equivalent (NE) equals 1 mg of niacin or 60 mg of dietary tryptophan.

■ It's the job of hospital dieticians to devise the large variety of diets needed by the patient population.

■ Many scientists urge women planning pregnancies to begin supplementing their diets with certain vitamins, like folate, *before* pregnancies.

The MDRs are just that, minimums. They are set very close to the levels at which actual signs of deficiencies occur. The RDAs are two to six times the MDRs. Just as individuals differ greatly in height, weight, and appearance, they also differ greatly in specific biochemical needs. Therefore an effort has been made to set the RDAs far enough above average requirements so that "practically all healthy people" will thrive. Most will receive more than they need; a few will not receive enough.

One of the controversies over the RDAs concerns the validity of the statistical research and analysis by which "average requirements" were determined. Moreover, some nutritionists insist that among healthy people the range of daily need for a particular nutrient can vary far more widely than the Food and Nutrition Board has determined.

3. *The RDAs do not define therapeutic nutritional needs.*
    People with chronic diseases, such as prolonged infections or metabolic disorders; people who take certain medications on a continuing basis; and prematurely born infants all require special diets. Recently the importance of certain vitamins to pregnant women, particularly in the first few weeks of pregnancy, has been established by controlled studies.

    The RDAs do cover people according to age, sex, and size, and they indicate special needs for pregnant and lactating women, but they do not include any other special needs. Therapeutic needs for water and salt increase during strenuous physical activity and prolonged exposure to high temperatures.

    In some areas of the United States and in many parts of the world, intestinal parasites are common. These organisms rob the affected people of some of their food intake each day, and such people also need special diets.

4. *The RDAs can (and ought to) be provided in the diet from a number of combinations and patterns of food.*
    No single food contains all nutrients. People take dangerous risks with their health when they go on fad diets limited to one particular food, like brown rice, gelatin, yogurt, or liquid protein. The ancient wisdom of a varied diet that includes meat, fruit, vegetables, grains, nuts, pulses (e.g., beans), and dairy products may seem to be supported solely by cultural and aesthetic factors. A varied diet, however, also assures us of getting any trace and needed nutrients that might not yet have been discovered. The U.S. Department of Agriculture uses a "food pyramid" for educating the public about how to organize personal nutrition to obtain all needed nutrients in healthy proportions through a varied diet (see Figure 29.1). Some objections to the food pyramid have been voiced by those who see it as reducing the value of meat in the diet.

FIGURE 29.1
The Food Group Pyramid—a device for educating the public about nutrition. (*Source:* U.S. Department of Agriculture.)

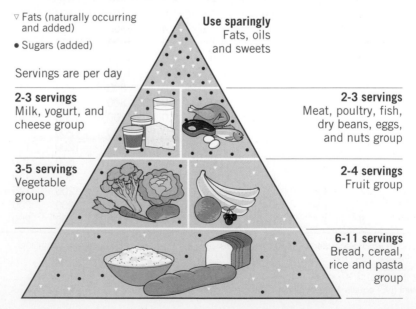

▽ Fats (naturally occurring and added)

● Sugars (added)

Servings are per day

Use sparingly
Fats, oils and sweets

2-3 servings
Milk, yogurt, and cheese group

2-3 servings
Meat, poultry, fish, dry beans, eggs, and nuts group

3-5 servings
Vegetable group

2-4 servings
Fruit group

6-11 servings
Bread, cereal, rice and pasta group

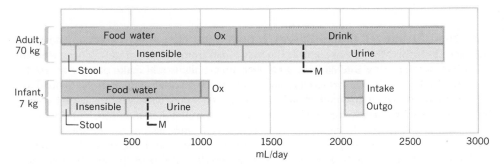

FIGURE 29.2
Water intake and outgo by the principal routes (excluding sensible perspiration or sweating). The dashed lines at **M** are the minimal volumes of urine at the maximal concentrations of solutes. "Ox" refers to water formed by the oxidation of foods. (*Source: Recommended Dietary Allowances,* 9th ed., 1980. National Academy of Sciences, Washington, D.C.)

Adults Have a Water Budget of 2.5 to 3 L per Day   Our water intake comes from the fluids we drink, the water in the foods we eat, and the water made by the oxidations of nutrients in cells. Most comes in response to the thirst mechanism.

Water leaves the body via the urine, the feces, exhaled air, and perspiration. Figure 29.2 gives the relative quantities by each route for adults and infants.

Not shown in Figure 29.2 is the water that can be lost by sensible perspiration (sweat). Any activity in the hot sun of an arid desert can cause the loss of as much as 10 L of water per day along with body salts. If these losses aren't made up at a reasonable rate, the results can be heat exhaustion, heat stroke, heat cramps, and even death.

■ No amount of physical training or willpower can condition anyone to go without water.

Energy Needs Determine Oxygen Requirements   Our daily oxygen needs vary with how much chemical energy we have to import each day to maintain metabolism under different activities. At rest, the largest user of oxygen in the body is the brain. Its mass is only about 2% of the body mass, but it consumes 20% of the oxygen used by the resting body. Much of the energy used in the brain goes to keep up concentration gradients of ions across the membranes of the billions of brain cells. In general, energy demands in the range of 2.0×10^3 to 3.0×10^3 kcal/day means a need for roughly 18 to 25 moles of oxygen per day. Table 29.2 gives data on energy demands for several activities.

TABLE 29.2
Daily Energy Expenditures of Adult Men and Women in Light Occupations

Activity Category	Time (h)	Man (70 kg)		Woman (58 kg)	
		Rate		Rate	
		kcal/min	Total (kcal)	kcal/min	Total (kcal)
Sleeping, reclining	8	1.0–1.2	540	0.9–1.1	440
Very light[a]	12	up to 2.5	1300	up to 2.0	900
Light[b]	3	2.5–4.9	600	2.0–3.9	450
Moderate[c]	1	5.0–7.4	300	4.0–5.9	240
Heavy[d]	0	7.5–12.0	——	6.0–10.0	——
Total	24		2740		2030

Source: J. V. G. A. Durin and R. Passmore, 1967. In *Recommended Dietary Allowances,* 9th ed., 1980. National Academy of Sciences, Washington, D.C.

[a] Seated and standing activities; driving cars and trucks; secretarial work; laboratory work; sewing; ironing, playing musical instruments.

[b] Walking (on the level, 2.5 to 3 mph); tailoring and pressing; carpentry; electrical trades; restaurant work; washing clothes; light recreation such as golf, table tennis, volleyball, sailing.

[c] Walking at 3.5 to 4 mph; garden work; scrubbing floors; shopping with heavy load; moderate sports such as skiing, tennis, dancing, bicycling.

[d] Uphill walking with a load; pick-and-shovel work; lumbering; heavy sports, such as swimming, climbing, football, basketball.

Fad Diets Meant to Control Energy Intake Can Cause Grave Harm The energy obtained from food should not come exclusively either from carbohydrates alone or from fats alone. If you went on a zero-carbohydrate diet, your body would make glucose to meet its needs. (The brain derives nearly all its energy from the breakdown of glucose.) If the body has to make glucose over a long period of time, say weeks, a slow buildup of toxic wastes would cause harm. On the other hand, if you were to go on a zero-fat diet, then some fatty acids that are important to the body would be unavailable, and your system would not absorb the fat-soluble vitamins very well. A daily diet that includes at least 15 to 25 g of food fat (the equivalent of 2 to 4 pats of margarine or butter) and 50 to 100 g of digestible carbohydrate prevents these problems.

■ High-protein diets, if eaten for an extended period, can make the kidneys overwork and become damaged.

Linoleic Acid Must Be in the Diet Linoleic acid is called an **essential fatty acid,** and it is particularly needed for the health and growth of infants.[1] This acid is so poorly supplied by the animal fats in the diet, including milk fat, that we rely almost totally on vegetable oils for it. Our bodies use arachidonic acid to make prostaglandins (Special Topic 20.1, page 568), and this acid is made in the body from linoleic acid. Animals on a diet free of linoleic acid show very poor growth, impaired healing of wounds, and skin problems, and they have shortened lives. Thanks to the widespread presence of linoleic acid in all edible vegetable oils, it is virtually impossible for an adult not to obtain enough. Dietary linoleic acid lowers the serum level of the *low-density lipoprotein complex* or LDL, the "bad cholesterol," which gives another reason for adults to have vegetable oils in the diet. (See Special Topic 27.1, page 750.)

■ *Hydrogenated* vegetable oils are, like animal fats, deficient in the important fatty acids because hydrogenation removes the carbon–carbon double bonds of linoleic, linolenic, and arachidonic acids.

29.2

PROTEIN REQUIREMENTS

A protein's biological value is determined largely by its digestibility and its ability to supply the essential amino acids.

The total protein requirement is 56 g/day for adult men and 46 g/day for adult women. Small as these numbers are, they are still more than the bare minimum. Their intent is to allow about 30% extra to cover a wide range of variations in the protein needs of individuals. Moreover, they assume that we digest and absorb into the bloodstream only 75% of the protein in the diet. (These allowances are considered too narrow by some nutritionists.)

■ This is only about 2 ounces per day.

Our Protein Requirement Actually Means a Requirement for Essential Amino Acids Proteins supply α-amino acids, and these are what protein requirements are all about. We particularly need daily intakes of a small number of specific amino acids whose presence in the diet is so critical that they are called the **essential amino acids.**

Of the 20 amino acids used to make proteins, adults can synthesize 12 from parts and pieces of other molecules, including other amino acids (as discussed in Section 28.1). This leaves eight essential amino acids, those that the body cannot synthesize, at least not at rates rapid enough to make much difference. They are listed in the margin.

Our total daily needs for the essential amino acids add up only to about 13 g, less than half an ounce. When a cell is actually making a protein, however, all of the amino acids needed for it must be available *at the same time,* even if only a trace of any one is required. The absence of lysine, for example, when a protein that includes it has to be made, would prevent the synthesis of the protein altogether.

■

Essential Amino Acids[a]	Daily Needs (g/day)
Isoleucine	1.4
Leucine	2.2
Lysine	1.6
Methionine	2.2
Phenylalanine	2.2
Threonine	1.0
Tryptophan	0.5
Valine	1.6

[a] Histidine is believed to be essential to infants. Data from F. E. Deatherage, *Food for Life,* Plenum Press, New York, 1975.

---

[1] All sources agree that linoleic acid is an *essential* fatty acid. Some add arachidonic acid to the list, but the body is able to make arachidonic acid from linoleic acid. Other lists include linolenic acid as an essential fatty acid, but the body can also make this acid from linoleic acid. No doubt in feeding studies, marked improvements in essential fatty acid deficiency occurs with giving any of the three—linoleic, linolenic, and arachidonic acids. It is arachidonic acid that is finally crucial internally, because many prostaglandins are made from it, and it is possible that infants (but not adults) lack the ability to make arachidonic acid from linoleic acid.

The Availability of the Amino Acids in Food Proteins Varies with the Digestibility of the Protein Not all proteins are easily and completely digested. The variations are expressed by a **coefficient of digestibility,** defined by Equation 29.1.

$$\text{Coefficient of digestibility} = \frac{(\text{N in food eaten}) - (\text{N in feces})}{(\text{N in food eaten})} \qquad (29.1)$$

The difference of terms in the numerator:

$$(\text{N in food eaten}) - (\text{N in feces})$$

is *the nitrogen that is actually absorbed by the bloodstream.*

Animal proteins have higher coefficients of digestibility than those of plants or fruits. For the average animal protein, this coefficient is 0.97, which means 97% digestible. The coefficients for the proteins in fruits and fruit juices average about 0.85; for whole wheat flour, 0.79; and for whole rye flour, 0.67. Milling improves the digestibility coefficients, raising them to 0.83 for all-purpose bread flour and 0.89 for cake flour.

Milling does the same to rice. The digestibility coefficient for brown rice is 0.75 but 0.84 for polished white rice. Milling, unfortunately, also reduces the quantities of vitamins, minerals, and fiber in the grain product. But these are usually put back, as in enriched flours, for example.

The average digestibility coefficient of the proteins in legumes and nuts is 0.78, and in vegetables the range is 0.65 to 0.74.

Proteins Low in Essential Amino Acids Have Low Nutritional Value The digestibility of a protein is just one factor in the quality of a protein source. If a protein is lacking or low in one or more of the essential amino acids, it is a low-quality protein in nutrition.

When the body gets amino acids from a low-quality protein, it simply will not be able to use the amino acids efficiently *to make protein.* When a low-quality protein diet fails to supply one essential amino acid, like lysine, the body still gets a load of other amino acids it can't use to make lysine-containing proteins. As we learned in Chapter 28, there is no way to store these excess amino acids until lysine arrives, and they cannot be excreted as amino acids. To get rid of them, the body has to break them down so their nitrogen can be excreted as urea. This leaves other compounds.

The remaining, nitrogen-free products can be used by the body for energy, as we'll study later. But if the body isn't demanding this energy by its level of activity, then it converts these breakdown products into fat. You can get fat on a high-protein, low-activity diet, and you strain your kidneys besides.

■ On a very-high-protein diet, the formation and removal of urea can place a strain on the liver and the kidneys.

One Measure of Protein Quality Is Called the Protein's *Biological Value* Measured when the body operates under the stress of receiving not quite enough protein overall, the percent of the nitrogen retained out of the total nitrogen eaten is called the **biological value** of the protein. It is a measure of the *efficiency* with which the body uses the nitrogen of *the actually absorbed amino acids* furnished by the protein.

The largest single factor in a protein's biological value is its amino acid composition. What most limits biological value is the extent to which the protein supplies the essential amino acids in sufficient quantities for human use. Human milk protein is the best of all proteins in terms of digestibility and biological value. Whole-egg protein is very close, and it is often taken as the reference for experimental work.

When the diet is well balanced in absorbable amino acids, the amount of nitrogen ingested equals the amount excreted. This condition is called **nitrogen balance.** Most 70-kg men could be in nitrogen balance by ingesting 35 g/day of the proteins in human milk, the value used as a reference standard in rating other proteins.

■ Infants and schoolchildren must have a positive nitrogen balance, meaning that they must ingest more nitrogen than they excrete, in order to grow.

The Essential Amino Acids Most Poorly Supplied Put an Upper Limit on a Protein's Biological Value Table 29.3 summarizes some information about most of the proteins, named in the first column, that are prominent in various diets of the world's peoples. We now study what the data in the other columns mean.

TABLE 29.3
Comparisons of Food Proteins with Human Milk Protein

Food	Limiting Amino Acid	Food's Protein Equivalent to 35 g Human Milk Protein (g)	Digestibility Coefficient of Food's Protein[a]	Amount of Food's Protein[b] (g)	Percentage of Protein in the Food	Amount of Food Needed[c]	Kilocalories of Food Received
Wheat	Lysine	80.6	79.0	102	13.3%	767 g (1.7 lb)	2560
Corn	Tryptophan and lysine	72.4	60.0	120	7.8%	1540 g (3.4 lb)	5660
Rice	Lysine	51.7	75.0	68.9	7.5%	919 g (2.0 lb)	3310
Beans	Valine	50.5	78.0	64.8	24.0%	270 g (0.59 lb)	913
Soybeans	Methionine and cysteine	43.8	78.0	56.2	34.0%	165 g (0.36 lb)	665
Potatoes	Leucine	71.6	74.0	96.7	2.1%	4600 g (10.1 lb)	3500
Cassava	Methionine and cysteine	82.4	60.0	137	1.1%	12,500 g (27.5 lb)	16,400
Eggs	Leucine	36.6	97.0	37.8	12.8%	295 g (0.65 lb)	477
Meat	Tryptophan	43.1	97.0	44.4	21.5%	206 g (0.45 lb)	295
Cow's milk	Methionine and cysteine	43.8	97.0	45.2	3.2%	1410 g (3.1 lb)	903

Source: Data from F. E. Deatherage, *Food for Life,* Plenum Press, New York, 1975. Used by permission.

[a] Expressed as percentages.

[b] The grams of protein that have to be obtained from each food source in order that the protein be nutritionally equivalent (with respect to essential amino acids) to 35 g of human milk protein, allowing for the poorer digestibility of that food's protein (i.e., its digestibility coefficient).

[c] The grams of each food that are equivalent in nutritional value (with respect to essential amino acids) to 35 g human milk protein allowing for the digestibility coefficient and the percent protein in the food.

Column 2   The essential amino acid that is most poorly supplied by a given protein is called its **limiting amino acid.** These are named in this column of Table 29.3.

Column 3   This column gives the number of grams of each food that a 70-kg man would have to digest *and absorb* per day to get the same amount of its limiting amino acid that is available from 35 g of human milk protein. Such an intake, of course, would also supply all the other needed but nonlimiting amino acids. For example, 80.6 g of wheat protein—not wheat, but wheat protein—would have to be digested and absorbed to obtain the lysine in 35 g of human milk protein. But this figure, 80.6 g, assumes 100% digestion, so we have to adjust for a lower percentage digestion.

Column 4   This column gives the digestibility coefficients of all the proteins listed. Wheat protein has a digestibility coefficient of 0.790 so, by a factor of 100/79.0, we need more wheat protein than 80.6 g in order to get the necessary lysine.

Column 5   Here is the result of multiplying 80.6 by (100/79.0). We need 102 g of wheat protein to get the necessary limiting amino acid, lysine. But remember, we're talking about wheat *protein,* not actual wheat, so we move on to the next column.

Column 6   Wheat, on the average, is only 13.3% protein, so we have to multiply our 102 g of wheat protein by the factor 100/13.3. The grim result, 767 g, is in the next column.

Column 7   A 70-kg man would have to eat 767 g (1.69 lb) of wheat if his daily needs for all amino acids are to be met by wheat alone. But all this wheat also has calories. The implication of a daily diet of 767 g of wheat is in the last column.

Column 8   Anyone eating 767 g of wheat also gets 2560 kcal of food energy. This is nearly as much total energy per day as a 70-kg adult should have, so that isn't prohibitive, but can you imagine a diet this boring? Just imagine his problems if he had to get all his amino acids from potatoes, 10.1 lb/day, or cassava, 27.5 lb/day!

The data in Table 29.3 are important to those concerned about both the protein and total calorie needs of the world's burgeoning population. Neither children nor adults can eat enough corn, rice, potatoes, or cassava per day to meet both their protein and energy needs.

Proteins That Provide a Balanced Supply of Essential Amino Acids Are Called Adequate Proteins   Proteins of eggs and meat, as you can see in Table 29.3, are particularly good. They are said to be **adequate proteins,** because they include all the essential amino acids in suitable proportions to make it possible to satisfy amino acid and total nitrogen needs without excess intake of calories.

Soybeans are the best of the nonanimal sources of proteins. Cassava, a root, is an especially inadequate protein, and corn (or maize) is also poor. Unhappily, huge numbers of the world's peoples, especially those in Africa, Central and South America, India, and the countries of the Middle and Far East, rely heavily on these two foods. Although they adequately provide energy needs, not enough of them can be eaten per day to give the essential amino acids, so there are widespread dietary deficiency diseases in these regions.

Variety in the Diet Offers Many Nutritional Advantages   The data in Table 29.3 also provide scientific support for the long-standing practices in all major cultures of including a wide variety of foods in the daily diet. Meat and eggs can ensure adequate protein while they leave room for an attractive variety that is important to good eating habits. Milk, soybeans, and other beans also leave room for variety.

With Planning and Good Timing, Vegetarians Obtain All Essential Amino Acids
When both rice (low in lysine) and beans (low in valine) are included in equal proportions, about 43 g of a rice–bean combination is equal in protein value to 35 g of human milk protein. Less of this combined diet is needed than rice alone or beans alone, which means that room is left for other foods that our taste buds crave.

The rice and beans should, of course, be eaten fairly closely together to ensure that all essential amino acids are simultaneously available during protein synthesis. Eating just the rice early in the morning and the beans in the evening works against efficient protein synthesis. All of us, including vegetarians, must also include vitamins, studied next.

29.3

VITAMINS

The vitamins essential to health must be in the diet because the body cannot make them.
The term **vitamin** applies to any compound or a closely related group of compounds satisfying the following criteria.

1. It is organic rather than inorganic or an element.
2. It cannot be synthesized at all (or at least in sufficient amounts) by the body and it must be in the diet.
3. Its absence causes a specific **vitamin deficiency disease.**
4. Its presence is essential to normal growth and health.
5. It is present in foods in *small* concentrations, and it is not a carbohydrate, a saponifiable lipid, an amino acid, or a protein.

Vitamins function in the body either as precursors for coenzymes or as coenzymes themselves. Most of the nutritionally required minerals function as cofactors to enzymes.

■ Vitamin = "vital amine," from an early belief that vitamins might all be amines.

Sometimes Any Member of a Set of Compounds Prevents a Vitamin Deficiency Disease

The members of a set of related compounds that prevent a specific vitamin deficiency disease are called *vitamers*. The body can convert any one of them into the active forms needed. Our needs for vitamin A, for example, are satisfied by several structurally related compounds, including a family of plant pigments called the *carotenoids*.

Some Individuals Require Greater Daily Vitamin Intakes because of Genetic Defects

In addition to vitamin deficiency diseases, there are at least 25 disorders classified as *vitamin-responsive inborn errors of metabolism*. These arise from genetic faults that can be partly and sometimes entirely overcome by the daily ingestion of 10 to 1000 times as much of a particular vitamin as normally should be present in the diet.

Genetic faults can compromise the body's use of a vitamin at several places along the vitamin's metabolic path. There might be impaired absorption of the vitamin, or impaired transport in the blood. Many vitamins are needed to provide a prosthetic group for an enzyme, so the genetic error could be a defect in building the vitamin molecule into the enzyme.

■ Dietary surpluses of *water-soluble* vitamins leave the body in the urine.

Vitamins Are Classified as Fat-Soluble or Water-Soluble

As we should now expect, fat-soluble vitamins are largely hydrocarbonlike, and water-soluble vitamins are polar or ionic. In this section we study what the vitamins are, their related deficiency diseases, and their sources.

The Fat-Soluble Vitamins Are A, D, E, and K

The fat-soluble vitamins occur in the fat fractions of living systems. As a group, the fat-soluble vitamins pose dangers when taken in excess. We all have at least some fatty tissue, and "like dissolves like." So when we ingest surpluses of the fat-soluble vitamins, fatty tissue absorbs some of the excess. Such accumulations can be dangerous, as we will see.

Because molecules of the fat-soluble vitamins have alkene double bonds or phenol rings, oxidizing agents readily attack them. These vitamins are destroyed, therefore, by prolonged exposures to air or to organic peroxides found in fats and oils that are turning rancid.

Vitamin A activity is given by several vitamers. In the body, the active forms—retinol, retinal, and retinoic acid—all have five alkene groups. β-Carotene is a source for all three because the liver has enzymes that can cleave its molecules to give the retinol skeleton. This is why β-carotene, the yellow pigment in many plants and available in liver and egg yolk, is an important source of vitamin A activity. It is an example of a *provitamin,* a compound that the body can change into an active vitamin.

■ Notice the alkene groups, which are easily oxidized and so can trap undesirable oxidizing agents, like peroxides.

Retinol, $G = CH_2OH$
Retinal, $G = CH = O$
Retinoic acid, $G = CO_2H$

β-Carotene

The deficiency disease for vitamin A is nyctalopia, or night blindness—impaired vision in dim light. Special Topic 13.2 (page 388) discussed the primary chemical event in vision and how it directly involves retinal. We also need vitamin A for healthy mucous membranes.

Several studies have shown that a high intake of β-carotene in a diet rich in vegetables and fruit is associated with a reduced risk of cancer of the epithelial cells in several locations. (Epithelial cell cancers, like colon cancer and lung cancer, account for over 90% of cancer deaths in the United States.) The ease of oxidation of the multiply unsaturated skeleton of β-carotene might account for this. Two kinds of oxidizing agents, excited (activated) oxygen and free radicals, have been implicated in the onset of cancer, and β-carotene might serve to trap and destroy them.

■ Epithelial cells make up the tissue that lines tubes and cavities.

As Special Topic 12.2 (page 372) described, free radicals are species having at least one unpaired electron. The hydroxy free radical is only one example and is represented as $HO\cdot$, where the dot signifies one unpaired electron. Free radicals lack outer octets and so are reactive and attack almost any organic substance with double bonds, including such crucial materials as genes, hormones, enzymes, and vitamins. Free radicals also attack the cholesterol present in low-density lipoprotein complexes, LDL. Partially oxidized cholesterol attracts white blood cells, which can lead to a build up of a plaque in a blood capillary more rapidly than usual. Thus anything that suppresses free radicals helps to provide protection against heart disease.

■ The cholesterol molecule has one double bond (page 576).

Carefully controlled studies in both Africa and India have shown that vitamin A supplements among peoples with chronically low availability of vitamin A significantly reduce infant mortality and morbidity. When given to children with severe measles, the mortality rate dropped appreciably.

■ *Morbidity* is the ratio of diseased to healthy individuals in a population.

Excess doses of vitamin A must not be taken because it is toxic to adults. The livers of polar bears and seals are particularly rich in vitamin A, and Eskimos are careful about eating these otherwise desirable foods. Siberian huskies also have very high levels of vitamin A in their livers, and early explorers, when driven to eating their sled dogs to survive, risked serious harm and death by eating too much husky liver. In early 1913 the Antarctic explorer X. Mertz almost certainly died in just this way, and his companion, Douglas Mawson, barely survived.

■ Those suffering from excess vitamin A recover quite quickly when they stop taking vitamin A.

Some and maybe all forms of vitamin A, when taken in excess, are also known to be teratogenic (they cause birth defects), so it is not recommended that vitamin supplements with extra vitamin A be taken by pregnant women.

Vitamin D also exists in a number of forms. Cholecalciferol (D_3) and ergocalciferol (D_2) are two that occur naturally.

Ergocalciferol (D_2)   Cholecalciferol (D_3)

These two are equally useful in humans, and either can be changed in the body to the slightly oxidized forms that the body uses as hormones to stimulate the absorption and uses of calcium ions and phosphate ions. Vitamin D is especially important during the years of early growth when bones and teeth, which need calcium and phosphate ions, are developing. Lack of vitamin D causes the deficiency disease known as rickets, a bone disorder.

■ Poor management of Ca^{2+} metabolism contributes to a disfiguring bone condition of old age called *osteoporosis*.

Eggs, butter, liver, fatty fish, and fish oils such as cod liver oil are good natural sources of vitamin D precursors. The most common source today is vitamin D fortified milk. We are able to make some vitamin D ourselves from certain steroids that we can manufacture, provided we get enough direct sunlight on the skin. Energy absorbed from sunlight converts these steroids to the vitamin. Youngsters who worked from dawn to dusk in dingy factories or mines during the early years of the industrial revolution were particularly prone to rickets simply because they saw little if any sun.

No advantage is gained by taking large doses of vitamin D, and in sufficient *excess* it is dangerous. Excesses promote an increase in the calcium ion level of the blood, and this damages the kidneys and causes soft tissue to calcify and harden.

Vitamin E needs are satisfied by any member of the tocopherol family. The most active member is α-tocopherol.

■ The benzene ring with the phenolic OH is easily oxidized, which makes the tocopherols effective in trapping unwanted oxidizing agents.

α-Tocopherol

Vitamin E occurs so widely in vegetable oils that it is almost impossible not to obtain enough of it. Vitamin E detoxifies peroxides, compounds which have the general formula R—O—O—H. These can form when oxygen attacks CH_2 groups that are adjacent to alkene double bonds, as in the many —CH_2—CH=CH— units in molecules of the unsaturated fatty acids. Because the acyl units of these acids are present in the phospholipids of which cell membranes are made, vitamin E is needed to protect all membranes. β-Carotene also protects cell membranes. Because β-carotene and vitamin E are both fat-soluble, they are actually able to enter the lipid-rich environment of the membrane.

In the absence of vitamin E, the activities of certain enzymes are reduced, and red blood cells hemolyze more readily. Anemia and edema are reported in infants whose feeding formulas are low in vitamin E.

Because vitamin E detoxifies peroxides it is called an *antioxidant*. Because vitamin E is a fat-soluble vitamin, some of it tends to be carried in the bloodstream within the lipid-rich environment of lipoprotein complexes, particularly the low-density lipoprotein complexes (LDL) described on page 749. Vitamin E molecules with their antioxidant properties are thus strategically positioned to inhibit what has lately come to be recognized as an important part of atherosclerosis: the oxidation of LDL. Vitamin E inhibits LDL oxidation and thus might act to thwart the onset of coronary disease. In two huge studies reported in 1993, both men and women who regularly took vitamin E supplements for more than two years experienced a significantly reduced risk of coronary disease. Whether the long-term intake (many years, up to a lifetime) of vitamin E supplements will cause toxic results is not known. Until this question is answered, specialists in coronary disease counsel caution in taking large doses of vitamin E.

■ Much of the vitamin K that we need is made for us by our own intestinal bacteria.

Vitamin K is the antihemorrhagic vitamin. It is present in green, leafy vegetables, and deficiencies are rare. It works as a cofactor in the blood-clotting mechanism. Sometimes women about to give birth and their newborn infants are given vitamin K to provide an extra measure of protection against possible hemorrhaging. Vitamin K also aids in the absorption of calcium ion into bone, so it works against the development of osteoporosis.

The Water-Soluble Vitamins Are Vitamin C, Choline, Thiamine, Riboflavin, Nicotinamide, Folate, Vitamin B_6, Vitamin B_{12}, Pantothenic Acid, and Biotin   These are the water-soluble vitamins recognized by the Food and Nutrition Board. It is widely believed that they are among the safest substances known, but this generalization is too broad, as we will see. Most water-soluble vitamins are now known to be essential components of cofactors for enzymes (page 615).

■ Collagen is the protein in bone that holds the minerals together, much as steel rods work in reinforced concrete.

Vitamin C, or ascorbic acid, prevents scurvy, a sometimes fatal disease in which collagen is not well made (see page 596). Whether it prevents other diseases is the subject of enormous controversy, speculation, and research. It is an antioxidant, much as is vitamin E. A variety of studies have found that vitamin C is involved in the metabolism of amino acids, in the synthesis of some adrenal hormones, and in the healing of wounds. There are probably millions of people who believe that vitamin C in sufficient dosages — up to several grams per day — acts to prevent or to reduce the severity of the common cold. The vitamin appears to be nontoxic at these high levels.

Vitamin C is also a destroyer of free radicals. Because it is water-soluble, it is able to do this work within a cell's cytosol. Where low levels of vitamin C exist in seminal fluid, high levels of oxidative damage to sperm have been observed, which may be a factor either in reduced fertility or in birth defects.

Vitamin C is present in citrus fruits, potatoes, leafy vegetables, and tomatoes. It is destroyed by extended cooking, heating over steam tables, or prolonged exposure to air or to ions of iron or copper. Even when vitamin C is kept in a refrigerator in well-capped bottles, it slowly deteriorates.

Ascorbic acid
(vitamin C)

$HOCH_2CH_2\overset{+}{N}(CH_3)$

Choline

Choline is needed to make complex lipids, as we discussed in Section 20.3. Acetylcholine, which is made from choline, is one of several substances that carry nerve signals from one nerve cell to another. The body can make choline, but the Food and Nutrition Board calls it a vitamin because there are ten species of higher animals that have dietary requirements for it. We make it too slowly to meet all our daily needs, so having it in the diet provides protection. It occurs widely in meats, egg yolk, cereals, and legumes. No choline deficiency disease has been demonstrated in humans, but in animals the lack of dietary choline leads to fatty livers and to hemorrhagic kidney disease.

■ Acetylcholine is one of several neurotransmitters (page 638).

Thiamine is needed for the breakdown of carbohydrates. Its deficiency disease is beriberi, a disorder of the nervous system. Good sources are lean meats, legumes, and whole (or enriched) grains. It is stable when dry but is destroyed by alkaline conditions or prolonged cooking. Thiamine is not stored, and excesses are excreted in the urine. One's daily thiamine requirement is proportional to the number of calories that are represented by the diet.

■ In rice, most of the thiamine is in the husk, which is lost when raw rice is milled.

Thiamine

Riboflavin

Riboflavin is required by a number of oxidative processes in metabolism. How riboflavin works was discussed in Special Topic 22.1 (page 617). Deficiencies lead to the inflammation and breakdown of tissue around the mouth and nose, as well as the tongue, a scaliness of the skin, and burning, itching eyes. Wound healing is impaired. The best source is milk, but certain meats (e.g., liver, kidney, and heart) also supply it. Cereals are poor sources unless they have been enriched. Little if any riboflavin is stored, and excesses in the diet are excreted. Alkaline substances, prolonged cooking, and irradiation by light destroy this vitamin.

Niacin, meaning both nicotinic acid and nicotinamide, is essential for nearly all biological oxidations. (Special Topic 22.1, page 617, described how niacin is built into NAD^+, $NADP^+$, and their reduced forms.) It's needed by every cell of the body every day. Its deficiency disease

■ Severe niacin deficiency can cause delerium and dementia.

is pellagra, a deterioration of the nervous system and the skin. Pellagra is particularly a problem where corn (or maize) is the major item of the diet. Corn (maize) is low in niacin and the essential amino acid tryptophan, from which we are able to make some of our own niacin. Where the diet is low in tryptophan, niacin must be provided in other foods, such as enriched grains. Prolonged cooking destroys niacin.

■ We can make about 1 mg of niacin for every 60 mg of dietary tryptophan—nearly all we need.

Nicotinic acid
(niacin)

Nicotinamide
(niacinamide)

■ Chronic alcoholism causes folacin deficiency.

Folate is the name used by the Food and Nutrition Board for folic acid and related compounds. Its deficiency disease is megaloblastic anemia. Several drugs, including alcohol, promote folic acid deficiency. The enzymes that use folic acid as a cofactor are largely involved in reactions that transfer one-carbon units, like $-CH_3$ and $-CH=O$. These reactions include the synthesis of nucleic acids and heme. Good sources are fresh, leafy green vegetables, asparagus, liver, and kidney. Folacin is relatively unstable to heat, air, and ultraviolet light, and its activity is often lost in both cooking and food storage.

Folic acid

In a careful study sponsored by the British Medical Research Council, it was found that supplementary folic acid (4 mg/day) prevented the recurrence of neural-tube defects (e.g., spina bifida) in 72% of the women who had a history of delivering babies with such defects. Its use during the first four weeks of pregnancy was particularly important. A large controlled study in Hungary showed that vitamin–mineral supplements, including folic acid, resulted in 50% fewer congenital malformations of all kinds, when taken prior to pregnancy.

Vitamin B_6 activity is supplied by pyridoxine, pyridoxal, or pyridoxamine. All can be changed in the body to the active form, pyridoxal phosphate. The activities of at least 60 enzymes involved in the metabolism of various amino acids depend on pyridoxal phosphate. One deficiency disease is hypochromic microcytic anemia, and disturbances in the central nervous system also occur. The vitamin is present in meat, wheat, yeast, and corn. It is relatively stable to heat, light, and alkali.

Pyridoxal phosphate

Pyridoxine

Pyridoxal

Pyridoxamine

Pyridoxine is widely used as a component of body-building diets and in the treatment of premenstrual syndrome. *Massive doses ("megavitamin doses") at levels of 500 mg/day and higher, however, can severely disable parts of the nervous system and should be avoided.* Pyridoxine is clearly not "among the safest substances known."

Vitamin B_{12}, or cobalamin, is a controlling factor for pernicious anemia. This deficiency disease, however, is very rare, because it is very difficult to design a diet that lacks this vitamin. Enzymes that use vitamin B_{12} are, like folate, involved in the transfers of one-carbon units. The synthesis of DNA, for example, depends on vitamin B_{12}.

TABLE 29.4
Estimated Safe and Adequate Daily Dietary Intakes of Biotin and Pantothenic Acid

Category	Age (years)	Biotin (μg)	Pantothenic acid (mg)
Infants	0–0.5	10	2
	0.5–1	15	3
Children and adolescents	1–3	20	3
	4–6	25	3–4
	7–10	30	4–5
	11+	30–100	4–7
Adults		30–100	4–7

[a] Food and Nutrition Board, National Academy of Sciences National Research Council, Revised 1989. Because less information is available for basing allowances, these are not in Table 29.1.

Animal products such as liver, kidney, and lean meats as well as milk products and eggs are good sources — and virtually the only sources. Thus most people who develop B_{12} deficiency are true vegetarians (or are infants born of true vegetarian mothers). The onset of any symptoms of B_{12} deficiency occurs slowly because the body stores it fairly well and because such minute traces are needed.

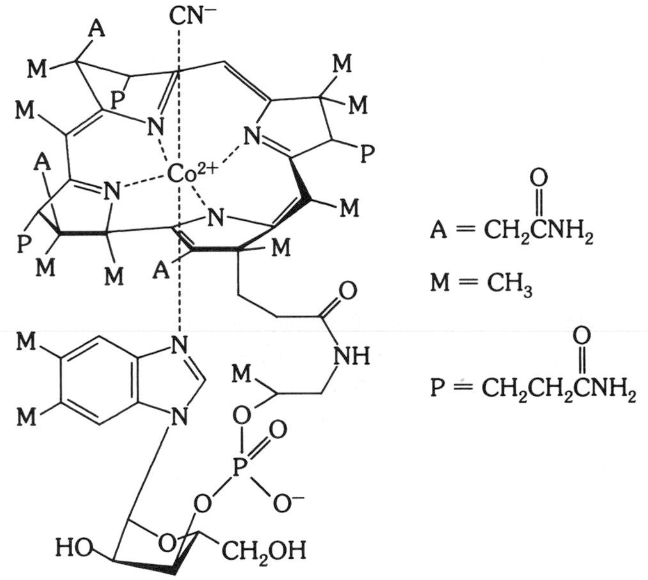

A = $CH_2\overset{O}{\overset{\|}{C}}NH_2$

M = CH_3

P = $CH_2CH_2\overset{O}{\overset{\|}{C}}NH_2$

Cyanocobalamin

■ The replacement of the CN group by other groups (e.g., OH or CH_3 and a few others) gives the members of a small family of compounds that function as cofactors for enzymes.

■ Dorothy Crowfoot Hodgkins, a British chemist, won the 1964 Nobel prize in chemistry for determining the structure of cyanocobalamin.

Pantothenic acid is used to make a coenzyme, coenzyme A (symbol: CoASH). As we learned in Chapter 27, the body needs coenzyme A to metabolize fatty acids. Signs of a deficiency disease for this vitamin have not been observed clinically in humans, but the deliberate administration of compounds that work to lower the availability of pantothenic acid in the body causes symptoms of cellular damage in vital organs. This vitamin is supplied by many foods, and especially by liver, kidney, egg yolks, and skim milk.

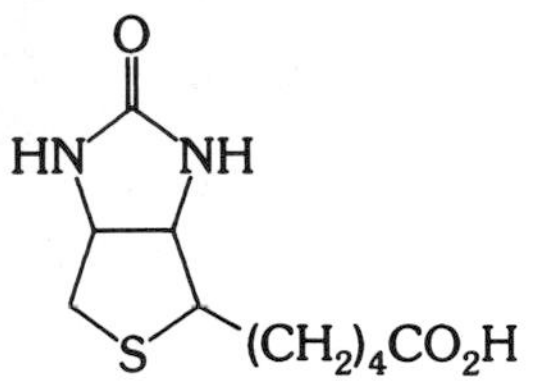

Biotin

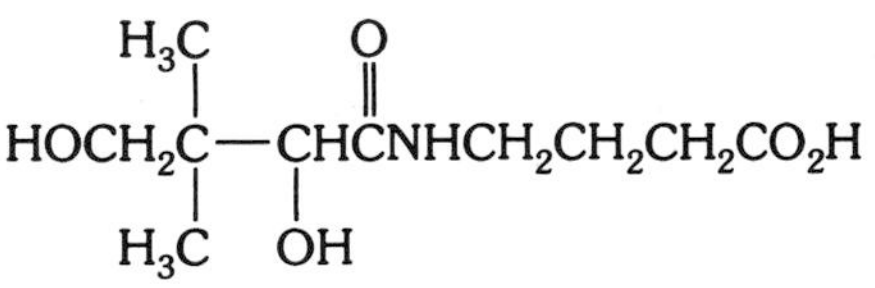

Pantothenic acid

Table 29.4 describes safe and adequate daily dietary intakes of biotin and pantothenic acid.

Biotin is required for all pathways in which carbon dioxide is temporarily used as a reactant, as in the synthesis of fatty acids (Section 27.4). Signs of biotin deficiency are hard to find, but when such a deficiency is deliberately induced, the individual experiences nausea, pallor, dermatitis, anorexia, and depression. When biotin is given again, the symptoms disappear. Our own intestinal microorganisms probably make biotin for us. Egg yolks, liver, tomatoes, and yeast are good sources.

29.4

MINERALS AND TRACE ELEMENTS IN NUTRITION

Many metal ions are necessary for the activities of enzymes.
Most of the minerals and trace elements needed in the diet are metal ions, but some anions are also required. The distinction between a *mineral* and a *trace element* is a matter of quantity.

Minerals Are Needed at Levels of 100 mg/Day or More   The dietary **minerals** are calcium (Ca^{2+}), phosphorus (phosphate ion, P_i), magnesium (Mg^{2+}), sodium (Na^+), potassium (K^+), and chlorine (Cl^-). As we learned in Section 23.2, these are the chief electrolytes in the body. Besides its involvement in bones, the calcium ion is an essential part of the pathways that operate the nervous system and that activate enzymes (Sections 22.6 and 22.7).

■ Selenocysteine is an amino acid with a CH_2SeH side chain in place of the CH_2SH side chain of normal cysteine.

■ 1 ppm $F^- = 1$ mg/L

Trace Elements Are Needed at Levels of 20 mg/Day or Less   The Food and Nutrition Board recognizes 17 **trace elements,** all of which have been found to have various biological functions in animals, and which therefore are quite likely used in the human body. Ten of them are *known* to be needed by humans. All are toxic in excessive amounts. They are fluorine (as F^-) and iodine (as I^-), and the ions of chromium, manganese, iron, cobalt, copper, zinc, and molybdenum; selenium occurs in selenocysteine. All occur widely in food and drink, but some are removed by food refining and processing. For trace elements not listed in Table 29.1, see Table 29.5 for estimated safe and adequate daily dietary intakes.

Fluorine (as fluoride ion) is essential to the growth and development of sound teeth, and the Food and Nutrition Board recommends that public water supplies be fluoridated at a level of about 1 ppm wherever natural fluoride levels are too low. Both the medical and dental associations in the United States strongly support this. Those in some other countries do not, France, West Germany, Denmark, for example. A controversy over the appropriateness of fluoridating drinking water supplies has gone on for nearly half a century.

TABLE 29.5
Estimated Safe and Adequate Daily Dietary Intakes of Selected Minerals[a]

Category	Age (years)	Copper (mg)	Manganese (mg)	Fluoride (mg)	Chromium (μg)	Molybdenum (μg)
Infants	0–0.5	0.4–0.6	0.3–0.6	0.1–0.5	10–40	15–30
	0.5–1	0.6–0.7	0.6–1.0	0.2–1.0	20–60	20–40
Children and	1–3	0.7–1.0	1.0–1.5	0.5–1.5	20–80	25–50
adolescents	4–6	1.0–1.5	1.5–2.0	1.0–2.5	30–120	30–75
	7–10	1.0–2.0	2.0–3.0	1.5–2.5	50–200	50–150
	11+	1.5–2.5	2.0–5.0	1.5–2.5	50–200	75–250
Adults		1.5–3.0	2.0–5.0	1.5–4.0	50–200	75–250

[a] The upper levels in this table should not be habitually exceeded because the toxic levels for many trace elements may be only several times the usual intakes. Data from Food and Nutrition Board, National Academy of Sciences National Research Council, Revised 1989. Because less information is available for basing allowances, these are not in Table 29.1.

Iodine (as iodide ion) is essential in the synthesis of certain hormones made by the thyroid gland (see page 332). Diets deficient in iodide ion cause an enlargement of the thyroid gland known as a goiter. It takes only about 1 μg (1×10^{-6} g) of I$^-$ per kilogram of body weight each day to prevent a goiter.

Seafood is an excellent source of iodide ion, but iodized salt with 75 to 80 μg of iodide ion equivalent per gram of salt is the surest way to obtain the iodine that is needed. Before the days of iodized salt, goiters were quite common, especially in regions where little if any fish or other seafood was in the diet.

Chromium (Cr^{3+}) is required for the work of insulin and normal glucose metabolism. Chromium that occurs naturally in foods is absorbed significantly more easily than chromium given as the simple salts added to vitamin–mineral supplements. Most animal proteins and whole grains supply this trace element. The daily intake should be 0.05 to 0.2 mg.

Manganese (Mn^{2+}) is required for normal nerve function, for the development of sound bones, and for reproduction. It is essential to the activities of certain enzymes in the metabolism of carbohydrates. Nuts, whole grains, fruits, and vegetables supply this element, but a recommended daily allowance has yet to be set by the Food and Nutrition Board. Some nutritionists recommend a daily intake of 2.5 to 5.0 mg/day.

Iron (Fe^{2+}) is an essential cofactor for heme, certain cytochromes, and the iron–sulfur proteins (Section 25.3). The intestines regulate how much dietary iron is absorbed, so a proper level of iron in circulation is maintained in this way. A high serum iron level apparently renders an individual more susceptible to infections. Pregnant women must have larger than usual amounts of iron to keep pace with the needs of fetal blood.

Cobalt (Co^{2+}) is part of the vitamin B$_{12}$ molecule (cyanocobalamin, page 795). Apparently there is no other use for it in humans. But without it there can be anemia and growth retardation.

Copper (Cu^{2+}) occurs in a number of proteins and enzymes, including certain cytochromes (Section 25.3). If deficient in copper, the individual synthesizes lower-strength collagen and elastin and will tend to suffer anemia, skeletal defects, and degeneration of the myelin sheaths of nerve cells. Ruptures and aneurysms of the aorta become more likely. The structure of hair is affected, and reproduction tends to fail.

Fortunately, copper occurs widely in foods, particularly in nuts, raisins, liver, kidney, certain shellfish, and legumes. An intake of just 2 mg/day assures a copper balance for nearly all people. Unfortunately, however, the copper contents of the foods in a typical American diet have declined over the last four decades. Some scientists believe that the increase in the incidence of heart disease in the United States during this period has been caused partly by diets low in copper. In one study that involved women, the lower the copper level in the blood the higher were their blood cholesterol levels. (We discussed cholesterol and heart disease in Chapter 27.)

Zinc (Zn^{2+}) is required for the activities of several enzymes. Without sufficient zinc in the diet, an individual will experience loss of appetite and poor wound healing. Insufficient zinc in an infant's diet, which is a chronic problem in the Middle East, causes dwarfism and poor development of the gonads. If too much zinc is in the diet, and too little copper is present, the extra zinc acts to inhibit the absorption of copper. The ratio of zinc to copper probably doesn't matter as long as sufficient copper is available. When enough zinc is present, it acts to inhibit the absorption of a rather toxic pollutant, cadmium (as Cd^{2+}), which is just below zinc in the periodic table.

Selenium is known to be essential in many animals, including humans. It is needed for thyroid hormone action. How much we need is not known, but probably we need 0.05 to 0.2 mg/day. *Too much is very toxic.*

A strong statistical correlation exists between high levels of selenium in livestock crops and low incidences of human deaths by heart disease. In the United States, those who live where selenium levels are high — the Great Plains between the Mississippi River and the eastern Rocky Mountains — have one-third the chance of dying from heart attack and strokes as those who live where levels are very low — the northeastern quarter of the United States, Florida, and the Pacific Northwest. Rats, lambs, and piglets on selenium-poor diets develop damage to heart tissue and abnormal electrocardiograms.

■ High-fiber diets can work against the absorption of some of the trace elements.

Selenium reduces the occurrence of certain cancers in animals, and it may possibly provide a similar benefit in humans; at excessive levels it causes cancer in experimental animals.

Molybdenum is in an enzyme required for the metabolism of nucleic acids as well as in some enzymes that catalyze oxidations. Deficiencies in humans are unknown, meaning that almost any reasonable diet furnishes enough.

Nickel, silicon, tin, vanadium, and boron are possibly trace elements for humans, because deficiency diseases for these elements have been induced in experimental animals.

SUMMARY

Nutrition Good nutrition entails the ingestion of all the substances needed for health—water, oxygen, food energy, essential amino acids and fatty acids, vitamins, minerals, and trace elements. Foods that best supply various nutrients have been identified. The Food and Nutrition Board of the National Academy of Sciences regularly publishes *recommended daily allowances* that are intended to be amounts that will meet the nutritional needs of practically all healthy people. Some people need more, most need less. But neither more nor less of anything is necessarily better. The RDAs should be obtained by a varied diet because it promotes good eating habits and because such a diet might supply a nutrient no one yet knows is essential.

Water needs The thirst mechanism leads to our chief source of water. Our principal routes of exporting water are the urine, perspiration (both sensible and insensible), and exhaled air. When excessive water losses occur during vigorous exercise, the body also loses electrolytes.

Energy Both carbohydrates and lipids should be in the diet as sources of energy. Without carbohydrates for several days, certain poisons build up in the blood as the body works to make glucose internally from amino acids. A zero-lipid diet means zero ingestion of essential fatty acids. Our oxygen requirements adjust to the caloric demands of our activities.

Protein in the diet What we need most from proteins are certain essential amino acids, and we need some extra nitrogen if we have to make the nonessential amino acids. When a diet excretes as much nitrogen as it takes in, the individual is in nitrogen balance. The most superior, balanced proteins—those that are highly digestible and that supply the essential amino acids in the right proportions—are proteins associated with animals such as the proteins in milk and whole eggs. (Human milk protein is the standard of excellence, and whole-egg protein is very close to it.)

The most important factor in the biological value of a protein is its limiting amino acid—the essential amino acid that it supplies in the lowest quantity. Another factor is the digestibility of the protein. For several reasons—poor source of an essential amino acid, low digestibility coefficients, low concentration of the protein—several foods cannot be used as exclusive or even major components of a healthy diet. Foods that have inadequate proteins include corn (maize), rice, potatoes, and cassava.

Vitamins Organic compounds, called vitamins, or sets of closely related compounds that satisfy the same need, must be in the diet in at least trace amounts or an individual will suffer from a vitamin deficiency disease. Vitamins can't be sufficiently made by the body. (Essential amino acids and essential fatty acids are generally not classified as vitamins, nor are carbohydrates, proteins, or triacylglycerols and other saponifiable lipids.) Excessive amounts of the fat-soluble vitamins (A, D, E, and K)—especially A and D—must be avoided. These vitamins accumulate in fatty tissue, and in excess they can cause serious trouble. Any excesses of the water-soluble vitamins are eliminated.

Minerals and trace elements The minerals are inorganic cations and anions that are needed in the diet in amounts in excess of 100 mg/day. The trace elements are needed at levels of 20 mg/day or less. The whole body quantities of the minerals are large relative to the trace elements. (More about minerals is given in other chapters.) The trace elements that are metal ions are essential to several enzyme systems. Two anions are trace elements, F^- and I^-. Fluoride ion is needed to make strong teeth, and iodide ion is needed to make thyroid hormones and to prevent goiter.

REVIEW EXERCISES

The answers to Review Exercises whose numbers are in color are found in Appendix C. The answers to the other Review Exercises are found in the Study Guide that accompanies this book. The more challenging questions are marked with asterisks.

Nutrition

29.1 What does the science of nutrition study?

29.2 What is meant by the term *nutrient*?

29.3 What is the relationship of nutrients to *foods*?

29.4 What is the relationship of the science of dietetics to nutrition?

29.5 Why are the recommended dietary allowances higher than minimum daily requirements?

29.6 What are seven situations that require special therapeutic diets?

29.7 Why should our diets be drawn from a variety of foods?

29.8 What is potentially dangerous about the following diets?
(a) a carbohydrate-free diet
(b) a lipid-free diet

29.9 All agree that one fatty acid is *essential*. What is its name and what does "essential" mean in this context?

29.10 What are two other fatty acids often cited as "essential" besides the one of Review Exercise 29.9?

29.11 Which fatty acid is needed to make prostaglandins and stands closest to prostaglandins in the synthetic pathway?

Protein and Amino Acid Requirements

29.12 Because the full complement of 20 amino acids is required to make all the body's proteins, why are fewer than half of this number considered essential amino acids?

29.13 If we are able to make all the nonessential amino acids, why should our daily protein intake include more than what is represented solely by the essential amino acids?

29.14 What does the body do with the amino acids that it absorbs from the bloodstream but doesn't use?

29.15 What is the equation that defines the coefficient of digestibility of a protein? What does the numerator in this equation stand for?

29.16 Which kind of protein is generally more fully digested by the body, the protein from an animal or a plant source?

29.17 What can be done to whole grains to improve the digestibility of their proteins? What also happens in this process that reduces the food value of the grains?

29.18 What is the most important factor in determining the biological value of a given protein?

29.19 Which specific protein has the highest coefficient of digestibility and the highest biological value for humans? Which protein comes so close to this on both counts that it is possible to use it as a substitute for research purposes?

29.20 What is meant by the *limiting* amino acid of a protein?

29.21 Why does the protein in corn have a lower biological value than the protein in whole eggs?

****29.22** In one variety of hybrid corn, the limiting amino acids are lysine and tryptophan. It takes 75 g of the protein of this corn to be equivalent to 35 g of human milk protein. The coefficient of digestibility of the protein in this corn is 0.59.
(a) In order to match the nutritional value of human milk protein, how many grams of the protein in this corn must be ingested?
(b) To get this much protein from this variety of corn, how many grams of the corn must be eaten? The corn is only 7.6% protein.

(c) How many kilocalories are also consumed with the quantity of corn calculated in part (b) if the corn has 360 kcal/100 g?
(d) Could a child eat enough corn per day to satisfy all its protein needs and still have room for any other food?

Vitamins

29.23 Make a table that lists each vitamin, at least one good source of each, and a serious consequence of a deficiency of each. (Use only the information available in this book.) Set up the table with the following column heads.

Vitamin	Source(s)	Problem(s) if Deficient

29.24 Why aren't the essential amino acids listed as vitamins?

29.25 On a strict vegetarian diet—no meat, eggs, and dairy products of any sort—which one vitamin is hardest to obtain?

29.26 Why should strict vegetarians use two or more different sources of proteins?

29.27 Which are the fat-soluble vitamins?

29.28 Which vitamin is activated when the skin is exposed to sunlight?

29.29 Which vitamin acts as a hormone in its active forms?

29.30 Which vitamin has been shown to be teratogenic when used in excess?

29.31 The yellow-orange pigment in carrots, β-carotene, can serve as a source of the activity of which vitamin?

29.32 The vitamin needed to participate in the blood-clotting mechanism is which one?

29.33 What is a free radical and why is this species dangerous in cells and their membranes?

29.34 What vitamins provide protection against free radicals within membranes of cells?

29.35 How might the attack of free radicals on the cholesterol in low-density lipoprotein complexes (LDL) accelerate the formation of a plaque in a capillary?

29.36 Name a vitamin that helps to destroy free radicals within the cytosol.

29.37 The Food and Nutrition Board of the National Academy of Sciences recognizes which substances as the water-soluble vitamins?

29.38 Scurvy is prevented by which vitamin?

29.39 Name at least two vitamins that tend to be destroyed by prolonged cooking.

29.40 Which vitamin prevents beriberi?

29.41 When corn (maize) is the chief food in the diet, which vitamin is likely to be in short supply because a raw material for making it is in short supply?

29.42 Which vitamin, when taken during (or before) pregnancies offers protection against neural tube defects?

Minerals and Trace Elements

29.43 What criterion makes the distinction between *minerals* and *trace elements?*

29.44 Name and give the chemical forms of the six minerals.

29.45 Make a table of the ten trace elements and at least one particular function of each.

29.46 What can be the body's response to an iodine-deficient diet?

Additional Exercises

°29.47 If an individual must take in 20.0 mol O_2 in one day to satisfy all needs, how many liters of *air* must be inhaled to supply this? (Air is 21.0% oxygen, on a volume/volume basis, and assume that the measurement is at STP.)

°29.48 The limiting amino acid in peanuts is lysine, and 62 g of protein obtained from peanuts is equivalent to 35 g of human milk protein. The coefficient of digestibility of peanut protein is 0.78. In order to match human milk protein in nutritional value,

(a) How many grams of peanut protein must be eaten?

(b) How many grams of peanuts must be eaten if peanuts are 26.2% protein?

(c) How many kilocalories are also ingested with this many grams of peanuts if peanuts have 282 kcal/100 g?

(d) Could a child eat enough peanuts per day to satisfy its protein needs and still have room for other foods?

MATHEMATICAL CONCEPTS

A.1

EXPONENTIALS

When numbers are either very large or very small, it's often more convenient to express them in what is called *exponential notation*. Several examples are given in Table A.1, which shows how multiples of 10, such as 10,000, and submultiples of 10, such as 0.0001, can be expressed in exponential notation.

Exponential notation expresses a number as the product of two numbers. The first is a digit between 1 and 10, and this is multiplied by the second, 10 raised to some whole-number power or exponent. For example, 55,000,000 is expressed in exponential notation as 5.5×10^7, in which 7 (meaning $+7$) is the exponent.

Exponents can be negative numbers, too. For example, 3.4×10^{-3} is a number with a negative exponent. Now let's learn how to move back and forth between the exponential and the expanded expressions.

Positive Exponents A positive exponent is a number that tells how many times the number standing before the 10 has to be multiplied by 10 to give the same number in its expanded form. For example:

$$5.5 \times 10^7 = \underbrace{5.5 \times 10 \times 10 \times 10 \times 10 \times 10 \times 10 \times 10}_{10^7}$$

$$6 \times 10^3 = 6 \times 10 \times 10 \times 10 = 6000$$
$$8.576 \times 10^2 = 8.576 \times 10 \times 10 = 857.6$$

The number before the 10 doesn't always have to be a number between 1 and 10. This is just a convention, which we sometimes might find useful to ignore. However, we can't ignore the rules of arithmetic in conversions from one form to another. For example,

$$0.00045 \times 10^5 = 0.00045 \times 10 \times 10 \times 10 \times 10 \times 10$$
$$= 45$$
$$87.5 \times 10^3 = 87.5 \times 10 \times 10 \times 10$$
$$= 87,500$$

In most problem-solving situations you find that the problem is given the other way around. You encounter a large number, and (after you've learned the usefulness of exponential notation) you know that the next few minutes of your life could actually be easier if you could quickly restate the number in exponential form. This is very easy to do. Just count the number of places that you have to move the decimal point to the *left* to put it right after the first digit of the given number. For example, you might have to work with a number such as 1500 (as in

TABLE A.1

Number	Exponential Form
1	1×10^0
10	1×10^1
100	1×10^2
1000	1×10^3
10,000	1×10^4
100,000	1×10^5
1,000,000	1×10^6
0.1	1×10^{-1}
0.01	1×10^{-2}
0.001	1×10^{-3}
0.0001	1×10^{-4}
0.00001	1×10^{-5}
0.000001	1×10^{-6}

1500 mL). You'd have to move the decimal point three places leftward from where it is (or is understood to be) to put it immediately after the first digit.

$$1 \underset{3}{\,5} \underset{2}{\,0} \underset{1}{\,0}$$

Each of these leftward moves counts as one unit for the exponent. Three leftward moves means an exponent of $+3$. Therefore 1500 can be rewritten as 1.500×10^3. A really large number and one that you'll certainly meet somewhere during the course is 602,000,000,000,000,000,000,000. It's called Avogadro's number, and you can see that manipulating it would be awkward. (How in the world does one even pronounce it?) In exponential notation, it's written simply as 6.02×10^{23}. Check it out. Do you have to move the decimal point leftward 23 places? (And now you could pronounce it: "six point oh two times ten to the twenty-third," but saying "Avogadro's number" is easier.) Do these exercises for practice.

EXERCISE A.1

Expand each of these exponential numbers.

(a) 5.050×10^6 (b) 0.0000344×10^8 (c) 324.4×10^3

EXERCISE A.2

Write each of these numbers in exponential form.

(a) 422,045   (b) 24,000,000,000,000,000,000   (c) 24.32

Answers to Exercises A.1 and A.2

A.1   (a) 5,050,000   (b) 3,440   (c) 324,400

A.2   (a) 4.22045×10^5 (b) 2.4×10^{19} (c) 2.432×10^1

Negative Exponents   A negative exponent is a number that tells how many times the number standing before the 10 has to be *divided* by 10 to give the number in its expanded form. For example:

$$1 \times 10^{-4} = 1 \div 10 \div 10 \div 10 \div 10$$

$$= \frac{1}{10 \times 10 \times 10 \times 10} = \frac{1}{10000} = \frac{1}{10^4}$$

$$= 0.0001$$

$$6 \times 10^{-3} = 6 \div 10 \div 10 \div 10$$

$$= \frac{6}{10 \times 10 \times 10} = \frac{6}{1000}$$

$$= 0.006$$

$$8.576 \times 10^{-2} = \frac{8.576}{10 \times 10} = \frac{8.576}{100} = 0.08576$$

You'll see negative exponents often when you study aqueous solutions that have very low concentrations.

Sometimes, you'll want to convert a very small number into its equivalent in exponential notation. This is also easy. This time we count *rightward* the number of times that you have to move the decimal point, one digit at a time, to place the decimal immediately to the right of the first nonzero digit in the number. For example, if the number is 0.00045, you have to move the decimal four times to the right to place it after the 4.

$$0.\underset{1}{\underbrace{0}}\,\underset{2}{\underbrace{0}}\,\underset{3}{\underbrace{0}}\,\underset{4}{\underbrace{4}}\,5$$

Therefore we can write $0.00045 = 4.5 \times 10^{-4}$. Similarly, we can write $0.0012 = 1.2 \times 10^{-3}$. And $0.00000000000000011 = 1.1 \times 10^{-16}$. Now try these exercises.

EXERCISE A.3

Write each number in expanded form.

(a) 4.3×10^{-2} (b) 5.6×10^{-10} (c) 0.00034×10^{-2} (d) 4523.34×10^{-4}

EXERCISE A.4

Write the following numbers in exponential forms.

(a) 0.115 (b) 0.00005000041 (c) 0.000000000000345

Answers to Exercises A.3 and A.4

A.3 (a) 0.043 (b) 0.00000000056 (c) 0.0000034 (d) 0.452334
A.4 (a) 1.15×10^{-1} (b) 5.000041×10^{-5} (c) 3.45×10^{-13}

Now that we can write numbers in exponential notation, let's learn how to manipulate them.

How to Add and Subtract Numbers in Exponential Notation We'll not spend too much time on this, because it doesn't come up very often. The only rule is that when you add or subtract exponentials, all of the numbers must have the same exponents of 10. If they don't, we have to reexpress them to achieve this condition. Suppose you want to add 4.41×10^3 and 2.20×10^3. The result is simply 6.61×10^3.

$$(4.41 \times 10^3) + (2.20 \times 10^3) = [4.41 + 2.20] \times 10^3$$
$$= 6.61 \times 10^3$$

However, we can't add 4.41×10^3 to 2.20×10^4 without first making the exponents equal. We can do this in either of the following ways. In one, we notice that $2.20 \times 10^4 = 2.20 \times 10 \times 10^3 = 22.0 \times 10^3$, so we have:

$$(4.41 \times 10^3) + (22.0 \times 10^3) = 26.41 \times 10^3 = 2.641 \times 10^4$$

Alternatively, we could notice that $4.41 \times 10^3 = 4.41 \times 10^{-1} \times 10^4 = 0.441 \times 10^4$, so we can do the addition as follows:

$$(0.441 \times 10^4) + (2.20 \times 10^4) = 2.641 \times 10^4$$

The result is the same both ways. The extension of this to subtraction should be obvious.[1]

How to Multiply Numbers Written in Exponential Form   Use the following two steps to multiply numbers that are expressed in exponential forms.

Step 1.   Multiply the numbers in front of the 10s.

Step 2.   *Add* the exponents of the 10s algebraically.

EXAMPLE A.1

$$(2 \times 10^4) \times (3 \times 10^5) = 2 \times 3 \times 10^{(4+5)}$$
$$= 6 \times 10^9$$

Usually, the problem you want to solve involves very large or very small numbers that aren't yet stated in exponential form. When this happens, convert the given numbers into their exponential forms first, and then carry out the operation. The next example illustrates this and shows how exponentials can make a calculation easier.

EXAMPLE A.2

$$6576 \times 2000 = (6.576 \times 10^3) \times (2 \times 10^3)$$
$$= 13.152 \times 10^6$$
$$= 1.3152 \times 10^7$$

EXERCISE A.5

Calculate the following products after you have converted large or small numbers to exponential forms.

(a) $6{,}000{,}000 \times 0.0000002$ **(b)** $10^6 \times 10^{-7} \times 10^8 \times 10^{-7}$

(c) $0.003 \times 0.002 \times 0.000001$ **(d)** $1{,}500 \times 3{,}000{,}000{,}000{,}000$

Answers

(a)   1.2   **(b)**   1   **(c)** 6×10^{-12} **(d)** 4.5×10^{15}

---

[1] Whenever these operations are with pure numbers and not with physical quantities obtained by measurements, we are not concerned about the numbers of significant figures in the answers.

How to Divide Numbers Written in Exponential Form To divide numbers expressed in exponential forms, use the following two steps.

Step 1. Divide the numbers that stand in front of the 10s.

Step 2. *Subtract* the exponents of the 10s algebraically.

EXAMPLE A.3

$$(8 \times 10^4) \div (2 \times 10^3) = (8 \div 2) \times 10^{(4-3)}$$
$$= 4 \times 10^1$$

EXAMPLE A.4

$$(8 \times 10^4) \div (2 \times 10^{-3}) = (8 \div 2) \times 10^{[4-(-3)]}$$
$$= 4 \times 10^7$$

EXERCISE A.6

Do the following calculations using exponential forms of the numbers.

(a) $6,000,000 \div 1500$

(b) $7460 \div 0.0005$

(c) $\dfrac{3\,000\,000 \times 6\,000\,000\,000}{20\,000}$

(d) $\dfrac{0.016 \times 0.0006}{0.000008}$

(e) $\dfrac{400 \times 500 \times 0.002 \times 500}{2\,500\,000}$

Answers

(a) 4×10^3 **(b)** 1.492×10^7 **(c)** 9×10^{11} **(d)** 1.2 **(e)** 8×10^{-2}

The Pocket Calculator and Exponentials The foregoing was meant to refresh your memory about exponentials, because you almost certainly studied them in any course in algebra or some earlier course. You probably own a good pocket calculator, at least one that can take numbers in exponential form. Go ahead and use it, but be sure that you understand exponentials well, first. Otherwise, there are many pitfalls.

Most pocket calculators have a key marked *EE* or *EXP*. This key is used to enter exponentials. Here is where an ability to *read* exponentials comes in handy. For example, the number 2.1×10^4 reads "two point one times ten to the fourth." The *EE* or *EXP* key on most calculators stands for ". . . times ten to the . . . ." Therefore to enter 2.1×10^4, punch the following keys.

$$\boxed{2}\boxed{.}\boxed{1}\boxed{EE}\boxed{4}$$

Try this on your own calculator, and be sure to see that the display is correct. If it isn't, you may have a calculator that works differently than most, so recheck your operations of entering and then check the owner's manual.

To enter an exponential with a negative exponent, you have to use one more key, the $\boxed{+/-}$ key. This switches a positive number to its negative, and you *must* use it rather than the $\boxed{-}$ key in this situation. Thus the number 2.1×10^{-5} enters as follows.

$$\boxed{2}\boxed{.}\boxed{1}\boxed{EE}\boxed{+/-}\boxed{5}$$

Try it and check the display. To see what happens if you use the $\boxed{-}$ key instead of the $\boxed{+/-}$ key, clear the display and enter this number only using the $\boxed{-}$ key instead of the $\boxed{+/-}$ key.

A.2

CROSS-MULTIPLICATION

In this section we will learn how to solve for x in such expressions as

$$\frac{12}{x} = \frac{16}{25} \quad \text{or} \quad \frac{32.0}{11.2} = \frac{6.15x}{13.1}$$

The operation is called *cross-multiplication,* and its object is to get x to stand alone, all by itself, on one side of the $=$ sign, and above any real or understood divisor line.

> To cross-multiply, move a number or a symbol both across the $=$ sign and across a divisor line, and then multiply.

EXAMPLE A.5

PROBLEM Solve for x in $\dfrac{25 \times 60}{12} = \dfrac{625}{x}$

SOLUTION To get x to stand alone, we carry out the following cross-multiplication.

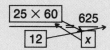

The result is:

$$x = \frac{625 \times 12}{25 \times 60}$$

$$= 5$$

EXERCISE A.7

Solve for x in the following.

(a) $\dfrac{12}{x} = \dfrac{16}{25}$ **(b)** $\dfrac{32.0}{11.2} = \dfrac{6.15x}{13.1}$

Answers

(a) $x = 18.75$ **(b)** $x = 6.085946574$

In Example 2.3 on page 39, the problem was to solve for Δt in the equation:

$$\frac{0.106 \text{ cal}}{\text{g } °\text{C}} = \frac{115 \text{ cal}}{25.4 \times \Delta t}$$

Here is a problem with both units and numbers, so now we have to add one more and very important principle. *We cross-multiply units as well as numbers.*

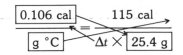

The result is the following, in which the cancel lines show how the units cancel.

$$\Delta t = \frac{(115 \; \cancel{cal}) \times (\cancel{g} \; ^\circ C)}{(25.4 \; \cancel{g}) \times 0.12 \; \cancel{cal}} = 38 \; ^\circ C$$

How to Do Chain Calculations with the Pocket Calculator Sometimes the steps in solving a problem lead to something like the following:

$$x = \frac{24.2 \times 30.2 \times 55.6}{2.30 \times 18.2 \times 4.44}$$

Many people will first calculate the value of the numerator and write it down. Then they'll compute the denominator and write it down. Finally, they'll divide the two results to get the final answer. There's no need to do this much work. All you have to do is enter the first number you see in the numerator, 24.2 in our example. Then use the $\boxed{\times}$ key for any number in the numerator and the $\boxed{\div}$ key for any number in the denominator. *Each number in the denominator is entered with the $\boxed{\div}$ key.* Any of the following sequences work. Try them.

$$24.2 \times 30.2 \times 55.6 \div 2.30 \div 18.2 \div 4.44 = 218.632...$$

Or

$$24.2 \div 2.30 \times 30.2 \div 18.2 \times 55.6 \div 4.44 = 218.632...$$

A.3

LOGARITHMS

The **common logarithm** or the **log** of a number N to the base 10 is the exponent to which 10 must be raised to give N. In other words, when

$$N = 10^x$$

then the log of N is simply x. For example, when

$N = 10$	$\log N = 1$ because	$10 = 10^1$
$= 100$	$= 2$	$100 = 10^2$
$= 1000$	$= 3$	$1000 = 10^3$
$= 0.1$	$= -1$	$0.1 = 10^{-1}$
$= 0.001$	$= -2$	$0.01 = 10^{-2}$
$= 0.0001$	$= -3$	$0.001 = 10^{-3}$

Usually, finding the value of $\log N$ is not this simple. Suppose, for example, that $N = 7.35$ rather than some simple multiple or submultiple of 10. By our definitions, to find the log of 7.35 (usually written log 7.35), we have to find the value of x in

$$7.35 = 10^x$$

The exponent x cannot now be a simple number. To analyze this we have to recognize that if

$$a = b$$

then it must be true that

$$\log a = \log b$$

So if

$$7.35 = 10^x$$

then taking the logarithms of both sides must give us

$$\log 7.35 = \log 10^x$$

But the definition of a log tells us that $\log 10^x = x$, so we can now write

$$\log 7.35 = x$$

There are two ways to determine x; use a table of logarithms or use a pocket calculator. A pocket calculator with a ⬚log key, in fact, has a built in table of logarithms. Just enter the number, 7.35, and hit the ⬚log key, and you will see 0.866287339 come up on the screen. This number has far more digits than we need. Let's round it to 0.866. Thus

$$\log 7.35 = 0.866$$

This result means that

$$7.35 = 10^{0.866}$$

Note that this makes some rough sense. If $10 = 10^1$ and log 10 is therefore 1, the log of a number slightly smaller than 10, like 7.35, should correspond to 10 raised to a power somewhat smaller than 1.

The only situation in this book in which logarithms must be used is in connection with pH or pK_a problems in Sections 9.2 and 9.5, and Henderson-Hasselbalch buffer problems in Section 9.7. For pH problems, Example 9.2 (page 259) gave detailed directions for using the pocket calculator to solve problems that require finding the log of a number. By definition, pH is given as follows.

$$pH = -\log[H^+]$$

If $[H^+] = 3.5 \times 10^{-8}$ mol/L, for example, then the log of this value is (correctly rounded) -7.46. The pH is the negative of this number, so the pH $= 7.46$.

If the pH of a solution is given and the question asks you to find the value of $[H^+]$ in the

solution, we can follow the directions of Example 9.3 (page 259). Now the alternative definition of pH is useful, which was given as Equation 9.4, page 256.

$$[H^+] = 1 \times 10^{-pH}$$

Thus if an aqueous solution has a pH of 4.66, we enter 4.66 and change sign with the ⬚+/− key. We have now entered x for the equation

$$N = 10^x$$

where N is $[H^+]$ and $-x$ is the pH.

We now use the ⬚10^x key to find N. Correctly rounded,

$$N = 2.2 \times 10^{-5}$$

So

$$[H^+] = 2.2 \times 10^{-5} \text{ mol/L}$$

ELECTRON CONFIGURATIONS OF THE ELEMENTS

Atomic Number			Atomic Number			Atomic Number		
1	H	$1s^1$	38	Sr	[Kr] $5s^2$	75	Re	[Xe] $6s^24f^{14}5d^5$
2	He	$1s^2$	39	Y	[Kr] $5s^24d^1$	76	Os	[Xe] $6s^24f^{14}5d^6$
3	Li	[He] $2s^1$	40	Zr	[Kr] $5s^24d^2$	77	Ir	[Xe] $6s^24f^{14}5d^7$
4	Be	[He] $2s^2$	41	Nb	[Kr] $5s^14d^4$	78	Pt	[Xe] $6s^14f^{14}5d^9$
5	B	[He] $2s^22p^1$	42	Mo	[Kr] $5s^14d^5$	79	Au	[Xe] $6s^14f^{14}5d^{10}$
6	C	[He] $2s^22p^2$	43	Tc	[Kr] $5s^24d^5$	80	Hg	[Xe] $6s^24f^{14}5d^{10}$
7	N	[He] $2s^22p^3$	44	Ru	[Kr] $5s^14d^7$	81	Tl	[Xe] $6s^24f^{14}5d^{10}6p^1$
8	O	[He] $2s^22p^4$	45	Rh	[Kr] $5s^14d^8$	82	Pb	[Xe] $6s^24f^{14}5d^{10}6p^2$
9	F	[He] $2s^22p^5$	46	Pd	[Kr] $4d^{10}$	83	Bi	[Xe] $6s^24f^{14}5d^{10}6p^3$
10	Ne	[He] $2s^22p^6$	47	Ag	[Kr] $5s^14d^{10}$	84	Po	[Xe] $6s^24f^{14}5d^{10}6p^4$
11	Na	[Ne] $3s^1$	48	Cd	[Kr] $5s^24d^{10}$	85	At	[Xe] $6s^24f^{14}5d^{10}6p^5$
12	Mg	[Ne] $3s^2$	49	In	[Kr] $5s^24d^{10}5p^1$	86	Rn	[Xe] $6s^24f^{14}5d^{10}6p^6$
13	Al	[Ne] $3s^23p^1$	50	Sn	[Kr] $5s^24d^{10}5p^2$	87	Fr	[Rn] $7s^1$
14	Si	[Ne] $3s^23p^2$	51	Sb	[Kr] $5s^24d^{10}5p^3$	88	Ra	[Rn] $7s^2$
15	P	[Ne] $3s^23p^3$	52	Te	[Kr] $5s^24d^{10}5p^4$	89	Ac	[Rn] $7s^26d^1$
16	S	[Ne] $3s^23p^4$	53	I	[Kr] $5s^24d^{10}5p^5$	90	Th	[Rn] $7s^26d^2$
17	Cl	[Ne] $3s^23p^5$	54	Xe	[Kr] $5s^24d^{10}5p^6$	91	Pa	[Rn] $7s^25f^26d^1$
18	Ar	[Ne] $3s^23p^6$	55	Cs	[Xe] $6s^1$	92	U	[Rn] $7s^25f^36d^1$
19	K	[Ar] $4s^1$	56	Ba	[Xe] $6s^2$	93	Np	[Rn] $7s^25f^46d^1$
20	Ca	[Ar] $4s^2$	57	La	[Xe] $6s^25d^1$	94	Pu	[Rn] $7s^25f^6$
21	Sc	[Ar] $4s^23d^1$	58	Ce	[Xe] $6s^24f^15d^1$	95	Am	[Rn] $7s^25f^7$
22	Ti	[Ar] $4s^23d^2$	59	Pr	[Xe] $6s^24f^3$	96	Cm	[Rn] $7s^25f^76d^1$
23	V	[Ar] $4s^23d^3$	60	Nd	[Xe] $6s^24f^4$	97	Bk	[Rn] $7s^25f^9$
24	Cr	[Ar] $4s^13d^5$	61	Pm	[Xe] $6s^24f^5$	98	Cf	[Rn] $7s^25f^{10}$
25	Mn	[Ar] $4s^23d^5$	62	Sm	[Xe] $6s^24f^6$	99	Es	[Rn] $7s^25f^{11}$
26	Fe	[Ar] $4s^23d^6$	63	Eu	[Xe] $6s^24f^7$	100	Fm	[Rn] $7s^25f^{12}$
27	Co	[Ar] $4s^23d^7$	64	Gd	[Xe] $6s^24f^75d^1$	101	Md	[Rn] $7s^25f^{13}$
28	Ni	[Ar] $4s^23d^8$	65	Tb	[Xe] $6s^24f^9$	102	No	[Rn] $7s^25f^{14}$
29	Cu	[Ar] $4s^13d^{10}$	66	Dy	[Xe] $6s^24f^{10}$	103	Lr	[Rn] $7s^25f^{14}6d^1$
30	Zn	[Ar] $4s^23d^{10}$	67	Ho	[Xe] $6s^24f^{11}$	104	Unq	[Rn] $7s^25f^{14}6d^2$
31	Ga	[Ar] $4s^23d^{10}4p^1$	68	Er	[Xe] $6s^24f^{12}$	105	Unp	[Rn] $7s^25f^{14}6d^3$
32	Ge	[Ar] $4s^23d^{10}4p^2$	69	Tm	[Xe] $6s^24f^{13}$	106	Unh	[Rn] $7s^25f^{14}6d^4$
33	As	[Ar] $4s^23d^{10}4p^3$	70	Yb	[Xe] $6s^24f^{14}$	107	Uns	[Rn] $7s^25f^{14}6d^5$
34	Se	[Ar] $4s^23d^{10}4p^4$	71	Lu	[Xe] $6s^24f^{14}5d^1$	108	Uno	[Rn] $7s^25f^{14}6d^6$
35	Br	[Ar] $4s^23d^{10}4p^5$	72	Hf	[Xe] $6s^24f^{14}5d^2$	109	Une	[Rn] $7s^25f^{14}6d^7$
36	Kr	[Ar] $4s^23d^{10}4p^6$	73	Ta	[Xe] $6s^24f^{14}5d^3$			
37	Rb	[Kr] $5s^1$	74	W	[Xe] $6s^24f^{14}5d^4$			

ANSWERS TO PRACTICE EXERCISES AND SELECTED REVIEW EXERCISES

If you have computed an answer that differs in a small way with the answer given, the difference most likely is caused by the timing of the rounding off of a calculated result after one or more intermediate steps. The answers given here were in almost every instance obtained by chain calculations. When working with atomic masses, remember the rule to round an atomic mass to the first decimal place *before* using it (except round hydrogen's atomic mass to 1.01).

CHAPTER 1

Practice Exercises, Chapter 1

1. 310 K
2. (a) 5.45×10^8 (b) 5.67×10^{12}
   (c) 6.454×10^3 (d) 2.5×10^1
   (e) 3.98×10^{-5} (f) 4.26×10^{-3}
   (g) 1.68×10^{-1} (h) 9.87×10^{-12}
3. (a) 10^{-6} (b) 10^{-9} (c) 10^{-6} (d) 10^3
4. (a) mL   (b) μL   (c) dL   (d) mm
   (e) cm   (f) kg   (g) μg   (h) mg
5. (a) kilogram   (b) centimeter
   (c) deciliter   (d) microgram
   (e) milliliter   (f) milligram
   (g) millimeter   (h) microliter
6. (a) 1.5 Mg   (b) 3.45 μL   (c) 3.6 mg
   (d) 6.2 mL   (e) 1.68 kg   (f) 5.4 dm
7. (a) 275 kg   (b) 62.5 μL
   (c) 82 nm or 0.082 μm
8. (a) 95   (b) 11.36   (c) 0.0263
   (d) 1.3000   (e) 16.1   (f) 3.8×10^2
   (g) 9.31   (h) 9.1×10^2
9. (a) $\dfrac{1 \text{ g}}{1000 \text{ mg}}$ or $\dfrac{1000 \text{ mg}}{1 \text{ g}}$

   (b) $\dfrac{1 \text{ kg}}{2.205 \text{ lb}}$ or $\dfrac{2.205 \text{ lb}}{1 \text{ kg}}$
10. 0.324 g of aspirin
11. (a) 324 mg of aspirin   (b) 3.28×10^4 ft
    (c) 18.5 mL   (d) 17.72 g
    (e) $4.78 \times 10^3 \mu$L

12. 40 °C
13. 59 °F (quite cool)
14. 20.7 mL
15. 32.1 g

Review Exercises, Chapter 1

1.3   At the *molecular* level of life *molecules* are exchanged.
1.7   To construct a testable hypothesis (or a diagnosis)
1.9   The better question is (a). Choice (b) begins with a bias toward the hypothesis being true, not false.
1.12   To construct hypotheses and theories from facts and observations
1.15   The observation of a *chemical* property necessarily converts the substance into a different substance. The observation or measurement of a physical property does not do this.
1.16   A physical *quantity* gives a description of a physical property in terms of a number and a unit.
1.22   A base unit
1.29   (a) meter   (b) inch
      (c) ounce   (d) centimeter
      (e) kilogram   (f) ton
      (g) liter   (h) milliliter
      (i) pound   (j) kilogram
1.40 -40 °F
1.41   41 °F (5 °C) (too cold without a coat or heavy sweater for most people)
1.42   160 °C
1.44   104 °F (The patient has a *fever*.)
1.47   (a) 5.230×10^3 g   (b) 4.50×10^{-2} L
      (c) 1.562×10^3 m   (d) 9.3×10^{-6} g

1.49 (a) 5.230 kg   (b) 4.50 cL
(c) 1.562 km   (d) 9.3 μg

1.53 (a) A, B, D   (b) E, F, G   (c) C, H, I

1.55 (a) 2.00×10^5 (b) 1.999×10^5
(c) 1.9990×10^5 (d) 1.99899×10^5
(e) 1.9989891×10^5 (f) 1.99898909×10^5

1.57 (a) 1.5×10^{-4} (b) 3.5×10^4
(c) 2.50×10^{24} (d) 6.65×10^1
(e) 2.8   (f) 4.5025×10^1
(g) 3.0×10^1 (h) 1×10^{-1}
(i) 2.00

1.59 (a) They are accurate; the average is 59.84. They are close to the true value.
(b) Evidently the numbers could be read to within one unit of the second decimal place, so the uncertainty is small.

1.63 (a) 200 mL   (b) 1.6 km
(c) 0.25 in.   (d) 1 L

1.65 (a) 50.6 kg   (b) 75.6 in.

1.67 10.0 liquid ounces

1.69 No. The vehicle has a mass of 2.0×10^3 kg, too much for the bridge.

1.71 3 tablets

1.73 7.00 mL

1.75 6194.2 m

1.77 0.500 g

1.79 Density = 0.787 g/cm³ (based on a mass of 1.70×10^3 g and a volume of 2.16×10^3 cm³). 24.8 kg (if lead) or 54.7 lb. (Answers can vary slightly depending on the timing of rounding.)

1.82 0.911 g/mL. 2.7×10^2 mL

CHAPTER 2

Practice Exercises, Chapter 2

1. Na_2S

2. 2 atoms potassium to 1 atom carbon to 3 atoms oxygen

3. (a) $\dfrac{1 \text{ cal}}{4.184 \text{ J}}$ and $\dfrac{4.184 \text{ J}}{1 \text{ cal}}$
(b) 333 J or 3.33×10^{-1} kJ

4. The temperature *increase* is 4.53 °C, so the final temperature is 24.5 °C.

5. In terms of *kilocalories* equation 2.3 becomes

$$\text{Heat capacity (in kcal/°C)} = \frac{\text{kcal}}{\Delta t}$$

so, 50 kcal/°C $= \dfrac{1200 \text{ kcal}}{\Delta t}$. Therefore, $\Delta t = 24$ °C.

6. 5.8×10^2 kcal

7. 2.99×10^4 cal or 29.9 kcal

8. 8.86 kcal

Review Exercises, Chapter 2

2.3 The fundamental particles of water (its molecules) are identical in all three states.

2.4 In frozen water, the molecules are much more strongly attracted to each other than they are at a higher temperature, where water is in the liquid state.

2.7 Metal, mostly because it is a conductor, but it is unusual for a metal to be a liquid at room temperature. (The element is mercury.)

2.9 (a) chemical   (b) physical
(c) physical   (d) chemical

2.21 A *mixture* consists of two or more *substances* (elements or compounds) combined physically in no particular ratio by mass.

2.24 Law of definite proportions

2.26 Law of definite proportions

2.28 Their relative masses would be equal.

2.29 Law of multiple proportions

2.31 According to the law of multiple proportions, the answer has to be 15.8731, because only this number, when divided by 7.93655, gives a whole number, 2.

$$\frac{15.8731}{7.93655} = 2.00000$$

whereas

$$\frac{9.45386}{7.93655} = 1.19118 \quad \text{and} \quad \frac{13.0521}{7.93655} = 1.64446$$

2.39 One atom of iron

2.43 (a) $Ca + S \rightarrow CaS$
(b) $2Na + S \rightarrow Na_2S$

2.49 (a) $kg \times \dfrac{m^2}{s^2}$
(b) 6.25×10^5 J
(c) 1.49×10^5 cal or 1.49×10^2 kcal
(d) 35.4 m/s (about 80 mi/hr) (Notice that the velocity does not have to double to double the associated kinetic energy.)

2.50 (a) 67.1 mi/hr
(b) 22.4 times the speed of walking
(c) 31.5 kJ in the vehicle versus 6.29×10^{-2} kJ for walking

2.53 As chemical energy in the substances of the muscle

2.55 (a) From A to B
(b) B is either melting or boiling.

2.60 The specific heat of granite as well as the mass of the sample

2.63 3.00 kcal. 12.6 kJ

2.73 199 °C ($\Delta t = 174$ °C)

2.75 14 °C ($\Delta t = -23$ °C; *minus* because the temperature is lowered)

CHAPTER 3

Practice Exercises, Chapter 3

1. 8.68×10^{22} atoms of gold per ounce

2. (a) 180   (b) 58.3   (c) 858.6

3. 408 g of NH_3

4. 0.0380 mol of aspirin

5. $3O_2 \rightarrow 2O_3$

6. $4Al + 3O_2 \rightarrow 2Al_2O_3$

7. (a) $2Ca + O_2 \rightarrow 2CaO$
   (b) $2KOH + H_2SO_4 \rightarrow 2H_2O + K_2SO_4$
   (c) $Cu(NO_3)_2 + Na_2S \rightarrow CuS + 2NaNO_3$
   (d) $2AgNO_3 + CaCl_2 \rightarrow 2AgCl + Ca(NO_3)_2$
   (e) $2Al + 3H_2SO_4 \rightarrow Al_2(SO_4)_3 + 3H_2$
   (f) $CH_4 + 2O_2 \rightarrow 2H_2O + CO_2$

8. 0.500 mol of H_2O. 500 mmol of H_2O

9. 4.20 mol of N_2 and 4.20 mol of O_2

10. 450 mol of H_2 and 150 mol of N_2

11. 11.5 g of O_2

12. 18.4 g of Na

13. (a) 2.45 g of H_2SO_4 (b) 9.01 g of $C_6H_{12}O_6$

14. 156 mL of 0.800 M Na_2CO_3 solution

15. 9.82 mL of 0.112 M H_2SO_4, when calculated step-by-step with rounding after each step. 9.84 mL of 0.112 M H_2SO_4, when found by a chain calculation.

16. Dilute 20.0 mL of 0.200 M $K_2Cr_2O_7$ to a final volume of 100 mL.

17. Dilute 14 mL of 18 M H_2SO_4 to a final volume of 250 mL.

Review Exercises, Chapter 3

3.5 The ratio by atoms is 2 Al atoms to 3 O atoms; the ratio by moles is 2 mol Al to 3 mol O.

3.10 They exhibit the same chemical reactions.

3.13 (a) Neither
   (b) The isotope with the larger mass
   (c) 23.0 (the average of 24.0 and 22.0 can be taken because of the ratio of 1:1)

3.17 6.02×10^{22} atoms of cobalt

3.20 (a) 36.5   (b) 56.1   (c) 184.1
   (d) 63.0   (e) 84.0   (f) 261.3
   (g) 132.1   (h) 158.2   (i) 180.2

3.23 The reactants of a reaction interact in whole number ratios by *formula units*, but we cannot count these units directly. By taking substances in ratios by *moles*, which can be weighed as an equivalent number of grams, we get them in the same ratios by formula units, so mole calculations let us count particles indirectly.

3.27 (a) 4.56 g HCl   (b) 7.01 g KOH
   (c) 23.0 g $MgBr_2$ (d) 7.88 g HNO_3
   (e) 10.5 g $NaHCO_3$ (f) 32.7 g $Ba(NO_3)_2$
   (g) 16.5 g $(NH_4)_2HPO_4$
   (h) 19.8 g $Ca(C_2H_3O_2)_2$
   (i) 22.5 g $C_6H_{12}O_6$

3.29 (a) 1.37 mol HCl   (b) 0.891 mol KOH
   (c) 0.272 mol $MgBr_2$ (d) 0.794 mol HNO_3
   (e) 0.595 mol $NaHCO_3$ (f) 0.191 mol $Ba(NO_3)_2$
   (g) 0.379 mol $(NH_4)_2HPO_4$
   (h) 0.316 mol $Ca(C_2H_3O_2)_2$
   (i) 0.277 mol $C_6H_{12}O_6$

3.31 2.15×10^{22} molecules N_2

3.33 6×10^{15} molecules O_3

3.36 Two molecules of nitrogen monoxide react with one molecule of oxygen to give two molecules of nitrogen dioxide.

3.38 (a) $N_2 + O_2 \rightarrow 2NO$
   (b) $MgO + 2HNO_3 \rightarrow Mg(NO_3)_2 + H_2O$
   (c) $CaBr_2 + 2AgNO_3 \rightarrow Ca(NO_3)_2 + 2AgBr$
   (d) $2HI + Mg(OH)_2 \rightarrow MgI_2 + 2H_2O$
   (e) $CaCO_3 + 2HBr \rightarrow CaBr_2 + CO_2 + H_2O$

3.40 (a) $CH_4 + 2O_2 \rightarrow CO_2 + 2H_2O$
   (b) $\dfrac{1 \text{ mol } CH_4}{2 \text{ mol } O_2}$ $\dfrac{2 \text{ mol } O_2}{1 \text{ mol } CH_4}$

   $\dfrac{1 \text{ mol } CH_4}{1 \text{ mol } CO_2}$ $\dfrac{1 \text{ mol } CO_2}{1 \text{ mol } CH_4}$

   $\dfrac{1 \text{ mol } CH_4}{2 \text{ mol } H_2O}$ $\dfrac{2 \text{ mol } H_2O}{1 \text{ mol } CH_4}$

3.42 (a) $\dfrac{2 \text{ mol Al}}{3 \text{ mol } H_2SO_4}$ $\dfrac{3 \text{ mol } H_2SO_4}{2 \text{ mol Al}}$
   (b) $\dfrac{2 \text{ mol Al}}{3 \text{ mol } H_2}$ $\dfrac{3 \text{ mol } H_2}{2 \text{ mol Al}}$
   (c) $\dfrac{3 \text{ mol } H_2SO_4}{1 \text{ mol } Al_2(SO_4)_3}$ $\dfrac{1 \text{ mol } Al_2(SO_4)_3}{3 \text{ mol } H_2SO_4}$

3.44 (a) $C_6H_{12}O_6 + 6O_2 \rightarrow 6CO_2 + 6H_2O$
   (b) $\dfrac{1 \text{ mol } C_6H_{12}O_6}{6 \text{ mol } O_2}$ $\dfrac{6 \text{ mol } O_2}{1 \text{ mol } C_6H_{12}O_6}$

3.46 (a) 0.417 mol O_2 (b) 0.278 mol Fe_2O_3

3.48 (a) $2Al_2O_3 \rightarrow 4Al + 3O_2$
   (b) 52.9 g Al   (c) 47.1 g O_2 (d) 100.0 g (Al + O_2), which is identical to the initial mass of Al_2O_3 used. Law of conservation of mass in chemical reactions

3.50 (a) 0.740 mg H_2S (b) 5.38 mg Ag_2S

3.52 32.4 g Na_2CO_3

3.62 (a) 0.0625 mol or 3.66 g NaCl
   (b) 0.0250 mol or 4.50 g $C_6H_{12}O_6$
   (c) 0.0250 mol or 2.45 g H_2SO_4
   (d) 0.0625 mol or 6.63 g Na_2CO_3

3.64 66.7 mL

3.66 400 mL

3.68 0.300 mol Na_2CO_3

3.70 79.4 mL

3.72 22.8 mL

3.74 28.6 mL of 0.150 M Na_2SO_4 solution and 42.9 mL of 0.100 M $Ba(NO_3)_2$ solution

3.76 33.5 mL of 0.500 M Na_2CO_3; 2.22 g of Na_2CO_3

3.78 1.04 L of 0.120 M Na_2CO_3

3.80 3.00 M NH_3

3.82 1.00 mol iodine (I)

3.84 (a) Yes; 0.307 mol HNO_3 was spilled and half this or 0.154 mol of Na_2CO_3 is needed. But 0.189 mol of Na_2CO_3 is provided in the 20.0 g of Na_2CO_3.
   (b) HNO_3 (the reactant that can be all used up)
   (c) 0.035 mol of Na_2CO_3 remained
   (d) 0.154 mol or 6.78 g of CO_2 produced

CHAPTER 4

Practice Exercises, Chapter 4

1. (a) Sn   (b) Cl   (c) Rb   (d) Mg   (e) Ar

2. (a) $1s^2 2s^2 2p^6 3s^2 3p_x{}^1$
(b) $1s^2 2s^2 2p^6 3s^2 3p_x{}^2 3p_y{}^2 3p_z{}^1$
(c) $1s^2 2s^2 2p^6 3s^2 3p_x{}^1 3p_y{}^1$
(d) $1s^2 2s^2 2p^6 3s^2 3p^6 4s^2$

3. (a) 1   (b) 6   (c) 5   (d) 7

Review Exercises, Chapter 4

4.4 (a) KH   (b) CaH_2 (c) GaH_3 (d) GeH_4
(e) AsH_3 (f) SeH_2 (usually written as H_2Se)
(g) BrH (usually written as HBr)

4.8 (a) IA
(b) base
(c) $XOH + HBr \rightarrow XBr + H_2O$
(d) acid–base neutralization

4.12 (a) 1+   (b) 1−
(c) Attract; they have opposite charges.
(d) X, 23; Y, 35

4.14 The protons. Whole numbers of protons, each with a whole-number charge of 1+, must result in a whole number for the nuclear charge.

4.16 Electrons are confined in atoms in certain allowed energy states. As long as electrons remain in their allowed states, the atom neither gives off nor absorbs energy. These two postulates are still true.

4.18 The light emerged only with certain values of energy, not with all possible values.

4.24

Principal Energy Level Number	Number of Sublevels	Number of Orbitals
1	1	1
2	2	4
3	3	9
4	4	16

4.25 (a) $1s$ (b) $2s$ (c) $2p$ (d) $3p$

4.27 (a) It is the volume of space near the atomic nucleus in which the probability of finding the $1s$ electron is very high.
(b) The probability of finding the $1s$ electron at any one place on the surface is identical to the probability of finding it anywhere else on the surface. (The surface is an equiprobability surface.)
(c) Yes, but the probability of finding the $1s$ electron outside the $1s$ sphere becomes increasingly small as the distance outward increases.

4.31 (a) Level 1, 2 electrons. Level 2, 8 electrons. Level 3, 18 electrons.
(b) At both levels 2 and 3, the p orbitals can hold up to six electrons, two in each of the three p orbitals.
(c) At any principal energy level, the s orbital can hold up to two electrons.
(d) The d orbitals can hold a maximum of 10 electrons.
(e) The f orbitals can hold a maximum of 14 electrons.

4.35 (a) 8. oxygen   (b) 13. aluminum

4.37 (a) $1s^2 2s^2 2p^6 3s^2 3p_x{}^1 3p_y{}^1 3p_z{}^1$
(b) $[Ne] 3s^2 3p_x{}^1 3p_y{}^1 3p_z{}^1$

4.39 (a) 30; its number of electrons, 30, equals its number of protons, which is its atomic number.

(b) Yes, it has 18 electrons.
(c) No, all occupied orbitals are *filled* orbitals (each with two electrons), and the Pauli exclusion principle tells us that orbitals can hold two electrons only if their spins are opposite or paired.
(d) $[Ar] 4s^2 3d^{10}$. (Argon has no $3d$ electrons so the $3d^{10}$ electrons must be shown as part of the condensed electron configuration.)
(e) $[Ar]\ 4s^2 3d^{10} 4p_x{}^2 4p_y{}^1 4p_z{}^1$. Atomic number = 34. Atomic symbol = Se.
(f) Three of the four $4p$ electrons spread out among the three $4p$ orbitals.
(g) The fourth $4p$ electron pairs up with another electron in the $4p_x$ orbital, and we assume that the spins of these two electrons are opposite.

4.41 (a) The same element; they have identical numbers of electrons and so identical atomic numbers.
(b) Configuration **1**; its electrons are in the lowest available energy states.
(c) Configuration **2**; configuration **2** has an electron in the $4s$ orbital while having empty $3p$ orbitals.
(d) The atom has to be given energy to change it from **1** to **2**. (The change from **2** to **1** would release the identical amount of energy.)

4.43 (a) $1s^2 2s^2 2p_x{}^1$ or $[He]\ 2s^2 2p_x{}^1$
(b) $1s^2 2s^2 2p_x{}^1 2p_y{}^1 2p_z{}^1$ or $[He]\quad 2s^2 2p_x{}^1 2p_y{}^1 2p_z{}^1$ (or $[He]\ 2s^2 2p^3$)
(c) $1s^2 2s^2 2p^6 3s^2 3p_x{}^1 3p_y{}^1$ or $[Ne]\ 3s^2 3p_x{}^1 3p_y{}^1$
(d) $1s^2 2s^2 2p^6 3s^2 3p_x{}^2 3p_y{}^2 3p_z{}^1$ or $[Ne]\ 3s^2 3p_x{}^2 3p_y{}^2 3p_z{}^1$ (or $[Ne]\ 3s^2 3p^5$)

4.45 (a) IVA. All inner levels, 1–3, are full and the outer level (5) has four electrons, like all elements in group IVA.
(b) $[Kr]\ 5s^2 4d^{10} 5p2$
(c) $[Kr]\ 5s^2 4d^{10} 5p^3$
(d) $[Kr]\ 5s^2 4d^{10} 5p^1$
(e) $[Ar]\ 4s^2 3d^{10} 4p^2$ (Notice that only the numbers of the principal energy levels change in shifting from element 50 to the element immediately above it in the periodic table.)

4.46 (a) 13, 14, 15, 16, 17
(b) 6, 14, 32
(c) 9, 17, 35
(d)

IIIA	IVA	VA	VIA	VIIA
5	6	7	8	9

(e) 2 (The element would be in group IIA, so has two outside shell electrons.)
(f) 6 [The element is in the same family as (d), group VIA, so it has six outside shell electrons, just like (d).]
(g) XH_3 and ZH_3
(h) 9 (It stands high up in a group with more than four outside level electrons.)

CHAPTER 5

Practice Exercises, Chapter 5

1. (a) AgBr   (b) Na_2O (c) Fe_2O_3 (d) $CuCl_2$

2. (a) copper(II) sulfide (cupric sulfide)
(b) sodium fluoride
(c) iron(II) iodide (ferrous iodide)

(d) zinc bromide

(e) copper(I) oxide (cuprous oxide)

3. (a) 3+   (b) 2+   (c) 2+

4. (a) $1s^22s^22p^63s^23p^64s^1$. 1+ charge on the ion
   (b) $1s^22s^22p^63s^23p_x^23p_y^13p_z^1$. 2− charge on the ion
   (c) $1s^22s^22p^63s^23p_x^13p_y^1$. No ion is predicted (or exists).

5. (a) $1s^22s^22p^63s^23p^6$
   (b) $1s^22s^22p^63s^23p_x^23p_y^23p_z^2$
   (c) No ion exists.

6. (a) Cs^+ (b) F^-
   (c) No ion is predicted.   (d) Sr^{2+}

7. (a) Mg is oxidized; S is reduced. Mg is the reducing agent; S is the oxidizing agent.
   (b) Zn is oxidized; Cu^{2+} is reduced. Zn is the reducing agent; Cu^{2+} is the oxidizing agent.

8. Na· ·Mg· ·Äl· ·S̈i· ·P̈· ·S̈· :C̈l· : Är:

9. ·S̈b·

10. ·Ca· + ·Ö· → Ca^{2+} + $\left[:Ö:\right]^{2-}$

11. H—S̈i—H (with H above and H below Si)

12. (a) $KHCO_3$ (b) Na_2HPO_4 (c) $(NH_4)_3PO_4$

13. (a) sodium cyanide
    (b) potassium nitrate
    (c) sodium hydrogen sulfite
    (d) ammonium carbonate
    (e) sodium acetate

14. H O S O H   H O P O H
           O   O
                                H

15. 32

16. H O Cl O   26 valence electrons

    H—Ö—C̈l—Ö: (with :Ö: above and :O: below Cl)

17. :Ö:S̈:Ö:

18. (a) $\left[\begin{array}{c} H \\ H:Ö:H \end{array}\right]^+$ (b) $\left[H:Ö:\right]^-$

19. (a) $\left[:Ö:C:Ö:\right]^{2-}$ (b) $\left[:Ö:N:Ö:\right]^-$

20.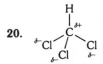

5.14 (a) MgF_2 (b) Li_2O
     (c) CuS (d) $FeCl_3$
     (e) $NaBr$ (f) CaO

5.16 (a) sodium fluoride
     (b) lithium oxide
     (c) copper(II) bromide, cupric bromide
     (d) magnesium chloride
     (e) zinc oxide
     (f) iron(III) bromide, ferric bromide

5.20 Ionic compounds. To be an electrolyte, the compound has to furnish ions, and these preexist in ionic compounds.

5.22

$\left(\begin{array}{c} 20\,p^+ \\ 20\,n \end{array}\right)$ $1s^22s^22p^63s^23p^64s^2$

Calcium atom

+

$2\left(\begin{array}{c} 17\,p^+ \\ 18\,n \end{array}\right)$ $1s^22s^22p^63s^23p_x^23p_y^23p_z^1$

Two chlorine atoms (^{35}Cl isotope)

↓

$\left[\left(\begin{array}{c} 20\,p^+ \\ 20\,n \end{array}\right) 1s^22s^22p^63s^23p^6\right]^{2+}$

Calcium ion

+

$2\left[\left(\begin{array}{c} 17\,p^+ \\ 18\,n \end{array}\right) 1s^22s^22p^63s^23p_x^23p_y^23p_z^2\right]^-$

Two chloride ions

5.24 IIA

5.26 (a) 1+   (b) 2+   (c) 3+
     (d) 1−   (e) 2−

5.28 Yes, 1−

5.30 Yes, 2−

5.32 No. The atom has an outer octet.

5.34 (a) M^+ $1s^22s^22p^63s^23p^6$
     (b) Q^+ $1s^2$
     (c) Z^{2-} $1s^22s^22p^63s^23p^6$

5.37 (a) 3+   (b) 3+   (c) 3+
     (d) 4+   (e) 4+   (f) 5+

5.39 W_2O_5

5.41 (a) Ca   (b) Cl_2 (c) Cl_2 (d) Ca

5.44 An ion and a molecule are both small particles, but the ion is electrically charged and a molecule is electrically neutral. (It is possible for ions to have more than one nucleus; molecules *always* do.)

5.46 Molecular compounds consist of electrically neutral particles, molecules, each capable of independent existence. Ionic compounds consist of oppositely charged ions.

5.50 Helium has atomic number 2. Its electron configuration is therefore $1s^2$, and it already has one of the two conditions of extraordinary stability specified by the octet rule. Its outside level, level 1, can hold no more than two electrons.

5.51 (a) Rb· (b) ·Sr· (c) ·G̈a· (d) ·T̈e·

5.54 In ethylene. There is twice as much electron density between the nuclei, so they are more strongly attracted toward this region and thus closer together.

5.57
(a) K_3PO_4 (b) Na_2CO_3
(c) $CaSO_4$ (d) NH_4CN
(e) $LiNO_2$ (f) $NaHSO_3$
(g) $CaCr_2O_7$ (h) $Mg(C_2H_3O_2)_2$

5.59
(a) sodium carbonate (b) ammonium nitrate
(c) magnesium hydroxide (d) barium sulfate
(e) potassium bicarbonate (f) calcium acetate
(g) sodium nitrite
(h) ammonium phosphate

5.61 (a) 14 (b) 22 (c) 15

5.63 49

5.65 HNO_3. Hydrogen ion, H^+, and nitrate ion, NO_3^-

5.67 Hydrogen ion, H^+, and chloride ion, Cl^-

5.71

$$H \overset{..}{\underset{..}{O}} H + H^+ \longrightarrow \left[H \overset{..}{O} H \atop H \right]^+$$

5.73

(a)
```
        Cl
Cl   Pb   Cl
        Cl
```

(b) F O F

(c)
```
        Cl
Cl   N   Cl
```

(d)
```
        H
H   N   H
```

(e)
```
        O
H   O   S   O
        O
```

(f) H O S O

5.74
(a) 32 (b) 20 (c) 26
(d) 8 (e) 32 (f) 26

5.75

(a)
```
       :C̈l:
        |
:C̈l—Pb—C̈l:
        |
       :C̈l:
```

(b) :F̈—Ö—F̈:

(c)
```
       :C̈l:
        |
:C̈l—N—C̈l:
```

(d)
```
        H
        |
H—N—H
   ..
```

(e)
$$\left[H—\overset{..}{\underset{..}{O}}—\overset{\overset{:\ddot{O}:}{|}}{\underset{\underset{:\ddot{O}:}{|}}{S}}—\overset{..}{\underset{..}{O}}:—H \right]^-$$

(f)
$$\left[H—\overset{..}{\underset{..}{O}}—\overset{\overset{:\ddot{O}:}{|}}{S}—\overset{..}{\underset{..}{O}}:—H \right]^-$$

5.77 (a) NO_2

(b)
```
     :Ö: :Ö:
:Ö:N : N :Ö:
```

5.82 The same as in methane, CH_4, 109.5°

5.84 X has the higher electronegativity *because* it has the larger nuclear charge. (Being in the same period, X and Y have the same inner level electrons that screen their nuclei, so the atom with the higher nuclear charge has a higher *unscreened* ability to draw electrons of a covalent bond toward it.)

5.85 (a) yes (b) $\delta+$ is near X and $\delta-$ is near Y.

5.88 Both kinds of orbitals are regions of space in which electrons can be, and both kinds can hold a maximum of two electrons (and then only if the spins are opposite).

5.91 By the partial overlapping of two $3p_z$ atomic orbitals, each with one electron and one such atomic orbital from each of the two chlorine atoms.

5.94 The bond angle would be 90° if this occurred, but the true bond angle (104°) is closer to the tetrahedral angle (109.5°).

5.97 Each pair is in an sp^3 hybrid orbital of the central atom (N or O).

5.112
(a) $1s^2 2s^2 2p_x^1$
(b) The axes of the three sp^2 orbitals would lie in a plane and make angles of 120° with each other.
(c) no
(d) All atoms lie in the same plane, and all Cl—B—Cl angles are 120°.

CHAPTER 6

Practice Exercises, Chapter 6

1. 2.6 L
2. 547 mL
3. 794 mm Hg
4. 214 mL
5. 98.1 g O_2
6. 159 mm Hg
7. 52 mm Hg
8. $P_{nitrogen} = 732$ mm Hg. 283 mL at STP

Review Exercises, Chapter 6

6.9 The column of air supported by the earth is shorter at the mountain top, so it exerts less weight on a unit of the earth's area.

6.11 The weight doubles; the pressure stays the same because pressure is the *ratio* of weight to area.

6.12 Although the weight of mercury in a larger diameter tube will be greater than that in a tube of smaller diameter, the weight *per unit area,* the pressure, exerted by the mercury stays the same. The *height* of the column of mercury in a Torricelli barometer thus depends solely on how much air pressure is available to hold the mercury up.

6.14 33.9 ft

6.18 0.329 atm. 250 torr. 33.3 kPa

6.20 666 mm Hg

6.26 By $\dfrac{745}{730}$. Boyle's law tells us that if we *reduce* the pressure (at constant temperature), the volume *increases,* so we must use the ratio of pressures that is greater than 1.

6.28 16.1 L

6.30 The gas pressure had to change, to a new value of 799 mm Hg.

6.32 241 mL

6.34 12.0 atm

6.35 740 mm Hg

6.37 257 mL

6.39 668 mm Hg

6.41 24.1 L

6.43 64.9 mol O_2; 2.08 kg O_2

6.45 4.74 kg N_2

6.47 Standard temperature and pressure; 273.15 K (which we round to 273 K) and 1 atm (760 mm Hg)

6.49 30.2 mmol O_2

6.51 (a) 0.0354 mol of the gas   (b) 32.0
(c) oxygen   (d) 0.0708 mol H_2
(e) 0.143 g H_2

6.53 (a) 9.82×10^{-3} mol CO_2
(b) 9.82×10^{-3} mol $CaCO_3$
(c) 0.983 g $CaCO_3$
(d) 0.551 g CaO

6.56 601 mm Hg (This is what *partial* pressure means, the pressure that a component of a gas mixture would exert if it were all alone in the container and all other gases were removed.)

6.58 20 °C (At this temperature, the vapor pressure of water is 17.5 mm Hg, which corresponds to the "missing" mm Hg of pressure when the water vapor is removed from the water-saturated gas mixture.)

6.61 It obeys all of the gas laws exactly.

6.71 Octane; its molecules are larger than those of butane, but both are compounds of carbon and hydrogen and so have roughly the same (very small) permanent polarity.

6.73 (a) Less volatile. With a higher boiling point than water, DMSO will at any given temperature have a lower vapor pressure than that of water. Lower vapor pressure, in this context, means a lower volatility.
(b) A $\delta+$ on S and a $\delta-$ on O. O is more electronegative than S.

(c) Dipole–dipole attraction

6.74 (a) B   (b) B   (c) A
(d) C   (e) A, B   (f) 100 °C
(g) Their rates are equal.   (h) Equilibrium

6.78 (a) Condensation of alcohol vapor to alcohol liquid occurs.
(b) Forward change
(c) Shift to the left
(d) Shift to the right; this supplies vapor to compensate for the loss of vapor and tends to restore equilibrium.
(e) Yes, the equilibrium will shift to the left because this releases heat and so tends to compensate for the heat loss caused by the cooling.

6.80 (a) Forward. $2NO_2 \rightleftharpoons N_2O_4 +$ heat
(b) Forward. The forward reaction is the conversion of 2 mol of gas (NO_2) to 1 mol (N_2O_4). Increasing the pressure favors this volume-reducing effect.

6.84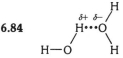

6.86 (a) The boiling point of ammonia is much less than that of water. Because the sizes of ammonia and water molecules are about the same, we cannot explain the difference in boiling points by differences in London forces between the molecules. Instead, there must be differences in the sizes of partial charges that exist more permanently than those of the London forces.
(b) Both $\delta+$ and $\delta-$ are larger in the water molecule than in the ammonia molecule.
(c) Oxygen is more electronegative than nitrogen.

6.96 $CO_2(s) +$ heat $\rightleftharpoons CO_2(g)$

6.100 737 mm Hg. 260 mL

6.102 1.85 g $CaCO_3$; 4.63 mL HCl

CHAPTER 7

Practice Exercises, Chapter 7
1. 1.47 mg N_2/100 g H_2O
2. 10.2 g of 96% H_2SO_4
3. 1.25 g of glucose and 499 g of water (rounded from 498.75, and assuming that the density of water is 1.00 g/mL)
4. 1.67×10^4 mm Hg
5. (a) 0.020 Osm   (b) 0.015 Osm   (c) 0.100 Osm
(d) 0.150 Osm

Review Exercises, Chapter 7
7.2 For measurements of physical properties, the sample of the mixture would have to include enough molecules so that the average composition would be consistent with the experimentally determined ratio of the various isotopes. For studies of chemical properties (of the type we are interested in), the isotope composition is unimportant because isotopes have the same chemical properties.

7.16 Water molecules are strongly attracted to each other, and they find nothing in molecules of CF_4 with which to establish forces of attraction of comparable strength, so the water molecules stay together.

7.18 (a) $KBr(s) \rightarrow K^+(aq) + Br^-(aq)$

(b) The rates at which solid dissolves and at which solid reforms from the ions are equal.

$$KBr(s) \rightleftharpoons K^+(aq) + Br^-(aq)$$

7.20 The dissolving of a solid in a liquid is usually an endothermic change, so heat shifts the following equilibrium to the right, in favor of more solute being in solution.

$$NH_4Cl(s) + heat \rightleftharpoons NH_4^+(aq) + Cl^-(aq)$$

The removal of heat (by lowering the temperature) shifts the equilibrium to the left, and more solid NH_4Cl forms.

7.21 $MgSO_4 \cdot 7H_2O(s) \rightleftharpoons MgSO_4(s) + 7H_2O(g)$

7.23 $Na_2SO_4(s) + 10H_2O \rightarrow Na_2SO_4 \cdot 10H_2O(s)$

7.26 $y = 3$. The formula is $Y \cdot 3H_2O$

7.28 0.035 g/L

7.34 The first region (where the gas tension is 79 mm Hg)

7.39 (a) 2.25 g NaCl
(b) 16.1 g $NaC_2H_3O_2$
(c) 8.44 g NH_4Cl
(d) 3.13 g Na_2CO_3

7.41 (a) 31.3 g NaCl
(b) 7.50 g KBr
(c) 1.12 g $CaCl_2$
(d) 4.50 g NaCl

7.43 100 mL ethyl alcohol

7.45 (a) 75.0 mL 4.00% NaOH solution
(b) 3.25 mL 10% Na_2CO_3 solution
(c) 6.38 mL 4.00% glucose solution
(d) 115 mL 4.00% NaOH solution
(e) 901 mL 4.00% glucose solution

7.47 16.3 g of the trihydrate

7.49 62.5 g 10.0% NaOH solution

7.51 (a) 35.9% (w/w) HCl
(b) 88.1 mL (104 g) of 35.9% (w/w) HCl (88.5 mL by a chain calculation)

7.54 The solute in the second solution breaks up (ionizes) into two ions per formula unit when it dissolves, so the effective concentration is 2.00 mol/1000 g H_2O.

7.60 0.080 M Na_2SO_4 solution, because it is 3×0.080 $M =$ 0.240 M in its osmolarity.

7.62 Solution A. While the osmolarities of A and B are identical with respect to their dissolved salts and sugars, A has the higher concentration of the colloidally dispersed starch.

7.64 457 mm Hg

7.66 (a) hypertonic
(b) (1) hemolysis (2) crenation

7.77 7.12 M H_2SO_4

7.79 0.33 osmol (0.12 osmol NaCl + 0.060 osmol $NaHCO_3$ + 0.040 osmol KCl + 0.11 osmol glucose)

CHAPTER 8

Practice Exercises, Chapter 8

1.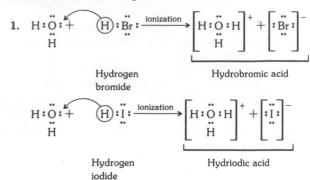

Hydrogen bromide   Hydrobromic acid

Hydrogen iodide   Hydriodic acid

2. $H_2SO_3(aq) \rightleftharpoons H^+(aq) + HSO_3^-(aq)$
$HSO_3^-(aq) \rightleftharpoons H^+(aq) + SO_3^{2-}(aq)$

3. $HNO_3(aq) + KOH(aq) \rightarrow KNO_3(aq) + H_2O$
$H^+(aq) + NO_3^-(aq) + K^+(aq) + OH^-(aq) \rightarrow$
$\qquad\qquad\qquad\qquad K^+(aq) + NO_3^-(aq) + H_2O$
$H^+(aq) + OH^-(aq) \rightarrow H_2O$

4. $Mg(OH)_2(s) + 2HCl(aq) \rightarrow MgCl_2(aq) + 2H_2O$
$Mg(OH)_2(s) + 2H^+(aq) \rightarrow Mg^{2+}(aq) + 2H_2O$

5. K_2SO_4

6. $2NaHCO_3(aq) + H_2SO_4(aq) \rightarrow$
$\qquad\qquad\qquad Na_2SO_4(aq) + 2CO_2(g) + 2H_2O$
$2Na^+(aq) + 2HCO_3^-(aq) + 2H^+(aq) + SO_4^{2-}(aq) \rightarrow$
$\qquad\qquad 2Na^+(aq) + SO_4^{2-}(aq) + 2CO_2(g) + 2H_2O$
$HCO_3^-(aq) + H^+(aq) \rightarrow CO_2(g) + H_2O$

7. $K_2CO_3(aq) + H_2SO_4(aq) \rightarrow K_2SO_4(aq) + CO_2(g) + H_2O$
$2K^+(aq) + CO_3^{2-}(aq) + 2H^+(aq) + SO_4^{2-}(aq) \rightarrow$
$\qquad\qquad 2K^+(aq) + SO_4^{2-}(aq) + CO_2(g) + H_2O$
$CO_3^{2-}(aq) + 2H^+(aq) \rightarrow CO_2(g) + H_2O$

8. $MgCO_3(s) + 2HNO_3(aq) \rightarrow Mg(NO_3)_2(aq) + CO_2(g) + H_2O$
$MgCO_3(s) + 2H^+(aq) + 2NO_3^-(aq) \rightarrow$
$\qquad\qquad Mg^{2+}(aq) + 2NO_3^-(aq) + CO_2(g) + H_2O$
$MgCO_3(s) + 2H^+(aq) \rightarrow Mg^{2+}(aq) + CO_2(g) + H_2O$

9. (a) $NH_3(aq) + HBr(aq) \rightarrow NH_4Br(aq)$
$NH_3(aq) + H^+(aq) \rightarrow NH_4^+(aq)$
(b) $2NH_3(aq) + H_2SO_4(aq) \rightarrow (NH_4)_2SO_4(aq)$
$NH_3(aq) + H^+(aq) \rightarrow NH_4^+(aq)$

10. $Mg(s) + 2HCl(aq) \rightarrow MgCl_2(aq) + H_2(g)$
$Mg(s) + 2H^+(aq) \rightarrow Mg^{2+}(aq) + H_2(g)$

11. (a) HNO_3 (b) HSO_3^- (c) HCO_3^-
(d) HSO_4^- (e) HCl   (f) H_3O^+
(g) H_2O

12. (a) CO_3^{2-} (b) PO_4^{3-} (c) HSO_4^-
(d) SO_4^{2-} (e) Br^- (f) H_2O
(g) OH^-

13. (a) weak acid   (b) weak acid   (c) weak acid

14. (a) weak base   (b) weak base
(c) strong base   (d) strong base

15. $NO_2^-(aq) + H^+(aq) \rightleftharpoons HNO_2(aq)$. The product is favored because we know that HNO_2 is a weak acid (not being on the list of strong acids).

16. $Na_2S(aq) + Cu(NO_3)_2(aq) \rightarrow CuS(s) + 2NaNO_3(aq)$
$S^{2-}(aq) + Cu^{2+}(aq) \rightarrow CuS(s)$

17. The acetate ion combines with (neutralizes) the hydrogen ion. In the following equilibrium, the product is favored.

$$H^+(aq) + C_2H_3O_2^-(aq) \rightleftharpoons HC_2H_3O_2(aq)$$

18. (a) AgCl precipitates. $Ag^+(aq) + Cl^-(aq) \rightarrow AgCl(s)$
    (b) CO_2 evolves. $CaCO_3(s) + 2H^+(aq) \rightarrow Ca^{2+}(aq) + CO_2(g) + H_2O$
    (c) No reaction occurs.

Review Exercises, Chapter 8

8.9 Ethyl alcohol does not furnish any ions of any kind in water.

8.11 It is a molecular compound. It does not consist of ions in the liquid state.

8.12 (a) $Na^+(l) + e^- \rightarrow Na(l)$
     (b) $2Cl^-(l) \rightarrow Cl_2(g) + 2e^-$
     (c) $2Na^+(l) + 2Cl^-(l) \rightarrow 2Na(l) + Cl_2(g)$

8.14 (a) No
     (b) Yes, any polyatomic ion such as SO_4^{2-}
     (c) No
     (d) No

8.16 (a) acidic   (b) neutral   (c) basic

8.27 $HClO_4(aq) + H_2O \rightarrow H_3O^+(aq) + ClO_4^-(aq)$

8.29 Both H's in H_2SO_4 can react with a base, but only one H in $HC_2H_3O_2$ can neutralize a base.

8.31 With greater difficulty. The second H^+ ion has to pull away from a particle that has more negative charge. Unlike charges attract, so the more unlike the charges are, the more energy is needed to separate particles bearing them.

8.34 Sulfuric acid. The extra oxygen atom provides more electron-withdrawing action, and this weakens the H—O bonds more.

8.42 $CO_2(aq) + NaOH(aq) \rightarrow NaHCO_3(aq)$

8.47 (a) $HNO_3(aq) + KOH(aq) \rightarrow KNO_3(aq) + H_2O$
        $H^+(aq) + OH^-(aq) \rightarrow H_2O$
     (b) $HCl(aq) + NaHCO_3(aq) \rightarrow$
                $NaCl(aq) + CO_2(g) + H_2O$
        $H^+(aq) + HCO_3^-(aq) \rightarrow CO_2(g) + H_2O$
     (c) $2HBr(aq) + MgCO_3(s) \rightarrow$
                $MgBr_2(aq) + CO_2(g) + H_2O$
        $2H^+(aq) + MgCO_3(s) \rightarrow Mg^{2+}(aq) + CO_2(g) + H_2O$
     (d) $HNO_3(aq) + KHCO_3(aq) \rightarrow$
                $KNO_3(aq) + CO_2(g) + H_2O$
        $H^+(aq) + HCO_3^-(aq) \rightarrow CO_2(g) + H_2O$
     (e) $HBr(aq) + NH_3(aq) \rightarrow NH_4Br(aq)$
        $H^+(aq) + NH_3(aq) \rightarrow NH_4^+(aq)$
     (f) $2HNO_3(aq) + Ca(OH)_2(s) \rightarrow Ca(NO_3)_2(aq) + 2H_2O$
        $2H^+(aq) + Ca(OH)_2(s) \rightarrow Ca^{2+}(aq) + 2H_2O$
     (g) $2HCl(aq) + Mg(s) \rightarrow MgCl_2(aq) + H_2(g)$
        $2H^+(aq) + Mg(s) \rightarrow Mg^{2+}(aq) + H_2(g)$

8.49 (a) $OH^-(aq) + H^+(aq) \rightarrow H_2O$
     (b) $HCO_3^-(aq) + H^+(aq) \rightarrow CO_2(g) + H_2O$
     (c) $CO_3^{2-}(aq) + 2H^+(aq) \rightarrow CO_2(g) + H_2O$
     (d) $NH_3(aq) + H^+(aq) \rightarrow NH_4^+(aq)$

8.51 $M(OH)_2(s) + 2H^+(aq) \rightarrow M^{2+}(aq) + 2H_2O$

8.53 (a) To be higher in the activity series means, in a qualitative sense, that the atoms of the element have a greater tendency to become ions than do those of elements lower in the series.

(b) The lower ionization energies of sodium and potassium also indicate this greater tendency to become ions.

8.56 0.512 mol of NaOH

8.58 6.34 g $CaCO_3$

8.60 0.989 g K_2CO_3

8.62 40.9 mL KOH solution

8.64 $Na_2CO_3(s) + 2HCl(aq) \rightarrow 2NaCl(aq) + CO_2(g) + H_2O$
     $CO_3^{2-}(aq) + 2H^+(aq) \rightarrow CO_2(g) + H_2O$
     23.9 L of CO_2; 101 g Na_2CO_3

8.67 (a) H_2SO_3 (b) HBr
     (c) H_3O^+ (d) $HC_2H_3O_2$

8.69 (a) NH_2^- (b) NO_2^-
     (c) SO_3^{2-} (d) HSO_3^-

8.71 (a) NH_2^- (b) OH^- (c) S^{2-}

8.73 (a) HCl   (b) H_2O (c) HSO_4^-

8.75 The reactants are favored because they include the weaker acid and the weaker base.

8.77 An aqueous solution of NaOH or KOH, when added to the solution to be tested, releases ammonia, which has a characteristic odor, if the solution in the test tube is ammonium chloride.

$$NH_4^+(aq) + OH^-(aq) \rightarrow NH_3(aq) + H_2O$$

8.79 Potassium hydroxide, KOH
     $KOH(aq) + HCl(aq) \rightarrow KCl(aq) + H_2O$
     Potassium bicarbonate, $KHCO_3$
     $KHCO_3(aq) + HCl(aq) \rightarrow KCl(aq) + CO_2(g) + H_2O$
     Potassium carbonate, K_2CO_3
     $K_2CO_3(aq) + 2HCl(aq) \rightarrow 2KCl(aq) + CO_2(g) + H_2O$

8.81 Compounds c and d are insoluble in water.

8.83 Compounds a and d are insoluble in water.

8.85 (a) $Cl^-(aq) + Ag^+(aq) \rightarrow AgCl(s)$
     (b) no reaction
     (c) $OH^-(aq) + H^+(aq) \rightarrow H_2O$
     (d) $Pb^{2+}(aq) + 2Cl^-(aq) \rightarrow PbCl_2(s)$
     (e) no reaction
     (f) $S^{2-}(aq) + Cu^{2+}(aq) \rightarrow CuS(s)$
     (g) $SO_4^{2-}(aq) + Ba^{2+}(aq) \rightarrow BaSO_4(s)$
     (h) $OH^-(aq) + H^+(aq) \rightarrow H_2O$
     (i) $S^{2-}(aq) + Ni^{2+}(aq) \rightarrow NiS(s)$
     (j) $Ag^+(aq) + Br^-(aq) \rightarrow AgBr(s)$
     (k) $HCO_3^-(aq) + H^+(aq) \rightarrow CO_2(g) + H_2O$
     (l) $Mg^{2+}(aq) + 2OH^-(aq) \rightarrow Mg(OH)_2(s)$

8.108 (a) We look for the salt with the anion having the highest charge to be least soluble. This salt would involve the strongest attraction between cation and anion and so would be least able to release the ions from the crystal.
      (b) $Ca(H_2PO_4)_2$
      (c) $Ca_3(PO_4)_2$

8.110 Make a solution of the unknown. If it is K_2O, you will notice that the temperature increases, because K_2O reacts exothermically with water to give KOH ($K_2O + H_2O \rightarrow 2KOH$). The solution can be tested with litmus. If the unknown is K_2O, red litmus will turn blue. KNO_3 will not do this.

8.112 0.232 g K_2CO_3

CHAPTER 9

Practice Exercises, Chapter 9

1. (a) 2.5×10^{-6} mol OH^-/L. Basic
   (b) 9.1×10^{-8} mol OH^-/L. Acidic
   (c) 1.1×10^{-7} mol OH^-/L. Basic

2. (a) 1.60   (b) 10.40   (c) 10.70

3. 7.14. Basic

4. (a) 4.6×10^{-7} mol/L, acidic
   (b) 1.3×10^{-8} mol/L, basic

5. 5.2×10^{-8} mol/L

6. 3.6×10^{-10} mol/L

7. $HC_2H_3O_2(aq) \rightleftharpoons H^+(aq) + C_2H_3O_2^-(aq)$

$$K_a = \frac{[H^+][C_2H_3O_2^-]}{[HC_2H_3O_2]}$$

8. $HCO_3^-(aq) \rightleftharpoons H^+(aq) + CO_3^{2-}(aq)$

$$K_a = \frac{[H^+][CO_3^{2-}]}{[HCO_3^-]}$$

9. $NH_4^+(aq) \rightleftharpoons H^+(aq) + NH_3(aq)$

$$K_a = \frac{[H^+][NH_3]}{[NH_4^+]}$$

10. ascorbic acid

11. cyanide ion

12. (a) $CO_3^{2-}(aq) + H_2O \rightleftharpoons HCO_3^-(aq) + OH^-(aq)$

$$K_b = \frac{[HCO_3^-][OH^-]}{[CO_3^{2-}]}$$

   (b) $C_2H_3O_2^-(aq) + H_2O \rightleftharpoons HC_2H_3O_2(aq) + OH^-(aq)$

$$K_b = \frac{[HC_2H_3O_2][OH^-]}{[C_2H_3O_2^-]}$$

   (c) $NH_3(aq) + H_2O \rightleftharpoons NH_4^+(aq) + OH^-(aq)$

$$K_b = \frac{[NH_4^+][OH^-]}{[NH_3]}$$

13. (a) Yes, basic   (b) Yes, basic
    (c) Yes, basic   (d) Yes, acidic
    (e) Yes, basic   (f) Yes, basic

14. basic

15. acidic

16. Yes, the pH is decreased because NH_4^+ hydrolyzes to give some H^+.

17. 10.33

18. acetic acid

19. 6.80

20. NH_3, $K_b = 1.8 \times 10^{-5}$

21. 4.8
    $CN^-(aq) + H_2O \rightleftharpoons HCN(aq) + OH^-(aq)$

$$K_b = \frac{[HCN][OH^-]}{[CN^-]}$$

22. 3.89

23. 3.86

24. (a) $pH = pK_a + 1$
    (b) $pH = pK_a - 1$

25. (a) 7.00
    (b) 7.18
    (c) Buffer *capacity* depends on the *absolute* values of [anion] and [acid].

26. 7.75. No

27. 0.105 *M* NaOH

28. 0.125 *M* H_2SO_4

Review Exercises, Chapter 9

9.3 The nature of the reactants; the concentrations of the reactants; the temperature of the reaction mixture; and the presence or absence of a catalyst. In a homogeneous reaction all substances are in the *same* phase (gas or liquid).

9.5 C and D

9.9 increase concentrations

9.10 (a) no effect   (b) reduces it
     (c) no effect   (d) increases

9.12 Catalase: $2H_2O_2 \xrightarrow{\text{catalase}} 2H_2O + O_2$

Carbonic anhydrase: $CO_2 + OH^- \xrightarrow{\text{carbonic anhydrase}} HCO_3^-$

9.15 Collision frequency refers to the *total* number of collisions (of every amount of collisional energy) per unit of volume per second. The rate of a reaction is the frequency of *successful* collisions per unit of volume per second.

9.17 (a) E_{act}
     (b) heat of reaction
     (c) If the reaction has both a large E_{act} and a large (evolving) heat of reaction
     (d) The reaction is instantaneous (an explosion).
     (e) E_{act}

9.19 (a) CO and O_2 (b) CO_2
     (c) C   (d) B
     (e) Exothermic. The energy of the product D is less than that of the reactants at A and the difference in energy *evolves* as the heat of reaction.
     (f) E

9.21

9.24 (a) $K_{eq} = \dfrac{[NH_4^+][OH^-]}{[NH_3][H_2O]}$

     (b) $K_{eq} = \dfrac{[H_2O]^2}{[H_2]^2[O_2]}$

     (c) $K_{eq} = \dfrac{[NH_3]^2}{[N_2][H_2]^3}$

9.29 2.43×10^{-14}. Neutral, because $[H^+] = [OH^-]$

9.32 acidic

9.34 3.0

9.37 4.5×10^{-6} mol/L. Slightly acidic

9.39 Strong. At a pH of 2, the value of $[H^+]$ is 1×10^{-2} mol/L, which is numerically identical to the initial concentration of the acid, so all of the acid is ionized.

9.41 basic (pH = 7.76)

9.43 $HNO_2(aq) \rightleftharpoons H^+(aq) + NO_2^-(aq)$

$$K_a = \frac{[H^+][NO_2^-]}{[HNO_2]}$$

9.45 $NH_3(aq) \rightleftharpoons NH_2^-(aq) + H^+(aq)$

$$K_a = \frac{[NH_2^-][H^+]}{[NH_3]}$$

9.47 ammonium ion

9.49 $NO_2^-(aq) + H_2O \rightleftharpoons HNO_2(aq) + OH^-(aq)$

$$K_b = \frac{[HNO_2][OH^-]}{[NO_2^-]}$$

9.51 ammonia

9.53 The sulfide ion reacts with water as follows, which produces an excess of hydroxide ions in the solution.

$$S^{2-}(aq) + H_2O \longrightarrow HS^-(aq) + OH^-(aq)$$

9.55 (a) neutral (b) basic (c) basic
(d) acidic (e) basic

9.57 basic

9.59 (a) 10.85 (b) 10.25

9.61 (a) 7.1×10^{-4}, 3.15 (b) 1.8×10^{-4}, 3.75

9.63 N is the stronger base, and it has the weaker conjugate acid.

9.67 More alkaline. Alkalosis

9.69 pH = 7.18. This means acidosis because 7.18 is less (so more acidic) than 7.35, the normal pH of blood.

9.80 (a) alkalosis
(b) For each CO_2 lost at the lungs, one H^+ ion is permanently neutralized. The loss of H^+ ion, of course, means an increase in the value of the pH of the blood.

9.83 If HA is the acid component of the buffer and A^- its conjugate base, then the presence of A^- from the *salt* used to supply the base suppresses the ionization of HA. Thus $[HA]_{eq} = [HA]_{initial} = [\text{acid}]$. However, because so little A^- is lost because of hydrolysis, we can set $[A^-]_{eq} = [A^-]_{initial} = [\text{anion}]$.

9.85 pH = 7.26

9.86 (a) pH = 5.10 (b) pH = 4.96 (c) pH = 5.24
(d) After the addition of acid to 500 mL water, pH = 1.40
After the addition of base to 500 mL water, pH = 12.60

9.90 (a) pH = 6.1 (b) pH = 7.1
(c) Hyperventilation expels additional CO_2.
(d) A net synthesis of HCO_3^- for the blood is performed by the kidneys.

9.94 The titration of any strong acid (e.g., HCl) with any strong base (e.g., NaOH) produces a salt whose ions do not hydrolyze.

9.96 The titration of a strong acid against aqueous ammonia would produce an ammonium salt whose cation, NH_4^+, would hydrolyze to give a slightly acidic pH. The anion would not hydrolyze.

9.98 (a) 0.3407 M HI (b) 0.2000 M H_2SO_4

9.100 64.00

9.102 (a) 1.279 g HI (b) 14.71 g H_2SO_4
(c) 59.62 g Na_2CO_3

9.104 (a) 0.2240 M NaOH (b) 8.96 g NaOH/L

9.106 K_a 1.10×10^{-4}; $pK_a = 3.96$

9.108 pH = 1.93

9.112 10.65

9.114 0.055 mol acetic acid

CHAPTER 10

Practice Exercises, Chapter 10

1. (a) H, +1 (b) S, +4 (c) H, +1
   S, −2   O, −2   O, −2
         S, +4
   (d) O, −2 (e) H, +1 (f) O, −2
   S, +6   O, −2   S, +6
         S, +6

2. (a) H, +1 (b) H, +1 (c) H, +1
   C, −4   C, −2   C, 0
         O, −2   O, −2
   (d) H, +1 (e) H, +1
   C, +2   C, +4
   O, −2   O, −2

3. (a) −1 (b) −2 (c) −3. Reduction

4. $Cu(s) + 4H^+(aq) + 2NO_3^-(aq) \rightarrow Cu^{2+}(aq) + 2NO_2(g) + 2H_2O$

5. $3C_2H_6O + 4MnO_4^-(aq) \rightarrow 3C_2H_3O_2^-(aq) + 4MnO_2(s) + OH^-(aq) + 4H_2O$

6. $Zn(s) + Fe^{2+}(aq) \rightarrow Zn^{2+}(aq) + Fe(s)$

7. No, the reduction potential for $2H^+/H_2$ is not more positive that that of Cu^{2+}/Cu.

8. $E_{cell}^{\circ} = +0.29$ V. $K_{eq} = 6.3 \times 10^9$

Review Exercises, Chapter 10

10.2 (a) N, −3; H, +1 (b) N, +3; F, −1
(c) N, +4; O, −2 (d) N, +3; O, −2
(e) N, +5; O, −2

10.4 (a) H, +1; S, −2 (b) H, +1; S, +4; O, −2
(c) H, +1; S, +6; O, −2

10.7 (a) $MnO_4^- + 8H^+ + 5e^- \rightarrow Mn^{2+} + 4H_2O$ (reduction)
(b) $2Fe^{2+} + 3H_2O \rightarrow Fe_2O_3 + 6H^+ + 2e^-$ (oxidation)
(c) $H_2C_2O_4 \rightarrow 2CO_2 + 2H^+ + 2e^-$ (oxidation)
(d) $2NO_2 + 8H^+ + 8e^- \rightarrow N_2 + 4H_2O$ (reduction)

10.9 (a) $2I^- + 2HNO_2 + 2H^+ \rightarrow I_2 + 2NO + 2H_2O$
(b) $3Sn + 4NO_3^- + 4H^+ \rightarrow 3SnO_2 + 4NO + 2H_2O$
(c) $Mg + SO_4^{2-} + 4H^+ \rightarrow Mg^{2+} + SO_2 + 2H_2O$
(d) $PbO_2 + 4Cl^- + 4H^+ \rightarrow PbCl_2 + Cl_2 + 2H_2O$

10.11 (a) $4MnO_4^- + 3CH_2O \rightarrow 4MnO_2 + 3CO_2 + 4OH^- + H_2O$

(b) $2CrO_4^{2-} + 3S^{2-} + 4H_2O \rightarrow 2CrO_2^- + 3S + 8OH^-$

10.14 (a) $Ca^{2+}(aq) + 2e^- \rightleftharpoons Ca(s)$ (b) lesser (c) H^+

10.16 (a) $I_2 < Br_2 < Cl_2 < F_2$

(b) $F^- < Cl^- < Br^- < I^-$

10.18 Au. Gold, the metal with the highest reduction potential, would have the lowest potential for becoming oxidized to a metal ion and so is the least reactive of all metals.

10.20 $4I^- + O_2 + 4H^+ \rightarrow 2I_2 + 2H_2O$

10.23 (a) no (b) no

10.24 1.51 V

10.26 1.0×10^{51}

10.28 When the temperature is 25 °C, all concentrations are 1 M, and all gas pressures are 1 atm.

10.38 19.1 g Na_2SO_3

10.40 (a) $4Au + 8CN^- + O_2 + 2H_2O \rightarrow 4Au(CN)_2^- + 4OH^-$

(b) $4ClO_3^- + 3N_2H_4 \rightarrow 4Cl^- + 6NO + 6H_2O$

CHAPTER 11

Practice Exercises, Chapter 11
1. $^{131}_{53}I \rightarrow ^{131}_{54}Xe + ^{0}_{-1}e + ^{0}_{0}\gamma$
2. $^{239}_{94}Pu \rightarrow ^{235}_{92}U + ^{4}_{2}He + ^{0}_{0}\gamma$
3. 10,000 units
4. 8.5 m

Review Exercises, Chapter 11
11.9 Isotope. A radionuclide is a radioactive *isotope* that consists of *atoms*.

11.11 (a) $^{211}_{83}Bi$ (b) $^{216}_{84}Po$ (c) $^{140}_{57}La$

11.13 (a) $^{144}_{60}Nd \rightarrow ^{140}_{58}Ce + ^{4}_{2}He$

(b) $^{40}_{19}K \rightarrow ^{40}_{20}Ca + ^{0}_{-1}e$

(c) $^{149}_{62}Sm \rightarrow ^{149}_{63}Eu + ^{0}_{-1}e$

(d) $^{251}_{98}Cf \rightarrow ^{247}_{96}Cm + ^{4}_{2}He + ^{0}_{0}\gamma$

11.17 1.500 ng

11.26 A factor of 4 (meaning that the intensity at the greater distance is one-fourth the intensity at the lesser distance).

11.30 15.1 m

11.33 The sample is undergoing 1.5×10^{-3} Ci $\times (3.7 \times 10^{10}$ disintegrations/Ci) $= 5.6 \times 10^7$ disintegrations per second.

11.46 Iron-55. $^{55}_{25}Mn + ^{1}_{1}p \rightarrow ^{1}_{0}n + ^{55}_{26}Fe$

11.48 Neutrons. $^{113}_{49}In \rightarrow ^{111}_{49}In + 2^{1}_{0}n$

11.51 $^{10}_{5}B + ^{4}_{2}He \rightarrow ^{13}_{7}N + ^{1}_{0}n$

11.53 A lithium-6 nucleus. $^{27}_{13}Al + ^{6}_{3}Li \rightarrow ^{32}_{15}P + ^{1}_{1}p$

11.57 A product produced by the ionizing radiation that is not present in the food or produced either by cooking or digestion.

11.60 botulinum bacillus

11.63 It emits only gamma radiation, and it has a shorter half-life.

11.65 As phosphate ion, because the hydroxyapatite in bone contains this ion.

11.68 Each fission event produces more neutron initiators than were needed to cause the fission event.

11.72 Unless the heat is carried away, the reactor will melt.

11.75 Strontium-90 is a bone seeker. Iodine-131 is taken up by the thyroid gland. Cesium-137 gets as widely distributed as the sodium ion.

11.77 Some radioactive isotopes in wastes have very long half-lives.

11.100 The atomic number is reduced by one unit.

$$^{50}_{23}V + ^{0}_{-1}e \xrightarrow{\text{electron capture}} ^{50}_{22}Ti$$

CHAPTER 12

Practice Exercises, Chapter 12
1. (a) $CH_3-CH_2-CH_3$ (b) $CH_3-CH-CH_3$ with CH_3

(c) structure

2. (a) $CH_3CH_2CH_3$ (b) CH_3CHCH_3 with CH_3

(c) structure

3. (a)–(c) structures

4. Structures (b) and (c) violate the tetravalences of carbon at one point.

5. (a)

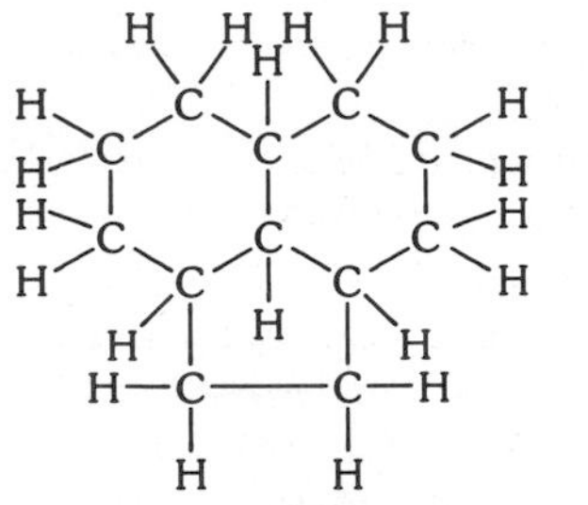

(b)

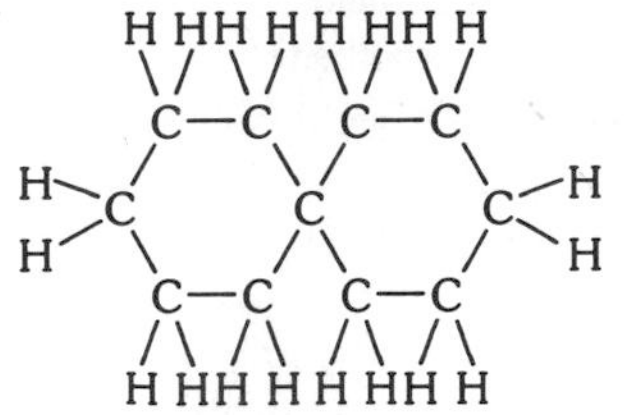

6. 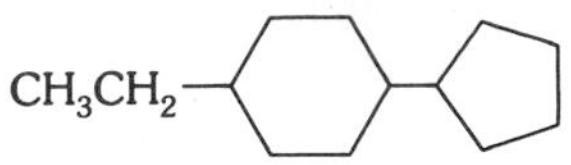

7. (a) identical (b) isomers
(c) identical (d) isomers
(e) different in another way

8. 2-methyl-1-butanol (less polar)

9. (a) 3-methylhexane
(b) 4-ethyl-2,3-dimethylheptane
(c) 5-ethyl-2,4,6-trimethyloctane

10. (a) 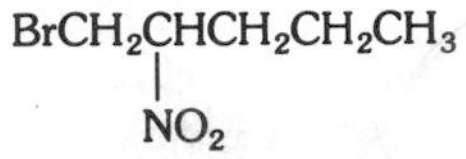

(b) 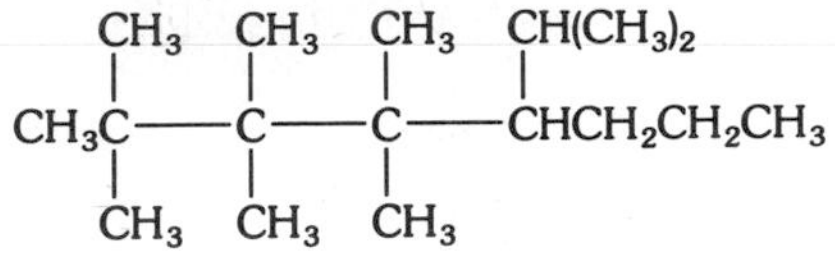

(c) 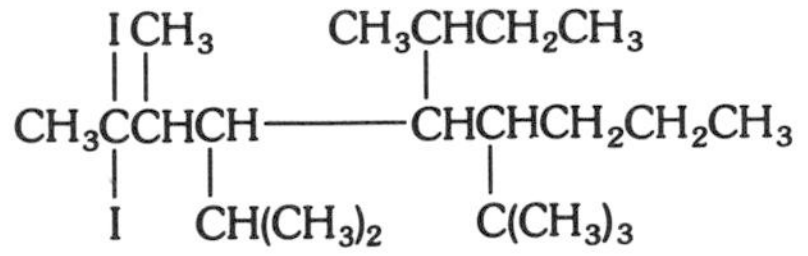

(d) 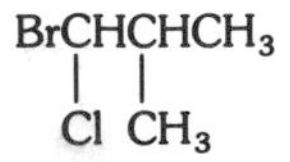

(e) 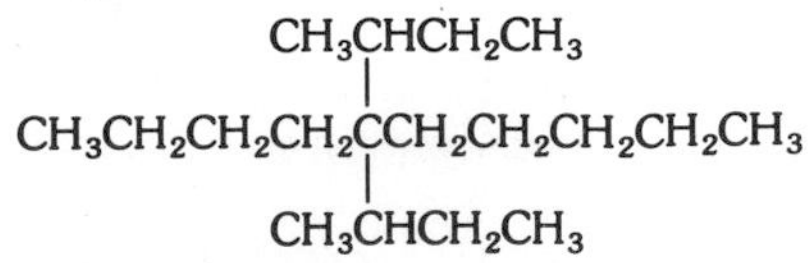

11. 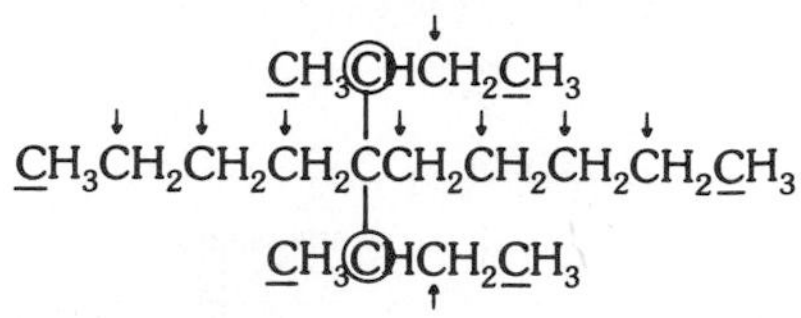

12. (a) ethyl chloride (b) butyl bromide
(c) isobutyl chloride (d) *t*-butyl bromide

13. (a) butyl chloride, 1-chlorobutane
sec-butyl chloride, 2-chlorobutane
(b) isobutyl chloride, 2-methyl-1-chloropropane
t-butyl chloride, 2-methyl-2-chloropropane

Review Exercises, Chapter 12

12.6 Compounds b, d, and e are considered to be inorganic.

12.9 (a) Molecular; low melting and flammable
(b) Ionic; water soluble (and likely a carbonate or a bicarbonate)
(c) Molecular; no ionic compound is a gas (or a liquid) at room temperature.
(d) Ionic; high melting and nonflammable
(e) Molecular; most liquid organic compounds are insoluble in water but will burn.
(f) Molecular; no ionic compound is a liquid at room temperature.

12.11 Compounds a, d, and e are possible.

12.12 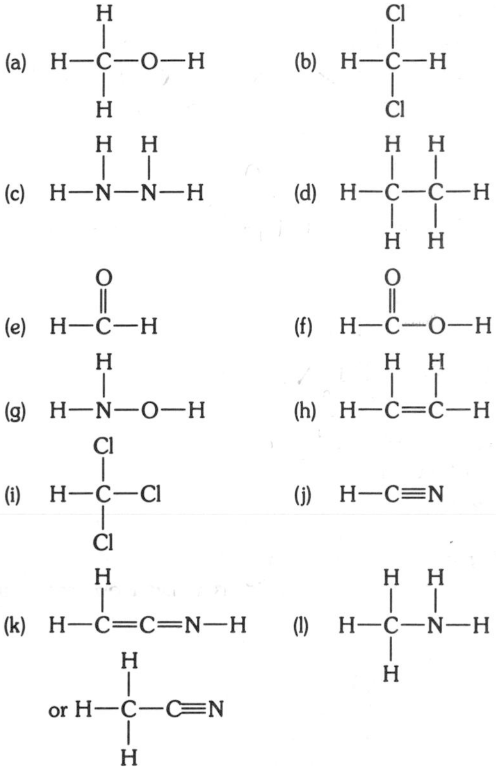

12.19 Unsaturated compounds are (b), (c), and (d).

12.21 Identical compounds are a, b, e, f, and m. Isomers are d, g, h, i, j, k, and l. Unrelated structures occur at c and n.

12.22 (a) alkane (b) alcohol
(c) thioalcohol (d) alkyne
(e) aldehyde (f) ester
(g) ester (h) ketone
(i) amine (j) ether

12.23 (a) alcohol
(b) alcohol
(c) both thioalcohols
(d) first, alkene; second, cycloalkane
(e) ketone
(f) alkane
(g) both amines
(h) first, carboxylic acid; second, alcohol + ketone
(i) first, ester; second, carboxylic acid (Why is an ester the first and not an aldehyde, will be explained later.)

(j) first, ester + alcohol; second, alcohol + carboxylic acid
(k) first, ether + ketone; second, ester
(l) both alkanes
(m) ketone

12.25 B. Its molecules have water-like OH groups.

12.27 Add a drop of the liquid to water. If it dissolves, it is methyl alcohol.

12.31 $CH_3CH_2CH_2CH_2CH_2CH_2CH_3$ heptane

CH_3 (on C-2)
$CH_3CHCH_2CH_2CH_2CH_3$ 2-methylhexane

CH_3 (on C-3)
$CH_3CH_2CHCH_2CH_2CH_3$ 3-methylhexane

CH_2CH_3 (on C-3)
$CH_3CH_2CHCH_2CH_3$ 3-ethylpentane

CH_3, CH_3 (on C-2)
$CH_3CCH_2CH_2CH_3$ 2,2-dimethylpentane

CH_3, CH_3 (on C-3)
$CH_3CH_2CCH_2CH_3$ 3,3-dimethylpentane

H_3C CH_3 (on C-2, C-3)
$CH_3CHCHCH_2CH_3$ 2,3-dimethylpentane

CH_3 CH_3 (on C-2, C-4)
$CH_3CHCH_2CHCH_3$ 2,4-dimethylpentane

H_3C, CH_3 (on C-2), CH_3 (on C-3)
$CH_3C—CHCH_3$ 2,2,3-trimethylbutane

12.32 (a) $CH_3CH_2CH_2CH_2CH_2CH_2CH_3$
(b) CH_3 (on C-2)
$CH_3CHCH_2CH_2CH_2CH_3$

12.33 (a) 5-*sec*-butyl-5-ethyl-2,3,3,9-tetramethyldecane
(b) 7-*t*-butyl-5-isobutyl-2-methyl-6-propyldecane

12.35 (a) $CH_3CH_2CH_2Cl$ (b) CH_3 (on C-2) CH_3CHCH_2I
(c) CH_3, CH_3 (on C) CH_3CBr (d) CH_3CH_2Br

12.37 (a) CH_3 (on C-2)
CH_3CHCH_2Cl
1-chloro-2-methylpropane

(b) 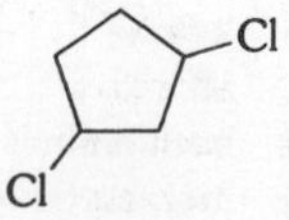
1,3-dichlorocyclopentane

(c) CH_2CH_3 (on C-3)
$CH_3CHCH_2CH_3$
3-methylpentane

(d) 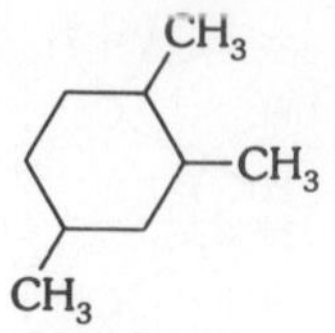
1,2,4-trimethylcyclohexane

12.39 $CH_3CH_2OH + 3O_2 \rightarrow 2CO_2 + 3H_2O$

12.41 CH_3CHCl_2, 1,1-dichloroethane
$ClCH_2CH_2Cl$, 1,2-dichloroethane

12.44 Petroleum is a mixture of liquid (crude oil) and gas.

12.51 (a) $H_2 + Cl_2 \rightarrow 2HCl$
(b) Cl_2 + UV or heat $\rightarrow 2Cl\cdot$
$Cl\cdot + H_2 \rightarrow HCl + H\cdot$
$H\cdot + Cl_2 \rightarrow HCl + Cl\cdot$

12.54 No reaction with any of the given reactants.

12.56 11.1 g chlorocyclopentane

CHAPTER 13

Practice Exercises, Chapter 13

1. (a) 2-methylpropene (or 2-methyl-1-propene)
(b) 4-isobutyl-3,6-dimethyl-3-heptene
(c) 1-chloropropene (or 1-chloro-1-propene)
(d) 3-bromopropene (or 3-bromo-1-propene)
(e) 4-methyl-1-hexene
(f) 4-methylcyclohexene

2. (a) CH_3 (on C-4)
$CH_3CH{=}CHCHCH_3$
(b) $CH_2CH_2CH_3$ (on C-3)
$CH_2{=}CHCHCH_2CH_2CH_2CH_3$
(c) CH_3, CH_3 (on C-3)
$CH_2{=}CHCCH_2Cl$
(d) H_3C CH_3 (on C-2, C-3)
$CH_3C{=}CCH_3$

3. Cis–trans isomerism is possible for (a) and (b). For part (a), cis versus trans is based on the way the main chain passes through the double bond.

(a) 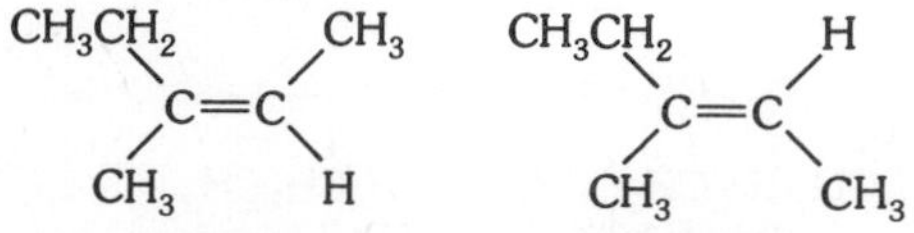
cis isomer trans isomer

(b) 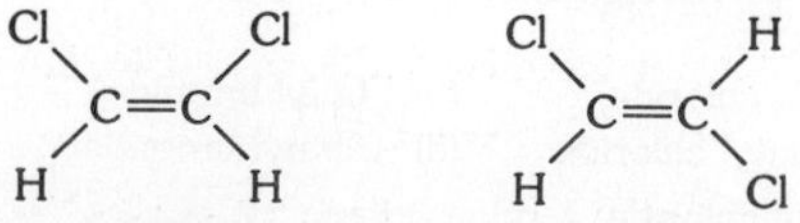
cis isomer trans isomer

4. (a) $CH_3CH_2CH_3$
(b) no reaction

(c)

(d) $CH_3(CH_2)_{16}CO_2H$

5. (a)
$$CH_3\overset{\overset{\displaystyle CH_3}{|}}{\underset{\underset{\displaystyle Br}{|}}{C}}CH_2Br$$

(b)   no reaction (in the absence of heat or UV radiation)

(c) $ClCH_2\overset{\underset{\displaystyle |}{}}{C}HCH_2CH_3$
$$\underset{Cl}{|}$$

(d) $CH_3CH_2CH_2CH_3$

6. (a) $CH_3\overset{}{C}HCH_2CH_3$
$$\underset{Cl}{|}$$

(b) $CH_3\overset{\overset{\displaystyle CH_3}{|}}{\underset{\underset{\displaystyle Br}{|}}{C}}CH_3$

(c) $CH_3CH_2\overset{\overset{\displaystyle CH_3}{|}}{\underset{\underset{\displaystyle OH}{|}}{C}}$ —cyclohexyl

(d)

(e)   no reaction

7. (a) $CH_3CH_2CH_2\overset{+}{C}H_2$ and $\boxed{CH_3CH_2\overset{+}{C}HCH_3}$

$$CH_3CH_2\underset{\underset{\displaystyle Cl}{|}}{C}HCH_3$$

(b) $CH_3\overset{\overset{\displaystyle CH_3}{|}}{\overset{+}{C}}HCH_2$ and $\boxed{CH_3\overset{\overset{\displaystyle CH_3}{|}}{\underset{+}{C}}CH_3}$ $CH_3\overset{\overset{\displaystyle CH_3}{|}}{\underset{\underset{\displaystyle Cl}{|}}{C}}CH_3$

(c)   cyclohexyl—$\overset{+}{C}H_2$ and $\boxed{\text{+-cyclohexyl—}CH_3}$

(d)   Only one carbocation is possible:

$\boxed{CH_3CH_2\overset{+}{C}HCH_3}$ $CH_3CH_2\overset{\underset{\displaystyle |}{}}{C}HCH_3$
$$\underset{Cl}{}$$

(e) $CH_3\overset{+}{C}HCH_2CH_2CH_3$ $CH_3\overset{\underset{\displaystyle |}{}}{C}HCH_2CH_2CH_3$
$$\underset{Cl}{}$$

and

$$CH_3CH_2\overset{+}{C}HCH_2CH_3 \qquad CH_3CH_2\overset{\underset{\displaystyle |}{}}{C}HCH_2CH_3$$
$$\underset{Cl}{}$$

(Both carbocations are 2°, so both are equally possible and both form. Thus two isomeric chloropentanes form.)

(f) $CH_3\overset{\underset{\displaystyle |}{}}{C}HCH_2CH_2CH_3$ and $CH_3CH_2\overset{\underset{\displaystyle |}{}}{C}HCH_2CH_3$
$$\underset{OH}{}\underset{OH}{}$$

Two 2° carbocations can form, so two isomeric pentanols form. Only one carbocation can form from propene, namely, the more stable isopropyl carbocation.

Review Exercises, Chapter 13

13.1 (a)   A, B, E
(b)   C
(c)   E (Structure C has "-pent-" in "cyclopentane.")
(d)   All are insoluble in water.
(e)   B, and C. No, C is not an alkene.
(f)   A, D, and E. Only D is an alkyne.
(g)   Not well; the presence of rings or multiple bonds can reduce the ratio of carbon atoms to hydrogen atoms of a hydrocarbon.

13.4 (a)   1-octene
(b)   1-bromo-3-methyl-1-butene
(c)   4-methyl-2-propyl-1-hexene
(d)   4,4-dimethyl-2-hexene

13.5 $CH_2{=}CHCH_2CH_2CH_3$ 1-pentene

cis-2-pentene

trans-2-pentene

$$\overset{\overset{\displaystyle CH_3}{|}}{CH_2{=}CCH_2CH_3}$$ 2-methyl-1-butene

$$\overset{\overset{\displaystyle CH_3}{|}}{CH_2{=}CHCHCH_3}$$ 3-methyl-1-butene

$$\overset{\overset{\displaystyle CH_3}{|}}{CH_3C{=}CHCH_3}$$ 2-methyl-2-butene

13.7 $HC{\equiv}CCH_2CH_3$ 1-butyne
$CH_3C{\equiv}CCH_3$ 2-butyne

13.9 $CH_2{=}C{=}CHCH_2CH_3$ 1,2-pentadiene
$CH_2{=}CHCH{=}CHCH_3$ 1,3-pentadiene
$CH_2{=}CHCH_2CH{=}CH_2$ 1,4-pentadiene
$CH_3CH{=}C{=}CHCH_3$ 2,3-pentadiene

$$\overset{\overset{\displaystyle CH_3}{|}}{CH_2{=}CCH{=}CH_2}$$ 2-methyl-1,3-butadiene

13.13 (a)   identical
(b)   identical
(c)   isomers
(d)   identical
(e)   identical

13.15 (a)

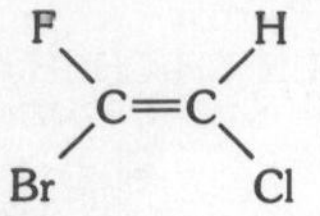

and

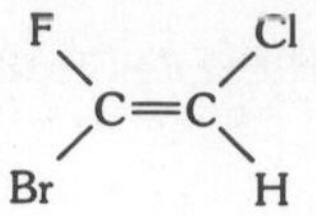

(b) No geometric isomers.

(c)

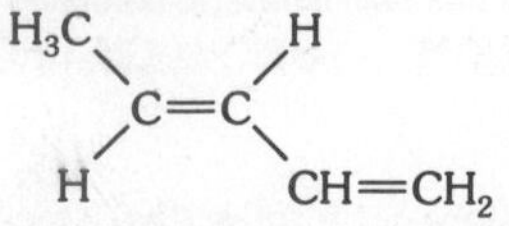

and

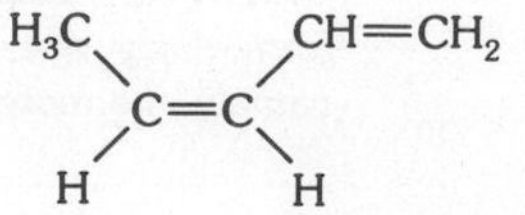

13.17 Compound A could be either of the following:

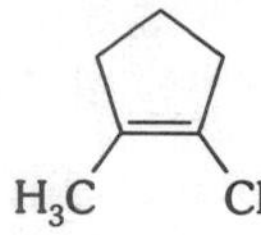

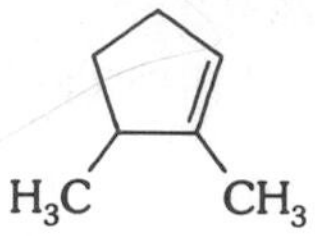

13.19 (a) 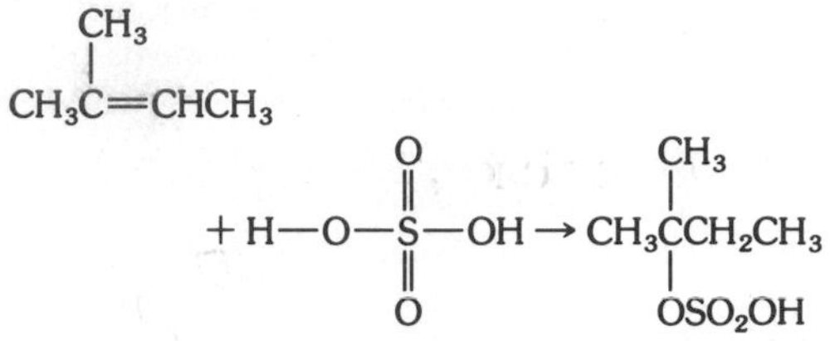 $CH_3\overset{\overset{\displaystyle CH_3}{|}}{C}=CHCH_3$

(b) $CH_3\overset{\overset{\displaystyle CH_3}{|}}{C}=CHCH_3 + H_2 \xrightarrow{\text{catalyst}} CH_3\overset{\overset{\displaystyle CH_3}{|}}{C}HCH_2CH_3$

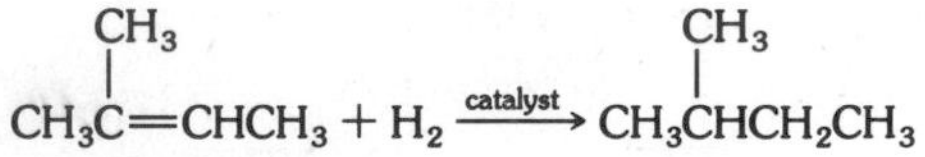

(c) $CH_3\overset{\overset{\displaystyle CH_3}{|}}{C}=CHCH_3 + H_2O \xrightarrow{H^+} CH_3\overset{\overset{\displaystyle CH_3}{|}}{\underset{\underset{\displaystyle OH}{|}}{C}}CH_2CH_3$

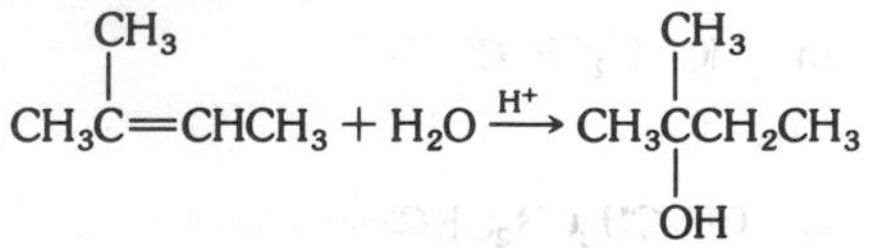

(d) 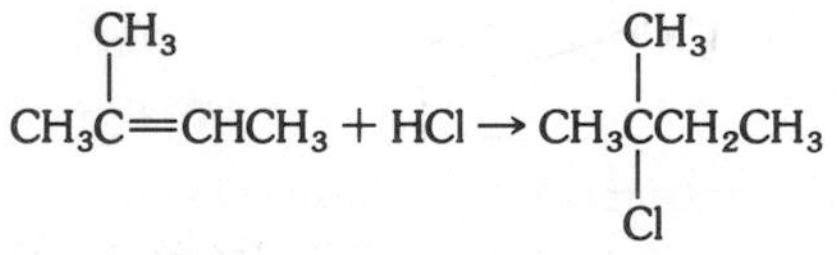 $CH_3\overset{\overset{\displaystyle CH_3}{|}}{C}=CHCH_3 + HCl \rightarrow CH_3\overset{\overset{\displaystyle CH_3}{|}}{\underset{\underset{\displaystyle Cl}{|}}{C}}CH_2CH_3$

(e) 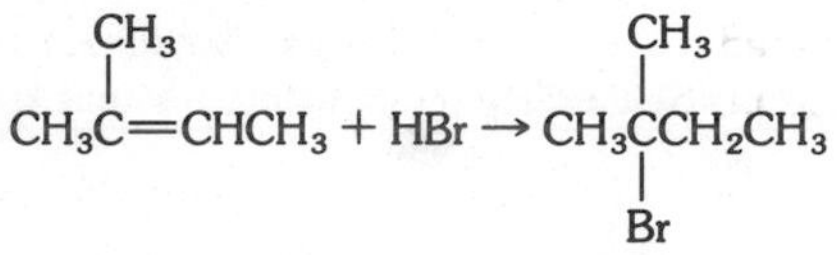 $CH_3\overset{\overset{\displaystyle CH_3}{|}}{C}=CHCH_3 + HBr \rightarrow CH_3\overset{\overset{\displaystyle CH_3}{|}}{\underset{\underset{\displaystyle Br}{|}}{C}}CH_2CH_3$

(f) 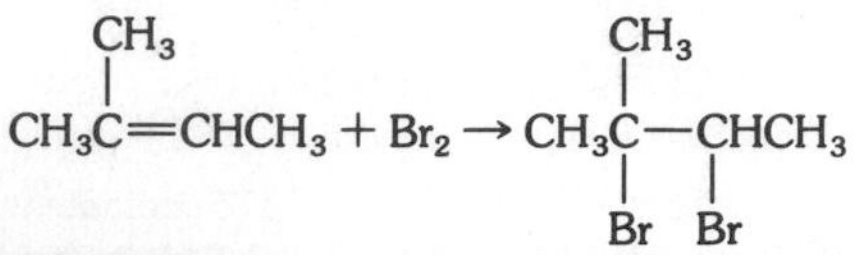 $CH_3\overset{\overset{\displaystyle CH_3}{|}}{C}=CHCH_3 + Br_2 \rightarrow CH_3\overset{\overset{\displaystyle CH_3}{|}}{\underset{\underset{\displaystyle Br}{|}}{C}}-\overset{}{\underset{\underset{\displaystyle Br}{|}}{C}}HCH_3$

13.22 Cycloalkanes with the formula C_5H_{10} have no alkene groups and so cannot react with the given reactants. Possible structures are

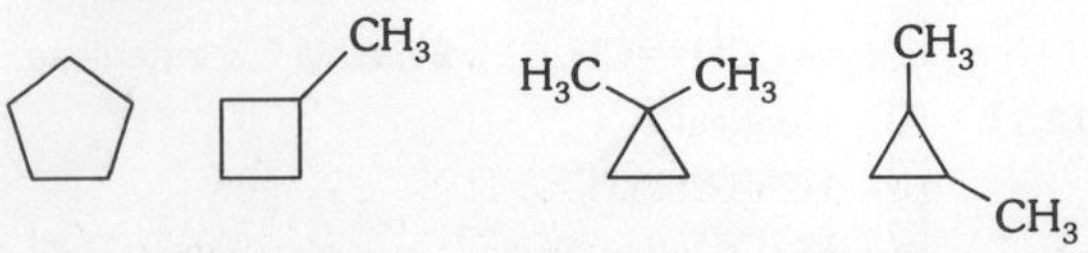

13.24 11.1 g $K_2C_6H_8O_4$

13.27 (a)

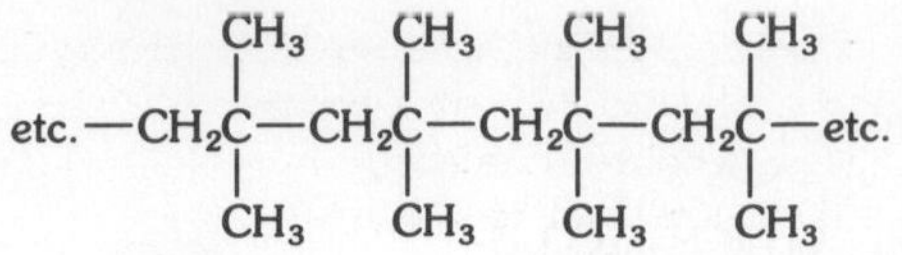

(b)

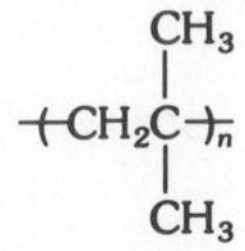

13.31 Dipentene has no benzene ring.

13.33 The products of the reactions are the following, where C_6H_5 is the phenyl group.

(a) $C_6H_5SO_2OH$ (or $C_6H_5SO_3H$, benzenesulfonic acid) $+ H_2O$

(b) $C_6H_5NO_2 + H_2O$

(c) no reaction

(d) no reaction

(e) no reaction (No iron or iron salt catalyst is specified.)

(f) no reaction

(g) $C_6H_5Br + HBr$

13.35

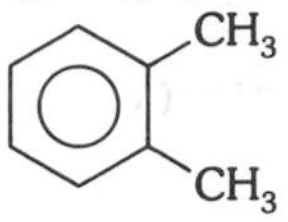

13.52 (a) $CH_3CH_2CH_2CH_3$

(b)

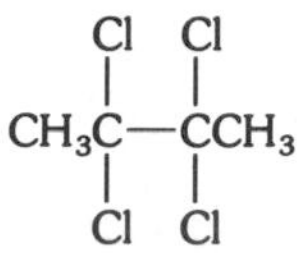

(c)

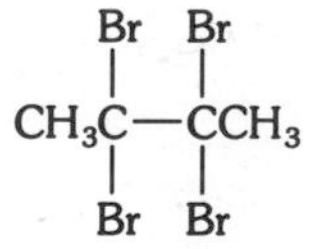

13.56 (a) 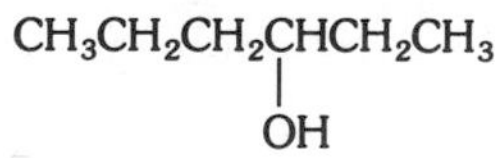 $CH_3CH_2CH_2\overset{}{\underset{\underset{\displaystyle OH}{|}}{C}}HCH_2CH_3$

(b) 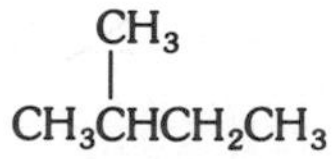 $CH_3\overset{\overset{\displaystyle CH_3}{|}}{C}HCH_2CH_3$

(c) $C_6H_5Br + HBr$

(d) no reaction

(e)

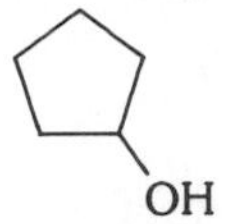

(f) $CH_3CH_2CH_2CH_2CH_3$

(g) no reaction

(h) 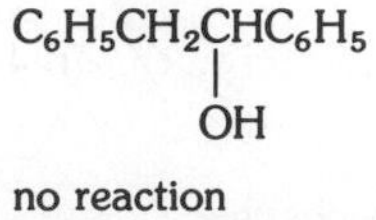 $C_6H_5CH_2\overset{}{\underset{\underset{\displaystyle OH}{|}}{C}}HC_6H_5$

(i) no reaction

(j) $C_5H_{12} + 8O_2 \rightarrow 5CO_2 + 6H_2O$

(k) no reaction

CHAPTER 14

Practice Exercises, Chapter 14

1. (a) monohydric, secondary

    (b) monohydric, secondary

(c) dihydric, unstable (two OH groups on the same carbon)
(d) dihydric, both are secondary
(e) monohydric, primary
(f) monohydric, primary
(g) monohydric, tertiary
(h) monohydric, secondary
(i) trihydric, unstable (three OH groups on the same carbon)

2. (a) alcohol
    (b) phenol
    (c) carboxylic acid
    (d) alcohol (but unstable; an enol)
    (e) alcohol
    (f) alcohol

3. (a) 4-methyl-1-pentanol
    (b) 2-methyl-2-propanol
    (c) 2-ethyl-2-methyl-1-pentanol
    (d) 2-methyl-1,3-propanediol

4. In 1,2-propanediol. Its boiling point is 189 °C, much higher than that of 1-butanol (bp 117 °C). 1,2-Propanediol is more soluble in water.

5. (a) $CH_3CH{=}CH_2$ (b) $CH_3CH{=}CH_2$

    (c) $CH_2{=}\overset{\underset{|}{CH_3}}{C}CH_3$ (d) [cyclohexene with CH_3]

6. (a) $CH_3\overset{\underset{|}{CH_3}}{C}HCH{=}O$ then $CH_3\overset{\underset{|}{CH_3}}{C}HCO_2H$
    (b) $C_6H_5CH{=}O$ then $C_6H_5CO_2H$

7. (a) $CH_3\overset{O}{\overset{||}{C}}CH_2CH_3$ (b) $C_6H_5\overset{O}{\overset{||}{C}}CH_3$ (c) [cyclopentanone]

8. (a) $CH_3\overset{O}{\overset{||}{C}}CH_2CH{=}O$ and $CH_3\overset{O}{\overset{||}{C}}CH_2CO_2H$
    (b) no reaction
    (c) $CH_3\overset{\underset{|}{CH_3}}{\underset{|}{\underset{CH_3}{C}}}CH{=}O$ and $CH_3\overset{\underset{|}{CH_3}}{\underset{|}{\underset{CH_3}{C}}}CO_2H$
    (d) $CH_3\overset{\underset{|}{CH_3}}{C}H\overset{O}{\overset{||}{C}}CH_3$

9. (a) CH_3OCH_3 (b) $CH_3CH_2CH_2OCH_2CH_2CH_3$
    (c) [cyclohexyl-O-cyclohexyl ether]

10. (a) $2CH_3SH$ (b) $CH_3\overset{\underset{|}{CH_3}}{C}HSS\overset{\underset{|}{CH_3}}{C}HCH_3$
    (c) [cyclohexane with two SH groups]   (d) [cyclopentyl-S-S-cyclopentyl disulfide]

Review Exercises, Chapter 14

14.1 1, 1°alcohol; 2, ketone; 3, 2°alcohol; 4, ketone; 5, alkene

14.3 (a) $CH_3\overset{\underset{|}{CH_3}}{C}HCH_2OH$ (b) $CH_3\overset{\underset{|}{OH}}{C}HCH_3$
      (c) $CH_3CH_2CH_2OH$ (d) $HOCH_2\overset{\underset{|}{OH}}{C}HCH_2OH$

14.8 (a) 2-methyl-1-propanol   (b) 2-propanol
      (c) 1-propanol   (d) 1,2,3-propanetriol

14.12 Hydrogen bonds of the following type form.

$CH_3CH_2{-}\overset{\cdot\cdot}{\underset{\cdot\cdot}{O}}:$

14.16 (a) $HOCH_2CH_2CH_3$ or $CH_3\overset{\underset{|}{OH}}{C}HCH_3$
       (b) [cyclopentane with OH]
       (c) $CH_3\overset{\underset{|}{CH_2OH}}{C}HCH_3$ or $CH_3\overset{\underset{|}{CH_3}}{\underset{|}{\underset{OH}{C}}}CH_3$
       (d) [cyclohexane with CH_3 and OH] or [cyclohexane with OH and CH_3]

14.17 (a) $HOCH_2CH_2CH_2CH_3$
       (b) $CH_3CH_2CH_2\overset{\underset{|}{OH}}{C}HCH(CH_3)_2$
       (c) [cyclopentane with HOCH_2]
       (d) $C_6H_5CH_2OH$

14.19 A reacts with aqueous NaOH; **B** does not. **B** can be dehydrated to an alkene; **A** cannot. (Both react with oxidizing agents, but in different ways.)

14.22 (a) $CH_3CH_2CH_2OH$ (b) $CH_3\overset{\underset{|}{CH_3}}{C}HCH_2OH$
       (c) [cyclohexane with OH]   (d) [cyclopentane with OH]

14.24 No reaction occurs. Ethers are stable in base.

14.26 (a) $CH_3CH_2CH_2SSCH_2CH_2CH_3$
       (b) $(CH_3)_2CHCH_2SH$
       (c)
       (d) $(CH_3)_2CHSSCH(CH_3)_2$

14.27 Hydrogen bonding in the thioalcohol family does exist, but the hydrogen bonds are not as strong as those in the alcohol family.

14.43 (a)    (b) $CH_3CH_2OCH_2CH_3$

(c) $CH_3\overset{O}{\overset{\|}{C}}CH_2CH_3$ (d) 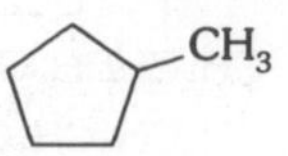

(e) no reaction   (f) no reaction

(g) no reaction   (h) $CH_3CH_2\underset{CH_3}{\underset{|}{C}}HCH_3$

(i) 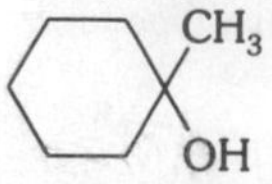   (j) 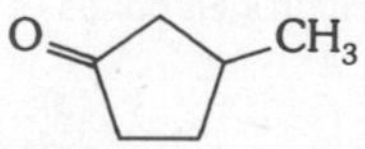

14.46 (a) $3C_3H_8O + 2Cr_2O_7^{2-} + 16H^+ \rightarrow 3C_3H_6O_2 + 4Cr^{3+} + 11H_2O$
(b) $3C_3H_8O + 2K_2Cr_2O_7 + 16HCl \rightarrow 3C_3H_6O_2 + 4CrCl_3 + 4KCl + 11H_2O$
(c) 2/3 mol $K_2Cr_2O_7$
(d) $K_2Cr_2O_7 \cdot 2H_2O$; 2/3 mol
(e) 15.3 g propanoic acid
(f) 53.1 g $K_2Cr_2O_7 \cdot 2H_2O$

CHAPTER 15

Practice Exercises, Chapter 15

1. (a) 2-methylpropanal   (b) 3-bromobutanal
   (c) 4-ethyl-2,4,6-trimethylheptanal
2. 2-Isopropylpropanal would have the structure:

$$CH_3\underset{CH_3\underset{}{C}HCH_3}{\underset{|}{C}H}\overset{O}{\overset{\|}{C}}H$$

and it should be named 2,3-dimethylbutanal.

3. (a) 2-butanone   (b) 6-methyl-2-heptanone
   (c) 2-methylcyclohexanone
4. (a) $CH_3CH_2\overset{O}{\overset{\|}{C}}\underset{CH_3}{\underset{|}{C}}HCH_3$
   (b) $CH_3\overset{O}{\overset{\|}{C}}C_6H_5$
   (c) $CH_3CH_2CH_2\overset{O}{\overset{\|}{C}}CH_2CH_2CH_3$
   (d) $(CH_3)_3C\overset{O}{\overset{\|}{C}}C(CH_3)_3$
5. (a) $CH_3CH_2\overset{OH}{\overset{|}{C}}HCH_3$ (b) $CH_3\underset{CH_3}{\underset{|}{C}}HCH_2CH_2OH$
   (c) 
6. (a) not a hemiacetal   (b) not a hemiacetal
   (c) $HOCH_2OCH_2CH_3$ (↑ hemiacetal position)   (d) hemiketal position 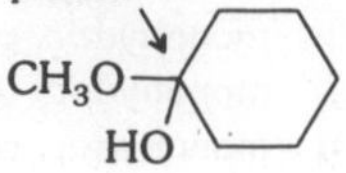
7. (a) $CH_3\overset{OH}{\overset{|}{C}}HOCH_3$
   (b) $CH_3CH_2CH_2\overset{OH}{\overset{|}{C}}HOCH_2CH_3$
   (c) $C_6H_5\overset{OH}{\overset{|}{C}}HOCH_2CH_2CH_3$
   (d) $HOCH_2OCH_3$
8. (a) $CH_3CH_2\overset{O}{\overset{\|}{C}}H + HOCH_3$
   (b) $CH_3CH_2OH + H\overset{O}{\overset{\|}{C}}CH_2CH_3$
9. (a) neither an acetal nor a ketal
   (b) A ketal. The carbon atom that holds two oxygen atoms is a keto group carbon. The breakdown products are:

$$2CH_3CH_2OH + CH_3\overset{O}{\overset{\|}{C}}CH_3$$

10. (a) $2CH_3OH + H\overset{O}{\overset{\|}{C}}H$ (b) no reaction
    (c) $2CH_3OH + CH_3\overset{H_3C}{\overset{|}{C}}H\overset{O}{\overset{\|}{C}}CH_3$

Practice Exercises, Chapter 15

15.2 $CH_3CH_2\overset{O}{\overset{\|}{C}}H$ (aldehyde) $CH_3\overset{O}{\overset{\|}{C}}CH_3$ (ketone) $CH_3CH_2\overset{O}{\overset{\|}{C}}OH$ (carboxylic acid) $CH_3\overset{O}{\overset{\|}{C}}OCH_3$ (ester)

15.3 (a) $CH_3CH_2\underset{CH_3}{\underset{|}{C}}HCHO$
(b) 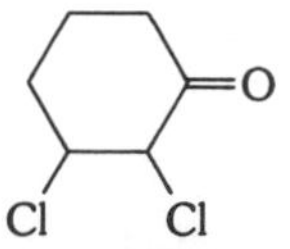
(c) $C_6H_5\overset{O}{\overset{\|}{C}}CH_3$
(d) $(CH_3)_2CHCH_2\overset{O}{\overset{\|}{C}}CH_2CH(CH_3)_2$
(e) $CH_3\overset{O}{\overset{\|}{C}}CH_2CH_2\overset{O}{\overset{\|}{C}}CH_3$

15.5 (a) 2-methylcyclohexanone   (b) propionic acid
(c) 2-pentanone   (d) propanal
(e) 2-methylpentanal

15.7 5-ketohexanal

15.9 (a) 3-methylbutanal
(b) 3-methylpentanal
(c) 3-ethyl-5-methylhexanal
(d) 4-methyl-2-pentanone
(e) 3-propylcyclopentanone

15.11 valeraldehyde

15.13 B < A < D < C

15.15 B < A < D < C

15.17

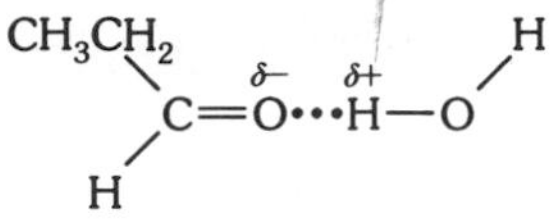

$$CH_3CH_2 \quad C=\overset{\delta-}{O}\cdots\overset{\delta+}{H}-O\quad ^H$$

15.19 (a) $CH_3CH_2CCH_3$
2-butanone

(b) $CH_3CHCH_2CH_2CHO$ (with CH_3 branch)
4-methylpentanal

(c) H_3C— (3-methylcyclopentanone structure) =O
3-methylcyclopentanone

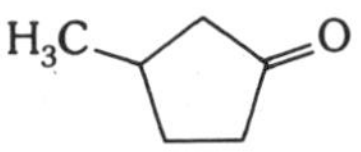

(d) $C_6H_5CH_2CHCHO$ (with CH_3 branch)
2-methyl-3-phenylpropanal

15.21 C_3H_6O is $CH_3CH_2CH=O$; $C_3H_6O_2$ is $CH_3CH_2CO_2H$.

15.23 Positive Benedict's tests are given by (a) and (b).

15.25 (a) The cations of transition metals
(b) Electron-rich particles, whether electrically neutral or negatively charged.
(c) F^-, Cl^-, Br^-, and I^- are four examples in the same family.
(d) H_2O, which forms $Cu(H_2O)_4^{2+}$. NH_3, which forms $Cu(NH_3)_4^{2+}$

15.30 $CH_3CHCO_2^-$ (with OH branch)

15.34 (a) $CH_3CH_2O^-$
(b) $CH_3CH_2O^- + H_2O \rightarrow CH_3CH_2OH + OH^-$
(c) ethanol

15.36 $\overset{+}{H_3N}CHCO_2^-$ (with $CH_2CH_2O^-$ branch) **A** $\overset{+}{H_3N}CHCO_2^-$ (with CH_2CH_2OH branch) **B**

15.38 (a) $CH_3CCH_2CH_3$ (ketone, O double bond)
(b) $HCCHCH_2CH_3$ (with OCH_3 branch, O=)
(c) cyclohexanone =O
(d) CH_3—(ring)—CH (O=)

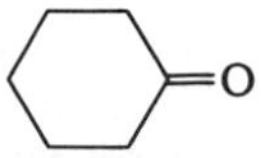

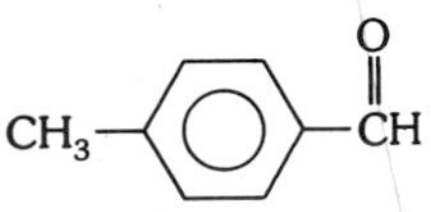

15.40 (a) hemiacetal   (b) acetal
(c) something else (a 1,2-diether)   (d) ketal

15.42 (a) $CH_3CH_2CHOCH_3$ (with OH above), $CH_3CH_2CHOCH_3$ (with OCH_3 above)
(b) $CH_3CH_2CHOCH_2CH_3$ (with OH above), $CH_3CH_2CHOCH_2CH_3$ (with OCH_2CH_3 above)

15.44 (ring structure with CH_3, CH—OH, CH_2, $CH=O$, CH_2—CH_2)

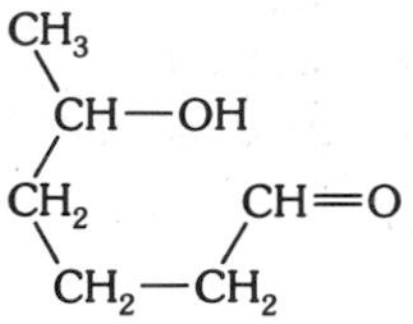

15.46 (ring structure with OH, CH_2, CH_3, $C=O$, CH_2—CH_2)

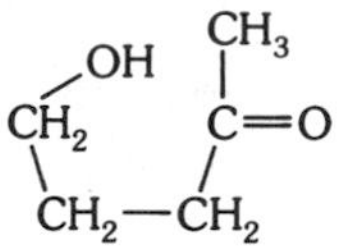

15.48 (a) $CH_3CH_2CHO + 2CH_3OH$
(b) no reaction
(c) $CH_3CCH_3 + 2CH_3CH_2OH$ (O=)
(d) cyclopentane =O + $2CH_3OH$

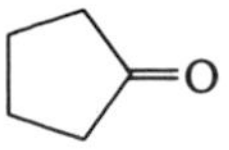

15.53 (a) CH_3CHCH_2OH (with CH_3 branch)
(b) $(CH_3)_2CHCCH_3$ (O=)
(c) no reaction
(d) $CH_3CH_2CH_2CH_2CH_3$
(e) $CH_3CH_2CHOCH_3$ (with OH above)
(f) $CH_3CH_2OH + Mtb^+$
(g) $CH_3CHOCH_2CH_3$ (with OCH_2CH_3 above)
(h) $CH_3CHO + 2CH_3OH$
(i) cyclopentane—CO_2H
(j) no reaction

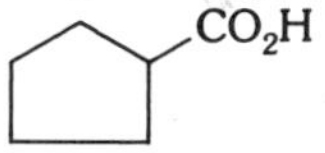

15.55 CH_3CH_2CH (O=) $CH_3CH_2CH_2OH$ $CH_3CH=CH_2$
A   **B**   **C**

CH_3CHCH_3 (with OH above) CH_3CCH_3 (O=)
D   **E**

15.57 (a) 0.173 mol butanal
(b) 1.23 mol CH_3OH
(c) Yes, 11.1 g CH_3OH needed but 39.4 g taken
(d) 3.11 g H_2O obtained
(e) To ensure that the equilibria involved in the reaction are all shifted as much as possible to the right, in favor of the products.

CHAPTER 16

Practice Exercises, Chapter 16

1. (a) 2,2-dimethylpropanoic acid
(b) 5-ethyl-5-isopropyl-3-methyloctanoic acid
(c) sodium ethanoate
(d) 5-chloro-3-methylheptanoic acid

2.   pentanedioic acid

3.   9-octadecenoic acid

4.   (a) $CH_3CH_2CO_2^-$ (b) $CH_3O-\langle\bigcirc\rangle-CO_2^-$

(c) $CH_3CH=CHCO_2^-$

5.   (a) $CH_3O-\langle\bigcirc\rangle-CO_2H$ (b) $CH_3CH_2CO_2H$

(c) $CH_3CH=CHCO_2H$

6.   (a) $CH_3\overset{O}{\overset{\|}{C}}CH_3$ (b) $CH_3\overset{O}{\overset{\|}{C}}CH_2CH_2CH_3$

(c) $CH_3\overset{O}{\overset{\|}{C}}\overset{CH_3}{\overset{|}{O}}CHCH_3$

7.   (a) $H\overset{O}{\overset{\|}{C}}OCH_2CH_3$ (b) $CH_3CH_2\overset{O}{\overset{\|}{C}}OCH_2CH_3$

(c) $C_6H_5\overset{O}{\overset{\|}{C}}OCH_2CH_3$

8.   (a) methyl propanoate
     (b) propyl 3-methylpentanoate

9.   (a) t-butyl acetate   (b) ethyl butyrate

10. (a) $CH_3OH + CH_3CO_2H$
     (b) $(CH_3)_2CHOH + CH_3CH_2CO_2H$
     (c) $CH_3CH_2CH_2OH + (CH_3)_2CHCO_2H$

11. (a) $C_6H_5OH + CH_3CO_2^-$

(b) $CH_3OH + {}^-O_2C-\langle\bigcirc\rangle-OCH_3$

Review Exercises, Chapter 16

16.1 (a) **B**   (b) **A**   (c) **B**   (d) **C**

16.6 (a) 2-methylpropanoic acid
      (b) 2,2-dimethylbutanoic acid
      (c) sodium 3-chloropentanoate
      (d) potassium benzoate

16.8 (a) sodium 3-hydroxybutanoate
      (b) sodium β-hydroxybutyrate

16.11

16.12 **C < A < B**

16.13 **B < C < D < A**

16.15 (a) $HCO_2H + H_2O \rightleftharpoons HCO_2^- + H_3O^+$
       (b) Toward the formate ion. The added OH⁻ (from NaOH) neutralizes H_3O^+ and so reduces the concentration of H_3O^+ in the equilibrium. The equilibrium thus must shift to the right to replace the lost H_3O^+.
       (c) $K_a = \dfrac{[HCO_2^-][H^+]}{[HCO_2H]}$
       (d) stronger

16.17 **B < A < C < D**

16.19 (a) $HO_2CCH_2CH_2CO_2H + 2OH^- \rightarrow$
            $^-O_2CCH_2CH_2CO_2^- + 2H_2O$
       (b) $HOCH_2CH_2CH_2CO_2H + OH^- \rightarrow$
            $HOCH_2CH_2CH_2CO_2^- + H_2O$
       (c) $H\overset{O}{\overset{\|}{C}}CH_2CH_2CH_2CO_2H + OH^- \rightarrow$
            $H\overset{O}{\overset{\|}{C}}CH_2CH_2CH_2CO_2^- + H_2O$
       (d) $O=\langle\bigcirc\rangle-CO_2H + OH^- \rightarrow$
            $O=\langle\bigcirc\rangle-CO_2^- + H_2O$

16.21 **A**, an ionic compound, is much more soluble in water than the molecular compound, **B**.

16.23 (a) $HOCH_2CH_2CO_2^- + H^+ \rightarrow HOCH_2CH_2CO_2H$
       (b) no reaction
       (c) $C_6H_5O^- + H^+ \rightarrow C_6H_5OH$

16.24 (a) $CH_3CH_2\overset{O}{\overset{\|}{C}}Cl$ (b) $CH_3CH_2\overset{O}{\overset{\|}{C}}O\overset{O}{\overset{\|}{C}}CH_2CH_3$

(c) $CH_3CH_2\overset{O}{\overset{\|}{C}}OH$

16.26 (a) $CH_3\overset{O}{\overset{\|}{C}}OCH_2CH_3$
       (b) The electronegativities of O and Cl place a relatively large $\delta+$ charge on the carbon atom of the carbonyl group in acetyl chloride. This charge is able quite strongly to attract the $\delta-$ charge on the O atom of the alcohol molecule. In addition, the Cl⁻ ion is a very stable leaving group and so quite readily leaves the carbonyl carbon atom of the acetyl chloride molecule when the alcohol molecule attacks.

16.28 (a) $CH_3CH_2\overset{O}{\overset{\|}{C}}OCH_2CH_3 + H_2O$
       (b) $(CH_3)_2CH\overset{O}{\overset{\|}{C}}OCH_2CH_3 + H_2O$
       (c) $O_2N-\langle\bigcirc\rangle-\overset{O}{\overset{\|}{C}}OCH_2CH_3 + H_2O$
       (d) $CH_3CH_2O\overset{O}{\overset{\|}{C}}-\langle\bigcirc\rangle-\overset{O}{\overset{\|}{C}}OCH_2CH_3$

16.31 A much stronger attraction can exist between the OH⁻ ion and the $\delta+$ charge on the carbonyl carbon atom of the ester, the specific site attacked in both hydrolysis and saponification, than between this $\delta+$ site and the $\delta-$ charge on a water molecule.

16.33 The acid chloride has a more stable leaving group (the weakly basic Cl^- ion) than the ester, for which the leaving group is a very strongly basic anion of an alcohol.

16.34 (a) $H\overset{O}{\overset{\|}{C}}OCH_2CH_3$ (b) 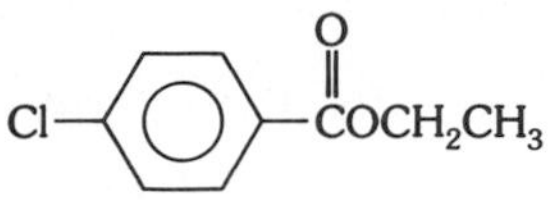

16.36 C < A < B < D

16.38 (a) $CH_3\overset{O}{\overset{\|}{C}}OCH_2\overset{CH_3}{\overset{|}{C}}HCH_3 + H_2O \xrightarrow{H^+}$

$$CH_3\overset{O}{\overset{\|}{C}}OH + HOCH_2\overset{CH_3}{\overset{|}{C}}HCH_3$$

(b) 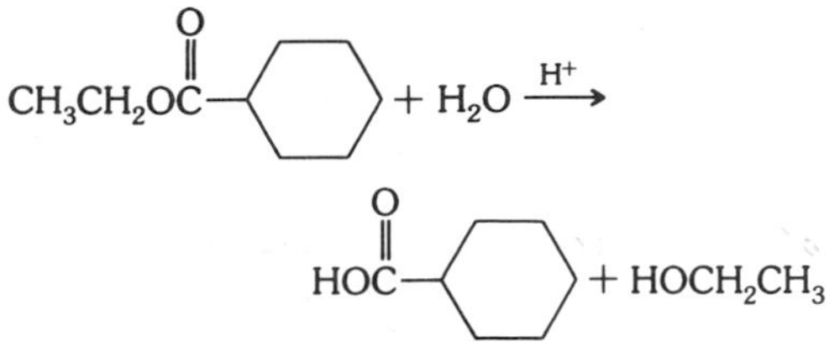

(c) no reaction (d) no reaction

16.40 $HOCH_2\underset{OH}{\underset{|}{C}}HCH_2OH + CH_3(CH_2)_{12}CO_2H +$

$$CH_3(CH_2)_{14}CO_2H + CH_3(CH_2)_{10}CO_2H$$

16.42 (a) $CH_3\overset{O}{\overset{\|}{C}}O^-Na^+ + HOCH_2\overset{CH_3}{\overset{|}{C}}HCH_3$

(b) 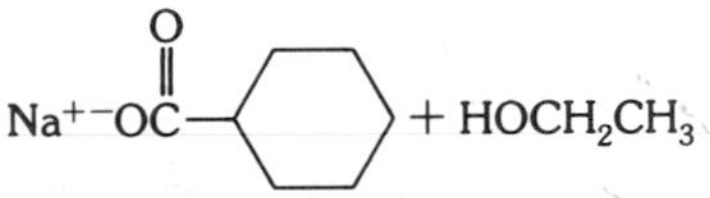

(c) no reaction (d) no reaction

16.44 $HOCH_2\underset{OH}{\underset{|}{C}}HCH_2OH + CH_3(CH_2)_{12}CO_2^-Na^+$

$$+ CH_3(CH_2)_{14}CO_2^-Na^+ + CH_3(CH_2)_{10}CO_2^-Na^+$$

16.46 If a large molar excess of ethyl alcohol is used, the following equilibrium will lie so much on the side of the products that essentially all of the expensive acid will be converted to the ester.

$$RCO_2H + CH_3CH_2OH \rightleftharpoons RCO_2CH_2CH_3 + H_2O$$

16.48 (a) $CH_3O{-}\underset{OH}{\underset{|}{\overset{O}{\overset{\|}{P}}}}{-}OH$

(b) $CH_3CH_2O{-}\underset{OH}{\underset{|}{\overset{O}{\overset{\|}{P}}}}{-}O{-}\underset{OH}{\underset{|}{\overset{O}{\overset{\|}{P}}}}{-}OH$

(c) $CH_3CH_2CH_2O{-}\underset{OH}{\underset{|}{\overset{O}{\overset{\|}{P}}}}{-}O{-}\underset{OH}{\underset{|}{\overset{O}{\overset{\|}{P}}}}{-}O{-}\underset{OH}{\underset{|}{\overset{O}{\overset{\|}{P}}}}{-}OH$

16.60 (a) 6.29 g methyl benzoate

(b) 1.48 g CH_3OH; 1.88 mL CH_3OH

(c) The reaction involves an equilibrium. By using a large excess of methyl alcohol, the equilibrium shifts in accordance with Le Châtelier's principle so that essentially all of the benzoic acid is converted to the ester.

16.61 (a) $CH_3\underset{CH_3O}{\underset{|}{C}}HCO_2^-Na^+$

(b) $CH_3CH_2CO_2H + CH_3OH$

(c) CH_3CO_2H

(d) 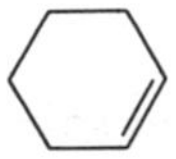

(e) $(CH_3)_2CHCH_2CO_2^-Na^+ + CH_3OH$

(f) $(CH_3)_2CH\overset{O}{\overset{\|}{C}}CH_3$

(g) $CH_3CH_2\overset{O}{\overset{\|}{C}}OCH_2CH_3$

(h) $CH_3CH_2CHO + 2CH_3OH$

(i) no reaction

(j) $C_6H_5\overset{O}{\overset{\|}{C}}OCH_2CH_3$

(k) $CH_3\underset{Cl}{\underset{|}{C}}HCH_2CH_3$

(l) no reaction

CHAPTER 17

Practice Exercises, Chapter 17

1. (a) isopropyldimethylamine
(b) cyclohexylamine
(c) *t*-butylisobutylamine

2. (a) $(CH_3)_3CNH\underset{CH_3}{\underset{|}{C}}HCH_2CH_3$

(b) 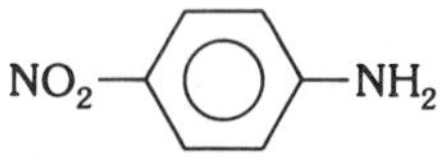

(c) 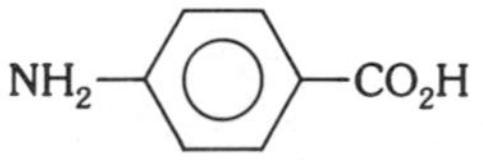

3. (a) $C_6H_5NH_3^+$ (b) $(CH_3)_3NH^+$
(c) $^+NH_3CH_2CH_2NH_3^+$

4. (a) 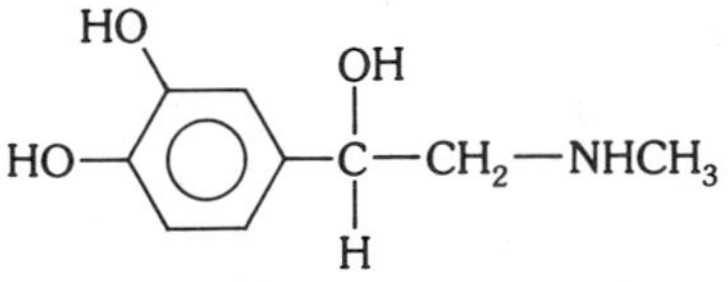

(b) 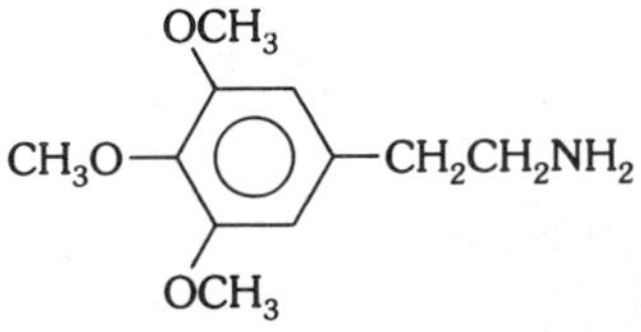

5. (a) 4-methylhexanamide
   (b) 2-ethylbutanamide

6. (a) $(CH_3)_2CHCNHCH_3$ (b) $CH_3CNHC_6H_5$

   (c) No amide forms.   (d) No amide forms.

7. (a) $C_6H_5CO_2H + NH_2CH_3$
   (b) No hydrolysis occurs.
   (c) $C_6H_5NH_2 + HO_2CCH_3$
   (d) $NH_2CH_2CH_2NH_2 + 2CH_3CO_2H$

8. $NH_2CH_2CO_2H + NH_2CHCO_2H +$
   $\qquad\qquad\qquad\qquad CH_3$

   $\qquad NH_2CHCO_2H + NH_2CHCO_2H$
   $\qquad\quad CH_3CH \qquad\qquad CH_2SH$
   $\qquad\qquad CH_3$

Review Exercises, Chapter 17

17.1 (a) aliphatic amide + ether group
   (b) aliphatic amine + ester group
   (c) aliphatic, heterocyclic amide
   (d) Aromatic *compound* overall because of the benzene ring, but the amine is an *aliphatic* amine. (The amino group is not attached directly to the ring.)

17.3 (a) amine; heterocyclic
   (b) 1, amine; 2, ester; 3, amine (aromatic)
   (c) 1, heterocyclic amine; 2, amine (heterocyclic)
   (d) 1, 2° alcohol; 2, amine

17.5 (a) isopropylpropylamine
   (b) ethylmethylpropylamine
   (c) *p*-bromoaniline
   (d) dipropylamine

17.7 (a) $CH_3CH_2CH_2NH_3^+$ (b) $CH_3CH_2CH_2NH_2$
   (c) no reaction   (d) no reaction

17.9 **A** is the stronger base; it is an amine (plus a ketone), and **B** is an amide.

17.11 (a) butanamide   (b) 3-methylbutanamide

17.13 $C_6H_5CON(CH_3)_2$

17.17 $CH_3CCl + 2NH_3 \rightarrow CH_3CNH_2 + NH_4^+Cl^-$

   $CH_3COCCH_3 + 2NH_3 \rightarrow CH_3CNH_2 + NH_4^+ \ ^-OCCH_3$

17.19 (a) $NH_2CH_2CNHCHC-$ (b) two
   $\qquad\qquad\qquad\qquad CH_3$

17.21 (a) $CH_3CH_2NH_2 + CH_3CO_2H$
   (b) $(CH_3)_2CHNH_2 + CH_3CH_2CO_2H$
   (c) $CH_3NH_2 + (CH_3)_2CHCO_2H$
   (d) Does not hydrolyze

17.33 (a) Water reacts quantitatively with alkenes, acetals or ketals, esters, and amides. (Not shown are the hydrolyses of acid chlorides, acid anhydrides, and esters of phosphoric acid.) The R groups can be alike or different. (H)R means that the group can be H or R.

$$\ce{>C=C< + H2O ->[H+] -C-C-} \quad \text{H OH}$$

Alkene $\qquad\qquad\qquad$ Alcohol

$(H)RC(OR)_2 + H_2O \xrightarrow{H^+} (H)RCR(H) + 2HOR$

Acetal or $\qquad\qquad$ Aldehyde $\quad$ Alcohol
ketal $\qquad\qquad\qquad$ or ketone

$(H)RCOR + H_2O \xrightarrow{H^+} (H)RCOH + HOR$

Ester $\qquad\qquad$ Carboxylic $\quad$ Alcohol
$\qquad\qquad\qquad\quad$ acid

$(H)RCNH_2 + H_2O \xrightarrow{H^+} (H)RCOH + NH_3$

Amide $\qquad\qquad$ Carboxylic $\quad$ Ammonia
$\qquad\qquad\qquad\quad$ acid

(The H's on N of the amide can be replaced by one or two alkyl groups.)

(b) The groups that can be hydrogenated are alkenes, aldehydes and ketones, and disulfides.

$$\ce{>C=C< + H2 ->[catalyst] -C-C-} \quad \text{H H}$$

Alkene $\qquad\qquad\qquad$ Alkane

$(H)RCR(H) + H_2 \xrightarrow{catalyst} (H)RCHR(H)$

Aldehyde $\qquad\qquad\qquad$ Alcohol
or ketone

$RSSR + H_2 \xrightarrow{catalyst} 2RSH$

Disulfide $\qquad\qquad\qquad$ Thioalcohol

(c) Oxidizable groups in our study are 1° and 2° alcohols, aldehydes, and thioalcohols.

$RCH_2OH + (O) \longrightarrow RCH=O$

1° Alcohol $\qquad\qquad$ Aldehyde

$RCHR + (O) \longrightarrow RCR$
OH $\qquad\qquad\qquad\qquad$ O

2° Alcohol $\qquad\qquad$ Ketone

$RCHO + (O) \longrightarrow RCO_2H$

Aldehyde $\qquad\qquad$ Carboxylic acid

$2RSH + (O) \longrightarrow RSSR$

Thioalcohol $\qquad\qquad$ Disulfide

17.35 (a) 1.66 g benzoic acid   (b) 28.3 mL 0.482 *M* HCl
17.36 (a) $CH_3CO_2H + CH_3OH$
   OH
   (b) $CH_3CHCH_2CH_3$

(c)   no reaction

(d) $CH_3CH_2\overset{\overset{\displaystyle O}{\|}}{C}NH_2$

(e)   no reaction
(f)   no reaction
(g) $CH_3CHO + 2HOCH_2CH_3$
(h) $CH_3CH_2CO_2H$
(i)   no reaction

(j) $CH_3CH_2\overset{\overset{\displaystyle O}{\|}}{C}OCH_3$

(k) $CH_3CH_2CO_2^-Na^+ + NH_3$

(l) $CH_3CH_2\overset{\overset{\displaystyle OCH_3}{|}}{C}HOCH_3$

(m) $CH_3CH_2SSCH_2CH_3$
(n) $C_6H_5CO_2^-Na^+ + HOCH_2CH(CH_3)_2$
(o) $Cl^- \, {}^+NH_3CH_2CH_2CH(CH_3)_2$
(p) $CH_3CH_2CH_2CH_2OCH_3$

17.39 (a)   sodium benzoate   (b)   propylamine
(c)   aniline   (d)   propionic acid
(e)   butyraldehyde   (f)   isobutyl alcohol
(g)   phenol   (h)   diethyl ether
(i)   ethyl butyrate   (j)   acetamide
(k)   acetone

17.40 **B.** It is a carboxylic acid that will become an anion at the basic pH and so more soluble in water. (**A** is an ester and **C** is an amine.)

CHAPTER 18

Practice Exercises, Chapter 18

1.   (a) $HO\text{—}\overset{\overset{\displaystyle HO}{|}}{\underset{}{\bigcirc}}\text{—}\overset{\overset{\displaystyle CH_3}{|}}{\underset{\displaystyle *}{C}}HCH_2NHCH_3$

(b) $CH_3\overset{*}{C}H\overset{\overset{}{}}{C}O_2H$
          $\overset{|}{OH}$

(c) $CH_3\overset{*}{C}H\overset{*}{C}HCO_2^-$
          $\overset{|}{HO}\;\overset{|}{NH_3^+}$

(d) $HOCH_2\overset{*}{C}H\text{—}\overset{*}{C}H\text{—}\overset{*}{C}H\overset{\overset{\displaystyle O}{\|}}{C}H$
              $\overset{|}{HO}\quad\overset{|}{OH}\quad\overset{|}{OH}$

2.   (a)   3   (b)   8   (c)   4

3. $CH_3\overset{*}{C}H\overset{*}{C}HCH_3$
          $\overset{|}{HO}\;\overset{|}{OH}$

The two tetrahedral stereocenters are identical; they hold identical sets of four different groups, CH_3, OH, H, and $CH_3CH(OH)$.

Review Exercises, Chapter 18
18.2 $CH_3CH_2CH_2OH$ and $(CH_3)_2CHOH$

18.5   (a)   stereoisomers   (b)   constitutional

18.7   (a) $HOCH_2\text{—}\overset{*}{C}H\text{—}\overset{*}{C}H\text{—}\overset{*}{C}H\text{—}\overset{*}{C}H\text{—}CH{=}O$
                  $\overset{|}{OH}\;\;\overset{|}{OH}\;\;\overset{|}{OH}\;\;\overset{|}{OH}$

(b)   All are different   (c)   16   (d)   8

18.8   No; glycine has no tetrahedral stereocenter.

18.10   148.5 °C. The designations (+) and (−) placed before otherwise identical names tell us that the two compounds are enantiomers, and enantiomers have identical physical properties.

18.14   −0.375°

18.18   3.01 g/100 mL

18.20   Strychnine. The calculated specific rotation for the sample is −139°, which corresponds to the value for strychnine, not for brucine.

18.22   (a)   They are not related as object to mirror image.
(b)   Because they are stereoisomers of each other
(c)   What makes them different is not a lack of free rotation.
(d)   diastereomers

18.23   Methane lacks a tetrahedral stereocenter and is not a member of a *set* of stereoisomers.

18.26   (a)   2-butanol   (b)   3-methylhexane

CHAPTER 19

Practice Exercises, Chapter 19
1.   (a)   a = b
          c = e
(b)   Compounds a and d are enantiomers.
(c)   Compound c (or e) is a meso compound.

2.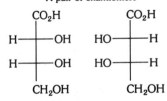
A pair of enantiomers

A pair of enantiomers

Review Exercises, Chapter 19
19.5   (a)   C   (b)   D   (c)   A   (d)   B and D

19.7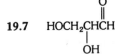
glyceraldehyde

19.9   polysaccharide

19.11 (a) 2 (b) trisaccharide

19.13

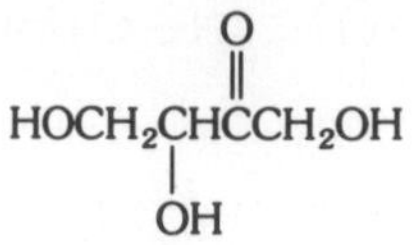

19.16 (a) 2

(b)

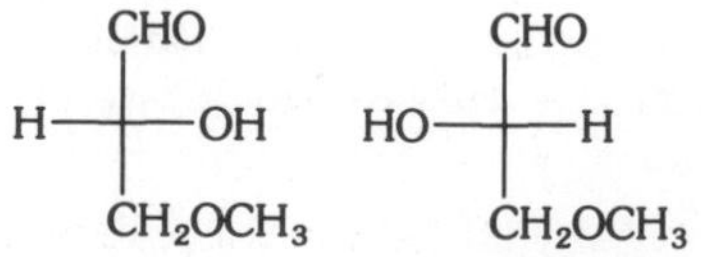

(c) D L

19.17 It has no tetrahedral stereocenter.

19.19

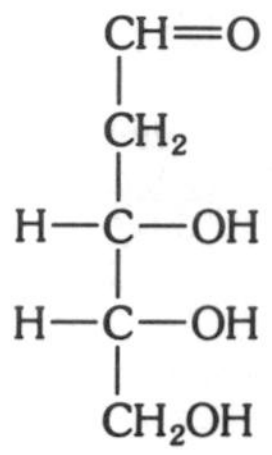

19.21 (a) L-family

(b)

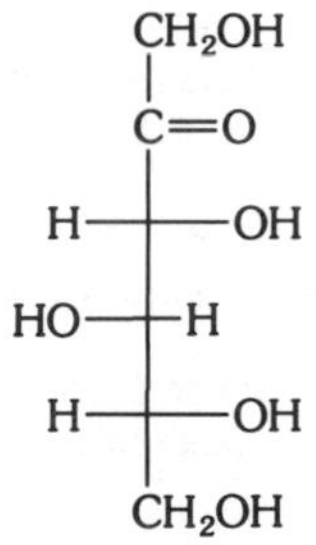

(c)

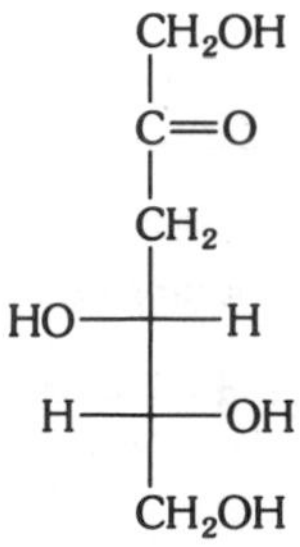

19.23

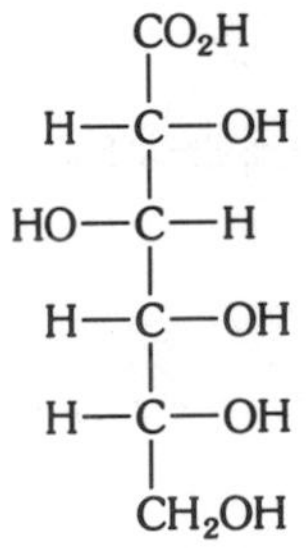

19.24 (a)

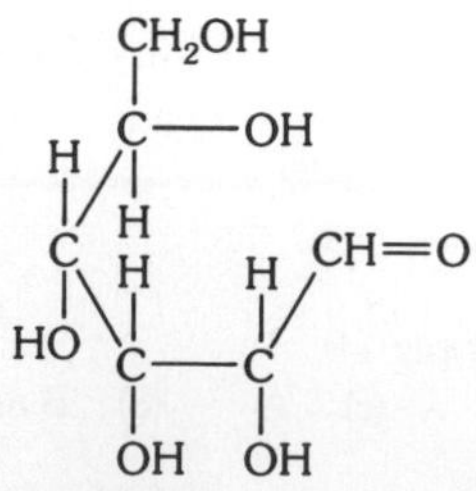

(b) at carbon 3

(c) D-family. The relative positions of the CH_2OH group and the O atom of the ring tell us that the compound is in the D-family.

(d) D-allose

(e) an epimer

19.26

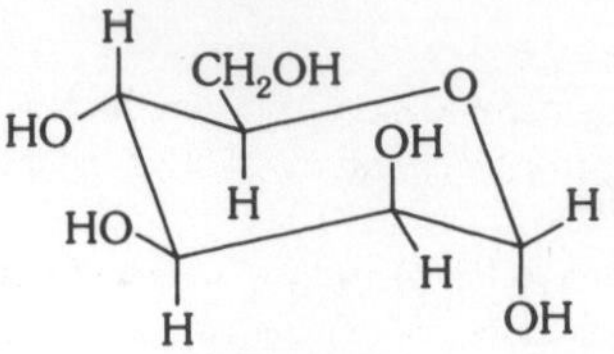

α-mannose

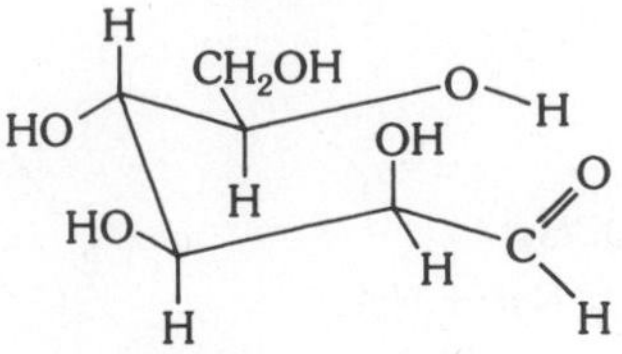

open form of mannose

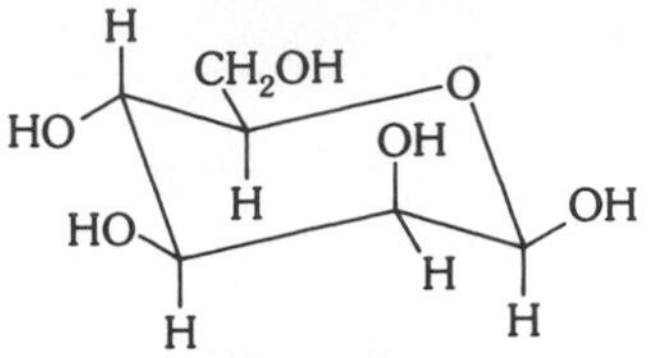

β-mannose

19.29 As the beta form is used, molecules of the other forms continuously change into it as the equilibria shift.

19.31 Something else: β-2-deoxyfructose

19.33 No, an OH group is required at position 4 to make possible the formation of the five-membered ring.

19.36

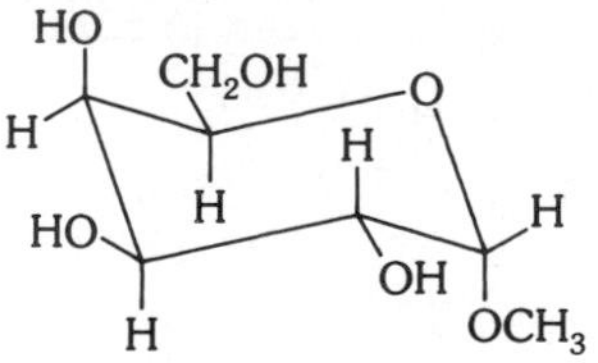

methyl α-galactoside

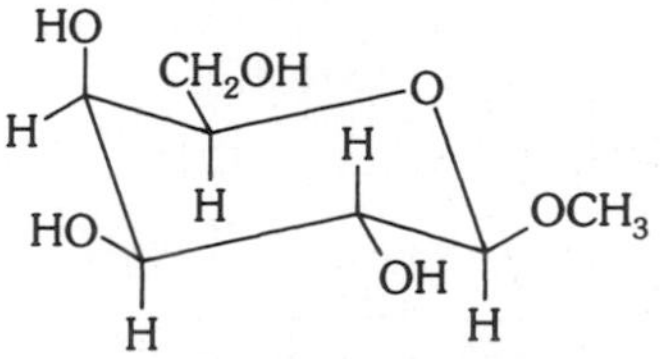

methyl β-galactoside

These are cis–trans isomers in the sense that we have used this concept, because the OCH_3 group is on opposite sides of the rings. However, the term is just not used in connection with glycosides.

19.40 (a) yes, see arrow

(b) yes, see enclosure

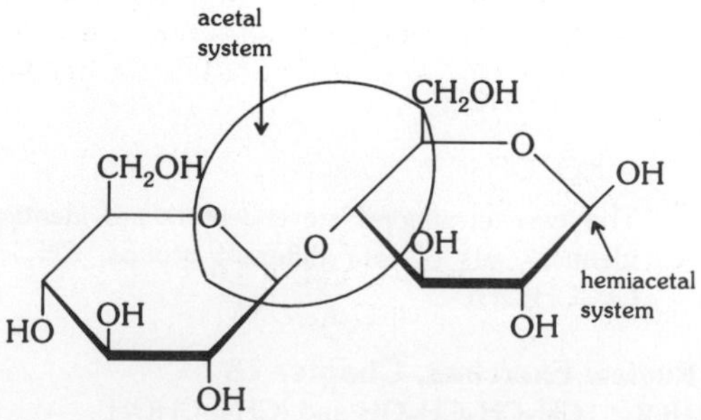

(c) a $\beta(1{\rightarrow}4)$ bridge

(d) Yes, it has the hemiacetal system, so the open form of the corresponding ring (on the right) has an aldehyde group.

(e) Maltose has an $\alpha(1{\rightarrow}4)$ bridge between the two rings.

(f) two glucose molecules

19.42

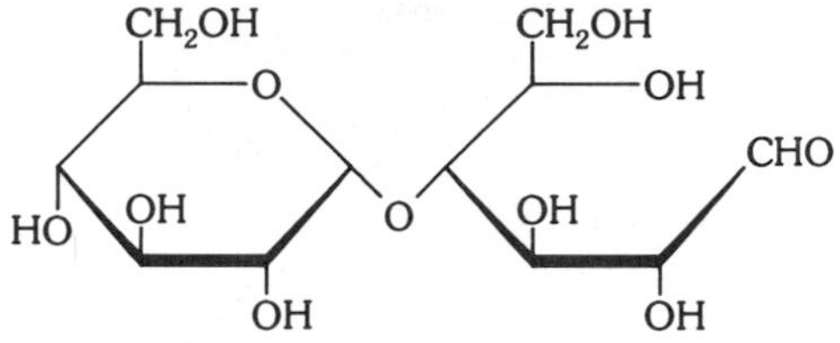

19.59 (a)

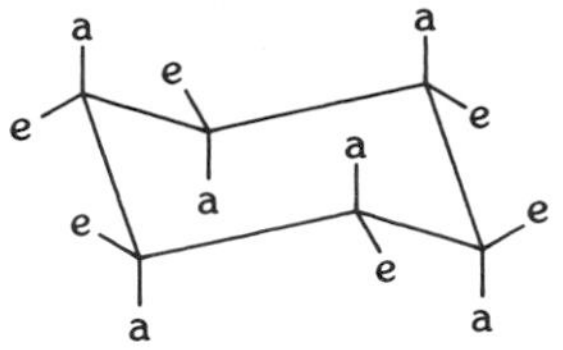

(b)

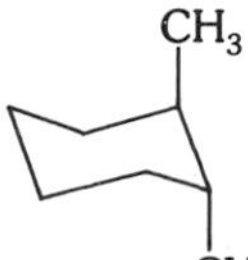

(c)

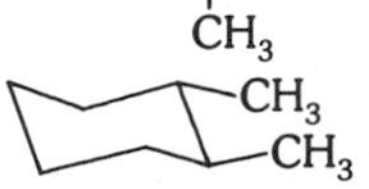

(d)

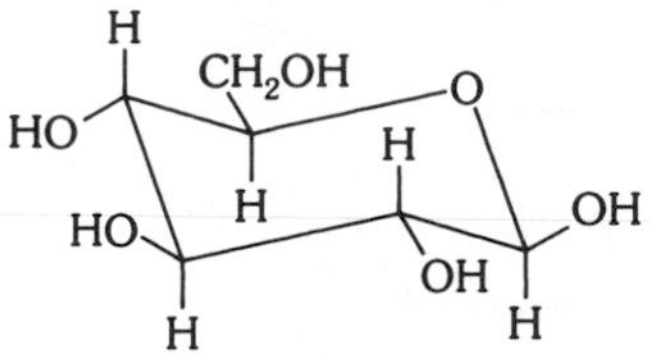

β-glucose
(all substituents are equatorial)

(d)

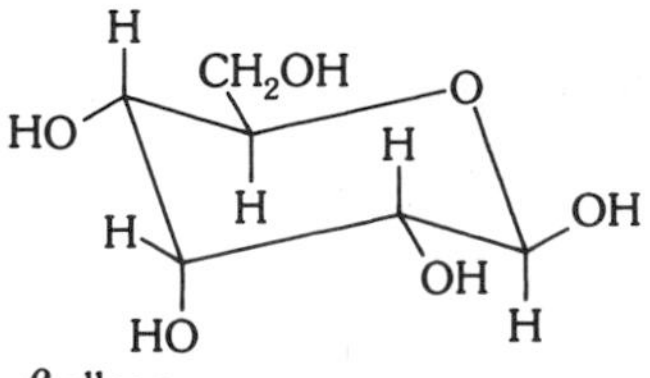

β-allose
(one substituent axial)

19.61 An enzyme (amylase) in the saliva catalyzes the hydrolysis of enough of the starch so that the resulting solution fails to give the iodine test.

19.63 To form a cyclic hemiacetal, the ring would be limited to four atoms. Although four-membered rings are known, they are difficult to form because of the unfavorable bond angle ($90°$ as compared to the normal angle of $109.5°$).

CHAPTER 20

Practice Exercises, Chapter 20

1.

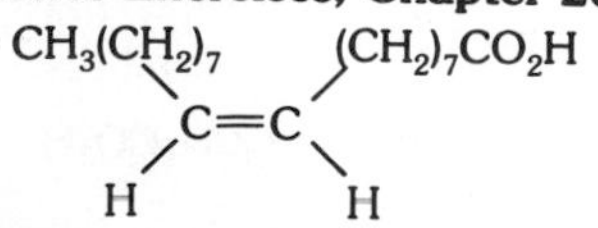

2. $CH_3(CH_2)_{26}CO_2(CH_2)_{25}CH_3$

3. $I + 3NaOH \longrightarrow$

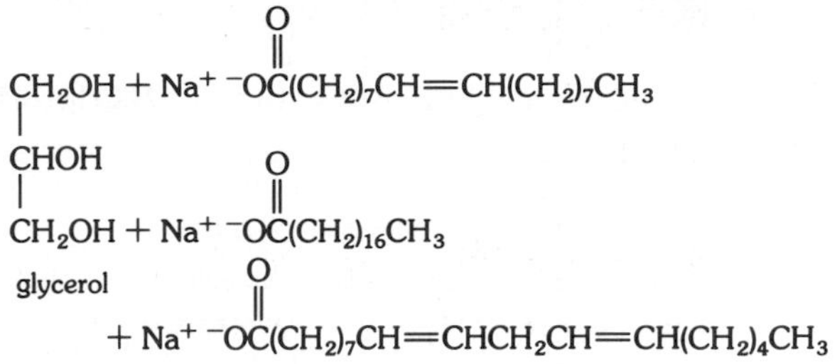

4.

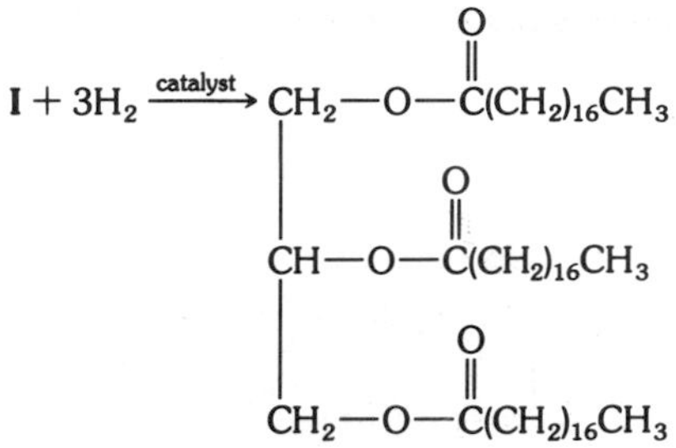

Review Exercises, Chapter 20

20.2 It is extractable from animal and plant sources by relatively nonpolar solvents.

20.4 It is present in undecomposed plant or animal materials and is extractable by relatively nonpolar solvents.

20.8 The organic products of the reactions are the following.

(a) $CH_3(CH_2)_7CH{-}CH(CH_2)_7CO_2H$
$\qquad\qquad\quad | \qquad |$
$\qquad\qquad\ Br \quad\ Br$

(b) $CH_3(CH_2)_7CH{=}CH(CH_2)_7CO_2^{-}K^{+}$

(c) $CH_3(CH_2)_{16}CO_2H$

(d) $CH_3(CH_2)_7CH{=}CH(CH_2)_7CO_2CH_2CH_3$

20.11

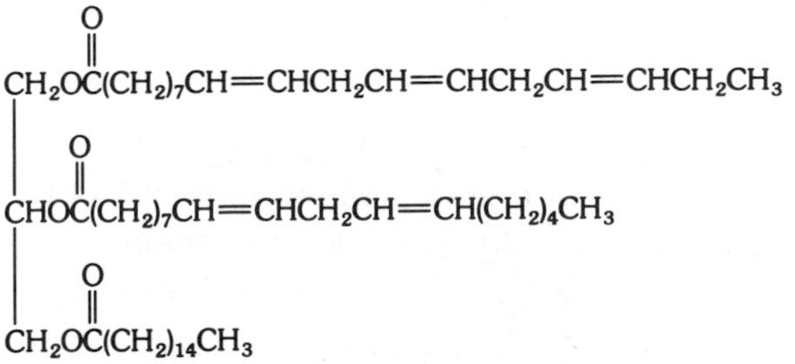

20.13 $HOCH_2CHCH_2OH$
$\qquad\qquad\ \ |$
$\qquad\qquad\ OH$

$\qquad + HO_2C(CH_2)_7CH{=}CHCH_2CH{=}CH(CH_2)_4CH_3$
$\qquad + HO_2C(CH_2)_{12}CH_3$
$\qquad + HO_2C(CH_2)_7CH{=}CH(CH_2)_7CH_3$

20.15 More than one structure is possible because the three different acyl groups can be joined in different orders to the glycerol unit. One possible structure is

$$CH_2-O-\overset{\overset{\displaystyle O}{\|}}{C}(CH_2)_{10}CH_3$$

$$CH-O-\overset{\overset{\displaystyle O}{\|}}{C}(CH_2)_7CH=CHCH_2CH=CH(CH_2)_4CH_3$$

$$CH_2-O-\overset{\overset{\displaystyle O}{\|}}{C}(CH_2)_7CH=CH(CH_2)_7CH_3$$

20.22 C. A is ruled out because *both* the acid and alcohol portions of the wax molecule are usually long chain. B is ruled out because *both* of these portions are likely to have an even number of carbons.

20.23 The molecules of both types give glycerol and phosphoric acid plus a fatty acid when fully hydrolyzed.

20.25 Molecules of both types are derivatives of sphingosine (rather than glycerol). Sphingomyelin molecules have a phosphate ester unit, but those of the cerebrosides have a monosaccharide unit instead.

20.27 Their molecules bear electrical charges at different locations.

20.29 sphingomyelins and cerebrosides

20.31

(a)
$$CH_2O\overset{\overset{\displaystyle O}{\|}}{C}(CH_2)_7CH=CHCH_2CH=CHCH_2CH=CHCH_2CH_3$$

$$\overset{\bullet}{C}HO\overset{\overset{\displaystyle O}{\|}}{C}(CH_2)_7CH=CH(CH_2)_7CH_3$$

$$CH_2O\overset{\displaystyle}{P}OCH_2CH_2\overset{+}{N}(CH_3)_3$$
$$\underset{\displaystyle O^-}{|}$$

(b) A glycerophospholipid, because it is based on glycerol, not sphingosine

(c) Yes, the asterisk in the structure of part (a) marks the tetrahedral stereocenter.

(d) A lecithin, because its hydrolysis would give 2-(trimethylamino)ethanol

20.40 The hydrophobic tails intermesh with each other between the two layers of the bilayer.

20.42 The water-avoiding properties of the hydrophobic units and the water-attracting properties of the hydrophilic units

20.54 (a) yes (b) yes (c) ester groups

(d)
$$CH_2-O-\overset{\overset{\displaystyle O}{\|}}{C}(CH_2)_{17}CH_3$$

$$CH-O-\overset{\overset{\displaystyle O}{\|}}{C}(CH_2)_{11}CH_3$$

$$CH_2-O-\overset{\overset{\displaystyle O}{\|}}{C}(CH_2)_{17}CH_3$$

(e) No. Its fatty acid units have odd numbers of carbon atoms.

CHAPTER 21

Practice Exercises, Chapter 21

1. glycine $^+NH_3CH_2CO_2^-$

   alanine $^+NH_3\overset{\displaystyle}{C}HCO_2^-$
   $$\underset{\displaystyle CH_3}{|}$$

   lysine $^+NH_3\overset{\displaystyle}{C}HCO_2^-$
   $$\underset{\displaystyle CH_2CH_2CH_2CH_2NH_2}{|}$$

   glutamic acid $^+NH_3\overset{\displaystyle}{C}HCO_2^-$
   $$\underset{\displaystyle CH_2CH_2CO_2H}{|}$$

2. (a) $^+NH_3\overset{\overset{\displaystyle O}{\|}}{C}HCO^-$ (b) $^+NH_3\overset{\overset{\displaystyle O}{\|}}{C}HCO^-$
   $$\underset{\displaystyle CH_2CO_2^-}{|} \qquad \underset{\displaystyle CH_2CONH_2}{|}$$

3. $^+NH_3\overset{\overset{\displaystyle O}{\|}}{C}HCO^- \qquad \overset{+}{N}H_2$
   $$\underset{\displaystyle CH_2CH_2CH_2NH\overset{\overset{\displaystyle}{\|}}{C}NH_2}{|}$$

4. Hydrophilic; neutral (The side chain has an amide group, not an amino group.)

5. $^+NH_3\overset{\overset{\displaystyle O}{\|}}{C}HC-NH\overset{\overset{\displaystyle O}{\|}}{C}HCO^-$ $\qquad$ $^+NH_3\overset{\overset{\displaystyle O}{\|}}{C}HC-NH\overset{\overset{\displaystyle O}{\|}}{C}HCO^-$

   $$\underset{\displaystyle CH_3}{|} \quad \underset{\displaystyle CH_2}{|} \qquad\qquad \underset{\displaystyle CH_2}{|} \quad \underset{\displaystyle CH_3}{|}$$
   $$\qquad \underset{\displaystyle CH_2}{|} \qquad\qquad\qquad \underset{\displaystyle CH_2}{|}$$
   $$\qquad \underset{\displaystyle CO_2H}{|} \qquad\qquad\qquad \underset{\displaystyle CO_2H}{|}$$

   Ala-Glu $\qquad\qquad\qquad$ Glu-Ala

Review Exercises, Chapter 21

21.1 B. Its NH_3^+ group is not on the same carbon that holds the CO_2^- group.

21.3 $^+NH_3CH_2CO_2H$

21.5 $NH_2\overset{\displaystyle}{C}HCO_2CH_2CH_3$
$$\underset{\displaystyle CH_3}{|}$$

The polarity of this molecule is much, much less than the polarity of the dipolar ionic form of alanine; thus, the ester molecules stick together with forces weaker than those present between alanine molecules, and the ester has a lower melting point than alanine.

21.7 A. It has amine-like groups that can both donate hydrogen bonds to water molecules and accept them. (B has an alkyl group side chain, which is hydrophobic.)

21.11 oxidizing agent

21.12 In the presence of additional acid, the following equilibrium shifts to the right in accordance with Le Châtelier's principle. This neutralizes the extra acid.

$$^+NH_3CH_2CO_2^- + H^+ \rightleftharpoons {}^+NH_3CH_2CO_2H$$

In the presence of additional base, the following equilibrium shifts to the right in accordance with Le Châtelier's principle. This neutralizes the extra base.

$$^+NH_3CH_2CO_2^- + OH^- \rightleftharpoons NH_2CH_2CO_2^- + H_2O$$

21.14 **A** has an amide bond not to the amino group of the α-position of an amino acid unit but to an amino group of a side chain (that of lysine). **B** has a proper peptide bond.

21.17 Lys-Glu-Cys Glu-Cys-Lys Cys-Lys-Glu
Lys-Cys-Glu Glu-Lys-Cys Cys-Glu-Lys

21.20

21.22 (a) **A**
(b) **B**; it has only hydrophilic side chains. All those in **A** are hydrophobic.

21.24 Gly-Cys-Ala
            |
        Gly-Cys-Ala

21.30 It forms *after* a polypeptide with a cysteine side chain has been put together, so it forms after the primary structure has become set.

21.32 reduce

21.34 A left-handed helix structure

21.36 It consists of three left-handed collagen helices wound together as a right-handed, cablelike triple helix. Between the strands occur molecular "bridges." A fibril forms when individual triple helices overlap lengthwise.

21.37 Covalent linkages fashioned from lysine side chains

21.39 No, they represent portions of the secondary structure of a polypeptide and often both features are present.

21.44 (a) Myoglobin is single stranded; hemoglobin has four subunits.
(b) Myoglobin is in muscle tissue; hemoglobin is in red cells.
(c) Both have the heme unit.
(d) Myoglobin accepts and stores O_2 molecules carried into tissue by hemoglobin molecules.

21.45

21.52 cell fluid

21.58 An oligosaccharide is made from more than two monosaccharide units, and it never has the thousands of such units commonly present in polysaccharide molecules.

21.60 D-glucosamine

21.63 The resiliency of ground substance depends on the hydrogen bonds increasing the "stickiness" of the molecules of ground substance and their abilities to hold large amounts of water as water of hydration.

CHAPTER 22

Practice Exercises, Chapter 22

1. (a) sucrose   (b) glucose
   (c) protein   (d) an ester

2. feedback inhibition

Review Exercises, Chapter 22

22.5 It catalyzes the rapid reestablishment of the equilibrium after it has been disturbed.

22.9 $H^+ + NADH + FAD \rightarrow NAD^+ + FADH_2$

22.11 (a) an oxidation
(b) the transfer of a methyl group
(c) a reaction with water
(d) an oxidation–reduction equilibrium

22.13 Hydrolysis is a kind of reaction catalyzed by a hydrolase enzyme.

22.20 (a) $V \propto [E_0]$
(b) $V \propto [S]$

22.22 At the another site. *Allosteric* describes an action induced at a site on an enzyme molecule at some distance from the active site.

22.24 As the substrate binds to one active site it induces changes in the shape of the enzyme that enable all active sites to become active, so the rate of the reaction suddenly increases rapidly.

22.26 Calmodulin and troponin, the latter being in muscle cells

22.28 At higher Ca^{2+} concentration, $Ca_3(PO_4)_2$ would precipitate.

22.30 Ca^{2+} converts them to activated effectors.

22.32 When a zymogen is cleaved properly, an active enzyme emerges. Trypsinogen is the zymogen for trypsin.

22.43 The levels of these enzymes increase in blood as the result of a disease or injury to particular tissues, which causes tissue cells to release their enzymes.

22.45 CK(*MM*).

22.54 They are primary chemical messengers.

22.56 cyclic AMP and inositol phosphate

22.58 When activated, adenylate cyclase catalyzes the formation of cyclic AMP (from ATP), which then activates an enzyme inside the target cell.

22.60 The cyclic AMP is hydrolyzed to AMP.

22.62 One helps keep the cellular glucose level high and the other helps to bring the Ca^{2+} level of the cytosol up.

22.64 They are hydrocarbon-like and so slip through a hydrocarbon-like lipid bilayer.

22.67 It is hydrolyzed back to acetic acid and choline. The enzyme is cholinesterase. Nerve poisons inactivate this enzyme.

22.69 It blocks the receptor protein for acetylcholine.

22.71 They catalyze the deactivation of neurotransmitters such as norepinephrine and thus reduce the level of signal-sending activity that depends on such neurotransmitters.

22.73 They inhibit the reabsorption of norepinephrine by the presynaptic neuron and thus reduce the rate of its deactivation by the monoamine oxidases.

22.75 dopamine

22.77 They accelerate the release of dopamine from the presynaptic neuron.

22.83 By reducing the flow of Ca^{2+} into cells of heart muscles, the heart beats with reduced vigor.

22.85 They consist of very tiny, gaseous molecules that easily slip through cell membranes.

22.102 (a) The lock-and-key theory, perhaps as modified by induced fit.
   (b) Add water to the alkene group and then oxidize the resulting 2° alcohol to a ketone. (An enzyme would have to guide the addition of the water molecule to give the specific 2° alcohol needed.)

22.104 165.1 mg of isoleucine

CHAPTER 23

Review Exercises, Chapter 23

23.2 Saliva, gastric juice, pancreatic juice, and intestinal juice

23.4 Molecules of a competitive inhibitor occupy the active sites of an enzyme. Cimetidine shuts down the K^+–H^+ pump and prevents the secretion of gastric juice, which gives the ulcer time to heal in a relatively acid-free environment.

23.6 (a) α-amylase
   (b) pepsinogen and gastric lipase
   (c) α-amylase, lipase, nuclease, trypsinogen, chymotrypsinogen, procarboxypeptidase, and proelastase
   (d) no enzymes
   (e) amylase, aminopeptidase, sucrase, lactase, maltase, lipase, nucleases, enteropeptidase

23.8 (a) amino acids
   (b) glucose, fructose, and galactose
   (c) fatty acids and monoacylglycerols (plus some diacylglycerols)

23.10 It catalyzes the conversion of trypsinogen to trypsin. Then trypsin catalyzes the conversion of other zymogens to chymotrypsin, carboxypeptidase, and elastin. Thus enteropeptidase turns on enzyme activity for three major protein-digesting enzymes.

23.12 They are surface-active agents that help to break up lipid globules, wash lipids from the particles of food, and aid in the absorption of fat-soluble vitamins.

23.14 (a) HCl   (b) enteropeptidase
   (c) trypsin   (d) trypsin
   (e) trypsin

23.16 This enzyme is inactive at the high acidity of the digesting mixture in the adult stomach, but the mixture's acidity in the infant's stomach is less.

23.20 The flow of bile normally delivers colored breakdown products from hemoglobin in the blood, and these products give the normal color to feces. When no bile flows, no colored products are available to the feces.

23.21 Cholesterol has no hydrolyzable groups.

23.23 serum-soluble proteins (albumins, mostly)

23.36 hypercalcemia

23.38 chloride ion, Cl^-

23.41 Blood pressure and osmotic pressure. The natural return of fluids to the blood on the venous side from the interstitial compartment is not balanced by the now reduced blood pressure on the venous side, so fluids return to the blood from which they left on the arterial side.

23.43 (a) Blood proteins leak out, which allows water to leave the blood and enter interstitial spaces throughout various tissues.
   (b) Blood proteins are lost to the blood by being consumed, which also leads to the loss of water from the blood and its appearance in interstitial compartments.
   (c) Capillaries are blocked at the injured site, reducing the return of blood in the veins, so fluids accumulate at the site.

23.46 The first oxygen molecule to bind changes the shapes of other parts of the hemoglobin molecule and makes it much easier for the remaining three oxygen molecules to bind. This ensures that all four oxygen-binding sites of each hemoglobin molecule will leave the lungs fully loaded with oxygen.

23.47 $HHb + O_2 \rightleftharpoons HbO_2^- + H^+$
   (a) to the left   (b) to the left
   (c) to the right   (d) to the left
   (e) to the left   (f) to the right

23.49 It generates H^+ needed to convert HCO_3^- to CO_2 and H_2O and to convert $HbCO_2^-$ to HHb and CO_2.

23.51 It helps to shift the following equilibrium to the left:

$$HHb + O_2 \rightleftharpoons HbO_2^- + H^+$$

23.56 For oxygenation:

$$HHb-BPG + O_2 + HCO_3^- \rightarrow$$
$$HbO_2^- + BPG + CO_2 + H_2O$$

For deoxygenation:

$$HbO_2^- + BPG + CO_2 + H_2O \rightarrow$$
$$HHb-BPG + O_2 + HCO_3^-$$

23.58 Oxygen affinity is lowered. Where the partial pressure of CO_2 is relatively high (as in actively metabolizing tissue) there is a need for oxygen, so the lowering effect of CO_2 on oxygen affinity helps to release O_2 precisely where O_2 is most needed.

23.63 The pH of the blood decreases in both, but both pCO_2 and $[HCO_3^-]$ increase in respiratory acidosis and both decrease in metabolic acidosis.

23.65 Hypoventilation is observed in metabolic alkalosis, and isotonic ammonium chloride can be given to neutralize the excess base. Involuntary hypoventilation is observed in respiratory acidosis, and isotonic sodium bicarbonate might be given to neutralize excess acid.

23.67 In respiratory alkalosis. The involuntary loss of CO_2 reduces the level of H_2CO_3 in the blood and thereby reduces the level of H^+.

23.69 In respiratory acidosis

23.71 (a) respiratory alkalosis   (b) metabolic alkalosis
       (c) respiratory acidosis   (d) respiratory acidosis
       (e) metabolic acidosis   (f) respiratory alkalosis
       (g) metabolic acidosis   (h) metabolic alkalosis
       (i) respiratory acidosis   (j) respiratory acidosis

23.72 (a) hyperventilation   (b) hypoventilation
       (c) hypoventilation   (d) hypoventilation
       (e) hyperventilation   (f) hyperventilation
       (g) hyperventilation   (h) hypoventilation
       (i) hypoventilation   (j) hypoventilation

23.74 Hypoventilation in emphysema lets the blood retain carbonic acid, and the pH decreases.

23.76 The loss of alkaline fluids from the duodenum and lower intestinal tract leads to a loss of base from the bloodstream, too. The result is a decrease in the blood's pH and thus acidosis.

23.78 hypercapnia

23.80 It acts to prevent the loss of water via the urine by letting the hypophysis secrete vasopressin, whose target cells are in the kidneys. The retention of water helps to keep the blood's osmotic pressure from increasing further. The thirst mechanism is also activated to bring in more water to dilute the blood.

23.82 The kidneys secrete a trace of renin into the blood. This catalyzes the conversion of angiotensinogen to angiotensin I, which catalyzes the formation of angiotensin II, a neurotransmitter and powerful vasoconstrictor. With constricting of the capillaries, the blood pressure has to increase to keep the delivery of blood going.

23.84 They transfer hydrogen ions into the urine and put bicarbonate ions into the bloodstream.

23.86 Without the bile salts in the bile, normally obtained from the gallbladder, there is less emulsifying action to aid in the digestion of triacylglycerols with the usual long fatty acyl side chains. The shorter chain molecules are a little more soluble in the digestive medium.

23.88 These animals can store more oxygen in heart muscle, which helps them to go longer without breathing.

CHAPTER 24

Practice Exercises, Chapter 24

1. (a) proline   (b) arginine
    (c) glutamic acid   (d) lysine

2. (a) serine   (b) CT(chain termination)
    (c) glutamic acid   (d) isoleucine

Review Exercises, Chapter 24

24.1 (a) cytosol   (b) protoplasm
      (c) cytoplasm   (d) ribosome
      (e) mitochondrion   (f) deoxyribonucleic acid

      (g) chromatin   (h) histone
      (i) gene

24.6 nucleotides

24.9 The main chains all have the same phosphate–deoxyribose–phosphate–deoxyribose repeating system.

24.10 In the sequence of bases attached to the deoxyribose units of the main chain

24.13 A and T pair to each other, so they must be in a 1:1 ratio regardless of the species. Similarly, G and C pair to each other and must be in a 1:1 ratio.

24.15 Hydrophobic interactions stabilize the helices. (Hydrogen bonds hold two helices together.)

24.19 The introns are b, d, and f, because they are the longer segments.

24.26 ATA. A codon is RNA material, and T does not occur in RNA.

24.28 Writing them in the 5′ to 3′ direction:
    (a) AAA   (b) GGA
    (c) UGU   (d) AUC

$$5' \rightarrow 3'$$

24.29 (a) UUUCUUAUAGAGUCCCCAACAGAU

$$5' \rightarrow 3'$$

    (b) UUUUCCACAGAU

    (c) Phe-Ser-Thr-Asp

24.32 (a) A large number of sequences are possible because three of the specific amino acid residues are coded by more than one codon. The possibilities are indicated by

           Met—Ala—Trp—Ser—Tyr
           AUG GCU UGG UCU UAU   (5′ → 3′)
                 GCC   UCC UAC
                 GCA   UCA
                 GCG   UCG

    (b) CAU (5′ → 3′) or, if (3′ → 5′), then UAC

24.34 How polypeptide synthesis can be controlled by the use of repressors

24.40 The codons specify the same amino acids in all organisms. No.

24.47 RNA replicase is able to direct the synthesis of RNA from the "directions" encoded on RNA. A normal host cell does not contain RNA replicase, so the virus either must bring it along or it must direct its synthesis inside the host cell.

24.49 (+)mRNA

24.51 RNA replicase

24.53 A silent gene is a unit provided by a virus particle to a host cell but which does not take over the genetic apparatus of the cell until activated.

24.57 The theory is that AZT molecules will bind to reverse transcriptase and inhibit the work of this enzyme in HIV.

24.59 restriction enzyme

24.63 a defect in a gene

24.64 The synthesis of a transmembrane protein that lets chloride ion pass through the membranes of mucous cells in the lungs and the digestive tract. The reduced movement of

Cl^- out of the cell means that less water is outside of the cell, and the mucous is thereby thickened and made more viscous. Breathing is thereby impaired.

24.66 Retroviruses can bring about the insertion of DNA into a chromosome of a host cell.

24.70 It is the synthesis of clones of DNA. DNA samples at crime scenes generally involve tiny amounts of DNA, and DNA typing depends on making enough cloned DNA for the next steps of the procedure.

24.76 (a) by one methyl group
(b) no
(c) Both uracil and thymine can form a base pair with adenine.

CHAPTER 25

Practice Exercise, Chapter 25
1. (a) yes (b) yes (c) no

Review Exercises, Chapter 25
25.1 All of them, but chiefly fatty acids and carbohydrates

25.3 Combustion produces just heat. The catabolism of glucose uses about half of the energy to make ATP, and the remainder appears as heat.

25.5

$$\text{Adenosine} - O - \overset{\displaystyle O}{\underset{\displaystyle O^-}{\overset{\|}{P}}} - O - \overset{\displaystyle O}{\underset{\displaystyle O^-}{\overset{\|}{P}}} - O - \overset{\displaystyle O}{\underset{\displaystyle O^-}{\overset{\|}{P}}} - O^-$$

25.7 The singly and doubly ionized forms of phosphoric acid, $H_2PO_4^- + HPO_4^{2-}$.

25.9 The first two have phosphate group transfer potentials equal to or higher than that of ATP, whereas this potential is lower than that of ATP for glycerol-3-phosphate.

25.11 (a) yes (b) yes (c) no (d) no

25.13 It stores phosphate group energy and transfers phosphate to ADP to remake the ATP consumed by muscular work.

25.15 (a) The aerobic synthesis of ATP
(b) The synthesis of ATP when a tissue operates anaerobically
(c) The supply of metabolites for the respiratory chain
(d) The supply of metabolites for the respiratory chain and for the citric acid cycle

25.17 An increase in its supply of ADP. The need to convert ADP back to ATP is met by metabolism, which requires oxygen.

25.19 (a) $E < B < C < A < D$ (b) $B < C < A < D$

25.23 When a cell is temporarily low on oxygen for running the respiratory chain (as a source of ATP), the cell can continue (for a while) to make ATP anyway.

25.25 An acetyl unit:

$$\overset{\displaystyle O}{\overset{\|}{CH_3C}}$$

25.27 two

25.30 9

25.32 An enzyme like isocitrate dehydrogenase, because the reaction is a dehydrogenation (not an addition of water to a double bond, catalyzed by an enzyme like aconitase)

25.35

$$\underset{\displaystyle \overset{\displaystyle OH}{|}}{CH_3CHCO_2^-} + NAD^+ \rightarrow \underset{\displaystyle \overset{\displaystyle O}{\|}}{CH_3CCO_2^-} + NAD\!:\!H + H^+$$

(a) $\underset{\displaystyle \overset{\displaystyle OH}{|}}{CH_3CHCO_2^-}$ (b) NAD^+

25.36

$$^-O_2CH_2CH_2CO_2^- \quad \diagdown \quad FAD$$
$$^-O_2CCH\!=\!CHCO_2^- \quad \diagup\!\!\!\!\longleftrightarrow \quad FADH_2$$

25.40 Across the inner membrane of the mitochondrion. The value of $[H^+]$ is higher on the outer side of the inner membrane than in the mitochondrial matrix.

25.42 If the membrane is broken, then the simple process of diffusion defeats any mitochondrial effort to set up a gradient of H^+ ions across the membrane, but the chain itself can still operate.

25.44 Oxidative phosphorylation is the kind made possible by the energy released from the operation of the respiratory chain. Substrate phosphorylation arises from the direct transfer of a phosphate unit from a higher to a lower energy phosphate.

25.46 An enzyme catalyzes the formation of ATP from ADP and P_1, but the new ATP sticks tightly to the enzyme. But this enzyme is at the end of the channel for protons in the inner mitochondrial membrane, and as protons flow through they change the enzyme so that it expels the ATP.

25.48

$$\underset{\displaystyle \overset{\displaystyle CH_3}{|}}{CH_3CH_2\overset{\displaystyle \overset{\displaystyle O}{\|}}{C}HCO_2CH_2CH_3}$$

25.51 $CH_3CH\!=\!O + NAD^+ + H_2O \rightarrow C_2H_3O_2^- + NADH + 2H^+$; $E'^\circ_{cell} = +0.28$ V. The spontaneous reaction is the oxidation of ethanal.

25.55 (a) $CH_3CH\!=\!O$
(b) acetic acid, CH_3CO_2H (or acetate ion, $CH_3CO_2^-$)
(c) α-ketoglutarate

25.57 (a) 4
(b) The nucleus of a hydrogen atom, because $H\!:\!^-$ transfers
(c) dehydrogenase

CHAPTER 26

Review Exercises, Chapter 26
26.5 99 mg/dL

26.8 The lack of glucose means the lack of the one nutrient most needed by the brain.

26.12 $D < C < E < A < B$

26.14 Glucose might be changed back to glycogen as rapidly as it is released from glycogen, and no glucose would be made available to the cell.

26.16 Glucose-1-phosphate is the end product, and phosphoglucomutase catalyzes its change to glucose-6-phosphate.

26.19 Glucagon, because it works better at the liver than epinephrine in initiating glycogenolysis, and when glycogenolysis occurs at the liver there is a mechanism for releasing glucose into circulation.

26.23 Too much insulin leads to a sharp decrease in the blood sugar level and therefore a decrease in the supply of glucose, the chief nutrient for the brain.

26.26 By its conversion to glucose-6-phosphate, which either enters a pathway that makes ATP or is converted to glycogen for storage

26.31 An overrelease of epinephrine (as in a stressful situation) that induces an overrelease of glucose

26.32 $C_6H_{12}O_6 + 2ADP + 2P_i \rightarrow 2C_3H_5O_3^- + 2H^+ + 2ATP$

26.34 Glyceraldehyde-3-phosphate is in the direct pathway of glycolysis, so changing dihydroxyacetone phosphate into it ensures that all parts of the original glucose molecule are used in glycolysis.

26.36 It is reoxidized to pyruvate, which then undergoes oxidative decarboxylation to the acetyl group in acetyl CoA. This enters the citric acid cycle.

26.38 NADPH forms, and the body uses it as a reducing agent to make fatty acids.

26.40 All are catabolized, and parts of some of their molecules are used to make fatty acids and, thence, fat.

26.42 α-Ketoglutarate is an intermediate in the citric acid cycle (Fig. 25.2) and so is changed eventually to oxaloacetate (normally the acceptor of acetyl units at the start of the citric acid cycle). Gluconeogenesis also can use oxaloacetate to make "new" glucose.

26.43 Succinyl units are in the citric acid cycle, which ends with the formation of oxaloacetate, and the latter can be used in gluconeogenesis.

26.57 (a) $\underset{\text{OH}}{\text{CH}_3\text{CHCH}_2}\underset{\text{O}}{\text{CH}}$

(b) $\text{CH}_3\text{CH}_2\underset{\text{OH}}{\text{CHCHCH}}\underset{\text{O}}{} $
        CH_3

(c) $\text{CH}_3\underset{\text{OH}}{\text{CCH}_2}\underset{\text{O}}{\text{CCH}_3}$
        CH_3

26.58 $C_6H_5CH_2\overset{O}{\overset{\|}{C}}H$

26.59 (a)   Yes, either pyruvate or lactate containing carbon-13 may reenter circulation.

(b)   Yes, either pyruvate or lactate containing carbon-13 might be absorbed by the liver from the bloodstream.

(c)   Yes, glucose with carbon-13 atoms might be made via gluconeogenesis from either pyruvate or lactate containing carbon-13.

26.60 (a)   15 ATP

(b)   3 ATP

(c)   3 glucose (because 6 ATP are needed to make each molecule of glucose by gluconeogenesis)

CHAPTER 27

Review Exercises, Chapter 27

27.2   They become reconstituted into triacylglycerols.

27.4   They unload some of their triacylglycerol.

27.6   Some cholesterol has originated in the diet and some has been synthesized in the liver.

27.8   triacylglycerol

27.10   the liver

27.12   The synthesis of steroids and the fabrication of cell membranes

27.14   When the receptor proteins are reduced in number, the liver cannot remove cholesterol from the blood, so the blood cholesterol level increases.

27.19   fasting and diabetes

27.21 $E < D < A < C < B$

27.23   The citric acid cycle processes the acetyl units manufactured by the β-oxidation pathway and so fuels the respiratory chain.

27.25   An increase in the blood sugar level triggers the release of insulin which inhibits the release of fatty acids from adipose fat.

27.28 $CH_3CH_2CH_2\overset{O}{\overset{\|}{C}}SCoA + FAD \rightarrow$

$CH_3CH=CH\overset{O}{\overset{\|}{C}}SCoA + FADH_2$

$CH_3CH=CH\overset{O}{\overset{\|}{C}}SCoA + H_2O \rightarrow CH_3\underset{OH}{CH}CH_2\overset{O}{\overset{\|}{C}}SCoA$

$CH_3\underset{OH}{CH}CH_2\overset{O}{\overset{\|}{C}}SCoA + NAD^+ \rightarrow$

$CH_3\overset{O}{\overset{\|}{C}}CH_2\overset{O}{\overset{\|}{C}}SCoA + NAD\!:\!H + H^+$

$CH_3\overset{O}{\overset{\|}{C}}CH_2\overset{O}{\overset{\|}{C}}SCoA + CoASH \rightarrow 2CH_3\overset{O}{\overset{\|}{C}}SCoA$

No more turns of the β-oxidation pathway are possible.

27.30   NADH passes its hydrogen into the respiratory chain and is changed back to NAD^+.

27.32   (a)   inside mitochondria   (b)   cytosol

27.34   The pentose phosphate pathway of glucose catabolism

27.36   Cholesterol inhibits the synthesis of HMG–CoA reductase.

27.40   A proton or hydrogen ion, H^+. If the level of hydrogen ion increases, the problem is acidosis.

27.47   acetoacetic acid

27.49   Their *over*production leads to acidosis.

27.51   Each Na^+ ion that leaves corresponds to the loss of one HCO_3^- ion, the true base, because HCO_3^- neutralizes acid generated as the ketone bodies are made. And for every negative ion that leaves with the urine a positive ion, mostly Na^+, has to leave to ensure electrical neutrality.

27.63   (a)   7   (b)   6   (c)   6   (d)   112

CHAPTER 28

Review Exercises, Chapter 28

28.3 infancy

28.5 glutamic acid (glutamate)

28.7 $C_6H_5CH_2CCO_2H$ (with O double-bonded above the second C)

28.9 The body can use it to make glucose by gluconeogenesis.

28.11 ketogenic

28.13 $4 < 1 < 5 < 2 < 3$

28.15 Yes: Glucose $\xrightarrow[\text{glycolysis}]{\text{aerobic}}$ pyruvate $\xrightarrow{\text{transamination}}$ alanine

28.17 HO—⬡—CH$_2$CH$_2$NH$_2$

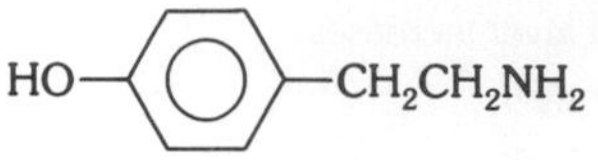

tyramine

28.19 To synthesize glucose by means of gluconeogenesis

28.20 (a) originally, the amino groups of amino acids
(b) carbon dioxide

28.23 Hyperammonemia. Step 2 consumes carbamoyl phosphate, which is made using ammonia. If carbamoyl phosphate levels rise, a backup occurs to cause ammonia levels to increase.

28.26 $3 < 2 < 1 < 4 < 5 < 7 < 6$

28.31 The exclusively ketogenic amino acids, those not also glucogenic, cannot produce net extra oxaloacetate.

CHAPTER 29

Review Exercises, Chapter 29

29.3 Foods are complex mixtures of nutrients.

29.5 To allow for individual differences among people and to ensure that practically all people can thrive

29.7 No food contains all of the essential nutrients, and there might still be nutrients yet to be discovered but which are routinely provided by a varied diet.

29.9 Linoleic acid. "Essential" means that it must be provided in the diet. If linoleic acid is absent in the diet, the prostaglandins are not made at a sufficient rate.

29.12 The body can make several amino acids itself.

29.14 It breaks them down and eliminates the products.

29.16 From an animal source

29.18 The proportions of essential amino acids available from it

29.20 The essential amino acid most poorly supplied by the protein

29.22 (a) 127 g
(b) 1.7×10^3 g
(c) 6.1×10^3 kcal
(d) Very likely not, since 1.7×10^3 g is nearly 3 lb.

29.24 They are needed in much more than trace amounts, and they come from proteins.

29.26 No single vegetable source has a balanced supply of essential amino acids.

29.28 vitamin D

29.30 vitamin A

29.32 vitamin K

29.34 vitamins C and E

29.36 vitamin C

29.38 vitamin C

29.40 thiamin

29.42 folate

29.43 The quantity needed per day. Minerals are needed in the amount of more than 100 mg/day and trace elements in the amount of less than 20 mg/day.

29.47 2.1×10^3 L air

GLOSSARY[1]

Absolute Configuration The actual arrangement in space about each tetrahedral stereocenter in a molecule. (19.3)

Absolute Zero The coldest temperature attainable; 0 K or −273.15 °C. (6.2)

Accuracy In science, the degree of conformity to some accepted standard or reference; freedom from error or mistake; correctness. (1.6)

Acetal Any organic compound in which two ether-like linkages extend from one CH unit. (15.5)

Acetyl Coenzyme A The molecule from which acetyl groups are transferred into the citric acid cycle or into the synthesis of fatty acids. (25.1)

Achiral Not possessing chirality; that quality of a molecule (or other object) that allows it to be superimposed on its mirror image. (18.2)

Acid *Brønsted theory:* Any substance that can donate a proton (H^+). (4.1, 5.5, 8.2)

Acid Anhydride In organic chemistry, a compound formed by splitting water out between two OH groups of the acid functions of two organic acids. (16.3)

Acid–Base Indicator (See *Indicator*)

Acid–Base Neutralization The reaction of an acid with a base. (8.2)

Acid Chloride A derivative of a carboxylic acid in which the OH group of the acid has been replaced by Cl. (16.3)

Acid Derivative Any organic compound that can be made from an organic acid or that can be changed back to the acid by hydrolysis. (Examples are acid chlorides, acid anhydrides, esters, and amides.) (16.3)

Acidic Solution A solution in which the molar concentration of hydronium ions is greater than that of hydroxide ions. (8.2, 9.5)

Acid Ionization Constant (K_a) A modified equilibrium constant for the following equilibrium:

$$HA + H_2O \rightleftharpoons H_3O^+ + A^-$$

$$K_a = \frac{[H_3O^+][A^-]}{[HA]} \quad (9.6)$$

Acidosis A condition in which the pH of the blood is below normal. *Metabolic acidosis* is brought on by a defect in some metabolic pathway. *Respiratory acidosis* is caused by a defect in the respiratory centers or in the mechanisms of breathing. (9.9, 23.4, 27.6)

Acid Rain Rain made acidic by the presence of air pollutants such as oxides of sulfur and nitrogen. (Special Topic 9.3)

Active Transport The movement of a substance through a biological membrane against a concentration gradient and caused by energy-consuming chemical changes that involve parts of the membrane. (7.5, 21.8)

Activity Series A list of elements (or other substances) in the order of the ease with which they release electrons under standard conditions and become oxidized. (8.3)

Acyl Group

$$R-\overset{\displaystyle O}{\overset{\displaystyle \|}{C}}- \quad (16.3)$$

Acyl group

Acyl Group Transfer Reaction Any reaction in which an acyl group transfers from a donor to an acceptor. (16.3)

Addition Reaction Any reaction in which two parts of a reactant molecule add to a double or a triple bond. (13.4)

Adenosine Diphosphate (ADP) A high-energy diphosphate ester obtained from adenosine triphosphate (ATP) when part of the chemical energy in ATP is tapped for some purpose in a cell. (25.1)

Adenosine Monophosphate (AMP) A low-energy phosphate ester that can be obtained by the hydrolysis of ATP or ADP; a monomer for the biosynthesis of nucleic acids. (25.1)

---

[1] The entries in this Glossary include the terms that appear in boldface within the chapters, including the margin comments, as well as several additional entries. The numbers in parentheses following the definitions are the section numbers (or Special Topics) where the entry was introduced or discussed.

Adenosine Triphosphate (ATP) A high-energy triphosphate ester used in living systems to provide chemical energy for metabolic needs. (25.1)

Adequate Protein A protein that, when digested, makes available all of the essential amino acids in suitable proportions to satisfy both the amino acid and total nitrogen requirements of good nutrition without providing excessive calories. (29.2)

ADP (See *Adenosine Diphosphate*)

Aerobic Sequence An oxygen consuming sequence of catabolism that starts with glucose and proceeds through glycolysis, the citric acid cycle, and the respiratory chain. (25.1)

Agonist A compound whose molecules can bind to a receptor on a cell membrane and cause a response by the cell. (22.7)

Albumin One of a family of globular proteins that tend to dissolve in water, and that in blood contribute to the blood's colloidal osmotic pressure and aid in the transport of metal ions, fatty acids, cholesterol, triacylglycerols, and other water insoluble substances. (21.9, 23.2)

Alcohol Any organic compound whose molecules have the OH group attached to a saturated carbon; ROH. (14.1)

Alcohol Group The OH group when it is joined to a saturated carbon. (14.1)

Aldehyde An organic compound that has a carbonyl group joined to H on one side and C on the other. (15.1)

Aldehyde Group $-CH=O$ (15.1)

Aldohexose A monosaccharide whose molecules have six carbon atoms and an aldehyde group. (19.2)

Aldose A monosaccharide whose molecules have an aldehyde group. (19.2)

Aldosterone A steroid hormone, made in the adrenal cortex, secreted into the bloodstream when the sodium ion level is low, and that signals the kidneys to leave sodium ions in the bloodstream. (23.5)

Aliphatic Compound Any organic compound whose molecules lack a benzene ring or a similar structural feature. (12.5)

Alkali Metals The elements of Group IA of the periodic table—lithium, sodium, potassium, rubidium, cesium, and francium. (4.1)

Alkaline Earth Metals The elements of Group IIA of the periodic table—beryllium, magnesium, calcium, strontium, barium, and radium. (4.1)

Alkaloid A physiologically active, heterocyclic amine isolated from plants. (17.2)

Alkalosis A condition in which the pH of the blood is above normal. *Metabolic alkalosis* is caused by a defect in metabolism. *Respiratory alkalosis* is caused by a defect in the respiratory centers of the brain or in the apparatus of breathing. (23.4)

Alkane A saturated hydrocarbon, one that has only single bonds. A *normal* alkane is any whose molecules have straight chains. (12.5, 13.1)

Alkene A hydrocarbon whose molecules have one or more double bonds. (12.5, 13.1)

Alkyl Group A substituent group that is an alkane minus one H atom. (12.6)

Alkyne A hydrocarbon whose molecules have triple bonds. (12.5, 13.1)

Allosteric Activation The activation of an enzyme's catalytic site by the binding of some molecule at a position elsewhere on the enzyme. (22.4)

Allosteric Inhibition The inhibition of the activity of an enzyme caused by the binding of an inhibitor molecule at some site other than the enzyme's catalytic site. (22.4)

Alloy A mixture of two or more metals made by mixing them in their molten states. (2.1)

Alpha (α) Particle The nucleus of a helium atom; $_2^4$He. (11.1)

Alpha (α) Radiation A stream of high-energy alpha particles. (11.1)

Amide An organic compound whose molecules have a carbonyl–nitrogen single bond. (16.3, 17.1)

Amide Bond The single bond that holds the carbonyl group to the nitrogen atom in an amide. (17.1)

Amine An organic compound whose molecules have a trivalent nitrogen atom, as in $R-NH_2$, $R-NH-R$, or R_3N. (17.1)

Amine Salt An organic compound whose molecules have a positively charged, tetravalent, protonated nitrogen atom, as in RNH_3^+, $R_2NH_2^+$, or R_3NH^+. (17.2)

Amino Acid Any organic compound whose molecules have both an amino group and a carboxyl group. (21.1)

Amino Acid Residue A structural unit in a polypeptide,

$$-NH-CH-\overset{\overset{\displaystyle O}{\|}}{C}-$$
$$\underset{R}{|}$$

furnished by an α-amino acid, where R is the side chain group of a particular amino acid. (21.1)

Aminoacyl Group

$$NH_2-CH-\overset{\overset{\displaystyle O}{\|}}{C}-$$
$$\underset{R}{|}$$

where R is one of the amino acid sidechains. (21.1)

AMP (See *Adenosine Monophosphate*)

Amphipathic Compound A substance whose molecules have both hydrophilic and hydrophobic groups. (20.5)

Anaerobic Sequence The oxygen-independent catabolism of glucose to lactate ion. (25.1, 26.3)

Anhydrous Without water. (7.2)

Anion A negatively charged ion. (5.1)

Anion Gap

$$\text{anion gap} = \frac{\text{meq of Na}^+}{L} - \left(\frac{\text{meq of Cl}^-}{L} + \frac{\text{meq of HCO}_3^-}{L} \right)$$

(Special Topic 8.5)

Anode The positive electrode to which negatively charged ions (anions) are attracted during electrolysis. (8.1)

Anoxia A condition of a tissue in which it receives no oxygen. (23.4)

Antagonist A compound that can bind to a membrane receptor but not cause any response by the cell. (22.7)

Antibiotics Antimetabolites made by bacteria and fungi. (22.4)

Anticodon A sequence of three adjacent sidechain bases on a molecule of tRNA that is complementary to a codon and that fits to its codon on an mRNA chain during polypeptide synthesis. (24.3)

Antimetabolite A substance that inhibits the growth of bacteria. (22.4)

Apoenzyme The wholly polypeptide part of an enzyme. (22.1)

Aromatic Compound Any organic compound whose molecules have a benzene ring (or a feature very similar to this). (12.5, 13.7)

Atmosphere, Standard (See Standard Atmosphere.)

Atom A small particle with one nucleus and zero charge; the smallest particle of a given element that bears the chemical properties of the element. (2.2, 3.1)

Atomic Mass The average mass, in atomic mass units (u), of the atoms of the isotopes of a given element as they occur naturally. (3.2)

Atomic Mass Number (See Mass Number)

Atomic Mass Unit (u) $1.6605665 \times 10^{-24}$ g. A mass very close to that of a proton or a neutron. (3.2)

Atomic Number The positive charge on an atom's nucleus; the number of protons in an atom's nucleus. (3.2, 4.2, 4.3)

Atomic Orbital A region in space close to an atom's nucleus in which one or two electrons can reside. (4.4)

Atomic Symbol A one- or two-letter symbol for an element or one of its atoms. (2.2)

ATP (See Adenosine Triphosphate)

Aufbau Principle A principle regarding the construction of electron configurations: As each additional proton is located in an atomic nucleus, an electron enters whichever of the available orbitals corresponds to the lowest energy. Hund's rule and the Pauli exclusion principle govern the term "available orbitals of lowest energy." (4.5)

Avogadro's Number 6.02×10^{23}. The number of formula units in one mole of any element or compound. (3.1)

Avogadro's Principle Equal volumes of gases contain equal numbers of moles when they are compared at identical temperatures and pressures. (6.3)

Background Radiation Cosmic rays plus the natural atomic radiation emitted by the traces of radioactive isotopes in soils and rocks plus any radiation that escapes from the operations of nuclear facilities. (11.2)

Balanced Equation (See Equation, Balanced)

Barometer An instrument for measuring atmospheric pressure. (6.1)

Basal Activities The minimum activities of the body needed to maintain muscle tone, control body temperature, circulate the blood, handle wastes, breathe, and carry out other essential activities. (Special Topic 2.1)

Basal Metabolic Rate The rate at which energy is expended to maintain basal activities. (Special Topic 2.1)

Basal Metabolism The total of all of the chemical reactions that support basal activities. (Special Topic 2.1)

Base *Brønsted theory:* A proton acceptor; a compound that neutralizes hydrogen ions. (4.1, 5.5, 8.2)

Base, Heterocyclic A heterocyclic amine obtained from the hydrolysis of nucleic acids: adenine, thymine, guanine, cytosine, or uracil. (24.2)

Base Ionization Constant (K_b) For the equilibrium (where B is some base)

$$B + H_2O \rightleftharpoons BH^+ + OH^-$$

$$K_b = \frac{[BH^+][OH^-]}{[B]} \quad (9.7, 17.2)$$

Base Pairing In nucleic acid chemistry, the association by means of hydrogen bonds of two heterocyclic, sidechain bases —adenine with thymine (or uracil) and guanine with cytosine. (24.2)

Base Quantity A fundamental quantity of physical measurement such as mass, length, and time; a quantity used to define derived quantities such as mass/volume for density. (1.4)

Base Unit A fundamental unit of measurement for a base quantity—such as the kilogram for mass, the meter for length, the second for time, the kelvin for temperature degree, and the mole for quantity of chemical substance; a unit to which derived units of measurement are related. (1.4)

Basic Solution A solution in which the molar concentration of hydroxide ions is greater than that of hydronium ions. (8.2, 9.5)

Becquerel (Bq) The SI unit for the activity of a radioactive source; one nuclear disintegration (or other transformation) per second. (11.3)

Benedict's Reagent A solution of copper(II) sulfate, sodium citrate, and sodium carbonate that is used in the Benedict's test. (15.3)

Benedict's Test The use of Benedict's reagent to detect the presence of any compound whose molecules have easily oxidized functional groups — α-hydroxyaldehydes and α-hydroxyketones — such as those present in monosaccharides. In a positive test the intensely blue color of the reagent disappears and a reddish precipitate of copper(I) oxide separates. (15.3)

Beta Oxidation The catabolism of a fatty acid by a series of repeating steps that produce acetyl units (in acetyl CoA); the fatty acid cycle of catabolism. (27.3)

Beta (β) Particle A high-energy electron emitted from a nucleus, $_{-1}^{0}e$. (11.1)

Beta (β) Radiation A stream of high-energy electrons. (11.1)

Bile   A secretion of the gallbladder that empties into the upper intestine and furnishes bile salts; a route of excretion for cholesterol and bile pigments.   (23.1)

Bile Pigment   Colored products of the partial catabolism of heme that are transferred from the liver to the gallbladder for secretion via the bile.   (28.4)

Bile Salts   Steroid-based detergents in bile that emulsify fats and oils during digestion.   (23.1)

Bilin   The brownish pigment that is the end product of the catabolism of heme and that contributes to the characteristic colors of feces and urine.   (28.4)

Bilinogen   A product of the catabolism of heme that contributes to the characteristic colors of feces and urine and some of which is oxidized to bilin.   (28.4)

Bilirubin   An reddish-orange substance that forms from biliverdin during the catabolism of heme and which enters the intestinal tract via the bile and is eventually changed into bilinogen and bilin.   (28.4)

Biliverdin   A greenish pigment that forms when partly catabolized hemoglobin (as verdohemoglobin) is further broken down, and which is changed in the liver to bilirubin.   (28.4)

Binary Compound   A compound made from two elements.   (4.1)

Biochemistry   The study of the structures and properties of substances found in living systems.   (19.1)

Biological Value   In nutrition, the percentage of the nitrogen of ingested protein that is absorbed from the digestive tract and retained by the body when the total protein intake is less than normally required.   (29.2)

Biotin   A water-soluble vitamin needed to make enzymes used in fatty acid synthesis.   (29.3)

2,3-Bisphosphoglycerate (BPG)   An organic ion that nestles within the hemoglobin molecule in deoxygenated blood but is expelled from the hemoglobin molecule during oxygenation.   (23.3)

Blood Sugar   The carbohydrates — mostly glucose — that are present in blood.   (19.2)

Blood Sugar Level   The concentration of carbohydrate — mostly glucose — in the blood; usually stated in units of mg/dL.   (26.1)

Boat Form   A conformation of a six-membered ring that resembles a boat.   (Special Topic 19.2)

Bohr Effect   The stimulation of hemoglobin to bind oxygen caused by the removal (neutralization) of the hydrogen ion released by oxygen binding.   (23.3)

Bohr Model of the Atom   The solar system model of the structure of an atom, proposed by Niels Bohr, that pictures the electrons circling the nucleus in discrete energy states called orbits.   (4.3)

Boiling   The turbulent behavior in a liquid when its vapor pressure equals the atmospheric pressure and when the liquid absorbs heat while experiencing no increase in temperature.   (6.7)

Boiling Point, Normal   The temperature at which a substance boils when the atmospheric pressure is 760 mm Hg (1 atm).   (2.4, 6.7)

Bond, Chemical   A net electrical force of attraction that holds atomic nuclei near each other within compounds.   (5.1)

Boron Family   Group IIIA of the periodic table: boron, aluminum, gallium, indium, and thallium.   (4.1)

Boyle's Law   (See *Pressure–Volume Law*)

BPG   (See *2,3-Bisphosphoglycerate*)

Branched Chain   A sequence of atoms to which additional atoms are attached at points other than the ends.   (12.2)

Brønsted Theory   An acid is a proton donor and a base is a proton acceptor.   (8.2)

Brownian Movement   The random, chaotic movements of particles in a colloidal dispersion that can be seen with a microscope.   (7.1)

Buffer   A combination of solutes that holds the pH of a solution relatively constant even if small amounts of acids or bases are added.   (9.9)

Butyl Group $CH_3CH_2CH_2CH_2-$ (12.6)

sec-**Butyl Group** $CH_3CH_2CH(CH_3)-$ (12.6)

t-**Butyl Group** $(CH_3)_3C-$ (12.6)

Calorie   The amount of heat that raises the temperature of 1 g of water by 1 degree Celsius from 14.5 °C to 15.5 °C.   (2.4)

Carbaminohemoglobin   Hemoglobin that carries chemically bound carbon dioxide.   (23.3)

Carbocation   Any cation in which a carbon atom has just six outer level electrons.   (13.5)

Carbohydrate   Any naturally occurring substance whose molecules are polyhydroxyaldehydes or polyhydroxyketones or can be hydrolyzed to such compounds.   (19.2)

Carbonate Buffer   A mixture or a solution that includes bicarbonate ions and dissolved carbon dioxide in which the bicarbonate ion can neutralize added acid and carbon dioxide can neutralize added base.   (9.9)

Carbon Family   The group IVA elements in the periodic table — carbon, silicon, germanium, tin, and lead.   (4.1)

Carbonyl Group   The atoms carbon and oxygen joined by a double bond, $C=O$.   (15.1)

Carboxylic Acid   A compound whose molecules have the carboxyl group, CO_2H.   (16.1)

Carcinogen   A chemical or physical agent that induces the onset of cancer or the formation of a tumor that might or might not become cancerous.   (11.2)

Catabolism   The reactions of metabolism that break molecules down.   (25.1)

Catalysis   The phenomenon of an increase in the rate of a chemical reaction brought about by a relatively small amount of a chemical — the catalyst — that is not permanently changed by the reaction.   (9.1)

Catalyst   A substance that is able, in relatively low concentrations, to accelerate the rate of a chemical reaction without itself being permanently changed. (In living systems, the catalysts are called enzymes.)   (9.1)

Cathode   The negative electrode to which positively charged ions — cations — are attracted during electrolysis.   (8.1)

Cation   A positively charged ion.   (5.1)

Cell Potential, Standard ($E°_{cell}$)   The difference between the standard reduction potentials of the two half-cell reactions of a redox reaction.   (10.4)

Centimeter (cm)   A length equal to one-hundredth of the meter.

$$1 \text{ cm} = 0.01 \text{ m} = 0.394 \text{ in.}   (1.4)$$

Chair Form   A conformation of a six-membered ring that resembles a chair.   (19.4, Special Topic 19.2)

Charles' Law   (See *Temperature–Volume Law*)

Chemical Bond   (See *Bond, Chemical*)

Chemical Energy   The potential energy that substances have because their arrangements of electrons and atomic nuclei are not as stable as are alternative arrangements that become possible in chemical reactions.   (2.3)

Chemical Equation   A shorthand representation of a chemical reaction that uses formulas instead of names for reactants and products; that separates reactant formulas from product formulas by an arrow; that separates formulas on either side of the arrow by plus signs; and that expresses the mole proportions of the chemicals by simple numbers (coefficients) placed before the formulas.   (2.2, 3.3)

Chemical Property   Any chemical reaction that a substance can undergo and the ability to undergo such a reaction.   (1.3)

Chemical Reaction   Any event in which substances change into different chemical substances.   (1.3)

Chemiosmotic Theory   An explanation of how oxidative phosphorylation is related to a flow of protons in a proton gradient established by the respiratory chain, a gradient that extends across the inner membrane of a mitochondrion.   (25.3)

Chemistry   The study of the compositions and structures of substances and their ability to change into other substances.   (1.3)

Chiral   Having handedness in a molecular structure. (See also *Chirality*)   (18.2)

Chiral Carbon   (See *Tetrahedral Stereocenter*)

Chirality   The quality of handedness that a molecular structure has that prevents this structure from being superimposable on its mirror image.   (18.2)

Chloride Shift   An interchange of chloride ions and bicarbonate ions between a red blood cell and the surrounding blood serum.   (23.3)

Choline   A compound needed to make complex lipids and acetylcholine; classified as a vitamin.   (29.3)

Chromosome   Small threadlike bodies in a cell nucleus that carry genes in a linear array and that are microscopically visible during cell division.   (24.1)

Citric Acid Cycle   A series of reactions that dismantle acetyl units and send electrons (and protons) into the respiratory chain; a major source of metabolites for the respiratory chain.   (25.1, 25.2)

Codon   A sequence of three adjacent sidechain bases in a molecule of mRNA that codes for a specific amino acid residue when the mRNA participates in polypeptide synthesis.   (24.3)

Coefficient of Digestibility   The proportion of an ingested protein's nitrogen that enters circulation rather than elimination (in feces); the difference between the nitrogen ingested and the nitrogen in the feces divided by the nitrogen ingested.   (29.2)

Coefficients   Numbers placed before formulas in chemical equations to indicate the mole proportions of reactants and products.   (2.2, 3.3)

Coenzyme   An organic compound needed to make a complete enzyme from an apoenzyme.   (22.1)

Cofactor   A nonprotein compound or ion that is an essential part of an enzyme.   (22.1)

Collagen   The fibrous protein of connective tissue that changes to gelatin in boiling water.   (21.9)

Colligative Property   A property of a solution that depends only on the concentrations of the solute and the solvent and not on their chemical identities (e.g., osmotic pressure).   (7.5)

Collision Theory   A theory about the rates of chemical reactions that postulates collisions between reacting particles.   (9.2)

Colloidal Dispersion   A relatively stable, uniform distribution in some dispersing medium of colloidal particles—those with at least one dimension between 1 and 1000 nm.   (7.1)

Colloidal Osmotic Pressure   The contribution made to the osmotic pressure of a solution by substances colloidally dispersed in it.   (7.5)

Common Ion Effect   The reduction in the solubility of a salt in some solution by the addition of another solute that furnishes one of the ions of this salt.   (8.5)

Competitive Inhibition   The inhibition of an enzyme by the binding of a molecule that can compete with the substrate for the occupation of the catalytic site.   (22.4)

Complex   (See *Complex Ion*)

Complex Ion   A combination of a metal ion with one or more ligands—negatively charged or neutral electron-rich species.   (15.3)

Compound   A substance made from the atoms of two or more elements that are present in a definite proportion by mass and by atoms.   (2.1)

Concentration   The quantity of some component of a mixture in a unit of volume or a unit of mass of the mixture.   (3.5)

Condensation   The physical change of a substance from its gaseous state to its liquid state.   (2.4, 6.7)

Condensed Structure   (See *Structural Formula*)

Conformation   One of the infinite number of contortions of a molecule that are permitted by free rotations around single bonds.   (12.2)

Conjugate Acid–Base Pair   Two particles whose formulas differ by only one H^+, such as NH_4^+ and NH_3, or HCl and Cl^-.   (8.4)

Constitutional Isomerism   The existence of two or more compounds with identical molecular formulas but different atom-to-atom sequences.   (18.1)

Constitutional Isomers   Compounds with identical molecular formulas but different atom-to-atom sequences. (12.3, 18.1)

Conversion Factor A fraction that expresses a relationship between quantities that have different units, such as 2.54 cm/in. (1.7)

Coordinate Covalent Bond A covalent bond in which both of the electrons of the shared pair originated from one of the atoms involved in the bond. (5.5)

Cori Cycle The sequence of chemical events and transfers of substances in the body that describes the distribution, storage, and mobilization of blood sugar, including the reconversion of lactate to glycogen. (26.2)

Cosmic Radiation A stream of ionizing radiations from the sun and outer space that consists mostly of protons but also includes alpha particles, electrons, and the nuclei of atoms up to atomic number 28. (Special Topic 11.1)

Covalence Number The number of covalent bonds that an atom can have in a molecule. (5.4)

Covalent Bond The net force of attraction that arises as two atomic nuclei share a pair of electrons. One pair is shared in a single bond, two pairs in a double bond, and three electrons pairs are shared in a triple bond. (5.3)

Crenation The shrinkage of red blood cells when they are in contact with a hypertonic solution. (7.6)

Curie (Ci) A unit of activity of a radioactive source.

$$1 \text{ Ci} = 3.70 \times 10^{10} \text{ disintegrations/s} \quad (11.3)$$

Dalton (D) A unit for a formula mass; one atomic mass unit. (3.2)

Dalton's Law (See *Law of Partial Pressures*)

Dalton's Theory A theory that accounts for the laws of chemical combination by postulating that matter consists of indestructible atoms; that all atoms of the same element are identical in mass and other properties; that the atoms of different elements are different in mass and other properties; and that in the formation of a compound atoms join together in definite, whole-number ratios. (2.2)

Deamination The removal of an amino group from an amino acid. (28.2)

Decarboxylation The removal of a carboxyl group. (28.2)

Degree Celsius One one-hundredth (1/100) of the interval on a thermometer between the freezing point and the boiling point of water. (1.4)

Degree Fahrenheit One one-hundred-and-eightieth (1/180) of the interval on a thermometer between the freezing point and the boiling point of water. (1.4)

Deliquescence The ability of a substance to attract water vapor to itself to form a concentrated solution. (7.2)

Denatured Protein A protein whose molecules have suffered the loss of their native shape and form as well as their ability to function biologically. (21.2)

Density The ratio of the mass of an object to its volume; the mass per unit volume. Density = mass/volume (usually expressed in g/mL). (1.8)

Deoxyribonucleic Acid (DNA) The chemical of a gene; one of a large number of polymers of deoxyribonucleotides and whose sequences of sidechain bases constitute the genetic messages of genes. (24.2)

Derived Quantity A quantity based on a relationship that involves one or more base quantities of measurement such as volume (length³) or density (mass/volume). (1.4)

Desiccant A substance that combines with water vapor to form a hydrate and thereby reduces the concentration of water vapor in the air space around the substance. (7.2)

Detergent A surface-active agent; a soap. (Special Topic 20.3)

Dextrorotatory That property of an optically active substance by which it can cause the plane of plane-polarized light to rotate clockwise. (18.3)

D-Family; L-Family The names of the two optically active families to which substances can belong when they are considered solely according to one kind of molecular chirality (molecular handedness) or the other. (19.3)

Diabetes Mellitus A disease in which there is an insufficiency of effective insulin and an impairment of glucose tolerance. (26.2)

Dialysis The passage through a dialyzing membrane of water and particles in solution, but not of particles that have colloidal size. (7.5)

Dialyzing Membrane A membrane permeable to solvent and small ions or molecules but impermeable to colloidal sized particles. (7.5)

Diastereomers Stereoisomers whose molecules are not related as an object is to its mirror image. (18.2)

Diatomic Molecule A molecule made of two atoms. (3.1)

Dietetics The application of the findings of the science of nutrition to the feeding of individual humans, whether well or ill. (29.1)

Diffusion A physical process whereby particles, by random motions, intermingle and spread out so as to erase concentration gradients. (6.1)

Digestive Juice A secretion into the digestive tract that consists of a dilute aqueous solution of digestive enzymes (or their zymogens) and inorganic ions. (23.1)

Dihydric Alcohol An alcohol with two OH groups; a glycol. (14.1)

Dipeptide A compound whose molecules have two α-amino acid residues joined by a peptide (amide) bond. (21.3)

Dipolar Ion A molecule that carries one plus charge and one minus charge, such as an α-amino acid. (21.1)

Dipole, Electrical A pair of equal but opposite (and usually partial) electrical charges separated by a small distance in a molecule. (5.8)

Dipole–Dipole Attraction The electrical force of attraction between $\delta+$ and $\delta-$ sites of polar molecules. (6.6)

Diprotic Acid An acid with two protons available per molecule to neutralize a base, e.g., H_2SO_4. (8.2)

Disaccharide A carbohydrate that can be hydrolyzed into two monosaccharides. (19.2)

Dissociation The separation of preexisting ions from one another as an ionic compound dissolves or melts. (7.2, 8.1)

Disulfide Link The sulfur–sulfur covalent bond in polypeptides. (21.1)

Disulfide System S—S as in R—S—S—R. (14.6)

DNA (See *Deoxyribonucleic Acid; see Double Helix DNA Model*)

Double Bond A covalent bond in which two pairs of electrons are shared. (5.4)

Double Helix DNA Model A spiral arrangement of two intertwining DNA molecules held together by hydrogen bonds between sidechain bases. (24.2)

Double Replacement A reaction in which a compound is made by the exchange of partner ions between two salts. (8.5)

Duplex DNA DNA in its double-stranded form. (24.2)

Dynamic Equilibrium (See *Equilibrium, Dynamic*)

Edema The swelling of tissue caused by the retention of water. (23.2)

Effector A chemical other than a substrate that can allosterically activate an enzyme. (22.4)

Elastin The fibrous protein of tendons and arteries. (21.9)

Electrical Balance The condition of a net ionic equation wherein the algebraic sum of the positive and negative charges of the reactants equals that of the products. (8.3)

Electrode A metal object, usually a wire, suspended in an electrically conducting medium through which electricity passes to or from an external circuit. (8.1)

Electrolysis A procedure in which an electrical current is passed through a solution that contains ions, or through a molten salt, for the purpose of bringing about a chemical change. (8.1)

Electrolyte Any substance whose solution in water conducts electricity; or the solution itself of such a substance. (8.1)

Electrolytes, Blood The ionic substances dissolved in the blood. (23.2)

Electron A subatomic particle that bears one unit of negative charge and has a mass that is 1/1836th the mass of a proton. (4.2)

Electron Cloud A mental model that views the one or two rapidly moving electrons of an orbital as creating a cloud-like distribution of negative charge. (4.4)

Electron Configuration The most stable arrangement (that is, the arrangement of lowest energy) of the electrons of an atom, ion, or molecule. (4.3)

Electron Dot Structure A Lewis structure of a molecule in which all valence shell electrons, whether shared or unshared, are shown either by dots or by lines. (5.4)

Electronegativity The ability of an atom joined to another by a covalent bond to attract the electrons of the bond toward itself. (5.8)

Electron Sharing The joint attraction of two atomic nuclei toward a pair of electrons situated between the nuclei and between which, therefore, a covalent bond exists. (5.3)

Electron Shell An alternative name for *principal energy level*. (4.3)

Electron Volt (eV) A very small unit of energy used to describe the energy of a radiation. (11.3)

$$1 \text{ eV} = 1.6 \times 10^{-19} \text{ joule}$$
$$1 \text{ eV} = 3.8 \times 10^{-20} \text{ calorie}$$
$$1000 \text{ eV} = 1 \text{ keV (1 kiloelectron volt)}$$
$$1000 \text{ keV} = 1 \text{ MeV (1 megaelectron volt)}$$

Element A substance that cannot be broken down into anything that is both stable and more simple; a substance in which all of the atoms have the same atomic number and the same electron configuration; one of the three broad kinds of matter, the others being compounds and mixtures. (2.1)

Emulsion A colloidal dispersion of tiny microdroplets of one liquid in another liquid. (7.1)

Enantiomers Stereoisomers whose molecules are related as an object is related to its mirror image but that cannot be superimposed. (18.2)

Endothermic Describing a change that needs a constant supply of heat energy to happen. (2.4, 6.7)

End Point The stage in a titration when the operation is stopped. (9.11)

Energy A capacity to cause a change that can, in principle, be harnessed for useful work. (2.3)

Energy Density The energy per gram of stored glycogen or fat. (27.2)

Energy Level A principal energy state in which electrons of an atom can be. (4.3)

Energy of Activation The minimum energy that must be provided by the collision between reactant particles to initiate the rearrangement of electrons relative to nuclei that must happen if the reaction is to occur. (9.2)

Enzyme A catalyst in a living system. (9.1, 22.1)

Enzyme Induction The chemical process whereby the synthesis of an enzyme is prompted. (24.4)

Enzyme–Substrate Complex The temporary combination that an enzyme must form with its substrate before catalysis can occur. (22.2)

Epinephrine A hormone of the adrenal medulla that activates the enzymes needed to release glucose from glycogen. (26.1)

Equation, Balanced A chemical equation in which all of the atoms represented in the formulas of the reactants are present in identical numbers among the products, and in which any net electrical charge provided by the reactants equals the same charge indicated by the products. (See also *Chemical Equation*) (2.2, 3.3)

Equation of State for an Ideal Gas (See *Ideal Gas Law*)

Equilibrium, Dynamic A situation in which two opposing events occur at identical rates so that no net change happens. (6.7)

Equilibrium Constant The value that the mass action expression has when a chemical system is at equilibrium. (9.3)

Equilibrium Equation A chemical equation in which oppositely pointing arrows separate reactants and products that are in equilibrium. (6.7)

Equilibrium Law The mathematical equation that describes the interrelationships among the molar concentrations of reactants and products when equilibrium exists. (9.3)

Equivalence Point The stage in a titration when the reactants have been mixed in the exact molar proportions represented by the balanced equation; in an acid–base titration, the stage when the moles of hydrogen ions furnished by the acid matches the moles of hydroxide ions (or other proton acceptor) supplied by the base. (9.11)

Equivalent (eq) For an ion, usually its mass in grams divided by the amount of its electrical charge. (8.5)

Error In a measurement, the difference between the measured value and the correct value of a physical quantity. (1.6)

Erythrocyte A red blood cell. (23.3)

Essential Amino Acid An α-amino acid that the body cannot make from other amino acids and that must be supplied by the diet. (29.2)

Essential Fatty Acid A fatty acid that must be supplied by the diet. (29.1)

Ester A derivative of an acid and an alcohol that can be hydrolyzed to these parent compounds. Esters of carboxylic acids and phosphoric acid occur in living systems. (16.3)

$$\underset{\text{Carboxylic acid ester}}{(H)R-\overset{\overset{O}{\|}}{C}-O-R'} \qquad \underset{\text{Phosphoric acid ester}}{RO-\overset{\overset{O}{\|}}{\underset{\underset{OH}{|}}{P}}-OH}$$

Esterification The formation of an ester. (16.3)

Ether An organic compound whose molecules have an oxygen attached by single bonds to separate carbon atoms neither of which is a carbonyl carbon atom: R—O—R′. (14.5)

Ethyl Group CH_3CH_2- (12.6)

Evaporation The conversion of a substance from its liquid to its vapor state. (2.4, 6.7)

Exon A segment of a DNA strand that eventually becomes expressed as a corresponding sequence of aminoacyl residues in a polypeptide. (24.2)

Exothermic Describing a change by which heat energy is released from the system. (2.4, 6.7)

Extensive Property Any property whose value is directly proportional to the size of the sample, such as volume or mass. (1.8)

Extracellular Fluids Body fluids that are outside of cells. (23.1)

Factor-Label Method A strategy for solving computational problems that uses conversion factors and the cancellation of the units of physical quantities as an aid in working toward the solution. (1.7)

Fatty Acid Any carboxylic acid that can be obtained by the hydrolysis of animal fats or vegetable oils. (16.1, 20.1)

Fatty Acid Cycle (See *Beta Oxidation*)

Favored Side (of an equilibrium) That side of the double arrows of an equilibrium equation whose chemical symbols are those of the species in higher concentrations at equilibrium. (7.3)

Feedback Inhibition The competitive inhibition of an enzyme by a product of its own action. (22.4)

Fibrin The fibrous protein of a blood clot that forms from fibrinogen during clotting. (21.9, 23.2)

Fibrinogen A protein in blood that is changed to fibrin during clotting. (21.9, 23.2)

Fibrous Proteins Water-insoluble proteins found in fibrous tissues. (21.9)

Fischer Projection Structure A two-dimensional representation, prepared according to rules, of the configuration at a tetrahedral stereocenter. (19.3)

Fission The splitting of the nucleus of a heavy atom approximately in half and that is accompanied by the release of one or a few neutrons and energy. (11.7)

Folate A vitamin supplied by folic acid or pteroylglutamic acid and that is needed to prevent megaloblastic anemia. (29.3)

Food A material that supplies one or more nutrients without contributing materials that, either in kind or quantity, would be harmful to most healthy people. (29.1)

Formula, Chemical A shorthand representation of a substance that uses atomic symbols and following subscripts to describe the elemental composition and the mole ratios in which the atoms of the elements are combined. (2.2)

Formula, Empirical A chemical symbol for a compound that gives just the ratios of the atoms and not necessarily the composition of a complete molecule. (2.2, 5.1)

Formula, Molecular A chemical symbol for a substance that gives the composition of a complete molecule. (5.4)

Formula, Structural A chemical symbol for a substance that uses atomic symbols and lines to describe the pattern in which the atoms are joined together in a molecule. (5.4)

Formula Mass The sum of the atomic masses of the atoms represented in a chemical formula. (3.2)

Formula Unit A small particle—an atom, a molecule, or a set of ions—that has the composition given by the chemical formula of the substance. (2.2)

Forward Reaction In a chemical equilibrium, the reaction whereby substances to the left of the double arrows are changed to the products shown on the right-hand side of the arrows. (6.7)

Free Rotation The absence of a barrier to the rotation of two groups with respect to each other when they are joined by a single, covalent bond. (12.2)

Functional Group An atom or a group of atoms in a molecule that is responsible for the particular set of reactions that all compounds with this group have. (12.4)

Gamma Radiation A natural radiation similar to but more powerful than X rays. (11.1)

Gap Junctions Tubules made of membrane-bound proteins that interconnect one cell to neighboring cells and through which materials can pass directly. (21.8)

Gas Any substance that must be contained in a wholly closed space and whose shape and volume are determined entirely by the shape and volume of its container; a state of matter.   (2.1, 6.1)

Gas Constant, Universal (R) The ratio of PV to nT for a gas, where P = the gas pressure, V = volume, n = number of moles, and T = the Kelvin temperature. When P is in mm Hg and V is in mL,

$$R = 6.24 \times 10^4 \text{ mm Hg mL/mol K} \quad (6.3)$$

Gas Tension The partial pressure of a gas over its solution in some liquid when the system is in equilibrium.   (7.3)

Gastric Juice The digestive juice secreted into the stomach and that contains pepsinogen, hydrochloric acid, and gastric lipase.   (23.1)

Gay-Lussac's Law (See *Pressure–Temperature Law*)

Gel A colloidal dispersion of a solid in a liquid that has adopted a semisolid form.   (7.1)

Gene A unit of heredity carried on a cell's chromosomes and consisting of DNA.   (24.1, 24.2)

Genetic Code The set of correlations that specify which codons on mRNA chains are responsible for which amino acyl residues when the latter are steered into place during the mRNA-directed synthesis of polypeptides.   (24.3)

Genetic Engineering The use of recombinant DNA to manufacture substances or to repair genetic defects.   (24.6)

Genome The entire complement of genetic information of a species; all the genes of an individual.   (24.7)

Geometric Isomerism Stereoisomerism caused by restricted rotation that gives different geometries to the same structural organization; cis–trans isomerism.   (13.3)

Geometric Isomers Stereoisomers whose molecules have identical atomic organizations but different geometries; cis–trans isomers.   (13.3)

Globular Proteins Proteins that are soluble in water or in water that contains certain dissolved salts.   (21.9)

Globulins Globular proteins in the blood that include γ-globulin, an agent in the body's defense against infectious diseases.   (21.9, 23.2)

Glucagon A hormone, secreted by the α-cells of the pancreas in response to a decrease in the blood sugar level that stimulates the liver to release glucose from its glycogen stores.   (26.1)

Gluconeogenesis The synthesis of glucose from compounds with smaller molecules or ions.   (26.4)

Glucose Tolerance The ability of the body to manage the intake of dietary glucose while keeping the blood sugar level from fluctuating widely.   (26.2)

Glucose Tolerance Test A series of measurements of the blood sugar level after the ingestion of a considerable amount of glucose, used to obtain information about an individual's glucose tolerance.   (26.2)

Glucoside An acetal formed from glucose (in its cyclic, hemiacetal form) and an alcohol.   (19.5)

Glucosuria The presence of glucose in urine.   (26.1)

Glycerophospholipid A hydrolyzable lipid that has an ester linkage between glycerol and one phosphoric acid unit (this, in turn, forming another ester link to a small molecule). In *phosphatides,* the remaining two OH units of glycerol are esterified with fatty acids. In *plasmalogens,* one OH is esterified with a fatty acid and the other is joined by an ether link to a long-chain unsaturated alcohol.   (20.3)

Glycogenesis The synthesis of glycogen.   (26.1)

Glycogenolysis The breakdown of glycogen to glucose.   (26.1)

Glycol A dihydric alcohol.   (14.1)

Glycolipid A lipid whose molecules include a glucose unit, a galactose unit or some other carbohydrate unit.   (20.3, 21.8)

Glycolysis A series of chemical reactions that break down glucose or glucose units in glycogen until pyruvate remains (when the series is operated aerobically) or lactate forms (when the conditions are anaerobic).   (25.1, 26.3)

Glycoprotein A protein, often membrane-bound, that is joined to a carbohydrate unit.   (21.8)

Glycoside An acetal or a ketal formed from the cyclic form of a monosaccharide and an alcohol.   (19.5)

Gradient The presence of a change in value of some physical quantity with distance, as in a *concentration* gradient in which the concentration of a solute is different in different parts of the system.   (7.3, 21.8)

Gram (g) A mass equal to one-thousandth of the kilogram mass, the SI standard mass.

$$1 \text{ g} = 0.001 \text{ kg} = 1000 \text{ mg}; 1 \text{ lb} = 453.6 \text{ g} \quad (1.4)$$

Gray (Gy) The SI unit of absorbed dose of radiation equal to one joule of energy absorbed per kilogram of tissue.   (11.3)

Greenhouse Effect The entrapment of heat radiating from the earth by the greenhouse gases in the atmosphere (principally carbon dioxide, water, methane, and others).   (Special Topic 8.4)

Ground Substance A gel-like material present in cartilage and other extracellular spaces that gives flexibility to collagen and other fibrous proteins.   (21.8)

Group A vertical column in the periodic table; a family of elements.   (4.1)

Half-Life The time needed for half of the atoms in a sample of a particular radioactive isotope to undergo radioactive decay.   (11.1)

Half-Reaction A net ionic equation that includes electrons either as reactants or as products and used to describe either the reduction or the oxidation part of a redox reaction.   (10.2)

Halogens The elements of group VIIA of the periodic table—fluorine, chlorine, bromine, iodine, and astatine.   (4.1)

Hard Water Water that contains one or more of the metallic ions Mg^{2+}, Ca^{2+}, Fe^{2+}, or Fe^{3+}. The negative ions present are usually Cl^- and SO_4^{2-}. If HCO_3^- is the chief negative ion, the water is said to be *temporary hard water;* otherwise it is *permanent hard water.*   (Special Topic 8.2)

Heat The form of energy that transfers between two objects in contact that have initially different temperatures. (2.4)

Heat Capacity The quantity of heat that a given object can absorb (or release) per degree Celsius change in temperature:

$$\text{heat capacity} = \text{heat}/\Delta t$$

where Δt is the change in temperature. (2.4)

Heat of Fusion The quantity of heat that one gram of a substance absorbs when it changes from its solid to its liquid state at its melting point. (2.4)

Heat of Reaction The net energy difference between the reactants and the products of a reaction. (9.2)

Heat of Vaporization The quantity of heat that one gram of a substance absorbs when it changes from its liquid to its gaseous state. (2.4)

Heisenberg Uncertainty Principle It is impossible simultaneously to determine with precision and accuracy both the position and the velocity of an electron. (4.3)

α-Helix One kind of secondary structure of a polypeptide in which its molecules are coiled. (21.4)

Heme The deep-red, iron-containing prosthetic group in hemoglobin and myoglobin. (21.5, 23.3)

Hemiacetal Any compound whose molecules have both an OH and an OR group coming to a CH unit. (15.5)

Hemiketal Any compound whose molecules have both an OH and an OR group coming to a carbon that otherwise bears no H atoms. (15.5)

Hemoglobin The oxygen-carrying protein in red blood cells. (21.5, 23.3)

Hemolysis The bursting of a red blood cell. (7.6)

Henderson-Hasselbalch Equation An equation used in buffer calculations when the buffer is prepared using a weak acid of some known pK_a and the anion of the weak acid:

$$pH = pK_a + \log \frac{[\text{anion}]}{[\text{acid}]}$$

where [anion] is the *initial* molar concentration of the anion component of a buffer pair and [acid] is the *initial* molar concentration of the acid component. (9.10)

Henry's Law See (*Pressure–Solubility Law*)

Heterocyclic Compound An organic compound with a ring in which an atom other than carbon takes up at least one position in the ring. (12.2)

Heterogeneous Mixture A mixture in which the composition of one small portion is not identical with that of another. (7.1)

Heterogeneous Nuclear RNA (hnRNA) RNA made directly at the guidance of DNA and from which messenger RNA (mRNA) is made. (Formerly called primary transcript RNA, ptRNA.) (24.3)

High-Energy Phosphate An organophosphate with a phosphate group transfer potential equal to or higher than that of ADP or ATP. (25.1)

Homeostasis The response of an organism to a stimulus such that the organism is restored to its prestimulated state. (22.4)

Homogeneous Mixture A mixture in which the composition and properties are uniform throughout. (7.1)

Hormone A primary chemical messenger made by an endocrine gland and carried by the bloodstream to a target organ where a particular chemical response is initiated. (22.6)

Human Growth Hormone One of the hormones that affects the blood sugar level; a stimulator of the release of the hormone glucagon. (26.1)

Hund's Rule Electrons become evenly distributed among *different* orbitals of the same sublevel insofar as there is room. (4.5)

Hybrid Orbital An atomic orbital obtained by mixing two or more pure orbitals (those of the *s, p, d,* or *f* types). (5.9)

Hydrate A compound in which intact molecules of water are held in a definite molar proportion to the other components. (7.2)

Hydrated Ion An ion around which molecules of water have been drawn by ion–dipole attractions. (7.2)

Hydration The association of water molecules with dissolved ions or polar molecules. (7.2)

Hydrocarbon An organic compound that consists entirely of carbon and hydrogen. (12.5)

Hydrogen Bond The force of attraction between a $\delta+$ on a hydrogen held by a covalent bond to oxygen or nitrogen (or fluorine) and a $\delta-$ charge on a nearby atom of oxygen or nitrogen (or fluorine). (6.8)

Hydrolase An enzyme that catalyzes a hydrolysis reaction. (22.1)

Hydrolysis of Anions Reactions in which anions (other than OH^-) react with water and increase the pH of a solution. (9.7)

Hydrolysis of Cations Reactions in which cations (other than H_3O^+) react with water and decrease the pH of the solution. (9.6)

Hydrolyzable Lipid A lipid that can be hydrolyzed or saponified. (Formerly called a saponifiable lipid.) (20.1)

Hydronium Ion H_3O^+ (8.2)

Hydrophilic Group Any part of a molecular structure that attracts water molecules; a polar or ionic group such as OH, CO_2^-, NH_3^+, or NH_2. (20.5)

Hydrophobic Group Any part of a molecular structure that has no attraction for water molecules; a nonpolar group such as any alkyl group. (20.5)

Hydrophobic Interaction The water avoidance by nonpolar groups or sidechains that is partly responsible for the shape adopted by a polypeptide or nucleic acid molecule in an aqueous environment. (21.1, 24.2)

Hydroxide Ion OH^- (8.2)

Hygroscopic Describing a substance that can reduce the concentration of water vapor in the surrounding air by forming a hydrate. (7.2)

Hyperammonemia   An elevated level of ammonium ion in the blood.   (28.3)

Hypercapnia   An elevated level of carbon dioxide in the blood as indicated by a partial pressure of CO_2 in venous blood above 50 mm Hg.   (23.4)

Hyperglycemia   An elevated level of glucose in the blood — above 110 mg/dL in whole blood.   (26.1)

Hyperkalemia   An elevated level of potassium ion in blood — above 5.0 meq/L.   (23.2)

Hypernatremia   An elevated level of sodium ion in blood — above 145 meq/L.   (23.2)

Hyperthermia   A condition of a core body temperature above normal.   (Special Topic 2.1)

Hypertonic   Having an osmotic pressure greater than some reference; having a total concentration of all solute particles higher than that of some reference.   (7.6)

Hyperventilation   Breathing considerably faster and deeper than normal.   (9.9)

Hypocapnia   A condition of a below-normal concentration of carbon dioxide in the blood as indicated by a partial pressure of CO_2 in venous blood of less than 35 mm Hg.   (23.4)

Hypoglycemia   A low level of glucose in blood — below 65 mg/dL of whole blood.   (26.1)

Hypokalemia   A low level of potassium ion in blood — below 3.5 meq/L.   (23.2)

Hyponatremia   A low level of sodium ion in blood — below 135 meq/L.   (23.2)

Hypothermia   A low body temperature.   (Special Topic 2.1)

Hypothesis   A conjecture, subject to being disproved, that explains a set of facts in terms of a common cause and that serves as the basis for the design of additional tests or experiments.   (1.2)

Hypotonic   Having an osmotic pressure less than some reference; having a total concentration of dissolved solute particles less than that of some reference.   (7.6)

Hypoventilation   Breathing more slowly and less deeply than normal; shallow breathing.   (9.9)

Hypoxia   A condition of a low supply of oxygen.   (23.4)

Ideal Gas   A hypothetical gas that obeys the gas laws exactly.   (6.2)

Ideal Gas Law $PV = nRT$ (6.3)

Indicator   A dye that has one color in solution below a measured pH range and a different color above this range.   (8.2, 9.11)

Induced Fit Model   Many enzymes are induced by their substrate molecules to modify their shapes to accommodate the substrate.   (22.2)

Inducer   A substance whose molecules remove repressor molecules from operator genes and so open the way for structural genes to direct the overall syntheses of particular polypeptides.   (24.4)

Inertia   The resistance of an object to a change in its position or its motion.   (1.4)

Inhibitor   A substance that interacts with an enzyme to prevent its acting as a catalyst.   (22.4)

Inner Transition Elements   The elements of the lanthanide and actinide series of the periodic table.   (4.1)

Inorganic Compound   Any compound that is not an organic compound.   (8.2, 12.1)

Insulin   A protein hormone made by the pancreas, released in response to an increase in the blood sugar level, and used by certain tissues to help them take up glucose from circulation.   (26.1)

Intensive Property   Any property whose value is independent of the size of the sample, such as temperature and density.   (1.8)

Internal Environment   Everything enclosed within an organism.   (23.1)

International System of Units (SI)   The successor to the metric system with new reference standards for the base units but with the same names for the units and the same decimal relationships.   (1.4)

International Union of Pure and Applied Chemistry System (IUPAC System)   A set of systematic rules for naming compounds and designed to give each compound one unique name and for which only one structure can be drawn.   (12.6)

Interstitial Fluids   Fluids in tissues but not inside cells or the blood.   (23.1)

Intestinal Juice   The digestive juice that empties into the duodenum from the intestinal mucosa and whose enzymes also work within the intestinal mucosa as molecules migrate through.   (23.1)

Intracellular Fluids   Fluids inside cells.   (23.1)

Intron   A segment of a DNA strand that separates exons and that does not become expressed as a segment of a polypeptide.   (24.2)

Inverse Square Law   The intensity of radiation varies inversely with the square of the distance from its source.   (11.2)

Invert Sugar   A 1 : 1 mixture of glucose and fructose.   (19.5)

In Vitro   Occurring in laboratory vessels.   (14.3)

In Vivo   Occurring within a living system.   (14.3)

Iodine Test   A test for starch by which a drop of iodine reagent produces an intensely purple color if starch is present.   (19.6)

Ion   An electrically charged, atomic or molecular-sized particle; a particle that has one or a few atomic nuclei and either one or two (seldom, three) too many or too few electrons to render the particle electrically neutral.   (5.1)

Ion – Dipole Attraction   The attraction between a partially charged site of a polar molecule and a fully charged ion.   (7.2)

Ionic Bond   The force of attraction between oppositely charged ions in an ionic compound.   (5.1)

Ionic Compound   A compound that consists of an orderly aggregation of oppositely charged ions that assemble in whatever ratio ensures overall electrical neutrality.   (5.1)

Ionic Equation   A chemical equation that explicitly shows all of the particles—ions, atoms, or molecules—that are involved in a reaction even if some are only spectator particles. (See also *Net Ionic Equation; Equation, Balanced*)   (8.3)

Ionization   A change, usually involving solvent molecules, whereby molecules change into ions.   (8.1)

Ionizing Radiation   Any radiation that can create ions from molecules within the medium that it enters, such as alpha, beta, gamma, X, and cosmic radiation.   (11.2)

Ion Product Constant (K_{sp})   Regarding the solubility equilibrium involving a relatively insoluble salt in a saturated solution, the product of the molar concentrations of the ions, each concentration raised to a power equaling the number of ions obtained from one formula unit of the salt.   (Special Topic 8.3)

Ion Product Constant of Water (K_w)   The product of the molar concentrations of hydrogen ions and hydroxide ions in water at a given temperature.

$$K_w = [H^+][OH^-]$$
$$= 1.0 \times 10^{-14} \text{ (at 25 °C)}   (9.4)$$

Isobutyl Group $(CH_3)_2CHCH_2-$ (12.6)

Isoelectric Molecule   A molecule which has an equal number of positive and negative sites.   (21.1)

Isoelectric Point (pI)   The pH of a solution in which a specified amino acid or a protein is in an isoelectric condition; the pH at which there is no net migration of the amino acid or protein in an electric field.   (21.1)

Isoenzymes   Enzymes that have identical catalytic functions but which are made of slightly different polypeptides.   (22.1)

Isohydric Shift   In actively metabolizing tissue, the use of a hydrogen ion released from newly formed carbonic acid to react with and liberate oxygen from oxyhemoglobin; in the lungs, the use of hydrogen ion released when hemoglobin oxygenates to combine with bicarbonate ion and liberate carbon dioxide for exhaling.   (23.3)

Isomerase   An enzyme that catalyzes the conversion of a compound into one of its isomers.   (22.1)

Isomerism   The phenomenon of the existence of two or more compounds with identical molecular formulas but different structures.   (12.3)

Isomers   Compounds with identical molecular formulas but different structures.   (12.3)

Isopropyl Group $(CH_3)_2CH-$ (12.6)

Isotonic   Having an osmotic pressure identical to that of a reference; having a concentration equivalent to the reference with respect to the ability to undergo osmosis.   (7.6)

Isotope   A substance in which all of the atoms are identical in atomic number, mass number, and electron configuration.   (3.2)

Isozyme   (See *Isoenzyme*)

IUPAC System   (See *International Union of Pure and Applied Chemistry System*)

Joule (J)   The SI derived unit of energy.   (2.3)

K_a (See *Acid Dissociation Constant*)

K_b (See *Base Dissociation Constant*)

K_{sp} (See *Ion Product Constant*)

K_w (See *Ion Product Constant of Water*)

Kelvin   The SI unit of temperature degree and equal to 1/100th of the interval between the freezing point and the boiling point of water when measured under standard conditions.   (1.4)

Kelvin Scale   The scale of absolute temperatures expressed in kelvins beginning with 0 K for the coldest temperature attainable.   (1.4)

Keratin   The fibrous protein of hair, fur, fingernails, and hooves.   (21.9)

Ketal   A substances whose molecules have two OR groups joined to a carbon that also holds two hydrocarbon groups.   (15.5)

Ketoacidosis   The acidosis caused by untreated ketonemia.   (27.6)

Keto Group   The carbonyl group when it is joined on each side to carbon atoms.   (15.1)

Ketohexose   A monosaccharide whose molecules contain six carbon atoms and have a keto group.   (19.2)

Ketone   Any compound with a carbonyl group attached to two carbon atoms, as in $R_2C=O$.   (15.1)

Ketone Bodies   Acetoacetate, β-hydroxybutyrate—or their parent acids—and acetone.   (27.6)

Ketonemia   An elevated concentration of ketone bodies in the blood.   (27.6)

Ketonuria   An elevated concentration of ketone bodies in the urine.   (27.6)

Ketose   A monosaccharide whose molecules have a ketone group.   (19.2)

Ketosis   The combination of ketonemia, ketonuria, and acetone breath.   (27.6)

Kilocalorie (kcal)   The quantity of heat equal to 1000 calories.   (2.3)

Kilogram (kg)   The SI base unit of mass; 1000 g; 2.205 lb.   (1.4)

Kinase   An enzyme that catalyzes the transfer of a phosphate group.   (22.1)

Kinetic Energy (KE)   The energy of an object by virtue of its motion.

$$KE = \tfrac{1}{2} \text{ mass} \times \text{velocity}^2   (2.3)$$

Kinetics   The field of chemistry that deals with the rates of chemical reactions.   (9.1)

Kinetic Theory of Gases   A set of postulates about the nature of an ideal gas: that it consists of a large number of very small particles in constant, random motion; that in their collisions the particles lose no frictional energy; that between collisions the particles neither attract nor repel each other; and that the motions and collisions of the particles obey all the laws of physics.   (6.5)

Krebs' Cycle   (See *Citric Acid Cycle*)

Law of Conservation of Energy   Energy can be neither created nor destroyed but only transformed from one form to another.   (2.3)

Law of Conservation of Mass   Matter is neither created nor destroyed in chemical reactions; the masses of all products equals the masses of all reactants.   (2.1)

Law of Definite Proportions   The elements in a compound occur in definite proportions by mass.   (2.1)

Law of Mass Action (Law of Guldberg and Waage)   The molar proportions of the interacting substances in a chemical equilibrium are related by the following equation (in which the ratio on the right is called the *mass action expression* for the system).

$$K_{eq} = \frac{[C]^c[D]^d}{[A]^a[B]^b}$$

The symbols refer to the following generalized equilibrium:

$$aA + bB \rightleftharpoons cC + dD$$

and the brackets, [   ], denote molar concentrations. When an equilibrium involves additional substances, the equation for the equilibrium constant is adjusted accordingly.   (9.3)

Law of Multiple Proportions   When two elements can combine to form more than one compound, the different masses of the first that can combine with the same mass of the second are in the ratio of small whole numbers.   (2.2)

Law of Partial Pressures (Dalton's Law)   The total pressure of a mixture of gases is the sum of their individual partial pressures.   (6.4)

Le Châtelier's Principle   If a system is in equilibrium and a change is made in its conditions, the system will change in whichever way most directly restores equilibrium.   (6.7)

Length   The base quantity for expressing distances or how long something is.   (1.4)

Levorotatory   The property of an optically active substance that causes a counterclockwise rotation of the plane of plane-polarized light.   (18.3)

Lewis Structure   A structural formula of a particle (atom, ion, or molecule) that shows all valence shell electrons in their correct places.   (5.4)

Ligand   An electron-rich species, either negatively charged or electrically neutral, that binds with a metal ion to form a complex ion.   (15.3)

Ligase   An enzyme that catalyzes the formation of bonds at the expense of triphosphate energy.   (22.1)

Like-Dissolves-Like Rule   Polar solvents dissolve polar or ionic solutes and nonpolar solvents dissolve nonpolar or weakly polar solutes.   (12.5)

Limiting Amino Acid   The essential amino acid most poorly provided by a dietary protein.   (29.2)

Limiting Reactant   The reactant that is completely consumed while one or more other reactants is not used up.   (3.3)

Lipid   A plant or animal product that tends to dissolve in such nonpolar solvents as ether, carbon tetrachloride, and benzene.   (20.1)

Lipid Bilayer   The sheet-like array of two layers of lipid molecules, interspersed with molecules of cholesterol and proteins, that make up the membranes of cells in animals.   (20.5)

Lipoprotein Complex   A combination of lipid and protein molecules that serves as the vehicle for carrying the lipid in the bloodstream.   (27.1)

Liquid   A state of matter in which a substance's volume but not its shape is independent of the shape of its container.   (2.1, 6.6)

Liter (L)   A volume equal to 1000 cm³ or 1000 mL or 1.057 liquid quart.   (1.4)

Lock-and-Key Theory   The specificity of an enzyme for its substrate is caused by the need for the substrate molecule to fit to the enzyme's surface much as a key fits to and turns only one tumbler lock.   (22.2)

London Force   A net force between molecules that arises from temporary polarities induced in the molecules by collisions or near-collisions with neighboring molecules.   (6.6)

Lyase   An enzyme that catalyzes an elimination reaction to form a double bond.   (22.1)

Macromolecule   Any molecule with a very high formula mass, generally several thousand or more.   (7.1)

Manometer   A device for measuring gas pressure.   (6.4)

Markovnikov's Rule   In the addition of an unsymmetrical reactant to an unsymmetrical double bond of a simple alkene, the positive part of the reactant molecule (usually H^+) goes to the carbon that has the greater number of hydrogen atoms and the negative part goes to the other carbon of the double bond.   (13.4)

Mass   A quantitative measure of inertia based on an artifact at Sèvres, France, called the standard kilogram mass; a measure of the quantity of matter in an object relative to this reference standard.   (1.4)

Mass Number   The sum of the numbers of protons and neutrons in one atom of an isotope.   (3.2)

Material Balance   The condition of a chemical equation in which all of the atoms present among the reactants are also found in the products.   (8.3)

Matter   Anything that occupies space and has mass.   (2.1)

Measurement   An operation that obtains a value for a physical quantity by the use of an instrument.   (1.3)

Melting Point   The temperature at which a solid changes into its liquid form; the temperature at which equilibrium exists between the solid and liquid forms of a substance.   (2.4, 6.9)

Mercaptan   A thioalcohol; R—S—H.   (14.6)

Meso Compound   One of a set of optical isomers whose own molecules are not chiral and which, therefore, is optically inactive.   (Special Topic 18.1)

Messenger RNA (mRNA)   RNA that carries the genetic code in the form of a specific series of codons for a specific polypeptide from the cell's nucleus to the cytoplasm.   (24.3)

Metabolism The sum total of all of the chemical reactions that occur in an organism. (2.4)

Metal Any substance, usually an element, that is shiny, conducts electricity well, and (if a solid) can be hammered into sheets and drawn into wires. (2.1)

Metalloids Elements that have some metallic and some nonmetallic properties. (4.1)

Metathesis Reaction (See *Double Replacement*)

Meter (m) The base unit of length in the International System of Units (SI).

$$1 \text{ m} = 100 \text{ cm} = 39.37 \text{ in.} = 3.280 \text{ ft} = 1.093 \text{ yd} \quad (1.4)$$

Methyl Group CH_3— (11.2)

Metric System A decimal system of weights and measures in which the conversion of a base unit of measurement into a multiple or a submultiple is done by moving the decimal point; the predecessor to the International System of Units (SI). (1.4)

Micelle A globular arrangement of the molecules of an amphipathic compound in water in which their hydrophobic parts intermingle inside the globule and their hydrophilic parts are exposed to the water. (20.5)

Microgram (μg) A mass equal to one-thousandth of a milligram.

$$1 \ \mu\text{g} = 0.001 \text{ mg} = 1 \times 10^{-6} \text{ g} \quad (1.4)$$

Microliter (μL) A volume equal to one-thousandth of a milliliter.

$$1 \ \mu\text{L} = 0.001 \text{ mL} = 1 \times 10^{-6} \text{ L} \quad (1.4)$$

Milliequivalent (meq) A quantity of substance equal to one-thousandth of an equivalent. (8.5)

Milligram (mg) A mass equal to one-thousandth of a gram.

$$1 \text{ mg} = 0.001 \text{ g}; 1000 \text{ mg} = 1 \text{ g} \quad (1.4)$$

Milliliter (mL) A volume equal to one-thousandth of a liter.

$$1 \text{ mL} = 0.001 \text{ L} = 1 \text{ cm}^3$$
$$1 \text{ liquid ounce} = 29.57 \text{ mL}$$
$$1 \text{ liquid quart} = 946.4 \text{ mL} \quad (1.4)$$

Millimeter (mm) A length equal to one-thousandth of a meter.

$$1 \text{ mm} = 0.001 \text{ m} = 0.0394 \text{ in.} \quad (1.4)$$

Millimeter of Mercury (mm Hg) A unit of pressure equal to 1/760 atm. (6.1)

Millimole (mmol) One-thousandth of a mole.

$$1000 \text{ mmol} = 1 \text{ mol} \quad (3.2)$$

Minerals, Dietary Ions that must be provided in the diet at levels of 100 mg/day or more; Ca^{2+}, Mg^{2+}, Na^+, K^+, Cl^-, and phosphate. (29.4)

Mitochondrion A unit inside a plant or animal cell in which the machinery for making high-energy phosphates by oxidative phosphorylation is located. (24.1)

Mixture One of the three kinds of matter (together with elements and compounds); any substance made up of two or more elements or compounds combined physically in no particular proportion by mass and separable into its component parts by physical means. (2.1, 7.1)

Model, Scientific A mental construction, often involving pictures or diagrams, that is used to explain a number of facts. (4.3)

Moderate Acid An acid with a K_a in the range of 1 to 10^{-3}. (9.6)

Molar Concentration (M) A solution's concentration in units of moles of solute per liter of solution; molarity. (3.5)

Molar Volume, Standard The volume occupied by one mole of a gas under standard conditions of temperature and pressure; 22.4 L at 273 K and 1 atm. (6.3)

Molarity (See *Molar Concentration*)

Mole (mol) A mass of a compound or element that equals its formula mass in grams; Avogadro's number of a substance's formula units. (3.1, 3.2)

Molecular Compound A compound whose smallest representative particle is a molecule; a covalent compound. (5.3)

Molecular Equation An equation that shows the complete formulas of all of the substances present in a mixture undergoing a reaction. (See also *Net Ionic Equation; Equation, Balanced*) (8.3)

Molecular Formula (See *Formula, Molecular*)

Molecular Mass The formula mass of a substance. (3.2)

Molecular Orbital A region in the space that envelopes two (or sometimes more) atomic nuclei where a shared pair of electrons of a covalent bond resides. (5.9)

Molecule An electrically neutral (but often polar) particle made up of the nuclei and electrons of two or more atoms and held together by covalent bonds; the smallest representative sample of a molecular compound. (3.1, 5.3)

Monatomic Ion An ion possessing only one atomic nucleus. (5.1)

Monoamine Oxidase An enzyme that catalyzes the inactivation of neurotransmitters or other amino compounds of the nervous system. (22.7)

Monohydric Alcohol An alcohol whose molecules have one OH group. (14.1)

Monomer A compound that can be used to make a polymer. (13.6)

Monoprotic Acid An acid with one proton per molecule that can neutralize a base. (8.2)

Monosaccharide A carbohydrate that cannot be hydrolyzed. (19.2)

Mucin A viscous glycoprotein released in the mouth and the stomach that coats and lubricates food particles and protects the stomach from the acid and pepsin of gastric juice. (23.1)

Mutagen A chemical or physical agent that can induce the mutation of a gene without preventing the gene from replicating. (11.2)

Mutarotation The gradual change in the specific rotation of a substance in solution but without a permanent, irreversible chemical change occurring. (19.4)

Myosins Proteins in contractile muscle. (21.9)

Native Protein A protein whose molecules are in the configuration and shape they normally have within a living system. (21.2)

Net Ionic Equation A chemical equation in which all spectator particles are omitted so that only the particles that participate directly are represented. (8.3)

Neurotransmitter A substance released by one nerve cell to carry a signal to the next nerve cell. (22.6)

Neutralization, Acid–Base A reaction between an acid and a base. (4.1)

Neutralizing Capacity The capacity of a solution or a substance to neutralize an acid or a base — expressed as a molar concentration. (9.11)

Neutral Solution A solution in which the molar concentration of hydronium ions exactly equals the molar concentration of hydroxide ions. (8.2, 9.5)

Neutron An electrically neutral subatomic particle with a mass of 1 u. (4.2)

Niacin A water-soluble vitamin needed to prevent pellagra and essential to the coenzymes in NAD^+ and $NADP^+$; nicotinic acid or nicotinamide. (29.3)

Nitrogen Balance A condition of the body in which it excretes as much nitrogen as it receives in the diet. (29.2)

Nitrogen Family The elements of group VA of the periodic table: nitrogen, phosphorus, arsenic, antimony, and bismuth. (4.1)

Nitrogen Pool The sum total of all nitrogen compounds in the body. (28.1)

Noble Gases The elements of group 0 of the periodic table: helium, neon, argon, krypton, xenon, and radon. (4.1)

Noble Gas Rule (See *Octet Rule*)

Nomenclature The system of names and the rules for devising such names, given structures, or for writing structures, given names. (12.6)

Nonelectrolyte Any substance that cannot furnish ions when dissolved in water or when melted. (8.1)

Nonfunctional Group A section of an organic molecule that remains unchanged during a chemical reaction at a functional group. (12.4)

Nonhydrolyzable Lipid Any lipid, such as the steroids, that cannot be hydrolyzed or similarly broken down by aqueous alkali. (20.1)

Nonmetal Any element that is not a metal. (See *Metal*) (2.1)

Nonvolatile Liquid Any liquid with a very low vapor pressure at room temperature and that does not readily evaporate. (6.7)

Normal Fasting Level The normal concentration of something in the blood, such as blood sugar, after about four hours without food. (26.1)

Nuclear Chain Reaction The mechanism of nuclear fission by which one fission event makes enough fission initiators (neu-

trons) to cause more than one additional fission event. (11.7)

Nuclear Equation A representation of a nuclear transformation in which the chemical symbols of the reactants and products include mass numbers and atomic numbers. (11.1)

Nucleic Acid A polymer of nucleotides in which the repeating units are pentose phosphate diesters, each pentose unit bearing a sidechain base (one of five heterocyclic amines); polymeric compounds that are involved in the storage, transmission, and expression of genetic messages. (24.2)

Nucleotide A monomer of a nucleic acid that consists of a pentose phosphate ester in which the pentose unit carries one of five heterocyclic amines as a sidechain base. (24.2)

Nucleus In chemistry and physics, the subatomic particle that serves as the core of an atom and that is made up of protons and neutrons. (4.2) In biology, the organelle in a cell that houses DNA. (24.1)

Nutrient Any one of a large number of substances in food and drink that is needed to sustain growth and health. (29.1)

Nutrition The science of the substances of the diet that are necessary for growth, operation, energy, and repair of body tissues. (29.1)

Octet, Outer A condition of an atom or ion in which its highest occupied energy level has eight electrons — a condition of stability. (5.2)

Octet Rule (Noble Gas Rule) The atoms of a reactive element tend to undergo those chemical reactions that most directly give them the electron configuration of the noble gas that stands nearest the element in the periodic table (all but one of which have outer octets). (5.2)

Olefin An alkene. (13.6)

One-Substance–One-Structure Rule If two samples of matter have identical physical and chemical properties, they have identical molecules. (18.1)

Optical Isomer One of a set of compounds whose molecules differ only in their chiralities. (18.3)

Optically Active The ability of a substance to rotate the plane of polarization of plane-polarized light. (18.3)

Optical Rotation The degrees of rotation of the plane of plane-polarized light caused by an optically active solution; the observed rotation of such a solution. (18.3)

Orbital (See *Atomic Orbital*)

Orbital Hybridization The mixing of two or more ordinary atomic orbitals to give an equal number of modified atomic orbitals, called *hybrid orbitals,* each of which possesses some of the characteristics of the originals. (5.9)

Orbital Overlap The interpenetration of one atomic orbital by another from an adjacent atom to form a molecular orbital. (5.9)

Organic Chemistry The chemistry of carbon compounds. (12.1)

Organic Compounds Compounds of carbon other than those related to carbonic acid and its salts, or to the oxides of carbon, or to the cyanides. (8.2, 12.1)

Osmolarity The molar concentration of all osmotically active solute particles in a solution. (7.5)

Osmosis The passage of water only, without any solute, from a less concentrated solution (or pure water) to a more concentrated solution when the two solutions are separated by a semipermeable membrane. (7.5)

Osmotic Membrane A semipermeable membrane that permits only osmosis, not dialysis. (7.5)

Osmotic Pressure The pressure that would have to be applied to a solution to prevent osmosis if the solution were separated from water by an osmotic membrane. (7.5)

Outer Octet (See *Octet, Outer*)

Outside Shell Electrons Electrons occupying any of the available orbitals at the highest occupied principal energy level. (4.5)

Oxidase (See *Oxidoreductase*)

Oxidation A reaction in which the oxidation number of one of the atoms of a reactant becomes more positive; in organic chemistry, the loss of hydrogen or the gain of oxygen. (5.2, 10.1)

Oxidation Number For simple monatomic ions, the quantity and sign of the electrical charge on the ion. (5.2, 10.1)

Oxidation–Reduction Reaction A reaction in which oxidation numbers change. (5.2, 10.1)

Oxidative Deamination The change of an amino group to a keto group with loss of nitrogen. (28.2)

Oxidative Phosphorylation The synthesis of high-energy phosphates such as ATP from lower energy phosphates and inorganic phosphate by the reactions that involve the respiratory chain. (25.1, 25.3)

Oxidizing Agent A substance that can cause an oxidation. (5.2)

Oxidoreductase An enzyme that catalyzes the formation of an oxidation–reduction equilibrium. (22.1)

Oxygen Affinity The percentage to which all of the hemoglobin molecules in the blood are saturated with oxygen molecules. (23.3)

Oxygen Debt The condition in a tissue when anaerobic glycolysis has operated and lactate has been excessively produced. (26.3)

Oxygen Family The elements in group VIA of the periodic table: oxygen, sulfur, selenium, tellurium, and polonium. (4.1)

Oxyhemoglobin Hemoglobin carrying its capacity of oxygen. (23.3)

P_i Inorganic phosphate ion(s) of whatever mix of PO_4^{3-}, HPO_4^{2-}, $H_2PO_4^-$, and even traces of H_3PO_4 that is possible at the particular pH of the system, but almost entirely $HPO_4^{2-} + H_2PO_4^-$. (16.6)

Pancreatic Juice The digestive juice that empties into the duodenum from the pancreas. (23.1)

Pantothenic Acid A water-soluble vitamin needed to make coenzyme A. (29.3)

Partial Pressure The pressure contributed to the total pressure by an individual gas in a mixture of gases. (6.4)

Parts per Billion (ppb) The number of parts in a billion parts. (Two drops of water in a railway tank car that holds 34,000 gallons of water correspond roughly to 1 ppb.) (7.4)

Parts per Million (ppm) The number of parts in a million parts. (Two drops of water in a large 32-gallon trash can correspond roughly to 1 ppm.) (7.4)

Pascal (Pa) The SI derived unit of pressure.

$$133.3224 \text{ Pa} = 1 \text{ mm Hg} \quad (6.1)$$

Pauli Exclusion Principle No more than two electrons can occupy the same orbital at the same time, and two can be present only if they have opposite spin. (4.4)

Pentose Phosphate Pathway The synthesis of NADPH that uses chemical energy in glucose-6-phosphate and that involves pentoses as intermediates. (26.3)

Peptide Bond The amide linkage in a protein; a carbonyl–nitrogen bond. (21.1, 21.3)

Percent (%) A measure of concentration.
 Vol/vol (v/v) percent: The number of volumes of solute in 100 volumes of solution.
 Wt/wt (w/w) percent: The number of grams of solute in 100 g of the solution.
 Wt/vol (w/v) percent: The number of grams of solute in 100 mL of the solution.
 Milligram percent: The number of milligrams of the solute in 100 mL of the solution. (7.4)

Period A horizontal row in the periodic table. (4.1)

Periodic Law Many properties of the elements are periodic functions of their atomic numbers. (4.1)

Periodic Table A display of the elements that emphasizes the family relationships. (4.1)

pH The negative power to which the base 10 must be raised to express the molar concentration of hydrogen ions in an aqueous solution.

$$[H^+] = 1 \times 10^{-pH}$$
$$-\log [H^+] = pH \quad (9.5)$$

Phenol Any organic compound whose molecules have an OH group attached to a benzene ring. (14.4)

Phenyl Group The benzene ring minus one H atom; C_6H_5. (13.7)

Phosphate Buffer Usually a mixture or a solution that contains dihydrogen phosphate ions ($H_2PO_4^-$) to neutralize OH^- and monohydrogen phosphate ions (HPO_4^{2-}) to neutralize H^+. (9.9)

Phosphate Group Transfer Potential The relative ability of an organophosphate to transfer a phosphate group to some acceptor. (25.1)

Phosphatide A glycerophospholipid whose molecules are esters between glycerol, two fatty acids, phosphoric acid, and a small alcohol. (20.3)

Phosphoglyceride (See *Glycerophospholipid*)

Phospholipid Lipids such as the glycerophospholipids (phosphatides and plasmalogens) and the sphingomyelins whose molecules include phosphate ester units. (20.3)

Photon A package of energy released when an electron in an atom moves from a higher to a lower energy state; a unit of light energy.   (4.3)

Photosynthesis The synthesis in plants of complex compounds from carbon dioxide, water, and minerals with the aid of sunlight captured by the plant's green pigment, chlorophyll.   (2.4, 19.2)

Physical Property Any observable characteristic of a substance other than a chemical property, such as color, density, melting point, boiling point, temperature, and quantity.   (1.3)

Physical Quantity A property of something to which we assign both a numerical value and a unit, such as mass, volume, or temperature.

$$\text{physical quantity} = \text{number} \times \text{unit.}   \text{(1.3)}$$

Physiological Saline Solution A solution of sodium chloride with an osmotic pressure equal to that of blood.   (7.6)

pI (See *Isoelectric Point*)

Pi Bond (π Bond) A covalent bond formed when two electrons fill a molecular orbital created by the side-to-side overlap of two p orbitals.   (5.9)

Pi (π) Electrons The pair of electrons in a pi bond.   (5.9)

pK_a p$K_a = -\log K_a$ (9.8)

pK_b p$K_b = -\log K_b$ (9.8)

Plane-Polarized Light Light whose electrical field vibrations are all in the same plane.   (18.3)

Plasmalogens Glycerophospholipids whose molecules include an unsaturated fatty alcohol unit.   (20.3)

Plasmid A circular molecule of supercoiled DNA in a bacterial cell.   (24.6)

β-Pleated Sheet A secondary structure for a polypeptide in which the molecules are aligned side by side in a sheet-like array with the sheet partially pleated.   (21.4)

pOH The negative power to which the base 10 must be raised to express the concentration of hydroxide ions in an aqueous solution in mol/L.

$$[OH^-] = 1 \times 10^{-pOH}$$

At 25 °C, pH + pOH = 14.00.   (9.5)

Poison A substance that reacts in some way in the body to cause changes in metabolism that threaten health or life.   (22.4)

Polar Bond A bond at which we can write $\delta+$ at one end and $\delta-$ at the other end, the end that has the more electronegative atom.   (5.8)

Polarimeter An instrument for detecting and measuring optical activity.   (18.3)

Polar Molecule A molecule that has sites of partial positive and partial negative charge and a permanent electrical dipole.   (5.8)

Polyatomic Ion A ion made from two or more atoms, such as OH^-, SO_4^{2-}, and CO_3^{2-}.   (5.5)

Polymer Any substance with a very high formula mass whose molecules have a repeating structural unit.   (13.6)

Polymerization A chemical reaction that makes a polymer from a monomer.   (13.6)

Polypeptide A polymer with repeating α-aminoacyl units joined by peptide (amide) bonds.   (21.1)

Polysaccharide A carbohydrate whose molecules are polymers of monosaccharides.   (19.2)

Potential Energy Stored or inactive energy.   (2.3)

PP$_i$ Inorganic diphosphate ion(s).   (25.1)

Ppb (See *Parts per Billion*)

Ppm (See *Parts per Million*)

Precipitate A solid that separates from a solution as the result of a chemical reaction.   (3.4)

Precipitation The formation and separation of a precipitate.   (3.4)

Precision The fineness of a measurement or the degree to which successive measurements agree with each other when several are taken one after the other. (See also *Accuracy*)   (1.6)

Pressure Force per unit area.   (6.1)

Pressure–Solubility Law (Henry's Law) The concentration of a gas in a liquid at any given temperature is directly proportional to the partial pressure of the gas on the solution.   (7.3)

Pressure–Temperature Law (Gay-Lussac's Law) The pressure of a gas is directly proportional to its Kelvin temperature when the gas volume is constant.   (6.2)

Pressure–Volume Law (Boyle's Law) The volume of a gas is inversely proportional to its pressure when the temperature is constant.   (6.2)

Primary Alcohol An alcohol in whose molecules an OH group is attached to a primary carbon, as in RCH_2OH.   (14.1)

Primary Carbon In a molecule, a carbon atom that is joined directly to just one other carbon, such as the end carbons in $CH_3CH_2CH_3$.   (12.6)

Primary Structure The sequence of amino acyl residues held together by peptide bonds in a polypeptide.   (21.2)

Primary Transcript RNA (ptRNA) [See *Heterogeneous Nuclear RNA (hnRNA)*]

Principal Energy Level A space near an atomic nucleus where there are one or more sublevels and orbitals in which electrons can reside; an electron shell.   (4.3)

Product A substance that forms in a chemical reaction.   (2.1)

Proenzyme (See *Zymogen*)

Property A characteristic of something by means of which we can identify it.   (1.3)

Propyl Group $CH_3CH_2CH_2-$ (11.6)

Prosthetic Group A nonprotein molecule joined to a polypeptide to make a biologically active protein.   (21.5)

Protein A naturally occurring polymeric substance made up wholly or mostly of polypeptide molecules.   (21.1)

Proton A subatomic particle that bears one unit of positive charge and has a mass of 1 u.   (4.2)

Proton-Pumping ATPase The enzyme on the matrix side of

the inner mitochondrial membrane that catalyzes the formation of ATP from ADP and P_i under the influence of a flow of protons across this membrane. (25.3)

Quantum A quantity of energy possessed by a photon. (4.3)

Quaternary Structure An aggregation of two or more polypeptide strands each with its own primary, secondary, and tertiary structure. (21.2, 21.6)

Racemic Mixture A 1:1 mixture of enantiomers and therefore optically inactive. (18.3)

Rad One rad equals 100 ergs (1×10^{-5} J) of energy absorbed per gram of tissue as a result of ionizing radiation. (11.3)

Radiation In atomic physics, the emission of some ray such as an alpha, beta, or gamma ray; any of the rays themselves. (11.1)

Radiation Sickness The set of symptoms that develops following exposure to heavy doses of ionizing radiations. (11.2)

Radical A particle with one or more unpaired electrons. (11.2)

Radioactive The property of unstable atomic nuclei whereby they emit alpha, beta, or gamma rays. (11.1)

Radioactive Decay The change occurring to a radioactive isotope by which it emits alpha rays, beta rays, or gamma rays. (11.1)

Radioactive Disintegration Series A series of isotopes selected and arranged such that each isotope except the first is produced by the radioactive decay of the preceding isotope and the last isotope is nonradioactive. (11.1)

Radioactivity The ability to emit atomic radiations. (2.1, 11.1)

Radiomimetic Substance A substance whose chemical effect in a cell mimics the effect of ionizing radiation. (24.4)

Radionuclide A radioactive isotope. (11.1)

Rate of Reaction The number of successful (product-forming) collisions that occur each second in each unit of volume of a reacting mixture. (9.2)

Reactant One of the substances that reacts in a chemical reaction. (2.1)

Reaction, Chemical An event in which chemical substances change into other substances. (2.1)

Reagent Any mixture of chemicals, usually a solution, that is used to carry out a chemical test. (7.4)

Receptor Molecule A molecule of a protein built into a cell membrane that can accept a molecule of a hormone or a neurotransmitter. (21.8, 22.6)

Recombinant DNA DNA made by combining the natural DNA of plasmids in bacteria or the natural DNA in yeasts with DNA from external sources, such as the DNA for human insulin, and made as a step in a process that uses altered bacteria or yeasts to make specific proteins (e.g., interferons, human growth hormone, and insulin). (24.6)

Recommended Dietary Allowance (RDA) The level of intake of a particular nutrient as determined by the Food and Nutrition Board of the National Research Council of the National Academy of Sciences to meet the known nutritional needs of most healthy individuals. (29.1)

Redox Reaction Abbreviation of *reduction–oxidation;* a reaction in which oxidation numbers change. (5.2, 10.1)

Reducing Agent A substance that can cause another to be reduced. (5.2)

Reducing Carbohydrate A carbohydrate that gives a positive Benedict's test. (19.2)

Reductase An enzyme that catalyzes a reduction. (See *Oxidoreductase*) (22.1)

Reduction A reaction in which the oxidation number of an atom of one reactant becomes less positive or more negative; in organic chemistry, the gain of hydrogen or the loss of oxygen. (5.2)

Reduction Potential The quantitative measure of a given half-reaction to proceed as a reduction relative to the standard hydrogen half-reaction. (10.3)

Reduction Potential, Standard ($E°$) The reduction potential under standard conditions, 25 °C, 1 atm, and concentrations of 1 M. (10.3)

Reductive Amination The conversion of a keto group to an amino group by the action of ammonia and a reducing agent. (28.1)

Rem One rem is the quantity of a radiation that produces the same effect in humans as one roentgen of X rays or gamma rays. (11.3)

Renal Threshold That concentration of a substance in blood above which it appears in the urine. (26.1)

Replication The reproductive duplication of a DNA double helix. (24.1)

Representative Element Any element in any A-group of the periodic table; any element in groups IA–VIIA and those in group 0. (4.1)

Repressor A substance whose molecules can bind to a gene and prevent the gene from directing the synthesis of a polypeptide. (24.4)

Respiration The intake and chemical use of oxygen by the body and the release of carbon dioxide. (23.3)

Respiratory Chain The reactions that transfer electrons from the intermediates made by other pathways to oxygen; the mechanism that creates a proton gradient across the inner membrane of a mitochondrion and that leads to ATP synthesis; the enzymes that handle these reactions. (25.1, 25.2)

Respiratory Enzymes The enzymes of the respiratory chain. (25.3)

Respiratory Gases Oxygen and carbon dioxide. (23.3)

Reverse Reaction The reaction that undoes the effect of the forward reaction of an equilibrium. (See *Forward Reaction*) (6.7)

Riboflavin A B vitamin needed to give protection against the breakdown of tissue around the mouth, the nose, and the tongue, as well as to aid in wound healing. (29.3)

Ribonucleic Acids (RNA) Polymers of nucleotides made using ribose that participate in the transcription and the translation of the genetic messages into polypeptides. (See also *Heterogeneous Nuclear RNA, Messenger RNA, Ribosomal RNA, and Transfer RNA*) (24.2)

Ribosomal RNA (rRNA)   RNA that is incorporated into cytoplasmic bodies called ribosomes.   (24.3)

Ribosome   A granular complex of rRNA that becomes attached to a mRNA strand and that supplies some of the enzymes for mRNA-directed polypeptide synthesis.   (24.1)

Ribozyme   An enzyme whose molecules consist of ribonucleic acid, not polypeptide.   (24.1)

Ring Compound   A compound whose molecules contain three or more atoms joined in a ring.   (12.2)

RNA   (See *Ribonucleic Acid*)

Roentgen   One roentgen is the quantity of X rays or gamma radiation that generates ions with an aggregate of 2.1×10^9 units of charge in 1 mL of dry air at normal pressure and temperature.   (11.3)

Saliva   The digestive juice secreted in the mouth whose enzyme, amylase, catalyzes the partial digestion of starch.   (23.1)

Salt   Any crystalline compound that consists of oppositely charged ions (other than H^+, OH^-, or O^{2-}).   (8.2)

Salt Bridge   A force of attraction between (+) and (−) sites on polypeptide molecules.   (21.5)

Saponifiable Lipid   (See *Hydrolyzable Lipid*)

Saponification   The reaction of an ester with base to give an alcohol and the salt of an acid.   (16.5)

Saturated Compound   A compound whose molecules have only single bonds.   (12.2)

Scientific Method   A method of solving a problem that uses facts to devise a hypothesis to explain the facts and to suggest further tests or experiments designed to discover if the hypothesis is true or false.   (1.2)

Scientific Notation   The method of writing a number as the product of two numbers, one being 10^x, where x is some positive or negative whole number.   (1.5)

Second(s)   The SI unit of time; 1/60th minute.   (1.4)

Secondary Alcohol   An alcohol in whose molecules an OH group is attached to a secondary carbon atom; R_2CHOH.   (14.1)

Secondary Carbon   Any carbon atom in an organic molecule that has two and only two bonds to other carbon atoms, such as the middle carbon atom in $CH_3CH_2CH_3$.   (12.6)

Secondary Structure   A shape, such as the α-helix or a unit in a β-pleated sheet, that all or a large part of a polypeptide molecule adopts under the influence of hydrogen bonds, salt bridges, and hydrophobic interactions after its peptide bonds have been made.   (21.2)

Semipermeable   Descriptive of a membrane that permits only certain kinds of molecules to pass through and not others.   (7.5)

Shock, Traumatic   A medical emergency in which relatively large volumes of blood fluid leave the vascular compartment and enter the interstitial spaces.   (23.2)

Sievert (Sv)   The SI unit of radiation dose equivalent.   (11.3)

Sigma Bond (σ Bond)   A covalent bond associated with a molecular orbital whose shape is symmetrical about the bonding axis.   (5.9)

Significant Figures   The number of digits in a numerical measurement or in the result of a calculation that are known with certainty to be accurate plus one more digit.   (1.6)

Simple Lipid   (See *Triacylglycerol*)

Simple Salt   A salt that consists of only one kind of cation and one kind of anion.   (8.2)

Simple Sugar   Any monosaccharide.   (19.2)

Single Bond   A covalent bond involving one shared pair of electrons.   (5.4)

Soap   A detergent that consists of salts of long-chain fatty acids.   (Special Topic 20.3)

Soft Water   Water with little if any of the hardness ions — Mg^{2+}, Ca^{2+}, Fe^{2+}, or Fe^{3+}.   (Special Topic 8.2)

Sol   A colloidal dispersion of tiny particles of a solid in a liquid.   (7.1)

Solid   A state of matter in which the visible particles of the substance have both definite shapes and definite volumes.   (2.1, 6.9)

Solubility   The extent to which a substance dissolves in a fixed volume or mass of a solvent at a given temperature.   (3.4)

Solute   The component of a solution that is understood to be dissolved in or dispersed in a continuous solvent.   (3.4)

Solution   A homogeneous mixture of two or more substances that are at the smallest levels of their states of subdivision — at the ion, atom, or molecule level.   (3.4, 7.1)

Solution, Aqueous   A solution in which water is the solvent.   (3.4)

Solution, Concentrated   A solution with a high ratio of solute to solvent.   (3.4)

Solution, Dilute   A solution with a low ratio of solute to solvent.   (3.4)

Solution, Saturated   A solution into which no more solute can be dissolved at the given temperature; a solution in which dynamic equilibrium exists between the dissolved and the undissolved solute.   (3.4, 7.2)

Solution, Supersaturated   An unstable solution that has a higher concentration of solute than that of the saturated solution.   (3.4)

Solution, Unsaturated   A solution into which more solute could be dissolved without changing the temperature.   (3.4)

Solvent   That component of a solution into which the solutes are considered to have dissolved; the component that is present as a continuous phase.   (3.4)

Somatostatin   A hormone of the hypothalamus that inhibits or slows the release of glucagon and insulin from the pancreas.   (26.1)

Specific Gravity   The ratio of the density of an object to the density of water.   (Special Topic 1.1)

Specific Heat   The amount of heat that one gram of a substance can absorb per degree Celsius increase in temperature:

$$\text{Specific heat} = \frac{\text{heat}}{\text{g } \Delta t}$$

where Δt = the change in temperature.   (2.4)

Specific Rotation [α] The optical rotation of a solution per unit of concentration per unit of path length in decimeters:

$$[\alpha] = \frac{\alpha}{cl}$$

where α = observed rotation; c = concentration in g/mL, and l = path length (in dm). (18.3)

Sphingolipid A lipid that, when hydrolyzed, gives spingosine instead of glycerol, plus fatty acids, phosphoric acid, and a small alcohol or a monosaccharide; sphingomyelins and cerebrosides. (20.3)

sp^2 Hybrid Orbital A hybrid orbital made by mixing one s orbital with two p orbitals to form three new, identical orbitals whose axes are in one plane and point to the corners of an equilateral triangle. (5.9)

sp^3 Hybrid Orbital One of four equivalent hybrid orbitals formed by the mixing of one s orbital and three p orbitals and whose axes point to the corners of a regular tetrahedron. (5.9)

Standard A physical description or embodiment of a base unit of measurement, such as the standard meter or the standard kilogram mass. (1.4)

Standard Atmosphere (atm) The pressure that supports a column of mercury 760 mm high when the mercury has a temperature of 0 °C. (6.1)

Standard Conditions of Temperature and Pressure (STP) 0 °C (or 273 K) and 1 atm (or 760 mm Hg). (6.3)

Standard Solution Any solution for which the concentration is accurately known. (9.11)

States of Matter The three possible physical conditions of aggregation of matter — solid, liquid, and gas. (2.1)

Stereoisomers Isomers whose molecules have the same atom-to-atom sequences but different geometric arrangements; a geometric (cis–trans) or optical isomer. (18.1)

Stereoisomerism The existence of stereoisomers. (18.1)

Steroids Nonhydrolyzable lipids such as cholesterol and several sex hormones whose molecules have the four fused rings of the steroid nucleus. (20.4)

Stoichiometry The branch of chemistry that deals with the mole proportions of chemicals in reactions. (3.1)

Straight Chain A continuous, open sequence of covalently bound carbon atoms from which no additional carbon atoms are attached at interior locations of the sequence. (12.2)

Stress In equilibrium chemistry, anything that upsets an equilibrium. (6.7)

Strong Acid An acid with a high percentage ionization and a high value of acid ionization constant, K_a. (8.2)

Strong Base A metal hydroxide with a high percentage ionization in solution. (8.2)

Strong Brønsted Acid Any species, molecule or ion, that has a strong tendency to donate a proton to some acceptor. (8.4)

Strong Brønsted Base Any species, molecular or ionic, that binds an accepted proton strongly. (8.4)

Strong Electrolyte Any substance that has a high percentage ionization in solution. (8.1)

Structural Formula A formula that uses lines representing covalent bonds to connect the atomic symbols in the pattern that occurs in one molecule of a compound. (5.4, 12.2)

Structural Isomer One of a set of isomers whose molecules differ in their atom-to-atom sequence. (5.4)

Structure Synonym for structural formula. (See *Structural Formula*) (12.2)

Subatomic Particle An electron, a proton, or a neutron; the atomic nucleus as a whole is also a subatomic particle. (4.2)

Sublimation The change from the solid state directly to the gaseous state. (6.9)

Subscripts Numbers placed to the right and a half space below the atomic symbols in a chemical formula. (2.2)

Subshell An energy sublevel; a region that makes up part (sometimes all) of a principal energy level and that can itself be subdivided into individual orbitals. (4.4)

Substance, Pure An element or a compound but not a mixture. (2.1)

Substitution Reaction A reaction in which one atom or group replaces another atom or group in a molecule. (12.7)

Substrate The substance on which an enzyme performs its catalytic work. (18.2)

Substrate Phosphorylation The direct transfer of a phosphate unit from an organophosphate to a receptor molecule. (25.1)

Superimposition A chirality testing operation to see if one molecular model can be made to blend simultaneously at exactly every point with another model. (18.2)

Supersaturated Describing an unstable condition of a solution in which more solute is in solution than could be if there were equilibrium between the undissolved and dissolved states of the solute. (3.4)

Surface-Active Agent (See *Surfactant*)

Surface Tension The quality of a liquid's surface by which it behaves as if it were a thin, invisible, elastic membrane. (6.8)

Surfactant A substance, such as a detergent, that reduces the surface tension of water. (6.8)

Suspension A mixture in which the particles of at least one component have average diameters greater than 1000 nm. (7.1)

Synapse The fluid-filled gap between the end of the axon of one nerve cell and the next nerve cell. (22.7)

Target Cell A cell at which a hormone molecule finds a site where it can become attached and then cause some action that is associated with the hormone. (22.6)

Target Tissue The organ whose cells are recognizable by the molecules of a particular hormone. (22.6)

Temperature The measure of the hotness or coldness of an object. *Degrees* of temperature, such as those of the Celsius, Fahrenheit, or Kelvin scales, are intervals of equal separation on the thermometer. (1.4)

Temperature–Volume Law (Charles' Law) The volume of a gas is directly proportional to its Kelvin temperature when the pressure is kept constant. (6.2)

Teratogen A chemical or physical agent that can cause birth defects in a fetus other than inherited defects. (11.2)

Tertiary Alcohol An alcohol in whose molecules an OH group is held by a carbon from which three bonds extend to other carbon atoms; R_3COH. (14.1)

Tertiary Carbon A carbon in an organic molecule that has three and only three bonds to adjacent *carbon* atoms. (12.6)

Tertiary Structure The shape of a polypeptide molecule that arises from further folding or coiling of secondary structures. (21.2, 21.5)

Tetrahedral Descriptive of the geometry of bonds at a central atom in which the bonds project to the corners of a regular tetrahedron. (5.7)

Tetrahedral Stereocenter An atom in a molecule with four single bonds arranged tetrahedrally and holding four different atoms or groups. (18.2)

Theory An explanation for a large number of facts, observations, and hypotheses in terms of one or a few fundamental assumptions of what the world (or some small part of the world) is like. (1.2)

Thiamine A B vitamin needed to prevent beriberi. (29.3)

Thioalcohol A compound whose molecules have the SH group attached to a saturated carbon atom; a mercaptan. (14.6)

Threshold Exposure The level of exposure to some toxic agent below which no harm is done. (11.2)

Time A period during which something endures, exists, or continues. (1.4)

Titration An experimental procedure for mixing two solutions using a buret in order to compare the concentration of one of the solutions with that of the other, the standard solution. (9.11)

Tollens' Reagent A slightly alkaline solution of the diammine complex of the silver ion, $Ag(NH_3)_2^+$, in water. (15.3)

Tollens' Test The use of Tollens' reagent to detect an easily oxidized group such as the aldehyde group. (15.3)

Torr A unit of pressure; 1 torr = 1 mm Hg; 1 atm = 760 torr. (6.1)

Trace Element, Dietary Any element that the body needs each day in an amount of no more than 20 mg. (29.4)

Transamination The transfer of an amino group from an amino acid to a receiver with a keto group such that the keto group changes to an amino group. (28.1)

Transcription The synthesis of messenger RNA under the direction of DNA. (24.3)

Transferase An enzyme that catalyzes the transfer of some group. (22.1)

Transfer RNA (tRNA) RNA that serves to carry an amino acyl group to a specific acceptor site of an mRNA molecule at a ribosome where the amino acyl group is placed into a growing polypeptide chain. (24.3)

Transition Elements The elements between those of group IIA and group IIIA in the long periods of the periodic table; a metallic element other than one in group IA or IIA or in the actinide or lanthanide families. (4.1)

Translation The synthesis of a polypeptide under the direction of messenger RNA. (24.4)

Transmutation The change of an isotope of one element into an isotope of a different element. (11.1)

Triacylglycerol A lipid that can be hydrolyzed to glycerol and fatty acids; a triglyceride; sometimes, simply called a glyceride or a simple lipid. (20.1)

Tricarboxylic Acid Cycle (See *Citric Acid Cycle*)

Triglyceride (See *Triacylglycerol*)

Trihydric Alcohol An alcohol with three OH groups per molecule. (14.1)

Triple Bond A covalent bond involving the sharing of three pairs of electrons. (5.9)

Triple Helix The quaternary structure of tropocollagen in which three polypeptide chains are coiled together. (21.6)

Triprotic Acid An acid that can supply three protons per molecule. (8.2)

Tyndall Effect The scattering of light by colloidal sized particles in a colloidal dispersion. (7.1)

Uncertainty The estimate of how finely a number can be read from a measuring instrument. (1.6)

Universal Gas Law (See *Ideal Gas Law*)

Unsaturated Compound Any compound whose molecules have a double or a triple bond. (12.2)

Unshared Pairs Pairs of valence-shell electrons not involved in covalent bonds. (5.4)

Urea Cycle The reactions by which urea is made from amino acids. (28.3)

Vacuum An enclosed space in which there is no matter. (6.1)

Valence Shell The highest energy level of an atom that is occupied by electrons; the outside shell. (5.2)

Valence-Shell Electron-Pair Repulsion Theory (VSEPR) Bond angles at a central atom are caused by the repulsions of the electron clouds of valence-shell electron pairs. (5.7)

Vapor Pressure The pressure exerted by the vapor that is in equilibrium with its liquid state at a given temperature. (6.4)

Vaporization The change of a liquid into its vapor. (2.4)

Vascular Compartment The entire network of blood vessels and their contents. (23.2)

Vasopressin A hypophysis hormone that acts at the kidneys to help regulate the concentrations of solutes in the blood by instructing the kidneys to retain water (if the blood is too concentrated) or to excrete water (if the blood is too dilute). (23.5)

Ventilation The movement of air into and out of the lungs by breathing. (9.9)

Virus One of a large number of substances that consist of nucleic acid surrounded by a protein overcoat and that can enter host cells, multiply, and destroy the host. (24.5)

Vital Force Theory A discarded theory that organic compounds could be made in the laboratory only if the chemicals possessed a vital force contributed by some living thing. (12.1)

Vitamin An organic substance that must be in the diet, whose absence causes a deficiency disease, which is present in foods in trace concentrations, and that isn't a carbohydrate, lipid, protein, or amino acid. (29.3)

Vitamin A Retinol; a fat-soluble vitamin in yellow-colored foods and needed to prevent night blindness and certain conditions of the mucous membranes. (29.3)

Vitamin B_6 Pyridoxine, pyridoxal, or pyridoxamine; a vitamin needed to prevent hypochromic microcytic anemia and used in enzymes of amino acid catabolism. (29.3)

Vitamin B_{12} Cobalamin; a vitamin needed to prevent pernicious anemia. (29.3)

Vitamin C Ascorbic acid; a vitamin needed to prevent scurvy. (29.3)

Vitamin D Cholecalciferol (D_3) or ergocalciferol (D_2); a fat-soluble vitamin needed to prevent rickets and to ensure the formation of healthy bones and teeth. (29.3)

Vitamin Deficiency Diseases Diseases caused not by bacteria or viruses but by the absence of specific vitamins, such as pernicious anemia (B_{12}), hypochromic microcytic anemia (B_6), pellagra (niacin), the breakdown of certain tissues (riboflavin), megaloblastic anemia (folate), beriberi (thiamin), scurvy (C), hemorrhagic disease (K), rickets (D), and night blindness (A). (29.3)

Vitamin E A mixture of tocopherols; a fat-soluble vitamin apparently needed for protection against edema and anemia (in infants) and possibly against dystrophy, paralysis, and heart attacks. (29.3)

Vitamin K The antihemorrhagic vitamin that serves as a cofactor in the formation of a blood clot. (29.3)

Volatile Liquid A liquid that has a high vapor pressure and readily evaporates at room temperature. (6.7)

Volt (V) The SI unit of electrical potential, the force that drives a flow of electrons when a current of electricity flows. (10.3)

Volume The capacity of an object to occupy space. (1.4)

Water of Hydration Water molecules held in a hydrate in some definite mole ratio to the rest of the compound. (7.2)

Wax A lipid whose molecules are esters of long-chain monohydric alcohols and long-chain fatty acids. (20.1)

Weak Acid An acid with a low percentage ionization in solution and with a small acid ionization constant, K_a. (8.2, 9.6)

Weak Base A base with a low percentage ionization in solution and a small base ionization constant. (8.2, 9.6)

Weak Brønsted Acid Any species, molecule or ion, that has a weak tendency to donate a proton and poorly serves as a proton donor. (8.4)

Weak Brønsted Base Any species, molecule or ion, that weakly holds an accepted proton and poorly serves as a proton acceptor. (8.4)

Weak Electrolyte Any electrolyte that has a low percentage ionization in solution. (8.1)

Weight The gravitational force of attraction on an object as compared to that of some reference. (1.4)

Zwitterion (See *Dipolar Ion*)

Zymogen A polypeptide that is changed into an enzyme by the loss of a few amino acid residues or by some other change in its structure; a proenzyme. (22.4)

PHOTO CREDITS

Capece. *Page 526:* Michael Watson. *Figure 18.7:* Courtesy of Polaroid Corporate Archives.

Chapter 19
Opener: Kathleen Hanzel/AllStock, Inc. *Page 539:* Paul Barton/ The Stock Market. *Figure 19.2 and Page 558:* Andy Washnik.

Chapter 20
Opener: Comstock, Inc. *Pages 566, 567, and 570:* Tripos Associates, *Pages 574, 575, and 576:* Courtesy of Richard Pastor, FDA.

Chapter 21
Opener: Irving Geiss/Peter Arnold. *Special Topic 21.1:* Bill Longcore/Photo Researchers.

Chapter 22
Opener: Darrell Jones/AllStock, Inc. *Page 630:* Courtesy Boehringer Mannheim Diagnostics.

Chapter 23
Opener: Galen Rowell/Peter Arnold.

Chapter 24
Opener: Werner H. Muller/Peter Arnold. *Special Topic 24.1, Figure 1:* Courtesy Lifecodes Corporation.

Chapter 25
Opener: Duomo. *Figure 25.3(a):* Courtesy of Dr. Keith R. Porter. *Figure 25.6(a):* From Parsons, D. F., *Science* 140, 985 (1973).

Chapter 26
Opener: Bruce Wilson/AllStock, Inc.

Chapter 27
Opener: Tim Davis/AllStock, Inc. *Special Topic 27.1, Figure 1:* W. Ober/Visuals Unlimited.

Chapter 28
Opener: FPG International.

Chapter 29
Opener: Barry L. Runk/Grant Heilman Photography.

INDEX

Phosphoenolpyruvate, *707, 741*
  in gluconeogenesis, 743
  in glycolysis, 740
Phosphofructokinase, 738
Phosphoglucomutase, 729
Phosphogluconate pathway, 742
Phosphoglucose isomerase, 738
2-Phosphoglycerate, 741
  in gluconeogenesis, 743
3-Phosphoglycerate, 740
  in gluconeogenesis, 743
Phosphoglycerate kinase, 740
Phosphoglycerate mutase, 740
Phosphoglycerides, *see* Glycerophospholipids
3-Phosphohydroxypyruvate, 769
Phosphoinositol cascade, *see* Inositol phosphate system
Phospholipase C, 635, 637
Phospholipids, 573
  in cell membranes, 578, 635
Phosphoric acid, 211, *212, 229*
  esters, 490
  ionization, 211
  K_a, 263
  from nucleic acids, 676
  from phospholipids, 574
Phosphoric anhydride system, 491
Phosphorus-32, 336
Phosphorus, dietary need, 796
  RDA, *783*
Phosphorus pentachloride, 130
  VSEPR theory and, 132
Phosphorylase, 599
Phosphorylase kinase, 729, 731
Phosphorylation:
  and enzyme activation, 729
  oxidative, 709, 721, *722*
  respiratory chain, 709, 721
  substrate, 709, 710, 740, 741
Photon, 86
Photosynthesis, 44, 238, 360, 539, 540
Physical property, 4
Physical quantity, 4
Physiological saline solution, 197
P_i, 493
p*I*, *586*, 587
Pi (π) bond, 141, 386
  in benzene, 407
  in carbonyl group, 447
Pint, *8*
PIP₂, 635
Pipet, 9
p*K'*, 283
pK_a, 271
pK_b, 271
pK_w, 273
PKU disease, *see* Phenylketonuria
Plane polarized light, 531
Plasmalogens, 575

Plasmids, 696
Plasmin, 631
Plasminogen, 631
  activator, 631
Plaster of paris, *186, 187, 230*
Platelets, 652
β-Pleated sheet, 590
Plutonium-239, 337
Pneumonia, acidosis from, 665
pOH, 257
Poisons, heavy metal ions as, 627
Polar bonds, 134
Polarimeter, 533
Polarity, London forces and, 166
Polarization, 166
Polarized light, 531
Polar molecule, 132, 135
Polaroid film, 531
Poliomyelitis, acidosis from, 665
Polio virus, 694
Polonium-210, alpha rays from, *323*
Polyamide, 510
Polyatomic ions, 123, *124*, 125
Polyester, 484
Polyethylene, 345, 402
Polymer, 401
Polymerase chain reaction, 682, 696
Polymerization, 402
Polyolefins, 402
Polypeptides, 584, 591. *See also* Proteins
  C-terminal, 593
  digestion, 601, 650
  α-helix form, 594
  hydrolysis, 601
  as neurotransmitters, 637, 640
  N-terminal, 593
  pleated sheet form, 596
  primary structure, 593
  quaternary structure and, 590
  mRNA-directed synthesis, 683, 688
  secondary structure, 590, 594
  tertiary structure, 590, 597
Polypropylene, 403
Polysaccharides, 541
Polysomes, 685
Polystyrene, *404*
Polyuria, 764
Polyvinyl chloride, *404*
Positron, 334
  and PET scan, 334
Potassium bicarbonate, 216
Potassium chlorate, 248, 252
Potassium dichromate, 395
Potassium hydroxide, 212, 574
  safety in handling, 213
  solubility, *213*
Potassium ion, *103*
  active transport, 604
  and alkalosis, 665
  in cells, *604*

dietary, 796
  and gastric juice, 649
  in plasma, *604*
Potassium permanganate, reaction:
  with alcohols, 428
  with alkenes, 395
  with alkylbenzenes, 408
Potassium–proton pump, 649
Potential energy, 36, 251
Pound, *8*
Precipitate, 65
Precipitation, 65
Precision, 14
Pressure, 148, 164
  boiling points and, 169
  gas solubility and, 187
  Henry's law and, 188
  osmotic, 194
  partial, 159
  units, 149
  vapor, 160
Pressure–solubility law, 188
  and decompression sickness, 189
Pressure–temperature law, 154, 165
Pressure–volume law, 150, 164
Primary carbon, 365
Primary structure, 590, 593
Primary transcript RNA, *see* Ribonucleic acid, heterogeneous nuclear
Principal energy levels, 87
Principal quantum number, 87
Procaine, 639
Procarboxypeptidase, 650
Procardia, 641
Products, 30
Proelastase, 650
Proenzyme, 626
Progesterone, *577, 605, 637*
Progestins, synthetic, *578*
Progress of reaction diagrams, 251
Proinsulin, 626
Proline, *586*
  biosynthesis, 769
  catabolism, 773
  codons, *687*
  hydroxylation, 594
Promoter, 402
Propanal, 445
Propanamide, *509*
Propane, *362*
  chlorination, 374
  combustion, 371
1,2-Propanediol, *419*, 420
  vapor pressure, 167
1-Propanethiol, 437
1,2,3-Propanetriol, *see* Glycerol
Propanoic acid, *see* Propionic acid
1-Propanol, *419*
  oxidation, 429
2-Propanol, *419*, 420

Student Guide

THIRD EDITION

KENDALL/HUNT PUBLISHING COMPANY
4050 Westmark Drive   Dubuque, Iowa 52002

A TIMS® Curriculum
University of Illinois at Chicago

MATH TRAILBLAZERS®

Dedication

This book is dedicated to
the children and teachers who
let us see the magic in their classrooms
and to our families who wholeheartedly
supported us while we searched for
ways to make it happen.

The TIMS Project

The University of Illinois
at Chicago

The original edition was based on work supported by the National Science Foundation under grant No. MDR 9050226 and the University of Illinois at Chicago. Any opinions, findings, and conclusions or recommendations expressed in this publication are those of the authors and do not necessarily reflect the views of the granting agencies.

Acknowledgments

Teaching Integrated Mathematics and Science (TIMS) Project Directors
Philip Wagreich, Principal Investigator
Joan L. Bieler
Howard Goldberg (emeritus)
Catherine Randall Kelso

Principal Investigators

| First Edition | Philip Wagreich |
| | Howard Goldberg |

Directors

| Third Edition | Joan L. Bieler |
| Second Edition | Catherine Randall Kelso |

Senior Curriculum Developers

First Edition	Janet Simpson Beissinger	Carol Inzerillo
	Joan L. Bieler	Andy Isaacs
	Astrida Cirulis	Catherine Randall Kelso
	Marty Gartzman	Leona Peters
	Howard Goldberg	Philip Wagreich

Curriculum Developers

Third Edition	Janet Simpson Beissinger	Philip Wagreich
	Lindy M. Chambers-Boucher	
Second Edition	Lindy M. Chambers-Boucher	Jennifer Mundt Leimberer
	Elizabeth Colligan	Georganne E. Marsh
	Marty Gartzman	Leona Peters
	Carol Inzerillo	Philip Wagreich
	Catherine Randall Kelso	
First Edition	Janice C. Banasiak	Jenny Knight
	Lynne Beauprez	Sandy Niemiera
	Andy Carter	Janice Ozima
	Lindy M. Chambers-Boucher	Polly Tangora
	Kathryn Chval	Paul Trafton
	Diane Czerwinski	

Illustrator

| | Kris Dresen |

Editorial and Production Staff

Third Edition	Kathleen R. Anderson	Anne Roby
	Lindy M. Chambers-Boucher	
Second Edition	Kathleen R. Anderson	Georganne E. Marsh
	Ai-Ai C. Cojuangco	Cosmina Menghes
	Andrada Costoiu	Anne Roby
	Erika Larson	
First Edition	Glenda L. Genio-Terrado	Sarah Nelson
	Mini Joseph	Biruté Petrauskas
	Lynelle Morgenthaler	

Acknowledgments

TIMS Professional Developers

	Barbara Crum	Cheryl Kneubuhler
	Catherine Ditto	Lisa Mackey
	Pamela Guyton	Linda Miceli

TIMS Director of Media Services

Henrique Cirne-Lima

TIMS Research Staff

	Stacy Brown	Catherine Ditto
	Reality Canty	Catherine Randall Kelso

TIMS Administrative Staff

	Eve Ali Boles	Enrique Puente
	Kathleen R. Anderson	Alice VanSlyke
	Nida Khan	

Research Consultant

First Edition   Andy Isaacs

Mathematics Education Consultant

First Edition   Paul Trafton

National Advisory Committee

First Edition	Carl Berger	Mary Lindquist
	Tom Berger	Eugene Maier
	Hugh Burkhardt	Lourdes Monteagudo
	Donald Chambers	Elizabeth Phillips
	Naomi Fisher	Thomas Post
	Glenda Lappan	

TIMS Project Staff, July 2005

Table of Contents

Additional student pages may be found in the *Discovery Assignment Book, Adventure Book,* or the *Unit Resource Guide.*

Table of Contents

Additional student pages may be found in the *Discovery Assignment Book, Adventure Book,* or the *Unit Resource Guide.*

Table of Contents

Additional student pages may be found in the *Discovery Assignment Book, Adventure Book,* or the *Unit Resource Guide.*

Table of Contents

Additional student pages may be found in the *Discovery Assignment Book, Adventure Book,* or the
Unit Resource Guide.

Letter to Parents

Dear Parents,

Math Trailblazers® is based on the ideas that mathematics is best learned through solving many different kinds of problems and that all children deserve a challenging mathematics curriculum. The program provides a careful balance of concepts and skills. Traditional arithmetic skills and procedures are covered through their repeated use in problems and through distributed practice. *Math Trailblazers,* however, offers much more. Students using this program will become proficient problem solvers, will know when and how to apply the mathematics they have

learned, and will be able to clearly communicate their mathematical knowledge. Computation, measurement, geometry, data collection and analysis, estimation, graphing, patterns and relationships, mental arithmetic, and simple algebraic ideas are all an integral part of the curriculum. They will see connections between the mathematics learned in school and the mathematics used in everyday life. And, they will enjoy and value the work they do in mathematics.

The *Student Guide* is only one component of *Math Trailblazers.* Additional material and lessons are contained in the *Discovery Assignment Book*, the *Adventure Book,* and in the teacher's *Unit Resource Guides.* If you have questions about the program, we encourage you to speak with your child's teacher.

This curriculum was built around national recommendations for improving mathematics instruction in American schools and the research that supported those recommendations. The first edition was extensively tested with thousands of children in dozens of classrooms over five years of development. In preparing the second and third editions, we have benefited from the comments and suggestions of hundreds of teachers and children who have used the curriculum. *Math Trailblazers* reflects our view of a complete and well-balanced mathematics program that will prepare children for the 21st century—a world in which mathematical skills will be important in most occupations and mathematical reasoning will be essential for acting as an informed citizen in a democratic society. We hope that you enjoy this exciting approach to learning mathematics and that you watch your child's mathematical abilities grow throughout the year.

Philip Wagreich

Philip Wagreich
Professor, Department of Mathematics, Statistics, and Computer Science
Director, Institute for Mathematics and Science Education
Teaching Integrated Mathematics and Science (TIMS) Project
University of Illinois at Chicago

Room 204
Bessie Coleman
Elementary
School

Unit 1

Data About Us

	Student Guide	Discovery Assignment Book	Adventure Book	Unit Resource Guide*
Lesson 1				
Getting to Know Room 204	●			
Lesson 2				
Getting to Know Room 204 a Little Better	●			
Lesson 3				
An Average Activity	●			
Lesson 4				
The Four Servants			●	
Lesson 5				
Arm Span vs. Height	●	●		●
Lesson 6				
Solving Problems About Room 204	●			

Unit Resource Guide pages are from the teacher materials.

Getting to Know Room 204

The class picture at the beginning of this unit shows Mrs. Dewey's fourth-grade class. Their classroom is Room 204 in the Bessie Coleman Elementary School in Chicago, Illinois.

Mrs. Dewey's class wants to share information about themselves with their pen pals at Westmont School in Phoenix, Arizona. They also want to get to know their pen pals.

Mrs. Dewey asked her class what they would like to know about their new pen pals. Tanya said, "I'd like to know what they are interested in. I love to read. I can recommend some really good books!"

Jessie said, "I wonder if they have some really cool trails in Arizona! I love to ride my bike and roller blade!"

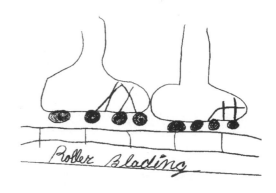

"I'll probably ask my pen pal to tell me about her favorite food, her favorite holiday, and her favorite color," said Keenya. "I'll also ask if she plays an instrument. I'm learning to play a keyboard."

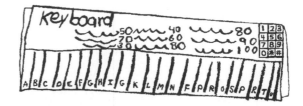

"I'll ask my pen pal about the color of his hair and eyes, and whether he is short or tall," said Jerome.

John said, "I'll ask my pen pal what his favorite sport is. I like soccer and basketball."

Values and Variables

The students in Room 204 decided to collect data they would like to share about their class. They started by making a list of things they wanted to learn about one another.

They made the following chart of the variables they chose to study. They included possible values for each variable. A **variable** is an attribute or quantity that may have one or many different values. The possible outcomes for each variable are called **values.**

Variables and Possible Values

Variable	Possible Values
Interests	Sports, Reading, Outdoor Activities, Playing Games, Music, Animals, etc.
Eye color	Blue, Hazel, Green, Brown
Favorite food	Pizza, Tacos, Liver, Chicken

Mrs. Dewey's class decided to collect data about the variable "main interest." Some students chose reading as their main interest. Others chose animals, sports, music, outdoor activities, or playing games. These possible outcomes are the values of the variable.

Discuss

1. What would you like to learn about the students in your class? Make a table like the one above. List variables you can study and possible values for each of the variables.

Room 204's Main Interests

Name	Main Interest
Linda	Animals
John	Sports
Tanya	Reading
Shannon	Reading
Jerome	Games
Romesh	Animals
Ana	Outdoors
Jackie	Sports
Nicholas	Animals
Ming	Sports
Luis	Music
Jacob	Reading
Jessie	Sports
Keenya	Music
Nila	Sports
Michael	Reading
Roberto	Outdoors
Irma	Reading
Maya	Sports
Lee Yah	Sports
Frank	Music
Grace	Sports

In order for the students in Phoenix to compare their class to Room 204, Mrs. Dewey's students organized their data. As the class read the main interest of each child from the large data table, Maya placed a tally mark next to his or her main interest. They then used the data in the table to make a bar graph.

Main Interests Data

Main Interest	Tally	Number of Students
Animals	|||	3
Music	|||	3
Reading	|||||	5
Outdoor Activities	||	2
Playing Games	|	1
Sports	|||| |||	8

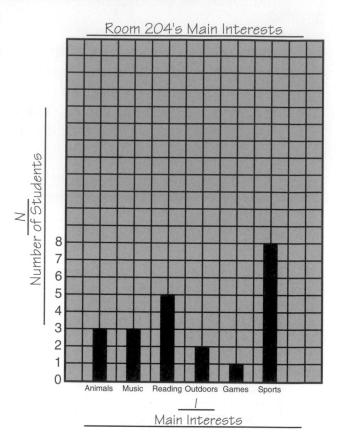

2. As a class, choose a variable to study. Choose from those variables you listed in Question 1.

Collect the data and organize it in a class data table.

Graph the data. Be sure you title your graph and label the axes.

3. What variable is on the horizontal axis (⟵⟶) on your graph?

4. What variable is on the vertical axis (↕) on your graph?

5. A. Which bar is the tallest on your graph?
   B. What does the tallest bar represent?

6. A. Which bar is the shortest on your graph?
   B. What does the shortest bar represent?

7. What else does your graph tell you about your class?

8. A. Look back at Room 204's graph. Which variable did they graph on the horizontal axis?
   B. Which variable did Room 204 graph on the vertical axis?

9. What does the tallest bar on Room 204's graph represent?

10. How many more students in Room 204 prefer to read than play games?

Homework

You will need a sheet of *Centimeter Graph Paper* to complete Question 6.

1. Mrs. Cook's fourth-grade class at Bessie Coleman School collected data to share with their pen pals. What variable did they study?

2. What does the tallest bar on their graph represent?

3. What do the shortest bars on Room 206's graph represent?

4. How many more students speak Spanish as their primary language than Assyrian?

5. What else does this graph tell you about Mrs. Cook's class?

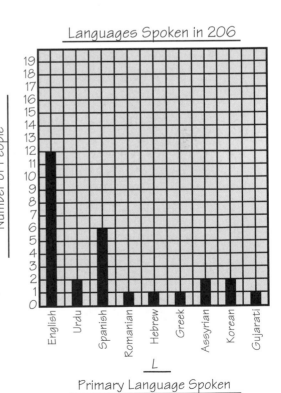

Languages Spoken in 206

N — Number of People

Primary Language Spoken — L

English, Urdu, Spanish, Romanian, Hebrew, Greek, Assyrian, Korean, Gujarati

6. Room 204's pen pals in Phoenix sent back data on their favorite subjects. Use the data to create a bar graph. Remember to label the axes and title your graph.

7. **A.** What variable is on the horizontal axis?

   B. What variable is on the vertical axis?

8. Answer the following questions based on the graph you created in Question 6.

   A. What is the most common favorite subject?

   B. What is the least common favorite subject?

   C. Which of the subjects in the data table is your favorite?

   D. How many Phoenix pen pals have the same favorite subject as you?

9. How many students are in the Phoenix class? Explain how you know.

Favorite Subjects

Favorite Subject	Number of Students
Writing	2
Social Studies	3
Math	8
Reading	2
Spelling	6
Science	7

Thinking of the sun rising over the horizon helps me remember which axis is the horizontal axis.

Getting to Know Room 204 a Little Better

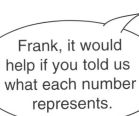

Here is what I am going to tell my pen pal about me. 57, 3, 1, 5, 4, 2

Frank, it would help if you told us what each number represents.

Frank continued to explain. "Well, I am 57 inches tall. I have 3 brothers and 1 sister. I live 5 blocks from school. I have moved 4 times. I have 2 pets."

Maya said, "I am 51 inches tall. I have 1 brother and no sisters. I live 4 blocks from school. I've only moved once. I have 1 pet."

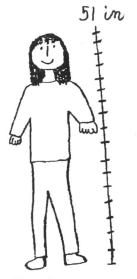

51 in

In Lesson 1, Room 204 discussed what they would like to know about their pen pals. They listed **categorical** variables such as interests, favorite food, and eye color.

In this lesson, Frank discussed **numerical** variables: height, number of brothers, number of sisters, number of blocks from school, number of times moved, and number of pets. Frank and Maya each had different values for the variables.

Numerical variables have values that are numbers; **categorical variables** do not.

Numerical Variables and Possible Values

Variable	Possible Values
Height	51 inches, 57 inches, etc.
Number of brothers	0, 1, 2, 3, etc.
Blocks from school	1, 2, 3, 4, etc.

Discuss

1. Think of numerical variables you would like to study about your classmates. Add these variables to the table you created in Lesson 1. List possible values for each variable.

Mrs. Dewey's class wanted to know how far their classmates lived from school. They chose to study the variable "number of blocks from school." First they recorded and organized their data in data tables.

Number of Blocks We Live from School

Number of Blocks	Tally	Number of Students
1	///	3
2	//// //	7
3	////	4
4	//	2
5	/	1
6		0
7	/	1
8	////	4

Room 204's Data

Name	Number of Blocks
Linda	2
John	1
Tanya	3
Shannon	2
Jerome	2
Romesh	3
Ana	8
Jackie	1
Nicholas	2
Ming	8
Luis	2
Jacob	3
Jessie	7
Keenya	1
Nila	2
Michael	8
Roberto	3
Irma	2
Maya	4
Lee Yah	4
Frank	5
Grace	8

Then, Mrs. Dewey's class graphed the data in a bar graph.

2. With your class, choose a numerical variable to study that helps you describe your class. Choose from the variables you listed in Question 1.

Room 204:
Number of Blocks We Live from School

Number of Blocks from School

Collect the data and organize it in a class table.

Graph the data in a bar graph. Be sure to title your graph and label the axes.

3. A. Which variable is on the horizontal axis on your graph?
   B. Is this variable a categorical or a numerical variable? How do you know?

4. A. Which variable is on the vertical axis on your graph?
   B. Is this variable a categorical or a numerical variable? How do you know?

5. A. Which bar is the tallest on your graph?
   B. What does the tallest bar represent?

6. A. Which bar is the shortest on your graph?
   B. What does the shortest bar represent?

7. **A.** Look back at Room 204's graph called Number of Blocks We Live from School. Is the variable graphed on the horizontal axis numerical or categorical?

   B. Is the variable graphed on the vertical axis numerical or categorical?

8. **A.** Would Room 204's graph be as easy to read if the numbers (values) on the horizontal axis were not in order? Explain.

   B. Does it matter in what order you label the horizontal axis when the variable is categorical? Refer back to Room 204's Main Interests graph in Lesson 1.

9. What story does the graph tell you about the students in Room 204?

10. **A.** How many students in Room 204 live 3 blocks or less from school?

    B. Is this more or less than half the class?

Homework

You will need one sheet of *Centimeter Graph Paper* to complete this homework.

Number of Times Families Moved

1. Room 204's Phoenix pen pals sent back the following data on the number of times their families have moved. Use the data to create a bar graph. Remember to label the axes and title your graph.

2. Answer the following questions using the bar graph you drew in Question 1.

   A. Is the variable on your horizontal axis numerical or categorical?

   B. Is the variable on your vertical axis numerical or categorical?

   C. Which is the tallest bar on the graph? What does it tell you?

   D. What is the most number of times any student has moved?

   E. Describe the shape of your graph.

Number of Times Moved	Number of Students
0	0
1	3
2	7
3	7
4	3
5	2
6	2
7	2
8	1
9	1
10	0

3. What story does the graph tell you about the Phoenix pen pals?

4. Decide whether each variable below is numerical or categorical. Then name three possible values for each variable.

A. ice cream flavors

B. number of telephones in homes

C. heights of tables at home

D. favorite kind of movie

E. weights of newborn babies

F. foot size

G. types of vehicles

An Average Activity

Mrs. Dewey's class was collecting data on the heights of fourth graders. She asked a doctor to talk to the class about how children grow and develop. One of the things Dr. Solinas talked about was the average height of ten-year-olds. What does "average" mean?

- "It was just an average day."
- "The doctor said my height is above average for kids my age."
- "We really need rain. Rainfall this year has been well below average."
- "My average grade in spelling is 75 percent."
- "Our soccer team averages about three goals per game."

Each sentence describes what is usual or typical for the situation.

You can calculate the average value for any set of numbers, such as the average height of fourth graders, in more than one way. In this lesson you will learn about one kind of average: the median. You can find the median of a set of numbers easily and use it to describe the data you collect. Later this year, you will learn to calculate another kind of average, the mean.

Finding Medians

Mrs. Dewey asked five students to stand in front of the room to show the class how to find medians. Jerome, Ana, Grace, Roberto, and Shannon stood in a line from shortest to tallest. The **median** is the number that is exactly in the middle of the data.

Discuss

1. **A.** Which student has the median height in Jerome's group?

    B. Does it make sense to say that this student's height is the "typical" height for this group? Why or why not?

2. **A.** Use the information in the table to find the median height for Keenya's group. Put the numbers in order from smallest to largest. The median height will be in the middle of the data.

    B. When you find the median, look back at the data. An **average** is one number that can be used to represent all the data. Does your answer make sense?

Keenya's Group: Our Heights

Name	Height in inches
Keenya	55 in
Nila	50 in
John	57 in
Michael	54 in
Luis	58 in
Jackie	54 in
Lee Yah	52 in

3. Keenya, Maya, Jessie, and Shannon all walk to school together.

- Jessie lives 7 blocks from school.
- Shannon lives 2 blocks from school.
- Maya lives 4 blocks from school.
- Keenya lives 1 block from school.

Jessie said, "The median number of blocks that the four of us walk to school is 3 blocks." Is she correct? Why or why not?
(*Hint:* Find the number halfway between the middle two values.)

4. Use the information in the data table to the right. Find the median number of blocks the students in Linda's group live from school.

Linda's Group: Number of Blocks We Live from School

Name	Number of Blocks from School
Linda	2
John	1
Tanya	3
Michael	8
Frank	5
Luis	2

Room 204: Number of Blocks We Live from School

Number of Blocks	Tally	Number of Students
1	///	3
2	//// //	7
3	////	4
4	//	2
5	/	1
6		0
7	/	1
8	////	4

5. Look at Room 204's data and graph.

A. Find the median number of blocks from school.

B. Explain how you found the median.

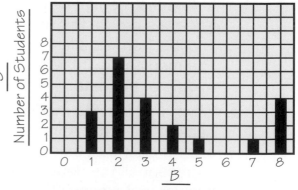

Room 204:
Number of Blocks We Live from School

6. The students in Room 204 collected data on the number of times their families moved. The data for Ming's group is in the table to the right. Find the median number of times the students in his group have moved.

7. Jerome's baseball team has played eight games. Here are the number of runs they scored: 1, 3, 5, 3, 2, 7, 2, 4. Find the median number of runs they scored.

8. John, Shannon, and Tanya made paper airplanes. They had a contest to see who made the best airplane. Each airplane was flown three times.

   A. Find the median distance for each student's airplane.

   B. Who do you think should win the contest? Why?

Ming's Group: Number of Times We Moved

Name	Number of Times Moved
Ming	2
Irma	1
Nicholas	5
Romesh	0
Linda	3

Name	Distance in cm			
	Trial 1	Trial 2	Trial 3	Median
John	410 cm	390 cm	640 cm	
Shannon	250 cm	230 cm	290 cm	
Tanya	420 cm	590 cm	600 cm	

1. Romesh took a survey on his block and recorded his data in the table shown to the right. Find the median number of pets on his block.

2. Ana and her two brothers play soccer. They all play on different teams. Find the median number of goals for each team.

  A. Ana's team has played 6 games. Here are the number of goals her team has scored in the six games: 4, 4, 0, 3, 2, 5.

  B. David's team has played 5 games. Here are the number of goals his team has scored: 2, 1, 3, 2, 3.

  C. Tony's team has played 4 games. Here are the number of goals his team has scored: 1, 0, 3, 6.

  D. Ana claims that her team is the best. Do you agree? Why or why not?

3. A. Make a data table like the one shown below. Measure the length of your hand and the hands of your family and friends. Measure at least five hands including your own. Carefully measure from the wrist to the end of the longest finger. Measure to the nearest centimeter.

  B. Record your data in the data table.

  C. Find the median value for your hand length data.

Pets Survey

Family	Number of Pets
Bailey	2
Johnson	0
Cruz	5
Kanno	3
Holt	4
Elkins	1
Roberts	2

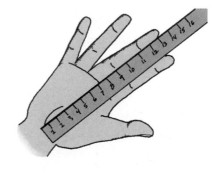

Hand Length Data

Name	Hand Length in cm

Arm Span vs. Height

The TIMS Laboratory Method

Irma and Jerome noticed that the Adventure Book story *The Four Servants* took place in China. All the measurements were of Chinese people. They wondered if the four servants would have found the same results if they had measured the people in Irma and Jerome's neighborhood.

Irma and Jerome used the TIMS Laboratory Method to help them solve problems involving hand length and height in their neighborhood. The four servants used this four-step method. First, Irma and Jerome **drew a picture** of the steps they would follow in the experiment. Irma's picture is shown below.

Notice that Irma showed the names of the people in her household, how she was going to measure hand length, and how she was going to measure height. Irma labeled the variables in her experiment, Hand Length and Height.

Irma and Jerome then **collected and organized** the data in a data table. Below is the data Jerome collected from his family:

Jerome's Family Table

Name	Hand Length (in cm)	Height (in cm)
Peter	12 cm	102 cm
Abby	13 cm	110 cm
Timothy	15 cm	124 cm
Jerome	14 cm	127 cm
Jenny	16 cm	147 cm
Mom		

Jerome **graphed** his family's data as a point graph:

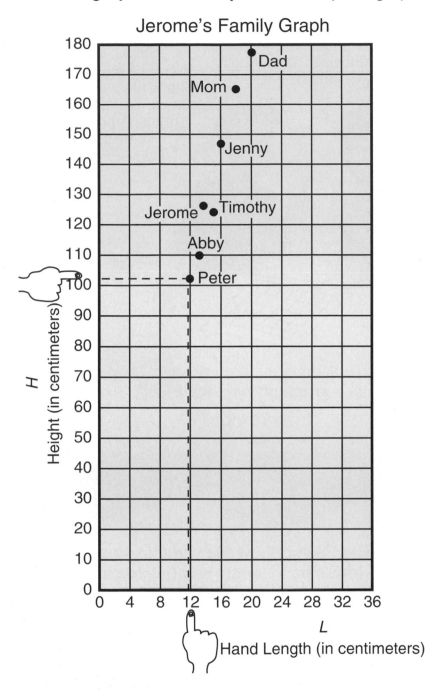

Jerome's Family Graph

Jerome graphed hand length on the horizontal axis and height on the vertical axis. To plot a point for his brother Peter's data, he first located his hand length, 12 cm, on the horizontal axis. Then he found his height, 102 centimeters, on the vertical axis.

- Make sure you see how Jerome graphed his brothers' and sisters' data.
- Use the graph to find the hand length and height of Jerome's parents.

Irma and Jerome measured their classmates. They also collected data from other families, schoolchildren, and even teachers. When they finished graphing the data, the graph represented Hand Length vs. Height for their neighborhood.

Finally, when their graph was finished, they **analyzed and discussed** their results.

Irma and Jerome chose to investigate the two variables, hand length and height. You also will investigate two variables that describe your class. The arm span and height of each student in your class will be measured. Your job is to find out whether you can predict a fourth-grade student's height if you know his or her arm span.

We never measured the principal's hand length and height.

It looks like she's in a hurry. Let's ask her if we can measure her hand. We can use our graph to predict her height.

Begin by drawing a picture of what you will do in the experiment. Then collect and organize data in a table. Next, make a graph of the data. Finally, explore the data by looking for patterns.

Draw

Draw a picture of the setup for your experiment. Show the variables Arm Span (*S*) and Height (*H*) in your picture. Use Irma's hand length and height picture to help you draw a picture of your *Arm Span vs. Height* experiment. Remember to label the variables.

Discuss

1. **A.** Is arm span a categorical or a numerical variable?

   B. Is height a categorical or a numerical variable? Explain how you know.

2. What is the same about all the people you measured for this experiment?

Collect

Measure the arm span and height of each person in your group to the nearest inch. Record your group's data in a data table like the one at the right. Discuss with your group what the letters *S* and *H* stand for.

Discuss any patterns you see in the data table.

Arm Span vs. Height Data Table

Name	S Arm Span (in inches)	H Height (in inches)

Graph

- Graph your group's data. Plot arm span on the horizontal axis and height on the vertical axis. Scale your horizontal axis to at least 75 inches and the vertical axis to at least 100 inches. Remember to label each axis.

- A class graph of *Arm Span vs. Height* will provide more data for you to analyze. Plot one point, your own data, for arm span and height on the class graph.

Explore

Use your class data and graphs to help you answer the following questions. Include units with your answers. Be ready to share your answers with the entire class.

3. **A.** Describe your group's graph. What do you notice about the points?

   B. Describe the class graph. What do you notice about the points?

4. Compare your group's graph and the class graph. How are they alike? How are they different?

5. **A.** If you measured a new classmate's arm span and height, where do you think his or her data would lie?

   B. If a fourth grader from another classroom had an arm span of 53 inches, what would you predict about his or her height?

6. **A.** In which part of the graph would first-grade data cluster in comparison to fourth-grade data—in the area marked A, B, or C?

   B. In which cluster would a kangaroo's data fall—in the area marked A, B, or C?

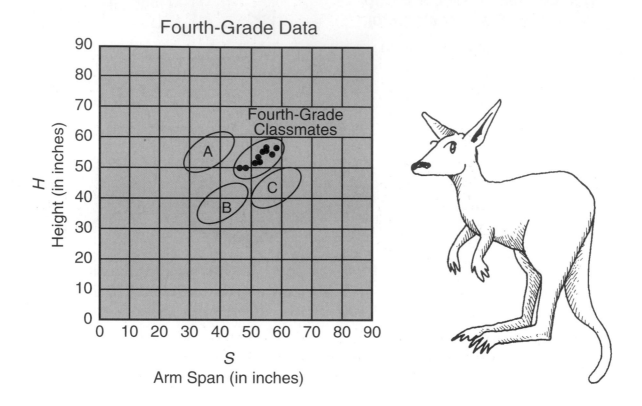

Use your class graph to discuss the following.

7. Use your graph to estimate the average arm span of your classmates. (*Hint:* This is a number that represents all the arm spans in your classroom.)

8. Use your graph to estimate the average height of your classmates.

9. **A.** Find the median height in your class.

   B. Find the median arm span in your class.

   C. Compare with your estimate. Were you close?

10. **A.** Use a red pen or marker to plot the data point on your graph for the median height and arm span.

    B. Where is the data point for the median values compared to the other data points on the graph?

Homework

1. The data table shows data for three groups of students in Room 204. Graph *Arm Span vs. Height* for these groups on a sheet of *Centimeter Graph Paper*. Title the graph so you know it is not your class data. Plot arm span on the horizontal axis and height on the vertical axis. Remember to label your axes and include units.

2. **A.** Estimate the average arm span of the groups using the graph.

   B. Estimate the average height using the graph.

   C. How does the groups' data compare to your class data?

3. If a new fourth grader who entered Mrs. Dewey's classroom had an arm span of 54 inches, what would you predict about the student's height?

4. If you measured the arm spans and heights of the parents of classmates in Mrs. Dewey's classroom, where would the data cluster? Show your answer on your graph of *Arm Span vs. Height* for the groups in Room 204.

Room 204 Arm Span and Height Data Table

Name	S Arm Span (in inches)	H Height (in inches)
Linda	51	51
Romesh	52	53
Nicholas	56	54
Jerome	49	50
Keenya	54	55
Frank	59	57
Luis	58	58
Roberto	55	57
Ana	52	52
Jacob	56	56
Grace	55	55
Lee Yah	53	52

Solving Problems About Room 204

Mrs. Dewey asked her students to solve some problems comparing Bessie Coleman School to Westmont School. She asked them to check their calculations by estimating the answer to each problem. She reminded them that sometimes using a convenient number that is easier to calculate in your head helps to make a quick estimate.

Solve the following problems Mrs. Dewey gave her class. Show how you solved each problem. Be ready to explain how you estimated your answers.

1. There are 396 students at Bessie Coleman School. There are 509 students at Westmont School in Phoenix. How many more students go to Westmont School?

2. There are 22 students in Mrs. Dewey's fourth-grade class at Bessie Coleman School. The fourth-grade class at Westmont School has 28 students.

   A. On the first day of school, Mrs. Dewey gave each student five textbooks. How many textbooks did she pass out?

   B. If each student in fourth grade at Westmont has 5 textbooks, how many textbooks do the Westmont fourth graders have in their classroom?

   C. Look back at your answers. Explain why you think they are reasonable.

3. Milk at both schools costs 25¢.

   A. If 20 students in Room 204 at the Bessie Coleman School bought milk, how much money did Mrs. Dewey collect?

   B. In the Westmont fourth-grade class, the total cost for milk was $5.25. How many students bought milk in this fourth-grade class?

4. Each of the 22 students in Mrs. Dewey's class sent a letter to a pen pal at Westmont School. Find the cost of the stamps.

5. In music class at Bessie Coleman School, Lee Yah, Roberto, Grace, and Luis demonstrated a folk dance. They lined up across the front of the room. They began with their arms outstretched and their fingers just touching. They could just reach across the room. Since the average arm span of the students in Room 204 is 54 inches, about how wide is the room?

6. Bessie Coleman School begins its school day at 8:30 A.M. and ends at 3:00 P.M. Jerome stays for lunch. How long is he at school?

7. The distance between Phoenix and Chicago is 1816 miles.

   A. If you traveled from Chicago to Phoenix and back again, how many miles would you travel?

   B. Make sure your answer in Question 7A is correct. Explain how you can make a quick estimate of the total distance to see if your answer is reasonable.

Unit 2

Geometric Investigations: A Baseline Assessment Unit

	Student Guide	Discovery Assignment Book	Adventure Book	Unit Resource Guide*
Lesson 1				
Investigating Perimeter and Area	●	●		
Lesson 2				
Perimeter vs. Length	●			
Lesson 3				
Letter to Myrna				●
Lesson 4				
Helipads for Antopolis	●			
Lesson 5				
Portfolios	●			
Lesson 6				
Angles	●			●
Lesson 7				
Angles in Pattern Blocks	●	●		

Unit Resource Guide pages are from the teacher materials.

Investigating Perimeter and Area

The ants have built a beautiful fountain in the Antopolis town square. In the evening, the ants come out to stroll around the perimeter of the fountain.

This diagram shows the Antopolis Town Square with the ants walking along the perimeter of the fountain.

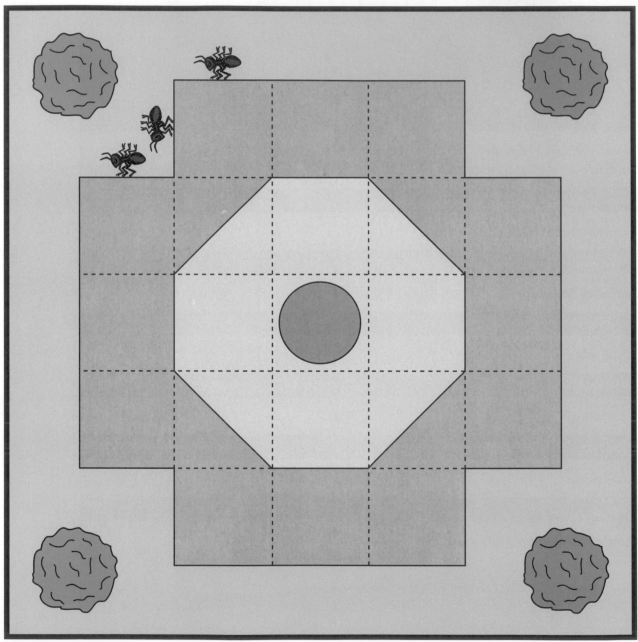

The **perimeter** is the distance around a two-dimensional shape.

Here is a picture of a one square-inch tile.

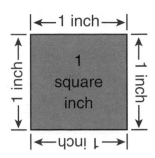

Each side of this tile is one inch long.
The **perimeter** of the tile is 4 inches.

The **area** of the tile is one **square inch.**

Area is the amount of surface needed to cover something. The size of an object or shape is often measured by counting the unit squares needed to cover the top of the shape. This is called the area of the object.

A grid of square-inch tiles has been drawn on the diagram of the fountain. You can use it to measure the distance the ants walked to travel the perimeter of the fountain. The grid of tiles can also help you measure the square inches needed to cover the area of the fountain.

1. Find the perimeter of the fountain.

2. Find the area of the fountain.

3. Find the area of the fountain that is covered by water.

4. Find the area of the fountain that is not covered by water.

The ants of Antopolis want to build a playground with an area of 8 square inches. They have ordered enough material to build a fence around the playground that is 14 inches long.

5. Draw a design for a playground for the ants that has an area of 8 square inches and a perimeter of 14 inches. (They want the little ants inside to be able to walk from one corner of the playground to any other corner.)

6. Write a paragraph that explains how you solved the problem. Use the Student Rubric: *Telling* to help you write your paragraph.

Dear Family Member:

Your child is learning to find the perimeter and area of shapes. In this assignment, help your child add the measurements marked around the shape to find the perimeter. The area can be found by counting the number of square centimeters that would fit inside the shape. The grid drawn on shape A will help your child see that the marked measurements can help find the number of squares needed to fill the shape.

Thank you for your help.

Here are some other shapes. Their edges have already been measured for you. Calculate their perimeters.

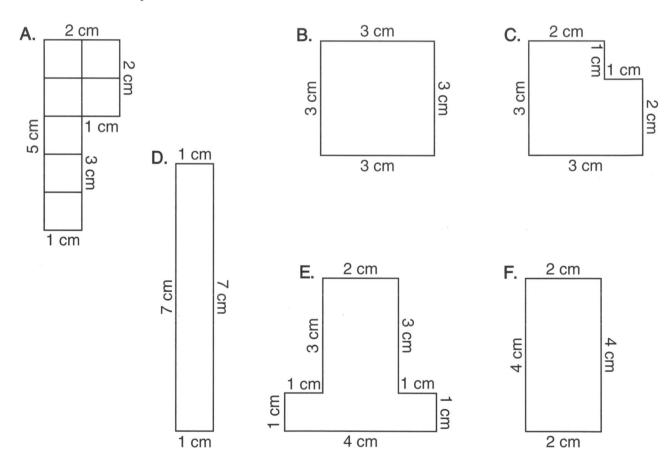

After you find the perimeter of each shape, find the area for 3 of the shapes.

Perimeter vs. Length

Myrna Myrmidon is helping to plan a new airport for Antopolis. Myrna is in charge of the runways.

The airport plans are not done yet, so Myrna doesn't know how long the runways will be. Myrna does know what kinds of airplanes will use the airport: light planes, commuter planes, short-haul jets, long-haul jets, and heavy-transport planes. She also knows that bigger airplanes need wider runways.

The light planes are the smallest airplanes that will use the new Antopolis airport. They need runways that are 1 inch wide.

Light Plane Runway

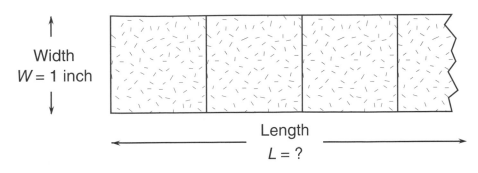

Width
$W = 1$ inch

Length
$L = ?$

The commuter planes need runways that are 2 inches wide.

Commuter Plane Runway

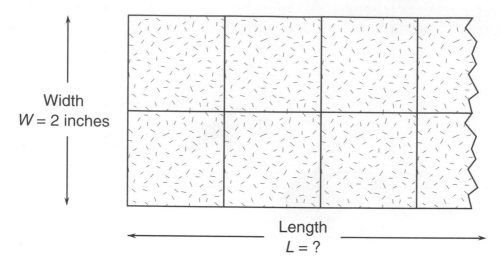

Width
$W = 2$ inches

Length
$L = ?$

The short-haul jets need 3-inch-wide runways.

Short-Haul Jet Runway

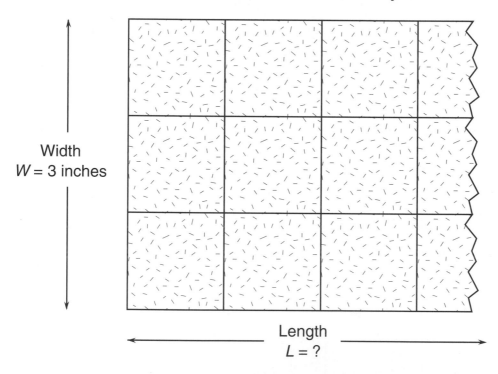

Width
$W = 3$ inches

Length
$L = ?$

The other planes need even wider runways. The long-haul jets need 4-inch wide runways and the heavy-transport planes need 5-inch wide runways.

Every runway will have lights all around it. Myrna has to know how much wire is needed to connect these lights. She must find the **perimeter** of the runways.

For example, a runway for a light plane that is 5 inches long needs 12 inches of wire for the perimeter lights.

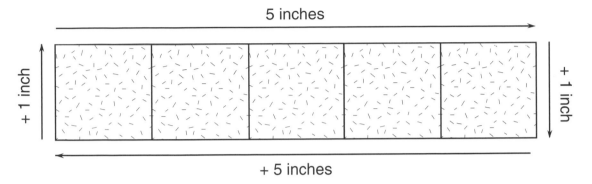

5 inches

+ 1 inch

+ 1 inch

+ 5 inches

A runway with a perimeter of 12 inches

Since Myrna doesn't know how long the runways will be, she doesn't know how much wire she needs. She has to find the amount of wire needed for any kind of runway, no matter how long.

Use the TIMS Laboratory Method to help Myrna. You will work on runways for only one kind of plane. Your teacher will help you choose.

You will use square-inch tiles to make several runways for your kind of plane. For each runway, you will record the length (L) and the perimeter (P) in a data table. Then you will graph your data and look for patterns.

1. Draw a picture of the lab.
   • Be sure to show the variables: length (L), perimeter (P), and width (W).
   • Show your kind of airplane and at least one of your runways. Show how wide your runways will be.

2. **A.** Which variable—length (L), width (W), or perimeter (P)—will stay the same for all your runways?
   B. Which two variables will change from runway to runway?

Collect

Use square-inch tiles to make 4–5 runways for your kind of airplane. Decide how long to make your runways. You might make one runway 1 inch long, another 2 inches long, and another 4 inches long. Or, you might make runways 2, 4, and 8 inches long. Do not make runways that are longer than 12 inches or you may have trouble graphing your data. Discuss with your group how long your runways should be.

3. Make your runways. Find the length and perimeter of each runway. Keep track of your data in a table like this:

W Width of Runway (in inches)	L Length of Runway (in inches)	P Perimeter of Runway (in inches)

Graph

4. Draw a point graph for your data. Put length (*L*) on the horizontal axis and perimeter (*P*) on the vertical axis. (Remember to include a title for the graph, label the axes, and include units.)

5. Look at your points on the graph. Describe your points.

6. If your points form a line, use a ruler to draw a line through your data points. Extend the line in both directions.

Discuss

Questions 7–12 are for runways the same width as yours.

7. How wide are your runways?

8. What is the perimeter of a runway that is 4 inches long?

9. Use your graph to find the perimeter of a runway that is 10 inches long. Use dotted lines on your graph to show your work.

10. What is the perimeter of a runway that is 100 inches long? Explain how you found your answer.

11. Give a rule for finding the perimeter of a runway for your type of plane, no matter what the length.

12. Plot your data on the class graph and draw the line. Label the line with your type of airplane. Then compare the different lines your class drew, and answer the following questions:

   A. What is similar about the lines?

   B. How do the lines differ?

Dear Family Member:

Your child is learning about perimeters in class. Encourage your child to use a yardstick, tape measure, ruler, or string for these problems. Your child can use the 6-inch line below to measure 36 inches of string and then use the string like a tape measure. Thank you.

_____

A 6-inch line

Bedroom Perimeter

1. Estimate, in inches, the perimeter of the room where you sleep. Explain how you made your estimate.

2. Measure the perimeter of the room where you sleep. Explain how you made your measurement. Draw a picture to help make your explanation clear.

3. What room in your home has a perimeter that is larger than your bedroom's perimeter? How did you decide?

4. What room in your home has a perimeter that is smaller than your bedroom's perimeter? How did you decide?

Runways

Use the graph to answer the following questions.

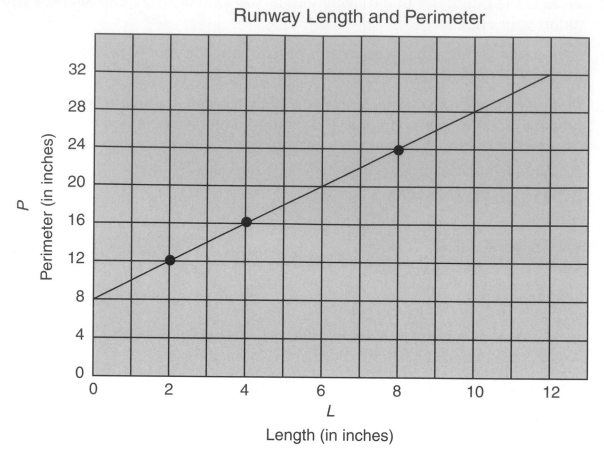

Runway Length and Perimeter

Length (in inches)

5. If the length of the runway is 6 inches, what is the perimeter?

6. If the length of the runway is 3 inches, what is the perimeter?

7. If the perimeter is 28 inches, what is the length?

8. If the perimeter is 18 inches, what is the length?

9. What is the width of the runways? Explain your answer.

Use what you have learned about lengths, widths, and perimeters of runways to answer the following questions.

10. If the length of a runway is 6 inches and the width is 3 inches, what is the perimeter? Draw a picture to help you.

11. If the length of a runway is 10 inches and the perimeter is 22 inches, how wide is the runway? Draw a picture to help you. Explain how you found your answer.

Helipads for Antopolis

Myrna Myrmidon's Aunt Penny likes to fly helicopters. When she flew to Ladybug Airport, she landed her helicopter on a helipad. The helipad at Ladybug Airport is 4 inches long and 2 inches wide.

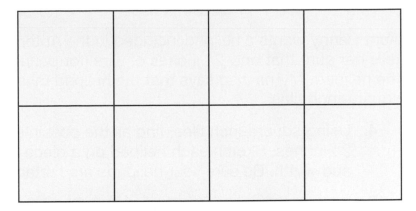

Discuss

1. Ladybug Airport is building a new helipad. They are also buying new perimeter lights. The lights are attached to a wire that goes around the entire helipad. How long does the wire need to be to fit around this helipad?

2. What is the area of this helipad?

While the construction crew at Ladybug Airport waited for the lights to be delivered, they discussed the new helipad. They knew that the perimeter of the helipad could not be changed since the wire had already been ordered. However, they could change the helipad's area. The head construction worker reminded them that the helipad needed to be a rectangle made with square-inch tiles.

3. Help the construction crew.
   A. Are there other rectangles besides the one shown above that have the same perimeter as the current helipad at Ladybug Airport? Use square-inch tiles to help you.
   B. What is the area of each rectangle you found?
   C. Which rectangle would you recommend for the new helipad? Why?

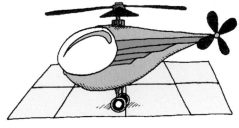

Helipads for Antopolis

Aunt Penny wants a helipad included in the Antopolis airport. Myrna agrees, but tells her aunt that only 24 inches of wire (for perimeter lights) can be allowed for the helipad. Myrna also says that the helipad must be a rectangle built with square-inch tiles.

4. Using square-inch tiles, find all the possible helipads with a perimeter of 24 inches. Sketch each helipad on a piece of paper, showing the length and width. Be sure your helipads are rectangles.

5. What is the area of each of your helipads?

6. Penny wants the helipad to be as big as possible. Design a helipad with a perimeter of 24 inches with the largest possible area.

7. Explain why you think your helipad has the largest possible area.

Portfolios

How do you set up a portfolio? First, you will need a collection folder. After you finish a piece of work, write the date on it and then put it in the collection folder. Not everything that goes in the collection folder will go in the portfolio. From time to time, you will pick pieces from your collection folder to put in your portfolio.

Mrs. Dewey's students began organizing portfolios in their mathematics class. They decided to organize the portfolios around the idea of communication. Communication has many different meanings. This is one of the reasons the class chose it. Jackie chose a picture to include in her portfolio. Frank chose some tables and graphs.

Students in Mrs. Dewey's class chose work to include in their portfolios. They explained why each piece of work was chosen and how each showed communication. The students also wrote what they learned and what they liked about their pieces.

To keep track of all the pieces, the students made a Table of Contents. Jackie's Table of Contents is shown:

Jackie's Portfolio
Table of Contents

Item	Description	Date
P vs. L	I made runways of different lengths for Antopolis Airport.	September 20
Myrna Let.	I wrote about the width, length, and perimeter of runways.	October 1
Journal	My favorite mathematics activity in third grade.	October 3

Your teacher will help you select pieces for the portfolio. After you start your portfolio, begin a Table of Contents. The Table of Contents should include the name of each piece of work, a short description, and the date it was finished. Add to the Table of Contents as you add to your portfolio.

Choose one of the following questions to write about. This work will be included in the portfolio.

1. What was your favorite mathematics activity last year? Why?

2. What would you like to learn in mathematics this year? Why?

Angles

Myrna's aunt, Cassandra, is in charge of designing a map for the runways at Antopolis Airport. She must think about such things as how many runways are needed, how many planes will be landing at one time, and how many runways will fit in the area that is planned for the airport.

At first, Cassandra thinks the runways could be lined up in rows:

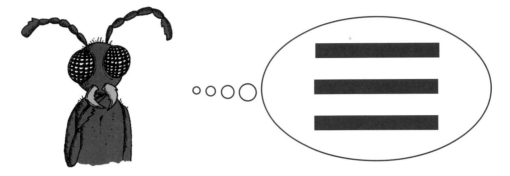

Then she changes her mind because she knows the planes will be flying in from many different directions.

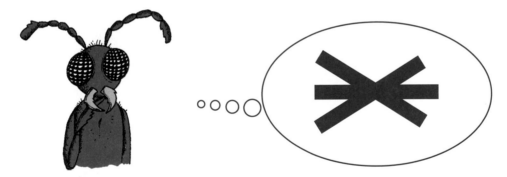

Cassandra decides to visit the airport at the nearby town, Anthill, to ask for some help. The manager of the airport asks, "Do you have any experience with angles?" Cassandra replies, "What are angles and how will they help me design the map for the runways at our airport?"

Here is what Cassandra learned.

What Is an Angle?

An **angle** is the amount of turning between two lines. The hands of a clock show turning.

The minute hand starts here:

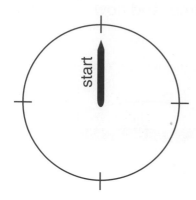

The hand turns to here:

The amount of turning from start to finish shows the angle. The inside of the angle is shaded to show turning:

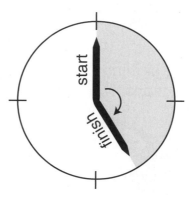

The size of the angle depends on the amount of turning. The more you turn, the larger the angle.

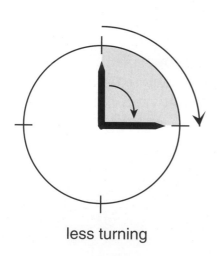

less turning

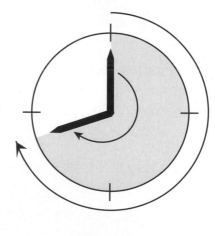

more turning

1. Compare the following pairs of shaded angles. For each pair, decide which shaded angle (the first or the second) has more turning.

   A.

    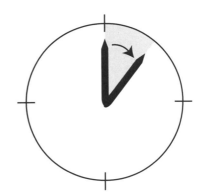

   B.

    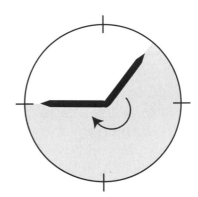

   C.

2. Mrs. Dewey's class used angle circles to make angles. For each angle circle, decide whether the white angle or the green angle has more turning.

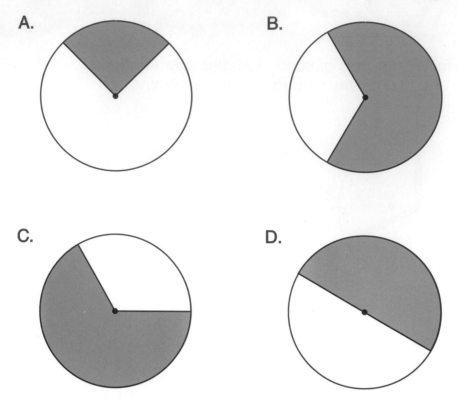

A.

B.

C.

D.

An angle with short sides can have more turning than another angle with longer sides. When you look at an angle, do not look at how long the sides are. Look at the amount of turning, or the inside of the angle. The angle on the right is bigger because it has more turning.

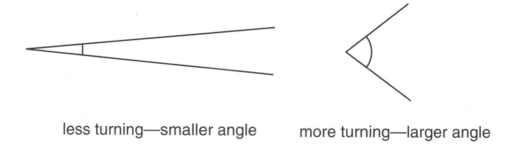

less turning—smaller angle   more turning—larger angle

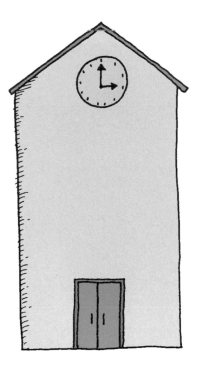

The hands of a clock on a building are much bigger than the hands of an alarm clock, but they can show the same amount of turning.

Angles in Shapes

Many shapes have angles. Look at the cover of a book. You should see four angles on the front cover.

ANGLES
iN OUR
WORLD

The figure below has five angles inside the shape.

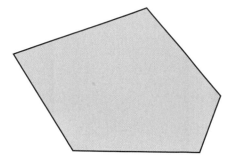

3. Look at the following shapes. How many angles do you see inside each shape?

A.

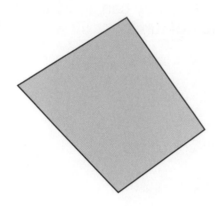

B.

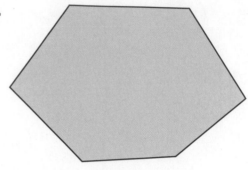

4. Look at the following shapes. How many angles do you see inside each shape?

A.

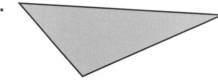

B.

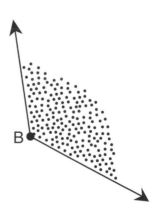

5. Look around the classroom. Find two angles in the room. Decide which of the two angles is larger. Remember, the angle with more turning is the larger angle.

Parts of an Angle

Every angle has two **sides.** We can think of the sides of an angle as the hands of a clock. The two sides of an angle meet at a point called the **vertex.** Sometimes an angle is named by its vertex.

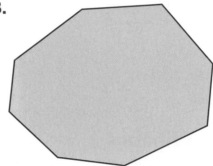

side

vertex

A

side

B

6. Use your angle circles to make a white angle larger than Angle A.

7. Use your angle circles to make a white angle smaller than Angle B.

Use a ruler to draw the following angles.

8. Draw an angle larger than Angle A. Label this Angle C.

9. Draw an angle smaller than Angle B. Label this Angle D.

10. Find an angle in the classroom that is larger than Angle A but smaller than Angle B.

Degrees

We measure the amount of turning in angles by **degrees.** Larger angles have larger degree measures. If you turn all the way around in a circle, you have turned 360° (read as 360 degrees). The little circle (°) stands for the word degree.

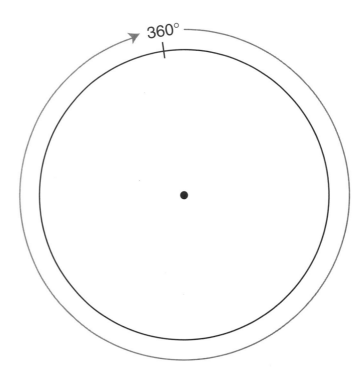

Use your calculator if needed to answer Questions 11 and 12.

11. How many degrees have you turned when you turn halfway around a circle?

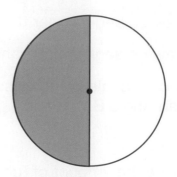

12. How many degrees have you turned when you turn one-quarter of the way around a circle?

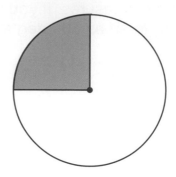

Right Angles

An angle that is easily recognized is a 90° angle or quarter-turn. Corners of books, papers, floors, and desks are usually right angles. A 90° angle is called a **right angle.** Since a right angle forms a square corner, a "box" is often drawn at its vertex.

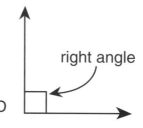

right angle

D

13. Make a white 90° angle with your angle circles.

14. Make a list of some right angles in your classroom.

15. Tear off one square corner from a piece of scrap paper. Draw a box at the vertex to show a right angle.

16. If a square corner of paper fits exactly in an angle, the angle is 90°. Use your piece of paper to decide if the following angles are right angles or not.

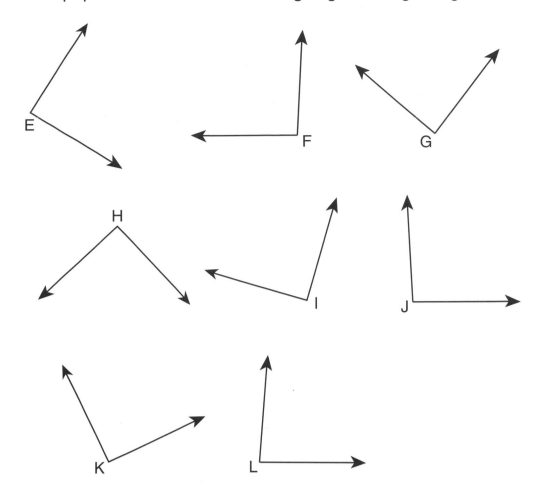

17. Use a ruler to draw an angle smaller than a right angle. Label the vertex with a letter. How can you show that this angle is less than 90°?

18. Use a ruler to draw an angle larger than a right angle. Label the vertex with a letter. How can you show that this angle is more than 90°?

19. This triangle contains a right angle. Which angle is the right angle?

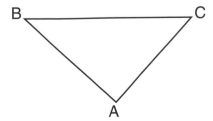

Acute Angles

These angles are **acute** angles:

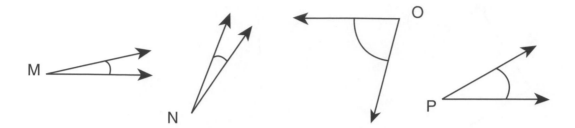

These angles are **not** acute angles:

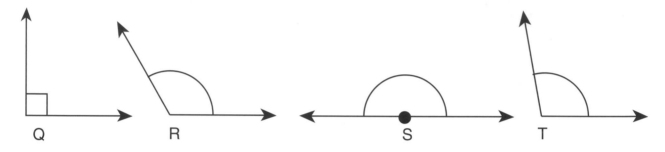

20. **A.** What can you say about the size of all the acute angles?

    B. How would you describe an acute angle?

    C. Acute means sharp. Why is that a good name?

    D. Make an acute white angle with your angle circles.

21. Which angles in the figure below are right angles?
    Which angles are acute angles?

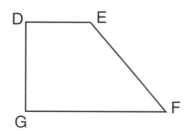

22. Tear off one square corner from a piece of scrap paper. Draw a box at the vertex to show the right angle. Fold the angle in half. What is the degree measure of the new angle? You may use your calculator if needed.

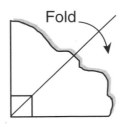

Fold

23. Use your new angle to estimate the measures of the following angles.

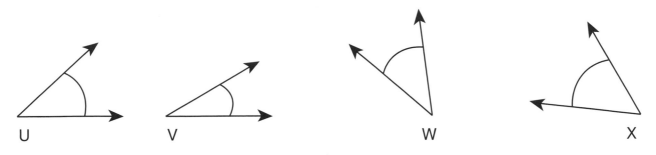

U   V   W   X

Obtuse Angles

These angles are acute angles:

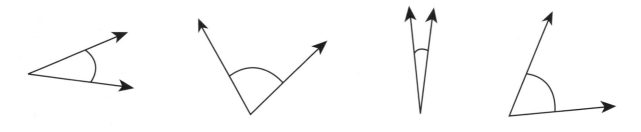

These angles are **obtuse** angles:

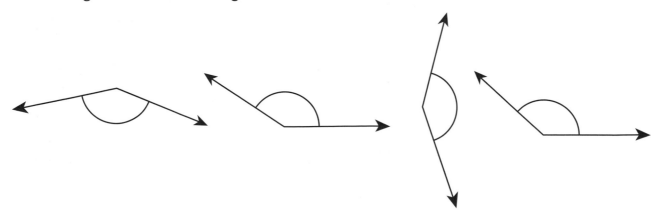

24. A. What can you say about the size of all of the obtuse angles?

B. How would you describe an obtuse angle?

C. Obtuse means dull. Why is that a good name?

D. Make an obtuse white angle with your angle circles.

25. This triangle has an obtuse angle and two acute angles. Which is the obtuse angle?

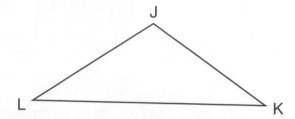

26. Here are some runway plans for Cassandra. For each plan, find the number of acute angles, obtuse angles, and right angles.

A.

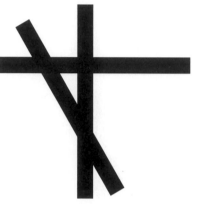

B.

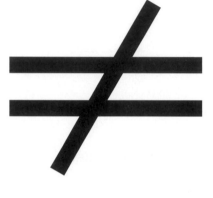

For each pair, which is the larger angle?

1.

A.

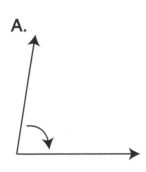

B.

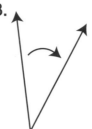

2.

A.

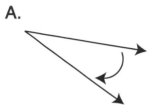

B.

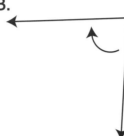

3.

A.

B.

4.

A.

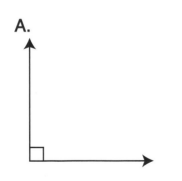

B.

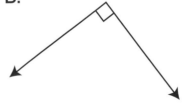

Name all the right angles, acute angles, and obtuse angles in the figures below.

5.

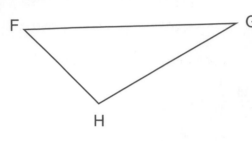

6.

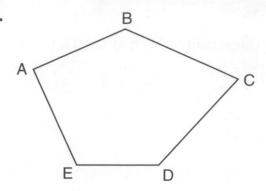

7. Draw two acute angles.

8. Draw two obtuse angles.

9. How many right angles do you have to put together to make one full turn?

10. Using a ruler, draw a shape with 4 angles.

11. Using a ruler, draw a shape with 5 angles.

Here are some more runway plans. For each plan, find the number of acute angles, obtuse angles, and right angles.

12.

13.

Angles in Pattern Blocks

Here are six pattern blocks and their names.

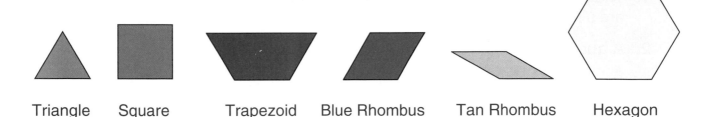

Triangle   Square   Trapezoid   Blue Rhombus   Tan Rhombus   Hexagon

1. Which shape has the smallest angle?

2. Which shape has the largest angle?

3. Which shapes have acute angles?

4. Which shapes have obtuse angles?

5. Which shapes have right angles?

6. Which shapes have both acute and obtuse angles?

7. **A.** Find angles in the pattern blocks that can be put together to make a right angle. Draw a picture of these pattern blocks. You may want to start by tracing a right angle. A right angle is shown here.

   B. Can you make a right angle with pattern blocks another way?

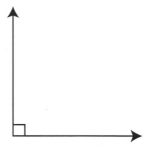

In class you explored pattern blocks and completed the table *Angle Measures in Pattern Blocks.* Use the table to answer the following questions.

1. The square has all 90° angles. Which other pattern blocks have angles that are all the same measurement?

2. A **quadrilateral** is a shape that has four sides (four angles).

    A. Make a list of the pattern blocks that are quadrilaterals.

    B. Find the sum of the angles of each of the quadrilaterals. For example, the sum of the angles for the square is 90° + 90° + 90° + 90° = 360°.

    C. What do you notice about the sum of the angles of the quadrilaterals in Question 2B?

Unit 3

Numbers and Number Operations

	Student Guide	Discovery Assignment Book	Adventure Book	Unit Resource Guide*
Lesson 1				
Multiplying and Dividing with 5s and 10s	●	●		●
Lesson 2				
Roman Numerals	●			
Lesson 3				
Place Value				●
Lesson 4				
The TIMS Candy Company	●			
Lesson 5				
Addition and Subtraction	●			●
Lesson 6				
What's Below Zero?	●	●		●
Lesson 7				
At the Hardware Store	●			

*Unit Resource Guide pages are from the teacher materials.

Multiplying and Dividing with 5s and 10s

Using Fact Families — Multiplying and Dividing with 5s and 10s

Jackson's Hardware Store donated 30 basketballs to the schools in the neighborhood.

John and his father went to pick up the basketballs for Bessie Coleman school. When they arrived at the store, there were people from four other schools waiting to pick up their basketballs. John helped divide the thirty basketballs into five groups.

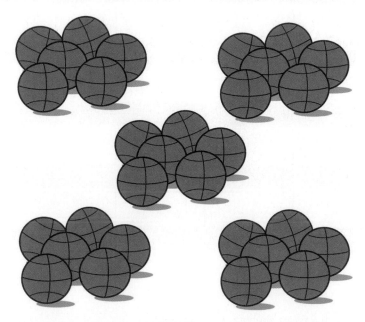

Each of the 5 schools got 6 new basketballs.

The **division sentence** for this is $30 \div 5 = 6$. The answer to a division problem is called the **quotient.** In this sentence the quotient is six. Thirty, or the number to be divided, is the **dividend.** The **divisor** is five.

Then John told everyone that he would label all the basketballs with the correct school name. Everyone brought the new basketballs back to him for labeling, one school at a time. John added 6 + 6 + 6 + 6 + 6. He knew this was the same as five groups of six or 5 times 6, or 30 basketballs in all.

John knew that 5 × 6 = 30 is related to the division sentence 30 ÷ 5 = 6. There are two more sentences that are related: 6 × 5 = 30 and 30 ÷ 6 = 5. We call all four of these sentences together a **fact family**:

5 × 6 = 30, 6 × 5 = 30, 30 ÷ 6 = 5, 30 ÷ 5 = 6

Discuss

1. Jackson's Hardware also gave away 30 soccer balls. Each school received a crate of six balls.

   A. How many schools got soccer balls? Write a number sentence to describe this.

   B. What does each number in the sentence represent?

2. John found he had 30 marbles at home. He decided to give an equal number of marbles to each of his three sisters. How many marbles did John give to each sister? Draw a picture for this problem and describe it using a division sentence. Write another number sentence that is in the same fact family.

3. Nila wrote a division story for 20 ÷ 5. Nila drew a picture for her story.

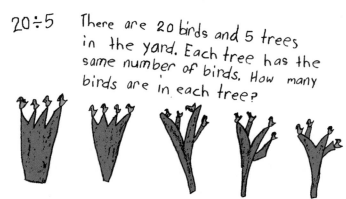

20 ÷ 5 There are 20 birds and 5 trees in the yard. Each tree has the same number of birds. How many birds are in each tree?

   A. What is another number sentence that is in the same fact family as 20 ÷ 5?

   B. Write a division story for 50 ÷ 10. Draw a picture for your story and write a number sentence. Write three more sentences that are in the same fact family.

4. Maya baked chocolate chip cookies. She counted out 45 cookies and put an equal number in each of 9 bags. Then she gave one bag of cookies to 9 friends.

   A. How many cookies did she give each friend? Write a number sentence for this story.

   B. Write a multiplication number sentence in the same fact family. What do the numbers in the multiplication sentence represent?

5. Which of the following number sentences are in the same fact family as $5 \times 8 = 40$?

   a) $40 \div 10 = 4$ **b)** $40 \div 5 = 8$ **c)** $8 \times 4 = 32$ **d)** $8 \times 5 = 40$

Solve Questions 6–15. Use fact families, manipulatives, or other strategies. Write a number sentence for each problem. Then write the other three sentences in the fact family.

6. How many dimes are in 80 cents?

7. How many nickels are in 35 cents?

8. How many nickels are in 15 cents?

9. How many nickels are in 40 cents?

10. How many dimes are in 60 cents?

11. How many nickels are in 20 cents?

12. How many dimes are in 40 cents?

13. Maya gets paid for helping her neighbor with her baby one afternoon each week. She saves all the money she gets. After five weeks, she has $25. How much money does Maya get paid each week? Write a number sentence.

14. How many weeks will Maya have to help her neighbor to make $45? Write a number sentence.

15. John lives 4 blocks from school. It takes him 20 minutes to walk to school. If John walks steadily, how long does it take John to walk one block? Write a number sentence.

Multiplying with 0 and 1

16. Think about multiplication as repeated addition of groups. Three groups of five makes fifteen: $3 \times 5 = 15$. Now think about what happens when you multiply with 1 or 0. How many groups do you have? How many are in each group? Try the following problems. You may use your calculator.

   A. $5 \times 0 =$ **B.** $5 \times 1 =$

   C. $10 \times 0 =$ **D.** $1 \times 10 =$

   E. $0 \times 98 =$ **F.** $98 \times 1 =$

   G. $0 \times 5348 =$ **H.** $1 \times 5348 =$

17. **A.** What can you say about multiplying numbers by 0? Explain.

   B. What can you say about multiplying numbers by 1? Explain.

Multiplication Facts and *Triangle Flash Cards*

Work with a partner. Use the directions below and your *Triangle Flash Cards: 5s* and *Triangle Flash Cards: 10s* to practice the multiplication facts.

- One partner covers the shaded number, the largest number on the card. This number will be the answer to the multiplication problem. It is called the **product.**

- The second person multiplies the two uncovered numbers (one in a circle, one in a square). These are the two **factors.** It does not matter which of the factors is said first: 4×5 and 5×4 both equal 20. $4 \times 5 = 20$ and $5 \times 4 = 20$ are called **turn-around facts.**

$5 \times 4 = ?$

$4 \times 5 = ?$

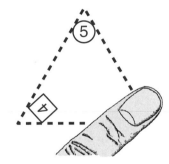

- Divide the cards into three piles: those facts you know and can answer quickly, those you can figure out with a strategy, and those you need to learn.

- Discuss how you can figure out facts that you do not recall right away. Share your strategies with your partner.

- Practice the last two piles again. Then make a list of the facts you need to practice at home.

- Circle the facts you know quickly on your *Multiplication Facts I Know* chart. Remember that if you know one fact, you also know its turn-around fact. Circle both on your chart.

- Review your answers to Question 17.

- You will continue to use *Triangle Flash Cards* to study other groups of facts. If you know one or two of the multiplication facts in a fact family, you can use those facts to help you learn the division facts.

Multiplication Facts I Know

×	0	1	2	3	4	5	6	7	8	9	10
0	0	0	0	0	0	0	0	0	0	0	0
1	0	1	2	3	4	5	6	7	8	9	10
2	0	2	4	6	8	10	12	14	16	18	20
3	0	3	6	9	12	15	18	21	24	27	30
4	0	4	8	12	16	(20)	24	28	32	36	40
5	0	5	10	15	(20)	25	30	35	40	45	50
6	0	6	12	18	24	30	36	42	48	54	60
7	0	7	14	21	28	35	42	49	56	63	70
8	0	8	16	24	32	40	48	56	64	72	80
9	0	9	18	27	36	45	54	63	72	81	90
10	0	10	20	30	40	50	60	70	80	90	100

Dear Family Member:

Your child is reviewing the multiplication facts and learning the division facts for fives and tens by studying fact families. For example, 5 × 4 = 20, 4 × 5 = 20, 20 ÷ 5 = 4, and 20 ÷ 4 = 5 is a fact family. Once a student learns the multiplication facts, learning the division facts becomes easier. Using fact families is a good strategy for solving most of the problems below. Remind your child to bring home the flash cards for the fives and tens. Help him or her study these facts.

Thank you for your cooperation.

1. How many dimes in 90 cents?

2. How many nickels in 30 cents?

3. Banks wrap dimes into packs of 50 dimes. If Nila takes 70 dimes to the bank to be wrapped, how many packs will she get? How many dollars will this be? How many dimes will be left over?

4. There are 20 Little League teams in the city. The League places 10 teams in a division. How many divisions will there be in the city? Write a number sentence for this story and all the number sentences in this fact family.

5. Write a story to show 45 ÷ 9. Draw a picture to go with your story, and write a number sentence. Write the other number sentences in this fact family.

6. Show two ways you can have 25 cents if you have only dimes and nickels.

7. Show three ways you can have 40 cents if you have only dimes and nickels.

8. Chewy Candies come in packs of five. Irma has 3 packs, Michael has 5 packs, Romesh has 1 pack, and Jessie has no packs.

   A. How many candies does each student have? Write a number sentence for each student.

   B. How many candies do they have altogether?

9. Jacob has 60 cents and needs $1.00 for a show. How many dimes does he need to make $1.00?

10. A pack of Chewy Candies costs 15 cents. How many packs can you buy with $1.00? Explain your solution.

Roman Numerals

One day, Mrs. Dewey started math class by asking, "How many of you speak another language besides English?"

Many hands went up. Roberto and Ana speak Spanish, Ming speaks Chinese, and Nila speaks Arabic. Linda speaks Tagalog, a language spoken in the Philippines, and Nicholas speaks Russian.

"That's wonderful," said Mrs. Dewey. "Did you know that there are more than 220 major languages in the world? Many cultures have their own language, but almost all share the same number system. They use the digits 0, 1, 2, 3, 4, 5, 6, 7, 8, 9 to count and compute."

The number system we and many other people use is called the Hindu-Arabic number system. It was invented around the 9th century AD. Many cultures had invented number systems long before this. One that is still used sometimes in our culture is the **Roman numeral** system. You can sometimes find Roman numerals on clocks and buildings and in movies. Sometimes the pages in the preface of a book are numbered using Roman numerals. Often, when people want to be a little bit fancy, they use Roman numerals. For example, the Super Bowls of the National Football League are all numbered with Roman numerals.

The Romans used a system that was based on counting groups similar to ours. Romans made groups of ten and also groups of five. Archaeologists have found evidence that their system was being used around 260 BCE. No one is really certain about where the Roman symbols for the numerals came from. They may have come from finger counting used in the market or the tally marks used to keep track of things like farm animals and soldiers. Some of the modern Roman numerals we use now came from the words Romans used.

Roman Numerals and Symbols

Early Roman Numerals	Modern Roman Numerals	Hindu-Arabic Numbers	Possible Origins	
			Finger Counting or Latin Words	Tally Marks
I	I	1		I
V	V	5		V
X	X	10		X
	L	50		
C	C	100	The Latin word for one hundred is *centum*	(X)
Ɔ	D	500		
Φ	M	1000	The Latin word for one thousand is *mille*	<I>

To write a number like 123, Romans put together their symbols for one hundred, two tens, and three ones to write the number like this:

CXXIII

At first, the Romans just used the symbols for ones, tens, hundreds, and thousands. So a number like 876 might have looked like this:

CCCCCCCCXXXXXXXIIIIII

To help make these numbers shorter, Romans began using symbols for half of ten, half of one hundred, and half of one thousand. Then they could write 876 as:

DCCCLXXVI

1. Use the chart to help you write these Roman numerals as Hindu-Arabic numbers:
   A. LXXXVIII
   B. MDCCLXXVII
   C. MMMMMMMMCCL

2. Write the following Hindu-Arabic numbers as Roman numerals:
   A. 68
   B. 108
   C. 286

3. In the earliest Roman times, four was written as "IIII" and nine was written as "VIIII." Later, a shorter way of writing numbers was invented. Study the table below to find the pattern for the shortcut. Copy the tables, then fill in the empty boxes in the top and bottom rows.

   A.

I	II	III	IV					IX	
1	2	3		5	6	7	8		10

   B.

	XII			XV		XVII		XIX	
11	12	13	14	15	16	17	18		20

   C.

X	XX		XL		LX	LXX		XC	C
10	20	30		50			80		100

   D. What patterns did you find?
   E. Can you give a rule for the shortcuts in the table?

This pattern or rule is called the **subtractive principle.** The symbol for a smaller number is placed before a symbol for the larger number. This indicates that the smaller number should be subtracted from the larger. **I** can come in front of only the **V** or **X**. **X** can be subtracted only from **L** or **C**. **C** can be subtracted only from **D** or **M**.

4. Use this pattern to write the Hindu-Arabic numbers for the following Roman numerals.

    A. XXXIV   **B.** XLIV

    C. CMXCIX   **D.** MCMXLVIII

5. Write the Roman numerals for the following Hindu-Arabic numbers in more than one way.

    A. 54   **B.** 47

    C. 192   **D.** 1996

6. Ana said, "I can think of one number we can't write in Roman numerals." What number is Ana thinking about?

7. Where do you see Roman numerals today?

Homework

Dear Family Member:

Your child is learning how to translate our numbers (Hindu-Arabic) into Roman numerals. Ask your child about Roman numerals that he or she may have seen. Then help your child with the translations. Encourage your child to use the Roman Numerals and Symbols chart as a guide.

Thank you for your help.

1. Write these numbers using Roman numerals.

    A. 12   **B.** 74

    C. 126   **D.** 239

2. Write these Roman numerals as Hindu-Arabic numbers.

    A. XIV   **B.** DXLV

    C. DCCXXIII   **D.** CMXCVIII

The TIMS Candy Company

Mr. and Mrs. Haddad own a chocolate factory that makes Chocos. The name of their company is the TIMS Candy Company. They use base-ten pieces to keep track of how much candy they make.

They use a **bit** for each Choco.

Whenever there are 10 bits, they can be packed together to make a **skinny.**

When there are 10 skinnies, they can be packaged together to make a **flat.**

A group of 10 flats makes a **pack.**

Mr. and Mrs. Haddad use a **Base-Ten Board** to show the bits, skinnies, flats, and packs. They also write the amounts in numbers on the **Recording Sheet.** For example, one day the company made 236 Chocos. This is how they recorded the candy:

Base-Ten Board

Recording Sheet

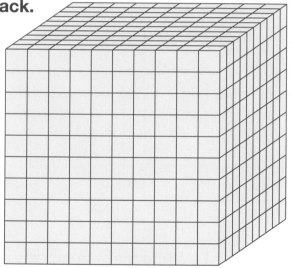

	2	3	6

Rhonda and Joe work for the TIMS Candy Company.

1. Rhonda has 3 flats, 1 skinny, and 5 bits. How many pieces of candy is that?

Base-Ten Board

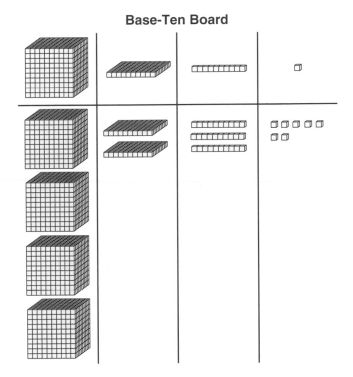

2. Joe has 4 packs, 2 flats, 3 skinnies, and 7 bits. How many pieces of candy is that?

Base-Ten Board

3. Rhonda has 15 bits, 3 flats, and 3 skinnies. She says this is 345 pieces of candy. Is she correct? If not, how many pieces of candy does she really have?

Base-Ten Board

4. Joe has 5 flats, 12 skinnies, and 8 bits. He says this is 528 pieces of candy. Is he correct? If not, how many pieces of candy does he really have?

Base-Ten Board

Sometimes Rhonda and Joe do not use the Base-Ten Boards. They put the blocks on a table.

5. Rhonda has 14 flats, 15 bits, and 4 skinnies. She says this is 1456 pieces of candy. Is she correct? If not, how many pieces of candy does she really have?

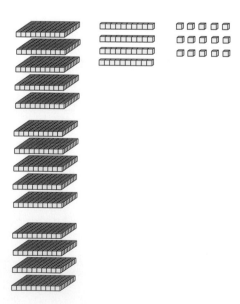

6. Joe has 2 packs, 5 skinnies, 4 flats, and 3 bits. He says this is 2543 pieces of candy. Is he correct? If not, how many pieces of candy does he really have?

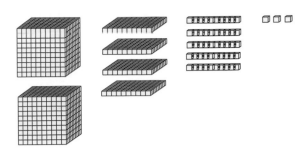

Base-Ten Shorthand

Sometimes it is useful to record your work with the base-ten pieces. Other times, drawing a picture of the base-ten pieces is helpful. Mr. Haddad decided to use a shorthand for the base-ten pieces.

• = Bit   / = Skinny   ☐ = Flat   ▱ = Pack

7. Joe says there are often several ways to show an amount of candy on the Base-Ten Board. For example, 26 can be shown as:

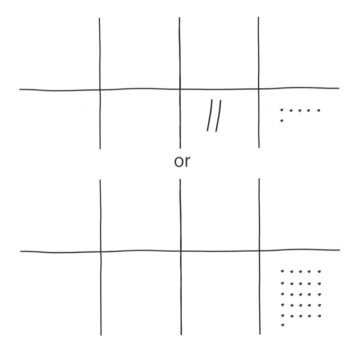

or

There is one more way 26 can be shown on the Base-Ten Board. What is this third way? Use base-ten shorthand to sketch your answer.

8. Use base-ten shorthand to show the number of candies Rhonda and Joe had in Questions 1–6.

9. Joe showed several ways of putting 32 Chocos on the Base-Ten Board by using base-ten shorthand. Some of Joe's work was erased. Fill in the missing pieces. Use base-ten shorthand to sketch your answer.

A. Here is one way to show 32.

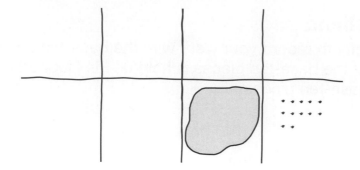

B. Here is another way to show 32.

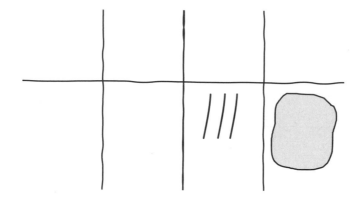

C. Here is a third way to show 32.

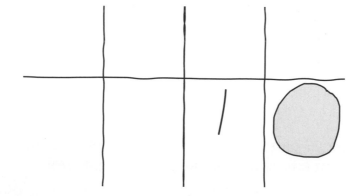

10. Rhonda made 267 pieces of candy. Fill in the missing information using base-ten shorthand and the Recording Sheet.

A. Here is one way to make 267.

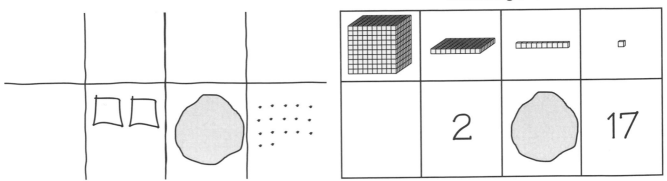

Recording Sheet

		2	17

B. Here is another way to show 267.

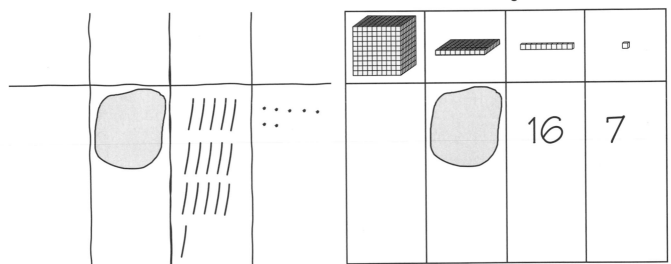

Recording Sheet

		16	7

C. Here is a third way to show 267.

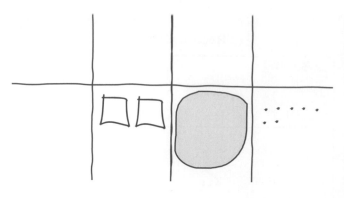

Recording Sheet

D. Here is yet another way to show 267.

Recording Sheet

The Fewest Pieces Rule

The TIMS Candy Company decided that the best way to record the amount of candy it makes is to use the smallest possible number of base-ten pieces. The company calls this the **Fewest Pieces Rule.** For example, the best way to record 32 candies is to use 3 skinnies and 2 bits.

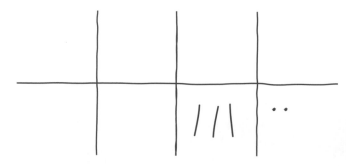

Recording Sheet

The best way to record 457 candies is to use 4 flats, 5 skinnies, and 7 bits.

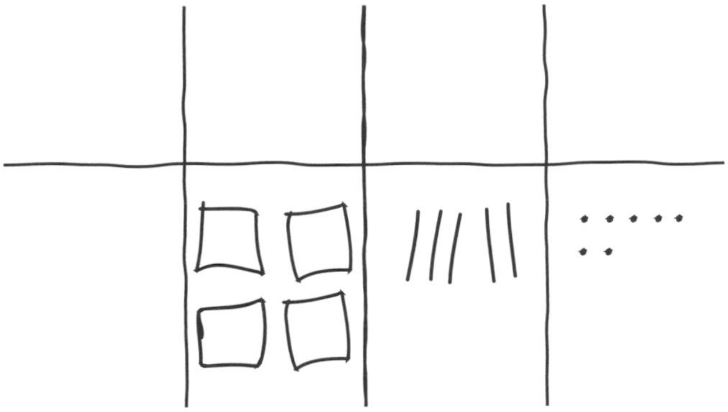

Recording Sheet

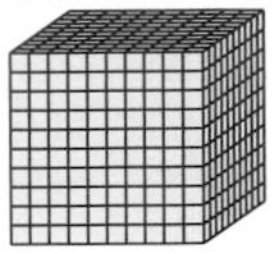	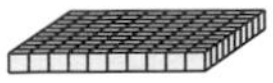		
	4	5	7

11. Show each number, using the Fewest Pieces Rule. Record your answer by using base-ten shorthand. You do not have to sketch the columns.

A. 236   **B.** 507   **C.** 5235   **D.** 6008

In problems 12–15:

A. Write how many Chocos were made.

B. Then check if the Fewest Pieces Rule is followed. If it is not, use base-ten shorthand to show the candy using the fewest pieces possible.

12.

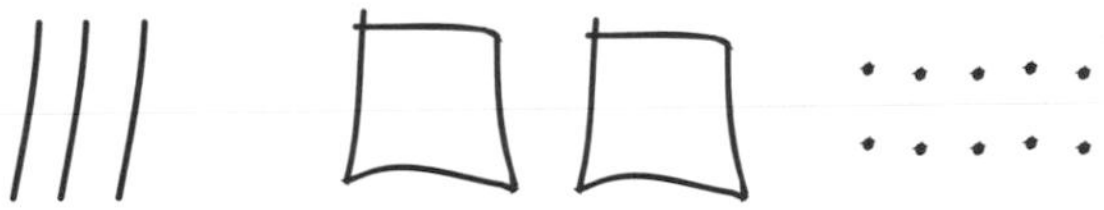

13.

14.

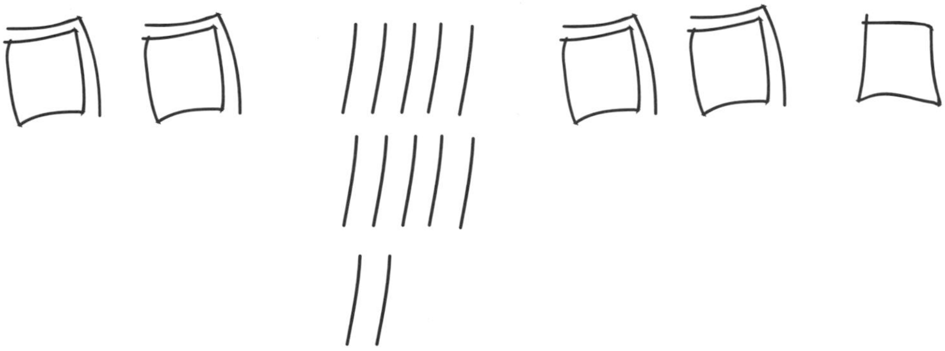

15.

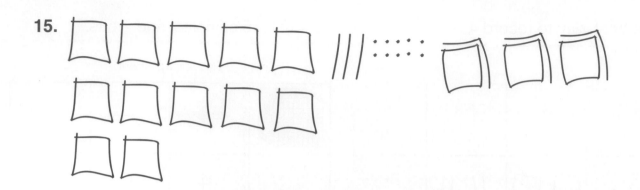

16. The number **157** has 3 digits: the 1, the 5, and the 7. Explain the value of each digit. Do you have the same amount if you mix up the digits and write **571**? Explain why or why not.

Dear Family Member:

Your child is reviewing place value—the idea that the value of a digit in a number depends upon where it is placed. For example, the 2 in 329 stands for 2 tens but the 2 in 7293 is 2 hundreds.

In class your child uses base-ten pieces to represent numbers. When the pieces are not available, students are encouraged to draw pictures of the base-ten pieces. We call these drawings of the base-ten pieces base-ten shorthand. To help your child with homework Questions 1–11, you may wish to review the Base-Ten Shorthand section on the previous pages.

Thank you for your help.

The sketches below show the number of Chocos made by workers at the TIMS Candy Company. Write the amount of candy using numbers.

1. │││ · · · · · · ·   **2.** ▢ │ · · · ·

3. ▢▢▢ ││ · ·   **4.** ▢▢ ▢

5. ▢ │││││││││   **6.** ▢▢▢ ·

The workers at the TIMS Candy Company recorded in numbers the amount of candy they made. Sketch each amount, using base-ten shorthand.

7. 356   **8.** 4206

9. 240   **10.** 3005

11. One way to show 352, using base-ten shorthand, is:

Sketch 352 two other ways, using base-ten shorthand.

Addition and Subtraction

Addition

One day, Rhonda made 326 pieces of candy. She used the base-ten pieces to show her work. She recorded 3 flats, 2 skinnies, and 6 bits. Joe made 258 candies, which he recorded as 2 flats, 5 skinnies, and 8 bits. Mrs. Haddad wanted to know how much candy they made altogether. She recorded her addition like this:

Base-Ten Board

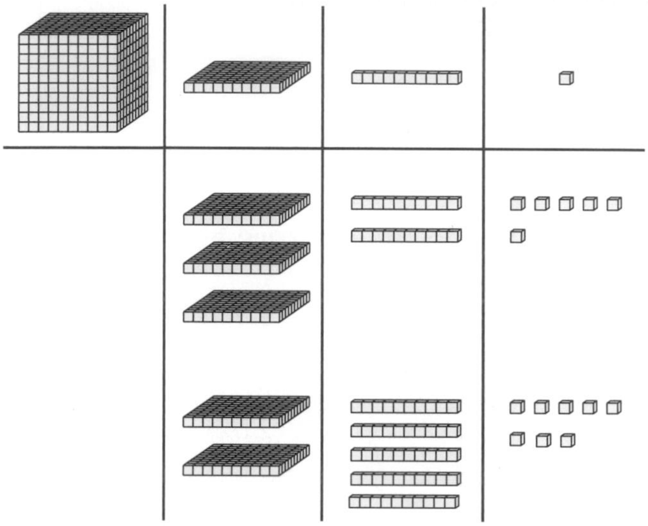

Recording Sheet

3	2	6
2	5	8
5	7	14

Mrs. Haddad saw that she was not using the fewest base-ten pieces possible. Since there are 14 bits, she can make 1 more skinny with 4 bits left over. Mrs. Haddad recorded her work like this:

Recording Sheet

+	3	2	6
	2	5	8
	5	7¹	̶1̶4̶
	5	8	4

Addition and Subtraction

1. On another day, Rhonda made 1326 candies and Joe made 575. They recorded their work by sketching the base-ten pieces using base-ten shorthand. Use your base-ten pieces to solve this problem.

Joe remembered the Fewest Pieces Rule and wrote:

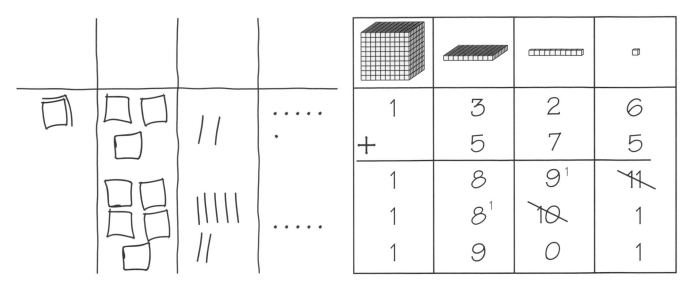

Mrs. Haddad noticed that drawing columns on the **Recording Sheet** was not necessary if she always used the Fewest Pieces Rule. Mrs. Haddad called this the **quick paper-and-pencil method for addition.** She wrote the problem like this:

$$1\overset{1}{3}\overset{1}{2}6$$
$$+\ \ 575$$
$$\overline{1901}$$

At the end of one day, Rhonda had made 1046, Joe had made 878, and Sam had made 767 candies. Mrs. Haddad found their total using the quick paper-and-pencil method:

$$\overset{1}{1}\overset{1}{0}\overset{2}{4}6$$
$$878$$
$$+\ \ 767$$
$$\overline{2691}$$

2. Explain Mrs. Haddad's method for adding the three numbers.

3. Dominque has 325 baseball cards. Her sister Rosie has 416. About how many baseball cards do the two girls have altogether?

One way to estimate is to think about base-ten pieces. The number of baseball cards Dominque has is 3 flats and some more. The number of baseball cards Rosie has is 4 flats and some more. Together they have 7 flats and some more—or more than 700 baseball cards.

I have about 300 baseball cards. You have about 400 baseball cards.

That means we have about 700 baseball cards.

Subtraction

Next to the factory, Mr. and Mrs. Haddad have a store where they sell their Chocos. They keep track of how much candy is sold, using the base-ten pieces. Sometimes they have to break apart skinnies, flats, or packs to keep track of how much candy they have in the store.

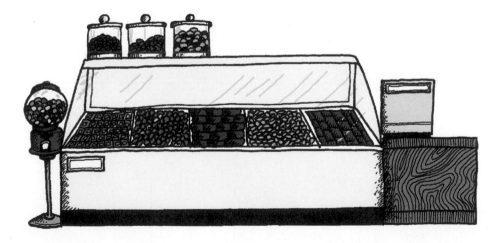

One morning, there were 3 flats, 6 skinnies, and 4 bits worth of candy in the store.

Base-Ten Board

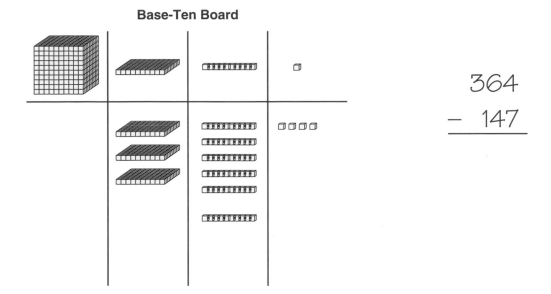

$$364$$
$$- \ 147$$

A customer came in and bought 147 pieces of candy. To find how much candy was left, Mrs. Haddad did the following:

Since 7 bits cannot be taken away from 4 bits, a skinny must be broken apart. Then there are 3 flats, 5 skinnies, and 14 bits. Now 1 flat, 4 skinnies, and 7 bits can be taken away.

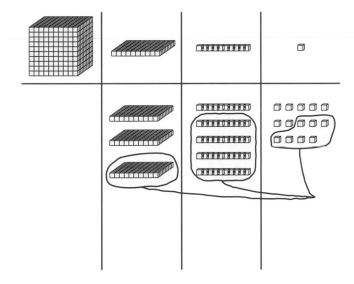

There are 2 flats, 1 skinny, and 7 bits left. Mrs. Haddad said she knew a different method to figure out how much candy is left. This is what Mrs. Haddad did:

$$\begin{array}{r} 3\overset{5}{\cancel{6}}\overset{1}{4} \\ -1\ 4\ 7 \\ \hline 2\ 1\ 7 \end{array}$$

4. Explain Mrs. Haddad's method in your own words.

Another day there were 1237 pieces of candy in the store. The store sold 459 pieces of candy that day. To find how much was left, Rhonda used Mrs. Haddad's method. Rhonda called this the **quick paper-and-pencil method for subtraction.**

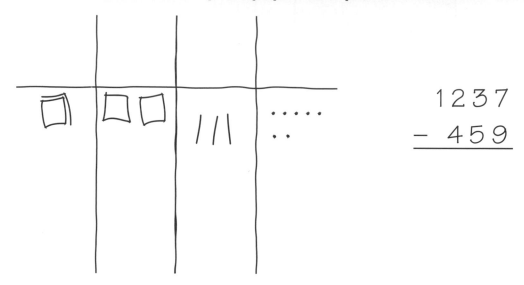

$$\begin{array}{r} 1\ 2\ 3\ 7 \\ -\ \ 4\ 5\ 9 \\ \hline \end{array}$$

Joe saw that he had to trade 1 skinny for 10 bits to subtract 9 bits.

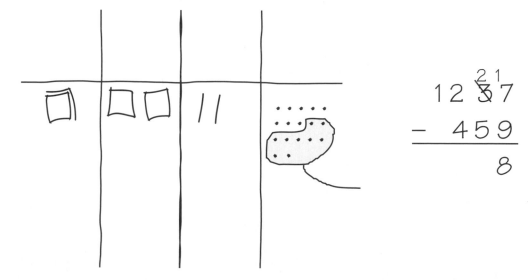

$$\begin{array}{r} 1\ 2\ \overset{2}{\cancel{3}}\overset{1}{7} \\ -\ \ 4\ 5\ 9 \\ \hline 8 \end{array}$$

Addition and Subtraction

Joe then broke up one flat so that he had 12 skinnies and was able to subtract.

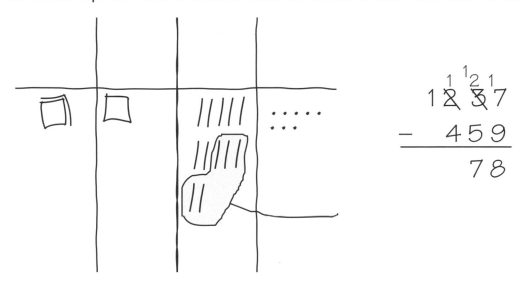

At the next step, Joe broke up his only pack so that he had 11 flats. Joe found that there were 778 pieces of candy left in the store.

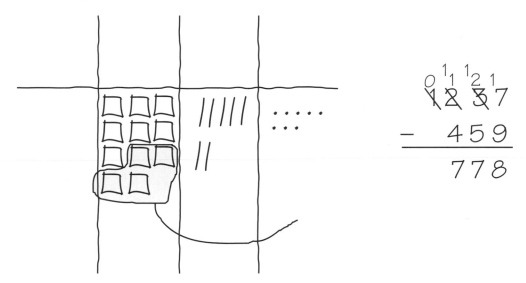

For problems 5–8:

A. **Use base-ten pieces or base-ten shorthand to solve the problem.**

B. **Then do the problem using paper and pencil or mental math.**

5. There were 578 pieces of candy in the store (5 flats, 7 skinnies, and 8 bits). The store sold 349 pieces of candy. How many pieces of candy were left?

6. Another day there were 4443 pieces of candy, and 1718 of them were sold. How many pieces of candy were left?

7. There are 2075 Chocos. The store sold 1539. How many are left?

8. There are 5204 Chocos in the store. A customer came in and bought 565. Another customer came in and wanted to buy 4859 pieces of candy. Was there enough candy in the store?

Homework

In Questions 1 through 3, draw a picture of the problem using base-ten shorthand. Then solve the problem using the picture to help you.

1. 364
      + 125

2. 1078
      + 2451

3. 1837
      + 2548

On Monday, Tuesday, and Wednesday, Rhonda and Joe were very busy and did not have time to compute their totals for the day. Help Rhonda and Joe compute their totals. Estimate to make sure your answer is reasonable.

Name	Monday	Tuesday	Wednesday
Rhonda	478	1003	576
Joe	589	1947	1756

4. How much candy was made on Monday?

5. How much candy was made on Tuesday?

6. Explain how you can compute in your head the amount of candy made on Tuesday.

7. How much candy was made on Wednesday?

8. How much candy did Rhonda make altogether on all three days?

9. How much candy did Joe make altogether on all three days?

10. How much candy did Rhonda and Joe make altogether on Monday, Tuesday, and Wednesday?

Solve the following problems. You may use any method you wish. Check your answer to make sure it is reasonable. Use base-ten shorthand when you need to.

11. 2357
       − 528

12. 2001
       − 432

13. 678
       + 1546

14. 1239
       − 643

15. The students from Livingston School and Stanley School are going on an outing. There are 765 students at Livingston School and 869 students at Stanley School. How many students are going on the outing?

16. To get free playground equipment, Livingston School needs to collect 4000 soup can labels by the end of the school year. During the first four months of school, they collected 487 soup labels. By the end of the first semester, they collected 752 more labels. How many more do they need?

17. A high school has 2456 student desks. The principal decided to replace 548 of these desks because they are not safe. How many old desks will be kept by the high school?

18. At Livingston School, Mr. Jones gave his class the following problem: Maya had 4006 stamps in her stamp collection. She sold 1658 of them. How many stamps does she have left? How would you solve this problem?

Maya thought about the base-ten pieces to solve Question 18.

John thought, "I can count up and do it in my head: 1658-2658-3658. That's 2000. Then 658-758-858-958. That's 300. Then 58-68-78-88-98 is 40. So far 2340. I have 8 more to go, so 2348 is the answer."

19. Think about John's method. Find another way to do this problem. Describe your method.

What's Below Zero?

Professor Peabody, Rhonda, and Joe work for the TIMS Candy Company in Arizona. They decided to visit the TIMS Candy Company in Minnesota to check on the production of Chocos. Before traveling to Minnesota, they checked the weather report for Minnesota.

Professor Peabody, Rhonda, and Joe expect to arrive in Minnesota in approximately 6 hours.

Professor Peabody, Joe, and Rhonda decided to use a number line to check. The current temperature in Minnesota is 20°F. For the next 6 hours, the temperature will drop 5° each hour. The change in temperature is shown below on a number line.

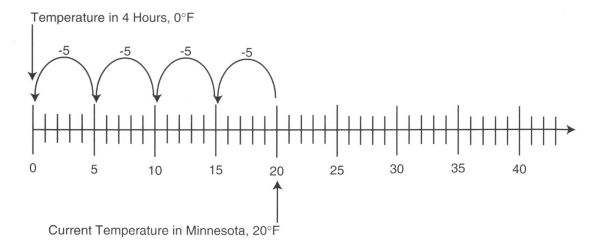

Temperature in 4 Hours, 0°F

Current Temperature in Minnesota, 20°F

Rhonda and Joe are puzzled by the number line and the weather forecaster's prediction.

After four hours, according to our number line, there is no more temperature!

That can't be true. We need to extend the number line. The temperature is still going down according to the weather forecaster.

Rhonda and Joe need to extend the number line beyond zero. There are numbers that go to the left of zero on the number line. They are called **negative numbers.**

Historical Note: It was not until the 16th century AD that it was understood that there would be numbers less than zero. Before that, if an answer was negative, it was regarded as absurd.

Negative numbers count back from zero just as **positive numbers** count up from zero. A number line with negative numbers and positive numbers is shown below.

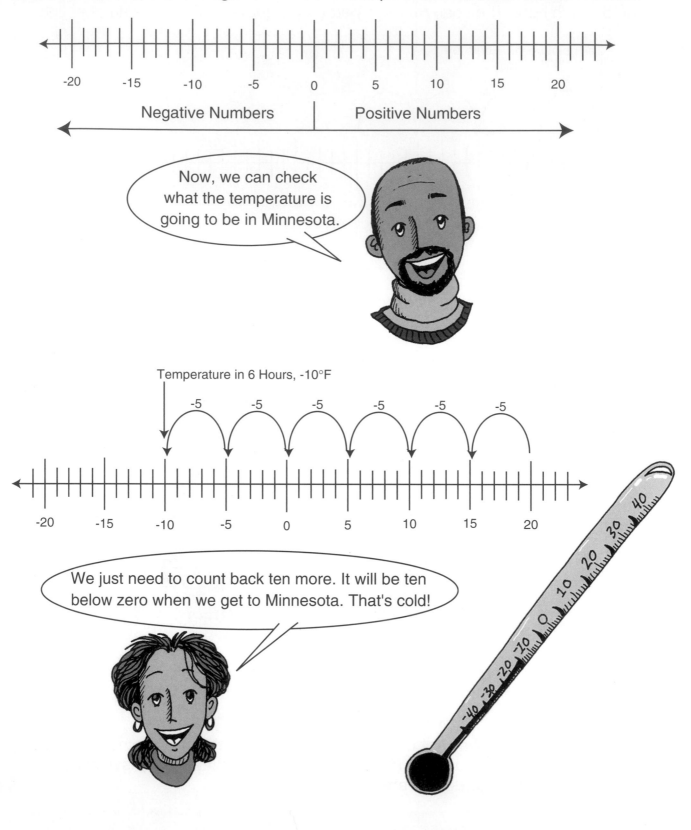

What's Below Zero?

I must have forgotten my negative sign earlier when I said that the temperature would be 10°F.

It's a good thing we checked!

Discuss

You can tell the difference between negative and positive numbers by the way they are written. "Negative five" is written as "-5."

1. Check Joe and Rhonda's calculations for the temperature when they arrive in Minnesota. Use your calculator. Skip count backwards by subtracting 5. Start at 20 and press ⬜−⬜ ⬜5⬜ six times.

2. Describe what happened in your calculator's window as you subtracted 5, six times.

Your calculator should show that the temperature in Minnesota in six hours will be -10°.

Joe, Rhonda, and Professor Peabody arrive in Minnesota and find that the temperature is -10°F. They watch the weather report and see that the temperature will be twenty degrees warmer on Monday morning.

3. **A.** Use your number line to find the temperature on Monday morning if the current temperature is -10°F.

   B. Use your calculator to verify your work on the number line.

Joe, Rhonda, and Professor Peabody stay in Minnesota for a week. Professor Peabody recorded the temperature each morning and the change in temperature throughout the day.

This is shown in the data table below:

Day of Week	Temperature in Morning	Change in Temperature	Temperature in Evening
Monday	10°	drops 15°	
Tuesday	-5°	rises 35°	
Wednesday	-20°	drops 10°	
Thursday	15°	rises 20°	
Friday	20°	drops 20°	

4. Tell the temperature in the evening for each day. Use your number line and check your answer with a calculator.

5. On Friday evening, Joe, Rhonda, and Professor Peabody return to Arizona, where the temperature is 85°F. If the temperature was -5°F when they left Minnesota, what is the change in temperature from Minnesota to Arizona? Explain.

Temperature Time

Use your number line or a calculator to find the final temperature for each of the temperature changes below.

1. It was -30°F at 10:00 this morning. The high temperature for the day was 10° higher than the temperature at 10:00 this morning. What was the high temperature for the day?

2. It was 20°F at 8:00 last night. The temperature dropped 20° overnight to reach the day's low temperature. What was the low temperature for the day?

3. The weather forecaster said that the current temperature is -15°F. The temperature is expected to drop 10 more degrees to reach the low temperature for the day. What is the low temperature expected to be?

Flying

Use a calculator to solve each of the following problems. List the keystrokes you used to solve each problem.

4. Chris was flying his plane at an altitude of 1000 feet. He decreased his altitude by 500 feet. At what altitude was Chris then flying?

5. Walt took off from an airport where the altitude is 100 feet below sea level. He then flew his plane 600 feet up. At what altitude was Walt then flying?

At the Bank

Show your work as you solve each of the following problems.

6. Sean opened a savings account with $50.00. He withdrew $45.00 and put it into a checking account. How much money was left in Sean's savings account?

7. Cindy opened a checking account with $5.00. She then wrote a check for $10.00. What is the balance in Cindy's account?

At the Hardware Store

Answer the following questions about the hardware store. Make sure you estimate to check if your answer is reasonable.

1. The hardware store has 576 drywall nails and 852 wood nails. How many drywall and wood nails does the store have altogether?

2. The hardware store had 217 cans of varnish. The store sold 89. How many are left?

3. The hardware store sells seed packets. It has 1145 vegetable seed packets and 2356 flower seed packets.

   A. Estimate the total number of seed packets in the store.

   B. Find the exact number of seed packets in the store.

4. On Monday morning, the hardware store had 2675 flower seed packets. By Friday, 1005 of these packets had sold.

   A. How many flower seed packets were left?

   B. Explain a way to do this problem in your head.

5. The hardware store has 672 gallon cans of white indoor paint and 743 gallon cans of white outdoor paint. How many cans of white paint does the store have altogether?

6. The hardware store has 260 gallon cans of glossy white paint and 240 quart cans of glossy white paint.

   A. How many gallons is 240 quarts? Remember, there are 4 quarts in one gallon.

   B. What is the total amount in gallons of glossy white paint in the store?

7. Frank buys a 2025-foot roll of string. He cuts off 215 feet for his kite. How many feet are left on the roll?

8. A gardener goes to the hardware store to buy some bags of grass seed. He buys 2 large bags and 1 small bag. A large bag of grass seed covers about 3275 square feet. A small bag covers about 1770 square feet. About how many square feet will the bags cover?

9. You need to cover 11,000 square feet of lawn with grass seed.

   A. How many bags of grass seed of each size do you need?

   B. If a large bag costs $5.00 and a small bag costs $2.50, what does it cost to cover 11,000 square feet?

10. Frank broke open his piggy banks to go to the toy section of the hardware store. In one piggy bank, he had 237 pennies. In another piggy bank, Frank had 522 pennies. In a sock in his drawer, he had 89 pennies.

    A. Explain a way to estimate the number of pennies Frank has altogether.

    B. Find the number of pennies Frank has.

    C. How much money does Frank have?

    D. A badminton set costs $9.51 with tax. How much more money does Frank need to buy the badminton set?

11. The hardware store sold 25 bags of seed on Monday, 32 on Tuesday, 11 on Wednesday, 9 on Thursday, and 41 on Friday. The hardware store then sold 197 bags of seed over the weekend.

    A. How many bags of seed did the hardware store sell from Monday to Friday?

    B. How many bags of seed did the hardware store sell all week?

    C. Did the hardware store sell more seeds from Monday to Friday or over the weekend?

    D. How many more bags of seed did the hardware store sell over the weekend than during the week?

Unit 4

Products and Factors

	Student Guide	Discovery Assignment Book	Adventure Book	Unit Resource Guide*
Lesson 1				
Multiplication and Rectangles	●	●		
Lesson 2				
Factors	●			
Lesson 3				
Floor Tiler	●	●		
Lesson 4				
Prime Factors	●			●
Lesson 5				
Product Bingo	●	●		
Lesson 6				
Multiplying to Solve Problems	●			

Unit Resource Guide pages are from the teacher materials.

Multiplication and Rectangles

Making Rectangles

I put these 12 tiles in 2 rows of 6 tiles, so $12 = 2 \times 6$.

I wonder if there are other ways to arrange 12 tiles.

1. Make as many different rectangles as you can, using 12 square-inch tiles. Then complete a table like the one below to record your rectangles.

Rectangles Possible with 12 Tiles

Number of Rows	Number in Each Row	Multiplication Sentence
2	6	$2 \times 6 = 12$

2. Draw your rectangles on *Square-Inch Grid Paper* and cut them out. Write a multiplication number sentence on each rectangle to match. Some rectangles like the 1×12 and 12×1 rectangles shown here have the same shape when you turn them sideways. You only have to cut out one of these.

Multiplication and Rectangles

3. Find all the rectangles you can make with 18 tiles. Follow the directions from Questions 1 and 2, using 18 tiles.

4. Your teacher will assign your group some of the numbers from 1 to 25.
   - For each number, arrange that many tiles into rectangles in as many ways as you can.
   - Then draw the rectangles on *Square-Inch Grid Paper* and cut them out. Write number sentences on each rectangle to match.

5. Put your rectangles on a class chart.

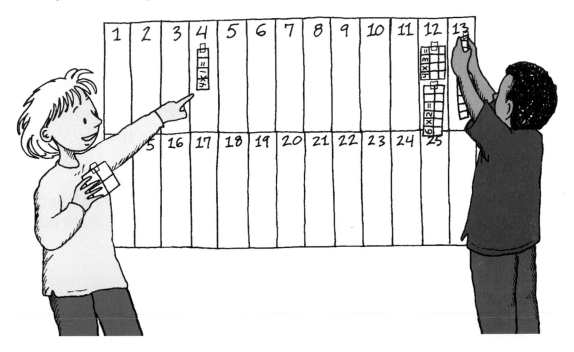

6. Use the class chart to make a table of the multiplication sentences for each of the numbers from 1–25. (See the example to the right.) Use the *Rectangles* Activity Page in the *Discovery Assignment Book.*

1	2	3	4	5
1 × 1	2 × 1	3 × 1	4 × 1 2 × 2	5 × 1

Multiples

Use the rectangles the class made to answer the following questions:

7. Which numbers have rectangles with 2 rows? List them from smallest to largest.

8. Which numbers have rectangles with 3 rows? List them from smallest to largest.

A number is a **multiple** of 2 if it equals 2 times another whole number. If you can make a rectangle with 2 rows for a number, then it is a multiple of 2.

Numbers that are multiples of two (2, 4, 6, 8, etc.) are called **even numbers.** Numbers that are not multiples of 2 (1, 3, 5, 7, etc.) are called **odd numbers.**

When you skip count, you say the multiples of a number. For example, skip counting by 3 gives the multiples of 3. The multiples of 3 are 3, 6, 9, 12, and so on. They are all the numbers that have rectangles with 3 rows.

9. Which numbers on the chart are multiples of 4 (have a rectangle with 4 rows)? List them from smallest to largest.

10. Which numbers on the chart are multiples of 5? List them from smallest to largest.

Prime Numbers

11. **A.** How many different rectangles can you make with 5 tiles?

    B. How many with 7 tiles?

Numbers that are larger than one and have only one rectangle have a special name. They are called **prime numbers.** For example, 5 and 7 are prime numbers.

12. List the prime numbers between 1 and 25.

13. Are all odd numbers prime? Explain.

Square Numbers

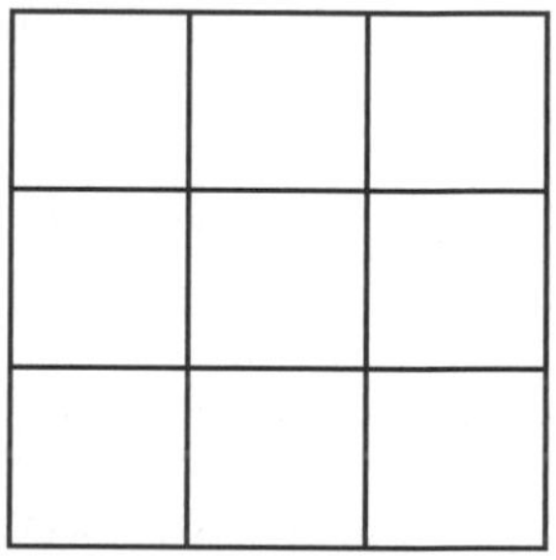

The number nine is special because it has a rectangle that is a square that has three rows and three columns.

14. Which other numbers have rectangles that are squares? These numbers are called **square numbers.**

15. Find the next largest square number after 25.

16. Another way mathematicians write 3×3 is 3^2. This is read "three to the second power" or "three squared." The raised 2 is called an **exponent.** Here are some more examples:

$$1^2 = 1 \times 1 = 1$$
$$2^2 = 2 \times 2 = 4$$
$$3^2 = 3 \times 3 = 9$$
$$4^2 = 4 \times 4 = 16$$

    A. What is 5^2?

    B. What is 6^2?

Arrays and Fact Families

17. My rectangle has a total of 18 square tiles. It has 3 rows of tiles.

   A. How many tiles are in each row? Write a number sentence for this rectangle.

   B. What are the other three sentences that are in this same family? Explain how all these sentences fit with this rectangle. Use the total number in each column in your explanation.

18. Another rectangle has 3 rows of tiles and a total of 24 square tiles.

   A. Write a number sentence to fit this rectangle.

   B. What are all the other number sentences in the same fact family?

19. **A.** Write a multiplication number sentence for a rectangle with 4 tiles in all and 2 tiles in each row.

   B. Can you write a different multiplication sentence for this rectangle? Why or why not?

   C. Write a division sentence for this rectangle.

   D. Can you write a different division sentence for this number?

20. A rectangle is made of 9 tiles and has 3 tiles in each row.

   A. How many different number sentences can you write for this rectangle?

   B. Look at the rectangles for the other square numbers. How many facts are in their fact families?

21. **A.** Write all the number sentences in the fact family for 5×2.

   B. Write all the number sentences in the fact family for 5^2.

You can draw pictures of rectangles on *Square-Inch Grid Paper* to help you solve these problems.

1. John built rectangles with 20 tiles, but some of his work was erased. Help John fill in the missing numbers.

Rectangles Possible with 20 Tiles

Number of Rows	Number in Each Row	Multiplication Sentence
1		$1 \times ? = 20$
	10	$? \times 10 = 20$
4		$4 \times ? = 20$
5		$5 \times ? = 20$
	2	$? \times 2 = 20$
20		$20 \times ? = 20$

2. **A.** Is 36 an even number? How do you know?

   B. Is 36 a square number? How do you know?

3. Find multiples by skip counting.

   A. Multiples of 2: Start at zero and skip count by 2s to 50.

   B. Multiples of 3: Start at zero and skip count by 3s to 48.

   C. Multiples of 5: Start at zero and skip count by 5s to 50.

   D. Multiples of 6: Start at zero and skip count by 6s to 48.

4. Tell whether the following numbers are even or odd.

   A. 10   **B.** 17   **C.** 21   **D.** 44

5. Jane says that any number that ends in 2, like 12, 72, and 102, is an even number. What other digits can even numbers end in?

6. **A.** Which of the following are multiples of 5?

20   34   45   56   60   73   35

B. Can you tell whether a number is a multiple of 5 by looking at the last digit? If so, tell what digits the multiples of 5 end in.

7. **A.** Which number in each of the following pairs is a multiple of 3?

11   21 (last digit 1)
12   22 (last digit 2)
23   33 (last digit 3)
14   24 (last digit 4)
15   25 (last digit 5)
16   36 (last digit 6)
17   27 (last digit 7)
18   28 (last digit 8)
39   19 (last digit 9)

B. Can you tell whether a number is a multiple of 3 by looking at its last digit? If so, tell what digits multiples of 3 end in.

8. Write the following multiplication problems using exponents. Then multiply.

A. 2×2 **B.** 5×5 **C.** 7×7 **D.** 10×10

9. Rewrite the following without using exponents. Then multiply.

A. 8^2 **B.** 3^2 **C.** 9^2

10. Ming has 32 rocks in his rock collection. He wants to buy a rectangular display box with one square compartment for each rock. At the store, he found boxes with the following designs:

- 6 rows and 6 columns
- 8 rows and 4 columns
- 2 rows and 16 columns
- 3 rows and 10 columns

Which boxes will hold his collection with no empty compartments?

Factors

Tile Problems

Use *Square-Inch Grid Paper* or tiles to help you solve these problems. Write a number sentence to go with each problem.

1. Shannon made a rectangle with 32 tiles. If there were 4 rows, how many tiles were in each row?

2. Roberto made a rectangle with 6 rows and 5 tiles in each row. How many tiles did he use?

3. Jackie made a rectangle with 21 tiles. There were 7 tiles in each row. How many rows were there?

4. A rectangle of 12 tiles has 3 different colors. There is an equal number of each color. How many of each color are there?

Finding Factors

Jacob wondered whether he could arrange 24 tiles into 5 rows. He used his calculator to divide 24 into 5 groups.

Hmm. $24 \div 5$ equals 4.8, so that means $4\frac{8}{10}$ tiles would go in each row. That's not possible, since I can't cut the tiles.

The **factors** of a number are the whole numbers that can be multiplied to get the number. For example, $3 \times 8 = 24$, so 3 and 8 are factors of 24. All the factors of 24 are 1, 2, 3, 4, 6, 8, 12, and 24, because we can multiply pairs of numbers to get 24 in the following ways: 1×24, 2×12, 3×8, and 4×6.

The factors of a number can also be described as the whole numbers that divide the number evenly. Two is a factor of 24 because $24 \div 2 = 12$. But 5 is not a factor of 24 because $24 \div 5 = 4.8$, which is not a whole number.

The factors of a number tell us which numbers of rows are possible in rectangles made with that number of tiles. Jacob couldn't make a rectangle with 5 rows and 24 tiles because 5 is not a factor of 24.

5. **A.** Is it possible to make a rectangle with 24 tiles and 6 rows? If so, how many tiles would be in each row? Use your calculator to check.

   B. Is it possible to make a rectangle with 24 tiles and 7 rows? Use your calculator to check. Explain.

6. **A.** Is 5 a factor of 38? Why or why not?

   B. Is it a factor of 35? Why or why not?

7. **A.** Is 8 a factor of 32? Why or why not?

   B. Is it a factor of 36? Why or why not?

8. The band leader at Coleman School wants to arrange the 48 members of its marching band into rows. He wants an equal number of students in each row.

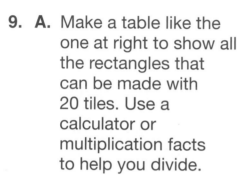

    A. Can he arrange them into 6 rows? Is 6 a factor of 48?

    B. Can he arrange them into 7 rows? Is 7 a factor of 48?

    C. Can he arrange them into 8 rows? Is 8 a factor of 48?

9. **A.** Make a table like the one at right to show all the rectangles that can be made with 20 tiles. Use a calculator or multiplication facts to help you divide.

    B. Use your table to help you list the factors of 20.

Rectangles Possible with 20 Tiles

Number of Rows	Number in Each Row	Division Sentence
1	20	$20 \div 1 = 20$
2	10	$20 \div 2 = 10$

    C. Look back at Question 1 in the Homework section on page 100. What do you notice about that table and this table? What is the same? What is different?

10. **A.** Make a table like the one in Question 9. Show all the rectangles that can be made with 36 tiles.

    B. Use your table to help you list the factors of 36.

11. Find the factors of:

    A. 12   **B.** 16   **C.** 18

A **prime number** is any number greater than one that has only two factors—itself and one. Thirteen is a prime number because its only factors are 13 and 1. Fourteen is not a prime number because it has four factors: 1, 2, 7, and 14.

12. Which of the following are prime numbers? Explain.

    A. 35   **B.** 27   **C.** 41

More Tile Problems

Use *Square-Inch Grid Paper* to help you solve these problems. Write a number sentence to go with each problem.

1. Irma made a rectangle with 28 tiles. If there were 7 rows, how many tiles were in each row?

2. Keenya made a rectangle with 8 rows and 5 tiles in each row. How many tiles did she use?

3. Romesh made a rectangle with 42 tiles. There were 6 tiles in each row. How many rows were there?

4. A rectangle of 18 tiles has 3 different colors. There is an equal number of each color. How many of each color are there?

More Finding Factors Problems

5. **A.** Make a table like the one at right. Show the rectangles that can be made with 28 tiles. You can use a calculator, multiplication facts, or *Square-Inch Grid Paper* to help you divide.

   B. Use the table to help you list the factors of 28.

Rectangles Possible with 28 Tiles

Number of Rows	Number in Each Row	Division Sentence
1	28	28 ÷ 1 = 28
2	14	28 ÷ 2 = 14

6. **A.** Make a table similar to the one above. Show the rectangles that can be made with 40 tiles.

   B. List the factors of 40.

7. **A.** Is 3 a factor of 27? How do you know?

   B. Is 7 a factor of 32? How do you know?

8. List all the factors of:

   A. 6

   B. 30

   C. 32

---

9. Help the Sunny Fruit Company design a rectangular-shaped box for shipping four dozen oranges. How many oranges are in four dozen? How many layers will your box have? How many rows of oranges will be in each layer? How many oranges will be in each row? (There is more than one way to pack four dozen oranges.)

10. Which of the following are prime numbers? How do you know?

    A. 39

    B. 51

    C. 67

11. Challenge question: Find all the prime numbers between 25 and 50. Explain what you did to find your answer.

Floor Tiler

Players

This is a game for two or more players.

Materials

- $\frac{1}{2}$ sheet of *Centimeter Grid Paper*
- *Spinners 1–4 and 1–10* Activity Page
- a clear plastic spinner or a paper clip and pencil
- a crayon or marker for each player

Rules

1. The first player spins twice so that he or she has two numbers. The player
    may either spin one spinner twice or spin each spinner once.

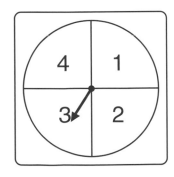 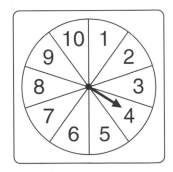

2. The player must then find the **product** of the two numbers he or she spun.
    For example, $3 \times 4 = $ **12**; 12 is the product. The product is the answer to a
    multiplication problem.

---

3. After finding the product, the player colors in a rectangle that has the same number of grid squares on the grid paper. For example, he or she might color in 3 rows of 4 squares for a total of 12 squares. But the player could also color in 2 rows of 6 squares or 1 row of 12 squares. (Remember, the squares colored in must connect so that they form a rectangle.)

4. Once the player has made his or her rectangle, the player draws an outline around it. He or she then writes its number sentence inside. For example, a player who colored in 3 rows of 4 squares would write "3 × 4 = 12." A player who colored in 2 rows of 6 squares would write "2 × 6 = 12."

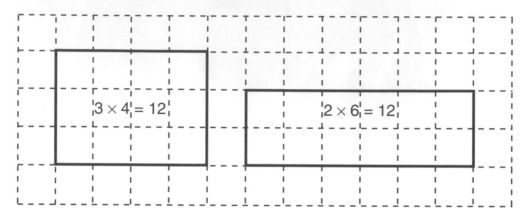

5. Players take turns spinning and filling in their grids.

6. If a player is unable to fill in a rectangle for his or her spin, that player loses the turn, and the next player can play.

7. The first player to fill in his or her grid paper completely wins the game.

Prime Factors

Multiplying Three Factors

Mrs. Dewey asked her class to multiply $3 \times 4 \times 5$.

David multiplied 3×4 first.

$$③ \times ④ \times 5$$
$$12 \times 5$$
$$60$$

Arti multiplied 4×5 first.

$$3 \times ④ \times ⑤$$
$$3 \times 20$$
$$60$$

Lin multiplied 3×5 first.

$$③ \times 4 \times ⑤$$
$$15$$
$$4 \times 15$$
$$60$$

1. When you find a product like $3 \times 4 \times 5$, you can only multiply 2 numbers at a time. It does not matter which two you multiply first. Multiply the following at least two different ways.

   A. $2 \times 2 \times 3 =$

   B. $2 \times 3 \times 3 =$

2. Find the products. You may use mental math, a calculator, or your multiplication tables.

   A. $2 \times 2 \times 5 =$ **B.** $2 \times 3 \times 5 =$

   C. $3 \times 3 \times 3 =$ **D.** $2 \times 5 \times 5 =$

   E. $4 \times 2 \times 5 =$ **F.** $4 \times 3 \times 2 =$

   G. $5 \times 6 \times 3 =$ **H.** $3 \times 6 \times 2 =$

Finding Prime Factors

Mrs. Dewey asked her class to write 24 as the product of at least three factors. Students had many different answers:

Luis:	$24 = 3 \times 4 \times 2$
Nicholas:	$24 = 3 \times 2 \times 2 \times 2$
Michael:	$24 = 6 \times 2 \times 2$
Nila:	$24 = 2 \times 3 \times 2 \times 2$

Which of these four ways show factors for 24 that are all prime?

The class noticed that all the factors Nicholas found were prime numbers. They are called the **prime factors** of 24. Nila found the same prime factors, but wrote them in a different order. There are many ways to factor a number, but only one way (not counting the order in which they are written) in which all the factors are prime numbers.

Nicholas showed how he found the prime factors of 24. He wrote: $24 = 3 \times 8$.

Then, he replaced the 8 with 4×2: $24 = 3 \times \mathbf{4} \times \mathbf{2}$.

Next, he replaced the 4 with 2×2: $24 = 3 \times \mathbf{2} \times \mathbf{2} \times 2$.

He stopped because none of the factors could be replaced. They were all prime.

Mrs. Dewey showed another way to write Nicholas's solution. She used a **factor tree.**

She factored 24 into 3 × 8. She circled the 3 because it cannot be factored anymore (it is prime). She factored 8 into 4 × 2 and circled the 2 because it is prime. She factored 4 into 2 × 2 and circled the 2s. She multiplied the circled numbers and got the same answer as Nicholas: 24 = 3 × 2 × 2 × 2.

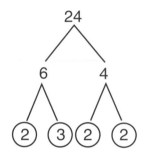

Nila decided to use a factor tree to show her solution. She factored 24 into 6 × 4. She decided not to write the multiplication signs in her factor tree. That was okay.

She factored 6 into 2 × 3 and 4 into 2 × 2. She circled the 2s and the 3 because they were prime; they could not be factored anymore.

She multiplied the prime numbers she had circled and got: 24 = 2 × 3 × 2 × 2.

Nila's answer was the same as Nicholas's, even though her factor tree was different.

3. John started the following factor tree for 24. Continue building his tree until all the numbers are prime. What factorization of 24 does your tree give you?

---

4. Complete the following factor trees for 36. Write 36 as a product of its prime factors.

A.
```
      36
     /  \
   18    2
```

B.
```
      36
     /  \
   12    3
```

C.
```
      36
     /  \
    9    4
```

D.
```
      36
     /  \
    6    6
```

5. Use factor trees to factor each of the following numbers into primes. Write number sentences to show your answers.

    A. 18

    B. 12

    C. 56

    D. 90

6. I am a prime number between 10 and 20. I am one more than a square number. What number am I?

7. I am a multiple of 3. I am a square number. I am less than 20. What number am I?

8. I am a multiple of 5. Two is not one of my factors. I am not prime, and I am not square. I am less than 30. What number am I?

Did You Know?

Some mathematicians study ways to factor large numbers. This is part of a branch of mathematics called number theory. Mathematicians study number theory because it is fun and interesting. Many of the discoveries that mathematicians made about number theory later turned out to be very useful. For example, factoring is important in making and breaking secret codes.

Exponents

Lee Yah factored 45 into prime factors. She wrote $45 = 5 \times 3 \times 3$. Linda found the same prime factors of 45, but she wrote them using an exponent: $45 = 5 \times 3^2$.

John factored 32. He wrote $32 = 2 \times 2 \times 2 \times 2 \times 2$. He wondered whether there was a shortcut for writing this. Mrs. Dewey showed how exponents can be used as a shortcut for writing products of the same factor:

$2 \times 2 \times 2 = 2^3$ (We read this as "2 cubed" or "two to the third power.")

Three is the **exponent.** Two is the **base.** The exponent tells us to multiply by 2 three times.

$2 \times 2 \times 2 \times 2 = 2^4$ (We read this as "two to the fourth power.")

$2 \times 2 \times 2 \times 2 \times 2 = 2^5$ (We read this as "two to the fifth power.")

How would you use this shortcut to write $3 \times 3 \times 3 \times 3$?

Jerome factored 72. He wrote $72 = 3 \times 3 \times 2 \times 2 \times 2$. Then, he wrote this with exponents: $72 = 3^2 \times 2^3$.

9. Rewrite the following factorizations using exponents:

   A. $600 = 2 \times 2 \times 2 \times 3 \times 5 \times 5$

   B. $378 = 2 \times 3 \times 3 \times 3 \times 7$

   C. $250 = 2 \times 5 \times 5 \times 5$

   D. $99 = 3 \times 3 \times 11$

10. Use exponents to rewrite each factorization you found in Question 5.

11. Write each of the following products without using exponents. Then multiply.

    A. $2^3 \times 3$

    B. $3^2 \times 5$

    C. 2×5^2

    D. $7^2 \times 2$

12. Find the prime factors of each of the following numbers. Write your answers, using exponents.

    A. 100

    B. 40

    C. 80

    D. 500

Homework

Dear Family Member:

This Homework section contains work with factors and multiplication. Background information about factors, prime numbers, and factor trees can be found in the previous pages of this section.

1. Find the products. You may use mental math, a calculator, or your multiplication table.

   A. $2 \times 3 \times 3 =$ **B.** $3 \times 2 \times 3 =$

   C. $2 \times 2 \times 2 =$ **D.** $3 \times 3 \times 5 =$

   E. $3 \times 3 \times 3 =$ **F.** $2 \times 2 \times 5 =$

   G. $3 \times 2 \times 5 =$ **H.** $7 \times 3 \times 2 =$

   I. $5 \times 4 \times 3 =$ **J.** $3 \times 4 \times 2 =$

2. Determine which of the following are prime numbers. If a number is prime, tell how you know. If it is not prime, write it as the product of prime factors.
   A. 6 **B.** 17
   C. 12 **D.** 39

3. Use factor trees to factor each of the following numbers into primes. Write multiplication sentences to show your answers.
   A. 20 **B.** 28 **C.** 60 **D.** 48
   E. 54 **F.** 72 **G.** 100 **H.** 42

4. I am a multiple of 2. I am not a multiple of 3. I am greater than 10 but less than 20. I am not a square number. What number am I?

5. I am between 6 and 35. I am one more than a square number. Five is one of my factors. What number am I?

6. I am the smallest square number that has the factors 2 and 3. What number am I?

7. Write your own number puzzle, similar to the ones in Questions 4–6. Use some of the following words: multiple, factor, prime number, square number.

8. Rewrite the following factorizations, using exponents:
   A. $200 = 2 \times 5 \times 5 \times 2 \times 2$
   B. $600 = 2 \times 3 \times 2 \times 5 \times 2 \times 5$
   C. $1200 = 2 \times 2 \times 3 \times 2 \times 2 \times 5 \times 5$
   D. $1500 = 2 \times 5 \times 2 \times 5 \times 5 \times 3$

9. Write each of the following products without exponents. Then multiply.
   A. $2^2 \times 3^2$ **B.** $3^3 \times 4^2$
   C. $2^2 \times 5^2$ **D.** $2^4 \times 3$

10. Find the prime factorizations of each of the following numbers. Write your answers, using exponents.
    A. 50 **B.** 66
    C. 96 **D.** 300

Product Bingo

Players

This is a game for five players.

Materials

- a clear plastic spinner or a pencil and a paper clip
- beans or other items to use as markers

Rules

1. One player is the Caller. The other players each choose a game board from the *Product Bingo Game Boards* Game Page in the *Discovery Assignment Book.*

2. The Caller spins twice. If the product of the spun digits is on your game board, then put a marker on it. The Caller should keep track of all digits spun by writing multiplication sentences on a piece of paper.

3. The first player with four markers in a row or a marker in each corner is the winner. (The **P** space, for **P**roduct, is a free space.)

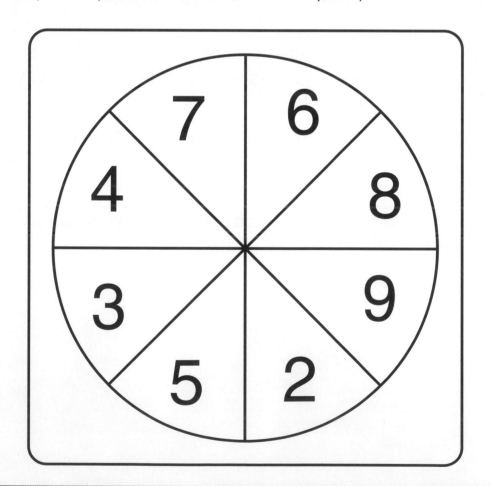

Answer these questions after you have played _Product Bingo._

1. Which game board is the best?

2. Which game board is the worst?

3. Why is the best game board better than the worst one?

4. Design your own _Product Bingo_ game board.

    A. First, draw an empty game board:

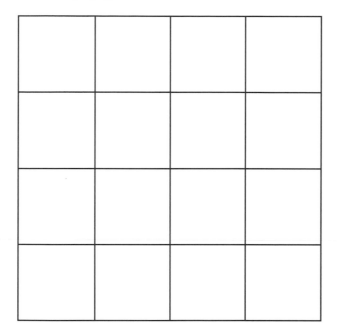

    B. Second, write products of the numbers 2 through 9 in all the squares except one.

    C. Finally, put a **P** in the last square.

5. Play _Product Bingo,_ using your game board. Did you win or not? Explain why.

Multiplying to Solve Problems

Solve the following problems. You may use paper and pencil, calculators, your multiplication tables, mental math, or manipulatives.

1. A movie theater has 20 rows with 10 seats in each row. How many seats are in the theater?

2. Twelve yards of fabric were used to make 4 dresses. How many such dresses can be made with 15 yards of fabric?

3. Jacob and his sister gave their mother a box of candy for Mother's Day. The box had 2 layers. Each layer had 5 rows with 8 pieces in each row. How many pieces of candy were in the box?

4. Jessie went to the Post Office to buy a sheet of 30 stamps. (Stamps usually come in rectangular sheets.) What are the possible rectangles of a sheet of 30 stamps? Tell how many rows and how many stamps are in each row.

5. Maya bought a box of paint at the hardware store. The box had 3 levels of gallons of paint with 3 rows of 4 gallons of paint in each row. How many gallons of paint were in the box?

6. **A.** In the morning, 80 new math books were delivered to the school in 10 identical boxes. How many books were in each box if each box had the same number of books?

   B. In the afternoon, 48 more math books were delivered in the same size boxes. How many boxes of books were delivered in the afternoon?

Multiplying to Solve Problems

Unit 5

Using Data to Predict

	Student Guide	Discovery Assignment Book	Adventure Book	Unit Resource Guide*
Lesson 1				
Predictions from Graphs	●	●		
Lesson 2				
Another Average Activity	●			
Lesson 3				
The Meaning of the Mean	●			●
Lesson 4				
Bouncing Ball	●			
Lesson 5				
Two Heads Are Better Than One			●	
Lesson 6				
Professor Peabody Invents a Ball				●
Lesson 7				
Speeds at the Indianapolis 500	●			
Lesson 8				
Midterm Test				●

Unit Resource Guide pages are from the teacher materials.

Predictions from Graphs

Graphs can tell stories. The following graph tells a story about the men's long jump competition in the Olympics. Contestants in the long jump try to jump as far as possible with a running start.

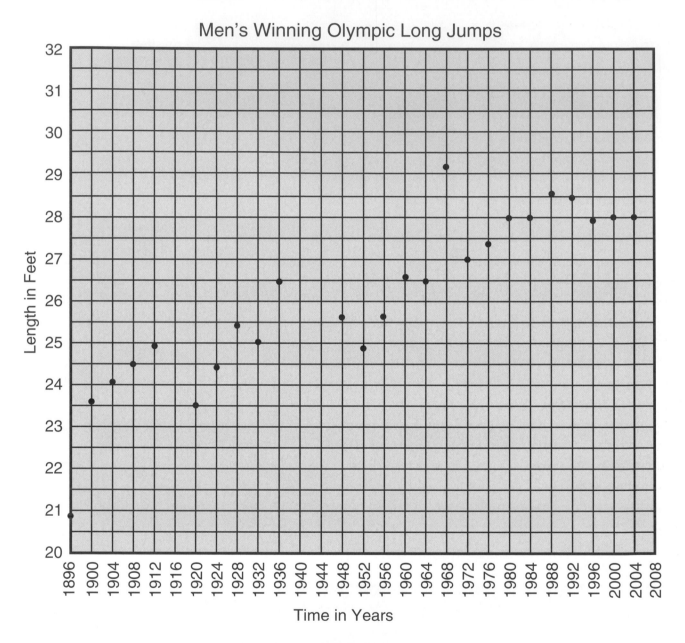

Men's Winning Olympic Long Jumps

1. **A.** What variable is on the horizontal axis?

   B. What variable is on the vertical axis?

2. Jesse Owens won the long jump competition in 1936.

   A. How far did he jump?

   B. Is the distance Jesse Owens jumped longer or shorter than the length of your classroom?

   C. How many years passed before someone jumped farther than Jesse Owens in the Olympics?

3. A. Describe the graph. What does it look like?

   B. If you read the graph from left to right, do the points tend to go uphill or downhill?

   C. What does the graph tell you about the winning long jumps in the Olympics?

4. In 1968, Bob Beamon of the United States won the long jump competition.

   A. How far did Beamon jump?

   B. What is unusual about this point on the graph?

   C. Do you think the winner in 2008 will jump as far as Bob Beamon jumped in 1968? Why or why not?

Here is another graph. It shows the history of the mile run in college championship races. Runners do not run the mile anymore in these track meets because the distances are measured using the metric system. Contestants now run 1500 meters, which is a little shorter than a mile.

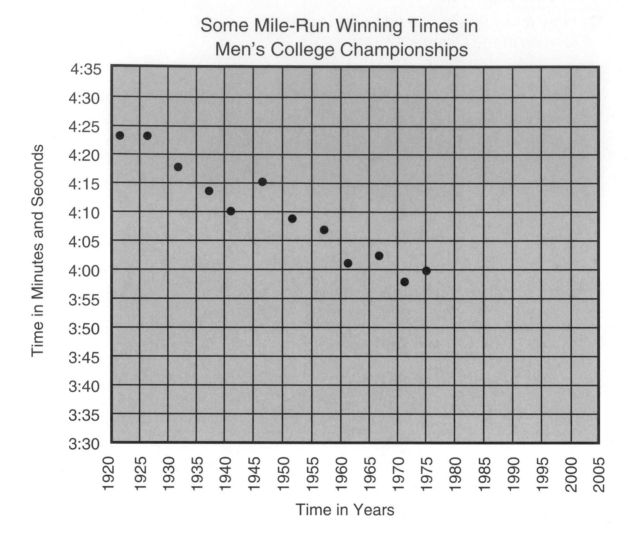

Some Mile-Run Winning Times in Men's College Championships

5. **A.** What variable is on the horizontal axis?

   B. What variable is on the vertical axis?

6. What was the winning time for running the mile in 1941?

7. Find the data point that shows a time for the mile race that is less than 4 minutes. What is the year for this data point?

8. **A.** Describe the graph. What does it look like?

   B. If you read the graph from left to right, do the points tend to go uphill or downhill?

   C. What does the graph tell you about the winning times for the mile run?

If the points on a graph lie close to a line, you can draw a line to help you make predictions. This line is called the **best-fit line.**

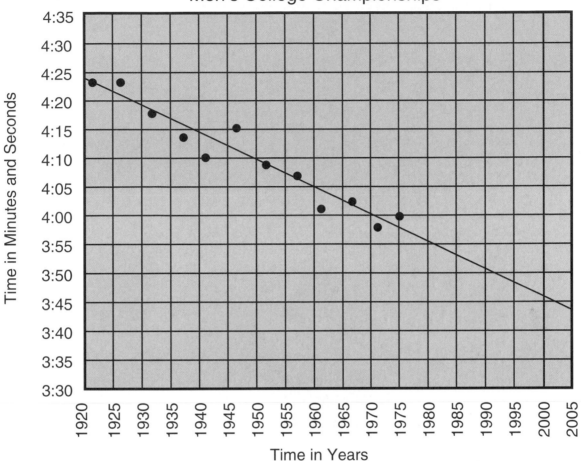

Some Mile-Run Winning Times in Men's College Championships

Time in Minutes and Seconds (y-axis)

Time in Years (x-axis)

9. **A.** Why do you think the line drawn on the graph is called a best-fit line?

    B. How many points on the graph are above the line?

    C. How many points are on the line?

    D. How many points are below the line?

10. Use this graph to estimate the winning time for the mile run in 1955.

11. Predict the winning time for the mile run in the year 2005. Explain how you made your prediction.

Using the graph to estimate the distances between two points on the graph is called **interpolation.** "Inter" means between points.

Using the graph to predict the distances beyond the data points on the graph is called **extrapolation.** "Extra" means beyond or outside the points on the graph.

12. **A.** Did you use interpolation or extrapolation to estimate the winning time in 1955?

  B. Did you use interpolation or extrapolation to predict the winning time in 2005?

  C. Which is more accurate? Explain.

Predictions from Graphs

Another Average Activity

Every week the students in Room 204 take a spelling test of 10 words. Each student records the number of words he or she spells correctly in a table. Mrs. Dewey reports the average score to the parents. Here are Ming's scores:

Ming's Scores

Test	Words Correct
Test #1	7
Test #2	9
Test #3	7
Test #4	7
Test #5	10

My median score is 7 words correct, so that is my average score. I guess you could say that 7 is a typical score, but it seems that the 9 and 10 should help bring my grades up.

Discuss

1. **A.** Do you agree that 7 words correct is Ming's median score?

   B. How do you find the median of a set of numbers?

2. In Unit 1, you learned that an **average** is one number that represents a set of data. For this data, the median number of words correct is 7. Is 7 a good number to represent all of Ming's scores? Why or why not?

3. Averages can also be used to make predictions. Do you think 7 words correct is a good prediction for the typical score on Ming's next five spelling tests? Why or why not?

The **median** is a useful average because it is often easy to find. Since it is the number that is exactly in the middle of the data, it can be used to describe what is normal or typical for that data. However, we can also use another kind of average to represent a set of data. This average is called the **mean.**

Mrs. Dewey showed Ming how to use the mean to average his spelling scores. She used connecting cubes. She said, "Each cube represents one spelling word. Make a tower of cubes to represent each of your spelling test scores. For example, the first tower will have 7 cubes because you spelled 7 words correctly on that test."

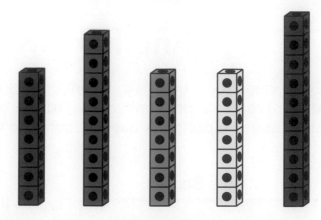

After Ming made these five towers, Mrs. Dewey said, "Use just the cubes in your towers. Even them out so that each of the five towers has the same number of cubes."

When the towers are evened out, the number of cubes in each of your towers is the mean.

Then my mean score is 8, and I can say that my average score is 8.

4. Is 8 a good number to represent all of Ming's scores? Why or why not?

5. How did the scores of 9 and 10 affect the mean score for Ming's tests?

Another Average Activity

Mrs. Dewey showed the students in Room 204 how to use connecting cubes to find the median and the mean. Students worked in pairs to complete the activity. Irma and Tanya's work is described on the following pages. Work with a partner to follow their example.

Irma and Tanya took turns pulling a handful of cubes from a paper bag and then building towers with the cubes. Together, they made the five towers shown here.

6. With your partner, build towers with the same numbers of cubes as in the picture.

Mrs. Dewey asked students to use one number, an average, to describe all the towers. One way to do this is to find the median. To do this, the girls lined up their towers from shortest to tallest.

7. Line up your towers as shown in the picture.

8. The number of cubes in the middle tower is the median. What is the median number of cubes?

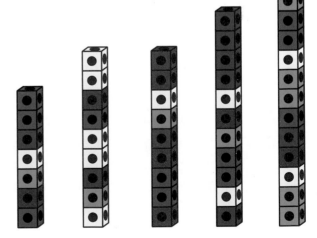

To find the mean, the girls tried to "even out" the towers so that they all had the same number of cubes.

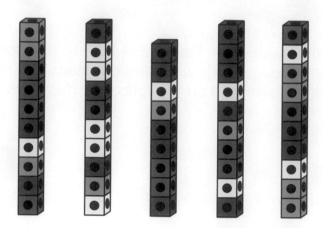

9. Even out your towers. (*Hint:* You can only use the cubes that are already in your towers, and you must keep the same number of towers.)

10. Now each tower has about the same number of cubes. This number is the mean. What is the mean?

11. The median number of cubes in Tanya and Irma's towers was nine. Do you agree that nine cubes is a normal handful for Irma and Tanya? Why or why not?

12. The mean number of cubes in the girls' towers was ten. Do you agree that this number can also describe a normal handful? Why or why not?

13. Scientists use averages to make predictions. Predict the number of cubes Tanya or Irma would pull, if they pulled another handful.

14. With your partner, complete the same activity that Irma and Tanya did.

   A. Return the cubes to your bag. Be sure they are all separated.

   B. Pull out one handful of cubes and build a tower with those cubes. Take turns with your partner until you have built five towers.

   C. Find the median number of cubes in your towers.

   D. Draw a picture of your towers. Record the median on your drawing.

   E. Find the mean number of cubes. Record the mean on your drawing. (Remember, you must keep the same number of towers. Do not add any more cubes from the bag or put any cubes back in the bag.)

Using Cubes to Solve Problems

Solve the following problems. Use towers of cubes to help you.

15. Jacob surveyed five families on his block. He filled in the following data table. Jacob used towers to find the median and the mean number of people in a household.

    A. Jacob built five towers. Why?

    B. What did each tower stand for?

    C. What did each cube stand for?

    D. Use towers to find the median.

    E. Use towers to find the mean. (*Hint:* Use the closest number for your answer.)

Jacob's Data

Family	Number of People in Household
Scott-Haines	2
Thomas	6
Molina	3
Chang	5
Green	3

16. When Mrs. Dewey was in the fourth grade, she took a math quiz each week. Every quiz had ten problems. She got 10 problems right the first week, 7 problems right the second week, 8 problems right the third week, and 4 problems right the fourth week. Use towers to find the median and mean.

    A. How many towers will you build?

    B. What does each tower stand for?

    C. What does each cube stand for?

    D. Find the median. (*Hint:* What number is halfway between the number of cubes in the middle two towers?)

    E. Even out the towers to find the mean. (You must use the same number of towers as in 16A.)

17. When Rita, a new Girl Scout leader, was introduced to her group, she asked the girls their ages. Keenya, Shannon, Ana, Grace, and Maya all said they were 10 years old.

    A. What one number can be used to describe the age of the girls?

    B. What is the median age for this group of Girl Scouts?

    C. What is the mean?

18. How are the two kinds of averages alike? (*Hint:* What do they tell us?)

19. How are the mean and the median different? (*Hint:* How do you find each one?)

20. Which average is easier to find?

21. Roberto rolled a toy car down a ramp and measured the distance it rolled. The first time it rolled 13 cm; the second time it rolled 14 cm; the third time it rolled 18 cm.

    A. Find the median distance the car rolled.

    B. Find the mean.

    C. Which number, the mean or the median, better describes the distance the car rolls? Explain your answer.

Dear Family Member:

In class, students used towers of cubes to learn how to find two kinds of averages: the mean and the median. You can look back at the previous pages in this section to see how this is done. In the next lesson, students will learn how to compute an average using calculators.

Use pennies or small building blocks to build towers to solve the following problems. You will need about 30 pennies or blocks.

1. Linda counted the number of plants her mom has in each room in the house. She filled in the following data table.

    A. Find the median number of plants in the house.

    B. Find the mean.

Linda's Data

Room	Number of Plants
Kitchen	5
Living Room	8
Family Room	7
Linda's Room	1
Bathroom	2
Dining Room	2
Mom's Room	3

2. John wanted to see how many free-throws he could make in one minute. The first minute he made 6 baskets. The second minute he made only 3. The third minute he made 6 baskets again. The fourth minute he made 9.

  A. Find the median number of baskets.

  B. Find the mean.

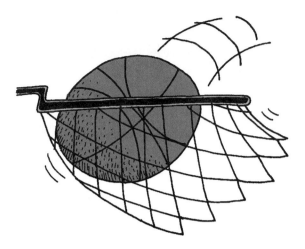

3. The students in Mrs. Dewey's class record the number of books they read each week. Here is Jerome's data for five weeks.

Jerome's Data

Week	Number of Books
Week #1	3
Week #2	2
Week #3	5
Week #4	2
Week #5	2

  A. Find the median.

  B. Use towers to find the mean. Give your answer to the nearest whole book.

  C. Which average (the median or the mean) do you think better represents the number of books that Jerome read? Why?

The Meaning of the Mean

The students in Mrs. Dewey's class are preparing for the Bessie Coleman School Olympic Day. They want to wear sweatbands during the races. To order the sweatbands, they need to know the average head circumference of students in the class.

The class first worked in groups of four students. They measured the distance around each person's head. Then they found the mean circumference for each group.

Finding the Mean Circumference

Work with a group of four people to complete the same activity:

Measure head circumference with strips of paper:

- For each group member, wrap adding machine tape around the student's head.
- Mark the distance around the head with a crayon.
- Cut the adding machine tape at the crayon mark to make a strip that is the same length as the circumference of the student's head.
- Check the strip. Wrap it around the student's head again. The ends of the strip should just touch each other.

- Measure to the nearest centimeter.
- Make a data table like the one shown.

Group 1's Data

Name	Circumference of Head (in cm)
Michael	52 cm
Ming	56 cm
Shannon	50 cm
Roberto	54 cm

Even out the strips to find the mean:

- Tape the four strips together end-to-end to make one long strip. Be careful not to let the strips overlap.

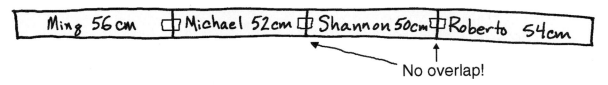

No overlap!

- Fold the long strip into four equal parts. (Fold the strip in half, then in half again.)

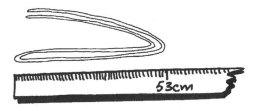

- Measure the length of one-fourth of the long strip to the nearest centimeter. This length is the mean.
- Write the mean for your group at the bottom of your data table.

The students in Group 1 finished evening out the length of their strips. They found that one-fourth of the long strip measured 53 cm. This is the mean length of their strips. They reported to the class that, on average, the circumference of their heads is 53 cm.

Discuss

1. Report to the class the average circumference of the heads of the students in your group. Give the mean.

2. Compare the means for all the groups. What can you say about them?

3. Estimate the average circumference for the whole class, using the means for each group.

4. **A.** How would you find the mean if there were five people in your group?
   B. Three people?

5. How could you use a calculator to find the mean? (*Hint:* What steps did you go through when using the strips to find the mean?)

Using a Calculator to Find the Mean

Michael's group used a chart to think through its answer to Question 5.

Finding the Mean

Steps with adding machine tape	Steps on the calculator
1. We taped the strips together.	1. Add the lengths of the strips together. 56 + 52 + 50 + 54 = 212 cm
2. We folded the long strip into 4 equal parts.	2. Divide the total length by the number of people in our group. 212 ÷ 4 = 53 cm
3. The length of one-fourth of the long strip is 53 cm.	3. The mean is 53 cm. On average, the circumference of our heads is 53 cm.

These are the keystrokes that Michael's group used. Try them on your calculator.

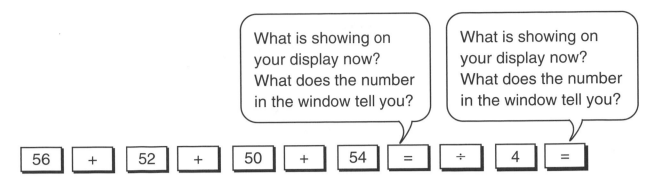

What is showing on your display now? What does the number in the window tell you?

What is showing on your display now? What does the number in the window tell you?

56 + 52 + 50 + 54 = ÷ 4 =

Michael's group found the mean by adding the values for each head circumference and dividing by the number of students in the group. The **mean** for any data set is an average that is found by adding the values in the set of data and dividing by the number of values.

6. The data for Group 2 is shown below.

Group 2's Data

Name	Circumference of Head (in cm)
Jacob	60 cm
Tanya	57 cm
Maya	56 cm
Jessie	54 cm

They used a calculator to find the mean for their data. The display on their calculator read 56.75.

A. Write calculator keystrokes for finding the mean for Group 2.

B. Is 56.75 cm closer to 56 cm or 57 cm?

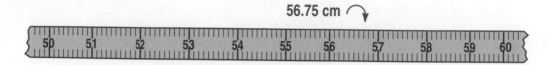

56.75 cm

C. Give the mean to the nearest whole centimeter.

7. A. Use your data table and your calculator to find the mean circumference for your group. Give your answer to the nearest centimeter. (*Hint:* Use a meterstick to help you find the nearest centimeter.)

B. Compare the number you found for the mean on your calculator to the number your group found by measuring the folded strip. Are the numbers close?

8. Groups 3 and 4 only had three students each.

A. Find the mean for each data set. Give your answer to the nearest centimeter.

B. Show the calculator keystrokes you used to find the mean.

C. Find the median for each set of data.

Group 3's Data

Name	Circumference of Head (in cm)
John	58 cm
Jerome	59 cm
Roberto	54 cm

Group 4's Data

Name	Circumference of Head (in cm)
Keenya	53 cm
Grace	53 cm
Ana	51 cm

9. Each day for a week, students in Room 204 recorded the outside temperature at noon.

   A. Find the mean temperature. Show your calculator keystrokes. Give your answer to the nearest whole degree.

   B. Find the median temperature.

Temperature Data

Day	Temperature at Noon in °F
Monday	47°
Tuesday	38°
Wednesday	37°
Thursday	43°
Friday	46°

10. In the first six soccer games of the season, Jackie's team scored 2, 3, 4, 0, 1, and 2 goals.

    A. Find the mean number of goals.

    B. Find the median number of goals.

    C. Look back. Do your answers make sense? Are the averages you found typical scores for Jackie's team?

11. Each week, a fourth-grade class has a test on 20 spelling words. A student got 13 right the first week, 19 right the second week, 12 right the third week, 20 right the fourth week, and 11 right the fifth week.

    A. On average, how many words did the student get right?

    B. Did you use the median or the mean? Why?

You will need a calculator to complete this homework.

1. The data table for a group of students from the experiment *Arm Span vs. Height* is shown below.

   A. Find the mean arm span for Jerome's group to the nearest inch. Show your calculator keystrokes.

   B. Find the median arm span.

   C. Find the mean height to the nearest inch.

   D. Find the median height.

   E. Look back at your answers. Do they make sense? Are these typical arm spans and heights for fourth graders?

Jerome's Group

Name	S Arm Span (in inches)	H Height (in inches)
Jerome	49	50
Keenya	54	55
Frank	59	57
Luis	58	58
Roberto	55	57

2. The students in Room 204 collected data on the number of times students have moved. Here is the data that one group collected: Shannon has moved 7 times, Linda has moved 3 times, Grace has moved 0 times, and Romesh has moved 2 times. Shannon said, "The average number of times we have moved is 12 times." Is Shannon correct? Why or why not?

3. John, Jackie, Nicholas, Irma, Maya, and Michael all walk to school together.

   John lives 1 block from school.
   Nicholas lives 2 blocks from school.
   Maya lives 4 blocks from school.

   Jackie lives 1 block from school.
   Irma lives 2 blocks from school.
   Michael lives 8 blocks from school.

   A. Find the median number of blocks the students live from school.

   B. Find the mean number of blocks.

   C. Michael lives much farther away from school than the other children. How does that affect the mean?

Bouncing Ball

Ana and Tanya wanted to play jacks at lunch time. Tanya brought jacks from home, but forgot the ball. Mrs. Dewey told the girls that they could look for a ball in the closet where the class keeps playground equipment. They found an old tennis ball and a Super Ball.

Ana said, "I'll use the Super Ball, and you can use the tennis ball."

"No, that wouldn't be fair," said Tanya. "The balls won't bounce the same. We have to use the same ball."

"Oh, you're right. Which one should we use?" Ana began to bounce the balls to try them out. "They both bounce pretty well when I drop them from here; but when we play jacks, we are sitting down."

Mrs. Dewey said, "You girls better go outside and get your game started or lunch will be over. Experiment with both of the balls before you start to play. You've given me an idea for an experiment we can do in class."

After lunch, Mrs. Dewey asked Tanya and Ana to let her have the tennis ball and the Super Ball. She asked the class, "Can you predict how high the tennis ball will bounce if I drop it from 1 meter?"

Students answered:

"One meter."

"Yeah, one meter."

"Half as high. 50 centimeters."

"Higher than a meter."

Mrs. Dewey said, "Are you guessing? Or can you give a reason for your answers? How can we make accurate predictions?"

Jessie said, "We'd have to try it out. We'd have to write down where we want to drop the ball. Then drop the ball and measure how high it bounces."

Mrs. Dewey said, "Here is the challenge: If I give you a drop height, can you predict the bounce height? Can you make a prediction that is close to the actual bounce height?"

You will use the TIMS Laboratory Method to carry out two experiments—one with a tennis ball and one with a Super Ball. Using the results of your experiments, you should be able to accurately predict the bounce height of either ball when you know the drop height.

Identifying the Variables

One of the first tasks in setting up an experiment is identifying the variables. The two main variables in these two experiments are the drop height (D) and the bounce height (B).

We have special names for the two main variables in any experiment. The variable with values we know at the beginning of the experiment is called the **manipulated variable.** We can often choose the values of the manipulated variable. The variable with values we learn by doing the experiment is called the **responding variable.**

The values for the manipulated and responding variables change during an experiment. In these experiments, good values for the drop height (D) are 40 cm, 80 cm, and 120 cm. Each time you change the drop height, you measure the bounce height (B).

Each experiment should tell you how much the bounce height changes when you change the drop height. Usually there are other variables involved in an experiment. Look at the experimental setup. Try bouncing a ball on the floor a few times. Other than the drop height, what could change the bounce height? List some variables.

The variables in your list should remain the same during an experiment, so the only thing that affects the bounce height is the drop height. These variables are called **fixed variables.** The results of a carefully controlled experiment will help you make accurate predictions.

1. What is the manipulated variable in both the tennis ball experiment and the Super Ball experiment?

2. What is the responding variable?

3. What are the fixed variables in each of the experiments?

4. **A.** Is the bounce height a categorical or numerical variable?

   B. Is the drop height a categorical or numerical variable?

   C. Is the type of ball a categorical or numerical variable?

5. Mrs. Dewey's class dropped the ball three times from each drop height. They measured the bounce height each time. Why is it a good idea to do three trials?

Draw a picture of the lab. Show the tools you will use. Be sure to label the two main variables. A student from another class should be able to look at your picture and know what you are going to do during the lab.

Work with your group to collect the data for each experiment. You will need two data tables, one for each type of ball.

- Tape two metersticks to the wall. Your teacher will show you how.
- Fill in the values for the drop height before starting. Follow the example.
- Do three trials for each drop height. Find the average bounce height for each trial. Record the average in the data table.

Tennis Ball

D Drop Height (in cm)	B Bounce Height (in cm)			
	Trial 1	Trial 2	Trial 3	Average
40				
80				
120				

- Make a point graph of your data for each experiment. Use two pieces of graph paper.
- Put the drop height (D) on the horizontal axis. Put the bounce height (B) on the vertical axis.
- Use the same scales on both graphs. Leave room for extrapolation.
- Remember to title the graphs, label the axes, and include units.
- Plot the average bounce height for each drop height.

6. **A.** If the drop height were 0 cm, what would the bounce height be?

    B. Put this point on your graphs.

7. Describe your graphs. Do the points lie close to a straight line? If so, use a ruler to draw best-fit lines.

8. Suppose you drop your tennis ball from 60 cm.

    A. Use your graph to predict how high it will bounce. $D = 60$ cm, predicted $B = ?$ Show your work using dotted lines on your graph.

    B. Did you use interpolation or extrapolation to find your answer?

    C. Check your prediction by dropping the tennis ball from 60 cm. What is the actual bounce height? $D = 60$ cm, actual $B = ?$

    D. Is your prediction close to the actual bounce height?

9. Suppose you want your tennis ball to bounce 75 cm.

    A. From what height should you drop it? $B = 75$ cm, predicted $D = ?$

    B. Did you use interpolation or extrapolation to find your answer?

    C. Check your prediction by dropping the tennis ball from your predicted drop height. What is the actual bounce height?

    D. Was the actual bounce height close to 75 cm?

10. Suppose you drop your tennis ball from 180 cm.

    A. Predict the bounce height. $D = 180$ cm, predicted $B = ?$

    B. How did you make your prediction?

    C. Check your prediction by dropping the tennis ball from 180 cm. What is the actual bounce height? $D = 180$ cm, actual $B = ?$

    D. Is your prediction close to the actual bounce height?

11. Suppose you drop your Super Ball from 1 meter.

   A. Use your graph to predict the bounce height. $D = 1$ m, predicted $B = ?$

   B. Did you use interpolation or extrapolation to find your answer?

   C. Check your prediction by dropping the Super Ball from 1 m. What is the actual bounce height? $D = 1$ m, actual $B = ?$

   D. Is your prediction close to the actual bounce height?

12. Suppose you want your Super Ball to bounce exactly 2 m. From what height should you drop the ball? Explain how you found your answer.

Discuss

13. Compare the graph for the tennis ball with the graph for the Super Ball. How are they alike? How are they different?

14. You find a strange ball on the playground. Because you have been investigating bouncing balls, you drop the ball from a height of 50 cm. It bounces back to a height of 18 cm. Is it more like the tennis ball or the Super Ball? How did you find your answer?

15. Maya brings in a ball which is not as bouncy as a tennis ball. Is the line for Maya's ball Line X or Line Y?

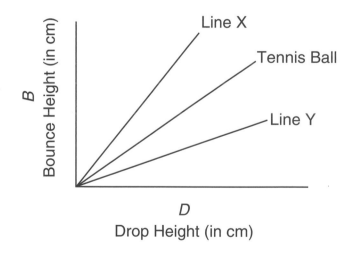

You will need a sheet of *Centimeter Graph Paper* **and a ruler to complete this homework.**

Here are the results of an experiment using a ball Frank found on the way to school:

1. Make a point graph of this data. Put the drop height (*D*) on the horizontal axis. Put the bounce height (*B*) on the vertical axis. The scale on the horizontal axis should go to at least 150 cm. The scale on the vertical axis should go to at least 100 cm.

2. **A.** If the drop height were 0 cm, what would be the bounce height?

   B. Put this point on your graph.

3. Draw a best-fit line.

4. Frank dropped his ball from 150 cm.

   A. Use your graph to predict the bounce height of the ball. Show how you found your answer on your graph. *D* = 150 cm, predicted *B* = ?

   B. Did you use interpolation or extrapolation to find your answer?

5. Frank dropped his ball and it bounced 25 cm.

   A. From what height was it dropped? Show how you found your answer on your graph. *B* = 25 cm, predicted *D* = ?

   B. Did you use interpolation or extrapolation to find your answer?

6. Frank wants his ball to bounce to a height of 100 cm. From what height should he drop the ball? Explain how you found your answer.

Frank's Data

D Drop Height in cm	B Bounce Height in cm
30	11
60	18
120	44

Speeds at the Indianapolis 500

The Indianapolis 500 is a famous car race that takes place every Memorial Day weekend in Indianapolis, Indiana. The cars race around an oval track until they have gone 500 miles. The graph shows the average speed in miles per hour for the winner of each race.

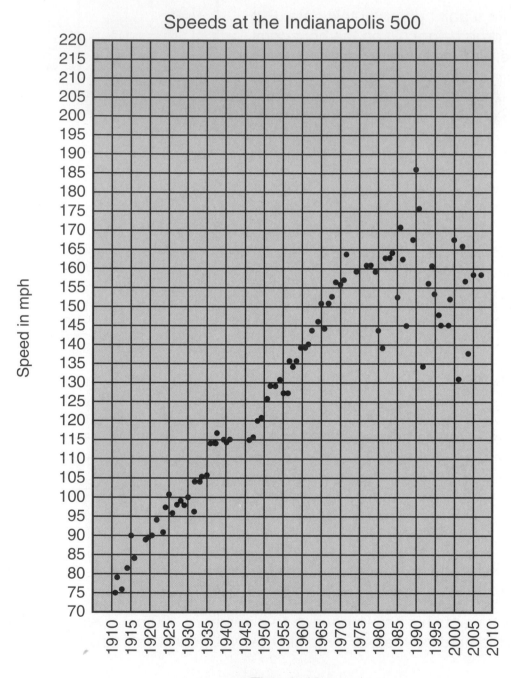

Speeds at the Indianapolis 500

The drivers go as fast as possible unless they are given a yellow light. This is a signal that the track is dangerous due to an accident or rain. The cars must slow down and maintain their race position until the track is safe again. When the drivers must drive slower, their average race speeds go down.

1. In 1920 the winner of the race, Gaston Chevrolet, won the race with a winning speed of 89 miles per hour. Fifty years later, Al Unser won with a winning speed of 156 miles per hour. What is the difference in the two speeds?

2. In 1993 the winning speed was 157 miles per hour.
   A. In 1911 the winning speed was 75 miles per hour. About how many times faster did the winner drive in 1993 than in 1911?
   B. The speed limit on freeways in cities is usually 55 miles per hour. About how many times faster did the winning 1993 car travel during the race than a car travels on a freeway?

3. When Jessie's family went on a trip, they drove about 50 miles each hour. How long did it take Jessie's family to drive 500 miles?

4. There were no races in 1917 or 1918 during World War I.
   A. Can you use the graph to estimate the winning speed if there had been a race in 1917? If so, what is your estimate?
   B. Did you use interpolation or extrapolation to make your estimate?

5. There were no races from 1942–1945 during World War II.
   A. Can you use the graph to estimate the winning speed if there had been a race in 1943? If so, what is your estimate?
   B. Did you use interpolation or extrapolation to make your estimate?

6. Can you use the graph to make an accurate prediction about the winning speed in 2010? Why or why not?

   A. If so, what is your prediction?

   B. Did you use interpolation or extrapolation?

7. Write a short paragraph that tells the story of the graph. In your paragraph, describe the graph. What does the graph tell you about the speeds of the winning cars over the years?

8. Here is part of the graph. This part shows the winning speeds from 1980 to 2005.

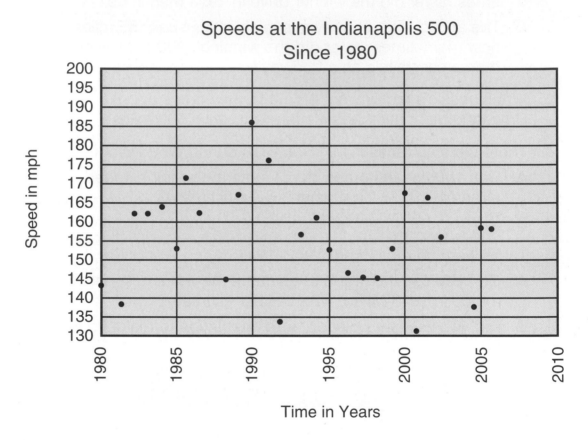

   A. Describe this part of the graph.

   B. Can you make predictions about the speed of the winner using only this data? Why or why not?

Place Value Patterns

	Student Guide	Discovery Assignment Book	Adventure Book	Unit Resource Guide*
Lesson 1				
Newswire	●	●		
Lesson 2				
Doubles	●			
Lesson 3				
Big Base-Ten Pieces	●	●		
Lesson 4				
News Number Line	●			
Lesson 5				
Close Enough	●	●		
Lesson 6				
Using Estimation	●			●
Lesson 7				
9 to 5 War	●	●		

Unit Resource Guide pages are from the teacher materials.

Newswire

Mrs. Dewey's class used a place value chart to compare numbers they found in newspaper articles. Using the place value chart helped them put the numbers in order on a wire.

Shannon found two large numbers. She wrote the numbers on her place value chart.

Millions' Period			Thousands' Period			Ones' Period		
4	2	3	1	7	6	3	2	1
	4	5	2	1	8	7	0	3

Then she read the first number out loud, "Four hundred twenty-three million, one hundred seventy-six thousand, three hundred twenty-one." Shannon noticed that there is a repeating pattern. Can you describe the pattern?

Each repeating core pattern is called a **period** on the *Place Value Chart.*

- The first three-digit group on the right is the **ones' period.** It is made up of ones, tens, and hundreds.

- The second group is the **thousands' period.** It is made up of thousands, ten thousands, and hundred thousands.

- The third group is the **millions' period.** It is made up of millions, ten millions, and hundred millions.

When you say a number, you say each period just the way you do when you read a number in the ones' period and then you add the name of the period. For example, to read Shannon's second number, say, "Forty-five million, two hundred eighteen thousand, seven hundred three." When writing numbers, place a comma or a space between each period to make reading easier: 45,218,703.

Jacob found these numbers in the *National Parks Gazette*:

National Parks Numbers

Crater Lake in Oregon **183,224 acres**	Kobuk Valley National Park in Alaska **1,750,736 acres**	Glacier National Park in Montana **1,013,572 acres**	Canyonlands National Park in Utah **337,570 acres**	Voyageurs National Park in Minnesota **218,200 acres**
Extraordinary blue lake in crater of extinct volcano encircled by lava walls 500 to 2000 feet high.	Caribou and black bears; archaeological sites indicate that humans have lived there for more than 10,000 years.	Superb Rocky Mountain scenery, numerous glaciers, and glacial lakes.	At junction of Colorado and Green Rivers; extensive evidence of prehistoric Indians.	Abundant lakes, forests, wildlife, and canoeing.

1. Put the numbers in bold in order from smallest to largest on your *Place Value Chart*. Then write each number in word form on a separate sheet of paper.

   For example, the smallest National Park in this table is Crater Lake in Oregon.

Millions			Thousands			Ones		
			1	8	3	2	2	4

Crater Lake National Park has one hundred eighty-three thousand, two hundred twenty-four acres.

2. Which national park in the table National Parks Numbers has an area closest to 1 million acres?

3. Rhode Island has an area of about 777,000 acres.

   A. Which national park has an area that is about half the area of Rhode Island?

   B. Which national park has an area that is about twice the area of Rhode Island?

To complete the homework, you need to take home your *Place Value Chart*.

Nila found these numbers in the *National Parks Gazette*.

More National Parks Numbers

Yellowstone National Park in Idaho, Montana, and Wyoming	Olympic National Park in Washington	Gates of the Arctic National Park in Alaska	Glacier Bay National Park in Alaska	Big Bend National Park in Texas
2,219,791 acres	**922,651 acres**	**7,523,898 acres**	**3,224,840 acres**	**801,163 acres**
The first national park; has the world's greatest geyser area, spectacular waterfalls, impressive canyons, and bears and moose.	Mountain wilderness containing the finest remnant of Pacific Northwest rain forest, active glaciers, Pacific shoreline, and rare elk.	The largest national park. Huge tundra wilderness, with rugged peaks and steep valleys.	Rugged mountains, with glaciers, lakes, sheep, bears, and bald eagles.	Desert land amid rugged Chisos mountains on the Rio Grande River; dinosaur fossils.

1. Put the numbers in bold in order from smallest to largest. Then write each number in word form. Use your *Place Value Chart* as a guide if you need it.

2. Which two of these national parks have areas closest to 1 million acres?

3. Find one number written in symbols (not words) greater than 1000 in a newspaper or magazine. Cut out the paragraph that contains the number. Highlight or circle the number. Bring the paragraph with the number to school and place it on the class newswire.

> **Acre.** An acre is a unit of measure for area. We use acres to measure land area. An acre is equal to 43,560 square feet. There are 640 acres in a square mile. A football field is a little larger than one acre.

Doubles

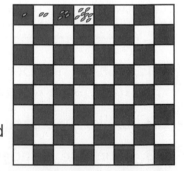

A Double Reward

There is an old story about the man who invented the game of chess. His name was Sissa Ben Dahir (da-here), and he was the Grand Vizier (viz-ear) of King Sirham of India. King Sirham was so pleased when Sissa Ben Dahir showed him his new game that he offered a great reward. "Choose your prize," said the king after he had played his first game of chess.

The vizier's reply seemed very foolish to the king: "Today, place one grain of wheat on the first square of the chessboard. Tomorrow, place two grains of wheat on the second square. On the third day, place four grains on the third square. Continue the pattern by doubling the number of grains each day. In this manner, give me enough grains to cover all 64 squares."

The king was glad to grant what he considered to be a small request, so he ordered a bag of wheat brought to the vizier.

However, when the counting began, it became clear that one bag of wheat was not nearly enough. On the first day, one grain of wheat was placed on the chessboard, two grains on the second day, four on the third day, and so forth. As the days passed, the king realized that he could not possibly keep his promise.

Some say Sissa Ben Dahir did not insist on receiving his full reward, and the king was again greatly impressed by his wisdom. Others say that the king was so angry that he could not fulfill the request that he cut off Sissa Ben Dahir's head.

1. Estimate how much wheat Sissa Ben Dahir asked for. More than 100 grains? More than 1000 grains? More than 1,000,000 grains?

2. **A.** One way to find out how much he requested is to make a data table. Copy the following data table in your journal or on a piece of paper. Continue the data table for eight days.

    B. Look for patterns.

Writing numbers using exponents in the second column may help you see more patterns. Each of the numbers in the second column of the data table are **powers of two.** For example, $2 \times 2 \times 2 = 2^3$ is read "two to the third power." We say that 2^3 is the "third power of two." Follow the examples to write the powers of two, using exponents in your data table. Use a calculator to help you. (*Hint:* You may need to stop writing $2 \times 2 \times 2 \ldots$ after several rows.)

Doubling Data Table

D Time in Days	N Number of Grains of Wheat Added	T Total Number of Grains of Wheat
1	1	1
2	$2 \times 1 = 2$	3
3	$2 \times 2 = 2^2 = 4$	7
4	$2 \times 2 \times 2 = 2^3 = 8$	15

3. Describe any patterns you see.

4. **A.** How many grains of wheat will be added on the eighteenth day?

   B. How many total grains of wheat are needed by the eighteenth day?

5. Use the patterns to help you predict when the total number of grains of wheat on the chessboard will reach 1 million.

6. Check your prediction. Continue your data table until the total number of grains of wheat reaches a million.

7. Make a point graph for the first two columns in your data table on *Centimeter Graph Paper*. Put the time in days (*D*) on the horizontal axis. Scale the horizontal axis by ones. Put the number of grains of wheat added each day (*N*) on the vertical axis. Scale the vertical axis by fours.

   A. Do the points form a straight line? If so, draw a best-fit line through the points.

   B. If the points do not form a line, describe the shape of the graph.

Use the *Solving* and the *Telling* Rubrics to help you organize your work for the following problems. Explain your strategies and show your work.

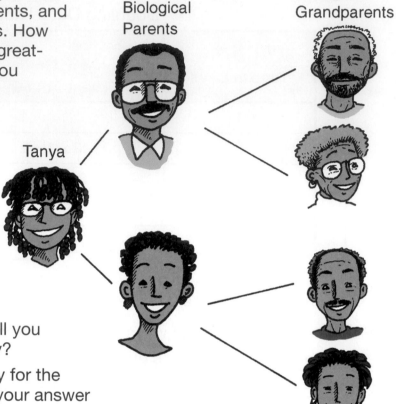

Biological Parents

Biological Grandparents

Tanya

1. You have two biological parents, four biological grandparents, and eight great-grandparents. How many great-great-great-great-great-grandparents do you have? Write this number using an exponent.

2. Suppose your pay for a job is one penny on the first day you work, two pennies on the second day, four pennies on the third day, eight pennies on the fourth day, etc.

   A. How much money will you earn on the tenth day?

   B. What is your total pay for the first ten days? (Give your answer in dollars and cents.)

   C. How long will you have to work to make a total of $1000?

3. You won the lottery! The lottery committee has given you a choice: get paid a cool $1 million in cash or get paid 1 cent on the first day, two on the second, four on the third, and so on for 1 month (30 days). What is your choice? Explain why you made the choice you did.

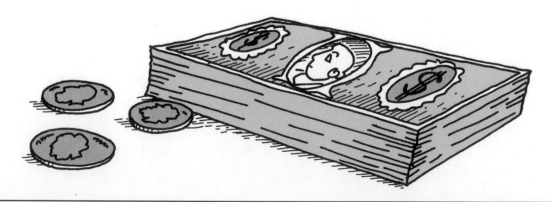

Big Base-Ten Pieces

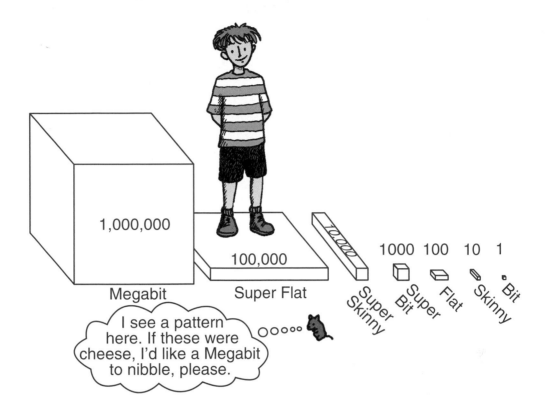

I see a pattern here. If these were cheese, I'd like a Megabit to nibble, please.

Patterns in the Base-Ten Pieces

The base-ten pieces are one model of the place value system. You have built models of base-ten pieces for numbers up to 1,000,000. We have given special names to the base-ten pieces to help us talk about the patterns in our base-ten number system. Starting in the ones' place we use the names: bit, skinny, flat, super bit, super skinny, super flat, and megabit.

1. What patterns do you see in the shapes of the base-ten pieces?

2. The sizes of the pieces also form a pattern.
   A. How many bits make a skinny?
   B. How many skinnies make a flat?
   C. How many flats make a super bit?
   D. Describe the pattern. Do all the pieces follow the pattern?

We can write the value of each piece, using the powers of 10. For example, $100 = 10 \times 10$, and it can be written as 10^2. This is read as 10 to the second power or 10 squared. The number $1000 = 10 \times 10 \times 10$, and it can be written as 10^3. This is read as 10 to the third power. The following chart helps to show these patterns.

3. Draw the chart on your paper and fill in the missing spaces.

Base-Ten Chart

Base-Ten Piece	Written as a Power of 10	Value
Bit	1	1
Skinny	$1 \times 10 = 10^1$	
Flat		100
Super Bit	$10 \times 10 \times 10 = 10^3$	
Super Skinny		10,000
Super Flat		
Megabit		

Each repeating core pattern is called a period on the *Place Value Chart.* The bit-skinny-flat group makes up the **ones'** period. The super bit-super skinny-super flat group makes up the **thousands'** period. The megabit begins the **millions'** period.

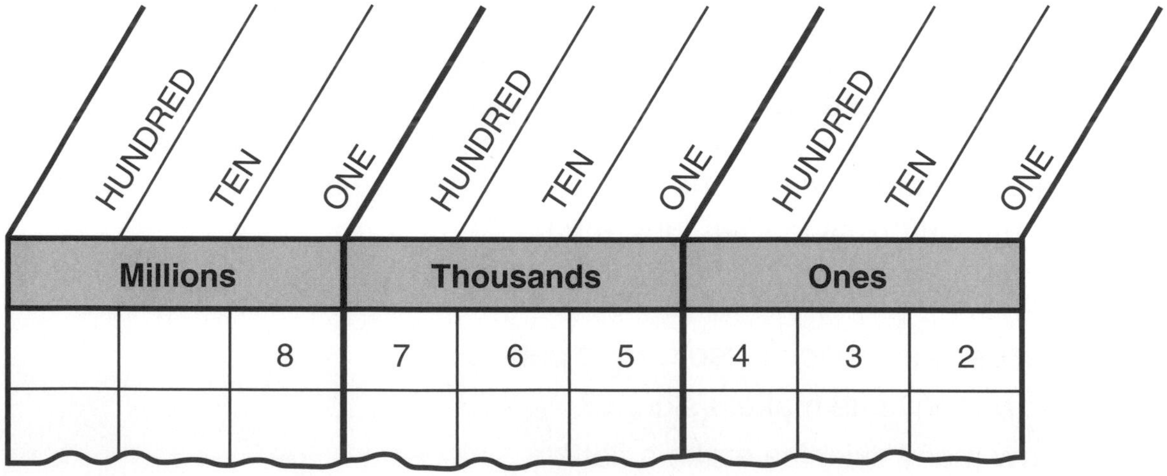

Millions			Thousands			Ones		
		8	7	6	5	4	3	2

Each period takes its name from the number that the cube represents in that period. In Lesson 1, you learned that a comma or space is placed between each period to make reading easier. Remember, the comma or space alerts you to say the period name. For instance: 8,765,432 is read as eight **million,** seven hundred sixty-five **thousand,** four hundred thirty-two.

Draw, Place, and Read

Play *Draw, Place, and Read*. Follow the directions that are written on the game page in the *Discovery Assignment Book*.

Tanya and Irma played *Draw, Place, and Read*. After all seven digit cards had been drawn, Tanya's number looked like this: 5,369,210. Irma's number looked like this: 6,935,021. Read each number.

4. Which girl recorded the larger number?

Play *Draw, Place, and Read* at home with your family.

News Number Line

How Big Is One Million?

1. Use your calculator to count by 10,000s to 1 million. If you count out loud, how many numbers will you say? Begin this way, "ten thousand, twenty thousand, thirty thousand, . . ."

2. Use your calculator to count by 100,000s to 1 million. If you count out loud, how many numbers will you say? Begin this way, "one hundred thousand, two hundred thousand, three hundred thousand, . . ."

3. A pencil manufacturer donates 1,000,000 pencils to your school. Your principal divides the pencils evenly among the students. How many pencils will each student get?

   A. What do you need to know before you can solve the problem?

   B. Use your calculator to find the answer.

Taking Attendance

Museum Attendance in 2001 and 2002

Museum	Attendance in 2001 and 2002	
	2001	2002
Science Museum	1,890,227	1,985,609
Institute of Art	1,338,266	1,295,321
Children's Museum	1,295,755	1,227,062
Aquarium and Oceanarium	1,858,766	1,789,222
Museum of the Stars	577,997	458,156

1. Put the 2001 attendance numbers in order from smallest to largest.

2. Name the most popular museum in 2001.

3. Which museum's attendance was closest to 1,000,000 in 2001?

4. Which museum had fewer visitors in 2001 than it had in 2002?

5. Put the 2002 attendance numbers in order from smallest to largest.

6. Name the least popular museum in 2002.

7. Write the 2002 attendance for each museum in words.

Making Mystery Jars

The next three problems prepare for the next lesson, *Close Enough*. If your class is not going to do Lesson 5, you may skip these questions.

8. Here is a picture Linda made using a computer.

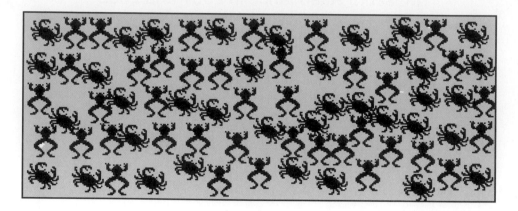

There are about 15 animals in the small picture at the right. Use this number as a reference number to help you estimate the number of animals in the picture above.

In the next lesson, you will use the same idea you used in Question 8 to estimate the number of objects in jars.

9. Bring in a mystery jar of objects to use in the activity.

- Find a clear, clean jar with a lid.

- Fill (or partly fill) the jar with one kind of object such as beans, pasta, or small building blocks. The objects should be small and about the same size.

- Count the objects in your jar. Write the number of objects on a small piece of paper. Tape the paper to the inside of the lid. Put the lid on the jar so that no one can see the number.

- Put your name on the outside of the lid.

One way to make good estimates is to compare the mystery jar to another jar or bag with a known number of objects. The picture below shows Ana's mystery jar of marbles and a bag with 50 marbles. Other students can use the 50 marbles as a reference to estimate the number of marbles in the mystery jar.

10. Bring in a number of objects in another jar or in a plastic bag. Your classmates will use this number as a reference to help them estimate the number of objects in your mystery jar.

- Use the same objects that are in your mystery jar. Count out a convenient number of objects—10, 25, 50, or 100 objects work well.

- Place them in another jar or a clear plastic bag.

- Label the jar (or bag) with your name and the number of objects so your classmates can see this number.

Close Enough

Look at the picture of the basketball game below. Do you think there are more or less than 100 people in the crowd? Do you think there are more or less than 500 people? more or less than 1000 people? more or less than 2000 people?

Use the small picture on the left to help you estimate. There are about 80 people in the small picture.

1. Estimate the number of people watching the game. How did you decide on your estimate?

Students in Mrs. Dewey's room made mystery jars for homework and brought them to class. Three jars the students brought to class are shown in the picture. Mrs. Dewey made a mystery jar of centimeter connecting cubes.

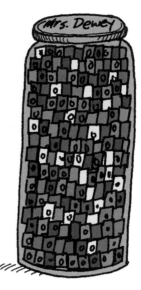

Linda and Romesh estimated the number of objects in each jar and recorded their estimates in data tables. Here are their estimates:

Linda's Data Table

Object	Estimate
marbles	130
marshmallows	50
beans	350
cubes	500

Romesh's Data Table

Object	Estimate
marbles	120
marshmallows	35
beans	375
cubes	600

Discuss

2. The actual number of marbles is 142. The actual number of marshmallows is 38.

   A. What is the difference between Linda's estimate of the number of marbles and the actual number?

B. What is the difference between Linda's estimate of the number of marshmallows and the actual number?

C. Is Linda's estimate for the number of marbles better than her estimate for the number of marshmallows? Why, or why not?

Mrs. Dewey said: "Let's say that an estimate is 'close enough' if it is within ten percent. Ten percent (10%) means 10 out of every 100. That's the same as 1 out of every 10 or $\frac{1}{10}$. So, to find out which numbers are within 10% of 142, we have to find $\frac{1}{10}$ of 142. How can we do that?"

Linda chose to find $\frac{1}{10}$ of 142 by dividing the marbles into 10 equal groups. When she finished, each group had 14 marbles, and two marbles were left over. She decided that $\frac{1}{10}$ of 142 is about 14.

Romesh used a calculator to find $\frac{1}{10}$ of 142. He divided 142 by 10. Since the display read 14.2, he agreed with Linda: 10% of 142 is about 14.

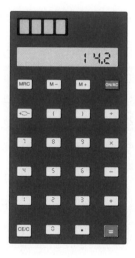

Any prediction in the range between 128 and 156 is within 10% of 142, since:

$$142 - 14 = 128$$
$$\text{and}$$
$$142 + 14 = 156.$$

3. A. Look back at Linda's and Romesh's estimates listed in the data tables. Which of the estimates for the jar of marbles is within 10% of 142?

B. Do you agree that this estimate is close enough?

4. A. Find 10% of 38. Check your answer by finding 10% of 38 in a different way.

B. Look back at Linda's and Romesh's estimates listed in the data tables. Which of the estimates for the jar of marshmallows is within 10%?

5. Draw the following data table on your paper. Complete the table.

Object	N Actual Number	N ÷ 10	10% of the Number	Range
marbles	142	14.2	About 14	128–156
marshmallows	38			
beans	351			
cubes	526			

6. **A.** Is Linda's estimate for the number of beans within 10%? Why or why not?

   B. Is the estimate Romesh made for the number of beans within 10%? Why or why not?

7. **A.** Is Linda's estimate for the number of cubes within 10%? Why or why not?

   B. Is the estimate Romesh made for the number of cubes within 10%? Why or why not?

8. Estimate the height of the door to your classroom to the nearest cm.

   A. Record your estimate and share it with your class.

   B. Measure the height of the door to the nearest cm.

   C. Which estimates are within 10% of the actual measurement?

9. A carpenter is making a door. The opening is 75 cm wide. If the carpenter measures the width of the door to within 10%, will the measurement be close enough? Why or why not?

Homework

Dear Family Member:

The students have learned to find 10% of various objects by finding one-tenth of the number. They can divide the number of objects into ten equal groups and count the number of objects in one group, or they can use a calculator to divide the number by ten. Ask your child to describe how he or she finds 10% of a number.

Thank you.

1. Here is part of the *10% Chart* made by students in Mrs. Dewey's class. Complete the table.

Object	N Actual Number	N ÷ 10	10% of the Number	Range
blocks	51			
macaroni	632			
pennies	198			
Super Balls	15			

2. **A.** Is an estimate of 60 blocks within 10% of the actual number? Show how you know.

   B. Is an estimate of 650 macaroni pieces within 10% of the actual number? Show how you know.

   C. Is an estimate of 175 pennies within 10% of the actual number? Show how you know.

   D. Is an estimate of 18 Super Balls within 10% of the actual number? Show how you know.

3. Tanya is working on the *Bouncing Ball* experiment. She predicts that the tennis ball will bounce to a height of 50 cm if it is dropped from a height of 100 cm. She checks her prediction, and the ball actually bounces to a height of 54 cm. Is her prediction of 50 cm within 10% of the actual bounce height of 54 cm? Why, or why not?

4. The average height of students in Room 204 is 55 inches. Nila is 48 inches tall. Is her height within 10% of the average height?

5. Keenya goes to the store with $32 for groceries. As she shops, she estimates the cost of the groceries so that she will have enough money when she goes to the cash register. She estimates that the groceries in her cart will cost about $32.

   A. The actual cost of the groceries is $34.52. Is her estimate within 10% of the actual cost?

   B. Is her estimate close enough? Explain.

Using Estimation

It's About . . .

Jerome brought in an article for the number newswire. Mrs. Dewey asked him to share the numbers he found. "My article says that 407,997 people visited the planetarium during 2001 and 458,156 people visited during 2002," said Jerome.

"Can anyone estimate the total number of people who visited the planetarium during these two years?" asked Mrs. Dewey.

Round numbers are often used when estimating because they are convenient to think about. Round numbers such as tens, hundreds, or thousands end in zeros. They are one type of convenient numbers. A number line can help you when you round a number.

1. **A.** Jerome estimated where 407,997 would be on the number line. He knew that it would be between 400,000 and 500,000 so he chose these two numbers as his **benchmarks.** Locate the mark Jerome made on the number line showing 407,997.

   B. Is 407,997 closer to 400,000 or 500,000?

   C. Round 407,997 to the nearest 100,000.

2. **A.** Using the same two benchmarks, Jerome estimated where 458,156 is on the number line. Find the mark Jerome made for 458,156 on the number line.

   B. Is it closer to 400,000 or 500,000?

   C. Round 458,156 to the nearest 100,000.

3. Ana did not round 458,156 to the nearest hundred thousand. She rounded 458,156 in two other ways. She used these two number lines.

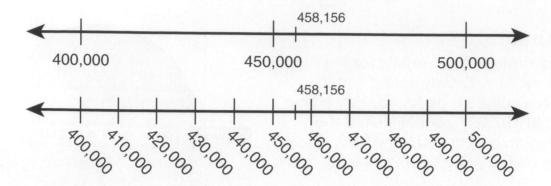

Using these two number lines, give two ways to round 458,156.

4. **A.** Ana used the second number line in Question 3 to round 407,997 to the nearest 10,000. What is her estimate?

   B. What benchmarks did she use?

5. Jerome estimated the total number of people who visited the planetarium over the last two years as 900,000. Ana's estimate was 870,000 people.

   A. Write a number sentence to show how Jerome found his estimate.

   B. Write a number sentence to show how Ana found her estimate.

   C. Which student is correct?

6. The museum estimated how much the attendance grew from 2001 to 2002. Use rounded numbers to estimate the increase in attendance.

Practice rounding the numbers in each problem below.

7. **A.** Use this number line to round 8207 to the nearest thousand.

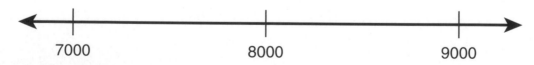

   B. What two benchmarks did you use?

8. **A.** Round 8207 to the nearest hundred using this number line.

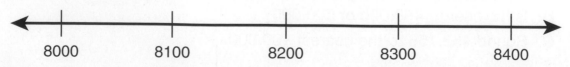

B. What two benchmarks did you use?

C. Compare this rounded number with the rounded number you found in Question 7. Which one is closer to the exact number?

D. Last year, 8207 people attended a high school play. The school play committee is expecting about the same number to attend this year's show. If the committee members are trying to plan refreshments for this year's show, which rounded number would be best to use?

9. **A.** Use this number line to round 36,736 to the nearest thousand.

B. What two benchmarks did you use?

C. Round 36,736 to the nearest 10,000. Draw a number line showing the benchmarks you would use.

D. Last year 36,736 tickets were sold at a fun fair during a summer festival. The planning committee is getting ready to order tickets for this summer's fun fair. Tickets are sold in rolls of 1000. If the planning committee expects about the same number of tickets to be sold, which rounded number should they use when ordering the tickets?

Estimating Sums and Differences

Ming and Keenya were researching the national parks in the United States. They found that Yellowstone National Park, established in 1872, was the world's first national park. Since then, more than 50 national parks have been set aside by the American government.

Ming used a table to organize some of the national park information they found.

National Parks

National Park	State	Established	Area
Acadia	Maine	1929	46,051 acres
Badlands	South Dakota	1978	242,756 acres
Carlsbad Caverns	New Mexico	1930	46,776 acres
Denali	Alaska	1980	4,740,911 acres
Everglades	Florida	1934	1,398,902 acres
Grand Canyon	Arizona	1919	1,217,403 acres
Mammoth Cave	Kentucky	1941	52,830 acres
Mesa Verde	Colorado	1906	52,122 acres
Petrified Forest	Arizona	1962	93,533 acres
Rocky Mountains	Colorado	1915	265,769 acres
Wind Cave	South Dakota	1903	28,295 acres

10. **A.** Ming estimated that about 1,300,000 acres of land have been set aside in Arizona as national park land. Ming wrote this number sentence showing the convenient numbers he chose:

$$1,200,000 + 100,000 = 1,300,000 \text{ acres of land set aside}$$

Explain how Ming arrived at his estimate.

B. Find another estimate for the amount of land set aside in Arizona as national park land. Explain your thinking.

Use Ming and Keenya's table to answer Questions 11–13. Write a number sentence showing the convenient numbers you chose to use.

11. Estimate the amount of land set aside in Colorado as national park land.

12. Which state, Arizona or Colorado, has more national park land? Estimate the difference.

13. Mammoth Cave is the longest known cave network in the world. Estimate the difference in size between Mammoth Cave National Park and Wind Cave National Park.

Making Money

The Parent-Teacher Committee at Bessie Coleman School wants to purchase computer hardware for the school computer lab over a two-year period. The committee's goal is to purchase hardware for 25 computer lab stations. Committee members made a table to show what they wanted to buy:

Quantity	Item	Total Cost ($)
25	14-inch color monitor	7469.00
25	extended keyboards	3127.00
25	personal computers	30,716.00
25	color printers	14,054.00

14. Use estimation to set a goal for the amount of money the Parent-Teacher Committee needs to earn. Write a paragraph explaining how you arrived at your goal. Use the Student Rubric: *Knowing* to help you write your paragraph.

1. Find at least two ways to round each of these numbers.
   - Draw a number line for each rounded number.
   - Label your line with the benchmarks you chose.
   - Estimate the location of the actual number on the number line.

   A. 5599   **B.** 24,681   **C.** 18,260

   D. 764,296   **E.** 206,492   **F.** 6,847,000

Use this number line to answer Questions 2–4.

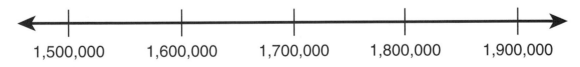

2. According to an article Linda hung on the newswire, 1,858,766 people visited the aquarium last year. Use the number line to round 1,858,766 to the nearest hundred thousand.

---

3. In one year, 1,510,063 people visited the Sears Tower Skydeck. Use the number line to round 1,510,063 to the nearest hundred thousand.

4. During one season, a total of 1,697,398 people attended home games for the Chicago White Sox. Use the number line to round 1,697,398 to the nearest hundred thousand.

Addition and Subtraction Practice with Paper and Pencil

Solve the following problems using paper and pencil. Find exact answers. Show all your work. Use estimation to look back and see if your answers are reasonable.

5. 9436
   + 4831

6. 4302
   + 3005

7. 7407
   − 3822

Estimate the answers to the following problems. Show the round numbers you used.

8. 23,065
   − 9,638

9. 94,378
   − 76,893

10. 80,025
    − 9,559

11. The United States has 12,383 miles of coastline along four different oceans. The Atlantic coast is 2069 miles long, the Arctic coast is 1060 miles long, and the coast of the Gulf of Mexico is 1631 miles long. About how long is the Pacific coast of the United States?

12. The United States has a total area of 3,787,319 square miles. Water covers 251,041 square miles. About how much of the United States area is land?

9 to 5 War

Players

This is a card game for two players.

Materials

- one pile per player of 9s and 5s cut from the two *9 to 5 War Cards* Activity Pages in the *Discovery Assignment Book*
- one pile per player of 20 other cards. This can be made from *Digit Cards (0–9)* or from a deck of playing cards with face cards removed (1–10, with aces representing 1s).

Rules

1. Players place their two piles face down in front of them.

2. Each player turns over two cards, one from the 9s and 5s pile, and one from the other pile.

3. Each player should say a number sentence that tells the product of his or her two cards. Whoever has the greater product wins all four cards.

4. If there is a tie, then each player turns over two more cards. The player with the greater product of the second pairs wins all eight cards.

5. Play for ten minutes or until the players run out of cards. The player with more cards at the end is the winner.

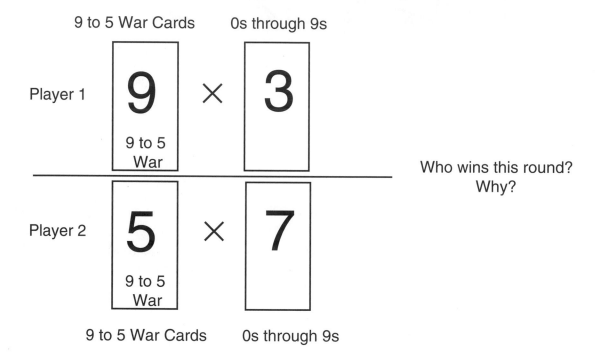

Who wins this round?
Why?

Variations

1. Whoever has the *smaller* product takes the cards.

2. Play with more than two players.

3. Each player is given only one pile of cards (playing cards with face cards removed or *Digit Cards*). Each player takes the top two cards from his or her pile and multiplies the numbers. The player with the larger product wins all four cards. This game practices all the facts—it does not just focus on the 9s and 5s.

Patterns in Multiplication

	Student Guide	Discovery Assignment Book	Adventure Book	Unit Resource Guide*
Lesson 1				
Order of Operations	●			
Lesson 2				
Divisibility Rules	●			●
Lesson 3				
Oh, No! My Calculator Is Broken	●			●
Lesson 4				
Multiplying by 10s	●			
Lesson 5				
Multiplication	●			
Lesson 6				
Estimation	●			
Lesson 7				
Multiplying Round Numbers	●			●
Lesson 8				
A Camping Trip	●			

Unit Resource Guide pages are from the teacher materials.

Order of Operations

An operation is work that someone or something does. When a surgeon removes a person's appendix, he or she performs an *operation*. People who use cranes and bulldozers *operate* heavy machinery. If a pop machine is broken, we say it is not *operating*.

Things that are done to numbers are also called **operations.** The four basic operations are addition, subtraction, multiplication, and division. People have agreed on a certain **order of operations** for arithmetic problems. The order in which you perform operations in an arithmetic problem is very important. You can see why in the problem below.

Mrs. Dewey brought raisins to the class picnic. There are six boxes of raisins in each snackpack.

After the picnic, there were 3 individual boxes of raisins and two unopened snackpacks left over. How many boxes of raisins were left?

Jessie said there were 3 boxes and 2 snackpacks left over. She wrote this number sentence: $3 + 2 \times 6$. But she didn't know whether to multiply or add first. Try both ways.

If you add $3 + 2$ first, the answer is 30 boxes of raisins. If you multiply 2×6 first, the answer is 15 boxes of raisins. Which is the correct answer?

There were three boxes and two snackpacks of raisins left over. This amount is shown below.

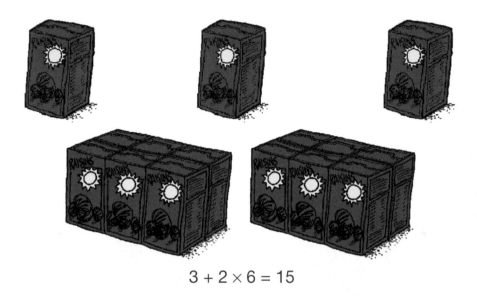

$$3 + 2 \times 6 = 15$$

If you count up all the boxes, you will see that the correct answer is 15 boxes of raisins left over. So, you need to multiply first in the problem: $3 + 2 \times 6$.

You cannot simply work from left to right when you solve problems like $3 + 2 \times 6$. Mathematicians have agreed on the following rules for the order of operations: First, do all the multiplications and divisions. If there are several multiplications and divisions, solve them from left to right.

When you finish all the multiplying and dividing, then do the additions and subtractions. Work from left to right. Some examples are shown below.

Do the multiplication first. Then, add.

$7 + (3 \times 4) = ?$
$7 + \quad 12 \quad = 19$

Do the division and multiplication first. Then, add.

$(6 \div 2) + (8 \times 2) = ?$
$\quad 3 \quad + \quad 16 \quad = 19$

Calculator Order of Operations

For each of these problems, first find the answer. Then, if possible, check your answer using a calculator that follows the order of operations.

1. $4 - 2 + 1$

2. $15 - 8 + 6 - 4$

3. $4 + 3 \times 2$

4. $4 + 9 - 3 \times 2$

5. $5 \times 2 + 3$

6. $3 + 5 \times 2$

7. $5 + 2 \times 3$

8. $6 \div 3 \times 2$

9. $4 \times 4 - 4 \div 4$

10. $4 + 4 \times 4 - 4$

11. $4 \times 4 - 4 - 4$

12. $4 \times 4 \div 4 - 4$

13. $10 + 6 \times 8$

14. $10 - 24 \div 6$

15. $8 \times 7 - 6$

Operation Target

Players

This is a cooperative contest for two or three people.

Materials

- paper
- pencil
- calculator

Rules

The goal is to use four digits and four operations (+, −, ×, and ÷) to make as many different whole numbers as you can.

You must use each of the four digits just once. You can use operations more than once or not at all. (All division operations must give whole numbers. For example, $9 \div 2 = 4.5$ is not allowed.)

16. Use 9, 5, 2, and 1 and +, −, ×, and ÷ to make as many whole numbers as you can. For example, $9 + 5 \times 2 − 1 = 18$. List the numbers you make and show how you made them.

 A. What is the largest whole number you can make?

 B. What is the smallest whole number you can make?

 C. How many whole numbers less than 10 can you make? Write number sentences for each number.

 D. What whole numbers can you make in more than one way? Show at least two number sentences for each.

17. Pick four different digits. Make as many whole numbers as you can using your four new digits and +, −, ×, and ÷. List the numbers you make and show how you made them.

18. Nila used 1, 2, 3, and 4 to make 10. How do you think she did it? Can you think of another way?

19. Luis used 1, 2, 3, and 4 to make 24. How could he have done it?

20. Romesh used 1, 3, 5, and 7 to make 8. How could he have done it?

21. Make up your own problem like those in Questions 18, 19, and 20.

Divisibility Rules

Is It Divisible by 2?

Shannon, Roberto, and Ming are talking about good books to read. Shannon shows Roberto and Ming the book she plans to read this weekend.

Shannon checked Ming's answer by multiplying 159 by 2 on the calculator. She got 318 as her answer: $159 \times 2 = 318$. Since 2 is a **factor** of 318, 318 is **divisible** by 2. A number is divisible by 2 if 2 divides it evenly—that is, if the answer is a whole number.

Shannon plans to read 159 pages each day. What if the book had 319 pages? On your calculator, press: | 319 | | ÷ | | 2 | | = |

Your calculator window should show:

$$159.5$$

The ".5" in the calculator window tells us there is a remainder. If the book had 319 pages, Shannon could read 159 whole pages each day—Saturday and Sunday. But if she did that, she would only read 159×2 or 318 pages. There would be a remainder of one page. Shannon could divide the leftover page in half. (The decimal .5 is the same as $\frac{1}{2}$.) She could read $\frac{1}{2}$ page on Saturday and $\frac{1}{2}$ page on Sunday.

Since 2 does not divide 319 evenly (we do not get a whole number answer), 319 is not divisible by 2.

1. On a copy of the *100 Chart*, use a blue crayon or pencil to circle all the **multiples** of 2. Then describe any patterns you see. Save your copy of the *100 Chart* for later use.

Discuss

2. Shannon's book is 318 pages long. In which column would 318 be if the 100 chart kept going beyond 100?

3. **A.** Which of the following numbers are divisible by 2? Why do you think so? Use a calculator to check your predictions.

| 109 | 213 | 216 | 275 | 784 |
| 1000 | 1358 | 2462 | 6767 | 8091 |

   B. Write a division sentence and a multiplication sentence for each number that is divisible by 2. For example: 216 ÷ 2 = 108, and 108 × 2 = 216.

4. How can you tell if a number is divisible by 2?

Is It Divisible by 3?

12 is divisible by 3.	21 is divisible by 3.	30 is divisible by 3.
$4 \times 3 = 12$	$7 \times 3 = 21$	$10 \times 3 = 30$
$12 \div 3 = 4$	$21 \div 3 = 7$	$30 \div 3 = 10$

The number 3 is a factor of 12, 21, and 30. A factor of a number can be divided evenly into the number—that is, the answer (or quotient) is a whole number. Since 12, 21, and 30 can be divided by 3 evenly, we say that 12, 21, and 30 are divisible by 3.

5. Use your copy of the *100 Chart* that you used earlier (the multiples of 2 should be circled in blue). Using a red crayon or pencil, mark all the multiples of 3 with an "X." Your *100 Chart* should look like the one below.

1	②	X	④	5	⑥	7	⑧	X	⑩
11	⑫	13	⑭	X	⑯	17	⑱	19	⑳
X	㉒	23	㉔	25	㉖	X	㉘	29	㉚

6. Describe any patterns you see.

7. Use your copy of the *100 Chart* and your calculator to help you answer the following questions:

   A. Is 27 a multiple of 3? Write a multiplication sentence.

   B. Is 27 divisible by 3? Write a division sentence.

   C. Is 51 a multiple of 3? Write a multiplication sentence.

   D. Is 51 divisible by 3? Write a division sentence.

8. **A.** Is 14 a multiple of 3? How do you know?

   B. If 14 is divided by 3, what is the remainder? Write a multiplication or division sentence. Remember to include the remainder.

   C. Use your calculator. Press: $\boxed{14}\ \boxed{\div}\ \boxed{3}\ \boxed{=}$
   How does your calculator show whether 14 is divisible by 3?

9. Is 74, 75, or 76 divisible by 3? Use your copy of the *100 Chart* or a calculator to decide. Write a division sentence showing which number is divisible by 3.

10. Look carefully at your *100 Chart*. Write in more numbers below it if you need to.

    A. Predict: Is 101 divisible by 3? Check your prediction with a calculator.

    B. Predict: Is 102 divisible by 3? Check your prediction with a calculator.

    C. Predict: Which of the following is divisible by 3: 116, 117, or 118? Why do you think so? Check your prediction with a calculator.

11. Mrs. Dewey started listing numbers from the *100 Chart* that were divisible by 3. Ming saw a pattern. Do you? Explain.

Number	Sum of Digits
18	1 + 8 = **9**
42	4 + 2 = **6**
51	5 + 1 = **6**
84	8 + 4 = **12**
99	9 + 9 = **18**

Explore

12. Name a number greater than 125 that is divisible by 3. Check your prediction.

13. Name a number greater than 200 that is divisible by 3. Check your prediction.

14. Use the numbers below to make predictions for Questions 14A–14C. Then check your predictions with a calculator.

126	209	342	177	1664
1002	991	297	8770	8775

    A. Which numbers are divisible by 3?

    B. Which are divisible by 2?

    C. Which are divisible by 2 and 3?

15. Is 12,345,678 divisible by 2? Divisible by 3? Use your calculator to check.

Is It Divisible by 6?

Discuss

16. **A.** Find out which numbers in Question 14 are divisible by 6. Use a calculator.

    B. How can you determine if a number is divisible by 6? Find the multiples of 6 by skip counting by 6s on your *100 Chart*. What do you notice?

    C. Based on the patterns you see, predict whether 12,345,678 (from Question 15) is divisible by 6. Check your prediction.

17. **A.** Give a number greater than 150 that is divisible by 6.

    B. Give a number greater than 225 that is divisible by 6. Explain how you found your number.

Is It Divisible by 9?

18. Copy and complete the list of facts for 9. Then write the products in a column, one on each line.

$$1 \times 9 =$$
$$2 \times 9 =$$
$$3 \times 9 =$$
$$4 \times 9 =$$
$$5 \times 9 =$$
$$6 \times 9 =$$
$$7 \times 9 =$$
$$8 \times 9 =$$
$$9 \times 9 =$$
$$9 \times 10 =$$

19. What patterns do you see? What can you say about the sum of the digits of the products?

20. Use your calculator to find more multiples of 9. Find the products below.

A. $9 \times 634 =$ **B.** $9 \times 23 =$

C. $9 \times 37 =$ **D.** $9 \times 73 =$

E. $9 \times 143 =$ **F.** $9 \times 444 =$

G. $9 \times 754 =$ **H.** $9 \times 4421 =$

21. Go back and add the digits of each product in Question 20.
For example: $9 \times 634 = 5706$
Add the digits in 5706: $5 + 7 + 0 + 6 = 18$
Now, add the digits in 18: $1 + 8 = 9$
Describe what happens when you add the digits of a multiple of 9.

22. A. Predict which numbers below are divisible by 9. Show how you decided.

B. Then check using a calculator.

C. Finally, write a multiplication and division sentence for each multiple of 9 you identify.

172	144	743	747	1007
2556	4906	8721	9908	12,987

You will need a clean copy of the *100 Chart* to complete this homework.

1. Which numbers below are divisible by 2? Tell how you decided.

345	980	1369	1197	3288
9036	2273	1035	8665	2073

2. Which numbers are divisible by 3? Tell how you decided.

3. Which numbers are divisible by 6? Tell how you decided.

4. Which numbers are divisible by 9? Tell how you decided.

5. Are any of the numbers divisible by 2, 3, 6, and 9? Which one(s)?

6. Use a clean copy of a *100 Chart* and a calculator to explore the following.

   A. Skip count by 5s on the *100 Chart*. Mark each multiple of 5. Then skip count by 10s and mark the multiples of 10.

   B. Describe how you know a number is divisible by 5. In your description, include:
      - Examples of numbers that are divisible by 5.
      - Multiplication sentences and division sentences for those numbers.
      - Descriptions of patterns you see on the *100 Chart*.
      - Any keystrokes you used on the calculator.

   C. Describe how you know a number is divisible by 10. In your description, include:
      - Examples of numbers that are divisible by 10.
      - Multiplication sentences and division sentences for those numbers.
      - Descriptions of patterns you see on the *100 Chart*.
      - Any keystrokes you used on the calculator.

Use the order of operations.

7. $6 \times 7 - 4 \times 8 =$

8. $1 + 48 \div 8 =$

9. $4 + 7 \times 8 =$

10. $4 \times 7 - 24 \div 6 =$

Oh, No! My Calculator Is Broken

Multiplication

What's the matter with this calculator?

Discuss

Mrs. Dewey passed out calculators to her class. She asked the class to use the calculators to do 6 × 8. John pressed: $\boxed{6}\ \boxed{\times}\ \boxed{8}\ \boxed{=}$.

He soon realized the $\boxed{8}$ key on his calculator was broken. Both Keenya and Grace tried their calculators. They each found that the $\boxed{\times}$ keys on their calculators were broken. In fact, all of the calculators in the class had keys that didn't work. Mrs. Dewey said, "Some of the keys on these calculators have been disconnected. Think about how you can use your calculator to multiply 6 × 8."

John knew that he could break apart 8 into 5 + 3. He could then multiply 6 × 5 and 6 × 3 and add the two products together to get 6 × 8. To show his solution, John recorded the following keystrokes:

$$\boxed{6}\ \boxed{\times}\ \boxed{5}\ \boxed{+}\ \boxed{6}\ \boxed{\times}\ \boxed{3}\ \boxed{=}$$

Since the $\boxed{\times}$ key was broken on Keenya's calculator, she decided to turn 6×8 into an addition problem. Keenya recorded the following keystrokes:

$$\boxed{8}\ \boxed{+}\ \boxed{8}\ \boxed{+}\ \boxed{8}\ \boxed{+}\ \boxed{8}\ \boxed{+}\ \boxed{8}\ \boxed{+}\ \boxed{8}\ \boxed{=}$$

Grace knew that 6×8 would be twice the answer to 3×8. "Since $3 \times 8 = 24$, I can use my calculator to add $24 + 24$ to find the answer for 6×8." She recorded these keystrokes:

$$\boxed{24}\ \boxed{+}\ \boxed{24}\ \boxed{=}$$

With your calculator, try each of the strategies suggested by Grace, John, and Keenya. Do you get the same answer with each strategy?

1. **A.** Pretend you are using Keenya's calculator and the $\boxed{\times}$ key is broken. Give a different list of keystrokes to find 6×8. Record your keystrokes on your own paper.

   B. Pretend you are using John's calculator and the $\boxed{8}$ key is broken. Give a different list of keystrokes to find 6×8. Record your keystrokes on your own paper.

Use a calculator to do the following problems two different ways. For each problem, record the keys you pressed.

2. Imagine that the $\boxed{6}$ key on your calculator is broken.

   A. $6 \times 7 =$ **B.** $6 \times 4 =$

3. Imagine that the $\boxed{7}$ key on your calculator is broken.

   A. $7 \times 8 =$ **B.** $6 \times 7 =$

4. Imagine that the $\boxed{2}$ key on your calculator is broken. Use your calculator to find the answer to this problem in two ways: $4 \times 2 \times 7 =$. Record your keystrokes.

5. Imagine that the $\boxed{\times}$ key on your calculator is broken. Use your calculator to find the answer to each problem in two ways. Record your keystrokes.

   A. $9 \times 6 =$ **B.** $9 \times 7 =$

There are many strategies for doing multiplication problems. Some strategies are easier to use without a calculator.

Maya found a strategy for multiplying 9×4. She said, "First I will multiply 10×4. Then if I subtract 4 from 40, I will get 9×4."

$$10 \times 4 = 40$$
$$40 - 4 = ?$$

6. **A.** Why is Maya multiplying 10×4?

   B. Why is she subtracting 4?

7. How would you explain Maya's strategy to a friend?

8. Use Maya's strategy to solve 9×7. Explain your thinking.

9. What are some strategies you could use to solve 12×8?

Addition and Subtraction

Shannon found an old calculator. When she tried the calculator, she found that only the clear key and these keys worked:

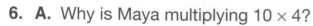

She found that she could get the number 2 on her display by pressing:
. She wondered if she could figure out a way to get the number 1 on her display.

10. Help Shannon think of a strategy she could use to get a 1 on her calculator display.

11. What keystrokes can Shannon use to get the number 55 on her calculator display? Remember, you can only use the 6 keys that work.

Multiplying by 10s

On Monday, ten of the candy machines at the TIMS Candy Company were working. Each machine can make four Chocos in one minute. On Monday, how many Chocos were made every minute at the company?

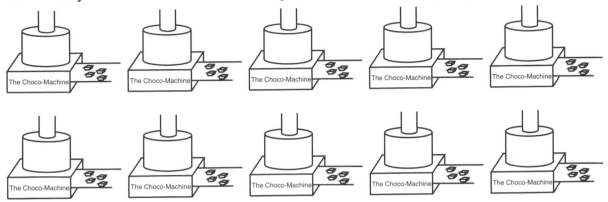

A number sentence for this question is $4 + 4 + 4 + 4 + 4 + 4 + 4 + 4 + 4 + 4 = n$.

Another way to write this is $10 \times 4 = n$. What number must n be to make the sentences true?

The value of n is 40 (or $n = 40$) since 40 is 10 groups of 4.

$4 + 4 + 4 + 4 + 4 + 4 + 4 + 4 + 4 + 4 = 40$ and $10 \times 4 = 40$.

We can also say this as $4 \times 10 = 40$. This is the same as 4 groups of 10.

Complete Questions 1 and 2. You may use your calculator.

1. There are 10 polar bears at the Greenville Zoo. Each bear eats 6 pounds of fish a day. How many pounds of fish are eaten by polar bears at the Greenville Zoo every day? Write an addition sentence for this problem and solve. Write a multiplication sentence for the problem and solve.

2. Tanya has 10 friends who wrote her letters when she was away at camp. Each friend wrote Tanya 3 letters. How many letters did Tanya receive? Draw a picture of the problem, and write a multiplication sentence.

3. Luis says that multiplication by multiples of 10 is easy.

   A. Do the following problems. You may use your calculator if necessary.

10	20	40	30	70	40	80	20	90
×6	×3	×3	×7	×2	×6	×2	×9	×8

   B. Do you notice a pattern in multiplying by multiples of 10? Describe it.

4. Do the following problems. You may use your multiplication table.

60	70	40	20	50	50	90
×4	×5	×7	×2	×6	×5	×3

5. Frank said one of the problems in Question 4 was tricky. He almost got it wrong. Which one do you think Frank is talking about, and why is it tricky?

6. Irma found some more patterns on her calculator. Describe the patterns.

   $7 \times 1 = 7$ $7 \times 4 = 28$
   $7 \times 10 = 70$ $7 \times 400 = 2800$
   $7 \times 100 = 700$ $7 \times 40 = 280$
   $7 \times 1000 = 7000$ $7 \times 40000 = 280,000$
   $7 \times 10,000 = 70,000$ $7 \times 4000 = 28,000$

7. Find the products. You may use a calculator if necessary.

   A. 6×1
   6×10
   6×100
   6×1000
   $6 \times 10,000$

   B. 6×3
   6×30
   6×300
   6×3000
   $6 \times 30,000$

   C. 6×5
   6×50
   6×500
   6×5000
   $6 \times 50,000$

8. Make a prediction of what you think the answer will be to each problem. Then do the problem on your calculator to check.

   A. 8×30　　　**B.** 40×5　　　**C.** 200×3　　　**D.** 5×600

   E. 7×2000　　　**F.** 90×4　　　**G.** 600×2

9. Predict what number n must be to make the number sentence true. Check your work on a calculator.

   A. $200 \times 5 = n$　　　**B.** $60 \times n = 120$　　　**C.** $5000 \times n = 15{,}000$

   D. $n \times 40 = 80$　　　**E.** $n \times 700 = 4900$　　　**F.** $6 \times n = 6000$

   G. $n \times 6 = 3600$　　　**H.** $2 \times n = 1400$

10. Can you find a rule that makes multiplying numbers that end in zeros easy?

Homework

Make a prediction of what you think the answer will be to each problem below. Then do the problem on your calculator to check.

1. 8×3
　　8×30
　　8×300
　　8×3000
　　$8 \times 30{,}000$

2. 9×6
　　9×600
　　9×60
　　$9 \times 60{,}000$
　　9×6000

3. 7×6
　　7×60
　　7×6000
　　7×600
　　$7 \times 60{,}000$

4. 30×7

5. 500×8

6. 300×9

7. 7×200

8. 4×6000

9. $\begin{array}{r} 3000 \\ \times\,4 \\ \hline \end{array}$　　**10.** $\begin{array}{r} 5000 \\ \times\,7 \\ \hline \end{array}$　　**11.** $\begin{array}{r} 100 \\ \times\,9 \\ \hline \end{array}$　　**12.** $\begin{array}{r} 400 \\ \times\,3 \\ \hline \end{array}$　　**13.** $\begin{array}{r} 600 \\ \times\,2 \\ \hline \end{array}$

14. A zoo has 10 displays of turtles. Each display has 4 turtles. How many turtles are at the zoo? Draw a picture. Write an addition sentence and a multiplication sentence. Solve.

15. There are 10 elevators in an office building. There can be up to 9 people in an elevator at one time. How many people can ride the elevators at the same time?

16. There are 3 juice boxes in a juice pack. If Roberto's mother buys 10 juice packs, how many juice boxes is that?

17. A juice pack costs $2.00. How much will 8 juice packs cost? How much will 10 juice packs cost?

18. There are 30 desks in every fourth-grade classroom at Holmes School. If there are 5 fourth-grade classes, how many desks are there for fourth graders?

19. There are 8 granola bars in a package. A school buys 40 packages. How many granola bars is that?

20. A large bottle of ketchup costs $3.00. The head cook at Stanley School buys 40 bottles. How much money does he spend on ketchup?

Predict what number _n_ must be to make the number sentence true. Check your work on a calculator.

21. $200 \times n = 600$

22. $n \times 2 = 800$

23. $2 \times n = 600$

24. $2 \times n = 20{,}000$

25. $n \times 7000 = 42{,}000$

26. $n \times 800 = 3200$

27. $60 \times 7 = n$

28. $n \times 8 = 320$

29. $n \times 400 = 2800$

Multiplication

Mrs. Haddad noticed that 4 of the workers at the TIMS Candy Company each made 32 Chocos in one hour. She wanted to find the total number of Chocos made. Mrs. Haddad used base-ten pieces and wrote 32 + 32 + 32 + 32. She said this could be an addition problem or a multiplication problem.

Base-Ten Board

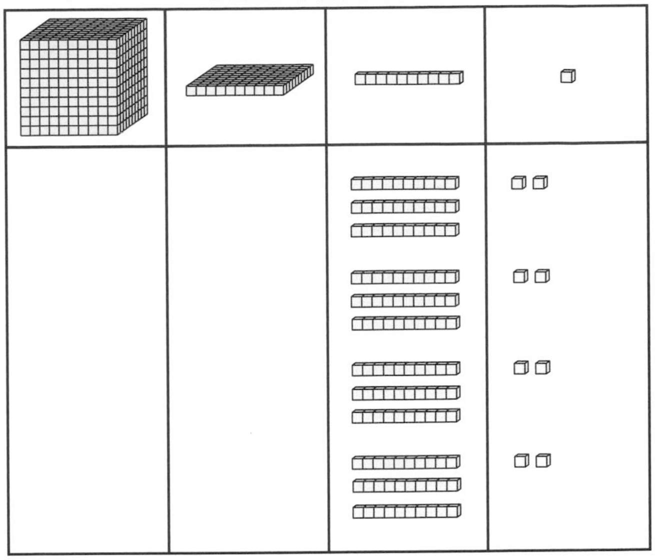

Mrs. Haddad used the **all-partials method** to solve the problem by multiplication. Since there are 4 groups of 2 bits, there are 8 bits total.

Recording Sheet

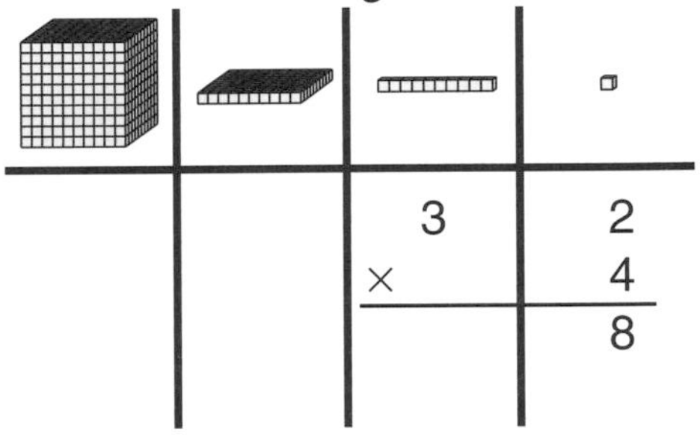

```
        3   2
    ×       4
            8
```

Since there are 4 groups of 3 skinnies, this makes 12 skinnies or 1 flat, 2 skinnies, and 0 bits, or 120 bits. The workers made a total of 128 Chocos in 1 hour.

Recording Sheet

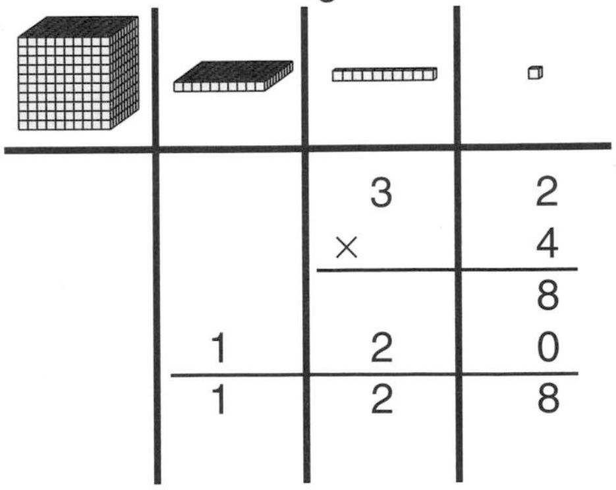

```
            3   2
        ×       4
                8
    1       2   0
    1       2   8
```

Multiplication

One of the employees at the TIMS Candy Company solved the problem a little differently. His work is shown here. He first found that 4 groups of 3 skinnies makes 120 bits. Then he found that 4 groups of 2 bits makes 8 bits total. The answer matches Mrs. Haddad's answer—128 Chocos.

Recording Sheet

	3	2
	×	4
1	2	0
		8
1	2	8

1. On another day, 3 workers at the TIMS Candy Company each made 26 pieces of candy in one hour.

Base-Ten Board

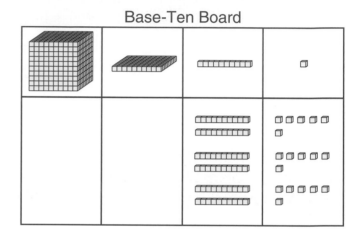

A. Fill in the missing numbers. Since there are 3 groups of 6 bits, there are ____ bits or ____ skinny and ____ bits.

We write this as shown on the *Recording Sheet* to the right.

Recording Sheet

	2	6
	×	3
	1	8

B. Since there are 3 groups of 2 skinnies, there are ____ skinnies. This is the same as ____ skinnies and ____ bits.

We write this as shown on the *Recording Sheet* to the right.

Recording Sheet

		2	6
	×		3
		1	8
		6	0
		7	8

C. What is the total number of candies made by the workers at the TIMS Candy Company?

Use base-ten pieces and the all-partials method to do the following problems:

2. 12
$\times 3$

3. 61
$\times 4$

4. 26
$\times 4$

5. 57
$\times 4$

6. 83
$\times 9$

Homework

Solve the following problems. You may use your multiplication table. You may also check your work on a calculator.

1. 20   50   90   60   40   70   80
$\times 8$ $\times 3$ $\times 4$ $\times 7$ $\times 9$ $\times 6$ $\times 6$

2. 100   30,000   700   200   40,000   700
$\times 5$ $\times 7$ $\times 5$ $\times 4$ $\times 6$ $\times 8$

3. Find *n* for each problem.

A. $5 \times 60 = n$

B. $70 \times 2 = n$

C. $n = 3 \times 600$

D. $n = 100 \times 5$

E. $n \times 70 = 140$

F. $500 \times n = 4000$

G. $80 = n \times 20$

H. $700 = 7 \times n$

I. $7 \times 800 = n$

Write a number sentence for each problem and solve. You may also want to draw a picture.

4. A sailboat can travel about 30 miles in 1 hour. How far can the sailboat travel in 6 hours?

5. There are about 50 crackers in a box. If 5 children share the box, how many crackers does each child get?

6. Mr. Thoms drove for 3 hours without stopping. He drove about 50 miles every hour (50 miles per hour or 50 mph). About how far did he drive?

Fill in the missing numbers in the multiplication problems. You may use base-ten shorthand if it is helpful. You may also use your multiplication table.

7. 13
    × 7
    ───
    21
    70
    ───
    ◯

8. 42
    × 6
    ───
    ◯
    240
    ───
    252

9. 51
    × 4
    ───
    4
    ◯
    ───
    204

10. Solve the problems. You may use base-ten shorthand if it is helpful. You may also use your multiplication tables.

    A. 14 × 2   **B.** 21 × 6   **C.** 52 × 3   **D.** 25 × 3   **E.** 41 × 6

    F. 65 × 6   **G.** 83 × 7   **H.** 76 × 9   **I.** 78 × 6   **J.** 67 × 4

11. There are 32 students in Miguel's class. For his birthday, Miguel's mother baked cookies to bring to school. If he gave each student 4 cookies, how many cookies did his mother bake?

12. There are 3 fourth-grade classes at Livingston School. If there are 26 students in each class, how many fourth graders are there at Livingston School?

13. A zoo has 22 lions. Each lion eats 8 pounds of meat a day. How much meat must the zookeeper bring to the lion exhibit each day?

14. The Rodriguez family is having a big party. Mrs. Rodriguez knows she should have 70 cans of soda. Ana buys 3 cases of soda. Each case contains 24 cans of soda. Will this be enough soda? Why or why not?

15. The array below has 6 rows of 13 tiles. How many tiles are in the array?

16. An array has 7 rows of 46 tiles. How many tiles in all are in the array?

17. An array has 9 rows of 58 tiles. How many tiles in all are in the array?

18. An array has 7 rows of 99 tiles. How many tiles? Explain a way to do this problem in your head.

Estimation

Nicholas wants to buy two books that cost $5.79 each. He **estimates** the cost of the two books to get a good idea of how much he will have to spend. In order to find the estimate, Nicholas uses a **convenient number.** He thinks, "$5.79 is close to $6.00, but $6.00 is easier to work with than $5.79." He multiplies 2 × $6 in his head and comes up with an estimate of $12.00.

As Nicholas waited in line at the cashier, he shared his estimate with Lee Yah. She estimated the total, too, and then said, "Your estimate is close to mine. I doubled $5.80. I knew 2 × $5 is $10 and 2 × 80¢ = $1.60. My estimate is $11.60."

Nicholas chose the number 6 when estimating because it was a convenient number to work with in his head. He rounded $5.79 to the nearest whole dollar—$6. Rounded numbers are one type of convenient number. Lee Yah chose the convenient number $5.80 because she could double $5.80 in her head more easily than $5.79. Remember, a number that is convenient for one person might not be convenient for another.

Discuss

1. **A.** Keenya has a bag of 25 oatmeal cookies to share with her friends. There are 6 girls altogether. How might Keenya decide the number of cookies each girl gets?

   B. Keenya thinks, "I have about 24 cookies. Since 6 × 4 is 24, each of us gets 4 cookies." What is Keenya's convenient number? How did she pick it?

2. Jackie's Girl Scout troop is planning a hike. The troop needs to know how heavy the supplies will be. Jackie weighs a bottle of water and finds that it weighs 4.1 kg. About how much will 3 bottles of water weigh?

We find estimates when we need to have a good idea about how big or small a number is, but we do not need to know exactly. Sometimes it is impossible to know an exact answer. We use estimates in many different situations.

For example:
You can estimate how far you live from Boston or Chicago.
You can estimate how many people are watching a Little League game.
You can estimate how much you will pay for a full cart of groceries.

When we estimate, we find a number or answer that is reasonably close to the actual number. It may be bigger or smaller than the actual number, depending on the problem. To make an estimate, we sometimes need to do some number operations in our head. To make these computations easy to do, we often choose **convenient numbers.** Convenient numbers are really estimates as well.

Which of the answers in Questions 3 and 4 are reasonable estimates? There may be more than one reasonable answer. When is one estimate better than another? Which answers are unreasonable?

3. Mrs. Borko buys 4 jackets for her children. Each jacket costs $47. About how much will all 4 jackets cost?

   A. $100   **B.** $20   **C.** $160   **D.** $2000   **E.** $200

4. There are about 27 students in each fourth-grade class at Hill Street Elementary School. If there are 6 fourth-grade classes, how many fourth-graders are there at the school?

   A. 120   **B.** 1200   **C.** 150   **D.** 180   **E.** 300

5. Keenya, her sister, and her parents went to a concert in the park. People sat in rows on benches. Keenya counted 12 people in the first row. There were 9 rows.

   A. Keenya thought, "There are about 10 people in each of the 9 rows. There are about . . ." Finish Keenya's statement using 10 as a convenient number.

   B. Keenya's sister thought, "There are about 10 rows and about 10 people in each row. I'd say there are about . . ." Finish her statement. What convenient numbers did Keenya's sister use to make her estimate?

6. At the grocery store, Jackie and her brother pick some grapes and weigh them. The grapes weigh 3 pounds and 7 ounces. Grapes are on sale for 49¢ a pound (1 pound = 16 ounces). Jackie and her brother estimate the price of the grapes.

   A. Jackie thinks, "3 pounds and 7 ounces is about 4 pounds. 4 pounds of grapes, at 50¢ a pound, would be . . ." Finish Jackie's statement. Will the actual price be higher or lower than this estimate?

Estimation

B. Jackie's brother thinks, "3 pounds of grapes cost 3 × 50¢ or $1.50. 7 ounces is about $\frac{1}{2}$ pound. If 1 pound costs 50¢, $\frac{1}{2}$ pound costs about 25¢. The grapes we picked should cost about . . ." Finish his statement.

C. If Jackie and her brother want to be sure they have enough money for the grapes, whose estimate would be better to use? Why do you think so?

7. Michael's father travels 36 miles to work every day.

   A. About how many miles does he travel to and from work in one week (5 days)? Solve this problem in your head. Then, explain how you solved it. Be sure to tell what convenient numbers you used.

   B. Share your method with a classmate. What convenient numbers did your classmate choose?

Estimating Mileage

Jerome and his family are thinking about taking a vacation. Jerome found a mileage chart.

	Chicago	Indianapolis	Louisville	St. Louis
Chicago	×	179	294	302
Indianapolis	179	×	112	257
Louisville	294	112	×	275
St. Louis	302	257	275	×

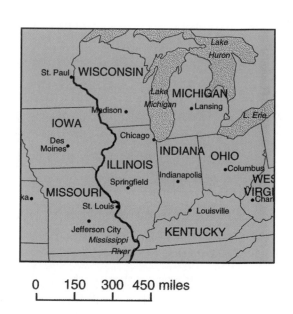

0   150   300   450 miles

8. **A.** How far is it from Chicago to Indianapolis?

   B. Give two possible convenient numbers for this distance.

9. About how many miles will they travel if Jerome's family drives from Chicago to Indianapolis and then back home to Chicago?

10. **A.** Which drive is longer: the drive from Chicago to Indianapolis or the drive from Indianapolis to St. Louis?

    B. About how much longer?

11. First, Jerome's family plans on taking the following trip: Chicago to Louisville, Louisville to St. Louis, and St. Louis back home to Chicago. About how many miles is this trip?

12. Jerome's family changes plans. They decide on the following trip: Chicago to Indianapolis, Indianapolis to Louisville, and Louisville back home to Chicago. About how many miles will they drive altogether?

13. Their car goes about 21 miles on one gallon of gasoline. At the start of the trip, their gas tank has about 12 gallons of gas.

    A. About how many miles can they travel on 12 gallons of gas?

    B. Will they have to get more gas before they reach Indianapolis? Before they reach Louisville?

14. At the time they were leaving, gasoline cost about $1.69 per gallon in Chicago. They estimated they would need about 28 gallons of gas in all to make the trip. If gas costs about the same in other cities as in Chicago, about how much money will they spend for gas?

15. Jerome's parents averaged 48 miles per hour on the entire trip. By the time they got back to Chicago, did the family drive for more or less than 8 hours? How did you decide?

Estimate answers to these problems. Be ready to tell how you used convenient numbers.

1. Ana found 12 tomatoes on one tomato plant. Ana has 11 tomato plants. All the plants have about the same number of tomatoes. About how many tomatoes do the plants have altogether?

2. A box of crackers weighs 269 grams. About how much do 5 boxes of crackers weigh together?

3. A pizza costs $3.75, a bag of apples $2.50, and a quart of milk $.80. About how much money is needed to buy one of each?

4. If a pizza costs $3.75, about how much will 3 pizzas cost?

5. Jacob's brother works at a fast-food restaurant and earns $5.15 an hour. If he worked 20 hours one week, about how much money did he earn?

6. Tanya earned $5.45 baby-sitting on Monday and $8.70 on Tuesday. She spent $6.65 of this money on Wednesday. About how much money does she have left?

7. Mount McKinley in Alaska is the highest mountain in the United States. It is 20,320 feet above sea level. The highest mountain in the 48 contiguous states (all the states except Alaska and Hawaii) is Mount Whitney in California. Mount Whitney is 14,494 feet above sea level. About how much higher is Mount McKinley than Mount Whitney?

Use the map below to answer Questions 8–13.

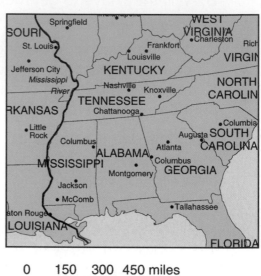

	Montgomery	Atlanta	Nashville	Jackson
Montgomery	×	163	286	247
Atlanta	163	×	250	381
Nashville	286	250	×	416
Jackson	247	381	416	×

0   150   300 450 miles

8. **A.** How far is it from Montgomery, Alabama, to Jackson, Mississippi?

B. What is a convenient number to use in place of this number?

9. **A.** How far is it from Nashville, Tennessee, to Montgomery, Alabama?

B. What is a convenient number to use in place of this number?

10. About how many miles is a drive from Nashville, Tennessee, to Montgomery, Alabama, and back home to Nashville?

11. **A.** Which drive is longer: a drive from Montgomery to Nashville or a drive from Montgomery to Jackson?

B. About how much longer?

12. About how long is the following trip: from Jackson to Montgomery, from Montgomery to Nashville, and Nashville back home to Jackson?

13. About how long is the following trip: from Atlanta to Montgomery, Montgomery to Jackson, and Jackson back home to Atlanta?

Multiplying Round Numbers

The Beautiful Blooms Garden Store sells many trays of flowers. A tray of flowers has 6 rows. There are 8 flowers in each row.

1. How many flowers are in each tray?

Ana and Grace went to the Beautiful Blooms Garden Store. They noticed that the trays of flowers were stored on a shelf. Ana counted 28 trays on a shelf.

Ana said, "I wonder how many flowers are on one of these shelves."

Grace replied, "We can estimate the number of flowers on a shelf: 48 flowers is about 50 and 28 trays is about 30 trays. So what is 50 × 30?"

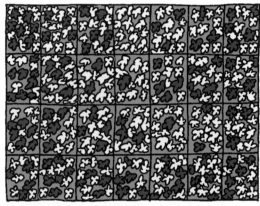

28 trays of flowers on a shelf

2. Ana learned that patterns often help us multiply. Find the following products, using a calculator if needed. Describe any patterns you see.

A. 5 × 3 =

B. 5 × 30 =

C. 50 × 3 =

D. 50 × 30 =

E. 50 × 300 =

F. 500 × 30 =

G. 500 × 300 =

3. About how many flowers did Ana and Grace see on a shelf?

4. Discuss other ways to compute 50 × 30.

---

5. Find the following pairs of products in your head. Check your work on a calculator, if needed.

A. 80   80
     ×2 ×20

B. 20   20
     ×4 ×40

C. 50   50
     ×7 ×70

D. 90   90
     ×7 ×70

E. 70   70
     ×1 ×10

F. 30   30
     ×6 ×60

G. 90 × 20 =

H. 40 × 50 =

I. 60 × 40 =

6. A. Ana and Grace counted 10 shelves of marigolds. If there are 28 trays on each shelf, about how many trays of marigolds are there?

Grace said, "Wow, there's a lot of marigolds here. There are 10 shelves and about 30 trays on each shelf. Since 10 × 30 = 300, I estimated 300 trays of marigolds."

28 trays on each of 10 shelves is about 10 × 30 or 300 trays.

300 trays with about 50 flowers on each is 50 × 300 or 15,000 marigolds.

B. Ana said, "If each tray has about 50 flowers and there are about 300 trays, that means there are 50 × 300 = 15,000 marigolds. That's amazing!"

What is another way to estimate the total number of flowers?

Find the following products with your calculator. Look for patterns.

7. 4 × 7 =
40 × 7 =
4 × 70 =
40 × 70 =
400 × 70 =
40 × 700 =
400 × 700 =

8. 6 × 7 =
60 × 7 =
6 × 70 =
60 × 70 =
600 × 70 =
60 × 700 =
600 × 700 =

9. 8 × 5 =
80 × 5 =
8 × 50 =
80 × 50 =
800 × 50 =
80 × 500 =
800 × 500 =

10. Ana says she can multiply 40×40 in her head easily. What method do you think Ana is using? What is 40×40?

11. Ana saw that for every zero in the factors, there is a zero in the product. Do you agree? Explain.

12. Grace says multiplying 400×50 is tricky. What is 400×50? Why is it tricky?

In Question 6, Ana and Grace used convenient numbers to estimate the number of flowers they saw.

For Questions 13–18, complete steps A and B.

   A. Estimate the following products in your head by finding convenient numbers. Be prepared to share your strategies.

   B. Use a calculator to compute the product. Then use your estimate to see if your answer is reasonable.

13. $32 \times 6 =$ **14.** $4 \times 67 =$ **15.** $8 \times 99 =$

16. $30 \times 41 =$ **17.** $40 \times 49 =$ **18.** $300 \times 24 =$

19. Describe a way to compute the exact product in your head in Question 15 without using a calculator or pencil and paper.

Find the products using mental computation.

1. 40 $\times 70$	**2.** 60 $\times 60$	**3.** 500 $\times 60$	**4.** 800 $\times 30$	**5.** 300 $\times 30$
6. 100 $\times 100$	**7.** 600 $\times 40$	**8.** 400 $\times 200$	**9.** 2000 $\times 800$	**10.** 6000 $\times 700$

11. Explain how to multiply two numbers that end in zeros.

For Questions 12–24, estimate the products using convenient numbers.

12. 42
$\times 8$ **13.** 76
$\times 4$ **14.** 69
$\times 7$

15. Describe a way to compute the exact product in Question 14 using mental math.

16. $60 \times 34 =$ **17.** $50 \times 79 =$ **18.** $320 \times 70 =$ **19.** $496 \times 90 =$

The Bessie Coleman Parent-Teacher Committee (PTC) decided to plant a garden by the school.

20. If there are about 38 daisies in a flat and 23 families each donate a flat of daisies to the garden, about how many daisy plants were donated?

21. The PTC bought 32 trays of assorted flowers. Each tray contains 48 flowers. About how many plants did the PTC buy?

22. A rose bush costs $6.89. The PTC bought 18 bushes. About how much did the PTC spend on rose bushes?

23. The PTC bought 35 bags of top soil. Each bag weighs 40 pounds. About how many pounds of top soil did they buy?

24. The PTC held a fun fair to help with the cost of the garden. Adult tickets cost $5.75 and children's tickets cost $2.25. The PTC sold 89 adult tickets and 112 children's tickets. About how much money did they make?

A Camping Trip

Estimate, then solve each problem. Use your estimate to check whether your solution is reasonable.

Ana and her family are going camping at the Potawatomi (POD-A-WAD-TO-MI) State Park in Wisconsin. The fee to camp in the park is $12.00 a night for families who live in Wisconsin and $16.00 a night for people who do not live in Wisconsin.

1. Ana's family is arriving on Tuesday and staying until Sunday afternoon. Ana's family lives in Illinois. How much does Ana's family have to pay?

2. Nadia, a fourth grader from Milwaukee, Wisconsin, is also camping from Tuesday until Sunday. How much does Nadia's family pay to stay at the park?

3. How much more does Ana's family pay than Nadia's?

A campground is a big area where many people can pitch their tents. A campground is divided into campsites. Each family that camps at the park gets a campsite.

4. There are 4 campgrounds in the park. Each campground has 120 campsites. How many families can stay at the park at one time?

5. Ana, who is 10 years old, Felipe, her 12-year-old brother, and Dalia, her 5-year-old sister, want to explore the park. Their father and mother ask them to help put up the tent. Dalia looked at her watch when they began. It was 2:06. They finished setting up the tent at 2:52. How long did it take them to set up the tent?

6. One evening, Ana's family ate dinner at a restaurant in a nearby town. The restaurant had an "all-you-can-eat" fish dinner. The cost is $7.00 for adults, $5.00 for children ages 6–11, and $2.50 for children ages 3–5. If the whole family orders the fish dinner, how much will the total bill be?

Ana's 11-year-old cousin Roberto, his mother, and his little sister Angela come to the park on Friday. They set up their tent at a nearby campsite. The families decide to go canoeing together. It costs $8.00 to rent a canoe for 1 hour. Each canoe holds 3 people.

7. Every person in a boat needs a life jacket. How many adult life jackets do they need? (Adults are ages 12 and up.) How many children's life jackets do they need?

8. How many canoes will they need to rent? There must be at least one adult in every canoe. (Adults are ages 12 and up.) Draw a picture of how Ana's and Roberto's families can seat themselves in the canoes.

9. How much will they have to pay if they canoe for 1 hour?

10. How much will they have to pay if they canoe for 2 hours?

11. **A.** After canoeing, Ana's and Roberto's families decide to stop for frozen yogurt at the snack shop. A single-dip cone is 69¢. The price includes tax. Estimate how much money is needed to buy a cone for every person in both families.

B. Ana gives the cashier $10.00 to pay for all the cones. About how much change will she get back?

12. The families decide to go on a long hike Saturday morning. Ana's father brought ingredients to make gorp. Gorp is a high-energy snack that is easy to take on a hike. Here is his recipe for 4 servings of gorp:

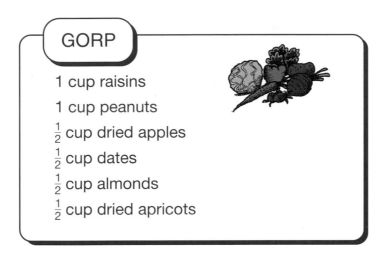

GORP

1 cup raisins

1 cup peanuts

$\frac{1}{2}$ cup dried apples

$\frac{1}{2}$ cup dates

$\frac{1}{2}$ cup almonds

$\frac{1}{2}$ cup dried apricots

Ana's father wants to make a serving of gorp for everybody in the group. How many servings of gorp does he need to make?

13. How many cups of peanuts should he use?

14. How many cups of dried apricots should he use?

15. Driving home from their vacation, Roberto's mother says they will travel about 50 miles in 1 hour. It takes them about 6 hours to get home. About how far is the park from their home?

16. Make up a story for the multiplication problems below and solve the problems. Your story can be about camping or anything else.

$$\begin{array}{ccc} 12 & 8 & \$1.75 \\ \times 8 & \times 6 & \times 8 \end{array}$$

Unit 8

Measuring Up: An Assessment Unit

	Student Guide	Discovery Assignment Book	Adventure Book	Unit Resource Guide*
Lesson 1				
Volume	●	●		
Lesson 2				
Fill It First	●	●		
Lesson 3				
Volume vs. Number	●	●		
Lesson 4				
Review Problems	●			
Lesson 5				
Hour Walk				●
Lesson 6				
Midyear Test				●
Lesson 7				
Midyear Experiment and Portfolio Review	●	●		
Lesson 8				
Facts I Know: Multiplication and Division Facts	●	●		●

Unit Resource Guide pages are from the teacher materials.

Volume

The Crow and the Pitcher

This is a very old story of a very thirsty crow. The crow, ready to die of thirst, flew with joy to a pitcher that he saw some distance away. When he came to the pitcher, he found water in it; but it was so near the bottom he was not able to drink. He tried to knock over the pitcher so he might at least get a little of the water. But he did not have enough strength for this. At last, seeing some pebbles nearby, he dropped them one by one into the pitcher. Little by little, he raised the water to the very brim and satisfied his thirst.

1. Why did the water in the pitcher rise?

2. Do you think the water in the pitcher rose the same amount each time a pebble was dropped in? Why or why not?

The **volume** of a rock is the amount of space it takes up. The volume of the pitcher is the amount of space inside it.

We measure volume in cubic units. A **cubic centimeter** (cc) is the amount of space taken up by a cube that is one centimeter long on each side.

1 cubic centimeter

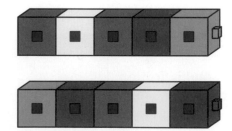

What is the total volume of these centimeter connecting cubes?

We can also measure the volume of an object using a graduated cylinder. This method is called **measuring volume by displacement** because you find out how much water the object displaces or pushes away.

3. Look carefully at the scale of the graduated cylinder before the cubes are added.

   A. How much water is in this graduated cylinder?

   B. How much water did the cubes displace or push away?

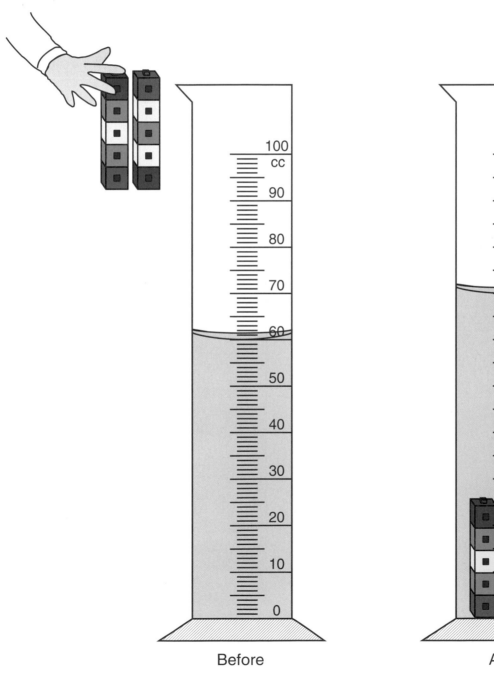

Before

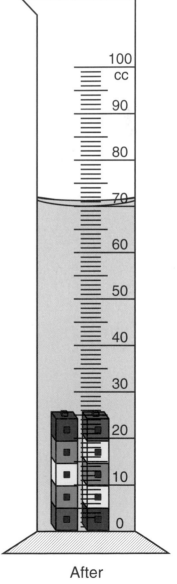

After

4. We can estimate the volume of a rock by making a model of the rock using centimeter connecting cubes and then counting the cubes. Estimate the volume of the rock using the picture of the cubes.

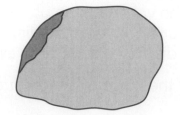 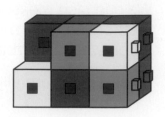

We can find a more exact measure of the volume of the rock by putting it into a graduated cylinder. The volume of the rock is the amount of water it displaces or pushes away.

5. A. Look carefully at the scale of the graduated cylinder before the rock is added. How much water is in the graduated cylinder?

B. Look at the scale after the rock has been added. What is the volume of the rock?

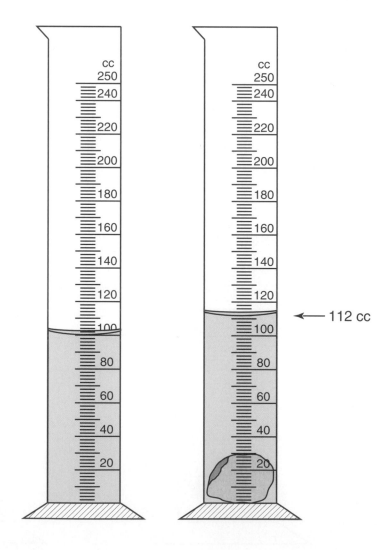

← 112 cc

Units of Volume

We can find the volume of a pitcher using a graduated cylinder, also. We sometimes measure liquid volume in **milliliters (ml)** or **liters (l).**

One milliliter is the same as one cubic centimeter.

1 cc = 1 ml

One liter is 1000 milliliters.

Jackie put water in a graduated cylinder until it reached the 250-cc mark. She emptied the cylinder into a pitcher. She did this four times until the pitcher was full.

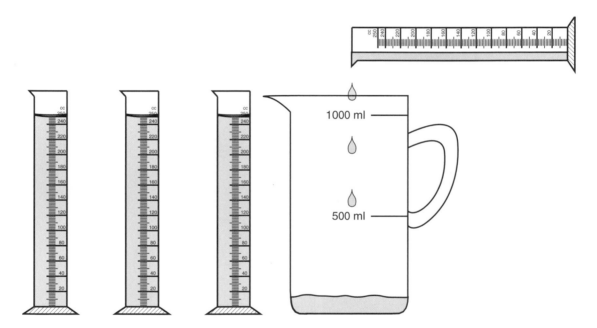

6. What is the volume of the pitcher? Give your answer in cubic centimeters.

7. How many milliliters does the pitcher hold?

8. Give the volume of the pitcher in liters.

9. A cubic foot is the amount of space taken up by a cube that is one foot long on each side. The volume of a 12-pack of soda is about half a cubic foot. What is the largest number of 12-packs that will fit into a refrigerator that can hold 15 cubic feet?

Fill It First

In this game, teams predict the rise in water level as they add marbles to a graduated cylinder. The winner is the team with the highest score after each team has added a total of 20 cc of marbles.

Players

This is a game for two teams of two students.

Materials for Each Team

- 1 100-cc graduated cylinder
- pitcher of water
- eyedropper
- score sheet
- spinner
- marbles

Rules

1. Fill your graduated cylinder with water to the 50-cc mark:

   - Add water to the cylinder until it almost reaches the 50-cc mark.
   - Use an eyedropper to carefully add the last few drops.
   - To read the water level correctly, put your eyes at the level of the water.
   - When water creeps up the sides of a cylinder, it forms a **meniscus.** This makes it look as though there are two lines. Read the lower line of the meniscus.

---

2. Each team makes a score sheet like the one shown here:

Fill It First Score Sheet

Predicted Volume in cc	Actual Volume in cc	Points

3. Team 1 spins the spinner to find out whether to add one, two, three, or four marbles to the water in their graduated cylinder.

4. Before adding the marbles, Team 1 predicts the volume of the marbles; they record the prediction on the score sheet.

5. Team 1 carefully slides the marble(s) into the water and calculates the volume of the marbles. If their prediction is correct, the team scores one point for each marble added. Both teams must agree on the actual volume of the marbles. Then Team 2 takes a turn.

6. When it is Team 1's turn again, they spin to find the number of marbles they will add to their cylinder. Before adding the marbles, team members predict the **total** volume of all the marbles in the cylinder after the new marbles are added. Then they slide the marbles into the cylinder. If they predicted correctly, they earn a point for each new marble added.

7. The first team to have a total of 20 cc of marbles in their cylinder scores an extra three points. The team then waits for the other team to have a total of 20 cc of marbles.

8. The team with the highest score wins.

Order of Operations Review

Solve the problems. Remember to use the correct order of operations. First, do all the multiplications and divisions, working from left to right. Then, do all the additions and subtractions, working from left to right.

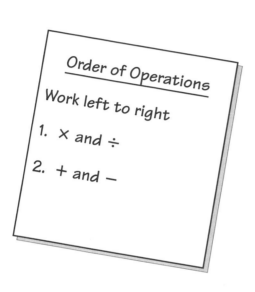

If you are using a calculator from home, check to see if it follows the correct order of operations. Try the first example. If the answer in the display is 10, then the calculator follows the order of operations. If the calculator gives an answer of 1, then it does not follow the order of operations.

Here are two examples:

A. $6 \times 2 - 8 \div 4 = ?$ (First multiply 6 times 2. Also, divide 8 by 4.)
$12 - 2 = ?$ (Then subtract 2 from 12.)
$12 - 2 = 10$

B. $8 + 4 \times 6 - 1 = ?$ (First multiply: $4 \times 6 = 24$.)
$8 + 24 - 1 = ?$ (Then add and subtract from left to right.)
$32 - 1 = 31$

1. $6 \times 4 \div 8 + 1 = ?$ **2.** $6 \times 4 \div 1 + 8 = ?$

3. $6 \times 4 \div 8 \times 1 = ?$ **4.** $8 + 16 \div 4 = ?$

Operation Target

Operation Target is a game that can be played many ways. One set of rules for the game is in Unit 7 Lesson 1. Here is another way to play:

- Use the four digits 1, 4, 6, and 8 and four operations (+, −, ×, and ÷).
- You must use each of the four digits just once.
- You can use each operation more than once or not at all.
- You can make two-digit numbers by putting two digits together. For example, you can use the numbers 14 or 68.
- No operation should give you a fraction, a decimal, or a negative number.

5. Here is a way to make the number 1:
$4 - 18 \div 6 = ?$ (First divide: $18 \div 6 = 3$)
$4 - 3 = 1$ (Then subtract 3 from 4.)

Find another way to make the number 1, following the new rules.

6. Make at least five numbers using these rules.

7. What is the largest number you can make?

8. What is the smallest number you can make?

Challenge: Make all the numbers from 0 to 9.

Volume vs. Number

Planning the Experiment

Mrs. Dewey's class is playing the game *Fill It First,* using marbles. Frank and Nicholas are on one team; Jackie and Maya are on the other team. Jackie and Maya are winning. They have been able to correctly predict the volume of the marbles more often than Frank and Nicholas.

"We lost again. They sure are lucky!" said Frank. "Their predictions are right almost every time."

"It may not be luck," said Nicholas. "They may have a way to figure out what the volume will be."

"How do they do it?" asked Frank.

"I don't know," said Nicholas. "Maybe they are following a pattern."

"I sure can't see a pattern on our score sheet. The numbers jump around too much," said Frank. "Before we play again, let's make a table that can help us make better predictions. We can find the volume of different numbers of marbles. Then we can look for a pattern that can help us."

"Won't that be cheating?" asked Nicholas.

Frank and Nicholas's Score Sheet

Predicted Volume in cc	Actual Volume in cc	Points
10 cc	6 cc	0
12 cc	11 cc	3
15 cc	13 cc	0

"Not if we figure it out for ourselves," said Frank. "We need to do something. Mrs. Dewey said that the next time we play the game, we are going to use a different-size marble. We'll have to think about the numbers of marbles, volume, and which size marble we're using. We'll never be able to keep all that straight."

"Okay. Let's figure something out and ask Mrs. Dewey if we can work on it before we play the game again," said Nicholas. "Whatever we do, we'll have to do two experiments—one for each size marble."

Here are Frank and Nicholas's data tables:

Small Marbles

N Number of Marbles	V Volume in cc
2	
4	
8	

Large Marbles

N Number of Marbles	V Volume in cc

Mrs. Dewey thought Frank and Nicholas had a good idea. She encouraged all the teams to design experiments to help them make predictions: "If you use the TIMS Laboratory Method, it will help you organize your thinking."

You are going to do the experiments Mrs. Dewey's class is working on. Your results should help you make predictions when you play the game *Fill It First*. Here are some things to think about as you plan your experiments:

Discuss

1. What are the two main variables in the experiments?

2. **A.** Which of the two main variables is the manipulated variable? Justify your answer.

   B. Which is the responding variable? Justify your answer.

   C. What letters will you use to stand for the two main variables?

3. The boys want to look for patterns in the data to help them make predictions. They want to predict the volume of the marbles when they know the number of marbles. What important variable should be held fixed in each experiment so they can do this?

4. A. What values for the number of small marbles did Frank and Nicholas choose?

  B. How will these values help them see patterns?

  C. What values will you choose?

Conducting the Experiment

You will collect and organize data. The data will help you make predictions about the volume of marbles when you know the number of marbles. You will investigate two sizes of marbles, so you will need to do two experiments.

Make a plan for your experiments. Draw a picture of your plan. Be sure to identify the variables in your picture. Label the main variables with letters.

- Label the columns in each of the *Two-column Data Tables* with the variables. Include units where needed.
- Your teacher will help you choose at least three values for the manipulated variable. The largest value will be no more than 8 marbles.

Volume vs. Number Data Tables

Experiment 1 Small Marbles	

Experiment 2 Large Marbles	

- Use a 250-cc graduated cylinder to measure the volume of the marbles.

- Choose a convenient amount of water to use in your cylinder. Use at least 140 cc.
- Collect the data.

Graph your data on a sheet of *Centimeter Graph Paper*.

- Label each axis, and write in the units.
- The scale on the horizontal axis should go to 15 or more.
- The scale on the vertical axis should go to 40 or more.

5. Plot your data points for both sizes of marbles on a single sheet of graph paper.

6. When the number of marbles equals 0, what is the volume of the marbles? Add this point to your graph for both sets of data.

7. **A.** If your points for the small marbles lie close to a line, use a ruler to draw a best-fit line.

   B. If your points for the large marbles lie close to a line, use a ruler to draw a best-fit line.

8. Why does the water level rise when you add marbles to the graduated cylinder?

9. Compare the line for the larger marble to the line for the smaller marble. How are they alike? How are they different?

10. **A.** Use your graph to estimate the volume of 7 small marbles. Show your work on your graph. Record your estimate.

    B. Did you use interpolation or extrapolation to find your answer?

    C. Check your estimate. Measure the volume of 7 small marbles. Record the volume.

D. How close is your estimate to the measured volume? Is it within 1 or 2 cc?

E. If your estimate was not close, you may need to correct your data on your graph before answering any more questions. Your teacher can help you decide.

11. **A.** Predict the volume of 15 small marbles using your graph. Show how you made your prediction. Record your prediction.

B. Did you use interpolation or extrapolation to find your answer?

C. Check your prediction by measuring the volume of 15 small marbles. Record the measured volume.

D. How close is your prediction to the measured volume? Is it within 3 cc?

12. **A.** Estimate the volume of 24 small marbles. Explain how you made your prediction and record it.

B. Check your prediction by measuring the volume of 24 small marbles. Record the measured volume.

C. How close is your prediction to the measured volume? Is it within 4 cc or 5 cc?

13. Irma and Jessie are playing *Fill It First* with the same large marbles you used. When they add the marbles for their turn, they will have a total of 9 marbles. Help them predict the volume of the 9 marbles. Show how you made your prediction.

14. Irma and Jessie have 150 cc of water in their graduated cylinder. They have added large marbles until the water level is 168 cc. How many large marbles are in the cylinder? Explain how you know.

15. Two students brought marbles from home. Keenya did the experiment with her marbles, and Jacob did the experiment with his. They graphed their data on the same graph. Which line (A or B) did Keenya draw? Explain.

Keenya's Marbles

Jacob's Marbles

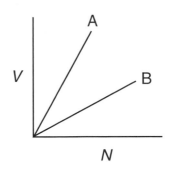

Review Problems

You will need a ruler and a calculator to complete the problems.

1. Make a factor tree to find the prime factors for the following numbers.

   A. 28   **B.** 124   **C.** 125

2. Solve the following problems without a calculator.

   A. $\begin{array}{r} 1267 \\ +1499 \\ \hline \end{array}$ **B.** $\begin{array}{r} 17{,}146 \\ -\quad 459 \\ \hline \end{array}$ **C.** $\begin{array}{r} 3000 \\ \times 9 \\ \hline \end{array}$

   D. $\begin{array}{r} 49 \\ \times 7 \\ \hline \end{array}$ **E.** $\begin{array}{r} 500 \\ \times 8 \\ \hline \end{array}$ **F.** $\begin{array}{r} 70 \\ \times 60 \\ \hline \end{array}$

3. Write the following numbers using base-ten shorthand.

   A. 467   **B.** 7615   **C.** 1042

4. Tell what the circled digit stands for in each of the following problems. The circled digit in the example stands for 30 or 3 tens.

   Example:
   $$\begin{array}{r} 1\,③\,2 \\ +7\,9 \\ \hline \end{array}$$

   A. $\begin{array}{r} 20{,}004 \\ \times 7 \\ \hline 2\,⑧ \end{array}$ **B.** $\begin{array}{r} 4236 \\ +2\,④\,9 \\ \hline \end{array}$ **C.** $\begin{array}{r} ① \\ 1278 \\ +5053 \\ \hline 1 \end{array}$

5. Find the median for this set of numbers: 12, 15, 17, 13, 9, 25, 12, 17. Then use a calculator to find the mean.

6. Write each of the following numbers in words.

   A. 1214   **B.** 77,589   **C.** 134,121

7. Write the following words as numbers.

   A. two thousand twenty-four

   B. forty-four thousand, three hundred sixty-nine

   C. two hundred sixty-five thousand, three hundred twenty-eight

8. Label the following angles as acute, right, or obtuse.

A.
B.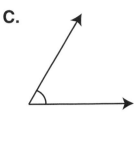
C.

9. Find the area of the following shape in square inches.

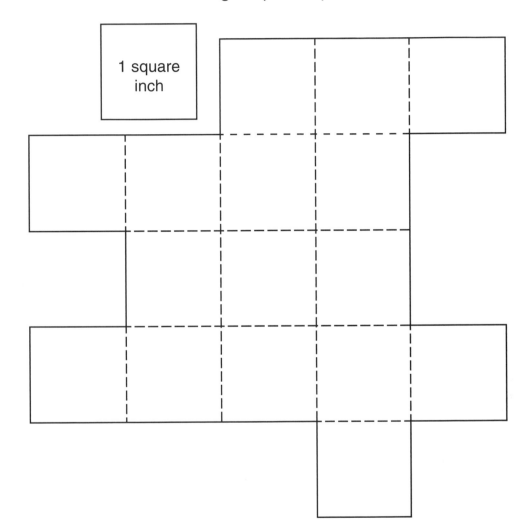

1 square inch

10. Find the perimeter of the shape.

11. Solve the given fact. Then write all the other number sentences in the same fact family.

A. $28 \div 4 =$
B. $7 \times 7 =$

Midyear Experiment and Portfolio Review

This is a good time to review your portfolio. Reviewing your work can help you see how much you have improved in math since the beginning of the year.

Experiment Review

The students in Mrs. Dewey's room were looking back through the labs in their collection folders. Ming found the survey he completed in the beginning of the school year that studied his classmates' main interests. He decided to compare this work with other labs he had completed during the first half of the year. He organized his work on an *Experiment Review Chart* in the *Discovery Assignment Book.*

Reviewing the labs you did this year is a good way to help you choose work for your portfolio.

1. Look through your *Student Guide* and your collection folder. Help your class make a list of the labs you have completed so far this year.

2. Use the following questions as you review a lab. Record your answers on the *Experiment Review Chart.*

   A. What variables did you study in this lab?

   B. Did you have to keep any variables fixed so the experiment would be fair? If so, which ones?

   C. Did you measure anything? If so, what did you measure? What units did you use?

D. How many trials did you do?

   E. Describe the shape of your graph.

   F. What were the most important problems you solved using your data and your graph?

Portfolio Review

3. If you have not done so recently, choose items from your collection folder to add to your portfolio.

4. Your *Experiment Review Chart* is a good choice for your portfolio.

5. Choose one or two pieces of work from this unit to include in your portfolio. Select pieces that are like other pieces of work that you put in your portfolio earlier in the year. For example, if you already have a lab in your portfolio, add the lab *Volume vs. Number*. Or, if you included a written solution to a problem like *Helipads for Antopolis* from Unit 2 or *Professor Peabody Invents a Ball* from Unit 5, then add *Hour Walk* to your portfolio now. Your teacher may help you make your choices.

6. Add to your Table of Contents. The Table of Contents should include the name of each piece of work, a short description of the work, and the date it was finished.

7. Write a paragraph comparing two pieces of work in your portfolio that are alike in some way. For example, you can compare two labs or your solutions to two problems. One piece should be new, and one should be from earlier in the year. Here are some questions to think about as you write your paragraph:

   • Which two pieces did you choose to compare?
   • How are they alike? How are they different?
   • Do you see any improvement in the newest piece of work as compared to the older work? Explain.
   • If you could redo the older piece of work, how would you improve it?
   • How could you improve the newer piece of work?

8. Write about your favorite piece of work in your portfolio. Tell why you like it. Explain what you learned from it.

Facts I Know: Multiplication and Division Facts

Picturing Fact Families

1. The picture below represents the following problem: If a rectangle has a total of 20 squares in 4 rows, how many squares are in each row?

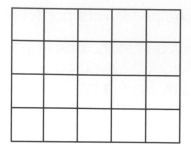

   What division sentence describes this problem?

2. The picture below represents the following problem: If a rectangle has a total of 20 squares in 5 rows, how many squares are in each row?

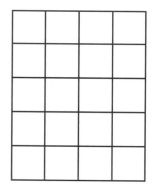

   A. What division sentence describes this problem?

   B. These two division sentences are members of the same **fact family.** What are the other number sentences in this same fact family?

3. Solve the given fact. Then name other facts in the same fact family.

   A. $9 \times 7 = ?$ **B.** $6 \times 4 = ?$ **C.** $7 \times 8 = ?$

Division Facts and *Triangle Flash Cards*

4. The directions that follow tell you how to use your *Triangle Flash Cards* to practice the division facts. Work with a partner. Use your *Triangle Flash Cards: 5s* and *10s.*

A. One partner covers the number in the square. This number will be the answer to a division problem. The answer to a division problem is called the **quotient.** The number in the circle is the **divisor.** The divisor is the number that divides the largest number on the flash card. The second person solves a division fact with the two uncovered numbers as shown below.

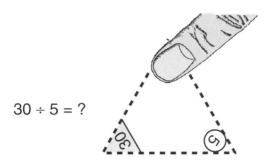

$30 \div 5 = ?$

B. Divide the cards into three piles: those facts you know and can answer quickly, those you can figure out with a strategy, and those you need to learn.

C. Begin your *Division Facts I Know* chart. Circle the facts you know well and can answer quickly.

For example, Jacob knew $30 \div 5 = 6$. The 5 is the divisor, so Jacob circled the 30 in the row for a divisor of 5.

Division Facts I Know

×	0	1	2	3	4	5	6	7	8	9	10
0	0	0	0	0	0	0	0	0	0	0	0
1	0	1	2	3	4	5	6	7	8	9	10
2	0	2	4	6	8	10	12	14	16	18	20
3	0	3	6	9	12	15	18	21	24	27	30
4	0	4	8	12	16	20	24	28	32	36	40
5	0	5	10	15	20	25	(30)	35	40	45	50
6	0	6	12	18	24	30	36	42	48	54	60
7	0	7	14	21	28	35	42	49	56	63	70
8	0	8	16	24	32	40	48	56	64	72	80
9	0	9	18	27	36	45	54	63	72	81	90
10	0	10	20	30	40	50	60	70	80	90	100

(Divisor — label along left side of table)

Recording $30 \div 5 = 6$ as a Fact I Know.

D. Sort the 5s and 10s cards again. This time your partner covers the number in the circle. The number in the square is now the **divisor.** The covered number in the circle is the answer to the division problem—the **quotient.** If we use the same example, 6 is now the **divisor.** Jacob knew this division problem also, 30 ÷ 6 = 5; so he drew a circle around the 30 in the row for a divisor of 6 on his *Division Facts I Know* chart. He circled 30 twice on his chart.

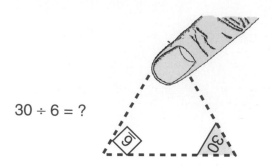

30 ÷ 6 = ?

E. Update your *Division Facts I Know* chart each time you go through the set of *Triangle Flash Cards.* Circle the facts you know well and can answer quickly.

F. Discuss how you can figure out facts you do not recall right away. Share your strategies with your partner.

G. Practice the last two piles at home for homework—the facts you can figure out with a strategy and those you need to learn. Make a list of these facts.

5. As you practice the division facts and update your *Division Facts I Know* chart, compare it to your *Multiplication Facts I Know* chart. Look for facts in the same fact family. Do you know any complete fact families? Which family or families? Explain.

You will continue to use *Triangle Flash Cards* to study all the groups of division facts in the units to come. You will update your *Division Facts I Know* chart each time you go through the cards. If you know one or two of the facts in a fact family, use those facts to help you learn the others.

Unit 9

Shapes and Solids

	Student Guide	Discovery Assignment Book	Adventure Book	Unit Resource Guide*
Lesson 1				
Lines	●			
Lesson 2				
What's Your Angle?	●	●		
Lesson 3				
Symmetry	●	●		
Lesson 4				
Journey to Flatopia			●	
Lesson 5				
Prisms	●	●		
Lesson 6				
Finding the Volume of a Prism	●			
Lesson 7				
Building an Octahedron		●		
Lesson 8				
Constructing a Prism				●

Unit Resource Guide pages are from the teacher materials.

Lines

TIMSville

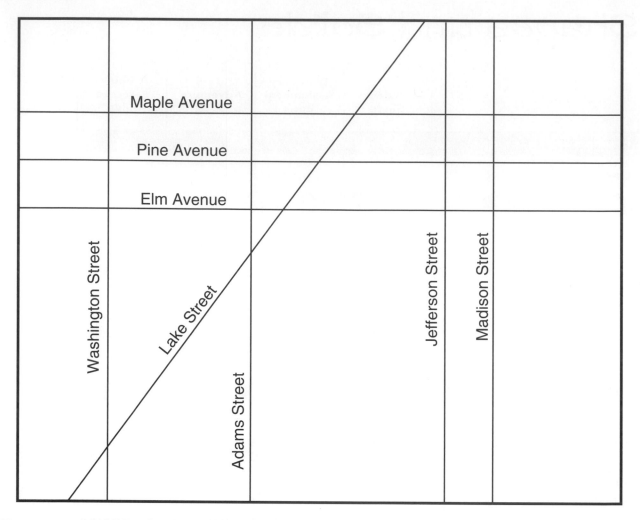

The map of TIMSville shows that Maple Avenue and Pine Avenue go in the same direction. We say Maple Avenue and Pine Avenue are **parallel.** These streets always stay the same distance apart. They will never **intersect** (meet) even if they go on, in both directions, forever.

Lake Street and Madison Street are not parallel. If you continued them, still going straight, they would intersect.

Pine Avenue and Jefferson Street are **perpendicular.** This means they form right angles where they intersect.

Washington Street and Lake Street intersect. However, they are not perpendicular.

1. Name a street parallel to Elm Avenue.

2. Name a street parallel to Madison Street.

3. Name a street perpendicular to Maple Avenue.

4. Name a street perpendicular to Washington Street.

5. Name two streets that Lake Street intersects.

On our map of TIMSville, we have drawn straight lines for each street. In mathematics "**line**" means "straight line." So, for the rest of this unit, we will always mean straight line when we say line.

Streets always have a beginning and an end. In mathematics, a line goes on forever in both directions. It has no beginning or end. What we have drawn on our maps are called line segments. A **line segment** has two endpoints. To draw a picture of a line, we draw a line segment and put arrows at the ends to show that it keeps going on infinitely (forever).

Lines are often named by two points on the line. This is line AC.

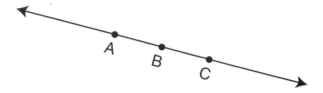

We write $\overleftrightarrow{AC}$. We can also call it $\overleftrightarrow{AB}$ or $\overleftrightarrow{BC}$.

The line segment AB is the part of the line that starts at A and ends at B. We write $\overline{AB}$ to show that we mean this segment.

6. Name two other line segments that are part of $\overleftrightarrow{AC}$.

7. This is line RT. ($\overleftrightarrow{RT}$)

A. What are two other names for this line?
B. Name at least 2 line segments that are part of $\overleftrightarrow{RT}$.

Rays are similar to lines, but they go on forever in only one direction. A ray has one endpoint.

8. This is ray XY. We can write ray XY like this: $\overrightarrow{XY}$. A ray is named by its endpoint first, followed by another point on the ray. What is another name for this ray?

9. Name the two rays you see here that make up angle R.

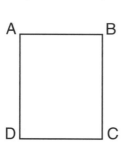

Talking about Shapes

ABCD is a rectangle. A rectangle has 4 line segments. Each line segment is part of a line. For example, $\overline{AB}$ is part of $\overleftrightarrow{AB}$.

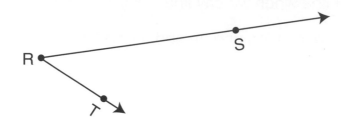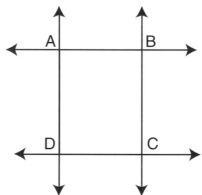

10. **A.** Name the 4 line segments that make up rectangle ABCD.

 B. $\overline{AD}$ is part of what line?

 C. $\overline{DC}$ is part of what line?

 D. There appear to be two pairs of parallel line segments. Name them.

 E. The angles of a rectangle are right angles. Name a line segment that is perpendicular to $\overline{AB}$.

 F. Name a line segment that is perpendicular to $\overline{CD}$.

 G. Name all the pairs of parallel lines.

 H. Name a line perpendicular to $\overleftrightarrow{AB}$.

11. Quadrilateral HIJK is a special quadrilateral called a **parallelogram.** A parallelogram is a quadrilateral with two pairs of opposite sides that are parallel.

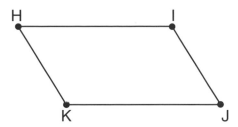

   A. Name two pairs of parallel line segments.

   B. Name two pairs of parallel lines.

   C. Measure all the line segments. What do you notice?

12. Is quadrilateral LMNO a parallelogram? Explain.

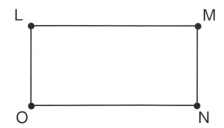

An **equilateral triangle** is a triangle that has three sides of equal length. All the triangles in this picture are equilateral triangles.

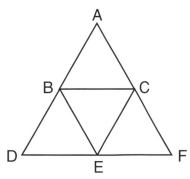

13. A. Name all the equilateral triangles you see in the figure.

   B. Name all the segments that appear to be parallel in the figure.

1. Name the line segments that form the sides of the parallelogram.

2. Name three points on the line segment NO.

3. Name the parallel line segments.

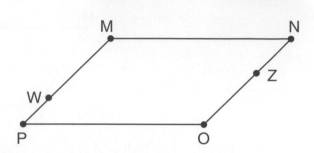

The figure below shows many lines, line segments, and rays.

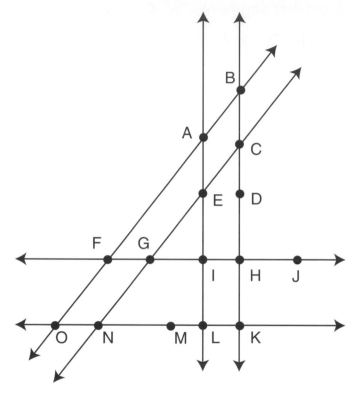

4. One triangle in this figure is triangle GEI. Name at least three other triangles in the figure.

5. Name three pairs of lines that appear to be parallel.

6. Name a pair of lines that appear to be perpendicular.

7. Name three shapes that appear to be parallelograms.

8. Which of the following pairs intersect (meet)?

   A. $\overrightarrow{LN}$ and $\overrightarrow{EC}$

   B. $\overrightarrow{AF}$ and $\overrightarrow{HI}$

   C. $\overline{IH}$ and $\overline{EC}$

   D. $\overleftrightarrow{LK}$ and $\overleftrightarrow{AB}$

What's Your Angle?

Looking at Angles

1. If both airplanes fly at the same speed, which airplane will be higher after 40 seconds? Why?

The paths of these two airplanes form angles with the ground. Different planes take off at different angles. A small plane has a climbing angle of about 10°. A jet can climb at a 30° angle. When a pilot is taking off, he or she needs to think about the climbing angle. Angles are also very important for drawing, building, and finding direction.

Recall that the sides of the angle meet at a point called the **vertex.** The sides of an angle are rays that have the same endpoint. Points are usually named with capital letters. We can call the angle here angle A and write ∠A as shorthand. We can also use 3 letters to name an angle. The angle here can be called ∠DAB or ∠BAD.

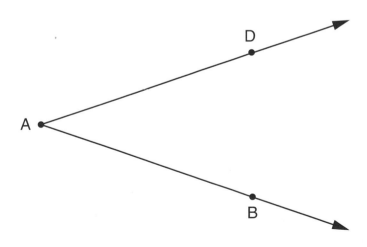

Measuring Angles

We measure the opening between the two sides of the angle in degrees. Protractors are used to measure angles. The measure of the angle shown here lies between 70° and 80°. Angle Z measures 73°. We write ∠Z = 73°. Notice where the vertex of the angle is placed.

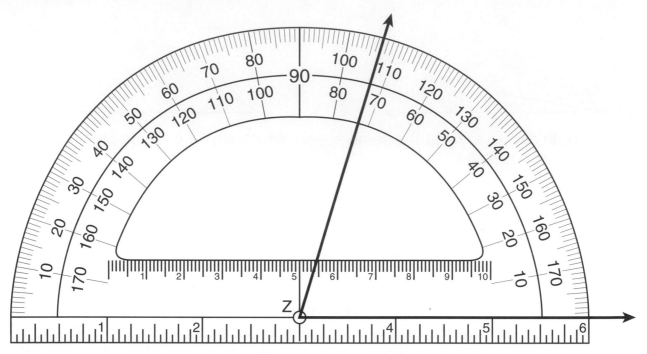

2. Is the angle shown above an acute or an obtuse angle?

3. **A.** Is the measure of this angle less than or greater than 90°?

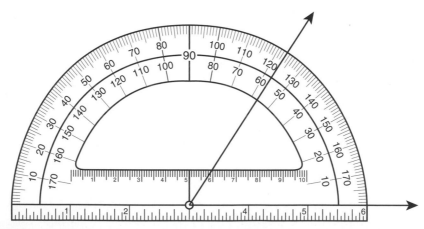

   B. Between which two numbers on the protractor does the measure of this angle lie?

   C. What is the measure of the angle?

4. A. Is the measure of this angle less than or greater than 90°?

B. Between which two numbers on the protractor does the measure of the angle lie?

C. Which scale do you use to read this angle measure— the inside scale on the protractor or the outside scale? How do you know?

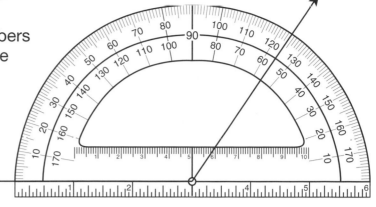

D. What is the measure of the angle?

Using Angles

5. How many acute angles do you see in the drawing? How could you name them?

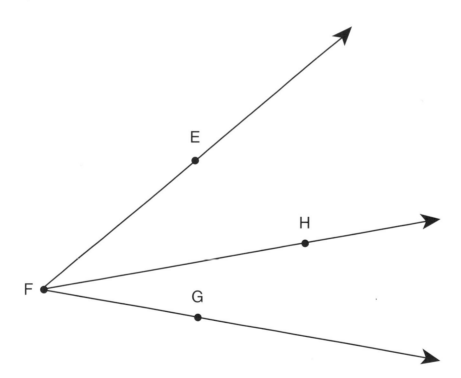

In the drawing, there are three angles that could be called ∠F. In order not to confuse them, we use three letters to name each angle. Find the measure of each angle.

6. ∠EFH =   **7.** ∠HFG =   **8.** ∠EFG =

The park district for the town of TIMSville has decided to build a new playground. One of the features will be a sandbox for little children. The town is asking for designs for the sandbox. A design must have all straight sides. The sandbox will be located between two sidewalks that meet at a 50° angle. So, one of the angles of the sandbox must be 50°.

Here is the design Ana submitted. She used the scale 1 cm = 1 foot.

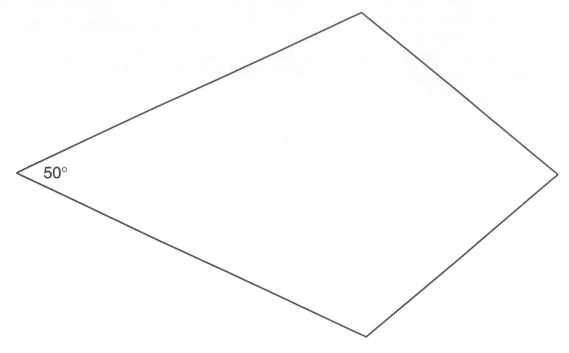

9. Design a sandbox for TIMSville's new playground. Your sandbox can have a different number of sides.

10. Design another sandbox for the playground. Remember, one of the angles must be 50°.

Shapes like the one Ana designed for the playground are called polygons. A **polygon** is a figure whose sides are line segments that are all connected. Every endpoint of a side meets the endpoint of exactly one other side and no sides overlap. The word *polygon* comes from the prefix *poly* (many) and suffix *gon* (angles). So, a polygon has many angles (or you could say many sides).

These shapes are polygons:

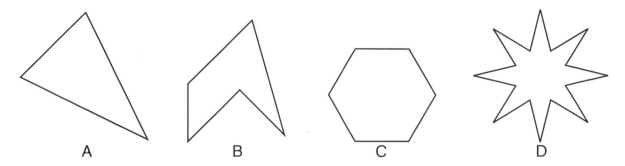

A   B   C   D

These shapes are not polygons:

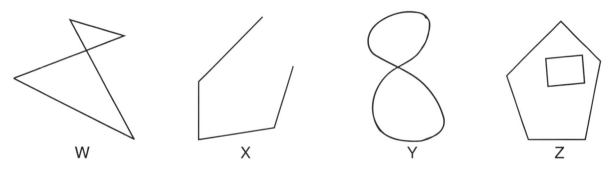

W   X   Y   Z

11. Explain why each shape above is not a polygon.

We often name polygons with letters. When we say the name, we go around the shape. For example, the shape Ana drew is a quadrilateral. A **quadrilateral** is a polygon with 4 sides (or we can say 4 angles).

There are many ways we can say the name of this quadrilateral. Three ways are: ABCD, CBAD, and DABC.

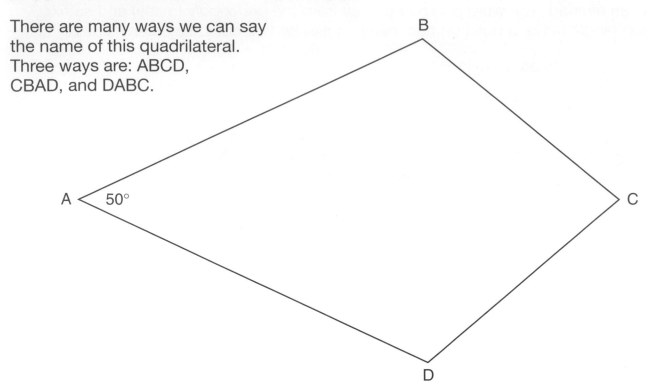

12. Find measures of all the angles in Ana's quadrilateral. Use a ruler to trace the design on another sheet of paper first.

   A. ∠A =   **B.** ∠B =   **C.** ∠C =   **D.** ∠D =

Homework

You will need a protractor to complete this homework.

1. Nila's design for a sandbox is shown here. She labeled the angles A, B, C, D, and E. The sandbox needed to have a 130° angle.

   A. Which angle do you think is 130°?

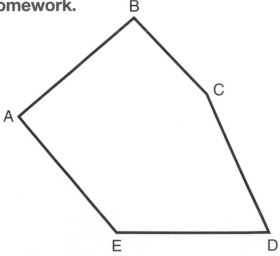

B. Trace the design on another paper. Extend the sides with the edge of your protractor. Then measure the angles of Nila's design.

∠A =

∠B =

∠C =

∠D =

∠E =

C. List 5 ways to name Nila's design.

2. Design a 3-sided sandbox for a playground. All the sides must be straight, and there must be a 45° angle in the design.

3. Design a 6-sided sandbox for a playground. All the sides must be straight, and there must be a 90° angle in the design.

4. A. Use a ruler to trace this figure first. Then find the measure of the angles.

∠RST = ?

∠TSU = ?

∠RSU = ?

B. What is the sum of the measures of ∠RST and ∠TSU? How does the sum compare to the measure of ∠RSU?

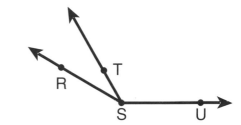

5. James works for the Sparkling Clean Window Washing company. His boss told him that his ladder must make an angle with the ground between 65° and 75° to be safe to climb. The picture here shows his ladder leaning against a building.

A. What is the measure of the angle between the ladder and the ground?

B. Will this ladder be safe to climb?

C. What is the measure of the angle between the ladder and the building (the top angle)?

D. What is the measure of the angle between the house and the ground?

6. The picture below shows many angles. Describe 5 angles you see in the picture.

7. Below is a side view of a playground slide. Measure the angles.

∠A =
∠B =
∠C =
∠D =

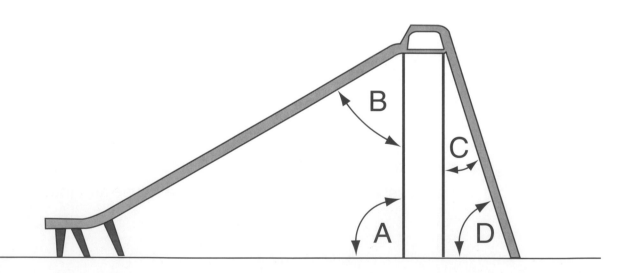

Symmetry

Turn and Line Symmetry

The picture below has a pattern that repeats as you turn it around the point in the center. This picture has **turn symmetry.** The **center of turning** is the point in the middle. The picture has $\frac{1}{6}$-**turn symmetry:** if you turn it $\frac{1}{6}$ of a full 360° turn, it lines back up with itself.

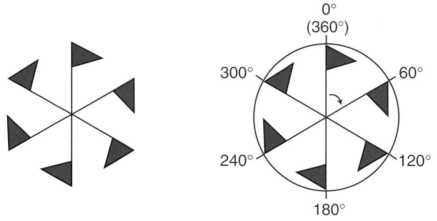

The heart shown here does not have turn symmetry. The heart has a different kind of symmetry called **line symmetry.** The heart has one line of symmetry. If you fold the heart along the **line of symmetry,** the two pieces fit exactly on top of one another.

The square below has both turn and line symmetry.

 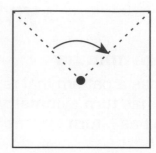

If you traced the figure, marked the center, and turned the square $\frac{1}{4}$ of a turn (90°) around the center, the traced figure would lie on the original. The square could be turned 4 times like this before it made a complete circle. The square has $\frac{1}{4}$-turn symmetry.

The square also has 4 lines of symmetry. You can fold the square on those lines and the halves of the square will match exactly.

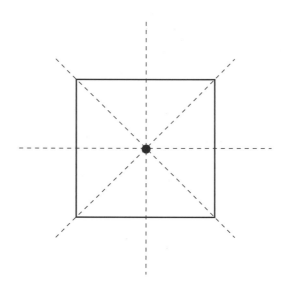

Discuss

1. Look at each shape below. Answer the following questions for each. You can use the shapes on the *Pattern Block Shapes* Activity Page in the *Discovery Assignment Book* to help you.

   - Does the shape have turn symmetry?
   - If the shape has turn symmetry, tell what kind of turn symmetry.
   - Does the shape have line symmetry?
   - If the shape has line symmetry, tell how many lines of symmetry.

A.

B.

C.

D.

E.

F.

G.

2. Many cultures use symmetry in their designs. Describe the symmetries you see in the following designs.

A. A design from ancient Pompeii (Roman Empire)

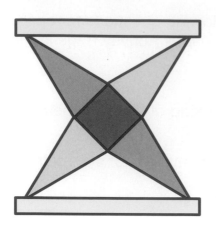

B. An Arabian design

C. An Apache design

D. A Latvian design

Making Spinners

The idea of turn symmetry is seen in many places. One example is a spinner used for games.

The workers at the TIMS Toy Company are designing spinners for board games. Here is a spinner that is divided into 3 equal pieces. When children play the game, there is an equal chance of the spinner landing on any of the 3 areas.

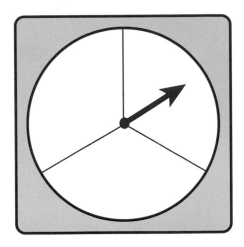

For Questions 3 and 4, you will need the *Blank Spinners* Activity Page in the *Discovery Assignment Book,* a protractor, and a calculator. Answer these questions after creating your spinners:

- **How big is the angle in each part of the spinner? Use your protractor to measure.**
- **What is the sum of all the angles? Does this make sense? Explain.**

3. Design a spinner that is divided into 4 equal pieces. Explain your work.

4. Design a spinner that is divided into 5 equal pieces. Explain your work.

Prisms

Mrs. Dewey's class is studying three-dimensional shapes. Another name for a three-dimensional shape is a **solid.** The class is thinking of ways to describe and record their shapes. Linda has a cereal box.

The sides of the box are called **faces.**

1. How many faces does Linda's cereal box have?

Two faces intersect in an **edge.**

2. How many edges does Linda's cereal box have?

The corners are called **vertices.**

3. How many vertices does Linda's cereal box have?

Linda wanted to draw her box. First she drew a rectangle for the front face.

Linda then drew parallel line segments, all the same length.

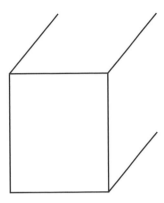

Then, Linda added two more edges:

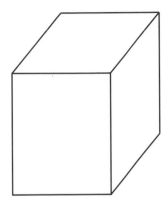

Grace showed Linda how to draw dotted lines to show the edges that cannot be seen. This helps us remember which edges are in the back, from our point of view.

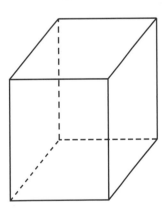

4. Do you know another way to draw a box? If you do, draw a box using your method. Otherwise, use Linda's method to draw a box. You may wish to look at a box as you draw.

Making Nets from Boxes

Linda was interested in the way boxes are made. She decided to flatten out her box. Linda cut along three of the top edges and four of the vertical edges. This is what she got.

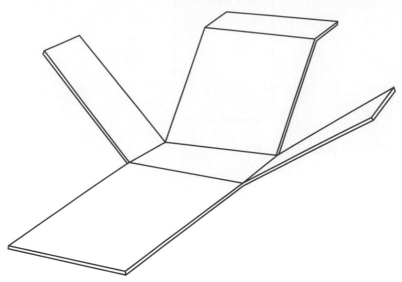

Linda flattened her box and turned it over. This is what it looked like:

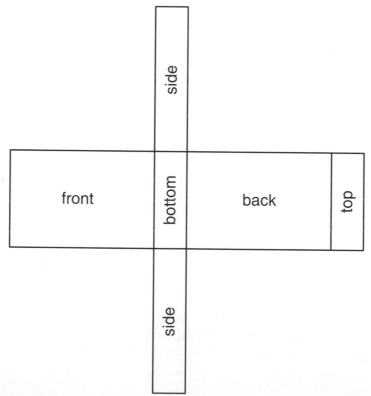

This is what we call a **net** for the box.

Prisms

5. There are other ways to flatten a box. Work with your group to make as many different nets of a box as you can. Each person in the group should have a box to cut. Make one net at a time. Discuss before you cut so that you do not repeat a net.

6. Draw a picture of your nets. Label the rectangles that were the bottom, top, and sides of the box.

Different Kinds of Prisms

Boxes are examples of special kinds of three-dimensional shapes. These shapes are prisms.

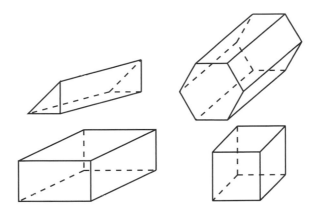

These shapes are not prisms.

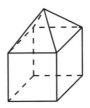

A **prism** is a three-dimensional shape. Prisms have two identical faces called bases. The bases are parallel to each other. The other faces are parallelograms.

One way to sketch a prism is to draw the two bases and then connect the matching vertices. You may want to trace the bases from a picture, a pattern block, or some other object. Here is a way to draw a pentagonal prism:

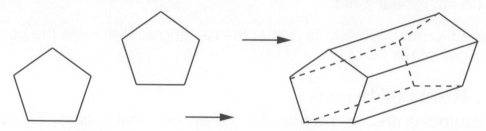

You may wish to use dotted lines to show the edges that are in the back of the figure. Prisms can be drawn with or without the dotted lines. Compare the drawing of the following pentagonal prism with the one above. Which one do you prefer?

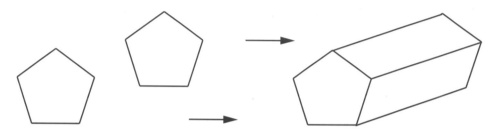

7. How many faces does a pentagonal prism have? How many edges? How many vertices?

Boxes

1. Find a box at home. Make a sketch of the box showing its length, width, and height. You can use centimeters or inches to measure.

For Questions 2–5, decide whether the figure is a net of a cube. You may need to trace the figures on a separate sheet of paper and cut them out.

2.

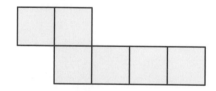

3.

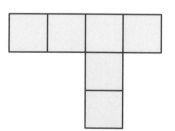

4.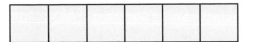

5.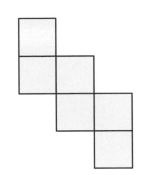

For Questions 6–12:

- **Decide whether the shape is a prism.**
- **If the shape is a prism, describe the bases.**
- **How many faces does the shape have?**

6.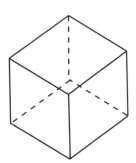

7.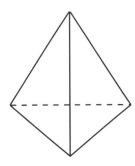

8.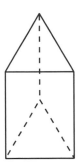

9.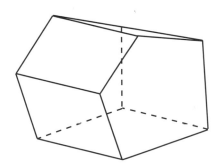

10.

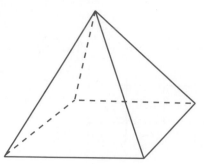

11.

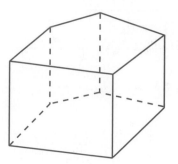

12.

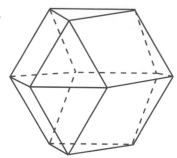

Finding the Volume of a Prism

Volumes of Prisms

Discuss

1. Use centimeter connecting cubes to build the prism shown in the picture at the right.

   A. How many centimeters tall is the prism? This is the prism's height.

   B. How many square centimeters cover the base of the prism? This is the area of the base.

   C. Count the number of cubic centimeters in the prism. This is the prism's volume.

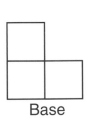

Base

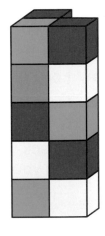

2. Build each of the three prisms shown below.

   A.

   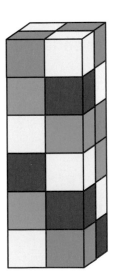

   B.

   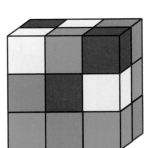

   C.

   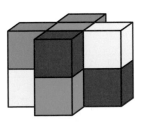

3. Copy the data table on a sheet of paper. Find the area of the base, the height, and the volume of each of the prisms in Question 2. Complete the table.

Prism	Area of Base in sq cm	Height in cm	Volume in cc
A			
B			
C			

4. **A.** Look at the information in the table. Describe any patterns you see.

   B. Describe a method for finding the volume of a prism without counting cubes.

   C. Check your method. Use it to find the volume of the prism in Question 1. Did you find the same volume using your new method as you did when you counted cubes?

Explore

One way to think about volume is to ask how many cubes (or parts of cubes) fit into the object.

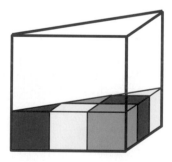

5. Jackie got a box of chocolates for her birthday. The box is a triangular prism.

Chocolate Triangles

Finding the Volume of a Prism

A. Jackie traced the base of the box on a sheet of *Centimeter Grid Paper* as shown here. What is the area of the base?

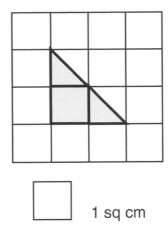

1 sq cm

B. The height (or length) of the box is 10 cm. Find the volume of Jackie's box. Use your method for finding the volume of a prism.

Juice Box Volume

6. About how much is a ml of juice—a cupful, a spoonful, or less than a spoonful? (Remember, 1 ml = 1 cc.)

7. Is your juice box a prism?

8. Find the volume of your juice box. Explain the steps you used. Give the volume of your juice box in ml.

9. Look at the label of your juice box. Was the volume you calculated close to the volume printed on the box? Explain.

You can find volumes of other prisms using the same method: Find the area of the base and then multiply by the height.

10. Find the volume of the right triangular prism you built in Lesson 5.

11. A. What is the area of the base of the prism shown in the picture below?

   B. What is the volume of the prism?

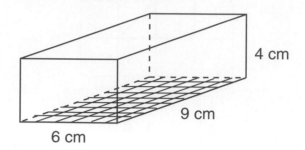

Homework

For Questions 1–4, a picture of a box is shown for each. Find the volume of the box. You can use a calculator to help you multiply.

1.

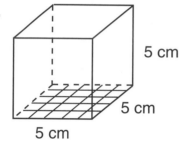

2.

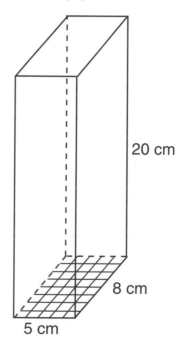

3. The area of the triangular base of this prism is 15 sq cm.

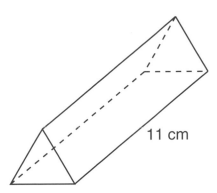

11 cm

4. The area of the hexagonal base of this prism is about 26 sq cm.

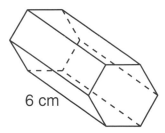

6 cm

5. Ana found that the area of the bottom (the base) of her purse is 90 sq cm. To find the volume, she measured the height of her purse. It is 15 cm tall. What is the volume of Ana's purse?

Finding the Volume of a Prism

Unit 10

Using Decimals

	Student Guide	Discovery Assignment Book	Adventure Book	Unit Resource Guide*
Lesson 1				
m, dm, cm, mm	●	●		
Lesson 2				
Tenths	●	●		
Lesson 3				
Hundredths	●	●		●
Lesson 4				
Downhill Racer	●			●
Lesson 5				
Decimal Hex		●		
Lesson 6				
Alberto in TenthsLand			●	

Unit Resource Guide pages are from the teacher materials.

m, dm, cm, mm

Meters (m)

Your teacher has made marks 1 and 2 **meters** above the floor. Use these marks to help you answer the questions below.

1. Are you more than 1 meter tall?

2. Are you more than 2 meters tall?

3. Are you closer to 1 meter or 2 meters?

4. As a class, measure objects around the classroom to the nearest whole meter. The symbol for meter is **m.** Keep track of the data on the *Class Measurement Tables* in the *Discovery Assignment Book.*

Class Measurement Table

Object	Measurement (nearest m)
Height of door	
Width of classroom	
Length of classroom	
Width of board	
Length of paper clip	
Length of pencil	

5. Jackie measured her calculator to the nearest meter and found it to be 0 meters long. What does this measurement (0 m) tell you? What unit would give you a better measurement for this length?

6. Sometimes it is good enough to measure to the nearest whole meter. Sometimes it is not. Usually, it does not make sense to measure a calculator to the nearest meter. List two things that probably should not be measured to the nearest meter.

7. List two things that you would measure to the nearest meter.

Decimeters (dm)

8. How many skinnies can you line up on a meterstick? Line them up along a meterstick to find out.

9. The length of one skinny is a **deci**meter. The symbol for decimeter is **dm.** How many dm are in 1 meter? How many dm are in $\frac{1}{2}$ meter?

10. A decimeter is $\frac{1}{10}$ of a meter. What do you think **deci-** means?

11. As a class, measure the objects in the table from Question 4 again. This time, measure to the nearest whole decimeter. Make a new table for this data like the one shown below.

Class Measurement Table

Object	Measurement (nearest dm)
Height of door	
Width of classroom	
Length of classroom	

12. John measured a paper clip to the nearest decimeter. He found it to be 0 decimeters long. What information does this measurement tell you? What unit would give you a better measurement for a paper clip?

13. Sometimes it is good enough to measure to the nearest whole decimeter. Sometimes it is not. List two things that should be measured to the nearest decimeter.

Centimeters (cm)

A **cent**ury is 100 years; a **cent**ennial is a 100-year anniversary; a **cent**ipede is said to have 100 legs. What do you think **cent-** means?

A **cent**imeter is $\frac{1}{100}$ of a meter and a **cent** is $\frac{1}{100}$ of a dollar. In other words, there are 100 centimeters in a meter and 100 cents in a dollar.

14. How many bits can you line up along a meterstick? How did you decide?

15. How long is a bit?

16. As a class, measure the objects from Question 4 again. This time, measure to the nearest whole centimeter. Make a new table for this data like the one shown below.

17. List two things it makes sense to measure to the nearest whole centimeter. List two things it does not make sense to measure in centimeters.

18. Find an object that measures less than 1 centimeter.

Class Measurement Table

Object	Measurement (nearest cm)
Height of door	
Width of classroom	
Length of classroom	

Millimeters (mm)

If you look at a meterstick very carefully, you will see little spaces between short lines. Each one of these spaces is a millimeter.

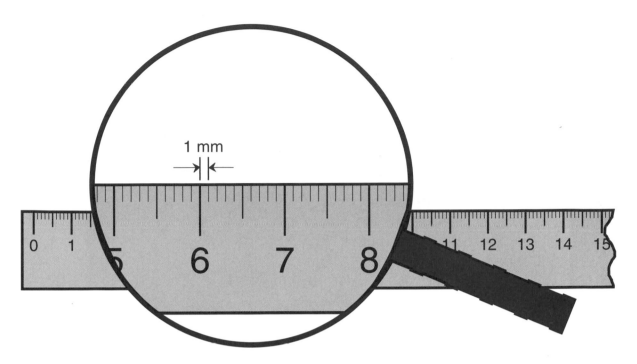

19. There are 1000 **millimeters** in a meter. What do you think **milli-** means?

20. How many millimeters are in a centimeter?

21. How many millimeters are in a decimeter? How did you decide?

22. People generally use millimeters to measure very short lengths. Give an example of when it would make sense to measure in millimeters.

Measuring with Metersticks, Skinnies, and Bits

John measured the length of the board in his classroom. First, he used two metersticks. He saw that less than half of a third meterstick would fit. John said, "To the nearest whole meter, this length is 2 meters."

Next, John put down four skinnies. John said, "Each meterstick is ten decimeters, and each skinny is one decimeter long. I have two metersticks and four skinnies. So, to the nearest decimeter, this length is 24 whole decimeters."

John still had a little space left. He put down one bit. One bit is one centimeter long.

Mrs. Dewey complimented John's work: "John, you have done a terrific job. You have found the length of the board using the fewest pieces. You can write down this measurement using decimals. How many metersticks did you use?" "Two," said John.

Mrs. Dewey wrote "2." on the board. "Each decimeter is one-tenth of a meter. How many skinnies did you use?" "Four," said John.

Mrs. Dewey wrote "2.4" on the board. "Each centimeter is one-hundredth of a meter. How many bits did you use?" "One," said John.

Mrs. Dewey wrote "2.41" on the board. "This number is read two and forty-one hundredths." John showed 2.41 m with two metersticks, four skinnies, and 1 bit.

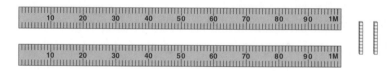

John continued to measure objects around the room to the nearest hundredth of a meter. For example, Mrs. Dewey had a life-size poster of a professional basketball player on the wall. John decided to measure the height of this player. He used 1 meterstick, 9 skinnies, and 8 bits. John wrote "1.98 meters" for this measurement. That told him the player's height was 1 meter, 9 decimeters, and 8 centimeters.

23. John wrote the following measurements on his paper. He used the fewest pieces for each measurement. How many metersticks, skinnies, and bits did John use for each measurement?

   A. 3.45 m

   B. 0.59 m

   C. 2.70 m

   D. 2.07 m

Dear Family Member:

In everyday life in the United States, we are slowly moving toward regular use of metric measurements. In scientific life, however, the metric system is already here. To succeed in a technological world, students need to know the metric system. This homework assignment will help your child become aware of the increasing use of metric units.

Thank you for your cooperation.

We often use customary units of measure (inches, pints, pounds, and so on) in everyday life. And there are many times we use metric units of measure (centimeters, liters, grams, and so on).

1. Look for metric units in the newspaper, on labels, and on other household items. Make a list showing what the unit is and what is being measured. If you can, bring in the item with the measurement on it.

2. **A.** Cut a piece of string 1 meter long.

   B. Carry your meter string with you all the time for one week.

   C. Estimate the length of various objects to the nearest meter. Then use your meter string to measure the objects to the nearest meter. Make a table showing the objects, your estimates, and your measurements.

3. Go on a measure hunt in your home. Look for objects that are between the specified lengths. Complete a data table like this one. (*Hint*: Half of your string is 0.5 m.)

Rule	Object
Between 1 and 2 m	
Between 1 and 1.5 m	
Between 0.5 and 1 m	

Tenths

A New Rule for Base-Ten Pieces

In the last lesson, Lee Yah used skinnies to measure to the nearest decimeter. She lined up skinnies along a meterstick. She learned that a skinny is one decimeter long and that a decimeter is one tenth of a meter.

1. **A.** How many skinnies can you lay along the edge of your meterstick?

   B. The length of one skinny is what fraction of a meterstick?

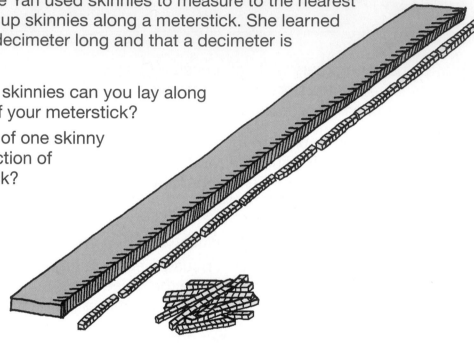

The fraction for one-tenth can be written as a common fraction ($\frac{1}{10}$). The **denominator** (the number on the bottom) tells us that the meterstick is divided into ten equal parts. The **numerator** (the number on the top) tells us that a skinny is one of these parts.

The fraction for one-tenth can also be written as a decimal fraction (0.1). The decimal point tells us that the numbers to the right of the decimal point are smaller than 1.

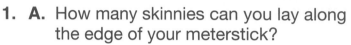
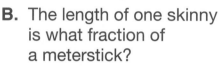

$$\frac{1}{10} \quad\begin{array}{l}\leftarrow \text{numerator} \\ \leftarrow \text{denominator}\end{array} \qquad\qquad \begin{array}{c}0.1 \\ \uparrow \\ \text{decimal point}\end{array}$$

2. Place 3 skinnies along the edge of the meterstick.

   A. The length of 3 skinnies is what fraction of a meter?

   B. Write this fraction as a common fraction and as a decimal fraction.

3. Place 5 skinnies along the edge of the meterstick.

   A. The length of 5 skinnies is what fraction of a meter?

   B. Write this fraction as a decimal fraction.

   C. Write this fraction as a common fraction in two different ways.

Doing mathematics is sometimes like playing a game. Just as you cannot play a game without rules, you cannot do mathematics without rules. But, just as people sometimes change game rules, we sometimes change rules in mathematics. And, just as we can still play a game if everyone agrees to the new rules, we can still do mathematics if everyone agrees to the new rules.

Now we are going to change a rule for base-ten pieces. When you worked with base-ten pieces before, usually a bit was one whole. When a bit is the unit, then a skinny is 10 units, a flat is 100 units, and a pack is 1000 units. Now we are going to change which piece is the whole. **For now, a flat will be one whole.**

4. Use skinnies to cover a flat.

   A. How many skinnies did you use?

   B. If a flat is 1 unit, then what fraction is a skinny?

5. A. Place 6 skinnies on your flat. Skip count by tenths as you place each skinny. Start like this: one-tenth, two-tenths, three-tenths. . . .

   B. What fraction of the flat is 6 skinnies?

   C. Write this fraction as a common fraction and a decimal fraction.

Tenths   SG • Grade 4 • Unit 10 • Lesson 2   **277**

6. **A.** Nicholas placed 4 skinnies on his flat. Put 4 skinnies on your flat. Skip count by tenths as you place each skinny.

  B. What fraction of the flat is 4 skinnies?

  C. Write this fraction as a common fraction and a decimal fraction.

7. **A.** Linda placed 10 skinnies on her flat. Put 10 skinnies on your flat. Skip count by tenths as you place each skinny.

  B. How many tenths is 10 skinnies?

  C. Linda noticed that 10 skinnies covered one whole. She recorded this 3 ways: $\frac{10}{10}$, 1, and 1.0. Explain how each of these represents the same number.

You can use the *Tenths Helper* to show how many tenths are in one whole and two wholes.

8. **A.** Cover your *Tenths Helper* with flats. How many flats did you use?

  B. What number does this represent?

9. **A.** Place 10 skinnies on your *Tenths Helper.* Count by tenths as you place each skinny on the chart.

  B. When you are skip counting by tenths, what number comes after 9 tenths? (*Hint*: There is more than one answer to this question.)

  C. Continue placing skinnies on your chart. What number will you say as you place the eleventh skinny on the chart? (*Hint*: There is more than one answer to this question.)

  D. How many skinnies does it take to fill the *Tenths Helper?*

  E. How many tenths are in two wholes?

Keenya uses a *Tenths Helper* to show tenths. First she places 12 skinnies on the chart. She counts by tenths as she places each skinny. Then she records the value of 12 skinnies on the chart.

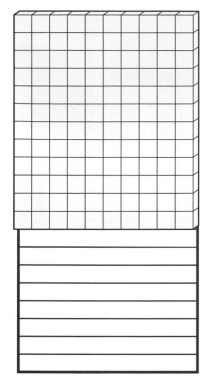

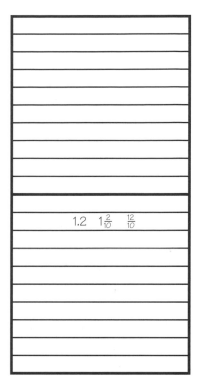

Place 12 skinnies on your chart.

Then record the value in more than one way.

Use the *Tenths Helper* to complete Questions 10–11. Use the skinnies to show each number. Then record its value on the chart in more than one way.

10. **A.** 17 skinnies

   B. 4 skinnies

   C. 15 skinnies

   D. 9 skinnies

   E. 1 skinny

   F. 13 skinnies

   G. 18 skinnies

11. Fill in all of the values in your *Tenths Helper*. If needed, use skinnies to build each number.

More Tenths

12. Romesh used base-ten pieces to make the following number. If a flat is one whole, what number did Romesh make?

13. If a flat is one whole, then what is the value of a pack? (*Hint*: how many flats does it take to make 1 pack?)

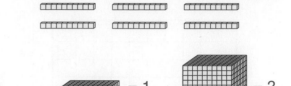

 = 1

= ?

If a flat is 1, the pieces below show a number a little more than 21.

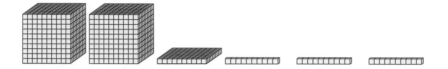

If a flat is 1, then these pieces are twenty-one and three-tenths. This can be written either 21.3 or $21\frac{3}{10}$. We can record this using base-ten shorthand.

$21\frac{3}{10}$ or 21.3

Explore

14. Use base-ten pieces to make these numbers. Then use base-ten shorthand to show what pieces you used. (Remember: the flat is one whole.)

   A. 1.7   **B.** 3.4   **C.** 0.6   **D.** 13.2   **E.** 10.1

15. Give decimals and common fractions for the base-ten shorthand below:

   A.

   B.

   C.

   D.

16. Use base-ten pieces to make a number between 2 and 3. Use base-ten shorthand to show the pieces you used. Then write your number.

17. Use base-ten pieces to make a number between 2.5 and 3. Use base-ten shorthand to show the pieces you used. Then write your number.

The Fewest Pieces Rule

The Fewest Pieces Rule says that you should trade ten base-ten pieces for the next size up whenever you can. Ten skinnies should be traded for a flat, and ten flats should be traded for a pack. For example, you might have these base-ten pieces:

Using the Fewest Pieces Rule, you should trade ten skinnies for a flat. You would then have these pieces:

For Questions 18–20, the flat is one whole. Follow these directions:

    A. Use base-ten shorthand to show how to make the number using the fewest base-ten pieces.

    B. Then write a decimal for the number.

First make each number with base-ten pieces. Then trade to get the fewest pieces. Finally, write the number.

18. ||||| |||||

19. ▢ ||||| ||||| ||||| ||

20. ▢ ▢ ||||| ||||| ||||| |||

Showing Numbers in Several Ways

Numbers can be shown in more than one way using base-ten pieces. Suppose a flat is one whole. Then, 2.3 can be shown in these three ways:

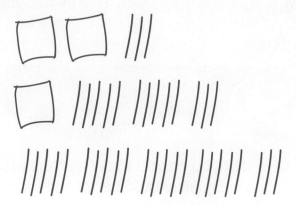

21. Use base-ten pieces to make 3.2 in several ways. Use base-ten shorthand to show each way you find. Circle the solution that uses the fewest pieces.

22. Work with a partner to practice making numbers with base-ten pieces. Write them, using decimals and common fractions. One person should lay out packs, flats, and skinnies. The other person should write numbers for the pieces. Keep track of your work in a table like this:

Tenths Data Table

Base-Ten Shorthand	Fraction of a Flat	
	Common	Decimal

Hundredths

Exploring Hundredths

Jackie is counting the pennies in her piggy bank to see how many dollars she has. She puts the pennies into piles of 100 since she knows that 100 pennies equals 1 dollar. Jackie knows that 1 penny is one-hundredth of a dollar. You can write the fraction for one-hundredth as a common fraction or as a decimal fraction:

$\frac{1}{100}$ 0.01

common fraction   decimal fraction

1. **A.** What does the denominator mean in the fraction $\frac{1}{100}$?

   B. What does the numerator mean in the fraction $\frac{1}{100}$?

   C. In the decimal fraction 0.01, what do the zeros mean?

After Jackie finished putting her pennies into piles of 100, she found that she had 28 pennies (28¢) left over. These 28 pennies are a fraction of a dollar. We can write it as a common fraction ($\frac{28}{100}$) or as a decimal fraction (0.28). We say: twenty-eight hundredths.

2. **A.** Jackie found 14 pennies in her desk drawer. What fraction of a dollar does this represent?

   B. Write this fraction as both a decimal fraction and as a common fraction.

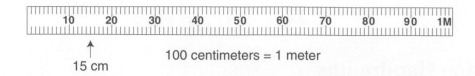

100 centimeters = 1 meter

15 cm

3. **A.** Frank knows that there are 100 centimeters in a meter. That means that the length of one centimeter is 0.01 or $\frac{1}{100}$ of a meter. Frank's pencil is 15 cm long. What fraction of a meter is the length of the pencil?

   B. Write this fraction as both a decimal fraction and as a common fraction.

Irma wanted to use base-ten pieces to show hundredths. She learned in the last lesson that if a flat is 1 whole, then a skinny is 0.1 and a pack is 10.

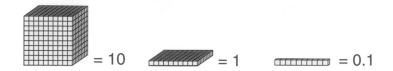

= 10   = 1   = 0.1

4. Which base-ten piece should Irma use to show one-hundredth? Explain why you chose the piece you did.

5. **A.** How many hundredths does a skinny represent? Write this number as a common fraction and as a decimal fraction.

   B. How many hundredths does a flat represent? Write this number as a common fraction and as a decimal fraction.

Nicholas used the following base-ten pieces to show a number. If a flat is one whole, then what number do these pieces represent?

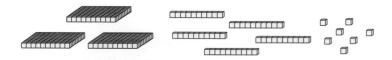

These pieces show 3 wholes, 5 tenths, and 7 hundredths. We can write $3\frac{57}{100}$ or 3.57 for this number. We read both $3\frac{57}{100}$ and 3.57 as "three and fifty-seven hundredths."

6. **A.** Irma placed the following base-ten pieces on her desk. If a flat is one whole, then what number do these pieces represent?

   B. Write this number as a common fraction and as a decimal fraction.

7. Mrs. Dewey showed the following base-ten pieces to the class. She asked each student to record the number for these pieces.

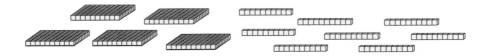

Romesh recorded 5.80, and Jessie recorded 5.8. Explain why both students are correct.

8. Get a handful of mixed skinnies and bits and count them by hundredths. Count the skinnies first (ten-hundredths, twenty-hundredths, thirty-hundredths, etc.), and then count on for the bits. When you finish, write the number for your handful.

Tanya used the following base-ten pieces to show the number 3.67.

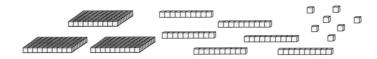

She recorded her work using base-ten shorthand.

9. Use base-ten pieces to make these numbers. Then, use base-ten shorthand to show what pieces you used. A flat is one whole.

    A. 2.34   **B.** 0.08   **C.** 0.15   **D.** 13.42   **E.** 3.04

10. Give a decimal number and a common fraction for the base-ten shorthand below:

    A.   **B.**

    C.   **D.**

11. **A.** Use base-ten pieces to make 0.1 and 0.2.

    B. Use base-ten pieces to make a number more than 0.1 but less than 0.2. Use base-ten shorthand to show the pieces you used. Then write your number.

12. **A.** Use base-ten pieces to make 2.5 and 2.6.

    B. Use base-ten pieces to make a number more than 2.5 but less than 2.6. Use base-ten shorthand to show the pieces you used. Then write your number.

For Questions 13–15, use base-ten shorthand to show how to make the number using the fewest base-ten pieces. Then write a decimal for the number.

13.

14.

15.

16. If you do not follow the fewest base-ten pieces rule, then the same number can be shown in several ways. Use base-ten pieces to make 0.42 in several ways. Use base-ten shorthand to show each way you find.

Homework

Playing *Hundredths, Hundredths, Hundredths*

Complete these questions after playing *Hundredths, Hundredths, Hundredths*.

1. Lee Yah and Jerome were playing *Hundredths, Hundredths, Hundredths*. Jerome tried to trick Lee Yah by making this number:

For her fractions Lee Yah wrote $\frac{23}{100}$ and 0.023 and said, "Twenty-three hundredths." Lee Yah said she should earn 3 points.

Jerome thought that Lee Yah was wrong, but he couldn't explain why. What do you think?

2. Luis and Ana were playing *Hundredths, Hundredths, Hundredths.* Ana made the following number.

Write the common fraction and the decimal fraction for Ana's number.

3. When it was his turn, Luis made the following number.

Write the common fraction and the decimal fraction for Luis's number.

4. Jessie and Roberto were playing *Hundredths, Hundredths, Hundredths.* Roberto made the following number.

Write the common fraction and the decimal fraction for Roberto's number.

5. Jessie wanted to make 6.48. Use base-ten shorthand to show Jessie's number.

6. Roberto wanted to build the number 9.06. Use base-ten shorthand to show what pieces he should use.

7. Jessie wrote nine and six-hundredths like this: 9.6. Explain why this is incorrect.

Downhill Racer

Jackie and her brother Derrick want to play with their toy cars. They set up a ramp using the steps in front of their apartment building and a thick board.

Jackie suggested using a meterstick to find how far each car traveled. Derrick wondered, "Will the cars roll different distances if we put the ramp on different steps?"

I think the cars will roll farther with the ramp on a higher step.

1. Do you think a car will roll farther when the ramp is set up on a higher step? Explain why you think so.

Jackie and Derrick did an experiment to find out how far each car rolled when the ramp was put on steps of different heights. First, they talked about how to be sure that the experiment was fair. Derrick suggested that the starting line on the ramp should stay the same.

2. What other variables should not change during the experiment? Why?

3. Jackie and Derrick decided to run three trials for each different height. Why was this a good idea?

Use the TIMS Laboratory Method to do an experiment like Jackie and Derrick's. Use a car and a ramp to study the relationship between the height of the ramp (*H*) and the distance your car will roll on the ground (*D*).

Use blocks or books to change the height of the ramp. The height (*H*) is the height of the blocks (or books). The blocks should touch the ramp at the same place for the entire experiment. The distance (*D*) should be measured from the bottom of the ramp to the back wheels of the car.

4. What is the manipulated variable?

5. What is the responding variable?

Draw a picture of the lab. Be sure to show the two main variables, Height (*H*) and Distance (*D*). Also show the length of your ramp and where your starting line is.

6. Later, you will make and check predictions about how far your car rolls. Unless you are careful now, you may not be able to check your predictions later. Write a paragraph that describes exactly how you set up your lab, so that later you can set it up again in exactly the same way. (*Hint:* Look at your answer to Question 2.)

---

7. **A.** Work with your group to do the experiment.

- Discuss what values for the height (*H*) you will use.
- Measure the height in centimeters.
- Measure the distance (*D*) the car rolls to the nearest hundredth of a meter. Use decimals to record your measurements of this distance.
- Do three trials for each height. Average the three distances for each height by finding the median distance.
- Keep track of your data in a table like this one:

H Ramp Height (in cm)	D Distance Rolled (in m)			
	Trial 1	Trial 2	Trial 3	Average

B. Why is it a good idea to find the average distance?

8. **A.** Plot your data points on *Centimeter Graph Paper.* Put the manipulated variable on the horizontal axis. Put the responding variable on the vertical axis. Remember to title your graph, label axes, and record units. Before you scale your axes, discuss with your group how much room you need on your graph for extrapolation. (*Hint:* Look at Questions 9–12.)

B. Look at your points on the graph. Do the points lie close to a straight line? If so, use a ruler to fit a line to the points. Extend the line in both directions.

Answer the following questions using your graph:

9. If the height of the ramp were 10 cm,

    A. How far would your car roll? H = 10 cm; Predicted D = ?

    B. Did you interpolate or extrapolate?

    C. Check your prediction. H = 10 cm; Actual D = ?

    D. Was your predicted distance close to the actual distance?

10. If the height of the ramp were 16 cm,

    A. How far would your car roll? H = 16 cm; Predicted D = ?

    B. Did you interpolate or extrapolate?

    C. Check your prediction. H = 16 cm; Actual D = ?

    D. Was your predicted distance close to the actual distance?

11. **A.** Predict how high the ramp should be if you want the car to roll 1.5 m. Explain how you found your answer.

    B. Check your prediction. How close to 1.5 m did your car roll?

12. Imagine doing the experiment again. Suppose you let the car go from a lower starting point on the ramp. Would your new line look like Line A or Line B? Explain why you think so.

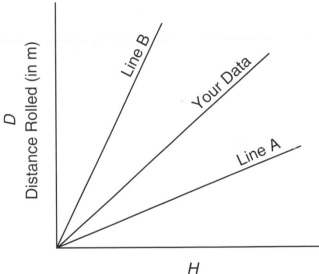

13. Sometimes knowing one variable helps in predicting another.

    A. Does knowing the height of the ramp (H) help you predict what the distance rolled (D) will be?

    B. Does knowing the distance rolled (D) help you predict what the height of the ramp (H) was?

    C. As the height of the ramp (H) increases, how does the distance (D) change?

You will need a calculator, a ruler, and a sheet of graph paper to solve Questions 14–19.

14. After Mrs. Dewey's class finished the experiment *Downhill Racer*, they did another experiment using ramps. In the new experiment, each group kept the height of the ramp the same and used the same car all the time. They chose three distances from the end of the ramp for starting points. Then, they rolled the cars down the ramp from each starting point and measured the distance the car rolled in centimeters. Here is one group's picture of the experiment.

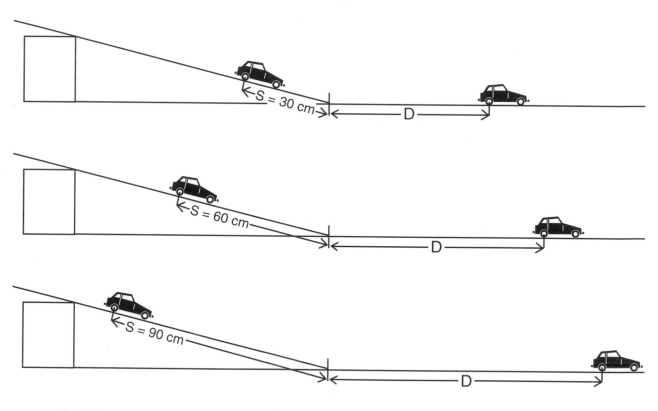

   A. What is the manipulated variable in this experiment?

   B. What is the responding variable in this experiment?

   C. Give one fixed variable.

   D. Is the manipulated variable numerical or categorical?

15. John's group rolled its car down the ramp three times from the starting point that was 60 cm from the end of the ramp. Here are the distances the car rolled for three trials: 83 cm, 84 cm, and 89 cm.

   A. Find the mean distance for these trials. Give your answer to the nearest centimeter.

   B. Find the median distance.

16. Nila's group chose to do four trials. Here are the distances the car rolled from a starting point that was 120 cm from the end of the ramp: 189 cm, 177 cm, 186 cm, and 188 cm.

   A. Find the mean distance for these trials.

   B. Find the median distance. Give your answer to the nearest centimeter.

17. Here is Shannon's data. Make a graph of the data.

Shannon's Data

H Starting Distance From End of Ramp (in cm)	D Distance Rolled (in cm)			
	Trial 1	Trial 2	Trial 3	Average
30	48	47	47	47
60	87	84	86	86
90	144	142	145	144

18. **A.** Shannon uses the same lab setup. She starts to roll the car down the ramp 45 cm from the end of the ramp. Use your graph to predict the distance the car will roll.

B. Did you use interpolation or extrapolation to find your answer?

19. **A.** Shannon uses the same lab setup. She starts to roll the car down the ramp 120 cm from the end of the ramp. Use your graph to predict the distance the car will roll.

B. Did you use interpolation or extrapolation to find your answer?

Miss Take made the following graph for Roberto's data. She has made many mistakes. How many can you find? Write a letter to Miss Take explaining as many errors as you can find.

Name: _____Miss Take_____

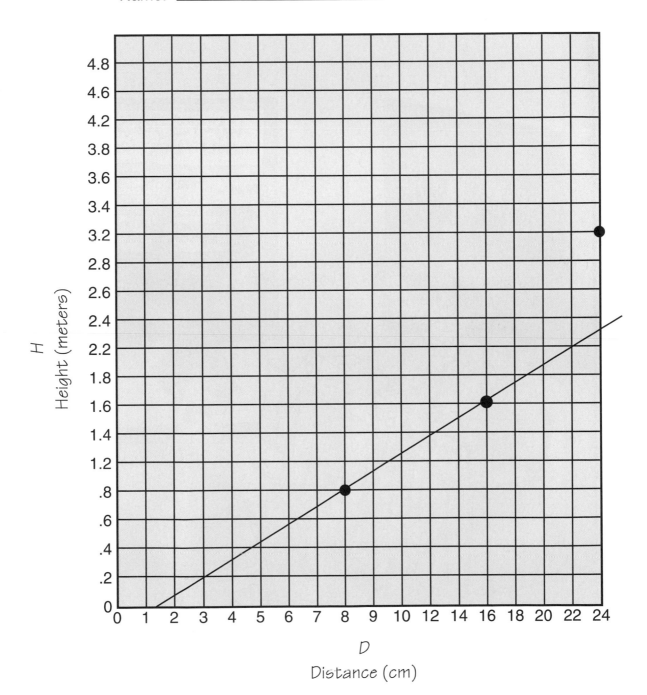

Multiplication

	Student Guide	Discovery Assignment Book	Adventure Book	Unit Resource Guide*
Lesson 1				
Modeling Multiplication	●			
Lesson 2				
More Multiplication	●			
Lesson 3				
Compact Multiplication	●			
Lesson 4				
All-Partials Revisited	●			●
Lesson 5				
More Compact Multiplication	●			
Lesson 6				
Phil and Howard's Excellent Egyptian Adventure			●	
Lesson 7				
Visiting Egypt	●			

Unit Resource Guide pages are from the teacher materials.

Modeling Multiplication

Break-Apart Products

There are many different ways to model multiplication problems. One way is to use an array. Another way is to use base-ten pieces. In this lesson and the other lessons in this unit, a bit has a value of 1 whole. Both pictures below model 3×12.

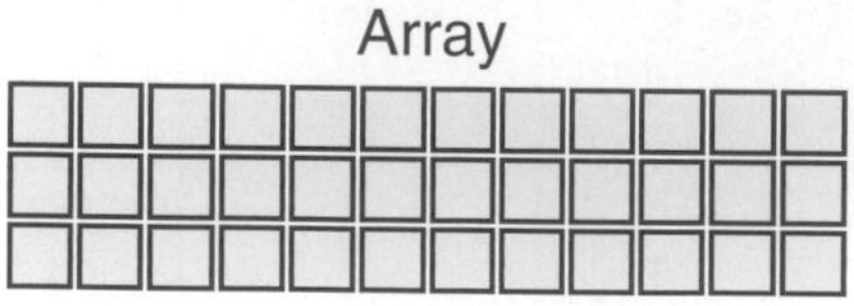

Array

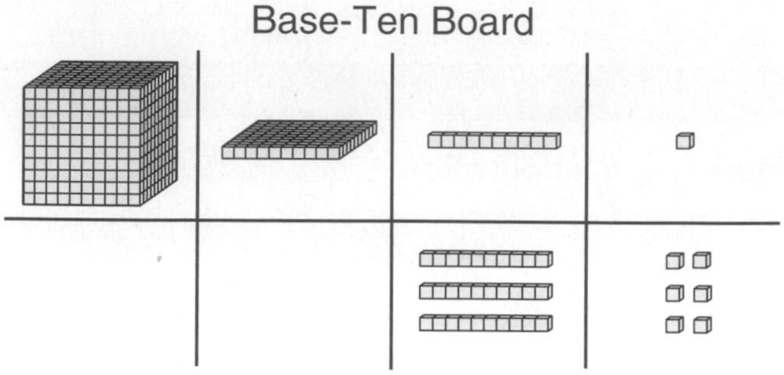

Base-Ten Board

For Questions 1–4, give a multiplication sentence for each picture.

1.

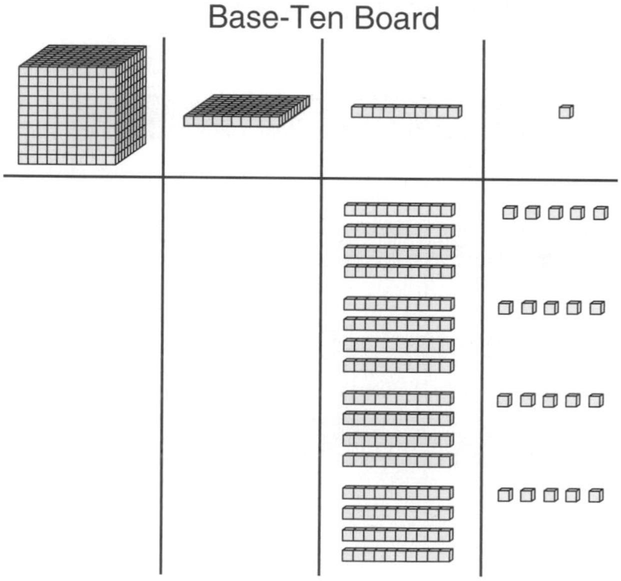

Base-Ten Board

2.

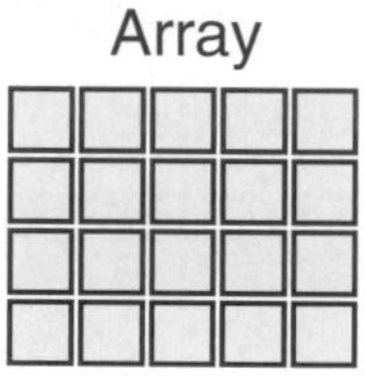

Array

3.

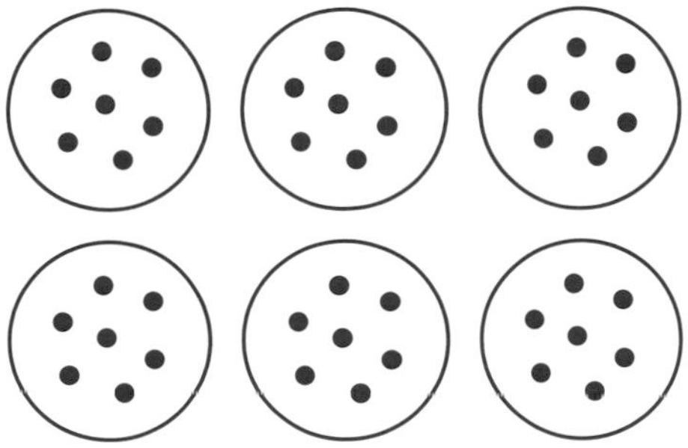

4.

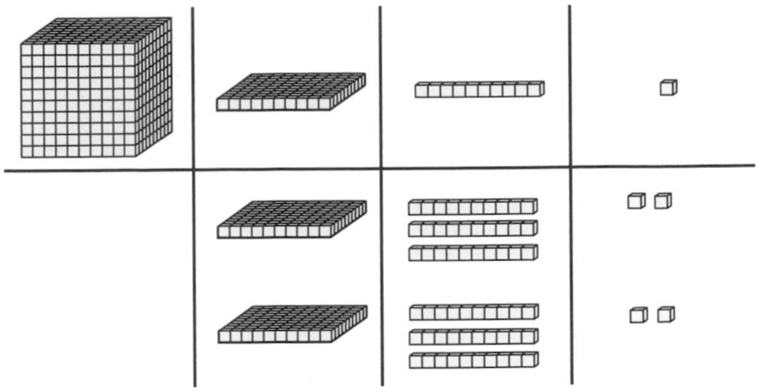

Base-Ten Board

For Questions 5 and 6, model the multiplication problem by drawing a picture.

5. 4×21

6. 5×32

For Questions 7 and 8, complete the break-apart number sentences.

7. $4 \times 45 = ? \times 40 + ? \times 5$
$= ? + ?$
$= ?$

8. $7 \times 28 = 7 \times ? + ? \times 8$
$= ? + ?$
$= ?$

9. When we multiply 6×34 using the all-partials method, we can think about the base-ten pieces.

$$\begin{array}{r} 34 \\ \times\ 6 \\ \hline 24 \\ 180 \\ \hline 204 \end{array}$$

$\leftarrow 6 \times 4$: 6 groups of 4 bits
$\leftarrow 6 \times 30$: 6 groups of 3 skinnies

Write a break-apart number sentence for 6×34.

10. When we multiply 7×27 using the all-partials method, we can think about the base-ten pieces.

$$\begin{array}{r} 27 \\ \times\ 7 \\ \hline 49 \\ 140 \\ \hline 189 \end{array}$$

$\leftarrow 7 \times 7$: 7 groups of 7 bits
$\leftarrow 7 \times 20$: 7 groups of 2 skinnies

Write a break-apart number sentence for 7×27.

Multiplying on the Base-Ten Board

11. This shows 23 in base-ten shorthand.

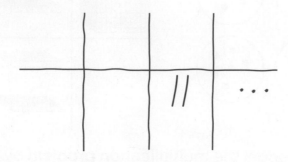

A. There are 2 _____?_____ and 3 _____?_____ .

This shows the answer when you multiply 10 × 23 = 230 in base-ten shorthand.

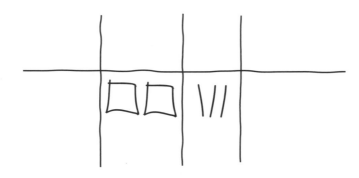

B. There are 2 _____?_____ and 3 _____?_____ .

C. Describe how the base-ten pieces change when you multiply a number by 10.

12. This shows 23 in base-ten shorthand.

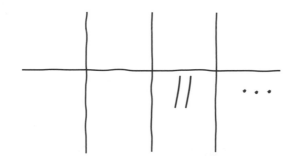

This shows the answer when you multiply $100 \times 23 = 2300$ in base-ten shorthand.

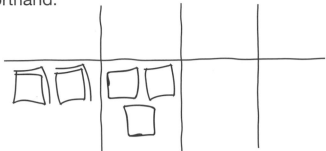

Explain how the base-ten pieces change when you multiply a number by 100.

Show just the answers to the following multiplication problems in base-ten shorthand. Draw in the column markings to help you.

13. **A.** 2×23

    B. 20×23

    C. Compare the shorthand for parts A and B. What is an easy way to think about 20×23?

14. **A.** 3×45

    B. 30×45

    C. Compare the shorthand for parts A and B. What is an easy way to think about 30×45?

15. **A.** 4×17

    B. 40×17

    C. 400×17

    D. Compare the shorthand for parts A, B, and C. What is an easy way of thinking about 40×17 and 400×17?

Do the following groups of problems and watch for patterns. Use a calculator when needed to check your work.

16.	17.	18.
2×3	4×8	5×4
2×30	4×80	5×40
20×3	40×8	50×4
20×30	40×80	50×40
20×300	40×800	50×400
200×3000	400×8000	500×4000

19. Explain how you can quickly compute products that end in zeros.

Compute the following sets of problems. Watch for patterns.

20. **A.** 31
     ×3

B. 310
     ×3

C. 3100
     ×3

D. 31,000
     ×3

21. **A.** 70×14 **B.** 70×140 **C.** 70×1400 **D.** $70 \times 14,000$

22. **A.** 4×25 **B.** 4×250 **C.** 4×2500 **D.** $4 \times 25,000$

Homework

For Questions 1–6, solve each two ways:

    A. Write a break-apart number sentence using multiples of 10 if possible.

    B. Use the all-partials method or facts you know to solve the problem.

1. $6 \times 8 =$ **2.** $7 \times 6 =$ **3.** $8 \times 7 =$

4. $5 \times 24 =$ **5.** $6 \times 12 =$ **6.** $36 \times 4 =$

Find the following products.

7. $15 \times 5 =$ **8.** $40 \times 6 =$ **9.** $8 \times 22 =$

10. $10 \times 5 =$ **11.** $7 \times 100 =$ **12.** $40 \times 10 =$

13. $21 \times 3 =$ **14.** $5 \times 30 =$ **15.** $60 \times 40 =$

16. $42 \times 10 =$ **17.** $12 \times 20 =$ **18.** $30 \times 50 =$

19. $17 \times 20 =$

20. Nila and her sister put together 5 jigsaw puzzles. If each puzzle has 32 pieces, how many pieces do the 5 jigsaw puzzles have?

21. Jerome practices piano for 30 minutes every day after school. He has two weeks to practice before the recital. How many minutes will he practice on school nights from now until the recital? How many hours?

22. Shannon is in a bowling league after school on Wednesday. This week she bowled an 86 in each game of a three-game series. What was her total score for the series?

23. Jesse sold Chocos for the Little League fund-raiser. One hundred twenty families each bought $5 worth of Chocos. How much money did Jesse raise?

24. Lee Yah watched 45 minutes of TV each day for 4 days. How many minutes did she watch in all?

For Questions 25–32, find the missing numbers. Some may have more than one answer.

25. $8 \times 60 = n$ **26.** $7 \times n = 140$ **27.** $50 \times n = 10{,}000$

28. $70 \times n = 490{,}000$ **29.** $n \times 200 = 4200$ **30.** $n \times 12 = 360{,}000$

31. $n \times m = 80$ **32.** $n \times m = 240{,}000$

Solve Questions 33–38 using either paper and pencil or mental math. You may use the all-partials method, base-ten pieces, or base-ten shorthand. Estimate to make sure your answer is reasonable.

33.   57 $\times 5$ **34.**   96 $\times 6$ **35.**   75 $\times 7$

36.   43 $\times 5$ **37.**   62 $\times 4$ **38.**   81 $\times 9$

More Multiplication

1. Pilot Jones flies her airplane about 324 minutes each week. Estimate the number of minutes Pilot Jones flies in a month. Assume there are 4 weeks in a month.

2. Use any method you wish. Calculate the exact number of minutes Pilot Jones flies in a month. Explain your work with numbers, words, pictures, or base-ten pieces.

You can find the total number of minutes by modeling 4 × 324 using the base-ten pieces.

Base-Ten Board

Another way of solving the problem is to use the all-partials method of multiplication.

$$
\begin{array}{r}
324 \\
\times\,4 \\
\hline
16 \\
80 \\
1200 \\
\hline
1296
\end{array}
$$

3. Explain where the partial product 80 came from.

4. Which base-ten pieces model 80?

5. Explain where the partial product 1200 came from.

6. Which base-ten pieces model 1200?

Do the following problems using the all-partials method.

7. $\begin{array}{r} 132 \\ \times\,3 \\ \hline \end{array}$

8. $\begin{array}{r} 3624 \\ \times\,2 \\ \hline \end{array}$

9. $\begin{array}{r} 1205 \\ \times\,4 \\ \hline \end{array}$

10. Grace notices that many of the partial products end in zeros. Circle the zeros in the partial products in Questions 7–9.

11. Explain where the zeros came from in Questions 7–9.

12. Professor Peabody always leaves out digits when he uses the all-partials method to multiply. Copy the problems, and find the missing digits as you find the products. Be careful to line up the columns.

   A. $\begin{array}{r} 24 \\ \times\,3 \\ \hline 12 \\ 6\square \\ \hline \end{array}$

   B. $\begin{array}{r} 734 \\ \times\,8 \\ \hline 32 \\ 24\square \\ 56\square\square \\ \hline \end{array}$

   C. $\begin{array}{r} 4251 \\ \times\,7 \\ \hline 7 \\ 35\square \\ 14\square\square \\ 28\square\square\square \\ \hline \end{array}$

13. Explain patterns you see in Question 12.

14. Does leaving out the missing digits in Question 12 affect the answer to the multiplication problems? Why or why not?

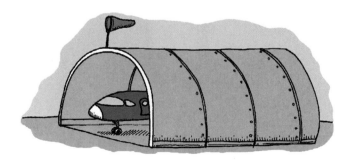

15. Grace computed 3 × 306 like this:

$$
\begin{array}{r}
306 \\
\times\,3 \\
\hline
18 \\
900 \\
\hline
918
\end{array}
$$

A. Is Grace correct?

B. Why are there only two partial products in this problem?

Do the following pairs of problems. Use the answer from the first problem to help solve the second problem in each pair without computing.

1. A.
$$\begin{array}{r} 32 \\ \times\,6 \\ \hline \end{array} \qquad \begin{array}{r} 320 \\ \times\,6 \\ \hline \end{array}$$

B.
$$\begin{array}{r} 403 \\ \times\,9 \\ \hline \end{array} \qquad \begin{array}{r} 4030 \\ \times\,9 \\ \hline \end{array}$$

C.
$$\begin{array}{r} 275 \\ \times\,3 \\ \hline \end{array} \qquad \begin{array}{r} 2,750,000 \\ \times\,3 \\ \hline \end{array}$$

Do the following problems. First make a mental estimate of the answer. Then do the problem. Compare your estimate with the answer.

2. Remember, there are 24 hours in a day. How many hours are there in a week?

3. One type of airplane can carry 229 passengers. North-South Airlines flies its airplane from Minneapolis to Ft. Lauderdale five times a week. How many people can travel every week from Minneapolis to Ft. Lauderdale on North-South Airlines?

4. An airplane has a cruising speed of about 558 miles per hour. How far can the aircraft travel in 3 hours?

5. Another airplane has a cruising speed of 1336 miles per hour. How far can this aircraft travel in 3 hours?

Practice Problems

Estimate your answer before solving the following problems.

6. 6×45 **7.** 8×32 **8.** 5×3049

9. 6×421 **10.** 30×312 **11.** 6×7213

12. 40×2301 **13.** 3×3008 **14.** 60×4250

15. 8×3453 **16.** 7×3024 **17.** 38×100

18. 124×200 **19.** 300×47 **20.** 94×400

21. 329×500 **22.** 417×60 **23.** 30×865

24. Explain your estimation strategy for Question 9.

25. Explain a mental math strategy for solving Question 13.

Did You Know?

Mrs. Dewey's classroom, Room 204, is in Bessie Coleman School. Bessie Coleman was the world's first African-American female aviator.

When Bessie's brother returned to America after World War I, he told Bessie that French women could fly airplanes. At that time, Bessie worked as a manicurist in a Chicago barber shop. Hearing this news, Bessie decided she too could learn to fly. She went to school to learn French. Then, she went to France. In 1921, she earned her pilot's license from the Federation Aeronautique Internationale. When she returned to Chicago, she became an air circus performer. A street and a library in Chicago are named after Bessie Coleman.

Compact Multiplication

Kris enjoys looking at travel brochures. He dreams about seeing all 50 states and other parts of the world, as well. Kris thinks it would be great to be a travel agent.

Travel agents need to think about transporting large groups of people. If an airplane holds 123 passengers, how many people will 4 flights of that airplane carry?

Jacob showed Kris his method for solving the problem. He first estimated the product. Since 123 is about 100 and $4 \times 100 = 400$, Jacob knows the product will be more than 400. Since 23 is about 25 and $4 \times 25 = 100$, Jacob thinks the product will be about 500 since $400 + 100 = 500$.

Jacob solved the problem by using the all-partials method of multiplication. Nicholas said he had a shortcut way of computing the product. Compare Jacob's method to Nicholas's method.

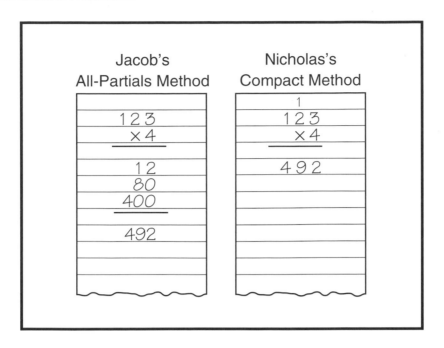

Review Jacob's method. Can you see how he found his answer?

Nicholas began by multiplying $3 \times 4 = 12$. Nicholas knows this is 1 ten and 2 ones.

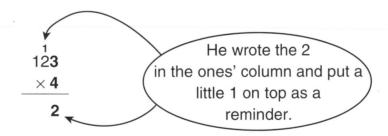

The 1 Nicholas wrote above the problem as a reminder is called a **carry.** Nicholas then multiplied 4×2 tens $= 8$ tens and added the extra ten to get 9 tens.

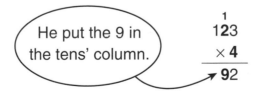

Last, Nicholas multiplied 4 × 1 hundred = 4 hundreds.

$$\begin{array}{r} 1 \\ 123 \\ \times\ 4 \\ \hline 492 \end{array}$$

1. Which method do you prefer: Jacob's method or Nicholas's method? Why?

He put the 4 in the hundreds' column.

One type of airplane can carry up to 376 passengers. How many people can 4 flights of the aircraft carry?

Nicholas did the problem using the compact method:

$$\begin{array}{r} 3\ 2 \\ 376 \\ \times\ 4 \\ \hline 1504 \end{array}$$

2. Why did Nicholas place a 2 above the problem? What does this 2 mean?

3. How did Nicholas get the 0 in the tens' column of the answer?

4. Why did Nicholas place a 3 above the problem? What does this 3 mean?

5. How did Nicholas get the 5 in the hundreds' column of the answer?

6. How did Nicholas get the 1 in the thousands' column of the answer?

Homework

1. Jerome wanted to check his answer to the problem 227 × 4. He knew 908 was a reasonable answer. However, he wasn't sure about the 0 in the tens' place. He laid down the base-ten pieces as shown.

Base-Ten Board

$$\begin{array}{r} 1\ 2 \\ 227 \\ \times\ 4 \\ \hline 908 \end{array}$$

A. How many bits are there?

B. How did Jerome record these bits using the compact method?

C. There are 8 skinnies shown on the base-ten board. Was Jerome correct in placing a 0 in the tens' column instead of an 8? Why or why not?

D. Would the answer 808 make sense? How did Jerome get the 9 in the hundreds' place?

Compute the following products. Try to use the compact method or mental math. Estimate your product first. Use your estimate to make sure your answer is reasonable.

2. 52 ×3	**3.** 325 ×2	**4.** 280 ×5	**5.** 307 ×3
6. 49 ×2	**7.** 43 ×5	**8.** 416 ×6	**9.** 401 ×8
10. 3210 ×5	**11.** 4200 ×6	**12.** 8007 ×9	**13.** 6018 ×7

14. Grace went to the airport to meet her grandmother, who was coming to Chicago from Rochester, New York. In the car, Grace's grandmother commented on the trip.

A. On the flight to Chicago, there were 139 passengers. They were served twice during the flight. If each passenger drank a total of two cans of juice or soda, how many cans would the flight attendants need?

B. Is this more or less than 20 twelve-packs?

15. Grace's grandmother told her that the plane she took to Chicago seemed small compared to the one she took to Hawaii. The main cabin of the larger plane was split into two parts. One part had 14 rows of 9 seats. The other part had 11 rows of 9 seats. The flight to Hawaii was filled to capacity.

A. How many seats did the flight to Hawaii have?

B. How many more passengers were on the flight to Hawaii than on the flight to Chicago?

16. Grace's grandmother met Pilot Johnson. He told her that he has flown round trip from Rochester to Chicago nine times in the past week. It is 528 miles one way from Rochester to Chicago. How many miles has Pilot Johnson traveled in all on the Rochester-Chicago route?

17. When they were picking up Grandma's luggage, an announcement was made for the returning flight to Rochester. Passengers could check in 3 bags and carry on 2 bags. If the returning flight to Rochester has 148 passengers, what is the largest number of pieces of luggage the airline will allow for this flight?

18. Nila gave the answer of 1000 for the following problem. Explain why Nila is wrong. Tell how to find the correct answer.

$$500 \times 20 =$$

19. $650 \times 100 =$

20. $760 \times 50 =$

21. $1350 \times 20 =$

　　　　　　　　　　　　　　　Compact Multiplication

All-Partials Revisited

TIMSville built a new school. The new school has 14 classrooms. Each classroom has 24 desks. How many desks are in the new school?

1. A. Estimate the number of desks in the new school. Then compare your estimate to Ana's solution.

Ana decided to model this multiplication problem using base-ten shorthand. She split the 14 classrooms into a group of 10 classrooms and a group of 4 classrooms.

10 classrooms with
24 desks in each

4 classrooms with
24 desks in each

There are 14 classrooms with
24 desks in each.

Ana computed the product using the all-partials method:

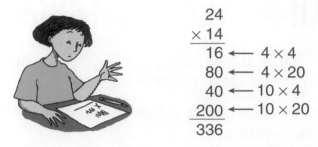

$$
\begin{array}{r}
24 \\
\times\,14 \\
\hline
16 \;\longleftarrow\; 4\times4 \\
80 \;\longleftarrow\; 4\times20 \\
40 \;\longleftarrow\; 10\times4 \\
200 \;\longleftarrow\; 10\times20 \\
\hline
336
\end{array}
$$

She multiplied $4 \times 4 = 16$ and $4 \times 20 = 80$. Then she multiplied $10 \times 4 = 40$ and $10 \times 20 = 200$. Her total is 336, so there are enough desks for 336 students.

B. Match the partial products with the base-ten shorthand.

2. Jessie also used the all-partials method. She multiplied 10×20 first.

$$
\begin{array}{r}
24 \\
\times\,14 \\
\hline
200 \;\longleftarrow\; 10\times20 \\
40 \;\longleftarrow\; 10\times4 \\
80 \;\longleftarrow\; 4\times20 \\
16 \;\longleftarrow\; 4\times4 \\
\hline
336
\end{array}
$$

She liked this method because the first product she found was the most important. Why might Jessie think her first product was the most important?

3. The new school has an auditorium with 32 rows of seats. Each row has 18 seats.

A. Estimate the number of seats in the auditorium.

B. Ana found the exact number of seats by using the all-partials method. Copy the problem, and fill in the missing numbers.

$$
\begin{array}{r}
32 \\
\times\,18 \\
\hline
?? \\
240 \\
?? \\
300 \\
\hline
???
\end{array}
$$

C. How did Ana get the partial product 240?

D. How did Ana get the partial product 300?

4. Compute 43×27 by using the all-partials method.

1. Irma used the all-partials method to do the problem 87 × 32. Copy the problem, and fill in the missing numbers in her work. Complete the problem.

$$
\begin{array}{r}
32 \\
\times\, 87 \\
\hline
?? \\
2?? \\
?6? \\
\underline{?40?}
\end{array}
$$

2. Roberto solved the problem 47 × 52 using the all-partials method.

$$
\begin{array}{r}
47 \\
\times\, 52 \\
\hline
14 \\
80 \\
35 \\
\underline{200} \\
\cancel{329}
\end{array}
$$

   A. Is Roberto's answer of 329 reasonable? Explain why or why not.

   B. What did Roberto do wrong?

   C. Find the correct answer.

For Questions 3–7, decide whether the estimate is a "could be" or "crazy" estimate. If the estimate seems appropriate, record your answer as "could be." If the estimate is too high or too low, record your answer as "crazy." Explain how you decided.

3. Tanya said, "76 × 42 is close to 280."

4. Romesh said, "35 × 35 is between 900 and 1600, so 1200 is my estimate."

5. Luis said, "The answer to 17 × 34 is less than 400."

6. Jessie said, "57 × 26 is less than 1000."

7. Shannon said, "A good estimate for 11 × 55 is 550."

For Questions 8–15, find the products using the all-partials method. Remember to estimate to see if your answer is reasonable. Choose two problems, and explain your estimation strategies for them.

8. 14×15

9. 45×12

10. 37×23

11. 64×78

12. 56×18

13. 93×44

14. 81×29

15. 62×65

16. Jacob's older sister Cara uses a graphing calculator. Her classroom has a set of 25. If one calculator costs $97, how much did the classroom set cost?

17. Smackin Good Apple Company shipped 14 small boxes of apples to Martha's Market. Each box has 48 apples in it. How many apples did Martha's Market receive?

18. Grace is starring in the school play. Her parents purchased tickets in advance. They got seat numbers 211 and 212. Grace looks for her parents from the stage. She knows there are 18 seats in a row. She also knows seat number 1 is in the first row. Where should Grace look to find her parents? About which row? (*Hint:* Use the picture to help you.)

19. The auditorium at Bessie Coleman School has 18 seats in each row and 33 rows. The movie theater has 28 seats in each row and 28 rows. Which has more seats? How many more?

More Compact Multiplication

A grocer received a shipment of eggs. The box contained 24 cartons of eggs. Each carton contained 12 eggs. How many eggs did the grocer receive?

1. Estimate the number of eggs by thinking about the base-ten pieces.

Jacob and Nicholas solved the problem using their paper-and-pencil methods.

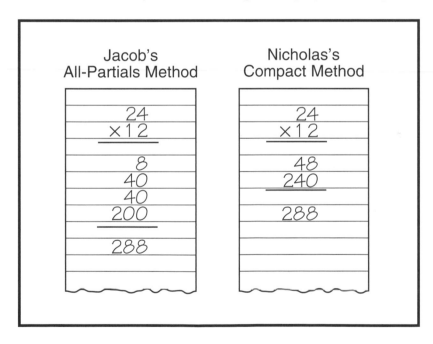

2. Identify the base-ten pieces that match each of the partial products in Jacob's method.

Nicholas showed how to do 12 × 24 using the compact method. He first multiplied 2 × 24. To do this, he multiplied 2 × 4 to get 8 and wrote the 8 in the ones' column. He then multiplied 2 × 20 to get 40 and wrote 4 in the tens' column.

$$\begin{array}{r} 24 \\ \times\,12 \\ \hline 48 \end{array}$$

For the next part of the problem, Nicholas multiplied 10 × 24. He first multiplied 10 × 4 to get 40.

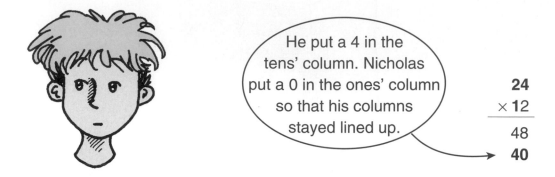

He put a 4 in the tens' column. Nicholas put a 0 in the ones' column so that his columns stayed lined up.

$$\begin{array}{r} 24 \\ \times\,12 \\ \hline 48 \\ 40 \end{array}$$

Then he multiplied 10 × 20 to get 200.

He put a 2 in the hundreds' column.

$$\begin{array}{r} 24 \\ \times\,12 \\ \hline 48 \\ 240 \end{array}$$

Nicholas then added to get 288.

$$\begin{array}{r} 24 \\ \times\,12 \\ \hline 48 \\ 240 \\ \hline 288 \end{array}$$

3. Compute 74×23 using the all-partials method.

To compute 74×23 using the compact method, Nicholas first multiplied $4 \times 3 = 12$. He put a 2 in the ones' column and a 1 above the problem as a reminder. He then multiplied $4 \times 20 = 80$. He now has 8 tens and the extra 1 ten, so he has 9 tens altogether. Nicholas wrote a 9 in the tens' column.

$$
\begin{array}{r}
1 \\
23 \\
\times\, 74 \\
\hline
92
\end{array}
$$

Nicholas then multiplied $70 \times 3 = 210$. He put a 0 in the ones' column and a 1 in the tens' column. He put a 2 above the problem as a reminder. He crossed out the 1 above the problem since he had taken care of it.

$$
\begin{array}{r}
2 \\
\not{1} \\
23 \\
\times\, 74 \\
\hline
92 \\
10
\end{array}
$$

Nicholas then multiplied $70 \times 20 = 1400$. This means he has 4 hundreds and the 2 extra hundreds, or 6 hundreds total. Nicholas put a 6 in the hundreds' column and a 1 in the thousands' column. He didn't have to carry the 1 because he had no more partial products to compute.

$$
\begin{array}{r}
2 \\
\not{1} \\
23 \\
\times\, 74 \\
\hline
92 \\
1610 \\
\hline
1702
\end{array}
$$

Nicholas found the product $74 \times 23 = 1702$.

Here is another problem that Nicholas did using the compact method.

$$
\begin{array}{r}
\overset{5}{\cancel{}} \\
49 \\
\times\,64 \\
\hline
196 \\
\underline{2940} \\
3136
\end{array}
$$

4. Why did Nicholas put a 3 above the problem?

5. How did Nicholas get a 9 in the tens' column of the first partial product?

6. How did Nicholas get the 4 in the tens' column and the 0 in the ones' column of the second partial product?

7. Why did Nicholas put a 5 above the problem?

8. How did Nicholas get a 9 in the hundreds' column in the second partial product?

Compute the problems using the compact method.

9. 34×79 **10.** 27×82 **11.** 42×28

Homework

1. Jackie solved the problem 47×52 using the all-partials method. Then she tried using the compact method.

$$
\begin{array}{r}
54 \\
\times\,47 \\
\hline
28 \\
350 \\
160 \\
\underline{2000} \\
2538
\end{array}
\qquad
\begin{array}{r}
\overset{1}{\cancel{}} \\
54 \\
\times\,47 \\
\hline
378 \\
\underline{2160} \\
2538
\end{array}
$$

A. Why did Jackie place a 1 above the problem in the compact method?

B. How did Jackie get a 6 in the tens' place in the second partial product in the compact method?

C. Where is Jackie's record of multiplying 7×50 in the compact method?

D. How did Jackie get 2160 when using the compact method?

2. Which of the following are the four partial products that make up the problem below using the all-partials method?

$$\begin{array}{r} 27 \\ \times\, 69 \\ \hline \end{array}$$

6×20	60×7	2×90	9×7
60×20	90×20	20×9	9×20

Find the following products.

3. $25 \times 15 =$ 4. $46 \times 34 =$ 5. $58 \times 76 =$ 6. $95 \times 64 =$

7. $70 \times 23 =$ 8. $68 \times 90 =$ 9. $52 \times 55 =$ 10. $76 \times 33 =$

11. The Victory Videos store was selling *Math and the Music Factory* videos for $14. Mrs. Dewey bought a dozen of these videos for the school library. How much did the videos cost in all?

12. On Monday night, only 17 movies were rented. On Saturday night, 16 times as many movies were rented. How many movies were rented on Saturday night?

13. A newly released movie was available for sale. Last weekend alone, Victory Videos sold 98 copies. If one copy costs $24, how much money did Victory Videos receive for this movie?

14. Victory Videos has 36 different sections. Each section has 6 shelves. Each shelf has about 10 movies on it. About how many videos can fill the video store?

15. Victory's ad in the paper reads, "We have 5 times as many videos as Urban Video." Urban Video has about 6000 videos. About how many videos does Victory Videos have?

16. There are 28 Victory Videos stores throughout the city. On average, each store has 28 employees. How many people work for Victory Videos?

Visiting Egypt

Solve the following problems about a fictional king in Egypt. Use paper and pencil or mental math. Estimate to be sure your answer is reasonable.

1. King Omar has eight stables full of camels. Each stable has 25 camels. How many camels does King Omar have?

2. Each camel has three riders. How many camel riders does King Omar have?

3. The camels can each carry five bundles of dates to King Omar. How many bundles of dates can they bring back to the king?

4. Half of King Omar's camels became ill and could only carry two bundles of dates instead of five bundles each. How many bundles came back to the king?

5. One night, 25 camels ran away.

   A. How many camels are left?

   B. How many bundles of dates can the remaining camels bring to King Omar if none are ill?

6. King Omar sent 30 groups of laborers to build a pyramid. Each group has 50 laborers. How many laborers did King Omar send?

7. Each group of laborers can build one small pyramid every year. How many small pyramids can they build for King Omar each year?

8. How many small pyramids can they build in 15 years?

9. How many small pyramids can they build in 25 years?

10. **A.** $37 \times 45 =$

    B. $2340 \times 9 =$

    C. $80 \times 53 =$

    D. $93 \times 70 =$

    E. $79 \times 46 =$

    F. $50 \times 600 =$

    G. $25 \times 40 =$

    H. $44 \times 44 =$

    I. $307 \times 7 =$

11. Explain your estimation strategy for Question 10H.

12. Describe a mental math strategy for Question 10I.

Unit 12

Exploring Fractions

	Student Guide	Discovery Assignment Book	Adventure Book	Unit Resource Guide*
Lesson 1				
Fraction Strips	●	●		
Lesson 2				
Adding and Subtracting with Fraction Strips	●			
Lesson 3				
Comparing Fractions	●			
Lesson 4				
Frabble Game	●	●		
Lesson 5				
Equivalent Fractions	●			
Lesson 6				
Pattern Block Fractions	●			●
Lesson 7				
Solving Problems with Pattern Blocks	●			
Lesson 8				
Fraction Puzzles	●			●
Lesson 9				
Midterm Test				●

Unit Resource Guide pages are from the teacher materials.

Fraction Strips

Lee Yah demonstrated folding fourths for her class.

"First, I folded my strip into two equal pieces. I folded my strip by matching the edges and then making a crease in the middle. Next, I kept it folded and then I folded it in half again. When I unfold the strip, it is divided into 4 equal pieces. Since the 4 parts are all the same size, each piece is $\frac{1}{4}$ of the strip."

She showed three of the pieces to show $\frac{3}{4}$ of the strip.

In a fraction, the bottom number is the **denominator.** This number tells us how many equal pieces the whole is divided into. The top number, the **numerator,** tells us how many of the pieces we are concerned with.

$$\frac{3}{4} \quad \longleftarrow \text{numerator}$$
$$\longleftarrow \text{denominator}$$

1. In the fraction $\frac{3}{4}$, what information does the denominator give us?

2. What information does the numerator give us in the fraction $\frac{3}{4}$?

3. What happens to the size of the fractional parts as the denominator gets bigger?

Halves, Fourths, and Eighths

1. Tanya showed the following fraction using her fraction strips. What fraction is she showing?

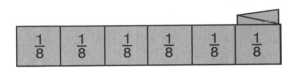

2. Luis showed the following fraction using his fraction strips. What fraction is he showing?

3. Ming showed the following fraction using his fraction strips. What fraction is he showing?

4. Write a fraction for each fraction strip.

A.

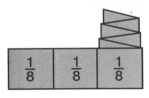

B.

C.

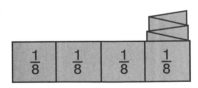

D.

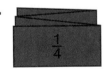

---

5. Roberto has 8 buttons. Three of the buttons are red, three of the buttons are green, and two of the buttons are blue.

   A. What fraction of the buttons are red?

   B. What fraction of the buttons are green?

   C. What fraction of the buttons are red or blue?

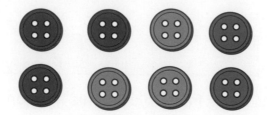

6. Jessie, Nila, Irma, and Tanya are standing together.

   A. What fraction of the girls are wearing red?

   B. What fraction of the girls are wearing glasses?

7. Linda folded her eighths strip to show $\frac{3}{8}$.

   A. In the fraction $\frac{3}{8}$, what information does the denominator give us?

   B. In the fraction $\frac{3}{8}$, what information does the numerator give us?

Thirds, Sixths, Ninths, and Twelfths

8. Romesh made the following fraction using his fraction strips. What fraction did he make?

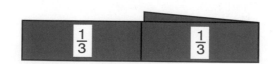

9. Jessie made the following fraction using her fraction strips. What fraction did she make?

10. Nicholas made the following fraction using his fraction strips. What fraction did he make?

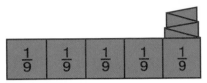

11. Write a fraction for each fraction strip.

A.

B.

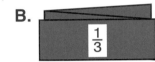

C.

D.

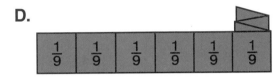

E.

12. Use the strips below to write two fractions equal to $\frac{4}{6}$.

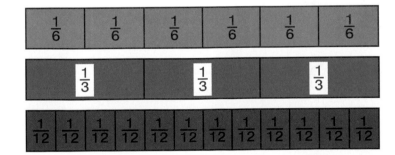

13. Use the strips below to find two fractions equal to $\frac{4}{12}$.

Fifths and Tenths

14. Michael showed the following fraction with his fraction strips. What fraction did he show?

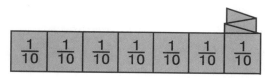

15. Roberto showed the following fraction with his fraction strip. What fraction did he show?

16. Lee Yah showed the following fraction with her fraction strip. What fraction did she show?

Fraction Strips

17. Write a fraction for each fraction strip.

A.
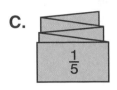

B.
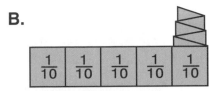

C.
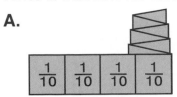

D.

| $\frac{1}{5}$ | $\frac{1}{5}$ | $\frac{1}{5}$ | $\frac{1}{5}$ | $\frac{1}{5}$ |

E.
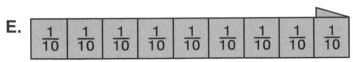

18. Use the fraction strips below to find a fraction equal to $\frac{4}{5}$.

| $\frac{1}{5}$ | $\frac{1}{5}$ | $\frac{1}{5}$ | $\frac{1}{5}$ | $\frac{1}{5}$ |

| $\frac{1}{10}$ | $\frac{1}{10}$ | $\frac{1}{10}$ | $\frac{1}{10}$ | $\frac{1}{10}$ | $\frac{1}{10}$ | $\frac{1}{10}$ | $\frac{1}{10}$ | $\frac{1}{10}$ | $\frac{1}{10}$ |

19. Use the fraction strips below to find five fractions equal to $\frac{1}{2}$. What do you notice about each of the denominators?

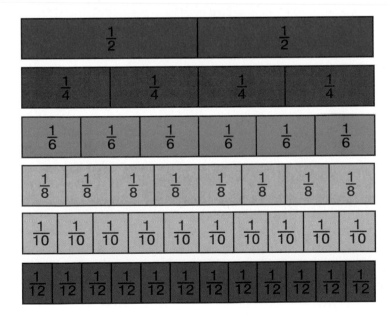

20. Use the fraction strips below to find three fractions equal to $\frac{2}{3}$.

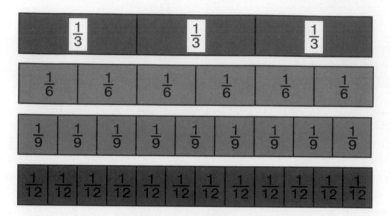

    A. What do you notice about each of the denominators?

    B. What do you notice about each of the numerators?

21. Use the fraction strips below to find two fractions equal to $\frac{3}{4}$.

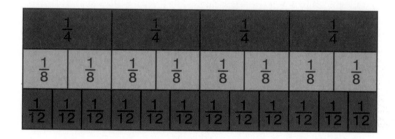

    A. What do you notice about each of the numerators?

    B. What do you notice about each of the denominators?

Adding and Subtracting with Fraction Strips

Discuss

Mrs. Dewey asked her class to use their fraction strips to add $\frac{1}{4}$ and $\frac{2}{4}$. Keenya explained her solution to the class:

"First, I folded my strip that shows fourths so that 1 piece or $\frac{1}{4}$ of the strip was showing. Then, I added $\frac{2}{4}$ of the strip by unfolding 2 more pieces. I ended up with $\frac{3}{4}$ of my strip showing. So, $\frac{1}{4} + \frac{2}{4} = \frac{3}{4}$."

Next, Mrs. Dewey asked the class to use their fraction strips to subtract $\frac{3}{8}$ from $\frac{7}{8}$.

Jacob explained his solution to the class:

"I started with my strip that is divided into eighths. I folded it so that $\frac{7}{8}$ of the strip was showing. Then I folded back $\frac{3}{8}$ of the strip or 3 more pieces since I was subtracting. This left me with $\frac{4}{8}$ of the strip showing, so $\frac{7}{8} - \frac{3}{8} = \frac{4}{8}$."

Adding and Subtracting with Fraction Strips

Work with a partner to solve the following problems. You will need to use two sets of fraction strips. Write a number sentence for each problem.

1. Maya has $\frac{5}{8}$ of a yard of fabric. She needs $\frac{3}{8}$ of a yard of fabric for a craft project. How much fabric will she have left over after she completes her project?

2. Frank is baking a cake. The recipe calls for $\frac{1}{4}$ cup of oil and $\frac{3}{4}$ cup of water. How much liquid will Frank add to the cake mix?

3. Jessie used $\frac{5}{12}$ of a board for a sign. What fraction of the board is left for another project?

4. There was $\frac{5}{6}$ of a pie on the counter when Luis got home from school.

   A. Luis ate $\frac{2}{6}$ of the pie. How much of the pie is left?

   B. Luis's sister ate another $\frac{1}{6}$ of the pie. Now how much of the pie is left?

   C. Use your fraction strips to find another fraction that is equal to your answer to Question 4B.

5. Ming rode his bike $\frac{8}{10}$ mile to Frank's house. He then rode $\frac{8}{10}$ mile back home. How far did Ming ride altogether?

6. Irma must finish her homework and practice piano before she can go outside to play. It takes her $\frac{3}{4}$ hour to do her homework, and she practices piano for $\frac{2}{4}$ hour. How long does she have to wait before going outside to play?

7. Use your fraction strips to complete the following number sentences.

   A. $\frac{3}{8} + \frac{2}{8} =$ **B.** $\frac{7}{10} + \frac{5}{10} =$ **C.** $\frac{3}{6} + \frac{3}{6} =$

   D. $\frac{11}{12} - \frac{4}{12} =$ **E.** $\frac{3}{5} - \frac{1}{5} =$ **F.** $\frac{7}{8} - \frac{3}{8} =$

Use your fraction strips to complete the following problems. Write a number sentence for each problem.

1. Grace needed $\frac{5}{8}$ of a yard of ribbon to decorate the outside edge of a picture frame. She needed another $\frac{3}{8}$ of a yard of ribbon to decorate the inside edge of her frame. How much ribbon did she need to finish the frame?

2. **A.** On Monday, John ate $\frac{1}{5}$ of a box of cookies. On Tuesday, he ate another $\frac{2}{5}$ of the cookies. What fraction of the cookies did he eat altogether?

   B. What fraction of the cookies is left?

3. **A.** Jerome lives $\frac{7}{10}$ of a mile from school. If he has already walked $\frac{3}{10}$ of a mile, how much farther does he have to go before he gets to school?

   B. Use your fraction strips to find another fraction that is equal to your answer.

4. Tanya and Nila used their fraction strips to add fractions. Look at their work. Write a number sentence to show what they did.

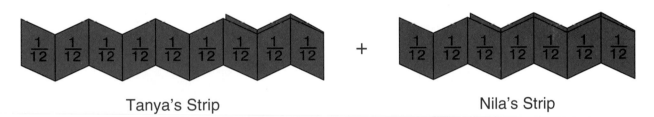

Tanya's Strip   Nila's Strip

5. Use your fraction strips to complete the following number sentences.

   A. $\frac{1}{12} + \frac{4}{12} =$ **B.** $\frac{7}{10} - \frac{5}{10} =$ **C.** $\frac{5}{8} + \frac{3}{8} =$

6. Maya and Jerome used their fraction strips to show the following addition problem. Write a number sentence for their work.

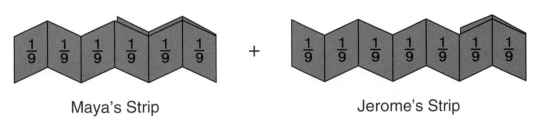

Maya's Strip   Jerome's Strip

7. Michael and his brother shared a pizza.

   A. Michael ate $\frac{2}{8}$ of a whole pizza. How much pizza was left?

   B. His brother ate another $\frac{3}{8}$ of the whole pizza. How much pizza was left?

   C. How much pizza did Michael and his brother eat altogether?

Comparing Fractions

Fraction Chart

Whole

| $\frac{1}{2}$ | $\frac{1}{2}$ |

| $\frac{1}{3}$ | $\frac{1}{3}$ | $\frac{1}{3}$ |

| $\frac{1}{4}$ | $\frac{1}{4}$ | $\frac{1}{4}$ | $\frac{1}{4}$ |

| $\frac{1}{5}$ | $\frac{1}{5}$ | $\frac{1}{5}$ | $\frac{1}{5}$ | $\frac{1}{5}$ |

| $\frac{1}{6}$ | $\frac{1}{6}$ | $\frac{1}{6}$ | $\frac{1}{6}$ | $\frac{1}{6}$ | $\frac{1}{6}$ |

| $\frac{1}{8}$ | $\frac{1}{8}$ | $\frac{1}{8}$ | $\frac{1}{8}$ | $\frac{1}{8}$ | $\frac{1}{8}$ | $\frac{1}{8}$ | $\frac{1}{8}$ |

| $\frac{1}{9}$ | $\frac{1}{9}$ | $\frac{1}{9}$ | $\frac{1}{9}$ | $\frac{1}{9}$ | $\frac{1}{9}$ | $\frac{1}{9}$ | $\frac{1}{9}$ | $\frac{1}{9}$ |

| $\frac{1}{10}$ | $\frac{1}{10}$ | $\frac{1}{10}$ | $\frac{1}{10}$ | $\frac{1}{10}$ | $\frac{1}{10}$ | $\frac{1}{10}$ | $\frac{1}{10}$ | $\frac{1}{10}$ | $\frac{1}{10}$ |

| $\frac{1}{12}$ | $\frac{1}{12}$ | $\frac{1}{12}$ | $\frac{1}{12}$ | $\frac{1}{12}$ | $\frac{1}{12}$ | $\frac{1}{12}$ | $\frac{1}{12}$ | $\frac{1}{12}$ | $\frac{1}{12}$ | $\frac{1}{12}$ | $\frac{1}{12}$ |

Use your fraction chart to complete the following questions.

1. Grace and her little sister each ordered a personal pizza for dinner. Grace ate $\frac{3}{4}$ of her pizza. Her little sister ate $\frac{1}{2}$ of her pizza. Who ate more pizza?

2. Roberto walks $\frac{7}{10}$ of a mile to get to school. Keenya walks $\frac{2}{3}$ of a mile to get to school. Who lives closer to the school, Keenya or Roberto?

3. Use your chart to find the fractions that are equal to $\frac{1}{2}$. Make a list of these fractions.

4. Use your fraction chart to compare the following pairs of fractions. Write number sentences using <, >, or =. For example, $\frac{1}{2} > \frac{1}{3}$.
   A. $\frac{1}{4}, \frac{1}{2}$ B. $\frac{2}{3}, \frac{1}{2}$ C. $\frac{1}{2}, \frac{2}{5}$
   D. $\frac{1}{2}, \frac{3}{6}$ E. $\frac{5}{12}, \frac{1}{2}$ F. $\frac{6}{9}, \frac{1}{2}$

5. Use $\frac{1}{2}$ as a benchmark or your fraction chart to compare the following pairs of fractions. Write number sentences using <, >, or =.
   A. $\frac{3}{4}, \frac{1}{3}$ B. $\frac{2}{5}, \frac{7}{10}$ C. $\frac{5}{8}, \frac{5}{12}$
   D. $\frac{2}{4}, \frac{6}{12}$ E. $\frac{3}{9}, \frac{2}{3}$ F. $\frac{3}{5}, \frac{1}{4}$

6. Use your fraction chart to put the following fractions in order from smallest to largest.
   A. $\frac{1}{3}, \frac{1}{6}, \frac{1}{2}$ B. $\frac{3}{5}, \frac{3}{4}, \frac{3}{12}$ C. $\frac{2}{10}, \frac{2}{4}, \frac{2}{9}$

7. Look at your answers for Question 6. Use them to help you answer this question: if two or more fractions have the same numerator, how can you tell which one is smallest?

8. Put the following fractions in order from smallest to largest.
   A. $\frac{4}{6}, \frac{1}{3}, \frac{1}{2}$ B. $\frac{7}{9}, \frac{4}{10}, \frac{3}{4}$ C. $\frac{3}{5}, \frac{5}{6}, \frac{1}{4}$ D. $\frac{5}{6}, \frac{5}{12}, \frac{5}{8}$

9. Explain your strategies for Questions 8A and 8D.

Complete the following questions. You may use your fraction chart to help you.

1. Find all the fractions equal to $\frac{1}{4}$ on your chart. Make a list of these fractions.

2. Jackie needs $\frac{5}{8}$ of a yard of fabric for a pillow. Luis needs $\frac{3}{4}$ of a yard of fabric for a banner. Who needs more fabric, Jackie or Luis?

3. Jessie's mom brought a pie to the potluck dinner. It was cut into 6 pieces. Romesh's dad also brought a pie, but it was cut into 12 pieces. At the end of the night, $\frac{1}{6}$ of Jessie's pie was left and $\frac{3}{12}$ of Romesh's pie was left. If the pies were the same size, who had more left-over pie, Jessie's mom or Romesh's dad?

4. Nila practiced her flute for $\frac{1}{2}$ hour on Monday, $\frac{3}{4}$ hour on Tuesday, and $\frac{1}{3}$ hour on Wednesday.

   A. On which day did she practice the longest period of time?

   B. On which day did she practice the shortest period of time?

5. Use your fraction chart to compare the following pairs of fractions. Write a number sentence for each one using <, >, or =.

   A. $\frac{3}{10}, \frac{1}{2}$

   B. $\frac{4}{8}, \frac{1}{2}$

   C. $\frac{1}{2}, \frac{2}{12}$

6. Use $\frac{1}{2}$ as a benchmark or your fraction chart to compare the following pairs of fractions. Write a number sentence for each one using <, >, or =.

   A. $1, \frac{1}{10}$

   B. $\frac{6}{9}, \frac{5}{12}$

   C. $\frac{3}{8}, \frac{3}{5}$

7. Use your fraction chart to put the following fractions in order from smallest to largest.

   A. $\frac{4}{8}, \frac{4}{6}, \frac{4}{10}$

   B. $\frac{3}{5}, \frac{3}{10}, \frac{3}{8}$

   C. $\frac{4}{8}, \frac{4}{12}, \frac{4}{6}$

   D. If two fractions have the same numerator, how can you tell which one is smaller?

8. Put the following fractions in order from smallest to largest. Be prepared to explain your strategies.

   A. $\frac{7}{12}, \frac{1}{3}, \frac{3}{8}$

   B. $\frac{3}{5}, \frac{5}{12}, \frac{1}{2}$

   C. $\frac{2}{3}, \frac{3}{4}, \frac{1}{6}$

   D. $\frac{1}{5}, \frac{1}{4}, \frac{1}{6}$

   E. $\frac{7}{12}, \frac{1}{12}, \frac{5}{12}$

   F. $\frac{1}{2}, \frac{3}{4}, \frac{2}{9}$

Frabble Game

Players

This game can be played in groups of two, three, or four players.

Materials

- one Fraction Chart from Lesson 3 for each student
- one deck of *Standard Frabble Cards* (24 cards) for each group
- two wild cards for each player
- pencil and paper for scoring

Rules

1. Shuffle the *Standard Frabble Cards,* and divide them evenly among all players.

2. Give each player two wild cards.

3. Players place all their cards face up in front of them.

4. The player with the Start card begins the game by placing this card in the center of the playing area.

5. The player to the left of the starting player takes the next turn. The player adds one card to the game, following these rules.

   - Cards with smaller fractions are placed to the left.

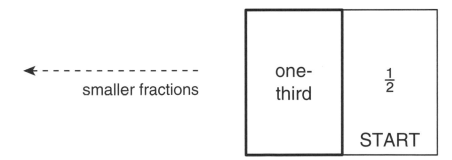

   - Cards with larger fractions are placed to the right.

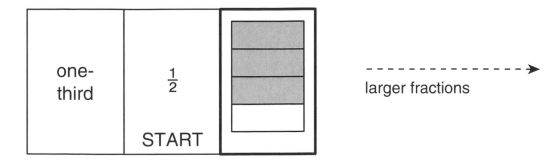

- Cards with equal fractions are placed above or below.

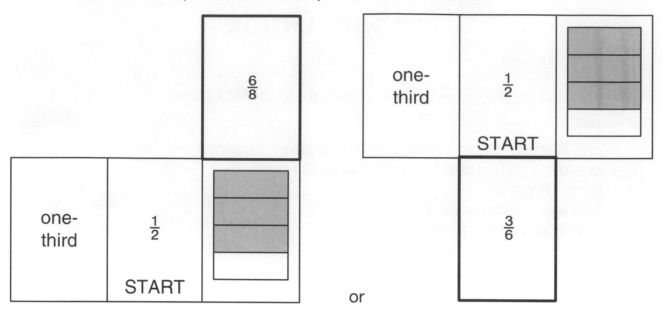

or

- You can place a wild card at any time. To place a wild card, you must give the card a name that is not already on the board. For example, a wild card named $\frac{2}{6}$ can be placed below the one-third card.

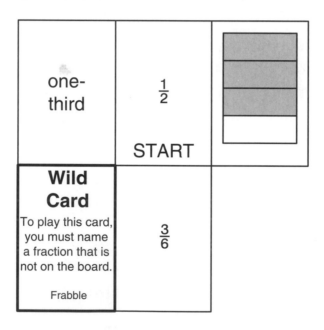

- You may not move any card that has already been played. For example, you may not move the $\frac{3}{4}$ card over in order to place a $\frac{2}{3}$ card between the $\frac{1}{2}$ and the $\frac{3}{4}$ card in the example shown above.

6. If you do not have a card that can be placed on the board, you lose a turn. The game ends when no player can take a turn. The player with the most points wins the game.

Scoring

One player is chosen to write down each person's points at the end of each turn. Points are scored by these rules:

- The player who places the "Start" card earns one point for placing one card on the board.
- Each player earns:
  1 point for each card played
  1 point for each card connected to the new card in the row
  1 point for each card on the board that has a fraction equal to the card played

Example 1: The player who places the "one" card earns four points—one point for playing the card and three points for the three cards connected to the new card in the same row.

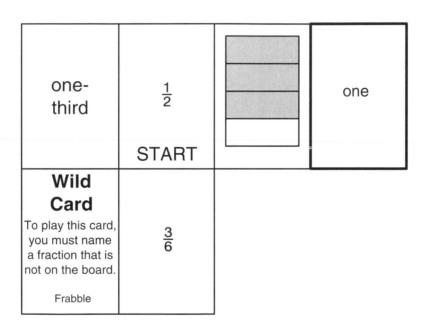

Example 2: The player who adds the $\frac{1}{1}$ card earns two points—one point for playing the card ($\frac{1}{1}$) and one point for the equal fraction (one).

one-third	$\frac{1}{2}$ START		one
Wild Card To play this card, you must name a fraction that is not on the board. Frabble	$\frac{3}{6}$		$\frac{1}{1}$

Example 3: The player who adds the $\frac{6}{8}$ card earns a total of five points—one point for the card added ($\frac{6}{8}$); three points for the cards connected in the row (wild card, $\frac{3}{6}$, and $\frac{1}{1}$); and one point for the equal fraction (picture of $\frac{3}{4}$).

one-third	$\frac{1}{2}$ START		one
Wild Card To play this card, you must name a fraction that is not on the board. Frabble	$\frac{3}{6}$	$\frac{6}{8}$	$\frac{1}{1}$

Equivalent Fractions

Discuss

Irma wants to bake some cookies. Her recipe calls for $\frac{3}{4}$ cup sugar. Irma can only find a $\frac{1}{8}$-cup measure. She needs to know how many eighths of a cup of sugar is the same as $\frac{3}{4}$ cup. She knows that two $\frac{1}{8}$-cup measures hold the same amount of sugar as a $\frac{1}{4}$-cup measure. She knows that she needs enough sugar to fill three $\frac{1}{4}$-cup measures because she needs $\frac{3}{4}$ cup. She reasons that she must fill the $\frac{1}{8}$ cup twice as many times, or six times. Irma also remembers what she learned in math class: if you multiply (or divide) the numerator and the denominator of a fraction by the same number, you will get an equal or equivalent fraction. **Equivalent fractions** are fractions that have the same value.

To solve this problem, Irma can use this number sentence: $\frac{3}{4} = \frac{?}{8}$.

1. **A.** Help Irma solve this problem. Think of a strategy she can use.

   B. Irma knows that $4 \times 2 = 8$. She multiplied 4 times 2 to find the new denominator. So she also must multiply 3 times 2 in order to find the missing numerator. Complete this number sentence for Irma: $\frac{3}{4} = \frac{?}{8}$.

Explore

2. Romesh is helping his father pack a box of key chains for a fundraiser. The box holds $\frac{1}{2}$ pound of merchandise. Each key chain weighs $\frac{1}{16}$ of a pound. Romesh must decide how many key chains he can fit in the box.

   A. Help Romesh by completing this number sentence: $\frac{1}{2} = \frac{?}{16}$.

   B. How many key chains can Romesh pack in the box?

---

3. **A.** Use your fraction chart to find three fractions that are equivalent to $\frac{3}{9}$. Write number sentences to record the equivalent fractions.

   B. Find three other fractions that are equivalent to $\frac{3}{9}$. Write number sentences to record the equivalent fractions.

   C. Explain the strategy you used to find the equivalent fractions.

4. Complete the number sentence: $\frac{4}{8} = \frac{?}{12}$. Explain how you know.

5. **A.** Use your fraction chart to find a fraction that is equivalent to $\frac{3}{5}$. Write a number sentence to record the equivalent fractions.

   B. Find three other fractions that are equivalent to $\frac{3}{5}$. Write number sentences to record the equivalent fractions.

   C. Explain the strategy you used to find the equivalent fractions.

6. Complete the number sentences below. Use your fraction chart.

   A. $\frac{3}{4} = \frac{?}{8}$ **B.** $\frac{1}{2} = \frac{?}{10}$ **C.** $\frac{2}{3} = \frac{?}{9}$ **D.** $\frac{6}{9} = \frac{?}{12}$

   E. $\frac{1}{2} = \frac{4}{?}$ **F.** $\frac{6}{10} = \frac{?}{5}$ **G.** $\frac{8}{12} = \frac{2}{?}$ **H.** $\frac{3}{12} = \frac{?}{8}$

Homework

1. Complete the number sentences to make each fraction equivalent to $\frac{1}{2}$.

   A. $\frac{1}{2} = \frac{3}{?}$ **B.** $\frac{1}{2} = \frac{?}{18}$ **C.** $\frac{1}{2} = \frac{12}{?}$

   D. $\frac{1}{2} = \frac{?}{60}$ **E.** $\frac{1}{2} = \frac{50}{?}$ **F.** $\frac{1}{2} = \frac{?}{?}$

2. Write 5 fractions equivalent to $\frac{2}{3}$.

3. Romesh is packing a box filled with plastic cars for his father. The box holds $\frac{3}{4}$ pound of merchandise. Each plastic car weighs $\frac{1}{16}$ pound.

   A. Complete this number sentence to help Romesh decide how many sixteenths of a pound is equivalent to $\frac{3}{4}$ pound. $\frac{3}{4} = \frac{?}{16}$.

   B. How many plastic cars can Romesh pack in the box?

   C. What is another name for $\frac{1}{16}$ of a pound?

Equivalent Fractions

4. Write 5 fractions equivalent to $\frac{2}{5}$.

5. Shannon wants to purchase $\frac{1}{3}$ yard of ribbon. There are 36 inches in a yard.

    A. Complete the following number sentence to help the clerk decide how many inches of ribbon she must cut: $\frac{1}{3} = \frac{?}{36}$.

    B. How many inches of ribbon should she cut?

6. Use the Fraction Chart to complete the number sentence: $\frac{6}{8} = \frac{?}{12}$.

Complete the following number sentences.

    7. $\frac{1}{2} = \frac{?}{12}$ **8.** $\frac{3}{4} = \frac{?}{16}$ **9.** $\frac{4}{6} = \frac{?}{9}$

  10. $\frac{3}{5} = \frac{?}{20}$ **11.** $\frac{10}{16} = \frac{?}{8}$ **12.** $\frac{8}{24} = \frac{?}{3}$

  13. $\frac{10}{15} = \frac{?}{3}$ **14.** $\frac{1}{5} = \frac{?}{100}$ **15.** $\frac{1}{5} = \frac{?}{20}$

  16. $\frac{75}{100} = \frac{?}{4}$ **17.** $\frac{2}{4} = \frac{?}{6}$ **18.** $\frac{20}{24} = \frac{5}{?}$

Use <, >, or = to write number sentences to compare the following pairs of numbers.

  19. $\frac{5}{9}, \frac{1}{2}$ **20.** $\frac{3}{4}, \frac{30}{40}$ **21.** $\frac{72}{100}, \frac{7}{10}$

Pattern Block Fractions

When Are Halves Different?

Jacob and Jerome looked at their data for the *Bouncing Ball* lab. They wondered what would happen if they dropped a tennis ball from a tall building. Jacob said, "Every time we dropped a ball during the lab, it bounced back about half of the drop height. Think how high a ball would bounce if we dropped it from the top of the Sears Tower in Chicago! That's one of the tallest buildings in the world."

Jerome said, "The CN Tower in Toronto is even taller. If we dropped the ball from the top of it, the ball would bounce even higher!"

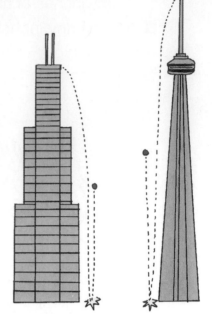

Discuss

1. If each ball bounces one-half the distance of the drop height, will the bounce heights be the same? Why or why not?

2. When are halves different?

Jerome and Jacob used fractions to estimate the bounce height. In this activity, you will study fractions using pattern blocks.

Exploring Pattern Blocks

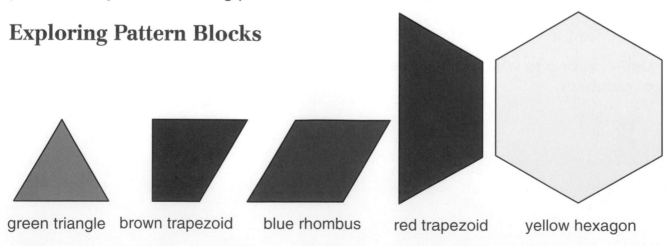

green triangle   brown trapezoid   blue rhombus   red trapezoid   yellow hexagon

Use the blocks to help you answer the following questions:

3. How many brown trapezoids equal one yellow hexagon?

4. How many brown trapezoids equal one red trapezoid?

5. How many red trapezoids equal one yellow hexagon?

6. One brown trapezoid is (less than, greater than, or equal to) one red trapezoid.

7. How many brown trapezoids equal three red trapezoids?

8. One brown trapezoid is (less than, greater than, or equal to) one green triangle.

9. How many green triangles equal one yellow hexagon?

10. One yellow hexagon equals two brown trapezoids plus how many green triangles?

11. How many green triangles equal two brown trapezoids?

12. How many green triangles equal two blue rhombuses?

13. Two blue rhombuses are (less than, greater than, or equal to) one brown trapezoid.

Exploring Pattern Block Fractions

14. Each of these figures shows thirds using pattern blocks. Build these figures with pattern blocks. Place three blue rhombuses on a yellow hexagon. Place three green triangles on a red trapezoid.

    A. If the red trapezoid is one whole, which block shows $\frac{1}{3}$?

    B. If the blue rhombus is $\frac{1}{3}$, which block shows one whole?

    C. If the red trapezoid is one whole, show $\frac{2}{3}$.

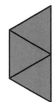

15. A. Show halves using pattern blocks in as many ways as you can.

    B. If the yellow hexagon is one whole, which block shows $\frac{1}{2}$?

    C. If the green triangle is $\frac{1}{2}$, which block is one whole?

    D. If the brown trapezoid is $\frac{1}{2}$, which block is one whole?

16. A. If the yellow hexagon is one whole, which block shows $\frac{1}{4}$?

    B. If the yellow hexagon is one whole, show $\frac{3}{4}$.

    C. If the yellow hexagon is one whole, show $\frac{5}{4}$.

17. A. If the green triangle is $\frac{1}{6}$, which block is one whole?

    B. If the yellow hexagon is one whole, show $\frac{3}{6}$.

    C. If the yellow hexagon is one whole, show $\frac{5}{6}$.

18. If the red trapezoid is one whole, name each of the following fractions:

    A. one green triangle

    B. two green triangles

    C. one blue rhombus

    D. one brown trapezoid

    E. three brown trapezoids

    F. five green triangles

one whole

19. If the yellow hexagon is one whole, name each of the following fractions:

    A. one red trapezoid

    B. one brown trapezoid

    C. two brown trapezoids

    D. one blue rhombus

    E. two green triangles

    F. two blue rhombuses

    G. three red trapezoids

one whole

Fraction Sentences

For Questions 20–26, the yellow hexagon is one whole. The red trapezoid is $\frac{1}{2}$. We can show $\frac{1}{2}$ using brown blocks. Since 1 red trapezoid equals 2 brown trapezoids, then $\frac{1}{2} = \frac{2}{4}$ or $\frac{1}{2} = \frac{1}{4} + \frac{1}{4}$.

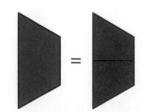

$\frac{1}{2} = \frac{2}{4}$

or

$\frac{1}{2} = \frac{1}{4} + \frac{1}{4}$

20. Show $\frac{1}{2}$ using green blocks. (Cover a red trapezoid with green blocks.) Write a number sentence to represent this figure.

21. The blue rhombus is $\frac{1}{3}$. Show $\frac{1}{3}$ using green blocks, and write a number sentence to represent this figure.

We can show 1 whole with two or more colors and write a number sentence to represent the figure.

$$1 = \frac{2}{4} + \frac{1}{6} + \frac{1}{3}$$

22. Show 1 whole another way using two or more colors. Write a number sentence for your figure.

For Questions 23–26, show each fraction using two or more colors. Write a number sentence for each figure.

23. Show $\frac{1}{2}$.

24. Show $\frac{3}{4}$.

25. Show $\frac{2}{3}$.

26. Show $\frac{3}{2}$.

Homework

Use your fraction chart from Lesson 3 or imagine pattern blocks to help you solve these problems.

1. Michael used $\frac{1}{2}$ yard of ribbon to decorate a gift for his mother. Irma used $\frac{2}{3}$ yard for her mother's present. Who used more ribbon?

2. Lee Yah drank $\frac{1}{3}$ cup of juice, and Roberto drank $\frac{1}{2}$ cup. Who drank more juice?

3. Put these fractions in order from smallest to largest: $\frac{5}{6}, \frac{1}{4}, \frac{1}{2}$.

4. Put these fractions in order from smallest to largest: $\frac{1}{2}, \frac{1}{3}, \frac{3}{4}$.

5. Put these fractions in order from smallest to largest: $\frac{2}{3}, \frac{1}{2}, \frac{1}{6}$. Explain your strategy.

6. Add or subtract.

A. $\frac{2}{6} + \frac{3}{6} =$

B. $\frac{1}{4} + \frac{2}{4} =$

C. $\frac{1}{3} + \frac{2}{3} =$

D. $\frac{3}{4} - \frac{1}{4} =$

E. $\frac{5}{6} - \frac{2}{6} =$

F. $\frac{3}{3} - \frac{1}{3} =$

7. Write three equivalent fractions for $\frac{3}{4}$.

Solving Problems with Pattern Blocks

You may use pattern blocks or your fraction chart to help you solve these problems.

1. Wednesday is pizza day at Bessie Coleman School. Each table in the lunchroom gets one pizza to share fairly among the students at the table. There are three students at Table A and four students at Table B.

   A. What fraction of the pizza will each student at Table A get?

   B. What fraction of the pizza will each student at Table B get?

   C. Who gets to eat more pizza, the students at Table A or the students at Table B?

   D. Which fraction is larger, $\frac{1}{3}$ or $\frac{1}{4}$? Explain how you know.

2. The cook made three small fruit pies that are all the same size. She divided the apple pie into 12 pieces, the cherry pie into six pieces, and the peach pie into four pieces. John ate two pieces of apple pie, Shannon ate two pieces of cherry pie, and Brandon ate two pieces of peach pie.

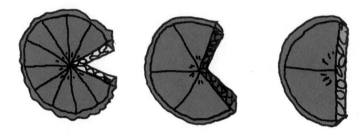

   A. What fraction of the apple pie did John eat?

   B. What fraction of the cherry pie did Shannon eat?

   C. What fraction of the peach pie did Brandon eat?

   D. Who ate the most pie? Tell how you know.

   E. Who ate the least pie?

3. One whole is divided into eight zax. Each zax is the same size. The same size whole is divided into ten snarks. Each snark is the same size.

   A. What fraction of the whole is one zax?

   B. What fraction of the whole is one snark?

   C. Which is larger, one zax or one snark? Explain.

4. Put each group of fractions in order from smallest to largest.

    A. $\frac{1}{2}, \frac{1}{6}, \frac{1}{3}, \frac{1}{4}, \frac{1}{12}$

    B. $\frac{2}{6}, \frac{2}{3}, \frac{2}{4}, \frac{2}{12}$

    C. $\frac{1}{10}, \frac{1}{8}, \frac{1}{5}$

    D. $\frac{3}{10}, \frac{3}{8}, \frac{3}{5}$

5. Describe a strategy for ordering fractions if the numerators are the same.

To solve the problems in Questions 6–8, you may use any tools such as pattern blocks, the Fraction Chart in Lesson 3, or pictures. Write number sentences to record your solutions.

6. Each of the following pairs of students shared a pizza. How much of the whole pizza did each pair eat?

    A. Manny ate $\frac{1}{2}$ of a pizza and Ming ate $\frac{1}{4}$ of it.

    B. Michael ate $\frac{3}{8}$ of a pizza and Frank ate $\frac{5}{8}$.

    C. Felicia ate $\frac{1}{3}$ of a pizza. Linda ate $\frac{1}{6}$ of it.

    D. Lee Yah ate $\frac{5}{12}$ and David ate $\frac{2}{12}$.

7. A. Four students each ate $\frac{1}{2}$ of a muffin. How many muffins did they eat altogether?

    B. Five students each ate $\frac{1}{2}$ of a muffin. How many muffins did they eat altogether?

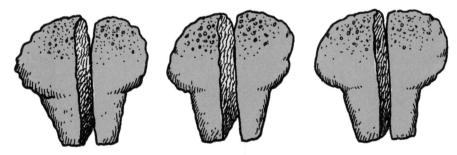

8. A. Eight students each ate $\frac{1}{4}$ of an apple. How many apples did they eat altogether?

    B. Three students each ate $\frac{1}{4}$ of an apple. How many apples did they eat altogether?

    C. Six students each ate $\frac{1}{4}$ of an apple. How many apples did they eat altogether?

1. Write these fractions in order from smallest to largest.

   A. $\frac{3}{12}, \frac{3}{4}, \frac{3}{3}, \frac{3}{6}$

   B. $\frac{3}{5}, \frac{3}{10}, \frac{3}{8}, \frac{3}{4}$

   C. $\frac{3}{8}, \frac{1}{8}, \frac{5}{8}, \frac{8}{8}$

   D. $\frac{1}{2}, \frac{1}{12}, \frac{5}{6}$

2. On Sunday, Shannon's family ate $\frac{5}{12}$ of a casserole. On Monday, they ate $\frac{3}{12}$ of the casserole. How much of the casserole did they eat? How much is left over?

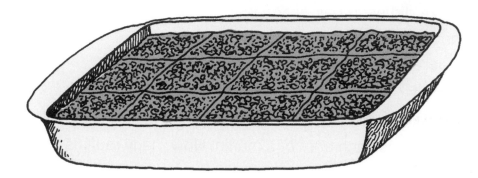

3. On Friday, a worker painted $\frac{3}{8}$ of a fence. On Saturday, he painted another $\frac{3}{8}$ of the fence.

   A. How much of the fence did he paint on the two days?

   B. How much more of the fence does he have left to paint?

4. Complete the following number sentences.

   A. $\frac{3}{4} = \frac{?}{20}$ **B.** $\frac{3}{6} = \frac{5}{?}$ **C.** $\frac{2}{3} = \frac{6}{?}$

   D. $\frac{3}{8} = \frac{?}{16}$ **E.** $\frac{1}{2} = \frac{?}{6}$ **F.** $\frac{30}{100} = \frac{3}{?}$

5. **A.** Four children each ate $\frac{1}{3}$ of a large cookie. How many cookies did they eat altogether?

   B. Six children each ate $\frac{1}{3}$ of a large cookie. How many cookies did they eat altogether?

Fraction Puzzles

Discuss

"Today in math, you will work in groups of four to solve puzzles using pattern blocks," said Mrs. Dewey. "Each puzzle has four clues to help you find a solution. Each group member will receive one of the clues. A clue can only be shared with your group by reading it. Once all the clues have been read, the group's job is to find a solution that meets all the guidelines given in the clues."

Roberto, Nicholas, Linda, and Jackie formed a group. Each received a clue for the first puzzle. "My clue says we need to use 2 or 3 blocks," said Nicholas as he reached to get some blocks.

"Wait," said Linda, "I think we should each read our clues to the group before we start building. That way we will have all of the information we need to start."

"Great idea! Go ahead and read your clue again, Nicholas, and then we will each read ours," said Jackie.

My clue says, "Use 2 or 3 blocks."

Mine says, "The yellow hexagon is equal to one whole."

Mine is, "Use at least 1 green triangle, but not all green triangles."

My clue is, "Make a shape with a value of $\frac{2}{3}$."

"Now that we have all the clues, let's get started," exclaimed Nicholas.

After some work, the students found this solution:

They wrote this number sentence to represent their solution: $\frac{1}{2} + \frac{1}{6} = \frac{2}{3}$.

1. Look back at their clues, and see if this solution fits all the clues they were given.

2. One of the other groups found this solution to the same puzzle.

They wrote this number sentence: $\frac{1}{6} + \frac{1}{6} + \frac{1}{6} + \frac{1}{6} = \frac{2}{3}$. Look back at the clues. Does this solution fit all the clues that were given? Why or why not?

3. Find another solution to this puzzle. Use the clues provided to make sure your solution fits all the clues. Draw a picture of your solution, and write a number sentence to represent your solution.

Homework

Solve the following problems. You may use your fraction chart to help you.

1. Put the following fractions in order from smallest to largest.
   A. $\frac{1}{5}, \frac{1}{9}, \frac{1}{8}, \frac{1}{3}$ **B.** $\frac{3}{10}, \frac{3}{4}, \frac{3}{8}, \frac{3}{5}$ **C.** $\frac{2}{3}, \frac{3}{4}, \frac{5}{8}, \frac{1}{2}$ **D.** $\frac{2}{5}, \frac{3}{8}, \frac{5}{12}, \frac{1}{4}$

2. Put the following fractions in order from smallest to largest.
   A. $\frac{1}{3}, \frac{1}{5}, \frac{1}{2}, \frac{1}{8}$ **B.** $\frac{2}{6}, \frac{2}{4}, \frac{2}{5}, \frac{2}{10}$ **C.** $\frac{4}{5}, \frac{4}{12}, \frac{4}{8}, \frac{4}{6}$ **D.** $\frac{3}{8}, \frac{3}{10}, \frac{3}{5}, \frac{3}{4}$

   E. Explain a strategy for putting fractions in order when the numerators are all the same.

3. Write a number sentence for each pair of fractions. Use the symbols <, >, or = in each sentence.

A. $\frac{6}{8}, \frac{3}{4}$ B. $\frac{3}{5}, \frac{3}{8}$ C. $\frac{1}{3}, \frac{3}{6}$

D. $\frac{1}{2}, \frac{5}{10}$ E. $\frac{4}{5}, \frac{5}{12}$ F. $\frac{3}{9}, \frac{1}{3}$

4. Frank and Jerome each ordered a small cheese pizza for lunch. Frank's pizza was cut into 6 pieces. Jerome's pizza was cut into 8 pieces. Frank ate 2 pieces of his pizza. Jerome ate 3 pieces of his pizza. Which boy ate more pizza? How do you know?

5. Nila and Tanya shared a sandwich for lunch. Nila ate $\frac{1}{2}$ of the sandwich, and Tanya ate $\frac{1}{4}$ of the sandwich. What fraction of the whole sandwich did the two girls eat? Explain how you found your answer.

6. Lee Yah, Luis, John, and Shannon solved a fraction puzzle. Their solution is below. If a yellow hexagon is one whole, write a number sentence for their solution.

7. Frank, Jacob, Irma, and Maya solved a fraction puzzle. Their solution is shown on the right. Does their solution fit the clues? Explain your thinking.

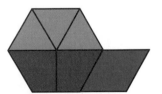

Clue 1: The red trapezoid is equal to 1 whole.
Clue 2: Make a shape with a value of $\frac{8}{3}$.
Clue 3: Use at least two brown trapezoids.
Clue 4: Do not use any blue rhombuses.

Unit 13

Division

	Student Guide	Discovery Assignment Book	Adventure Book	Unit Resource Guide*
Lesson 1				
TV Survey	●	●		●
Lesson 2				
Division	●			
Lesson 3				
More Division	●	●		
Lesson 4				
Solving Problems Using Multiplication and Division	●			●
Lesson 5				
Plant Growth	●			

Unit Resource Guide pages are from the teacher materials.

TV Survey

Mrs. Dewey shared with her class an article about the amount of television an average fourth-grader watches. Studies show that children between the ages of 6 and 11 watch an average of 3 hours of TV every day. Mrs. Dewey's class decided to find out how much TV they watch. They will record the number of minutes of television they watch for four days. Then the class will analyze the results.

Conduct a survey to find out how much television you watch. A *survey* is an investigation made by collecting information and then analyzing it. Collect data for four days. Record the number of minutes of television you watch each day.

1. Record your data for a four-day period. Use the four tables on the *Daily TV Time* Activity Pages in the *Discovery Assignment Book.* One table should be used for each day. Irma's table for the fourth and final day is shown below.

Daily TV Time

Program	Starting Time	Ending Time	Minutes Watched
Exploring Space	3:00	4:00	60
Milo and Bob	4:00	4:30	30
Power Troopers	6:00	6:30	30
Take a Chance	6:30	7:00	30
Treasure Search	8:00	8:45	45

(left side label: Thursday / Day)

Average Number of Minutes per Day ___157.5___

Average Number of Hours per Day _____

Total TV Minutes ___195___

Total Number of Minutes for 4 Days ___630___

2. Find and record the following information on your copy of the *Daily TV Time* Activity Page.

   A. The total number of minutes of TV you watched over the four days

   B. The average (mean) number of minutes of TV per day

Converting Average Number of Minutes to Average Number of Hours

3. Irma calculated that her average number of minutes of TV time per day is 157.5 minutes.

   A. Is her average daily TV time between 1 and 2 hours, or between 2 and 3 hours? How did you decide?

   B. How many whole hours of TV, on average, did she watch per day?

Irma finds her average daily TV time in hours. She knows there are 60 minutes in 1 hour. She uses the calculator to do the following problem:

| 1 | 5 | 7 | . | 5 | ÷ | 6 | 0 | = | .

This is what the calculator window displayed: 2.625

4. A. How many full hours did Irma watch on average per day?

  B. What does the ".625" tell you?

  C. Does the answer in the calculator display agree with your estimate in Question 3A?

5. Look at your average number of minutes of TV per day.

  A. Is your average daily TV time more or less than 1 hour? More or less than 2 hours? More or less than 3 hours?

  B. Calculate your average number of hours of TV per day. Record it on your copy of the *Daily TV Time* Activity Page.

6. Compare your average daily TV time to the data reported in the newspaper article. (The newspaper article reported that students between the ages of 6 and 11 watch 3 hours of TV every day.) Did you watch more or less than 3 hours of TV each day of the survey?

Making a Class Data Table

7. Share your data with the class by recording it in a class data table like the one below.

TV Survey: Class Data

Name	Total TV Time for 4 Days in Minutes	Average TV Time per Day in Minutes	Average TV Time per Day in Hours
John	480	120	2
Irma	630	157.5	2.625
Roberto	165	41.25	0.6875

Use your class data to answer Questions 8–10.

8. A. Who watched the most TV in your class?

  B. How many minutes did he or she watch TV?

9. A. Who watched the least TV in your class?

  B. How many minutes did he or she watch TV?

10. What can you learn from the class data table?

11. Look at the last column in your class data table. How many students' average daily TV time was between 0 and 1 hour? Between 1 and 2 hours? Between 2 and 3 hours, and so forth? Complete a table like the one below.

How Much TV Do We Watch?

T Average TV Time per Day (in hours)	S Number of Students	
	Tallies	Total
0 – .99	~~IIII~~	
1.00 – 1.99	III	
2.00 – 2.99	II	
3.00 – 3.99	IIII	
4.00 – 4.99	I	

12. What are the two main variables we are studying?

13. Make a bar graph on *Centimeter Graph Paper* using your class television data. Label the horizontal axis with *T*, TV Time in Hours, and the vertical axis with *S*, Number of Students.

14. How many students in your class averaged between 0 and 1 hour of TV per day?

15. What average TV time is most common?

16. Describe the graph. What does the graph tell you about the amount of TV the students in your class watched?

Extension

Follow-up Survey. One month from now, collect data over four days one more time. Make a data table and graph. Compare your data to the first TV Survey you completed. See if the TV Survey helped to change your habits.

TV Time vs. Reading Time. Collect data on the number of minutes you watch TV and the number of minutes you read every day. Graph and discuss your results. Give up television for four days, and collect data on your reading habits for those four days. Discuss your results.

You will need a calculator, a *Three-Column Data Table,* and *Centimeter Graph Paper* to complete the homework.

On Monday, Mrs. Dewey's class began their data collection for TV Survey. The data collection continued through Thursday evening. On Friday, the students shared their data by combining it on one large data table. A section of their class data table is shown below.

TV Survey: Class Data

Name and Group Number		Average TV Time per Day in Minutes	Average TV Time per Day in Hours
Ming	(Group 7)	240	
Shannon	(Group 7)	135	
Ana	(Group 7)	390	
Luis	(Group 7)	195	
Nicholas	(Group 8)	45	
Lee Yah	(Group 8)	210	
Jacob	(Group 8)	105	
Nila	(Group 8)	180	

Mrs. Dewey's students sit in groups of four. The students listed in the table above are in Groups 7 and 8.

1. Who watched more television, Group 7 or Group 8? How did you decide?

2. How many minutes of television did Group 7 watch in all?

3. How many minutes of television did Group 8 watch in all?

4. How many more minutes of television did Ana watch than Shannon?

5. Lee Yah watched about twice as much television as another student. Who? How did you decide?

6. Copy the previous data table onto a *Three-column Data Table*. Find the number of hours each student watched TV. Fill in the third column with your findings.

7. Each student in Groups 1 through 6 came to the overhead. The student placed a tally for himself or herself as shown in the following data table. Copy this data table onto a copy of a *Three-column Data Table*. Add the data for the students in Groups 7 and 8 to the table, using tallies. Then total the tallies.

How Much TV Do We Watch?

T Average TV Time per Day (in hours)	S Number of Students	
	Tallies	Total
0 – .99	ⅢⅢ	
1.00 – 1.99	IIII	
2.00 – 2.99	III	
3.00 – 3.99		
4.00 – 4.99	I	
5.00 – 5.99		
6.00 – 6.99	I	
7.00 – 7.99		

8. What amount of TV viewing is most common in Mrs. Dewey's class?

9. Graph all of Room 204's data on *Centimeter Graph Paper*.

10. Describe the graph. What does it tell you about the amount of TV the students in Room 204 watched?

Division

Modeling Division

Explore

Ana and Roberto were given a box of 75 marbles to share.

1. Can Ana and Roberto divide up the marbles evenly? How do you know?

2. Estimate how many marbles each will get.

3. Discuss how Ana and Roberto can decide how many marbles each of them will get.

One way to divide the marbles is to pass them out, keeping track of how many each person gets.

Another way is to first decide how many marbles each person will get and then pass them out. Base-ten pieces can be used to model the problem. Ana and Roberto use these base-ten pieces to represent 75 marbles. A bit is one.

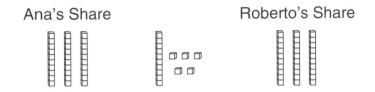

4. Since there are 7 skinnies, Ana and Roberto get 3 skinnies each. How many marbles do 3 skinnies represent?

Ana's Share   Roberto's Share

There is 1 skinny left over. This skinny can be traded for 10 bits, so there are 15 bits left. Ana and Roberto get 7 bits each, with 1 left over.

So, each child receives 37 marbles.

Ana's Share   Roberto's Share

There are several ways to write 75 divided by 2 using math symbols. Here are 2 ways:

$$2\overline{)75} \qquad 75 \div 2$$

Write a story for Questions 5 and 6, and then find the answer. Be sure to explain what any remainders mean.

5. $6\overline{)78}$

6. $93 \div 4$

The Forgiving Method

Keenya said she had a third way of solving 75 divided by 2 using paper and pencil. We call this the **forgiving method.** Keenya wrote:

$$2\overline{)75}$$

Keenya then estimated how many marbles she thought each child would get. Her first estimate was 20 each. Since each gets 20, and $2 \times 20 = 40$, 40 marbles are taken care of. Keenya wrote:

$$
\begin{array}{r}
2\overline{)75} \\
-40 \quad | \ 20 \\
\hline
35 \qquad
\end{array}
$$

Keenya has 35 marbles left to pass out.

7. How many marbles should she give each child now? What is a good estimate?

Keenya decided to give each child 10 more marbles.

8. Why did Keenya know that giving 20 more marbles to each child is too much?

She wrote:

$$
\begin{array}{r}
2\overline{)75} \\
-40 \quad | \ 20 \\
\hline
35 \qquad \\
-20 \quad | \ 10 \\
\hline
15 \qquad
\end{array}
$$

Keenya then chose 7. Since $2 \times 7 = 14$, she wrote:

```
  2 ) 75
    - 40 | 20
      35
    - 20 | 10
      15
    - 14 |  7
       1
```

Keenya saw that she could not divide further. She added up $20 + 10 + 7 = 37$. This is the number of times 2 divides 75. She finished the problem by writing the quotient on top: 37 and remainder, 1.

```
       37 R1
  2 ) 75
    - 40 | 20
      35
    - 20 | 10
      15
    - 14 |  7
       1
```

Here is another way to do this problem using the forgiving method.

```
       37 R1
  2 ) 75
    - 60 | 30
      15
    - 14 |  7
       1
```

The Boys' Wilderness Club from Bessie Coleman School is going camping.

9. The younger boys will be sleeping in cabins. Each cabin holds 7 people. If there are 95 younger boys, how many cabins will be needed?

   A. Estimate the number of cabins. Will the number of cabins be more than 10? more than 20? more than 15?

   B. Divide 95 by 7.

   C. What does the quotient mean?

   D. What does the remainder mean?

   E. How many boys will sleep in each cabin?

   F. Is there another way to arrange the number of boys in each cabin?

10. The older boys in the Wilderness Club want to sleep in tents. The club has 9 tents. There are 43 older boys. How many boys can sleep in each tent so that no tent is overcrowded?

   A. Estimate the number of boys in each tent.

   B. Solve the problem, and explain your reasoning.

Solve the following problems using the forgiving method or mental math. Record both the quotient and the remainder for your final answer. Estimate to be sure your answer is reasonable.

11. $49 \div 3$ **12.** $92 \div 4$ **13.** $56 \div 7$ **14.** $89 \div 6$

Homework

Use base-ten shorthand to model the two story problems that follow. Remember to describe how to use the remainder.

1. A group of campers are setting up tents. Each tent needs 6 poles. There are 32 poles in the box. How many tents can be set up?

2. Some campers want to go boating. Each boat can safely hold 4 people. There are 6 boats and 22 campers.

  A. Can they all go boating?

  B. Will all the boats be full?

Write a story for each of the following problems. Then solve the problems. You may show your work using base-ten shorthand. Remember to talk about any remainder.

  3. $34 \div 7$ 4. $81 \div 3$

  5. $67 \div 6$ 6. $75 \div 4$

Use the forgiving method to solve the following division problems. Remember to record your final answer, the quotient, and the remainder. Estimate to make sure your answer is reasonable.

  7. $74 \div 2$ 8. $87 \div 3$

  9. $43 \div 3$ 10. $95 \div 8$

  11. $73 \div 6$ 12. $97 \div 4$

13. **A.** The fourth-grade students at Bessie Coleman School are going on a field trip to the museum. There are 66 students and 5 adults who will be riding the bus. Each seat on the bus will hold 3 people. Use the forgiving method or mental math to divide 71 by 3. What remainder do you get?

  B. What does this remainder mean?

  C. How many seats will the bus need to have?

14. **A.** Once the students arrive at the museum, each adult will take a group of students to see the exhibits. Use the forgiving method or mental math to divide 66 by 5. What remainder do you get?

    B. What does the remainder mean?

    C. Show how you can divide the 66 students into 5 groups.

15. **A.** Jacob took pictures at the museum. He is going to put them into a photograph album. He has 82 pictures. Each page of his album will hold 6 pictures. How many full pages will he have?

    B. Will there be any remaining pictures?

16. The museum runs a train through several exhibits. Each car holds 6 people. The train has 9 cars.

    A. Can the whole group (students and adults) ride the train at the same time?

    B. How many cars will be needed in all?

More Division

Nicholas watched 347 minutes of television in 5 days. If he watched about the same number of minutes each day, about how many minutes per day did he watch television?

1. Estimate how many minutes Nicholas watched television each day. Did he watch more than 100 minutes each day?

2. Model the problem using the base-ten pieces or base-ten shorthand.

3. Will 5 divide into 347 evenly? Why or why not?

4. What is the average number of minutes Nicholas watched television each day? Explain the remainder.

Keenya showed her method again for doing division.

How do you know what numbers to pick?

You don't always. Try to get a good estimate. The only time you have to erase is if your estimate is too big. I try to choose easy numbers at the beginning. There are many ways of getting the correct answer.

Here is one way to do the problem using paper and pencil.

```
           69 R2
      5 ) 347
        - 200      40
          147
        - 100      20
           47
        -  45       9
            2
```

Keenya found that 347 divided by 5 is 69 with remainder 2. So, the answer is a little more than 69 minutes.

---

5. Mr. Haddad of the TIMS Candy Company has 398 Chocos to divide evenly among 6 orders.

   A. Estimate the number of Chocos each order will receive.

   B. Model the problem using base-ten shorthand.

   C. How many Chocos will be in each order?

   D. How many Chocos will be left?

   E. Do the problem using the forgiving method.

6. Joe made 282 Chocos. Mrs. Haddad had to divide them among 5 orders.

   A. Estimate the number of Chocos each order will receive.

   B. Model the problem using base-ten shorthand.

   C. How many Chocos will be in each order?

   D. How many Chocos will be left?

   E. Do the problem using the forgiving method.

7. One of the stores decided to sell Chocos in small bags. Each bag will have 6 Chocos. Mrs. Haddad has 96 Chocos. How many bags will she use?

   A. Model the problem using base-ten shorthand.

   B. How many bags will she use?

   C. Do the problem using the forgiving method.

Homework

Model Questions 1–4 using base-ten shorthand. Then solve them using the forgiving method.

1. 643 ÷ 5   2. 852 ÷ 3   3. 1533 ÷ 8   4. 2835 ÷ 9

Solve Questions 5–8 using mental math or pencil and paper.

5. 9076 ÷ 7   6. 2412 ÷ 3   7. 1889 ÷ 2   8. 3600 ÷ 4

Solve the following problems using paper-and-pencil methods or mental math. Remember to record your final answer. Also, if there is a remainder, describe how it is used in each problem.

9. Mr. Haddad's company packages damaged Chocos in packages called "Handful Packs." At the end of the day, Carmen, an employee who packages candy, has been given 132 damaged pieces of candy. How many Handful Packs can Carmen make if each pack must contain 8 Chocos?

10. The TIMS Candy Company has 176 employees. Mr. Haddad evenly divides his staff into four categories: management, candy making, packaging, and distribution. How many employees are in each category?

11. A. Hank works 5 hours a day. He gets paid $7 an hour. How much does Hank earn in one work week (5 days)?

    B. How much does he earn in four weeks?

12. James's candy-wrapping machine grabs 6 Chocos at a time and wraps them in about 1 second. James is helping to fill an order for 4568 Chocos.

    A. How many times must the candy-wrapping machine run in order to wrap the Chocos individually?

    B. Will it take more or less than 30 minutes to wrap all these candies? How did you decide?

13. The TIMS Candy Company holds an annual company picnic. One hundred eighteen employees came to the picnic. Sixty-seven employees brought one guest. Forty-three employees brought two guests. The rest of the employees brought three guests. How many people attended the picnic in all?

14. Bob transports boxes of nuts for the Crunchy Nut factory. Nine individual boxes of Crunchy Nuts fit inside a carton. Bessie Coleman School needs 3100 boxes for their fund-raiser. How many cartons of Crunchy Nuts will Bob need to deliver?

15. A. If the 3100 boxes of Crunchy Nuts are divided evenly among 7 classrooms, how many boxes should each classroom get?

    B. Mrs. Randall's classroom has 32 students. If each child sells about 14 boxes of Crunchy Nuts, will they sell all the boxes of Crunchy Nuts Mrs. Randall's class needs to sell?

16. In the end, Mrs. Randall's class raised $1835. How many boxes did Mrs. Randall's class sell, if each box cost $5?

Solving Problems Using Multiplication and Division

TV Survey

Mrs. Dewey's class learned that children between the ages of six and eleven watch an average of three hours of TV each day. Use this fact to answer Questions 1–3. Practice using mental math or a paper-and-pencil method as you solve the problems.

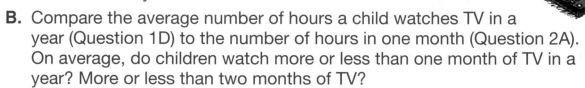

1. A. On average, how many hours of television does a child in this age group watch in 10 days?

    B. How many hours in 30 days?

    C. How many in 100 days?

    D. How many in one year (365 days)?

2. A. How many hours are there in a month that has 30 days?

    B. Compare the average number of hours a child watches TV in a year (Question 1D) to the number of hours in one month (Question 2A). On average, do children watch more or less than one month of TV in a year? More or less than two months of TV?

3. A. If a child watches 3 hours of TV each day, how many days would it take a child to watch 500 hours of television?

    B. How many weeks?

    C. How many months?

Reading Survey

For Questions 4–7, choose an appropriate method to solve the problem. For some questions, you may need to find an exact answer. For other questions, you may only need an estimate. For each question, you may choose to use paper and pencil, mental math, or a calculator. Be prepared to tell the class how you solved each problem.

4. Mrs. Dewey's class took a survey of the amount of time students read for pleasure in the evening. For the reading survey, they collected data for seven days. Here is Shannon's data for the seven days: 30 min, 75 min, 60 min, 45 min, 75 min, 90 min, and 45 min. Shannon averaged her data in two ways.

    A. Find the median number of minutes Shannon read for pleasure during that week.

    B. Find the mean.

5. Using the class data, they found that, on average, a student in Mrs. Dewey's class reads 45 minutes for pleasure each evening.

   A. On average, how many minutes does a student in Mrs. Dewey's class read in four days?

   B. How many hours?

6. A. On average, how many minutes does a student in Mrs. Dewey's class read in 10 days?

   B. How many minutes in 30 days?

   C. How many in 100 days?

7. A. On average, about how many minutes does a student in Mrs. Dewey's class read for pleasure in one year?

   B. About how many hours?

   C. Is the amount of time a student in Mrs. Dewey's class reads for pleasure in one year more or less than the number of hours in a month? (*Hint:* See Question 2A.)

Zeros and Division

Tanya and Frank were studying their division facts. They began with 24 ÷ 4.

Frank wrote "24 ÷ 4 = 7."

Tanya wrote "24 ÷ 4 = 6."

She said, "One of us must be wrong. There can't be two different answers to the same division problem."

Mrs. Dewey said, "That's right, Tanya. Each division problem has a unique solution. Work together to find the correct answer. Try using fact families."

   8. Write the fact family for 24 ÷ 4. Who is correct, Tanya or Frank?

Tanya said, "To find the answer to 24 ÷ 4, I look for the only number that makes 4 × ? = 24 true. Since 4 × 6 = 24, then 24 ÷ 4 = 6."

"That's good thinking," said Mrs. Dewey. "Let's use your reasoning to think about division and zero. Find 0 ÷ 24."

Tanya replied, "To find 0 ÷ 24, I find the only number that makes 24 × ? = 0 true. Since any number times zero is zero, 24 × 0 = 0 and 0 ÷ 24 = 0."

   9. Use Tanya's reasoning to find 0 ÷ 5.

Mrs. Dewey said, "Tanya, now try, 24 ÷ 0."

Tanya began, "To find 24 ÷ 0, I find the number that makes $0 \times ? = 24$. But, no number makes this number sentence true. What do I do?"

"Since there is no solution for $0 \times ? = 24$, we say that 24 ÷ 0 is undefined. In fact, if you use your reasoning with any number divided by zero, you will find the same thing. So, mathematicians say that division by zero is **undefined**."

10. Use Tanya's reasoning to find 5 ÷ 0.

"Now, think about 0 ÷ 0," said Mrs. Dewey.

This time Frank began, "To think about 0 ÷ 0, I try to find the only number that makes $0 \times ? = 0$ true. But, any number works. $0 \times 5 = 0$ and $0 \times 24 = 0$. Zero times any number is zero. Mrs. Dewey, I thought you said there is just one right answer. I remember you said, 'a unique solution.'"

"That's right, Frank," Mrs. Dewey replied. "Since there is not a unique solution, mathematicians say that 0 ÷ 0 is undefined as well."

For each statement below, find one number that will make it true. If there is no such number, say so.

11. **A.** 8 ÷ 4 = _____ , since 4 × _____ = 8

B. 42 ÷ 7 = _____ , since 7 × _____ = 42

C. 5 ÷ 1 = _____ , since 1 × _____ = 5

D. 0 ÷ 3 = _____ , since 3 × _____ = 0

E. 28 ÷ 7 = _____ , since 7 × _____ = 28

F. 2 ÷ 0 = _____ , since 0 × _____ = 2

G. 36 ÷ 6 = _____ , since 6 × _____ = 36

H. 0 ÷ 0 = _____ , since 0 × _____ = 0

Solve the following problems. When necessary, use "undefined." Justify your reasoning, using related multiplication sentences.

12. **A.** 35 ÷ 7 =   **B.** 0 ÷ 7 =

C. 7 ÷ 0 =   **D.** 0 ÷ 0 =

13. Do the division problems in Question 12 on a calculator. Explain what happens.

Solving Problems Using Multiplication and Division

A Saturday Visit

Nicholas's cousin Stan came to visit on a Saturday afternoon. Solve the following problems that describe their day.

- **Show your calculations using a paper-and-pencil method or explain a mental math strategy.**
- **Use estimation when appropriate.**
- **If the answer includes a remainder, explain how the remainder is used.**

1. It took Stan 4 hours to get to Nicholas's house. Stan and his mother took the freeway. If Stan's mother drove 55 miles per hour, how many miles away does Stan live?

2. Stan plans to surprise Nicholas with three tickets to see the Silver Blades hockey team. If Stan's dad paid $54 for three tickets, how much did one ticket cost?

3. When Stan arrived, Nicholas was just finishing a book. Nicholas said, "This book has 273 pages. It took me four days to read it." On average, how many pages did Nicholas read each day?

4. **A.** The boys' mothers talked about their exercise routines while they enjoyed a cup of tea. Stan's mother burns 10 calories in 1 minute on her bike. How many calories does she burn in 20 minutes?

Solving Problems Using Multiplication and Division   SG • Grade 4 • Unit 13 • Lesson 4   **377**

B. Nicholas's mother burns about 9 calories per minute on the Super Step machine. How many calories does she burn in 24 minutes?

C. How many minutes does Nicholas's mother need to use the Super Step machine to burn off a 340-calorie dessert?

5. A. When they got back to Nicholas's house, the boys played a board game. While playing, Nicholas had a chance to double his winnings of $2972. How much would Nicholas have if he doubled his money?

B. Instead, Nicholas landed on "Donate your earnings to 4 of your favorite charities." If Nicholas shares his earnings of $2972 equally, how much money will each charity receive?

6. Later in the evening, Nicholas and Stan played a video game involving a skyscraper. In this game, if you answer all the questions in any one round correctly, you go up 18 floors. If you have a perfect score after 12 rounds, you reach the top of the building. How many stories are in the building in this video game?

7. $1632 \div 8$ **8.** $976 \div 4$ **9.** $2832 \div 5$

Plant Growth

Introduction to Plant Growth

The students in Mrs. Dewey's class are studying plants and how they grow. To help the students see how plants grow, Mrs. Dewey suggested they design an experiment to do in their classroom.

"Why don't we each plant a seed and then measure how tall our plants grow each day?" suggested Grace.

"That's a great idea," echoed Jerome and Maya. "Then, we can graph our data and see if all our plants have the same growth pattern."

"That would make a good experiment," said Mrs. Dewey.

1. **A.** What are the two main variables in this experiment?
   B. What unit of measurement should students use to measure plant height? Explain your thinking.
   C. How should students measure time in this experiment: days, minutes, hours, etc.? Explain your thinking.
   D. Which variable is the manipulated variable?
   E. Which variable is the responding variable?

"During our experiment, we will use scientific time instead of calendar time," said Mrs. Dewey. "Scientific time always starts at $T = 0$. So, in our experiment, we will start at $T = 0$ days. You will each start your scientific clock when your plant first pushes out of the soil. Let's say that this happens on March 21. We would then call this date $T = 0$ days. March 22 would then be $T = 1$ day."

March

S	M	T	W	Th	F	Sa
		1	2	3	4	5
6	7	8	9	10	11	12
13	14	15	16	17	18	19
20	(21)	22	23	24	25	26
27	28	29	30	31		

2. How is scientific time different from calendar time?

"To keep our experiment fair, there are variables that need to be held fixed," continued Mrs. Dewey.

3. A. What variables should be held fixed in this experiment?

    B. Why is it important to hold these variables fixed?

The students in Mrs. Dewey's class decided on the following setup for their experiment. Your class may need to use a different setup.

- Each student will plant four bean seeds in a clean $\frac{1}{2}$-pint milk carton saved from the lunchroom.
- Each student will plant their seeds $\frac{1}{2}$-inch deep.
- Each student will choose one of his or her plants to measure during the experiment, cutting off any other plants that grow.
- Students will measure their plants on Monday, Wednesday, and Friday mornings. They will record their measurements on a data table.
- Plants will be watered with the same amount of water after they are measured on Mondays and Fridays.

Your class will now complete a plant growth experiment like the one described by the students in Mrs. Dewey's class. Draw a picture of your experimental setup. Be sure to label the variables using symbols.

4. What variables will you measure in this experiment?

Plant your seeds. As soon as the first sprout breaks through the soil, begin your scientific clock.

- Measure the height of your first sprout to the base of the first set of leaves several times each week.
- Measure to the nearest half centimeter.
- Use a data table similar to the one shown here to record your data.
- Continue to collect data for at least 21 days.

H=5.5cm

Plant Growth

Date	T in Days	H in cm
March 21	0	0 cm
March 23	2	1 cm
March 25	4	3 cm

You will graph and analyze your data from this experiment in Unit 15 Lesson 1.

Chancy Predictions: An Introduction to Probability

	Student Guide	Discovery Assignment Book	Adventure Book	Unit Resource Guide*
Lesson 1				
Chance Discussions	●			
Lesson 2				
Bean Counter's Game	●			
Lesson 3				
Rolling a Number Cube	●			●
Lesson 4				
From Number Cubes to Spinners	●	●		
Lesson 5				
Exploring Spinners	●	●		●
Lesson 6				
Make Your Own Spinners	●	●		
Lesson 7				
Prob_e Quest			●	

Unit Resource Guide pages are from the teacher materials.

Chance Discussions

Sure Things?

Some things are **impossible.**
They cannot happen. What is
impossible about the picture below?

Other things are **unlikely.** They can happen,
but they won't happen very often. It is unlikely
that you will grow to a height of 7 feet.

Still other things are **likely.** They probably will happen, but they might not. It is likely that you will grow to a height of at least 4 feet.

Finally, some things are **certain.** They are sure to happen. It is certain that you will grow to a height of more than 1 inch.

Probability is a measure of how likely things are to happen. Events that are impossible have probability 0%. Events that are certain have probability 100%. The larger the probability, the more likely an event is to happen.

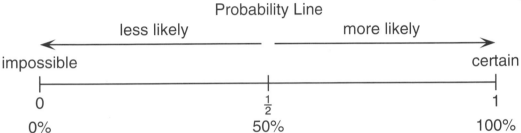

Draw a probability line like the one above. Where would each of the following events appear on the line? Use the letters to place each event on the line.

    A. It will be cold tomorrow at the South Pole.

    B. There is fruit in your refrigerator at home.

    C. Elvis Presley will be the next President of the United States.

    D. You will receive mail today.

    E. The Chicago Cubs will win the Super Bowl next year.

    F. You will fly to the moon tomorrow.

    G. A penny will show heads when flipped once.

    H. When flipping a penny, you get heads ten times in a row.

    I. Monday will follow Sunday.

    J. You will have homework tonight.

    K. A newborn baby will be a girl.

    L. It is going to snow tomorrow.

    M. You will be sick tomorrow.

The letters A, B, C, D, and E on the probability line below represent five different probability categories: 0%, not likely, 50%, likely, and 100%. Look around your home. For each letter on the line, give an example of an event with that probability. Record your examples on a separate sheet of paper.

Probability Line

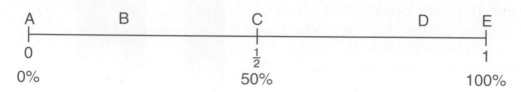

Bean Counter's Game

Players

This is a game for 2–3 players.

The object of this game is to be the first player to eliminate all twelve beans from his or her number line.

Materials

- scratch paper
- twelve beans or game markers per player
- one number cube

Rules

1. Each player first draws a line on a piece of scratch paper and then draws the six faces of a number cube as shown.

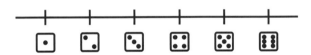

2. Each player distributes twelve beans any way he or she chooses above the drawings of the cube faces. Here are two examples of ways to spread the beans on the number line.

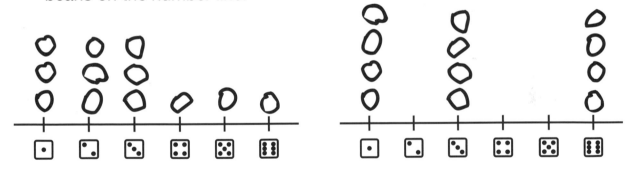

3. Decide which player will roll the number cube for the whole game. That player rolls the cube and reads the number that is face up.

4. For each roll, each player removes a bean above the matching face on the number line. If there is no bean above that face, the player removes nothing.

5. Continue rolling the number cube and removing beans. The first player to remove all the beans from his or her board is the winner.

Rolling a Number Cube

In this lab, you and a partner will study what happens when a number cube is rolled many times. The main variables are *F,* the faces on the cube, and *T,* the number of times each face comes up.

1. Which variable is the manipulated variable?

2. Which variable is the responding variable?

3. With your partner, roll the number cube 60 times. Tally the number of times each face comes up, using a table like this:

Group's Data Table

F Face	*T* Number of Times Face Appeared	
	Tallies	**Total**
⚀		
⚁		
⚂		
⚃		
⚄		
⚅		

Group Grand Total _____

4. **A.** Make a bar graph of your results.

   B. Describe your graph. For example, what can you say about the heights of the bars?

5. Combine your data with other groups to get class data. Record the class data on a table like this:

Class's Data Table

F Face	*T* Total Number of Times Face Appeared
⚀	
⚁	
⚂	
⚃	
⚄	
⚅	

Class Grand Total _____

6. **A.** Make a bar graph of the class data.

   B. Describe the class graph. What can you say about the heights of the bars?

   C. How is your graph of the class data different from your 60-roll group graph? Compare their shapes.

7. A number cube is **fair** if each face is equally likely to come up.

    A. What is the total number of times your class rolled the number cubes? About how many times would you expect each face to come up if each face is equally likely?

    B. Look at your class data table and graph. In general, does the data show that your class used fair number cubes? Why or why not?

8. Suppose a number cube was rolled 1000 times and a 6 came up 500 times. Do you think it is a fair cube? Why or why not?

9. Think about the *Bean Counter's Game* in Lesson 2. Based on the results of this lab, describe a winning strategy for the game.

Number Cubes and Probability

Sometimes we can only estimate the probability of an event. We can say that the event is likely or unlikely, but we don't know its exact probability. However, sometimes we can find the probability exactly.

For example, we can find the **probability** of rolling a particular face of a number cube. A normal number cube has six faces, and only one of those faces is ⚅. So, there is just one chance out of six that ⚅ will show when the number cube is rolled. The probability of rolling ⚅ with a fair number cube is $\frac{1}{6}$. This means we can expect ⚅ to show about $\frac{1}{6}$ of the time.

What is the probability of rolling a number greater than 4? Two out of the six possible faces are greater than 4—namely 5 and 6. Therefore, the probability is $\frac{2}{6}$. If we rolled a number cube 600 times, we would expect to get a number greater than 4 about 200 times, since 200 is $\frac{2}{6}$ (or $\frac{1}{3}$) of 600.

10. Probability predicts that each face will come up *about* $\frac{1}{6}$ of the time when a number cube is rolled.

    A. Did this happen in your group's 60-roll data?

    B. Did this happen in your class's data?

11. Probability predicts that a number greater than 4 (a 5 or a 6) will come up $\frac{1}{3}$ $(\frac{2}{6})$ of the time when a number cube is rolled. Does your class's data agree with this?

Answer the following questions about rolling a number cube.

12. **A.** What is the probability of rolling a 4? (Express your answer as a fraction.)

   B. Where would you place your answer to 12A on a probability line—nearer "1" or nearer "0"?

13. **A.** What are the odd numbers on the cube?

   B. What is the probability of rolling an odd number? (Express your answer as a fraction.) Explain your thinking.

14. **A.** What is the probability of rolling a number less than 3? (Express as a fraction.)

   B. On the probability line, would you place your answer to 14A closer to 0%, 50%, or 100%?

15. **A.** If Jessie rolls a 3 with a fair number cube, which of the following describes the probability of rolling a 3 on her next roll: (a) a little less than $\frac{1}{6}$, (b) equal to $\frac{1}{6}$, or (c) a little more than $\frac{1}{6}$?

   B. Why do you think so?

16. If you rolled a number cube 1200 times, about how many fives do you think would come up? Why do you think so?

From Number Cubes to Spinners

Tanya and Grace want to play a game after school at Tanya's house. They read in the directions that they need a number cube to play the game, but they can't find one anywhere. Finally, Tanya had an idea, "Maybe we can make a spinner that will work the same way a number cube does when we play the game. All we need is a paper clip, a pencil, some paper, and a protractor."

1. What is the probability of rolling each face on a fair number cube?

2. What would a spinner look like that would give the same results as a number cube?

3. How many regions would be on such a spinner?

4. What would be the measure of each angle on the spinner?

5. Create a spinner that will give the same results as a number cube. Use the *A Spinner for a Number Cube* Activity Page in the *Discovery Assignment Book* and a protractor.

6. Test your spinner. Spin your spinner 60 times, and record the number of times the spinner lands in each region.

7. Share your data with the class. Record your data on a class data table.

8. **A.** Make a bar graph of the class data.

   B. Describe the graph.

9. Compare the graph of the class "spinners" data to the graph of the class "number cube" data in the lab in Lesson 3.

   A. How many bars does each graph have?

   B. Are the shapes of the graphs similar to one another?

10. Will your spinner give the same results as a number cube? Why or why not?

Exploring Spinners

Exploring Spinner 1

You and a partner are going to spin 40 times using Spinner 1. You will need a copy of Spinner 1 from the *Discovery Assignment Book* and a sheet of *Centimeter Graph Paper* to complete this part of the lesson.

Spinner 1

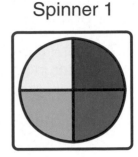

1. Draw the following data table on your own paper, and use it to record your data.

Spinner 1 Data

Region	Predicted Number of Times Spinner Will Land in Region (Out of 40)	Number of Times Spinner Landed in Region (Out of 40)	Class Data: Total Number of Times Spinner Landed in Region (Out of _____)
Yellow			
Red			
Blue			
Green			

    A. Before you begin, predict the number of times the spinner will land in each region. Record your predictions in the second column of your data table.

    B. Spin 40 times, and record your results in the third column of your data table.

2. Are your results for Spinner 1 close to your predictions? Why or why not?

3. Predict what a graph of all the class data will look like when each group adds their data to a class data table. Show your prediction using words or a drawing.

4. Add your group's results for 40 spins to the class data table. Then record the totals from the class data table in the last column of your data table.

5. Make a bar graph of the class data.

6. Is the shape of your predicted graph for the class data similar to the shape of the class graph?

7. Draw a probability line on your paper. Each of the following statements has a letter in front of it. Use the letter to place the statement correctly on the line.

   A. The probability of the spinner landing on yellow.

   B. The probability of the spinner landing on green.

   C. The probability of the spinner landing in a purple region.

   D. The probability of the spinner landing in a region other than yellow.

8. Choose the probability that will correctly complete this statement: The chance of spinning yellow is (a) one in four, (b) one in six, or (c) three in four.

9. Write the probability of spinning yellow as a fraction.

10. What is the probability of the spinner landing on yellow **or** green? The word "or" has a special meaning in math. This question asks for the probability of the spinner landing in the area covered by stripes in the picture.

11. Write the probability of the spinner landing on red, blue, or green.

Spinner 1

Exploring Spinner 2

You and a partner are going to spin 40 times using Spinner 2. You will need a copy of Spinner 2 and a sheet of _Centimeter Graph Paper_ to complete this part of the lesson.

12. Draw a data table for Spinner 2 just like the table for Spinner 1.

    A. Predict the number of times the spinner will land in each region. Record your predictions in your data table.

    B. Spin 40 times and record your results.

Spinner 2

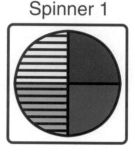

13. Predict what a graph of the class data will look like. Use words or a drawing to show your prediction.

14. Add your results to a class data table. Then record the class data in your table.

15. Make a bar graph of the class data.

Discuss

16. Is the shape of your predicted graph for the class data similar to the shape of the class graph?

17. Compare the class graph for Spinner 1 to the class graph for Spinner 2. How are they alike? How are they different?

18. **A.** The regions on Spinner 1 all have the same area. How does that affect the shape of the graph?

    B. The regions on Spinner 2 do not have the same area. How does that affect the shape of the graph?

Explore

19. Find the probability of Spinner 2 landing:

    A. On red

    B. On yellow

    C. On green

    D. On green or yellow

    E. On red or blue

    F. In an orange region

    G. In a region that is not red, yellow, or green

Probability Problems

20. A. When you spin this spinner, what is the probability it will land on A?

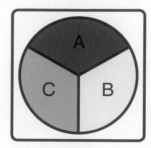

B. Danny spins the spinner 150 times. About how many times do you expect it to land on A?

21. Tim painted 3 sides of a cube red and 3 sides blue.

A. What is the probability that a red side will come up when the cube is rolled?

B. Tim rolls the cube 5 times, and it comes up red each time. What is the probability that it will come up red on the next roll?

22. A. What two ways can a nickel land when it is tossed in the air?

B. What is the probability that a nickel will land "heads up" when it is tossed in the air?

C. Jenny tosses a nickel 100 times. About how many times do you expect it to land heads up?

23. Abby wanted to invite a friend over to play, but she couldn't decide who to call. She put the names of Erin, Whitney, Tess, and Sam into a hat, closed her eyes, and pulled out a name. What is the probability that she chose Tess?

24. Peter, Nick, and Nate play a spinner game. Before playing, they agree that Peter will have Region A, Nick will have Region B, and Nate will have Region C. If the spinner lands in a player's region, he scores the point or points for that region. Do you think this is a fair game? That is, do you think each boy has an equal chance of winning? Explain your thinking.

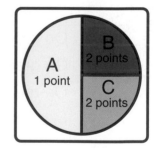

Make Your Own Spinners

Professor Peabody is inventing games for children to play. He needs a different spinner for each game. Each spinner will be divided into regions with different areas.

1. Design a spinner for Game 1. Professor Peabody wants the spinner to make data like this. Use the *Blank Spinners* Activity Pages from the *Discovery Assignment Book* to make your spinners.

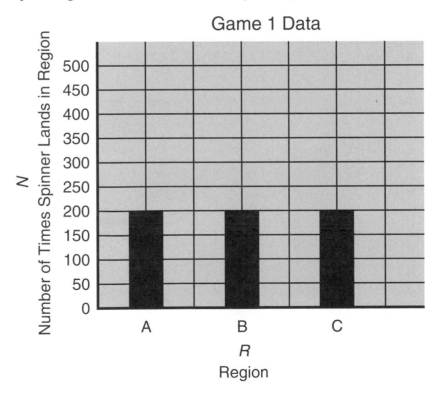

Game 1 Data

2. Design a spinner for Game 2 that you think will make data like this.

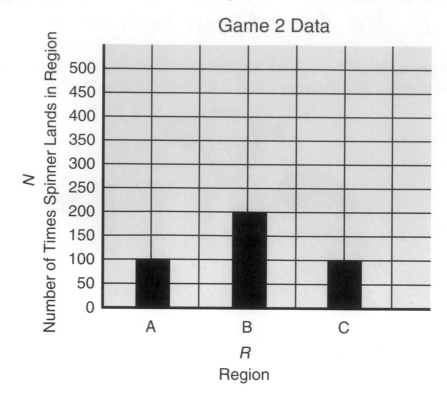

Game 2 Data

3. Design a spinner for Game 3 that you think will make data like this.

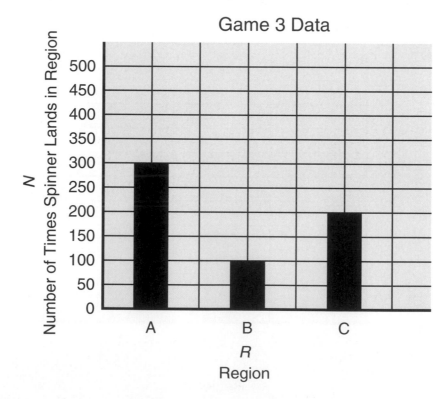

Game 3 Data

4. Design a spinner for Game 4 that you think will make data like this.

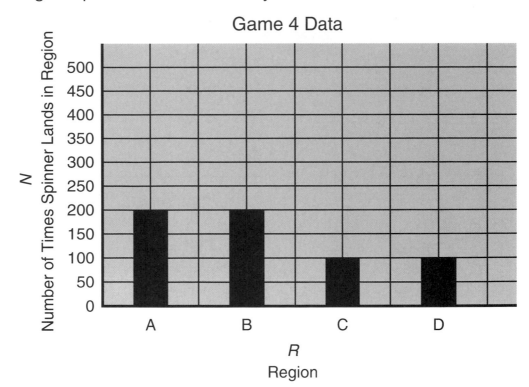

Game 4 Data

5. Choose one spinner. Test your spinner by collecting data and making a graph of the data.

   A. How many times will your group spin your spinner?

   B. Is the shape of your graph similar to the shape of the graph in the question for the spinner you have chosen? Why or why not?

Unit 15

Using Patterns

	Student Guide	Discovery Assignment Book	Adventure Book	Unit Resource Guide*
Lesson 1				
Plant Growth Conclusion	●	●		
Lesson 2				
In the Shade of the Old Meranpi Tree			●	
Lesson 3				
Planet Gzorp	●			●
Lesson 4				
Function Machines	●			
Lesson 5				
Taste of TIMS	●			●
Lesson 6				
Patterns and Problems	●			

Unit Resource Guide pages are from the teacher materials.

Plant Growth Conclusion

In Unit 13, you started the *Plant Growth* lab. When you finish collecting data, review the data for your plant. Here is a fourth grader's picture of the lab.

Use your *Plant Growth* picture to help you answer these questions.

1. What is the manipulated variable?

2. What is the responding variable?

3. What variables are held fixed?

4. Make a point graph of the data in your data table. Plot time on the horizontal axis and height on the vertical axis.

Use the graph to answer the following questions:

5. If there is a pattern to your data, fit a line or curve to the data points. Describe the shape of your graph.

6. How does your graph compare to those of your classmates?

7. Use your graph or data table. What was the height (H) of the lowest leaf at time $T = 5$ days?

8. Use your graph or data table. On what day was the height (H) of the lowest leaf equal to 10 cm?

9. What was the height of the lowest leaf at the end of your experiment?

10. **A.** What does your graph look like when your plant is growing fastest?

    B. What does it look like when it is growing slowest?

11. Write a paragraph that tells the story of your graph. Use your answers to Question 10 to help you. Include the following information: On which days did your plant grow the most? On which days did it grow the least? How did your plant grow in the beginning, middle, and end of your experiment?

12. What do you think the data would show if you continued to record the growth of the plant?

Discuss

13. Which graph, Graph A or Graph B, looks more like the graphs of the plants in your class?

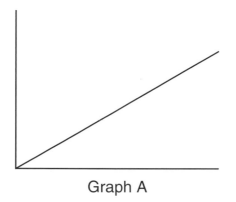

Graph A   Graph B

14. Compare your graph with some of the other graphs you have drawn this year. How are the graphs different? How are they similar?

15. Shannon and Jerome graphed the growth of their plants. What do the graphs tell you about how the plants grew? Using the graphs, tell as much as you can about how Shannon's and Jerome's plants grew. Compare the stories of the two graphs. How were they the same, and how were they different?

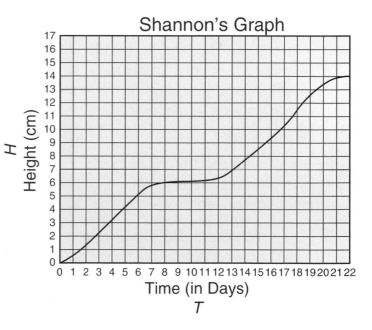

Shannon's Graph

H Height (cm)

Time (in Days)
T

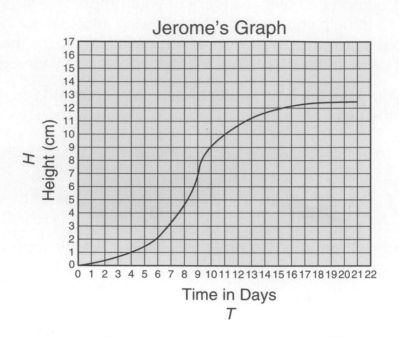

Jerome's Graph

Height (cm) **H**

Time in Days
T

🎒 Homework

You will need a sheet of *Centimeter Graph Paper*.

1. Grace planted a sunflower seed. She waited until it sprouted and measured its height every few days. She measured the plant at the same time each day—8 A.M. Here is her data table. Make a graph of Height (in cm) versus Time (in days) of her data on *Centimeter Graph Paper.*

2. What was the height of the plant at the end of the experiment?

3. On which days did the plant grow the most?

4. On which days is the graph the steepest?

Date	T Time in Days	H Height in cm
10/19	0	0
10/21	2	$\frac{1}{2}$
10/23	4	1
10/26	7	6
10/28	9	10
10/30	11	12
11/2	14	16
11/4	16	18
11/6	18	19
11/9	21	19

5. Here is the graph of Professor Peabody's plant growth data.

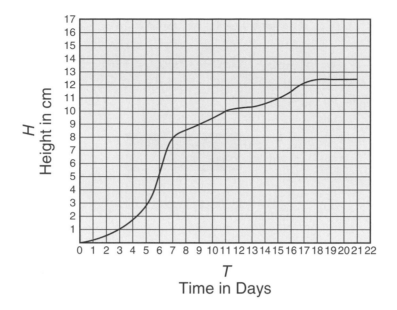

T
Time in Days

Professor Peabody's mouse, Milo, chewed up his data table. Make a data table like the one below, and fill in the missing information using the graph.

T Time in Days	H Height in cm
0	
2	0.5
3	
	3
7	
8	
9	
	10
	10.5
19	

Planet Gzorp

Far, far away, there is a planet called Gzorp. You can find many strange and beautiful things on Gzorp. The plants and animals on Gzorp are especially weird. Some of them are made of all squares!

This is an Add Three Gator.

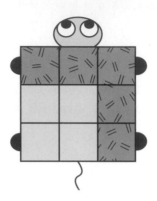

This is a Square Turtle.

And, this is a Triangle Fish.

Add Three Gator

The plants and animals that are made of squares grow by adding more squares. Different kinds of plants and animals add squares differently. For example, a 1-year-old Add Three Gator looks like this:

When it is 2 years old, it looks like this:

When it is 3 years old, it looks like this:

It keeps growing, getting three squares larger each year. The dark blue squares show the squares it grows each year.

1. Create a data table like the one below. Show the age in years and the size in squares for Add Three Gators between the ages of 1–7.

Add Three Gator Growth

Age in Years	Size in Squares
1	3
2	
3	
4	

2. Do you think an Add Three Gator that is as old as you is very big? Find out. Record the data for this Add Three Gator in your data table.

---

3. Use your data table to find out how many squares a 6-year-old Add Three Gator has.

4. How many squares does an 11-year-old Add Three Gator have? Tell how you solved this problem.

5. How old is an Add Three Gator that has 66 squares? Tell how you solved this problem.

6. If an Add Three Gator has 100 squares, about how old is it? Tell how you solved this problem.

Square Turtle

A Square Turtle grows into a bigger square each year. These are 1-, 2-, and 3-year-old Square Turtles.

7. **A.** Draw a 4-year-old Square Turtle.

   B. Write the number of squares it has in all.

   C. How many squares does it have on one side?

8. **A.** How many squares does a Square Turtle have on one side when it is 5 years old?

   B. How many squares does a 5-year-old Square Turtle have in all?

9. **A.** Create a data table for Square Turtles between the ages of 1–10.

   B. Describe patterns you see in the table.

Square Turtle Growth

Age in Years	Size in Squares
1	1
2	
3	
4	

10. How many squares would a 22-year-old Square Turtle have on one side?

11. **A.** If a Square Turtle has 36 squares, how old is it?

    B. If a Square Turtle has 169 squares, how old is it?

12. **A.** How many squares does a 10-year-old Square Turtle have?

    B. How many squares does a 75-year-old Square Turtle have? Tell how you know.

13. Estimate the age of a Square Turtle that has 1000 squares.

Triangle Fish

A Triangle Fish is a sea creature on Planet Gzorp. It lives in families. A family of Triangle Fish is shown below.

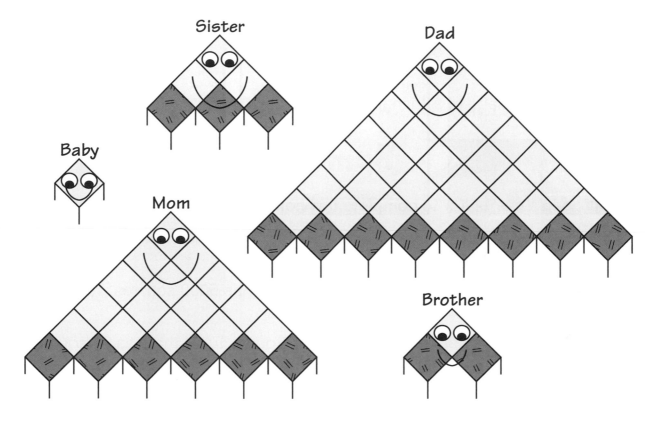

14. How old are the children in the Triangle Fish family?

15. How old are the parents?

16. **A.** Create a data table for Triangle Fish between the ages of 1–10. (*Hint:* the four-, five-, and seven-year-old Triangle Fish are not shown in the family.)

B. Describe in words how a Triangle Fish grows.

17. **A.** How many squares does a 10-year-old Triangle Fish have?

B. How many squares does a 15-year-old Triangle Fish have?

Triangle Fish Growth

Age in Years	Size in Squares
1	1
2	
3	
4	

18. **A.** If a Triangle Fish has 36 squares, how old is it?

B. If a Triangle Fish has 210 squares, how old is it?

C. If a Triangle Fish has 465 squares, how old is it?

19. How could you find out how many squares there are in a 48-year-old Triangle Fish? Estimate the answer.

20. Estimate the age of a Triangle Fish that has 20,100 squares.

Homework

You will need a calculator to complete this homework.

1. How many squares does a 30-year-old Add Three Gator have? Show your work.

2. Estimate the age of an Add Three Gator that has 5000 squares. Tell how you made your estimate.

3. How many squares are on one side of a Square Turtle that has 81 squares in all?

4. How many squares does a Square Turtle have on each side if it has 361 squares in all?

5. How many squares does a 30-year-old Square Turtle have? Show your work.

6. Estimate the age of a Square Turtle that has 7500 squares. Tell how you made your estimate.

7. Estimate the number of squares a 25-year-old Triangle Fish has.

Function Machines

Irma and Luis are exploring their Auntie Pat's attic one day.

Irma: Hey, Luis, look at this!

Luis: It looks like some kind of machine.

Irma: Yeah, but what is it?

Irma: It says Double Machine. What do you think that means?

Luis: Maybe it doubles stuff.

Irma: Look at all these cards with numbers on them. They look as though they fit in the slot.

Luis: Let's try it!

They put in a card with 25 on it and turn the crank. With a lot of coughing, sputtering, and choking, the machine spits out a card saying 50.

Irma: It really works!

Luis: This machine could help us double all kinds of numbers.

---

1. Irma and Luis tried the double machine on lots of numbers. They put their results in a data table like the one below. Make a data table like this one and fill in the blank spaces.

Doubling Machine

Input	Output
25	50
7	14
14	
	30
100	
	100
	7
N	$2 \times N$

Luis: Say, Irma, I found another machine over here, but I can't read the label.

Irma: Let's try it out and see if it works.

Luis: I put in 10 and out came 20.

Irma: It looks like another doubling machine. Let's put in 20.

Luis: Oh! 30 came out, so it can't be doubling.

2. Make a data table like the one below, and fill in the missing entries.

Mystery Machine

Input	Output
10	20
20	30
5	15
17	27
	45
25	
0	
	39
N	

3. What does the mystery machine do? There are many ways to answer this question. You can write the answer in words: the mystery machine is an "add ten machine."

You can write the answer in symbols. If we use N to stand for the Input number, then:

$$Output = N + 10$$

4. If you are given an output number, how can you find the input number?

Here are two more function machines. The machine in Question 5 multiplies the input number by 10 and then subtracts 5. The machine in Question 6 subtracts 5 from the input number. Set up two-column data tables like the ones below, and fill in the missing values.

5.

Input	Output
1	5
2	15
3	25
4	35
	55
15	
	205
100	
N	$10 \times N - 5$

6.

Input	Output
12	7
11	
	5
9	
	3
	2
6	
	0
N	$N - 5$

Guess My Rule

Players

This is a game for two or more players.

Materials

The players will need a *Two-column Data Table.* They can use calculators.

Rules

1. One player is the Function Machine. The player thinks of a rule and writes it down on a piece of paper, but doesn't tell it to the other players.

2. The other players take turns.

3. Each one gives the Function Machine an input number and writes it in the data table. The player who is the Function Machine tells the other players the output number and writes it in the data table.

4. A player may make one guess, describing the rule during his or her turn.

5. The first player to guess the rule is the winner. In the next round, the winner becomes the Function Machine.

In Questions 1–2, use the Input-Output Patterns to complete each data table.

1.

Input	Output
3	10
10	
	27
53	
	200
	1000
100	
N	N + 7

2.

Input	Output
1	30
2	
3	
4	
5	
10	120
	100
N	10 × N + 20

Here are data tables for two function machines. Find the Input-Output patterns. Use the patterns to fill in the blanks in each data table. Describe the Input-Output patterns, using words or symbols (or both).

3.

Input	Output
1	12
2	24
3	
4	48
5	
10	
	240
N	

4.

Input	Output
20	41
15	31
10	
	11
0	
2	5
	101
N	

5. Make a data table like this on your paper. Make up your own values for the input and output columns that follow a rule. Write the rule in symbols in the last row.

Input	Output
N	

6. Play *Guess My Rule* with a family member.

Function Machines

Taste of TIMS

Weight and Mass

Professor Peabody is having trouble telling the difference between weight and mass. He knows that **mass** is the amount of matter in an object and that **weight** is a measure of the pull of gravity. But, it's hard to tell the difference on Earth. So, he decides to travel to the moon to see what happens to his weight and mass.

Professor Peabody weighs 148 pounds on Earth. He discovers that his weight on the moon is less than his weight on Earth. Why do you think this happens?

The moon's gravity is weaker than Earth's. Because there is less gravity pulling on Professor Peabody, he has less weight.

Taste of TIMS

Now, he checks his mass.

His mass has stayed the same! Why do you think this happened? Why didn't Professor Peabody's mass change when the gravity changed?

Taste of TIMS

Shannon used the sandwich she brought for lunch to do an experiment. She placed her sandwich on a two-pan balance and used standard masses to find its mass. Then, she took a bite out of her sandwich and found the mass of the remaining sandwich. Shannon kept taking bites out of her sandwich, each time finding the mass, until her sandwich was gone.

Repeat Shannon's experiment using a sandwich of your own.

Draw a picture of the experiment.

1. What is the manipulated variable?

2. What is the responding variable?

3. What variable or variables are fixed during the experiment?

4. Collect data for 1, 2, and 4 bites. Record your data in a table like the one shown below.

N Number of Bites	M Mass in Grams
0	
1	

5. Make a point graph of the data in your table.

6. Do the points lie close to a straight line? If so, use a ruler to draw a best-fit line. If not, fit a curve to the data points.

7. A. Use your graph to predict the mass of the sandwich after your third bite. Write down your prediction.

    B. Did you use interpolation or extrapolation to make your prediction?

8. A. Use your graph to predict the mass of the sandwich after your sixth bite. Write down your prediction.

    B. Check your prediction. Is it close to the actual mass?

9. Predict how many bites it will take for you to eat your entire sandwich. Check your prediction and finish your sandwich.

10. Find the mean number of grams you ate with each bite.

11. Plot your partner's data on your graph. Compare your graph and data with your partner's. How are they the same? How are they different?

12. **A.** Who has a bigger bite size, you or your partner?

    B. Which student had the larger sandwich?

13. Two students' lines for the experiment are shown below. Tell a story for this graph. Include in your story which student had the sandwich with the most mass and which student took bigger bites.

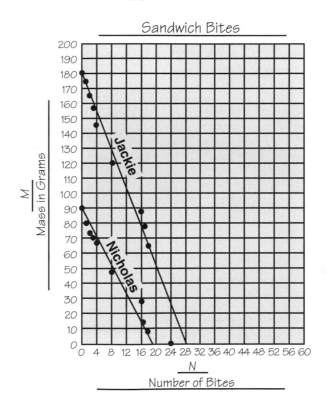

14. Three pairs of students do a similar experiment. They find the mass of a sandwich. Then they record the mass after eating one bite, two bites, and four bites.

A. Tell what is the same and different for each graph.

B. For each pair, what does the graph say about the mass of each sandwich?

C. For each pair, what does the graph say about the size of the students' bites?

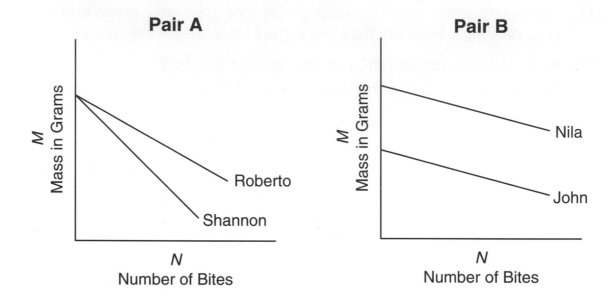

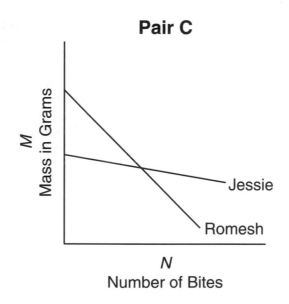

15. Romesh did the *Taste of TIMS* lab with an apple. He plotted his data in a graph. Tell a story about the graph.

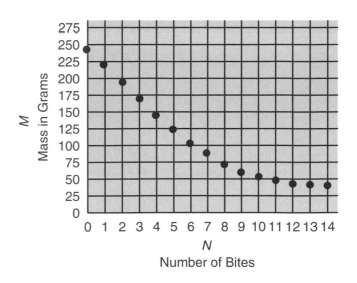

Homework

You will need a sheet of *Centimeter Graph Paper*.

1. Professor Peabody used the *Taste of TIMS* lab to see how long it took his mouse, Milo, to eat a dish of food. He recorded the data in a table. Plot a graph for Professor Peabody's data.

2. Predict the mass of the food left in the dish after 3 days.

3. Predict how many days it would take Milo to eat the entire dish of food.

4. What is the average amount of food eaten each day? Show how you got your answer.

N Number of Days	M Mass in Grams
0	114
1	103
2	91
4	67
6	43
8	22

Taste of TIMS

Patterns and Problems

You will need a sheet of *Centimeter Graph Paper* and a calculator to complete these problems.

1. Ming's function machine triples a number, then subtracts three. Jackie's function machine doubles a number, then subtracts two.

Machine X

Input	Output
1	0
2	3
3	6

Machine Y

Input	Output
1	0
2	2
3	4

   A. Which data table is Ming's?

   B. Which data table is Jackie's?

2. Maya's and Roberto's function machines have different rules.

   A. Help them complete their data tables for the numbers 0–5.

Double Plus Two

Input	Output
0	
1	
2	

Add 1, Then Double

Input	Output
0	
1	
2	

   B. What do you notice about Maya's and Roberto's data tables? Explain.

3. Jacob's function machine is missing its rule. Help Jacob find the rule for his function machine.

Input	Output
0	5
1	7
2	9
3	11
4	13
5	15

4. Nila's sandwich has a mass of 153 grams. She took one bite, and the mass of her sandwich is 122 grams.

   A. If each of Nila's bites has the same mass, what is the mass of two bites?

   B. How much mass will three bites have?

   C. How many bites can Nila take until her sandwich is gone?

5. John's sandwich has a mass of 139 grams. He took one bite, and the mass of his sandwich is 109 grams.

   A. If each of John's bites has the same mass, how many bites can John take until his sandwich is gone?

   B. Who has a bigger bite size, Nila or John?

6. Irma has organized her plant growth data in a table. Plot Irma's plant growth data on *Centimeter Graph Paper.*

Plant Growth

Day	Height (in centimeters)
0	0
4	2
5	4
7	8
8	12
11	14
15	15
17	15

A. How many centimeters did Irma's plant grow by Day 6? Show your work.

B. Use your graph to predict how many centimeters Irma's plant will grow by Day 18.

7. Frank organized his plant growth data in a table. Plot Frank's plant growth data on the same sheet of *Centimeter Graph Paper* as Irma's. Describe the differences and similarities in their graphs.

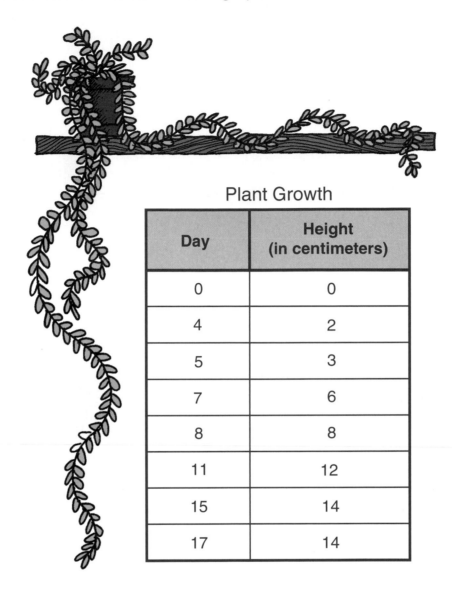

Plant Growth

Day	Height (in centimeters)
0	0
4	2
5	3
7	6
8	8
11	12
15	14
17	14

Unit 16

Assessing Our Learning

	Student Guide	Discovery Assignment Book	Adventure Book	Unit Resource Guide*
Lesson 1				
Experiment Review	●	●		
Lesson 2				
Problems and Practice	●			
Lesson 3				
Area vs. Length	●			
Lesson 4				
The Many-Eyed Dragonfly				●
Lesson 5				
End-of-Year Test				●
Lesson 6				
Portfolios	●			

*Unit Resource Guide pages are from the teacher materials.

Experiment Review

Professor Peabody loves to play with his toy cars. He likes to see how far each car can roll down ramps that are set at different heights.

Watching his toy cars roll down ramps reminds Professor Peabody of an experiment he did several months ago.

Discuss

1. Which lab do you think Professor Peabody is thinking about?

2. Answer the following questions about the lab in Question 1. You may look at the lab pages in the *Student Guide* or your portfolio to help you.

   A. What variables did you study in the lab?

   B. Did you have to keep any variables fixed, so the experiment would be fair? If so, which ones?

   C. Did you measure anything? If so, what did you measure? What units did you use?

   D. How many trials did you do? If you did more than one trial, tell why.

   E. Describe the shape of your graph. Use words or a sketch.

   F. What were the most important problems you solved using your data?

3. Look at the picture of Professor Peabody on the opening page of this unit. This picture, the labs in your portfolio, and the *Student Guide* can help your class make a list of the labs you completed. For each lab, answer each question in Question 2. Use the *Experiment Review Chart* in your *Discovery Assignment Book* to help organize the information.

Problems and Practice

Use appropriate tools such as paper and pencil, calculators, or pattern blocks to solve the following problems. For some problems, you need to find an exact answer. For others, you need only an estimate.

1. **A.** Tanya and her sister planted a rectangular flower garden. The garden plot is 4 feet wide and 6 feet long. What is the area of the garden plot?

   B. Tanya wants to put a fence around the garden. How many feet of fencing should she buy?

2. A rectangle is made of 36 square-inch tiles. Sketch all the possible rectangles. Write a multiplication sentence for the number of tiles in each rectangle.

3. Jessie recorded the time she spent watching television over a four-day period. Her data is recorded in the table below.

Day	Minutes of Television
Monday	240 minutes
Tuesday	210 minutes
Wednesday	255 minutes
Thursday	90 minutes

   A. Find the median number of minutes Jessie watched television during these four days.

   B. Find the mean number of minutes Jessie watched television during these four days. Give your answer to the nearest whole minute.

   C. What is the total number of minutes Jessie watched television during these four days?

   D. How many total hours of television did Jessie watch over the four days? Give your answer to the nearest hour.

4. Shannon helped her mom sell donuts at the Farmer's Market. Each donut costs $0.50. If Shannon sold 67 donuts and her mom sold 43 donuts, about how much money did they collect?

5. **A.** Show each of the decimals below using base-ten shorthand. The flat is one whole.

    0.45   0.68   1.04   0.1   0.05

    B. Arrange the decimals in order from least to greatest.

6. Ana and Nila are using this spinner to play a game.

    A. What fraction of the spinner is green?

    B. What fraction of the spinner is red?

    C. What fraction of the spinner is covered by blue or yellow?

    D. What is the probability of the spinner landing on the red region?

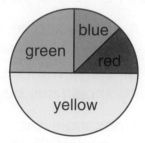

    E. What is the probability of the spinner landing on the red or the blue region?

    F. What is the probability of the spinner landing in a region other than the green region?

7. **A.** Irma and Maya are working with pattern blocks. They call the yellow hexagon one whole. Maya builds a shape using three green triangles, one red trapezoid, and one blue rhombus. Write a number for Maya's shape.

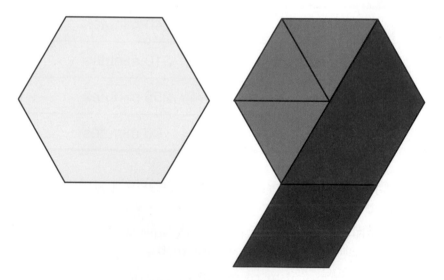

    B. Write a number sentence to show Maya's work.

8. Jerome, Shannon, and Nicholas each play on a different soccer team. They have played the same number of games. Jerome's team won $\frac{5}{6}$ of their games. Shannon's team won $\frac{3}{4}$ of their games. Nicholas's team won $\frac{7}{12}$ of their games.

   A. Put the fractions in order from smallest to greatest. You can use your Fraction Chart from Unit 12.

   B. Which team had the best record?

9. Linda made the following function machine. Copy the table. Help her complete her data table using her rule.

Double, then subtract 3

Input	Output
4	5
7	
8	13
	21
N	

10. **A.** Grace's sandwich has a mass of 142 grams. She took one bite and then found the mass of the remaining part of the sandwich to be 110 grams. If each of her bites is the same size, what will be the mass of the sandwich after Grace has taken a total of three bites?

   B. How many total bites will it take Grace to eat all of her sandwich?

11. Jackie and her sister are planning a party at a local indoor play park. They plan on inviting six people besides themselves. The total cost of the party is $96. How much will the party cost per person?

12. Michael, Jackie, Shannon, and Frank each brought a sandwich for the lab *Taste of TIMS.* They found that the total mass of all four sandwiches was 592 grams. What is the average mass of each sandwich?

13. Tanya's mom is working in an office supply store. Yesterday, a shipment of assorted notepads arrived at the store. There are 36 notepads in each box. The store received 28 boxes. How many notepads did they receive?

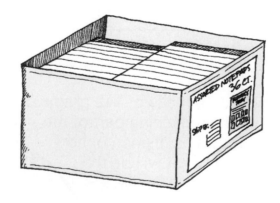

Area vs. Length

Michael and Jessie were looking through their portfolios. Michael pulled out the *Perimeter vs. Length* lab. Jessie did too. They compared the types of airplanes they chose to study. Michael studied the commuter plane, and Jessie studied the long-haul jet.

1. Look back at the *Perimeter vs. Length* lab in your portfolios or Unit 2 Lesson 2 in the *Student Guide.* What plane did you study?

2. What variables did you study?

In Unit 2, you helped Myrna find the perimeter of runways so that lights could be strung all around them. You chose one of the following planes to study:

- The light planes are the smallest airplanes at the Antopolis airport. They need runways that are 1 inch wide.
- The commuter planes need runways that are 2 inches wide.
- The short-haul jets need 3-inch-wide runways.
- The long-haul jets need 4-inch-wide runways.
- The heavy-transport planes need runways that are 5 inches wide.

Myrna needs your help again. The runways at Antopolis Airport need to be coated with a special paint to seal the cracks. To buy enough paint, Myrna needs to find the area of each runway—no matter how long.

Use the TIMS Laboratory Method to help Myrna. You will work on runways for only one kind of plane. Your teacher will help you choose.

You will use square-inch tiles to make several runways of different lengths for your kind of plane. For each runway, you will record the length and the area in a data table. Then you will graph your data and look for patterns.

3. Draw a picture of the lab. Be sure to show the two main variables, Length (*L*) and Area (*A*). Also show your kind of airplane and how wide your runways will be.

You will use square-inch tiles to make several runways for your kind of airplane. With your group, decide how long to make your runways. Do not make runways longer than 10 inches or you may have trouble graphing your data.

4. Make your runways. Record the length and area of each runway. Keep track of your data in a table like the one shown here.

Runway Data Table for _____

(Type of Plane)

Width = _____

W Width of Runway (in inches)	*L* Length of Runway (in inches)	*A* Area of Runway (in square inches)

5. **A.** What is the manipulated variable?

   B. What variable is the responding variable?

   C. What variable stays the same for all your runways?

6. Graph your data. Put Length (*L*) on the horizontal axis and Area (*A*) on the vertical axis.

7. Look at the points on the graph. If the points form a line, use a ruler to draw a line through the points. Extend your line in both directions.

Questions 8 to 12 are for runways the same width as yours.

8. How wide are your runways?

9. What is the area of a runway that is 4 inches long?

10. Find the area of a 12-inch-long runway. Show how you found your answer.

11. What is the area of a 100-inch-long runway? Explain how you found your answer.

12. Give a rule for finding the area of a runway for your type of airplane—no matter what the length.

13. Explain the difference between inches and square inches.

14. Professor Peabody tried to help Myrna. He worked hard and studied runways for three kinds of planes: light planes (1-inch-wide runways), short-haul jets (3-inch-wide runways), and heavy-transport planes (5-inch-wide runways). Unfortunately, Professor Peabody forgot which line on his graph he drew for each kind of runway. Which line did he draw for each kind of runway? Explain how you know.

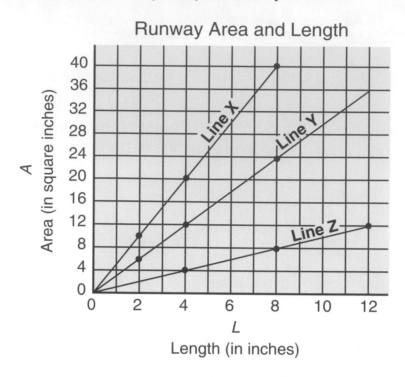

Runway Area and Length

15. Write a letter to Myrna about your kind of runway. Tell her what kind of plane your runways are for, and explain what you learned about the area of your runways. Explain how you can find the area for your kind of runway—no matter what the length. Use the Student Rubric: *Telling* to help you organize and write your letter.

Area vs. Length

Portfolios

Look at the work in your portfolios. Compare work from the beginning of the year to work you have just completed. Have you improved?

I have. I used to just write down answers. Now I show my work in lots of ways: I draw pictures, write sentences, make graphs…

I can't wait to show my portfolio to my mom and dad! They will be amazed.

Explore

1. Choose items from your collection folder to add to your portfolio. Choose items that are similar to the work that is already in your portfolio. Here are some examples of things you can choose:

   A. The *Experiment Review* from Lesson 1 of this unit.

   B. The solution to a problem you have solved. For example, you may choose to include your solutions to the problems in *The Many-Eyed Dragonfly*.

   C. A lab you have completed. *Area vs. Length* is a good example.

   D. Items your teacher recommends.

2. Add the name of each new piece to your Table of Contents. Include a short description of the work and the date it was finished.

3. Write a paragraph comparing two pieces of work in your portfolio that are alike in some way. For example, you can compare two labs or your solutions to two problems. One piece should be new, and one should be from the beginning of the year. Use these questions to help you write your paragraph:

   - Which two pieces did you choose to compare?
   - How are they alike? How are they different?
   - Do you see any improvement in the newest piece of work? Explain.
   - How could you improve your newest piece of work?

Student Rubric: Knowing

In My Best Work in Mathematics:

- I show that I understand the ideas in the problem.

- I show the same mathematical ideas in different ways. I use pictures, tables, graphs, and sentences when they fit the problem.

- I show that I can use tools and rules correctly.

- I show that I can use the mathematical facts that apply to the problem.

Student Rubric: Solving

In My Best Work in Mathematics:

- I read the problem carefully, make a good plan for solving it, and then carry out that plan.

- I use tools like graphs, pictures, tables, or number sentences to help me.

- I use ideas I know from somewhere else to help me solve a problem.

- I keep working on the problem until I find a good solution.

- I look back at my solution to see if my answer makes sense.

- I look back at my work to see what more I can learn from solving the problem.

Student Rubric: Telling

In My Best Work in Mathematics:

- I show all of the steps that I used to solve the problem. I also tell what each number refers to (such as 15 boys or 6 inches).

- I explain why I solved the problem the way I did so that someone can see why my method makes sense.

- If I use tools like pictures, tables, graphs, or number sentences, I explain how the tools I used fit the problem.

- I use math words and symbols correctly. For example, if I see "6 – 2," I solve the problem "six minus two," not "two minus six."

Index/Glossary

This index provides page references for the *Student Guide*. Definitions or explanations of key terms can be found on the pages listed in bold.

A

Addition
    with base-ten pieces, 78–80, 84–85
    multidigit, 174
    paper-and-pencil, 79–80, 84, 174
Angle, 41–56, **42,** 243–250
    acute, **50**–51, 54–55
    comparing, 43–44, 53
    drawing, 47, 54
    measuring, 243–250
    obtuse, **51,** 54–55
    in pattern blocks, 55–56
    right, **48**–49, 54–55
    in shapes, 45–46, 56
    vertex of, 243
Area, 28–38, **29**
    counting square units, 433–436
Arm span, 20–23
Average, 12, **13,** 22, **125**

B

Bar graph
    interpreting, 5, 9, 11, 14. *See also* Labs
    making, 4, 10. *See also* Labs

Base (with exponents), 113
Base-ten
    division, 364–378
    multiplication, 298–303
Base-Ten Board, 68
Base-ten pieces, 68–85
    bit, **68**
    flat, **68**
    pack, **68**
    skinny, **68**
Base-Ten Recording Sheet, 68
Base-ten shorthand, 71–85, 280–282, 285–287
Best-fit line, 123–124

C

Categorical variable, 7, 20
Certain event, 385
Center of turning, 251–254
Centimeter, 272
Circumference, 132
Comparing
    angles, 43–44, 53
    fractions, 351–352, 354
Convenient number, 202, 203
Cubic centimeter, 216
Cubic foot, 220

Odd number, 98
Order of operations, 180–183, **223**

Parallel lines, 238–242
Parallelogram, 241
Patterns and functions, 406–410
Percent, 10% as a measure of closeness, 166–168
Perimeter, 28–38, **29,** 31–38
Periods for place value, 150–153, **151**
Perpendicular lines, 238
Pictures, in Labs, 17. *See also* Labs
Place value, 68–85
    big base-ten pieces, 157–159, 159–161
    big numbers, 157–161
    decimals, 276–287
    Fewest Pieces Rule, **74**–75
    ordering numbers, 160–161
    periods, 150–153, **151,** 152–155, 158–159
    reading and writing numerals, 150, 152–155,
        157–159
Point graphs
    interpreting and making predictions, 120–124,
        146–148. *See also* Labs
    making, 157
    powers of two, 155
Polygon, 247
Portfolios, 39–40, 233, 439
Positive numbers, 88–91
Powers of two, 155–156, 156–158
Predictions, 143–145. *See also* Labs
    making, 120–124
    using graphs, 120–124
Prime number, 98, 104
Prism, 256–267, **259**
    drawing, 257
    volume of, 263–267
Probability, 384–399, **390**
    as a fraction, 390–391
    and spinners, 392–399
Probability line, 385–386
Product, 61, 107–108
Protractor, for measuring angles, 244–245

Quadrilateral, 56, 247
Quotient, 58, 235–236

Ray, 240
Responding variable, 140
Roman numerals, 64–67
    subtractive principle, **66**
Rounding, 170–171. *See also* Convenient number

Shapes, geometric, 240–241
Square inch, 29
Square number, 98
Student Rubric: *Knowing,* 441
Student Rubric: *Solving,* 442
Student Rubric: *Telling,* 443
Subtraction
    with base-ten pieces, 80–85
    multidigit, 174
    paper-and-pencil, 82–85, 174
Survey, 3
    hours of TV watched, 358–363
Symmetry, 251–255
    line, **251**–254
    turn, **251**–254

Temperature, 137
TIMS Laboratory Method, 17. *See also* Labs
Triangle Flash Cards, activities with, 61–62
Turn-around facts, multiplication, **61**

Unlikely event, 384

Value, 3, 7
Variable, 3
    categorical, **7,** 20
    fixed, **141**
    manipulated, **140.** *See also* Labs
    numerical, **7,** 20
    responding, **140.** *See also* Labs
Vertex, 46, 243, 256
    of a 3-dimensional shape, **256**–259
Vertical axis, 5–6
Volume, 216–222, 225
    cube models, **216**–218
    measure by displacement, **217**–218
    of a prism, 263–267

Weight, 417
Width, 31–38
Word problems, 24–25, 92–93, 118, 147–148,
    203–206, 211–213, 230–231, 332–334,
    374–375, 377–378, 424–427, 431–434.
    See also Labs

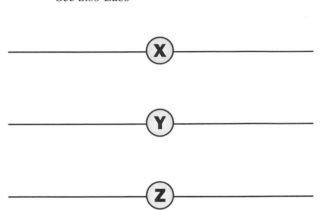